STUDY GUIDE AND SOLUTIONS MANUAL

PAULA YURKANIS BRUICE

University of California, Santa Barbara

ORGANIC CHEMISTRY

Eighth Edition

PEARSON

Editor in Chief: Jeanne Zalesky
Senior Acquisitions Editor: Chris Hess
Marketing Manager: Elizabeth Ellsworth
Project Manager: Brett Coker
Program Manager: Lisa Pierce
Editorial Assistant: Fran Falk
Marketing Assistant: Megan Riley
Executive Content Producer: Kristin Mayo
Media Producer: Lauren Layn
Director of Development: Jennifer Hart
Development Editor: Matt Walker
Program Management Team Lead: Kristen Flathman
Project Management Team Lead: David Zielonka
Production Management: GEX Publishing Services
Compositor: GEX Publishing Services
Art Specialist: Wynne Au Yeung
Illustrator: Imagineering
Text and Image Lead: Maya Gomez
Text and Image Researcher: Amanda Larkin
Design Manager: Derek Bacchus
Cover Designer: Tamara Newnam
Operations Specialist: Maura Zaldivar-Garcia
Cover Photo Credit: Olga Yakovenko/Shutterstock

4 18

www.pearsonhighered.com

ISBN 10: **0-13-406658-8;** ISBN 13: **978-0-13-406658-5**

CONTENTS

to my students

I am deeply grateful to Thomas Bertolini of the University of Southern California, who reworked all the problems to make this book as error free as possible and who gave me many important suggestions along the way. I am solely responsible for any errors that may remain. If you find any, please email me so they can be fixed in a future printing. I am also very grateful to Richard Morrison of the University of Georgia and Jess Jones of Saint Leo University who provided most of the spectroscopy problems.

Try to work as many problems as possible, so you can truly enjoy the wonderful world of organic chemistry.

Paula Yurkanis Bruice
pybruice@chem.ucsb.edu

Spectroscopy Problems

For problems that require specific chemical knowledge, the chapter number where the information can be found is given just after the problem number.

1. Determine the structure of the straight-chain five-carbon alcohol that produces the mass spectrum shown here.

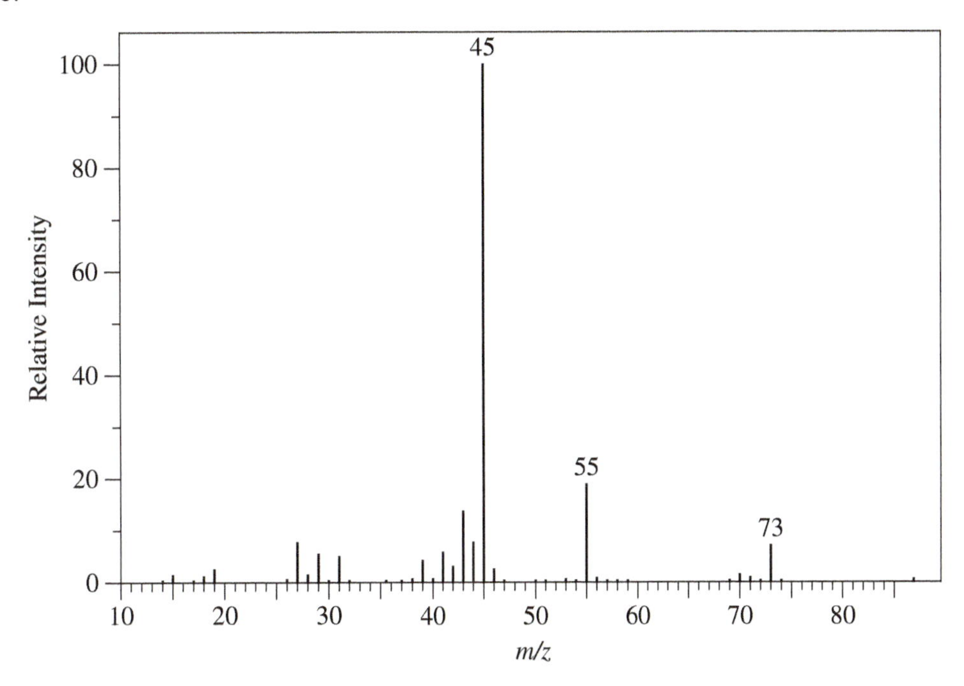

2. The mass spectrum of an ether is shown here. Determine the molecular formula of the ether that produces this spectrum and then draw possible structures for it.

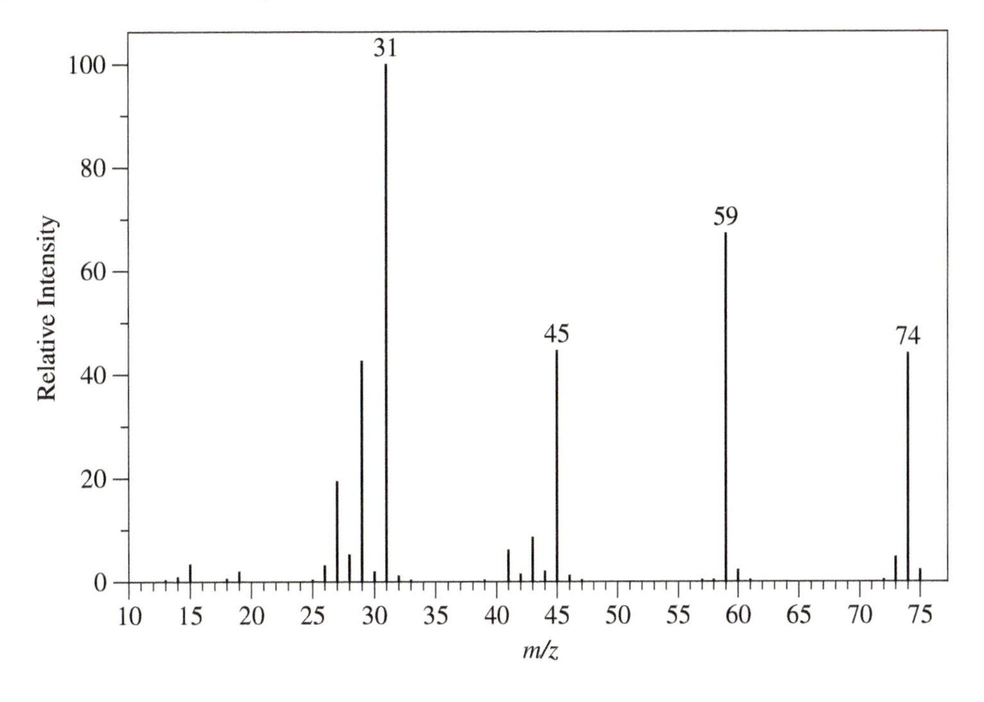

3. The mass spectra of pentane and isopentane are shown here. Determine which spectrum belongs to which compound.

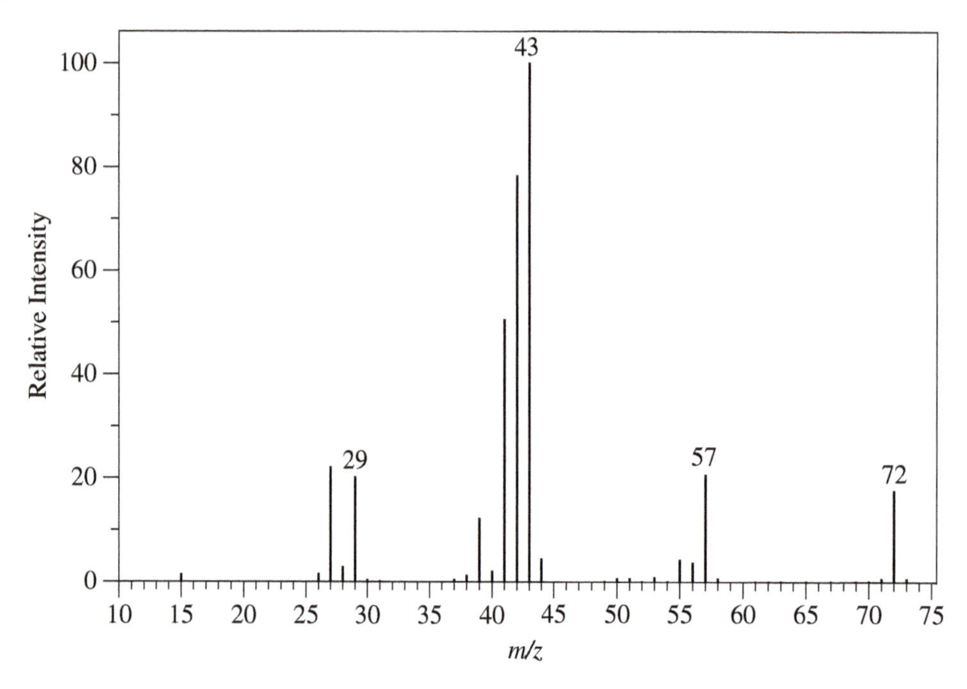

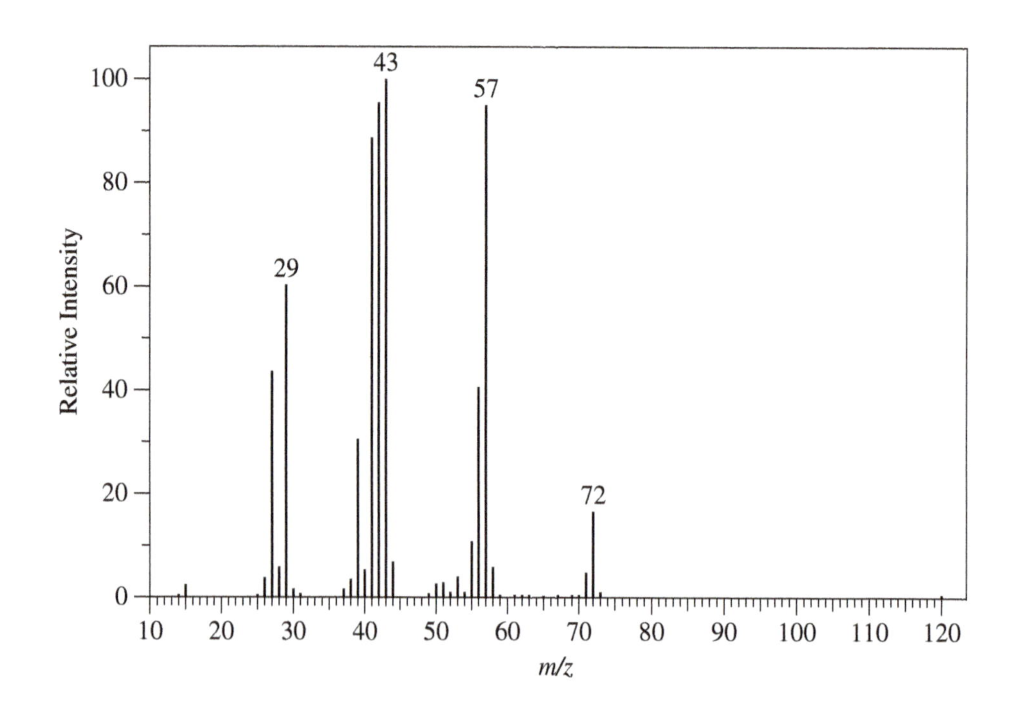

4. Which of the following compounds gives the mass spectrum shown here?

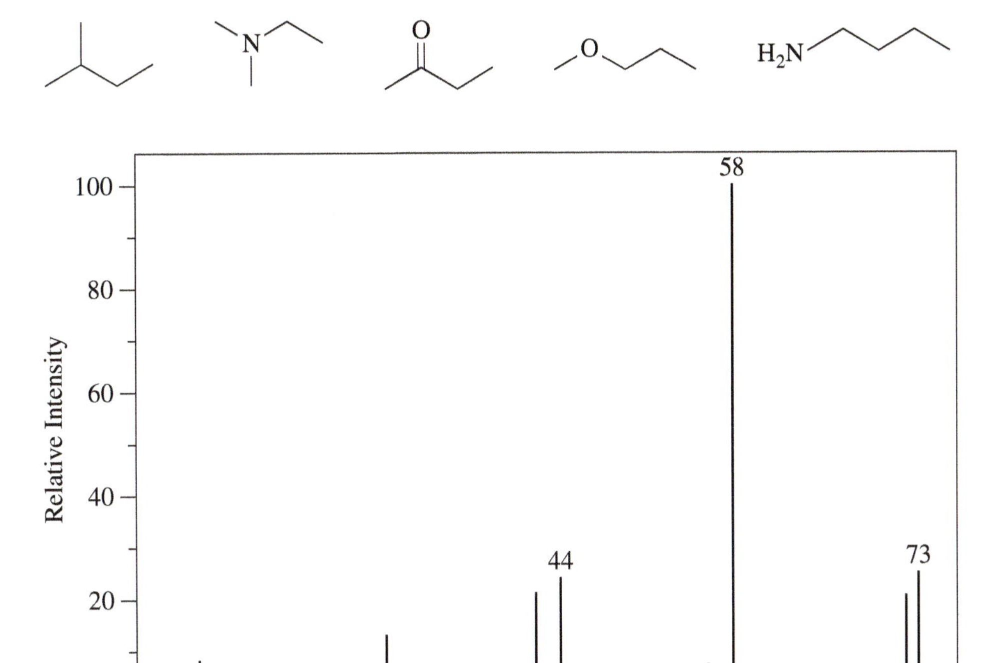

5. **Chapter 10**
An unknown acid reacted with 1-butanol. The product of the reaction gave the mass spectrum shown here. What is the product of the reaction, and what acid was used?

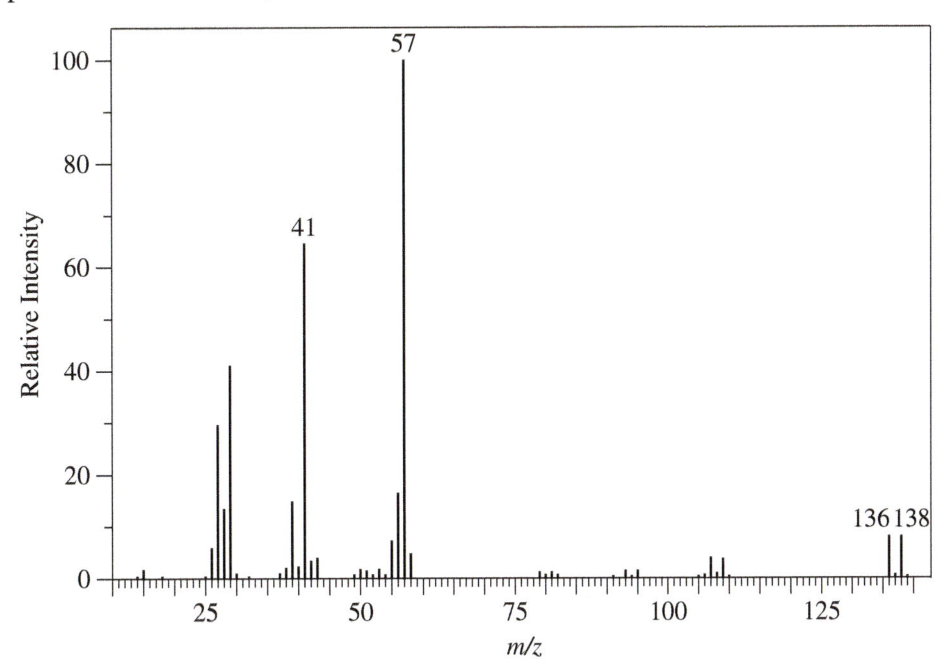

6. Identify the compound with molecular formula $C_9H_{10}O_3$ that gives the following IR and ^{1}H NMR spectra.

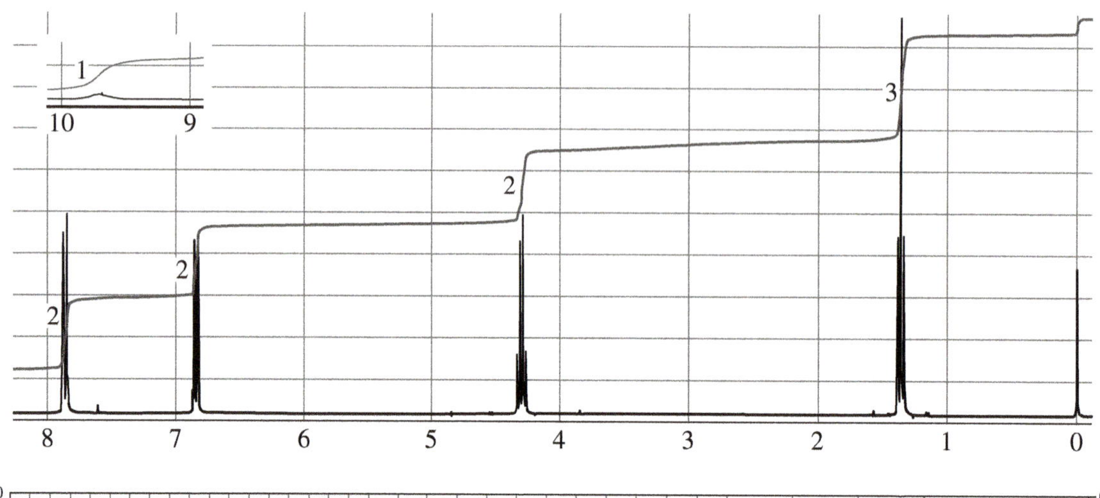

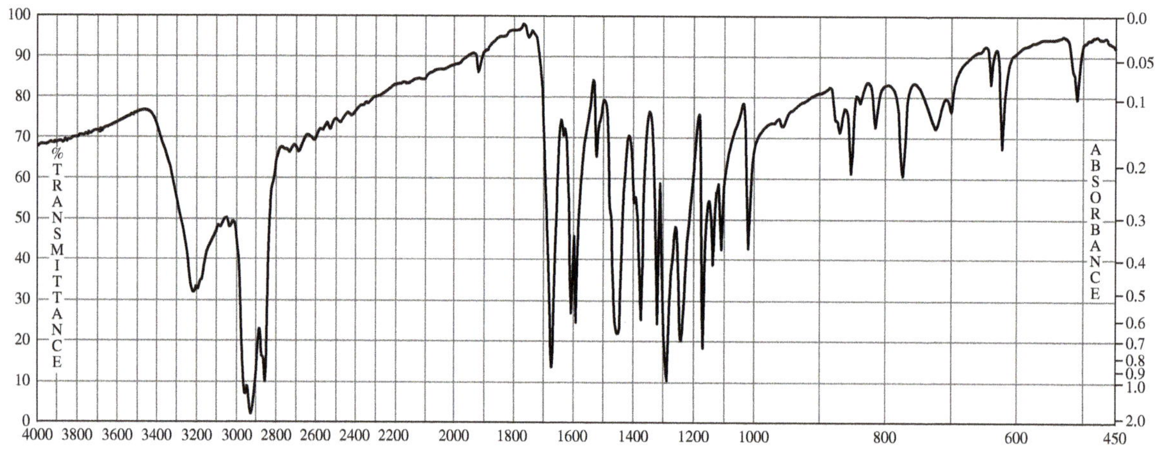

7. Identify the compound with molecular formula $C_5H_{11}Br$ that gives the following ^{1}H NMR spectrum.

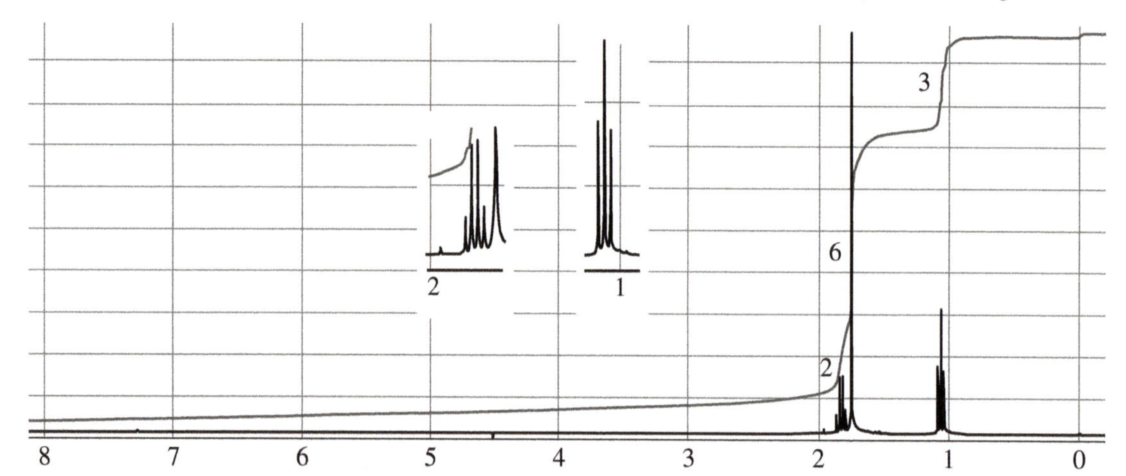

8. Identify the compound with molecular formula $C_6H_{12}O$ that gives the following IR and ^1H NMR spectra.

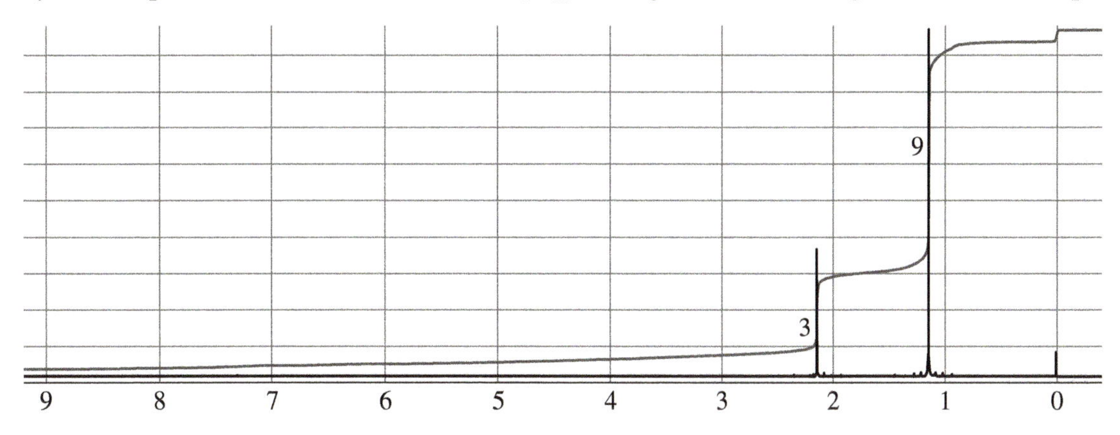

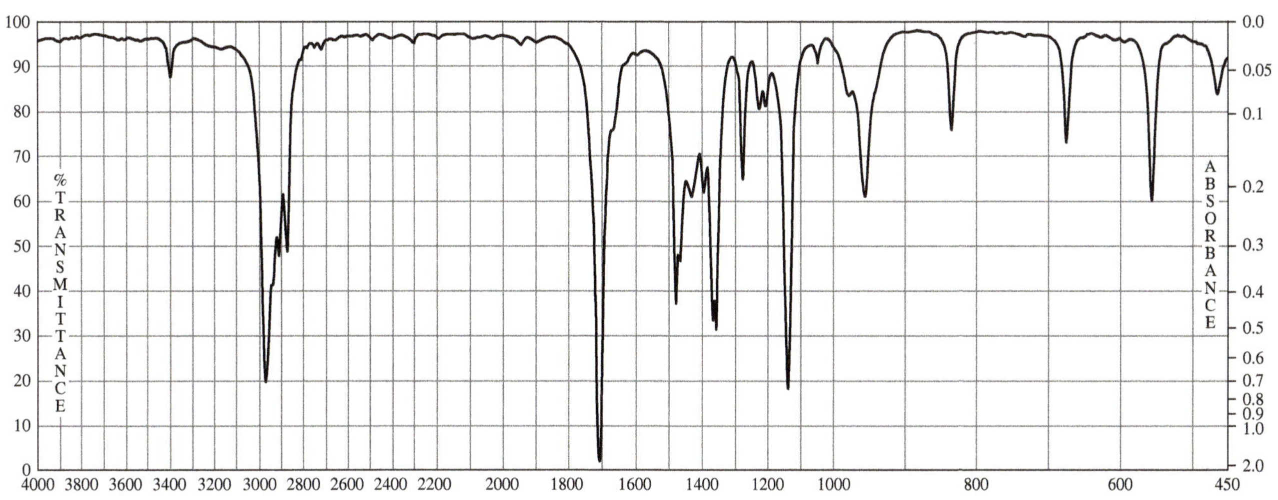

9. **Chapter 18**

A methyl-substituted benzene was treated with Cl_2 in the presence of $AlCl_3$. The ^1H NMR spectrum of one of the monochlorinated products is shown below. Identify the product.

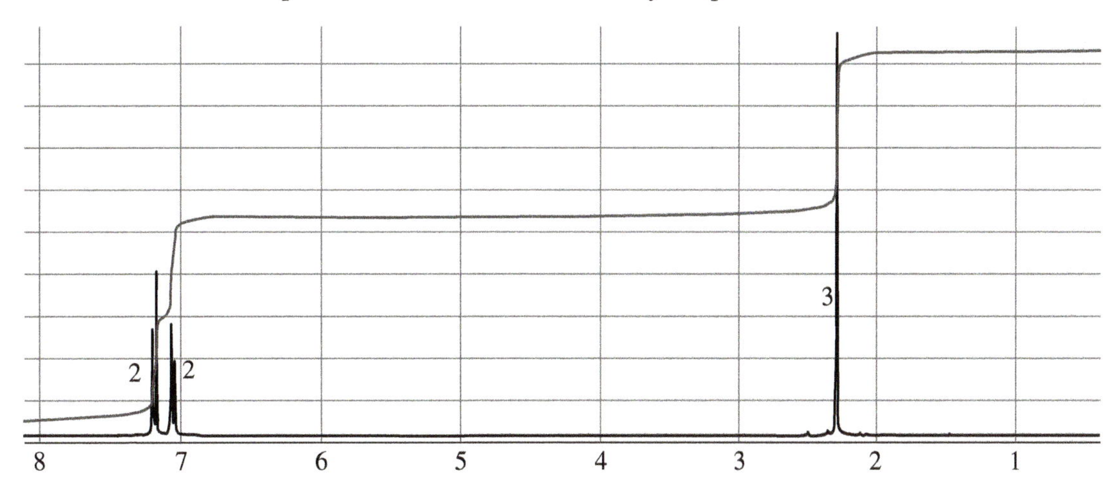

10. Identify the compound with molecular formula $C_6H_{14}O$ that gives the following IR and ^1H NMR spectra.

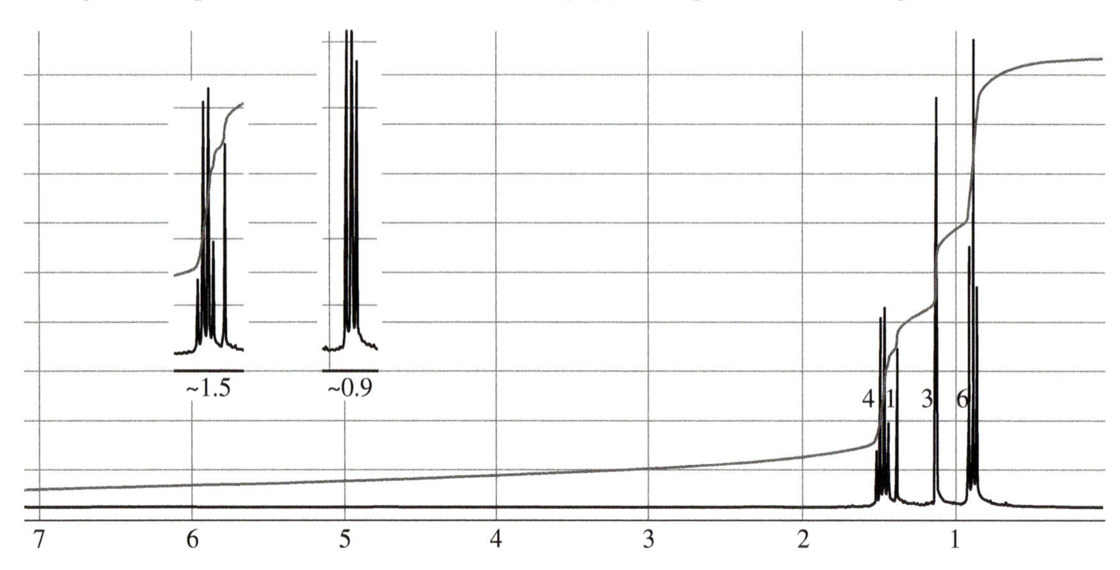

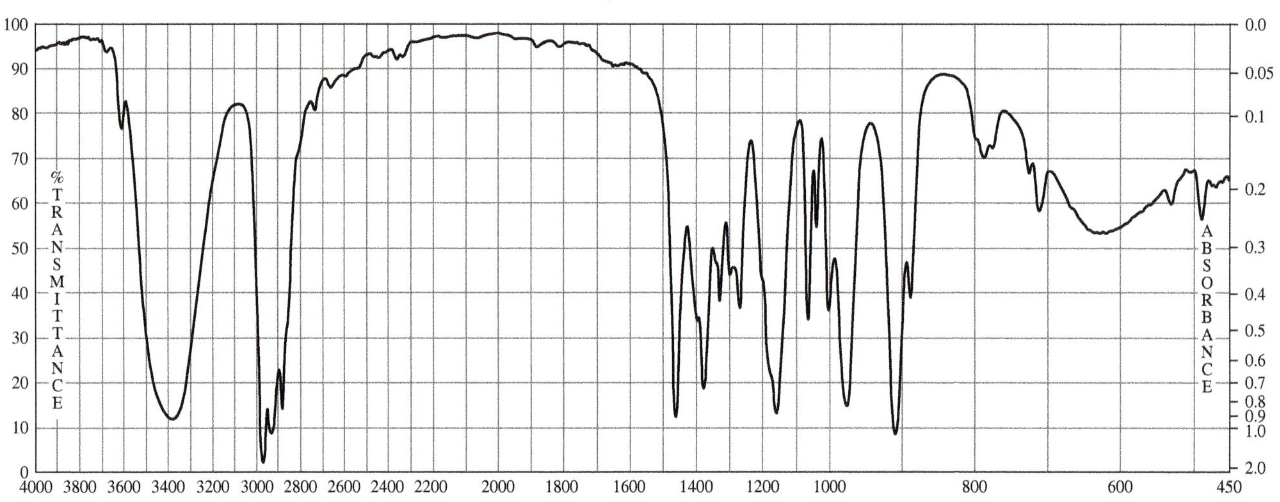

11. Identify the compound with molecular formula C_4H_9Br that gives the following ^1H NMR spectrum.

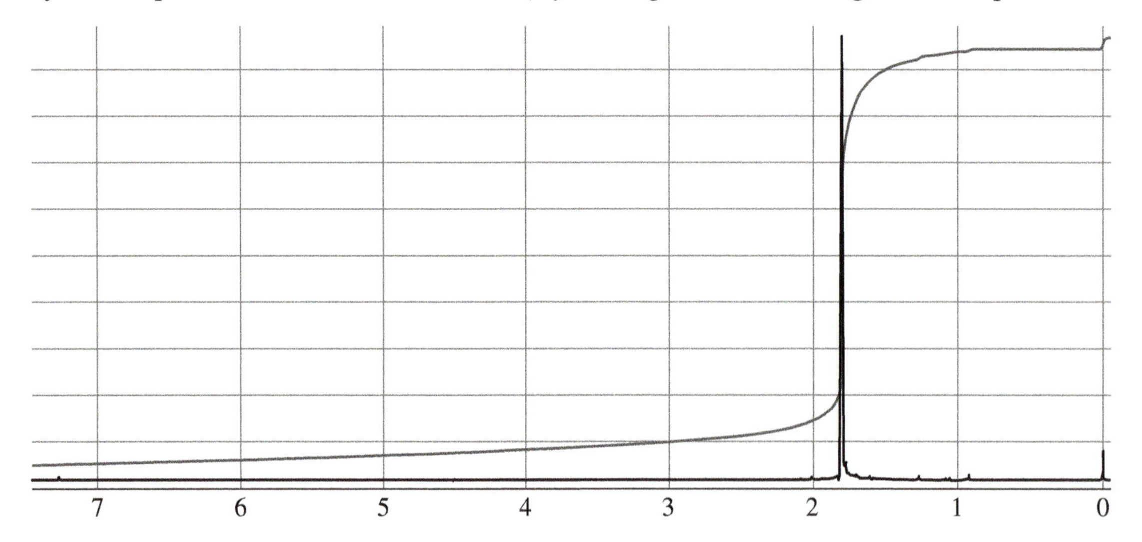

12. Identify the compound with molecular formula C_6H_{12} that gives the following IR and ^{1}H NMR spectra.

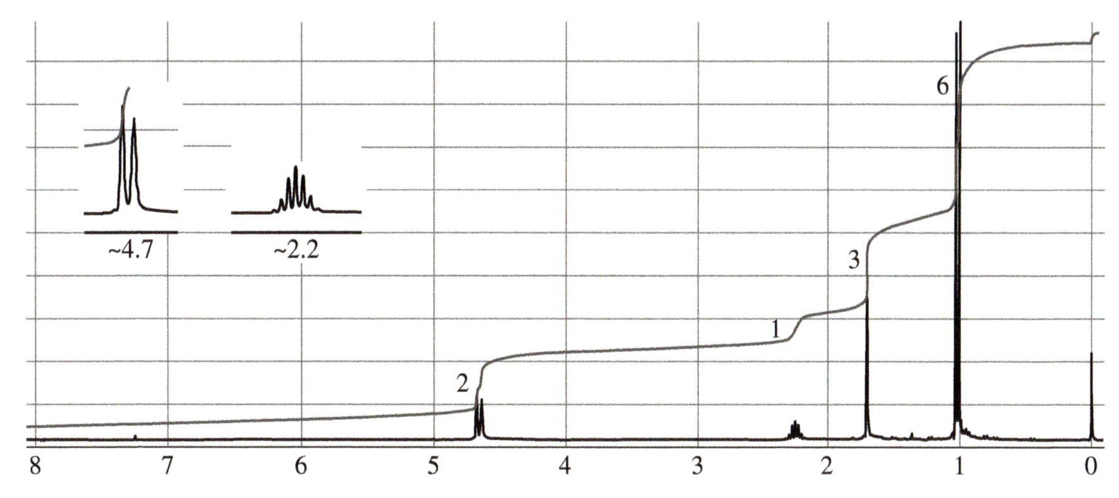

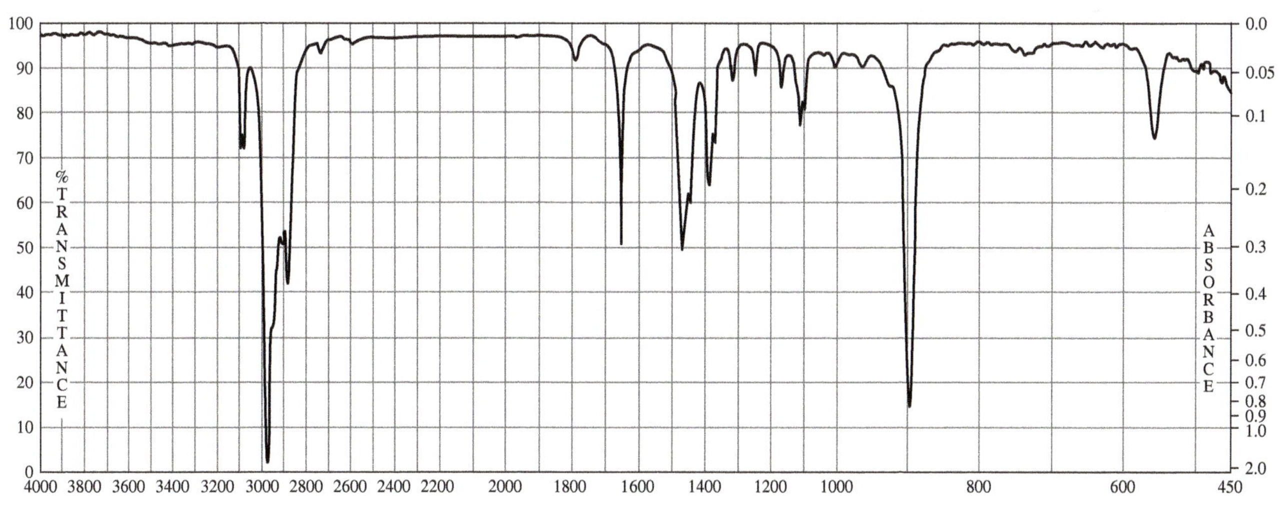

13. Identify the compound with molecular formula $C_8H_{10}O$ that gives the following ^{1}H NMR spectrum.

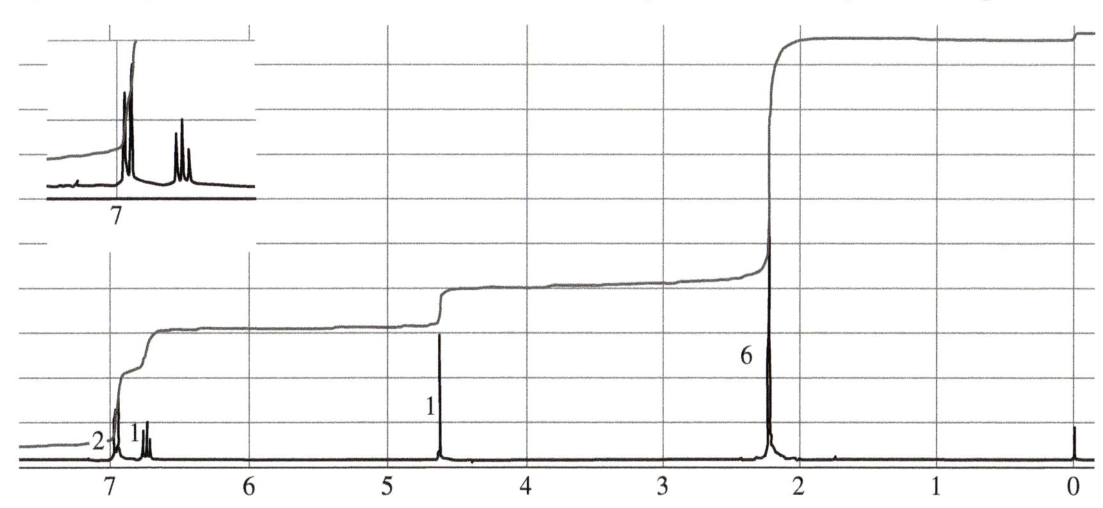

14. Identify the compound with molecular formula C_4H_7ClO that gives the following IR and ^1H NMR spectra.

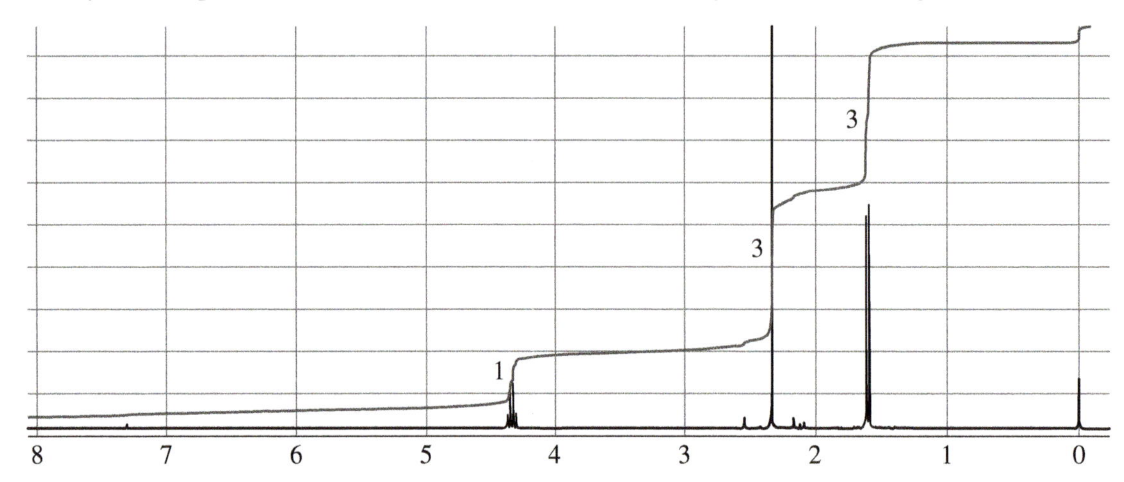

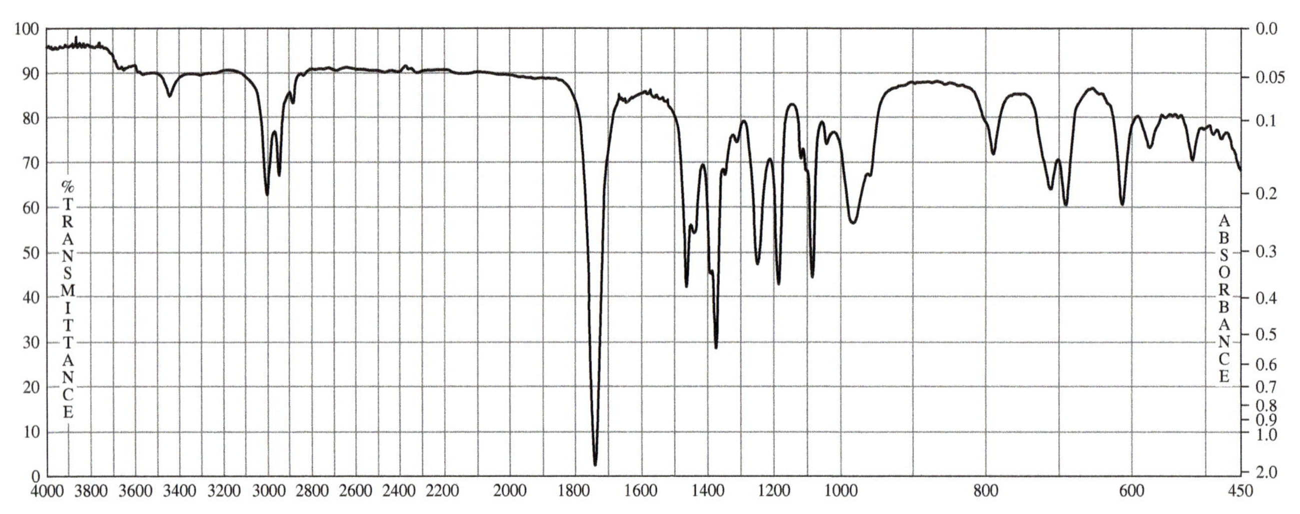

15. Identify the compound with molecular formula $C_8H_8Br_2$ that gives the following ^1H NMR spectrum.

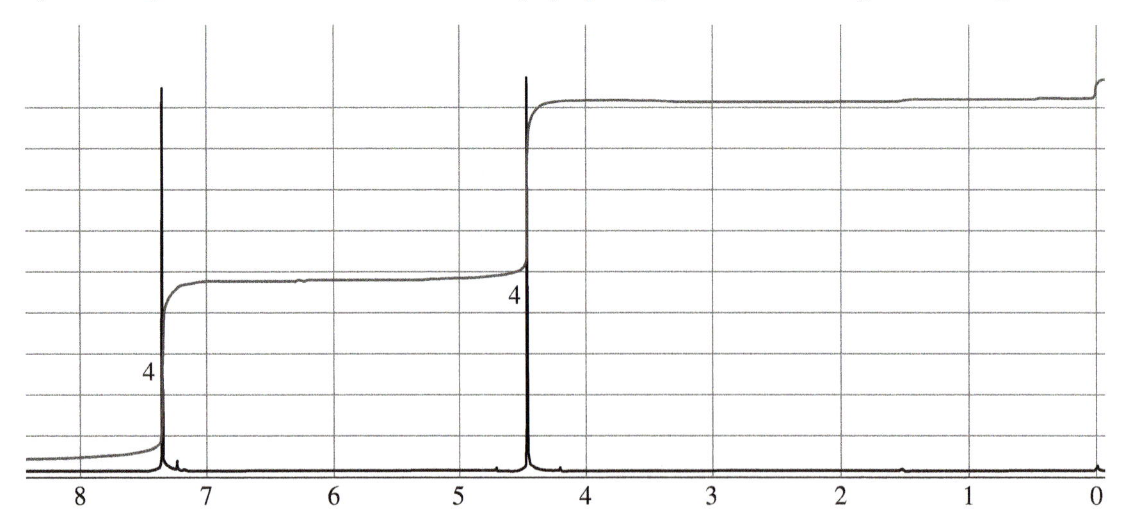

16. Identify the compound with molecular formula C_4H_6O that gives the following IR and ^1H NMR spectra.

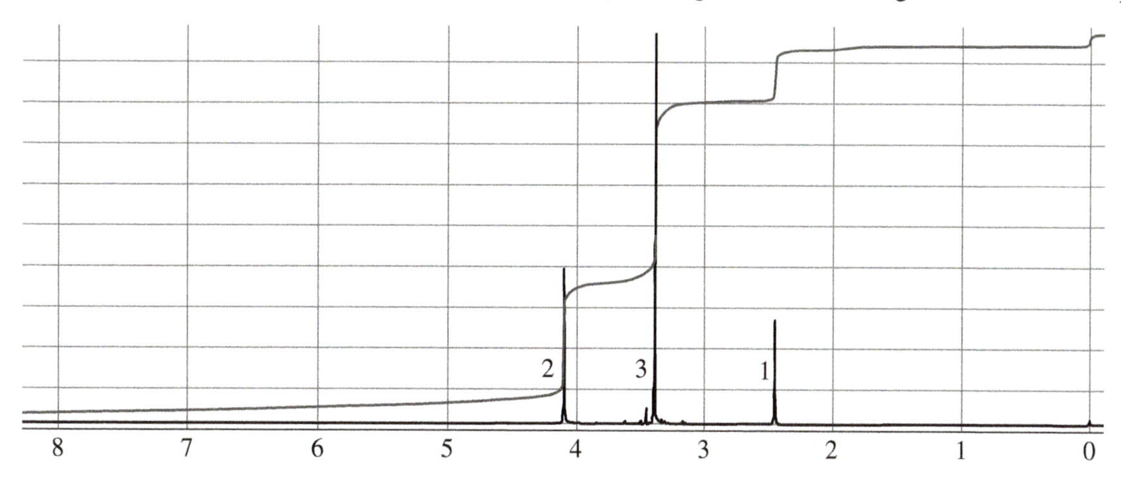

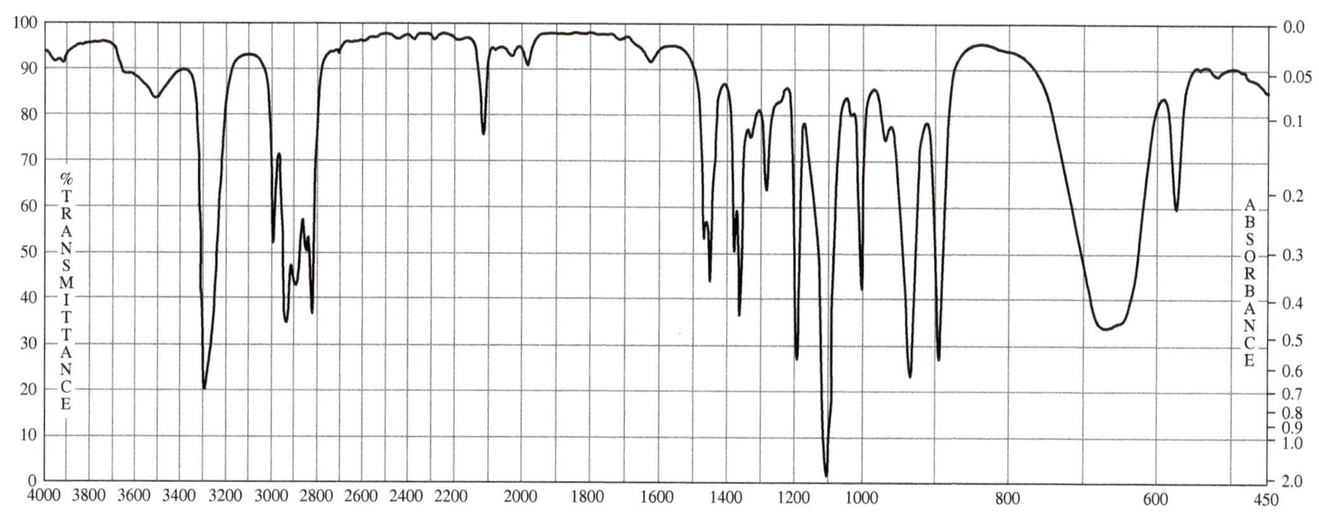

17. Identify the compound with molecular formula C_7H_8BrN that gives the following ^1H NMR spectrum.

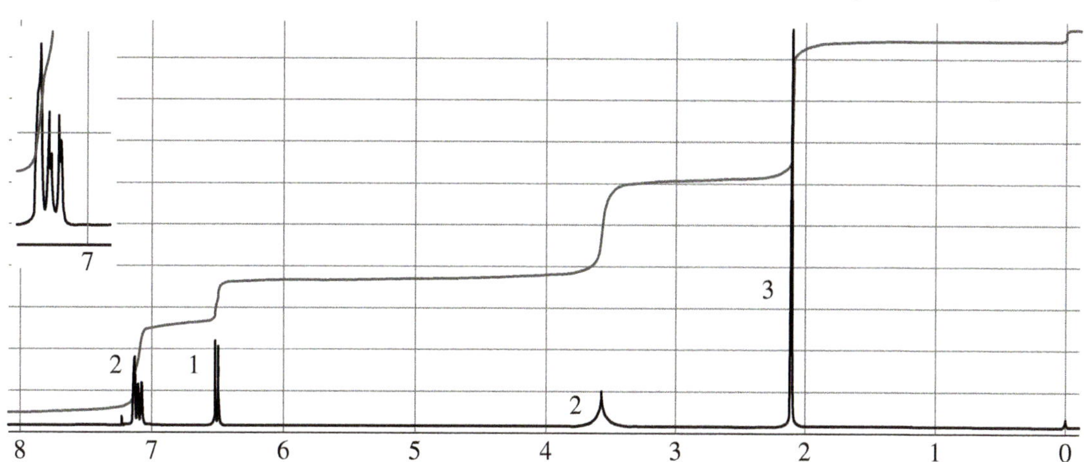

18. The ^{1}H NMR spectra of 1-chloro-3-iodopropane and 1-bromo-3-chloropropane are shown here. Which compound gives each spectrum?

a.

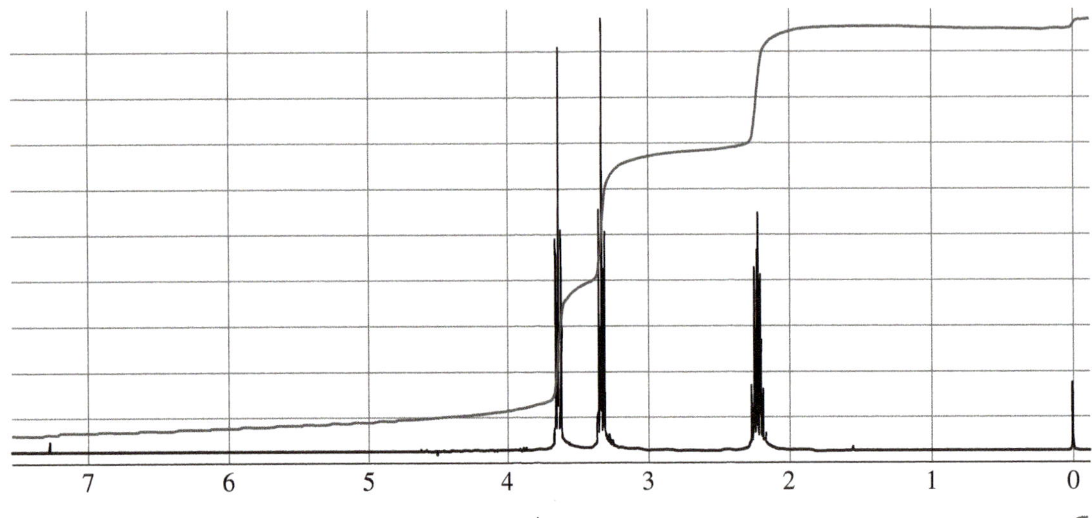

b.

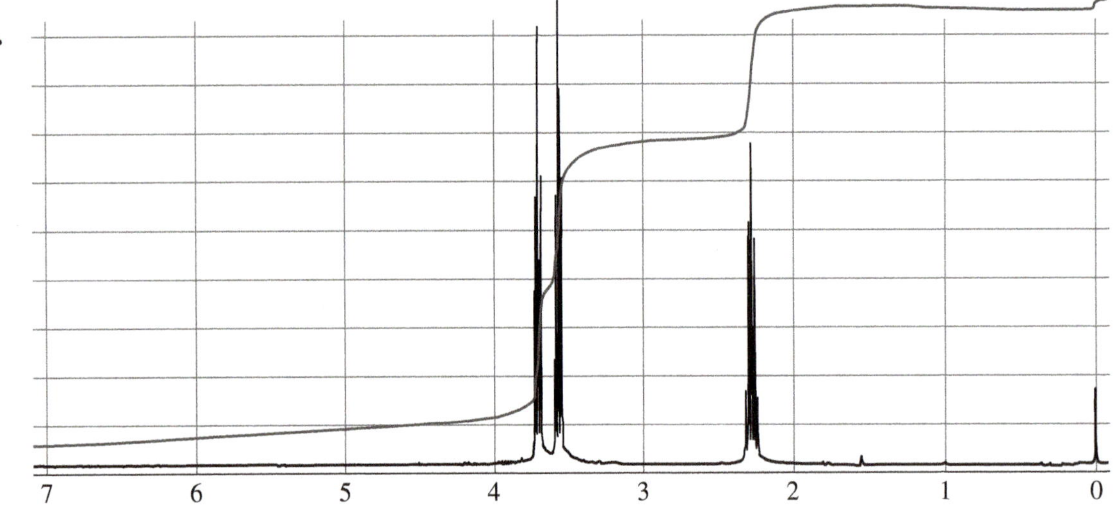

19. Identify the compound with molecular formula C_4H_8O that gives the following ^{1}H NMR spectrum.

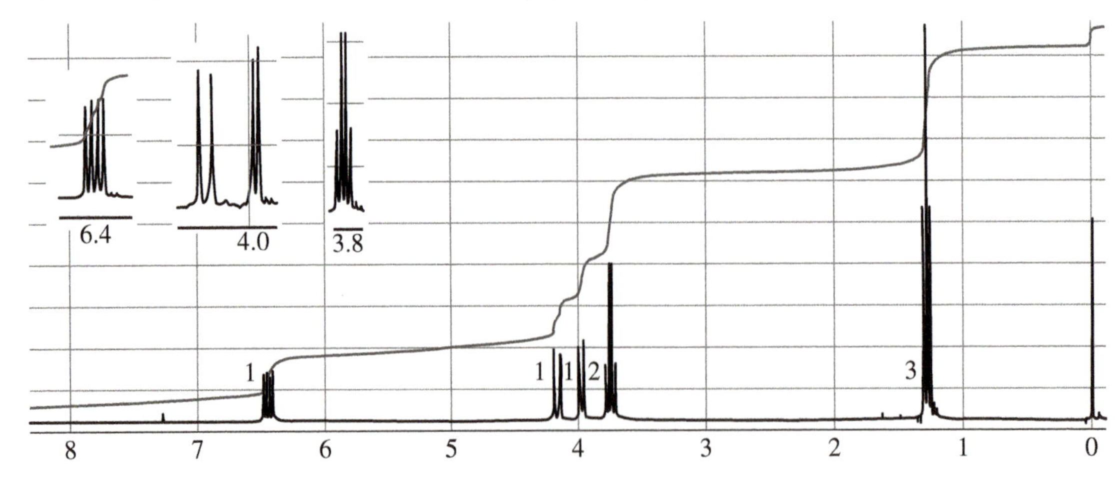

20. Identify the compound with molecular formula $C_5H_{10}O_2$ that gives the following ^{1}H NMR spectrum.

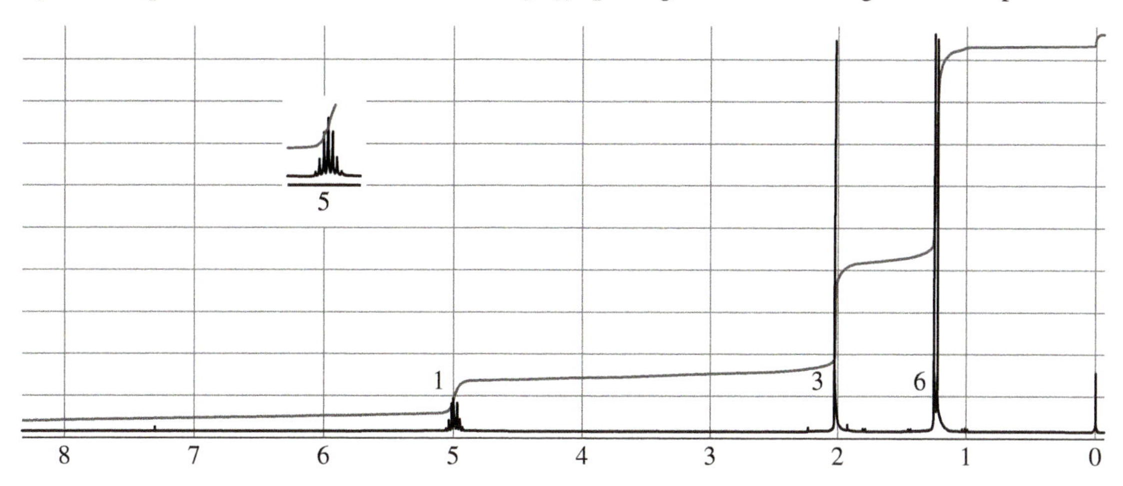

21. Identify the compound with molecular formula $C_8H_7O_2Br$ that gives the following IR and ^{1}H NMR spectra.

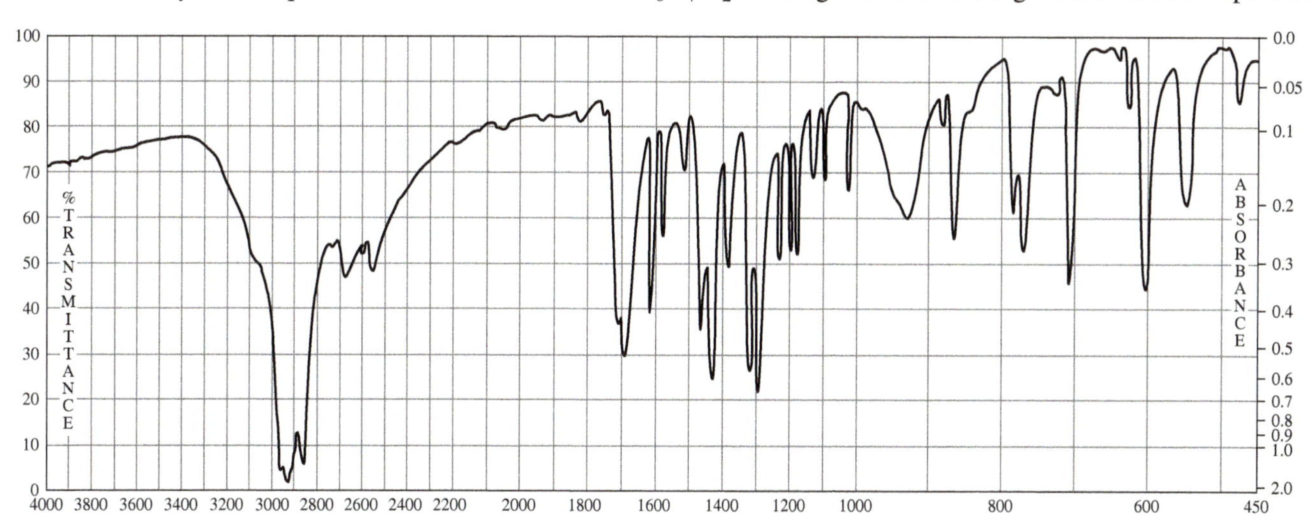

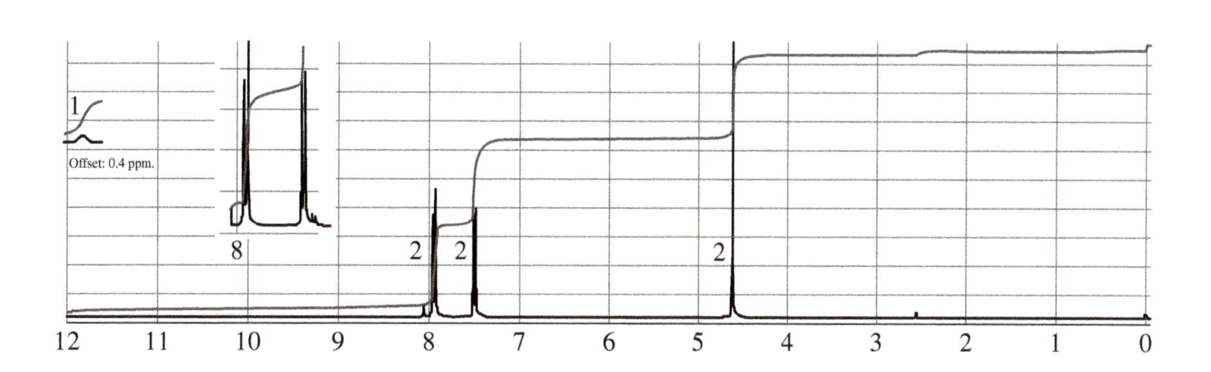

22. Identify the compound with molecular formula C_7H_6O that gives the following ^1H NMR spectrum.

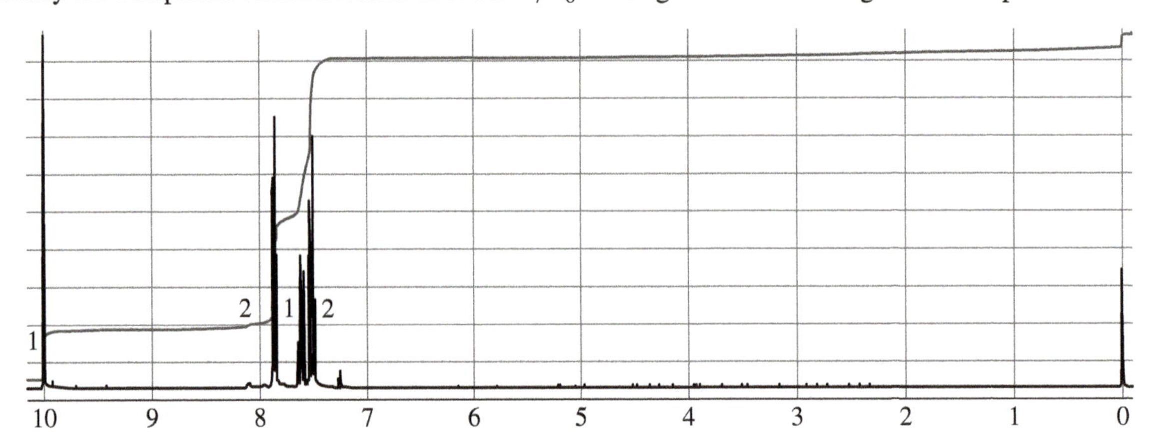

23. The two ^1H NMR spectra shown here are given by constitutional isomers with molecular formula C_3H_7Br. Identify each isomer.

a.

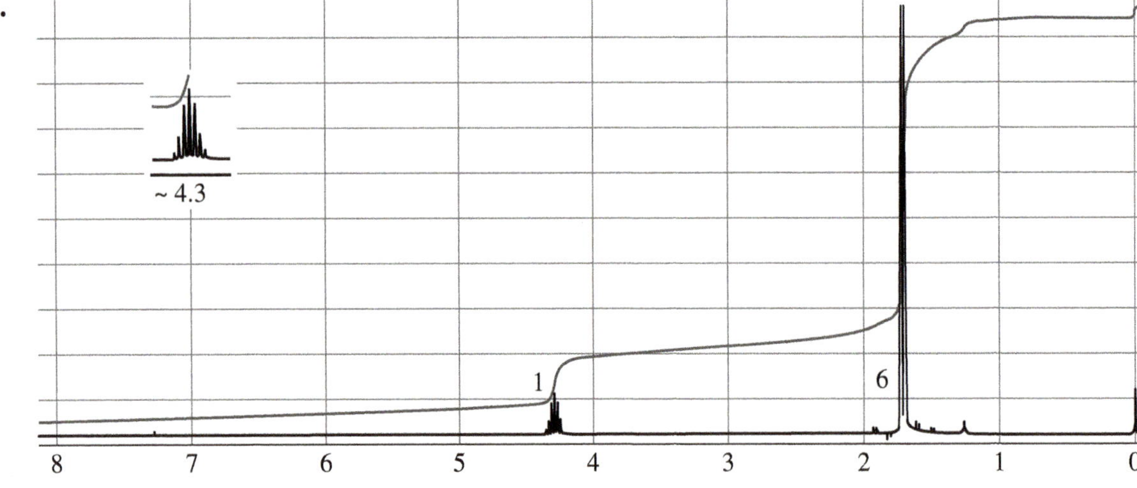

b.

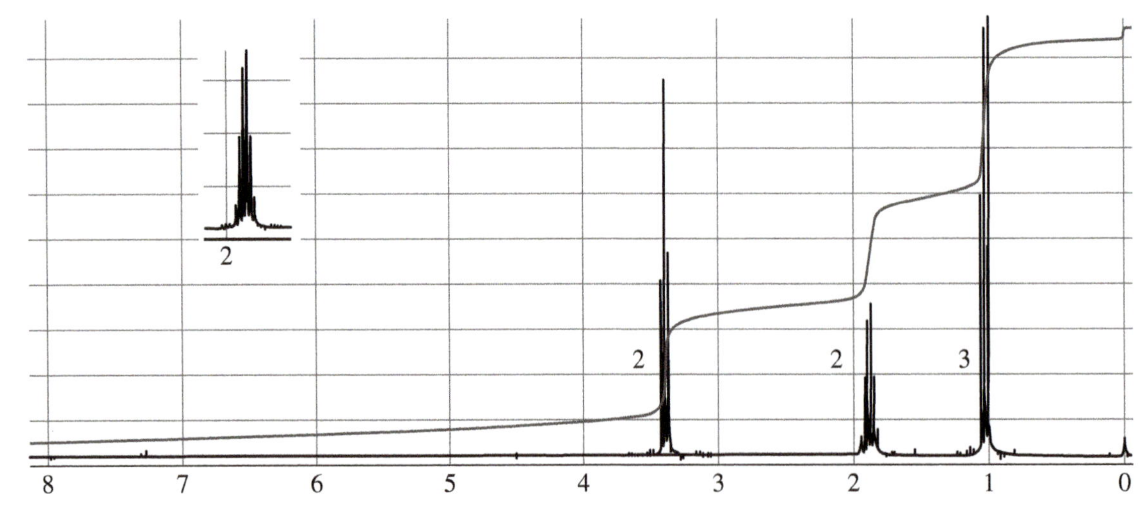

24. Chapter 6

An unknown alkene with molecular formula C_8H_{16} undergoes ozonolysis (Section 6.11 in the text). Only one product is formed. The IR and ^{1}H NMR spectra of the product are shown here. What is the product? What alkene produced this product?

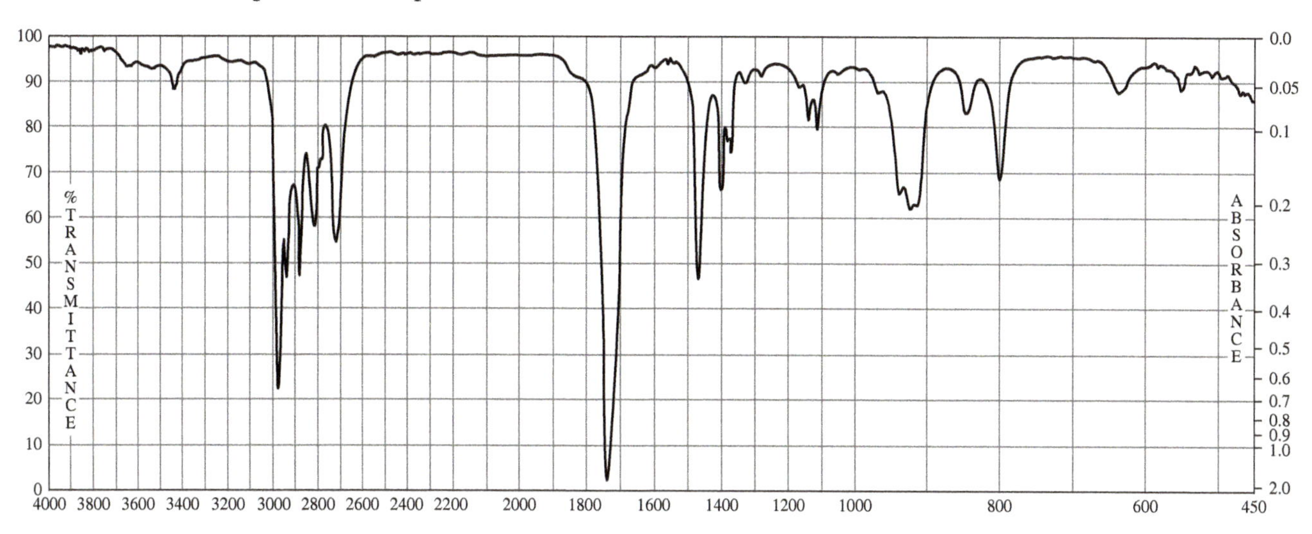

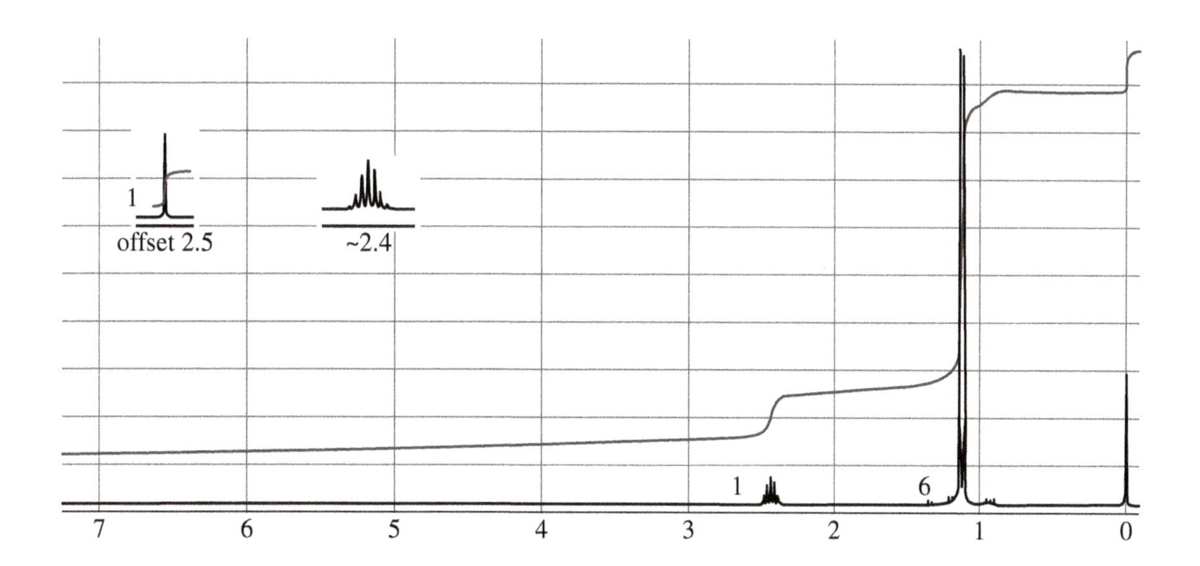

25. Identify the compound with molecular formula $C_4H_7BrO_2$ that gives the following IR and ^{1}H NMR spectra.

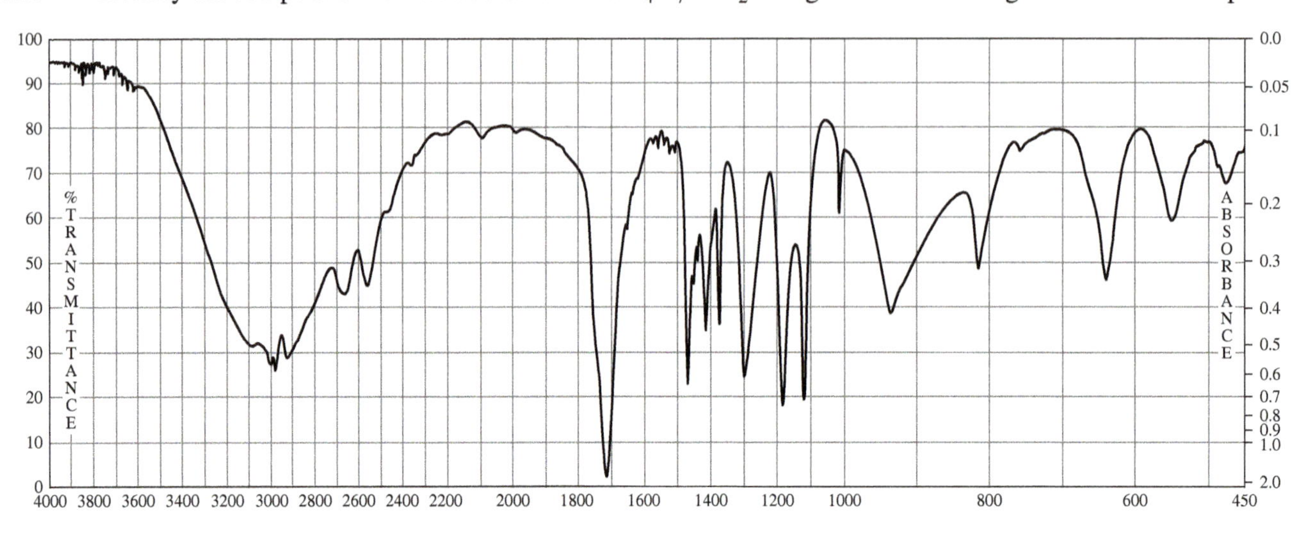

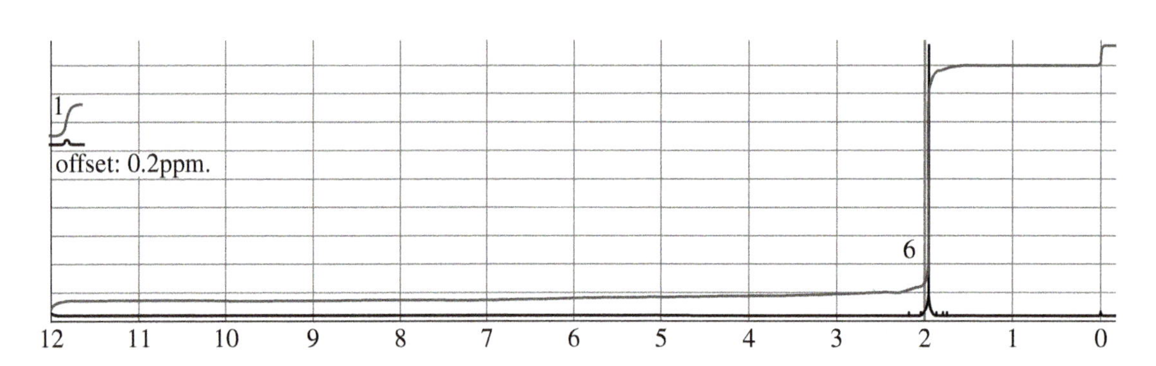

26. Identify the compound with molecular formula $C_3H_6Cl_2$ that gives the following ^{1}H NMR spectrum.

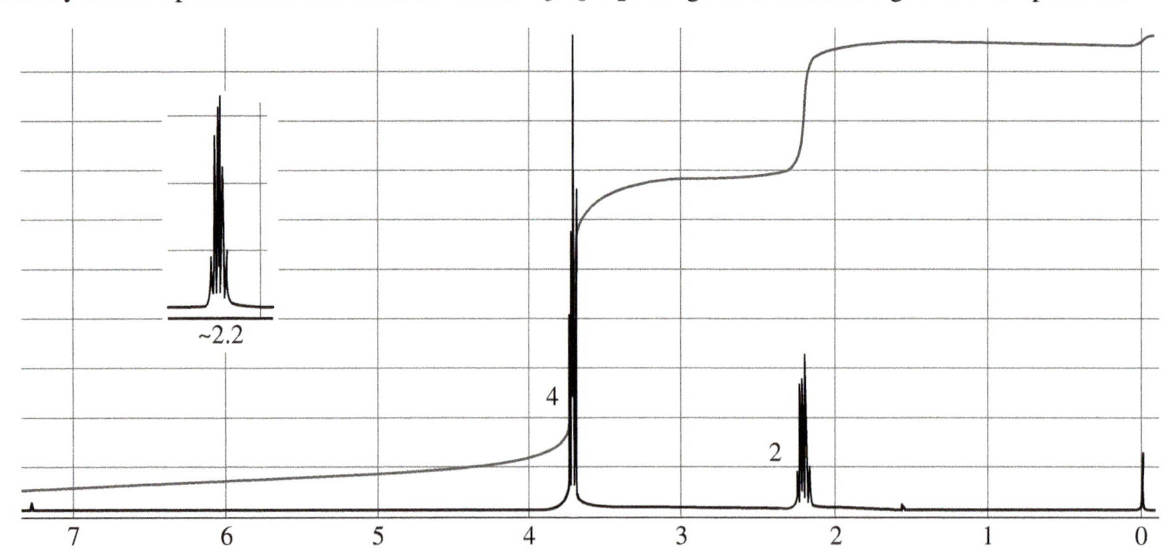

27. **Chapter 7**

A compound with the following IR spectrum was formed by a reaction with 1-propyne. If the number of carbons in the reactant and product is the same, what compound is formed and what reaction conditions produced this compound?

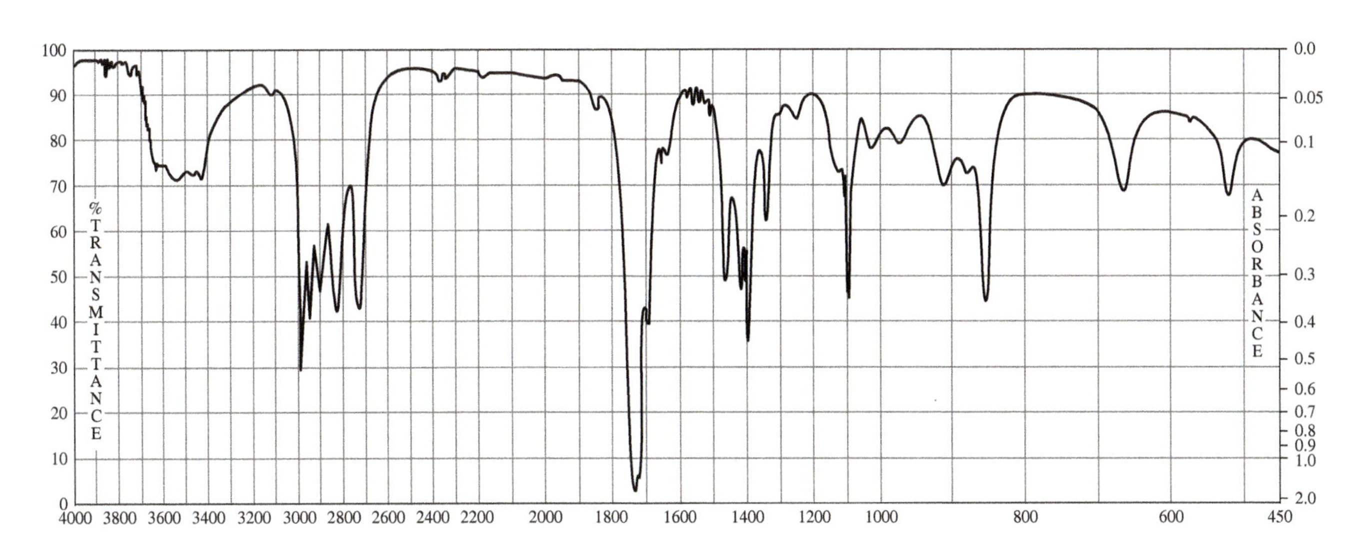

28. Identify the compound with molecular formula $C_4H_8O_2$ that gives the following ^1H and ^{13}C NMR spectra.

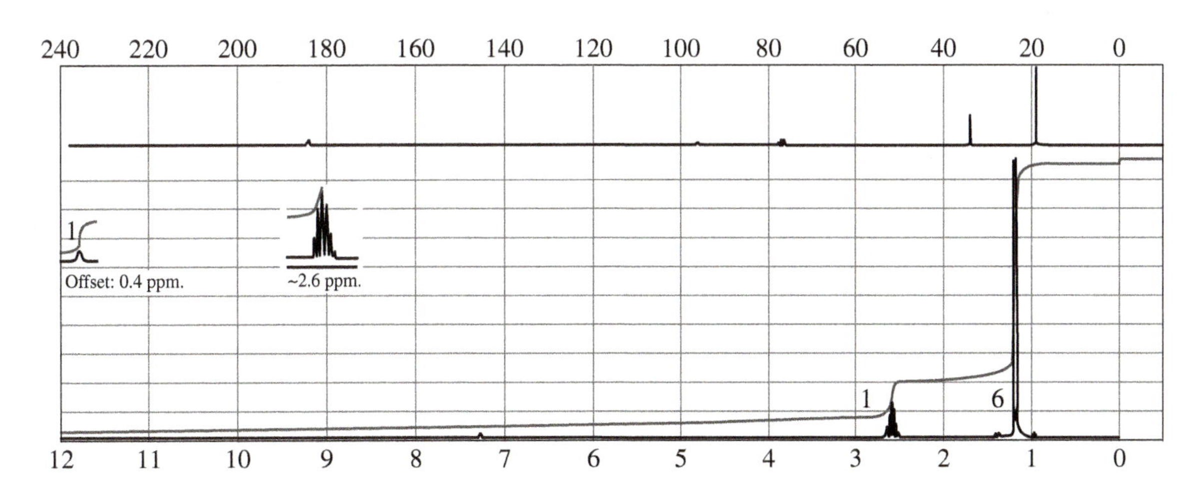

29. Identify the compound with molecular formula $C_4H_8O_2$ that gives the following ^1H NMR spectrum.

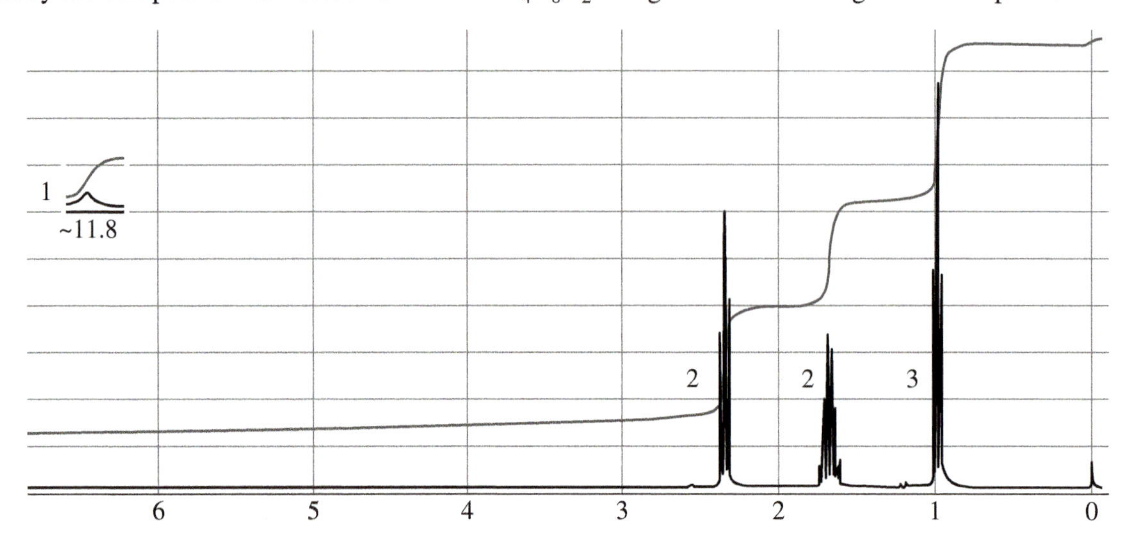

30. Identify the compound with molecular formula $C_9H_{11}NO$ that gives the following ^1H NMR spectrum.

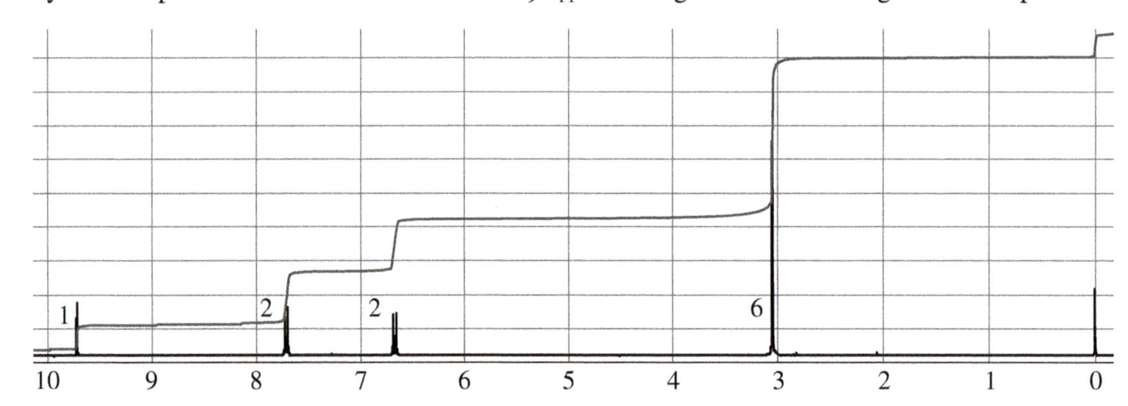

31. Identify the compound with molecular formula $C_9H_{10}O_2$ that gives the following ^1H NMR spectrum.

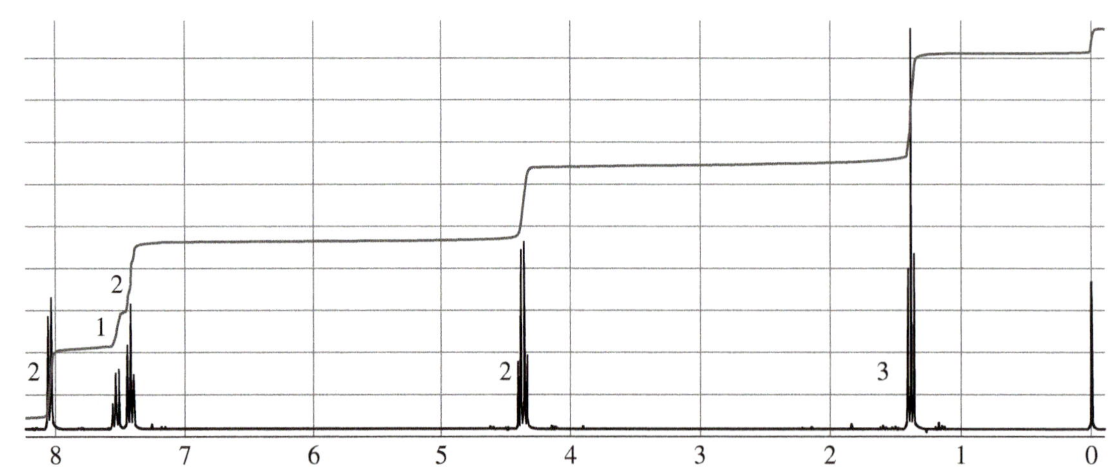

32. The ^{13}C NMR and ^{1}H NMR spectra of 1,2-, 1,3-, and 1,4-ethylmethylbenzene are shown here. Determine which spectrum belongs to which compound.

CH_3 / CH_2CH_3 CH_3 / CH_2CH_3 CH_3 / CH_2CH_3

a.

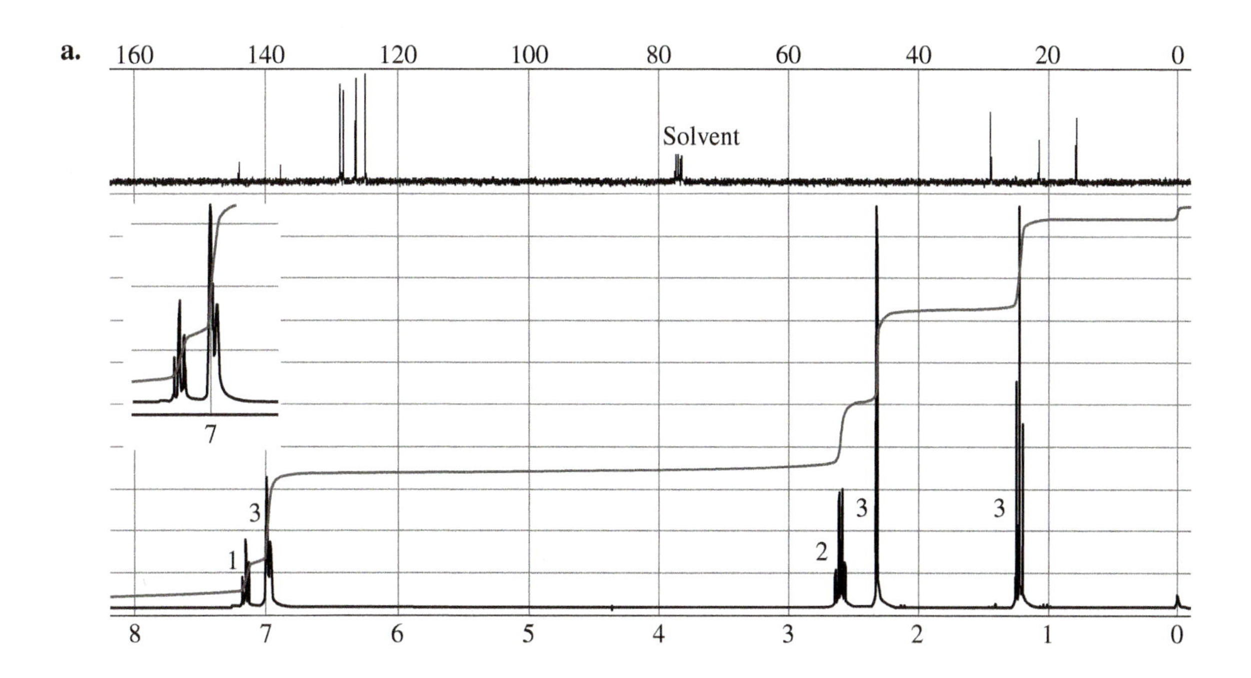

b.

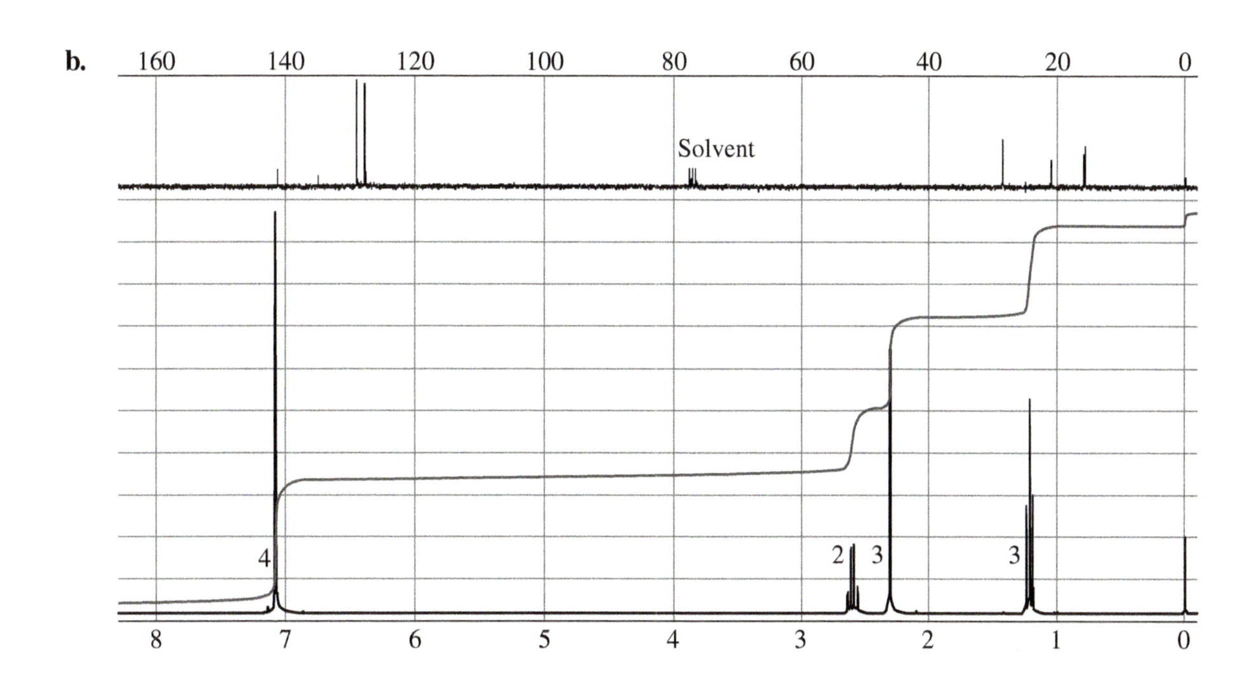

c.

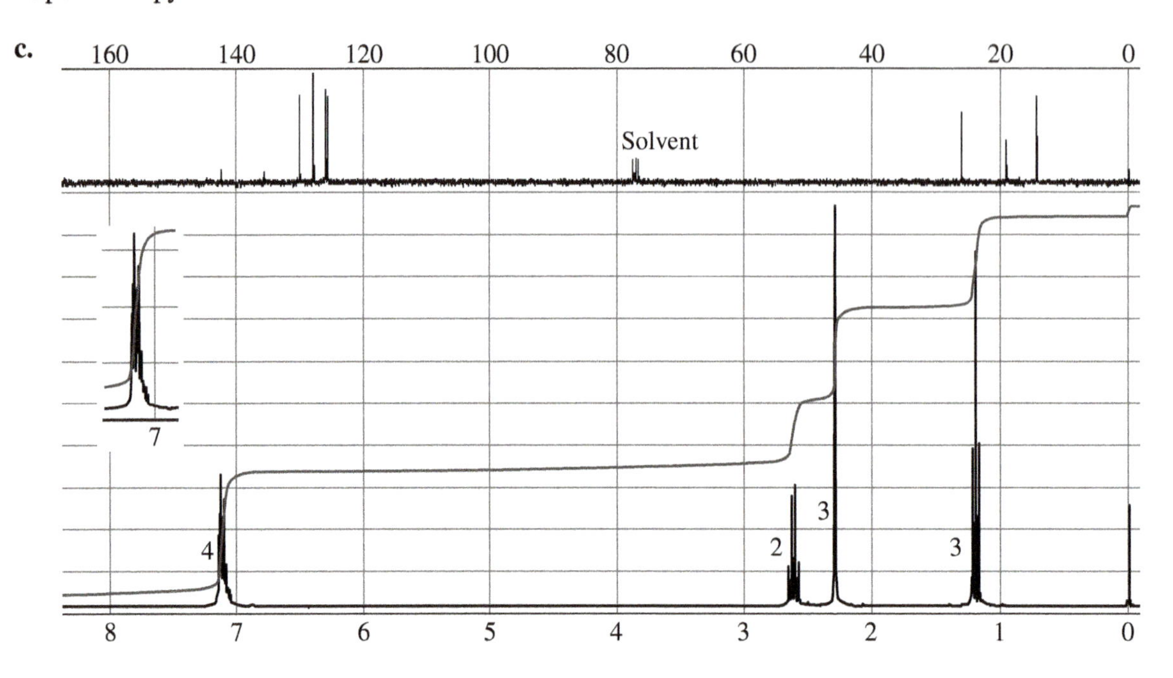

33. Identify the compound with molecular formula $C_7H_{14}O$ that gives the following ^1H NMR spectrum.

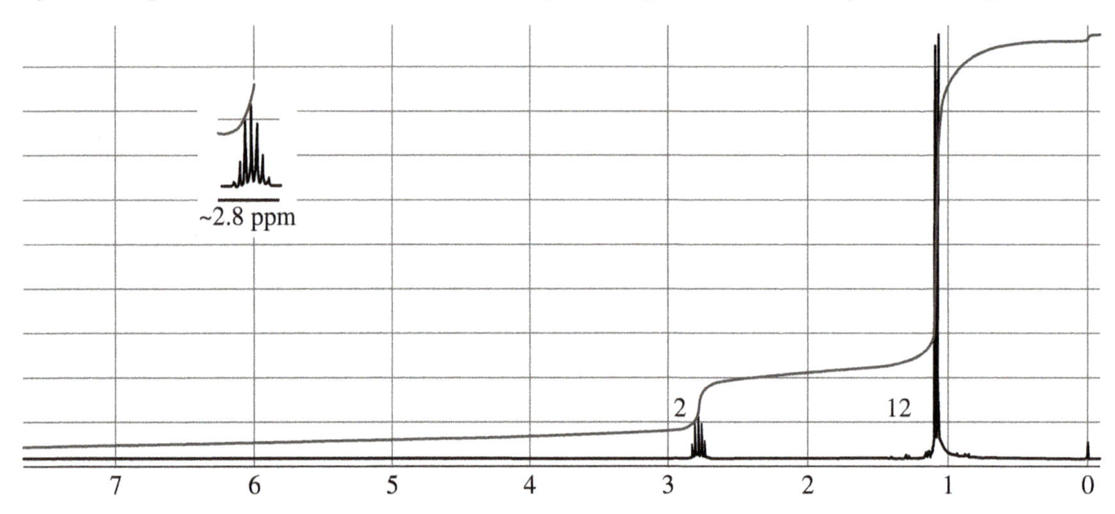

34. Identify the compound with molecular formula C_5H_{10} that gives the following ^1H NMR spectrum.

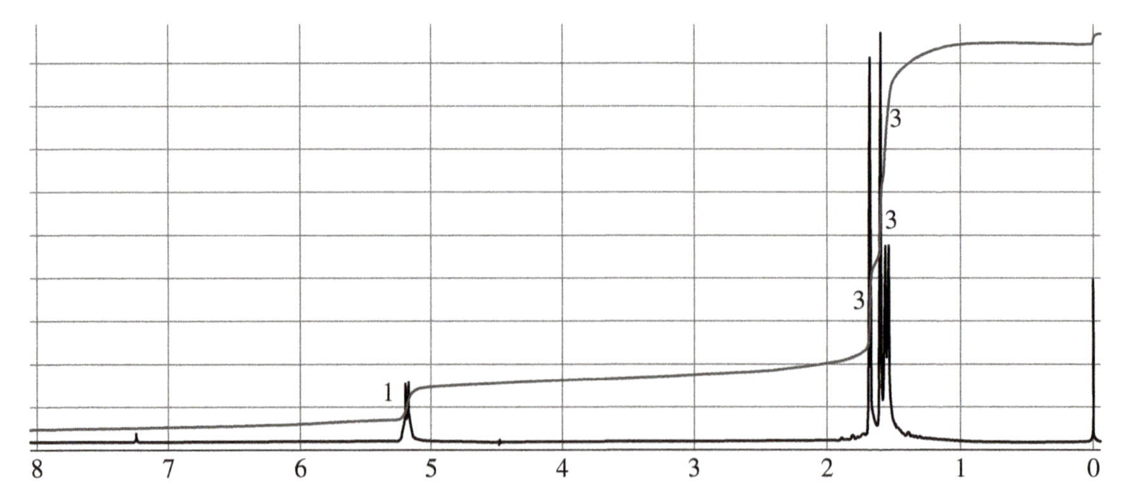

35. Identify the compound with molecular formula C_3H_4O that gives the following IR and ^1H NMR spectra.

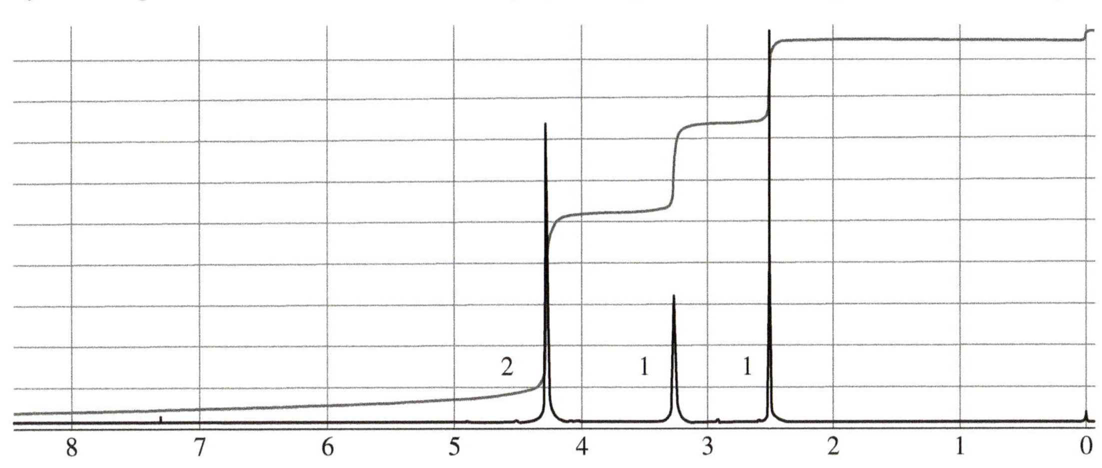

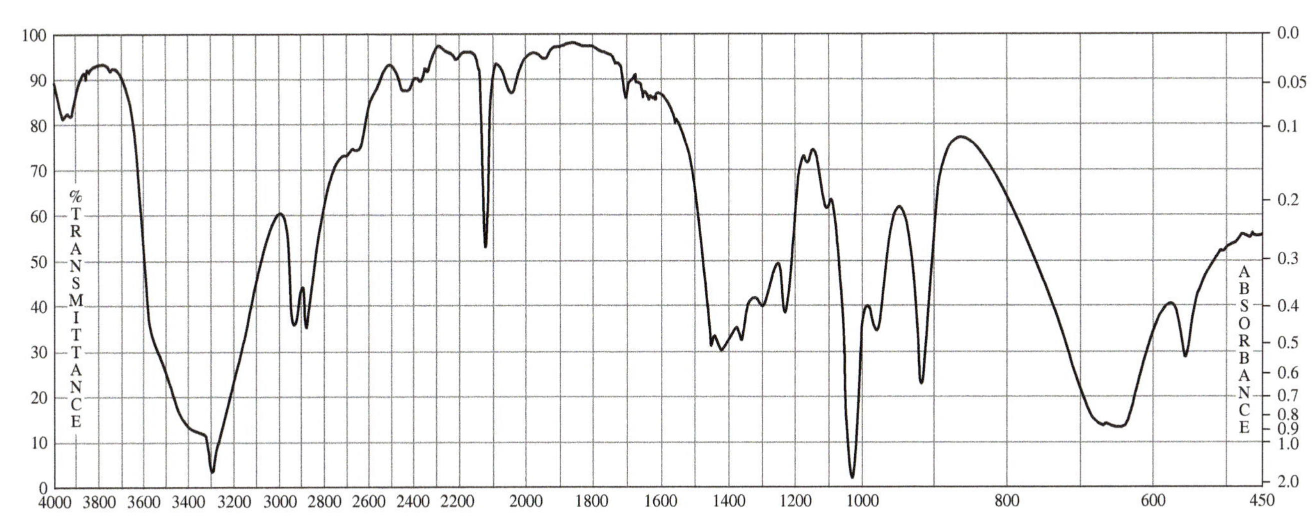

36. Identify the compound with molecular formula C_4H_7Cl that gives the following ^1H NMR spectrum.

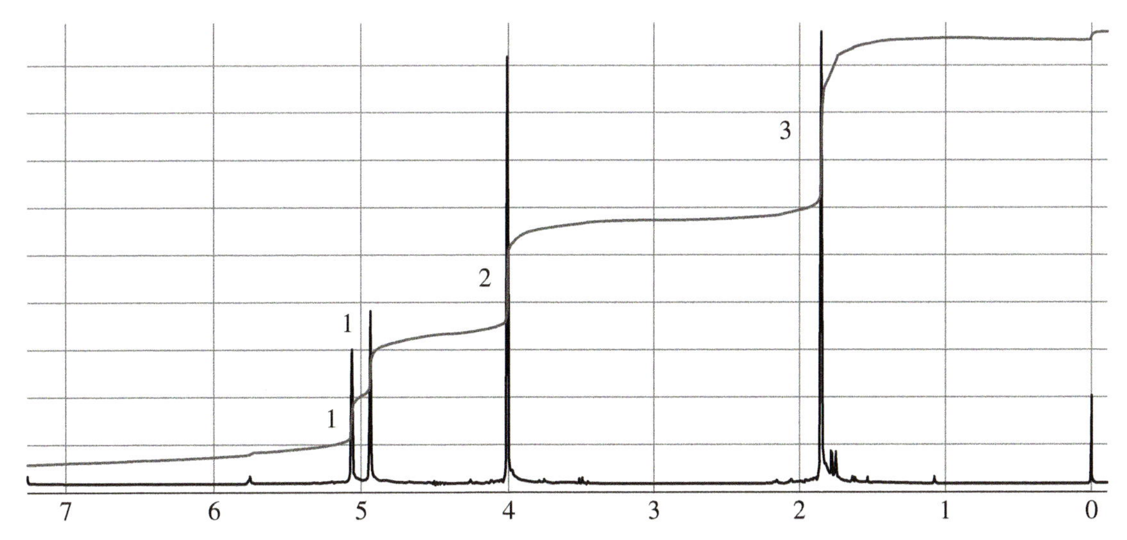

37. Identify the alcohol that gives the following ^{1}H NMR spectrum.

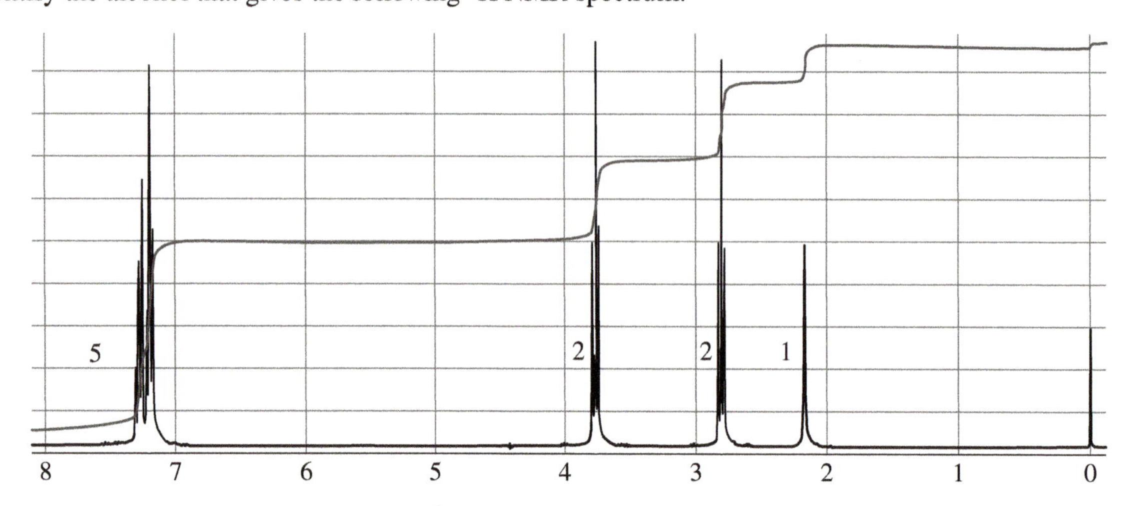

38. An unknown alcohol gives the following ^{1}H NMR spectrum. Identify the alcohol. (HINT: Because this is a spectrum of a pure alcohol, the OH proton is split by adjacent protons.)

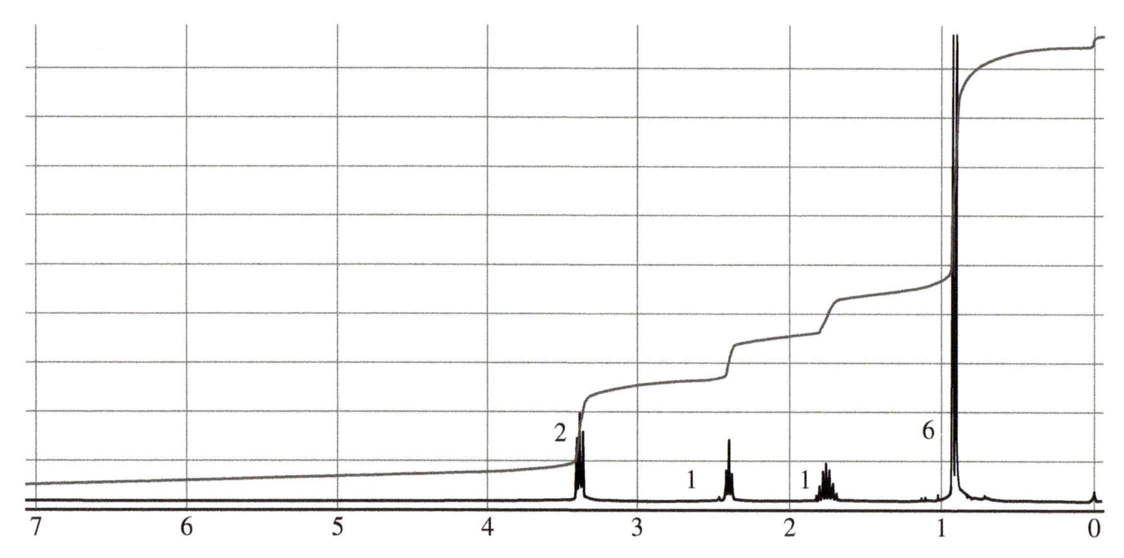

39. Identify the compound with molecular formula $C_5H_{12}O_2$ that gives the following ^{1}H NMR spectrum.

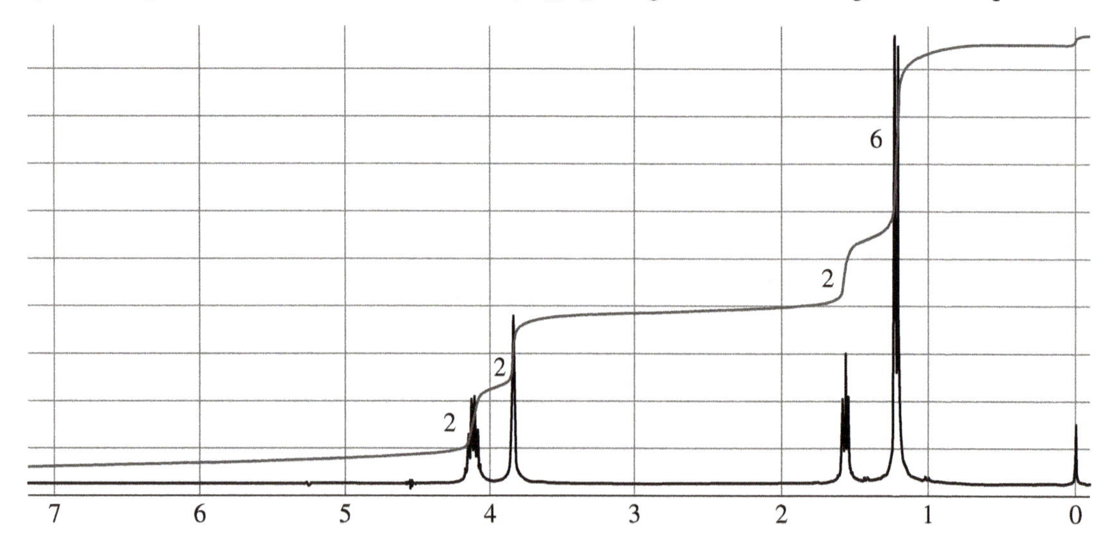

40. Identify the compound with molecular formula C_3H_7NO that gives the following IR and ^1H NMR spectra.

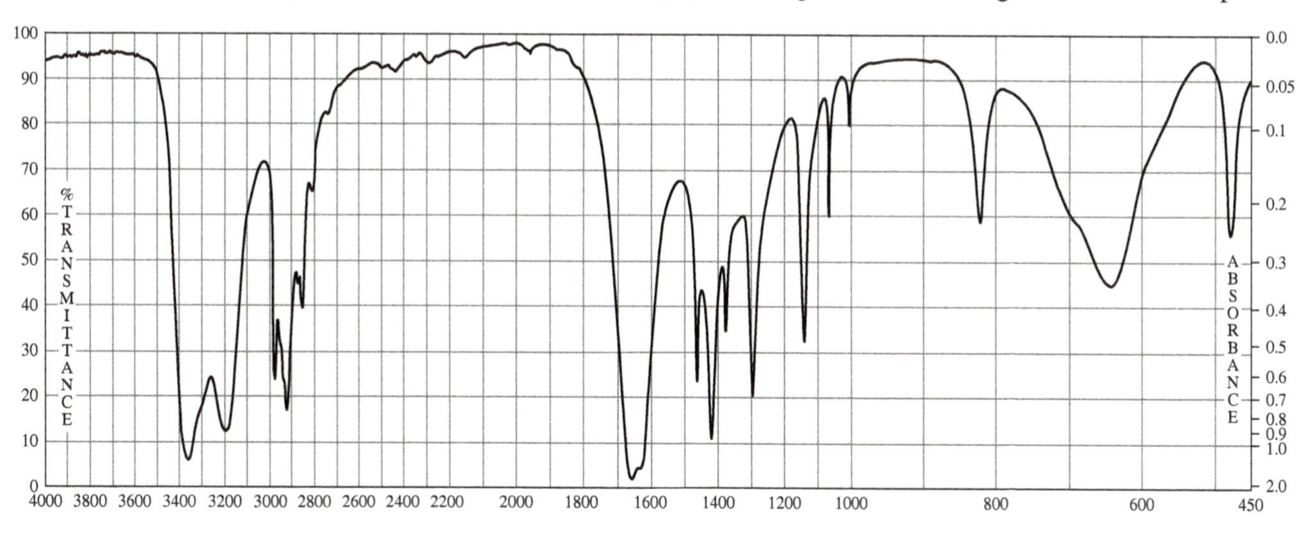

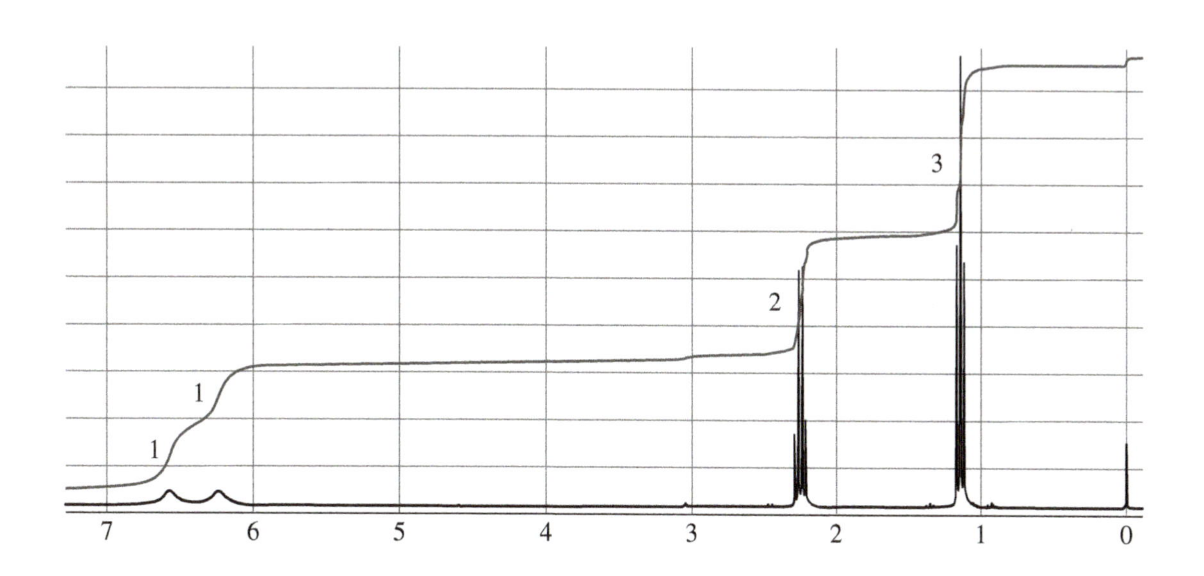

41. The ^{1}H NMR spectra shown here are given by constitutional isomers of propylamine (C_3H_9N). Identify the isomer that gives each spectrum.

a.

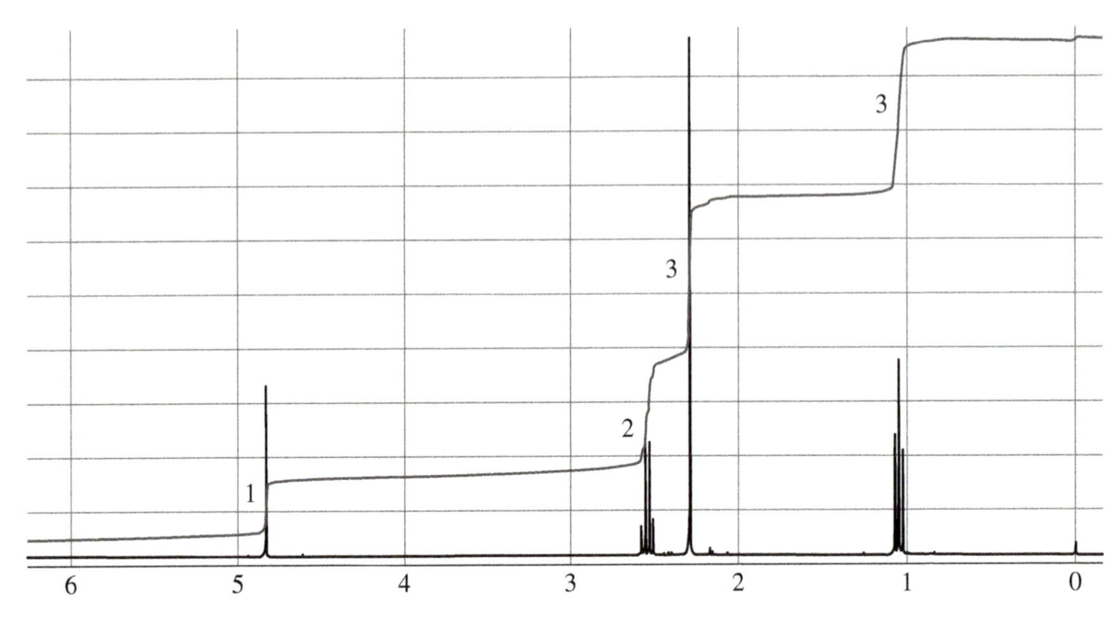

b.

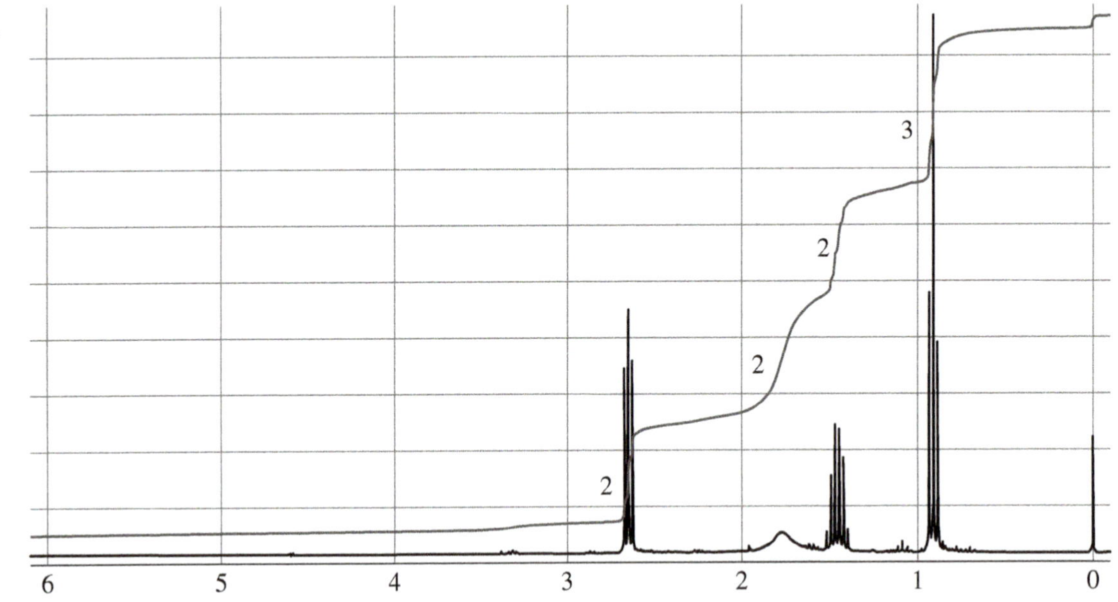

42. Identify the compound with molecular formula C_3H_7N that gives the following IR and ^1H NMR spectra.

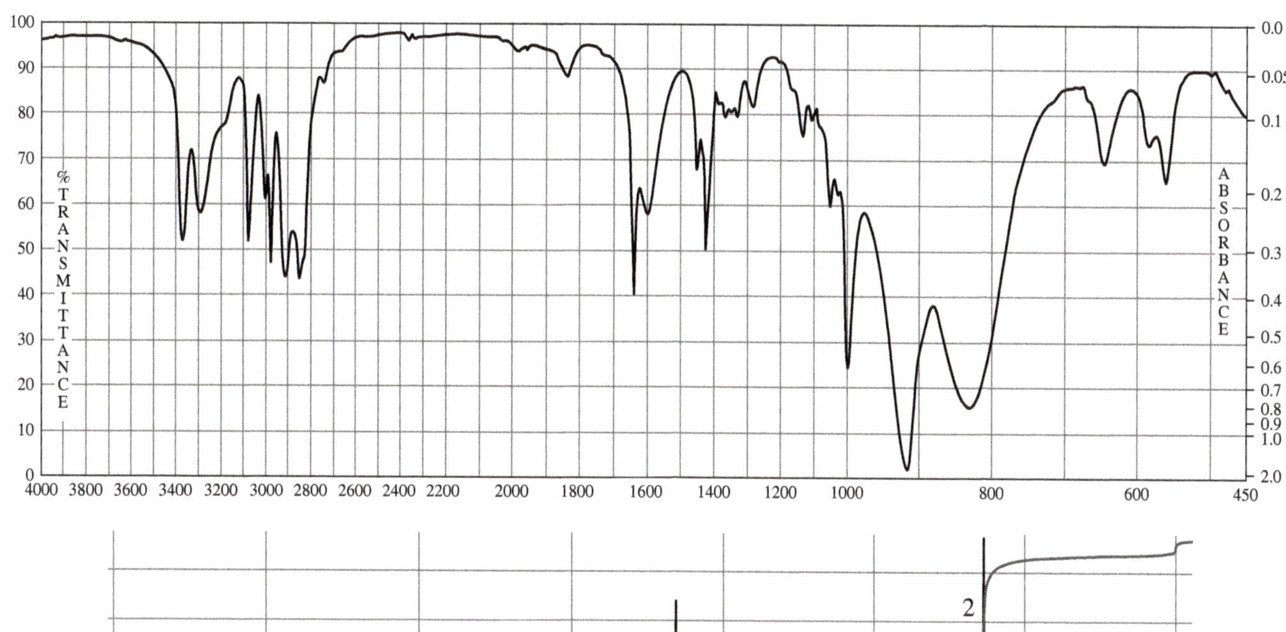

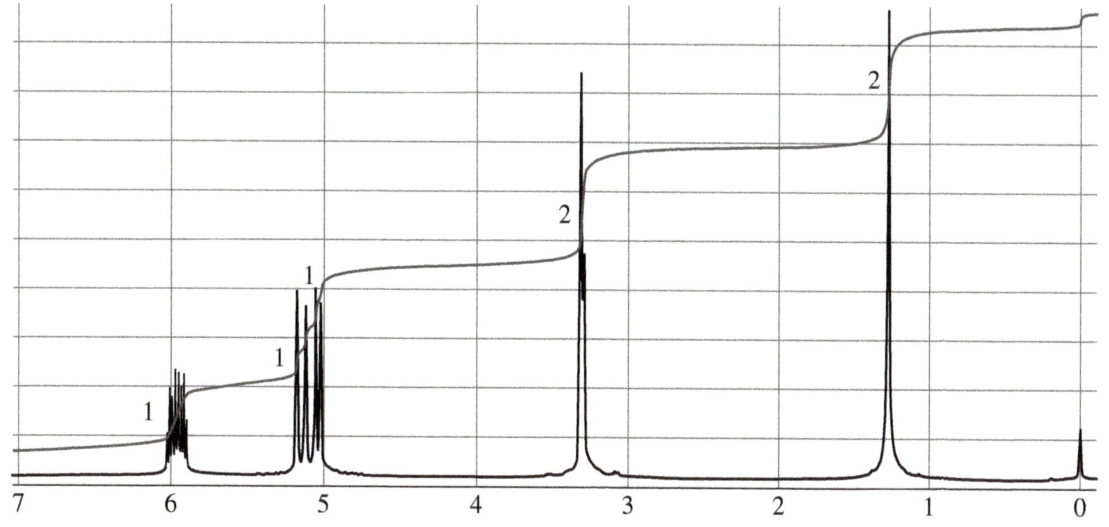

43. Identify the compound with molecular formula C_4H_9BrO that gives the following ^1H NMR spectrum.

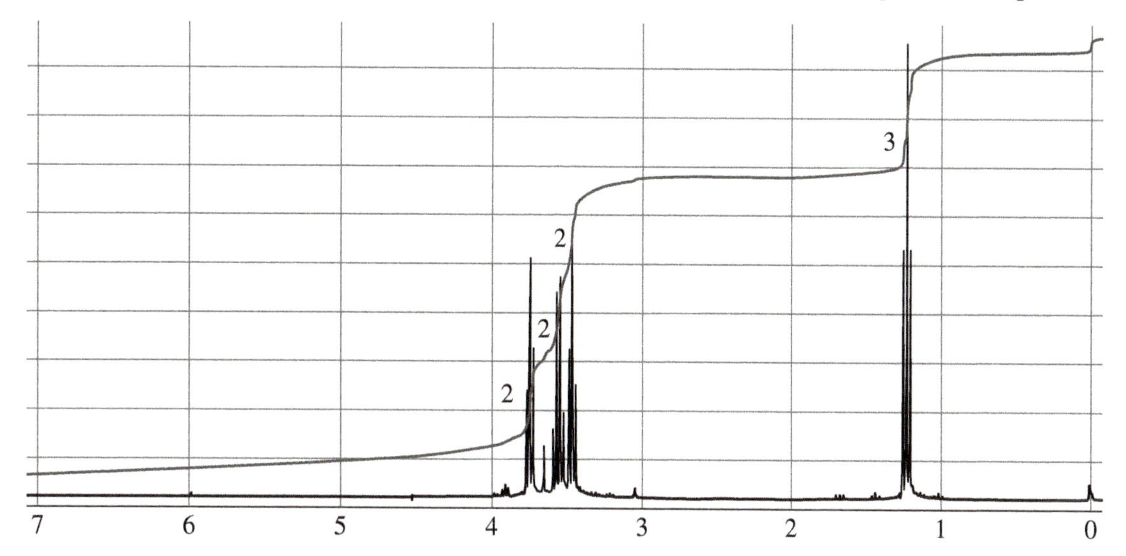

44. Identify the compound with molecular formula $C_7H_8O_2$ that gives the following IR and ^{1}H NMR spectra.

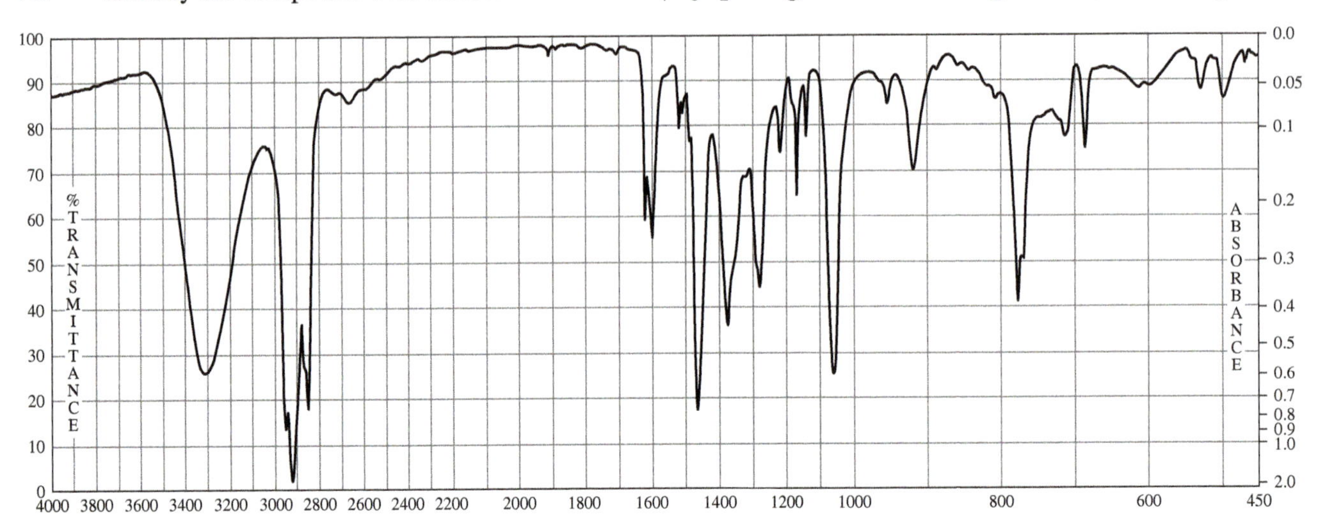

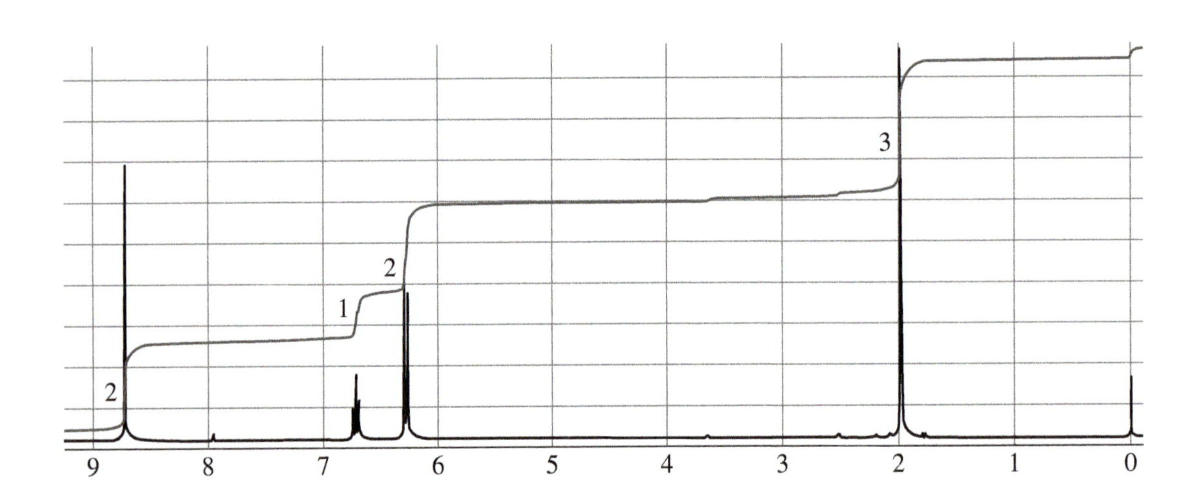

45. Identify the alcohol that gives the following ^{1}H NMR spectrum.

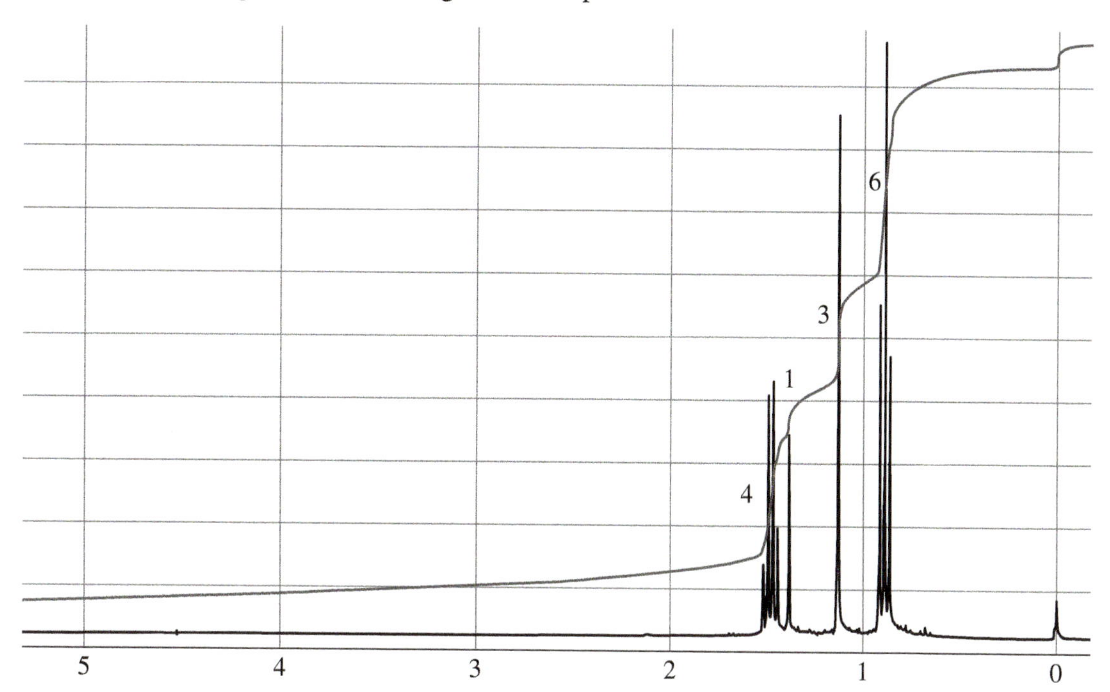

46. Identify the compound with molecular formula $C_6H_{12}O_2$ that gives the following ^{1}H NMR data. The number of hydrogens responsible for each signal is given in parentheses.

1.1 ppm (6H) doublet 2.2 ppm (2H) quartet

1.7 ppm (3H) triplet ~5 ppm (1H) septet

47. Identify the compound with molecular formula $C_9H_{10}O$ that gives the following ^{1}H NMR data. The number of hydrogens responsible for each signal is given in parentheses.

1.4 ppm (2H) multiplet 3.8 ppm (2H) triplet

2.5 ppm (2H) triplet 6.9–7.8 ppm (4H) multiplet

48. Identify the compound with molecular formula $C_{11}H_{22}O$ that gives the following ^{1}H NMR data.

1.1 ppm (18H) singlet 2.2 ppm (4H) singlet

49. Identify the compound with molecular formula C_4H_8O that gives the following ^{1}H NMR data.

1.6 ppm (4H) multiplet 3.8 ppm (4H) triplet

50. Identify the compound with molecular formula $C_{15}H_{14}O_3$ that gives the following ^{1}H NMR data.

3.8 ppm (6H) singlet 7.7 ppm (4H) doublet 7.3 ppm (4H) doublet

51. Propose structures for isomers with molecular formula C_3H_9NO that give the ^{1}H NMR spectra shown below.

A.

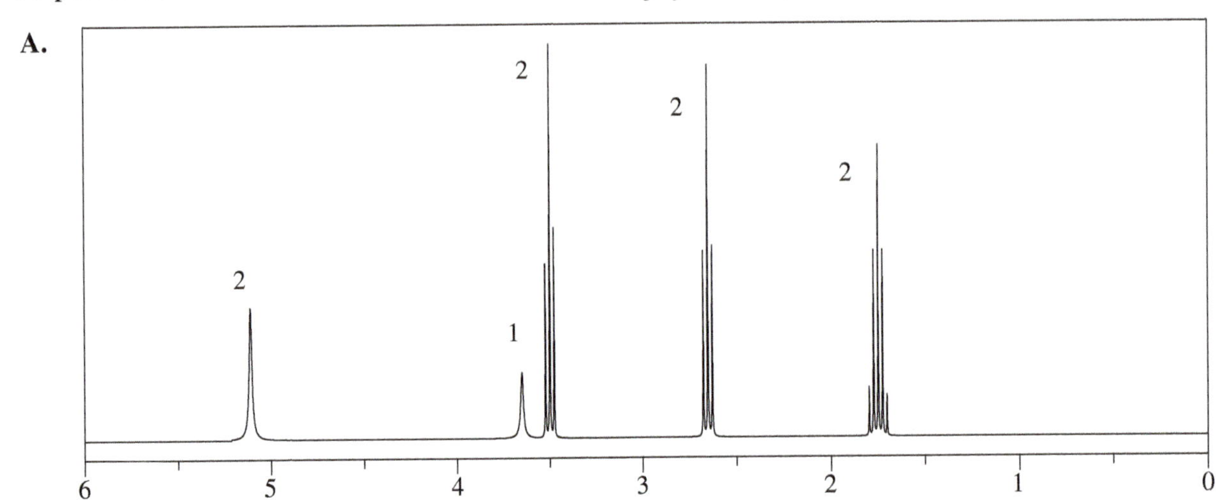

B.

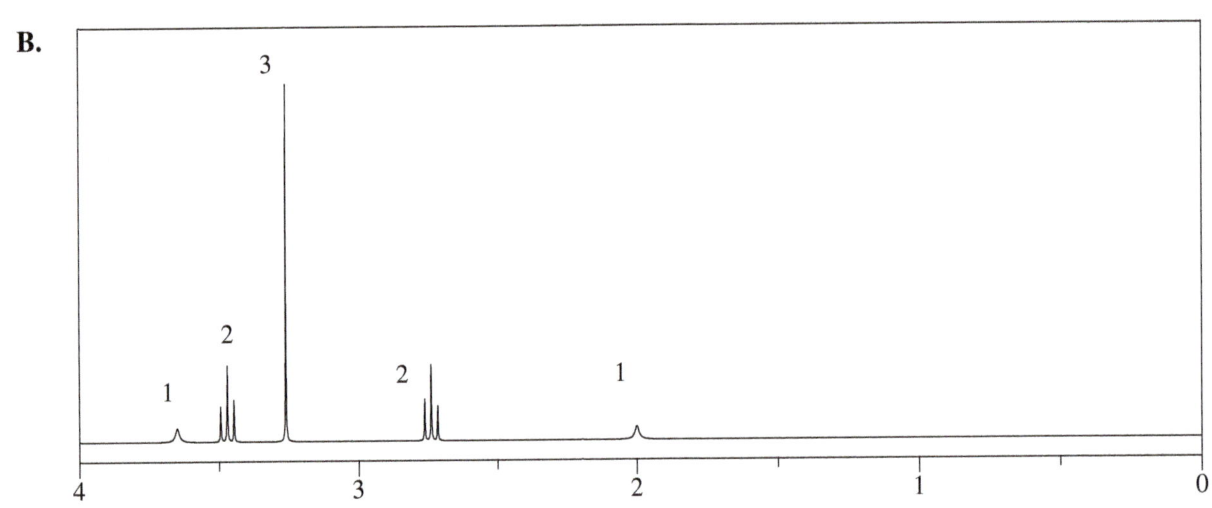

C.

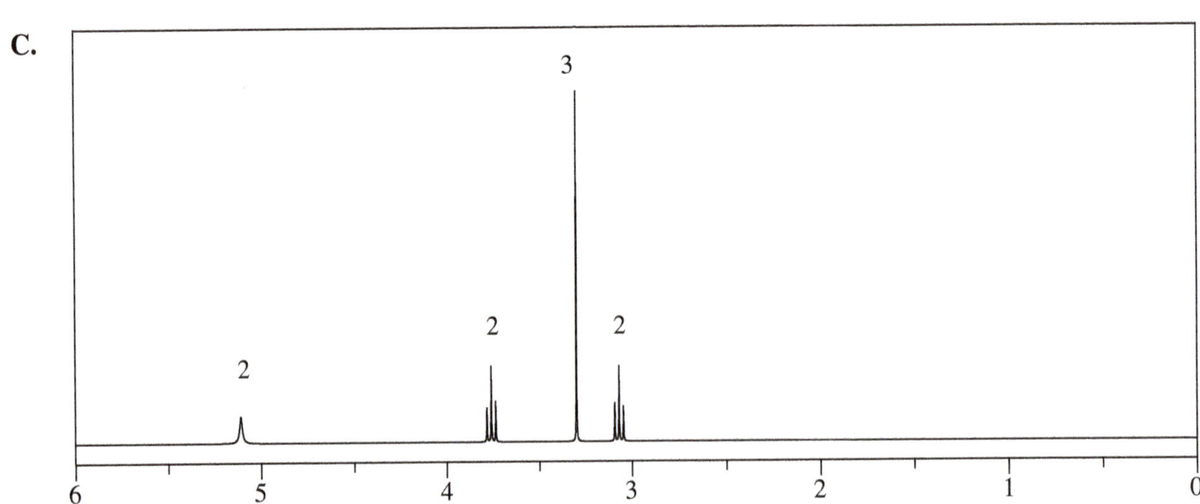

52. Match each of the four compounds to one of the IR spectra shown below.

A.

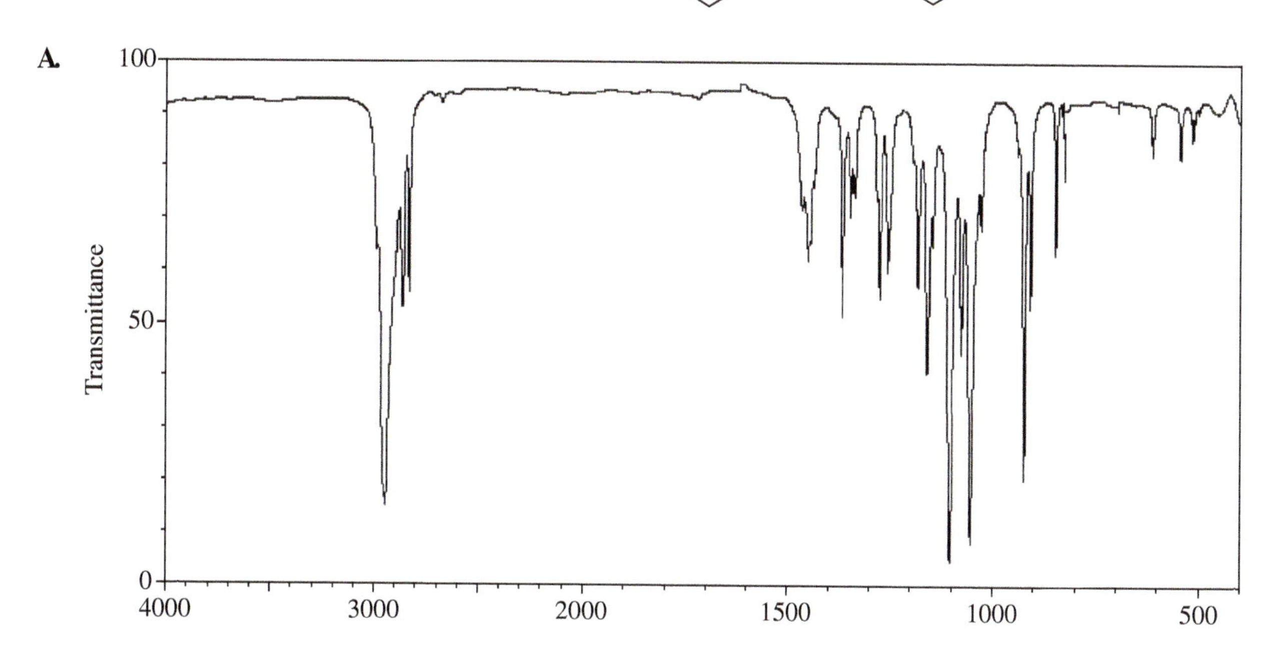

B.

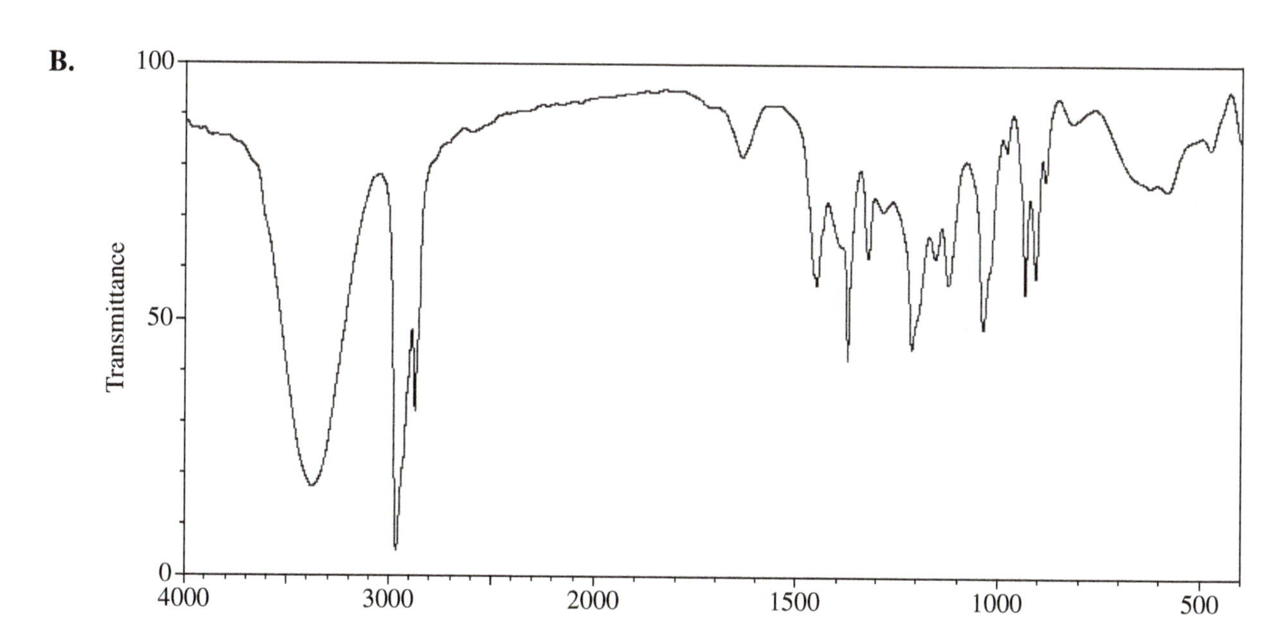

C.

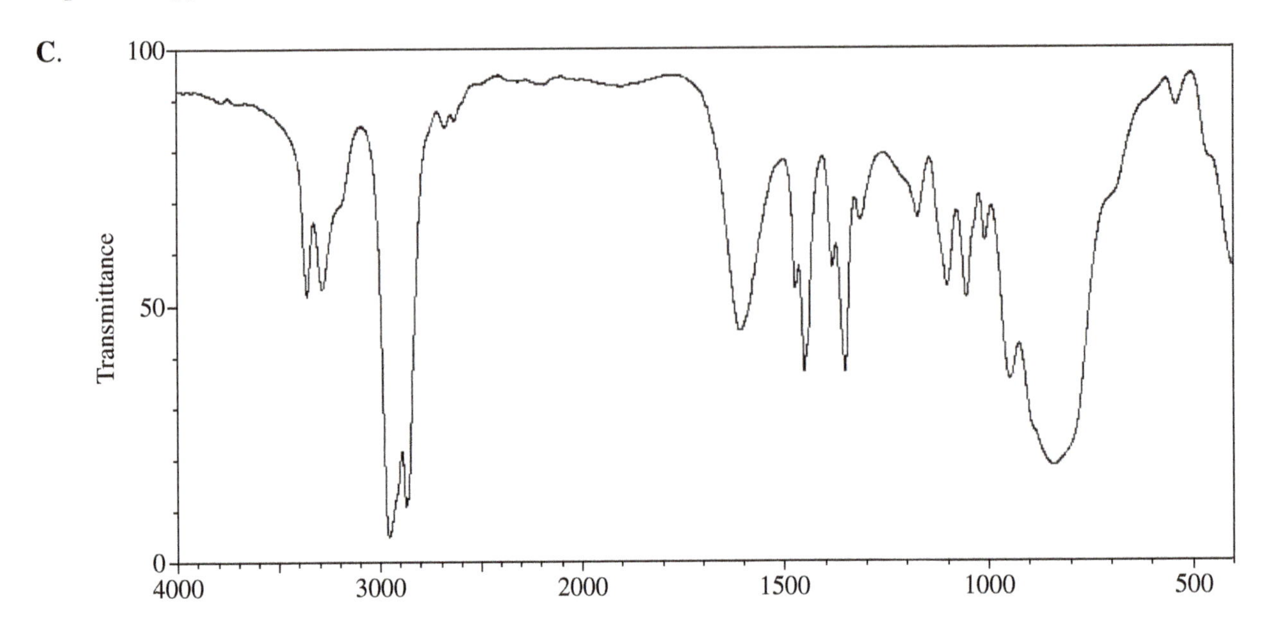
100
50
0
Transmittance
4000
3000
2000
1500
1000
500

D.

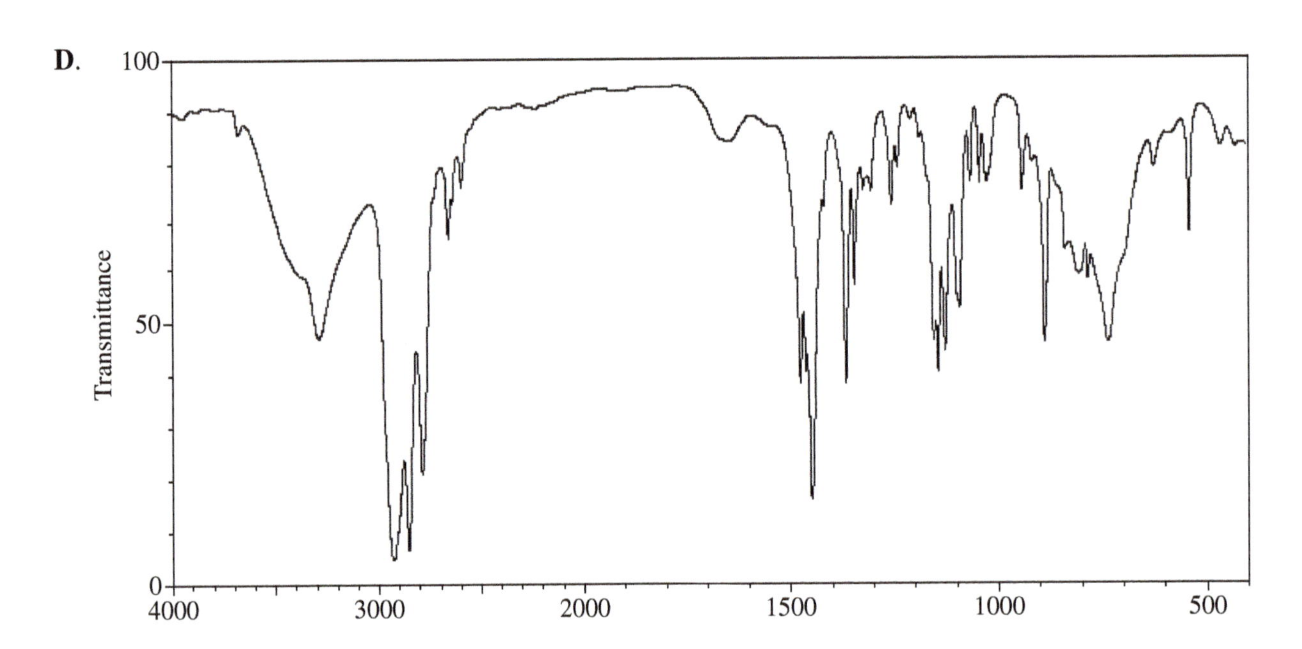
100
50
0
Transmittance
4000
3000
2000
1500
1000
500

53. Identify the compounds that give the IR and ^{1}H NMR spectra shown below. One has a molecular formula of $C_5H_{13}N$ and the other a molecular formula of $C_5H_{12}O$.

A.

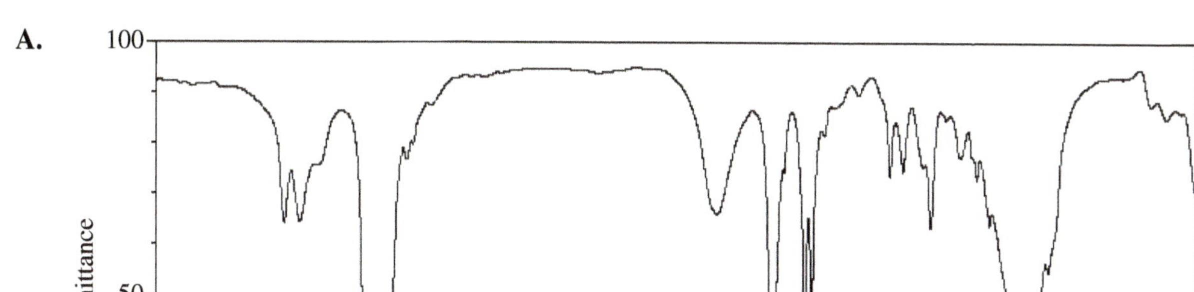

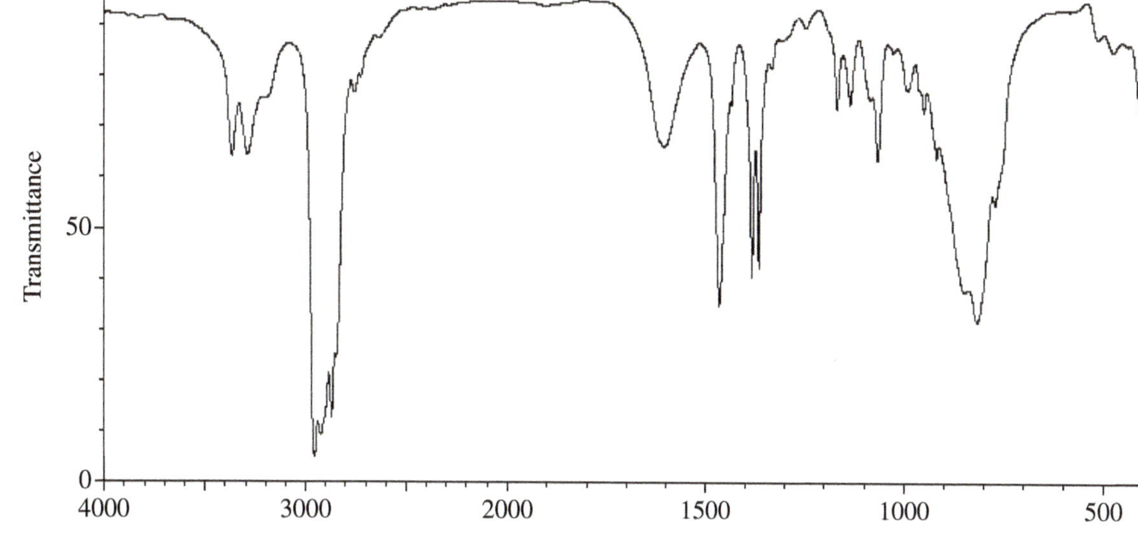

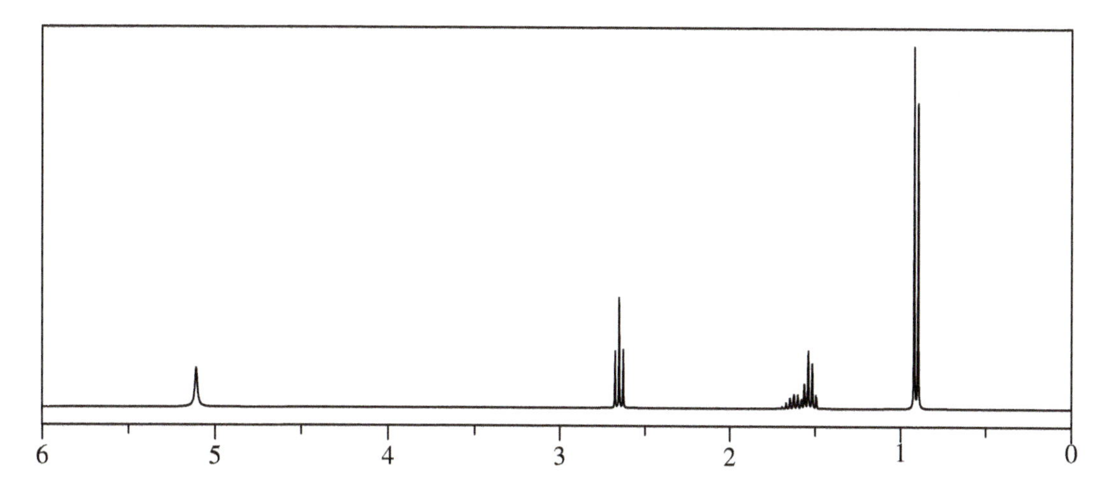

B.

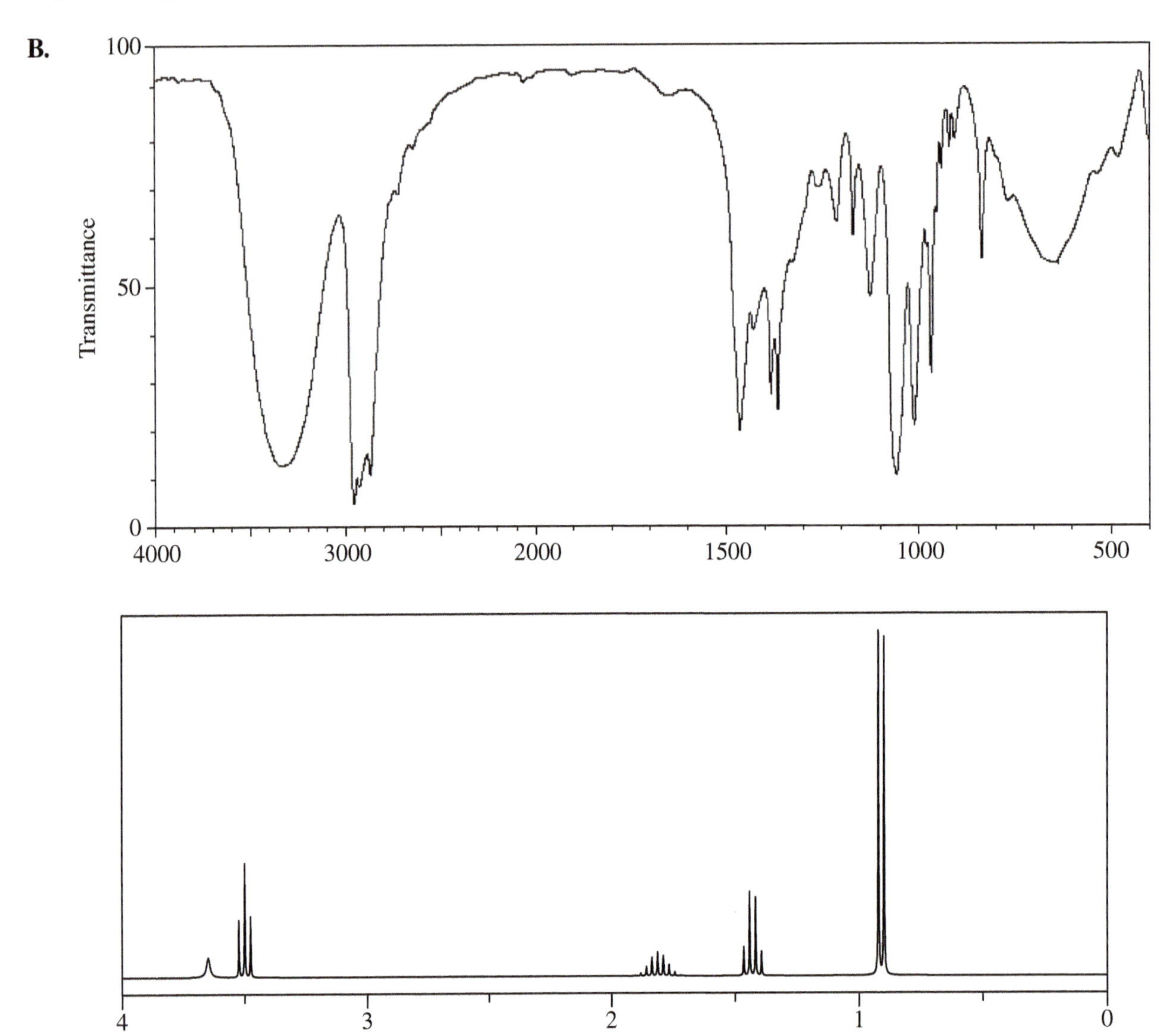

54. Identify the compounds that give the ^{1}H NMR spectra shown below. One has a molecular formula of $C_{10}H_{23}N$ and the other a molecular formula of $C_{10}H_{22}O$.

A.

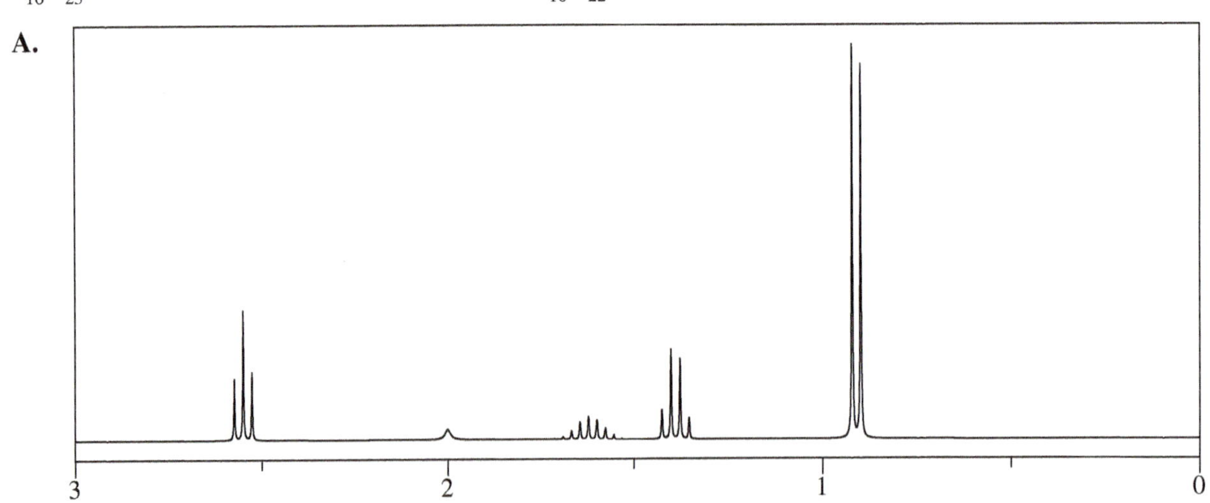

B.

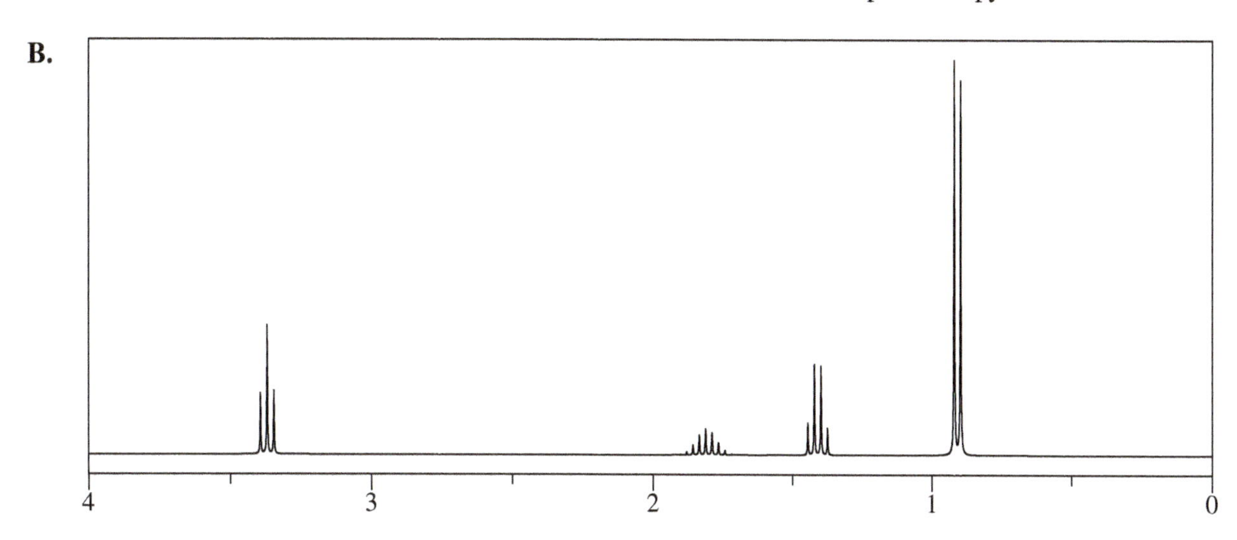

55. D.J., an undergraduate researcher, was asked to obtain a ^{1}H NMR spectrum for a sample of *cis*-1, 3-dibromocyclobutane and explain the results to his fellow group members. D.J. predicted that he would see two distinct signals in the ^{1}H NMR spetrum: a 4H triplet and a 2H quintet. Below is the ^{1}H NMR spectrum for his sample, which his advisor assured him was the correct spectrum for *cis*-1,3-dibromocyclobutane. How did D.J. rationalize this spectrum to his research group?

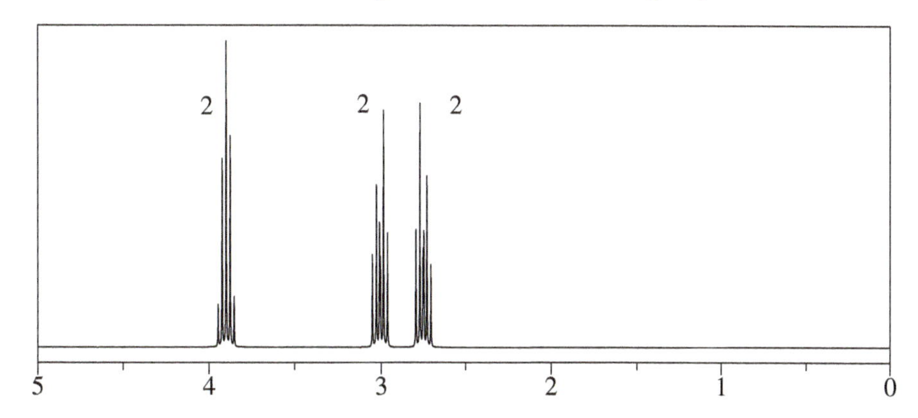

56. D.J., now more experienced in ^{1}H NMR interpretation, was asked to obtain a ^{1}H NMR spectrum for a sample of cyclopropanol. He predicted that he would see three distinct signals in the ^{1}H NMR spectrum: a 1H broad singlet, a 4H doublet, and a 1H quintet. Below is the ^{1}H NMR spectrum he obtained for his sample. How does he interpret this spectrum for his group members?

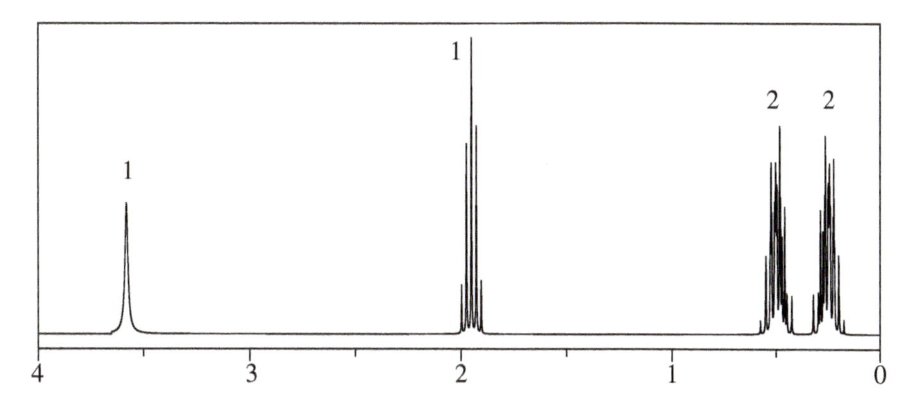

57. Chapter 6

A compound with molecular formula C_5H_{10} forms **A** and when it reacts with HBr and a minor amount of **B**. Identify the products from their ^1H NMR spectra. Write the reaction that forms **A** and **B**.

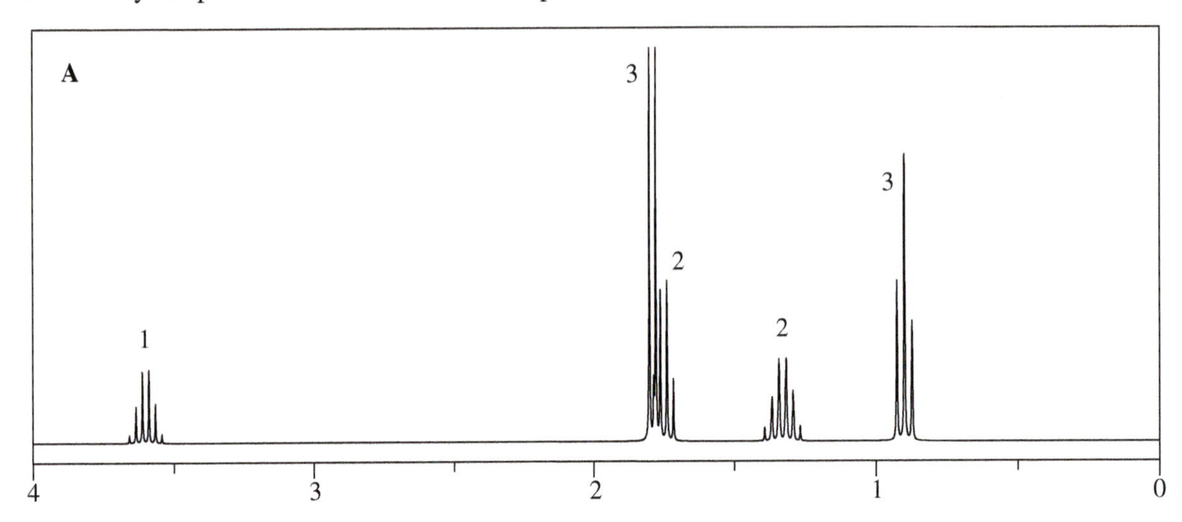

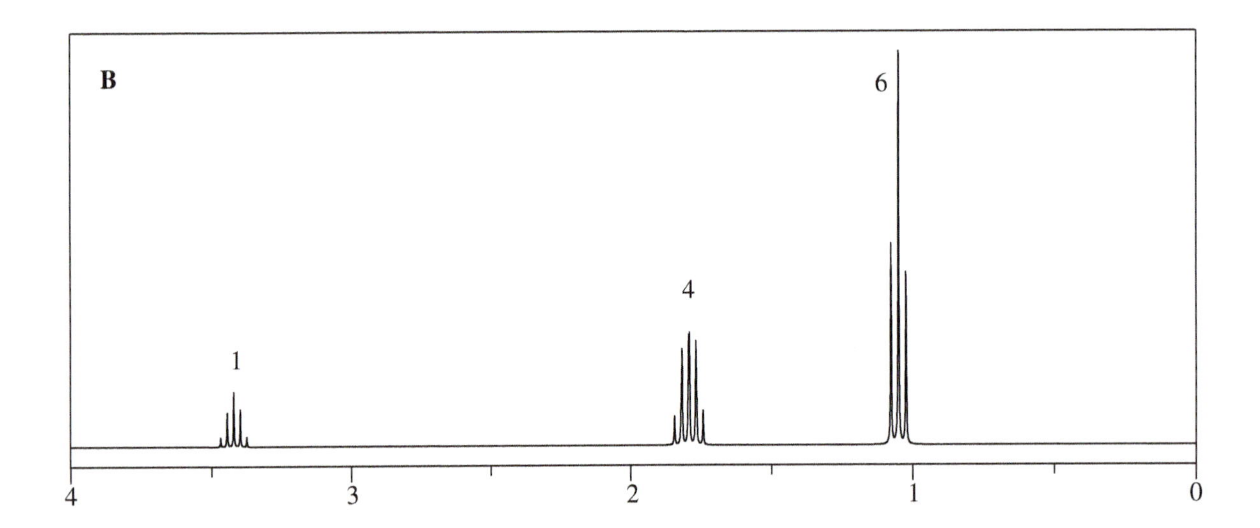

An isomer of the starting material of the previous problem forms **C** when it reacts with HBr and a minor amount of **D**. Identify **C** and **D** from their ^{1}H NMR spectra. Write the reaction that forms **C** and **D**.

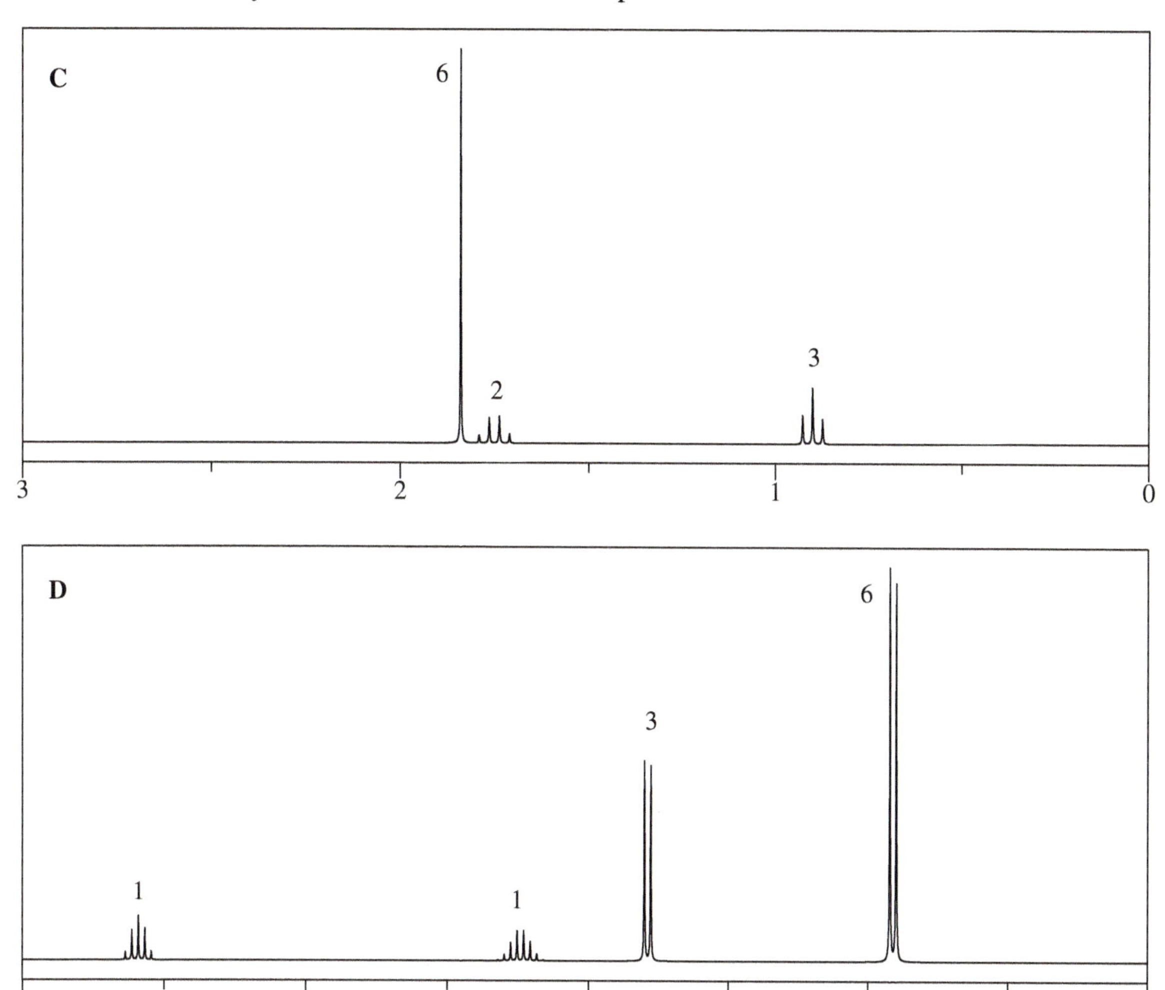

58. A compound with molecular formula C_4H_9ClO gives the IR and ^1H NMR spectra shown below. Identify the compound.

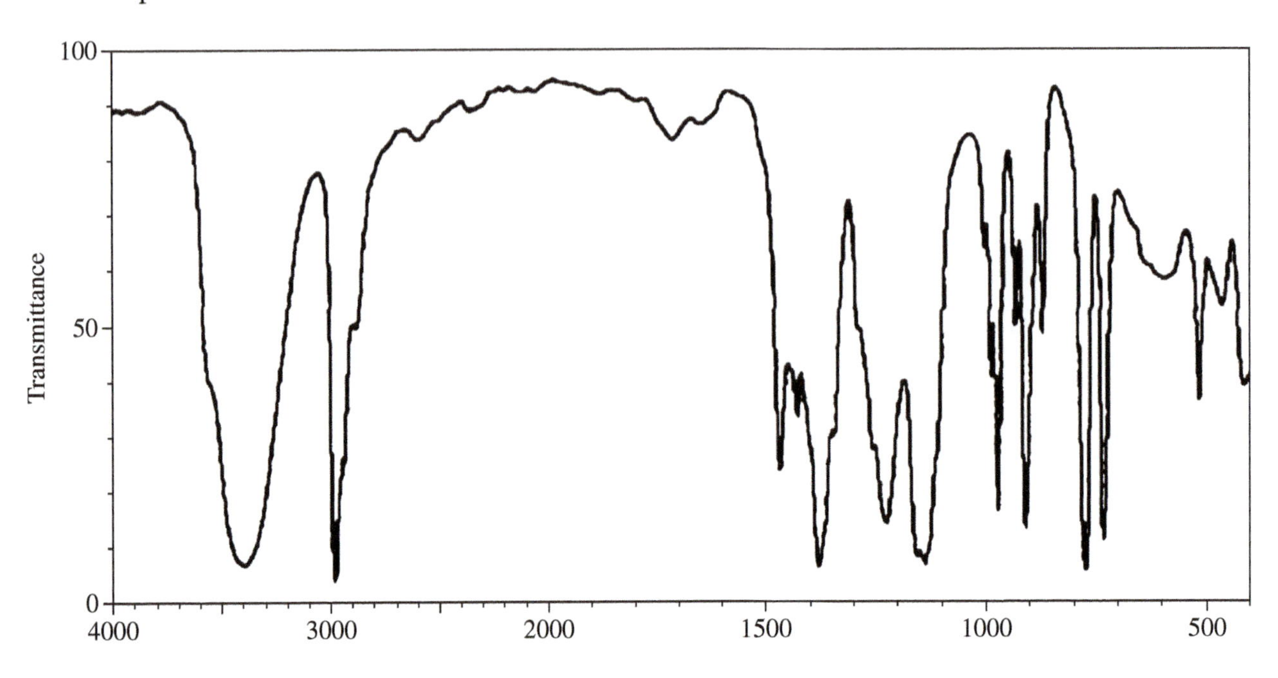

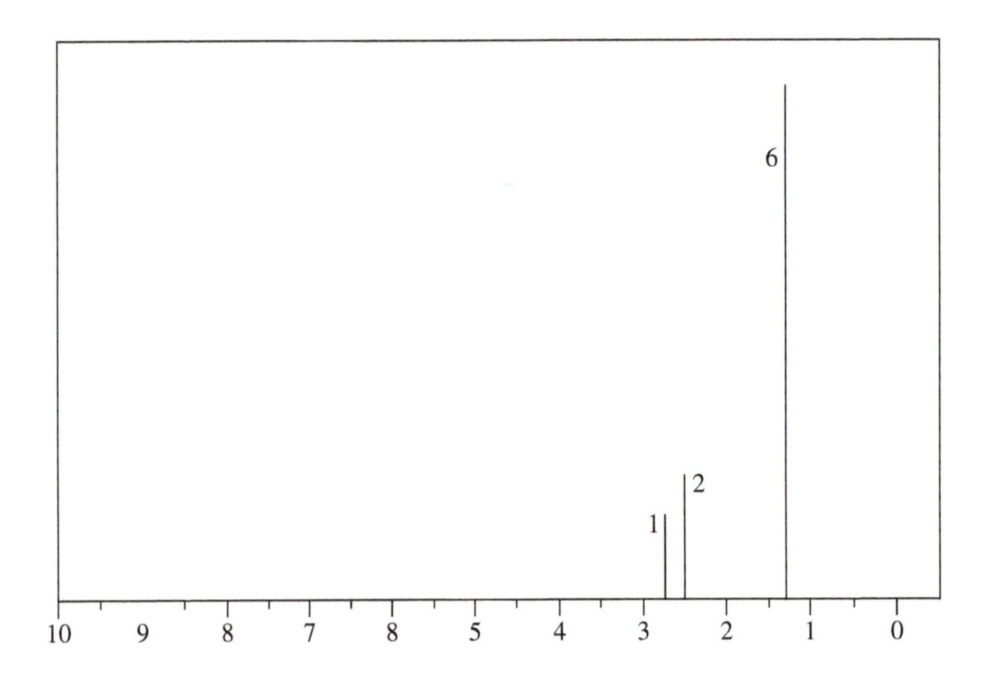

59. The mass spectrum for a compound with a molecular ion at $m/z = 102$ is shown below. The IR spectrum of the compound has a broad, strong absorption at 3600 cm^{-1} and a medium absorption at 1360 cm^{-1}.

a. Identify the compound. What fragments are responsible for the base peak at $m/z = 45$ and the *peak* at $m/z = 84$?

b. Explain the peak at $m/z = 84$ and draw a structure for the compound formed as a result.

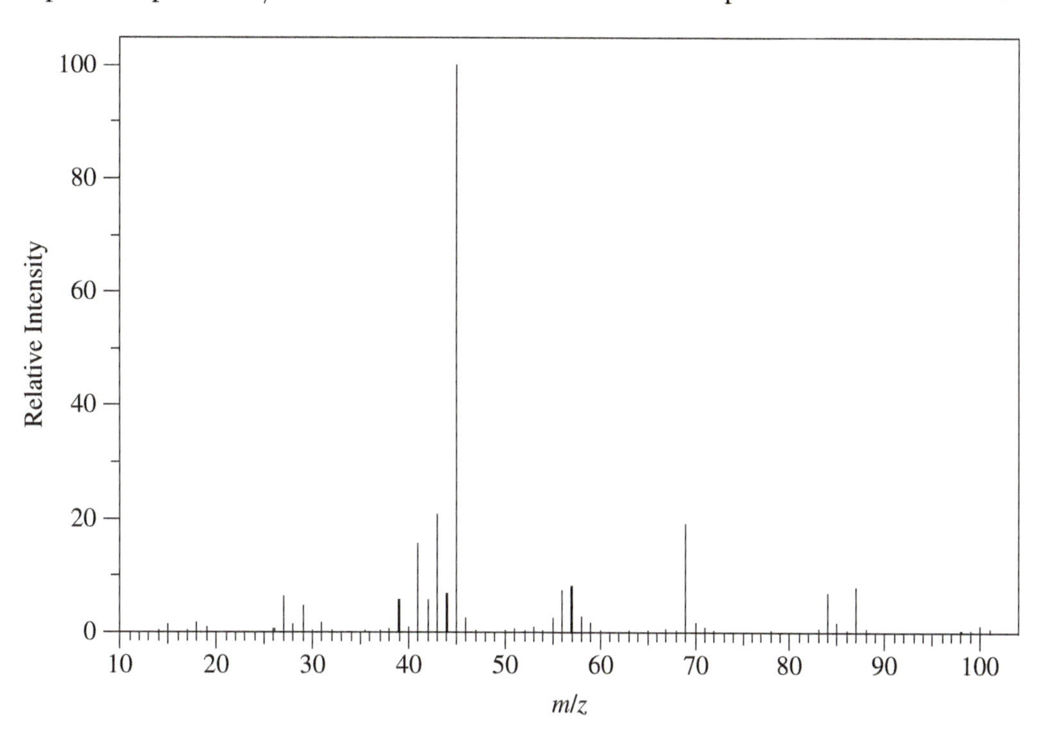

60. **Chapter 9**
Three isomeric bromobutanes (**A**, **B**, and **C**) were each treated with sodium hydroxide. Identify the bromobutane from the IR spectra of the product(s) it formed.

A IR absorption bands at 2960–2850 cm^{-1} and 1670 cm^{-1}

B IR absorption bands at 2960–2850 cm^{-1} and 3350 cm^{-1}

C IR absorption bands at 2960–2850 cm^{-1}, 3350 cm^{-1}, and 1670 cm^{-1}

61. IR spectra **A–E** are shown below for the following compounds. Match each compound to its IR spectrum.

O OH O C H OH COOH

1 2 3 4 5

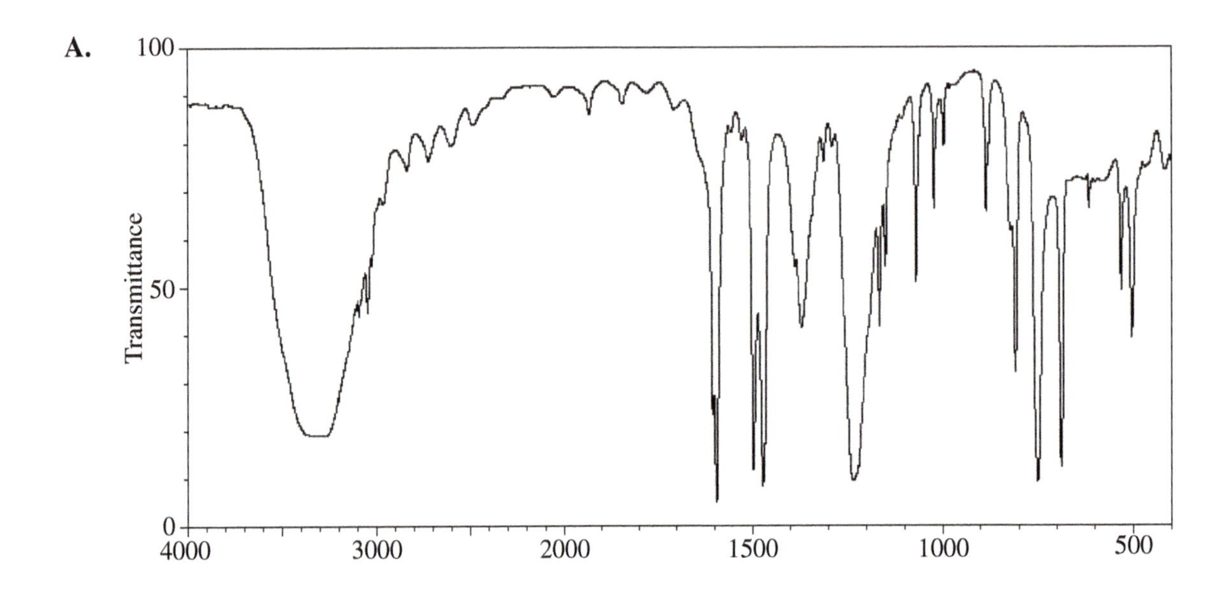

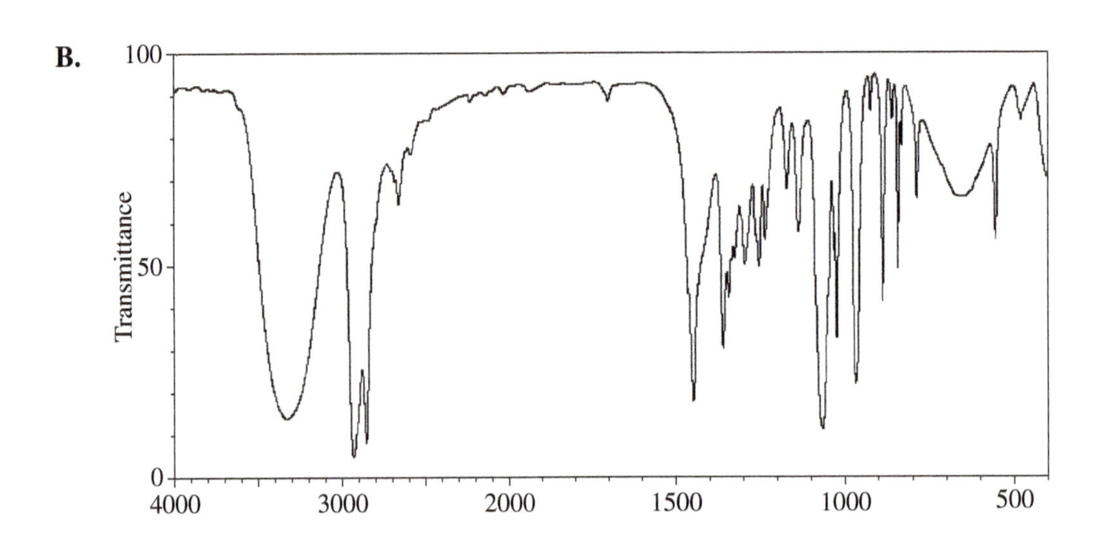

C.

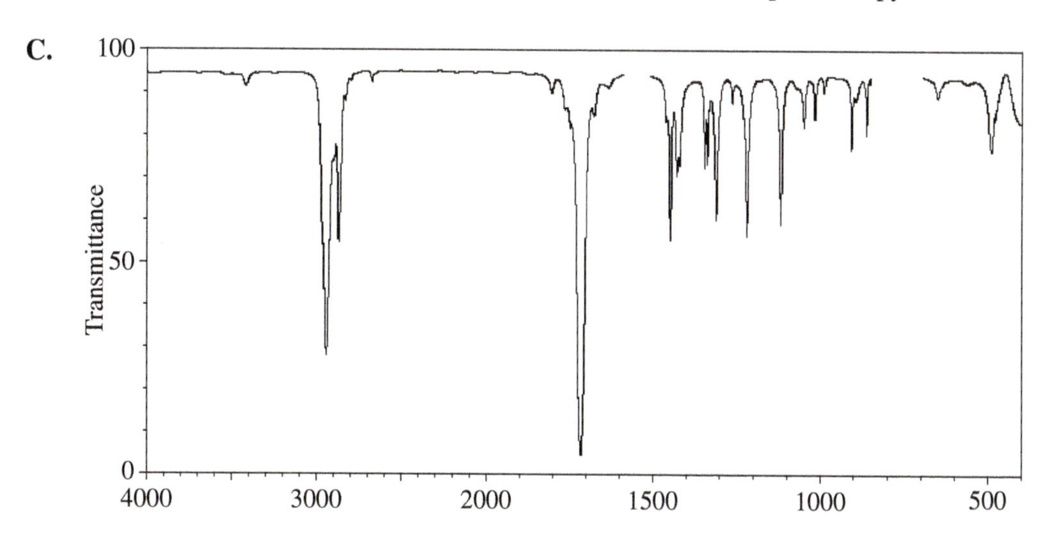
100
50
0
Transmittance
4000
3000
2000
1500
1000
500

D.

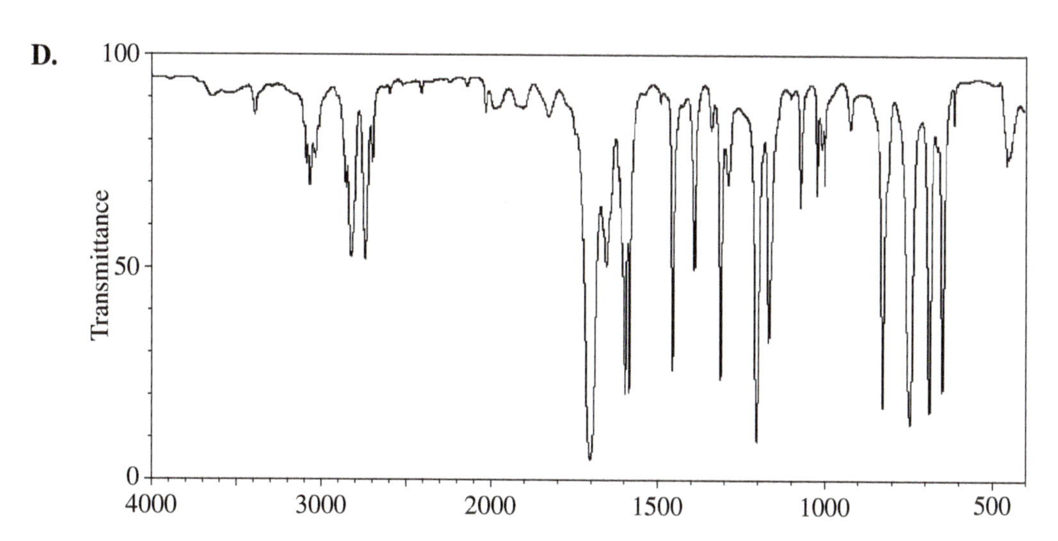
100
50
0
Transmittance
4000
3000
2000
1500
1000
500

E.

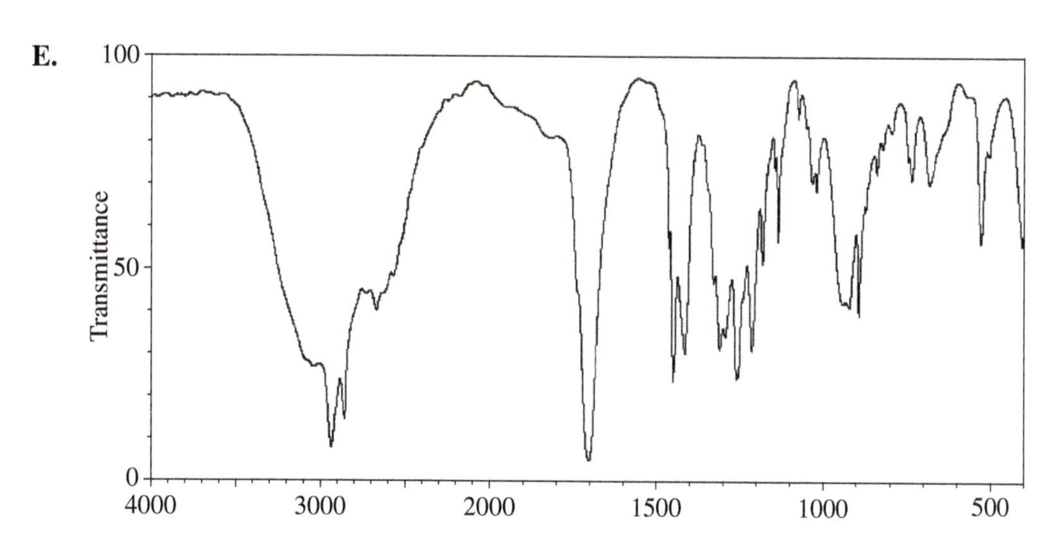
100
50
0
Transmittance
4000
3000
2000
1500
1000
500

62. Identify the compound with molecular formula C_4H_6O that gives the following ^1H NMR and ^{13}C NMR spectra.

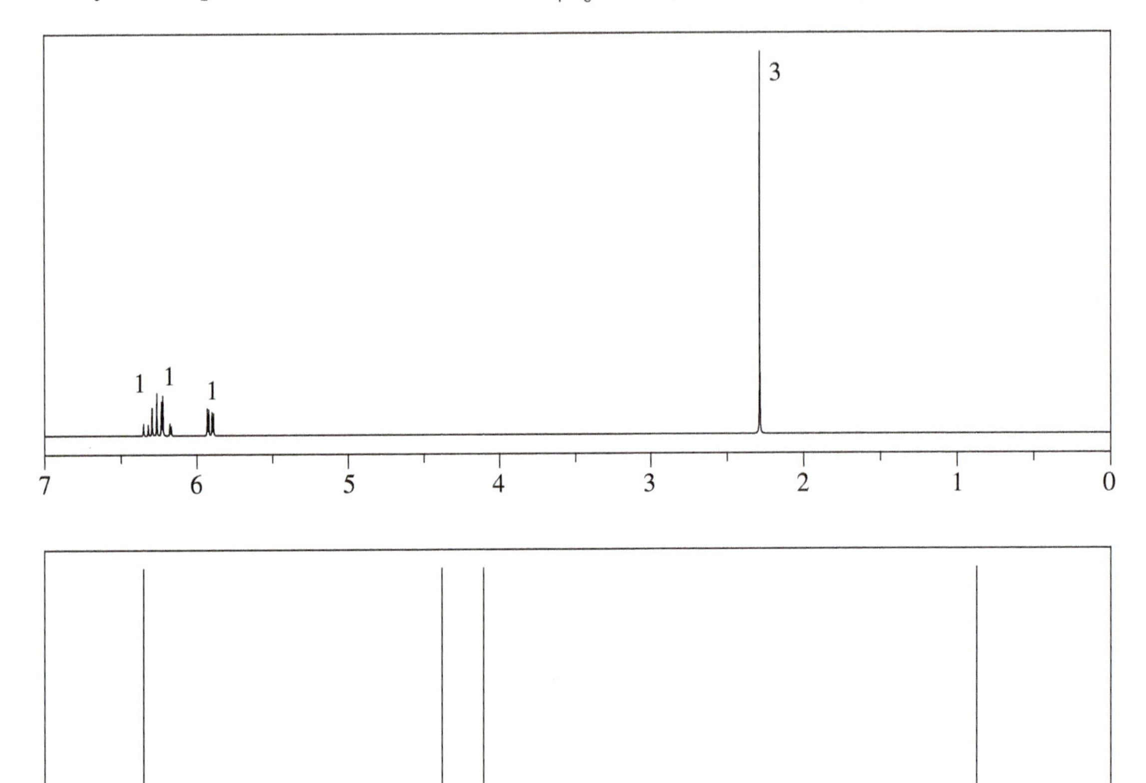

63. **Chapter 11**

The IR spectrum of a compound with molecular formula $C_6H_{14}O$ is shown below. When the compound is heated with a catalytic amount of H_2SO_4, it forms a product with molecular formula C_6H_{12} that gives the ^1H NMR and ^{13}C NMR spectra shown below. Identify the reactant and the product.

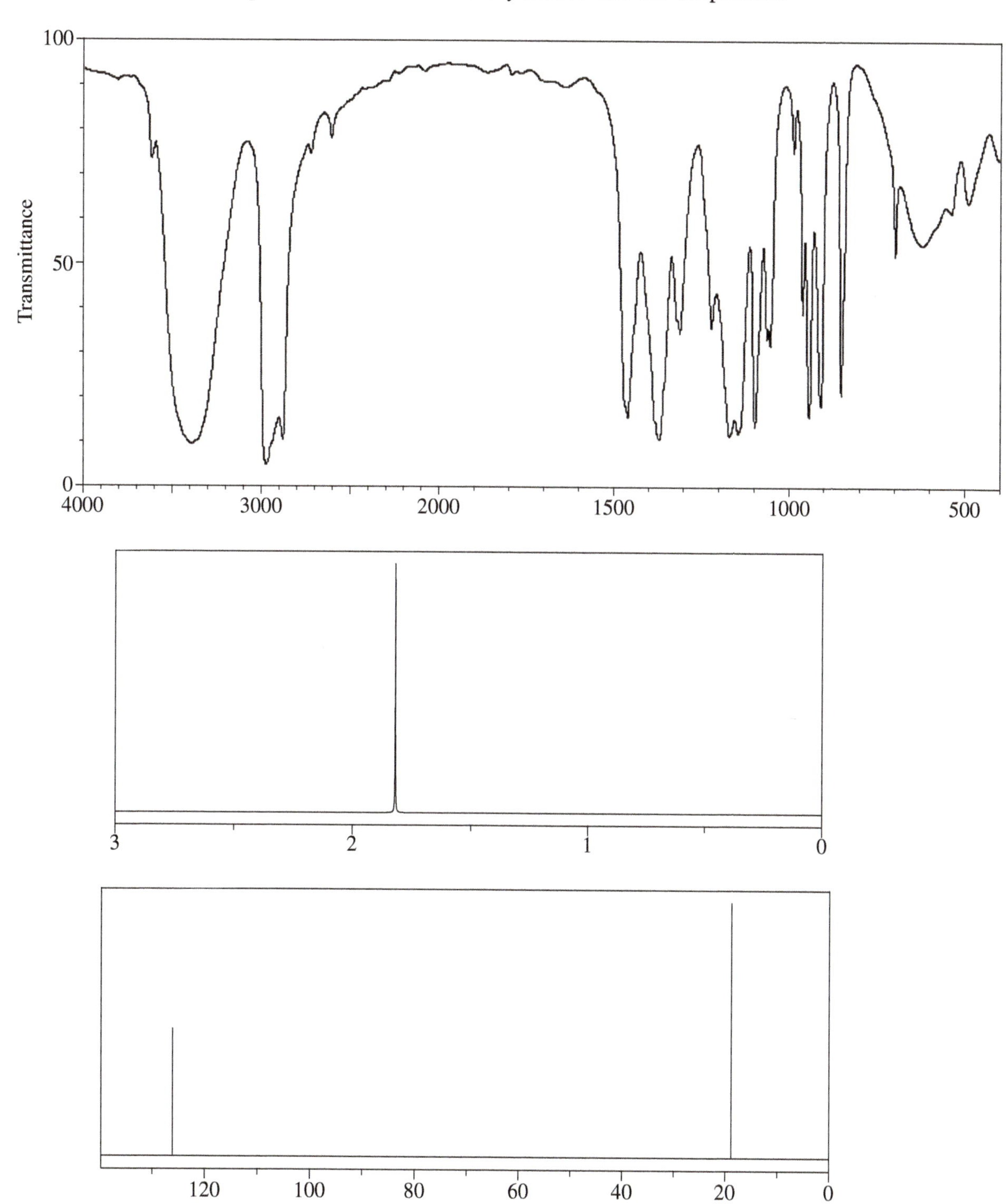

64. A compound with molecular formula $C_{14}H_{20}O_2$ gives the ^1H NMR spectrum shown below. Identify the compound. Important IR data are given below.

IR Data: 3030 cm^{-1} medium
1690 cm^{-1} strong
1600 cm^{-1} sharp, medium
1500 cm^{-1} sharp, medium
1380 cm^{-1} medium

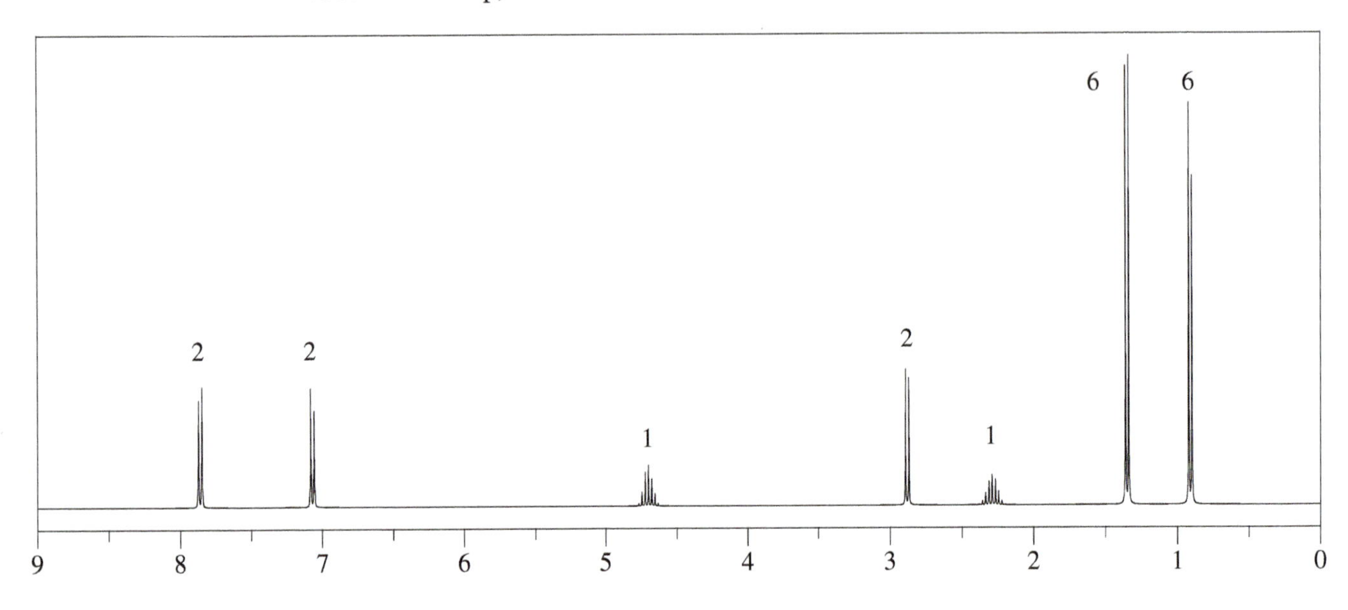

65. Compound A, with molecular formula $C_{10}H_{12}O$, gives the ^1H NMR spectrum shown below. Identify the compound. Important IR data are given below.

IR Data: 1703 cm^{-1} (strong absorption)
2700–2800 cm^{-1} (two week absorptions)
1500 and 1600 cm^{-1} (two medium absorptions)

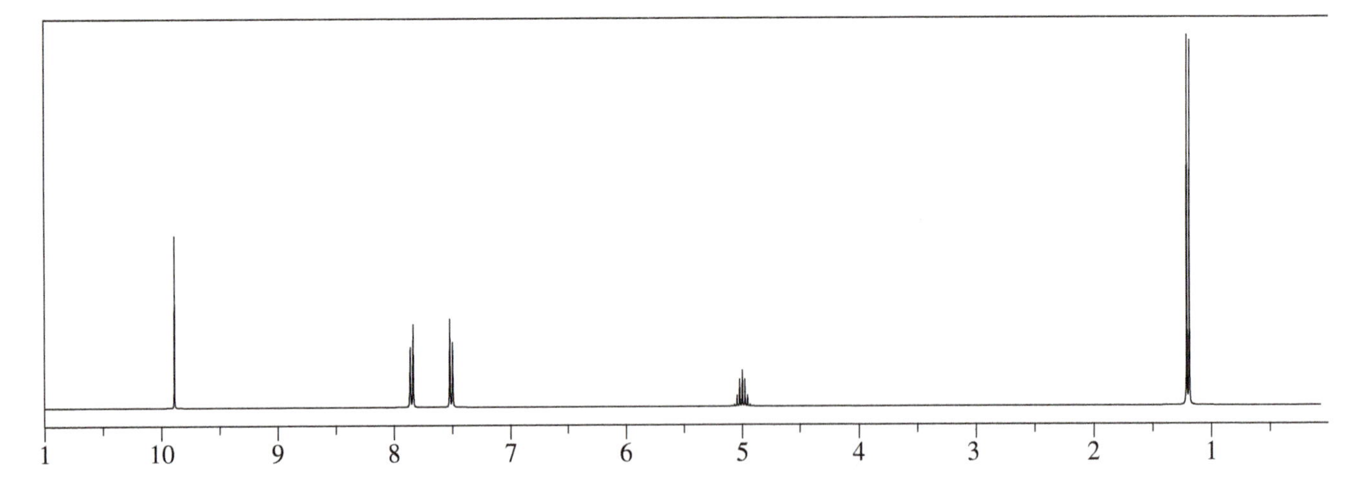

Answers to Spectroscopy Problems

1. There are three straight-chain pentanols: 1-pentanol, 2-pentanol, and 3-pentanol. Because the most stable fragment for an alcohol is the one formed by α-cleavage, we can see which of the alcohols forms the base peak shown in the spectrum (that is, a base peak with $m/z = 45$) as a result of α-cleavage.

For 1-pentanol, only one α-cleavage is possible. It forms a cationic fragment with $m/z = 31$.

$$CH_3CH_2CH_2CH_2{-}CH_2{-}OH \xrightarrow{\alpha\text{-cleavage}} CH_3CH_2CH_2\dot{C}H_2 + H\overset{+}{O}{=}CH_2 \quad (m/z = 31)$$

For 2-pentanol, two α-cleavages are possible. One forms a cationic fragment with $m/z = 73$ and a methyl radical. The second forms a cationic fragment with $m/z = 45$ and a propyl radical. Because a propyl radical is more stable than a methyl radical, the base peak is expected to have $m/z = 45$.

$$CH_3{-}CH(OH)CH_2CH_2CH_3 \xrightarrow{\alpha\text{-cleavage}} \dot{C}H_3 + H\overset{+}{O}{=}CHCH_2CH_2CH_3 \quad (m/z = 73)$$

$$CH_3CH(OH){-}CH_2CH_2CH_3 \xrightarrow{\alpha\text{-cleavage}} CH_3CH_2\dot{C}H_2 + H\overset{+}{O}{=}CHCH_3 \quad (m/z = 45)$$

For 3-pentanol, only one α-cleavage is possible because of the symmetry of the molecule. α-Cleavage forms a cationic fragment with $m/z = 59$.

$$CH_3CH_2{-}CH(OH){-}CH_2CH_3 \xrightarrow{\alpha\text{-cleavage}} CH_3\dot{C}H_2 + H\overset{+}{O}{=}CHCH_2CH_3 \quad (m/z = 59)$$

The base peak of the given mass spectrum has $m/z = 45$. Thus, the mass spectrum is that of **2-pentanol**. We also see a significant fragment at $m/z = 73$, the m/z value of the other α-cleavage product.

2. First, we must first identify the molecular ion. The molecular ion, the peak that represents the intact starting compound, has an $m/z = 74$. Now we can use the rule of 13 to determine the molecular formula.

$$\frac{74}{13} = 5 \text{ carbons with 9 left over}$$

From the rule of 13, we end up with a molecular formula of C_5H_{14}. Because the compound is an ether, we know that it has one oxygen, so we must add one O and subtract one C and four Hs from the molecular formula. The resulting molecular formula is:

$$C_4H_{10}O$$

Three ethers have this molecular formula: methyl propyl ether, diethyl ether, and isopropyl methyl ether.

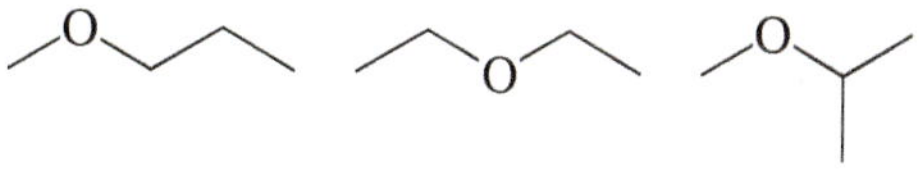

3. First, we need to determine the most abundant cationic fragments for each compound.

The possible fragments for **pentane** are:

1 $[CH_3{-}CH_2CH_2CH_2CH_3]^{\bullet+} \longrightarrow \cdot CH_3 + {}^{+}CH_2CH_2CH_2CH_3$ $m/z = 57$

2 $[CH_3CH_2{-}CH_2CH_2CH_3]^{\bullet+} \longrightarrow \cdot CH_2CH_3 + {}^{+}CH_2CH_2CH_3$ $m/z = 43$

3 $[CH_3CH_2{-}CH_2CH_2CH_3]^{\bullet+} \longrightarrow \cdot CH_2CH_2CH_3 + {}^{+}CH_2CH_3$ $m/z = 29$

4 $[CH_3{-}CH_2CH_2CH_2CH_3]^{\bullet+} \longrightarrow \cdot CH_2CH_2CH_2CH_3 + {}^{+}CH_3$ $m/z = 15$

The most abundant fragments result from bond cleavages that produce the most stable cations and radicals. Fragments from **2** and **3** are the most abundant because, in each case, a primary carbocation and a primary radical are formed.

2 is expected to give the base peak (the most stable fragment). The cation formed in **2** ($m/z = 43$) is more stable than the cation formed in **3** ($m/z = 29$), because the former is more stabilized by inductive electron donation from the alkyl group.

Fragments from **1** and **4** are expected to be less abundant. They each form one primary species, but the second species is a methyl fragment (either a radical or a carbocation), which is less stable than the second species formed in **2** and **3**.

Four sets of fragments are shown for **isopentane**. Fragmentations that result in a primary fragment and a methyl fragment have been excluded because they would be less abundant than those shown here.

1 $[CH_3CH(CH_3)CH_2CH_3]^{\bullet+} \longrightarrow \cdot CH_3 + CH_3\overset{+}{C}HCH_2CH_3$ $m/z = 57$

2 $[CH_3CH(CH_3)CH_2CH_3]^{\bullet+} \longrightarrow CH_3\dot{C}HCH_2CH_3 + {}^{+}CH_3$ $m/z = 15$

3 $[CH_3CH(CH_3){-}CH_2CH_3]^{\bullet+} \longrightarrow \cdot CH_2CH_3 + CH_3\overset{+}{C}H(CH_3)$ $m/z = 43$

4 $[CH_3CH(CH_3){-}CH_2CH_3]^{\bullet+} \longrightarrow CH_3\dot{C}H(CH_3) + {}^{+}CH_2CH_3$ $m/z = 29$

The four major fragments have the same m/z values (57, 43, 29, and 15) as those formed by pentane, but their relative intensities are different.

1 and 3 produce the most abundant fragments because they both form a secondary cation, and the stability of the cation is more important than the stability of the radical in determining the most abundant fragments. Therefore, we expect a base peak with $m/z = 43$ (because the secondary cation is accompanied by a primary radical) and a less intense peak with $m/z = 57$ (because the secondary cation is accompanied by a methyl radical).

Both spectra show a base peak at $m/z = 43$. The major difference in the two spectra is the intensity of the peak with $m/z = 57$. The spectrum of isopentane should show a more intense peak because it is due to a secondary cation, whereas the peak with $m/z = 57$ in the spectrum of pentane is due to a primary cation.

Thus, **pentane** gives the first mass spectrum and **isopentane** gives the second.

4. The molecular ion for this compound has an $m/z = 73$. The nitrogen rule states that if a molecular ion has an odd value, then the structure must have an odd number of nitrogens. Therefore, we can eliminate the alkane, the ketone, and the ether.

Now we can determine the molecular formula of the compound using the rule of 13. When we subtract 14 (the mass of nitrogen) from 73, we get 59.

$$\frac{59}{13} = 4 \text{ carbons with 7 left over}$$

Therefore, the molecular formula is $C_4H_{11}N$. Both amines given as possible structures have this formula.

To determine which of the amines is responsible for the spectrum, we can take clues from how ethers and alcohols cleave and apply them to amines. Oxygen-containing species undergo α-cleavage. If nitrogen behaved similarly, then we would expect a fragment to form by cleaving a C—C bond alpha to the nitrogen in each compound. Given the relative stability of this fragment, we can anticipate that it will be the base peak of the mass spectrum.

$$(H_3C)(CH_3)N{-}CH_2{-}CH_3 \xrightarrow{\alpha\text{-cleavage}} (H_3C)(CH_3)\overset{+}{N}{=}CH_2 + \cdot CH_3$$

$m/z = 58$

$$H_2N{-}CH_2{-}CH_2CH_2CH_3 \xrightarrow{\alpha\text{-cleavage}} H_2\overset{+}{N}{=}CH_2 + \cdot CH_2CH_2CH_3$$

$m/z = 30$

α-Cleavage of *N,N*-dimethylethylamine gives a cation with $m/z = 58$. α-Cleavage of butylamine gives a cation with $m/z = 30$. The spectrum shows a base peak with $m/z = 58$, indicating that the compound that gives the spectrum is ***N,N*-dimethylethylamine**.

N

5. The mass spectrum has two peaks with the same height with m/z values = 136 and 138, indicating the presence of bromine in the product. (Recall that bromine has two isotopes of equal abundance with weights of 79 and 81 amu.)

Now we need to think about the type of reaction that occurred.

Under acidic conditions, the starting material (1-butanol) will be protonated.

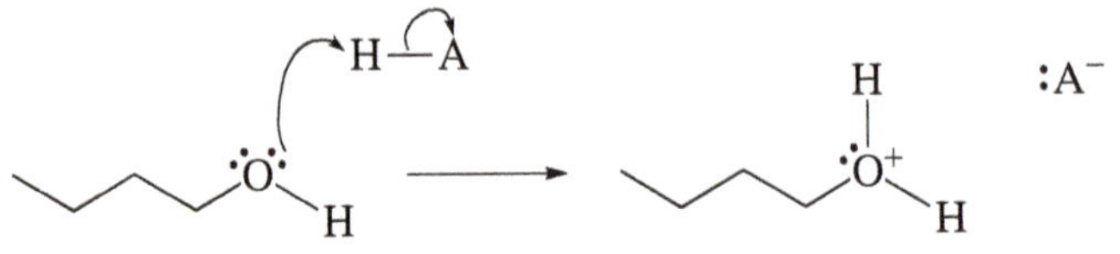

H—A H :A⁻ O H O+ H

The protonated alcohol now has a leaving group that can be replaced by a nucleophile. Because we know that bromine is present in the product, we can assume that bromide ion is the incoming nucleophile.

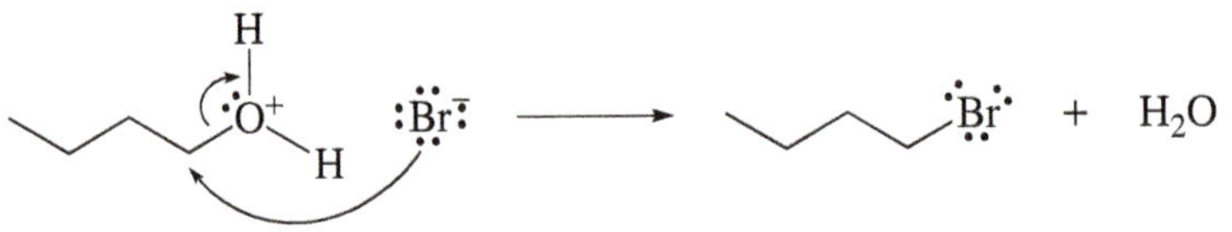

H O+ H :Br: Br + H_2O

We now know that the product of the reaction is **1-bromobutane**. The acid, which must be the source of the nucleophile, is **HBr**.

6. The two signals near 7 and 8 ppm are due to the hydrogens of a benzene ring. Because these signals integrate to 4 protons, the benzene ring must be disubstituted. The fact that both signals are doublets tells us that the protons that give each signal must be coupled to one proton ($N + 1 = 1 + 1 = 2$). Therefore, the substituents must be at the 1- and 4-positions.

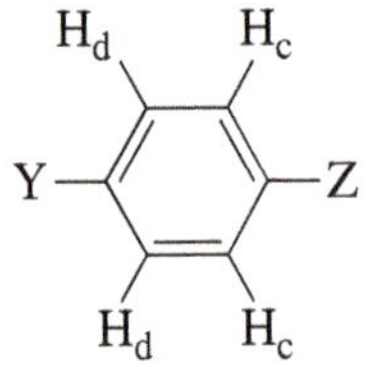

H_d H_c Y Z H_d H_c

By subtracting the six Cs and four Hs of the benzene ring from the molecular formula, we know that the two substituents contain three Cs, six Hs, and three Os ($C_9H_{10}O_3 - C_6H_4 = C_3H_6O_3$).

A triplet (1.4 ppm) that integrates to 3 protons and a quartet (4.2 ppm) that integrates to 2 protons are characteristic of an ethyl group. Because the signal for the CH_2 group of the ethyl substituent appears at a relatively high frequency, we know that it is attached to an electronegative atom (in this case, an O).

The presence of the CH_3CH_2O group consumes more of the remaining molecular formula ($C_3H_6O_3 - C_2H_5O = CHO_2$). There is one remaining NMR signal, a singlet (9.8 ppm) that integrates to 1 proton.

To help with the identification, we turn to the IR spectrum. The broad absorption near 3200 cm^{-1} indicates the O—H stretch of an alcohol; the proton of the OH group would give the broad NMR signal at 9.8 ppm. The strong absorption at 1680 cm^{-1} indicates the presence of a carbonyl C=O group.

Now that all the fragments of the compound have been identified, we can put them together. The compound is **ethyl 4-hydroxybenzoate**.

$$HO-C_6H_4-C(=O)OCH_2CH_3$$

7. The signals at 1.1 and 1.8 ppm have been magnified and are shown as insets on the spectrum (the **2** and **1** represent the ppm scale) so that you can better see the splitting. The triplet (1.1 ppm) that integrates to 3 protons and the quartet (1.8 ppm) that integrates to 2 protons are characteristic of an ethyl group. (The peak to the right of the quartet is actually the beginning of the adjacent signal that integrates to 6 protons.)

The singlet (1.7 ppm) that integrates to 6 protons indicates that there are two methyl groups in the same environment. Because the signal is a singlet, the carbon to which they are attached cannot be bonded to any hydrogens. The only atom not accounted for in the molecular formula is Br.

$$-C(CH_3)_2-$$

Therefore, the ethyl group and the bromine must be the two substituents that are attached to the carbon. Thus, the compound is **2-bromo-2-methylbutane**.

$$CH_3-C(CH_3)(Br)-CH_2CH_3$$

8. A major clue comes from the IR spectrum. The strong absorption at ~1710 cm^{-1} indicates the presence of a carbonyl (C=O) group. Because the compound has only one oxygen, we know that it must be an aldehyde or a ketone. The absence of absorptions at 2820 and 2720 cm^{-1} tells us that the compound is not an aldehyde.

The absorptions at 2880 and 2970 cm^{-1} are due to C—H stretches of hydrogens attached to sp^3 carbons.

The ^1H NMR spectrum has two unsplit signals. One integrates to 9 protons and the other to 3 protons. A signal that integrates to 9 protons suggests a *tert*-butyl group, and a signal that integrates to 3 protons suggests a methyl group. The fact that they are both singlets indicates that they are on either side of the carbonyl group. Therefore, the compound is **3,3-dimethyl-2-butanone**.

$$CH_3-C(CH_3)_2-C(=O)-CH_3$$

That the methyl group shows a signal at ~2.1 ppm reinforces this conclusion because that is where a methyl group attached to a carbonyl group is expected to occur.

9. From the reaction conditions provided, we know that the product is a monochlorinated toluene.

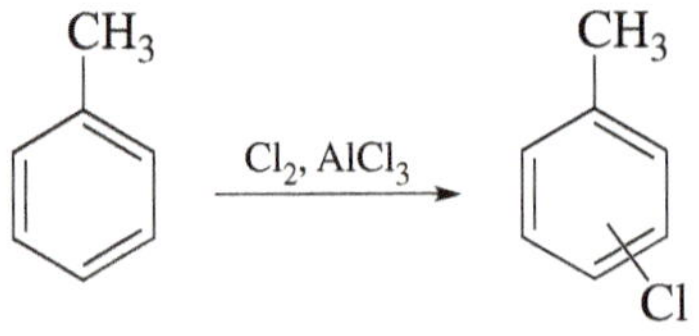

$$\text{C}_6\text{H}_5\text{CH}_3 \xrightarrow{\text{Cl}_2,\ \text{AlCl}_3} \text{ClC}_6\text{H}_4\text{CH}_3$$

The singlet (2.3 ppm) that integrates to 3 protons is due to the methyl group.

The signals in the 7–8 ppm region that integrate to 4 protons are due to the protons of a disubstituted benzene ring. Because both signals are doublets, we know that each proton is coupled to one adjacent proton. Thus, the compound has a 1,4-substituted benzene ring.

Therefore, the compound is **4-chloromethylbenzene**.

10. The strong and broad absorption in the IR spectrum at 3400 cm^{-1} indicates a hydrogen-bonded O—H group. The absorption bands between 2800 and 3000 cm^{-1} indicate hydrogens bonded to sp^3 carbons.

Only one signal in the ^{1}H NMR spectrum integrates to 1 proton, so it must be due to the hydrogen of the OH group. The singlet that integrates to 3 protons can be attributed to a methyl group that is attached to a carbon that is not attached to any hydrogens.

Because the other two signals show splitting, we know that they represent coupled protons (that is, protons on adjacent carbons). The quartet and triplet combination indicates an ethyl group. Because the quartet and triplet integrate to 6 and 4 protons, respectively, the compound must have two ethyl groups.

The identified fragments of the molecule are:

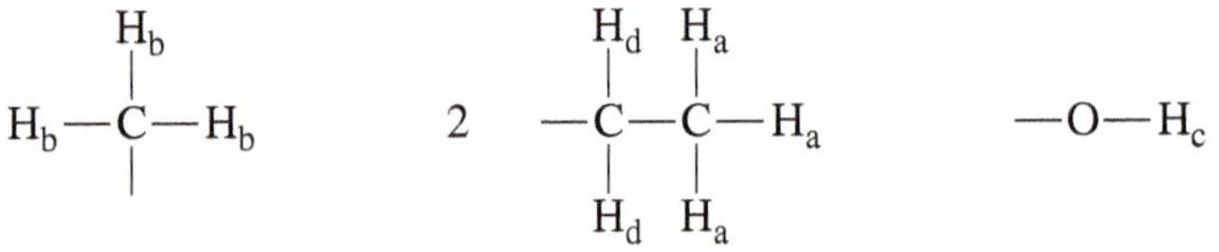

$$\text{H}_b\text{—C(H}_b\text{)(H}_b\text{)—} \qquad 2\ \text{—C(H}_d\text{)(H}_d\text{)—C(H}_a\text{)(H}_a\text{)—H}_a \qquad \text{—O—H}_c$$

When these fragments are subtracted from the molecular formula, only one carbon remains. Therefore, this carbon must connect the four identified fragments. The compound is **3-methyl-3-pentanol**.

$$(\text{CH}_3\text{CH}_2)_2\text{C(CH}_3\text{)OH}$$

11. The ^{1}H NMR spectrum contains only one signal, so only one type of hydrogen is present in the molecule. Because the compound has 4 carbons and 9 identical hydrogens, the compound must be ***tert*-butyl bromide**.

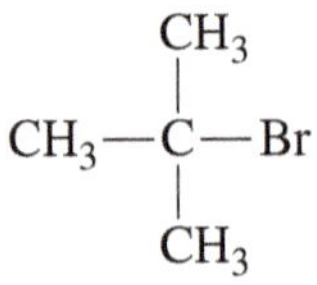

$$\text{CH}_3\text{—C(CH}_3\text{)(CH}_3\text{)—Br}$$

12. The molecular formula indicates that the compound is a hydrocarbon with one degree of unsaturation. The IR spectrum can tell us whether the degree of unsaturation is due to a cyclic system or a double bond. The absorption of moderate intensity near 1660 cm^{-1} indicates a C═C stretch. The absorption at ~3100 cm^{-1}, due to C—H stretches of hydrogens attached to sp^2 carbons, reinforces the presence of the double bond.

The two relatively high-frequency singlets (4.7 ppm) is given by vinylic protons. Because the signal integrates to 2 protons, we know that the compound has two vinylic protons. Because the signals are not split, the vinyl protons must not be on adjacent carbons. Thus, they must be on the same carbon.

The singlet (1.8 ppm) that integrates to 3 protons must be a methyl group. Because it is a singlet, the methyl group must be bonded to a carbon that is not attached to any protons.

The doublet (1.1 ppm) that integrates to 6 protons and the septet (2.2 ppm) that integrates to 1 proton is characteristic of an isopropyl group.

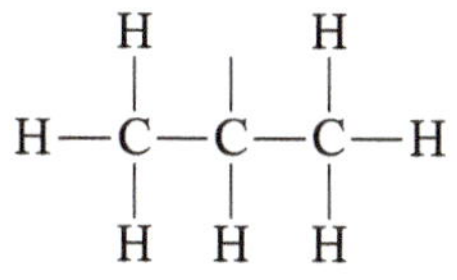

isopropyl group

Because the compound has a methyl group, an isopropyl group, and two vinylic hydrogens attached to the same carbon, we know that the compound must be **2,3-dimethyl-1-butene**.

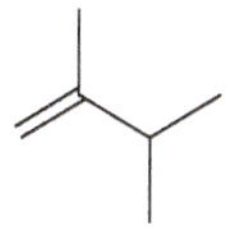

13. The signals in the ^{1}H NMR spectrum between 6.7 and 6.9 ppm indicate the presence of a benzene ring. Because the signals integrate to 3 protons, it must be a trisubstituted benzene ring.

The triplet (6.7 ppm) that integrates to 1 proton and the doublet (6.9 ppm) that integrates to 2 protons tell us that the three substituents are adjacent to one another. (The H_d protons are split into a doublet by the H_c proton, and the H_c proton is split into a triplet by the two H_d protons.)

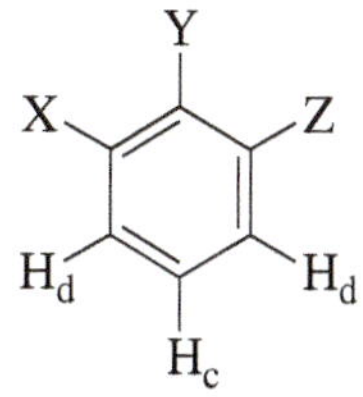

Subtracting the trisubstituted benzene (C_6H_3) from the molecular formula leaves C_2H_7O unaccounted for. The singlet (2.2 ppm) that integrates to 6 protons indicates that two methyl groups are in identical environments. Now only OH is left from the molecular formula. The singlet at 4.6 ppm is due to the proton of the OH group. The compound is **2,6-dimethylphenol**.

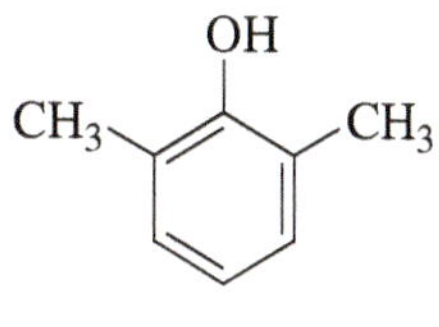

14. A major clue to the compound's structure comes from the IR spectrum. The strong absorption at ~1740 cm^{-1} indicates the presence of a carbonyl (C=O) group. Because the compound has only one oxygen, the compound must be an aldehyde or a ketone. The absence of absorptions at 2820 and 2720 cm^{-1} tells us that the compound is not an aldehyde.

The NMR spectrum shows a singlet (2.3 ppm) that integrates to 3 protons, indicating that it is due to a methyl group. The chemical shift of the signal (hydrogens attached to carbons adjacent to carbonyl carbons typically have shifts between 2.1 and 2.3 ppm) and the fact that the signal is a singlet suggest that the methyl group is attached directly to the carbonyl group.

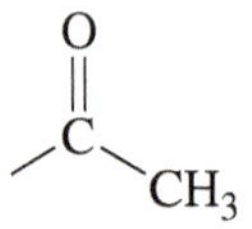

The two remaining signals are split, indicating that the protons that give these signals are attached to adjacent carbons. Because the signal at 4.3 ppm is a quartet, we know that the proton that gives this signal is bonded to a carbon that is attached to a methyl group. The other signal (1.6 ppm) is a doublet, so the proton that gives this signal is bonded to a carbon that is attached to one hydrogen.

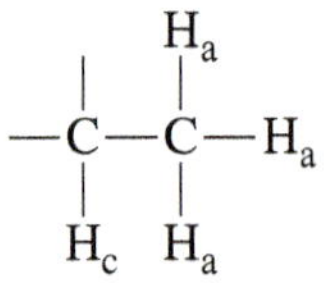

When these two fragments are subtracted from the molecular formula, only a Cl remains.

The only possible arrangement has the alkyl group bonded directly to the other side of the carbonyl group and the chlorine on the last available bond. The relatively high-frequency chemical shift of the quartet (4.3 ppm) reinforces this assignment because it must be attached to an electronegative atom. Thus, the compound is **3-chloro-2-butanone**.

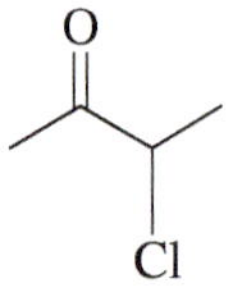

15. Given the simplicity of the ^{1}H NMR spectrum, the product must be highly symmetrical.

The singlet (7.4 ppm) that integrates to 4 protons is due to benzene-ring protons. Because there are four aromatic protons, we know that the benzene ring is disubstituted. Because the signal is a singlet, we know that the four protons are chemically equivalent. Therefore, the two substituents must be the same and they must be on the 1- and 4-positions of the benzene ring.

Subtracting the disubstituted benzene ring from the molecular formula, only $C_2H_4Br_2$ remains. Thus, each substituent must contain 1 carbon, 2 hydrogens, and 1 bromine. The compound that gives the spectrum, therefore, is the one shown here.

$BrCH_2$—C_6H_4—CH_2Br

16. The molecular formula indicates that the compound has two degrees of unsaturation. The weak absorption at ~2120 cm^{-1} is due to a carbon–carbon triple bond, which accounts for the two degrees of unsaturation. The intense and sharp absorption at 3300 cm^{-1} is due to the C—H stretch of a hydrogen attached to an *sp* carbon. The intensity and shape of this absorption distinguishes it from an alcohol (intense and broad) and an amine (weaker and broad). Thus, we know that the compound is a terminal alkyne.

The absorptions between 2800 and 3000 cm^{-1} are due to the C—H stretch of hydrogens attached to sp^3 carbons.

All three signals in the ^1H NMR spectrum are singlets, indicating that none of the protons that give these signals have neighboring protons. The singlet (2.4 ppm) that integrates to 1 proton is the proton of the terminal alkyne.

—C≡C—H

The two remaining signals (3.4 and 4.1 ppm) that integrate to 3 protons and 2 protons, respectively, can be attributed to a methyl group and a methylene group. When the alkyne fragment and the methyl and methylene groups are subtracted from the molecular formula, only an oxygen remains.

—O— —CH_2— —CH_3

The arrangement of these groups can be determined by the splitting and the chemical shift of the signals. Because each signal is a singlet, the methyl and methylene groups cannot be adjacent or they would split each other's signal. Because the terminal alkyne and the methyl group must be on the ends of the molecule, the only possible arrangement is shown below. Thus, the compound is **3-methoxy-1-propyne**.

$CH_3OCH_2C{\equiv}CH$

Notice that both the methyl and methylene groups show strong deshielding because of their direct attachment to the oxygen. The methylene hydrogens are also deshielded by the neighboring alkyne.

17. The signals in the ^1H NMR spectrum between 6.5 and 7.2 ppm indicate the presence of a benzene ring. Because the signals integrate to 3 protons, it must be a trisubstituted benzene ring.

The singlet (2.1 ppm) that integrates to three protons must be a methyl group; 2.1 ppm is characteristic of protons bonded to a benzylic carbon.

When the trisubstituted benzene ring (C_6H_3) and the methyl group (CH_3) are subtracted from the molecular formula, NH_2Br is all that remains. Thus, the three substituents must be a methyl group, bromine, and an amino group (NH_2). The amino group gives the broad singlet (3.6 ppm) that integrates to 2 protons. Hydrogens attached to nitrogens and oxygens typically give broad signals.

The substitution pattern for the trisubstituted benzene can be determined from the splitting patterns. Because the signal (6.5 ppm) that integrates to 1 proton is a doublet, we know that the proton that gives this signal has only one neighboring proton. Looking at the magnification of the signal at 7.1 ppm, we see that it is actually two separate signals. One is a singlet; therefore, it is attached to a carbon that is separated by substituents from the carbons that are attached to protons. The other signal is a doublet that integrates to 1 proton; because it gives a doublet, we know that it is next to the proton that gives the doublet at 6.5 ppm.

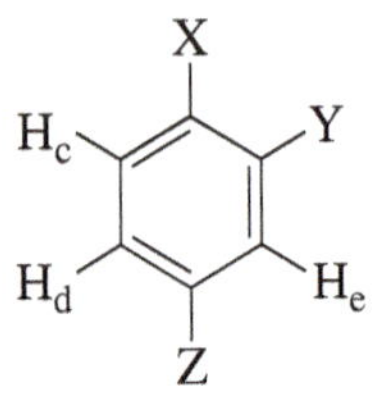

X, Y, Z, H_c, H_d, H_e

To determine the relative positions of the substituents, the chemical shifts must be analyzed. Bromine is the most electronegative substituent and, therefore, must be adjacent to the two protons that give signals at 7.1 ppm. Thus, Z is Br. The amino group donates its lone-pair electrons into the ring, so it shields benzene-ring protons. Thus, the signal at 6.5 ppm is from a proton in close proximity to the amino group. Therefore, X must be the amino group.

The compound that gives the spectrum is shown here.

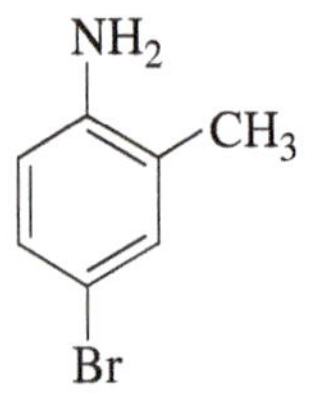

NH_2, CH_3, Br

18. The two compounds that produce the spectra have the following structures.

$ClCH_2CH_2CH_2Br$	$ClCH_2CH_2CH_2I$
1-bromo-3-chloropropane	1-chloro-3-iodopropane

The number of signals (three) and the splitting patterns are identical for each compound. The only difference is variations in the chemical shift due to the different electronegativities of bromine and iodine.

Because chlorine is more electronegative than bromine or iodine, the protons bonded to the carbon that is attached to chlorine has the most deshielded signal (that is, the signal that occurs at the highest frequency). This is the triplet that occurs at 3.7 ppm in both spectra.

The spectra differ in the signal that occurs at 3.4 ppm in the top spectrum and the signal that appears at 3.6 ppm in the second spectrum. Because bromine is more electronegative than iodine, the protons bonded to the carbon that is attached to bromine occurs at a higher frequency than the protons bonded to the carbon that is attached to iodine.

Thus, **1-chloro-3-iodopropane** gives the top spectrum, and **1-bromo-3-chloropropane** gives the bottom spectrum.

19. The molecular formula shows that the compound has one degree of unsaturation, indicating a cyclic compound, an alkene, or a carbonyl group.

A cyclic system containing an oxygen (a cyclic ether) would have the most deshielded signal at ~3.5 ppm, which would be due to the hydrogens attached to the carbon adjacent to the oxygen. Therefore, a cyclic ether would not give a signal at 6.4 ppm, so it can be ruled out.

Protons attached to a carbon adjacent to a carbonyl group show a signal at ~2.1 ppm. Because there is no signal in that region, a carbonyl group can also be ruled out.

Vinylic protons would account for the signals in the 3.9–4.2 ppm range that integrate to 2 protons, so we can conclude that the compound is an alkene. Because a highly deshielding oxygen is also present, the high-frequency signal (6.4 ppm) is not unexpected.

The triplet (1.3 ppm) that integrates to 3 protons and the quartet (3.8 ppm) that integrates to 2 protons indicate the presence of an ethyl group. The fact that the quartet is deshielded suggests that the ethyl's methylene group is attached to the oxygen.

The highly deshielded doublet of doublets (6.4 ppm) that integrates to 1 proton suggests that the proton that gives this signal is attached to an sp^2 carbon that is attached to the oxygen. The fact that the signal is a doublet of doublets indicates that it is split by each of two nonidentical protons on the adjacent carbon. Thus, the compound is **ethyl vinyl ether**.

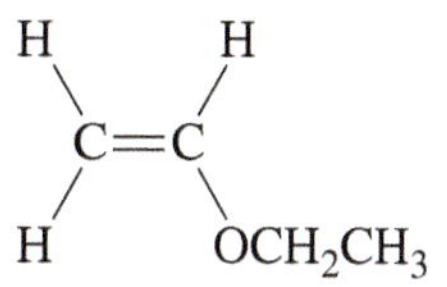

```
H       H
 \     /
  C=C
 /     \
H       OCH2CH3
```

The identification is confirmed by the two doublets (~4.0 and 4.2 ppm) that each integrate to 1 proton. When those signals are magnified, we can see that each is actually a doublet of doublets. The doublets of doublets are not well defined because of the small coupling constant (*J* value) for geminal coupling on sp^2 carbons.

20. The doublet (1.2 ppm) that integrates to 6 protons and the septet (5.0 ppm) that integrates to 1 proton are characteristic of an isopropyl group. (The two methyl groups are split by a single proton, and the single proton is split by six protons.)

The remaining signal (a singlet at a tiny bit more than 2.0 ppm) that integrates to 3 protons indicates an unsplit methyl group.

When the isopropyl and methyl groups are subtracted from the molecular formula, one carbon and two oxygens are left over. Thus, the compound has the following fragments:

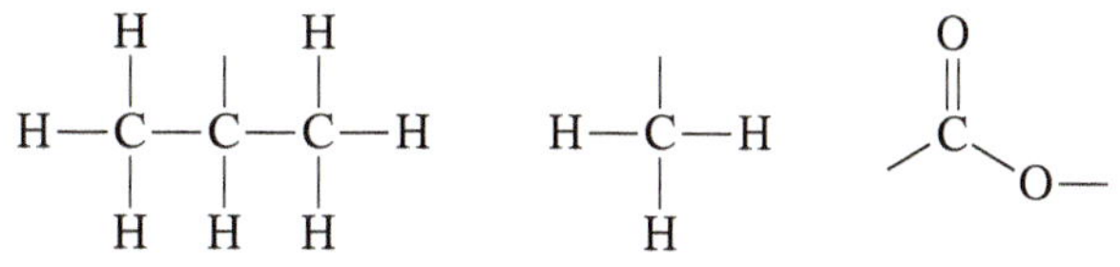

```
   H       H                        O
   |   |   |           |            ||
H—C—C—C—H        H—C—H          /C\
   |   |   |           |               O—
   H   H   H           H
```

These fragments can be pieced together two ways. Because the most deshielded signal in the spectrum (the one at 5.0 ppm) is the proton bonded to the central carbon of the isopropyl group, that carbon must be attached directly to the oxygen. Thus, the compound is **isopropyl acetate**.

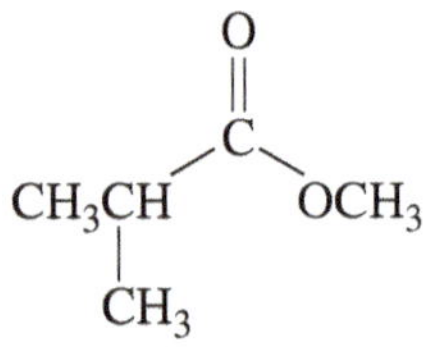

$CH_3CH(CH_3)C(=O)OCH_3$ — methyl 2-methylpropanoate

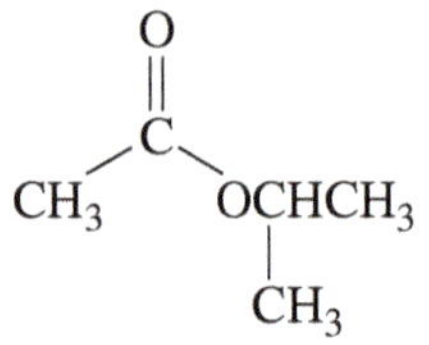

$CH_3C(=O)OCH(CH_3)CH_3$ — isopropyl acetate

21. The IR spectrum shows an absorption at ~1700 cm^{-1} for a C=O stretch and a very broad absorption (2300–3300 cm^{-1}) for an O—H stretch, indicating that the compound is a carboxylic acid. Intermolecular hydrogen bonding explains the broad nature of this peak as well as the broader-than-expected carbonyl peak absorption. The proton of the carboxylic acid gives a singlet at 12.4 ppm in the NMR spectrum.

The two doublets (7.5 and 7.9 ppm) that each integrate to 2 protons indicate a 1,4-disubstituted benzene ring.

Subtracting the disubstituted benzene ring and the COOH group from the molecular formula leaves CH_2Br. Therefore, we know that the second substituent is a bromomethyl group; it gives the singlet at ~4.7 ppm.

Therefore, the compound that gives the spectrum is the one shown here.

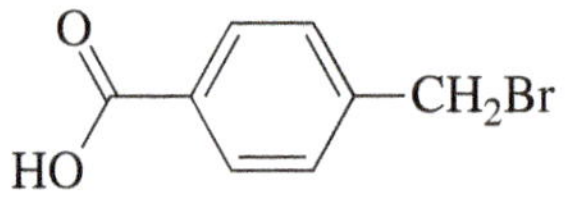

HO(O=)C–C_6H_4–CH_2Br

22. The signals with chemical shifts in the range of 7–8 ppm are due to benzene-ring protons. Because the three signals integrate to a total of 5 protons, we know that the benzene ring is monosubstituted.

The singlet at 10 ppm indicates the hydrogen of an aldehyde or a carboxylic acid. Because only one oxygen is in the molecular formula, we know that the compound is an aldehyde. Thus, the compound is **benzaldehyde**—a compound with a monosubstituted benzene ring and an attached aldehyde.

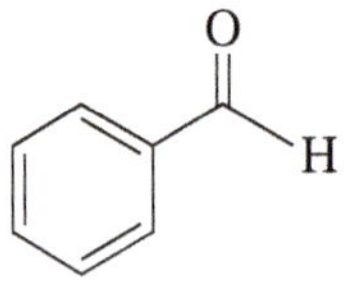

C_6H_5–C(=O)H

23. In the first spectrum, the doublet (~1.7 ppm) that integrates to 6 protons and the septet (~4.2 ppm) that integrates to 1 proton indicate an isopropyl group. When the isopropyl group is subtracted from the molecular formula, only a Br remains. Thus, the compound is **2-bromopropane**.

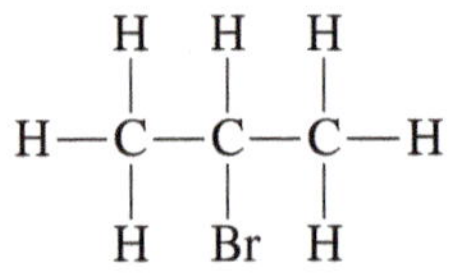

```
    H  H  H
    |  |  |
 H—C—C—C—H
    |  |  |
    H  Br H
```

In the second spectrum, the triplet (~ 1.0 ppm) that integrates to 3 protons is a methyl group that is attached to a methylene group. The triplet (~3.4 ppm) that integrates to 2 protons is a methylene group that is also attached to a methylene group; the highly deshielded nature of the signal indicates that the carbon is attached to an electronegative group. Thus, the compound is **1-bromopropane**.

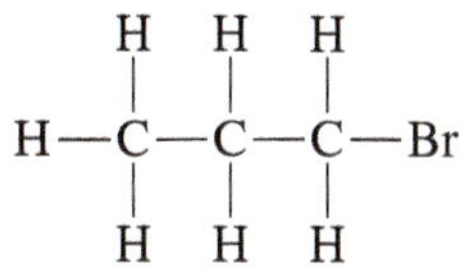

H H H
H—C—C—C—Br
H H H

The structure is confirmed by the multiplet (~ 1.8) that integrates to 2 protons; the signal is split by both the adjacent methyl and methylene groups.

Notice that the pattern of a triplet that integrates to 3 protons, a multiplet that integrates to 2 protons, and a triplet that integrates to 2 protons is characteristic of a propyl group.

24. A strong and sharp absorption in the IR spectrum at ~ 1730 cm^{-1} indicates a carbonyl (C=O) group. The two absorptions at 2710 and 2810 cm^{-1} tell us that the product of ozonolysis is an aldehyde. The aldehydic proton is also visible in the NMR as a singlet (9.0 ppm) that integrates to 1 proton.

The NMR spectrum has two additional signals. One is a doublet (1.1 ppm) that integrates to 6 protons, and the other is a septet (2.4 ppm) that integrates to 1 proton. This is characteristic of an isopropyl group. Therefore, we know that the product of ozonolysis is **2-methylpropanal**.

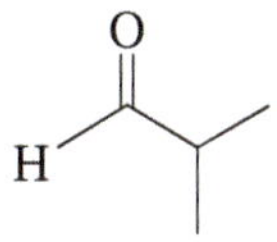

Because only one product is formed, we know that the alkene that formed the aldehyde must be symmetrical. The identification of the aldehyde also agrees with the molecular formula of the alkene that underwent ozonolysis—that is, an eight-carbon symmetrical alkene will form a four-carbon carbonyl compound.

Two symmetrical alkenes will form 2-methylpentanal—***trans*-2,5-dimethyl-3-hexene** and ***cis*-2,5-dimethyl-3-hexene**. We are not given any information that distinguishes between the two stereoisomers. Therefore, the unknown alkene can be either of the two stereoisomers.

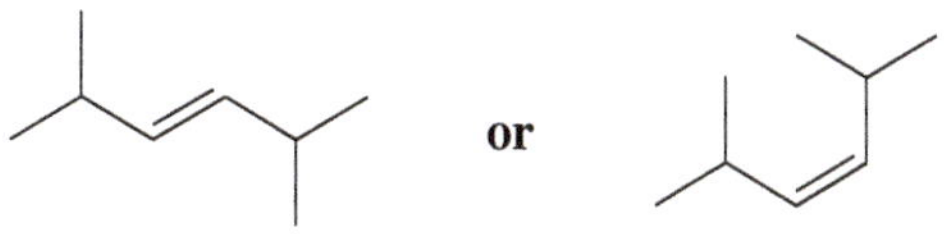

or

25. The strong and sharp absorption in the IR spectrum at ~ 1720 cm^{-1} indicates the presence of a carbonyl group. The broad absorption centered at 3000 cm^{-1} tells us that the carbonyl-containing compound is a carboxylic acid. The broad singlet (12.0 ppm) in the NMR spectrum (shown as offset by 0.2 ppm from where it is placed on the spectrum) confirms the presence of a carboxylic acid group.

The only other signal in the NMR spectrum is a singlet (2.0 ppm) that integrates to six protons, indicating two methyl groups in the same environment. Because the signal is a singlet, the methyl groups must be attached to a carbon that is not attached to a proton. Because we know that the compound has only four carbons and contains a bromine, the compound must be **2-bromo-2-methylpropanoic acid**.

$$(CH_3)_2CBr\text{—}COOH$$

26. The quintet (~2.2 ppm) that integrates to 2 protons indicates that the protons that give this signal have four identical neighboring protons. A carbon cannot be bonded to four protons and still be able to bond to anything else. Therefore, the two protons that give the quintet must be bonded to a carbon that is attached to two methylene groups in the same environment.

The triplet (~3.8 ppm) that integrates to 4 protons must be the signal for the four protons of the two methylene groups. The two methylene groups must be on either side of a carbon that is bonded to two protons (that is, the protons that give the quintet).

$$\text{—}CH_2\text{—}CH_2\text{—}CH_2\text{—}$$

Two bonds are left unaccounted for, so this is where the two chlorines shown in the molecular formula go. Therefore, the compound is **1,3-dichloropropane**. The highly deshielded nature of the signal at 3.8 ppm for the protons bonded to the carbons that are attached to chlorines is further evidence that the chlorines are attached to these carbons.

$$ClCH_2CH_2CH_2Cl$$

27. The IR spectrum shows a strong and sharp absorption at ~1720 cm^{-1}, indicating a carbonyl (C═O) group. The two absorptions at 2720 and 2820 cm^{-1} are characteristic of an aldehyde; they are due to the C—H stretch of the bond between the carbonyl carbon and the aldehydic hydrogen. Because the reactant has three carbons, the aldehyde that produces the IR spectrum must also have three carbons. Therefore, the product of the reaction is **propanal**.

$$CH_3CH_2\text{—}C(=O)H$$

Thus, the reaction that occurred was the conversion of 1-propyne to propanal. This reaction can occur by hydroboration–oxidation of the alkyne (Section 7.8 in the text).

$$H_3C\text{—}C\equiv CH \xrightarrow[\text{2. } H_2O_2,\ HO^-,\ H_2O]{\text{1. } R_2BH,\ THF} CH_3CH_2\text{—}C(=O)H$$

28. The short signal at ~185 ppm in the ^{13}C NMR spectrum suggests the presence of the carbonyl group of a carboxylic acid.

The broad singlet (12.2 ppm) in the ^{1}H NMR spectrum that integrates to 1 proton confirms that the compound contains a carboxylic acid group.

The doublet (1.2 ppm) that integrates to 6 protons and the septet (2.6 ppm) that integrates to 1 proton are characteristic of an isopropyl group.

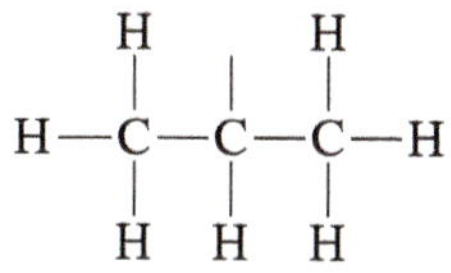

```
     H       H
     |   |   |
  H—C—C—C—H
     |   |   |
     H   H   H
```

Therefore, the compound is **2-methylpropanoic acid**.

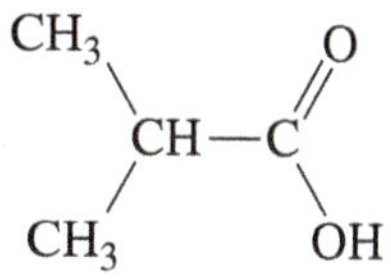

```
CH3       O
   \     //
    CH—C
   /     \
CH3       OH
```

29. The breadth of the singlet (11.8 ppm) that integrates to 1 proton indicates a hydrogen that is attached to an oxygen. The chemical shift of the signal indicates that it is due to the OH group of a carboxylic acid.

```
    O
    ||
   C
 /   \
      OH
```

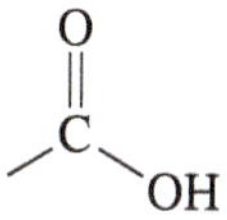

The triplet (~0.9 ppm) that integrates to 3 protons is a methyl group that is attached to a methylene group. The triplet (~2.3 ppm) that integrates to 2 protons indicates a methylene group that is also attached to a methylene group; the chemical shift of this signal indicates that the protons that give this signal are closest to the electron-withdrawing carboxylic acid group. The multiplet at 1.7 ppm that integrates to 2 protons is given by the two protons that split the other two signals into triplets.

We can conclude that the compound responsible for the spectrum is **butanoic acid**.

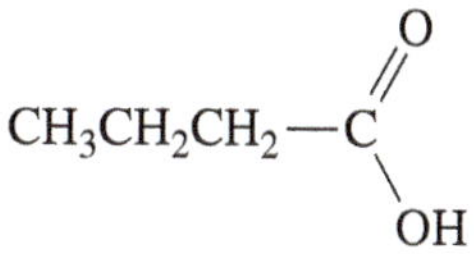

```
              O
             //
CH3CH2CH2—C
             \
              OH
```

30. The singlet (~9.7 ppm) that integrates to 1 proton and the molecular formula that contains one oxygen suggest that an aldehyde is present.

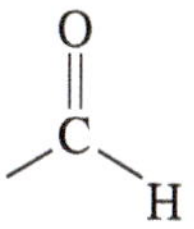

```
    O
    ||
   C
 /   \
      H
```

The signals at 7.7 and 6.7 ppm are due to benzene-ring protons. The fact that they are both doublets that integrate to 2 protons tells us that substituents are on the 1- and 4-positions of the benzene ring.

If the aldehyde group and the disubstituted ring are subtracted from the molecular formula, we find that the second substituent contains 2 carbons, 6 hydrogens, and 1 nitrogen. The remaining NMR signal (~3.0 ppm) is a singlet that integrates to 6 hydrogens. These must be due to two methyl groups in the same environment. The nitrogen must be between the two methyl group; otherwise, they would split each other's signals. The nitrogen causes the signal for the methyl groups to appear at a higher frequency than where methyl groups normally appear.

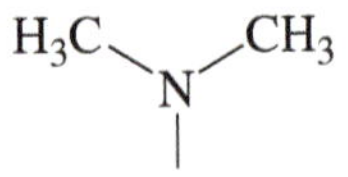

H_3C CH_3
N

Thus, the compound is **4-(dimethylamino)benzaldehyde**.

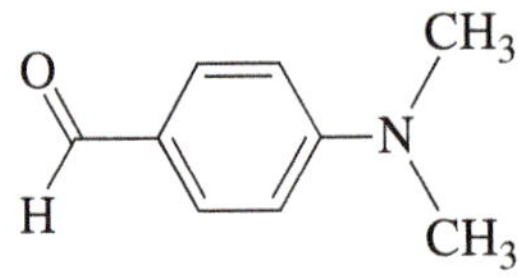

CH_3
O
N
H
CH_3

31. The three signals between 7.4 and 8.1 ppm that together integrate to 5 protons indicate a monosubstituted benzene ring. Subtracting the monosubstituted ring (C_6H_5) from the molecular formula leaves $C_3H_5O_2$ to be accounted for.

The two oxygens in the molecular formula tells us that the compound is an ester because a broad singlet between 10 and 12 ppm that would indicate a carboxylic acid is not present. The remainder of the molecule contains two carbons and five hydrogens.

The two remaining signals, a triplet (1.4 ppm) that integrates to 3 protons and a quartet (4.4 ppm) that integrates to 2 protons, are characteristic of an ethyl group. The three known segments can now be joined in one of two ways:

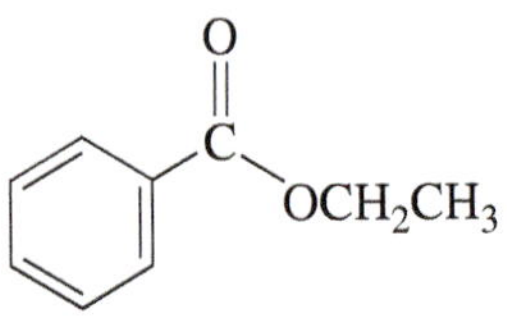

O
C
OCH_2CH_3

ethyl benzoate

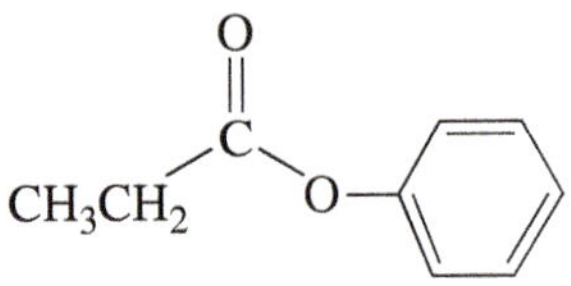

O
C
CH_3CH_2 O

phenyl propanoate

The choice between the two compounds can be made by looking at the chemical shift of the methylene protons. In the ethyl ester, the signal will be highly deshielded by the adjacent oxygen. In the phenyl ester, the signal will be at ~2.1 ppm because the methylene protons are next to the carbonyl group. Because the chemical shift of the methylene protons is 4.4 ppm, we know that the compound is **ethyl benzoate**.

32. Because all three spectra are given by ethylmethylbenzenes, the low-frequency signals in both the ^{1}H NMR and ^{13}C NMR spectra can be ignored because they belong to the methyl and ethyl substituents. The key to determining which spectrum belongs to which ethylmethylbenzene can be found in the aromatic region of the ^{1}H NMR and ^{13}C NMR spectra.

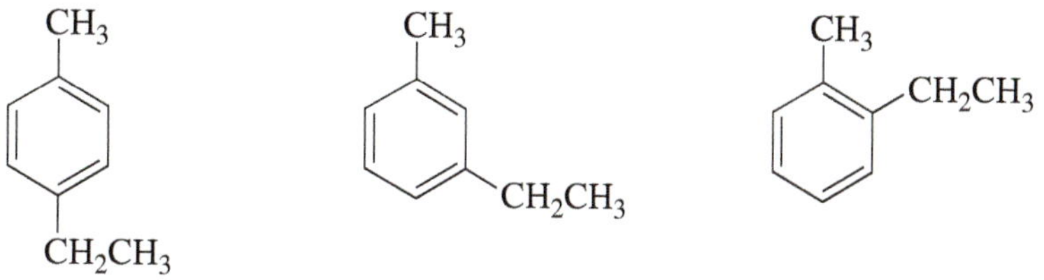

4-ethylmethylbenzene 3-ethylmethylbenzene 2-ethylmethylbenzene

The aromatic region of the ^{13}C NMR spectrum of 4-ethylmethylbenzene will show four signals because it has four different ring carbons.

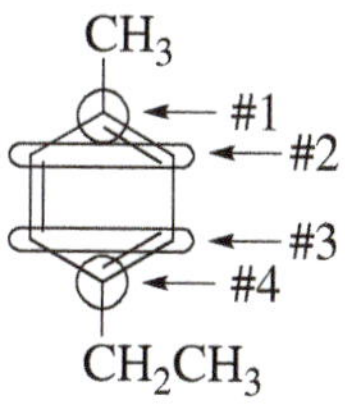

The aromatic region of the ^{13}C NMR spectrum of 3-ethylmethylbenzene will show six signals because it has six different ring carbons.

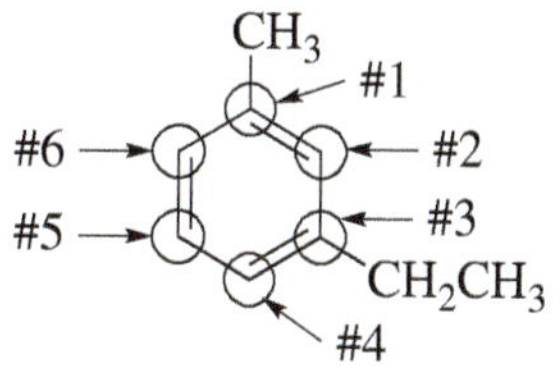

The aromatic region of the ^{13}C NMR spectrum of 2-ethylmethylbenzene will also show six signals because it has six different ring carbons.

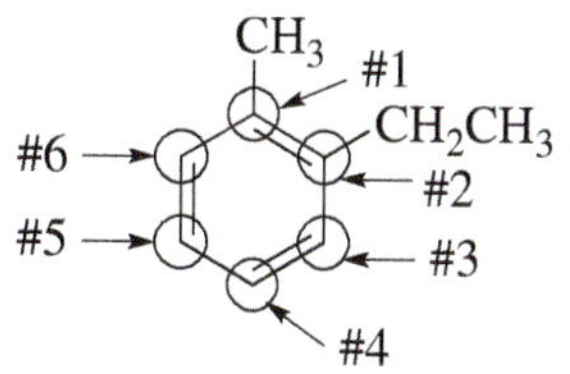

We now know that spectrum **(b)** is the spectrum of 4-ethylmethylbenzene because its ^{13}C NMR spectrum has four signals and the other two compounds will show six signals.

To distinguish between 2-ethylmethylbenzene and 3-ethylmethylbenzene, we need to look at the splitting patterns in the aromatic regions of the ^{1}H NMR spectra. Analysis of the aromatic region for spectrum **(c)** is difficult because the signals are superimposed. Analysis of the aromatic region for spectrum **(a)** provides the needed information. A triplet (7.2 ppm) that integrates to 1 proton is clearly present. This means that spectrum **(a)** is 3-ethylmethylbenzene because 2-ethylmethylbenzene would not show a triplet. Therefore, spectrum **(c)** is 2-ethylmethylbenzene.

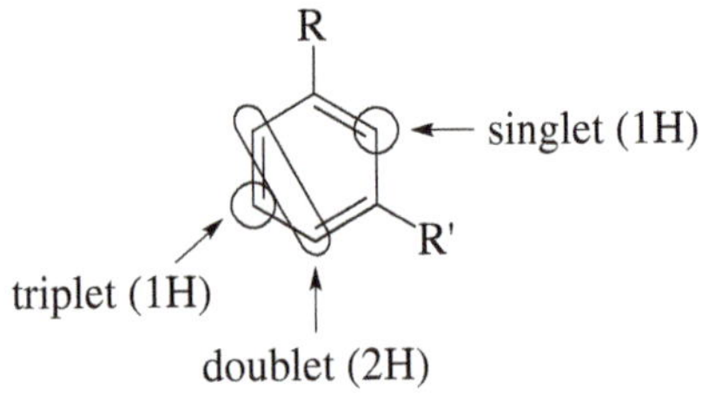

R
singlet (1H)
R'
triplet (1H)
doublet (2H)
splitting pattern for 3-methylethyl benzene

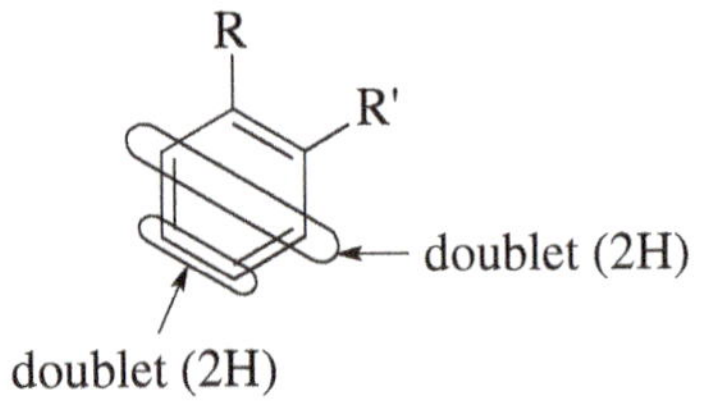

R
R'
doublet (2H)
doublet (2H)
splitting pattern for 2-methylethyl benzene

The final assignments are:

(a) 3-ethylmethylbenzene **(b) 4-ethylmethylbenzene** **(c) 2-ethylmethylbenzene**

33. The simplicity of the NMR spectrum of a compound with 7 carbons and 14 hydrogens indicates that the compound must be symmetrical. From the molecular formula, we see that it has one degree of unsaturation. The absence of signals near 5 ppm rules out an alkene. Because the compound has an oxygen, the degree of unsaturation may be due to a carbonyl group.

The doublet (1.1 ppm) that integrates to 12 protons and the septet (2.8 ppm) that integrates to 2 protons suggest the presence of two isopropyl groups.

If two isopropyl groups are subtracted from the molecular formula, we find that the remainder of the molecule is composed of one carbon and one oxygen. Thus, the compound is the one shown here.

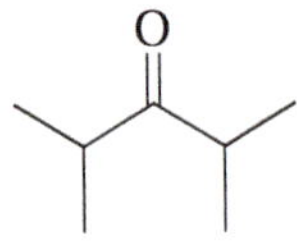

O

34. The molecular formula tells us that the compound has one degree of unsaturation. The multiplet (5.2 ppm) that integrates to 1 proton is due to a vinylic proton (that is, it is attached to an *sp*^{2} carbon). Thus, the degree of unsaturation is due to a carbon–carbon double bond. Because there is only one vinylic proton, we can assume that the alkene is trisubstituted.

Three additional signals are present that each integrate to 3 protons, suggesting that all three signals are due to methyl groups. The alkene, therefore, is **2-methyl-2-butene**.

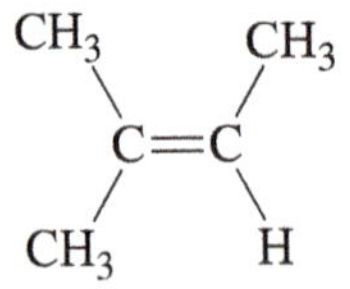

CH_3 CH_3
C=C
CH_3 H

Notice that two of the three signals given by the methyl groups are singlets and one is a doublet. The methyl group that gives the doublet is bonded to the carbon that is attached to the vinylic proton. The other two methyl groups are bonded to the other *sp*^{2} carbon.

35. A medium-intensity absorption at ~2120 cm^{-1} indicates the presence of a carbon–carbon triple bond. The sharp absorption at 3300 cm^{-1} is due to the C—H stretch of a hydrogen bonded to an *sp* carbon. Thus, the compound is a terminal alkyne.

$$—C \equiv C—H$$

The intense and broad peak centered at 3300 cm^{-1} is evidence of an O—H group.

The NMR spectrum can be used to determine the connectivity between the groups. The signal (2.5 ppm) that integrates to 1 proton is due to the proton of the terminal alkyne.

The singlet (3.2 ppm) that integrates to 1 proton must be due to the proton of the OH group.

The singlet (4.2 ppm) that integrates to 2 protons must be due to a methylene group that connects the triply bonded carbon to the OH group. The compound is **2-propyn-1-ol**.

$$H—C \equiv C—CH_2—OH$$

This arrangement explains the absence of any splitting and the highly deshielded nature of the signal for the methylene group.

36. The two singlets (4.9 and 5.1 ppm) that each integrate to 1 proton are vinylic protons. Therefore, we know that the compound is an alkene.

The singlet (2.8 ppm) that integrates to 3 protons is a methyl group. The deshielding results from its being attached to an sp^2 carbon.

If we subtract the two vinylic protons, the two sp^2 carbons of the alkene, and the methyl group from the molecular formula, we are left with CH_2Cl. Thus, a chloromethyl group is the fourth substituent of the alkene and gives the singlet (4.9 ppm) that integrates to 2 protons. Its deshielding is due to the proximity to the electronegative chlorine.

$$—CH_3 \qquad —CH_2Cl$$

Now we need to determine the substitution pattern of the alkene. The absence of splitting indicates that the two vinylic protons must be attached to the same carbon. If these protons were cis or trans to each other, they would give doublets with significant *J* values. Geminal protons attached to sp^2 carbons have very small *J* values, so splitting is typically not observed. (See Table 14.2 on page 644 of the text.)

Thus, the compound is **3-chloro-2-methyl-1-propene**.

$$ClCH_2—C(CH_3)=CH_2$$

37. The only signal that integrates to 1 proton is the singlet at 2.2 ppm. This must be due to the OH group of the alcohol.

The signals centered around 7.3 ppm are given by benzene-ring protons. Because they integrate to 5 protons, the benzene ring must be monosubstituted.

The two triplets (2.8 and 3.8 ppm) that each integrate to 2 protons suggest two adjacent methylene groups. Both signals are fairly deshielded, indicating an electronegative atom nearby.

The fragments identified at this point are a monosubstituted benzene ring, an OH group, and two adjacent methylene groups.

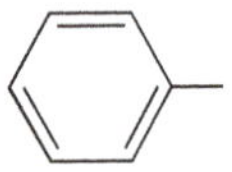

HO— $—CH_2CH_2—$

No other signals are in the NMR spectrum, so the compound must be the one shown here.

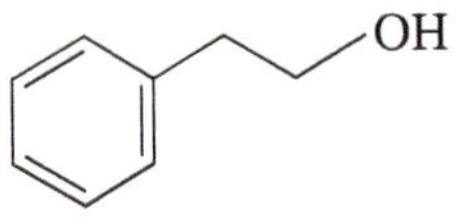

OH

38. The doublet (0.9 ppm) that integrates to 6 protons and the multiplet (1.8 ppm) that integrates to 1 proton suggest the presence of an isopropyl group.

Because we are told that the compound is an alcohol, the other signal that integrates to 1 proton (the triplet at 2.4 ppm) must be due to the OH proton. The fact that the signal is a triplet indicates that the OH group is probably attached to a methylene group.

The signal for the methylene group must be the remaining signal at 3.4 ppm because it integrates to 2 protons. The relatively high-frequency chemical shift confirms that the methylene group is attached to the oxygen.

Putting together the isopropyl group and the methylene group that is attached to an OH group identifies the compound as **2-methyl-1-propanol**.

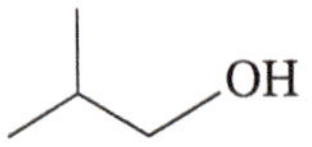

OH

39. The protons that are responsible for the doublet (1.2 ppm) that integrates to 6 protons must be adjacent to a carbon that is attached to only one proton. Because the spectrum does not have a signal that integrates to one proton, the compound must have two methyl groups in the same environment. Each methyl group must be adjacent to a carbon that is attached to one proton, and those two single protons must be in identical environments.

H—C(H)(H)—C(H)— —C(H)—C(H)(H)—H

Because the compound must be symmetrical, the two oxygens in the compound must be due to two OH groups in identical environments. The hydrogens of the OH groups give a singlet (3.8 ppm) that integrates to 2 protons.

```
    H  OH           OH H
    |  |            |  |
 H—C—C—          —C—C—H
    |  |            |  |
    H  H            H  H
```

The protons that give the triplet (2.6 ppm) must be bonded to a carbon that is adjacent to a total of two protons. Because the triplet integrates to 2 protons, it must be due to a methylene group that connects the two pieces.

```
    H  OH H  OH H
    |  |  |  |  |
 H—C—C—C—C—C—H
    |  |  |  |  |
    H  H  H  H  H
```

This structure is confirmed by the relatively high-frequency multiplet (4.2 ppm) that is given by the protons attached to the carbons that are attached to the OH groups. The signal for these protons is split by both the adjacent methyl group and the adjacent methylene group.

40. The absorption in the IR spectrum at ~1650 cm^{-1} could be due to either a carbonyl group or an alkene. Its strength and breadth tells us that it is probably due to a carbonyl (C═O) group. The strong and broad absorption at ~3300 cm^{-1} that contains two broad peaks suggests two N—H bonds; thus, an NH_2 group is present. When these two groups are subtracted from the molecular formula, all that is left is C_2H_5.

The triplet (~1.1 ppm) that integrates to 3 protons and the quartet (2.2 ppm) that integrates to 2 protons indicate the presence of an ethyl group; this accounts for the C_2H_5 fragment. Thus, all the fragments of the compound have been identified: C═O, NH_2, and CH_3CH_2. The compound, therefore, is **propanamide**.

O

NH_2

The presence of an amide explains the lower-than-normal frequency of the C═O stretch in the IR spectrum. The breadth of the N—H stretches confirms that these are amide N—H stretches and not amine N—H stretches. The broad singlets (6.2 and 6.6 ppm) in the NMR spectrum are given by the protons attached to the nitrogen. The protons resonate at different frequencies because the C—N bond has partial double-bond character, which causes the protons to be in different environments.

41. The singlet (2.3 ppm) in the first spectrum that integrates to 3 hydrogens must be due to an isolated methyl group.

The triplet (1.1 ppm) that integrates to 3 protons and the quartet (2.5 ppm) that integrates to 2 protons are characteristic of an ethyl group.

The singlet (4.8 ppm) that integrates to 1 proton must be due to a single hydrogen attached to nitrogen.

```
   H        H  H
   |        |  |
 —C—H    —C—C—H     —N—
   |        |  |        |
   H        H  H        H
```

Now that the three fragments have been identified, we know that the compound is **ethylmethylamine**.

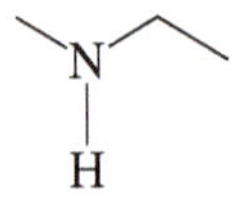

N
H

The second spectrum shows that a broad singlet (2.8 ppm) must be due to hydrogens that are attached to nitrogens. Because the signal integrates to 2 protons, we know that the compound is a primary amine.

The triplet (0.8 ppm) that integrates to 3 protons is due to a methyl group that is adjacent to a methylene group. The triplet (2.7 ppm) that integrates to 2 protons must also be adjacent to a methylene group. The multiplet (1.5 ppm) that integrates to 2 protons is the methylene group that splits both the methyl and methylene groups. (The two triplets and multiplet are characteristic of a propyl group.)

Therefore, the compound is **propylamine**.

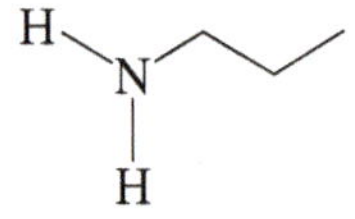

H
N
H

42. The relatively weak absorption in the IR spectrum at ~1650 cm^{-1} tells us that it is probably due to a carbon–carbon double bond. This is reinforced by the presence of absorptions at ~3080 cm^{-1}, indicating C—H bond stretches of hydrogens attached to sp^2 carbons.

The shape of the two absorptions at ~3300 cm^{-1} suggests the presence of an NH_2 group of a primary amine. (Compare these to the shape of the N—H stretches of an NH_2 group of an amide in Problem 40.)

The three signals in the NMR spectrum between 5.0 and 6.0 ppm that integrate as a group to 3 protons indicate that there are three vinylic protons. Therefore, we know that the alkene is monosubstituted.

The two remaining signals in the NMR spectrum are a doublet (3.3 ppm) and a singlet (1.3 ppm) that each integrate to 2 protons. Because splitting is not typically seen with protons attached to nitrogens, we can identify the singlet at 1.3 ppm as due to the two amine protons. The doublet must be due to a methylene group that is attached to an sp^2 carbon and split by a vinylic proton that is attached to the same carbon. The compound, therefore, is **allylamine**.

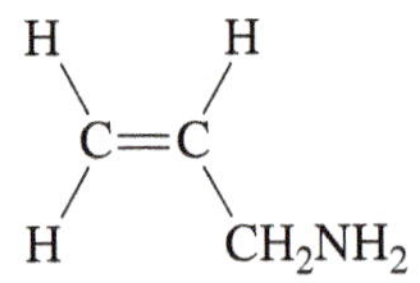

H H
C=C
H CH_2NH_2

Now we can understand why the signal at 5.9 ppm is a multiplet. This vinylic proton is split by the methylene group and two unique vinylic protons. The signals for the other two vinylic protons are doublets because each is split by the single proton attached to the adjacent sp^2 carbon. Notice that the higher-frequency doublet has the larger J value. This is the signal for the proton that is trans to its coupled proton.

43. The molecular formula tells us that the compound does not have any degrees of unsaturation. Therefore, the oxygen must be the oxygen of either an ether or an alcohol. Because there are no signals that integrate to one proton, we can conclude that the compound is an ether.

The triplet (1.2 ppm) that integrates to 3 protons and the quartet (3.5 ppm) that integrates to 2 protons suggests an ethyl group. The high-frequency chemical shift of the ethyl's methylene group and the fact that it shows splitting only by the three protons of the methyl group indicate that the ethyl group is next to the oxygen.

The two remaining signals are both triplets (3.4 and 3.5 ppm), and each integrates to 2 protons. Thus, the signals are due to two adjacent methylene groups. Because both signals occur at high frequencies, both must be attached to electron-withdrawing atoms.

Because the molecular formula tells us that the compound contains a bromine, we can conclude that the compound is **2-bromoethyl ethyl ether**.

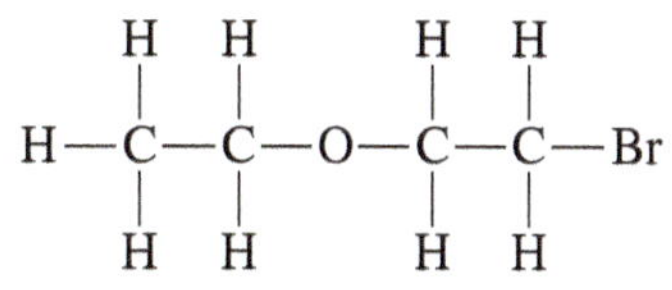

```
   H  H     H  H
   |  |     |  |
H—C—C—O—C—C—Br
   |  |     |  |
   H  H     H  H
```

44. The IR spectrum shows a strong and broad absorption at ~3300 cm^{-1}, indicating that the compound is an alcohol.

The signals in the NMR spectrum between 6 and 7 ppm indicate a benzene ring. Because these signals integrate to a total of 3 protons, the benzene ring must be trisubstituted.

Because the signal at 6.3 ppm is a doublet, it must be adjacent to one proton, and because the signal at 6.7 ppm is a triplet, it must be adjacent to two protons. Thus, the three benzene-ring protons must be adjacent to one other.

The singlet at 8.7 ppm is the only signal in the spectrum that can be attributed to the proton of the OH group. Because the signal integrates to 2 protons, the compound must have two OH groups in the same environment.

The singlet (2.0 ppm) that integrates to three protons indicates that the compound has a methyl group that is not adjacent to a carbon that is attached to any hydrogens.

Therefore, we know that the three substituents that are attached to adjacent carbons on the benzene ring are two OH groups and a methyl group. Because the OH groups are in the same environment, the compound must be the one shown here.

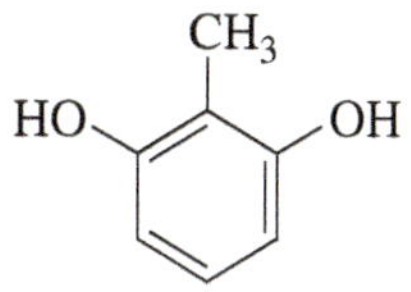

CH_3

HO OH

45. We are told that the compound is an alcohol. Because the singlet (1.4 ppm) is the only signal that integrates to 1 proton, it must be the signal given by the OH group.

We know that a triplet that integrates to 3 protons and a quartet that integrates to 2 protons are characteristic of an ethyl group. In this case, the triplet (0.8 ppm) integrates to 6 protons and the quartet integrates to 4 protons. Therefore, the compound must have two ethyl groups in identical environments.

The only other signal in the spectrum is the singlet (1.2 ppm) that integrates to 3 protons. This signal must be due to a methyl group that is bonded to a carbon that is not attached to any hydrogens.

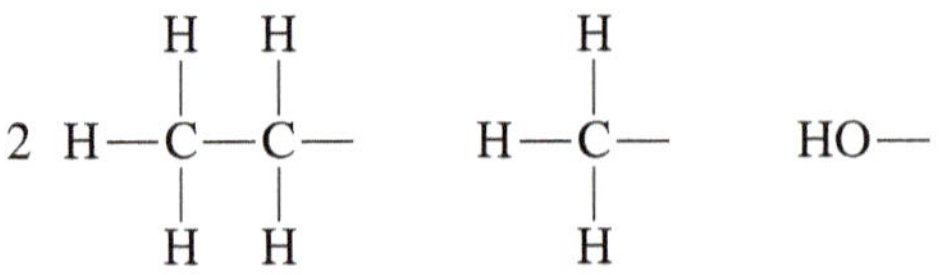

2 CH_3CH_2— H—C(H)(H)— HO—

Because the NMR spectrum does not show any additional signals, the compound must be **3-methyl-3-pentanol**.

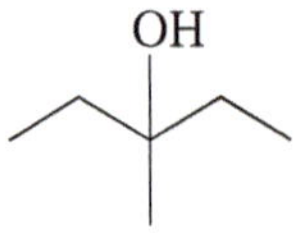

OH

46. The doublet (1.1 ppm) that integrates to 6 protons and the septet (~5 ppm) that integrates to 1 proton suggest an isopropyl group. The triplet (1.7 ppm) that integrates to 3 protons and the quartet (2.2 ppm) that integrates to 2 protons suggest an ethyl group. When these two groups are subtracted from the molecular formula, all that remains is CO_2. The splitting patterns tell us that the isopropyl and ethyl groups are isolated from one another. We can conclude then that the compound is an ester. There are two possibilities:

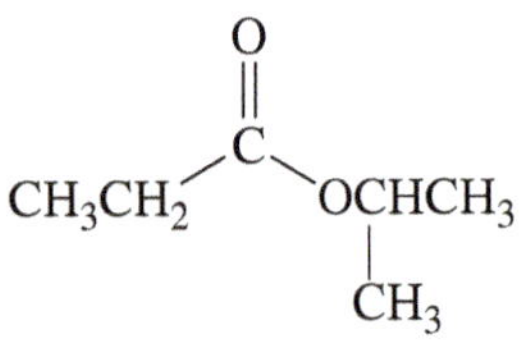

$CH_3CH_2C(=O)OCH(CH_3)CH_3$ **or** $CH_3CH(CH_3)C(=O)OCH_2CH_2$

isopropyl propanoate ethyl 2-methylpropanoate

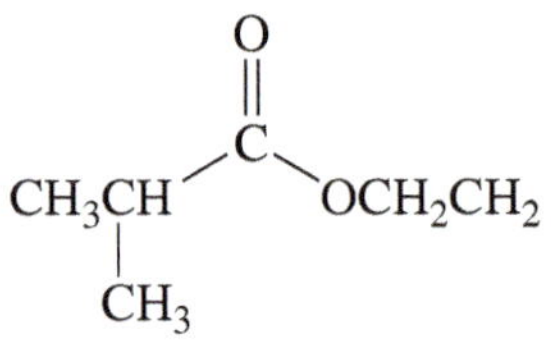

Because the highest-frequency signal (the septet) is given by the CH of the isopropyl group, we know that the CH is attached to an oxygen. Therefore, we know that the compound that gives the NMR data is **isopropyl propanoate**.

47. The multiplet (6.9–7.8 ppm) that integrates to 4 protons indicates a disubstituted benzene ring. Because the signal is a multiplet, we know that the substituents are on either the 1- and 2-positions or the 1- and 3-positions. If the substituents were on the 1- and 4-positions, either one singlet (if the two substituents are identical) or two doublets (if the substituents are not identical) would be observed.

Three signals (1.4, 2.5, 3.8 ppm) each integrate to 2 protons. The fact that the signals are two triplets and a multiplet suggests that the compound has three adjacent methylene groups. (The methylene groups on the ends will be triplets, and the one in the middle will be a multiplet.)

—$CH_2CH_2CH_2$—

From the molecular formula, we know that the compound has an oxygen. The triplet at 3.8 ppm indicates that that particular methylene group is next to the oxygen. The two fragment we have identified account for the entire molecular formula. Therefore, the compound must have the structure shown here.

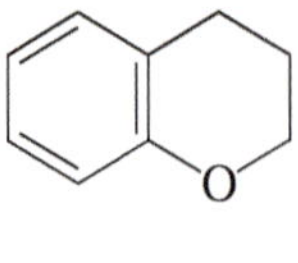

O

48. The molecular formula indicates that the compound has one degree of unsaturation. The NMR spectrum does not show any signals in the area expected for vinylic protons, so the compound must be either a ketone or a cyclic ether. A cyclic ether would be expected to have protons on adjacent carbons, so the signals would show splitting. Because the two signals in the spectrum are both singlets, the compound must be a ketone.

The fact that the compound has 11 carbons and 22 hydrogens but gives only two singlets in the NMR spectrum indicates that the compound must be symmetrical.

We know that a *tert*-butyl group gives a singlet that integrates to 9 protons. The symmetry of the molecule leads us to conclude that the singlet that integrates to 18 protons is due to two *tert*-butyl groups. We can then assume that the singlet that integrates to 4 protons is due to two nonadjacent methylene groups.

$$\overset{O}{\overset{\|}{C}} \qquad 2\ CH_3-\underset{CH_3}{\underset{|}{\overset{CH_3}{\overset{|}{C}}}}- \qquad 2\ -CH_2-$$

These fragments account for all atoms in the molecular formula. Therefore, the compound must be **2,2,4,4-tetramethyl-4-heptanone**.

$$CH_3\underset{CH_3}{\underset{|}{\overset{CH_3}{\overset{|}{C}}}}CH_2\overset{O}{\overset{\|}{C}}CH_2\underset{CH_3}{\underset{|}{\overset{CH_3}{\overset{|}{C}}}}CH_3$$

2,2,6,6-tetramethyl-4-heptanone

49. The molecular formula has one degree of unsaturation, so it must have a carbon–carbon double bond, a cyclic structure, or a carbonyl group.

The signal (3.8 ppm) that integrates to 4 hydrogens suggests the presence of two methylene groups in identical environments because a single carbon (other than the carbon in methane) cannot be attached to four hydrogens. The chemical shift suggests that each methylene group must be attached to an oxygen. Because the compound has only one oxygen, the two methylene groups must be attached to the same oxygen.

$$-CH_2-O-CH_2-$$

The signal (1.6 ppm) that integrates to 4 hydrogens also suggests the presence of two methylene groups in identical environments. The four methylene groups and the oxygen account for all the atoms in the molecular formula. Thus, the compound must be a cyclic ether.

[five-membered ring structure with O]

The fact that the signal at the higher frequency is a triplet and the other signal is a multiplet confirms this structure.

50. Each of the doublets (7.3 and 7.7 ppm) that integrates to 4 protons is given by benzene-ring protons. Because a benzene ring does not have 8 protons, there must be two benzene rings in the compound. The doublets indicate that the benzene rings have substituents at the 1- and 4-positions and, because each doublet integrates to 4 protons, the two substituents on each of the benzene rings must be the same.

The singlet (3.8 ppm) that integrates to 6 protons suggests the compound has two methyl groups in an identical environment. The chemical shift of the singlet indicates that each is attached to an electronegative atom. The molecular formula indicates that the electronegative atom is an oxygen.

$$2 \quad -OCH_3$$

When the two disubstituted benzene rings and the two CH_3O groups are subtracted from the molecular formula, all that remains is CO. Therefore, a carbonyl group must connect the two benzene rings.

CH_3O—C₆H₄—C(=O)—C₆H₄—OCH_3

51. **A.** The broad singlet at ~5.2 ppm that integrates to 2 protons indicates an NH_2 group. The broad singlet at ~3.7 ppm that integrates to 1 proton indicates an OH group. Subtracting NH_2 and OH from the molecular formula leaves C_3H_6. The two triplets and the quintet that each integrate to 2 hydrogens suggests a $-CH_2CH_2CH_2-$ unit with a group on either end that does not cause splitting. Therefore, this isomer is

HO—CH₂CH₂CH₂—NH_2

B. The broad singlet at ~3.7 ppm that integrates to 1 proton indicates an NH group. The broad singlet at ~2.0 ppm that integrates to 1 proton indicates an OH group. The 3H singlet at ~3.3 ppm indicates a methyl group attached to an electron-withdrawing group. Subtracting NH, OH, and CH_3 from the molecular formula leaves C_2H_4. The two triplets that each integrate to 2 hydrogens suggest a $-CH_2CH_2-$ unit with a group on either end that does not cause splitting. Therefore, this isomer is

HO—CH₂CH₂—NH—CH₃

C. The broad singlet at ~5.1 ppm that integrates to 2 protons indicates an NH_2 group. The singlet at ~3.3 that integrates to 3 protons indicates a methyl group attached to an electron-withdrawing group. The electron-withdrawing group must be the oxygen because the nitrogen has two protons attached to it, so it can be attached to only one other group. Subtracting NH_2 and OCH_3 from the molecular formula leaves C_2H_4. The two triplets that each integrate to 2 hydrogens suggest a $-CH_2CH_2-$ unit with a group on either end that does not cause splitting. Therefore, this isomer is

CH₃—O—CH₂CH₂—NH_2

52. **A.** The absence of an absorption band > 3000 cm^{-1} eliminates all the structures with an O—H or N—H bond. Therefore, the following compound is responsible for the spectrum.

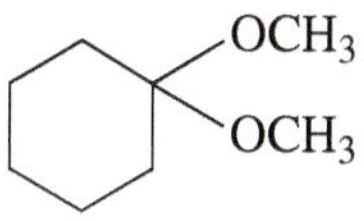

B. The broad absorption band at ~3400 cm^{-1} is due to an O—H bond. Therefore, the following compound is responsible for the spectrum.

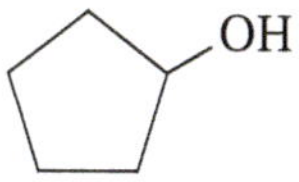

C. The two absorption bands at ~3400–3300 cm^{-1} indicate two N—H bonds. Therefore, the following compound is responsible for the spectrum.

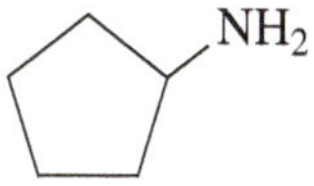

D. The absorption band at ~3400–3300 cm^{-1} is due to one N—H bond. Therefore, the following compound is responsible for the spectrum.

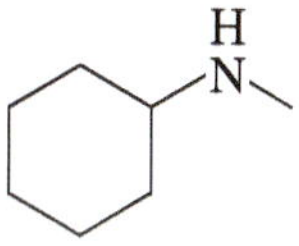

53. **A.** The two absorption bands at ~3400–3300 cm^{-1} indicate two N—H bonds. The doublet at 0.9 ppm indicates an isopropyl group. Subtracting NH_2 and $(CH_3)_2CH$ from the molecular formula leaves C_2H_4. Therefore, this compound is

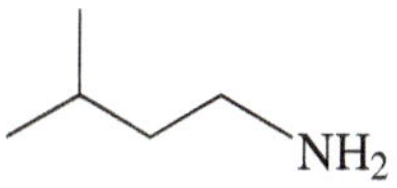

B. The broad absorption band at ~3400–3300 cm^{-1} is due to an O—H bond. The doublet at 0.9 ppm indicates an isopropyl group. Subtracting NH_2 and $(CH_3)_2CH$ from the molecular formula leaves C_2H_4. Therefore, this compound is

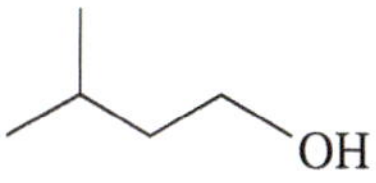

54. **A.** The broad singlet at ~2.0 ppm indicates an N—H bond. The compound has 23 hydrogens but only 4 additional signals, suggesting that it is a symmetrical compound. The doublet at 0.9 ppm indicates an isopropyl group. The triplet at ~2.6 ppm indicates a group split only by an adjacent CH_2 group. Therefore, this compound is

B. This spectrum is similar to the spectrum in part A. However, the signal indicating a hydrogen bonded to a nitrogen is missing, and the triplet that indicates a group split only by an adjacent CH_2 group is at a higher frequency (~3.4 ppm), indicating that it is adjacent to the oxygen. Therefore, this compound is

55. D.J. made the mistake of thinking that the H_a and H_b protons are equivalent. This made him conclude that there would be only 2 signals in the spectrum. However, the H_a and H_b protons are not in the same environment—H_a is trans to Br, and H_b is cis to Br. Therefore, there are 3 signals and each is a multiplet.

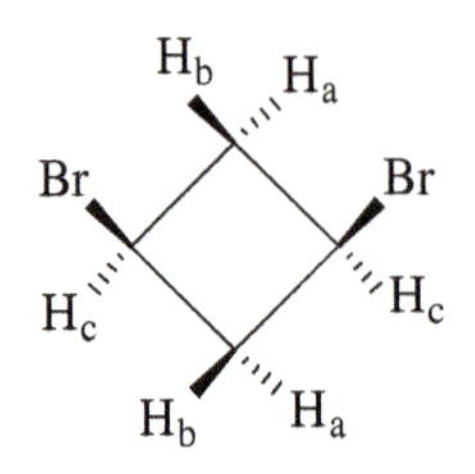

56. Although D.J. is now more experienced, he made the same mistake, still thinking that the H_a and H_b protons are equivalent. The H_a and H_b protons are not in the same environment —H_a is trans to OH, and H_b is cis to OH. Therefore, there are 3 signals (all multiplets) in addition to the signal for the H that is attached to the oxygen.

57. **A.** The highest-frequency signal is for the H that is attached to the same carbon that Br is attached to. The signal at ~0.9 ppm that integrates to 3 hydrogens is a methyl group that is attached to a CH_2 group. The signal at ~2.8 ppm that integrates to 3 hydrogens is a methyl group that is close to an electron-withdrawing group. Integration shows that the compound has two CH_2 groups. Thus, compound **A** is

B. The 6H triplet at ~1.1 ppm indicates two methyl groups in the same environment. There are also two CH_2 groups in the same environment. Thus, compound **B** is

$$CH_3CH_2\underset{\displaystyle Br}{\underset{|}{C}}HCH_2CH_3$$

The reaction that produces **A** and **B** is

$$CH_3CH{=}CHCH_2CH_3 \;+\; HBr \longrightarrow CH_3\underset{\displaystyle Br}{\underset{|}{C}}HCH_2CH_2CH_3 \;+\; CH_3CH_2\underset{\displaystyle Br}{\underset{|}{C}}HCH_2CH_3$$

C. The 6H singlet at ~2.8 ppm indicates 2 methyl groups that are attached to a carbon that is not attached to a hydrogen. Because the signal is at a higher frequency than expected for a methyl group, the carbon must be attached to an electron-withdrawing group (a Br). Integration shows that the compound has a CH_2 group and another methyl group. Therefore, compound **C** is

Br

D. The 6H doublet at 0.9 ppm indicates an isopropyl group. The compound has a third methyl group as well as two carbons that are attached to only one H. Therefore, compound **D** is

Br

The reaction that produces **C** and **D** is

+ HBr ⟶ Br + Br

58. The strong absorption band at ~3400 cm^{-1} is due to an OH group, which also gives the 1H singlet. The 6H singlet is due to two methyl groups that are attached to a carbon that is not attached to a hydrogen. This information and the molecular formula give two possible structures for the compound:

OH Cl **or** Cl OH

59. The absence of a molecular ion peak suggests that the compound might be an alcohol. Subtracting 84 from the molecular ion (102 – 84) = 18 shows that the peak at $m/z = 84$ results from loss of water from the molecular ion, confirming that the compound is an alcohol.

In order to lose water, the alcohol must have a γ-hydrogen. The rule of 13 gives a molecular formula of C_7H_{18}. Because we know that the compound is an alcohol, we must add an O and subtract a C and 4 Hs from the molecular formula, resulting in a molecular formula of $C_6H_{14}O$.

The molecular ions of three alcohols with molecular formula $C_6H_{14}O$ and a γ-hydrogen are shown below. α-Cleavage of the first alcohol results in a peak at $m/z = 31$; α-cleavage of the second alcohol results in peaks at $m/z = 45$ and 87; α-cleverage of the third alcohol results in a peak at $m/z = 73$.

$$CH_3CH_2CH_2CH_2CH_2CH_2\overset{+}{O}H \xrightarrow{\alpha\text{-cleavage}} CH_3CH_2CH_2CH_2\dot{C}H_2 + \underset{m/z\,=\,31}{CH_2{=}\overset{+}{O}H}$$

$$CH_3CH_2CH_2CH_2\underset{\overset{|}{{}^{+}OH}}{C}HCH_3 \xrightarrow{\alpha\text{-cleavage}} CH_3CH_2CH_2\dot{C}H_2 + \underset{m/z\,=\,45}{CH_3CH{=}\overset{+}{O}H}$$

$$\cdot CH_3 + \underset{m/z\,=\,87}{CH_3CH_2CH_2CH_2CH{=}\overset{+}{O}H}$$

$$CH_3CH_2CH_2\underset{\overset{|}{{}^{+}OH}}{C}HCH_2CH_2CH_3 \xrightarrow{\alpha\text{-cleavage}} CH_3CH_2\dot{C}H_2 + \underset{m/z\,=\,73}{CH_3CH_2CH_2CH{=}\overset{+}{O}H}$$

Therefore, the alcohol that gives the mass spectrum is

$$CH_3CH_2CH_2CH_2\underset{\overset{|}{OH}}{C}HCH_3$$

60. **A** shows an absorption for a double bond but not for an OH group. Therefore, **A** must be a tertiary alkyl halide (2-bromo-2-methylpropane) because tertiary alkyl halides undergo only elimination with a strong base.

B shows an absorption for an OH group but not for a double bond. Therefore, **B** must be a primary alkyl halide (1-bromobutane) because primary alkyl halides undergo primarily substitution.

C shows an absorption for a double bond and for an OH group. Therefore, **C** must be a secondary alkyl halide (2-bromobutane) because secondary alkyl halides undergo both substitution and elimination.

61. The strong and broad absorption bands at $\sim 3400\ cm^{-1}$ indicate that **A** and **B** are the spectra of alcohols. The absorptions in Spectrum **A** at a little $<3000\ cm^{-1}$ indicate hydrogens attached to sp^2 carbons, and the absorption at $1600\ cm^{-1}$ indicates a benzene ring. Spectrum **B** has neither of these absorptions. Therefore,

A is the spectrum of Compound **2**.
B is the spectrum of Compound **4**.

The strong absorption at $\sim 1700\ cm^{-1}$ indicates that **C**, **D**, and **E** are the spectra of compounds with a carbonyl group. The broad absorption at $\sim 3000\ cm^{-1}$ indicates that

E is the spectrum of a carboxylic acid (Compound **5**).

The absorption at $\sim 2700\ cm^{-1}$ indicates that

D is the spectrum of an aldehyde (Compound **3**). Therefore,
C is the spectrum of Compound **1**.

62. The three signals in the ^{1}H NMR spectrum at 5.9–6.3 ppm are due to three vinylic hydrogens. The 3H signal in the ^{1}H NMR spectrum at 2.3 ppm is due to a methyl group close to an electron-withdrawing group. The signal in the ^{13}C NMR spectrum at ~200 ppm indicates a carbonyl carbon. Putting the pieces together results in the following compound.

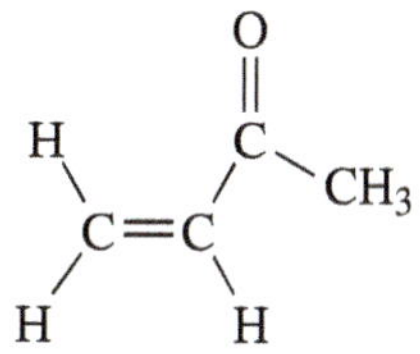

63. The strong and broad absorption band at ~3400 cm^{-1} indicates that the compound is an alcohol. The compound loses water when it is heated with H_2SO_4, forming a compound with six carbons that has only one kind of hydrogen and two kinds of carbon atoms. The reactant and product are shown below.

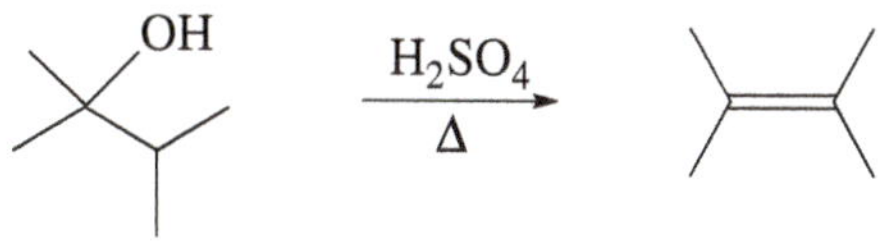

$$\xrightarrow[\Delta]{H_2SO_4}$$

64. The IR absorption bands at 1600 cm^{-1}, 1500 cm^{-1}, and 3030 cm^{-1} indicate a benzene ring. This is confirmed by the two doublets between 7.1 and 7.9 ppm in the ^{1}H NMR spectrum that indicate a 1,4-disubstituted benzene. The IR absorption band at 1690 cm^{-1} indicates a ketone carbonyl group with significant single-bond character. Therefore, the carbonyl group must be attached to the benzene ring. The two 6H doublets indicate two isopropyl groups. The septet at ~4.8 ppm indicates that one of the isopropyl groups is attached to an electron-withdrawing group. The structure of the compound is shown below.

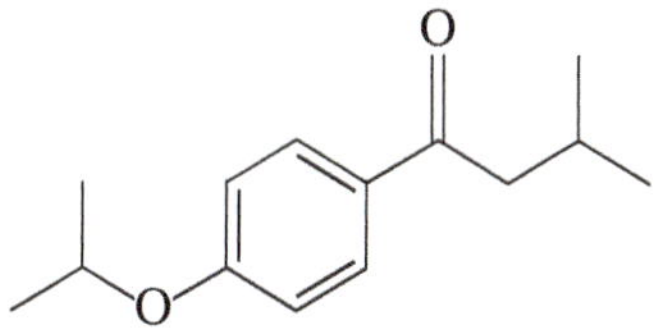

65. The IR absorption band at 1730 cm^{-1} indicates a carbonyl group, the IR absorption bands at 2700–2800 cm^{-1} indicate an aldehyde, and the IR absorption bands at 1600 cm^{-1} and 1500 cm^{-1} indicate a benzene ring. The two doublets between 7.5 and 7.9 ppm in the ^{1}H NMR spectrum indicate a 1,4-disubstituted benzene. The doublet at 1.2 ppm and the multiplet at 5.0 ppm indicate an isopropyl group. The structure of the compound is shown below.

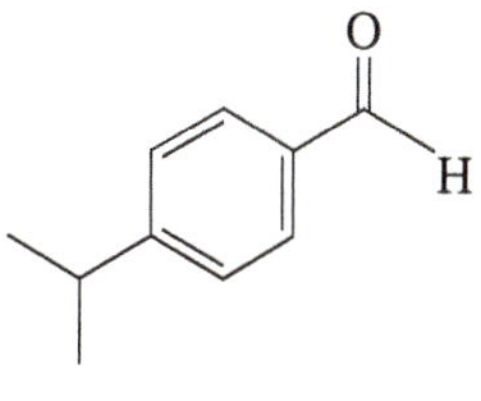

CHAPTER 1
Remembering General Chemistry: Electronic Structure and Bonding

Important Terms

antibonding molecular orbital	a molecular orbital that results when two atomic orbitals with opposite phases interact. Electrons in an antibonding orbital decrease bond strength.
atomic number	the number of protons (or electrons) that a neutral atom has.
atomic mass	the average mass of the atoms in the naturally occurring element.
atomic orbital	an orbital associated with an atom; the three-dimensional area around its nucleus where electrons are most likely to be found.
aufbau principle	the principle that states that an electron will always go into the available orbital with the lowest energy.
bond dissociation energy	the amount of energy required to break a bond in a way that allows each of the atoms to retain one of the bonding electrons; the amount of energy released when a bond is formed.
bonding molecular orbital	a molecular orbital that results when two atomic orbitals with the same phase interact. Electrons in a bonding orbital increase bond strength.
bond length	the internuclear distance between two atoms at minimum energy (maximum stability).
bond order	describes the number of covalent bonds shared by two atoms.
carbanion	a species containing a negatively charged carbon.
carbocation	a species containing a positively charged carbon.
condensed structure	a structure that does not show some (or all) of the covalent bonds.
core electrons	electrons in filled shells.
covalent bond	a bond created as a result of sharing electrons.
degenerate orbitals	orbitals that have the same energy.
dipole	a separation of positive and negative charges.
dipole moment (μ)	a measure of the separation of charge in a bond or in a molecule.
double bond	a bond composed of a sigma bond and a pi bond.

electronegative	describes an element that readily acquires an electron.
electronegativity	the tendency of an atom to pull electrons toward itself.
electrostatic attraction	an attractive force between opposite charges.
electrostatic potential map (potential map)	a map that shows how electrons are distributed in a molecule.
equilibrium constant	the ratio of products to reactants at equilibrium.
excited-state electronic configuration	the electronic configuration that results when an electron in the ground state has moved to a higher-energy orbital.
formal charge	the number of valence electrons − (the number of nonbonding electrons + the number of bonds).
free radical (radical)	a species with an unpaired electron.
ground-state electronic configuration	a description of the orbitals the electrons of an atom occupy when they are all in their lowest available energy orbitals.
Heisenberg uncertainty principle	a principle that states that both the precise location and the momentum of an atomic particle cannot be simultaneously determined.
Hund's rule	a rule that states that when there are degenerate orbitals, an electron will occupy an empty orbital before it will pair up with another electron.
hybrid orbital	an orbital formed by hybridizing (mixing) atomic orbitals.
hydride ion	a negatively charged hydrogen (a hydrogen atom with an extra electron).
hydrogen ion (proton)	a positively charged hydrogen (a hydrogen atom without its electron).
ionic compound	a compound composed of a positive ion and a negative ion held together by electrostatic attraction.
ionization energy	the energy required to remove an electron from an atom.
isotopes	atoms with the same number of protons but a different number of neutrons.
Kekulé structure	a model that represents the bonds between atoms as lines.
Lewis structure	a model that represents the bonds between atoms as lines or dots and the lone-pair electrons as dots.
lone-pair electrons (nonbonding electrons)	valence electrons not used in bonding.

mass number the number of protons plus the number of neutrons in an atom.

molecular mass the sum of the atomic masses of all the atoms in the molecule.

molecular orbital an orbital associated with a molecule that results from the combination of atomic orbitals.

molecular orbital (MO) theory a theory that describes a model in which the electrons occupy orbitals as they do in atoms but the orbitals extend over the entire molecule.

node a region within an orbital where there is zero probability of finding an electron.

nonbonding electrons valence electrons not used in bonding.

nonpolar covalent bond a bond formed between two atoms that share the bonding electrons equally.

octet rule a rule that states that an atom will give up, accept, or share electrons to achieve a filled outer shell (or an outer shell that contains eight electrons) and no electrons of higher energy. Because a filled second shell contains eight electrons, this is known as the octet rule.

orbital the volume of space around the nucleus where an electron is most likely to be found.

orbital hybridization mixing of atomic orbitals.

organic compound a compound that contains carbon.

Pauli exclusion principle a principle that states that no more than two electrons can occupy an orbital and that the two electrons must have opposite spin.

pi (π) bond a bond formed as a result of side-to-side overlap of *p* orbitals.

polar covalent bond a bond formed between two atoms that do not share the bonding electrons equally.

potential map (electrostatic potential map) a map that allows you to see how electrons are distributed in a molecule.

proton (hydrogen ion) a positively charged hydrogen ion.

quantum mechanics the use of mathematical equations to describe the behavior of electrons in atoms or molecules.

radical (free radical) a species with an unpaired electron.

sigma (σ) bond a bond with a symmetrical distribution of electrons about the internuclear axis.

single bond a pair of electrons shared between two atoms.

skeletal structure shows the carbon–carbon bonds as lines but does not show the carbons or the hydrogens that are bonded to the carbons.

tetrahedral bond angle the bond angle (109.5°) formed by an sp^3 hybridized atom that has no lone pairs.

tetrahedral carbon a carbon that forms covalent bonds using four sp^3 hybrid orbitals.

trigonal planar carbon an sp^2 hybridized carbon.

triple bond a bond composed of a sigma bond and two pi bonds.

valence electron an electron in an outermost shell.

valence-shell electron-pair repulsion (VSEPR) model a model for the prediction of molecular geometry based on the minimization of electron repulsion between bonding electrons and nonbonding electrons around an atom.

wave equation an equation that describes the behavior of each electron in an atom or a molecule.

wave functions a series of solutions to a wave equation.

Solutions to Problems

1. The atomic number = the number of protons.
The mass number = the number of protons + the number of neutrons.
All isotopes have the same atomic number; in the case of oxygen, it is 8. Therefore:

The isotope of oxygen with a mass number of 16 has 8 protons and 8 neutrons.
The isotope of oxygen with a mass number of 17 has 8 protons and 9 neutrons.
The isotope of oxygen with a mass number of 18 has 8 protons and 10 neutrons.

2. The number of protons an element has never changes. The number of electrons depends on the charge on the element.

a. **1.** 11 **2.** 18 **3.** 17
b. **1.** 10 **2.** 18 **3.** 18

3. (percentage of naturally occurring $^{35}Cl \times$ atomic mass of ^{35}Cl) +
(percentage of naturally occurring $^{37}Cl \times$ atomic mass of ^{37}Cl)

$(.7577 \times 34.969) + (.2423 \times 36.966)$
$(26.496 + 8.957) = 35.45$

4. All four atoms have 2 core electrons in their filled first shell. (Notice that because the four atoms in the question are in the same row of the periodic table, they have the same number of core electrons.) The electrons that are not in a filled shell are valence electrons.

a. 3 **b.** 5 **c.** 6 **d.** 7

5. **a.** Use the aufbau principle (electrons go into available orbitals with the lowest energy) and the Pauli exclusion principle (no more than two electrons are in each atomic orbital). The relative energies of the orbitals:

$$1s < 2s < 2p < 3s < 3p < 4s < 3d < 4p < 5s < 4d < 5p$$

Remember that each shell has one *s* atomic orbital and three degenerate *p* atomic orbitals. The third and fourth shells also have five degenerate *d* atomic orbitals.

Cl $1s^2, 2s^2\ 2p^6, 3s^2\ 3p^5$
Br $1s^2, 2s^2\ 2p^6, 3s^2\ 3p^6\ 3d^{10}, 4s^2\ 4p^5$
I $1s^2, 2s^2\ 2p^6, 3s^2\ 3p^6\ 3d^{10}, 4s^2\ 4p^6\ 4d^{10}, 5s^2\ 5p^5$

b. They each have 7 electrons in their outer shell; in each case, 2 are in an *s* orbital and 5 are in *p* orbitals. Notice that because the 3 elements all are in the same column of the periodic table, they have the same number of valence electrons.

6. The atomic numbers can be found in the periodic table on the last page of the text. Notice that elements in the same column of the periodic table have the same number of valence electrons and that their valence electrons are in similar orbitals.

a. carbon (atomic number = 6; 2 core, 4 valence): $1s^2\ 2s^2\ 2p^2$
silicon (atomic number = 14; 10 core, 4 valence): $1s^2\ 2s^2\ 2p^6\ 3s^2\ 3p^2$

b. oxygen (atomic number = 8; 2 core, 6 valence): $1s^2\,2s^2\,2p^4$
sulfur (atomic number = 16; 10 core, 6 valence): $1s^2\,2s^2\,2p^6\,3s^2\,3p^4$

c. nitrogen (atomic number = 7; 2 core, 5 valence): $1s^2\,2s^2\,2p^3$
phosphorus (atomic number = 15; 10 core, 5 valence): $1s^2\,2s^2\,2p^6\,3s^2\,3p^3$

d. magnesium (atomic number = 12; 10 core, 2 valence): $1s^2\,2s^2\,2p^6\,3s^2$
calcium (atomic number = 20; 18 core, 2 valence): $1s^2\,2s^2\,2p^6\,3s^2\,3p^6\,4s^2$

7. **a.** Potassium is in the first column of the periodic table; therefore, like lithium and sodium that are also in the first column, potassium has one valence electron.
b. It occupies a 4*s* orbital.

8. The polarity of a bond can be determined by the difference in the electronegativities (given in Table 1.3 on page 10 of the text) of the atoms sharing the bonding electrons. The greater the difference in electronegativity, the more polar the bond.

a. $Cl—CH_3$ **b.** $H—OH$ **c.** $H—F$ **d.** $Cl—CH_3$

9. The electronegativity differences in the four listed compounds are as follows:

$$\begin{aligned} KCl &\quad 3.0 - 0.8 = 2.2 \\ LiBr &\quad 2.8 - 1.0 = 1.8 \\ NaI &\quad 2.5 - 0.9 = 1.6 \\ Cl_2 &\quad 3.0 - 3.0 = 0 \end{aligned}$$

a. KCl has the most polar bond because its two bonded atoms have the greatest differences in electronegativity.
b. Cl_2 has the least polar bond because the two chlorine atoms share the bonding electrons equally.

10. Solved in the text.

11. To answer this question, compare the electronegativities of the two atoms sharing the bonding electrons using Table 1.3 on page 10 of the text.

a. $\overset{\delta-}{HO}—\overset{\delta+}{H}$ **b.** $\overset{\delta-}{F}—\overset{\delta+}{Br}$ **c.** $\overset{\delta+}{H_3C}—\overset{\delta-}{NH_2}$ **d.** $\overset{\delta+}{H_3C}—\overset{\delta-}{Cl}$

e. $\overset{\delta-}{HO}—\overset{\delta+}{Br}$ **f.** $\overset{\delta-}{H_3C}—\overset{\delta+}{Li}$ **g.** $\overset{\delta+}{I}—\overset{\delta-}{Cl}$ **h.** $\overset{\delta+}{H_2N}—\overset{\delta-}{OH}$

(Notice that if the two atoms being compared are in the same row of the periodic table, the atom farther to the right is the more electronegative atom; if the atoms being compared are in the same column, the one closer to the top of the column is the more electronegative atom.)

12. Solved in the text.

13. The dipole moment is the magnitude of the charge times the distance between the charges. Because fluorine is more electronegative than Cl, the charge on H and F in HF is larger than the charge on H and Cl in HCl. The larger charge on F compared to the charge on Cl is more than enough to make up for the fact that H—F is a shorter bond than H—Cl.

14. **a.** LiH and HF are polar (they have a red end and a blue end).

b. A potential map marks the edges of the molecule's electron cloud. The electron cloud is largest around the hydrogen in LiH, because that hydrogen has more electrons around it than do the hydrogens in the other molecules.

c. Because the hydrogen of HF is blue, we know that this compound has the most positively charged hydrogen and, therefore, will be most apt to attract a negatively charged species.

15. By answering this question, you will see that a formal charge is a bookkeeping device. It does *not necessarily* tell you which atom has the greatest electron density or is the most electron deficient.

a. oxygen **b.** oxygen (it is more red)

c. oxygen **d.** hydrogen (it is the deepest blue)

Notice that in the hydroxide ion, the atom with the formal negative charge **is** the atom with the greater electron density. In the hydronium ion, however, the atom with the formal positive charge **is not** the most electron-deficient atom.

16. formal charge = number of valence electrons − (number of lone-pair electrons + the number of bonds)

In all four structures, every H is singly bonded and thus has a formal charge $= 1 - (0 + 1) = 0$.
Similarly, all CH_3 carbon atoms have four bonds and a formal charge $= 4 - (0 + 4) = 0$.
The formal charges on the remaining atoms:

a. $CH_3-\ddot{O}^+-CH_3$ (with H on O)
formal charge on O
$6 - (2 + 3) = +1$

b. $H-\ddot{C}^--H$ (with H on C)
formal charge on C
$4 - (2 + 3) = -1$

c. $CH_3-N^+(CH_3)_2-CH_3$
formal charge on N
$5 - (0 + 4) = +1$

d. $H-N^+H_2-B^-H_2-H$
formal charge on
N: $5 - (0 + 4) = +1$
B: $3 - (0 + 4) = -1$

17. The bond between two atoms can be shown by a pair of dots or by a line, so there are two ways each of the answers can be written. Remember that all lone pairs have to be shown.

a. $:\ddot{O}:N^+:\ddot{O}:^-$ with $\ddot{O}:$ on N
or
$^-:\ddot{O}-N^+(=\ddot{O}:)-\ddot{O}:^-$

b. $:\ddot{O}::N^+::\ddot{O}:$
or
$:\ddot{O}=N^+=\ddot{O}:$

c. $H:\ddot{C}:\ddot{C}:^-$ (H H above, H H below)
or
$H-CH_2-\ddot{C}H_2^-$

d. $H:\ddot{C}:C^+:H$ (H above, H H below)
or
$H-CH_2-C^+H-H$

e. $H:\ddot{C}:\ddot{N}^+:H$ (H H above, H H below)
or
$H-CH_2-N^+H_2-H$

f. $Na^+\ ^-:\ddot{O}:H$
or
$Na^+\ ^-:\ddot{O}-H$

g. $^-:\ddot{O}:\ddot{C}:\ddot{O}:H$ with $\ddot{O}:$ on C
or
$^-:\ddot{O}-C(=\ddot{O}:)-\ddot{O}-H$

h. $H:\ddot{C}:\ddot{O}:^-$ with $\ddot{O}:$ on C
or
$H-C(=\ddot{O}:)-\ddot{O}:^-$

18. **a.**

```
 H H                 H   H
H:C:C:Ö:H    and   H:C:Ö:C:H
 H H                 H   H
```

or

```
   H   H                  H       H
   |   |                  |       |
H—C—C—Ö—H    and    H—C—Ö—C—H
   |   |                  |       |
   H   H                  H       H
```

b.

```
 H H H                 H H   H               H H H
H:C:C:C:Ö:H   and   H:C:C:Ö:C:H   and   H:C:C:C:H
 H H H                 H H   H               H:Ö:H
                                               H
```

or

```
   H   H   H                 H   H       H                 H   H   H
   |   |   |                 |   |       |                 |   |   |
H—C—C—C—Ö—H    and    H—C—C—Ö—C—H    and    H—C—C—C—H
   |   |   |                 |   |       |                 |   |   |
   H   H   H                 H   H       H                 H  :O:  H
                                                                |
                                                                H
```

19. Because the compounds are neutral, a halogen has 3 lone pairs, an oxygen has 2, a nitrogen has 1, and a carbon or a hydrogen has no lone pairs.

a. $CH_3CH_2\ddot{N}H_2$ **b.** $CH_3\ddot{N}HCH_3$ **c.** $CH_3CH_2\ddot{\underset{..}{O}}H$

d. $CH_3\ddot{\underset{..}{O}}CH_3$ **e.** $CH_3CH_2\ddot{\underset{..}{Cl}}:$ **f.** $H\ddot{\underset{..}{O}}\ddot{N}H_2$

20. **a.** $CH_3CH_2CH_2Cl$ **b.** $CH_3\overset{O}{\overset{\|}{C}}OCH_2CH_3$ **c.** $CH_3CH_2\overset{O}{\overset{\|}{C}}\underset{CH_3}{\underset{|}{N}}CH_2CH_3$ **d.** $CH_3CH_2C{\equiv}N$

21. **a.** the (green) chlorine atom
b. the (red) oxygen atoms
c. the (blue) nitrogen atoms
d. the (black) carbon atoms and (gray) hydrogen atoms

22.

a.

```
   H       H   H   H
   |       |   |   |
H—C—N—C—C—C—H
   |   |   |   |   |
   H   H   H   H   H
```

b.

```
   H           H
   |           |
H—C—————C—Cl
   |           |
   H       H—C—H
               |
               H
```

c.

```
                H
                |
      H   H—C—H
      |         |
   H—C———C—Br
      |         |
      H   H—C—H
                |
                H
```

d.

```
              H
              |
   H     H—C—H   H   H   H          O
   |          |       |   |   |          ‖
H—C————C————C—C—C————C—H
   |          |       |   |   |
   H     H—C—H   H   H   H
              |
              H
```

23. **a.** 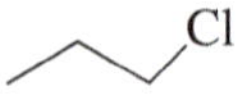 Cl **b.** 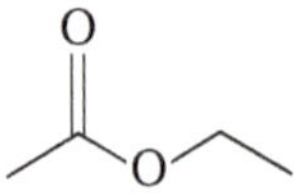 O O **c.** 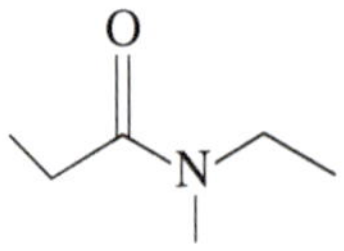 O N **d.** 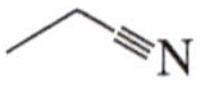 N

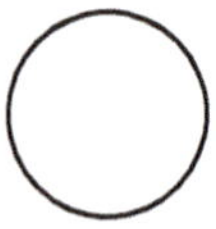

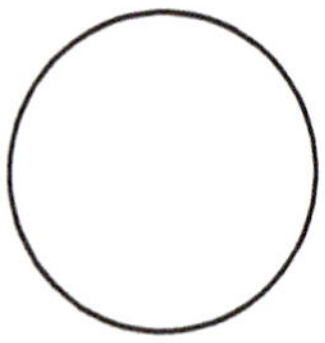

24. **a.** a ball larger than the ball that represents a 2*s* orbital **b.** a ball larger than the ball that represents a 3*s* orbital **c.**

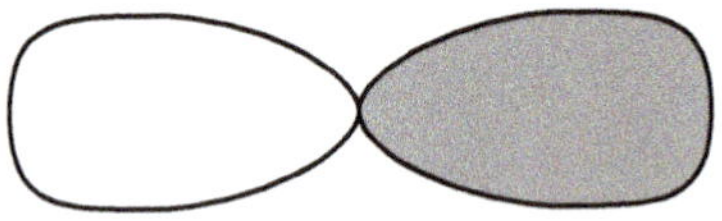

25. He_2^+ has three electrons. Using Figure 1.3 on page 23 of the text, two electrons will be in a bonding molecular orbital and one electron will be in an antibonding molecular orbital. Because there are more electrons in the bonding molecular orbital than in the antibonding molecular orbital, He_2^+ exists.

26. **a.** π^* This involves out-of-phase interaction of atomic orbitals (the interacting lobes have different colors), leading to an antibonding molecular orbital. Because this example involves the side-to-side overlap of *p* orbitals, it is a π^* antibonding molecular orbital.

b. π This involves in-phase overlap of atomic orbitals (the overlapping lobes have the same color), leading to a bonding molecular orbital. Because this example involves the side-to-side overlap of *p* orbitals, it is a π bonding molecular orbital.

c. σ^* This involves out-of-phase interaction of atomic orbitals (the interacting lobes have different colors), leading to an antibonding molecular orbital. Because this example involves the end-on overlap of atomic orbitals, it is a σ^* antibonding molecular orbital.

d. σ This involves in-phase overlap of atomic orbitals (the overlapping lobes have the same color), leading to a bonding molecular orbital. Because this example involves the end-on overlap of atomic orbitals, it is a σ bonding molecular orbital.

27. The 3 carbon–carbon bonds form as a result of sp^3–sp^3 overlap.
The 7 carbon–hydrogen bonds form as a result of sp^3–s overlap.

28. The electron density of the large lobe of an sp^3 orbital (the lobe that overlaps the *s* orbital) is greater than the electron density of a lobe of a *p* orbital. Therefore, the overlap of an *s* orbital with an sp^3 orbital forms a stronger bond than does the overlap of an *s* orbital with a *p* orbital.

29. Solved in the text.

30. **a.** **One** *s* orbital and **three** *p* orbitals form **four** sp^3 orbitals.
b. **One** *s* orbital and **two** *p* orbitals form **three** sp^2 orbitals.
c. **One** *s* orbital and **one** *p* orbital form **two** *sp* orbitals.

31. **a(1).** Solved in the text.
b(1). Solved in the text.

32. **a(1).** The first attempt at drawing a Lewis structure results in a carbon that does not have a complete octet and does not form the needed number of bonds.

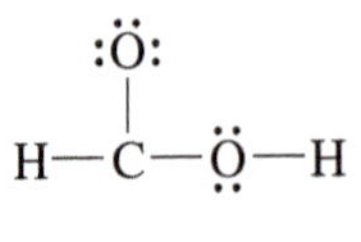

:Ö:
|
H—C—Ö—H

Using one of oxygen's lone pairs to put a double bond between the carbon and oxygen solves both problems.

H—C(=O)—O—H

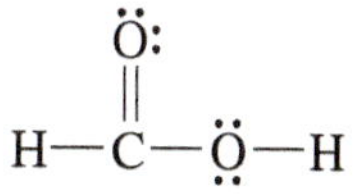

b(1). The sp^2 hybridized C=O carbon has 120° bond angles, uses sp^2 orbitals to form the three σ bonds, a p orbital to form the π bond, and has bond angles of 120°.

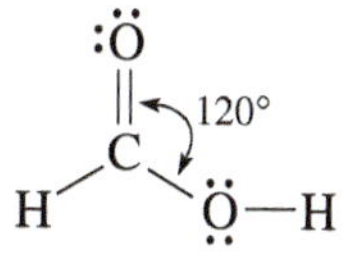

120°

a(2). In order to fill their octets and form the required number of bonds, carbon and nitrogen must form a triple bond.

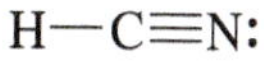

H—C≡N:

b(2). Because the carbon is sp hybridized, the carbon uses sp orbitals to form the two σ bonds and p orbitals to form the two π bonds. The bond angle is 180°.

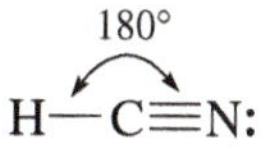

180°

H—C≡N:

a(3). The carbon forms four bonds, and each chlorine forms one bond.

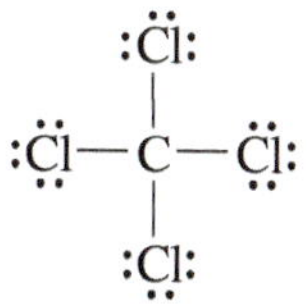

b(3). The carbon uses sp^3 orbitals to form the bonds with the chlorine atoms, so the bond angles are all 109.5°.

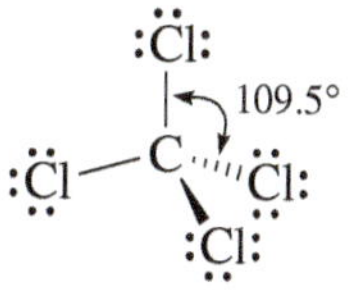

109.5°

a(4). The first attempt at drawing a Lewis structure (and remembering to avoid oxygen–oxygen single bonds) results in a carbon that does not have a complete octet and does not form the needed number of bonds.

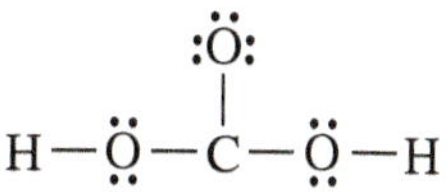

H—O—C(—O)—O—H

Using one of oxygen's lone pairs to put a double bond between the carbon and the oxygen solves both problems.

H—O—C(=O)—O—H

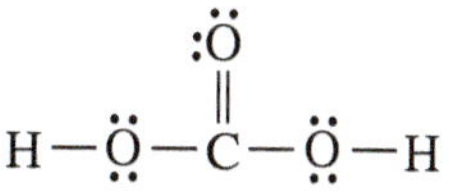

b(4). The carbon uses sp^2 orbitals to form the three σ bonds and a p orbital to form the π bond. The bond angles are 120°.

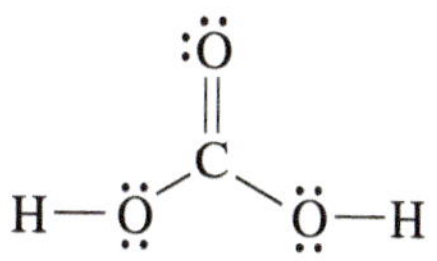

33. **a.** 120° **b.** 120°

c. Because the carbon is sp^3 hybridized and it has one lone pair, you can predict that the bond angle is similar to that in NH_3 (107.3°).

34. The nitrogen atom has the greatest electron density.
The hydrogens are the bluest atoms. Therefore, they have the least electron density. In other words, they have the most positive (least negative) electrostatic potential.

35. Water is the most polar—it has a deep red area and the most intense blue area.
Methane is the least polar—it is all nearly the same color (green) with no red or blue areas.

36. Solved in the text.

37. Electrons in atomic orbitals farther from the nucleus form **longer** bonds; they also form **weaker** bonds due to less electron density in the region of orbital overlap. Therefore:

a. **relative lengths** of the bonds in the halogens: $Br_2 > Cl_2$
relative strengths of the bonds: $Cl_2 > Br_2$

b. **relative lengths**: $CH_3—Br > CH_3—Cl > CH_3—F$
relative strengths: $CH_3—F > CH_3—Cl > CH_3—Br$

38. **a.** **longer:** **1.** C—I **2.** C—Cl **3.** H—Cl
b. **stronger:** **1.** C—Cl **2.** C—C **3.** H—F

39. **a.** CH_3O^-
The carbon in CH_3O^- is bonded to four atoms, so it uses four sp^3 orbitals.
Each carbon–hydrogen bond is formed by the overlap of an sp^3 orbital of carbon with the s orbital of hydrogen. The carbon–oxygen bond is formed by the overlap of an sp^3 orbital of carbon with an sp^3 orbital of oxygen. Because the four sp^3 orbitals of carbon orient themselves to get as far away from each other as possible, the bond angles are all **109.5°**.

O^- | C bonded to H, H, H

bond angles = 109.5°

b. CO_2
The carbon in CO_2 is bonded to two atoms, so it uses two sp orbitals. Each carbon–oxygen bond is a double bond. One of the bonds of each double bond is formed by the overlap of an sp orbital of carbon with an sp^2 orbital of oxygen. The second bond of the double bond is formed as a result of side-to-side overlap of a p orbital of carbon with a p orbital of oxygen. Because carbon's two sp orbitals orient themselves to get as far away from each other as possible, the bond angle in CO_2 is **180°**.

O=C=O
bond angle = 180°

c. H_2CO

The double-bonded carbon and the double-bonded oxygen in H_2CO both use sp^2 orbitals; thus, the bonds around the double-bonded carbon are all 120°. Each carbon–hydrogen bond is formed by the overlap of an sp^2 orbital of carbon with the *s* orbital of hydrogen.

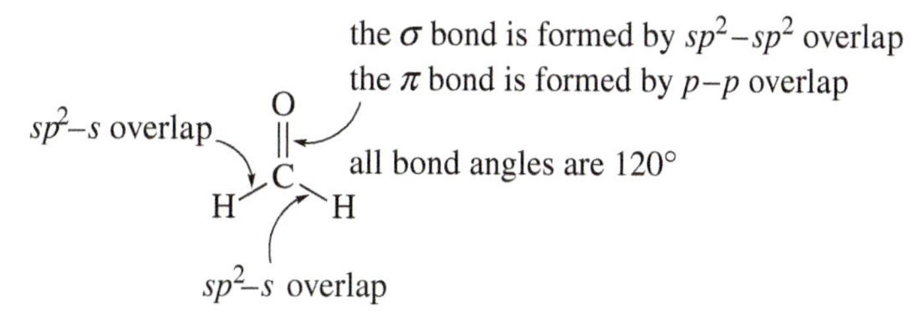

d. N_2

The triple bond consists of one σ bond and two π bonds. Each nitrogen has two *sp* orbitals; one is used to form the σ bond, and the other contains the lone pair. Each nitrogen has two *p* orbitals that are used to form the two π bonds. A bond angle is the angle formed by three atoms. Therefore, there are no bond angles in this two-atom containing compound.

:N≡N: the σ bond is formed by *sp–sp* overlap
each π bond is formed by *p–p* overlap

e. BF_3

Promotion gives boron three unpaired electrons, and hybridization gives it three sp^2 orbitals.

$1s$	$2s$	$2p_x$	$2p_y$		$1s$	$2s$	$2p_x$	$2p_y$		$1s$	$2sp^2$	$2sp^2$	$2sp^2$
↑↓	↑↓	↑		promotion →	↑↓	↑	↑	↑	hybridization →	↑↓	↑	↑	↑

Each sp^2 orbital of boron overlaps an sp^3 orbital of fluorine. The three sp^2 orbitals orient themselves to get as far away from each other as possible, resulting in bond angles of **120°**.

F–B(–F)–F

bond angles = 120°

40. Solved in the text.

41. We know that the σ bond is stronger than the π bond, because the σ bond in ethane has a bond dissociation energy of 90.2 kcal/mol, whereas the bond dissociation energy of the double bond ($\sigma + \pi$) in ethene is 174.5 kcal/mol, which is less than twice as strong.
Because the σ bond is stronger, we know that it has more effective orbital–orbital overlap.

42. Because electrons in an *s* orbital are closer on average to the nucleus than those in a *p* orbital, the greater the *s* character in the interacting orbitals, the stronger (and shorter) the bond. Therefore, the carbon–carbon σ bond formed by sp^2–sp^2 overlap is stronger (and shorter) than the carbon–carbon bond formed by sp^3–sp^3 overlap, because an sp^2 orbital has 33.3% *s* character, whereas an sp^3 orbital has 25% *s* character.

43.

44. **a.** $CH_3CHCH{=}CHCH_2C{\equiv}CCH_3$ (with CH_3 branch)

sp^3 *sp*^{2} *sp* *sp*^{3} *sp*^{3}

b.

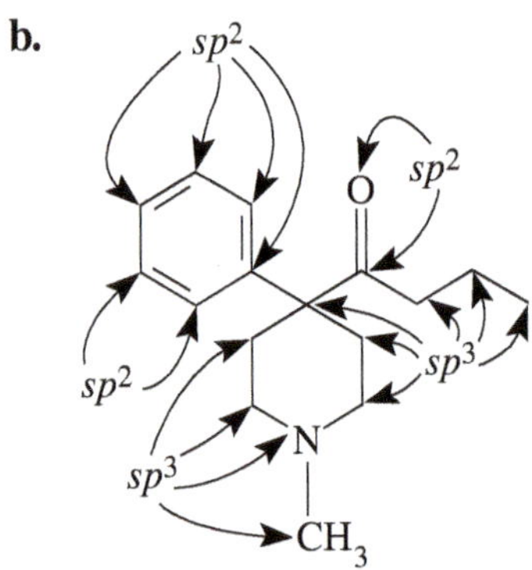

45. The bond angle depends on the central atom.

a. *sp*^{3} nitrogen with no lone pair: 109.5°

b. *sp*^{3} nitrogen with one lone pair: 107.3°

c. *sp*^{3} carbon with no lone pair: 109.5°

d. *sp*^{3} carbon with no lone pair: 109.5°

46. **a, e, g,** and **h** have a dipole moment of zero because they are symmetrical molecules.

$H_3C{-}CH_3$ $\quad$ $H_2C{=}CH_2$ $\quad$ $Cl{-}Be{-}Cl$ $\quad$ BF_3

47. The electrostatic potential map of ammonia is not symmetrical in the distribution of the charge—the nitrogen is more electron rich and, therefore, more red than the three hydrogens. Therefore, its shape, which indicates charge distribution, is not symmetrical.

The electrostatic potential map of the ammonium ion is symmetrical in the distribution of the charge, so its shape is symmetrical. Its symmetry results from the fact that nitrogen forms a bond with each of the four hydrogens and the four bonds point to the corners of a regular tetrahedron. The nitrogen in the ammonium ion has significantly lower electron density than the nitrogen in ammonia as a result of the lone pair having formed a bond to hydrogen.

48. The atom with the greater electronegativity will decrease the electron flow toward the electronegative F atom, giving the compound a smaller dipole moment. Since CH_3F has a smaller dipole moment than CD_3F, we know that hydrogen is more electronegative than deuterium.

49.
a. H:N̈:N̈:H (each N bearing one H below) or H—N̈—N̈—H (each N bearing one H below)

b. Lewis structure of the carbonate ion: one O above C, two O⁻ on either side (:Ö:C:Ö:⁻) or O=C with two single-bonded O⁻ ($^{-}$:Ö—C—Ö:$^{-}$, with :Ö: double-bonded above C)

c. H:N̈::N̈:H or H—N̈=N̈—H

d. Ö::C::Ö or Ö=C=Ö

e. H:Ö:C̈l: or H—Ö—C̈l:

50. **a.** CH_3NH_2, CH_3F, CH_3OH **b.** CH_3F

51. If the central atom is sp^3 hybridized, the bond angle will depend on the number of lone pairs it has: no lone pairs = 109.5°; one lone pair = 107.3°; two lone pairs = 104.5°.

a. sp^3, 107.3°
b. sp^2, 120°
c. sp^3, 107.3°
d. sp^2, 120°
e. sp^3, 109.5°
f. sp^2, 120°
g. sp, 180°
h. sp^3, 109.5°
i. sp^3, 107.3°
j. sp^2, 120°

52. **a.** $CH_3CH_2CH_3$ **b.** $CH_3CH{=}CH_2$ **c.** $CH_3C{\equiv}CCH_3$ or $CH_3CH_2C{\equiv}CH$

53. The hybridization of the central atom determines the bond angle. If the hybridization is sp^3, the number of lone pairs on the central atom determines the bond angle.

a. 109.5° **b.** 104.5°* **c.** 107.3° **d.** 107.3°

*104.5° is the correct prediction based on the bond angle in water.
However, the bond angle is actually somewhat larger (108.2°) because the bond opens up to minimize the interaction between the electron cloud of the relatively bulky CH_3 group.

54.

		$1s$	$2s$	$2p_x$	$2p_y$	$2p_z$	$3s$	$3p_x$	$3p_y$	$3p_z$	
a.	Mg	↿⇂	↿⇂	↿⇂	↿⇂	↿⇂	↿⇂				$1s^2\,2s^2\,2p^6\,3s^2$
b.	Ca^{2+}	↿⇂	↿⇂	↿⇂	↿⇂	↿⇂	↿⇂	↿⇂	↿⇂	↿⇂	$1s^2\,2s^2\,2p^6\,3s^2\,3p^6$
c.	Ar	↿⇂	↿⇂	↿⇂	↿⇂	↿⇂	↿⇂	↿⇂	↿⇂	↿⇂	$1s^2\,2s^2\,2p^6\,3s^2\,3p^6$
d.	Mg^{2+}	↿⇂	↿⇂	↿⇂	↿⇂	↿⇂					$1s^2\,2s^2\,2p^6$

55.
a. H:C:N̈:H (C bearing H above and below, N bearing H below) or H—C—N̈—H (C bearing H above and below, N bearing H below)

b. H—Ö—N̈=Ö

c. Na^+ $^{-}$:N̈:H (N bearing H below) or Na^+ $^{-}$:N̈—H (N bearing H below)

d. H:N̈:Ö:$^{-}$ (N bearing H below) or H—N̈—Ö:$^{-}$ (N bearing H below)

56.

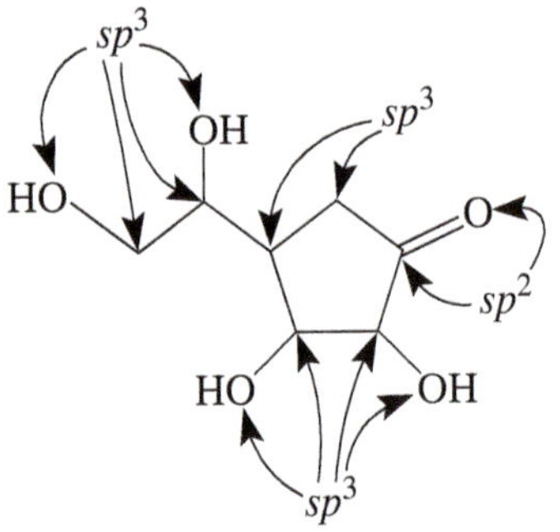

sp^3
OH
sp^3
HO
O
sp^2
HO
OH
sp^3

57. The greater the electronegativity difference between the two bonded atoms, the more polar the bond. (See Table 1.3 on page 10.)

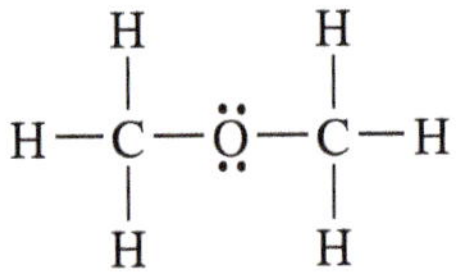

a. C—F > C—O > C—N

b. C—Cl > C—Br > C—I

c. H—O > H—N > H—C

d. C—N > C—H > C—C

58.

a. 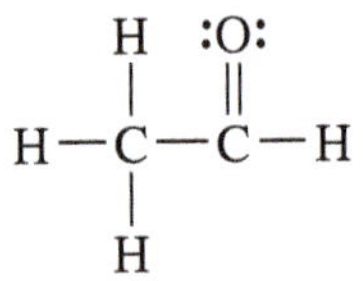
H :O:
H—C—C—H
H

b.
H H
H—C—Ö—C—H
H H

c.
H :O:
H—C—C—Ö—H
H

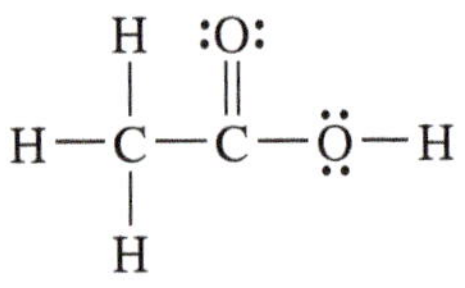

59. **a.** 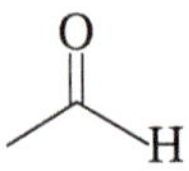 O, H **b.** 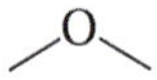 O **c.** 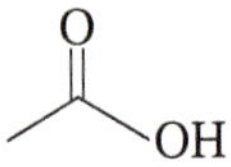 O, OH

60.

a. sp^2 ⇓ $CH_3CH{=}CH_2$

b. O ⇐ sp^2 CH_3CCH_3 (C=O)

c. sp^3 ⇓ CH_3CH_2OH

d. sp ⇓ $CH_3C{\equiv}N$

e. sp^2 ⇓ $CH_3CH{=}NCH_3$

f. sp^3 ⇓ $CH_3OCH_2CH_3$

61. **a.** 107.3° **b.** 109.5° **c.** 180° **d.** 109.5°

62.

a. H_3C—Br (⟼)

b. H_3C—Li (⟻)

c. HO—NH_2 (⟻)

d. I—Br (⟼)

e. H_3C—OH (⟼)

f. $(CH_3)_2N$—H (⟻)

63. formal charge = number of valence electrons
− (number of lone-pair electrons + the number of bonds)

In all four compounds, H has a single bond and is neutral and each C has four bonds and is neutral. Thus, the indicated formal charge is for O or N.

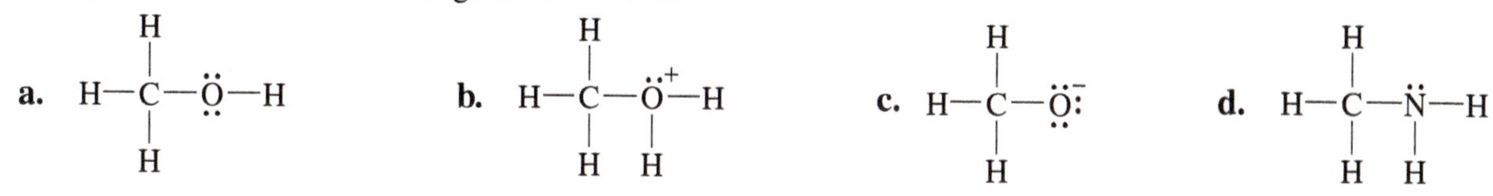

a. formal charge $6 - (4 + 2) = 0$

b. formal charge $6 - (2 + 3) = +1$

c. formal charge $6 - (6 + 1) = -1$

d. formal charge $5 - (2 + 3) = 0$

64. The open arrow in the structures points to the shorter of the two indicated bonds in each compound.

1. $CH_3CH{=}CHC{\equiv}CH$ (sp^3 sp^2 sp^2 sp sp)

2. CH_3CCH_2OH with C=O (sp^2, sp^3, sp^3, sp^3, sp^2)

3. $CH_3NHCH_2CH_2N{=}CHCH_3$ (sp^3 sp^3 sp^3 sp^3 sp^2 sp^2 sp^3)

4. $H_2C{=}CHC{\equiv}C{-}H$ (sp^2 sp^2 sp sp)

5. $H_2C{=}CHC{\equiv}C{-}C(CH_3)_2{-}H$ (sp^2 sp^2 sp sp sp^3 sp^3)

6. $Br{-}CH_2CH_2CH_2{-}Cl$ (sp^3 sp^3 sp^3)

For **1**, **2**, and **3**: A triple bond is shorter than a double bond, which is shorter than a single bond.

For **4** and **5**: The greater the *s* character in the hybrid orbital, the shorter the bond formed using that orbital, because an *s* orbital is closer than a *p* orbital to the nucleus. Therefore, the bond formed by a hydrogen and an *sp* carbon is shorter than the bond formed by a hydrogen and an sp^2 carbon, which is shorter than the bond formed by a hydrogen and an sp^3 carbon. (See Table 1.7 on page 42 of the text.)

For **6**: Cl forms a bond using a $3sp^3$ orbital, and Br forms a bond using a $4sp^3$ orbital. Therefore, the C—Cl bond is shorter.

65. **a.**

```
        H
  sp    |   180°
        C
       |||  180°
        C
sp3     |
      H'''C
        H   H
        109.5°
```

b.

```
          H     sp3
 109.5°   |
     H'''C   120°  H
        H    C=C    120°
            /    \
           H      H
             sp2
```

c.

```
     H   H   H
     |   |   |
 H — C — C — C — H
     |   |   |
     H   H   H
```

Each of the three carbons is sp^3 hybridized.
All the bond angles are 109.5°.

d.

```
 H        H
   \     /
    C = C         H
   /      \      /
 H         C = C
          /     \
         H       H
```

Each of the four carbons is sp^2 hybridized.
All the bond angles are 120°.

66.

a.

```
               H
               |
     H    H —  C  — H    H
     |         |         |
 H — C ——————— C ——————— C — H
     |         |         |
     H        :O:        H
               |
               H
```

b.

```
     H   H   H
     |   |   |
 H — C — C — C — C≡N:
     |   |   |
     H  :O:  H
         |
         H
```

c.

```
              H                          H
              |                          |
     H  H  —  C  — H  H        H  H   —  C  — H  H
     |        |       |        |         |       |
 H — C ——————— C ———— C ——————— C ——————— C ————— C — H
     |        |       |        |         |       |
     H        H H  —  C  — H  H H   —    C  — H H
                      |                  |
                      H                  H
```

67. **a.** OH **b.** N OH **c.**

68.

Cl Cl — 1 highest dipole moment

Cl — 2

Cl Br — 3

Cl Cl — 4 lowest dipole moment

69.

N̈ CH$_3$ sp^3

N̈ sp^2

70.

```
     H          H           H
     |          |           |
 H — C:⁻    H — C⁺      H — C•     H — C̈ — H    CH2=C̈H    CH2=C⁺H    CH2=Ċ H
     |          |           |
     H          H           H
```

71. In an alkene, six atoms are in the same plane: the two sp^2 carbons and the two atoms that are bonded to each of the two sp^2 carbons. The other atoms in the molecule will not necessarily be in the same plane with these six atoms.

If you put stars next to the six atoms that lie in a plane in each molecule, you will be able to see more clearly whether the indicated atoms lie in the same plane.

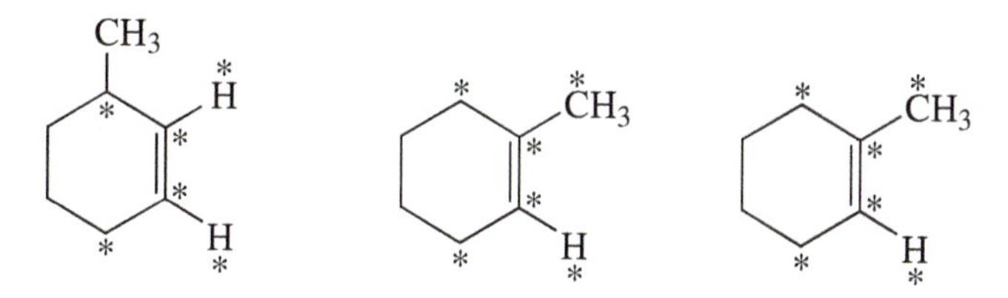

72. **a.** If the central atom is sp^3 hybridized and it does not have a lone pair, the molecule will have tetrahedral bond angles (109.5°). Therefore, only $^{+}NH_4$ has tetrahedral bond angles. The following species are close to being perfectly tetrahedral: H_2O, H_3O^+, NH_3, $^{-}CH_3$. However, they all have bond angles slightly smaller than 109.5°.

b. $^{+}CH_3$ and BF_3

73. CH_3CH_2Cl has the longer C—Cl bond because it is formed by the overlap of an sp^3 orbital of Cl with an sp^3 orbital of C, whereas the C—Cl bond in $CH_2{=}CHCl$ is formed by the overlap of an sp^3 orbital of Cl with an sp^2 orbital of C. (The more the *s* character, the shorter and stronger the bond.)

74. CH_2Cl_2 has the larger dipole moment because the two chlorines are withdrawing electrons in the same general direction, whereas in CH_3Cl, only one chlorine is withdrawing electrons.

75. The bond angles at the triple-bonded carbons, when the bonding orbitals overlap maximally, are 180°. These 180° angles cannot fit into the ring structure. Therefore, the overlap between the *sp* orbital and the adjacent sp^3 orbital becomes distorted from the ideal end-on overlap. This poor overlap causes the compound to be unstable. (Compare the structure shown here with Figure 3.8 on page 122 of the text.)

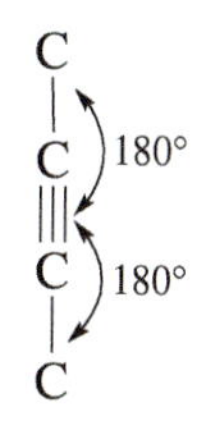

76. The dipole moment depends on the size of the charge and the distance between the bonded electrons. The longer C—Cl bond more than makes up for the greater charge on fluorine.

77. **a.** **1.** 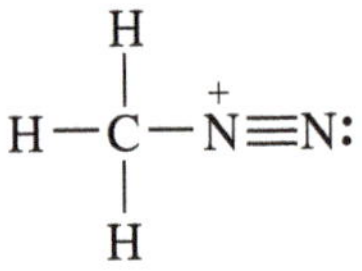 $H_3C-\overset{+}{N}\equiv N:$

3. 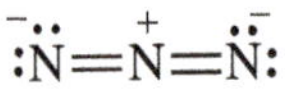 $:\overset{-}{\ddot{N}}=\overset{+}{N}=\overset{-}{\ddot{N}}:$

5. 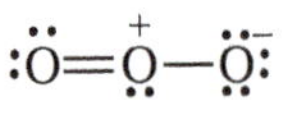 $:\ddot{O}=\overset{+}{\ddot{O}}-\ddot{\underset{..}{O}}:^-$

2. 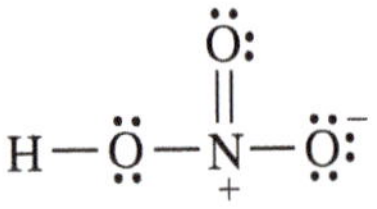 $H-\ddot{\underset{..}{O}}-\overset{+}{N}(=\ddot{O}:)-\ddot{\underset{..}{O}}:^-$

4. 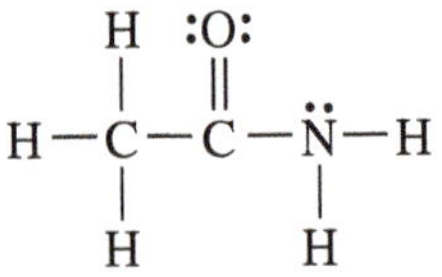 $H_3C-C(=\ddot{O}:)-\ddot{N}H_2$

b. **1.** $CH_3\overset{+}{N}\equiv N$

3. $\overset{-}{N}=\overset{+}{N}=\overset{-}{N}$

5. 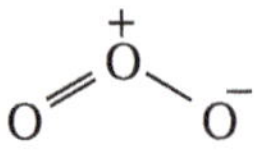 $O=\overset{+}{O}-O^-$

2. 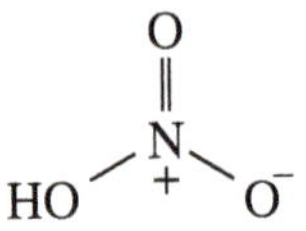 $HO-\overset{+}{N}(=O)-O^-$

4. 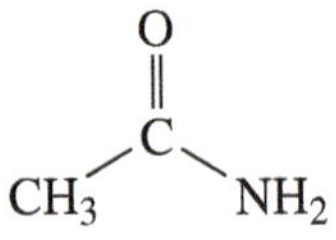 $CH_3C(=O)NH_2$

c. N_3^-

78. Only the first structure has no dipole moment.

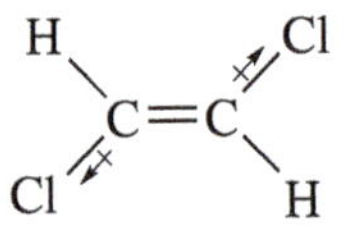

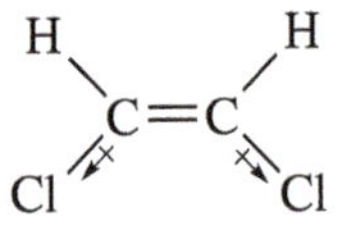

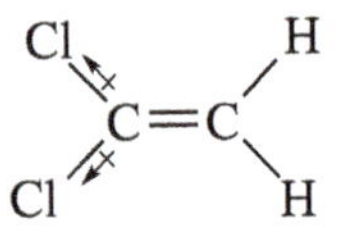

H Cl C=C Cl H

H H C=C Cl Cl

Cl H C=C Cl H

Chapter 1 Practice Test

1. Answer the following:

a. Which bond has a greater dipole moment, a carbon–oxygen bond or a carbon–fluorine bond?

b. If He_2^+ has three electrons in its molecular orbitals, how many electrons are in an antibonding molecular orbital?

c. Which is the longer bond, a carbon–hydrogen bond in ethene or a carbon–hydrogen bond in ethane?

d. Which is larger, the bond angle in water or the bond angle in ammonia?

2. What is the hybridization of the carbon atom in each of the following compounds?

$$^{+}CH_3 \qquad ^{-}CH_3 \qquad \cdot CH_3$$

3. Draw the Lewis structure for HCO_3^-.

4. Circle the compounds below that have a dipole moment = 0.

$$CH_2Cl_2 \qquad CH_3CH_3 \qquad CH_3Cl \qquad H_2C{=}O \qquad CCl_4$$

5. Which compound has greater bond angles, H_3O^+ or $^{+}NH_4$?

6. Draw the structure for each of the following:

a. a methyl cation

b. a hydride ion

c. a bromine radical

d. an alkane with only primary carbons

7. Draw the structure of a compound that contains only carbon and hydrogen, and that has five carbons, two of which are sp^2 hybridized and three of which are sp^3 hybridized.

8. What is the hybridization of each of the indicated atoms?

$$CH_3CH_2C{\equiv}N \qquad CH_3\underset{\displaystyle CH_3}{\underset{|}{C}}{=}NCH_3 \qquad CH_3\overset{\displaystyle O}{\overset{\|}{C}}CH_3 \qquad O{=}C{=}O$$

9. **a.** What orbitals do carbon's electrons occupy before promotion?

b. What orbitals do carbon's electrons occupy after promotion and before hybridization?

c. What orbitals do carbon's electrons occupy after hybridization?

10. Answer the following:

a. What is the H—C—O bond angle in CH_3OH?
b. What is the H—Be—H bond angle in BeH_2?
c. What is the H—B—H bond angle in BH_3?
d. What is the C—O—H bond angle in CH_3OH?

11. For each of the following compounds, indicate the hybridization of the atom to which the arrow is pointing.

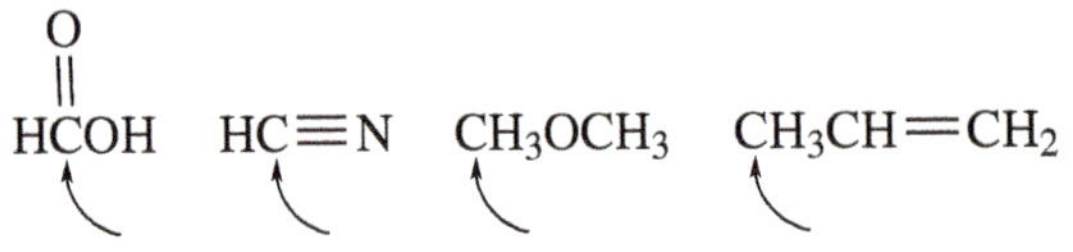

HCOH (with C=O) HC≡N CH_3OCH_3 $CH_3CH{=}CH_2$

12. Indicate whether each of the following statements is true or false.

a. A pi bond is stronger than a sigma bond. T F

b. A triple bond is shorter than a double bond. T F

c. The oxygen–hydrogen bonds in water are formed by the overlap of an sp^2 orbital of oxygen with an *s* orbital of hydrogen. T F

d. A double bond is stronger than a single bond. T F

e. A tetrahedral carbon has bond angles of 107.5°. T F

ANSWERS TO ALL THE PRACTICE TESTS CAN BE FOUND AT THE END OF THE SOLUTIONS MANUAL.

CHAPTER 2
Acids and Bases: Central to Understanding Organic Chemistry

Important Terms

acid (Brønsted acid)	a species that loses a proton.
acid–base reaction	a reaction of an acid with a base.
acid dissociation constant	a measure of the degree to which an acid dissociates.
acidity	a measure of how easily a compound gives up a proton.
base (Brønsted base)	a species that gains a proton.
basicity	a measure of the tendency of a compound to share its electrons with a proton.
Brønsted acid	a species that loses a proton.
Brønsted base	a species that gains a proton.
buffer solution	solution of a weak acid and its conjugate base.
conjugate acid	the species formed when a base gains a proton.
conjugate base	the species formed when an acid loses a proton.
delocalized electrons	electrons that are shared by three or more atoms (that is, do not belong to a single atom nor are they shared in a bond between two atoms).
equilibrium constant	the ratio of products to reactants at equilibrium.
Henderson–Hasselbalch equation	$pK_a = pH + \log[HA]/[A^-]$
inductive electron withdrawal	the pull of electrons through sigma bonds by an atom or by a group of atoms.
Lewis acid	a species that accepts a share in an electron pair.
Lewis base	a species that donates a share in an electron pair.
pH	the pH scale used to describe the acidity of a solution ($pH = -\log[H^+]$).
pK_a	a measure of the tendency of a compound to lose a proton ($pK_a = -\log K_a$, where K_a is the acid dissociation constant).
proton	a positively charged hydrogen ion.
proton transfer reaction	a reaction in which a proton is transferred from an acid to a base.
resonance	delocalized electrons.
resonance contributors	structures with localized electrons that together approximate the true structure of a compound with delocalized electrons.
resonance hybrid	the actual structure of a compound with delocalized electrons.

Solutions to Problems

1. CO_2 and CCl_4 are not acids because neither has a proton that it can lose.

2.
a. HBr
b. $^{-}C{\equiv}N$
c. Br^-
d. $HC{\equiv}N$
e. $HC{\equiv}N$
f. Br^-
g. $^{-}C{\equiv}N$
h. HBr

3.
a. $HCl + NH_3 \rightleftharpoons Cl^- + {}^{+}NH_4$
b. $H_2O + {}^{-}NH_2 \rightleftharpoons HO^- + NH_3$

4. The conjugate acid is obtained by adding an H^+ to the species.

a. (1) $^{+}NH_4$ (2) HCl (3) H_2O (4) H_3O^+

The conjugate base is obtained by removing an H^+ from the species.

b. (1) $^{-}NH_2$ (2) Br^- (3) NO_3^- (4) HO^-

5.
a. The lower the pK_a, the stronger the acid, so the compound with $pK_a = 5.2$ is the stronger acid.
b. The greater the dissociation constant, the stronger the acid, so the compound with a dissociation constant $= 3.4 \times 10^{-3}$ is the stronger acid.

6. Because we know that $K_a = K_{eq}[H_2O]$

$$K_{eq} = \frac{K_a}{[H_2O]} = \frac{4.53 \times 10^{-6}}{55.5} = 8.16 \times 10^{-8}$$

7. Its K_a value is 1.51×10^{-5}. It is a weaker acid than vitamin C whose K_a value was determined to be 6.8×10^{-5} in the Problem-Solving Strategy.

8.
a. $HO^- + H^+ \rightleftharpoons H_2O$
b. $HCO_3^- + H^+ \rightleftharpoons H_2CO_3 \longrightarrow H_2O + CO_2$
c. $CO_3^{2-} + 2\,H^+ \rightleftharpoons H_2CO_3 \longrightarrow H_2O + CO_2$

9. If the pH is <7, the body fluid is acidic; if the pH is >7, the body fluid is basic.

a. basic **b.** acidic **c.** basic

10. Remember that a proton can be picked up by an atom that has one or more lone pairs. Notice that two oxygens have lone pairs in part **e**. To see which one gets the proton, see Table 2.1 on page 57 of the text and the Problem-Solving Strategy on page 68.

a. $CH_3CH_2\overset{+}{O}H_2$ **b.** CH_3CH_2OH **c.** $CH_3C(=O)OH$ **d.** $CH_3CH_2\overset{+}{N}H_3$ **e.** $CH_3CH_2C(={}^{+}OH)OH$

11. In each of the following reactions, the position of equilibrium is established by analyzing the relative strengths of the acids on either side of the reaction—the equilibrium favors reaction of the stronger acid to form the weaker acid (see Section 2.5). (Note that HCl is a stronger acid than H_3O^+; see Appendix I in the text.)

If the lone pairs are not shown:

a. CH_3OH as an acid: $CH_3OH + NH_3 \rightleftharpoons CH_3O^- + \overset{+}{N}H_4$

CH_3OH as a base: $CH_3OH + HCl \rightleftharpoons CH_3\overset{+}{O}H(H) + Cl^-$

b. NH_3 as an acid: $NH_3 + CH_3O^- \rightleftharpoons {}^-NH_2 + CH_3OH$

NH_3 as a base: $NH_3 + HBr \rightleftharpoons \overset{+}{N}H_4 + Br^-$

If the lone pairs are shown:

a. CH_3OH as an acid: $CH_3\ddot{\underset{..}{O}}H + \ddot{N}H_3 \rightleftharpoons CH_3\ddot{\underset{..}{O}}:^- + \overset{+}{N}H_4$

CH_3OH as a base: $CH_3\ddot{\underset{..}{O}}H + H\ddot{\underset{..}{Cl}}: \rightleftharpoons CH_3\overset{+}{\ddot{O}}H(H) + :\ddot{\underset{..}{Cl}}:^-$

b. NH_3 as an acid: $\ddot{N}H_3 + CH_3\ddot{\underset{..}{O}}:^- \rightleftharpoons {}^-:\ddot{N}H_2 + CH_3\ddot{\underset{..}{O}}H$

NH_3 as a base: $\ddot{N}H_3 + H\ddot{\underset{..}{Br}}: \rightleftharpoons \overset{+}{N}H_4 + :\ddot{\underset{..}{Br}}:^-$

12. **a.** ~40 **b.** ~15 **c.** ~5 (Note that a $-C(=O)OH$ group can be written as —COOH) **d.** ~10

13. **a.** CH_3COO^- is the stronger base.
Because CH_3COOH is the weaker acid, it has the stronger conjugate base.

b. ${}^-NH_2$ is the stronger base.
Because NH_3 is the weaker acid, it has the stronger conjugate base.

c. H_2O is the stronger base.
Because H_3O^+ is the weaker acid, it has the stronger conjugate base.

14. The conjugate acids of the given bases have the following relative strengths:

$$CH_3\overset{+}{O}H_2 > CH_3C(=O)OH > CH_3\overset{+}{N}H_3 > CH_3OH > CH_3NH_2$$

The bases, therefore, have the following relative strengths, because the weakest acid has the strongest conjugate base.

$$CH_3\overset{-}{N}H > CH_3O^- > CH_3NH_2 > CH_3C(=O)O^- > CH_3OH$$

15. Methanol is the acid because it is a stronger acid (pK_a ~15) than methylamine (pK_a about 40).

16. Recall that the equilibrium favors reaction of the stronger acid to form the weaker acid. Because the pK_a values in part **a** are similar, there will be similar amounts of reactants and products at equilibrium.

a.

$$\underset{pK_a = 15.5}{CH_3OH} + HO^- \rightleftharpoons CH_3O^- + \underset{pK_a = 15.7}{H_2O}$$

$$CH_3OH + \underset{pK_a = -1.7}{H_3O^+} \rightleftharpoons \underset{pK_a = -2.5}{CH_3\overset{+}{O}H_2} + H_2O$$

$$\underset{pK_a = 4.8}{CH_3C(=O)OH} + HO^- \rightleftharpoons CH_3C(=O)O^- + \underset{pK_a = 15.7}{H_2O}$$

$$CH_3C(=O)OH + \underset{pK_a = -1.7}{H_3O^+} \rightleftharpoons \underset{pK_a = -6.1}{CH_3C(=\overset{+}{O}H)OH} + H_2O$$

$$\underset{pK_a = 40}{CH_3NH_2} + HO^- \rightleftharpoons CH_3\bar{N}H + \underset{pK_a = 15.7}{H_2O}$$

$$CH_3NH_2 + \underset{pK_a = -1.7}{H_3O^+} \rightleftharpoons \underset{pK_a = 10.7}{CH_3\overset{+}{N}H_3} + H_2O$$

b.

$$\underset{pK_a = -7}{HCl} + H_2O \rightleftharpoons \underset{pK_a = -1.7}{H_3O^+} + Cl^-$$

$$NH_3 + \underset{pK_a = 15.7}{H_2O} \rightleftharpoons \underset{pK_a = 9.4}{^+NH_4} + HO^-$$

17. Because a strong acid is more likely to lose a proton than a weak acid, the equilibrium favors loss of a proton from the strong acid and formation of the weak acid.

a. $\underset{pK_a = 25}{HC\equiv CH} + HO^- \rightleftharpoons HC\equiv C^- + \underset{pK_a = 15.7}{H_2O}$

b. $\underset{pK_a = 25}{HC\equiv CH} + {}^-NH_2 \rightleftharpoons HC\equiv C^- + \underset{pK_a = 36}{NH_3}$

c. ${}^-NH_2$ would be a better base because when it removes a proton, the equilibrium favors the products. When HO^- removes a proton, the equilibrium favors the reactants.

18. Each of the following bases will remove a proton from acetic acid in a reaction that favors products, because each of these bases forms an acid that is a weaker acid than acetic acid.

$$HO^- \qquad CH_3NH_2 \qquad HC\equiv C^-$$

The other three choices form an acid that is a stronger acid than acetic acid.

19. $pK_{eq} = pK_a\text{ (reactant acid)} - pK_a\text{ (product acid)}$

For **a**, the reactant acid is HCl and the product acid is H_3O^+.

For **b**, the reactant acid is CH_3COOH and the product acid is H_3O^+.

For **c**, the reactant acid is H_2O and the product acid is $CH_3\overset{+}{N}H_3$.

For **d**, the reactant acid is $CH_3\overset{+}{N}H_3$ and the product acid is H_3O^+.

a. $pK_{eq} = -7 - (-1.7) = -5.3$
$K_{eq} = 2.0 \times 10^5$

b. $pK_{eq} = 4.8 - (-1.7) = 6.5$
$K_{eq} = 3.2 \times 10^{-7}$

c. $pK_{eq} = 15.7 - (10.7) = 5.0$
$K_{eq} = 1.0 \times 10^{-5}$

d. $pK_{eq} = 10.7 - (-1.7) = 12.4$
$K_{eq} = 4.0 \times 10^{-13}$

20. Recall that the weakest acid has the strongest conjugate base.

$^-CH_3 > {}^-NH_2 > HO^- > F^-$

21. Again, the weakest acid has the strongest conjugate base.

$CH_3\bar{C}H_2 > H_2C{=}\bar{C}H > HC{\equiv}C^-$

22. The species on the right is the stronger acid because its hydrogen is attached to an sp^2 oxygen, which is more electronegative than the sp^3 oxygen to which the hydrogen in the protonated alcohol is attached.

23. **a.** **A** $HC{\equiv}CH + CH_3\bar{C}H_2 \rightleftharpoons HC{\equiv}C^- + CH_3CH_3$

B $H_2C{=}CH_2 + HC{\equiv}C^- \rightleftharpoons H_2C{=}\bar{C}H + HC{\equiv}CH$

C $CH_3CH_3 + H_2C{=}\bar{C}H \rightleftharpoons CH_3\bar{C}H_2 + H_2C{=}CH_2$

b. Only **A**, because only **A** has a reactant that is a stronger acid than the acid that is formed in the product.

24. Reaction **B**. The equilibrium constants for the three reactions are: $K_{eq}(A) = 10^{35}$; $K_{eq}(B) = 10^{-19}$; $K_{eq}(C) = 10^{-16}$.

25. The smaller the ion, the stronger it is as a base.

$F^- > Cl^- > Br^- > I^-$

26. **a.** oxygen **b.** H_2S **c.** CH_3SH

The size of an atom is more important than its electronegativity in determining stability. Therefore, even though oxygen is more electronegative than sulfur, H_2S is a stronger acid than H_2O and CH_3SH is a stronger acid than CH_3OH, because the sulfur atom is larger than the oxygen atom.

Because the sulfur atom is larger, the electrons in its conjugate base are spread out over a greater volume, which stabilizes it. The more stable the base, the stronger its conjugate acid.

27. The stronger acid has its proton attached to the more electronegative atom (if the atoms are about the same size) or to the larger atom (if the atoms are not the same size).

a. HBr **b.** $CH_3CH_2CH_2\overset{+}{O}H_2$ **c.** $CH_3CH_2CH_2OH$ **d.** $CH_3CH_2CH_2SH$

28. Remember that the stronger the acid, the weaker (or more stable) its conjugate base,

a. Because HI is the strongest acid, I^- is the most stable (weakest) base.
b. Because HF is the weakest acid, F^- is the least stable (strongest) base.

29. Compare the acid strengths of the conjugate acids, recalling that a weaker acid has a stronger conjugate base.

a. CH_3O^- because CH_3OH is a weaker acid than CH_3SH.

b. HO^- because H_2O is a weaker acid than H_3O^+.

c. NH_3 because $^+NH_4$ is a weaker acid than H_3O^+.

d. CH_3O^- because CH_3OH is a weaker acid than CH_3COOH.

30. **a.** $CH_3OCH_2CH_2OH$ because its conjugate base has its negative charge stabilized by electron withdrawal by the CH_3O group.

b. $CH_3CH_2CF_2CH_2\overset{+}{O}H_2$ because oxygen is more electronegative than nitrogen.

c. $CH_3CH_2OCH_2CH_2OH$ because the electron-withdrawing oxygen is closer to the OH group.

d. $CH_3CH_2\overset{O}{\overset{\|}{C}}OH$ because the electron-withdrawing C=O is closer to the OH group.

31. $\underset{F}{C}H_2\underset{F}{C}HCH_2COOH > CH_3\underset{F}{C}HCH_2COOH > \underset{F}{C}H_2CH_2CH_2COOH > CH_3CH_2CH_2COOH$

The first listed compound is the most acidic because it has two electron-withdrawing substituent that stabilize the conjugate base.

The second listed compound is a stronger acid than the third listed compound because the fluorine in the third compound is farther away from the O—H bond, so the electron-withdrawing group will not be as effective in stabilizing the conjugate base.

The compound on the far right does not have a substituent that withdraws electrons inductively, so it is the least acidic of the four compounds.

32. The weaker acid has a stronger conjugate base.

a. $CH_3\underset{Br}{C}H\overset{O}{\overset{\|}{C}}O^-$ **b.** $CH_3\underset{Cl}{C}HCH_2\overset{O}{\overset{\|}{C}}O^-$ **c.** $CH_3CH_2\overset{O}{\overset{\|}{C}}O^-$ **d.** $CH_3\overset{O}{\overset{\|}{C}}CH_2CH_2O^-$

33. Solved in the text.

34. **a.** 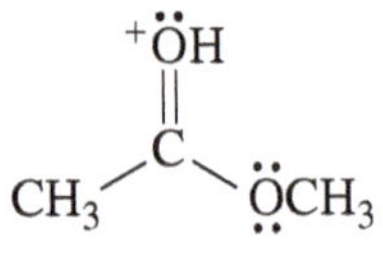 $CH_3C(=\overset{+}{O}H)\ddot{O}CH_3$ **b.** 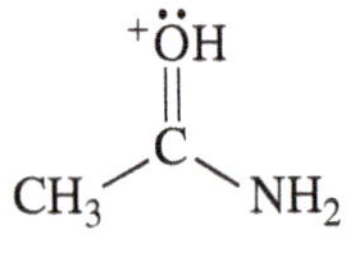 $CH_3C(=\overset{+}{O}H)NH_2$ **c.** 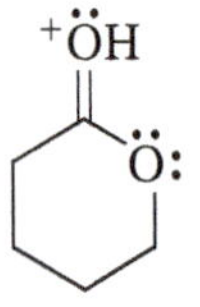 cyclic ester with $=\overset{+}{O}H$ and ring $\ddot{O}$: **d.** 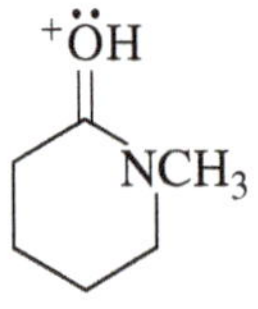 cyclic amide with $=\overset{+}{O}H$ and ring NCH_3

35. **a.** 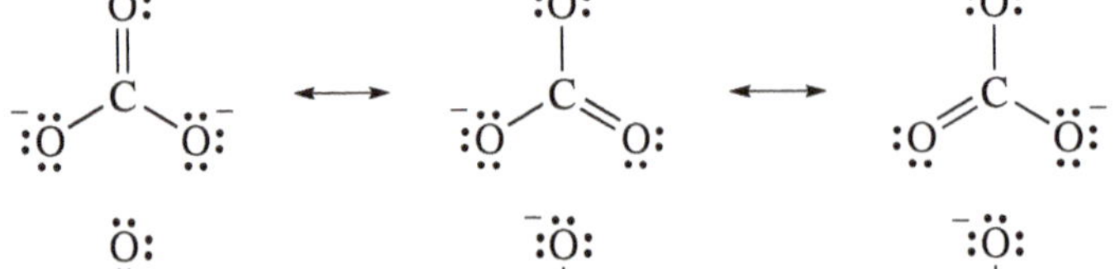 resonance contributors of carbonate: $C(=\ddot{O})(\ddot{O}^-)(\ddot{O}^-)$ ⟷ $C(\ddot{O}^-)(=\ddot{O})(\ddot{O}^-)$ ⟷ $C(\ddot{O}^-)(\ddot{O}^-)(=\ddot{O})$

b. 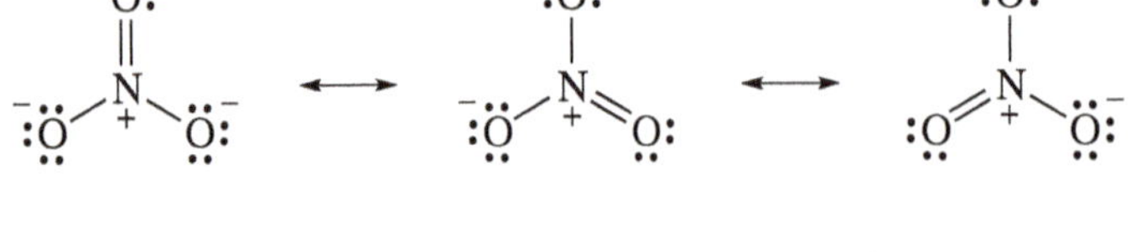 resonance contributors of nitrate: $\overset{+}{N}(=\ddot{O})(\ddot{O}^-)(\ddot{O}^-)$ ⟷ $\overset{+}{N}(\ddot{O}^-)(=\ddot{O})(\ddot{O}^-)$ ⟷ $\overset{+}{N}(\ddot{O}^-)(\ddot{O}^-)(=\ddot{O})$

36. When a sulfonic acid loses a proton, the electrons left behind are shared by three oxygens. In contrast, when a carboxylic acid loses a proton, the electrons left behind are shared by two oxygens. The sulfonate ion, therefore, is more stable than the carboxylate ion.

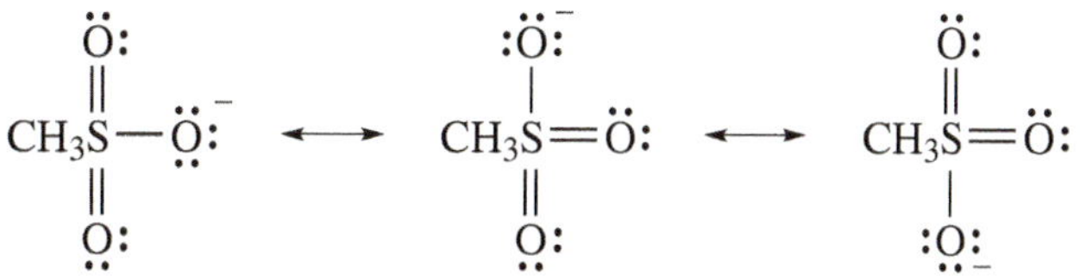

$$CH_3S(=O)_2-O^- \longleftrightarrow CH_3S(-O^-)(=O)_2 \longleftrightarrow CH_3S(=O)_2(-O^-)$$

a sulfonate ion

$$CH_3C(=O)-O^- \longleftrightarrow CH_3C(-O^-)=O$$

a carboxylate ion

The more stable the base, the stronger its conjugate acid.

Therefore, the sulfonic acid is a stronger acid than the carboxylic acid.

37. **a.** Because the atom (P) to which each of the OH groups is attached is also attached to two electronegative oxygens and when each OH group loses a proton, the electrons left behind can be shared by two oxygens.

b. The middle OH group is the weakest of the remaining acidic groups. It is an alcohol ($pK_a \sim 15$), and the atom (C) to which it is attached is not attached to any strongly electronegative atoms. (The protonated amino group has a pK_a value of ~10.)

38. Remember, the smaller the pK_a, the stronger the acid.

a. $CH_3C\equiv\overset{+}{N}H$

b. CH_3CH_3

c. 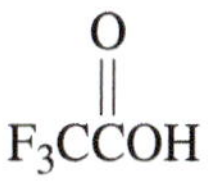 F_3CCOH (with C=O)

d. an sp^2 oxygen

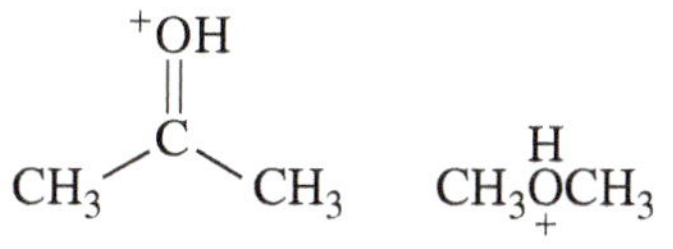

$CH_3C(=\overset{+}{O}H)CH_3$ $pK_a = -7.3$ $\quad CH_3\overset{+}{O}(H)CH_3$ $pK_a = -3.6$

e. 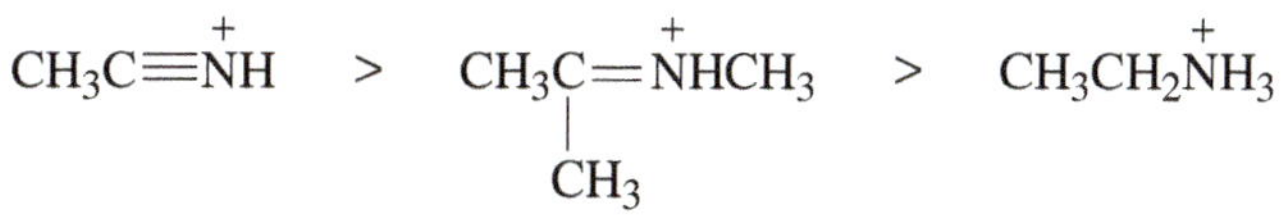 $CH_3C\equiv\overset{+}{N}H$ > $CH_3C(CH_3)=\overset{+}{N}HCH_3$ > $CH_3CH_2\overset{+}{N}H_3$

$pK_a = -10.1$ $\quad pK_a = 5.5$ $\quad pK_a = 11.0$

39. If the pH of the solution is less than the compound's pK_a value, the compound will be in its acidic form (with its proton).

If the pH of the solution is greater than the compound's pK_a value, the compound will be in its basic form (without its proton).

a. CH_3COO^-
b. $CH_3CH_2\overset{+}{N}H_3$
c. H_2O
d. Br^-
e. $^+NH_4$
f. $HC{\equiv}N$
g. NO_2^-
h. NO_3^-
i. $HO\overset{+}{N}H_3$

40. pH 10.4
(As long as the pH is greater than the pK_a of the compound, at least 50% of the compound will be in its basic form.)

41. Solved in the text.

42.
a. **1.** charged **2.** charged **3.** charged **4.** charged **5.** neutral **6.** neutral
b. **1.** neutral **2.** neutral **3.** neutral **4.** neutral **5.** neutral **6.** neutral

43.
a. The $^+NH_3$ group withdraws electrons, which increases the acidity of the COOH group.
b. Both the COOH group and the $^+NH_3$ group will be in their acidic forms, because the pH of the solution is less than both of their pK_a values.

$$CH_3CH(^+NH_3)C(=O)OH$$

c. The COOH group will be in its basic form because the pH of the solution is greater than its pK_a. The $^+NH_3$ group will be in its acidic form because the pH of the solution is less than its pK_a.

$$CH_3CH(^+NH_3)C(=O)O^-$$

d. Both the COOH group and the $^+NH_3$ group will be in their basic forms because the pH of the solution is greater than both of their pK_a values.

$$CH_3CH(NH_2)C(=O)O^-$$

e. No, alanine can never be without a charge. To be without a charge would require a group with a pK_a value of 9.69 to lose a proton before a group with a pK_a value of 2.34. This clearly cannot happen. A weak acid cannot have a greater tendency to lose a proton than a strong acid has.

f. As the pH becomes more basic than 2.34, the COOH group will become more negatively charged. As the pH becomes more acidic than 9.69, the $^{+}NH_3$ group will become more positively charged. Therefore, the amount of negative charge will be the same as the amount of positive charge at the pH that is equidistant from the two pK_a values.

$$\frac{2.34 + 9.69}{2} = 6.02$$

44. Solved in the text.

45.
a. 6.4 (two log units more acidic than the pK_a)
b. 7.3 (when the pH is equal to the pK_a)
c. 5.6 (one log unit more basic than the pK_a)

46.
a. **1.** pH = 4.9 — When the pH = pK_a, half the compound is in its acidic form (with its proton) and half is in its basic form (without its proton).
2. pH = 10.7

b. **1.** pH > 6.9 — Because the basic form is the form in which the compound is charged, the pH needs to be more than two units more basic than the pK_a value.
2. pH < 8.7 — Because the acidic form is the form in which the compound is charged, the pH needs to be more than two units more acidic than the pK_a value.

47. Solved in the text.

48.
a. For the carboxylic acid to dissolve in water, it must be charged (in its basic form), so the pH will have to be 6.8 or greater. For the amine to dissolve in ether, it must be neutral (in its basic form); the pH must be 12.7 or greater to have essentially all of it in the neutral form. Therefore, the pH of the water layer must be 12.7 or greater.
b. To dissolve in ether, the carboxylic acid has to be neutral; the pH must be 2.8 or lower to have essentially all the carboxylic acid in the acidic (neutral) form. To dissolve in water, the amine has to be charged; the pH must be 8.7 or lower to have essentially all the amine in the acidic form. Therefore, the pH of the water layer must be 2.8 or lower.

49.
a. The basic form of the buffer removes a proton from the solution.

$$CH_3COO^- + H^+ \rightleftharpoons CH_3COOH$$

b. The acidic form of the buffer donates a proton to remove hydroxide ion from the solution.

$$CH_3COOH + HO^- \rightleftharpoons CH_3COO^- + H_2O$$

50. Solved in the text.

51.
a. $ZnCl_2 + CH_3\ddot{O}H \rightleftharpoons {}^{-}ZnCl_2\text{—}H\overset{+}{O}CH_3$

b. $FeBr_3 + :\ddot{Br}:^- \rightleftharpoons Br\text{—}\bar{Fe}Br_3$

c. $AlCl_3 + :\ddot{Cl}:^- \rightleftharpoons Cl\text{—}\bar{Al}Cl_3$

52. **a, b, c,** and **h** are Brønsted acids (proton-donating acids). Therefore, they react with HO^- by giving a proton to it. **d, e, f,** and **g** are Lewis acids. They react with HO^- by accepting a pair of electrons from it.

a. $CH_3OH + HO^- \rightleftharpoons CH_3O^- + H_2O$

b. $^+NH_4 + HO^- \rightleftharpoons NH_3 + H_2O$

c. $CH_3\overset{+}{N}H_3 + HO^- \rightleftharpoons CH_3NH_2 + H_2O$

d. $BF_3 + HO^- \rightleftharpoons HO{-}\bar{B}F_3$

e. $^+CH_3 + HO^- \rightleftharpoons CH_3OH$

f. $FeBr_3 + HO^- \rightleftharpoons HO{-}\bar{F}eBr_3$

g. $AlCl_3 + HO^- \rightleftharpoons HO{-}\bar{A}lCl_3$

h. $CH_3COOH + HO^- \rightleftharpoons CH_3COO^- + H_2O$

53. The stronger base has the weaker conjugate acid.

a. HO^- **b.** $CH_3\bar{N}H$ **c.** CH_3O^- **d.** Cl^- **e.** CH_3COO^- **f.** $CH_3CHBrCOO^-$

54. **a.** $\ddot{N}H_3 + H{-}\ddot{\underset{..}{Cl}}: \rightleftharpoons {}^+NH_4 + :\ddot{\underset{..}{Cl}}:^-$

b. $H_2\ddot{O}: + FeBr_3 \rightleftharpoons H_2\overset{+}{O}{-}\bar{F}eBr_3$

c. $HC(=\ddot{O}:)OH + H{-}\ddot{\underset{..}{Cl}}: \rightleftharpoons HC(={}^+\ddot{O}H)OH + :\ddot{\underset{..}{Cl}}:^-$

55. **a.** $CCl_3CH_2OH > CHCl_2CH_2OH > CH_2ClCH_2OH$

b. The greater the number of electron-withdrawing chlorine atoms equidistant from the OH group, the stronger the acid. (Notice that the larger the K_a, the stronger the acid.)

56. **a.** $CH_3CH_2CH(Cl)COOH > CH_3CH(Cl)CH_2COOH > ClCH_2CH_2CH_2COOH > CH_3CH_2CH_2COOH$

b. An electron-withdrawing substituent makes the carboxylic acid more acidic, because it stabilizes its conjugate base by decreasing the electron density around the oxygen atom. (Remember that the larger the K_a, the stronger the acid.)

c. The closer the electron-withdrawing chloro substituent is to the acidic proton, the more it can decrease the electron density around the oxygen atom because it has to exert its effect through fewer bonds. Therefore, the closer it is, the more it stabilizes the conjugate base and increases the acidity of its conjugate acid.

57. **a.** $CH_3\overset{..+}{O}(CH_3){-}\bar{B}F_3$ **b.** $CH_3\overset{..+}{O}(CH_3){-}H + Cl^-$ **c.** $CH_3\overset{+}{N}H_2{-}\bar{A}lCl_3$

58. **a.** $HOCH_2CH_2CH_2\overset{+}{N}H_3$ **b.** $^-OCH_2CH_2CH_2NH_2$

59. O is more electronegative than N, which is more electronegative than C.
Therefore, the alcohol is more acidic than the amine, which is more acidic than the alkane.
S is larger than O, so CH_3CH_2SH is more acidic than CH_3CH_2OH.

$$CH_3CH_2SH > CH_3CH_2OH > CH_3CH_2NH_2 > CH_3CH_2CH_3$$

60. If the pH of the solution is less than the pK_a of the compound, the compound will be in its acidic form (with the proton).
If the pH of the solution is greater than the pK_a of the compound, the compound will be in its basic form (without the proton).

a. at pH = 3 CH_3COOH
at pH = 6 CH_3COO^-
at pH = 10 CH_3COO^-
at pH = 14 CH_3COO^-

b. at pH = 3 $CH_3CH_2\overset{+}{N}H_3$
at pH = 6 $CH_3CH_2\overset{+}{N}H_3$
at pH = 10 $CH_3CH_2\overset{+}{N}H_3$
at pH = 14 $CH_3CH_2NH_2$

c. at pH = 3 CF_3CH_2OH
at pH = 6 CF_3CH_2OH
at pH = 10 CF_3CH_2OH
at pH = 14 $CF_3CH_2O^-$

61. In all four reactions, the products are favored at equilibrium. (Recall that the equilibrium favors formation of the weaker acid.)

a. $CH_3COOH + CH_3O^- \rightleftharpoons CH_3COO^- + CH_3OH$

b. $CH_3CH_2OH + {}^-NH_2 \rightleftharpoons CH_3CH_2O^- + NH_3$

c. $CH_3COOH + CH_3NH_2 \rightleftharpoons CH_3COO^- + CH_3\overset{+}{N}H_3$

d. $CH_3CH_2OH + HCl \rightleftharpoons CH_3CH_2\overset{+}{O}H_2 + Cl^-$

62. **a.** $HC{\equiv}CCH_2OH > CH_2{=}CHCH_2OH > CH_3CH_2CH_2OH$

b. These three compounds differ only in the group that is attached to CH_2OH. The more electronegative the group attached to CH_2OH, the stronger the acid because inductive electron withdrawal stabilizes the conjugate base, and the more stable the base, the stronger its conjugate acid. An *sp* carbon is more electronegative than an sp^2 carbon, which is more electronegative than an sp^3 carbon.

63. The direction of the dipole will be toward the more electronegative of the two atoms that are sharing the bonding electrons.

a. $CH_3{-}C{\equiv}CH$ (dipole arrow pointing from CH_3 toward C) **b.** $CH_3{-}CH{=}CH_2$ (dipole arrow pointing from CH_3 toward CH)

64. In each compound, the nitrogen atom is the atom most apt to be protonated because it is the stronger base.

a. $CH_3{-}CH(OH){-}CH_2\overset{+}{N}H_3$ **b.** $CH_3{-}C(CH_3)(\overset{+}{N}H_3){-}OH$ **c.** $CH_3{-}C(CH_3)(\overset{+}{N}H_3){-}CH_2OH$

65. The log of $10^{-4} = -4$, the log of $10^{-5} = -5$, the log of $10^{-6} = -6$, and so on. Because the $pK_a = -\log K_a$, the pK_a of an acid with a K_a of 10^{-4} is $-(-4) = 4$. An acid with a K_a of 4.0×10^{-4} is a stronger acid than one with a K_a of 1.0×10^{-4}. Therefore, the pK_a can be estimated as being between 3 and 4.

a.
1. between 3 and 4
2. between −2 and −1
3. between 10 and 11
4. between 9 and 10
5. between 3 and 4
6. between −1 and 0

b.
1. $pK_a = 3.4$
2. $pK_a = -1.3$
3. $pK_a = 10.2$
4. $pK_a = 9.1$
5. $pK_a = 3.7$
6. $pK_a = -0.3$

c. Nitric acid (HNO_3) is the strongest acid because it has the lowest pK_a value. (The lower the pK_a value, the stronger the acid.)

66. The nitrogen on the right in the chain is the most basic. The nitrogen of the NH_2 group is less basic because its lone pair is delocalized onto the oxygen, so it is not available to be protonated.

$$H_2\ddot{N}-C(=O)- \longleftrightarrow H_2\overset{+}{N}=C(-O^-)-$$

67. **A** and **C** because in each case, the acid is stronger than the acid (H_2O) that is formed as a product.

68. The reaction with the more favorable equilibrium constant is the one with the smallest difference between the pK_a value of the reactant acid and the pK_a value of the product acid, because $pK_{eq} = pK_a$ (reactant acid) – pK_a (product acid) and the smaller the pK_{eq}, the larger the K_{eq}.

a. **1.** CH_3CH_2OH $pK_a = 15.9$; CH_3OH $pK_a = 15.5$; $^+NH_4$ $pK_a = 9.4$

$$\begin{aligned} pK_{eq} &= pK_a\,(\text{reactant acid}) - pK_a\,(\text{product acid}) \\ &= 15.9 - 9.4 = 6.5;\ K_{eq} = 3.2 \times 10^{-7} \\ &= 15.5 - 9.4 = 6.1;\ K_{eq} = 7.9 \times 10^{-7} \end{aligned}$$

Thus, the reaction of CH_3OH with NH_3 has the more favorable equilibrium constant.

2. CH_3CH_2OH $pK_a = 15.9$; $^+NH_4$ $pK_a = 9.4$; $CH_3\overset{+}{N}H_3$ $pK_a = 10.7$

$$\begin{aligned} pK_{eq} &= pK_a\,(\text{reactant acid}) - pK_a\,(\text{product acid}) \\ &= 15.9 - 9.4 = 6.5;\ K_{eq} = 3.2 \times 10^{-7} \\ &= 15.9 - 10.7 = 5.2;\ K_{eq} = 6.3 \times 10^{-6} \end{aligned}$$

Thus, the reaction of CH_3CH_2OH with CH_3NH_2 has the more favorable equilibrium constant.

b. Because the reaction of CH_3CH_2OH with CH_3NH_2 has the smallest difference between the pK_a values of the reactant and product acids, it has the most favorable equilibrium constant.

69. If the reaction is producing protons, the basic form of the buffer will pick up the protons. At the pH at which the reaction is carried out $(pH = 10.5)$, a protonated methylamine/methylamine buffer with a $pK_a = 10.7$ will have a larger percentage of the buffer in the needed basic form than will a protonated ethylamine/ethylamine buffer with a pK_a of 11.0.

70. **a.** $CH_2{=}CHCOOH$ because an sp^2 carbon is more electronegative than an sp^3 carbon

b. 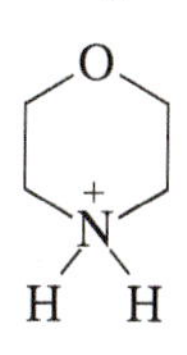
because an oxygen can withdraw electrons inductively

c. $HC{\equiv}CCOOH$ because an sp carbon is more electronegative than an sp^2 carbon

d. 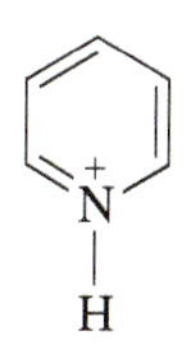
because an sp^2 nitrogen is more electronegative than an sp^3 nitrogen

71. **a.** between 9 and 10
b. between 0 and 1
c. between 2 and 3

a. 9.5
b. 0.08
c. 2.8

72. **a.** The first pK_a is lower than the pK_a of acetic acid because the middle COOH group of citric acid has additional oxygen-containing groups that acetic acid does not have that withdraw electrons inductively and thereby stabilize the conjugate base.

b. The third pK_a is greater than the pK_a of acetic acid because loss of the third proton puts a third negative charge on the molecule. Increasing the number of charges on a species destablizes it, and the less stable the base, the weaker its conjugate acid.

73.

$$K_a = \frac{[H^+][HO^-]}{[H_2O]}$$

Because $[H^+] = [HO^-]$, both must be 1×10^{-7} M

$$K_a = \frac{(1 \times 10^{-7})(1 \times 10^{-7})}{55.5}$$

$$K_a = 1.80 \times 10^{-16}$$

$$pK_a = -\log 1.80 \times 10^{-16}$$

$$pK_a = 15.7$$

The answer can also be obtained in the following way:

$$K_a = \frac{[H^+][HO^-]}{[H_2O]}$$

$$K_a[H_2O] = [H^+][HO^-]$$

take the log of both sides

$$\log K_a + \log[H_2O] = \log[H^+] + \log[HO^-]$$

multiply both sides by -1

$$-\log K_a - \log[H_2O] = -\log[H^+] - \log[HO^-]$$

$$pK_a - \log[H_2O] = pH + pOH$$

$$pK_a - \log[H_2O] = 14$$

$$pK_a = 14 + \log[H_2O]$$

$$pK_a = 14 + \log 55.5$$

$$pK_a = 14 + 1.7$$

$$pK_a = 15.7$$

74. Charged compounds dissolve in water, and uncharged compounds dissolve in ether.
The acidic forms of carboxylic acids and alcohols are neutral, and the basic forms are charged.
The acidic forms of amines are charged, and the basic forms are neutral.

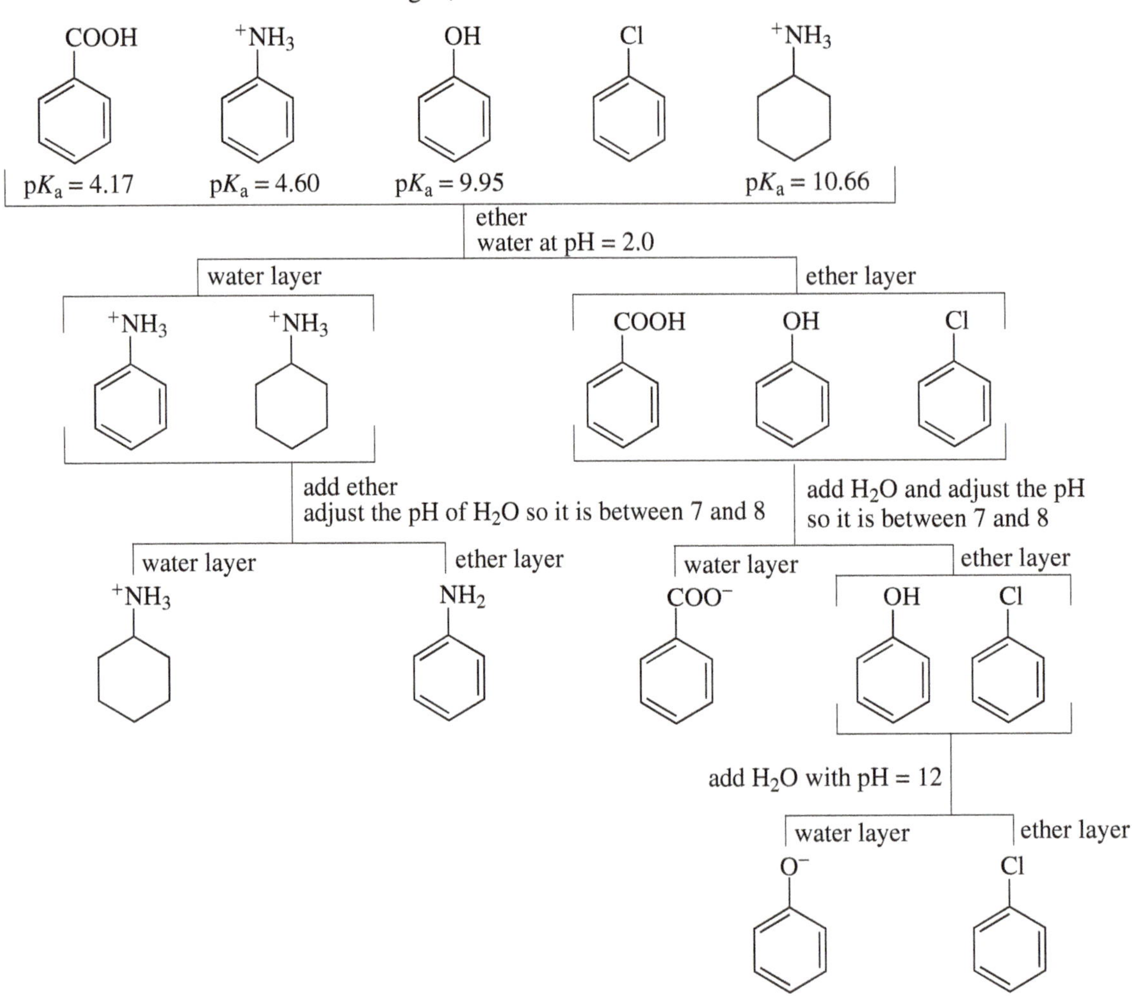

75. For a discussion of how to do problems such as Problems 68–70, see **Special Topic I** (pH, pK_a, and Buffers).

$$pK_a = pH + \log\frac{[HA]}{[A^-]}$$

The above equation, called the Henderson–Hasselbalch equation, shows that:

1. When the pH equals the pK_a, the concentration of buffer in the acidic form [HA] equals the concentration of buffer in the basic form $[A^-]$.
2. When the pH of the solution is less than the pK_a, more buffer species are in the acidic form than in the basic form.
3. When the pH of the solution is greater than the pK_a, more buffer species are in the basic form than in the acidic form.

Because the pH of the blood (~ 7.3) is greater than the pK_a of the buffer (6.1), more buffer species are in the basic form than in the acidic form. Therefore, the buffer is better at neutralizing excess acid.

76. **a.** fraction present in the acidic form $= \dfrac{\text{amount in the acidic form}}{\text{amount in the acidic form} + \text{amount in the basic form}}$

$$= \frac{[HA]}{[HA] + [A^-]}$$

Because there are two unknowns, we must define one in terms of the other.
By using the definition of the acid dissociation constant, we can define $[A^-]$ in terms of $[HA]$, $[H^+]$, and K_a, so we have only one unknown.

$$K_a = \frac{[H^+][A^-]}{[HA]}$$

$$[A^-] = \frac{K_a[HA]}{[H^+]}$$

Substituting the definition of $[A^-]$ into the equation for the fraction present in the acidic form gives:

$$\frac{[HA]}{[HA] + [A^-]} = \frac{[HA]}{[HA] + \dfrac{K_a[HA]}{[H^+]}} = \frac{1}{1 + \dfrac{K_a}{[H^+]}} = \frac{[H^+]}{[H^+] + K_a}$$

Therefore, the percentage that is present in the acidic form is given by:

$$\frac{[H^+]}{[H^+] + K_a} \times 100$$

Because the pK_a of the acid is given as 5.3, we know that K_a is 5.0×10^{-6} (because p$K_a = -\log K_a$). Because the pH of the solution is given as 5.7, we know that $[H^+]$ is 2.0×10^{-6} (because pH $= -\log [H^+]$).

Substituting into the equation for the percentage present in the acidic form gives:

$$\frac{[H^+]}{[H^+] + K_a} = \frac{2.0 \times 10^{-6}}{2.0 \times 10^{-6} + 5.0 \times 10^{-6}} \times 100$$

$$\frac{2.0 \times 10^{-6}}{7.0 \times 10^{-6}} \times 100 = 29\%$$

b. Fraction present in the acidic form $= \dfrac{[H^+]}{[H^+] + K_a} = 0.80$

$$[H^+] = 0.80([H^+] + K_a)$$
$$[H^+] = 0.80\,[H^+] + 0.80\,K_a$$
$$0.20\,[H^+] = 0.80\,K_a$$
$$[H^+] = 4\,K_a$$
$$[H^+] = 4 \times 5.0 \times 10^{-6}$$
$$[H^+] = 20 \times 10^{-6}$$
$$\text{pH} = 4.7$$

77. In each problem, we define $[H^+] = x$. Then $[A^-]$ is also x. In part **a**, because we have a 1.0 M solution, $[HA] = 1.0 - x$; in part **b**, because we have a 0.1 M solution, $[HA] = 0.1 - x$.

a.

$$K_a = \frac{[H^+][A^-]}{[HA]}$$

$$1.74 \times 10^{-5} = \frac{x^2}{1.0 - x}$$

$$1.74 \times 10^{-5} = x^2$$

$$x = 4.16 \times 10^{-3}$$

$$pH = 2.38$$

b.

$$K_a = \frac{[H^+][A^-]}{[HA]}$$

$$2.00 \times 10^{-11} = \frac{x^2}{0.1 - x}$$

$$2.00 \times 10^{-12} = x^2$$

$$x = 1.41 \times 10^{-6}$$

$$pH = 5.85$$

c. This question can be answered by plugging the given concentrations into the Henderson–Hasselbalch equation.

$$pK_a = pH + \log\frac{[acid]}{[base]}$$

$$3.76 = pH + \log\frac{0.3}{0.1}$$

$$3.76 = pH + \log 3$$

$$3.76 = pH + 0.48$$

$$pH = 3.76 - 0.48 = 3.28$$

Chapter 2 Practice Test

1. Which compound is the stronger acid?

   a. 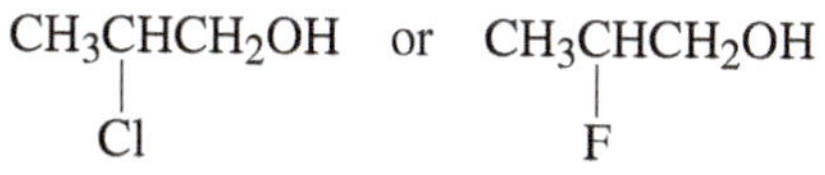 $CH_3CH(Cl)CH_2OH$ or $CH_3CH(F)CH_2OH$

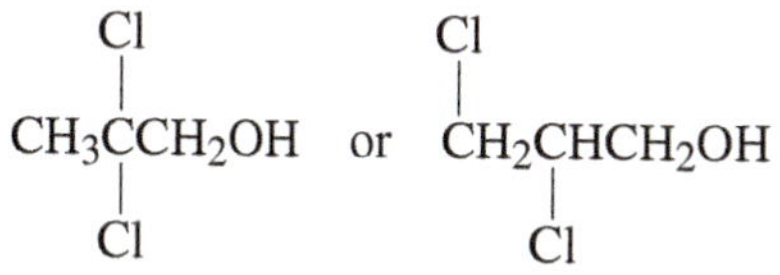

   b. HBr or HI

   c. $CH_3CCl_2CH_2OH$ or $CH_2ClCHClCH_2OH$

   d. CH_4 or NH_3

2. Which compound is the stronger base?

   a. $CH_3CH_2NH_2$ or CH_3CH_2OH **b.** F^- or I^-

3. Draw a circle around the strongest base and draw a square around the weakest base.

   $CH_3C(=O)O^-$ CH_3O^- CH_3OH $CH_3\bar{N}H$ CH_3NH_2

4. The following compounds are drawn in their acidic forms, and their pK_a values are given. Draw the form in which each compound will predominantly exist at pH = 8.

CH_3COOH	CH_3CH_2OH	$CH_3\overset{+}{O}H_2$	$CH_3CH_2\overset{+}{N}H_3$
pK_a = 4.8	pK_a = 15.9	pK_a = −2.5	pK_a = 11.2

5. **a.** Write the acid–base reaction that occurs when methylamine is added to water.

   b. Does the above reaction favor reactants or products?

6. **a.** What is the conjugate base of NH_3?

   b. What is its conjugate acid?

7. What is the pK_a of a compound that at pH = 7.2 has a 10 times greater concentration in its basic form than in its acidic form?

8. **a.** What products are formed from the following reaction?

   $$CH_3OH + {}^+NH_4 \rightleftharpoons$$

   b. Does the reaction favor reactants or products?

9. A compound has a $K_a = 6.3 \times 10^{-9}$. What is its approximate pK_a (that is, between what two integers)?

10. Label the compounds in order of decreasing acidity. (Label the most acidic compound #1.)

CH_3CH_2OH $CH_3CH_2NH_2$ CH_3CH_2SH $CH_3CH_2CH_3$

11. You are planning to carry out a reaction at pH = 4 that releases protons. Would it be better to use a 1.0 M formic acid/sodium formate buffer or a 1.0 M acetic acid/sodium acetate buffer? (The pK_a of formic acid is 3.75; the pK_a of acetic acid is 4.76.)

12. Indicate whether each of the following statements is true or false.

a. HO^- is a stronger base than $^-NH_2$. T F
b. A Lewis acid is a compound that accepts a share in a pair of electrons. T F
c. CH_3CH_3 is more acidic than $H_2C{=}CH_2$. T F
d. The weaker the acid, the more stable the conjugate base. T F
e. The larger the pK_a, the weaker the acid. T F
f. The weaker the base, the more stable it is. T F

ANSWERS TO ALL THE PRACTICE TESTS CAN BE FOUND AT THE END OF THE SOLUTIONS MANUAL.

SPECIAL TOPIC I

pH, pK_a, and Buffers

This is a continuation of the discussion on acids and bases found in Chapter 2 of the text. First, we will see how the pH of solutions of acids and bases can be calculated. We will look at three different kinds of solutions.

1. A solution made by dissolving a strong acid or a strong base in water.
2. A solution made by dissolving a weak acid or a weak base in water.
3. A solution made by dissolving a weak acid and its conjugate base in water. Such a solution is known as a **buffer solution**.

Before we start, we need to review a few terms.

An acid is a compound that loses a proton, and a base is a compound that gains a proton.

The degree to which an acid (HA) dissociates is described by its acid dissociation constant (K_a).

$$HA \rightleftharpoons H^+ + A^-$$

$$K_a = \frac{[H^+][A^-]}{[HA]}$$

The strength of an acid is indicated by its acid dissociation constant (K_a) or by its pK_a.

$$pK_a = -\log K_a$$

The stronger the acid, the **larger** its acid dissociation constant and the **smaller** its pK_a.

For example, an acid with an acid dissociation constant of 1×10^{-2} $(pK_a = 2)$ is a stronger acid than one with an acid dissociation constant of 1×10^{-4} $(pK_a = 4)$.

While pK_a is used to describe the strength of an acid, pH is used to describe the acidity of a solution. In other words, pH describes the concentration of hydrogen ions in a solution.

$$pH = -\log[H^+]$$

The smaller the pH, the more acidic the solution:
acidic solutions have pH values <7;
a neutral solution has a pH $= 7$;
basic solutions have pH values >7.

A solution with a pH $= 2$ is more acidic than a solution with a pH $= 4$.
A solution with a pH $= 12$ is more basic than a solution with a pH $= 8$.

Determining the pH of a Solution

To determine the pH of a solution, the concentration of hydrogen ion $[H^+]$ in the solution must be determined.

Strong Acids

A strong acid dissociates completely in solution. Strong acids have pK_a values <1.

Because a strong acid dissociates completely, the concentration of hydrogen ions is the same as the concentration of the acid: a 1.0 M HCl solution contains 1.0 M $[H^+]$; a 1.5 M HCl solution contains 1.5 M $[H^+]$.

Therefore, to determine the pH of a strong acid, the $[H^+]$ value does not have to be calculated; it is the same as the molarity of the strong acid.

Solution	$[H^+]$	pH
1.0 M HCl	1.0 M	0
1.0×10^{-2} M HCl	1.0×10^{-2} M	2.0
6.4×10^{-4} M HCl	6.4×10^{-4} M	3.2

Strong Bases

Strong bases are compounds such as NaOH or KOH that dissociate completely in water.

Because they dissociate completely, the concentration of hydroxide ion is the same as the concentation of the strong base.

pOH describes the basicity of a solution. The smaller the pOH, the more basic the solution; just like the smaller the pH, the more acidic the solution.

$$\text{pOH} = -\log[\text{HO}^-]$$

$[HO^-]$ and $[H^+]$ are related by the ionization constant for water (K_w).

$$K_w = [H^+][HO^-] = 10^{-14}$$
$$\text{pH} + \text{pOH} = 14$$

Solution	$[HO^-]$	pOH	pH
1.0 M NaOH	1.0 M	0	14.0 – 0 = 14.0
1.0×10^{-4} M NaOH	1.0×10^{-4} M	4.0	14.0 – 4.0 = 10.0
7.8×10^{-2} M NaOH	7.8×10^{-2} M	1.1	14.0 – 1.1 = 12.9

Weak Acids

A weak acid does not dissociate completely in solution. Therefore, $[H^+]$ must be calculated to determine the pH.

Acetic acid (CH_3COOH) is an example of a weak acid. It has an acid dissociation constant of 1.74×10^{-5} ($pK_a = 4.76$). The pH of a 1.00 M solution of acetic acid can be calculated as follows:

$$CH_3COOH \rightleftharpoons H^+ + CH_3COO^-$$

$$K_a = \frac{[H^+][CH_3COO^-]}{[CH_3COOH]}$$

Each molecule of acetic acid that dissociates forms one proton and one acetate ion. Therefore, the concentration of protons in solution equals the concentration of acetate ions. Each has a concentration that can be represented by x. The concentration of acetic acid, therefore, is the concentration we started with minus x.

$$1.74 \times 10^{-5} = \frac{(x)(x)}{1.00 - x}$$

The denominator ($1.00 - x$) can be simplified to 1.00 because 1.00 is much greater than x. (When we actually calculate the value of x, we see that it is 0.004. And $1.00 - 0.004 = 1.00$.)

$$1.74 \times 10^{-5} = \frac{x^2}{1.00}$$

$$x = 4.17 \times 10^{-3}$$

$$pH = -\log 4.17 \times 10^{-3}$$

$$pH = 2.38$$

Formic acid (HCOOH) has a pK_a value of 3.75. The pH of a 1.50 M solution of formic acid can be calculated as follows:

$$HCOOH \rightleftharpoons H^+ + HCOO^-$$

$$K_a = \frac{[H^+][HCOO^-]}{[HCOOH]}$$

A compound with a $pK_a = 3.75$ has an acid dissociation constant of 1.78×10^{-4}.

$$1.78 \times 10^{-4} = \frac{(x)(x)}{1.50 - x} = \frac{x^2}{1.50}$$

$$x^2 = 1.50(1.78 \times 10^{-4})$$

$$x^2 = 2.67 \times 10^{-4}$$

$$x = 1.63 \times 10^{-2}$$

$$pH = -\log(1.63 \times 10^{-2})$$

$$pH = 1.79$$

Weak Bases

When a weak base is dissolved in water, it accepts a proton from water, creating hydroxide ion.

Determining the concentration of hydroxide allows the pOH to be determined, and this, in turn, allows the pH to be determined.

The pH of a 1.20 M solution of sodium acetate can be calculated as follows:

$$CH_3COO^- + H_2O \rightleftharpoons CH_3COOH + HO^-$$

$$\frac{K_w}{K_a} = \frac{[HO^-][CH_3COOH]}{[CH_3COO^-]}$$

$$\frac{1.00 \times 10^{-14}}{1.74 \times 10^{-5}} = \frac{(x)(x)}{1.20 - x}$$

$$5.75 \times 10^{-10} = \frac{x^2}{1.20}$$

$$x^2 = 6.86 \times 10^{-10}$$

$$x = 2.62 \times 10^{-5} = [HO^-]$$

$$pOH = -\log 2.62 \times 10^{-5}$$

$$pOH = 4.58$$

$$pH = 14.00 - 4.58$$

$$pH = 9.42$$

Notice that by setting up the equation equal to K_w/K_a, we can avoid the introduction of a new term (K_b), because $K_w/K_a = K_b$.

Buffer Solutions

A buffer solution is a solution that maintains nearly constant pH in spite of the addition of small amounts of H^+ or HO^-. That is because a buffer solution contains both a weak acid and its conjugate base. The weak acid can give a proton to any HO^- added to the solution, and the conjugate base can accept any proton that is added to the solution, so the addition of small amounts of HO^- or H^+ does not significantly change the pH of the solution.

A buffer can maintain nearly constant pH in a range of one pH unit on either side of the pK_a of the conjugate acid. For example, an acetic acid/sodium acetate mixture can be used as a buffer in the pH range 3.76–5.76 because acetic acid has a $pK_a = 4.76$; methylammonium ion/methylamine can be used as a buffer in the pH range 9.7–11.7 because the methylammonium ion has a $pK_a = 10.7$.

The pH of a buffer solution can be determined from the Henderson–Hasselbalch equation. This equation comes directly from the expression defining the acid dissociation constant. Its derivation is found on pages 72–73 of the text.

Henderson–Hasselbalch equation

$$pK_a = pH + \log \frac{[HA]}{[A^-]}$$

The pH of an acetic acid/sodium acetate buffer solution (pK_a of acetic acid $= 4.76$) that is 1.00 M in acetic acid and 0.50 M in sodium acetate is calculated as follows:

$$\begin{aligned} pK_a &= pH + \log\frac{[HA]}{[A^-]} \\ 4.76 &= pH + \log\frac{1.00}{0.50} \\ 4.76 &= pH + \log 2 \\ 4.76 &= pH + 0.30 \\ pH &= 4.46 \end{aligned}$$

Remember from Section 2.10 that compounds exist primarily in their acidic forms in solutions that are more acidic than their pK_a values and primarily in their basic forms in solutions that are more basic than their pK_a values. Therefore, it could have been predicted that the above solution will have a pH less than the pK_a of acetic acid, because there is more conjugate acid than conjugate base in the solution.

There are three ways a buffer solution can be prepared:

1. Weak Acid and Weak Base

A buffer solution can be prepared by mixing a solution of a weak acid with a solution of its conjugate base.

The pH of a formic acid/sodium formate buffer (pK_a of formic acid $= 3.75$) solution prepared by mixing 25 mL of 0.10 M formic acid and 15 mL of 0.20 M sodium formate is calculated as follows:

The equation below shows that the number of millimoles (mmol) of each of the buffer components can be determined by multiplying the number of milliliters (mL) by the molarity (M).

$$M = \text{molarity} = \frac{\text{moles}}{\text{liters}} = \frac{\text{millimoles}}{\text{milliliters}}$$

Therefore:

$$\begin{aligned} 25\text{ mL} \times 0.10\text{ M} &= 2.5\text{ mmol formic acid} \\ 15\text{ mL} \times 0.20\text{ M} &= 3.0\text{ mmol sodium formate} \end{aligned}$$

Notice that in the following equation, we use mmol for both $[HA]$ and $[A^-]$ rather than molarity (mmol/mL) because both the acid and the conjugate base are in the same solution, so they have the same volume. Therefore, volumes cancel in the equation.

$$\begin{aligned} pK_a &= pH + \log\frac{[HA]}{[A^-]} \\ 3.75 &= pH + \log\frac{2.5}{3.0} \\ 3.75 &= pH + \log 0.83 \\ 3.75 &= pH - 0.08 \\ pH &= 3.83 \end{aligned}$$

It could have been predicted that the above solution would have a pH greater than the pK_a of formic acid, because there is more conjugate base than conjugate acid in the solution.

2. Weak Acid and Strong Base

A buffer solution can be prepared by mixing a solution of a weak acid with a solution of a strong base such as NaOH. The NaOH reacts completely with the weak acid, thereby creating the conjugate base needed for the buffer solution.

The pH of a solution prepared by mixing 10 mL of a 2.0 M solution of a weak acid with a pK_a of 5.86 with 5.0 mL of a 1.0 M solution of sodium hydroxide can be calculated as follows:

When the 20 mmol of HA and the 5.0 mmol of HO^- are mixed, the 5.0 mmol of strong base react with 5.0 mmol of HA, with the result that 5.0 mmol of A^- is formed and 15 mmol (20 mmol − 5.0 mmol) of HA is left unreacted.

$$10\text{ mL} \times 2.0\text{ M} = 20\text{ mmol HA} \longrightarrow 15\text{ mmol HA}$$

$$5.0\text{ mL} \times 1.0\text{ M} = 5.0\text{ mmol HO}^- \longrightarrow 5.0\text{ mmol A}^-$$

$$\begin{aligned} pK_a &= pH + \log\frac{[HA]}{[A^-]} \\ 5.86 &= pH + \log\frac{15}{5} \\ 5.86 &= pH + \log 3 \\ 5.86 &= pH + 0.48 \\ pH &= 5.38 \end{aligned}$$

3. Weak Base and Strong Acid

A buffer solution can be prepared by mixing a solution of a weak base with a strong acid such as HCl. The strong acid reacts completely with the weak base, thereby forming the conjugate acid needed for the buffer solution.

The pH of an ethylammonium ion/ethylamine buffer (pK_a of $CH_3CH_2\overset{+}{N}H_3$ = 11.0) prepared by mixing 30 mL of 0.20 M ethylamine with 40 mL of 0.10 M HCl can be calculated as follows:

$$30\text{ mL} \times 0.20\text{ M} = 6.0\text{ mmol RNH}_2 \longrightarrow 2.0\text{ mmol RNH}_2$$

$$40\text{ mL} \times 0.10\text{ M} = 4.0\text{ mmol H}^+ \longrightarrow 4.0\text{ mmol R}\overset{+}{N}H_3$$

Notice that 4.0 mmol H^+ reacts with 4.0 mmol RNH_2, forming 4.0 mmol $R\overset{+}{N}H_3$, and 2.0 mmol of RNH_2 is left unreacted.

$$\begin{aligned} pK_a &= pH + \log\frac{[HA]}{[A^-]} \\ 11.0 &= pH + \log\frac{4.0}{2.0} \\ 11.0 &= pH + \log 2.0 \\ 11.0 &= pH + 0.30 \\ pH &= 10.7 \end{aligned}$$

Fraction Present in the Acidic or the Basic Form

A common question asked is what fraction of a buffer will be in a particular form at a given pH—either what fraction will be in the acidic form or what fraction will be in the basic form.
This is an easy question to answer if you remember the following formulas that are derived at the end of this section:

$$\text{fraction present in the acidic form} = \frac{[H^+]}{K_a + [H^+]}$$

$$\text{fraction present in the basic form} = \frac{K_a}{K_a + [H^+]}$$

What fraction of an acetic acid/sodium acetate buffer (pK_a of acetic acid $= 4.76$; $K_a = 1.74 \times 10^{-5}$) is present in the acidic form at pH $= 5.20$; $[H^+] = 6.31 \times 10^{-6}$?

$$\frac{[H^+]}{K_a + [H^+]} = \frac{6.31 \times 10^{-6}}{(1.74 \times 10^{-5}) + (6.31 \times 10^{-6})}$$

$$= \frac{6.31 \times 10^{-6}}{(17.4 \times 10^{-6}) + (6.31 \times 10^{-6})}$$

$$= \frac{6.31 \times 10^{-6}}{23.7 \times 10^{-6}} = \frac{6.31}{23.7}$$

$$= 0.26$$

What fraction of a formic acid/sodium formate buffer (pK_a of formic acid $= 3.75$; $K_a = 1.78 \times 10^{-4}$) is present in the basic form at pH $= 3.90$; $[H^+] = 1.26 \times 10^{-4}$?

$$\frac{K_a}{K_a + [H^+]} = \frac{1.78 \times 10^{-4}}{(1.78 \times 10^{-4}) + (1.26 \times 10^{-4})}$$

$$= \frac{1.78 \times 10^{-4}}{3.04 \times 10^{-4}} = \frac{1.78}{3.04}$$

$$= 0.586$$

$$= 0.59$$

The formulas describing the fraction present in the acidic or basic form are obtained from the definition of the acid dissociation constant.

$$K_a = \frac{[H^+][A^-]}{[HA]}$$

To derive the equation for the fraction present in the acidic form, we need to define $[A^-]$ in terms of [HA], so we have only one unknown in the equation.

$$[A^-] = \frac{K_a[HA]}{[H^+]}$$

$$\text{fraction present in the acidic form} = \frac{[HA]}{[HA] + [A^-]} = \frac{[HA]}{[HA] + \dfrac{K_a[HA]}{[H^+]}} = \frac{1}{1 + \dfrac{K_a}{[H^+]}}$$

$$= \frac{[H^+]}{K_a + [H^+]}$$

To derive the equation for the fraction present in the basic form, we need to define [HA] in terms of $[A^-]$, so we can get rid of the [HA] term.

$$K_a = \frac{[H^+][A^-]}{[HA]}$$

$$[HA] = \frac{[H^+][A^-]}{K_a}$$

$$\text{fraction present in the basic form} = \frac{[A^-]}{[HA] + [A^-]} = \frac{[A^-]}{[A^-] + \dfrac{[H^+][A^-]}{K_a}}$$

$$= \frac{1}{1 + \dfrac{[H^+]}{K_a}}$$

$$= \frac{K_a}{K_a + [H^+]}$$

Preparing Buffer Solutions

The type of calculations just shown can be used to determine how to make a buffer solution.

For example, how can 100 mL of a 1.00 M buffer solution with a pH = 4.24 be prepared if you have 1.50 M solutions of acetic acid ($pK_a = 4.76$; $K_a = 1.74 \times 10^{-5}$), sodium acetate, HCl, and NaOH?

First, we need to determine what fraction of the buffer is present in each form at pH = 4.24; $[H^+] = 5.75 \times 10^{-5}$. We start by calculating the fraction of the buffer present in the acidic form.

$$\frac{[H^+]}{K_a + [H^+]} = \frac{5.75 \times 10^{-5}}{(1.74 \times 10^{-5}) + (5.75 \times 10^{-5})}$$

$$= \frac{5.75 \times 10^{-5}}{7.49 \times 10^{-5}}$$

$$= 0.77$$

If a 1.00 M buffer solution is desired, the buffer must be 0.77 M in acetic acid and 0.23 M in sodium acetate.

There are three ways to make such a buffer solution:

1. **By mixing the appropriate amounts of acetic acid and sodium acetate in water and adding water to obtain a final volume of 100 mL.**

The amount of acetic acid needed: $[CH_3COOH] = 0.77\ M$

$$M = \frac{mmol}{mL} = \frac{x\ mmol}{100\ mL} = 0.77\ M$$
$$x = 77\ mmol$$

Therefore, we need 77 mmol of acetic acid in the final solution.

To obtain 77 mmol of acetic acid from a 1.50 M solution of acetic acid:

$$\frac{77\ mmol}{y\ mL} = 1.50\ M$$
$$y = 51.3\ mL$$

Notice that the formula M = mmol/mL is used twice. The first time it is used to determine the number of mmol of acetic acid that is needed in the final solution. The second time it is used to determine how that number of mmol can be obtained from an acetic acid solution with a known concentration.

The amount of sodium acetate needed: $[CH_3COO^-] = 0.23\ M$

$$\frac{x\ mmol}{100\ mL} = 0.23$$
$$x = 23\ mmol$$

To obtain 23 mmol of sodium acetate from a 1.50 M solution of sodium acetate:

$$\frac{23\ mmol}{y\ mL} = 1.50\ M$$
$$y = 15.3\ mL$$

The desired buffer solution can be prepared using:

51.3 mL 1.50 M acetic acid
15.3 mL 1.50 M sodium acetate
33.4 mL H_2O
______
100.0 mL

2. By mixing the appropriate amounts of acetic acid and sodium hydroxide and adding water to obtain a final volume of 100 mL.

Sodium hydroxide is used to convert some of the acetic acid into sodium acetate. This means that acetic acid will be the source of both acetic acid and sodium acetate.

The concentrations needed: $[CH_3COOH] = 1.00\text{ M}$
$[NaOH] = 0.23\text{ M}$

The amount of acetic acid needed: $[CH_3COOH] = 1.00\text{ M}$

$$\frac{x\text{ mmol}}{100\text{ mL}} = 1.00\text{ M}$$
$$x = 100\text{ mmol}$$

To obtain 100 mmol of acetic acid from a 1.50 M solution of acetic acid:

$$\frac{100\text{ mmol}}{y\text{ mL}} = 1.50\text{ M}$$
$$y = 66.7\text{ mL}$$

The amount of sodium hydroxide needed: $[NaOH] = 0.23\text{ M}$

$$\frac{x\text{ mmol}}{100\text{ mL}} = 0.23\text{ M}$$
$$x = 23\text{ mmol}$$

To obtain 23 mmol of sodium hydroxide from a 1.50 M solution of NaOH:

$$\frac{23\text{ mmol}}{y\text{ mL}} = 1.50\text{ M}$$
$$y = 15.3\text{ mL}$$

The desired buffer solution can be prepared using:
66.7 mL 1.50 M acetic acid
15.3 mL 1.50 M NaOH
18.0 mL H_2O
100.0 mL

3. By mixing the appropriate amounts of sodium acetate and hydrochloric acid and adding water to obtain a final volume of 100 mL.

Hydrochloric acid is used to convert some of the sodium acetate into acetic acid.

This means that sodium acetate will be the source of both acetic acid and sodium acetate.

The concentrations needed: $[CH_3COONa] = 1.00\text{ M}$
$[HCl] = 0.77\text{ M}$

The amount of sodium acetate needed: $[CH_3COONa] = 1.00\text{ M}$

$$\frac{x\text{ mmol}}{100\text{ mL}} = 1.00\text{ M}$$

$$x = 100\text{ mmol}$$

To obtain 100 mmol of sodium acetate from a 1.50 M solution of sodium acetate:

$$\frac{100\text{ mmol}}{y\text{ mL}} = 1.50\text{ M}$$

$$y = 66.7\text{ mL}$$

The amount of hydrochloric acid needed:

$$[HCl] = 0.77\text{ M}$$

$$\frac{x\text{ mmol}}{100\text{ mL}} = 0.77\text{ M}$$

$$x = 77\text{ mmol}$$

To obtain 77 mmol of hydrochloric acid from a 1.50 M solution of HCl:

$$\frac{77\text{ mmol}}{y\text{ mL}} = 1.50\text{ M}$$

$$y = 51.3\text{ mL}$$

100 mL of a 1.00 M acetic acid/acetate buffer cannot be made from these reagents, because the volumes needed $(66.7\text{ mL} + 51.3\text{ mL})$ add up to more than 100 mL. To make this buffer using sodium acetate and hydrochloric acid, you need to use a more concentrated solution $(>1.50\text{ M})$ of sodium acetate and/or a more concentrated solution $(>1.50\text{ M})$ of HCl.

Problems on pH, pK_a, and Buffers

1. Calculate the pH of each of the following solutions:

   a. 1×10^{-3} M HCl

   b. 0.60 M HCl

   c. 1.40×10^{-2} M HCl

   d. 1×10^{-3} M KOH

   e. 3.70×10^{-4} M NaOH

   f. a 1.20 M solution of an acid with a $pK_a = 4.23$

   g. 1.60×10^{-2} M sodium acetate (pK_a of acetic acid $= 4.76$)

2. Calculate the pH of each of the following buffer solutions:

   a. A buffer prepared by mixing 20 mL of 0.10 M formic acid and 15 mL of 0.50 M sodium formate (pK_a of formic acid $= 3.75$).

   b. A buffer prepared by mixing 10 mL of 0.50 M aniline and 15 mL of 0.10 M HCl (pK_a of the anilinium ion $= 4.60$).

   c. A buffer prepared by mixing 15 mL of 1.00 M acetic acid and 10 mL of 0.50 M NaOH (pK_a of acetic acid $= 4.76$).

3. What fraction of a carboxylic acid with a $pK_a = 5.23$ will be ionized at pH $= 4.98$?

4. What will be the concentration of formic acid and sodium formate in a 1.00 M buffer solution with a pH $= 3.12$ (pK_a of formic acid $= 3.75$)?

5. You have found a bottle labeled 1.00 M RCOOH. You want to identify the carboxylic acid, so you decide to determine its pK_a value. How can you do this?

6. a. How can you prepare 100 mL of a buffer solution that is 0.30 M in acetic acid and 0.20 M in sodium acetate using a 1.00 M acetic acid solution and a 2.00 M sodium acetate solution?

   b. The pK_a of acetic acid is 4.76. Will the pH of the above solution be greater or less than 4.76?

7. You have 100 mL of a 1.50 M acetic acid/sodium acetate buffer solution that has a pH $= 4.90$. How can you change the pH of the solution to 4.50?

8. You have 100 mL of a 1.00 M solution of an acid with a $pK_a = 5.62$ to which you add 10 mL of 1.00 M sodium hydroxide. What fraction of the acid will be in the acidic form? How much more sodium hydroxide will you need to add so that 40% of the acid is in its acidic form (that is, with its proton)?

9. Describe three ways to prepare a 1.00 M acetic acid/sodium acetate buffer solution with a pH = 4.00.

10. You have available to you 1.50 M solutions of acetic acid, sodium acetate, sodium hydroxide, and hydrochloric acid. How can you make 50 mL of each of the buffers described in the preceding problem?

11. How can you make a 1.0 M buffer solution with a pH = 3.30?

12. You are planning to carry out a reaction that produces protons. In order for the reaction to take place at constant pH, it will be carried in a solution buffered at pH = 4.2. Would it be better to use a formic acid/formate buffer or an acetic acid/acetate buffer?

Solutions to Problems on pH, pK_a, and Buffers

1. **a.** $pH = -\log(1 \times 10^{-3})$
$pH = 3$

b. $pH = -\log 0.60$
$pH = 0.22$

c. $pH = -\log(1.40 \times 10^{-2})$
$pH = 1.85$

d. $pOH = -\log(1 \times 10^{-3})$
$pOH = 3$
$pH = 14 - 3 = 11$

e. $pOH = -\log(3.70 \times 10^{-4})$
$pOH = 3.43$
$pH = 10.57$

f. $pK_a = 4.23, \quad K_a = 5.89 \times 10^{-5}$

$$K_a = \frac{[H^+][A^-]}{[HA]}$$

$$5.89 \times 10^{-5} = \frac{x^2}{1.20}$$

$$x^2 = 7.07 \times 10^{-5}$$

$$x = 8.41 \times 10^{-3}$$

$$pH = 2.08$$

g. $$\frac{K_w}{K_a} = \frac{[HO^-][HA]}{[A^-]} \qquad (K_a = 10^{-4.76} = 1.74 \times 10^{-5})$$

$$\frac{1.0 \times 10^{-14}}{1.74 \times 10^{-5}} = \frac{x^2}{1.60 \times 10^{-2}}$$

$$5.75 \times 10^{-10} = \frac{x^2}{1.60 \times 10^{-2}}$$

$$x^2 = 9.20 \times 10^{-12}$$

$$x = 3.03 \times 10^{-6}$$

$$pOH = 5.52$$

$$pH = 14.00 - 5.52 = 8.48$$

2. **a.** formic acid: 20 mL × 0.10 M = 2.0 mmol
sodium formate: 15 mL × 0.50 M = 7.5 mmol

$$pK_a = pH + \log\frac{[HA]}{[A^-]}$$
$$3.75 = pH + \log\frac{2.0}{7.5}$$
$$3.75 = pH + \log 0.27$$
$$3.75 = pH + (-0.57)$$
$$pH = 4.32$$

b. aniline: 10 mL × 0.50 M = 5.0 mmol ⟶ 3.5 mmol aniline (RNH_2)
HCl: 15 mL × 0.10 M = 1.5 mmol ⟶ 1.5 mmol anilinium hydrochloride ($R\overset{+}{N}H_3$)

$$pK_a = pH + \log\frac{[HA]}{[A^-]}$$
$$4.60 = pH + \log\frac{1.5}{3.5}$$
$$4.60 = pH + \log 0.43$$
$$4.60 = pH + (-0.37)$$
$$pH = 4.97$$

c. acetic acid: 15 mL × 1.00 M = 15 mmol ⟶ 10 mmol acetic acid
NaOH: 10 mL × 0.50 M = 5.0 mmol ⟶ 5.0 mmol sodium acetate

$$pK_a = pH + \log\frac{[HA]}{[A^-]}$$
$$4.76 = pH + \log\frac{10}{5.0}$$
$$4.76 = pH + \log 2$$
$$4.76 = pH + 0.30$$
$$pH = 4.46$$

3. The ionized form is the basic form. Therefore, we need to use the equation that allows us to calculate the fraction present in the basic form.

$$\text{fraction of buffer in the basic form} = \frac{K_a}{K_a + [H^+]} = \frac{5.89 \times 10^{-6}}{(5.89 \times 10^{-6}) + (10.47 \times 10^{-6})} = \frac{5.89 \times 10^{-6}}{16.36 \times 10^{-6}}$$
$$= 0.36$$

4. fraction of buffer in the basic form $= \dfrac{K_a}{K_a + [H^+]} = \dfrac{1.78 \times 10^{-4}}{(1.78 \times 10^{-4}) + (7.59 \times 10^{-4})}$

$$= \frac{1.78 \times 10^{-4}}{9.37 \times 10^{-4}}$$

$$= 0.19$$

$$[\text{sodium formate}] = 0.19 \text{ M}$$
$$[\text{formic acid}] = 0.81 \text{ M}$$

5. From the Henderson–Hasselbalch equation, we see that the pH of the solution will be the same as the pK_a of the compound when the concentration of the compound in the acidic form is the same as the concentration of the compound in the basic form.

$$pK_a = pH + \log\frac{[HA]}{[A^-]}$$

$$\text{when } [HA] = [A^-],$$
$$pK_a = pH$$

Therefore, in order to have a solution in which the pH will be the same as the pK_a, the number of mmol of acid must equal the number of mmol of conjugate base.

Preparing a solution of x mmol of RCOOH and $1/2\,x$ mmol NaOH gives a solution in which $[RCOOH] = [RCOO^-]$.

For example: 20 mL of 1.00 M RCOOH = 20 mmol
10 mL of 1.00 M NaOH = 10 mmol

This gives a solution that contains 10 mmol RCOOH and 10 mmol $RCOO^-$.

The pH of this solution is the pK_a of RCOOH.

6. **a.** $\dfrac{x \text{ mmol}}{100 \text{ mL}} = 0.30 \text{ M}$

$x = 30$ mmol of acetic acid

$\dfrac{x \text{ mmol}}{100 \text{ mL}} = 0.20 \text{ M}$

$x = 20$ mmol of sodium acetate

$\dfrac{30 \text{ mmol}}{y \text{ mL}} = 1.00 \text{ M}$

$y = 30$ mL of 1.00 M acetic acid

$\dfrac{20 \text{ mmol}}{y \text{ mL}} = 2.00 \text{ M}$

$y = 10$ mL of 2.00 M acetic acid

The buffer solution can be prepared by mixing:
30 mL of 1.00 M acetic acid
10 mL of 2.00 M sodium acetate
60 mL of water
———
100 mL

b. Because the concentration of buffer in the acidic form (0.30 M) is greater than the concentration of buffer in the basic form (0.20 M), the pH of the solution will be less than 4.76.

7. **Original solution**

$$\text{fraction of buffer in the basic form} = \frac{K_a}{K_a + [H^+]} = \frac{1.74 \times 10^{-5}}{(1.74 \times 10^{-5}) + (1.26 \times 10^{-5})}$$

$$= \frac{1.74 \times 10^{-5}}{3.00 \times 10^{-5}}$$

$$= 0.58$$

$$0.58 \times 1.50 \text{ M} = 0.87 \text{ M}$$

$$[A^-] = 0.87 \text{ M}$$

$$[HA] = 0.63 \text{ M}$$

Desired solution

$$\text{fraction of buffer in the basic form} = \frac{K_a}{K_a + [H^+]} = \frac{1.74 \times 10^{-5}}{(1.74 \times 10^{-5}) + (3.16 \times 10^{-5})}$$

$$= \frac{1.74 \times 10^{-5}}{4.90 \times 10^{-5}}$$

$$= 0.35$$

$$0.35 \times 1.50 \text{ M} = 0.53 \text{ M}$$

$$[A^-] = 0.53 \text{ M}$$

$$[HA] = 0.97 \text{ M}$$

The original solution contains 87 mmol of A^- $(100 \text{ mL} \times 0.87 \text{ M})$.

The desired solution with a pH $= 4.50$ must contain 53 mmol of A^-.

Therefore, 34 mmol of A^- $(87 - 53 = 34)$ must be converted to HA.

This can be done by adding 34 mmol of HCl to the original solution.

If you have a 1.00 M HCl solution, you will need to add 34 mL of it to the original solution in order to change its pH from 4.90 to 4.50.

$$\frac{34 \text{ mmol}}{x \text{ mL}} = 1.00 \text{ M}$$

$$x = 34 \text{ mL}$$

Note that after adding HCl to the original solution, it will no longer be a 1.50 M buffer; it will be more dilute $(150 \text{ mmol}/134 \text{ mL} = 1.12 \text{ M})$.

The change in the concentration of the buffer solution will be less if a more concentrated solution of HCl is used to change the pH. For example, if you have a 2.00 M HCl solution:

$$\frac{34 \text{ mmol}}{x \text{ mL}} = 2.00 \text{ M}$$

$$x = 17 \text{ mL}$$

You will need to add 17 mL to the original solution, and the concentration of the buffer will be 1.28 M (150 mmol/117 mL = 1.28 M).

8. acid: 100 mL × 1.00 M = 100 mmol HA ⟶ 90 mmol HA

NaOH: 10 mL × 1.00 M = 10 mmol HO^- ⟶ 10 mmol A^-

Therefore, 90% is in the acidic form.

For 40% to be in the acidic form, you need:

40 mmol HA
60 mmol A^-

You need to have 60 mmol rather than 10 mmol in the basic form. To get the additional 50 mmol in the basic form, you need to add 50 mL of 1.0 M NaOH.

9.

$$\text{fraction of buffer in the basic form} = \frac{K_a}{K_a + [H^+]} = \frac{1.74 \times 10^{-5}}{(1.74 \times 10^{-5}) + (1.00 \times 10^{-4})} = \frac{1.74 \times 10^{-5}}{(1.74 \times 10^{-5}) + (10.00 \times 10^{-5})}$$

$$= \frac{1.74 \times 10^{-5}}{11.74 \times 10^{-5}}$$

$$= 0.15$$

$$[A^-] = 0.15\text{ M}$$
$$[HA] = 0.85\text{ M}$$

a. [acetic acid] = 0.85 M
[sodium acetate] = 0.15 M

b. [acetic acid] = 1.00 M
[NaOH] = 0.15 M

c. [sodium acetate] = 1.00 M
[HCl] = 0.85 M

10. **a.**

$$\frac{x\text{ mmol}}{50\text{ mL}} = 0.85\text{ M}$$

$$x = 42.5\text{ mmol of acetic acid}$$

$$\frac{42.5\text{ mmol}}{y\text{ mL}} = 1.50\text{ M}$$

$$y = 28.3\text{ mL of 1.50 M acetic acid}$$

$$\frac{x\text{ mmol}}{50\text{ mL}} = 0.15\text{M}$$

$$x = 7.5\text{ mmol of sodium acetate}$$

$$\frac{7.5\text{ mmol}}{y\text{ mL}} = 1.50\text{ M}$$

$$y = 5.0\text{ mL of 1.50 M sodium acetate}$$

28.3 mL of 1.50 M acetic acid
5.0 mL of 1.50 M sodium acetate
16.7 mL of H_2O
50.0 mL

b. $\dfrac{x \text{ mmol}}{50 \text{ mL}} = 1.00 \text{ M}$

$x = 50$ mmol of acetic acid

$\dfrac{50 \text{ mmol}}{y \text{ mL}} = 1.50 \text{ M}$

$y = 33.3$ mL of 1.50 M acetic acid

$\dfrac{x \text{ mmol}}{50 \text{ mL}} = 0.15 \text{ M}$

$x = 7.5$ mmol of NaOH

$\dfrac{7.5 \text{ mmol}}{y \text{ mL}} = 1.50 \text{ M}$

$y = 5.0$ mL of 1.50 M NaOH

33.3 mL of 1.50 M acetic acid
5.0 mL of 1.50 M NaOH
11.7 mL of H_2O
50.0 mL

c. $\dfrac{x \text{ mmol}}{50 \text{ mL}} = 1.00 \text{ M}$

$x = 50$ mmol of sodium acetate

$\dfrac{50 \text{ mmol}}{y \text{ mL}} = 1.50 \text{ M}$

$y = 33.3$ mL of 1.50 M sodium acetate

$\dfrac{x \text{ mmol}}{50 \text{ mL}} = 0.85 \text{ M}$

$x = 42.5$ mmol of HCl

$\dfrac{42.5 \text{ mmol}}{y \text{ mL}} = 1.5 \text{ M}$

$y = 28.3$ mL of 1.5 M HCl

We cannot make the required buffer with these solutions, because 33.3 mL + 28.3 mL > 50 mL.

11. Because formic acid has a pK_a = 3.75, a formic acid/formate buffer can be a buffer at pH = 3.30, because this pH is within one pH unit of the pK_a value.

$$\text{fraction of buffer in the basic form} = \frac{K_a}{K_a + [H^+]} = \frac{1.78 \times 10^{-4}}{(1.78 \times 10^{-4}) + (5.01 \times 10^{-4})}$$

$$= \frac{1.78 \times 10^{-4}}{6.79 \times 10^{-4}}$$

$$= 0.26$$

The solution must have [formic acid] = 0.74 M and [sodium formate] = 0.26 M.

12. The reaction to be carried out produces protons that will react with the basic form of the buffer in order to keep the pH constant.

Therefore, the better buffer is the one that has the larger percentage of the buffer in the basic form.

The pK_a of formic acid is 3.74. Because the pH of the solution (4.2) is greater than the pK_a of the compound, formic acid will be primarily in its basic form.

The pK_a of acetic acid is 4.76. Because the pH of the solution (4.2) is less than the pK_a of the compound, acetic acid will be primarily in its acidic form.

Therefore, the formate buffer is preferred, because it has a greater percentage of the buffer in the basic form.

CHAPTER 3
An Introduction to Organic Compounds
Nomenclature, Physical Properties, and Structure

Important Terms

alcohol	a compound with an OH group in place of one of the hydrogens of an alkane (ROH).
alkane	a hydrocarbon that contains only single bonds.
alkyl halide	a compound with a halogen in place of one of the hydrogens of an alkane.
alkyl substituent	a substituent formed by removing a hydrogen from an alkane.
amine	a compound in which one or more of the hydrogens of NH_3 are replaced by an alkyl substituent (RNH_2, R_2NH, R_3N).
angle strain	the strain introduced into a molecule as a result of its bond angles being distorted from their ideal values.
anti conformer	the staggered conformer in which the largest substituents bonded to the two carbons are opposite each other. It is the most stable of the staggered conformers.
axial bond	a bond of the chair conformer of cyclohexane that points directly up or directly down.
boat conformer	a conformer of cyclohexane that roughly resembles a boat.
boiling point	the temperature at which the vapor pressure of a liquid equals the atmospheric pressure.
chair conformer	a conformer of cyclohexane that roughly resembles a chair. It is the most stable conformer of cyclohexane.
cis fused	two rings fused together in such a way that if the second ring were considered to be two substituents of the first ring, the two substituents would be on the same side of the first ring.
cis isomer (for a cyclic compound)	the isomer with two substituents on the same side of the ring.
cis-trans stereoisomers	see the definition of "cis isomer" and "trans isomer."
common name	nonsystematic nomenclature.
conformation	the three-dimensional shape of a molecule at a given instant.
conformers	different conformations of a molecule.
constitutional isomers (structural isomers)	molecules that have the same molecular formula but differ in the way the atoms are connected.

cycloalkane an alkane with its carbon chain arranged in a closed ring.

1,3-diaxial interaction the interaction between an axial substituent and one of the other two axial substituents on the same side of a cyclohexane ring.

dipole–dipole interaction an interaction between the dipole of one molecule and the dipole of another.

eclipsed conformer a conformer in which the bonds on adjacent carbons are parallel to each other when viewed looking down the carbon–carbon bond.

equatorial bond a bond of the chair conformer of cyclohexane that juts out from the ring but does not point directly up or directly down.

ether a compound in which an oxygen is bonded to two alkyl substituents (ROR).

flagpole hydrogens the two hydrogens in the boat conformer of cyclohexane that are at the 1- and 4-positions of the ring.

functional group the center of reactivity of a molecule.

gauche conformer a staggered conformer in which the largest substituents bonded to the two carbons are gauche to each other; that is, their bonds have a dihedral angle of approximately 60°.

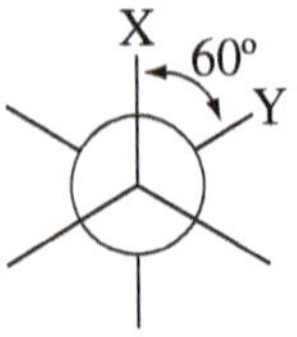

X 60° Y

The substituents are gauche to each other.

gauche interaction the interaction between two atoms or groups that are gauche to each other.

geometric isomers cis–trans isomers.

half-chair conformer the least stable conformer of cyclohexane.

homologue a member of a homologous series.

homologous series a family of compounds in which each member differs from the next by one methylene group.

hydrocarbon a compound that contains only carbon and hydrogen.

hydrogen bond an unusually strong dipole–dipole interaction between a hydrogen bonded to O, N, or F and the lone pair of a different O, N, or F.

hyperconjugation delocalization of electrons by the overlap of a σ orbital with an empty orbital.

induced-dipole–induced-dipole interaction an interaction between a temporary dipole in one molecule and the dipole that the temporary dipole induces in another molecule.

IUPAC nomenclature systematic nomenclature developed by the International Union of Pure and Applied Chemistry.

London dispersion forces induced-dipole–induced-dipole interactions.

melting point the temperature at which a solid becomes a liquid.

methylene group a CH_2 group.

Newman projection a way to represent the three-dimensional spatial relationships of atoms by looking down the length of a particular carbon–carbon bond.

packing a property that determines how well individual molecules fit into a crystal lattice.

parent hydrocarbon the longest continuous carbon chain in a molecule; if the molecule has a functional group, it is the longest continuous carbon chain that contains the functional group.

perspective formula a way to represent the three-dimensional spatial relationships of atoms using two adjacent solid lines, one solid wedge, and one hatched wedge.

polarizability the ease with which an electron cloud of an atom can be distorted.

primary alcohol an alcohol in which the OH group is bonded to a primary carbon.

primary alkyl halide an alkyl halide in which the halogen is bonded to a primary carbon.

primary amine an amine with one alkyl group bonded to the nitrogen.

primary carbon a carbon bonded to only one other carbon.

primary hydrogen a hydrogen bonded to a primary carbon.

quaternary ammonium salt a compound with four alkyl groups bonded to a nitrogen, plus an accompanying anion.

ring flip (chair-chair interconversion) the conversion of a chair conformer of cyclohexane into the other chair conformer; bonds that are axial in one chair conformer are equatorial in the other chair conformer.

sawhorse projection a way to represent the three-dimensional spatial relationships of atoms by looking at the carbon–carbon bond from an oblique angle.

secondary alcohol an alcohol in which the OH group is bonded to a secondary carbon.

secondary alkyl halide an alkyl halide in which the halogen is bonded to a secondary carbon.

secondary amine an amine with two alkyl groups bonded to the nitrogen.

secondary carbon a carbon bonded to two other carbons.

secondary hydrogen a hydrogen bonded to a secondary carbon.

solubility the extent to which a compound dissolves in a solvent.

solvation the interaction between a solvent and another molecule (or ion).

staggered conformer a conformer in which the bonds on one carbon bisect the bond angles on the adjacent carbon when viewed looking down the carbon-carbon bond.

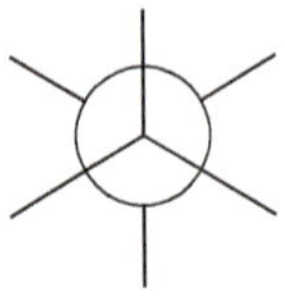

steric hindrance hindrance due to groups occupying a volume of space.

steric strain the repulsion between the electron cloud of an atom or group of atoms and the electron cloud of another atom or group of atoms.

straight-chain alkane an alkane in which the carbons form a continuous chain with no branches.

structural isomers (constitutional isomers) molecules that have the same molecular formula but differ in the way the atoms are connected.

symmetrical ether an ether with two identical alkyl substituents bonded to the oxygen.

systematic nomenclature a system of nomenclature based on rules.

tertiary alcohol an alcohol in which the OH group is bonded to a tertiary carbon.

tertiary alkyl halide an alkyl halide in which the halogen is bonded to a tertiary carbon.

tertiary amine an amine with three alkyl groups bonded to the nitrogen.

tertiary carbon a carbon bonded to three other carbons.

tertiary hydrogen a hydrogen bonded to a tertiary carbon.

trans-fused two rings fused together in such a way that if the second ring were considered to be two substituents of the first ring, the two substituents would be on opposite sides of the first ring.

trans isomer (for a cyclic compound) the isomer with two substituents on opposite sides of the ring.

twist-boat conformer one of the conformers of a cyclohexane ring.

unsymmetrical ether an ether with two different alkyl substituents bonded to the oxygen.

Solutions to Problems

1. **a.** C_nH_{2n+2} If there are 17 carbons, then there are 36 hydrogens.

   b. C_nH_{2n+2} If there are 74 hydrogens, then there are 36 carbons.

2. $CH_3CH_2CH_2CH_2CH_2CH_2CH_2CH_3$ — octane

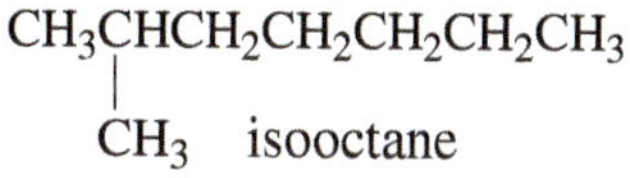

   $CH_3CH(CH_3)CH_2CH_2CH_2CH_2CH_3$ — isooctane

3. **a.** propyl alcohol **b.** butyl methyl ether **c.** propylamine

4. **a.** 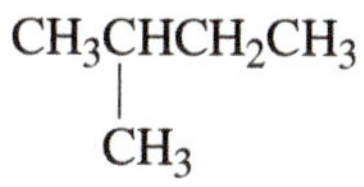 $CH_3CH(CH_3)CH_2CH_3$ — 2-methylbutane

   b. 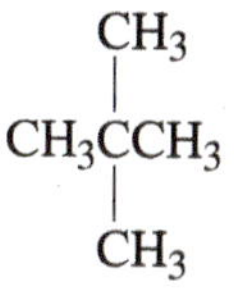 $CH_3C(CH_3)_2CH_3$ — 2,2-dimethylpropane

5. Notice that each carbon forms four bonds and each hydrogen and bromine forms one bond.

   $CH_3CH_2CH_2CH_2Br$ — *n*-butyl bromide or butyl bromide

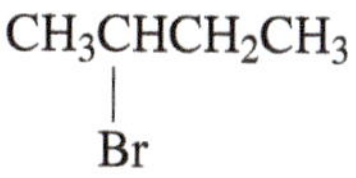

   $CH_3CH(Br)CH_2CH_3$ — *sec*-butyl bromide

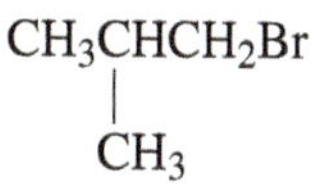

   $CH_3CH(CH_3)CH_2Br$ — isobutyl bromide

   $CH_3C(CH_3)(Br)CH_3$ — *tert*-butyl bromide

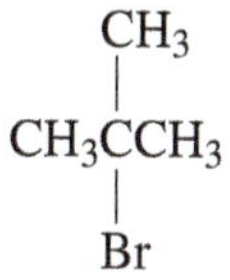

6. "Dibromomethane does not have constitutional isomers" proves that carbon is tetrahedral.

   If carbon were flat, rather than tetrahedral, dibromomethane would have constitutional isomers because the two structures shown below would be different since the bromines would be 90° apart in one compound and 180° apart in the other compound. Only because carbon is tetrahedral are the two structures identical.

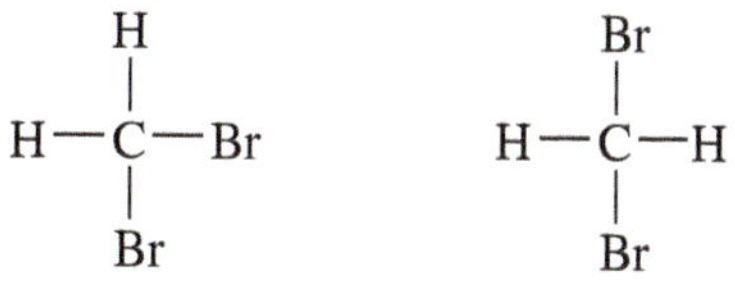

   H—C(H)(Br)—Br   H—C(Br)(Br)—H

7. **a.** 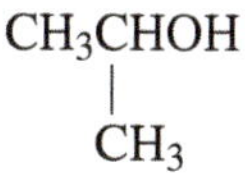 $CH_3CH(CH_3)OH$

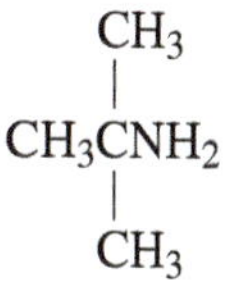

   b. 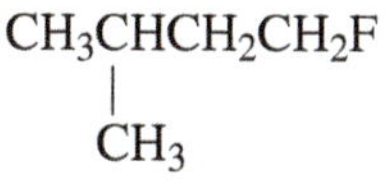 $CH_3CH(CH_3)CH_2CH_2F$

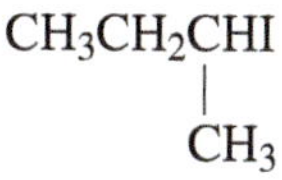

   c. $CH_3CH_2CH(CH_3)I$

   d. 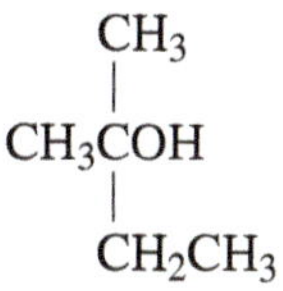 $CH_3C(CH_3)(CH_2CH_3)OH$

   e. $CH_3C(CH_3)_2NH_2$

   f. $CH_3CH_2CH_2CH_2CH_2CH_2CH_2CH_2Br$

8.
a. ethyl methyl ether
b. methyl propyl ether
c. *sec*-butylamine
d. butyl alcohol or *n*-butyl alcohol
e. isobutyl bromide
f. *sec*-butyl chloride

9.
a. 2,2,4-trimethylhexane
b. 2,2-dimethylbutane
c. 3-methyl-4-propylheptane
d. 2,2,5-trimethylhexane
e. 3,3-diethyl-4-methyl-5-propyloctane
f. 5-ethyl-4,4-dimethyloctane
g. 3,3-diethylhexane
h. 4-(1-methylethyl)octane or 4-isopropyloctane
i. 2,5-dimethylheptane

10. Solved in the text.

11. A substituent can be drawn pointing up from the chain or pointing down from the chain.

a.
CH_3
$CH_3CCH_2—CHCH_2CH_2CH_2CH_3$
CH_3 CH_3CHCH_3

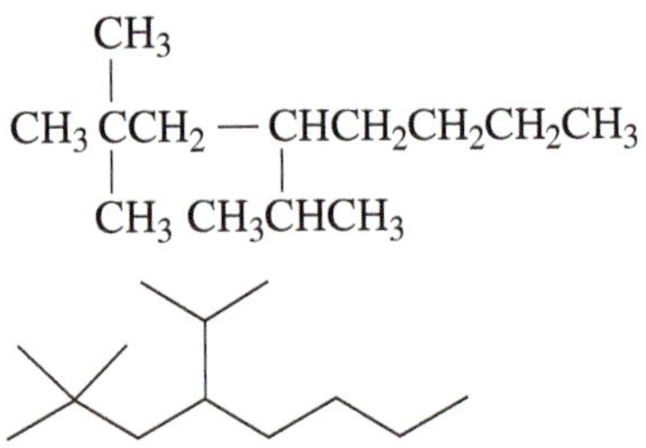

b.
CH_3
$CH_3CHCHCH_2CH_2CH_3$
CH_3

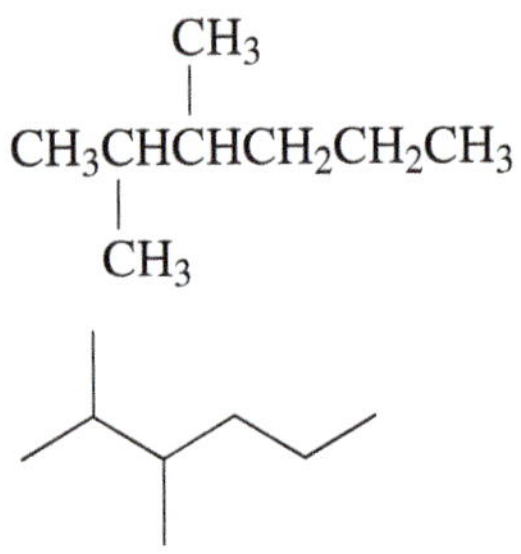

c.
CH_2CH_3
$CH_3CH_2CH_2CCH_2CH_2CH_2CH_2CH_2CH_3$
CH_2CH_3

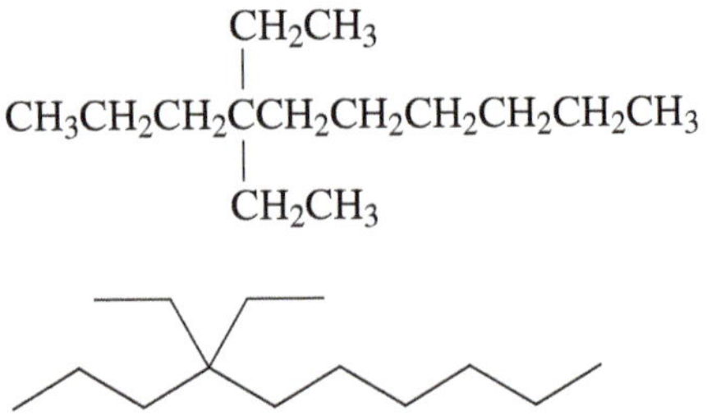

d.
CH_3 CH_3 CH_3
$CH_3CHCH_2C—CHCH_2CH_3$
CH_3CHCH_3

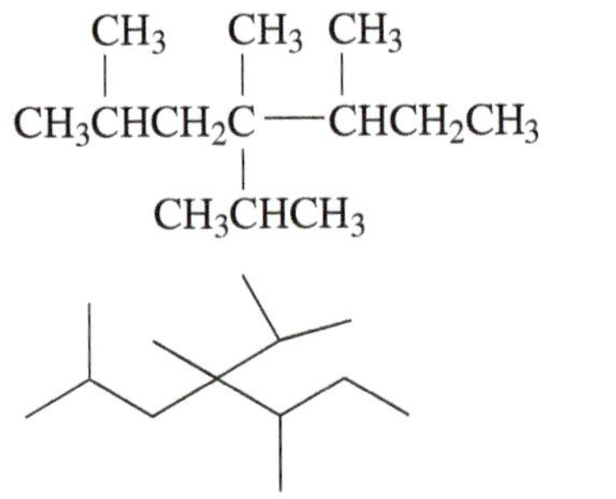

e.
CH_3 CH_3
$CH_3CHCH_2CHCHCH_2CH_2CH_3$
CH_2
CH_3CHCH_3

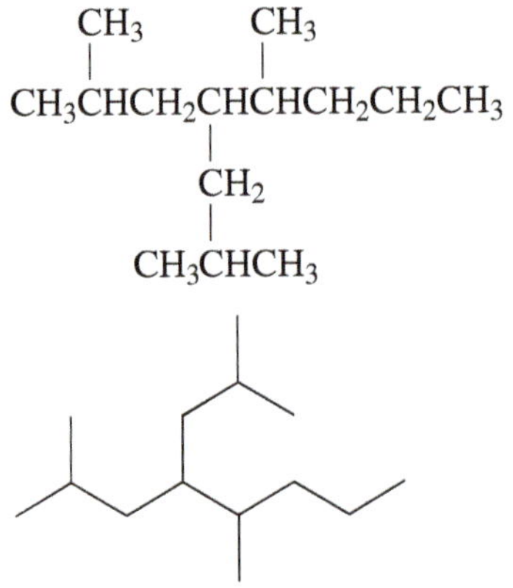

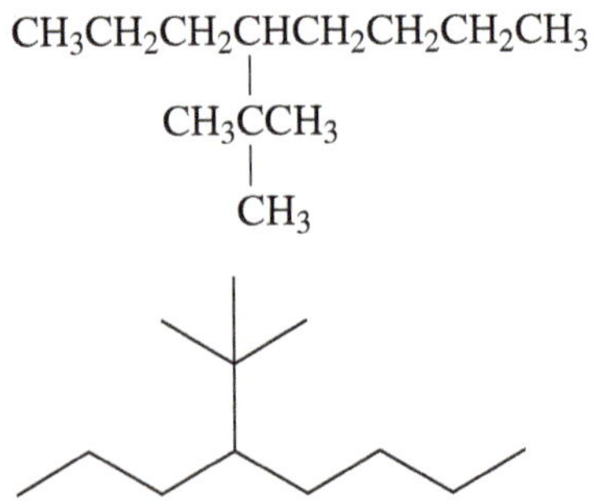

f.
$CH_3CH_2CH_2CHCH_2CH_2CH_2CH_3$
CH_3CCH_3
CH_3

12. **a.**

#1 $CH_3CH_2CH_2CH_2CH_2CH_2CH_2CH_3$
octane

#2 $CH_3CHCH_2CH_2CH_2CH_2CH_3$
CH_3
2-methylheptane

#3 $CH_3CH_2CHCH_2CH_2CH_2CH_3$
CH_3
3-methylheptane

#4 $CH_3CH_2CH_2CHCH_2CH_2CH_3$
CH_3
4-methylheptane

CH_3
#5 $CH_3CCH_2CH_2CH_2CH_3$
CH_3
2,2-dimethylhexane

CH_3
#6 $CH_3CH_2CCH_2CH_2CH_3$
CH_3
3,3-dimethylhexane

#7 $CH_3CH-CHCH_2CH_2CH_3$
CH_3 CH_3
2,3-dimethylhexane

CH_3 CH_3
#8 $CH_3CHCH_2CHCH_2CH_3$
2,4-dimethylhexane

#9 CH_3 CH_3
$CH_3CHCH_2CH_2CHCH_3$
2,5-dimethylhexane

CH_3 CH_3
#10 $CH_3CH_2CH-CHCH_2CH_3$
3,4-dimethylhexane

CH_3 CH_3
#11 $CH_3C-CHCH_2CH_3$
CH_3
2,2,3-trimethylpentane

CH_3 CH_3
#12 $CH_3CCH_2CHCH_3$
CH_3
2,2,4-trimethylpentane

CH_3 CH_3
#13 $CH_3CH-CCH_2CH_3$
CH_3
2,3,3-trimethylpentane

CH_3 CH_3
#14 $CH_3CH-CH-CHCH_3$
CH_3
2,3,4-trimethylpentane

CH_3 CH_3
#15 CH_3C-CCH_3
CH_3 CH_3
2,2,3,3-tetramethylbutane

#16 $CH_3CH_2CHCH_2CH_2CH_3$
CH_2CH_3
3-ethylhexane

CH_3
#17 $CH_3CH_2CHCHCH_3$
CH_2CH_3
3-ethyl-2-methylpentane

CH_3
#18 $CH_3CH_2CCH_2CH_3$
CH_2CH_3
3-ethyl-3-methylpentane

b. The systematic name is under each structure.
c. Only #1 (octane or *n*-octane) and #2 (isooctane) have common names.
d. #2, #7, #8, #9, #12, #13, #14, #17
e. #3, #8, #10, #11
f. #5, #11, #12, #15

13. **a.** isopropyl and (1-methylethyl)
b. *tert*-butyl and (1,1-dimethylethyl)
c. *sec*-butyl and (1-methylpropyl)
d. isobutyl and (2-methylpropyl)

14. **a.** $CH_3CH_2CH_2CH_2CH_3$

pentane

b. 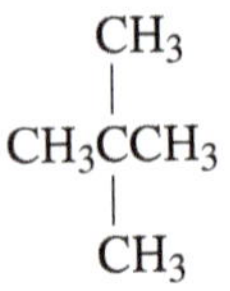 CH_3 | CH_3CCH_3 | CH_3

2,2-dimethylpropane

c. 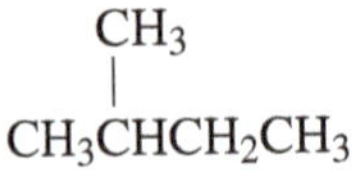 CH_3 | $CH_3CHCH_2CH_3$

2-methylbutane

d. 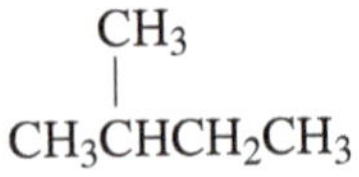 CH_3 | $CH_3CHCH_2CH_3$

2-methylbutane

15. **a.** 1-ethyl-2-methylcyclopentane
b. ethylcyclobutane
c. 4-ethyl-1,2-dimethylcyclohexane
d. 3,6-dimethyldecane
e. 2-cyclopropylhexane
f. 1-ethyl-3-isobutylcyclohexane or 1-ethyl-3-(2-methylpropyl)cyclohexane
g. 5-isopropylnonane or 5-(1-methylethyl)nonane
h. 1-*sec*-butyl-4-isopropylcyclohexane or 1-(1-methylethyl)-4-(1-methylpropyl)cyclohexane
i. 2,2,6-trimethylheptane

16.

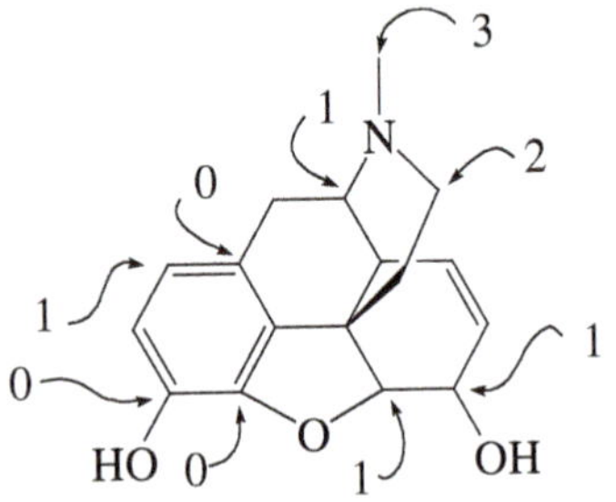

3 1 N 2 0 1 1 0 HO 0 O 1 OH

17. **a.**  OH

b.

c.

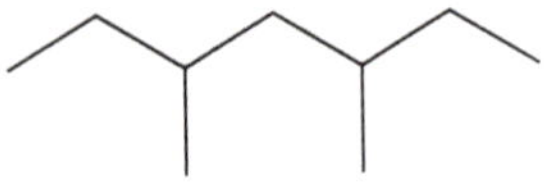

d.  O

e. 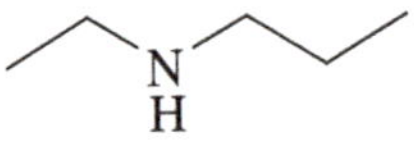 N H

f. 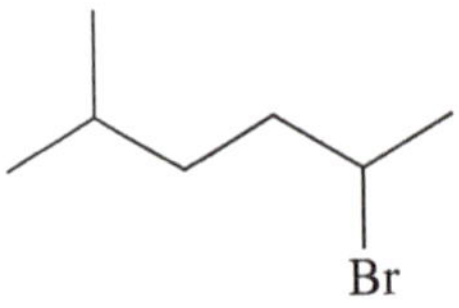 Br

18. a.

d. 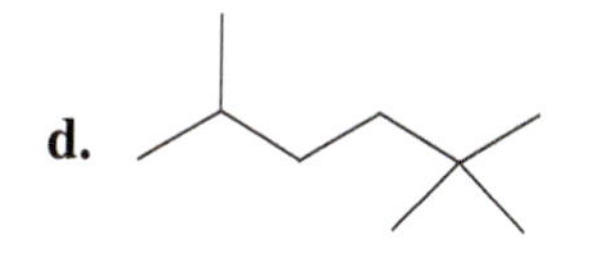

g.

b.

e.

h.

c.

f.

i.

19. a. $CH_3CH(CH_3)CH(CH_2CH_3)CH(CH_2CH_3)CH_2CH_2CH_3$

$$\begin{array}{c} \quad CH_3 \quad CH_2CH_3 \\ CH_3\overset{|}{C}H\underset{|}{C}H\overset{|}{C}HCH_2CH_2CH_3 \\ CH_2CH_3 \end{array}$$

b.

$$\begin{array}{c} CH_3 \\ CH_3\overset{|}{\underset{|}{C}}CH_2CH_2\underset{|}{C}HCH_3 \\ CH_3 \qquad\quad CH_3 \end{array}$$

20. a. *sec*-butyl chloride
2-chlorobutane
secondary

b. cyclohexyl bromide
bromocyclohexane
secondary

c. isohexyl chloride
1-chloro-4-methylpentane
primary

d. isopropyl fluoride
2-fluoropropane
secondary

21. a. CH_2Cl

chloromethylcyclohexane

Note that the name of a CH_2Cl sustituent is "chloromethyl," because a Cl is in place of one of the Hs of a methyl substituent.

b. Cl CH_3

1-chloro-1-methylcyclohexane

c. CH_3 Cl

1-chloro-2-methylcyclohexane

CH_3 Cl

1-chloro-3-methylcyclohexane

CH_3 Cl

1-chloro-4-methylcyclohexane

22. **a.** **1.** methoxyethane
2. ethoxyethane
3. 4-methoxyoctane
4. 1-isopropoxy-3-methylbutane
5. 1-propoxybutane
6. 2-isopropoxyhexane

b. No.

c. **1.** ethyl methyl ether
2. diethyl ether
3. no common name
4. isopentyl isopropyl ether
5. butyl propyl ether
6. no common name

23. **a.** 1-pentanol
primary

b. 5-chloro-2-methyl-2-pentanol
tertiary

c. 5-methyl-3-hexanol
secondary

d. 7-methyl-3,5-octanediol (Notice that because there are two OH groups, the suffix is "diol.")
both alcohol groups are secondary

24. CH_3OH
common = methyl alcohol
systematic = methanol

CH_3CH_2OH
common = ethyl alcohol
systematic = ethanol

$CH_3CH_2CH_2OH$
common = propyl alcohol or *n*-propyl alcohol
systematic = 1-propanol

$CH_3CH_2CH_2CH_2OH$
common = butyl alcohol or *n*-butyl alcohol
systematic = 1-butanol

$CH_3CH_2CH_2CH_2CH_2OH$
common = pentyl alcohol or *n*-pentyl alcohol
systematic = 1-pentanol

$CH_3CH_2CH_2CH_2CH_2CH_2OH$
common = hexyl alcohol or *n*-hexyl alcohol
systematic = 1-hexanol

25. $CH_3C(CH_3)(OH)CH_2CH_2CH_3$ $CH_3CH_2C(CH_3)(OH)CH_2CH_3$ $CH_3C(CH_3)(OH)—CH(CH_3)CH_3$

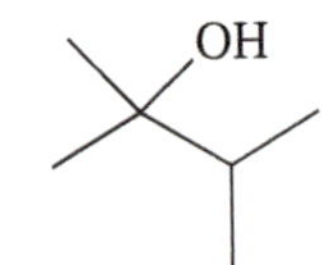

2-methyl-2-pentanol 3-methyl-3-pentanol 2,3-dimethyl-2-butanol

26. **a.** 4-chloro-3-ethylcyclohexanol
secondary

b. 7,8-dimethyl-3-nonanol
secondary

c. 1-bromo-5,5-dimethyl-3-heptanol
secondary

d. 4-methylcyclohexanol
secondary

27.

a. 1-hexanamine
hexylamine
primary

b. *N*-*sec*-butyl-4-methyl-1-pentanamine or
4-methyl-*N*-(1-methylpropyl)-1-pentanamine
sec-butylisohexylamine
secondary

c. *N*-ethyl-*N*-methylethanamine
diethylmethylamine
tertiary

d. *N*-propyl-1-butanamine
butylpropylamine
secondary

e. *N*,*N*-diethyl-1-propanamine
diethylpropylamine
tertiary

f. *N*-ethyl-3-methylcyclopentanamine
no common name
secondary

28. **a.** tertiary alkyl halide **b.** primary amine **c.** secondary alcohol **d.** secondary amine

29.

a. $CH_3CH_2CH_2NHCH_2CHCH_3$ (with CH_3 on the CH)

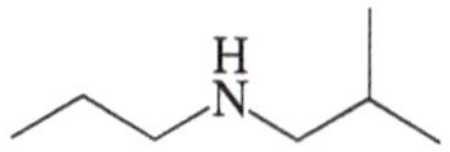

b. $CH_3CH_2NHCH_2CH_3$

c. $CH_3CHCH_2CH_2CH_2CH_2NH_2$ (with CH_3 on the CH)

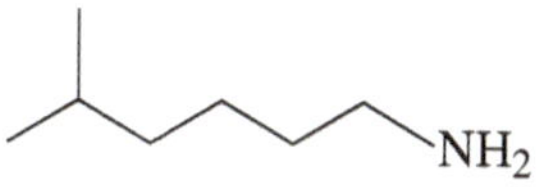

d. $CH_3CH_2CH_2NCH_2CH_2CH_3$ (with CH_3 on the N)

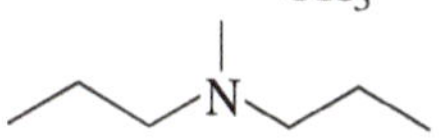

e. $CH_3CH_2CHCH_2CH_3$ (with $N(CH_3)_2$ on the CH)

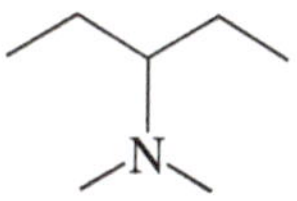

f. cyclohexyl–NCH_2CH_3 (with CH_3 on the N)

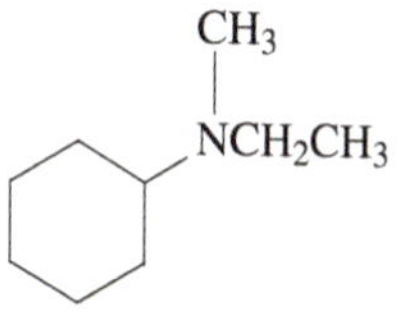

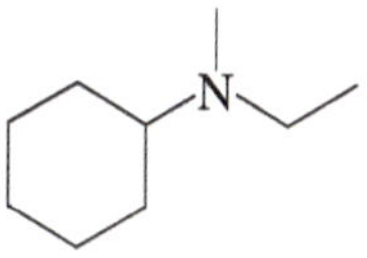

30.

a. 6-methyl-1-heptanamine
isooctylamine
primary

b. cyclohexanamine
cyclohexylamine
primary

c. 4-methyl-*N*-propyl-1-pentanamine
isohexylpropylamine
secondary

d. 2,5-dimethylcyclohexanamine
no common name
primary

31.

a. The bond angle is predicted to be similar to the bond angle in water (104.5°).
b. The bond angle is predicted to be similar to the bond angle in ammonia (107.3°).
c. The bond angle is predicted to be similar to the bond angle in water (104.5°).
d. The bond angle is predicted to be similar to the bond angle in the ammonium ion (109.5°).

32. To be a liquid at room temperature, the compound must have a boiling point that is greater than room temperature.

pentane

33. **a.** 1, 4, and 5
b. 1, 2, 4, 5, and 6

34. **a.** Each water molecule has two hydrogens that can form hydrogen bonds, whereas each alcohol molecule has only one hydrogen that can form a hydrogen bond. Therefore, there are more hydrogen bonds between water molecules than between alcohol molecules.

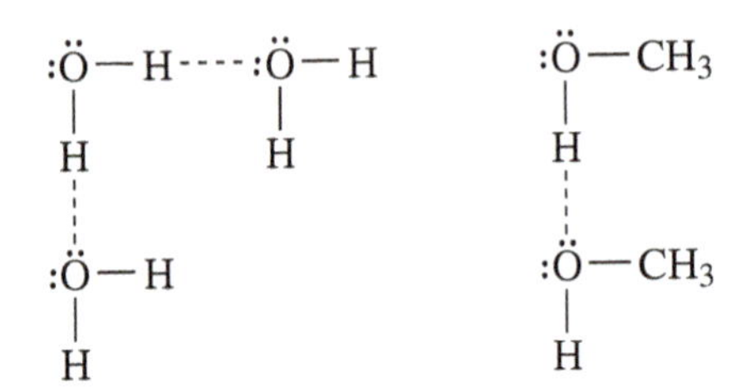

b. Each water molecule has two hydrogens that can form hydrogen bonds and two lone pairs that can accept hydrogen bonds.
Ammonia has three hydrogens that can form hydrogen bonds but only one lone pair that can accept hydrogen bonds.
Therefore, only one hydrogen of an ammonia molecule can engage in hydrogen bonding, so it will have a lower boiling point than water. In addition, oxygen forms stronger hydrogen bonds than nitrogen because oxygen is more electronegative than nitrogen.

c. Each water molecule has two hydrogens that can form hydrogen bonds and two lone pairs that can accept hydrogen bonds.
HF has three lone pairs that can accept hydrogen bonds but only one hydrogen that can form a hydrogen bond.
Therefore, only one lone pair of a HF molecule can engage in hydrogen bonding, so it will have a lower boiling point than water.

d. HF and ammonia can each form only one hydrogen bond, but HF has a higher boiling point because the hydrogen bond formed by HF is stronger since fluorine is more electronegative than nitrogen.

35. HO OH OH > OH OH > OH > NH_2 > >

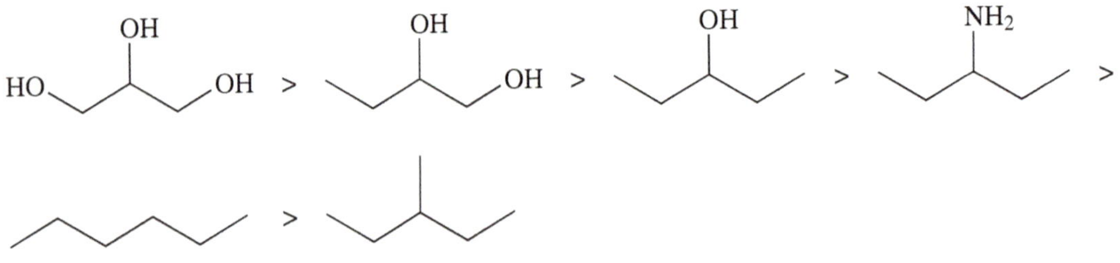

36. **a.** Br > Br > Br

b. > > >

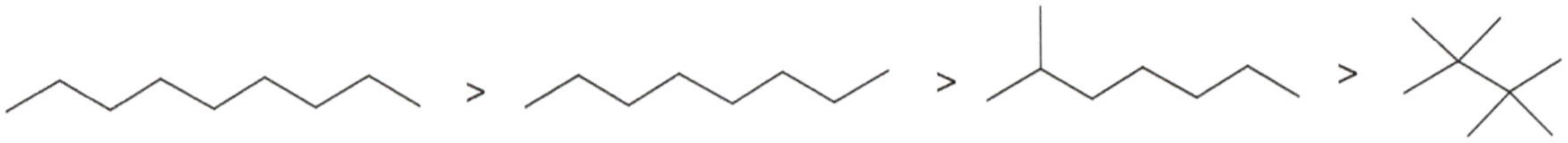

c. OH > OH > Cl >

37. Because cyclohexane is a nonpolar compound, it will have the lowest solubility in the most polar solvent, which, of the solvents given, is ethanol.

$CH_3CH_2CH_2CH_2CH_2OH$	$CH_3CH_2OCH_2CH_3$	CH_3CH_2OH	$CH_3CH_2CH_2CH_2CH_2CH_3$
1-pentanol	diethyl ether	ethanol	hexane

38. **a.** OH, OH, O > OH, O > OH >

b. $HOCH_2CH_2CH_2OH > CH_3CH_2CH_2CH_2OH > CH_3CH_2CH_2CH_2Cl$

39. Hexethal would be expected to be the more effective sedative because it is less polar than barbital since hexethal has a hexyl group in place of the ethyl group of barbital. Being less polar, hexethal will be better able to penetrate the nonpolar membrane of the cell.

40. Start with the least stable conformer and then obtain the others by keeping the front carbon constant and rotating the back carbon clockwise.

a.

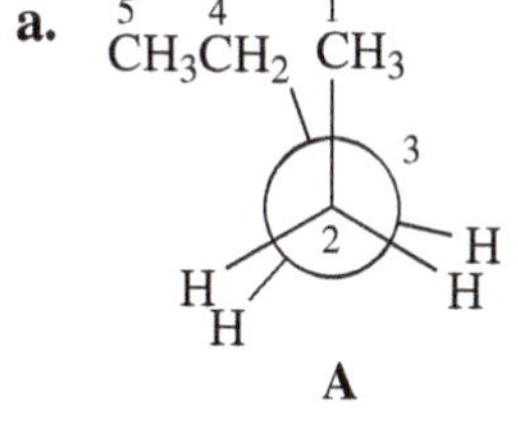

CH_3, H, CH_2CH_3, H, H, H — **B**

H, CH_3, CH_2CH_3, H, H, H — **C**

CH_3, H, H, H, H, CH_2CH_3 — **D**

H, CH_3, H, H, H, CH_3CH_2 — **E**

CH_3, CH_3CH_2, H, H, H, H — **F**

b.

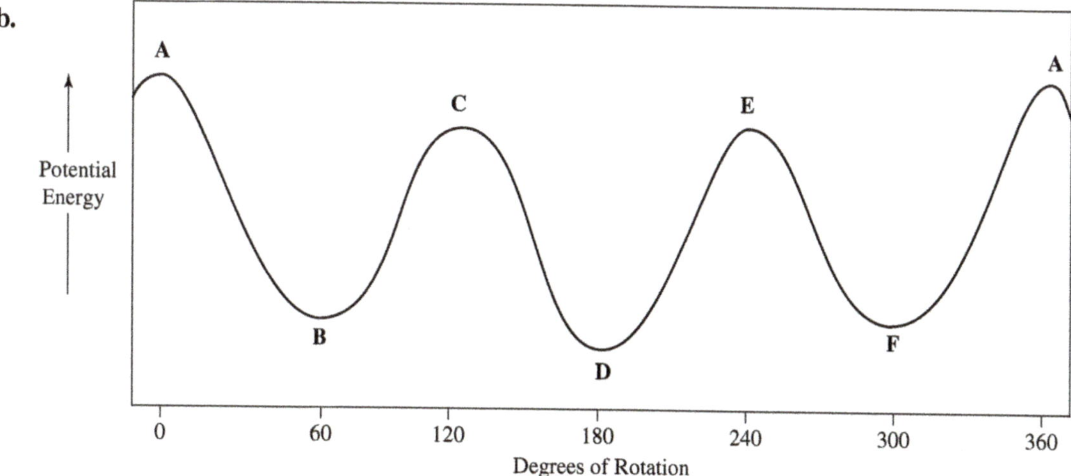

41. **a.** The Newman projection shows rotation about the C-2—C-3 bond.

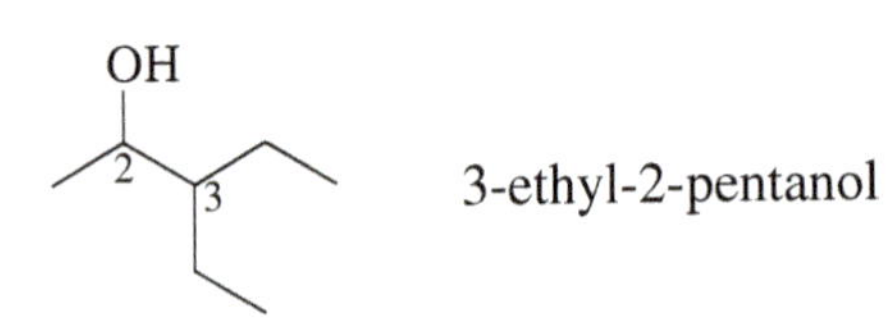

b. The Newman projection shows rotation about the C-2—C-3 bond.

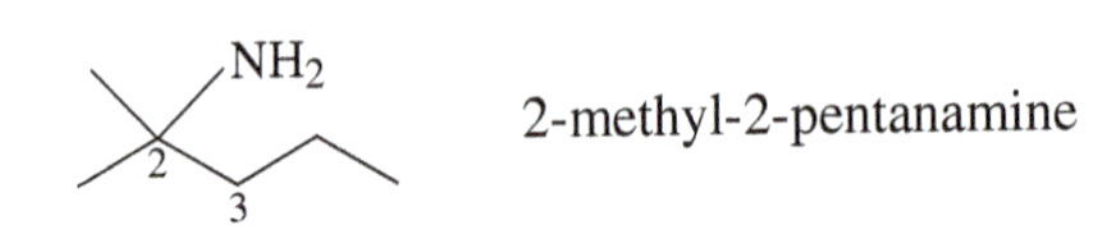

42. To draw the most stable conformer: put the largest group on the front carbon opposite the largest group on the back carbon.

a. (Newman projection: CH_3 front top; H, CH_3, H, H; CH_2CH_3 back bottom)

b. (Newman projection: CH_2CH_3 top; H, H, H, CH_3; CH_2CH_3 bottom)

c. (Newman projection: CH_2CH_3 top; H, H, CH_3, CH_3; CH_2CH_3 bottom)

43. **a.** $180° - \dfrac{360°}{8}$

$180° - 45° = 135°$

b. $180° - \dfrac{360°}{9}$

$180° - 40° = 140°$

44. You can get the total strain energy of cycloheptane by subtracting the strainless heat of formation from the actual heat of formation:

The "strainless" heat of formation of cycloheptane is $7\,(-4.92) = -34.4$ kcal/mol.
The actual heat of formation of cycloheptane is -28.2 kcal/mol (from Table 3.8 on page 124 of the text).

Therefore, the total strain energy of cycloheptane is $-28.2 - (-34.4) = 6.2$ kcal/mol.

45. **a.** (chair cyclohexane with Cl substituents: Cl, Cl, Cl, Cl, Cl, Cl)

b. (chair cyclohexane with Cl substituents: Cl, Cl, Cl, Cl, Cl, Cl)

46.

$$K_{eq} = \frac{[\text{equatorial conformer}]}{[\text{axial conformer}]} = \frac{5.4}{1}$$

$$\%\text{ of equatorial conformer} = \frac{[\text{equatorial conformer}]}{[\text{equatorial conformer}] + [\text{axial conformer}]} \times 100$$

$$= \frac{5.4}{5.4 + 1} \times 100 = \frac{5.4}{6.4} \times 100 = 84\%$$

47. Two 1,3-diaxial (gauche) interactions cause the chair conformer of fluorocyclohexane to be 0.25 kcal/mol less stable when the fluoro substituent is in the axial position than when it is in the equatorial position.

The gauche conformer of 1-fluoropropane has one gauche interaction (see Figure 3.15 on page 128 of the text). Therefore, the gauche conformer is $(0.25/2) = 0.13$ kcal/mol less stable than the anti conformer that has no gauche interactions.

48. If both substituents point downward or both point upward, it is a cis isomer.
If one substituent points upward and the other downward, it is a trans isomer.

a. cis **b.** cis **c.** cis **d.** trans **e.** trans **f.** trans

49. Both *trans*-1,4-dimethylcyclohexane and *cis*-1-*tert*-butyl-3-methylcyclohexane have a conformer with two substituents in the equatorial position and a conformer with two substituents in the axial position.

cis-1-*tert*-Butyl-3-methylcyclohexane will have a higher percentage of the diequatorial-substituted conformer because the bulky *tert*-butyl substituent will have a greater preference for the equatorial position than will a less bulky methyl substituent, since the larger substituent will have greater destabilizing 1,3-diaxial interactions when it is in an axial position.

50. **a.** CH_2CH_3 CH_3 **b.** CH_2CH_3 CH_3

c. *trans*-1-Ethyl-2-methylcyclohexane is more stable because both substituents can be in equatorial positions.

51.
a. one equatorial and one axial in each
b. both equatorial in one and both axial in the other
c. both equatorial in one and both axial in the other
d. one equatorial and one axial in each
e. one equatorial and one axial in each
f. both equatorial in one and both axial in the other

52. Solved in the text.

53. **a.**

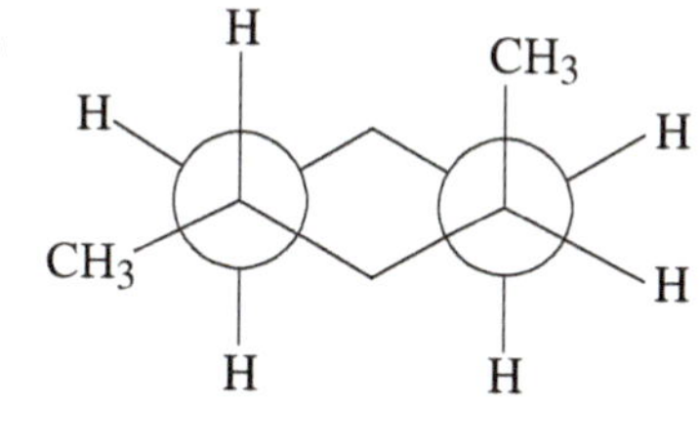

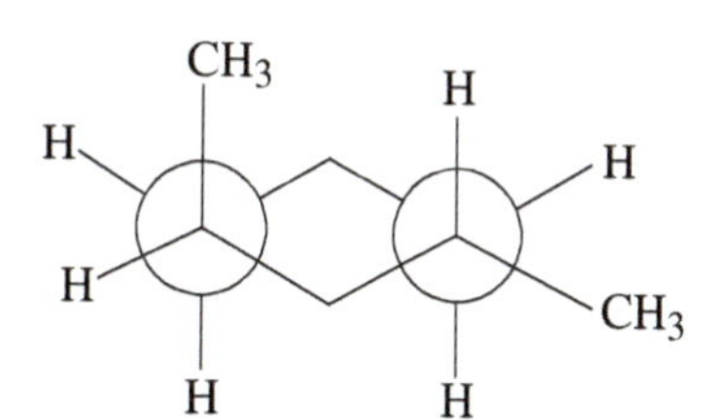

b. There will be equal amounts of the two conformers at equilibrium because they have the same stability—each one has one methyl group on an equatorial bond and one methyl group on an axial bond.

54. **a.** One chair conformer of *trans*-1,4-dimethylcyclohexane has both substituents in equatorial positions, so it does not have any 1,3-diaxial interactions. The other chair conformer has both substituents in axial positions. When a substituent is in an axial position, it experiences two 1,3-diaxial interactions, so this chair conformer has a total of four 1,3-diaxial interactions.

Because the 1,3-diaxial interaction between a methyl group and a hydrogen causes a strain energy of 0.9 kcal/mol, the chair conformer with both substituents in axial positions is $4 \times 0.9 = 3.6$ kcal/mol less stable than the chair conformer with both substituents in equatorial positions.

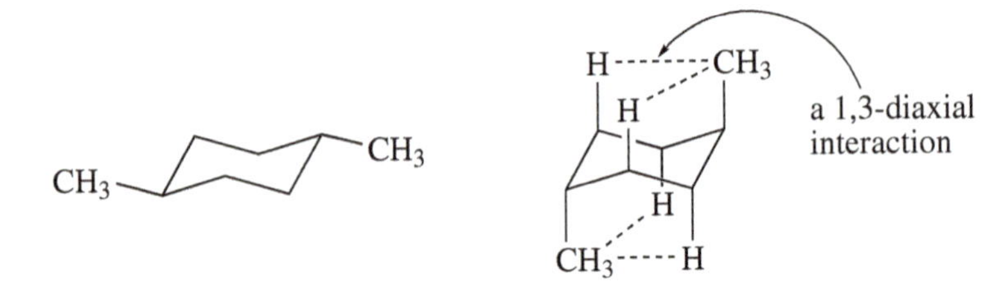

b. Each of the chair conformers of *cis*-1,4-dimethylcyclohexane has one substituent in an equatorial position and one in an axial position. Therefore, the two conformers are equally stable.

55. Both condensed and skeletal structures are shown.

a.
$$\begin{array}{l} \qquad\qquad\quad CH_3 \\ CH_3CH_2CHOCCH_3 \\ \qquad\quad\; CH_3\;\; CH_3 \end{array}$$

d.
$$\begin{array}{l} CH_3CHCH_2CH_2Br \\ \quad\; CH_3 \end{array}$$

Br

b.
$$\begin{array}{l} CH_3CHCH_2CH_2CH_2CH_2OH \\ \quad\; CH_3 \end{array}$$

OH

e.
$$\begin{array}{l} \qquad\qquad\qquad\; CH_3CHCH_3 \\ CH_3CH_2CH_2CH_2CHCH_2CH_2CH_2CH_3 \end{array}$$

c.
$$\begin{array}{l} CH_3CH_2CHNH_2 \\ \qquad\quad CH_3 \end{array}$$

NH_2

f.
$$\begin{array}{l} CH_3CH_2NCH_2CH_3 \\ \qquad\;\; CH_2CH_3 \end{array}$$

N

g. $CH_3CH_2CH_2CHCH_2CH_2CH_3$
CH_3CCH_3
CH_3

j. $CH_3CH_2CH_2CH_2CHCH_2CH_2CH_2CH_3$
$CHCH_3$
$CHCH_3$
CH_3

Br
h. $CH_3CHCH_2CH_2CCH_2CH_2CH_3$
CH_3 Br

Br Br

CH_3
k. $CH_3CH_2CHCHCH_2CH_2CH_2CH_3$
CH_3

CH_3
i. $CH_3CHCHCH_2CH_2CH_3$
OCH_2CH_3

O

CH_3CHCH_3
l. $CH_3CH_2CH_2CH_2CHCH_2CH_2CH_2CH_2CH_3$

56. OH O > OH > NH_2 > N H > O >

has two groups that form hydrogen bonds	O is more electronegative than N, so OH hydrogen bonds are stronger than NH hydrogen bonds	primary amines form stronger hydrogen bonds than do secondary amines	relatively weak hydrogen bonds	no hydrogen bonds; only dipole–dipole interactions

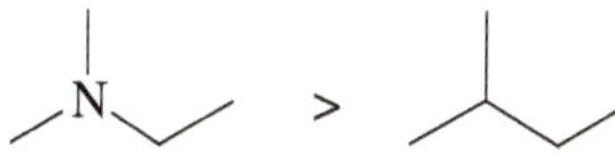

no hydrogen bonds; weaker dipole–dipole interactions than oxygen-containing compounds because N is less electro-negative than O

no dipole–dipole interactions

57. **a.**

1. 2,2,6-trimethylheptane
2. 5-bromo-2-methyloctane
3. 5-methyl-3-hexanol
4. 3,3-diethylpentane
5. 5-bromo-*N*-ethyl-1-pentanamine
6. 2,3,5-trimethylhexane
7. 3-ethoxyheptane
8. 1,3-dimethoxypropane
9. *N,N*-dimethylcyclohexanamine
10. 3-ethylcyclohexanol
11. 1-bromo-4-methylcyclohexane

b.

1.

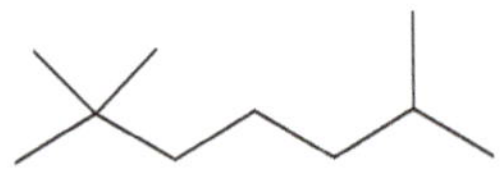

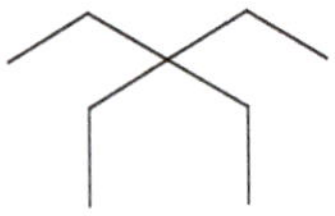

5. 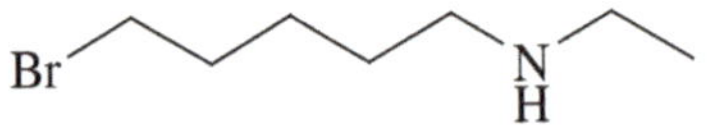 Br N H

9. 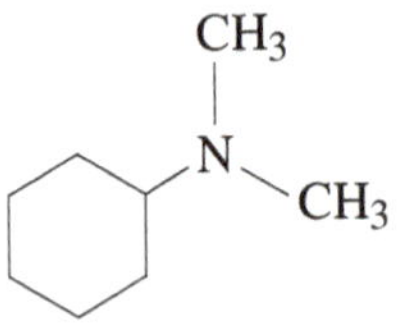 CH_3 N CH_3

2. CH_3 above; $CH_3CHCH_2CH_2CHCH_2CH_2CH_3$; Br below

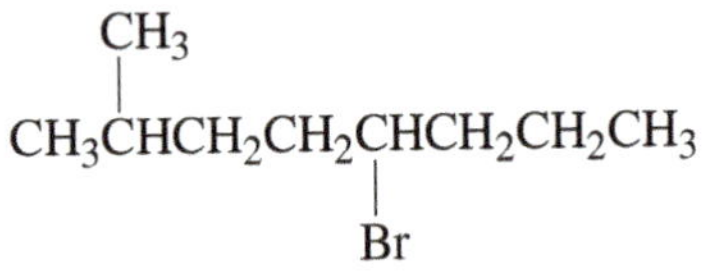

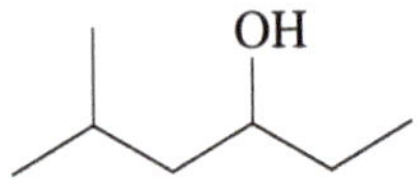

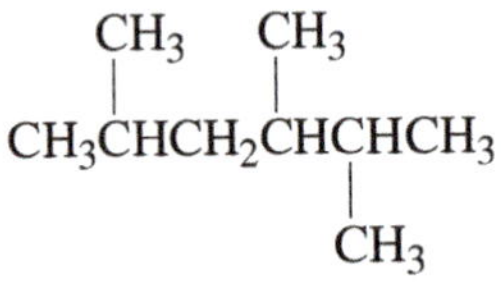

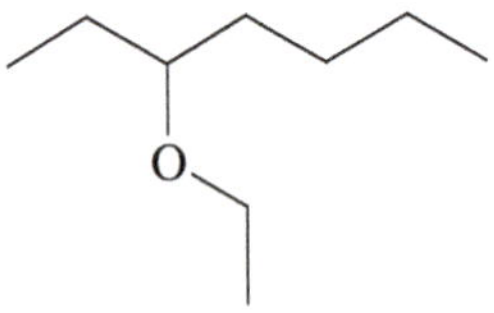

6. CH_3 CH_3 above; $CH_3CHCH_2CHCHCH_3$; CH_3 below

10. 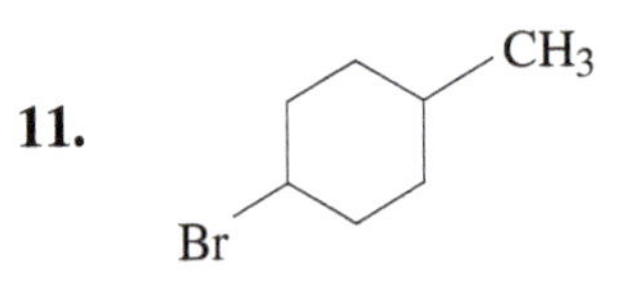

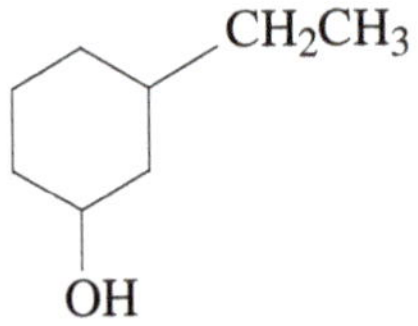

3. OH

7. O

11. CH_3 Br

4.

8. O O

58. **C** and **D** are cis isomers. (Both substituents are downward pointing in **C**; both substituents are upward pointing in **D**.)

59. **a.** **1.** 3 **2.** 4 **b.** **1.** 6 **2.** 5 **c.** **1.** 3 **2.** 4

60. The first conformer (**A**) is the most stable because the three substituents are more spread out, so its gauche interactions will not be as large. (The Cl in **A** is between a CH_3 and an H, whereas the Cl in **B** and **C** is between two CH_3 groups.)

61. **a.** 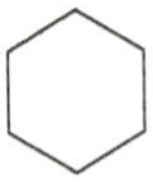 **b.** 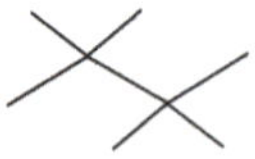 **c.**

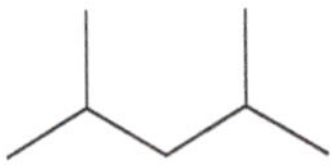

62.
- **a.** 2-butanamine
- **b.** 2-chlorobutane
- **c.** *N*-ethyl-2-butanamine
- **d.** 1-ethoxypropane
- **e.** 2-methylpentane
- **f.** 2-propanamine
- **g.** 2-bromo-2-methylbutane
- **h.** 4-methyl-1-pentanol
- **i.** bromocyclopentane
- **j.** cyclohexanol

63.
- **a.** 1-bromohexane (larger, so greater surface area)
- **b.** pentyl chloride (greater surface area than the branched compound)
- **c.** 1-butanol (fewer carbons)
- **d.** 1-hexanol (forms hydrogen bonds)
- **e.** hexane (greater area of contact)
- **f.** 1-pentanol (forms hydrogen bonds)
- **g.** 1-bromopentane (bromine larger and more polarizable)
- **h.** butyl alcohol (forms hydrogen bonds)
- **i.** octane (see Table 3.1)
- **j.** isopentyl alcohol (forms stronger hydrogen bonds)
- **k.** hexylamine (primary amines form stronger hydrogen bonds than do secondary amines)

64. **a.**

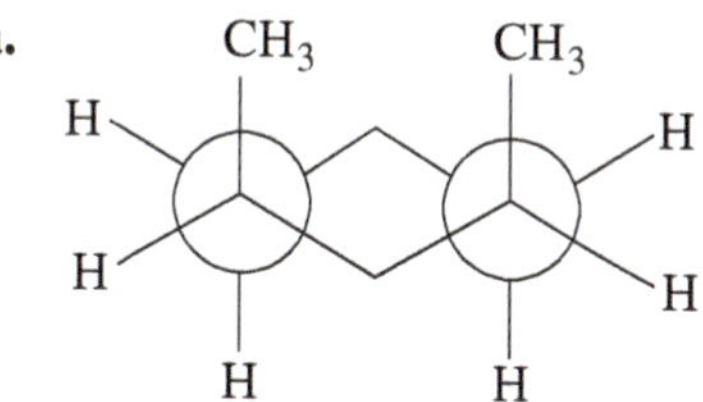

H H H H CH3 CH3 H H

b. The one on the right predominates at equilibrium, because it is more stable since both methyl groups are in equatorial positions.

c.

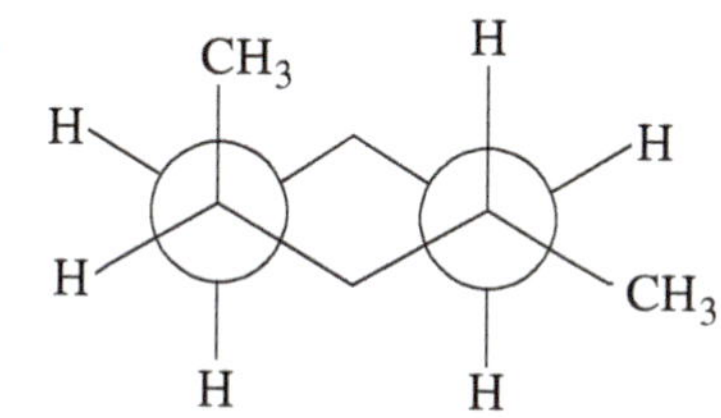

H CH3 H H CH3 H H H

d. There will be the same amount of each one at equilibrium, because they have the same stability since each conformer has one methyl group in an equatorial position and one methyl group in an axial position.

65. Ansaid is more soluble in water. It has a fluoro substituent that can form a hydrogen bond with water. Hydrogen bonding increases its solubility in water compared to Motrin.

66.

CH_3 CH_3 CH_3
H—O:--H—O:--H—O:

CH_3 CH_3 CH_3
H—O:--H—O:--H—O:

CH_3 CH_3 CH_3
H—O:--H—O:--H—O:

67. The student named only one compound correctly.

a. 2-bromo-3-pentanol
b. 4-ethyl-2,2-dimethylheptane
c. 3-methylcyclohexanol
d. 2,2-dimethylcyclohexanol
e. 5-(2-methylpropyl)nonane
f. 1-bromo-3-methylbutane
g. correct
h. 2,5-dimethylheptane
i. 5-bromo-2-pentanol
j. 3-ethyl-2-methyloctane
k. 2,3,3-trimethyloctane
l. *N,N*,5-trimethyl-3-hexanamine

68. All three compounds are diaxial-substituted cyclohexanes. **B** has the highest energy. Only **B** has a 1,3-diaxial interaction between CH_3 and Cl, which will be greater than a 1,3-diaxial interaction between CH_3 and H or between Cl and H.

69. The only one is 2,2,3-trimethylbutane.

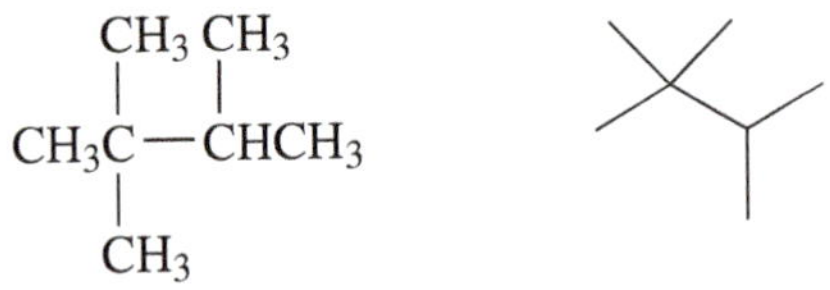

$$\begin{array}{c} CH_3\ CH_3 \\ |\quad\ | \\ CH_3C-CHCH_3 \\ | \\ CH_3 \end{array}$$

70. a. 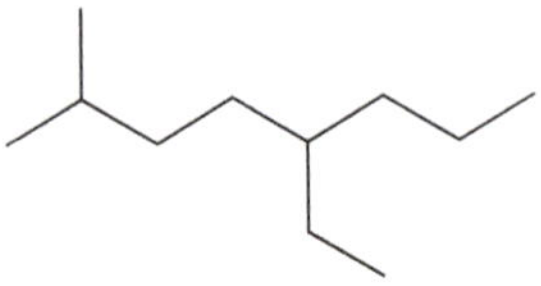 c. 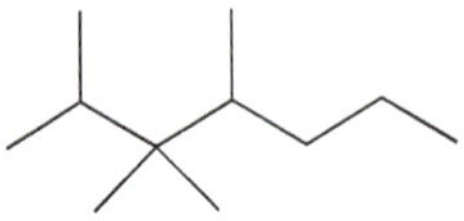 e.

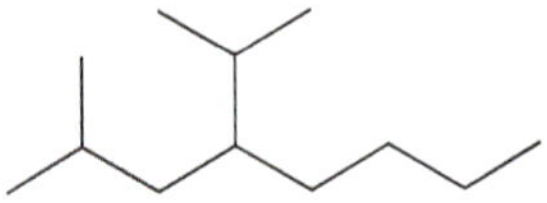

b. 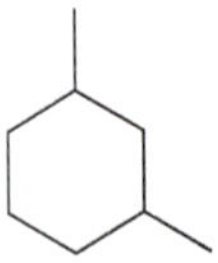 d. 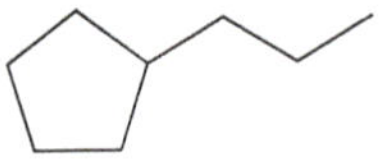 f.

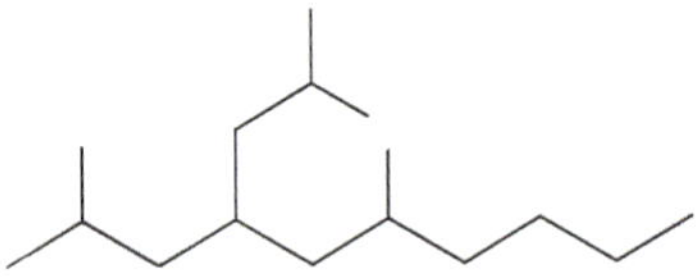

71. First draw the structure so that you know what groups to put on the bonds in the Newman projections.

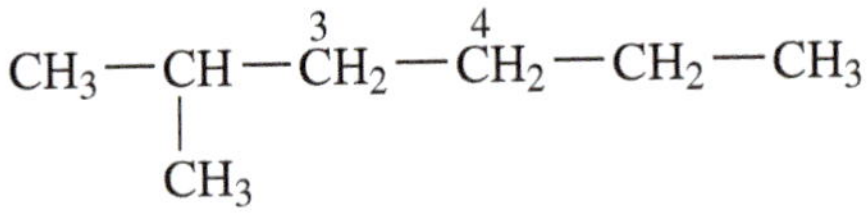

$$\begin{array}{l} \qquad\qquad\qquad\ \ 3\qquad\ \ 4 \\ CH_3-CH-CH_2-CH_2-CH_2-CH_3 \\ \qquad\quad\ | \\ \qquad\quad CH_3 \end{array}$$

a. CH_3 / $CHCH_3$ (front); H, H, H, H; CH_2CH_3

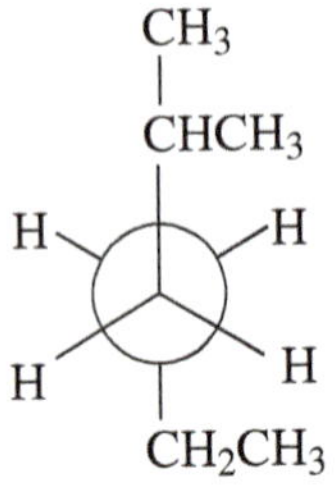

most stable

b. CH_3 / CH_3CH_2 $CHCH_3$; H, H, H, H

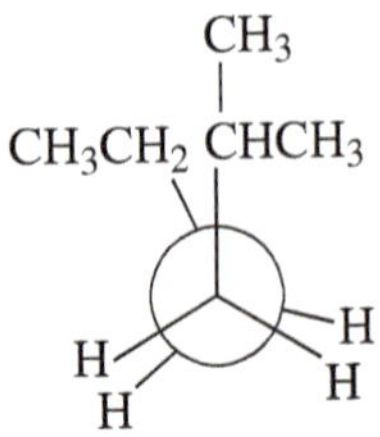

least stable

c. Rotation can occur about all the C—C bonds. There are six carbon–carbon bonds in the compound, so there are five other carbon–carbon bonds, in addition to the $C_3—C_4$ bond, about which rotation can occur.

d. Three of the carbon–carbon bonds have staggered conformers that are equally stable, because each is bonded to a carbon with three identical substituents.

$$CH_3{-}\underset{\displaystyle CH_3}{\underset{|}{CH}}{-}CH_2{-}CH_2{-}CH_2{-}CH_3$$

72.

$CH_3CH_2CH_2CH_2CH_2Br$	**a.** 1-bromopentane **b.** pentyl bromide	primary alkyl halide
$CH_3CH_2CH_2\underset{\displaystyle Br}{\underset{\vert}{C}}HCH_3$	**a.** 2-bromopentane **b.** no common name	secondary alkyl halide
$CH_3CH_2\underset{\displaystyle Br}{\underset{\vert}{C}}HCH_2CH_3$	**a.** 3-bromopentane **b.** no common name	secondary alkyl halide
$CH_3\overset{\displaystyle CH_3}{\overset{\vert}{C}}HCH_2CH_2Br$	**a.** 1-bromo-3-methylbutane **b.** isopentyl bromide	primary alkyl halide
$CH_3CH_2\overset{\displaystyle CH_3}{\overset{\vert}{C}}HCH_2Br$	**a.** 1-bromo-2-methylbutane **b.** no common name	primary alkyl halide
$CH_3CH_2\overset{\displaystyle Br}{\overset{\vert}{\underset{\displaystyle CH_3}{\underset{\vert}{C}}}}CH_3$	**a.** 2-bromo-2-methylbutane **b.** *tert*-pentyl bromide	tertiary alkyl halide
$CH_3\overset{\displaystyle Br}{\overset{\vert}{C}}H\underset{\displaystyle CH_3}{\underset{\vert}{C}}HCH_3$	**a.** 2-bromo-3-methylbutane **b.** no common name	secondary alkyl halide
$CH_3\overset{\displaystyle CH_3}{\overset{\vert}{\underset{\displaystyle CH_3}{\underset{\vert}{C}}}}CH_2Br$	**a.** 1-bromo-2,2-dimethylpropane **b.** no common name, but in older literature, the common name neopentyl bromide can be found.	primary alkyl halide

c. Four isomers are primary alkyl halides.
d. Three isomers are secondary alkyl halides.
e. One isomer is a tertiary alkyl halide.

73.
a. butane
b. 1-propanol
c. 5-propyldecane
d. 4-propyl-1-nonanol
e. 2-methyl-5-(1-methylethyl)octane or 5-isopropyl-2-methyloctane
f. 6-chloro-4-ethyl-3-methyloctane
g. 1-methoxy-5-methyl-3-propylhexane
h. 2,3-dimethyl-6-(2-methylpropyl)decane or 6-isobutyl-2,3-dimethyldecane
i. 8-methyl-4-decanamine
j. 1-methyl-2-(2-methylpropyl)cyclohexane or 1-isobutyl-2-methylcyclohexane

74. **a.**

more stable

CH_3 CH_2CH_3

b.

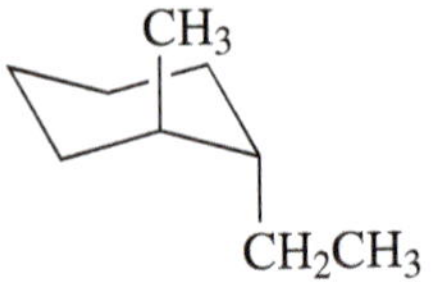

more stable

c.

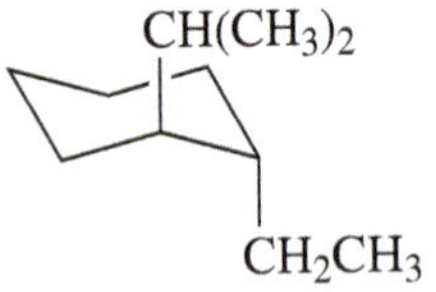

more stable

d.

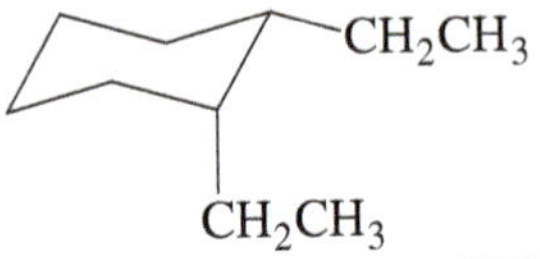

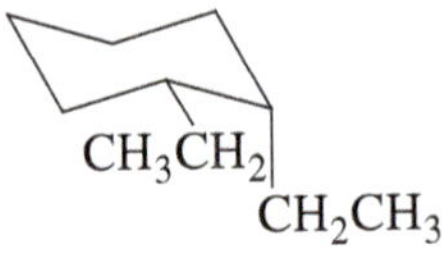

equally stable

e.

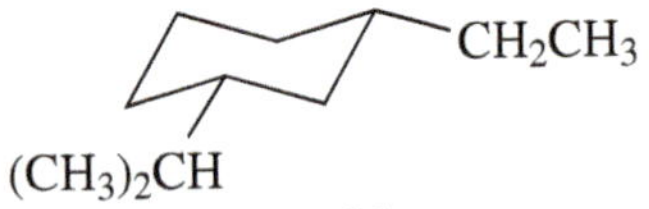

more stable

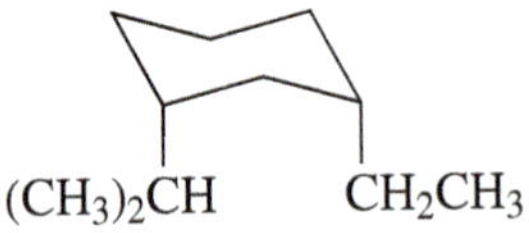

f.

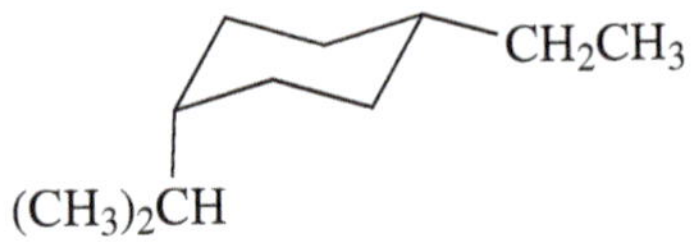

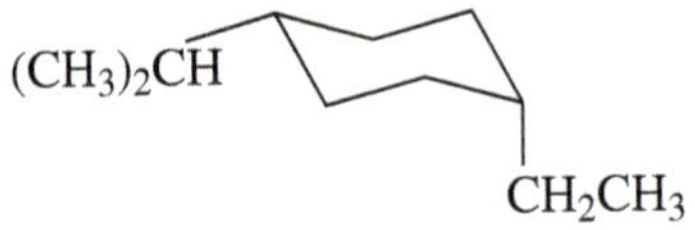

more stable

75. Alcohols with low molecular weights are more water soluble than alcohols with high molecular weights because, with fewer carbons, they have a smaller nonpolar component that has to be dragged into water.

76. **a.**

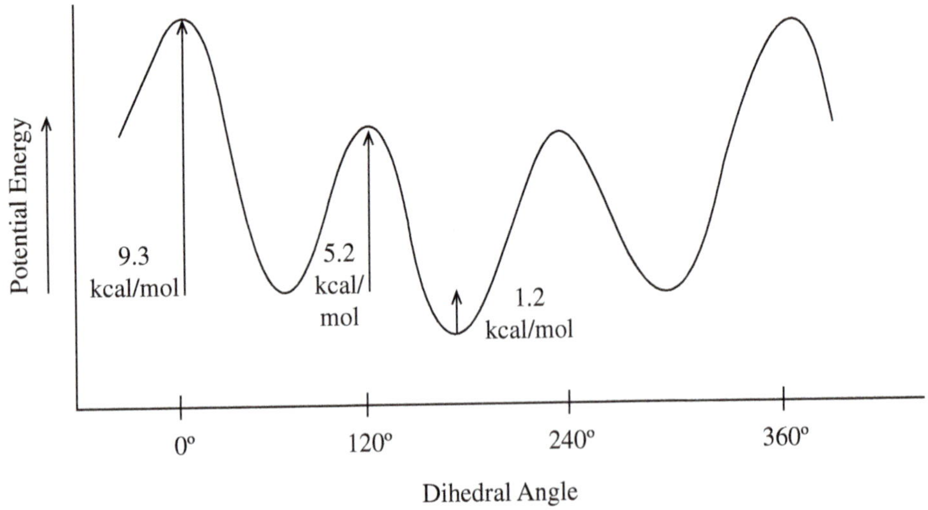

b.

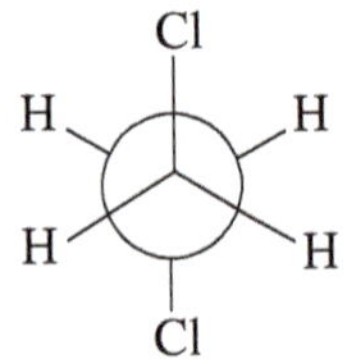

c. $1.2 + 5.2 = 6.4$ kcal/mol

d. $1.2 + 9.3 = 10.5$ kcal/mol

77. The more stable isomer is the one that has a conformer with both substituents in equatorial positions.

Using the following structure, you can determine easily the isomer that has both substituents in axial positions. That will be the more stable isomer because it will have a conformer with both groups in equatorial positions.

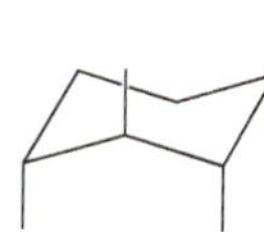

a. The cis isomer of a 1,3-disubstituted compound is the more stable isomer, because it has a conformer with both substituents in axial positions. Therefore, its other conformer has both groups in equatorial positions.

b. The trans isomer of a 1,4-disubstituted compound is the more stable isomer, because it has a conformer with both substituents in axial positions. Therefore, its other conformer has both groups in equatorial positions.

c. The trans isomer of a 1,2-disubstituted compound is the more stable isomer, because has a conformer with both substituents in axial positions. Therefore, its other conformer has both groups in equatorial positions.

78. Six ethers have a molecular formula of $C_5H_{12}O$.

$CH_3OCH_2CH_2CH_2CH_3$
1-methoxybutane
butyl methyl ether

$CH_3CH(OCH_3)CH_2CH_3$
2-methoxybutane
sec-butyl methyl ether

$CH_3CH_2OCH_2CH_2CH_3$
1-ethoxypropane
ethyl propyl ether

$CH_3CH(OCH_2CH_3)CH_3$
2-ethoxypropane
ethyl isopropyl ether

$CH_3C(CH_3)_2OCH_3$
2-methoxy-2-methylpropane
tert-butyl methyl ether

$CH_3CH(CH_3)CH_2OCH_3$
1-methoxy-2-methylpropane
isobutyl methyl ether

79. The most stable conformer has two CH_3 groups in equatorial positions and one in an axial position. (The other conformer would have two CH_3 groups in axial positions and one in an equatorial position.)

CH_3 CH_3 CH_3

80.

a. *N*,6-dimethyl-3-heptanamine
b. 3-ethyl-2,5-dimethylheptane
c. 1,2-dichloro-3-methylpentane
d. 2,3-dimethylpentane
e. 5-butyl-3,4-dimethylnonane
f. 5-butyl-3,3,9-trimethylundecane (undecane is an 11 carbon straight-chain hydrocarbon; Table 3.1 on page 89 of the text).

81. One chair conformer of *trans*-1,2-dimethylcyclohexane has both substituents in equatorial positions, so it does not have any 1,3-diaxial interactions. However, the figure on the right on the middle of page 130 of the text shows that the two methyl substituents are gauche to each other (as they would be in gauche butane; see Figure 3.15 on page 128), giving it a strain energy of 0.87 kcal/mol.

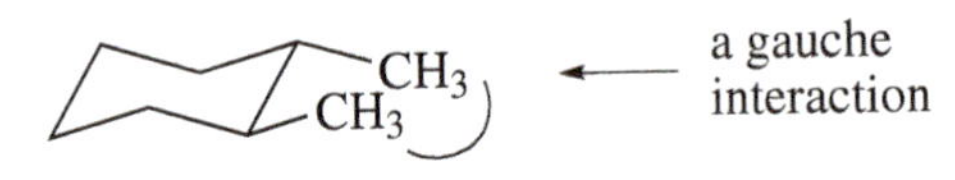

The other chair conformer of *trans*-1,2-dimethylcyclohexane has both substituents in axial positions. When a substituent is in an axial position, it experiences two 1,3-diaxial interactions. This chair conformer, therefore, has a total of four 1,3-diaxial interactions. Each diaxial interaction is between a CH_3 and an H, so each results in a strain energy of 0.87 kcal/mol. Therefore, this chair conformer has a strain energy of 3.48 kcal/mol ($4 \times 0.87 = 3.48$).

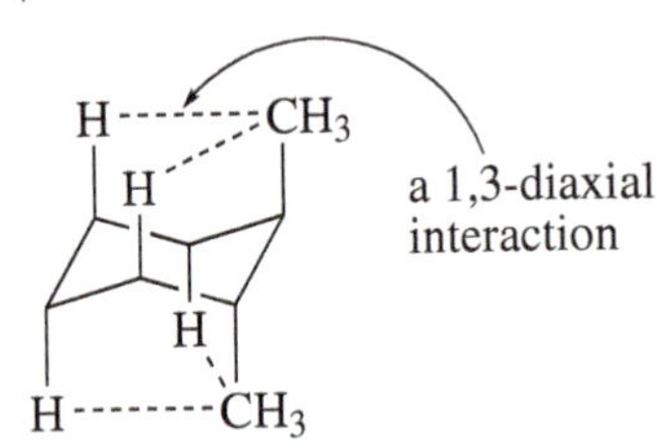

Therefore, one conformer is 2.61 kcal/mol (3.48 – 0.87) more stable than the other.

82.

CH₂OH, HO, O, HO, OH, OH

83.

a. 5-methyl-3-hexanol
b. 1-bromo-2-propylcyclopentane
c. 2-methyl-3-pentanol
d. 5-bromo-2-methyloctane
e. 1,5-hexanediol
f. 6-bromo-2-hexanol
g. 4-ethyl-3-methylcyclohexanol
h. 4-bromo-1-ethyl-2-methylcyclohexane
i. 3-butyl-4-methylcyclopentanamine

84.

a. 1-Hexanol has a higher boiling point than 3-hexanol because the alkyl group in 1-hexanol has stronger London dispersion forces, because the OH group of 3-hexanol makes it more difficult for its six carbons to lie close to the six carbons of another molecule of 3-hexanol.

b. The floppy ethyl groups in diethyl ether make it difficult for the water molecules to approach the oxygen in order to engage in hydrogen bonding. Therefore, it is less soluble in water than is tetrahydrofuran, in which the alkyl groups are pinned back in a ring.

85. One of the chair conformers of *cis*-1,3-dimethylcyclohexane has both substituents in equatorial positions, so there are no unfavorable 1,3-diaxial interactions. The other chair conformer has three 1,3-diaxial interactions, two between a CH_3 and an H and one between two CH_3 groups.

We know that a 1,3-diaxial interaction between a CH_3 and an H is 0.87 kcal/mol. Subtracting 1.7, for the two interactions between a CH_3 and an H, from 5.4 (the energy difference between the two conformers) results in a value of 3.7 kcal/mol for the 1,3-diaxial interaction between the two CH_3 groups.

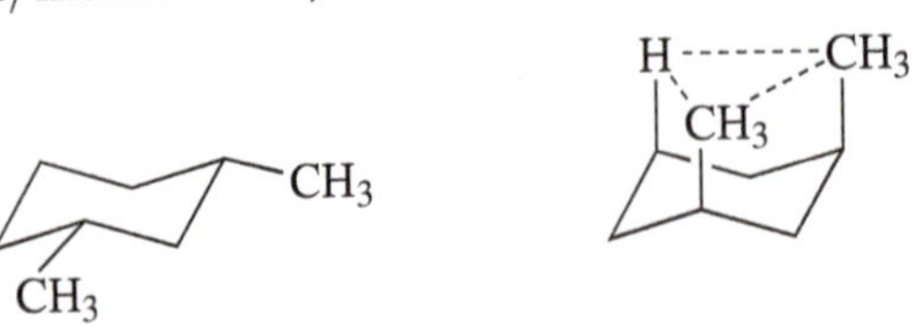

86. Because bromine has a larger diameter than chlorine, one would expect bromine to have a greater preference for the equatorial position that would be indicated by a larger $\Delta G°$. However, Table 3.9 on page 128 of the text shows that it has a smaller $\Delta G°$, indicating that it has less preference for the equatorial position than chlorine has.

The C—Br bond is longer than the C—Cl bond, which causes bromine to be farther away than chlorine from the other axial substituents. Apparently, the longer bond more than offsets the larger diameter.

87.
a. 7-bromo-6-ethyl-5-decanol
b. 5-chloro-3-ethyl-2,7-dimethylnonane
c. 7,7-dimethyl-3-nonanol

88. Problem 81 shows that the energy difference between the two chair conformers is 2.61 kcal/mol. To calculate the equilibrium constant needed to answer the question, see Problem 19 on page 204 of the textbook.

$$\Delta G° = -2.61 \text{ kcal/mol}$$
$$\Delta G° = -RT \ln K_{eq}$$
$$-2.61 \text{ kcal/mol} = -1.986 \times 10^{-3} \text{ kcal/mol K} \times 298 \text{ K} \times \ln K_{eq}$$
$$-2.61 \text{ kcal/mol} = -0.5918 \text{ kcal/mol} \ln K_{eq}$$
$$\ln K_{eq} = 4.41$$
$$K_{eq} = 82.3 = \frac{\text{both equatorial}}{\text{both axial}} = \frac{82.3}{1}$$

$$\text{percentage of molecule with both groups in equatorial positions} = \frac{\text{both equatorial}}{\text{both equatorial} + \text{both axial}} \times 100 = \frac{82.3}{82.3 + 1} \times 100 = 98.8\%$$

89. The conformer on the left has two 1,3-diaxial interactions between a CH_3 and an H (2×0.87 kcal/mol) for a total strain energy of 1.7 kcal/mol.

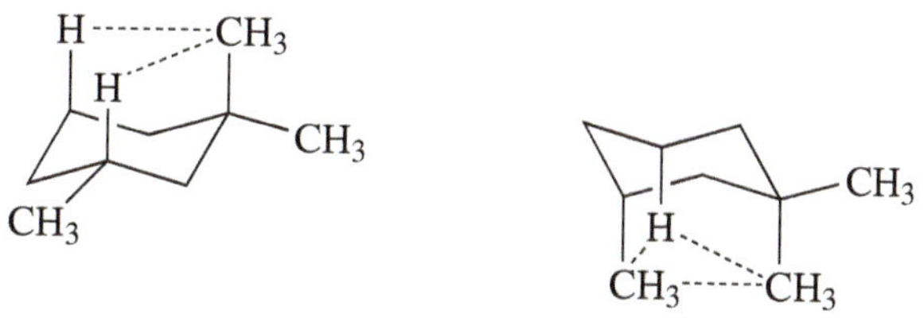

The conformer on the right has three 1,3-diaxial interactions, two between a CH_3 and an H (1.7 kcal/mol) and one between two CH_3 groups (3.7 kcal/mol; see Problem 85) for a total strain energy of 5.4 kcal/mol. Therefore, the conformer on the left predominates at equilibrium.

90.

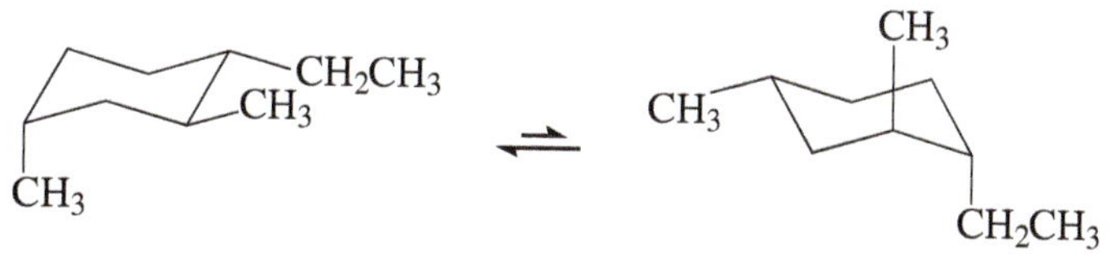

Because the ethyl and methyl substituents are on adjacent carbons, they experience a gauche interaction.
There are two 1,3-diaxial interactions between a CH_3 and an H.

$0.96 + 0.87 + 0.87 = 2.7$ kcal/mol

There are two 1,3-diaxial interactions between a CH_3 and an H and two 1,3-diaxial interactions between a CH_3CH_2 and an H.

$0.87 + 0.87 + 1.00 + 1.00 = 3.7$ kcal/mol

Chapter 3 Practice Test

1. Name the following compounds:

a.

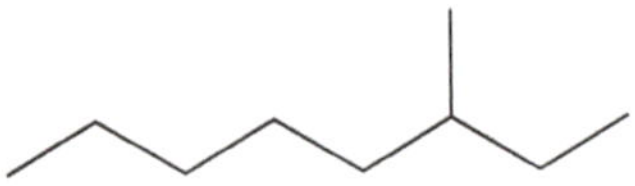

b. 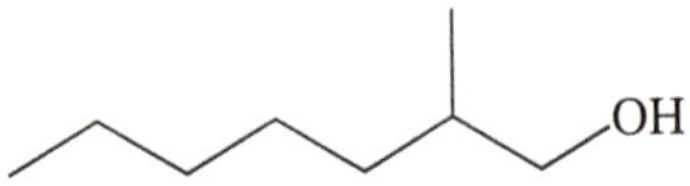 OH

c. 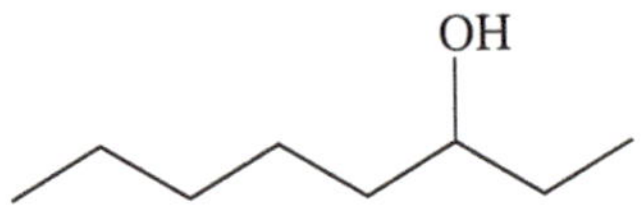 OH

2. Using Newman projections, draw the following conformers of hexane considering rotation about the $C_3—C_4$ bond:

a. the most stable of all the conformers

b. the least stable of all the conformers

c. a gauche conformer

3. What are the common and systematic names of the following compounds?

a. $CH_3CH_2CHCH_3$
    Cl

b. $CH_3CHCH_2CH_2CH_2OH$
    CH_3

c. 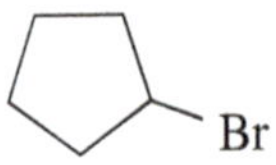 Br

4. Rank the three compounds in each set from highest boiling to lowest boiling.

a. $CH_3CH_2CH_2CH_2CH_2Br$ $CH_3CH_2CH_2Br$ $CH_3CH_2CH_2CH_2Br$

b. $CH_3CH_2CH_2CH_2CH_3$ $CH_3CH_2CH_2CH_2OH$ $CH_3CH_2CH_2CH_2Cl$

c.
    CH_3 CH_3
$CH_3C—CCH_3$
    CH_3 CH_3

$CH_3CH_2CH_2CH_2CH_2CH_2CH_2CH_3$

$CH_3CHCH_2CH_2CH_2CH_2CH_3$
    CH_3

5. Name the following compounds:

a. $CH_3CHCH_2CH_2CHCH_2CH_3$
    CH_3 OH

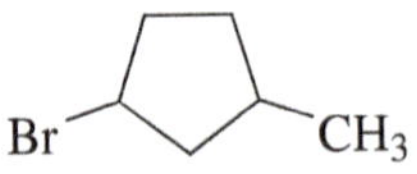

b. $CH_3CH_2CHOCH_2CH_3$
    $CH_2CH_2CH_2CH_3$

c. Br CH_3

d.
    Cl
$CH_3CHCHCH_2CH_2CH_2Cl$
    CH_2CH_3

6. Draw the other chair conformer for the following compound:

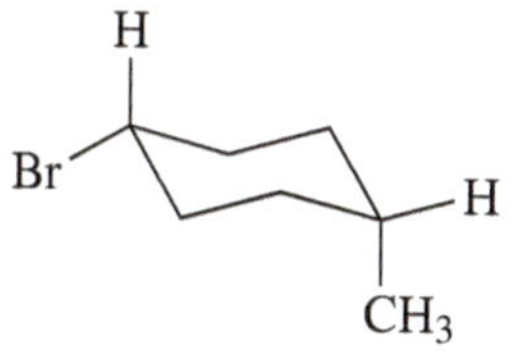

7. Which is more stable, *cis*-1-isopropyl-3-methylcyclohexane or *trans*-1-isopropyl-3-methylcyclohexane?

8. Which of the following has:

a. the higher boiling point: diethyl ether or butyl alcohol?

b. the greater solubility in water: 1-butanol or 1-pentanol?

c. the higher boiling point: hexane or isohexane?

d. the higher boiling point: pentylamine or ethylmethylamine?

e. the greater solubility in water: ethyl alcohol or ethyl chloride?

9. What are the common and systematic names of the following compounds?

a. $CH_3CHCH_2CH_2Br$
CH_3

b. $CH_3CHCH_2CH_2OH$
CH_3

c. $CH_3CHCH_2CH_2NH_2$
CH_3

10. Draw the more stable conformer of:

a. *cis*-1-*sec*-butyl-4-isopropylcyclohexane

b. *trans*-1-*sec*-butyl-4-isopropylcyclohexane

c. *trans*-1-*sec*-butyl-3-isopropylcyclohexane

11. Draw the structure for each of the following:

a. a secondary alkyl bromide that has three carbons

b. a secondary amine that has three carbons

c. an alkane with no secondary hydrogens

d. a constitutional isomer of butane

e. three compounds with molecular formula C_3H_8O

12. Name the following compounds:

a. $CH_3CHCH_2CH_2CHCH_3$
CH_3 CH_2CH_3

b. OH
$CH_3CHCH_2CH_2CH_2CH_2CH_2Br$

c. $CH_3CHCH_2CHCH_2CH_2CH_3$
Cl OH

d.

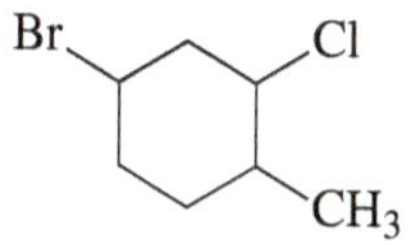

e. $CH_3CH_2CH_2CHCH_2OCH_2CH_2CH_2CH_3$
CH_3

f. $CH_3CH_2CH_2NHCH_2CH_2CHCH_2CH_3$
CH_3

CHAPTER 4
Isomers: The Arrangement of Atoms in Space

Important Terms

achiral	does not rotate the plane of polarization of plane-polarized light.
amine inversion	a process in which the lone pair of an sp^3 nitrogen of an amine migrates from one face of the atom to the other face. This causes the nitrogen's substituents to move like an umbrella inverting in a windstorm.
asymmetric center	an atom that is bonded to four different substituents.
chiral	rotates the plane of polarization of plane-polarized light.
chiral probe	something capable of distinguishing between enantiomers.
chromatography	a separation technique in which the mixture to be separated is dissolved in a solvent and the solution is passed through a column packed with an adsorbent stationary phase.
cis isomer	the isomer with substituents on the same side of a cyclic structure, or the isomer with the hydrogens on the same side of a double bond.
cis–trans isomers (*E,Z* isomers)	isomers that result from not being able to rotate about a carbon–carbon double bond.
configuration	the three-dimensional structure of a chiral compound. The configuration at a specific atom is designated by *R* or *S*.
configurational isomers	stereoisomers that cannot interconvert unless a covalent bond is broken. Cis–trans isomers and isomers with asymmetric centers are configurational isomers.
constitutional isomers	molecules that have the same molecular formula but differ in the way the atoms are connected.
dextrorotatory	the enantiomer that rotates the plane of polarization of plane-polarized light in a clockwise direction (+).
diastereomers	stereoisomers that are not enantiomers.
***E* isomer**	the isomer with the high-priority groups on opposite sides of the double bond.
enantiomerically pure	only one enantiomer is present in an enantiomerically pure sample.
enantiomeric excess (optical purity)	how much excess of one enantiomer is present in a mixture of a pair of enantiomers, expressed as a percentage.
enantiomers	nonsuperimposable mirror-image molecules.

erythro enantiomers the pair of enantiomers with similar groups on two asymmetric centers on the same side when drawn in a Fischer projection.

***E,Z* isomers (cis–trans isomers)** isomers that result from not being able to rotate about a carbon–carbon double bond.

Fischer projection a method of representing the spatial arrangement of groups bonded to an asymmetric center. The asymmetric center is the point of intersection of two perpendicular lines; the horizontal lines represent bonds that project out of the plane of the paper toward the viewer, and the vertical lines represent bonds that project back from the plane of the paper away from the viewer.

isomers nonidentical compounds with the same molecular formula.

levorotatory the enantiomer that rotates the plane of polarization of plane-polarized light in a counterclockwise direction $(-)$.

meso compound a compound that possesses asymmetric centers and a plane of symmetry; it is achiral, because it has a plane of symmetry.

observed rotation the amount of rotation observed in a polarimeter.

optically active rotates the plane of polarization of plane-polarized light.

optically inactive does not rotate the plane of polarization of plane-polarized light.

optical purity (enantiomeric excess) the amount of excess of one enantiomer present in a mixture of a pair of enantiomers.

perspective formula a method of representing the spatial arrangement of groups bonded to an asymmetric center. Two adjacent bonds are drawn in the plane of the paper; a solid wedge depicts a bond that projects out of the plane of the paper toward the viewer, and a hatched wedge depicts a bond that projects back from the paper away from the viewer.

plane-polarized light light that oscillates in a single plane.

plane of symmetry an imaginary plane that bisects a molecule so that the two halves are a pair of mirror images.

polarimeter an instrument that measures the rotation of the plane of polarization of plane-polarized light.

racemic mixture (racemic) a mixture of equal amounts of a pair of enantiomers.

***R* configuration** after assigning relative priorities to the four groups bonded to an asymmetric center, if the lowest priority group is on a vertical axis in a Fischer projection (or pointing away from the viewer in a perspective formula), the arrow drawn from the highest priority group to the next highest priority group and then to the next highest priority group is clockwise.

receptor a protein that binds a particular molecule.

resolution of a racemic mixture separation of a racemic mixture into the individual enantiomers.

***S* configuration** after assigning relative priorities to the four groups bonded to an asymmetric center, if the lowest priority group is on a vertical axis in a Fischer projection (or pointing away from the viewer in a perspective formula), the arrow drawn from the highest priority group to the next highest priority group and then to the next highest priority group is counterclockwise.

specific rotation the amount of rotation that will be observed for a compound with a concentration given in grams per 100 mL of solution (or g/mL if it is a pure liquid) in a sample tube 1.0 dm long.

stereocenter (stereogenic center) an atom at which the interchange of two groups produces a stereoisomer.

stereoisomers isomers that differ in the way the atoms are arranged in space.

threo enantiomers the pair of enantiomers with similar groups on two asymmetric centers on opposite sides when drawn in a Fischer projection.

trans isomer the isomer with substituents on the opposite sides of a cyclic structure, or the isomer with the hydrogens on the opposite sides of a double bond.

***Z* isomer** the isomer with the high-priority groups on the same side of the double bond.

Solutions to Problems

1. **a.** 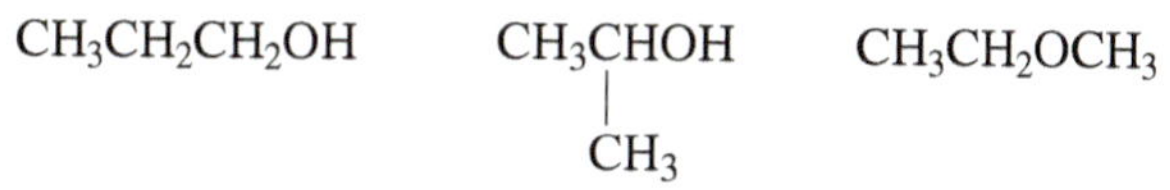 $CH_3CH_2CH_2OH$ $CH_3CH(CH_3)OH$ $CH_3CH_2OCH_3$

b. There are seven constitutional isomers with molecular formula $C_4H_{10}O$.

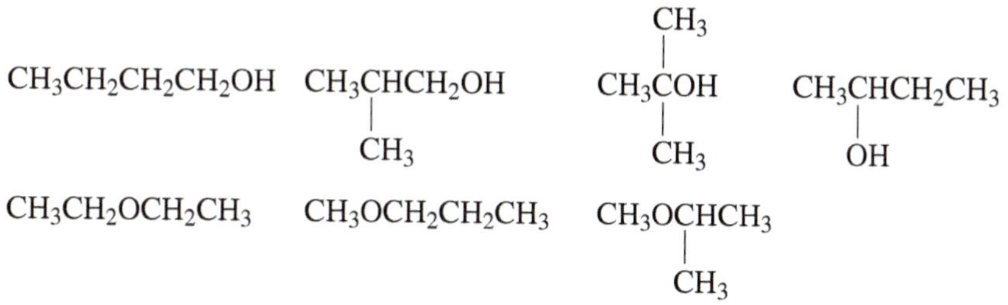

$CH_3CH_2CH_2CH_2OH$ $CH_3CH(CH_3)CH_2OH$ $CH_3C(CH_3)_2OH$ $CH_3CH(OH)CH_2CH_3$

$CH_3CH_2OCH_2CH_3$ $CH_3OCH_2CH_2CH_3$ $CH_3OCH(CH_3)CH_3$

2. **a.** 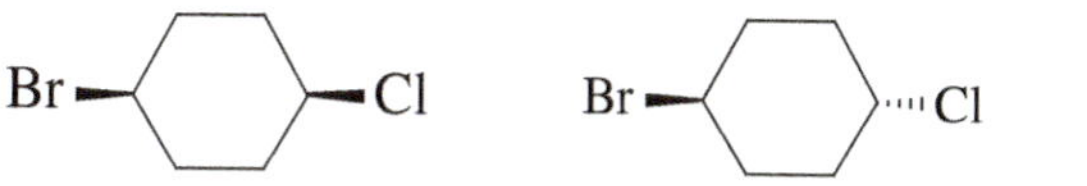 Br–(cyclohexane)–Cl (both wedged) Br–(cyclohexane)····Cl (wedge and hash) **b.** 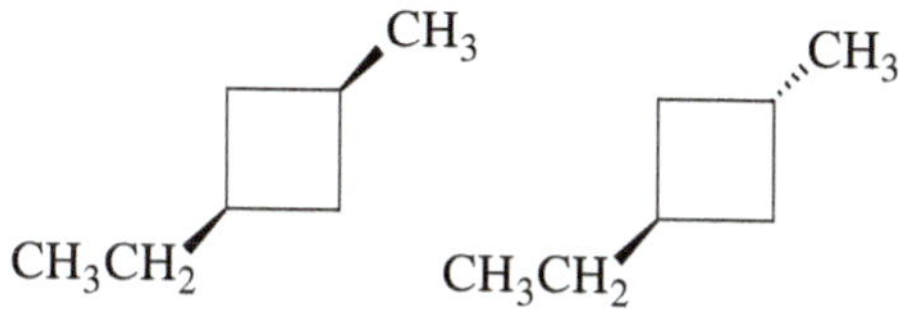 CH_3 / CH_3CH_2 cyclobutane (both wedged) CH_3 / CH_3CH_2 cyclobutane (hash and wedge)

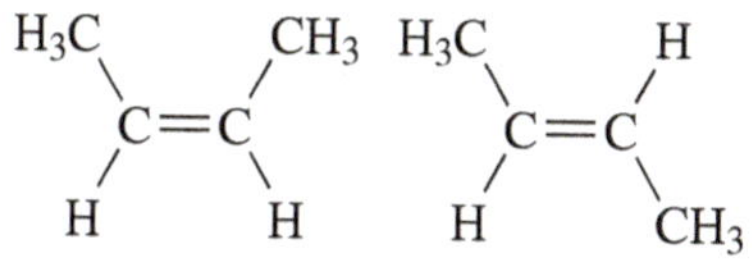

3. **a.** 1 and 3

b. **1.** 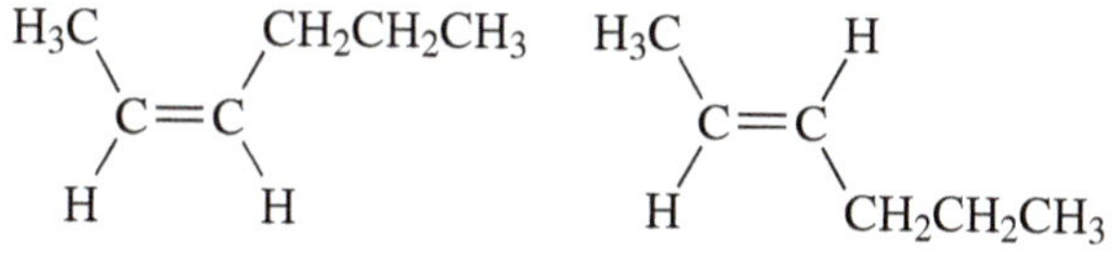 H_3C, H / C=C / H, $CH_2CH_2CH_3$ — cis; H_3C, H / C=C / H, $CH_2CH_2CH_3$ — trans

3. H_3C, H / C=C / H, CH_3 — cis; H_3C, H / C=C / H, CH_3 — trans

4. **1.**  trans cis **3.**  trans cis

2. 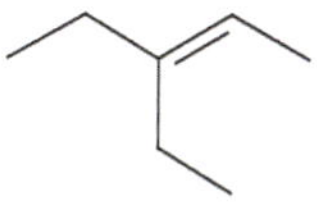 **4.**

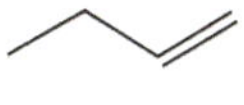

5. 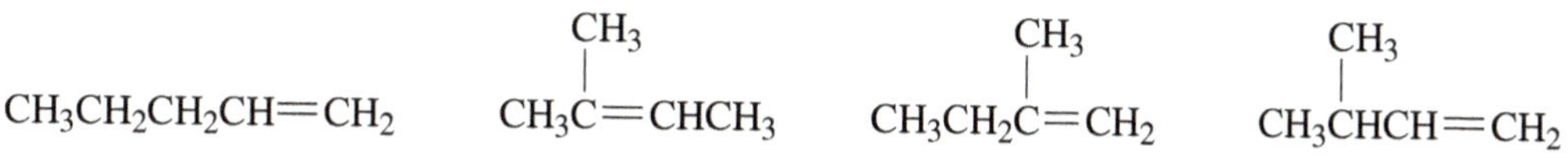 $CH_3CH_2CH_2CH{=}CH_2$ $CH_3C(CH_3){=}CHCH_3$ $CH_3CH_2C(CH_3){=}CH_2$ $CH_3CH(CH_3)CH{=}CH_2$

6. Only **C** has a dipole moment of zero, because the bond dipoles cancel since they are in opposite directions.

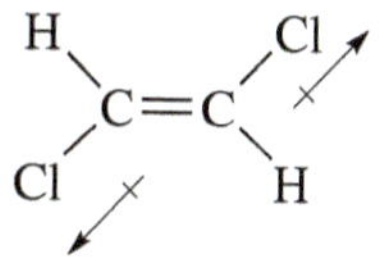

H, Cl / C=C / Cl, H

7. **a.**

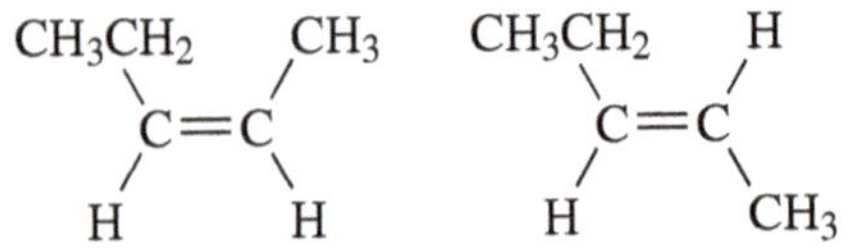

CH_3CH_2 CH_3 $C{=}C$ H H — ***Z***

CH_3CH_2 H $C{=}C$ H CH_3 — ***E***

b.

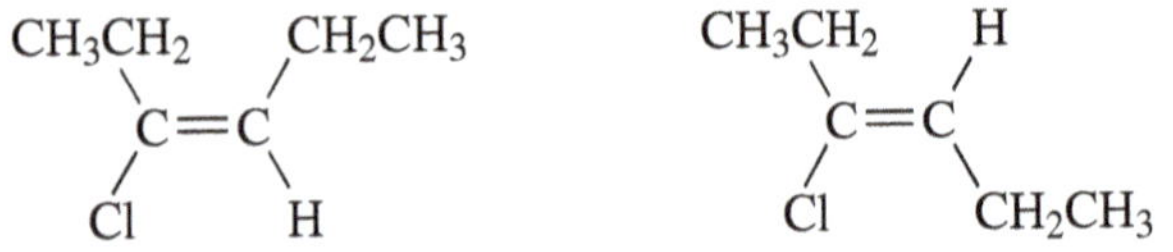

CH_3CH_2 CH_2CH_3 $C{=}C$ Cl H — ***E***

CH_3CH_2 H $C{=}C$ Cl CH_2CH_3 — ***Z***

c.

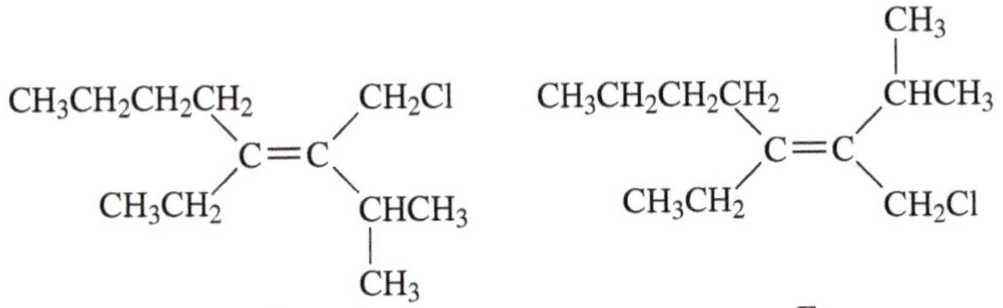

$CH_3CH_2CH_2CH_2$ CH_2Cl $C{=}C$ CH_3CH_2 $CHCH_3$ (CH_3) — ***Z***

$CH_3CH_2CH_2CH_2$ (CH_3) $CHCH_3$ $C{=}C$ CH_3CH_2 CH_2Cl — ***E***

d.

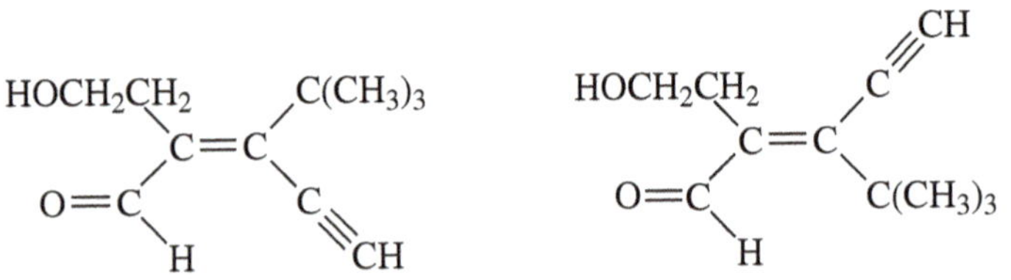

$HOCH_2CH_2$ $C(CH_3)_3$ $C{=}C$ $O{=}C$ H $C{\equiv}CH$ — ***Z***

$HOCH_2CH_2$ $C{\equiv}CH$ $C{=}C$ $O{=}C$ H $C(CH_3)_3$ — ***E***

8. **a.** $-I > -Br > -OH > -CH_3$

b. $-OH > -CH_2Cl > -CH{=}CH_2 > -CH_2CH_2OH$

9. The high-priority groups are on same side of the double bond, so tamoxifen has the *Z* configuration.

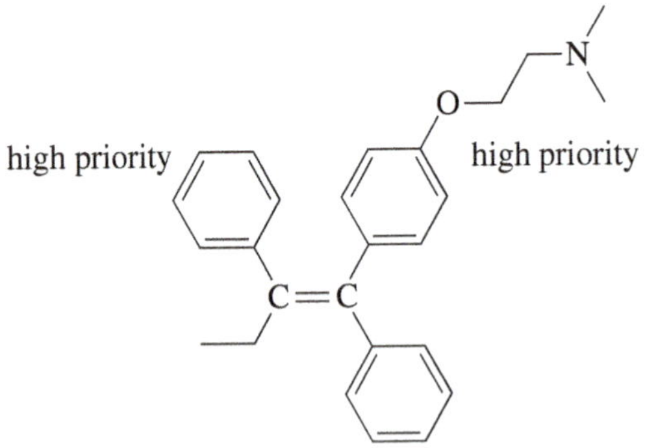

high priority C=C O N high priority

10. **a.**

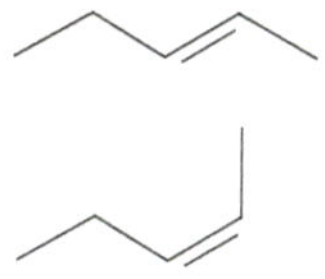

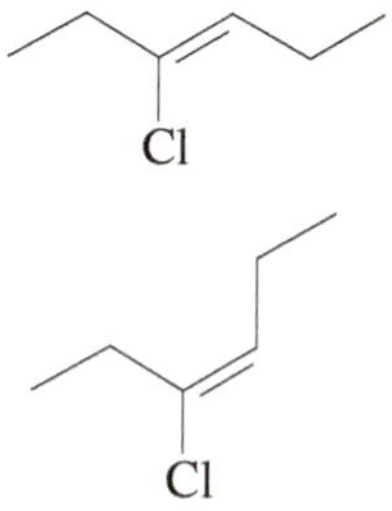

c. Cl

Cl

b. Cl

Cl

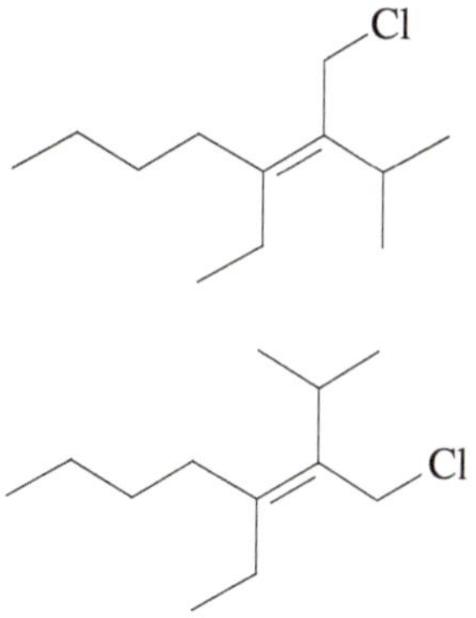

d. HO O H

HO O H

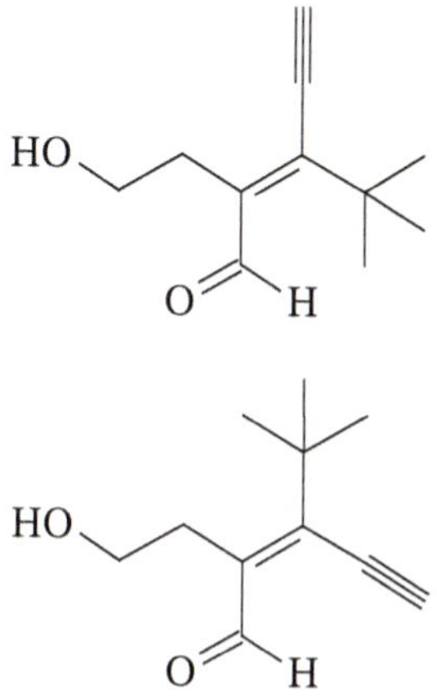

11. **a.** (*E*)-2-heptene **b.** (*Z*)-3,4-dimethyl-2-pentene **c.** (*Z*)-1-chloro-3-ethyl-4-methyl-3-hexene

12.

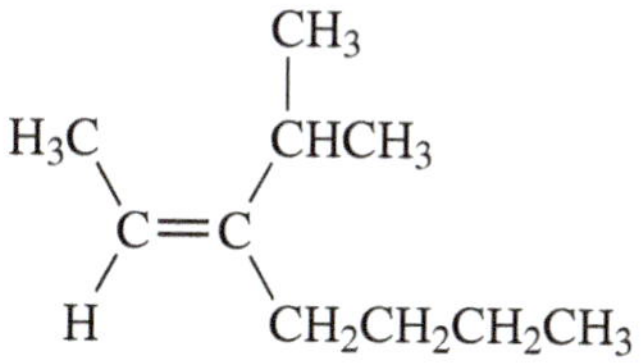

```
              CH3
              |
H3C           CHCH3
   \         /
    C=C
   /         \
  H           CH2CH2CH2CH3
```

13. **a**, **b**, **c**, **f**, and **h** are chiral.
d, **e**, and **g** are each superimposable on its mirror image. These, therefore, are achiral.

14. **a**, **c**, and **f** have asymmetric centers.

15. Solved in the text.

16. **a**, **c**, and **f**, because to be able to exist as a pair of enantiomers, the compound must have an asymmetric center (except in the case of certain compounds with unusual structures; see Problem 103).

17. **a.** It has one asymmetric center.
b. It has three stereocenters.

18. Draw the first enantiomer with the groups in any order you want. Then draw the second enantiomer by drawing the mirror image of the first enantiomer. Your answer might not look like the ones shown below because the first enantiomer can be drawn with the four groups on any of the four bonds. The next one is the mirror image of the first one.

a. 1. CH_3, Br, H, CH_2OH on C; mirror image: CH_3, $HOCH_2$, H, Br on C

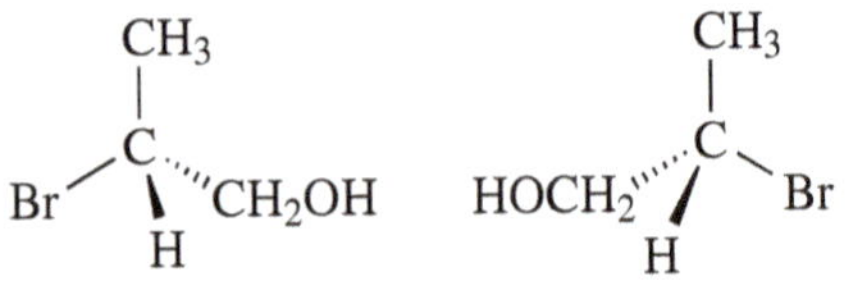

3. OH, $(CH_3)_2CH$, H, CH_3 on C; mirror image: OH, CH_3, H, $CH(CH_3)_2$ on C

2. CH_2CH_2Cl, CH_3CH_2, H, CH_3 on C; mirror image: CH_2CH_2Cl, CH_3, H, CH_2CH_3 on C

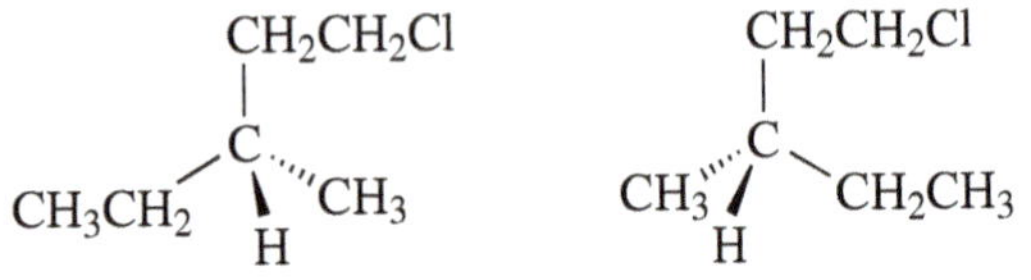

b. 1. Fischer projections: top CH_3, left Br, right H, bottom CH_2OH; mirror image: top CH_3, left H, right Br, bottom CH_2OH

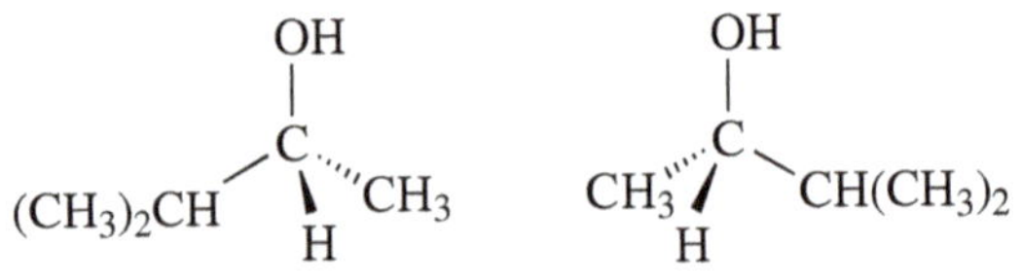

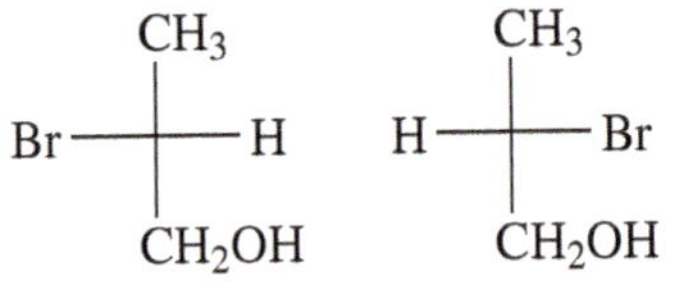

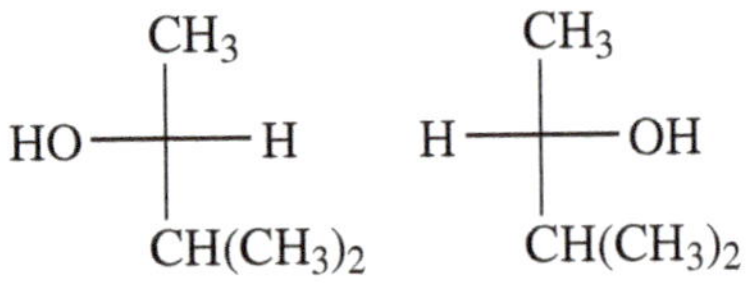

3. Top CH_3, left HO, right H, bottom $CH(CH_3)_2$; mirror image: top CH_3, left H, right OH, bottom $CH(CH_3)_2$

2. Top CH_2CH_2Cl, left H, right CH_3, bottom CH_2CH_3; mirror image: top CH_2CH_2Cl, left CH_3, right H, bottom CH_2CH_3

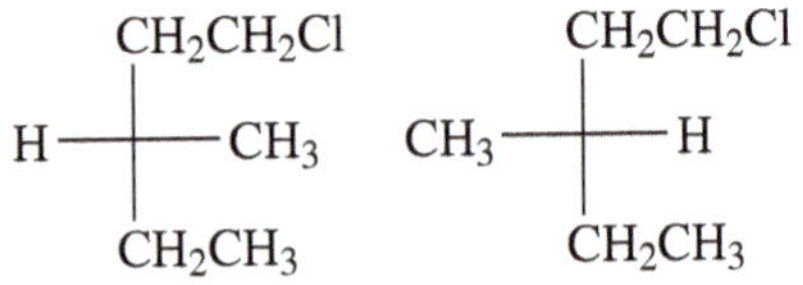

19. Solved in the text.

20.

a. $—CH_2OH$ (1) $—CH_3$ (3) $—H$ (4) $—CH_2CH_2OH$ (2)

b. $—CH_2Br$ (2) $—OH$ (1) $—CH_3$ (4) $—CH_2OH$ (3)

c. $—CH(CH_3)_2$ (2) $—CH_2CH_2Br$ (3) $—Cl$ (1) $—CH_2CH_2CH_2Br$ (4)

d. $—CH{=}CH_2$ (2) — forms 2 bonds to C, so considered to be attached to 2 Cs

$—CH_2CH_3$ (3) — attached to 1 C

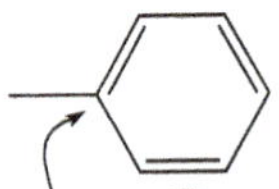

—phenyl (benzene ring) (1) — forms 3 bonds to C, so considered to be attached to 3 Cs

$—CH_3$ (4) — attached only to Hs

21. Solved in the text.

22. **a.** *R*

b. To determine the configuration, first add the fourth bond to the asymmetric center. Remember that it cannot be drawn between the two solid bonds. (It can be drawn on either side of the solid wedge.)

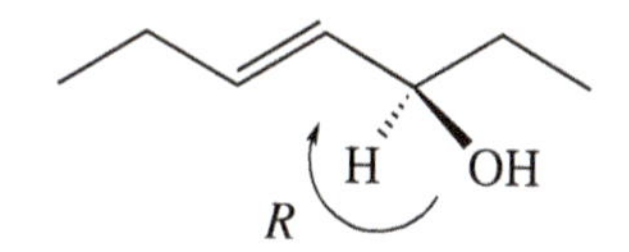

c. *R*.

23. **a.** *S* **b.** *R* **c.** *S* **d.** *S*

24. The easiest way to determine whether two compounds are identical or enantiomers is to determine their configurations: if both are *R* (or both are *S*), they are identical, one is *R* and the other is *S*, they are enantiomers.

a. identical **b.** enantiomers **c.** enantiomers **d.** enantiomers

25. **a.**

Cl
CH_3CH_2 — C ···H
CH_3

b.

Br
CH_3CH_2 — C ···H
CH_2Br

26. Solved in the text.

27.

COO^-
H — + — CH_3
CH_2CH_3

COO^-
CH_3 — C ···H
CH_2CH_3

28. **a.** levorotatory **b.** dextrorotatory

29. Solved in the text.

30. **a.** *R* **b.** *R* **c.** *S*

31. Solved in the text.

32. We see that the (*R*)-alkyl halide reacts with HO^- to form the (*R*)-alcohol. We are told that the product (the (*R*)-alcohol) is (−). We can, therefore, conclude that the (+)-alcohol has the *S* configuration.

33.

$$\text{specific rotation} = \frac{\text{observed rotation (degrees)}}{\text{concentration (g in 100 mL)} \times \text{length (dm)}}$$

There are 4 g in 100 mL, so

$$[\alpha] = \frac{+138^\circ}{4 \times 2} = \frac{+138^\circ}{8} = +1.7$$

34. **a.** −24 **b.** 0

35. **a.** 0 (It is a racemic mixture.)

b. 50% of the mixture is excess (+)-mandelic acid.

$$\text{optical purity} = 0.50 = \frac{\text{observed specific rotation}}{\text{specific rotation of the pure enantiomer}}$$

$$0.50 = \frac{\text{observed specific rotation}}{+158}$$

observed specific rotation = +79

c. 50% of the mixture is excess (−)-mandelic acid.
observed specific rotation = −79 (For the calculation, see part **b.**)

36. **a.** From the data given, you cannot determine the configuration of naproxen.
b. 97% of the commercial preparation is (+)-naproxen; 3% is a racemic mixture. Therefore, the commercial preparation forms 98.5% (+)-naproxen and 1.5% (−)-naproxen.

37. Solved in the text.

38. As a result of the double bond, the compound has a cis isomer and a trans isomer. Because the compound also has an asymmetric center, the cis isomer can exist as a pair of enantiomers and the trans isomer can exist as a pair of enantiomers.

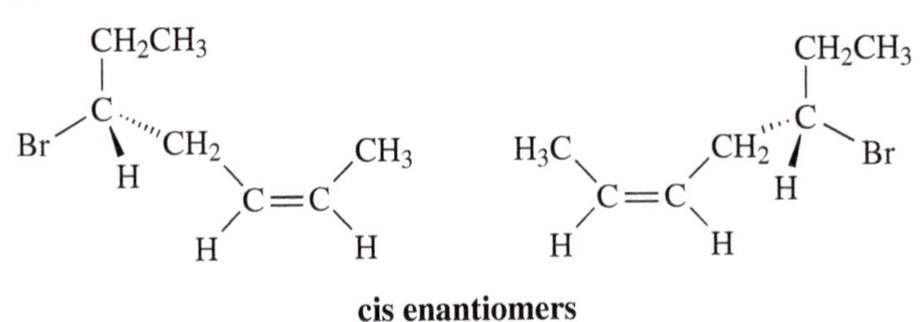

cis enantiomers

trans enantiomers

39. **a.** enantiomers
b. identical compounds (Therefore, they are not stereoisomers.)
c. diastereomers

40. **a.** Find the sp^3 carbons that are bonded to four different substituents; these are the asymmetric centers. Cholesterol has eight asymmetric centers. They are indicated by arrows.

b. $2^8 = 256$
Only the stereoisomer shown above is found in nature.

41. Your perspective formulas may not look exactly like the ones drawn here because you can draw the first one with the groups attached to any bonds you want. Just make certain that the second one is a mirror image of the first one.

a. Leucine has one asymmetric center, so it has two stereoisomers.

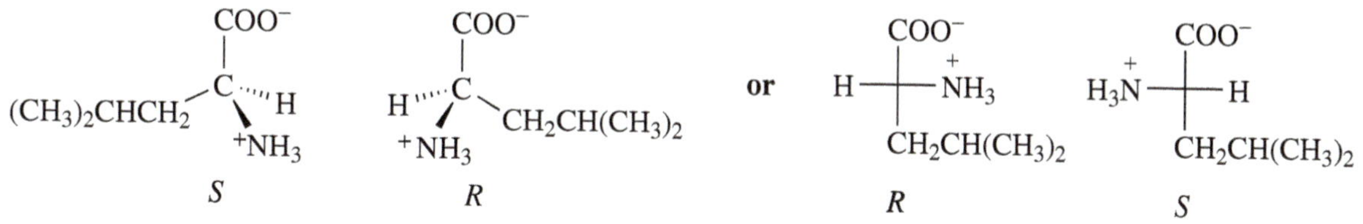

b. Isoleucine has two asymmetric centers, so it has four stereoisomers. Again, your perspective formulas may not look like the ones drawn here. To make sure you have all four, determine the configuration of each of the asymmetric centers. You should have *R,R*, *S,S*, *R,S*, and *S,R*. Notice that the asymmetric centers in mirror images have the opposite configurations.

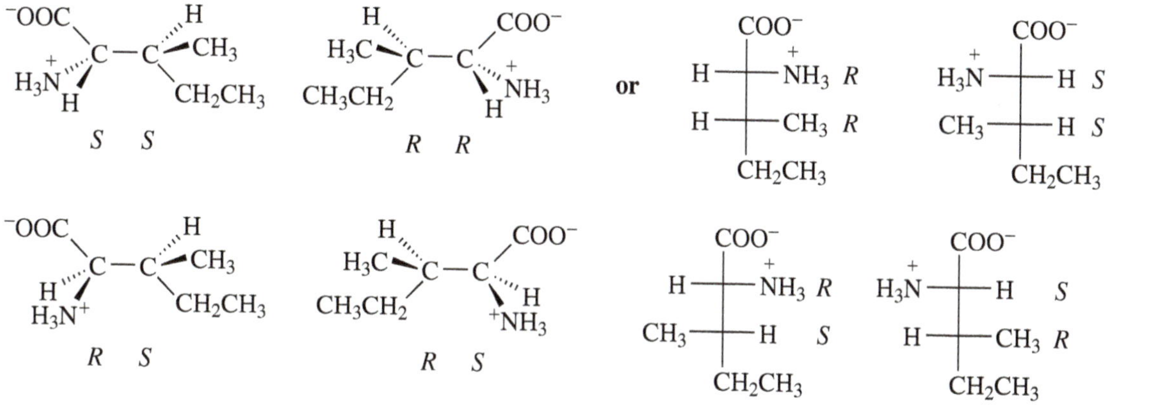

42.

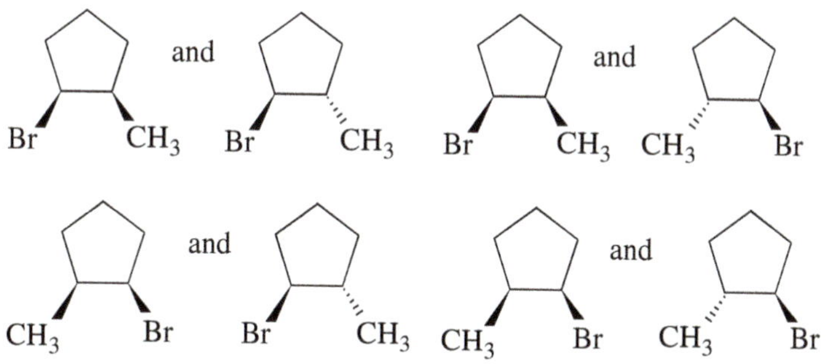

43. **B** and **D** have no asymmetric centers.

A and **C** each has one asymmetric center. **E** has two asymmetric centers.

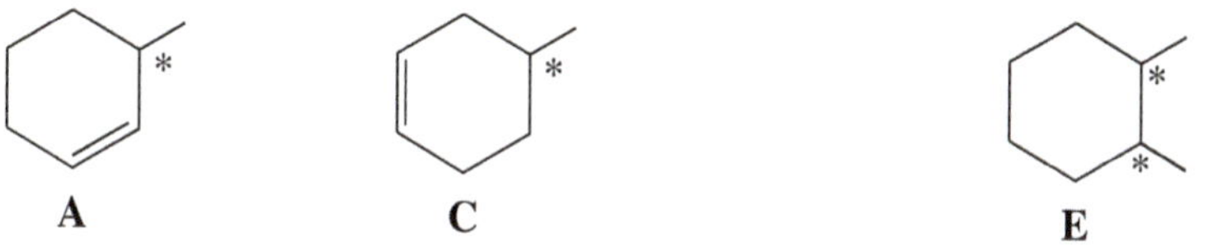

44. **a.**

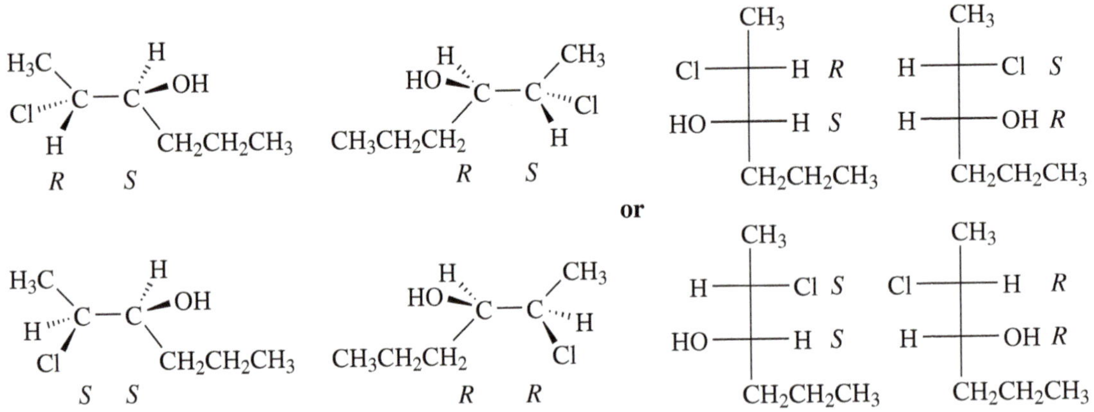

b.

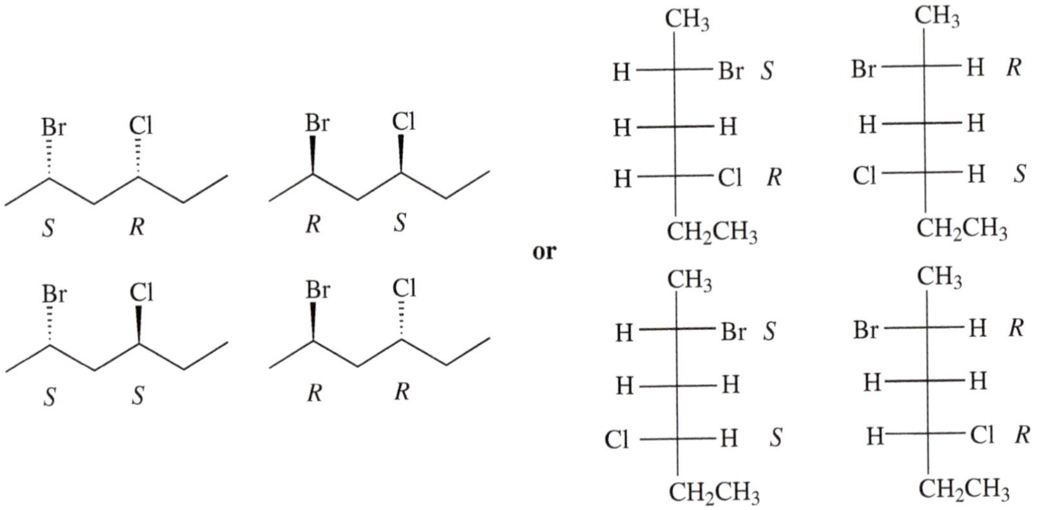

c.

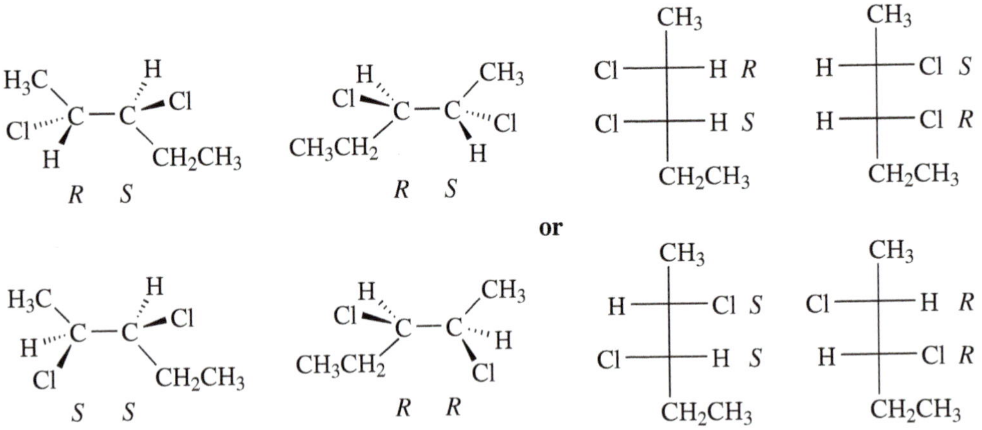

d.

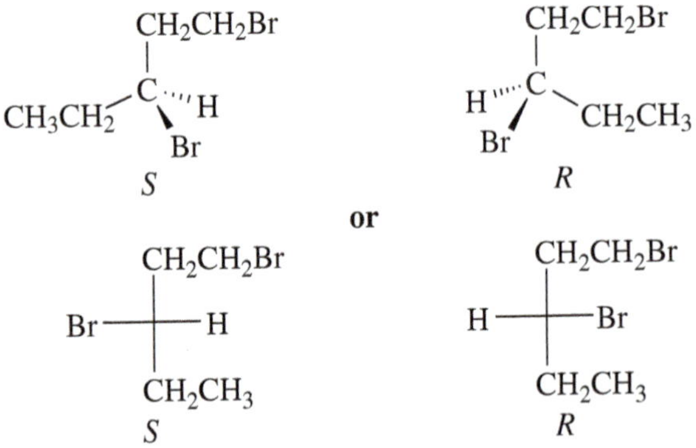

45.

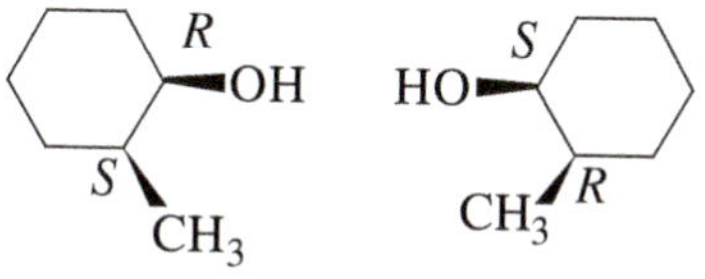

cis-2-methylcyclohexanol

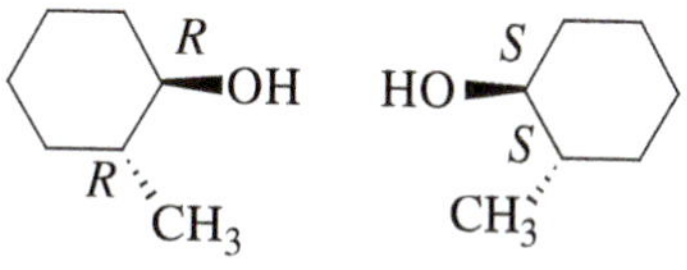

trans-2-methylcyclohexanol

46.

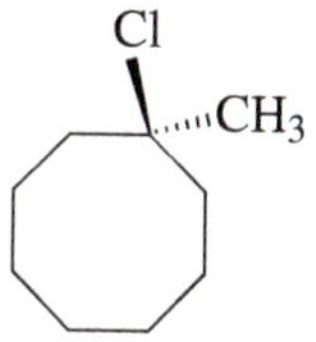

1-chloro-1-methylcyclooctane

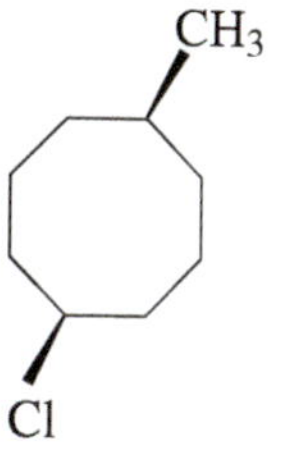

cis-1-chloro-5-methylcyclooctane

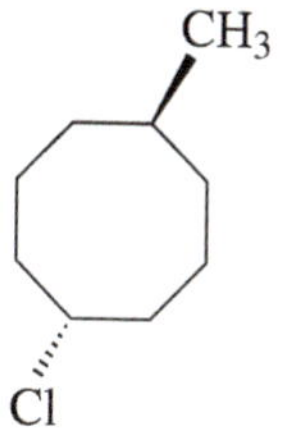

trans-1-chloro-5-methylcyclooctane

47. There is more than one diastereomer for **a, b,** and **d**; **c** has only one diastereomer.

To draw a diastereomer, switch any one pair of substituents bonded to one of the asymmetric centers. Because any one pair can be switched, your diastereomer may not be the one drawn here, unless you happened to switch the same pair that was switched here.

a.

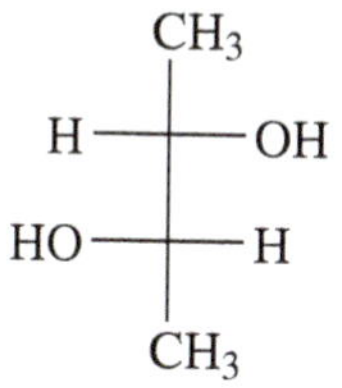

b.

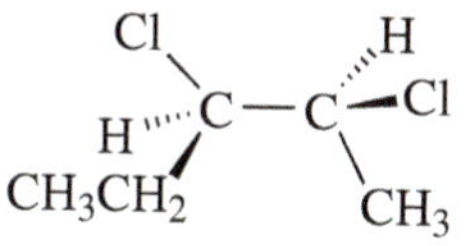

c.

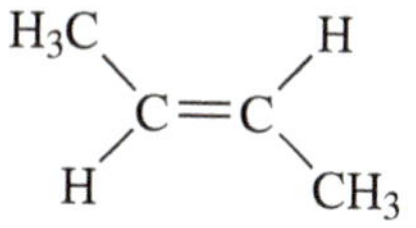

d.

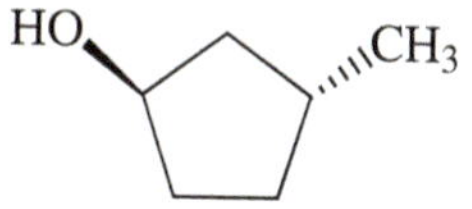

48. **A** = identical **B** = enantiomer **C** = diastereomer **D** = identical

49. **B**, **D**, and **F** because each has two asymmetric centers and the same four groups bonded to each of the asymmetric centers.
A has two asymmetric centers, but it does not have a stereoisomer that is a meso compound, because it does not have the same four groups bonded to each of the asymmetric centers.
C and **E** do not have a stereoisomer that is a meso compound, because they do not have asymmetric centers.

50. Solved in the text.

51. **a.**

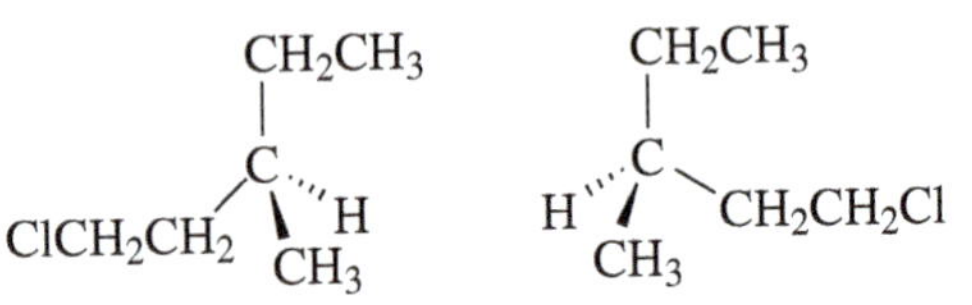

CH_2CH_3
C
$ClCH_2CH_2$ H CH_3

CH_2CH_3
C
H CH_2CH_2Cl CH_3

or

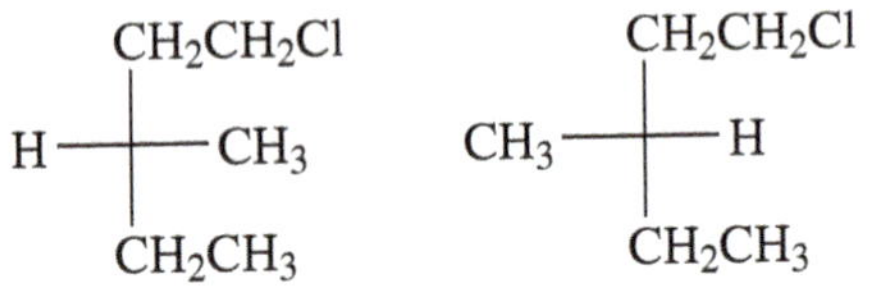

	CH_2CH_2Cl			CH_2CH_2Cl	
H	+	CH_3	CH_3	+	H
	CH_2CH_3			CH_2CH_3	

b. CH_3
CH_3CHCH_2OH No stereoisomers, because the compound does not have an asymmetric center.

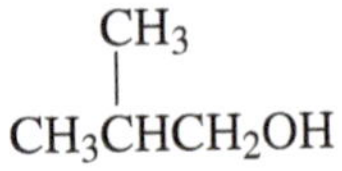

c.

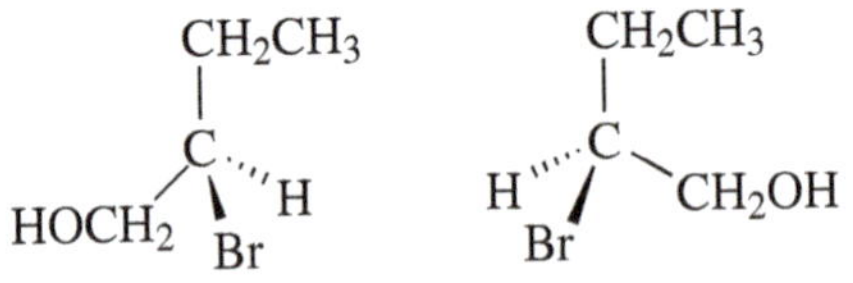

CH_2CH_3
C
$HOCH_2$ H Br

CH_2CH_3
C
H CH_2OH Br

or

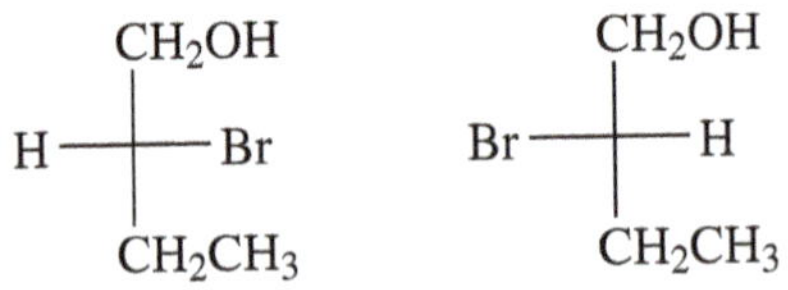

	CH_2OH			CH_2OH	
H	+	Br	Br	+	H
	CH_2CH_3			CH_2CH_3	

d.

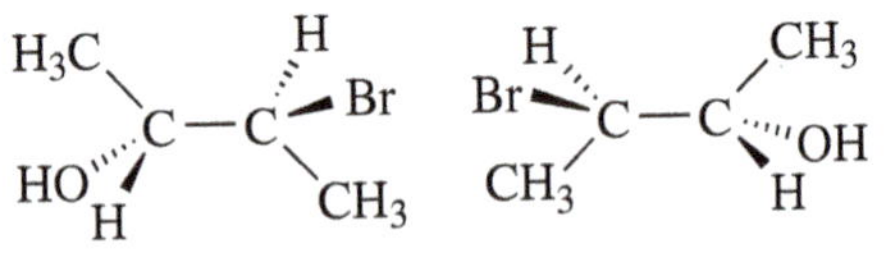

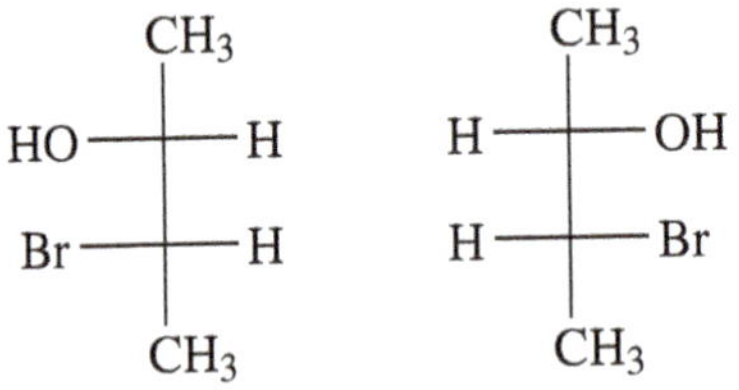

H_3C H
C—C Br
HO H CH_3

H CH_3
Br C—C OH
CH_3 H

	CH_3			CH_3	
HO	+	H	H	+	OH
Br	+	H	H	+	Br
	CH_3			CH_3	

or

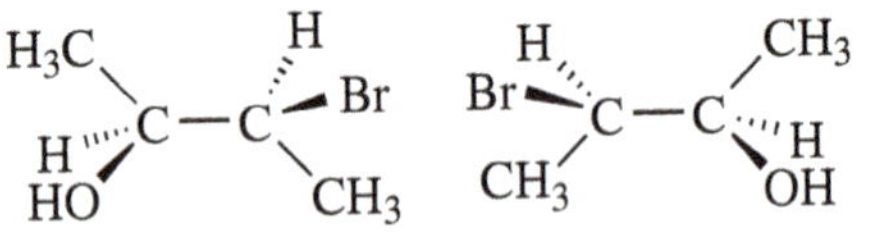

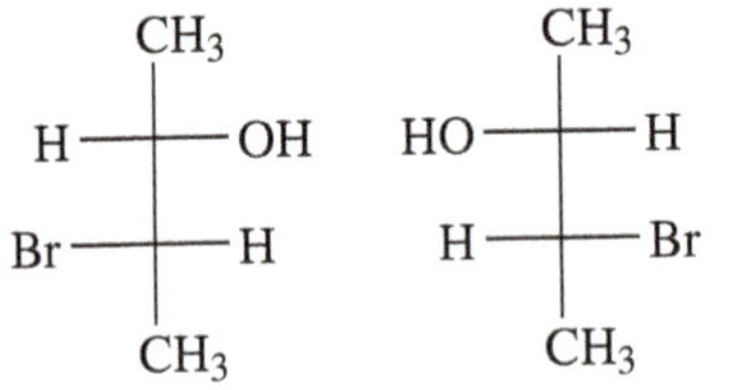

H_3C H
H C—C Br
HO CH_3

H CH_3
Br C—C H
CH_3 OH

	CH_3			CH_3	
H	+	OH	HO	+	H
Br	+	H	H	+	Br
	CH_3			CH_3	

e.

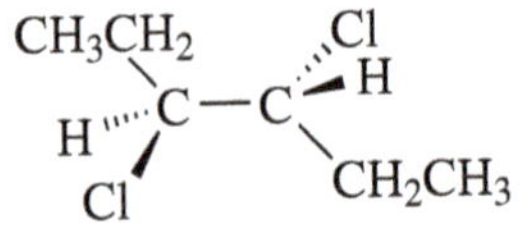

CH_3CH_2 Cl
H C—C H
Cl CH_2CH_3

a meso compound

or

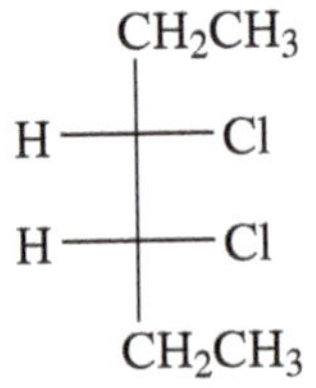

	CH_2CH_3	
H	+	Cl
H	+	Cl
	CH_2CH_3	

a meso compound

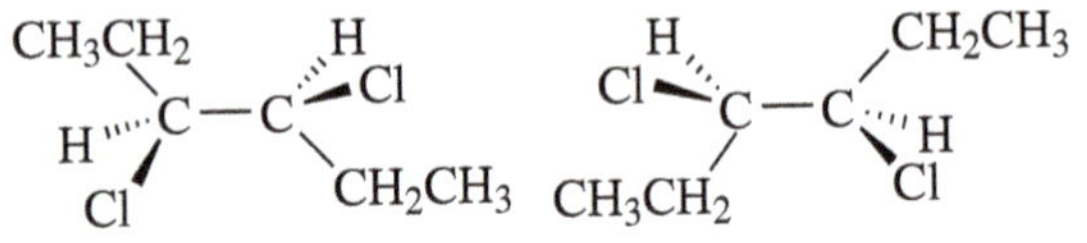

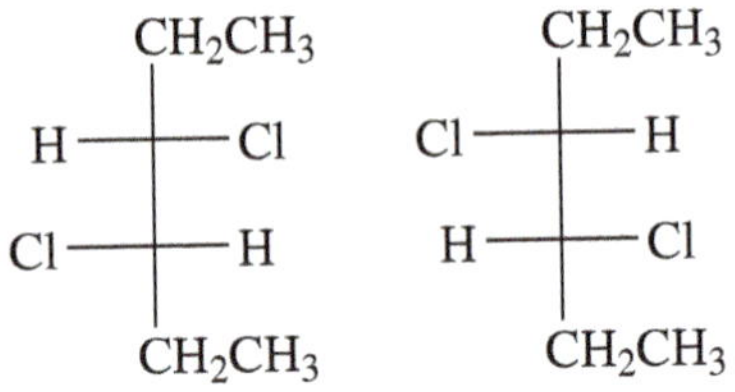

CH_3CH_2 H
H C—C Cl
Cl CH_2CH_3

H CH_2CH_3
Cl C—C H
CH_3CH_2 Cl

	CH_2CH_3			CH_2CH_3	
H	+	Cl	Cl	+	H
Cl	+	H	H	+	Cl
	CH_2CH_3			CH_2CH_3	

f.

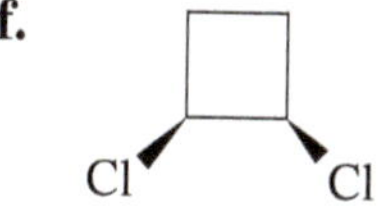

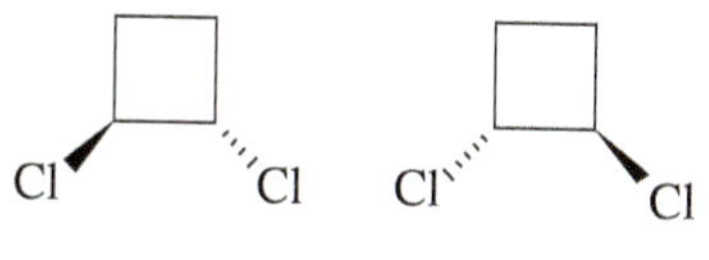

The cis stereoisomer is a meso compound.

g.

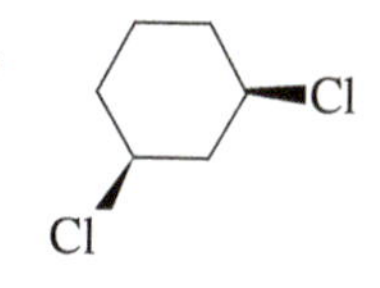

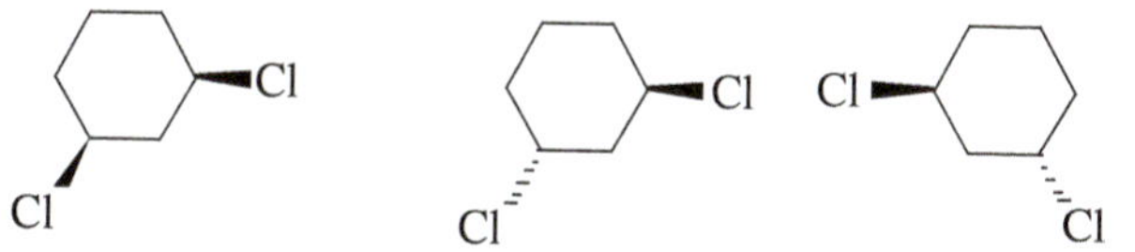

The cis stereoisomer is a meso compound.

h.

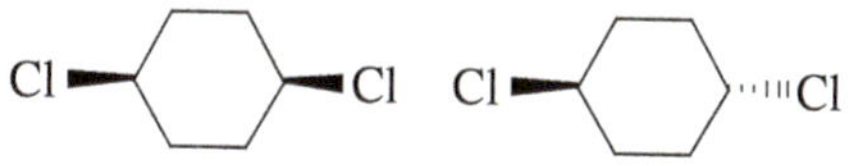

This compound does not have any asymmetric centers, so it has only cis–trans isomers.

i.

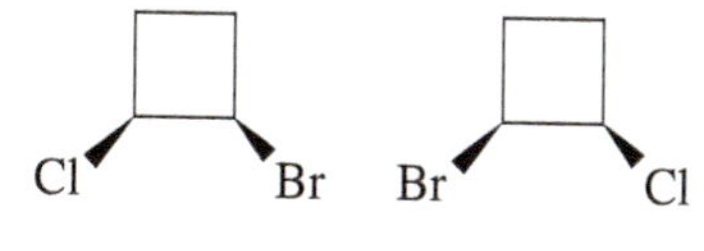

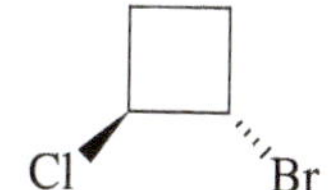

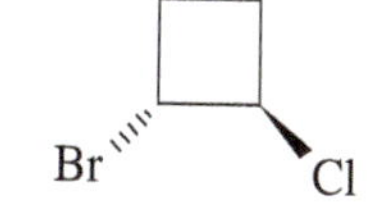

j.

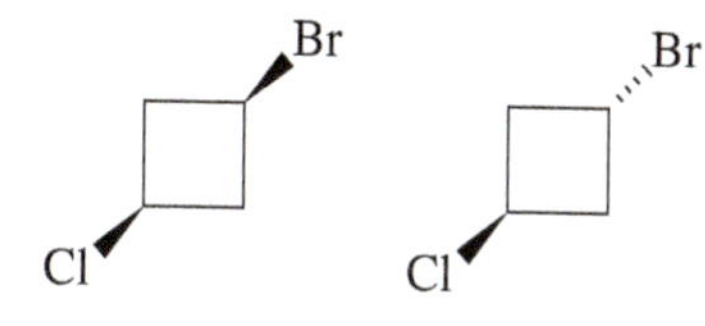

This compound does not have any asymmetric centers, so it has only cis–trans isomers.

52. and **53.** How to draw perspective formulas for compounds that have two asymmetric centers is described on page 175 of the text.

a.

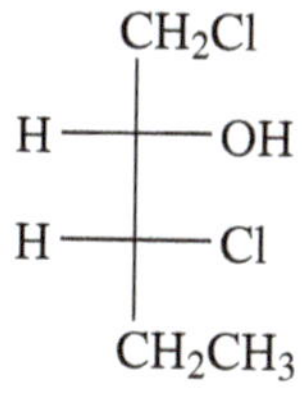

(2*S*,3*R*)-1,3-dichloro-2-pentanol

(2*R*,3*S*)-1,3-dichloro-2-pentanol

(2*S*,3*S*)-1,3-dichloro-2-pentanol

(2*R*,3*R*)-1,3-dichloro-2-pentanol

b.

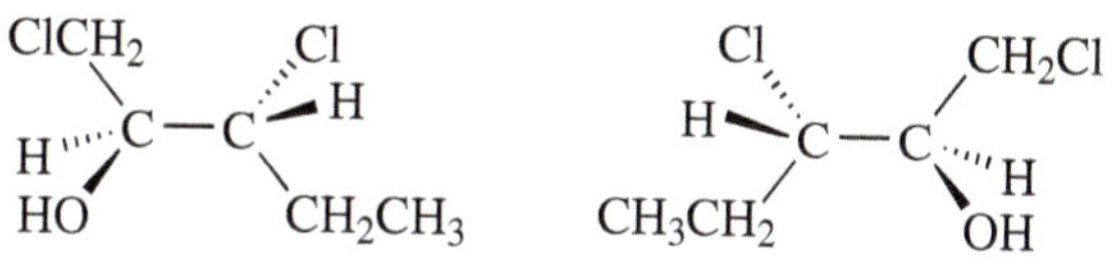

(2*S*,3*R*)-1,3-dichloro-2-pentanol

(2*R*,3*S*)-1,3-dichloro-2-pentanol

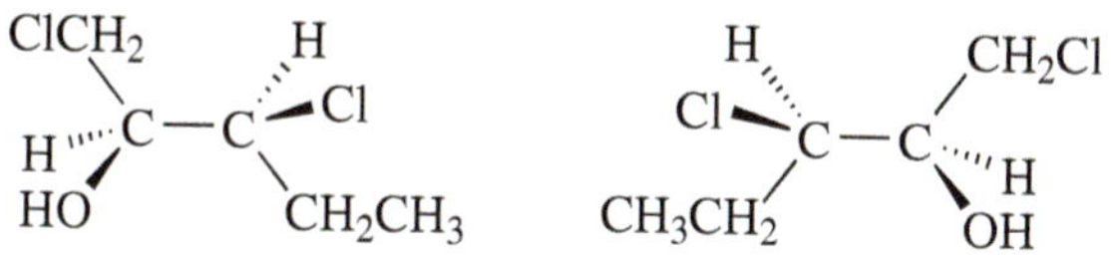

(2*S*,3*S*)-1,3-dichloro-2-pentanol

(2*R*,3*R*)-1,3-dichloro-2-pentanol

54.

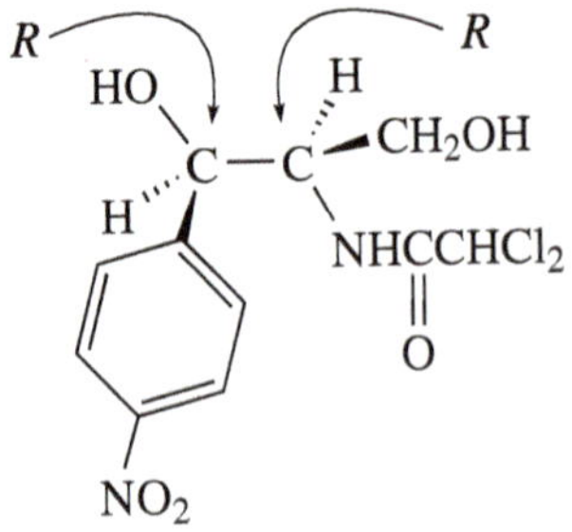

R *R* HO H CH_2OH H C—C $NHCCHCl_2$ O NO_2

55. Your answer might be correct yet not look like the answers shown here. If you can get the answer shown here by interchanging **two** pairs of groups bonded to an asymmetric center on the structure you drew, then your answer is correct. If you get the answer shown here by interchanging **one** pair of groups bonded to an asymmetric center, then your answer is not correct.

a. Cl C H $HOCH_2CH_2$ CH_2CH_3

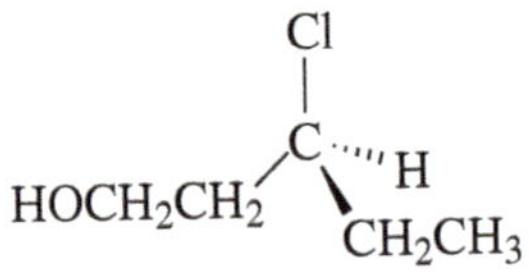

c. H_3C H CH_2CH_3 H C—C HO CH_3

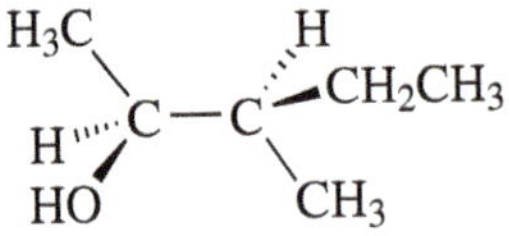

b. Br H CH_2CH_3 H C—C CH_3 Br

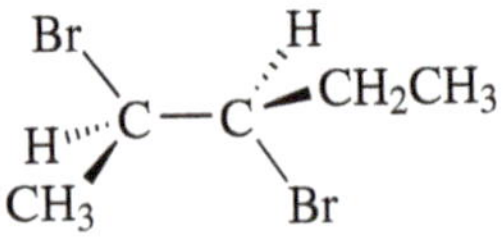

d. CH_2Br C H Br CH_2CH_3

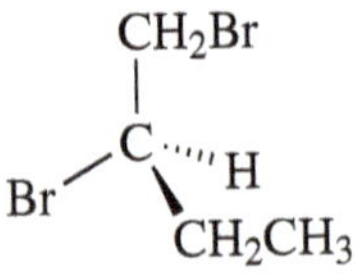

56.
a. (3*R*,4*S*)-3-chloro-4-methylhexane
b. (2*S*,3*S*)-2-bromo-3-chloropentane
c. (1*R*,3*S*)-3-bromocyclopentanol
d. (2*R*,3*R*)-2,3-dichloropentane

57. The first structure is 2*R*,3*S*. Therefore, naturally occurring threonine, with a configuration of 2*S*,3*R*, is the mirror image of the first structure. Thus, the second structure is naturally occurring threonine.

58. Solved in the text.

59.

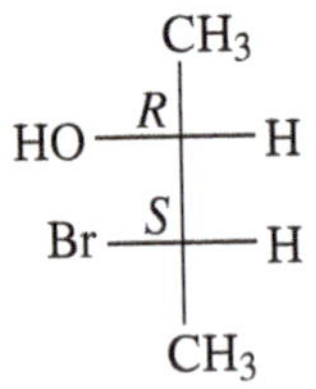

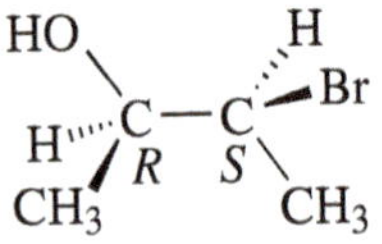

CH_3 HO *R* H Br *S* H CH_3

HO H Br H C—C *R* *S* CH_3 CH_3

60. Solved in the text.

61.

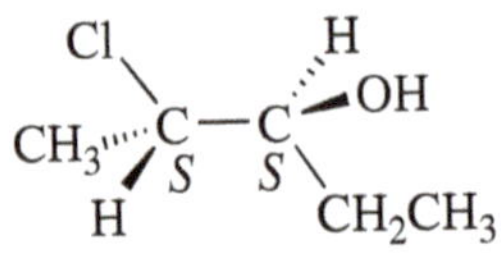

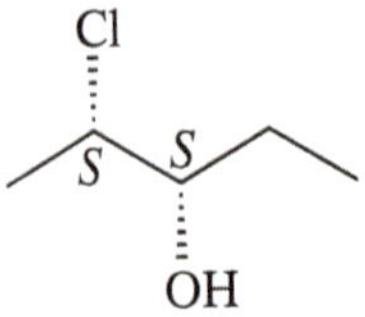

Cl H OH CH_3 C—C *S* *S* H CH_2CH_3

Cl *S* *S* OH

62. Start by naming the first stereosiomer. Finding that **A** is 2*R*,3*R* allows you to answer both questions.

a. **A** is D-erythrose.

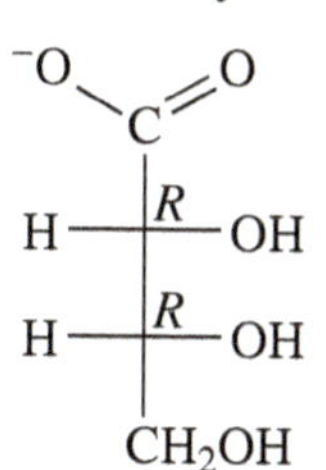

b. D-Threose has the opposite configuration at C-2 and the same configuration at C-3. Therefore, **C** is D-threose.

63. Compound **A** has two stereoisomers, because it has an asymmetric center (at N).

Compound **B** does not have stereoisomers, because it does not have an asymmetric center.

Compound **C** has an asymmetric center at N but, because of the lone pair, the two enantiomers rapidly interconvert, so it exists as a single compound.

64. (+)-Limonene has the *R* configuration, so it is the stereoisomer found in oranges and lemons.

65. $CH_3CH{=}CHCH_2CH_3$ 2 stereoisomers [cis and trans]

$CH_2{=}CHCH_2CH_2CH_3$ no stereoisomers

$CH_3C(CH_3){=}CHCH_3$ no stereoisomers

$CH_3CH(CH_3)CH{=}CH_2$ no stereoisomers

$CH_3CH_2C(CH_3){=}CH_2$ no stereoisomers

1,1-dimethylcyclopropane (CH_3, CH_3) no stereoisomers

ethylcyclopropane (CH_2CH_3) no stereoisomers

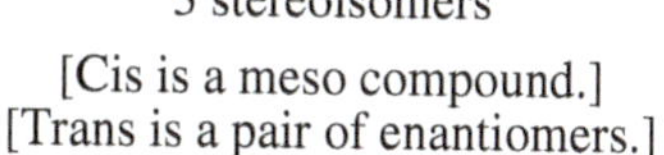

3 stereoisomers

[Cis is a meso compound.]
[Trans is a pair of enantiomers.]

methylcyclobutane (CH_3) no stereoisomers

cyclopentane no stereoisomers

66. **a.** Br Cl Cl Br Br Cl Cl Br

b. CH_3CHCH_2–C(Br)(H)(CH_3) with CH_3 branch + H–C(Br)(CH_3)–CH_2CHCH_3 with CH_3 branch **or** H–C(CH₃)(Br)–CH_2CHCH_3 + Br–C(CH₃)(H)–CH_2CHCH_3

c.

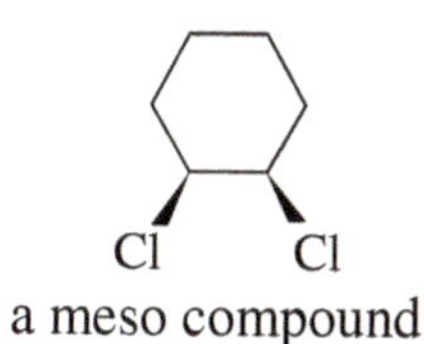

a meso compound

Cl Cl Cl Cl

d.

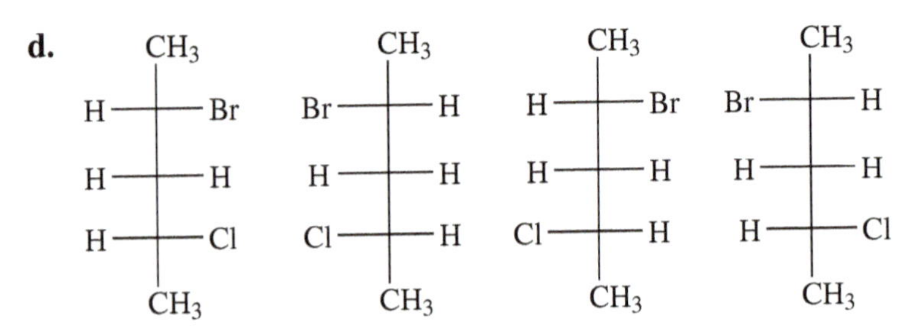

e. Br, Cl (cis); Br, Cl (trans)

cis and trans

f.

g.

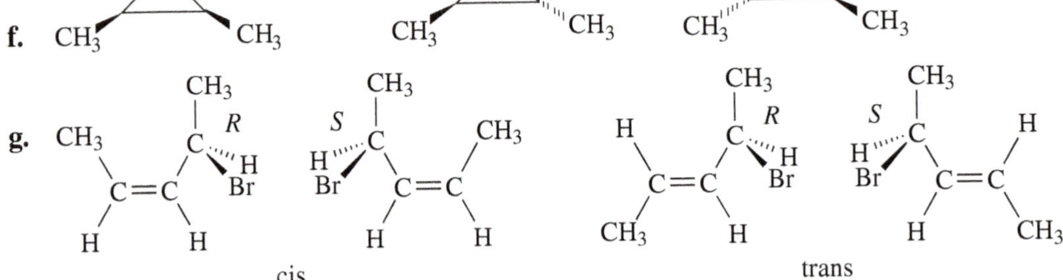

h. $CH_3CH_2C(CH_3)_2CH_2CH_3$ No isomers are possible for this compound, because it does not have an asymmetric center.

i.

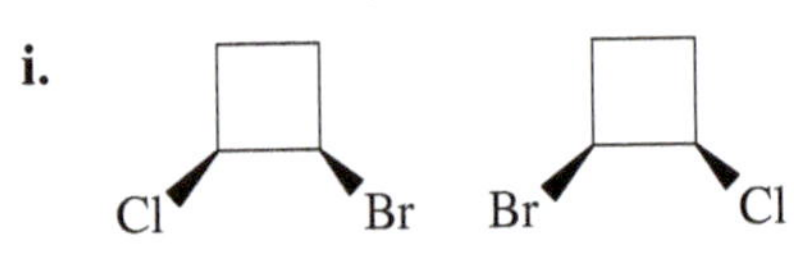

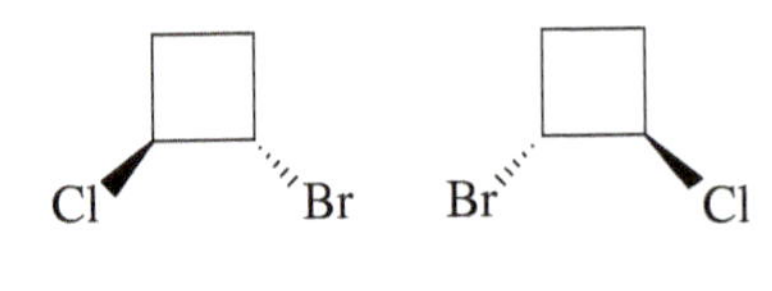

j.

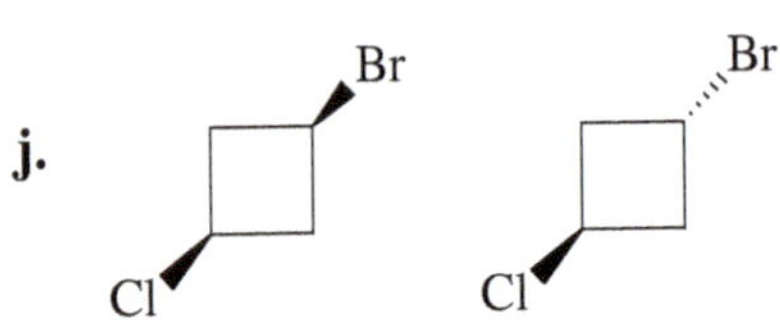

cis and trans only (no asymmetric centers)

67. Only the fourth one (CHFBrCl) has an atom with four different atoms attached to a carbon, so it is the only one that has an asymmetric center.

68. **a.** (2*R*,3*R*)-3-chloro-2-pentanol
b. (*S*)-2-methyl-1,2,5-pentanetriol
c. (2*S*,3*S*)-1,2-dibromo-2-methyl-3-pentanol

69. Mevacor has eight asymmetric centers, which are indicated by stars.

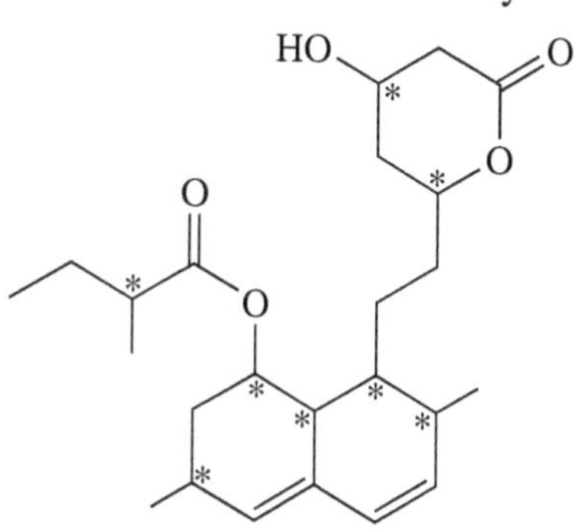

70. **a.** *Z* **b.** *E* **c.** *E* **d.** *Z* **e.** *E* **f.** *E*

71.
a. diastereomers (one asymmetric center has the same configuration in both compounds, and the other has the opposite configuration in both compounds)
b. enantiomers (they are mirror images)
c. constitutional isomers
d. diastereomers (the configuration of two asymmetric centers is the same, and the configuration of one asymmetric center is different)
e. diastereomers
f. identical
g. diastereomers
h. identical

72.
a. $—CH{=}CH_2$ > $—CH(CH_3)_2$ > $—CH_2CH_2CH_3$ > $—CH_3$

b. $—OH$ > $—NH_2$ > $—CH_2OH$ > $—CH_2NH_2$

c. $—Cl$ > $—C({=}O)CH_3$ > $—C{\equiv}N$ > $—CH{=}CH_2$

73.
a.
H
F—+—F
I—+—I
H

b.
H
F—+—I
F—+—I
H

c.
H
F—+—I
I—+—F
H

74. Only the compound on the far right is optically active, because it is the only one that has one or more asymmetric centers and no plane of symmetry.

75. **a.** Because there are two asymmetric centers, there are four stereoisomers.

b.
H
NHCH₃
HO C—C
H CH₃
R *S*

76. **a.** (R)-3-bromo-2,5-dimethylhexane

b. (2R,4R)-4-methyl-2-hexanol

c. (2R,4S,6S)-4-chloro-6-methyl-2-octanol

77. Compounds **A, D, E, F, I,** and **J** have a stereoisomer that is achiral.

Compounds **A, D, F, I,** and **J** have two asymmetric centers bonded to identical substituents.

Therefore, there are three stereoisomers, one of which is an achiral meso compound.

Compounds **E** and **H** do not have any asymmetric centers; they have cis and trans stereosiomers, so each is achiral and, therefore, each has an achiral stereoisomer.

Compounds **B** and **G** each have two asymmetric centers bonded to different substituents.

Therefore, there are four stereoisomers, all of which are chiral.

Compound **C** does not have any asymmetric centers.

Therefore, it is achiral and does not have any stereoisomers.

78. **a.** Cl Cl Cl Cl Cl Cl Cl Cl

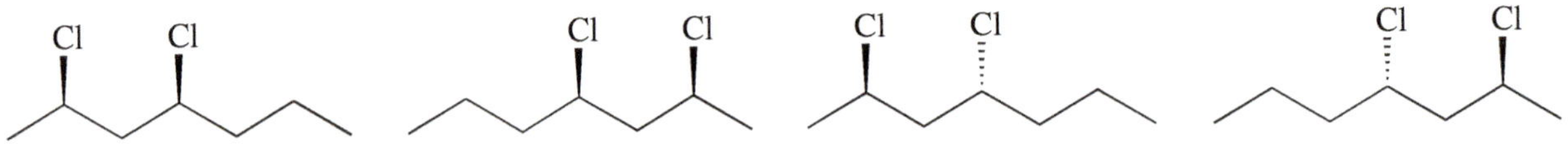

b. Cl Cl Cl Cl Cl Cl

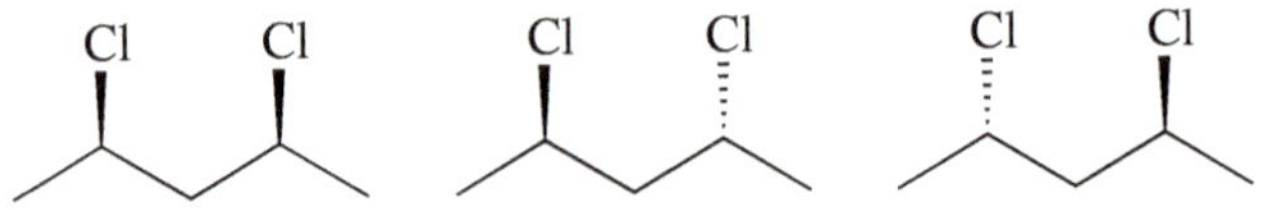

79. **a.** One asymmetric center has the same configuration in both compounds; the other asymmetric center has the opposite configuration in both. Therefore, the compounds are diastereomers.

b. Both asymmetric centers in one compound have the opposite configuration in the other, so the compounds are enantiomers.

c. They are identical because if one is flipped over, it will superimpose on the other.

d. They are constitutional isomers because the atoms are hooked up differently; one compound is 1-chloro-2-methylcyclopentane, and the other is 1-chloro-3-methylcyclopentane.

80. **a.**

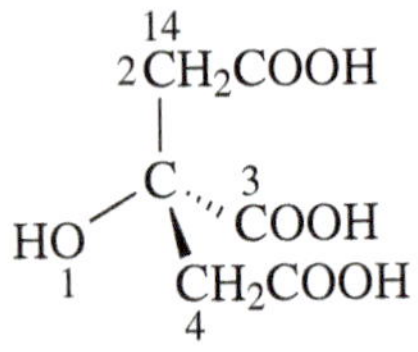

$$\overset{14}{2\,CH_2COOH}$$

HO (1) — C — COOH (3), CH_2COOH (4)

(S)-citric acid

b. The product of the reaction will be achiral, because if it does not have a ^{14}C label, the two CH_2COOH groups are identical, so it does not have an asymmetric center.

81. $$[\alpha] = \frac{\alpha}{l \times c} = \frac{-18^\circ}{[2.0 \text{ dm}][1.5 \text{ g in } 100 \text{ mL}]} = -6.0$$

82. **a.** identical

b. identical

c. enantiomers

d. constitutional isomers

e. constitutional isomers

f. diastereomers

g. constitutional isomers

h. enantiomers

83. optical purity $= \dfrac{+1.4}{+8.7} = 0.16 = 16\%$ excess *R* enantiomer

$100\% - 16\% = 84\%$ is a racemic mixture

R enantiomer $= 1/2\,(84\%) + 16\% = 42\% + 16\% = 58\%$

84. **a.** *R*

b. *S*

c. Cl = Cl H

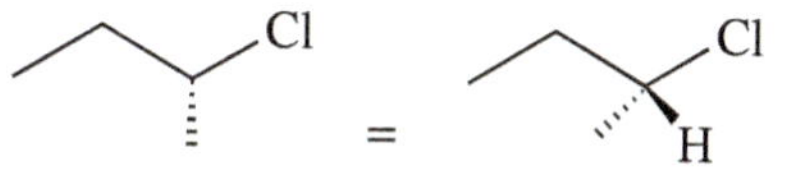

R

d. Cl = H Cl

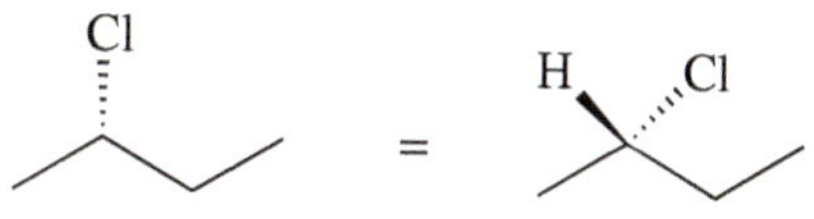

S

e. Cl, H, CH_3, CH_2CH_3

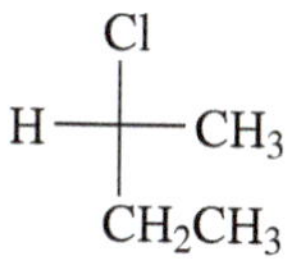

R

f. CH_3, H, Cl, CH_2CH_3

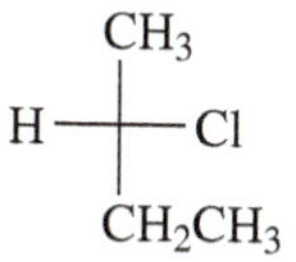

S

Fisher projections show the molecule with eclipsed bonds. Therefore, to answer parts **e** and **f**, first rotate the Newman projection so it is eclipsed. Then turn the Newman projection into a Fisher projection. (See page 188 in the text.)

85. Butaclamol has four asymmetric centers; three of them are carbons, and one is a nitrogen.

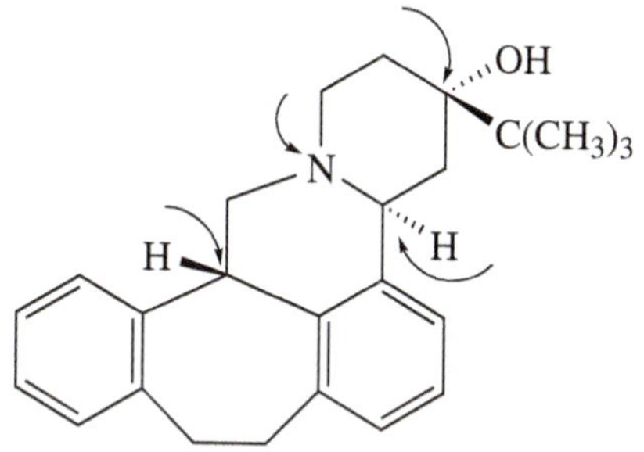

OH
$C(CH_3)_3$
N
H
H

86. The only way that *R* and *S* are related to (+) and (−) is that if the configuration of one enantiomer (for example, the *R* enantiomer) is (−), the the configuration of the other enantiomer is (+).

Because some compounds with the *R* configuration are (+) and some are (−), there is no way to determine whether a particular *R* enantiomer is (+) or (−) without putting the compound in a polarimeter or finding out whether someone else has previously determined how the compound rotates the plane of polarization of plane-polarized light.

87. First convert the staggered Newman projection to an eclipsed Newman projection, which can then be converted to a Fischer projection because that too is eclipsed (see page 188 of the text). Then name the Fischer projection.

a. Because one is *R* and the other is *S*, they are enantiomers

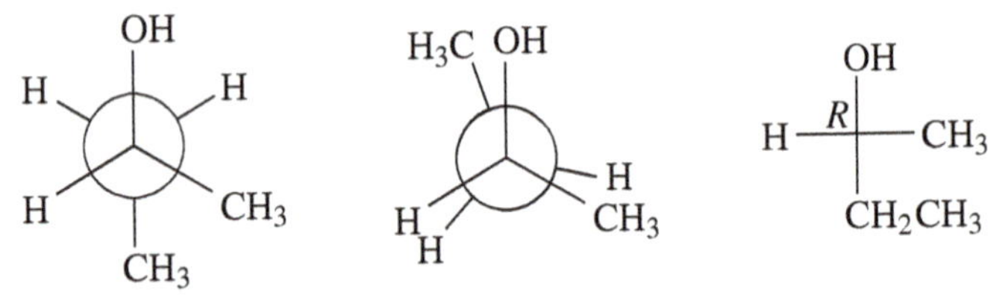

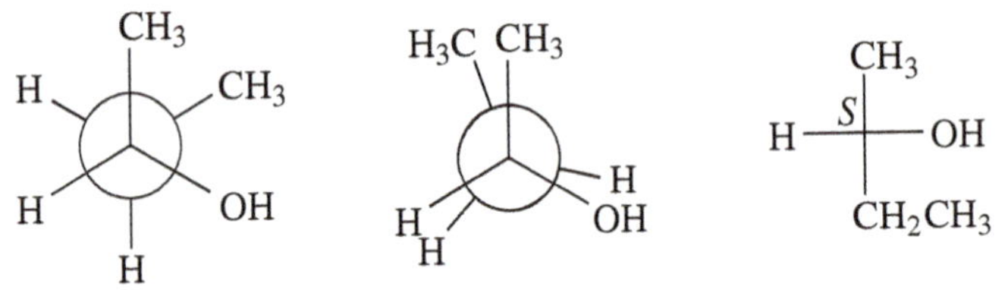

b. Because one is *R*,*R* and the other is *S*,*S*, they are enantiomers.

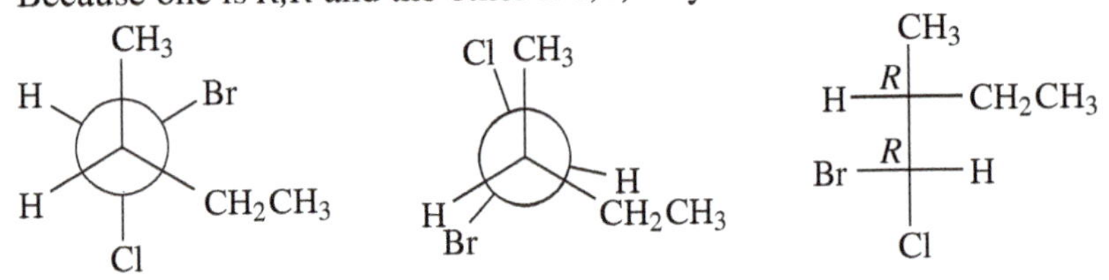

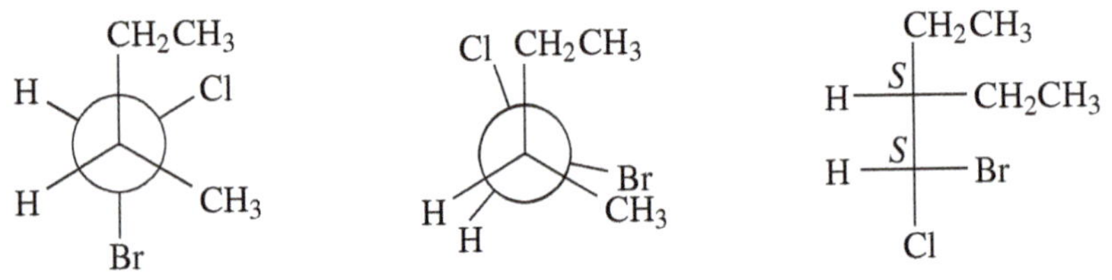

88. **a.** The compound has four stereoisomers.

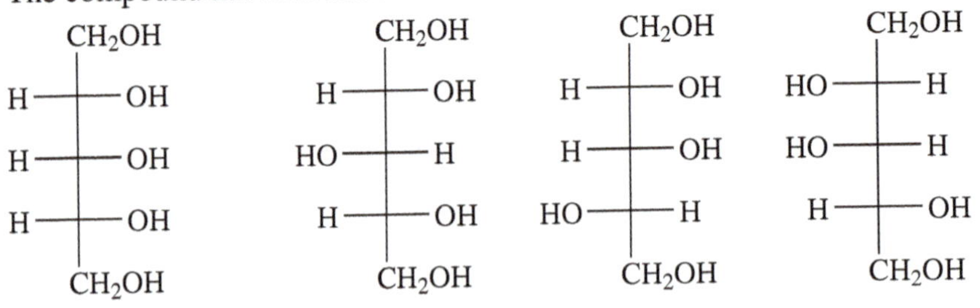

b. The first two stereoisomers are optically inactive because they are meso compounds. (Each has a plane of symmetry.)

89. **a.** **b.** 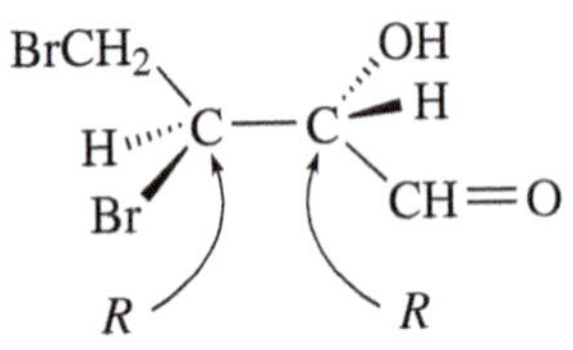 **c.**

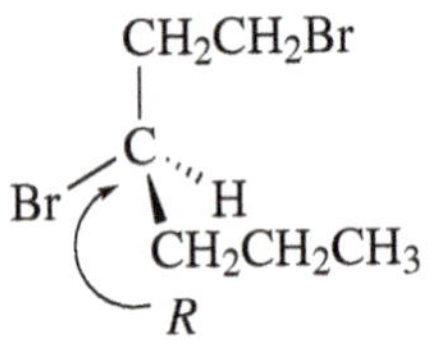

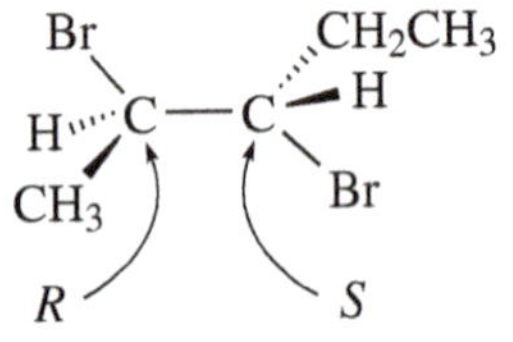

90. **a.** and **b.**

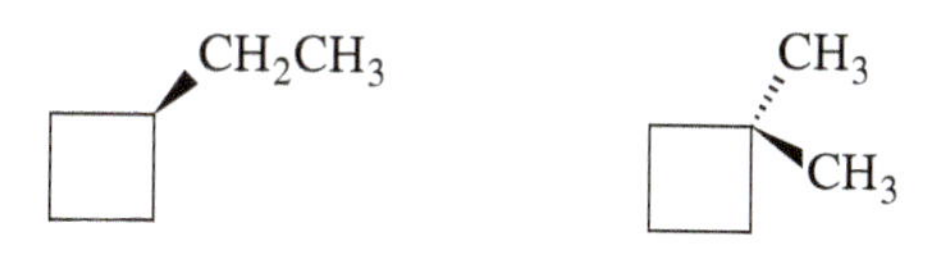

ethylcyclobutane 1,1-dimethylcyclobutane

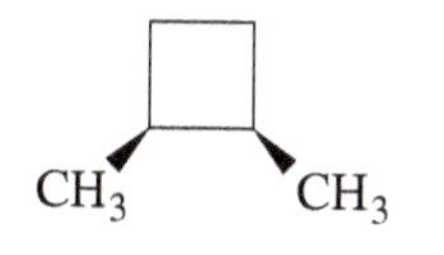

CH_3 CH_3 CH_3 CH_3

cis-1,2-dimethylcyclobutane *trans*-1,2-dimethylcyclobutane

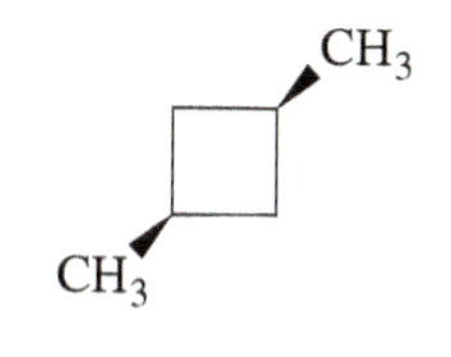

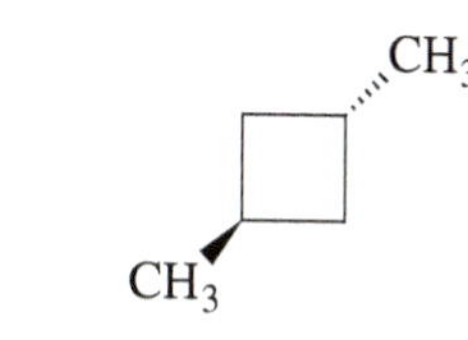

cis-1,3-dimethylcyclobutane *trans*-1,3-dimethylcyclobutane

c. **1.** ethylcyclobutane
1,1-dimethylcyclobutane
1,2-dimethylcyclobutane
1,3-dimethylcyclobutane

2. the three isomers of 1,2-dimethylcyclobutane
the two isomers of 1,3-dimethylcyclobutane

3. *cis*- and *trans*-1,2-dimethylcyclobutane
cis- and *trans*-1,3-dimethylcyclobutane

4. the two stereoisomers of *trans*-1,2-dimethylcyclobutane

5. all the isomers except the two stereoisomers of *trans*-1,2-dimethylcyclobutane

6. *cis*-1,2-dimethylcyclobutane
(Note: *cis*-1,3-dimethylcyclobutane is not a meso compound because it does not have any asymmetric centers.)

7. the two stereoisomers of *trans*-1,2-dimethylcyclobutane

8. *cis*-1,3-dimethylcyclobutane and *trans*-1,3-dimethylcyclobutane
cis-1,2-dimethylcyclobutane and either of the enantiomers of *trans*-1,2-dimethylcyclobutane

91. $$\text{observed specific rotation} = \frac{\text{observed rotation}}{\text{concentration} \times \text{length}}$$

$$\frac{-6.52°}{0.187 \text{ g in} 100 \text{ mL} \times 1 \text{ dm}} = -34.9$$

$$\% \text{ optical purity} = \frac{\text{observed specific rotation}}{\text{specific rotation of the pure enantiomer}} \times 100$$

$$= \frac{-34.9}{-39.0} \times 100$$

$$= 89.5\%$$

$$\% \text{ of the (+)-isomer} = \frac{100 - 89.5}{2} = 5.25\%$$

$$\% \text{ of the (−)-isomer} = 89.5 + 5.25 = 94.75\%$$

92. **a.** diastereomers [The configuration of all the symmetric centers is not the same in both (then the two would be the same) and not opposite in both (then the two would be enantiomers.)] Recall that diastereomers are stereoisomers that are not enantiomers.

b. identical (by rotating the first compound clockwise, you can see that it is superimposable on the other)

c. constitutional isomers

d. diastereomers (the configuration of all the stereoisomers is not the same in both and not opposite in both)

93. **a.** Br, Cl, H, $CH_2CH_2CH_3$ (wedge-dash structure) or Br, H, Cl, $CH_2CH_2CH_3$ (Fischer projection)

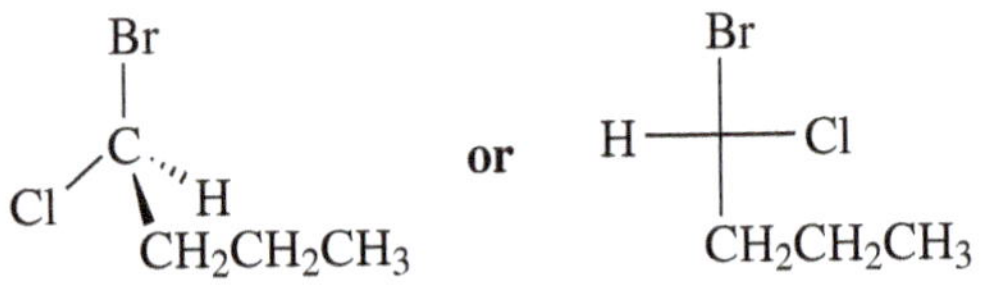

b. H, Cl, CH_3CH_2, CH_3, H, Cl (wedge-dash structure) or CH_3, Cl, H, H, Cl, CH_2CH_3 (Fischer projection)

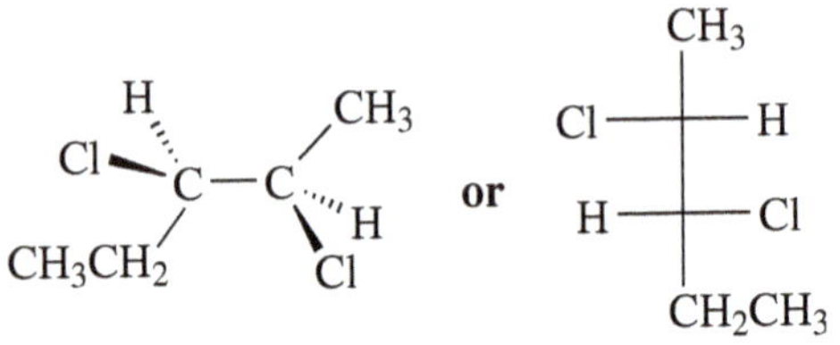

c. CH_3 CH_3

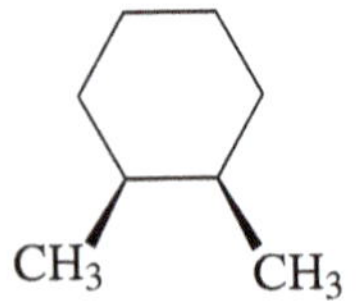

d. Br Br or Br Br

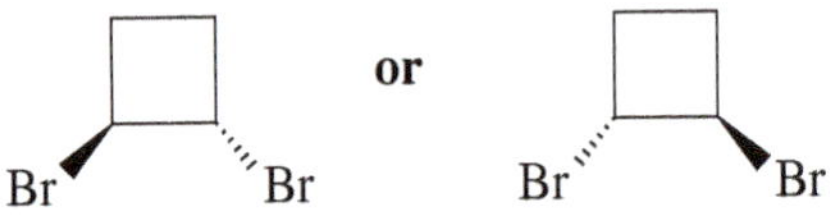

e. CH_2CH_3; H, CH_3; H, CH_3; H, CH_3; CH_2CH_3 and CH_2CH_3; CH_3, H; H, CH_3; CH_3, H; CH_2CH_3

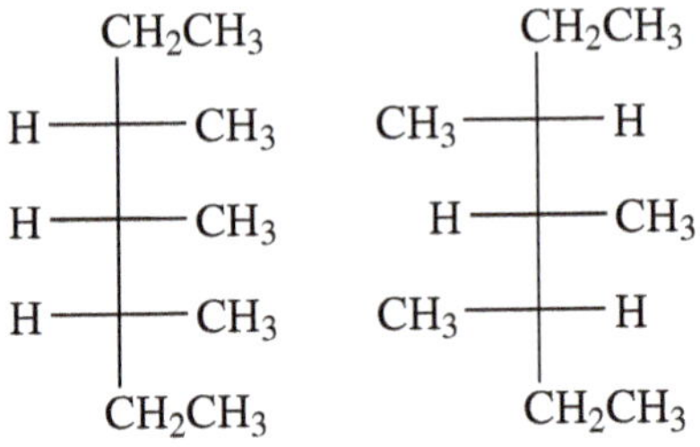

94. **a.**

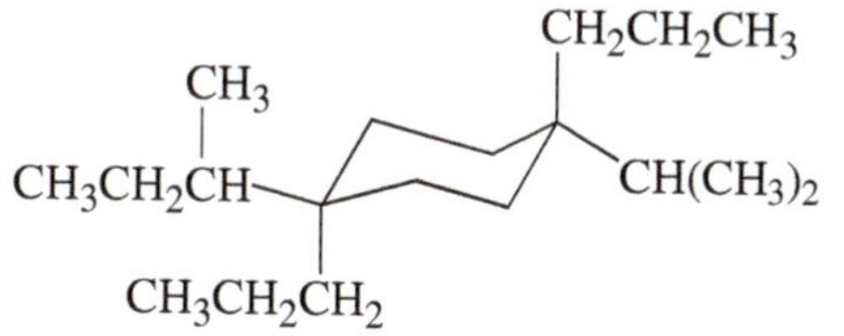

b.

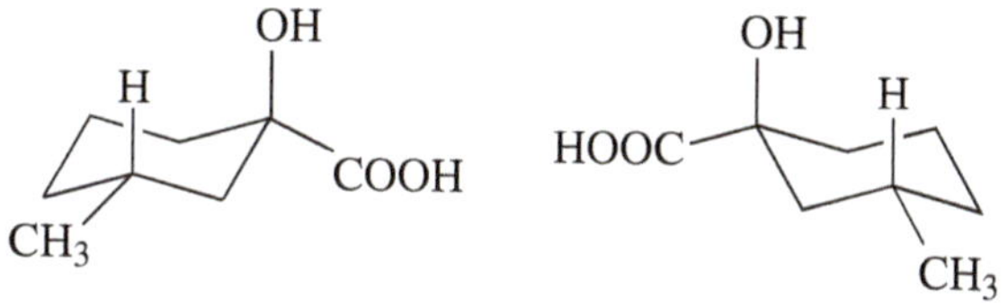

These represent the most stable conformer of each enantiomer.

95. In the transition state for amine inversion, the nitrogen atom is sp^2 hybridized, which means that it has bond angles of 120°. A nitrogen atom in a three-membered ring cannot achieve a 120° bond angle, so the amine inversion that would interconvert the enantiomers cannot occur. Therefore, the enantiomers can be separated.

96. The fact that the optical purity is 72% means that there is 72% enantiomeric excess of the *S* isomer and 28% racemic mixture. Therefore, the actual amount of the *S* isomer in the sample is $72\% + 1/2(28\%) = 86\%$. The amount of the *R* isomer in the sample is $1/2(28\%) = 14\%$ (or $100\% - 86\% = 14\%$).

97. **A** = a diastereomer **C** = a diastereomer **E** = a diastereomer
B = a diastereomer **D** = a diastereomer

98. **a.**

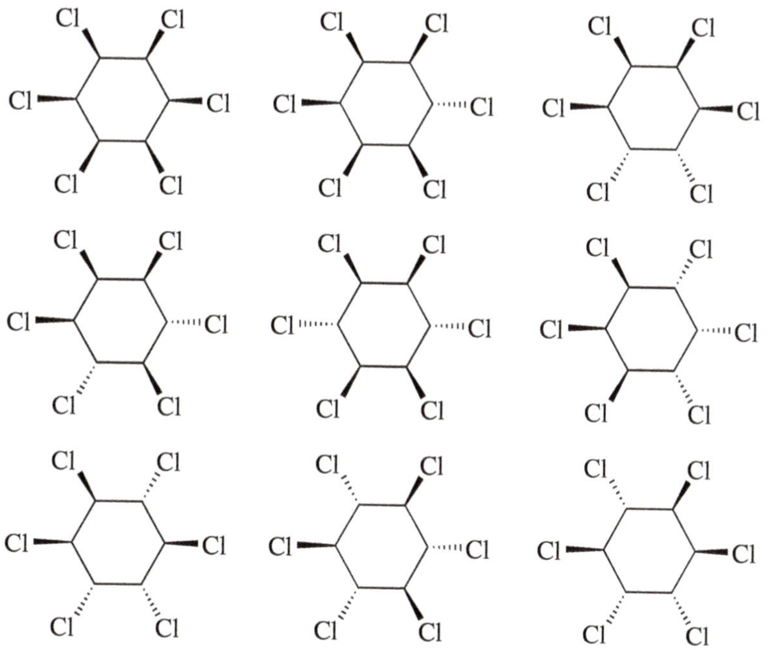

b.

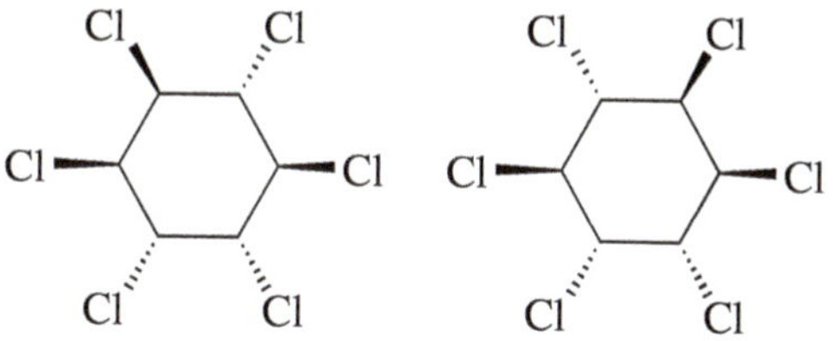

This is a pair of enantiomers because they are nonsuperimposable mirror images.

c.

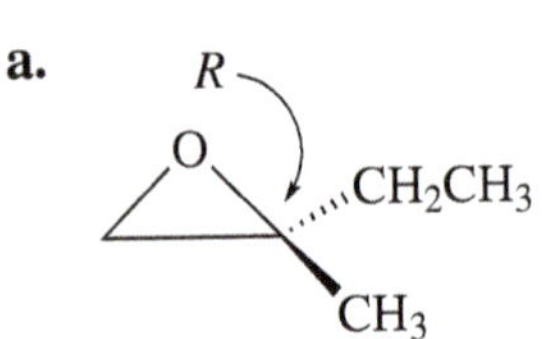

=

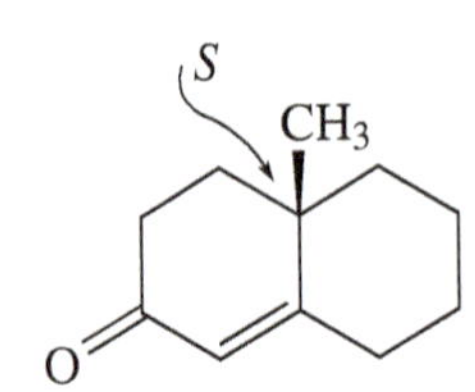

This is the most stable stereoisomer. Because the chloro substituents are all trans to each other, they can all be in the more stable equatorial position. (Recall that there is less steric strain when a substituent is in the equatorial position.)

99. **a.** **b.** **c.**

100.

101. Yes, as long as the Fischer projection is drawn with the #1 carbon at the top of the chain of carbons.

102. The trans compound exists as a pair of enantiomers.

As a result of ring flip, each enantiomer has two chair conformers. In each case, the more stable conformer is the one with the larger group (the *tert*-butyl group) in the equatorial position.

more stable

more stable

103. **a.** The compounds do not have any asymmetric centers.

b. **1.** It is not chiral.

2. It is chiral. Because of its unusual geometry, it is a chiral molecule, even though it does not have any asymmetric centers, because it cannot be superimposed on its mirror image. This will be easier to understand if you build models.

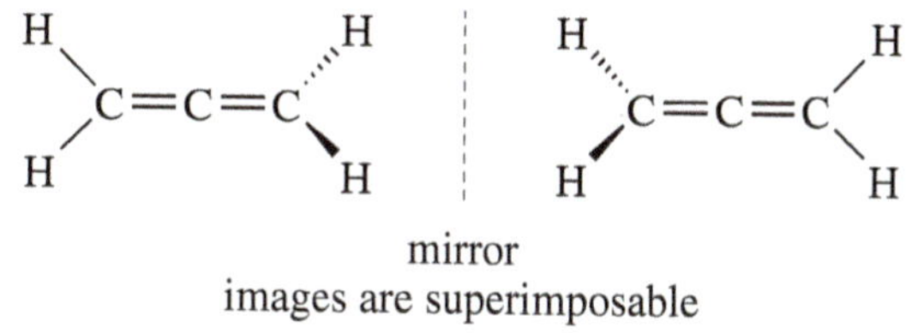

mirror

images are superimposable

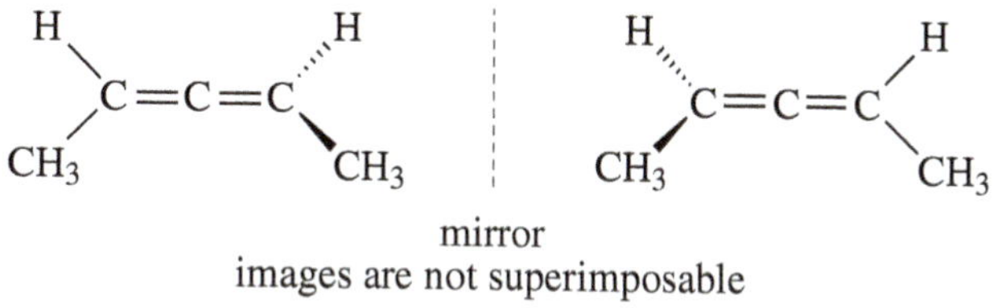

mirror

images are not superimposable

104. The compound is not optically active because it has a point of symmetry.

A point of symmetry is a point, and if a line is drawn to this point from an atom or a group and then extended an equal distance beyond the point, the line will touch an identical atom or group.

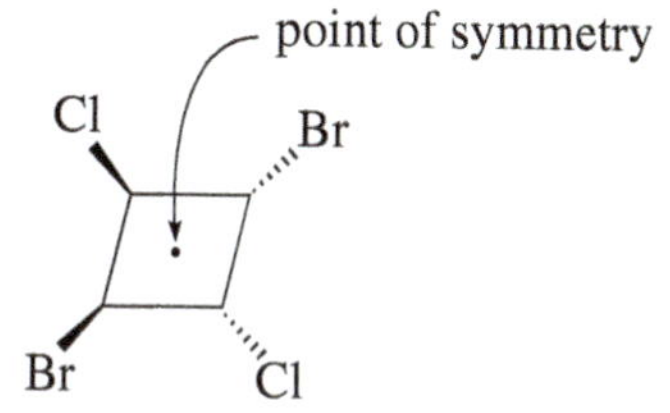

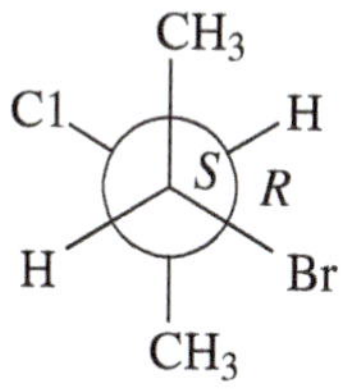

point of symmetry

105.

a. *R*

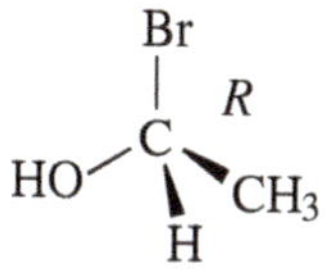

b. *R*

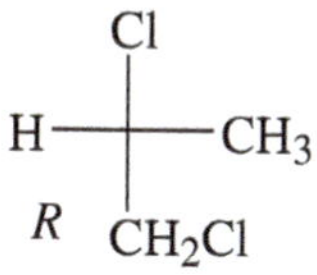

c. *R*

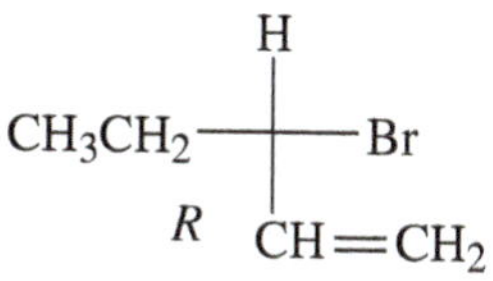

d. *S*, *R*, *R*

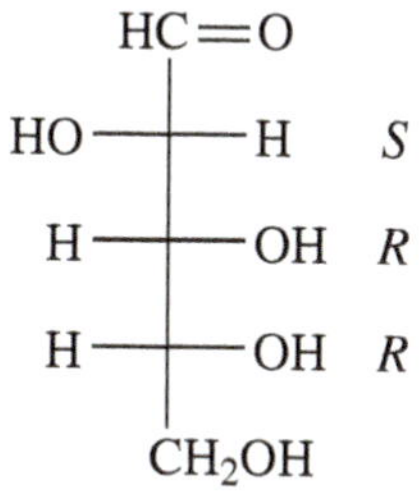

e. *R*, *S*

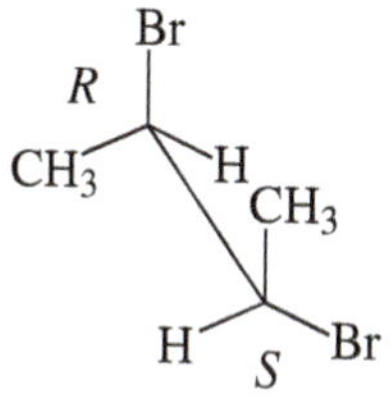

f. *S*, *R*

Chapter 4 Practice Test

1. Are the following pairs of compounds identical or a pair of enantiomers?

a. Fischer projection: H (top), CH_3CH_2 (left), CH_2OH (right), CH_3 (bottom) and Fischer projection: CH_3 (top), H (left), CH_2CH_3 (right), CH_2OH (bottom)

b. C with CH_2CH_3, CH_3, H, Cl and C with CH_3, Cl, H, CH_2CH_3

2. Label the following substituents in order from highest priority to lowest priority in the *E*,*Z* system of nomenclature.

$-\overset{O}{\overset{\|}{C}}CH_3$ $\quad -CH{=}CH_2 \quad -Cl \quad -C{\equiv}N$

3. 100 mL of a solution containing 0.80 g of a compound rotates the plane of polarization of plane-polarized light −4.8° in a polarimeter with a 2 dm sample tube. What is the specific rotation of the compound?

4. Which are meso compounds?

Fischer projections:
- CH_3 (top); H, Cl; H, Cl; CH_2CH_3 (bottom)
- CH_3 (top); H, Cl; Cl, H; CH_2CH_3 (bottom)
- CH_3 (top); Br, Cl; Br, Cl; CH_2Cl (bottom)
- CH_3 (top); H, Cl; H, Cl; CH_3 (bottom)
- CH_3 (top); Cl, H; H, Cl; CH_3 (bottom)

5. Draw all the constitutional isomers with molecular formula C_4H_9Cl.

6. Do the following compounds have the *E* or the *Z* configuration?

a. C=C with $CH_3\overset{CH_3}{CH}CH_2$ and $CH_3\underset{CH_3}{CH}$ on one carbon; $CH_2CH_2CH_2Br$ and CH_2OH on the other

b. C=C with $CH_3\overset{Cl}{CH}$ and $CH_3\underset{CH_3}{CH}$ on one carbon; $\overset{O}{\overset{\|}{C}}CH_3$ and CH_2OH on the other

7. Draw all the possible stereoisomers for each compound that has them.

a. cyclopentane with HO and OH

b. $CH_3\underset{Br}{CH}\underset{Br}{CH}CH_2CH_3$

c. $CH_3CH_2\underset{Cl}{CH}CH_2CH_3$

d. $CH_3CH_2\underset{Cl}{CH}CH_2CH_2Cl$

e. cyclohexane with Br and OH

f. HO–cyclohexane–CH_3

8. Which of the following three perspective formulas are the same as the Fischer projection shown here?

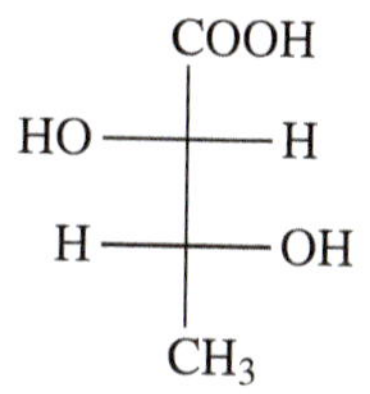

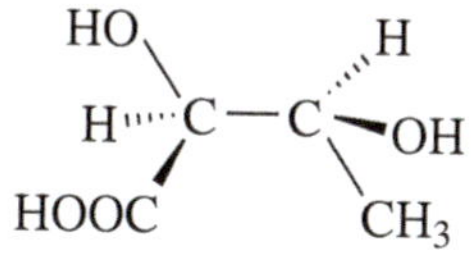

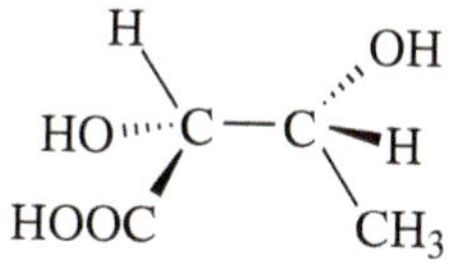

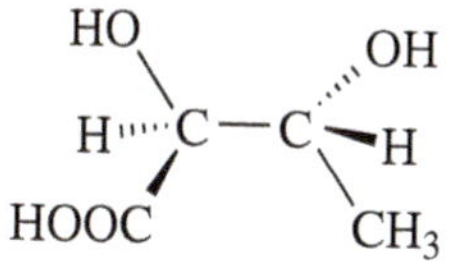

COOH
HO — H
H — OH
CH_3

HO, H, H····C—C, OH, HOOC, CH_3

H, OH, HO····C—C, H, HOOC, CH_3

HO, OH, H····C—C, H, HOOC, CH_3

9. (*R*)-(−)-2-Methyl-1-butanol can be oxidized to (+)-2-methylbutanoic acid without breaking any of the bonds to the asymmetric center. What is the configuration of (−)-2-methylbutanoic acid?

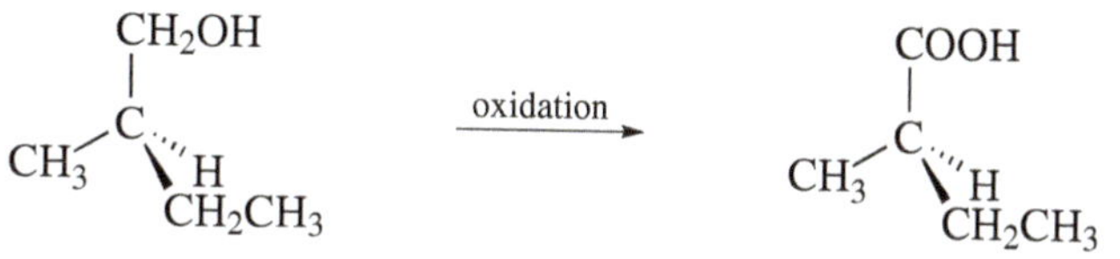

CH_2OH, CH_3, C, H, CH_2CH_3 —oxidation→ COOH, CH_3, C, H, CH_2CH_3

(*R*)-(−)-2-methyl-1-butanol (+)-2-methylbutanoic acid

10. (−)-Cholesterol has a specific rotation of −32. What would be the observed specific rotation of a solution that contains 25% (+)-cholesterol and 75% (−)-cholesterol?

11. Draw and label the *E* and *Z* stereoisomers of:

a. 1-bromo-2,3-dimethyl-2-pentene **b.** 2,3,4-trimethyl-3-hexene

12. Which of the following have the *R* configuration?

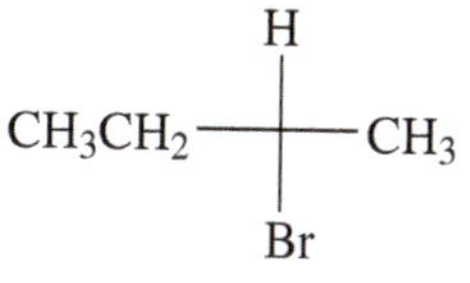

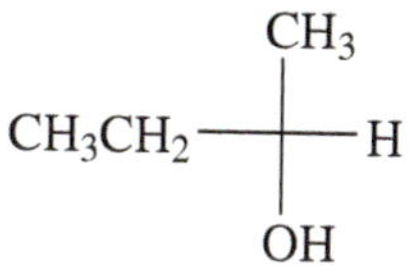

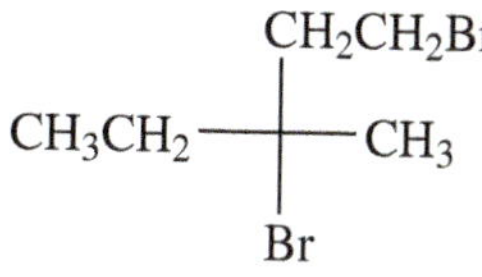

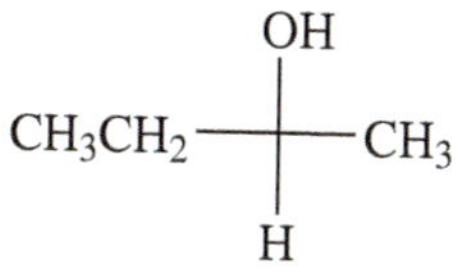

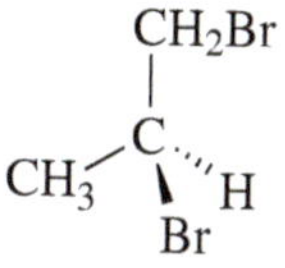

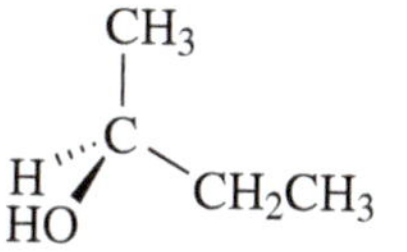

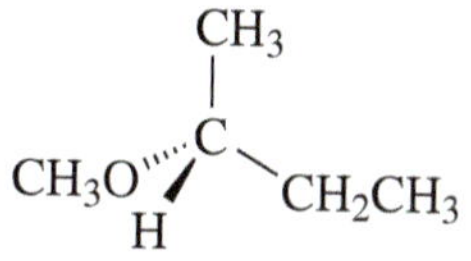

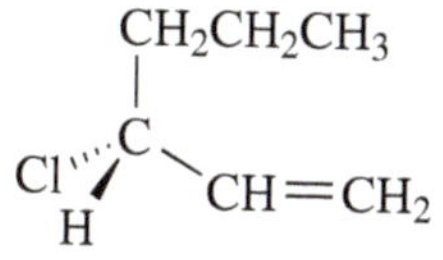

H; CH_3CH_2 — CH_3; Br

CH_3; CH_3CH_2 — H; OH

CH_2CH_2Br; CH_3CH_2 — CH_3; Br

OH; CH_3CH_2 — CH_3; H

CH_2Br, CH_3, C, H, Br

CH_3, H, C, HO, CH_2CH_3

CH_3, CH_3O, C, H, CH_2CH_3

$CH_2CH_2CH_3$, Cl, C, H, $CH{=}CH_2$

13. Answer the following:

a. Are the following compounds identical or a pair of enantiomers?

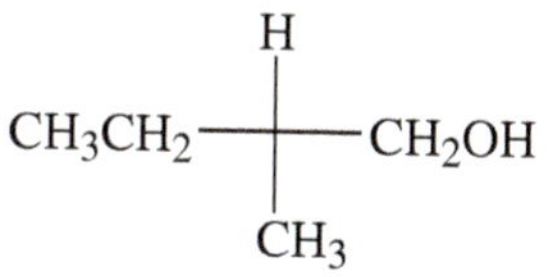

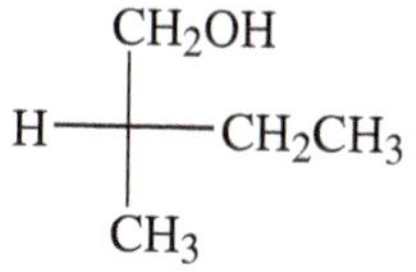

H; CH_3CH_2 — CH_2OH; CH_3

CH_2OH; H — CH_2CH_3; CH_3

b. Put the remaining groups on the structure so it represents (*R*)-2-butanol.

OH
|
C
H

c. Put the remaining groups on the Fischer projection so it represents the Newman projection shown.

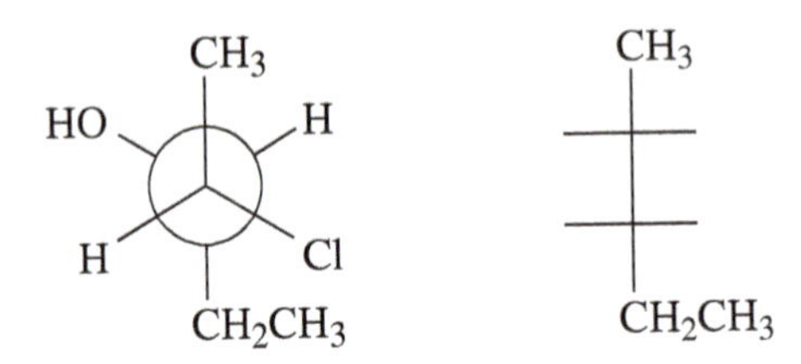

d. Draw a diastereomer for each of the following:

1. Cl Cl

2. CH_3; H—C—OH; H—C—Br; CH_3

14. Indicate whether each of the following statements is true or false:

a.	Diastereomers have the same melting points.	T	F
b.	3-Chloro-2,3-dimethylpentane has two asymmetric centers.	T	F
c.	Meso compounds do not rotate the plane of polarization of plane-polarized light.	T	F
d.	2,3-Dichloropentane has a stereoisomer that is a meso compound.	T	F
e.	All chiral compounds with the *R* configuration are dextrorotatory.	T	F
f.	A compound with three asymmetric centers can have a maximum of nine stereoisomers.	T	F

15. Which of the following have cis–trans isomers?

$CH_3CH_2CH_2CH{=}CH_2$ $\quad$ $CH_3CH_2\underset{\mid \atop CH_3}{C}HCH{=}CHCH_3$ $\quad$ $CH_3CH_2CH{=}CH\underset{\mid \atop Br}{C}HCH_3$ $\quad$ $CH_3CH_2CH_2CH{=}\underset{\mid \atop CH_3}{C}CH_3$

CHAPTER 5
Alkenes: Structure, Nomenclature, and an Introduction to Reactivity • Thermodynamics and Kinetics

Important Terms

active site — the pocket of an enzyme where all the bond-making and bond-breaking steps of an enzyme-catalyzed reaction occur.

acyclic — noncyclic.

addition reaction — a reaction in which atoms or groups are added to the reactant.

alkene — a hydrocarbon that contains a double bond.

allyl group — $CH_2{=}CHCH_2{-}$

allylic carbon — an sp^3 carbon adjacent to a vinyl carbon.

allylic hydrogen — a hydrogen bonded to an allylic carbon.

Arrhenius equation — an equation that relates the rate constant of a reaction to the energy of activation and the temperature at which the reaction is carried out ($k = Ae^{-E_a/RT}$).

catalyst — a species that increases the rate at which a reaction occurs without being consumed or changed in the reaction.

catalytic hydrogenation — the addition of hydrogen to a double or a triple bond with the aid of a metal catalyst.

coupled reactions — an endergonic reaction followed by an exergonic reaction.

degree of unsaturation — the sum of the number of π bonds and rings in a hydrocarbon.

electrophile — an electron-deficient atom or molecule.

electrophilic addition reaction — an addition reaction in which the first species that adds to the reactant is an electrophile.

endergonic reaction — a reaction with a positive $\Delta G°$; it consumes more energy than it releases.

endothermic reaction — a reaction with a positive $\Delta H°$.

enthalpy — the heat given off (if $\Delta H° < 0$) or the heat absorbed (if $\Delta H° > 0$) during the course of a reaction.

entropy — a measure of the freedom of motion in a system.

enzyme — a protein that is a biological catalyst.

exergonic reaction — a reaction with a negative $\Delta G°$; it releases more energy than it consumes.

exothermic reaction	a reaction with a negative $\Delta H°$.
experimental energy of activation ($E_a = \Delta H^{\ddagger} + RT$)	a measure of the approximate energy barrier to a reaction. (It is approximate because it does not contain an entropy component.)
free energy of activation ($\Delta G^{\ddagger}$)	the energy barrier to a reaction.
functional group	the center of reactivity of a molecule.
Gibbs free-energy change ($\Delta G°$)	the difference between the free energy of the products and the free energy of the reactants at equilibrium under standard conditions (1M, 25 °C, 1 atm).
heat of hydrogenation	the heat ($\Delta H°$) released in a hydrogenation reaction.
hydrogenation	addition of hydrogen.
intermediate	a species formed during a reaction that is not the final product of the reaction.
kinetics	the field of chemistry that deals with the rates of chemical reactions.
kinetic stability	indicated by $\Delta G^{\ddagger}$. If $\Delta G^{\ddagger}$ is large, the compound is kinetically stable (is not very reactive). If $\Delta G^{\ddagger}$ is small, the compound is kinetically unstable (is very reactive).
Le Châtelier's principle	a principle that states that if an equilibrium is disturbed, the components of the equilibrium will adjust to offset the disturbance.
mechanism of the reaction	a description of the step-by-step process by which reactants are changed into products.
metabolic pathway	a series of reactions that convert complex nutrient molecules to simple molecules.
molecular recognition	the ability of one molecule to recognize another as a result of intermolecular interactions.
nucleophile	an electron-rich atom or molecule.
pheromone	a chemical substance used for the purpose of communication.
rate constant	the proportionality constant in the rate law.
rate of a reaction	the speed at which the reactants are converted to products.
rate-determining step or rate-limiting step	the step in a reaction that has the transition state with the highest energy.

rate law	the relationship between the rate of a reaction and the concentration of the reactants.
reaction coordinate diagram	a diagram that describes the energy changes that take place during the course of a reaction.
saturated hydrocarbon	a hydrocarbon that is completely saturated with hydrogen (contains no double or triple bonds).
solvation	the interaction between a solvent and another molecule (or ion).
steric strain	the repulsion between the electron cloud of an atom or a group of atoms and the electron cloud of another atom or group of atoms.
substrate	the reactant of an enzyme-catalyzed reaction.
thermodynamic stability	thermodynamic stability is indicated by ΔG°. If ΔG° is negative, the product is thermodynamically stable compared to the reactant. If ΔG° is positive, the reactant is thermodynamically stable compared to the product.
thermodynamics	the field of chemistry that describes the properties of a system at equilibrium.
transition state	the energy maximum of a reaction step on a reaction coordinate diagram. In the transition state, bonds in the reactant that will break are partially broken and bonds in the product that will form are partially formed.
unsaturated hydrocarbon	a hydrocarbon that contains one or more double or triple bonds.
vinyl group	$CH_2{=}CH{-}$
vinylic carbon	a carbon that is doubly bonded to another carbon.
vinylic hydrogen	a hydrogen bonded to a vinylic carbon.

Solutions to Problems

1. Solved in the text.

2. **a.** C_4H_6 **b.** $C_{10}H_{16}$

3. Solved in the text.

4. **a.** 4 **b.** 1 **c.** 3 **d.** 13

5. **a.** degree of unsaturation = 1

$CH_3CH{=}CH_2$

b. degree of unsaturation = 2

$CH_3C{\equiv}CH$

$CH_2{=}C{=}CH_2$

c. degree of unsaturation = 2

$HC{\equiv}CCH_2CH_3$

$CH_3C{\equiv}CCH_3$

$CH_2{=}CHCH{=}CH_2$

$CH_2{=}C{=}CHCH_3$

6. A hydrocarbon with no rings and no double bonds would have a molecular formula of $C_{40}H_{82}$. $C_{40}H_{56}$ has 26 fewer hydrogens. Therefore, β-carotene has a total of 13 rings and double bonds. Because we know it has two rings, it has 11 double bonds.

7.
a. 4-methyl-2-pentene
b. 2-chloro-3,4-dimethyl-3-hexene
c. 1-bromocyclopentene
d. 1-bromo-4-methyl-3-hexene
e. 1,5-dimethylcyclohexene
f. 1-butoxy-2-butene
g. (1*E*,3*E*)-1-bromo-2-methyl-1,3-pentadiene
h. 8,8-dimethyl-1-nonene

8.
a. It has two vinylic hydrogens.
b. It has four allylic hydrogens.

9. **a.**

CH_3

CH_3

b.

CH_3

$CH_3C{=}CCH_2CH_2CH_2Br$

CH_3

c. $CH_3CH_2OCH{=}CH_2$

d. $CH_2{=}CHCH_2OH$

10. Solved in the text.

11. **a.** 4 **b.** 4 **c.** 6

12. **a.** (*E*)-2-methyl-2,4-hexadiene 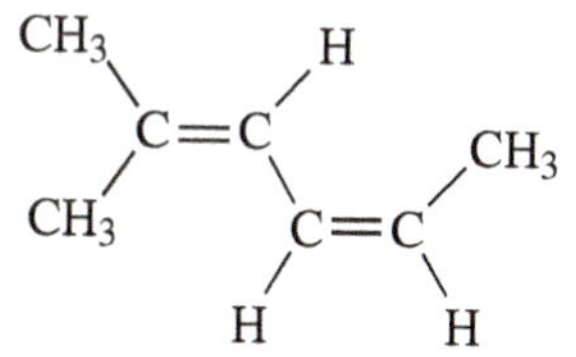 (*Z*)-2-methyl-2,4-hexadiene

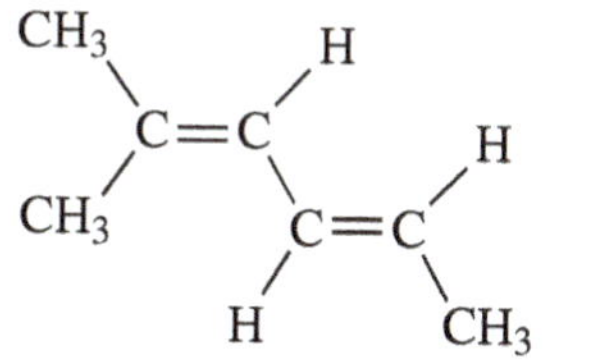

b. (2*E*,4*E*)-2,4-heptadiene 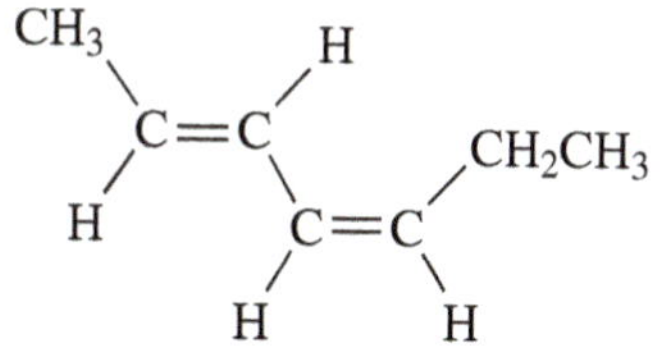 (2*E*,4*Z*)-2,4-heptadiene

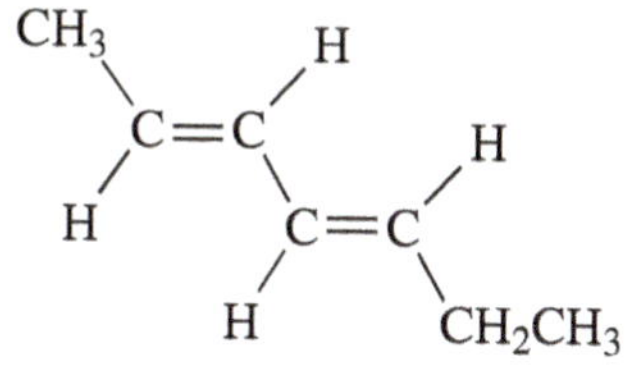

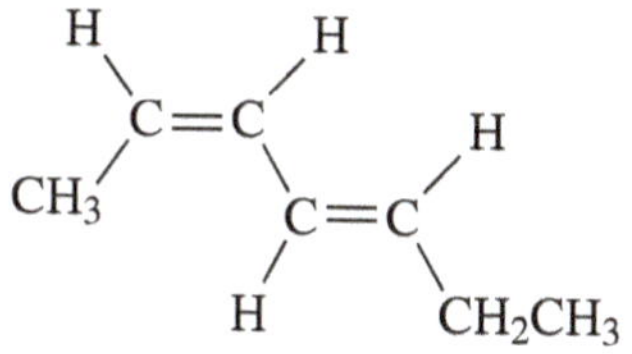

(2*Z*,4*E*)-2,4-heptadiene 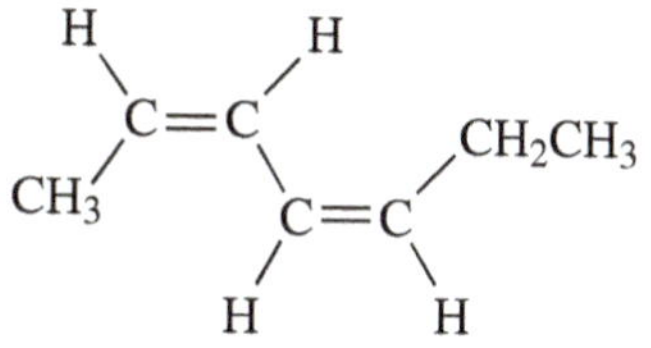 (2*Z*,4*Z*)-2,4-heptadiene

c. (*E*)-1,3-pentadiene 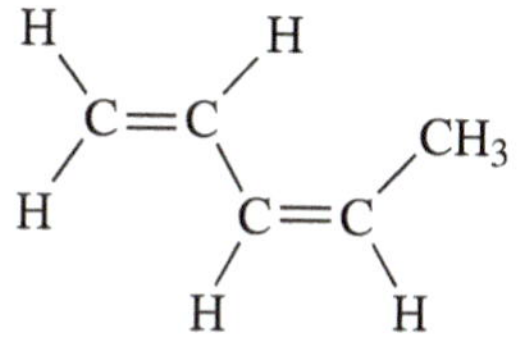 (*Z*)-1,3-pentadiene

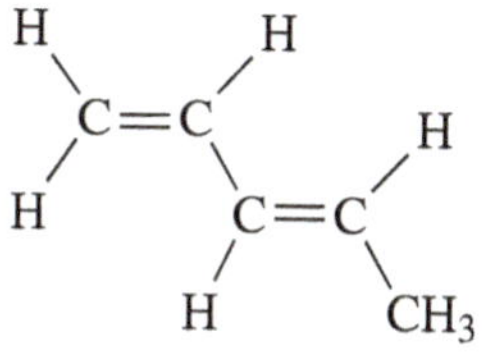

b has four stereoisomers because each double bond can have either the *E* or the *Z* configuration.

a and **c** have only two stereoisomers because, in each case, there are two identical substituents bonded to one of the sp^2 carbons, so only one of the double bonds can have either the *E* or the *Z* configuration.

13. nucleophiles: H^- CH_3O^- $CH_3C{\equiv}CH$ NH_3

electrophiles: $CH_3\overset{+}{C}HCH_3$

14. **a.** $AlCl_3$ is the electrophile, and NH_3 is the nucleophile.
b. The $H^{\delta+}$ of HBr is the electrophile, and HO^- is the nucleophile.

15. **a.** $CH_3C(=O)OH$ + H_3O^+ ⟶ $CH_3C(=\overset{+}{O}H)OH$ + H_2O

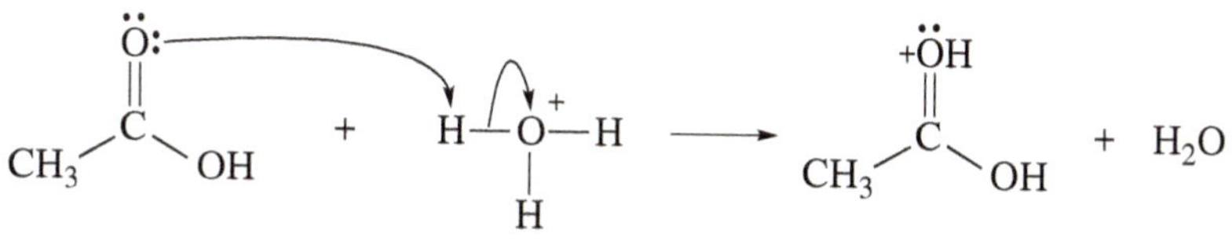

nucleophile electrophile

b. cyclohexene + Br^+ $\longrightarrow$ bromocyclohexyl cation (Br, +)

nucleophile electrophile

c. $CH_3C(=O)O{-}H$ + $H\ddot{O}:^-$ $\longrightarrow$ $CH_3C(=O)O^-$ + H_2O

electrophile nucleophile

d. $CH_3{-}\overset{CH_3}{\underset{CH_3}{C^+}}$ + $:\ddot{Cl}:^-$ $\longrightarrow$ $CH_3{-}\overset{CH_3}{\underset{CH_3}{C}}{-}Cl$

electrophile nucleophile

16. The labels are under the structures in Problem 15.

17. **1.**

$CH_3{-}\overset{:\ddot{O}:^-}{\underset{CH_3}{C}}{-}Br \longrightarrow CH_3{-}\overset{:\ddot{O}:^{2-}}{\underset{CH_3}{C}} + \overset{+}{\ddot{Br}}:$

Drawing the arrows incorrectly leads to a bromine with an incomplete octet and a positive charge as well as an oxygen with 10 valence electrons and 2– charge.

$CH_3{-}\overset{+}{\ddot{O}}(H){-}H \longrightarrow CH_3{-}\overset{2+}{\dot{O}}{-}H + H:^-$

Drawing the arrows incorrectly leads to an oxygen with an incomplete octet and a 2+ charge.

2. This one cannot be drawn because the arrow is supposed to show where the electrons move to, but there are no electrons on the H to go anywhere.

3. $CH_3\overset{:\ddot{O}}{\overset{\|}{C}}OCH_3$ + $H\ddot{O}:^-$ $\longrightarrow$?

The product cannot be drawn because the destination of the electrons in the breaking π bond is not clear.

4. This one cannot be drawn because the arrow is supposed to show where the electrons move to, but there are no electrons on the C to go anywhere.

18. **a.** Because the equilibrium constants for all the monosubstituted cyclohexanes in Table 3.9 on page 128 of the text are greater than 1, all of the equilibria have negative ΔG° values.

(Recall that $\Delta G^\circ = -RT \ln K_{eq}$.)

b. *tert*-butylcyclohexane

c. *tert*-butylcyclohexane, because it is the largest substituent

d.
$$\Delta G^\circ = -RT \ln K_{eq}$$
$$K_{eq} = 18$$
$$\Delta G^\circ = -1.986 \times 10^{-3} \text{ kcal/mol K} \times 298 \text{ K} \times \ln 18 \text{ (recall that } T = {}^\circ\text{C} + 273)$$
$$\Delta G^\circ = -0.59 \times \ln 18 \text{ kcal/mol}$$
$$\Delta G^\circ = -0.59 \times 2.89 \text{ kcal/mol}$$
$$\Delta G^\circ = -1.7 \text{ kcal/mol}$$

19. Solved in the text.

20. **a.**
$$\Delta G^\circ = -RT \ln K_{eq}$$
$$-2.1 = -1.986 \times 10^{-3} \times 298 \times \ln K_{eq}$$
$$\ln K_{eq} = 3.56$$
$$K_{eq} = 35$$
$$K_{eq} = \frac{[\text{ isopropylcyclohexane }]_{\text{equatorial}}}{[\text{ isopropylcyclohexane }]_{\text{axial}}} = \frac{35}{1}$$

$$\begin{aligned}\text{\% of equatorial isopropylcyclohexane} &= \frac{[\text{isopropylcyclohexane}]_{\text{equatorial}}}{[\text{isopropylcyclohexane}]_{\text{equatorial}} + [\text{isopropylcyclohexane}]_{\text{axial}}} \times 100 \\ &= \frac{35}{35 + 1} \times 100 \\ &= \frac{35}{36} \times 100 \\ &= 97\%\end{aligned}$$

b. Isopropylcyclohexane has a greater percentage of the conformer with the substituent in the equatorial position because the isopropyl substituent is larger than the fluoro substituent. The larger the substituent, the less stable is the conformer in which the substituent is in the axial position because of the 1,3-diaxial interactions.

21. ΔS° is more significant in reactions in which the number of reactant molecules and the number of product molecules are not the same.

a. **1.** $A + B \rightleftharpoons C$
2. $A \rightleftharpoons B + C$

b. Reaction **2.** has a positive ΔS°.
In order to have a positive ΔS°, the products must have greater freedom of motion than the reactants. (In other words, there must be more molecules of products than molecules of reactant.)

22. **a.** **1.**

$$\Delta G^\circ = \Delta H^\circ - T\Delta S^\circ$$

(recall that $T = {}^\circ C + 273$)

$$\Delta G^\circ = -12 - (273 + 30)(.01)$$

$$\Delta G^\circ = -12 - 3 = -15 \text{ kcal/mol}$$

$$\Delta G^\circ = -RT \ln K_{eq}$$

$$\Delta G^\circ = -(1.986 \times 10^{-3})(303) \ln K_{eq}$$

$$-15 = -0.60 \ln K_{eq}$$

$$\ln K_{eq} = 25$$

$$K_{eq} = 7.2 \times 10^{10}$$

2.

$$\Delta G^\circ = \Delta H^\circ - T\Delta S^\circ$$

$$\Delta G^\circ = -12 - (273 + 150)(.01)$$

$$\Delta G^\circ = -12 - 4 = -16 \text{ kcal/mol}$$

$$\Delta G^\circ = -(1.986 \times 10^{-3})(423) \ln K_{eq}$$

$$-16 = -0.84 \ln K_{eq}$$

$$\ln K_{eq} = 19$$

$$K_{eq} = 1.8 \times 10^{8}$$

b. For this reaction: the calculations show that increasing the temperature causes ΔG° to be more negative.

c. For this reaction: the calculations show that increasing the temperature causes K_{eq} to be smaller, because $\ln K_{eq} = -\Delta G^\circ/RT$.

23. The value for the π bond of ethene (62 kcal/mol) is given in the text on page 43.

a.

bonds broken		bonds formed		
π bond of ethene	62	$CH_3CH_2—H$	101	
H—Cl	103	$CH_3CH_2—Cl$	85	
	165 kcal/mol		186 kcal/mol	$\Delta H^\circ = 165 - 186 = -21$ kcal/mol

b.

bonds broken		bonds formed		
π bond of ethene	62	$CH_3CH_2—H$	101	
H—H	104	$CH_3CH_2—H$	101	
	166 kcal/mol		202 kcal/mol	$\Delta H^\circ = 166 - 202 = -36$ kcal/mol

c. Both are exothermic, because they both have a negative ΔH° value.

d. The ΔH° values of both reactions are sufficiently negative to allow you to expect that they will be exergonic as well.

24. **a.** $CH_2{=}CHCH_2CH_2CH_3$ **or** $CH_3CH{=}CHCH_2CH_3$

b. CH_2CH_3 **or** CH_2CH_3 **or** CH_2CH_3

or $CH{=}CH_2$ **or** $CHCH_3$

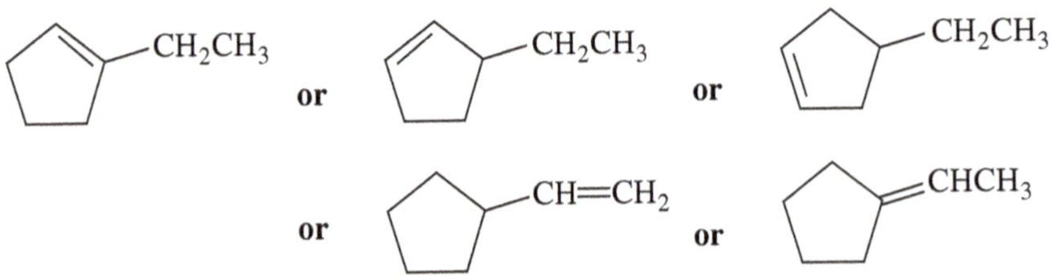

25. **a.** three alkenes: 1-butene, *cis*-2-butene, *trans*-2-butene

b. four alkenes: 3-methyl-1-pentene, (*E*)-3-methyl-2-pentene, (*Z*)-3-methyl-2-pentene, 2-ethyl-1-butene

c. five alkenes: 1-hexene, *cis*-2-hexene, *trans*-2-hexene, *cis*-3-hexene, *trans*-3-hexene

26. Because alkene **A** has the smaller heat of hydrogenation, it is more stable.

27.

a. CH_2CH_3 / CH_2CH_3

This alkene is the most stable because it has the greatest number of alkyl substituents bonded to the sp^2 carbons.

b. CH_2CH_3 / CH_2CH_3

This alkene is the least stable because it has the fewest number of alkyl substituents bonded to the sp^2 carbons.

c. CH_2CH_3 / CH_2CH_3

This alkene has the smallest heat of hydrogenation because it is the most stable of the three alkenes.

28. $(CH_3)(CH_3CH_2)C{=}C(CH_3)(CH_2CH_3)$ > $(H)(CH_3CH_2)C{=}C(CH_2CH_3)(H)$ > $(CH_3CH_2)(H)C{=}C(CH_2CH_3)(H)$ > $(CH_3{-}CH(CH_3))(H)C{=}C(CH(CH_3){-}CH_3)(H)$

4 alkyl substituents; 2 trans alkyl substituents; 2 cis alkyl substituents; 2 cis alkyl substituents that cause greater steric strain than those in *cis*-3-hexene

29.
a. **a** and **b**, because the product is more stable than the reactant.
b. **b** is the most kinetically stable product, because it has the smallest rate constant (greatest $\Delta G^\ddagger$) leading from the product to the transition state.
c. **c** is the least kinetically stable product, because it has the largest rate constant (smallest $\Delta G^\ddagger$) leading from the product to the transition state.

30.
a. A thermodynamically **unstable** product is less stable than the reactant.
A kinetically **unstable** product has a large rate constant (small $\Delta G^\ddagger$) for the reverse reaction.
b. A kinetically **stable** product has a small rate constant (large $\Delta G^\ddagger$) for the reverse reaction.

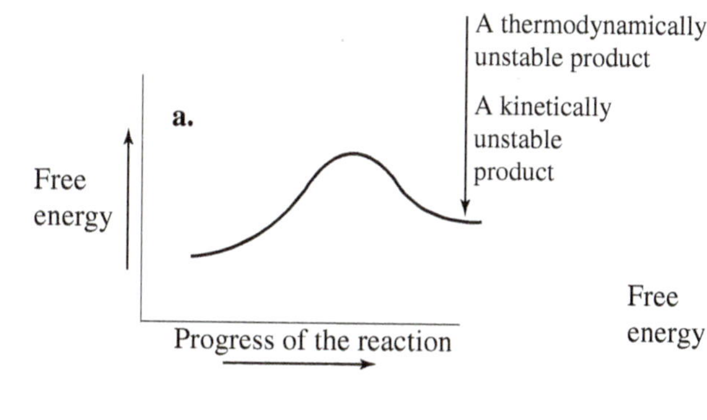

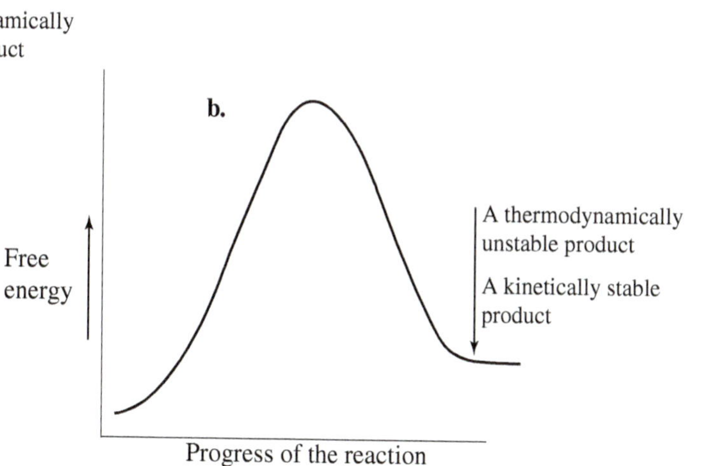

31. **a.** Solved in the text.

b. **1.** Decreasing the concentration of methyl chloride (by a factor of 10) decreases **the rate of the reaction** (by a factor of 10) to 1×10^{-8} M s^{-1}.

2. Decreasing the concentration of methyl chloride has no effect on **the rate constant** for the reaction.

32. The rate constant for a reaction can be increased by **decreasing** the stability of the reactant (increasing its energy) or by **increasing** the stability of the transition state (decreasing its energy).

33. Taking the logarithm of both sides of the Arrhenius equation gives the following equation (where k is the rate constant), which we can use to answer the questions:

$$\ln k = \ln A - \frac{E_a}{RT}$$

a. Increasing the experimental activation energy (E_a) decreases the rate constant of a reaction (causes the reaction to be slower).

b. Increasing the temperature (T) increases the rate constant of a reaction (causes the reaction to go faster).

34. **a.** The first stated reaction has the greater equilibrium constant:

$$K_{eq} = \frac{1 \times 10^{-3}}{1 \times 10^{-5}} = 1 \times 10^{2} \qquad K_{eq} = \frac{1 \times 10^{-2}}{1 \times 10^{-3}} = 10$$

b. Because both reactions start with the same concentration, the first stated reaction will form the most product when the reactions have reached equilibrium, because it has the greater equilibrium constant.

35.

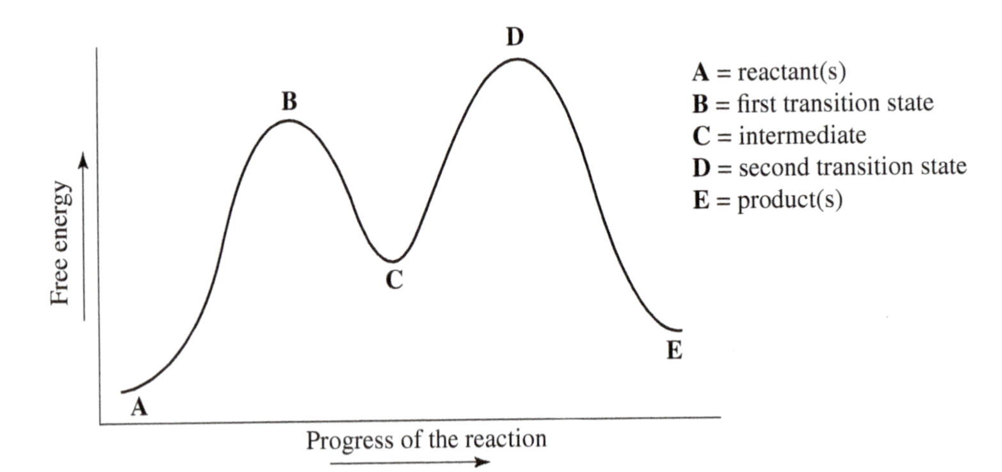

36. **a.** The first step in the forward direction (**A** → **B**) has the greatest free energy of activation.
b. The first-formed intermediate (**B**) is more apt to revert to reactants, because the free energy of activation for **B** to form **A** (the reactants) is less than the free energy of activation for **B** to form **C**.
c. The second step (**B** → **C**) is the rate-determining step because it has the transition state with the highest energy.

Notice that the second step is rate-determining even though the first step has the greater energy of activation (steeper hill to climb). That is because it is easier for the intermediate that is formed in the first step to go back to starting material than to undergo the second step of the reaction. So the second step is the rate-limiting step.

37.

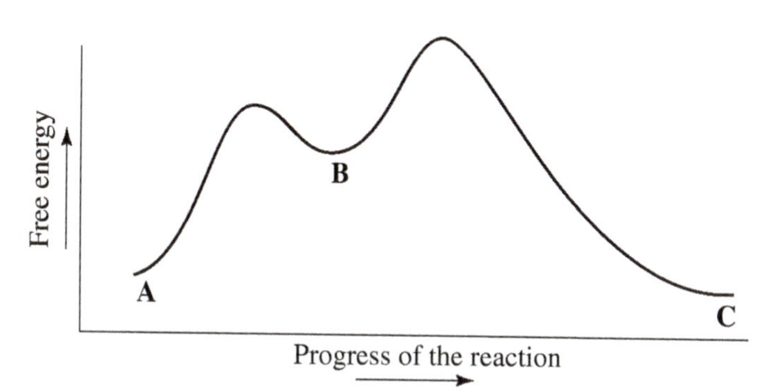

a. one (B)
b. two
c. the second step (k_2), B to form C (In this particular diagram, $k_2 > k_1$: if you had made the transition state for the second step a lot higher, you could have had a diagram in which $k_1 > k_2$.)
d. the second step in the reverse direction (k_{-1})
e. the second step in the reverse direction (k_{-1}), B to form A
f. B to C
g. C to B

38. A catalyst will change the energy difference between the reactants and the transition state, but it will not change the energy difference between the reactants and the products.

$$\Delta H^{\ddagger}, E_a, \Delta S^{\ddagger}, \Delta G^{\ddagger}, k$$

39. **a.** 3,8-dibromo-4-nonene
b. (*Z*)-4-ethyl-3,7-dimethyl-3-octene
c. 1,5-dimethylcyclopentene
d. 3-ethyl-2-methyl-2-heptene
e. 4-methylcyclohexene
f. 4-ethyl-5-methylcyclohexene

40. **a.** $CH_3C(CH_3){=}CHCH_2CH_3$ **b.** [skeletal structure] **c.** [1-methylcyclohexene structure with CH_3]

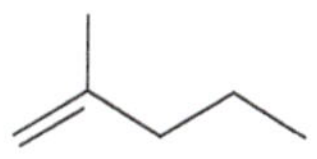

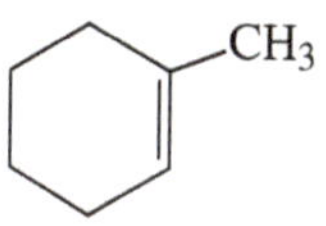

41. a. $CH_2{=}CHCH_2CH_2CH_2CH_3$

b. 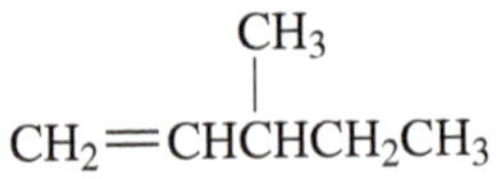 $CH_2{=}CHCH(CH_3)CH_2CH_3$

c. 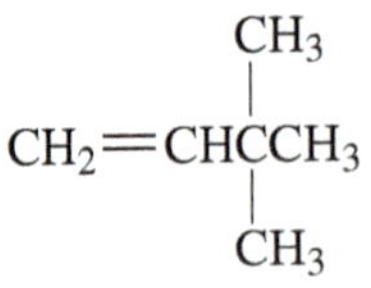 $CH_2{=}CHC(CH_3)_2CH_3$

42. a. 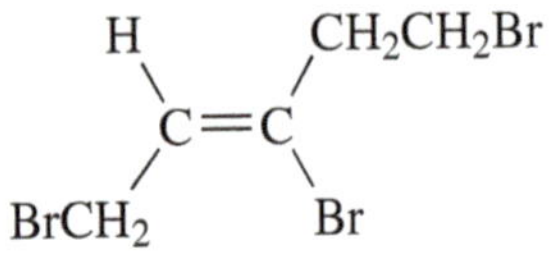 H, CH_2CH_2Br / C=C / $BrCH_2$, Br

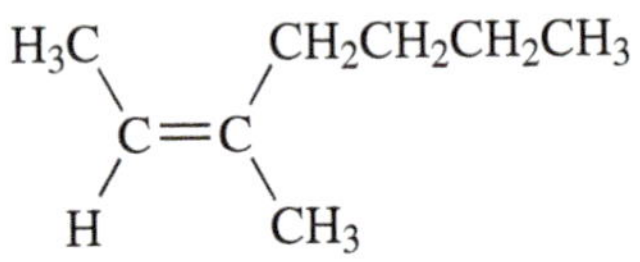

b. H_3C, $CH_2CH_2CH_2CH_3$ / C=C / H, CH_3

c. $BrCH_2$, $CH(CH_3)CH_3$ / C=C / Br, $CH_2CH_2CH_3$

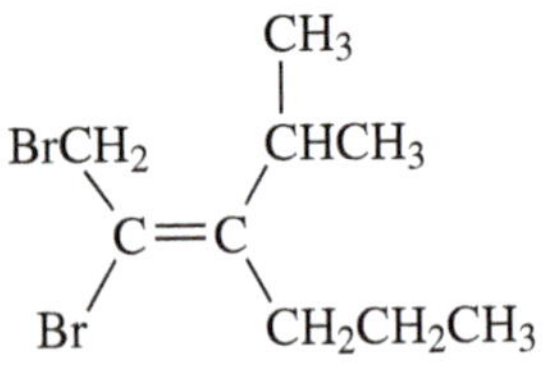

d. $CH_2{=}CHBr$

e. 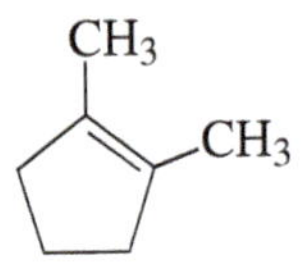 CH_3, CH_3 (1,2-dimethylcyclopentene ring)

f. $CH_2{=}CHCH_2NHCH_2CH{=}CH_2$

43. a. 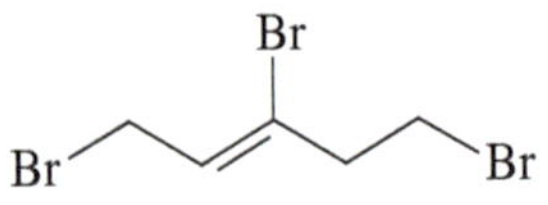 Br Br Br

b.

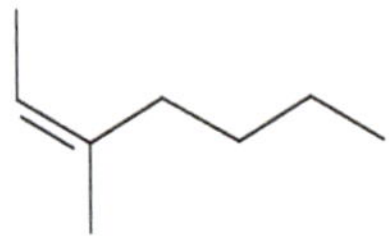

c. 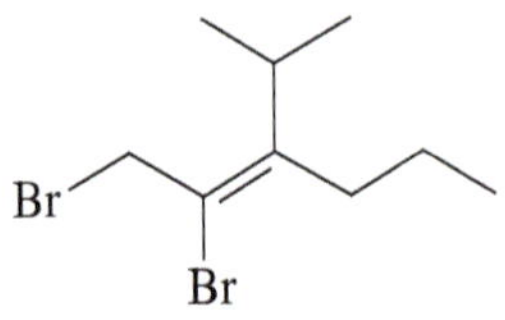 Br Br

d. 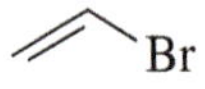 Br

e.

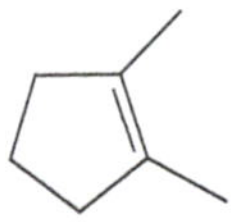

f. 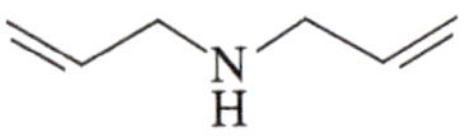 N H

44. **a.**

$CH_2{=}CHCH_2CH_2CH_2CH_3$
1-hexene

$CH_3CH{=}CHCH_2CH_2CH_3$
2-hexene

$CH_3CH_2CH{=}CHCH_2CH_3$
3-hexene

$CH_2{=}CCH_2CH_2CH_3$
CH_3
2-methyl-1-pentene

$CH_2{=}CHCHCH_2CH_3$
CH_3
3-methyl-1-pentene

$CH_2{=}CHCH_2CHCH_3$
CH_3
4-methyl-1-pentene

$CH_3C{=}CHCH_2CH_3$
CH_3
2-methyl-2-pentene

$CH_3CH{=}CCH_2CH_3$
CH_3
3-methyl-2-pentene

$CH_3CH{=}CHCHCH_3$
CH_3
4-methyl-2-pentene

CH_3
$CH_2{=}CCHCH_3$
CH_3
2,3-dimethyl-1-butene

CH_3
$CH_3CCH{=}CH_2$
CH_3
3,3-dimethyl-1-butene

$CH_3CH_2C{=}CH_2$
CH_2CH_3
2-ethyl-1-butene

CH_3
$CH_3C{=}CCH_3$
CH_3
2,3-dimethyl-2-butene

b. Of the compounds shown in part **a**, the following can have *E* and *Z* isomers:
2-hexene, 3-hexene, 3-methyl-2-pentene, 4-methyl-2-pentene

c.

CH_3
$CH_3C{=}CCH_3$
CH_3

This compound is the most stable.
It has four alkyl substituents bonded to the sp^2 carbons.

d.

$CH_2{=}CHCH_2CH_2CH_2CH_3$

$CH_2{=}CHCH_2CHCH_3$
CH_3

$CH_2{=}CHCHCH_2CH_3$
CH_3

CH_3
$CH_3CCH{=}CH_2$
CH_3

These compounds are the least stable.
Each has only one alkyl substituent bonded to the sp^2 carbons.

45. **a.** 2,3-Dimethyl-2-butene is the most stable because it has four alkyl groups attached to its sp^2 carbons.
b. 1-Hexene, 3-methyl-1-pentene, 4-methyl-1-pentene, and 3,3-dimethyl-1-butene are the least stable because they have one alkyl group attached to their sp^2 carbons.

46. **a.** (E)-3-methyl-3-hexene
b. *trans*-8-methyl-4-nonene or (E)-8-methyl-4-nonene
c. *trans*-9-bromo-2-nonene or (E)-9-bromo-2-nonene
d. 2,4-dimethyl-1-pentene
e. 2-ethyl-1-pentene
f. *cis*-2-pentene or (Z)-2-pentene

47. First draw the structures so you can see the number of alkyl groups bonded to the sp^2 carbons:

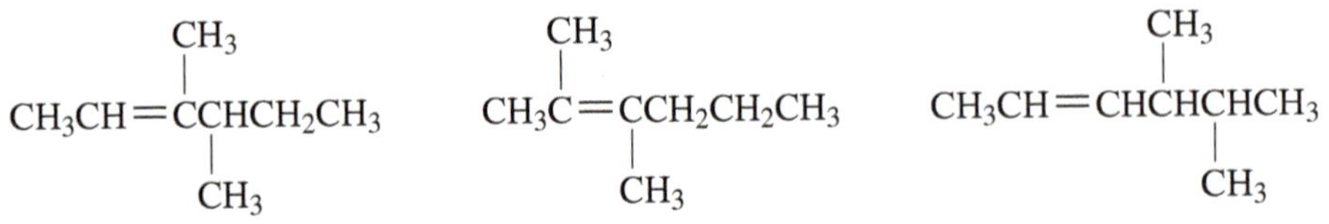

$CH_3CH{=}C(CH_3)CH(CH_3)CH_2CH_3$ 3,4-dimethyl-2-hexene

$CH_3C(CH_3){=}CCH_2CH_2CH_3$ (with CH_3 on C-3) 2,3-dimethyl-2-hexene

$CH_3CH{=}CHCH(CH_3)CH(CH_3)CH_3$ 4,5-dimethyl-2-hexene

a. 2,3-Dimethyl-2-hexene is the most stable of the three alkenes because it has the greatest number of alkyl substituents bonded to the sp^2 carbons.
b. 4,5-Dimethyl-2-hexene has the fewest alkyl substituents bonded to the sp^2 carbons, making it the least stable of the three alkenes. It, therefore, has the greatest heat of hydrogenation.
c. Because it is the most stable, 2,3-dimethyl-2-hexene has the smallest heat of hydrogenation.

48. $H{-}\ddot{O}{:}^- + H{-}CH_2{-}CH_2{-}Br \longrightarrow H_2C{=}CH_2 + H_2O + Br^-$

49.

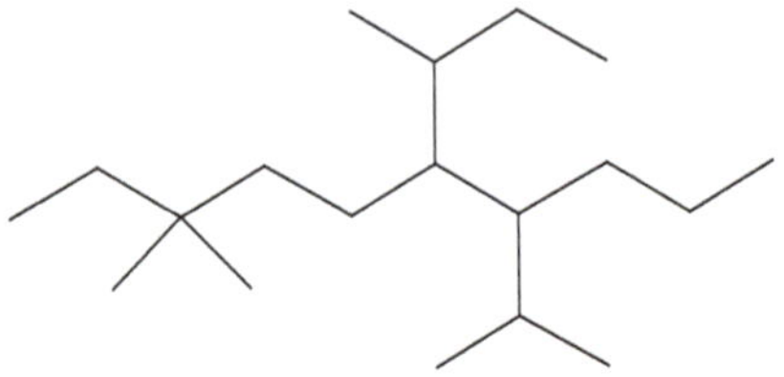

50.
$$\Delta G^\circ = -RT \ln K_{eq}$$
$$\ln K_{eq} = -\Delta G^\circ/RT$$
$$\ln K_{eq} = -\Delta G^\circ/0.59 \text{ kcal/mol}$$

a.
$$\ln K_{eq} = -2.72/0.59 \text{ kcal/mol}$$
$$\ln K_{eq} = -4.6$$
$$K_{eq} = [B]/[A] = 0.01$$

b.
$$\ln K_{eq} = -0.65/0.59 \text{ kcal/mol}$$
$$\ln K_{eq} = -1.10$$
$$K_{eq} = [B]/[A] = 0.33$$

c.
$$\ln K_{eq} = 2.72/0.59 \text{ kcal/mol}$$
$$\ln K_{eq} = 4.6$$
$$K_{eq} = [B]/[A] = 100$$

d.
$$\ln K_{eq} = 0.65/0.59 \text{ kcal/mol}$$
$$\ln K_{eq} = 1.10$$
$$K_{eq} = [B]/[A] = 3.0$$

51. **a.** The C—Cl bond is a stronger bond because Cl uses a $3sp^3$ orbital to overlap the $2sp^3$ orbital of carbon, whereas Br uses a $4sp^3$ orbital. A $4sp^3$ has a greater volume than a $3sp^3$ and, therefore, has less electron density in the region of orbital-orbital overlap, so it forms a weaker bond.
b. The Br—Br bond is a stronger bond because Br uses a $4sp^3$ orbital to overlap the $4sp^3$ orbital of the other bromine, whereas I uses a $5sp^3$ orbital. A $5sp^3$ has a greater volume than a $4sp^3$ and, therefore, has less electron density in the region of orbital–orbital overlap, so it forms a weaker bond.

52. If the number of carbons is 30, $C_nH_{2n+2} = C_{30}H_{62}$. A compound with molecular formula $C_{30}H_{50}$ is missing 12 hydrogens. Because it has no rings, squalene has 6 π bonds $(12/2 = 6)$.

53. **a.** (*E*)-2-methyl-2,4-hexadiene 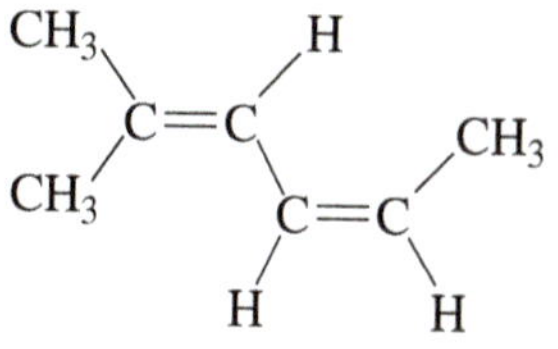   (*Z*)-2-methyl-2,4-hexadiene

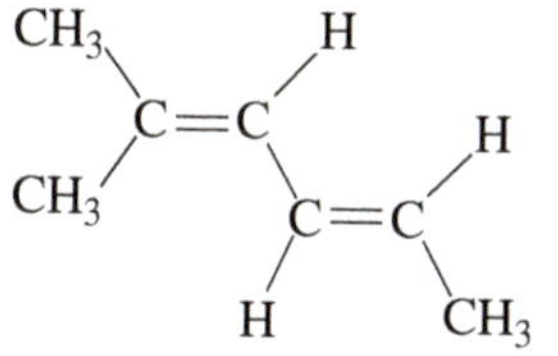

b. (*E*)-1,5-heptadiene 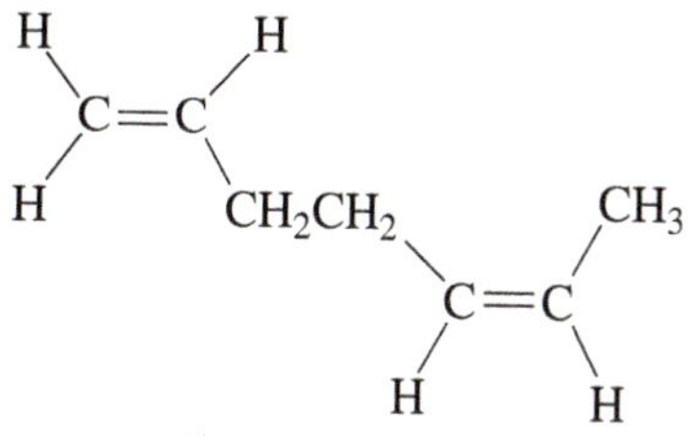   (*Z*)-1,5-heptadiene

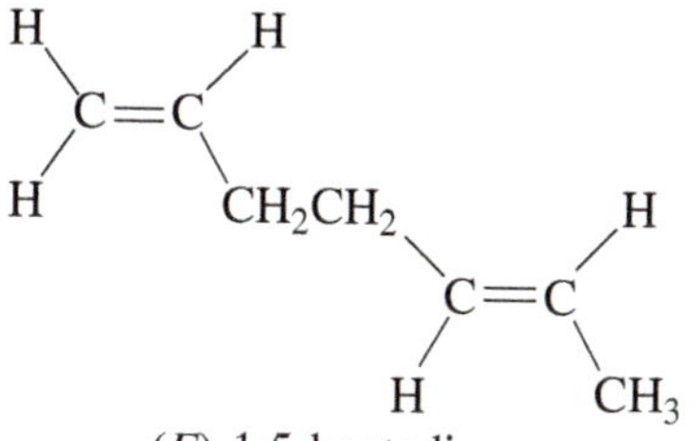

c. 1,4-pentadiene

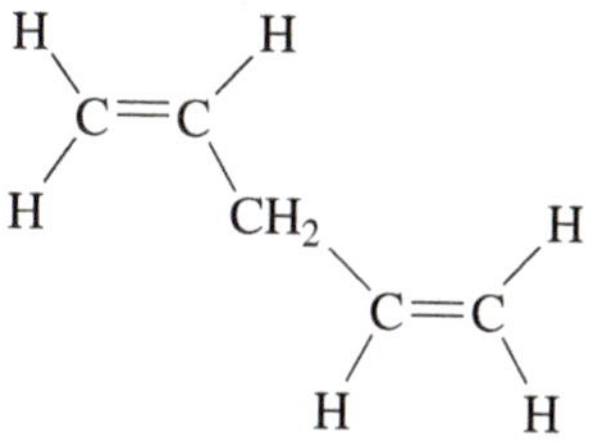

d. (2*E*,4*Z*)-3-methyl-2,4-hexadiene 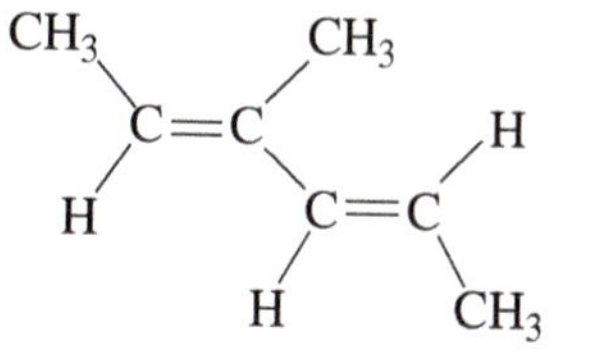   (2*E*,4*E*)-3-methyl-2,4-hexadiene

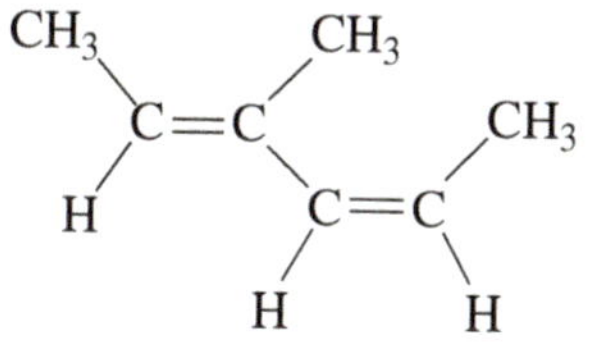

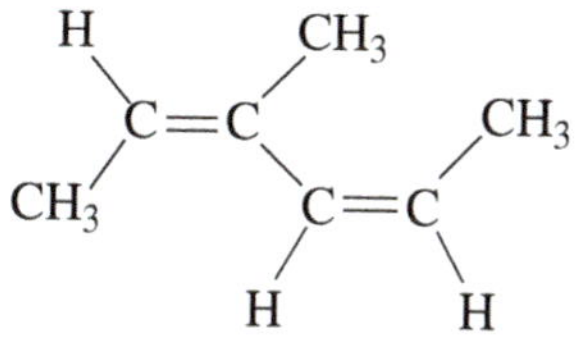

(2*Z*,4*Z*)-3-methyl-2,4-hexadiene 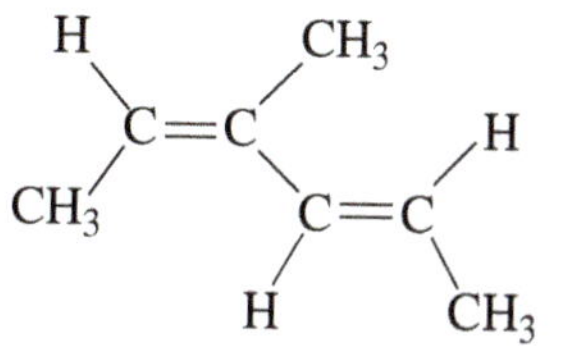   (2*Z*,4*E*)-3-methyl-2,4-hexadiene

54. **a.** $CH_3CH_2\overset{+}{N}H_3 + Br^-$

electrophile: ethyl chloride
nucleophile: ammonia

b. $CH_3-C(O^-)(OH)-CH_3$

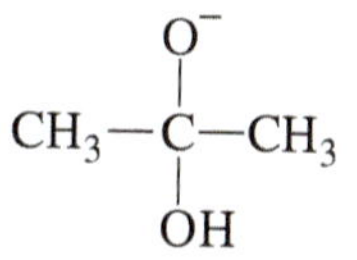

electrophile: the compound with the double-bonded carbon
nucleophile: hydroxide ion

c. $CH_3CH(CH_3)-\overset{+}{C}(CH_3)CH_3 + Cl^-$

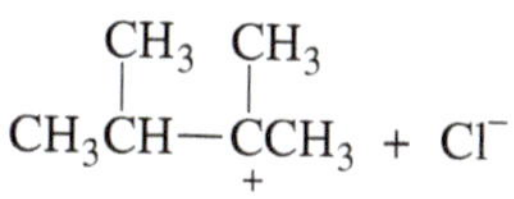

electrophile: the H of HCl
nucleophile: the alkene

55. Only one name is correct.

a. 2-pentene
b. correct
c. 3-methyl-1-hexene (the parent hydrocarbon is the longest chain that contains the functional group)

d. 3-heptene
e. 4-ethylcyclohexene
f. 2-chloro-3-hexene
g. 3-methyl-2-pentene
h. 2-methyl-1-hexene (it does not have *E*,*Z* isomers)
i. 1-methylcyclopentene

56. **a.** 4 alkenes

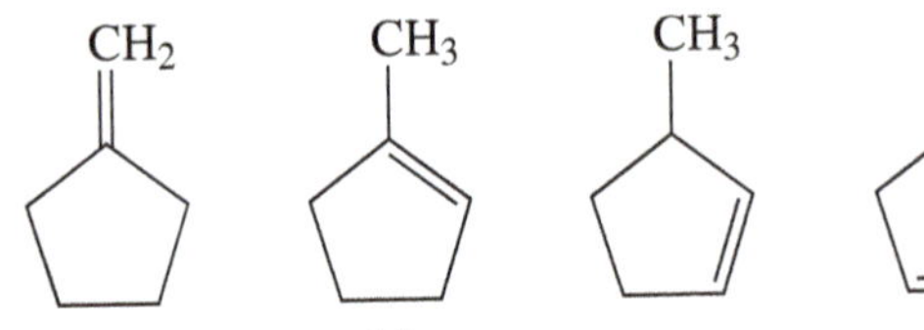

b. 1-Methylcyclopentene is the most stable.
c. Because 1-methylcyclopentene is the most stable, it has the smallest heat of hydrogenation.

57. **a.** $ClCH_2$, CH_3, C=C, H, H, C=C, H, CH_3

b. H, CH_3, C=C, CH_3CH_2, H, C=C, H, $CH_2CH_2CH_3$

c. H, CH_3, C=C, CH_3CH_2, $CH_2CH_2CH_3$, C=C, CH_3, H

d. Br, CH_3CH, H, C=C, H, CH_2CH_3, C=C, Br, H

58. **a.** 2
b. B, D, F
c. E to G (The fastest step has the smallest energy of activation to overcome.)
d. G
e. A
f. C
g. C
h. endergonic
i. exergonic
j. E to G (The largest rate constant corresponds to the smallest energy of activation.)
k. G to E (The smallest rate constant corresponds to the largest energy of activation.)

59. **a.** B will have the larger $\Delta S°$ value because, unlike A, the number of reactants is not the same as the number of products.
b. $\Delta S°$ = (the freedom of motion of the products) − (the freedom of motion of the reactants).
Because the three products have a greater freedom of motion than the two reactants, $\Delta S°$ is positive.

60.

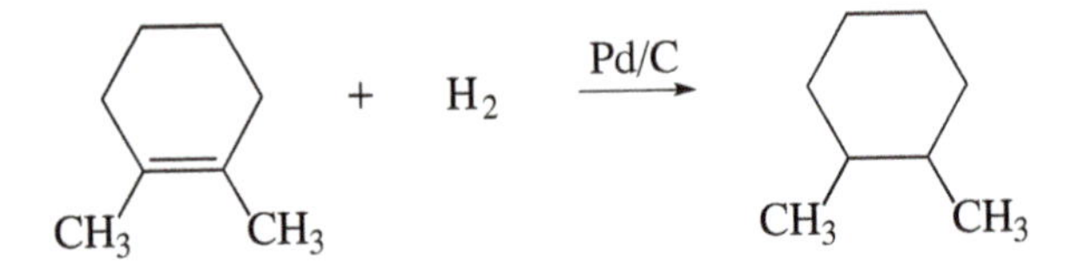

61. $\Delta G^\circ = \Delta H^\circ - T\Delta S^\circ$

a. $\Delta G^\circ = 20 - (298)(0.05)$
$\Delta G^\circ = 20 - 14.9 = 5.1 \text{ kcal/mol}$

$\Delta G^\circ = -RT \ln K_{eq}$
$5.1 = -1.986 \times 10^{-3} \times 298 \times \ln K_{eq}$
$5.1 = -0.59 \ln K_{eq}$
$-8.6 = \ln K_{eq}$
$K_{eq} = 1.8 \times 10^{-4}$

b. $\Delta G^\circ = 20 - (398)(0.05)$
$\Delta G^\circ = 20 - 20 = 0$
$K_{eq} = 1.0$

62.

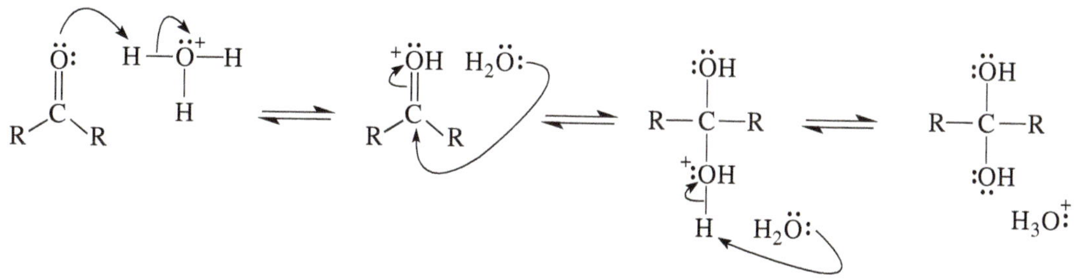

63. **a.** $\Delta G^\circ = -RT \ln K_{eq}$
$= -1.986 \times 10^{-3} \times 298 \times \ln 10^{-3}$
$= -2.72 \text{ kcal/mol}$

$\Delta G^\circ = -1.986 \times 10^{-2} \times 298 \times \ln 10^{-2}$
$= -1.36 \text{ kcal/mol}$

$\Delta\Delta G^\circ = -2.72 - (-1.36) = -1.36 \text{ kcal/mol}$
Thus, ΔG° must change by 1.36 kcal/mol.

b. $\Delta G^\circ = \Delta H^\circ - 0$
$-1.36 = \Delta H^\circ - 0$
$\Delta H^\circ = -1.36 \text{ kcal/mol}$

c. $\Delta G^\circ = 0 - T\Delta S^\circ$
$-1.36 = 0 - 298\Delta S^\circ$
$\Delta S^\circ = 1.36/298 = 4.56 \times 10^{-3} \text{ kcal/(mol deg)}$

64. $\Delta G^\circ = -RT \ln K_{eq}$
$\ln K_{eq} = -\Delta G^\circ/RT$
$\ln K_{eq} = -\Delta G^\circ/0.59 \text{ kcal/mol}$
$\ln K_{eq} = -5.3/0.59 \text{ kcal/mol}$
$\ln K_{eq} = -9.0$
$K_{eq} = [B]/[A] = 0.00013$
$[B]/[A] = 0.00013/1 = 0.13/1000$

The calculation shows that for every 1000 molecules in the chair conformation, there is 0.13 molecule in a twist-boat conformation.

This agrees with the statement on page 127 of the text that for every 10,000 chair conformers of cyclohexane, there is no more than one twist-boat conformer.

65. A step-by-step description of how to solve this problem is given in the box entitled "Calculating Kinetic Parameters" on page 224 of the text.

E_a can be determined from the Arrhenius equation ($\ln k = -E_a/RT$), because a plot of $\ln k$ versus $1/T$ gives a slope $= -E_a/R$.

$\ln 2.11 \times 10^{-5} = -10.77$	$T = 304$	$1/T = 3.29 \times 10^{-3}$
$\ln 4.44 \times 10^{-5} = -10.02$	$T = 313$	$1/T = 3.19 \times 10^{-3}$
$\ln 1.16 \times 10^{-4} = -9.06$	$T = 324.5$	$1/T = 3.08 \times 10^{-3}$
$\ln 2.10 \times 10^{-4} = -8.47$	$T = 332.8$	$1/T = 3.00 \times 10^{-3}$
$\ln 4.34 \times 10^{-4} = -7.74$	$T = 342.2$	$1/T = 2.92 \times 10^{-3}$

$$\text{slope} = -8290$$
$$E_a = -(\text{slope})\,R$$
$$E_a = -(-8290) \times 1.98 \times 10^{-3}\ \text{kcal/mol}$$
$$E_a = 16.4\ \text{kcal/mol}$$

To find $\Delta G°$:

$-\Delta G° = RT \ln kh/Tk_b$

From the graph used to determine E_a, one can find the rate constant (k) at 30°.
(It is $1.84 \times 10^{-5}\ \text{s}^{-1}$.)

$$-\Delta G° = 1.98 \times 10^{-3} \times 303 \ln (1.84 \times 10^{-5} \times 1.58 \times 10^{-31})/(303 \times 3.30 \times 10^{-19})$$
$$-\Delta G° = 1.98 \times 10^{-3} \times 303 \ln (2.90 \times 10^{-36})/(1.00 \times 10^{-16})$$
$$-\Delta G° = 1.98 \times 10^{-3} \times 303 \ln 2.90 \times 10^{-20}$$
$$-\Delta G° = 1.98 \times 10^{-3} \times 303 \times (-3.50)$$
$$\Delta G° = 2.10\ \text{kcal/mol}$$

To find $\Delta H°$:

$$\Delta H° = E_a - RT$$
$$\Delta H° = 16.4 - 1.98 \times 10^{-3} \times 303$$
$$\Delta H° = 16.4 - 0.6$$
$$\Delta H° = 15.8\ \text{kcal/mol}$$

To find $\Delta S°$:

$$\Delta S° = (\Delta H° - \Delta G°)/T$$
$$\Delta S° = (15.8 - 2.10)/303$$
$$\Delta S° = (-13.7)/303$$
$$\Delta S° = 0.045\ \text{kcal/(mol deg)} = 45\ \text{cal/(mol deg)}$$

Chapter 5 Practice Test

1. Name each of the following:

a. $CH_3CH_2CHCH_2CH{=}CH_2$ with CH_3 on C3

c. $CH_3CH_2CH{=}CHCH_2CH_2CHCH_3$ with CH_2CH_3 on the CH

b. (cyclopentene ring bearing Br)

d. (cyclohexene ring bearing Cl and CH_3)

2. Which member of each pair is more stable?

a. (cyclohexene with CH_3 on an sp³ carbon adjacent to the double bond) or (1-methylcyclohexene, CH_3 on the double-bond carbon)

b. H_3C(H)C=C(H)CH_3 (trans) or H_3C(H)C=C(H)CH_3 (cis)

3. Correct the incorrect names.

a. 3-pentene

b. 2-vinylpentane

c. 2-ethyl-2-butene

d. 2-methylcyclohexene

4. Indicate whether each of the following statements is true or false:

a. Increasing the energy of activation increases the rate of the reaction. T F

b. Decreasing the entropy of the products compared to the entropy of the reactants makes the equilibrium constant more favorable. T F

c. An exergonic reaction is one with a $-\Delta G°$. T F

d. An alkene is an electrophile. T F

e. The higher the energy of activation, the more slowly the reaction takes place. T F

f. Another name for *trans*-2-butene is (*Z*)-2-butene. T F

g. A reaction with a negative $\Delta G°$ has an equilibrium constant greater than one. T F

h. Increasing the free energy of the reactants increases the rate of the reaction. T F

i. Increasing the free energy of the products increases the rate of the reaction. T F

j. The magnitude of a rate constant is not dependent on the concentration of the reactants. T F

k. 2,3-Dimethyl-2-pentene is more stable than 3,4-dimethyl-2-pentene. T F

5. The addition of H_2 in the presence of Pd/C to alkenes **A** and **B** results in the formation of the same alkane. The addition of H_2 to alkene **A** has a heat of hydrogenation ($-\Delta H°$) of 29.7 kcal/mol, whereas the addition of H_2 to alkene **B** has a heat of hydrogenation of 27.3 kcal/mol. Which is the more stable alkene, **A** or **B**?

6. Draw structures for each of the following:

a. allyl alcohol **b.** 3-methylcyclohexene **c.** *cis*-3-heptene **d.** vinyl bromide

7. What is the total number of π bonds and rings in a hydrocarbon with a molecular formula of C_8H_8?

8. Using curved arrows, show the movement of electrons in the following mechanism:

$$CH_3CH{=}CH_2 + H{-}\ddot{\underset{..}{Cl}}: \rightleftharpoons CH_3\overset{}{\underset{+}{C}}H{-}CH_3 + :\ddot{\underset{..}{Cl}}:^- \longrightarrow CH_3\underset{\underset{:\underset{..}{Cl}:}{|}}{C}H{-}CH_3$$

9. A favorable (negative) $\Delta G°$ is given by:
a positive or negative $\Delta H°$, a positive or negative $\Delta S°$, a high or low temperature.

10. Which of the following has a more favorable equilibrium constant (that is, which reaction favors products more)?

a. A reaction with a $\Delta H°$ of 4 kcal/mol or a reaction with a $\Delta H°$ of 7 kcal/mol? (Assume a constant $\Delta S°$ value.)

b. A reaction with a positive $\Delta S°$ value that takes place at 25 °C or the same reaction that takes place at 35 °C?

c. A reaction in which two reactants form one product or a reaction in which one reactant forms two products? (Assume a constant $\Delta H°$ value.)

11. Draw a reaction coordinate diagram for a one-step reaction with a product that is thermodynamically unstable but kinetically stable.

CHAPTER 6
The Reactions of Alkenes: The Stereochemistry of Addition Reactions

Important Terms

acid-catalyzed reaction — a reaction catalyzed by an acid.

aldehyde — a compound with a carbonyl group that is bonded to an alkyl group and to a hydrogen (or bonded to two hydrogens).

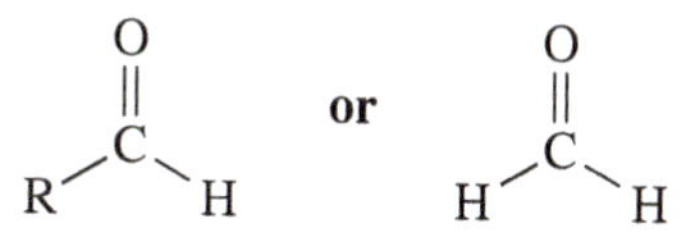

anti addition — an addition reaction in which the two added substituents add to opposite sides of the molecule.

biochemistry — the chemistry associated with living organisms.

carbocation rearrangement — the rearrangement of a carbocation to a more stable carbocation.

concerted reaction — a reaction in which all the bond-making and bond-breaking processes take place in a single step.

dimer — a molecule formed by joining two identical molecules.

electrophilic addition reaction — an addition reaction in which the first species that adds to the reactant is an electrophile.

enzyme — a protein that catalyzes a biological reaction.

epoxide (oxirane) — an ether in which the oxygen is incorporated into a three-membered ring.

halohydrin — an organic molecule that contains a halogen atom and an OH group.

Hammond postulate — states that the transition state will be more similar in structure to the species (reactants or products) that it is closer to energetically.

hydration — addition of water to a compound.

1,2-hydride shift — the movement of a hydride ion (a hydrogen with its pair of bonding electrons) from one carbon to an adjacent carbon.

hydroboration–oxidation — the addition of borane (or R_2BH) to a double or triple bond followed by reaction with hydrogen peroxide and hydroxide ion.

hyperconjugation — delocalization of electrons by overlap of carbon–hydrogen or carbon–carbon σ bonds with a *p* orbital on an adjacent carbon.

ketone a compound with a carbonyl group that is bonded to two alkyl groups.

$$R-\overset{\overset{O}{\|}}{C}-R$$

Markovnikov's rule the actual rule is as follows: "when a hydrogen halide adds to an asymmetrical alkene, the addition occurs such that the halogen attaches itself to the carbon atom of the alkene bearing the least number of hydrogen atoms." Chemists use the rule as follows: the hydrogen adds to the sp^2 carbon that is bonded to the most hydrogens.

Here is a more general rule: the electrophile adds to the sp^2 carbon that is bonded to the most hydrogens.

mechanism of the reaction a description of the step-by-step process by which reactants are changed into products.

1,2-methyl shift the movement of a methyl group with its bonding electrons from one carbon to an adjacent carbon.

oxidation reaction a reaction that decreases the number of C—H bonds in the reactant or increases the number of C—O, C—N, or C—X bonds (X denotes a halogen).

oxidative cleavage an oxidation reaction that cleaves the reactant into two or more compounds.

ozonolysis the reaction of an alkene with ozone.

primary carbocation a carbocation with a positive charge on a primary carbon.

reduction reaction a reaction that increases the number of C—H bonds in the reactant or decreases the number of C—O, C—N, or C—X bonds (X denotes a halogen).

regioselective reaction a reaction that leads to the preferential formation of one constitutional isomer over another.

secondary carbocation a carbocation with a positive charge on a secondary carbon.

stereocenter (stereogenic center) an atom at which the interchange of two groups produces a stereoisomer.

stereochemistry the field of chemistry that deals with the structure of molecules in three dimensions.

stereoisomers isomers that differ in the way the atoms are arranged in space.

stereoselective reaction a reaction that leads to the preferential formation of one stereoisomer over another.

stereospecific reaction a reaction in which the reactant can exist as stereoisomers and each stereoisomeric reactant leads to a different stereoisomeric product or products.

steric effect an effect due to the space occupied by a substituent.

steric hindrance a hindrance due to bulky groups at the site of a reaction that make it difficult for the reactants to approach one another.

syn addition an addition reaction in which the two added substituents add to the same side of the molecule.

tertiary carbocation a carbocation with a positive charge on a tertiary carbon.

vicinal refers to substituents on adjacent carbons.

Solutions to Problems

1. Cyclohexene + H—Cl → cyclohexyl cation + $:\ddot{Cl}:^-$ → chlorocyclohexane

2. **a.** The σ bond orbitals of the carbon adjacent to the positively charged carbon are available for overlap with the vacant p orbital. Because the methyl cation does not have a carbon adjacent to the positively charged carbon, no σ bond orbitals are available for overlap with the vacant p orbital.

b. An ethyl cation is more stable because the carbon adjacent to the positively charged carbon has three σ bond orbitals available for overlap with the vacant p orbital, whereas a methyl cation does not have any σ bond orbitals available for overlap with the vacant p orbital.

3. **a.** $CH_3—\overset{+}{C}(H)(CH_3)H_2$ — 3 $\quad$ $CH_3CH_2—\overset{+}{C}(H)(H)H_2$ — 3 $\quad$ $CH_3—\overset{+}{C}(H)(H)HC(H)(H)—H$ — 6 $\quad$ **b.** *sec*-butyl cation

4. **a.** $CH_3CH_2\overset{+}{C}(CH_3)CH_3 > CH_3CH_2\overset{+}{C}HCH_3 > CH_3CH_2CH_2\overset{+}{C}H_2$

b. The reason a halogen atom decreases the stability of the carbocation is because it is an electronegative atom and, therefore, withdraws electrons away from the positively charged carbon. This increases the concentration of positive charge on the carbocation, which makes it less stable.

$$CH_3CH(CH_3)CH_2\overset{+}{C}H_2 > CH_3CH(Cl)CH_2\overset{+}{C}H_2 > CH_3CH(F)CH_2\overset{+}{C}H_2$$

Because fluorine is more electronegative than chlorine and, therefore, withdraws electrons more strongly, the fluorine-substituted carbocation is less stable than the chlorine-substituted carbocation.

5. The transition state resembles the one (reactants or products) that it is closer to on the reaction coordinate diagram; that is, the one that it is closer to in energy.

a. products **b.** reactants **c.** reactants **d.** products

6. **a.** $CH_3CH_2C(CH_3)=CH_2$ $\quad$ In both **a** and **b**, the compound that is more highly regioselective is the one where the choice is between forming a tertiary carbocation and a primary carbocation.

b. Methylenecyclohexane (CH_2=cyclohexane) $\quad$ In both **a** and **b**, the less regioselective compound is the one in which the choice is between forming a tertiary carbocation and a secondary carbocation, because the difference in the stability of the two possible carbocations, and therefore the difference in the amount of product formed, is not as great as when the choice is between a tertiary and a primary carbocation.

7.

a. $CH_3CH_2CHCH_3$ with Br on C-2

b. $CH_3CH_2CCH_3$ with CH_3 and Br on C-2

c. 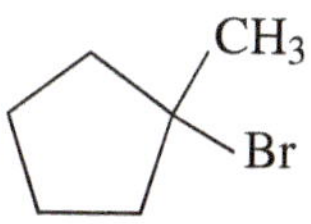 cyclopentane with CH_3 and Br on the same carbon

d. $CH_3CCH_2CH_2CH_3$ with CH_3 and Br on C-2

e. 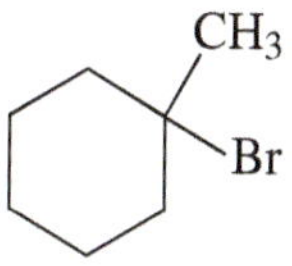 cyclohexane with CH_3 and Br on the same carbon

f. $CH_3CH_2CHCH_3$ with Br on C-2

8.

a. $CH_2{=}CCH_3$ with CH_3 on C-2

b. 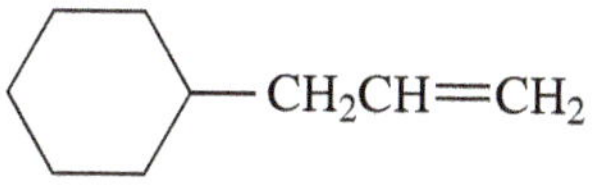 cyclohexyl–$CH_2CH{=}CH_2$

cyclohexyl–$CH{=}CHCH_3$

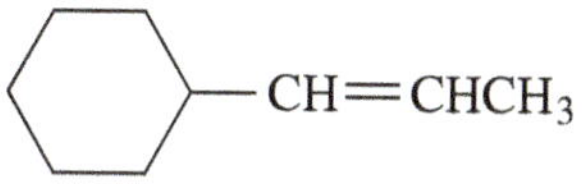

This is not a good choice, because it forms approximately equal amounts of two products.

c. 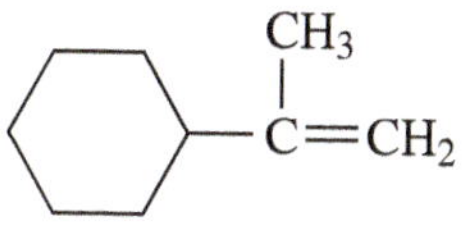 cyclohexyl–$C{=}CH_2$ with CH_3 on that C

d. 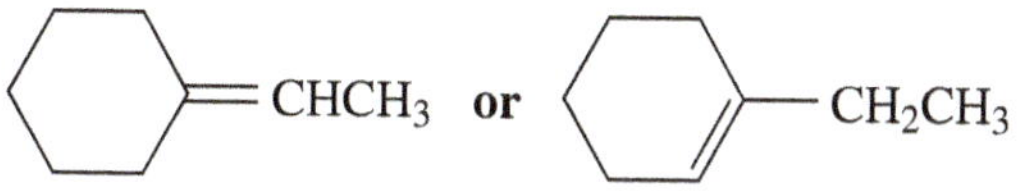 cyclohexylidene$=CHCH_3$ **or** 1-cyclohexene ring with CH_2CH_3 on the double-bond carbon

9. As long as the pH is greater than about −2.5 and less than about 15, more than 50% of 2-propanol will be in its neutral, nonprotonated form.

$$R\overset{+}{O}H_2 \underset{}{\overset{pK_a = -2.5}{\rightleftharpoons}} ROH \overset{pK_a = \sim 15}{\rightleftharpoons} RO^-$$

Recall that when the pH = pK_a, half the compound is in its acidic form and half is in its basic form. Therefore, at a pH less than about −2.5, more than half of the compound is in its positively charged protonated form. At a pH greater than about ~15, more than half of the compound exists as the negatively charged anion.

Therefore, at a pH between −2.5 and ~15, more than half of the compound exists in the neutral nonprotonated form.

10. **a.** three transition states **b.** two intermediates

c. The first step is the *slowest step*, so it has the *smallest rate constant*; the second step is fast because no bonds are being broken; the third step is fast because transfer of a proton from or to an O or an N is always a fast reaction.

11.

a. $CH_3CH_2CH_2CHCH_3$ with OH on C-2

b. 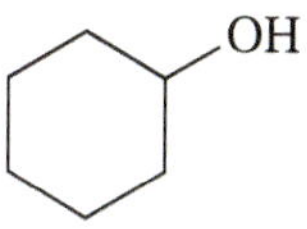 cyclohexane with OH

c. $CH_3CH_2CH_2CH_2CHCH_3$ with OH on C-2 and $CH_3CH_2CH_2CHCH_2CH_3$ with OH on C-3

approximately equal amounts are formed

d. 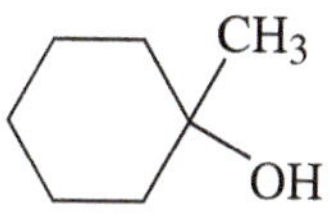 cyclohexane with CH_3 and OH on the same carbon

12. **a.** **1.** $CH_3C(CH_3)(Cl)CH_3$ **2.** $CH_3C(CH_3)(Br)CH_3$ **3.** $CH_3C(CH_3)(OH)CH_3$ **4.** $CH_3C(CH_3)(OCH_3)CH_3$

b. **1.** The first step in all the reactions is addition of an electrophilic proton (H^+) to the carbon of the CH_2 group.
2. A *tert*-butyl carbocation is formed as an intermediate in each of the reactions.

c. **1.** The nucleophile that adds to the *tert*-butyl carbocation is different in each reaction.
2. In reactions #3 and #4, there is a third step—a proton is lost from the group that was the nucleophile in the second step of the reaction.

3. $CH_3C(CH_3)(\overset{+}{O}H_2)CH_3 + H_2\ddot{O} \longrightarrow CH_3C(CH_3)(OH)CH_3 + H_3O^+$

4. $CH_3C(CH_3)(\overset{+}{O}(H)CH_3)CH_3 + CH_3\ddot{O}H \longrightarrow CH_3C(CH_3)(OCH_3)CH_3 + CH_3\overset{+}{O}H_2$

13. Solved in the text.

14. **a.** cyclopentene $+ H_2O \xrightarrow{H_2SO_4}$ cyclopentanol (OH)

b. $CH_2{=}C(CH_3)CH_3 + CH_3OH \xrightarrow{H_2SO_4} CH_3OC(CH_3)_2CH_3$

c. $CH_3CH{=}CHCH_3 + H_2O \xrightarrow{H_2SO_4} CH_3CH(OH)CH_2CH_3$

or

$CH_2{=}CHCH_2CH_3 + H_2O \xrightarrow{H_2SO_4} CH_3CH(OH)CH_2CH_3$

d. methylenecyclohexane (CH_2) $+ CH_3OH \xrightarrow{H_2SO_4}$ 1-methoxy-1-methylcyclohexane (CH_3, OCH_3)

1-Methylcyclohexene could also be used, but the reaction would form less of the desired product because it less highly regioselective.

e. $CH_2{=}C(CH_3)CH_2CH_3 + H_2O \xrightarrow{H_2SO_4} CH_3C(CH_3)(OH)CH_2CH_3$

2-Methyl-2-butene could also be used, but the reaction would form less of the desired product because it is less highly regioselective.

15.

$$CH_3C(CH_3){=}CH_2 + H{-}OSO_3H \rightleftharpoons CH_3\overset{+}{C}(CH_3)CH_3 + R\ddot{O}H \rightleftharpoons CH_3C(CH_3)(\overset{+}{O}(H)R)CH_3 \;[+\ R\ddot{O}H] \rightleftharpoons CH_3C(CH_3)(OR)CH_3 + R\overset{+}{O}H_2$$

16. Solved in the text.

17.

a. $CH_3CH(CH_3)CH{=}CH_2 \xrightarrow{HBr} CH_3CH(CH_3)\overset{+}{C}HCH_3$ (secondary) $\xrightarrow{\text{1,2-hydride shift}} CH_3\overset{+}{C}(CH_3)CH_2CH_3$ (tertiary) $\xrightarrow{Br^-} CH_3CBr(CH_3)CH_2CH_3$

secondary **tertiary**

b. $CH_3CH(CH_3)CH_2CH{=}CH_2 \xrightarrow{HBr} CH_3CH(CH_3)CH_2\overset{+}{C}HCH_3 \xrightarrow{Br^-} CH_3CH(CH_3)CH_2CHBrCH_3$

c. methylenecyclohexane ($=CH_2$) $\xrightarrow{HBr}$ 1-methylcyclohexyl cation (CH_3, +) $\xrightarrow{Br^-}$ 1-bromo-1-methylcyclohexane (Br, CH_3)

d. 1-methylcyclohexene (CH_3) $\xrightarrow{HBr}$ 1-methylcyclohexyl cation (CH_3, +) $\xrightarrow{Br^-}$ 1-bromo-1-methylcyclohexane (Br, CH_3)

e. $CH_2{=}CHC(CH_3)_2CH_3 \xrightarrow{HBr} CH_3\overset{+}{C}HC(CH_3)_2CH_3$ (secondary carbocation) $\xrightarrow{\text{1,2-methyl shift}} CH_3CH(CH_3)\overset{+}{C}(CH_3)CH_3$ (tertiary carbocation) $\xrightarrow{Br^-} CH_3CH(CH_3){-}CBr(CH_3)CH_3$

secondary carbocation **tertiary carbocation**

f. 4-methylcyclohexene (CH_3) $\xrightarrow{HBr}$ 3-methylcyclohexyl cation (CH_3, +) $\xrightarrow{Br^-}$ 1-bromo-3-methylcyclohexane (CH_3, Br)

4-methylcyclohexene (CH_3) $\xrightarrow{HBr}$ 4-methylcyclohexyl cation (CH_3, +) $\xrightarrow{Br^-}$ 1-bromo-4-methylcyclohexane (CH_3, Br)

Approximately equal amounts of 1-bromo-3-methylcyclohexane and 1-bromo-4-methylcyclohexane are obtained because in each case, the intermediate is a secondary carbocation. Therefore, the two compounds are formed at about the same rate. A carbocation rearrangement does not occur because it would just form another secondary carbocation.

18. The reaction with 9-BBN is more highly regioselective. 9-BBN is sterically hindered, so it will be more likely than BH_3 to add to the least sterically hindered carbon.

19. Either BH_3 or R_2BH can be used for the hydroboration reactions in this chapter.

a. $CH_3C(CH_3){=}CHCH_3 \xrightarrow[2.\ HO^-,\ H_2O_2,\ H_2O]{1.\ BH_3/THF} CH_3CH(CH_3)CH(OH)CH_3$

b. 1-methylcyclohexene $\xrightarrow[2.\ HO^-,\ H_2O_2,\ H_2O]{1.\ BH_3/THF}$ 2-methylcyclohexanol (ring bearing CH_3 and OH on adjacent carbons)

20. $CH_3CBr(CH_3)CH_2CH_3$

Addition of H^+ forms a carbocation that rearranges.

Addition of Br^+ forms a cyclic bromonium ion rather than a carbocation, so there is no carbocation rearrangement.

21.

a. The first step in the reaction of propene with Br_2 forms a cyclic bromonium ion, whereas the first step in the reaction of propene with HBr forms a carbocation.

b. If the bromide ion were to attack the positively charged bromine, a highly unstable compound (with a negative charge on carbon and a positive charge on bromine) would be formed.

$$\text{cyclic bromonium ion } (:\!Br\!:^+) + :\!\ddot{Br}\!:^- \longrightarrow \overset{-}{C}H_2{-}CH_2{-}\overset{+}{\ddot{Br}}{-}\ddot{Br}:$$

Notice that the electrostatic potential map of the cyclic bromonium ion on page 254 of the text shows that the ring carbons are the least electron dense (most blue) atoms in the intermediate and, therefore, are the ones most susceptible to nucleophilic attack.

22. Sodium and potassium achieve an outer shell of eight electrons by losing the single electron they have in the 3*s* (in the case of Na) or 4*s* (in the case of K) orbital, thereby becoming Na^+ and K^+.

In order to form a covalent bond, they would have to regain electrons in these orbitals, thereby losing the stability associated with having an outer shell of eight electrons and no extra electrons.

23. The nucleophile that is present in greater concentration is more apt to collide with the intermediate. For example, in part **a** the concentration of CH_3OH (the solvent) is much greater than the concentration of Cl^-; in part **b**, the nucleophile is most likely to be I^- because there are two equivalents of NaI and one equivalent of HBr.

a. $ClCH_2C(CH_3)(OCH_3)CH_3$ (**major**) and $ClCH_2C(CH_3)(Cl)CH_3$

b. $CH_3CH(I)CH_3$ (**major**) and $CH_3CH(Br)CH_3$

c. $CH_3CH_2CH(OH)CH_3$ (**major**) and $CH_3CH_2CH(Cl)CH_3$

d. $CH_3CH(Br)CH(OCH_3)CH_3$ (**major**) and $CH_3CH(Br)CH(Br)CH_3$

24. Because chlorine is more electronegative than iodine, iodine is the electrophile. Therefore, it ends up attached to the sp^2 carbon that is bonded to the most hydrogens.

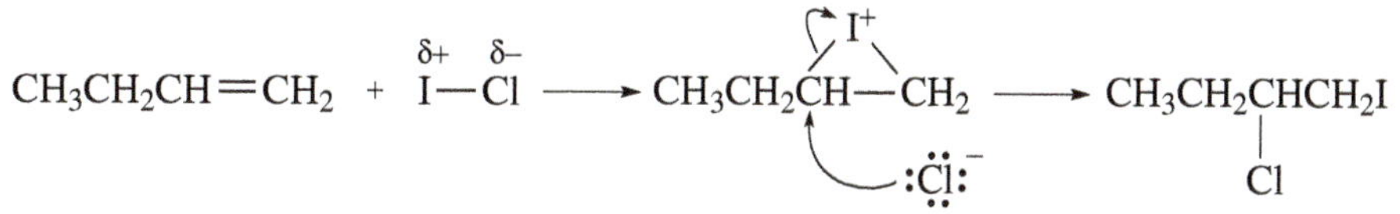

$$CH_3CH_2CH{=}CH_2 + \overset{\delta+}{I}{-}\overset{\delta-}{Cl} \longrightarrow CH_3CH_2\overset{I^+}{CH{-}CH_2} \xrightarrow{:\ddot{Cl}:^-} CH_3CH_2CH(Cl)CH_2I$$

25. **a.** 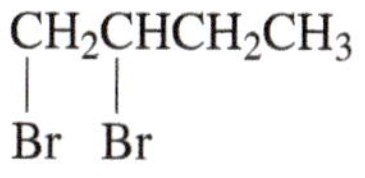 $CH_2(Br)CH(Br)CH_2CH_3$ **b.** 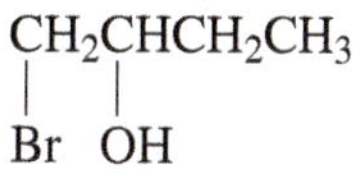 $CH_2(Br)CH(OH)CH_2CH_3$ **c.** 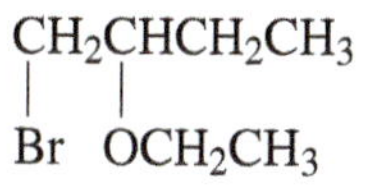 $CH_2(Br)CH(OCH_2CH_3)CH_2CH_3$ **d.** 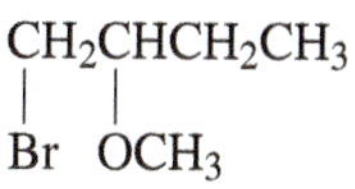 $CH_2(Br)CH(OCH_3)CH_2CH_3$

26. Look at the reagent (Cl_2) and remember what that reagent does when it reacts with an alkene (do not let the rest of the molecule confuse you): when Cl_2 adds to an alkene, it forms a cyclic chloronium ion intermediate. You know that the intermediate then reacts with a nucleophile.

There are two nucleophiles in the solution that can react with the intermediate, a Cl^- and the OH group at the end of the molecule. There is a greater probability that the OH group is the nucleophile that attacks the chloronium ion because, since it is attached to the reactant, it does not have to wander through the solution to find the chloronium ion as the Cl^- has to. Loss of a proton forms the six-membered ring ether.

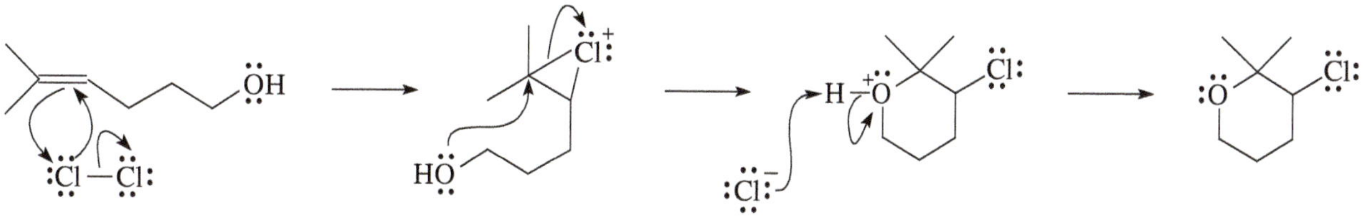

:Cl—Cl: ÖH → Cl⁺ HÖ → :Cl:⁻ H—O⁺ Cl: → :O Cl:

27. **a.** 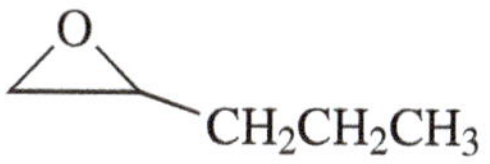 epoxide with $CH_2CH_2CH_3$ **b.** 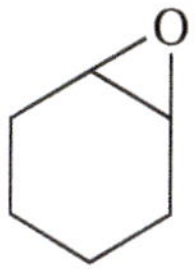 cyclohexene oxide (O) **c.** H_3C, H_3C, O, CH_3, CH_3 **d.** 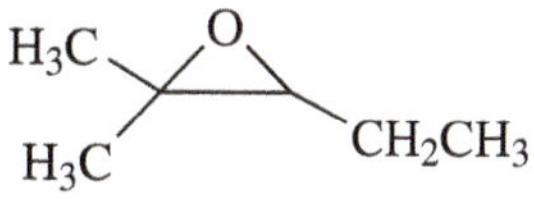 H_3C, H_3C, O, CH_2CH_3

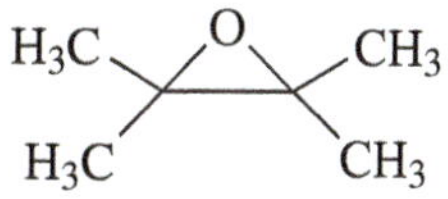

28. **a.** 1-pentene **b.** cyclohexene **c.** 2,3-dimethyl-2-butene **d.** 2-methyl-2-pentene

29. Solved in the text.

30. **a.** 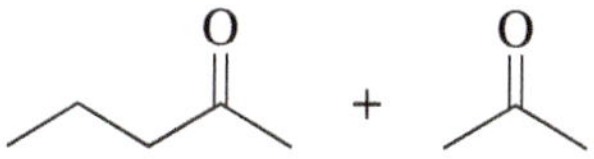 O + O
b. 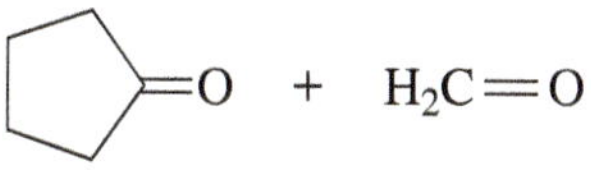 cyclopentanone =O + $H_2C{=}O$
c. 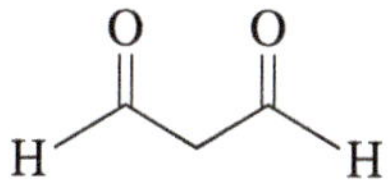 H, O, O, H
d. H, O, O, H + H, O, O
e. O, H, O + O, H, H, O
f. 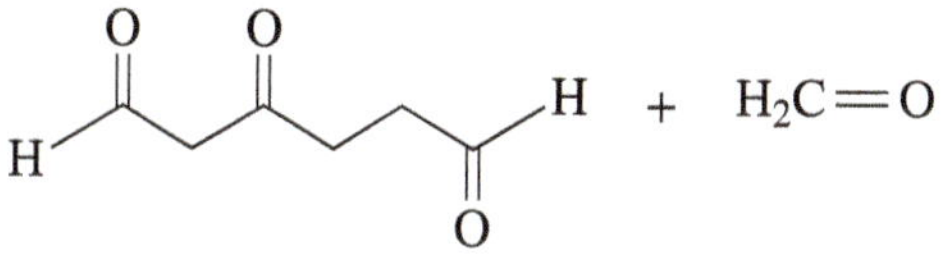 O, O, H, O, H + $H_2C{=}O$

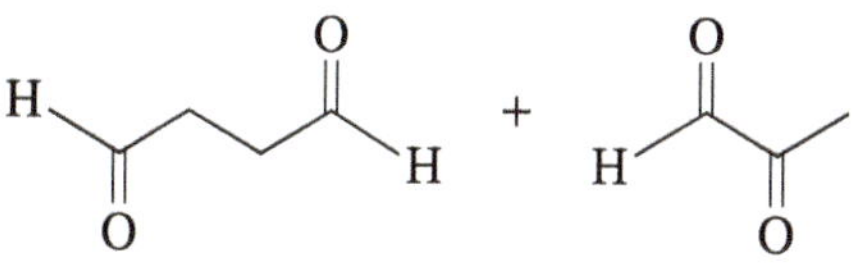

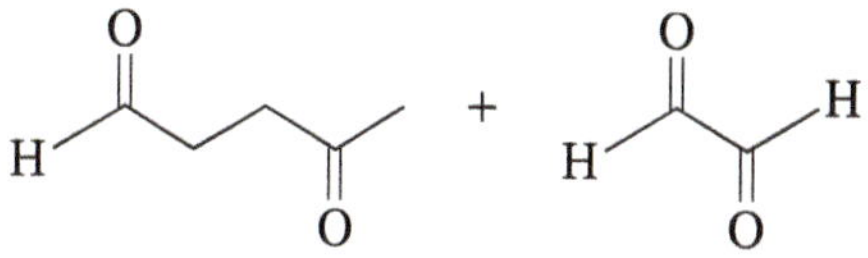

31. **a.** $(CH_3)_2C{=}C(CH_3)_2$

2,3-dimethyl-2-butene

b. $CH_3CH_2CH_2CH{=}CHCH_2CH_2CH_3$ (both H on same side)

cis-4-octene

or

$CH_3CH_2CH_2CH{=}CHCH_2CH_2CH_3$ (H on opposite sides)

trans-4-octene

or

$(CH_3)_2CHCH{=}CHCH(CH_3)_2$ (both H on same side)

cis-2,5-dimethyl-3-hexene

or

$(CH_3)_2CHCH{=}CHCH(CH_3)_2$ (H on opposite sides)

trans-2,5-dimethyl-3-hexene

32. It does not tell you whether the double bond has the *E* or *Z* configuration.

33. Solved in the text.

34. **a.** **b.** **c.**

35. The reactant must have *E,Z* stereoisomers or *R,S* stereoisomers.

36. **a.** No, because only one constitutional isomer can be formed as a product since 2-butene is a symmetrical alkene.

$CH_3CH_2CHCH_3$ (Br on C-2)

b. No, because it forms a racemic mixture.
c. No, because *cis*-butene and *trans*-butene form the same product.
d. Yes, because two constitutional isomers are possible but only one is formed.
e. No, because it forms a racemic mixture.
f. No, because 1-butene does not have stereoisomers.

37. Only the stereoisomers of the major product of each reaction are shown.

a. *S* and *R* enantiomers of $CH_3CHClCH_2CH_2CH_3$ — racemic mixture

or

Fischer projections: CH_3 top, $CH_2CH_2CH_3$ bottom; H left, Cl right *S*; Cl left, H right *R*

b. $CH_2CH_2CH_3$ / CH_3CH_2 / S C H / OH racemic mixture $CH_2CH_2CH_3$ / R C / H HO CH_2CH_3 or CH_2CH_3 / H—OH S / $CH_2CH_2CH_3$ CH_2CH_3 / HO—H R / $CH_2CH_2CH_3$

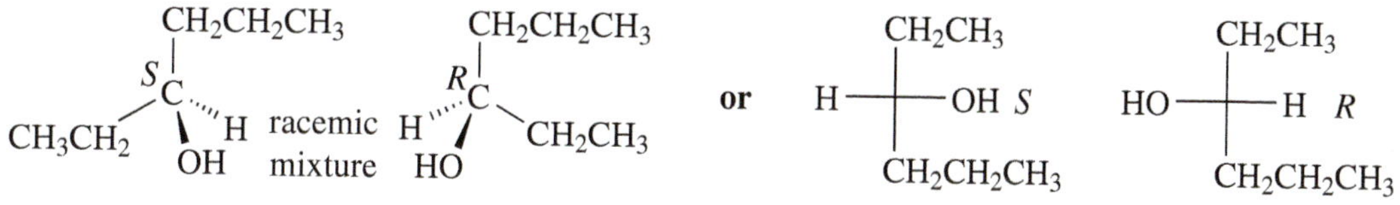

c. CH_3O CH_3

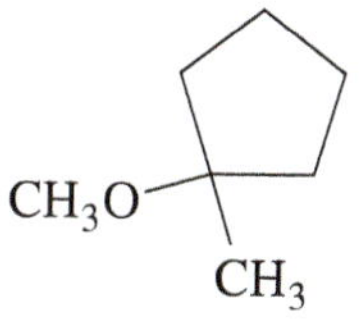

This compound does not have any stereoisomers, because it does not have any asymmetric centers.

d. $CH_3CH_2CCH_2CH_3$ with CH_2CH_3 and Br

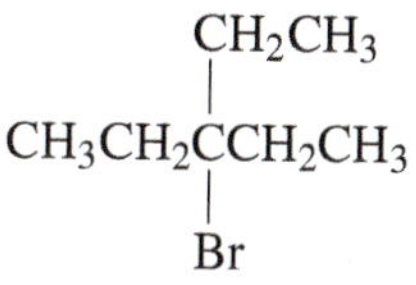

This compound does not have any stereoisomers, because it does not have an asymmetric center.

38. Solved in the text.

39. Solved in the text.

40. Solved in the text.

41. **a.** **1.** *trans*-3-heptene **2.** *cis*-3-heptene

b. and **c.** The enantiomer of each of the structures shown is also formed, because the peroxyacid can approach both the top and the bottom of the plane defined by the double bond.

1. O; S S; $CH_3CH_2CH_2$ C—C H; H CH_2CH_3

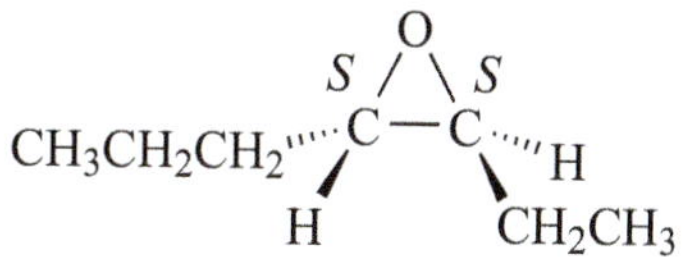

2. O; R S; H C—C H; $CH_3CH_2CH_2$ CH_2CH_3

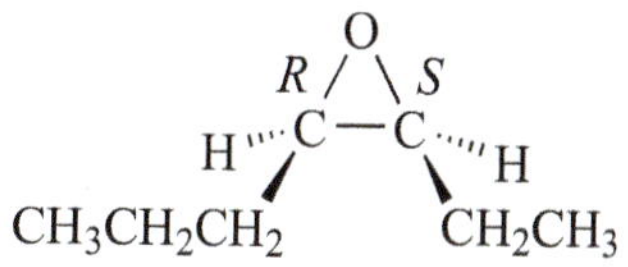

42. **a.** OH

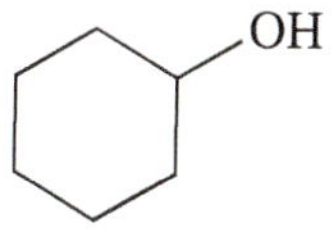

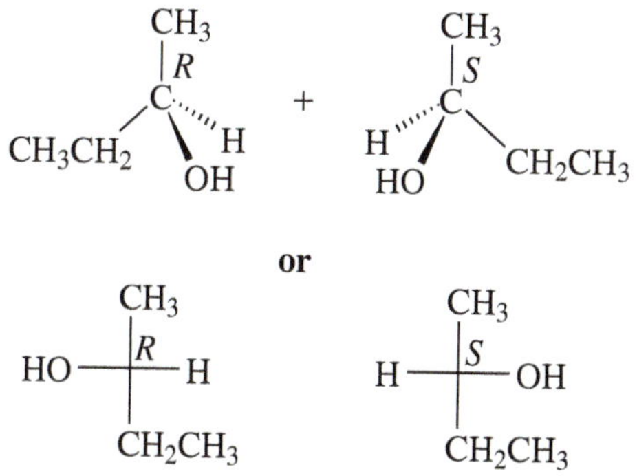

b. S S CH_3CH_2 OH + R R HO CH_2CH_3

c. CH_3 R C CH_3CH_2 H OH + CH_3 S C H HO CH_2CH_3

or

CH_3 / HO—H R / CH_2CH_3 CH_3 / H—OH S / CH_2CH_3

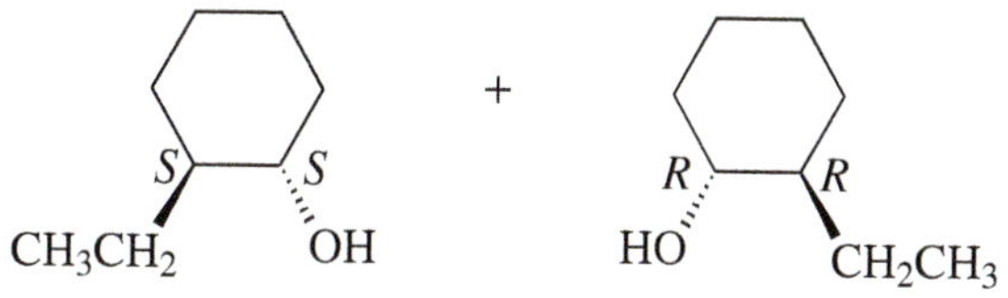

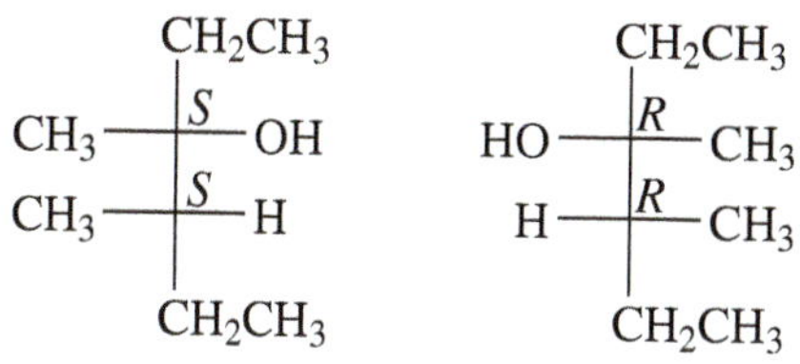

d. CH_2CH_3 / CH_3—OH S / CH_3—H S / CH_2CH_3 CH_2CH_3 / HO—CH_3 R / H—CH_3 R / CH_2CH_3

43.

$$CH_2{=}C(CH_2CH_3)CH_2CH_2CH_3 \xrightarrow{Br_2} BrCH_2\overset{*}{C}(CH_2CH_3)(Br)CH_2CH_2CH_3$$

$$CH_2{=}C(CH_2CH_3)CH_2CH_2CH_3 \xrightarrow[Pd/C]{H_2} CH_3\overset{*}{C}H(CH_2CH_3)CH_2CH_2CH_3$$

$$CH_2{=}C(CH_2CH_3)CH_2CH_2CH_3 \xrightarrow[2.\ HO^-,\ H_2O_2,\ H_2O]{1.\ BH_3} HOCH_2\overset{*}{C}H(CH_2CH_3)CH_2CH_2CH_3$$

(* indicates an asymmetric center)

Each of the reactions forms a compound with one asymmetric center from a compound with no asymmetric centers. Therefore, each of the products is a racemic mixture.

44. Solved in the text.

45. **a.** Wedge-dash structures: C bearing CH_2CH_3 (up), CH_3, H (hashed), Br (wedge) — *S* + C bearing CH_2CH_3 (up), H (hashed), Br (wedge), CH_3 — *R*

or Fischer projections: CH_3 (top), H (left), Br (right), CH_2CH_3 (bottom) — *S* + CH_3 (top), Br (left), H (right), CH_2CH_3 (bottom) — *R*

b.

$$CH_3CH_2C(CH_3)(Br)CH_2CH_3$$

This compound does not have an asymmetric center.

c. Same as **b**.

d. Wedge-dash structures: C bearing $CH_2CH_2CH_3$ (up), CH_3CH_2, H (hashed), Br (wedge) — *S* + C bearing $CH_2CH_2CH_3$ (up), H (hashed), Br (wedge), CH_2CH_3 — *R*

or Fischer projections: CH_2CH_3 (top), H (left), Br (right), $CH_2CH_2CH_3$ (bottom) — *S* + CH_2CH_3 (top), Br (left), H (right), $CH_2CH_2CH_3$ (bottom) — *R*

e. Wedge-dash structures: CH_3, H (hashed), Br (wedge) on C — C bearing H (hashed), Br (wedge), CH_2CH_3 — *S S* + Br (wedge), CH_3CH_2 on C — C bearing CH_3, H (hashed), Br (wedge); H on first C — *R R*

or Fischer projections: CH_3 (top); H (left), Br (right) *S*; Br (left), H (right) *S*; CH_2CH_3 (bottom) + CH_3 (top); Br (left), H (right) *R*; H (left), Br (right) *R*; CH_2CH_3 (bottom)

f. Wedge-dash structures: C bearing $CH_2CH_2CH_2CH_3$ (up), $BrCH_2$, H (hashed), Br (wedge) — *R* + C bearing $CH_2CH_2CH_2CH_3$ (up), H (hashed), Br (wedge), CH_2Br — *S*

or Fischer projections: CH_2Br (top), H (left), Br (right), $CH_2CH_2CH_2CH_3$ (bottom) — *R* + CH_2Br (top), Br (left), H (right), $CH_2CH_2CH_2CH_3$ (bottom) — *S*

46. Two different bromonium ions are formed because Br_2 can add to the double bond either from the top of the plane or from the bottom of the plane defined by the double bond; the two bromonium ions are formed in equal amounts. Attacking the less hindered carbon of one bromonium ion forms one stereoisomer, whereas attacking the less hindered carbon of the other bromonium ion forms the other stereoisomer. Because Br^- can attack the least sterically hindered carbon with equal ease from pathway **a** as from pathway **b**, equal amounts of the threo enantiomers are obtained. Of course, some reaction will occur at the more hindered end of the bromonium ion, but it will occur to the same extent in both pathways.

H_3C / $CH_2CH_2CH_2CH_3$ / C=C / H / H

pathway **a** Br_2 → bromonium ion (H_3C, $CH_2CH_2CH_2CH_3$, H, H, Br^+, $:Br:^-$)

Br_2 pathway **b** → bromonium ion (H_3C, $CH_2CH_2CH_2CH_3$, H, H, Br^+, $:Br:^-$)

47. **a.** HO, Br and Br, OH

The addition of Br and OH are anti, so these two substituents are trans to each other in the cyclic product.

b. <u>**mechanism of the reaction**</u>

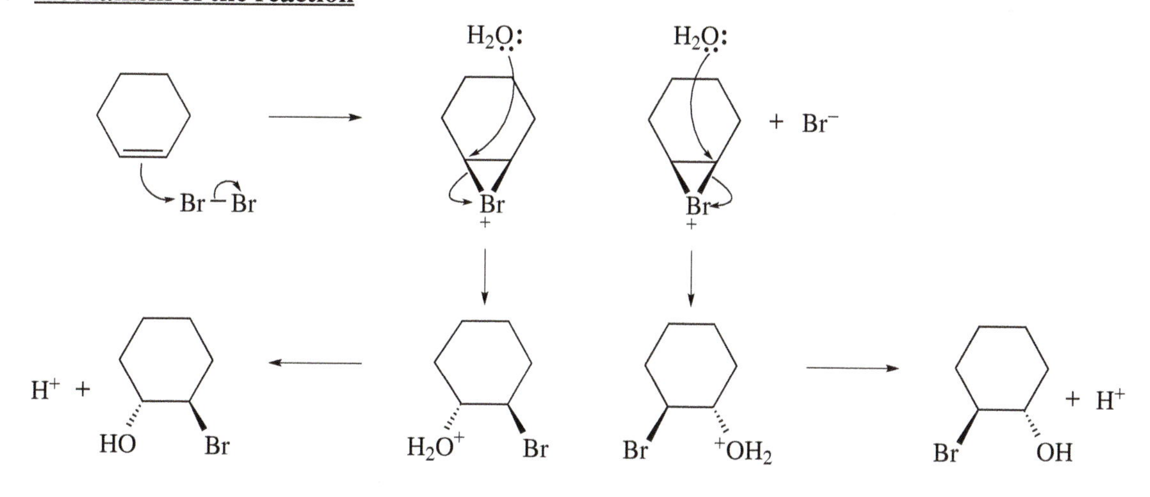

48. **a.** Only anti addition occurs. Because the reactant is trans, the product is expected to be the erythro pair of enantiomers. However, in this case, each asymmetric carbon is attached to the same four groups, so the product is a meso compound. Therefore, only one stereoisomer is obtained.

CH_2CH_3
CH_3 — Br *S*
CH_3 — Br *R*
CH_2CH_3

or

CH_3CH_2 Br
CH_3—C—C—CH_3
Br CH_2CH_3
S *R*

b. Only one asymmetric center is created in the product, so the product is a racemic mixture.

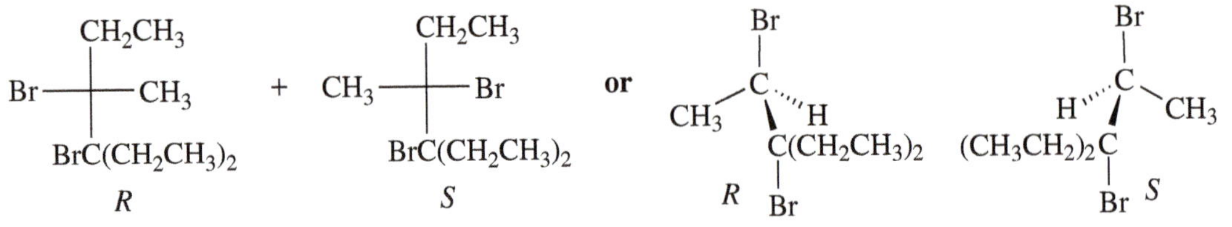

c. Only anti addition occurs. Because the reactant is cis (a cyclopentene ring cannot exist in a trans configuration), the product is the pair of enantiomers with the bromines on opposite sides of the ring.

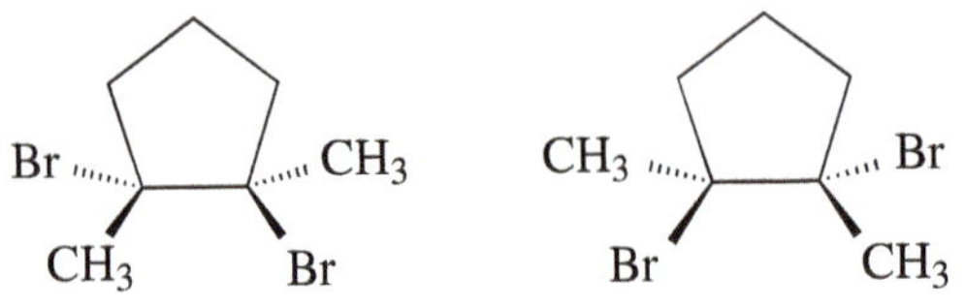

d. Only anti addition occurs. Because the reactant is cis, the product is the pair of enantiomers with the bromines on opposite sides of the ring.

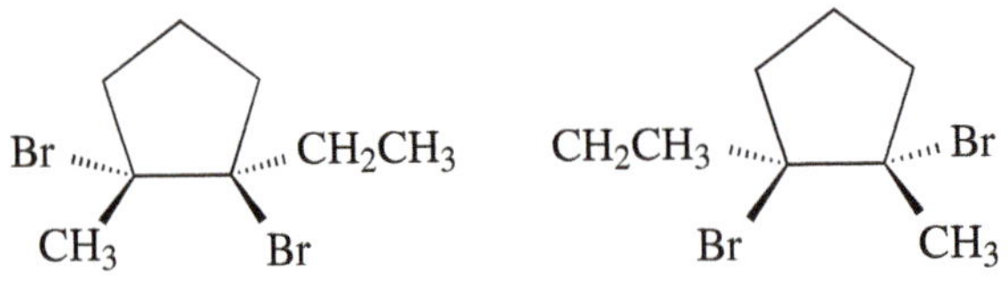

49. **a.** Only syn addition occurs. Because the reactant is trans, the product is the threo pair of enantiomers.

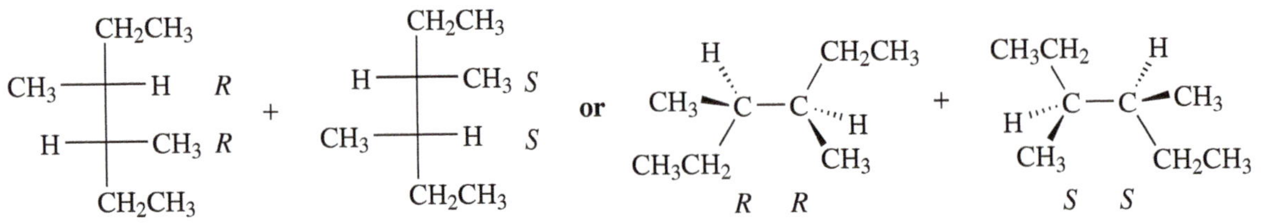

b. The product of the reactions does not have any asymmetric centers, so it does not have any stereoisomers.

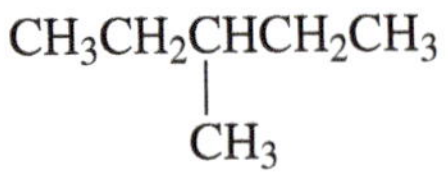

$CH_3CH_2CH(CH_3)CH_2CH_3$

c. Only syn addition occurs. Because the reactant is cis, the product would be the pair of enantiomers with the hydrogens on the same side of the ring, but in this case, the product is a meso compound, so only one stereoisomer is obtained.

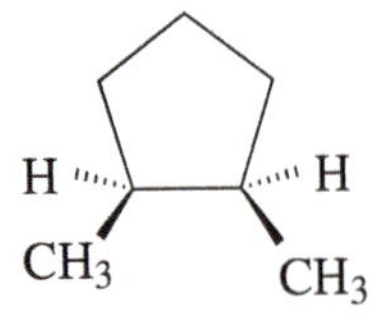

d. Only syn addition occurs. Because the reactant is cis, the product is the pair of enantiomers with the hydrogens on the same side of the ring.

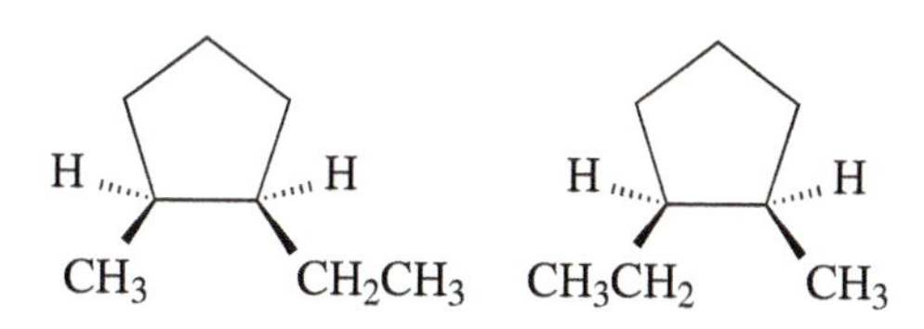

50. **a.** CH_3CHCH_2Br (with Cl on C-2)

b. Equal amounts of the *R* and *S* enantiomers are formed (a racemic mixture) because a reactant without an asymmetric center forms a product with one asymmetric center.

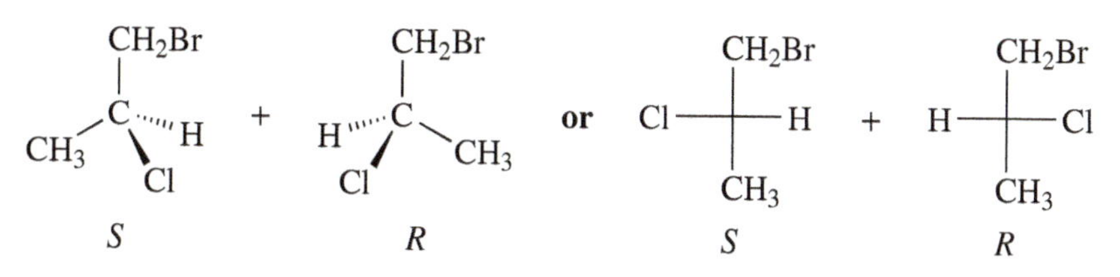

51. **a.** A racemic mixture of (*R*)-malate and (*S*)-malate. (A product with one asymmetric center is formed from a reactant with no asymmetric centers.)

b. A racemic mixture of (*R*)-malate and (*S*)-malate. In the absence of an enzyme, the reactions are neither stereoselective (part **a**) nor stereospecific (part **b**).

52. Solved in the text.

53. **a.** $CH_3CH{=}CH_2 \xrightarrow[CH_3OH]{H_2SO_4} CH_3CHCH_3$ (with OCH_3 on C-2)

b. + Br_2 $\xrightarrow{CH_2Cl_2}$ Br, Br

c. CH_2 $\xrightarrow[\text{2. } HO^-, H_2O_2, H_2O]{\text{1. } R_2BH/THF}$ CH_2OH

d. CH_2 or CH_3 $\xrightarrow[CH_3OH]{H_2SO_4}$ CH_3, OCH_3

Methylenecyclohexane is the better choice because the reaction is more highly regioselective.

e. CH_2 or CH_3 $\xrightarrow{HBr}$ CH_3, Br

Methylenecyclohexane is the better choice because the reaction is more highly regioselective.

f. + $CH_3CH_2CH_2OH$ $\xrightarrow{H_2SO_4}$ $OCH_2CH_2CH_3$

54. Less of the desired product would be formed from 3-methylcyclohexene than from 1-methylcyclohexene. Hydroboration–oxidation of 3-methylcyclohexene would form both 2-methylcyclohexanol (the desired product) and 3-methylcyclohexanol at approximately the same rate because the transition states formed when borane adds to the 1-position or the 2-position of the alkene have approximately the same stability.

CH3 (3-methylcyclohexene) —1. R_2BH/THF; 2. H_2O_2, HO^-, H_2O→ CH3, OH (2-methylcyclohexanol) + CH3, OH (3-methylcyclohexanol)

55. **a.** 1-bromo-1-ethylcyclohexane (Br, CH_2CH_3) **b.** $CH_3C(CH_3)(Br)CH_2CH_3$ **c.** 1-bromo-1-ethylcyclohexane (CH_2CH_3, Br) **d.** $CH_3CH_2C(CH_3)(Br)—CH(CH_3)CH_3$

56. **a.** addition of a hydrogen halide, acid-catalyzed addition of water, acid-catalyzed addition of an alcohol

b. hydroboration, epoxidation

c. addition of Br_2 or Cl_2

d. ozonolysis

57. **a.** electrophile nucleophile

$CH_3\overset{+}{C}HCH_3$ + $:\ddot{Cl}:^-$ ⟶ $CH_3CH(:\ddot{Cl}:)CH_3$

b. nucleophile electrophile

$CH_3CH{=}CH_2$ + H—Br ⟶ $CH_3\overset{+}{C}H—CH_3$ + Br^-

c. electrophile nucleophile

$CH_3CH{=}CH_2$ ⟶ $CH_3CH_2—CH_2BH_2$

+ H—BH_2

nucleophile electrophile

58. **a.** $CH_3C(CH_3){=}CHCH_3$ + HBr ⟶ $CH_3C(CH_3)(Br)—CH_2CH_3$

b. $CH_3C(CH_3){=}CHCH_3$ + HI ⟶ $CH_3C(CH_3)(I)—CH_2CH_3$

c. $$CH_3C(CH_3){=}CHCH_3 + Cl_2 \longrightarrow CH_3C(CH_3)(Cl){-}CH(Cl)CH_3$$

d. $$CH_3C(CH_3){=}CHCH_3 \xrightarrow[2.\ (CH_3)_2S]{1.\ O_3,\ -78\ ^{\circ}C} CH_3C(=O)CH_3 + CH_3C(=O)H$$

e. $$CH_3C(CH_3){=}CHCH_3 \xrightarrow[Pd/C]{H_2} CH_3CH(CH_3){-}CH_2CH_3$$

f. $$CH_3C(CH_3){=}CHCH_3 + RC(=O)OOH \longrightarrow CH_3CH{-}C(CH_3)CH_3 \ (\text{with O bridging the two carbons})$$

g. $$CH_3C(CH_3){=}CHCH_3 + H_2O \xrightarrow{H_2SO_4} CH_3C(CH_3)(OH){-}CH_2CH_3$$

h. $$CH_3C(CH_3){=}CHCH_3 + Br_2 \xrightarrow[CH_2Cl_2]{} CH_3C(CH_3)(Br){-}CH(Br)CH_3$$

i. $$CH_3C(CH_3){=}CHCH_3 + Br_2 \xrightarrow[H_2O]{} CH_3C(CH_3)(HO){-}CH(Br)CH_3$$

j. $$CH_3C(CH_3){=}CHCH_3 + Br_2 \xrightarrow[CH_3OH]{} CH_3C(CH_3)(CH_3O){-}CH(Br)CH_3$$

k. $$CH_3C(CH_3){=}CHCH_3 \xrightarrow[2.\ H_2O_2,\ HO^-,\ H_2O]{1.\ BH_3/THF} CH_3CH(CH_3){-}CH(OH)CH_3$$

59. **a.** 2,2-diethyl-3-isopropyloxirane or 3,4-epoxy-4-ethyl-2-methylhexane

b. 3-ethyl-2,2-dimethyloxirane or 2,3-epoxy-2-methylpentane

60. R_2BH is the electrophile, and H^- is the nucleophile.

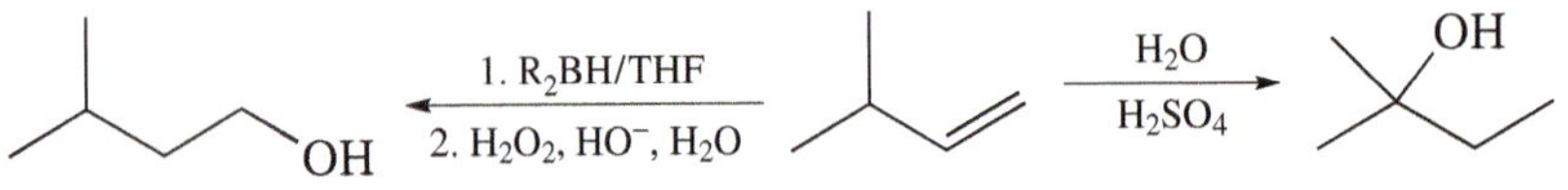

(alcohol, OH) ← 1. R_2BH/THF; 2. H_2O_2, HO^-, H_2O — (alkene) — H_2O / H_2SO_4 → (tertiary alcohol, OH)

forms a carbocation intermediate, so a carbocation rearrangement will occur

61. **a.** $CH_3CH_2C(=O)CH_3$ + $CH_3CH_2C(=O)H$

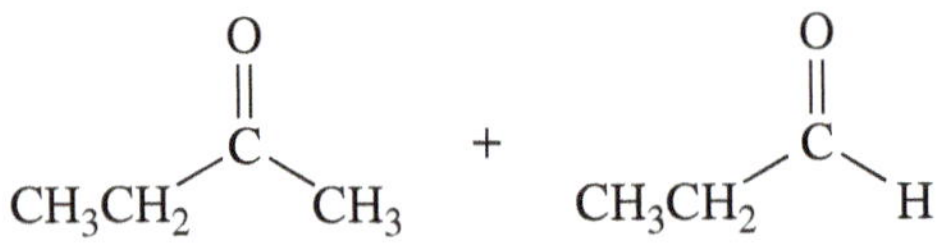

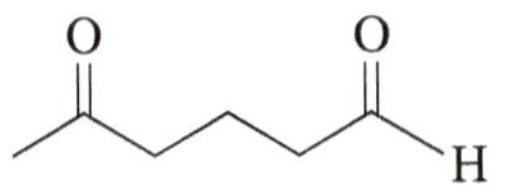

oxidation

b. $CH_3CH(CH_3)CH_2CH_2CH_3$

reduction

c. (keto aldehyde, O, O, H)

oxidation

62. $CH_3CH(CH_3)CH{=}CH_2$ + H—Br ⟶ $CH_3C(H)(CH_3)—\overset{+}{C}HCH_3$ —1,2-hydride shift→ $CH_3\overset{+}{C}(CH_3)CH_2CH_3$

$:\ddot{Br}:^-$ ↓ $:\ddot{Br}:^-$ ↓

$CH_3CH(CH_3)CH(Br)CH_3$ 2-bromo-3-methylbutane

$CH_3C(Br)(CH_3)CH_2CH_3$ 2-bromo-2-methylbutane

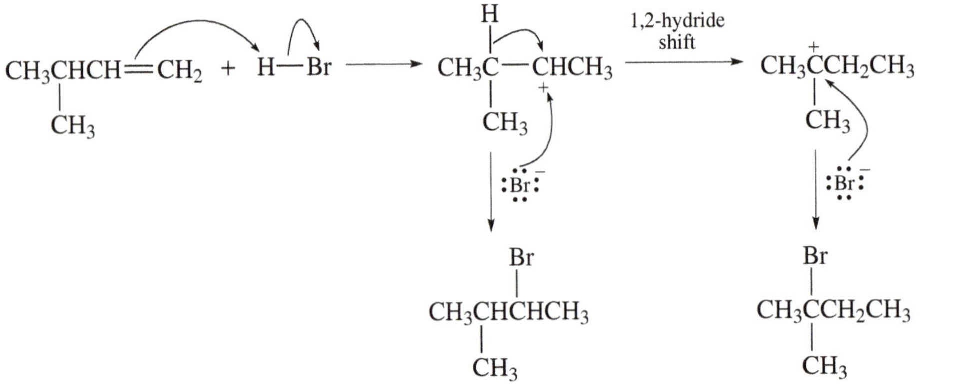

63. **a.** $CH_3—C(\ddot{O}:^-)(CH_3)—OCH_3$ ⟶ $CH_3—C(=\ddot{O}:)—CH_3$ + CH_3O^-

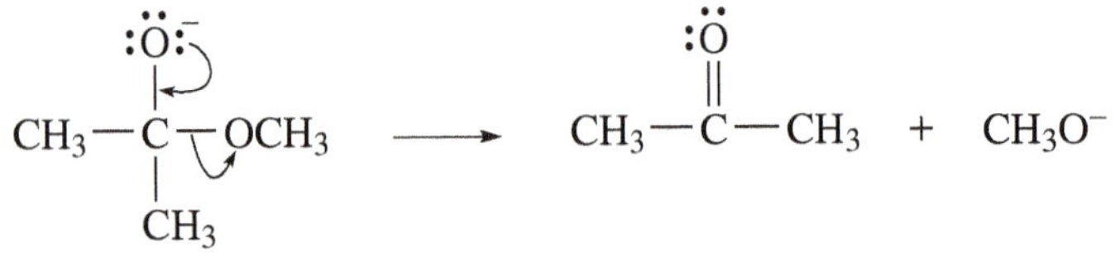

b. $CH_3C{\equiv}C—H$ + $:\ddot{N}H_2^-$ ⟶ $CH_3C{\equiv}C:^-$ + $:NH_3$

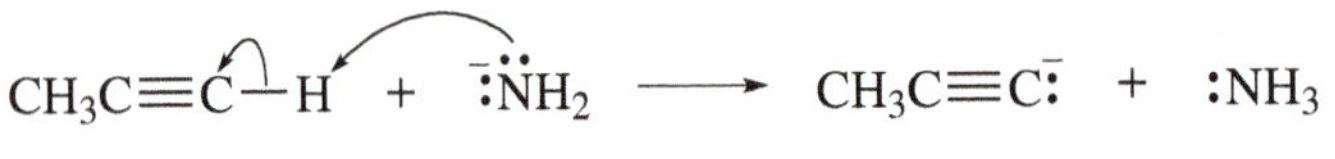

c. $CH_3CH_2—Br$ + $CH_3\ddot{O}:^-$ ⟶ $CH_3CH_2—\ddot{O}CH_3$ + Br^-

64.

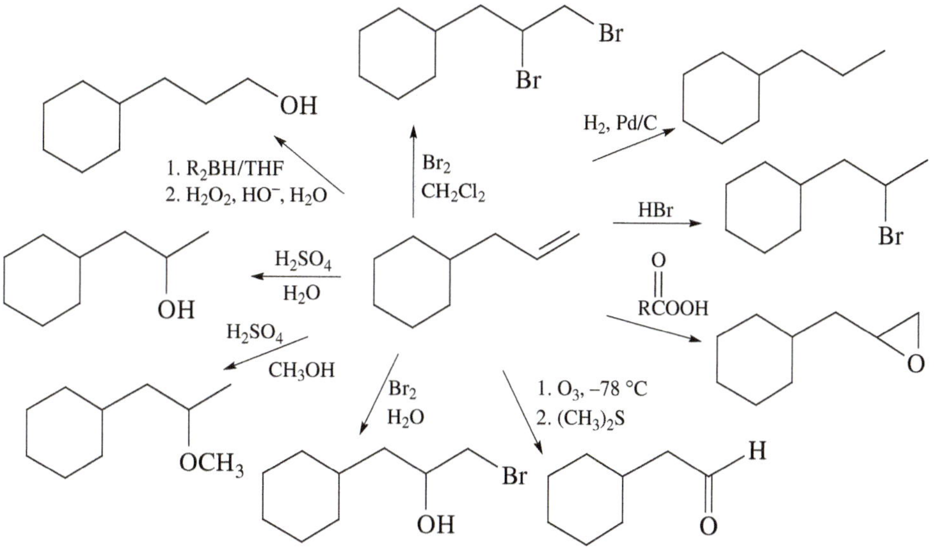

OH

1. R_2BH/THF
2. H_2O_2, HO^-, H_2O

Br

Br

Br_2
CH_2Cl_2

H_2, Pd/C

HBr

Br

H_2SO_4
H_2O

OH

O
‖
RCOOH

O

H_2SO_4
CH_3OH

OCH_3

Br_2
H_2O

Br

OH

1. O_3, −78 °C
2. $(CH_3)_2S$

H

O

65. **a.** 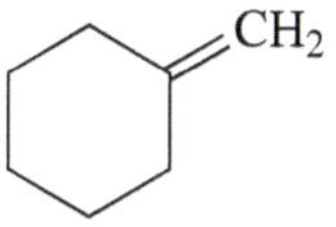 CH_2 and CH_3

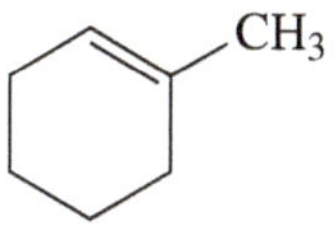

b. (Note: D stands for deuterium, an isotope of hydrogen; DBr reacts in a manner similar to HBr.) While HBr forms the same product when it reacts with the two alkenes, DBr forms different products. They are shown here.

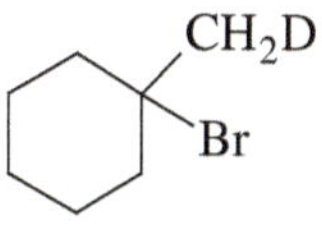

CH_2D, Br and D, CH_3, Br

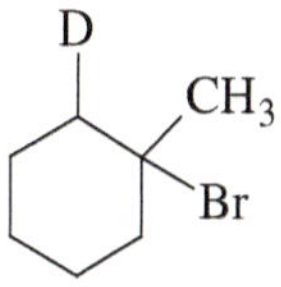

66. **a.** 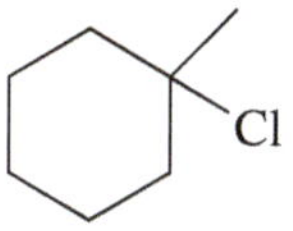 Cl

d. 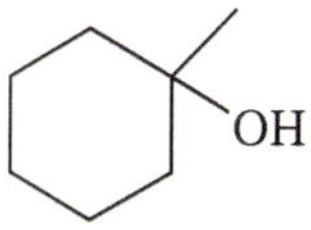 OH

g. 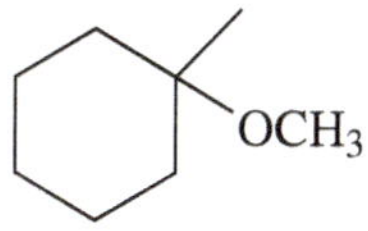 OCH_3

b. 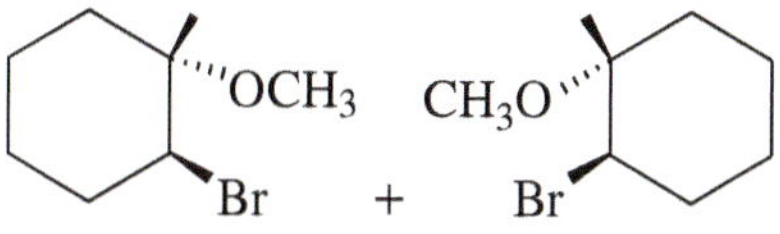 OCH_3, Br + CH_3O, Br

e. 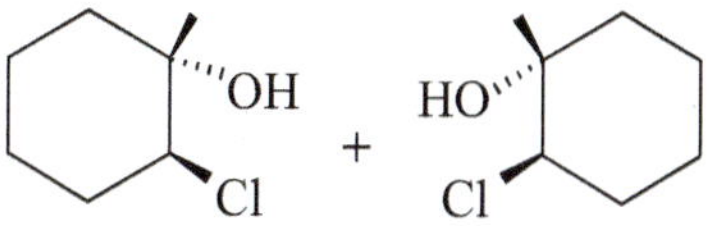 OH, Cl + HO, Cl

h. 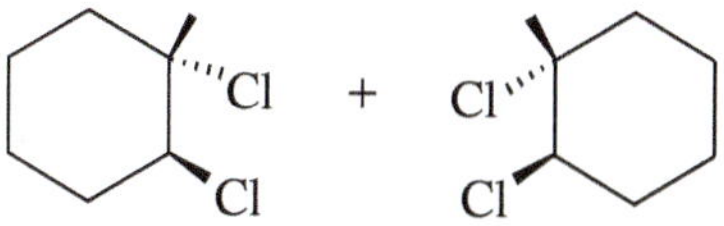 Cl, Cl + Cl, Cl

c. 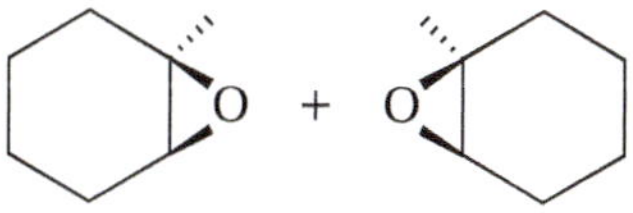 O + O

f. 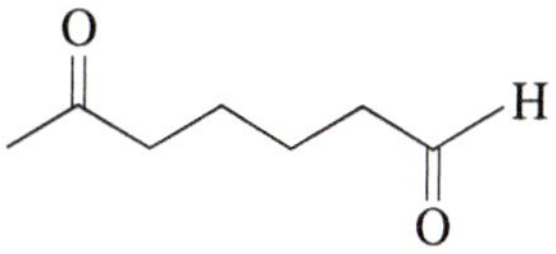 O, H, O

67. **a.** 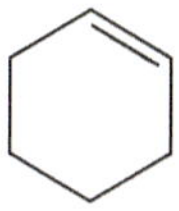 $\xrightarrow[\text{Pd/C}]{H_2}$

b. $CH_3CH_2CH_2CH{=}CH_2 \xrightarrow{HCl}$

c. CH_2 $\xrightarrow[\text{2. } H_2O_2,\ HO^-,\ H_2O]{\text{1. } R_2BH/THF}$

d. $CH_3CH_2CH{=}CHCH_2CH_3 \xrightarrow[H_2O]{Br_2}$

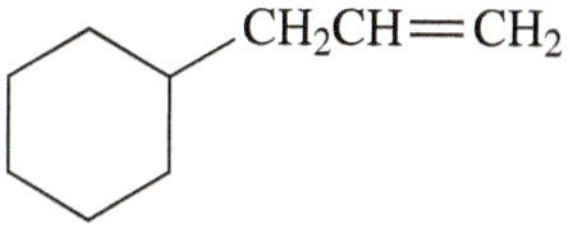

e. $CH_2CH{=}CH_2$ $\xrightarrow[H_2O]{H_2SO_4}$

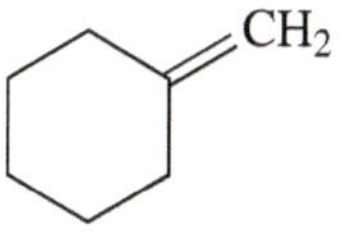

f. $CH_3CH_2CH{=}CHCH_2CH_3 \xrightarrow[\text{NaBr excess}]{Cl_2}$

or

$CH_3CH_2CH{=}CHCH_2CH_3 \xrightarrow[\text{NaCl excess}]{Br_2}$

68. **a.** CH_3 | $CH_3\overset{+}{C}CH_3$

Tertiary is more stable than secondary.

b. $CH_3\overset{+}{C}HCH_3$

The electron–withdrawing chlorine destabilizes the carbocation by increasing the amount of positive charge on the carbon.

c. +

Tertiary is more stable than secondary.

69. **a.** H H + H—Br ⟶ H + H + Br^- ⟶ H H + :Br: ⟶ H H Br

b.

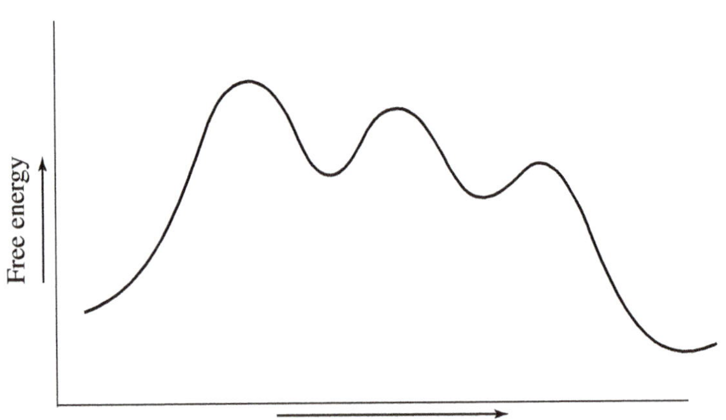

70. **a.** **1.** Both *cis*- and *trans*-2-butene form these products; in each case, a product with one asymmetric center is formed, so the product is a racemic mixture.

CH_2CH_3, C, CH_3, H, Cl — *S* + CH_2CH_3, C, H, Cl, CH_3 — *R* **or** CH_3, H—Cl, CH_2CH_3 — *S* + CH_3, Cl—H, CH_2CH_3 — *R*

2. Both *cis*- and *trans*-2-butene form these products; in each case, a product with one asymmetric center is formed, so the product is a racemic mixture.

CH_2CH_3, C, CH_3, H, OH — *S* + CH_2CH_3, C, H, HO, CH_3 — *R* **or** CH_3, H—OH, CH_2CH_3 — *S* + CH_3, HO—H, CH_2CH_3 — *R*

3. *cis*-2-Butene forms a meso compound; the product has two asymmetric centers, and only syn addition occurs.

trans-2-Butene forms a pair of enantiomers; the product has two asymmetric centers, and only syn addition occurs.

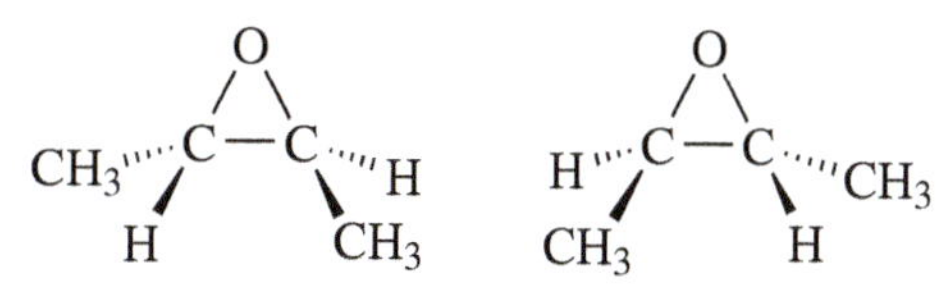

4. *cis*-2-Butene forms the threo pair of enantiomers; a product with two asymmetric centers is formed, and only anti addition of Br_2 occurs.

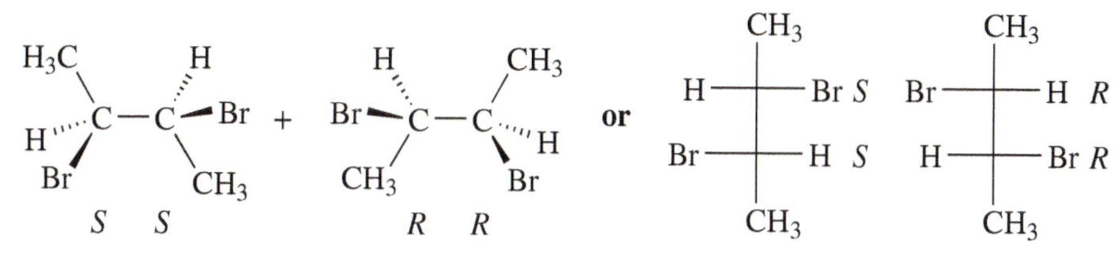

trans-2-Butene forms a meso compound; a product with two asymmetric centers is formed, and only anti addition of Br_2 occurs.

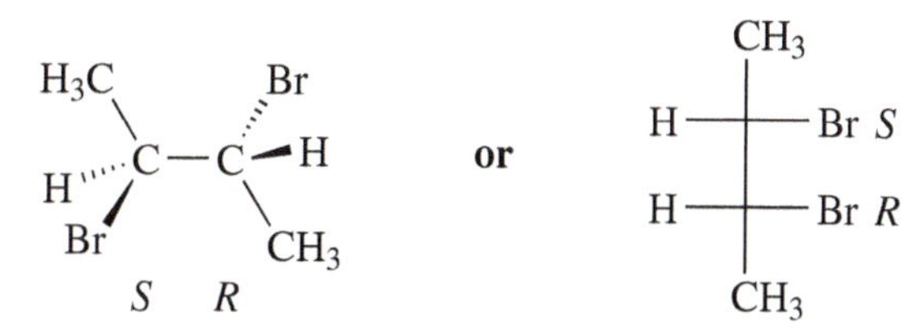

5. *cis*-2-Butene forms the threo pair of enantiomers; a product with two asymmetric centers is formed, and only anti addition of Br and OH occurs.

trans-2-Butene forms the erythro pair of enantiomers; a product with two asymmetric centers is formed, and only anti addition of Br and OH occurs.

6. Both *cis*- and *trans*-2-butene form this product; a product with no asymmetric centers is formed.

$$CH_3CH_2CH_2CH_3$$

7. Both *cis*- and *trans*-2-butene form these products; a product with one asymmetric center is formed, so the product is a racemic mixture.

CH_2CH_3, CH_3, H, OH — *S* + *R* or CH_3, H, OH, CH_2CH_3 — *S* + *R*

8. Both *cis*- and *trans*-2-butene form these products; a product with one asymmetric center is formed, so the product is a racemic mixture.

CH_2CH_3, CH_3, H, OCH_3 — *S* + *R* or CH_3, H, OCH_3, CH_2CH_3 — *S* + *R*

b. For *cis*- and *trans*-2-butene to form different products, (1) the reaction must form a product with two new asymmetric centers and (2) either syn or anti addition must occur (but not both). Therefore, the *cis*- and *trans*-2-butene form different products when they react with a peroxyacid, with Br_2, and with Br_2 in H_2O.

71. **a.** CH_3 :Br—Br → CH_3 Br^+ $CH_3\ddot{O}H$ → CH_3 OCH_3 Br + OCH_3 CH_3 Br

b. The addition of Br_2/CH_3OH is anti. In the second step of the reaction, CH_3OH can add to the cyclic bromonium ion from the top of the plane or from the bottom of the plane. The products are a pair of enantiomers.

72. **a.** To determine relative rates, the rate constant of each alkene is divided by the smallest rate constant of the series (3.51×10^{-8}).

		relative rates
propene	$= (4.95 \times 10^{-8})/(3.51 \times 10^{-8})$	$= 1.41$
(*Z*)-2-butene	$= (8.32 \times 10^{-8})/(3.51 \times 10^{-8})$	$= 2.37$
(*E*)-2-butene	$= (3.51 \times 10^{-8})/(3.51 \times 10^{-8})$	$= 1$
2-methyl-2-butene	$= (2.15 \times 10^{-4})/(3.51 \times 10^{-8})$	$= 6.12 \times 10^3$
2,3-dimethyl-2-butene	$= (3.42 \times 10^{-4})/(3.51 \times 10^{-8})$	$= 9.74 \times 10^3$

b. Both compounds form the same carbocation but, because (*Z*)-2-butene is less stable than (*E*)-2-butene, (*Z*)-2-butene has a smaller free energy of activation.

c. 2-Methyl-2-butene is more stable than (*Z*)-2-butene, and it forms a more stable carbocation intermediate (tertiary) and, therefore, a more stable transition state than does (*Z*)-2-butene (secondary). Knowing that 2-methyl-2-butene reacts faster tells us that the energy difference between the transition states is greater than the energy difference between the alkenes. This is what we would expect from the Hammond postulate, because the transition states look more like the carbocations than like the alkenes.

d. 2,3-Dimethyl-2-butene is more stable than 2-methylbutene, and both compounds form a tertiary carbocation intermediate. On this basis, you would predict that 2,3-dimethyl-2-butene would react more slowly than 2-methylbutane. However, 2,3-dimethyl-2-butene has two sp^2 carbons that can react with a proton to form the tertiary carbocation, whereas 2-methyl-2-butene has only one. The fact that 2,3-dimethyl-2-butene reacts faster in spite of being more stable tells us that the more important factor is the greater number of collisions with the proper orientation that lead to a productive reaction in the case of 2,3-dimethyl-2-butene.

73. **a.** $H(H)C{=}C(H)Cl$ has a greater dipole moment than $Cl(H)C{=}C(H)Cl$ where the C—Cl dipoles oppose each other

b. $Cl(H)C{=}C(H)CH_3$ has a greater dipole moment than

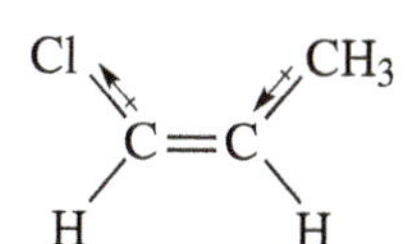

Recall that an sp^2 carbon is more electronegative than an sp^3 carbon. Therefore, the C—CH_3 and C—Cl dipoles reinforce each other, leading to a higher dipole, in the compound on the left and oppose each other in the compound on the right.

74. **a.** CH_3, Br^+, CH_2CH_3 $\xrightarrow{Br^-}$ CH_3, Br, Br, CH_2CH_3

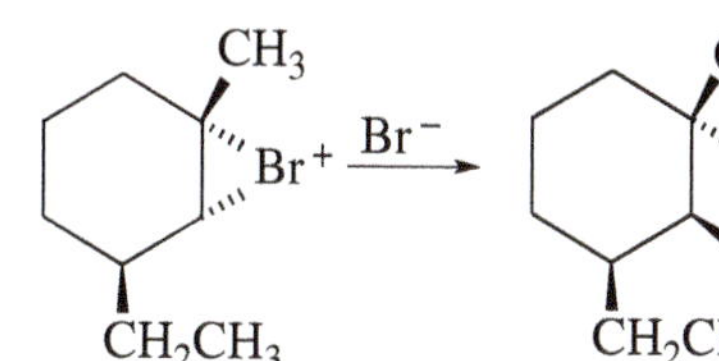

b. $CH_3CH_2CH_2C(CH_3)(H)OH$ + $H(HO)C(CH_3)CH_2CH_2CH_3$

c. CH_3, Br, Br, CH_3 CH_3, Br, Br, CH_3

d. $CH_3CH_2CH_2C(CH_2CH_3)(CH_3)Br$ + $CH_3C(CH_2CH_3)(Br)CH_2CH_2CH_3$

75. No, he should not follow his friend's advice. Adding the electrophile to the sp^2 carbon bonded to the most hydrogens forms a secondary carbocation in preference to a primary carbocation. However, in this case, the primary carbocation is more stable than the secondary carbocation. The electron-withdrawing fluorine substituents are closer to the positively charged carbon in the secondary carbon. Therefore, they will destabilize the secondary carbocation more than the primary carbocation. So the major product will be 1,1,1-trifluoro-3-iodopropane and not 1,1,1-trifluoro-2-iodopropane, the compound that would be predicted to be the major product by following the rule.

$F_3CCH_2—\overset{+}{C}H_2$ $F_3C\overset{+}{C}H—CH_3$

more stable   less stable because of the nearby electron-withdrawing fluorines

76. **a.** $CH_3CH_2CH{=}CH_2$ $\xrightleftharpoons{H—\overset{+}{O}(H)CH_3}$ $CH_3CH_2\overset{+}{C}HCH_3$ $\xrightleftharpoons{CH_3\ddot{O}H}$ $CH_3CH_2CH(\overset{+}{O}(H)CH_3)CH_3$ $\xrightleftharpoons{CH_3\ddot{O}H}$ $CH_3CH_2CH(OCH_3)CH_3$

b. the first step (See the answer to Problem 10.)
c. H^+
d. 1-butene
e. the *sec*-butyl cation
f. methanol

77. **a.** CH_2CH_3 (1-ethylcyclohexene) $\xrightarrow{HBr}$ CH_2CH_3, Br (1-bromo-1-ethylcyclohexane)

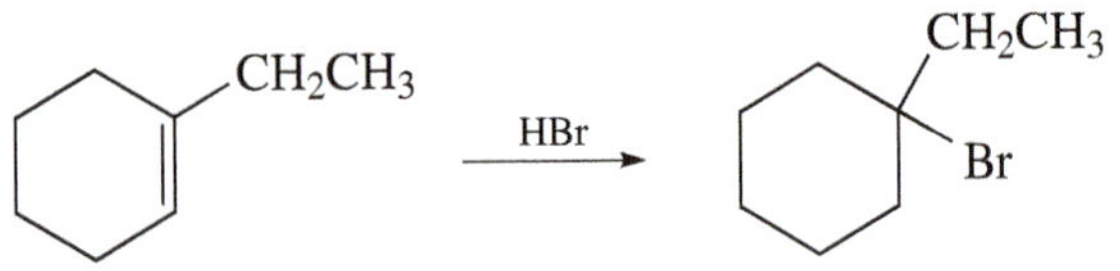

b. CH_2CH_3 $\xrightarrow[Pd/C]{H_2}$ CH_2CH_3 (ethylcyclohexane)

c. CH_2CH_3 $\xrightarrow[2.\ HO^-,\ H_2O_2,\ H_2O]{1.\ R_2BH/THF}$ CH_2CH_3, OH + CH_2CH_3, OH

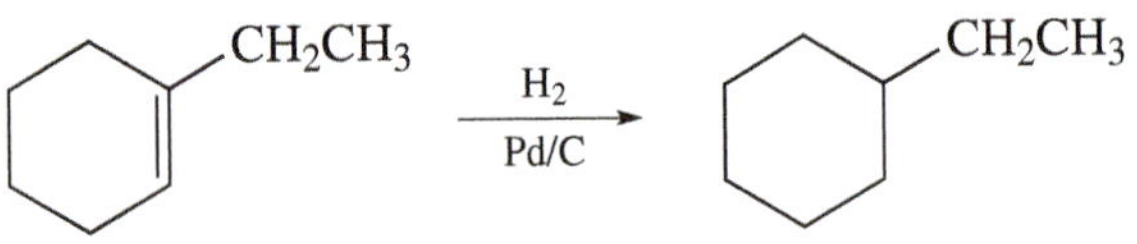

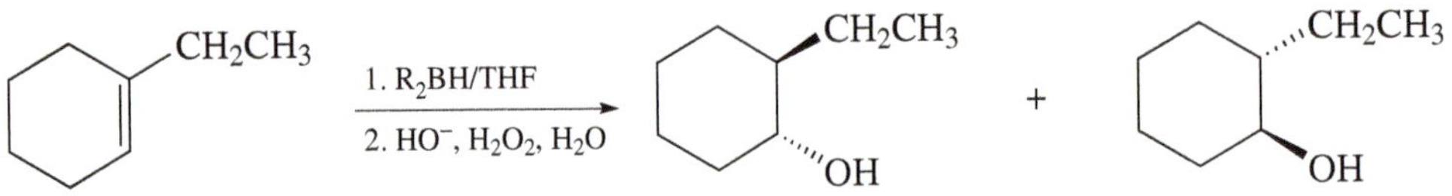

(The H and OH are added to the same side.)

d. CH_2CH_3 $\xrightarrow[CH_2Cl_2]{Br_2}$ CH_2CH_3, Br, Br CH_2CH_3, Br, Br

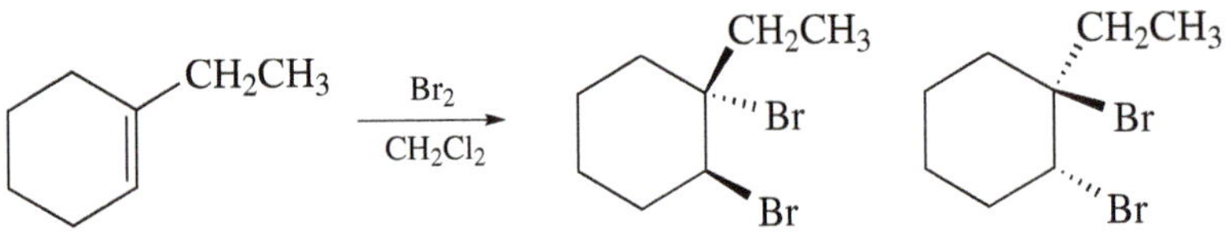

(The two bromines are added to opposite sides of the cyclohexane ring.)

78. CH_3CH_2, H, C=C, H, CH_2CH_3 $\xrightarrow{Br_2}$ CH_2CH_3; H—Br; H—Br; CH_2CH_3

trans-3-hexene

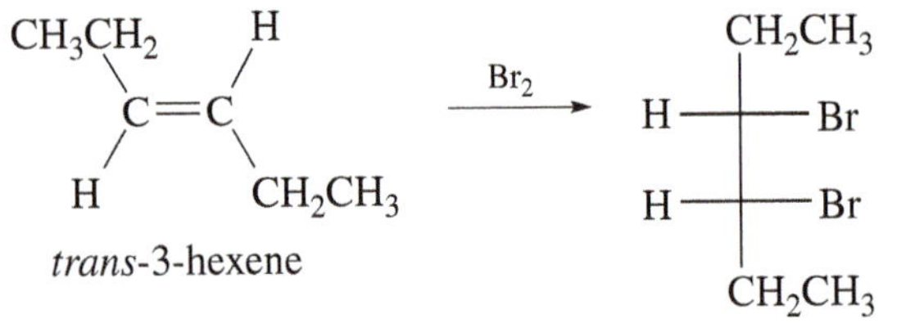

a meso compound

79. CH_3CH_2, CH_2CH_3, C=C, H, H $\xrightarrow{Br_2}$ CH_2CH_3; H—OH; Br—H; CH_2CH_3 + CH_2CH_3; HO—H; H—Br; CH_2CH_3

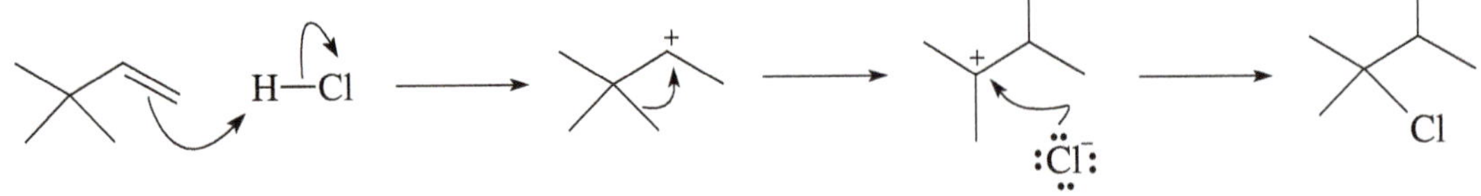

cis-3-hexene

(3*S*,4*S*)-4-bromo-3-hexanol (3*R*,4*R*)-4-bromo-3-hexanol

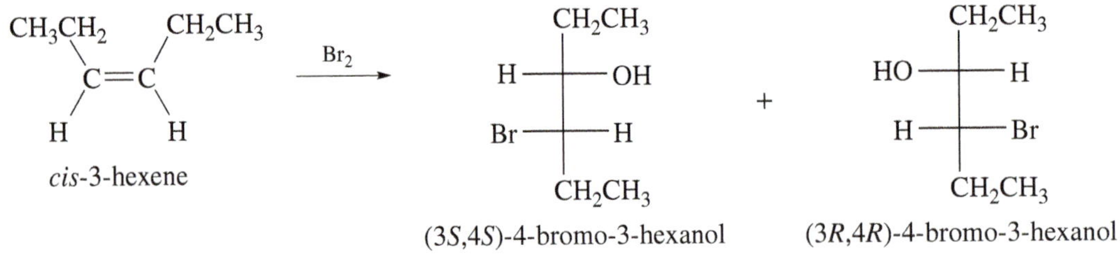

80. **a.** H—$\overset{+}{O}$H, H, $H_2\ddot{O}$:, $H_2\ddot{O}$:, H, $\overset{+}{:}\ddot{O}H$, $H_3\ddot{O}^+$, :$\ddot{O}$H

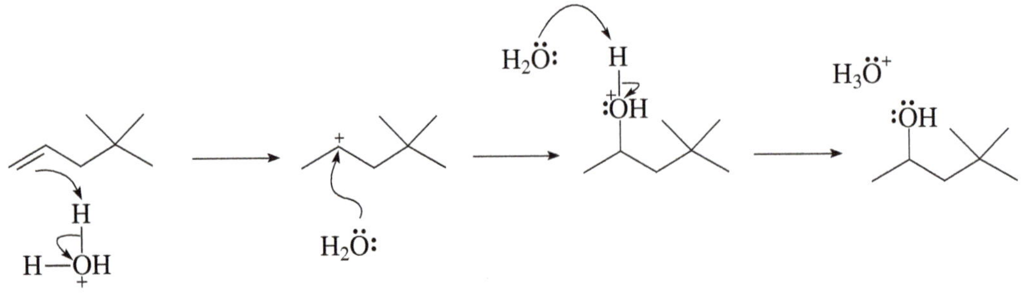

b. H—Cl, +, +, :$\ddot{Cl}^-$:, Cl

81. Five- and six-membered rings are more stable than smaller or larger rings.

a. Either a six-membered ring or a five-membered ring can be formed. However, because the more substituted three-membered ring carbon has the larger partial positive charge (it is secondary, and the less substituted one is primary), the three-membered ring breaks in the direction that forms the five-membered ring.

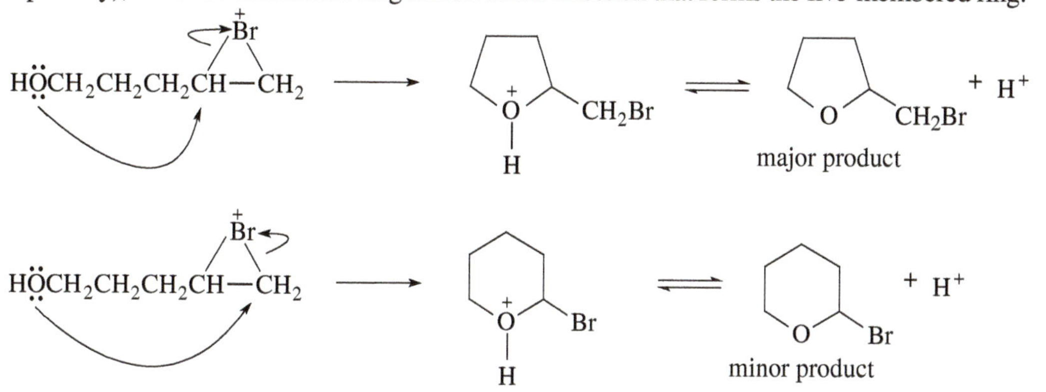

$HOCH_2CH_2CH_2CH_2CH{-}CH_2$ (bromonium ion, $\overset{+}{Br}$) → CH_2Br ⇌ CH_2Br + H^+

H

major product

$HOCH_2CH_2CH_2CH_2CH{-}CH_2$ (bromonium ion, $\overset{+}{Br}$) → Br ⇌ Br + H^+

H

minor product

b. A six-membered ring will form in preference to a less stable seven-membered ring. (In addition, formation of the six-membered ring involves the preferred attack on the more substituted ring carbon.)

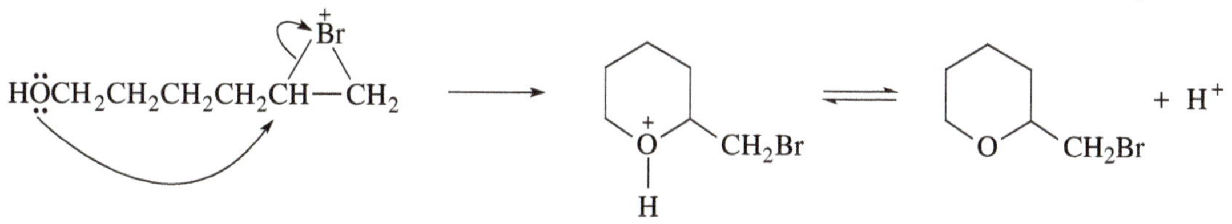

$HOCH_2CH_2CH_2CH_2CH_2CH{-}CH_2$ (bromonium ion, $\overset{+}{Br}$) → CH_2Br ⇌ CH_2Br + H^+

H

82. **a.** $CH_2CH_2CH_3$, CH_3, H, Cl (*S*) + $CH_2CH_2CH_3$, H, CH_3, Cl (*R*) + $CH_3CH_2CHCH_2CH_3$ (Cl)

or

CH_3, H—Cl, $CH_2CH_2CH_3$ (*S*) + CH_3, Cl—H, $CH_2CH_2CH_3$ (*R*) + $CH_3CH_2CHCH_2CH_3$ (Cl)

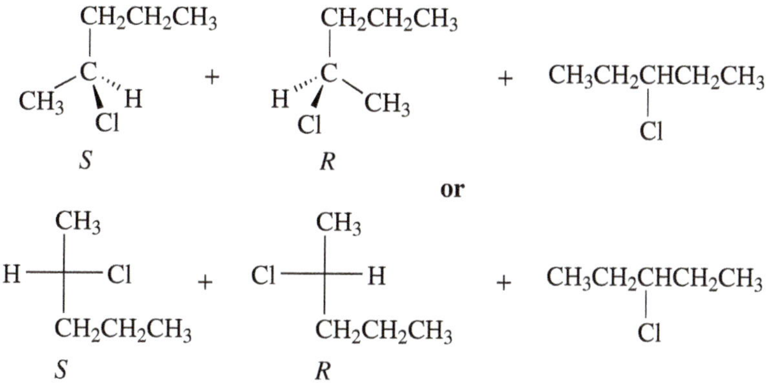

b. $CH_2CH_2CH_3$, CH_3, H, Cl (*S*) + $CH_2CH_2CH_3$, H, CH_3, Cl (*R*) + $CH_3CH_2CHCH_2CH_3$ (Cl)

or

CH_3, H—Cl, $CH_2CH_2CH_3$ (*S*) + CH_3, Cl—H, $CH_2CH_2CH_3$ (*R*) + $CH_3CH_2CHCH_2CH_3$ (Cl)

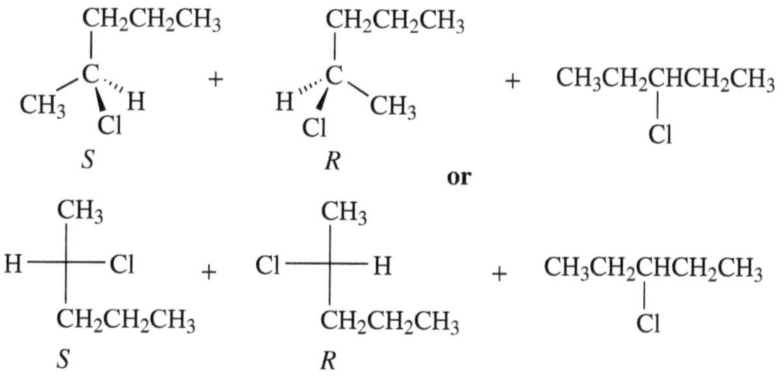

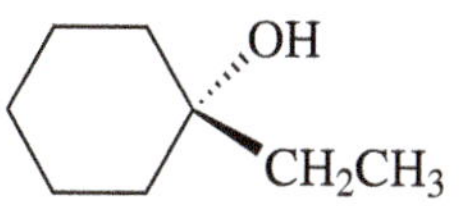

c. OH, CH_2CH_3

d. 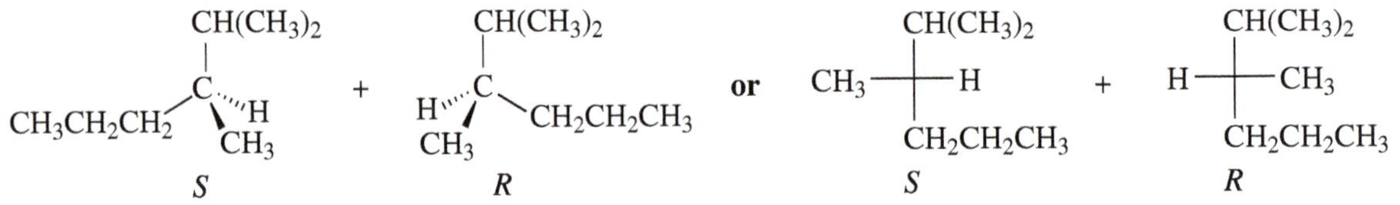

e. 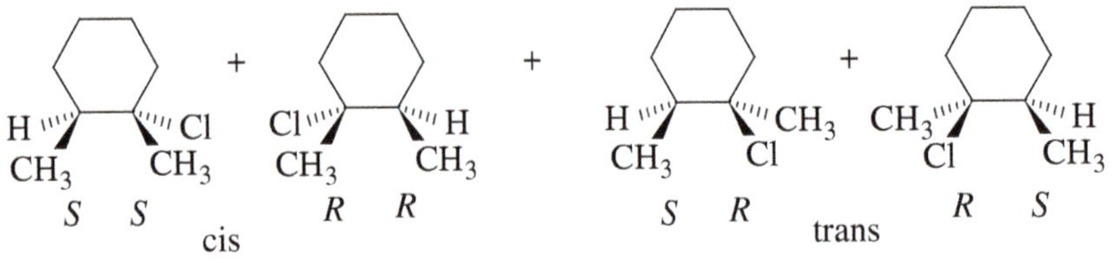

f. 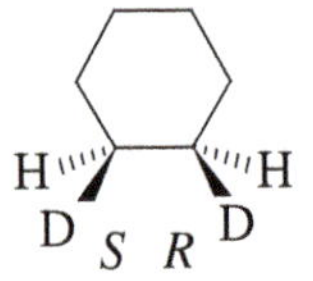

g. 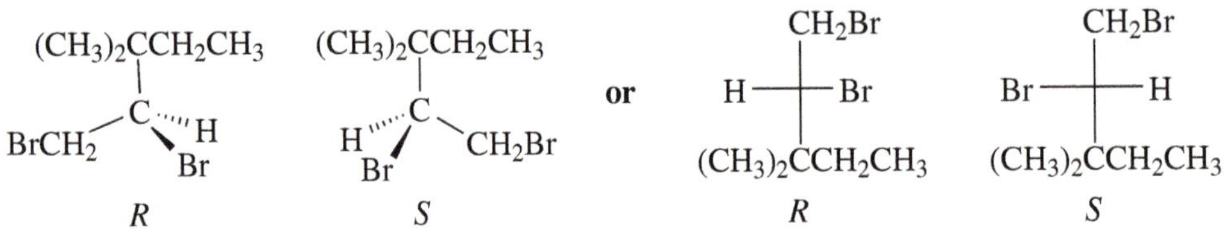

h. 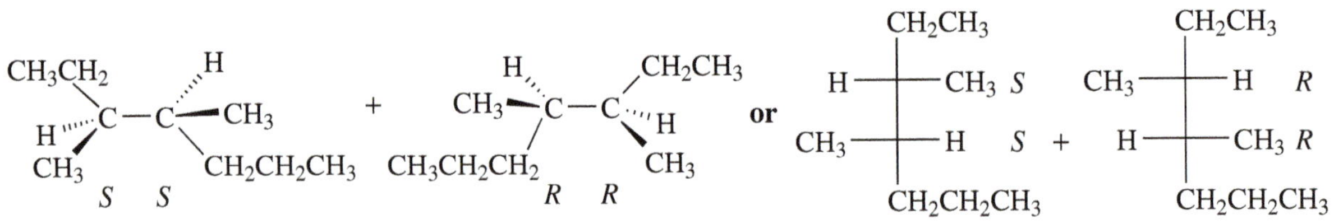

i. 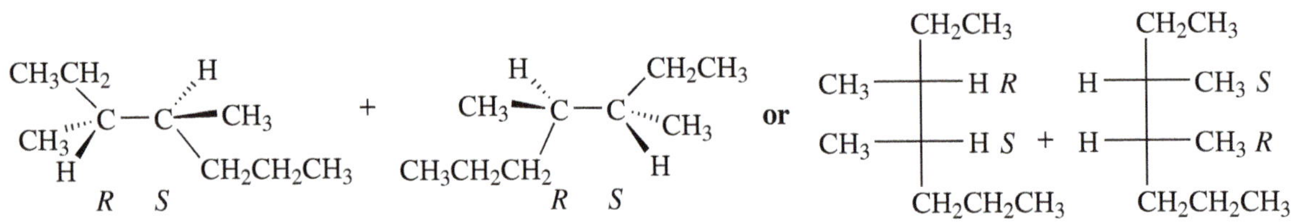

j. 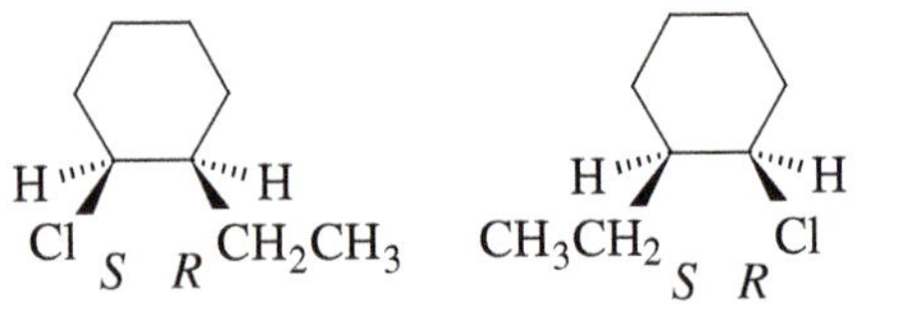

83. **a.** Both 1-butene and 2-butene react with HCl to form 2-chlorobutane.

b. Both alkenes form the same carbocation and, therefore, have transition states that are close in energy, but because 2-butene is more stable than 1-butene, 2-butene has the greater free energy of activation.

c. Both compounds form the same carbocation and, therefore, have transition states that are close in energy, but because (*Z*)-2-butene is less stable, it reacts more rapidly with HCl.

84. **a.** 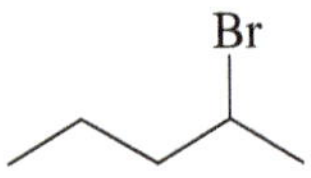 and 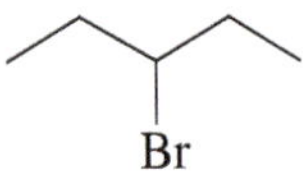 **b.** 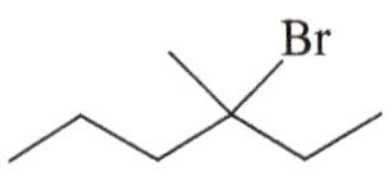

c. 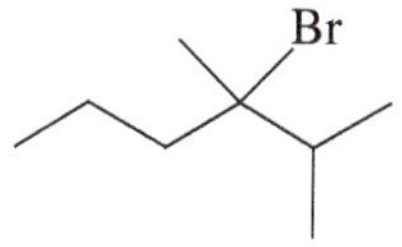 Br

d. 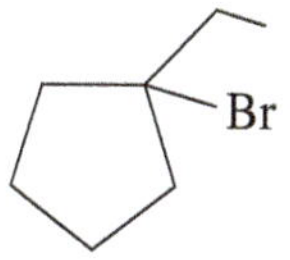 Br

85. a. 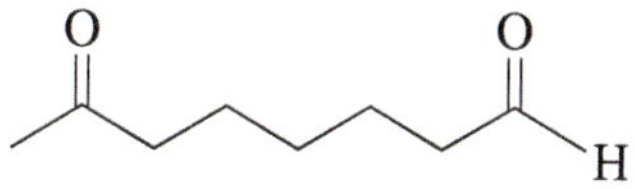 O O H

b. 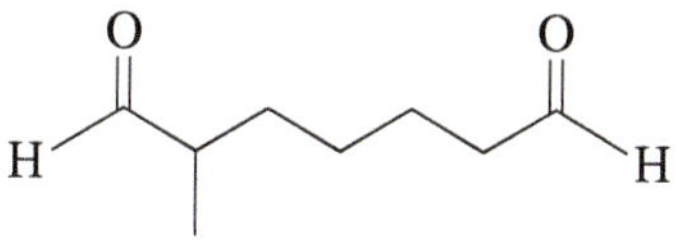 O O H H

c. $H_2C{=}O$ + H O O H O

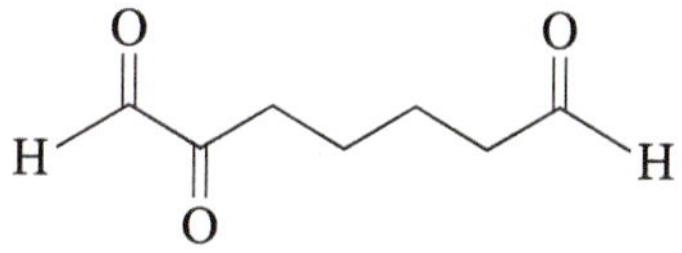

d. $H_2C{=}O$ + H O O H O + O O H H

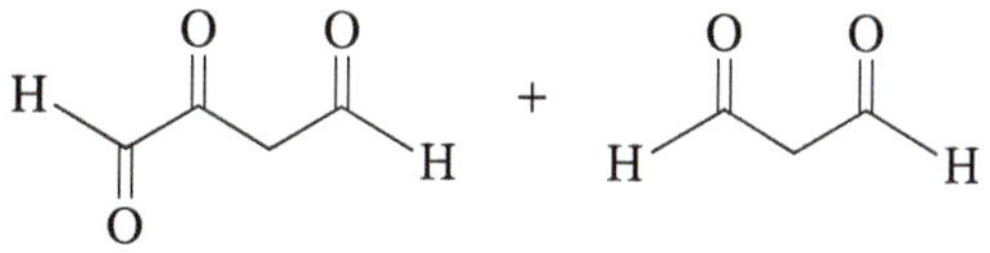

e. $H_2C{=}O$ + O O O H H + O H H O

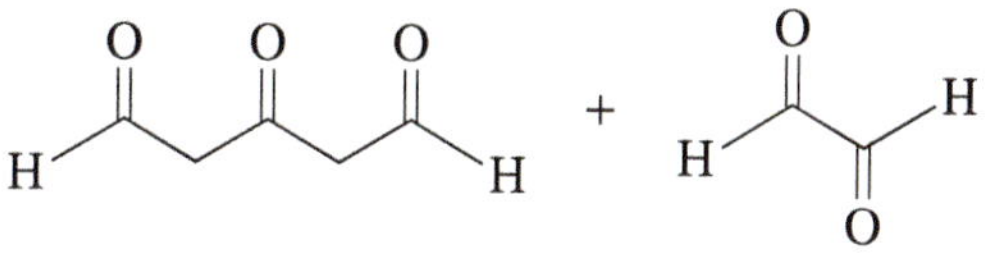

86. 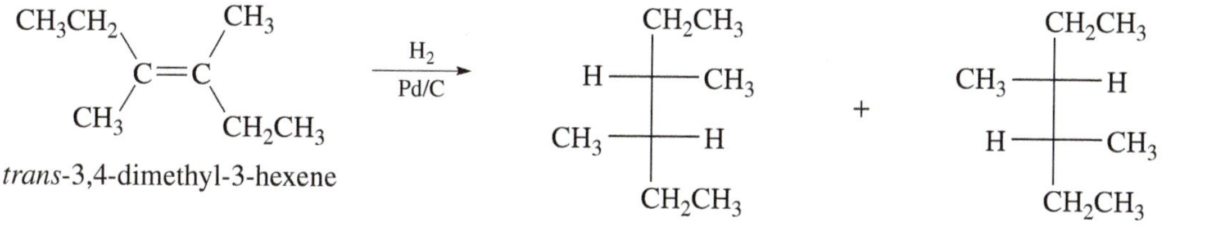 CH_3CH_2 CH_3 C=C CH_3 CH_2CH_3 $\xrightarrow[\text{Pd/C}]{H_2}$ CH_2CH_3 H — CH_3 CH_3 — H CH_2CH_3 + CH_2CH_3 CH_3 — H H — CH_3 CH_2CH_3

trans-3,4-dimethyl-3-hexene

(3*S*,4*S*)-3,4-dimethyhexane

(3*R*,4*R*)-3,4-dimethylhexane

87. a. 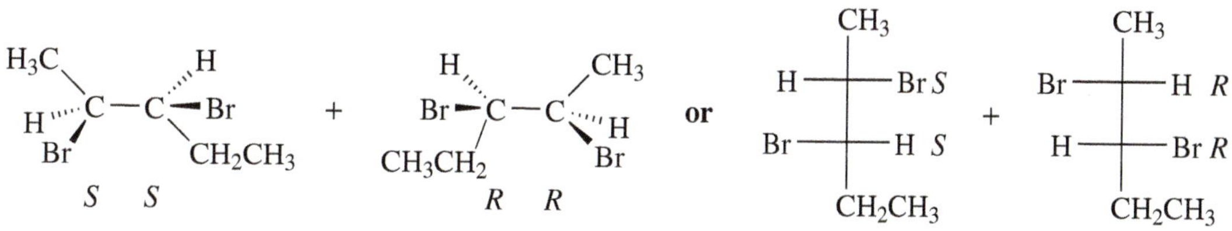 H_3C H H C—C Br Br CH_2CH_3 + H CH_3 Br C—C H CH_3CH_2 Br or CH_3 H — Br *S* Br — H *S* CH_2CH_3 + CH_3 Br — H *R* H — Br *R* CH_2CH_3

S *S* *R* *R*

b. H_3C H Br C—C Br H CH_2CH_3 + H CH_3 Br C—C Br CH_3CH_2 H or CH_3 Br — H *R* Br — H *S* CH_2CH_3 + CH_3 H — Br *S* H — Br *R* CH_2CH_3

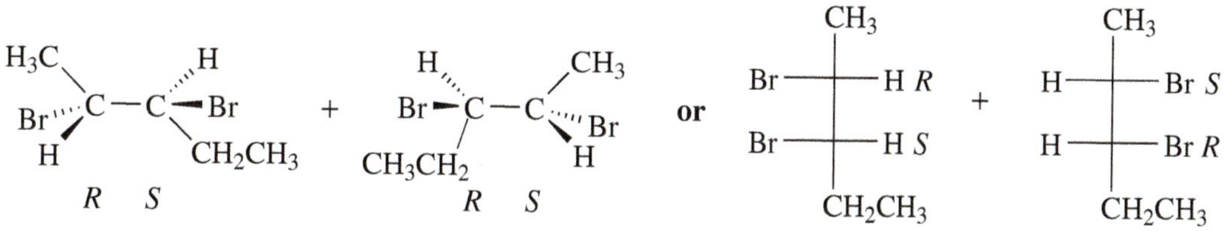

R *S* *R* *S*

c. 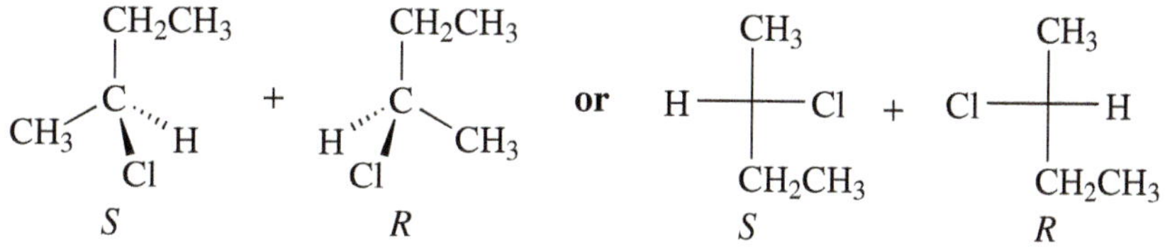 CH_2CH_3 C CH_3 H Cl + CH_2CH_3 C H CH_3 Cl or CH_3 H — Cl CH_2CH_3 + CH_3 Cl — H CH_2CH_3

S *R* *S* *R*

d. 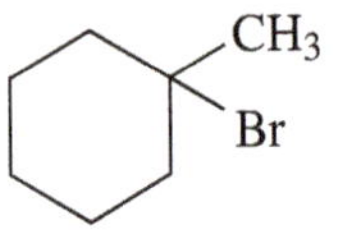 CH_3 Br

e. 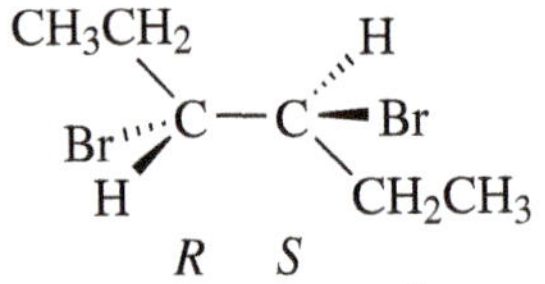 CH_3CH_2 H Br C—C Br H CH_2CH_3

R *S*

a meso compound

or

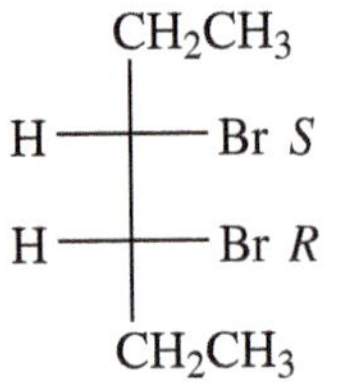

CH_2CH_3
H — Br *S*
H — Br *R*
CH_2CH_3

f. 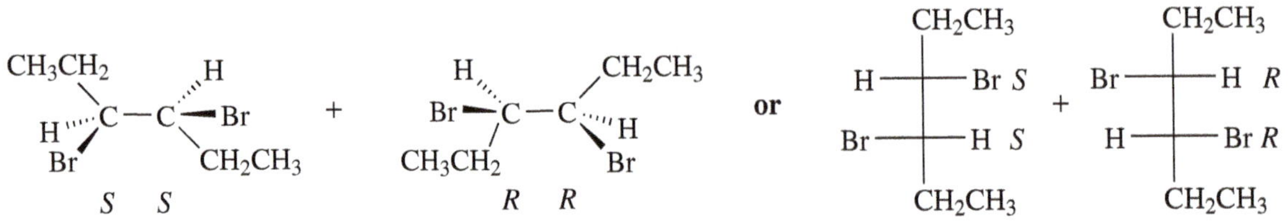 CH_3CH_2 H H C—C Br Br CH_2CH_3

S *S*

\+

H CH_2CH_3 Br C—C H CH_3CH_2 Br

R *R*

or

CH_2CH_3
H — Br *S*
Br — H *S*
CH_2CH_3

\+

CH_2CH_3
Br — H *R*
H — Br *R*
CH_2CH_3

g. The initially formed carbocation is secondary. It undergoes a 1,2-methyl shift to form a tertiary carbocation that forms the products shown below.

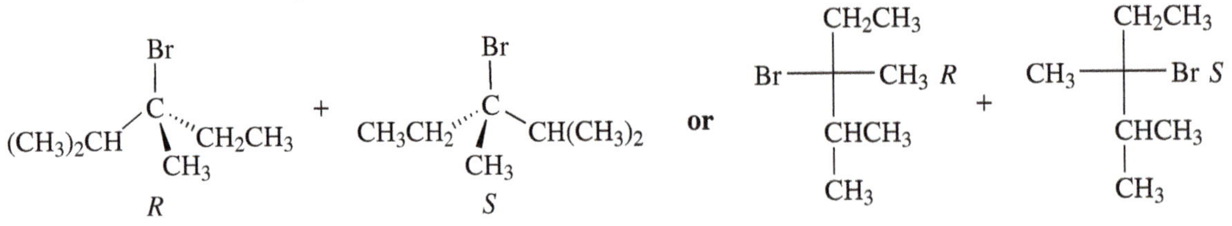

Br C $(CH_3)_2CH$ CH_2CH_3 CH_3

R

\+

Br C CH_3CH_2 $CH(CH_3)_2$ CH_3

S

or

CH_2CH_3
Br — CH_3 *R*
$CHCH_3$
CH_3

\+

CH_2CH_3
CH_3 — Br *S*
$CHCH_3$
CH_3

h. 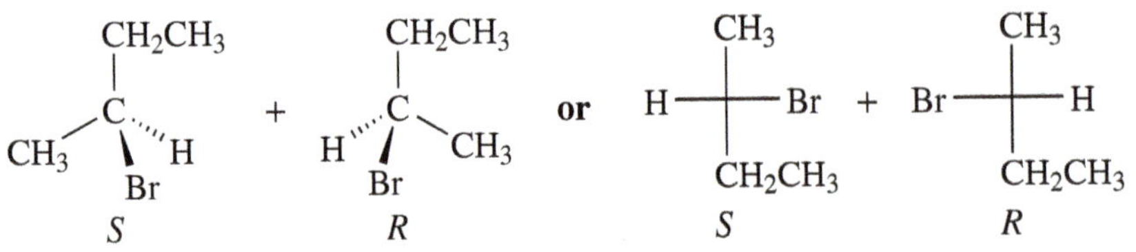 CH_2CH_3 C CH_3 H Br

S

\+

CH_2CH_3 C H CH_3 Br

R

or

CH_3
H — Br
CH_2CH_3
S

\+

CH_3
Br — H
CH_2CH_3
R

i. 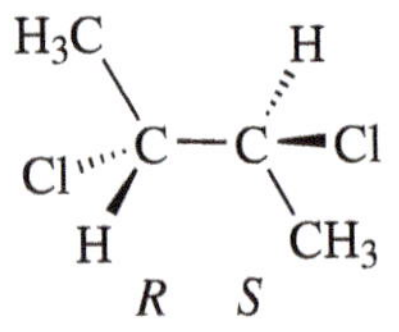 H_3C H Cl C—C Cl H CH_3

R *S*

a meso compound

or

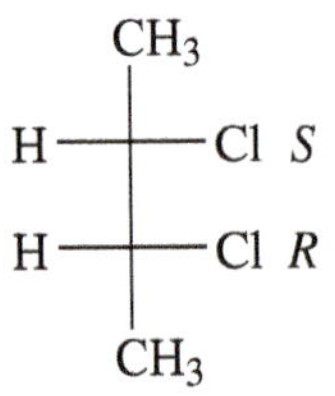

CH_3
H — Cl *S*
H — Cl *R*
CH_3

j. 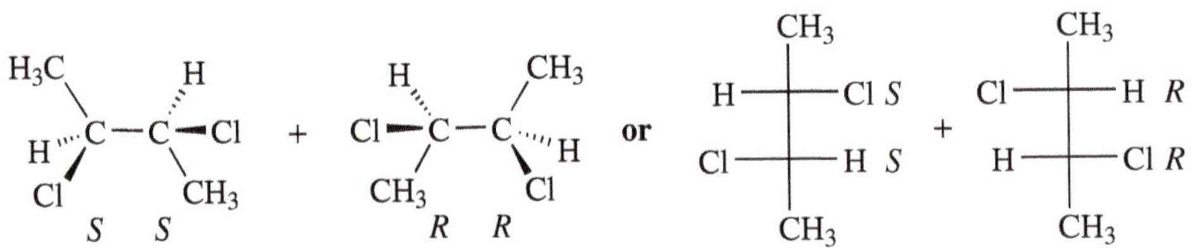 H_3C H H C—C Cl Cl CH_3

S *S*

\+

H CH_3 Cl C—C H CH_3 Cl

R *R*

or

CH_3
H — Cl *S*
Cl — H *S*
CH_3

\+

CH_3
Cl — H *R*
H — Cl *R*
CH_3

k. 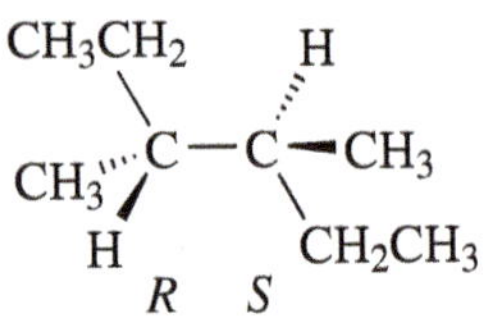 CH_3CH_2 H CH_3 C—C CH_3 H CH_2CH_3

R *S*

a meso compound

or

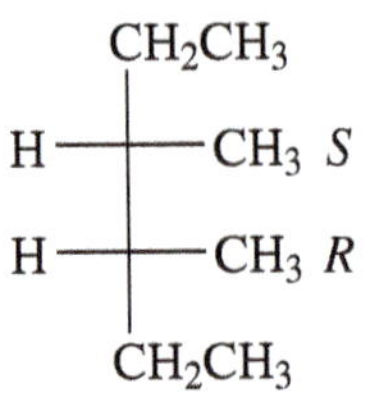

CH_2CH_3
H — CH_3 *S*
H — CH_3 *R*
CH_2CH_3

l. 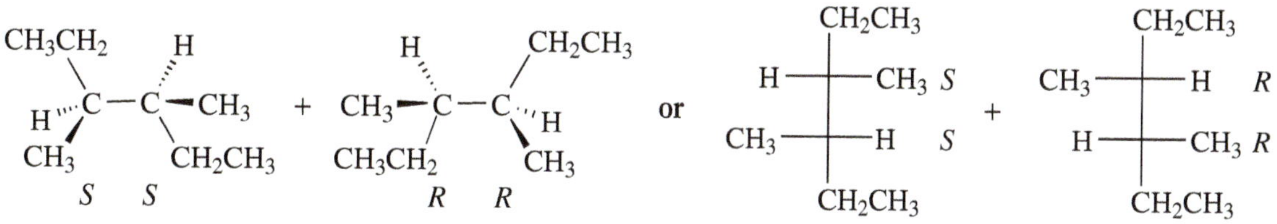 CH₃CH₂, H, C—C, CH₃ ... S S + R R or (Fischer projections: H—CH₃ S, CH₃—H S + CH₃—H R, H—CH₃ R)

88. The first product would not be formed, because none of the bonds attached to C-2 were broken during the reaction. Therefore, the configuration at C-2 cannot change.

89. **A** has two degrees of unsaturation. Because it has three methylene groups, it must be a methyl-substituted cyclopentene. Because **A** forms only one product when it reacts with aqueous acid, **A** must be 1-methylcyclopentene. Therefore, **A–F** have the following structures.

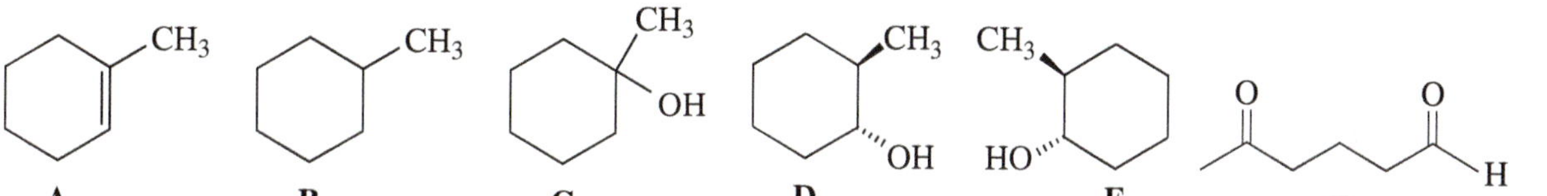

A **B** **C** **D** **E** **F**

90. Diazomethane is a very reactive compound because the triple-bonded nitrogen has a strong propensity to depart from the carbon to form a very stable molecule of nitrogen gas. As it departs, the nucleophilic alkene attacks the electrophilic carbon, and in the same step, the lone pair is the nucleophile that adds to the other sp^2 carbon of the alkene.

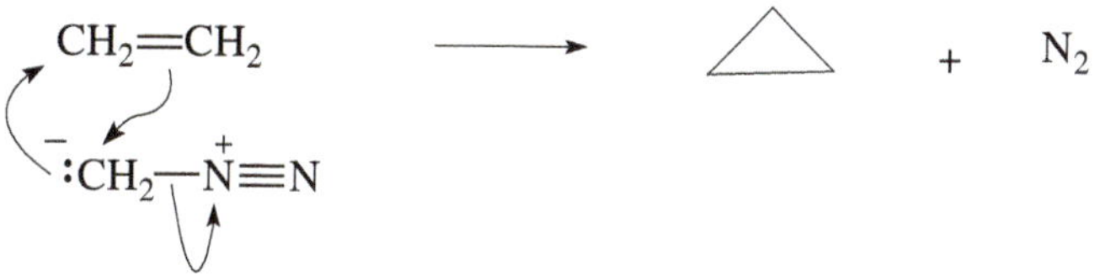

$CH_2{=}CH_2$ + $^{-}{:}CH_2{-}\overset{+}{N}{\equiv}N$ ⟶ (cyclopropane) + N_2

91. $I{-}\ddot{C}H_2^{-}$ reacts like Br_2. The CH_2 group is the electrophile that adds to the sp^2 carbon, and its lone pair is the nucleophile that adds to the other sp^2 carbon.

$I{-}CH_2{-}ZnI$ reacts as if it were $I{-}\ddot{C}H_2^{-}$ $\overset{+}{Z}nI$

$I{-}\ddot{C}H_2^{-}$ + $CH_2{=}CH_2$ ⟶ (cyclopropane) + I^-

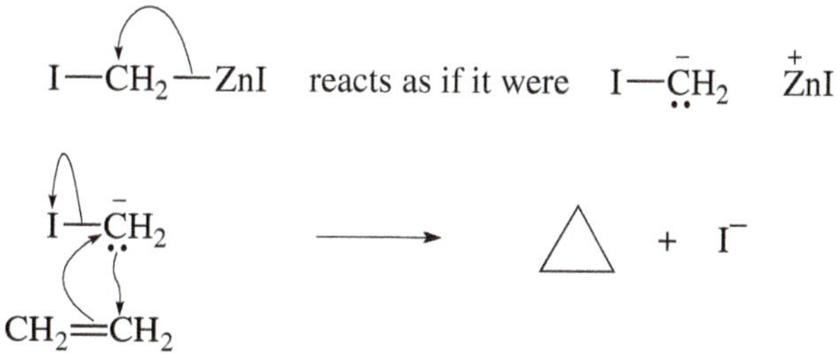

92. **a.** The base removes a proton. Then do what is needed to get the known product of the reaction.

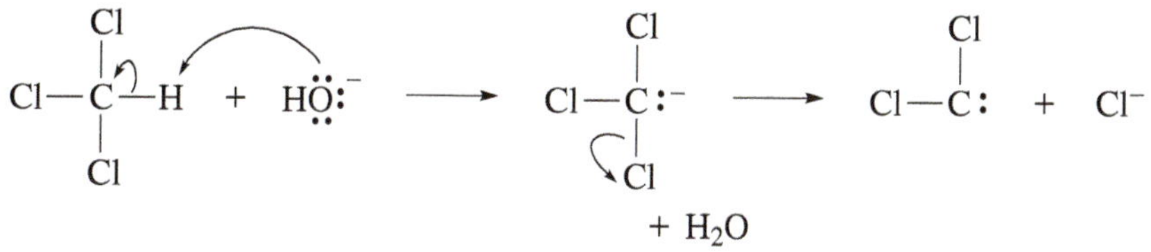

$Cl_3C{-}H$ + HO^- ⟶ $Cl_3C{:}^-$ + H_2O ⟶ $Cl_2C{:}$ + Cl^-

b. Notice that this reaction has the same mechanism as that in part **a** except instead of a base removing a proton and leaving behind its bonding electrons, heat causes CO_2 to be removed, leaving behind its bonding electrons.

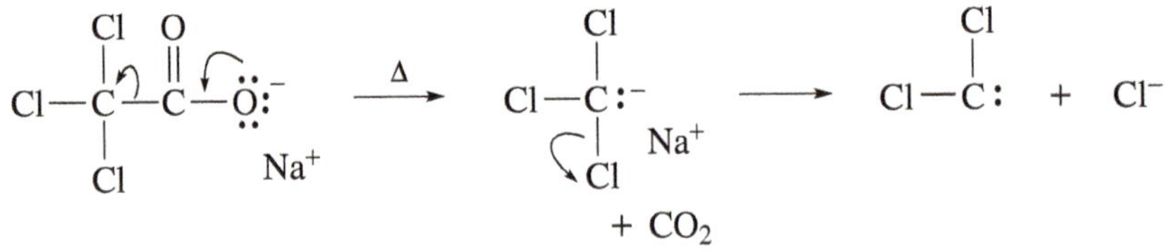

$$Cl_3C{-}CO_2^-\ Na^+ \xrightarrow{\Delta} Cl_3C{:}^-\ Na^+ + CO_2 \longrightarrow Cl_2C{:} + Cl^-$$

93. 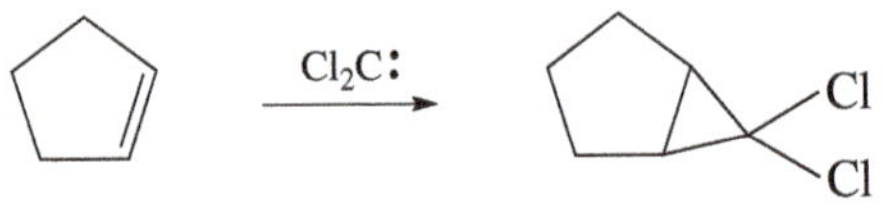 $\xrightarrow{Cl_2C:}$ Cl Cl

94. **a.** 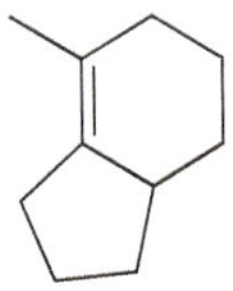 **b.** 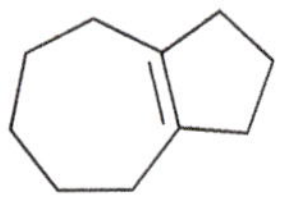 **c.**

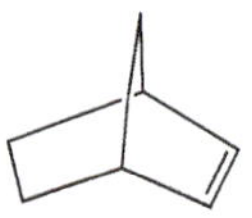

95. HBr; S; R; or; CH(CH3)2, Br, CH3, CH2CH3, S; CH(CH3)2, CH3, Br, CH2CH3, R

Br_2 / CH_3OH; S S; R R; or; H—Br S, CH3O—CH3 S; Br—H R, CH3—OCH3 R

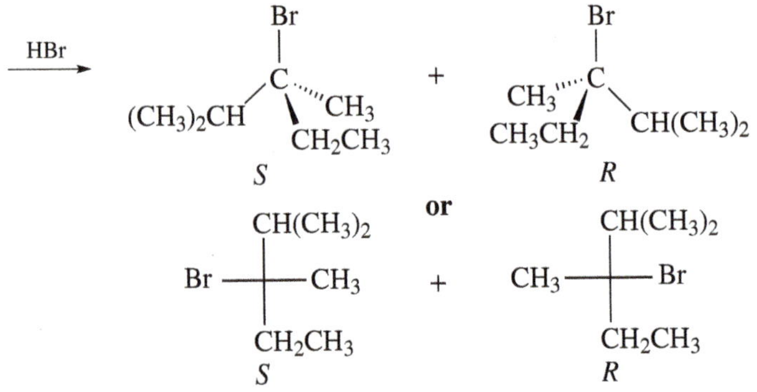

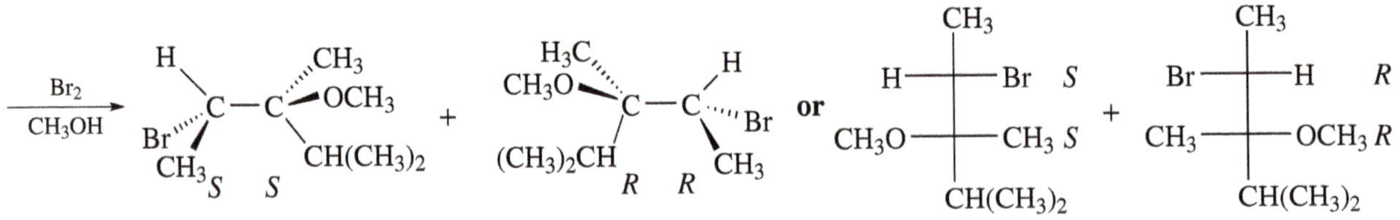

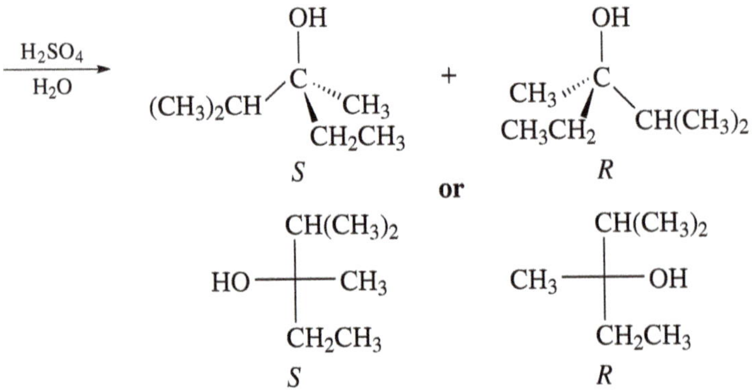

H_2SO_4 / H_2O; S; R; or; CH(CH3)2, HO, CH3, CH2CH3, S; CH(CH3)2, CH3, OH, CH2CH3, R

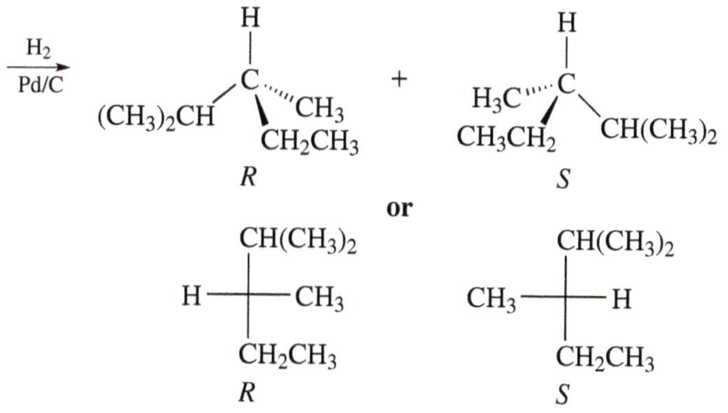

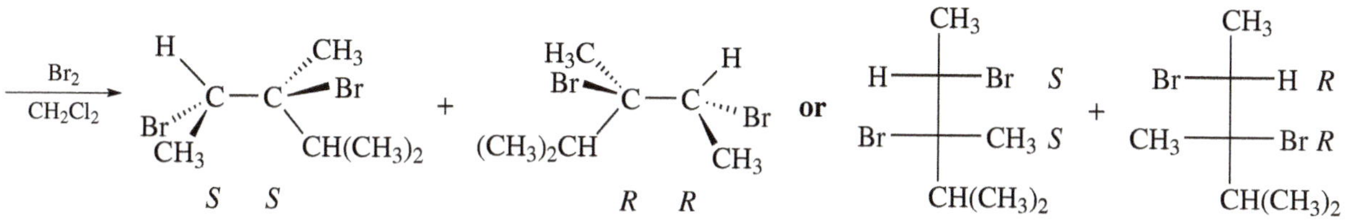

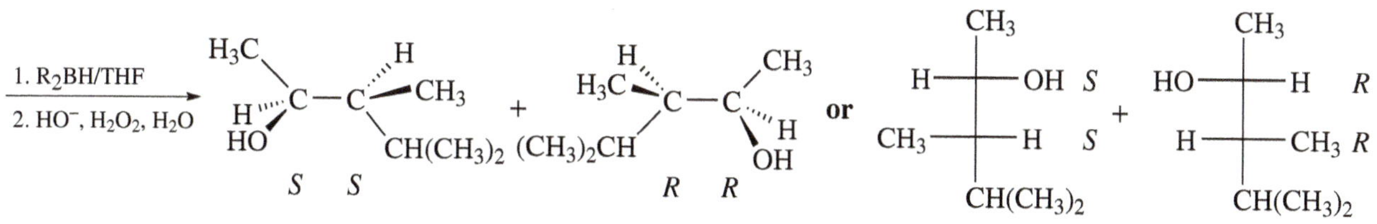

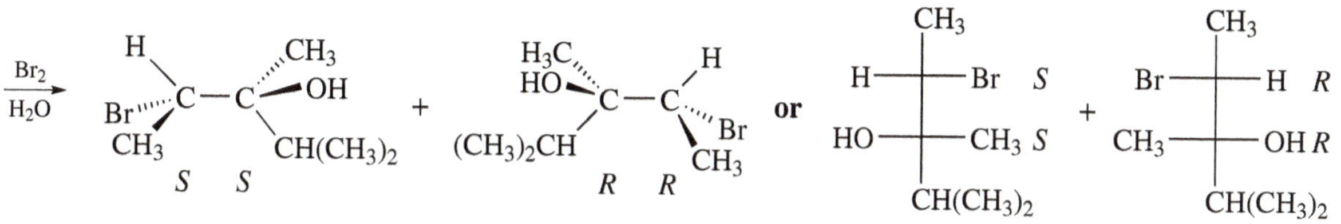

$\xrightarrow[\text{Pd/C}]{H_2}$ (R)-$(CH_3)_2CH$–CH(CH_3)(CH_2CH_3) + (S)-$(CH_3)_2CH$–CH(CH_3)(CH_2CH_3)

R *S*

or

Fischer projections: *R* (H left, CH_3 right; top $CH(CH_3)_2$, bottom CH_2CH_3) and *S* (CH_3 left, H right; top $CH(CH_3)_2$, bottom CH_2CH_3)

$\xrightarrow[CH_2Cl_2]{Br_2}$ *S S* + *R R* or

	CH_3	
H		Br *S*
Br		CH_3 *S*
	$CH(CH_3)_2$	

+

	CH_3	
Br		H *R*
CH_3		Br *R*
	$CH(CH_3)_2$	

$\xrightarrow[\text{2. HO}^-,\ H_2O_2,\ H_2O]{\text{1. }R_2BH/THF}$ *S S* + *R R* or

	CH_3	
H		OH *S*
CH_3		H *S*
	$CH(CH_3)_2$	

+

	CH_3	
HO		H *R*
H		CH_3 *R*
	$CH(CH_3)_2$	

$\xrightarrow[H_2O]{Br_2}$ *S S* + *R R* or

	CH_3	
H		Br *S*
HO		CH_3 *S*
	$CH(CH_3)_2$	

+

	CH_3	
Br		H *R*
CH_3		OH *R*
	$CH(CH_3)_2$	

96. **a.** H—Br; secondary; tertiary; $:\ddot{Br}:^-$; Br

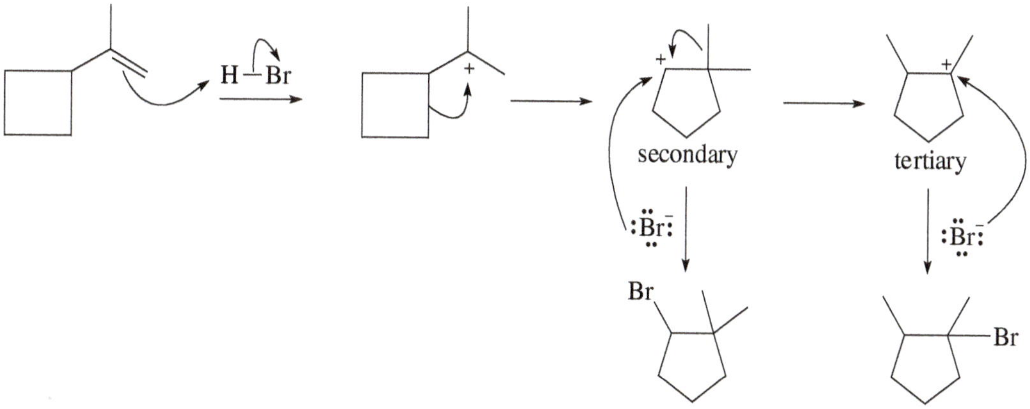

b. The initially formed carbocation is tertiary.

c. The rearranged carbocation is secondary, which undergoes another rearrangement to a more stable tertiary carbocation.

d. The initially formed carbocation rearranges in order to release the strain in the four-membered ring. (A tertiary carbocation with a strained four-membered ring is less stable than a secondary carbocation with an unstrained five-membered ring.)

97. 2-Methylpropene will be hydrated more rapidly.

1. It is more reactive than the chloro-substituted alkene, because the electron-withdrawing chlorine makes the alkene less nucleophilic.
2. The carbocation intermediate that 2-methylpropene forms (and, therefore, the transition state leading to its formation) is more stable, because the electron-withdrawing chlorine increases the amount of positive charge on the carbon.

98. It tells us that the first step of the mechanism is the slow step. If the first step is slow, the carbocation will react with water in a subsequent fast step, which means that the carbocation will not have time to lose a proton to reform the alkene, so all the deuterium atoms (D) will be retained in the unreacted alkene.

$$C_6H_5-CH{=}CD_2 \xrightarrow{H_2SO_4} C_6H_5-\overset{+}{C}H-CHD_2 \xrightarrow[\text{fast}]{H_2O} C_6H_5-CH(OH)-CHD_2$$

$$C_6H_5-\overset{+}{C}H-CHD_2 \not\longrightarrow C_6H_5-CH{=}CHD + D^+$$

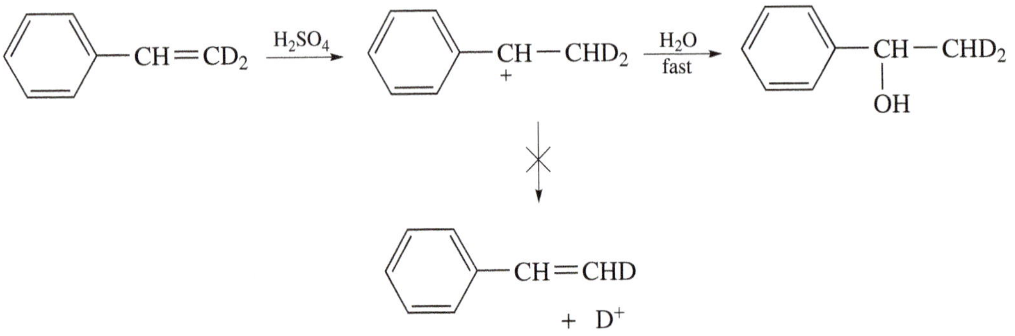

If the first step is not the slow step, an equilibrium will be set up between the alkene and the carbocation, and because the carbocation could lose either H^+ or D^+ when it reforms the alkene, all the deuterium atoms would not be retained in the unreacted alkene.

99. Because fumarate is the trans isomer and it forms an erythro product, the enzyme must catalyze the anti addition of D_2O. (Recall: CIS-SYN-ERYTHRO allows TRANS-ANTI-ERYTHRO but does not allow TRANS-SYN-ERYTHRO, because two terms must be changed.)

100. When (*S*)-3-methyl-1-pentene reacts with Cl_2, a compound with a new asymmetric center (*) is formed. The relative position of the groups around the asymmetric center does not change because no bonds to it are broken during the course of the reaction.

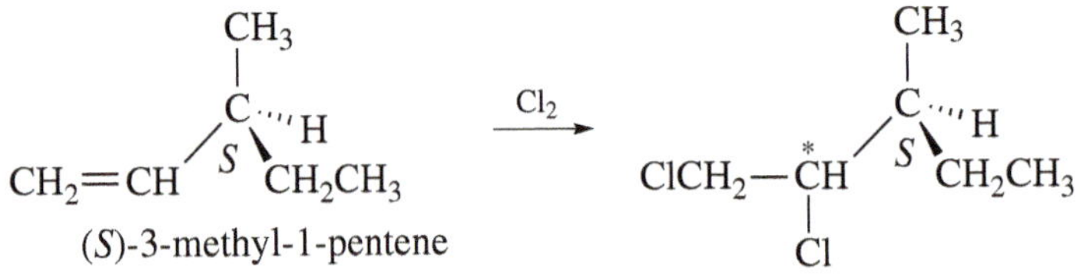

$$CH_2{=}CH-\underset{S}{C}(CH_3)(H)CH_2CH_3 \xrightarrow{Cl_2} ClCH_2-\overset{*}{C}H(Cl)-\underset{S}{C}(CH_3)(H)CH_2CH_3$$

(*S*)-3-methyl-1-pentene

The new asymmetric center can have either the *R* or *S* configuration. Therefore, a pair of diastereomers is obtained.

(2*R*,3*S*)-1,2-dichloro-3-methylpentane (2*S*,3*S*)-1,2-dichloro-3-methylpentane

101. **a.** A proton adds to the alkene, forming a secondary carbocation, which undergoes a ring-expansion rearrangement to form a more stable tertiary carbocation.

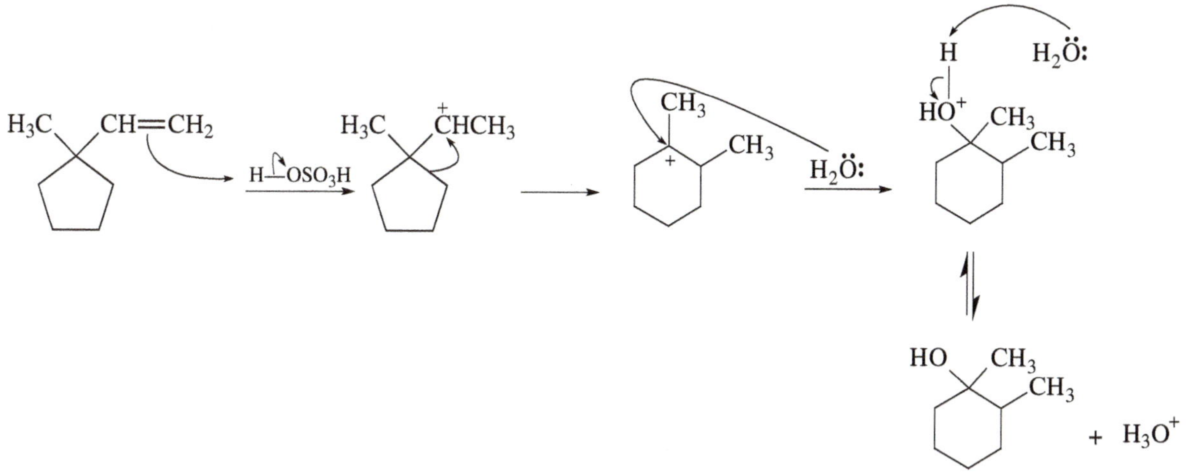

H_3C $CH{=}CH_2$ $H{-}OSO_3H$ H_3C $\overset{+}{C}HCH_3$ CH_3 + CH_3 $H_2\ddot{O}$: H $H_2\ddot{O}$: HO^+ CH_3 CH_3

HO CH_3 CH_3 + H_3O^+

102.

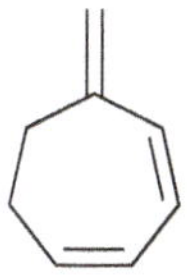

103. **a.** 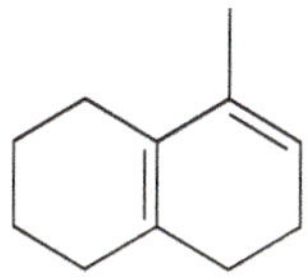 **b.**

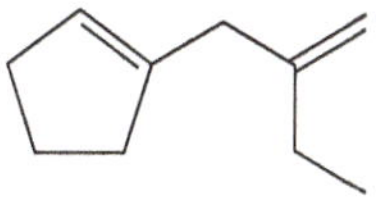

Chapter 6 Practice test

1. Which member of each pair is more stable?

a. $CH_3\overset{+}{C}HCH_2CH_3$ **or** $CH_3\overset{+}{C}(CH_3)CH_3$

b. $CH_3CH_2CH_2\overset{+}{C}H_2$ **or** $CH_3CH_2\overset{+}{C}HCH_3$

c. $CH_3CH_2\overset{+}{C}H_2$ **or** $ClCH_2CH_2\overset{+}{C}H_2$

2. Which is a better compound to use as a starting material for the synthesis of 2-bromopentane?

$CH_3CH_2CH_2CH{=}CH_2$ **or** $CH_3CH_2CH{=}CHCH_3$

3. What is the major product of each of the following reactions?

a. $CH_2{=}C(CH_3)CH_2CH_3 \;+\; HBr \longrightarrow$

b. $CH_3CH(CH_3)CH{=}CH_2 \;+\; HCl \longrightarrow$

c. $CH_3CH_2CH{=}CH_2 \;+\; Cl_2 \xrightarrow{H_2O}$

d. $CH_3C(CH_3)_2CH{=}CH_2 \;+\; HBr \longrightarrow$

e. $CH_3CH{=}CH_2 \;+\; RC(=O)OOH \longrightarrow$

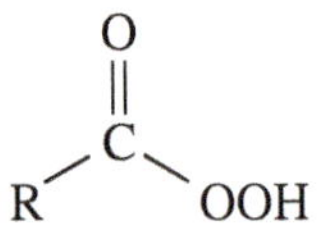

f. 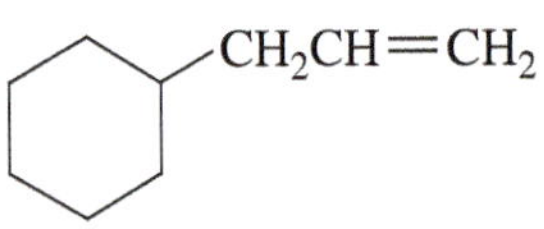 $C_6H_{11}CH_2CH{=}CH_2$ (cyclohexyl) $\xrightarrow[\text{2. } HO^-,\ H_2O_2,\ H_2O]{\text{1. } R_2BH/THF}$

4. What is the major product obtained from hydroboration–oxidation of the following alkenes?

a. 2-methyl-2-butene **b.** 1-ethylcyclopentene

5. Indicate how each of the following compounds can be synthesized using an alkene as one of the starting materials:

a. $\longrightarrow CH_3C(CH_3)_2CH_2CH_3$

b. $\longrightarrow$ 2-chlorocyclopentanol (cyclopentane ring bearing OH and Cl on adjacent carbons)

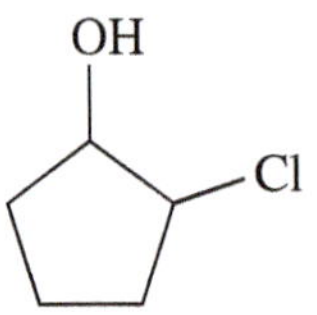

c. 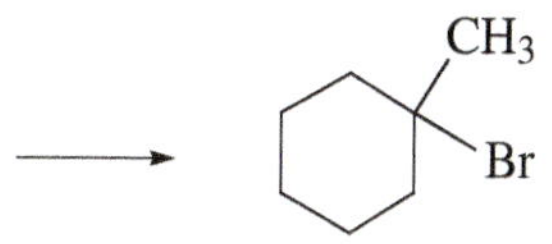 $\longrightarrow$ (1-bromo-1-methylcyclohexane: ring with CH_3 and Br on the same carbon)

d. 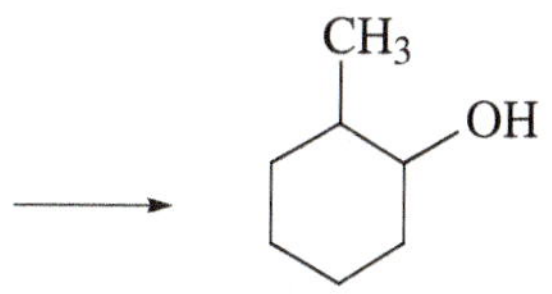 $\longrightarrow$ (2-methylcyclohexanol: ring with CH_3 and OH on adjacent carbons)

6. Indicate the carbocations that you would expect to rearrange to give a more stable carbocation.

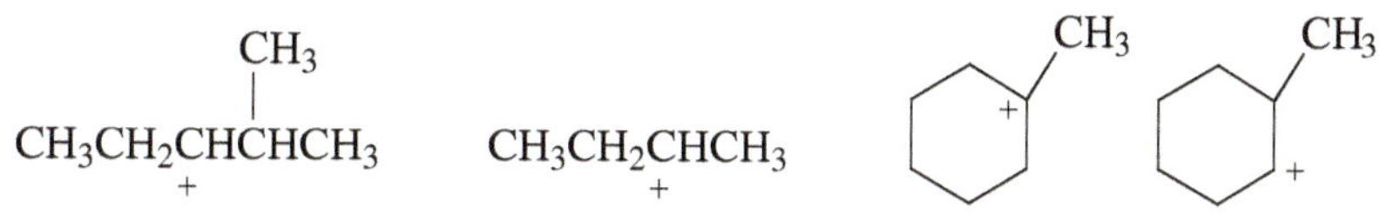

$CH_3CH_2CH\overset{+}{C}HCH_3$ (with CH_3 on the third carbon) $CH_3CH_2\overset{+}{C}HCH_3$ (cyclohexyl cation with CH_3 on the cationic carbon)   (cyclohexane with CH_3 on C1 and the positive charge on the adjacent carbon)

7. Indicate whether each of the following statements is true or false:

a. The addition of Br_2 to 1-butene to form 1,2-dibromobutane is a concerted reaction. T F

b. The reaction of 1-butene with HCl forms 1-chlorobutane as the major product. T F

c. The reaction of HBr with 3-methylcyclohexene is more highly regioselective than is the reaction of HBr with 1-methylcyclohexene. T F

d. The reaction of an alkene with a carboxylic acid forms an epoxide. T F

e. A catalyst increases the equilibrium constant of a reaction. T F

f. The addition of HBr to 3-methyl-2-pentene is a stereospecific reaction. T F

g. The addition of HBr to 3-methyl-2-pentene is a stereoselective reaction. T F

h. The addition of HBr to 3-methyl-2-pentene is a regioselective reaction. T F

8. Draw all the products that are obtained from each of the following reactions, indicating the stereoisomers that are formed:

a. 1-butene + HCl

b. 2-pentene + HBr

c. *trans*-3-hexene + Br_2

d. *trans*-3-heptene + Br_2

9. Draw the stereoisomers that are obtained from each of the following reactions:

a. 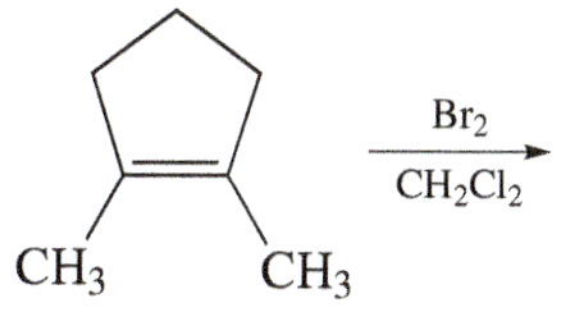 1,2-dimethylcyclopentene (substituents CH_3, CH_3) $\xrightarrow[CH_2Cl_2]{Br_2}$

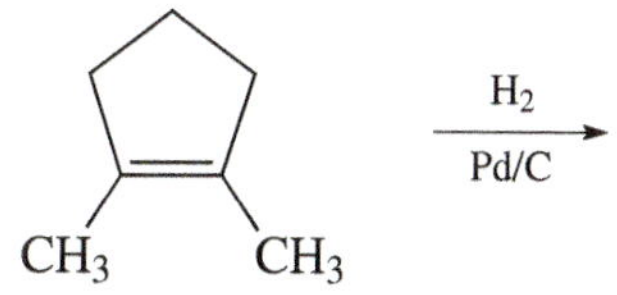

b. 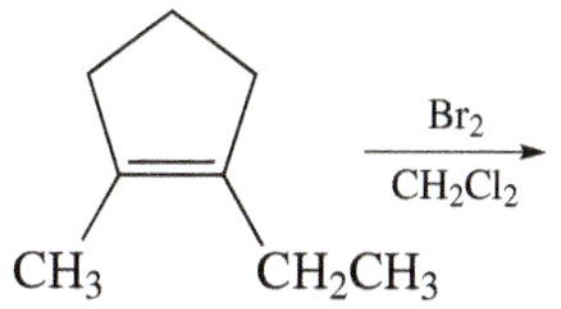 1-ethyl-2-methylcyclopentene (substituents CH_3, CH_2CH_3) $\xrightarrow[CH_2Cl_2]{Br_2}$

c. 1,2-dimethylcyclopentene (substituents CH_3, CH_3) $\xrightarrow[Pd/C]{H_2}$

d. 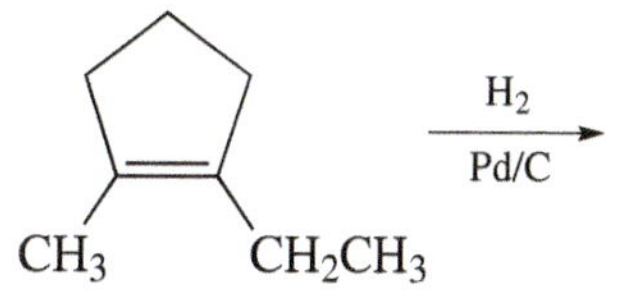 1-ethyl-2-methylcyclopentene (substituents CH_3, CH_2CH_3) $\xrightarrow[Pd/C]{H_2}$

The Reactions of Alkynes • An Introduction to Multistep Synthesis

Important Terms

acetylide ion — the conjugate base of a terminal alkyne.

$RC\equiv C^-$

aldehyde — a compound with a carbonyl group that is bonded to an alkyl group and to a hydrogen (or bonded to two hydrogens).

$RC(=O)H$ **or** $HC(=O)H$

alkylation reaction — a reaction that adds an alkyl group to a reactant.

alkyne — a hydrocarbon that contains a triple bond.

carbonyl group — a carbon doubly bonded to an oxygen.

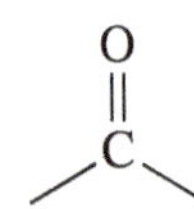

enol — an alkene with an OH group bonded to one of the sp^2 carbons.

$C=C(OH)$

geminal dihalide — a compound with two halogen atoms bonded to the same carbon.

internal alkyne — an alkyne with its triple bond not at the end of the carbon chain.

keto–enol tautomers — a ketone or an aldehyde and its isomeric enol. The keto and enol tautomers differ only in the location of a double bond and a hydrogen.

$RCH_2C(=O)R \rightleftharpoons RCH=C(OH)R$

keto — enol

ketone — a compound with a carbonyl group that is bonded to two alkyl groups.

$RC(=O)R$

π-complex — a complex formed between an electrophile and a triple bond.

radical anion a species with an atom that has a negative charge and an unpaired electron.

retrosynthetic analysis or retrosynthesis working backward (on paper) from a target molecule to available starting materials.

tautomerization interconversion of keto–enol tautomers.

tautomers constitutional isomers that are in rapid equilibrium; for example, keto and enol tautomers. The keto and enol tautomers differ only in the location of a double bond and a hydrogen.

terminal alkyne an alkyne with its triple bond at the end of the carbon chain.

vinylic cation a species with a positive charge on a vinylic carbon.

vinylic radical a species with an unpaired electron on a vinylic carbon.

Solutions to Problems

1. **a.** 5-bromo-2-pentyne
b. 6-bromo-2-chloro-4-octyne
c. 1-methoxy-2-pentyne
d. 3-ethyl-1-hexyne

2. **a.** 6-methyl-2-octyne **b.** 5-ethyl-4-methyl-1-heptyne **c.** 2-bromo-4-octyne

3. The general molecular formula of a noncyclic hydrocarbon is C_nH_{2n+2}.
Therefore, the molecular formula for a noncyclic hydrocarbon with 14 carbons is $C_{14}H_{30}$.

Because a compound has two fewer hydrogens for every ring and π bond, a compound with one ring and 4 π bonds (2 triple bonds) has 10 fewer hydrogens than the C_nH_{2n+2} formula.
Therefore, the molecular formula is $\mathbf{C_{14}H_{20}}$.

4. **a.** Cl $ClCH_2CH_2C{\equiv}CCH_2CH_3$

b. $C{\equiv}C$, CH_2, CH_2, CH_2, CH_2, $CH_2{-}CH_2$ (cyclooctyne)

c. $CH_3CHC{\equiv}CH$ with CH_3 on C-2

d. $CH_3CH_2CHC{\equiv}CCH_2CHCH_3$ with CH_3 on C-3 and C-7

e. $HC{\equiv}CCH_2CCH_3$ with two CH_3 on C-4

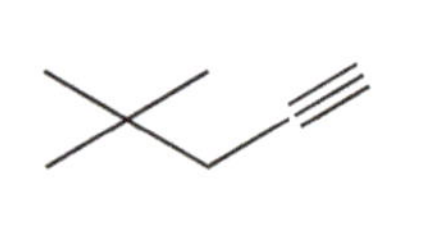

f. $CH_3C{\equiv}CCH_3$

5.

$HC{\equiv}CCH_2CH_2CH_2CH_3$
1-hexyne
butylacetylene

$CH_3C{\equiv}CCH_2CH_2CH_3$
2-hexyne
methylpropylacetylene

$CH_3CH_2C{\equiv}CCH_2CH_3$
3-hexyne
diethylacetylene

$CH_3CH_2CHC{\equiv}CH$ with CH_3 on C-3
3-methyl-1-pentyne
sec-butylacetylene

$CH_3CHCH_2C{\equiv}CH$ with CH_3 on C-4
4-methyl-1-pentyne
isobutylacetylene

$CH_3CHC{\equiv}CCH_3$ with CH_3 on C-4
4-methyl-2-pentyne
isopropylmethylacetylene

$CH_3CC{\equiv}CH$ with two CH_3 on C-3
3,3-dimethyl-1-butyne
tert-butylacetylene

6. **a.** 1-hepten-4-yne
b. 4-methyl-1,4-hexadiene
c. 5-vinyl-5-octen-1-yne
(One of the functional groups cannot be included in the parent hydrocarbon.)
d. 3-butyn-1-ol
e. 1,3,5-heptatriene
f. 2,4-dimethyl-4-hexen-1-ol

7. **a.** (E)-2-hepten-4-ol **b.** 1-hepten-5-yne **c.** (E)-4-hepten-1-yne

8. **a.** sp^2-sp^2 **d.** $sp-sp^3$ **g.** sp^2-sp^3
b. sp^2-sp^3 **e.** $sp-sp$ **h.** $sp-sp^3$
c. $sp-sp^2$ **f.** sp^2-sp^2 **i.** sp^2-sp

9. alkane = pentane alkene = 1-pentene alkyne = 1-pentyne

10. The cis isomer has a higher boiling point because it has a dipole moment, whereas the dipole moment of the trans isomer is zero.

11. Solved in the text.

12. **a.** $CH_2{=}CCH_3$ (Br on C2) **c.** $CH_3C{=}CCH_3$ (Br on C2, Br on C3) **e.** $CH_3CH_2CCH_3$ (Br, Br on C2)

b. CH_3CCH_3 (Br, Br on C2) **d.** $HC{-}CCH_3$ (Br, Br on C1; Br, Br on C2) **f.** $CH_3CCH_2CH_2CH_3$ (Br, Br on C2) + $CH_3CH_2CCH_2CH_3$ (Br, Br on C3)

13. **a.** $CH_3C{\equiv}CCH_3$ + :Br–Br: ⟶ $CH_3C{=}CCH_3$ (cyclic bromonium ion, $:\overset{+}{Br}:$ bridging C2–C3) + $:Br:^-$ ⟶ $CH_3C{=}CCH_3$ (Br on C2, Br on C3)

b. Only anti addition occurs because the intermediate is a cyclic bromonium ion. Therefore, the product has the E configuration.

$$\begin{array}{ccc} H_3C & & Br \\ & C{=}C & \\ Br & & CH_3 \end{array}$$

14. Because the alkyne is not symmetrical, two ketones are obtained.

$CH_3CH_2C(=O)CH_2CH_2CH_2CH_3$ and $CH_3CH_2CH_2C(=O)CH_2CH_2CH_3$

15. **a.** $CH_3C{\equiv}CH$ **b.** $CH_3CH_2C{\equiv}CCH_2CH_3$ **c.** $HC{\equiv}C$–cyclohexyl

The best answer for **b** is 3-hexyne, because it would form only the desired ketone.
2-Hexyne would form two different ketones, so only half of the product would be the desired ketone.

16. **a.** $CH_2{=}CCH_3$ (OH on C2) Because the ketone has identical substituents bonded to the carbonyl carbon, it has only one enol tautomer.

b. $CH_3CH{=}C(OH)CH_2CH_2CH_3$ and $CH_3CH_2C(OH){=}CHCH_2CH_3$

E and *Z* isomers are possible for each of these enols.

$(CH_3)(H)C{=}C(OH)(CH_2CH_2CH_3)$ — *E*

$(CH_3)(H)C{=}C(CH_2CH_2CH_3)(OH)$ — *Z*

$(HO)(CH_3CH_2)C{=}C(H)(CH_2CH_3)$ — *E*

$(HO)(CH_3CH_2)C{=}C(CH_2CH_3)(H)$ — *Z*

c. $CH_2{=}C(OH)$–cyclohexyl and $CH_3C(OH){=}$cyclohexylidene

Because each enol has identical groups bonded to one of its sp^2 carbons, *E* and *Z* isomers are not possible for either enol.

17. **a.** (1) $CH_3CH_2C(=O)CH_3$ **b.** (1) $CH_3CH_2C(=O)CH_3$ **c.** (1) $CH_3CH_2CH_2C(=O)CH_3$ and $CH_3CH_2C(=O)CH_2CH_3$

(2) $CH_3CH_2CH_2CH(=O)$ (2) $CH_3CH_2C(=O)CH_3$ (2) $CH_3CH_2CH_2C(=O)CH_3$ and $CH_3CH_2C(=O)CH_2CH_3$

18. Ethyne (acetylene)

An alkyne can form an aldehyde only if the OH group adds to a terminal *sp* carbon. In the acid-catalyzed addition of water to a terminal alkyne, the proton adds to the terminal *sp* carbon. Therefore, the only way the OH group can add to a terminal *sp* carbon under these conditions is if there are two terminal *sp* carbons in the alkyne. In other words, the alkyne must be ethyne.

19. **a.** $CH_3CH_2CH_2C{\equiv}CH$ (1-pentyne) **or** $CH_3CH_2C{\equiv}CCH_3$ (2-pentyne) $\xrightarrow[Pd/C]{H_2}$ $CH_3CH_2CH_2CH_2CH_3$

b. $CH_3C{\equiv}CCH_3$ (2-butyne) $\xrightarrow[\text{Lindlar catalyst}]{H_2}$ $(H_3C)(H)C{=}C(CH_3)(H)$ (cis)

c. $CH_3CH_2C{\equiv}CCH_3$ (2-pentyne) $\xrightarrow[NH_3(liq),\ -78\ °C]{Na}$ $(CH_3CH_2)(H)C{=}C(H)(CH_3)$ (trans)

d. $CH_3CH_2CH_2CH_2C{\equiv}CH$ (1-hexyne) $\xrightarrow[\text{Lindlar catalyst}]{H_2}$ $CH_3CH_2CH_2CH_2CH{=}CH_2$

$Na/NH_3(liq)$ cannot be used to reduce terminal alkynes because Na removes the hydrogen that is attached to the *sp* carbon of the terminal alkyne.

$$2\ RC{\equiv}CH + 2\ Na \longrightarrow 2\ RC{\equiv}C^- + 2\ Na^+ + H_2$$

20. **a.**  **b.**  **c.**  **d.** 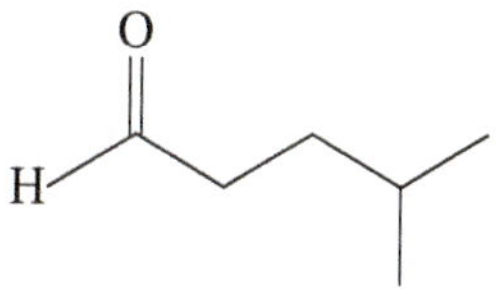 **e.** 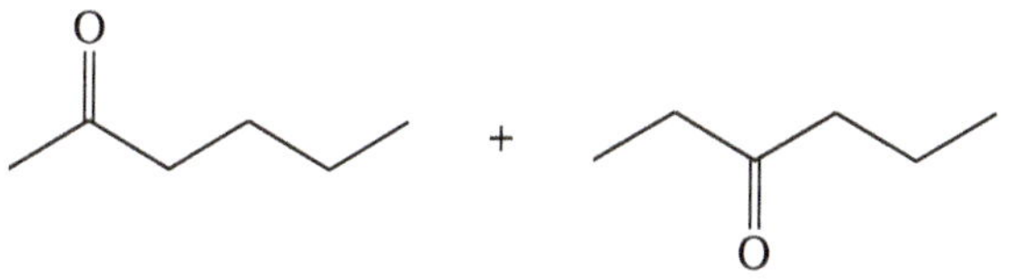 **f.**

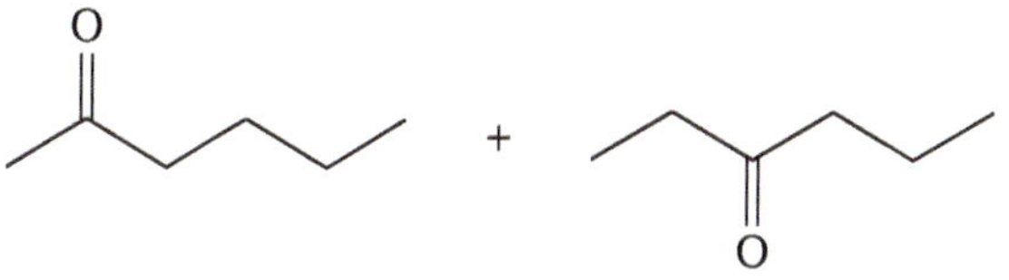

21. A terminal alkyne has a pK_a = 25. A base that removes a proton from a terminal alkyne in a reaction that favors products must have a conjugate acid that is a weaker acid than a terminal alkyne. That is, it must have a pK_a > 25. (Recall that the equilibrium favors formation of the weak acid.)

$CH_3\bar{C}H_2$ and $H_2C{=}\bar{C}H$ because the pK_a values of their conjugate acids are > 60 and 44, respectively.

The pK_a values of the conjugate acids of the other choices are all < 25.

22. The reaction of sodium amide with an alkane does not favor products because the acid that would be formed is a stronger acid than the alkane (the reactant acid). Recall that the equilibrium favors reaction of the strong acid (or strong base) and formation of the weak acid (or weak base); Section 2.5.

$$\underset{\substack{pK_a > 60 \\ \text{weaker acid}}}{CH_3CH_3} + \underset{\text{weaker base}}{^{-}NH_2} \rightleftharpoons \underset{\text{stronger base}}{CH_3\bar{C}H_2} + \underset{\substack{pK_a = 36 \\ \text{stronger acid}}}{NH_3}$$

23. The base used to remove a proton must be stronger than the base that is formed as a result of removing the proton. Therefore, the base used to remove a proton from a terminal alkyne must be a stronger base than the conjugate base of the terminal alkyne. A terminal alkyne has a pK_a ~ 25. In other words, any base whose conjugate acid has a pK_a greater than 25 can be used.

24. **a.** $CH_3CH_2CH_2\bar{C}H_2$ > $CH_3CH_2CH{=}\bar{C}H$ > $CH_3CH_2C{\equiv}C^-$

b. $^-NH_2$ > $CH_3C{\equiv}C^-$ > $CH_3CH_2O^-$ > F^-

25. Solved in the text.

26. **a.** $H_2C{=}\overset{+}{C}H$ **b.** $CH_3\overset{+}{C}H_2$

A triply bonded (*sp*) carbon is more electronegative than an sp^2 or sp^3 carbon. Therefore, a triply bonded carbon with a positive charge is less stable than a doubly bonded or singly bonded carbon with a positive charge. Thus, in **a**, the vinyl cation is more stable and in **b**, the ethyl cation is more stable.

27. Solved in the text.

28. **a.** $HC\equiv CH \xrightarrow[2.\ CH_3CH_2CH_2Br]{1.\ NaNH_2} CH_3CH_2CH_2C\equiv CH$

b. $HC\equiv CH \xrightarrow[2.\ CH_3CH_2Br]{1.\ NaNH_2} CH_3CH_2C\equiv CH \xrightarrow[2.\ CH_3CH_2Br]{1.\ NaNH_2}$

$CH_3CH_2C\equiv CCH_2CH_3 \xrightarrow[\text{Lindlar catalyst}]{H_2} \begin{array}{ccc} CH_3CH_2 & & CH_2CH_3 \\ & C=C & \\ H & & H \end{array}$

c. $HC\equiv CH \xrightarrow[2.\ CH_3Br]{1.\ NaNH_2} CH_3C\equiv CH \xrightarrow[\text{Lindlar catalyst}]{H_2} CH_3CH=CH_2$

d. $HC\equiv CH \xrightarrow[2.\ CH_3Br]{1.\ NaNH_2} CH_3C\equiv CH \xrightarrow{\text{excess HCl}} CH_3\overset{Cl}{\underset{Cl}{C}}CH_3$

e. product of **a** $\xrightarrow[2.\ HO^-,\ H_2O_2,\ H_2O]{1.\ R_2BH/THF} CH_3CH_2CH_2CH_2\overset{O}{\overset{\|}{C}}H$

f. product of **c** $\xrightarrow{HBr} CH_3\underset{Br}{C}HCH_3$

29. **a.** $CH_3CH_2\overset{Cl}{\underset{Cl}{C}}CH_3$

b. $CH_3CH_2CH_2\overset{Cl}{\underset{Cl}{C}}CH_2CH_3$

c. $CH_3CH_2CH_2\overset{Cl}{\underset{Cl}{C}}CH_2CH_2CH_3 + CH_3CH_2\overset{Cl}{\underset{Cl}{C}}CH_2CH_2CH_2CH_3$

equal amounts

30. **a.** $CH_3C\equiv CCH_2CH_2CH_3$

b. $CH_3CH_2C\equiv C\underset{CH_2CH_3}{C}HCH_2CH_2CH_3$

c. $CH_3C\equiv CH$

d. $CH_2=CHC\equiv CH$

e. $CH_3OC\equiv CH$

f. $CH_3\overset{CH_3}{\underset{CH_3}{C}}C\equiv C\overset{CH_3}{C}HCH_2CH_3$

g. $BrC\equiv CCH_2CH_2CH_3$

h. OH, CH_3

i. $CH_3CH_2C\equiv CCH_2CH_3$

j. $CH_3\overset{CH_3}{\underset{CH_3}{C}}C\equiv C\overset{CH_3}{\underset{CH_3}{C}}CH_3$

k. $-C\equiv CH$

l. $CH_3C\equiv CCH_2\overset{CH_3}{C}H\underset{CH_3}{C}HCH_3$

31. The student named only one correctly.

a. 4-methyl-2-hexyne **b.** 7-bromo-3-heptyne **c.** correct **d.** 2-pentyne

32.

$CH_3CH_2\overset{+}{C}(\text{:Br:})=CH$ + $:\ddot{Br}:^-$ ⟶ $CH_3CH_2C(Br)=CH(Br)$

electrophile nucleophile

$CH_3C\equiv C-H$ + $^-:\ddot{N}H_2$ ⟶ $CH_3C\equiv C:^-$ + $\ddot{N}H_3$

electrophile nucleophile

$CH_3C\equiv C:^-$ CH_3-Br ⟶ $CH_3C\equiv CCH_3$ + $:\ddot{Br}:^-$

nucleophile electrophile

(Methyl bromide is an electrophile because the carbon has a partial positive charge since bromine is more electronegative than carbon.)

33.

a. 5-bromo-2-hexyne **b.** 5-methyl-2-octyne **c.** 5,5-dimethyl-2-hexyne **d.** 6-chloro-2-methyl-3-heptyne **e.** 1,5-cyclooctadiene **f.** 1,6-dimethyl-1,3-cyclohexadiene

34.

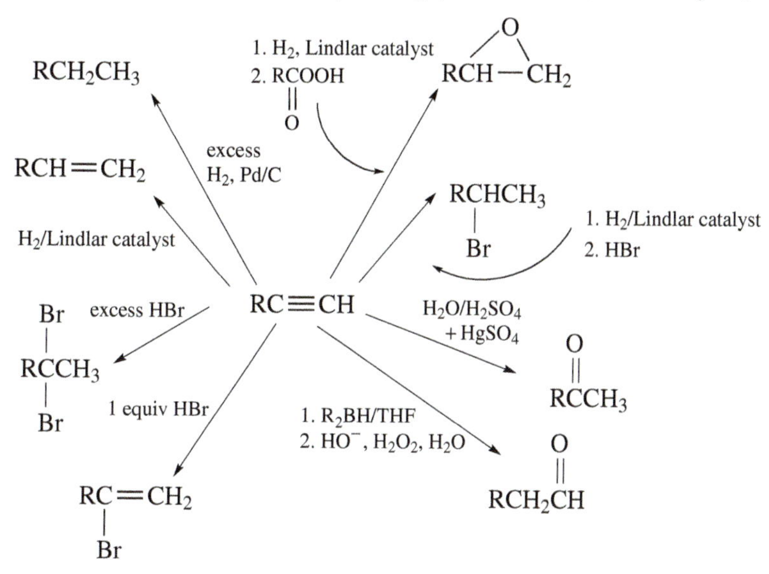

35. **a.** First draw the straight-chain compounds with seven carbons; then draw the straight-chain compounds with six carbons and one methyl group; then draw the straight-chain compounds with five carbons and two methyl groups (or with one ethyl group). Naming them will tell you if you have drawn one compound more than once because if two compounds have the same name, they are the same compound.

$HC\equiv CCH_2CH_2CH_2CH_2CH_3$
1-heptyne
pentylacetylene

$CH_3C\equiv CCH_2CH_2CH_2CH_3$
2-heptyne
butylmethylacetylene

$CH_3CH_2C\equiv CCH_2CH_2CH_3$
3-heptyne
ethylpropylacetylene

$CH_3CH_2CH_2\overset{*}{C}H(CH_3)C\equiv CH$
3-methyl-1-hexyne

$CH_3CH_2\overset{*}{C}H(CH_3)CH_2C\equiv CH$
4-methyl-1-hexyne

$CH_3CH(CH_3)CH_2CH_2C\equiv CH$
5-methyl-1-hexyne
isopentylacetylene

$CH_3CH_2\overset{*}{C}HC{\equiv}CCH_3$ with CH_3 on C4
4-methyl-2-hexyne
sec-butylmethylacetylene

$CH_3CHCH_2C{\equiv}CCH_3$ with CH_3 on C5
5-methyl-2-hexyne
isobutylmethylacetylene

$CH_3CHC{\equiv}CCH_2CH_3$ with CH_3 on C2
2-methyl-3-hexyne
ethylisopropylacetylene

$CH_3C(CH_3)_2C{\equiv}CCH_3$
4,4-dimethyl-2-pentyne
tert-butylmethylacetylene

$CH_3C(CH_3)_2CH_2C{\equiv}CH$
4,4-dimethyl-1-pentyne

$CH_3CH_2C(CH_3)_2C{\equiv}CH$
3,3-dimethyl-1-pentyne
tert-pentylacetylene

$CH_3CH_2CH(CH_2CH_3)C{\equiv}CH$
3-ethyl-1-pentyne

$CH_3CH(CH_3)\overset{*}{C}H(CH_3)C{\equiv}CH$
3,4-dimethyl-1-pentyne

b. There are 14 alkynes if stereoisomers are ignored. Four of the 14 alkynes have an asymmetric center, so each of these can have either the *R* or *S* configuration. Therefore, if stereoisomers are included, there would be 18 alkynes.

36. $CH_3CH_2CH_2C{\equiv}CCH + Br_2 \longrightarrow CH_3CH_2CH_2C{=}CH$ (cyclic bromonium ion, Br^+) $\longrightarrow$ attack by $H_2\ddot{O}$:

$CH_3CH_2CH_2C(Br){=}CH$ with $H{-}OH^+$ (deprotonated by $H_2\ddot{O}$:) $\longrightarrow CH_3CH_2CH_2C{=}CH(Br)$ with OH (enol) $\rightleftharpoons CH_3CH_2CH_2\overset{O}{\overset{\|}{C}}{-}CH_2Br$

37. **a.** $CH_3CH_2CH_2C{\equiv}CH \xrightarrow[\text{2. } HO^-,\ H_2O_2,\ H_2O]{\text{1. } R_2BH/THF} CH_3CH_2CH_2CH_2\overset{O}{\overset{\|}{C}}H$

b. $CH_3CH_2CH{=}CH_2 \xrightarrow[\text{2. } HO^-,\ H_2O_2,\ H_2O]{\text{1. } R_2BH/THF} CH_3CH_2CH_2CH_2OH$

c. $CH_3CH_2CH_2C{\equiv}CCH_2CH_2CH_3 + H_2O \xrightarrow{H_2SO_4} CH_3CH_2CH_2\overset{O}{\overset{\|}{C}}CH_2CH_2CH_2CH_3$

This symmetrical alkyne will give the greatest yield of the desired ketone.
Because the reactant is not a terminal alkyne, the reaction can take place without the mercuric ion catalyst.

38. **a.** H_2/Lindlar catalyst **b.** Na, NH_3(liq), −78 °C **c.** excess H_2, Pd/C

39.

a. 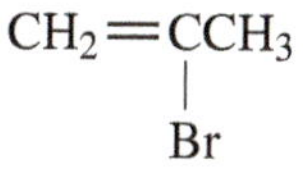 $CH_2{=}C(Br)CH_3$

b. $CH_3CBr_2CH_3$

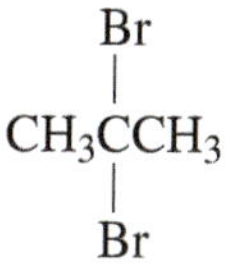

c. $(Br)(H)C{=}C(CH_3)(Br)$

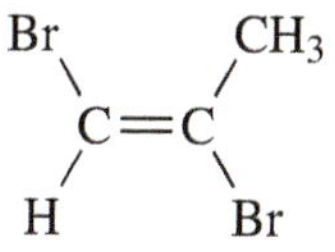

d. 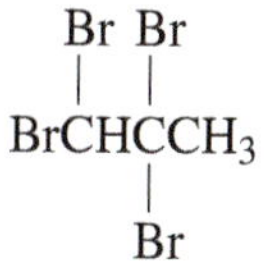 $BrCH(Br)CBr_2CH_3$

e. 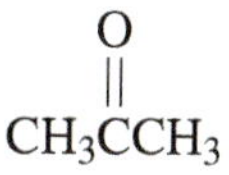 $CH_3C(=O)CH_3$

f. 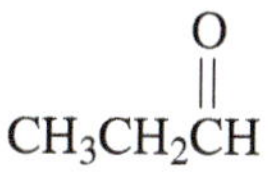 $CH_3CH_2CH{=}O$

g. $CH_3CH_2CH_3$

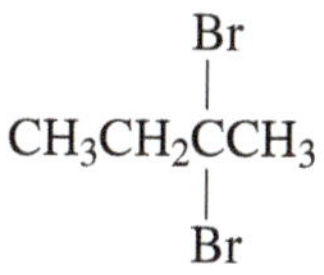

h. $CH_3CH{=}CH_2$

i. $CH_3C{\equiv}C^-$

j. $CH_3C{\equiv}CCH_2CH_2CH_3$

40.

a. 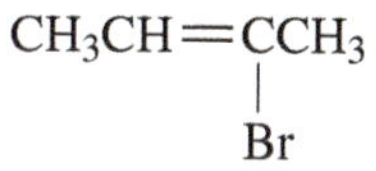 $CH_3CH{=}C(Br)CH_3$

b. $CH_3CH_2CBr_2CH_3$

c. $(Br)(CH_3)C{=}C(CH_3)(Br)$

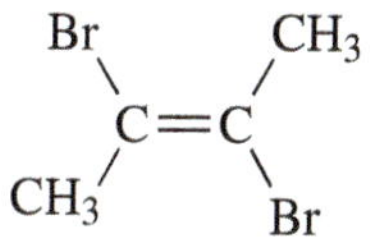

d. 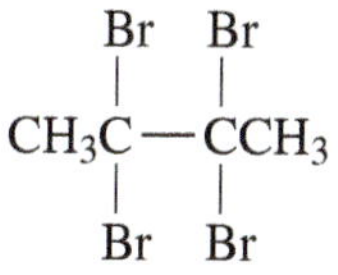 $CH_3CBr_2{-}CBr_2CH_3$

e. 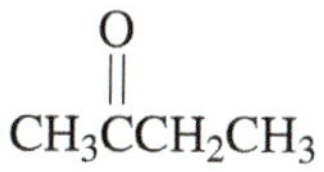 $CH_3C(=O)CH_2CH_3$

f. 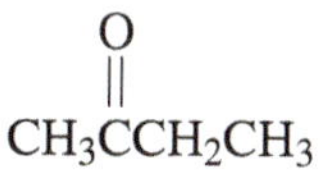 $CH_3C(=O)CH_2CH_3$

g. $CH_3CH_2CH_2CH_3$

h. $(CH_3)(H)C{=}C(CH_3)(H)$

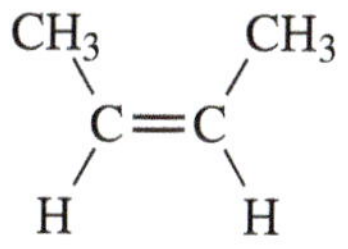

i. no reaction

j. no reaction

41.
a. 1-octen-6-yne
b. *cis*-3-hexen-1-ol or (*Z*)-3-hexen-1-ol
c. 1,5-octadiyne
d. 5-chloro-1,3-cyclohexadiene
e. 1-methyl-1,3,5-cycloheptatriene
f. 6-methyl-3-cyclohexenamine

42. The molecular formula of the hydrocarbon is $C_{32}H_{56}$.
With one triple bond, two double bonds, and one ring, the degree of unsaturation is 5.
Therefore, the compound is missing 10 hydrogens from $C_nH_{2n+2} = C_{32}H_{66}$.

43. **a.** **1.**

$$CH_3\underset{\displaystyle CH_3}{\underset{|}{C}}HC{\equiv}CH \xrightarrow[\text{Lindlar catalyst}]{H_2} CH_3\underset{\displaystyle CH_3}{\underset{|}{C}}HCH{=}CH_2 \xrightarrow[H_2O]{H_2SO_4} CH_3\underset{\displaystyle CH_3}{\underset{|}{\overset{\displaystyle H}{\overset{|}{C}}}}\overset{+}{C}HCH_3 \longrightarrow CH_3\underset{\displaystyle CH_3}{\underset{|}{\overset{+}{C}}}CH_2CH_3 \xrightarrow{H_2O} CH_3\underset{\displaystyle CH_3}{\underset{|}{\overset{\displaystyle OH}{\overset{|}{C}}}}CH_2CH_3$$

2.

$$CH_3\underset{\displaystyle CH_3}{\underset{|}{C}}HC{\equiv}CH \xrightarrow[\text{Lindlar catalyst}]{H_2} CH_3\underset{\displaystyle CH_3}{\underset{|}{C}}HCH{=}CH_2 \xrightarrow[\text{2. } H_2O_2,\ HO^-,\ H_2O]{\text{1. } R_2BH/THF} CH_3\underset{\displaystyle CH_3}{\underset{|}{C}}HCH_2CH_2OH$$

b. 3-Methyl-2-butanol would be a minor product obtained from both **1** and **2**.

$$CH_3\underset{\displaystyle CH_3}{\underset{|}{C}}H\overset{\displaystyle OH}{\overset{|}{C}}HCH_3$$

3-methyl-2-butanol

3-Methyl-2-butanol will be obtained from **1**, because occasionally water will attack the secondary carbocation before it has a chance to rearrange to the tertiary carbocation.
3-Methyl-2-butanol will be obtained from **2**, because in the second step of the synthesis, boron can also add to the other sp^2 carbon; it will be a minor product because the transition state for its formation is less stable than the transition state leading to the major product. Because a carbocation is not formed as an intermediate, a carbocation rearrangement cannot occur.

(The proton cannot add to the other sp^2 carbon in the second step of part **1** because that would form a primary carbocation. Primary carbocations are so unstable that they can never be formed.)

44. Three of the names are correct.

a. 3-heptyne
b. 5-methyl-3-heptyne
c. correct
d. 6,7-dimethyl-3-octyne
e. correct
f. correct

45. Only **c** and **e** are keto–enol tautomers. Notice that an enol tautomer has an OH group bonded to an sp^2 carbon. The structures in **d** are not enol tautomers, because they do not have the oxygen on the same carbon.

46. **a.**

$$HC{\equiv}CH \xrightarrow[HgSO_4]{H_2O,\ H_2SO_4} \underset{\text{enol}}{CH_2{=}CHOH} \rightleftharpoons CH_3\overset{\displaystyle O}{\overset{\|}{C}}H$$

b.

$$HC{\equiv}CH \xrightarrow[\text{2. } CH_3CH_2Br]{\text{1. } NaNH_2} CH_3CH_2C{\equiv}CH \xrightarrow[\text{Lindlar catalyst}]{H_2} CH_3CH_2CH{=}CH_2 \xrightarrow[CH_2Cl_2]{Br_2} CH_3CH_2\underset{\displaystyle Br}{\underset{|}{C}}HCH_2Br$$

c. $HC{\equiv}CH \xrightarrow[\text{2. } CH_3Br]{\text{1. } NaNH_2} HC{\equiv}CCH_3 \xrightarrow[HgSO_4]{H_2O,\ H_2SO_4} CH_3\overset{O}{\overset{\|}{C}}CH_3$

d.

$HC{\equiv}CH \xrightarrow[\text{2. } CH_3Br]{\text{1. } NaNH_2} HC{\equiv}CCH_3 \xrightarrow[\text{2. } CH_3CH_2Br]{\text{1. } NaNH_2} CH_3C{\equiv}CCH_2CH_3 \xrightarrow[NH_3\text{ (liq)}, -78\,°C]{Na}$

e. $HC{\equiv}CH \xrightarrow[\text{2. } CH_3Br]{\text{1. } NaNH_2} HC{\equiv}CCH_3 \xrightarrow[\text{2. } CH_3CH_2Br]{\text{1. } NaNH_2} CH_3C{\equiv}CCH_2CH_3 \xrightarrow[\text{Lindlar catalyst}]{H_2}$

f. $HC{\equiv}CH \xrightarrow[\text{2. } CH_3CH_2CH_2Br]{\text{1. } NaNH_2} CH_3CH_2CH_2C{\equiv}CH \xrightarrow[Pd/C]{\text{excess } H_2}$

47. The first equilibrium lies to the right because HOOH ($pK_a = 11.6$) is a stronger acid than H_2O ($pK_a = 15.7$).
The second equilibrium lies to the left because the alkyne ($pK_a = 25$) is a weaker acid than HOOH ($pK_a = 11.6$).

48.
a. (*Z*)-3,6-dimethyl-2-hepten-4-yne
b. 5-*tert*-butyl-2-methyl-3-octyne
c. 4,4-dimethyl-1-nonen-6-yn-3-ol
d. 4-(2-methylbutyl)-5-heptyn-3-amine

49. **a.** Syn addition of H_2 forms *cis*-2-butene; when Br_2 adds to *cis*-2-butene, the threo pair of enantiomers is formed.

CH₃ / H—Br / Br—H / CH₃ and CH₃ / Br—H / H—Br / CH₃ **or** CH₃, H, H, Br, Br, CH₃ **+** H, CH₃, Br, H, CH₃, Br

b. Reaction with sodium and liquid ammonia forms *trans*-2-butene; when Br_2 adds to *trans*-2-butene, a meso compound is formed.

CH₃ / H—Br / H—Br / CH₃ **or** CH₃, H, H, CH₃, Br, Br

c. Anti addition of Cl_2 forms *trans*-2,3-dichloro-2-butene; when Br_2 adds to *trans*-2,3-dichloro-2-butene, a meso compound is formed.

CH₃ / Cl—Br / Cl—Br / CH₃ **or** CH₃, Cl, Cl, CH₃, Br, Br

50. **a.** $HC{\equiv}CH \xrightarrow{NaNH_2} HC{\equiv}C^- \xrightarrow{CH_3Br} HC{\equiv}CCH_3 \xrightarrow{NaNH_2} {}^-C{\equiv}CCH_3 \xrightarrow{CH_3CH_2CH_2CH_2CH_2Br} CH_3CH_2CH_2CH_2CH_2C{\equiv}CCH_3 \xrightarrow[\text{Lindlar catalyst}]{H_2}$ cis-alkene (H_3C and $CH_2CH_2CH_2CH_2CH_3$ on the same side of $C{=}C$, each carbon bearing H)

b. $HC{\equiv}CH \xrightarrow{NaNH_2} HC{\equiv}C^- \xrightarrow{CH_3CH_2Br} HC{\equiv}CCH_2CH_3 \xrightarrow{NaNH_2} {}^-C{\equiv}CCH_2CH_3 \xrightarrow{CH_3CH_2CH_2Br} CH_3CH_2CH_2C{\equiv}CCH_2CH_3 \xrightarrow[NH_3(liq),\ -78\ °C]{Na}$ trans-alkene (CH_3CH_2 and $CH_2CH_2CH_3$ on opposite sides of $C{=}C$, each carbon bearing H)

51. **a.** $CH_3CH_2\overset{O}{\overset{\|}{C}}CH_3$ **b.** $CH_3CH_2CH_2\overset{O}{\overset{\|}{C}}CH_3$ **c.** cyclohexanone (cyclohexane ring $=O$) **d.** cyclohexyl$-\overset{O}{\overset{\|}{C}}H$

52. **a.** $HC{\equiv}CH \xrightarrow[2.\ CH_3CH_2CH_2CH_2Br]{1.\ NaNH_2} CH_3CH_2CH_2CH_2C{\equiv}CH \xrightarrow[HgSO_4]{H_2O,\ H_2SO_4} CH_3CH_2CH_2CH_2\overset{O}{\overset{\|}{C}}CH_3$

b. $HC{\equiv}CH \xrightarrow[2.\ CH_3CH_2Br]{1.\ NaNH_2} CH_3CH_2C{\equiv}CH \xrightarrow[\text{or Na, }NH_3(liq),\ -78\ °C]{H_2,\ \text{Lindlar catalyst}} CH_3CH_2CH{=}CH_2 \xrightarrow{HBr} CH_3CH_2\underset{Br}{\underset{|}{C}}HCH_3$

c. $HC{\equiv}CH \xrightarrow[2.\ CH_3CH_2CH_2Br]{1.\ NaNH_2} CH_3CH_2CH_2C{\equiv}CH \xrightarrow[\text{or Na, }NH_3(liq),\ -78\ °C]{H_2,\ \text{Lindlar catalyst}} CH_3CH_2CH_2CH{=}CH_2 \xrightarrow{H_2O,\ H_2SO_4} CH_3CH_2CH_2\underset{OH}{\underset{|}{C}}HCH_3$

d. cyclohexyl$-C{\equiv}CH \xrightarrow[2.\ HO^-,\ H_2O_2,\ H_2O]{1.\ R_2BH/THF}$ cyclohexyl$-CH_2\overset{O}{\overset{\|}{C}}H$

e. cyclohexyl$-C{\equiv}CH \xrightarrow[HgSO_4]{H_2O,\ H_2SO_4}$ cyclohexyl$-\overset{O}{\overset{\|}{C}}-CH_3$

f. phenyl$-C{\equiv}CCH_3 \xrightarrow[\text{Lindlar catalyst}]{H_2}$ cis-alkene (phenyl and CH_3 on the same side of $C{=}C$, each carbon bearing H)

53. The chemist can make 3-octyne by using 1-hexyne instead of 1-butyne. He would then need to use ethyl bromide (instead of butyl bromide) for the alkylation step:

$$CH_3CH_2CH_2CH_2C{\equiv}CH \xrightarrow[\text{2. } CH_3CH_2Br]{\text{1. } NaNH_2} CH_3CH_2CH_2CH_2C{\equiv}CCH_2CH_3$$

Or he could make the 1-butyne he needed by alkylating ethyne:

$$HC{\equiv}CH \xrightarrow[\text{2. } CH_3CH_2Br]{\text{1. } NaNH_2} CH_3CH_2C{\equiv}CH$$

54. a. Only one product is obtained from hydroboration–oxidation of 2-butyne because the alkyne is symmetrical. Two different products can be obtained from hydroboration–oxidation of 2-pentyne because the alkyne is not symmetrical.

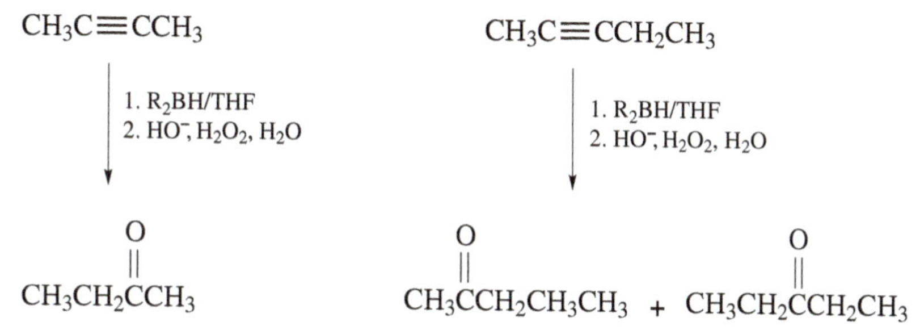

b. Only one product is obtained from hydroboration–oxidation of a symmetrical alkyne such as 3-hexyne or 4-octyne.

$CH_3CH_2C{\equiv}CCH_2CH_3$ (3-hexyne) $\quad$ $CH_3CH_2CH_2C{\equiv}CCH_2CH_2CH_3$ (4-octyne)

55. a. The first step forms a trans alkene. Syn addition to a trans alkene forms the threo pair of enantiomers.

Fischer projections: CH_2CH_3 (top); H–C–D; D–C–H; CH_2CH_3 (bottom) and CH_2CH_3 (top); D–C–H; H–C–D; CH_2CH_3 (bottom) **or** the corresponding wedge-dash structures (CH_3CH_2, H, D / H, D, CH_2CH_3 + H, D, CH_3CH_2 / CH_2CH_3, H, D)

b. The first step forms a cis alkene. Syn addition to a cis alkene forms the erythro pair of enantiomers, but because each asymmetric carbon is bonded to the same four groups, the product is a meso compound.

Fischer projection: CH_2CH_3 (top); H–C–D; H–C–D; CH_2CH_3 (bottom) **or** the corresponding wedge-dash structure (CH_3CH_2, H, D / D, H, CH_2CH_3)

56. If HO^- and HOOH are added at the same time as BH_3 (or R_2BH), HO^- is a better nucleophile than the alkyne (no bonds have to be broken when HO^- reacts with the electrophile), so BH_3 (or R_2BH) will react with HO^- instead of with the alkyne.

$$HO^- + BH_3 \rightleftharpoons H\overset{+}{O}—\overset{-}{B}H_3$$

57. **a.** $HC{\equiv}CH \xrightarrow{NaNH_2} HC{\equiv}C^- \xrightarrow{CH_3CH_2Br} HC{\equiv}CCH_2CH_3 \xrightarrow{NaNH_2} {}^-C{\equiv}CCH_2CH_3 \xrightarrow{CH_3CH_2Br} CH_3CH_2C{\equiv}CCH_2CH_3$

$CH_3CH_2C{\equiv}CCH_2CH_3 \xrightarrow[-78\ °C]{Na/NH_3(liq)}$ trans alkene (C=C with H and CH_2CH_3 on one carbon, CH_3CH_2 and H on the other) $\xrightarrow[H_2O]{Br_2}$

Fischer projections (top CH_2CH_3, bottom CH_2CH_3): H—OH, H—Br + HO—H, Br—H

(3*S*,4*R*)-4-bromo-3-hexanol + (3*R*,4*S*)-4-bromo-3-hexanol

b. $HC{\equiv}CH \xrightarrow{NaNH_2} HC{\equiv}C^- \xrightarrow{CH_3CH_2Br} HC{\equiv}CCH_2CH_3 \xrightarrow{NaNH_2} {}^-C{\equiv}CCH_2CH_3 \xrightarrow{CH_3CH_2Br} CH_3CH_2C{\equiv}CCH_2CH_3$

$CH_3CH_2C{\equiv}CCH_2CH_3 \xrightarrow[\text{Lindlar catalyst}]{H_2}$ cis alkene (C=C with CH_3CH_2 and H on one carbon, CH_2CH_3 and H on the other) $\xrightarrow[H_2O]{Br_2}$

Fischer projections (top CH_2CH_3, bottom CH_2CH_3): HO—H, H—Br + H—OH, Br—H

(3*R*,4*R*)-4-bromo-3-hexanol + (3*S*,4*S*)-4-bromo-3-hexanol

Because anti addition occurs in the last step, the threo enantiomers are formed.
If $Na/NH_3(liq)/-78$ °C is used instead of H_2/Lindlar catalyst in the fifth step, the trans alkene will be formed. Reaction of Br_2 and H_2O with the trans alkene will form the erythro enantiomers.

58. (3*E*,6*E*)-3,7,11-trimethyl-1,3,6,10-dodecatetraene

The configuration of the double bond at the 1-position and at the 10-position is not specified because isomers are not possible at those positions, because there are two hydrogens bonded to C-1 and two methyl groups bonded to C-11.

59. $HC{\equiv}CH \xrightarrow[\text{2. } CH_3CH_2Br]{\text{1. } \bar{N}H_2} HC{\equiv}CCH_2CH_3 \xrightarrow[\text{2. } CH_3Br]{\text{1. } \bar{N}H_2} CH_3C{\equiv}CCH_2CH_3$

$\xrightarrow{\text{Na, } NH_3\text{(liq)}, -78\ ^\circ C}$ H, CH_2CH_3, C=C, CH_3, H

$\xrightarrow{RCOOH}$ O, CH_3, CH_2CH_3 + O, CH_3CH_2, CH_3

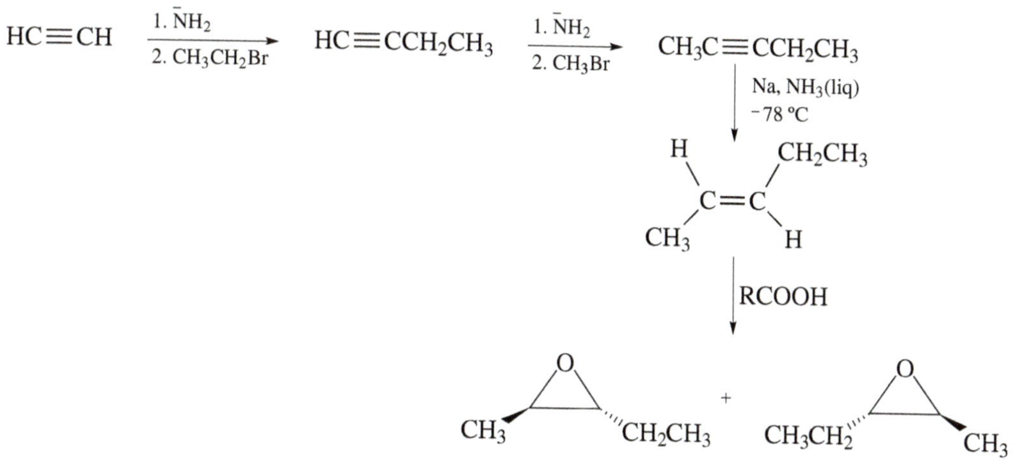

60.

a. Cl Cl Cl Cl

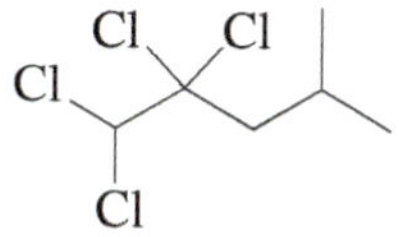

b. H Cl Cl

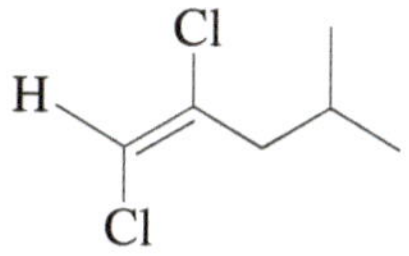

c. O

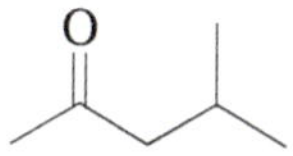

d. H O

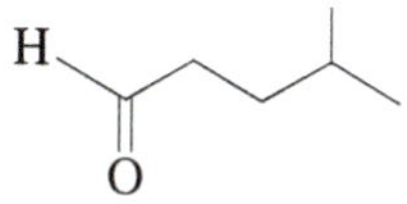

e.

f.

g.

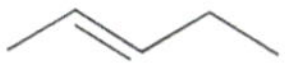

h. O + O

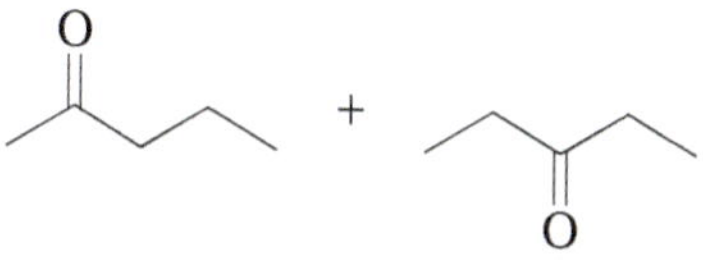

i. Br Br + Br Br

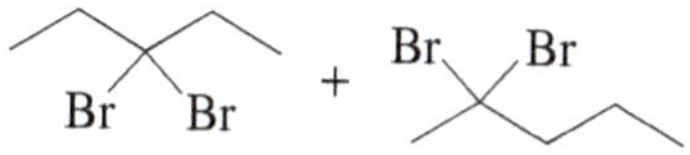

Chapter 7 Practice Test

1. What reagents can be used to convert the given starting material into the desired product?

a. 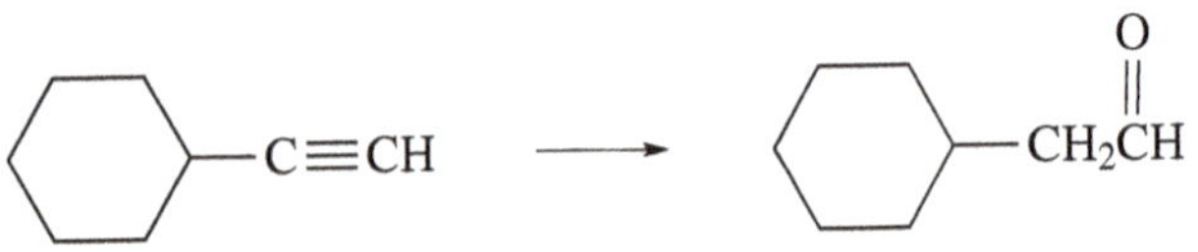 cyclohexyl–C≡CH ⟶ cyclohexyl–CH_2CH(=O)

b. 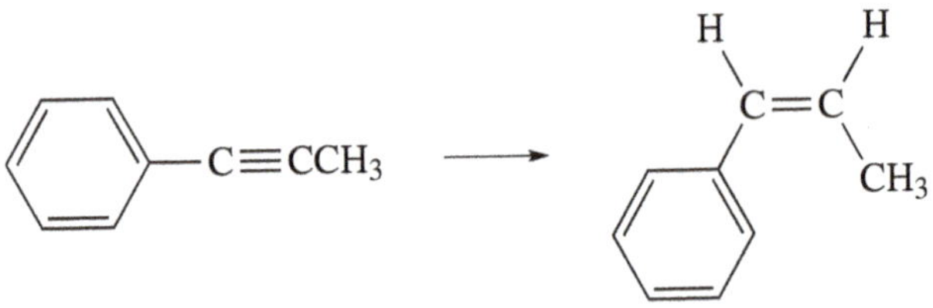 phenyl–C≡CCH_3 ⟶ (H)(phenyl)C=C(H)(CH_3)

2. Draw the enol tautomer(s) of the following ketones:

a. 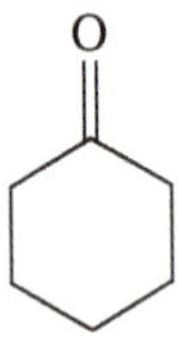 cyclohexanone

b. 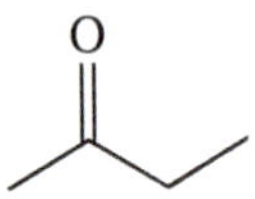 2-butanone

3. Draw the structure for each of the following:

a. *sec*-butylisobutylacetylene

b. 2-methyl-1,3-cyclohexadiene

4. Indicate whether each of the following statements is true or false:

a.	A terminal alkyne is more stable than an internal alkyne.	T	F
b.	Propyne is more reactive than propene toward reaction with HBr.	T	F
c.	1-Butyne is more acidic than 1-butene.	T	F
d.	An sp^2 carbon is more electronegative than an sp^3 carbon.	T	F
e.	The reactions of internal alkynes are more regioselective than the reactions of terminal alkynes.	T	F
f.	Alkenes are more reactive than alkynes.	T	F

5. What is each compound's systematic name?

a. $CH_3CH(CH_3)C{\equiv}CCH_2CH_2Br$

b. $CH_3CH(CH_3)C{\equiv}CCH_2CH_2OH$

6. What alkyne is the best reagent to use for the synthesis of each of the following ketones?

a. $CH_3CH_2CH_2C(=O)CH_3$

b. $CH_3CH_2C(=O)CH_2CH_2CH_3$

7. Rank the following compounds from most acidic to least acidic:

NH_3 $CH_3C{\equiv}CH$ CH_3CH_3 H_2O $CH_3CH{=}CH_2$

8. Give an example of a ketone that has two enol tautomers.

9. Show how the target molecules can be prepared from the given starting materials.

a. $CH_3CH_2C{\equiv}CH \longrightarrow CH_3CH_2CH_2CH_2CH_2CH_3$

b. $CH_3CH_2C{\equiv}CH \longrightarrow CH_3CH_2\underset{\displaystyle Br}{\underset{|}{C}}HCH_2CH_2CH_3$

c. $CH_3CH_2C{\equiv}CH \longrightarrow CH_3CH_2\overset{\displaystyle O}{\overset{\|}{C}}CH_2CH_2CH_3$

CHAPTER 8

Delocalized Electrons: Their Effect on Stability, pK_a, and the Products of a Reaction • Aromaticity and Electronic Effects: An Introduction to the Reactions of Benzene

Important Terms

1,2-addition (direct addition)	addition to the 1- and 2-positions of a conjugated system.
1,4-addition (conjugate addition)	addition to the 1- and 4-positions of a conjugated system.
aliphatic compound	an organic compound that does not contain an aromatic ring.
allene	a compound with two adjacent double bonds.
allylic carbon	a carbon adjacent to an sp^2 carbon of a carbon–carbon double bond.
allylic cation	a compound with a positive charge on an allylic carbon.
antiaromatic compound	a cyclic and planar compound with an uninterrupted cloud of π electrons containing an even number of pairs of π electrons.
antibonding molecular orbital	the molecular orbital formed when out-of-phase orbitals interact.
antisymmetric molecular orbital	a molecular orbital that does not have a plane of symmetry but would have one if half of the MO is turned upside down.
aromatic compound	a cyclic and planar compound with an uninterrupted cloud of π electrons containing an odd number of pairs of π electrons.
benzylic carbon	a carbon, joined to other atoms by single bonds, that is bonded to a benzene ring.
benzylic cation	a compound with a positive charge on a benzylic carbon.
bonding molecular orbital	the molecular orbital formed when in-phase orbitals overlap.
bridged bicyclic compound	a bicyclic compound in which the rings share two nonadjacent carbons.
common intermediate	an intermediate that reaction pathways have in common.
concerted reaction	a reaction in which all the bond-making and breaking processes occur in a single step.
conjugate addition	addition to the 1- and 4-positions of a conjugated system.
conjugated diene	a compound with two conjugated double bonds.

conjugated double bonds double bonds separated by one single bond.

contributing resonance structure a structure with localized electrons that approximates the true structure of a compound with delocalized electrons.

cumulated double bonds double bonds that are adjacent to one another.

cycloaddition reaction a reaction in which two π electron-containing reactants combine to form a single cyclic product.

[4+2] cycloaddition reaction a cycloaddition reaction in which six π electrons participate in the transition state with four π electrons coming from one reactant and two π electrons coming from the other reactant.

delocalization energy (resonance energy) the extra stability associated with a compound as a result of having delocalized electrons.

delocalized electrons electrons that result from a *p* orbital overlapping the *p* orbitals of two adjacent atoms; therefore, delocalized electrons are shared by three or more atoms.

Diels–Alder reaction a [4+2] cycloaddition reaction.

diene a hydrocarbon with two double bonds.

dienophile an alkene that reacts with a diene in a Diels–Alder reaction.

direct addition (1,2-addition) addition to the 1- and 2-positions of a conjugated system.

donation of electrons by resonance donation of electrons through π bonds.

electron delocalization the sharing of electrons by more than two atoms.

endo a substituent is endo if it and the bridge are on opposite sides of a bicyclic compound.

equilibrium control thermodynamic control.

exo a substituent is exo if it and the bridge are on the same side of a bicyclic compound.

fused rings rings that share two adjacent carbons.

highest occupied molecular orbital (HOMO) the highest-energy molecular orbital that contains electrons.

heteroatom an atom other than carbon.

heterocyclic compound (heterocycle) a cyclic compound in which one or more of the atoms of the ring is a heteroatom.

Hückel's rule or the $4n + 2$ rule	a rule that gives the number of π electrons a compound must have in its π cloud to be aromatic.
isolated double bonds	double bonds separated from one another by more than one single bond.
kinetic control	when a reaction is under kinetic control, the relative amounts of the products depend on the rates at which they are formed.
kinetic product	the product that is formed the fastest.
linear combination of molecular orbitals (LCAO)	the combination of atomic orbitals to produce molecular orbitals.
localized electrons	electrons that are restricted to a particular locality.
lowest unoccupied molecular orbital (LUMO)	the lowest-energy molecular orbital that does not contain electrons.
pericyclic reaction	a reaction that takes place in one step as a result of a cyclic reorganization of electrons.
phenyl group	C_6H_5—
polyene	a compound that has several double bonds.
polymer	a large molecule made by linking many small molecules together.
polymerization	the process of linking many small molecules to form a polymer.
proximity effect	an effect caused by one species being close to another.
resonance	electron delocalization.
resonance contributor (resonance structure)	a structure with localized electrons that approximates the true structure of a compound with delocalized electrons.
resonance electron donation	donation of electrons through π bonds.
resonance electron withdrawal	withdrawal of electrons through π bonds.
resonance energy (delocalization energy) (resonance stabilization energy)	the extra stability a compound possesses as a result of having delocalized electrons.

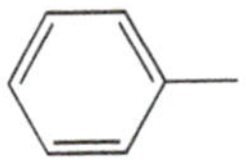

resonance hybrid the actual structure of a compound with delocalized electrons; it is represented by two or more resonance contributors with localized electrons.

***s*-cis-conformation** the conformation in which two double bonds of a conjugated diene are on the same side of a connecting single bond.

separated charges a positive and a negative charge that can be neutralized by the movement of electrons.

***s*-trans-conformation** the conformation in which two double bonds of a conjugated diene are on opposite sides of a connecting single bond.

symmetric molecular orbital an orbital with a plane of symmetry so that one half is the mirror image of the other half.

thermodynamic control when a reaction is under thermodynamic control, the relative amounts of the products depend on their stabilities.

thermodynamic product the most stable product.

withdrawal of electrons by resonance withdrawal of electrons through π bonds.

Solutions to Problems

1. **a.** **1.** If stereoisomers are not included, three different monosubstituted compounds are possible.

$BrC{\equiv}CC{\equiv}CCH_2CH_3$ $HC{\equiv}CC{\equiv}CCH(Br)CH_3$ $HC{\equiv}CC{\equiv}CCH_2CH_2Br$

If stereoisomers are included, four different monosubstituted compounds are possible, because the second listed compound has an asymmetric center so that it can have both the *R* and *S* configuration.

2. If stereoisomers are not included, two different monosubstituted compounds are possible.

$BrCH{=}CHC{\equiv}CCH{=}CH_2$ $CH_2{=}C(Br)C{\equiv}CCH{=}CH_2$

If stereoisomers are included, three different monosubstituted compounds are possible, because the first listed compound has a double bond that can have cis-trans isomers.

b. **1.** If stereoisomers are not included, five different disubstituted compounds are possible.

$HC{\equiv}CC{\equiv}CCH_2CHBr_2$ $HC{\equiv}CC{\equiv}CCBr_2CH_3$ $HC{\equiv}CC{\equiv}CCH(Br)CH_2Br$

$BrC{\equiv}CC{\equiv}CCH_2CH_2Br$ $BrC{\equiv}CC{\equiv}CCH(Br)CH_3$

2. If stereoisomers are not included, five different disubstituted compounds are possible.

$CH_2{=}C(Br)C{\equiv}CC(Br){=}CH_2$ $CH_2{=}CHC{\equiv}CC(Br){=}CHBr$ $CH_2{=}C(Br)C{\equiv}CCH{=}CHBr$

$CH_2{=}CHC{\equiv}CCH{=}CBr_2$ $BrCH{=}CHC{\equiv}CCH{=}CHBr$

c. **1.** If stereoisomers are included, seven different disubstituted compounds are possible, because two of the compounds have asymmetric centers so each can have either the *R* or *S* configuration.

2. If stereoisomers are included, nine different disubstituted compounds are possible, because the second and third compounds can have cis–trans isomers and the fifth compound can have cis–cis, trans–trans, and cis–trans isomers. (Note that cis–trans is the same as trans–cis.)

2. Ladenburg benzene is a better proposal. It would form one monosubstituted compound and three disubstituted compounds, in accordance with what early chemists knew about the structure of benzene.

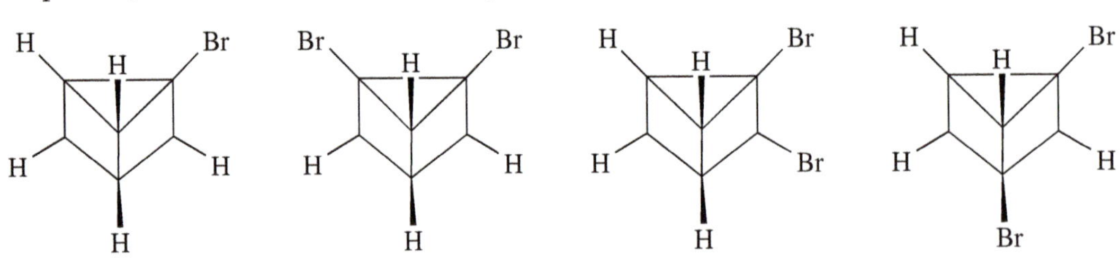

Dewar benzene is not in accordance with what early chemists knew about the structure of benzene, because it would form two monosubstituted compounds and six disubstituted compounds and it would undergo electrophilic addition reactions.

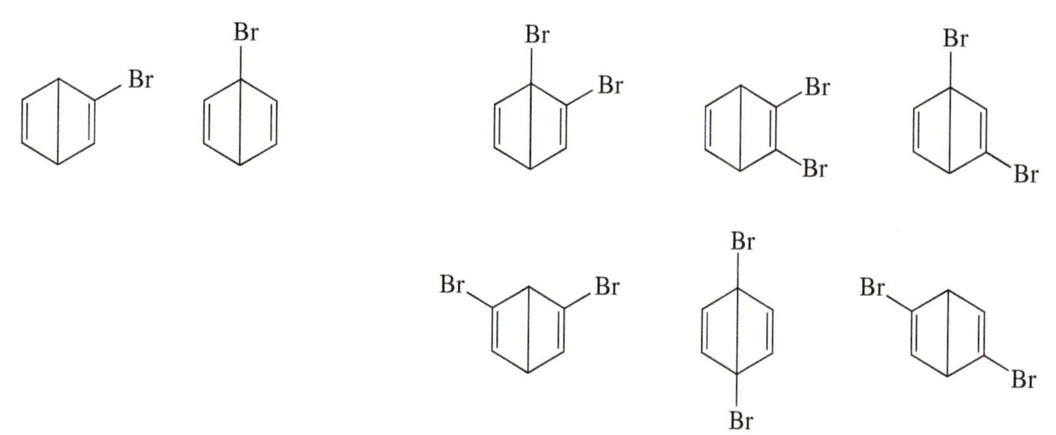

3. **a.** **2, 4,** and **5** have delocalized electrons.

1, 3, 6, and **7** do not have delocalized electrons, because electrons cannot be moved to an sp^3 carbon.

2. $CH_3CH{=}CH{-}CH{=}CH{-}\overset{+}{C}H_2$ ⟷ $CH_3CH{=}CH{-}\overset{+}{C}H{-}CH{=}CH_2$ ⟷ $CH_3\overset{+}{C}H{-}CH{=}CH{-}CH{=}CH_2$

4. N H ⟷ N H ⟷ N H ⟷ N H ⟷ N H

5. $\ddot{N}H_2$ ⟷ $=\overset{+}{N}H_2$

4. **a.** $CH_3CH_2\overset{CH_2}{\overset{\|}{C}}{-}\overset{+}{C}H_2$ ⟷ $CH_3CH_2\overset{\overset{+}{C}H_2}{\overset{|}{C}}{=}CH_2$

Less stable, because the positive charge is shared by two primary allylic carbons.

$CH_3CH_2CH{=}CH{-}\overset{+}{C}H_2$ ⟷ $CH_3CH_2\overset{+}{C}H{-}CH{=}CH_2$

More stable, because the positive charge is shared by a primary allylic and a secondary allylic carbon.

b. $CH_3C(=O)CH{=}CH_2 \longleftrightarrow CH_3C(O^-){=}CH{-}\overset{+}{C}H_2$

Less stable, because the positive charge is on a primary allylic carbon.

$CH_3C(=O)CH{=}CHCH_3 \longleftrightarrow CH_3C(O^-){=}CH{-}\overset{+}{C}HCH_3$

More stable, because the positive charge is on a secondary allylic carbon.

c. $CH_3CH(O^-){-}CH{=}CH_2$

Less stable, because the negative charge is not delocalized.

$CH_3C(\ddot{O}:^-){=}CHCH_3 \longleftrightarrow CH_3C(=O){-}\ddot{C}^-HCH_3$

More stable, because the negative charge is delocalized.

d. $CH_3C(={}^+OH)\ddot{N}H_2 \longleftrightarrow CH_3C(OH){=}\overset{+}{N}H_2$

Less stable, because the positive charge is shared by an O and an N.

$CH_3C(={}^+NH_2)\ddot{N}H_2 \longleftrightarrow CH_3C(NH_2){=}\overset{+}{N}H_2$

More stable, because the positive charge is shared by two nitrogens. Nitrogen is less electronegative than oxygen, so nitrogen is more comfortable with a positive charge.

5. Solved in the text.

6. The resonance contributor that makes the greatest contribution to the hybrid is labeled **A**. **B** contributes less to the hybrid than **A**, and **C** contributes less to the hybrid than **B**.

a. $CH_3C(=O)\ddot{O}CH_3$ (**A**) $\longleftrightarrow$ $CH_3C(O^-){=}\overset{+}{\ddot{O}}CH_3$ (**B**)

A is more stable than **B** because **B** has separated charges and has a positive charge on an oxygen.

$CH_3C(\overset{\delta-}{O})\overset{\delta+}{\ddot{O}}CH_3$

b. $CH_3\ddot{N}H-CH=CHCH_3 \longleftrightarrow CH_3\overset{+}{N}H=CH-\overset{-}{\ddot{C}}HCH_3$

A B

$CH_3\overset{\delta+}{N}H \text{---} CH \text{---} \overset{\delta-}{C}HCH_3$

A is more stable than **B** because **B** has separated charges and a positive charge on a nitrogen.

c. 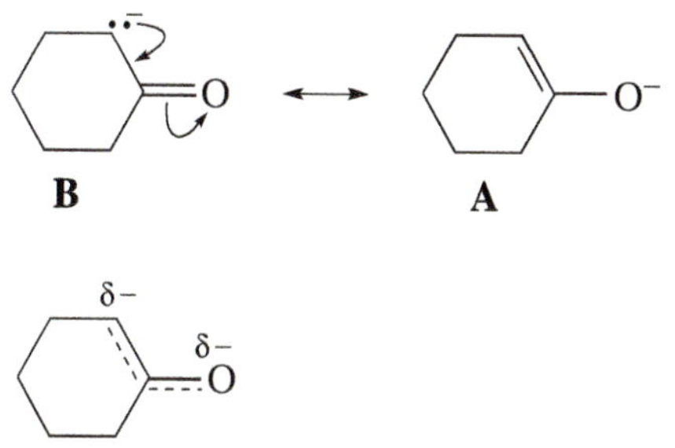 B ⟷ A

A is more stable than **B** because the negative charge in **A** is on an oxygen, whereas the negative in **B** is on carbon, which is less electronegative than oxygen.

δ− δ− O

d. 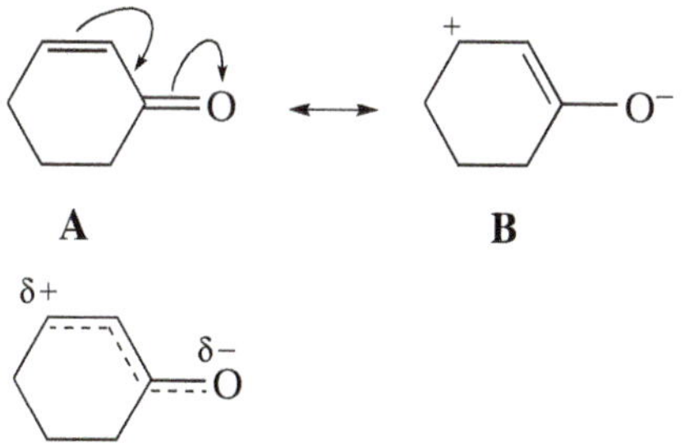 A ⟷ B

A is more stable than **B** because **A** does not have separated charges and **B** has an atom with an incomplete octet.

δ+ δ− O

e. 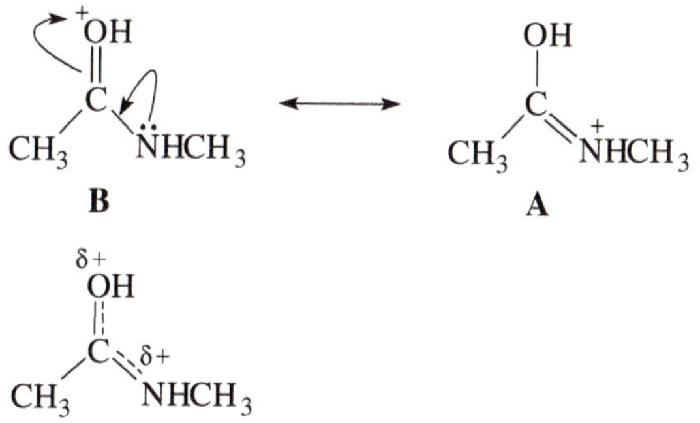 B ⟷ A

A is more stable than **B** because the positive charge in **A** is on a less electronegative atom. (N is less electronegative than O.)

δ+ OH, C, δ+ NHCH$_3$, CH$_3$

f. $CH_3\overset{+}{C}H-CH=CHCH_3 \longleftrightarrow CH_3CH=CH-\overset{+}{C}HCH_3$

A B

$CH_3\overset{\delta+}{C}H \text{---} CH \text{---} \overset{\delta+}{C}HCH_3$

A and **B** are equally stable.

7. **a.** All the carbon–oxygen bonds in the carbonate ion are the same length, because each carbon–oxygen bond is represented in one resonance contributor by a double bond and in two resonance contributors by a single bond.

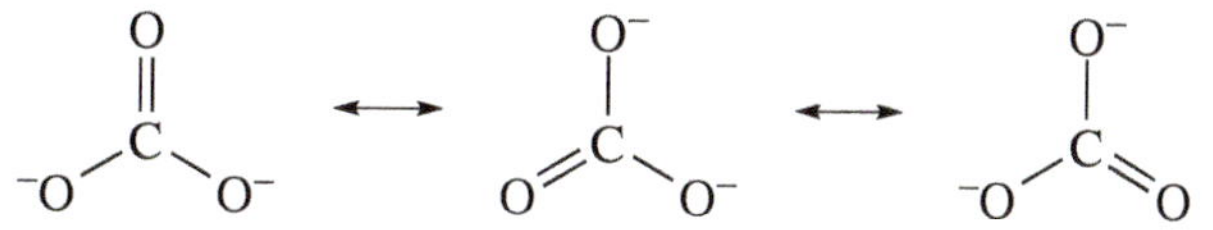

b. Because the two negative charges are shared equally by three oxygens, each oxygen has two-thirds of a negative charge.

8. $^{-}O{-}C({=}O){-}O^{-}$ > $H{-}C({=}O){-}O^{-}$ > $H{-}C({=}O){-}OH$

The dianion has the greatest delocalization energy; it has three relatively stable resonance contributors. (See the answer to Problem 7.)

The monoanion is the next most stable; it has two relatively stable resonance contributors.

The neutral species has the smallest delocalization energy; it has one relatively stable resonance contributor and one that has separated charges.

9. The carboxylate ion has the greater delocalization energy because it has two equivalent relatively stable resonance contributors.

$CH_3{-}C({=}O){-}O^{-} \longleftrightarrow CH_3{-}C(O^{-}){=}O$

In contrast, the diene has three resonance contributors but only one is relatively stable (see the top of page 330 in the text).

10. The compound on the left has the greater delocalization energy because the lone-pair electrons on oxygen are delocalized, whereas the lone-pair electrons on oxygen in the compound on the right are not delocalized.

11. The smaller the heat of hydrogenation (the positive value of ΔH°), the more stable the compound. Therefore, the relative stabilities of the dienes are

conjugated diene > isolated diene > cumulated diene

12. $CH_3C(CH_3){=}CHCH{=}C(CH_3)CH_3$ > $CH_3CH{=}CHCH{=}CHCH_3$ > $CH_3CH{=}CHCH{=}CH_2$ > $CH_2{=}CHCH_2CH{=}CH_2$

2,5-dimethyl-2,4-hexadiene 2,4-hexadiene 1,3-pentadiene 1,4-pentadiene

13. **a.** The compound with delocalized electrons is more stable than the compound in which all the electrons are localized.

$\overset{+}{C}HCH_3$ ⟷ $CHCH_3$ $\overset{+}{C}HCH_3$

electrons are delocalized electrons are localized

b. Because nitrogen is less electronegative than oxygen, it shares the positive charge better.

$CH_3\ddot{O}-\overset{+}{C}H_2 \longleftrightarrow CH_3\overset{+}{O}=CH_2$ $CH_3\ddot{N}H-\overset{+}{C}H_2 \longleftrightarrow CH_3\overset{+}{N}H=CH_2$

more stable

c. In order for electron delocalization to occur, the atoms that share the π electrons must be in the same plane so that their *p* orbitals can overlap. The two bulky *tert*-butyl groups do not allow enough room for the group with the positively charged carbon to be in the same plane as the benzene ring. Therefore, the carbocation cannot be stabilized by electron delocalization because the *p* orbital of the positively charged carbon cannot overlap the *p* orbitals of the benzene ring since they are not in the same plane.

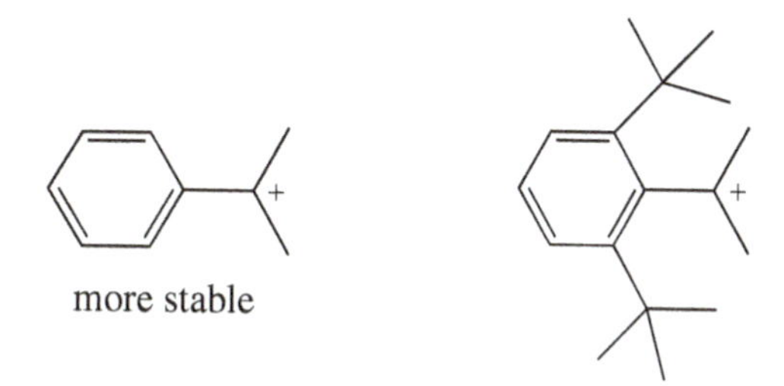

14. The ψ_3 molecular orbital of 1,3-butadiene has three nodes (two vertical and one horizontal).

The ψ_4 molecular orbital of 1,3-butadiene has four nodes (three vertical and one horizontal).

15.
a. ψ_1 and ψ_2 are bonding molecular orbitals, and ψ_3 and ψ_4 are antibonding molecular orbitals.
b. ψ_1 and ψ_3 are symmetric molecular orbitals, and ψ_2 and ψ_4 are antisymmetric molecular orbitals.
c. ψ_2 is the HOMO and ψ_3 is the LUMO in the ground state.
d. ψ_3 is the HOMO and ψ_4 is the LUMO in the excited state.
e. If the HOMO is symmetric, the LUMO is antisymmetric and vice versa.

16. Benzene is more stable than 1,3,5-hexatriene because the MO of benzene has six bonding interactions, whereas the MO of 1,3,5-hexatriene has only five bonding interactions.

17. In each case, the compound shown is the stronger acid because the negative charge that results when it loses a proton can be delocalized. Electron delocalization stabilizes the base, and the more stable the base, the more acidic its conjugate acid. Electron delocalization is not possible for the other compound in each pair.

a. $CH_3C(=O)OH \longrightarrow H^+ + CH_3C(=O)O^- \longleftrightarrow CH_3C(O^-)=O$

b. $CH_3CH=CH\overset{+}{N}H_3 \longrightarrow H^+ + CH_3CH=CHNH_2 \longleftrightarrow CH_3\bar{C}HCH=\overset{+}{N}H_2$

c. $CH_3CH=CHOH \longrightarrow H^+ + CH_3CH=CHO^- \longleftrightarrow CH_3\bar{C}HCH=O$

18. **a.** Ethylamine is a stronger base because when the lone pair on the nitrogen in aniline is protonated, it can no longer be delocalized into the benzene ring. Therefore, aniline is less apt to share its electrons with a proton.

b. Ethoxide ion is a stronger base because a negatively charged oxygen is a stronger base than a neutral nitrogen.

c. Ethoxide ion is a stronger base because when the phenolate ion is protonated, the pair of electrons that is protonated can no longer be delocalized into the benzene ring. Therefore, the phenolate ion is less apt to share its electrons with a proton.

d. Phenolate ion is the stronger base because its conjugate acid is a weaker acid than the conjugate acid of acetate ion.

19. The carboxylic acid is the most acidic because its conjugate base has greater delocalization energy than does the conjugate base of phenol. The alcohol is the least acidic because, unlike the negative charge on the conjugate base of phenol, the negative charge on the conjugate base of the alcohol cannot be delocalized.

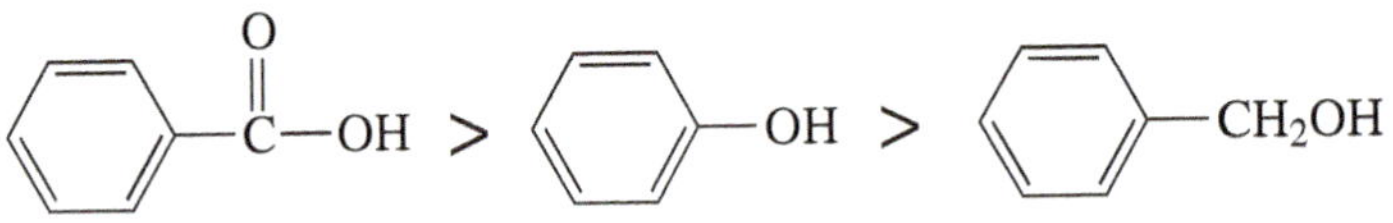

$$C_6H_5\text{—}\overset{O}{\overset{\|}{C}}\text{—OH} > C_6H_5\text{—OH} > C_6H_5\text{—}CH_2OH$$

20. **a.** donates electrons by resonance and withdraws electrons inductively
b. donates electrons by hyperconjugation
c. withdraws electrons by resonance and withdraws electrons inductively
d. donates electrons by resonance and withdraws electrons inductively
e. donates electrons by resonance and withdraws electrons inductively
f. withdraws electrons inductively

21. **a.** $ClCH_2COOH$

b. O_2NCH_2COOH
The closer the electron-withdrawing substituent is to the COOH group, the stronger the acid.

c. COOH (benzene ring, para) $CH_3C{=}O$

d. $H_3\overset{+}{N}CH_2COOH$

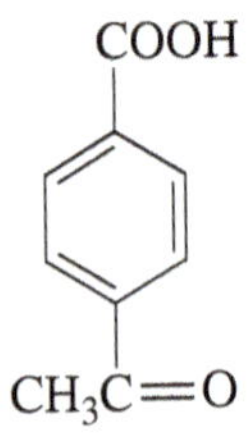

e. HCOOH
A hydrogen is electron-withdrawing compared to a methyl group, because a methyl group can donate electrons by hyperconjugation.

f. COOH (benzene ring, para) Cl

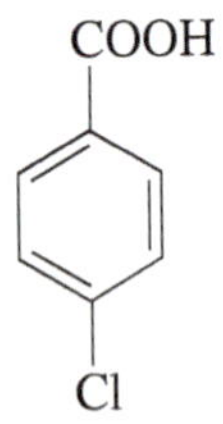

22. Solved in the text.

23. When *para*-nitrophenol loses a proton, the negative charge in the conjugate base can be delocalized onto the nitro substituent. Therefore, the *para*-nitro substituent decreases the pK_a both by resonance electron withdrawal and by inductive electron withdrawal.

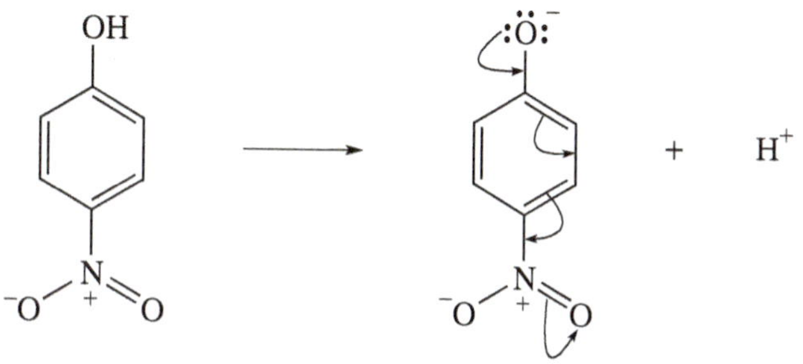

When *meta*-nitrophenol loses a proton, the negative charge in the conjugate base cannot be delocalized onto the nitro substituent. Therefore, the *meta*-nitro substituent can decrease the pK_a only by inductive electron withdrawal. Therefore, the para isomer has a lower pK_a.

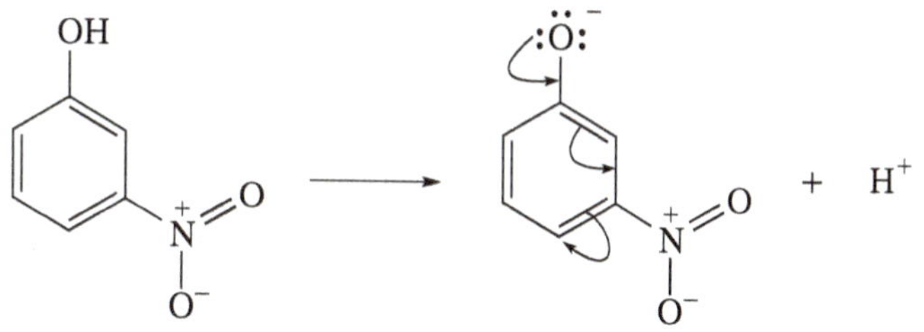

24. Recall that if a more stable carbocation can be formed as a result of carbocation rearrngement, rearrangement will occur.

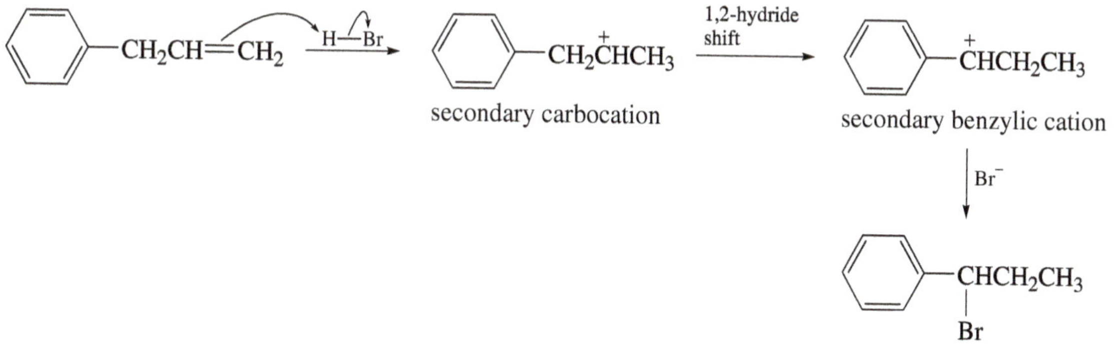

secondary carbocation

1,2-hydride shift

secondary benzylic cation

25. Solved in the text.

26. The resonance contributors show that one of the atoms has a partial negative charge. Therefore, that is the atom that is more apt to be protonated.

site of protonation

27. **a.** The more reactive double bond is the one that forms a tertiary carbocation.

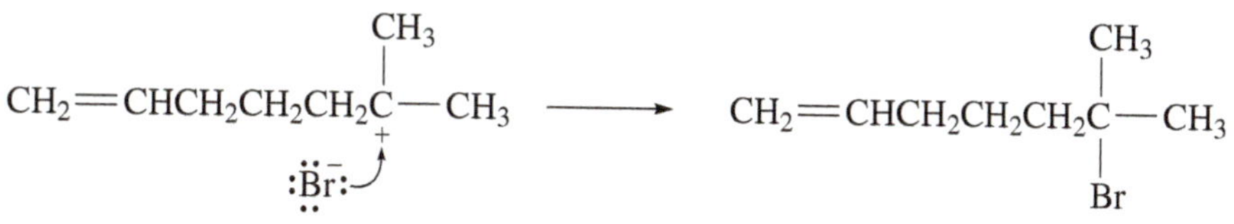

Instead of Br^- being the nucleophile that adds to the tertiary carbon, the π bond can be the nucleophile. In that case, a stable six-membered ring is formed. (See the Problem-Solving Strategy on page 257 of the text.) This is expected to be a minor product because, unlike the above reaction of the carbocation with Br^-, bond breaking is required to form the product.

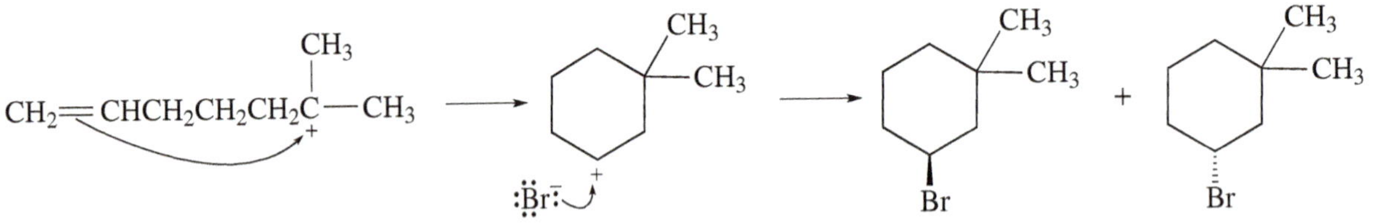

$$CH_2{=}CHCH_2CH_2CH_2\overset{+}{C}(CH_3){-}CH_3 \longrightarrow$$

CH3, CH3, :Br:⁻, +, CH3, CH3, Br, +, CH3, CH3, Br

b. The double bond is more reactive than the triple bond. The reaction forms a new asymmetric center, so a pair of enantiomers is formed.

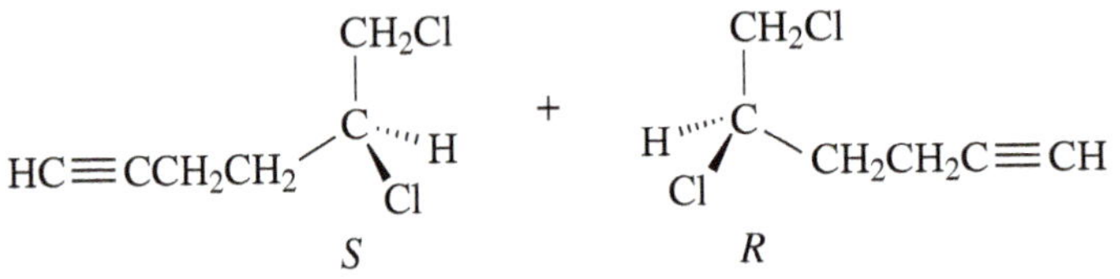

$$HC{\equiv}CCH_2CH_2C(CH_2Cl)(H)(Cl) \quad + \quad H(Cl)C(CH_2Cl)CH_2CH_2C{\equiv}CH$$

S *R*

It is unlikely that the triple bond will act as a nucleophile, because it would have to form an unstable vinylic cation intermediate.

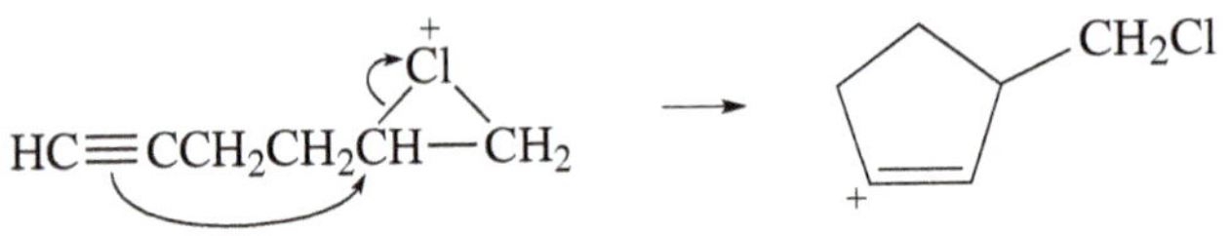

$$HC{\equiv}CCH_2CH_2\overset{\overset{+}{Cl}}{CH{-}CH_2} \longrightarrow$$

CH2Cl, +

c. The more reactive double bond is the one that forms a tertiary carbocation.

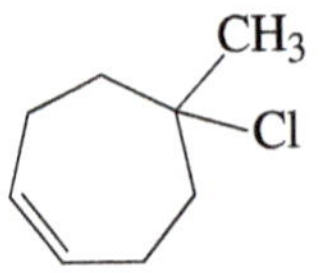

CH3, Cl

28. **a.** $CH_3CH{=}CH{-}CH{=}CHCH_3$

$\downarrow Cl_2$

$$CH_3CH{-}CH{-}CH{=}CHCH_3 \text{ (bridged } \overset{+}{Cl}\text{)} \longleftrightarrow CH_3\underset{Cl}{CH}{-}CH{=}CH{-}\overset{+}{C}HCH_3$$

$\downarrow Cl^-$ $\downarrow Cl^-$

$$CH_3\overset{Cl}{C}H\overset{Cl}{C}HCH{=}CHCH_3 \quad + \quad CH_3\overset{Cl}{C}HCH{=}CH\overset{Cl}{C}HCH_3$$

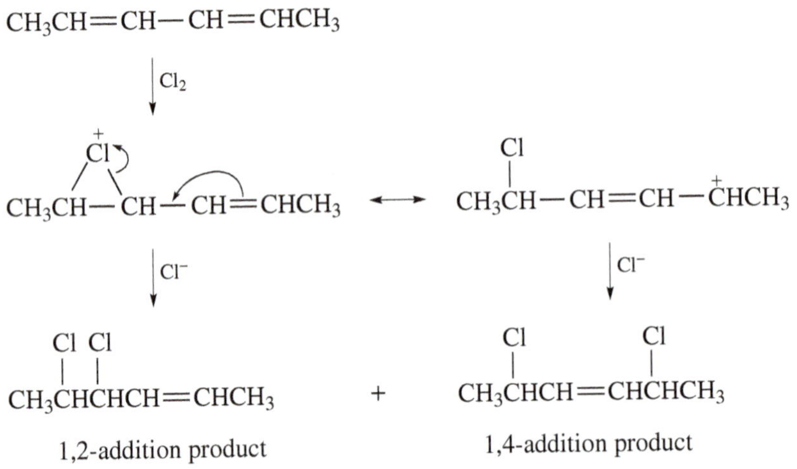

1,2-addition product 1,4-addition product

b. 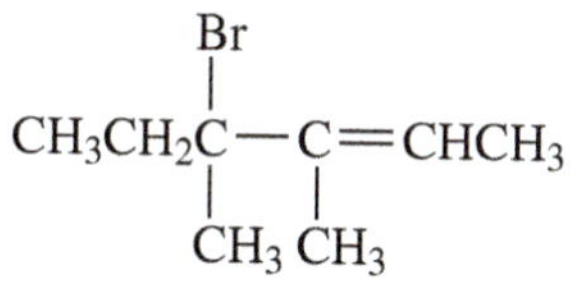 $CH_3CH_2C(Br)(CH_3){-}C(CH_3){=}CHCH_3$ + $CH_3CH_2C(CH_3){=}C(CH_3)CH(Br)CH_3$

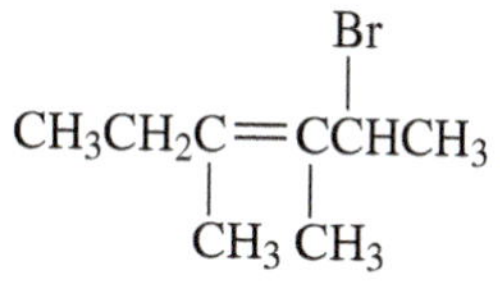

1,2-addition product 1,4-addition product

c. 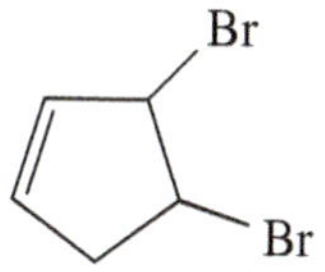 1,2-addition product + 1,4-addition product

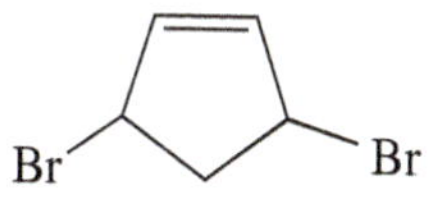

d.

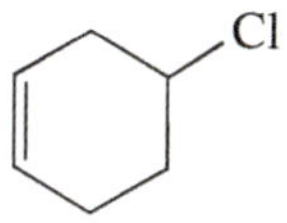

29. The indicated double bond is the most reactive in an electrophilic addition reaction with HBr because addition of an electrophile to this double bond forms the most stable carbocation (a tertiary allylic cation).

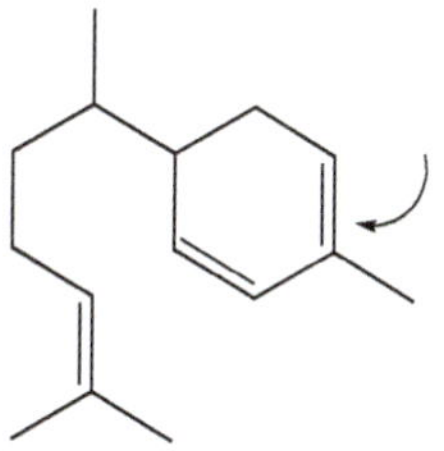

30. first reaction: $\underset{Cl}{CH_2}{-}\overset{*}{\underset{Cl}{CH}}{-}CH{=}CH_2$ $\underset{Cl}{CH_2}{-}CH{=}CH{-}\underset{Cl}{CH_2}$

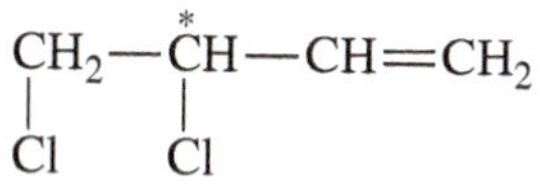

This compound has an asymmetric center, so both the *R* and *S* stereoisomers will be obtained. (Note that *E* and *Z* stereoisomers are not possbile for the double bond.)

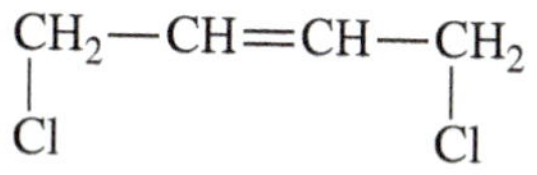

This compound has a double bond, so both the *E* and *Z* stereoisomers will be obtained.

second reaction: $CH_3\overset{*}{\underset{Br}{CH}}{-}CH{=}CH_2$ $CH_3CH{=}CH{-}\underset{Br}{CH_2}$

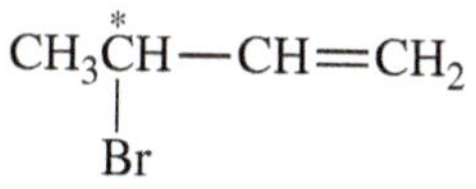

This compound has an asymmetric center, so both the *R* and *S* stereoisomers will be obtained. (Note that *E* and *Z* stereoisomers are not possbile for the double bond.)

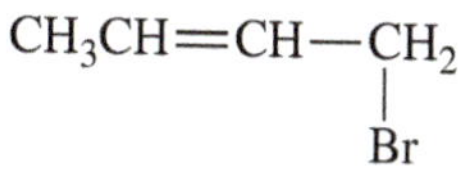

This compound has a double bond, so both the *E* and *Z* stereoisomers will be obtained.

31. $CH_2{=}CH{-}CH{=}CH{-}CH{=}CH_2 \xrightarrow{HBr} CH_3{-}CH(Br){-}CH{=}CH{-}CH{=}CH_2$

$CH_3{-}CH{=}CH{-}CH(Br){-}CH{=}CH_2$

$CH_3{-}CH{=}CH{-}CH{=}CH{-}CH_2Br$

32. **a.** The chlorine adds so that the positive charge in the resonance contributor is on a secondary allylic carbon. (If the chlorine had added to the other double bond, the positive charge would be on a primary allylic carbon.)

$CH_3CH{=}CH{-}C(CH_3){=}CH_2 \xrightarrow{Cl_2} CH_3CH{=}CH{-}\overset{+}{C}l\text{-bridged } C(CH_3){-}CH_2 \longleftrightarrow CH_3\overset{+}{C}H{-}CH{=}C(CH_3){-}CH_2Cl$

$\downarrow Cl^-$: $CH_3CH{=}CH{-}C(Cl)(CH_3){-}CH_2Cl$

$\downarrow Cl^-$: $CH_3CH(Cl){-}CH{=}C(CH_3){-}CH_2Cl$

b. The proton adds so that the positive charge in the carbocation is shared by a tertiary allylic and a secondary allylic carbon. (If the proton had added to the other double bond, the positive charge would be shared by two secondary allylic carbons.)

$CH_3CH{=}CH{-}C(CH_3){=}CHCH_3 \xrightarrow{HBr} CH_3CH{=}CH{-}\overset{+}{C}(CH_3){-}CH_2CH_3 \longleftrightarrow CH_3\overset{+}{C}H{-}CH{=}C(CH_3){-}CH_2CH_3$

$\downarrow Br^-$: $CH_3CH{=}CH{-}C(Br)(CH_3){-}CH_2CH_3$

$\downarrow Br^-$: $CH_3CH(Br){-}CH{=}C(CH_3){-}CH_2CH_3$

33. **a.** Addition at C-1 forms the more stable carbocation because the positive charge is shared by two secondary allylic carbons. If the deuterium had added to C-4, the positive charge would be shared by a secondary allylic and a primary allylic carbon.

b. DCl was used to cause the 1,2- and 1,4-products to be different. If HCl had been used, the 1,2- and 1,4-products would be the same.

34. She should follow her friend's advice. If she uses 2-methyl-1,3-cyclohexadiene, the fastest formed product will be 3-chloro-3-methylcyclohexene, both if the proximity effect controls which product is formed faster *and* if the more stable transition state controls which product is formed faster, because this product is formed through a transition state in which the positive charge is primarily on a tertiary allylic carbon. Therefore, the experiment will not be able to establish which of the two effects controls which product is formed faster.

CH_3 HCl Cl CH_3

3-chloro-3-methylcyclohexene

If she follows her friend's advice and uses 1-methyl-1,3-cyclohexadiene, the faster-formed product will be 3-chloro-1-methylcyclohexene only if the proximity effect controls which product is formed faster. The product will be 3-chloro-3-methylcyclohexene if the more stable transition state controls which product is formed faster, because only this product is formed through a transition state in which the positive charge is primarily on a tertiary allylic carbon.

CH_3 HCl CH_3 Cl + Cl CH_3

3-chloro-1-methylcyclohexene 3-chloro-3-methylcyclohexene

35. **a.** The rate-determining step is formation of the carbocation.

b. The product-determining step is reaction of the carbocation with the nucleophile.

36. Solved in the text.

37. **a.** CH_3 HCl Cl CH_3 + CH_3 Cl

kinetic product thermodynamic product

b. $CH{=}CHCH_3$ HCl $CH{=}CHCH_3$ Cl + $CH{-}CHCH_3$ Cl

kinetic product thermodynamic product

Notice that the 1,2-product is always the kinetic product.
The thermodynamic product is the product with the most highly substituted double bond.

38.

kinetic product
thermodynamic product

The first compound is the kinetic product because it is the 1,2-product.
The first compound is the thermodynamic product because it is more stable since the double bond is conjugated with the benzene ring.

39. **a.**

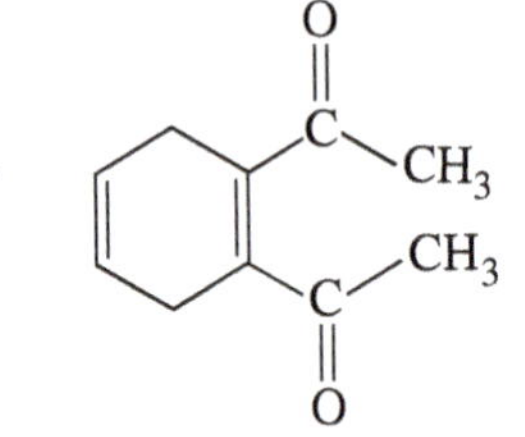

c.

b.

d.

40. First draw the resonance contributors to determine where the charges are on the reactants. The major product is obtained by joining the negatively charged carbon of the diene with the positively charged carbon of the dienophile.

diene

dienophile

Because the reaction creates an asymmetric center, the product will be a racemic mixture.

41. The resonance contributors show that if the electron-donating substituent is at the end of the conjugated system, it and the electron-withdrawing substituent of the dienophile will be adjacent to one another in the major product of the Diels–Alder reaction.

$\delta+$ OCH_3 $\delta-$ O C $\delta-$ $\delta+$ → OCH_3 O C

If the electron-donating substituent is not at the end of the conjugated system, it and the electron-withdrawing substituent of the dienophile will be opposite each other in the major product of the Diels–Alder reaction.

$\delta-$ O C $\delta+$ CH_3O $\delta-$ $\delta+$ → O C CH_3O

42. **a.** CH_3 C≡N + C≡N CH_3 **b.** C≡N CH_3 + CH_3 C≡N

43. **A** and **C** will not react, because they are both locked in an *s*-trans conformation.

D and **E** will react, because they are both locked in an *s*-cis conformation.

B and **F** will react, because they can rotate into an *s*-cis conformation.

44. **a.** C CH_3 O O C CH_3 **b.** C CH_3 O O C CH_3

45. Solved in the text.

46. **a.** It is not optically active, because it is a meso compound.
(It has two asymmetric centers and a plane of symmetry.)

Cl Cl

b. It is not optically active, because it is a racemic mixture. (Identical amounts of the enantiomers will be obtained.)

Cl Cl + Cl Cl

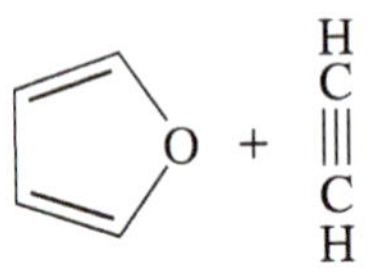

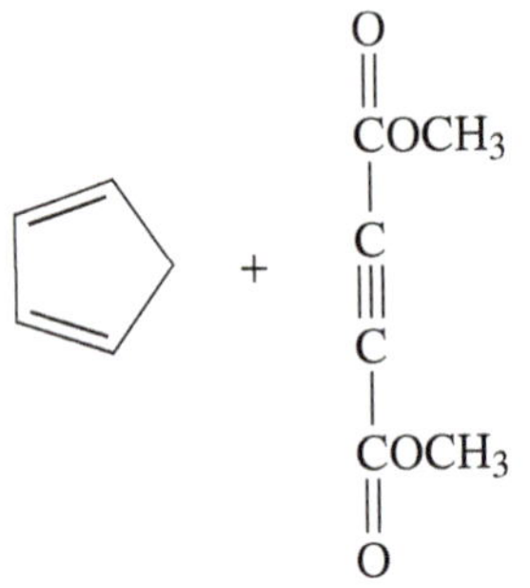

47. **a.** 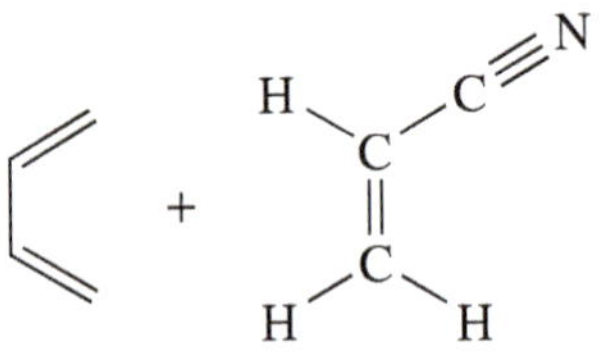 + H C≡N C H H **b.** O + HC≡CH **c.** + COCH$_3$ C≡C COCH$_3$ **d.** 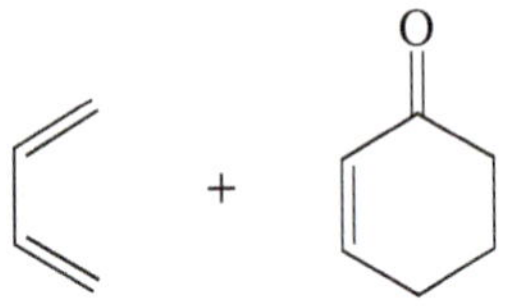 + O **e.** 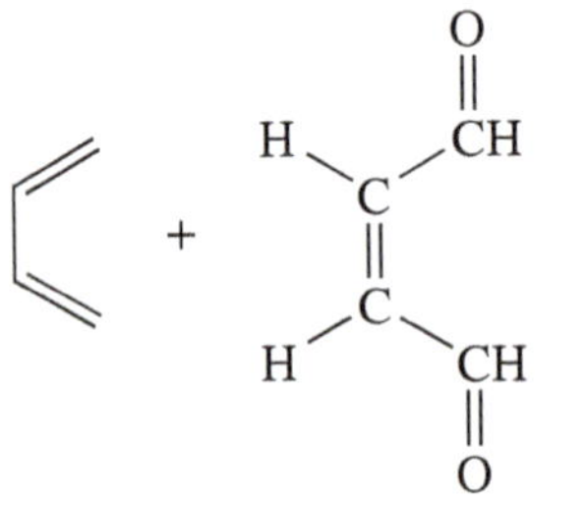 + H CH=O H CH=O **f.** 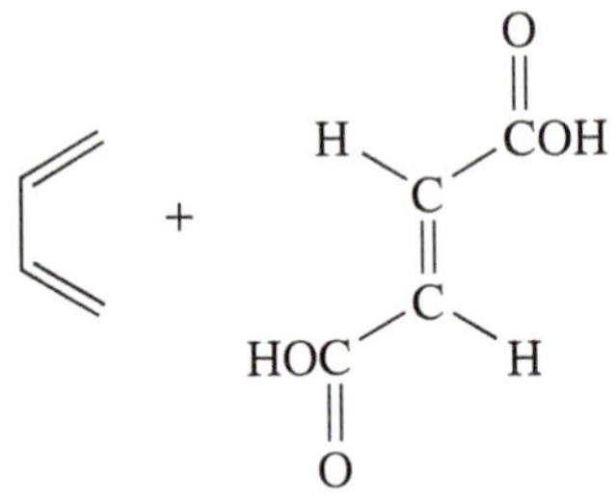 + H COOH HOOC H

48. None are aromatic.

c is not aromatic, because it has two pairs of π electrons.

a, **b**, **d**, and **e** are not aromatic, because each compound has two pairs of π electrons and every atom in the ring does not have a p orbital.

f is not aromatic, because it is not cyclic.

49. **a.** In the case of 9 pairs of π electrons, there are 18 electrons. Therefore, $4n + 2 = 18$ where $n = 4$.

b. Because it has an odd number of pairs of π electrons, it will be aromatic if it is cyclic and planar and if every atom in the ring has a p orbital.

50. **a.** 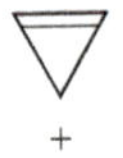 + This is the only one that is aromatic; it is cyclic, it is planar, every ring atom has a p orbital, and it has one pair of π electrons.

The first compound is not aromatic, because one of the atoms is sp^3 hybridized and, therefore, does not have a p orbital.

The third compound is not aromatic, because it has two pairs of π electrons.

b. 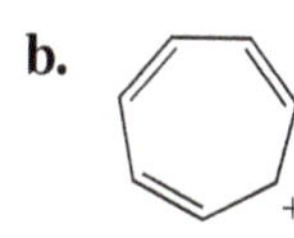

This is the only one that is aromatic; it is cyclic, it is planar, every ring atom has a p orbital, and it has three pairs of π electrons.

The first compound is not aromatic, because one of the atoms is sp^3 hybridized and, therefore, does not have a p orbital.

The third compound is not aromatic, because it has four pairs of π electrons.

51. Solved in the text.

52. **a.** Cyclopentadiene has a lower pK_a value. That is, it is a stronger acid. When cyclopentadiene loses a proton, a relatively stable aromatic compound is formed. When cycloheptatriene loses a proton, an unstable antiaromatic compound is formed (Section 8.19). Recall that the more stable the base, the stronger its conjugate acid.

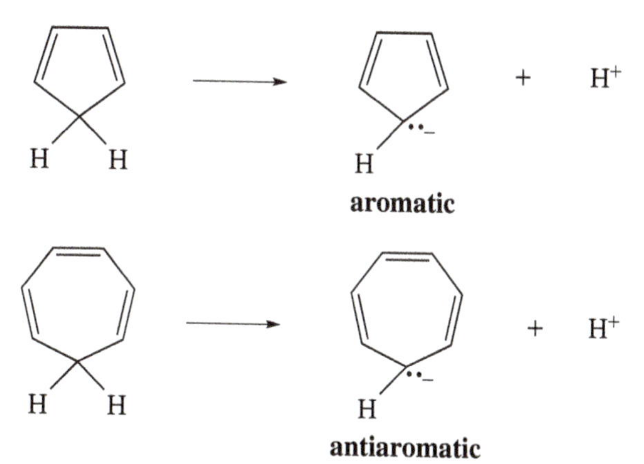

b. Cyclopropane has a lower pK_a value because a very unstable antiaromatic compound is formed when cyclopropene loses a proton.

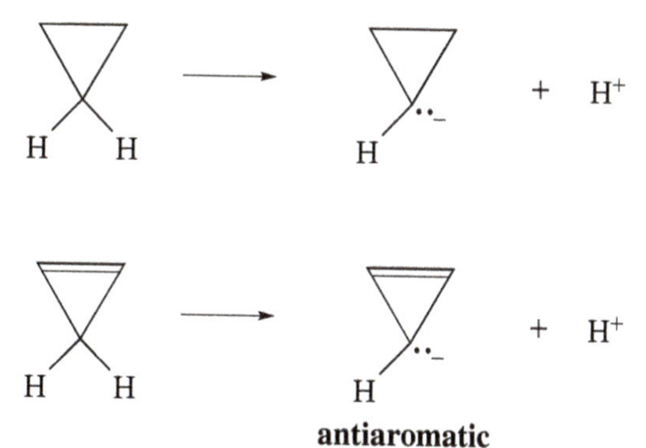

53. 3-Bromocyclopropene is more soluble in water because it is more apt to ionize since an aromatic compound is formed when its carbon–bromine bond breaks.

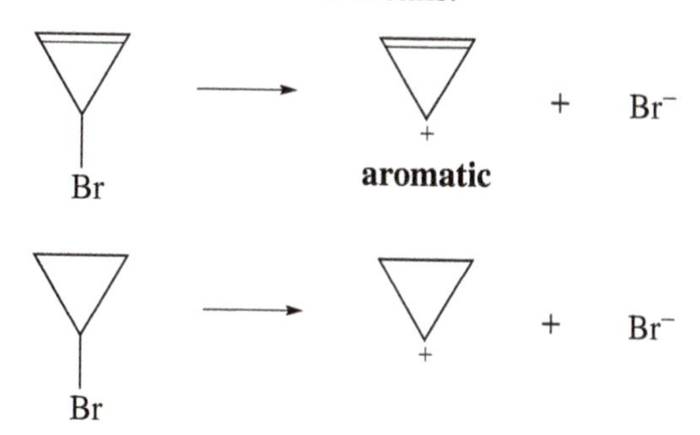

54. **a.** In fulvene, the electrons in the exocyclic double bond (a double bond attached to a ring) move toward the five-membered ring, because that results in a resonance contributor that is aromatic. Moving the electrons in the other direction would result in a resonance contributor with an antiaromatic ring.

CH_2 ⟷ $\overset{+}{C}H_2$

fulvene

b. In calicene, the electrons in the double bond between the two rings move toward the five-membered ring, because that results in a resonance contributor with two aromatic rings.

calicene

55. The Frost circles show that species with completely filled bonding MOs and with electrons in no other orbitals are aromatic (for example, the cycloheptatrienyl cation and the cyclopropenyl cation).

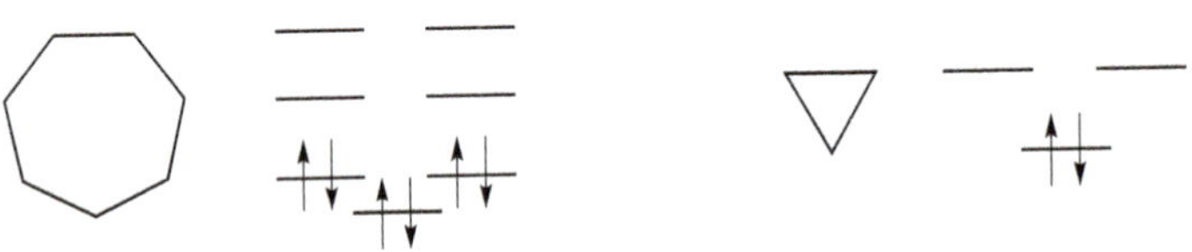

cycloheptatrienyl cation cyclopropenyl cation

The species with unpaired electrons in degenerate MOs is antiaromatic (the cycloheptatrienyl anion).

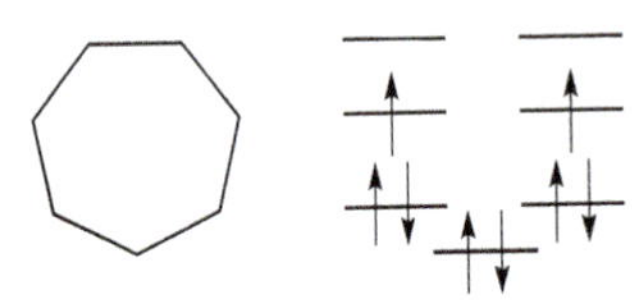

cycloheptatrienyl anion

56. **a.** $CH_3CH_2\ddot{N}H_2$ (sp^3) **b.** cyclohexyl$-CH{=}\ddot{N}CH_2CH_3$ (sp^2) **c.** $CH_3CH_2C{\equiv}N{:}$ (sp)

57. sp^2 p sp^2 p sp^2 sp^2 sp^2 p sp^2 sp^2

58. Solved in the text.

59. The lone-pair electrons on both nitrogens in the second compound are sp^2 electrons; therefore, they are not part of the π cloud and can be protonated without destroying the compound's aromaticity.

The lone-pair electrons on the nitrogen in the first compound are π electrons and are part of the π cloud. If these π electrons are protonated, they will no longer be part of the π cloud (they cannot be delocalized) and the compounds will not be aromatic.

60. **a.** The nitrogen atom (the atom at the bottom of the epm) in pyrrole has a partial positive charge because it donates electrons by resonance into the ring.

b. The nitrogen atom (the atom at the bottom of the epm) in pyridine cannot donate electrons by resonance; it withdraws electrons from the ring inductively because it is the most electronegative atom in the molecule. Thus, this nitrogen is electron-rich.

c. The relatively electronegative nitrogen atom in pyridine withdraws electrons from the ring.

61. **A, B, C, E, F, L, M, N, O**

62. **a** and **b.**

1.  $\ddot{C}H_2^-$—$\overset{+}{N}$≡N: ⟷ CH_2=$\overset{+}{N}$=$\ddot{N}$:$^-$

More stable, because the negative charge is on nitrogen rather than on carbon.

2.  :N≡$\overset{+}{N}$—$\ddot{O}$:$^-$ ⟷ $^-$:$\ddot{N}$=$\overset{+}{N}$=$\ddot{O}$:

More stable because the negative charge is on oxygen rather than on nitrogen (oxygen is more electronegative).

3.  $^-$:$\ddot{O}$—$\ddot{N}$=$\ddot{O}$: ⟷ :$\ddot{O}$=$\ddot{N}$—$\ddot{O}$:$^-$

Both are equally stable.

63. In each case, the proton adds to the sp^2 carbon that results in formation of a tertiary carbocation.

a. 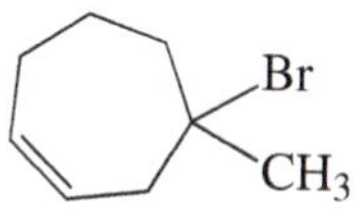 Br, CH_3

b. 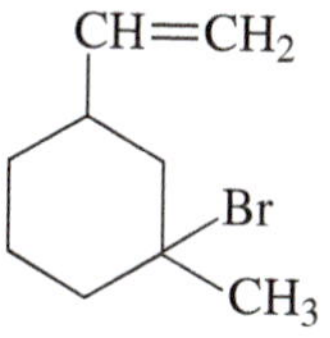 $CH=CH_2$, Br, CH_3

64. **a.** 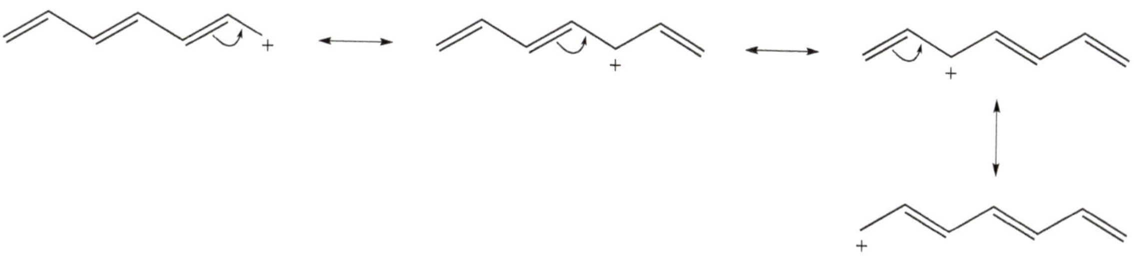 + ⟷ + ⟷ + ⟷ +

b.

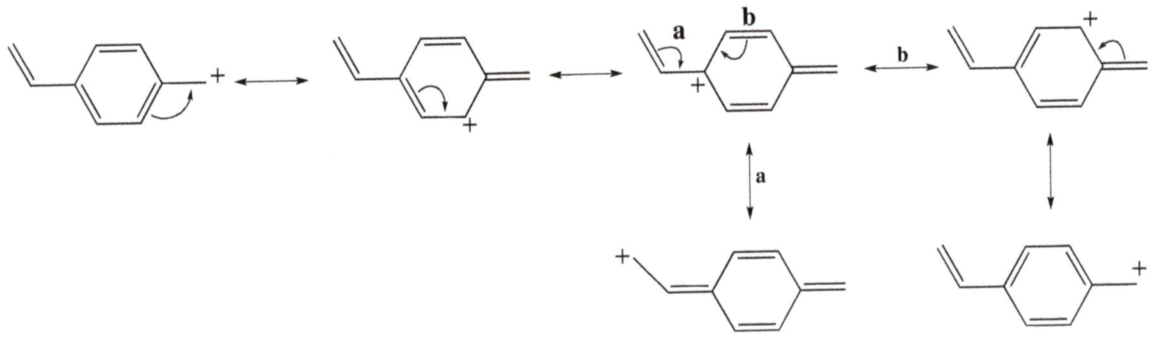

c.

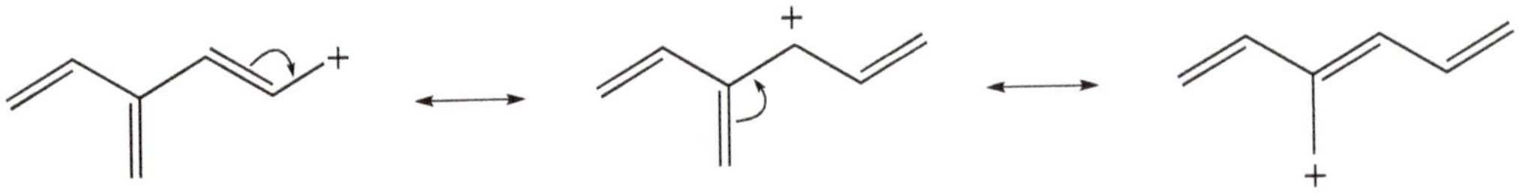

d.

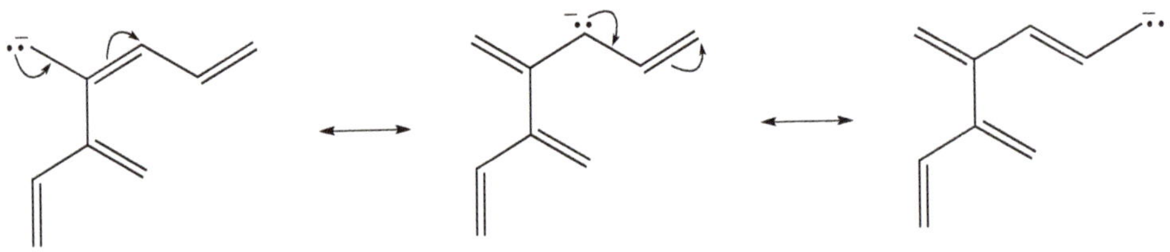

65. Draw the resonance contributors. Then have the Cl pointing up in all the products, with the OH pointing up in one product and down in the other. Notice that the products in row 4 are the mirror images of the products in row 1 and that the products in row 3 are the mirror images of the products in row 2. (The same products would be obtained if the Cl pointed down in all the products with the OH pointing up in one structure and down in the other.)

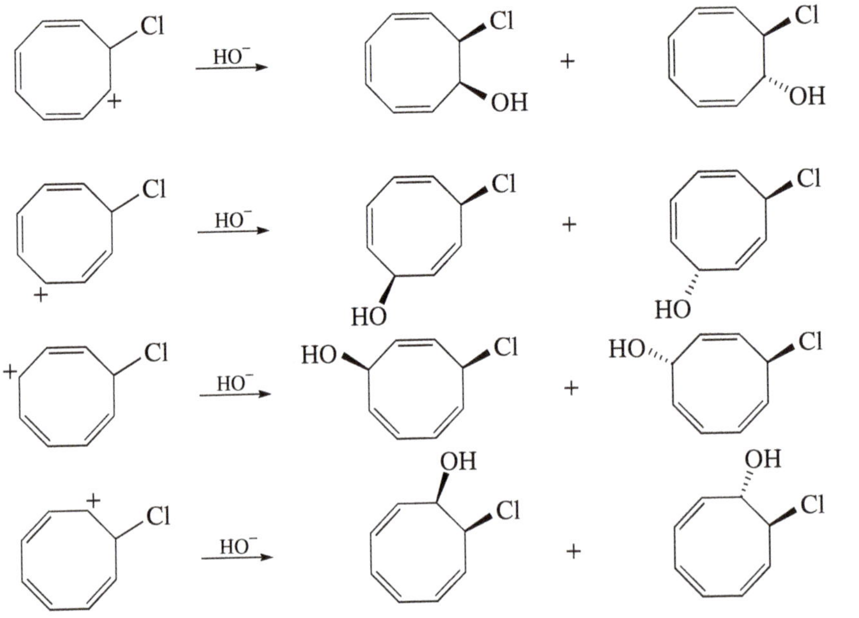

66. **a.** different compounds
b. different compounds
c. resonance contributors
d. resonance contributors
e. different compounds

Notice that in the structures that are different compounds, both atoms and electrons have changed their locations. In the structures that are resonance contributors, only the electrons have moved.

67. **a.** There are six linear dienes with molecular formula C_6H_{10}.

b. Two are conjugated dienes.
$CH_2{=}CHCH{=}CHCH_2CH_3$
$CH_3CH{=}CHCH{=}CHCH_3$

c. Two are isolated dienes.
$CH_2{=}CHCH_2CH{=}CHCH_3$
$CH_2{=}CHCH_2CH_2CH{=}CH_2$

d. Two are cumulated dienes.
$CH_2{=}C{=}CHCH_2CH_2CH_3$
$CH_3CH{=}C{=}CHCH_2CH_3$

68. **a.** **1.** $CH_3CH{=}CH{-}\ddot{O}CH_3$ ⟷ $CH_3\bar{C}H{-}CH{=}\overset{+}{O}CH_3$
major minor

2. (benzene ring)CH_2NH_2 ⟷ (benzene ring)CH_2NH_2

The two resonance contributors have the same stability and, therefore, contribute equally to the resonance hybrid.

3. (acetophenone resonance contributors: $C_6H_5C(=O)CH_3$ ⟷ three contributors with O^- and positive charge on the ring ⟷ $C_6H_5C(=O)CH_3$)
major minor minor minor major

4. $CH_3{-}\overset{+}{N}({=}O)({-}\ddot{O}{:}^-)$ ⟷ $CH_3{-}\overset{+}{N}({-}O^-)({=}O)$

The two resonance contributors have the same stability and, therefore, contribute equally to the hybrid.

5. $CH_3\bar{C}H{-}\overset{+}{N}({=}O)({-}\ddot{O}{:}^-)$ ⟷ $CH_3\ddot{\bar{C}}H{-}\overset{+}{N}({-}O^-)({=}O)$ ⟷ $CH_3CH{=}\overset{+}{N}(O^-)(O^-)$
minor minor major

6. $CH_3CH{=}CH{-}\overset{+}{C}H_2$ ⟷ $CH_3\overset{+}{C}H{-}CH{=}CH_2$
minor major

7.

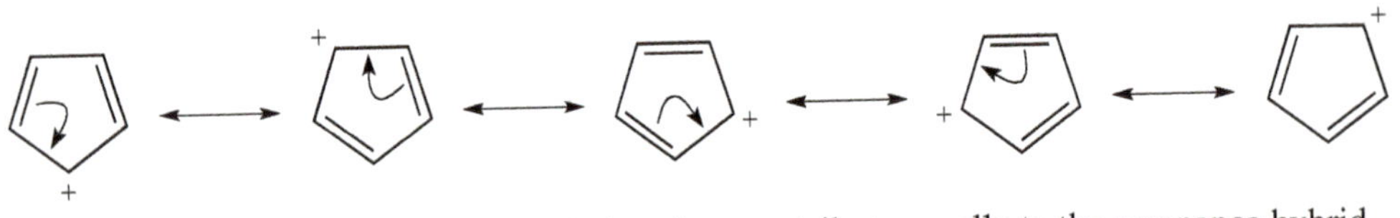

The five contributors are equally stable and, therefore, contribute equally to the resonance hybrid.

8. $CH_3CH_2C(=O)\ddot{O}CH_2CH_3$ ⟷ $CH_3CH_2C(O^-)=\overset{+}{O}CH_2CH_3$

major — minor

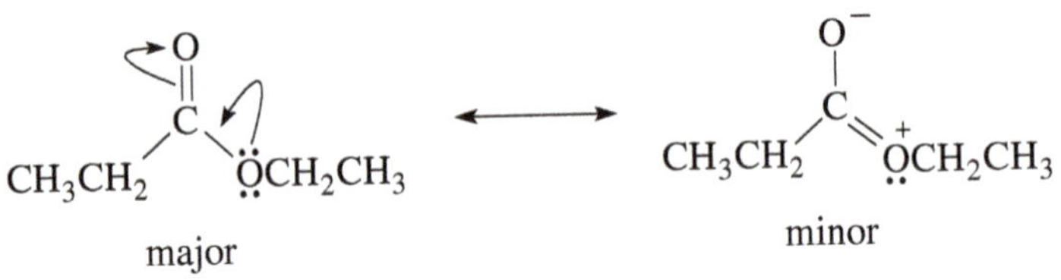

9. $CH_3CH{=}CH{-}CH{=}CH{-}\overset{+}{C}H_2$

minor
(the positive charge is on a primary allylic carbon)

$CH_3\overset{+}{C}H{-}CH{=}CH{-}CH{=}CH_2$

major
(the positive charge is on a secondary allylic carbon)

$CH_3CH{=}CH{-}\overset{+}{C}H{-}CH{=}CH_2$

major (the positive charge is on a secondary allylic carbon)

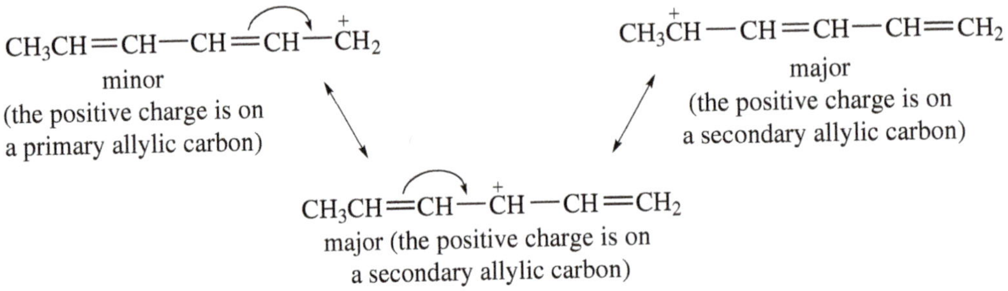

10. $^-\ddot{C}H_2C(=O)CH_2CH_3$ ⟷ $CH_2{=}C(O^-)CH_2CH_3$

minor — major

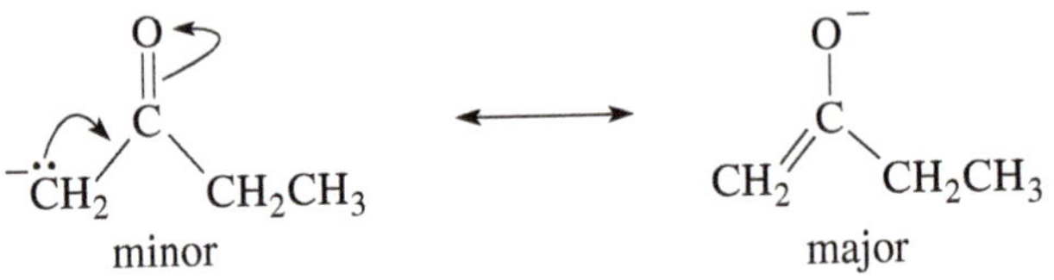

The major contributor has a negative charge on oxygen, which is more stable than a contributor with a negative charge on carbon.

11. $CH_3\overset{-}{\ddot{C}}H{-}C{\equiv}N$ ⟷ $CH_3CH{=}C{=}\ddot{N}^-$

minor — major

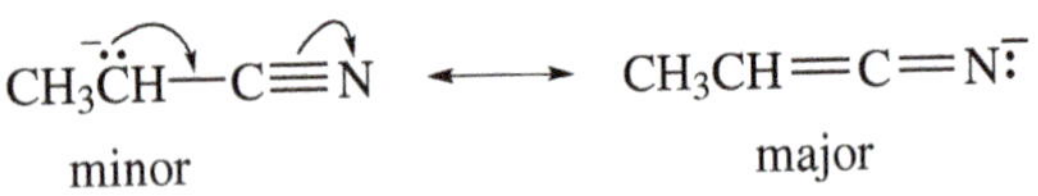

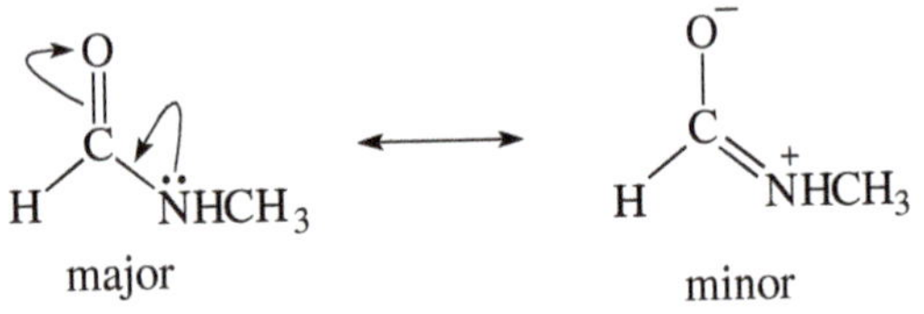

12. $\ddot{O}CH_3$ ⟷ $\overset{+}{O}CH_3$ ⟷ $\overset{+}{O}CH_3$ ⟷ $\overset{+}{O}CH_3$ ⟷ OCH_3

major — minor — minor — minor — major

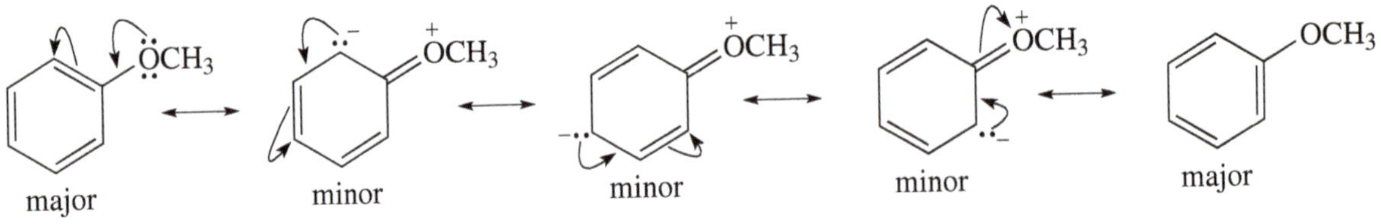

13. $HC(=O)\ddot{N}HCH_3$ ⟷ $HC(O^-)=\overset{+}{N}HCH_3$

major — minor

14. H–C(=O)–CH=CH–$\ddot{C}H_2^-$ ⟷ H–C(=O)–$\ddot{C}H^-$–CH=CH$_2$ ⟷ H–C(O$^-$)=CH–CH=CH$_2$

minor minor major

The major contributor has a negative charge on oxygen, which is more stable than a contributor with a negative charge on carbon.

15. Notice that the electrons on the center carbon can be delocalized onto both of the carbonyl oxygens.

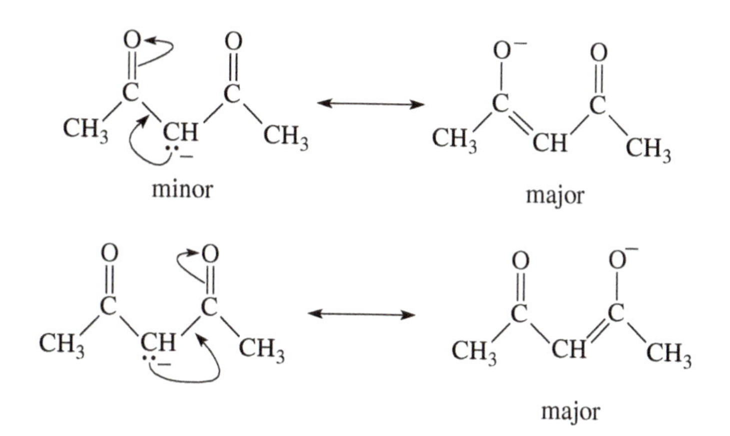

b. **2, 4,** and **7** have resonance contributors that all contribute equally to the hybrid.

69. **a.** It is aromatic. **b.** It is aromatic. **c.** It is aromatic. **d.** It is not antiaromatic.

70. Both compounds form the same product when they are hydrogenated, so the difference in the heats of hydrogenation depend only on the difference in the stabilities of the reactants. Because 1,2-pentadiene has cumulated double bonds and 1,4-pentadiene has isolated double bonds, 1,2-pentadiene is less stable and, therefore, has a greater heat of hydrogenation (a more negative $\Delta H°$).

$$CH_2{=}C{=}CHCH_2CH_3 \xrightarrow[Pd/C]{H_2} CH_3CH_2CH_2CH_2CH_3$$

1,2-pentadiene

$$CH_2{=}CHCH_2CH{=}CH_2 \xrightarrow[Pd/C]{H_2} CH_3CH_2CH_2CH_2CH_3$$

1,4-pentadiene

71. **a.** $CH_3\overset{+}{C}HCH{=}CH_2$
This makes the greater contribution because the positive charge is on a secondary allylic carbon.

b.

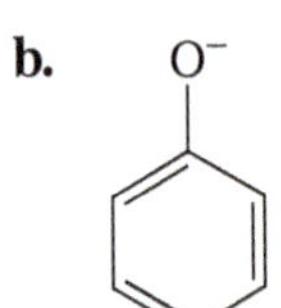

This makes the greater contribution because the negative charge is on an oxygen.

c.

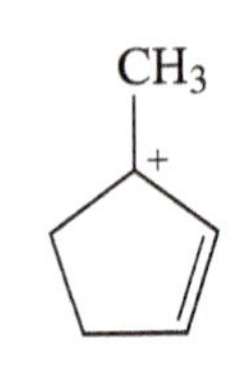

This makes the greater contribution because the positive charge is on a tertiary allylic carbon.

d. $\overset{+}{C}HCH_2CH_3$

This makes the greater contribution because a secondary benzylic cation is more stable than a secondary alkyl cation.

72.

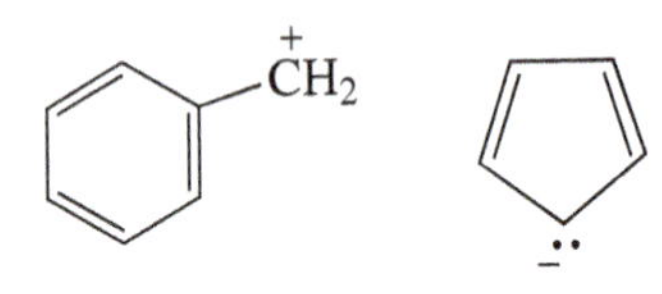

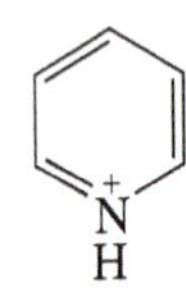

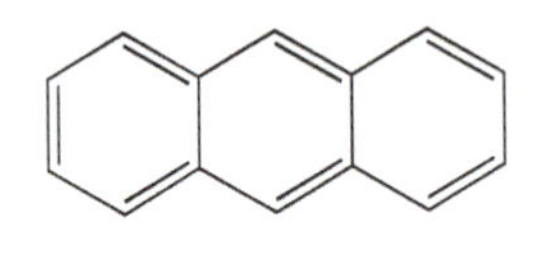

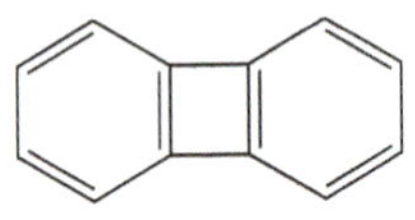

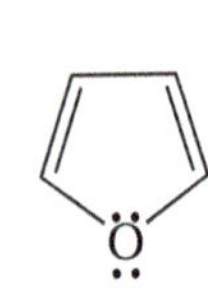

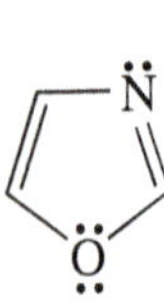

73. **a.** The resonance contributors show that the carbonyl oxygen has the greater electron density.

carbonyl oxygen

CH_3 $\ddot{O}CH_3$ ⟷ CH_3 $\overset{+}{O}CH_3$

b. The compound on the right has the greater electron density on its nitrogen, because the compound on the left has a resonance contributor with a positive charge on the nitrogen as a result of electron delocalization.

c. The compound with the cyclohexane ring has the greater electron density on its oxygen, because the lone pair on the nitrogen can be delocalized only onto the oxygen.

CH_3 ⟷ CH_3

There is less delocalization onto oxygen by the lone pair in the compound with the benzene ring (path **a**) because the lone pair can also be delocalized away from the oxygen into the benzene ring (path **b**).

74. Remember that an sp^2 nitrogen is more electronegative than an sp^3 nitrogen, and the more electronegative the atom to which a hydrogen is attached, the stronger the acid. The stronger the acid, the weaker its conjugate base.

This is the strongest base because the lone pair is localized and the nitrogen is sp^3 hybridized.

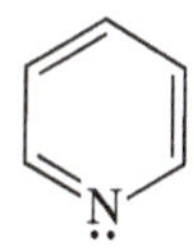

The lone pair is localized and the nitrogen is sp^2 hybridized, which is not as strong a base as an sp^3 hybridized nitrogen.

These are weak bases because the lone-pair electrons are delocalized.

75. The methyl group on benzene can lose a proton easier than the methyl group on cyclohexane because the electrons left behind on the carbon in the former can be delocalized into the benzene ring. In contrast, the electrons left behind in the other compound cannot be delocalized.

76. The triphenylmethyl carbocation is stable because the positive charge is shared by 10 carbons (the central carbon and three carbons of each of the three benzene rings) as a result of electron delocalization.

77. **a.** CH_3CH_2O + → CH_3CH_2O

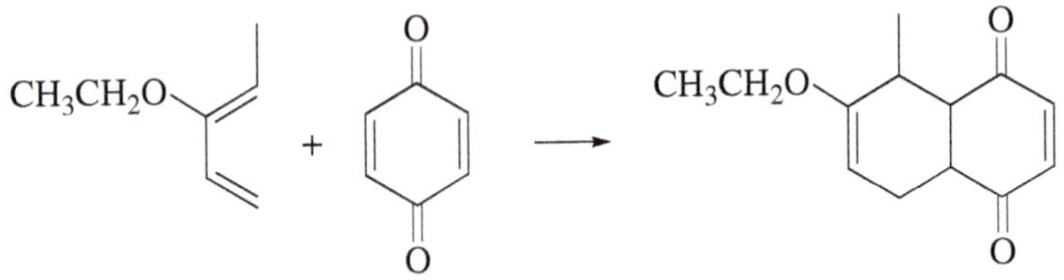

b. CH_3O + → CH_3O

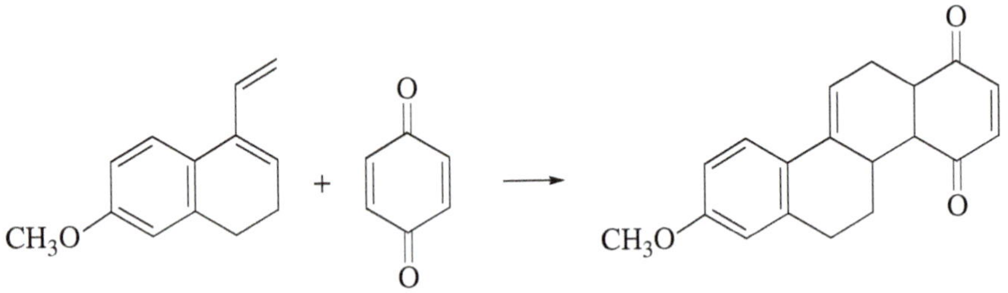

78. **a.** The structure shown below is the stronger acid because it has the weaker conjugate base. When this compound loses a proton, the electrons left behind can be delocalized onto six different carbons. The electrons left behind on the conjugate base of the other compound can be delocalized onto only three different carbons; they cannot be delocalized into the second benzene ring.

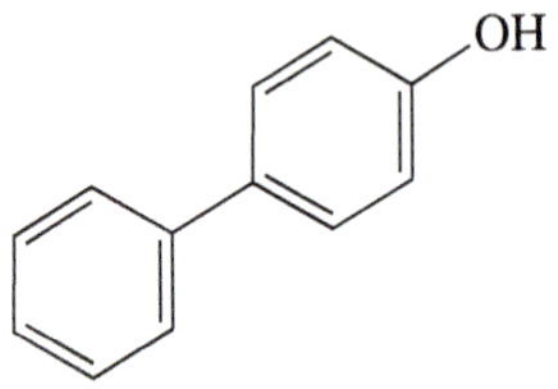

OH

b. **1.** The first compound has the greater electron density on its oxygen because the resonance contributor with a negative charge on oxygen is particularly stable since it is aromatic.

aromatic

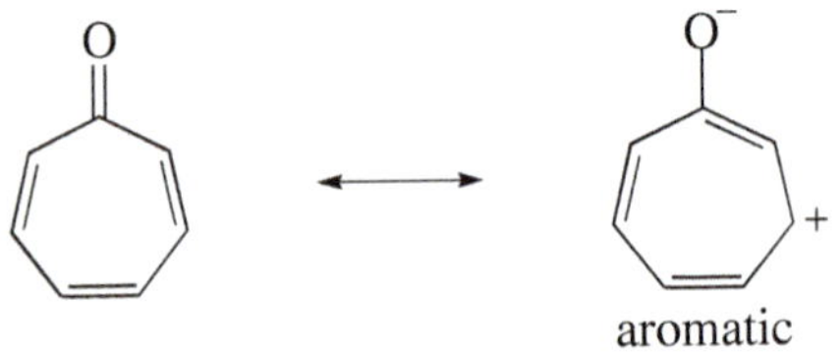

2. The first compound has the greater electron density on its oxygen because the resonance contributor with a negative charge on oxygen is particularly stable since it is aromatic.

aromatic

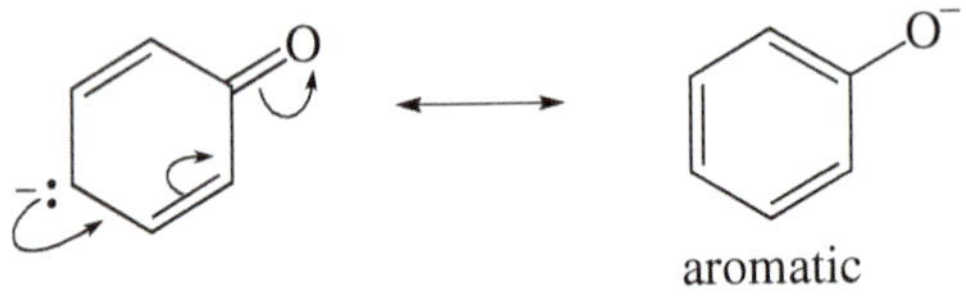

79. The first carbocation is the most stable because its positive charge can be shared by two other carbons; the positive charge on the next carbocation can be shared by one other carbon; the positive charge on the cation on the right cannot be shared by other carbons.

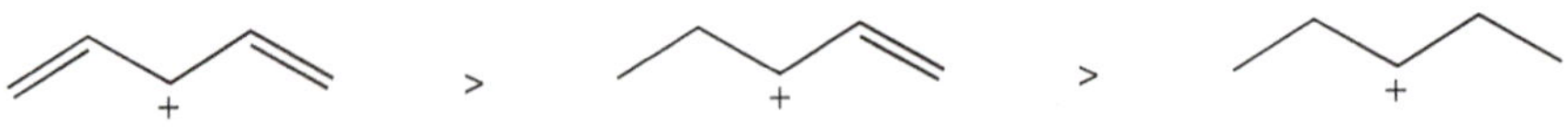

80. **a.** 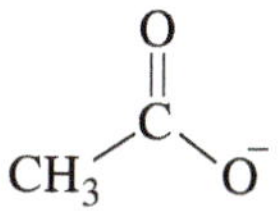 $CH_3C(=O)O^-$ The negative charge is shared by two oxygens.

b. 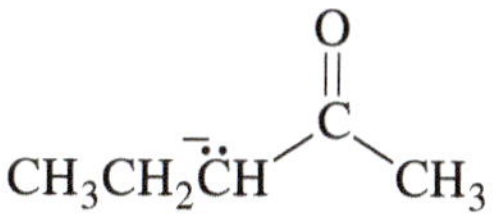 $CH_3CH_2\ddot{C}H^-C(=O)CH_3$ The negative charge is shared by a carbon and an oxygen.

c. 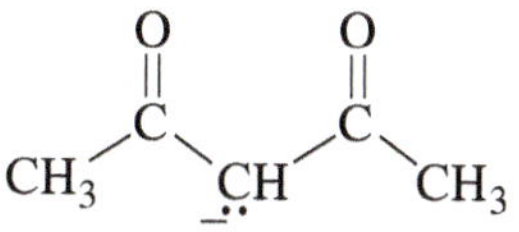 $CH_3C(=O)\ddot{C}H^-C(=O)CH_3$ The negative charge is shared by a carbon and two oxygens.

d. 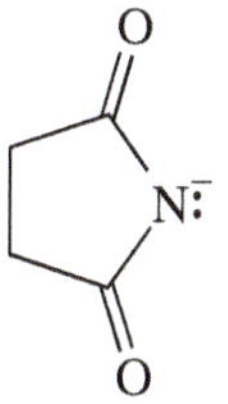 (five-membered ring with two C=O groups flanking $N\colon^-$) The negative charge is shared by a nitrogen and two oxygens.

81. The stronger base is the less stable base of each pair in Problem 80.

a. 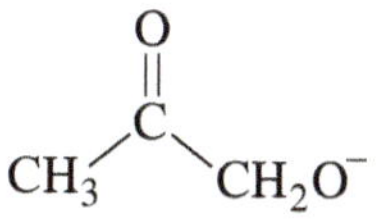 $CH_3C(=O)CH_2O^-$ Less stable because the negative charge cannot be delocalized.

b. 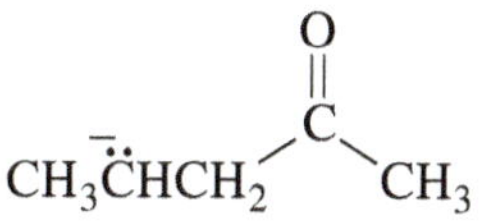 $CH_3\ddot{C}H^-CH_2C(=O)CH_3$ Less stable because the negative charge cannot be delocalized.

c. 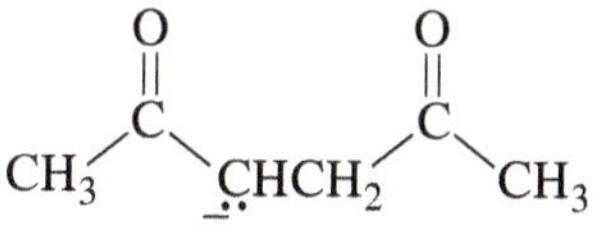 $CH_3C(=O)\ddot{C}H^-CH_2C(=O)CH_3$ Less stable because the negative charge can be delocalized onto only one oxygen.

d. 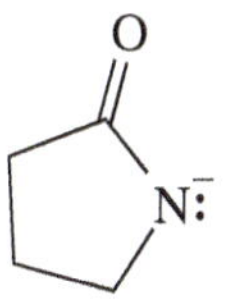 (five-membered ring with one C=O group adjacent to $N\colon^-$) Less stable because the negative charge can be delocalized onto only one oxygen.

82. The resonance contributor shown here indicates which nitrogen is most apt to be protonated (the one with the greatest negative charge) and which nitrogen is least apt to be protonated (the one with the greatest positive charge).

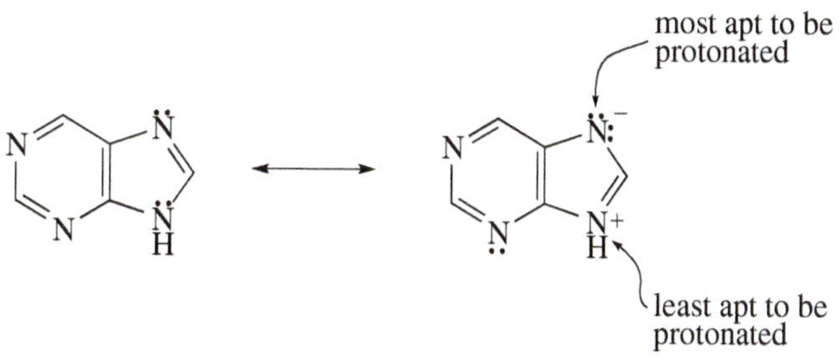

most apt to be protonated

least apt to be protonated

83. The following compound is the strongest acid, because it is the only one that forms a conjugate base that is aromatic. Recall that the more stable (weaker) the base, the stronger is its conjugate acid.

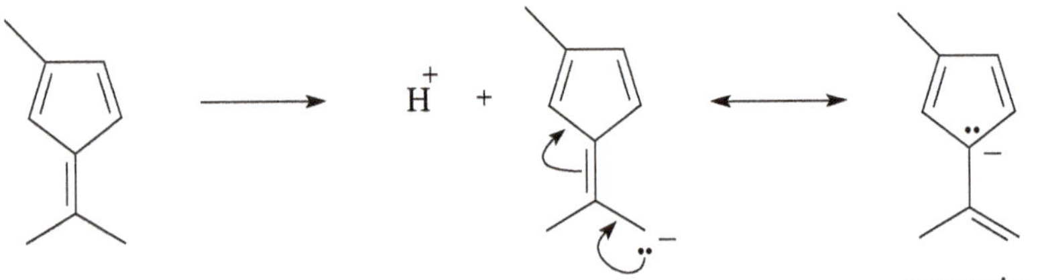

$\overset{+}{H}$ +

aromatic

84. The resonance contributors of pyrrole are more stable because the positive charge is on nitrogen. In furan, the positive charge is on oxygen, which, being more electronegative, is less stable with a positive charge.

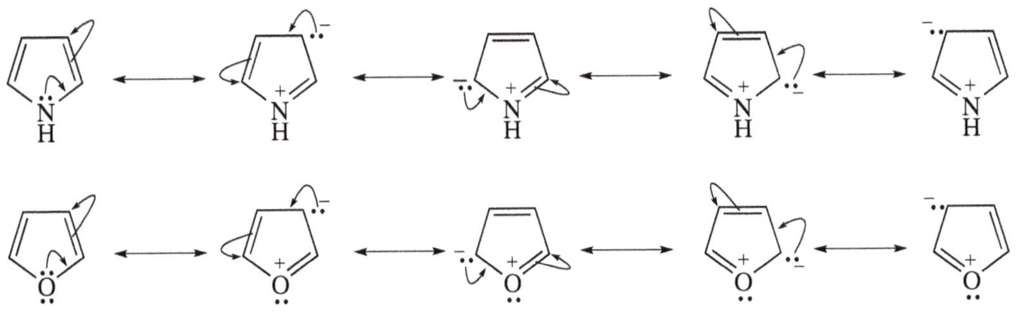

85. **A** is the most acidic because the electrons left behind when the proton is removed can be delocalized onto two oxygens.

B is the next most acidic because the electrons left behind when the proton is removed can be delocalized onto one oxygen.

C is the least acidic because the electrons left behind when the proton is removed cannot be delocalized.

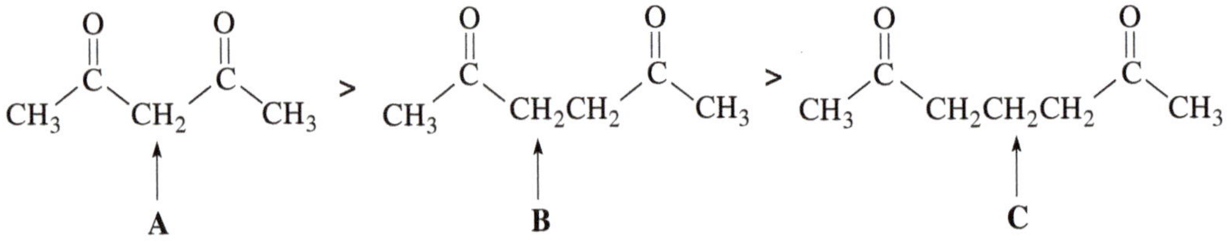

$CH_3C(=O)CH_2C(=O)CH_3$ > $CH_3C(=O)CH_2CH_2C(=O)CH_3$ > $CH_3C(=O)CH_2CH_2CH_2C(=O)CH_3$

A **B** **C**

86.
- **a.** It has eight molecular orbitals.
- **b.** ψ_1, ψ_2, ψ_3, and ψ_4 are bonding molecular orbitals; ψ_5, ψ_6, ψ_7, and ψ_8 are antibonding molecular orbitals.
- **c.** ψ_1, ψ_3, ψ_5, and ψ_7 are symmetric molecular orbitals; ψ_2, ψ_4, ψ_6, and ψ_8 are antisymmetric molecular orbitals.
- **d.** ψ_4 is the HOMO and ψ_5 is the LUMO in the ground state.
- **e.** ψ_5 is the HOMO and ψ_6 is the LUMO in the excited state.
- **f.** If the HOMO is symmetric, the LUMO is antisymmetric and vice versa.
- **g.** It has seven nodes between the nuclei. It also has one node that passes through the nuclei.

87. + $\xrightarrow{\Delta}$ $\xrightarrow[2.\ (CH_3)_2S]{1.\ O_3,\ -78\ °C}$

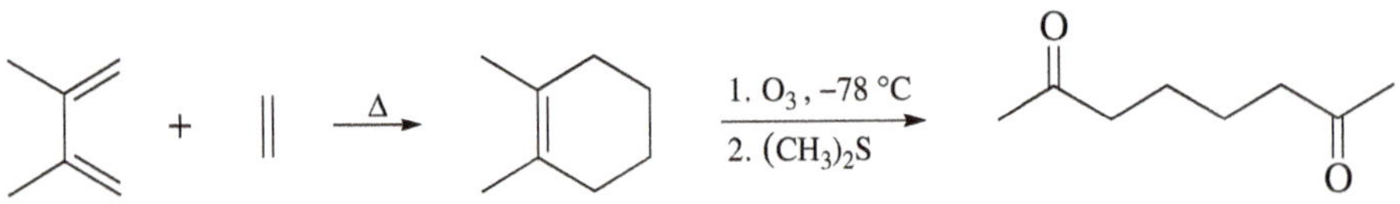

88. The reaction of 1,3-cyclohexadiene with Br_2 forms 3,4-dibromocyclohexene as the 1,2-addition product and 3,6-dibromocyclohexene as the 1,4-addition product. The reaction of 1,3-cyclohexadiene with HBr forms only 3-bromocyclohexene, so it is both the 1,2-addition product and the 1,4-addition product.

Br_2

Br Br

3,4-dibromocyclohexene
1,2-addition product

+

Br Br

3,6-dibromocyclohexene
1,4-addition product

1,3-cyclohexadiene

HBr

Br

3-bromocyclohexene
1,2-addition product
1,4-addition product

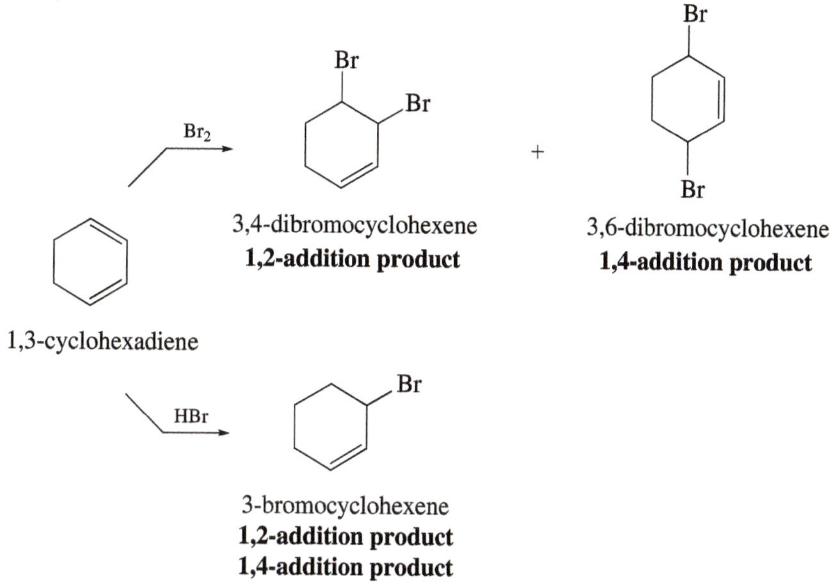

89. **a.** Only part **a** involves the reaction of two unsymmetrically substituted reactants. Therefore, only for part **a** do we need to look at the charge distribution in the reactants to determine the major product of the reaction.

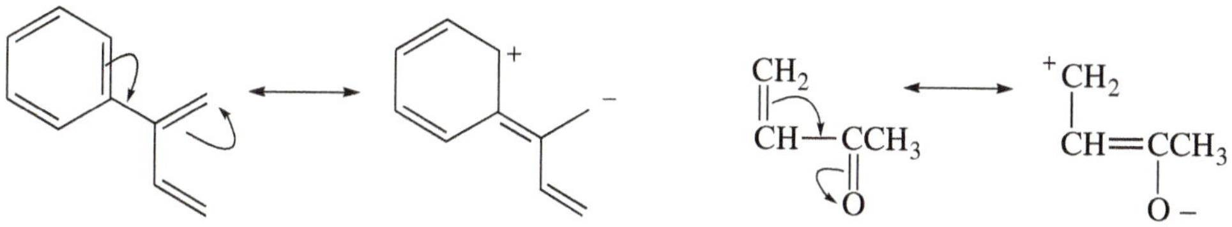

+ −

CH_2 = $CH-CCH_3$ (C=O) ⟷ $^+CH_2-CH=CCH_3$ (O^-)

We now need to join the negative end of the diene and the positive end of the dienophile.

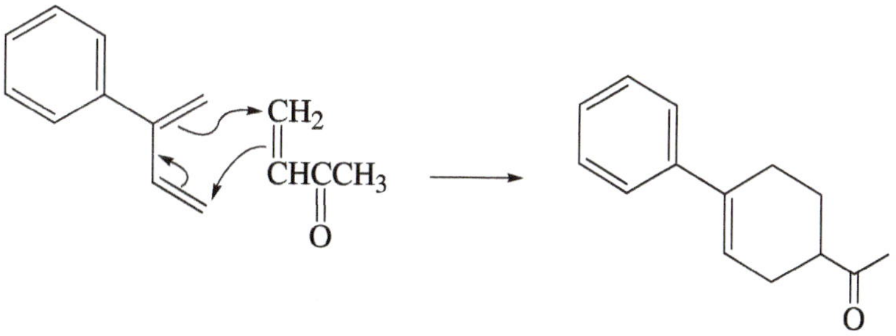

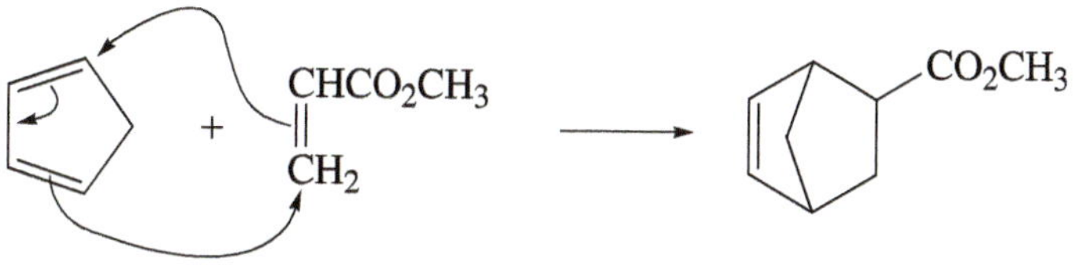

CH_2 = $CHCCH_3$ (C=O) ⟶

O

b. + $CHCO_2CH_3$ = CH_2 ⟶ CO_2CH_3

c. 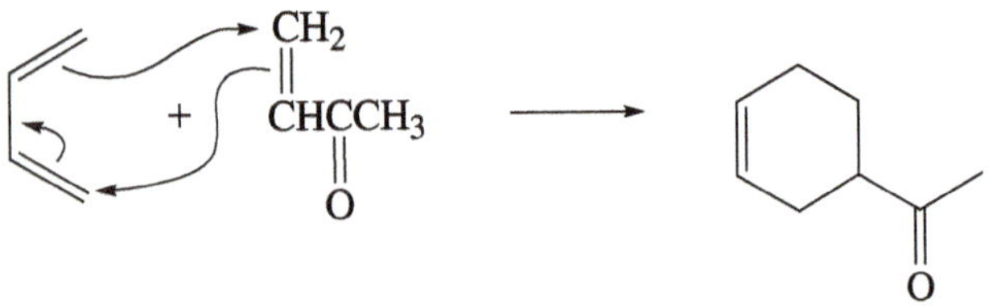 CH_2 + $CH_2{=}CHCCH_3$ (C=O) ⟶

d. 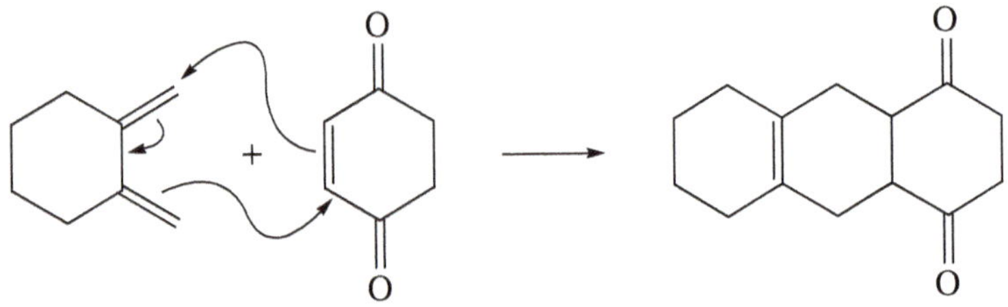 +

90. Numbering the carbons in the conjugated system will help you determine the 1,2- and 1,4-addition products.

a. 1. 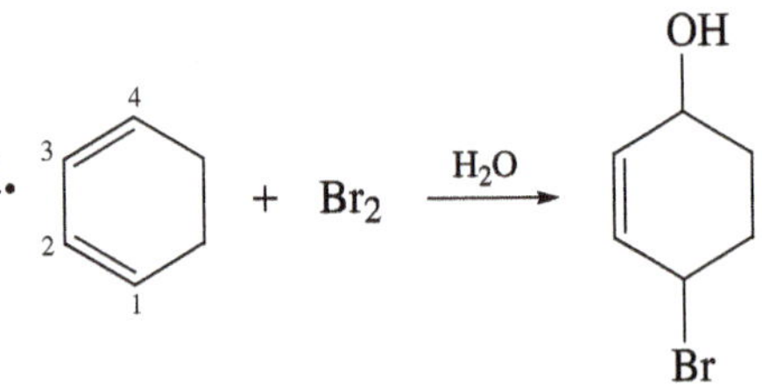 + Br_2 $\xrightarrow{H_2O}$ (OH, Br)

2. CH_3 + HBr ⟶ Br, CH_3

or

CH_2 + HBr ⟶ Br, CH_3

3. CH_3, CH_3 + H_2O $\xrightarrow{H_2SO_4}$ CH_3, HO, CH_3

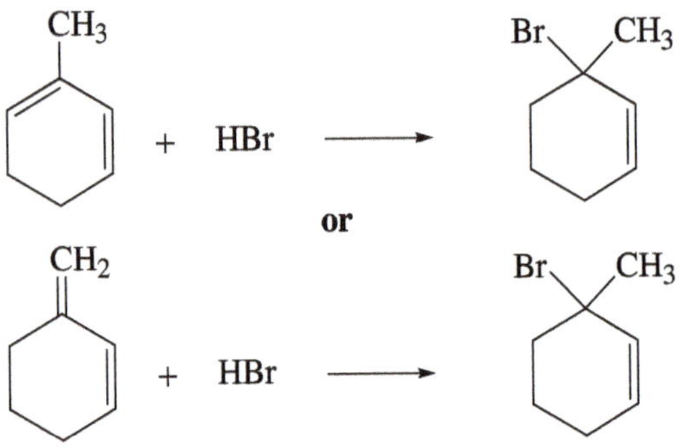

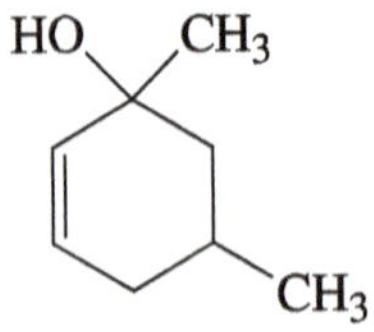

or

CH_3, CH_3 + H_2O $\xrightarrow{H_2SO_4}$ CH_3, HO, CH_3

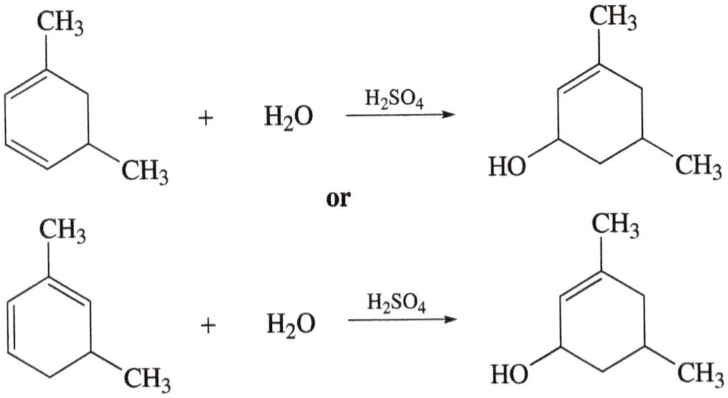

b. 1. 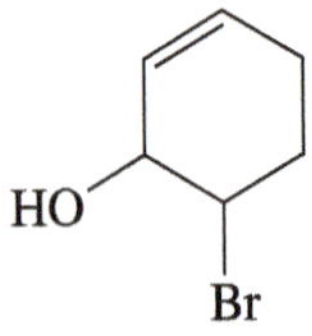 HO, Br

2. 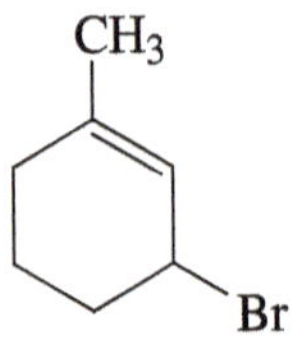 CH_3, Br

3. HO, CH_3, CH_3

91. **a.** $CH_2{=}CHCH{=}CHCH{=}CH_2 \xrightarrow{HBr} CH_3CH(Br)CH{=}CHCH{=}CH_2$ (A) +

1,3,5-hexatriene

$CH_3CH{=}CHCH(Br)CH{=}CH_2$ (B) + $CH_3CH{=}CHCH{=}CHCH_2Br$ (C)

b. **A** will predominate if the reaction is under kinetic control because it is the 1,2-product and, therefore, is the product formed most rapidly as a result of the proximity effect. Notice that **A** will be the 1,2-product regardless of which end of the conjugated system reacts with the electrophile.

c. **C** will predominate if the reaction is under thermodynamic control because it is the most stable diene. (It is the most substituted conjugated diene.)

92. The diene is the nucleophile, and the dienophile is the electrophile in a Diels–Alder reaction.

a. An electron-donating substituent in the diene will increase the rate of the reaction, because electron donation increases its nucleophilicity.

b. An electron-donating substituent in the dienophile will decrease the rate of the reaction, because electron donation decreases its electrophilicity.

c. An electron-withdrawing substituent in the diene will decrease the rate of the reaction, because electron withdrawal decreases its nucleophilicity.

93. **a.** Addition of an electrophile to C-1 forms a carbocation with two resonance contributors, a *tertiary allylic cation* and a *secondary allylic cation*. Addition of an electrophile to C-4 forms a carbocation with two resonance contributors, a *tertiary allylic cation* and a *primary allylic cation*. Therefore, addition to C-1 results in formation of the more stable carbocation intermediate, and the more stable intermediate leads to the major products.

$CH_2{=}C(CH_3){-}C(CH_3){=}CHCH_3 \xrightarrow{HCl} CH_3\overset{+}{C}(CH_3){-}C(CH_3){=}CHCH_3 \longleftrightarrow CH_3C(CH_3){=}C(CH_3){-}\overset{+}{C}HCH_3$

$\downarrow Cl^-$ (from each contributor)

$CH_3C(Cl)(CH_3){-}C(CH_3){=}CHCH_3$

kinetic product

$CH_3C(CH_3){=}C(CH_3){-}CH(Cl)CH_3$

thermodynamic product

b. Addition of an electrophile to C-1 forms a carbocation with two resonance contributors; both are *tertiary allylic* cations. Addition of an electrophile to C-4 forms a carbocation with two resonance contributors, a *secondary allylic* cation and a *primary allylic* cation. Therefore, addition to C-1 results in formation of the more stable carbocation. Only one product is formed, because the carbocation is symmetrical.

$CH_2{=}C(CH_3){-}CH{=}C(CH_3)CH_3 \xrightarrow{HCl} CH_3\overset{+}{C}(CH_3)CH{=}C(CH_3)CH_3 \longleftrightarrow CH_3C(CH_3){=}CH\overset{+}{C}(CH_3)CH_3$

$\xrightarrow{Cl^-} CH_3C(Cl)(CH_3)CH{=}C(CH_3)CH_3$

This is the only product because the carbocation is symmetrical.

94. **a** and **d.**

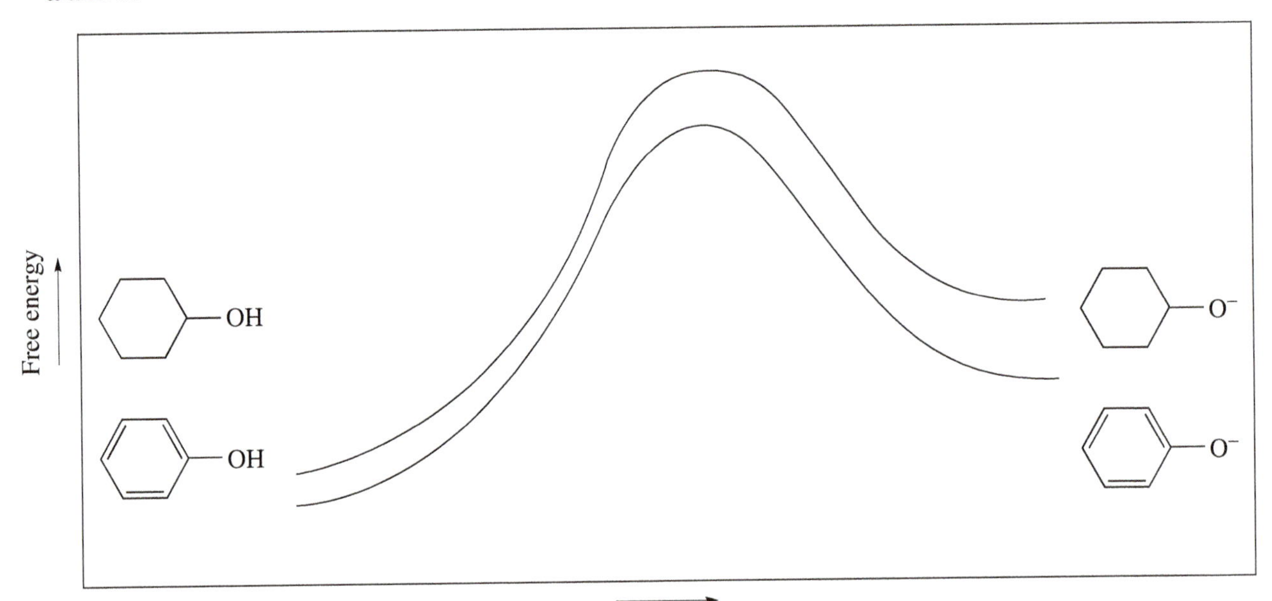

Progress of the reaction

b. [resonance contributors of phenol: $\ddot{O}H$ ⟷ $\overset{+}{\ddot{O}}H$ ⟷ $\overset{+}{\ddot{O}}H$ ⟷ $\overset{+}{\ddot{O}}H$ ⟷ $\ddot{O}H$]

c. [resonance contributors of phenolate ion: $\ddot{O}:^-$ ⟷ $\ddot{O}:$ ⟷ $\ddot{O}:$ ⟷ $\ddot{O}:$ ⟷ $\ddot{O}:^-$]

d. The 2nd, 3rd, and 4th resonance contributors in **c** are more stable than those resonance contributors in **b**, because in **b**, a positive charge is on the most electronegative atom (the oxygen) and there is charge separation. Therefore, the phenolate ion has greater electron delocalization than phenol. Thus, as shown in the energy diagram, the difference in energy between the phenolate ion and the cyclohexoxide ion is greater than the difference in energy between phenol and cyclohexanol.

e. Because of greater electron delocalization in the phenolate ion compared to that in phenol, phenol has a larger K_a than cyclohexanol.

f. Because it has a larger K_a (a lower pK_a), phenol is a stronger acid.

95.

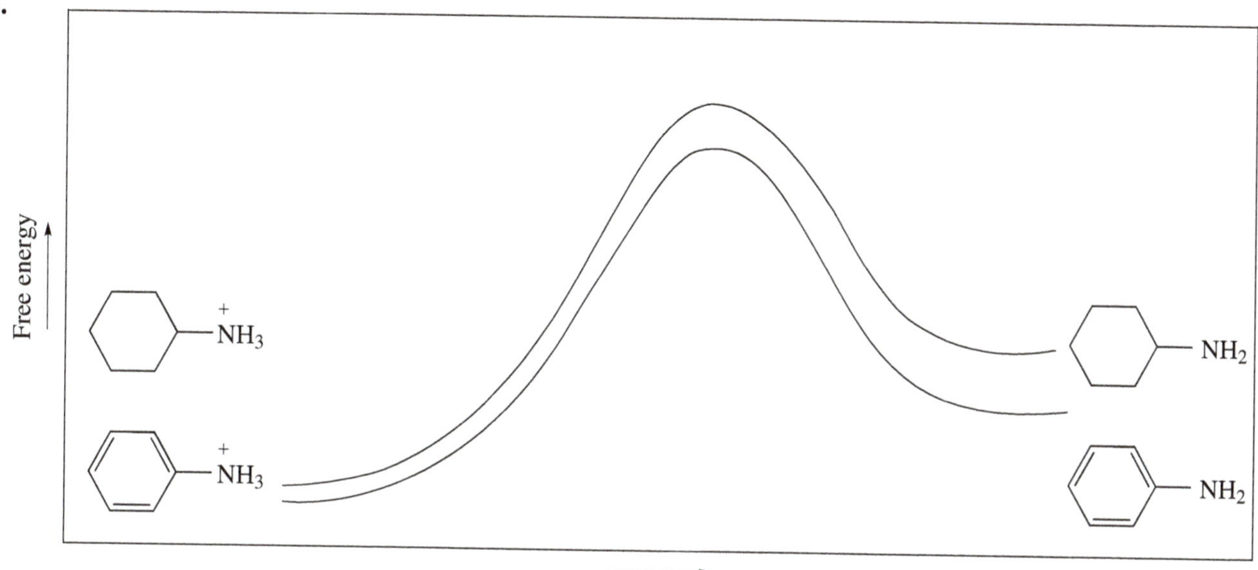

b. $\overset{+}{N}H_3$ ⟷ $\overset{+}{N}H_3$

c. $\ddot{N}H_2$ ⟷ $\overset{+}{N}H_2$ ⟷ $\overset{+}{N}H_2$ ⟷ $\overset{+}{N}H_2$ ⟷ $\ddot{N}H_2$

d. Aniline has greater electron delocalization than the anilinium ion. Therefore, in the energy diagram, the difference in energy between aniline and cyclohexylamine is greater than the difference in energy between the anilinium ion and the cyclohexylammonium ion.

e. Because of greater electron delocalization in aniline compared to that in the anilinium ion, the anilinium ion has a larger K_a than the cyclohexylammonium ion.

f. Because it has a larger K_a (a lower pK_a), the anilinium ion is a stronger acid than the cyclohexylammonium ion. Therefore, cyclohexylamine is a stronger base than aniline. (The stronger the acid, the weaker its conjugate base.)

96. **a.**

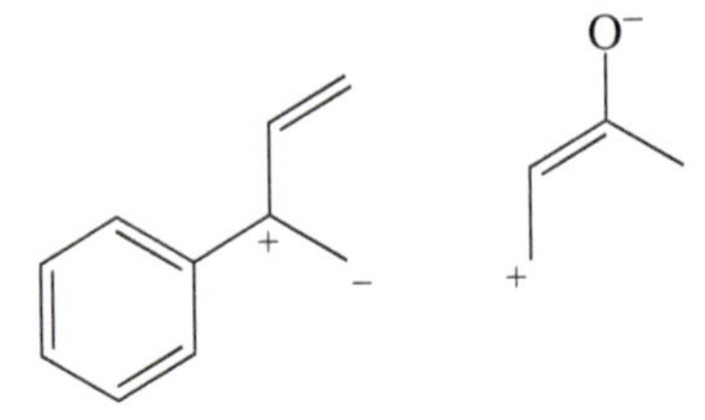

Because the reaction creates an asymmetric center in the product, the product will be a racemic mixture.

b.

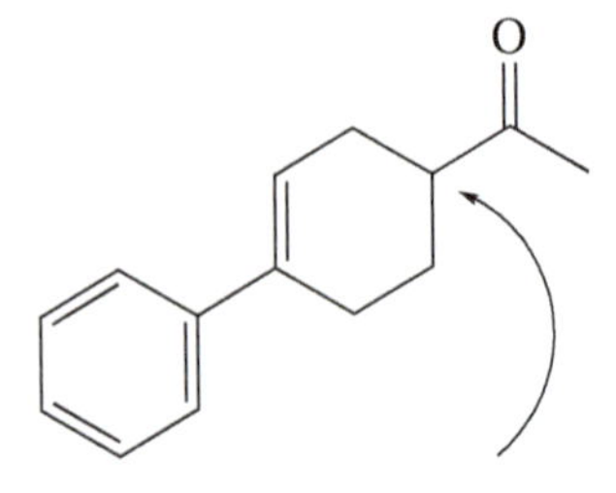

Even though both reactants are unsymmetrically substituted, they will be aligned primarily as shown, because of the relatively stable tertiary benzylic cation resonance contributor and delocalization of the π electrons of the dienophile onto the oxygen.

Because the reaction creates an asymmetric center, the product will be a racemic mixture.

97. The first pair is the preferred set of reagents because it has the more nucleophilic diene and the more electrophilic dienophile.

or

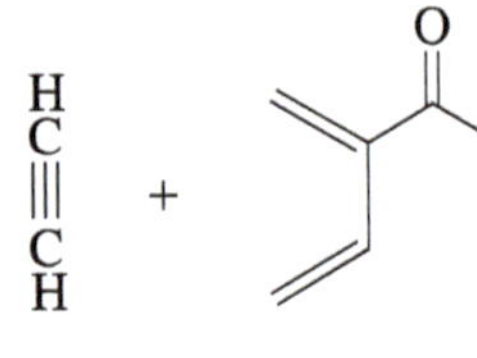

98. A Diels–Alder reaction is a reaction between a nucleophilic diene and an electrophilic dienophile.

a. The compound shown below is more reactive in both **1** and **2**, because electron delocalization increases the electrophilicity of the dienophile.

$$CH_2{=}CH{-}CH{=}O \longleftrightarrow \overset{+}{C}H_2{-}CH{=}CH{-}O^-$$

b. The compound shown below is more reactive, because electron delocalization increases the nucleophilicity of the diene.

$$CH_2{=}CH{-}CH{=}CH{-}\ddot{O}CH_3 \longleftrightarrow \bar{C}H_2{-}CH{=}CH{-}CH{=}\overset{+}{O}CH_3$$

99. **a.**

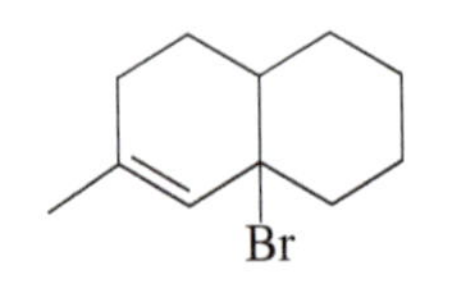

1,2-product
kinetic product

1,4-product

b.

1,2-product
kinetic product
thermodynamic product

1,4-product

Both have the same stability,
so neither exclusively is the thermodynamic product.
When the reaction is under thermodynamic control,
approximately equal amouts of the two products will be obtained.

100.

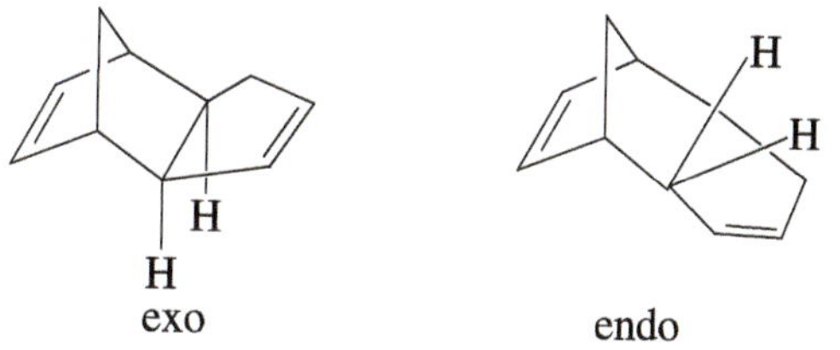

101. **a.**

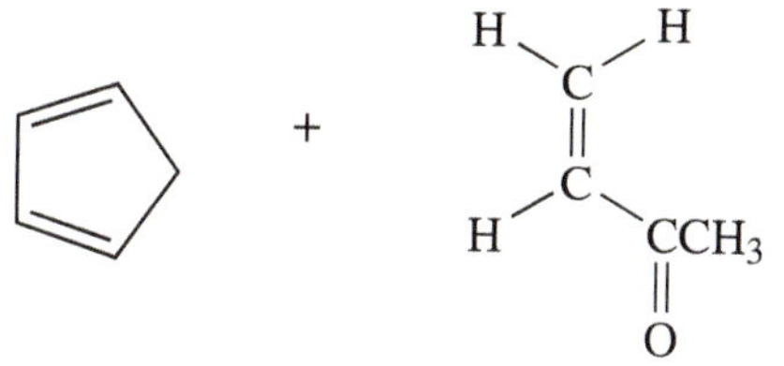

c.

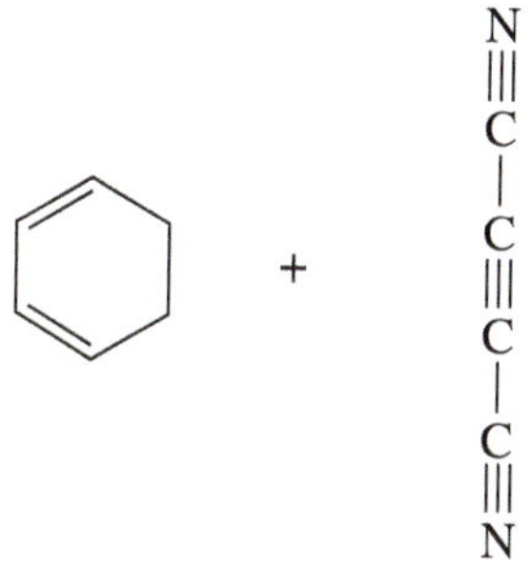

b.

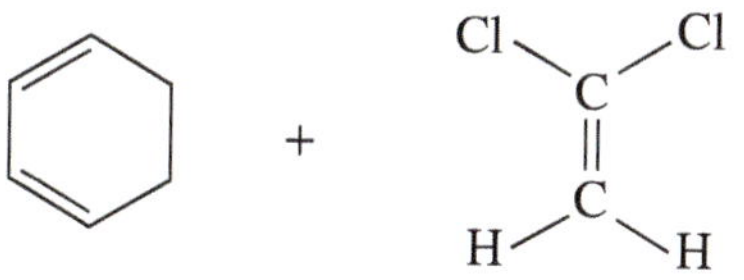

d.

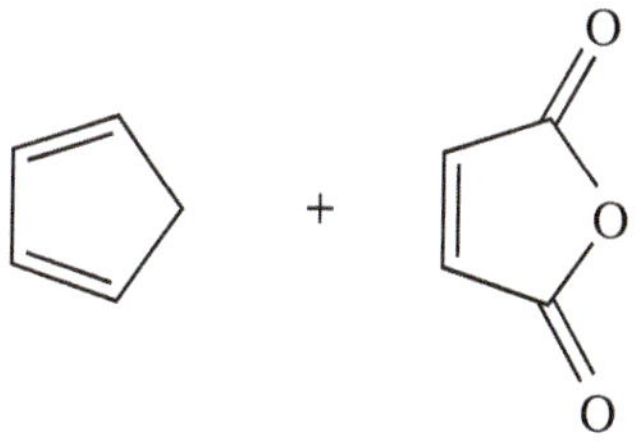

e.

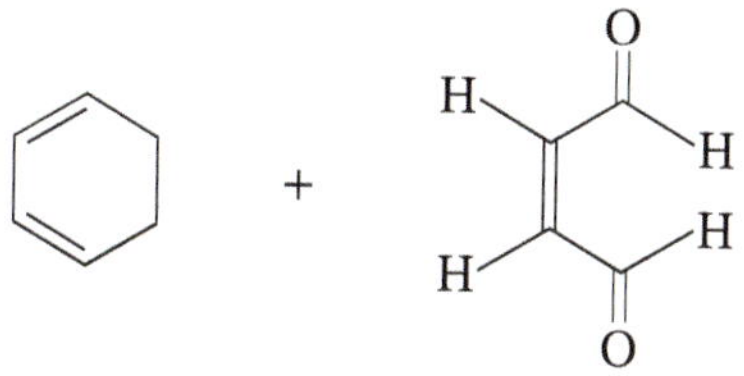

102. a.

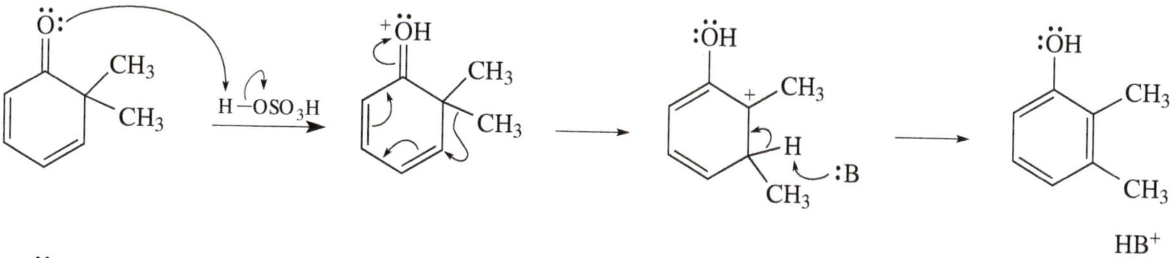

b.

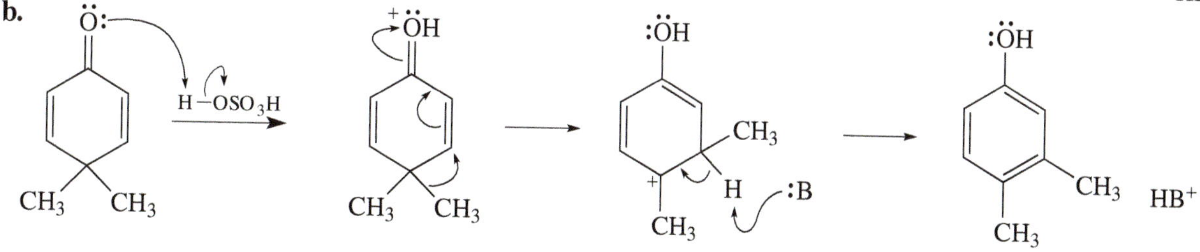

103. The electrophile can add to either end of the conjugated system. Adding the electrophile to one end forms **A** and **B**; adding the electrophile to the other end forms **C** and **B**.

a.

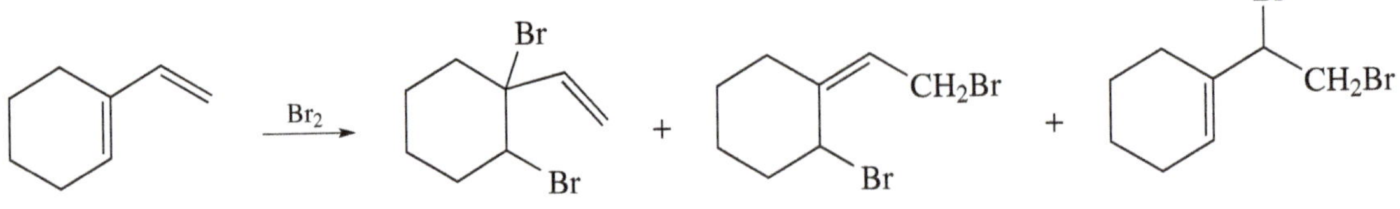

b. **A** has two asymmetric centers, but only two stereoisomers are obtained because only anti addition of Br_2 can occur.

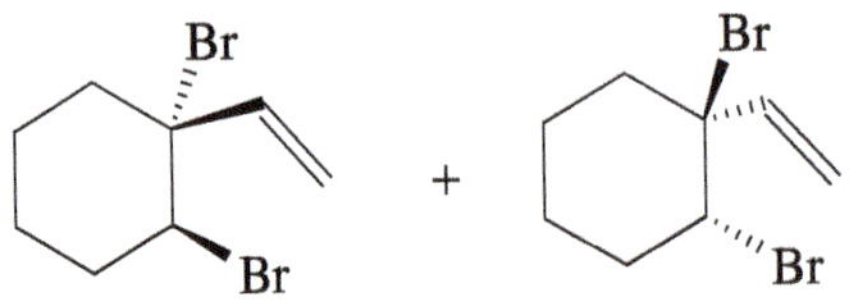

B has four stereoisomers because it has an asymmetric center and a double bond that can be in either the *E* or *Z* configuration.

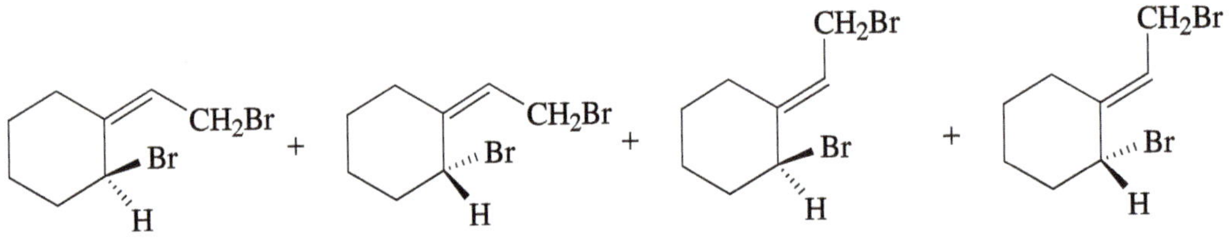

C has two stereoisomers because it has one asymmetric center.

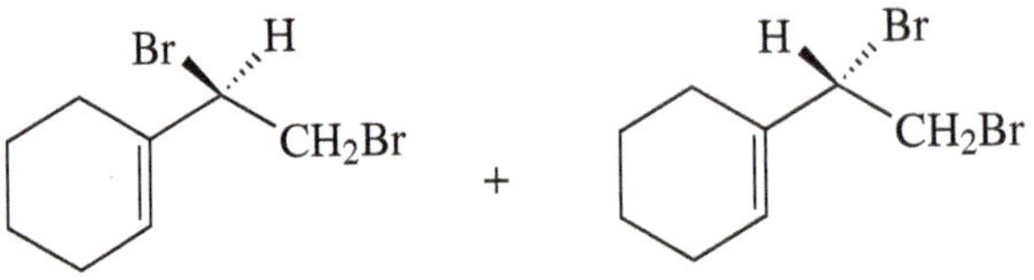

104. Nine of the compounds are shown below. Because each has one asymmetric center, each can have either the *R* or *S* configuration. Therefore, 18 different products can be obtained.

1,3-Butadiene is the electrophile.

The 3,4-bond of 2-methyl-1,3-butadiene is the electrophile.

The 1,2-bond of 2-methyl-1,3-butadiene is the electrophile.

1,3-Butadiene is the nucleophile.

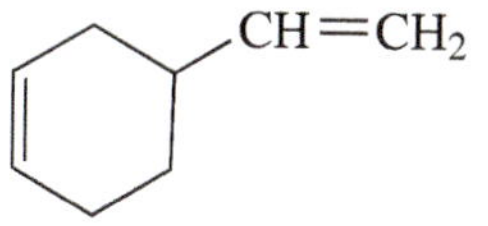

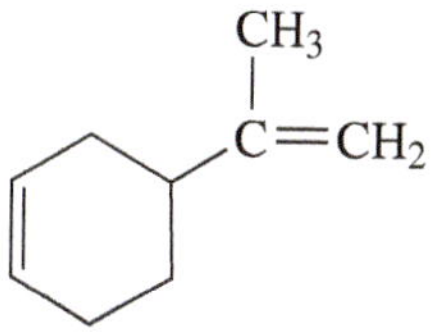

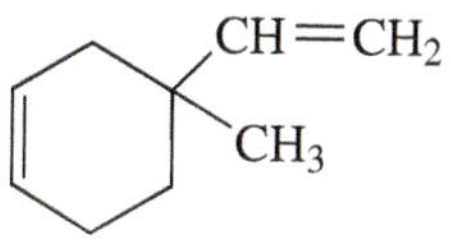

2-Methyl-1,3-butadiene is the nucleophile (the 1-position is on top).

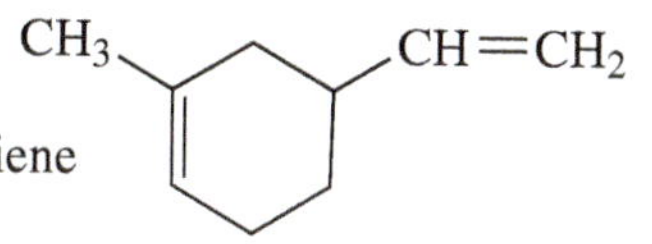

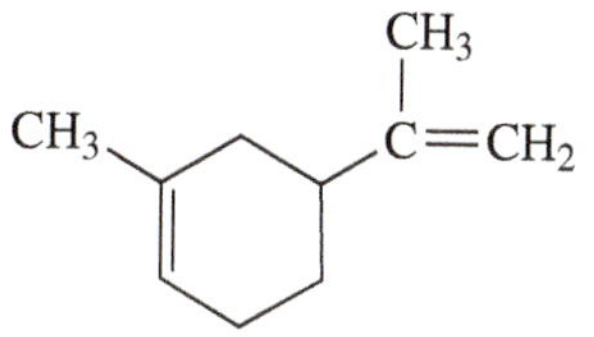

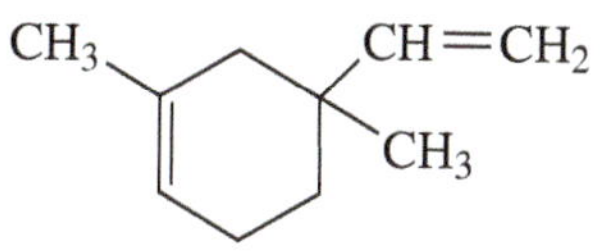

2-Methyl-1,3-butadiene is the nucleophile (the 4-position is on top).

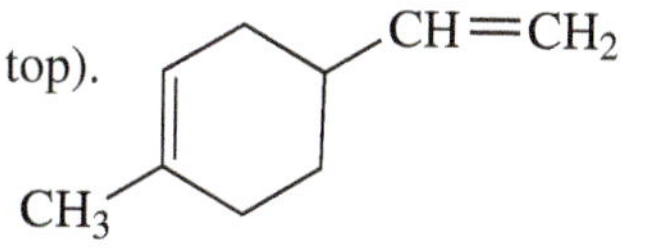

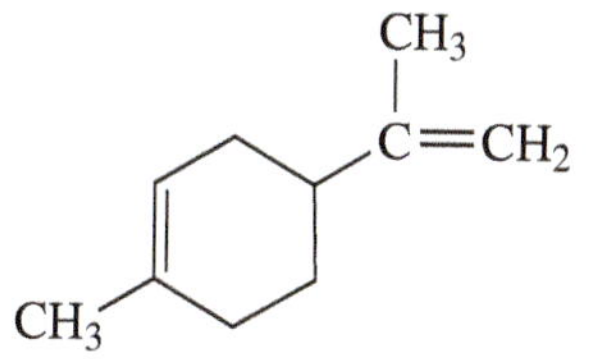

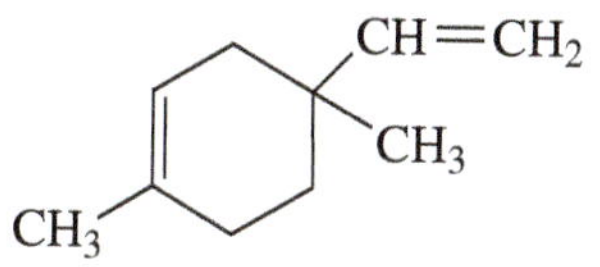

105. 2-Methyl-1,3-pentadiene (with conjugated double bonds) is more stable than 2-methyl-1,4-pentadiene (with isolated double bonds). The rate-limiting step of the reaction is formation of the carbocation intermediate. 2-Methyl-1,3-pentadiene forms a more stable carbocation intermediate than does 2-methyl-1,4-pentadiene.

Because the more stable reactant forms the more stable carbocation intermediate, the relative free energies of activation for the rate-limiting steps of the two reactions depend on whether the difference in the stabilities of the reactants is greater or less than the difference in the stabilities of the transition states leading to formation of the carbocation intermediates (which depend on the difference in stabilities of the carbocation intermediates).

There is a significant difference in the stabilities of the carbocation intermediates because one is stabilized by electron delocalization and one is not. The transitions states look more like the carbocation intermediates than like the alkenes.

Therefore, the difference in the stabilities of the reactants is less than the difference in the stabilities of the transition states, so the rate of reaction of HBr with 2-methyl-1,3-pentadiene is the faster reaction. (If the difference in the stabilities of the reactants had been greater than the difference in the stabilities of the transition states, the rate of reaction of HBr with 2-methyl-1,4-pentadiene would have been the faster reaction.)

$$CH_2{=}C(CH_3)CH{=}CHCH_3 \xrightarrow{HBr} CH_3\overset{+}{C}(CH_3)CH{=}CHCH_3 \longleftrightarrow CH_3C(CH_3){=}CH\overset{+}{C}HCH_3$$

2-methyl-1,3-pentadiene

$$CH_3\overset{+}{C}(CH_3)CH{=}CHCH_3 \xrightarrow{Br^-} CH_3C(CH_3)(Br)CH{=}CHCH_3$$

4-bromo-4-methyl-2-pentene

$$CH_3C(CH_3){=}CH\overset{+}{C}HCH_3 \xrightarrow{Br^-} CH_3C(CH_3){=}CHCH(Br)CH_3$$

4-bromo-2-methyl-2-pentene

$$CH_2{=}C(CH_3)CH_2CH{=}CH_2 \xrightarrow{HBr} CH_3\overset{+}{C}(CH_3)CH_2CH{=}CH_2 \xrightarrow{Br^-} CH_3C(CH_3)(Br)CH_2CH{=}CH_2$$

2-methyl-1,4-pentadiene — 4-bromo-4-methyl-1-pentene

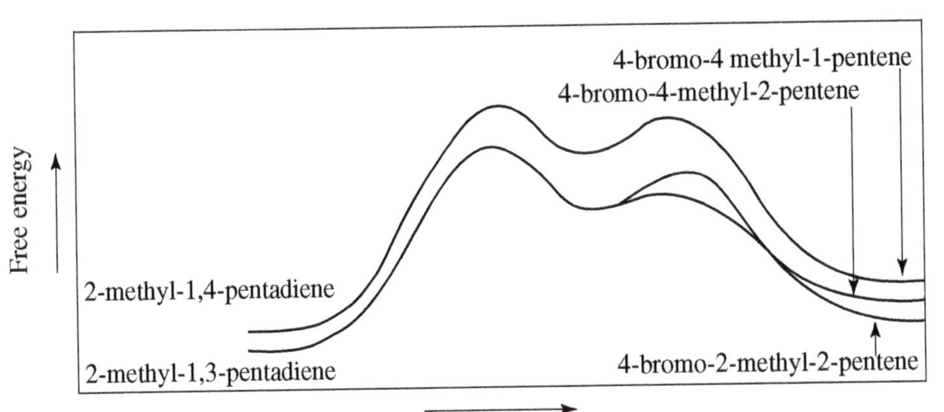

106. Her recrystallization was not successful. Because maleic anhydride is a dienophile, it reacts with cyclopentadiene in a Diels–Alder reaction.

cyclopentadiene + maleic anhydride ⟶ endo + exo

107. We saw in Problem 106 that maleic anhydride reacts with cyclopentadiene. The function of maleic anhydride in this reaction is to remove cyclopentadiene because removal of a product drives the equilibrium toward products. (See Le Châtelier's principle on page 205 of the text.)

108. The bridgehead carbon cannot have the 120° bond angle required for the sp^2 carbon of a double bond because, if it did, the compound would be too strained to exist.

109. **a.** Unless the reaction is being carried out under kinetic control, the amount of product obtained is not dependent on the rate at which the product is formed, so the relative amounts of products obtained will not tell you which product was formed faster.

b. In a thermodynamically controlled reaction, the product distribution depends on the relative stabilities of the products because the products come to equilibrium. Therefore, if the distribution of products does not reflect the relative stabilities of the products, the reaction must have been kinetically controlled.

110.

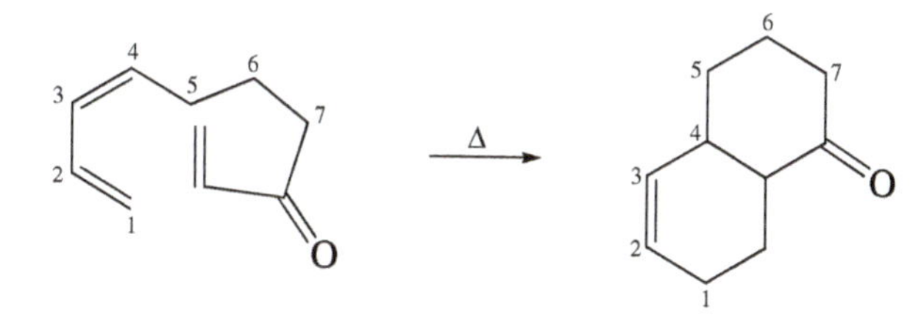

First line up the conjugated diene and the double bond of the dienophile in a way that prepares them to undergo the Diels–Alder reaction. Once they are lined up correctly, you can rearrange the electrons to determine the product of the reaction.

111. **a.** The three resonance contributors marked with an X are the least stable because in these contributors, the two negative charges are on adjacent carbons.

b. Because these contributors are the least stable, they make the smallest contribution to the hybrid.

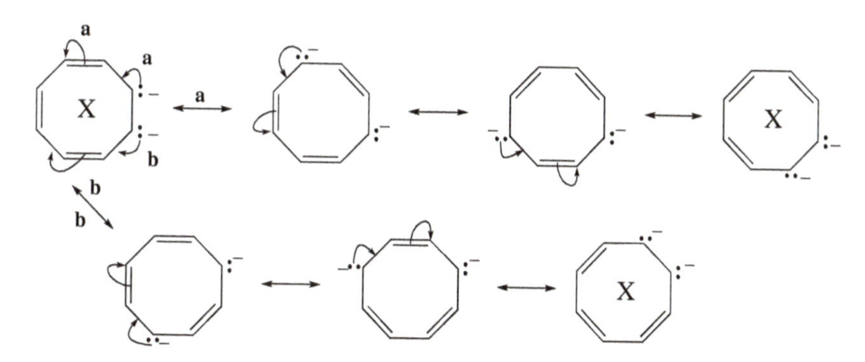

112. We have classified cyclobutadiene as antiaromatic. However, the recent observation that cyclobutadiene is rectangular and the observation that there are two different 1,2-dideuterio-1,3-cyclobutadienes both indicate that the π electrons are localized rather than delocalized. Localization of the π electrons prevents the compound from being antiaromatic. Apparently, the extreme instability associated with being antiaromatic causes cyclobutadiene to be rectangular.

Chapter 8 Practice Test

1. Indicate the more stable species in each pair:

a. 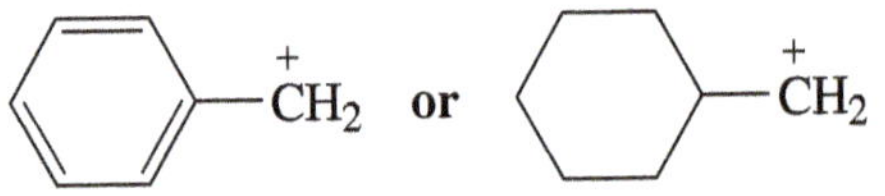 $C_6H_5\overset{+}{C}H_2$ (benzene ring) or $C_6H_{11}\overset{+}{C}H_2$ (cyclohexane ring)

b. $CH_3\bar{C}HCH_3$ or $CH_3\bar{C}HC{\equiv}CH$

c. $CH_3\bar{C}HCH_2\overset{O}{\overset{\|}{C}}CH_3$ or $CH_3\bar{C}H\overset{O}{\overset{\|}{C}}CH_3$

d. $CH_2{=}CH\overset{+}{C}H_2$ or $CH_2{=}CHCH_2\overset{+}{C}H_2$

e. 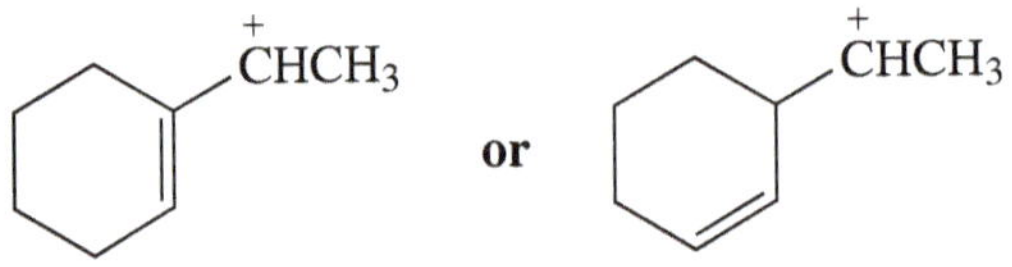 cyclohexenyl–$\overset{+}{C}HCH_3$ (double bond adjacent) or cyclohexenyl–$\overset{+}{C}HCH_3$ (double bond not adjacent)

2. Draw resonance contributors for each of the following:

a. $CH_3CH{=}CH{-}\ddot{\underset{..}{O}}CH_3$

b. $CH_3CH{=}CH{-}CH{=}CH{-}\overset{+}{C}H_2$

c. $\overset{-}{\ddot{C}}H_2{-}CH{=}CH{-}\overset{O}{\overset{\|}{C}}H$

3. Which compounds do not have delocalized electrons?

$CH_3CH_2NHCH{=}CHCH_3$ $\quad CH_3\overset{CH_3}{\overset{|}{\underset{+}{C}}}CH_2CH{=}CH_2$ $\quad CH_2{=}CHCH_2CH{=}CH_2$

$CH_2{=}CH\overset{O}{\overset{\|}{C}}CH_3$ $\quad CH_3CH_2NHCH_2CH{=}CHCH_3$ $\quad$ (cyclohexa-1,3-diene)

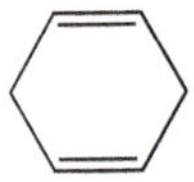

4. What are the products of the following reactions?

a. 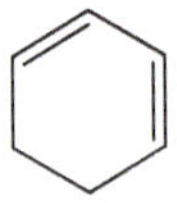 (cyclohexa-1,3-diene) + Br_2 ⟶

b. 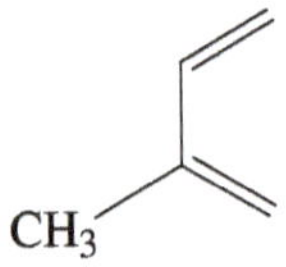 (CH_3-substituted 1,3-butadiene) + $\underset{CH_2}{\overset{\|}{C}H}\overset{O}{\overset{\|}{C}}CH_3$ ⟶

5. Which of the following pairs are resonance contributors?

a. CH_3CH_2OH and CH_3OCH_3

b. $CH_3\overset{O}{\overset{\|}{C}}OH$ and $CH_3\overset{O}{\overset{\|}{C}}O^-$

c. $CH_3\overset{O}{\overset{\|}{C}}OH$ and $CH_3\overset{O^-}{\overset{|}{C}}{=}\overset{+}{O}H$

d. $CH_3CH_2\overset{O}{\overset{\|}{C}}H$ and $CH_3CH{=}\overset{OH}{\overset{|}{C}}H$

6. Which of the following dienes can be used in a Diels–Alder reaction?

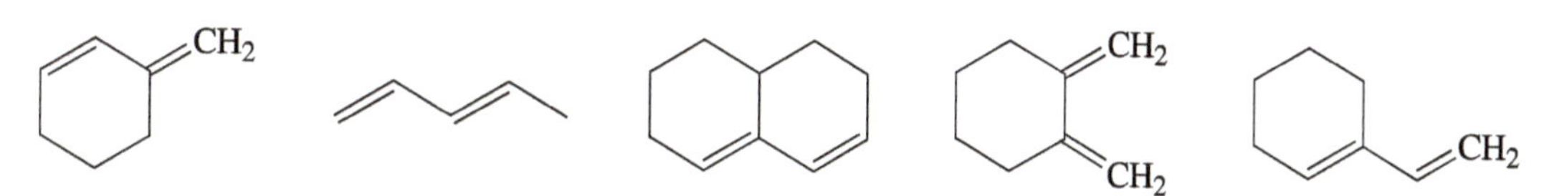

7. Which is a stronger base?

a.

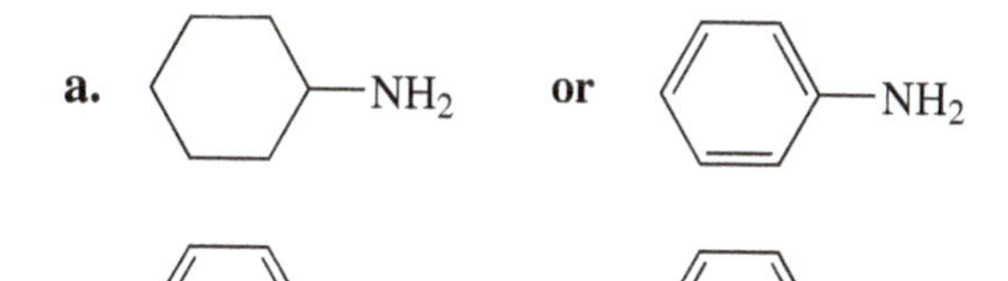

b. $-CH_2O^-$ or $-O^-$

8. Draw resonance contributors for each of the following:

a. $\ddot{N}H_2$

b. $\overset{+}{N}H_3$

c. $\ddot{O}:^-$

9. Which resonance contributor makes a greater contribution to the resonance hybrid?

a. O^- or O

b. CH_3 or CH_3

10. Indicate whether each of the following statements is true or false.

a. A compound with four conjugated double bonds has four molecular orbitals. T F

b. ψ_1 and ψ_2 are symmetric molecular orbitals. T F

c. If ψ_3 is the HOMO in the ground state, ψ_4 will be the HOMO in the excited state. T F

d. If ψ_3 is the LUMO, ψ_4 will be the HOMO. T F

e. If the ground-state HOMO is symmetric, the ground-state LUMO will be antisymmetric. T F

f. A single bond formed by an sp^2—sp^2 overlap is longer than a single bond formed by an sp^2—sp^3 overlap. T F

g. The thermodynamically controlled product is the major product obtained when the reaction is carried out under mild conditions. T F

h. 1,3-Hexadiene is more stable than 1,4-hexadiene. T F

11. Draw the four products that will be obtained from the following reaction. Ignore stereoisomers.

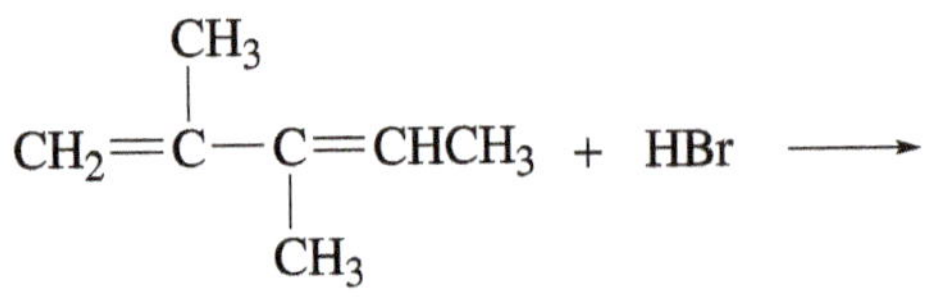

$$CH_2{=}\overset{\overset{CH_3}{|}}{C}{-}\underset{\underset{CH_3}{|}}{C}{=}CHCH_3 + HBr \longrightarrow$$

12. What reactants are necessary for the synthesis of the following compound via a Diels–Alder reaction?

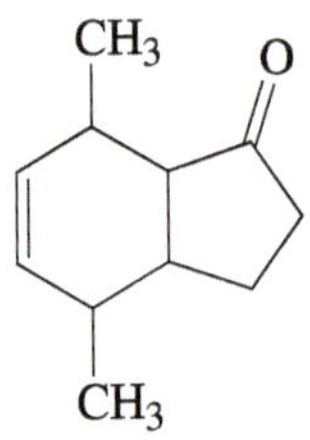

13. Rank the following carbocations from most stable to least stable:

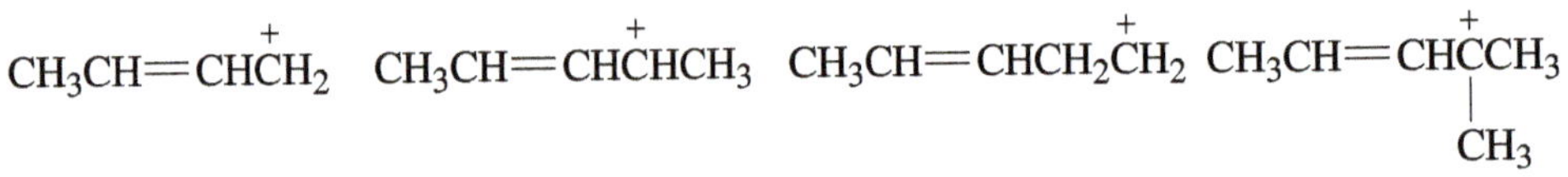

$$CH_3CH{=}CH\overset{+}{C}H_2 \quad CH_3CH{=}CH\overset{+}{C}HCH_3 \quad CH_3CH{=}CHCH_2\overset{+}{C}H_2 \quad CH_3CH{=}CH\underset{\underset{CH_3}{|}}{\overset{+}{C}}CH_3$$

14. **a.** Draw the predominant 1,2- and 1,4-products of the following reaction.

b. Which is the product of thermodynamic control?

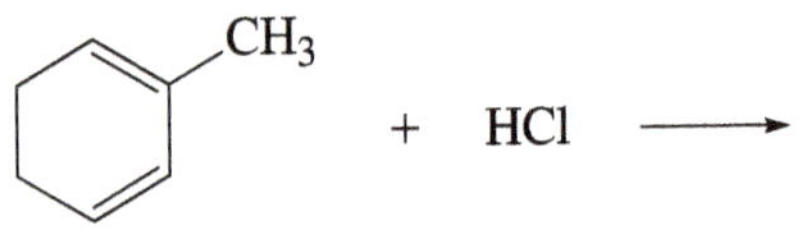

$$+ \ HCl \longrightarrow$$

15. What reagents can be used to convert the given starting material into the given product?

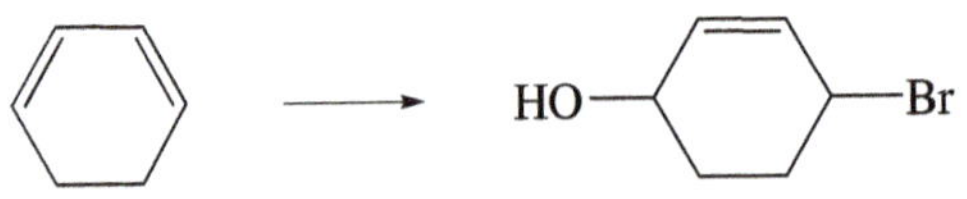

$$\longrightarrow \quad HO{-}\quad{-}Br$$

16. Draw the product of the following reaction, showing its configuration:

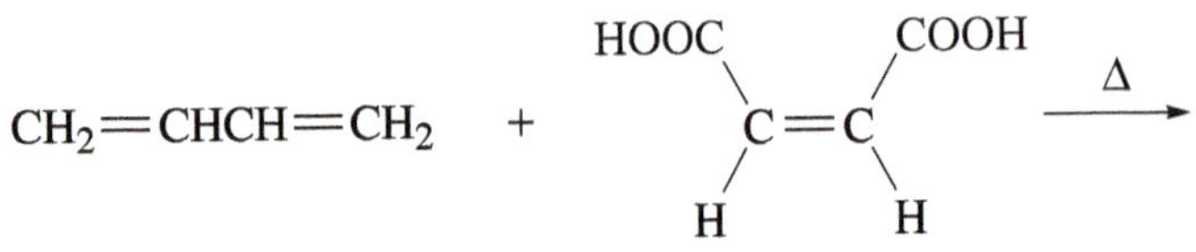

$$CH_2{=}CHCH{=}CH_2 \quad + \quad \begin{matrix} HOOC & & COOH \\ & C{=}C & \\ H & & H \end{matrix} \xrightarrow{\Delta}$$

17. For each of the following reactions, give the major 1,2- and 1,4-products. Label the product of kinetic control and the product of thermodynamic control.

a. $CH_2{=}CH{-}\overset{CH_3}{\overset{|}{C}}{=}CH_2$ + HCl ⟶

b. CH_2 (methylenecyclohexene) + HBr ⟶

18. Which are aromatic compounds?

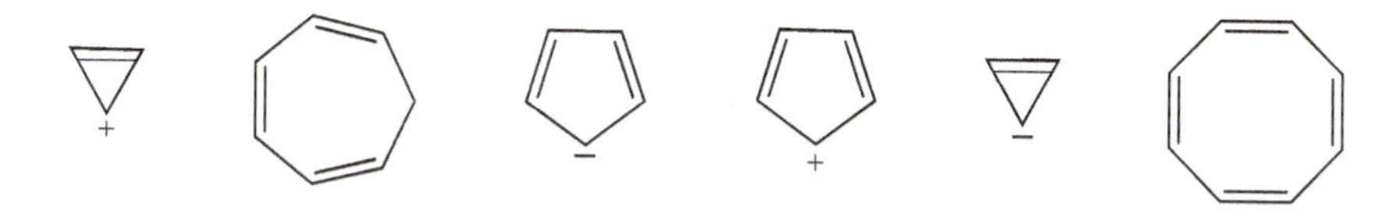

19. Which compound has the greater delocalization energy?

(cycloheptatrienyl cation) or (cycloheptatrienyl anion)

20. Which is a stronger acid?

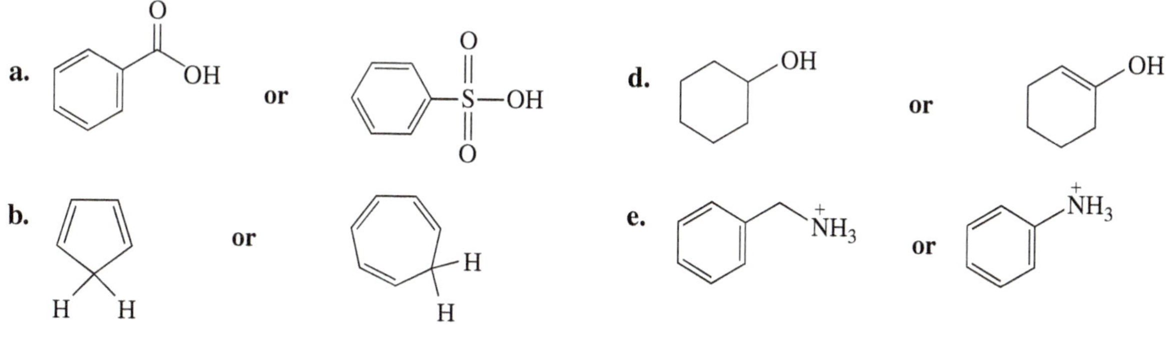

c. COOH or COOH CH_3O

f. OH or OH O_2N Cl

SPECIAL TOPIC II

Molecular Orbital Theory

Chemists use models to describe such things as the bonding in molecules, the stability of molecules, and the reactions between molecules. The model used is generally the one that provides the best description of the molecule under consideration. One very powerful model is molecular orbital theory.

You were introduced to molecular orbital (MO) theory in Section 1.6 where you saw that electrons are assigned to a volume of space called an orbital. According to MO theory, covalent bonds are formed when atomic orbitals combine to form molecular orbitals. Let's review some important principles:

1. Orbitals are conserved. In other words, the number of molecular orbitals formed must equal the number of atomic orbitals combined. For example, when two atomic orbitals interact, two molecular orbitals are formed—one lower in energy (a bonding MO) and one higher in energy (an antibonding MO, indicated by a *) than the interacting orbitals.

2. Electrons always occupy the available *atomic orbitals* with the lowest energy, and no more than two electrons can occupy an *atomic orbital*. Similarly, electrons always occupy the available *molecular orbitals* with the lowest energy, and no more than two electrons can occupy a *molecular orbital*.

3. The relative energies of the molecular orbitals are $\sigma < \pi < \pi^* < \sigma^*$.

4. The strongest covalent bonds are formed by electrons that occupy the molecular orbitals with the lowest energy. For example, the energy of a σ MO is lower than that of a π MO, and we have seen that a σ bond is stronger than a π bond (Section 1.15).

Atomic orbitals can overlap to form molecular orbitals in two ways; these are shown in Figures 1 and 2.

In Figure 1, each of the overlapping atomic orbitals contributes one electron to the bond.
In Figure 2, a filled atomic orbital (it has two electrons) overlaps an empty atomic orbital.

In each case, we see that the electrons are stabilized (are lower in energy) as a result of orbital overlap.

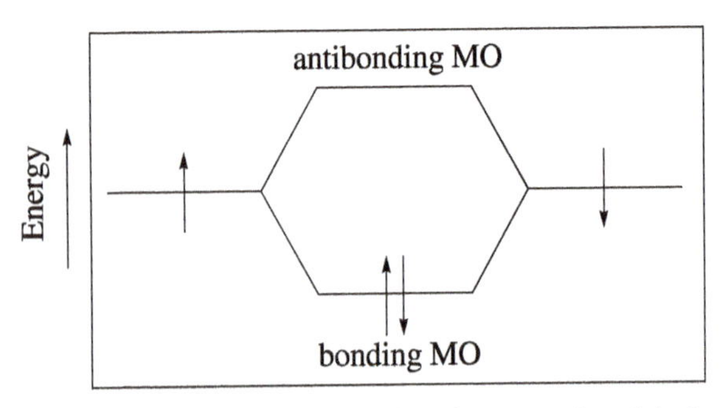

Figure 1. Each of the overlapping atomic orbitals contributes one electron to the bond.

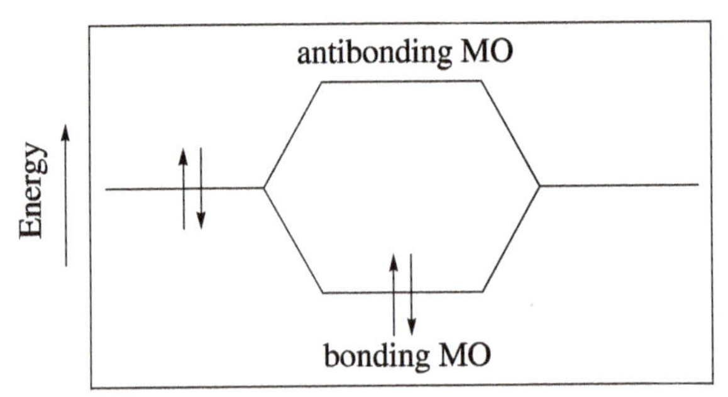

Figure 2. A bond is formed as a result of the overlap of a filled atomic orbital with an empty atomic orbital.

I. Using Molecular Orbital Theory to Describe Covalent Bond Formation

A. Each of the Overlapping Atomic Orbitals Contributes One Electron to the Bond

Take a look at Figure 1.3 on page 23 of the text. There you can see that an H—H bond is formed by the overlap of a 1*s* atomic orbital of a hydrogen atom with a 1*s* atomic orbital of another hydrogen atom; each of the atomic orbitals contributes one electron to the molecular orbital.

Figure 3 shows that the C—C bond in ethane is formed by the overlap of an sp^3 atomic orbital of carbon with an sp^3 atomic orbital of another carbon; again, each of the overlapping atomic orbitals contributes one electron to the bond.

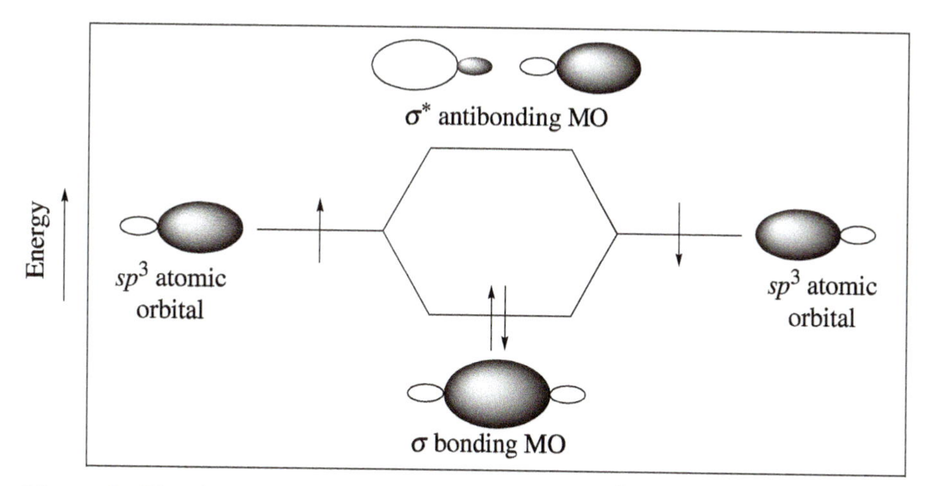

Figure 3. The C—C bond of ethane is formed by sp^3—sp^3 overlap.

Figure 4 shows that the C—H bond of ethane is formed by the overlap of an sp^3 atomic orbital of carbon with an *s* atomic orbital of hydrogen. Because an *s* atomic orbital of hydrogen is more stable than an sp^3 atomic orbital of carbon (see page 27 of the text), the MO formed by sp^3—*s* overlap is more stable than the MO formed by sp^3—sp^3 overlap. As a result, the C—H bond is stronger (and shorter) than the C—C bond.

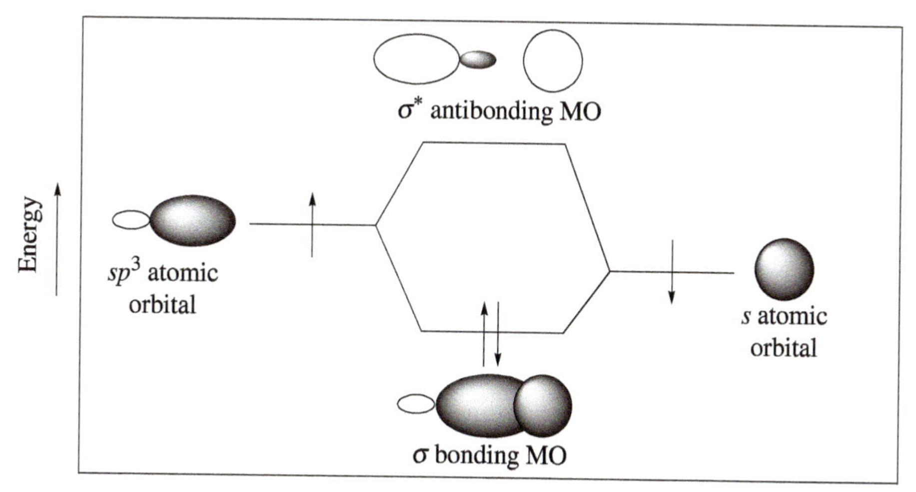

Figure 4. A C—H bond of ethane is formed by sp^3—s overlap.

Figure 5 shows that the two sp^3 atomic orbitals that overlap to form the C—O bond of an alcohol or of an ether do not have the same energy. An electron is more stable in the atomic orbital of the more electronegative atom. Thus, the C—O bond is a little stronger and shorter than the C—C bond.

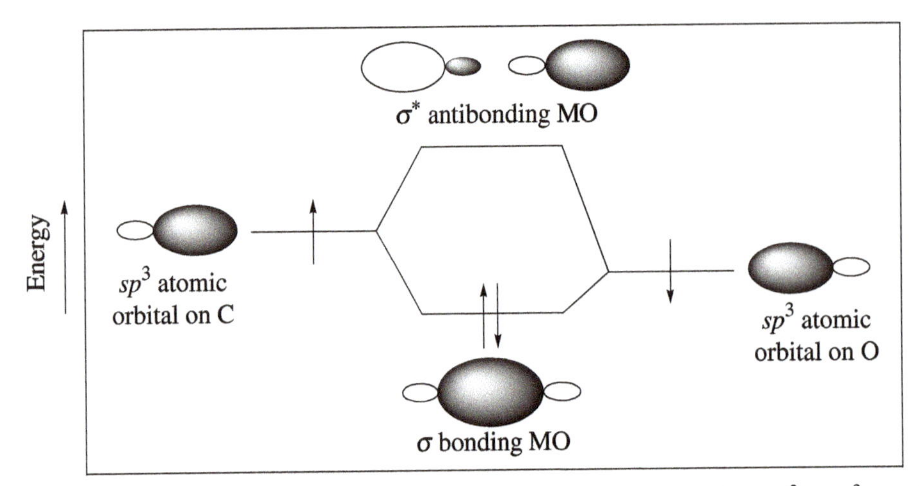

Figure 5. The C—O bond of an ether or an alcohol is formed by sp^3—sp^3 overlap.

Figure 6 shows that the C—C σ bond of ethene is formed by the overlap of an sp^2 atomic orbital of carbon with an sp^2 atomic orbital of another carbon. The π bond of ethene is formed by the side-to-side overlap of two p orbitals. (See Figure 1.4 on page 25 of the text.) A π molecular orbital is less stable than a σ molecular orbital. The π bond, therefore, is weaker than the σ bond.

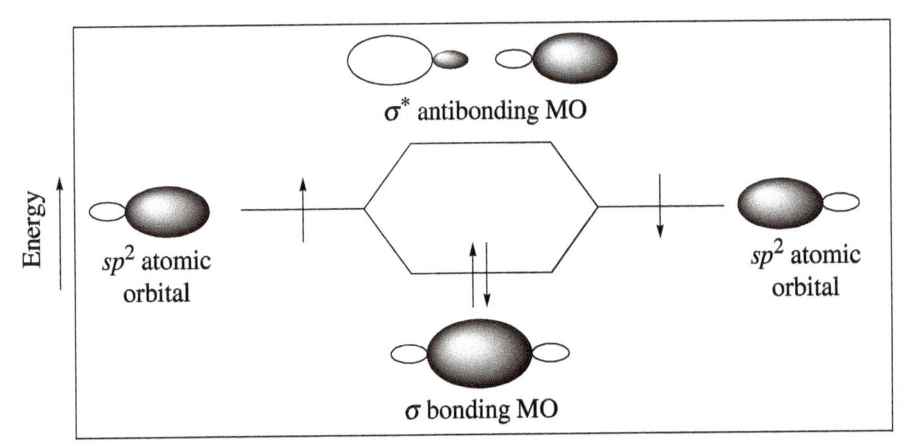

Figure 6. The C—C σ bond of ethene is formed by sp^2—sp^2 overlap.

B. A Filled Atomic Orbital Overlaps an Empty Atomic Orbital to Form a Bond

The overlap of a filled atomic orbital with an empty atomic orbital is the second way two atomic orbitals can overlap to form a bond (Figure 2).

For example, the bond formed between a Lewis base (such as NH_3) and a Lewis acid (such as $FeBr_3$) results from the base sharing a pair of electrons with the acid. Bond formation results from the overlap of a filled sp^3 orbital of nitrogen with an empty orbital of iron. This type of reaction is discussed in Section 2.12 of the text.

$$FeBr_3 + :NH_3 \longrightarrow Br_3\overset{-}{Fe}—\overset{+}{N}H_3$$

II. Using Molecular Orbital Theory to Describe Chemical Reactions

We have seen that most organic reactions involve the reaction of a nucleophile with an electrophile. Molecular orbital theory describes a reaction between a nucleophile and an electrophile as the result of the interaction of the HOMO (highest occupied MO) of the nucleophile with the LUMO (lowest unoccupied MO) of the electrophile, because the most stabilizing interaction is between orbitals closest in energy. Notice that in these examples, a filled orbital overlaps an empty orbital. The interaction of a filled orbital with an empty orbital is stabilizing, because the two electrons involved in bond formation end up in the lower-energy bonding MO and no electrons have to be placed in an antibonding MO (Figure 7).

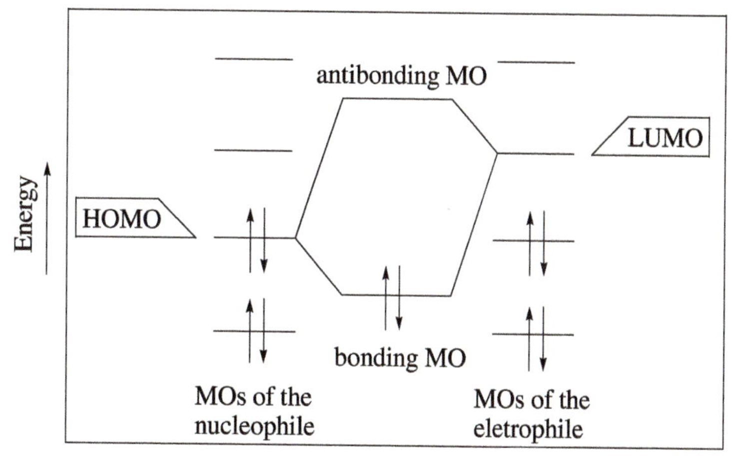

Figure 7. The interaction of the HOMO of the nucleophile with the LUMO of the electrophile.

We will start by looking at the reaction of 2-butene with HBr, an electrophilic addition reaction we first examined in Section 5.5.

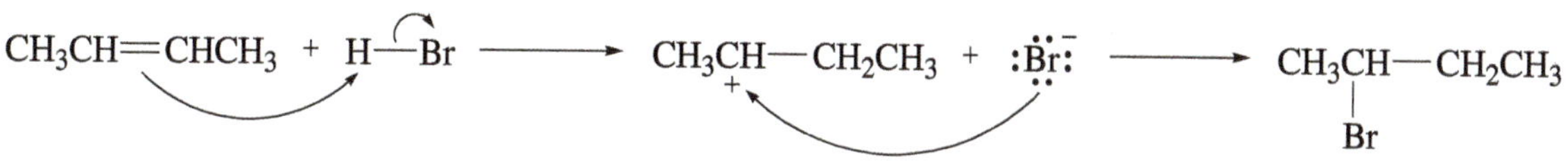

$$CH_3CH{=}CHCH_3 + H{-}Br \longrightarrow CH_3\overset{+}{C}H{-}CH_2CH_3 + :\ddot{Br}:^- \longrightarrow CH_3CH(Br){-}CH_2CH_3$$

In the first step of the reaction, the alkene is the nucleophile; the electrons of the π bond are in the π bonding MO; this is the HOMO. HBr is the electrophile. The electrons that form the H—Br bond are in a σ bonding MO. Therefore, the LUMO of HBr is the σ^* antibonding MO (Figure 8).

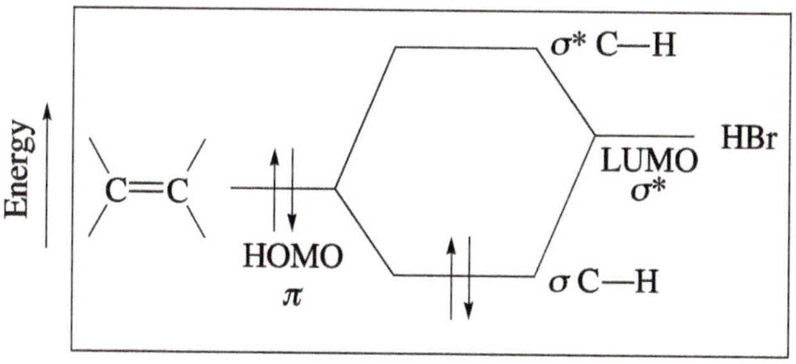

Energy; C=C; HOMO π; σ^* C—H; HBr; LUMO σ^*; σ C—H

Figure 8. Interaction of the HOMO of the alkene with the LUMO of HBr.

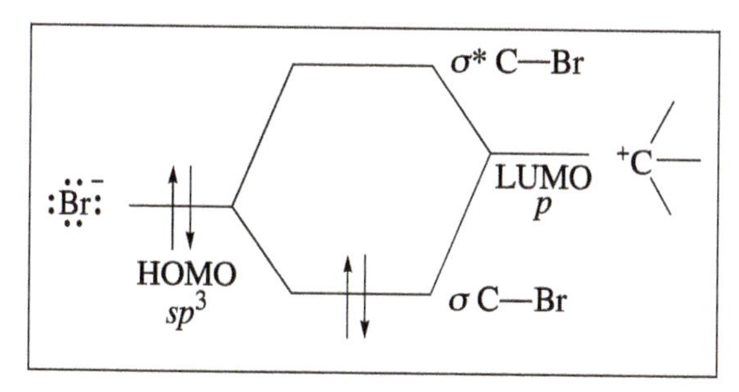

Figure 9. Interaction of the HOMO of Br^- with the LUMO of the carbocation.

The three p atomic orbitals of the three allyl carbons combine to produce three π molecular orbitals (Figure 10).

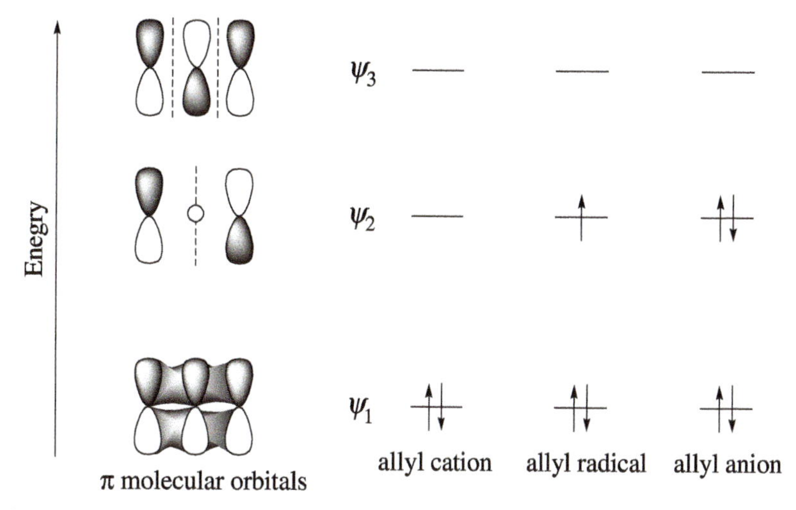

Figure 10. The distribution of the electrons in the molecular orbitals of the allyl cation, the allyl radical, and the allyl anion.

The bonding MO (ψ_1) encompasses the three carbons. In an acyclic system, the number of bonding MOs always equals the number of antibonding MOs. Therefore, when there is an odd number of MOs, one of them must be a nonbonding molecular orbital; ψ_2 is the nonbonding MO. We have seen (Section 8.8) that as the energy of the MO increases, the number of nodes increases. Consequently, ψ_2 must have a node—in addition to the one that ψ_1 has—that bisects the p orbitals. The only symmetrical position for a node to pass through in ψ_2 is through the middle carbon. You also know that it needs to pass through the middle carbon because that is the only way ψ_2 can be fully antisymmetric, which it must be since ψ_1 and ψ_3 are symmetric (recall that MOs alternate between being symmetric and antisymmetric; Section 8.8.)

You can see why ψ_2 is called a nonbonding molecular orbital—there is no overlap between the p orbital on the middle carbon and the p orbital on either of the end carbons. Notice that a nonbonding MO has the same energy as the isolated p atomic orbitals. The third MO (ψ_3) is an antibonding MO.

The two π electrons of the allyl cation are in the bonding MO, which means they are spread out over all three carbons. Consequently, the two carbon–carbon bonds are identical, with each having some double-bond character. The resonance contributors show that the positive charge is shared equally by the end carbon atoms, which is another way of showing that the stability of the allyl cation is due to electron delocalization:

$$CH_2{=}CH{-}\overset{+}{C}H_2 \longleftrightarrow \overset{+}{C}H_2{-}CH{=}CH_2$$

The contributing resonance structures show that when a nucleophile such as Br^- reacts with an allyl cation, the Br^- can bond to either of the end carbons—because they share the positive charge—but cannot bond to the middle carbon. Likewise, MO theory shows that only the end carbons have an empty orbital with which the filled orbital of Br^- can overlap. The central carbon has a node, so there can be no interaction with this carbon.

The allyl radical has two electrons in the bonding MO, so these electrons are spread over all three carbon atoms. The third electron is in the nonbonding MO. The MO diagram shows that the third electron is shared equally by the end carbons with none of the electron density residing on the middle carbon. This agrees with what the resonance contributors show:

$$CH_2{=}CH{-}\dot{C}H_2 \longleftrightarrow \dot{C}H_2{-}CH{=}CH_2$$

Finally, the allyl anion has two electrons in the nonbonding MO. These two electrons are shared equally by the end carbons. This, too, agrees with what the resonance contributors show:

$$CH_2{=}CH{-}\ddot{C}H_2^{-} \longleftrightarrow \ddot{C}H_2^{-}{-}CH{=}CH_2$$

We have seen that both molecular orbital theory and contributing resonance structures can be used to explain electron delocalization. The choice is a matter of preference. Hyperconjugation is probably best shown by molecular orbital theory, because the contributing resonance structures that describe hyperconjugation would require breaking σ bonds. Contributing resonance structures are sometimes preferred when one needs to see on which atoms charges reside.

CHAPTER 9
Substitution and Elimination Reactions of Alkyl Halides

Important Terms

β-elimination reaction or 1,2-elimination reaction — an elimination reaction where the groups being eliminated are bonded to adjacent carbons.

anti elimination — an elimination reaction in which the substituents being eliminated are removed from opposite sides of the molecule.

anti-periplanar — substituents are attached to parallel bonds on opposite sides of a molecule.

aprotic solvent — a solvent that does not have a hydrogen bonded to an oxygen or to a nitrogen; some aprotic solvents are polar; others are nonpolar.

back-side attack — nucleophilic attack on the side of the carbon opposite the side bonded to the leaving group.

base — a substance that gains a proton.

basicity — the tendency of a compound to share its electrons with a proton.

bifunctional molecule — a molecule with two functional groups.

bimolecular reaction — a reaction in which two molecules are involved in the transition state of the rate-determining step.

complete racemization — formation of a pair of enantiomers in equal amounts.

dehydrohalogenation — elimination of a proton and a halide ion.

E1 reaction — a unimolecular elimination reaction.

E2 reaction — a bimolecular elimination reaction.

elimination reaction — a reaction that removes atoms or groups from the reactant to form a π bond.

first-order reaction — a reaction whose rate is proportional to the concentration of one reactant.

intermolecular reaction — a reaction that takes place between two molecules.

intimate ion pair — an ion pair that results when the covalent bond that joined the cation and anion has broken but the cation and anion are still next to each other.

intramolecular reaction — a reaction that takes place within a molecule.

inversion of configuration turning the carbon inside out like an umbrella so that the resulting product has a configuration opposite that of the reactant.

ion-dipole interaction the interaction between an ion and the dipole of a molecule.

kinetics the field of chemistry that deals with the rates of chemical reactions.

leaving group the group that is displaced in a substitution reaction.

nucleophile an electron-rich atom or molecule.

nucleophilicity a measure of how readily an atom or a molecule with a lone pair attacks another atom.

nucleophilic substitution reaction a reaction in which a nucleophile substitutes for an atom or a group.

partial racemization formation of a pair of enantiomers in unequal amounts.

protic solvent a solvent that has a hydrogen bonded to an oxygen or to a nitrogen.

rate constant the constant of proportionality in the rate law for a reaction; it describes how difficult it is to overcome the energy barrier of a reaction.

rate law the equation that shows the relationship between the rate of a reaction and the concentration of the reactants.

regioselectivity the preferential formation of a constitutional isomer.

sawhorse projection a way to represent the three-dimensional spatial relationships of atoms by looking at the carbon–carbon bond from an oblique angle.

second-order reaction a reaction whose rate is dependent on the concentration of two reactants.

S_N1 reaction a unimolecular nucleophilic substitution reaction.

S_N2 reaction a bimolecular nucleophilic substitution reaction.

solvent-separated ion pair an ion pair that results when the cation and anion are separated by one or more solvent molecules.

solvolysis reaction with a solvent.

steric effects effects due to the fact that groups occupy a certain volume of space.

steric hindrance caused by bulky groups at the site of a reaction that make it difficult for the reactants to approach each other.

substitution reaction a reaction that exchanges one substituent of a reactant for another.

syn elimination — an elimination reaction in which substituents being eliminated are removed fromthe same side of the molecule.

syn-periplanar — substituents are attached to parallel bonds on the same side of a molecule.

target molecule — the desired product of a synthesis.

unimolecular reaction — a reaction in which only one molecule is involved in the transition state of the rate-determining step.

Williamson ether synthesis — formation of an ether from the reaction of an alkoxide ion with an alkyl halide.

Zaitsev's rule — the rule that states that the more stable alkene product of an elimination reaction is obtained by removing a proton from the β-carbon that is bonded to the fewest hydrogens.

Solutions to Problems

1. DEE is formed when HCl is eliminated from DDT. (See the box on page 392.)

DDE

2. Methoxychlor has methoxy groups in place of the chlorines on the benzene rings of DDT. The oxygen of the methoxy groups can form hydrogen bonds with water, making methoxychlor more soluble in water and, therefore, less soluble in fatty tissues.

3. $\text{rate} = k\,[\text{alkyl halide}][\text{nucleophile}]$

original: $\text{rate} = k\,[1.0][1.0]$

a. $\text{rate} = k\,[1.0][3.0] = 3.0$ The rate is tripled.

b. $\text{rate} = k\,[0.50][1.0] = 0.50$ The rate is cut in half.

c. $\text{rate} = k\,[0.5][2.0] = 1.0$ The rate is not changed.

4. Increasing the height of the energy barrier decreases the magnitude of the rate constant; this causes the reaction to be slower.

5. The closer the methyl group is to the site of nucleophilic attack, the greater the steric hindrance to nucleophilic attack and the slower the rate of the reaction.

$CH_3CH_2CH_2CH_2CH_2Br > CH_3CH(CH_3)CH_2CH_2Br > CH_3CH_2CH(CH_3)CH_2Br > CH_3CH_2CBr(CH_3)_2$

6. **a.** Solved in the text.

b. $CH_3CH_2C(CH_3)(H)Br$ $\xrightarrow{CH_3O^-}$ $CH_3CH_2C(CH_3)(H)OCH_3$

(*R*)-2-bromobutane → (*S*)-2-methoxybutane

c. $CH_3CH_2C(CH_2CH_2CH_3)(H)Cl$ $\xrightarrow{HO^-}$ $CH_3CH_2C(CH_2CH_2CH_3)(H)OH$

(*S*)-3-chlorohexane → (*R*)-3-hexanol

d. $CH_3CH_2CH(I)CH_2CH_3$ $\xrightarrow{HO^-}$ $CH_3CH_2CH(OH)CH_2CH_3$

3-iodopentane → 3-pentanol

7. Solved in the text.

8. **a.**

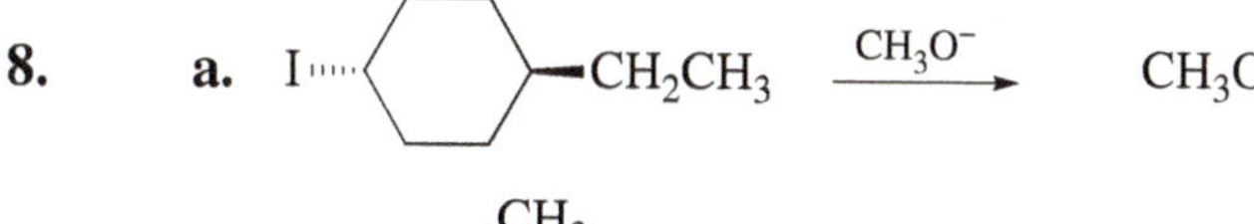

b. CH_3 Cl $\xrightarrow{CH_3CH_2O^-}$ CH_3 CH_3CH_2O

9. **a.** Br The primary alkyl halide is less sterically hindered than the secondary alkyl halide (the CH_3 group is farther away from the back side of the carbon attached to the Br).

b. Br Br is a weaker base; therefore, it is a better leaving group.

c. Br With one methyl and one ethyl group, this alkyl halide is less sterically hindered than the other alkyl halide that has two ethyl groups.

d. Br The primary alkyl halide is less sterically hindered than the secondary alkyl halide.

10. A protic solvent has a hydrogen bonded to an oxygen or to a nitrogen, whereas an aprotic solvent does not have a hydrogen bonded to an oxygen or to a nitrogen.

a. aprotic **b.** aprotic **c.** protic **d.** aprotic

11. **a.** RO^-, because ROH is a weaker acid than RSH since the hydrogen is attached to a smaller atom.

b. RS^-, because it is less well solvated by water and sulfur is more polarizable than oxygen.

c. RO^-, because, although they differ in size, they are in an aprotic solvent. Remember that the stronger base is always the better nucleophile in an aprotic solvent.

12. Remember that the stronger base is always the better nucleophile unless they differ in size *and* they are in a protic solvent.

a. They differ in size, and because they are in a protic solvent, the larger one (Br^-) is the better nucleophile.

b. They differ in size, and because they are in an aprotic solvent, the stronger base (Cl^-) is the better nucleophile.

c. Because the oxygen is negatively charged, CH_3O^- is the better nucleophile.

d. Because the oxygen is negatively charged, CH_3O^- is the better nucleophile.

e. Because H_2O is a stronger acid than NH_3, $^-NH_2$ is the stronger base and the better nucleophile.

f. Because H_2O is a stronger acid than NH_3, $^-NH_2$ is the stronger base and the better nucleophile.

g. They differ in size and, because they are in a protic solvent, the larger one (I^-) is the better nucleophile.

h. They differ in size and, because they are in an aprotic solvent, the stronger base (Br^-) is the better nucleophile.

13. Solved in the text.

14.

a. $CH_3CH_2Br + HO^-$ — HO^- is a better nucleophile than H_2O.

b. $CH_3CH(CH_3)CH_2Br + HO^-$ — This alkyl halide has less steric hindrance toward nucleophilic attack.

c. $CH_3CH_2Cl + CH_3S^-$ — CH_3S^- is a better nucleophile than CH_3O^- in a protic solvent (a solvent that can form hydrogen bonds).

d. $CH_3CH_2Br + I^-$ — Br^- is a weaker base than Cl^-, so Br^- is a better leaving group.

15. These are all S_N2 reactions.

a. $CH_3CH_2OCH_2CH_2CH_3$

b. $CH_3CH_2C{\equiv}CCH_3$

c. $CH_3CH_2\overset{+}{N}(CH_3)_3\ Br^-$

d. $CH_3CH_2SCH_2CH_3$

16. Solved in the text.

17.

a. Reaction of an alkyl halide with ammonia gives a low yield of primary amine, because as soon as the primary amine is formed, it can react with another molecule of alkyl halide to form a secondary amine; the secondary amine can react with the alkyl halide to form a tertiary amine, which can then react with an alkyl halide to form a quaternary ammonium salt. (See Problem 16 on page 404.)

b. The alkyl azide is not treated with hydrogen until after all the alkyl halide has reacted with azide ion. Therefore, when the primary amine is formed, there is no alkyl halide for it to react with to form a secondary amine.

18.

a. OCH₃

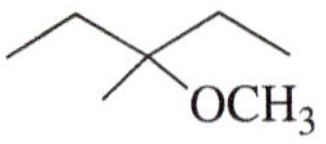

one product because the leaving group is not attached to an asymmetric center

b. OCH₃ + OCH₃

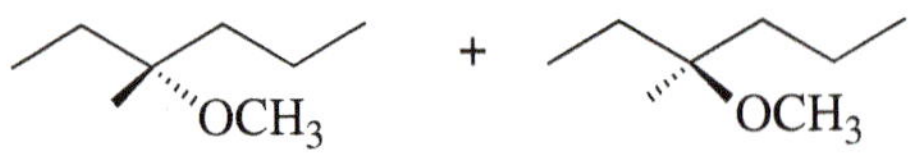

R and *S* because the leaving group is attached to an asymmetric center

19. I > Br > Cl > Cl

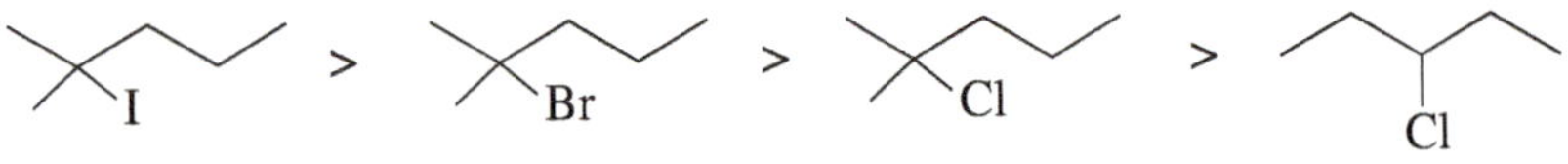

20.

a. CH_3, H, CH_2CH_3, CH_3CH_2O — C

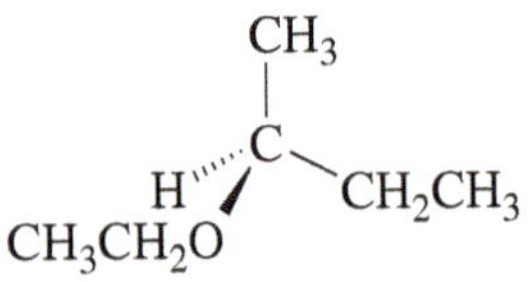

The product has the inverted configuration compared to that of the reactant.

b. CH_3CH_2–(cyclohexane)–CH_3/OCH_3 + CH_3CH_2–(cyclohexane)–OCH_3/CH_3

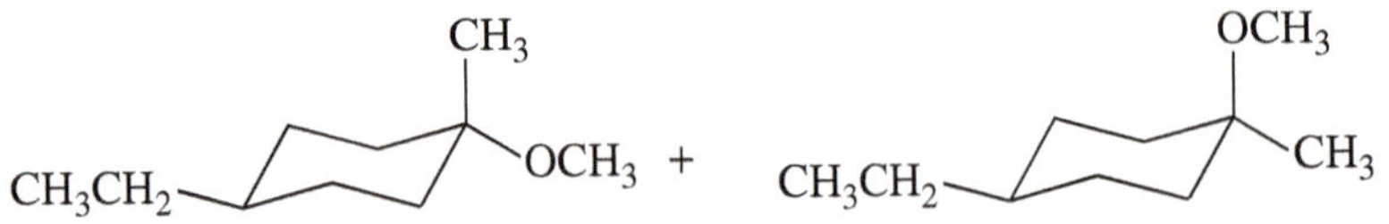

Once the tertiary carbocation forms, methanol can attack the sp^2 carbon from the top or bottom of the planar carbocation.

21. The rate of an S_N1 reaction is not affected by increasing the concentration of the nucleophile, whereas the rate of an S_N2 reaction is increased when the concentration of the nucleophile is increased. Therefore, we first have to determine whether the reactions are S_N1 or S_N2 reactions.

A is an S_N2 reaction because the reactant is a secondary alkyl halide (and the configuration of the product is inverted compared to that of the reactant).

B is an S_N2 reaction because the reactant is a primary alkyl halide.

C is an S_N1 reaction because the reactant is a tertiary alkyl halide.

Because they are S_N2 reactions, the rate of **A** and **B** increases if the concentration of the nucleophile is increased. Because it is an S_N1 reaction, the rate of **C** does not change if the concentration of the nucleophile is increased.

22. **a.** 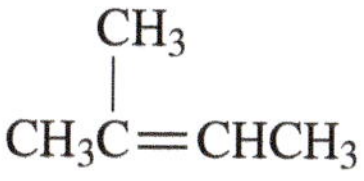 $CH_3C(CH_3){=}CHCH_3$ **b.** 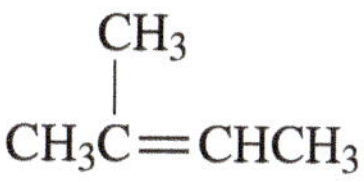 $CH_3C(CH_3){=}CHCH_3$ **c.** 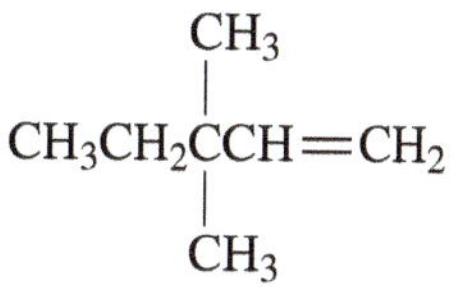 $CH_3CH_2C(CH_3)_2CH{=}CH_2$

23. **a.** $CH_3CH{=}CHCH_3$ — Removal of a hydrogen from the more substituted β-carbon forms the more stable "alkene-like" transition state.

b. 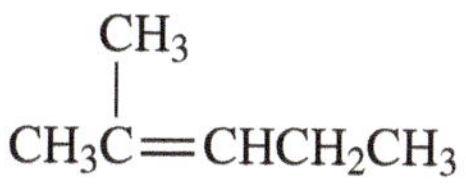 $CH_3C(CH_3){=}CHCH_2CH_3$ — Removal of a hydrogen from the more substituted β-carbon forms the more stable "alkene-like" transition state.

c. $CH_3CH{=}CHCH{=}CH_2$ — The hydrogen is removed from the β-carbon that leads to a conjugated alkene.

d. $CH_2{=}CHCH_2CH_3$ — Removal of a hydrogen from the less substituted β-carbon forms the more stable "carbanion-like" transition state.

e. 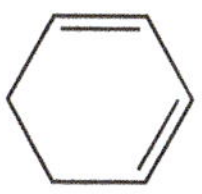 (1,3-cyclohexadiene structure) — The hydrogen is removed from the β-carbon that leads to a conjugated alkene.

f. 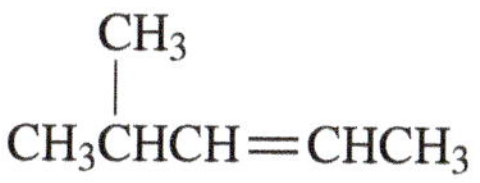 $CH_3CH(CH_3)CH{=}CHCH_3$ — Removal of a hydrogen from the less substituted β-carbon forms the more stable "carbanion-like" transition state.

24. **a.** 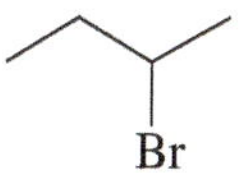 (2-bromobutane skeletal structure; Br) — The alkene (2-butene) that is formed is more stable than the alkene (1-butene) that is formed from the other alkyl halide.

b. 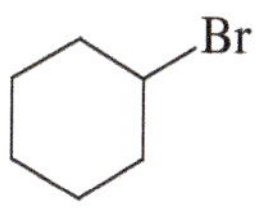 (bromocyclohexane skeletal structure; Br) — Br^- is a better leaving group (weaker base) than Cl^-.

c. 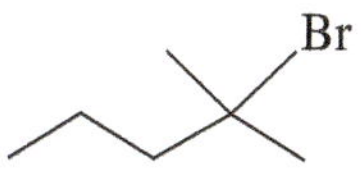 (2-bromo-2-methylpentane skeletal structure; Br) — The alkene (2-methyl-2-pentene) that is formed is more stable than the alkene (4-methyl-2-pentene) that is formed from the other alkyl halide.

d. 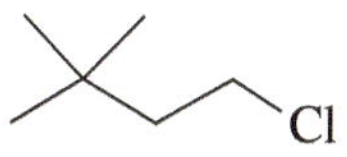 (1-chloro-3,3-dimethylbutane skeletal structure; Cl) — The other alkyl halide cannot undergo an E2 reaction, because it does not have any β-hydrogens.

25. **a.** 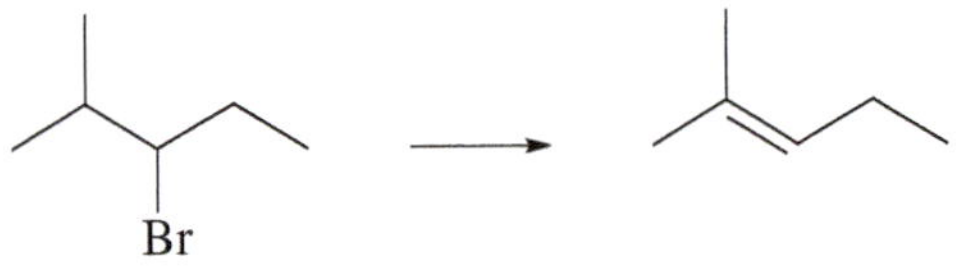 Br

It forms the more stable alkene (the alkene with the most substituents bonded to the sp^2 carbons), so it has the more stable transition state.

b. 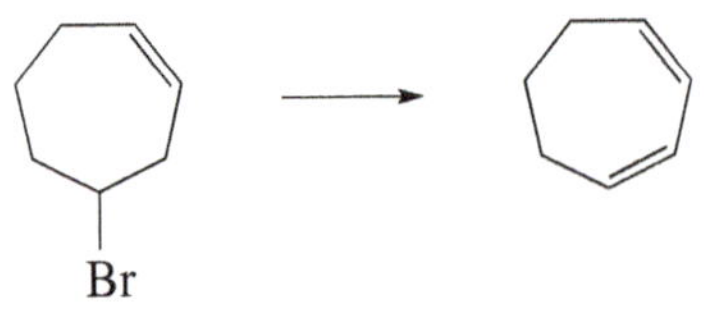 Br

It forms the more stable alkene (the double bonds are conjugated, so it has the more stable transition state.

c.  Br

It has four hydrogens that can be removed to form an alkene with two substituents on the sp^2 carbons, so it has a greater probability of having an effective collision with the nucleophile than the other alkyl halide that has only two such hydrogens.

d.  Br

It forms the more stable alkene (the new double bond is conjugated with the phenyl substituent), so it has the more stable transition state.

26.

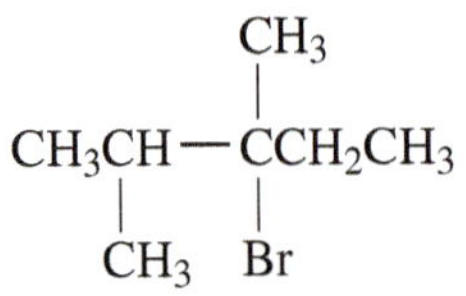

$$CH_3CH(CH_3)-C(CH_3)(Br)CH_2CH_3$$

3-bromo-2,3-dimethylpentane

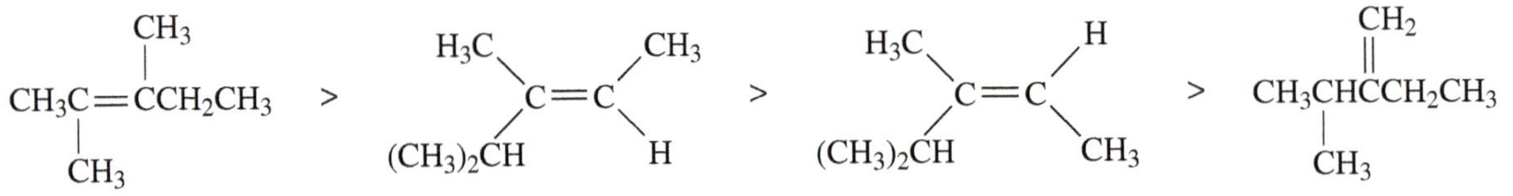

$CH_3C(CH_3){=}C(CH_3)CH_2CH_3$ > $(H_3C)((CH_3)_2CH)C{=}C(CH_3)(H)$ > $(H_3C)((CH_3)_2CH)C{=}C(H)(CH_3)$ > $CH_3CH(CH_3)C({=}CH_2)CH_2CH_3$

$CH_3C(CH_3){=}C(CH_3)CH_2CH_3$	H_3C, $(CH_3)_2CH$ / $C{=}C$ / CH_3, H	H_3C, $(CH_3)_2CH$ / $C{=}C$ / H, CH_3	$CH_3CHC(=CH_2)CH_2CH_3$ with CH_3
Four alkyl substituents are bonded to the sp^2 carbons.	Three alkyl substituents are bonded to the sp^2 carbons; the largest groups are on opposite sides of the double bond.	Three alkyl substituents are bonded to the sp^2 carbons; the largest groups are on the same side of the double bond.	Two alkyl substituents are bonded to the sp^2 carbons.

27. The major product is the one predicted by Zaitsev's rule, because the fluoride ion dissociates in the first step, forming a carbocation. Loss of a proton from the carbocation follows Zaitsev's rule, as it does in other E1 reactions.

28. **a.** B because it forms the more stable carbocation.
b. B because it forms the more stable alkene.
c. B because it forms the more stable carbocation.
d. A because it is less sterically hindered.

29. A tertiary carbocation with a strained four-membered ring is less stable than a secondary carbocation with an unstrained five-membered ring, so a carbocation rerrangement occurs. A second carbocation rearrangement forms a tertiary carbocation.

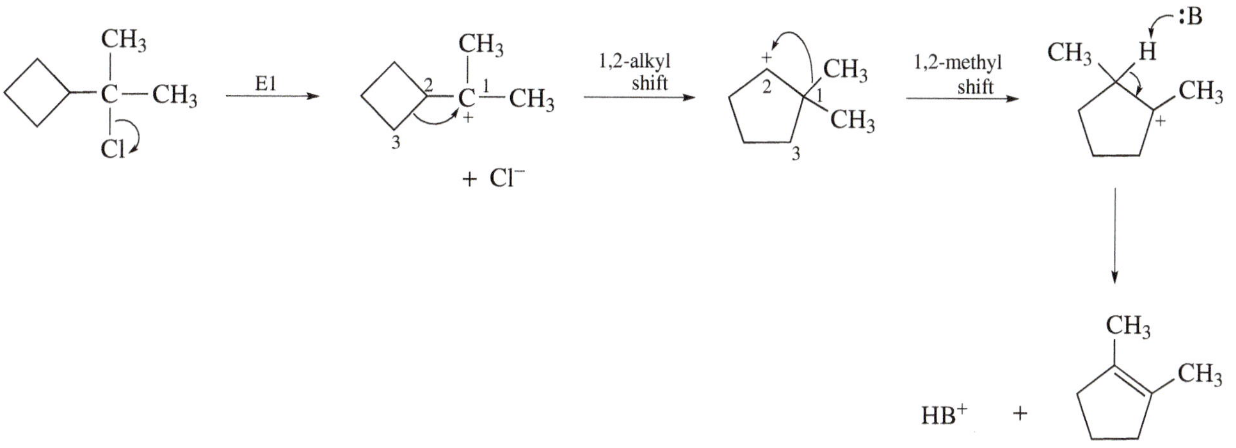

E1 → + Cl^- — 1,2-alkyl shift → — 1,2-methyl shift → (:B) ↓ HB^+ +

30. **a.** E2 $CH_3CH{=}CHCH_3$

b. E2 $CH_3CH_2CH{=}CH_2$

c. E1 $CH_3C(CH_3){=}CH_2$

d. E2 $CH_3C(CH_3){=}CH_2$

31. Solved in the text.

32.
$$\frac{E2}{E2 + E1} = \frac{7.1 \times 10^{-5} \times 2.5 \times 10^{-3}}{7.1 \times 10^{-5} \times 2.5 \times 10^{-3} + 1.50 \times 10^{-5}} = \frac{1.78 \times 10^{-7}}{1.78 \times 10^{-7} + 150 \times 10^{-7}} = \frac{1.78}{152} = 1.2\%$$

33. **a.** **1.** $CH_3CH_2CH{=}C(CH_3)CH_3$ — No stereoisomers are possible because there are two methyl groups on one of the sp^2 carbons.

2. $CH_3CH_2(H)C{=}C(H)CH{=}CH_2$ 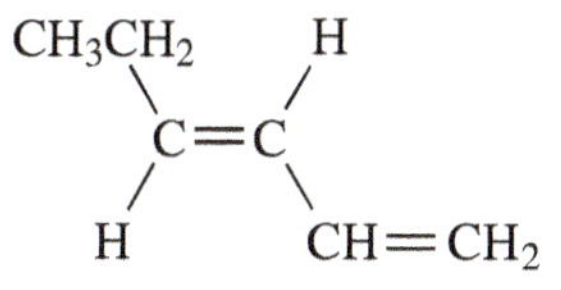 — The major product is the conjugated diene with the larger group bonded to one sp^2 carbon on the opposite side of the double bond from the larger group bonded to the other sp^2 carbon.

3. 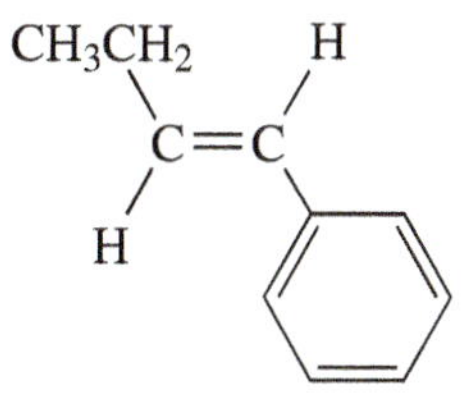 $CH_3CH_2(H)C{=}C(H)C_6H_5$ — The major product is the conjugated alkene with the larger group bonded to one sp^2 carbon on the opposite side of the double bond from the larger group bonded to the other sp^2 carbon.

b. In none of the reactions is the major product dependent on whether you start with the *R* or *S* enantiomer of the reactant.

34. Solved in the text.

35. **a.** $CH_3CH_2CH{=}C(CH_3)CH_3$

$$\begin{array}{c} CH_3 \\ | \\ CH_3CH_2CH{=}CCH_3 \end{array}$$

b. (Structure: phenyl and CH_3 on one sp^2 carbon, H and CH_3 on the other, C=C)

The larger substituent attached to one sp^2 carbon and the larger substituent attached to the other sp^2 carbon are on opposite sides of the double bond.

c. (Structure: 1-methylcyclohexene, CH_3)

36. E2 elimination reactions from six-membered rings occur only when the substituents to be eliminated are both in axial positions.

In the cis isomer, when Br is in an axial position, there is an axial hydrogen on each of the adjacent carbons. The one bonded to the same carbon as the ethyl group is more apt to be the one eliminated with Br because the product formed is more stable and, therefore, more easily formed than the product formed when the other H is eliminated with Br. (Recall that when there is a choice, a hydrogen is removed from the β-carbon bonded to the fewest hydrogens.)

In the trans isomer, when Br is in an axial position, there is an axial hydrogen on only one adjacent carbon, and it is not the carbon that is bonded to the ethyl group. Therefore, a different product is formed. (Notice in this case there is no choice; there is only one hydrogen bonded by an axial bond to a β-carbon.)

Br, H, CH_2CH_3, H, H — *cis*-1-bromo-2-ethylcyclohexane

Br, H, H, H, CH_2CH_3 — *trans*-1-bromo-2-ethylcyclohexane

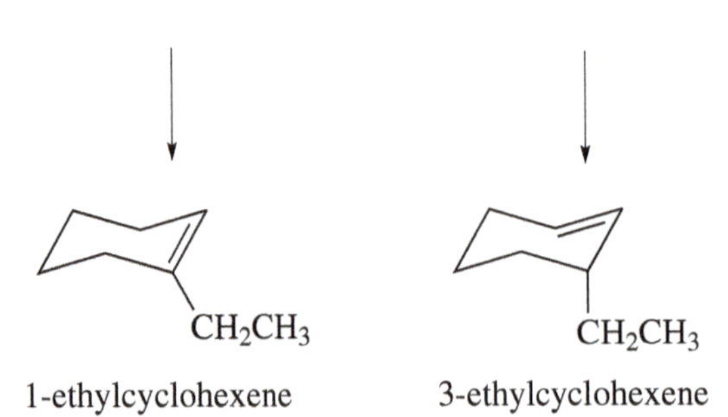

1-ethylcyclohexene 3-ethylcyclohexene

37. In order for a six-membered ring to undergo an E2 reaction, the substituents that are to be eliminated must both be in axial positions.

When bromine and an adjacent hydrogen are both in axial positions, the large *tert*-butyl substituent is in an equatorial position in the cis isomer and in an axial position in the trans isomer. The rate constant for the reaction is $k'K_{eq}$.

Because a large substituent is more stable in an equatorial position than in an axial position, elimination of the cis isomer occurs through its more stable chair conformer (K_{eq} is large; see page 427 of the text), whereas elimination of the trans isomer has to occur through its less stable chair conformer (K_{eq} is small). The cis isomer, therefore, reacts more rapidly in an E2 reaction.

cis-1-bromo-4-*tert*-butylcyclohexane *trans*-1-bromo-4-*tert*-butylcyclohexane

38. **a.** *trans*-1-Chloro-2-methylcyclohexane has two stereoisomers, and each forms a substitution product and an elimination product. Notice that the substitution products and the elimination products are enantiomers.

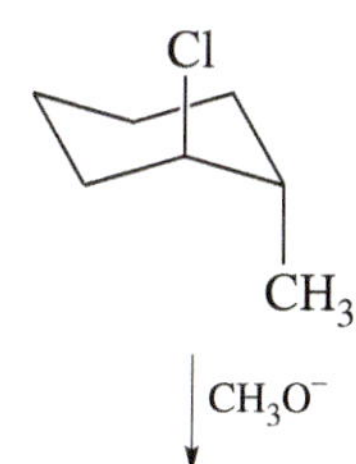

Elimination occurs through the less stable conformer.

Substitution occurs through back-side attack.

b. *cis*-l-Chloro-2-methylcyclohexane has two stereoisomers, and each forms a substitution product and an elimination product. Notice that the substitution products and the elimination products are enantiomers.

Elimination occurs through the more stable conformer.

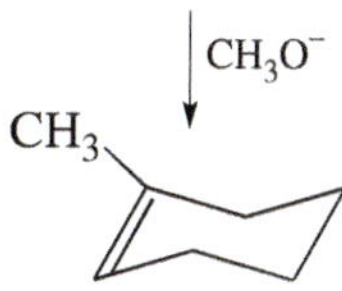

c. $\xrightarrow{CH_3O^-}$ only product

d. $\xrightarrow{CH_3OH}$ +

39. **a.** $CH_3CH_2CH_2Br$ This compound has less steric hindrance.

b. 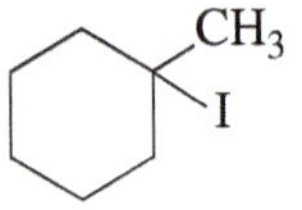 1-iodo-1-methylcyclohexane (structure: cyclohexane ring bearing CH_3 and I) I^- is a better leaving group (weaker base) than Br^-.

c. 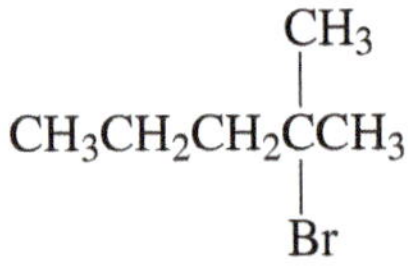 $CH_3CH_2CH_2C(CH_3)(Br)CH_3$ The tertiary alkyl halide because a secondary alkyl halide does not undergo S_N1 reactions.

40. The reaction of an alkyl halide with an acetylide ion is an S_N2 reaction. Methyl and primary alkyl halides work best because they have the least steric hindrance to back-side attack. In addition, primary alkyl halides form mainly the desired substitution product and methyl halides form only the desired substitution product.

41. Because CH_3S^- is a better nucleophile in the protic polar solvent and a weaker base than CH_3O^-, the ratio of substitution (where Y^- reacts as a nucleophile) to elimination (where Y^- reacts as a base) increases when the nucleophile is changed from CH_3O^- to CH_3S^-.

42. In order to undergo an E2 reaction, the substituents to be eliminated (H and Br) must both be in axial positions. Drawing the compound in the chair conformation shows that when Br is in an axial position, neither of the adjacent β-carbons has a hydrogen in an axial position, so an elimination reaction cannot take place.

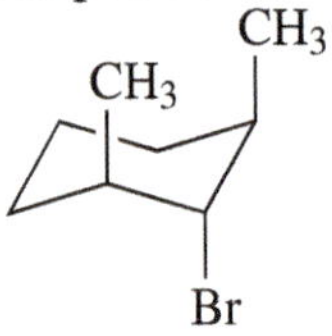

CH3 CH3 Br

43.

$CH_3C(CH_3)_2CH_2Br$

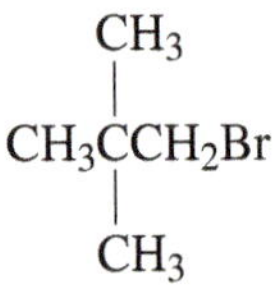

1-bromo-2,2-dimethylpropane

a. The bulky *tert*-butyl substituent blocks the back side of the carbon bonded to the bromine to nucleophilic attack, making an S_N2 reaction difficult. An S_N1 reaction cannot occur because it requires formation of an unstable primary carbocation.

b. It cannot undergo an E2 reaction, because the β-carbon is not bonded to a hydrogen.
It cannot undergo an E1 reaction, because that requires the formation of a primary carbocation.

44. **a.** *trans*-4-Bromo-2-hexene (the compound on the right) is more reactive, because the carbocation that is formed is stabilized by electron delocalization. (It is a secondary allylic cation.) The other alkyl halide is a secondary alkyl halide and does not undergo an S_N1 reaction.

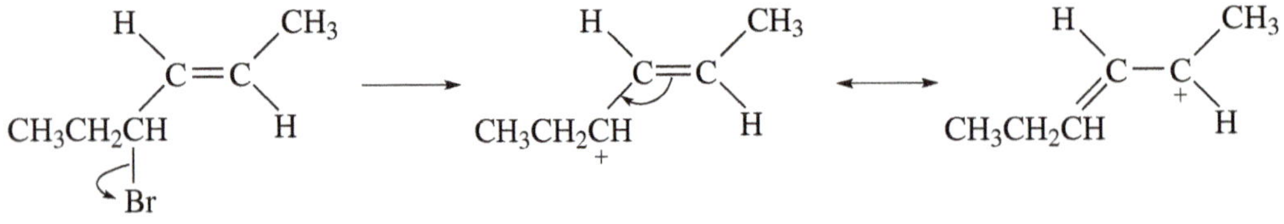

$CH_3CH_2CH(Br)CH{=}CHCH_3 \longrightarrow CH_3CH_2\overset{+}{C}HCH{=}CHCH_3 \longleftrightarrow CH_3CH_2CH{=}CH\overset{+}{C}HCH_3$

b.

$$\underset{\underset{\displaystyle OCH_2CH_3}{|}}{CH_3CH_2CH}\!-\!CH\!=\!CH\!-\!CH_3 \text{ (H and } CH_3 \text{ trans)} \qquad CH_3CH_2CH{=}CH\underset{\underset{\displaystyle OCH_2CH_3}{|}}{C}HCH_3$$

45. Conjugated double bonds are more stable and, therefore, are easier to form than isolated double bonds.

a. (1,3-cyclohexadiene) **b.** C_6H_5—CH=CHCH$_3$ **c.** $CH_2{=}CHCH{=}\overset{\overset{\displaystyle CH_3}{|}}{C}CH_3$

46. **a.** (1,3-cyclohexadiene) **c.** $CH_2{=}CHCH{=}\overset{\overset{\displaystyle CH_3}{|}}{C}CH_3 \;+\; CH_3CH{=}CH\overset{\overset{\displaystyle CH_3}{|}}{C}{=}CH_2$

b. The secondary alkyl halide does not undergo an El reaction.

47. **a.**

$$CH_3{-}\underset{\underset{\displaystyle Br}{|}}{\overset{\overset{\displaystyle CH_3}{|}}{C}}{-}CH{=}CH_2 + CH_3O^- \longrightarrow CH_2{=}\overset{\overset{\displaystyle CH_3}{|}}{C}{-}CH{=}CH_2 + CH_3OH + Br^-$$

b.

$$CH_3{-}\underset{\underset{\displaystyle Br}{|}}{\overset{\overset{\displaystyle CH_3}{|}}{C}}{-}CH{=}CH_2 \longrightarrow CH_3\overset{\overset{\displaystyle CH_3}{|}}{\underset{+}{C}}{-}CH{=}CH_2 \longleftrightarrow CH_3\overset{\overset{\displaystyle CH_3}{|}}{C}{=}CH{-}\underset{+}{C}H_2 \quad + \quad Br^-$$

$$\downarrow CH_3OH$$

$$CH_2{=}\overset{\overset{\displaystyle CH_3}{|}}{C}{-}CH{=}CH_2 \quad + \quad CH_3OH \quad + \quad HBr$$

Only one elimination product is formed because elimination from the other resonance contributor would form a cumulated diene, which is much less stable and, therefore, much harder to form than a conjugated diene.

48. **a.** (benzyl bromide, $C_6H_5CH_2Br$) An aryl halide cannot undergo an S_N1 reaction because an aryl cation is too unstable to form.

b. (4-bromo-2-pentene, skeletal structure with Br) A vinyl halide cannot undergo an S_N1 reaction because an vinyl cation is too unstable to form.

49. **a.** (benzyl bromide, $C_6H_5CH_2Br$) An aryl halide cannot undergo an S_N2 reaction because it cannot undergo back-side attack.

b. (4-bromo-2-pentene, skeletal structure with Br) A vinyl halide cannot undergo an S_N2 reaction because it cannot undergo back-side attack.

50. Because a cumulated diene is less stable than an alkyne, the transition state for its formation is less stable than that for the formation of the alkyne, so the cumulated diene is harder to make.

51. Solved in the text.

52. In the first elimination reaction, a hydrogen will be removed from the β-carbon bonded to the fewest hydrogens as expected. In the second elimination reaction, a hydrogen will be removed from the β-carbon that results in the formation of a conjugated double bond.

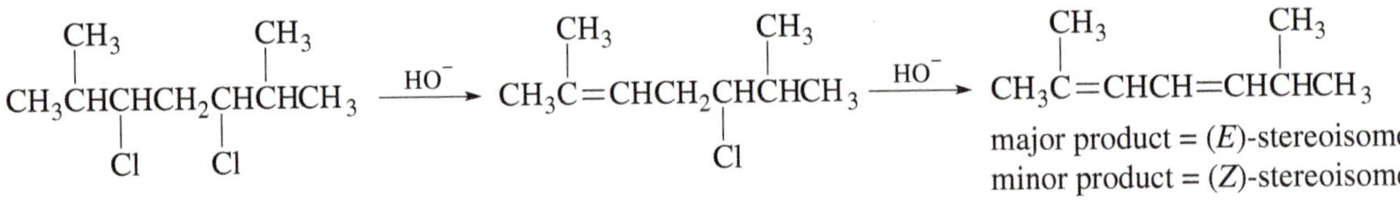

$$CH_3CH(CH_3)CH(Cl)CH_2CH(Cl)CH(CH_3)CH_3 \xrightarrow{HO^-} CH_3C(CH_3){=}CHCH_2CH(Cl)CH(CH_3)CH_3 \xrightarrow{HO^-} CH_3C(CH_3){=}CHCH{=}CHCH(CH_3)CH_3$$

major product = (*E*)-stereoisomer
minor product = (*Z*)-stereoisomer

53. **a.** [cyclohexene with OCH_3 (hashed wedge)] + [1,3-cyclohexadiene]

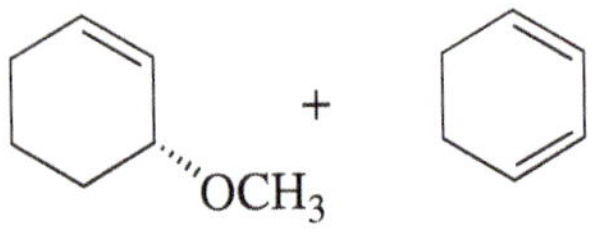

b. [cyclohexene with OCH_3 (solid wedge)] + [cyclohexene with OCH_3 (hashed wedge)] + [1,3-cyclohexadiene]

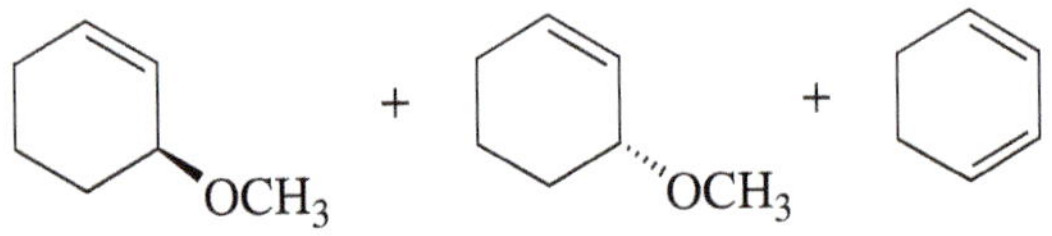

54. Because both reactants in the rate-limiting step are neutral, the reaction will be faster if the polarity of the solvent is increased.

55. **a.** Increasing the polarity decreases the rate of the reaction because the concentration of charge on the reactants is greater (the reactants are charged) than the concentration of charge on the transition state.

b. Increasing the polarity decreases the rate of the reaction because the concentration of charge on the reactants is greater (the reactants are charged) than the concentration of charge on the transition state.

c. Increasing the polarity increases the rate of the reaction because the concentration of charge on the reactants is less (the reactants are not charged) than the concentration of charge on the transition state.

56. **a.** $CH_3Br + HO^- \longrightarrow CH_3OH + Br^-$

HO^- is a better nucleophile than H_2O.

b. $CH_3I + HO^- \longrightarrow CH_3OH + I^-$

I^- is a better leaving group than Cl^-.

c. $CH_3Br + NH_3 \longrightarrow CH_3\overset{+}{N}H_3 + Br^-$

NH_3 is a better nucleophile than H_2O.

d. $CH_3Br + HO^- \xrightarrow{DMSO} CH_3OH + Br^-$

Unlike ethyl alcohol, DMSO does not stabilize the nucleophile (and, therefore, decrease the rate of the reaction) by hydrogen bonding.

e. $CH_3Br + NH_3 \xrightarrow{EtOH} CH_3\overset{+}{N}H_3 + Br^-$

A more polar solvent stabilizes the transition state more than it stabilizes the reactants. (EtOH is ethanol.)

57. Solved in the text.

58. Acetate ion is a better nucleophile in dimethyl sulfoxide because dimethyl sulfoxide does not stabilize the negatively charged nucleophile by ion–dipole interactions, whereas methanol does stabilize it by ion–dipole interactions.

59. The only way the product with retention of configuration can be obtained is via an S_N1 reaction. An S_N1 reaction will be faster in the more polar solvent. So the reaction should be carried out in the 50% water/50% ethanol mixture.

60. Because a strong base is used in the Williamson ether synthesis, the reaction is an S_N2 reaction, so a competing E2 reaction can also occur. The elimination product is a minor product because substitution is favored when the alkyl halide is primary.

a. $CH_3CH_2CH_2Br + CH_3CH_2CH_2CH_2O^- \longrightarrow CH_3CH_2CH_2CH_2OCH_2CH_2CH_3 + CH_3CH{=}CH_2$

butyl propyl ether; propene

b. $CH_3CH_2CH_2CH_2Br + CH_3CH_2CH_2O^- \longrightarrow CH_3CH_2CH_2CH_2OCH_2CH_2CH_3 + CH_3CH_2CH{=}CH_2$

butyl propyl ether; 1-butene

61. To maximize the amount of ether formed in the S_N2 reaction, make sure the less hindered group is provided by the alkyl halide. In order to convert the alcohol ($pK_a \sim 15$) to an alkoxide ion in a reaction that favors products (in parts **a**, **b**, and **c**), a strong base (H^-) is needed.

a. $CH_3CH_2CH(CH_3)OH \xrightarrow{NaH} CH_3CH_2CH(CH_3)O^- \xrightarrow{CH_3CH_2CH_2Br} CH_3CH_2CH(CH_3)OCH_2CH_2CH_3$

b. $CH_3CH_2CH_2CH(CH_3)CH_2OH \xrightarrow{NaH} CH_3CH_2CH_2CH(CH_3)CH_2O^- \xrightarrow{CH_3CH_2Br} CH_3CH_2CH_2CH(CH_3)CH_2OCH_2CH_3$

c. cyclohexyl–OH $\xrightarrow{NaH}$ cyclohexyl–O^- $\xrightarrow{CH_3Br}$ cyclohexyl–OCH_3

In part **d**, HO^- can be used to convert phenol ($pK_a \sim 15$) to phenoxide ion in a reaction that favors products, because phenol is a considerably stronger acid than an alcohol.

d. phenyl–OH $\xrightarrow{NaOH}$ phenyl–O^- $\xrightarrow{\text{phenyl–}CH_2Br}$ phenyl–CH_2O–phenyl

Remember that aryl halides cannot undergo S_N2 reactions.

62. $(CH_3)_3C{-}Br \xrightarrow[H_2O]{CH_3CH_2OH} (CH_3)_3C{-}OH + (CH_3)_3C{-}OCH_2CH_3 + CH_3C(CH_3){=}CH_2$

63. $CH_3CH_2C(CH_3)(Br)CH_2CH_3 \xrightarrow{HO^-}$ $(H_3C)(H)C{=}C(CH_3)(CH_2CH_3)$ (major) + $(H_3C)(H)C{=}C(CH_2CH_3)(CH_3)$ (minor)

A tertiary alkyl halide cannot undergo a substitution reaction under S_N2/E2 conditions.

$CH_3CH_2C(CH_3)(Br)CH_2CH_3 \xrightarrow{H_2O}$ $CH_3CH_2C(CH_3)(OH)CH_2CH_3$ + $(H_3C)(H)C{=}C(CH_3)(CH_2CH_3)$ + $(H_3C)(H)C{=}C(CH_2CH_3)(CH_3)$ (minor)

64. a. HO~~~~~Br because it forms a six-membered ring, whereas the other compound would form a seven-membered ring. A seven-membered ring is more strained than a six-membered ring, so the six-membered ring is formed more easily. (See Table 3.8 on page 124 of the text.)

b. HO~~~~Br because it forms a five-membered ring, whereas the other compound would form a four-membered ring. A four-membered ring is more strained than a five-membered ring, so the five-membered ring is formed more easily.

c. HO~~~~~~Br because it forms a seven-membered ring, whereas the other compound would form an eight-membered ring. An eight-membered ring is more strained than a seven-membered ring, so the seven-membered ring is formed more easily; also, the Br and OH in the compound that leads to the eight-membered ring are less likely to be in the proper position relative to each other for reaction because there are more bonds around which rotation to an unfavorable conformation can occur.

65. a. When hydride ion removes a proton from the OH group, the alkoxide ion cannot react in an intramolecular reaction with the alkyl chloride to form an epoxide, because it cannot reach the back side of the carbon attached to the chlorine. Therefore, the major product will result from an intermolecular reaction.

Cl, OH (cyclopentane) $\xrightarrow{NaH}$ Cl, O^- (cyclopentane) + H_2O ⟶ Cl, O–(cyclopentyl), OH

b. Hydride ion removes a proton from the OH group more rapidly than it attacks the alkyl chloride. Once the alkoxide ion is formed, it attacks the back side of the alkyl chloride, forming an epoxide. (Removing a proton from an oxygen is always a fast reaction.)

Cl, OH (cyclopentane) $\xrightarrow{NaH}$ Cl, :Ö:$^-$ (cyclopentane) ⟶ O (epoxide) + Cl^-

c. Hydride ion removes a proton from the OH group and the alkoxide ion attacks the back side of the carbon attached to the bromine, forming a six-membered ring ether.

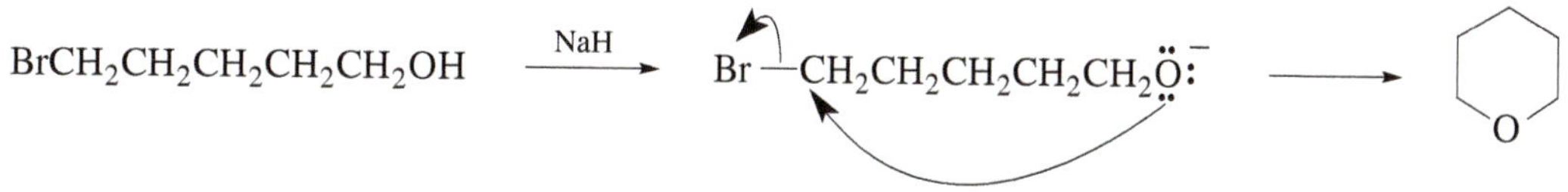

$$BrCH_2CH_2CH_2CH_2CH_2OH \xrightarrow{NaH} Br{-}CH_2CH_2CH_2CH_2CH_2\ddot{O}:^- \longrightarrow$$

O

d. Hydride ion removes a proton from the OH group and the alkoxide ion attacks the back side of the carbon attached to the chlorine, forming an epoxide.

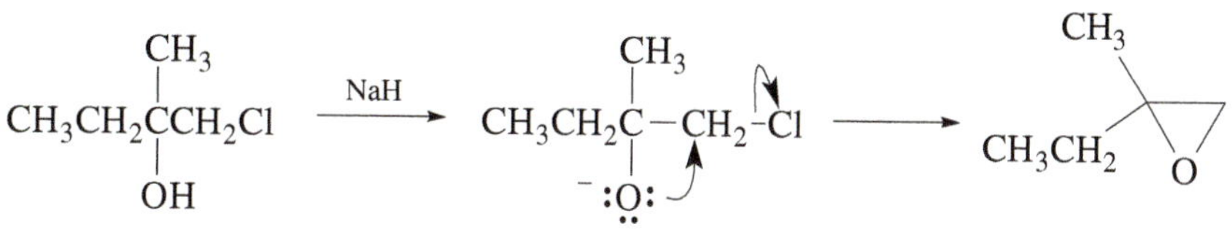

$$CH_3CH_2C(CH_3)(OH)CH_2Cl \xrightarrow{NaH} CH_3CH_2C(CH_3)(:\ddot{O}:^-){-}CH_2{-}Cl \longrightarrow$$

CH_3 CH_3CH_2 O

e. After the halohydrin is formed, hydride ion removes a proton from the OH group and the alkoxide ion forms an epoxide.

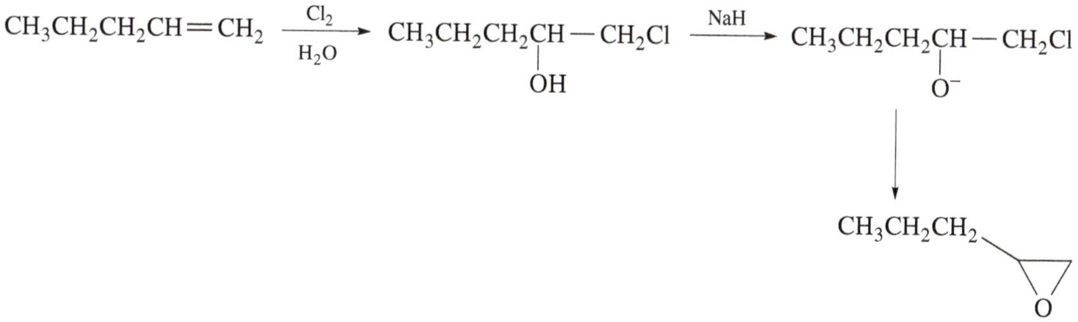

$$CH_3CH_2CH_2CH{=}CH_2 \xrightarrow[H_2O]{Cl_2} CH_3CH_2CH_2CH(OH){-}CH_2Cl \xrightarrow{NaH} CH_3CH_2CH_2CH(O^-){-}CH_2Cl \longrightarrow$$

$CH_3CH_2CH_2$ O

66. In parts **a** and **b**, a bulky base is used to encourage elimination over substitution.

a. Br $\xrightarrow{tert\text{-}BuO^-}$ $\xrightarrow{RCOOH}$ O

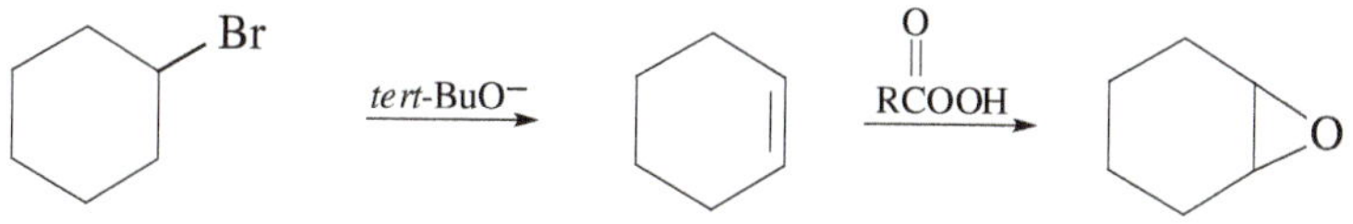

or

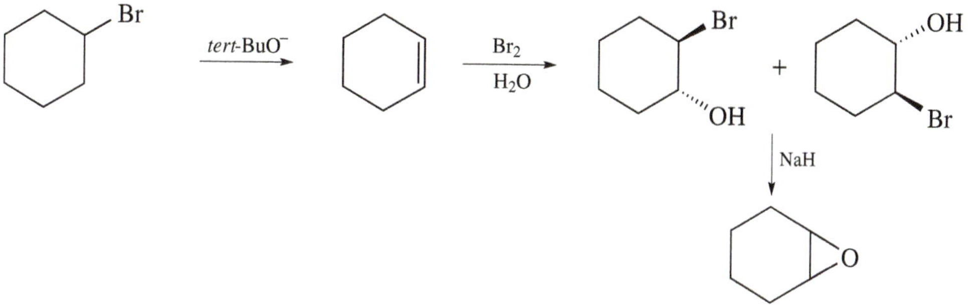

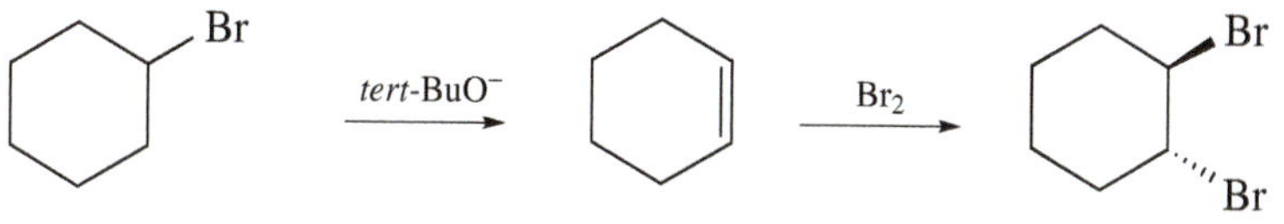

Br $\xrightarrow{tert\text{-}BuO^-}$ $\xrightarrow[H_2O]{Br_2}$ Br OH + OH Br $\xrightarrow{NaH}$ O

b. Br $\xrightarrow{tert\text{-}BuO^-}$ $\xrightarrow{Br_2}$ Br Br

c. cyclohexyl–$CH{=}CH_2$ $\xrightarrow[CH_2Cl_2]{Br_2}$ cyclohexyl–$\underset{Br}{CH}CH_2Br$ $\xrightarrow[\text{excess}]{^-NH_2}$ cyclohexyl–$C{\equiv}CH$ $\xrightarrow[\text{2. HO}^-,\ H_2O_2,\ H_2O]{\text{1. } R_2BH/THF}$ cyclohexyl–$CH_2\overset{O}{\overset{\|}{C}}H$

d. cyclohexyl–$CH{=}CH_2$ $\xrightarrow{Br_2}$ cyclohexyl–$\underset{Br}{CH}CH_2Br$ $\xrightarrow{2\ NaNH_2}$ cyclohexyl–$C{\equiv}CH$ $\xrightarrow[\text{2. } CH_3CH_2Br]{\text{1. } NaNH_2}$ cyclohexyl–$C{\equiv}CCH_2CH_3$ $\xrightarrow[Pd/C]{H_2}$ cyclohexyl–$CH_2CH_2CH_2CH_3$

67. a. $CH_3CH_2CH_2OH$ c. $CH_3CH_2CH_2SCH_3$ e. $CH_3CH_2CH_2OCH_3$

b. $CH_3CH_2CH_2NH_2$ d. $CH_3CH_2CH_2SH$ f. $CH_3CH_2CH_2\overset{+}{N}H_2CH_3$

(Notice that the product in part **c** is not protonated because its pK_a is ~−7; the product in **f** is protonated because its pK_a is ~11. In part **f**, $CH_3\overset{+}{N}H(CH_2CH_2CH_3)_2$ and $CH_3N(CH_2CH_2CH_3)_3$ can also be formed, depending on the concentration of 1-bromopropane; see Problem 16 on page 404 of the text.)

68. If the atoms are in the same horizontal row of the periodic table, the stronger base is the better nucleophile. If the atoms are in the same column, the larger atom is the better nucleophile in the protic polar solvent because the solvent forms stronger hydrogen bonds with the smaller atom.

a. HO^- c. H_2S e. I^-

b. NH_3 d. HS^- f. Br^-

69. The weaker base is the better leaving group.

a. H_2O c. H_2S e. I^-

b. H_2O d. HS^- f. Br^-

70. a. HO^- c. CH_3NH_2 e. $CH_3CH_2S^-$ g. $^-C{\equiv}N$

b. CH_3O^- d. HS^- f. $CH_3\overset{O}{\overset{\|}{C}}O^-$ h. $CH_3CH_2C{\equiv}C^-$

In part **c**, a tertiary amine and a quaternary ammonium ion can also form unless a large excess of CH_3NH_2 is used. (See Problem 16 on page 404 of the text.)

71.
- **a.** The rate will be increased nine-fold.
- **b.** The reaction will be slower because of the more polar solvent.
- **c.** The reaction will be slower because the leaving group will be poorer.
- **d.** The reaction will be slower because there will be more steric hindrance.

72.
- **a.** The reaction will be slower because the leaving group is poorer.
- **b.** The reaction will be slower because it will be an S_N2 reaction with a poor nucleophile, and the leaving group is poorer.

73.

a. cyclohexene $\xrightarrow{HBr}$ bromocyclohexane (Br) $\xrightarrow{CH_3O^-}$ methoxycyclohexane (OCH_3)

b. cyclohexene $\xrightarrow{HBr}$ bromocyclohexane (Br) $\xrightarrow{CH_3NH_2}$ $\overset{+}{N}H_2CH_3$ $\xrightarrow{HO^-}$ $NHCH_3$

(A large excess of methylamine has to be used in the second step to minimize the formation of a tertiary amine and a quarternary ammonium ion.)

c. Half of the cyclohexene is converted to bromocyclohexane, and half is converted to an alkoxide ion. The ether is formed from the reaction of bromocyclohexane with the alkoxide ion.

cyclohexene $\xrightarrow{HBr}$ bromocyclohexane (Br)

cyclohexene $\xrightarrow{H_2SO_4,\ H_2O}$ cyclohexanol (OH) $\xrightarrow{NaH}$ cyclohexoxide (O^-)

bromocyclohexane + cyclohexoxide $\longrightarrow$ dicyclohexyl ether (O)

74.

a. $CH_3CH_2S^- > CH_3CH_2O^- > CH_3\overset{O}{\overset{\|}{C}}O^-$

b. cyclohexyl–O^- > phenyl–O^-

c. $NH_3 > H_2O$

d. $I^- > Br^- > Cl^-$

75. The pK_a will increase (it will be a weaker acid) because of a decreased tendency to form a charged species in a less polar solvent. (See Problem 57 on page 442.)

76. **a.**

CH_3
CH_3CCH_3
Br
→
CH_3
$CH_3\overset{+}{C}CH_3$
Br^- Cl^-
$\xrightarrow[H_2O]{CH_3CH_2OH}$
CH_3
CH_3CCH_3
OCH_2CH_3
+
CH_3
CH_3CCH_3
OH

CH_3
CH_3CCH_3
Cl
→

b. The products are obtained as a result of the nucleophiles reacting with the carbocation. 2-Bromo-2-methylpropane and 2-chloro-2-methylpropane form the same carbocation, so both alkyl halides form the same products.

77. **a.** The S_N2 reaction takes place with inversion of configuration.

$CH_2CH_2CH_3$
$Br-C$ (H, CH_3)
$\xrightarrow{CH_3O^-}$
$CH_2CH_2CH_3$
(H, CH_3) $C-OCH_3$

(*R*)-2-bromopentane (*S*)-2-methoxypentane

or

CH_3
$Br-C-H$
$CH_2CH_2CH_3$
$\xrightarrow{CH_3O^-}$
CH_3
$H-C-OCH_3$
$CH_2CH_2CH_3$

b. The S_N1 reaction takes place with inversion of configuration.

CH_2CH_3
$CH_3CH_2CH_2CH_2-C$ (CH_3, Br)
$\xrightarrow{CH_3OH}$
CH_2CH_3
$CH_3CH_2CH_2CH_2-C$ (CH_3, OCH_3)
+
CH_2CH_3
(CH_3, CH_3O) $C-CH_2CH_2CH_2CH_3$

(*R*)-3-bromo-3-methylheptane (*R*)-3-methoxy-3-methylheptane (*S*)-3-methoxy-3-methylheptane

or

CH_2CH_3
$Br-C-CH_3$
$CH_2CH_2CH_2CH_3$
$\xrightarrow{CH_3OH}$
CH_2CH_3
$CH_3O-C-CH_3$
$CH_2CH_2CH_2CH_3$
+
CH_2CH_3
$CH_3-C-OCH_3$
$CH_2CH_2CH_2CH_3$

c. $C_6H_5-CH_2Cl \xrightarrow{CH_3CH_2OH} C_6H_5-CH_2OCH_2CH_3$

d. $CH_2{=}CHCH_2Cl \xrightarrow{CH_3OH} CH_2{=}CHCH_2OCH_3$

e. $CH_3CH{=}CHCH_2Br \xrightarrow{CH_3O^-} CH_3CH{=}CHCH_2OCH_3$

f. $CH_3CH{=}CHCH_2Br \xrightarrow{CH_3OH} CH_3CH{=}CHCH_2OCH_3$ + $CH_3CHCH{=}CH_2$
OCH_3

78. a. Br → + → CH_3CH_2OH → OCH_2CH_3 +

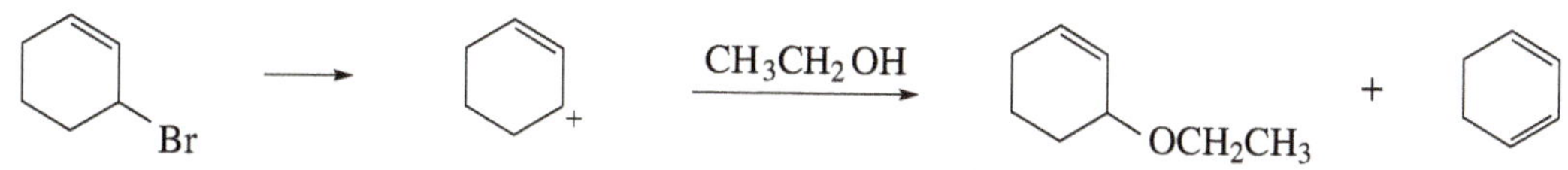

Both the *R* and *S* stereoisomers are obtained.

b. CH_2Br → $\overset{+}{C}H_2$ + Br^- ⟷ CH_2 +

CH_3CH_2OH CH_3CH_2OH

$CH_2OCH_2CH_3$ + CH_2 OCH_2CH_3 + CH_2

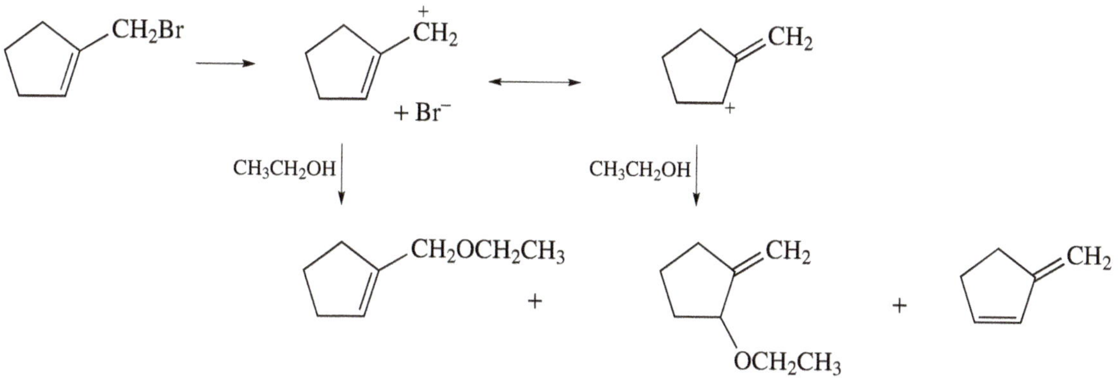

Both the *R* and *S* stereoisomers are obtained.

c. Br → + Br^- ⟷ +

CH_3CH_2OH CH_3CH_2OH

+ OCH_2CH_3 OCH_2CH_3 +

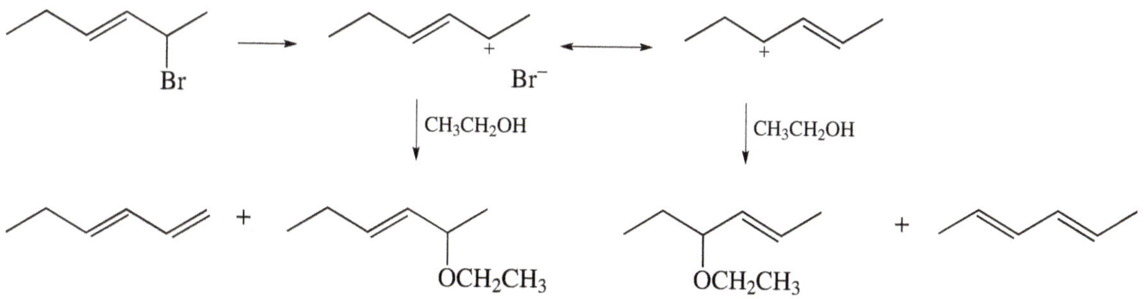

This compound has an asymmetric center, so both the *R* and *S* stereoisomers are obtained.

This compound has an asymmetric center, so both the *R* and *S* stereoisomers are obtained.

79. Methoxide ion is a better nucleophile in DMSO, because DMSO cannot stabilize the ion by ion–dipole interactions.

80. a. Cl $\xrightarrow{CH_3S^-}$ S + Cl^-

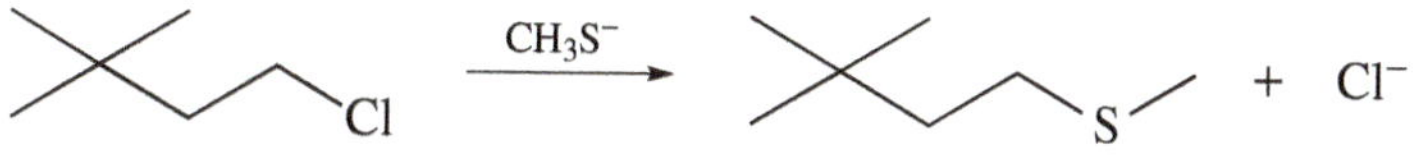

The nucleophile is less sterically hindered.

b. O Cl $\xrightarrow[H_2O]{HO^-}$ O OH + Cl^-

The electron-withdrawing oxygen increases the electrophilicity of the carbon that the nucleophile attacks.

c. Cl $\xrightarrow{H_2O}$ OH + H_3O^+ + Cl^-

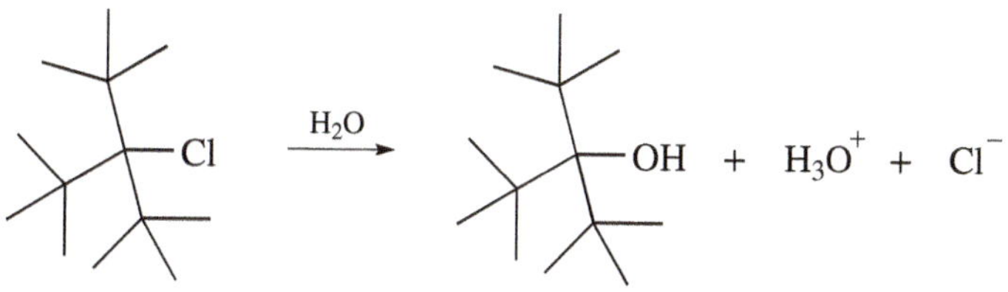

Steric strain is decreased when the alkyl halide dissociates to form the carbocation because the hybridization of the carbon changes from sp^3 to sp^2, allowing the bond angle between the bulky groups to increase from 109.5° to 120°, which allows more room for the bulky groups. Relieving steric strain causes the carbocation to be formed more rapidly.

d. $(CH_3)_3CBr \xrightarrow{H_2O} (CH_3)_3COH + HBr$

Because the reactants are neutral, the reaction is faster in the more polar solvent.

81. 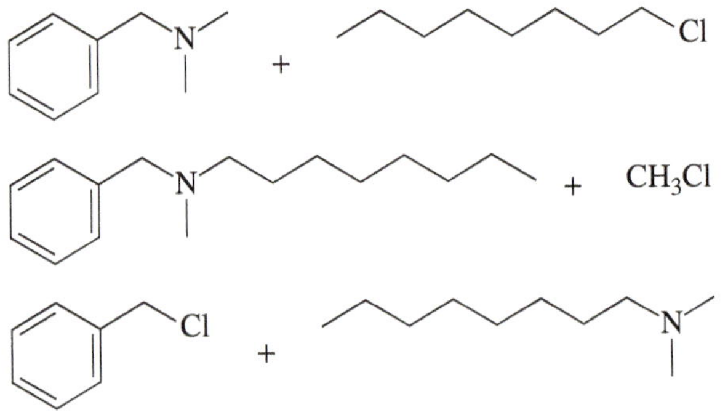 N,N-dimethylbenzylamine + 1-chlorooctane

N-benzyl-N-methyloctylamine + CH_3Cl

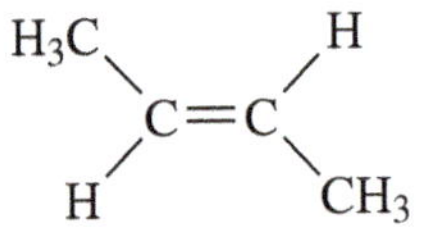

benzyl chloride + N,N-dimethyloctylamine

82. **a.** CH_3O^- **b.** 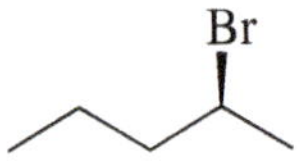 2-bromobutane (Br) **c.** 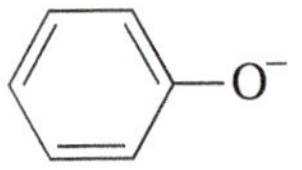 phenoxide ($-O^-$) **d.** 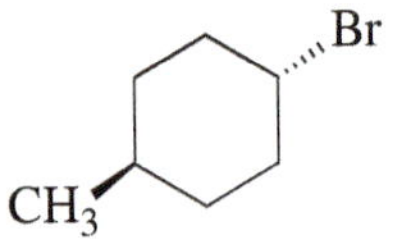 bromomethylcyclohexane with Br and CH_3

83. **a.** 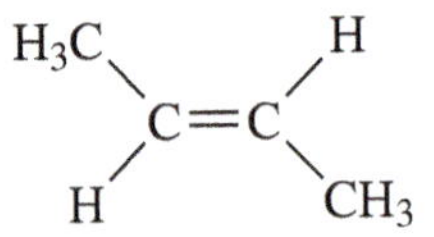 H_3C, H / C=C / H, CH_3 **c.** H_3C, H / C=C / H, CH_3 **e.** 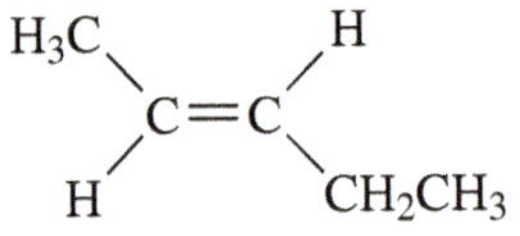 H_3C, H / C=C / H, CH_2CH_3

b. 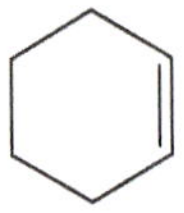 cyclohexene **d.** 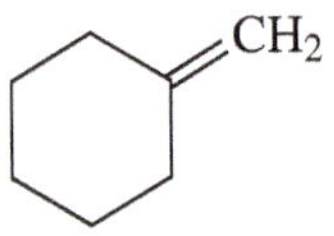 methylenecyclohexane (CH_2) **f.** 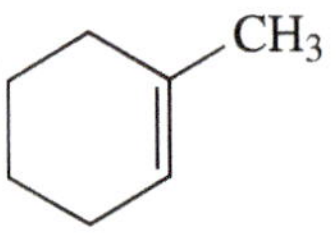 1-methylcyclohexene (CH_3)

84. **f.** This is the only one that can undergo an E1 reaction. 1-chloro-1-methylcyclohexane (CH_3, Cl) → 1-methylcyclohexene (CH_3)

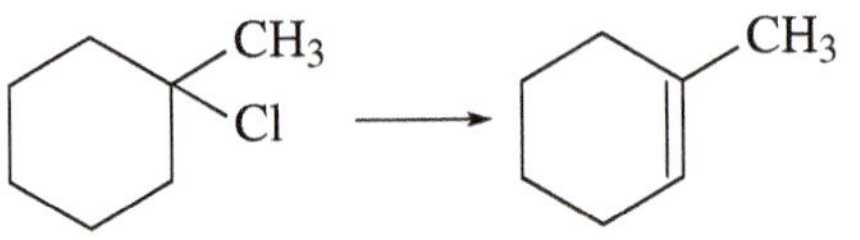

85. **a.** **1.** An E1 reaction is not affected by the strength of the base but, if the reaction can take place by both an E1 and E2 reaction, a weak base will favor the E1 reaction by disfavoring an E2 reaction.

2. An E1 reaction is not affected by the concentration of the base but, if the reaction can take place by both an E1 and E2 reaction, a low concentration of a base will favor the E1 reaction by disfavoring an E2 reaction.

3. If the reactant is charged, an E1 reaction will be favored by the least polar solvent that will dissolve the reactant (generally an aprotic polar solvent). If the reactant is not charged, an E1 reaction will be favored by a protic polar solvent.

b. **1.** A strong base favors an E2 reaction.

2. A high concentration of a base favors an E2 reaction.

3. If one of the reactants is charged, an E2 reaction will be favored by the least polar solvent that will dissolve the reactant (generally an aprotic polar solvent). If neither of the reactants is charged, an E2 reaction will be favored by a protic polar solvent.

86. a. $CH_3CH_2CH_2\bar{C}H_2$

b. $CH_3\overset{+}{C}HCH_2CH_3$

c. $CH_3\bar{C}HCH{=}CH_2$
stabilized by electron delocalization (resonance)

d. $CH_3CH_2C(CH_3){=}CH_2$

e. $CH_3\overset{+}{C}HCH{=}CH_2$
stabilized by electron delocalization (resonance)

f. $CH_3CH(CH_3)\bar{C}HCH_3$

g. $CH_3\overset{+}{C}(CH_3)CH_2CH_3$

87. He obtained only the elimination product because a tertiary alkyl halide does not undergo an S_N2 reaction. Because of steric hindrance, only an E2 reaction occurs.

$$CH_3C(CH_3)(Cl)CH_3 + CH_3CH_2O^- \longrightarrow CH_3C(CH_3){=}CH_2$$

predominant product

Rather than a tertiary alkyl halide and a primary alkoxide ion, he should have used a primary alkyl halide and a tertiary alkoxide ion. They will react in an S_N2 reaction to form the desired ether.

$$CH_3CH_2Cl + CH_3C(CH_3)_2O^- \longrightarrow CH_3C(CH_3)_2OCH_2CH_3 + Cl^-$$

Although the Williamson ether synthesis (an S_N2 reaction) is the preferred way to synthesize an ether because it gives higher yields, the ether also could be synthesized using an S_N1 reaction.

$$CH_3C(CH_3)_2Br + CH_3CH_2OH \longrightarrow CH_3C(CH_3)_2OCH_2CH_3 + HBr$$

88. a. $$(CH_3)_3CBr \xrightarrow[H_2O]{HO^-} (CH_3)_2C{=}CH_2 + Br^-$$

because Br^- is a better leaving group than Cl^-

b. [cyclohexane chair: H, H, CH_3, H, CH_3, Br] This compound is the only one that can undergo an E2 reaction because the other compound does not have an axial hydrogen bonded to a β-carbon.

[cyclohexane chair: CH_3, CH_3, H, H, H, Br] This compound will not be able to undergo an E2 reaction because it does not have an adjacent H and Br that are both attached to axial bonds.

89. The very minor products that are obtained from "anti-Zaitsev" elimination (that is, the less substituted alkenes) are not shown.

a.

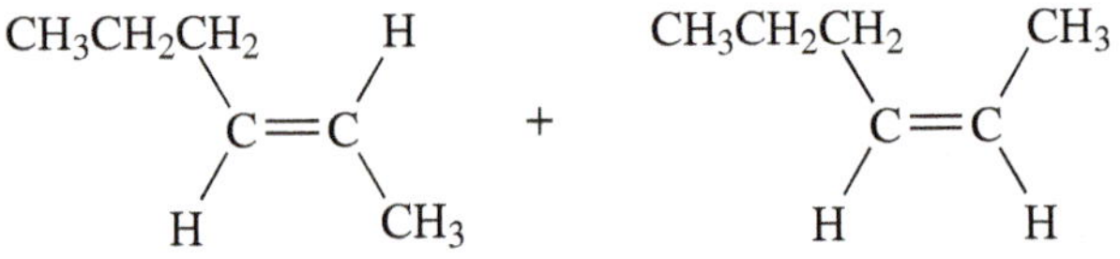

$CH_3CH_2CH_2$ H
C=C
H CH_3

major

\+

$CH_3CH_2CH_2$ CH_3
C=C
H H

minor

b. The reactant has two β-carbons that are attached to two hydrogens, so the reaction is not regioselective. Therefore, two constitutional isomers can be formed. Each constitutional isomer has *E* and *Z* stereoisomers.

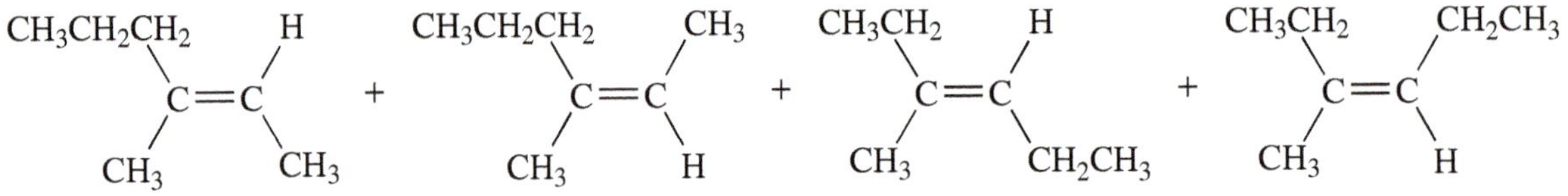

$CH_3CH_2CH_2$ H
C=C
CH_3 CH_3

major

\+

$CH_3CH_2CH_2$ CH_3
C=C
CH_3 H

minor

\+

CH_3CH_2 H
C=C
CH_3 CH_2CH_3

major

\+

CH_3CH_2 CH_2CH_3
C=C
CH_3 H

minor

c. *trans*-1-Chloro-2-methylcyclohexane has two stereoisomers.
A hydrogen cannot be removed from the β-carbon bonded to the fewest hydrogens because that hydrogen is not attached to an axial bond.

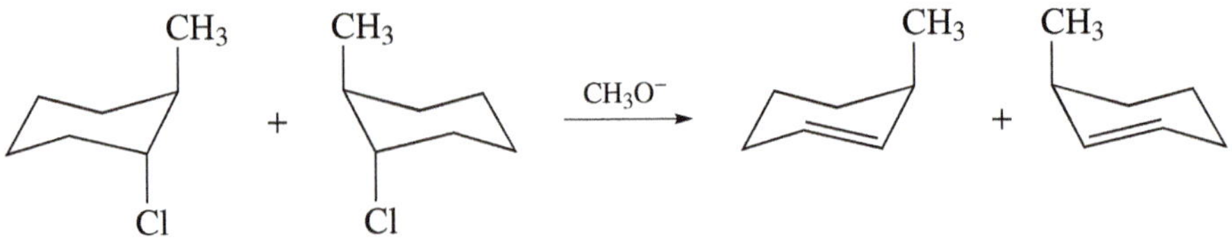

CH_3 / Cl + CH_3 / Cl $\xrightarrow{CH_3O^-}$ CH_3 + CH_3

d. *trans*-1-Chloro-3-methylcyclohexane has two stereoisomers, and each can form two elimination products.

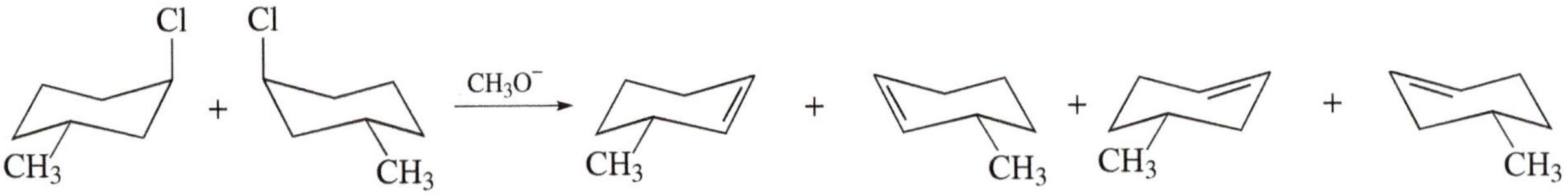

Cl / CH_3 + Cl / CH_3 $\xrightarrow{CH_3O^-}$ CH_3 + CH_3 + CH_3 + CH_3

e.

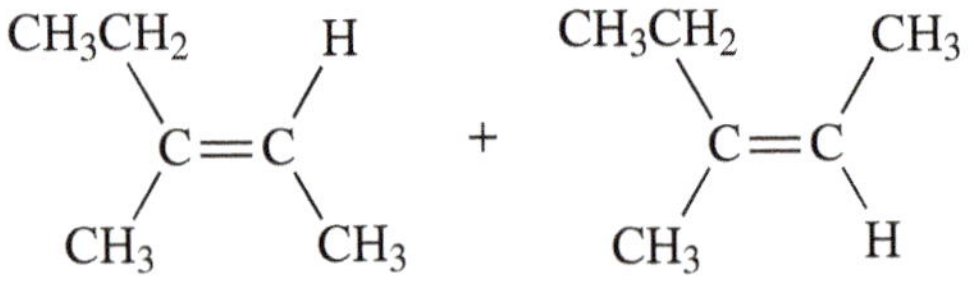

CH_3CH_2 H
C=C
CH_3 CH_3

major

\+

CH_3CH_2 CH_3
C=C
CH_3 H

minor

f.

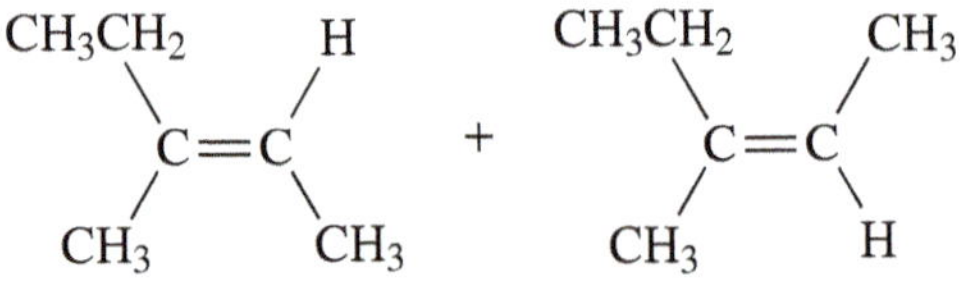

CH_3CH_2 H
C=C
CH_3 CH_3

major

\+

CH_3CH_2 CH_3
C=C
CH_3 H

minor

90. **a.** 3-Bromocyclohexene forms 1,3-cyclohexadiene; bromocyclohexane forms cyclohexene. 3-Bromocyclohexene reacts faster in an E2 reaction, because a conjugated double bond is more stable than an isolated double bond. So the transition state leading to formation of the conjugated double bond is more stable and, therefore, the conjugated double bond is easier to form.

b. 3-Bromocyclohexene, because it forms a relatively stable secondary allylic cation. The other compound is a secondary alkyl halide, so it does not undergo an E1 reaction.

91. That fact that the change from hydrogen to deuterium affects the rate of the reaction tells us that the C–H (or C–D) bond is broken in the rate-limiting step. This is consistent with the mechanism for an E2 reaction but not for the mechanism for an E1 reaction.

92. Alkyl chlorides and alkyl iodides could also be used. Do not use alkyl fluorides because they have the poorest leaving groups and cannot undergo S_N2 reactions.

a. $CH_3CH(O^-)CH_2CH_3 \xrightarrow{CH_3Br} CH_3CH(OCH_3)CH_2CH_3$

b. $CH_3CH_2CH_2CH_2Br \xrightarrow{CH_3O^-} CH_3CH_2CH_2CH_2OCH_3$

c. $CH_3CH_2CH_2CH_2Br \xrightarrow{CH_3NH_2} CH_3CH_2CH_2CH_2\overset{+}{N}H_2CH_3 \xrightarrow{HO^-} CH_3CH_2CH_2CH_2NHCH_3$

93. **a.** ethoxide ion, because elimination is favored by bulky bases and *tert*-butoxide ion is bulkier than ethoxide ion

b. ^-SCN because elimination is favored by strong bases and ^-OCN is a stronger base than ^-SCN

c. Br^- because elimination is favored by strong bases and Cl^- is a stronger base than Br^-

d. CH_3S^- because elimination is favored strong bases and CH_3O^- is a stronger base than CH_3S^-

94. The first compound listed below is the most reactive compound because it has two axial hydrogens attached to β-carbons. The second compound has one axial hydrogen attached to a β-carbon, but it cannot form the more substituted (more stable) alkene that can be formed by the first compound. The last compound cannot undergo an E2 reaction because it does not have an axial hydrogen attached to a β-carbon.

95. Draw the carbocation that each compound forms in an S_N1 reaction. Because carbocation formation is the rate-limiting step, the more stable the carbocation, the faster is the rate of the S_N1 reaction.

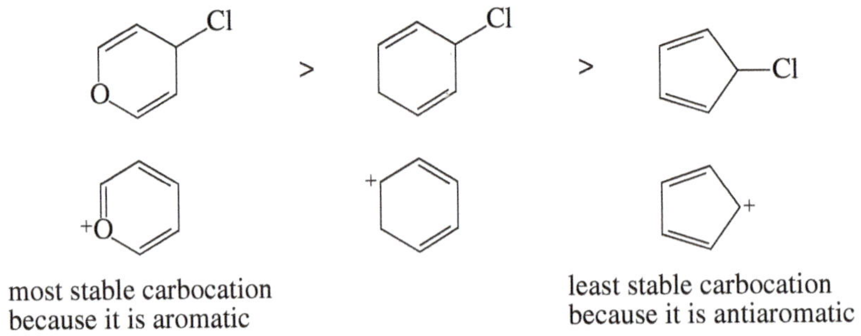

Cl > Cl > Cl

most stable carbocation because it is aromatic

least stable carbocation because it is antiaromatic

96. **a.** The stereoisomer formed in greatest yield is the one in which the larger group attached to one sp^2 carbon and the larger group attached to the other sp^2 carbon are on opposite sides of the double bond.

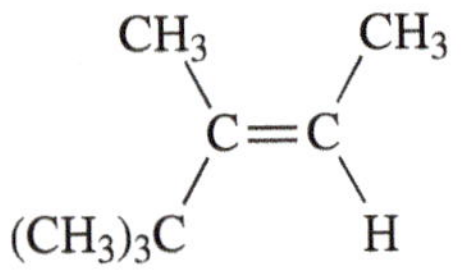

CH_3 CH_3

C=C

$(CH_3)_3C$ H

b. No stereoisomers are possible for this compound because one of the sp^2 carbons is bonded to two hydrogens.

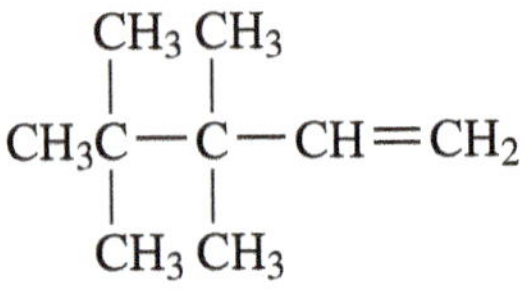

CH_3 CH_3

$CH_3C-C-CH=CH_2$

CH_3 CH_3

c. No stereoisomers are possible for this compound because one of the sp^2 carbons is bonded to two methyl groups.

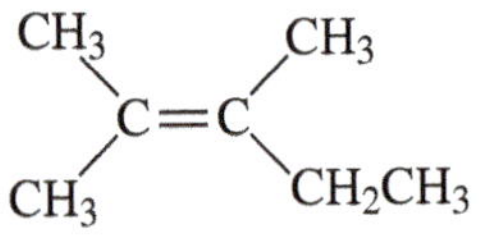

CH_3 CH_3

C=C

CH_3 CH_2CH_3

d. Because it is an E2 reaction and only one hydrogen is attached to the β-carbon, the stereoisomer formed in greater yield depends on the configuration of the reactant. The reactant can have four different configurations: *S,S*; *S,R*; *R,R*; and *R,S*. To determine the product of the reaction:

1. Draw the skeleton of a perspective formula, putting the groups to be eliminated on the solid lines. Notice that on each carbon, the solid wedge is below the hatched wedge.

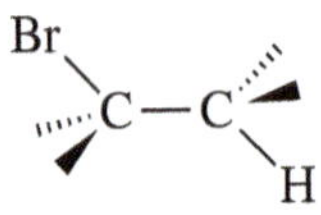

Br

C—C

H

2. Add the remaining groups to the structure in a way that gives the asymmetric centers the desired configurations. For example, the structure below is (3*S*,4*S*)-3-bromo-3,4-dimethylhexane.

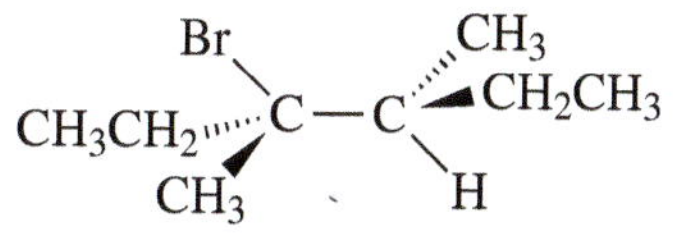

Br, CH_3CH_2, CH_3 on C—C with CH_3, CH_2CH_3, H

(3*S*,4*S*)-3-bromo-3,4-dimethylhexane

3. Because the groups to be eliminated are both attached to solid lines, they are anti to each other.

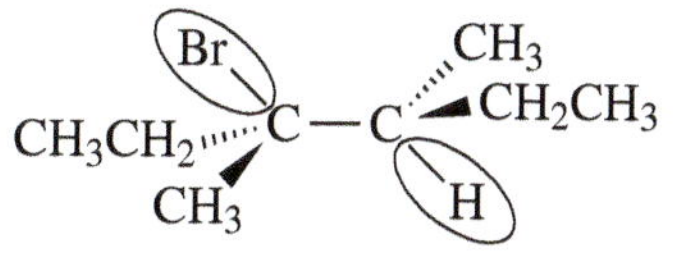

Br, CH_3CH_2, CH_3 on C—C with CH_3, CH_2CH_3, H

4. Once the groups are eliminated, you can draw the structure of the alkene product. (The groups attached to the solid wedges are on the same side of the double bond, and the groups attached by hatched wedges are on the other side of the double bond.)

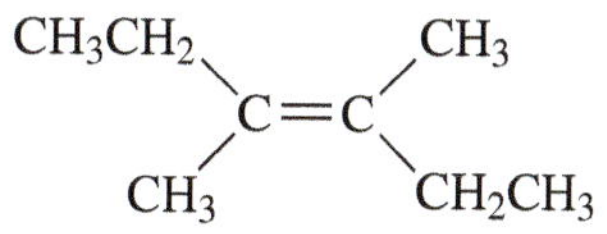

CH_3CH_2, CH_3 C=C CH_3, CH_2CH_3

(*E*)-3,4-dimethyl-3-hexene

The configuration of the elimination product obtained from each of the other three stereoisomers can be determined in the same way.

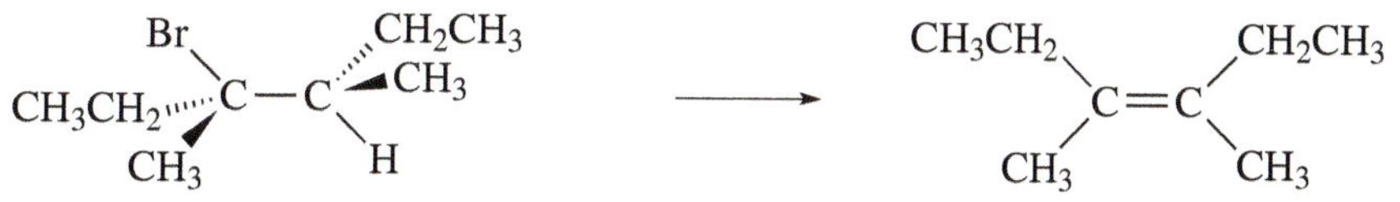

Br, CH_3CH_2, CH_3 C—C CH_2CH_3, CH_3, H → CH_3CH_2, CH_2CH_3 C=C CH_3, CH_3

(3*S*,4*R*)-3-bromo-3,4-dimethylhexane → (*Z*)-3,4-dimethyl-3-hexene

Br, CH_3, CH_2CH_2 C—C CH_2CH_3, CH_3, H → CH_3CH_2, CH_3 C=C CH_3, CH_2CH_3

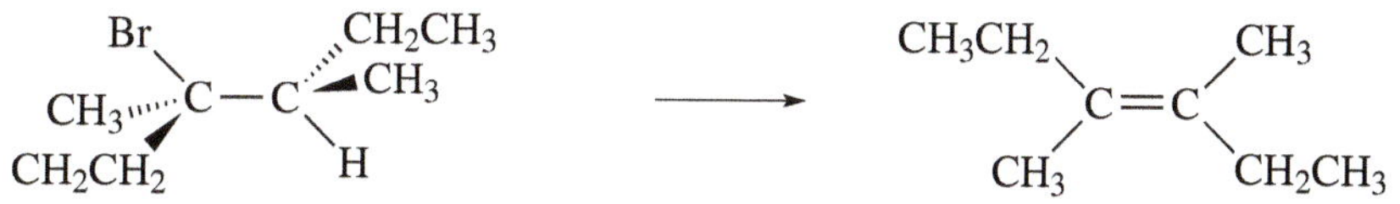

(3*R*,4*R*)-3-bromo-3,4-dimethylhexane → (*E*)-3,4-dimethyl-3-hexene

Br, CH_3, CH_2CH_2 C—C CH_3, CH_2CH_3, H → CH_3, CH_3 C=C CH_2CH_2, CH_2CH_3

(3*R*,4*S*)-3-bromo-3,4-dimethylhexane → (*Z*)-3,4-dimethyl-3-hexene

97. 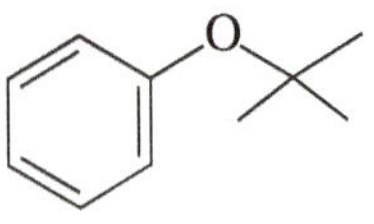 This ether cannot be made by a Williamson ether synthesis.

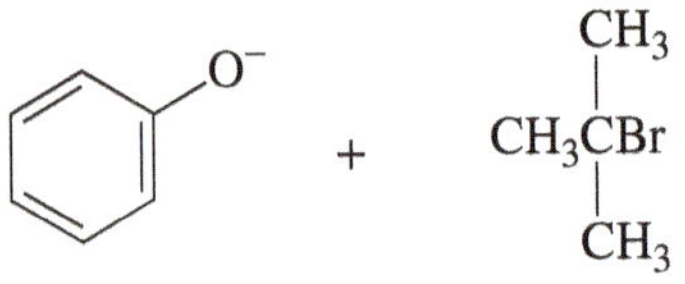

$$C_6H_5O^- \quad + \quad CH_3\overset{CH_3}{\underset{CH_3}{C}}Br$$

These reagents cannot be used because the tertiary alkyl halide undergoes only an elimination reaction.

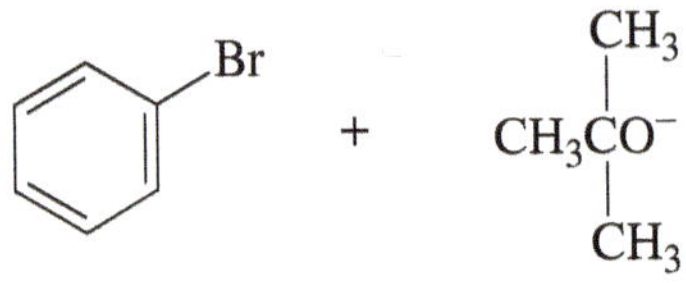

$$C_6H_5Br \quad + \quad CH_3\overset{CH_3}{\underset{CH_3}{C}}O^-$$

These reagents cannot be used because the aryl halide cannot undergo a substitution reaction.

98.

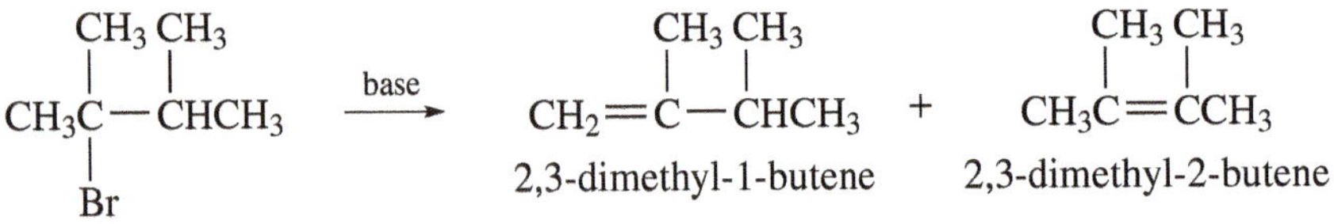

$$CH_3\overset{CH_3}{\underset{Br}{C}}-\overset{CH_3}{C}HCH_3 \xrightarrow{\text{base}} CH_2=\overset{CH_3}{C}-\overset{CH_3}{C}HCH_3 + CH_3\overset{CH_3}{C}=\overset{CH_3}{C}CH_3$$

2,3-dimethyl-1-butene   2,3-dimethyl-2-butene

a. 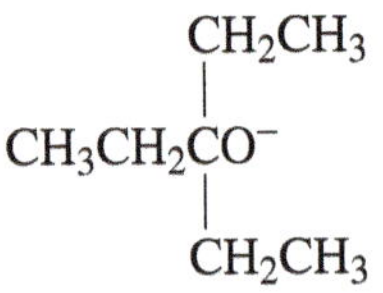 $CH_3CH_2\overset{CH_2CH_3}{\underset{CH_2CH_3}{C}}O^-$

Because it is the most sterically hindered base, **B** gives the highest percentage of the less stable 1-alkene because it is easier for it to remove the most accessible hydrogen.

b. $CH_3CH_2O^-$

Because it is the least sterically hindered base, **C** gives the highest percentage of the more stable 2-alkene.

99.

a. 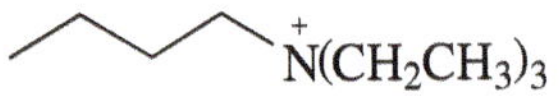 $\overset{+}{N}(CH_2CH_3)_3$

b. 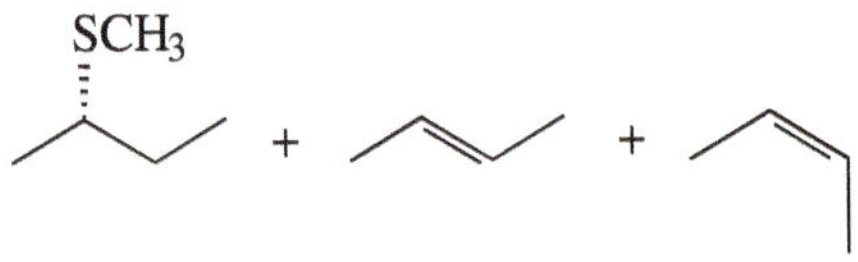 SCH_3 + +

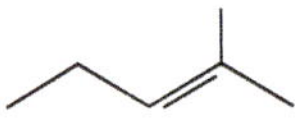

c. 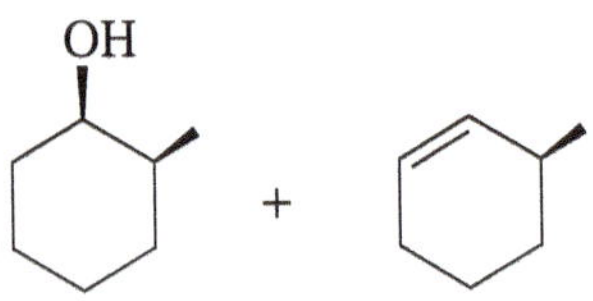 OH +

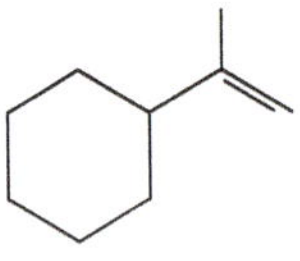

d. 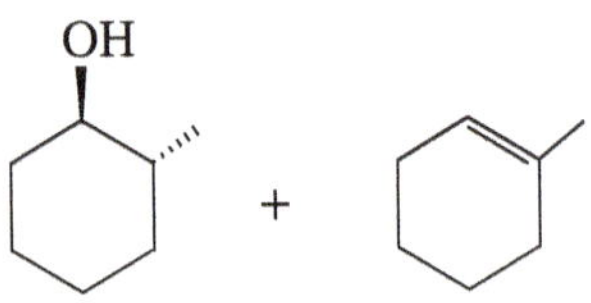 OH +

e. OCH_3

f.

g.

h. 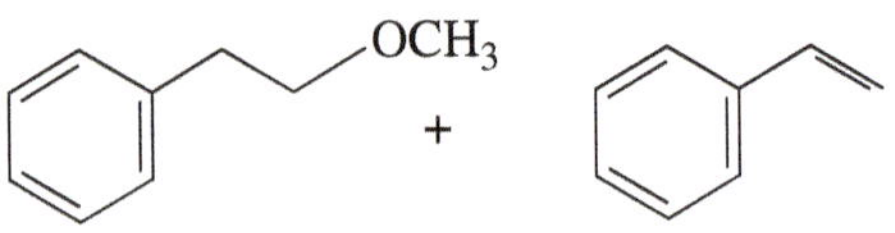 OCH_3 +

This primary alkyl halide forms more than the usual amout of elimination product because the new double bond is conjugated with the benzene ring. Its greater stability makes it easier to form.

100. **a.** These are $S_N2/E2$ reactions, because the alkyl halide is secondary.

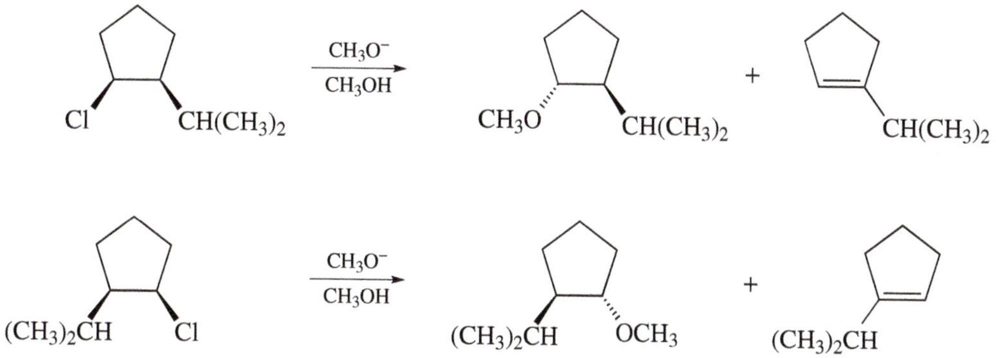

CH_3O^- / CH_3OH

Cl, $CH(CH_3)_2$ → CH_3O, $CH(CH_3)_2$ + $CH(CH_3)_2$

$(CH_3)_2CH$, Cl → $(CH_3)_2CH$, OCH_3 + $(CH_3)_2CH$

b. Only the substitution products are optically active; the elimination product does not have an asymmetric center.

c.

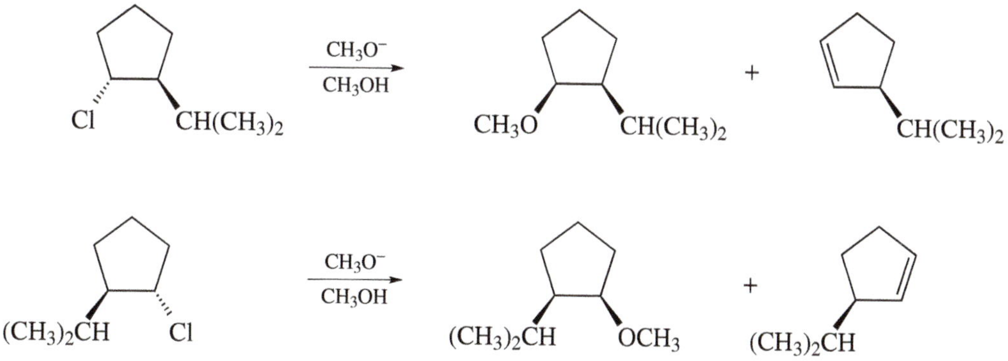

CH_3O^- / CH_3OH

Cl, $CH(CH_3)_2$ → CH_3O, $CH(CH_3)_2$ + $CH(CH_3)_2$

$(CH_3)_2CH$, Cl → $(CH_3)_2CH$, OCH_3 + $(CH_3)_2CH$

All the products are optically active.

d. The cis enantiomers form the substitution products more rapidly, because there is less steric hindrance from the adjacent substituent to back-side attack by the nucleophile.

e. The cis enantiomers form the elimination products more rapidly, because the alkenes formed from the cis enantiomers are more substituted and, therefore, more stable. The more stable the alkene, the lower the energy of the transition state leading to its formation and the more rapidly it is formed.

101. CH_2Br → $\overset{+}{C}H_2$ ↔ CH_2 (+) → CH_2

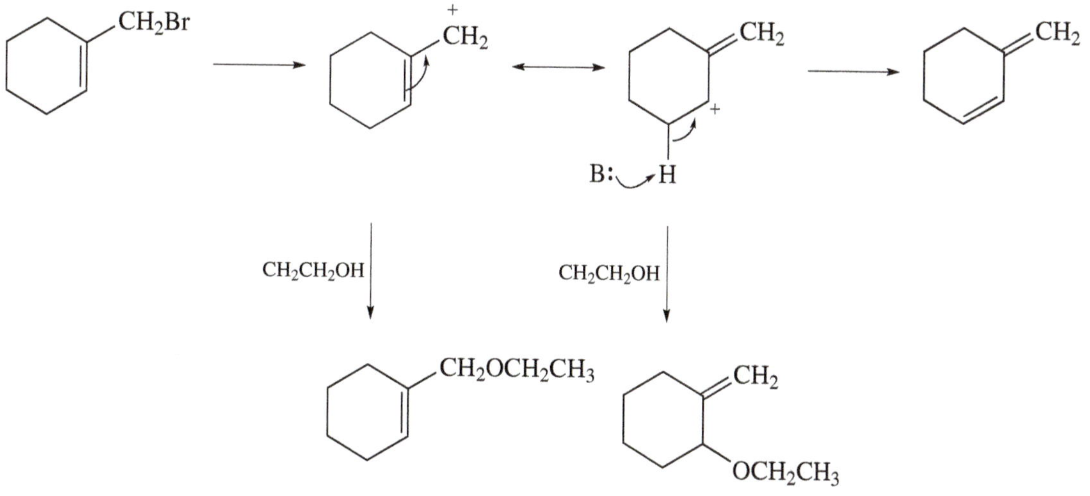

B: H

CH_2CH_2OH ↓ $CH_2OCH_2CH_3$

CH_2CH_2OH ↓ CH_2, OCH_2CH_3

102. In an E2 reaction, both groups to be eliminated must be in axial positions.

When the bromine is in the axial position in the cis isomer, the *tert*-butyl substituent is in the more stable equatorial position.

When the bromine is in the axial position in the trans isomer, the *tert*-butyl substituent is in the less stable axial position.

Therefore, elimination takes place via the most stable conformer in the cis isomer and via the less stable chair conformer in the trans isomer, so the cis isomer undergoes elimination more rapidly.

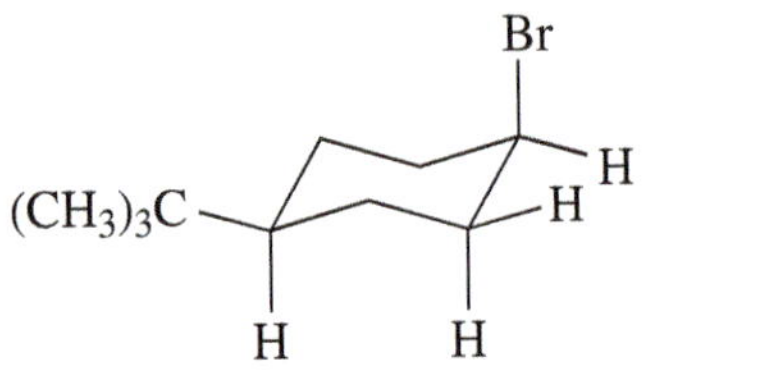

cis-1-bromo-4-*tert*-butylcyclohexane

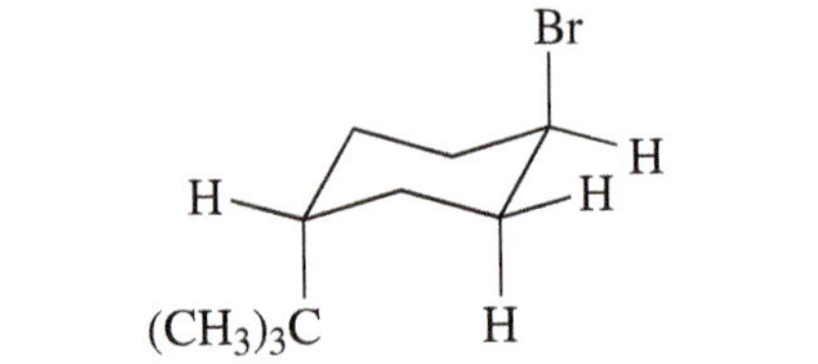

trans-1-bromo-4-*tert*-butylcyclohexane

103. **a.** minor + major (OCH_3)

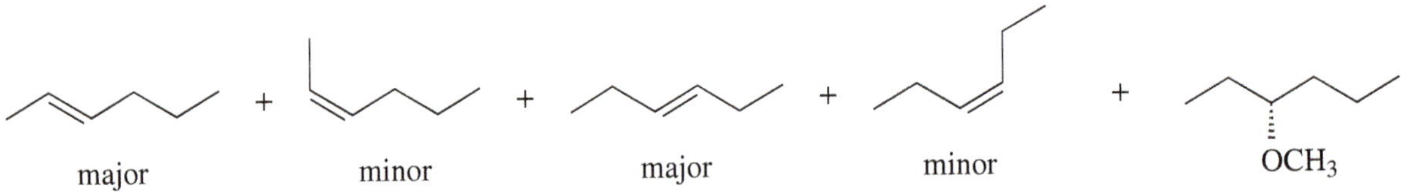

b. major + minor + major + minor + (OCH_3)

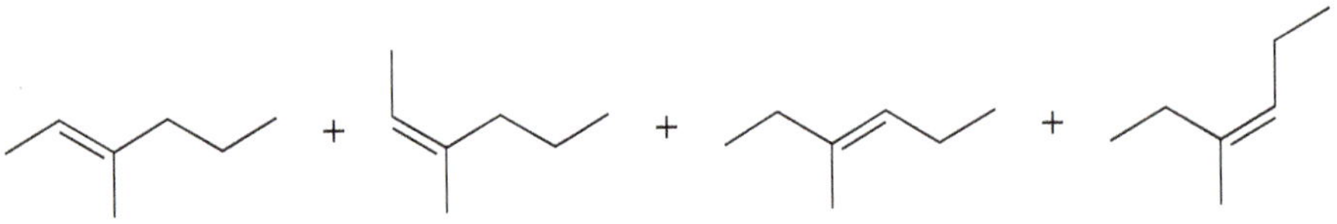

c. major + minor + major + minor

d. major + minor + major + minor + (OCH_3) + (CH_3O)

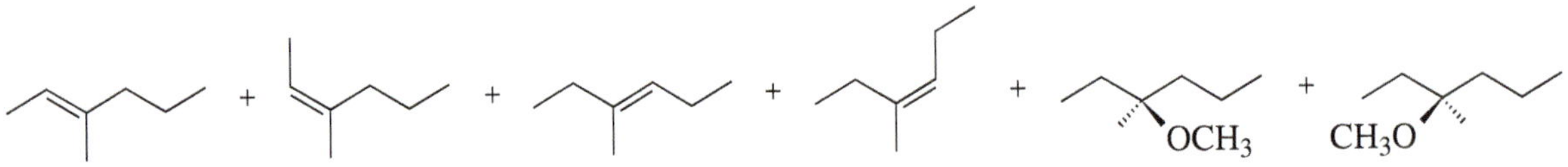

e. + (OCH_3)

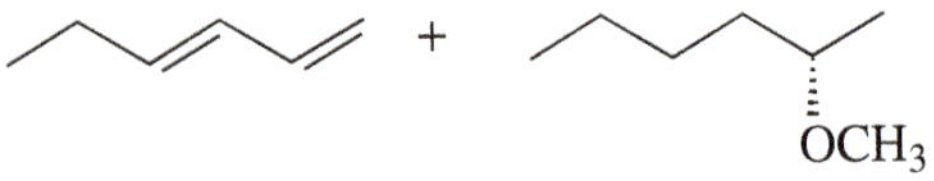

f. no substitution or elimination reaction

104. The reactants are neutral. Therefore, increasing the polarity of the solvent increases the rate of the reactions. Therefore, the reaction is faster in formic acid, the more polar of the two solvents.

105. **a.** We can predict that this is an S_N1 reaction because acetate ion is a relatively poor nucleophile.

$$CH_3C(CH_3)(Cl)CH{=}CHCH_2CH_3 \longrightarrow CH_3\overset{+}{C}(CH_3)CH{=}CHCH_2CH_3 \longleftrightarrow CH_3C(CH_3){=}CH\overset{+}{C}HCH_2CH_3 + Cl^-$$

$$\xrightarrow{CH_3CO_2^-} CH_3C(CH_3)(OCOCH_3)CH{=}CHCH_2CH_3 + CH_3C(CH_3){=}CHCH(OCOCH_3)CH_2CH_3$$

b. We can predict that this is an S_N1 reaction because acetate ion is a relatively poor nucleophile.

3-bromo-1-methylcyclohexene → allylic carbocation ⟷ tertiary carbocation + Br^-

$\xrightarrow{CH_3CO_2^-}$ 3-methylcyclohex-2-enyl acetate ($OCCH_3$ with C=O) + 1-methylcyclohex-2-enyl acetate (CH_3CO, CH_3 on same carbon)

106. **a.** $CH_3CH_2CH_2Br + HC{\equiv}C^- \longrightarrow CH_3CH_2CH_2C{\equiv}CH \xrightarrow[Pd/C]{H_2} CH_3CH_2CH_2CH_2CH_3$

b. $CH_3CH_2CH_2Br + HC{\equiv}C^- \longrightarrow CH_3CH_2CH_2C{\equiv}CH \xrightarrow[H_2O]{H_2SO_4,\ HgSO_4} CH_3CH_2CH_2\overset{O}{\overset{\|}{C}}CH_3$

c. $CH_3CH_2CH_2Br + HC{\equiv}C^- \longrightarrow CH_3CH_2CH_2C{\equiv}CH \xrightarrow[2.\ HO^-,\ H_2O_2,\ H_2O]{1.\ R_2BH/THF} CH_3CH_2CH_2CH_2\overset{O}{\overset{\|}{C}}H$

d. $CH_3CH_2CH_2Br + HC{\equiv}C^- \longrightarrow CH_3CH_2CH_2C{\equiv}CH \xrightarrow[\text{Lindlar catalyst or } Na/NH_3(liq)]{H_2} CH_3CH_2CH_2CH{=}CH_2$

$\xrightarrow{RCOOH\ (\text{peroxyacid, } RCO_3H)}$ $CH_3CH_2CH_2\overset{O}{CH{-}CH_2}$ (epoxide)

107. The equilbrium contstant is given by the relative stabilities of the products and reactants. Therefore, any factor that stabilizes the products increases the equilibrium constant.

$$K_{eq} = \frac{[\text{products}]}{[\text{reactants}]}$$

Ethanol will stabilize the charged products more than will diethyl ether because ethanol is a more polar solvent. Therefore, the equilibrium will lie farther to the right (toward products) in ethanol.

108. **a.** The reaction with quinuclidine had the larger rate constant because quinuclidine is less sterically hindered as a result of the substituents on the nitrogen being pulled back into a ring structure.

b. The reaction with quinuclidine had the larger rate constant for the same reason given in part **a**.

c. Isopropyl iodide exhibits the larger difference in rate constants and, therefore, the larger $k_{quinuclidine}/k_{triethylamine}$ ratio. Because it is more sterically hindered than methyl iodide, it is more affected by differences in the amount of steric hindrance in the nucleophile.

109. Because methanol is a poor nucleophile, it is an S_N1 reaction. The bromine that departs is the one that forms a secondary benzylic cation and not the one that would form a secondary carbocation. The nucleophile can approach from the top or bottom of the planar carbocation.

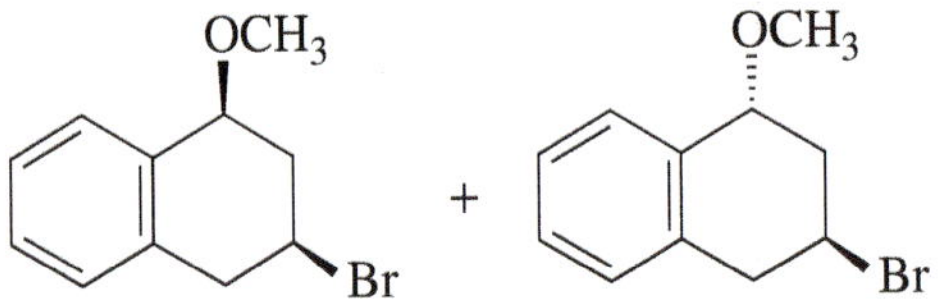

OCH_3 Br + OCH_3 Br

110. Because all the reactions are S_N2 reactions, the configuration of the asymmetric center attached to the Br in the reactant is inverted in the product.

a. 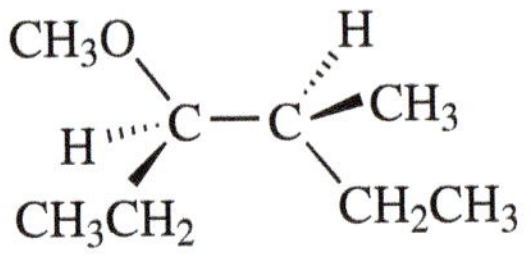 CH_3O, H, CH_3CH_2 — C—C — H, CH_3, CH_2CH_3 or 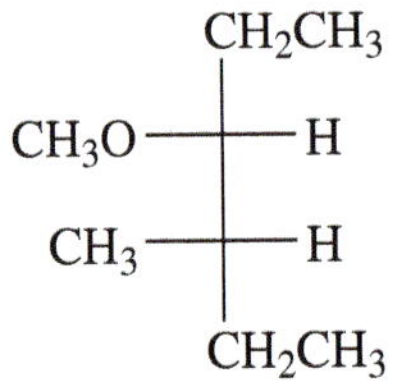 (3*R*,4*S*)

Fischer projection: CH_2CH_3 (top); CH_3O—H; CH_3—H; CH_2CH_3 (bottom)

b. 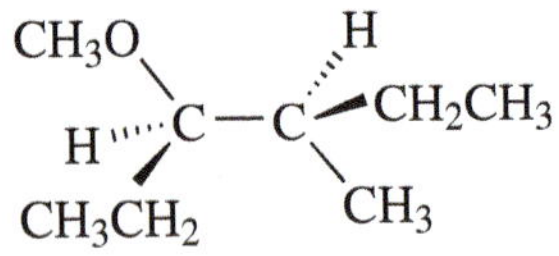 CH_3O, H, CH_3CH_2 — C—C — H, CH_2CH_3, CH_3 or 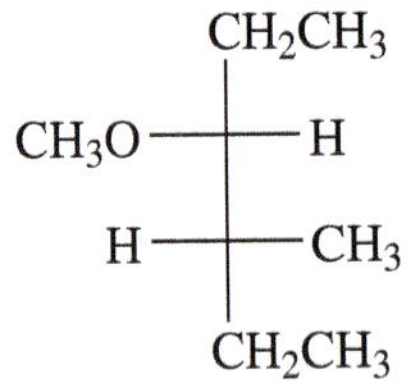 (3*R*,4*R*)

Fischer projection: CH_2CH_3 (top); CH_3O—H; H—CH_3; CH_2CH_3 (bottom)

c. 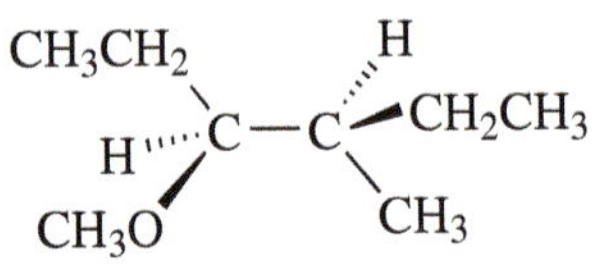 CH_3CH_2, H, CH_3O — C—C — H, CH_2CH_3, CH_3 or 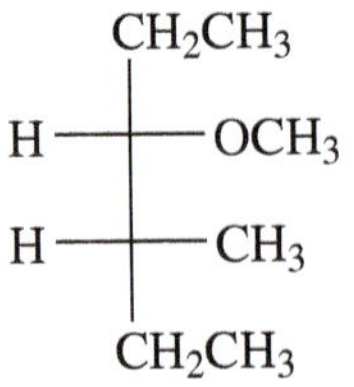 (3*S*,4*R*)

Fischer projection: CH_2CH_3 (top); H—OCH_3; H—CH_3; CH_2CH_3 (bottom)

d. 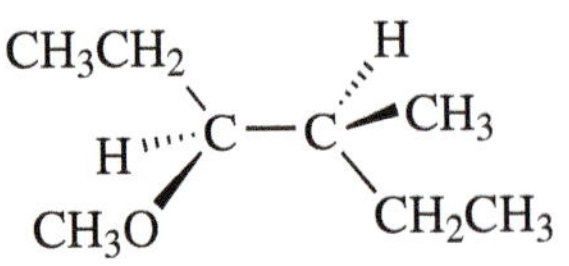 CH_3CH_2, H, CH_3O — C—C — H, CH_3, CH_2CH_3 or 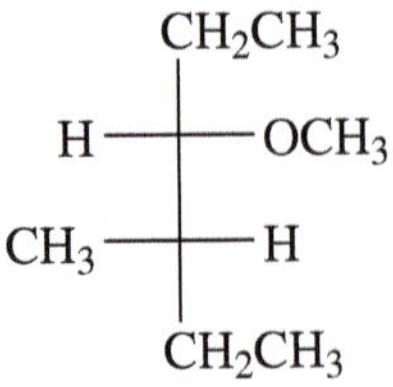 (3*S*,4*S*)

Fischer projection: CH_2CH_3 (top); H—OCH_3; CH_3—H; CH_2CH_3 (bottom)

111. In reactions **a–d**, the β-carbon from which the hydrogen is to be removed is bonded to only one hydrogen. Therefore, the configuration of the reactant determines the configuration of the product of the E2 reaction.

To determine the configuration of the product, see the instructions in the Problem-Solving Strategy on page 425.

a. (2*S*,3*S*)-2-chloro-3-methylpentane → (*E*)-3-methyl-2-pentene

b. (2*S*,3*R*)-2-chloro-3-methylpentane → (*Z*)-3-methyl-2-pentene

c. (2*R*,3*S*)-2-chloro-3-methylpentane → (*Z*)-3-methyl-2-pentene

d. (2*R*,3*R*)-2-chloro-3-methylpentane → (*E*)-3-methyl-2-pentene

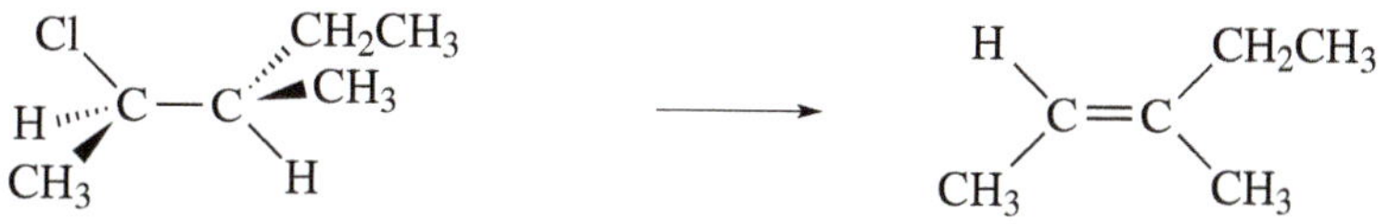

e. This reactant has two hydrogens bonded to the β-carbon, so both *Z* and *E* stereoisomers are formed.

CH_3CH_2, $(CH_3)_3C$ / C=C / CH_3, H — major product

+

CH_3CH_2, $(CH_3)_3C$ / C=C / H, CH_3 — minor product

112. The two tertiary alkyl halides form different products when they react with a strong base (an E2 reaction) because the H and Br attached to the ring can be eliminated only if they are anti to each other. Because the H and Br are on the same side of the ring in **b**, a hydrogen has to be removed from the other β-carbon (that is, from the methyl group).

a. →

b. →

The two tertiary alkyl halides form the same product when they react with a weak base (an E1 reaction) because both form the same carbocation.

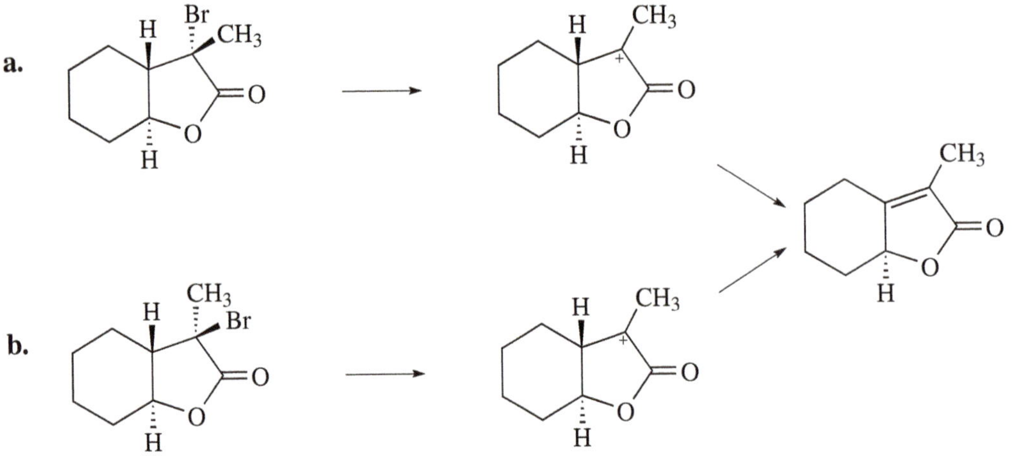

a.

b.

113. In order to undergo an E2 reaction, a chlorine and a hydrogen bonded to an adjacent carbon must be trans to each other so they can both be in the required axial positions. Every Cl in the following compound has a Cl trans to it, so no Cl has a hydrogen trans to it. Therefore, it is the least reactive of the isomers; it cannot undergo an E2 reaction.

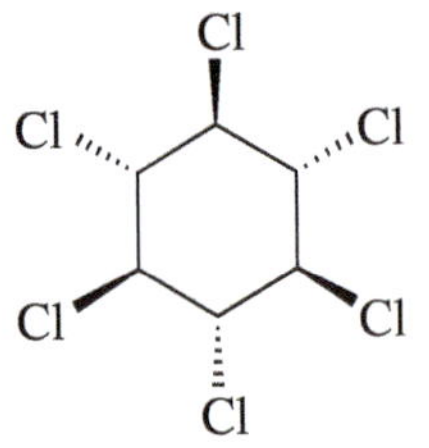

114. The silver ion increases the ease of departure of the halogen atom by sharing one of bromine's lone pairs, which weakens the carbon–bromine bond.

$$CH_3CH{=}CHCH_2{-}\ddot{B}r{:} + Ag^+ \longrightarrow CH_3CH{=}CHCH_2{-}\overset{+}{Br}{-}Ag \longrightarrow CH_3CH{=}CH\overset{+}{C}H_2 + AgBr$$

115. For a description of how to do this problem, see the Problem-Solving Strategy on page 425.

a. 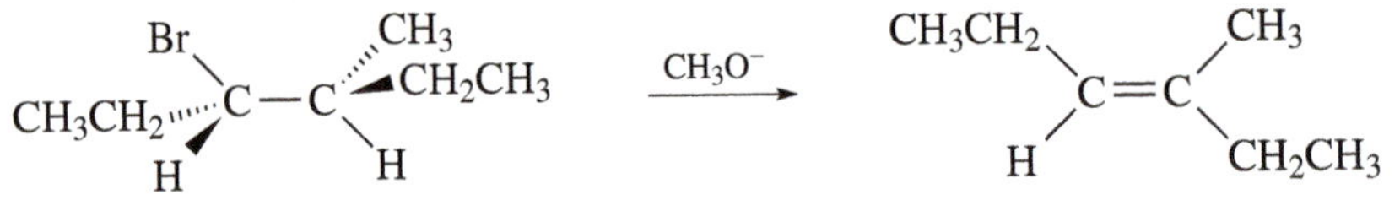 (3*S*,4*S*)-3-bromo-4-methylhexane $\xrightarrow{CH_3O^-}$ (*E*)-3-methyl-2-pentene + substitution product (CH_3O replaces Br)

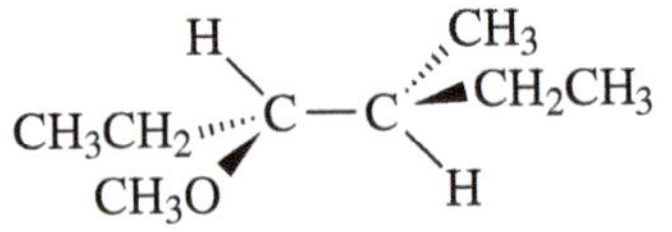

b. 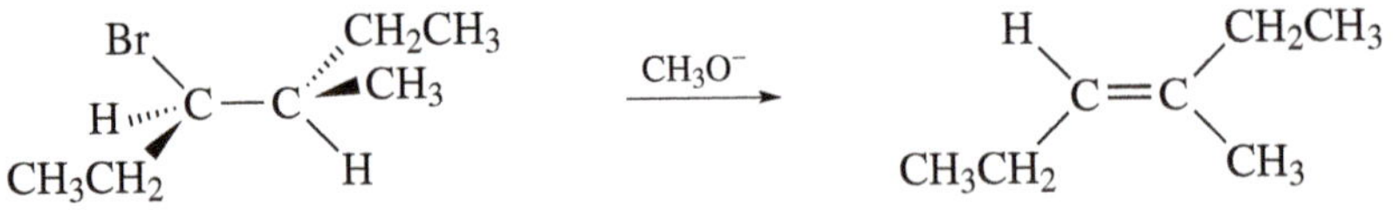 (3*R*,4*R*)-3-bromo-4-methylhexane $\xrightarrow{CH_3O^-}$ (*E*)-3-methyl-2-pentene + substitution product (CH_3O replaces Br)

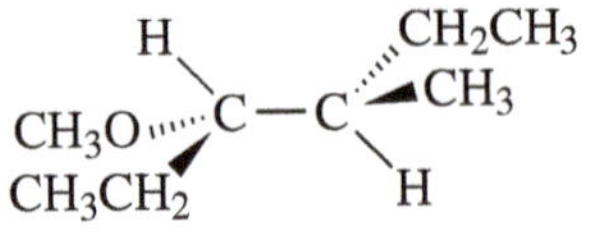

c. (3*S*,4*R*)-3-bromo-4-methylhexane $\xrightarrow{CH_3O^-}$ (*Z*)-3-methyl-2-pentene + [substitution product: CH_3CH_2, H, CH_3O on C3; CH_2CH_3, CH_3, H on C4]

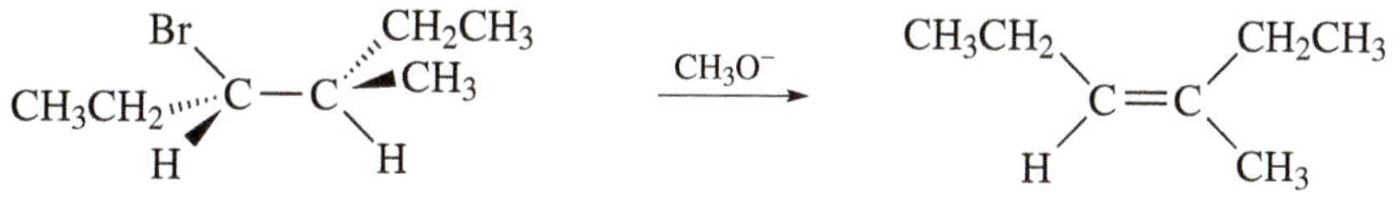

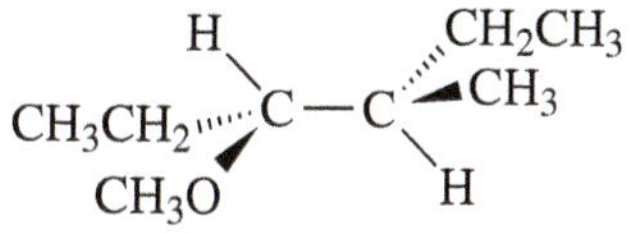

d. 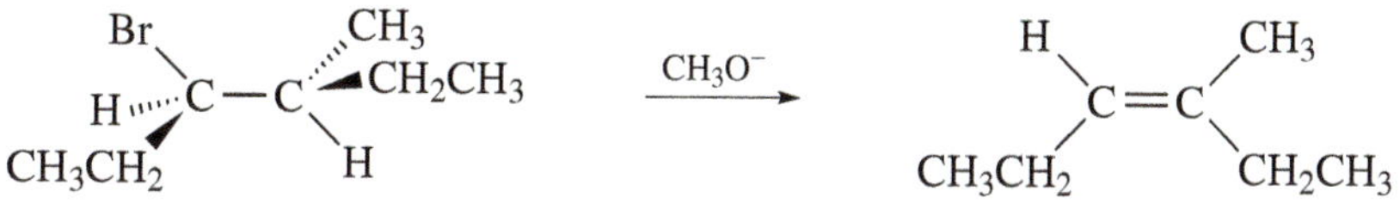 (3*R*,4*S*)-3-bromo-4-methylhexane $\xrightarrow{CH_3O^-}$ (*Z*)-3-methyl-2-pentene + [substitution product: H, CH_3O, CH_3CH_2 on C3; CH_3, CH_2CH_3, H on C4]

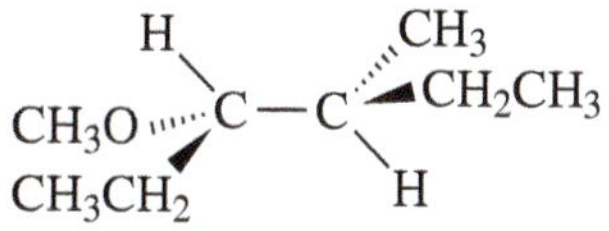

116. **a.** $CH_3CH_2CD{=}CH_2$ and $CH_3CH_2CH{=}CH_2$

b. The deuterium-containing compound results from elimination of HBr, whereas the non-deuterium-containing compound results from elimination of DBr. The deuterium-containing compound is obtained in greater yield, because a C—H bond is easier to break than a C—D bond.

117. [decalin with CH_3 and OCH_3] + [decalin with OCH_3 and CH_3] + [octalin with CH_3 on the double bond]

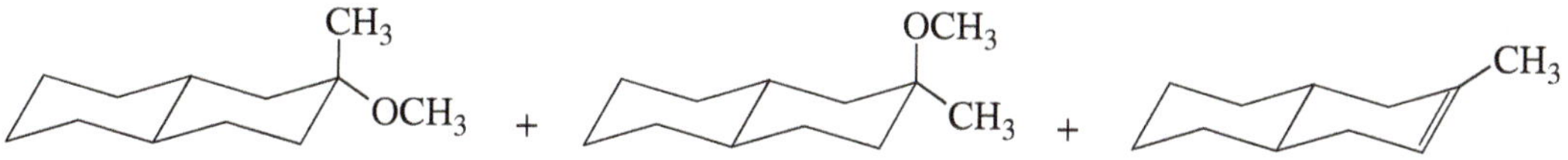

118. **a.** $CH_3CH_2CH_2CH_2Br \xrightarrow{tert\text{-}BuO^-} CH_3CH_2CH{=}CH_2 \xrightarrow{Br_2} CH_3CH_2CHBrCH_2Br \xrightarrow{2\ ^-NH_2} CH_3CH_2C{\equiv}CH \xrightarrow{^-NH_2} CH_3CH_2C{\equiv}C^- \xrightarrow{CH_3CH_2Br} CH_3CH_2C{\equiv}CCH_2CH_3 \xrightarrow[H_2SO_4]{H_2O} CH_3CH_2C({=}O){-}CH_2CH_2CH_3$

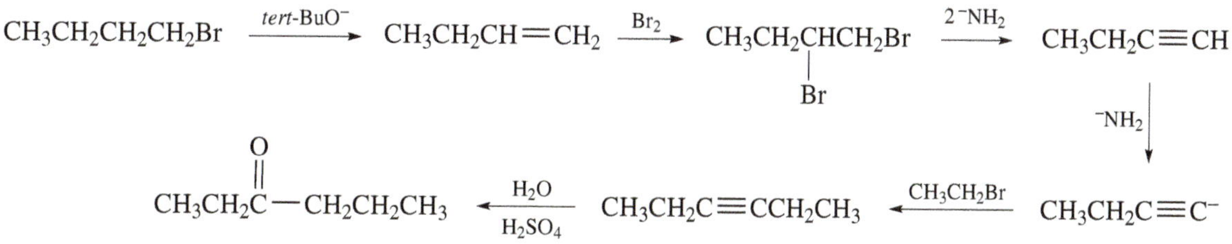

b. Br(CH₂)₄Br $\xrightarrow[\text{excess}]{tert\text{-}BuO^-}$ 1,3-butadiene (two equivalents) $\xrightarrow[\text{Diels-Alder reaction}]{\Delta}$ 4-vinylcyclohexene $\xrightarrow[Pd/C]{H_2}$ ethylcyclohexane

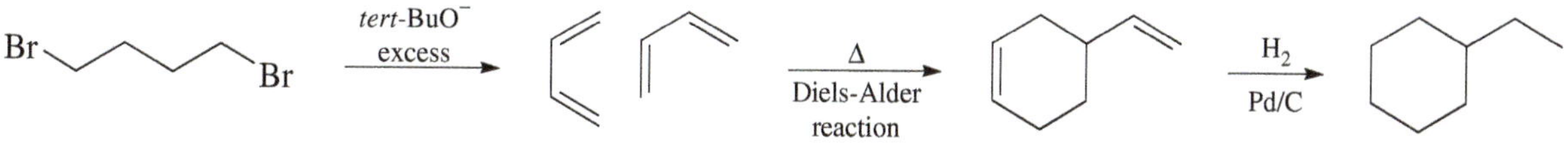

One equivalent reacts in the Diels–Alder reaction as a diene and the other as a dienophile.

119. **a.** Both compounds form the same elimination products because they both have hydrogens bonded to the same β-carbons that are anti to the bromine.

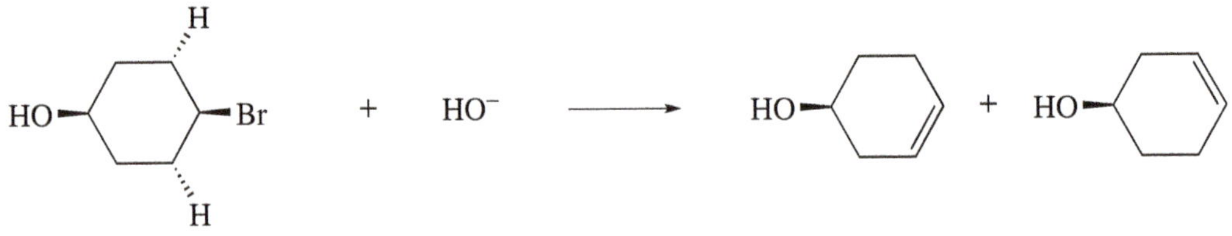

HO–(cyclohexane)–Br + HO^- ⟶ HO–(cyclohexene) + HO–(cyclohexene)

cis-4-bromocyclohexanol

HO–(cyclohexane)····Br + HO^- ⟶ HO–(cyclohexene) + HO–(cyclohexene)

trans-4-bromocyclohexanol

b. Only the trans isomer can undergo an intramolecular substitution reaction because the S_N2 reaction requires back-side attack.

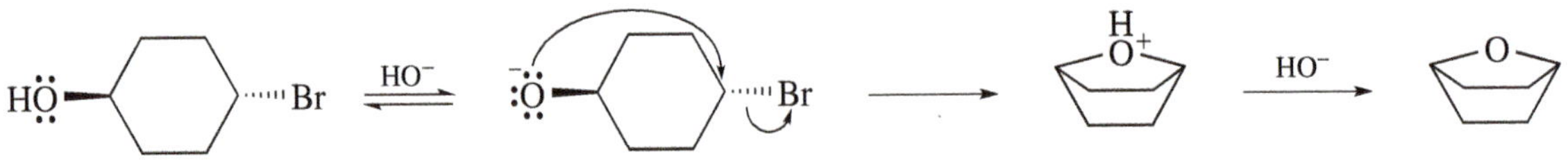

$H\ddot{O}$–(cyclohexane)····Br $\overset{HO^-}{\rightleftharpoons}$ $:\ddot{O}^-$–(cyclohexane)····Br ⟶ (bicyclic $\overset{+}{O}H$) $\xrightarrow{HO^-}$ (bicyclic ether)

trans-4-bromocyclohexanol

The cis isomer can undergo only an intermolecular substitution reaction.

HO–(cyclohexane)–Br + $H\ddot{O}:^-$ ⟶ HO–(cyclohexane)····OH

cis-4-bromocyclohexanol

c. The elimination reaction forms a pair of enantiomers because the reaction creates an asymmetric center in the product. Both substitution reactions form a single stereoisomer, because the reaction does not create an asymmetric center in the product.

120. **a.** **1.** $CH_3CH_2C{\equiv}CCH_2CH_3$ **2.** $CH_3CH_2CH_2OCH_3$

b. $CH_3CH_2C{\equiv}C^-$ + $CH_3CH_2CH_2CH_2Br$ **or** $CH_3CH_2CH_2CH_2C{\equiv}C^-$ + CH_3CH_2Br

c. $CH_3CH_2O^-$ + $CH_3CH(CH_3)CH_2CH_2Br$ **or** $CH_3CH(CH_3)CH_2CH_2O^-$ + CH_3CH_2Br

121. A tertiary carbocation with a strained four-membered ring is less stable than a secondary carbocation with an unstrained five-membered ring, so a carbocation rerrangement occurs. A second carbocation rearrangement forms a tertiary carbocation.

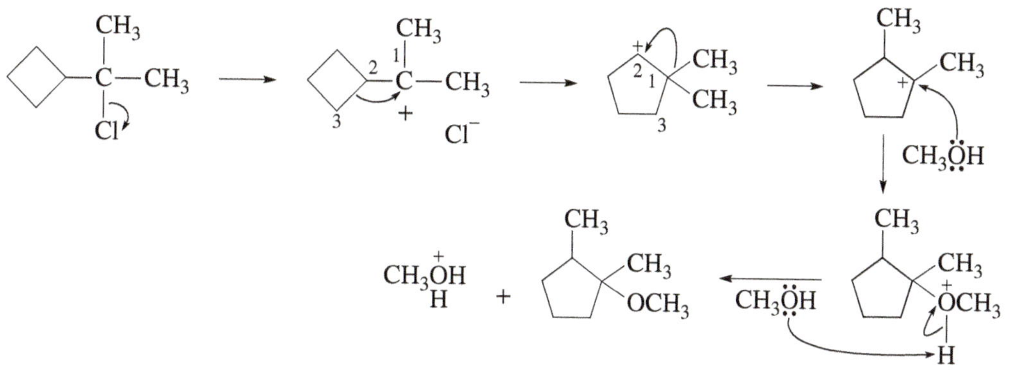

122. Tetrahydrofuran can solvate a charge better than diethyl ether can, because the floppy ethyl substituents of diethyl ether provide steric hindrance, making it difficult for the nonbonding electrons of the oxygen to approach the positive charge that is to be solvated.

123. $CH_3CH_2Cl + K^+I^- \xrightarrow{\text{acetone}} CH_3CH_2I + K^+Cl^-$

K^+Cl^- precipitates out in acetone, which drives the reaction to the right.

124. a.

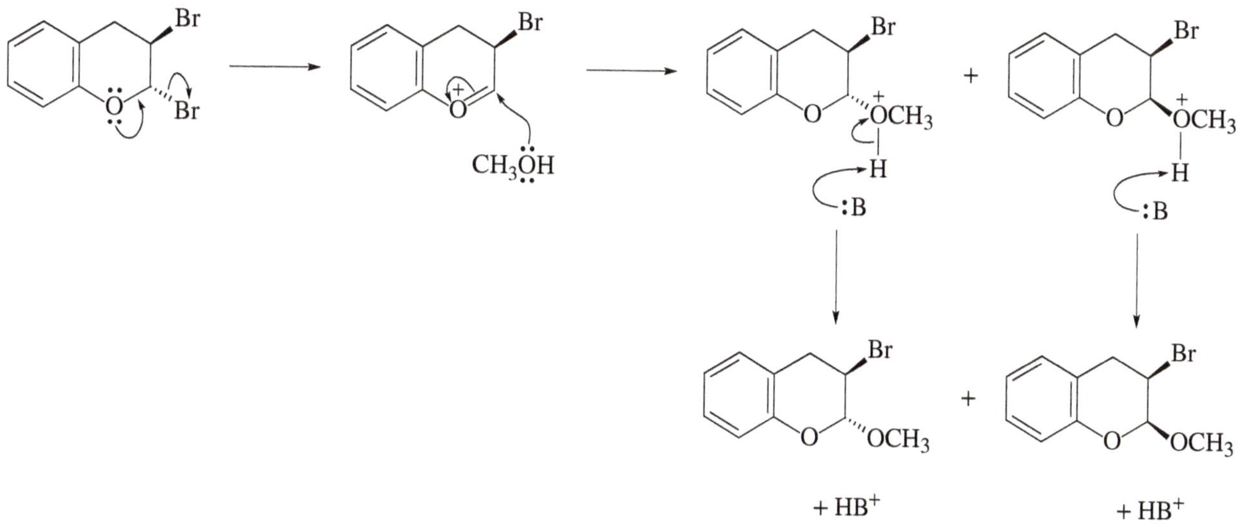

+ HB^+ + HB^+

b. Two products are obtained because methanol can add to the top or bottom of the planar double bond.

c. One bromine is eliminated with the help of one of oxygen's lone pairs, forming a species in which the positive charge is shared by a carbon and an oxygen. The oxygen cannot help eliminate the other bromine.

125.

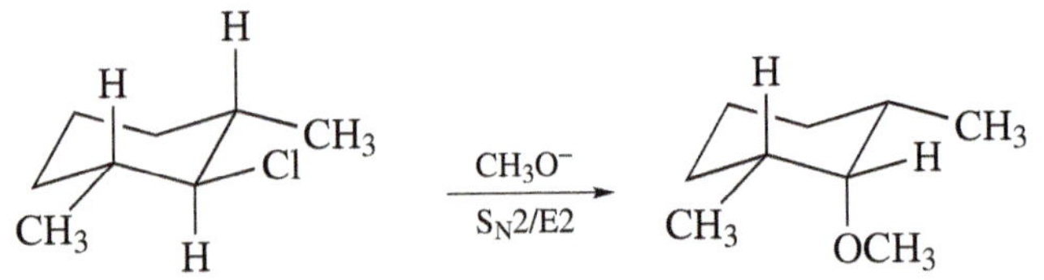

H, H, CH_3, Cl, CH_3, H $\xrightarrow[S_N2/E2]{CH_3O^-}$ H, CH_3, H, CH_3, OCH_3

There is only a substitution product.
The reactant does not undergo elimination because when Cl is in an axial position, neither of the β-carbons is bonded to an axial hydrogen.

126.

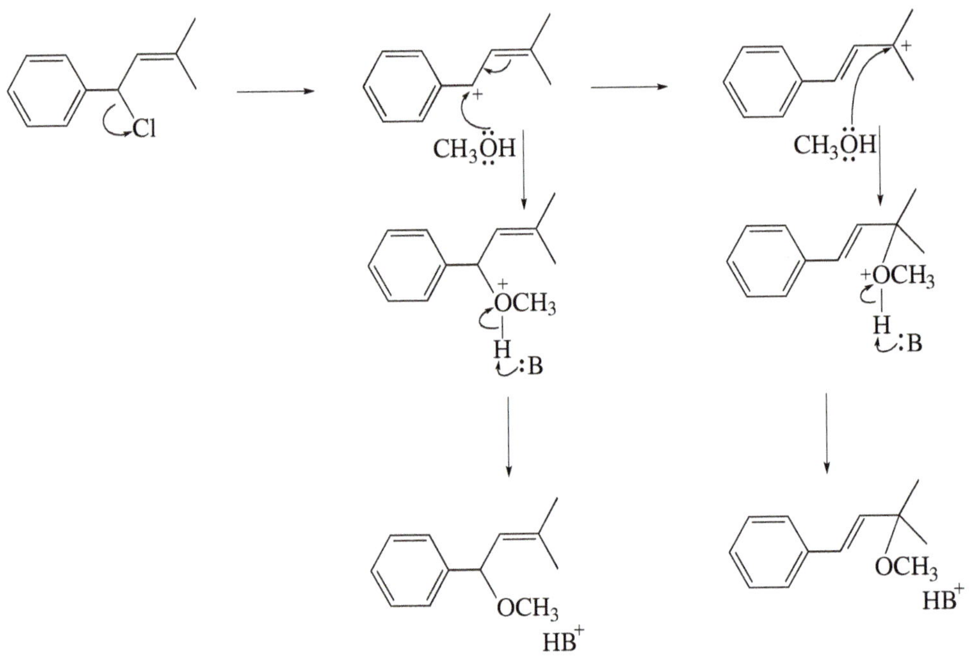

Cl → + → +

$CH_3\ddot{O}H$ $CH_3\ddot{O}H$

$\overset{+}{O}CH_3$ H :B $\overset{+}{O}CH_3$ H :B

OCH_3 HB^+ OCH_3 HB^+

127. a.

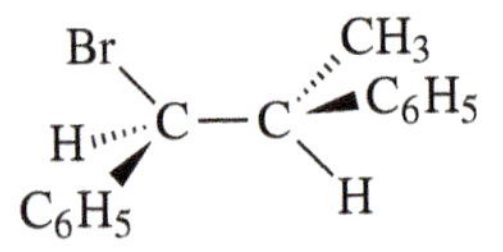

Br, H, C_6H_5, C—C, CH_3, C_6H_5, H → H, CH_3, C=C

(1*S*,2*S*)-1-bromo-1,2-diphenylpropane

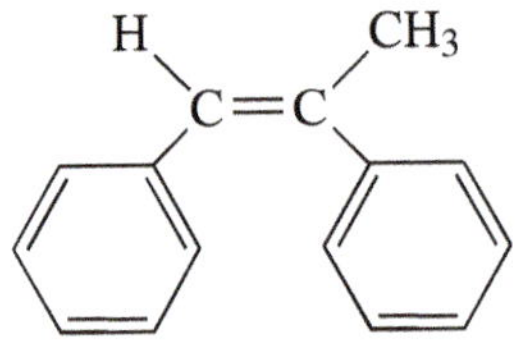

(*Z*)-3-methyl-2-pentene

b.

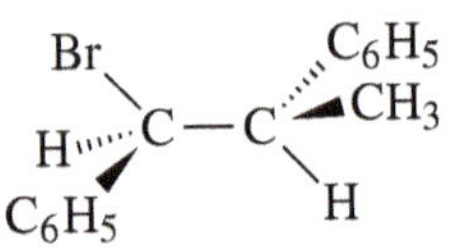

Br, H, C_6H_5, C—C, C_6H_5, CH_3, H → H, C=C, CH_3

(1*S*,2*R*)-1-bromo-1,2-diphenylpropane

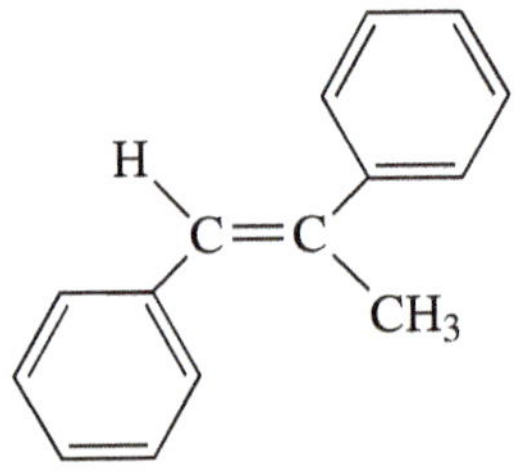

(*E*)-3-methyl-2-pentene

128. A Diels–Alder reaction between hexachlorocyclopentadiene and 1,4-dichlorocyclopentene forms chlordane.

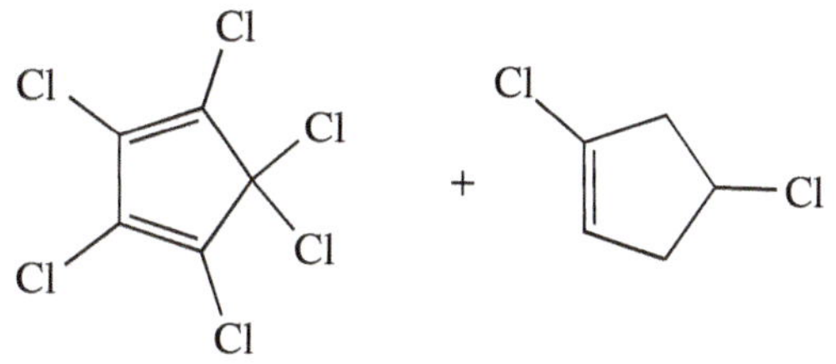

129. I^- is a good nucleophile in a polar solvent such as methanol, so it reacts rapidly with methyl bromide, causing the concentration of I^- to decrease rapidly.

$$:\ddot{\underset{..}{I}}:^- + CH_3-Br \xrightarrow{\text{fast}} CH_3-I + :\ddot{\underset{..}{Br}}:^-$$

I^- is a good leaving group, so methyl iodide undergoes an S_N2 reaction with methanol.
Methanol is a poor nucleophile, so the S_N2 reaction is slow. Therefore, iodide ion returns slowly to its original concentration.

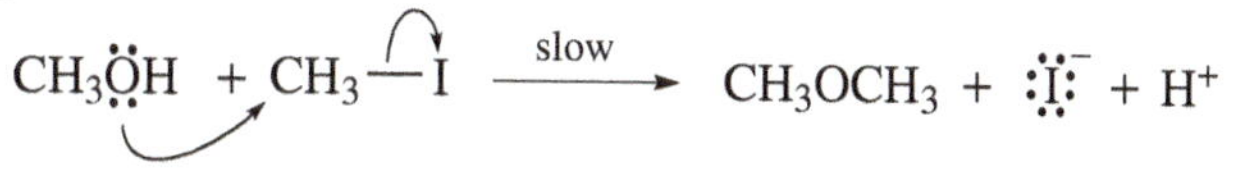

$$CH_3\ddot{\underset{..}{O}}H + CH_3-I \xrightarrow{\text{slow}} CH_3OCH_3 + :\ddot{\underset{..}{I}}:^- + H^+$$

130. It does not undergo an S_N2 reaction, because of steric hindrance to back-side attack.
It does not undergo an S_N1 reaction, because the carbocation that would be formed is unstable; the ring structure prevents it from achieving the 120° bond angles required for an sp^2 carbon.

131. The equation needed to calculate K_{eq} from the change in free energy is given in Section 5.6.

$$\ln K_{eq} = \frac{-\Delta G^\circ}{RT}$$

$$\ln K_{eq} = \frac{-(-21.7 \text{ kcal mol}^{-1})}{0.001986 \text{ kcal mol}^{-1}\text{K}^{-1} \times 303 \text{ K}} = \frac{21.7}{0.60}$$

$$\ln K_{eq} = 36.1$$

$$K_{eq} = 4.8 \times 10^{15}$$

As expected, this highly exergonic reaction has a very large equilibrium constant.

Chapter 9 Practice Test

1. Which of the following is more reactive in an S_N1 reaction?

a. 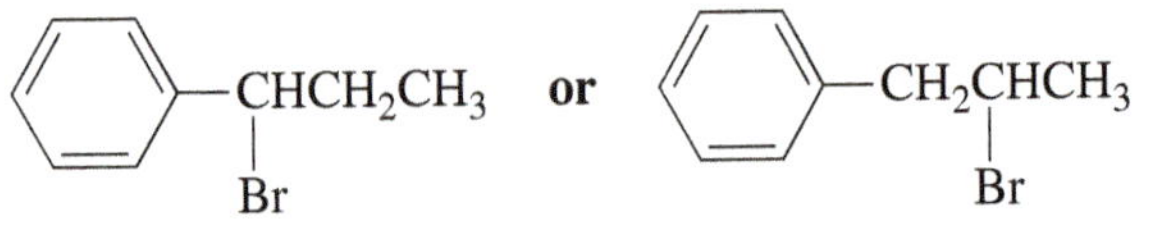 $C_6H_5CH(Br)CH_2CH_3$ or $C_6H_5CH_2CH(Br)CH_3$

b. 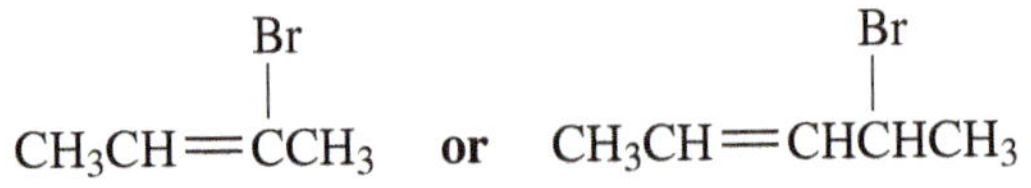 $CH_3CH{=}C(Br)CH_3$ or $CH_3CH{=}CHCH(Br)CH_3$

2. Which of the following is more reactive in an S_N2 reaction?

a. 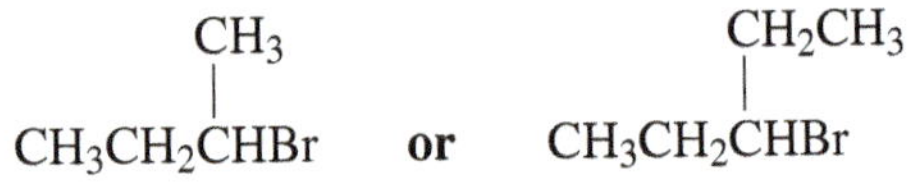 $CH_3CH_2CH(CH_3)Br$ or $CH_3CH_2CH(CH_2CH_3)Br$

b. 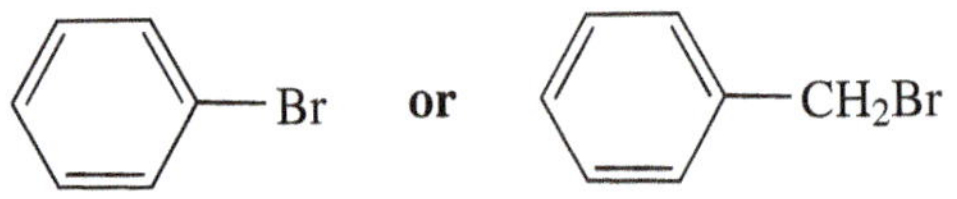 C_6H_5Br or $C_6H_5CH_2Br$

3. Draw the product(s) of the following reactions, showing the stereoisomers that are formed:

a. 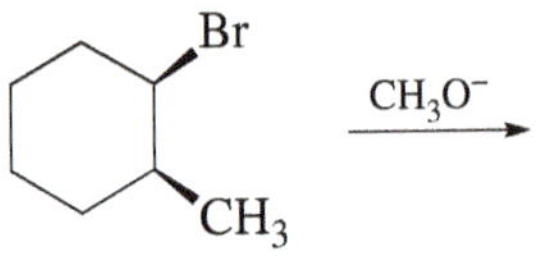 1-bromo-2-methylcyclohexane (Br and CH_3 on wedges) $\xrightarrow{CH_3O^-}$

b. $C_6H_5CH(Br)CH_2CH_3 \xrightarrow{CH_3OH}$

c. Fischer projection: CH_3 (top), H (left), Br (right), CH_2CH_3 (bottom) $\xrightarrow{HO^-}$

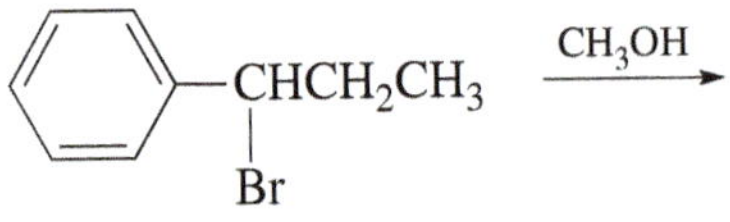

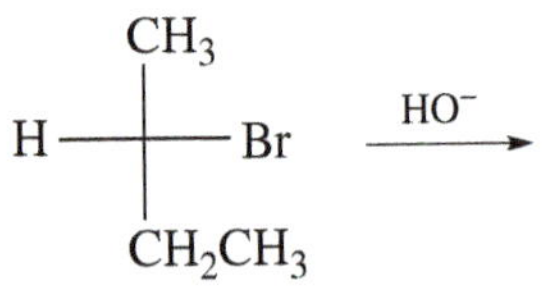

d. 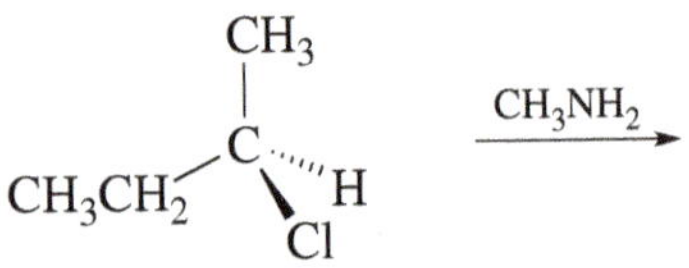 CH_3 (top), CH_3CH_2, H (hashed), Cl (wedge) on C $\xrightarrow{CH_3NH_2}$

e.  $CH_3CH{=}CHCH_2Br \xrightarrow{CH_3OH}$

4. Indicate whether each of the following statements is true or false:

a. Increasing the concentration of the nucleophile favors an S_N1 reaction over an S_N2 reaction. T F

b. Ethyl iodide is more reactive than ethyl chloride in an S_N2 reaction. T F

c. In an S_N1 reaction, the product with the retained configuration is obtained in greater yield. T F

d. The rate of a substitution reaction in which none of the reactants is charged will increase if the polarity of the solvent is increased. T F

e. An S_N2 reaction is a two-step reaction. T F

f. The pK_a of a carboxylic acid is greater in water than it is in a less polar solvent. T F

g. 4-Bromo-1-butanol forms a cyclic ether faster than does 3-bromo-1-propanol. T F

5. Answer the following:

a. Which is a stronger base, CH_3O^- or CH_3S^-?

b. Which is a better nucleophile in an aqueous solution, CH_3O^- or CH_3S^-?

6. For each of the following pairs of S_N2 reactions, indicate the one that occurs with the greater rate constant (that is, occurs faster):

a. $CH_3CH_2CH_2Cl + HO^-$ **or** $CH_3CHCH_3 + HO^-$ (with Cl on the central carbon)

b. $CH_3CH_2CH_2Cl + HO^-$ **or** $CH_3CH_2CH_2I + HO^-$

c. $CH_3CH_2CH_2Br + HO^-$ **or** $CH_3CH_2CH_2Br + H_2O$

d. CH_3CHCH_3 (with Br on the central carbon) $\xrightarrow[H_2O/CH_3OH]{CH_3O^-}$ **or** CH_3CHCH_3 (with Br on the central carbon) $\xrightarrow[CH_3OH]{CH_3O^-}$

e. $BrCH_2CH_2CH_2CH_2NHCH_3$ **or** $BrCH_2CH_2CH_2NHCH_3$

7. Circle the aprotic solvents.

a. dimethyl sulfoxide **b.** diethyl ether **c.** ethanol **d.** hexane

8. How does increasing the polarity of the solvent affect the following?

a. the rate of the S_N2 reaction of methylamine with 2-bromobutane

b. the rate of the S_N1 reaction of methylamine with 2-bromo-2-methylbutane

c. the rate of the S_N2 reaction of methoxide ion with 2-bromobutane

d. the pK_a of acetic acid

e. the pK_a of phenol

9. Which of the following is more reactive in an E2 reaction?

a. 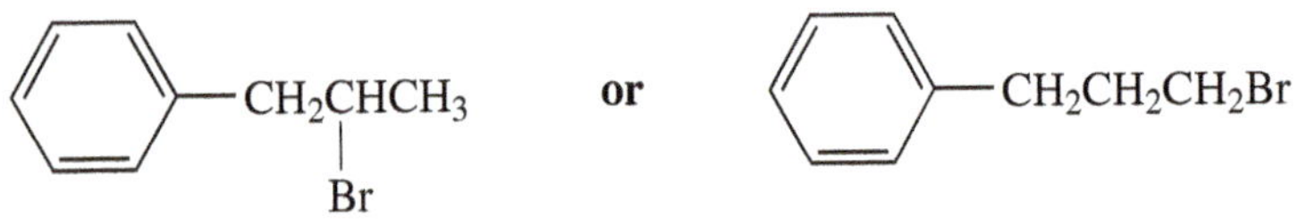 C_6H_5—CH_2CHCH_3 (with Br on CH) or C_6H_5—$CH_2CH_2CH_2Br$

b. 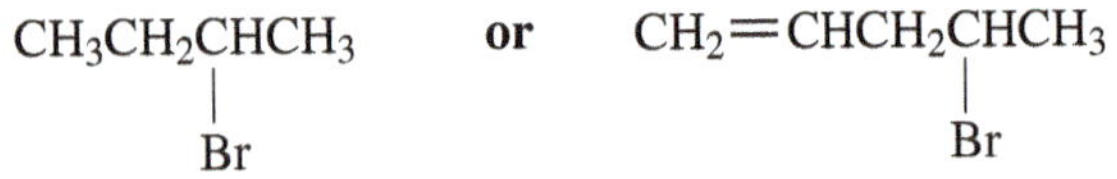 $CH_3CH_2CHCH_3$ (with Br on CH) or $CH_2{=}CHCH_2CHCH_3$ (with Br on CH)

10. Which of the following gives the greater amount of substitution product when it reacts with a good nucleophile/strong base?

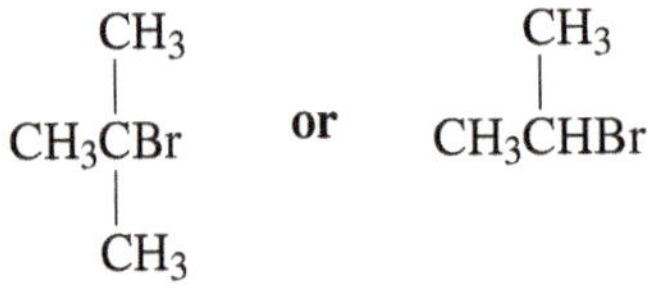

CH_3CBr (with CH_3 above and CH_3 below C) or CH_3CHBr (with CH_3 above C)

11. What products are obtained when (*R*)-2-bromobutane reacts with CH_3O^- / CH_3OH under conditions that favor S_N2/E2 reactions? Include the configuration of the products.

12. What alkoxide ion and what alkyl bromide should be used to synthesize the following ethers?

a. 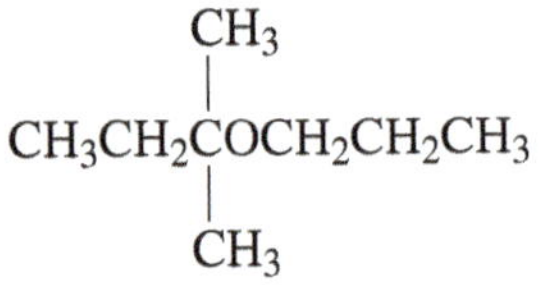 $CH_3CH_2COCH_2CH_2CH_3$ (with CH_3 above and CH_3 below C)

b. 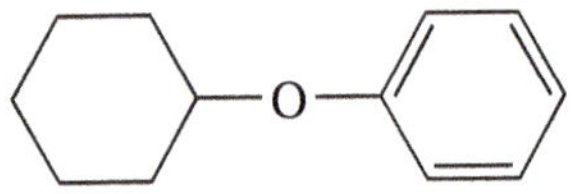 cyclohexyl—O—phenyl

c. $CH_3CH_2CH_2OCH_3$

13. For each of the following pairs of E2 reactions, indicate the one that occurs with the greater rate constant:

a. 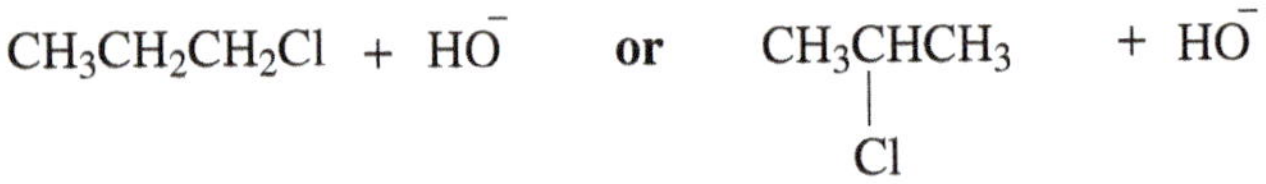 $CH_3CH_2CH_2Cl + HO^-$ or CH_3CHCH_3 (with Cl on CH) $+ HO^-$

b. $CH_3CH_2CH_2Cl + HO^-$ or $CH_3CH_2CH_2I + HO^-$

c. $CH_3CH_2CH_2Br + HO^-$ or $CH_3CH_2CH_2Br + H_2O$

d. 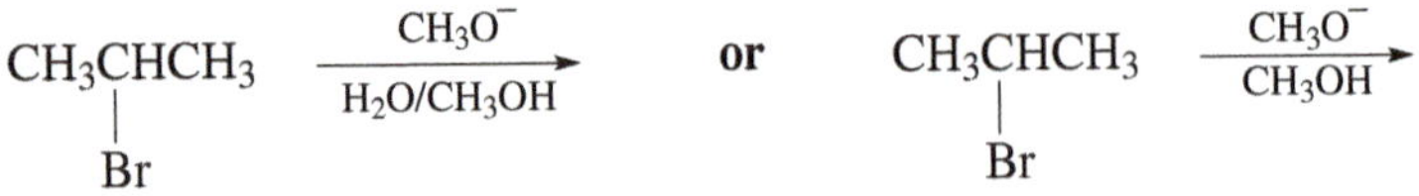 CH_3CHCH_3 (with Br on CH) $\xrightarrow[H_2O/CH_3OH]{CH_3O^-}$ or CH_3CHCH_3 (with Br on CH) $\xrightarrow[CH_3OH]{CH_3O^-}$

e. 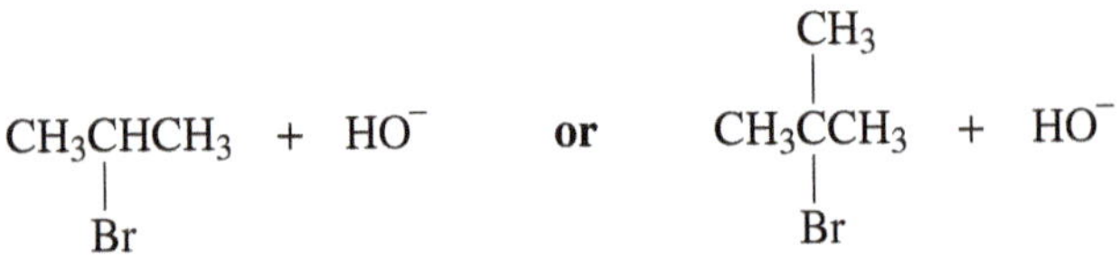 CH_3CHCH_3 (with Br on CH) $+ HO^-$ or CH_3CCH_3 (with CH_3 above and Br below C) $+ HO^-$

14. What is the major product obtained from the E2 reaction of each of the following compounds with hydroxide ion?

a. 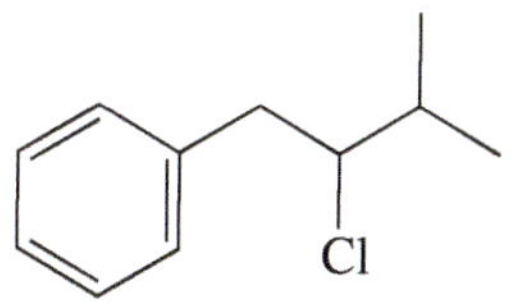 Cl

b. 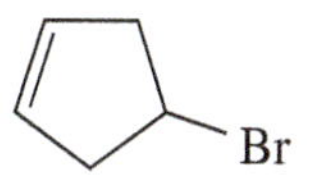 Br

c. 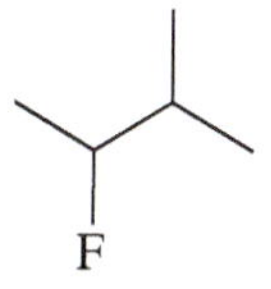 F

d. 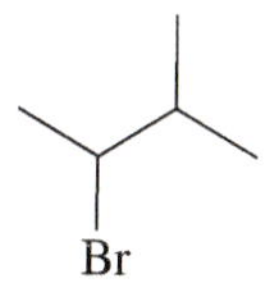 Br

15. Which is more reactive in an E2 reaction, *cis*-1-bromo-2-methylcyclohexane or *trans*-1-bromo-2-methylcyclohexane?

CHAPTER 10
Reactions of Alcohols, Ethers, Epoxides, Amines, and Sulfur-Containing Compounds

Important Terms

alcohol an alkane with an OH group in place of an H (ROH).

alkaloid amines found in the leaves, bark, roots, or seeds of plants.

alkyl tosylate an ester of *para*-toluenesulfonic acid.

crown ether a cyclic molecule that possesses several ether linkages around a central cavity.

dehydration loss of water.

epoxide an ether in which the oxygen is incorporated into a three-membered ring.

ether a compound containing an oxygen bonded to two carbons (ROR).

inclusion compound the complex formed when a crown ether specifically binds a metal ion or an organic molecule.

Hofmann elimination reaction elimination of a proton and a tertiary amine from a quaternary ammonium hydroxide.

lead compound a prototype in a search for other physiologically active compounds.

mercapto group an SH group.

molecular modification changing the structure of a lead compound.

quaternary ammonium ion a cation containing a nitrogen bonded to four alkyl groups (R_4N^+).

sulfide (thioether) the sulfur analogue of an ether (RSR).

sulfonate ester the ester of a sulfonic acid (RSO_2OR).

sulfonium salt $R_3S^+X^-$

thioether (sulfide) the sulfur analogue of an ether (RSR).

thiol (mercaptan) the sulfur analogue of an alcohol (RSH).

Solutions to Problems

1. They no longer have a lone pair of electrons.

2. Solved in the text.

3. The leaving group of $CH_3\overset{+}{O}H_2$ is H_2O; the pK_a of its conjugate is -1.7.

The leaving group of CH_3OH is HO^-; the pK_a of its conjugate acid is 15.7.

Because H_3O^+ is a much stronger acid than H_2O, H_2O is a much weaker base than HO^- and, therefore, is the better leaving group. Therefore, $CH_3\overset{+}{O}H_2$ is more reactive than CH_3OH.

4. Solved in the text.

5. **a.** 1-butanol (OH) $\xrightarrow[\text{2. } CH_3CH_2\overset{O}{\overset{\|}{C}}O^-]{\text{1. HBr, } \Delta}$ butyl propanoate

b. 1-butanol (OH) $\xrightarrow[\text{2. } CH_3CH_2N(CH_3)_2]{\text{1. HBr, } \Delta}$ butylethyldimethylammonium ($\overset{+}{N}$)

c. $CH_3CH_2CH_2CH_2OH \xrightarrow[\text{2. } ^-C\equiv N]{\text{1. HBr, } \Delta} CH_3CH_2CH_2CH_2C\equiv N$

6. All six alcohols undergo an S_N1 reaction, because they are either secondary or tertiary alcohols. The arrows are shown for the first protonation step in part **a**, but are not shown for that step in parts **b**, **c**, **d**, **e**, and **f**.

a. $CH_3CH_2\underset{:\ddot{O}H}{\underset{|}{C}}HCH_3 + H{-}Br \rightleftharpoons CH_3CH_2\underset{:\overset{+}{O}H_2}{\underset{|}{C}}HCH_3 \rightleftharpoons CH_3CH_2\overset{+}{C}HCH_3 + H_2O \xrightarrow{Br^-} CH_3CH_2\underset{Br}{\underset{|}{C}}HCH_3$

b. 1-methylcyclopentanol (CH3, OH) $\underset{\text{HCl}}{\rightleftharpoons}$ protonated alcohol (CH3, $\overset{+}{O}H$, H) $\rightleftharpoons$ 1-methylcyclopentyl cation (CH3, +) + H_2O $\xrightarrow{Cl^-}$ 1-chloro-1-methylcyclopentane (CH3, Cl)

The carbocations that are initially formed in **c** and **d** rearrange to form more stable carbocations.

c. CH_3 / $CH_3C—CHCH_3$ / CH_3 OH $\xrightleftharpoons{HBr}$ CH_3 / $CH_3C—CHCH_3$ / CH_3 $\overset{+}{O}H$ H $\rightleftharpoons$ CH_3 / $CH_3C—\overset{+}{C}HCH_3$ / CH_3 + H_2O

secondary carbocation

1,2-methyl shift

CH_3 / $CH_3\overset{+}{C}—CHCH_3$ / CH_3

tertiary carbocation

$\xrightarrow{Br^-}$ CH_3 / $CH_3C—CHCH_3$ / Br CH_3

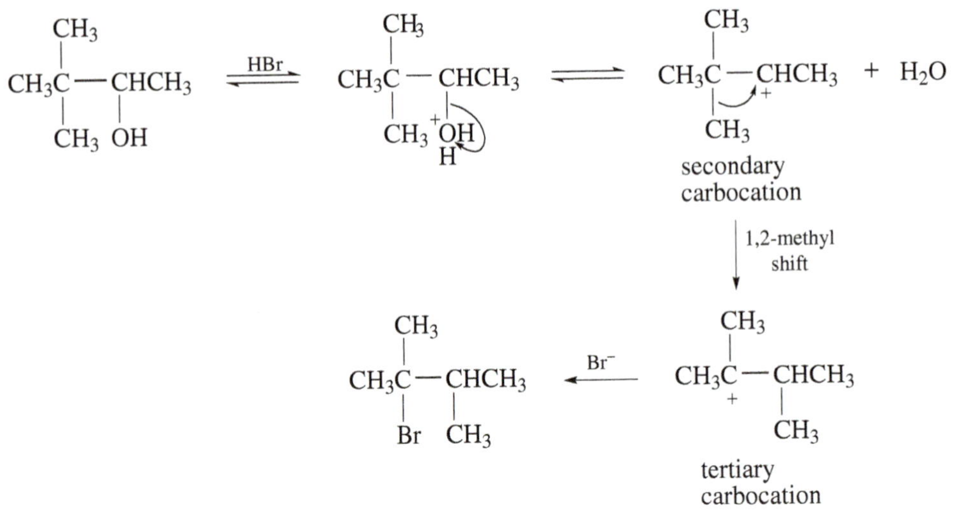

d. OH $\xrightleftharpoons{HBr}$ $\overset{+}{O}H$ H $\rightleftharpoons$ H + $\xrightarrow{\text{1,2-hydride shift}}$ + :Br:$^-$ → Br

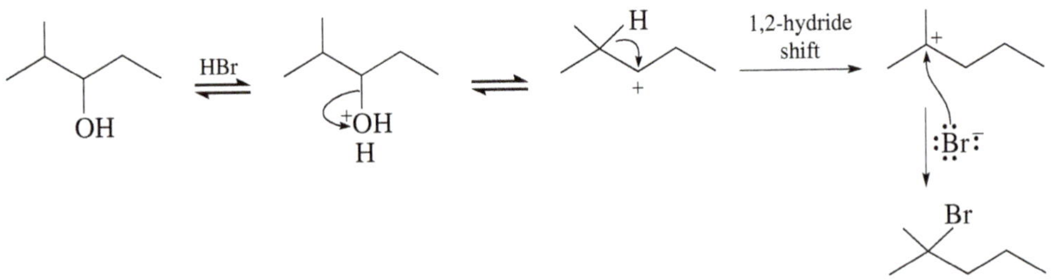

e. CH_3 OH $\xrightleftharpoons{HCl}$ CH_3 $\overset{+}{O}H$ H $\rightleftharpoons$ CH_3 H + $\xrightarrow{\text{1,2-hydride shift}}$ + CH_3 :Cl:$^-$ → CH_3 Cl

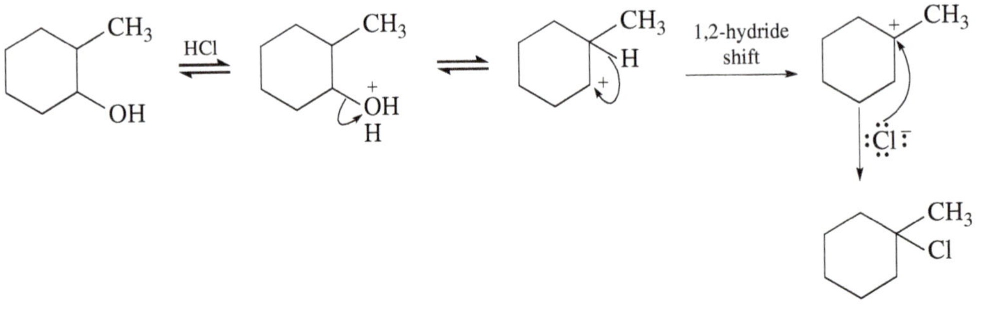

f. CH_3 OH $CHCH_3$ $\xrightleftharpoons{HCl}$ CH_3 H $\overset{+}{O}H$ $CHCH_3$ $\rightleftharpoons$ CH_3 $\overset{+}{C}HCH_3$ H + H_2O

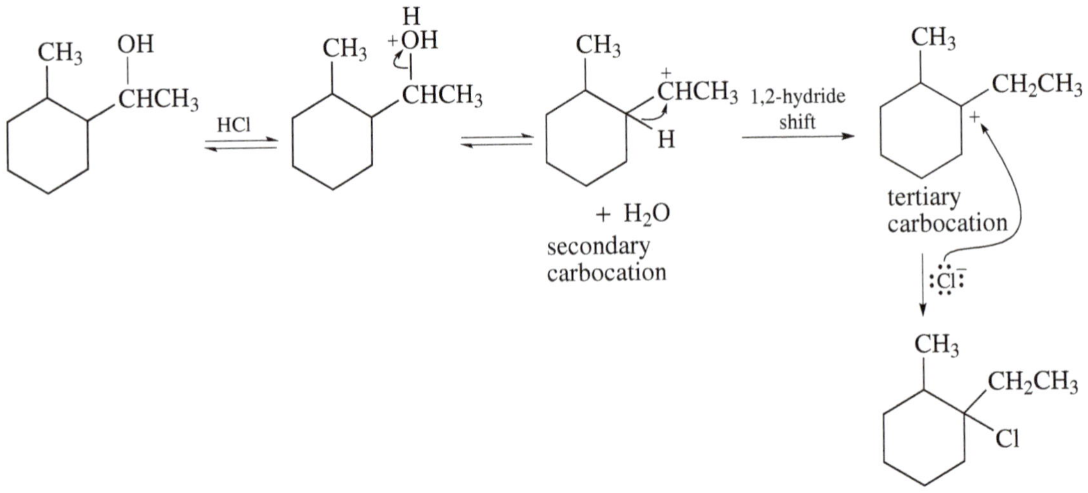

secondary carbocation

$\xrightarrow{\text{1,2-hydride shift}}$ CH_3 CH_2CH_3 +

tertiary carbocation

:Cl:$^-$ → CH_3 CH_2CH_3 Cl

7. Solved in the text.

8.

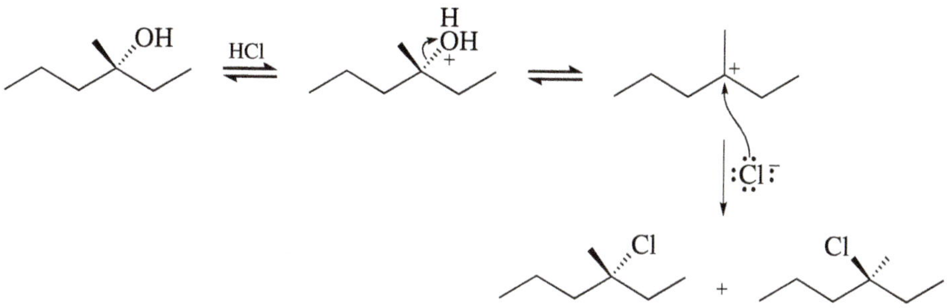

OH

HCl

H

OH

+

+

:Cl:⁻

Cl

Cl

+

9. **a.**

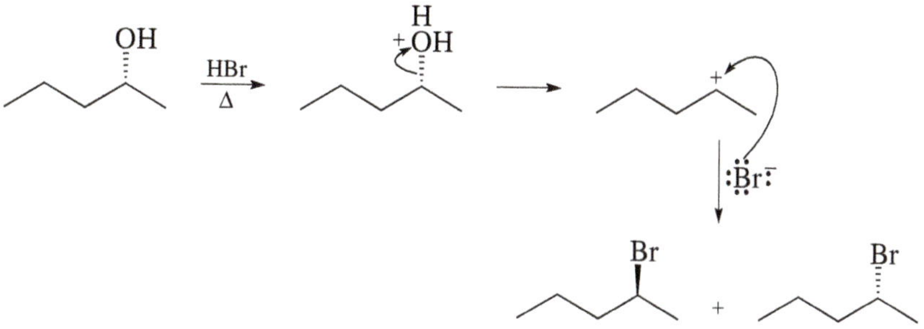

OH

HBr

Δ

H

+OH

+

:Br:⁻

Br

Br

+

b.

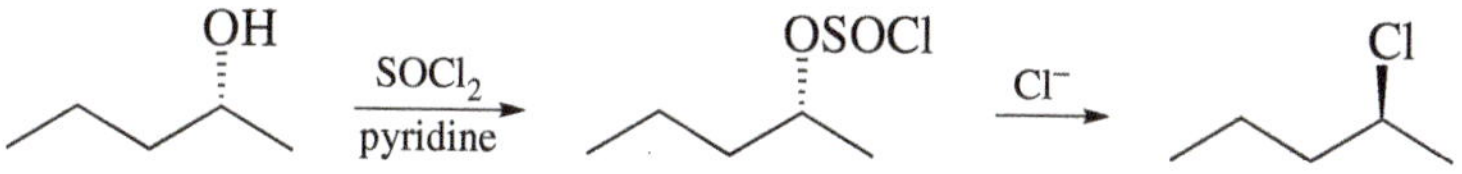

OH

$SOCl_2$

pyridine

OSOCl

Cl^-

Cl

c.

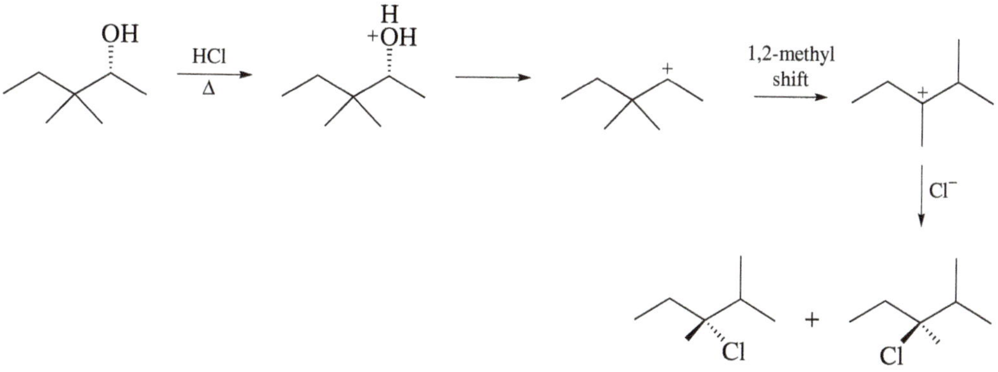

OH

HCl

Δ

H

+OH

+

1,2-methyl shift

+

Cl^-

Cl

+

Cl

d.

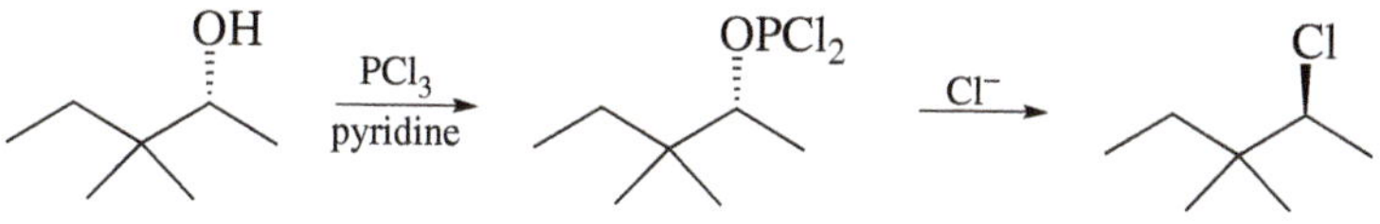

OH

PCl_3

pyridine

$OPCl_2$

Cl^-

Cl

10. Solved in the text.

11. **a.**

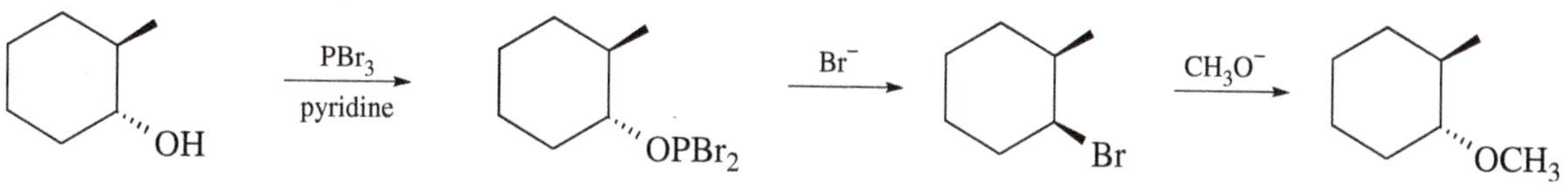

OH

PBr_3

pyridine

$OPBr_2$

Br^-

Br

CH_3O^-

OCH_3

b. (cyclohexanol with methyl) $\xrightarrow[\Delta]{HBr}$ (protonated alcohol, $\overset{+}{O}H$ H) $\longrightarrow$ (secondary cation) $\xrightarrow{\text{1,2-hydride shift}}$ (tertiary cation) $\longrightarrow$ (1-bromo-1-methylcyclohexane, Br)

c. (OH) $\xrightarrow[\text{pyridine}]{SOCl_2}$ (OSOCl) $\xrightarrow{Cl^-}$ (Cl) $\xrightarrow{CH_3O^-}$ (OCH_3)

d. (OH) $\xrightarrow[\text{pyridine}]{TsCl}$ (OTs) $\xrightarrow{CH_3O^-}$ (OCH_3)

12. **a.** $CH_3CH_2CH_2OH \xrightarrow[\text{2. } CH_3CH_2S^-]{\text{1. TsCl/pyridine}} CH_3CH_2CH_2SCH_2CH_3$

b. $CH_3CH_2CH_2OH \xrightarrow[\text{2. } CH_3CH(CH_3)CH_2O^-]{\text{1. TsCl/pyridine}} CH_3CH_2CH_2OCH_2CH(CH_3)CH_3$

13. **D** because it forms a very stable tertiary allylic cation intermediate.

14. **a.** $CH_3CH_2C(CH_3)(OH)CHCH_3(CH_3) \underset{}{\overset{H_2SO_4}{\rightleftharpoons}} CH_3CH_2C(CH_3)(\overset{+}{O}H_2)CHCH_3(CH_3) \rightleftharpoons CH_3CH_2\overset{+}{C}(CH_3)CHCH_3(CH_3) + H_2O$

$\rightleftharpoons H_3O^+ + (H_3C)(CH_3CH_2)C{=}C(CH_3)(CH_3)$

b. (cyclohex-2-enol, OH) $\underset{\Delta}{\overset{H_2SO_4}{\rightleftharpoons}}$ (protonated, $\overset{+}{O}H$ H) $\rightleftharpoons$ (cation, +) $\underset{}{\overset{\text{1,2-hydride shift}}{\rightleftharpoons}}$ (allylic cation, H, $H_2\ddot{O}:$) $\rightleftharpoons$ (cyclohexadiene) $+ H_3O^+$

c. $CH_3CH_2CH(OH)C(CH_3)_2CH_3 \xrightarrow[\Delta]{H_2SO_4} CH_3CH_2CH(\overset{+}{O}H_2)C(CH_3)_2CH_3 \rightleftharpoons CH_3CH_2\overset{+}{C}HC(CH_3)_2CH_3 + H_2O$

1,2-methyl shift

$CH_3CH_2CH(CH_3)\overset{+}{C}(CH_3)CH_3 \longrightarrow (H_3C)(CH_3CH_2)C{=}C(CH_3)_2 + H_3O^+$

d. $CH_3CH_2CH_2CH_2CH_2OH \xrightarrow[\Delta]{H_2SO_4} CH_3CH_2CH_2CH_2CH_2\overset{+}{O}H_2 \longrightarrow CH_3CH_2CH_2CH{=}CH_2 + H_3O^+$

$\rightleftharpoons CH_3CH_2CH_2\overset{+}{C}HCH_3 + H_2O \rightleftharpoons$ (CH₃CH₂)(H)C=C(CH₃)(H) (cis) + (CH₃CH₂)(H)C=C(H)(CH₃) (trans) + H_3O^+

major product

e. $CH_2{=}CHCH_2CH_2OH \xrightarrow[\Delta]{H_2SO_4} CH_2{=}CHCH_2CH_2\overset{+}{O}H_2 \rightleftharpoons CH_2{=}CHCH{=}CH_2 + H_3O^+$

f. In **d** and **f**, the reactant is a primary alcohol. Therefore, elimination of water takes place via an E2 reaction. Because the dehydration reaction is being carried out in an acidic solution, the alkene that is formed initially is protonated to form a carbocation. The proton that is then lost from the carbocation results in formation of the most stable alkene.

cyclohexyl-$CH_2OH \xrightarrow{H_2SO_4}$ cyclohexyl-$CH_2\overset{+}{O}H_2$ ($H_2\ddot{O}$: removes H) $\rightleftharpoons$ methylenecyclohexane ($=CH_2$) $+ H_3O^+$ $\longrightarrow$ 1-methylcyclohexyl cation (CH_3, $H_2\ddot{O}$: removes H) $\rightleftharpoons$ 1-methylcyclohexene (CH_3) $+ H_3O^+$

15. a. To synthesize an unsymmetrical ether (ROR′) by this method, two different alcohols (ROH and R′OH) have to be heated with sulfuric acid. Therefore, three different ethers would be obtained as products. Consequently, less than half of the total amount of ether that is synthesized would be the desired ether.

$$ROH + R'OH \xrightarrow[\Delta]{H_2SO_4} ROR + ROR' + R'OR'$$

(ROR′ = target molecule)

b. It could be synthesized by a Williamson ether synthesis. (See Section 9.15 on page 443 of the text.)

$$CH_3CH_2CH_2OH \xrightarrow{NaH} CH_3CH_2CH_2O^- \xrightarrow{CH_3CH_2Br} CH_3CH_2CH_2OCH_2CH_3 + Br^-$$

16. a. $H{-}OSO_3H$ CH_3 CH_3 $\ddot{O}H$ CH_3 CH_3 $\overset{+}{O}H$ H CH_3 CH_3 $+ H_2O$ CH_3 H $H_2\ddot{O}:$ CH_3 $H_3O^+ +$ CH_3 CH_3

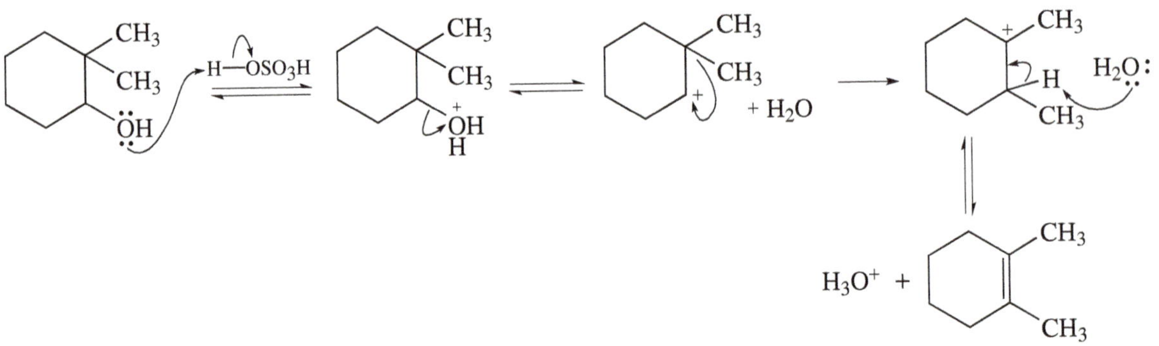

b. Because a primary carbocation cannot be formed, the dehydration is an E2 reaction. The alkene that results is protonated, and the proton that is removed results in formation of the most stable alkene.

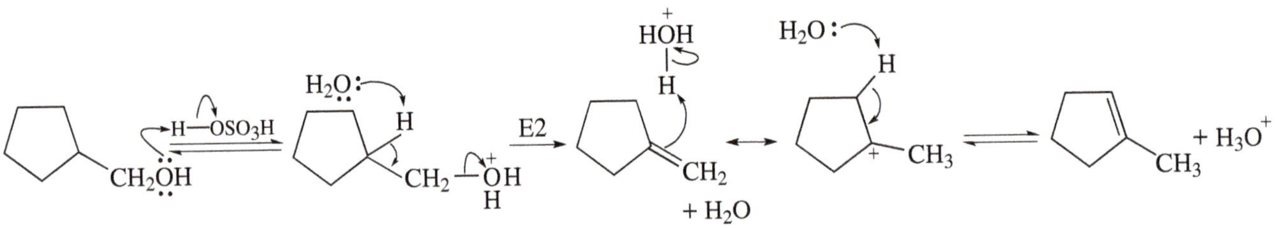

$H{-}OSO_3H$ $CH_2\ddot{O}H$ $H_2\ddot{O}:$ H $CH_2{-}\overset{+}{O}H$ H E2 $\overset{+}{HOH}$ H CH_2 $+ H_2O$ $H_2O:$ H CH_3 CH_3 $+ H_3O^+$

17. The strained three-membered ring causes a tertiary carbocation with a three-membered ring to be less stable than a secondary carbocation with a less strained four-membered ring. The secondary carbocation then rearranges to a tertiary carbocation.

a. CH_3 $\ddot{O}H$ CH_3 $H{-}Br$ CH_3 H $\overset{+}{O}H$ CH_3 Br^- 3 2 1 4 CH_3 CH_3 $+ H_2O$ 3 2 4 1 CH_3 CH_3 CH_3 CH_3 $:\ddot{B}r:^-$ CH_3 CH_3 Br

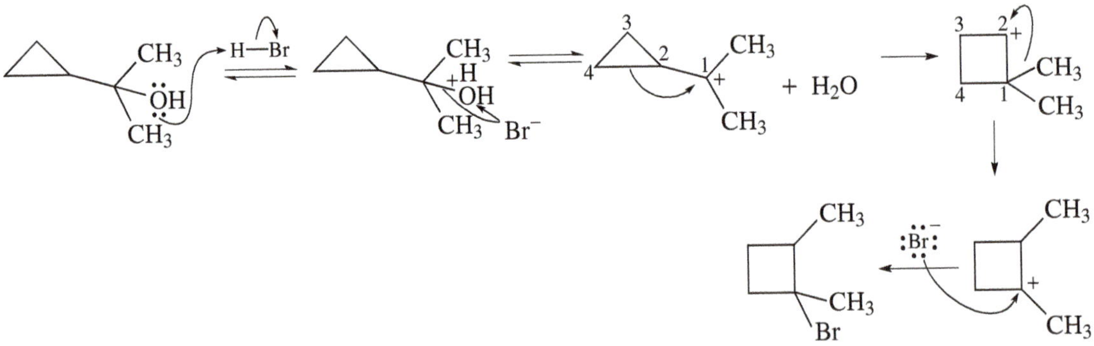

b. $:\ddot{O}H$ $CHCH_2CH_3$ $H{-}OSO_3H$ H $\overset{+}{O}H$ $CHCH_2CH_3$ $\overset{+}{C}HCH_2CH_3$ $H_2\ddot{O}:$ H CH_2CH_3 $H_3O^+ +$ CH_2CH_3

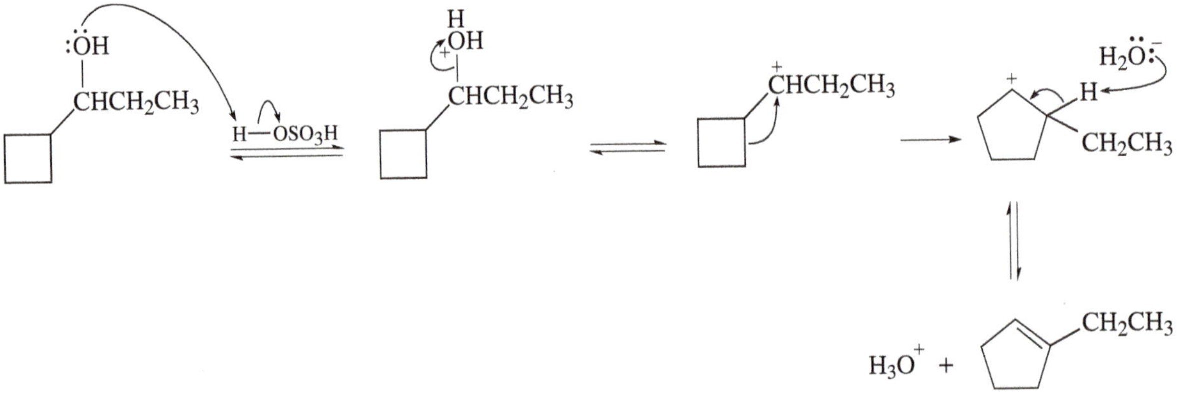

18. Both alcohols form the same carbocation, so they both form the same alkenes.

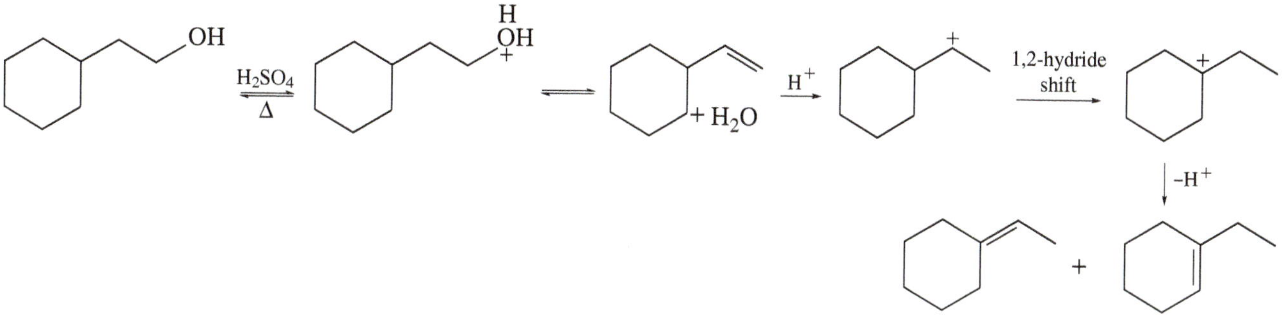

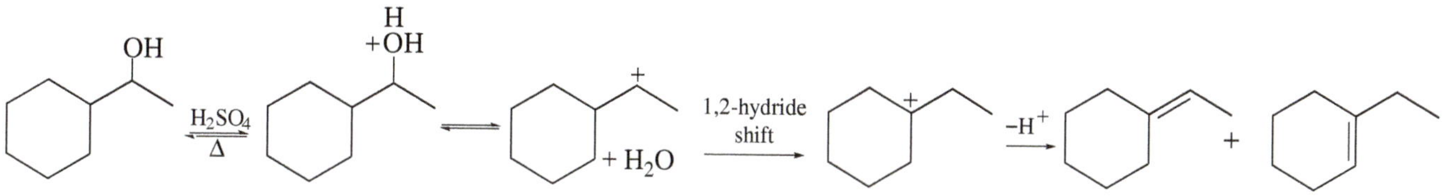

OH $\xrightleftharpoons[\Delta]{H_2SO_4}$ H OH + $\rightleftharpoons$ + H_2O $\xrightarrow{H^+}$ + $\xrightarrow{\text{1,2-hydride shift}}$ + $\xrightarrow{-H^+}$ +

OH $\xrightleftharpoons[\Delta]{H_2SO_4}$ H +OH $\rightleftharpoons$ + + H_2O $\xrightarrow{\text{1,2-hydride shift}}$ + $\xrightarrow{-H^+}$ +

19. **a.** H, CH_3 C=C H, CH_2CH_3 (major) + H, CH_3 C=C H, CH_2CH_3 (minor)

b. CH_3, CH_3CH_2 C=C CH_2CH_3, CH_3 (major) + CH_3, CH_3CH_2 C=C CH_3, CH_2CH_3 (minor)

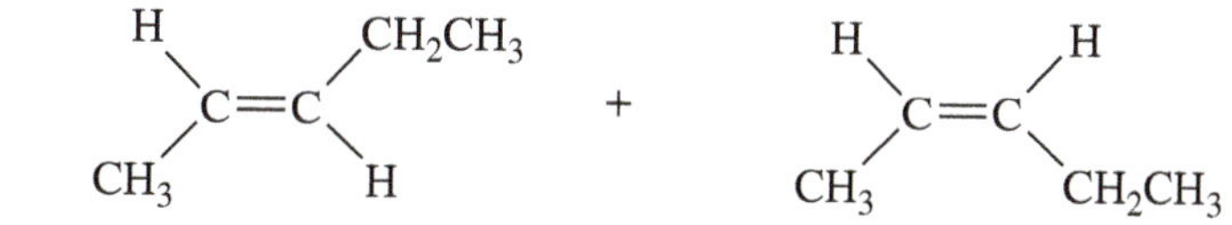

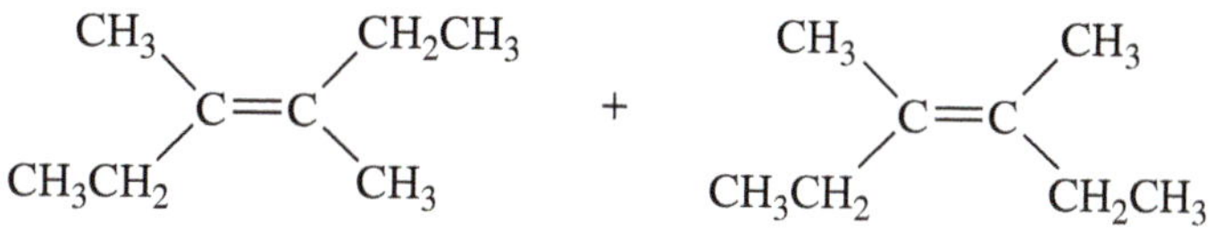

20. **a.** $CH_3CH_2C{=}CCHCH_3$ (with CH_3 and CH_2CH_3 on the alkene carbons, CH_3 on CH) **b.** CH_3, CH_2CH_3 C=C $CHCH_3$ (CH_3), CH_2CH_3

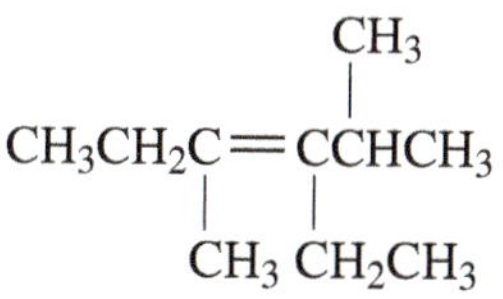

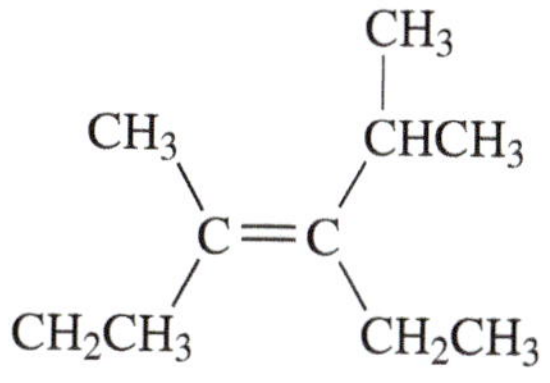

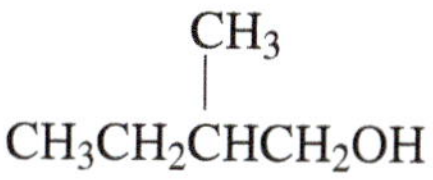

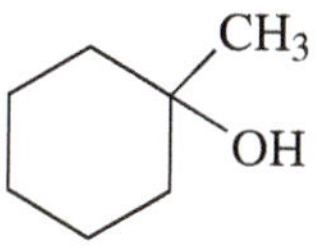

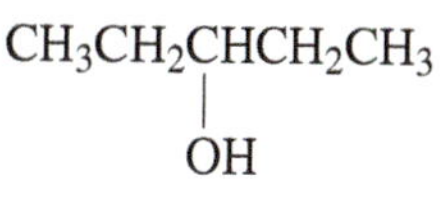

21. **a.** $CH_3CH_2CHCH_2OH$ (CH_3 on CH) **b.** CH_3, OH (on cyclohexane) **c.** $CH_3CH_2CHCH_2CH_3$ (OH on CH) **d.** CH_2OH (on cyclohexane)

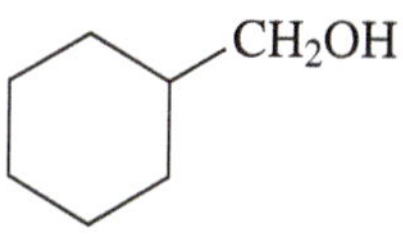

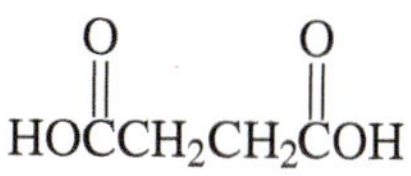

22. **a.** The answers to **1**, **3**, **4**, and **5** are the same as in part **b**.

2. $CH_3CH_2CH_2CH_2\overset{O}{\overset{\|}{C}}OH$ **6.** $HO\overset{O}{\overset{\|}{C}}CH_2CH_2\overset{O}{\overset{\|}{C}}OH$

b. **1.** 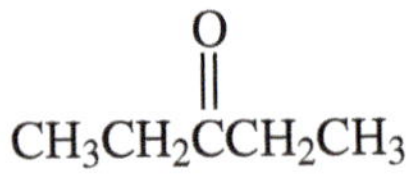 $CH_3CH_2\overset{O}{\overset{\|}{C}}CH_2CH_3$

2. $CH_3CH_2CH_2CH_2\overset{O}{\overset{\|}{C}}H$

3. A tertiary alcohol is not oxidized to a carbonyl compound.

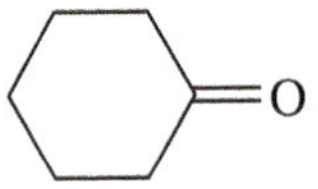

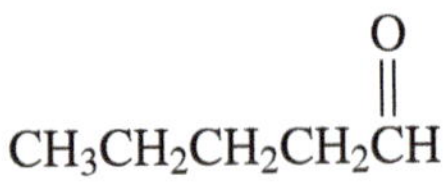

4. 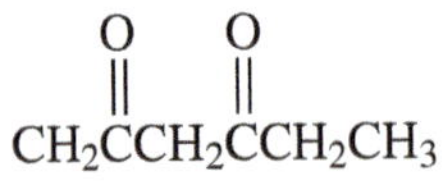 $CH_3\overset{O}{\overset{\|}{C}}CH_2\overset{O}{\overset{\|}{C}}CH_2CH_3$

5. (cyclohexane ring)=O

6. 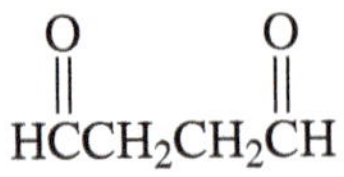 $H\overset{O}{\overset{\|}{C}}CH_2CH_2\overset{O}{\overset{\|}{C}}H$

c. The same products as formed in **b.**

23. Solved in the text.

24. **a.** Cleavage occurs by an S_N1 pathway because the allyl cation is relatively stable.

$$CH_2{=}CHCH_2\overset{+}{O}(H)CH_2CH_3 \longrightarrow CH_2{=}CH\overset{+}{C}H_2 + HOCH_2CH_3$$

$$CH_2{=}CH\overset{+}{C}H_2 \xrightarrow{I^-} CH_2{=}CHCH_2I$$

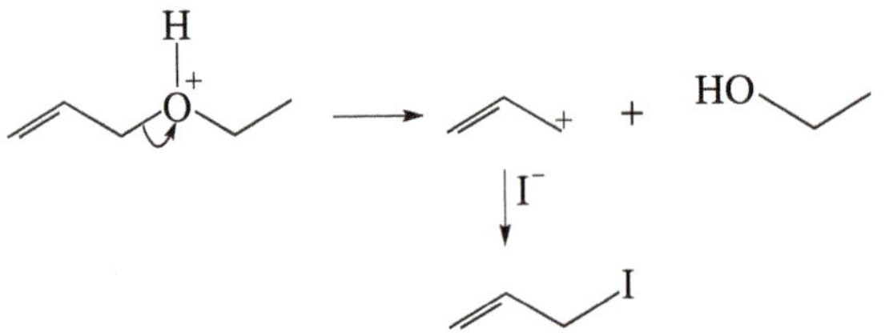

b. Cleavage occurs by an S_N2 pathway because a primary carbocation is too unstable to be formed.

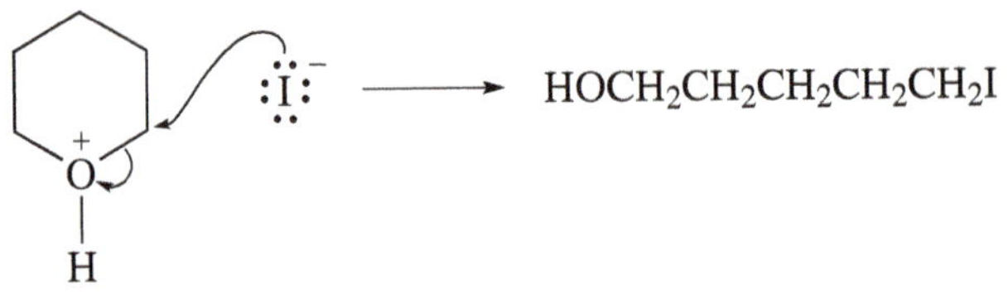

$$\overset{+}{O}{-}H \text{ (ring)} + :\ddot{I}:^- \longrightarrow HOCH_2CH_2CH_2CH_2CH_2I$$

c. Cleavage occurs by an S_N1 pathway because the benzyl cation is relatively stable.

$$C_6H_5{-}CH_2{-}\overset{+}{O}(H){-}C_6H_5 \longrightarrow C_6H_5{-}\overset{+}{C}H_2 + HO{-}C_6H_5$$

$$C_6H_5{-}\overset{+}{C}H_2 \xrightarrow{I^-} C_6H_5{-}CH_2I$$

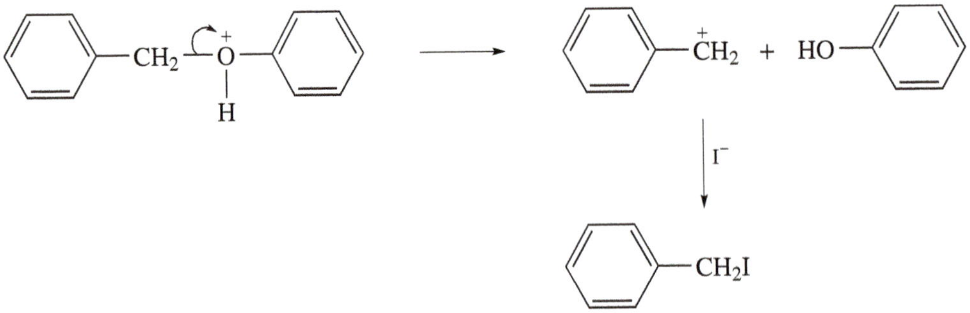

d. Cleavage occurs by an S_N1 pathway because the benzyl cation is relatively stable.

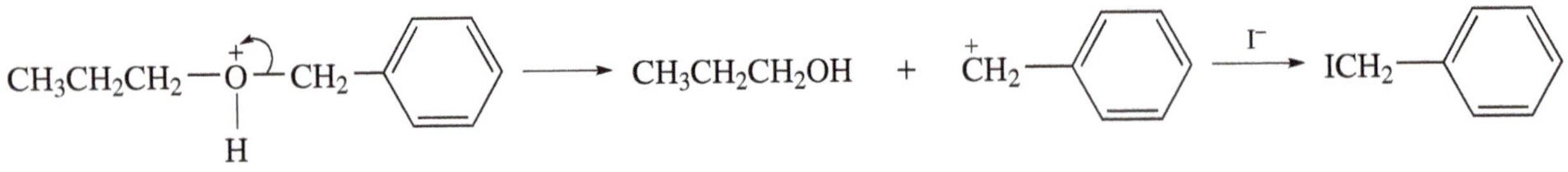

$$CH_3CH_2CH_2{-}\overset{+}{O}(H){-}CH_2{-}C_6H_5 \longrightarrow CH_3CH_2CH_2OH + \overset{+}{C}H_2{-}C_6H_5 \xrightarrow{I^-} ICH_2{-}C_6H_5$$

e. Cleavage occurs by an S_N2 pathway because a primary carbocation or a vinylic cation is too unstable to be formed.

$$\text{cyclohexenyl-}\overset{+}{O}(H)\text{-}CH_3 \quad :\ddot{I}:^- \longrightarrow CH_3I + \text{cyclohexenol (enol)} \xrightleftharpoons{\text{tautomerization}} \text{cyclohexanone (ketone)}$$

enol ketone

f. Cleavage occurs by an S_N1 pathway, because the tertiary carbocation is relatively stable.

$$\text{2,2-dimethyltetrahydrofuran-}\overset{+}{O}H \longrightarrow HOCH_2CH_2CH_2\overset{+}{C}(CH_3)CH_3 \xrightarrow{I^-} HOCH_2CH_2CH_2C(CH_3)(I)CH_3$$

25. Solved in the text.

26. We saw that HCl does not cleave ethers because Cl^- is not a good enough nucleophile. F^- is an even poorer nucleophile, so HF cannot cleave ethers. Therefore, ethers can be cleaved only with HBr or HI.

27. **a.** $HOCH_2C(OCH_3)(CH_3)CH_3$ **b.** $CH_3OCH_2C(OH)(CH_3)CH_3$ **c.** $CH_3C(OH)(CH_3)\text{—}C(OCH_3)(CH_3)CH_3$ **d.** $CH_3C(OCH_3)(CH_3)\text{—}C(OH)(CH_3)CH_3$

28. The reactivity of tetrahydrofuran is more similar to that of a noncyclic ether, because the five-membered ring does not have the strain that makes the epoxide reactive.

29. **a.**

$$CH_3C(CH_3)_2CH{=}CH_2 \xrightleftharpoons{H_2SO_4} CH_3C(CH_3)_2\overset{+}{C}H\text{—}CH_3 \xrightarrow{\text{1,2-methyl shift}} CH_3\overset{+}{C}(CH_3)CH(CH_3)\text{—}CH_3 \xrightarrow{H_2O} CH_3C(OH)(CH_3)\text{—}CH(CH_3)CH_3 + H^+$$

b.

$$CH_3C(CH_3)_2CH{=}CH_2 \xrightarrow{\text{MCPBA}} CH_3C(CH_3)_2\text{—}\overset{O}{CH\text{—}CH_2} \xrightarrow{H^-} CH_3C(CH_3)_2\text{—}CH(O^-)CH_3 \xrightarrow{HCl} CH_3C(CH_3)_2\text{—}CH(OH)CH_3$$

30. **a.** 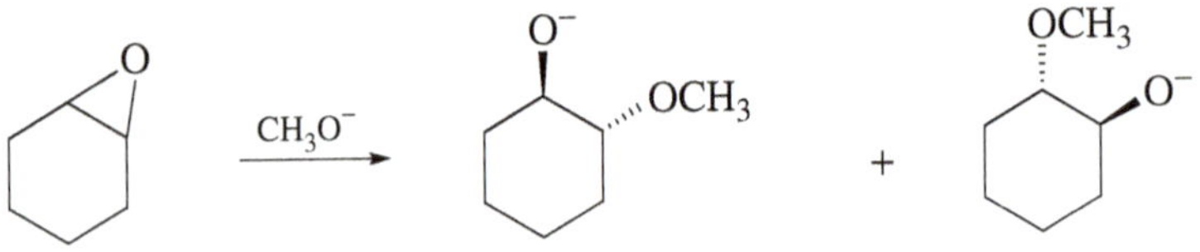 $\xrightarrow{CH_3O^-}$ (trans-2-methoxycyclohexanolate) + (enantiomer)

The nucleophile can attack the back side of either of the carbons in the three-membered ring.

b. 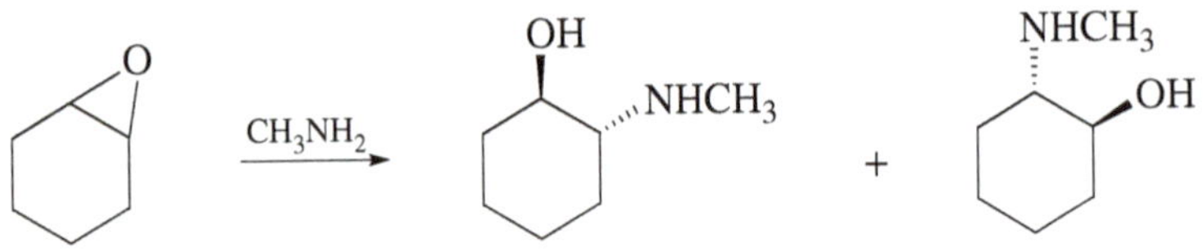 $\xrightarrow{CH_3NH_2}$ (trans-2-(methylamino)cyclohexanol) + (enantiomer)

31. **a.** 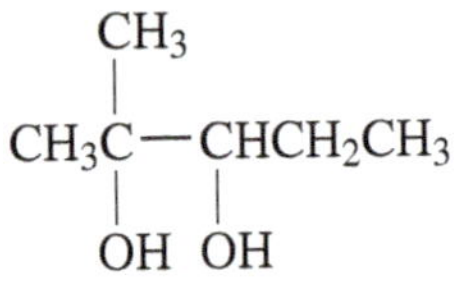 $CH_3C(CH_3)(OH)-CH(OH)CH_2CH_3$ **b.** 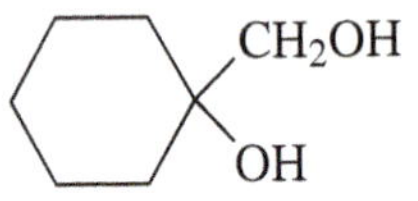 1-(hydroxymethyl)cyclohexanol (CH_2OH, OH)

32. **a.** The reaction of an alkene with OsO_4 is a syn addition reaction. Syn addition to the trans isomer forms the threo pair of enantiomers. (See Section 6.13, particularly page 268.)

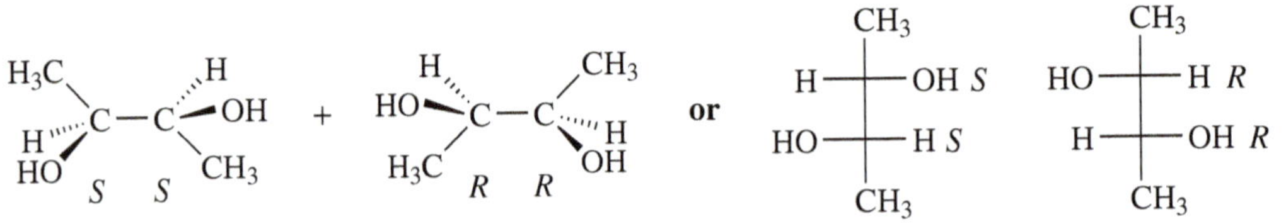

S S + *R R* or

	CH_3	
H		OH *S*
HO		H *S*
	CH_3	

	CH_3	
HO		H *R*
H		OH *R*
	CH_3	

b. Syn addition to the cis isomer forms the erythro pair of enantiomers. In this case, the product is a meso compound because each asymmetric center is bonded to the same four substituents. Therefore, only one stereoisomer is formed.

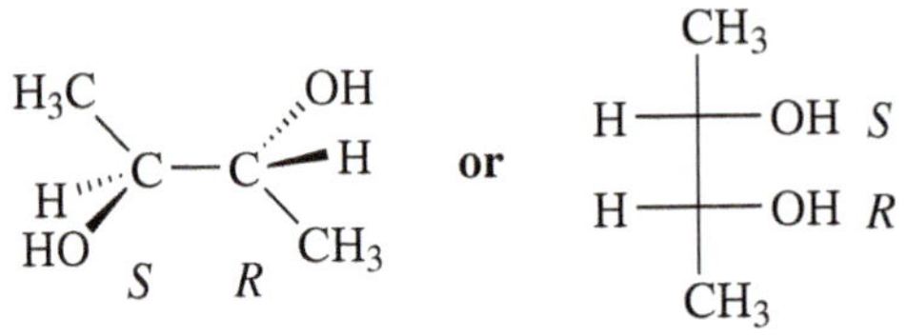

S R or

	CH_3	
H		OH *S*
H		OH *R*
	CH_3	

c. Syn addition to the cis isomer forms the erythro pair of enantiomers.

S R + *S R* or

	CH_3	
H		OH *S*
H		OH *R*
	CH_2CH_3	

	CH_3	
HO		H *R*
HO		H *S*
	CH_2CH_3	

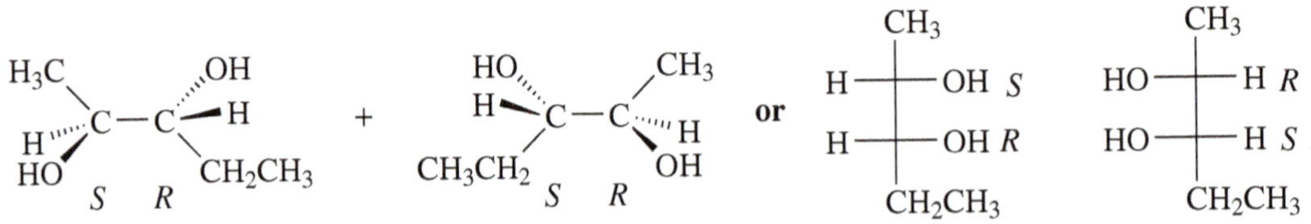

d. Syn addition to the trans isomer forms the threo pair of enantiomers.

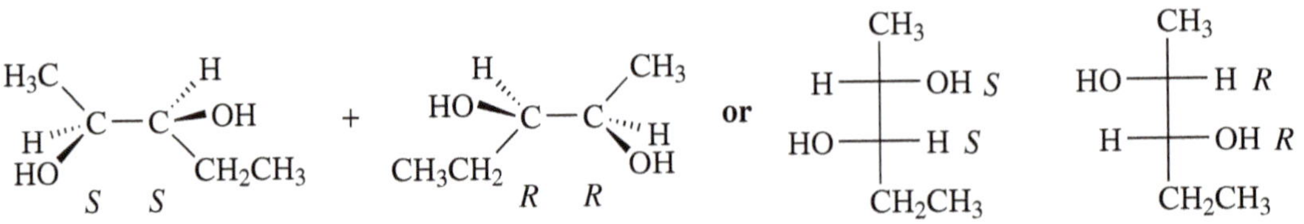

S S + *R R* or

	CH_3	
H		OH *S*
HO		H *S*
	CH_2CH_3	

	CH_3	
HO		H *R*
H		OH *R*
	CH_2CH_3	

33. **a.** Reaction with a peroxyacid forms a product with two new asymmetric centers. Because it is a syn addition reaction, the trans alkene forms the two trans products.

Hydroxide ion can attack either of the two asymmetric centers in the epoxide in an S_N2 reaction. The asymmetric center that is attacked will have its configuration inverted. The configuration of the other asymmetric center will not change.

Therefore, when hydroxide ion attacks the asymmetric center on the left of epoxide **A**, the *S,R* stereoisomer is formed; when it attacks the asymmetric center on the right of epoxide **A**, the *R,S* stereoisomer is formed. The *S,R* and *R,S* stereoisomers are identical; it is a meso compound.

Attack of hydroxide ion on epoxide **B** forms the same meso compound.

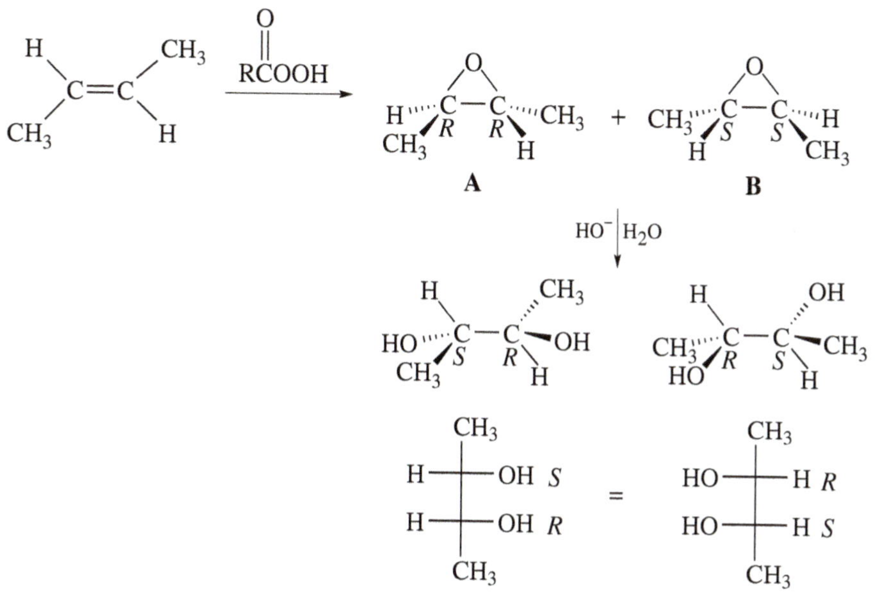

a meso compound

b. Reaction of the cis alkene with a peroxyacid forms only one peroxide (a meso compound). The meso compound has the *R,S* configuration, so when it reacts with hydroxide ion, the *S,S* and *R,R* products are formed.

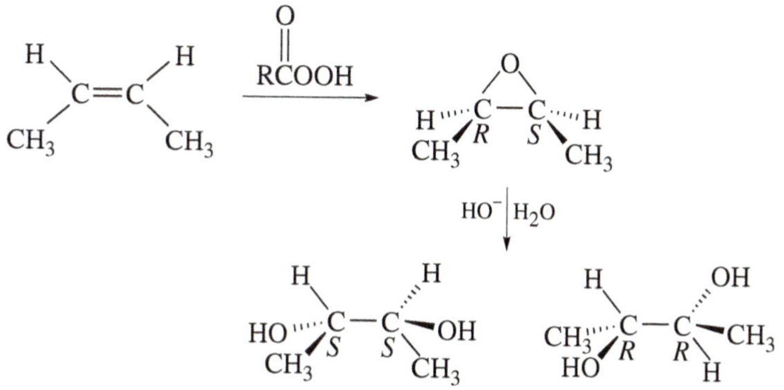

c. Reaction of the cis alkene with a peroxyacid forms the two cis isomers. Hydroxide ion preferentially attacks the least sterically hindered carbon of each epoxide. Therefore, the *S,S* and *R,R* products are formed.

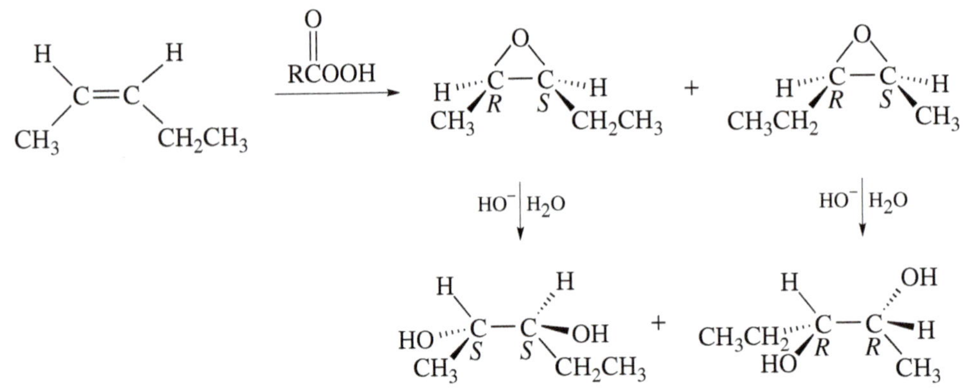

d. Reaction of the trans alkene with a peroxyacid forms the two trans isomers. Hydroxide ion preferentially attacks the least sterically hindered carbon of each epoxide. Therefore, the *S,R* and *R,S* products are formed.

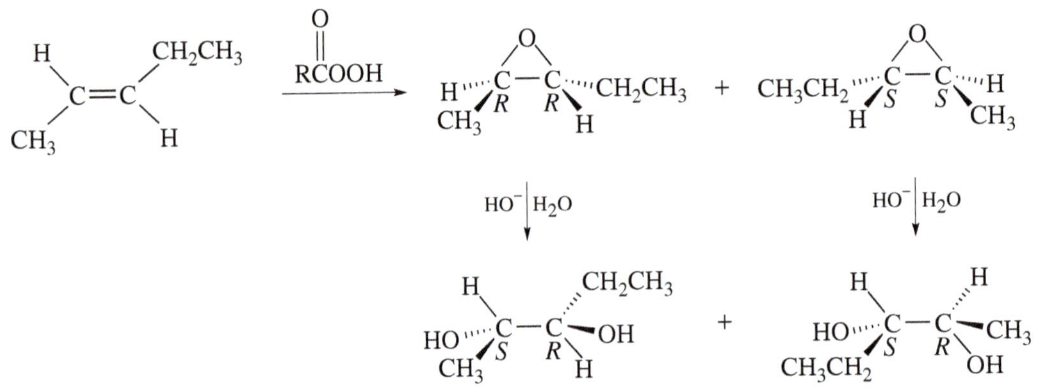

34.

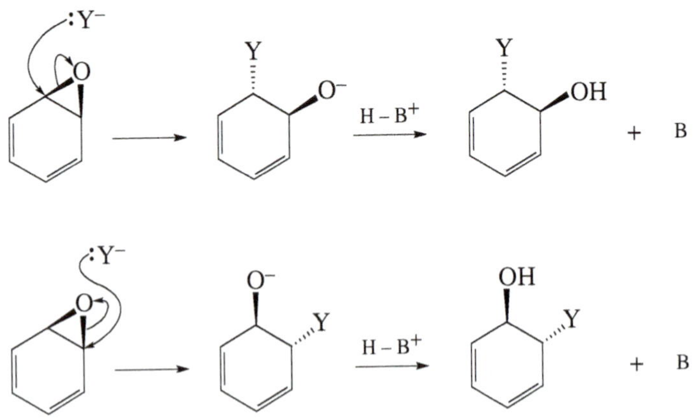

35. The carbocation leading to 1-naphthol can be stabilized by electron delocalization without destroying the aromaticity of the intact benzene ring. The carbocation leading to 2-naphthol can be stabilized by electron delocalization only by destroying the aromaticity of the intact benzene ring. Therefore, the carbocation leading to 1-naphthol is more stable.

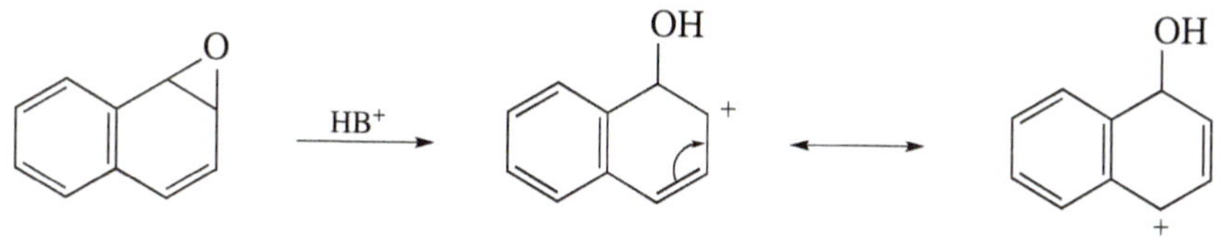

HB^+

carbocation that leads to 1-naphthol

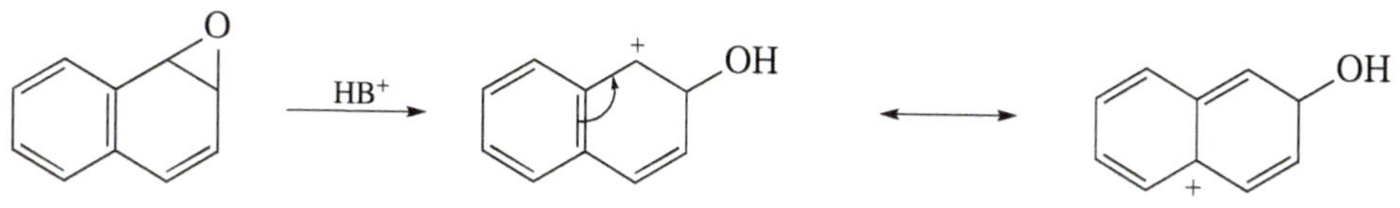

HB^+

carbocation that leads to 2-naphthol

36. **a.** **with an NIH shift**

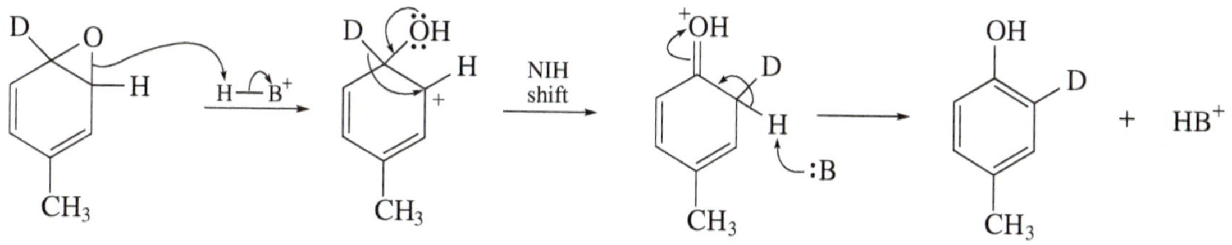

NIH shift

$+ \; HB^+$

b. **without an NIH shift**

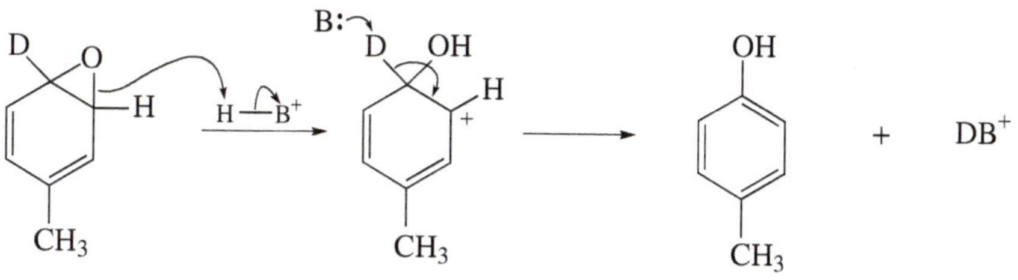

$+ \; DB^+$

37. The two arene oxides form the same major and minor products.
Each epoxide opens in the direction that forms the most stable carbocation. The carbocation undergoes an NIH shift, and as a result of the NIH shift, both reactants form the same protonated ketone intermediate. Because they form the same intermediate, they form the same products. The deuterium-containing product is the major product, because in the last step of the reaction, it is easier to break a carbon–hydrogen bond than a carbon–deuterium bond.

H O D $\xrightarrow{HB^+}$ H OH D + $\xrightarrow{\text{NIH shift}}$ $^+$OH H D

D O H $\xrightarrow{HB^+}$ D OH H + $\xrightarrow{\text{NIH shift}}$

:B

BH^+ + OH D + OH H + BD^+

major product minor product

38. Solved in the text.

39. Each arene oxide opens in the direction that forms the more stable carbocation.
Therefore, the methoxy-substituted arene oxide opens so that the positive charge can be stabilized by electron donation from the methoxy group.

O :$\ddot{O}CH_3$ $\xrightarrow{H_2O}$ OH + :$\ddot{O}CH_3$ more stable; + OH :$\ddot{O}CH_3$ less stable

The nitro-substituted arene oxide opens in the direction that forms the carbocation intermediate that has its positive charge farther away from the electron-withdrawing NO_2 group.

O $^-O-N^+=O$ $\xrightarrow{H_2O}$ + OH $^-O-N^+=O$ more stable; OH + $^-O-N^+=O$ less stable

40. The compound without the double bond in the second ring is more apt to be carcinogenic. It opens to form a less stable carbocation than the other compound, because the carbocation can be stabilized by electron delocalization only if the aromaticity of the benzene ring is destroyed. Because the carbocation is less stable, it is formed more slowly, giving the carcinogenic pathway a better chance to compete with ring-opening.

less stable carbocation

more stable carbocation

41. **a.** Notice that a bond shared by two rings cannot be epoxidized.

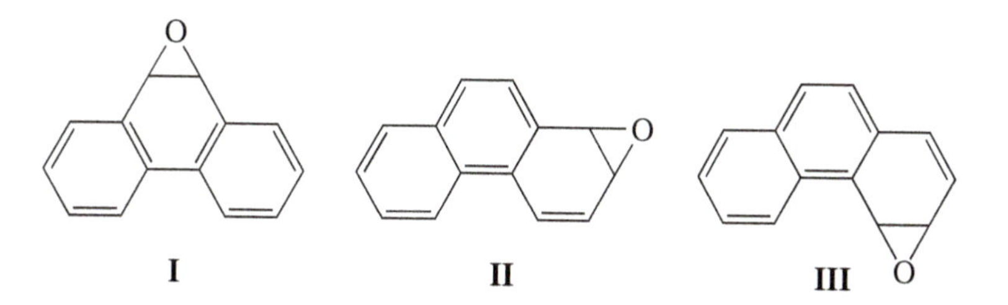

b. The epoxide ring in phenanthrene oxides II and III can open in two different directions to form two different carbocations and, therefore, two different phenols.

c. The two different carbocations formed by phenanthrene oxides II and III differ in stability. One carbocation is more stable than the other because it can be stabilized by electron delocalization without disrupting the aromaticity of the adjacent ring. The more stable carbocation leads to the major product.

II → [carbocation with OH, +] + [carbocation with OH, +]

major product minor product

III → [carbocation with HO, +] + [carbocation with OH, +]

major product minor product

d. Phenanthrene oxide I is the most carcinogenic, because it is the only one that opens to form a carbocation that cannot be stabilized without disrupting the aromaticity of the other ring(s).

42. The drugs are metabolized as a result of reacting with water. Water adds to the electrophilic carbon of the carbonyl group. The resonance contributor on the right (the one with separated charges) decreases the electrophilicity of the carbonyl carbon, which makes it less reactive.

Xylocaine's resonance contributor with separated charges is more stable than Novocaine's resonance contributor with separated charges because its positive charge is on nitrogen, which is a less electronegative atom than oxygen. Because it is more stable, it makes a greater contribution to the hybrid. Therefore, Xylocaine is less reactive and, consequently, has a longer half-life.

$H_2\ddot{O}$: → δ^-O=Cδ^+–O— ⟷ O^-–C=$\overset{+}{O}$— Novocaine

$H_2\ddot{O}$: → δ^-O=Cδ^+–NH— ⟷ O^-–C=$\overset{+}{N}$H— Xylocaine

43. The strong base immediately removes a proton from the protonated amine, and then there is no base to carry out the elimination reaction.

44. The major products result from removing a proton from the β-carbon that is bonded to the most hydrogens.

$$CH_3\underset{\displaystyle CH_3}{\underset{|}{C}}HCH_2\overset{\displaystyle CH_3}{\overset{|}{\underset{\displaystyle CH_3}{\underset{|}{N}}}} \quad + \quad CH_2{=}CHCH_3$$

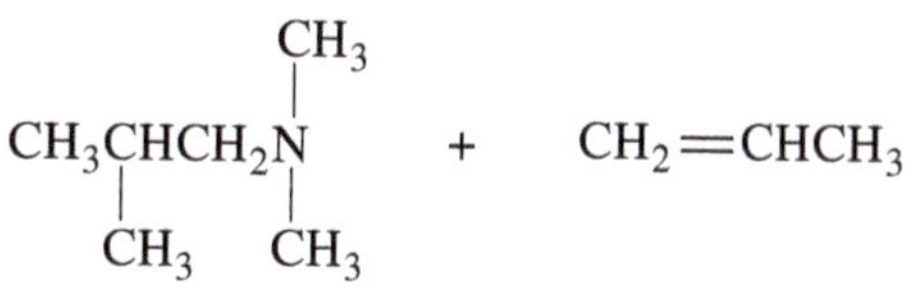

45. The minor products result from removing a proton from the β-carbon that is bonded to the fewest hydrogens.

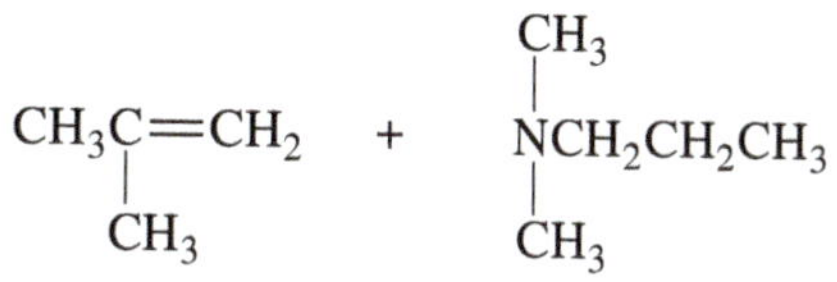

$$CH_3\underset{\displaystyle CH_3}{\underset{|}{C}}{=}CH_2 \quad + \quad \overset{\displaystyle CH_3}{\overset{|}{\underset{\displaystyle CH_3}{\underset{|}{N}}}}CH_2CH_2CH_3$$

46. The only difference is the leaving group.

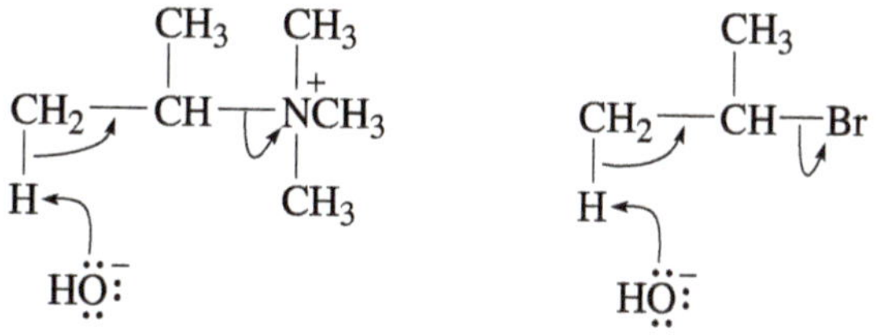

$$\underset{\underset{\displaystyle H\overset{\frown}{O}^-}{\displaystyle H}}{CH_2}{-}\overset{\displaystyle CH_3}{\overset{|}{CH}}{-}\overset{\displaystyle CH_3}{\overset{|}{\underset{\displaystyle CH_3}{\underset{|}{\overset{+}{N}}}}}CH_3 \qquad \underset{\underset{\displaystyle HO^-}{\displaystyle H}}{CH_2}{-}\overset{\displaystyle CH_3}{\overset{|}{CH}}{-}Br$$

47. In each case, a proton is removed from the β-carbon that is bonded to the most hydrogens.

a. 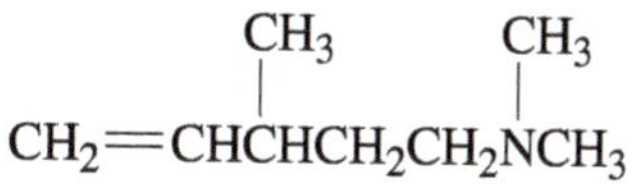 $CH_2{=}CH\overset{\displaystyle CH_3}{\overset{|}{C}}HCH_2CH_2\overset{\displaystyle CH_3}{\overset{|}{N}}CH_3$

c. $CH_3\underset{\displaystyle CH_3}{\underset{|}{N}}CH_2\overset{\displaystyle CH_3}{\overset{|}{C}}HCH_2CH{=}CH_2$

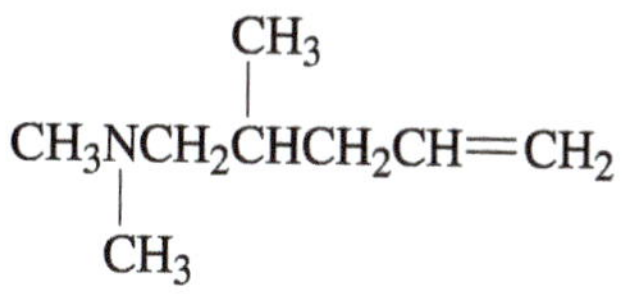

b. 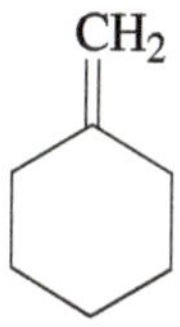 methylenecyclohexane (CH_2 on the ring) + $N(CH_3)_3$

48. Solved in the text.

49. **a.** $CH_3CH_2CH_2CHCH_3$ (Br) $\xrightarrow{(CH_3)_3N}$ $CH_3CH_2CH_2CHCH_3$ ($CH_3-\overset{+}{N}(CH_3)-CH_3$ Br^-) $\xrightarrow[H_2O]{Ag_2O}$ $CH_3CH_2CH_2CHCH_3$ ($CH_3-\overset{+}{N}(CH_3)-CH_3$ HO^-) $\xrightarrow{\Delta}$ $(CH_3)_3N$ + $CH_3CH_2CH_2CH{=}CH_2$

b. pyrrolidine (N–H) $\xrightarrow[K_2CO_3]{CH_3I \text{ excess}}$ N,N-dimethylpyrrolidinium I^- $\xrightarrow[H_2O]{Ag_2O}$ N,N-dimethylpyrrolidinium HO^- $\xrightarrow{\Delta}$ $CH_2{=}CHCH_2CH_2N(CH_3)_2$ $\xrightarrow{CH_3I}$ $CH_2{=}CHCH_2CH_2\overset{+}{N}(CH_3)_3$ I^- $\xrightarrow[H_2O]{Ag_2O}$ $CH_2{=}CHCH_2CH_2\overset{+}{N}(CH_3)_3$ HO^- $\xrightarrow{\Delta}$ $(CH_3)_3N$ + $CH_2{=}CHCH{=}CH_2$

50. **a.** $CH_3CH_2CH_2SH$ $\xrightarrow{HO^-}$ $CH_3CH_2CH_2S^-$ + H_2O $\xrightarrow{CH_3CH_2CH_2Br}$ $CH_3CH_2CH_2SCH_2CH_2CH_3$ + Br^-

b. The synthesis must be done with *tert*-butylthiol and a methyl halide. It cannot be done with methanethiol and *tert*-butyl bromide because a tertiary alkyl halide cannot undergo an S_N2 reaction. Therefore, it would form an elimination product rather than a substitution product.

$(CH_3)_3CSH$ $\xrightarrow{HO^-}$ $(CH_3)_3CS^-$ + H_2O $\xrightarrow{CH_3Br}$ $(CH_3)_3CSCH_3$ + Br^-

c. The highest yield of the target molecule is obtained by having the less substituted of the two R groups of the thioether be the alkyl halide and the more substituted be the thiol.

$CH_2{=}CH\overset{CH_3}{C}HSH$ $\xrightarrow{HO^-}$ $CH_2{=}CH\overset{CH_3}{C}HS^-$ + H_2O $\xrightarrow{CH_3CH_2Br}$ $CH_2{=}CH\overset{CH_3}{C}HSCH_2CH_3$ + Br^-

d. The synthesis must be done with these reagents because the sp^2 carbon of the benzene ring cannot undergo back-side attack.

C_6H_5SH $\xrightarrow{HO^-}$ $C_6H_5S^-$ + H_2O $\xrightarrow{C_6H_5CH_2Br}$ $C_6H_5SCH_2C_6H_5$ + Br^-

51. The first compound is too insoluble.

The second compound is used clinically; it has fewer carbons than the first compound, so it is more soluble in water.

The third compound is less reactive than the second compound because the lone pair on the nitrogen can be delocalized into the benzene ring. Therefore, the lone pair is less apt to displace a chloride ion and form the three-membered ring that is needed for the compound to be an alkylating agent.

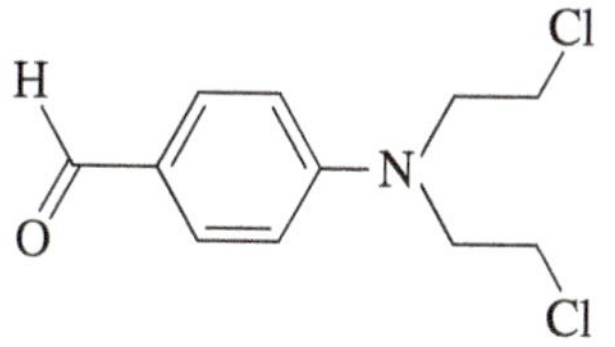

52. Melphalan is a good alkylating reagent because the group on the side chain makes the compound water soluble. The group can delocalize electrons into the benzene ring so that it is not as reactive as the too reactive compound in Problem 51.

53. Cyclophosphamide, carmustine, and chloroambucil are less reactive than mechlorethamine because, unlike the lone-pair electrons on the nitrogen of mechlorethamine, the lone-pair electrons on the nitrogen of the other three compounds can be delocalized. Therefore, these compounds are less apt to form the three-membered ring that is needed for the compound to be an alkylating agent.

54.

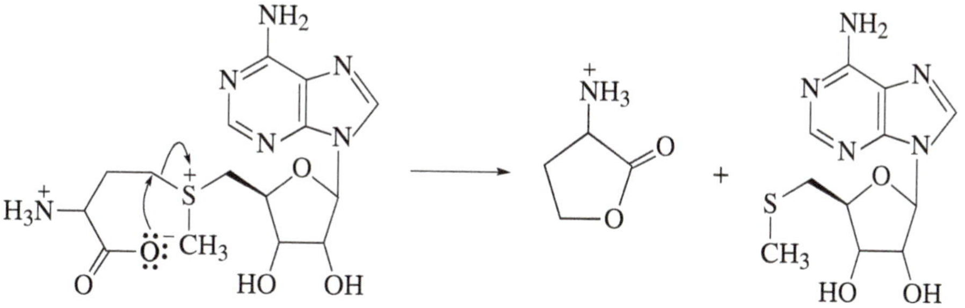

55.

a. $CH_3CH_2CH_2OC(=O)CH_3$

b. $CH_3CH_2CH_2CH_2Br$

c. $CH_3CH(CH_3)CH_2CH_2O{-}C_6H_5$

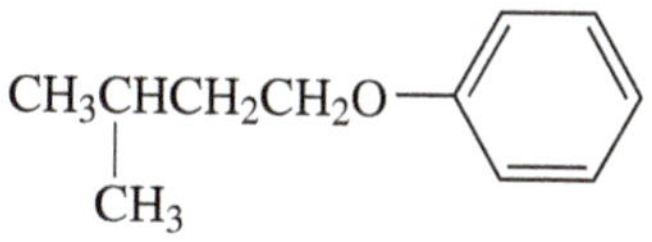

d. $CH_3CH{=}CHCH_3$

e. $CH_3CH_2CH(OCH_3){-}C(CH_3)(OH)CH_3$

f. $CH_3CH_2CH(OH){-}C(CH_3)(OCH_3)CH_3$

g. $CH_3CH(CH_3)CH_2CH_2Cl$

56. **a.** 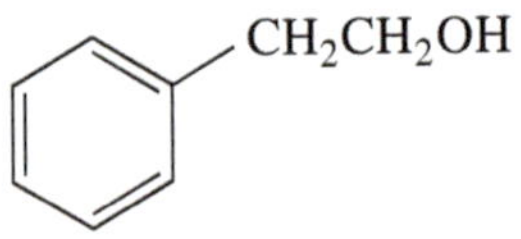 CH_2CH_2OH (attached to benzene ring)

The other alcohol cannot undergo dehydration to form an alkene, because its β-carbon is not bonded to a hydrogen.

b. 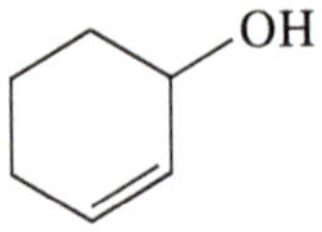 OH

The rate-limiting step in a dehydration is carbocation formation. A secondary allylic cation is more stable than a secondary alkyl carbocation and therefore is formed more rapidly.

c. 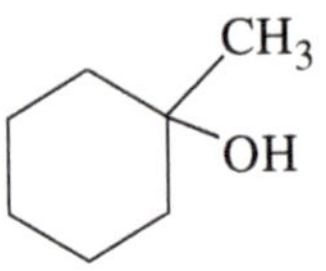 CH_3 OH

A tertiary alkyl carbocation is more stable than a secondary alkyl carbocation.

d. 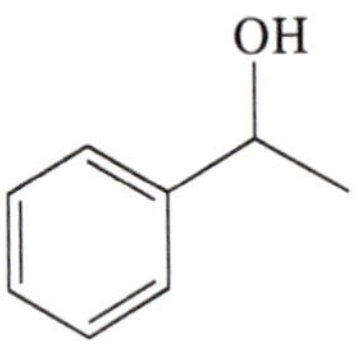 OH

A secondary benzylic cation is more stable than a secondary alkyl carbocation.

e. 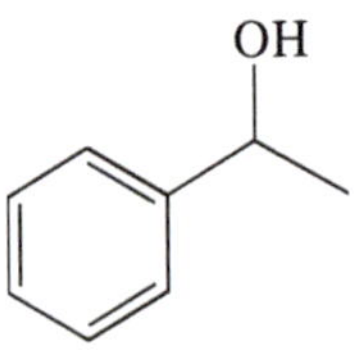 OH

A secondary benzylic cation is more stable than a primary alkyl carbocation.

57. 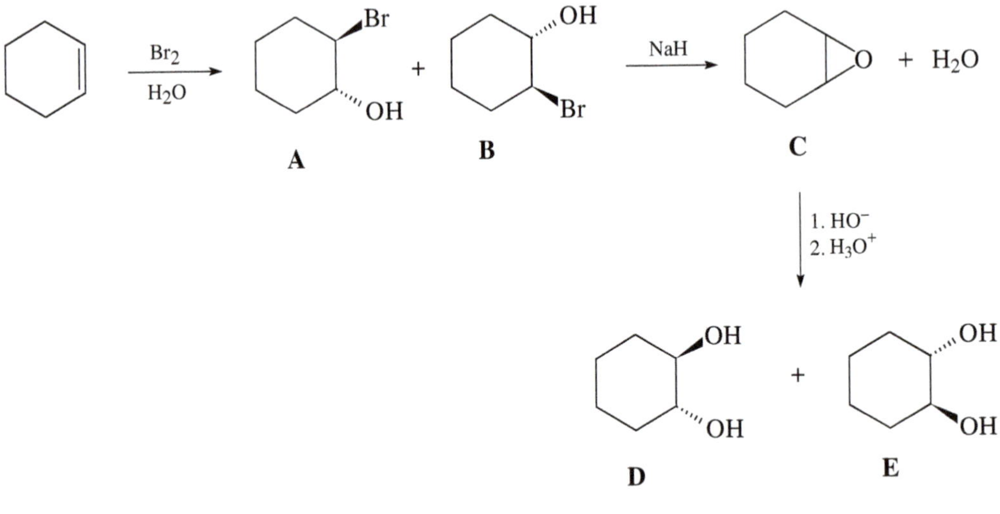 $\xrightarrow[H_2O]{Br_2}$ Br, OH (**A**) + OH, Br (**B**) $\xrightarrow{NaH}$ O (**C**) + H_2O

1. HO^-
2. H_3O^+

OH, OH (**D**) + OH, OH (**E**)

58. Only one S_N2 reaction takes place in **a** and **b**, so the product has the inverted configuration compared to the configuration of the reactant.

a. CH_3CH_2–C(D)(H)(OH) + R–SO_2–Cl $\xrightarrow{\text{pyridine}}$ CH_3CH_2–C(D)(H)(OSO_2R) $\xrightarrow[S_N2]{HO^-}$ CH_3CH_2–C(OH)(H)(D)

(*R*)-1-deuterio-1-propanol (*S*)-1-deuterio-1-propanol

or

HO–C(CH₂CH₃)(D)(H) + R–SO_2–Cl $\xrightarrow{\text{pyridine}}$ RSO_2O–C(CH_2CH_3)(D)(H) $\xrightarrow[S_N2]{HO^-}$ D–C(CH_2CH_3)(OH)(H)

(*R*)-1-deuterio-1-propanol (*S*)-1-deuterio-1-propanol

b. CH_3CH_2–C(D)(H)(OH) + R–SO_2–Cl $\xrightarrow{\text{pyridine}}$ CH_3CH_2–C(D)(H)(OSO_2R) $\xrightarrow[S_N2]{CH_3O^-}$ CH_3CH_2–C(OCH_3)(H)(D)

(*R*)-1-deuterio-1-propanol (*S*)-1-deuterio-1-methoxypropane

or

HO–C(CH_2CH_3)(D)(H) + R–SO_2–Cl $\xrightarrow{\text{pyridine}}$ RSO_2O–C(CH_2CH_3)(D)(H) $\xrightarrow[S_N2]{CH_3O^-}$ D–C(CH_2CH_3)(OCH_3)(H)

(*R*)-1-deuterio-1-propanol (*S*)-1-deuterio-1-methoxypropane

c. Because the desired product has the same configuration as the starting material, it can be synthesized using reactions that do not break any of the bonds to the asymmetric center.

CH_3CH_2–C(D)(H)(OH) $\xrightarrow{\text{NaH}}$ CH_3CH_2–C(D)(H)(O^-) $\xrightarrow{CH_3I}$ CH_3CH_2–C(D)(H)(OCH_3)

(*R*)-1-deuterio-1-propanol (*R*)-1-deuterio-1-methoxypropane

It can also be synthesized using two consecutive S_N2 reactions. Because each one involves inversion of configuration, the final product will have the same configuration as the starting material. For example, treating the starting material with PBr_3 forms (*S*)-1-bromo-1-deuteriopropane. Treating (*S*)-1-bromo-1-deuteriopropane with methoxide ion forms (*R*)-1-deuterio-1-methoxypropane.

CH_3CH_2–C(D)(H)(OH) $\xrightarrow[S_N2]{PBr_3/\text{pyridine}}$ H–C(D)(Br)–CH_2CH_3 $\xrightarrow[S_N2]{CH_3O^-}$ CH_3CH_2–C(D)(H)(OCH_3)

(*R*)-1-deuterio-1-propanol (*S*)-1-bromo-1-deuteriopropane (*R*)-1-deuterio-1-methoxypropane

59. 2,3-Dimethyl-2-butanol dehydrates faster because it forms a tertiary carbocation, whereas 3,3-dimethyl-2-butanol initially forms a secondary carbocation.

$CH_3CH(CH_3)C(CH_3)(OH)CH_3$ — 2,3-dimethyl-2-butanol, **a tertiary alcohol**

$CH_3CH(OH)C(CH_3)_2CH_3$ — 3,3-dimethyl-2-butanol, **a secondary alcohol**

60.

2-ethyloxirane (epoxide ring with CH_2CH_3 substituent)

a. $CH_3CH_2CH(OH)CH_2OH$

0.1 M HCl is a dilute solution of HCl in water.

b. $CH_3CH_2CH(OCH_3)CH_2OH$

c. $CH_3CH_2CH(OH)CH_2OH$

d. $CH_3CH_2CH(OH)CH_2OCH_3$

61.

Cyclohexanol $\xrightarrow{NaOCl,\ CH_3COOH,\ 0\ °C}$ cyclohexanone

Cyclohexanol $\xrightarrow{H_2SO_4/\Delta \text{ or } 1.\ HBr/\Delta\ 2.\ tert\text{-}BuO^-}$ cyclohexene $\xrightarrow{1.\ OsO_4\ 2.\ H_2O_2,\ H_2O}$ cis-1,2-cyclohexanediol

Cyclohexanol $\xrightarrow{TsCl,\ pyridine}$ cyclohexyl-OTs $\xrightarrow{CH_3O^-}$ cyclohexyl-OCH_3

Cyclohexanol $\xrightarrow{HBr/\Delta}$ bromocyclohexane $\xrightarrow{CH_3O^-}$ cyclohexyl-OCH_3

Bromocyclohexane $\xrightarrow{tert\text{-}BuO^-}$ cyclohexene

Cyclohexene $\xrightarrow{MCPBA \text{ or } 1.\ Cl_2/H_2O\ 2.\ NaH}$ cyclohexene oxide $\xrightarrow{HO^-}$ trans-1,2-cyclohexanediol + trans-1,2-cyclohexanediol (enantiomer)

62. 2-hexene and 3-hexene

$$CH_3CH_2CH_2CH_2CH_2CH_2OH \xrightarrow[\Delta]{H_2SO_4} CH_3CH_2CH_2CH_2CH{=}CH_2 + H_2O$$

$$CH_3CH_2CH_2CH_2CH{=}CH_2 \xrightarrow{H^+} CH_3CH_2CH_2CH_2\overset{+}{C}HCH_3$$

$$H^+ + CH_3CH_2CH_2CH{=}CHCH_3 \rightleftharpoons CH_3CH_2CH_2CH_2\overset{+}{C}HCH_3$$

2-hexene

$$H^+ + CH_3CH_2CH_2CH{=}CHCH_3 \rightleftharpoons CH_3CH_2CH_2\overset{+}{C}HCH_2CH_3$$

$$CH_3CH_2CH_2\overset{+}{C}HCH_2CH_3 \longrightarrow CH_3CH_2CH{=}CHCH_2CH_3 + CH_3CH_2CH_2CH{=}CHCH_3 + H^+$$

3-hexene (cis and trans); 2-hexene (cis and trans)

63.

a. 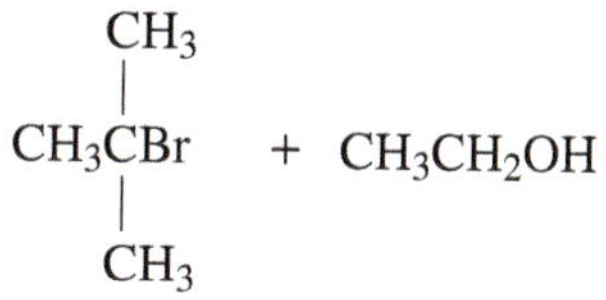 CH_3 / CH_3CBr / CH_3 + CH_3CH_2OH

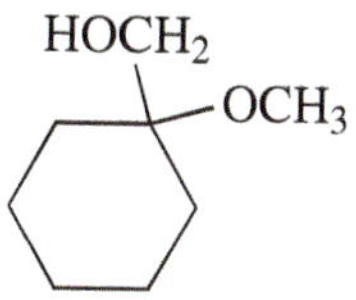

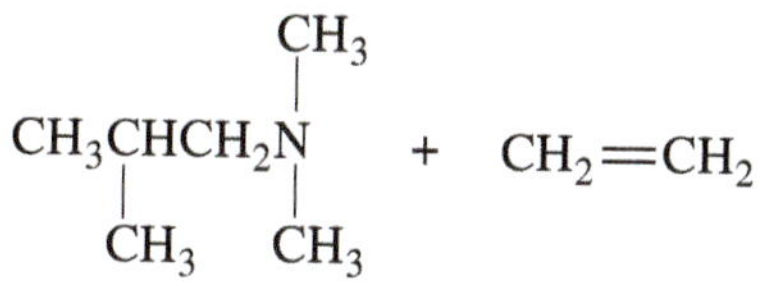

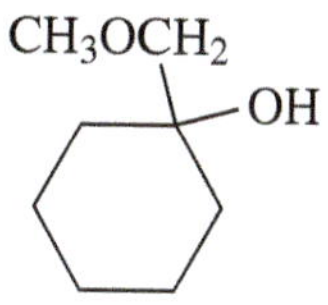

b. 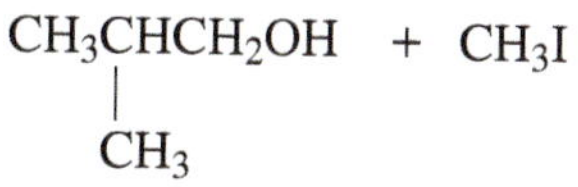 CH_3CHCH_2OH / CH_3 + CH_3I

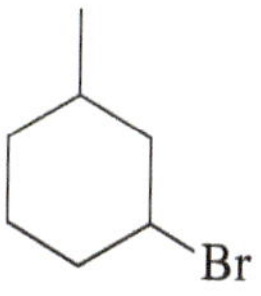

c. 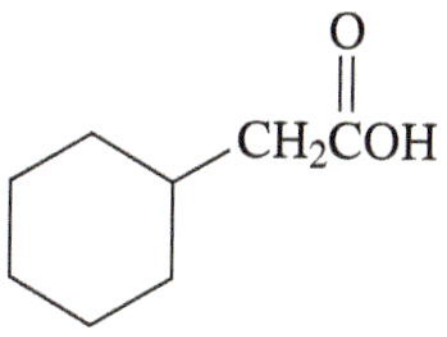 O / CH_2COH

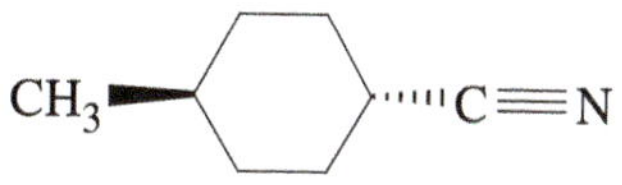

d.

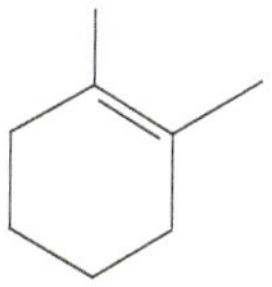

e. $HOCH_2$ / OCH_3

f. CH_3OCH_2 / OH

g. CH_3 C≡N

h. 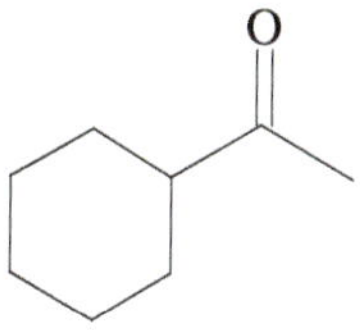 O

i. CH_3 / CH_3CHCH_2N / CH_3 CH_3 + $CH_2{=}CH_2$

j. Br

64. a. 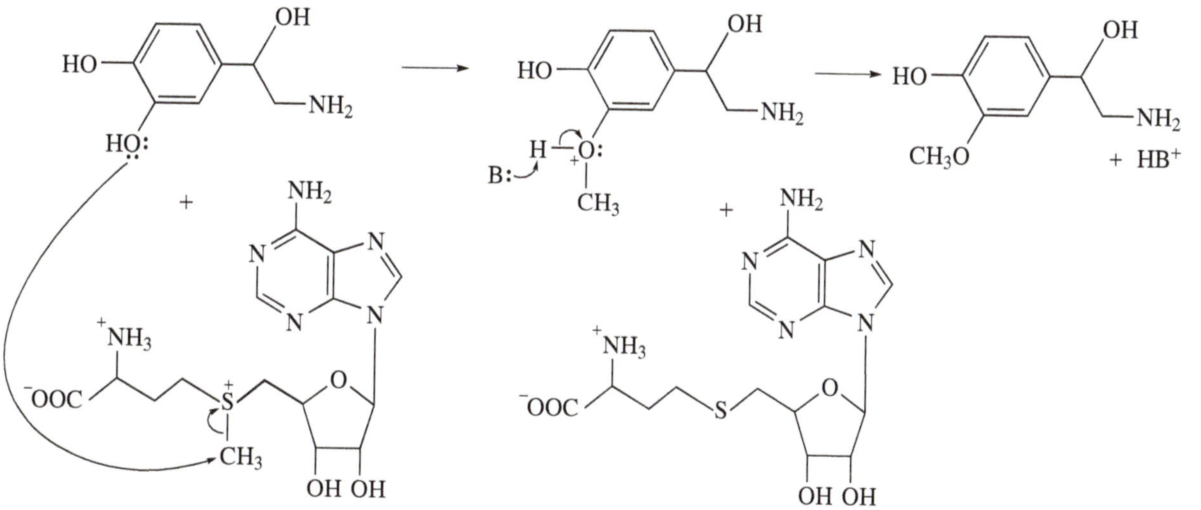 HO / OH / NH_2 / HO: + NH_2 / N / N / N / N / $\overset{+}{N}H_3$ / ^-OOC / $\overset{+}{S}$ / CH_3 / O / OH OH → HO / OH / NH_2 / B: / $H-\overset{+}{O}:$ / CH_3 + NH_2 / N / N / N / N / $\overset{+}{N}H_3$ / ^-OOC / S / O / OH OH → HO / OH / NH_2 / CH_3O + HB^+

b. Methylation on nitrogen, because nitrogen is the stronger base.

65. **a.** If an NIH shift occurs, both carbocations will form the same intermediate ketone. Because it is about four times easier to break a C—H bond (k_3) compared with a C—D bond (k_4), about 80% of the deuterium will be retained.

k_1 H_2O k_2 H_2O

NIH shift NIH shift

k_3 k_4

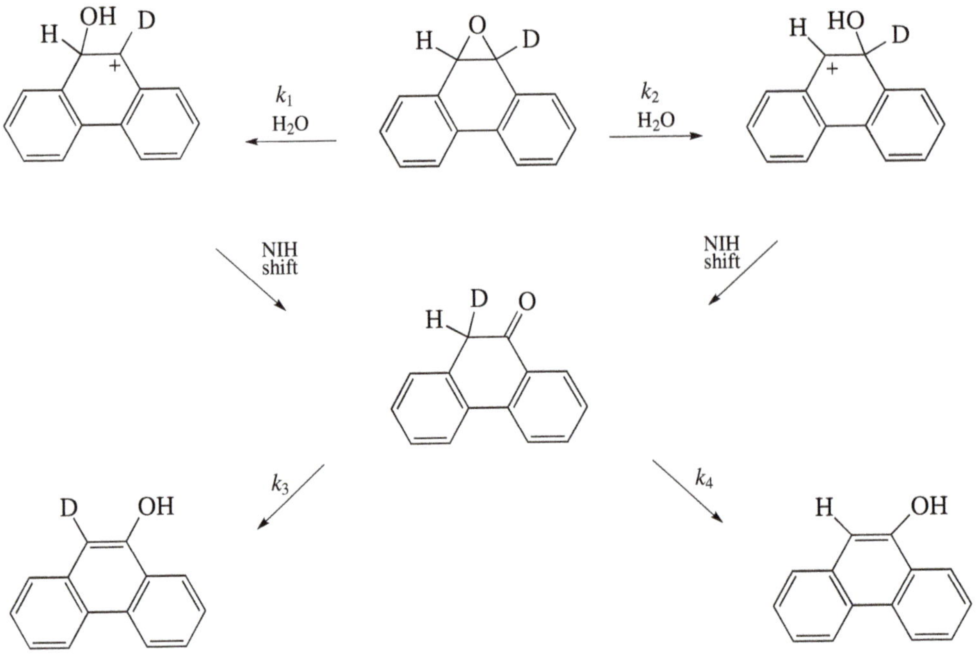

b. If an NIH shift does not occur, 50% of the deuterium will be retained, because the epoxide can open equally easily in either direction (k_1 is equal to k_2) and subsequent loss of H^+ or D^+ is fast.

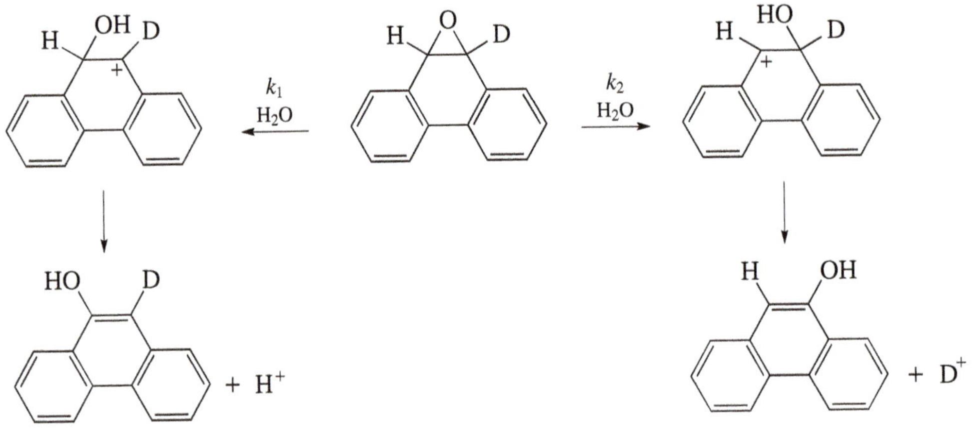

k_1 H_2O k_2 H_2O

+ H^+ + D^+

66. H_2SO_4, Δ

unknown alcohol

major product

minor product

NaOCl CH_3COOH 0 °C

1. R_2BH/THF
2. H_2O_2, HO^-, H_2O

original unknown alcohol

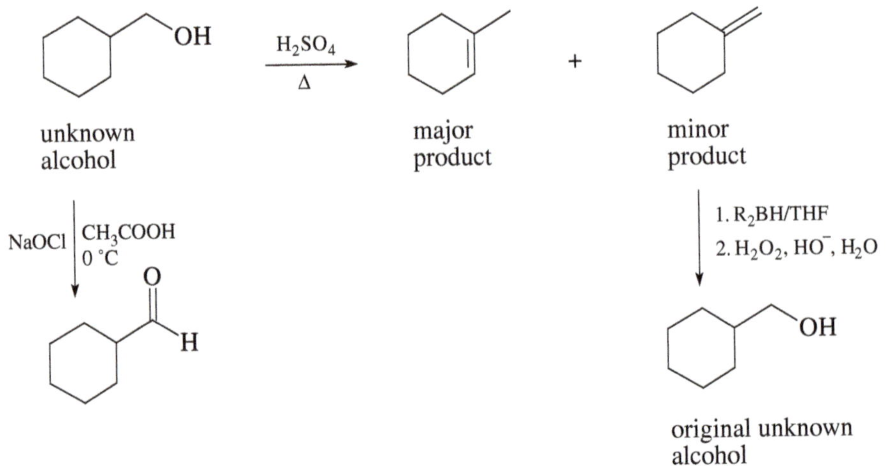

67. The product of each reaction is an alkene.

$$CH_3CH_2CHCH_3(OH) \underset{\Delta}{\overset{H_2SO_4}{\rightleftharpoons}} CH_3CH_2CHCH_3(\overset{+}{O}H_2) \rightleftharpoons CH_3CH_2\overset{+}{C}HCH_3 + H_2O \rightleftharpoons CH_3CH{=}CHCH_3 + H_3O^+$$

$$CH_3CH_2CHCH_3(Br) \xrightarrow{HO^-} CH_3CH{=}CHCH_3 + H_2O + Br^-$$

Recall that alkenes undergo electrophilic addition reactions, and the first step in an electrophilic addition reaction is addition of an electrophile to the alkene.

The acid-catalyzed dehydration reaction is reversible, because the electrophile (H^+) needed to react with the alkene in the first step of the reverse reaction is available.

The base-promoted elimination reaction of a hydrogen halide is not reversible, because an electrophile is not available to react with the alkene in the first step of the reverse reaction.

68. When (S)-2-butanol loses water as a result of being heated with sulfuric acid, the asymmetric center in the reactant becomes a planar sp^2 carbon. Therefore, the chirality is lost. When water attacks the carbocation, it can attack from either side of the planar carbocation, forming (S)-2-butanol and (R)-2-butanol with equal ease.

(S)-2-butanol $\underset{\Delta}{\overset{H_2SO_4}{\rightleftharpoons}}$ protonated (S)-2-butanol $\rightleftharpoons$ H_2O + $CH_3CH_2\overset{+}{C}HCH_3$ $\rightleftharpoons$ $CH_3CH{=}CHCH_3 + H_3O^+$

H^+ + (S)-2-butanol + (R)-2-butanol $\rightleftharpoons$ protonated (S)-2-butanol + protonated (R)-2-butanol $\rightleftharpoons$ $CH_3CH_2\overset{+}{C}HCH_3$ + H_2O

69. **a.** $CH_3CH_2CH{=}CH_2 \xrightarrow[\text{2. } H_2O_2,\ HO^-,\ H_2O]{\text{1. } R_2BH/THF} CH_3CH_2CH_2CH_2OH$

$CH_3CH_2CH_2CH_2OH \underset{\Delta}{\overset{H_2SO_4}{\rightleftharpoons}} CH_3CH{=}CHCH_3 + H_2O$

$CH_3CH{=}CHCH_3 \xrightarrow[\text{2. } (CH_3)_2S]{\text{1. } O_3,\ -78\ °C} 2\ CH_3CHO$

b. 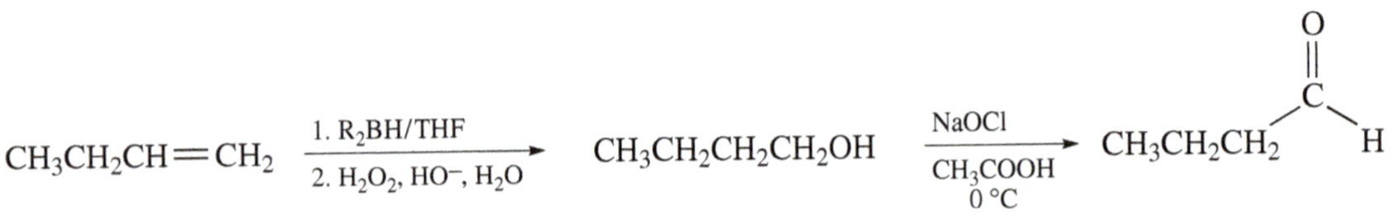 $CH_3CH_2CH{=}CH_2 \xrightarrow[2.\ H_2O_2,\ HO^-,\ H_2O]{1.\ R_2BH/THF} CH_3CH_2CH_2CH_2OH \xrightarrow[CH_3COOH,\ 0\ °C]{NaOCl} CH_3CH_2CH_2C(=O)H$

c. cyclohexene $\xrightarrow{HBr}$ bromocyclohexane (Br) $\xrightarrow{HO^-}$ cyclohexanol (OH) $\xrightarrow[CH_3COOH,\ 0\ °C]{NaOCl}$ cyclohexanone (O)

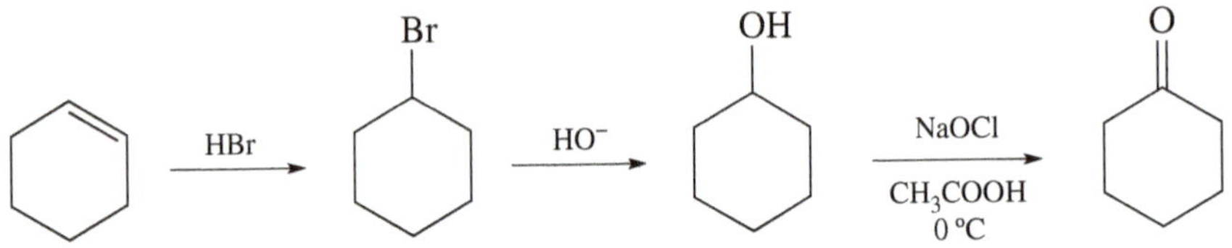

70.

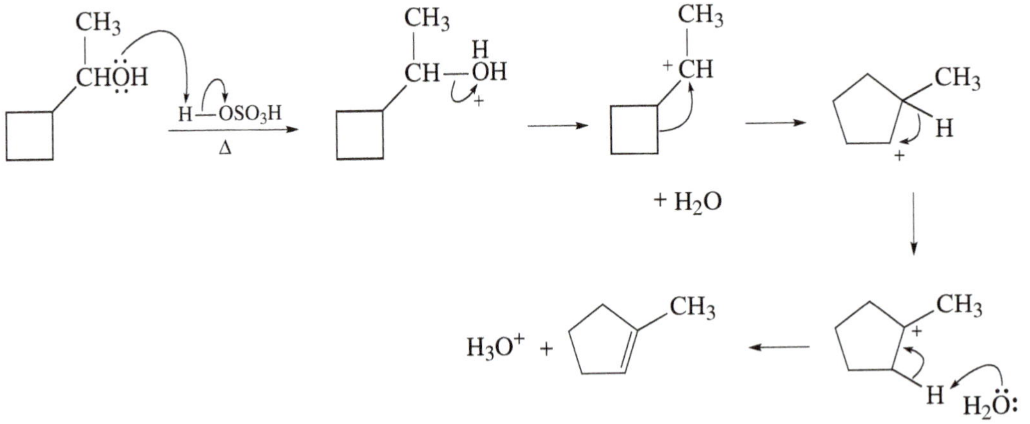

CH_3, $\ddot{C}HOH$ (cyclobutyl) $\xrightarrow[\Delta]{H{-}OSO_3H}$ CH_3, $CH{-}\overset{+}{O}H$, H $\longrightarrow$ CH_3, ^+CH + H_2O $\longrightarrow$ CH_3, H, + $\longrightarrow$ CH_3, +, H, $H_2\ddot{O}$: $\longrightarrow$ H_3O^+ + CH_3

71.

CH_3, $\ddot{C}HOH$ (cyclobutyl) $\xrightarrow[\Delta]{H{-}Br}$ CH_3, $CH{-}\overset{+}{O}H$, H $\longrightarrow$ CH_3, ^+CH $\longrightarrow$ CH_3, H, + $\longrightarrow$ CH_3, +, $:\ddot{Br}:^-$ $\longrightarrow$ CH_3, Br

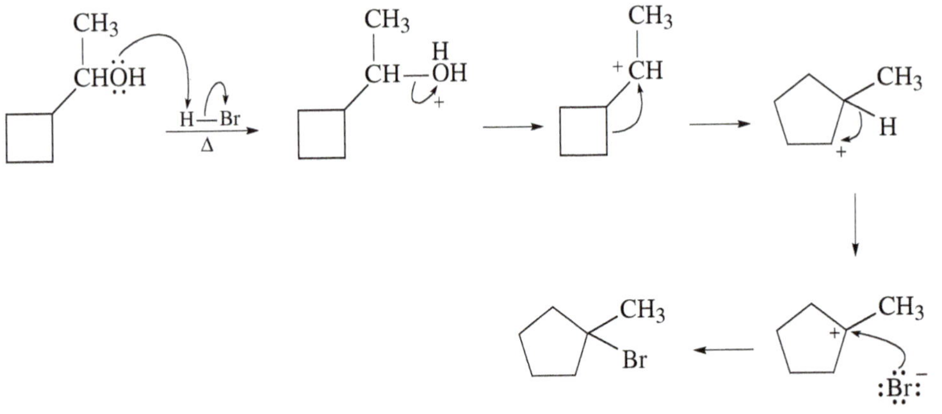

72. **Diethyl ether** is the ether that would be obtained in greatest yield, because it is a symmetrical ether. Because it is symmetrical, only one alcohol is used in its synthesis, so only one ether is formed.

In contrast, the synthesis of an unsymmetrical ether requires two different alcohols, so three different ethers are formed.

73. **a.** cyclohexanol $\xrightarrow[\Delta]{H_2SO_4}$ cyclohexene $\xrightarrow[Pd/C]{H_2}$ cyclohexane

b. $CH_3CH_2CH_2CH_2Br \xrightarrow{HO^-} CH_3CH_2CH_2CH_2OH \xrightarrow[\Delta]{H_2CrO_4} CH_3CH_2CH_2\overset{O}{\overset{\|}{C}}OH$

74. *N*-Methylpiperidine forms 1,4-pentadiene.

N-methylpiperidine $\xrightarrow[2.\ Ag_2O,\ H_2O]{1.\ CH_3I}$ N,N-dimethylpiperidinium HO^- $\xrightarrow{\Delta}$ $CH_2{=}CHCH_2CH_2CH_2N(CH_3)CH_3$

$\downarrow$ 1. CH_3I 2. Ag_2O, H_2O 3. Δ

$CH_2{=}CHCH_2CH{=}CH_2$

2-Methylpiperidine forms 1,5-hexadiene.

2-methylpiperidine $\xrightarrow[2.\ Ag_2O,\ H_2O]{1.\ CH_3I\ \text{excess}\ K_2CO_3}$ 1,1,2-trimethylpiperidinium HO^- $\xrightarrow{\Delta}$ $CH_2{=}CHCH_2CH_2CH_2CH_2N(CH_3)CH_3$

β-carbon bonded to the greatest number of hydrogens

$\downarrow$ 1. CH_3I 2. Ag_2O, H_2O 3. Δ

$CH_2{=}CHCH_2CH_2CH{=}CH_2$

3-Methylpiperidine forms 2-methyl-1,4-pentadiene.

3-methylpiperidine $\xrightarrow[2.\ Ag_2O,\ H_2O]{1.\ CH_3I\ \text{excess}\ K_2CO_3}$ 1,1,3-trimethylpiperidinium HO^- $\xrightarrow{\Delta}$ $CH_2{=}CHCH_2CH(CH_3)CH_2N(CH_3)CH_3$

$\downarrow$ 1. CH_3I 2. Ag_2O, H_2O 3. Δ

$CH_2{=}CHCH_2C(CH_3){=}CH_2$

4-Methylpiperidine forms 3-methyl-1,4-pentadiene.

4-methylpiperidine $\xrightarrow[2.\ Ag_2O,\ H_2O]{1.\ CH_3I\ \text{excess}\ K_2CO_3}$ 1,1,4-trimethylpiperidinium HO^- $\xrightarrow{\Delta}$ $CH_2{=}CHCH(CH_3)CH_2CH_2N(CH_3)CH_3$

$\downarrow$ 1. CH_3I 2. Ag_2O, H_2O 3. Δ

$CH_2{=}CHCH(CH_3)CH{=}CH_2$

75. 3-Methyl-2-butanol is a secondary alcohol and, therefore, undergoes an S_N1 reaction. The secondary carbocation intermediate rearranges to a more stable tertiary carbocation.

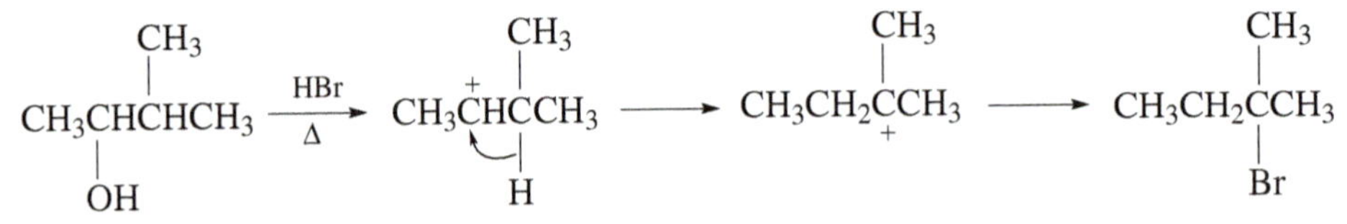

$$CH_3CH(OH)CH(CH_3)CH_3 \xrightarrow[\Delta]{HBr} CH_3\overset{+}{C}HC(H)(CH_3)CH_3 \longrightarrow CH_3CH_2\overset{+}{C}(CH_3)CH_3 \longrightarrow CH_3CH_2C(CH_3)(Br)CH_3$$

3-methyl-2-butanol

2-Methyl-1-propanol is a primary alcohol and, therefore, undergoes an S_N2 reaction. Because carbocations are not formed in S_N2 reactions, a carbocation rearrangement cannot occur.

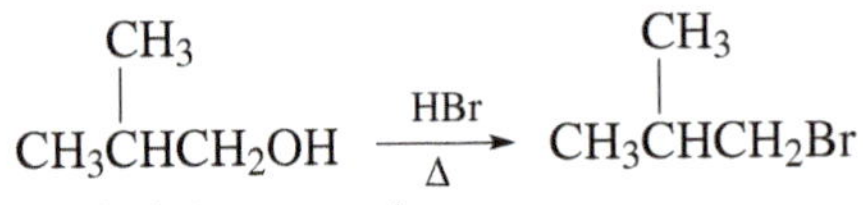

$$CH_3CH(CH_3)CH_2OH \xrightarrow[\Delta]{HBr} CH_3CH(CH_3)CH_2Br$$

2-methyl-1-propanol

76. OH → A (Cl) → B → C (OH) → D → F (O) + E (O, H)

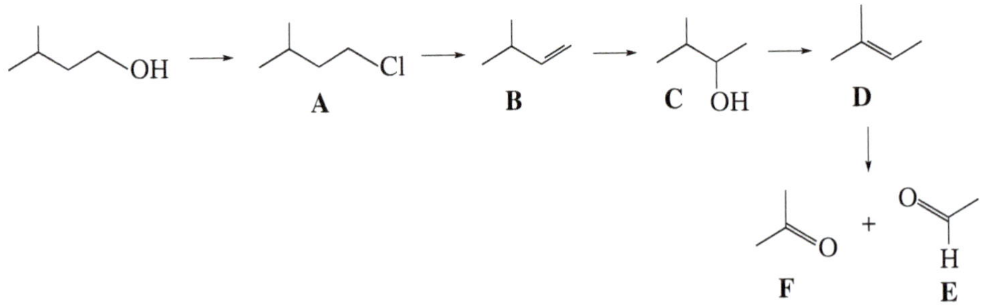

77. **a.** $CH_3CH(Cl)CH{-}CH_2$ (epoxide, O) + $CH_3\ddot{O}{:}^-$ ⟶ $CH_3CH(Cl)CH(:\ddot{O}:^-){-}CH_2OCH_3$ ⟶ $CH_3CH{-}CHCH_2OCH_3$ (epoxide, O) + Cl^-

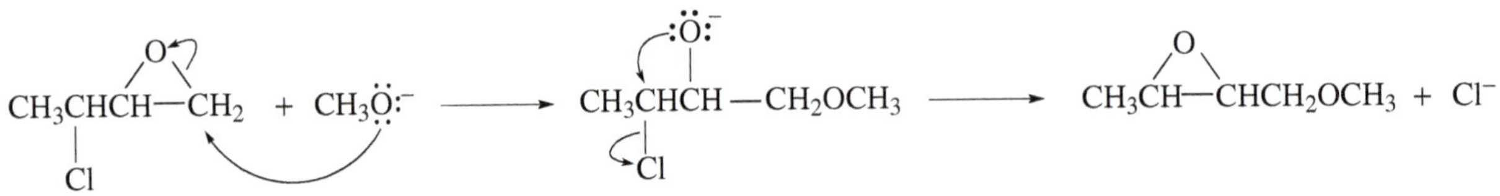

b. OH, $\ddot{O}H$ + $H{-}\ddot{O}SO_3H$ $\xrightarrow{\Delta}$ OH, $\overset{+}{O}H$ H ⟶ $\ddot{O}H$, + ↓ $H_2\ddot{O}$, $\overset{+}{O}$–H ⟶ H_3O^+ + O

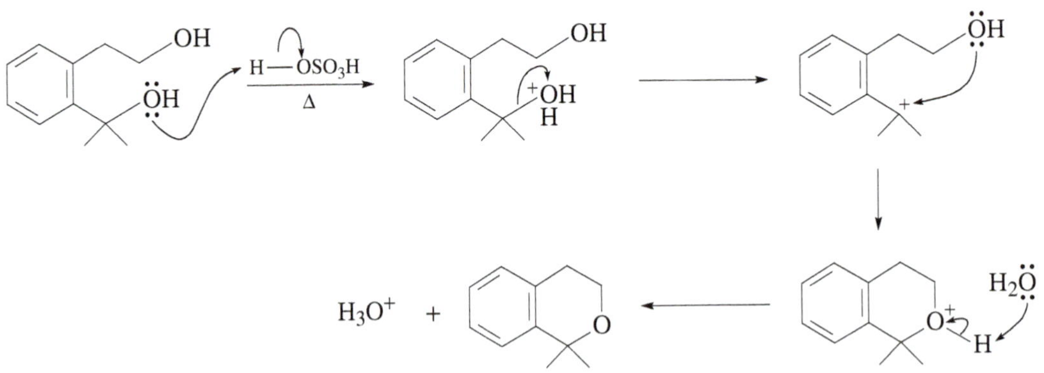

78.

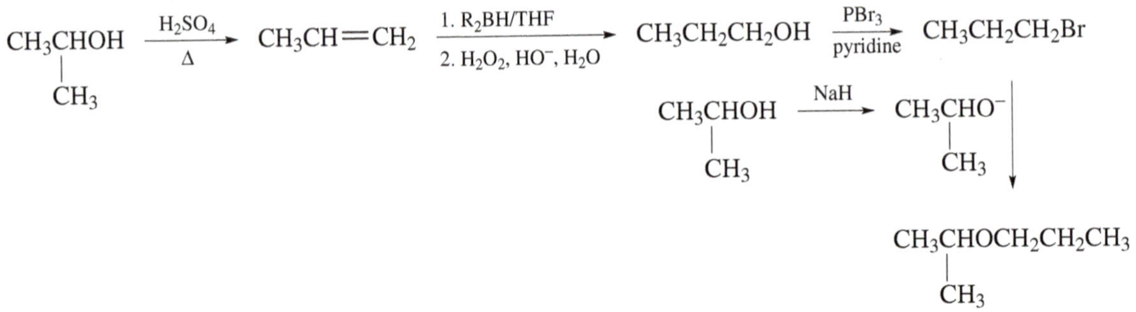

$$CH_3CH(CH_3)OH \xrightarrow[\Delta]{H_2SO_4} CH_3CH{=}CH_2 \xrightarrow[2.\ H_2O_2,\ HO^-,\ H_2O]{1.\ R_2BH/THF} CH_3CH_2CH_2OH \xrightarrow[pyridine]{PBr_3} CH_3CH_2CH_2Br$$

$$CH_3CH(CH_3)OH \xrightarrow{NaH} CH_3CH(CH_3)O^- \longrightarrow CH_3CH(CH_3)OCH_2CH_2CH_3$$

79. Ethyl alcohol is not obtained as a product because it reacts with the excess HI, forming ethyl iodide.

$$CH_3CH_2OCH_2CH_3 \xrightarrow[\Delta]{HI} CH_3CH_2I + CH_3CH_2OH \xrightarrow[\Delta]{HI} CH_3CH_2I + H_2O$$

80.

CH3 CH3 CH3 OH $\xrightarrow[\Delta]{H_2SO_4}$ CH3 CH3 CH3 + → CH3 + CH3 CH3 → CH3 CH3 + CH3

CH3 CH3 CH3 + H+   CH3 CH2 CH3 + H+   CH3 CH3 CH3 + H+

81. Cyclopropane does not react with HO^- because cyclopropane does not contain a leaving group; a carbanion is far too basic to serve as a leaving group. Ethylene oxide reacts with HO^- because ethylene oxide contains an RO^- leaving group.

82. a. OH $\xrightarrow[\Delta]{H_2SO_4}$ → 1. R_2BH, THF 2. HO^-, H_2O_2, H_2O → OH → NaOCl, CH_3COOH, 0 °C → O

b. N $\xrightarrow{CH_3I}$ N+ I⁻ $\xrightarrow[H_2O]{Ag_2O}$ N+ HO⁻ $\xrightarrow{\Delta}$ → 1. OsO_4 2. H_2O_2, H_2O → OH OH

83. a. HO ŎH H—OSO_3H → HO H OH + → + O H H_2O: → O + H_3O^+

b. [Mechanism: tetrahydropyran oxygen is protonated by H–Br; :Br:⁻ attacks to give HO(CH₂)₅Br; the OH is protonated by H–Br and :Br:⁻ displaces water]

H_2O + Br(CH₂)₅Br

84. H_2C—CH_2 (epoxide) + HO^- ⟶ ^-O–CH_2CH_2OH ⟶ (+ epoxide) $HOCH_2CH_2OCH_2CH_2O^-$ ⟶ (+ epoxide) $HOCH_2CH_2OCH_2CH_2OCH_2CH_2O^-$

$HOCH_2CH_2OCH_2CH_2OCH_2CH_2O^-$ $\xrightarrow{H_2O}$ $HOCH_2CH_2OCH_2CH_2OCH_2CH_2OH$ + HO^-

85. a. Epoxide–$CH_2CH_2CH_2Br$ + CH_3O^- ⟶ $CH_3OCH_2CH(O^-)CH_2CH_2CH_2$–Br ⟶ 2-($CH_2OCH_3$)tetrahydrofuran + Br^-

b. Epoxide–$CH_2CH_2CH_2Br$ + CH_3O^- ⟶ $^-OCH_2CH(OCH_3)CH_2CH_2CH_2$–Br ⟶ 3-($OCH_3$)tetrahydropyran + Br^-

c. The six-membered ring is formed by attack on the more sterically hindered carbon of the epoxide. Attack on the less sterically hindered carbon that leads to the five-membered ring is preferred.

86. :B is any base in the solution (HSO_4^-, H_2O, ROH).

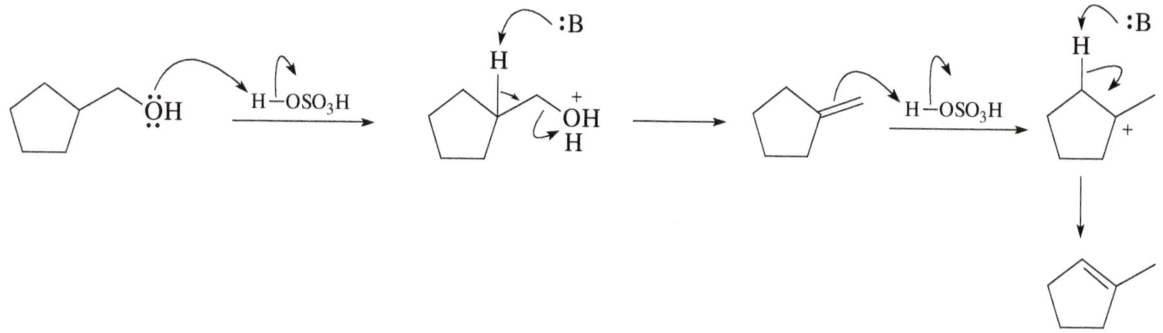

87. **a.** The primary amine is 2-methyl-3-pentanamine.

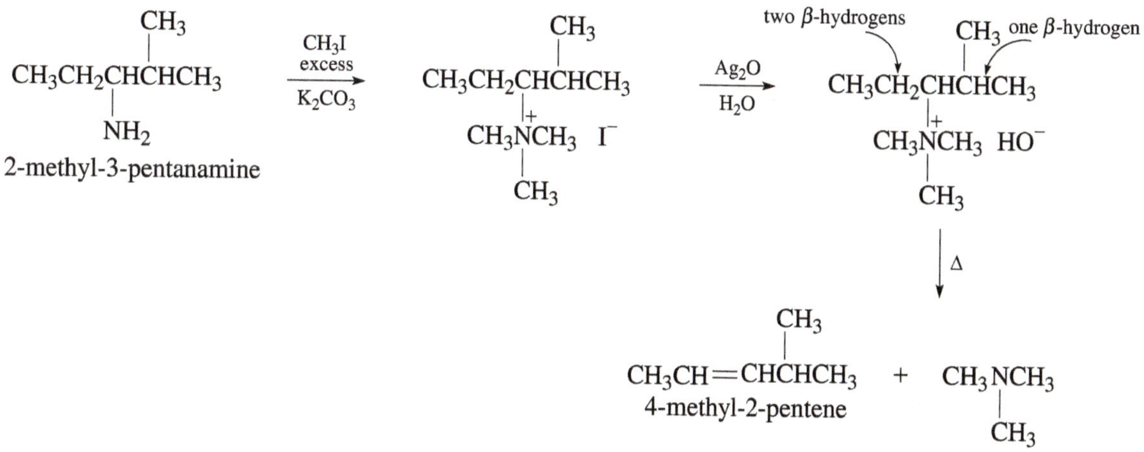

2-methyl-3-pentanamine

CH_3I excess, K_2CO_3; Ag_2O, H_2O; Δ

two β-hydrogens; one β-hydrogen

4-methyl-2-pentene

b. The primary amine is 3-methyl-2-butanime.

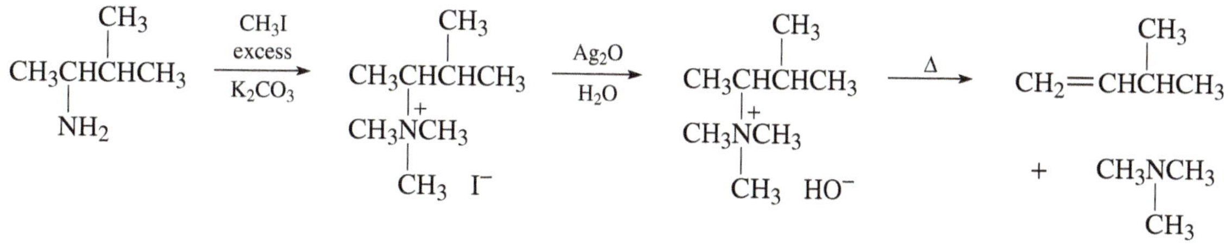

c. The secondary amine is 3-methylpyrrolidine.

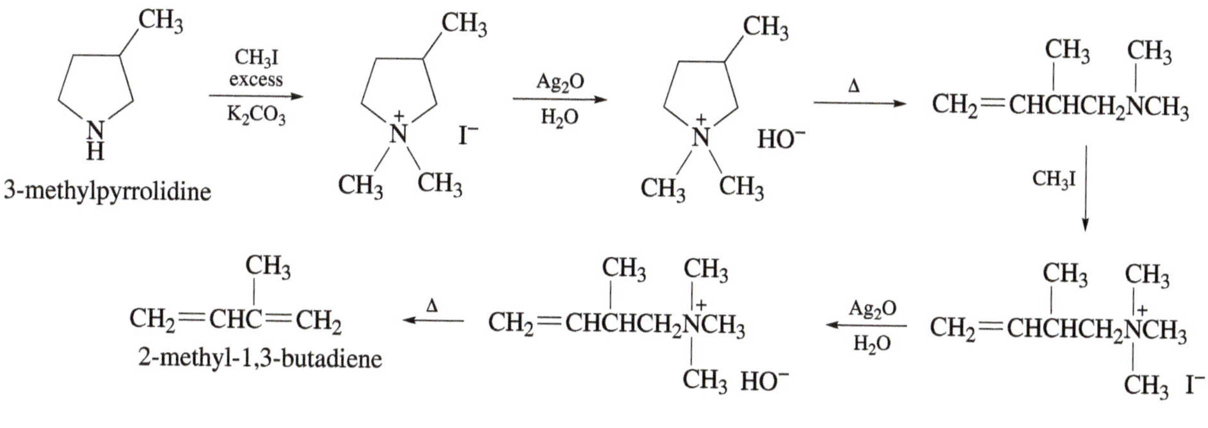

3-methylpyrrolidine

2-methyl-1,3-butadiene

88. **A** is the substitution product that forms when methoxide ion attacks a carbon of the three-membered ring and eliminates the amino group, thereby opening the ring. It does not discolor Br_2 because it does not have a double bond to which Br_2 can add.

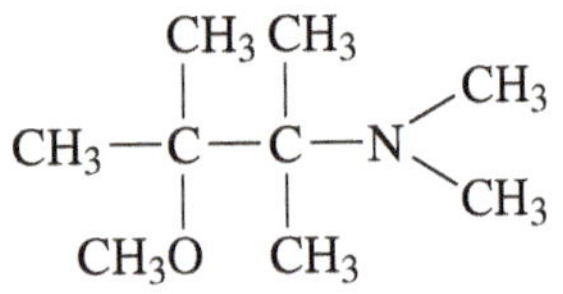

$$\underset{\textbf{A}}{CH_3-\underset{CH_3O}{\overset{CH_3}{C}}-\underset{CH_3}{\overset{CH_3}{C}}-N\begin{matrix}CH_3\\CH_3\end{matrix}} \qquad \underset{\textbf{B}}{CH_2=\underset{CH_3}{C}-\underset{CH_3}{\overset{CH_3}{C}}-N\begin{matrix}CH_3\\CH_3\end{matrix}}$$

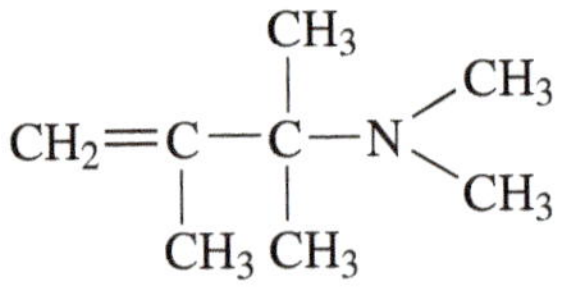

B is the product of a Hofmann elimination reaction: methoxide ion removes a proton from a methyl group bonded to a ring carbon and eliminates the amino group. The red color disappears when Br_2 is added to **B**, because Br_2 adds to the double bond.

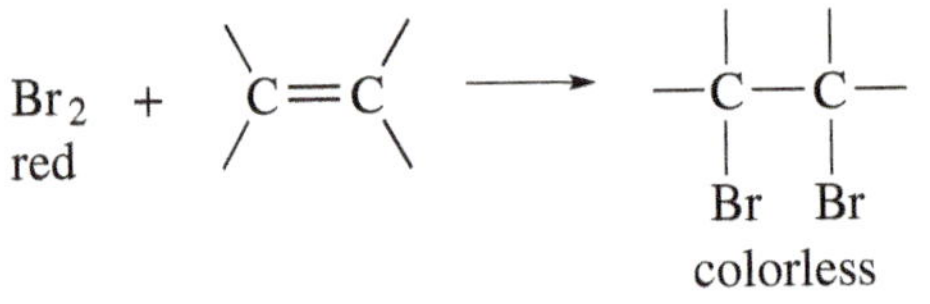

$$\underset{\text{red}}{Br_2} + \;>C=C<\; \longrightarrow \underset{\text{colorless}}{-\underset{Br}{\overset{|}{C}}-\underset{Br}{\overset{|}{C}}-}$$

When the aziridinium ion reacts with methanol, only **A**, the substitution product, is formed.

1. $\overset{+}{N}$ (aziridinium ion) + $CH_3\ddot{O}:^-$ ⟶ N, OCH_3 — substitution product

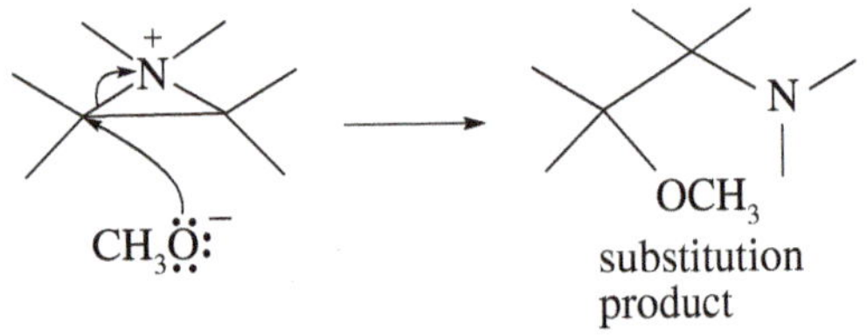

2. $\overset{+}{N}$ (aziridinium ion), H + $CH_3\ddot{O}:^-$ ⟶ N — eliminiation product + Br_2 (red) ⟶ N, Br, Br — colorless

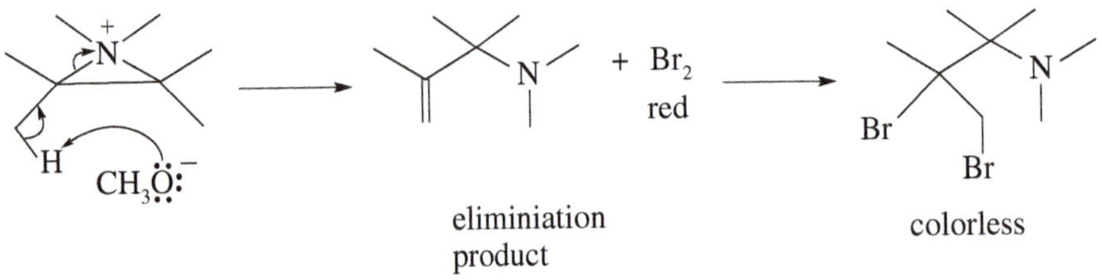

89. **a.** The reaction of 2-chlorobutane with HO^- is an intermolecular reaction, so the two compounds have to find each other in the solution.

$$CH_3\underset{Cl}{CH}CH_2CH_3 + HO^- \longrightarrow CH_3\underset{OH}{CH}CH_2CH_3 + Cl^-$$

The following reaction takes place in two steps. The first is an intramolecular S_N2 reaction; the reaction is much faster than the above reaction because the two reactants are in the same molecule and can find each other relatively easily.

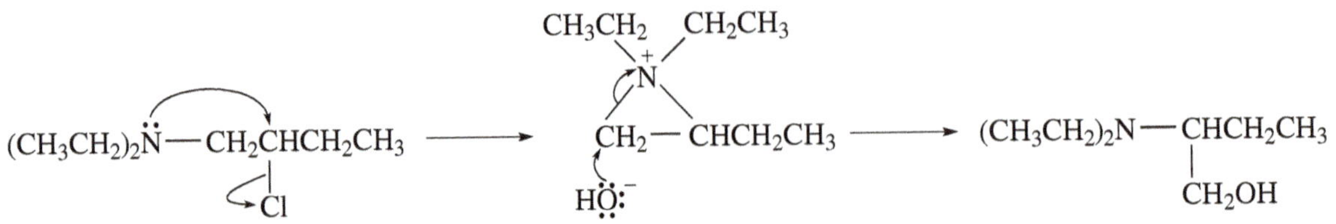

$$(CH_3CH_2)_2\ddot{N}-CH_2\underset{Cl}{CH}CH_2CH_3 \longrightarrow \overset{CH_3CH_2\;\;CH_2CH_3}{\overset{+}{N}}\;\; CH_2-CHCH_2CH_3 \;\; (H\ddot{O}:^-) \longrightarrow (CH_3CH_2)_2N-\underset{CH_2OH}{CH}CH_2CH_3$$

The second reaction is also an S_N2 reaction and is fast because the strain of the three-membered ring and the positive charge on the nitrogen make the amine a very good leaving group.

b. The HO group is bonded to a different carbon because HO^- attacks the least sterically hindered carbon of the three-membered ring.

90. **B** is the fastest reaction; **A** is the slowest reaction.

To form the epoxide, the alkoxide ion must attack the back side of the carbon that is bonded to Br. This means that the OH and Br substituents must both be in axial positions. To be 1,2-diaxial, they must be trans to each other.

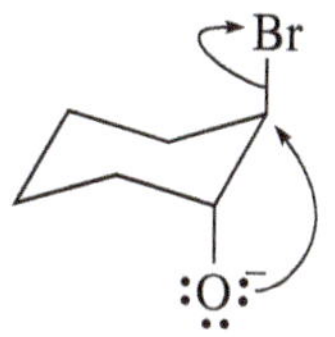

A does not form an epoxide, because the OH and Br substituents are cis to each other.
B and **C** can form epoxides because the OH and Br substituents are trans to each other.

The rate of formation of the epoxide is given by $k'K_{eq}$, where k' is the rate constant for the substitution reaction and K_{eq} is the equilibrium constant for the [equatorial]/[axial] conformers.

When the OH and Br substituents are in the required diaxial position, the large *tert*-butyl substituent is in the equatorial position in **B** and in the axial position in **C**.

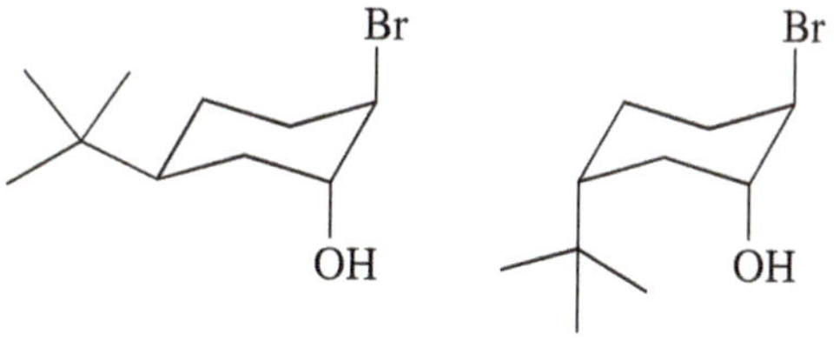

Because the more stable conformer has the large *tert*-butyl group in the equatorial position, the OH and Br substituents are in the required diaxial position in the more stable conformer of **B** (K_{eq} is large), whereas the OH and Br substituents are in the required diaxial position in the less stable conformer of **C** (K_{eq} is small). Therefore, **B** reacts faster than **C**.

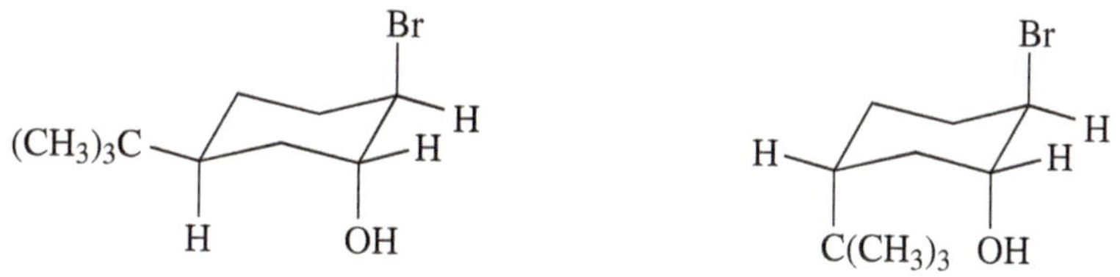

91. **a.**

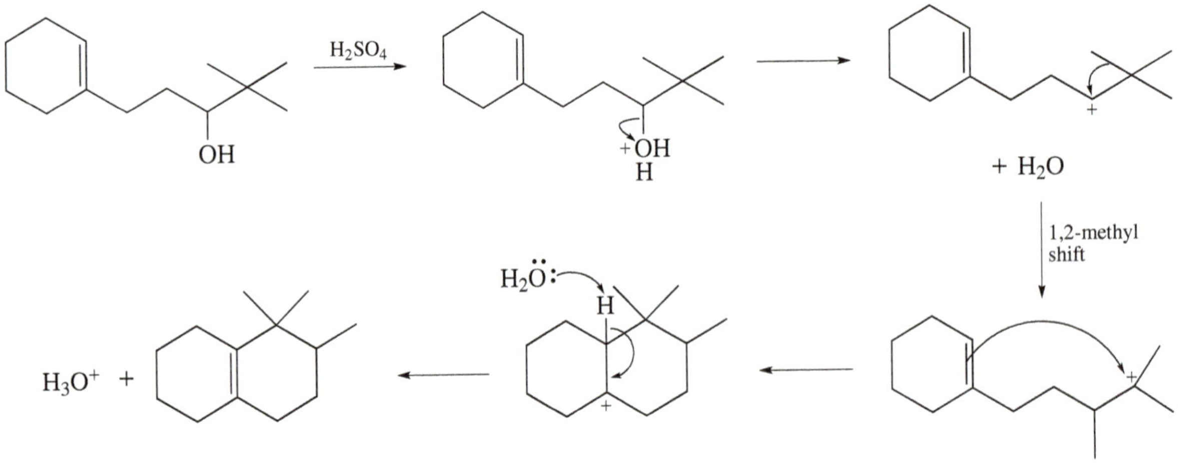

H$_2$SO$_4$

OH

+OH
H

+ H_2O

1,2-methyl shift

$H_2\ddot{O}$:

H

H_3O^+ +

b.

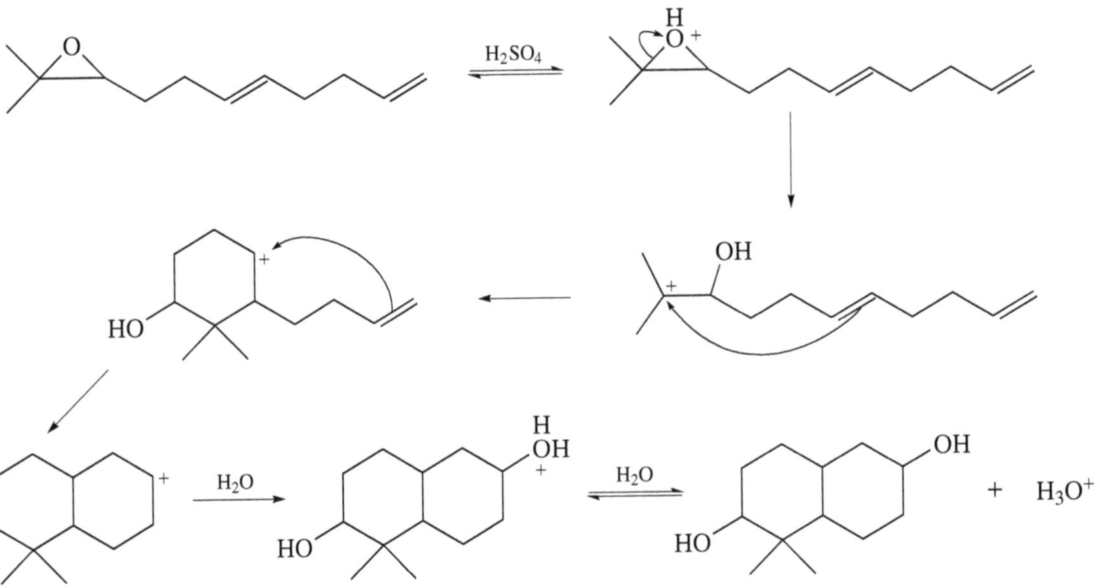

O

H$_2$SO$_4$

H
O+

OH

HO

HO

H$_2$O

H
OH
+

HO

H$_2$O

OH

HO

+ H_3O^+

c.

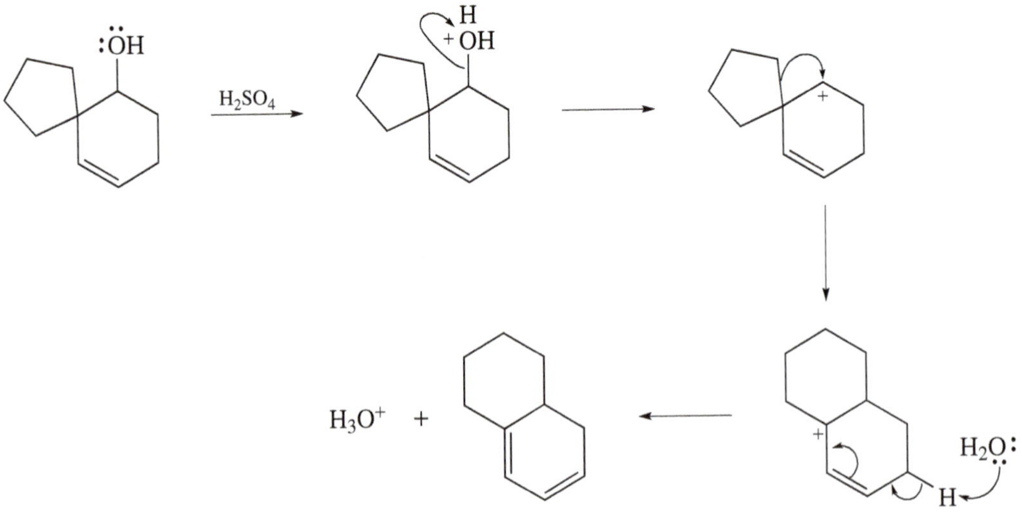

:ÖH

H$_2$SO$_4$

H
+OH

H_3O^+ +

H_2O:

H

92. $CH_3C(OH)(CH_3)-C(OH)(CH_3)-CH_3$ $\xrightarrow{H_2SO_4}$ $CH_3C(\overset{+}{O}H_2)(CH_3)-C(OH)(CH_3)-CH_3$ $\longrightarrow$ $CH_3\overset{+}{C}(CH_3)-C(OH)(CH_3)-CH_3 + H_2O$ $\rightleftharpoons$ $CH_3C(CH_3)_2-C(=\overset{+}{O}H)-CH_3$ $\rightleftharpoons$ $H_3O^+ + CH_3C(CH_3)_2-C(=O)-CH_3$

93. **a.** $CH_3C(CH_3)(OH)-CH_2OH$ $\xrightarrow{H_2SO_4}$ $CH_3C(CH_3)(\overset{+}{O}H_2)-CH_2OH$ $\longrightarrow$ $CH_3\overset{+}{C}(CH_3)-CH_2OH + H_2O$ $\longrightarrow$ $CH_3C(CH_3)=CHOH + H_3O^+$ $\xrightleftharpoons{\text{keto-enol interconversion}}$ $CH_3CH(CH_3)-CH=O$

b. Dehydration of the primary alcohol group cannot occur, because it cannot lose water via an E1 pathway, since a primary carbocation cannot be formed. It cannot lose water via an E2 pathway because the β-carbon is not bonded to a hydrogen. However, dehydration of the tertiary alcohol group can occur. The product is an enol that tautomerizes to an aldehyde.

94. OH, OH $\xrightleftharpoons[\Delta]{H_2SO_4}$ OH, $\overset{+}{O}H$ H $\longrightarrow$ $\ddot{O}H$, + , $+ H_2O$ $\longrightarrow$ ^{+}OH $\rightleftharpoons$ $H_3O^+ +$ O

95. Cyclopentene oxide undergoes back-side attack by the nucleophile, so the two substituents in the products are trans to each other. Therefore, both the *R,R*-isomer and the *S,S*-isomer are formed.

cyclopentene oxide + CH_3NHCH_3 (dimethylamine) $\longrightarrow$ OH, $N(CH_3)_2$ (*R,R*-isomer) + $N(CH_3)_2$, OH (*S,S*-isomer)

96. Notice that the initially generated carbocation can undergo either a 1,2-methyl shift or a 1,2-alkyl shift.

a.

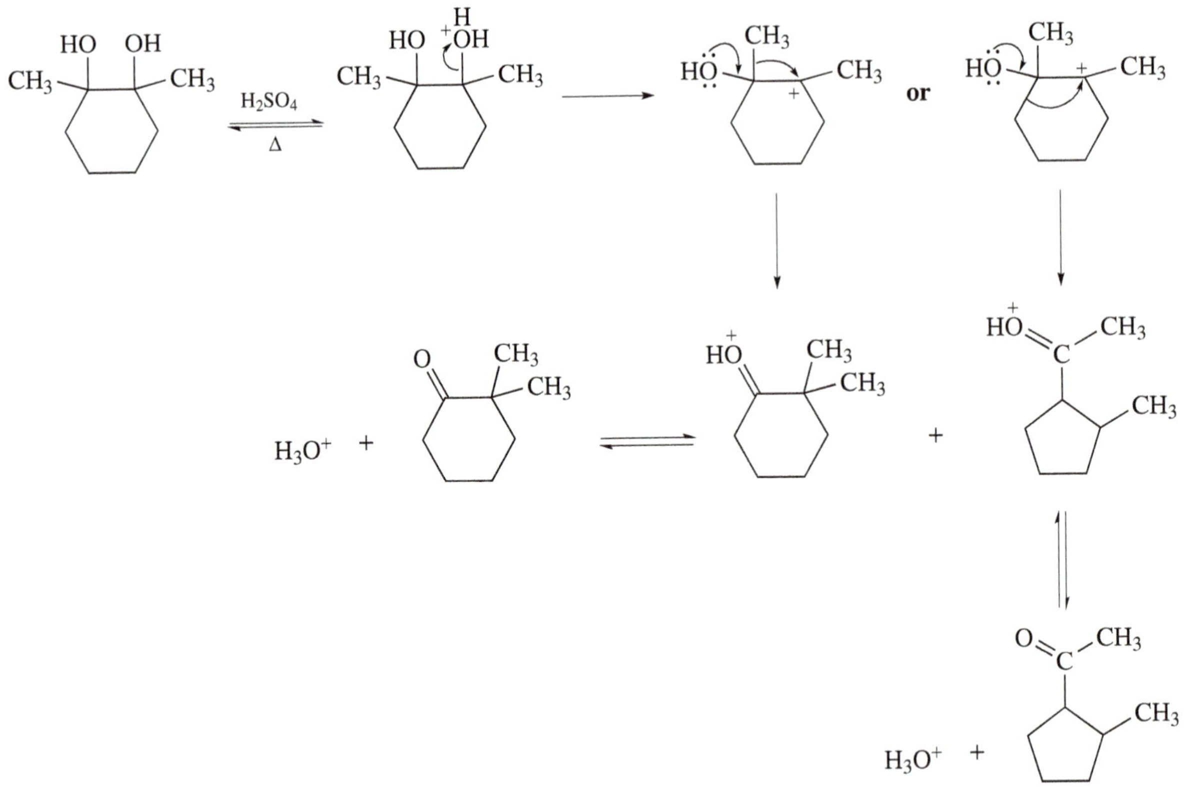

b.

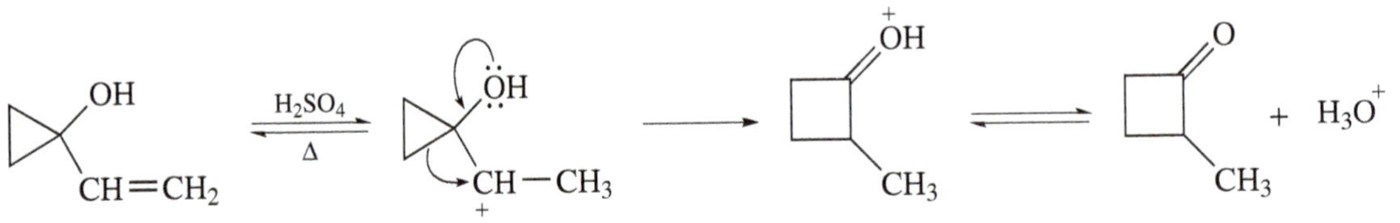

97. **a.** A nitrogen is a stronger base than an oxygen, so unlike an epoxide that can be opened without the oxygen being protonated, the three-membered nitrogen-containing ring has to be protonated to improve the leaving propensity of the group.

b. A nucleophile such as an NH_2 group on a chain of DNA can react with the three-membered ring. If a nucleophile on another chain of DNA reacts with another of the three-membered rings in this compound, the two DNA chains will be cross-linked.

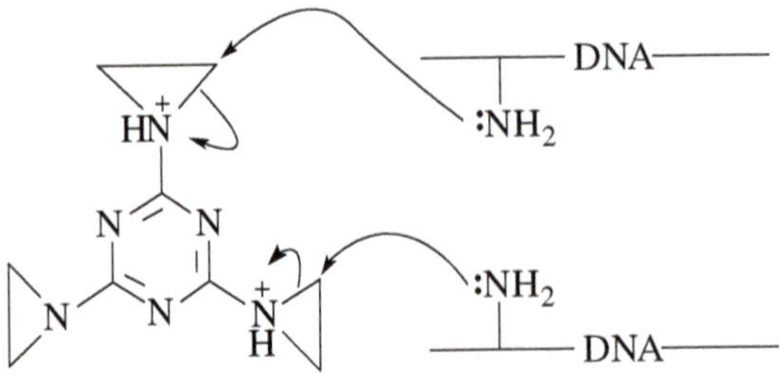

Chapter 10 Practice Test

1. Which of the following reagents is the best one to use to convert methyl alcohol into methyl bromide?

Br^- HBr Br_2 NaBr Br^+

2. Which of the following reagents is the best one to use to convert methyl alcohol into methyl chloride?

Cl_2/CH_2Cl_2 Cl_2/Δ Cl^- $SOCl_2$ NaCl

3. a. What is the major product obtained from the reaction of the epoxide shown below in methanol containing 0.1 M HCl?

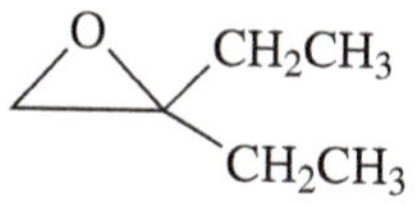

CH_2CH_3

CH_2CH_3

b. What is the major product obtained from the reaction of the epoxide in methanol containing 0.1 M $NaOCH_3$?

4. Draw the major product(s) of each of the following reactions:

a. 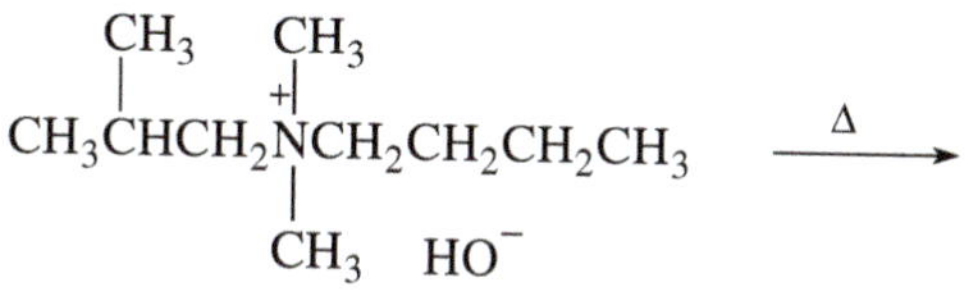 $CH_3CH(CH_3)CH_2\overset{+}{N}(CH_3)_2CH_2CH_2CH_2CH_3$ HO^- $\xrightarrow{\Delta}$

b. 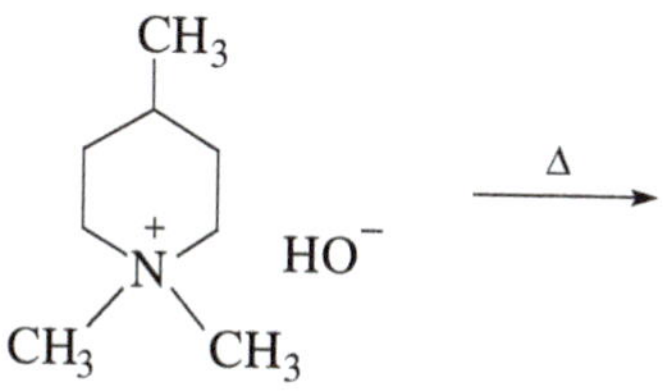 CH_3 $\overset{+}{N}$ HO^- CH_3 CH_3 $\xrightarrow{\Delta}$

c. 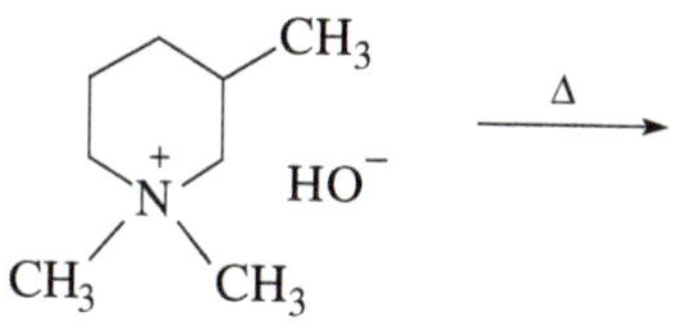 CH_3 $\overset{+}{N}$ HO^- CH_3 CH_3 $\xrightarrow{\Delta}$

5. Draw the major elimination product that is obtained when each of the following alcohols is heated in the presence of H_2SO_4:

a. 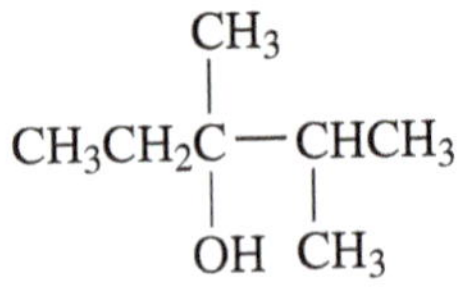 $CH_3CH_2C(CH_3)(OH)-CH(CH_3)CH_3$

c. $CH_3CH_2CH_2CH_2CH_2OH$

b. 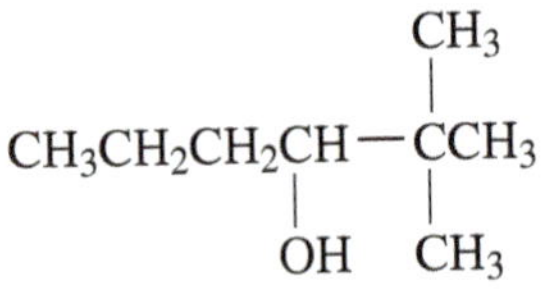 $CH_3CH_2CH_2CH(OH)-C(CH_3)_2CH_3$

d. 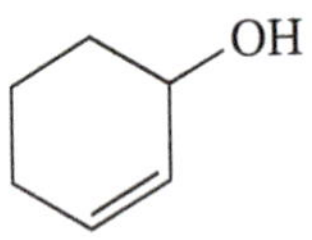 (cyclohex-2-en-1-ol: cyclohexene ring with OH)

6. Indicate whether each of the following statements is true or false:

a.	Tertiary alcohols are easier to dehydrate than secondary alcohols.	T	F
b.	Alcohols are more acidic than thiols.	T	F
c.	Alcohols have higher boiling points than thiols.	T	F
d.	The acid-catalyzed dehydration of a primary alcohol is an S_N1 reaction.	T	F
e.	The Hofmann elimination reaction is an E2 reaction.	T	F

7. What products are obtained when the following ethers are heated with one equivalent of HI?

a. 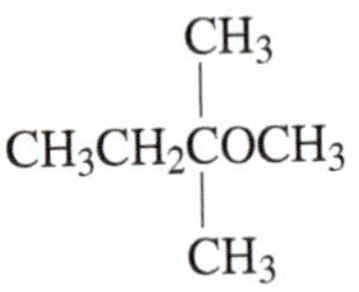 $CH_3CH_2C(CH_3)_2OCH_3$

b. 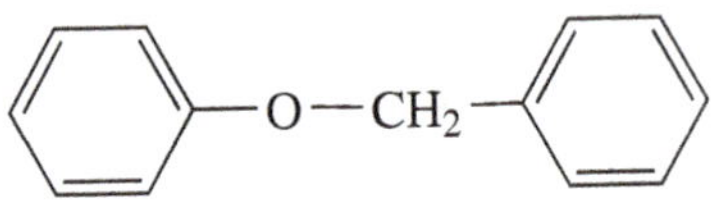 $C_6H_5-O-CH_2-C_6H_5$

8. Draw the major product of each of the following reactions:

a. 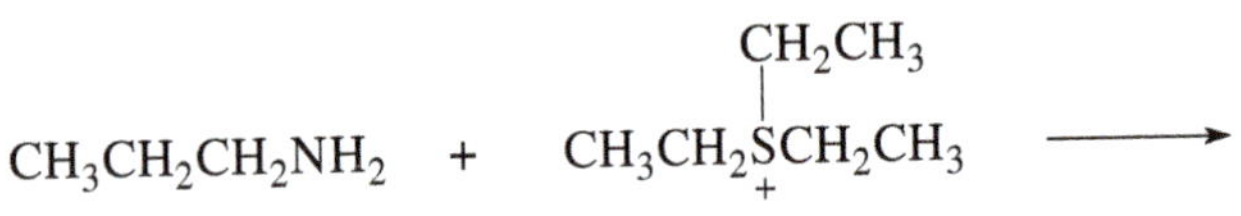 $CH_3CH_2CH_2NH_2 \quad + \quad CH_3CH_2\overset{+}{S}(CH_2CH_3)CH_2CH_3 \longrightarrow$

b. $CH_3CH_2CH_2OH \quad + \quad SOCl_2 \xrightarrow[\text{pyridine}]{}$

c. $CH_3CH_2CH_2CH_2CH_2CH_2OH \xrightarrow[\Delta]{H_2SO_4}$

d. $CH_3CH_2CH_2CH_2OH \xrightarrow[\substack{CH_3COOH \\ 0\ °C}]{NaOCl}$

e. $CH_3CH_2CH_2CH_2OH \xrightarrow{H_2CrO_4}$

CHAPTER 11
Organometallic Compounds

Important Terms

alkene metathesis (olefin metathesis) — a reaction that breaks the double bond of an alkene and then rejoins the fragments to form a new double bond between two sp^2 carbons that were not previously bonded.

alkyne metathesis — a reaction that breaks the triple bond of an alkyne and then rejoins the fragments to form a new triple bond between two *sp* carbons that were not previously bonded.

coupling reaction — a reaction that joins two groups with a carbon–carbon bond.

dialkyl lithium cuprate — $(R)_2CuLi$; prepared by treating an organolithium reagent with cuprous iodide.

Gilman reagent (organocuprate) — a dialkyl lithium cuprate used to replace a halogen in an alkyl, aryl, or vinylic halide with an alkyl group.

Grignard reagent (organomagnesium compound) — RMgBr or RMgCl; prepared by adding an alkyl halide to magnesium shavings.

Heck reaction — a reaction that couples an aryl halide or a vinylic halide with an alkene.

olefin metathesis (alkene metathesis) — a reaction that breaks the double bond of an alkene and then rejoins the fragments to form a new double bond between two sp^2 carbons that were not previously bonded.

organoboron compound — an alkyl-organoboron compound, an alkenyl-organoboron compound, or an aryl-organoboron compound: $R'B(OR)_2$.

organocuprate (Gilman reagent) — a dialkyl lithium cuprate used to replace a halogen in an alkyl, aryl, or vinylic halide with an alkyl group.

organolithium compound — RLi; prepared by adding lithium to an alkyl halide.

organomagnesium compound (Grignard reagent) — RMgBr or RMgCl; prepared by adding an alkyl halide to magnesium shavings.

organometallic compound — a compound with a carbon–metal bond.

organopalladium compound — a compound with a palladium–carbon bond.

oxidative addition — insertion of a metal between two atoms.

reductive elimination — elimination of a metal from between two carbons, resulting in formation of a carbon–carbon bond.

Suzuki reaction — a reaction that couples an aryl halide or vinylic halide with an organoboron reagent.

transmetallation — the exchange of two groups (often alkyl and halide) between two metals.

Solutions to Problems

1. All the reactions occur because in each case, the reactant acid is a stronger acid than the product acid (methane, pK_a ~ 60).

2. The greater the polarity of the carbon–metal bond, the more reactive the organometallic compound. Polarity depends on the difference in electronegativity between the atoms forming the bond; the greater the difference in electronegativity, the more polar the bond.

C—Li	C—Na
2.5–1.0 = 1.5	2.5–0.9 = 1.6

The electronegativity difference between carbon and sodium is greater than that between carbon and lithium. Therefore, organosodium compounds (with a more polar carbon–metal bond) are more reactive than organolithium compounds.

3. Transmetallation will occur if the new metal's electronegativity is closer to that of carbon. Because gallium (1.8) is more electronegative than magnesium (1.2) and, therefore, closer to the electronegativity of carbon, transmetallation occurs. Notice that after transmetallation, the number of alkyl groups attached to gallium is the same as the number of chorines that were attached to it in the reactant.

$$3\ CH_3MgCl + GaCl_3 \longrightarrow (CH_3)_3Ga + 3\ MgCl_2$$

4. Solved in the text.

5. The Br is replaced by the alkyl group of the organocuprate.

$$2\ CH_3(CH_2)_3CH_2Br \xrightarrow[\text{hexane}]{\text{2 Li}} 2\ CH_3(CH_2)_3CH_2Li \xrightarrow[\text{THF}]{\text{CuI}} (CH_3CH_2CH_2CH_2CH_2)_2CuLi$$

$CH_3(CH_2)_7CH{=}CH(CH_2)_7CH_2Br$ (cis; H, H on same side) or $BrCH_2(CH_2)_2CH{=}CH(CH_2)_{12}CH_3$ (cis; H, H on same side)

$\downarrow$

$CH_3(CH_2)_7CH{=}CH(CH_2)_{12}CH_3$ (cis; H, H on same side)

6. Solved in the text.

7. a. bromobenzene (benzene ring with Br) b. cyclohexyl–CH=CH–CH$_2$Br c. cyclohexyl–CH$_2$–CH=CH–Br

8. a. $CH_3CH_2CH_2CH_2CH_2OH$ b. $CH_3CH{=}CHCH_2CH_2OH$ c. C_6H_5–$CH_2CH_2CH_2OH$

9. **a.** $CH_3CH{=}CH_2CH_2C(OH)(CH_3)CH_3$ **b.** $CH_3CH_2CH(CH_3)CH(OH)CH_3$

10. **a.** cyclohexene $\xrightarrow{RCOOH\ (peroxyacid)}$ cyclohexene oxide $\xrightarrow{(CH_3CH_2)_2CuLi}$ trans-2-ethylcyclohexanolate (O^-, CH_2CH_3) + enantiomer (CH_2CH_3, O^-) $\xrightarrow{HCl}$ trans-2-ethylcyclohexanol (OH, CH_2CH_3) + enantiomer (CH_2CH_3, OH)

b. cyclohexene $\xrightarrow{RCOOH}$ cyclohexene oxide $\xrightarrow{(CH_3CH_2)_2CuLi}$ 2-ethylcyclohexanolate (O^-, CH_2CH_3) $\xrightarrow{HCl}$ 2-ethylcyclohexanol (OH, CH_2CH_3) $\xrightarrow{H_2SO_4,\ \Delta}$ 1-ethylcyclohexene (CH_2CH_3)

c. cyclohexene $\xrightarrow{RCOOH}$ cyclohexene oxide $\xrightarrow{(CH_3)_2CuLi}$ 2-methylcyclohexanolate (O^-, CH_3) $\xrightarrow{HCl}$ 2-methylcyclohexanol (OH, CH_3) $\xrightarrow{H_2SO_4,\ \Delta}$ 1-methylcyclohexene (CH_3) $\xrightarrow{HBr}$ 1-bromo-1-methylcyclohexane (CH_3, Br)

d. cyclohexene $\xrightarrow{RCOOH}$ cyclohexene oxide $\xrightarrow{1.\ (C_6H_5)_2CuLi\quad 2.\ HCl}$ phenylcyclohexanol (OH) $\xrightarrow{1. H_2SO_4, \Delta\quad 2. H_2, Pd/C}$ phenylcyclohexane

11. Solved in the text.

12. Solved in the text.

13. **a.** **b.**

14. Add the number of carbons in the chain in the starting material to the number of carbons in the organoboron compound. It should be the same as the number of carbons in the chain in the product.

a. Br **b.** Br **c.** Br

15. An alkenylboron compound is prepared from an alkyne.

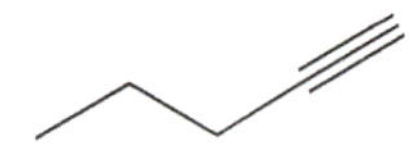

1-pentyne

16.

17. **a.** C≡N **c.** CH_3O

b. O NH$_2$ **d.** O

18. Solved in the text.

19. **a.** Br + O **b.** Br +

20. **Suzuki reactions**

a. (bromobenzene) + (alkenyl boronate, B(OR)$_2$) $\xrightarrow[HO^-]{PdL_2}$ (product)

$\uparrow$ $HB(OR)_2$

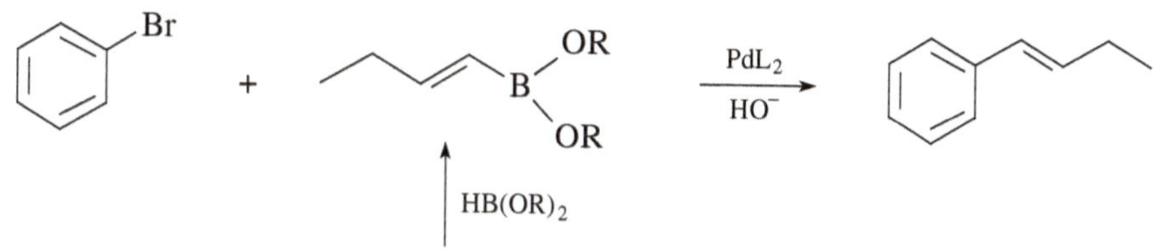

$CH_3CH_2C{\equiv}CH$

b. (bromobenzene) + (alkyl boronate, B(OR)$_2$) $\xrightarrow[HO^-]{PdL_2}$ (product)

$\uparrow$ $HB(OR)_2$

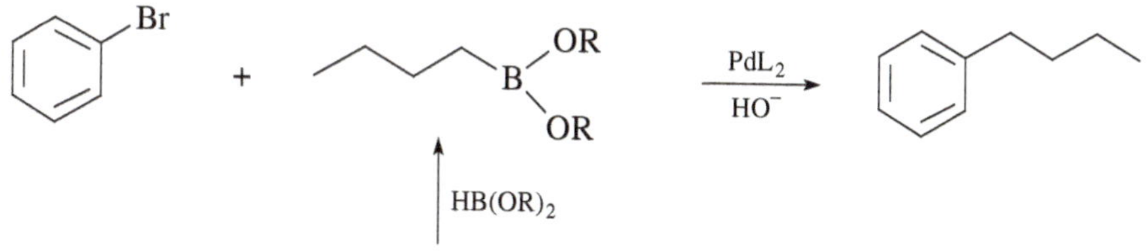

$CH_3CH_2CH{=}CH_2$

Heck reactions

a. (bromobenzene) + (1-butene) $\xrightarrow[(CH_3CH_2)_3N]{PdL_2}$ (product)

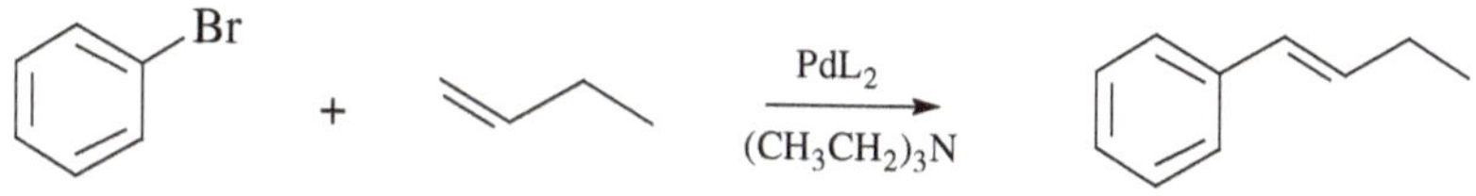

b. (bromobenzene) + (1-butene) $\xrightarrow[(CH_3CH_2)_3N]{PdL_2}$ (product) $\xrightarrow[H_2]{Pd/C}$ (product)

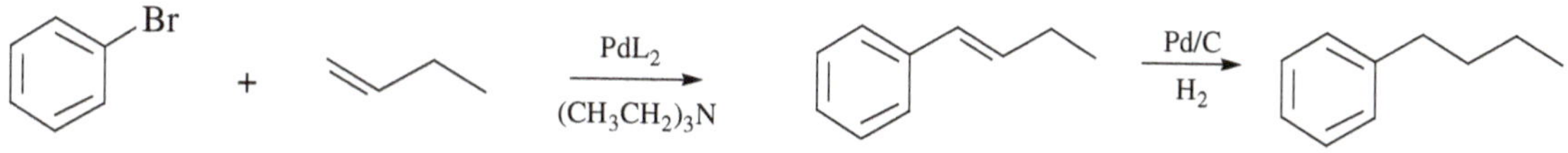

Notice that only a Suzuki reaction can be used to replace a halogen with an alkyl group. A Heck reaction replaces a halogen wih an alkenyl group. However, the alkenyl group can subsequently be reduced to an alkyl group.

21. $CH_3\overset{O}{\overset{\|}{C}}$–C$_6H_4$–Br + CH_3O–C$_6H_4$–$CH{=}CH_2$

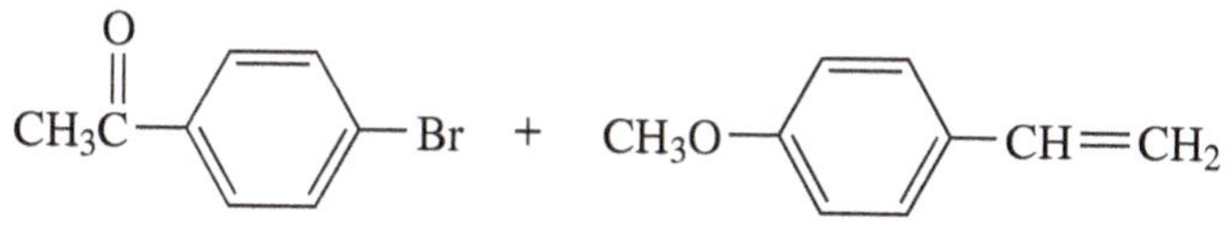

or

CH_3O–C$_6$H$_4$–Br + $CH_3\overset{O}{\overset{\|}{C}}$–C$_6H_4$–$CH{=}CH_2$

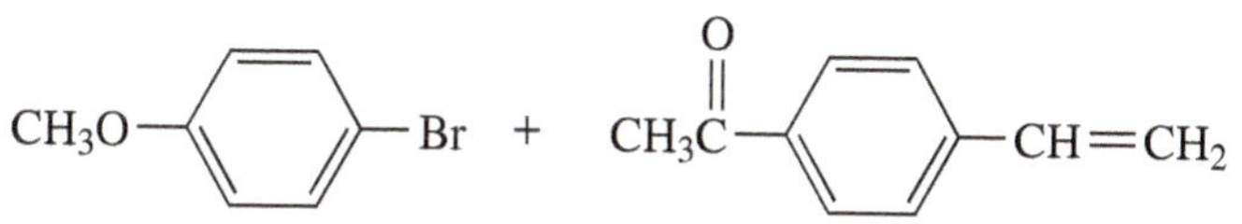

22. **a.** $CH_2{=}CH_2$ + $CH_3CH_2CH{=}CHCH_2CH_3$

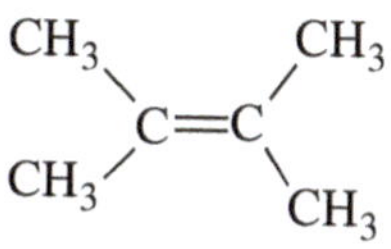

E and *Z*

b. $CH_2{=}CH_2$ + (alkene) + (alkene)

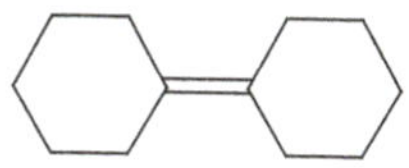

c. $CH_2{=}CH_2$ + $(CH_3)_2C{=}C(CH_3)_2$

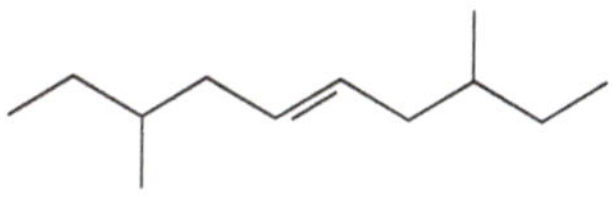

d. $CH_2{=}CH_2$ + (cyclohexylidenecyclohexane)

23. To determine the size of the ring that is formed in ring-closing metathesis, subtract 2 from the number of carbons in the reactant.

a. $CH_2{=}CH_2$ +

c. $CH_2{=}CH_2$ +

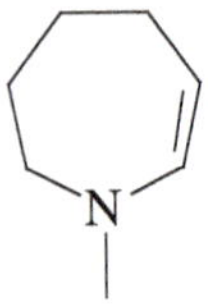

N

b. $CH_2{=}CH_2$ +

OCH_3

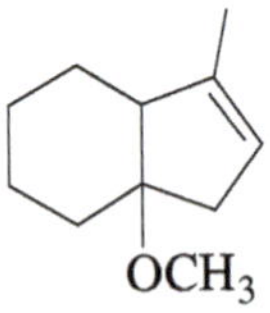

24. Solved in the text.

25. a.

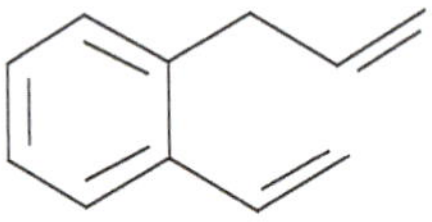

b.

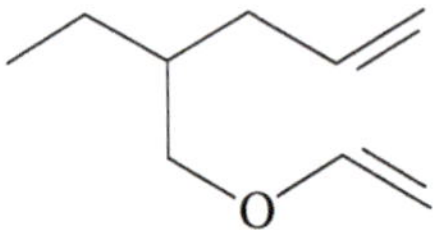

O

26. 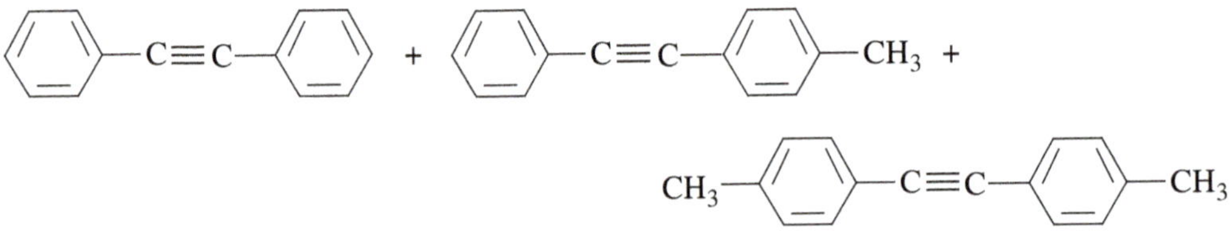 C≡C + C≡C CH_3 +

CH_3 C≡C CH_3

27. a.

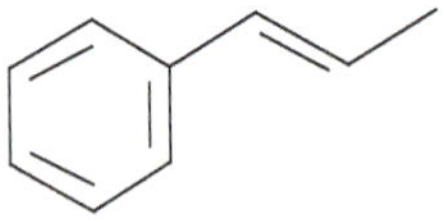

d.

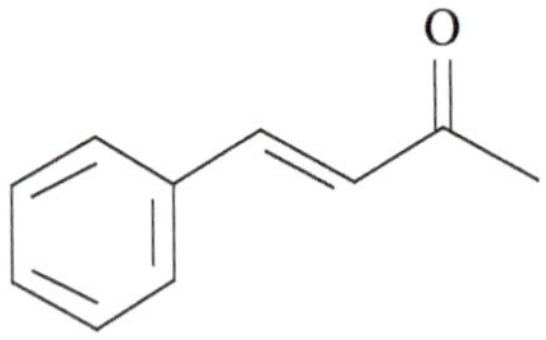

O

b. 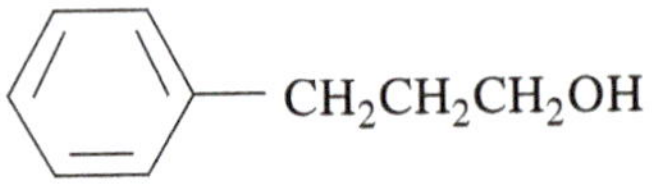 $CH_2CH_2CH_2OH$

e.

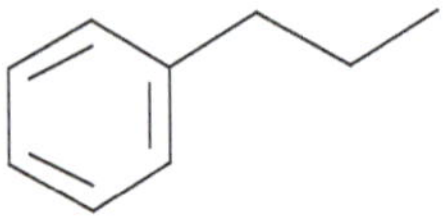

c.

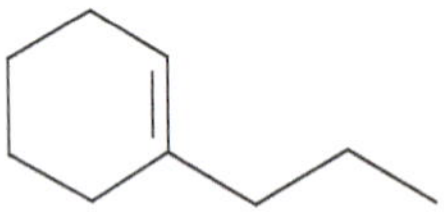

f.

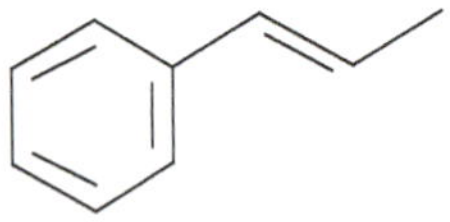

28. **C** is the only one that can be used to form a Grignard reagent.
The Grignard reagent formed from **A** will be destroyed immediately by reacting with the proton of the alcohol group.
The Grignard reagent formed from **B** will be destroyed immediately by reacting with the proton of the carboxylic acid group.
The Grignard reagent formed from **D** will be destroyed immediately by reacting with the proton of the amino group.

29. 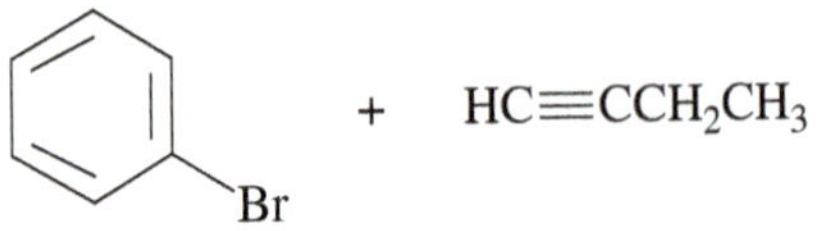 Br + $HC{\equiv}CCH_2CH_3$

30. **A** = Li **D** = ethylene oxide **G** = ethylene oxide
B = CuI **E** = H^+ **H** = H^+
C = $(CH_3)_2CuLi$ **F** = NaH

31. **a.** bromobenzene + $CH_2{=}CH_2$ $\xrightarrow[(CH_3CH_2)_3N]{PdL_2}$ C₆H₅–$CH{=}CH_2$ $\xrightarrow[2.\ H_2O_2,\ HO^-,\ H_2O]{1.\ R_2BH/THF}$ C₆H₅–CH_2CH_2OH

or

bromobenzene + $(CH_2{=}CH)_2CuLi$ → C₆H₅–$CH{=}CH_2$

or

bromobenzene $\xrightarrow[2.\ CuI]{1.\ Li}$ $(C_6H_5)_2CuLi$ $\xrightarrow[2.\ HCl]{1.\ \text{ethylene oxide}}$ C₆H₅–CH_2CH_2OH

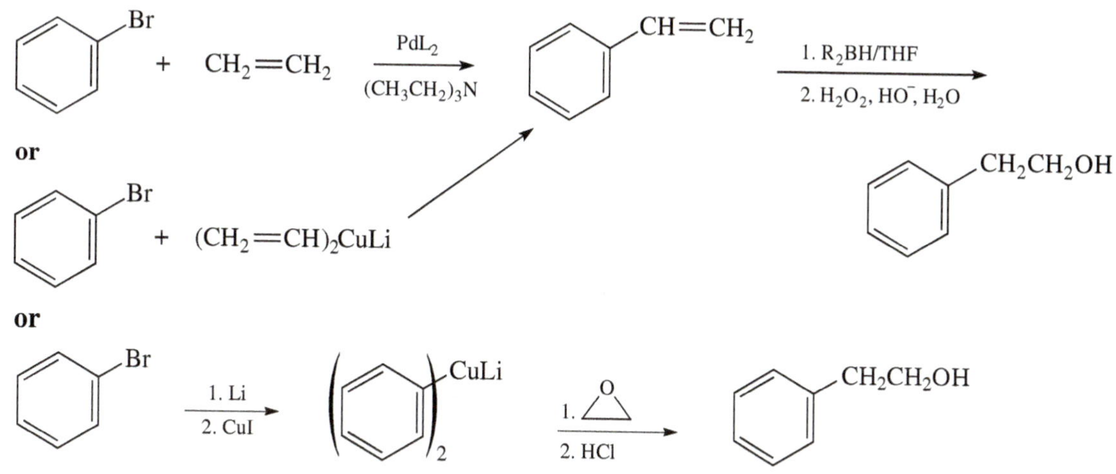

b. $CH_3CH(CH_3)CH_2OH$ $\xrightarrow[\Delta]{HBr}$ $CH_3CH(CH_3)CH_2Br$ $\xrightarrow[2.\ CuI]{1.\ Li}$ $(CH_3CH(CH_3)CH_2)_2CuLi$ $\xrightarrow[2.\ HCl]{1.\ \text{ethylene oxide}}$ $CH_3CH(CH_3)CH_2CH_2CH_2OH$

c. $CH_3CH_2C{\equiv}CH$ $\xrightarrow{NaNH_2}$ $CH_3CH_2C{\equiv}C^-$ $\xrightarrow{CH_3CH_2Br}$ $CH_3CH_2C{\equiv}CCH_2CH_3$

d. bromobenzene (Br) + methyl vinyl ketone (O) $\xrightarrow[(CH_3CH_2)_3N]{PdL_2}$ C₆H₅–CH=CH–C(=O)CH₃

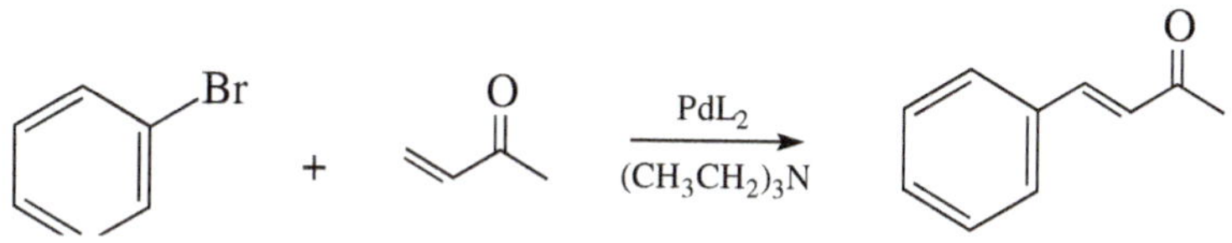

e. 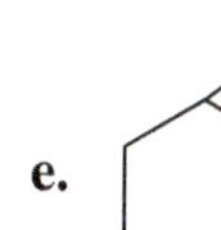

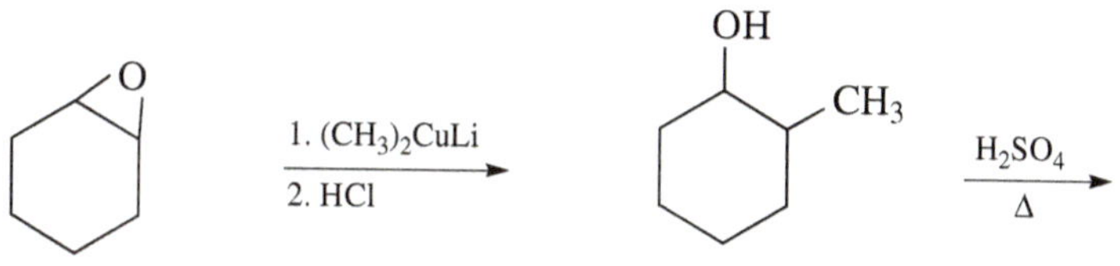

$\xrightarrow[2.\ HCl]{1.\ (CH_3)_2CuLi}$ 2-methylcyclohexanol (OH, CH_3) $\xrightarrow[\Delta]{H_2SO_4}$ 1-methylcyclohexene (CH_3)

f. ethanol (OH) $\xrightarrow[pyridine]{SOCl_2}$ chloroethane (Cl) $\xrightarrow[hexane]{Li}$ ethyllithium (Li) $\xrightarrow[THF]{CuI}$ $(CH_3CH_2)_2CuLi$ $\xrightarrow[2.\ HCl]{1.\ \text{ethylene oxide}}$ 1-butanol (OH)

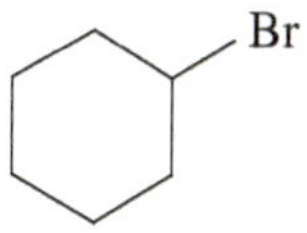

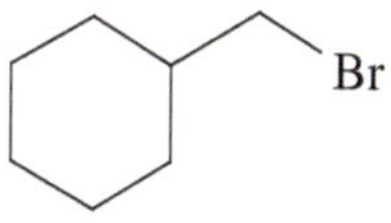

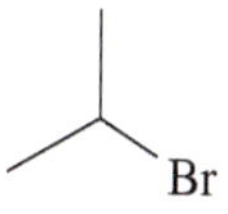

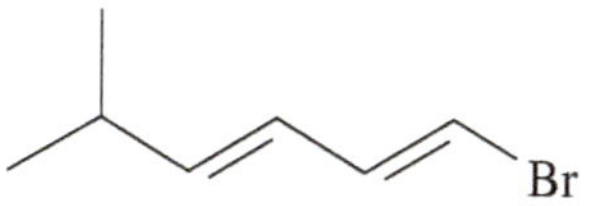

32. **a.** bromocyclohexane (Br) **b.** (bromomethyl)cyclohexane (Br) **c.** 2-bromopropane (Br) **d.** 1-bromo-3-methyl-1-butene (Br)

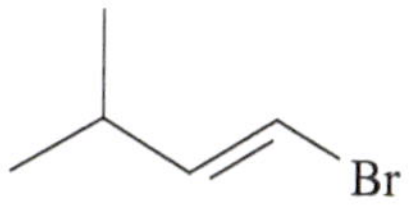

e. 1-bromo-5-methyl-1,3-hexadiene (Br)

33. a.

b.

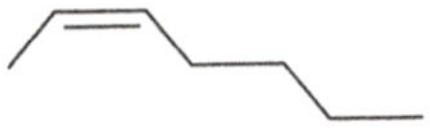

c.

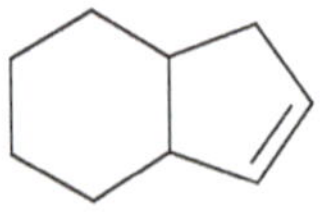

d.

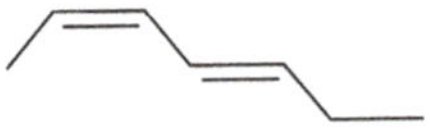

e.

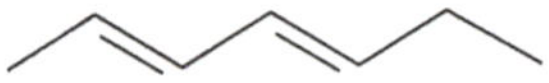

f.

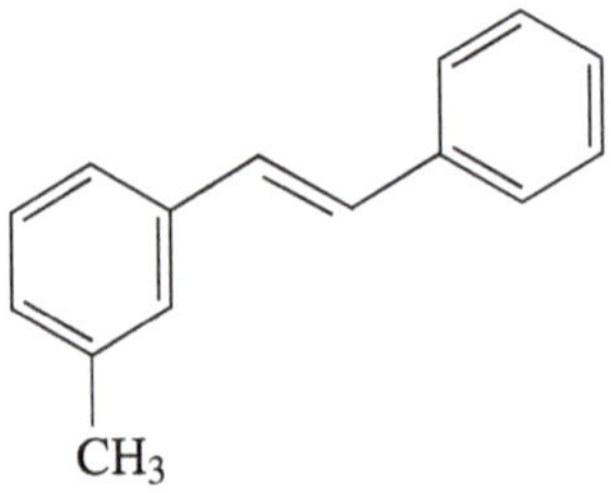

CH_3

34. a. 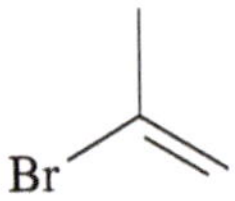 Br

b. 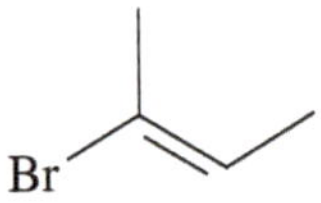 Br

c. 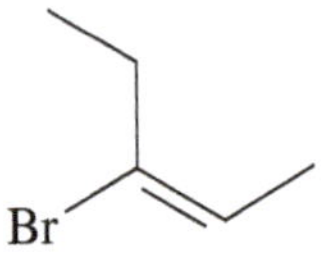 Br

35. 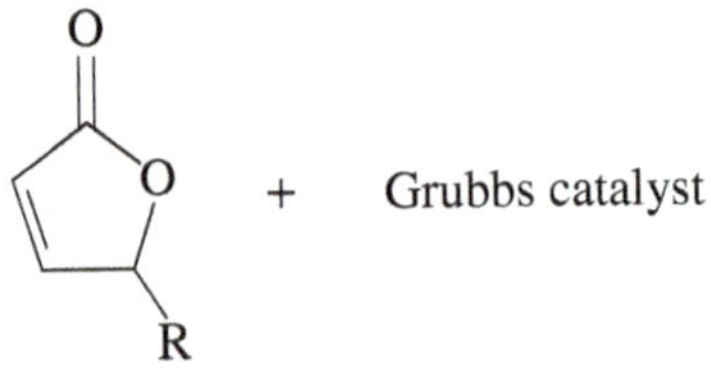 O O R + Grubbs catalyst

36. Remember that the double bond that comes from the alkenyl-organoboron compound always has the *E* (trans) configuration.

a. 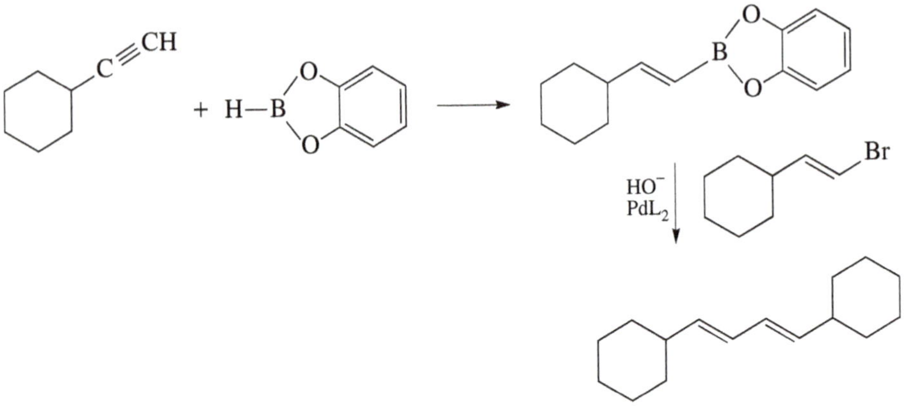 C≡CH + H—B(O)(O) ⟶ B(O)(O)

HO^- PdL_2 Br

b. 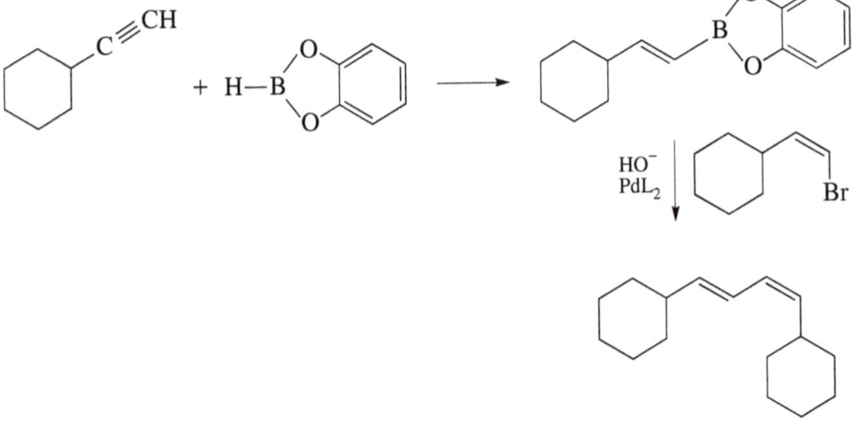 C≡CH + H—B(O)(O) ⟶ B(O)(O)

HO^- PdL_2 Br

c.

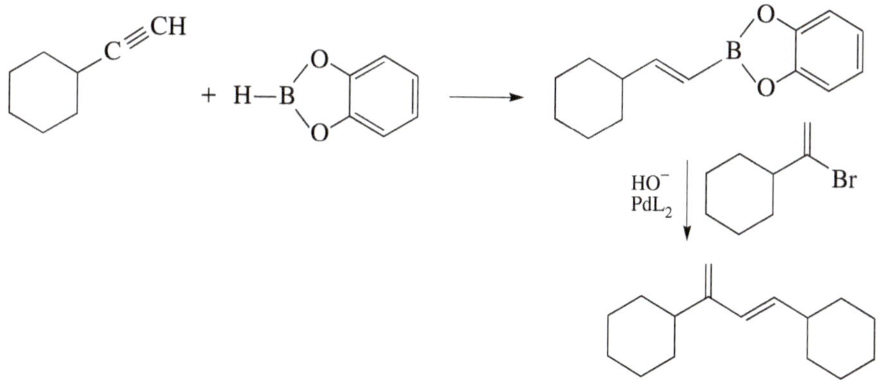
C≡CH
+ H—B
O
O
B
HO⁻
PdL2
Br

37. a.

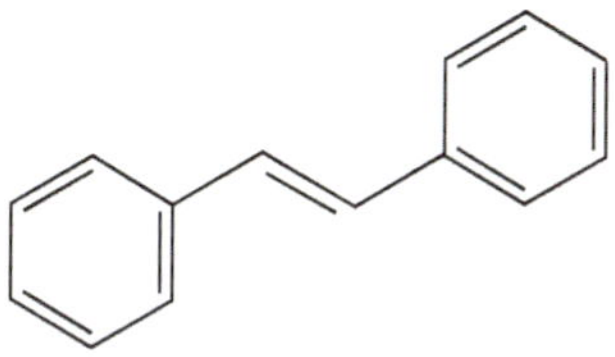

b.

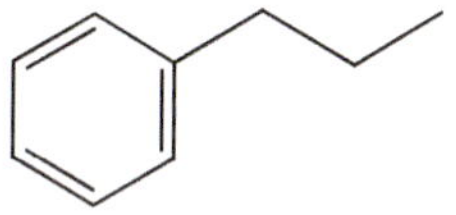

c.

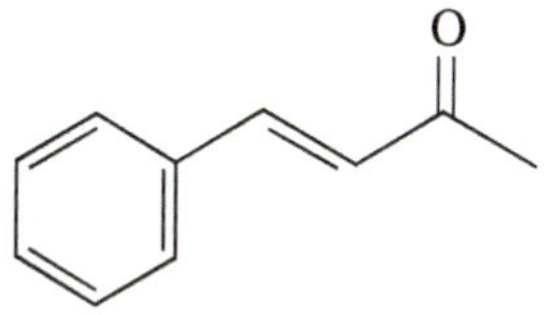
O

d.

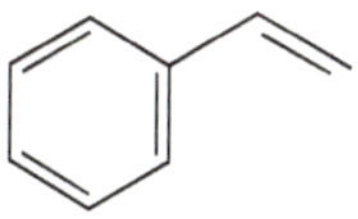

38. a.

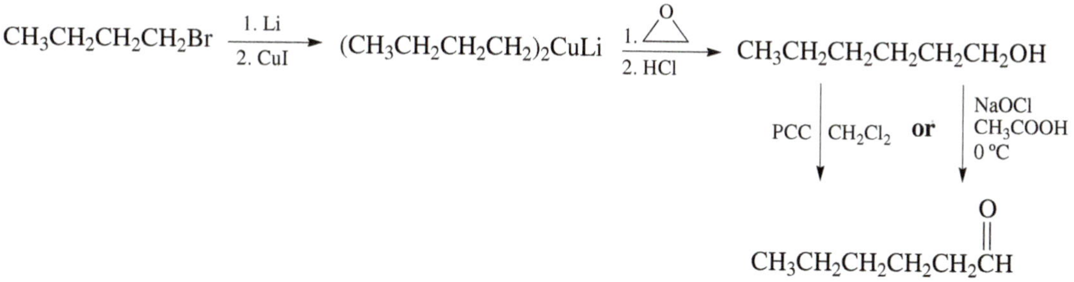
CH3CH2CH2CH2Br
1. Li
2. CuI
(CH3CH2CH2CH2)2CuLi
1.
O
2. HCl
CH3CH2CH2CH2CH2CH2OH
PCC
CH2Cl2
or
NaOCl
CH3COOH
0 °C
O
CH3CH2CH2CH2CH2CH

b.

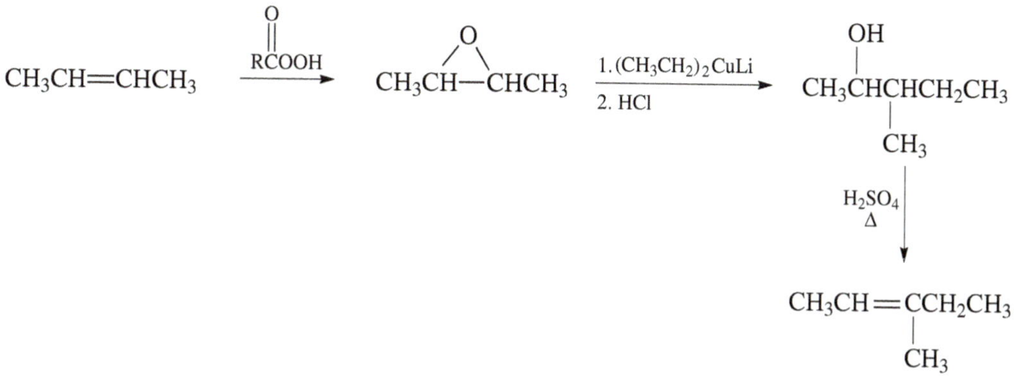
CH3CH=CHCH3
O
RCOOH
O
CH3CH—CHCH3
1. (CH3CH2)2CuLi
2. HCl
OH
CH3CHCHCH2CH3
CH3
H2SO4
Δ
CH3CH=CCH2CH3
CH3

c. $CH_3CH{=}CHCH_3$ $\xrightarrow{RCOOH\ (\text{peroxyacid})}$ $CH_3CH{-}CHCH_3$ (epoxide, O bridging) $\xrightarrow[2.\ HCl]{1.\ (CH_3)_2CuLi}$ $CH_3CH(OH)CH(CH_3)CH_3$ $\xrightarrow{NaOCl,\ CH_3COOH,\ 0\ °C}$ $CH_3C(=O)CH(CH_3)CH_3$

39. Once some bromocyclohexane has been converted to a Grignard reagent, the Grignard reagent reacts with the bromocyclohexane that has not been converted to a Grignard reagent.

Cyclohexyl–Br $\xrightarrow[Et_2O]{Mg}$ Cyclohexyl–MgBr + Br–Cyclohexyl ⟶ Cyclohexyl–Cyclohexyl + $MgBr_2$

40. The student did not get any of the expected product, because the Grignard reagent removed a proton from the alcohol group. Addition of HCl/H_2O protonated the alkoxide ion and opened the epoxide ring.

CH_3MgBr + (H$_3$C, O epoxide cyclohexane with OH) ⟶ CH_4 + (H$_3$C, O epoxide cyclohexane with O^-) $\xrightarrow[H_2O]{HCl}$ (H$_3$C, OH, OH cyclohexane with OH)

41. **a.** Cyclohexyl–OH $\xrightarrow[pyridine]{PBr_3}$ Cyclohexyl–Br $\xrightarrow[2.\ CuI]{1.\ Li}$ (Cyclohexyl)$_2$CuLi $\xrightarrow[2.\ HCl]{1.\ \text{ethylene oxide}}$ Cyclohexyl–CH_2CH_2OH

b. (OH) $\xrightarrow[pyridine]{PBr_3}$ (Br) $\xrightarrow[Et_2O]{Mg}$ (MgBr) $\xrightarrow{D_2O}$ (D)

42. Remember that the double bond that comes from the alkenyl-organoboron compound always has the *E* (trans) configuration.
Both double bonds in **C** have the *Z* configuration.

a. **C** cannot be prepared by a Heck reaction because the double bond in the alkenyl-organoboron compound reactant of a Heck reaction always has the *E* configuration in the product.

A cannot be *directly* synthesized by a Heck reaction. A Heck reaction couples a vinylic halide and an alkene, so the product contains at least two double bonds. However, **A** can be made by a Heck reaction followed by reduction of the double bonds.

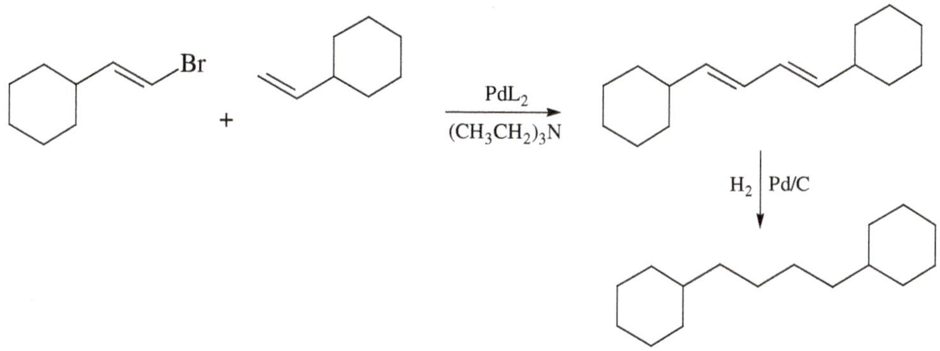

Br + $\xrightarrow[(CH_3CH_2)_3N]{PdL_2}$

H_2 | Pd/C

b. The starting materials for the synthesis of **A** are shown above. Those for the synthesis of **B** are shown here.

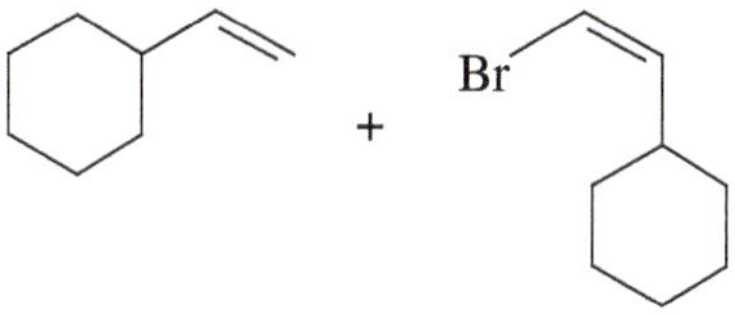

+ Br

43. $CH_3CH{=}CHCH_2CH_3$
2-pentene

A = $CH_3CH{=}$ **B** = $CH_3CH_2CH{=}$

A = a piece containing two carbons
B = a piece containing three carbons
The two pieces can be put together in four different ways.

A—A B—B A—B B—A

Two of the four pieces (**A—B** and **B—A**) represent 2-pentene. Therefore, the maximum yield of 2-pentene is 50%.

44. Bombykol can be prepared using a Suzuki reaction. First make the organoboron compound. Draw the vinylic bromide so that you can see what product will be obtained from the coupling reaction.

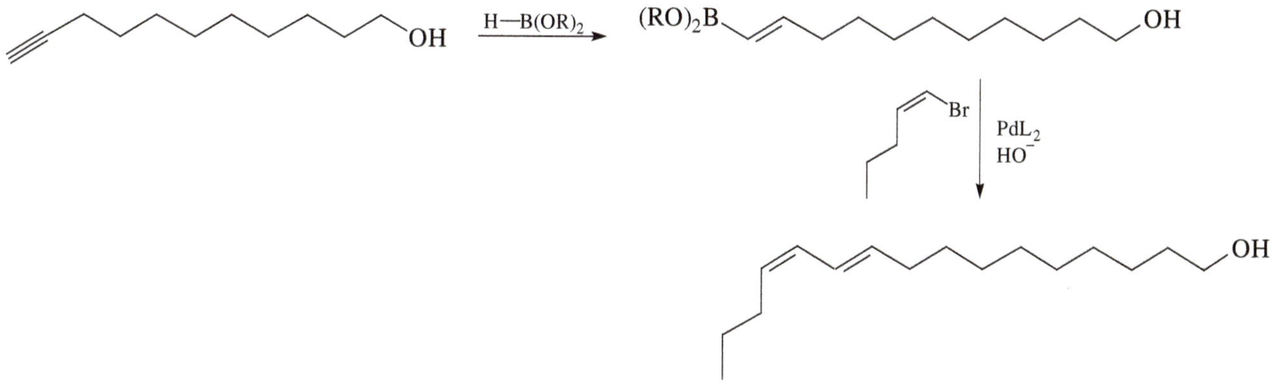

OH $\xrightarrow{H{-}B(OR)_2}$ $(RO)_2B$ OH

Br PdL_2 HO^-

OH

Bombykol can also be prepared using a Heck reaction.

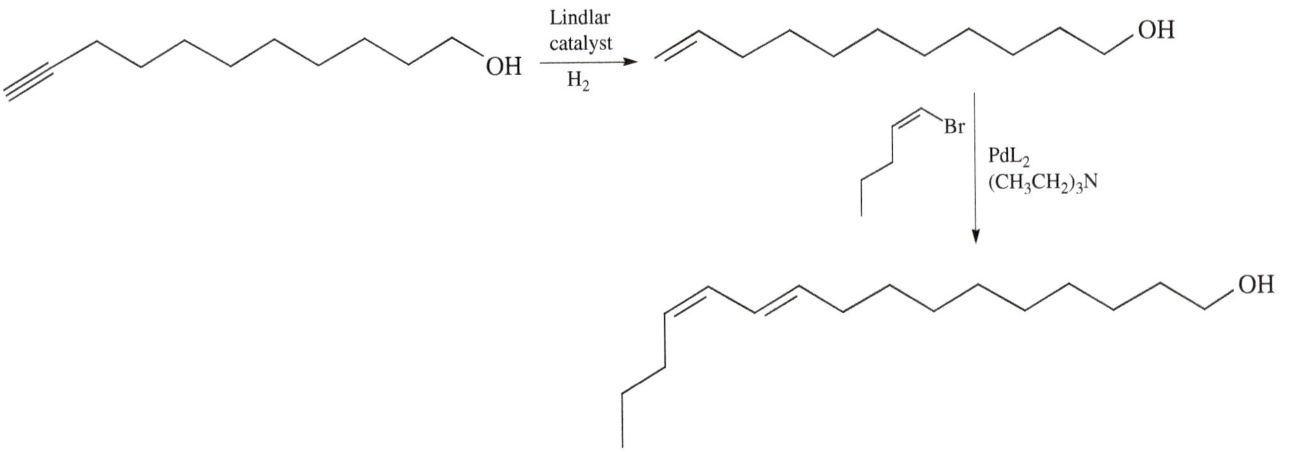

Lindlar catalyst, H_2

OH

OH

Br

PdL_2
$(CH_3CH_2)_3N$

OH

45. **a.** 2-butene and 3-hexene
b. Two symmetrical alkenes. The only alkenes that will be synthesized other than the target alkene are the reactant alkenes.

46.

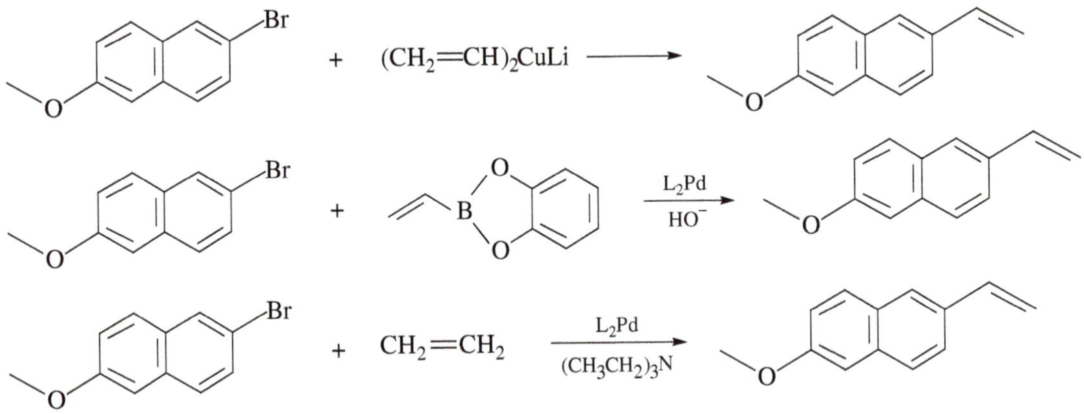

Br + $(CH_2{=}CH)_2CuLi$ →

O

Br + B, O, O → L_2Pd, HO^-

O

Br + $CH_2{=}CH_2$ → L_2Pd, $(CH_3CH_2)_3N$

O

47. There are three possibilities: **A**, **B**, and **C**. Only the bromine bonded to the benzylic carbon can be replaced by hydroxide ion via a nucleophilic substitution reaction. Both bromines in each compound can react with magnesium to form a Grignard reagent. When the Grignard reagents are treated with acid, a H replaces each of the bromines, so all three compounds form toluene.

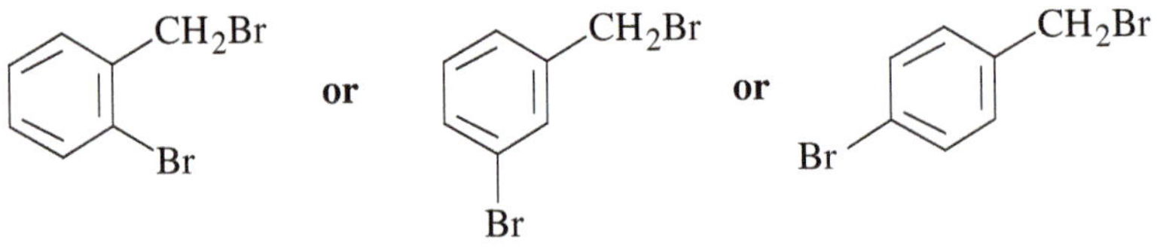

CH_2Br, Br **A** or CH_2Br, Br **B** or CH_2Br, Br **C**

Compound **A**, for example, reacts as follows:

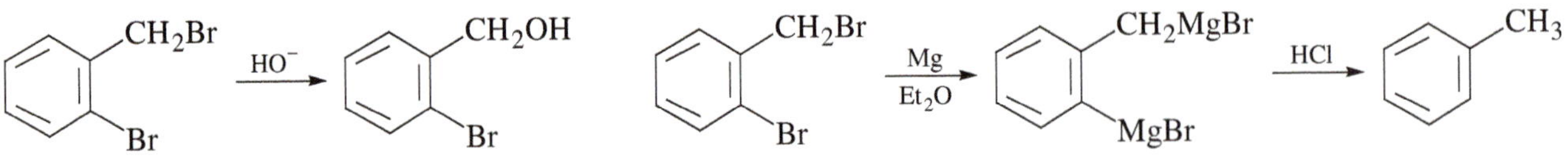

CH_2Br, Br → (HO^-) → CH_2OH, Br

CH_2Br, Br → (Mg, Et_2O) → CH_2MgBr, MgBr → (HCl) → CH_3

48. **a.** 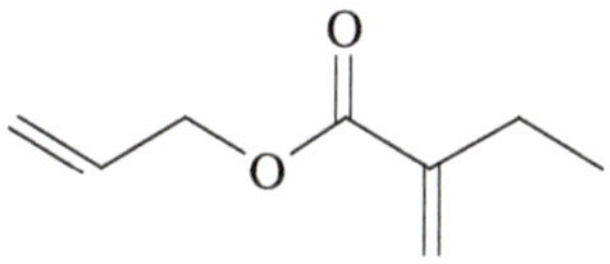 **b.** 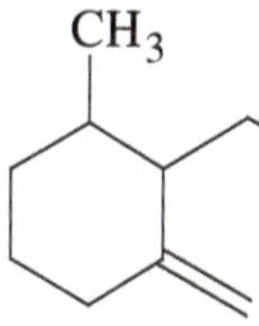 CH_3 **c.**

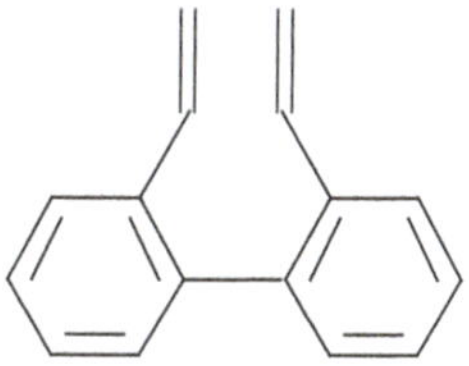

49. Some alkenes can undergo ring-opening metathesis to form high-molecular-weight polymers. Relieving the strain of a ring that contains a double bond (recall that sp^2 carbons have 120° bond angles) causes the reaction to go to completion.

a. First draw the product obtained by breaking the double bond and forming new double bonds. Seeing that six-membered rings are formed allows you to draw a reasonable structure for the polymer.

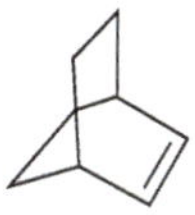

$$\longrightarrow -CH{=}CHCHCH_2CH_2CHCH{=}CHCHCH_2CH_2CHCH{=}CH-$$

CH=CH CH=CH CH=CH

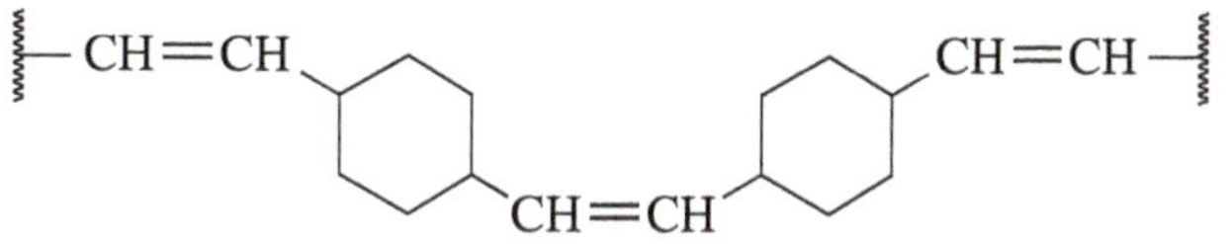

b.

$$\longrightarrow -CH{=}CHCHCH_2CH_2CHCH{=}CHCHCH_2CH_2CHCH{=}CH-$$

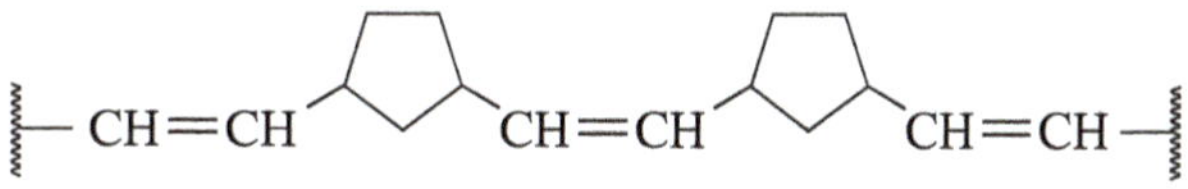

CH=CH CH=CH CH=CH

Chapter 11 Practice Test

1. Which is more reactive, an organocadmium compound (electronegativity of Cd = 1.5) or an organolead compound (electronegativity of lead = 1.6)?

2. Which of the following alkyl halides could be used to form a Grignard reagent?

3. What are the products of the following reactions?

a. (bromobenzene) + (alkene) $\xrightarrow[(CH_3CH_2)_3N]{PdL_2}$

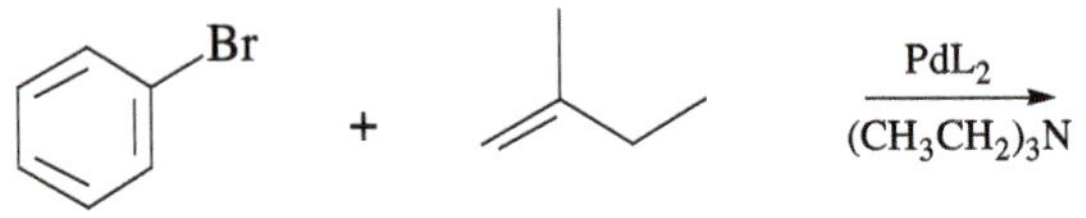

b. (bromobenzene) + $(CH_3CH_2CH_2)_2CuLi$ $\longrightarrow$

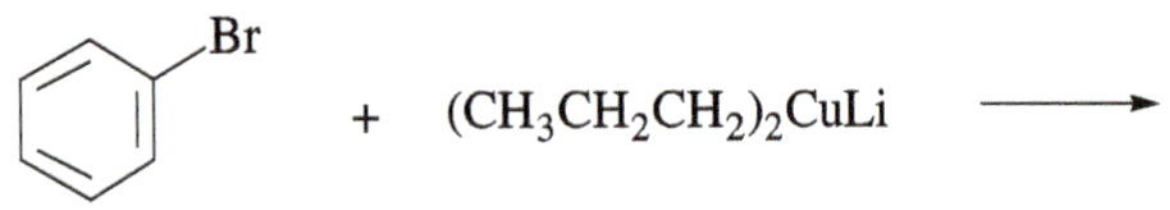

c. (bromobenzene) + (alkenyl) $B(OR)_2$ $\xrightarrow[HO^-]{PdL_2}$

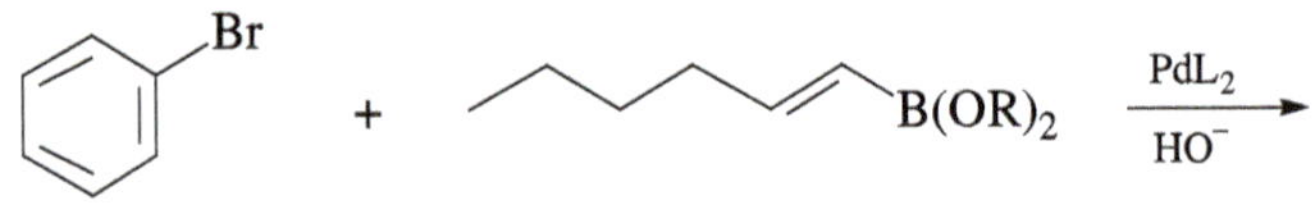

4. What are the products of the following reactions?

a. (vinyl bromide) Br + (alkene) $\xrightarrow[(CH_3CH_2)_3N]{PdL_2}$

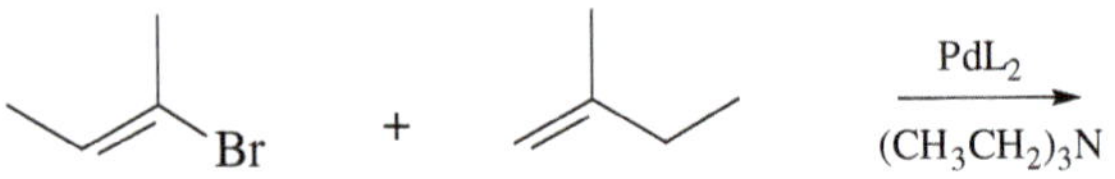

b. (vinyl bromide) Br + $(CH_3CH_2CH_2)_2CuLi$ $\longrightarrow$

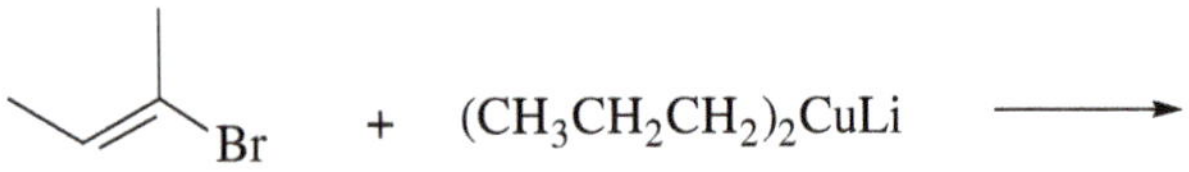

c. (vinyl bromide) Br + (alkenyl) $B(OR)_2$ $\xrightarrow[HO^-]{PdL_2}$

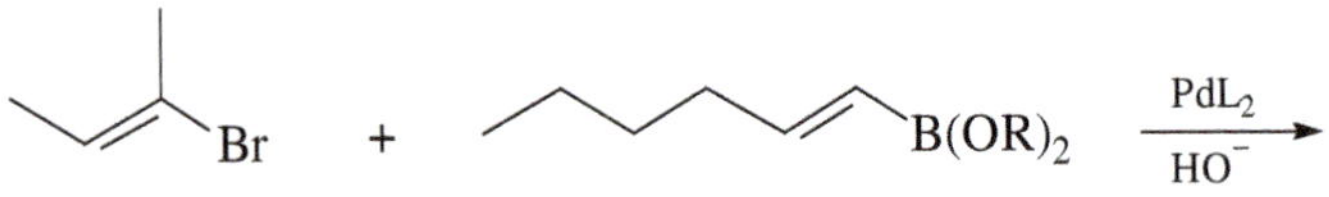

5. What are the products of the following reactions?

a. (alkene) $\xrightarrow{\text{Grubbs catalyst}}$

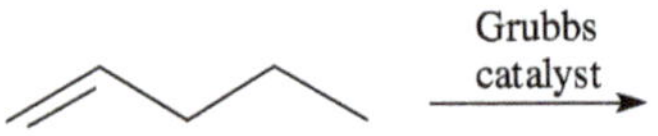

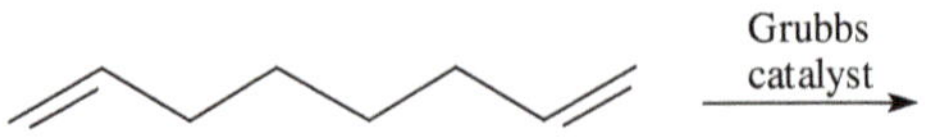

b. (diene) $\xrightarrow{\text{Grubbs catalyst}}$

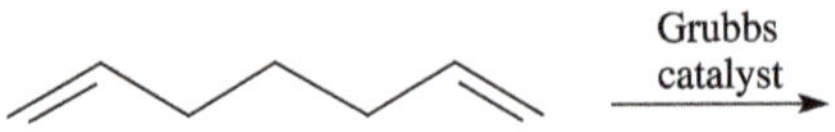

c. (diene) $\xrightarrow{\text{Grubbs catalyst}}$

6. Indicate how each of the following compounds can be prepared using the given starting material:

a. 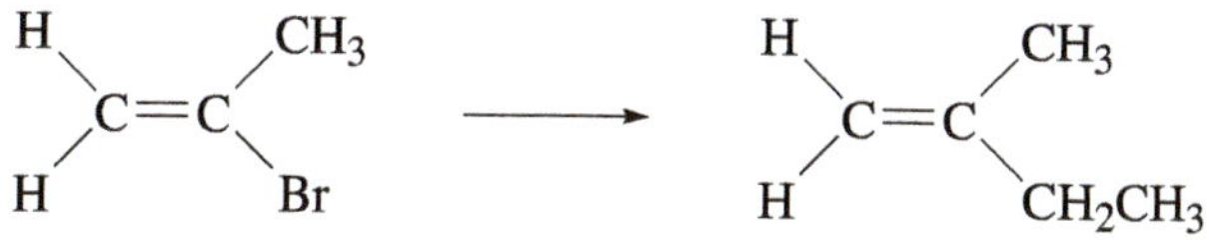 $H_2C{=}C(CH_3)Br \longrightarrow H_2C{=}C(CH_3)CH_2CH_3$

b. 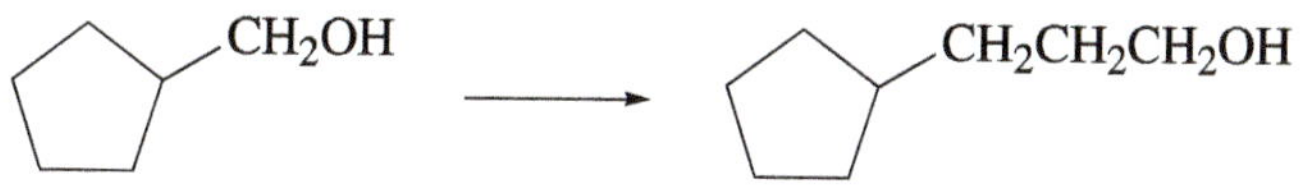 cyclopentyl–$CH_2OH \longrightarrow$ cyclopentyl–$CH_2CH_2CH_2OH$

7. What alcohols are formed from the reaction of ethylene oxide with the following organocuprates?

a. $(CH_3CH_2CH_2)_2CuLi$

b. 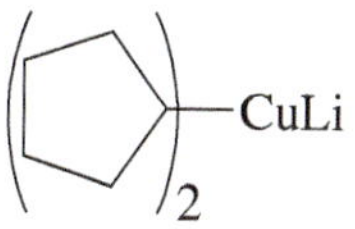 (cyclopentyl)$_2$CuLi

8. Which of the following can be used as a reactant in a Heck or Suzuki reaction?

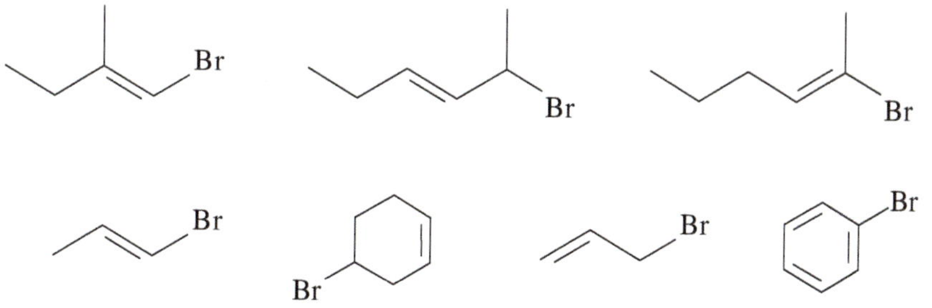

Br Br Br

Br Br Br Br

9. Show how the following compounds can be prepared by:

a. a Heck reaction b. a Suzuki reaction

1. 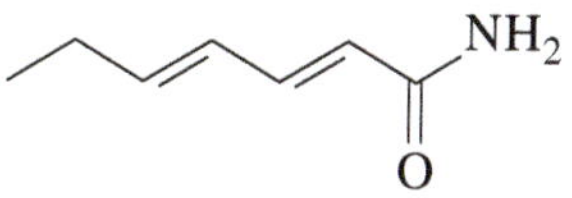 NH_2 O

2.

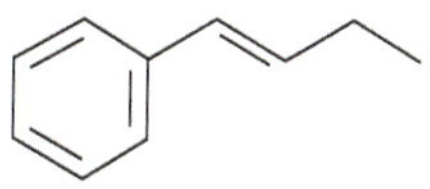

10. What are the products of the following reactions?

a. 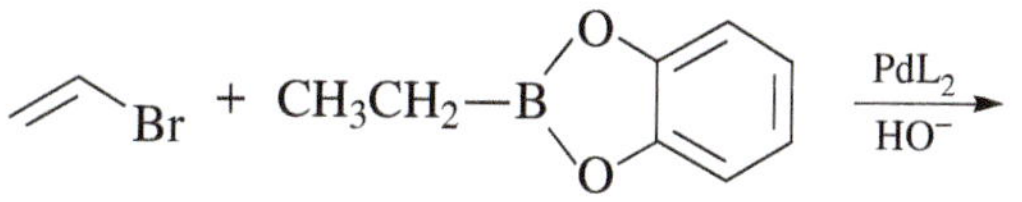 Br + CH_3CH_2—B(O,O-catechol) $\xrightarrow[HO^-]{PdL_2}$

b. Br + $\xrightarrow[(CH_3CH_2CH_2)_3N]{PdL_2}$

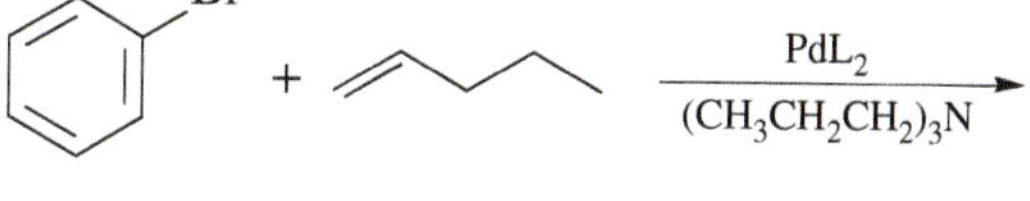

11. Using any needed reagents, show how the following synthesis can be carried out:

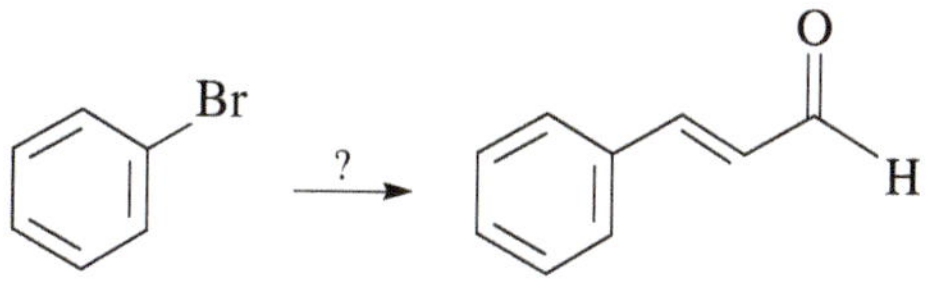

Br $\xrightarrow{?}$ O H

CHAPTER 12
Radicals

Important Terms

alkane a hydrocarbon that contains only single bonds.

alkoxy radical an alkoxy group with an unpaired electron on oxygen (RO·).

allylic radical a species with an unpaired electron on an allylic carbon.

benzylic radical a species with an unpaired electron on a benzylic carbon.

combustion a reaction with oxygen that takes place at high temperatures and converts alkanes (or other organic compounds) to carbon dioxide and water.

free radical (radical) an atom or a molecule with an unpaired electron.

halogenation reaction the reaction of an alkane (or other organic compounds) with a halogen.

heterolytic bond cleavage (heterolysis) breaking a bond with the result that both bonding electrons stay with one of the previously bonded atoms.

homolytic bond cleavage (homolysis) breaking a bond with the result that each of the atoms that formed the bond gets one of the bonding electrons.

initiation step the step in which radicals are created and/or the step in which the radical needed for the first propagating step is created.

peroxide a compound with an O—O bond.

peroxide effect peroxide causes the initial step in the addition of HBr to an alkene to be addition of a bromine radical instead of a proton.

primary alkyl radical an alkyl radical with the unpaired electron on a primary carbon.

propagation step in the first of a pair of propagation steps, a radical reacts to produce another radical that reacts in the second propagation step to produce the radical that was the reactant in the first propagation step.

radical (often called a free radical) an atom or a molecule with an unpaired electron.

radical addition reaction an addition reaction in which the first species that adds is a radical.

radical chain reaction a reaction in which radicals are formed and react in repeating propagating steps.

radical inhibitor a compound that traps radicals.

radical initiator a compound that creates radicals.

radical substitution reaction a substitution reaction that has a radical intermediate.

reactivity-selectivity principle the greater the reactivity of a species, the less selective it will be.

saturated hydrocarbon a hydrocarbon that contains only single bonds (it is saturated with hydrogen).

secondary alkyl radical an alkyl radical with the unpaired electron on a secondary carbon.

termination step two radicals combine to produce a molecule in which all the electrons are paired.

tertiary alkyl radical an alkyl radical with the unpaired electron on a tertiary carbon.

Solutions to Problems

1.

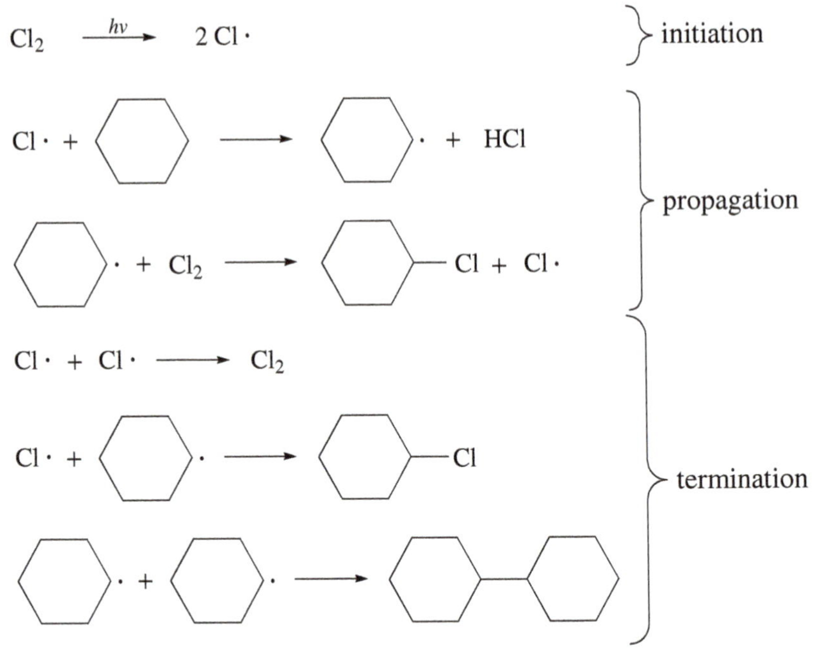

$$Cl_2 \xrightarrow{h\nu} 2\,Cl\cdot$$ — initiation

$$Cl\cdot + C_6H_{12} \longrightarrow C_6H_{11}\cdot + HCl$$

$$C_6H_{11}\cdot + Cl_2 \longrightarrow C_6H_{11}Cl + Cl\cdot$$

— propagation

$$Cl\cdot + Cl\cdot \longrightarrow Cl_2$$

$$Cl\cdot + C_6H_{11}\cdot \longrightarrow C_6H_{11}Cl$$

$$C_6H_{11}\cdot + C_6H_{11}\cdot \longrightarrow C_6H_{11}\text{–}C_6H_{11}$$

— termination

2.

$$Cl_2 \xrightarrow{h\nu} 2\,Cl\cdot$$

$$Cl\cdot + CH_4 \longrightarrow \cdot CH_3 + HCl$$

$$\cdot CH_3 + Cl_2 \longrightarrow CH_3Cl + Cl\cdot$$

$$Cl\cdot + CH_3Cl \longrightarrow \cdot CH_2Cl + HCl$$

$$\cdot CH_2Cl + Cl_2 \longrightarrow CH_2Cl_2 + Cl\cdot$$

$$Cl\cdot + CH_2Cl_2 \longrightarrow \cdot CHCl_2 + HCl$$

$$\cdot CHCl_2 + Cl_2 \longrightarrow CHCl_3 + Cl\cdot$$

$$Cl\cdot + CHCl_3 \longrightarrow \cdot CCl_3 + HCl$$

$$\cdot CCl_3 + Cl_2 \longrightarrow CCl_4 + Cl\cdot$$

$$Cl\cdot + \cdot CCl_3 \longrightarrow CCl_4$$

3. **a.** $CH_3CH_2CH(CH_3)CH_2C(CH_3)_2CH_2CH_3$ (arrow pointing to the CH carbon bearing the CH_3 group)

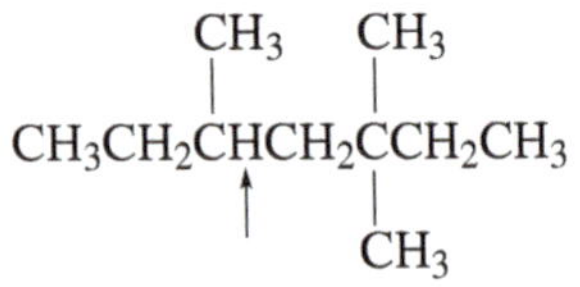

b. six secondary hydrogens

4.

a. 3 **b.** 3 **c.** 5 **d.** 1 **e.** 5 **f.** 6 **g.** 2 **h.** 4 **i.** 4

5. Solved in the text.

6. **a.** 4 One of the 3 alkyl halides has an asymmetric center, so it has *R* and *S* stereoisomers.

b. 5 Two of the 3 alkyl halides has an asymmetric center, so each has *R* and *S* stereoisomers.

c. 8 Three of the 5 alkyl halides have an asymmetric center, so each has *R* and *S* stereoisomers.

d. 1

e. 12 Two of the 5 alkyl halides have two asymmetric asymmetric centers, so each has 4 stereoisomers. One of the 5 alkyl halides has cis–trans stereoisomers.

f. 14 Two of the 6 alkyl halides have two asymmetric asymmetric centers, so each has 4 stereoisomers. One of the 6 alkyl halides has one asymmetric center, so it has *R* and *S* stereoisomers. One of the 6 alkyl halides has cis–trans stereoisomers.

g. 2

h. 6 Two of the 4 alkyl halides have an asymmetric center, so they have *R* and *S* stereoisomers.

i. 6 Two of the 4 alkyl halides have an asymmetric center, so they have *R* and *S* stereoisomers.

7. Solved in the text.

8. **a.** The major product is obtained from removal of a secondary hydrogen.

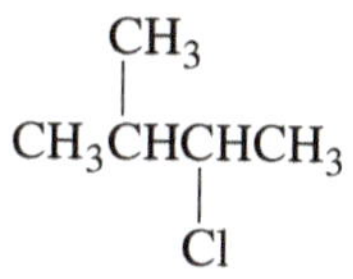

$$CH_3\overset{CH_3}{\overset{|}{C}}H\underset{\underset{Cl}{|}}{C}HCH_3$$

b. The relative amounts of the four possible products are:

$6 \times 1 = 6 \qquad 1 \times 5 = 5 \qquad 2 \times 3.8 = 7.6 \qquad 3 \times 1 = 3$

$$\text{percentage of major product} = \frac{7.6}{21.6} \times 100 = 35\%$$

9. **a.** Chlorination, because the halogen is substituting for a primary hydrogen.
b. Bromination, because the halogen is substituting for a tertiary hydrogen.
c. Because the molecule has only one kind of hydrogen, both chlorination and bromination will form only one monohalogenated product.

10. Solved in the text.

11. **a.** 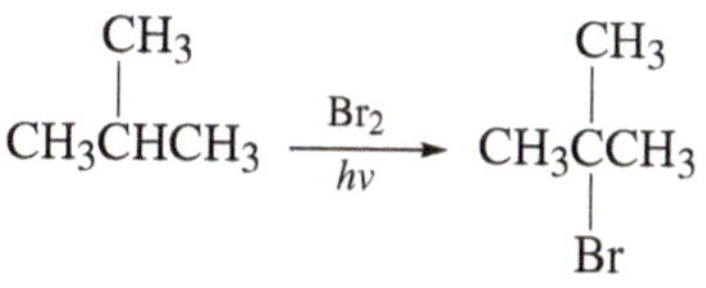 $$CH_3\overset{CH_3}{\overset{|}{C}}HCH_3 \xrightarrow[h\nu]{Br_2} CH_3\underset{\underset{Br}{|}}{\overset{CH_3}{\overset{|}{C}}}CH_3$$

b. $$CH_3\overset{CH_3}{\overset{|}{C}}HCH_3 \xrightarrow[h\nu]{Br_2} CH_3\underset{\underset{Br}{|}}{\overset{CH_3}{\overset{|}{C}}}CH_3 \xrightarrow{tert\text{-}BuO^-} CH_2{=}\overset{CH_3}{\overset{|}{C}}CH_3$$

c. $$CH_3\overset{CH_3}{\overset{|}{C}}HCH_3 \xrightarrow[h\nu]{Br_2} CH_3\underset{\underset{Br}{|}}{\overset{CH_3}{\overset{|}{C}}}CH_3 \xrightarrow{tert\text{-}BuO^-} CH_2{=}\overset{CH_3}{\overset{|}{C}}CH_3 \xrightarrow{HI} CH_3\underset{\underset{I}{|}}{\overset{CH_3}{\overset{|}{C}}}CH_3$$

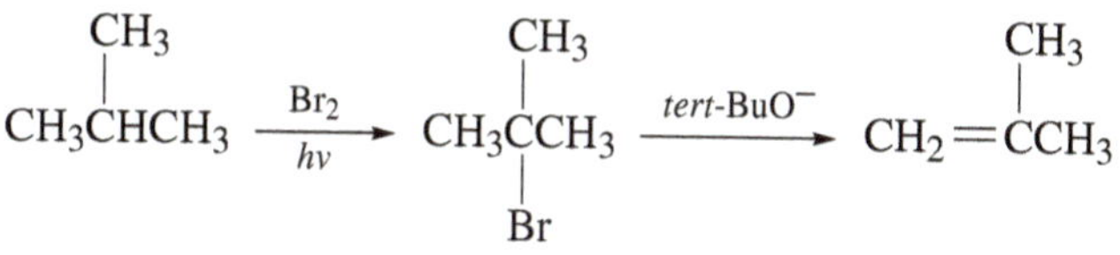

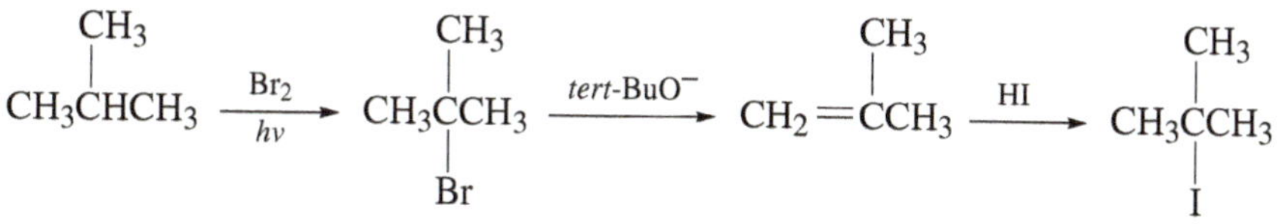

12. To start peroxide formation, the chain-initiating radical removes a hydrogen atom from an α-carbon of the ether.

a. **D** is most apt to form a peroxide, because removal of a hydrogen from an α-carbon forms a secondary radical.

b. **B** is least apt to form a peroxide, because it does not have any hydrogens bonded to its α-carbons.

13.

CH_3 (1-methylcyclohexene) + Br• ⟶ CH_3, Br (radical on ring carbon bearing CH_3, Br on adjacent carbon)

CH_3, Br (radical) + HBr ⟶ CH_3, Br (1-bromo-2-methylcyclohexane) + Br•

14.

a. $CH_3\overset{\displaystyle CH_3}{\overset{|}{C}}{=}CHCH_3$ + HBr ⟶ $CH_3\underset{\displaystyle Br}{\overset{\displaystyle CH_3}{C}}{-}CH_2CH_3$

b. $CH_3\overset{\displaystyle CH_3}{\overset{|}{C}}{=}CHCH_3$ + HCl ⟶ $CH_3\underset{\displaystyle Cl}{\overset{\displaystyle CH_3}{C}}{-}CH_2CH_3$

c. $CH_3\overset{\displaystyle CH_3}{\overset{|}{C}}{=}CHCH_3$ + HBr $\xrightarrow{\text{peroxide}}$ $CH_3\overset{\displaystyle CH_3}{CH}{-}\underset{\displaystyle Br}{CH}CH_3$

d. $CH_3\overset{\displaystyle CH_3}{\overset{|}{C}}{=}CHCH_3$ + HCl $\xrightarrow{\text{peroxide}}$ $CH_3\underset{\displaystyle Cl}{\overset{\displaystyle CH_3}{C}}{-}CH_2CH_3$

15.

a. $CH_3\overset{\displaystyle CH_3}{CH}CH_3$ $\xrightarrow[h\nu]{Cl_2}$ $CH_3\overset{\displaystyle CH_3}{CH}CH_2Cl$ achiral + $CH_3\underset{\displaystyle Cl}{\overset{\displaystyle CH_3}{C}}CH_3$ achiral

b. $CH_3CH_2CH_2CH_3$ $\xrightarrow[h\nu]{Cl_2}$ $CH_3CH_2CH_2CH_2Cl$ achiral + CH_2CH_3, CH_3, C, H, Cl (chiral) + CH_2CH_3, H, C, Cl, CH_3 (chiral)

chiral chiral — enantiomers

16. The major monobromination product results from bromination at a tertiary carbon. Bromination can occur at either of the two tertiary carbons. Because the radical intermediate is sp^2 hybridized and, therefore, planar, the incoming bromine can add to a tertiary radical from either the top or bottom of the plane. Notice that the product (1-bromo-1,3-dimethylcyclohexane) has two asymmetric centers; therefore, four stereoisomers are formed: *R,R, R,S, S,S,* and *S,R*.

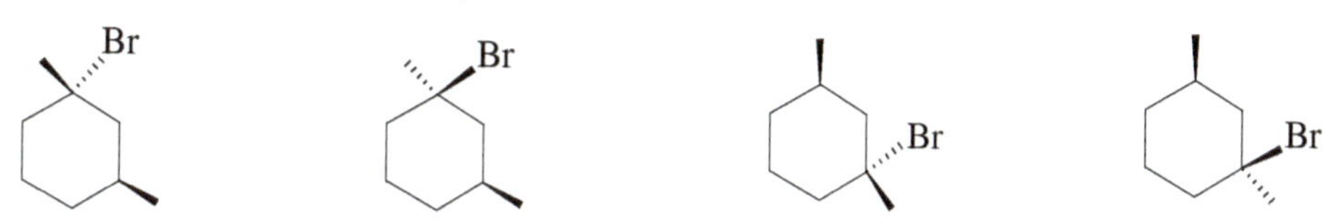

17. Solved in the text.

18. [Reaction scheme: methylenecyclohexane + Br• → allylic radical ↔ resonance radical + HBr; Br₂ → 2-bromo-1-methylenecyclohexane (two stereoisomers) and 1-(bromomethyl)cyclohexene]

19. The product of the reaction in Problem 17 has one asymmetric center. Therefore, two allylic substituted bromoalkenes are obtained, one with the *R* configuration at the asymmetric center and one with the *S* configuration.

(*R*)-4-bromo-2-pentene (*S*)-4-bromo-2-pentene

20. **a.** Two stereoisomers are formed because the reaction forms a compound with an asymmetric center.

[Cyclohexene —NBS, Δ, peroxide→ 3-bromocyclohexene (two stereoisomers)]

b. The reaction forms two constitutional isomers, and each constitutional isomer has two stereoisomers.

[3-methylcyclohexene —NBS, Δ, peroxide→ radical ↔ radical + HBr; Br₂ → 3-bromo-3-methylcyclohexene (two stereoisomers) + 5-bromo-1-methylcyclohexene (two stereoisomers)]

21. a.

b.

c.

d.

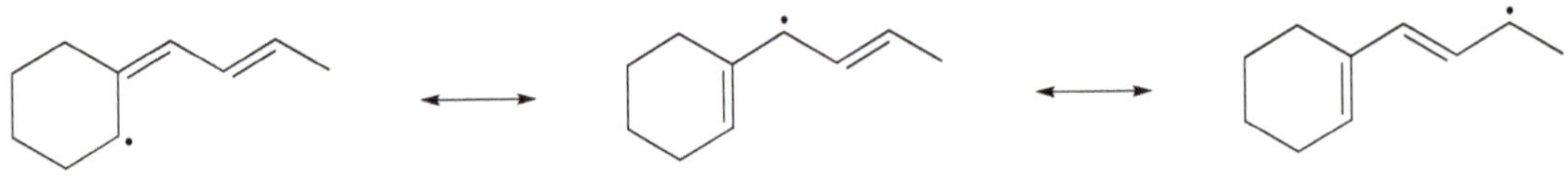

22. a. 1. There are two sets of secondary allylic hydrogens, **a** and **b** (plus a less reactive set of primary allylic hydrogens on the methyl group).
First, we need to determine which allylic hydrogen is the easiest one to remove. Removing one of the **a** allylic hydrogens forms an intermediate in which the unpaired electron is shared by two secondary allylic carbons. Removing one of the **b** allylic hydrogens forms an intermediate in which the unpaired electron is shared by a tertiary allylic carbon and a secondary allylic carbon.

CH_3 a b — removing an H from a → CH_3 secondary ⟷ CH_3 secondary

CH_3 — removing an H from b → CH_3 secondary ⟷ CH_3 tertiary

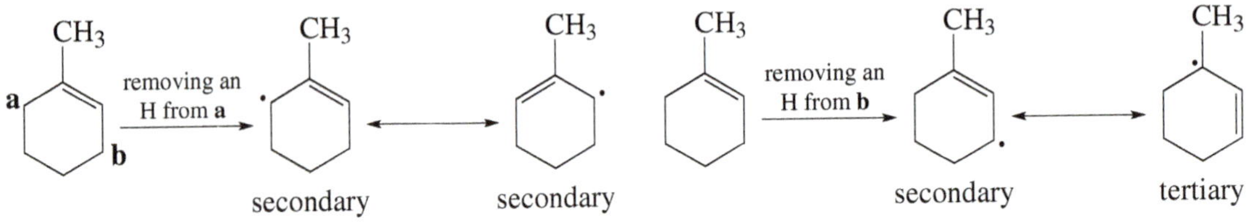

Therefore, the major products are obtained by removing a **b** allylic hydrogen.

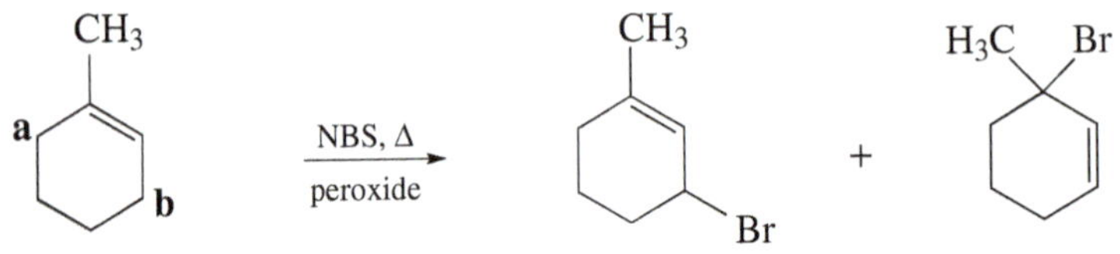

CH_3 a b (1-methylcyclohexene) — NBS, Δ, peroxide → CH_3 Br + H_3C Br

2. 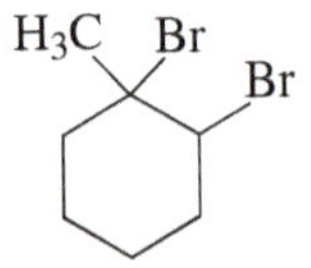 H_3C Br Br

3. H_3C Br

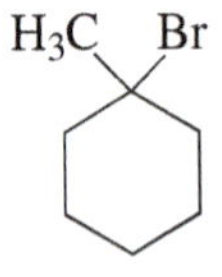

4. 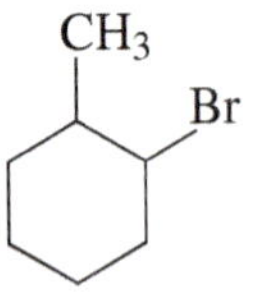 CH_3 Br

b. 1. Each of the products has one asymmetric center.
The *R* and the *S* stereoisomers are obtained for each product.

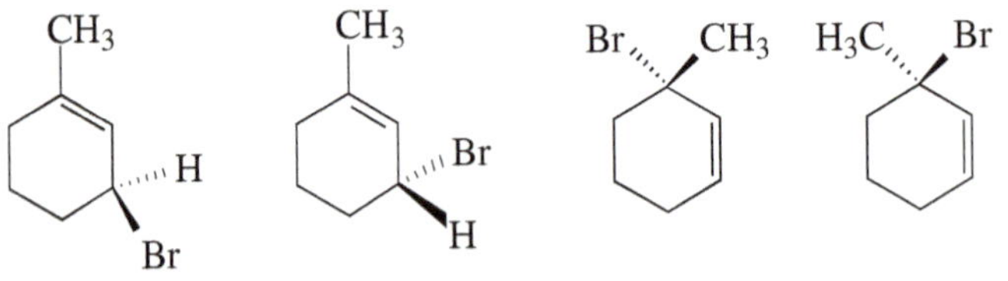

CH_3 H Br CH_3 Br H   Br CH_3 H_3C Br

2. The product has two asymmetric centers.
Because addition of Br_2 is anti, only the stereoisomers with the bromine atoms on opposite sides of the ring are obtained.

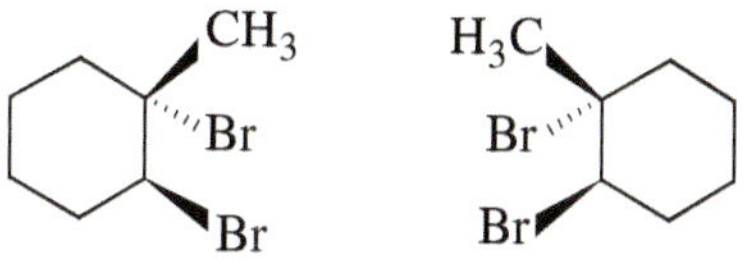

CH_3 Br Br H_3C Br Br

3. The product does not have an asymmetric center, so it does not have stereoisomers.

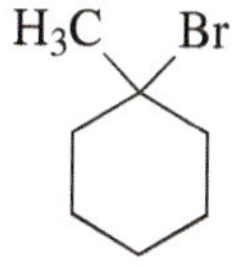

H_3C Br

4. The product has two asymmetric centers.
Because radical addition of HBr can be either syn or anti, four stereoisomers are obtained as products.

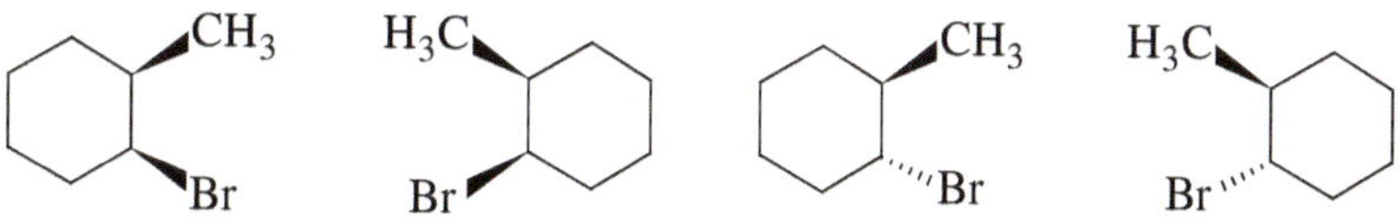

CH_3 Br H_3C Br CH_3 Br H_3C Br

23. **a.** $\xrightarrow[h\nu]{Br_2}$ Br $\xrightarrow{tert\text{-}BuO^-}$ $\xrightarrow{RCOOH}$ O

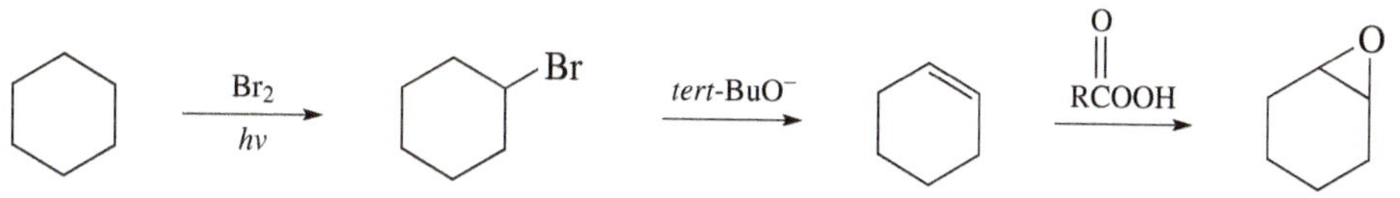

b. CH_3 $\xrightarrow[h\nu]{Br_2}$ CH_2Br $\xrightarrow{HO^-}$ CH_2OH $\xrightarrow[CH_3COOH, 0\ °C]{NaOCl}$ C(=O)H

c. $\xrightarrow[h\nu]{Br_2}$ Br $\xrightarrow{HO^-}$ OH $\xrightarrow[CH_3COOH, 0\ °C]{NaOCl}$ O

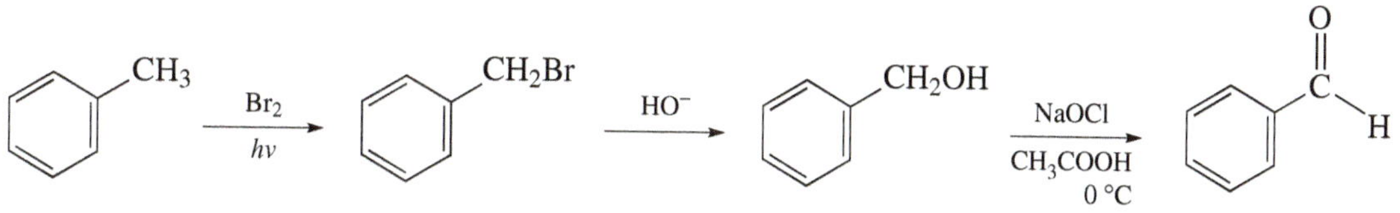

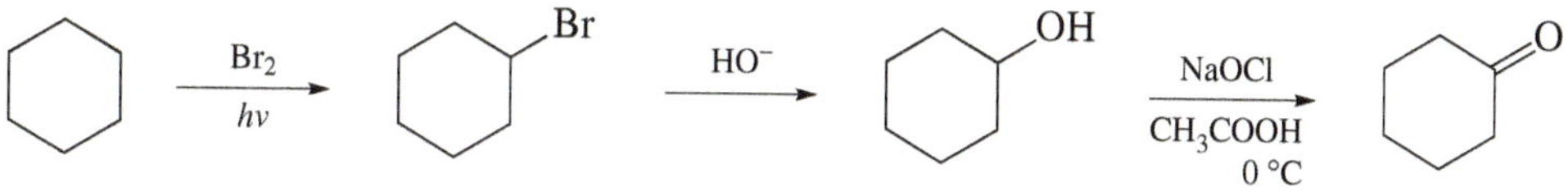

d. $\xrightarrow[h\nu]{Br_2}$ Br $\xrightarrow{tert\text{-}BuO^-}$ $\xrightarrow{RCOOH}$ O $\xrightarrow[H_2O]{HO^-}$ OH OH

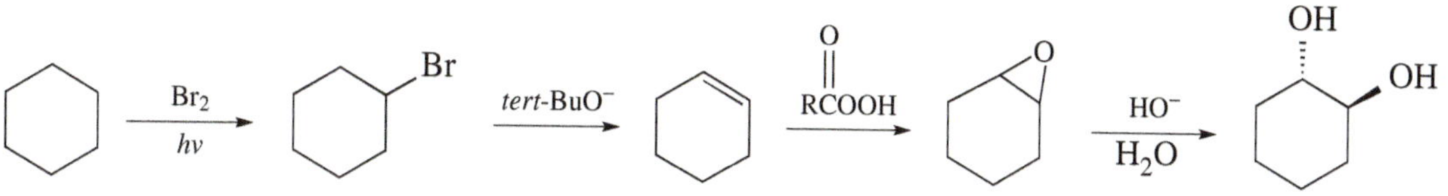

24. Four atoms (three carbons and one oxygen) share the unpaired electrons.

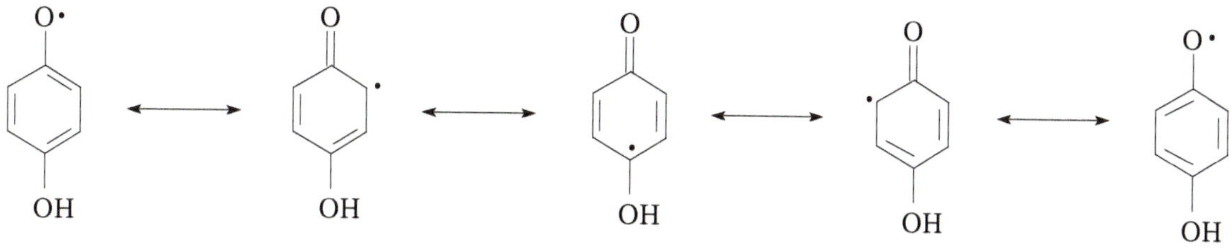

O• OH ⟷ O OH ⟷ O OH ⟷ O OH ⟷ O• OH

25. Antioxidants are radical inhibitors. That is, they react with radicals and, thereby, prevent radical chain reactions. There are several OH groups in a catechin that upon losing a hydrogen atom as a result of reacting with a reactive radical, form a radical that is stabilized by electron delocalization. Because this highly stabilized radical is sufficiently unreactive, it cannot damage cells by reacting with them. An example of one of the stabilized radicals is shown below.

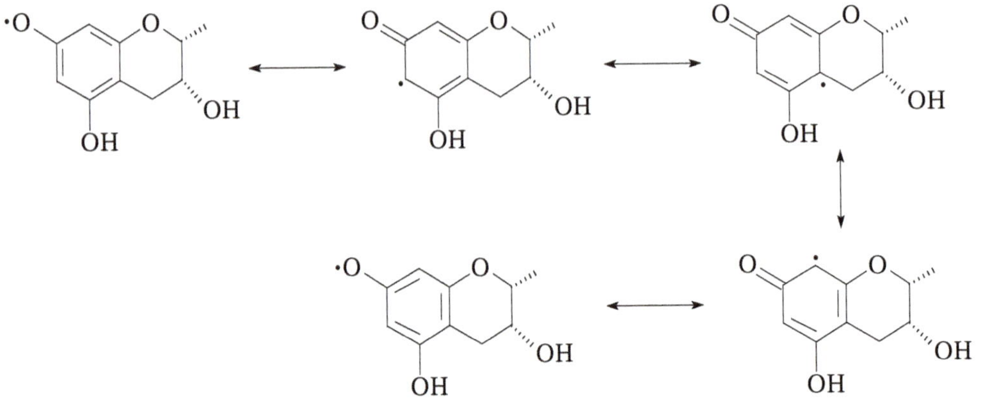

•O OH O OH ⟷ O OH OH ⟷ O OH OH
↕
•O OH OH ⟷ O OH OH

26. **a.** $CH_2{=}CH\dot{C}HCH_2CH_3$ ⟷ $\dot{C}H_2CH{=}CHCH_2CH_3$

$\downarrow Br_2$ $\downarrow Br_2$

$CH_2{=}CHCH(Br)CH_2CH_3$ + $BrCH_2CH{=}CHCH_2CH_3$

b. In the radical intermediate that leads to the major products, the unpaired electron is shared by a primary carbon and a tertiary carbocation. In the radical intermediate that leads to the minor products, the unpaired electron is shared by a primary and a secondary carbocation. It, therefore, is not as stable as the intermediate that leads to the major products.

$CH_3\overset{\delta\cdot}{C}(CH_3){\cdots}CH{\cdots}\overset{\delta\cdot}{C}H_2$ $\overset{\delta\cdot}{C}H_2{\cdots}C(CH_3){\cdots}\overset{\delta\cdot}{C}HCH_3$

$CH_3C(CH_3){=}CHCH_2Br$ + $CH_3C(CH_3)(Br)CH{=}CH_2$ $BrCH_2C(CH_3){=}CHCH_3$ + $CH_2{=}C(CH_3)CH(Br)CH_3$

major products minor products

In bromination, selectivity is more important than probability. Therefore, even though twice as many hydrogens are available for removal by a bromine radical that leads to the minor products, they will still be minor products because the easier-to-remove hydrogens lead to the major products.

c. Br

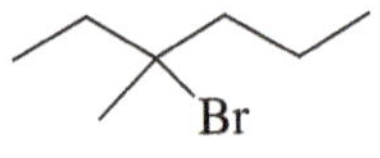

major product

d. Cl

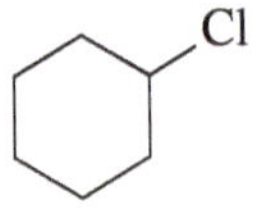

e. no reaction
(There is no light or heat, so radicals cannot be formed.)

f. CH_2Cl + CH_3 Cl + CH_3 Cl + CH_3 Cl

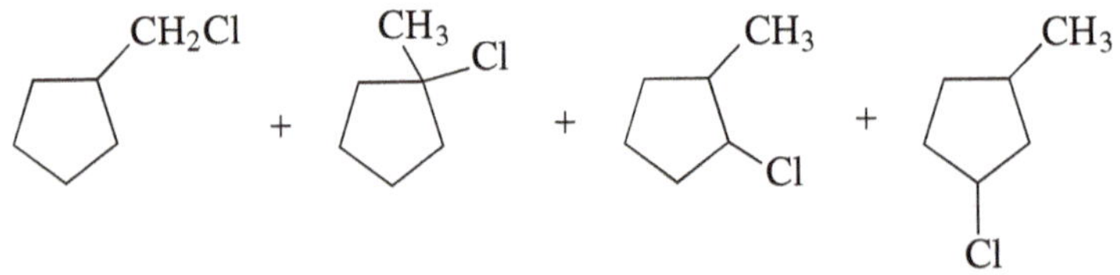

27. **a.** $CH_3C(CH_3)_2CH_3$

dimethylpropane

b. $CH_3CH_2CH(CH_3)CH_2CH_2CH_3$

3-methylhexane

28. Removing a hydrogen atom from ethane by an iodine radical is a highly endothermic reaction ($\Delta H° = 101 - 71 = 30$ kcal/mol; see Table 5.1 on page 206 of the text), so the iodine radicals will reform I_2 rather than remove a hydrogen atom.

$$\underset{101\ \text{kcal/mol}}{CH_3CH_3} + I\cdot \not\longrightarrow CH_3\dot{C}H_2 + \underset{71\ \text{kcal/mol}}{HI}$$

29. **a.** CH3, CH3, Cl

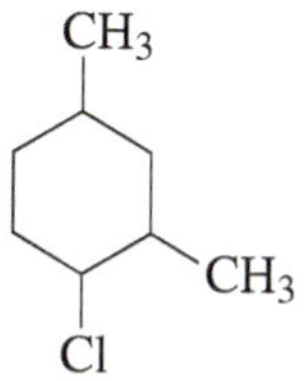

Four secondary hydrogens can be removed to form this product.

b. CH3, Cl, CH3

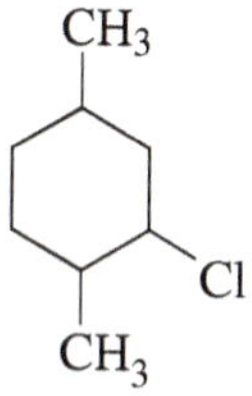

Eight secondary hydrogens can be removed to form this product.

c. H3C, CH3, Cl, CH3

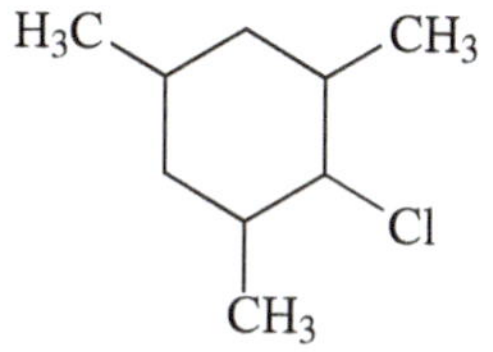

Six secondary hydrogens can be removed to form this product.

30. Because a bromine radical is more selective than a chlorine radical, the bromine radical will remove a tertiary hydrogen to form a tertiary radical.

a. CH3, CH3, Br

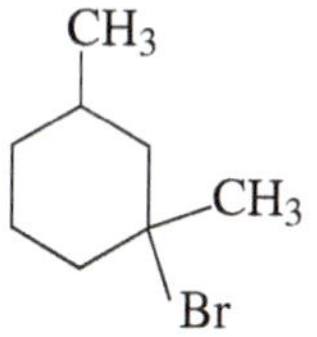

b. Br, CH3, CH3

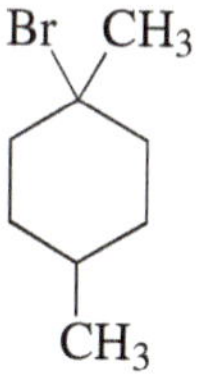

c. H3C, CH3, Br, CH3

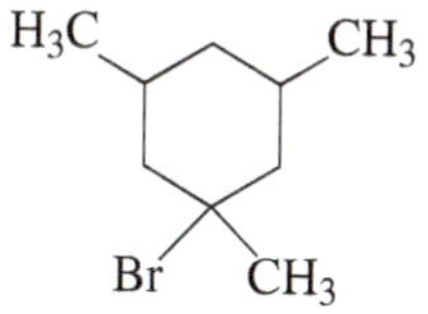

31. **a.** Br

b. Br

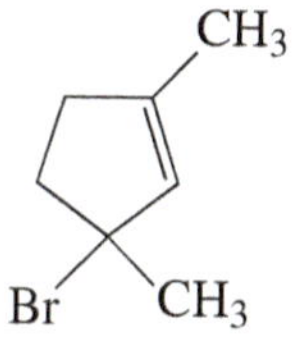

c. CH3, CHBr

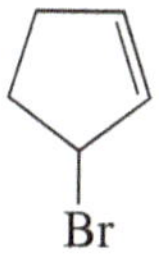

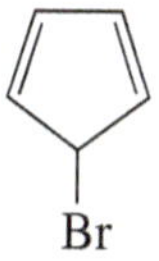

d. CH3, Br, CH3 + Br, CH3, CH3

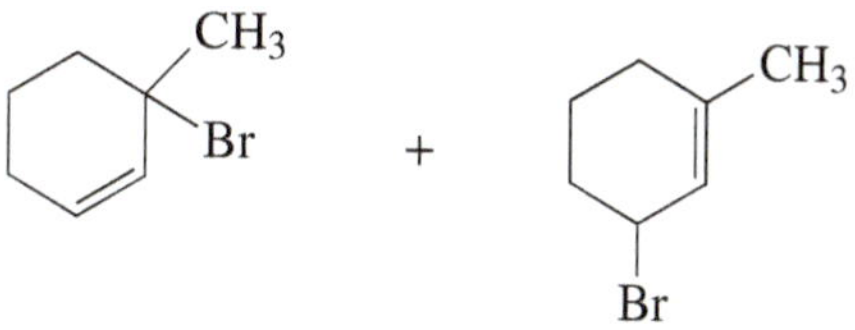

e. CH3, Br, CH3

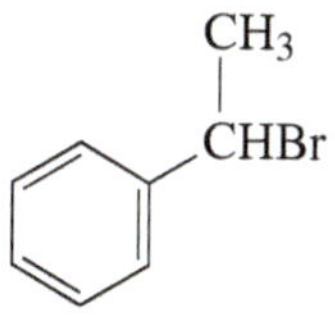

f. CH3, Br + CH3, Br

In **d** and **f**, because one product is under kinetic control and the other is under thermodynamic control, the major product depends on the conditions under which the reaction is carried out.

32. $CH_3C(CH_3)(Cl)CH_3$ (2-chloro-2-methylpropane): $1 \times x = x$

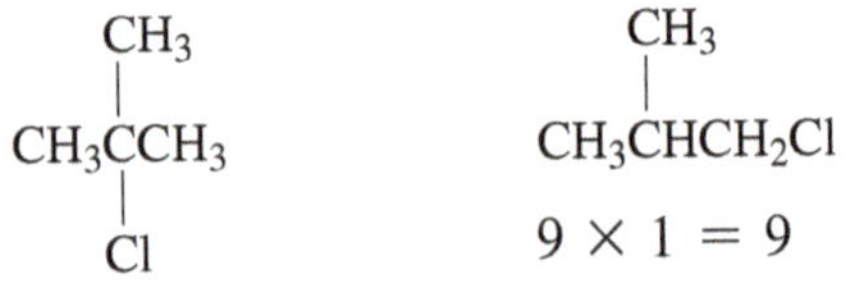

$CH_3CH(CH_3)CH_2Cl$ (1-chloro-2-methylpropane): $9 \times 1 = 9$

$$\text{fraction of the total that is 1-chloro-2-methylpropane} = \frac{\text{1-chloro-2-methylpropane}}{\text{1-chloro-2-methylpropane} + \text{2-chloro-2-methylpropane}} = 0.64$$

$$\frac{9}{9 + x} = 0.64$$

$$9 = 0.64\,(9 + x)$$

$$9 = 5.76 + 0.64x$$

$$3.24 = 0.64x$$

$$x = 5$$

We have found that it is five times easier for a chlorine radical to remove a hydrogen atom from a tertiary carbon than from a primary carbon.

33.

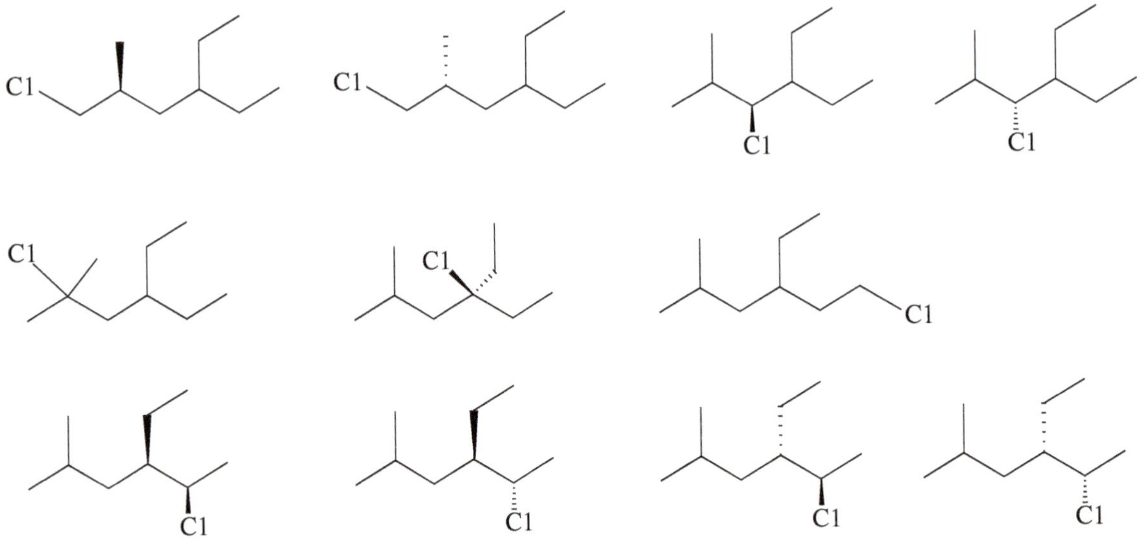

Cl Cl Cl Cl

Cl Cl Cl

Cl Cl Cl Cl

34. + HBr

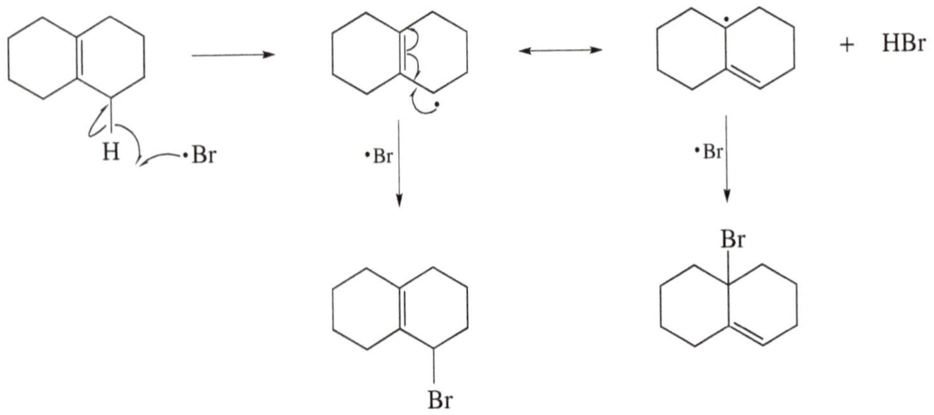

•Br •Br •Br

Br Br

35. It is easier to break a C—H bond than a C—D bond.
Because a bromine radical is less reactive (and more selective) than a chlorine radical, a bromine radical has a greater preference for the more easily broken C—H bond. Bromination, therefore, would have a greater deuterium kinetic isotope effect than chlorination.

36.

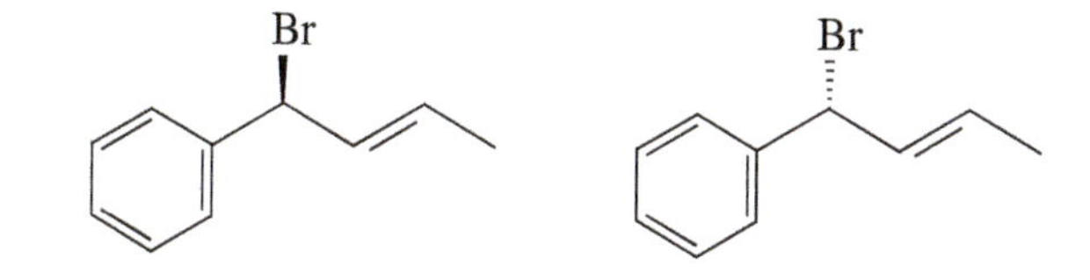

37. **a.** Five monochlorination products are possible.

CH_3 $\xrightarrow[hv]{Cl_2}$ CH_2Cl Cl CH_3 CH_3 Cl CH_3 Cl CH_3 Cl

b. The two compounds below are obtained in the greatest yield because each one can be formed by removing one of four hydrogens to form a secondary radical. More of the compound on the left is obtained because the methyl group in the compound on the right provides some steric hindrance to the approach of the chlorine atom.

CH_3 Cl CH_3 Cl

$4 \times 3.8 = 15.2$ $4 \times 3.8 = 15.2$

c. The number of possible stereoisomers for each compound is indicated below each structure.

CH_3 $\xrightarrow[hv]{Br_2}$ CH_2Cl Cl CH_3 CH_3 Cl CH_3 Cl CH_3 Cl

1 1 — These have 0 asymmetric centers.

4 4 — These have 2 asymmetric centers.

2 — This does not have an asymmetric center, but has cis–trans isomers.

Therefore, a total of 12 stereoisomers can be obtained.

38. For an alkene to form the same product whether it reacts with HBr in the presence or absence of peroxide, it must be symmetrical about the double bond.

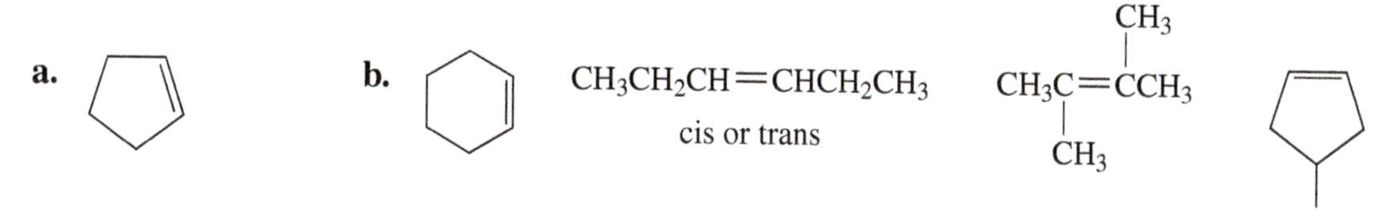

39.

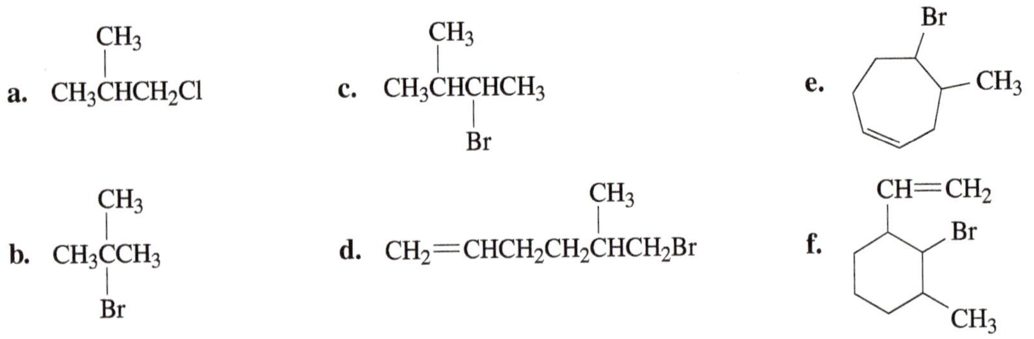

a. $CH_3CH(CH_3)CH_2Cl$

b. $CH_3C(CH_3)(Br)CH_3$

c. $CH_3CH(CH_3)CH(Br)CH_3$

d. $CH_2{=}CHCH_2CH_2CH(CH_3)CH_2Br$

e. Br, CH_3

f. $CH{=}CH_2$, Br, CH_3

40. a. $\xrightarrow[hv]{Br_2}$ Br $\xrightarrow{tert\text{-}BuO^-}$

b. $\xrightarrow[hv]{Br_2}$ Br $\xrightarrow{tert\text{-}BuO^-}$ $\xrightarrow[\text{peroxide}]{\text{NBS, }\Delta}$ Br

c. $\xrightarrow[hv]{Br_2}$ Br $\xrightarrow{tert\text{-}BuO^-}$ $\xrightarrow[\text{peroxide}]{\text{NBS, }\Delta}$ Br $\xrightarrow{CH_3O^-}$ OCH_3

d. $\xrightarrow[hv]{Br_2}$ Br $\xrightarrow{tert\text{-}BuO^-}$ $\xrightarrow[\text{peroxide}]{\text{NBS, }\Delta}$ Br $\xrightarrow{(CH_3)_2CuLi}$ CH_3

e. $\xrightarrow[hv]{Br_2}$ Br $\xrightarrow{tert\text{-}BuO^-}$ $\xrightarrow{RCOOH}$ (O=C) O $\xrightarrow[\text{2. HCl}]{\text{1. } CH_3O^-}$ OCH_3, OH + OH, OCH_3

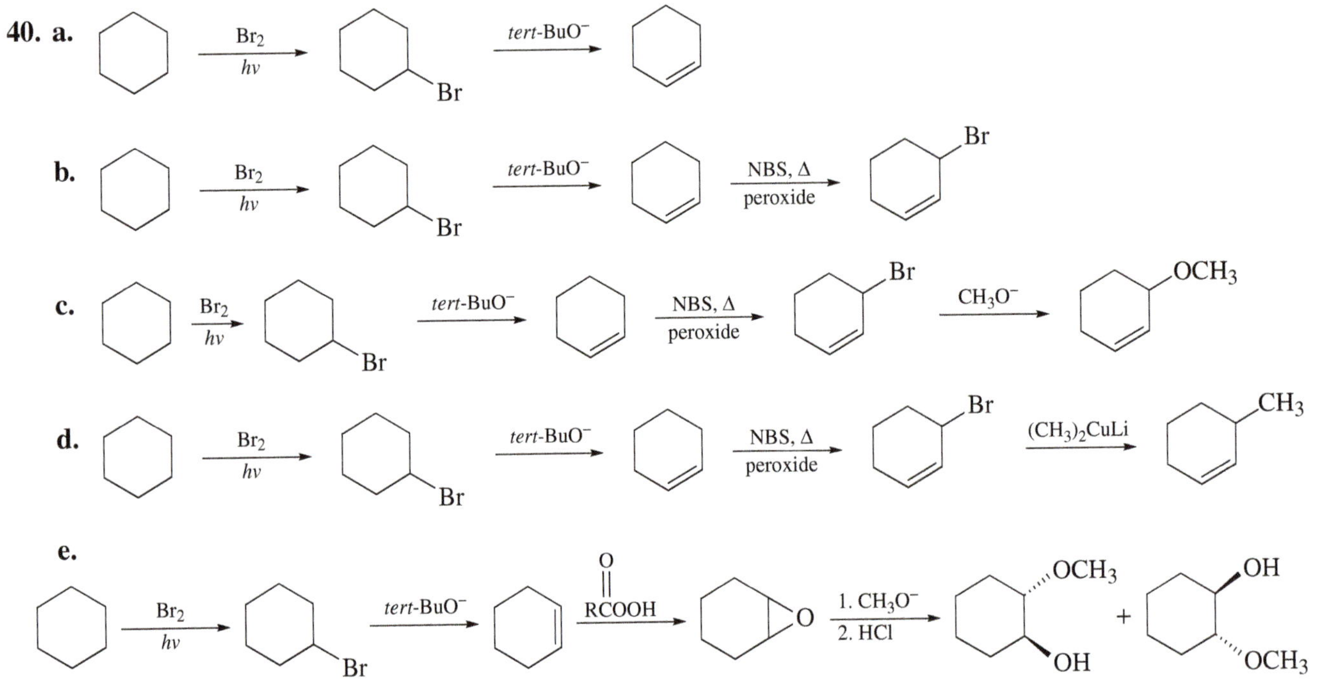

41. **a.** This reaction has two sets of propagation steps because two different radicals are generated in the initiation step. This reaction also has several more termination steps because of the two different radicals generated in the initiation step. (Bond dissociation enthalpies can be found in Table 5.1 on page 206 of the text.)

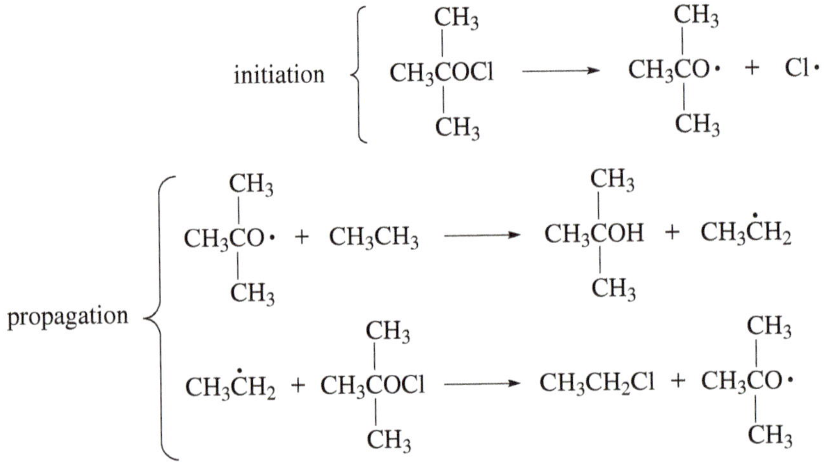

initiation

$$(CH_3)_3COCl \longrightarrow (CH_3)_3CO\cdot + Cl\cdot$$

propagation

$$(CH_3)_3CO\cdot + CH_3CH_3 \longrightarrow (CH_3)_3COH + CH_3\dot{C}H_2$$

$$CH_3\dot{C}H_2 + (CH_3)_3COCl \longrightarrow CH_3CH_2Cl + (CH_3)_3CO\cdot$$

propagation

$$Cl\cdot + CH_3CH_3 \longrightarrow HCl + CH_3\dot{C}H_2$$

$$CH_3\dot{C}H_2 + CH_3C(CH_3)_2OCl \longrightarrow CH_3CH_2Cl + CH_3C(CH_3)_2O\cdot$$

This species enters the first propagation cycle.

termination

$$CH_3\dot{C}H_2 + Cl\cdot \longrightarrow CH_3CH_2Cl$$

$$CH_3C(CH_3)_2O\cdot + CH_3C(CH_3)_2O\cdot \longrightarrow CH_3C(CH_3)_2O{-}OC(CH_3)_2CH_3$$

$$CH_3C(CH_3)_2O\cdot + Cl\cdot \longrightarrow CH_3C(CH_3)_2OCl$$

$$CH_3C(CH_3)_2O\cdot + CH_3\dot{C}H_2 \longrightarrow CH_3C(CH_3)_2OCH_2CH_3$$

$$Cl\cdot + Cl\cdot \longrightarrow Cl_2$$

$$CH_3\dot{C}H_2 + CH_3\dot{C}H_2 \longrightarrow CH_3CH_2CH_2CH_3$$

b. ΔH° = bonds broken − bonds formed

Let x = the bond dissociation enthalpy of the O—Cl bond.

$$-42\ \text{kcal/mol} = [101\ \text{kcal/mol} + x\ \text{kcal/mol}] - [85\ \text{kcal/mol} + 105\ \text{kcal/mol}]$$

$$-143\ \text{kcal/mol} = x\ \text{kcal/mol} - 190\ \text{kcal/mol}$$

$$x = 47\ \text{kcal/mol}$$

42. The reaction forms a product with two new asymmetric centers.

$$CH_3CH_2CH(CH_3){-}C(CH_3)(Br)CH_2CH_3$$

Because the reaction involves both syn and anti addition, four stereoisomers are obtained.

(Fischer projections, top to bottom: CH_2CH_3 / first carbon / second carbon / CH_2CH_3)

1. H—C—CH_3; Br—C—CH_3
2. CH_3—C—H; CH_3—C—Br
3. H—C—CH_3; CH_3—C—Br
4. CH_3—C—H; Br—C—CH_3

43. **a.** HO $\xrightarrow[\text{peroxide}]{\text{HBr}}$ HO Br $\xrightarrow{\text{NaH}}$ $^{-}$O Br $\longrightarrow$ O

b. $\xrightarrow[\text{peroxide}]{\text{HBr}}$ Br $\xrightarrow{CH_3O^-}$ OCH_3

c. OH $\xrightarrow[\Delta]{H_2SO_4}$ $\xrightarrow[\text{peroxide}]{\text{HBr}}$ Br

d. $\xrightarrow{Br_2}$ Br Br $\xrightarrow{HO^-}$ Br

PdL_2, $(CH_3CH_2)_3N$

$\xleftarrow[\text{Pd/C}]{H_2}$

44.

$$\underset{\displaystyle CH_3}{CH_3\overset{|}{C}HCH_2CH_3}$$

2-methylbutane

2-Methylbutane forms two primary alkyl halides, one secondary alkyl halide, and one tertiary alkyl halide. First, we need to calculate how much alkyl halide is formed from each primary hydrogen available, from each secondary hydrogen available, and from each tertiary hydrogen available.

a primary alkyl halide

$ClCH_2CHCH_2CH_3$ (with CH_3 on C-2)

Substitution of any one of six hydrogens leads to this product.
percentage of this product that is formed = 36%
percentage formed per hydrogen available = 36/6 = 6%

$CH_3CHCH_2CH_2Cl$ (with CH_3 on C-2)

Substitution of any one of three hydrogens leads to this product.
percentage of this product that is formed = 18%
percentage formed per hydrogen available = 18/3 = 6%

a secondary alkyl halide

$CH_3CHCHCH_3$ (with Cl on C-3 and CH_3 on C-2)

Substitution of any one of two hydrogens leads to this product.
percentage of this product that is formed = 28%
percentage formed per hydrogen available = 28/2 = 14%

a tertiary alkyl halide

$CH_3CCH_2CH_3$ (with Cl and CH_3 on C-2)

Substitution of the one tertiary hydrogen leads to this product.
percentage of this product that is formed = 18%
percentage formed per hydrogen available = 18/1 = 18%

From the above calculations, we see that at 300 °C, the relative rates of removal of a hydrogen atom from a tertiary, secondary, and primary carbocation are:

$$18 : 14 : 6 = 3.0 : 2.3 : 1$$

In Section 12.4, we saw that at room temperature, the relative rates are:

$$5.0 : 3.8 : 1$$

Therefore, we can conclude that at higher temperatures, the radical is less selective about which hydrogen atom it removes.

45. The chlorine radical is more reactive at the higher temperature (600 °C), so it is less selective.

46. **a.** 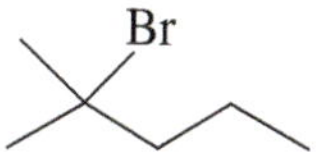 Br

b. 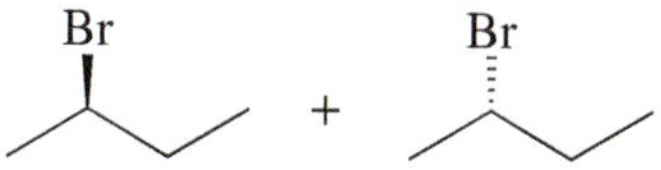 Br + Br

c. 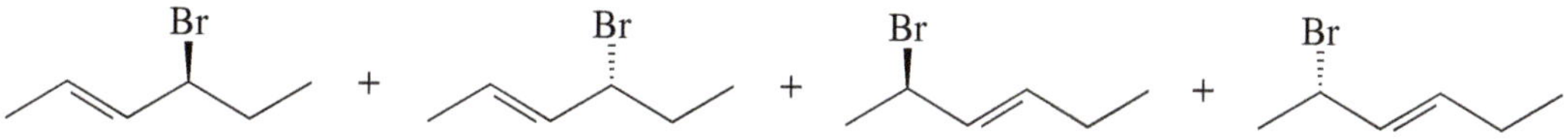 Br + Br + Br + Br

d. 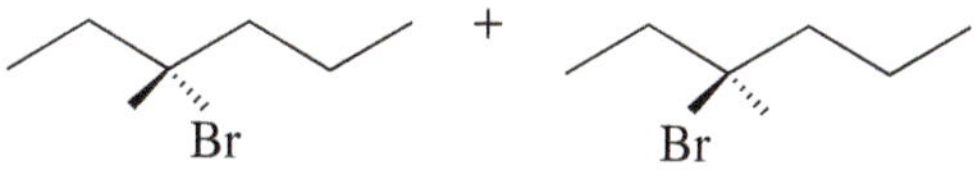 Br + Br

e. 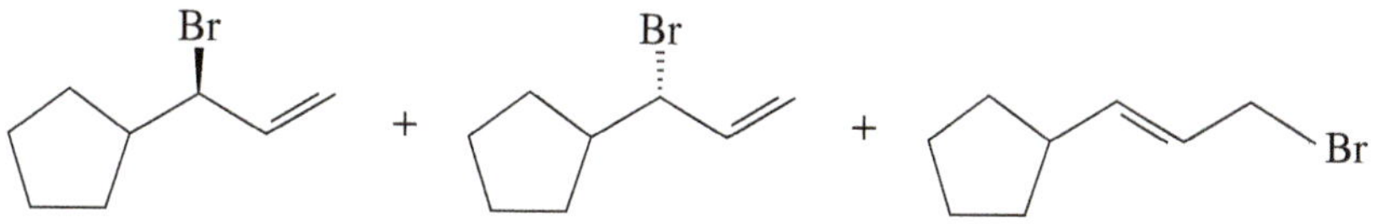 Br + Br + Br

f. 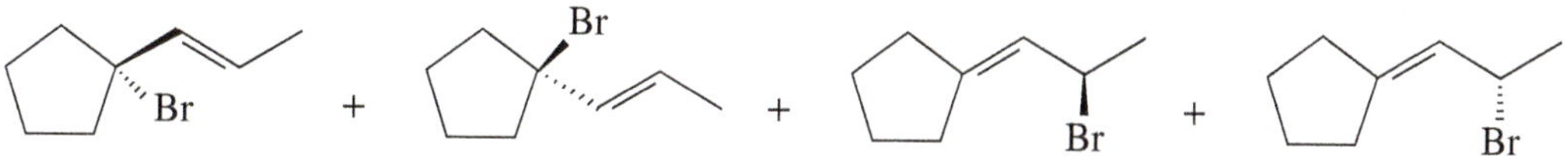 Br + Br + Br + Br

47. **a.** CH_3—H + Cl—Cl ⟶ CH_3—Cl + H—Cl

105 58 84 103

ΔH° = bonds broken − bonds formed

$\Delta H^\circ = [105 \text{ kcal/mol} + 58 \text{ kcal/mol}] - [84 \text{ kcal/mol} + 103 \text{ kcal/mol}]$

$\Delta H^\circ = 163 - 187 = -24 \text{ kcal/mol}$

b. CH_3—H + $\dot{Cl}$ ⟶ $\dot{C}H_3$ + H—Cl $\quad \Delta H^\circ = 105 - 103 = 2 \text{ kcal/mol}$

$\dot{C}H_3$ + Cl—Cl ⟶ CH_3—Cl + $\dot{Cl}$ $\quad \Delta H^\circ = 58 - 84 = -26 \text{ kcal/mol}$

Overall $\Delta H^\circ = 2 + (-26) = -24 \text{ kcal/mol}$

c. If you cancel the elements that are the same on opposite sides of the equations in part **b** and then add the two equations, you are left with the equation in part **a**.

CH_3—H + $\dot{Cl}$ ⟶ $\dot{C}H_3$ + H—Cl

$\dot{C}H_3$ + Cl—Cl ⟶ CH_3—Cl + $\dot{Cl}$

CH_4 + Cl_2 ⟶ CH_3Cl + HCl

48. The overall $\Delta H°$ value is the same for both mechanisms. However, the mechanism that occurs is controlled by the first propagation step. The first propagation step of the alternative mechanism is very endothermic, so it would not be able to compete with the first propagation step of the mechanism shown in Problem 47b.

$$CH_3{-}H + \cdot Cl \longrightarrow CH_3{-}Cl + \cdot H \qquad \Delta H° = 105 - 84 = 21 \text{ kcal/mol}$$

$$H\cdot + Cl{-}Cl \longrightarrow H{-}Cl + \cdot Cl \qquad \Delta H° = 58 - 103 = -45 \text{ kcal/mol}$$

$$\text{Overall } \Delta H° = 21 + (-45) = -24 \text{ kcal/mol}$$

49.

$$RO{-}OR \longrightarrow 2\ RO\cdot$$

$$RO\cdot + HCCl_3 \longrightarrow ROH + \cdot CCl_3$$

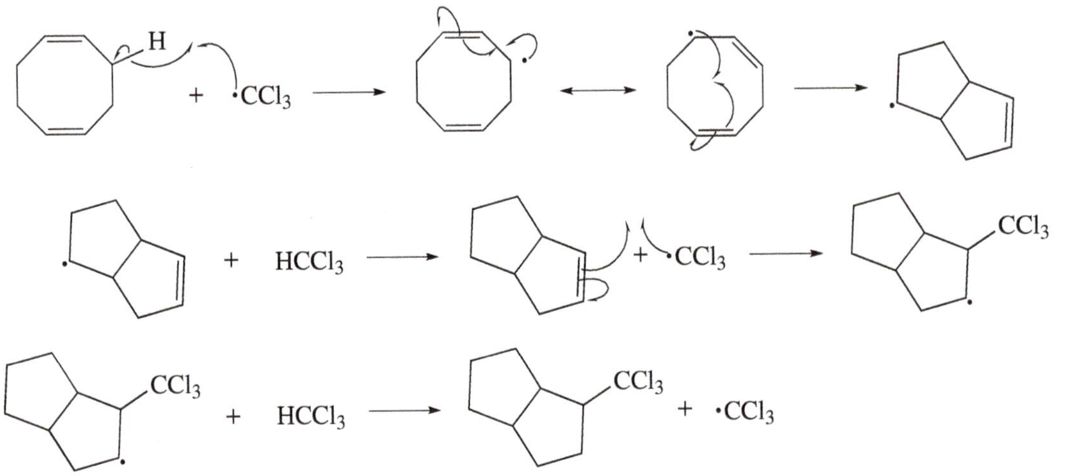

H + $\cdot CCl_3$ ⟶ ⟷ ⟶

+ $HCCl_3$ ⟶ + $\cdot CCl_3$ ⟶ CCl_3

CCl_3 + $HCCl_3$ ⟶ CCl_3 + $\cdot CCl_3$

50. The methyl radical that is created in the first propagation step reacts with Br_2, forming bromomethane.

$$\cdot Br + CH_4 \longrightarrow \cdot CH_3 + HBr$$
$$\cdot CH_3 + Br_2 \longrightarrow CH_3Br + \cdot Br$$

If HBr is added to the reaction mixture, the methyl radical that is created in the first propagation step of the bomination of methane can react with Br_2 or with the added HBr. Because only reaction with Br_2 forms bromomethane (reaction with HBr re-forms methane), the overall rate of formation of bromomethane is decreased.

$$\cdot Br + CH_4 \longrightarrow \cdot CH_3 + HBr$$

$$\cdot CH_3 + Br_2 \longrightarrow CH_3Br + \cdot Br$$

$$\cdot CH_3 + HBr \longrightarrow CH_4 + \cdot Br$$

51.

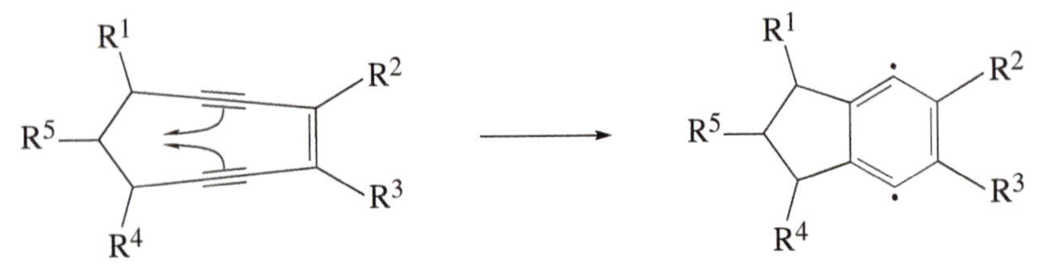

R^1 R^2 R^5 R^3 R^4 ⟶ R^1 R^2 R^5 R^3 R^4

Chapter 12 Practice Test

1. How many monochlorinated products would be obtained from the monochlorination of the following alkanes? (Ignore stereoisomers.)

a.

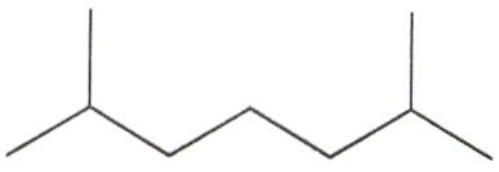

d.

b.

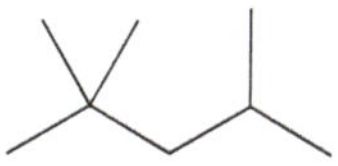

e.

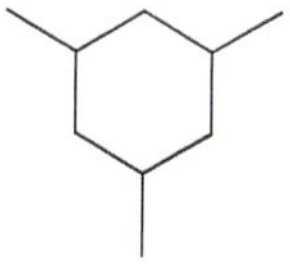

c.

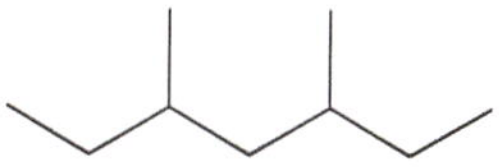

2. What is the first propagation step in the monochlorination of ethane?

3. When (*S*)-2-bromopentane is brominated, two 2,3-dibromopentanes are formed. Which of the following 2,3-dibromopentanes are **not** formed?

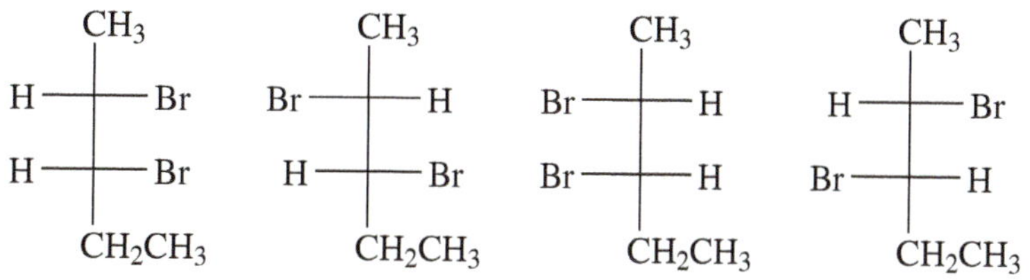

CH_3	CH_3	CH_3	CH_3
H—Br	Br—H	Br—H	H—Br
H—Br	H—Br	Br—H	Br—H
CH_2CH_3	CH_2CH_3	CH_2CH_3	CH_2CH_3

4. Determine the $\Delta H°$ of the two propagation steps in the monochlorination of ethane, using Table 5.1 on page 206 of the text.

5. Rank the radicals from most stable to least stable.

$\dot{C}H_2CH_2CH_2CH{=}CH_2$ $CH_3CH_2\dot{C}HCH{=}CH_2$

$CH_3CH_2CH_2CH{=}\dot{C}H$ $CH_3\dot{C}HCH_2CH{=}CH_2$

6. Draw the product(s) of each of the following reactions, ignoring stereoisomers:

a. 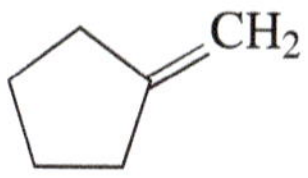 (cyclopentylidene)CH_2 + NBS $\xrightarrow[\text{peroxide}]{\Delta}$

b. $CH_3CH_2CH{=}CH_2$ + NBS $\xrightarrow[\text{peroxide}]{\Delta}$

7. **a.** Draw the products that are obtained from the monochlorination of the following alkane at room temperature. (Ignore stereoisomers.)

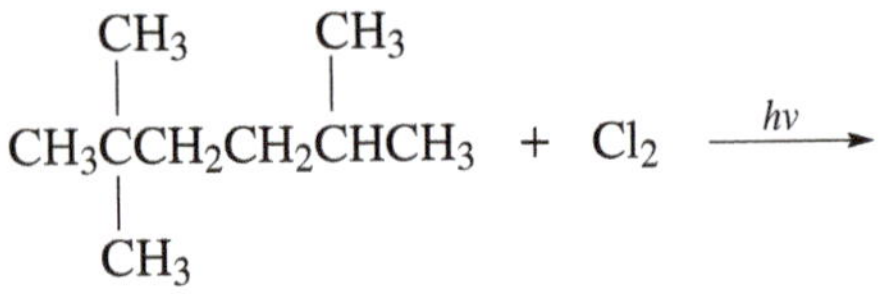

$$CH_3\overset{\displaystyle CH_3}{\underset{\displaystyle CH_3}{C}}CH_2CH_2\overset{\displaystyle CH_3}{C}HCH_3 + Cl_2 \xrightarrow{h\nu}$$

b. Which product is obtained in greatest yield?

c. Which product would be obtained in greatest yield if the alkane is brominated instead of chlorinated?

8. Calculate the yield of 2-chloro-2-methylbutane (as a percentage of all the monochlorinated products obtained) when 2-methylbutane is chlorinated in the presence of light at room temperature.

9. What is the major product of each of the following reactions? Ignore stereoisomers.

a. 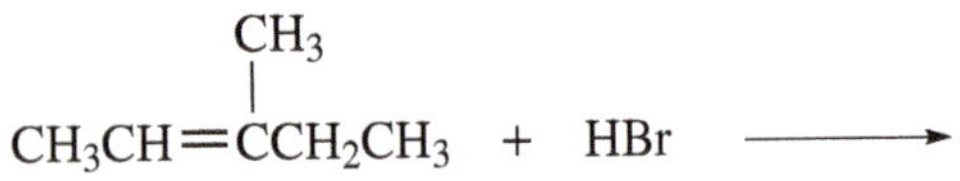 $CH_3CH{=}\overset{\displaystyle CH_3}{C}CH_2CH_3 + HBr \longrightarrow$

b. $CH_3CH{=}\overset{\displaystyle CH_3}{C}CH_2CH_3 + HBr \xrightarrow{\text{peroxide}}$

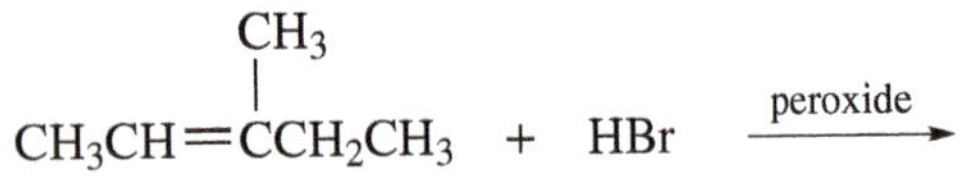

c. $CH_3CH{=}\overset{\displaystyle CH_3}{C}CH_2CH_3 + HCl \longrightarrow$

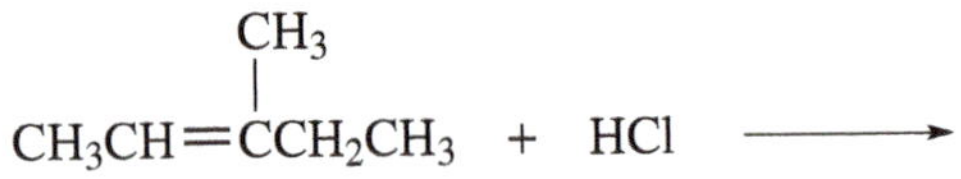

d. 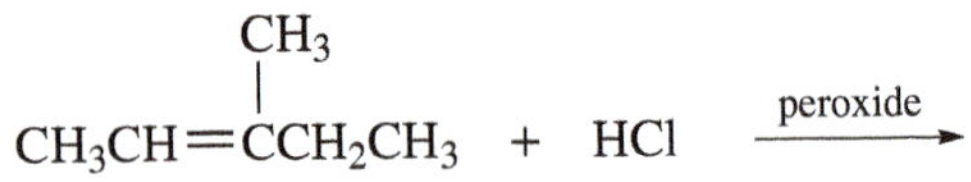 $CH_3CH{=}\overset{\displaystyle CH_3}{C}CH_2CH_3 + HCl \xrightarrow{\text{peroxide}}$

10. Which of the above reactions forms more than one stereoisomer?

CHAPTER 13
Mass Spectrometry; Infrared Spectroscopy; Ultraviolet/Visible Spectroscopy

Important Terms

absorption band	a signal in a spectrum that occurs as a result of absorption of energy.
auxochrome	a substituent that when attached to a chromophore alters the λ_{max} and intensity of absorption of UV/Vis radiation.
base peak	the peak in a mass spectrum with the greatest intensity.
Beer–Lambert law	an equation that states the relationship between the absorbance of UV/Vis light, the concentration of the sample, the length of the light path, and the molar absorptivity.
bending vibration	a vibration that does not occur along the line of the bond.
chromophore	the part of a molecule responsible for a UV or visible spectrum.
α-cleavage	homolytic cleavage of an alpha substituent.
electromagnetic radiation	radiant energy that displays wave properties.
electronic transition	promotion of an electron from its HOMO to its LUMO.
fingerprint region	the right-hand third of an IR spectrum (1400–600 cm^{-1}), where the absorption bands are characteristic of the compound as a whole.
fragment ion peak	a positively charged fragment of a molecular ion.
frequency	the velocity of a wave divided by its wavelength.
functional group region	the left-hand two-thirds of an IR spectrum (4000–1400 cm^{-1}), where most functional groups show absorption bands.
highest occupied molecular orbital (HOMO)	the highest energy molecular orbital that contains electrons.
Hooke's law	an equation that describes the motion of a vibrating spring.
infrared radiation	electromagnetic radiation familiar to us as heat.
infrared spectroscopy	spectroscopy that uses infrared radiation to provide a knowledge of the functional groups in a compound.
infrared spectrum (IR spectrum)	a plot of relative absorption versus wavenumber (or wavelength) of absorbed infrared radiation.
λ_{max}	the wavelength at which there is maximum UV/Vis absorbance.

lowest unoccupied molecular orbital (LUMO) the lowest energy molecular orbital that does not contain electrons.

mass spectrometry an instrumental technique that provides a knowledge of the molecular weight and certain structural features of a compound.

mass spectrum a plot of the relative abundance of the positively charged fragments produced in a mass spectrometer versus their *m/z* values.

McLafferty rearrangement rearrangement of the molecular ion of a ketone that contains a γ-hydrogen; the bond between the α- and β-carbons breaks, and a γ-hydrogen migrates to the oxygen.

molar absorptivity the absorbance obtained from a 1.00 M solution in a cell with a 1.00 cm light path.

molecular ion (M) the radical cation formed by removing one electron from a molecule.

nominal molecular mass mass to the nearest whole number.

radical cation a species with a positive charge and an unpaired electron.

rule of 13 a rule that allows possible molecular formulas to be determined from the *m/z* value of the molecular ion.

spectroscopy study of the interaction of matter and electromagnetic radiation.

stretching vibration a vibration occurring along the line of the bond.

ultraviolet light electromagnetic radiation with wavelengths ranging from 180 to 400 nm.

UV/Vis spectroscopy the absorption of electromagnetic radiation that is useful in determining information about conjugated systems.

visible light electromagnetic radiation with wavelengths ranging from 400 to 780 nm.

wavelength distance from any point on one wave to the corresponding point on the next wave.

wavenumber the number of waves in 1 cm.

Solutions to Problems

1. Only positively charged fragments are accelerated through the analyzer tube.

$$CH_3CH_2\overset{+}{C}H_2 \qquad [CH_3CH_2CH_3]^{+\bullet} \qquad \overset{+}{C}H_2CH{=}CH_2$$

2. The peak at $m/z = 57$ is more intense for 2,2-dimethylpropane than for isopentane or pentane. The peak at $m/z = 57$ is due to loss of a methyl radical: loss of a methyl radical from 2,2-dimethylpropane forms a tertiary carbocation, whereas loss of a methyl radical from isopentane or pentane forms a less stable secondary and primary carbocation, respectively.

$$[(CH_3)_3CCH_3]^{+\bullet} \longrightarrow CH_3\overset{+}{C}(CH_3)CH_3 + \dot{C}H_3$$

2,2-dimethylpropane; $m/z = 57$, a tertiary carbocation

$$[CH_3CH(CH_3)CH_2CH_3]^{+\bullet} \longrightarrow CH_3\overset{+}{C}HCH_2CH_3 + \dot{C}H_3$$

2-methylbutane; $m/z = 57$, a secondary carbocation

$$[CH_3CH_2CH_2CH_2CH_3]^{+\bullet} \longrightarrow CH_3CH_2CH_2\overset{+}{C}H_2 + \dot{C}H_3$$

pentane; $m/z = 57$, a primary carbocation

Notice that the mass spectrum of isopentane can be distinguished from those of the other isomers by the peak at $m/z = 43$. The peak at $m/z = 43$ is intense for isopentane because such a peak is due to loss of an ethyl radical, which forms a secondary carbocation. Pentane gives a less intense peak at $m/z = 43$ because loss of an ethyl radical from pentane forms a primary carbocation. 2,2-Dimethylpropane does not show a peak at $m/z = 43$ because it does not have an ethyl group.

$$[CH_3CH(CH_3)CH_2CH_3]^{+\bullet} \longrightarrow CH_3\overset{+}{C}HCH_3 + CH_3\dot{C}H_2$$

$m/z = 43$, a secondary carbocation

$$[CH_3CH_2CH_2CH_2CH_3]^{+\bullet} \longrightarrow CH_3CH_2\overset{+}{C}H_2 + CH_3\dot{C}H_2$$

$m/z = 43$, a primary carbocation

3. Intense peaks are expected at $m/z = 57$ for loss of an ethyl radical ($86 - 29$) and at $m/z = 71$ for loss of a methyl radical ($86 - 15$).

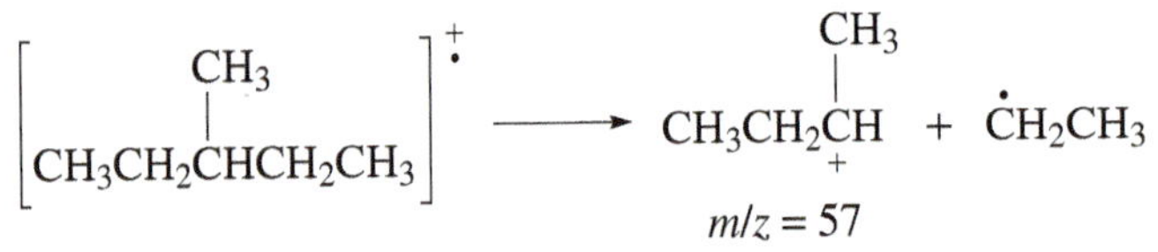

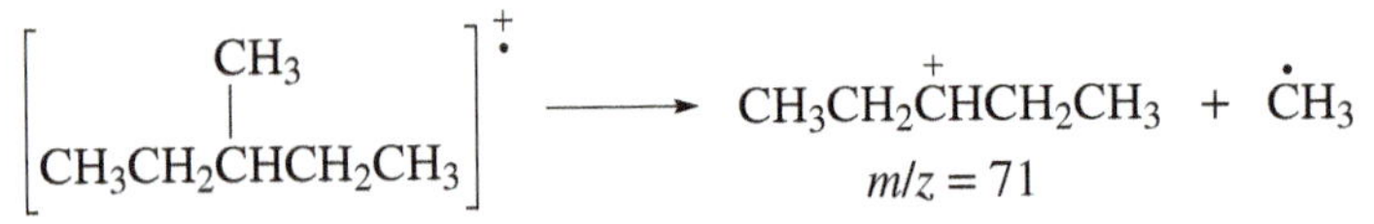

$$\left[CH_3CH_2\overset{\overset{CH_3}{|}}{C}HCH_2CH_3\right]^{+\cdot} \longrightarrow CH_3CH_2\overset{\overset{CH_3}{|}}{\underset{+}{C}H} + \dot{C}H_2CH_3 \quad m/z = 57$$

$$\left[CH_3CH_2\overset{\overset{CH_3}{|}}{C}HCH_2CH_3\right]^{+\cdot} \longrightarrow CH_3CH_2\overset{+}{C}HCH_2CH_3 + \dot{C}H_3 \quad m/z = 71$$

A secondary carbocation is formed in each case. Because an ethyl radical is more stable than a methyl radical and there is more than one way to produce this peak, the base peak is most likely at $m/z = 57$.

4. Solved in the text.

5.

a. Dividing 72 by 13 gives 5 with 7 left over, giving a base value of C_5H_{12}. Because the compound contains only carbons and hydrogens, we know that the base value is also the molecular formula of the compound.

b. Dividing 100 by 13 gives 7 with 9 left over, giving a base value of C_7H_{16}. Because the compound contains one oxygen, an O must be added to the base value and one C and four Hs must be subtracted. Therefore, the molecular formula is $C_6H_{12}O$.

c. Dividing 102 by 13 gives 7 with 11 left over, giving a base value of C_7H_{18}. Because the compound contains two oxygens, two Os must be added to the base value and two Cs and eight Hs must be subtracted. Therefore, the molecular formula is $C_5H_{10}O_2$.

d. Dividing 115 by 13 gives 8 with 11 left over, giving a base value of C_8H_{19}. Because the compound contains one oxygen, an O must be added to the base value and one C and four Hs must be subtracted. Because the compound contains an N, an N must be added to the base value and one C and two Hs must be subtracted. Therefore, the molecular formula is $C_6H_{13}NO$.

6.

a. Dividing 86 by 13 gives 6 with 8 left over, giving a base value of C_6H_{14}.
If the compound contains only carbons and hydrogens, the base value is also the molecular formula of the compound. Some possible structures are:

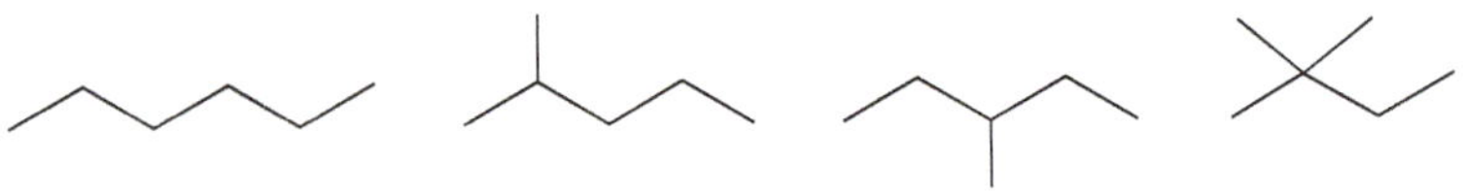

If the compound contains one oxygen, the molecular formula is $C_5H_{10}O$. Some possible structures are:

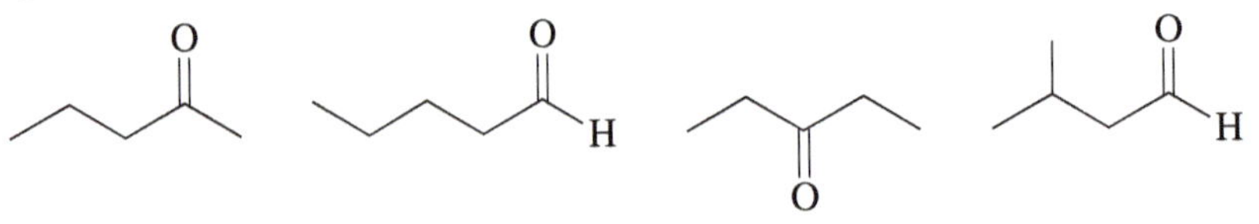

O O O
H H

If the compound contains two oxygens, the molecular formula is $C_4H_6O_2$. Some possible structures are:

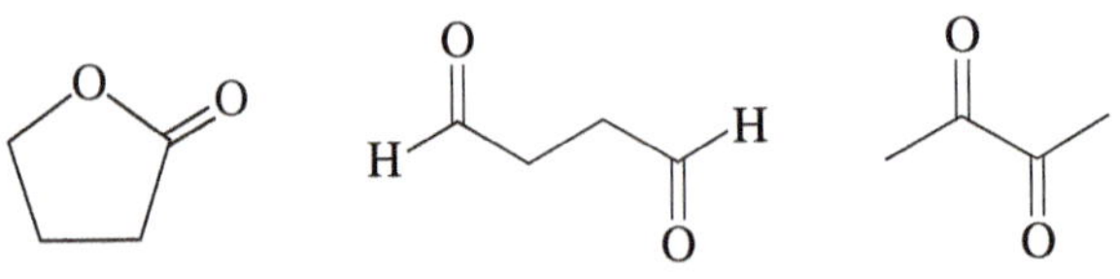

O O O O
H H O O

b. Because the compound has an even-numbered mass, we know that it does not contain one nitrogen atom (see Problem 7). It could, however, contain two nitrogen atoms ($C_4H_{10}N_2$). A possible structure is:

$$H_2NCH_2CH{=}CHCH_2NH_2$$

7. **a.** **1.** $15 + (3 \times 14) + 16 = 73$ **2.** $16 + (3 \times 14) + 16 = 74$

b. An alkane has an even-mass molecular ion. If a CH_2 group (14) of an alkane is replaced by an NH group (15) or if a CH_3 group (15) of an alkane is replaced by an NH_2 group (16), the molecular ion will have an odd mass.

A second nitrogen in the molecular ion will cause it to have an even mass.

Thus, for a molecular ion to have an odd mass, it must have an odd number of nitrogens.

c. An even-mass molecular ion either has no nitrogens or has an even number of nitrogens.

8. A hydrocarbon with molecular formula C_9H_{20} has a molecular mass of 128.
Because $C_9H_{20} = C_nH_{2n+2}$, we know that the hydrocarbon has no rings and no π bonds.
The hydrocarbon is **2,6-dimethylheptane**.

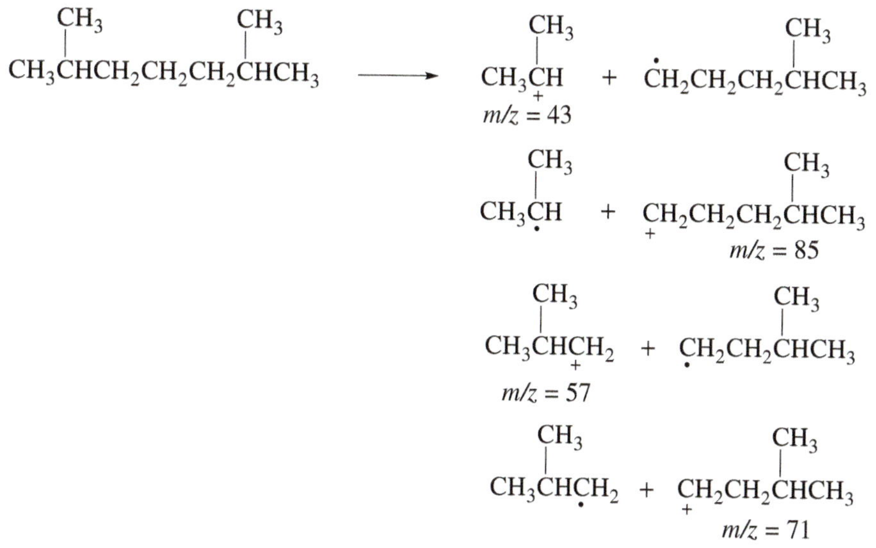

$$CH_3CH(CH_3)CH_2CH_2CH_2CH(CH_3)CH_3 \longrightarrow CH_3\overset{+}{C}H(CH_3) + \dot{C}H_2CH_2CH_2CH(CH_3)CH_3$$
$m/z = 43$

$$CH_3\dot{C}H(CH_3) + \overset{+}{C}H_2CH_2CH_2CH(CH_3)CH_3$$
$m/z = 85$

$$CH_3CH(CH_3)\overset{+}{C}H_2 + \dot{C}H_2CH_2CH(CH_3)CH_3$$
$m/z = 57$

$$CH_3CH(CH_3)\dot{C}H_2 + \overset{+}{C}H_2CH_2CH(CH_3)CH_3$$
$m/z = 71$

2-Methyloctane is also expected to give a base peak of $m/z = 43$ because it, too, forms a secondary (isopropyl carbocation) together with a primary radical, and all other cleavages that form primary radicals form primary carbocations. However, we would expect fragments with $m/z = 29$ and 99 to be present to the same extent as those with $m/z = 57$, 85, and 71. Because fragments with $m/z = 29$ and 99 are not mentioned, we can conclude that the hydrocarbon is more likely to be 2,6-dimethylheptane than 2-methyloctane.

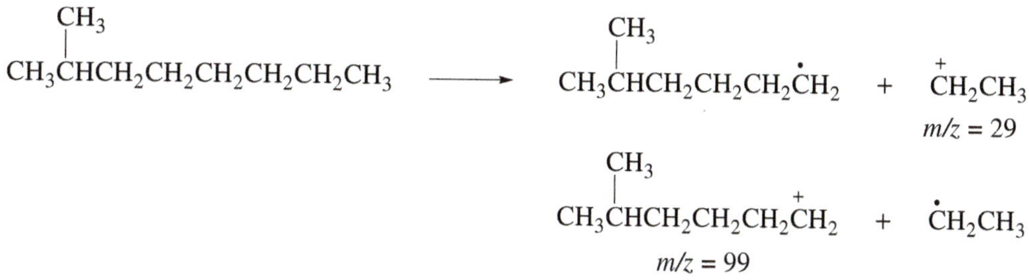

$$CH_3CH(CH_3)CH_2CH_2CH_2CH_2CH_2CH_3 \longrightarrow CH_3CH(CH_3)CH_2CH_2CH_2\dot{C}H_2 + \overset{+}{C}H_2CH_3$$
$m/z = 29$

$$CH_3CH(CH_3)CH_2CH_2CH_2\overset{+}{C}H_2 + \dot{C}H_2CH_3$$
$m/z = 99$

9. The ratio is 1:2:1.
To get the M peak, both Br atoms must be ^{79}Br. To get the M+4 peak, they both must be ^{81}Br. To get the M + 2 peak, the first Br atom can be ^{79}Br and the second ^{81}Br, or the first can be ^{81}Br and the second ^{81}Br. So the relative intensity of the M+2 peak will be twice that of the others.

M	M+2	M+4
$^{79}Br\ ^{79}Br$	$^{79}Br\ ^{81}Br$ $^{81}Br\ ^{79}Br$	$^{81}Br\ ^{81}Br$

10. The calculated exact masses show that only C_6H_{14} has an exact mass of 86.10955.

$$C_6H_{14} \quad \begin{aligned} 6(12.00000) &= 72.00000 \\ 14(1.007825) &= \underline{14.10955} \\ &\ 86.10955 \end{aligned}$$

$$C_4H_6O_2 \quad \begin{aligned} 4(12.00000) &= 48.00000 \\ 6(1.007825) &= 6.04695 \\ 2(15.9949) &= \underline{31.9898} \\ &\ 86.03675 \end{aligned}$$

$$C_4H_{10}N_2 \quad \begin{aligned} 4(12.00000) &= 48.00000 \\ 10(1.007825) &= 10.07825 \\ 2(14.0031) &= \underline{28.0064} \\ &\ 86.08465 \end{aligned}$$

11. **a.** A low-resolution spectrometer cannot distinguish between them because they both have the same molecular mass (29).
b. A high-resolution spectrometer can distinguish between these ions because they have different exact molecular masses; one has an exact molecular mass of 29.039125, and the other an exact molecular mass of 29.002725.

12. Because the compound contains chlorine, the M+2 peak is one-third the size of the M peak. Breaking the weak C—Cl bond heterolytically and, therefore, losing a chlorine atom from either the M+2 peak (80 − 37) or the M peak (78 − 35) gives the base peak with $m/z = 43$ ($[CH_3CH_2CH_2]^{+\bullet}$).

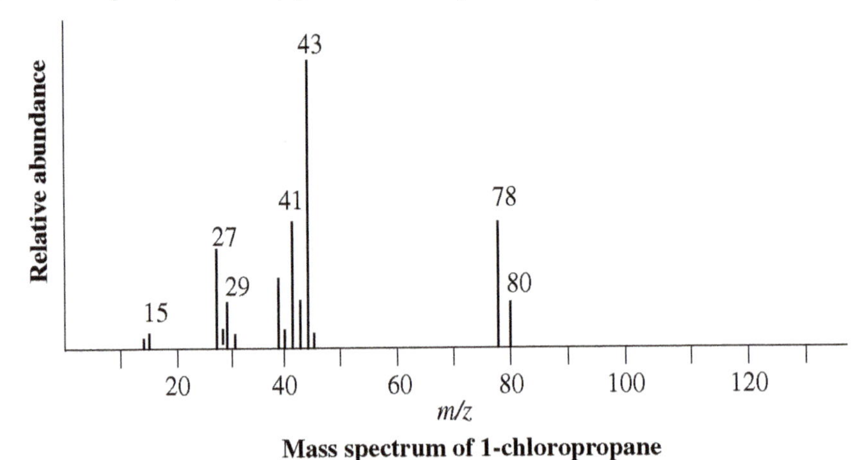

Mass spectrum of 1-chloropropane

13. The dominant fragmentation pathway in each case is loss of an alkyl radical via α-cleavage to form a cation in which all the atoms have complete octets.

The base peak at $m/z = 73$ (88 − 15), due to loss of a methyl radical, indicates that **a.** is the mass spectrum of **2-methoxy-2-methylpropane**.

The base peak at $m/z = 59$ (88 − 29), due to loss of an ethyl radical, indicates that **b.** is the mass spectrum of **2-methoxybutane**.

The base peak at $m/z = 45$ (88 − 43), due to loss of a propyl radical, indicates that **c.** is the mass spectrum of **1-methoxybutane**.

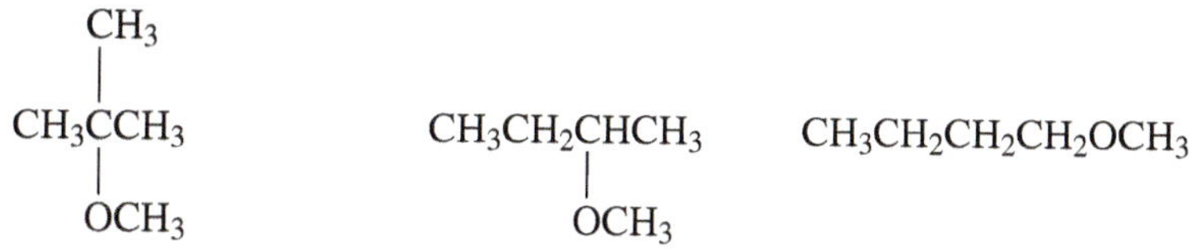

$CH_3C(CH_3)_2OCH_3$ — 2-methoxy-2-methylpropane

$CH_3CH_2CH(OCH_3)CH_3$ — 2-methoxybutane

$CH_3CH_2CH_2CH_2OCH_3$ — 1-methoxybutane

14. a. $CH_2{=}\overset{+}{O}H$

$m/z = 31$

b. $R{-}CH_2{-}\ddot{O}H \xrightarrow{-e^-} R{-}CH_2{-}\overset{+}{\dot{O}}H \longrightarrow R\cdot + CH_2{=}\overset{+}{O}H$

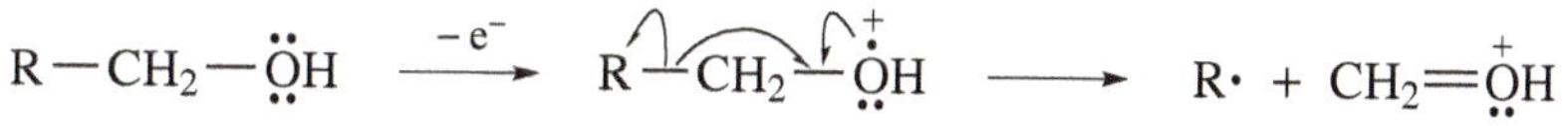

15. The molecular ions with $m/z = 86$ indicate that both ketones have the molecular formula $C_5H_{10}O$. Spectrum **a.** shows a base peak at $m/z = 43$ for loss of a propyl (or isopropyl) radical (86 − 43), indicating that it is the mass spectrum of either ketone A or ketone B because each of these has a propyl or isopropyl group. The fact that the spectrum has a peak at $m/z = 58$, indicating loss of ethene (86 − 28), indicates that the compound has a γ-hydrogen that enables it to undergo a McLafferty rearrangement. Therefore, the ketone must be **A** because **B** does not have a γ-hydrogen.

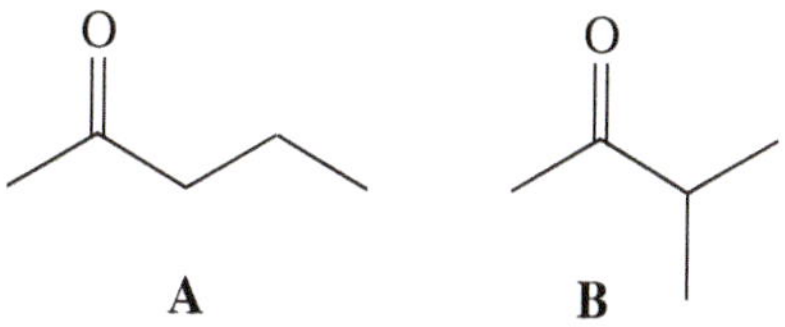

A **B**

Spectrum **b.** shows a base peak at $m/z = 57$ for loss of an ethyl radical (86 − 29), indicating that it is the mass spectrum of A.

16. All three ketones have a molecular ion with $m/z = 86$.

$86 - 29 = 57$ $86 - 43 = 45$ $86 - 43 = 45$

The first one will have a base peak at $m/z = 57$, whereas the other two ketones will have base peaks at $m/z = 43$.

The second one will have a peak at $m/z = 58$ due to a McLafferty rearrangement.

The third one does not have any γ-hydrogens. Therefore, it cannot undergo a McLafferty rearrangement, so it will not have a peak at $m/z = 58$.

17. **a.** $CH_3CH_2CH_2CH_2CH_2\ddot{O}H$

$\downarrow -e^-$

$CH_3CH_2CH_2CH_2—CH_2—\dot{O}^+H$ $\xrightarrow{\alpha\text{-cleavage}}$ $CH_3CH_2CH_2\dot{C}H_2 + CH_2=\overset{+}{O}H$

$m/z = 88$ $m/z = 31$

$CH_3CH_2CH(H)CH_2CH_2—\overset{+}{O}H \longrightarrow CH_3CH_2\dot{C}HCH_2\overset{+}{C}H_2 + H_2O$

$m/z = 70$

b. $CH_3CH_2CH(\ddot{O}H)CH_2CH_2CH_2CH_3$

$\downarrow -e^-$

$CH_3CH_2—CH(\dot{O}^+H)CH_2CH_2CH_2CH_3$ $\xrightarrow{\alpha\text{-cleavage}}$ $CH_3CH_2CH_2CH_2CH=\overset{+}{O}H + CH_3\dot{C}H_2$

$m/z = 116$ $m/z = 87$

$CH_3CH_2CH(\dot{O}^+H)—CH_2CH_2CH_2CH_3$ $\xrightarrow{\alpha\text{-cleavage}}$ $CH_3CH_2CH=\overset{+}{O}H + CH_3CH_2CH_2\dot{C}H_2$

$m/z = 59$

$CH_2(H)CH_2CH(\dot{O}^+H)CH_2CH_2CH_2CH_3 \longrightarrow \dot{C}H_2CH_2\overset{+}{C}HCH_2CH_2CH_2CH_3 + H_2O$

$m/z = 98$

$CH_3CH_2CH(\dot{O}^+H)CH_2CH(H)CH_2CH_3 \longrightarrow CH_3CH_2\overset{+}{C}HCH_2\dot{C}HCH_2CH_3 + H_2O$

$m/z = 98$

c.

$$CH_3CH_2-\ddot{O}-C(CH_2CH_3)(CH_3)CH_2CH_2CH_3 \xrightarrow{-e^-} CH_3CH_2-\dot{O}^{+}-C(CH_2CH_3)(CH_3)CH_2CH_2CH_3$$

$$\underset{m/z = 144}{CH_3CH_2-\dot{O}^{+}-C(CH_2CH_3)(CH_3)CH_2CH_2CH_3} \longrightarrow CH_3CH_2-\dot{O}: + \underset{m/z = 99}{{}^{+}C(CH_2CH_3)(CH_3)CH_2CH_2CH_3}$$

$$CH_3CH_2-\dot{O}^{+}-C(CH_2CH_3)(CH_3)CH_2CH_2CH_3 \xrightarrow{\alpha\text{-cleavage}} \underset{m/z = 129}{CH_3CH_2-\overset{+}{O}=C(CH_2CH_3)CH_2CH_2CH_3} + \dot{C}H_3$$

$$CH_3CH_2-\dot{O}^{+}-C(CH_2CH_3)(CH_3)CH_2CH_2CH_3 \xrightarrow{\alpha\text{-cleavage}} \underset{m/z = 115}{CH_3CH_2-\overset{+}{O}=C(CH_3)CH_2CH_2CH_3} + CH_3\dot{C}H_2$$

$$CH_3CH_2-\dot{O}^{+}-C(CH_2CH_3)(CH_3)-CH_2CH_2CH_3 \xrightarrow{\alpha\text{-cleavage}} \underset{m/z = 101}{CH_3CH_2-\overset{+}{O}=C(CH_3)CH_2CH_3} + CH_3CH_2\dot{C}H_2$$

$$CH_3-CH_2-\dot{O}^{+}-C(CH_2CH_3)(CH_3)-CH_2CH_2CH_3 \xrightarrow{\alpha\text{-cleavage}} \underset{m/z = 129}{CH_2=\overset{+}{O}-C(CH_2CH_3)(CH_3)CH_2CH_2CH_3} + \dot{C}H_3$$

d.

$$CH_3\overset{\overset{\ddot{O}:}{\|}}{C}CH_2CH_2CH_2CH_3 \xrightarrow{-e^-} CH_3-\overset{\overset{\dot{O}^{+}:}{\|}}{C}CH_2CH_2CH_2CH_3$$

$m/z = 100$

$$CH_3-\overset{\overset{\dot{O}^{+}:}{\|}}{C}CH_2CH_2CH_2CH_3 \xrightarrow{\alpha\text{-cleavage}} CH_3CH_2CH_2CH_2C{\equiv}\overset{+}{\ddot{O}} + \dot{C}H_3$$

$m/z = 85$

$$CH_3\overset{\overset{:\dot{O}^{+}}{\|}}{C}-CH_2CH_2CH_2CH_3 \xrightarrow{\alpha\text{-cleavage}} CH_3C{\equiv}\overset{+}{\ddot{O}} + CH_3CH_2CH_2\dot{C}H_2$$

$m/z = 43$

$$CH_3\overset{\overset{:\dot{O}^{+}}{\|}}{C}CH_2-CH_2-\overset{H}{C}HCH_3 \xrightarrow[\text{rearrangement}]{\text{McLafferty}} CH_3\overset{\overset{:\overset{+}{O}H}{\|}}{C}\dot{C}H_2 + CH_2{=}CHCH_3$$

$m/z = 58$

e.

$$CH_3CH_2\underset{CH_3}{\underset{|}{C}}H\ddot{\underset{..}{Cl}}: \xrightarrow{-e^-} CH_3CH_2\underset{CH_3}{\underset{|}{C}}H-\dot{\underset{..}{Cl}}^{+}: \longrightarrow CH_3CH_2\underset{CH_3}{\underset{|}{\overset{+}{C}}}H + :\dot{\underset{..}{Cl}}:$$

$m/z = 92$ and 94 $\qquad$ $m/z = 57$

$$CH_3CH_2\underset{CH_3}{\underset{|}{C}}H-\dot{\underset{..}{Cl}}^{+}: \xrightarrow{\alpha\text{-cleavage}} CH_3CH_2CH{=}\overset{+}{\underset{..}{Cl}}: + \dot{C}H_3$$

$m/z = 77$ and 79

$$CH_3CH_2-\underset{CH_3}{\underset{|}{C}}H-\dot{\underset{..}{Cl}}^{+}: \xrightarrow{\alpha\text{-cleavage}} CH_3CH{=}\overset{+}{\underset{..}{Cl}}: + CH_3\dot{C}H_2$$

$m/z = 63$ and 65

f.

$$CH_3\underset{CH_3}{\underset{|}{\overset{CH_3}{\overset{|}{C}}}}-\ddot{\underset{..}{Br}}: \xrightarrow{-e^-} CH_3\underset{CH_3}{\underset{|}{\overset{CH_3}{\overset{|}{C}}}}-\dot{\underset{..}{Br}}^{+}: \longrightarrow CH_3\underset{CH_3}{\underset{|}{\overset{CH_3}{\overset{|}{C}}}}^{+} + :\dot{\underset{..}{Br}}:$$

$m/z = 136$ and 138 $\qquad$ $m/z = 57$

18. We know from Section 6.5 that when (*Z*)-2-pentene reacts with water and an acid catalyst, 3-pentanol and 2-pentanol are formed. Both alcohols have a molecular weight of 88. (Notice that the first step in solving this problem is to use chemical knowledge to identify the products.) The absence of a molecular ion peak is consistent with the fact that the compounds are alcohols.

$$\underset{(Z)\text{-2-pentene}}{(H_3C)(H)C{=}C(CH_2CH_3)(H)} + H_2O \xrightarrow{H_2SO_4} \underset{\text{3-pentanol}}{CH_3CH_2\underset{OH}{\underset{|}{C}}HCH_2CH_3} + \underset{\text{2-pentanol}}{CH_3\underset{OH}{\underset{|}{C}}HCH_2CH_2CH_3}$$

Spectrum **a.** shows a base peak at $m/z = 59$ due to loss of an ethyl radical (88 − 29), indicating that it is the spectrum of 3-pentanol.

Spectrum **b.** shows a base peak at $m/z = 45$ due to loss of a propyl radical (88 − 43), indicating that it is the spectrum of 2-pentanol.

19. **a.** 2000 cm^{-1} (The larger the wavenumber, the higher the energy.)
b. 850 nm (The shorter the wavelength, the higher the energy.)

20. The wavelength is the distance from the top of one wave to the top of the next wave. We see that **A** has a longer wavelength than **B**.
Infrared radiation has longer wavelengths than visible light because infrared radiation is lower in energy. Therefore, **A** depicts infrared radiation and **B** depicts visible light.

21. **a.**

1.	C≡C stretch	A triple bond is stronger than a double bond, so it takes more energy to stretch a triple bond.
2.	C—H stretch	It requires more energy to stretch a given bond than to bend it.
3.	C=N stretch	A double bond is stronger than a single bond, so it takes more energy to stretch a double bond.
4.	C=O stretch	A double bond is stronger than a single bond, so it takes more energy to stretch a double bond.

b.

1.	C—O	Vibrations of lighter atoms occur at larger wavenumbers.
2.	C—C	Vibrations of lighter atoms occur at larger wavenumbers.

22. **a.** The carbon–oxygen stretch of phenol because it has partial double-bond character as a result of electron delocalization.

ÖH ⟷ $\overset{+}{O}H$

b. The carbon–oxygen double-bond stretch of a ketone because it has more double-bond character. The double-bond character of the carbonyl group of an amide is reduced by electron delocalization.

:Ö: C R N̈H_2 ⟷ :Ö:⁻ C R $\overset{+}{N}H_2$

c. The C—N stretch of aniline because it has partial double-bond character.

N̈H_2 ⟷ $\overset{+}{N}H_2$

23. A carbonyl group bonded to an sp^3 carbon exhibits an absorption band at a larger wavenumber because a carbonyl group bonded to an sp^2 carbon of an alkene has greater single-bond character as a result of electron delocalization.

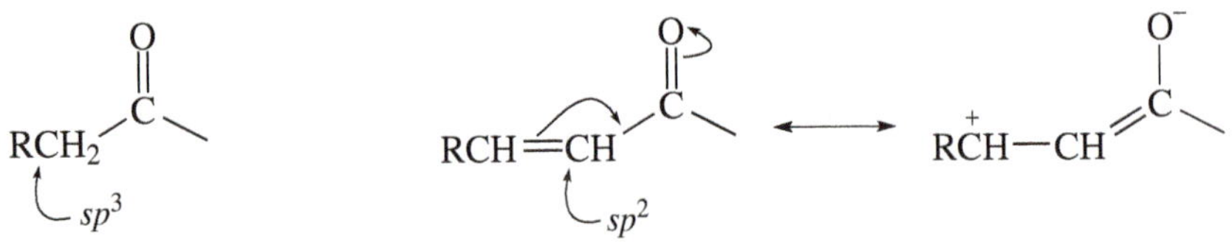

RCH_2–C(=O)– (sp^3) RCH=CH–C(=O)– (sp^2) ⟷ $R\overset{+}{C}H$–CH=C(–O^-)–

24. The C—O bond of the alcohol is a pure single bond. In contrast, the C—O bond of the carboxylic acid has double-bond character, so it is a stronger bond and, therefore, takes more energy to stretch it.

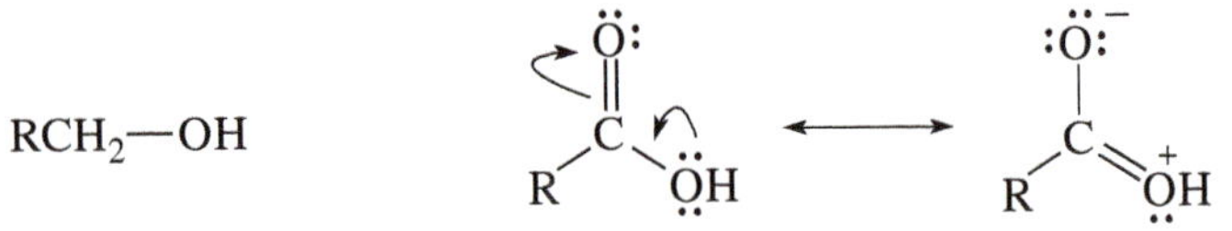

RCH_2—OH R–C(=O)–OH ⟷ R–C(–O^-)=$\overset{+}{O}$H

25. **a.** The C=O absorption band of an ester occurs at the largest wavenumber because the carbonyl group of an ester has the most double-bond character, since the predominant effect of the ester oxygen atom is inductive electron withdrawal.

The C=O absorption band of an amide occurs at the smallest wavenumber because the carbonyl group of an amide has the least double-bond character, since the predominant effect of the amide nitrogen atom is electron donation by resonance.

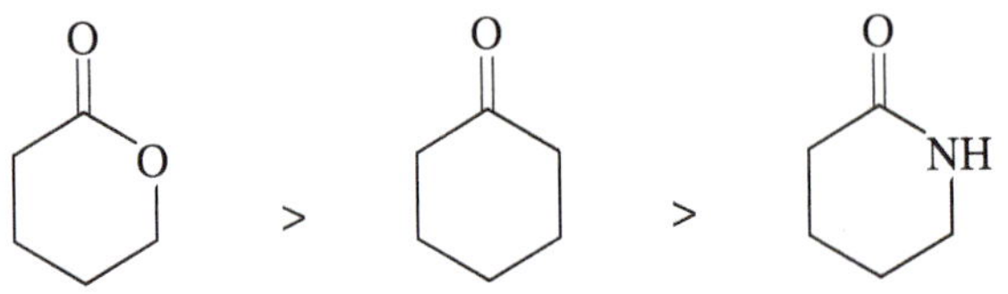

(six-membered lactone) > (cyclohexanone) > (six-membered lactam, NH)

b. The C=O absorption of the four-membered ring lactone occurs at the highest wavenumber because it is the least able to accommodate double-bond character in the ring, which is required if the carbonyl group is to have any single-bond character.

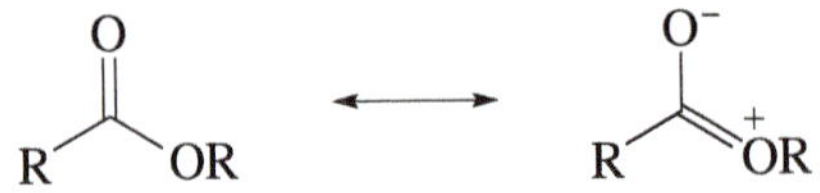

R–C(=O)–OR ⟷ R–C(–O^-)=$\overset{+}{O}$R

The C=O absorption of the six-membered ring lactone occurs at the lowest wavenumber because it is the most able to accommodate double-bond character in the ring.

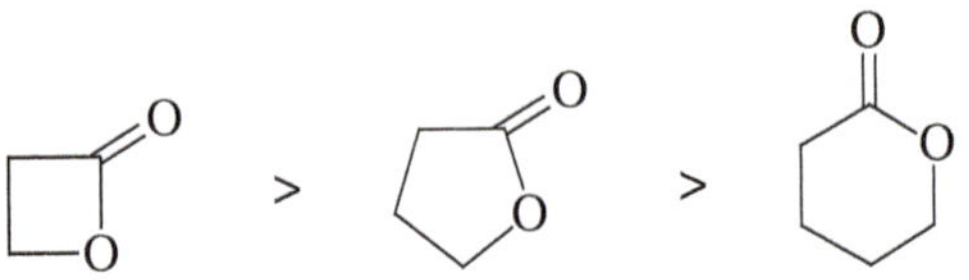

(four-membered lactone) > (five-membered lactone) > (six-membered lactone)

26. Ethanol dissolved in carbon disulfide shows an oxygen–hydrogen stretch at a larger wavenumber. There is extensive hydrogen bonding in the undiluted alcohol and an oxygen–hydrogen bond is easier to stretch if it is hydrogen bonded. Therefore, the O—H stretch will be at a smaller wavenumber.

27. The absorption band at 1100 cm^{-1} would be less intense if it were due to a C—N bond because a smaller change in dipole moment is associated with the stretch of a C—N bond compared to the change in dipole moment associated with the stretch of a C—O bond.

28. **a.** The absorption band at 1700 cm^{-1} indicates that the compound has a carbonyl group.

The absence of an absorption band at 3300 cm^{-1} indicates that the compound is not a carboxylic acid.

The absence of an absorption band at 2700 cm^{-1} indicates that the compound is not an aldehyde.

The absence of an absorption band at 1100 cm^{-1} indicates that the compound is not an ester or an amide. The compound, therefore, must be a **ketone**.

b. The absence of an absorption band at 3400 cm^{-1} indicates that the compound does not have an N—H bond.

The absence of absorption between 2260 – 2220 cm^{-1} indicates that the compound does not have a C≡N bond.

The absence of a carbonyl absorption band between 1700 cm^{-1} and 1600 cm^{-1} indicates that the compound is not an amide. The compound, therefore, must be a **tertiary amine** or a **quaternary ammonium salt**.

29. **a.** An aldehyde would show absorption bands at 2820 and 2720 cm^{-1}. A ketone would not have these absorption bands.

b. An open-chain ketone would have a methyl substituent and, therefore, an absorption band at 1385–1365 cm^{-1} that a cyclic ketone would not have.

c. Cyclohexene would show an sp^3 C—H stretch slightly to the right of 3000 cm^{-1}. Benzene would not show an absorption band in this region.

d. The cis isomer would show a carbon–hydrogen bending vibration at 730–675 cm^{-1}, whereas the trans isomer would show a carbon–hydrogen bending vibration at 960–980 cm^{-1}.

e. Cyclohexene would show a carbon–carbon double-bond stretching vibration at 1680–1600 cm^{-1} and an sp^2 carbon–hydrogen stretching vibration at 3100–3020 cm^{-1}. Cyclohexane would not show these absorption bands.

f. A primary amine would show a nitrogen–hydrogen stretch at 3500–3300 cm^{-1}, and a tertiary amine would not have this absorption band.

30. **a.** An absorption band at 1150–1050 cm^{-1} due to a C—O stretching vibration would be present for the ether and absent for the alkane.

b. An absorption band at 3300–2500 cm^{-1} due to an O—H stretching vibration would be present for the carboxylic acid and absent for the ester.

c. An absorption band at 1385–1365 cm^{-1} due to C—H bending vibrations of a methyl group would be present for methylcyclohexane and absent for cyclohexane.

d. Only the terminal alkyne would show an absorption band at 3300 cm^{-1} due to an *sp* C—H stretching vibration.

e. An absorption band at 1780–1650 cm^{-1} due to a C=O stretching vibration would be present for the carboxylic acid and absent for the alcohol.

f. An absorption band at 2960–2850 cm^{-1} due to a sp^3 C—H stretching vibration would be present for the compound with the methyl group and absent for benzene.

31. 2-butyne, H_2, Cl_2, and ethene because they are symmetrical molecules.

32. The absorption bands in the vicinity of 3000 cm^{-1} indicate that the compound has hydrogens attached to both sp^2 and sp^3 carbons. The absence of absorptions at 1600–1800 cm^{-1} and the absence of broad absorptions between 2500 and 3650 cm^{-1} rules out compounds containing C=O, N—H, and O—H groups. The lack of absorption at 1600 cm^{-1} and 1500 cm^{-1} indicates that the compound does not have a benzene ring. The sp^2 hydrogens, therefore, must be those of an **alkene**.

The lack of significant absorption at 1600 cm^{-1} indicates that the compound must be an alkene with a relatively small (if any) dipole moment change when the vibration occurs. The absorption band at 965 cm^{-1} indicates that the compound is a ***trans*-alkene**.

The molecular ion with $m/z = 84$ suggests that the compound has a molecular formula of C_6H_{12}. The base peak with $m/z = 55$ indicates that the group that the molecular ion most easily loses most easily is an ethyl radical ($84 - 29 = 55$). Therefore, the ethyl group must be attached to an allylic carbon. The compound, therefore, is ***trans*-2-hexene**.

allylic carbon

H, $CH_2—CH_2CH_3$ / C=C / CH_3, H

trans-2-hexene

33. The absorption band at ~1700 cm^{-1} indicates that the compound has a carbonyl group, and the absorption band at ~1600 cm^{-1} indicates that the compound has a carbon–carbon double bond. The absorption bands in the vicinity of 3000 cm^{-1} indicate that the compound has hydrogens attached to both sp^2 and sp^3 carbons. The absorption band at ~1380 cm^{-1} indicates that the compound has a methyl group. Because the compound has only four carbons and one oxygen, it must be **methyl vinyl ketone**. Notice that the carbonyl stretch is at a lower frequency (1700 cm^{-1}) than expected for a ketone (1720 cm^{-1}) because the carbonyl group has partial single-bond character due to electron delocalization.

$CH_3C(=O)CH=CH_2$

34. $A = c\,l\,\varepsilon$

$$c = \frac{A}{l\varepsilon}$$

$$c = \frac{0.52}{12{,}600} = 4.1 \times 10^{-5}\text{ M}$$

35. $A = c\,l\,\varepsilon$

$$c = \frac{A}{l\varepsilon}$$

$$\varepsilon = \frac{0.40}{4.0 \times 10^{-5}} = 10{,}000\ \mathrm{M^{-1}\,cm^{-1}}$$

36. The compound has the same chromophore as methyl vinyl ketone. So it must have approximately the same value for its λ_{max}, which is 219 nm.

37. **a.** C_6H_5–CH=CH–C_6H_5 > C_6H_5–C_6H_5 > C_6H_5–CH=CH_2 > C_6H_6

b. C_6H_5–$N(CH_3)_2$ > C_6H_5–$\overset{+}{N}(CH_3)_3$ > cyclohexadienyl–$N(CH_3)_2$ > cyclohexenyl–$N(CH_3)_2$

38. **a.** Blue results from absorption of light that has a longer wavelength than the light that produces purple when absorbed. The compound on the right has two $N(CH_3)_2$ auxochromes that cause it to absorb light with a longer wavelength than the compound on the left, which has only one $N(CH_3)_2$ auxochrome. Therefore, the compound on the right is the blue compound.

b. They will be the same color at pH = 3 because the $N(CH_3)_2$ groups will be protonated and, therefore, will not possess the lone pair that causes the compound to absorb light of a longer wavelength.

39. NADH is formed as a product; it absorbs light at 340 nm. Therefore, the rate of the oxidation reaction can be determined by monitoring the increase in absorbance at 340 nm as a function of time.

$$CH_3CH_2OH + NAD^+ \xrightarrow{\text{alcohol dehydrogenase}} CH_3CHO + NADH + H^+$$

40. The Henderson–Hasselbalch equation (Section 2.10) shows that when the pH of the solution equals the pK_a of the compound, the concentration of the species compound in the acidic form is the same as the concentration of the species compound in the basic form.

From the data given, we see that the absorbance of the acid is 0. We also see that the absorbance ceases to increase with increasing pH after the absorbance reaches 1.60. That means that all of the compound is in the basic form when the absorbance is 1.60. Therefore, when the absorbance is half of 1.60 (or 0.80), half of the compound is in the basic form and half is in the acidic form; in other words, the concentration in the acidic form is the same as the concentration in the basic form. We see that the absorbance is 0.80 at pH = 5.0. Therefore, the pK_a of the compound is 5.0.

41. The molecular ion peak for these compounds is $m/z = 86$; the peak at $m/z = 57$ is due to loss of an ethyl radical (86 − 29), and the peak at $m/z = 71$ is due to loss of a methyl radical (86 − 15).

a. 3-Methylpentane is more apt to lose an ethyl radical (forming a secondary carbocation and a primary radical) than a methyl radical (forming a secondary carbocation and a methyl radical). In addition, 3-methylpentane has two pathways to lose an ethyl radical. Therefore, the peak at $m/z = 57$ is more intense than the peak at $m/z = 71$.

$CH_3CH_2CHCH_2CH_3$ (with CH_3 on the central carbon)

3-methylpentane

b. 2-Methylpentane has two pathways to lose a methyl radical (forming a secondary carbocation and a methyl radical in each pathway), and it cannot form a secondary carbocation by losing an ethyl radical. (Loss of an ethyl radical would form a primary carbocation and a primary radical.) Therefore, it is more apt to lose a methyl radical than an ethyl radical, so the peak at $m/z = 71$ is more intense than the peak at $m/z = 57$.

$CH_3CHCH_2CH_2CH_3$ (with CH_3 on C-2)

2-methylpentane

42. **1.** the change in the dipole moment when the bond stretches or bends

2. the number of bonds that cause the absorption band

3. the concentration of the sample

43. Dividing 128 by 13 gives 9 with 11 left over, giving a base value of C_9H_{20}. Because the compound is a saturated hydrocarbon, we know that the base value is also the molecular formula of the compound. Some possible structures are:

44. The more conjugated double bonds in a compound, the greater its λ_{max}.

C, with 2 conjugated double bonds, has the smallest λ_{max}.

D, with 5 conjugated double bonds, has the greatest λ_{max}.

C < B < A <

D

45. **a.** An absorption band at ~1250 cm^{-1} due to a C—O stretching vibration will be present for the ester and absent for the ketone.

b. An absorption band at 720 cm^{-1} due to in-phase rocking of the five adjacent methylene groups will be present for heptane and absent for methylcyclohexane.

c. An absorption band at 3650–3200 cm^{-1} due to an O—H stretching vibration will be present for the alcohol and absent for the ether.

d. An absorption band at 3500–3300 cm^{-1} due to an N—H stretching vibration will be present for the amide and absent for the ester.

e. The secondary alcohol will have an absorption band at 1385–1365 cm^{-1} for the methyl group. The primary alcohol does not have a methyl group, so it will not have this absorption band.

f. The trans isomer will have a C—H bending absorption band at 980–960 cm^{-1}, whereas the cis isomer will have the absorption band at 730–675 cm^{-1}. In addition, a weak C═C absorption band at 1680–1600 cm^{-1} will be present for the cis isomer and absent for the trans isomer because only the trans isomer has no dipole moment.

g. The C═C absorption band will be at a larger wavenumber for the ester (1740 cm^{-1}) than for the ketone (1720 cm^{-1}).

h. The C═O absorption band will be at a larger wavenumber for the β,γ-unsaturated ketone (1720 cm^{-1}) than for the α,β-unsaturated ketone (1680 cm^{-1}), since the double bonds in the latter are conjugated.

i. The alkene will have absorption bands at 1680–1600 cm^{-1} due to a C═C stretching vibration and at 3100–3020 cm^{-1} due to an sp^2 C—H stretching vibration that the alkyne will not have. The alkyne will have an absorption band at 2260–2100 cm^{-1} that the alkene will not have.

j. An absorption band at ~2820 and ~2720 cm^{-1} due to the aldehyde C—H stretching vibration will be present for the aldehyde and absent for the ketone.

k. Absorption bands at 1600 cm^{-1} and 1500 cm^{-1} (aromatic ring stretching vibrations) and at 3100–3020 cm^{-1} (sp^2 C—H stretching vibration) will be present for the compound with the benzene ring and absent for the compound with the cyclohexane ring. An absorption band at 2960–2850 cm^{-1} due to an sp^3 C—H stretching vibration will be present for the compound with the cyclohexane ring and absent for the compound with the benzene ring.

l. Absorption bands at 990 cm^{-1} and 910 cm^{-1} due to an sp^2 C—H bending vibration will be present for the terminal alkene and absent for the internal alkene.

46. **a.** If the reaction had occurred, the intensity of the absorption bands at ~1700 cm^{-1} (due to the carbonyl group) and at ~2700 cm^{-1} (due to the aldehyde C—H bond) of the reactant would have decreased. If all the aldehyde had reacted, these absorption bands would have disappeared.

b. If all the NH_2NH_2 had been removed, there would be no N—H absorption at ~3400 cm^{-1}.

47. If the force constants are approximately the same, the lighter atoms absorb at higher frequencies.

$$\text{C—C} \quad > \quad \text{C—N} \quad > \quad \text{C—O}$$

48. Enovid would have its carbonyl stretch at a higher frequency. The carbonyl group in Norlutin has some single-bond character because of electron delocalization as a result of having conjugated double bonds. The single-bond character causes the carbon–oxygen bond to be easier to stretch than the carbon–oxygen bond in Enovid, which has isolated double bonds and, therefore, no electron delocalization.

Norlutin

49.

$3600\ cm^{-1}$ — 3000	3000 — 1800	1800 — 1400	1400 — 1000
OH 3300–3000	sp^3 CH 2950	C=O 1700	C—O 1250–1050
NH 3600–3200	RC(=O)H 2700	C=C 1600	C—N 1230–1030
sp CH 3300	C≡C 2100	benzene ring 1500	
sp^2 CH 3050	C≡N 2250		

50. The molecular weight of each of the alcohols is 158. The peak at $m/z = 140$ is due to loss of water $(158 - 18)$; each of the alcohols will show such a peak. The peaks at $m/z = 87$, 115, and 143 are due to loss of a group with five carbons (C_5H_{11}), a group with three carbons (C_3H_7), and a methyl group, respectively. Only 2,2,4-trimethyl-4-heptanol can lose all three groups via α-cleavage.

51. $CH_2{=}CHCH_2CH_2CH{=}CH_2$ — 1,5-hexadiene; $CH_3CH{=}CHCH{=}CHCH_3$ — 2,4-hexadiene

One way to distinguish the two compounds is by the presence or absence of an absorption band at $\sim 1370\ cm^{-1}$ due to the methyl group that 2,4-hexadiene has but that 1,5-hexadiene does not have. In addition, 2,4-hexadiene has conjugated double bonds; therefore, its double bonds have some single-bond character due to electron delocalization. Consequently, they are easier to stretch than the isolated double bonds of 1,5-hexadiene. Therefore, the carbon–carbon double-bond stretch of 2,4-hexadiene will be at a smaller wavenumber than the carbon–carbon double-bond stretch of 1,5-hexadiene.

52. The fact that the abundance of the M+2 peak is 30% of the abundance of the M peak indicates that the compound has one chlorine atom. The peak at $m/z = 77$ is due to loss of the chlorine atom $(112 - 35 = 77)$. The fact that the peak at $m/z = 77$ does not fragment indicates that it is a phenyl cation. Therefore, the compound is chlorobenzene.

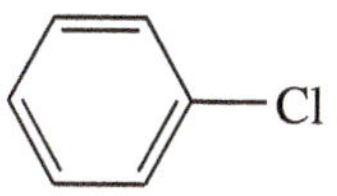

53. Dividing 112 by 13 gives 8 with 8 left over, giving a base value of C_8H_{16}, and because the compound is a hydrocarbon, this is also its molecular formula. The molecular formula indicates that it has one degree of unsaturation, which is accounted for by the fact that we know it has a six-membered ring. Possible structures are shown here. Possible stereoisomers are not shown: the second and third structures have three stereoisomers, and the fourth structure has two stereoisomers.

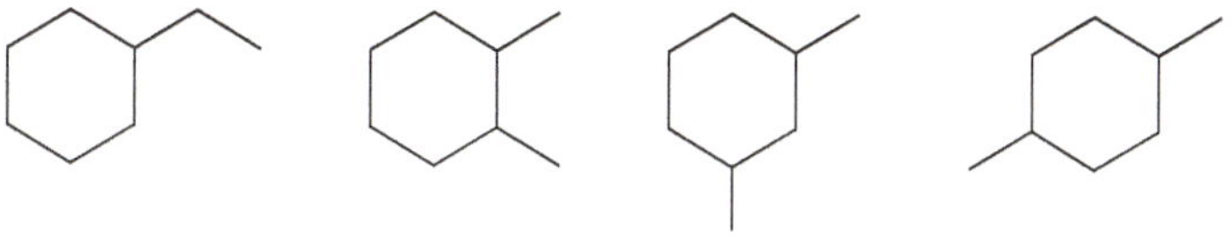

54. The absorption band at ~1740 cm^{-1} indicates that the compound has a carbonyl group, and the absence of an absorption band at ~1380 cm^{-1} indicates that it has no methyl groups. The absence of an absorption band at ~1600 cm^{-1} indicates that the compound does not have a carbon–carbon double bond, and the absence of an absorption band at ~3050 cm^{-1} indicates that the compound does not have hydrogens bonded to sp^2 carbons.

From the molecular formula, you can deduce that the compound is **cyclopentanone**.

55. Hydrogens are more electron-withdrawing than alkyl groups. Therefore, the carbonyl group bonded to two relatively electron-withdrawing hydrogens has the largest wavenumber for its C=O absorption band, whereas the carbonyl group bonded to two alkyl groups has the lowest wavenumber.

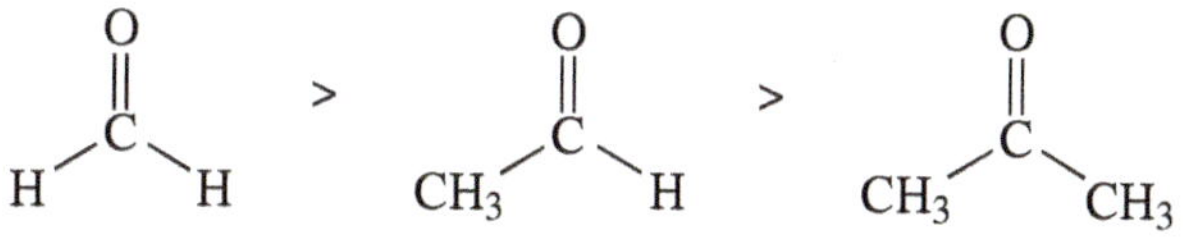

56. The C=O absorption band of the three compounds decreases in the following order.

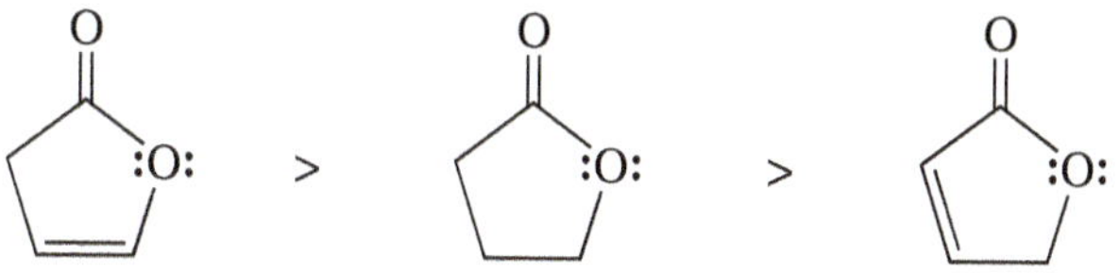

The first compound has the C=O absorption band at the largest wavenumber because a lone pair on the ring oxygen can be delocalized onto two different atoms; therefore, it is less apt than the lone pair in the other compounds to be delocalized onto the carbonyl oxygen atom.

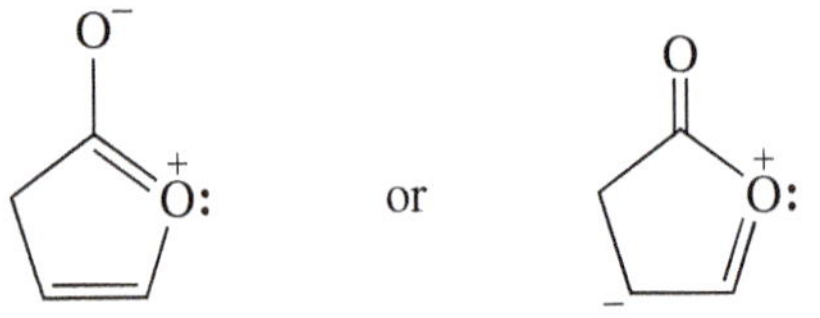

The third compound has the C=O absorption band at the smallest wavenumber because its carbonyl group has more single-bond character due to contributions from two other resonance contributors.

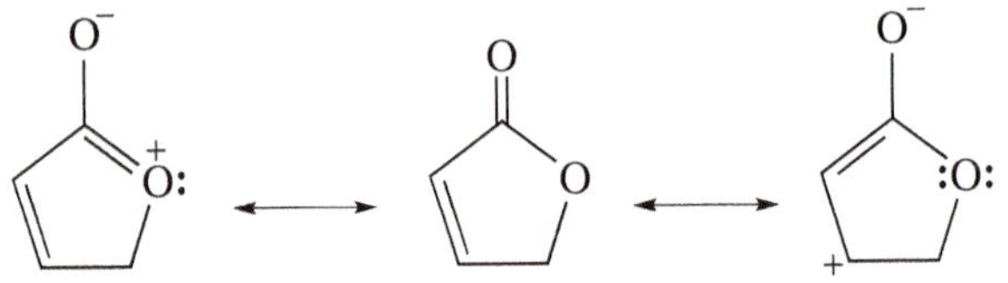

57. The ketone shown below will show peaks at $m/z = 85$ (loss of a methyl group) and at $m/z = 43$ (loss of an isobutyl group) and a peak at $m/z = 58$ due to a McLafferty rearrangement.

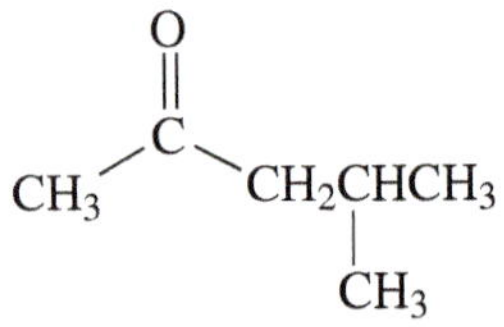

$CH_3C(=O)CH_2CH(CH_3)CH_3$

The ketone shown below will show peaks at $m/z = 71$ (loss of an ethyl group) and at $m/z = 57$ (loss of an isopropyl group). Because it does not have a γ-hydrogen, it cannot undergo a McLafferty rearrangement. Therefore, it will not have a peak at $m/z = 58$.

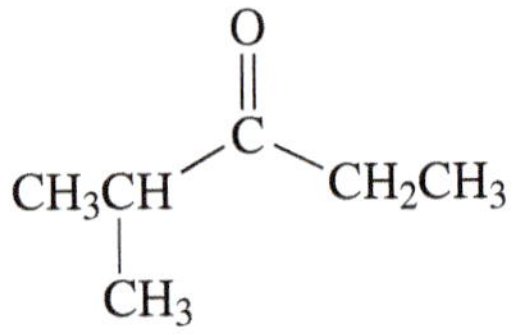

$CH_3CH(CH_3)C(=O)CH_2CH_3$

58. **a.** The tiny molecular ion peak at 102 and the broad absorption at 3600 cm^{-1} indicate that the compound is an alcohol. The base peak at $m/z = 45$ indicates that the OH group is on the second carbon. The absence of significant peaks an $m/z = 29$ and 27 indicates that the compound does not have an ethyl group that can be cleaved from the molecule. The compound is 4-methyl-2-pentanol.

$CH_3CH=\overset{+}{O}H$

$m/z = 45$

b. 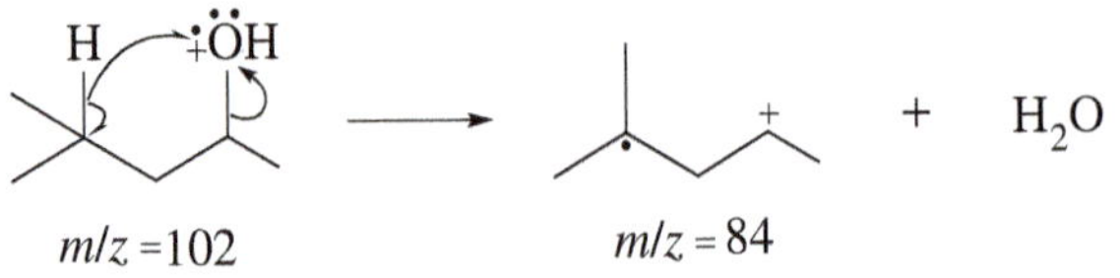 $m/z = 102$ ⟶ $m/z = 84$ + H_2O

59. The broad absorption band at ~3300 cm^{-1} indicates that the compound has an OH group. The absence of absorbance at ~1700 cm^{-1} shows that the compound does not have a carbonyl group. The absence of absorption at ~2950 cm^{-1} indicates that the compound does not have any hydrogens bonded to sp^3 carbons. Therefore, the compound is **phenol**.

60.

$CH_3CH(OH)CH_2CH_2CH_3$ (**A**) $\xrightarrow{\text{oxidation}}$ $CH_3C(=O)CH_2CH_2CH_3$ (**B**)

A $\xrightarrow{H_2SO_4,\ \Delta}$ $CH_2{=}CHCH_2CH_2CH_3$ (**C**) + $CH_3CH{=}CHCH_2CH_3$ (**D**)

C $\xrightarrow{1.\ O_3,\ -78\,°C;\ 2.\ (CH_3)_2S}$ $HC(=O)H$ + $HC(=O)CH_2CH_2CH_3$ (**E**)

D $\xrightarrow{1.\ O_3,\ -78\,°C;\ 2.\ (CH_3)_2S}$ $CH_3CH_2C(=O)H$ (**F**) + $CH_3C(=O)H$ (**G**)

61. **a.** The absorption band at ~2100 cm^{-1} indicates a carbon–carbon triple bond, and the absorption band at ~3300 cm^{-1} indicates a hydrogen bonded to an *sp* carbon.

$$CH_3CH_2CH_2CH_2C{\equiv}CH$$

b. The absence of an absorption band at ~2700 cm^{-1} indicates that the compound is not an aldehyde, and the absence of a broad absorption band in the vicinity of 3000 cm^{-1} indicates that the compound is not a carboxylic acid. The ester and the ketone can be distinguished by the absorption band at ~1200 cm^{-1} that indicates the carbon–oxygen single bond of an ester.

$$CH_3CH_2C(=O)OCH_2CH_3$$

c. The absorption band at ~1360 cm^{-1} indicates the presence of a methyl group.

$$C_6H_5C(CH_3)_3$$

62. **a.** The broad absorption band at ~3300 cm^{-1} is characteristic of the oxygen–hydrogen stretch of an alcohol; the absence of absorption bands at ~1600 cm^{-1} and ~3100 cm^{-1} indicates that it is not the alcohol with a carbon–carbon double bond.

$$CH_3CH_2CH_2CH_2OH$$

b. The absorption band at ~1685 cm^{-1} indicates a carbon–oxygen double bond. The absence of a strong and broad absorption band at ~3000 cm^{-1} rules out the carboxylic acid, and the absence of an absorption band at ~2700 cm^{-1} rules out the aldehyde. Therefore, it must be one of the ketones. The ketone with the conjugated carbonyl group would be expected to show a C=O stretch at ~1685 cm^{-1}, whereas the ketone with the isolated carbonyl group would show a C=O stretch at ~1720 cm^{-1}. Thus, the compound is the ketone with the conjugated carbonyl group.

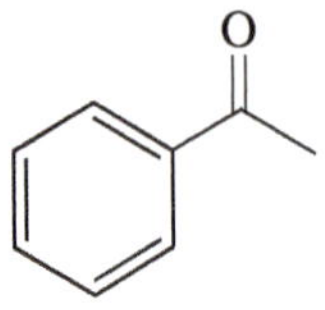

c. The absorption band at ~1700 cm^{-1} indicates a carbon–oxygen double bond. The absence of an absorption band at ~1600 cm^{-1} rules out the ketones with the benzene or cyclohexene rings. The absence of absorption bands at ~2100 cm^{-1} and ~3300 cm^{-1} rules out the ketone with the carbon–carbon triple bond. Therefore, it must be 4-ethylcyclohexanone.

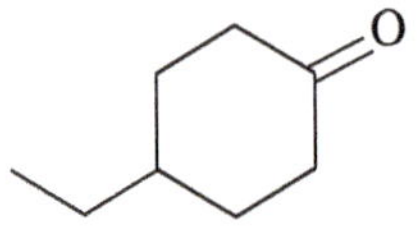

63. The absorption bands at ~2700 cm^{-1} for the aldehyde hydrogen and at ~1380 cm^{-1} for the methyl group will distinguish the compounds.

A will have the band at ~2700 cm^{-1} but not the one at ~1380 cm^{-1}.

B will have the band at ~1380 cm^{-1} but not the one at ~2700 cm^{-1}.

C will have the band at both ~ 2700 cm^{-1} and ~ 1380 cm^{-1}.

64. 1-Hexyne will show absorption bands at ~3300 cm^{-1} for a hydrogen bonded to an *sp* carbon and at ~2100 cm^{-1} for the triple bond.

$CH_3CH_2CH_2CH_2C{\equiv}CH$

1-hexyne

2-Hexyne will show the absorption band at ~2100 cm^{-1} but not the one at ~3300 cm^{-1}.

$CH_3CH_2CH_2C{\equiv}CCH_3$

2-hexyne

3-Hexyne will not show the absorption band at either ~3300 cm^{-1} or ~2100 cm^{-1} (there is no change in dipole moment when the C≡C stretches).

$CH_2CH_2C{\equiv}CCH_2CH_3$

3-hexyne

65. Dividing 116 by 13 gives 8 with 12 left over, giving a base value of C_8H_{20}. Because the compound contains two oxygens, two Os must be added to the base value and two Cs and eight Hs must be subtracted. Therefore, the molecular formula is $C_6H_{12}O_2$. Some possible structures are:

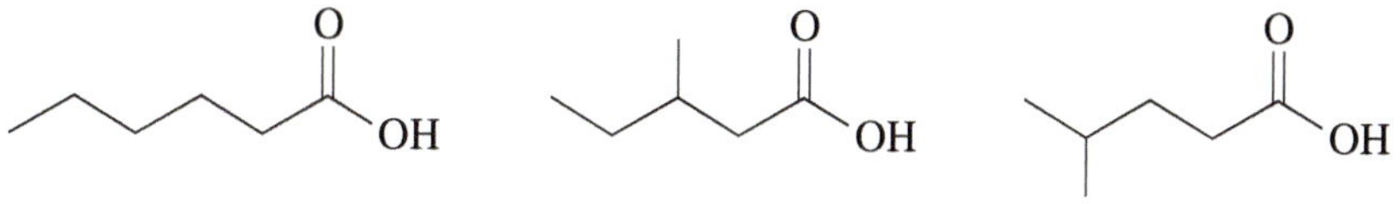

O OH O OH O OH

66. The concentration of benzene can be determined using the Beer–Lambert law, because only benzene absorbs light of 260 nm and the length of the light path of the cell (1.0 cm) and the molar absorptivity of benzene at 260 nm (ε) are known.

$$\text{Beer–Lambert law: } A = c\,l\,\varepsilon$$

$$\text{Therefore, the concentration of benzene} = \frac{A}{l\varepsilon}$$

67.

a. O=C–H 2700 cm^{-1}

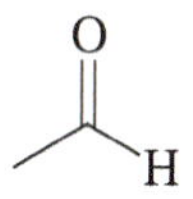

C=O 1700 cm^{-1}

C=C 1600 cm^{-1}

b. C=O 1700 cm^{-1}

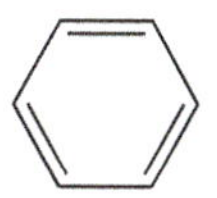

benzene ring 1500 cm^{-1}, 1600 cm^{-1}

C—O ~1250 cm^{-1}, ~1050 cm^{-1}

c. C—O ~1050 cm^{-1}

O—H 3600–3200 cm^{-1} (broad)

d. C=O 1700 cm^{-1}

C—N ~1030 cm^{-1}

N—H 3500–3300 cm^{-1}

e. *sp* CH 3300 cm^{-1} (narrow)

C≡C 2100 cm^{-1}

f. C=O 1700 cm^{-1}

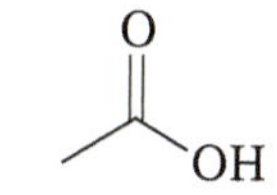

O OH ~3000 cm^{-1} (broad)

C—O ~1250 cm^{-1}

68. Calculating the term in Hooke's law that depends on the masses of the atoms joined by a bond for a C—H bond and for a C—C bond shows why stretching vibrations for smaller atoms occur at larger wavenumbers.

for a carbon–hydrogen bond

$$\frac{m_1 + m_2}{m_1 \times m_2} = \frac{12 + 1}{12 \times 1} = \frac{13}{12} = 1.08$$

for a carbon–hydrogen bond

$$\frac{m_1 + m_2}{m_1 \times m_2} = \frac{12 + 12}{12 \times 12} = \frac{24}{144} = 0.17$$

69. The broad absorption band at ~3300 cm^{-1} indicates that the compound has an OH group. The absorption bands at ~2900 cm^{-1} indicate that the compound has hydrogens attached to an sp^3 carbon. The compound, therefore, is **benzyl alcohol**.

70. In an acidic solution, the three benzene rings are isolated from one another, so phenolphthalein is colorless. In a basic solution, loss of the proton from one of the OH groups causes the five-membered ring to open. As a result, the number of conjugated double bonds increases, causing the solution to become colored.

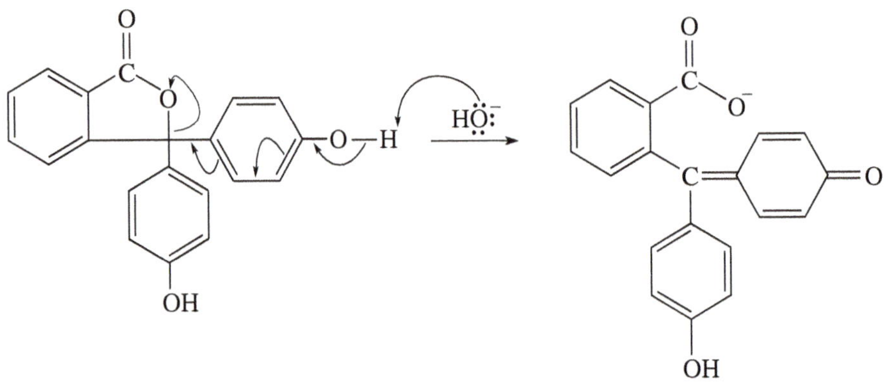

71. **a.** The absorption bands at 1720 cm^{-1} and ~2700 cm^{-1} (C—H of an aldehyde) indicate that the compound is an aldehyde. The absence of an absorption band at ~1600 cm^{-1} rules out the aldehyde with the benzene ring. Therefore, it must be the other aldehyde.

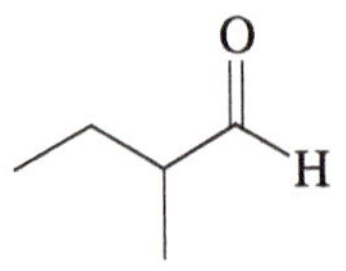

b. The absorption bands at ~3350 cm^{-1} and ~3200 cm^{-1} indicate that the compound is an amide (nitrogen–hydrogen stretch). The absence of an absorption band at ~3050 cm^{-1} indicates that the compound does not have hydrogens bonded to sp^2 carbons. Therefore, it is not the amide that has a benzene ring. Thus, it must be the other amide.

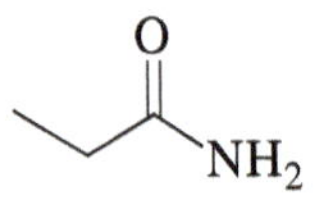

c. The absence of absorption bands at ~1600 cm^{-1} and ~1500 cm^{-1} indicates that the compound does not have a benzene ring. Therefore, it must be the ketone. This is confirmed by the absence of an absorption band at ~1380 cm^{-1}, indicating that the compound does not have a methyl group.

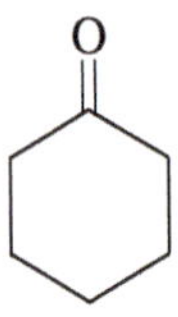

72. **a.**

and

The λ_{max} will be at a longer wavelength, because there are three conjugated double bonds.

b. OH and OCH_3

The λ_{max} of the phenolate ion is at a longer wavelength than the λ_{max} of phenol. Because the pK_a of phenol is ~10, the λ_{max} will be at a longer wavelength at pH = 11 than at pH = 7.

The λ_{max} is pH-independent, so it will be the same at pH 7 and 11.

c. O and O

The λ_{max} will be at a longer wavelength because the carbonyl group is conjugated with the benzene ring.

73.

$$\tilde{\nu} = \frac{1}{2\pi c}\left[\frac{f(m_1 + m_2)}{m_1 m_2}\right]^{1/2}$$

$$\tilde{\nu} = \frac{1}{2 \times 3.1416 \times 3 \times 10^{10}\ \text{cm s}^{-1}}\sqrt{\frac{10 \times 10^5\ \text{g s}^{-2}\left(\frac{12}{6.02} + \frac{12}{6.02}\right) \times 10^{-23}\ \text{g}}{\frac{12}{6.02} \times 10^{-23}\ \text{g} \times \frac{12}{6.02} \times 10^{-23}\ \text{g}}}$$

$$\tilde{\nu} = \frac{1}{18.85 \times 10^{10}}\sqrt{10.0 \times 10^{28}}$$

$$\tilde{\nu} = \frac{1}{18.85 \times 10^{10}} \times 3.16 \times 10^{14}$$

$$\tilde{\nu} = 1676\ \text{cm}^{-1}$$

74. **a.** The IR spectrum indicates that the compound is an aliphatic ketone with at least one methyl group. The M peak at $m/z = 100$ indicates that the ketone is a hexanone. The peak at 43 (100 − 57) for loss of a butyl radical and the peak at 85 for loss of a methyl radical (100 − 15) suggest that the compound is **2-hexanone**.

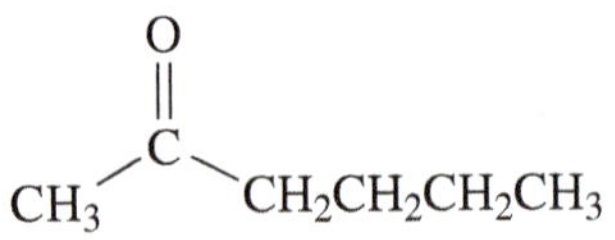

$$CH_3C(=O)CH_2CH_2CH_2CH_3$$

This is confirmed by the peak at 58 for loss of propene (100 − 42) as a result of a McLafferty rearrangement.

$$CH_3\overset{\overset{+\bullet}{:O}}{\overset{\|}{C}}CH_2{-}CH_2{-}\overset{H}{CH}CH_3 \xrightarrow[\text{rearrangement}]{\text{McLafferty}} CH_3\overset{\overset{+}{:OH}}{\overset{\|}{C}}\dot{C}H_2 + CH_2{=}CHCH_3$$

$m/z = 58$

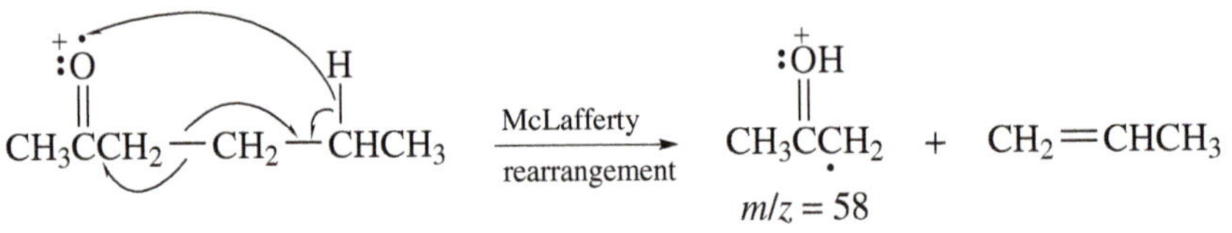

b. The equal heights of the M and M+2 peaks at 162 and 164 indicate that the compound contains bromine. The peak at $m/z = 83$ (162 − 79) is for the carbocation that is formed when the bromine atom is eliminated. The IR spectrum does not indicate the presence of any functional groups, and it shows that no methyl groups are present. The m/z peak $= 83$ indicates a carbocation with a formula of C_6H_{11}. The fact that the compound does not contain a methyl group indicates that the compound is **bromocyclohexane**.

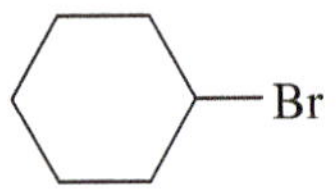

$$C_6H_{11}{-}Br$$

bromocyclohexane

c. The absorption bands at ~1700 cm^{-1} and ~2700 cm^{-1} indicate that the compound is an aldehyde. The molecular ion peak at $m/z = 72$ indicates that the aldehyde contains four carbons (C_4H_8O). The peak at $m/z = 44$ for loss of a group with molecular weight 28 indicates that ethene has been lost as a result of a McLafferty rearrangement.

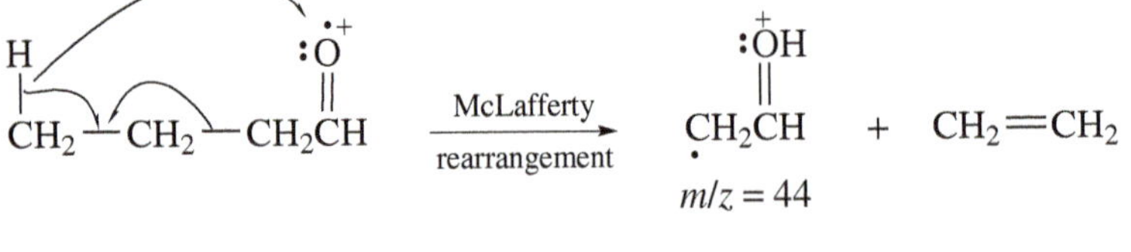

$$\overset{H}{CH_2}{-}CH_2{-}CH_2\overset{\overset{\bullet +}{:O}}{\overset{\|}{C}}H \xrightarrow[\text{rearrangement}]{\text{McLafferty}} \dot{C}H_2\overset{\overset{+}{:OH}}{\overset{\|}{C}}H + CH_2{=}CH_2$$

$m/z = 44$

A McLafferty rearrangement can occur only if the aldehyde has a γ-hydrogen. The only four-carbon aldehyde that has a γ-hydrogen is **butanal**.

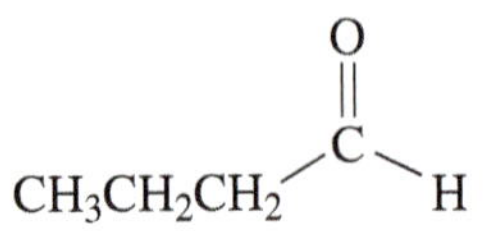

$$CH_3CH_2CH_2C(=O)H$$

Chapter 13 Practice Test

1. Give one IR absorption band that can be used to distinguish the following pairs of compounds. Indicate the compound for which the band would be present.

a. 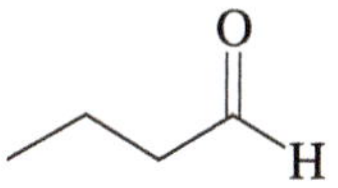 O, H and O

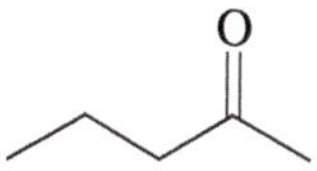

b. 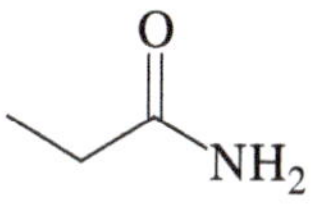 O, NH_2 and O, OCH_3

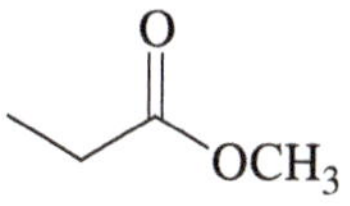

c. $CH_3CH_2CH_2CH_2OH$ and $CH_3CH_2CH_2OCH_3$

d. 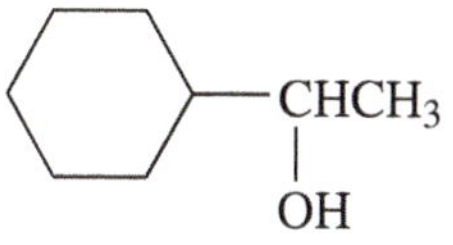 $CHCH_3$, OH and CH_2CH_2OH

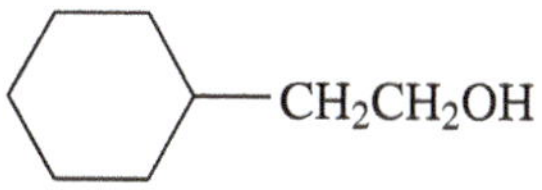

e. 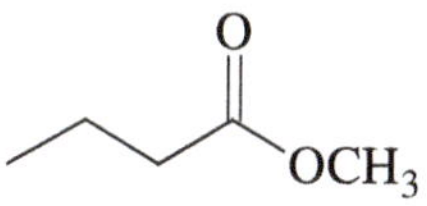 O, OCH_3 and O

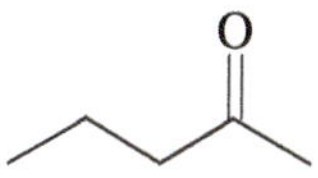

f. $CH_3CH_2CH{=}CHCH_3$ and $CH_3CH_2C{\equiv}CCH_3$

g. $CH_3CH_2C{\equiv}CH$ and $CH_3CH_2C{\equiv}CCH_3$

2. Indicate whether each of the following statements is true or false:

a. The O—H stretch of a concentrated solution of an alcohol occurs at a higher frequency than the O—H stretch of a dilute solution. T F

b. Light of 280 nm is of higher energy than light of 320 nm T F

c. It takes more energy for a bending vibration than for a stretching vibration. T F

d. Propyne will not have an absorption band at 3100 cm^{-1} because there is no change in the dipole moment. T F

e. The M+2 peak of an alkyl chloride is half the height of the M peak. T F

3. The major peaks shown in the mass spectrum of a tertiary alcohol are at m/z = 73, 87, 98, and 101. Identify the alcohol.

4. A 3.8×10^{-4} M solution of cyclohexanone shows an absorbance of 0.75 at 280 nm in a 1.00 cm cell. What is the molar absorptivity of cyclohexanone at 280 nm?

5. How can you distinguish between the IR spectra of the following compounds?

a.

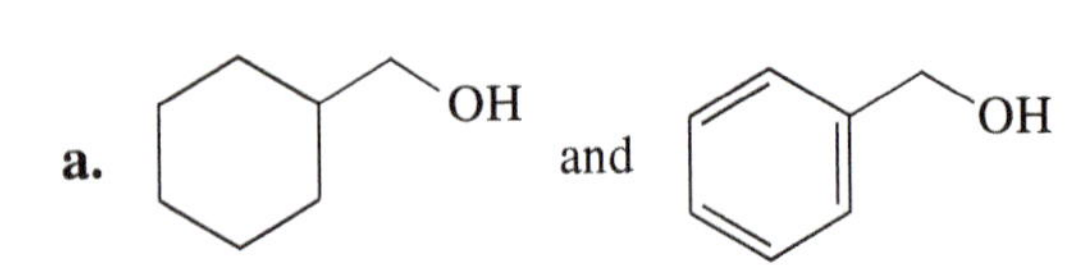

d.

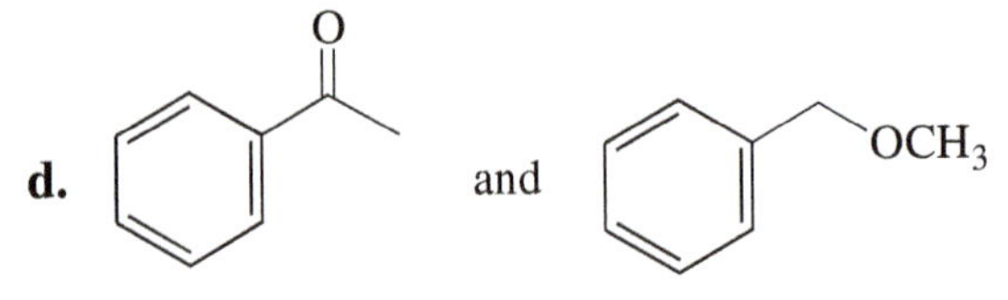

b.

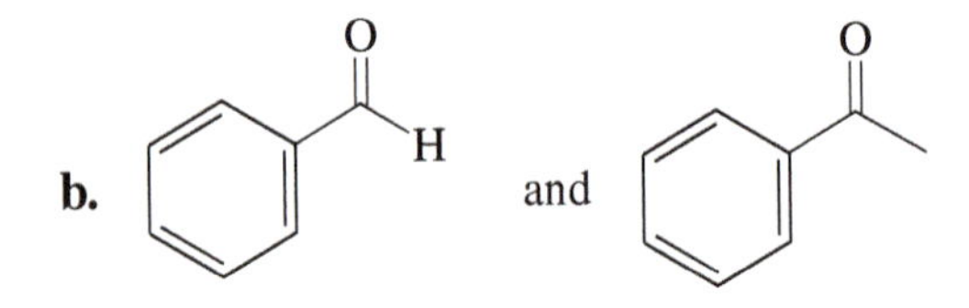

e.

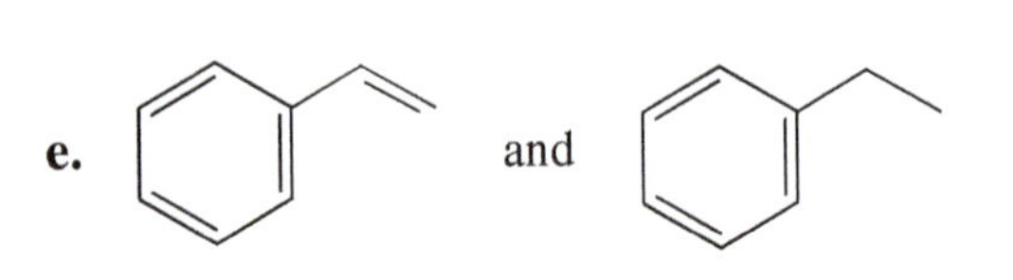

c. O OH and O OCH_3

6. Which compound has the greater λ_{max}? (300 nm is a greater λ_{max} than 250 nm.)

a.

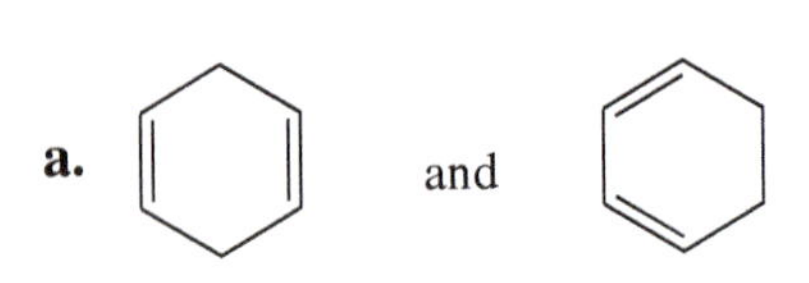

c.

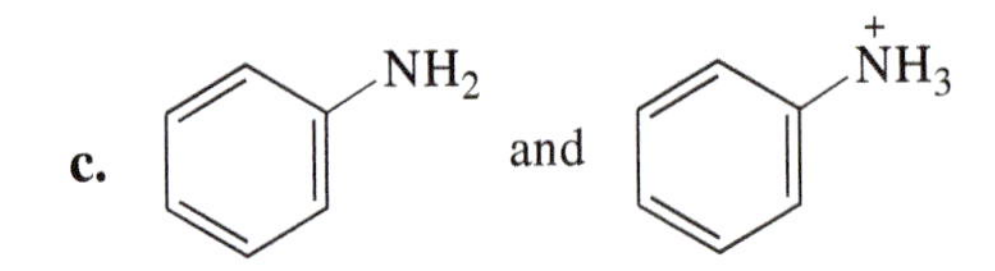

b. H N and NH_2

7. A solution of a compound with a molar absorptivity of 1200 $M^{-1}cm^{-1}$ at 297 nm gives an absorbance of 0.76 at that wavelength in a 1.0 cm quartz cell. What is the concentration of the solution?

8 Draw possible structures for an alcohol that has a molecular ion with an m/z value of 60.

9. A bond between a carbon and an atom of similar electronegativity breaks ____________, whereas a bond between a carbon and a more electronegative atom breaks ____________.

CHAPTER 14
NMR Spectroscopy

Important Terms

applied magnetic field — the externally applied magnetic field.

chemically equivalent protons — protons with the same connectivity relationship to the rest of the molecule.

chemical shift — location of a signal occurring in an NMR spectrum. It is measured downfield from a reference compound (most often TMS).

^{13}C NMR — nuclear magnetic resonance from carbon (^{13}C) nuclei.

COSY spectrum — a 2-D NMR spectrum showing ^{1}H-^{1}H correlations.

coupled protons — protons that split each other's signals. Coupled protons have the same coupling constant.

coupling constant — the distance (in hertz) between two adjacent peaks of a split NMR signal.

DEPT ^{13}C NMR spectrum — a group of four ^{13}C NMR spectra that distinguish CH_3, CH_2, and CH groups.

diamagnetic anisotropy — the term used to describe the greater freedom of π electrons to move in response to a magnetic field as a consequence of their greater polarizability compared with σ electrons.

diamagnetic shielding — shielding by the local magnetic field that opposes the applied magnetic field.

diastereotopic hydrogens — two hydrogens bonded to the same carbon that will result in a pair of diastereomers when each of them is replaced in turn by deuterium.

2-D NMR — two-dimensional nuclear magnetic resonance.

doublet — an NMR signal that is split into two peaks.

doublet of doublets — an NMR signal that is split into four peaks of approximately equal height. A doublet of doublets is caused by splitting a signal into a doublet by one hydrogen and into another doublet by another (nonequivalent) hydrogen.

downfield — farther to the left-hand side of the spectrum.

effective magnetic field — the magnetic field that a nucleus "senses" through the surrounding cloud of electrons.

enantiotopic hydrogens — two hydrogens bonded to a carbon that is bonded to two other groups that are nonidentical.

Fourier transform NMR (FT-NMR) — a technique in which all the nuclei are excited simultaneously by an rf pulse, their relaxation monitored, and the data mathematically converted to a frequency spectrum.

geminal coupling the mutual splitting of two nonidentical protons bonded to the same carbon.

gyromagnetic ratio the ratio of the magnetic moment of a rotating charged particle to its angular momentum.

HETCOR spectrum a 2-D NMR spectrum showing ^{13}C-^{1}H correlations.

high-resolution NMR spectroscopy NMR spectroscopy that uses a spectrometer with a high operating frequency.

^{1}H NMR nuclear magnetic resonance from hydrogen nuclei.

long-range coupling splitting by a proton more than three σ bonds away.

magnetic resonance imaging (MRI) NMR used in medicine. The difference in the way water is bound in different tissues produces the signal variation between organs as well as between healthy and diseased states.

methine hydrogen a tertiary hydrogen.

MRI scanner an NMR spectrometer used in medicine for whole-body NMR.

multiplet an NMR signal split by two (or more) nonequivalent sets of protons.

multiplicity the number of peaks in an NMR signal.

$N + 1$ rule a rule that states that an ^{1}H NMR signal for a hydrogen with N equivalent hydrogens bonded to an adjacent carbon is split into $N + 1$ peaks; a proton-coupled ^{13}C NMR signal for a carbon bonded to N hydrogens is split into $N + 1$ peaks.

NMR spectroscopy the absorption of rf radiation by nuclei in an applied magnetic field to determine the structural features of an organic compound. In the case of ^{1}H NMR spectroscopy, it reveals the carbon–hydrogen framework.

operating frequency the frequency at which an NMR spectrometer operates.

prochiral carbon a carbon (bonded to two hydrogens) that will become an asymmetric center if one of the hydrogens is replaced by deuterium.

pro-*R*-hydrogen replacing this hydrogen with deuterium creates an asymmetric center with the *R* configuration.

pro-*S*-hydrogen replacing this hydrogen with deuterium creates an asymmetric center with the *S* configuration.

proton-coupled ^{13}C NMR spectrum a ^{13}C NMR spectrum in which each signal for a carbon is split by the hydrogens bonded to that carbon.

proton exchange the transfer of a proton from one molecule to another.

quartet an NMR signal that is split into four equally spaced peaks with an integral ratio of 1:3:3:1.

reference compound a compound added to the sample whose NMR spectrum is to be taken. The position of the signals in the NMR spectrum are measured from the position of the signal given by the reference compound.

rf radiation radiation in the radiofrequency region of the electromagnetic spectrum.

shielding the electrons around a proton shield the proton from the full effect of the applied magnetic field. The more a proton is shielded, the farther to the right its signal appears in an NMR spectrum.

singlet an unsplit NMR signal.

spin-spin coupling the splitting of a signal in an NMR spectrum described by the $N + 1$ rule.

α-spin state nuclei in this spin state have their magnetic moments oriented in the same direction as the applied magnetic field.

β-spin state nuclei in this spin state have their magnetic moments oriented opposite the direction of the applied magnetic field.

splitting diagram (splitting tree) a diagram that describes the splitting of a set of protons.

triplet an NMR signal that is split into three equally spaced peaks with an integral ratio of 1:2:1.

upfield farther to the right-hand side of the spectrum.

X-ray crystallography a technique used to determine the arrangement of atoms within a crystal.

X-ray diffraction a technique used to obtain images that are used to determine the electron density within a crystal.

Solutions to Problems

1. $$\nu = \frac{\gamma}{2\pi}B_0$$
$$= \frac{26.75 \times 10^7 \text{ rad T}^{-1}\text{ sec}^{-1} \times 1.0\text{ T}}{2(3.1416)\text{ rad}}$$
$$= 43 \times 10^6 \text{ sec}^{-1}$$
$$= 43 \times 10^6 \text{ Hz} = 43 \text{ MHz}$$

2. a. $$\nu = \frac{\gamma}{2\pi}B_0$$
$$B_0 = \frac{\nu \times 2\pi}{\gamma}$$
$$B_0 = \frac{360 \times 10^6 \times 2(3.1416)}{26.75 \times 10^7}$$
$$B_0 = \frac{226.2}{26.75}$$
$$B_0 = 8.46\text{ T}$$

b. $$B_0 = \frac{\nu \times 2\pi}{\gamma}$$
$$B_0 = \frac{500 \times 10^6 \times 2(3.1416)}{26.75 \times 10^7}$$
$$B_0 = \frac{314.2}{26.75}$$
$$B_0 = 11.75\text{ T}$$

From these calculations, you can see that the greater the operating frequency of the instrument (360 MHz versus 500 MHz), the more powerful the magnet (8.46 T versus 11.75 T) required to operate it.

3. 3 5 4 3 2

4. a. 2 b. 1 c. 1 d. 4 e. 1 f. 4 g. 3 h. 3 i. 5 j. 3 k. 4 l. 3 m. 3 n. 2 o. 3

5. **A** would give two signals, **B** would give one signal, and **C** would give three signals.

6. a.
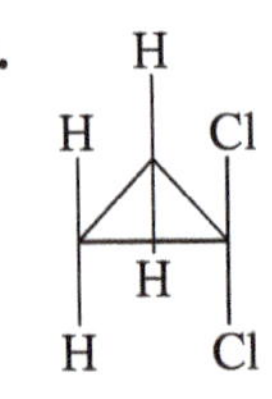

All the Hs are equivalent.

b.
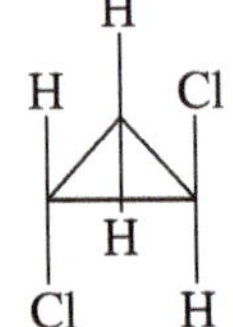

The Hs attached to the front of the molecule are equivalent, and the methylene Hs are equivalent.

c.
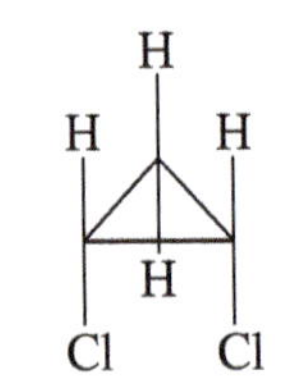

The Hs attached to the front of the molecule are equivalent, and the methylene Hs are not equivalent.

7. $$\text{chemical shift in ppm} = \frac{\text{downfield from TMS in Hertz}}{\text{operating frequency in Megahertz}}$$

a. $$2 \text{ ppm} = \frac{x \text{ Hz}}{300 \text{ MHz}}$$
$$x = 600 \text{ Hz}$$

b. $$2 \text{ ppm} = \frac{x \text{ Hz}}{500 \text{ MHz}}$$
$$x = 1000 \text{ Hz}$$

8. **a.** $$\frac{600 \text{ Hz}}{300 \text{ MHz}} = 2.0 \text{ ppm}$$

b. The answer is still 2.0 ppm because the chemical shift is independent of the operating frequency of the spectrometer.

c. $$\frac{x \text{ Hz}}{500 \text{ MHz}} = 2.0 \text{ ppm}$$
$$x = 1000$$

1000 Hz downfield from TMS

9. **a.** The chemical shift is independent of the operating frequency. Therefore, if the two signals differ by **1.5 ppm** in a 300 MHz spectrometer, they differ by **1.5 ppm** in a 100 MHz spectrometer.

b. $$\frac{\text{Hz}}{\text{MHz}} = \text{ppm}$$
$$\frac{90 \text{ Hz}}{300 \text{ MHz}} = 0.3 \text{ ppm}$$
$$\frac{x \text{ Hz}}{500 \text{ MHz}} = 0.3 \text{ ppm}$$
$$x = 150 \text{ Hz}$$

They differ by 150 Hz.

10. Magnesium is less electronegative than silicon. (See Table 11.1 on page 509 of the text.) Therefore, the peak for $(CH_3)_2Mg$ would be upfield from the TMS peak.

11. **a.** and **b.**

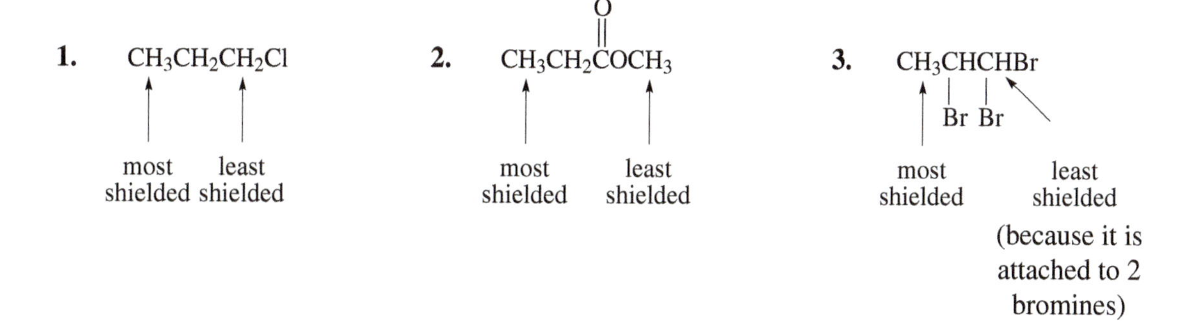

12. Recall that the highest frequency signal is the one that is farthest to the left on the spectrum. The proton(s) that is underlined in the answer gives the higher frequency signal.

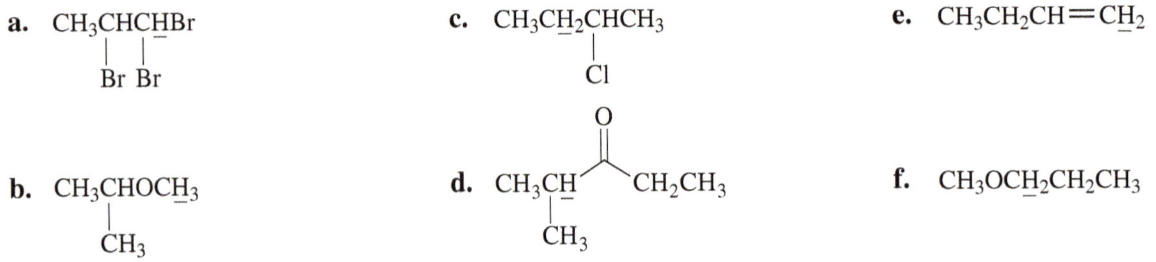

a. $CH_3CHC\underline{H}Br$
with Br Br on the second and third carbons

b. $CH_3CHOC\underline{H_3}$
CH_3

c. $CH_3C\underline{H_2}CHCH_3$
Cl

d. $CH_3C\underline{H}C(=O)CH_2CH_3$
CH_3

e. $CH_3CH_2CH{=}C\underline{H_2}$

f. $CH_3OC\underline{H_2}CH_2CH_3$

13. **a.** $CH_3CH_2C\underline{H_2}Cl$ **b.** $CH_3CH_2C\underline{H}CH_3$ with Cl **c.** $CH_3CH_2C(=O)\underline{H}$

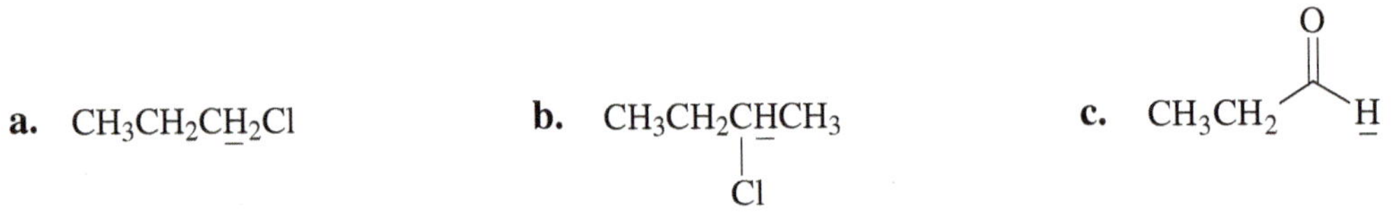

14.

a. $CH_3CH_2C(=O)H$ (a b c)

b. $CH_3CH_2CHCH_3$ (a c e b)
OCH_3 (d)

c. $ClCH_2CH_2CH_2Cl$ (b a b)

d. $CH_3CH_2CHCH_2CH_3$ (a b d b a)
OCH_3 (c)

e. $CH_3CH_2CH_2C(=O)OCH_3$ (a b c d)

f. $CH_3CH_2CH_2OCHCH_3$ (a c d e b)
CH_3 (b)

g. $CH_3CH_2CH_2C(=O)CH_3$ (a b d c)

h. $CH_3CHCH_2OCH_3$ (a b d c)
CH_3 (a)

i. $CH_3CHCHCH_3$ (a c d b), with CH_3 (a) on C-2
Cl

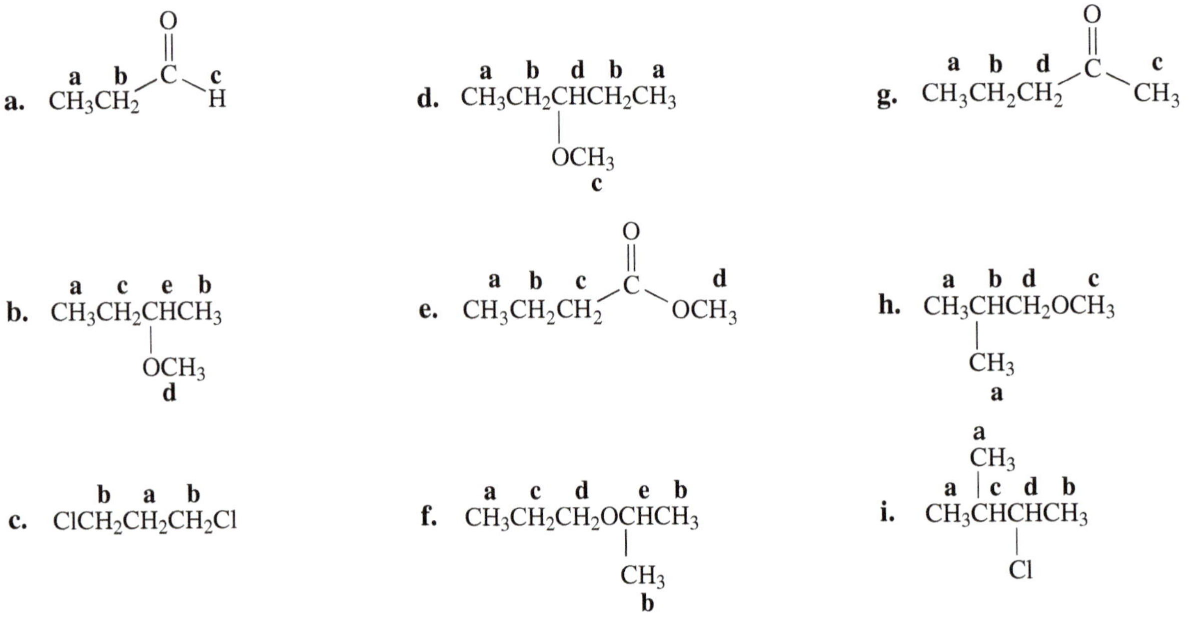

15. From the direction of the electron flow around the benzene ring pictured in Figure 14.6 on page 631 of the text, you can see that the magnetic field induced in the region of the hydrogens that protrude out from the compound in this problem is in the same direction as the applied magnetic field. However, the magnetic field induced in the region of the hydrogens that protrude into the center of the compound is in the opposite direction of the applied magnetic field.

Therefore, the signal at 9.25 ppm is for the hydrogens that protrude out because they need a higher frequency to come into resonance due to the fact that they sense a larger effective magnetic field since the induced and applied magnetic fields are in the same direction. The signal at −2.88 ppm is for the hydrogens that protrude inward because a smaller frequency is necessary since the induced magnetic field and the applied magnetic field are in opposite directions.

16. Each of the compounds would show two signals, but the ratio of the integrals for the two signals will be different for each of the compounds. The ratio of the integrals for the signals given by the first compound will be 2:9 (or 1:4.5), the ratio of the integrals for the signals given by the second compound will be 1:3, and the ratio of the integrals for the signals given by the third compound will be 1:2.

17. Solved in the text.

18. The heights of the integrals for the signals in the spectrum are about 3.5 and 5.2. The ratio of the integrals, therefore, is $5.2/3.5 = 1.5$. This matches the ratio of the integrals calculated for **B**. (Later we will see that a signal at ~7 ppm is characteristic of a benzene ring.)

$HC{\equiv}C$–C_6H_4–$C{\equiv}CH$ $\quad \frac{4}{2} = 2$

CH_3–C_6H_4–CH_3 $\quad \frac{6}{4} = 1.5$

$ClCH_2$–C_6H_4–CH_2Cl $\quad \frac{4}{4} = 1.0$

Br_2CH–C_6H_4–$CHBr_2$ $\quad \frac{4}{2} = 2$

19. The highest frequency signal in both spectra is the signal for the hydrogens bonded to the carbon that is also bonded to the halogen. Because chlorine is more electronegative than iodine, that signal should be at a higher frequency in the ^1H NMR spectrum for 1-chloropropane than in the ^1H NMR spectrum for 1-iodopropane. Therefore, the **first spectrum** is the ^1H NMR spectrum for **1-iodopropane**, and the **second spectrum** is the ^1H NMR spectrum for **1-chloropropane**.

20. **C** is easiest to distinguish because it will have **two** signals, whereas **A** and **B** will each have **three** signals.

A and **B** can be distinguished by looking at the splitting of their signals.

The signals in the ^1H NMR spectrum of **A** will be (left to right across the spectrum): **triplet**, **triplet**, **multiplet**.

The signals in the ^1H NMR spectrum of **B** will be (left to right across the spectrum): **doublet**, **multiplet**, **doublet**.

21. From the molecular formula and the splitting patterns of the signals, the spectra can be identified as the ^1H NMR spectrum of:

a. $CH_3CH(Cl)COOH$ **b.** $ClCH_2CH_2COOH$

22. **a.** A triplet is caused by coupling to two equivalent protons on an adjacent carbon. The two protons can be aligned in three different ways: both with the field, one with the field and one against the field, or both against the field. That is why the signal is a triplet. There is only one way to align two protons that are with the field or two protons that are against the field. However, there are two ways to align two protons if one is with and one is against the field: with and against or against and with. Consequently, the peaks in a triplet have relative intensities of 1:2:1.

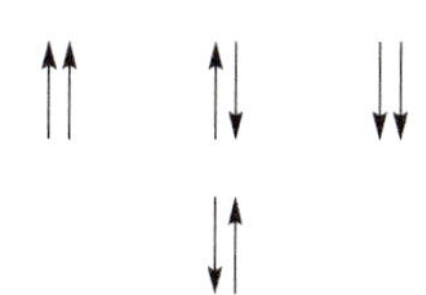

b. A quintet is caused by coupling to four equivalent protons on adjacent carbons. The four protons can be aligned in five different ways. The following possible arrangements for the alignment of four protons explain why the relative intensities of a quintet are 1:4:6:4:1.

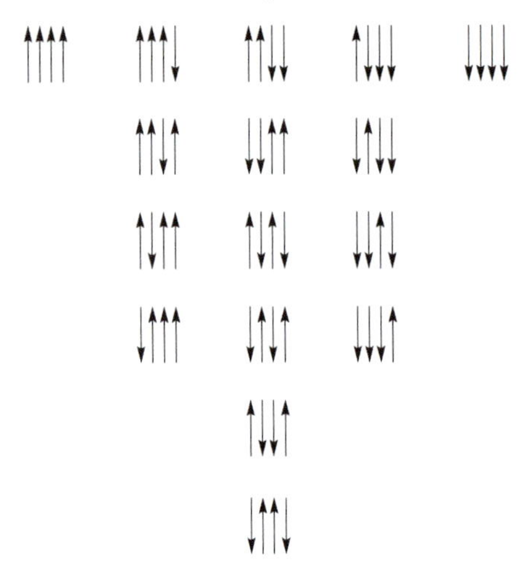

23. **a.** three signals

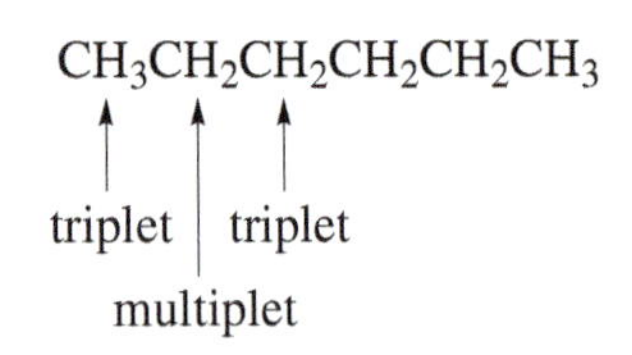

b. three signals

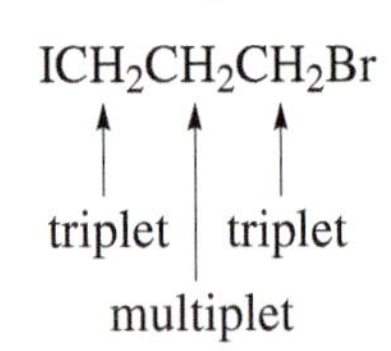

c. two signals

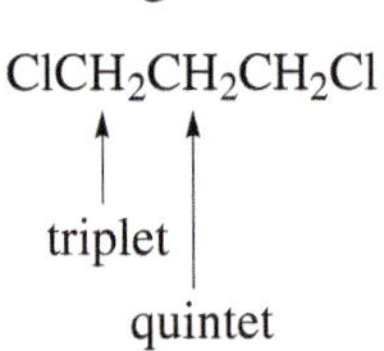

d. three signals

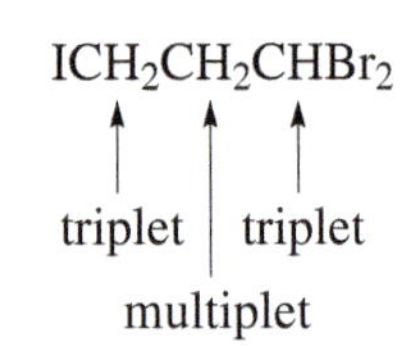

24. The nitro group withdraws electrons inductively and by resonance.

The ***a*** protons are at the lowest frequency because the nitro group withdraws electrons only inductively from the carbons to which they are attached.

The ***b*** and ***c*** protons are at a higher frequency because the nitro group withdraws electrons both inductively and by resonance from the carbons to which they are attached.

The ***c*** protons are at the highest frequency because the nitro group withdraws electrons inductively more strongly from the carbons to which the ***c*** protons are attached since those carbons are closest to the nitro group.

25. Each compound will have two doublets. In addition, **A** will have two doublet of doublets, **B** will have one doublet of doublets and a singlet, and **C** will have no other signals.

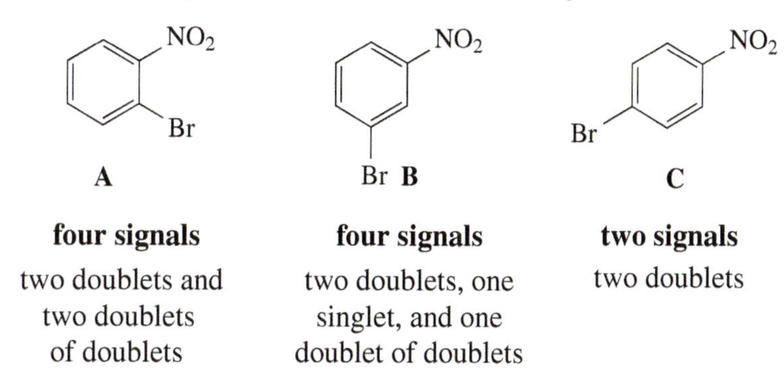

26. **a.** The signal at ~7.2 ppm indicates the presence of a benzene ring. The integrations shows that the ring has a single substituent; thus, the ring has a formula of C_6H_5. Subtracting C_6H_5 from the molecular formula of the compound ($C_9H_{12} - C_6H_5$) gives a substituent with a formula of C_3H_7. The triplet at ~0.9 ppm that integrates to 3 protons indicates a methyl group adjacent to a methylene group. The identical integration of the signals at 1.6 ppm and 2.6 ppm indicates two methylene groups. Thus, the compound is propylbenzene.

C_6H_5—$CH_2CH_2CH_3$

b. The triplet that integrates to 3 protons and the quartet that integrates to 2 protons indicate a CH_3CH_2 group bonded to an atom that is not bonded to any hydrogens. The molecular formula of $C_5H_{10}O$ indicates that the compound is diethyl ketone.

$CH_3CH_2C(=O)CH_2CH_3$

c. The signals between ~7 and 8 ppm that integrate to 5 protons indicate the presence of a monosubstituted benzene ring. Subtracting C_6H_5 from the molecular formula of the compound ($C_9H_{10}O_2 - C_6H_5$) shows that the substituent has a formula of $C_3H_5O_2$.

The triplet that integrates to 3 protons and the quartet that integrates to 2 protons suggest a CH_3CH_2 group bonded to an atom that is not bonded to any hydrogens. Therefore, we can conclude that the compound is one of the following:

C_6H_5—$C(=O)OCH_2CH_3$ (**A**) or C_6H_5—$OC(=O)CH_2CH_3$ (**B**)

The substituent attached to the benzene ring in compound **A** withdraws electrons from the ring, whereas the substituent attached to the benzene ring in compound **B** donates electrons into the benzene ring. That one of the peaks of the benzene ring signal is at a higher frequency (farther downfield) than usual (>8 ppm) suggests that an electron-withdrawing substituent is present on the benzene ring. This is confirmed by the signal for the CH_2 group at 4.3 ppm, indicating that the methylene group is adjacent to an oxygen. Thus, the spectrum is that of compound **A**.

27. s = singlet, d = doublet, t = triplet, q = quartet, quin = quintet, d of d = doublet of doublets, and m = multiplet.

a. 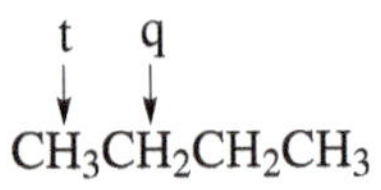 $CH_3CH_2CH_2CH_3$ (t, q)

b. 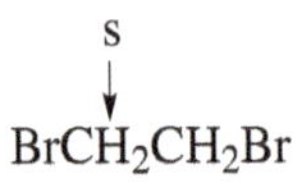 $BrCH_2CH_2Br$ (s)

c. 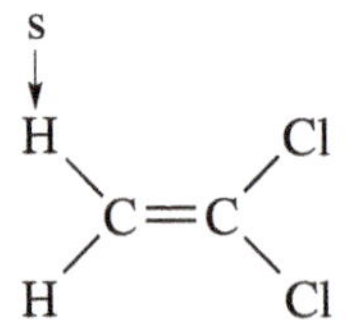 H(s), H / C=C / Cl, Cl

(Remember that equivalent hydrogens do not split each other's signal.)

d. 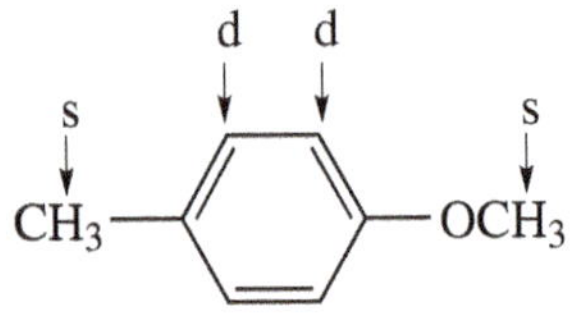 CH_3(s)–C₆H₄(d, d)–OCH_3(s)

e. 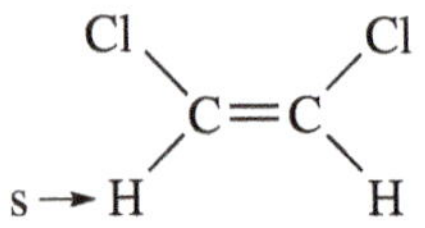 Cl, Cl / C=C / H(s), H

g. 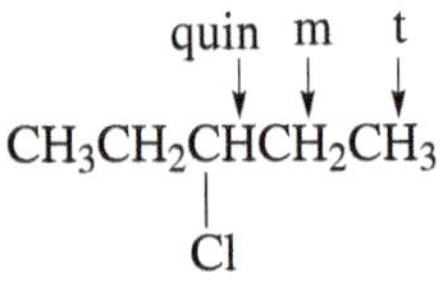 $CH_3CH_2CHCH_2CH_3$ with Cl on C-3 (quin, m, t)

h. $CH_3CHCH_2CHCH_3$ with CH_3, CH_3 (d, m, t)

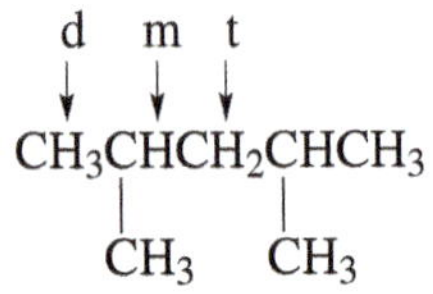

f. 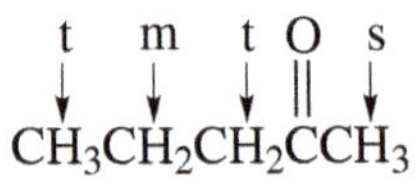 $CH_3CH_2CH_2CCH_3$ with O (t, m, t, s)

i. 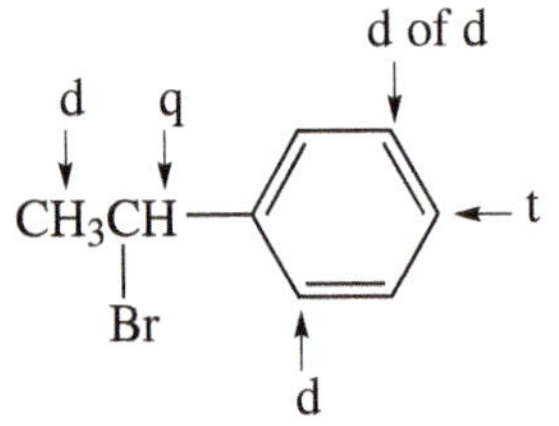 CH_3CH with Br–benzene (d, q; ring: d of d, t, d)

j. 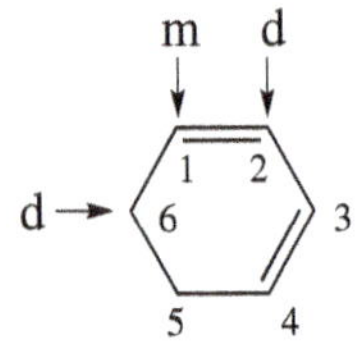 (ring labeled 1–6; m, d at C-1, C-2; d at C-6)

The H on C-2 will not be split by the H on C-3 because they are equivalent. The Hs on C-6 will not be split by the Hs on C-5 because the Hs on C-5 and C-6 are equivalent.

k. 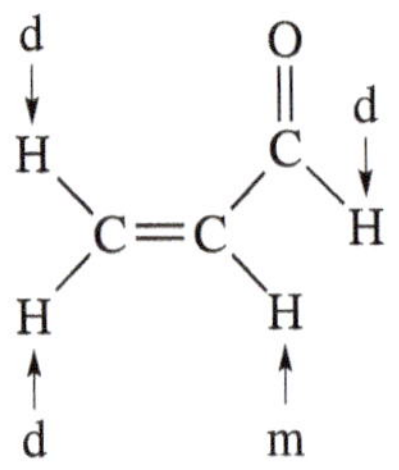 H(d), H(d) / C=C / C(=O)H(d), H(m)

l. 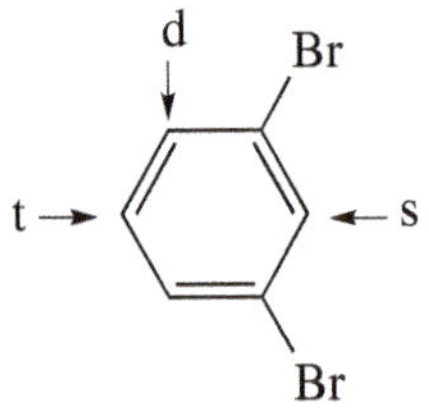 benzene with Br, Br (d, t, s)

m. 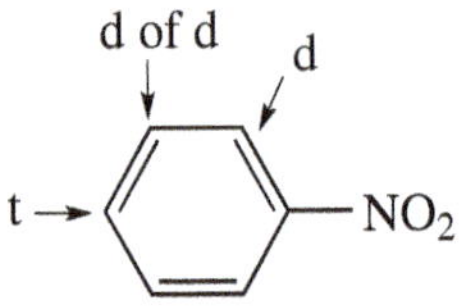 benzene–NO_2 (d of d, d, t)

n. 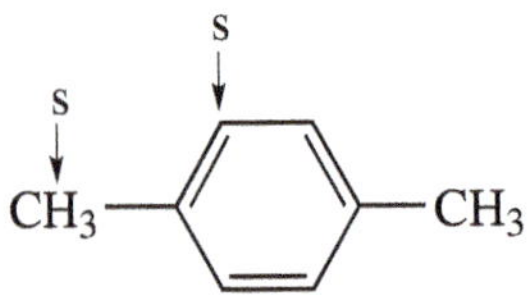 CH_3(s)–C₆H₄(s)–CH_3

o. Cl, CH_3(d) / C=C / H(d), H(d)

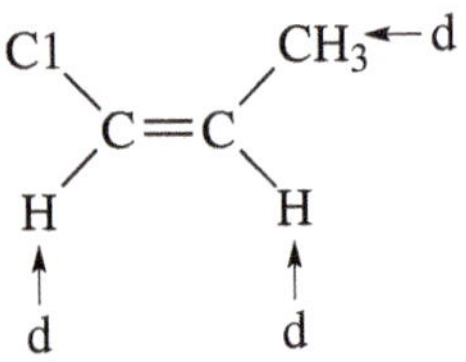

28. Each spectrum is described going from left to right.

a. $BrCH_2CH_2CH_2CH_2Br$

triplet triplet (This triplet will not be split by the adjacent equivalent protons.)

b. two triplets (close to each other) singlet multiplet

(The table "^{1}H NMR Chemical Shifts" in Appendix V indicates that a methylene group adjacent to an RO and a methylene group adjacent to a Br appear at about the same place.)

c. singlet (Equivalent Hs do not split each other's signals.)

d. quartet singlet triplet

e. three singlets

f. three doublets of doublets

g. quartet triplet

h. singlet quartet triplet

i. doublet multiplet doublet

j. triplet quintet

k. doublet doublet of septets doublet

l. triplet singlet quintet

m. singlet

n. singlet (Equivalent Hs do not split each other's signals.)

o. singlet

p. quintet, multiplet, multiplet

29. **a.** $CH_3OCH_2CH_2OCH_3$

b. There are three possibilities:

$CH_3\overset{O}{\overset{\|}{C}}CH_2CH_2\overset{O}{\overset{\|}{C}}CH_3$

$CH_3OCH_2C{\equiv}CCH_2OCH_3$

(a ring with two O atoms: 2,3-dimethyl-1,4-dioxene, ring carbons bearing CH_3 and CH_3)

c. $H\overset{O}{\overset{\|}{C}}-C_6H_4-\overset{O}{\overset{\|}{C}}H$

30. There is no coupling between H_a and H_c because they are separated by four σ bonds. (We will see in Section 14.17 that unless the sample is pure and dry, a proton bonded to an oxygen is not split and does not split other protons.)

There is no coupling between H_b and H_c because they are separated by five σ bonds.

31. The two singlets in the ^1H NMR spectrum that each integrate to 3 protons suggest two methyl groups, one of which is adjacent to an oxygen. That the benzene ring protons (6.7−7.1 ppm) consist of two doublets (each integrating to two hydrogens) indicates a 1,4-disubstituted benzene ring. The IR spectrum also indicates a benzene ring (1600 and 1500 cm^{-1}) with hydrogens bonded to sp^2 carbons and no carbonyl group. The absorption bands in the 1250–1000 cm^{-1} region suggest that there are two C—O single bonds, one with no double-bond character and one with some double-bond character.

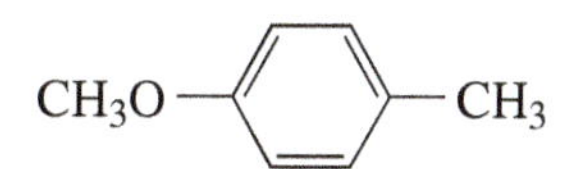

32. **a.**

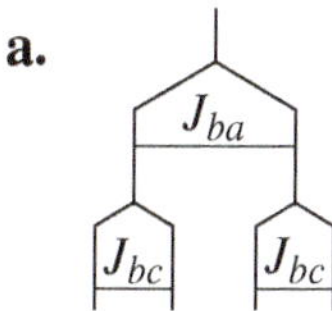

b.

J_{ba}, J_{bc}, J_{bc}

33. **a.** **A** **b.** **B** and **D**

34. Solved in the text.

35.

$CH_3C(Br)_2CH_3$ (a, a)	$BrCH_2CH_2CH_2Br$ (b a b)	CH_3CH_2CHBr (a b c)	$CH_3CH(Br)CH_2Br$ (a d b, c)
one signal	**two signals**	**three signals**	**four signals**
a singlet	a quintet (a) and a triplet (b)	two triplets (a and c) and a multiplet (b)	a doublet (a), two doublet of doublets (b and c), and a multiplet (d)

36. Solved in the text.

37. The OH proton of a carboxylic acid is more deshielded than the OH proton of an alcohol as a result of electron delocalization that decreases the electron density around the OH proton. In addition, the extent of hydrogen bonding affects the chemical shift of a proton bonded to an oxygen. Carboxylic acids exist as tightly hydrogen-bonded dimers (see page 693 in the text); this strongly deshields the OH proton.

$R\overset{O}{\overset{\|}{C}}\ddot{O}-H \longleftrightarrow R\overset{O^-}{\overset{|}{C}}{=}\overset{+}{\ddot{O}}-H$ versus $R-O-H$

38. The greater the extent of hydrogen bonding, the greater the chemical shift. Therefore, the ^{1}H NMR spectrum of pure ethanol would have the signal for the OH proton at a greater chemical shift because it would be hydrogen bonded to a greater extent.

39. $R\ddot{O}-H + {:}\ddot{O}H^- \rightleftharpoons R\ddot{O}{:}^- + H-\ddot{O}H$

40. It is the ^{1}H NMR spectrum of propanamide. Notice that the signals for the N—H protons are unusually broad. Because of the partial double-bond character of the C—N bond, there is no free rotation about the C—N bond, so the two N—H protons are not chemically equivalent. The quartet and triplet are characteristic of an ethyl group.

$CH_3CH_2C(=O)NH_2 \longleftrightarrow CH_3CH_2C(-O^-)=N^+H_2$

41. **a.** **1.** 3 **2.** 3 **3.** 2 **4.** 4 **5.** 3 **6.** 4 **7.** 3 **8.** 3 **9.** 2

b. An arrow is drawn to the carbon that gives the signal at the lowest frequency.

1. $CH_3CH_2CH_2Br$

2. $(H_3C)_2C=CH_2$

3. CH_3CHCH_3 (Br)

4. $CH_3CH_2COOCH_3$

5. $CH_3CH(CH_3)CHO$

6. chlorobenzene

7. $(H_3C)_2C(CH_3)OCH_3$

8. $CH_3COCH_2CH_2COCH_3$

9. $H_2C=CHBr$

42. Each spectrum is described going from left to right:

1. triplet triplet quartet

3. doublet quartet

5. doublet doublet quartet

43.

NO_2 NO_2 NO_2 (ortho-, meta-, and para-dinitrobenzene structures)

a.	^{1}H NMR	two signals	three signals	one signal
b.	^{13}C NMR	three signals	four signals	two signals

44. **a.** The signal at 210 ppm is for the carbonyl carbon of a ketone. There are 10 other carbons in the compound but only 5 other signals. This suggests that the compound is a symmetrical ketone with identical five-carbon alkyl groups.

$$CH_3CH_2CH_2CH_2CH_2C(=O)CH_2CH_2CH_2CH_2CH_3$$

b. Because there are only four signals for the six carbons of the benzene ring (the signals between 110 and 117 ppm), the compound must be a 1,4-disubstituted benzene with two different substituents.

$$Br{-}C_6H_4{-}CH_2CH_3$$

c. The signal at 212 ppm is for the carbonyl carbon of a ketone. There are five other carbons in the compound but only three other signals. This suggests that the compound is a symmetrical cyclic ketone.

(cyclohexanone structure, =O)

d. The molecular formula indicates that the compound has one double bond or one ring. The presence of a signal at 130 ppm indicates the presence of sp^2 carbons; thus, the compound must have a double bond. Each of the two sp^2 carbons must be bonded to the same groups because the six carbons exhibit only three signals. Whether the compound is *cis*-3-hexene or *trans*-3-hexene cannot be determined from the spectrum.

$$CH_3CH_2CH{=}CHCH_2CH_3$$

45. If the triangles shown below are drawn on the spectrum, you will be able to identify the coupled protons.

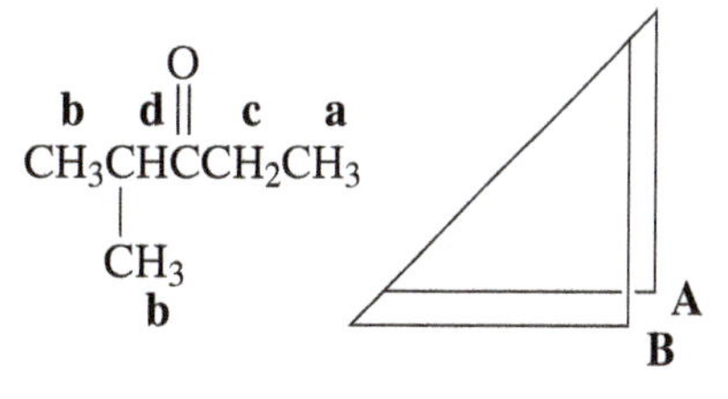

Point **A** shows that the **a** protons are split by the **c** protons.
Point **B** shows that the **b** protons are split by the **d** protons.

46. Cross peak X tells you that the hydrogens that produce the signal at ~1.0 ppm in the ^{1}H NMR spectrum are bonded to the carbon that produces the signal at ~19 ppm in the ^{13}C NMR spectrum.

47. **a.** **1.** 5 **2.** 5 **3.** 4 **4.** 2 **5.** 3 **6.** 3

b. **1.** 7 **2.** 7 **3.** 5 **4.** 2 **5.** 2 **6.** 4

48. The H_b proton will be split into a quartet by the H_a protons and into a doublet by the H_c protons.

a.

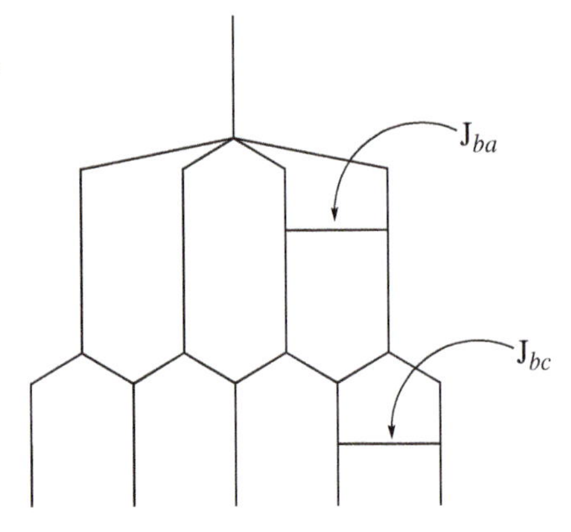

five peaks (a quintet)

b.

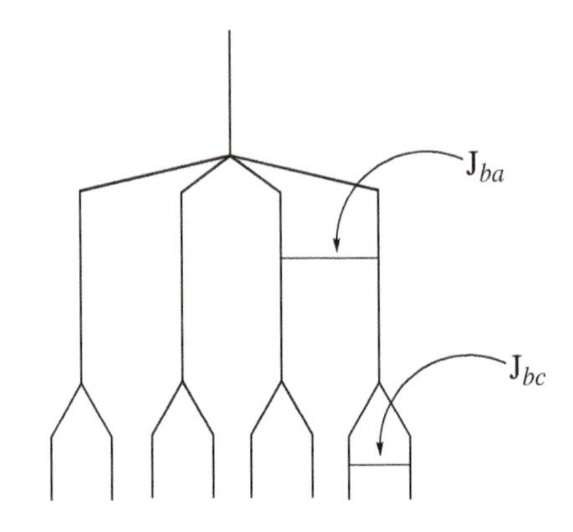

eight peaks (a quartet of doublets)

49. **a.**

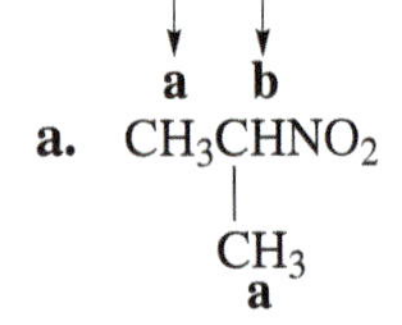

b. $CH_3CH_2CH_2OCH_3$ — a: t; b: m; d: t; c: s

c. $CH_3CH(CH_3)C(=O)CH_2CH_2CH_3$ — a: t; b: d; c: m; d: t; e: septet

d. $CH_3CH_2CH_2C(=O)CH_2Cl$ — a: t; b: m; c: t; d: s

e. $ClCH_2C(CH_3)_2CHCl_2$ — a: s; b: s; c: s

f. $ClCH_2CH_2CH_2CH_2CH_2Cl$ — c: t; b: m; a: quintet

50. By dividing the value of the integration steps by the smallest one, the ratios of the hydrogens are found to be 3:2:1:9.

$$\frac{40.5}{13} = 3.1 \qquad \frac{27}{13} = 2.1 \qquad \frac{13}{13} = 1 \qquad \frac{118}{13} = 9.1$$

Because the ratios are given in the highest frequency to lowest frequency direction, a possible compound is the following ester:

$$(CH_3)_3C{-}CH(CH_3){-}C(=O)OCH_3$$

51. **a.**  four signals and  two signals

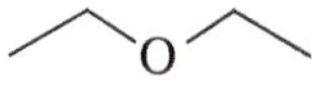

b. Br⁄⁄⁄Br two signals and Br⁄⁄⁄NO_2 three signals

c. 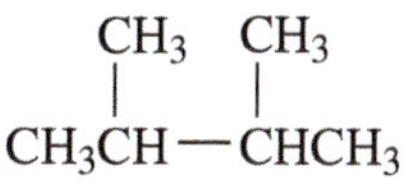 $CH_3CH(CH_3)—CH(CH_3)CH_3$ two signals and $CH_3C(CH_3)_2CH_2CH_3$ three signals

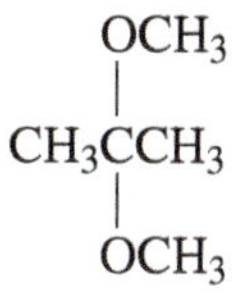

d. 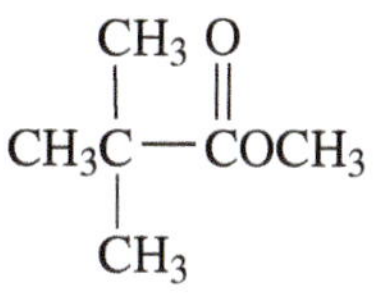 $CH_3C(CH_3)_2—COCH_3$ (C=O) two signals with integration 3:1 and $CH_3C(OCH_3)_2CH_3$ two signals with integration 1:1

e. 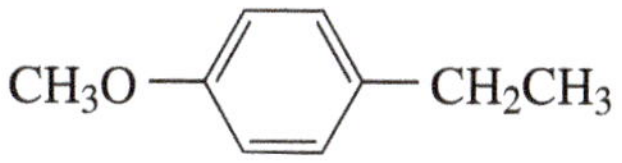 $CH_3O—C_6H_4—CH_2CH_3$ and $CH_3—C_6H_4—OCH_2CH_3$

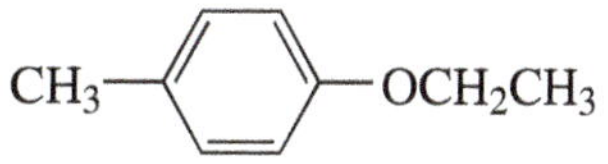

The singlet will be at a higher frequency than the quartet.

The quartet will be at a higher frequency than the singlet.

f. 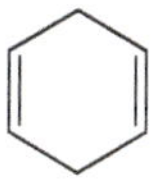 and

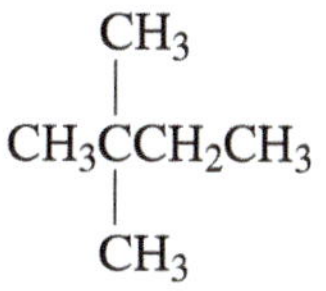

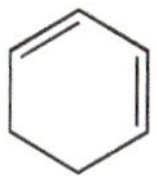

two signals three signals

g. 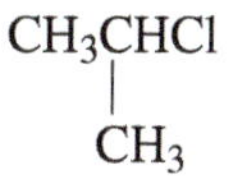 $CH_3CHCl(CH_3)$ and $CH_3CDCl(CH_3)$

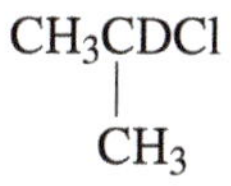

two signals one signal

h. 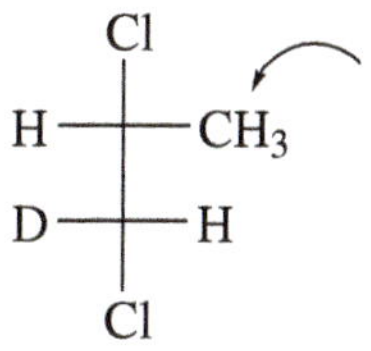 and

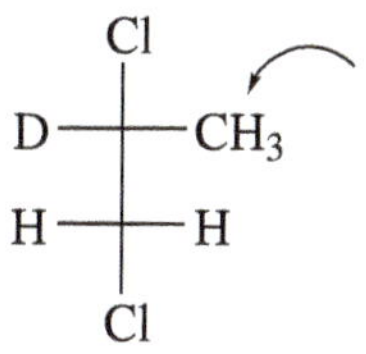

The lowest frequency signal will be a doublet. The lowest frequency signal will be a singlet.

i. 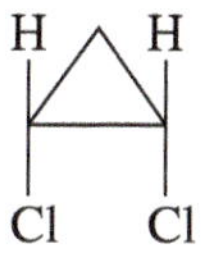 and

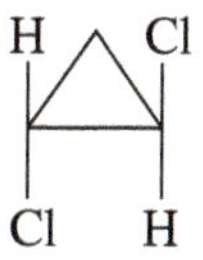

three signals two signals

j. CH_3–(benzene ring)–$C(CH_3)_2CH_3$ and (benzene ring)–$CH_2C(CH_3)_2CH_3$

the signals for the benzene ring protons plus two signals with integration 3:1

the signals for the benzene ring protons plus two signals with integration 4.5:1

52. **a.** Chemical shift in ppm is independent of the operating frequency.
b. Chemical shift in hertz is proportional to the operating frequency.
c. The coupling constant in hertz is independent of the operating frequency.
d. The frequency required for NMR spectroscopy is lower than that required for IR spectroscopy, which is lower than that required for UV/Vis spectroscopy.

53. **a.** The spectrum must be that of **2-bromopropane**, because the NMR spectrum has two signals—the lowest frequency signal is a doublet, and the other signal is given by a single hydrogen.

$CH_3CH(CH_3)Br$

b. The spectrum must be that of **1-nitropropane**, because the NMR spectrum has three signals—both the lowest frequency and highest frequency signals are triplets.

$CH_3CH_2CH_2NO_2$

c. The spectrum must be that of **ethyl methyl ketone**, because the NMR spectrum has three signals—a triplet, a singlet, and a quartet.

$CH_3CH_2C(=O)CH_3$

54. **a.** $CH_3CH_2CH(CH_3)Br$ **b.** $CH_3CH_2CH_2CH_2Br$ **c.** $CH_3CH(CH_3)CH_2Br$

55. **a.** $CH_3C(CH_3)(Br)CH_2Br$ **b.** $C_6H_5CH(Br)CH_3$ **c.** $CH_3CH_2C(=O)OCH_2CH_3$

56. The singlet at 210 ppm indicates the carbonyl group of a ketone. The splitting of the other two signals indicates an isopropyl group. The molecular formula indicates that the compound must have two isopropyl groups.

$CH_3CH(CH_3)C(=O)CH(CH_3)CH_3$

57.

$CH_3C(CH_3)(Cl)CH_3$ — *tert*-butyl chloride, Compound **A**

$CH_3CH(Cl)CH_2CH_3$ — *sec*-butyl chloride, Compound **B**

58.

$CH_3CH_2CH{=}CH_2$ — 1-butene

$H_3C(H)C{=}C(H)CH_3$ — *cis*-2-butene

$CH_3C(CH_3){=}CH_2$ — 2-methylpropene

It would be better to use ^{13}C NMR because you would have to look only at the number of signals in each spectrum: 1-butene will show four signals, *cis*-2-butene will show two signals, and 2-methylpropene will show three signals. (In the ^1H NMR spectrum, 1-butene will show five signals, and *cis*-2-butene and 2-methylpropene will both show two signals.)

59.

A	**B**	**C**	**D**
$CH_3CH_2C(=O)OCH_3$	$CH_3C(=O)OCH_2CH_3$	$HC(=O)OCH_2CH_2CH_3$	$HC(=O)OCH(CH_3)CH_3$
three signals	**three signals**	**four signals**	**three signals**
singlet, quartet, triplet (singlet farthest downfield)	singlet, quartet, triplet (quartet farthest downfield)	singlet, triplet multiplet, triplet	singlet, septet doublet

C can be distinguished from **A, B,** and **D** because **C** has four signals and the others have three signals.

D can be distinguished from **A** and **B** because the three signals of **D** are a singlet, a septet, and a doublet, whereas the three signals of **A** and **B** are a singlet, a quartet, and a triplet.

A and **B** can be distinguished because the highest frequency signal in **A** is a singlet, whereas the highest frequency signal in **B** is a quartet.

60. It is the ^1H NMR spectrum of 2,3-dimethylbutane.

$CH_3CH(CH)CH(CH_3)CH_3$

61. It is the ^1H NMR spectrum of *tert*-butyl methyl ether.

$$CH_3Br + CH_3C(CH_3)_2{-}O^- \longrightarrow CH_3C(CH_3)_2{-}OCH_3$$

tert-butyl methyl ether

62. **a.** $CH_3C(=O)CH_2C(CH_3)_2CH_3$ **b.** $CH_3CH(CH_3)C(=O)CH_2CH_2CH_3$ **c.** $CH_3CH(CH_3)C(=O)CH(CH_3)CH_3$

63. **a.** $CH_3CH_2CH_2CH_2OCH_3$ **c.** $CH_3CH(Cl)C(=O)OCH_3$

b. $CH_3C(CH_3)_2C(=O)OCH_3$ **d.** $CH_3CH_2CH_2C(=O)OH$

64. **a.** 19, 19, 30, 70, OH **b.** 24, 26, 36, 71, OH

The numbers indicate the ppm value of the signal given by that carbon.

65. The broad signal at ~2.9 ppm is for the H_a proton that is bonded to the oxygen. The signal for the H_b protons at ~4.2 ppm is split into a doublet by the H_e proton. The H_c proton and the H_d proton are each split by the H_e proton; the coupling constant is greater for the trans protons. Thus, the H_c proton is at ~5.1 ppm and the H_d proton is at ~5.3 ppm. The H_e proton at ~6.0 ppm is split by the H_b protons and by the H_c proton and the H_d proton.

c: H, e: H, C=C, d: H, CH_2OH (b a)

66. Each compound will have two doublets in the 6.5–8.1 ppm region. In addition, in that region, **A** will have two doublet of doublets, **B** will have one doublet of doublets and a singlet, and **C** will have no other signals.

OCH_3, NO_2 — **A**

OCH_3, NO_2 — **B**

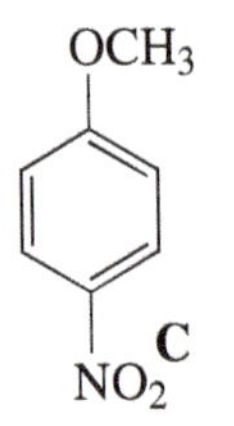

In the 6.5–8.1 ppm region:	**four signals** two doublets and two doublets of doublets	**four signals** one singlet, two doublets, and one doublet of doublets	**two signals** two doublets

67. The signals at ~7.2 ppm indicate that the compounds whose spectra are shown in parts **a** and **b** contain a benzene ring. From the molecular formula, you know that each compound has five additional carbons.

a. The integration of the benzene ring protons indicate that the ring is monosubstituted. The hydrogens bonded to the five additional carbons in part **a** must all be accounted for by two singlets with integral ratios of ~2:9, indicating that the compound is **2,2-dimethyl-1-phenylpropane**.

2,2-dimethyl-1-phenylpropane

b. The integration of the benzene ring protons indicate that the ring is 1,4-disubstituted. The two singlets in the spectrum in part **b** have integral ratios of ~1:3, indicating a methyl substituent and a *tert*-butyl substituent. Therefore, the compound is **1-*tert*-butyl-4-methylbenzene**.

1-*tert*-butyl-4-methylbenzene

68.

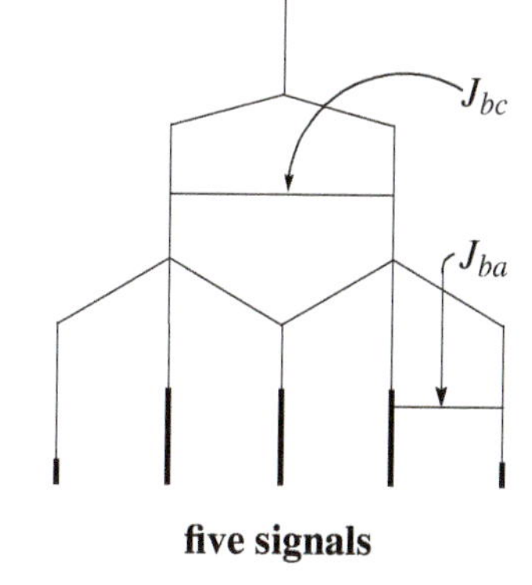

69. Oxygen is more electronegative than chlorine. (See Table 1.3 on page 10 of the text.)

a.

b a c
$ClCH_2CH_2OH$

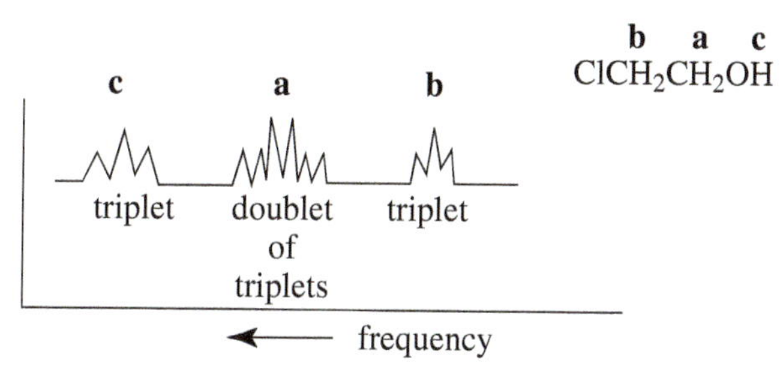

b.

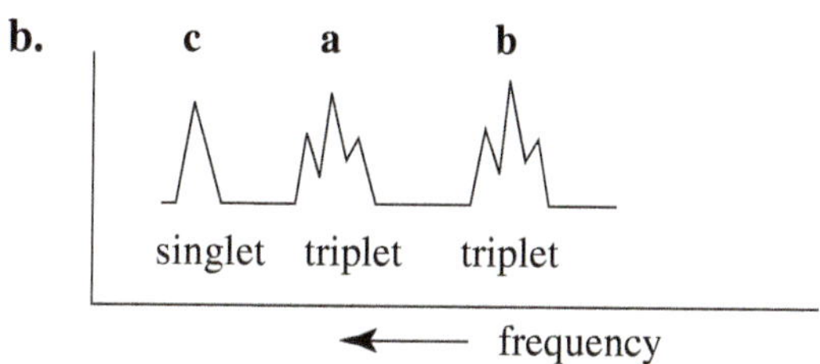

c.

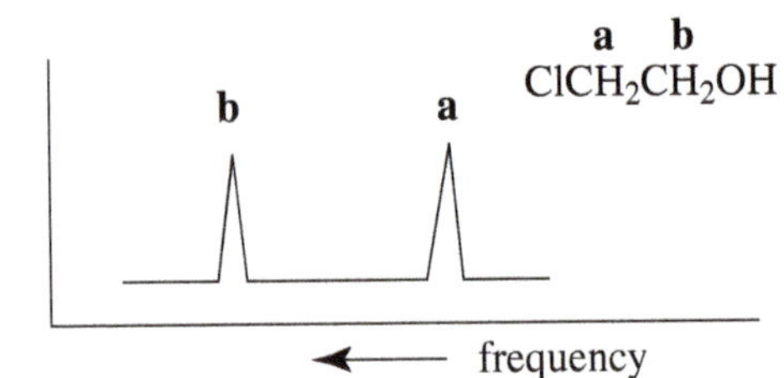

d.

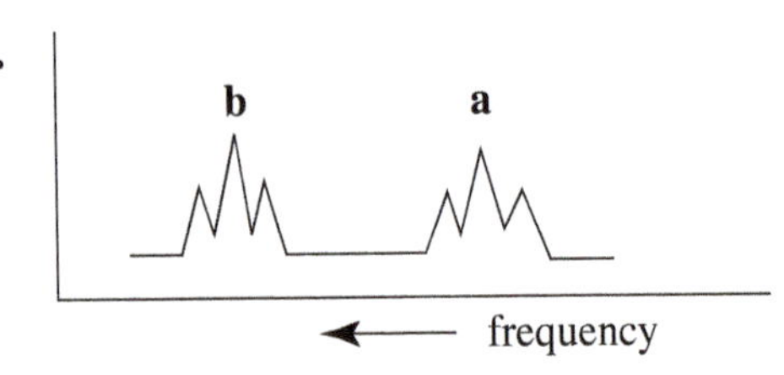

e.

CH_3 carbons

CH_2 carbons

CH carbons

all carbons

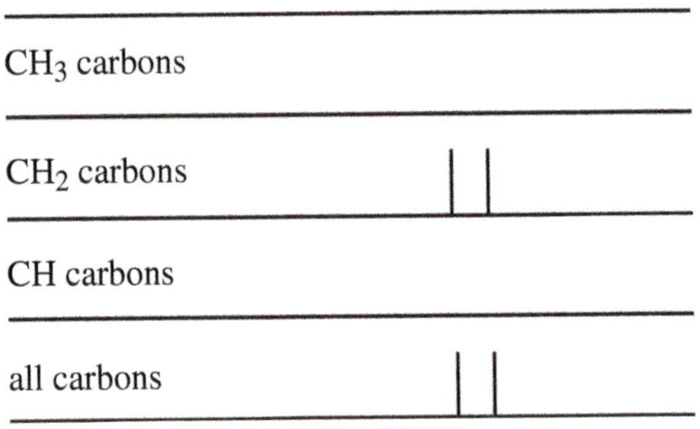

70. If addition of HBr to propene follows the rule that says that the electrophile adds to the sp^2 carbon that is bonded to the most hydrogens, the product of the reaction will give an NMR spectrum with two signals (a doublet and a septet). If addition of HBr does not follow the rule, the product will give an NMR spectrum with three signals (two triplets and a multiplet).

$CH_3CH{=}CH_2$ + HBr ⟶ $CH_3CHBrCH_3$ (Br on the central carbon) — follows the rule — **two signals**

$CH_3CH_2CH_2Br$ — does not follow the rule — **three signals**

71. a. 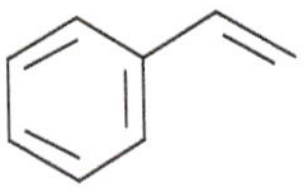 b. 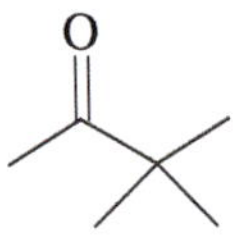 c. 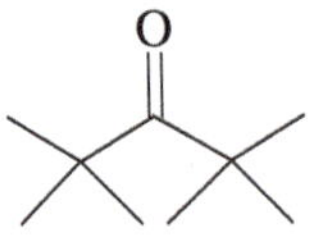 d.

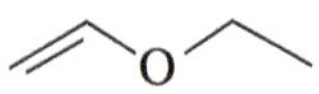

72. Bromomethane gives a signal at 2.7 ppm for its three methyl protons, and 2-bromo-2-methylpropane gives a signal at 1.8 ppm for its nine methyl protons. If equal amounts of each were present in a solution, the ratio of the hydrogens (and, therefore, the ratio of the relative integrals) would be 3:1. Because the relative integrals are 1:6, there must be an 18-fold greater concentration of bromomethane in the mixture.

$$3:x = 1:6$$
$$x = 18$$

73. Using the formula given on page 621 of the text:

$$\Delta E = h\nu = \frac{h\gamma}{2\pi}B_0$$

$$\begin{aligned}
\text{Given: Planck's constant} = h &= 6.626 \times 10^{-34}\ \text{Js} \\
1\ \text{cal} &= 4.184\ \text{J} \\
300\ \text{MHz} &= 7.046\ \text{T} \qquad \text{(page 622 of the text)} \\
\gamma \text{ for } {}^1\text{H} &= 2.675 \times 10^{8}\ \text{T}^{-1}\text{s}^{-1}
\end{aligned}$$

$$\begin{aligned}
\Delta E &= \frac{h\gamma}{2\pi}B_0 \\
&= \frac{6.626 \times 10^{-34}\ \text{Js} \times 2.675 \times 10^{8}\ \text{T}^{-1}\text{s}^{-1}}{2(3.1416)} \times 1.4092\ \text{T} \times \frac{1\ \text{cal}}{4.184\ \text{J}} \\
&= 9.50 \times 10^{-26}\ \text{cal}
\end{aligned}$$

74. All four spectra show a singlet at ~2.0 ppm, suggesting that they are all esters with a methyl group attached to the carbonyl group. Now the problem becomes determining the nature of the group attached to the oxygen.

$CH_3C(=O)O{-}R$

a. The highest frequency signal in the first spectrum is a triplet that integrates to 2 protons, indicating that a CH_2 group is attached to the oxygen and is bonded to another CH_2 group. The lowest frequency signal is a triplet that integrates to 3 protons, indicating a methyl group that is attached to a CH_2 group. The presence of two multiplets that each integrate to 2 protons confirms the structure.

$CH_3C(=O)O{-}CH_2CH_2CH_2CH_3$

b. The highest frequency signal in the second spectrum is a multiplet that integrates to 1 proton, indicating that the carbon attached to the oxygen is attached to one proton and two nonequivalent carbons bonded to hydrogens. The lowest frequency signal is a triplet that integrates to 3 protons, indicating a methyl group attached to a CH_2 group. The doublet at ~1.2 ppm that integrates to 3 protons is due to a methyl group attached to a carbon bonded to one hydrogen.

$CH_3C(=O)O{-}CH(CH_3)CH_2CH_3$

c. The highest frequency signal is a doublet that integrates to 2 protons, indicating that the methylene group that is attached to the oxygen is attached to a carbon bonded to one hydrogen. The lowest frequency signal that integrates to 6 protons indicates two equivalent methyl groups that are attached to a carbon bonded to one hydrogen.

$CH_3C(=O)O{-}CH_2CH(CH_3)CH_3$

d. The group attached to the oxygen in the fourth spectrum has only one kind of hydrogen. The carbon attached to the oxygen is not bonded to any hydrogens, because there is no signal at ~4.0.

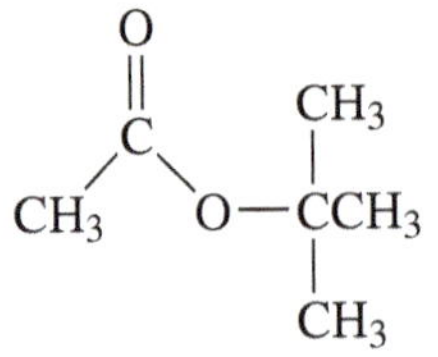

$CH_3C(=O)-O-C(CH_3)_3$

75.

$CH_3CH(OH)CH(CH_3)CH_3 \xrightarrow[\Delta]{HBr} CH_3CH_2CBr(CH_3)CH_3$

compound **A** → compound **B**

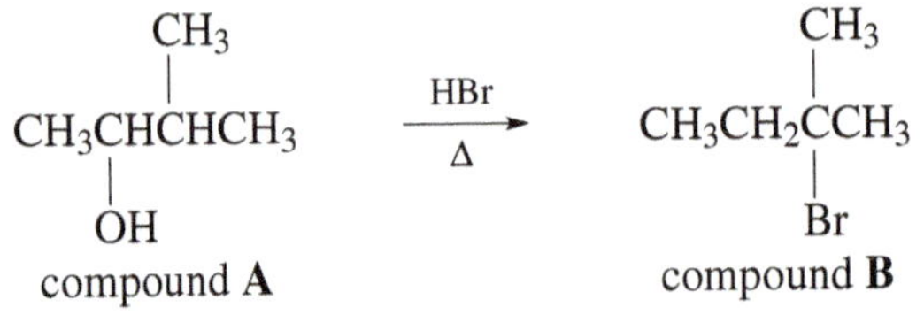

76. The DEPT ^{13}C NMR spectrum indicates that the compound has six carbons. The molecular formula shows that the compound has six carbons and an oxygen. The signal at 220 ppm suggests a ketone. The following ketones have the single CH_3 group, the single CH group, and the three CH_2 groups indicated by the DEPT spectrum. The highest frequency signals are a CH_2 group and a CH group, indicating that these are the groups attached to the carbonyl carbon. Thus, the compound on the right is responsible for the spectrum.

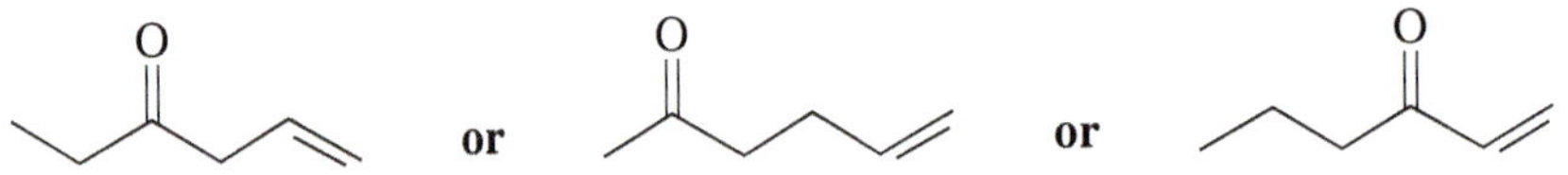

O or O or O

77. **a.** The IR spectrum indicates that the compound is a ketone. The molecular formula shows that there are 12 hydrogens in the compound. Therefore, the signals in the NMR spectrum are due to 2, 4, and 6 protons. The doublet in the NMR spectrum at ~0.9 ppm that integrates to 6 protons suggests an isopropyl group. There is a singlet at ~2.1 ppm on top of a multiplet, which is due to the methine proton of an isopropyl group plus a three-hydrogen singlet for a methyl group adjacent to the carbonyl carbon. The doublet at ~2.2 ppm that integrates to 2 protons suggests a CH_2 group adjacent to the carbonyl group that couples to an adjacent methine proton. Knowing that the compound contains six carbons helps in the identification.

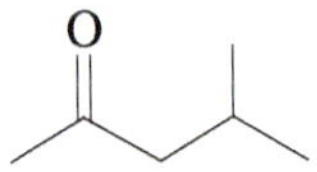

O

b. The IR spectrum indicates that this oxygen-containing compound is not a carbonyl compound or an alcohol; the absorption band at ~1000 cm^{-1} suggests that it is an ether. The doublet at ~1.1 ppm in the ^1H NMR spectrum that integrates to 6 protons and the septet that integrates to 1 proton indicate an isopropyl group. Because there are no other signals in the NMR spectrum, the compound must be a symmetrical ether. The compound is **diisopropyl ether**.

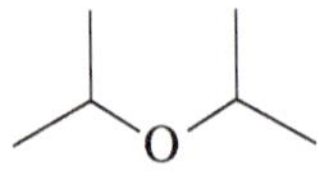

O

c. The IR absorption band at ~3400 cm^{-1} indicates that the compound is an alcohol. The two doublets in the ^1H NMR spectrum at 7.3 and 8.1 ppm that integrate to 4 protons indicate a 1,4-disubstituted benzene ring with a strongly electron-withdrawing substituent. The two triplets that each integrate to 2 protons and the two multiplets that each integrate to 2 protons indicate that the four-carbon substituent is not branched. The broad signal at 2.1 ppm is due to the OH proton.

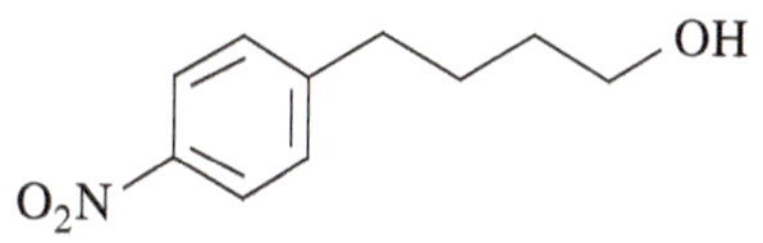

d. The IR absorption bands at ~1700 and 2700 cm^{-1} indicate that the compound is an aldehyde. The two doublets at ~7.0 and 7.8 ppm in the ^1H NMR spectrum indicate a 1,4-disubstituted benzene ring. That none of the remaining NMR signals is a doublet suggests that the aldehyde group is attached directly to the benzene ring. The two triplets and two multiplets indicate an unbranched substituent. The triplet at ~4.0 ppm indicates that the group giving this signal is next to an electron-withdrawing group.

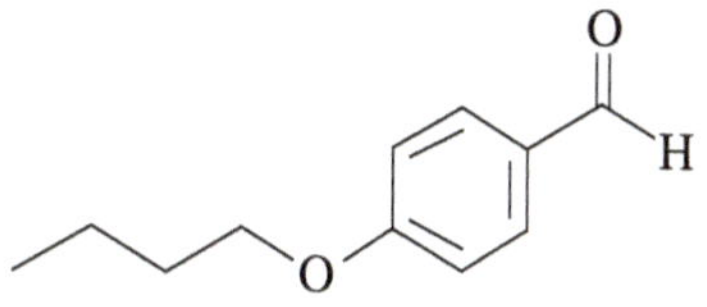

78.

a.	**1.** 7	**2.** 3	**3.** 4	**4.** 3	**5.** 4	**6.** 4
b.	**1.** 5	**2.** 4	**3.** 3	**4.** 3	**5.** 3	**6.** 3

79. **a.** The IR absorption band at ~1730 cm^{-1} and the absence of an aldehyde C—H stretch suggest that the compound is a ketone. The signals in the ^1H NMR spectrum are consistent with a $CH_3CH_2CH_2(C{=}O)$ group. That there are no other signals in the ^1H NMR spectrum suggests that it is a symmetrical ketone. The mass spectrum shows a molecular ion with $m/z = 114$; this is consistent with the ketone shown below. Also, the large fragments at $m/z = 71$ (M − propyl) and at $m/z = 43$ (a propyl cation) are consistent with the expected α-cleavage fragmentation. The fact that there are only three signals in the NMR spectrum suggests that it is a symmetrical ketone. The splitting pattern confirms the structure.

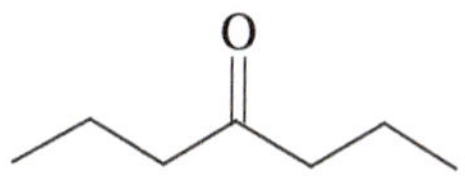

b. The M+2 peak at $m/z = 156$ in the mass spectrum indicates that the compound contains chlorine; the IR spectrum indicates that it is a ketone; the NMR spectrum indicates that it has a monosubstituted benzene ring. The singlet at ~4.7 ppm indicates that the CH_2 group giving this signal is in a strongly electron-withdrawing environment. The major fragment ions at $m/z = 105$(M − CH_2Cl) and at $m/z = 77$ ($C_6H_5^+$) are consistent with the structure shown below.

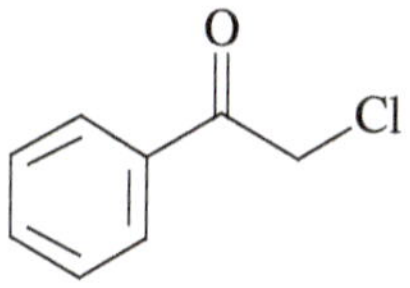

Chapter 14 Practice Test

1. How many signals would you expect to see in the ^{1}H NMR spectrum of each of the following compounds?

$CH_3CH_2CH_2C(=O)CH_3$ $CH_3CH_2CH(Cl)CH_2CH_3$ $CH_2{=}CHC(=O)H$

$C_6H_5NO_2$ (nitrobenzene) m-dichlorobenzene $CH_3CH(CH_3)CH_2CH(CH_3)CH_3$

2. Indicate the multiplicity of each of the indicated sets of protons. (That is, indicate whether it is a singlet, doublet, triplet, quartet, quintet, multiplet, or doublet of doublets.)

$CH_3CH_2C(=O)CH_3$ H–C_6H_4–NO_2 $CH_3CH_2C(=O)OCH_2CH_3$

$H_2C{=}CHCl$ $CH_3CH(CH_3)CH_2Cl$ $ClCH_2CH_2CH_2OCH_3$

$CH_3OCH_2CH_2CH_2OCH_3$ $BrCH_2CH_2Br$

3. How can you distinguish the following compounds using ^{1}H NMR spectroscopy?

$CH_3C(=O)OCH_2CH_3$ $CH_3CH_2C(=O)OCH_3$ $HC(=O)OCH_2CH_2CH_3$

4. Indicate whether each of the following statements is true or false:

a. The signals on the right of an NMR spectrum are deshielded compared to the signals on the left. T F

b. Dimethyl ketone has the same number of signals in its ^{1}H NMR spectrum as in its ^{13}C NMR spectrum. T F

c. In the ^1H NMR spectrum of the compound shown below, the lowest frequency signal (the one farthest upfield) is a singlet and the highest frequency signal (the one farthest downfield) is a doublet. T F

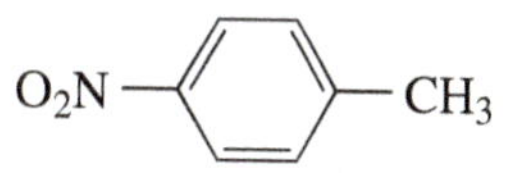

O_2N—(benzene ring)—CH_3

d. The greater the frequency of the signal, the greater its chemical shift in ppm. T F

5. For each compound:

a. Indicate the number of signals you would expect to see in its ^1H NMR spectrum.
b. Indicate the hydrogen or set of hydrogens that will give the highest frequency signal.
c. Indicate the multiplicity of that signal.

1. $CH_3CH_2CH_2Cl$ **2.** $CH_3CH_2COCH_3$ (C=O) **3.** CH_3CHCH_3 (Br on the central carbon)

6. For each compound in Problem 5:

a. Indicate the number of signals you would expect to see in its ^{13}C NMR spectrum.
b. Indicate the carbon that would give the highest frequency signal.
c. Indicate the multiplicity of that signal in a proton-coupled ^{13}C NMR spectrum.

CHAPTER 15
Reactions of Carboxylic Acids and Carboxylic Acid Derivatives

Important Terms

acid anhydride $R{-}C(=O){-}O{-}C(=O){-}R$

acyl adenylate a carboxylic acid derivative with an AMP leaving group.

acyl group $R{-}C(=O){-}$

acyl halide $R{-}C(=O){-}Cl$ or $R{-}C(=O){-}Br$

acyl phosphate a carboxylic acid derivative with a phosphate leaving group.

acyl transfer reaction a reaction that transfers an acyl substituent from one group to another.

alcoholysis a reaction with an alcohol that converts one compound into two compounds.

amide $R{-}C(=O){-}NH_2$ or $R{-}C(=O){-}NHR$ or $R{-}C(=O){-}NR_2$

amino acid an α-amino carboxylic acid.

aminolysis a reaction with an amine that converts one compound into two compounds.

biosynthesis synthesis that occurs in a biological system.

carbonyl carbon the carbon of a C=O group.

carbonyl compound a compound that contains a C=O group.

carbonyl group a carbon doubly bonded to an oxygen (C=O).

carbonyl oxygen the oxygen of a C=O group.

carboxyl group ${-}C(=O){-}OH$ or $-COOH$ or $-CO_2H$

carboxylic acid $R{-}C(=O){-}OH$

carboxylic acid derivative a compound that is hydrolyzed to a carboxylic acid.

carboxyl oxygen the single-bonded oxygen of a carboxylic acid or ester.

catalyst a species that increases the rate of a reaction without being consumed or changed in the reaction.

ester $R{-}C({=}O){-}OR'$

Fischer esterification reaction a reaction of a carboxylic acid with excess alcohol and an acid catalyst.

Gabriel synthesis a method used to convert an alkyl halide into a primary amine, involving S_N2 attack of phthalimide ion on an alkyl halide followed by hydrolysis.

hydrolysis a reaction with water that converts one compound into two compounds.

imide a compound with two acyl groups bonded to a nitrogen.

lactam a cyclic amide.

lactone a cyclic ester.

mixed anhydride an acid anhydride with two different R groups.

nitrile a compound that contains a carbon–nitrogen triple bond.

$$R{-}C{\equiv}N$$

nucleophilic acyl substitution reaction a reaction in which a group bonded to an acyl group is substituted by another group.

nucleophilic addition–elimination reaction another name for a nucleophilic acyl substitution reaction, which emphasizes the two-step nature of the reaction: a nucleophile adds to the carbonyl carbon in the first step, and a group is eliminated in the second step.

symmetrical anhydride an acid anhydride with identical R groups.

tetrahedral intermediate the intermediate formed in an adddition–elimination (or a nucleophilic acyl substitution) reaction.

thioester the sulfur analog of an ester.

$$R{-}C({=}O){-}SR'$$

transesterification reaction the reaction of an ester with an alcohol to form a different ester.

Solutions to Problems

1. **a.** benzyl acetate **b.** isopentyl acetate **c.** methyl butyrate

2. **a.** potassium butanoate
potassium butyrate

b. isobutyl butanoate
2-methylpropyl butanoate
isobutyl butyrate

c. *N,N*-dimethylhexanamide
N,N-dimethylcaproamide

d. pentanoyl chloride
valeryl chloride

e. 5-methylhexanoic acid
δ-methylcaproic acid

f. propanamide
propionamide

g. γ-butyrolactam or
2-azacyclopentanone

h. cyclopentanecarboxylic acid

i. β-methyl-δ-valerolactone or
5-methyl-2-oxacyclohexanone

3. **a.** 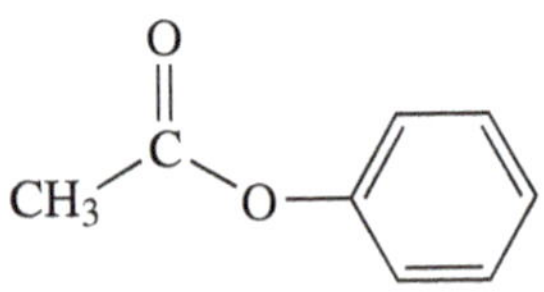 $CH_3C(=O)O-C_6H_5$

b. (5-ethyl-2-pyrrolidinone: five-membered ring with C=O and NH, CH_2CH_3 on the carbon adjacent to NH)

c. $HC(=O)O^- Na^+$

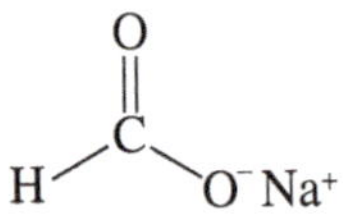

d. 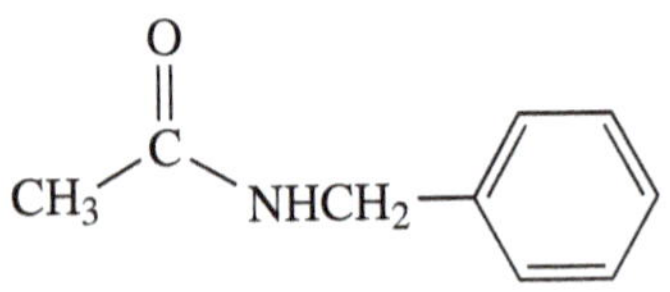 $CH_3C(=O)NHCH_2-C_6H_5$

e. 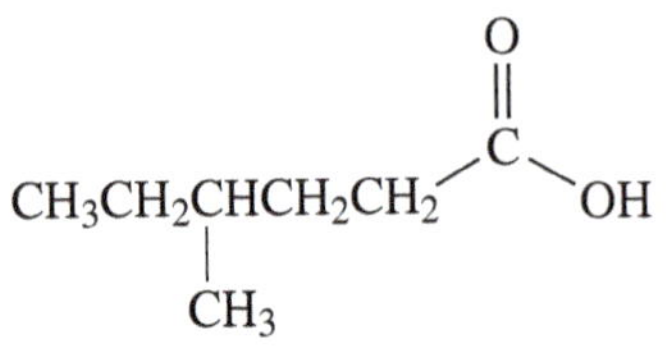 $CH_3CH_2CH(CH_3)CH_2CH_2C(=O)OH$

f. 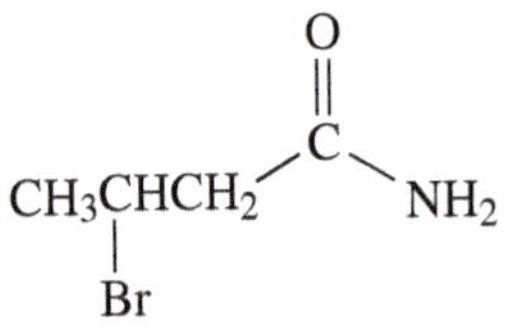 $CH_3CH(Br)CH_2C(=O)NH_2$

g. 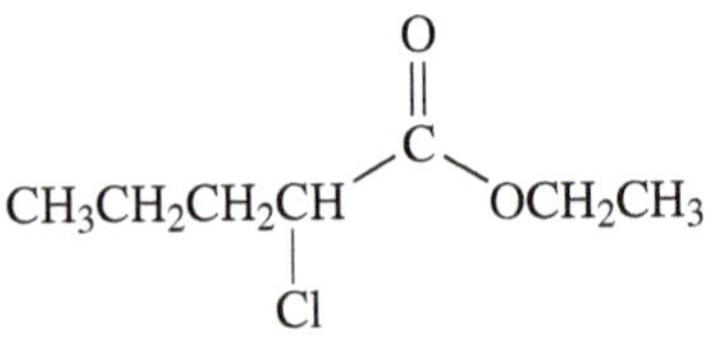 $CH_3CH_2CH_2CH(Cl)C(=O)OCH_2CH_3$

h. 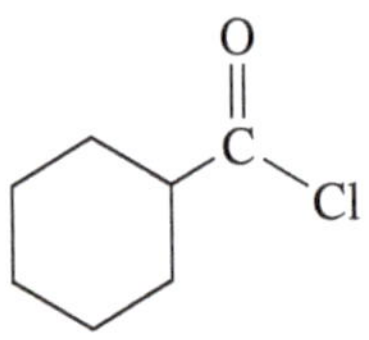 (cyclohexyl–C(=O)Cl)

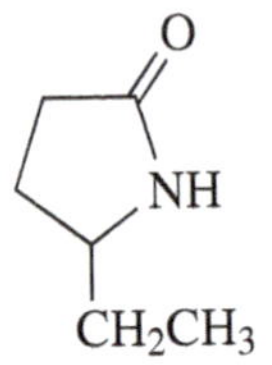

i. 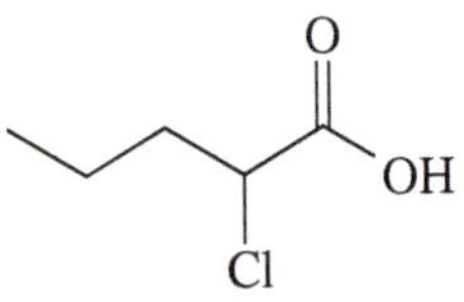 (skeletal: 2-chloropentanoic acid, with Cl and OH labels)

4. The carbon–oxygen single bond in an alcohol is longer because, as a result of electron delocalization, the carbon–oxygen single bond in a carboxylic acid has some double-bond character.

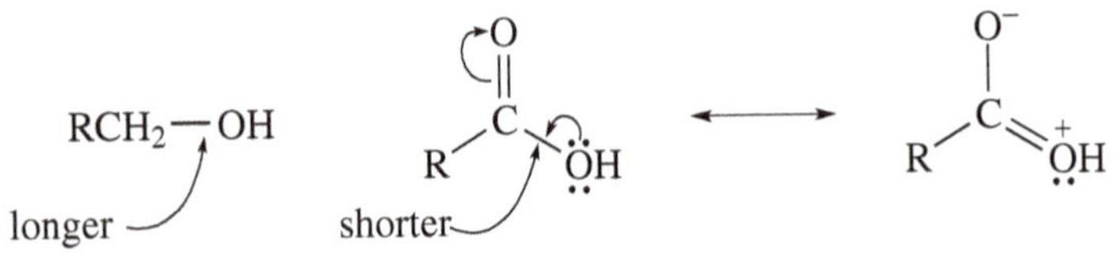

RCH_2-OH (longer) $RC(=O)\ddot{O}H$ (shorter) ⟷ $RC(O^-)=\overset{+}{O}H$

5. **a.** The bond between oxygen and the methyl group is the longest, because it is a pure single bond, whereas the other two carbon–oxygen bonds have some double-bond character.

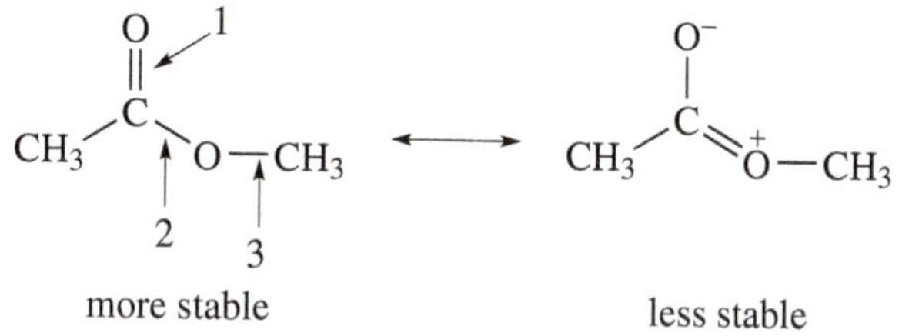

more stable　　　less stable

3 = longest　1 = shortest

The bond between carbon and the carbonyl oxygen is the shortest, because it has the most double-bond character.

b. Notice that the longer the bond, the lower its IR stretching frequency.

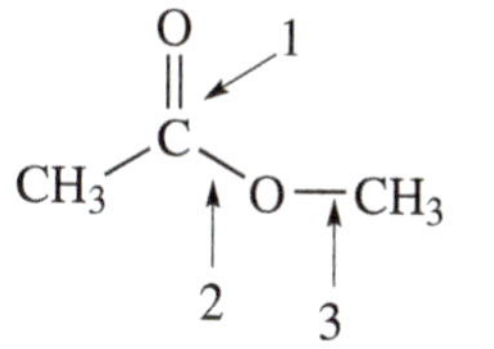

1 = highest frequency
3 = lowest frequency

6. **B** is a correct statement. The delocalization energy (resonance energy) is greater for the amide than for the ester because the second resonance contributor of the amide has a greater predicted stability and so contributes more to the overall structure of the amide. (Recall that nitrogen is less electronegative than oxygen, so it is more stable with a positive charge.)

7. **a.** Because HCl is a stronger acid than H_2O, Cl^- is a weaker base than HO^-.
Therefore, Cl^- will be eliminated from the tetrahedral intermediate, so the product of the reaction will be acetic acid. Because the solution is basic, acetic acid will be in its basic form as a result of losing a proton.

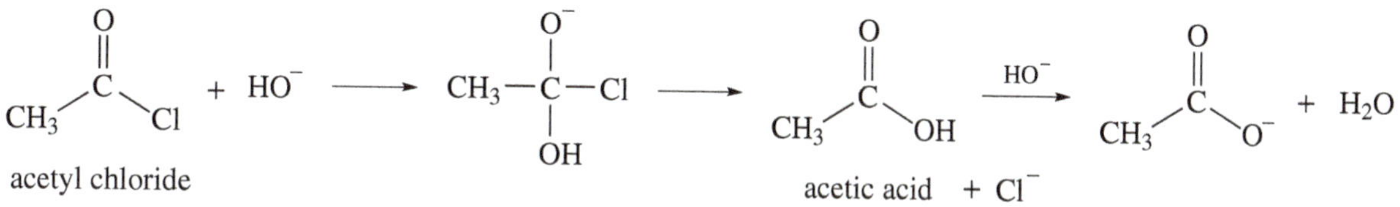

$$CH_3COCl + HO^- \longrightarrow CH_3-C(O^-)(OH)-Cl \longrightarrow CH_3COOH + Cl^- \xrightarrow{HO^-} CH_3COO^- + H_2O$$

acetyl chloride　　　acetic acid

b. Because H_2O is a stronger acid than NH_3, HO^- is a weaker base than $^-NH_2$.
Therefore, HO^- will be eliminated from the tetrahedral intermediate, so the reactant will reform. In other words, no reaction will take place.

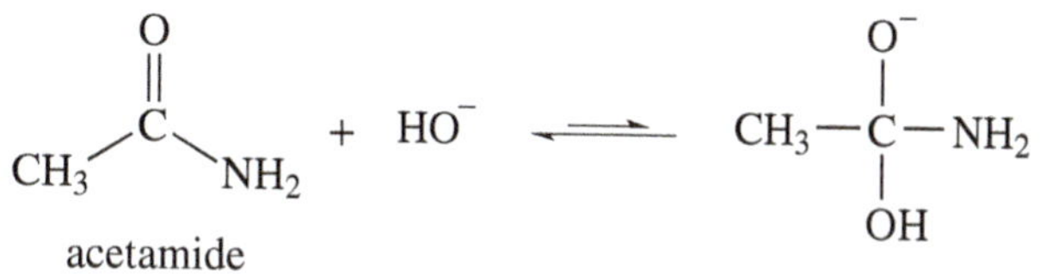

$$CH_3CONH_2 + HO^- \rightleftharpoons CH_3-C(O^-)(OH)-NH_2$$

acetamide

8. **a.** a new carboxylic acid derivative
b. no reaction
c. a mixture of two carboxylic acid derivatives

9. **a.** Acetyl chloride has the stretching vibration for its carbonyl group at the highest frequency because it has the most C=O double-bond character, since it has the smallest contribution from the resonance contributor with a positive charge on Y.

$$CH_3C(=O)\ddot{Y} \longleftrightarrow CH_3C(O^-)=\overset{+}{Y}$$

b. Acetamide has the stretching vibration for its carbonyl group at the lowest frequency because it has the least C=O double-bond character, since it has the largest contribution from the resonance contributor with a positive charge on Y.

10. **a.** CH_3COOCH_3 + NaCl ⟶ no reaction, because Cl^- is weaker base than CH_3O^-

b. CH_3COCl + NaOH ⟶ CH_3COOH + NaCl $\xrightarrow{HO^-}$ CH_3COO^- + H_2O

c. CH_3CONH_2 + NaCl ⟶ no reaction, because Cl^- is a weaker base than $^-NH_2$

d. CH_3CONH_2 + NaOH ⟶ no reaction, because HO^- is a weaker base than $^-NH_2$

11. It is a true statement.

If the nucleophile is the stronger base, it will be harder to eliminate the nucleophile from the tetrahedral intermediate (**B**) than the group attached to the acyl group in the reactant. In other words, the hill that has to be climbed from the intermediate back to the reactants (**B** to **A**) is higher than the hill that has to be climbed from the intermediate to the products (**B** to **C**). We know that the first step is the rate-limiting step because it has the transition state with the greatest energy.

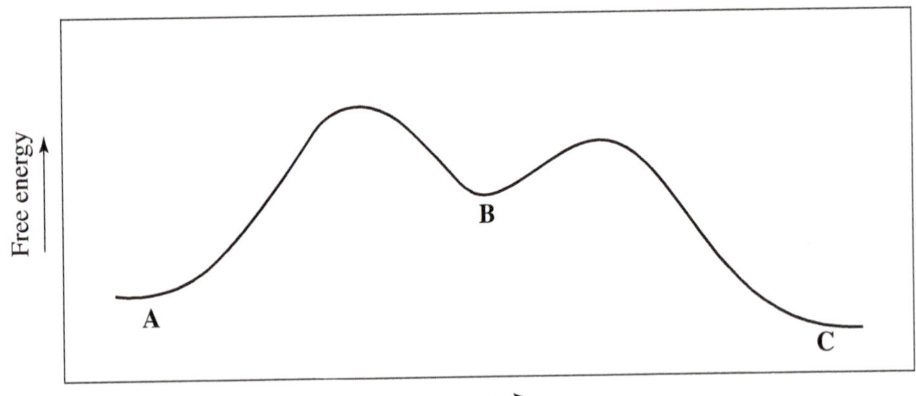

12. **a.** H_2O **c.** $CH_3CH_2CH_2OH$ **e.** HO–cyclohexyl

b. NH_3 **d.** $(CH_3)_2NH$ **f.** HO–C_6H_4–NO_2

13. **a.** $CH_3COCl + H_2O \rightleftharpoons CH_3C(O^-)(Cl)\overset{+}{O}H_2$ (B: removes H) $\rightleftharpoons CH_3C(O^-)(Cl)OH + HB^+ \rightarrow CH_3COOH + Cl^-$

b. $C_6H_5COCl + CH_3NH_2 \rightleftharpoons C_6H_5C(O^-)(Cl)\overset{+}{N}H_2CH_3$ (CH_3NH_2 removes H) $\rightleftharpoons C_6H_5C(O^-)(Cl)NHCH_3 \rightarrow C_6H_5CONHCH_3 + Cl^-$

14. Solved in the text.

15. **a.** ethanol **b.** ammonia **c.** phenol **d.** benzyl alcohol

16. A protonated amine has a $pK_a \sim 11$. Therefore, the neutral amine will be protonated by the acid that is produced in the reaction, and a protonated amine is not a nucleophile. Excess amine is used to have some unprotonated amine available to react as a nucleophile.

A protonated alcohol has a $pK_a \sim -2$. Therefore, the alcohol will not be protonated by the acid that is produced in the reaction.

17. **a.**

$$CH_3CH_2C(=O)OCH_3 + H_2\ddot{O}: \rightleftharpoons CH_3CH_2-C(O^-)(OCH_3)-\overset{+}{O}H_2 \xrightarrow{B:} CH_3CH_2-C(O^-)(OCH_3)-OH + HB^+$$

CH_3O^- (path a) and HO^- (path b) have the same leaving propensity, so either one can be eliminated.

Path a: $CH_3CH_2C(=O)OH + CH_3O^- \rightarrow CH_3CH_2C(=O)O^- + CH_3OH$

Path b: $CH_3CH_2C(=O)OCH_3 + HO^-$

b.

$$HC(=O)OC_6H_5 + CH_3\ddot{N}H_2 \rightleftharpoons H-C(O^-)(OC_6H_5)-\overset{+}{N}H_2CH_3 \xrightarrow{B:} H-C(O^-)(OC_6H_5)-NHCH_3 + HB^+ \rightarrow HC(=O)NHCH_3 + {}^-OC_6H_5$$

18. Solved in the text.

19. **a.** phenyl acetate because phenoxide is a weaker base than methoxide
b. phenyl acetate because phenoxide is a weaker base than the conjugate base of benzyl alcohol

20. **a.** **1.** The carbonyl group of an ester is a weak electrophile.
2. Water is a weak nucleophile.
3. $^-OCH_3$ is a strong base and, therefore, a poor leaving group.
b. Aminolysis is faster because an amine is a better nucleophile than water.

21. **a.** Any species with an acidic proton can be represented by HB^+.

H_3O^+ $\quad CH_3\overset{+}{O}H_2$ $\quad CH_3C(=O)OH$ $\quad CH_3C(=\overset{+}{O}H)OH$ $\quad CH_3C(OH)(OCH_3)(\overset{+}{O}H_2)$ $\quad CH_3C(OH)_2(\overset{+}{O}HCH_3)$

b. Any species with a lone pair can be represented by :B.

H_2O $\quad CH_3OH$ $\quad CH_3C(=O)OH$ $\quad CH_3C(OH)_2OCH_3$

c. H_3O^+

d. H_3O^+ if excess water is used; $CH_3\overset{+}{O}H_2$ if excess water is not used.

22. The mechanism for the acid-catalyzed reaction of acetic acid and methanol is the exact reverse of the mechanism for the acid-catalyzed hydrolysis of methyl acetate.

:Ö: H—B⁺ CH₃ C OH ⇌ ⁺O H CH₃ C OH + CH₃ÖH ⇌ OH CH₃—C—OH ⁺OCH₃ B: H ⇌ OH H—B⁺ CH₃—C—ÖH OCH₃

HB⁺ O CH₃ C OCH₃ ⇌ ⁺O H :B CH₃ C OCH₃ + H₂O ⇌ :ÖH H CH₃—C—OH ⁺ OCH₃

23. **a.** C₆H₅–C(=O)–OH + CH_3CH_2OH

b. $CH_3CH_2CH_2$–C(=O)–OH + CH_3OH

c. $HOCH_2CH_2CH_2CH_2$–C(=O)–OH

The mechanism for the hydrolysis of a cyclic ester is the same as that for an acyclic ester.

24. HO ... C(=Ö:) OH H—B⁺ ⇌ HÖ ... C(=⁺OH) OH ⇌ HO OH ⁺:O—H :B ⇌ HO ÖH O H—B⁺ ⇌ HÖ ⁺ÖH H O

O O + HB⁺ ⇌ ⁺:O—H :B O + H₂O ⇌

25. These are transesterification reactions.

a. $C_6H_5COOCH_2CH_3 + CH_3CH(OH)CH_3 \underset{}{\overset{HCl}{\rightleftharpoons}} C_6H_5COOCH(CH_3)CH_3 + CH_3CH_2OH$

b. $CH_3COOC_6H_5 + CH_3CH_2OH \underset{}{\overset{HCl}{\rightleftharpoons}} CH_3COOCH_2CH_3 + C_6H_5OH$

26. $CH_3C(=O)OCH_2CH_3 \overset{H-B^+}{\rightleftharpoons} CH_3C(=O^+H)OCH_2CH_3 + CH_3\ddot{O}H \rightleftharpoons CH_3-C(OH)(OCH_2CH_3)-\overset{+}{O}(H)CH_3 \rightleftharpoons CH_3-C(OH)(OCH_3)-\ddot{O}CH_2CH_3 \rightleftharpoons CH_3-C(\ddot{O}H)(OCH_3)-\overset{+}{O}(H)CH_2CH_3 \rightleftharpoons CH_3C(=O^+H)OCH_3 + CH_3CH_2OH \rightleftharpoons CH_3C(=O)OCH_3 + HB^+$

27. The mechanism is the same as the one on page 705 of the text except that the nucleophile is methanol rather than water.

$RC(=O)OC(CH_3)_3 \overset{H-B^+}{\rightleftharpoons} RC(=O^+H)OC(CH_3)_3 \rightleftharpoons RC(=O)OH + {}^+C(CH_3)_3$

${}^+C(CH_3)_3 \xrightarrow{CH_3\ddot{O}H} CH_3\overset{+}{O}(H)-C(CH_3)_3 \xrightarrow{B:} CH_3O-C(CH_3)_3 + HB^+$

28. **a.** The conjugate base ($CH_3CH_2CH_2O^-$) of the reactant alcohol ($CH_3CH_2CH_2OH$) can be used to increase the rate of the reaction because it is the nucleophile that we want to become attached to the carbonyl carbon.

b. If H^+ is used as a catalyst, the amine will be protonated in the acidic solution and, therefore, will not be able to react as a nucleophile.

If HO^- is used as a catalyst, HO^- will be the best nucleophile in the solution. Therefore, it will add to the ester, and the product of the reaction will be a carboxylate ion rather than an amide.

If RO^- is used as a catalyst, RO^- will be the nucleophile, and the product of the reaction will be an ester rather than an amide.

29. **a.** The alcohol (CH_3CH_2OH) contained the ^{18}O label.

b. The carboxylic acid would have contained the ^{18}O label.

30. Solved in the text.

31. **a.**

$$CH_3CH_2CH_2C(=O)O^- \xrightarrow{CH_3Br} CH_3CH_2CH_2C(=O)OCH_3$$

or

$$CH_3CH_2CH_2C(=O)OH \xrightarrow[HCl]{CH_3OH \text{ excess}} CH_3CH_2CH_2C(=O)OCH_3$$

b.

$$CH_3C(=O)O^- \xrightarrow{CH_3(CH_2)_7Br} CH_3C(=O)O(CH_2)_7CH_3$$

or

$$CH_3C(=O)OH \xrightarrow[HCl]{CH_3(CH_2)_7OH \text{ excess}} CH_3C(=O)O(CH_2)_7CH_3$$

32. In the first step, protonation occurs on the carbon that results in formation of the most stable carbocation (a tertiary carbocation). In the second step, the pair of electrons of the second π bond provides the nucleophile that adds to the carbocation because, as a result of that reaction, a stable six-membered ring is formed.

$CH_2{=}CHCH_2CH_2CH{=}C(CH_3)CH_3$ $\xrightarrow{H-B^+}$ $CH_2{=}CHCH_2CH_2CH_2-\overset{+}{C}(CH_3)CH_3$ (carbons numbered 6, 5, 4, 3, 2, 1) → cyclohexyl cation with + at C5, 1,1-dimethyl (ring positions 1–6) $\xrightarrow{CH_3C(=O)\ddot{O}H}$ $CH_3C(=\overset{+}{O}-H)O$-(3,3-dimethylcyclohexyl) + :B → $CH_3C(=O)O$-(3,3-dimethylcyclohexyl) + HB^+

33. **a.** $CH_3CH_2CH_2C(=O)Cl + 2\ CH_3CH_2NH_2$

b. $C_6H_5C(=O)Cl + 2\ CH_3NHCH_3$

34. Only **B** and **E** will form amides. **A** will form an amide if it is heated to about 225 °C.

35.

$$\mathrm{RC(=O)NH_2} \underset{}{\overset{\mathrm{H{-}B^+}}{\rightleftharpoons}} \mathrm{RC(={}^{+}OH)NH_2} + \mathrm{CH_3OH} \rightleftharpoons \mathrm{R{-}C(OH)(NH_2){-}{}^{+}O(H)CH_3} \;(:B) \rightleftharpoons \mathrm{R{-}C(OH)(NH_2){-}OCH_3} \;(\mathrm{H{-}B^+})$$

$$\rightleftharpoons \mathrm{R{-}C(OH)({}^{+}NH_3){-}OCH_3} \rightleftharpoons \mathrm{RC(={}^{+}OH)OCH_3} + \mathrm{NH_3} \rightleftharpoons \mathrm{RC(=O)OCH_3} + {}^{+}\mathrm{NH_4}$$

36. The relative reactivities of the amides depend on the basicities of their leaving groups: the weaker the base, the more reactive the amide. (*para*-Nitroanilinium ion has a pK_a value of 0.98, the anilinium ion has a pK_a value of 4.58, and the cyclohexylammonium ion has a pK_a value of 11.2.) Recall that the stronger the acid, the weaker its conjugate base and that the weaker the base, the better it is as a leaving group.

$$\underset{\mathbf{B}}{\mathrm{CH_3C(=O)NH{-}C_6H_4{-}NO_2}} > \underset{\mathbf{C}}{\mathrm{CH_3C(=O)NH{-}C_6H_5}} > \underset{\mathbf{A}}{\mathrm{CH_3C(=O)NH{-}C_6H_{11}}}$$

37. Because the acid–base reaction below favors the products. Recall that the equilibrium favors reaction of the strong acid (H_2O) and formation of the weak acid (NH_3) (Section 2.5).

$$^{-}NH_2 + H_2O \rightleftharpoons NH_3 + HO^{-}$$

38. **a.** pentyl bromide **b.** isohexyl bromide **c.** benzyl bromide **d.** cyclohexyl bromide

39. The reaction of an alkyl halide with ammonia leads to primary, secondary, and tertiary amines and even quaternary ammonium ions. Thus, the yield of primary amine would be relatively low.

$$\mathrm{RBr} \xrightarrow{\mathrm{NH_3}} \mathrm{R\overset{+}{N}H_3} + \mathrm{Br^-} \overset{\mathrm{NH_3}}{\rightleftharpoons} \mathrm{RNH_2} + {}^{+}\mathrm{NH_4} \xrightarrow{\mathrm{RBr}} \mathrm{R_2\overset{+}{N}H_2} + \mathrm{Br^-} \overset{\mathrm{NH_3}}{\rightleftharpoons} \mathrm{R_2NH} + {}^{+}\mathrm{NH_4}$$

$$\mathrm{R_2NH} \xrightarrow{\mathrm{RBr}} \mathrm{R_3\overset{+}{N}H} + \mathrm{Br^-} \overset{\mathrm{NH_3}}{\rightleftharpoons} \mathrm{R_3N} + {}^{+}\mathrm{NH_4} \xrightarrow{\mathrm{RBr}} \mathrm{R_4\overset{+}{N}} + \mathrm{Br^-}$$

In contrast, the Gabriel synthesis forms only primary amines. The reaction of an alkyl halide with azide ion also forms only primary amines because the compound formed from the initial reaction of the two reagents is not nucleophilic, so polyalkylation does not occur.

40. **a.** butanenitrile
butyronitrile
propyl cyanide

b. 4-methylpentanenitrile
γ-methylvaleronitrile
isopentyl cyanide

41. Notice that the alkyl halide has one less carbon than the target carboxylic acid, because the alkyl halide will obtain a carbon from the cyanide ion.

a. $CH_3CH_2CH_2Br$ **b.** $CH_3CH(CH_3)CH_2Br$ **c.** cyclohexyl–Br

42. Solved in the text.

43. **a.** $CH_3C(=O)OC(=O)CH_3 + H_2\ddot{O}: \rightleftharpoons CH_3C(O^-)(\overset{+}{O}H_2)-O-C(=O)CH_3 \rightleftharpoons CH_3C(O^-)(OH)-O-C(=O)CH_3 + HB^+$

$\downarrow$

$CH_3C(=O)OH + HOC(=O)CH_3 + :B$

b. The mechanisms are exactly the same.

$CH_3C(=O)OC(=O)CH_3 + CH_3\ddot{O}H \rightleftharpoons CH_3C(O^-)(\overset{+}{O}HCH_3)-O-C(=O)CH_3 \rightleftharpoons CH_3C(O^-)(OCH_3)-O-C(=O)CH_3 + HB^+$

$\downarrow$

$CH_3C(=O)OCH_3 + HOC(=O)CH_3 + :B$

44. $RC(=O)Cl + CH_3C(=O)\ddot{O}:^- \rightleftharpoons R-C(O^-)(Cl)-O-C(=O)CH_3 \longrightarrow RC(=O)OC(=O)CH_3 + Cl^-$

45. When an acid anhydride reacts with an amine, the tetrahedral intermediate does not have to lose a proton before it eliminates the carboxylate ion, because the carboxylate ion is a weaker base (pK_a of its conjugate acid is ~5) than the amine (pK_a of its conjugate acid is ~10). Therefore, the carboxylate ion is the better leaving group.

$CH_3C(=O)OC(=O)CH_3 + CH_3\ddot{N}H_2 \rightleftharpoons CH_3C(O^-)(\overset{+}{N}H_2CH_3)-O-C(=O)CH_3$

46. **a.** At neutral pH, both carboxyl groups of succinic acid are in their basic form.

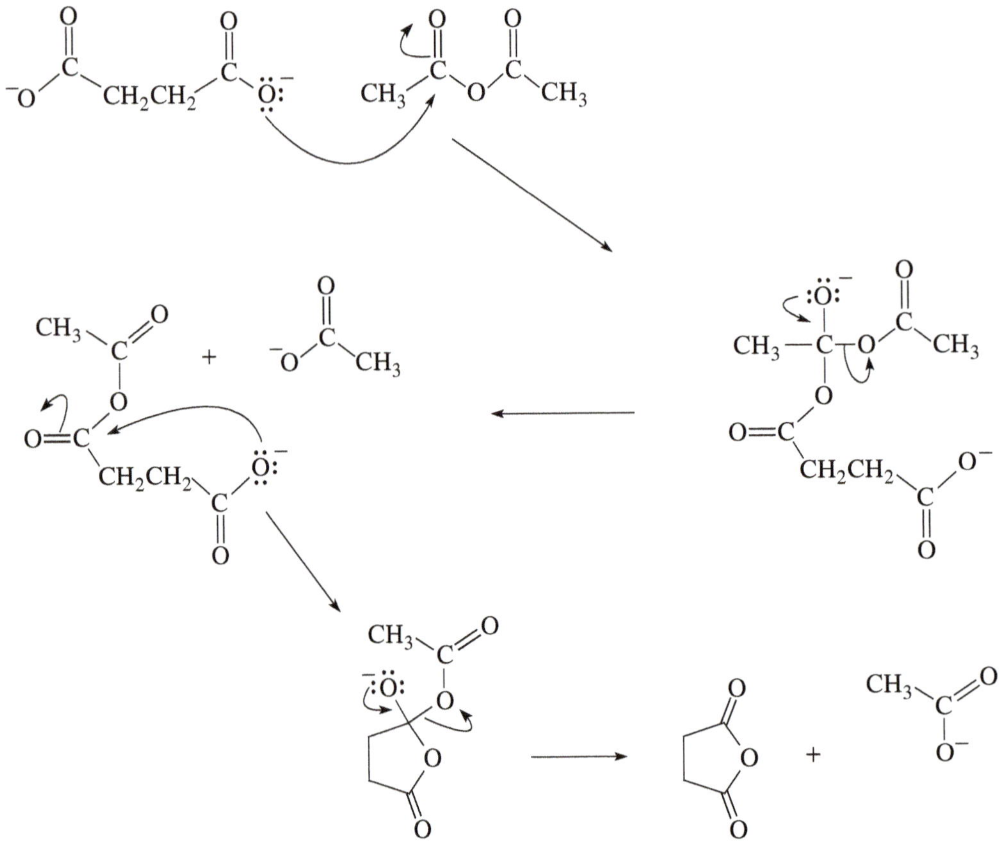

b. Without acetic anhydride, the leaving group would be hydroxide ion. Acetic anhydride causes the reaction to take place via two successive nucleophilic acyl substitution reactions. In both reactions, the leaving group is acetate ion, which is much less basic than hydroxide ion and therefore a better leaving group.

47. **a.** $CH_3CH_2COOH \xrightarrow{SOCl_2} CH_3CH_2COCl \xrightarrow{C_6H_5OH} CH_3CH_2COOC_6H_5$

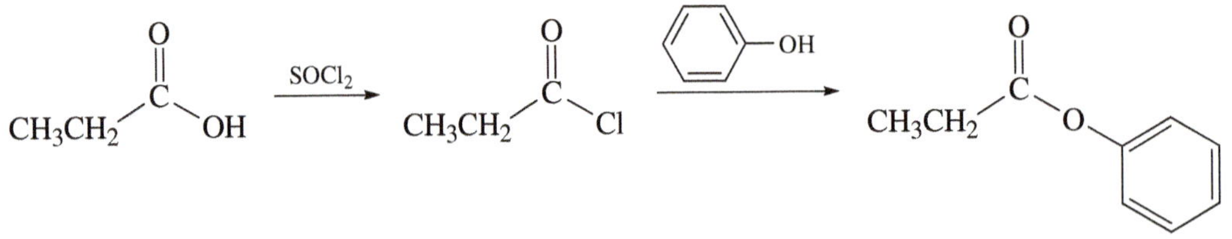

b. $C_6H_5COOH \xrightarrow{SOCl_2} C_6H_5COCl \xrightarrow{2\ CH_3CH_2NH_2} C_6H_5CONHCH_2CH_3$

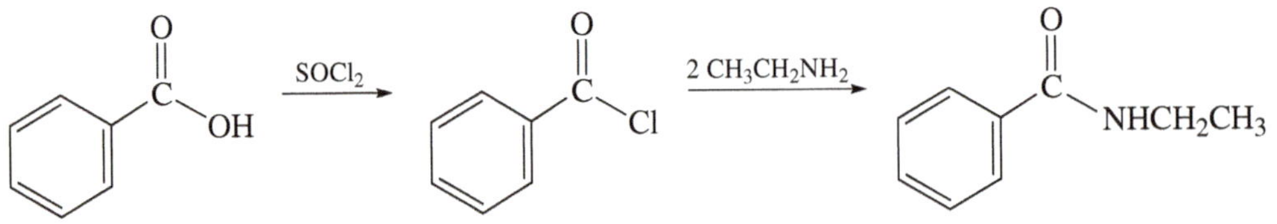

48.

a. 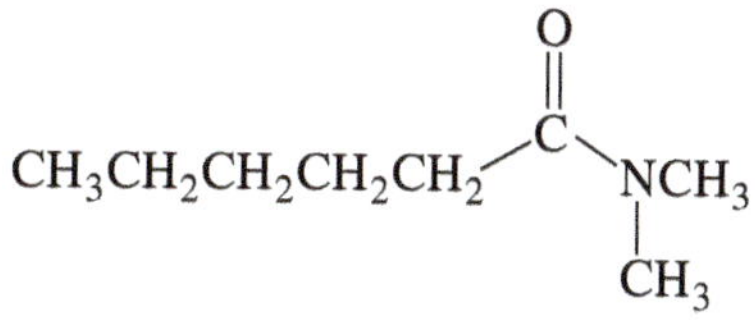 $CH_3CH_2CH_2CH_2CH_2C(=O)N(CH_3)CH_3$

b. 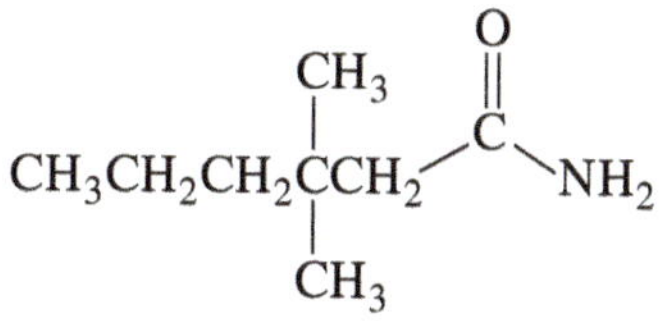 $CH_3CH_2CH_2C(CH_3)_2CH_2C(=O)NH_2$

c. 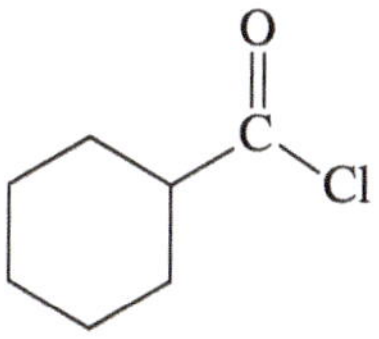 cyclohexyl–$C(=O)Cl$

d. $CH_3CH_2C{\equiv}N$

e. 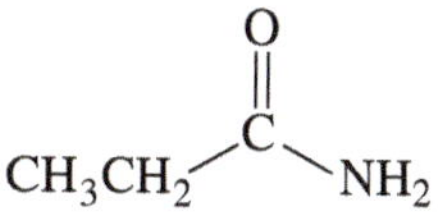 $CH_3CH_2C(=O)NH_2$

f. 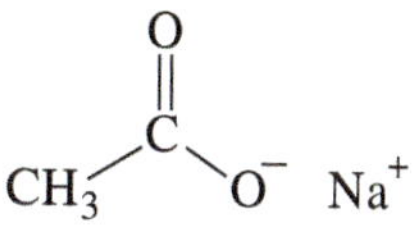 $CH_3C(=O)O^-\ Na^+$

g. 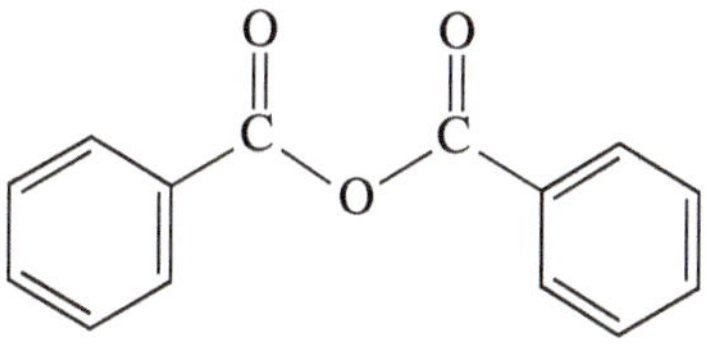 $C_6H_5C(=O)OC(=O)C_6H_5$

h. 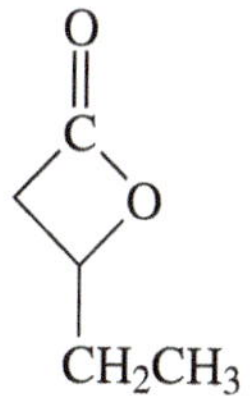 four-membered lactone ring with $C=O$ and O, bearing CH_2CH_3

i. 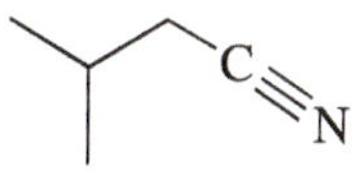 $(CH_3)_2CHCH_2C{\equiv}N$

j. 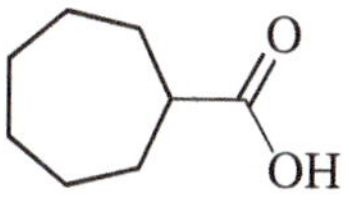 cycloheptyl–$C(=O)OH$

k. 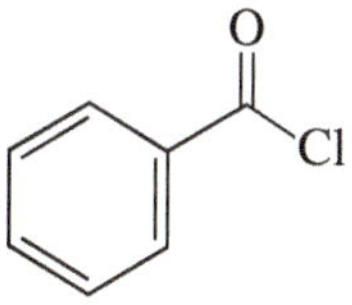 $C_6H_5C(=O)Cl$

49.

a. pentanoyl chloride
valeryl chloride

b. *N,N*-dimethylbutanamide
N,N-dimethylbutyramide

c. (*S*)-3-methylpentanoic acid
(*S*)-β-methylvaleric acid

d. propyl propanoate
propyl propionate

e. acetic benzoic anhydride
benzoic ethanoic anhydride

f. propanoic anhydride
propionic anhydride

g. 5-ethylheptanoic acid

h. (*R*)-3-methylhexanenitrile

50.

a. 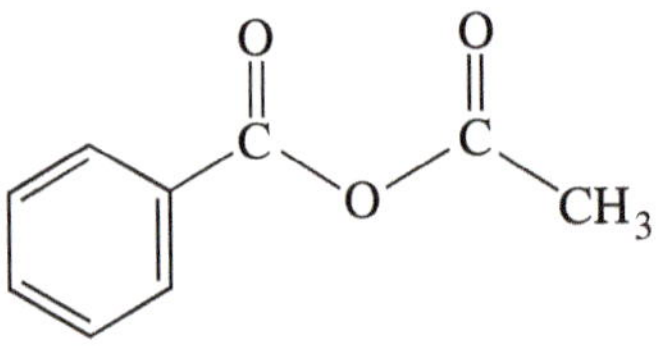 $C_6H_5-C(=O)-O-C(=O)-CH_3$

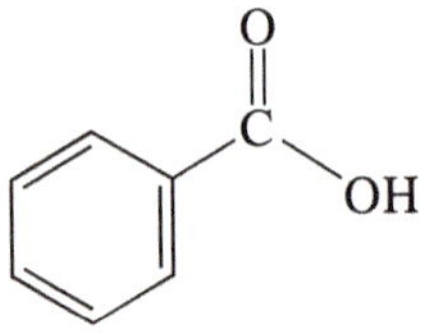

b. $C_6H_5-C(=O)-OH$

c. $C_6H_5-C(=O)-N(CH_3)-CH_3$

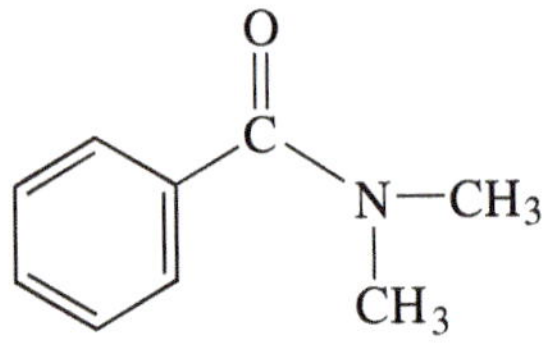

d. 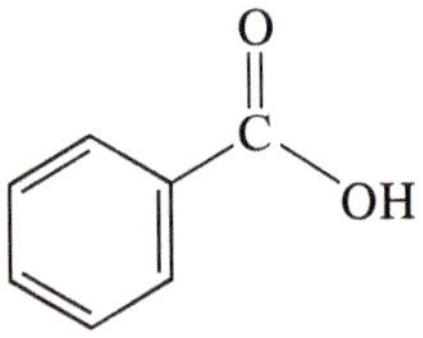 $C_6H_5-C(=O)-OH$

e. 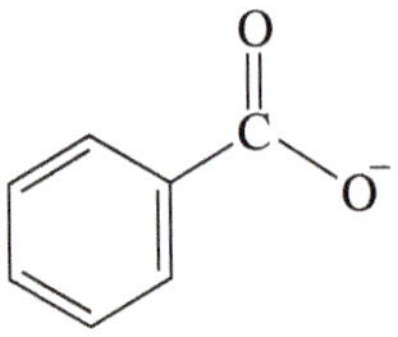 $C_6H_5-C(=O)-O^-$

A carboxylic acid is in its basic form in a basic solution.

f. 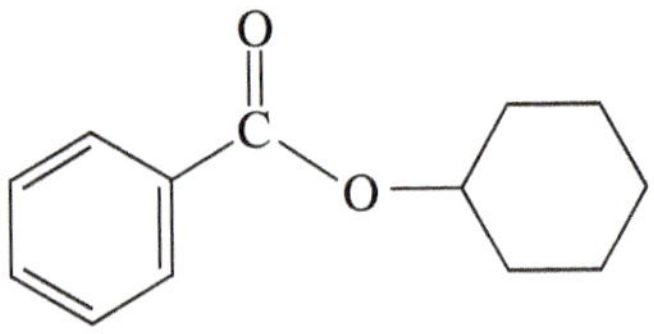 $C_6H_5-C(=O)-O-C_6H_{11}$

g. 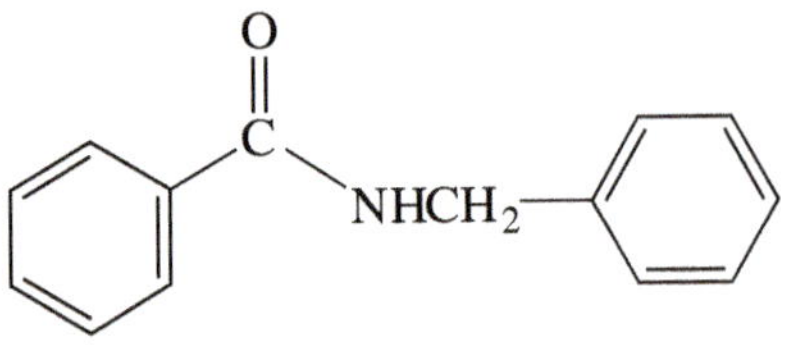 $C_6H_5-C(=O)-NHCH_2-C_6H_5$

h. 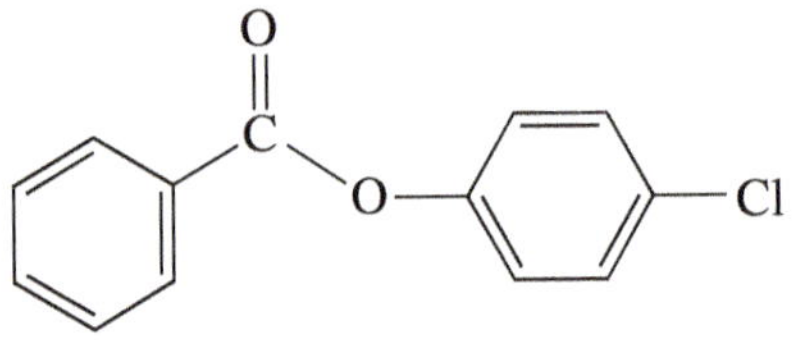 $C_6H_5-C(=O)-O-C_6H_4-Cl$

i. 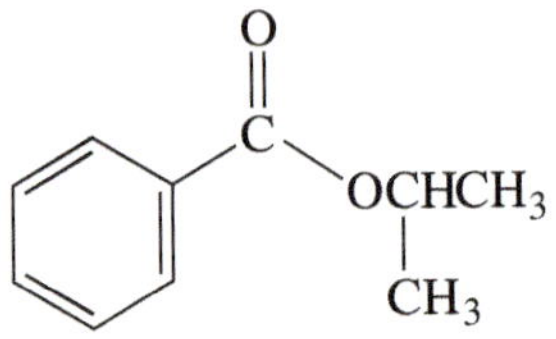 $C_6H_5-C(=O)-OCHCH_3$ (with CH_3 on the CH)

j. 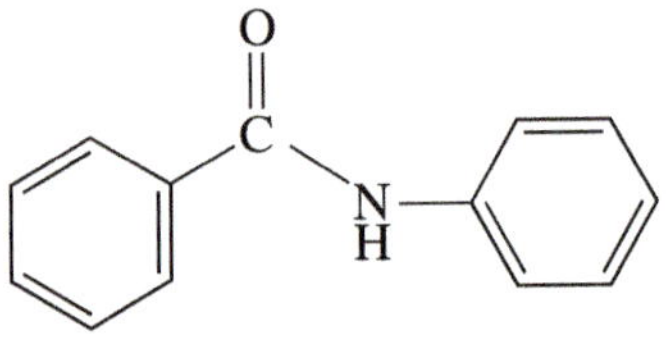 $C_6H_5-C(=O)-NH-C_6H_5$

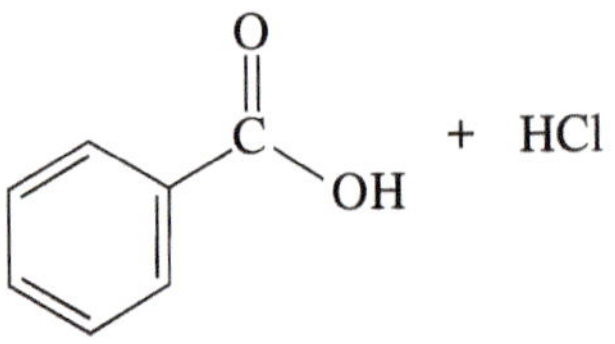

k. 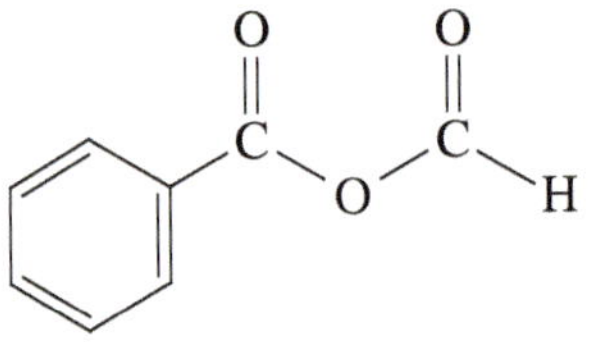 $C_6H_5-C(=O)-O-C(=O)-H$

51.

a. 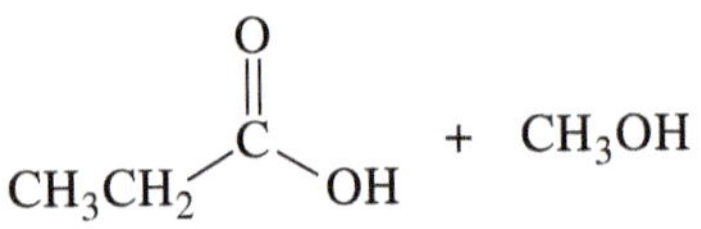 $CH_3CH_2-C(=O)-OH$ + CH_3OH

b. $C_6H_5-C(=O)-OH$ + HCl

c. $C_6H_5-C(=O)-OH$ + $CH_3CH_2\overset{+}{N}H_3$ Cl^-

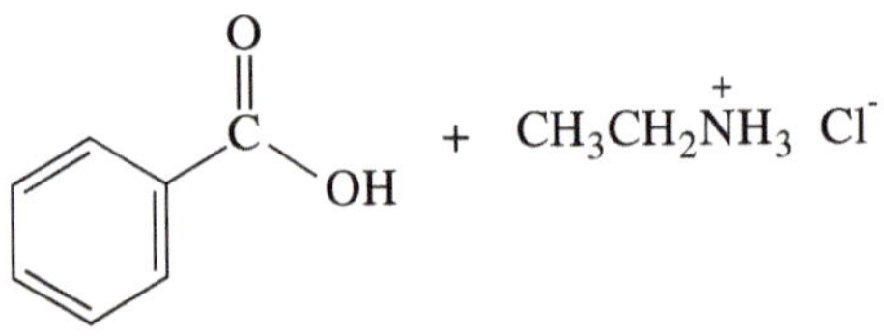

d. 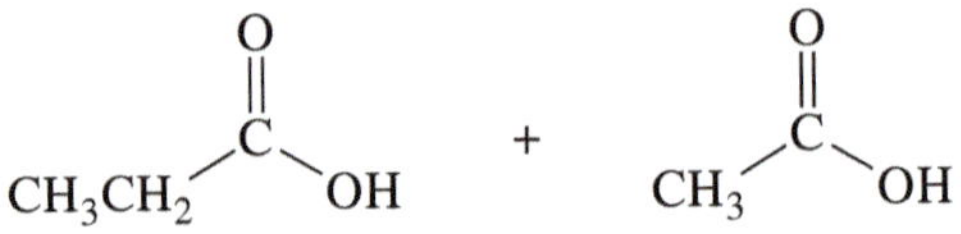 $CH_3CH_2-C(=O)-OH$ + $CH_3-C(=O)-OH$

52. **a.** The weaker the base attached to the acyl group, the stronger its electron-withdrawing ability; therefore, the easier it is to form the tetrahedral intermediate. *para*-Chlorophenol is a stronger acid than phenol, so the conjugate base of *para*-chlorophenol is a weaker base than the conjugate base of phenol.

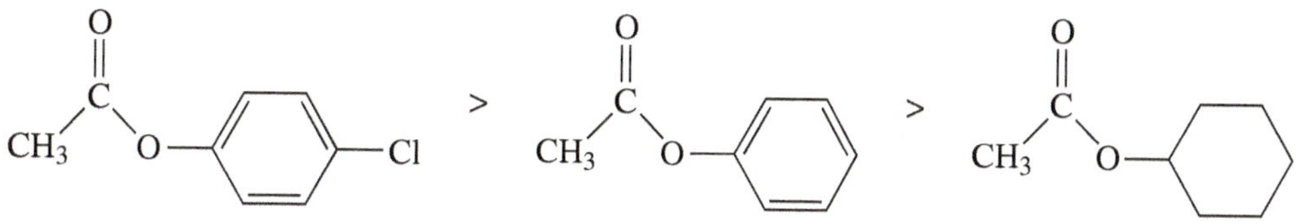

C > A > B

b. The tetrahedral intermediate collapses by eliminating the OR group. The weaker the OR group is as a base, the easier it is to eliminate.

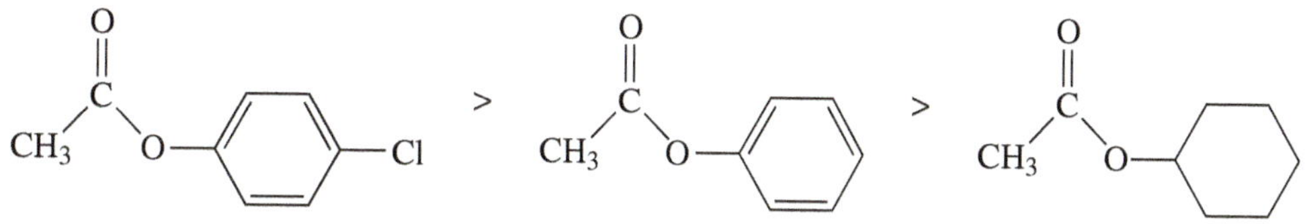

Thus, the rate of both formation of the tetrahedral intermediate and collapse of the tetrahedral intermediate is decreased by increasing the basicity of the OR group.

53. Because acetate ion is a weak base, the S_N2 reaction forms only a substitution product. (See page 430 of the text.) The product of the S_N2 reaction is an ester which, when hydrolyzed, forms cyclohexanol (the target molecule) and acetic acid.

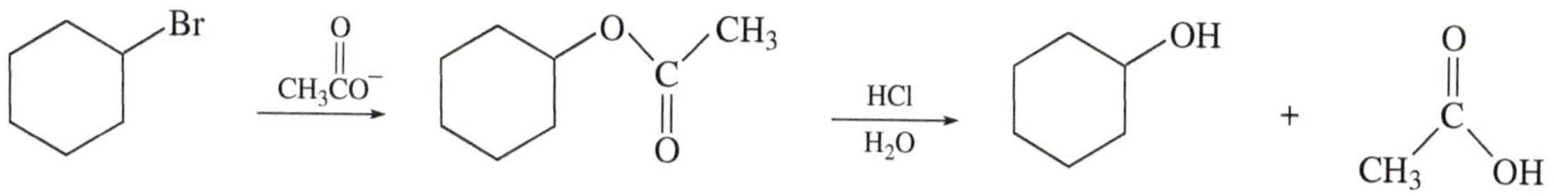

CH_3CO^- ; HCl, H_2O

54. **a.** Methyl acetate has a resonance contributor that butanone does not. This resonance contributor causes methyl acetate to be more polar than butanone. Because methyl acetate is more polar, it has the greater dipole moment.

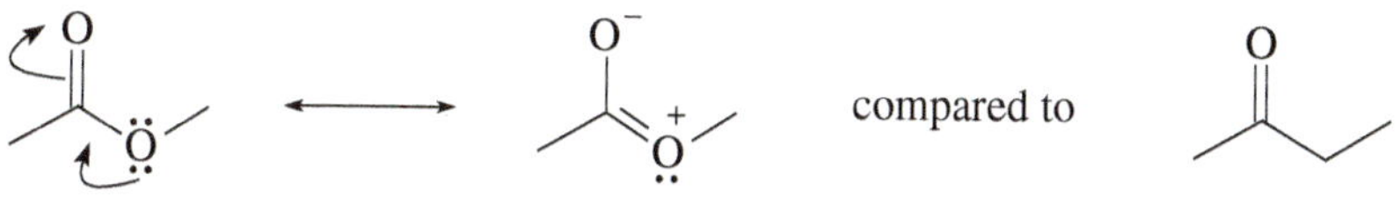

compared to

b. Because it is more polar, the intermolecular forces holding methyl acetate molecules together are stronger, so we expect methyl acetate to have a higher boiling point.

55. Propyl formate is easy to distinguish because it is the only ester that shows four signals. The other three esters each show three signals.

Isopropyl formate can be distinguished by its unique splitting pattern: a singlet, a doublet, and a septet. The splitting patterns of the other two esters are the same: a singlet, a triplet, and a quartet. They can be distinguished because the highest frequency signal is due to the protons attached to the carbon that is next to the oxygen. For ethyl acetate, it is a quartet, whereas for methyl propionate, it is a singlet.

$HC(=O)OCH_2CH_2CH_3$

4 signals

$HC(=O)OCH(CH_3)CH_3$ — **3 signals**; singlet, doublet, septet

$CH_3C(=O)OCH_2CH_3$ — **3 signals**; singlet, triplet, quartet; The highest frequency signal is a quartet.

$CH_3CH_2C(=O)OCH_3$ — **3 signals**; singlet, triplet, quartet; The highest frequency signal is a singlet.

56. The carbonyl IR absorption band decreases in the order:

$CH_3C(=O)Cl > CH_3C(=O)OCH_3 > CH_3C(=O)H > CH_3C(=O)NH_2$

The carbonyl IR absorption band of the acyl chloride occurs at the highest frequency because there is essentially no electron delocalization from a lone pair on chlorine and the chlorine withdraws electrons inductively. The carbonyl IR absorption band of the ester is next because the predominant effect of the oxygen of an ester is inductive electron withdrawal because resonance electron donation from the oxygen is minimal since it puts a positive charge on the electronegative oxygen. Therefore, it takes more energy to stretch the bond. (See page 592 of the text.)

$RC(=\ddot{O}:)\ddot{Y}$

The carbonyl group of the amide stretches at the lowest frequency because the nitrogen is less electronegative than oxygen, so it is better able to accommodate a positive charge. Therefore, the resonance contributor on the right makes a significant contribution to the overall structure. The larger the contribution from the resonance contributor on the right, the greater the single bond character of the C=O group.

$RC(=\ddot{O}:)\ddot{Y} \longleftrightarrow RC(-\ddot{O}:^-)=\ddot{Y}^+$

57. **a.** 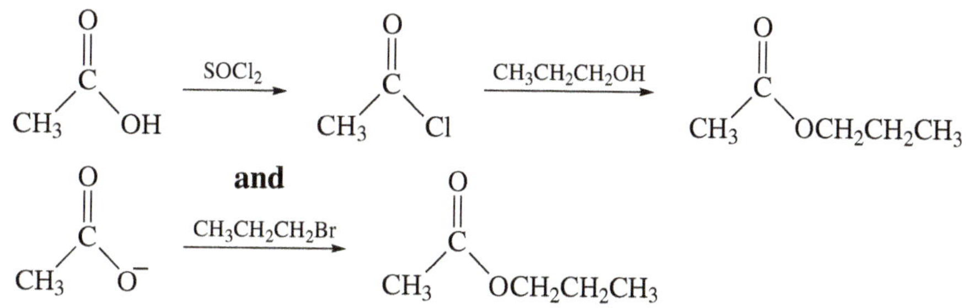 $CH_3COOH \xrightarrow{SOCl_2} CH_3COCl \xrightarrow{CH_3CH_2CH_2OH} CH_3COOCH_2CH_2CH_3$

and

$CH_3COO^- \xrightarrow{CH_3CH_2CH_2Br} CH_3COOCH_2CH_2CH_3$

b. 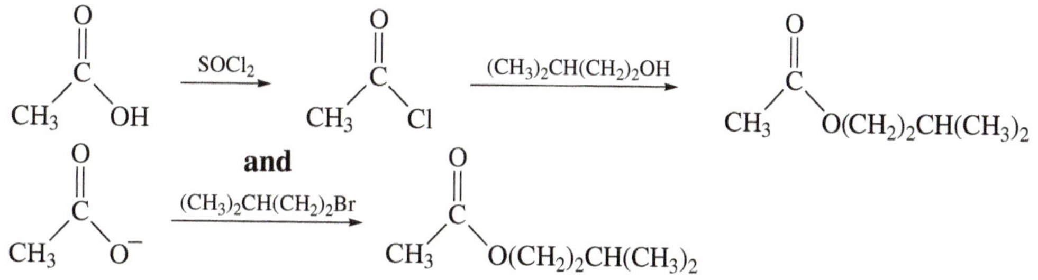 $CH_3COOH \xrightarrow{SOCl_2} CH_3COCl \xrightarrow{(CH_3)_2CH(CH_2)_2OH} CH_3COO(CH_2)_2CH(CH_3)_2$

and

$CH_3COO^- \xrightarrow{(CH_3)_2CH(CH_2)_2Br} CH_3COO(CH_2)_2CH(CH_3)_2$

c. 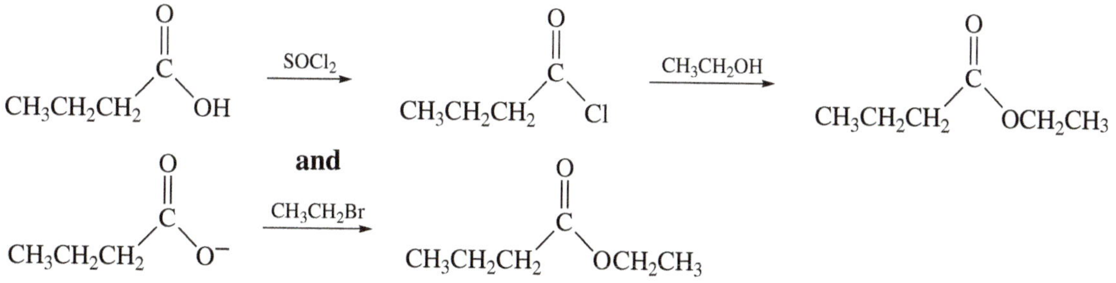 $CH_3CH_2CH_2COOH \xrightarrow{SOCl_2} CH_3CH_2CH_2COCl \xrightarrow{CH_3CH_2OH} CH_3CH_2CH_2COOCH_2CH_3$

and

$CH_3CH_2CH_2COO^- \xrightarrow{CH_3CH_2Br} CH_3CH_2CH_2COOCH_2CH_3$

d. 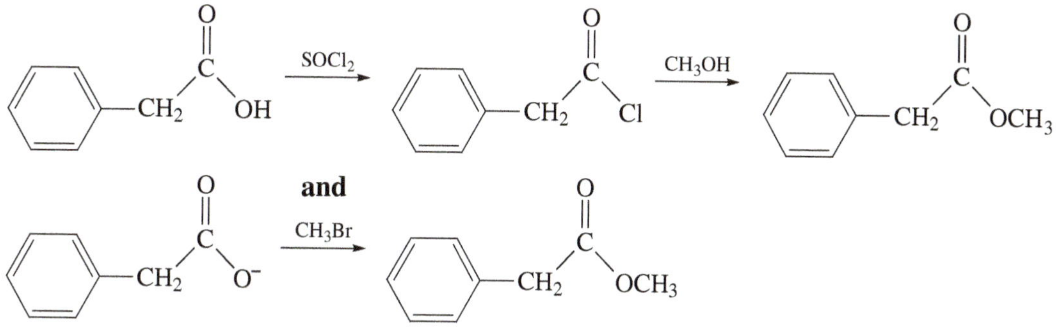 $C_6H_5CH_2COOH \xrightarrow{SOCl_2} C_6H_5CH_2COCl \xrightarrow{CH_3OH} C_6H_5CH_2COOCH_3$

and

$C_6H_5CH_2COO^- \xrightarrow{CH_3Br} C_6H_5CH_2COOCH_3$

58. **a.** isopropyl alcohol and HCl

b. aqueous sodium hydroxide

c. ethylamine

d. water and HCl

59. **a.** 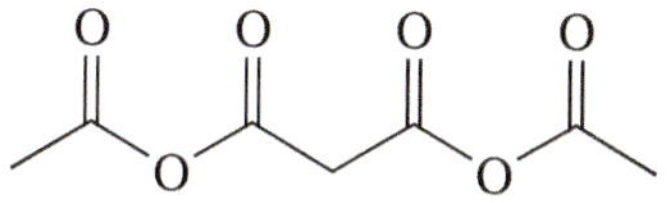 $CH_3COOCOCH_2COOCOCH_3$

b. 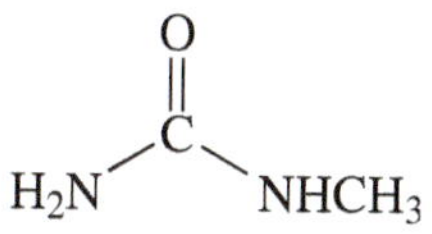 $H_2NCONHCH_3$

c. no reaction; an amide will not react with water without a catalyst

d. 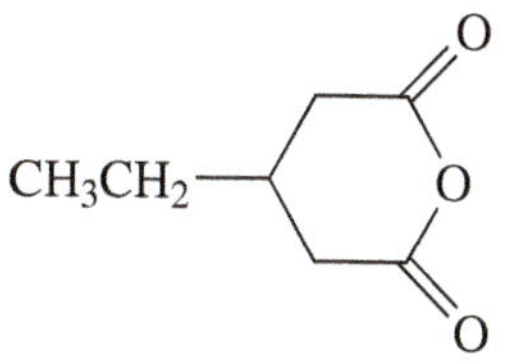 CH_3CH_2– (cyclic anhydride, six-membered ring with O and two C=O)

60. The offset in the NMR spectrum shows that there is a broad signal at ~ 10 ppm, which is characteristic of a COOH group. The two triplets and the multiplet are characteristic of a propyl group.

$$\underset{t}{CH_3}-\underset{m}{CH_2}-\underset{t}{CH_2}-$$

Therefore, the product of the hydrolysis reaction is butanoic acid.

$CH_3CH_2CH_2C(=O)OCH_3$ (reactant) $CH_3CH_2CH_2C(=O)OH$ (hydrolysis product)

The molecular formula shows that the reactant has one more carbon than butanoic acid; the IR spectrum shows that it is an ester. Because butanoic acid is formed from acid hydrolysis of the ester, the reactant must be the methyl ester of butanoic acid.

61. Aspartame has an amide group and an ester group that will be hydrolyzed in an aqueous solution of HCl. Because the hydrolysis is carried out in an acidic solution, the carboxylic acid groups and the amino groups in the products will be in their acidic forms.

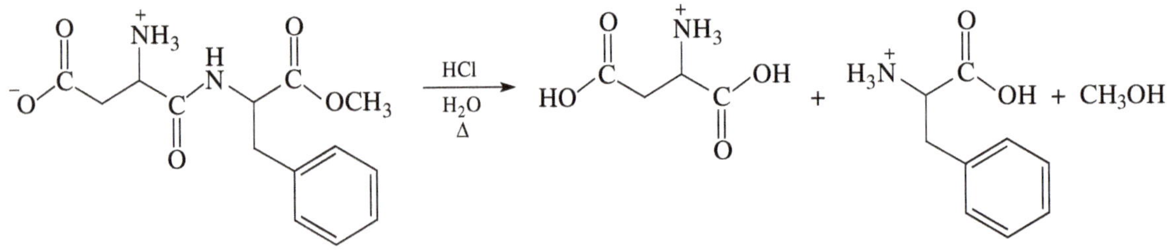

$$\text{aspartame} \xrightarrow[H_2O,\ \Delta]{HCl} \text{HOOC-CH}_2\text{-CH}(\overset{+}{N}H_3)\text{-COOH} + H_3\overset{+}{N}\text{-CH(CH}_2C_6H_5)\text{-COOH} + CH_3OH$$

62. If it were an S_N2 reaction, the ester would be the only species that would be isotopically labeled and there would be no unlabeled ester because the carbonyl group would not have participated in the reaction.

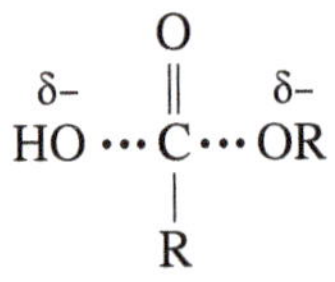

$$\overset{\delta-}{HO}\cdots \overset{O}{\overset{\|}{\underset{R}{\underset{|}{C}}}}\cdots \overset{\delta-}{OR}$$

If the reaction formed a tetrahedral intermediate, some of the ester would not be isotopically labeled because some of the label would have been transferred to hydroxide ion.

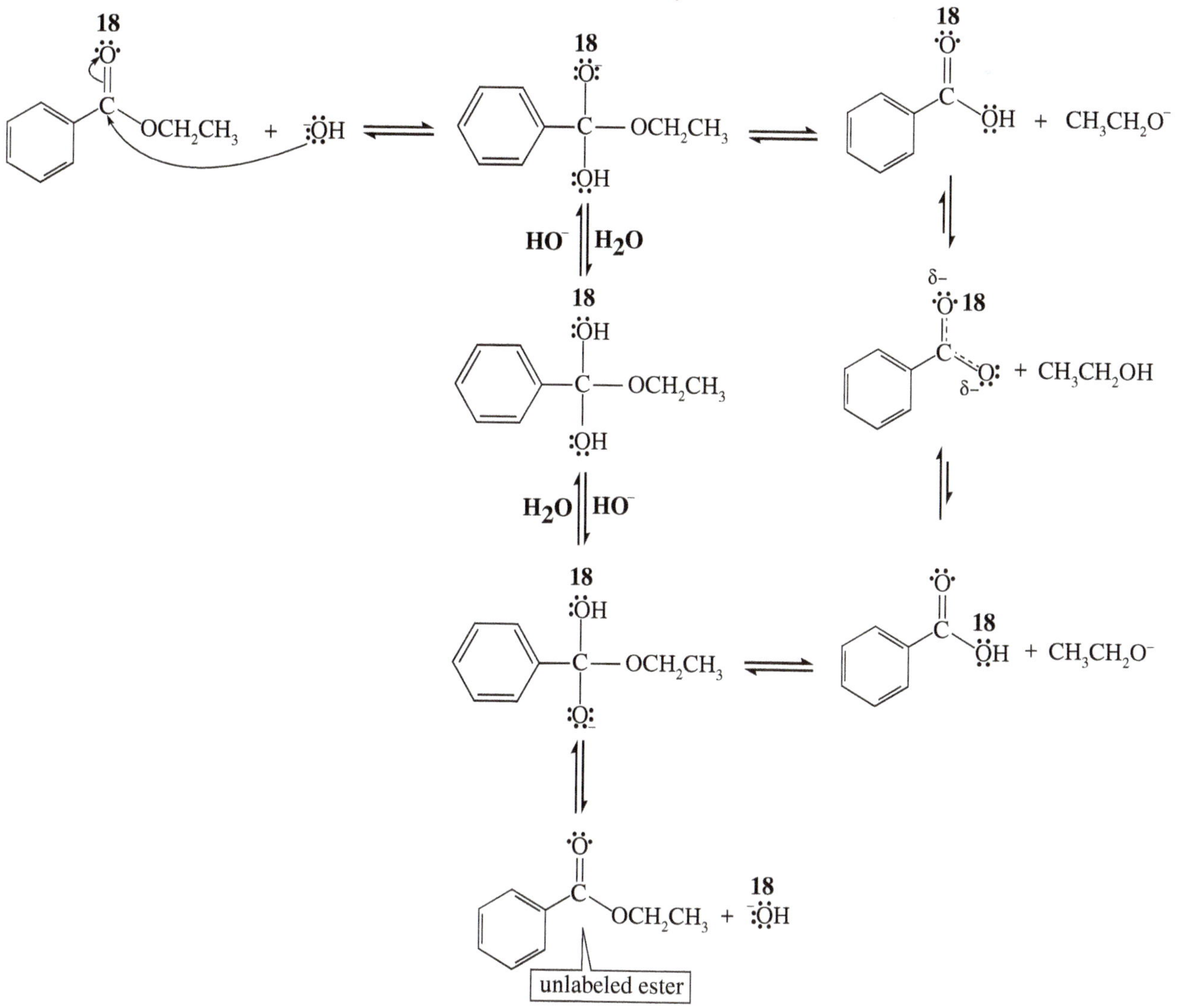

18 O, OCH_2CH_3 + :ÖH ⇌ 18 O⁻, C—OCH_2CH_3, :ÖH ⇌ 18 O, C, ÖH + $CH_3CH_2O^-$

HO⁻ | H_2O

18 :ÖH, C—OCH_2CH_3, :ÖH

δ– Ö 18, C, O: δ– + CH_3CH_2OH

H_2O | HO⁻

18 :ÖH, C—OCH_2CH_3, :O:⁻ ⇌ Ö, C, 18 ÖH + $CH_3CH_2O^-$

Ö, C, OCH_2CH_3 + 18 :ÖH⁻

unlabeled ester

63. **a.** 1, 3, 4, 6, 7, 9 will not form the indicated products under the given conditions.
b. 9 will form the product shown in the presence of an acid catalyst.

64. Notice that in order to maximize the substitution product, a poor nucelophile is used for the substitution reaction of the alkyl halide.

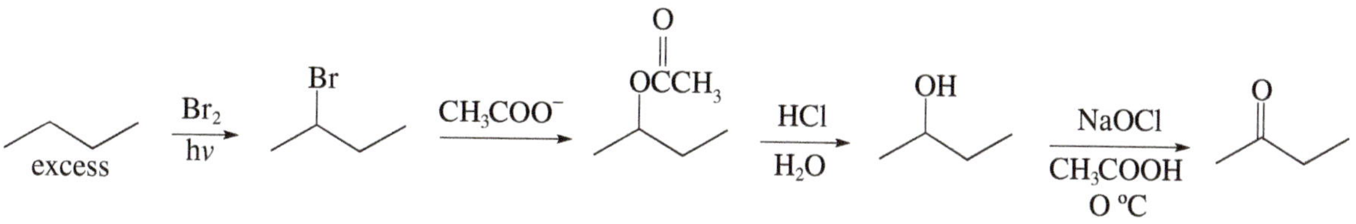

excess $\xrightarrow[h\nu]{Br_2}$ Br $\xrightarrow{CH_3COO^-}$ $OCCH_3$ (O) $\xrightarrow[H_2O]{HCl}$ OH $\xrightarrow[CH_3COOH, 0\ °C]{NaOCl}$ O

65. The tertiary amine is a better nucleophile than the alcohol,
so formation of the amide with a positively charged nitrogen will be faster than formation of the new ester would have been.
The positively charged amide is more reactive than an ester,

so formation of the new ester by reaction of the alcohol with the charged amide will be faster than formation of the ester by reaction of the alcohol with the starting ester would have been.
In other words, both reactions that occur in the presence of the tertiary amine are faster than the single reaction that occurs in the absence of the tertiary amine.

$$CH_3C(=O)OCH_3 + :N(CH_2CH_2)_3N: \rightleftharpoons CH_3C(=O)\overset{+}{N}(CH_2CH_2)_3N: + CH_3OH \xrightleftharpoons{ROH} CH_3C(=O)OR + :N(CH_2CH_2)_3N:$$

66. **a.** $CH_3COOH + CH_3CH_2OH \rightleftharpoons CH_3CO_2CH_2CH_3 + H_2O$

$$K_{eq} = \frac{[CH_3CO_2CH_2CH_3]\,[H_2O]}{[CH_3COOH]\,[CH_3CH_2OH]}$$

$$4.02 = \frac{x^2}{(1-x)^2}$$

take the square root of both sides

$$2.00 = \frac{x}{(1-x)}$$

$$2 - 2x = x$$

$$2 = 3x$$

$$x = 0.667$$

[ethyl acetate] = 0.667 times the concentration of acetic acid used

b. $$4.02 = \frac{x^2}{(10-x)\,(1-x)}$$

$$4.02 = \frac{x^2}{x^2 - 11x + 10}$$

$$4.02\,(x^2 - 11x + 10) = x^2$$

$$3.02x^2 - 44.22x - 4.02 = 0$$

solving for x using the quadratic formula:

$$x = 0.974$$

[ethyl acetate] = 0.974 times the concentration of acetic acid used

c. $$4.02 = \frac{x^2}{(100-x)\,(1-x)}$$

$$x = 0.997$$

[ethyl acetate] = 0.997 times the concentration of acetic acid used

67. 1-Bromobutane undergoes an S_N2 reaction with NH_3 to form butylamine (**A**). Butylamine can then react with 1-bromobutane to form dibutylamine (**B**). The amines each form an amide upon reaction with acetyl chloride. The IR spectrum of **C** exhibits an NH stretch at about ~3300 cm^{-1}, whereas the IR spectrum of **D** does not exhibit this absorption band because the nitrogen in **D** is not bonded to a hydrogen.

$$CH_3CH_2CH_2CH_2Br \xrightarrow{NH_3} \underset{\textbf{A}}{CH_3CH_2CH_2CH_2NH_2} \xrightarrow{CH_3CH_2CH_2CH_2Br} \underset{\textbf{B}}{(CH_3CH_2CH_2CH_2)_2NH}$$

$$CH_3C(=O)Cl + \underset{\textbf{A}}{CH_3CH_2CH_2CH_2NH_2} \longrightarrow \underset{\textbf{C}}{CH_3C(=O)NHCH_2CH_2CH_2CH_3}$$

$$CH_3C(=O)Cl + \underset{\textbf{B}}{(CH_3CH_2CH_2CH_2)_2NH} \longrightarrow \underset{\textbf{D}}{CH_3C(=O)N(CH_2CH_2CH_2CH_3)_2}$$

68. a. $CH_3C(=O)F$

b. $HOC(=O)CH_2CH_2CH_2\overset{+}{N}H_3\ Cl^-$

c. $C_6H_5C(=O)NHCH_3$

d. $HOC(=O)CH_2CH_2C(=O)OH$

e. benzene ring fused to a five-membered cyclic carbonate (O–C(=O)–O)

f. $HOC(=O)CH_2CH_2CH_2OH$

g. $CH_3C(=O)CH_2OH + CH_3OC(=O)CH_3$

h. benzene ring fused to a six-membered cyclic anhydride (ring with CH_2, C=O, O, C=O)

i. benzene ring with ortho substituents $C(=O)NH_2$ and $C(=O)O^-$

j. benzene ring with ortho substituents $C(=O)OCH_3$ and $CH_2C(=O)OCH_3$

69. **a.** 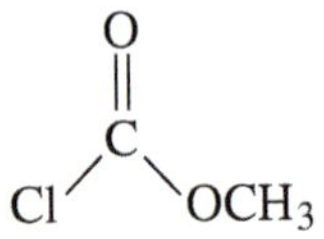 $Cl-C(=O)-OCH_3$

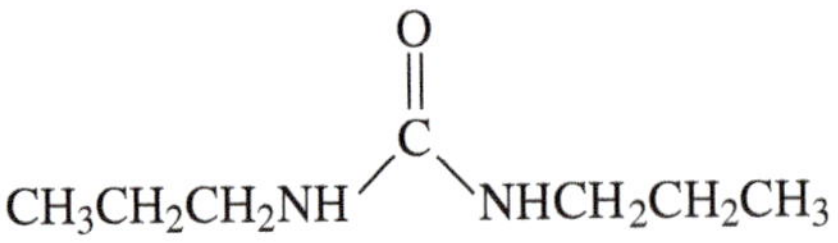

b. 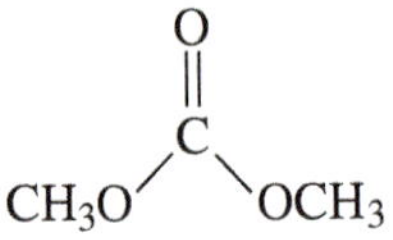 $CH_3O-C(=O)-OCH_3$

c. $CH_3CH_2CH_2NH-C(=O)-NHCH_2CH_2CH_3$

d.  $HO-C(=O)-OH \rightleftharpoons CO_2 + H_2O$

see page 77 of the text

70. This is the reaction of an acyl chloride or an anhydride with an alcohol.

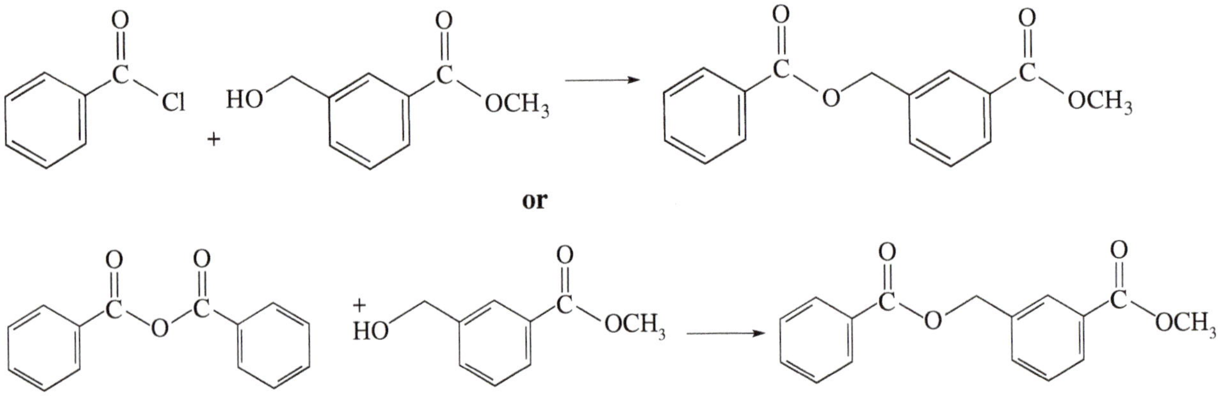

$C_6H_5C(=O)Cl$ + HO–CH_2–C_6H_4–$C(=O)OCH_3$ ⟶ $C_6H_5C(=O)O$–CH_2–C_6H_4–$C(=O)OCH_3$

or

$C_6H_5C(=O)–O–C(=O)C_6H_5$ + HO–CH_2–C_6H_4–$C(=O)OCH_3$ ⟶ $C_6H_5C(=O)O$–CH_2–C_6H_4–$C(=O)OCH_3$

71. 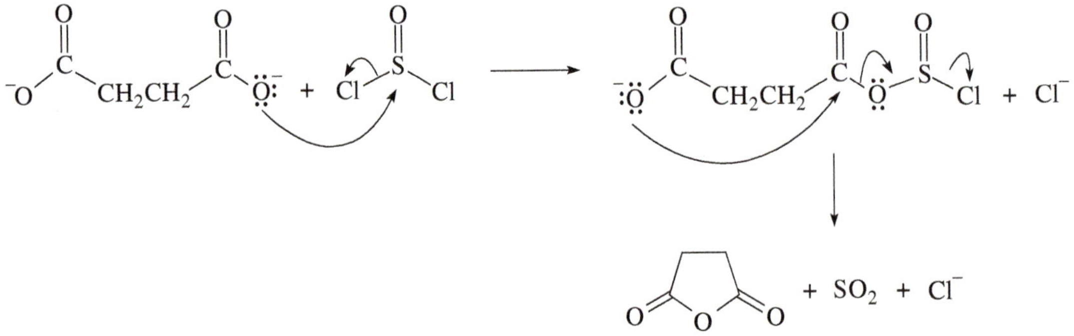 $^{-}O-C(=O)-CH_2CH_2-C(=O)-\ddot{O}:^{-}$ + $Cl-S(=O)-Cl$ ⟶ $:\ddot{O}^{-}-C(=O)-CH_2CH_2-C(=O)-\ddot{O}-S(=O)-Cl$ + Cl^{-}

↓

(cyclic anhydride) + SO_2 + Cl^{-}

72. 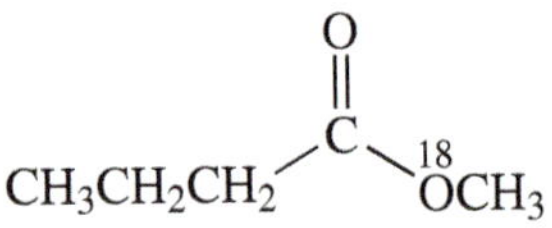 $CH_3CH_2CH_2-C(=O)-{}^{18}OCH_3$ and $CH_3{}^{18}OH$

73. 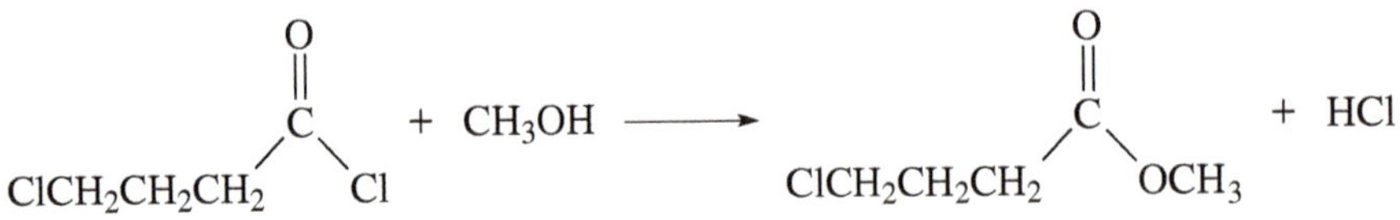 $ClCH_2CH_2CH_2-C(=O)-Cl$ + CH_3OH ⟶ $ClCH_2CH_2CH_2-C(=O)-OCH_3$ + HCl

Compound **A**

74. **a.** The amino alcohol has two groups that can be acetylated by acetic anhydride, the NH_2 group and the OH group.

$$2\ CH_3C(=O)OC(=O)CH_3 + CH_3CH(NH_2)CH_2CH_2OH \longrightarrow CH_3C(=O)NHCH(CH_3)CH_2CH_2OC(=O)CH_3 + 2\ CH_3C(=O)O^-$$

b. Because the NH_2 group is a better nucleophile than the OH group, it will be the first group that is acetylated. Therefore, if the reaction is stopped before it is half over, the product will have only its NH_2 group acetylated.

$$CH_3C(=O)NHCH(CH_3)CH_2CH_2OH$$

75. If the amine is tertiary, the nitrogen in the amide product cannot get rid of its positive charge by losing a proton. An amide with a positively charged nitrogen is very reactive toward nucleophilic acyl substitution because the positively charged nitrogen-containing group is an excellent leaving group. Therefore, if water is added, it will immediately react with the amide and, because the $^+NR_3$ group is a better leaving group than the OH group, the $^+NR_3$ group will be eliminated. The product will be a carboxylic acid, which will lose its acidic proton to the amine.

$$CH_3C(=O)Cl \xrightarrow{2\ NR_3} CH_3C(=O)\overset{+}{N}R_3 + H\overset{+}{N}R_3\ Cl^- \xrightarrow{H_2O} CH_3C(=O)OH + NR_3 \rightleftharpoons CH_3C(=O)O^- + H\overset{+}{N}R_3$$

76. The reaction of methylamine with propionyl chloride generates a proton that will protonate unreacted amine, thereby destroying the amine's nucleophilicity. If two equivalents of CH_3NH_2 are used, one equivalent will remain unprotonated and be able to react as a nucleophile with propionyl chloride to form *N*-methylpropanamide.

$$CH_3CH_2C(=O)Cl + CH_3NH_2 \longrightarrow CH_3CH_2C(=O)NHCH_3 + H^+ + Cl^-$$

$$H^+ \xrightarrow{CH_3NH_2} CH_3\overset{+}{N}H_3$$

77. **a.**

Ö: / CH_3–C–OH + H–Cl ⇌ $\overset{+}{O}$H / CH_3–C–OH + $H_2\overset{18}{O}$: ⇌ CH_3—C(OH)—OH with $^{18}\overset{+}{O}H$ (H :B) ⇌ CH_3—C(OH)—ÖH with ^{18}OH (H–B^+) ⇌ CH_3—C(:ÖH)—$\overset{+}{O}$H(H) with ^{18}OH

⇌ $\overset{+}{O}H$ / CH_3–C–^{18}OH + $H_2\ddot{O}$:^{18}

⇌ CH_3—C(OH)—^{18}OH with $^{18}\overset{+}{O}H$ (H :B) ⇌ CH_3—C(:ÖH)—^{18}OH with ^{18}OH (H–B^+) ⇌ CH_3—C($\overset{+}{O}$H H)—^{18}ÖH with ^{18}OH ⇌ $^{18}\overset{+}{O}$–H (:B) / CH_3–C–^{18}OH ⇌ ^{18}O / CH_3–C–^{18}OH

b. The carboxyl oxygen will be labeled.
Only one isotopically labeled oxygen can be incorporated into the ester because the bond between the methyl group and the labeled oxygen does not break. Therefore, there is no way for the carbonyl oxygen to become labeled.

cannot become labeled

O / CH_3–C–^{18}O–CH_3

bond does not break

c. In the presence of an acid catalyst, the ester will be hydrolyzed to a carboxylic acid and an alcohol. Both oxygen atoms of the carboxylic acid will be labeled for the same reason both oxygen atoms of the carboxylic acid are labeled in part **a**. The alcohol will not contain any label, because the bond between the methyl group and the oxygen does not break.

78. **a.** The steric hindrance provided by the methyl groups prevents methyl alcohol from attacking the carbonyl carbon.

COOH
H_3C
CH_3
CH_3

b. No, because there would be no steric hindrance to nucleophilic attack.

79. The amine is a better nucleophile than the alcohol but, because the acyl chloride is very reactive, it can react easily with both nucleophiles. Therefore, steric hindrance is the most important factor in determining the rate of formation of the products. The amino group is less sterically hindered than the alcohol group, so that will be the group most easily acetylated.

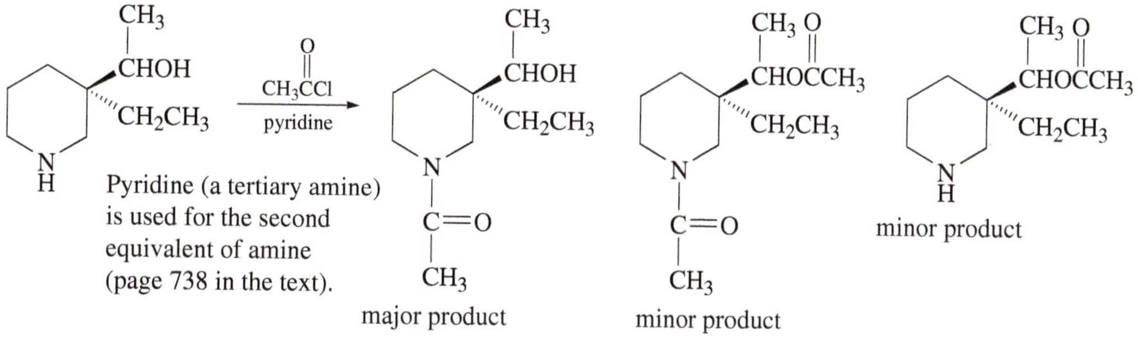

Pyridine (a tertiary amine) is used for the second equivalent of amine (page 738 in the text).

major product

minor product

minor product

80. The spectrum shows that the compound has four different kinds of hydrogens with relative ratios 2:1:1:6. The doublet at 0.9 ppm that integrates to 6 protons suggests that the compound has an isopropyl group. The signal at 3.4 ppm that integrates to 2 hydrogens indicates a methylene group that is attached to an oxygen. We can conclude that the spectrum is that of isobutyl alcohol. The molecular formula indicates that the compound that undergoes hydrolysis is an ester. Subtracting the atoms due to the isobutyl group lets us identify the ester as isobutyl benzoate.

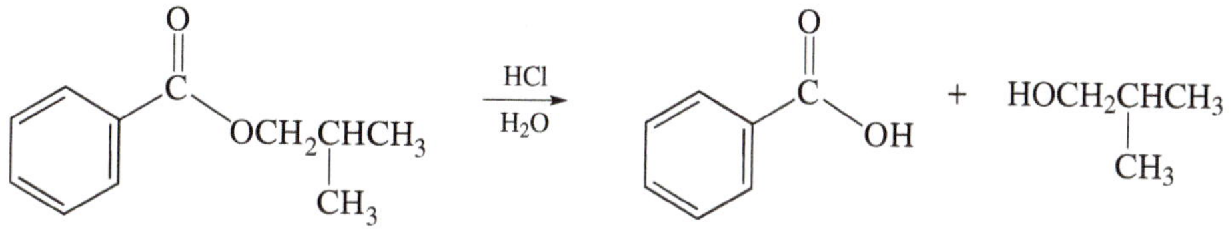

81. **a.** K_2CO_3 causes the solution to be at a pH that will cause the reactant to lose one of its protons. The electron-withdrawing carbonyl group makes its α-carbon more susceptible to nucleophilic attack than its β-carbon.

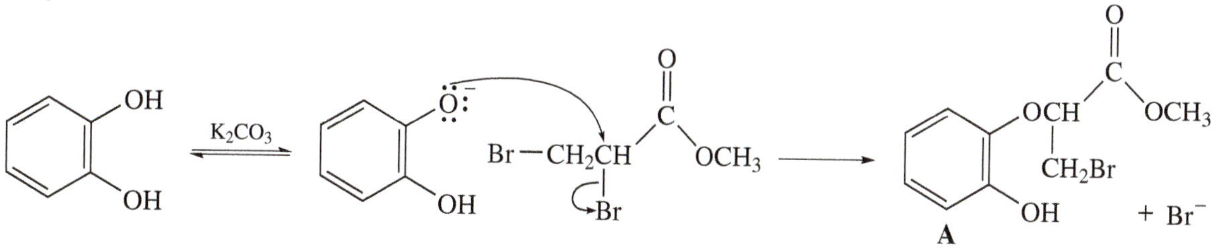

b.

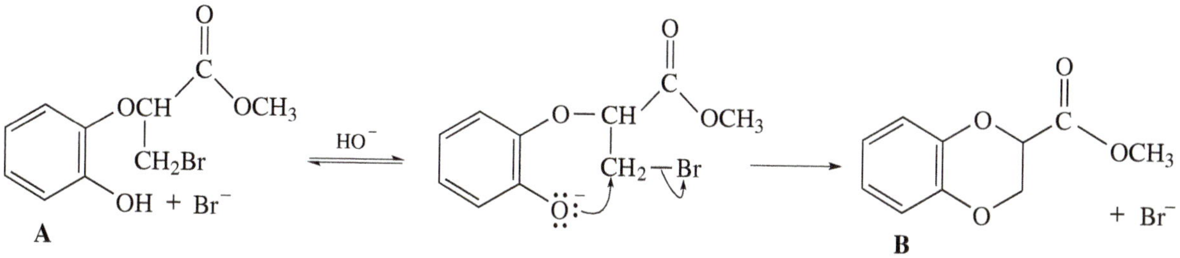

The *intramolecular* formation of **B** occurs more rapidly than the *intermolecular* formation of **A** because in the reaction that forms **B**, the reagents do not have to wander through the solution to find one another.

82. **a.** R′—ÖH → (H—B⁺) → R′—$\overset{+}{O}$H(H) → R′⁺ + R—C≡N: → R—C≡$\overset{+}{N}$R′ + H_2O: → ... R—C(=NR′)—$\overset{+}{O}$H—H + :B ⇌ R—C(=NR′)—OH + H—B⁺ ⇌ R—C(=$\overset{+}{N}$HR′)—:ÖH ⟷ R—C(NHR′)=$\overset{+}{O}$—H + :B ⇌ R—C(=O)—NHR′ + HB^+

b. The Ritter reaction does not work with primary alcohols because primary alcohols do not form carbocations.

c. The only difference in the two reactions is the electrophile that attaches to the nitrogen of the nitrile: it is a carbocation in the Ritter reaction and a proton in the acid-catalyzed hydrolysis of a nitrile.

83. Before the bond breaks, the three rings are not conjuated. The rings are conjugated in the carbocation. Therefore, the carbocation is a highly colored species.

H^+ ⇌ HO^- ; COOH

84. Each of the NMR spectra has signals between about 7 and 8 ppm (integrating to 5H), indicating that the compound has a benzene ring with one substituent. There is one additional signal that is a singlet. From the molecular formulas, it can be determined that the esters have the following structures. The singlet in each spectrum is due to a methyl group. Because the methyl group is at a higher frequency in the lower spectrum, it is the spectrum of the compound on the right, since its methyl group is adjacent to an electron-withdrawing oxygen. The compound on the left will be hydrolyzed more rapidly because its OR group is a weaker base. (The pK_a of phenol is ~10; the pK_a of methanol is ~16.)

CH_3—C(=O)—O—C_6H_5 C_6H_5—C(=O)—OCH_3

85. **a.** $CH_3CH_2C(=O)NH_2 \xrightarrow[H_2O,\ \Delta]{HO^-} CH_3CH_2C(=O)O^- \xrightarrow{SOCl_2} CH_3CH_2C(=O)Cl$

b. $CH_3CH_2CH_2CH_2OH \xrightarrow[pyridine]{SOCl_2} CH_3CH_2CH_2CH_2Cl \xrightarrow{^-C{\equiv}N} CH_3CH_2CH_2CH_2C{\equiv}N \xrightarrow{H_2O\ |\ HCl,\ \Delta} \overset{+}{N}H_4 + CH_3CH_2CH_2CH_2C(=O)OH$

c. $C_6H_5CH_3 \xrightarrow[2.\ HO^-]{1.\ H_2CrO_4/\Delta} C_6H_5C(=O)O^- \xrightarrow{SOCl_2} C_6H_5C(=O)Cl \xrightarrow{excess\ CH_3NH_2} C_6H_5C(=O)NHCH_3$

d. $C_6H_5CH_3 \xrightarrow[peroxide]{NBS,\ \Delta} C_6H_5CH_2Br \xrightarrow{^-C{\equiv}N} C_6H_5CH_2C{\equiv}N \xrightarrow[HCl,\ \Delta]{H_2O} C_6H_5CH_2C(=O)OH$

86. The acid-catalyzed hydrolysis of acetamide forms acetic acid and ammonium ion. It is an irreversible reaction because the pK_a of an acetic acid is less than the pK_a of the ammonium ion. Therefore, it is impossible to have the carboxylic acid in its reactive acidic form and ammonia in its reactive basic form.

If the solution is sufficiently acidic to have the carboxylic acid in its acidic form, ammonia will also be in its acidic form, so it will not be a nucleophile.

$CH_3C(=O)OH + \overset{+}{N}H_4 \longrightarrow$ no reaction
The ammonium ion is not nucleophilic.

If the pH of the solution is sufficiently basic to have ammonia in its nucleophilic basic form, the carboxylic acid will also be in its basic form; a negatively charged carboxylate ion cannot be attacked by nucleophiles.

$CH_3C(=O)O^- + NH_3 \longrightarrow$ no reaction
The carboxylate ion is not attacked by nucleophiles.

87. **a.** (six-membered lactone with CH_2CH_3 substituent)

b. (bicyclic lactone fused to cyclopentane, H, H) $+ CH_3CH_2OH$

88.

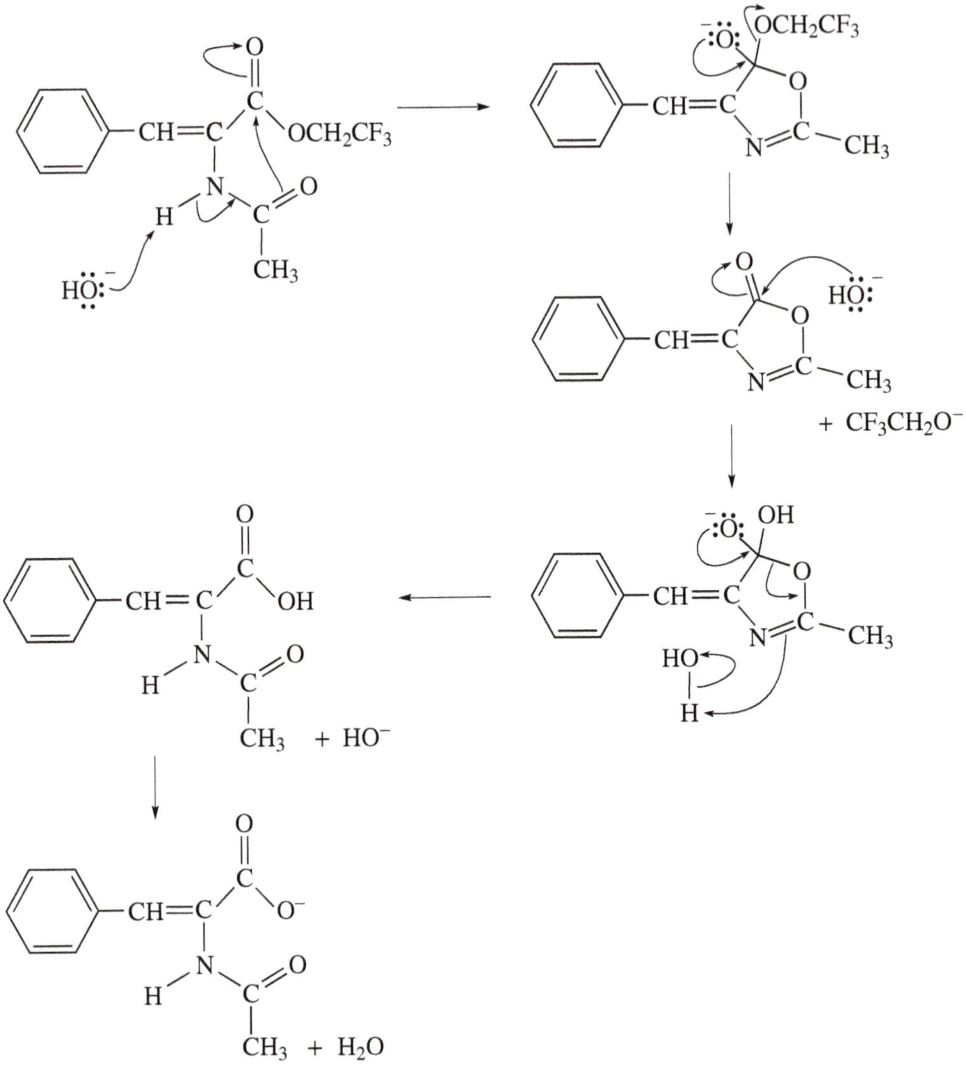

$$\text{C}_6\text{H}_5\text{–CH=C(NHCOCH}_3\text{)–COOCH}_2\text{CF}_3 \longrightarrow \text{oxazolone tetrahedral intermediate} \longrightarrow \text{oxazolone} + CF_3CH_2O^- \xrightarrow{HO^-} \text{tetrahedral intermediate} \longrightarrow \text{C}_6\text{H}_5\text{–CH=C(NHCOCH}_3\text{)–COOH} + HO^- \longrightarrow \text{C}_6\text{H}_5\text{–CH=C(NHCOCH}_3\text{)–COO}^- + H_2O$$

89. An enzyme called penicillinase provides resistance to penicillin using its CH_2OH group to open the four-membered ring. (See page 714 of the text.)

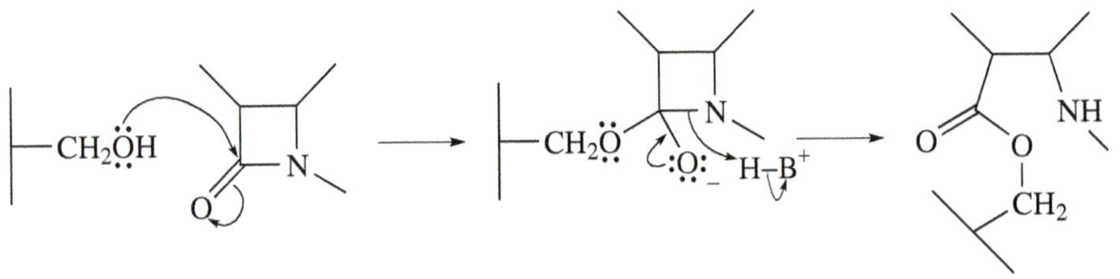

$$\text{enzyme–}CH_2OH + \beta\text{-lactam} \longrightarrow \text{tetrahedral intermediate} \xrightarrow{H\text{–}B^+} \text{ring-opened ester (enzyme–}CH_2\text{–O–C(=O)–)} + \text{NH}$$

The inhibitor of penicillinase is a phosphate ester with an excellent leaving group, so it readily reacts with nucleophiles. When the CH_2OH group of penicillinase attacks the inhibitor, a relatively stable phosphonyl-enzyme is formed, so its OH group is no longer available to react with penicillin. (The two sets of double arrows indicate formation and collapse of the tetrahedral intermediate.)

$CH_2\ddot{O}H$ H N P O O O O^- O^- H–B^+ → H N P O CH_2 O O^-

a phosphonyl-enzyme

+ HO– COO^-

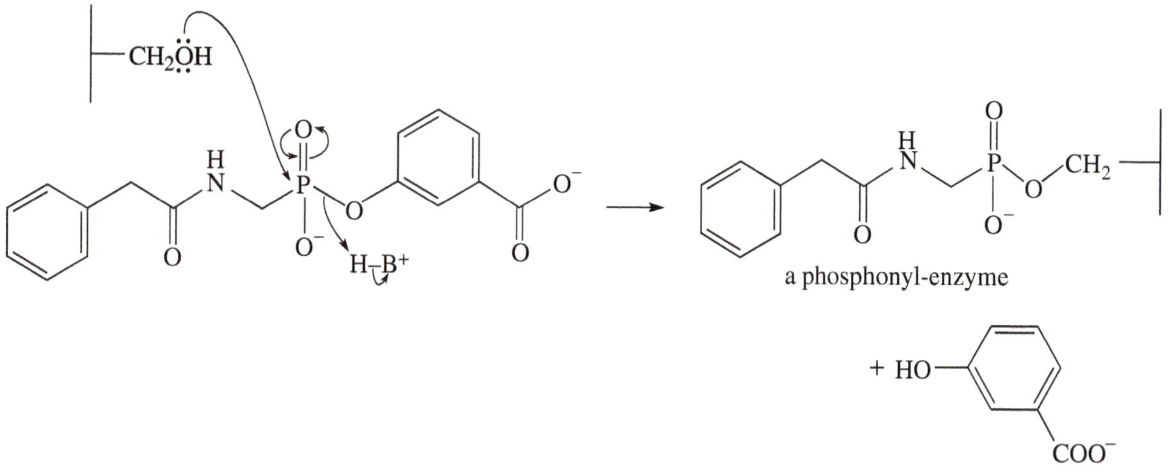

Hydroxylamine (H_2NOH) reactivates penicillinase by liberating the enzyme from the phosphonyl-enzyme. (Unlike alkylation reactions with hydroxylamine where nitrogen is the nucleophile, it is known that oxygen is the nucleophile when hydroxylamine reacts with phosphate esters.) Again, the two sets of double arrows indicate formation and collapse of the tetrahedral intermediate.

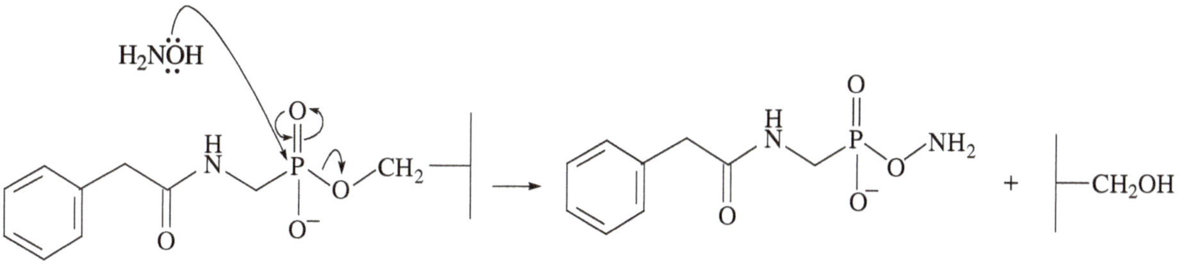

$H_2N\ddot{O}H$ H N O P O CH_2 O O^- → H N O P O NH_2 O O^- + CH_2OH

90. $H\ddot{O}$ $\ddot{N}HCO_2CH_3$ $\underset{\text{HCl}}{\rightleftharpoons}$ H $H\ddot{O}^+$ $\ddot{N}HCO_2CH_3$ →

CO_2CH_3 $^+\ddot{N}$ H :B ⇌ N–CO_2CH_3 HB^+

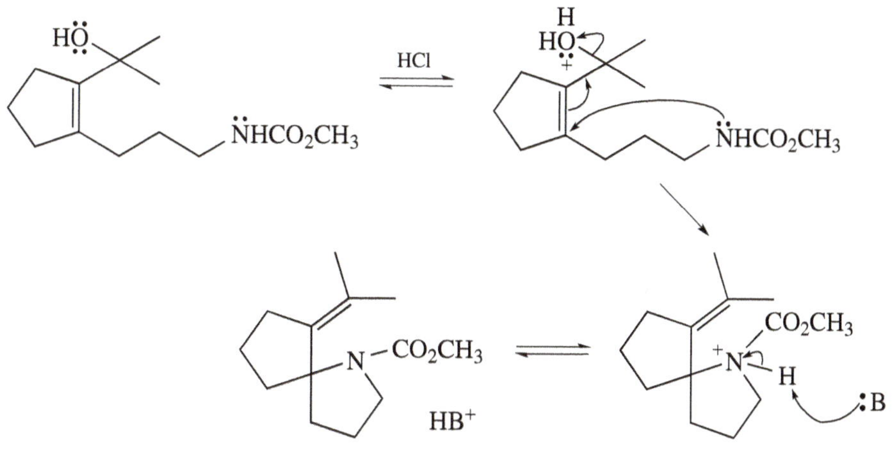

91. **a.** HO, O, δ−, HO, δ−, O, C, O, NO_2 HO, O, δ−, HO, O, C, δ−, O, NO_2

transition state if formation of the tetrahedral intermediate is rate-limiting

transition state if collapse of the tetrahedral intermediate is rate-limiting

b. $\overset{+}{N}$, O, C, O, NO_2 $\xrightarrow{H_2O}$ $\overset{+}{N}$, O, C, OH + HO, NO_2

c. HO, O, O, H N, $^-$O, O, P, O, NO_2

or NH

92. Because electron-withdrawing substituents have positive substituent constants and electron-donating substituents have negative substituent constants, a reaction with a positive ρ value is one in which compounds with electron-withdrawing substituents react more rapidly than compounds with electron-donating substituents, and a reaction with a negative ρ value is one in which compounds with electron-donating substituents react more rapidly than compounds with electron-withdrawing substituents.

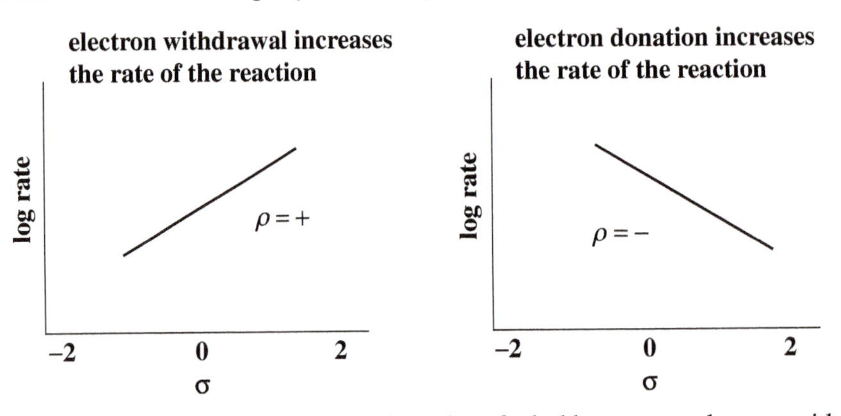

a. In the hydroxide-ion-promoted hydrolysis of a series of ethyl benzoates, electron-withdrawing substituents increase the rate of the reaction by increasing the amount of positive charge on the carbonyl carbon, thereby making it more readily attacked by hydroxide ion. The ρ value for this reaction is, therefore, positive.

O, C, OCH_2CH_3, Y, $H\ddot{O}:^-$ ⟶

In amide formation with a series of anilines, electron donation increases the rate of the reaction by increasing the nucleophilicity of the aniline. The ρ value for this reaction is, therefore, negative.

O
C
Cl
$H_2\ddot{N}$
Y

b. Because ortho substituents are close to the site of the reaction, they introduce steric factors into the rate constant for the reaction. In other words, the presence of an ortho substituent can slow a reaction down not because it can donate or withdraw electrons, but because it can get in the way of the nucleophile. Therefore, any change in the rate is due to a combination of steric effects and the electron-donating or electron-withdrawing ability of the substituent.
Because the change in rate cannot be attributed solely to the electron-donating or electron-withdrawing ability of the substituent, ortho-substituted compounds were not included in the study.

c. An electron-withdrawing substituent will make it easier for benzoic acid to lose a proton, so ionization of a series of substituted benzoic acids will show a positive ρ value.

Chapter 15 Practice Test

1. Circle the compound in each pair that is more reactive toward nucleophilic acyl substitution.

a. 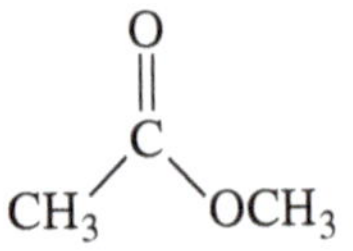 $CH_3C(=O)OCH_3$ **or** $CH_3C(=O)NHCH_3$

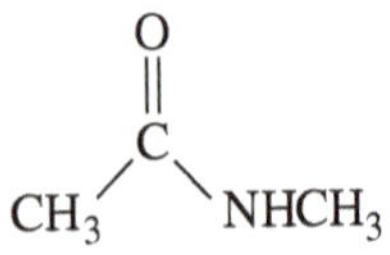

b. 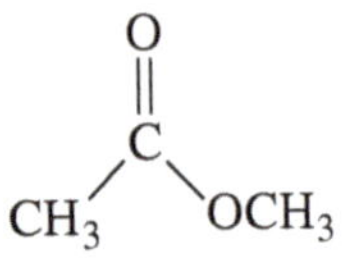 $CH_3C(=O)OCH_3$ **or** $CH_3C(=O)O{-}C_6H_5$

c. $CH_3C(=O)O{-}C_6H_4{-}OCH_3$ **or** $CH_3C(=O)O{-}C_6H_4{-}NO_2$

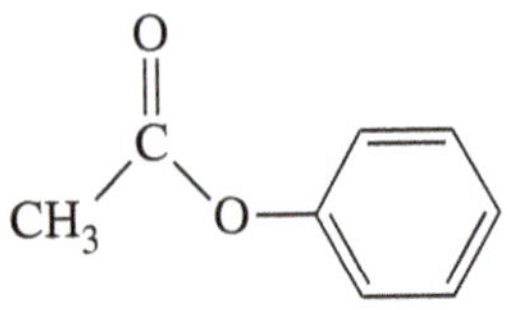

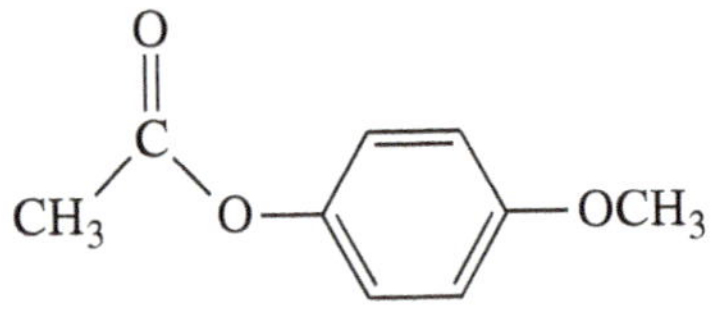

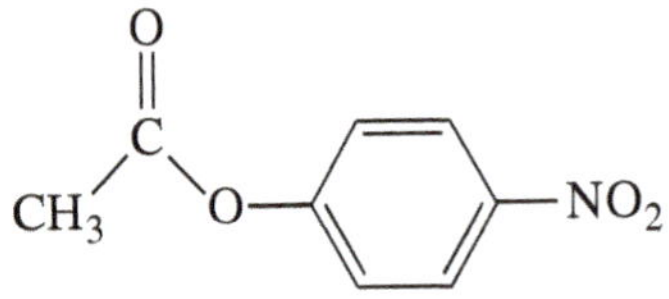

d. 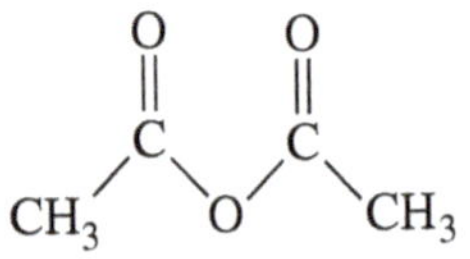 $CH_3C(=O)OC(=O)CH_3$ **or** $CH_3C(=O)Cl$

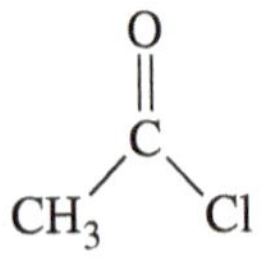

2. What is each compound's systematic name?

a. 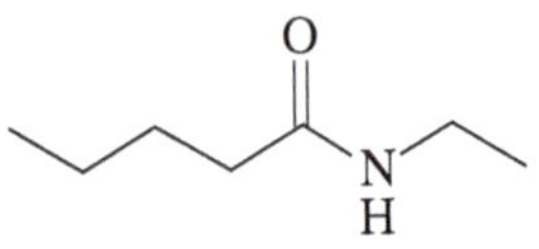 (skeletal structure: C=O, N, H)

c. (skeletal structure: benzene ring, O, OCH_3)

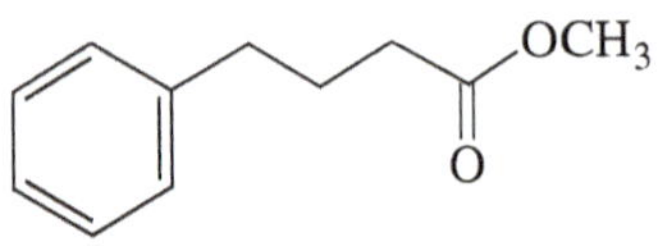

b. 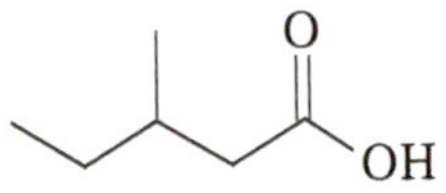 (skeletal structure: O, OH)

d. 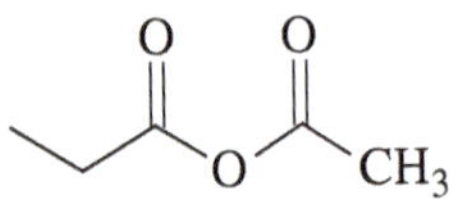 (skeletal structure: O, O, O, CH_3)

3. Give an example of each of the following:

a. a symmetrical anhydride

b. a hydrolysis reaction

c. a transesterification reaction

d. aminolysis of an ester

4. What carbonyl compound will be obtained from collapse of each of the following tetrahedral intermediates?

a. 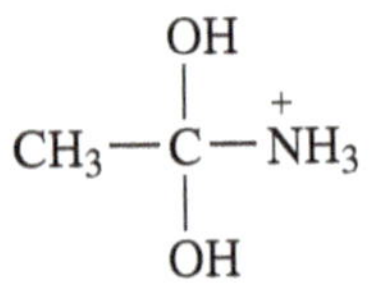 $CH_3{-}C(OH)(OH){-}\overset{+}{N}H_3$

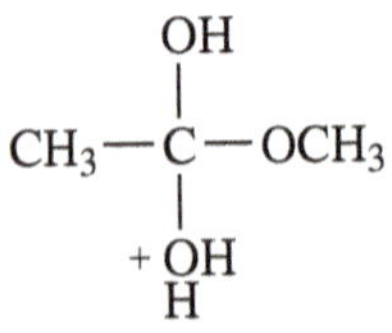

b. $CH_3{-}C(OH)(\overset{+}{O}H_2){-}OCH_3$

c. $CH_3{-}C(O^-)(OH){-}NH_2$

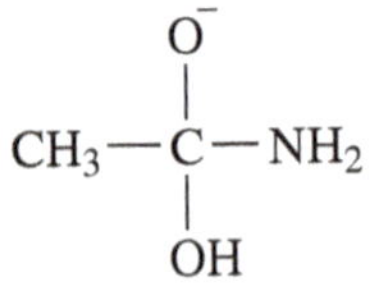

d. 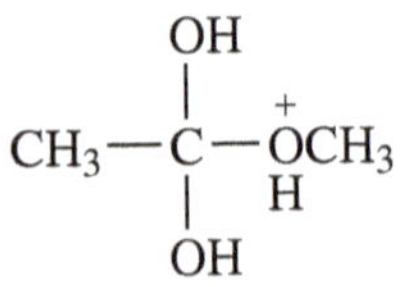 $CH_3{-}C(OH)(OH){-}\overset{+}{O}(H)CH_3$

5. What are the products of the following reactions?

a. $CH_3CH_2C{\equiv}N + H_2O \xrightarrow[\Delta]{HCl}$

b. 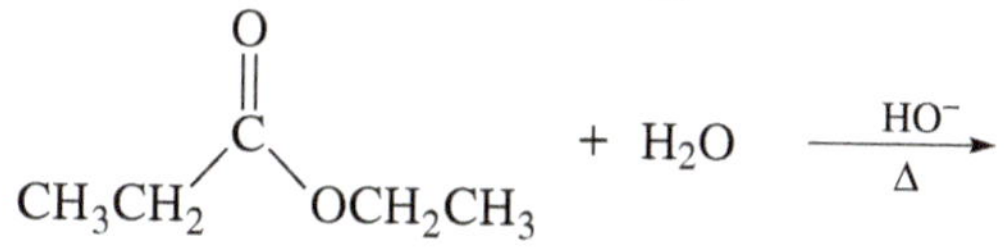 $CH_3CH_2C(=O)OCH_2CH_3 + H_2O \xrightarrow[\Delta]{HO^-}$

c. 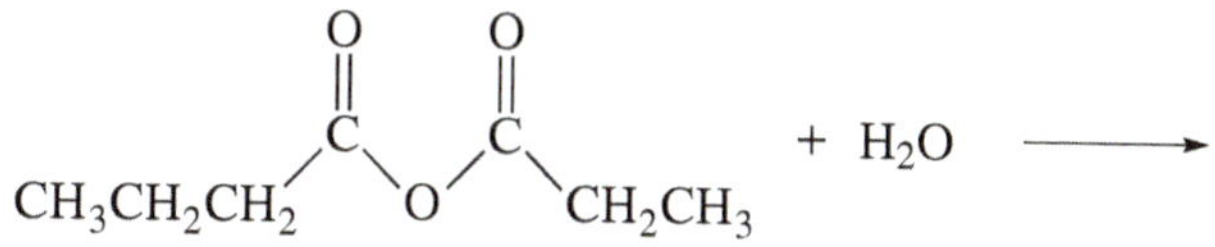 $CH_3CH_2CH_2C(=O)OC(=O)CH_2CH_3 + H_2O \longrightarrow$

d. 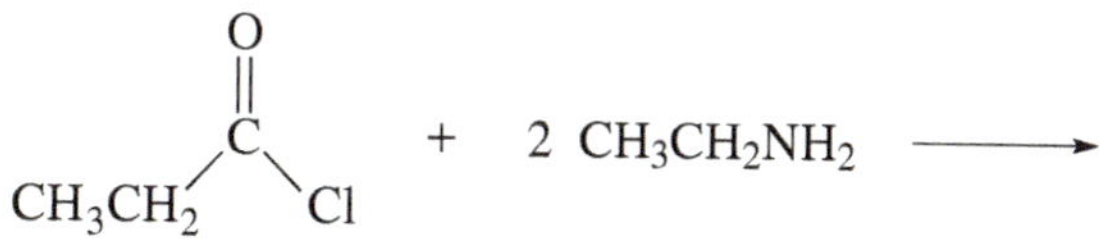 $CH_3CH_2C(=O)Cl + 2\ CH_3CH_2NH_2 \longrightarrow$

e. 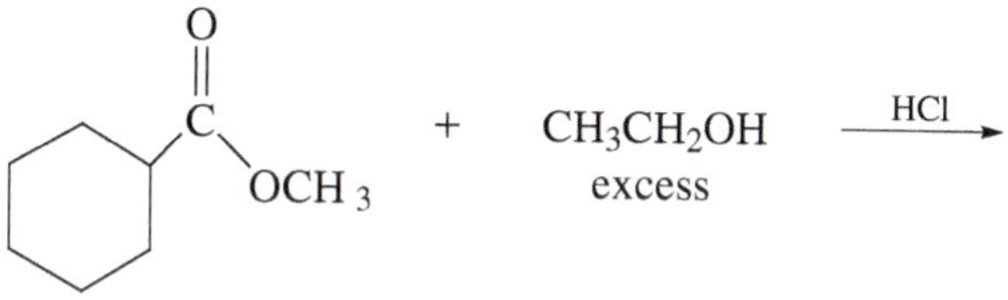 $C_6H_{11}C(=O)OCH_3 + \underset{\text{excess}}{CH_3CH_2OH} \xrightarrow{HCl}$

6. Show how the following compounds can be prepared from the given starting material:

a. $CH_3CH_2CH_2Br \longrightarrow CH_3CH_2CH_2CH_2NH_2$

b. 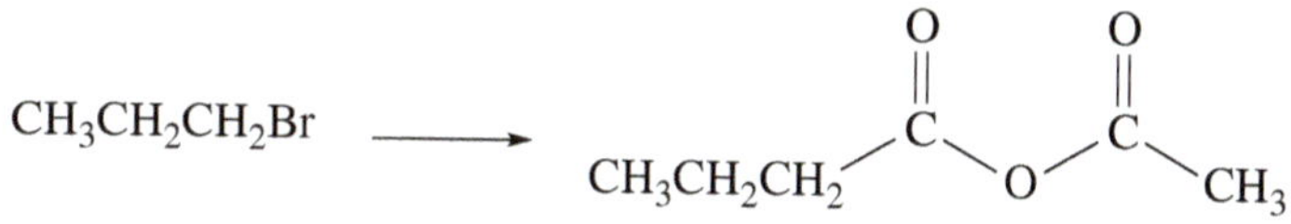 $CH_3CH_2CH_2Br \longrightarrow CH_3CH_2CH_2C(=O)OC(=O)CH_3$

7. What are the products of the following reactions?

a. 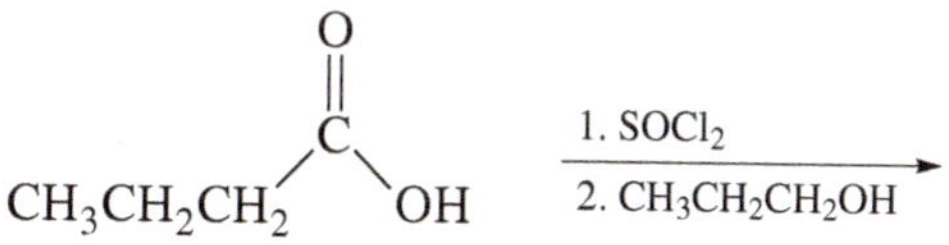 $CH_3CH_2CH_2C(=O)OH \xrightarrow[\text{2. } CH_3CH_2CH_2OH]{\text{1. } SOCl_2}$

b. 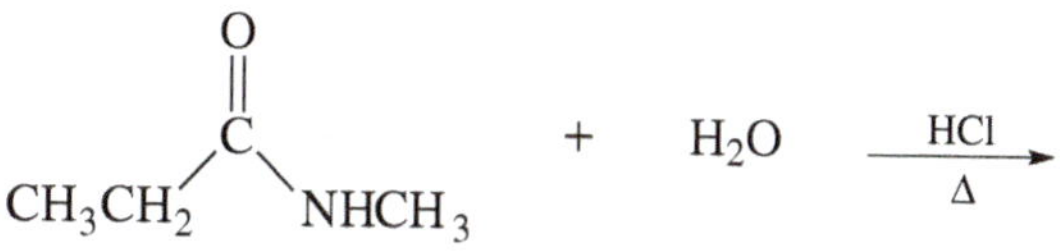 $CH_3CH_2C(=O)NHCH_3 + H_2O \xrightarrow[\Delta]{HCl}$

c. 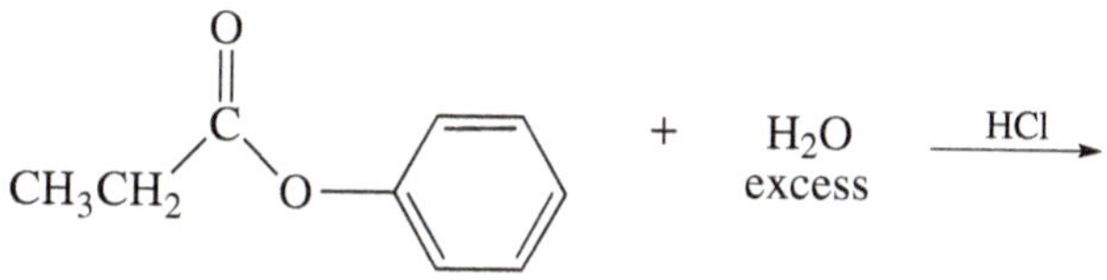 $CH_3CH_2C(=O)OC_6H_5 + \underset{\text{excess}}{H_2O} \xrightarrow{HCl}$

d. 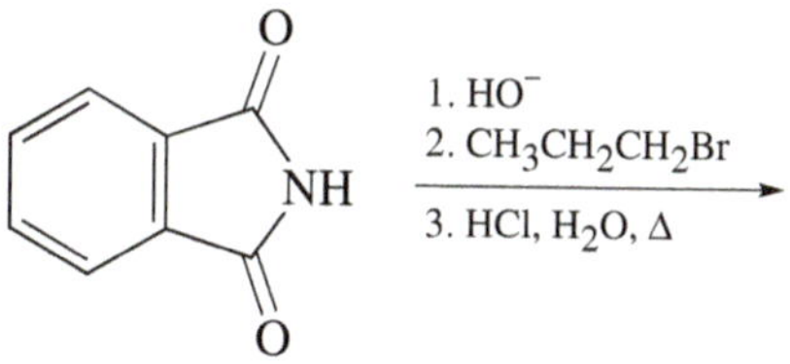 phthalimide (C=O, NH, C=O) $\xrightarrow[\text{3. } HCl,\ H_2O,\ \Delta]{\text{1. } HO^-\ \ \text{2. } CH_3CH_2CH_2Br}$

e. 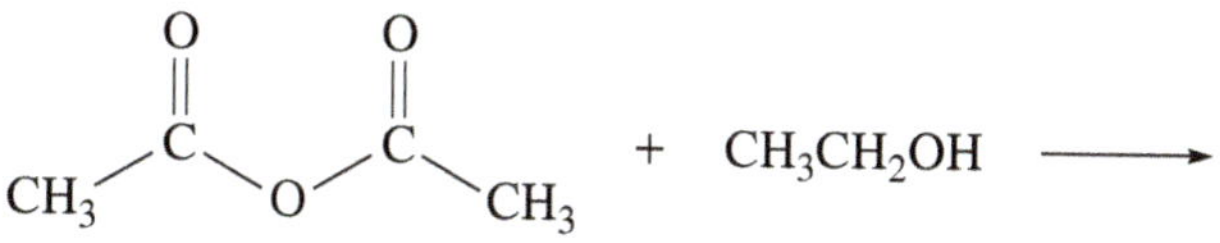 $CH_3C(=O)OC(=O)CH_3 + CH_3CH_2OH \longrightarrow$

CHAPTER 16
Reactions of Aldehydes and Ketones • More Reactions of Carboxylic Acid Derivatives

Important Terms

acetal $R{-}C(OR)(OR){-}H$ or $R{-}C(OR)(OR){-}R$

aldehyde $R{-}C({=}O){-}H$

catalytic hydrogenation addition of hydrogen to a double or triple bond in the presence of a metal catalyst.

chemoselective reaction a reaction in which a reagent reacts with one functional group in preference to another.

conjugate addition (1,4-addition) nucleophilic addition to the β-carbon of an α,β-unsaturated carbonyl compound.

cyanohydrin $R{-}C(OH)(R'(H)){-}C{\equiv}N$

deoxygenation removal of an oxygen from a reactant.

direct addition (1,2-addition) nucleophilic addition to the carbonyl carbon of an α,β-unsaturated carbonyl compound.

disconnection the breaking of a bond to carbon in a retrosynthetic analysis to give a simpler molecule.

dissolving metal reduction a reduction using sodium or lithium metal dissolved in liquid ammonia.

enamine an α,β-unsaturated tertiary amine.

***gem*-diol (hydrate)** a molecule with two OH groups on the same carbon.

hemiacetal $R{-}C(OH)(OR){-}H$ or $R{-}C(OH)(OR){-}R$

hydrate a molecule with two OH groups on the same carbon.

hydrazone

$$\ce{R(R')C=N-NH2}$$

(R' = H or R)

imine (Schiff base)

$$\ce{R(R')C=N-R}$$

(R' = H or R)

ketone

$$\ce{R-C(=O)-R}$$

nucleophilic addition reaction — a reaction that involves the addition of a nucleophile to the carbonyl carbon of an aldehyde or a ketone.

nucleophilic addition–elimination reaction — a reaction involving nucleophilic addition to a carbonyl group to form a tetrahedral intermediate, which then undergoes elimination of a leaving group. Imine formation is an example: an amine adds to the carbonyl carbon, and water is eliminated.

oxime

$$\ce{R(R')C=N-OH}$$

(R' = H or R)

pH-rate profile — a plot of the rate constant of a reaction versus the pH of the reaction mixture.

phenylhydrazone

$$\ce{R(R')C=N-NHC6H5}$$

(R' = H or R)

protecting group — a group that protects a functional group from a synthetic operation that it would otherwise not survive.

reduction reaction — in the case of an organic molecule, a reaction in which the number of C—H bonds is increased or the number of C—O, C—N, or C—X (X = halogen) bonds is decreased.

reductive amination — a reaction of an aldehyde or a ketone with ammonia or with a primary amine in the presence of a reducing agent.

synthetic equivalent — the reagent actually used as the source of a synthon.

synthon — a fragment of a disconnection.

Wittig reaction — a reaction of an aldehyde or a ketone with a phosphonium ylide, resulting in the formation of an alkene.

ylide — a compound with opposite charges on adjacent covalently bonded atoms with complete octets.

Solutions to Problems

1.
a. 3-methylpentanal, β-methylvaleraldehyde
b. 4-heptanone, dipropyl ketone
c. 2-methyl-4-heptanone, isobutyl propyl ketone
d. 4-phenylbutanal, γ-phenylbutyraldehyde
e. 4-ethylhexanal, γ-ethylcaproaldehyde
f. 1-hepten-3-one, butyl vinyl ketone

2. If the carbonyl group were anywhere else in these compounds, they would not be ketones (they would be aldehydes) and, therefore, would not have the "one" suffix.

3. **a.** 6-hydroxy-3-heptanone **b.** 2-oxocyclohexylmethanenitrile **c.** 4-formylhexanamide

4. **a.** 2-Heptanone is more reactive because it has less steric hindrance. There is little difference in the amount of steric hindrance provided by the propyl and the pentyl group at the carbonyl carbon (the site of nucleophilic addition) because they differ at a point somewhat removed from the site of nucleophilic addition. However, there is a significant difference in size between a methyl group and a propyl group at the site of nucleophilic addition.

O
‖
CH_3–C–$CH_2CH_2CH_2CH_2CH_3$
2-heptanone

O
‖
$CH_3CH_2CH_2$–C–$CH_2CH_2CH_3$
4-heptanone

b. Chloromethyl phenyl ketone is more reactive because chlorine is more strongly electron withdrawing than bromine since chlorine is more electronegative. Withdrawing electrons inductively away from the carbonyl group makes the carbonyl carbon more electrophilic and, therefore, more reactive toward a nucleophile.

O
‖
C–CH_2Br
bromomethyl phenyl ketone

O
‖
C–CH_2Cl
chloromethyl phenyl ketone

5. **a.** $CH_3CH_2CHCH_3$ (OH on C-2)

b. $CH_3CH_2CH_2CCH_3$ (OH and CH_3 on C-2)

c. HO, CH_3 (1-methylcyclohexanol)

6.

O
‖
CH_3–C–CH_2CH_3 + $CH_3CH_2CH_2MgBr$ **or**

O
‖
CH_3CH_2–C–$CH_2CH_2CH_3$ + CH_3MgBr

7. **a.** Two stereoisomers are obtained, because the reaction creates an asymmetric center in the product.

$$CH_3C(=O)CH_2CH_2CH_3 \xrightarrow[\text{2. } H_3O^+]{\text{1. } CH_3CH_2MgBr} CH_3C(OH)(CH_2CH_3)CH_2CH_2CH_3$$

(*S*)-3-methyl-3-hexanol (*R*)-3-methyl-3-hexanol

b. Only one compound is obtained, because the product does not have an asymmetric center.

$$CH_3C(=O)CH_2CH_2CH_3 \xrightarrow[\text{2. } H_3O^+]{\text{1. } CH_3MgBr} CH_3C(OH)(CH_3)CH_2CH_2CH_3$$

8. **a.** Solved in the text.

b. **A** Solved in the text.

B $CH_3C(=O)OR$ + 2 CH_3MgBr

D $CH_3C(=O)OR$ + 2 CH_3CH_2MgBr

F $CH_3C(=O)OR$ + 2 C_6H_5MgBr

9. If a secondary alcohol is formed from the reaction of a formate ester with excess Grignard reagent, the two alkyl substituents of the alcohol will be identical because they both come from the Grignard reagent. Therefore, only the following two alcohols (**B** and **D**) can be prepared that way.

$$CH_3CH(OH)CH_3 \qquad CH_3CH_2CH(OH)CH_2CH_3$$

10.

$$CH_3C(=O)Cl + CH_3CH_2{-}MgBr \longrightarrow CH_3{-}C(O^-\,\overset{+}{M}gBr)(CH_2CH_3){-}Cl \longrightarrow CH_3C(=O)CH_2CH_3$$

$$CH_3C(=O)CH_2CH_3 \xrightarrow{CH_3CH_2{-}MgBr} CH_3{-}C(O^-\,\overset{+}{M}gBr)(CH_2CH_3){-}CH_2CH_3 \xrightarrow{H_3O^+} CH_3{-}C(OH)(CH_2CH_3){-}CH_2CH_3$$

11. **A** and **C** will not undergo nucleophilic addition with a Grignard reagent. **A** has an H bonded to a nitrogen, and **C** has an H bonded to an oxygen; these acidic hydrogens will react rapidly with the Grignard reagent, converting it to an alkane before the Grignard reagent has a chance to react as a nucleophile.

B will undergo a nucleophilic addition–elimination reaction with the Grignard reagent to form a ketone. This will be followed by a nucleophilic addition reaction of the ketone with another equivalent of the Grignard reagent. The product will be a tertiary alcohol.

12. **a.** OH; NaOCl, CH_3COOH, 0 °C; O; 1. CH_3MgBr 2. H_3O^+; OH; H_2SO_4, Δ

b. OH; NaOCl, CH_3COOH, 0 °C; O; 1. CH_3CH_2MgBr 2. H_3O^+; OH; H_2SO_4, Δ; H_2, Pd/C

c. OH; HBr, Δ; Br; Mg, Et_2O; MgBr; 1. $CH_3\overset{O}{\parallel}CH$ 2. H_3O^+; OH

13. **a.** 1. $HC\equiv CH \xrightarrow{NaNH_2} HC\equiv C^- \xrightarrow{CH_3CH_2\overset{O}{\parallel}CH} HC\equiv C\overset{O^-}{|}CHCH_2CH_3 \xrightarrow{\text{pyridinium chloride}} HC\equiv C\overset{OH}{|}CHCH_2CH_3$

2. $HC\equiv CH \xrightarrow{NaNH_2} HC\equiv C^- \xrightarrow{CH_3Br} HC\equiv CCH_3 \xrightarrow{NaNH_2} {}^-C\equiv CCH_3$

$\downarrow$ $C_6H_5\overset{O}{\parallel}CH$

$C_6H_5\overset{O^-}{|}CHC\equiv CCH_3 \xrightarrow{\text{pyridinium chloride}} C_6H_5\overset{OH}{|}CHC\equiv CCH_3$

Pyridinium chloride is used to protonate the alkoxide ion in the last step because it will not add to the triple bond like HCl will.

3. $HC{\equiv}CH \xrightarrow{NaNH_2} HC{\equiv}C^- \xrightarrow{CH_3CH_2Br} HC{\equiv}CCH_2CH_3 \xrightarrow{NaNH_2} {}^-C{\equiv}CCH_2CH_3$

$$\xrightarrow{CH_3\overset{O}{\overset{\|}{C}}CH_3} CH_3\underset{CH_3}{\underset{|}{\overset{O^-}{\overset{|}{C}}}}C{\equiv}CCH_2CH_3 \xrightarrow{\text{pyridinium chloride}} CH_3\underset{CH_3}{\underset{|}{\overset{OH}{\overset{|}{C}}}}C{\equiv}CCH_2CH_3$$

b. Ethyne must be alkylated before it reacts as a nucleophile with the carbonyl compound. If ethyne is alkylated after nucleophilic addition to the ketone, alkylation can occur on both carbon and oxygen.

$$^-C{\equiv}CH \xrightarrow{CH_3\overset{O}{\overset{\|}{C}}CH_3} CH_3\underset{CH_3}{\underset{|}{\overset{O^-}{\overset{|}{C}}}}C{\equiv}CH \xrightarrow{NaNH_2} CH_3\underset{CH_3}{\underset{|}{\overset{O^-}{\overset{|}{C}}}}C{\equiv}C^- \xrightarrow{CH_3CH_2Br} CH_3\underset{CH_3}{\underset{|}{\overset{OCH_2CH_3}{\overset{|}{C}}}}C{\equiv}CCH_2CH_3$$

14. Similar to the way a Grignard reagent reacts with an ester, an acetylide ion first reacts with the ester in a nucleophilic acyl substitution reaction to form a ketone; this is followed by a nucleophilic addition reaction to form a tertiary alcohol.

$$R\overset{O}{\overset{\|}{C}}OR \xrightarrow{RC{\equiv}C^-} R-\underset{C{\equiv}CR}{\underset{|}{\overset{O^-}{\overset{|}{C}}}}-OR \longrightarrow \underset{\text{a ketone}}{R\overset{O}{\overset{\|}{C}}C{\equiv}CR + RO^-} \xrightarrow[\text{2. pyridinium chloride}]{\text{1. } RC{\equiv}C^-} \underset{\text{a tertiary alcohol}}{R-\underset{C{\equiv}CR}{\underset{|}{\overset{OH}{\overset{|}{C}}}}-C{\equiv}CR}$$

Notice that the alkoxide ion is protonated by pyridinium chloride rather than by HCl or H_3O^+ because these acids would add to the triple bonds.

15. The reaction is carried out with excess cyanide ion in order to have some unprotonated cyanide ion to act as a nucleophile. HCl is a strong acid, so it dissociates completely and protonates the cyanide ion. Therefore, there is no HCl in the reaction mixture, so HCN is the only acid available to protonate the alkoxide ion.

$$HCl + {}^-C{\equiv}N \longrightarrow HC{\equiv}N + Cl^-$$

16. No, an acid must be present in the reaction mixture to protonate the oxygen of the cyanohydrin. Otherwise, the cyano group will be eliminated and the reactants will reform.

17. Strong acids like HCl and H_2SO_4 have very weak conjugate bases (Cl^- and HSO_4^-), which are excellent leaving groups. When these bases add to the carbonyl group, they are readily eliminated, reforming the starting materials. Cyanide ion is a strong enough base not to be eliminated unless the oxygen in the product is negatively charged.

18. Solved in the text.

19.

1. 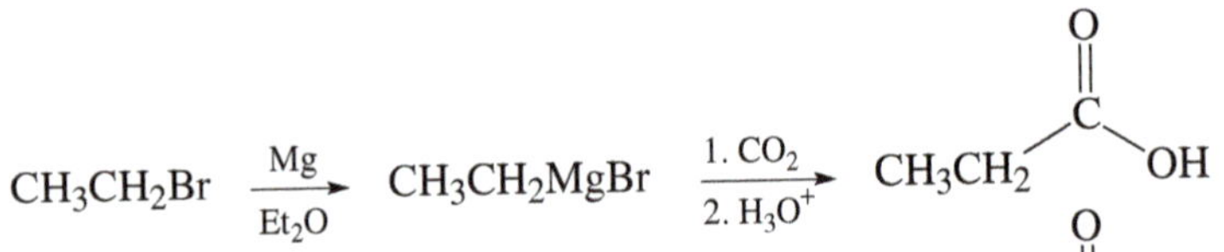 $CH_3CH_2Br \xrightarrow[Et_2O]{Mg} CH_3CH_2MgBr \xrightarrow[2.\ H_3O^+]{1.\ CO_2} CH_3CH_2C(=O)OH$

2. $CH_3CH_2Br \xrightarrow{^-C\equiv N} CH_3CH_2C\equiv N \xrightarrow[\Delta]{HCl,\ H_2O} CH_3CH_2C(=O)OH$

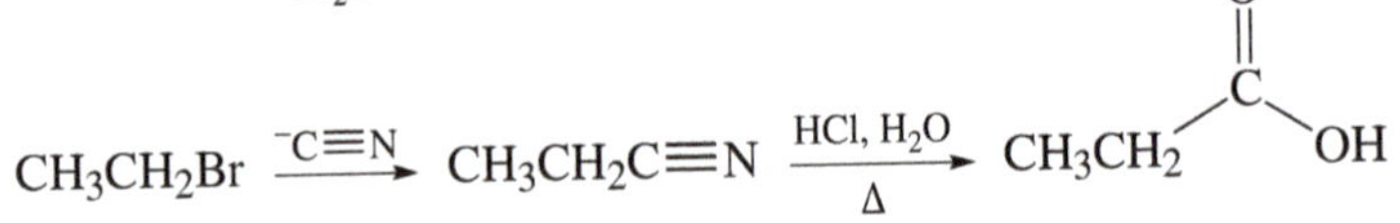

20. **a.** $CH_3CH(CH_3)CH_2OH$ **b.** 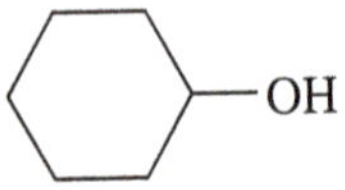 cyclohexanol (cyclohexane–OH) **c.** 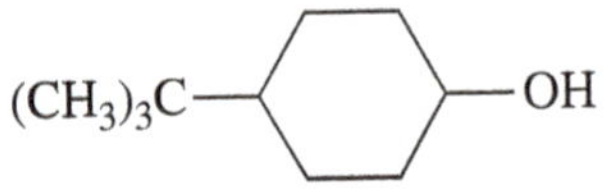 $(CH_3)_3C$–cyclohexane–OH **d.** 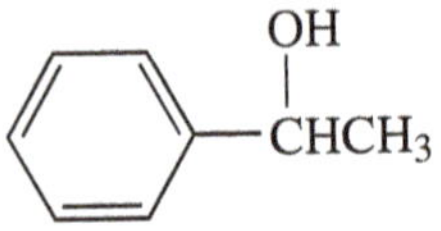 benzene–$CH(OH)CH_3$

21. **a.** $CH_3CH_2CH_2CH_2OH$ + CH_3CH_2OH

b. 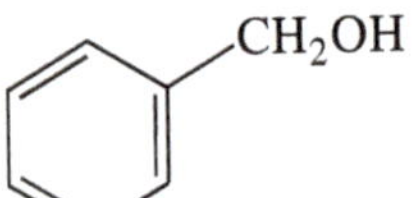 benzene–CH_2OH

c. 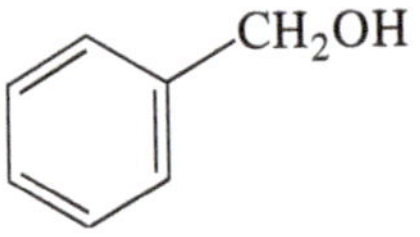 benzene–CH_2OH + CH_3OH

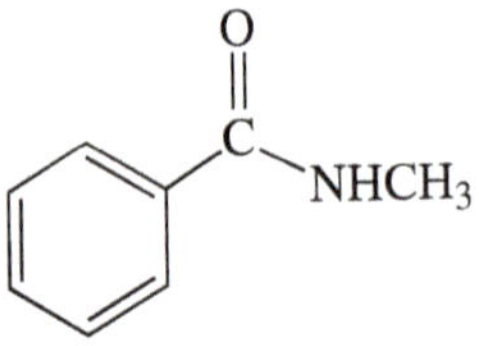

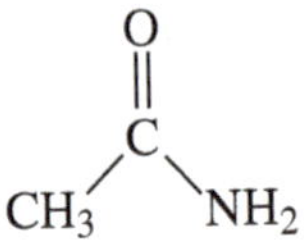

d. $CH_3CH_2CH_2CH_2CH_2OH$

22. **a.** 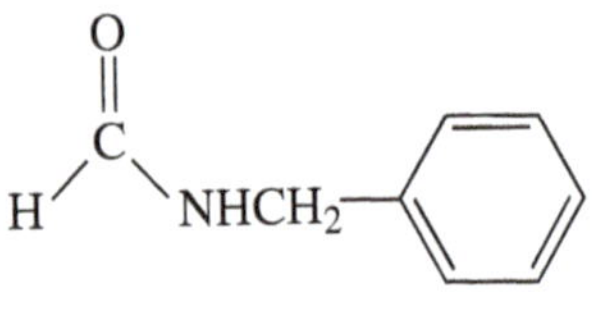 $HC(=O)NHCH_2$–benzene **or** benzene–$C(=O)NHCH_3$

b. $CH_3C(=O)NH_2$

c. $CH_3C(=O)NHCH_2CH_3$

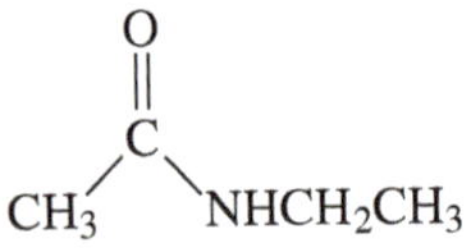

d. 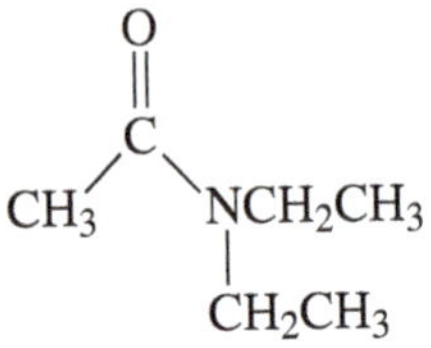 $CH_3C(=O)N(CH_2CH_3)CH_2CH_3$

23. **a.** **1.** $LiAlH_4$ **2.** H_2O

b. HCl, H_2O, Δ

c. **1.** HCl, H_2O, Δ **2.** $LiAlH_4$ **3.** HCl

24. **a.** 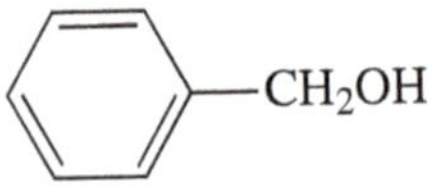 benzene–CH_2OH

b. $CH_3CH_2CH_2CH_2NH_2$

c. 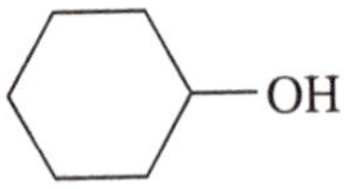 cyclohexane–OH

d. no reaction

25.

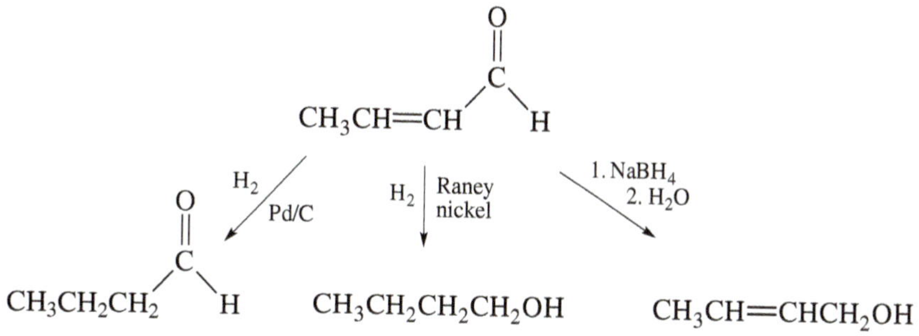

$$\mathrm{CH_3CH{=}CHCH{=}O}$$

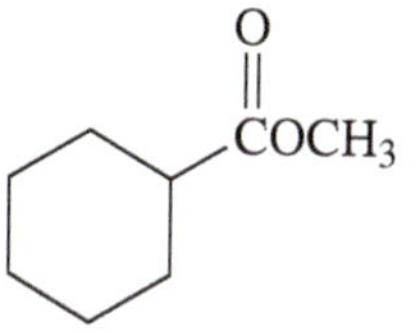

- H_2, Pd/C → $\mathrm{CH_3CH_2CH_2CH{=}O}$
- H_2, Raney nickel → $\mathrm{CH_3CH_2CH_2CH_2OH}$
- 1. $NaBH_4$ 2. H_2O → $\mathrm{CH_3CH{=}CHCH_2OH}$

26.

a. cyclohexene ring bearing $\mathrm{CH(OH)CH_3}$

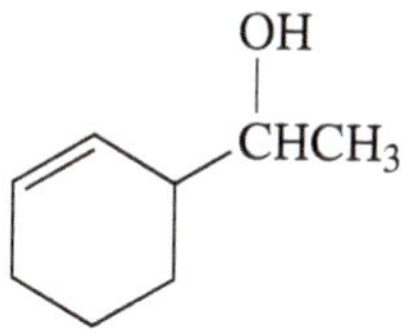

b. cyclohexane ring bearing $\mathrm{C(=O)OCH_3}$ (COCH$_3$ with C=O)

c. cyclohexane ring bearing OH and $\mathrm{CH_2C(=O)OCH_3}$

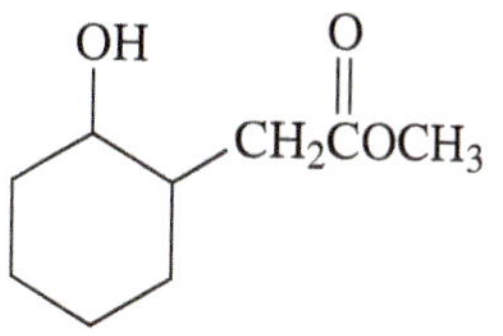

d. cyclohexane ring bearing OH and $\mathrm{CH_2CH_2OH}$ + CH_3OH

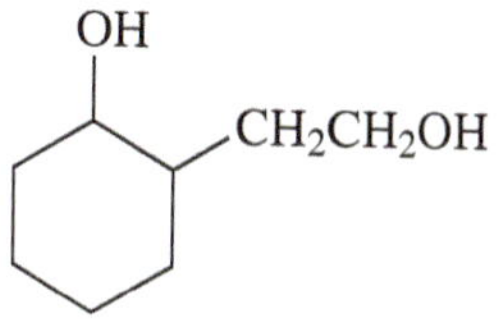

27. The nitrile reacts with the Grignard reagent to form an imine, which can then be hydrolyzed to a ketone.

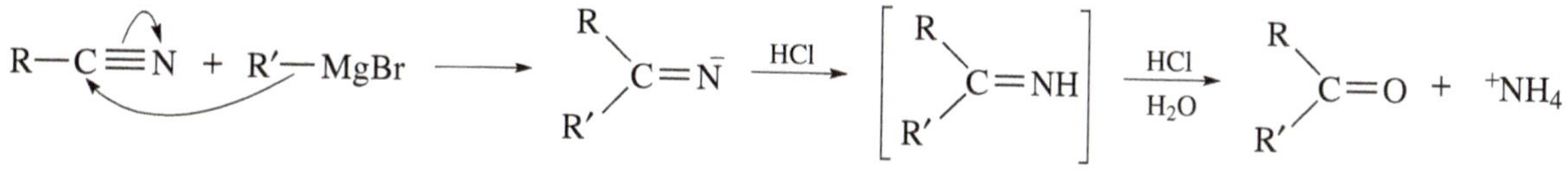

$$\mathrm{R{-}C{\equiv}N + R'{-}MgBr \longrightarrow RR'C{=}\bar{N} \xrightarrow{HCl} [RR'C{=}NH] \xrightarrow[H_2O]{HCl} RR'C{=}O + {}^{+}NH_4}$$

28. The OH group withdraws electrons inductively, and this makes the ammonium ion a stronger acid.

$$\mathrm{HO{-}\overset{+}{N}H_3} \qquad \mathrm{CH_3{-}\overset{+}{N}H_3}$$

29. Figure 16.2 for imine formation when the amine is hydroxylamine (pK_a of its conjugate acid $= 6.0$) shows that the maximum rate is obtained when the pH is about 1.5 units lower than the pK_a of the amine's conjugate acid. Therefore, for an amine such as ethylamine whose conjugate acid has a $pK_a \sim 11$, imine formation should be carried out at about pH $= 9.5$.

30. For the derivations of the equations used to calculate the amount of a compound that is present in either its acidic or basic form, refer to Special Topic I in this *Study Guide and Solutions Manual.*

a. fraction present in the acidic form $= \dfrac{[H^+]}{K_a + [H^+]}$

$$\frac{[H^+]}{K_a + [H^+]} = \frac{3.2 \times 10^{-5}}{3.2 \times 10^{7} + 3.2 \times 10^{-5}}$$

$$= \frac{3.2 \times 10^{-5}}{3.2 \times 10^{7}}$$

$$= 1 \times 10^{-12}$$

b. fraction present in the acidic form $= \dfrac{[H^+]}{K_a + [H^+]}$

$$\frac{[H^+]}{K_a + [H^+]} = \frac{3.2 \times 10^{-2}}{3.2 \times 10^{7} + 3.2 \times 10^{-2}}$$
$$= \frac{3.2 \times 10^{-2}}{3.2 \times 10^{7}}$$
$$= 1 \times 10^{-9}$$

c. fraction present in the basic form $= \dfrac{K_a}{K_a + [H^+]}$

$$\frac{K_a}{K_a + [H^+]} = \frac{1.0 \times 10^{-6}}{1.0 \times 10^{-6} + 3.2 \times 10^{-2}}$$
$$= \frac{1.0 \times 10^{-6}}{3.2 \times 10^{-2}}$$
$$= 3.1 \times 10^{-5}$$

31. **a.** C=N̈ with H and NH_2 (phenyl on C) **b.** C=N̈ with OH on N, CH_3CH_2 and phenyl on C

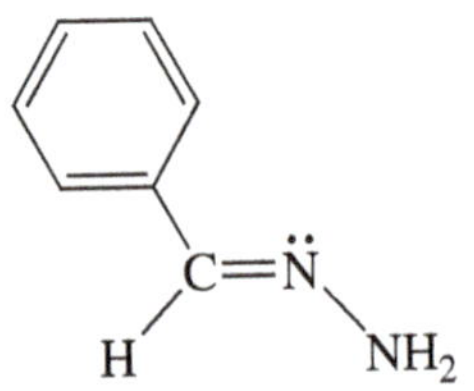

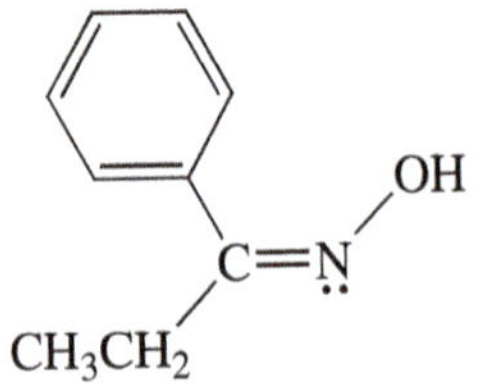

32. **a.** Notice that both of these mechanisms have a feature that we have seen in many mechanisms—formation of three tetrahedral intermediates: a protonated tetrahedral intermediate, then a neutral tetrahedral intermediate, and then a second protonated tetrahedral intermediate.

1. R, H—B⁺, + H_2O:, NHR, $\overset{+}{O}H$, H, B:, protonated tetrahedral intermediate, H—B⁺, NHR, OH, neutral tetrahedral intermediate, B:, $\overset{+}{N}H_2R$, OH, protonated tetrahedral intermediate, O: + $R\overset{+}{N}H_3$, $\overset{+}{O}H$ + $R\ddot{N}H_2$

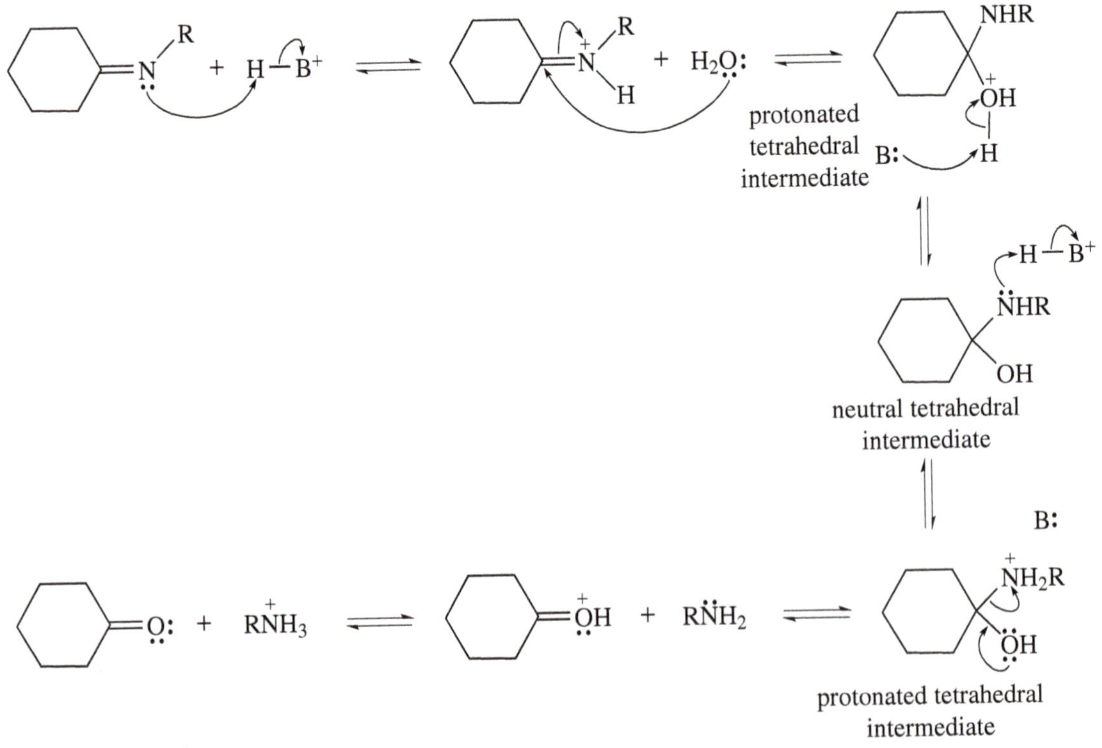

2.

$$\text{1-cyclohexenyl-}\ddot{N}R_2 + H{-}B^+ \rightleftharpoons \text{cyclohexylidene}{=}\overset{+}{N}R_2 + H_2\ddot{O}: \rightleftharpoons \text{1-}NR_2\text{-1-}\overset{+}{O}H_2\text{-cyclohexane} \; (B:)$$

β-carbon

$$\rightleftharpoons \text{1-}\ddot{N}R_2\text{-1-OH-cyclohexane} + H{-}B^+ \rightleftharpoons \text{1-}\overset{+}{N}HR_2\text{-1-}\ddot{O}H\text{-cyclohexane} \; (B:)$$

$$\rightleftharpoons \text{cyclohexyl}{=}\overset{+}{\ddot{O}}H + R_2\ddot{N}H \rightleftharpoons \text{cyclohexanone}{=}\ddot{O}: + R_2\overset{+}{N}H_2$$

b. The only difference is the first step of the mechanism: in imine hydrolysis, the acid protonates the nitrogen; in enamine hydrolysis, the acid protonates the β-carbon.

33. **a.** $$\text{cyclopentanone} + CH_3CH_2NH_2 \xrightleftharpoons{\text{trace } H^+} \text{cyclopentylidene}{=}NCH_2CH_3 + H_2O$$

b. $$\text{cyclopentanone} + CH_3CH_2NHCH_2CH_3 \xrightleftharpoons{\text{trace } H^+} \text{1-cyclopentenyl-}N(CH_2CH_3)_2 + H_2O$$

c. $$C_6H_5C(CH_3){=}O + CH_3(CH_2)_5NH_2 \xrightleftharpoons{\text{trace } H^+} C_6H_5C(CH_3){=}N(CH_2)_5CH_3 + H_2O$$

d. $$C_6H_5C(CH_3){=}O + C_6H_{11}NH_2 \xrightleftharpoons{\text{trace } H^+} C_6H_5C(CH_3){=}N{-}C_6H_{11} + H_2O$$

34. A tertiary amine will be obtained because the primary amine synthesized by reductive animation will react with the excess carbonyl compound, forming an imine that will be reduced to a secondary amine. The secondary amine will then react with the carbonyl compound, forming an enamine that will be reduced to a tertiary amine.

$$R_2C{=}O + NH_3 \xrightarrow{NaBH_3CN} \underset{\text{primary amine}}{R_2CHNH_2} \xrightarrow[NaBH_3CN]{R_2C{=}O} \underset{\text{secondary amine}}{R_2CHNHCHR_2} \xrightarrow[NaBH_3CN]{R_2C{=}O} \underset{\text{tertiary amine}}{(R_2CH)_3N}$$

35. **a.** 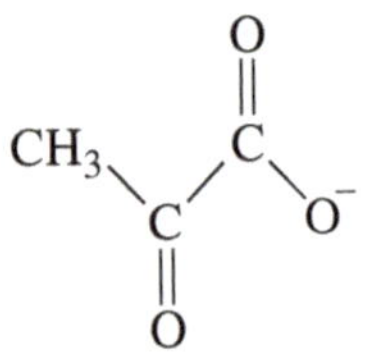 CH_3COCOO^- **b.** 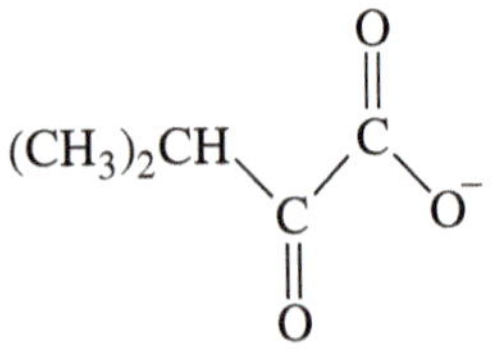 $(CH_3)_2CHCOCOO^-$

36. 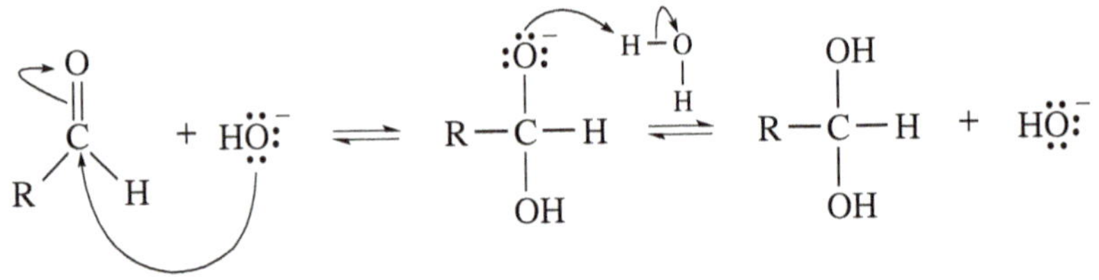 $RCHO + HO^- \rightleftharpoons R-CH(O^-)-OH \xrightleftharpoons{H_2O} R-CH(OH)-OH + HO^-$

37. Because an electron-withdrawing substituent decreases the stability of a ketone and increases the stability of a hydrate, the compound with the electron-withdrawing nitro substituents has the largest equilibrium constant for addition of water.

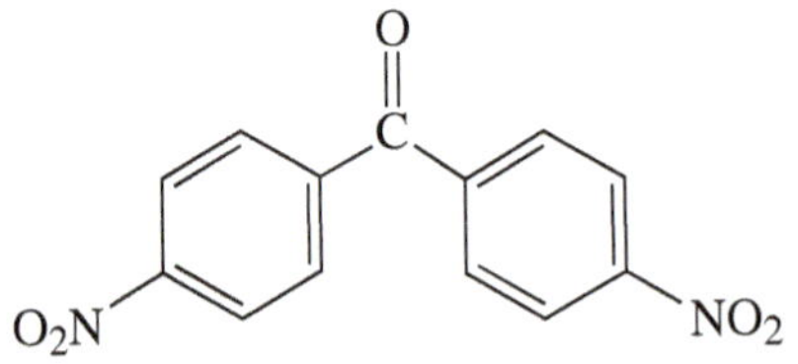

$O_2N-C_6H_4-CO-C_6H_4-NO_2$

38. The electron-withdrawing chlorines decrease the stability of the aldehyde (and, therefore, decrease its concentration), increasing the equilibrium constant for hydrate formation compared to, for instance, acetaldehyde.

$$K_{eq} = \frac{[\text{hydrate}]}{[\text{aldehyde}][H_2O]}$$

the concentration is decreased

39. **a.** hemiacetals: 1, 7, 8

b. acetals: 2, 3, 5

c. hydrates: 4, 6

40. **a.** Hemiacetals are unstable in basic solution because the base can remove a proton from an OH group, thereby providing an oxyanion that can expel the OR group (since the charge that develops on the oxygen in the transition state is negative) and form a stable aldehyde.

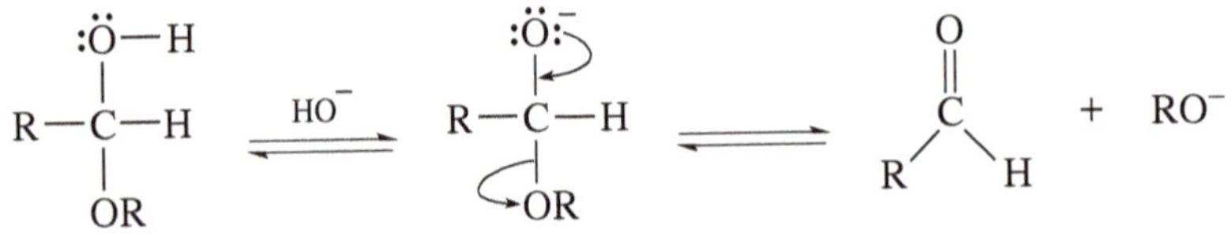

$R-CH(OH)-OR \xrightleftharpoons{HO^-} R-CH(O^-)-OR \rightleftharpoons RCHO + RO^-$

b. In order for an acetal to form, the CH_3O group in the hemiacetal must eliminate an OH group. Hydroxide ion is too basic to be eliminated by a CH_3O group (because the charge that develops on the oxygen in the transition state is positive), but water, a much weaker base, can be eliminated by a CH_3O group. In other words, the OH group must be protonated before it can be eliminated by a CH_3O group. Therefore, acetal formation must be carried out in an acidic solution.

$$R-C(H)(\ddot{O}CH_3)(OH) \not\longrightarrow$$

$$R-C(H)(\ddot{O}CH_3)(\overset{+}{O}H_2) \rightleftharpoons RCH{=}\overset{+}{\ddot{O}}CH_3 + H_2O$$

c. Hydrate formation can be catalyzed by hydroxide ion because a group does not have to be eliminated after hydroxide ion attacks the aldehyde or ketone.

$$RCH{=}O \underset{}{\overset{HO^-}{\rightleftharpoons}} R-C(H)(O^-)(OH) \overset{H_2O}{\rightleftharpoons} R-C(H)(OH)(OH)$$

41. When a tetrahedral intermediate collapses, the intermediate that is formed is unstable because of the positive charge on the sp^2 oxygen atom. In the case of an acetal, the only way to form a neutral species is to reform the acetal.

$$\underset{\text{acetal}}{R-C(H)(\ddot{O}R)(\ddot{O}R)} \overset{H-B^+}{\rightleftharpoons} R-C(H)(\overset{+}{O}HR)(\ddot{O}R) \overset{:B}{\rightleftharpoons} \underset{\text{unstable}}{RCH{=}\overset{+}{O}R} \quad (\text{attacked by } RO\text{H})$$

In the case of a hydrate, a neutral species can be formed by loss of a proton.

$$\underset{\text{hydrate}}{R-C(H)(\ddot{O}H)(\ddot{O}H)} \overset{H-B^+}{\rightleftharpoons} R-C(H)(\overset{+}{O}H_2)(\ddot{O}H) \overset{:B}{\rightleftharpoons} \underset{\text{unstable}}{RCH{=}\overset{+}{O}H} + H_2O \rightleftharpoons RCH{=}\ddot{O}: + HB^+$$

42. **a.** (cyclohexane with OH and CH_2OH substituents) both groups would be reduced

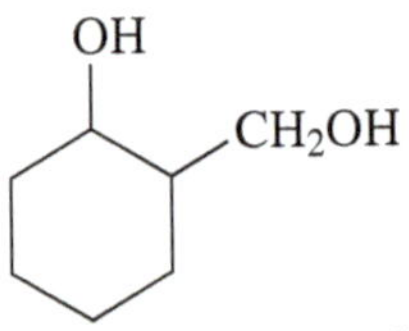

b. $NaBH_4$ because it is not strong enough a reducing agent to reduce the less reactive ester

43. An acetal has a very poor leaving group (CH_3O^-).

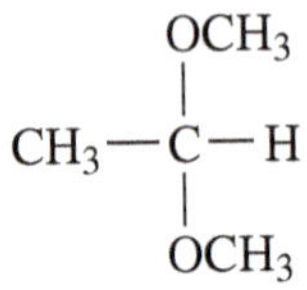

$$CH_3-\underset{OCH_3}{\overset{OCH_3}{C}}-H$$

44. If the OH group of the carboxylic acid is not protected, both OH groups will be converted to COOH groups.

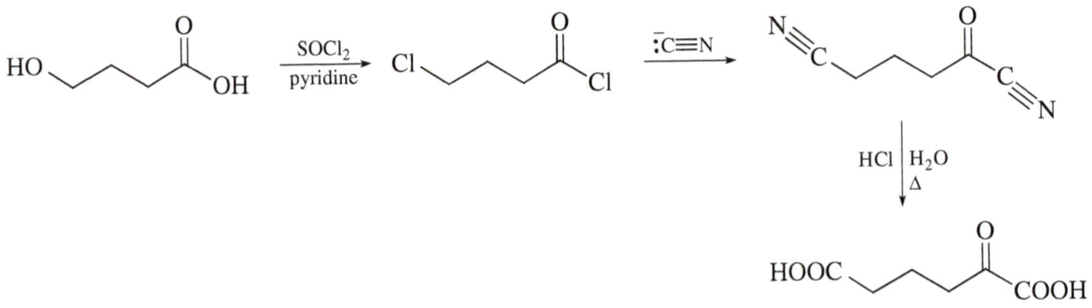

HO–$CH_2CH_2CH_2$–C(=O)OH $\xrightarrow[\text{pyridine}]{SOCl_2}$ Cl–$CH_2CH_2CH_2$–C(=O)Cl $\xrightarrow{^-:C\equiv N}$ N≡C–$CH_2CH_2CH_2$–C(=O)–C≡N $\xrightarrow[\Delta]{HCl,\ H_2O}$ HOOC–$CH_2CH_2CH_2$–C(=O)–COOH

45. If the yield of each step is 80%, the yield of **B** is 80%; the yield of **C** from **B** is 80% times the yield of **B**; the yield of **D** from **C** is 80% times the yield of **C**; and so on.

a.

A	→ B	→ C	→ D	→ E	→ F	→ G
	80%	0.80×0.80 64%	0.80×0.64 51%	0.80×0.51 41%	0.80×0.41 33%	0.80×0.33 26%

b.

G	→ H	→ I
26%	0.80×0.26 21%	0.80×0.21 17%

46. a. HO~~~Br + $(CH_3)_3CSi(CH_3)_2Cl$ $\xrightarrow{\text{imidazole}}$ $(CH_3)_3CSi(CH_3)_2O$~~~Br

$\xrightarrow{\text{Mg / Et}_2\text{O}}$ $(CH_3)_3CSi(CH_3)_2O$~~~MgBr

$\xrightarrow{CH_3CH{=}O}$ $(CH_3)_3CSi(CH_3)_2O$~~~CH(O^-)CH$_3$

$\xrightarrow{H_3O^+}$ $(CH_3)_3CSi(CH_3)_2O$~~~CH(OH)CH$_3$

$\xrightarrow{(CH_3CH_2CH_2CH_2)_4N^+F^-}$ HO~~~CH(OH)CH$_3$

b. HO~~C(=O)CH$_3$ + $(CH_3)_3CSi(CH_3)_2Cl$ $\xrightarrow{\text{imidazole}}$ $(CH_3)_3CSi(CH_3)_2O$~~C(=O)CH$_3$

$\xrightarrow{CH_3MgBr}$ $(CH_3)_3CSi(CH_3)_2O$~~C(O^-)(CH$_3$)$_2$

$\xrightarrow{H_3O^+}$ $(CH_3)_3CSi(CH_3)_2O$~~C(OH)(CH$_3$)$_2$

$\xrightarrow{(CH_3CH_2CH_2CH_2)_4N^+F^-}$ HO~~C(OH)(CH$_3$)$_2$

c. The carbonyl group has to be protected to prevent it from reacting with the Grignard reagent. This synthesis is easier to understand if the CO_2 needed in the third step is written as O=C=O.

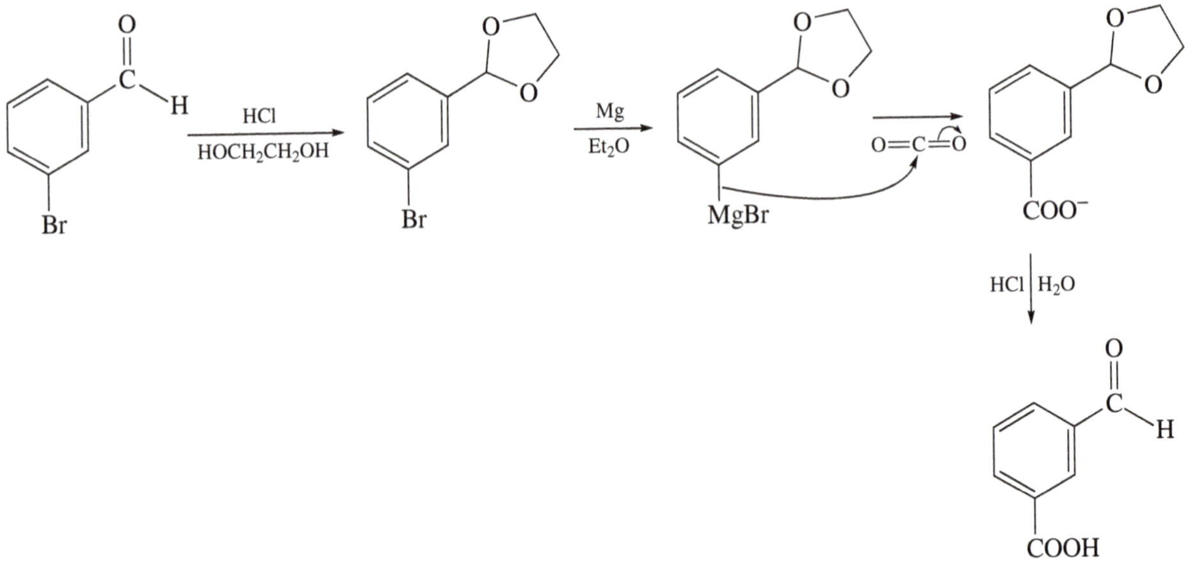

HCl, $HOCH_2CH_2OH$; Mg, Et_2O; O=C=O; HCl, H_2O

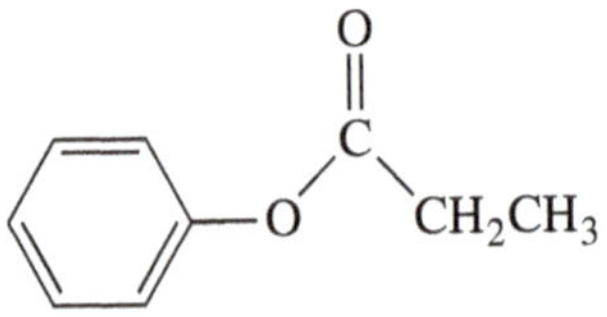

Br; Br; MgBr; COO^-; COOH

47. a. CH_2CH_3 c. CH_3

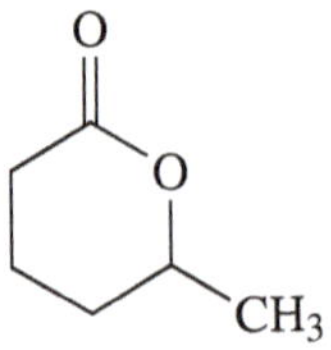

b. OH d.

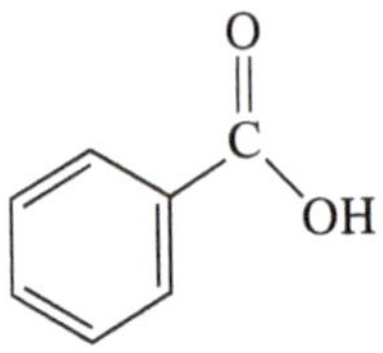

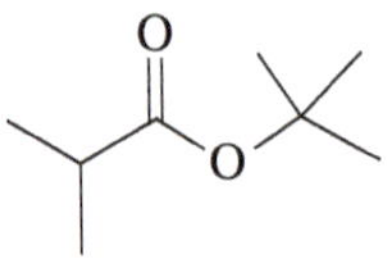

48. Solved in the text.

49. a.

1. 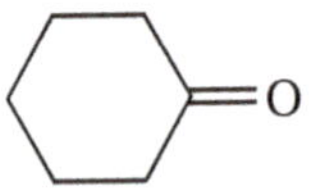 =O + $CH_3CH_2CH{=}P(C_6H_5)_3$ or $=P(C_6H_5)_3$ + $CH_3CH_2CH{=}O$

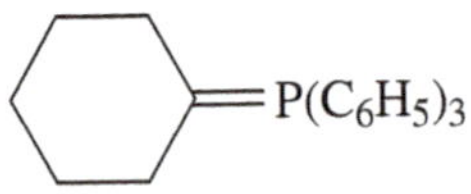

2. $(C_6H_5)_2C{=}O$ + $CH_3CH{=}P(C_6H_5)_3$ or $(C_6H_5)_2C{=}P(C_6H_5)_3$ + $CH_3CH{=}O$

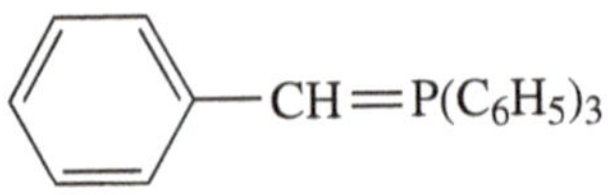

3. 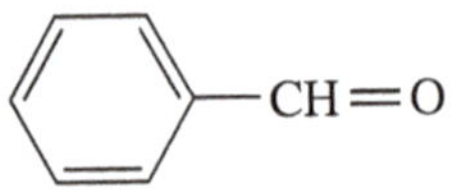 $-CH{=}O$ + $CH_2{=}P(C_6H_5)_3$ or $-CH{=}P(C_6H_5)_3$ + $CH_2{=}O$

b.

1. $CH_3CH_2CH_2Br$ **or** Br (cyclohexyl bromide)

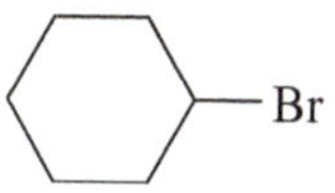

2. CH_3CH_2Br **or** $(C_6H_5)_2CHBr$

3. CH_3Br **or** CH_2Br (benzyl bromide)

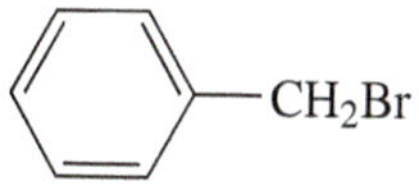

c. The best set of reagents is the pair that has the ylide synthesized from the less sterically hindered alkyl halide.

1. $CH_3CH_2CH{=}P(C_6H_5)_3$ + 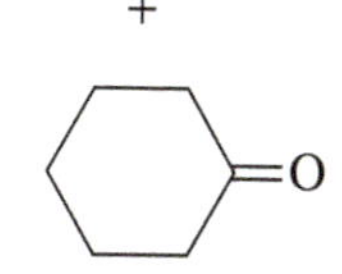

2. $CH_3CH{=}P(C_6H_5)_3$ + $(C_6H_5)_2C{=}O$

3. $CH_2{=}P(C_6H_5)_3$ + $CH{=}O$ (benzaldehyde)

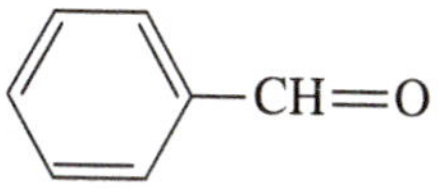

50. a. 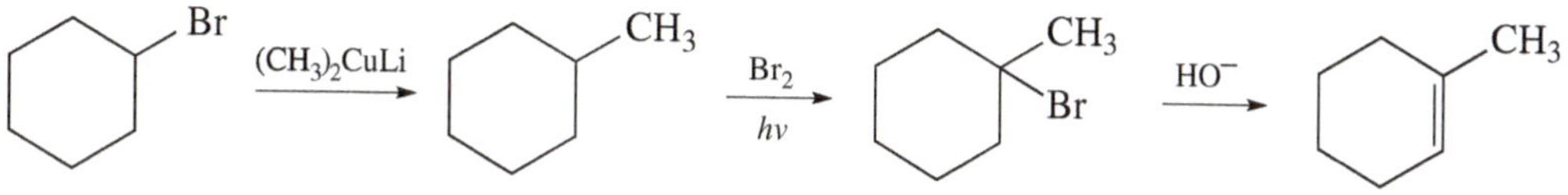 Br $\xrightarrow{(CH_3)_2CuLi}$ CH_3 $\xrightarrow[h\nu]{Br_2}$ CH_3, Br $\xrightarrow{HO^-}$ CH_3

b. 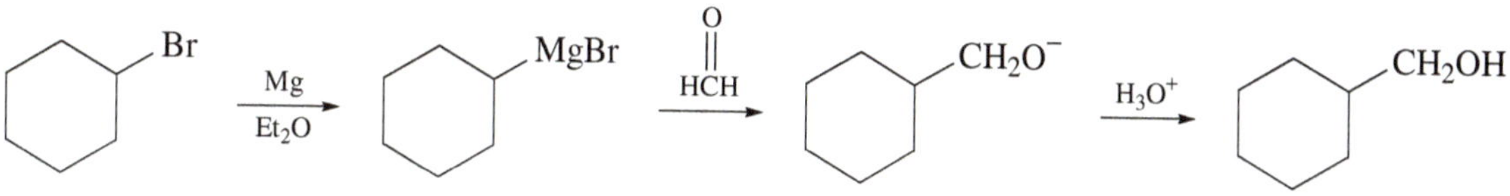 Br $\xrightarrow[Et_2O]{Mg}$ MgBr $\xrightarrow{HCH \text{ (O)}}$ CH_2O^- $\xrightarrow{H_3O^+}$ CH_2OH

c. 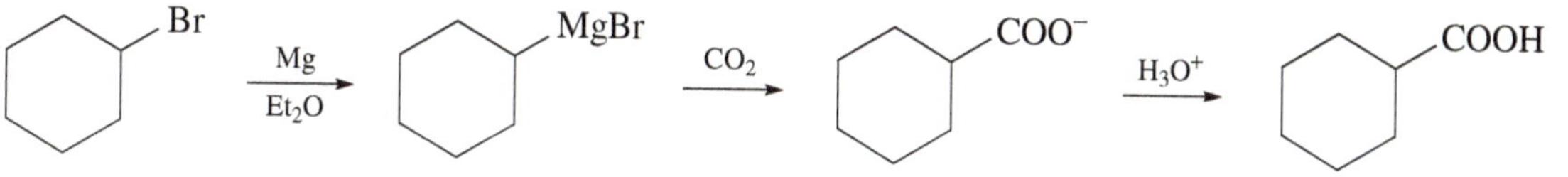 Br $\xrightarrow[Et_2O]{Mg}$ MgBr $\xrightarrow{CO_2}$ COO^- $\xrightarrow{H_3O^+}$ COOH

or

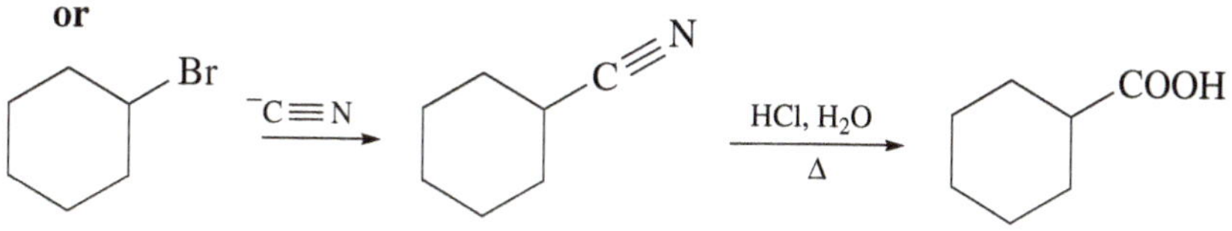

Br $\xrightarrow{^-C{\equiv}N}$ $C{\equiv}N$ $\xrightarrow[\Delta]{HCl, H_2O}$ COOH

d. 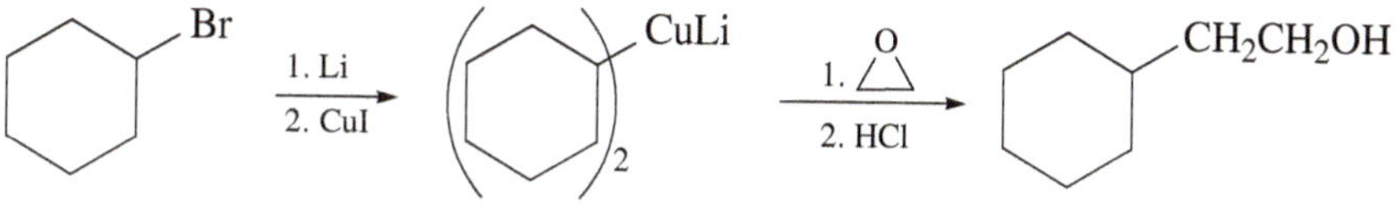 Br $\xrightarrow[2.\ CuI]{1.\ Li}$ $(\ )_2$CuLi $\xrightarrow[2.\ HCl]{1.\ \text{(O) ethylene oxide}}$ CH_2CH_2OH

e. Bromocyclohexane $\xrightarrow{P(C_6H_5)_3}$ cyclohexyl–$\overset{+}{P}(C_6H_5)_3$ $\xrightarrow{CH_3(CH_2)_3Li}$ cyclohexyl carbanion–$\overset{+}{P}(C_6H_5)_3$ $\xrightarrow{CH_3COCH_3}$ isopropylidenecyclohexane (cyclohexane=C(CH_3)CH_3) + $(C_6H_5)_3P{=}O$

f. Bromocyclohexane $\xrightarrow[Et_2O]{Mg}$ cyclohexyl–MgBr $\xrightarrow{CH_3CH{=}O}$ cyclohexyl–$\underset{O^-}{CH}CH_3$ $\xrightarrow[\Delta]{H_2SO_4}$ ethylidenecyclohexane (cyclohexane=CHCH_3) + H_2O

$\xrightarrow{H^+}$ cyclohexyl cation (+ on ring carbon bearing CH_2CH_2) $\xrightarrow{-H^+}$ 1-ethylcyclohexene (ring with CH_2CH_3) + H_2O $\xrightarrow{HCl}$ 1-chloro-1-ethylcyclohexane (ring with Cl and CH_2CH_3)

To see why 1-ethylcyclohexene is the final product of the dehydration, see page 470 in the text.

51.

a. Bicyclic ketone with C≡N at the ring-fusion carbon (=O on ring)

weak base favors conjugate addition

b. Bicyclic alkene with OH

steric hindrance is greater at the β-carbon than at the carbonyl carbon, so direct addition is favored

c. $CH_3\underset{CH_3}{C}{=}CH\overset{OH}{\underset{CH_3}{C}}CH_3$

strong base favors direct addition

d. $CH_3CH{=}CHCH_2OH$

strong base favors direct addition

52. Conjugate addition occurs in **a** and **c** because the nucleophile is a relatively weak base. Nucleophilic addition–elimination occurs in **b** because the carbonyl group is very reactive. Because excess base is used in part **d**, both nucleophilic addition–elimination and conjugate addition occur.

a. $CH_3\underset{Br}{CH}CH_2C({=}O)OCH_3$

c. $CH_3\underset{SCH_3}{CH}CH_2C({=}O)OCH_3$

b. $CH_3CH{=}CHC({=}O)OCH_3$

d. $CH_3\underset{NH_2}{CH}CH_2C({=}O)NH_2$

53.

a. 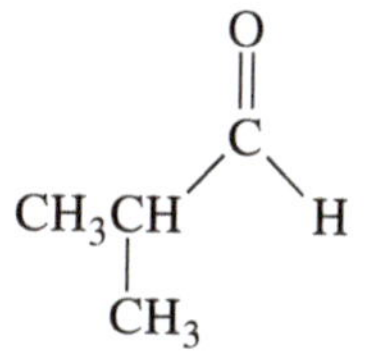 O ‖ C — CH_3CH (with CH_3 below) — H

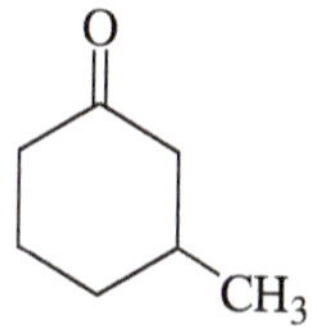

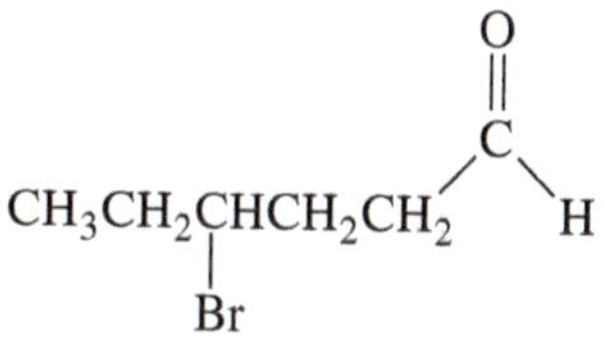

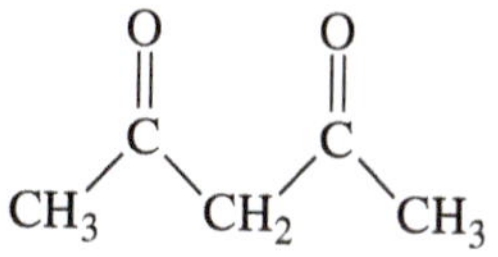

b. 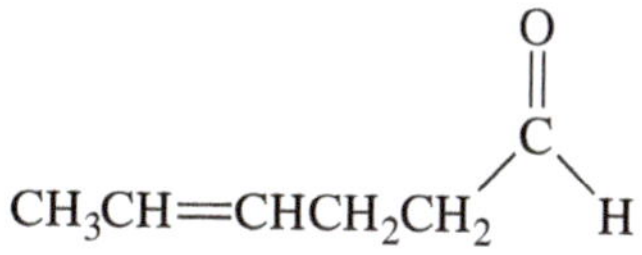 $CH_3CH{=}CHCH_2CH_2$ — $C(=O)$ — H

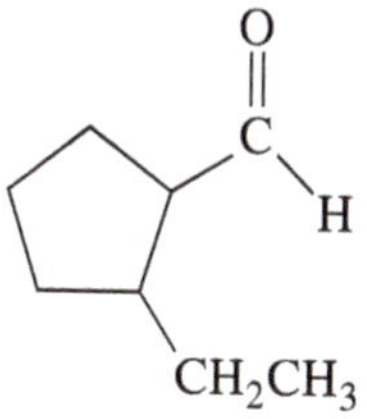

c. 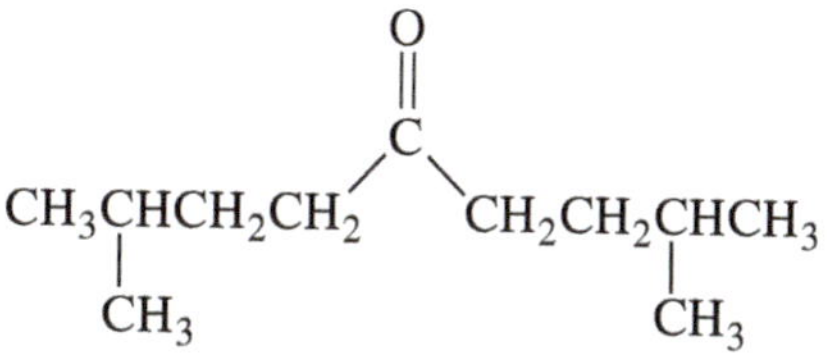 $CH_3CHCH_2CH_2$ (with CH_3 below) — $C(=O)$ — $CH_2CH_2CHCH_3$ (with CH_3 below)

d. O (cyclohexanone ring) CH_3

e. CH_3 — $C(=O)$ — CH_2 — $C(=O)$ — CH_3

f. 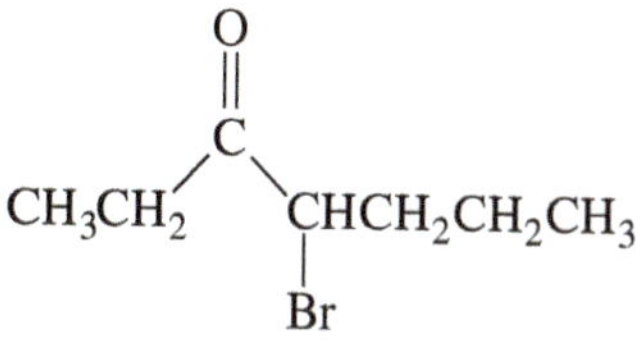 CH_3CH_2 — $C(=O)$ — $CHCH_2CH_2CH_3$ (with Br below)

g. $CH_3CH_2CHCH_2CH_2$ (with Br below) — $C(=O)$ — H

h. cyclopentane ring — $C(=O)$ — H; CH_2CH_3

i. 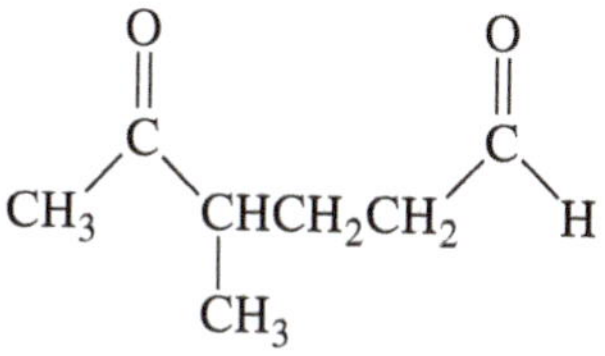 CH_3 — $C(=O)$ — $CHCH_2CH_2$ (with CH_3 below) — $C(=O)$ — H

54.

a. 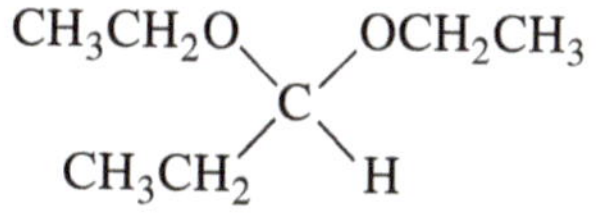 CH_3CH_2O, OCH_2CH_3, C, CH_3CH_2, H

b. (benzene ring) — $C(=O)$ — $CH_2CH_2CH_3$

c. $CH_3CH_2CHCH_3$ (with OH above)

d. $CH_3CH_2CCH_2CH_3$ (with OH above and $C{\equiv}N$ below)

e. $CH_3CH_2CH_2CH_2OH$ + CH_3CH_2OH

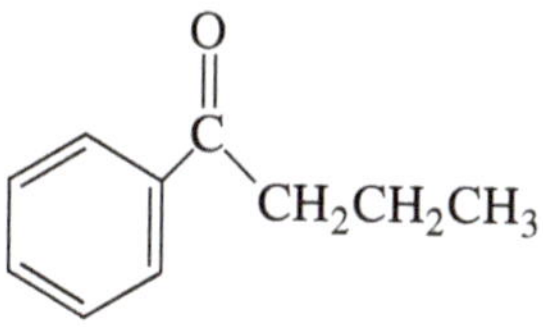

f. 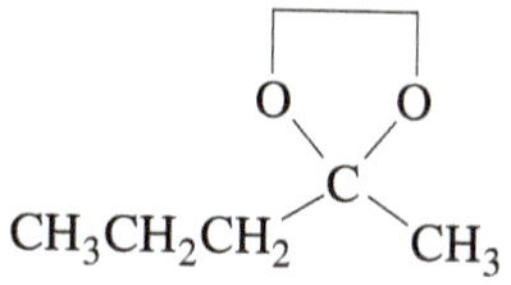 O O (ring); $CH_3CH_2CH_2$ — C — CH_3

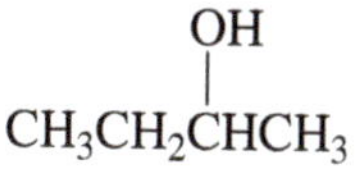

g. 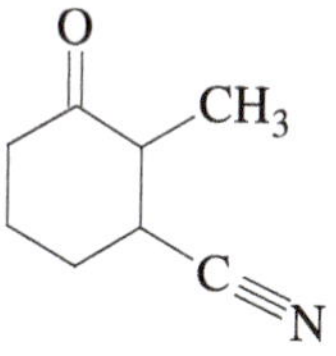 O (cyclohexanone ring) CH_3; $C{\equiv}N$

h. 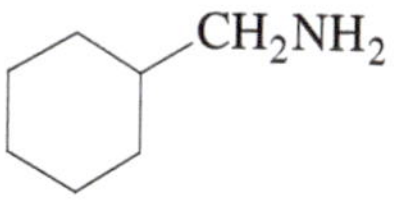 (cyclohexane ring) CH_2NH_2

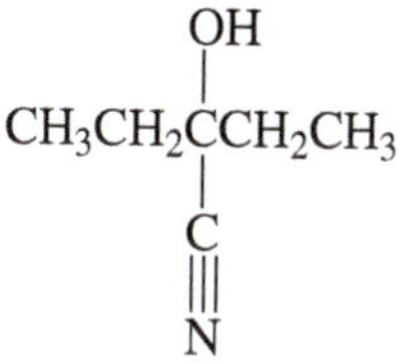

55. The greater the steric hindrance at the site of nucleophilic addition, the less reactive the carbonyl compound.

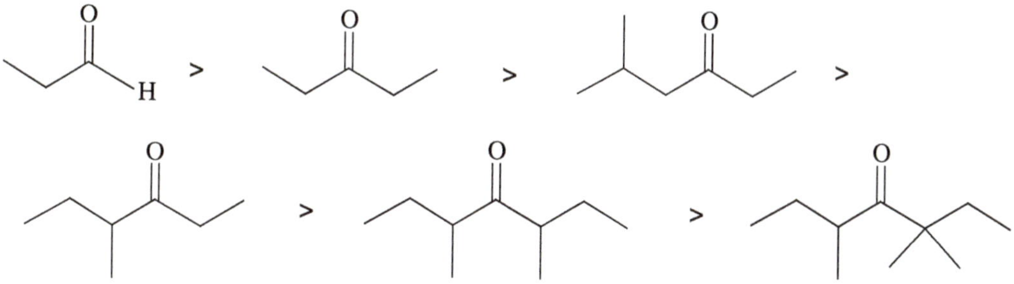

O, H > O > O >

O > O > O

56. $CH_3CH_2CH_2C(=O)OCH_2CH_2CH_3$ **and** $CH_3CH_2C(=O)OCH_2CH_2CH_2CH_3$

57. **a.**

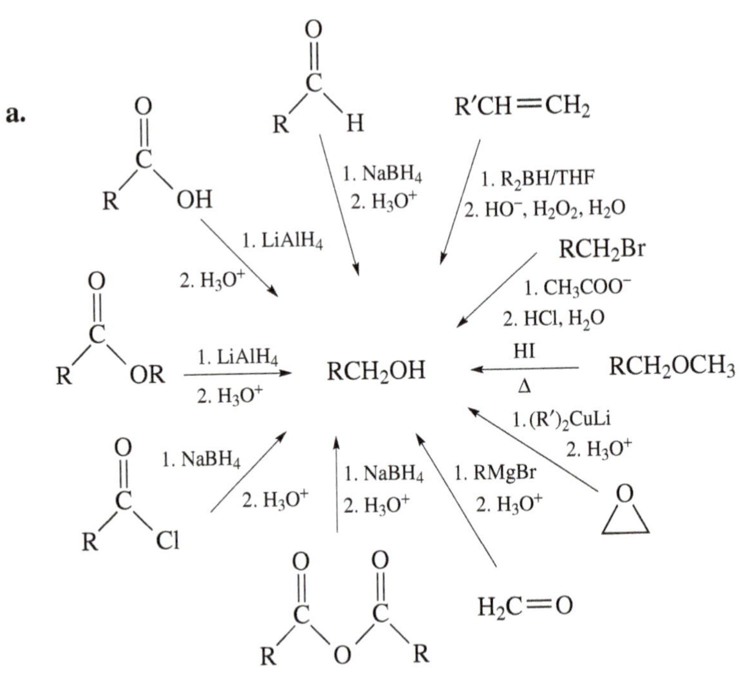

b. The reaction of an organocuprate with ethylene oxide leads to $R'CH_2CH_2OH$. Because this alcohol does not have a methyl group on the β-carbon, this method cannot be used to synthesize isobutyl alcohol.

$R'CH_2CH_2OH$ cannot lead to $CH_3CH(CH_3)CH_2OH$

58. **a.** $CH_3CH_2CH_2CH{=}CHCH_3$ (trans) reduction

b. $CH_3CH_2CH_2NHCH_3$ reduction

c. $C_6H_5CH_2CH_2CH_3$ reduction

d. $C_6H_5CH_2OH$ reduction

e. $C_6H_5CH_2OH$ + CH_3OH reduction

f. C_6H_5COOH oxidation

59. **a.** 6-methyl-4-hepten-3-one
b. 6-ethoxy-5-ethyl-2,2-dimethyl-4-heptanol
c. (*S*)-4-methylhexanal
d. 5-amino-6-hydroxy-2-heptanone

60. **a.** O → (1. $NaBH_4$; 2. H_3O^+) → OH

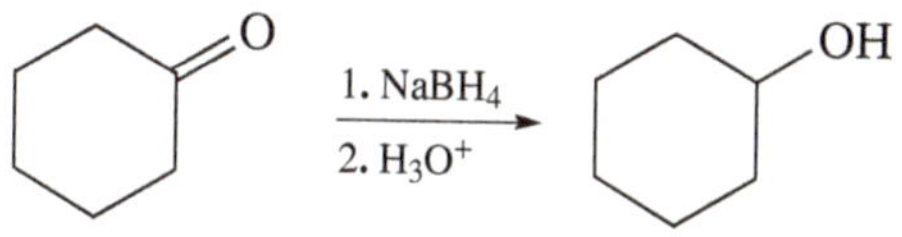

b. O → (1. $NaBH_4$; 2. H_3O^+) → OH → (H_2SO_4, Δ)

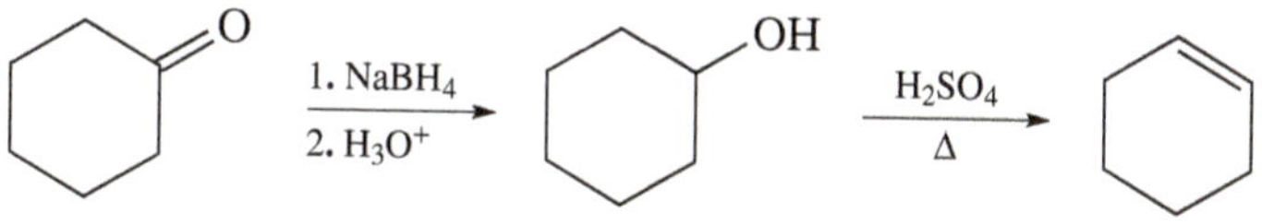

c. O → (1. $NaBH_4$; 2. H_3O^+) → OH → (PBr_3, pyridine) → Br

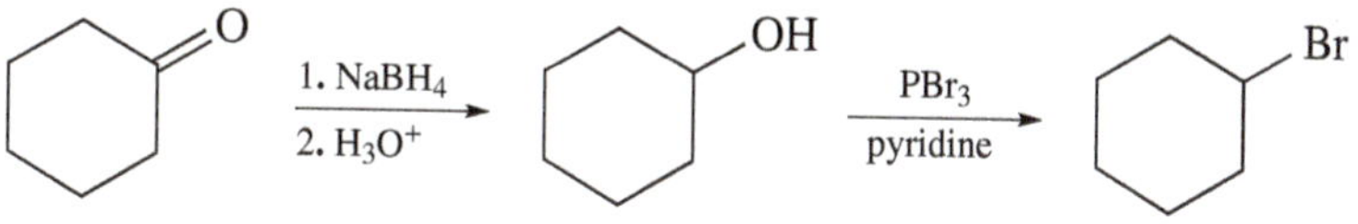

d. O → (1. trace acid, excess NH_3; 2. H_2, Pd/C) → NH_2

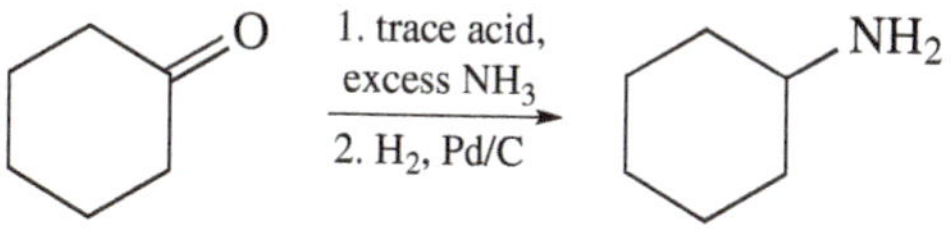

e. Br (product of **c**) → ($^-C{\equiv}N$) → $C{\equiv}N$ → (H_2, Raney Ni) → CH_2NH_2

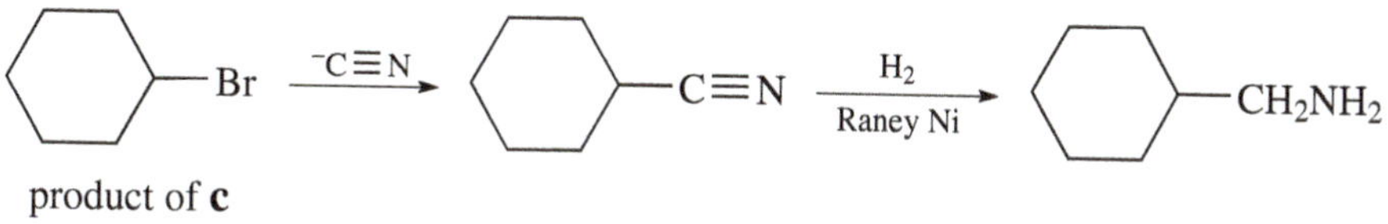

f. O + CH_3NH (CH_3) → (trace acid) → $N(CH_3)_2$ → ($NaBH_3CN$) → $N(CH_3)_2$

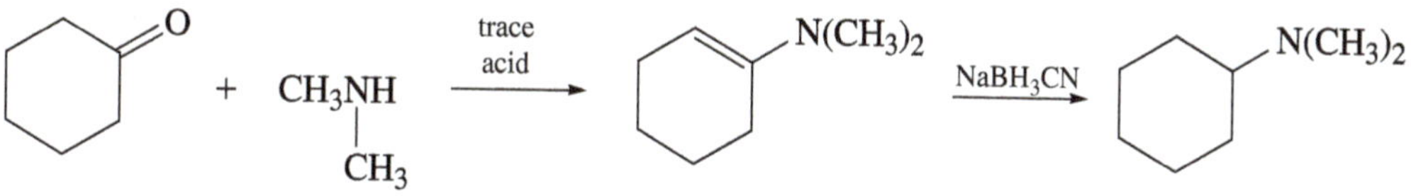

g. Br (product of **c**) + $(CH_2{=}CH)_2CuLi$ → $CH{=}CH_2$

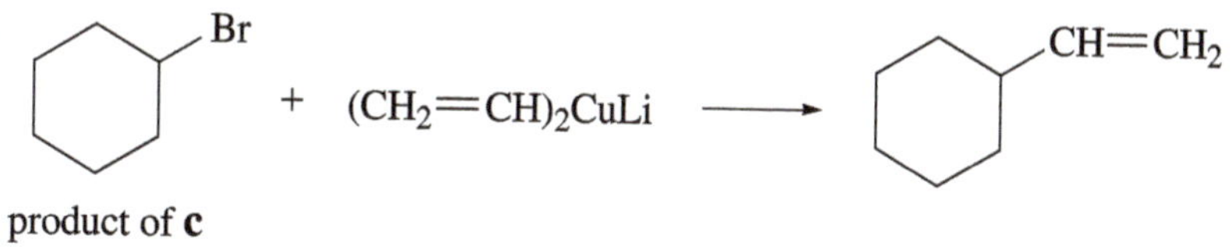

or

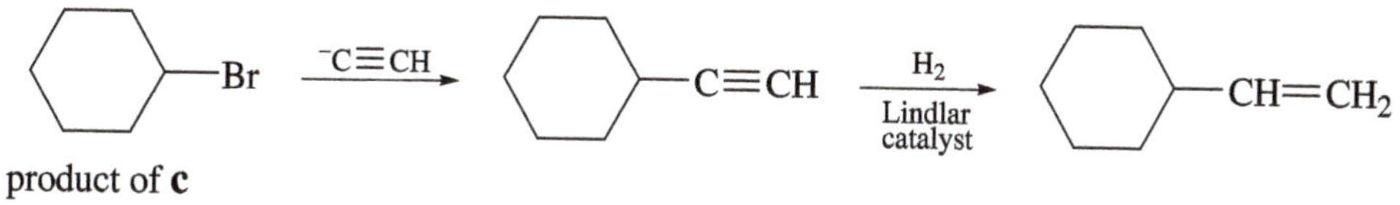

Br (product of **c**) → ($^-C{\equiv}CH$) → $C{\equiv}CH$ → (H_2, Lindlar catalyst) → $CH{=}CH_2$

h. O → (1. $HSCH_2CH_2CH_2SH/HCl$; 2. H_2, Raney Ni) →

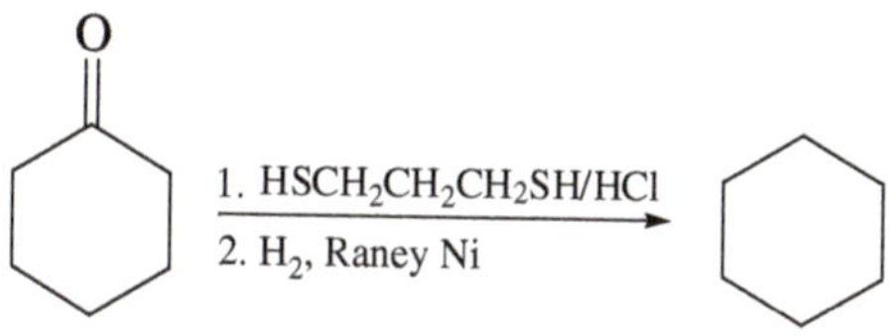

or

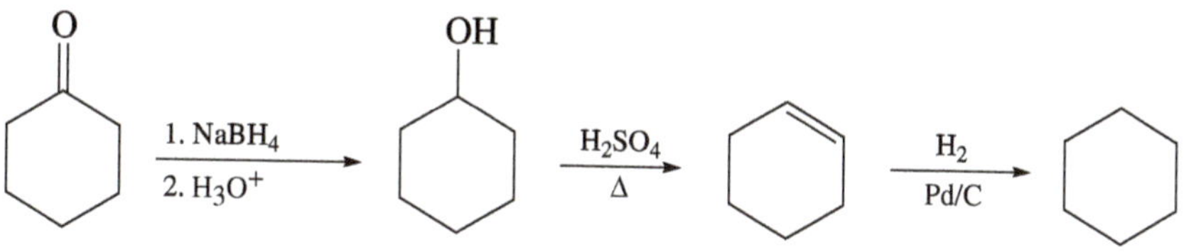

O → (1. $NaBH_4$; 2. H_3O^+) → OH → (H_2SO_4, Δ) → (H_2, Pd/C) →

i.

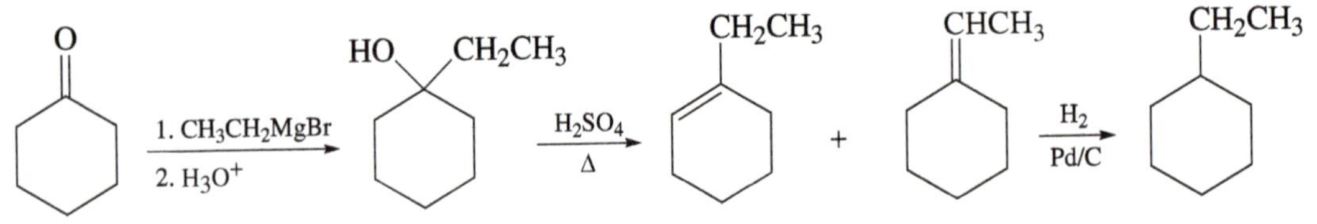

or

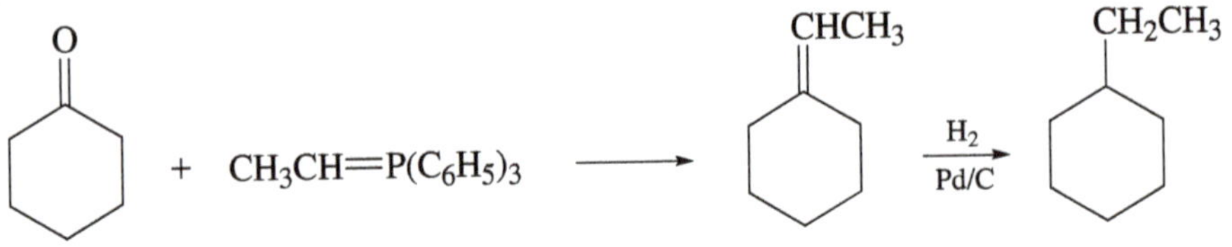

61. a.

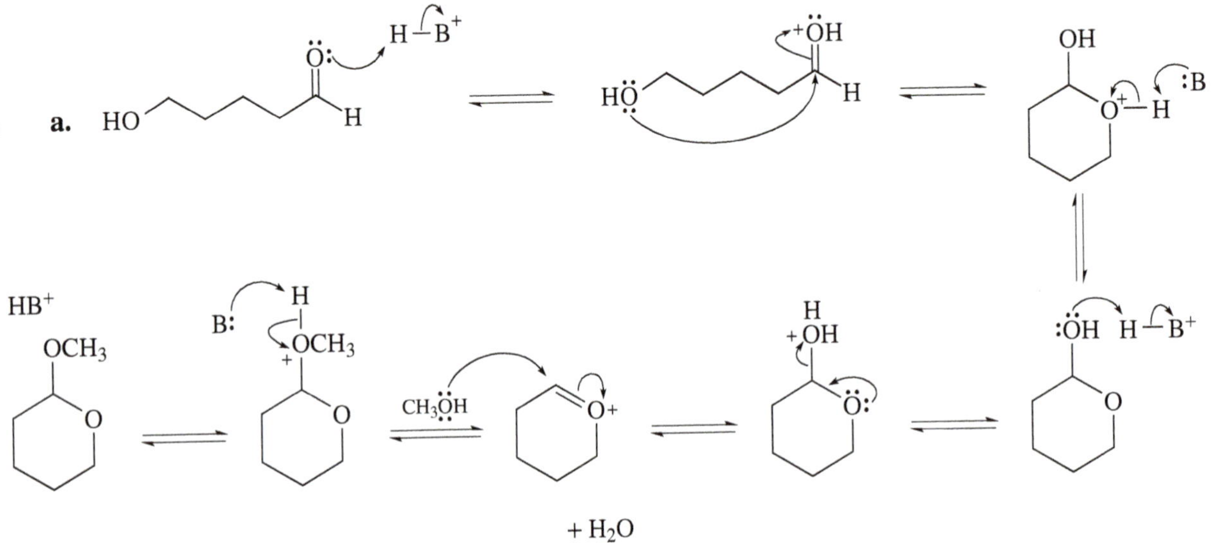

b.

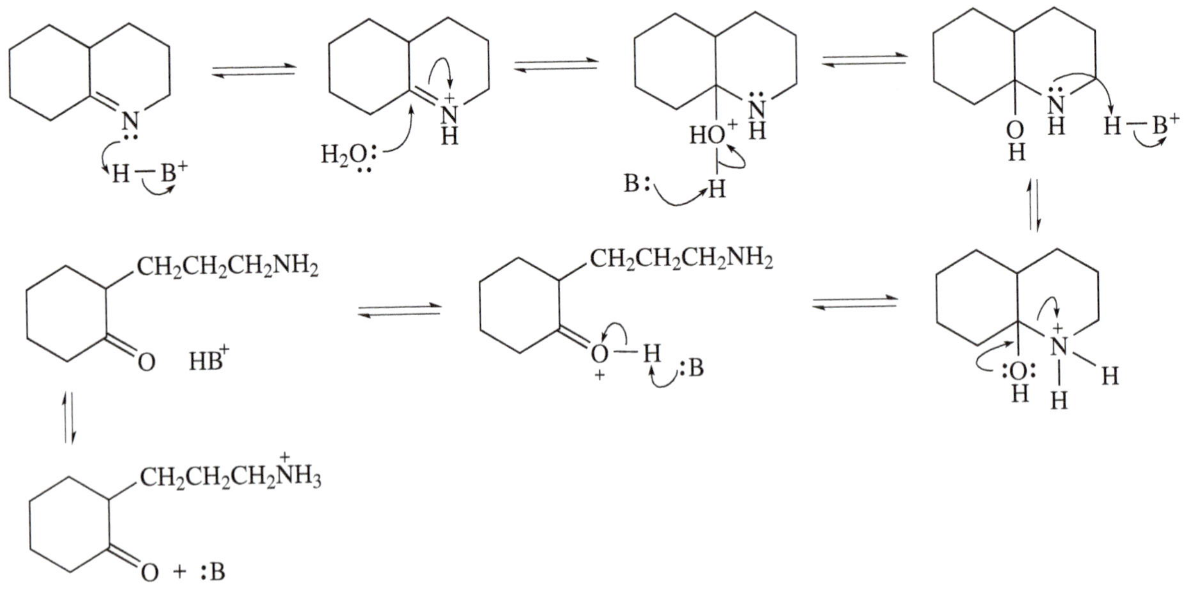

62. **a.** O C H $\xrightarrow{H_2CrO_4}$ O C OH $\xrightarrow[\text{pyridine}]{SOCl_2}$ O C Cl

1. $NaBH_4$ 2. H_3O^+ → $CH_3CH_2CH_2OH$ ↓

O C O

b. OH $\xrightarrow[\text{CH}_3\text{COOH, 0 °C}]{NaOCl}$ O C H $\xrightarrow[2.\ H_3O^+]{1.\ CH_3CH_2MgBr}$ OH

NaOCl | CH_3COOH, 0 °C ↓

O

c. OH $\xrightarrow[\text{CH}_3\text{COOH, 0 °C}]{NaOCl}$ O $\xrightarrow[2.\ NaBH_3CN]{1.\ CH_3NH_2,\ \text{trace acid}}$ $NHCH_3$

d. O $\xrightarrow[2.\ H_3O^+]{1.\ NaBH_4}$ OH $\xrightarrow[\Delta]{H_2SO_4}$ $\xrightarrow[2.\ (CH_3)_2S]{1.\ O_3,\ -78\ °C}$ H O H O

↓ H_2CrO_4

HO O OH O

e. O $\xrightarrow[2.\ HCl]{1.\ CH_3MgBr}$ CH_3 OH $\xrightarrow[\Delta]{H_2SO_4}$ CH_3 $\xrightarrow[2.\ (CH_3)_2S]{1.\ O_3,\ -78\ °C}$ O H O

↓ H_2CrO_4

O OH O

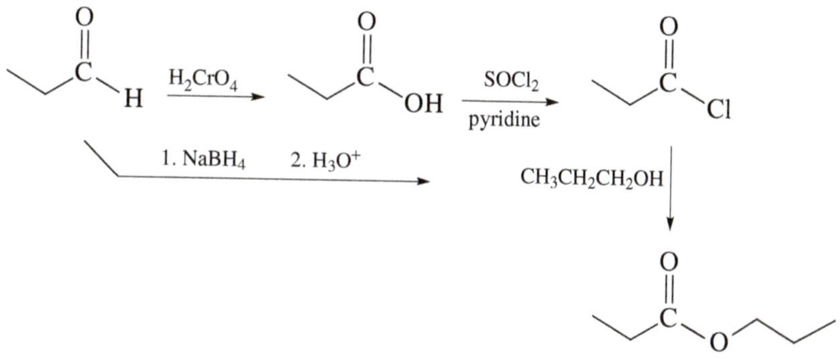

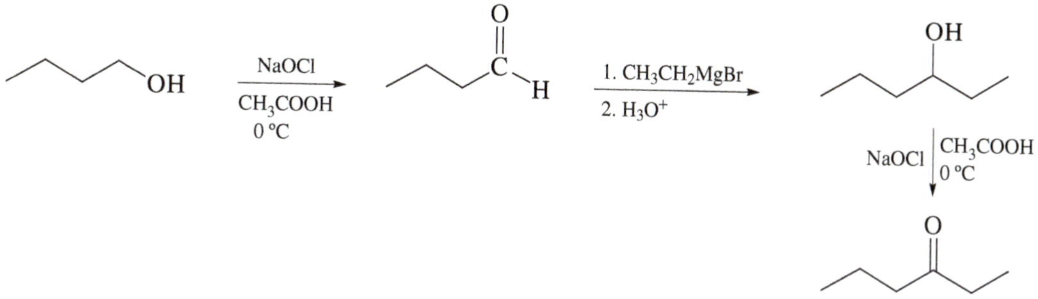

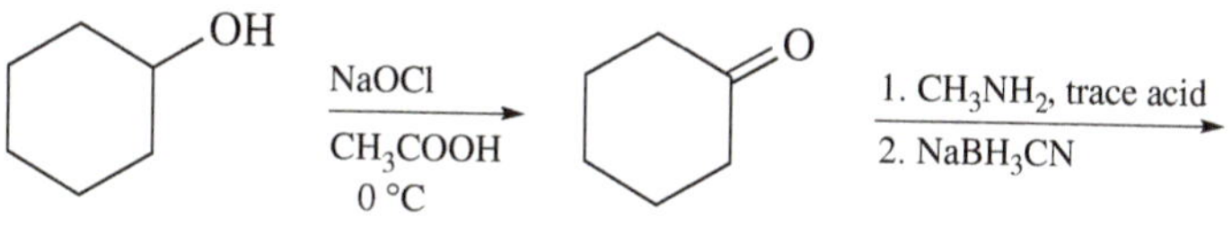

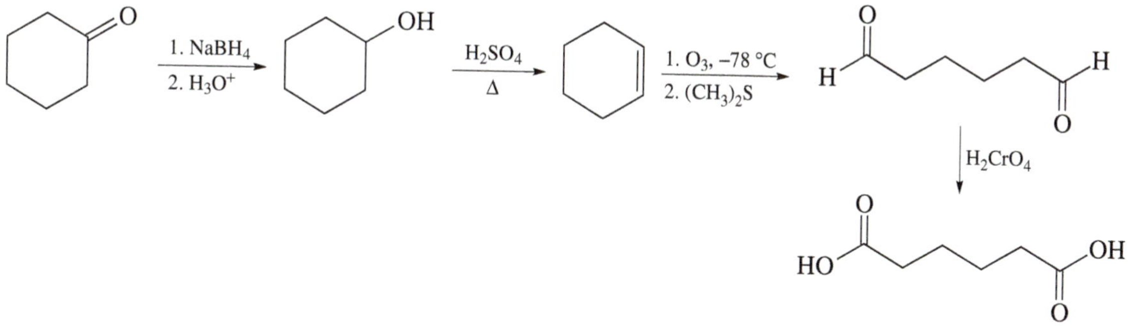

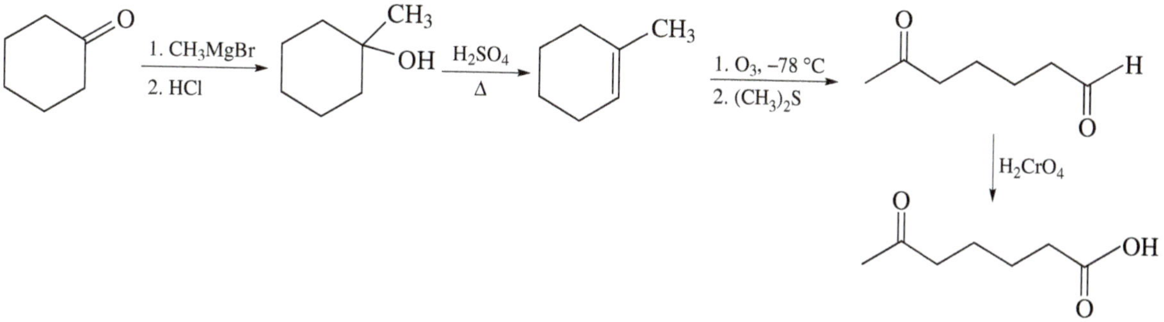

f. cyclohexyl–CH_2Br $\xrightarrow[Et_2O]{Mg}$ cyclohexyl–CH_2MgBr $\xrightarrow{CO_2}$ cyclohexyl–$CH_2CO_2^-$

$\xrightarrow{SOCl_2}$ cyclohexyl–CH_2COCl $\xrightarrow{CH_3CH_2NH_2}$ cyclohexyl–$CH_2C(=O)NHCH_2CH_3$

or

cyclohexyl–CH_2Br $\xrightarrow{^-C{\equiv}N}$ cyclohexyl–$CH_2C{\equiv}N$ $\xrightarrow[2.\ HO^-]{1.\ HCl,\ H_2O,\ \Delta}$ cyclohexyl–$CH_2CO_2^-$

$\xrightarrow{SOCl_2}$ cyclohexyl–CH_2COCl $\xrightarrow{CH_3CH_2NH_2}$ cyclohexyl–$CH_2C(=O)NHCH_2CH_3$

63. a. $CH_3OH \xrightarrow[pyridine]{PBr_3} CH_3Br \xrightarrow[2.\ CuI]{1.\ Li} (CH_3)_2CuLi \xrightarrow[2.\ HCl]{1.\ H_2C{=}O} CH_3CH_2OH$

b. $\underset{excess}{CH_4} \xrightarrow[h\nu]{Br_2} CH_3Br \xrightarrow[2.\ CuI]{1.\ Li} (CH_3)_2CuLi \xrightarrow[2.\ HCl]{1.\ ethylene\ oxide} CH_3CH_2CH_2OH$

64. a. $(H_2N)_2C{=}\ddot{S}{:}$ $\xrightarrow{CH_3CH_2-Br}$ $(H_2N)_2C{=}\overset{+}{S}-CH_2CH_3$ $\underset{H\ddot{O}{:}^-}{\overset{HO^-}{\rightleftharpoons}}$ $H_2N-C(OH)(-\ddot{S}-CH_2CH_3)-NH_2$

$\rightleftharpoons$ $(H_2N)_2C{=}\overset{+}{O}H$ + $CH_3CH_2S^-$ $\rightleftharpoons$ $(H_2N)_2C{=}O$ + CH_3CH_2SH

b. $CH_3CH_2CH_2CH_2CH_2SH$

65.

CH_3 (methylcyclohexane) $\xrightarrow[hv]{Br_2}$ **A** (1-bromo-1-methylcyclohexane; CH_3, Br) $\xrightarrow{tert\text{-}BuO^-}$ **B** (1-methylcyclohexene; CH_3) $\xrightarrow{Br_2}$ **C** (CH_3, Br, Br) + **D** (CH_3, Br, Br)

B $\xrightarrow{1.\ R_2BH/THF;\ 2.\ H_2O_2,\ HO^-,\ H_2O}$ (CH_3, OH) + **E** (CH_3, OH)

B $\xrightarrow{H_2SO_4,\ H_2O}$ **M** (CH_3, OH) $\xrightarrow{HBr}$ **N** (CH_3, Br) $\xrightarrow{1.\ Li;\ 2.\ CuI;\ 3.\ \text{ethylene oxide};\ 4.\ HCl}$ **O** (CH_3, CH_2CH_2OH)

B $\xrightarrow{1.\ O_3,\ -78\ °C;\ 2.\ (CH_3)_2S}$ **I** $CH_3C(=O)CH_2CH_2CH_2CH_2C(=O)H$

I $\xrightarrow{H_2CrO_4}$ **J** $CH_3C(=O)CH_2CH_2CH_2CH_2C(=O)OH$

J $\xrightarrow{SOCl_2}$ **K** $CH_3C(=O)CH_2CH_2CH_2CH_2C(=O)Cl$

K $\xrightarrow{CH_3OH}$ **L** $CH_3C(=O)CH_2CH_2CH_2CH_2C(=O)OCH_3$

E $\xrightarrow{HOCl}$ **F** (2-methylcyclohexanone; CH_3, O)

F $\xrightarrow{CH_3CH_2OH\ \text{excess},\ HCl}$ **G** (CH_3, OCH_2CH_3, OCH_2CH_3)

F $\xrightarrow{\text{trace acid},\ CH_3CH_2NH_2}$ **H** (CH_3, NCH_2CH_3)

66.

a. $C_6H_5C(=O)CH_2CH_3$ + $CH_3CH_2\overset{+}{N}H_3$

b. $CH_3CH_2C(OH)(CH_2CH_3)CH_3$

c. ethylidenecyclopentane ($CHCH_3$)

d. $CH_3CH_2C(OH)(CH_2CH_3)CH_2CH_3$

e. HO, OCH_3 (phenyl, propyl)

f. pyrrolidine (N–H)

g. cyclohexylidene–NCH_2CH_3

h. CH_3CH_2–N–CH_2CH_3 on cyclohexenyl

i. 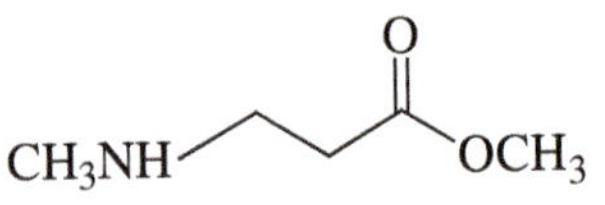 CH_3NH ... OCH_3 (O)

j. 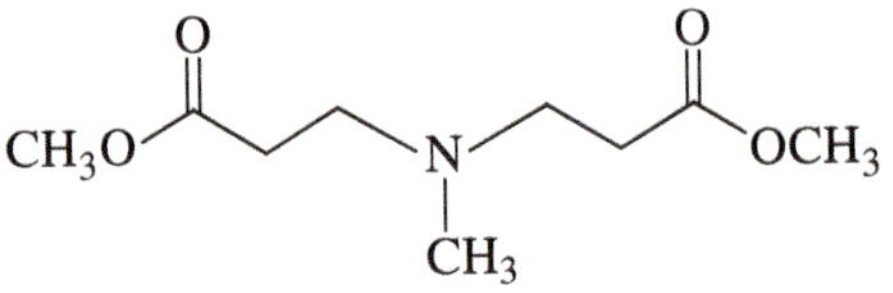 CH_3O ... N (CH_3) ... OCH_3 (O, O)

67. The offset shows that there is a signal at ~10.0 ppm, indicating an aldehyde. The ^{1}H NMR spectrum is that of benzaldehyde. Phenylmagnesium bromide, therefore, must react with a compound with one carbon to form an alcohol that can be oxidized to benzaldehyde. Therefore, Compound **Z** must be formaldehyde.

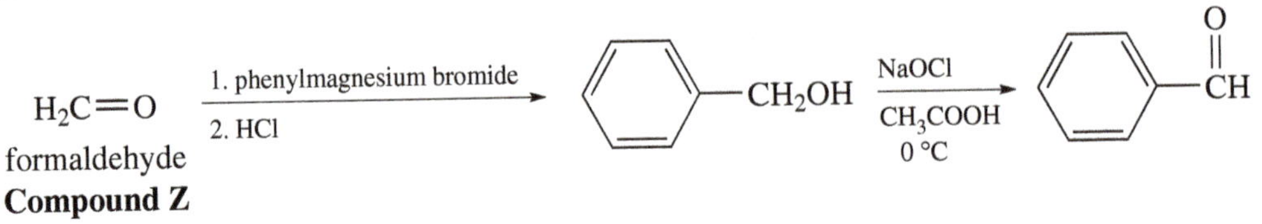

$H_2C{=}O$ formaldehyde **Compound Z** → (1. phenylmagnesium bromide; 2. HCl) → $-CH_2OH$ → (NaOCl, CH_3COOH, 0 °C) → $-\overset{O}{\overset{\|}{C}}H$

68. **a.** 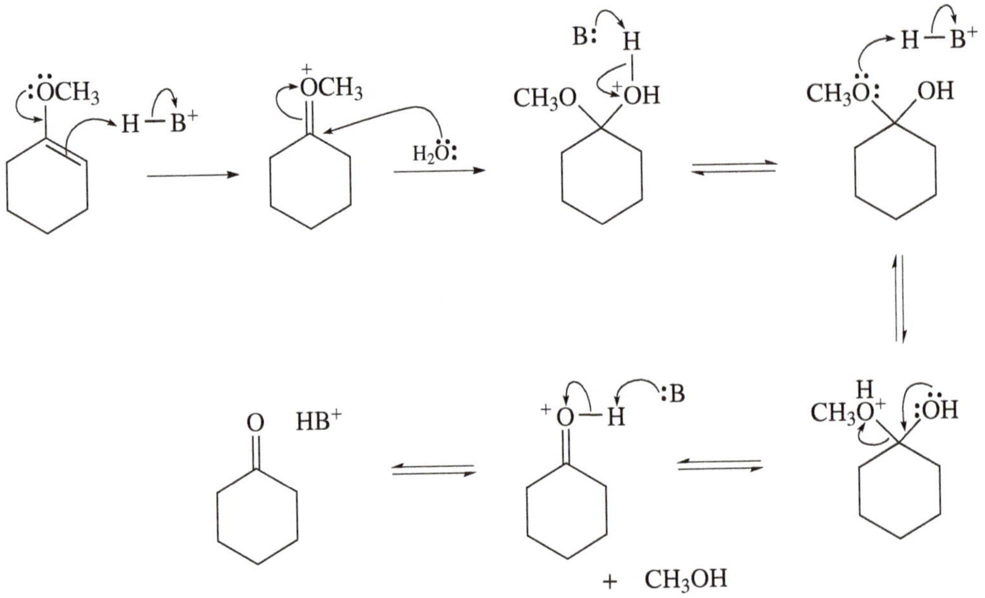 $:\ddot{O}CH_3$, $H-B^+$ → $\overset{+}{O}CH_3$ → $H_2\ddot{O}:$ → B:, H, CH_3O, $\overset{+}{O}H$ ⇌ $H-B^+$, $CH_3\ddot{O}:$, OH ⇌ H, $CH_3\overset{+}{O}$, $:\ddot{O}H$ ⇌ $\overset{+}{O}-H$, :B, + CH_3OH ⇌ O, HB^+

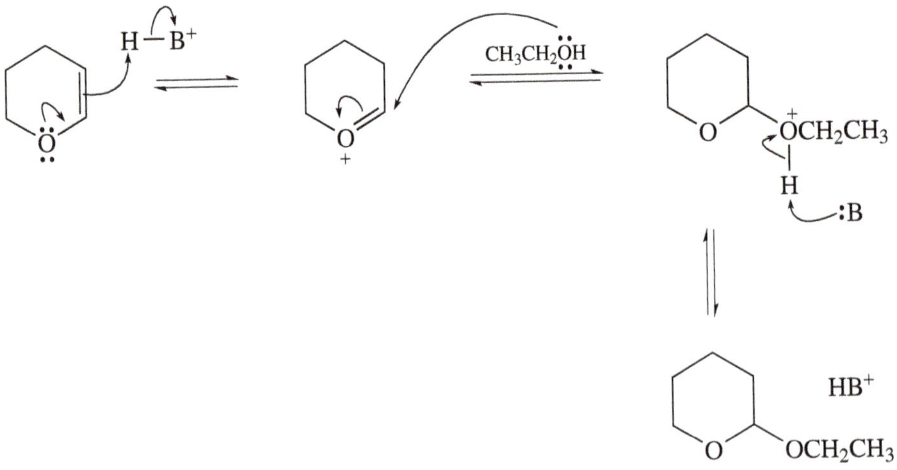

b. $H-B^+$ ⇌ $\overset{+}{O}$ ⇌ ($CH_3CH_2\ddot{O}H$) O, $\overset{+}{O}CH_2CH_3$, H, :B ⇌ O, OCH_2CH_3, HB^+

69. **a.** three signals in its ^{1}H NMR spectrum **b.** three signals in its ^{13}C NMR spectrum

$$CH_3C(=O)CH_2CH_2C(=O)OCH_3 \xrightarrow[2.\ HCl]{1.\ excess\ CH_3MgBr} CH_3C(OH)(CH_3)CH_2CH_2C(OH)(CH_3)CH_3$$

70.

$$CH_3CH_2C(=O)H \xrightarrow[2.\ HCl]{1.\ CH_3MgBr} CH_3CH_2CH(OH)CH_3 \xrightarrow[CH_3COOH,\ 0\ °C]{NaOCl} CH_3CH_2C(=O)CH_3 \xrightarrow[CH_3OH\ excess]{HCl} CH_3CH_2C(OCH_3)_2CH_3$$

71. Electron withdrawal from the benzene ring increases the electrophilicity of the carbonyl carbon, which makes it more reactive.
Electron donation into the benzene ring decreases the electrophilicity of the carbonyl carbon, which makes it less reactive.

The first compound is the most reactive toward nucleophilic addition because the nitro substituent withdraws electrons from the benzene ring inductively and by resonance.
The third compound is less reactive than the compound without a substituent on the benzene ring because the isopropyl substituent donates electrons into the benzene ring by hyperconjugation.
The fourth compound is least reactive because the ethoxy substituent donates electrons strongly into the benzene ring by resonance.

$$p\text{-}O_2N\text{-}C_6H_4CHO > C_6H_5CHO > p\text{-}(CH_3)_2CH\text{-}C_6H_4CHO > p\text{-}CH_3CH_2O\text{-}C_6H_4CHO$$

72. **a.**

$$C_6H_5C(=O)NHCH_3 \xrightarrow[2.\ H_2O]{1.\ LiAlH_4} C_6H_5CH_2NHCH_3$$

b.

$$C_6H_5C(=O)NHCH_3 \xrightarrow[H_2O,\ \Delta]{HCl} C_6H_5C(=O)OH$$

c. $C_6H_5C(=O)NHCH_3$ $\xrightarrow[H_2O, \Delta]{HCl}$ $C_6H_5C(=O)OH$ $\xrightarrow{SOCl_2}$ $C_6H_5C(=O)Cl$ $\xrightarrow{CH_3OH}$ $C_6H_5C(=O)OCH_3$

d. $C_6H_5C(=O)NHCH_3$ $\xrightarrow[H_2O, \Delta]{HCl}$ $C_6H_5C(=O)OH$ $\xrightarrow[2.\ H_3O^+]{1.\ LiAlH_4}$ $C_6H_5CH_2OH$

73. In parts **a**, **b**, and **d**, a new asymmetric center is formed in the product, so a racemic mixture is formed. In part **c**, both *E* and *Z* stereoisomers are formed.

a. O, CH₃ (wedge) ; O, CH₃ (hash)

c. H_3C, H, C=C, N (pyrrolidine), phenyl ; H_3C, H, C=C, phenyl, N (pyrrolidine)

b. HO, CH_3 ; HO, CH_3

d. $CH_2CH_2CH_2CH_3$, CH_3CH_2, C, H, OH ; $CH_2CH_2CH_2CH_3$, H, HO, C, CH_2CH_3

74. **a.** $CH_3CH_2C(=O)CH_2CH_2CH_2CH_3$ + C_6H_5MgBr

$C_6H_5C(=O)CH_2CH_3$ + $CH_3CH_2CH_2CH_2MgBr$

$C_6H_5C(=O)CH_2CH_2CH_2CH_3$ + CH_3CH_2MgBr

b. 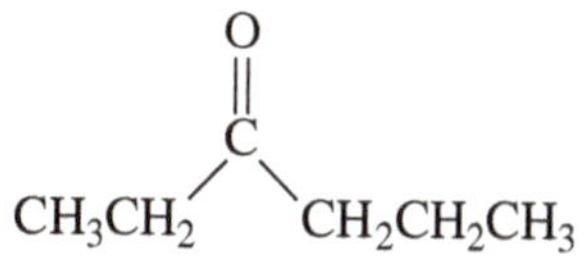 $CH_3CH_2C(=O)CH_2CH_2CH_3$ + CH_3CH_2MgBr

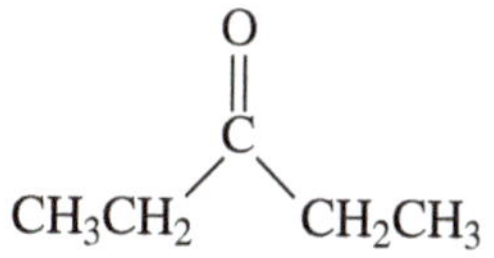

$CH_3CH_2C(=O)CH_2CH_3$ + $CH_3CH_2CH_2MgBr$

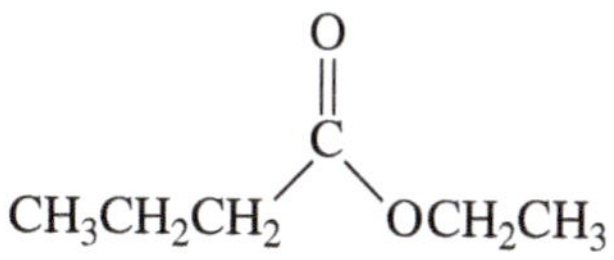

$CH_3CH_2CH_2C(=O)OCH_2CH_3$ + 2 CH_3CH_2MgBr

75. **a.** 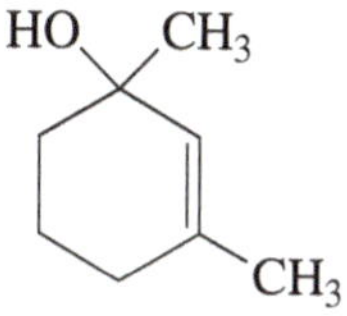 HO, CH_3, CH_3 **b.** 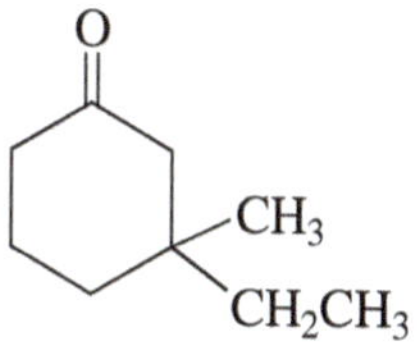 O, CH_3, CH_2CH_3 **c.** 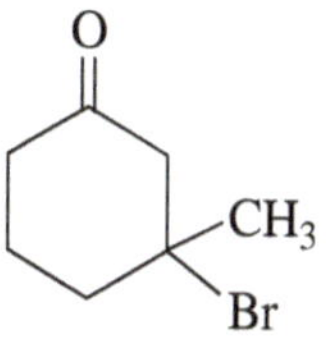 O, CH_3, Br **d.** 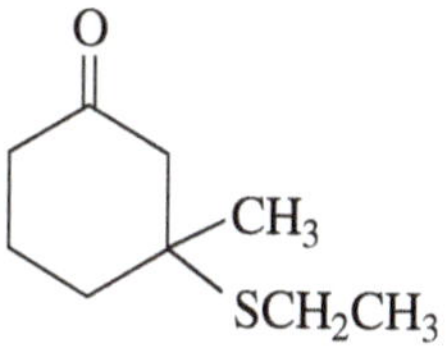 O, CH_3, SCH_2CH_3

76. Notice that ketal formation can be used to protect either ketones or 1,2-diols. Thus, in part **a**, the 1,2-diol is protected by acetone, whereas in part **b**, the ketone is protected by 1,2-ethanediol.

a. 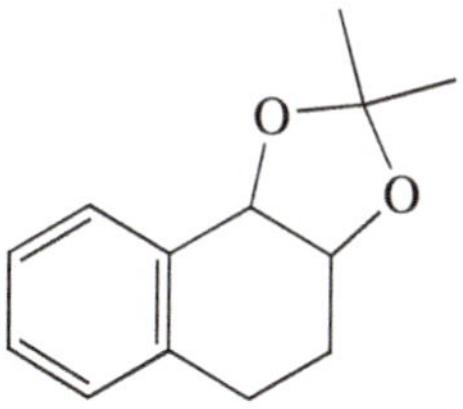 O, O **b.** 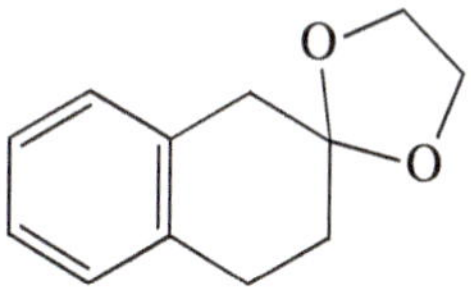 O, O

77. **a.** 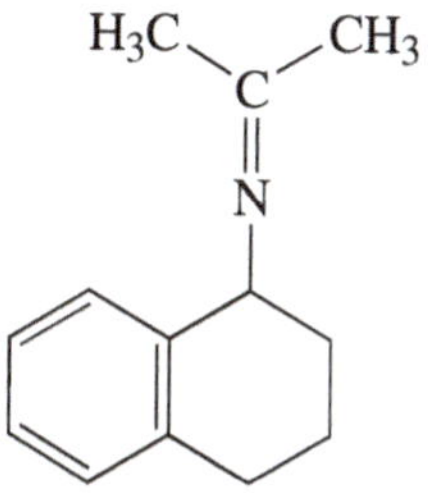 H_3C, CH_3, C, N **c.** O, NH

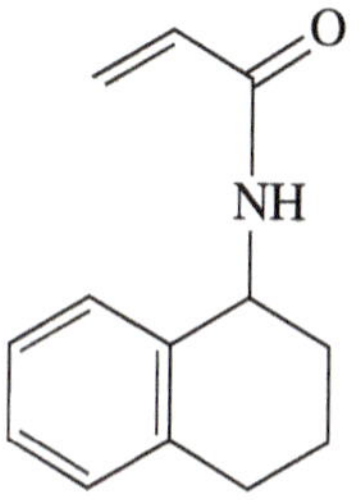

b. 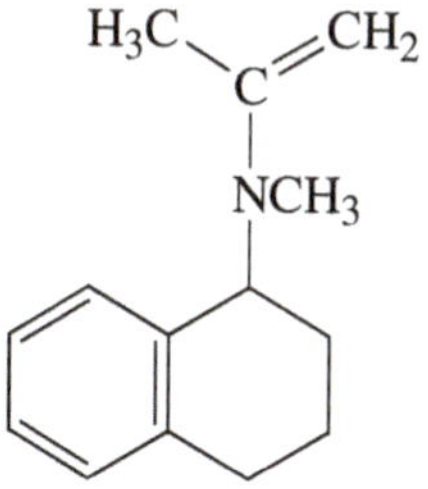 H_3C, CH_2, C, NCH_3 **d.** 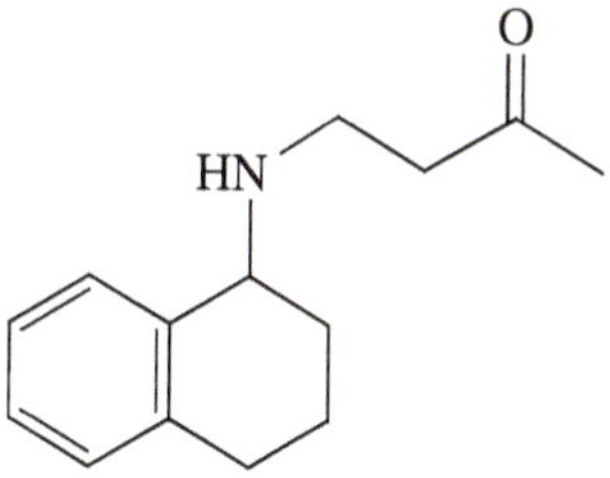 O, HN

78. **a.**

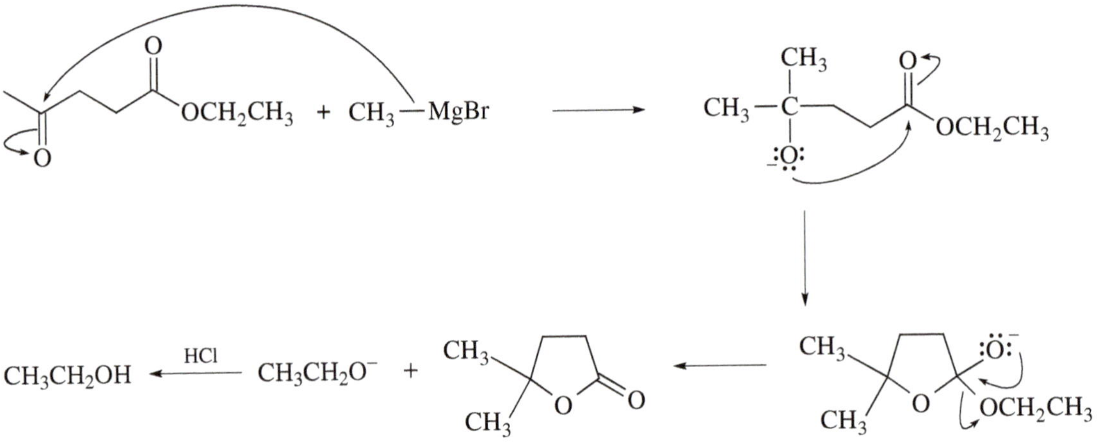

b.

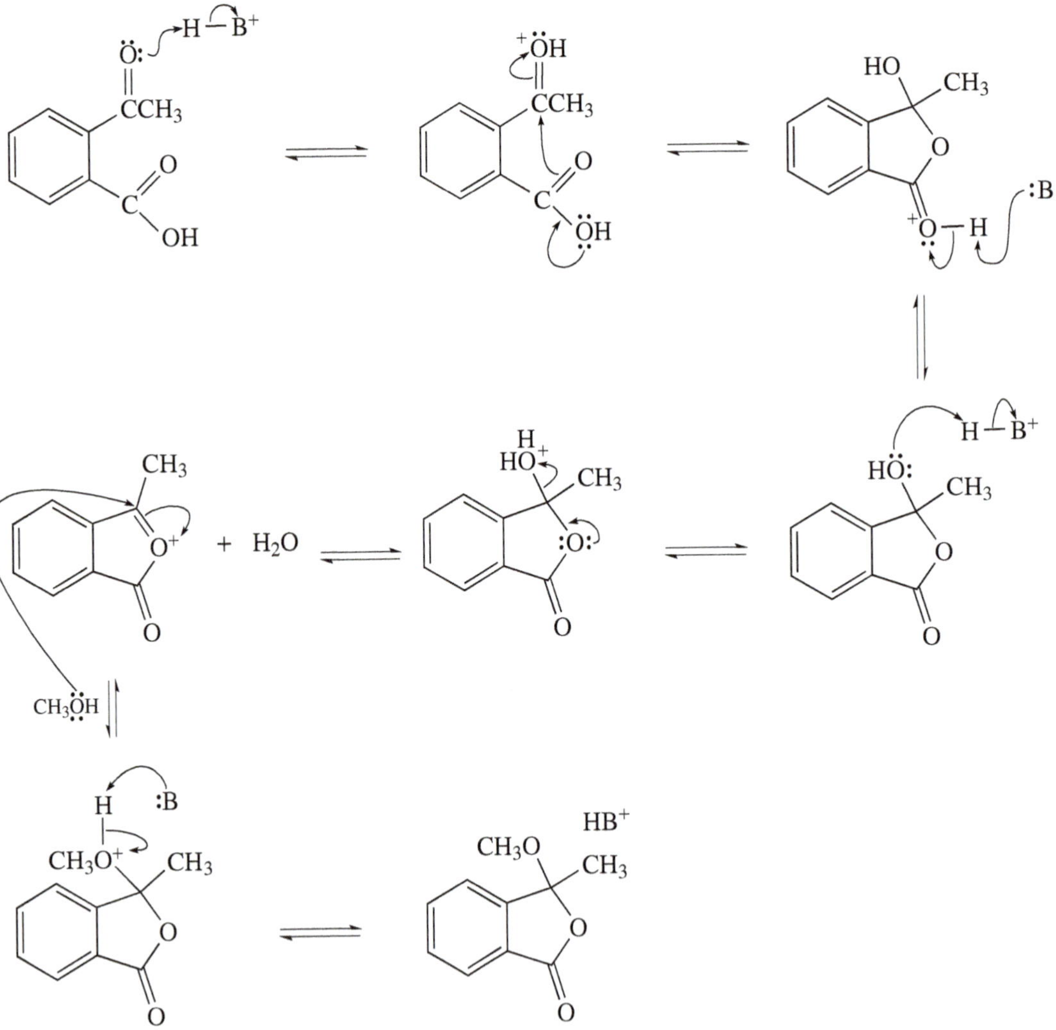

79. a.

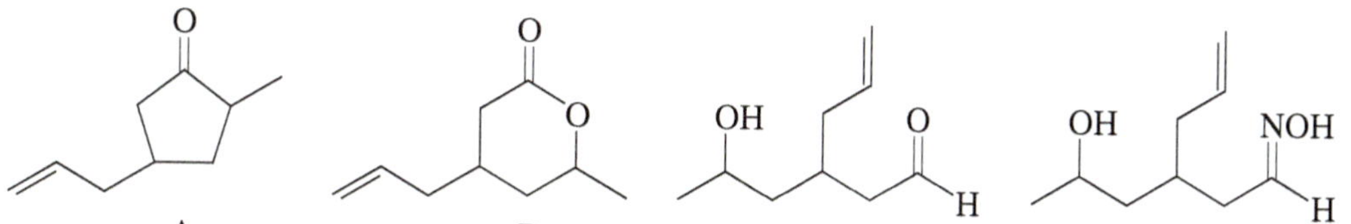

A B C D

b.

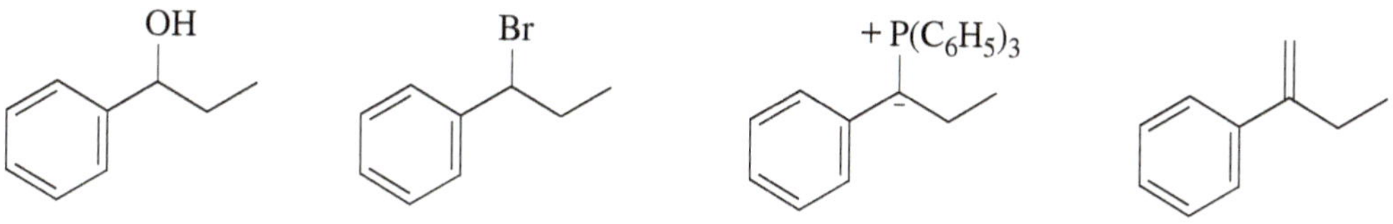

A B C D

80.

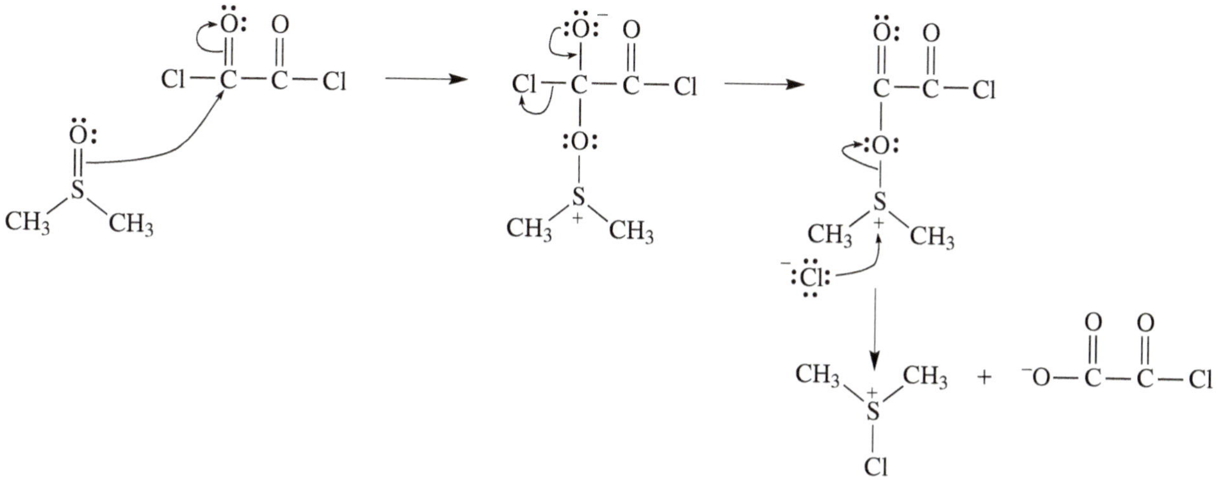

dimethylchlorosulfonium ion

81. In the first step of each reaction, cyanide is eliminated because it is a weak base. Conjugate addition occurs in the second step, again, because cyanide ion is a weak base.

a. $CH_2{=}CHCH(OH)C{\equiv}N \longrightarrow CH_2{=}CHCH{=}O + {}^{-}{:}C{\equiv}N \longrightarrow N{\equiv}CCH_2CH_2CH{=}O$

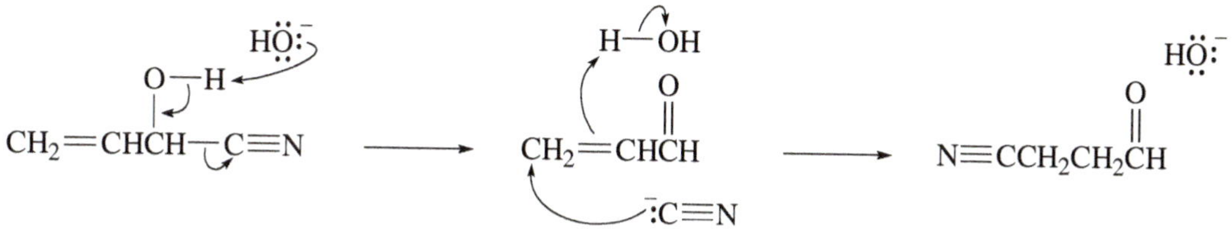

b. $CH_3C(OH)(C{\equiv}N)CH{=}CH_2 \longrightarrow CH_3C(=O)CH{=}CH_2 + {}^{-}{:}C{\equiv}N \longrightarrow CH_3C(=O)CH_2CH_2C{\equiv}N$

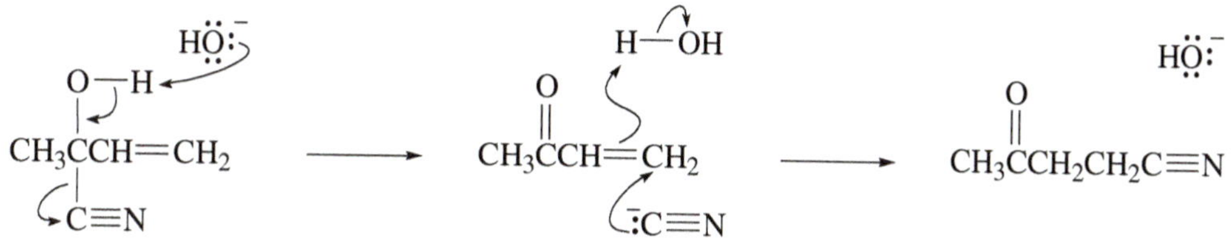

82. We have seen that the Wittig reaction proceeds via a concerted [2+2] cycloaddition reaction.

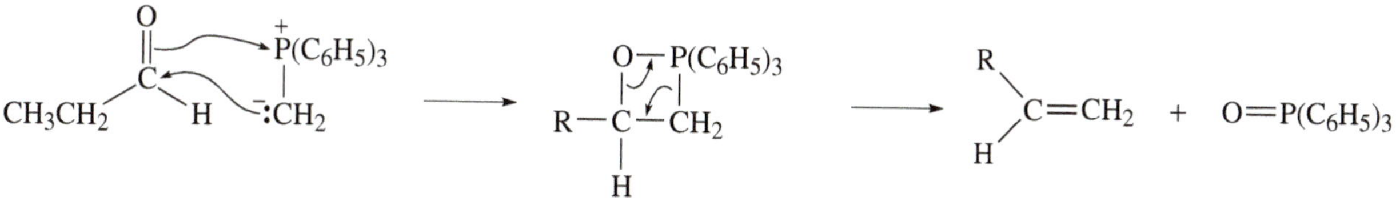

$CH_3CH_2CHO + (C_6H_5)_3\overset{+}{P}{-}\overset{-}{C}H_2 \longrightarrow$ R—C(H)—CH₂ / O—P(C₆H₅)₃ $\longrightarrow$ RHC=CH₂ + O=P(C₆H₅)₃

In contrast, the sulfonium ylide adds to the carbonyl group to form the intermediate shown here. The intermediate has a nucleophile (an oxyanion) that attacks the methylenecarbon because $(CH_3)_2S$ is a weak base and, therefore, a very good leaving group.

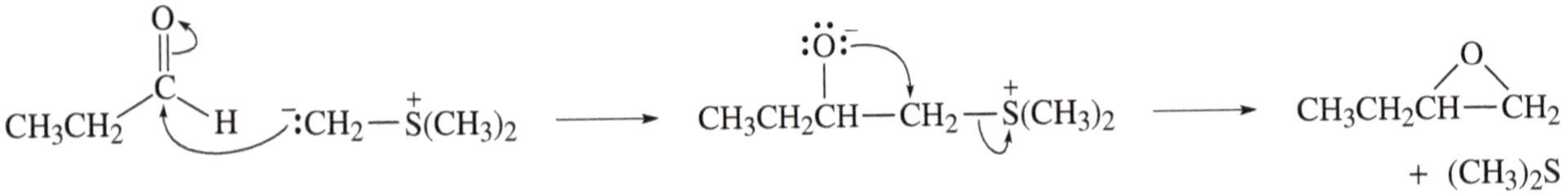

$$CH_3CH_2C(=O)H + :\bar{C}H_2-\overset{+}{S}(CH_3)_2 \longrightarrow CH_3CH_2CH(\ddot{O}:^-)-CH_2-\overset{+}{S}(CH_3)_2 \longrightarrow CH_3CH_2\overset{O}{CH-CH_2} + (CH_3)_2S$$

83. The absorption bands at 1600 cm^{-1}, 1500 cm^{-1}, and >3000 cm^{-1} in the IR spectrum indicate that the compound has a benzene ring. The absorption band at 1720 cm^{-1} suggests that it is an aldehyde or a ketone and the carbonyl group is not conjugated with the benzene ring. The absorption bands at 2720 cm^{-1} and 2820 cm^{-1} indicate that the compound is an aldehyde.

The broad signal at 1.8 ppm in the ^1H NMR spectrum of the product indicates an OH group. We also see that there are three different kinds of hydrogens (other than the OH group and the benzene ring hydrogens). The doublet at ~1.2 ppm that integrates to 3 protons indicates a methyl group adjacent to a carbon bonded to one hydrogen. The doublet at 3.6 ppm that integrates to 2 protons is due to a methylene group bonded to a CH group. The chemical shift of this signal suggests that the electron-withdrawing OH group is also bonded to this methylene group.

The IR spectrum is the spectrum of 2-phenylpropanal, and the ^1H NMR spectrum is the spectrum of 2-phenyl-1-propanol.

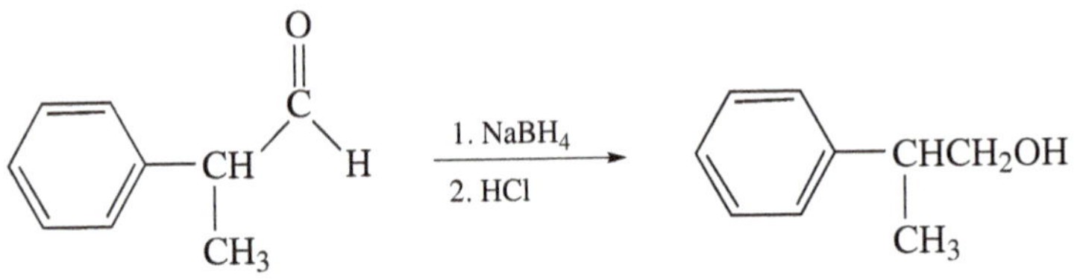

$$C_6H_5CH(CH_3)CHO \xrightarrow[\text{2. HCl}]{\text{1. NaBH}_4} C_6H_5CH(CH_3)CH_2OH$$

84. a.

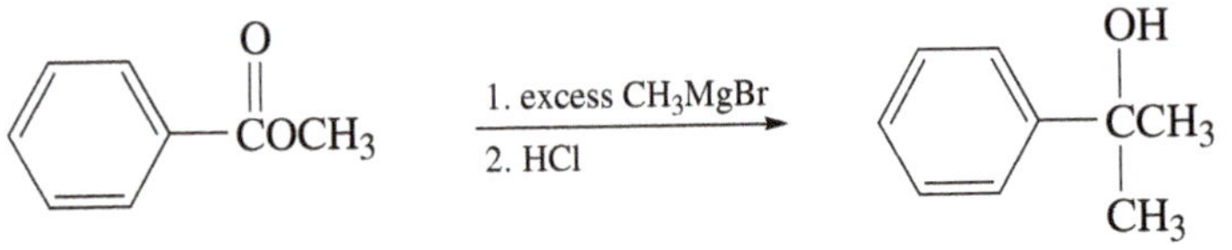

$$C_6H_5COCH_3 \xrightarrow[\text{2. HCl}]{\text{1. excess CH}_3\text{MgBr}} C_6H_5C(OH)(CH_3)CH_3$$

b.

$$C_6H_5COOCH_3 \xrightarrow[\text{2. HCl}]{\text{1. LiAlH}_4} C_6H_5CH_2OH \xrightarrow[\text{CH}_3\text{COOH, 0 °C}]{\text{NaOCl}} C_6H_5CHO$$

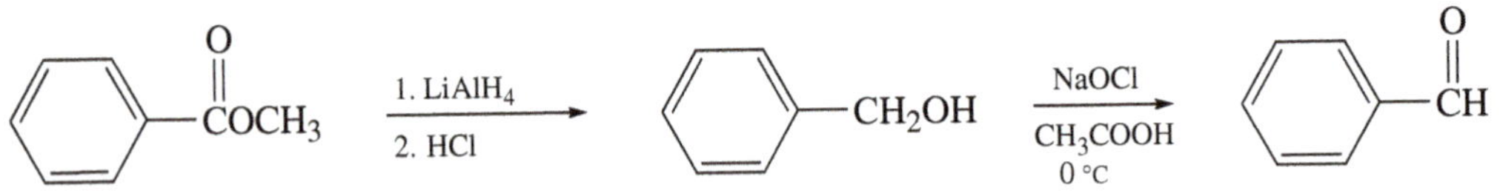

The synthesis can also be carried out in one step.

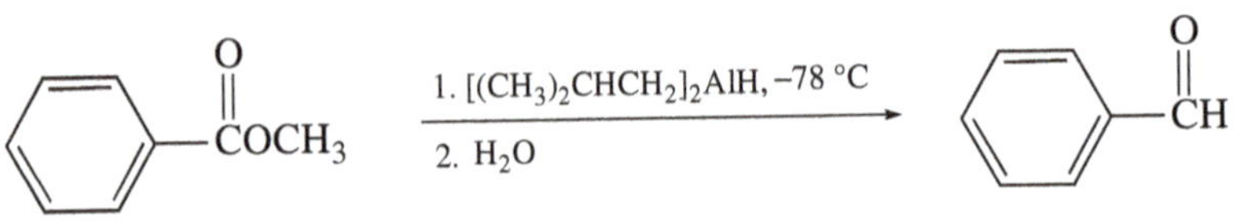

$$C_6H_5COOCH_3 \xrightarrow[\text{2. H}_2\text{O}]{\text{1. [(CH}_3)_2\text{CHCH}_2]_2\text{AlH, −78 °C}} C_6H_5CHO$$

c. 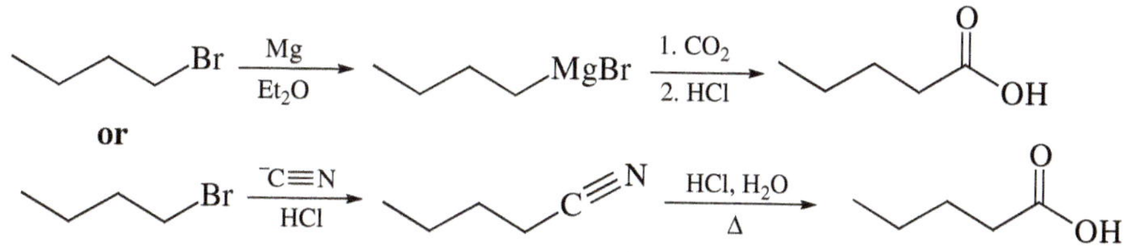

d. 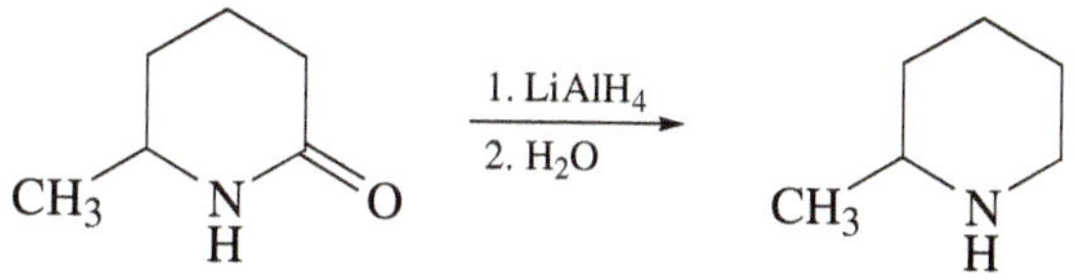

e. 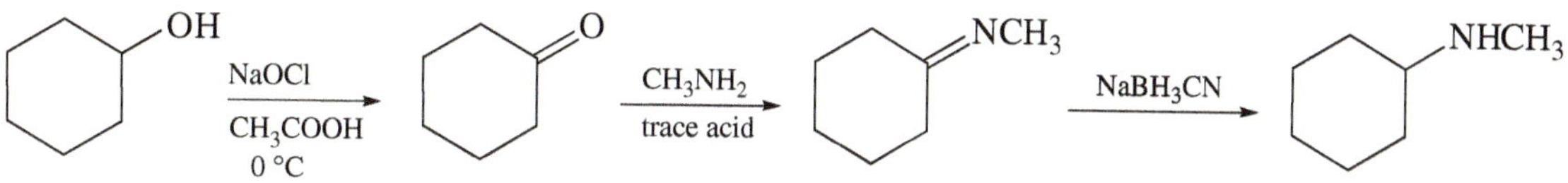

f. 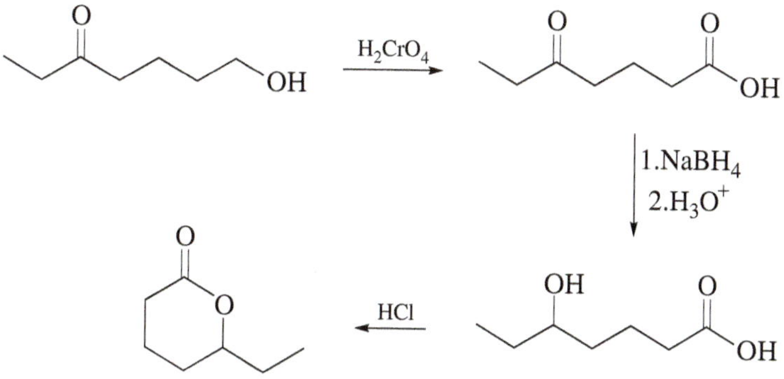

85. a. 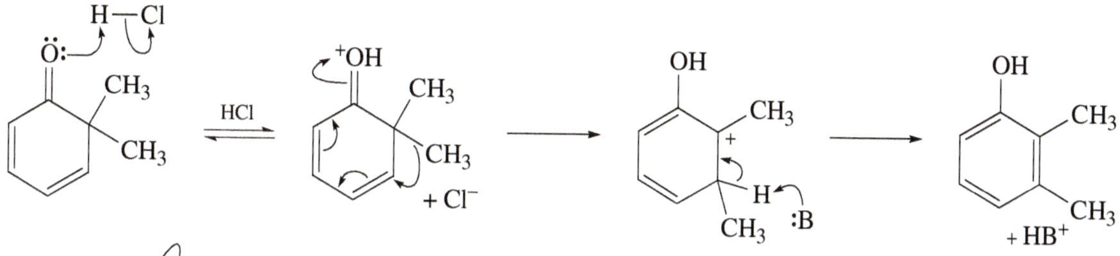

b. 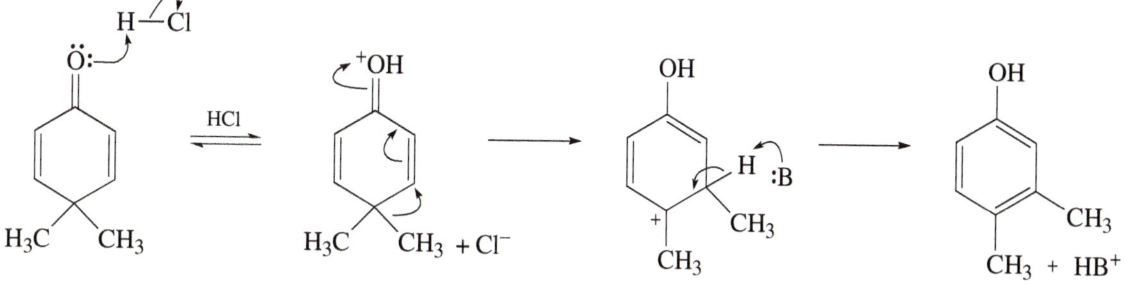

86. **a.** The OH group on C-5 of glucose reacts with the aldehyde group in an intramolecular reaction, forming a cyclic hemiacetal. Because the reaction creates a new asymmetric center, two cyclic hemiacetals are formed, one with the *R* configuration at the new asymmetric center and one with the *S* configuration.

HOCH$_2$ O OH new asymmetric center HC=O H OH HO H H OH H ÖH 6CH$_2$OH = HOCH$_2$ ÖH H OH C O HO OH HO OH OH HOCH$_2$ O OH OH HO OH

b. The two products can be drawn in their chair conformations by putting the largest group (CH_2OH) in the equatorial position and then putting the other groups in axial or equatorial positions depending on whether they are cis or trans to one another. For example, the OH attached to the carbon that is next to the carbon attached to the CH_2OH group is trans to the CH_2OH group. The hemiacetal on the left has all but one of its OH groups in the equatorial position, whereas the hemiacetal on the right has all of its OH groups in the equatorial position. Because a compound can avoid unfavorable 1,3-diaxial interactions by having its substituents in equatorial positions, the hemiacetal on the right is more stable.

CH_2OH HO O HO OH OH hemiacetal — less stable

CH_2OH HO O HO OH OH hemiacetal — more stable

87. The alkyl bromide is 1-bromo-2-phenylethane. The molecular formula of the product of the Wittig reaction indicates that the ketone that reacts with the phosphonium ylide has three carbons (that is, the ketone is acetone).

$C_6H_5CH_2CH_2Br \xrightarrow{P(C_6H_5)_3} C_6H_5CH_2CH_2\overset{+}{P}(C_6H_5)_3 \xrightarrow{CH_3(CH_2)_3Li} C_6H_5CH_2\overset{-}{C}H\overset{+}{P}(C_6H_5)_3$

$\xrightarrow{O=C(CH_3)_2} C_6H_5CH_2CH=C(CH_3)CH_3$

88.

$$CH_3CH{=}O + H{-}B^+ \rightleftharpoons CH_3CH{=}\overset{+}{O}H \rightleftharpoons CH_3{-}CH(OH){-}\overset{+}{O}{=}CHCH_3 \;\; (+\; CH_3CH{=}O) \longrightarrow \ldots$$

$$HB^+ + \text{2,4,6-trimethyl-1,3,5-trioxane} \rightleftharpoons \text{protonated trioxane} + B{:}$$

89. **a.** $C_6H_5{-}CH{=}O + (C_6H_5)_3P{=}CHCH_2CH_2CH_3$ **or** $C_6H_5{-}CH{=}P(C_6H_5)_3 + O{=}CHCH_2CH_2CH_3$

b. cyclopentanone $+ (C_6H_5)_3P{=}CHCH_2CH_3$ **or** cyclopentylidene${=}P(C_6H_5)_3 + O{=}CHCH_2CH_3$

c. $C_6H_5{-}CH{=}O + (C_6H_5)_3P{=}CH{-}C_6H_5$

d. cyclohexenone ${=}O + (C_6H_5)_3P{=}CH_2$ **or** cyclohexenylidene${=}P(C_6H_5)_3 + O{=}CH_2$

90. The first step is conjugate addition of an organocuprate, and the second step is direct addition of a Grignard reagent.

$$\underset{\textbf{A}}{CH_3C(CH_3){=}CHCH{=}O} \xrightarrow[\text{THF}]{(CH_2{=}CH)_2CuLi} \underset{\textbf{B}}{CH_2{=}CHC(CH_3)_2{-}CH_2CH{=}O} \xrightarrow[2.\ H_3O^+]{1.\ CH_3MgBr} CH_2{=}CHC(CH_3)_2CH_2CH(OH)CH_3$$

91. cyclohexanone + $\overset{-}{C}H_2-\overset{+}{N}\equiv N$ ⟶ (tetrahedral intermediate: $:\ddot{O}:^-$ and $CH_2-\overset{+}{N}\equiv N$ on the ring carbon) ⟶ cycloheptanone + N_2

92. The compound that gives the ^{1}H NMR spectrum is 2-phenyl-2-butanol.

$$C_6H_5-C(OH)(CH_3)CH_2CH_3$$

Therefore, the compound that reacts with methylmagnesium bromide is 1-phenyl-1-propanone.

$$C_6H_5-\overset{O}{\overset{\|}{C}}CH_2CH_3 \xrightarrow[2.\ HCl]{1.\ CH_3MgBr} C_6H_5-C(OH)(CH_3)CH_2CH_3$$

93. **a.** $CH_3\underset{OH}{CH}CH_2\overset{O}{\overset{\|}{C}}OCH_3 \xrightarrow[\text{imidazole}]{(CH_3)_3Si(CH_3)_2Cl} CH_3\underset{OSi(CH_3)_3}{CH}CH_2\overset{O}{\overset{\|}{C}}OCH_3 \xrightarrow[2.\ HCl]{1.\ CH_3MgBr\ \text{excess}} CH_3\underset{OH}{CH}CH_2\overset{OH}{\underset{CH_3}{C}}CH_3$

b. 2-chlorocyclohexanol (Cl, OH) $\xrightarrow[\text{imidazole}]{(CH_3)_3Si(CH_3)_2Cl}$ cyclohexane with Cl and $OSi(CH_3)_3$ $\xrightarrow[Et_2O]{Mg}$ cyclohexane with MgCl and $OSi(CH_3)_3$ $\xrightarrow{CO_2}$ cyclohexane with COO^- and $OSi(CH_3)_3$ $\xrightarrow[2.\ (CH_3CH_2CH_2CH_2)_4NF]{1.\ HCl}$ cyclohexane with COOH and OH

c. 3-bromophenyl $-\overset{O}{\overset{\|}{C}}-CH_3$ $\xrightarrow[HOCH_2CH_2OH]{HCl}$ 3-bromophenyl ketal (1,3-dioxolane with CH_3) $\xrightarrow{1.\ Li;\ 2.\ CuLi;\ 3.\ \text{ethylene oxide}}$ ketal with $CH_2CH_2O^-$ $\xrightarrow{HCl,\ H_2O}$ $3\text{-}(CH_2CH_2OH)C_6H_4-\overset{O}{\overset{\|}{C}}-CH_3$

94. Cyclohexane $\xrightarrow[h\nu]{Br_2}$ bromocyclohexane $\xrightarrow[2.\ HCl,\ H_2O]{1.\ CH_3CO^-}$ cyclohexanol $\xrightarrow[CH_3COOH,\ 0\ °C]{NaOCl}$ cyclohexanone $\xrightarrow[2.\ HCl]{1.\ CH_3CH_2MgBr}$ 1-ethylcyclohexanol (OH, CH_2CH_3)

or

bromocyclohexane $\xrightarrow{1.\ (C_6H_5)_3P \mid 2.\ Bu^-\ Li^+}$ $=P(C_6H_5)_3$ $\xrightarrow{CH_3CH{=}O}$ $=CHCH_3$ $\xrightarrow[Pd/C]{H_2}$ OH, CH_2CH_3

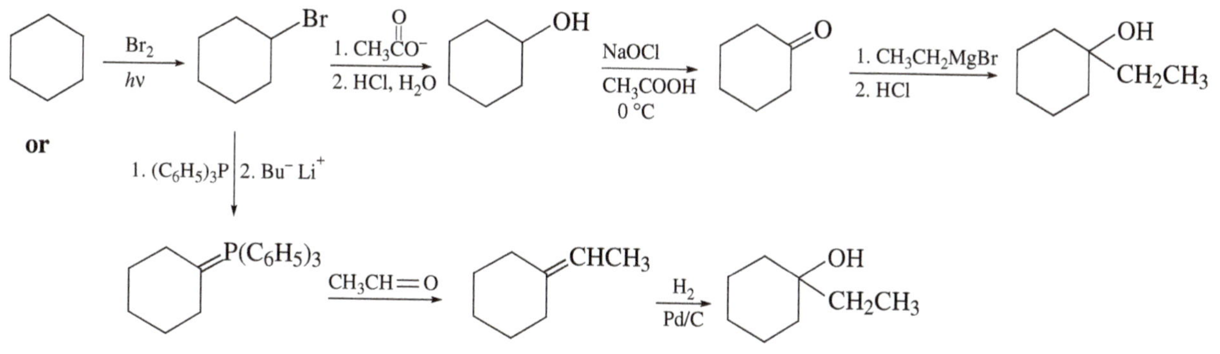

95. **a.** This carboxyl group is more acidic, because its conjugate base is more stable since it is closest to the electron-withdrawing keto group. (Recall that the more stable the base, the stronger its conjugate acid.)

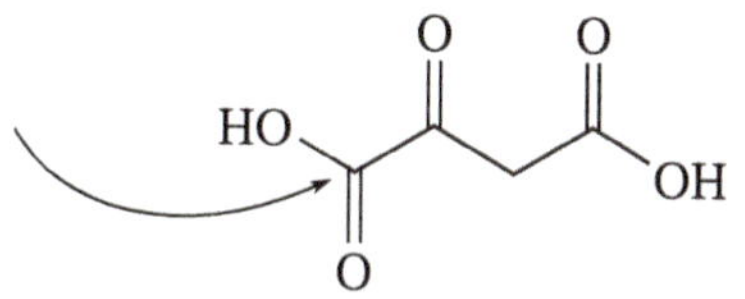

HO, O, O, O, OH

b. The data show that the amount of hydrate decreases with increasing pH until about pH = 6 and that increasing the pH beyond 6 has no effect on the amount of hydrate.

A hydrate is stabilized by electron-withdrawing groups. A COOH group is electron withdrawing, but a COO^- group is less so. In acidic solutions, where both carboxylic acid groups are in their acidic (COOH) forms, the compound is essentially all hydrate. As the pH of the solution increases and the COOH groups become COO^- groups, the amount of hydrate decreases. Above pH = 6, where both carboxyl groups are fully in their basic (COO^-) forms, there is no change in the amount of hydrate.

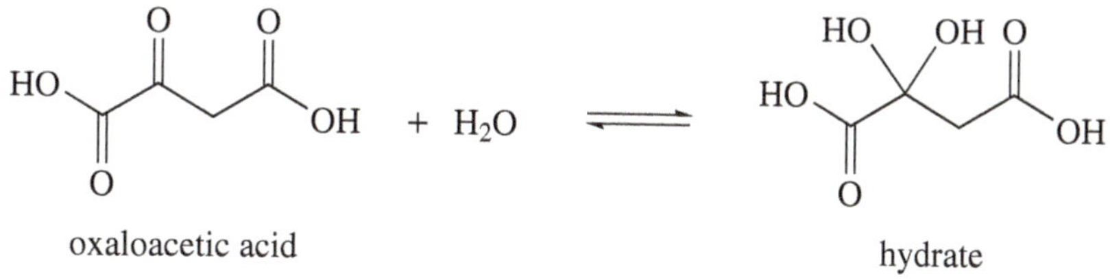

oxaloacetic acid + H_2O ⇌ hydrate

96.

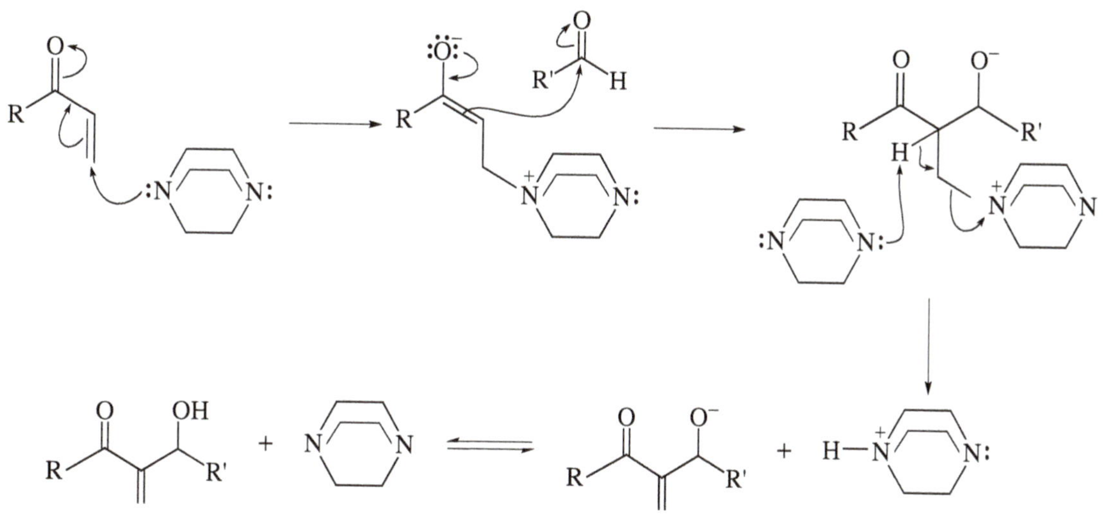

R, O, :N, N: → :Ö:, R, R', O, H, N+, N: → O, O⁻, R, R', H, :N, N:, N+, N

R, O, OH, R' + N, N ⇌ R, O, O⁻, R' + H—N+, N:

97. **a.** The negative ρ value obtained when hydrolysis is carried out in a basic solution indicates that electron-donating substituents increase the rate of the reaction. This means that the rate-determining step must be protonation of the carbon (the first step), because the more electron donating the substituent is, the greater the partial negative charge on the carbon is and the easier it will be to protonate.

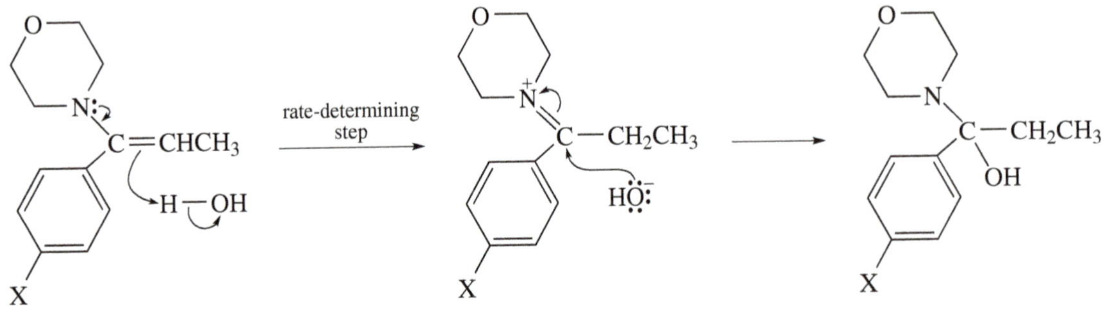

rate-determining step

$C{=}CHCH_3$ H—OH → $C-CH_2CH_3$ HÖ:⁻ → $C-CH_2CH_3$ OH

b. The positive ρ value obtained when hydrolysis is carried out in an acidic solution indicates that electron-withdrawing substituents increase the rate of the reaction. This means that the rate-determining step must be nucleophilic addition of water to the iminium carbon to form the tetrahedral intermediate (the second step), because electron withdrawal increases the electrophilicity of the iminium carbon, making it more susceptible to nucleophilic addition.

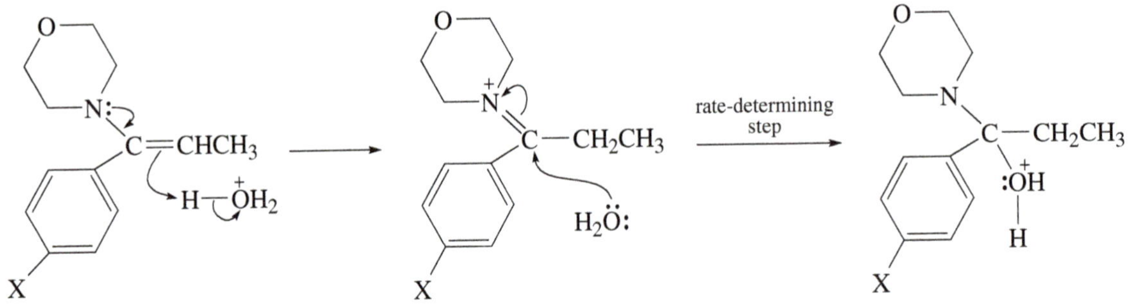

$C{=}CHCH_3$ $H-\overset{+}{O}H_2$ → $C-CH_2CH_3$ $H_2\ddot{O}$: → rate-determining step → $C-CH_2CH_3$ $:\overset{+}{O}H$ H

98. Notice that in the mechanism in part **a**, there is an equilibrium between a protonated intermediate, a neutral intermediate, and a second protonated intermediate, just as we saw in many previous acid-catalyzed mechanisms in this chapter and in Chapter 15.

a.

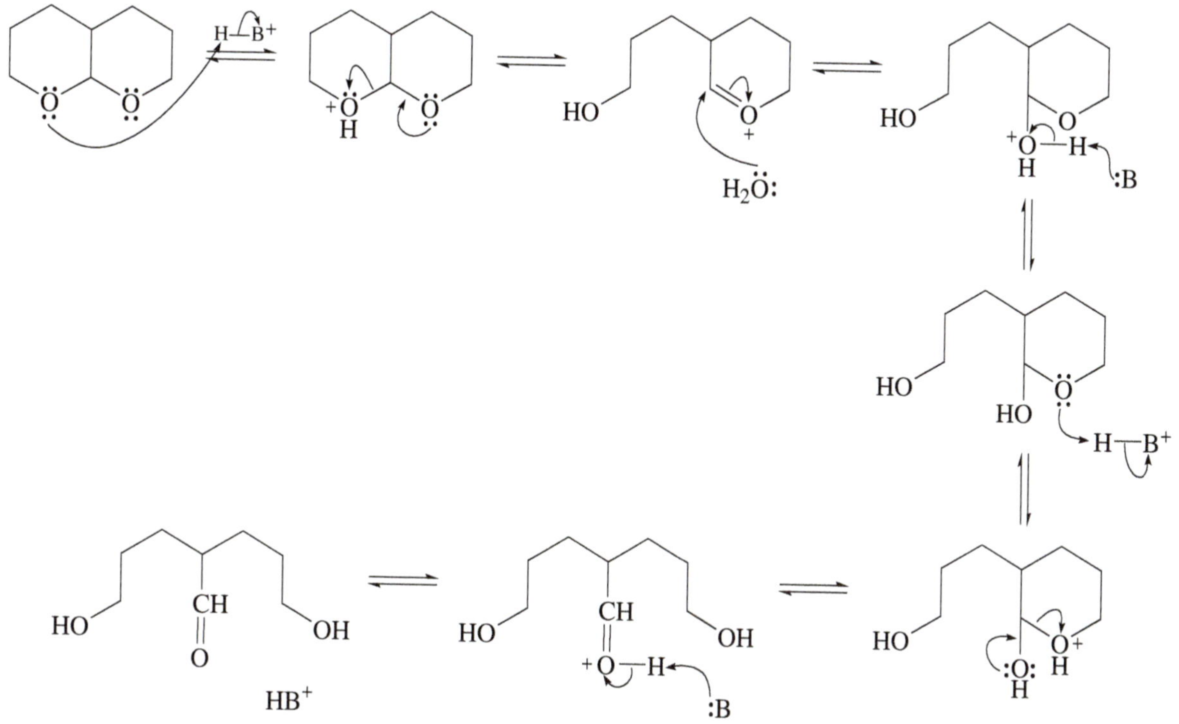

H—B+ HO $H_2\ddot{O}$: HO H :B HO HO H—B+ HO CH O OH HB+ HO CH +O—H OH :B HO :O: H H O+

b.

O O H—B+ $CH_3CH_2\ddot{S}H$ O H :B SCH_2CH_3 OH :O:− H—B+ SCH_2CH_3 OH OH :B SCH_2CH_3 OH

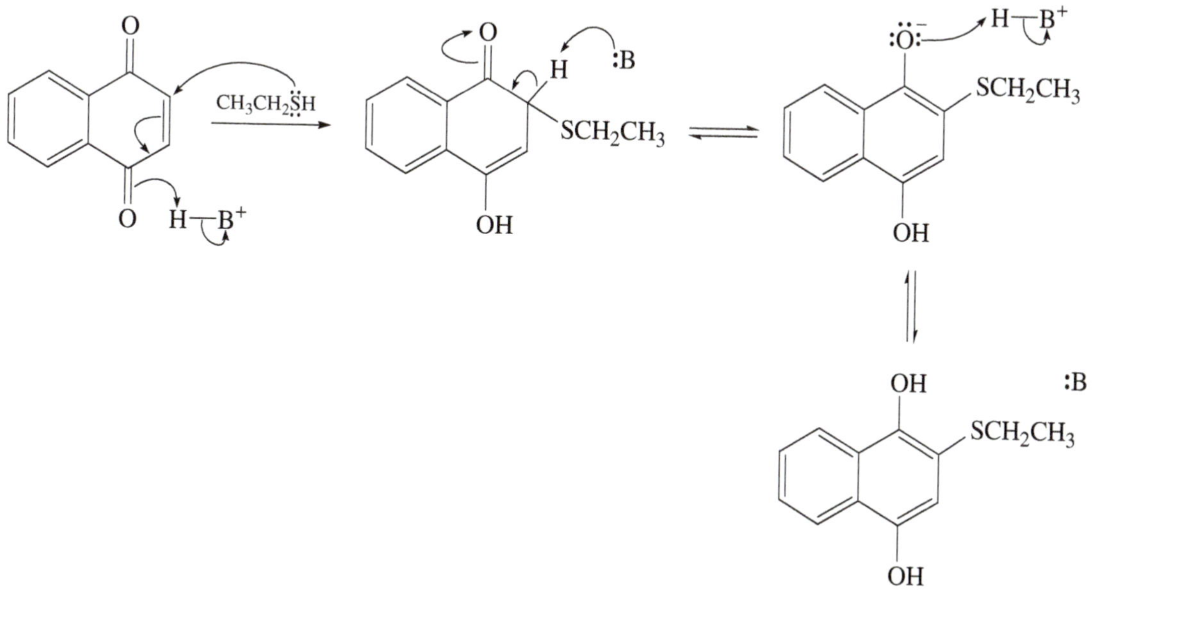

Chapter 16 Practice Test

1. What is the product of the following reactions?

a. 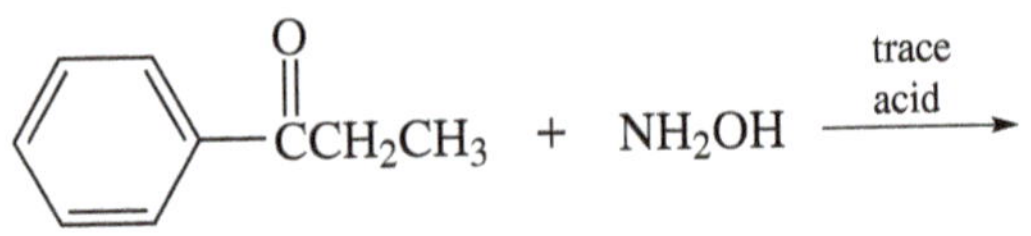 $C_6H_5\overset{O}{\overset{\|}{C}}CH_2CH_3 + NH_2OH \xrightarrow{\text{trace acid}}$

b. 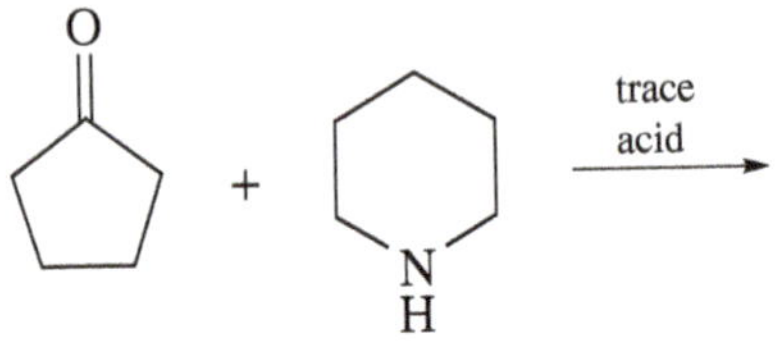 cyclopentanone + piperidine $\xrightarrow{\text{trace acid}}$

c. $CO_2 \xrightarrow[\text{2. HCl, } H_2O]{\text{1. } CH_3CH_2CH_2MgBr}$

d. 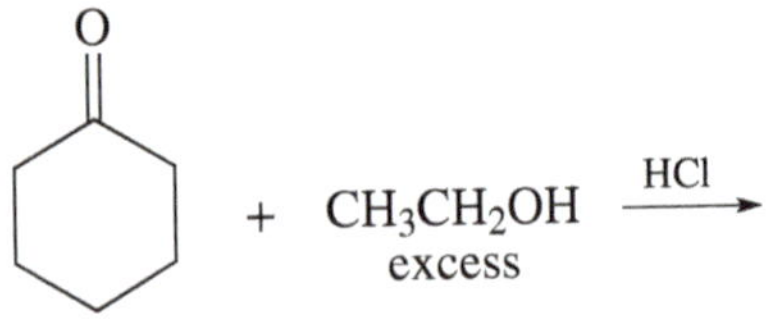 cyclohexanone + CH_3CH_2OH (excess) $\xrightarrow{HCl}$

e. $CH_3CH_2\overset{O}{\overset{\|}{C}}CH_2CH_3 \xrightarrow[\text{2. HCl}]{\text{1. } CH_3MgBr}$

f. 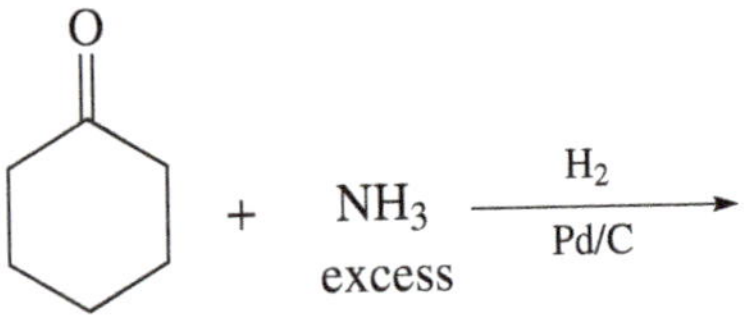 cyclohexanone + NH_3 (excess) $\xrightarrow[Pd/C]{H_2}$

g. $CH_3CH_2\overset{O}{\overset{\|}{C}}CH_2CH_3 \xrightarrow[HCl]{^-C{\equiv}N}$

h. $CH_3CH{=}CH\overset{O}{\overset{\|}{C}}CH_3 + CH_3SH \longrightarrow$

i. 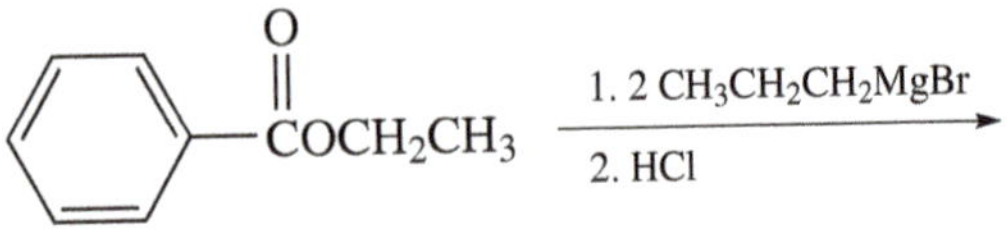 $C_6H_5\overset{O}{\overset{\|}{C}}OCH_2CH_3 \xrightarrow[\text{2. HCl}]{\text{1. 2 } CH_3CH_2CH_2MgBr}$

2. Which of the following alcohols cannot be prepared by the reaction of an ester with excess Grignard reagent?

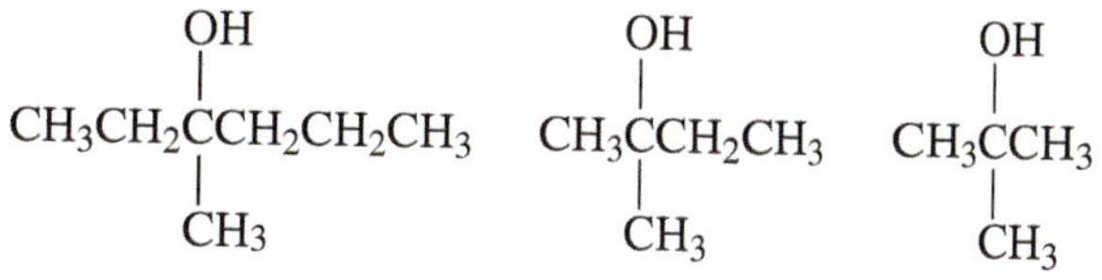

$CH_3CH_2C(OH)(CH_3)CH_2CH_2CH_3$ $CH_3C(OH)(CH_3)CH_2CH_3$ $CH_3C(OH)(CH_3)CH_3$

3. Name the following:

a. 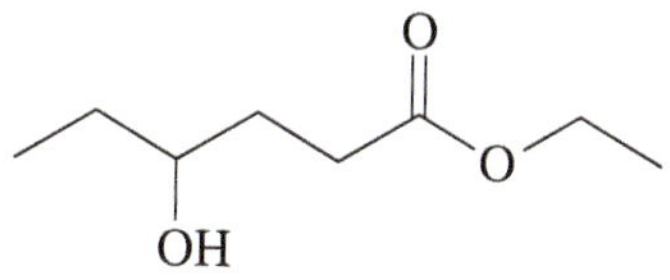 (skeletal structure: ester with OH substituent, O, OH)

b. 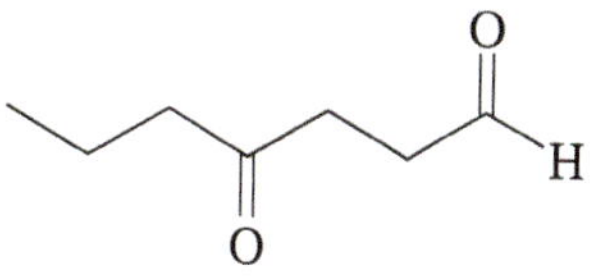 (skeletal structure: keto aldehyde, O, O, H)

c. 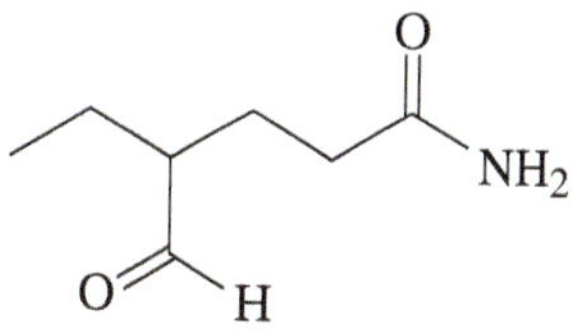 (skeletal structure: amide with CHO substituent, O, NH_2, O, H)

4. Give an example of each the following:

a. an enamine

b. an acetal

c. an imine

d. a hemiacetal

e. a phenylhydrazone

5. Which is more reactive toward nucleophilic addition?

a. butanal or methyl propyl ketone

b. 2-heptanone or 2-pentanone

6. Indicate how the following compounds can be prepared using the given starting material:

a. $CH_3CH_2CH_2Br \longrightarrow CH_3CH_2CH_2C(=O)OCH_2CH_3$

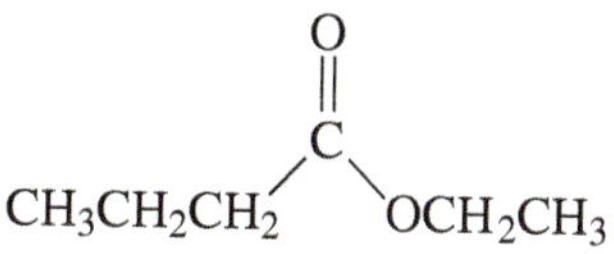

b. 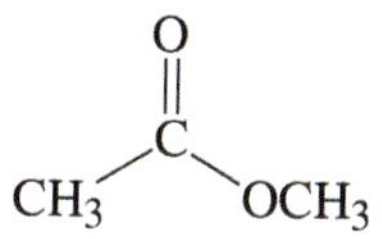 $CH_3C(=O)OCH_3 \longrightarrow CH_3C(OH)(CH_3)CH_3$

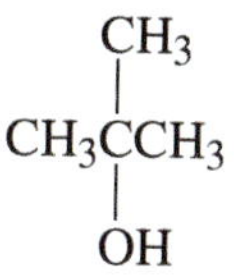

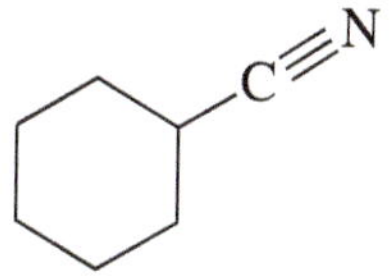

c. $CO_2 \longrightarrow CH_3CH_2OH$

d. 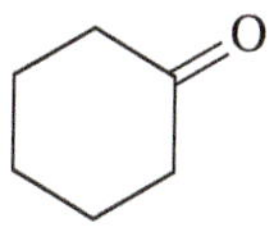 cyclohexanone (C=O) $\longrightarrow$ cyclohexyl–C≡N

CHAPTER 17
Reactions at the α-Carbon

Important Terms

acetoacetic ester synthesis synthesis of a methyl ketone using ethyl acetoacetate as the starting material.

aldol addition a reaction between two molecules of an aldehyde (or two molecules of a ketone) that connects the α-carbon of one with the carbonyl carbon of the other.

aldol condensation an aldol addition followed by elimination of water.

annulation reaction a ring-forming reaction.

α-carbon a carbon adjacent to a carbonyl carbon.

carbon acid a compound that contains a carbon bonded to a relatively acidic hydrogen.

Claisen condensation a reaction between two molecules of an ester that connects the α-carbon of one with the carbonyl carbon of the other and eliminates an alkoxide ion.

condensation reaction a reaction combining two molecules while removing a small molecule (usually water or an alcohol).

crossed aldol addition an aldol addition using two different aldehydes or ketones.

crossed Claisen condensation a Claisen condensation using two different esters.

decarboxylation loss of carbon dioxide.

Dieckmann condensation an intramolecular Claisen condensation.

β-diketone a ketone with a second ketone carbonyl group at the β-position.

E1cB a two-step elimination reaction that proceeds via a delocalized carbanion intermediate.

enolization keto–enol interconversion.

gluconeogenesis the synthesis of D-glucose from pyruvate.

glycolysis the breakdown of D-glucose into two molecules of pyruvate.

haloform reaction the conversion of a methyl ketone to a carboxylic acid and haloform using Cl_2 (or Br_2 or I_2) and HO^-.

Hell–Volhard–Zelinski (HVZ) reaction the conversion of a carboxylic acid into an α-bromocarboxylic acid using $PBr_3 + Br_2$.

α-hydrogen	a hydrogen bonded to the carbon adjacent to a carbonyl carbon.
keto–enol tautomerism (keto–enol interconversion)	interconversion of keto and enol tautomers.
β-keto ester	an ester with a ketone carbonyl group at the β-position.
Kolbe–Schmidt carboxylation reaction	a reaction of a phenolate ion with carbon dioxide under pressure.
malonic ester synthesis	synthesis of a carboxylic acid using diethyl malonate as the starting material.
Michael reaction	the addition of an α-carbanion to the β-carbon of an α, β-unsaturated carbonyl compound.
Robinson annulation	a Michael reaction followed by an intramolecular aldol condensation.
α-substitution reaction	a reaction that puts a substituent on an α-carbon in place of an α-hydrogen.
tautomers	constitutional isomers that are in rapid equilibrium (for example, keto and enol tautomers). The keto and enol tautomers differ only in the location of a double bond and a hydrogen.

Solutions to Problems

1. The electrons left behind when a base removes a proton from propene are delocalized over three carbons. In contrast, the electrons left behind when a base removes a proton from an alkane are localized—they belong to a single carbon. Electron delocalization stabilizes the base, and the more stable the base, the stronger its conjugate acid. Therefore, propene is a stronger acid than an alkane.

$$B:^- + CH_2{=}CHCH_3 \rightleftharpoons CH_2{=}CH\ddot{C}H_2^- \longleftrightarrow {}^-\ddot{C}H_2CH{=}CH_2 \quad + HB$$

$$B:^- + CH_3CH_2CH_3 \rightleftharpoons {}^-\ddot{C}H_2CH_2CH_3 + HB$$

Propene, however, is not as acidic as the carbon acids in Table 17.1, because the electrons left behind when a base removes a proton from these carbon acids are delocalized onto an oxygen or a nitrogen, which are more electronegative than carbon and, therefore, are better able to accommodate the electrons.

$$B:^- + RCH_2{-}\overset{\overset{\ddot{O}:}{\|}}{C}{-}R \longrightarrow R\ddot{C}H^-{-}\overset{\overset{\ddot{O}:}{\|}}{C}{-}R + HB \longleftrightarrow RCH{=}\overset{\overset{:\ddot{O}:^-}{|}}{C}{-}R$$

2.

$$R{-}\overset{\overset{O}{\|}}{C}{-}CH_2C{\equiv}N$$
a β-keto nitrile

$$RO{-}\overset{\overset{O}{\|}}{C}{-}CH_2{-}\overset{\overset{O}{\|}}{C}{-}OR$$
a β-diester

$$R{-}\overset{\overset{O}{\|}}{C}{-}CH_2{-}\overset{\overset{O}{\|}}{C}{-}H$$
a β-keto aldehyde

3. A proton cannot be removed from the α-carbon of *N*-methylethanamide or ethanamide because these compounds have a hydrogen bonded to the nitrogen and this hydrogen is more acidic than the one attached to the α-carbon. Therefore, a base will remove the hydrogen attached to the nitrogen. In the case of *N,N*-dimethylethanamide, there is no N—H proton, so a proton can be removed from the α-carbon.

$$CH_3{-}\overset{\overset{O}{\|}}{C}{-}NHCH_3$$
N-methylethanamide

$$CH_3{-}\overset{\overset{O}{\|}}{C}{-}NH_2$$
ethanamide

$$CH_3{-}\overset{\overset{O}{\|}}{C}{-}N(CH_3)CH_3$$
N,N-dimethylethanamide

The following resonance contributors show why the hydrogen attached to the nitrogen is more acidic (the nitrogen has a partial positive charge) than the hydrogen attached to the α-carbon.

$$CH_3{-}\overset{\overset{O}{\|}}{C}{-}\ddot{N}H_2 \longleftrightarrow CH_3{-}\overset{\overset{O^-}{|}}{C}{=}\overset{+}{N}H_2$$

4. Electron delocalization of the lone pair on nitrogen or oxygen competes with electron delocalization of the electrons left behind on the α-carbon when it loses a proton. The lone pair on nitrogen is more delocalized than the lone pair on oxygen, because nitrogen is better able to accommodate a positive charge since it is less electronegative than oxygen. Therefore, the amide competes better with the carbanion for electron delocalization, so the α-hydrogen is less acidic.

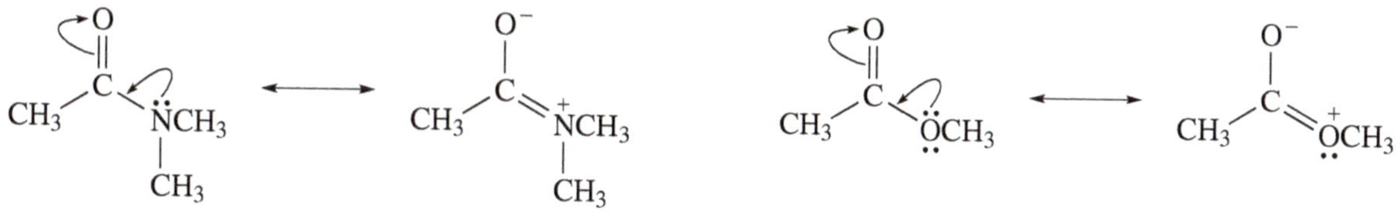

$CH_3C(=O)\ddot{N}(CH_3)CH_3 \longleftrightarrow CH_3C(O^-)=\overset{+}{N}(CH_3)CH_3$ $CH_3C(=O)\ddot{O}CH_3 \longleftrightarrow CH_3C(O^-)=\overset{+}{O}CH_3$

5. **a.** 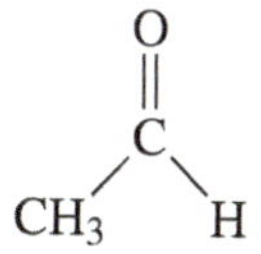 $CH_3C(=O)H > HC{\equiv}CH > CH_2{=}CH_2 > CH_3CH_3$

b. 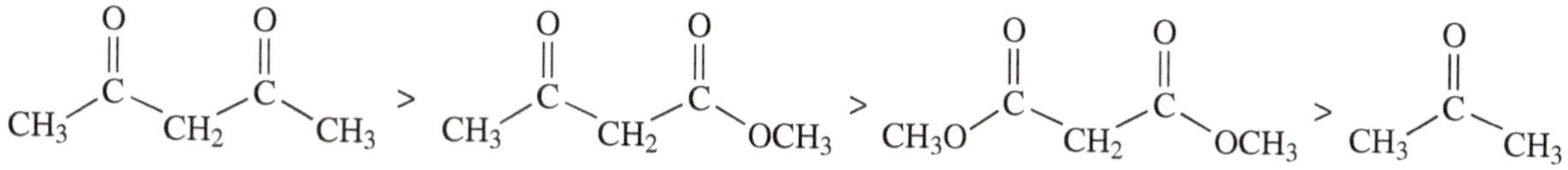 $CH_3C(=O)CH_2C(=O)CH_3 > CH_3C(=O)CH_2C(=O)OCH_3 > CH_3OC(=O)CH_2C(=O)OCH_3 > CH_3C(=O)CH_3$

c. 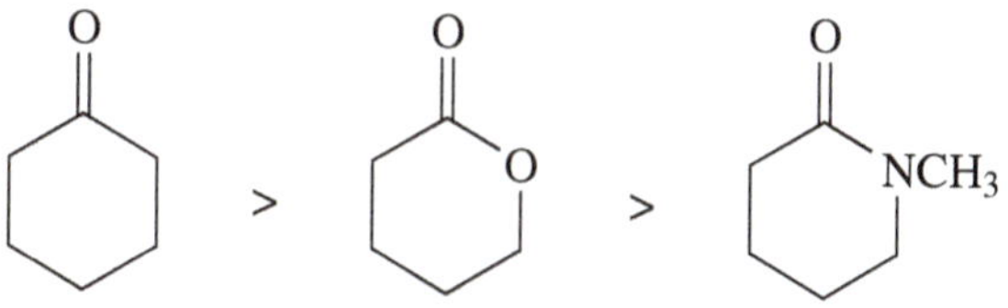 cyclohexanone > six-membered lactone (ring O) > six-membered N-methyl lactam (NCH_3)

Problem 4 explains why the ester is more acidic than the amide.

The ketone is a stronger acid than the ester or the *N*-alkylated amide, because there is no competition for delocalization of the electrons that are left behind when the α-hydrogen is removed.

6. Both the keto and enol tautomers of 2,4-pentanedione can form hydrogen bonds with water. Neither the keto nor the enol tautomer can form hydrogen bonds with hexane, but the enol tautomer can form an intramolecular hydrogen bond, which stabilizes it. Therefore, the enol tautomer is more stable relative to the keto tautomer in hexane than in water and, therefore, more of the enol tautomer is present in hexane than in water.

$CH_3C(=O)CH_2C(=O)CH_3$ keto tautomer $CH_3C(=O)CH{=}C(OH)CH_3$ (O···H–O intramolecular hydrogen bond) enol tautomer

2,4-pentanedione

7.

a. 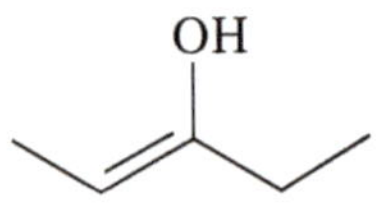 OH

b. OH

c. OH

d. 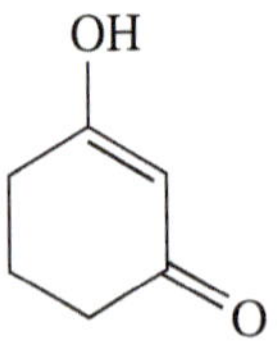 OH and OH

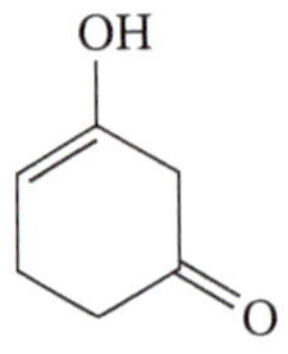

more stable, because the double bonds are conjugated

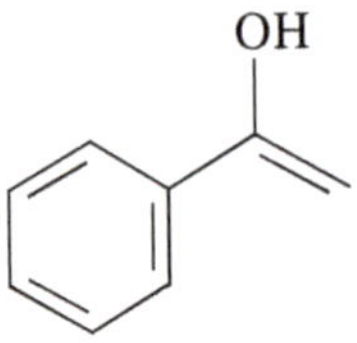

e. 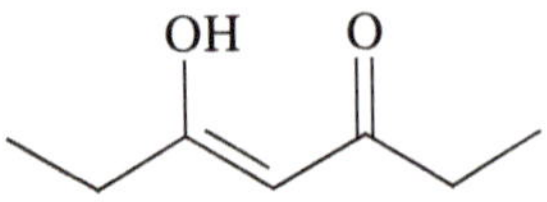 OH O OH O

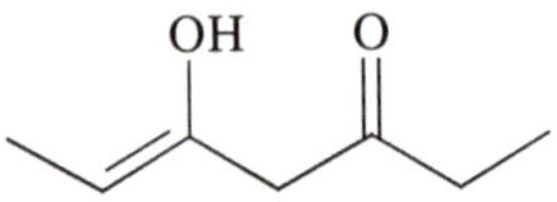

more stable, because the double bonds are conjugated

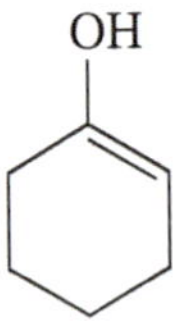

f. 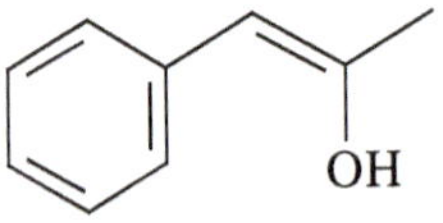 OH OH

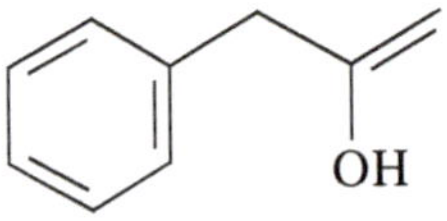

more stable, because the double bond is conjugated with the benzene ring

8. The methyl hydrogens can be removed by a base (^{-}OD) and then are reprotonated by D_2O. The aldehyde hydrogen cannot be removed by a base because the electrons left behind if it were to be removed cannot be delocalized.

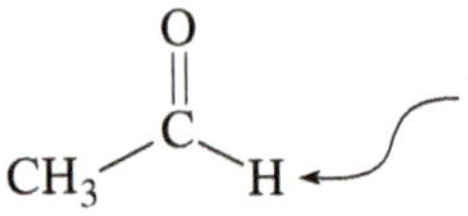

$CH_3C(=O)H$ — The aldehyde hydrogen is not acidic.

9. In an acidic or basic solution, the aldehyde is in equilibrium with its enol. When the ketone enolizes, the asymmetric center is lost. When the enol reforms the ketone, the proton can add to the sp^2 carbon from above or below the plane of the double bond defined by the sp^2 carbons. As a result, equal amounts of the *R* and *S* ketones are formed.

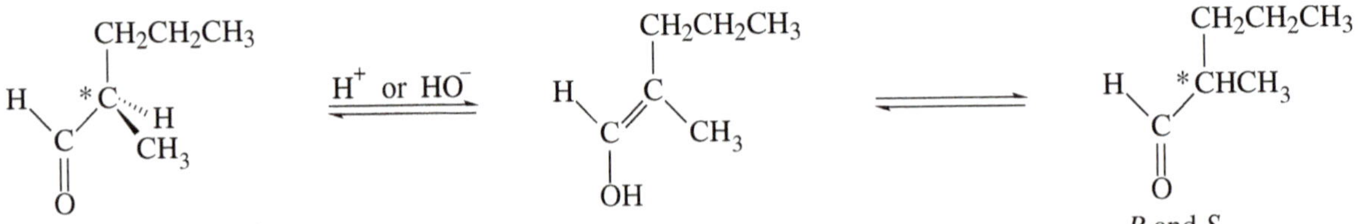

$CH_2CH_2CH_3$ / H–C(=O)–*C(H)(CH_3) $\underset{}{\overset{H^+ \text{ or } HO^-}{\rightleftharpoons}}$ $CH_2CH_2CH_3$ / H–C(OH)=C–CH_3 $\rightleftharpoons$ $CH_2CH_2CH_3$ / H–C(=O)–*$CHCH_3$

(*R*)-2-methylpentanal

R and *S*

10. A Br—Br bond is weaker and easier to break than a Cl—Cl bond, which in turn is weaker and easier to break than a D—O bond. Because the experimentally determined rates of bromination, chlorination, and deuterium exchange are about the same, you know that breaking the Br—Br, Cl—Cl, or D—O bond, which occur at different rates, takes place after the rate-determining step. Therefore, the rate-determining step must be removal of the proton from the α-carbon of the ketone.

$$RCH_2C(=O)R + H_3O^+ \rightleftharpoons RCH(H)C(={}^+OH)R \xrightarrow[\text{rate-determining step}]{H_2\ddot{O}:} RCH{=}C(\ddot{O}H)R + H_3O^+ \xrightarrow[\text{rate of this step varies}]{X-Y} RCH(Br)C(={}^+OH)R + X^- \text{ or } {}^-OD$$

$$[X-Y = Br-Br,\ Cl-Cl,\ D-OD]$$

$$RCH(Br)C(={}^+OH)R \xrightleftharpoons{H_2O} H_3O^+ + RCH(Br)C(=O)R$$

11. **a.** $CH_3CH_2C(=O)O^- \xrightarrow{1.\ PBr_3,\ Br_2;\ 2.\ H_2O} CH_3CH(Br)C(=O)O^- \xrightarrow{CH_3SH} CH_3CH(SCH_3)C(=O)O^- + H^+$

b. $CH_3C(=O)O^- \xrightarrow{1.\ PBr_3,\ Br_2;\ 2.\ H_2O} CH_2(Br)C(=O)O^- \xrightarrow{(CH_3)_2CuLi} CH_3CH_2C(=O)O^-$

12. cyclohexanone $\xrightarrow[D_2O]{^-OD}$ 2,2,6,6-tetradeuteriocyclohexanone

13. Alkylation of an alpha carbon is an S_N2 reaction. S_N2 reactions work best with primary alkyl halides because a primary alkyl halide has less steric hindrance than a secondary alkyl halide. S_N2 reactions do not work at all with tertiary alkyl halides because they are the most sterically hindered of the alkyl halides. Therefore, in the case of tertiary alkyl halides, the S_N2 reaction cannot compete with the E2 elimination reaction.

14. **a.** O; 1. LDA/THF; 2. $ICH_2CH{=}CH_2$; O

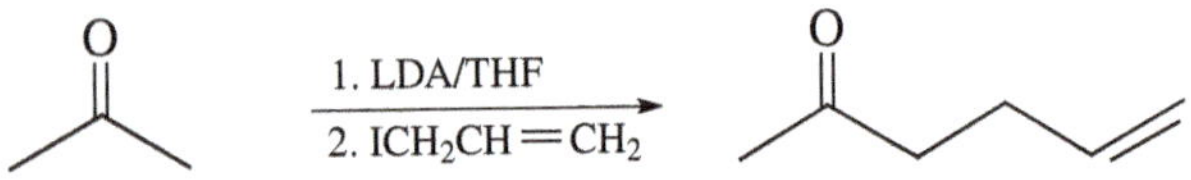

b. O; 1. LDA/THF; 2. CH_2Br; O

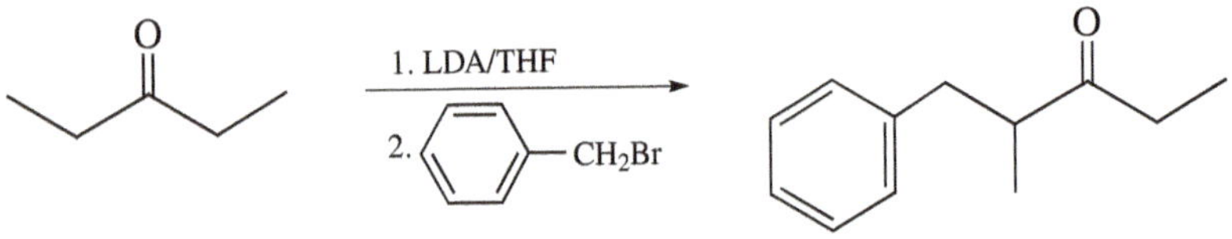

15. The compound formed in **14a** has no stereoisomers.
The compound formed in **14b** has a new asymmetric center, so both the R and S stereoisomers are obtained.

16. **a.** O; 1. LDA/THF; 2. Br; O

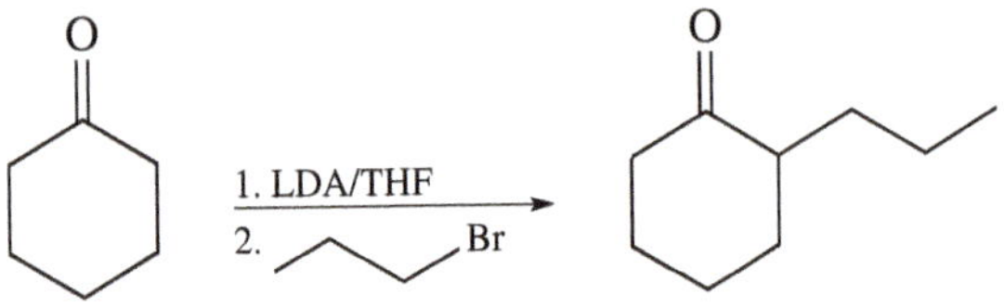

b. O; 1. LDA/THF; 2. Br; O

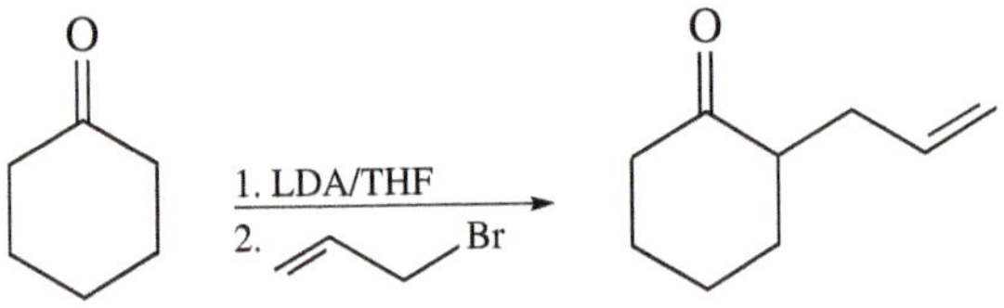

c. O; 1. LDA/THF; 2. Br; O

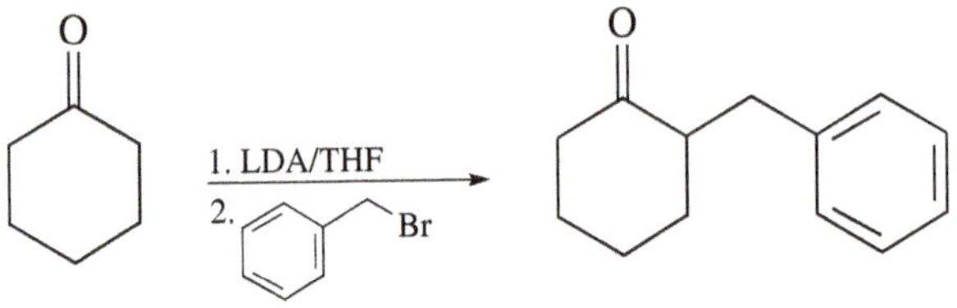

17. **a.** O + N H; trace acid; N; $CH_3CH_2CH_2Br$; $\overset{+}{N}$ Br^- $CH_2CH_2CH_3$; HCl | H_2O; $\overset{+}{N}$ H H Br^- + O $CH_2CH_2CH_3$

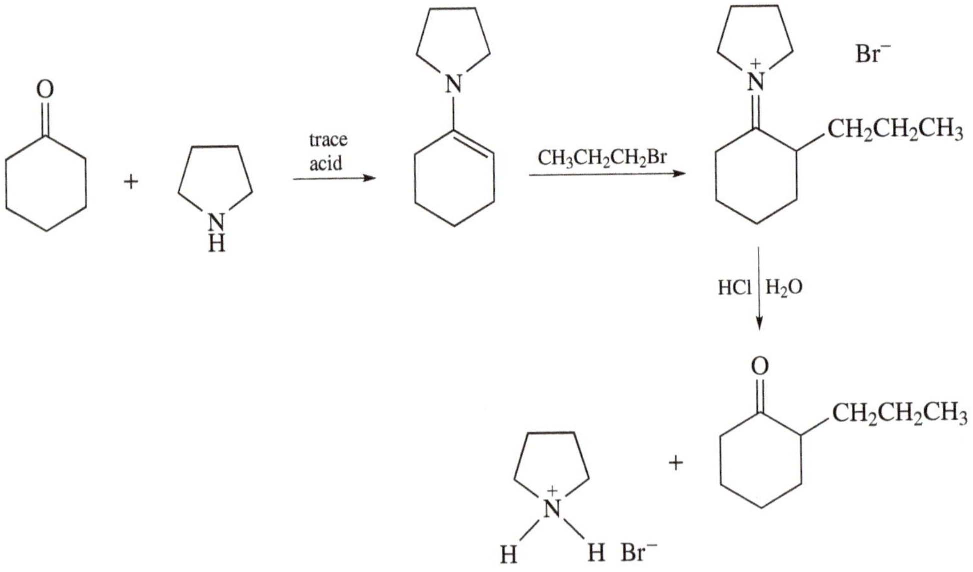

b. 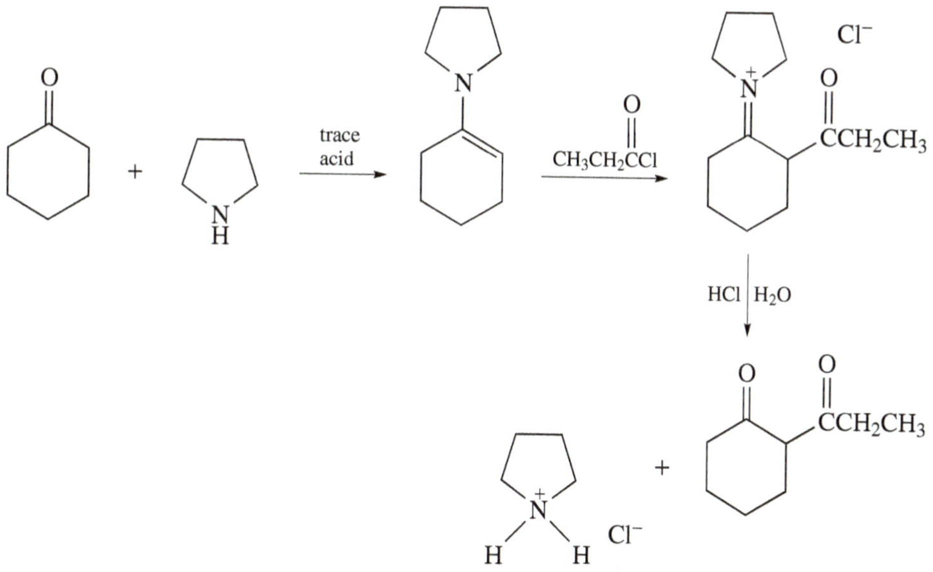

18. **a.** 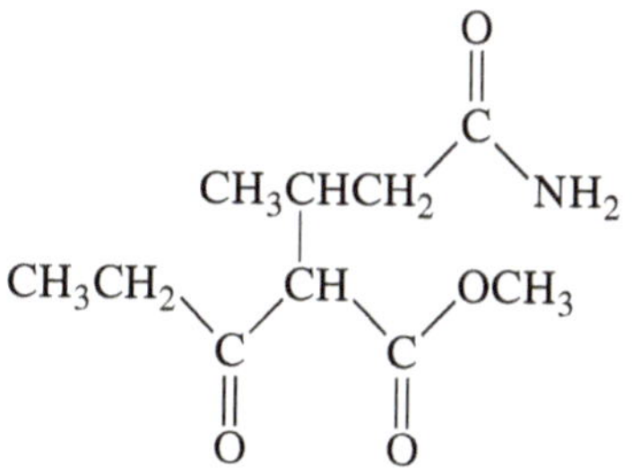

b. 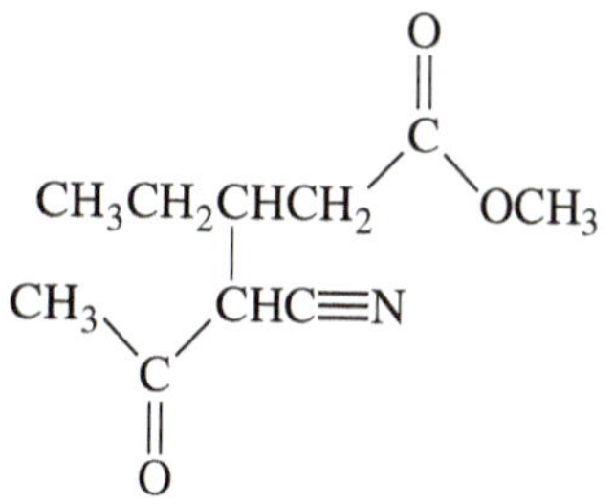

19. **a.** 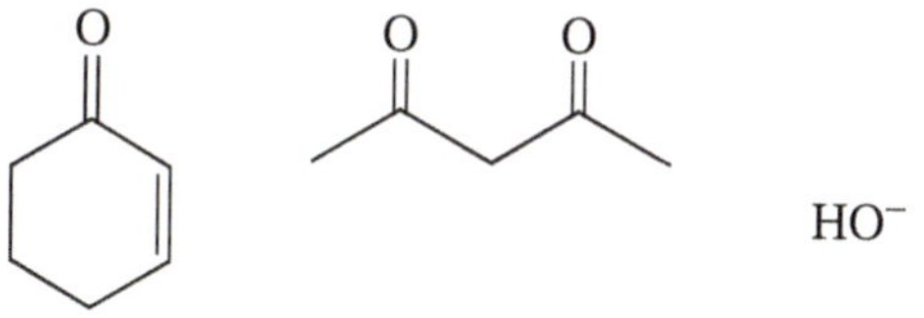

b. 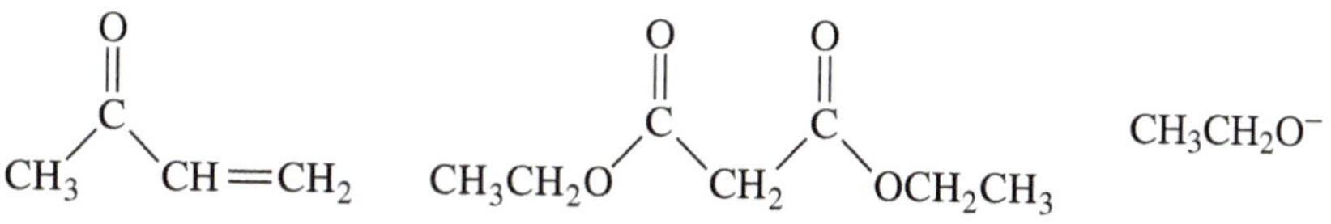

c. 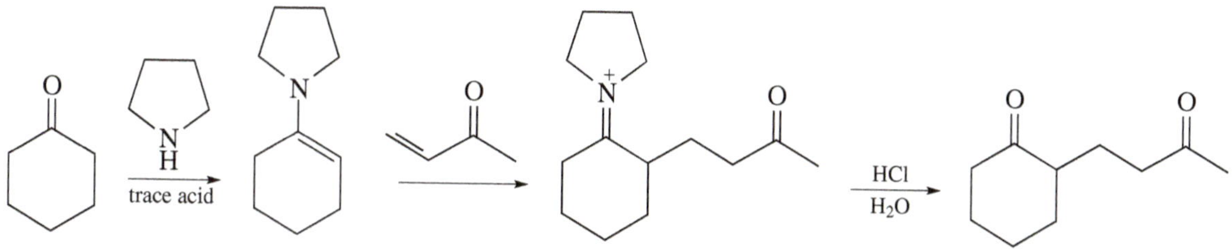

20. **a.** 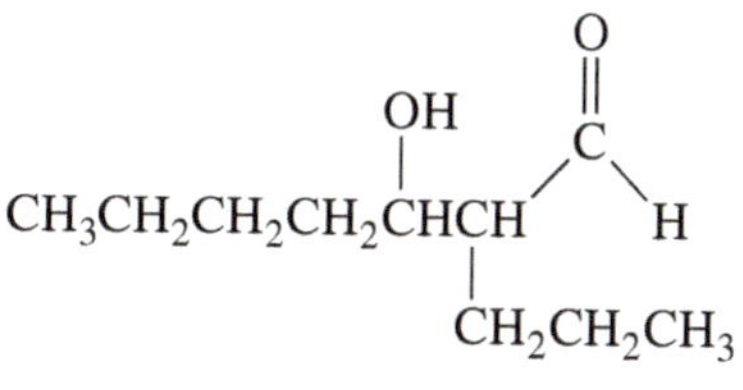

b. 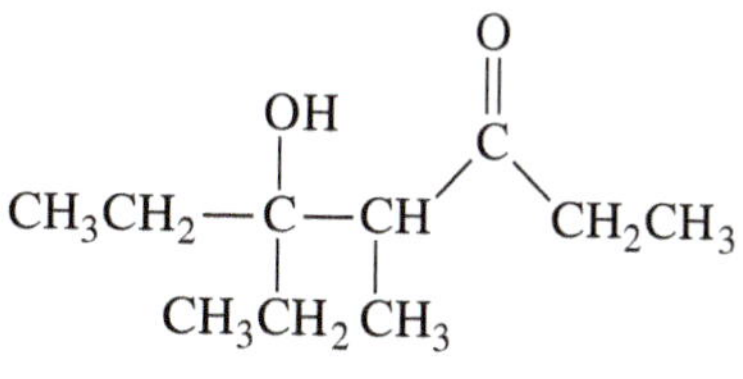

c. 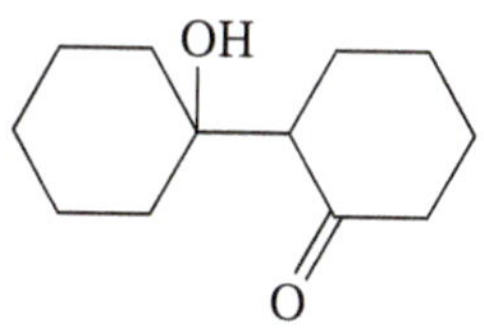

21. **a.** 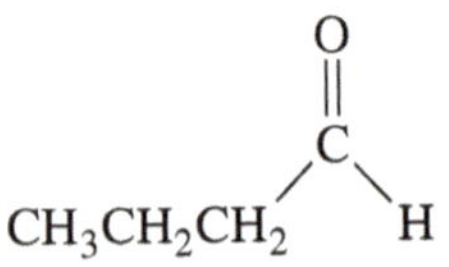 $CH_3CH_2CH_2$–C(=O)–H **b.** 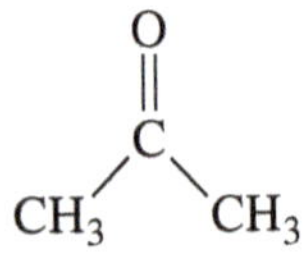 CH_3–C(=O)–CH_3 **c.** 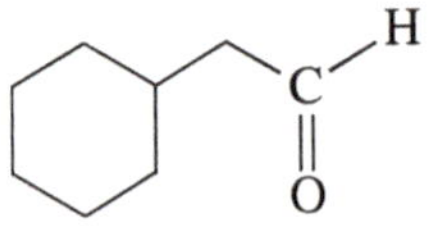 **d.** 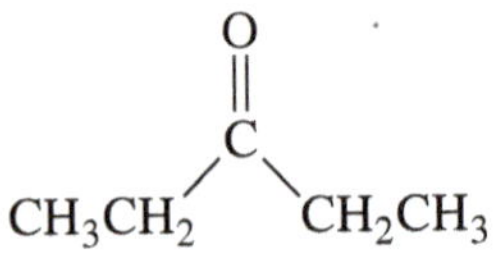 CH_3CH_2–C(=O)–CH_2CH_3

22. 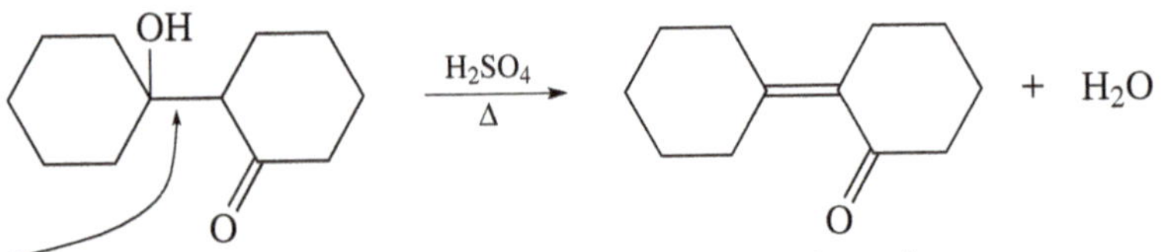 OH $\xrightarrow[\Delta]{H_2SO_4}$ + H_2O

aldol addition aldol condensation

A bond has formed between the α-carbon of one molecule and the carbonyl carbon of another.

23. Solved in the text.

24. 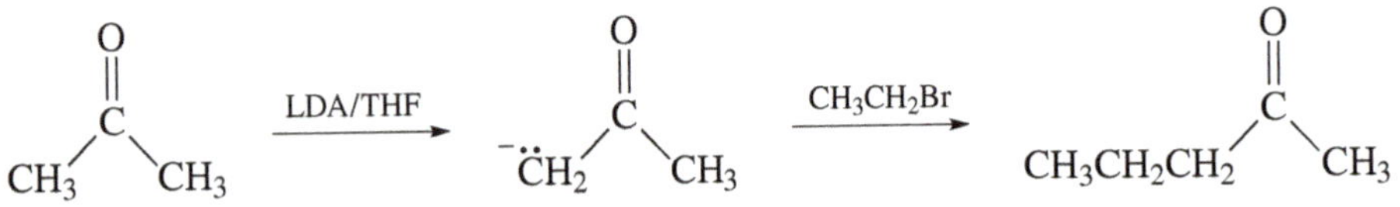 CH_3–C(=O)–CH_3 $\xrightarrow{\text{LDA/THF}}$ $^{-}\ddot{C}H_2$–C(=O)–CH_3 $\xrightarrow{CH_3CH_2Br}$ $CH_3CH_2CH_2$–C(=O)–CH_3

25. **a.** 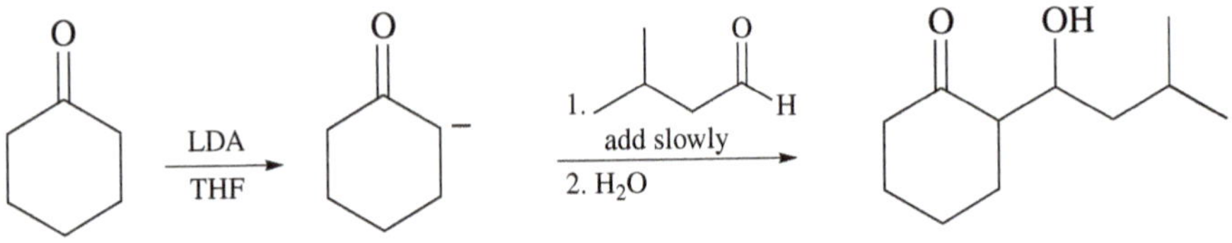 LDA / THF; 1. add slowly 2. H_2O

b. 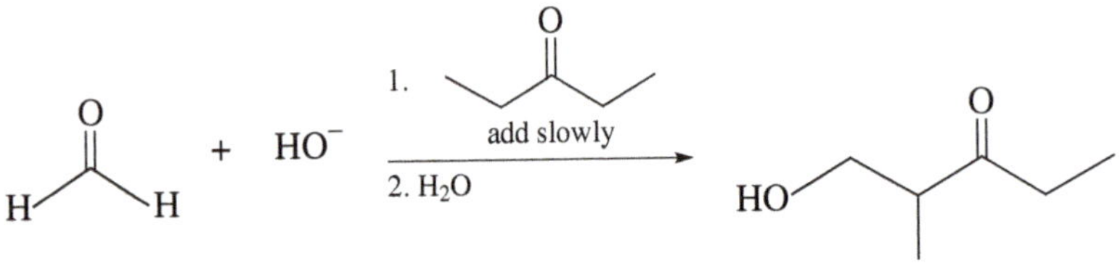 + HO^-; 1. add slowly 2. H_2O; HO

c. 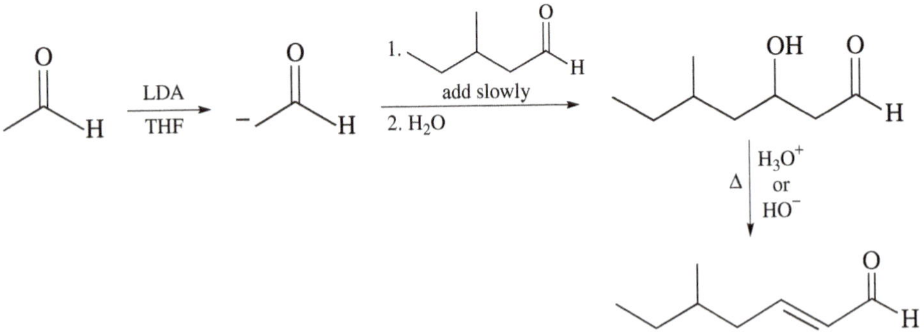 LDA / THF; 1. add slowly 2. H_2O; OH O H; Δ H_3O^+ or HO^-

d. 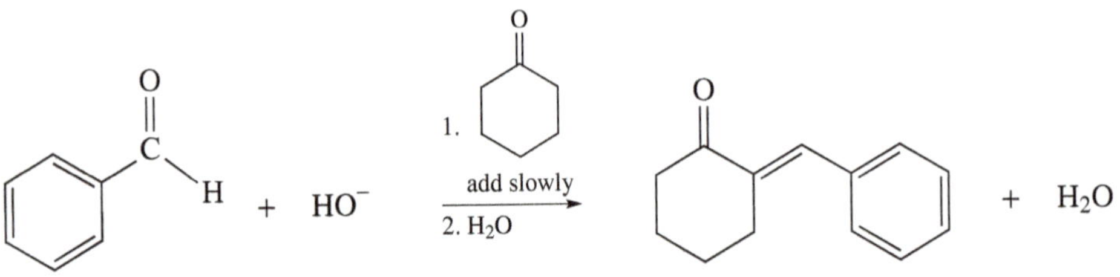 + HO^-; 1. add slowly 2. H_2O; + H_2O

26.

27.

28. **a.**

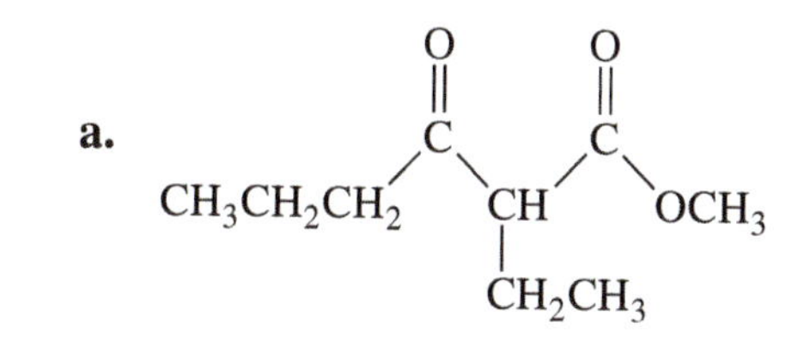

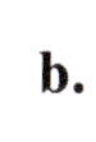

b.

29. **A**, **B**, and **D** cannot undergo a Claisen condensation.
A cannot, because a proton cannot be removed from an sp^2 carbon.
B and **D** cannot, because they do not have an α-hydrogen.

30. **a.** $CH_3CH_2O-C(=O)-CH(CH_3)-C(=O)-OCH_2CH_3$

b. $H-C(=O)-CH(CH_2CH_3)-C(=O)-OCH_3$

31. **a.** Only one carbonyl compound has α-hydrogens, so the one with α-hydrogens is added slowly to a basic solution of the one without α-hydrogens.

$CH_3O-C(=O)-OCH_3 + CH_3O^-$ → (1. $C_6H_5C(=O)CH_3$, add slowly; 2. HCl) → $C_6H_5-C(=O)-CH_2-C(=O)-OCH_3$

b. Both carbonyl compounds have α-hydrogens, so LDA is used to form the enolate, and the other carbonyl compound is added slowly.

$C_6H_5-C(=O)-CH_3$ → (LDA, THF) → $C_6H_5-C(=O)-\bar{C}H_2$ → (1. $CH_3C(=O)OCH_3$, add slowly; 2. HCl) → $C_6H_5-C(=O)-CH_2-C(=O)-CH_3$

c. Only one carbonyl compound has α-hydrogens, so the one with α-hydrogens is added slowly to a basic solution of the one without α-hydrogens.

$H-C(=O)-OCH_3 + CH_3O^-$ → (1. $C_6H_5C(=O)CH_3$, add slowly; 2. HCl) → $C_6H_5-C(=O)-CH_2-C(=O)-H$

32. a 1,7-diester + $CH_3\ddot{O}^-$ ⇌ enolate + CH_3OH ⇌ tetrahedral intermediate (cyclohexane with COCH₃, OCH₃, O⁻) ⇌ 2-(methoxycarbonyl)cyclohexanone ($COCH_3$, =O) + CH_3O^-

a 1,7-diester

33. **a.** 1-methyl-2-acetylcyclobutanol (OH, CH_3, CCH_3 with =O)

b. cyclooctanone bearing OH and CH_3 on the adjacent ring carbon (OH, CH_3, O)

34. No, because an intramolecular reaction would lead to a strained four-membered ring. Therefore, an intermolecular reaction is preferred; the reaction invloves removal of a proton from the more acidic central carbon.

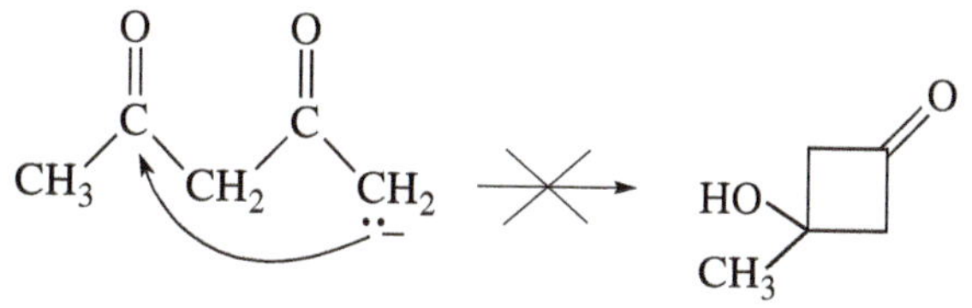

intramolecular product

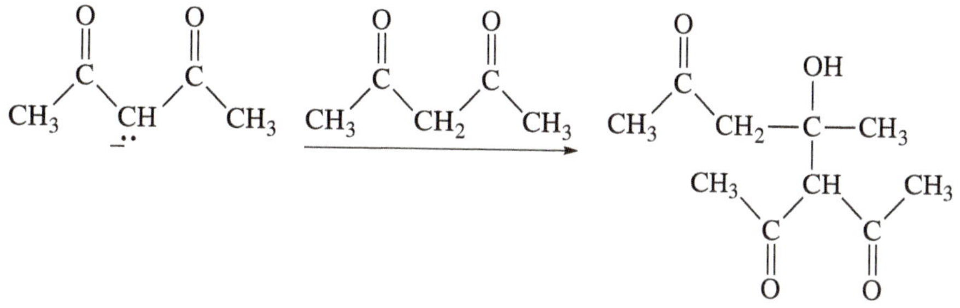

intermolecular product

35. a. 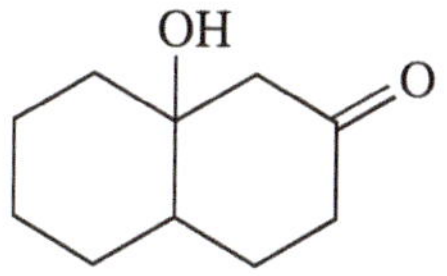 b. 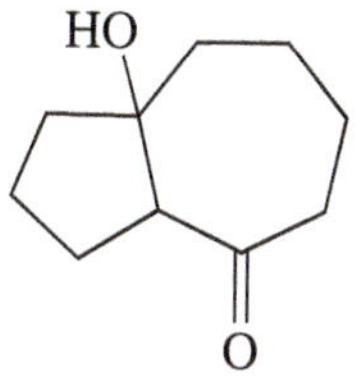 c. 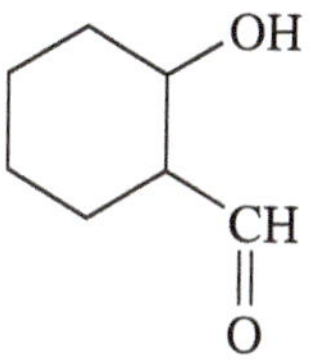 d.

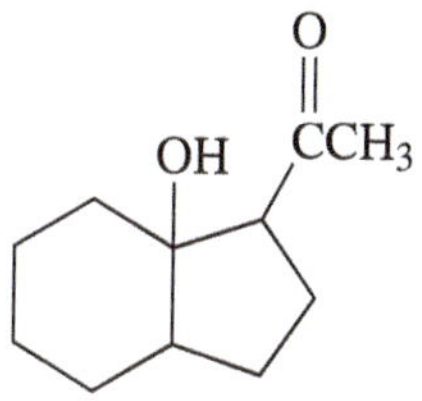

36. a. 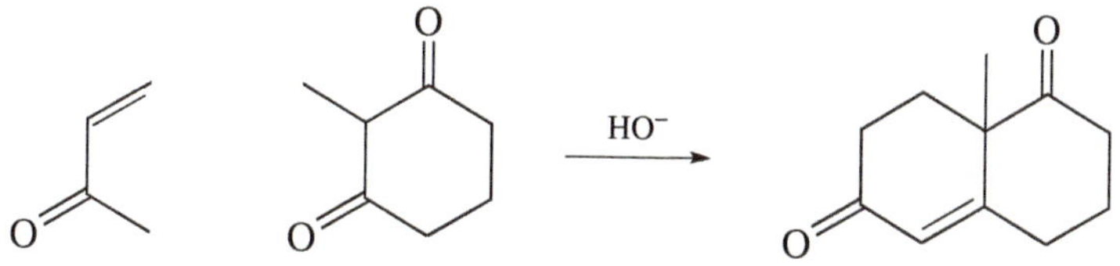 b.

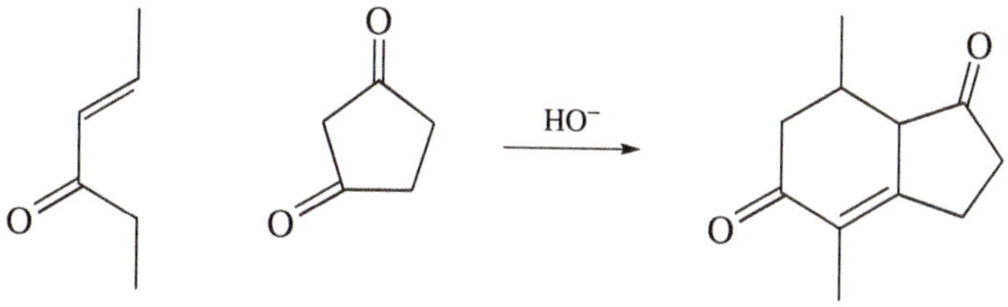

37. **a.**

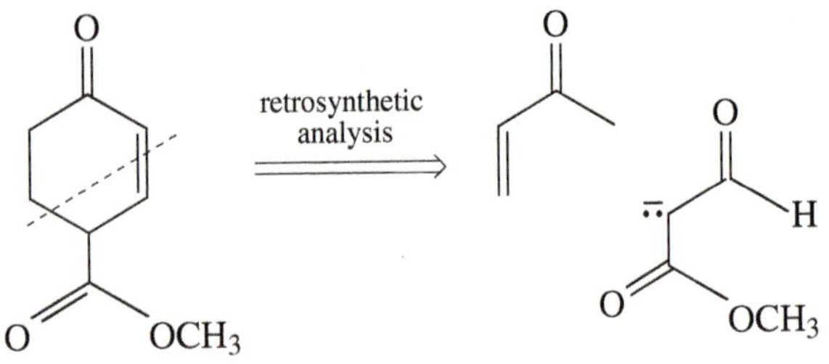

retrosynthetic analysis

O, O, H, O, OCH_3

synthesis

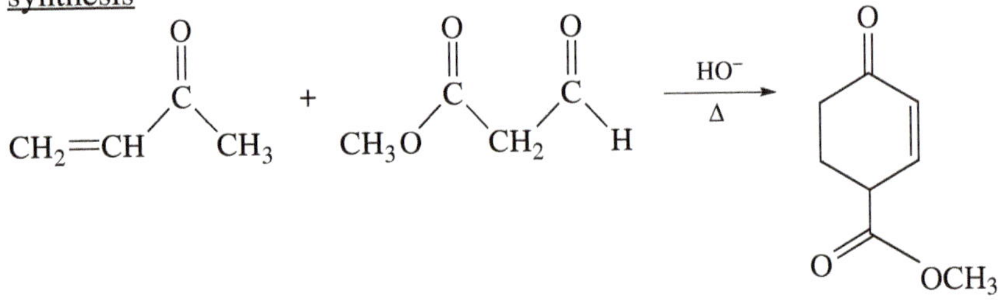

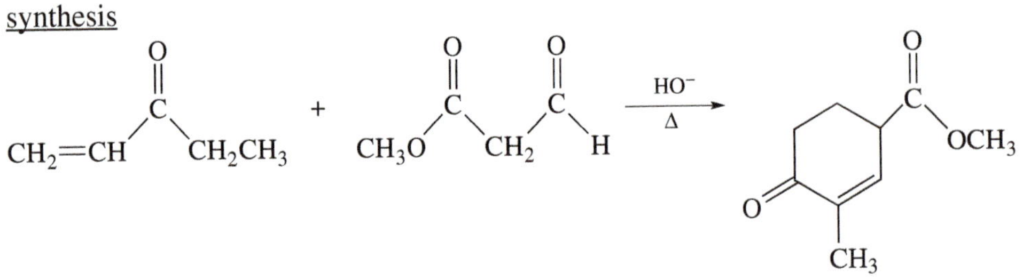

$CH_2{=}CH{-}C(=O){-}CH_3$ + $CH_3O{-}C(=O){-}CH_2{-}C(=O){-}H$ $\xrightarrow[\Delta]{HO^-}$ (product with O, OCH_3)

b.

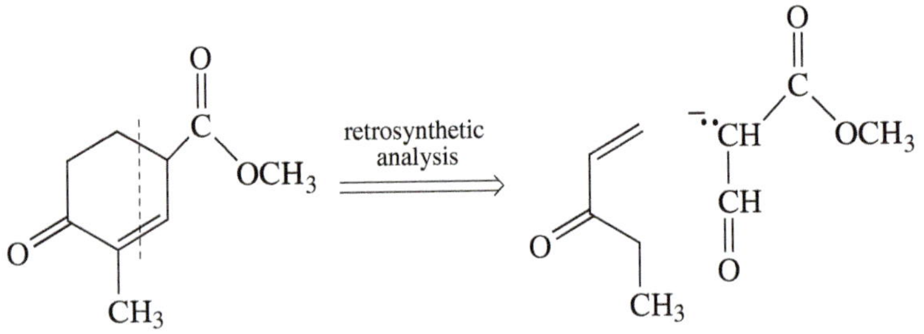

OCH_3, CH_3

retrosynthetic analysis

CH_3, $\ddot{C}H^-$, OCH_3, CH, O

synthesis

$CH_2{=}CH{-}C(=O){-}CH_2CH_3$ + $CH_3O{-}C(=O){-}CH_2{-}C(=O){-}H$ $\xrightarrow[\Delta]{HO^-}$ (product with O, OCH_3, CH_3)

c.

H_3C, O, O, CH_3

retrosynthetic analysis

O, CH_3, H_3C, O, O

$CH_2{=}CH{-}C(=O){-}CH_2CH_3$ + (H_3C, O, O) $\xrightarrow[\Delta]{HO^-}$ (product with H_3C, O, O, CH_3)

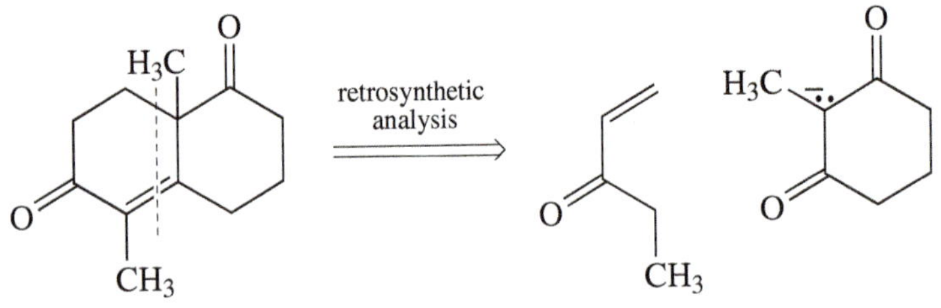

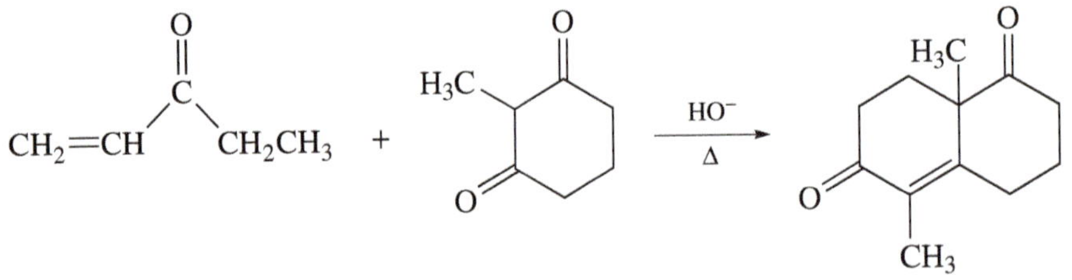

38. **A, D,** and **E** are β-keto acids, so they can be decarboxylated on heating.
B cannot be decarboxylated because it does not have a carboxyl group.
The electrons left behind if **C** were decarboxylated cannot be delocalized onto an oxygen.

39. **a.** methyl bromide **c.** benzyl bromide
b. methyl bromide (twice) **d.** isobutyl bromide

40. **a.** An S_N2 reaction cannot be done on bromobenzene. (See page 436 of the text.)
b. An S_N2 reaction cannot be done on vinyl bromide. (See page 436 of the text.)
c. An S_N2 reaction cannot be done on a tertiary alkyl halide. (Only elimination occurs; see Table 9.7 on page 432 of the text.)

41. **a.** ethyl bromide **b.** pentyl bromide **c.** benzyl bromide

42. Solved in the text.

43. **a.** CH_3 — NH, trace acid → CH_2 — CH_3CH_2—Br → Br^- $CH_2CH_2CH_3$ — HCl, H_2O → Cl^- + $CH_2CH_2CH_3$

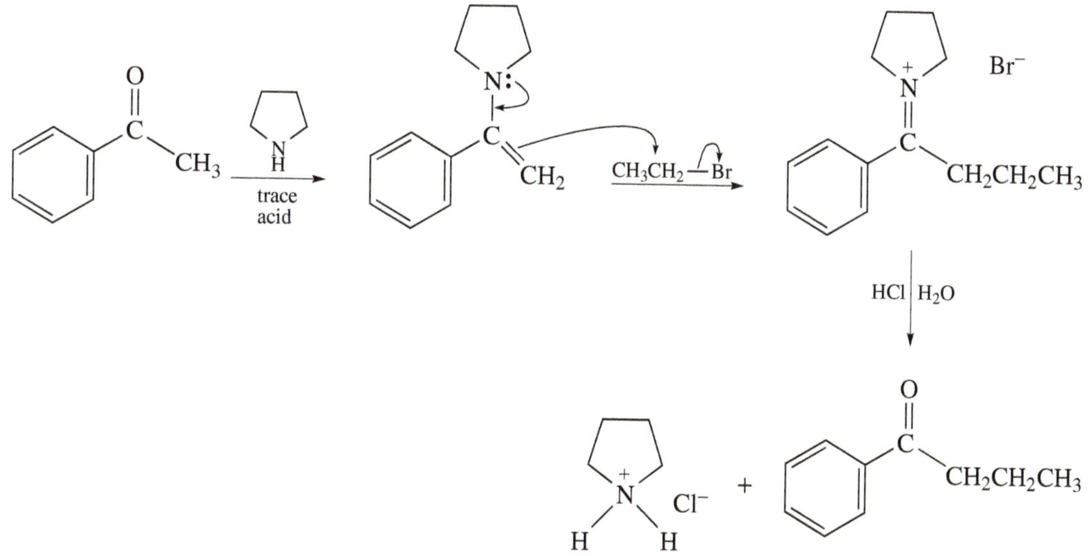

or CH_3 — LDA / THF → CH_2 — CH_3CH_2—Br → $CH_2CH_2CH_3$

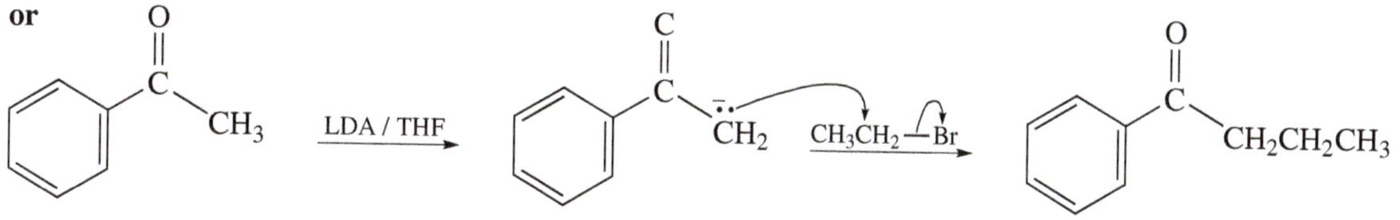

b. CH_3O $(CH_2)_5$ OCH_3 — CH_3O^- → $COCH_3$ — 1. CH_3O^- 2. CH_3CH_2Br → CH_2CH_3 $COCH_3$ — HCl, H_2O, Δ → CH_2CH_3 + CO_2

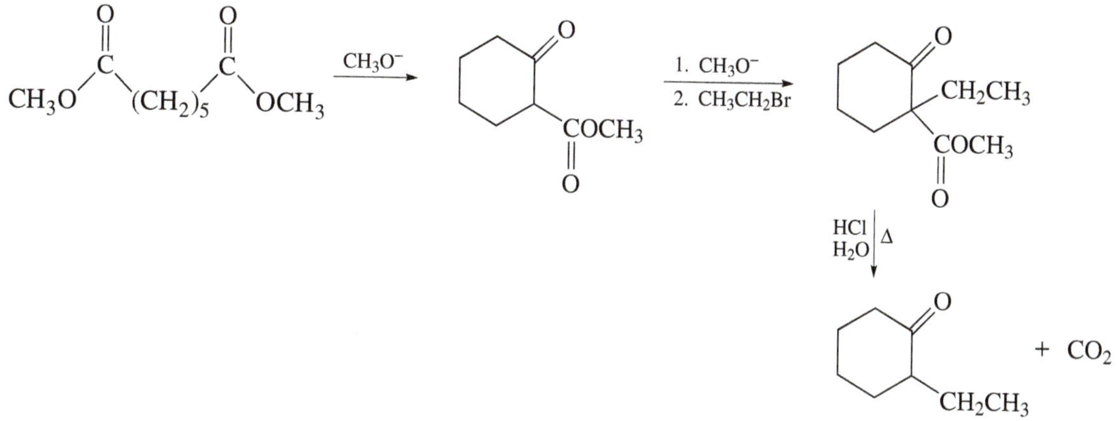

c. Benzaldehyde + HO^- $\xrightarrow[\text{2. HCl}]{\text{1. } CH_3CHO \text{ add slowly}}$ $C_6H_5CH(OH)CH_2CHO$ $\longrightarrow$ $C_6H_5CH{=}CHCHO + H_2O$

$C_6H_5CH{=}CHCHO$ $\xrightarrow{\text{NaOCl} \mid CH_3COOH,\ 0\ °C}$ $C_6H_5CH{=}CHCOOH$

d. $CH_3OC(O)CH_2C(O)OCH_3$ $\xrightarrow[\text{2. } Br(CH_2)_4Br]{\text{1. } CH_3O^-}$ $CH_3OC(O)CH(CH_2(CH_2)_3Br)C(O)OCH_3$ $\xrightarrow{CH_3O^-}$ cyclopentane-1,1-$(COOCH_3)_2$

$\xrightarrow{HCl,\ H_2O,\ \Delta}$ $2\ CH_3OH + CO_2 +$ cyclopentanecarboxylic acid

44. Because the catalyst is a hydroxide ion rather than an enzyme, four stereoisomers will be formed since two asymmetric centers are created in the product.

$CH_2OPO_3^{2-}$–C(=O)–CH_2OH (H removed by HO^-) $\rightleftharpoons$ $CH_2OPO_3^{2-}$–C(=O)–$\ddot{C}H$–OH $\rightleftharpoons$ (+ glyceraldehyde 3-phosphate: O=CH–C(H)(OH)–$CH_2OPO_3^{2-}$) $\rightleftharpoons$ $CH_2OPO_3^{2-}$–C=O–CHOH–CHO^-–C(H)(OH)–$CH_2OPO_3^{2-}$ $\xrightleftharpoons{H-OH}$ $CH_2OPO_3^{2-}$–C=O–*CHOH–*CHOH–C(H)(OH)–$CH_2OPO_3^{2-}$

45. **Seven** moles. The first two carbons in the fatty acid come from acetyl CoA.
Each subsequent two-carbon piece comes from malonyl CoA.
Because this amounts to 14 carbons for the synthesis of the 16-carbon fatty acid, seven moles of malonyl CoA are required.

46. a. **Three** deuteriums would be incorporated into palmitic acid, because only one CD_3COSR is used in the synthesis.

b. **Seven** deuteriums would be incorporated into palmitic acid, because seven $^{-}OOCCD_2COSR$ are used in the synthesis (for a total of 14 Ds), and each $^{-}OOCCD_2COSR$ loses one deuterium in the dehydration step (14 Ds − 7 Ds = 7 Ds).

47. It tells you that an imine is formed as an intermediate. Because an imine is formed, the only source of oxygen for acetone formation is $H_2^{18}O$.

$CH_3C(=\overset{+}{N}HR)CH_2COO^- \rightarrow CH_3C(NHR)=CH_2 + CO_2 \xrightarrow{H-{}^{18}O-H} CH_3C(=\overset{+}{N}HR)CH_3 \xrightarrow{H_2^{18}O} R\overset{+}{N}H_3 + CH_3C(={}^{18}O)CH_3$

If decarboxylation occurred without imine formation, most of the acetone would contain ^{16}O. There would be some ^{18}O incorporated into acetone because acetone would react with $H_2^{18}O$ and form a hydrate. The ^{18}O would become incorporated into acetone when the hydrate reforms acetone.

$CH_3C(=O)CH_2COO^- \rightarrow CH_3C(O^-)=CH_2 + CO_2 \xrightarrow{H-{}^{18}O-H} CH_3C(=O)CH_3 \xrightarrow{H_2^{18}O} CH_3-C(OH)({}^{18}OH)-CH_3 \rightleftharpoons H_2O + CH_3C(={}^{18}O)CH_3$

48. a. $CH_3C(=O)CH_2C(=O)OCH_2CH_3$

b. $HOC(=O)CH(CH_3)C(=O)OH$

c. $RC(=O)CH_2C(=O)OR'$

d. 1-hydroxycyclopentene (cyclopentene with OH on the double-bond carbon)

e. $CH_3CH_2CH_2CH_2C(=O)OH$

49. **a.** 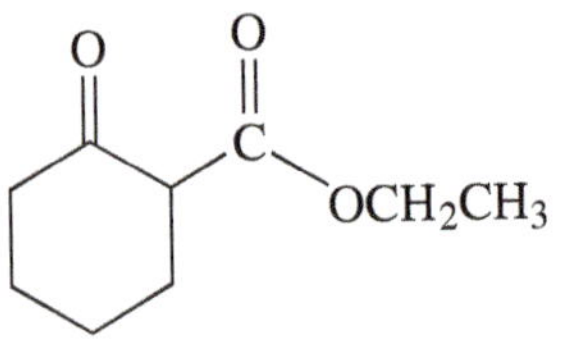 O O C OCH_2CH_3

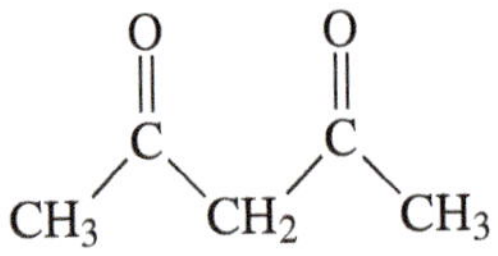

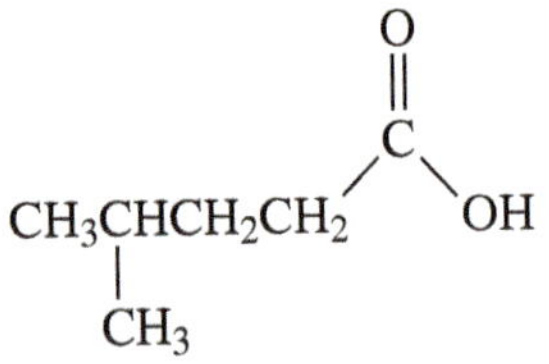

b. 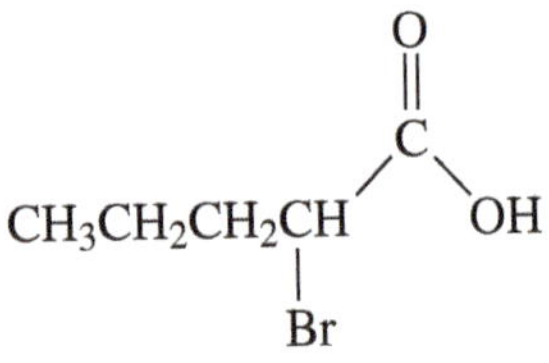 O C $CH_3CH_2CH_2CH$ OH Br

c. O O C C CH_3 CH_2 CH_3

d. 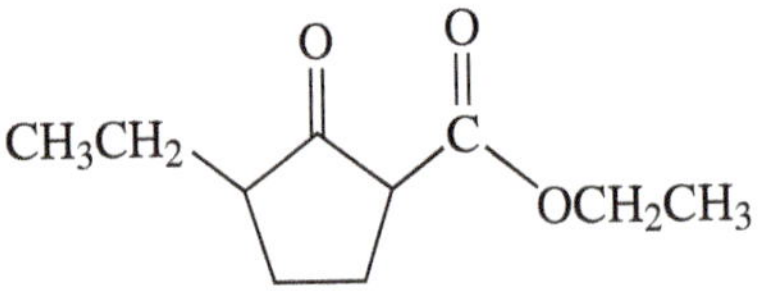 CH_3CH_2 O O C OCH_2CH_3

e. O C $CH_3CHCH_2CH_2$ OH CH_3

f. 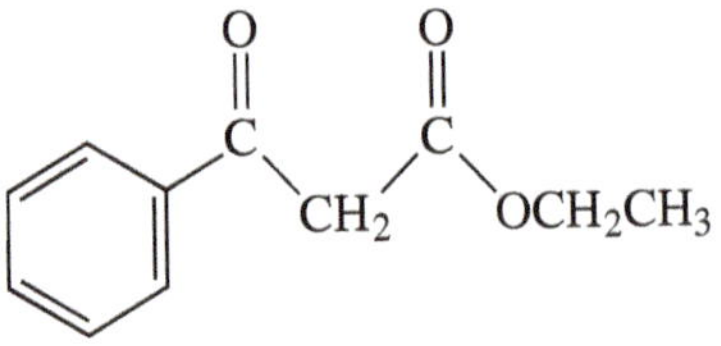 O O C C CH_2 OCH_2CH_3

50. 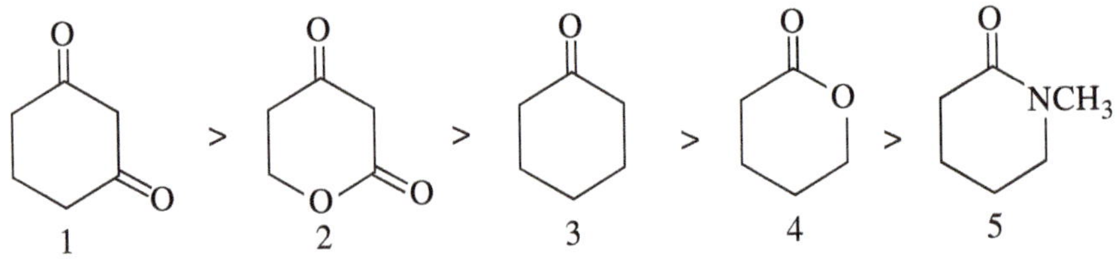 O O > O O O > O > O O > O NCH_3

1 2 3 4 5

51. The electron-withdrawing nitro group causes the signal to occur at a higher frequency (has a greater chemical shift). The more acidic the α-hydrogen, the greater the chemical shift.

CH_3NO_2	$CH_2(NO_2)_2$	$CH(NO_2)_3$
δ 4.33	δ 6.10	δ 7.52

52. Compound **C** loses CO_2 at the lowest temperature because it is a β-keto acid. Therefore, the electrons left behind when CO_2 is removed can be delocalized onto oxygen.

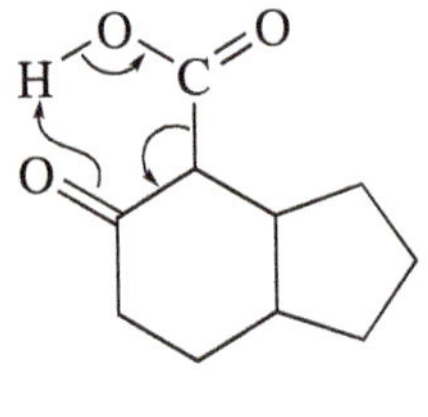

H O C=O O

53. **a.** 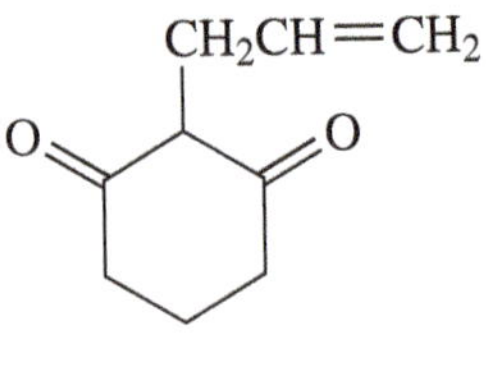 $CH_2CH{=}CH_2$ O O

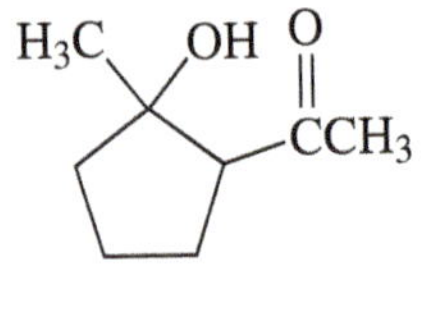

b. 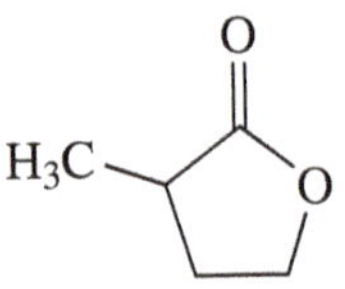 H_3C O O

c. H_3C OH O CCH_3

d. 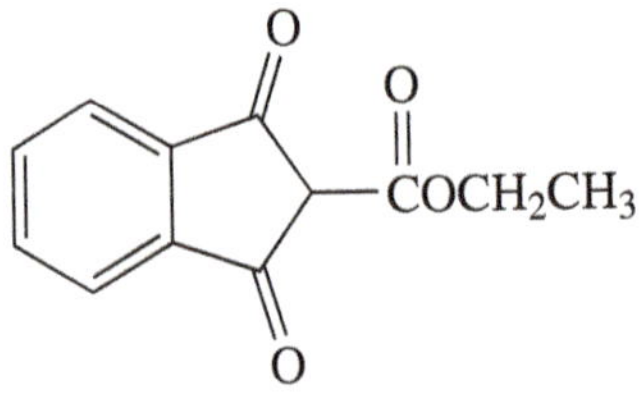 O O O $COCH_2CH_3$

54. When the ketone enolizes, the asymmetric center is lost. When the enol reforms the ketone, an asymmetric center is created; the *R* and *S* enantiomers can be formed equally easily, so a racemic mixture is obtained.

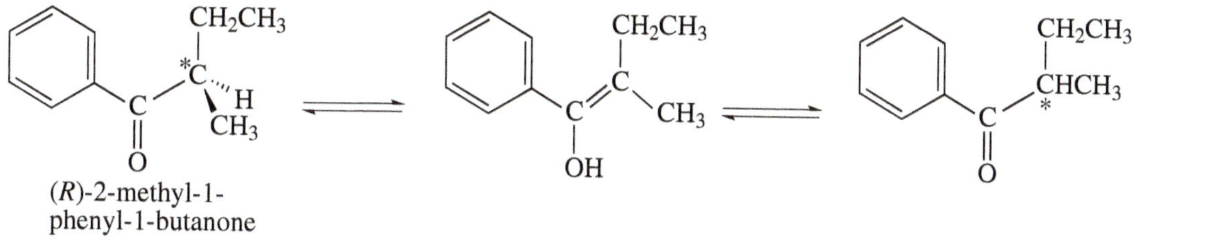

(*R*)-2-methyl-1-phenyl-1-butanone

You need a ketone with an α-carbon that is an asymmetric center. Racemization occurs when an α-hydrogen is removed from the asymmetric center.

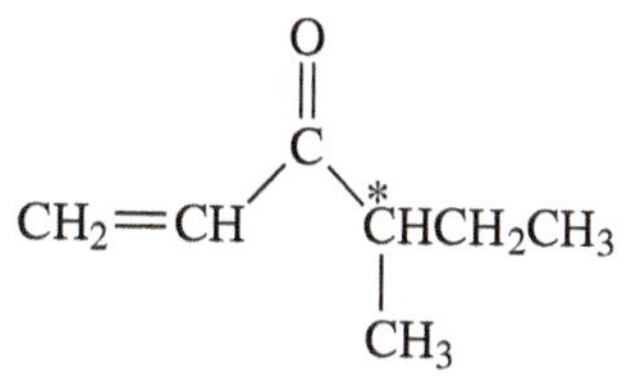

$CH_2{=}CH{-}C(=O){-}\overset{*}{C}H(CH_3)CH_2CH_3$

55. At 0 °C, enolate ion formation is reversible, so the major product of the reaction is the thermodynamic enolate—the one obtained by removing a proton from the more substituted α-carbon. Therefore, the deuterium will substitute for a hydrogen attached to the more substituted α-carbon.

LDA/THF

D_2O

thermodynamic enolate

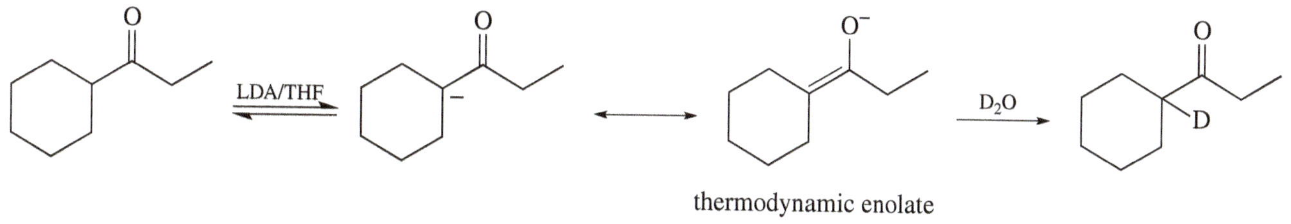

At −78 °C, enolate ion formation is irreversible, so the major product of the reaction is the kinetic enolate—the one obtained by removing a proton from the less substituted α-carbon. Therefore, the deuterium will substitute for a hydrogen attached to the less substituted α-carbon.

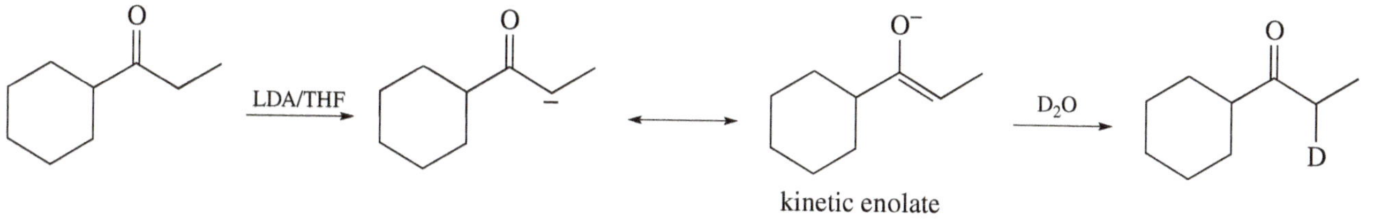

LDA/THF

D_2O

kinetic enolate

56. Two aldol additions occur. The initially formed addition product loses water immediately because the new double bonds are particularly stable (and therefore easy to form) since each is conjugated with a benzene ring.

OH O OH ⇌ O + 2 H_2O

57. $CH_3CH_2CH(=\ddot{O}:)$ ⇌ (H—B⁺) $CH_3CH-CH(=\overset{+}{O}H)$, H, B: ⇌ $CH_3CH=CH(-\ddot{O}H)$ + $CH_3CH_2CH(=\overset{+}{O}H)$, HB⁺

$CH_3CH_2CHCH(CH_3)CH$ with OH, O (HB⁺) ⇌ $CH_3CH_2CHCH(CH_3)CH$ with OH, $\overset{+}{O}$—H :B

58. The haloform reaction requires that a group be created that is a weaker base than hydroxide ion so that hydroxide ion is not the group eliminated from the tetrahedral intermediate. For an alkyl group to be the weaker base, the α-carbon must be bonded to three halogen atoms. The only alkyl group that can become bonded to three halogen atoms at the α-carbon is a methyl group (because it is bonded to three hydrogens).

$RC(=O)CI_3$ $\xrightarrow{HO^-}$ $R-C(-\ddot{O}:^-)(OH)-CI_3$ ⟶ $RC(=O)OH$ + $^-CI_3$ ⟶ $RC(=O)O^-$ + HCI_3

59. $CH_3C(=O)CH_2C(=O)OCH_3$ (**A**, $C_5H_8O_3$) $\xrightarrow{HCl,\ H_2O,\ \Delta}$ $CH_3C(=O)CH_3$ (**B**) $\xrightarrow{HO^-,\ H_2O}$ $CH_3C(OH)(CH_3)CH_2C(=O)CH_3$ (**C**) $\xrightarrow{HCl,\ \Delta}$ $CH_3C(CH_3){=}CHC(=O)CH_3$ (**D**)

A $\xrightarrow{1.\ CH_3O^-\ \ 2.\ CH_3Br}$ $CH_3C(=O)CH(CH_3)C(=O)OCH_3$ (**E**)

E $\xrightarrow{HCl,\ H_2O,\ \Delta}$ $CH_3C(=O)CH_2CH_3$ (**H**) $\xrightarrow{HO^-\ \text{excess},\ I_2\ \text{excess}}$ $CH_3CH_2C(=O)O^-$ (**I**) $\xrightarrow{SOCl_2}$ $CH_3CH_2C(=O)Cl$ (**J**)

E $\xrightarrow{1.\ CH_3O^-\ \ 2.\ CH_3Br}$ $CH_3C(=O)C(CH_3)_2C(=O)OCH_3$ (**F**) $\xrightarrow{HCl,\ H_2O,\ \Delta}$ $CH_3C(=O)CHCH_3$ with CH_3 on CH (**G**)

J $\xrightarrow{CH_3OH}$ $CH_3CH_2C(=O)OCH_3$ (**K**) $\xrightarrow{1.\ CH_3O^-\ \ 2.\ HCl}$ $CH_3CH_2C(=O)CH(CH_3)C(=O)OCH_3$ (**L**)

60. **a.** cyclopentanone ⇌ cyclopent-1-en-1-ol (OH)

b. 2-(1-hydroxycyclopentyl)cyclopentanone (OH, O)

c. 2-cyclopentylidenecyclopentanone (O)

61. $CH_3CH_2C(=O)CH_2CH_3$ $\xrightarrow{1.\ LDA/THF\ \ 2.\ CH_3CH_2Br}$ $CH_3CH_2C(=O)CHCH_3$ with CH_2CH_3 on CH $\xrightarrow{1.\ NaBH_4\ \ 2.\ H_3O^+}$ $CH_3CH_2CH(OH)CHCH_3$ with CH_2CH_3 on CH

62.

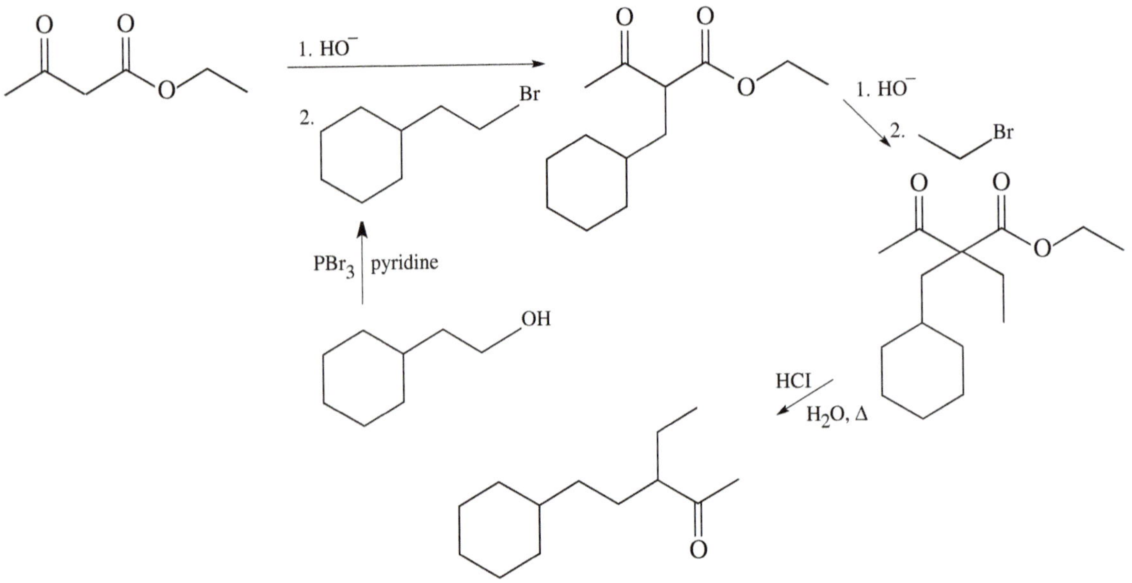

1. HO^-
2. Br

Br

PBr_3 pyridine

OH

1. HO^-
2. Br

HCl
H_2O, Δ

O

63.

1. $CH_3\overset{O}{\overset{\|}{C}}CH_2\overset{O}{\overset{\|}{C}}OCH_3$ + CH_3O^-
2. HCl, H_2O, Δ

or

1. (pyrrolidine enamine)
2. HCl, H_2O

H_2SO_4
Δ

OH

1. Br_2, H_3O^+
2. HO^-

1. Br_2, H_3O^+
2. $CH_3CO_2^-$
3. HCl, H_2O

$CH_2CH_2CH_3$

1. LDA/THF
2. $CH_3CH_2CH_2\overset{O}{\overset{\|}{C}}OCH_3$ add slowly

CH_2 CH_3

1. LDA/THF
2. $CH_3CH_2CH_2Br$ add slowly

$CH_2CH_2CH_3$

1. $CH_3O\overset{O}{\overset{\|}{C}}OCH_3$ + CH_3O^-
2. HCl, H_2O

1. LDA/THF
2. $BrCH_2$– (phenyl) add slowly

OH

CH_2

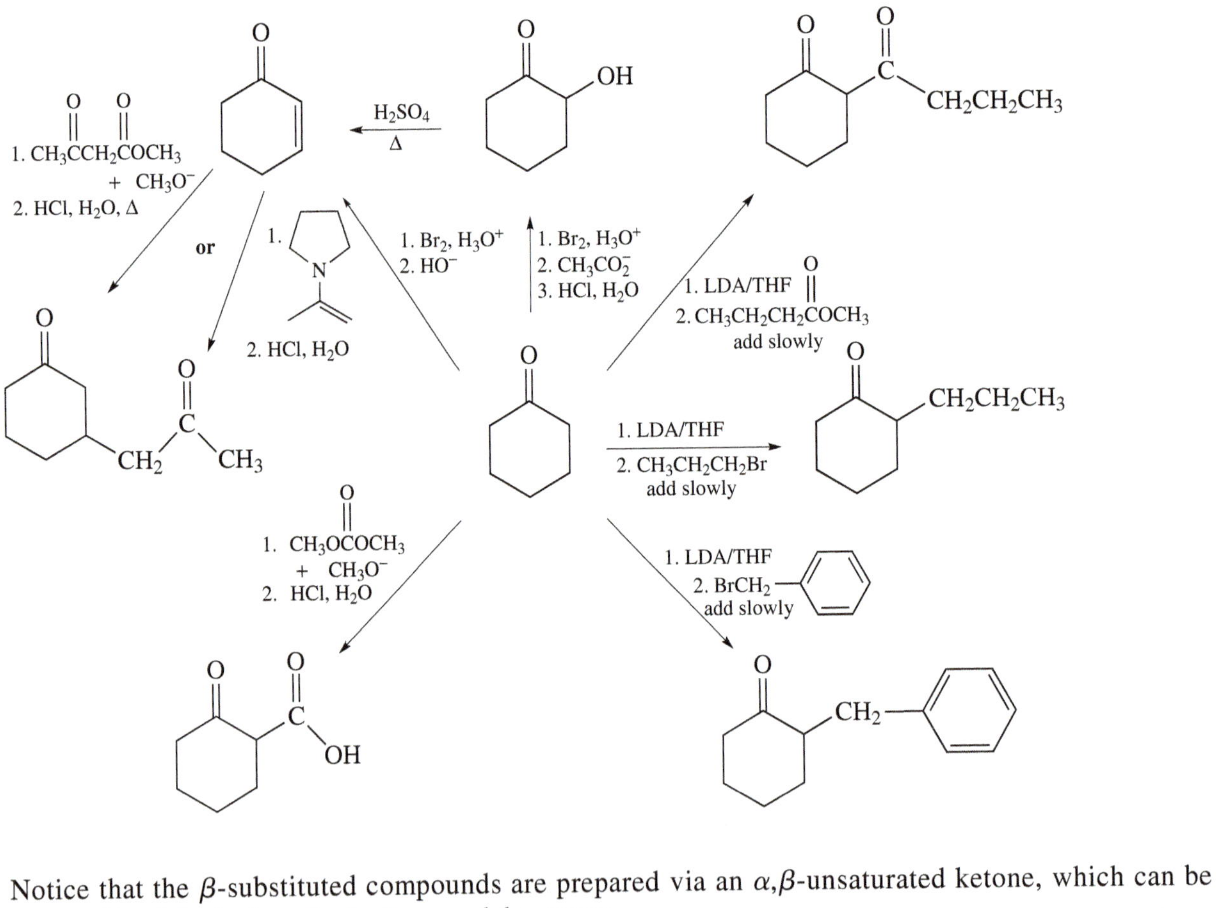

Notice that the β-substituted compounds are prepared via an α,β-unsaturated ketone, which can be prepared by dehydration of α-hydroxycyclohexanone.

64. Remember that there are no positively charged organic reactants, intermediates, or products in a basic solution and no negatively charged organic reactants, intermediates, or products in an acidic solution.

a. HO^- ; $H-OH$; $+ HO^-$

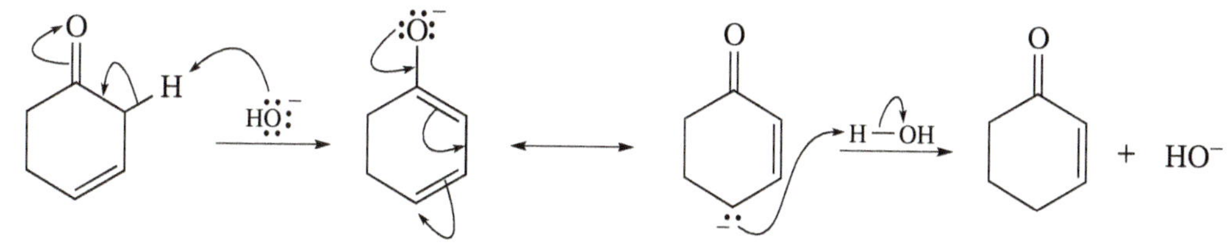

b. HOH (protonated) ; H_2O: ; HOH (protonated) ; H_2O: ; $+ H_3O^+$

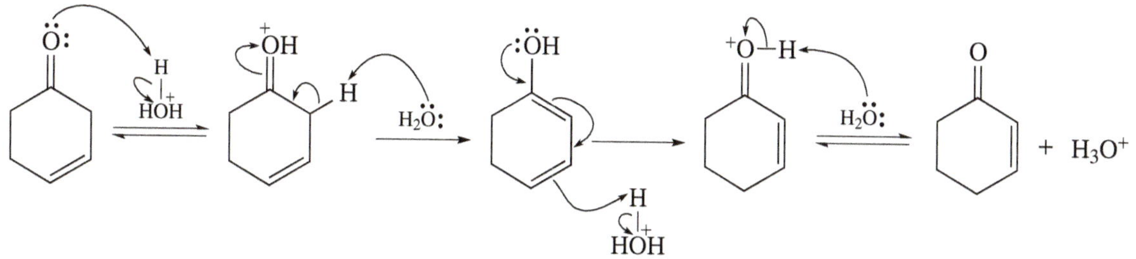

65. a. H + CH_2 ... O ... CH_3 → OH, CHCH$_2$, O, CH_3

CH_3CO^-

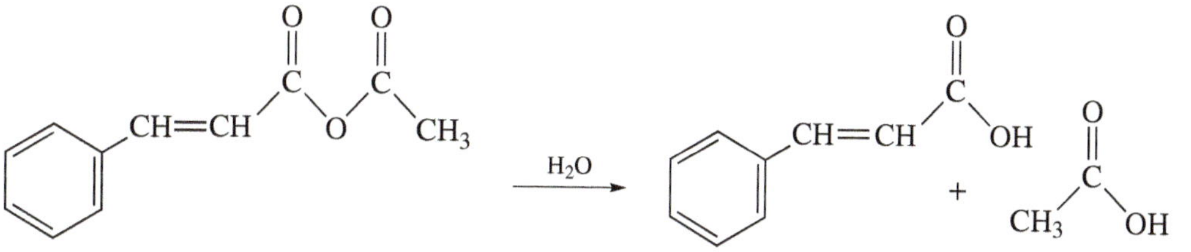

CH_3 O CH_3

CH=CH O CH_3 + H_2O

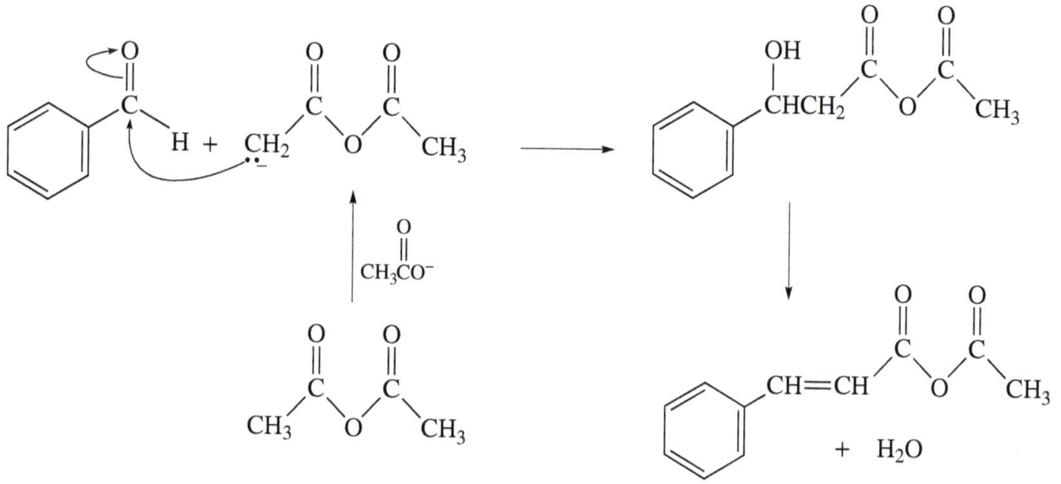

b. CH=CH O CH_3 $\xrightarrow{H_2O}$ CH=CH OH + CH_3 OH

c.

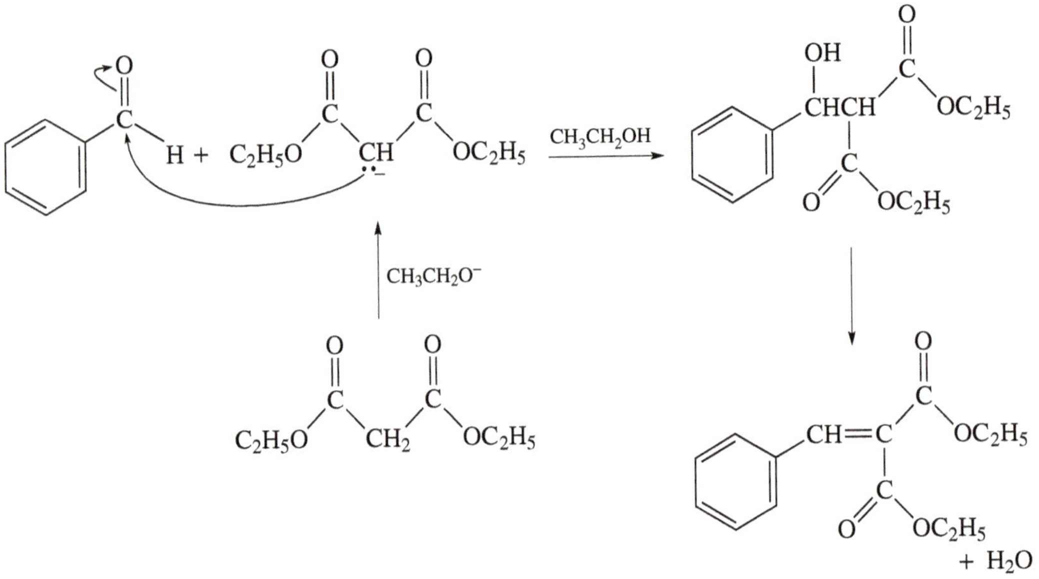

$C_6H_5-C(=O)-H$ + $C_2H_5O-C(=O)-\ddot{C}H^{-}-C(=O)-OC_2H_5$ $\xrightarrow{CH_3CH_2OH}$ $C_6H_5-CHCH(OH)$ with $C(=O)OC_2H_5$ and $C(=O)OC_2H_5$

(O, O, C, C, OH, CHCH, OC_2H_5, OC_2H_5)

$\uparrow$ $CH_3CH_2O^-$

$C_2H_5O-C(=O)-CH_2-C(=O)-OC_2H_5$

$\downarrow$

$C_6H_5-CH=C(-C(=O)OC_2H_5)-C(=O)OC_2H_5$ + H_2O

d.

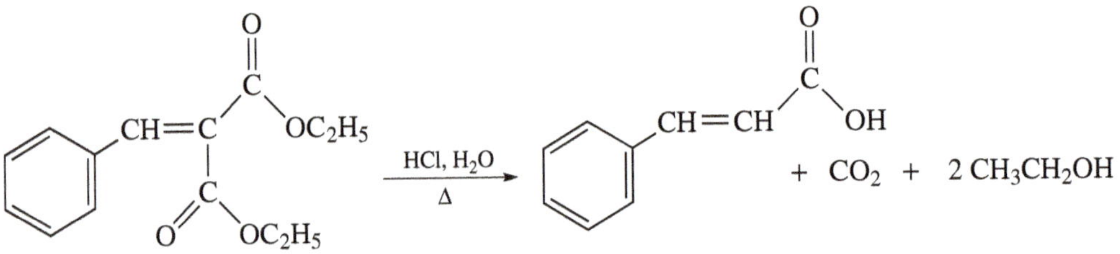

$C_6H_5-CH=C(-C(=O)OC_2H_5)-C(=O)OC_2H_5$ $\xrightarrow[\Delta]{HCl,\ H_2O}$ $C_6H_5-CH=CH-C(=O)-OH$ + CO_2 + 2 CH_3CH_2OH

66. a.

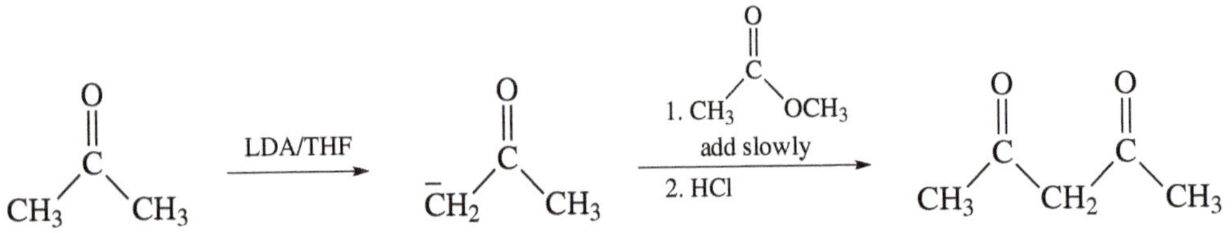

$CH_3-C(=O)-CH_3$ $\xrightarrow{LDA/THF}$ $\bar{C}H_2-C(=O)-CH_3$ $\xrightarrow[2.\ HCl]{1.\ CH_3-C(=O)-OCH_3,\ \text{add slowly}}$ $CH_3-C(=O)-CH_2-C(=O)-CH_3$

b.

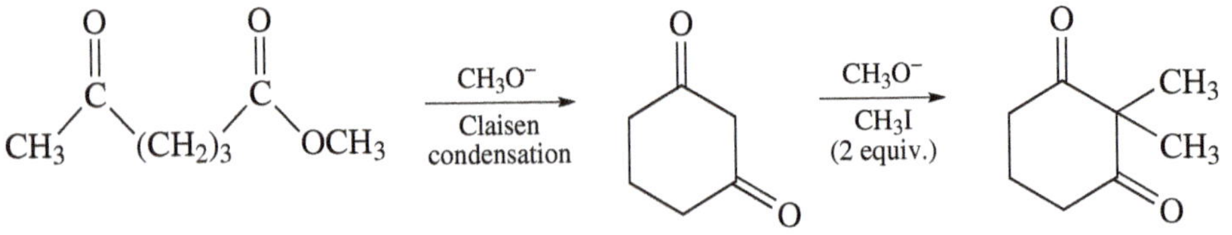

$CH_3-C(=O)-(CH_2)_3-C(=O)-OCH_3$ $\xrightarrow[\text{Claisen condensation}]{CH_3O^-}$ (cyclohexane-1,3-dione: O, O) $\xrightarrow[CH_3I\ (2\ equiv.)]{CH_3O^-}$ (2,2-dimethylcyclohexane-1,3-dione: O, CH_3, CH_3, O)

c. $CH_3C(=O)CH_2C(=O)OCH_2CH_3$ $\xrightarrow[\text{2. cyclopentyl bromide}]{\text{1. } CH_3CH_2O^-}$ $CH_3C(=O)CH(C_5H_9)C(=O)OCH_2CH_3$

$\xrightarrow{\Delta \text{ HCl, } H_2O}$ CO_2 + CH_3CH_2OH + $CH_3C(=O)CH_2$–cyclopentyl

d. $CH_3CH_2OC(=O)(CH_2)_4C(=O)OCH_2CH_3$ $\xrightarrow[\text{Claisen condensation}]{CH_3CH_2O^-}$ 2-oxocyclopentane–$C(=O)OCH_2CH_3$

$\xrightarrow{\Delta \text{ HCl, } H_2O}$ cyclopentanone + CO_2 + CH_3CH_2OH

cyclopentanone $\xrightarrow[\text{2. } CH_3CH_2CH_2Br]{\text{1. LDA/THF}}$ 2-($CH_2CH_2CH_3$)cyclopentanone

67. The Reformatsky reaction requires an organozinc reagent that is obtained by treating an α-bromo ester with zinc.

a. $CH_3CH_2CH_2C(=O)H$ + $BrCH_2C(=O)OCH_3$ $\xrightarrow[\text{2. } H_2O]{\text{1. Zn}}$ $CH_3CH_2CH_2CH(OH)CH_2C(=O)OCH_3$

b. $CH_3CH_2C(=O)H$ + $CH_3CH_2CH(Br)C(=O)OCH_3$ $\xrightarrow[\text{2. } H_2O]{\text{1. Zn}}$ $CH_3CH_2CH(OH)CH(CH_2CH_3)C(=O)OCH_3$

$\xrightarrow{\text{HCl, } H_2O \text{ excess}}$ $CH_3CH_2CH(OH)CH(CH_2CH_3)C(=O)OH$

c. CH_3CH_2CHO + $CH_3CH(Br)COOCH_3$ $\xrightarrow[2.\ H_2O]{1.\ Zn}$ $CH_3CH_2CH(OH)CH(CH_3)COOCH_3$

$\xrightarrow{HCl,\ H_2O\ \text{excess}}$ $CH_3CH_2CH(OH)CH(CH_3)COOH$ $\xrightarrow{H_2SO_4,\ \Delta}$ $CH_3CH_2CH{=}C(CH_3)COOH$

d. $CH_3CH_2COCH_2CH_3$ + $BrCH_2COOCH_3$ $\xrightarrow[2.\ H_2O]{1.\ Zn}$ $CH_3CH_2C(OH)(CH_2CH_3)CH_2COOCH_3$

68. The ketone is 2-hexanone. Therefore, the alkyl halide is a propyl halide (propyl bromide, propyl chloride, or propyl iodide).

$CH_3COCH_2CH_2CH_2CH_3$

This part comes from acetoacetic ester. ($CH_3C(=O)CH_2$–)

69. **a.** cyclohexanone $\xrightarrow[2.\ CH_3CH_2CH_2CH_2Br]{1.\ LDA/THF}$ 2-butylcyclohexanone (ring–$CH_2CH_2CH_2CH_3$)

b. **1.** cyclohexanone $\xrightarrow{1.\ \text{pyrrolidine, trace acid};\ 2.\ CH_3CH_2CH_2COCl;\ 3.\ HCl,\ H_2O}$ 2-(butanoyl)cyclohexanone (ring–$C(=O)CH_2CH_2CH_3$)

2. cyclohexanone $\xrightarrow{1.\ LDA/THF;\ 2.\ CH_3CH_2CH_2COCl\ \text{add slowly}}$ 2-(butanoyl)cyclohexanone (ring–$C(=O)CH_2CH_2CH_3$)

c. $CH_3CH_2OC(=O)OCH_2CH_3$ + $CH_3CH_2O^-$ $\xrightarrow[\text{2. HCl}]{\text{1. cyclohexanone, add slowly}}$ 2-(ethoxycarbonyl)cyclohexanone (OCH_2CH_3)

d. $HC(=O)OCH_2CH_3$ + $CH_3CH_2O^-$ $\xrightarrow[\text{2. HCl}]{\text{1. cyclohexanone, add slowly}}$ 2-formylcyclohexanone (H)

e. cyclohexane-1,3-dione + $CH_3CH{=}CHCCH_3$ (with C=O) $\xrightarrow[\text{a Robinson annulation}]{HO^-}$ bicyclic hydroxy diketone (OH) $\xrightarrow[\Delta]{H_2SO_4}$ enone $\xrightarrow{H_2\ Pd/C}$ saturated bicyclic diketone

70. The positive iodoform test indicates that compound **B** is a methyl ketone. The two singlets with a ratio of 3:1 indicate that if the methyl group represents the "3", there is only one other hydrogen in the compound or if the methyl group represent the "1", there are nine other hydrogens. The former does not agree with the molecular formula. Therefore, the compound must have three identical methyl groups that are attached to a carbon that is not attached to a hydrogen.
3,3-Dimethyl-1-butyne will show two singlets with a ratio of 9:1. When water is added, a methyl ketone with two singlets with a ratio of 3:1 is formed.

$$(CH_3)_3C{-}C{\equiv}CH \xrightarrow[H_2O,\ HgSO_4]{H_2SO_4} (CH_3)_3C{-}\overset{O}{\overset{\|}{C}}{-}CH_3$$

two singlets (9:1) two singlets (3:1)

71. **a.** $C_2H_5OC(=O)CH_2C(=O)OC_2H_5$ $\xrightarrow[\text{2. } Br(CH_2)_5Br]{\text{1. } CH_3CH_2O^-}$ $C_2H_5OC(=O)CH(CH_2CH_2CH_2CH_2CH_2Br)C(=O)OC_2H_5$

$\xrightarrow{CH_3CH_2O^-}$ cyclohexane-1,1-di($C{-}OC_2H_5$, C=O)

$\xrightarrow[\Delta]{HCl,\ H_2O}$ $2\ CH_3CH_2OH + CO_2 +$ cyclohexane–C(=O)OH

b. $C_2H_5OC(=O)CH_2C(=O)OC_2H_5$ $\xrightarrow[\text{2. } Br(CH_2)_5Br]{\text{1. } CH_3CH_2O^-}$ $C_2H_5OC(=O)CH(CH_2CH_2CH_2CH_2CH_2Br)C(=O)OC_2H_5$

$\xrightarrow[CH_3CH_2O^-]{\text{malonic ester}}$ $C_2H_5OC(=O)CH(CH_2CH_2CH_2CH_2CH_2CH(C(=O)OC_2H_5)_2)C(=O)OC_2H_5$

$\xrightarrow[\Delta]{HCl,\ H_2O}$ $HOC(=O)(CH_2)_7C(=O)OH$ $+ 2\ CO_2 + 4\ CH_3CH_2OH$

72. **a.** The first reaction is an intermolecular Claisen condensation. An intramolecular condensation reaction forms the final product. Notice that the most acidic hydrogen is the one removed for the intramolecular condensation.

b. The reaction involves two successive aldol condensations.

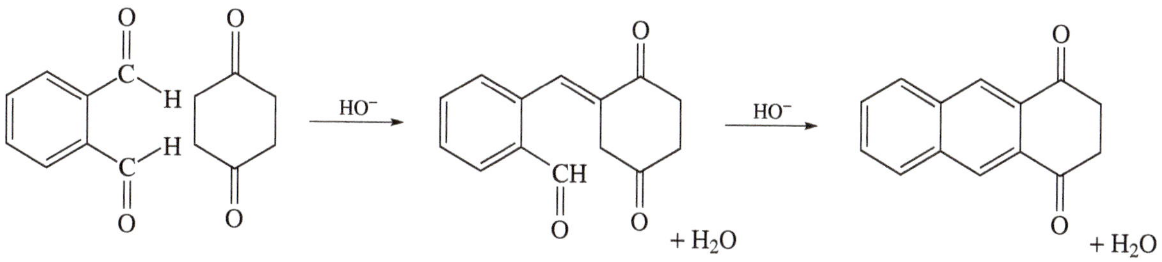

HO⁻ HO⁻ CH + H_2O + H_2O

c. The base removes a proton from the carbon that is flanked by two carbonyl groups; this is followed by an intramolecular S_N2 reaction.

HO⁻ Br H Br H + Br^-

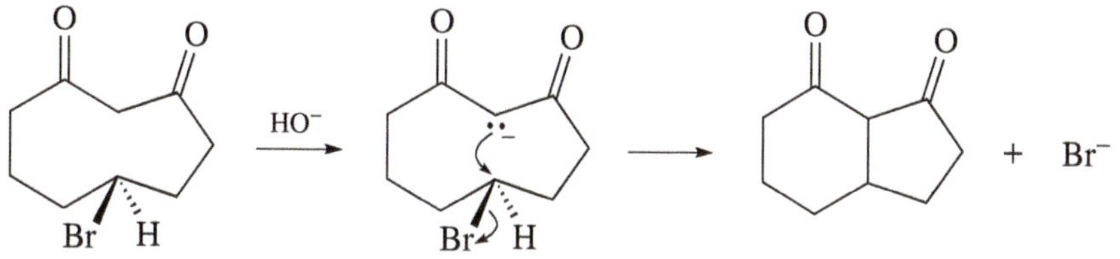

d. A proton is removed from the α-carbon that is flanked by a carbonyl group and a nitrile; then an intramolecular S_N2 reaction occurs.

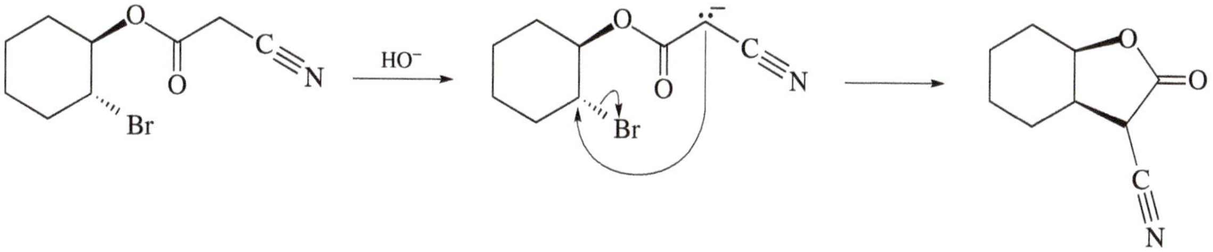

HO⁻ Br Br C≡N

73. LDA/THF add slowly HCl tautomerization

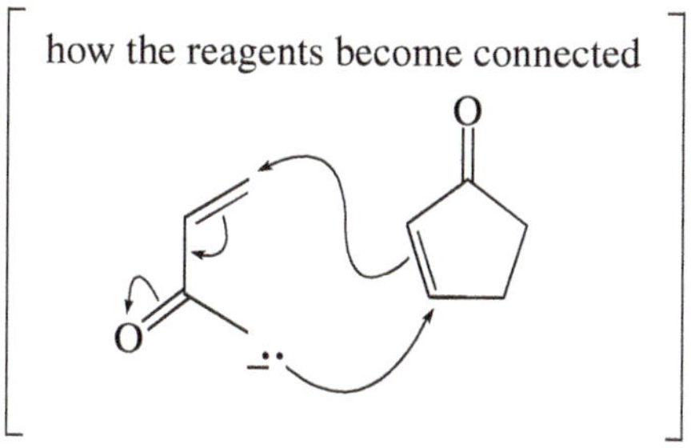

how the reagents become connected

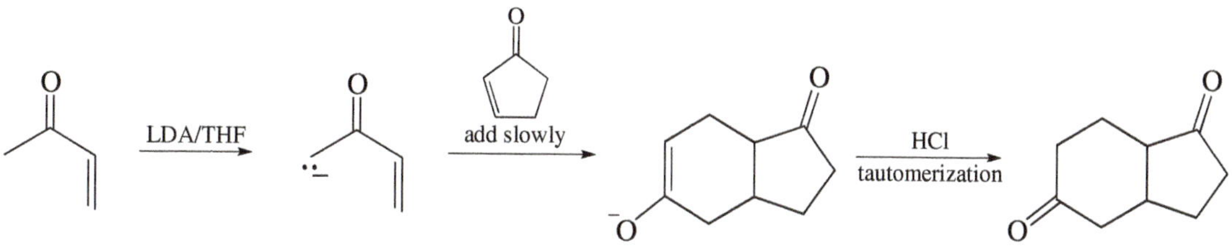

74. a. CH_3CH_2COOH 1. Br_2, PBr_3 2. H_2O → $CH_3CH(Br)COOH$ excess NH_3 → $CH_3CH(\overset{+}{N}H_3)COO^-$ $\overset{+}{N}H_4$

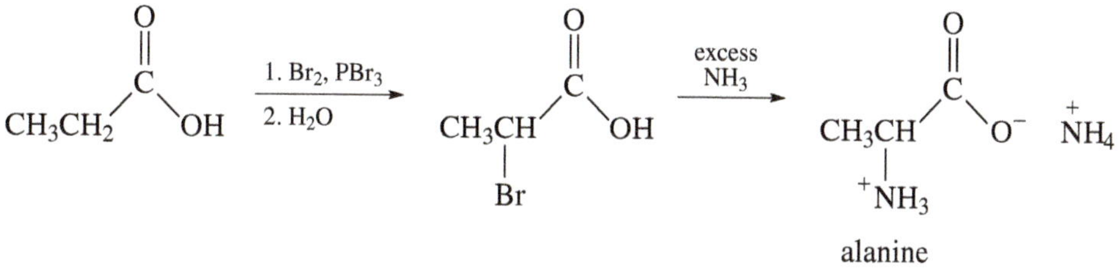

alanine

b. 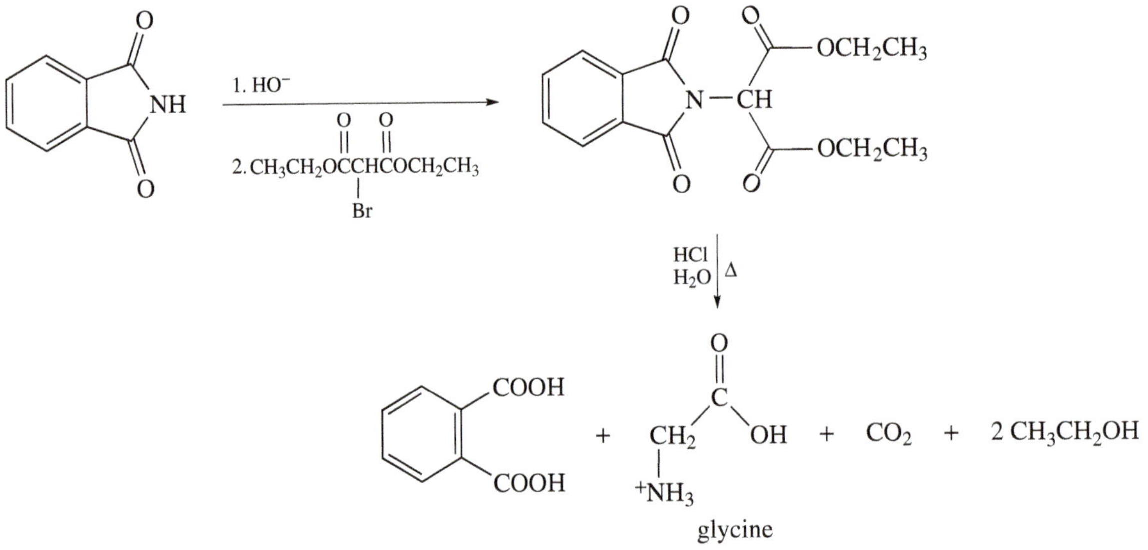
1. HO⁻
2. CH3CH2OCCHCOCH2CH3
Br
N—CH
OCH2CH3
OCH2CH3
HCl
H2O
Δ
COOH
COOH
CH2
+NH3
OH
CO2
2 CH3CH2OH
glycine

75. a. 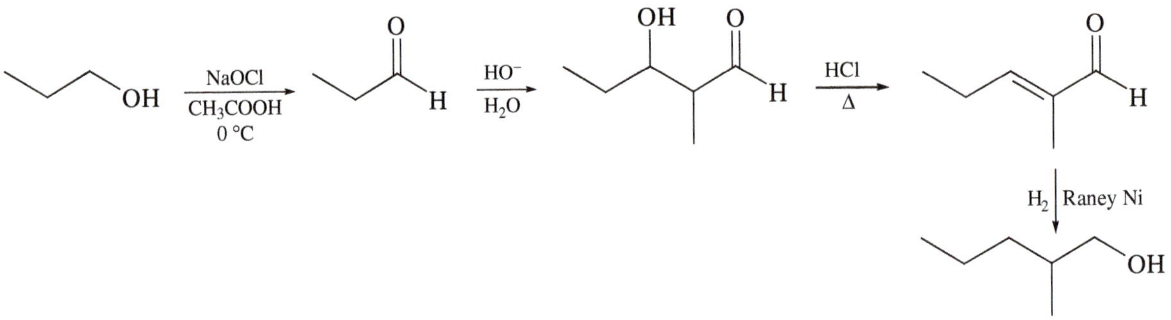
OH
NaOCl
CH3COOH
0 °C
H
HO⁻
H2O
HCl
Δ
H2
Raney Ni

b. 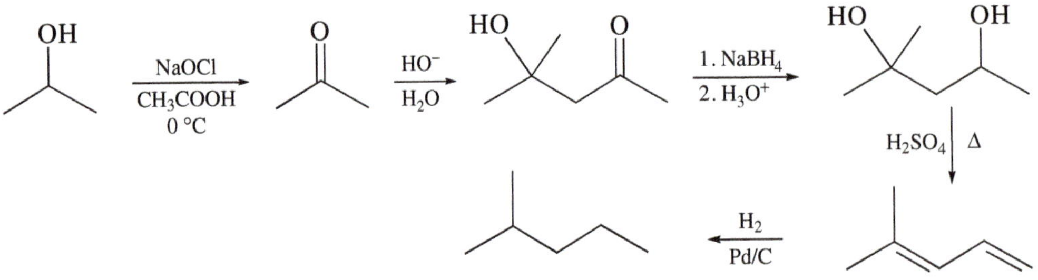
OH
NaOCl
CH3COOH
0 °C
HO⁻
H2O
HO
1. NaBH4
2. H3O+
H2SO4
Δ
H2
Pd/C

c. 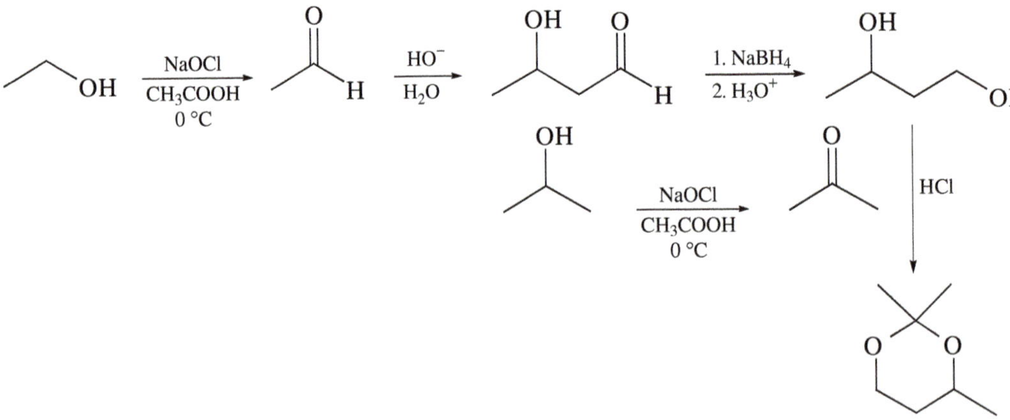
OH
NaOCl
CH3COOH
0 °C
H
HO⁻
H2O
1. NaBH4
2. H3O+
HCl

76. **A** can be prepared by adding the compound with α-hydrogens slowly to a basic solution of the compound without α-hydrogens, followed by dehydration.

$$HCHO + HO^- \xrightarrow[\text{add slowly}]{CH_3COCH_3} {}^-OCH_2CH_2COCH_3 \xrightarrow[\Delta]{H_2SO_4} CH_2{=}CHCOCH_3$$

Or **A** can be prepared by removing an α-hydrogen from acetone with LDA and then adding formaldehyde, followed by dehydration.

$$CH_3COCH_3 \xrightarrow{LDA/THF} \ddot{C}H_2^-COCH_3 \xrightarrow{HCHO} {}^-OCH_2CH_2COCH_3 \xrightarrow[\Delta]{H_2SO_4} CH_2{=}CHCOCH_3$$

B can be prepared by removing an α-hydrogen from propionaldehyde with LDA and then slowly adding acetaldehyde, followed by dehydration.

$$CH_3CH_2CHO \xrightarrow{LDA/THF} CH_3\ddot{C}H^-CHO \xrightarrow[\text{add slowly}]{CH_3CHO} CH_3CH(O^-)CH(CH_3)CHO \xrightarrow[\Delta]{H_2SO_4} CH_3CH{=}C(CH_3)CHO$$

C can be prepared by removing an α-hydrogen from 3-hexanone with LDA and then adding formaldehyde, followed by dehydration. The yield is poor because 3-hexanone is an asymmetrical ketone; therefore, two different α-carbanions are formed that lead to two different α,β-unsaturated ketones.

$$CH_3CH_2COCH_2CH_2CH_3 \xrightarrow{LDA/THF} CH_3\ddot{C}H^-COCH_2CH_2CH_3 \xrightarrow{HCHO} {}^-OCH_2CH(CH_3)COCH_2CH_2CH_3 \xrightarrow[\Delta]{H_2SO_4} CH_2{=}C(CH_3)COCH_2CH_2CH_3$$

$$CH_3CH_2COCH_2CH_2CH_3 \xrightarrow{LDA/THF} CH_3CH_2CO\ddot{C}H^-CH_2CH_3 \xrightarrow{HCHO} CH_3CH_2COCH(CH_2O^-)CH_2CH_3 \xrightarrow[\Delta]{H_2SO_4} CH_3CH_2COC({=}CH_2)CH_2CH_3$$

D can be prepared by removing an α-hydrogen from acetaldehyde with LDA and then slowly adding acetone, followed by dehydration.

$$CH_3CHO \xrightarrow{\text{LDA/THF}} {}^{-}\!:CH_2CHO \xrightarrow[\text{add slowly}]{CH_3COCH_3} CH_3C(O^-)(CH_3)CH_2CHO \xrightarrow[\Delta]{H_2SO_4} CH_3C(CH_3){=}CHCHO$$

E can be prepared via an intramolecular aldol condensation using 5-oxooctanal.

$$CH_3CH_2CH_2C(=O)(CH_2)_3CHO \xrightarrow{HO^-} CH_3CH_2\ddot{C}H^-C(=O)(CH_2)_3CHO \longrightarrow \text{2-ethyl-3-oxidocyclohexanone} \xrightarrow[\Delta]{H_2SO_4} \text{2-ethylcyclohex-2-enone}$$

F can be prepared via an intramolecular aldol condensation using 1,7-heptanedial.

$$HC(=O)(CH_2)_4CH_2CHO \xrightarrow{HO^-} HC(=O)(CH_2)_4\ddot{C}H^-CHO \longrightarrow \text{2-oxidocyclohexanecarbaldehyde} \xrightarrow[\Delta]{H_2SO_4} \text{cyclohex-1-enecarbaldehyde}$$

A small amount of **G** can be formed via an intramolecular aldol condensation using 7-oxooctanal, but this compound can form two different α-carbanions that can react with a carbonyl group to form a six-membered ring. Because an aldehyde is more reactive than a ketone toward nucleophilic addition, the target molecule will be a minor product.

$$\underset{\textbf{minor product}}{\text{2-methylcyclohex-1-enecarbaldehyde}} \longleftarrow CH_3C(=O)CH_2CH_2CH_2CH_2CH_2CHO \longrightarrow \underset{\textbf{major product}}{\text{1-acetylcyclohexene}}$$

77. At low temperatures (−78 °C), the proton will be more apt to be removed from the methyl group, because its hydrogens are the most accessible and are slightly more acidic (that is, the kinetically controlled product will be formed).

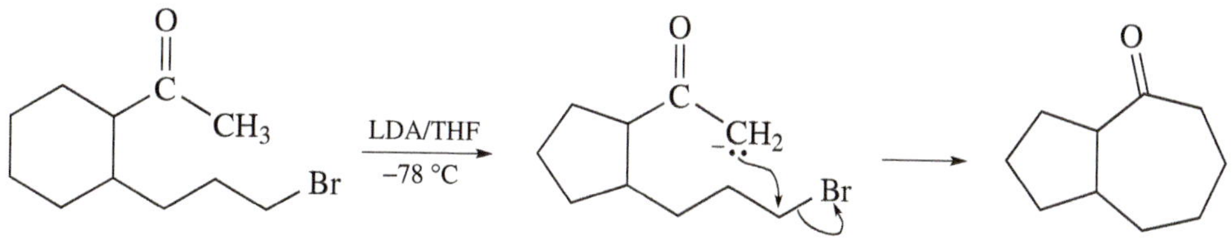

LDA/THF
−78 °C

At higher temperatures (0 °C), the proton will be more apt to be removed from the more substituted α-carbon because in that way, the more stable enolate is formed (the one with the more stable double bond). Therefore, the thermodynamically controlled product is formed.

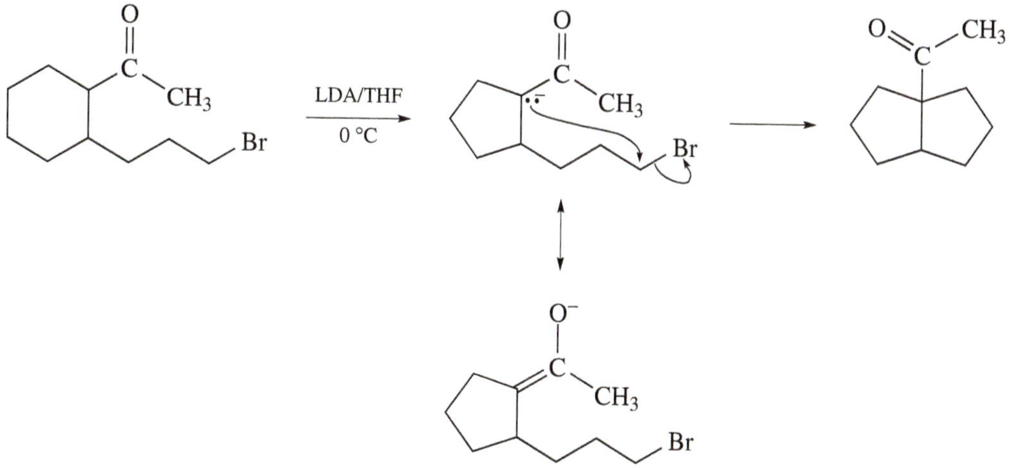

LDA/THF
0 °C

78.

$+ HB^+$

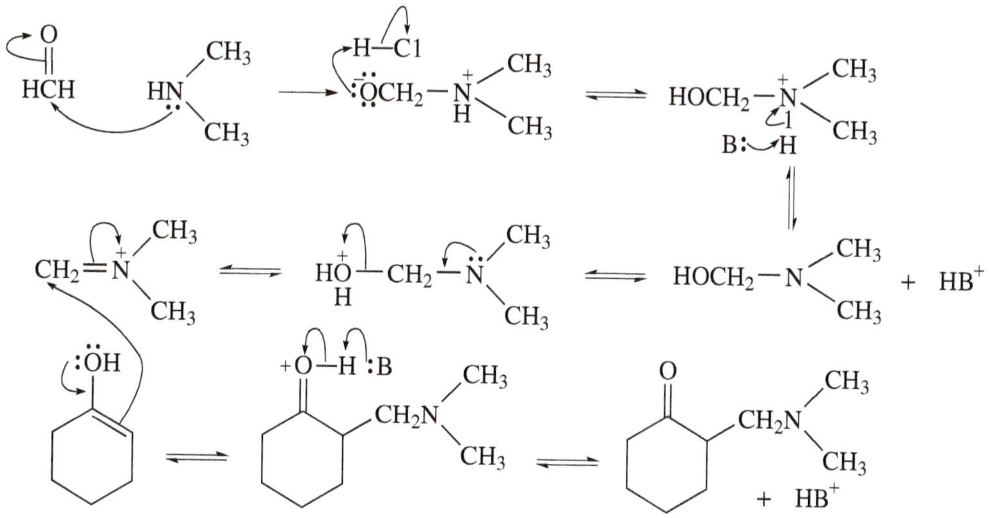

79.

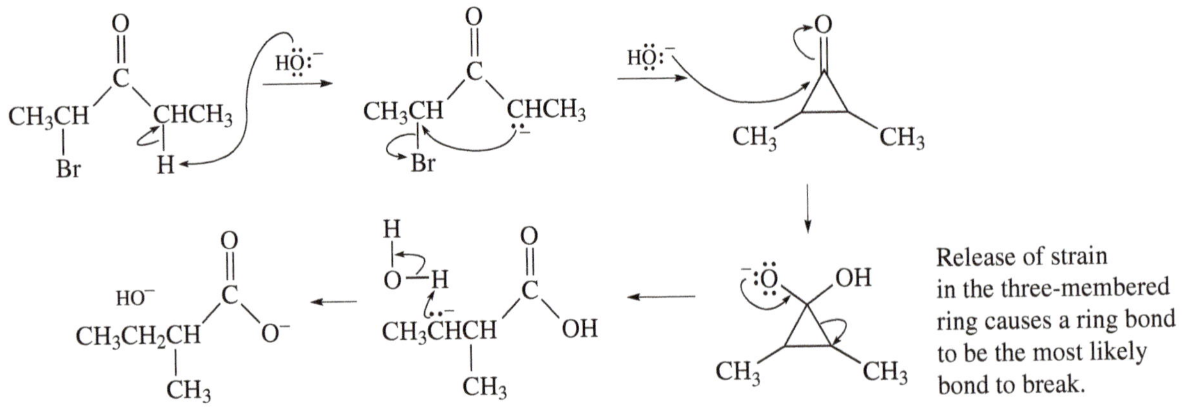

$$CH_3CH(Br)\overset{O}{\overset{\|}{C}}CHCH_3 \xrightarrow{HO^-} CH_3CH(Br)\overset{O}{\overset{\|}{C}}\overset{-}{C}HCH_3 \rightarrow \text{cyclopropanone}(CH_3)_2 \xrightarrow{HO^-}$$

Release of strain in the three-membered ring causes a ring bond to be the most likely bond to break.

$$\rightarrow CH_3\overset{-}{C}HCH(CH_3)COOH \rightarrow CH_3CH_2CH(CH_3)COO^- + HO^-$$

80.

Cyclohexanone $\xrightarrow{LDA/THF}$ enolate carbanion + PhSe–Br → 2-(phenylseleno)cyclohexanone $\xrightarrow{HO-OH}$ selenonium hydroxide ⇌ selenoxide (with H^+) → 2-cyclohexenone + PhSeOH

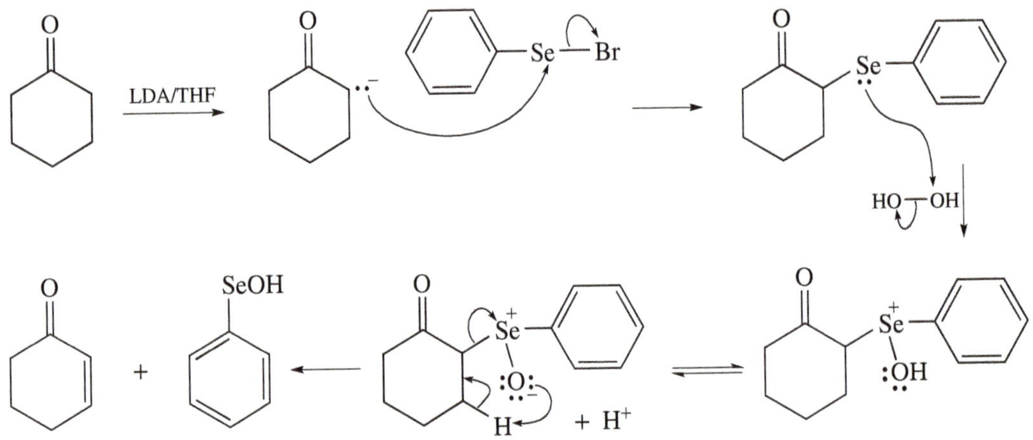

81. The compound that gives the ^1H NMR spectrum is 4-phenyl-3-buten-2-one. The singlet at 2.3 ppm that integrates to 3 protons is the methyl group. Because the benzene ring protons that show signals between 7 and 8 ppm apparently integrate to 6 protons, we know that the signals include one of the vinylic protons of the double bond. The other vinyl proton is the doublet (6.7 ppm) that integrates to 1 proton. The compounds that would form this compound (via an aldol condensation) are benzaldehyde and acetone.

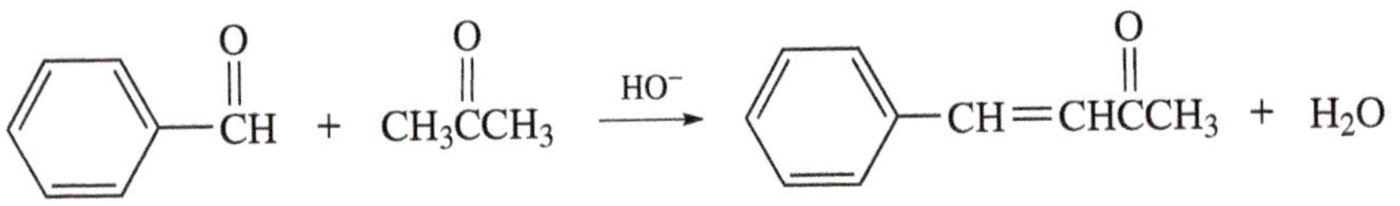

$$C_6H_5\overset{O}{\overset{\|}{C}}H + CH_3\overset{O}{\overset{\|}{C}}CH_3 \xrightarrow{HO^-} C_6H_5CH{=}CH\overset{O}{\overset{\|}{C}}CH_3 + H_2O$$

4-phenyl-3-buten-2-one

82. The middle carbonyl group is hydrated, because the hydrate is stabilized by the electron-withdrawing carbonyl groups on either side of it.

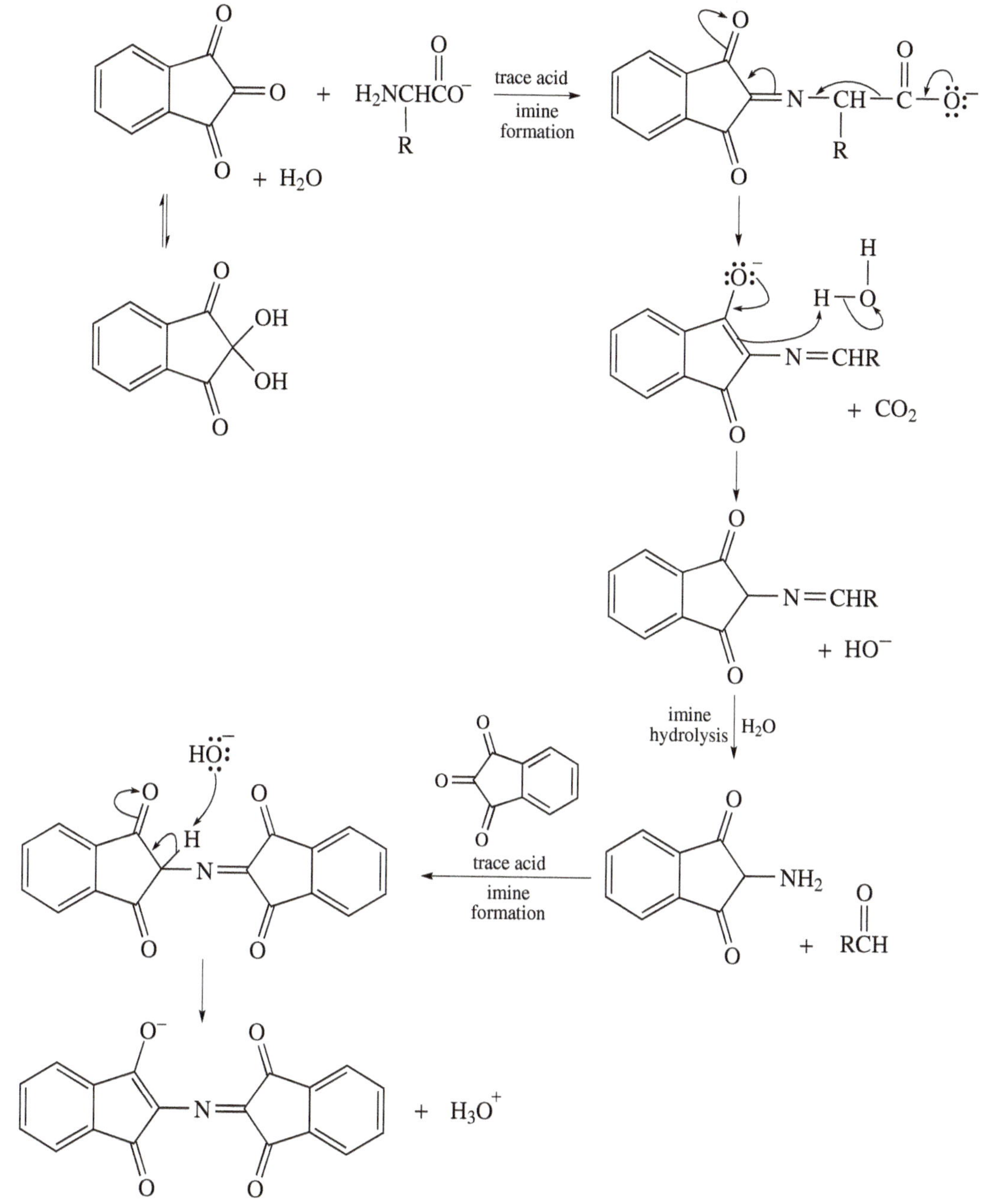

trace acid

imine formation

H_2NCHCO^- (R) + H_2O

$N=CHR$ + CO_2

$N=CHR$ + HO^-

imine hydrolysis H_2O

NH_2 + RCH(=O)

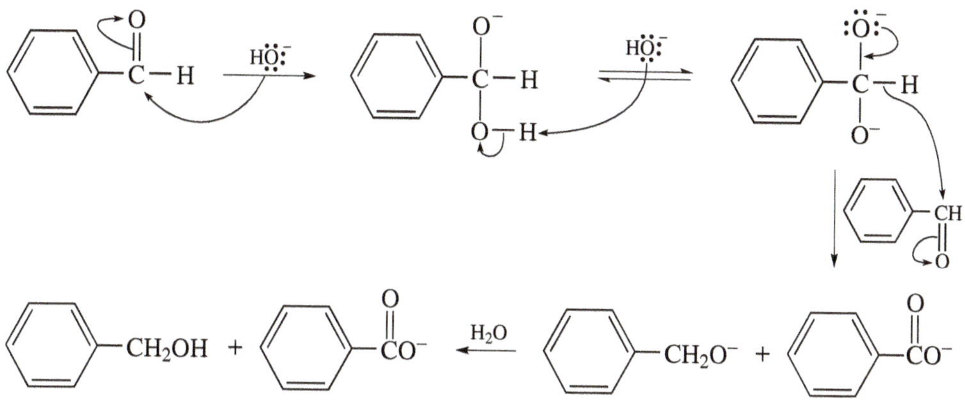

trace acid

imine formation

HO^-

+ H_3O^+

83. HO^- ... HO^- ... H_2O

CH_2OH + CO^- ← CH_2O^- + CO^-

84. **a.** The mechanism starts with an isomerization that converts isolated double bonds to conjugated double bonds.

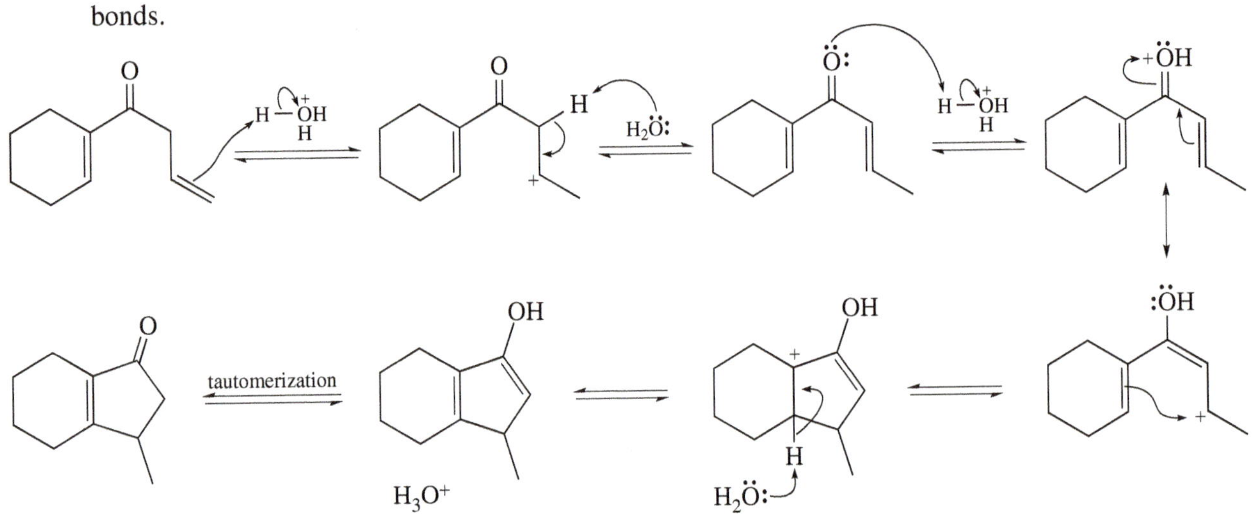

$H-\overset{+}{O}H$ H

H

$H_2\ddot{O}$:

$\ddot{O}$:

$H-\overset{+}{O}H$ H

$+\ddot{O}H$

$:\ddot{O}H$

OH

OH

O

tautomerization

H_3O^+

$H_2\ddot{O}$:

H

b. The carbonyl group on the left is protonated because electron delocalization causes it to be more basic than the other carbonyl group.

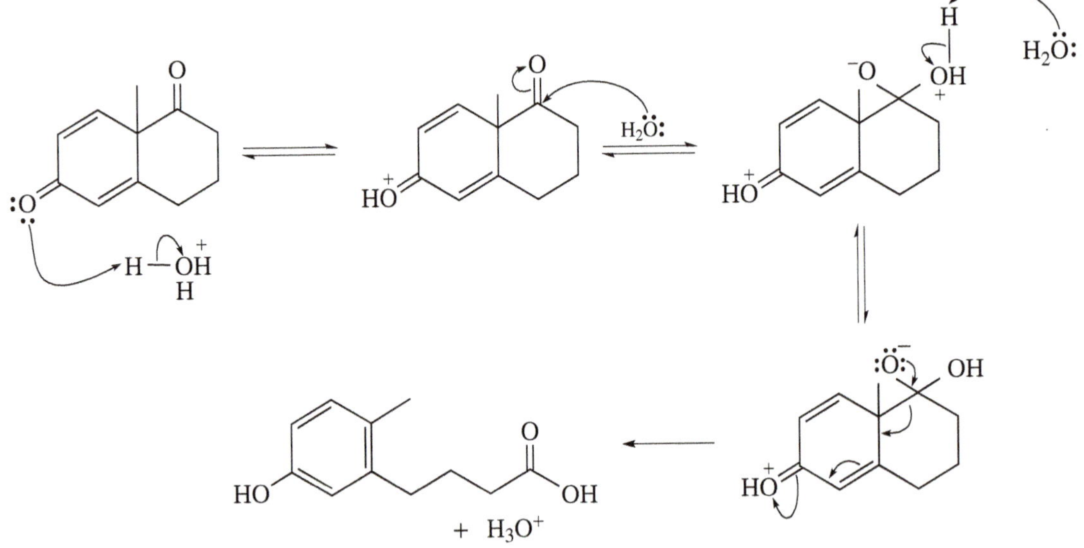

O

:O

$H-\overset{+}{O}H$ H

O

$H\overset{+}{O}$

$H_2\ddot{O}$:

^-O

$\overset{+}{O}H$

H

$H_2\ddot{O}$:

$H\overset{+}{O}$

$:\ddot{O}:^-$

OH

$H\overset{+}{O}$

HO

O

OH

+ H_3O^+

85. The benzoin condensation requires the aldehyde hydrogen to be removed. This hydrogen cannot be removed unless the electrons left behind can be delocalized onto an electronegative atom. The nitrogen of the cyano group serves that purpose. Therefore, the reaction will not occur if hydroxide ion is used instead of cyanide ion because the electrons left behind if the hydrogen were to be removed cannot be delocalized.

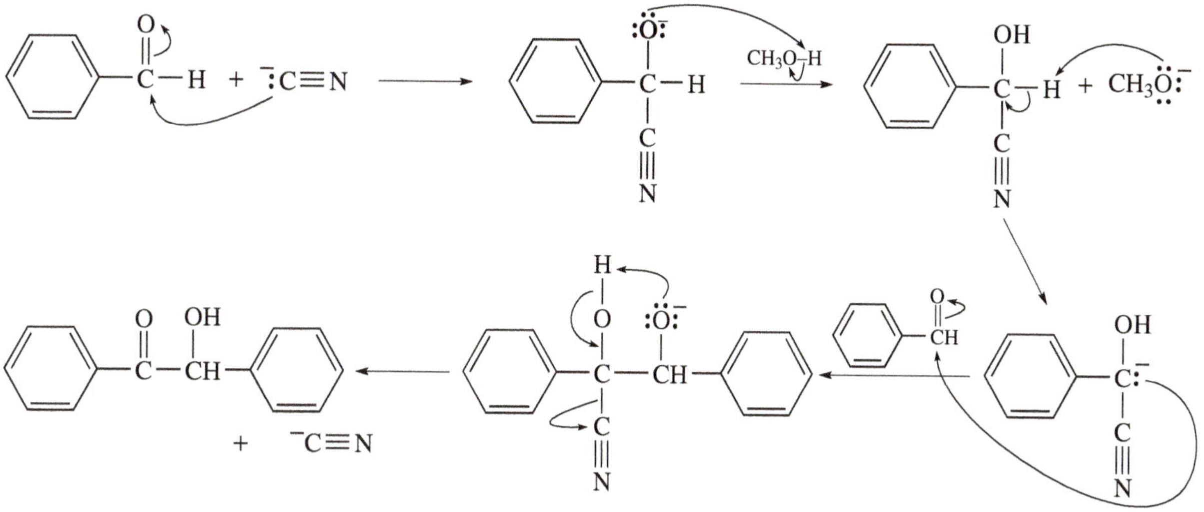

86. To arrive at the final product, three equivalents of malonyl thioester are needed. (See page 840 in the text.) Because only one acetate ion was used, only one carbon is labeled in the product.

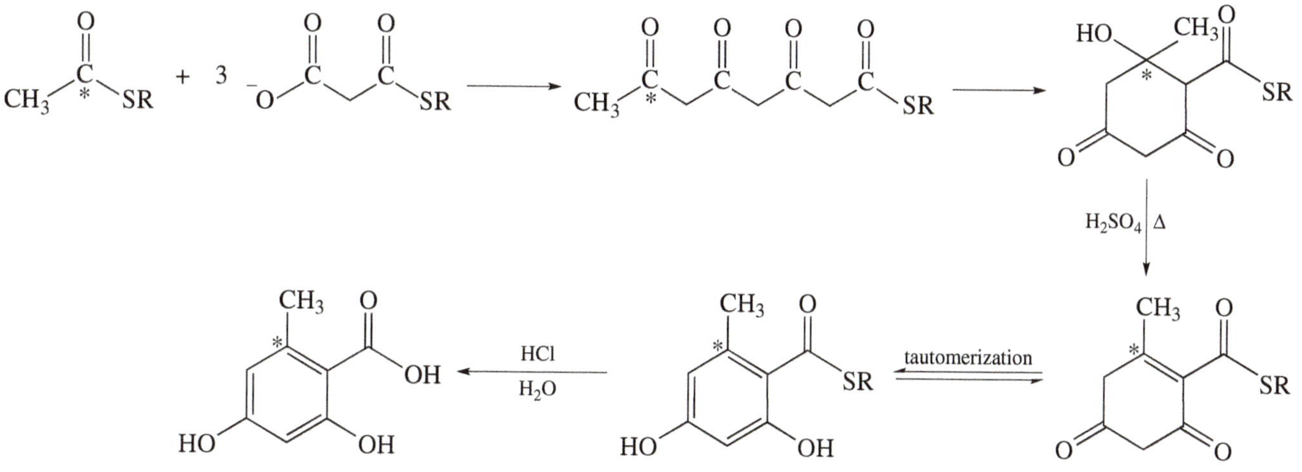

87. **a.**

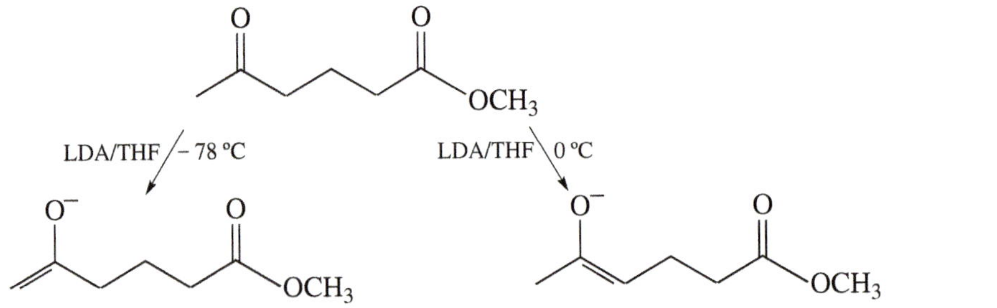

LDA/THF / − 78 °C

LDA/THF \ 0 °C

OCH_3

At −78 °C, the kinetic enolate is formed (the hydrogen is removed from the less substituted α-carbon).

At 0 °C, the thermodynamic enolate is formed (the hydrogen is removed from the more substituted α-carbon).

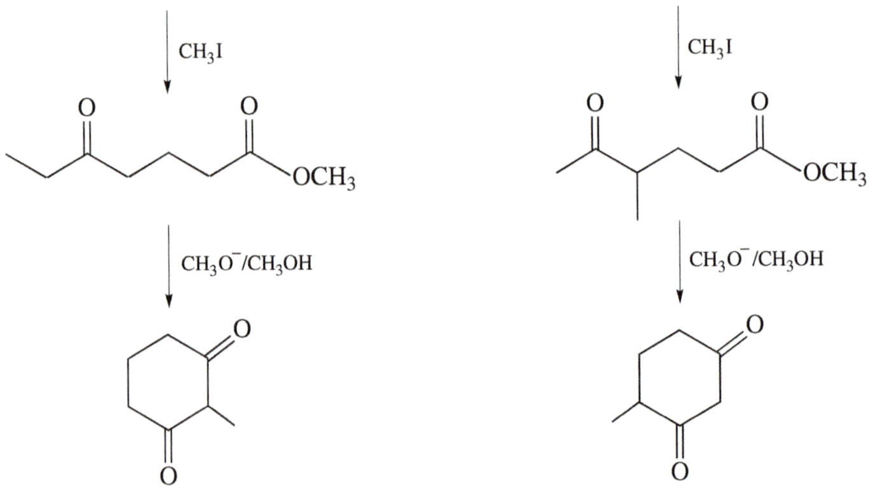

CH_3I

CH_3I

OCH_3

OCH_3

CH_3O^-/CH_3OH

CH_3O^-/CH_3OH

b.

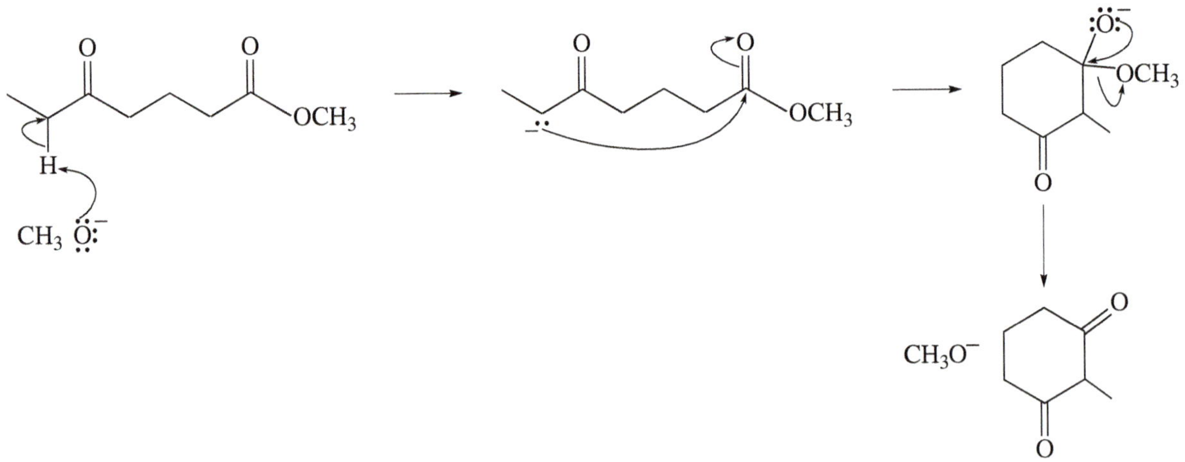

OCH_3

H

$CH_3\ddot{O}:^-$

OCH_3

OCH_3

CH_3O^-

88.

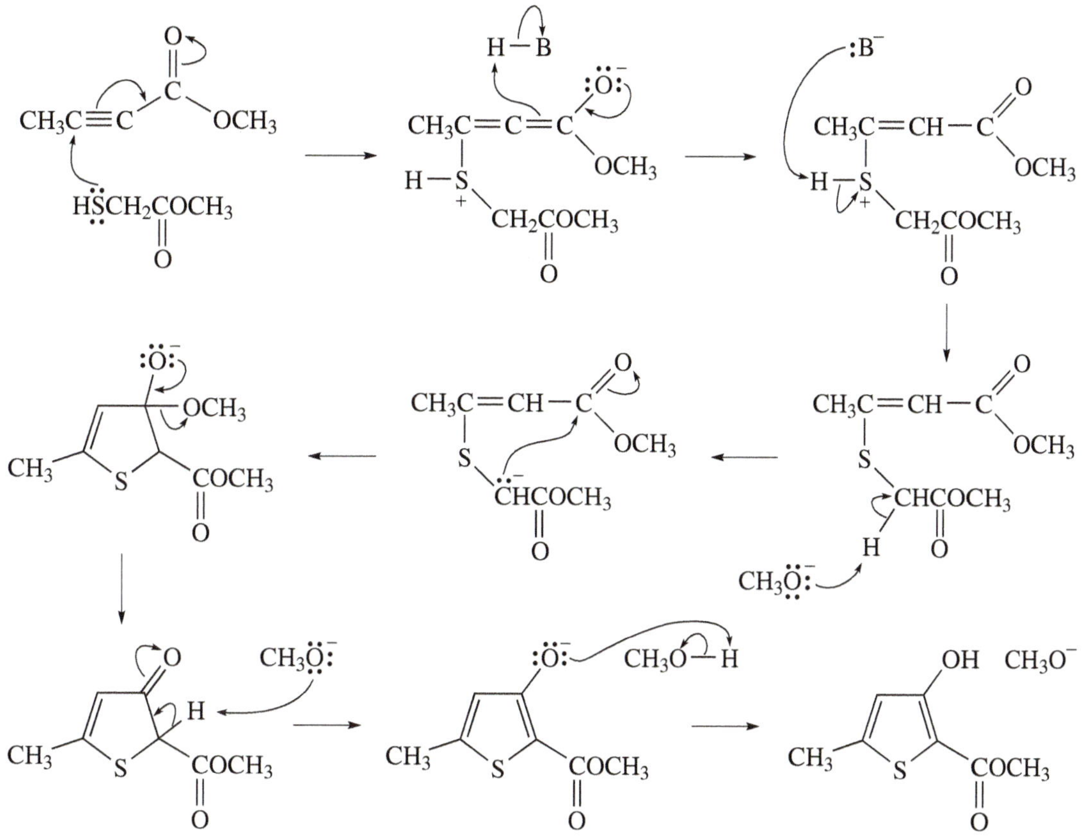

$CH_3C{\equiv}C-C(=O)OCH_3$ + $HSCH_2COCH_3$ (with C=O) → $CH_3C(=C=C(O^-)OCH_3)-S^+(H)CH_2COCH_3$ (with C=O); H—B → $CH_3C=CH-C(=O)OCH_3$, $H-S^+-CH_2COCH_3$ (with C=O); :B$^-$ ↓

$CH_3C=CH-C(=O)OCH_3$, S—$CHCOCH_3$ (with C=O; H); CH_3O^- → $CH_3C=CH-C(=O)OCH_3$, S—$\ddot{C}HCOCH_3$ (with C=O) → CH_3, S, O^-, OCH_3, $COCH_3$ (with C=O) ↓

CH_3, S, O, H, $COCH_3$ (with C=O); CH_3O^- → CH_3, S, O^-, $COCH_3$ (with C=O); CH_3O-H → CH_3, S, OH, $COCH_3$ (with C=O) + CH_3O^-

89.

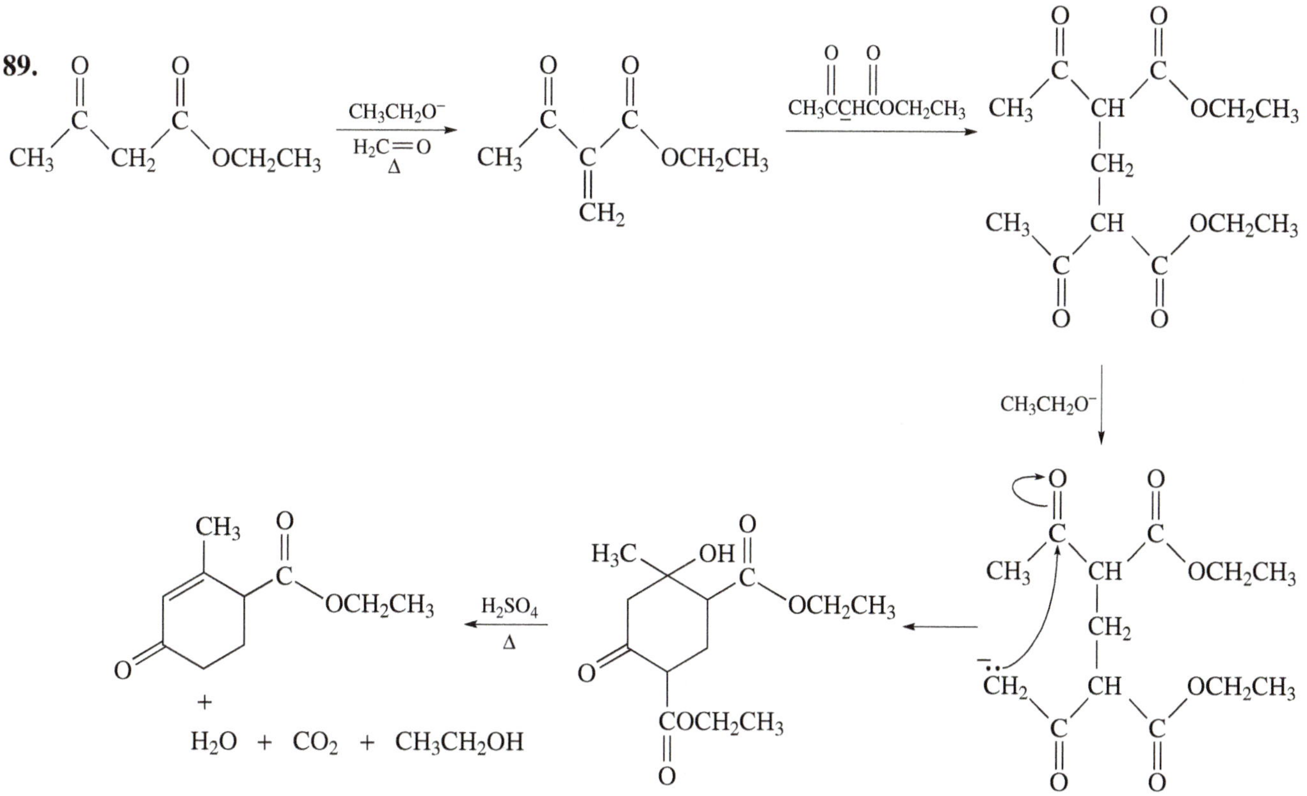

$CH_3C(=O)CH_2C(=O)OCH_2CH_3$ —($CH_3CH_2O^-$; $H_2C{=}O$; Δ)→ $CH_3C(=O)C(=CH_2)C(=O)OCH_2CH_3$ —($CH_3C(=O)CHC(=O)OCH_2CH_3$)→ $CH_3C(=O)CH(C(=O)OCH_2CH_3)-CH_2-CH(C(=O)CH_3)C(=O)OCH_2CH_3$

↓ $CH_3CH_2O^-$

$CH_3C(=O)CH(C(=O)OCH_2CH_3)-CH_2-CH(C(=O)\ddot{C}H_2^-)C(=O)OCH_2CH_3$ → H_3C, OH, O, $C(=O)OCH_2CH_3$, $COCH_2CH_3$ (with C=O) —(H_2SO_4, Δ)→ CH_3, O, $C(=O)OCH_2CH_3$

+

H_2O + CO_2 + CH_3CH_2OH

90. $CH_3CH_2OC(=O)CH_2C(=O)OCH_2CH_3$ $\xrightarrow[\text{2. } CH_3CH_2Br]{\text{1. } CH_3CH_2O^-}$ $CH_3CH_2OC(=O)CH(CH_2CH_3)C(=O)OCH_2CH_3$

$\xrightarrow[\text{2. } CH_3CHCH_2CH_2Br \text{ (with } CH_3 \text{ on CH)}]{\text{1. } CH_3CH_2O^-}$ $CH_3CH_2OC(=O)C(CH_2CH_3)(CH_2CH_2CH(CH_3)CH_3)C(=O)OCH_2CH_3$

$\xrightarrow[\Delta]{H_2NC(=O)NH_2}$ cyclic product (HN, N–H, three C=O; ring carbon bearing $CH_2CH_2CHCH_3$ with CH_3, and CH_2CH_3)

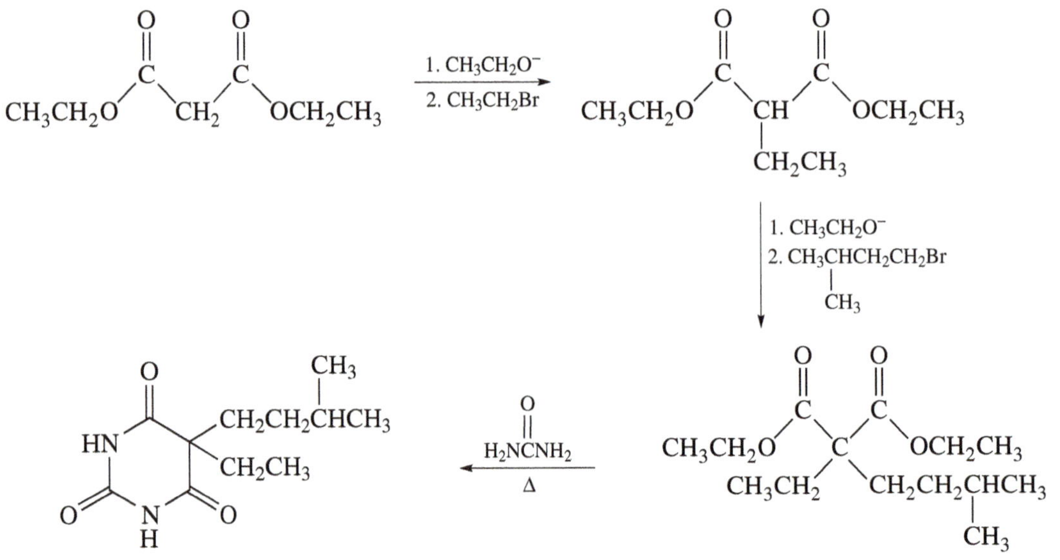

Notice that a diester is converted to a diamide.

91. $CH_3C(=O)CH_2C(=O)OCH_2CH_3$ + $CH_3CH_2\ddot{O}{:}^-$ $\xrightarrow{CH_3CH_2O^-}$ $CH_3C(=O)\bar{C}HC(=O)OCH_2CH_3$ + epoxide bearing CH_2CH_3 $\longrightarrow$ $CH_3C(=O)CH(CH_2CH(O^-)CH_2CH_3)C(=O)OCH_2CH_3$

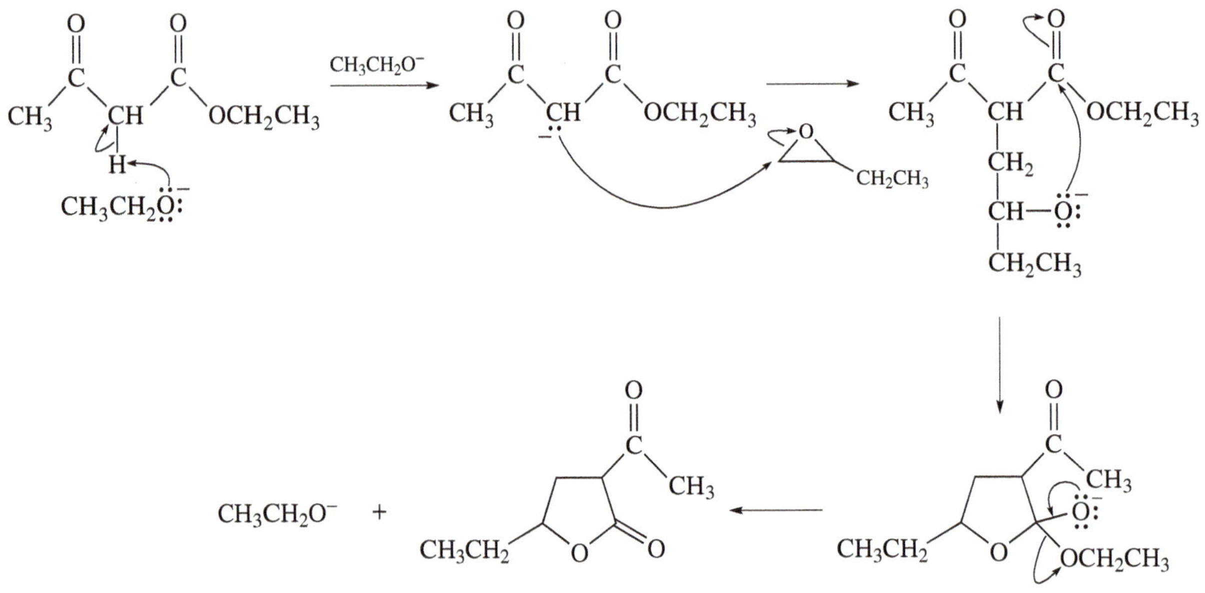

$\longrightarrow$ cyclic tetrahedral intermediate (CH_3CH_2 on ring, ring O, C bearing O^- and OCH_2CH_3, ring carbon bearing $C(=O)CH_3$) $\longrightarrow$ $CH_3CH_2O^-$ + lactone (CH_3CH_2, ring O, C=O, $C(=O)CH_3$)

Chapter17 Practice Test

1. Rank the following compounds from most acidic to least acidic:

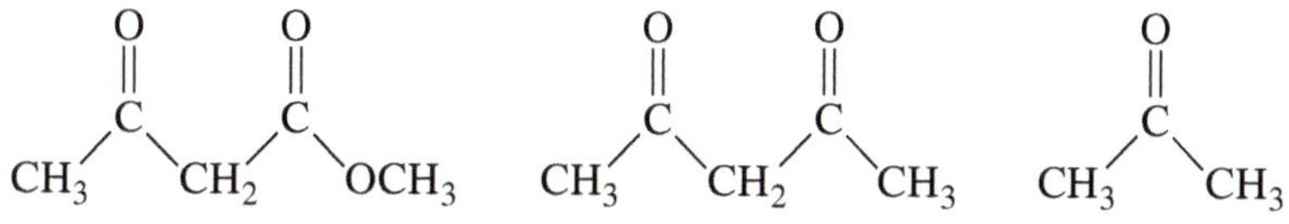

$CH_3C(=O)CH_2C(=O)OCH_3$ $CH_3C(=O)CH_2C(=O)CH_3$ $CH_3C(=O)CH_3$

2. Draw a structure for each of the following:

a. the most stable enol tautomer of 2,4-pentanedione

b. a β-keto ester

3. Draw the product of the following reactions:

a. $CH_3CH_2C(=O)CH_2C(=O)OH \xrightarrow{\Delta}$

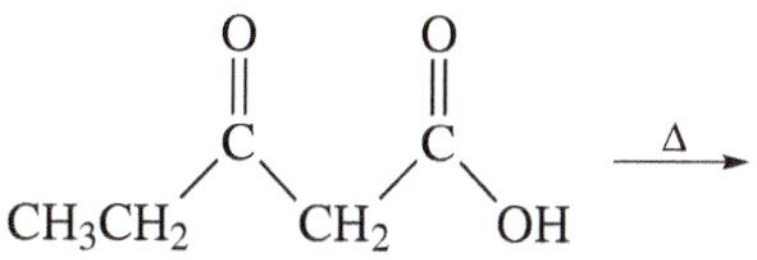

b. $CH_3CH_2CH_2C(=O)H \xrightarrow{Br_2,\ HO^-}$

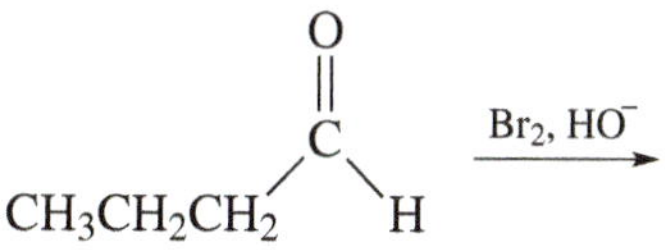

c. $CH_3CH_2CH_2C(=O)OCH_3 \xrightarrow[2.\ HCl]{1.\ CH_3O^-}$

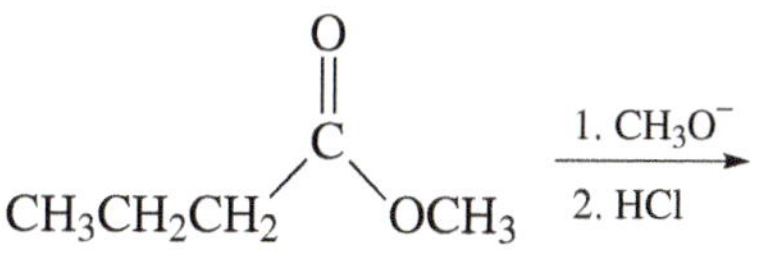

d. $CH_2{=}CHC(=O)CH_3 + CH_3C(=O)CH_2C(=O)CH_3 \xrightarrow[H_2O]{HO^-}$

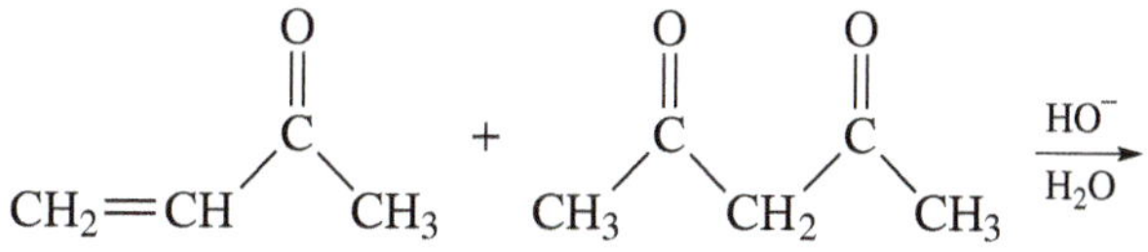

e. cyclopentanone $+ Br_2 \xrightarrow{H_3O^+}$

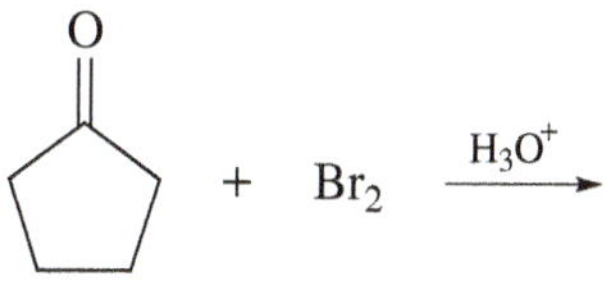

f. cyclohexanone $+ Cl_2 \xrightarrow{HO^-}$

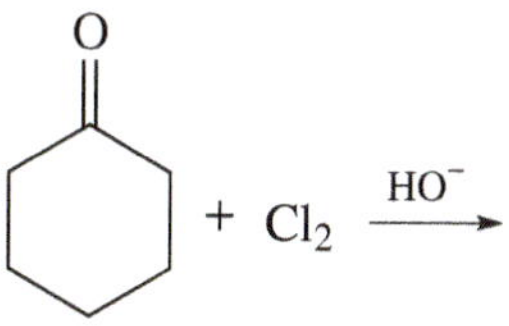

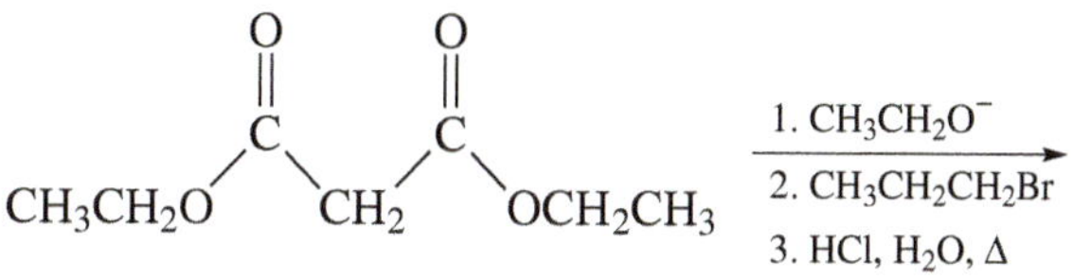

g. $CH_3CH_2OC(=O)CH_2C(=O)OCH_2CH_3$ $\xrightarrow{\text{1. } CH_3CH_2O^- \text{ 2. } CH_3CH_2CH_2Br \text{ 3. } HCl,\ H_2O,\ \Delta}$

4. Give an example of each of the following:

a. an aldol addition

b. an aldol condensation

c. a Claisen condensation

d. a Dieckmann condensation

e. a malonic ester synthesis

f. an acetoacetic ester synthesis

5. Draw the products of the following crossed aldol addition:

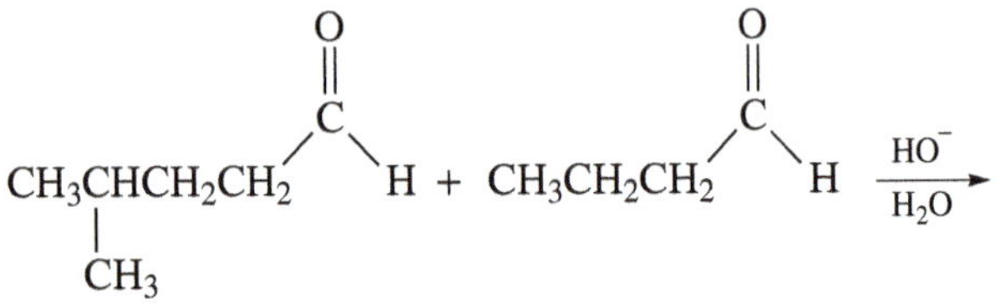

$$CH_3CH(CH_3)CH_2CH_2C(=O)H + CH_3CH_2CH_2C(=O)H \xrightarrow[H_2O]{HO^-}$$

6. What ester is required to prepare each β-ketoester?

a. $CH_3CH_2CH_2C(=O)CH(CH_2CH_3)C(=O)OCH_3$

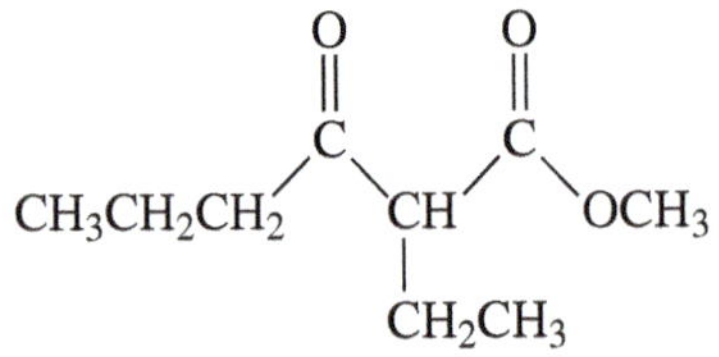

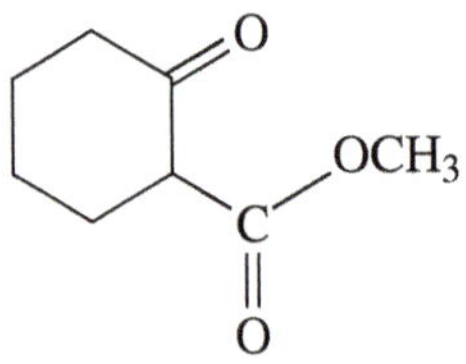

b. (2-oxocyclohexane with $C(=O)OCH_3$ substituent)

CHAPTER 18
Reactions of Benzene and Substituted Benzenes

Important Terms

activating substituent	a substituent that increases the reactivity of an aromatic ring. Electron-donating substituents activate aromatic rings toward electrophilic attack, and electron-withdrawing substituents activate aromatic rings toward nucleophilic attack.
arenediazonium salt	$Ar\overset{+}{N}\equiv N\ X^-$
aromatic compound	a cyclic and planar compound with an uninterrupted cloud of electrons containing an odd number of pairs of π electrons.
azo linkage	a $-N{=}N-$ bond.
benzyl group	$C_6H_5-CH_2-$
deactivating substituent	a substituent that decreases the reactivity of an aromatic ring. Electron-withdrawing substituents deactivate aromatic rings toward electrophilic attack, and electron-donating substituents deactivate aromatic rings toward nucleophilic attack.
donate electrons by resonance (resonance electron donation)	donation of electrons through p orbital overlap with neighboring π bonds.
electrophilic aromatic substitution reaction	a reaction in which an electrophile substitutes for a hydrogen of an aromatic ring.
Friedel–Crafts acylation	an electrophilic aromatic substitution reaction that puts an acyl group on an aromatic ring.
Friedel–Crafts alkylation	an electrophilic aromatic substitution reaction that puts an alkyl group on an aromatic ring.
Gatterman–Koch reaction	a reaction that uses a high-pressure mixture of carbon monoxide and HCl to form benzaldehyde.
halogenation	reaction with a halogen (Br_2, Cl_2, I_2).
inductive electron withdrawal	withdrawal of electrons through a σ bond.

Meisenheimer complex a resonance-stabilized complex formed by addition of a nucleophile to a benzene ring.

meta director a substituent that directs an incoming substituent meta to an existing substituent.

nitration substitution of a nitro group (NO_2) for a hydrogen of an aromatic ring.

nitrosamine (*N*-nitroso compound) an amine with a nitroso (N=O) substituent bonded to its nitrogen atom.

nucleophilic aromatic substitution (S_NAr) reaction a reaction in which a nucleophile substitutes for a halo-substituent on a benzene ring.

ortho-para director a substituent that directs an incoming substituent ortho and para to an existing substituent.

phenyl group C_6H_5—

Sandmeyer reaction the reaction of an arenediazonium ion with a cuprous salt.

Schiemann reaction the reaction of an arenediazonium ion with HBF_4.

S_NAr reaction a nucleophilic aromatic substitution reaction.

sulfonation substitution of a hydrogen of an aromatic ring with a sulfonic acid group (SO_3H).

Suzuki reaction a reaction that replaces the R group of a vinylic or aryl halide with an alkyl, alkenyl, or aryl substituent.

withdraw electrons by resonance (resonance electron withdrawal) withdrawal of electrons through *p* orbital overlap with neighboring π bonds.

Wolff–Kishner reduction a reaction that reduces the carbonyl group of a ketone to a methylene group using NH_2NH_2/HO^-, Δ.

Solutions to Problems

1. a. 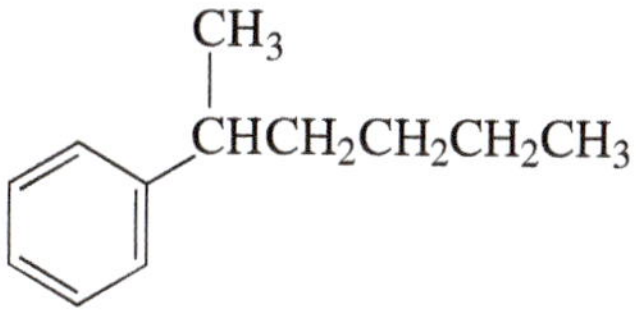 CH_3 over $CHCH_2CH_2CH_2CH_3$ attached to a benzene ring

b. 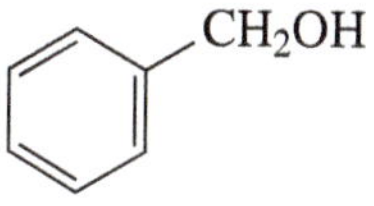 benzene ring–CH_2OH

c. 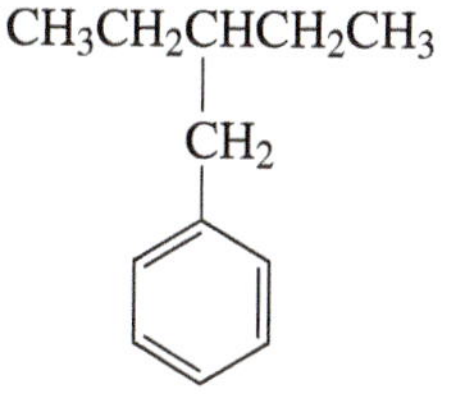 $CH_3CH_2CHCH_2CH_3$ with CH_2–benzene ring on the middle carbon

d. 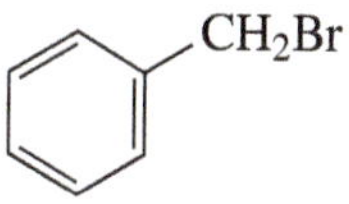 benzene ring–CH_2Br

2. Ferric bromide activates Br_2 for nucleophilic attack by accepting a pair of electrons from it. Hydrated ferric bromide cannot do this because it has already accepted a pair of electrons from water.

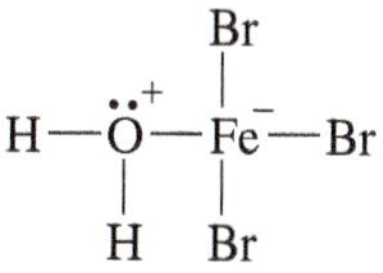

$$H-\overset{+}{\ddot{O}}(H)-\overset{-}{Fe}Br_3$$

3. Solved in the text.

4. 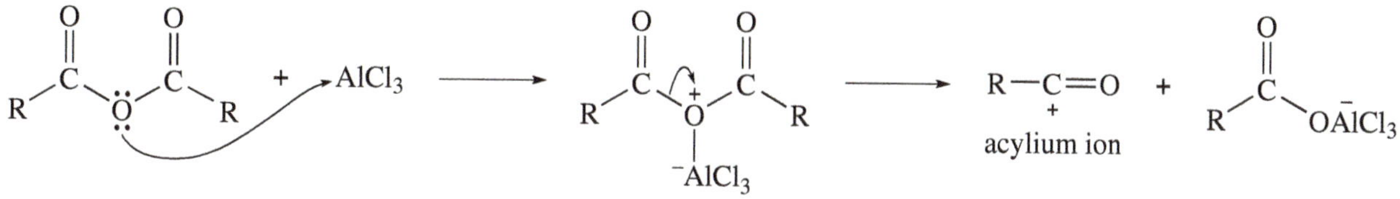 $RC(=O)\ddot{O}C(=O)R + AlCl_3 \longrightarrow RC(=O)\overset{+}{O}(^{-}AlCl_3)C(=O)R \longrightarrow R-\overset{+}{C}=O + RC(=O)O\overset{-}{Al}Cl_3$

acylium ion

5. A carbocation rearrangement occurs in **b.** and **e.**

a. 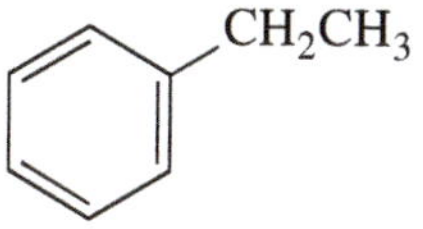 benzene ring–CH_2CH_3

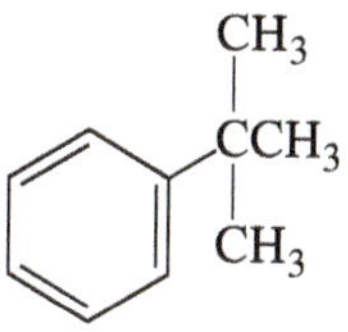

b. 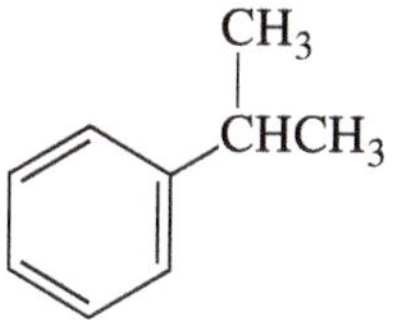 benzene ring–$CHCH_3$ with CH_3

c. benzene ring–$CHCH_2CH_3$ with CH_3

d. 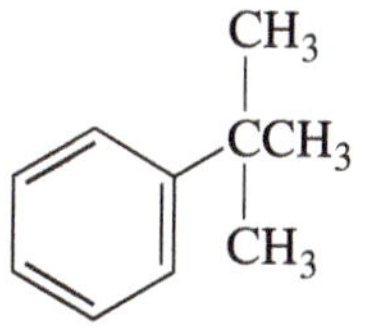 benzene ring–CCH_3 with two CH_3

e. benzene ring–CCH_3 with two CH_3

f. 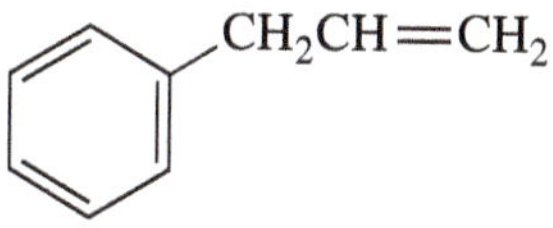 benzene ring–$CH_2CH=CH_2$

6. a. 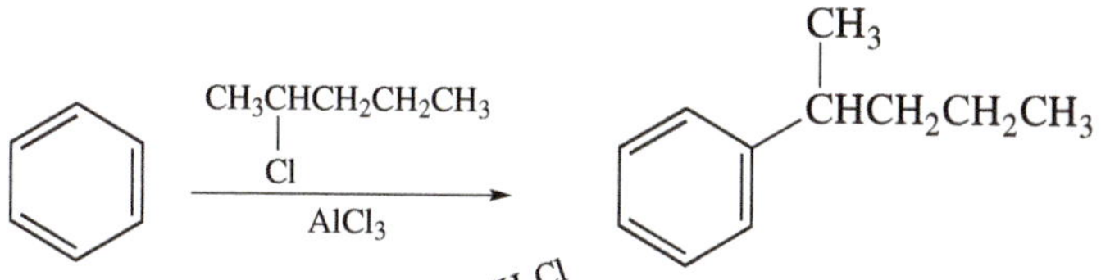 benzene $\xrightarrow[AlCl_3]{CH_3CHClCH_2CH_2CH_3}$ benzene ring–$CH(CH_3)CH_2CH_2CH_3$

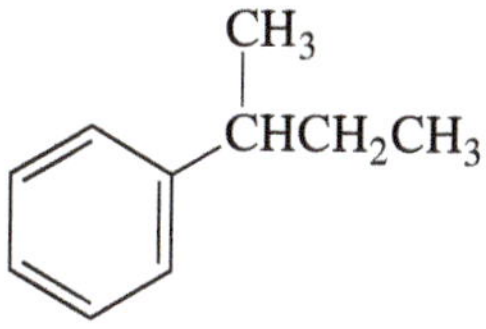

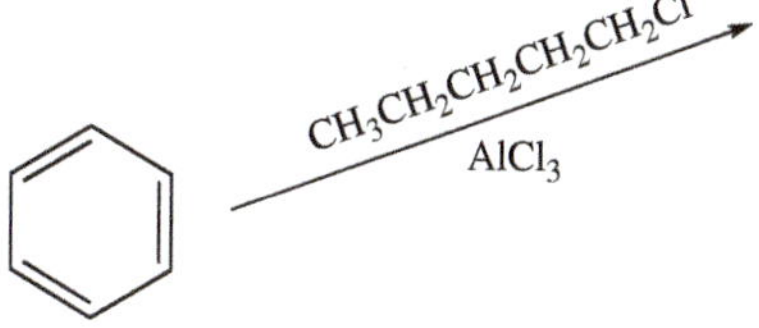

benzene $\xrightarrow[AlCl_3]{CH_3CH_2CH_2CH_2CH_2Cl}$ benzene ring–$CH(CH_3)CH_2CH_2CH_3$

b.

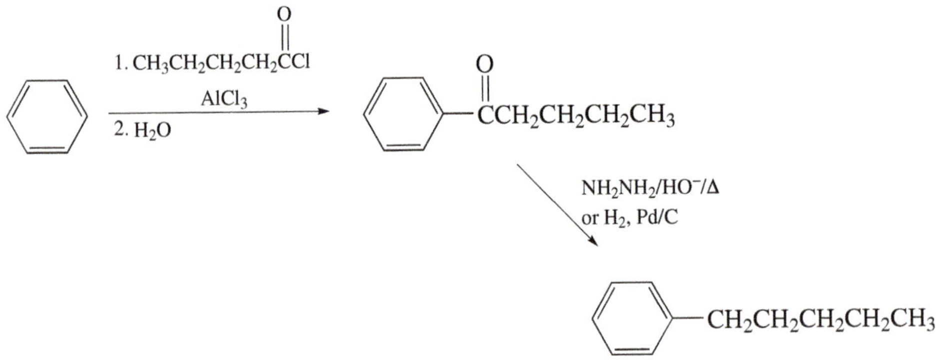

$$C_6H_6 \xrightarrow[\text{2. } H_2O]{\text{1. } CH_3CH_2CH_2CH_2CCl \text{ (O)},\ AlCl_3} C_6H_5\text{—}\overset{O}{\overset{\|}{C}}CH_2CH_2CH_2CH_3$$

$$\xrightarrow{NH_2NH_2/HO^-/\Delta \text{ or } H_2, Pd/C} C_6H_5\text{—}CH_2CH_2CH_2CH_2CH_3$$

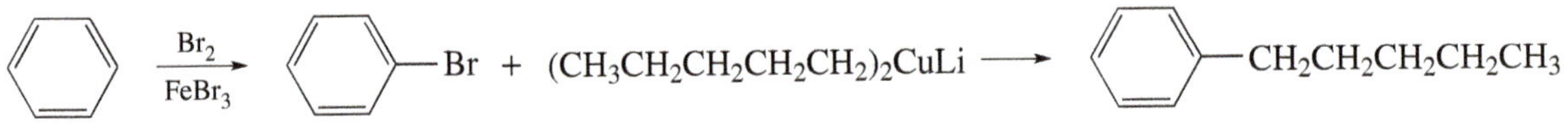

or

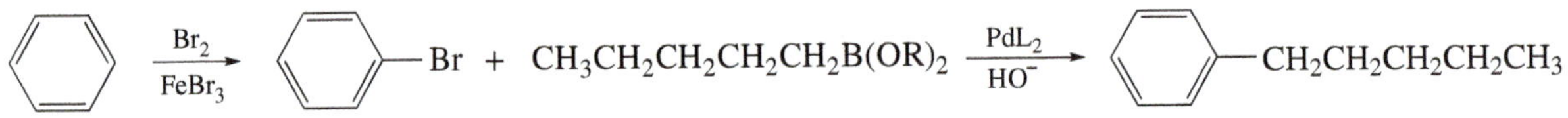

$$C_6H_6 \xrightarrow[FeBr_3]{Br_2} C_6H_5\text{—}Br + (CH_3CH_2CH_2CH_2CH_2)_2CuLi \longrightarrow C_6H_5\text{—}CH_2CH_2CH_2CH_2CH_3$$

or

$$C_6H_6 \xrightarrow[FeBr_3]{Br_2} C_6H_5\text{—}Br + CH_3CH_2CH_2CH_2CH_2B(OR)_2 \xrightarrow[HO^-]{PdL_2} C_6H_5\text{—}CH_2CH_2CH_2CH_2CH_3$$

7.

a. 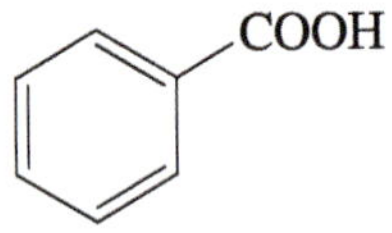 $C_6H_5\text{—}COOH$

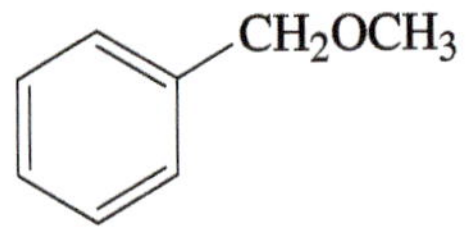

b. $C_6H_4(COOH)_2$ (COOH groups meta)

c. $C_6H_5\text{—}CH_2OCH_3$

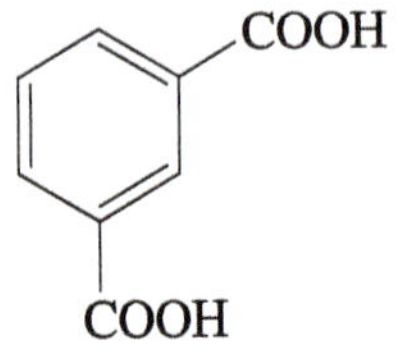

d. $C_6H_5\text{—}CH_2CH_2NH_2$

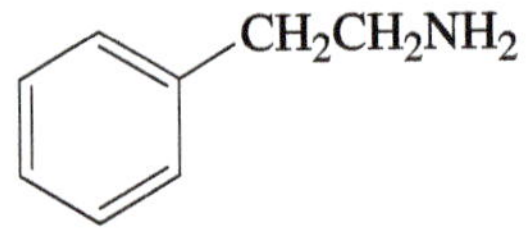

8.

a. Solved in the text.

b.

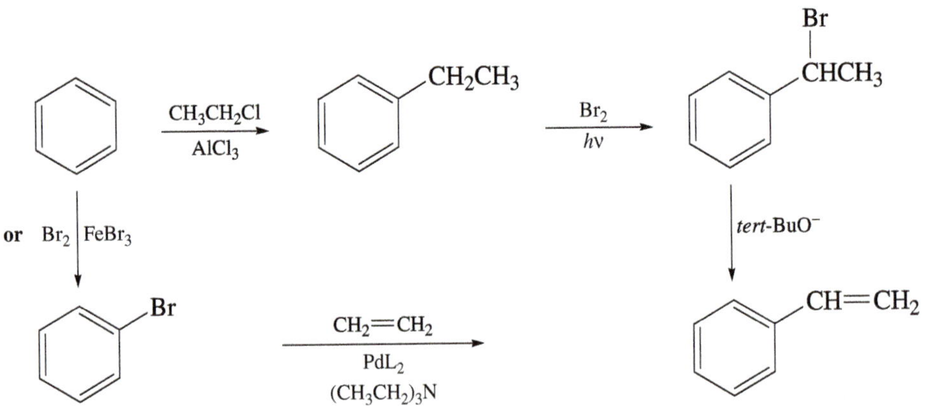

$$C_6H_6 \xrightarrow[AlCl_3]{CH_3CH_2Cl} C_6H_5\text{—}CH_2CH_3 \xrightarrow[h\nu]{Br_2} C_6H_5\text{—}\overset{Br}{\overset{|}{C}}HCH_3 \xrightarrow{tert\text{-}BuO^-} C_6H_5\text{—}CH{=}CH_2$$

or

$$C_6H_6 \xrightarrow{Br_2,\ FeBr_3} C_6H_5\text{—}Br \xrightarrow[PdL_2,\ (CH_3CH_2)_3N]{CH_2{=}CH_2} C_6H_5\text{—}CH{=}CH_2$$

c.

$$C_6H_5\text{—}CH{=}CH_2 \xrightarrow[\text{peroxide}]{HBr} C_6H_5\text{—}CH_2CH_2Br$$

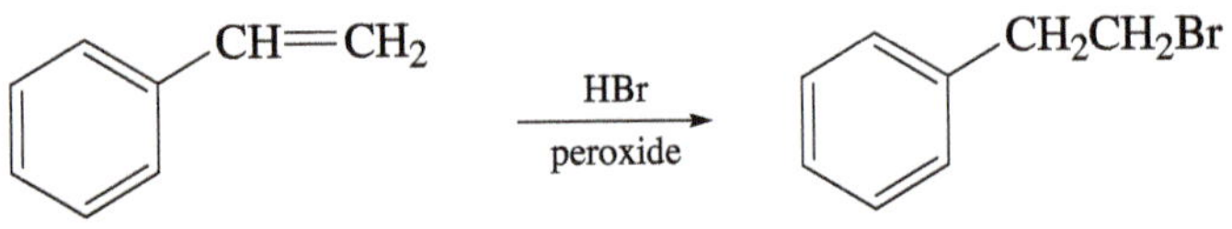

product of **b**

d. $CH{=}CH_2$ (product of **b**) $\xrightarrow[\text{2. HO}^-,\ H_2O_2,\ H_2O]{\text{1. }R_2BH/THF}$ CH_2CH_2OH

e.

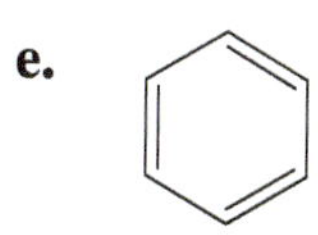

$\xrightarrow[H_2SO_4]{HNO_3}$ NO_2 $\xrightarrow[Pd/C]{H_2}$ NH_2

f.

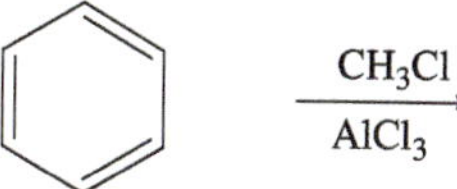

$\xrightarrow[AlCl_3]{CH_3Cl}$ CH_3 $\xrightarrow[\Delta]{H_2CrO_4}$ COOH

9.
a. *ortho*-ethylphenol or 2-ethylphenol
b. *meta*-bromochlorobenzene or 1-bromo-3-chlorobenzene
c. *meta*-bromobenzaldehyde or 3-bromobenzaldehyde
d. *ortho*-ethylmethylbenzene or 1-ethyl-2-methylbenzene

10.

a.

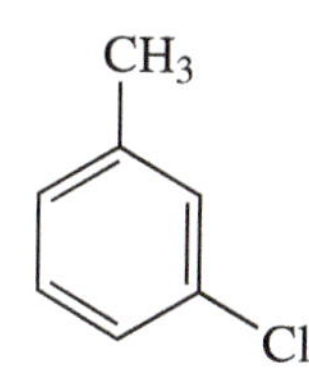

c. NH_2, NO_2

e. Cl, Cl

b.

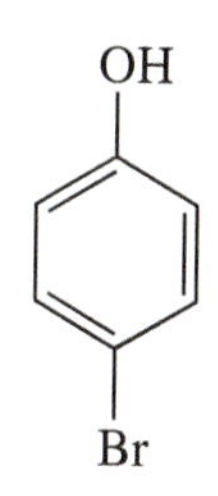

d.

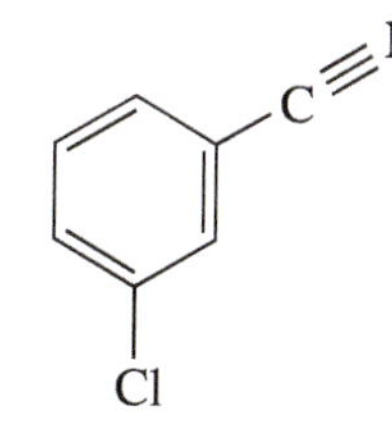

f. CH_3, CH_3

11. **a.** **b.** 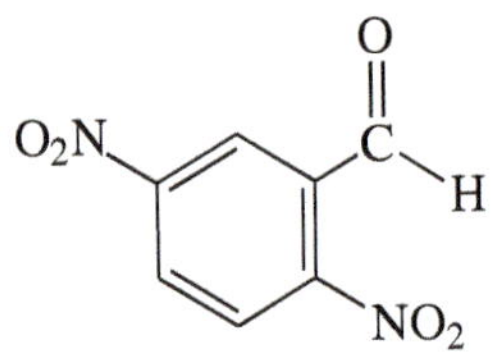 **c.**

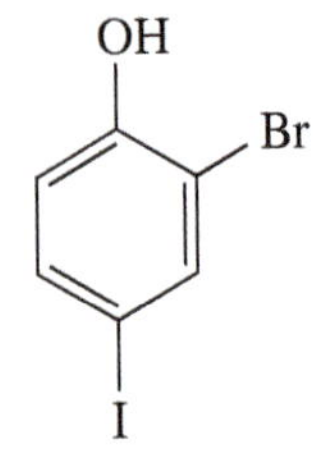

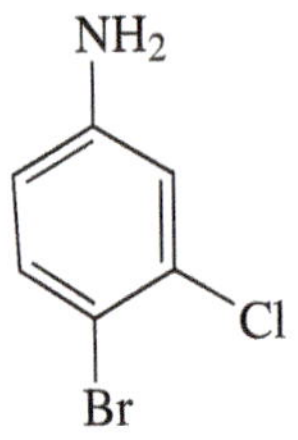

12.
a. 1,3,5-tribromobenzene
b. *meta*-nitrophenol or 3-nitrophenol
c. *para*-bromomethylbenzene or 1-bromo-4-methylbenzene
d. *ortho*-dichlorobenzene or 1,2-dichlorobenzene
e. *meta*-bromomethylbenzene or 1-bromo-3-methylbenzene
f. 2-ethyl-4-iodoaniline

13. Solved in the text.

14. When a halogen is attached to a benzene ring, it can donate electrons by resonance and withdraw electrons inductively. We saw in Problem 13 that fluorobenzene is more reactive than chlorobenzene because fluorine is better at donating electrons by resonance. A halogen attached to a methyl group can only withdraw electrons inductively. Because fluorine is more electronegative than chlorine, it is better at withdrawing electrons as well. Therefore, chloromethylbenzene is more reactive than fluoromethylbenzene.

15.
a. phenol > toluene > benzene > bromobenzene > nitrobenzene
b. toluene > chloromethylbenzene > dichloromethylbenzene > difluoromethylbenzene

16. **a.** **b.**

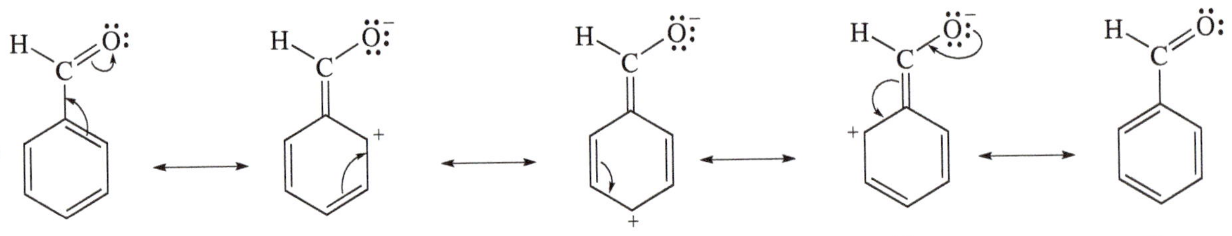

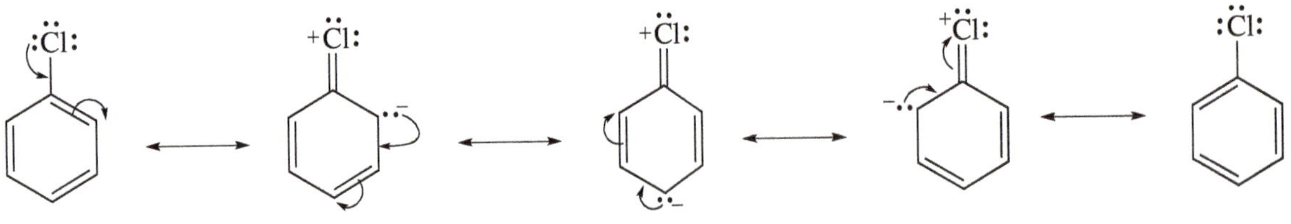

17. **a.** (p-nitropropylbenzene, $CH_2CH_2CH_3$ / NO_2) + (o-nitropropylbenzene, $CH_2CH_2CH_3$ / NO_2)

c. (m-nitrobenzaldehyde: $C(=O)H$ / NO_2)

e. (m-nitrobenzenesulfonic acid: SO_3H / NO_2)

b. (p-bromonitrobenzene: Br / NO_2) + (o-bromonitrobenzene: Br / NO_2)

d. (m-nitrobenzonitrile: $C\equiv N$ / NO_2)

f. (p-nitrocyclohexylbenzene / NO_2) + (o-nitrocyclohexylbenzene / NO_2)

18. They are all meta directors:

a. This group withdraws electrons by resonance from the ring. The relatively electronegative nitrogen atom causes it to also withdraw electrons inductively from the ring.

$$C_6H_5-CH{=}CH-C{\equiv}N$$

b. NO_2 withdraws electrons inductively and withdraws electrons by resonance from the ring.

c. CH_2OH withdraws electrons inductively from the ring.

d. COOH withdraws electrons inductively and withdraws electrons by resonance from the ring.

$$C_6H_5-C(=O)-OH$$

e. CF_3 withdraws electrons inductively from the ring.

f. N=O withdraws electrons inductively and withdraws electrons by resonance from the ring.
You can draw resonance contributors for electron donation into the ring by resonance. However, the most stable resonance contributors are obtained by electron flow out of the benzene ring toward oxygen because oxygen is the most electronegative atom in the compound.

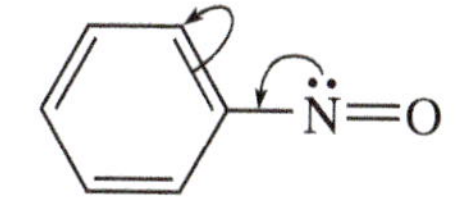

resonance electron donation into the ring

$$C_6H_5-\ddot{N}{=}O$$

resonance electron withdrawal out of the ring

19. Solved in the text.

20. For each compound, determine which benzene ring is more highly activated. The more highly activated ring is the one that undergoes electrophilic aromatic substitution.

a. Because the left-hand ring is weakly activated and the right-hand ring is strongly activated, electrophilic aromatic substitution occurs on the right-hand ring. Because a catalyst is employed, monobrominted and dibrominated compounds can be obtained. However, because there is not enough bromine to brominate all the reactive positions, a mixture of unreacted starting material and the two products shown below will be obtained.

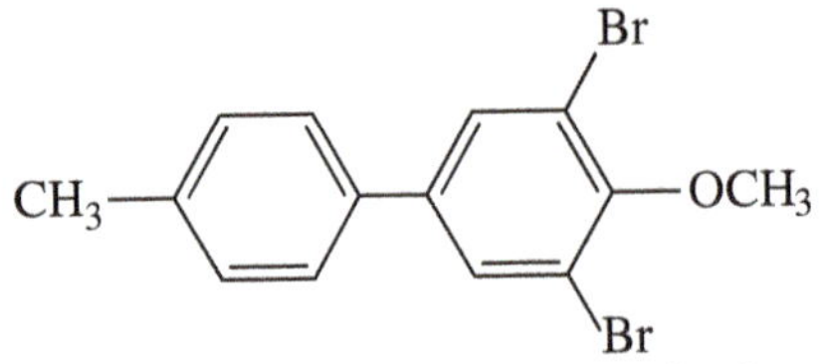

Br

CH_3 — — OCH_3 CH_3 — — OCH_3

Br Br

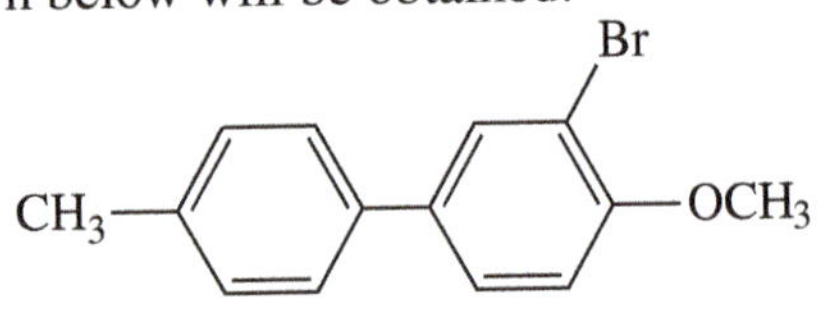

b. Because the left-hand ring is activated and the right-hand ring is deactivated, electrophilic aromatic substitution occurs on the left-hand ring.

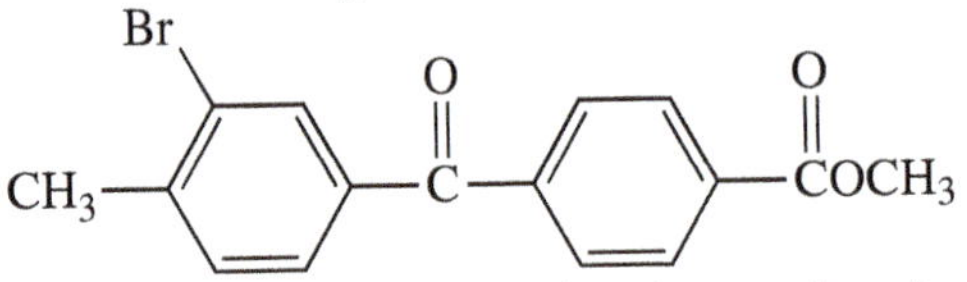

Br O O

CH_3 — C — $COCH_3$

c. Because the right-hand ring is strongly activated and a catalyst is employed, monobrominated, dibrominated, and tribrominated compounds can be obtained. However, because only one equivalent of Br_2 is available, there is not enough bromine to brominate all the reactive positions. Therefore, a mixture of unreacted starting material and the four products shown below will be obtained.

Br

CH_2O — Br CH_2O —

Br

CH_2O — Br CH_2O — Br

Br Br

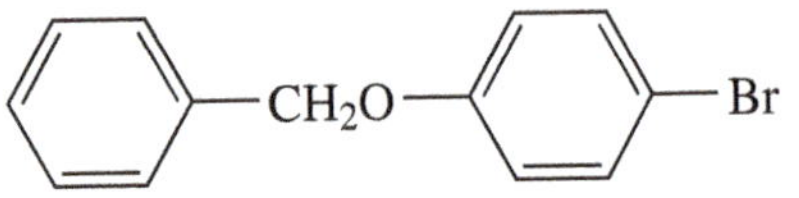

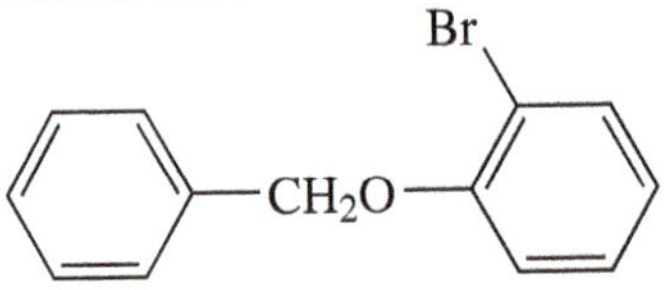

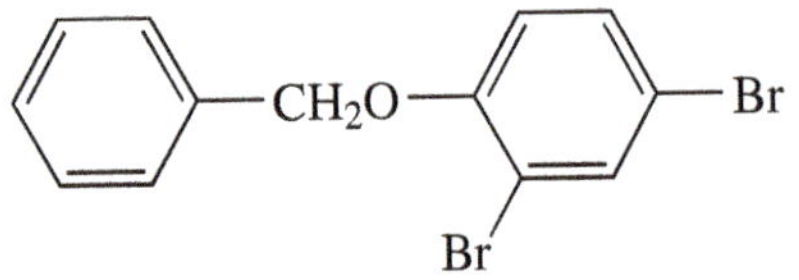

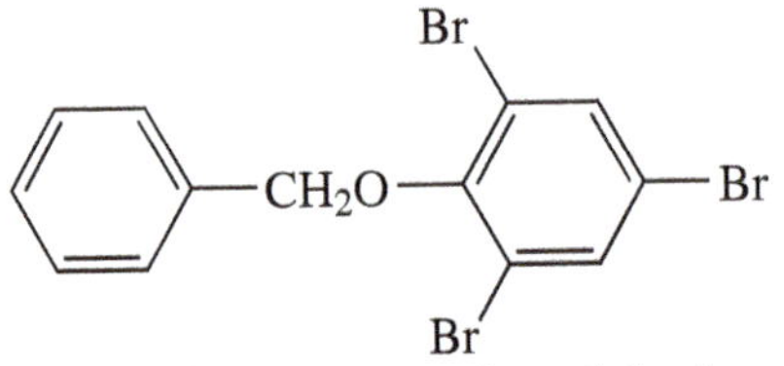

d. Because the left-hand ring is strongly activated and a catalyst is employed, both ortho positions can be brominated. Because not enough Br_2 is available to brominate both reactive positions, a mixture of unreacted starting material and the two products shown below will be obtained.

Br Br

CH_3O — — NO_2 CH_3O — — NO_2

Br

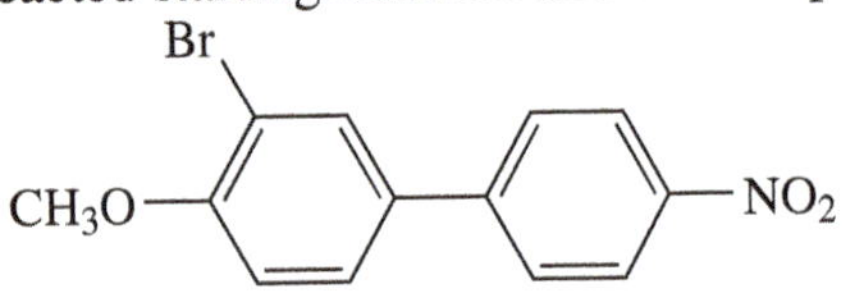

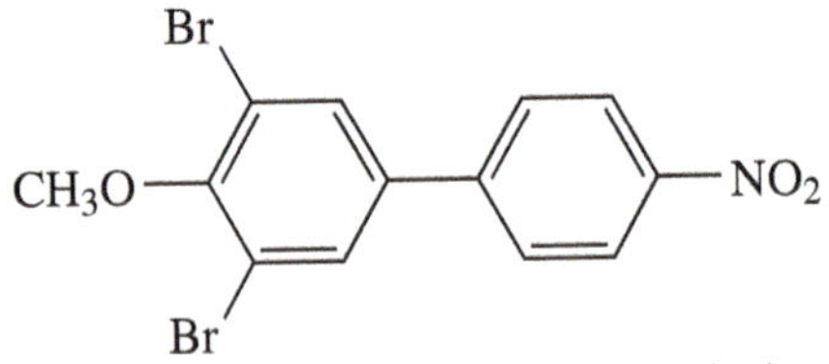

21. No reaction will occur in **a** and **c**, because a Friedel–Crafts reaction cannot be carried out on a ring that possesses a meta director.

a. no reaction

b. OH_2 Br + OH_2 Br

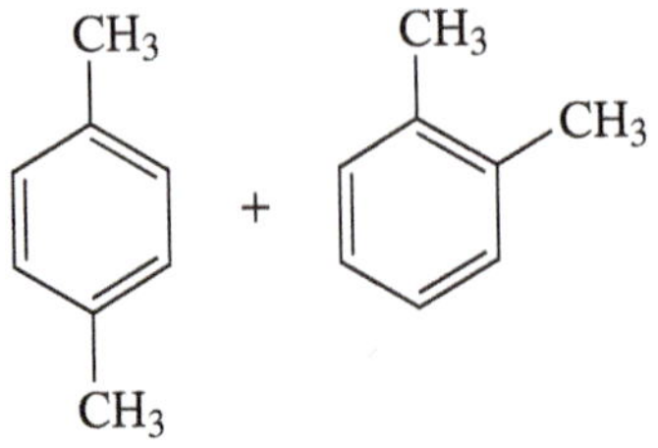

c. no reaction

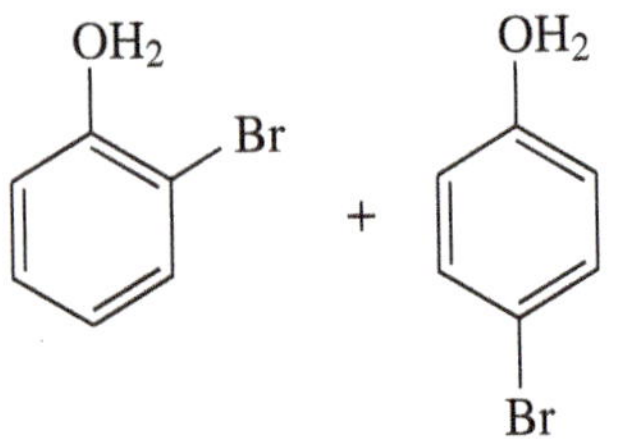

d. CH_3 CH_3 + CH_3 CH_3

22. **a.** Unlike *m*-dipropylbenzene that requires a coupling reaction to be used in its synthesis, a coupling reaction does not have to be used in the synthesis of *p*-dipropylbenzene. The propyl group is an ortho/para director, so propylbenzene can undergo a Friedel–Crafts reaction.

1. $CH_3CH_2\overset{O}{\overset{\|}{C}}Cl/AlCl_3$ 2. H_2O ; CH_2CH_3 ; H_2 Pd/C ; $CH_2CH_2CH_3$

1. $CH_3CH_2\overset{O}{\overset{\|}{C}}Cl/AlCl_3$ 2. H_2O

$CH_2CH_2CH_3$; H_2 Pd/C ; $CH_2CH_2CH_3$; CH_3CH_2 ; $CH_3CH_2CH_2$

b. A coupling reaction can be used in the synthesis of *p*-dipropylbenzene. Notice that bromination occurs after the reduction of the carbonyl group. In contrast, bromination occurs before the reduction reaction in the synthesis of *m*-dipropylbenzene.

1. $CH_3CH_2\overset{O}{\overset{\|}{C}}Cl/AlCl_3$ 2. H_2O ; CH_2CH_3 ; H_2 Pd/C ; $CH_2CH_2CH_3$

Br_2 $FeBr_3$

$CH_2CH_2CH_3$; $CH_3CH_2CH_2B(OR)_2$ PdL_2, NaOH ; $CH_2CH_2CH_3$; Br ; $CH_3CH_2CH_2$

or

$(CH_3CH_2CH_2)_2CuLi$

23. Notice that in all three syntheses, the Friedel–Crafts reaction has to be done first. Both NO_2 and SO_3H are meta directors, and a Friedel–Crafts reaction cannot be carried out if a meta director is on the ring.

a. 1. $CH_3\overset{O}{\overset{\|}{C}}Cl/AlCl_3$ 2. H_2O ; CH_3 ; HNO_3 H_2SO_4 ; CH_3 ; NO_2

b. Benzene $\xrightarrow[\text{2. } H_2O]{\text{1. } CH_3CH_2\overset{O}{\overset{\|}{C}}Cl/AlCl_3}$ $C_6H_5\overset{O}{\overset{\|}{C}}CH_2CH_3$ $\xrightarrow{H_2SO_4}$ m-(SO_3H)$C_6H_4\overset{O}{\overset{\|}{C}}CH_2CH_3$

c. Benzene $\xrightarrow[AlCl_3]{CH_3Cl}$ $C_6H_5CH_3$ $\xrightarrow[\Delta]{H_2CrO_4}$ C_6H_5COOH $\xrightarrow[H_2SO_4]{HNO_3}$ m-(NO_2)C_6H_4COOH

24.

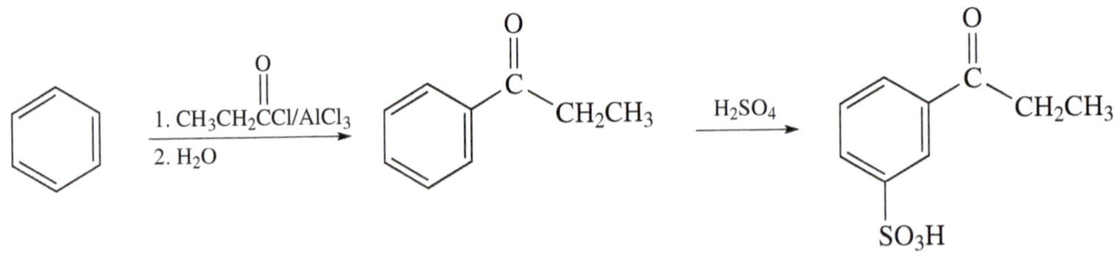

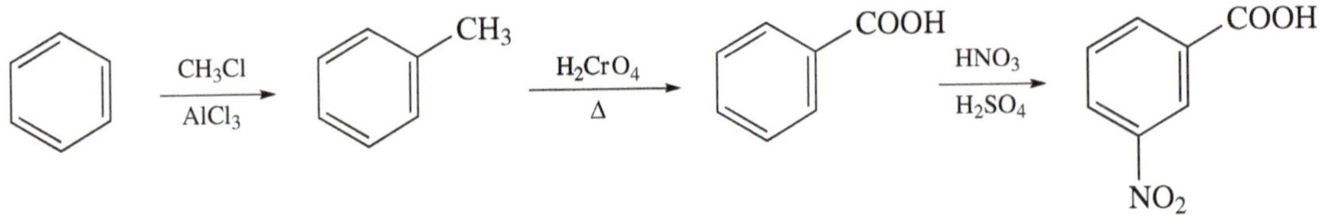

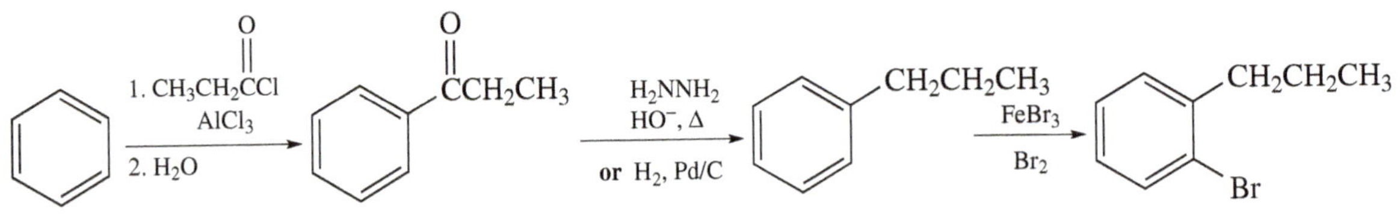

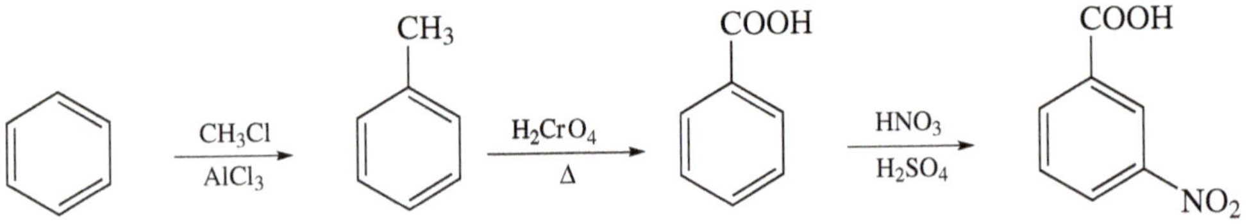

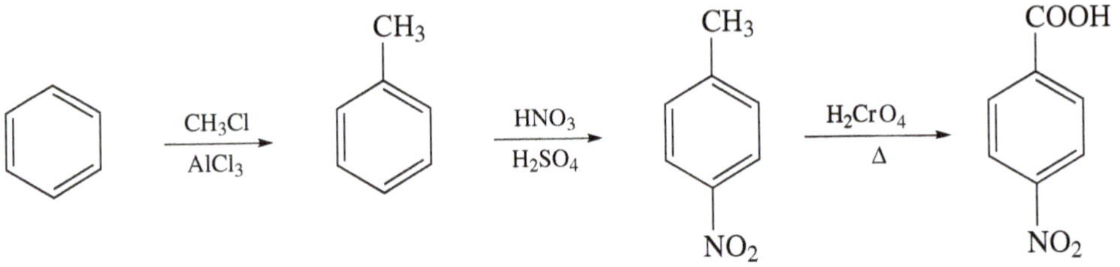

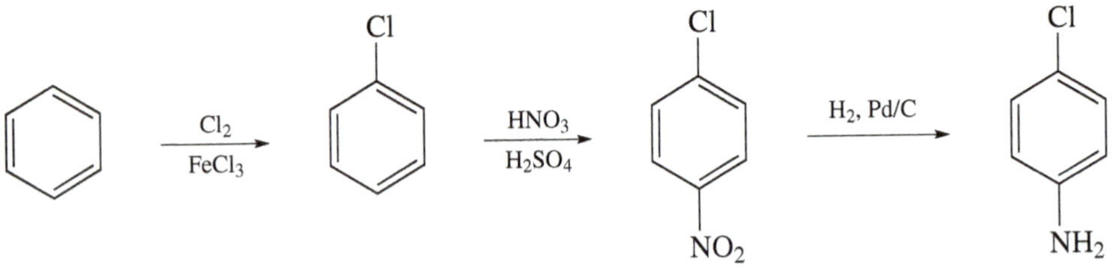

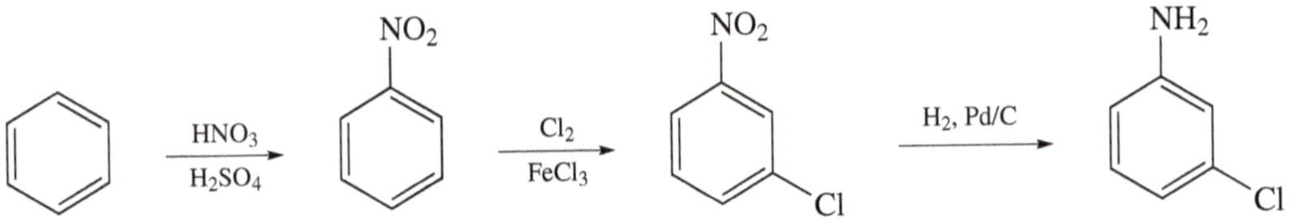

a. Benzene $\xrightarrow[\text{2. } H_2O]{\text{1. } CH_3CH_2\overset{O}{\overset{\|}{C}}Cl \; AlCl_3}$ $C_6H_5\overset{O}{\overset{\|}{C}}CH_2CH_3$ $\xrightarrow[\textbf{or } H_2,\ Pd/C]{H_2NNH_2 \; HO^-,\ \Delta}$ $C_6H_5CH_2CH_2CH_3$ $\xrightarrow[Br_2]{FeBr_3}$ o-Br$C_6H_4CH_2CH_2CH_3$

b. Benzene $\xrightarrow[AlCl_3]{CH_3Cl}$ $C_6H_5CH_3$ $\xrightarrow[\Delta]{H_2CrO_4}$ C_6H_5COOH $\xrightarrow[H_2SO_4]{HNO_3}$ m-(NO_2)C_6H_4COOH

c. Benzene $\xrightarrow[AlCl_3]{CH_3Cl}$ $C_6H_5CH_3$ $\xrightarrow[H_2SO_4]{HNO_3}$ p-(NO_2)$C_6H_4CH_3$ $\xrightarrow[\Delta]{H_2CrO_4}$ p-(NO_2)C_6H_4COOH

d. Benzene $\xrightarrow[FeCl_3]{Cl_2}$ C_6H_5Cl $\xrightarrow[H_2SO_4]{HNO_3}$ p-(NO_2)C_6H_4Cl $\xrightarrow{H_2,\ Pd/C}$ p-(NH_2)C_6H_4Cl

e. Benzene $\xrightarrow[H_2SO_4]{HNO_3}$ $C_6H_5NO_2$ $\xrightarrow[FeCl_3]{Cl_2}$ m-Cl$C_6H_4NO_2$ $\xrightarrow{H_2,\ Pd/C}$ m-Cl$C_6H_4NH_2$

f. Note that benzaldehyde is formed by the Gatterman–Koch reaction (page 877 in the text).

Benzene $\xrightarrow[\text{high pressure, } AlCl_3,\ CuCl]{CO,\ HCl}$ benzaldehyde (C_6H_5CHO) $\xrightarrow[FeBr_3]{Br_2}$ m-bromobenzaldehyde $\xrightarrow[\text{or } H_2,\ Pd/C]{H_2NNH_2,\ HO^-,\ \Delta}$ m-bromotoluene $\xrightarrow{(CH_3)_2CuLi \text{ or } CH_3B(OR)_2,\ PdL_2,\ NaOH}$ m-xylene

g. Benzene $\xrightarrow[AlCl_3]{CH_3CH(CH_3)Cl}$ $C_6H_5CH(CH_3)_2$ $\xrightarrow[\text{peroxide}]{NBS,\ \Delta}$ $C_6H_5CBr(CH_3)_2$ $\xrightarrow{tert\text{-}BuO^-}$ $C_6H_5C(CH_3){=}CH_2$

h. Benzene $\xrightarrow[2.\ H_2O]{1.\ CH_3CH_2COCl,\ AlCl_3}$ $C_6H_5COCH_2CH_3$ $\xrightarrow[FeBr_3]{Br_2}$ m-$BrC_6H_4COCH_2CH_3$ $\xrightarrow[\text{or } H_2,\ Pd/C]{H_2NNH_2,\ HO^-,\ \Delta}$ m-$BrC_6H_4CH_2CH_2CH_3$

i. Notice that in the second step of the synthesis, epoxidation followed by the addition of hydride ion (page 483) is used to add water to the double bond in order to avoid the carbocation rearrangement that would occur with the acid-catalyzed addition of water.

Benzene $\xrightarrow[AlCl_3]{CH_2{=}CHCH_2Cl}$ $C_6H_5CH_2CH{=}CH_2$ $\xrightarrow{MCPBA}$ $C_6H_5CH_2CH{-}CH_2$ (epoxide, O bridging) $\xrightarrow[2.\ HCl]{1.\ NaH}$ $C_6H_5CH_2CH(OH)CH_3$

25. **a.** 4-methylbenzoic acid (CH_3 para to COOH) $\xrightarrow[FeBr_3]{Br_2}$ product with Br meta to COOH and ortho to CH_3

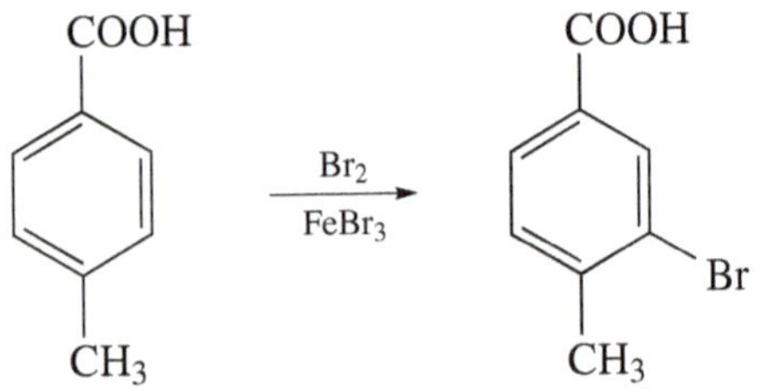

COOH is a meta director, and CH_3 is an ortho-para director, so both direct to the same position.

b. Benzene-1,2-dicarboxylic acid (two adjacent COOH groups) $\xrightarrow[FeCl_3]{Cl_2}$ product with Cl para to the first COOH + product with Cl next to the second COOH

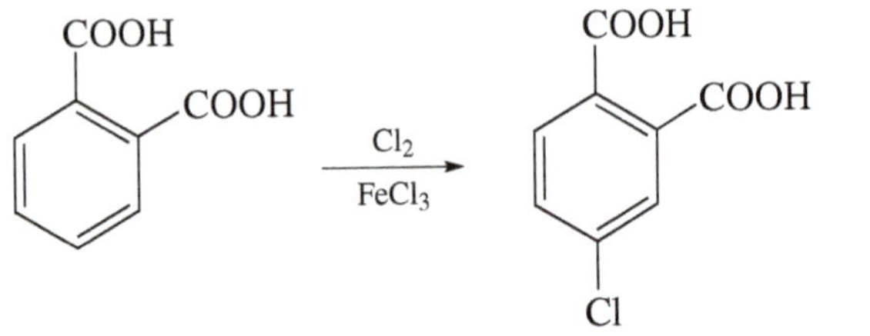

COOH directs to the meta position. The same product will be obtained regardless of which COOH is the director.

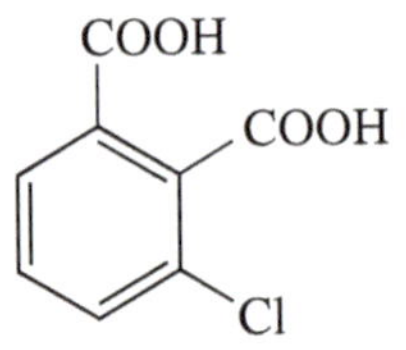

Less of this compound will be obtained because of steric hindrance.

c. 4-chlorobenzoic acid (Cl para to COOH) $\xrightarrow[FeBr_3]{Br_2}$ product with Br meta to COOH and ortho to Cl

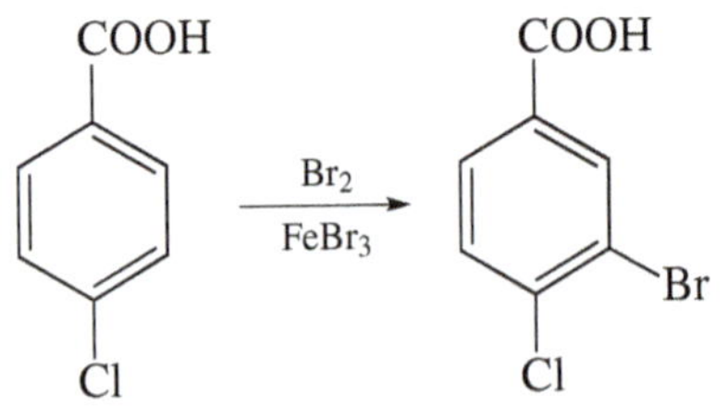

COOH directs to its meta position, and Cl directs to its ortho position, so they both direct to the same position on the ring.

d. 4-fluoroanisole (OCH_3 para to F) $\xrightarrow[H_2SO_4]{HNO_3}$ product with NO_2 ortho to OCH_3

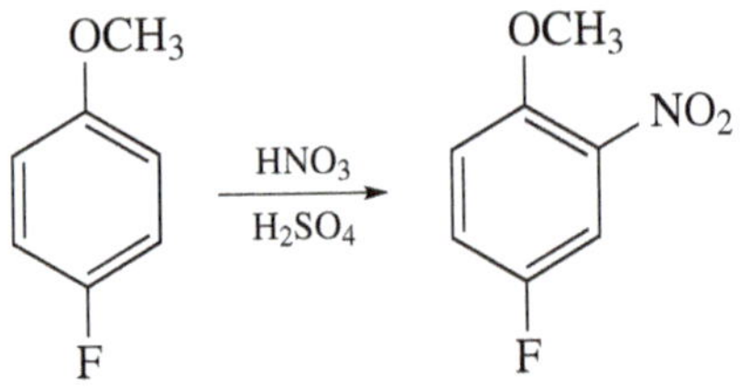

A methoxy substituent is strongly activating, and a fluorine substituent is deactivating, so the methoxy substituent will do the directing.

e. 4-methoxybenzaldehyde (HC=O para to OCH_3) $\xrightarrow[H_2SO_4]{HNO_3}$ product with NO_2 meta to the aldehyde and ortho to OCH_3

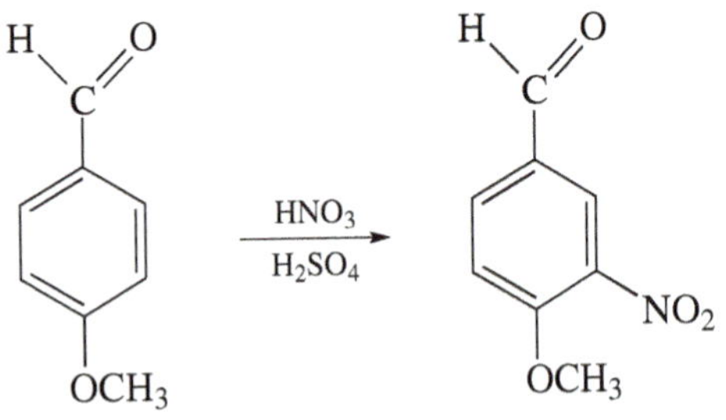

The aldehyde group is a meta director, and the methoxy group is an ortho-para director, so both direct to the same position.

f. 4-tert-butyltoluene (CH_3 para to CH_3CCH_3 with a third CH_3) $\xrightarrow[H_2SO_4]{HNO_3}$ product with NO_2 ortho to CH_3 + product with NO_2 ortho to the tert-butyl group

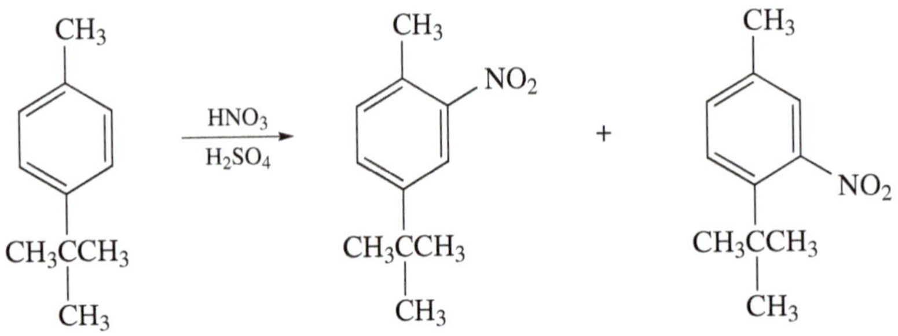

Less of this product will be obtained because of steric hindrance.

26. Yes, the advice is sound.

The para isomer will form one product, because the formyl and ethyl groups both direct to the same positions and both positions result in the same product.

The ortho isomer will form two products, because the formyl and ethyl groups both direct to the same positions but different products are obtained from each position.

The meta isomer will form four products, because the formyl and ethyl groups direct to four different positions and a different product is obtained from each position.

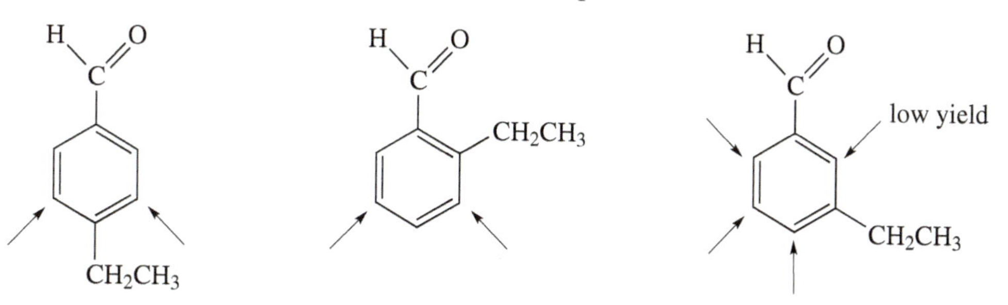

27. Solved in the text.

28. $FeBr_3$ will complex with the amino group, converting it into a meta director. The NH_2 group is a strongly activating substituent, so a Lewis acid is not needed.

$\ddot{N}H_2$ + $FeBr_3$ ⟶ ${}^{+}NH_2$–${}^{-}FeBr_3$

29. Because a diazonium ion is electron withdrawing, it deactivates the benzene ring toward electrophilic aromatic substitution. A deactivated benzene ring is too unreactive to undergo an electrophilic substitution reaction at the cold temperature necessary to keep the benzenediazonium ion from decomposing.

30. benzene $\xrightarrow[H_2SO_4]{HNO_3}$ NO_2 $\xrightarrow{H_2,\ Pd/C}$ NH_2 $\xrightarrow[HCl,\ 0\ °C]{NaNO_2}$ $\overset{+}{N}{\equiv}N$ Cl^-

31. a. benzene $\xrightarrow[AlCl_3]{CH_3Cl}$ CH_3 $\xrightarrow[H_2SO_4]{HNO_3}$ CH_3, O_2N $\xrightarrow[\Delta]{H_2CrO_4}$ $COOH$, O_2N

b.

$$\text{benzene} \xrightarrow[H_2SO_4]{HNO_3} \text{nitrobenzene} \xrightarrow{FeBr_3,\ Br_2} m\text{-bromonitrobenzene} \xrightarrow[Pd/C]{H_2} m\text{-bromoaniline} \xrightarrow[HCl,\ 0\ ^\circ C]{NaNO_2} m\text{-bromobenzenediazonium } Br^- \xrightarrow[Cu(NO_3)_2,\ H_2O \text{ or } H_3O^+,\ \Delta]{Cu_2O} m\text{-bromophenol}$$

c. The first approach is longer, but it will generate a higher yield of the ortho isomer.

$$\text{benzene} \xrightarrow[H_2SO_4]{HNO_3} \text{nitrobenzene} \xrightarrow{H_2,\ Pd/C} \text{aniline} \xrightarrow[HCl,\ 0\ ^\circ C]{NaNO_2} \text{benzenediazonium } Cl^- \xrightarrow[Cu(NO_3)_2,\ H_2O \text{ or } H_3O^+,\ \Delta]{Cu_2O} \text{phenol}$$

$$\text{phenol} \xrightarrow[\Delta]{H_2SO_4} p\text{-}HO_3S\text{-phenol} \xrightarrow[FeCl_3]{Cl_2} \text{2-chloro-4-}HO_3S\text{-phenol} \xrightarrow[\Delta]{H_3O^+} o\text{-chlorophenol}$$

or

$$\text{benzene} \xrightarrow[FeCl_3]{Cl_2} \text{chlorobenzene} \xrightarrow[H_2SO_4]{HNO_3} o\text{-chloronitrobenzene} \xrightarrow{H_2,\ Pd/C} o\text{-chloroaniline} \xrightarrow[HCl,\ 0\ ^\circ C]{NaNO_2} o\text{-chlorobenzenediazonium } Cl^- \xrightarrow[Cu(NO_3)_2,\ H_2O \text{ or } H_3O^+,\ \Delta]{Cu_2O} o\text{-chlorophenol}$$

d. The nitro group cannot be placed on the benzene ring first, because a Friedel–Crafts reaction cannot be carried out on a ring with a meta director. Because formyl chloride is too unstable to be purchased, benzaldehyde is prepared by the Gatterman–Koch reaction (page 884 in the text).

+ CO + HCl $\xrightarrow[\text{high pressure}]{AlCl_3/CuCl}$ C(=O)H $\xrightarrow[H_2SO_4]{HNO_3}$ C(=O)H, NO_2 $\xrightarrow{H_2NNH_2,\ HO^-,\ \Delta}$ CH_3, NO_2

e. $\xrightarrow[AlCl_3]{CH_3Cl}$ CH_3 $\xrightarrow[H_2SO_4]{HNO_3}$ CH_3, O_2N $\xrightarrow{H_2,\ Pd/C}$ CH_3, H_2N $\xrightarrow[HCl]{NaNO_2,\ 0\ °C}$ CH_3, $\overset{+}{N}\equiv N$ Cl^- $\xrightarrow{CuC\equiv N}$ CH_3, $N\equiv C$

f. $\xrightarrow[AlCl_3]{CH_3Cl}$ CH_3 $\xrightarrow[h\nu]{Br_2}$ CH_2Br $\xrightarrow{HO^-}$ CH_2OH $\xrightarrow[CH_3COOH]{NaOCl,\ 0\ °C}$ C(=O)H $\xrightarrow[FeCl_3]{Cl_2}$ C(=O)H, Cl

or

$\xrightarrow[\text{high pressure, } AlCl_3,\ CuCl]{CO,\ HCl}$ C(=O)H $\xrightarrow[FeCl_3]{Cl_2}$ C(=O)H, Cl

32. Because the para position is occupied, the electrophile adds to the ortho position. (Attack at the ortho position is slower than attack at the para position because of steric hindrance, but since the para position is not available, the slower reaction prevails.)

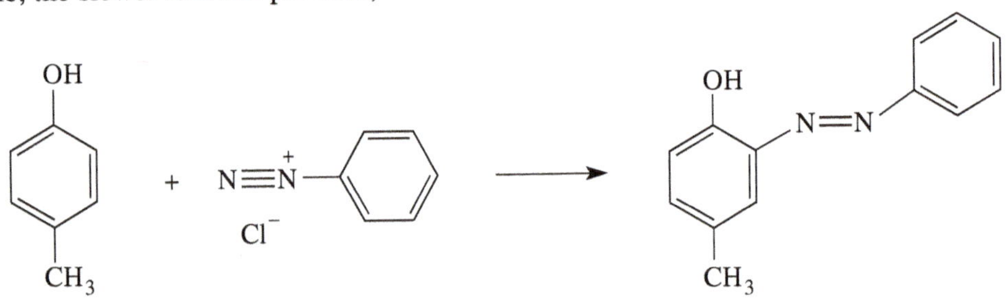

OH, CH_3 + $N\equiv\overset{+}{N}$ Cl^- → OH, N=N, CH_3

33. You can see why nucleophilic attack occurs on the neutral nitrogen if you compare the products of nucleophilic attack on the two nitrogens. Nucleophilic attack on the neutral nitrogen forms a stable product, whereas nucleophilic attack on the positively charged nitrogen would form an unstable compound with two charged nitrogen atoms.

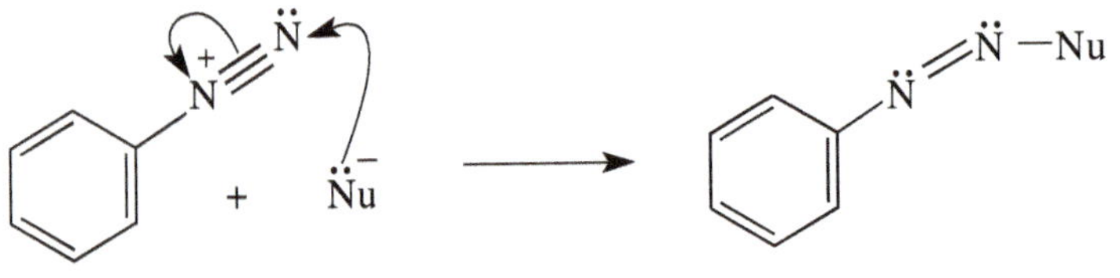

+ $\ddot{N}u^-$ → $\ddot{N}-Nu$

compared to

+ $\ddot{N}u^-$ → Nu

The terminal nitrogen is electrophilic because of electron withdrawal by the positively charged nitrogen. If you draw the resonance contributors, you can see that the "neutral" nitrogen is electron deficient.

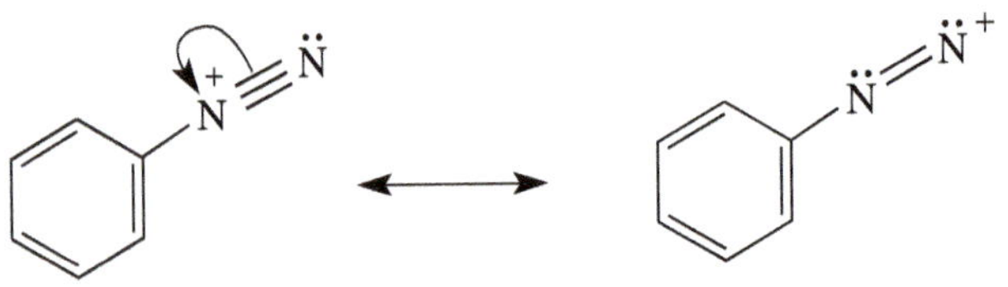

34. **a.** +

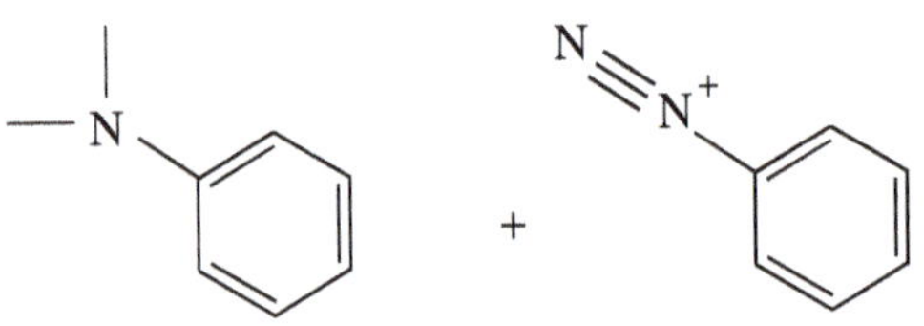

activated ring diazonium ion

b. + SO_3^-

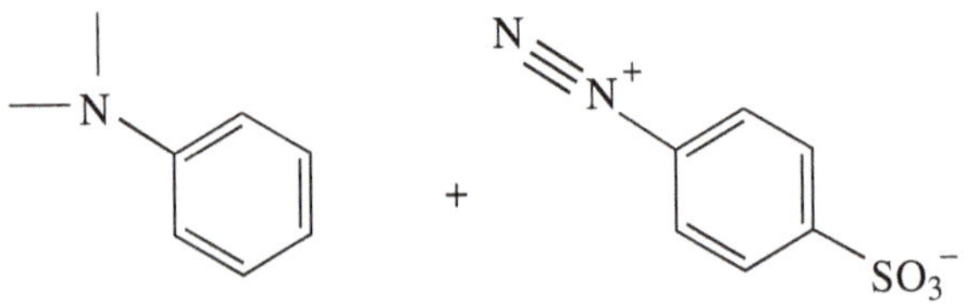

activated ring diazonium ion

35. Immediately after hydrolysis of the amide bond, carbon dioxide is lost. We know then that the indicated amide bond is the one that is hydrolyzed because carbon dioxide can then be lost since the electrons left behind can be delocalized onto the carbonyl oxygen.

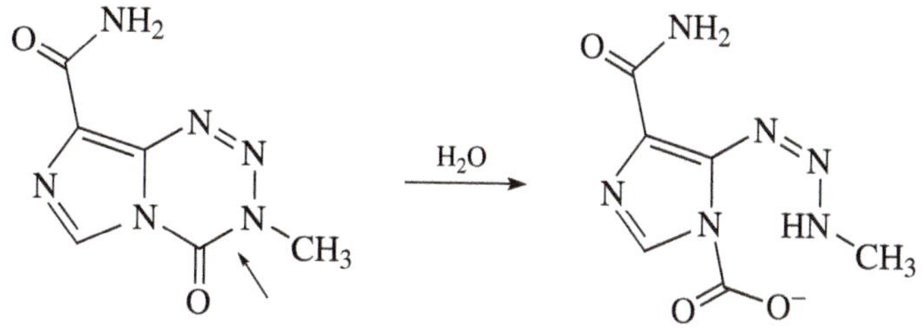

$\xrightarrow{H_2O}$

36. The nitrosamine formed from a secondary amine cannot form a diazonium ion when the nitrogen–nitrogen double bond is formed because the nitrogen cannot lose its positive charge by losing a proton. Therefore, the reaction stops at the nitrosamine.

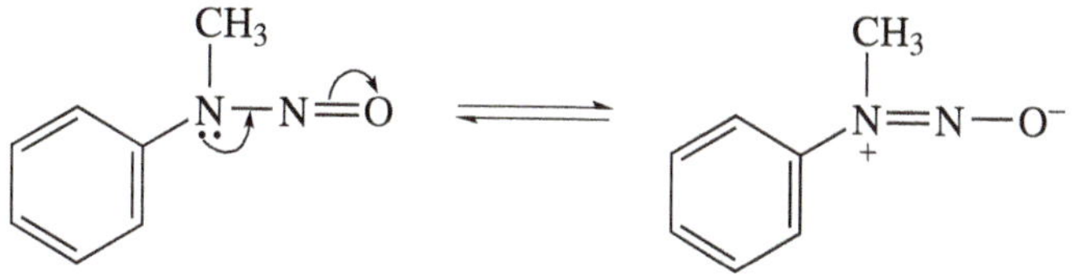

nitrosamine

37. Note: Diazomethane is both explosive and toxic, so it should be synthesized only in small amounts by experienced laboratory workers.

The first step of the reaction is formation of the methyldiazonium ion as a result of removal of a proton from the carboxylic acid by diazomethane.

In the second step of the reaction, the carboxylate ion displaces nitrogen gas (N_2) from the methyldiazonium ion in an S_N2 reaction. High yields are obtained because the only side product is a gas.

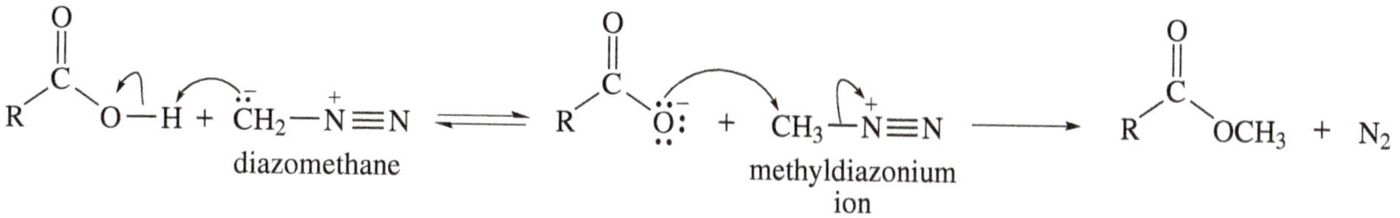

$RCOOH + \ddot{C}H_2{-}\overset{+}{N}{\equiv}N \rightleftharpoons RCOO^- + CH_3{-}\overset{+}{N}{\equiv}N \longrightarrow RCOOCH_3 + N_2$

diazomethane methyldiazonium ion

38. From the resonance contributors, you can see that the reason that *meta*-chloronitrobenzene does not react with hydroxide ion is because the negative charge that is generated on the benzene ring cannot be delocalized onto the nitro substituent.

Electron delocalization onto the nitro substituent can occur only if the nitro substituent is ortho or para to the site of nucleophilic attack.

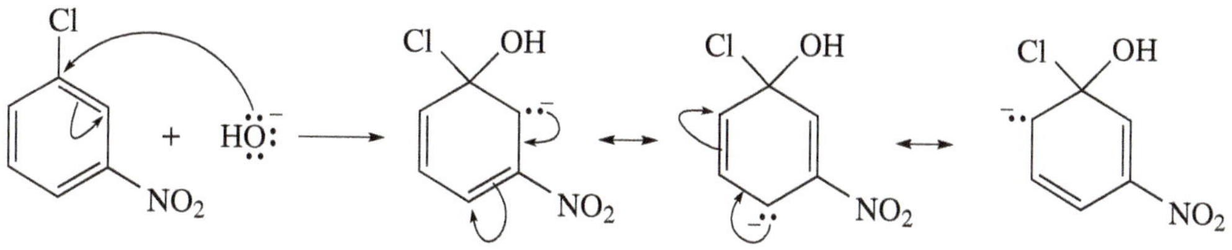

39. **a.** The more nitro substituents ortho and para to the halogen, the faster the rate of nucleophilic aromatic substitution.

1-chloro-2,4-dinitrobenzene > *p*-chloronitrobenzene > chlorobenzene

b. The fewer nitro substituents attached to the benzene ring, the faster the rate of electrophilic aromatic substitution.

chlorobenzene > *p*-chloronitrobenzene > 1-chloro-2,4-dinitrobenzene

40. **a.** benzene $\xrightarrow[FeBr_3]{Br_2}$ bromobenzene (Br) $\xrightarrow[H_2SO_4]{HNO_3}$ (Br, NO_2) $\xrightarrow[\Delta]{HO^-}$ (OH, NO_2)

b. benzene $\xrightarrow[FeBr_3]{Br_2}$ (Br) $\xrightarrow[H_2SO_4]{HNO_3}$ (Br, NO_2) $\xrightarrow[\Delta]{NH_3}$ (NH_2, NO_2)

c. benzene $\xrightarrow[FeBr_3]{Br_2}$ (Br) $\xrightarrow[H_2SO_4]{HNO_3}$ (Br, NO_2) $\xrightarrow[\Delta]{CH_3O^-}$ (OCH_3, NO_2) $\xrightarrow{H_2, Pd/C}$ (OCH_3, NH_2) $\xrightarrow[HCl,\ 0\ °C]{NaNO_2}$ (OCH_3, $N^+{\equiv}N\ Cl^-$) $\xrightarrow{CuBr}$ (OCH_3, Br)

d. This synthesis would be carried out exactly as in **c** (above) except for the second to last step which would be:

(OCH_3, $N^+{\equiv}N\ Cl^-$) $\xrightarrow{H_3PO_2}$ (OCH_3)

41. **a.** $BrCH_2CH_2CH_2CHCH_2CH_3$ (with OH on the CH) $\xrightarrow{NaH}$ (tetrahydrofuran ring with O) CH_2CH_3

b. (benzene ring)$CH_2CH_2CH_2CHCH_3$ (with Cl on the CH) $\xrightarrow{AlCl_3}$ (tetralin with) CH_3

c. In this synthesis of **c** and **d**, a primary alkyl halide cannot be used in the first step because the carbocation will rearrange to a secondary carbocation, which will result in the formation of a methyl-substituted five-membered ring. A primary allyl halide can be used because the initially formed allyl cation is stabilized by electron delocalization.

(benzene ring)$CH_2CH{=}CHCH_2Cl$ $\xrightarrow{AlCl_3}$ $\xrightarrow[\text{peroxide}]{NBS,\ \Delta}$ Br $\xrightarrow{H_2,\ Pd/C}$ Br $\xrightarrow{\textit{tert}\text{-}BuO^-}$

d.

(benzene ring)$CH_2CH{=}CHCH_2Cl$ $\xrightarrow{AlCl_3}$ $\xrightarrow[\text{peroxide}]{NBS,\ \Delta}$ Br $\xrightarrow{H_2,\ Pd/C}$ Br

Br $\xrightarrow{^-C{\equiv}N}$ $C{\equiv}N$ $\xrightarrow[\Delta]{H_2O,\ HCl}$ COOH

Br $\xrightarrow[Et_2O]{Mg}$ MgBr $\xrightarrow[2.\ HCl]{1.\ CO_2}$ COOH

e. $BrCH_2CH_2CH_2CHCH_2CH{=}CH_2$ (with OH on the CH) $\xrightarrow{NaH}$ (tetrahydrofuran ring with O) $CH_2CH{=}CH_2$

f. 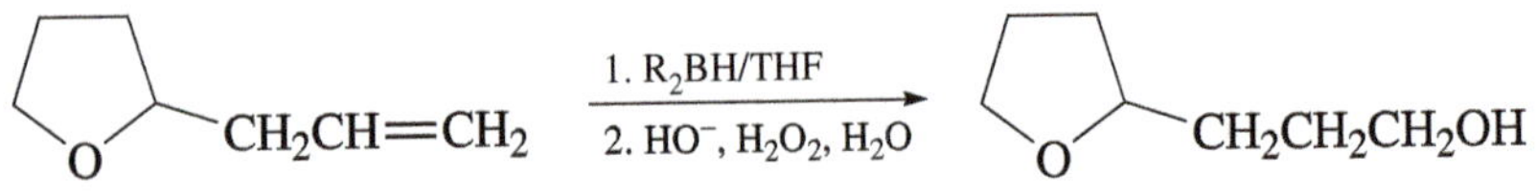 product of **e** $\xrightarrow[\text{2. } HO^-, H_2O_2, H_2O]{\text{1. } R_2BH/THF}$ (tetrahydrofuran-2-yl)$CH_2CH_2CH_2OH$

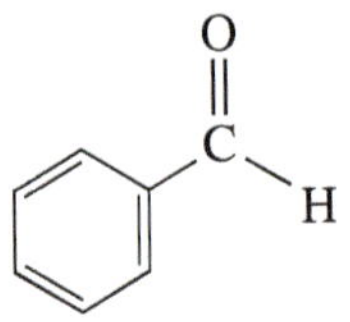

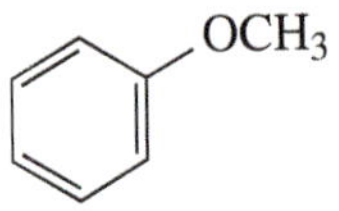

(Starting material: tetrahydrofuran ring bearing $CH_2CH{=}CH_2$)

42.

a. 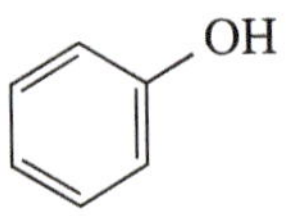 benzene ring with OH

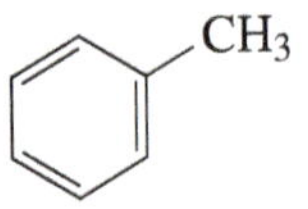

b. 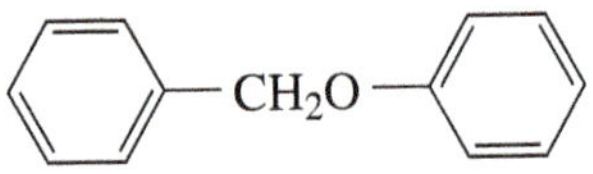 benzene ring—CH_2O—benzene ring

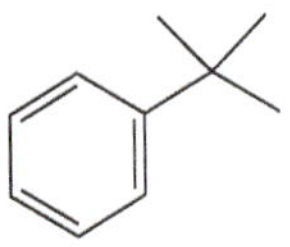

c. 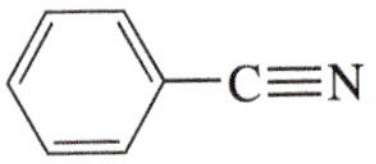 benzene ring—$C{\equiv}N$

d. benzene ring with C(=O)H

e. benzene ring with OCH_3

f. 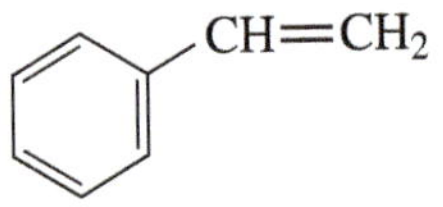 benzene ring with $CH{=}CH_2$

g. benzene ring with CH_3

h. benzene ring with tert-butyl group

i. 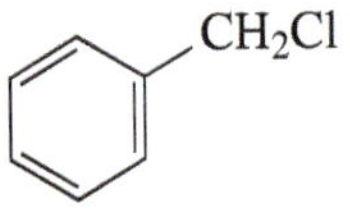 benzene ring with CH_2Cl

43.

a. *m*-bromobenzoic acid or 3-bromo-benzoic acid
b. 1,2,4-tribromobenzene
c. 2,6-dimethylphenol
d. *p*-nitrostyrene or 4-nitrostyrene
e. *m*-ethylanisole or 3-ethylanisole
f. 3,5-dichlorobenzenesulfonic acid
g. *o*-bromom ethylbenzene or 1-bromo-2-methylbenzene
h. *p*-cyclohexylmethylbenzene or 1-cyclohexyl-4-methylbenzene
i. 2-chloro-4-ethylaniline

44.

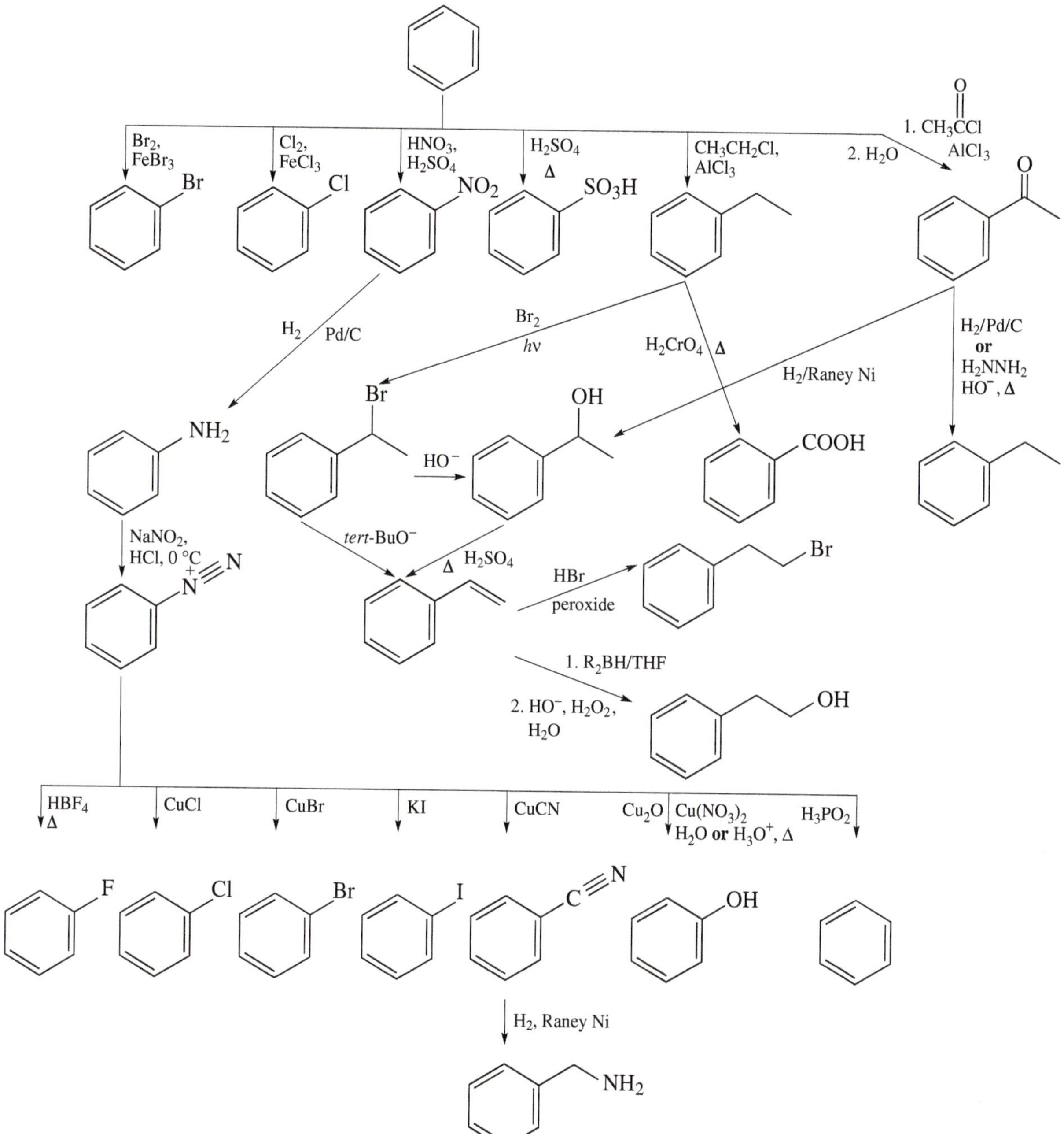

Br_2, $FeBr_3$ — Br

Cl_2, $FeCl_3$ — Cl

HNO_3, H_2SO_4 — NO_2

H_2SO_4, Δ — SO_3H

CH_3CH_2Cl, $AlCl_3$

1. CH_3CCl (O), $AlCl_3$ 2. H_2O

H_2 Pd/C — NH_2

Br_2, hν — Br

H_2CrO_4, Δ — COOH

H_2/Raney Ni — OH

H_2/Pd/C **or** H_2NNH_2, HO^-, Δ

HO^-

$NaNO_2$, HCl, 0 °C — $\overset{+}{N}{\equiv}N$

tert-BuO^-

Δ H_2SO_4

HBr peroxide — Br

1. R_2BH/THF 2. HO^-, H_2O_2, H_2O — OH

HBF_4 Δ — F

CuCl — Cl

CuBr — Br

KI — I

CuCN — $C{\equiv}N$

Cu_2O, $Cu(NO_3)_2$, H_2O **or** H_3O^+, Δ — OH

H_3PO_2

H_2, Raney Ni — NH_2

45.

a. 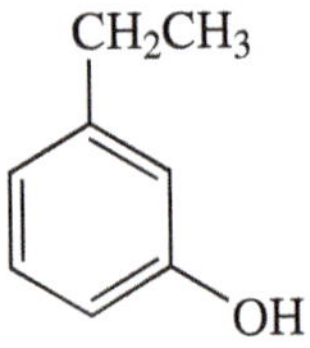 CH_2CH_3 OH

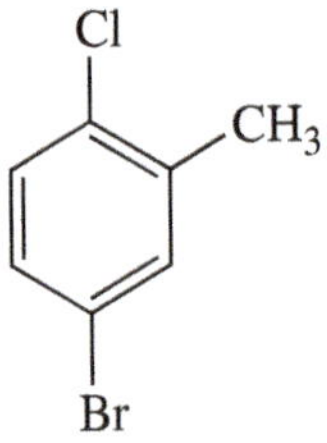

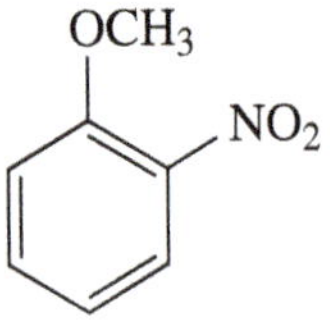

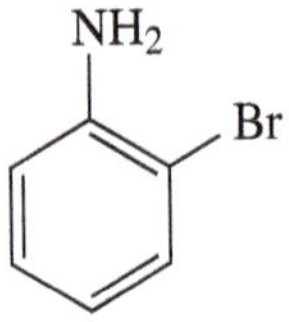

b. 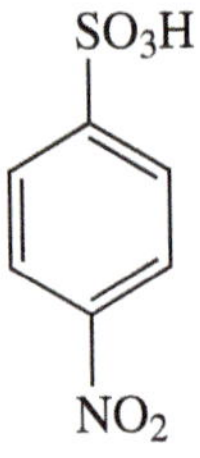 SO_3H NO_2

c. 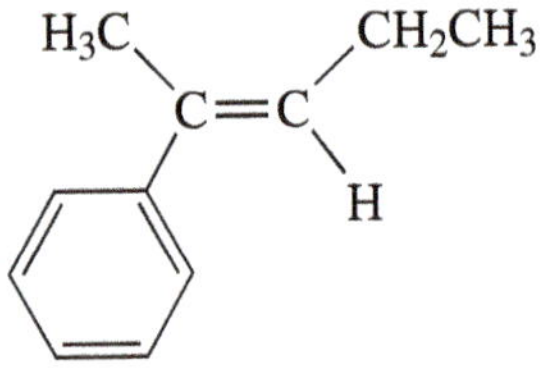 H_3C CH_2CH_3 C=C H

d. NH_2 Br

e. Cl CH_3 Br

f. 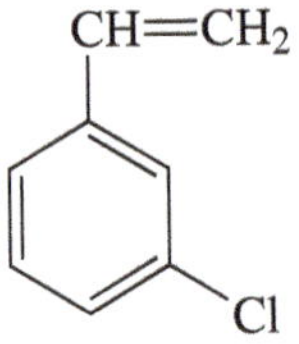 CH=CH_2 Cl

g. OCH_3 NO_2

h. CH_3 Cl Cl

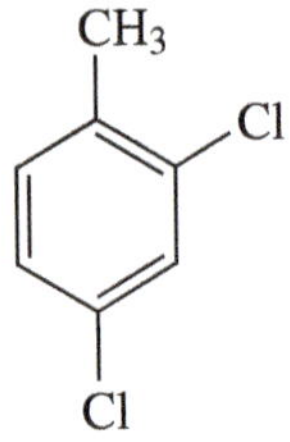

i. 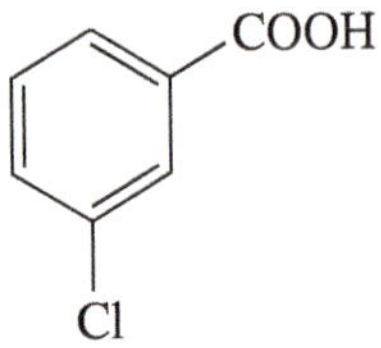 COOH Cl

46.

a. CH_2CH_3 donates electrons by hyperconjugation and does not donate or withdraw electrons by resonance.

b. NO_2 withdraws electrons inductively and withdraws electrons by resonance.

c. Br deactivates the ring and directs ortho/para.

d. OH withdraws electrons inductively, donates electrons by resonance, and activates the ring.

e. $^{+}NH_3$ withdraws electrons inductively and does not donate or withdraw electrons by resonance.

47. a. 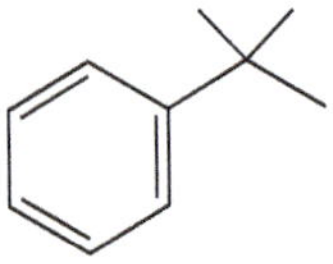 b. 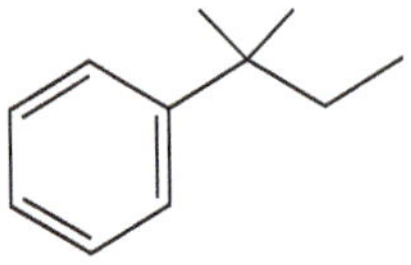 c.

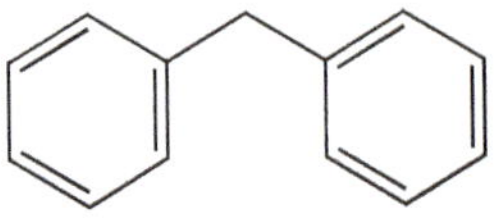

48.

a. 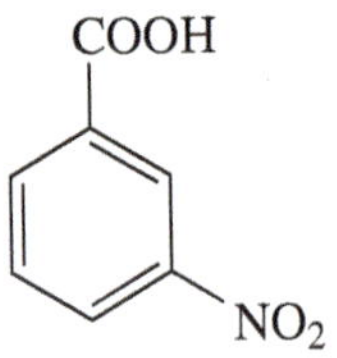 COOH NO_2

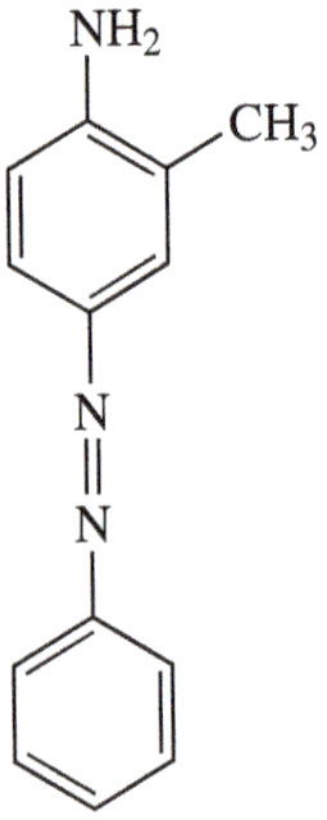

b. 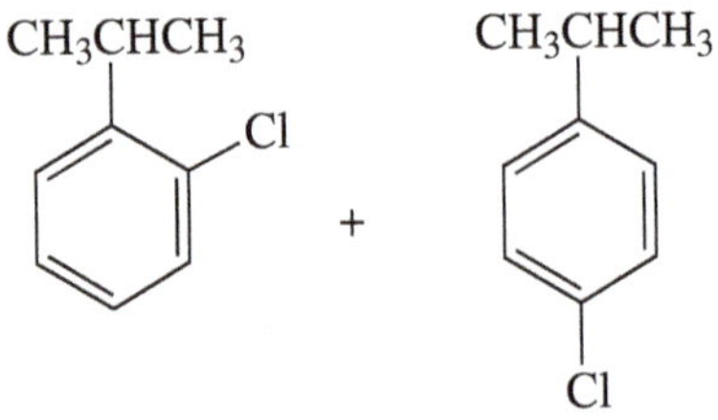 CH_3CHCH_3 Cl + CH_3CHCH_3 Cl

c. 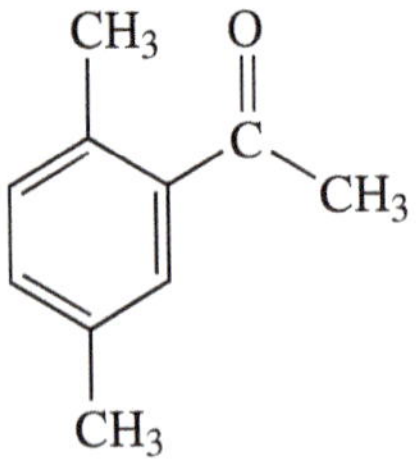 CH_3 O C CH_3 CH_3

d. NH_2 CH_3 N=N

e.

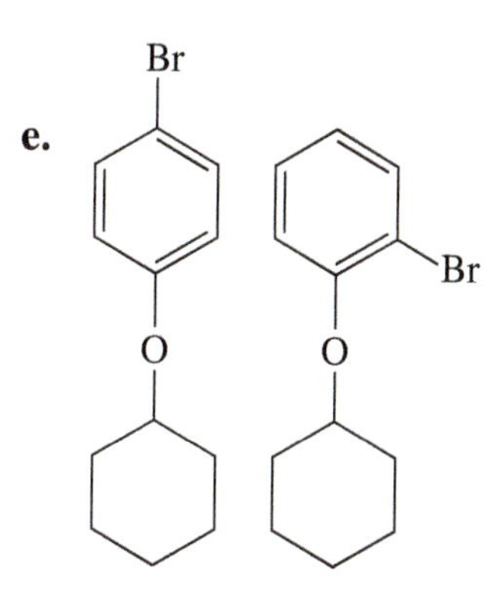

g.

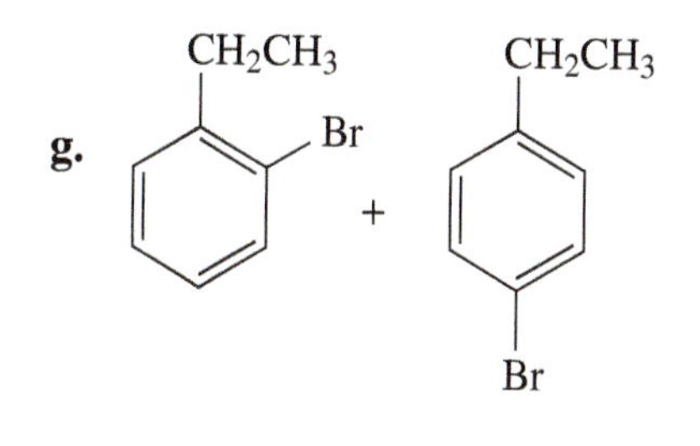

f.

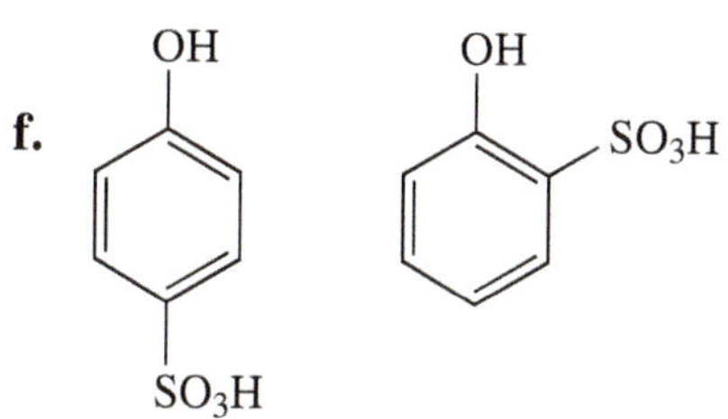

h.

COOH

COOH

49. The more activating the substituent, the more basic the NH_2 group.

CH_3O–C_6H_4–NH_2 > CH_3–C_6H_4–NH_2 > Br–C_6H_4–NH_2 > $CH_3\overset{O}{\overset{\|}{C}}$–$C_6H_4$–$NH_2$

50. The least reactive compound yields the highest percentage of meta product.

1. CH_3 CHF_2 CF_3

most reactive

least reactive
highest % meta product

2. $\overset{+}{N}(CH_3)_3$ $CH_2\overset{+}{N}(CH_3)_3$ $CH_2CH_2\overset{+}{N}(CH_3)_3$

least reactive
highest % meta product

most reactive

3. OCH_2CH_3 CH_2OCH_3 $\overset{O}{\overset{\|}{C}}OCH_3$

most reactive

least reactive
highest % meta product

51. **a.** These are intramolecular Friedel–Crafts alkylation reactions. Drawing the mechanism allows you to determine the product of the reaction. Notice that a carbocation rearrangement occurs in part **a**.

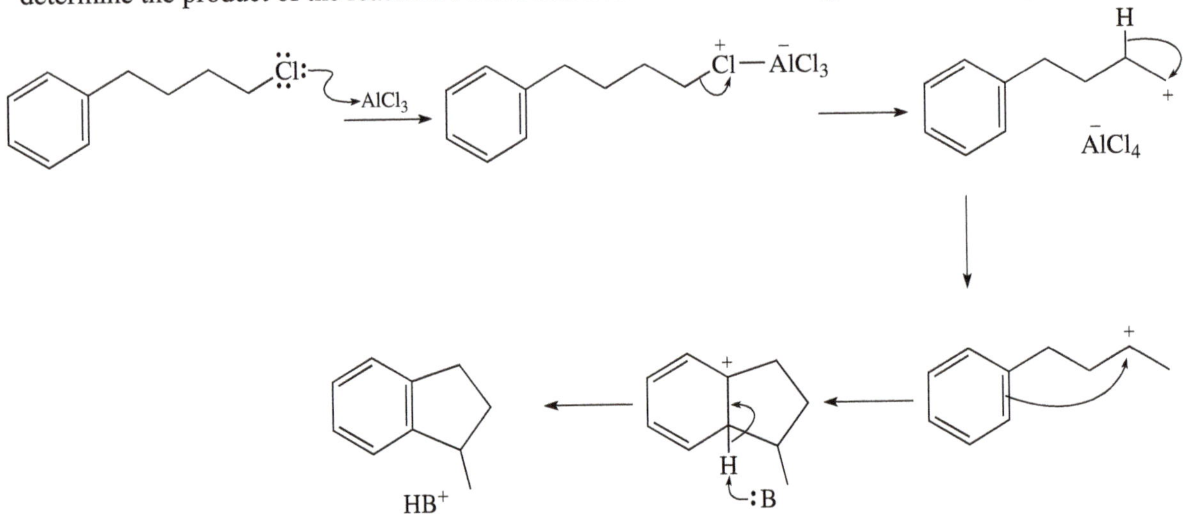

$AlCl_3$ $\overset{+}{Cl}-\overset{-}{Al}Cl_3$ H + $\overset{-}{Al}Cl_4$

+ H :B HB^+

b. Now that you understand how the product in part **a** is formed, you can determine the product of this reaction without drawing the mechanism.

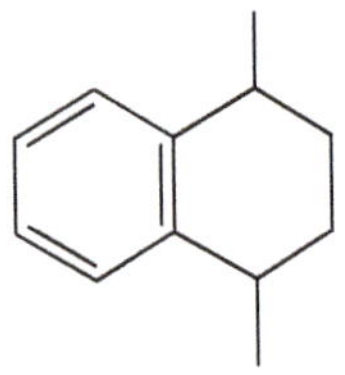

52. **a.** $\xrightarrow[\Delta]{H_2SO_4}$ SO_3H $\xrightarrow[FeCl_3]{Cl_2}$ SO_3H Cl

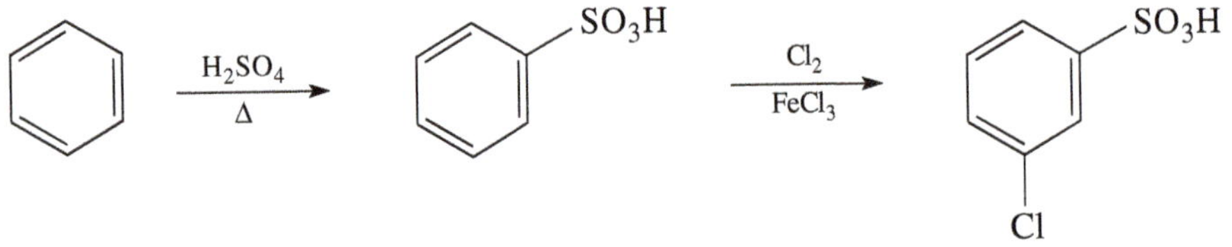

b. $\xrightarrow[2.\ H_2O]{1.\ CH_3\overset{O}{\overset{\|}{C}}Cl,\ AlCl_3}$ O C CH_3 $\xrightarrow[FeCl_3]{Cl_2}$ O C CH_3 Cl $\xrightarrow[\text{or } H_2,\ Pd/C]{H_2NNH_2,\ HO^-,\ \Delta}$ CH_2CH_3 Cl

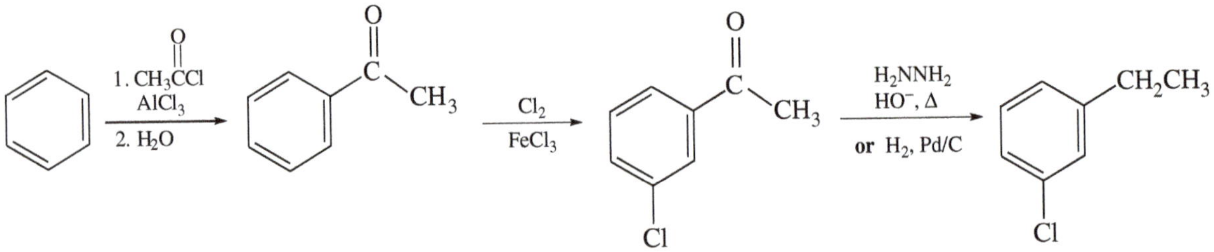

c. benzene $\xrightarrow[H_2SO_4]{HNO_3}$ NO_2-benzene $\xrightarrow[FeBr_3]{Br_2}$ m-bromonitrobenzene (NO_2, Br) $\xrightarrow{H_2, Pd/C}$ m-bromoaniline (NH_2, Br)

$\xrightarrow[0\ °C]{NaNO_2, HCl}$ m-bromobenzenediazonium chloride ($\overset{+}{N}\equiv N$ Cl^-, Br) $\xrightarrow{CuC\equiv N}$ m-bromobenzonitrile ($C\equiv N$, Br)

d.

benzene $\xrightarrow[2.\ H_2O]{1.\ CH_3CH_2CH_2CH_2\overset{O}{\overset{\|}{C}}Cl,\ AlCl_3}$ $C_6H_5\overset{O}{\overset{\|}{C}}CH_2CH_2CH_2CH_3$ $\xrightarrow[\textbf{or}\ H_2,\ Pd/C]{H_2NNH_2,\ HO^-,\ \Delta}$ $C_6H_5CH_2CH_2CH_2CH_2CH_3$

e. benzene $\xrightarrow[AlCl_3]{CH_3Cl}$ $C_6H_5CH_3$ $\xrightarrow[\Delta]{H_2CrO_4}$ C_6H_5COOH $\xrightarrow[FeBr_3]{Br_2}$ m-bromobenzoic acid (COOH, Br)

f. benzene $\xrightarrow[AlCl_3]{CH_3Cl}$ $C_6H_5CH_3$ $\xrightarrow[\Delta]{H_2CrO_4}$ C_6H_5COOH $\xrightarrow[H_2SO_4]{HNO_3}$ m-nitrobenzoic acid (COOH, NO_2)

$\xrightarrow{H_2, Pd/C}$ m-aminobenzoic acid (COOH, NH_2) $\xrightarrow[HCl,\ 0\ °C]{NaNO_2}$ m-carboxybenzenediazonium (COOH, $\overset{+}{N}\equiv N$) $\xrightarrow[Cu(NO_3)_2,\ H_2O\ \textbf{or}\ H_3O^+,\ \Delta]{Cu_2O}$ m-hydroxybenzoic acid (COOH, OH)

g. benzene $\xrightarrow[AlCl_3]{CH_3Cl}$ toluene (CH_3) $\xrightarrow[H_2SO_4]{HNO_3}$ (O_2N, CH_3) $\xrightarrow{H_2,\ Pd/C}$ (H_2N, CH_3)

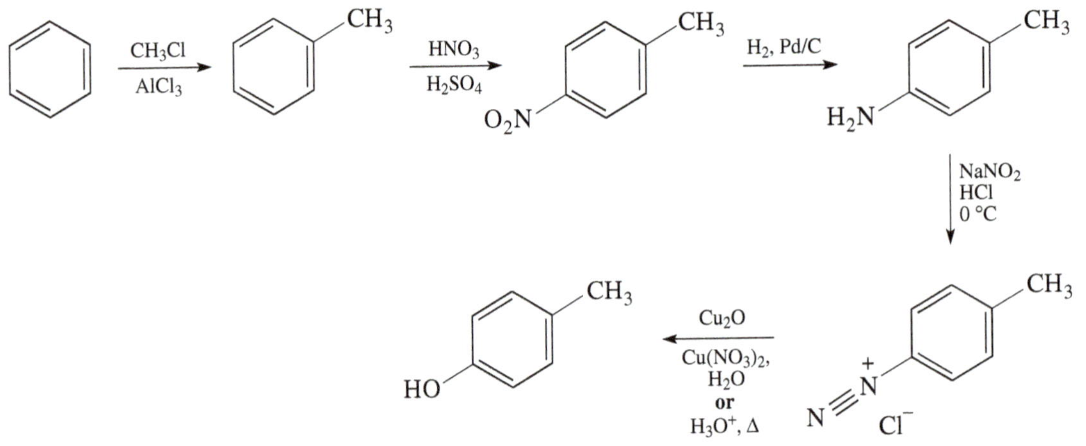

$\downarrow$ $NaNO_2$, HCl, 0 °C

($N{\equiv}\overset{+}{N}$, Cl^-, CH_3) $\xrightarrow{Cu_2O,\ Cu(NO_3)_2,\ H_2O\ \textbf{or}\ H_3O^+,\ \Delta}$ (HO, CH_3)

h. benzene $\xrightarrow[AlCl_3]{CH_3Cl}$ (CH_3) $\xrightarrow[h\nu]{Br_2}$ (CH_2Br) $\xrightarrow{HO^-}$ (CH_2OH)

or benzene $\xrightarrow[AlCl_3,\ CuCl]{CO,\ HCl,\ \text{high pressure}}$ ($C(=O)H$) $\xrightarrow[2.\ H_3O^+]{1.\ NaBH_4}$ (CH_2OH)

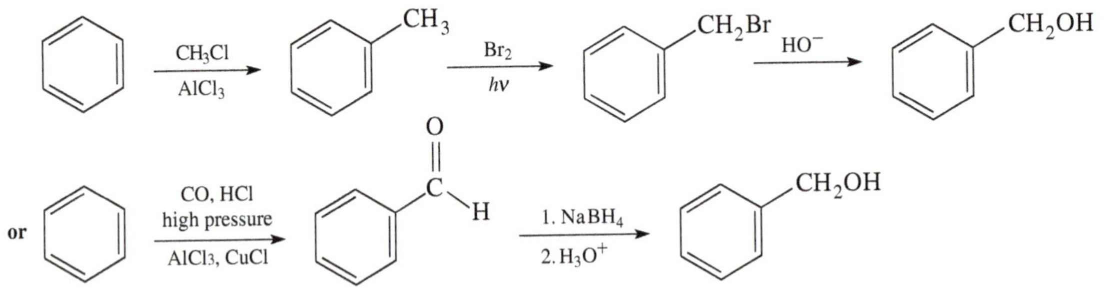

i. benzene $\xrightarrow[H_2SO_4]{HNO_3}$ (NO_2) $\xrightarrow[Pd/C]{H_2}$ (NH_2) $\xrightarrow[HCl,\ 0\ °C]{NaNO_2}$ ($\overset{+}{N}{\equiv}N$) $\xrightarrow{CuCN}$ ($C{\equiv}N$)

$\downarrow$ H_2, Raney Ni

(CH_2NH_2)

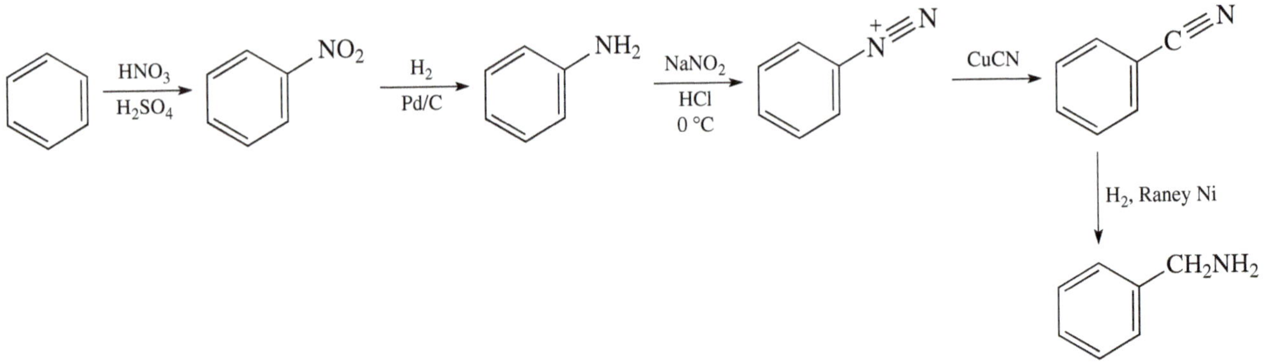

53. Anisole undergoes electrophilic aromatic substitution (D^+ is the electrophile) primarily at the ortho and para positions.

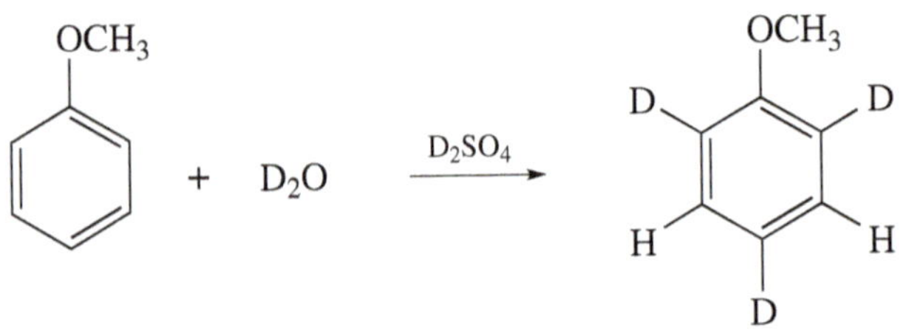

(OCH_3) + D_2O $\xrightarrow{D_2SO_4}$ (OCH_3; D, D, H, H, D)

54.  CH_3O–(benzene ring)–OCH_3

55. **a.** anisole > ethylbenzene > benzene > chlorobenzene > nitrobenzene
b. 2,4-dinitrophenol > 1-methyl-2,4-dinitrobenzene > 1-chloro-2,4-dinitrobenzene
c. *p*-cresol > *p*-xylene > toluene > benzene
d. phenol > propylbenzene > benzene > benzoic acid
e. *p*-chloromethylbenzene > *p*-methylnitrobenzene > 2-chloro-1-methyl-4-nitrobenzene > 1-methyl-2,4-dinitrobenzene
f. fluorobenzene > chlorobenzene > bromobenzene > iodobenzene

56. **a.** Two products are obtained. Less of the product on the right is obtained because of steric hindrance.

$\xrightarrow[FeCl_3]{Cl_2}$

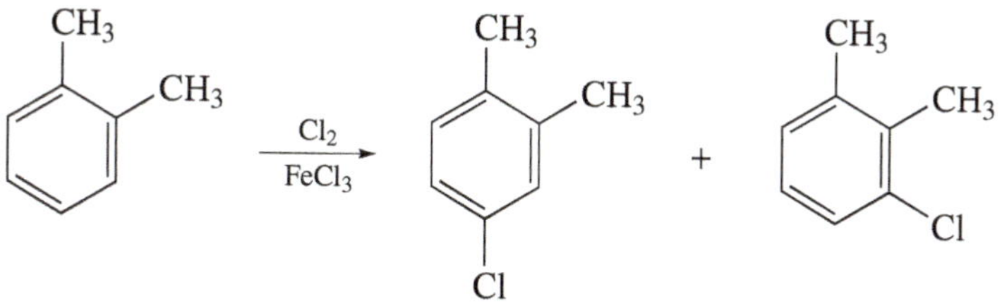

b. One product is obtained. (Notice that the four positions are equivalent.)

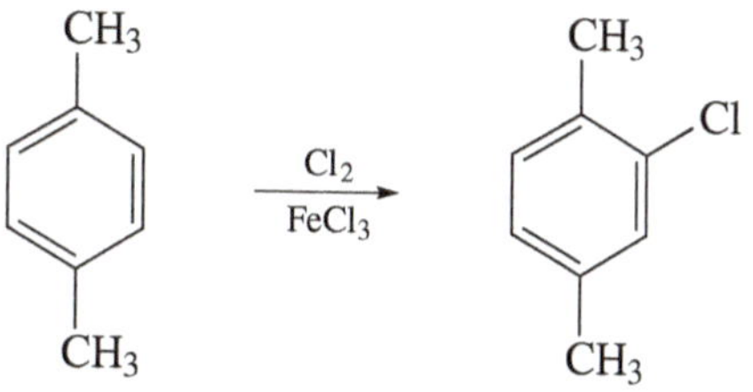

$\xrightarrow[FeCl_3]{Cl_2}$

c. Two products can be obtained, but little, if any, of one of the products is obtained because of steric hindrance.

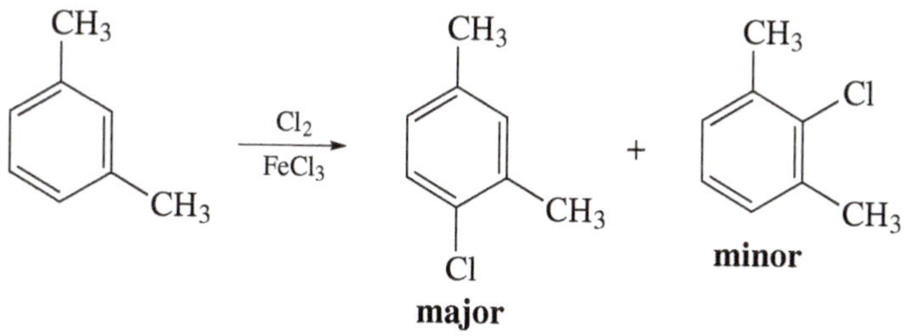

$\xrightarrow[FeCl_3]{Cl_2}$

major **minor**

57. **a.** 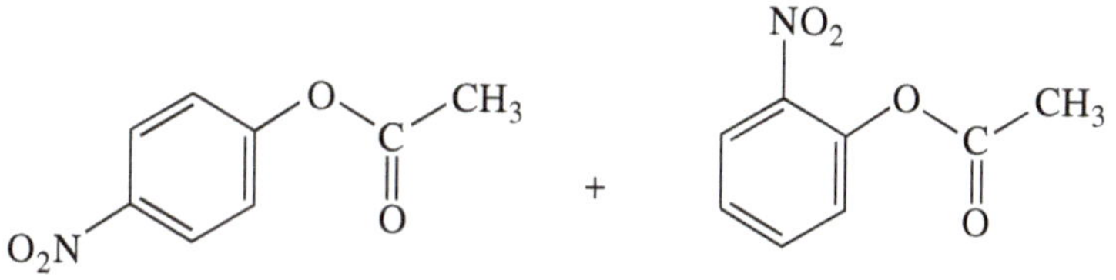 + **b.**

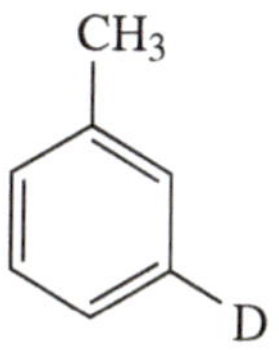

c. +

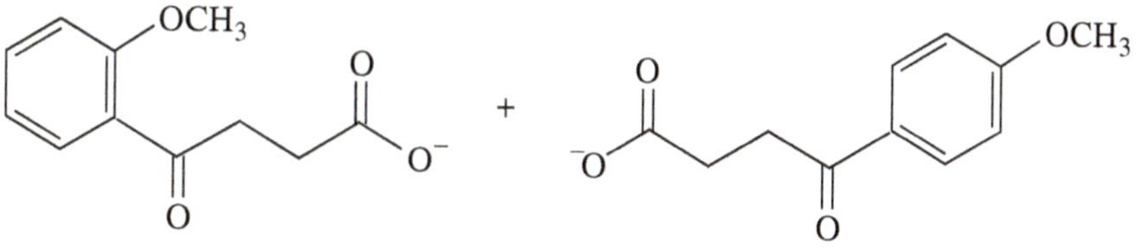

d. 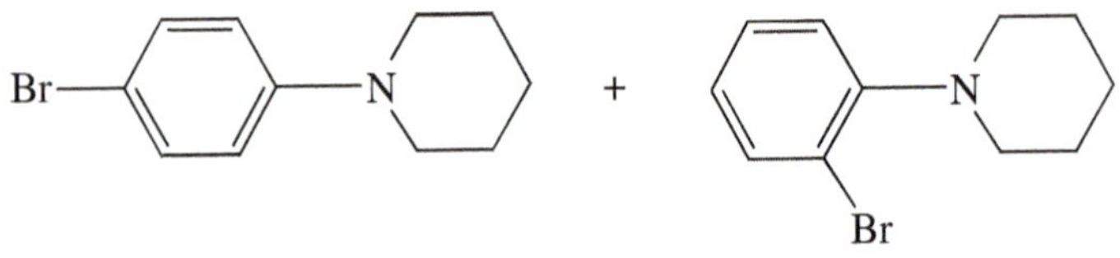 Br (para-bromophenyl piperidine) + (ortho-bromophenyl piperidine) Br

e. 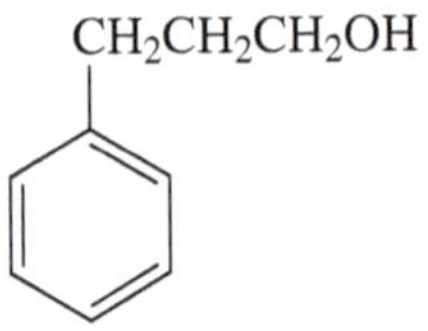 $CH_2CH_2CH_2OH$

f. 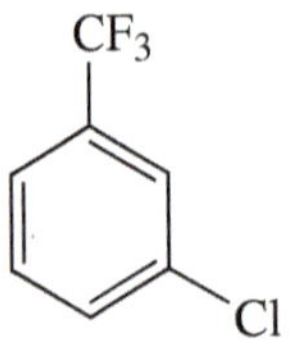 CF_3 Cl

58. **a.** 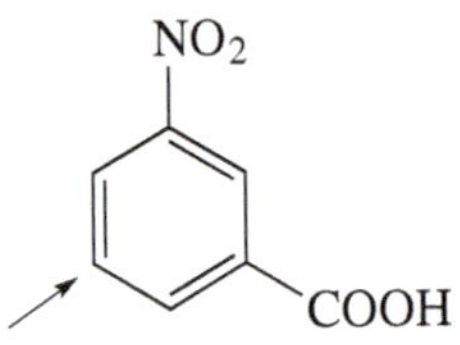 NO_2 COOH

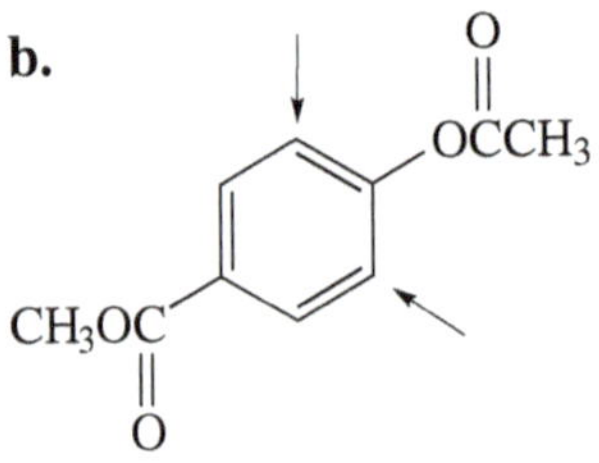

c. 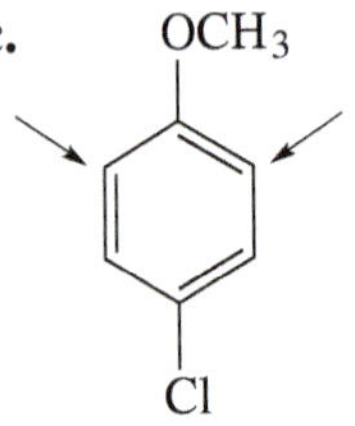 OCH_3 Cl

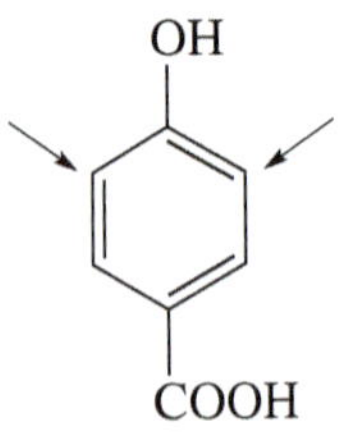

e. 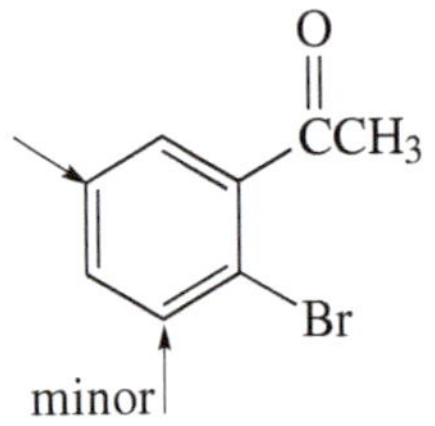 O ‖ CCH_3 Br minor

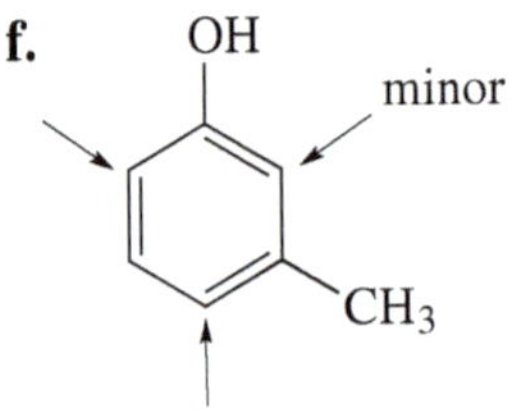

g. 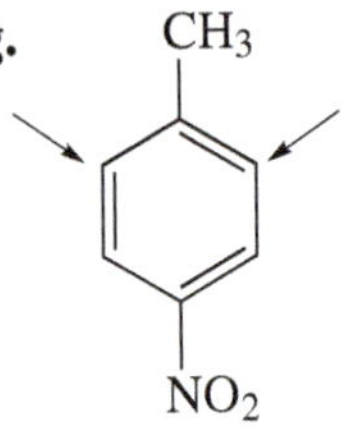 CH_3 NO_2

b. O ‖ $OCCH_3$ CH_3OC ‖ O

d. OH COOH

f. OH minor CH_3

h. 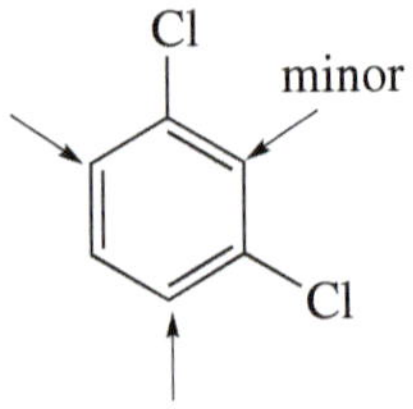 Cl minor Cl

59. **1.** 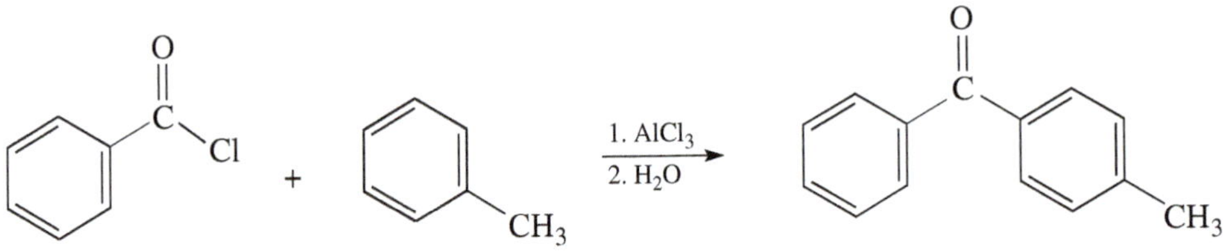 O ‖ C Cl + CH_3 $\xrightarrow[\text{2. } H_2O]{\text{1. } AlCl_3}$ O ‖ C CH_3

2. + Cl C ‖ O CH_3 $\xrightarrow[\text{2. } H_2O]{\text{1. } AlCl_3}$ O ‖ C CH_3

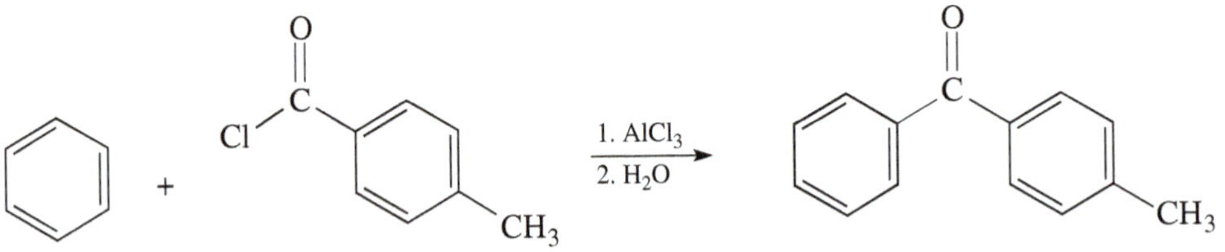

60. Because the 2*p* orbital of oxygen overlaps the 2*p* orbital of carbon better than does the 3*p* orbital of sulfur, oxygen is better than sulfur at donating electrons by resonance. Therefore, the benzene ring of anisole is more activated toward electrophilic aromatic substitution than is the benzene ring of thioanisole.

61. Three ways to synthesize anisole from benzene are shown below.

1. benzene $\xrightarrow[H_2SO_4]{HNO_3}$ nitrobenzene (NO_2) $\xrightarrow{H_2, Pd/C}$ aniline (NH_2) $\xrightarrow[NaNO_2, 0\ °C]{HCl}$ benzenediazonium chloride ($\overset{+}{N}\equiv N$ Cl^-) $\xrightarrow{CH_3OH\ |\ HCl, \Delta}$ anisole (OCH_3)

2. benzene $\xrightarrow[H_2SO_4]{HNO_3}$ nitrobenzene (NO_2) $\xrightarrow{H_2, Pd/C}$ aniline (NH_2) $\xrightarrow[NaNO_2, 0°\ C]{HCl}$ benzenediazonium chloride ($\overset{+}{N}\equiv N$ Cl^-) $\xrightarrow{Cu_2O,\ Cu(NO_3)_2,\ H_2O \text{ or } H_3O^+, \Delta}$ phenol (OH) $\xrightarrow{1.\ NaH \quad 2.\ CH_3I}$ anisole (OCH_3)

3. benzene $\xrightarrow[FeCl_3]{Cl_2}$ chlorobenzene (Cl) $\xrightarrow[H_2SO_4]{HNO_3}$ 1-chloro-4-nitrobenzene (Cl, O_2N) $\xrightarrow[\Delta]{CH_3O^-}$ 4-nitroanisole (OCH_3, O_2N) $\xrightarrow{H_2, Pd/C}$ 4-methoxyaniline (OCH_3, H_2N) $\xrightarrow[NaNO_2, 0\ °C]{HCl}$ 4-methoxybenzenediazonium chloride (OCH_3, $N\equiv\overset{+}{N}$ Cl^-) $\xrightarrow{H_3PO_2}$ anisole (OCH_3)

62. The compound with the methoxy substituent is the more reactive because it forms the more stable carbocation intermediate. The carbocation intermediate is stabilized by resonance electron donation.

$CH_3\ddot{\underset{\cdot\cdot}{O}}$–$C_6H_4$–$\overset{+}{C}H$–$CH_3$ (curved arrows showing electron donation from the methoxy oxygen through the ring to the carbocation)

63. The rate-determining step in the S_N1 reaction is formation of the tertiary carbocation. An electron-donating substituent stabilize the carbocation and causes it to be more easily formed. An electron-withdrawing substituent will destabilizes the carbocation and causes it to be less easily formed.

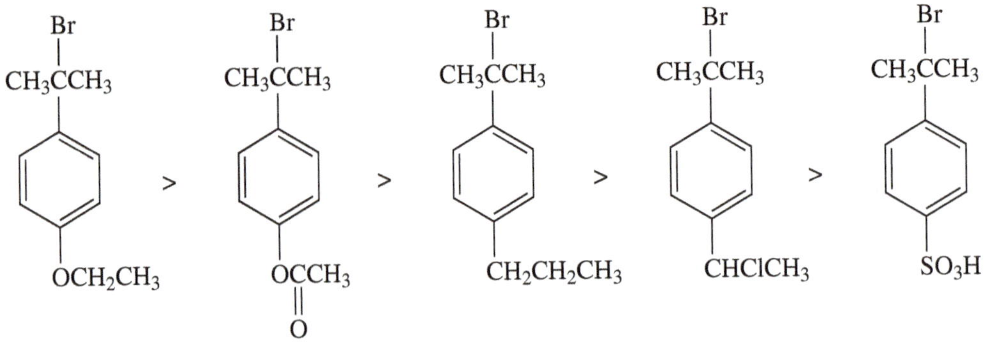

Br, CH_3CCH_3, OCH_2CH_3 > Br, CH_3CCH_3, $OCCH_3$ (=O) > Br, CH_3CCH_3, $CH_2CH_2CH_3$ > Br, CH_3CCH_3, $CHClCH_3$ > Br, CH_3CCH_3, SO_3H

64. The signal at ~7 ppm that integrates to 5H indicates a monosubstituted benzene ring. A monosubstituted benzene ring contains six carbons and five hydrogens. Subtracting these atoms from the molecular formula of $C_{13}H_{20}$ gives us a substituent with 7 carbons and 15 hydrogens. The compound shown below has the correct number of carbons and hydrogens that will give two singlets, with one (9H) having 1.5 times the area of the second (6H).

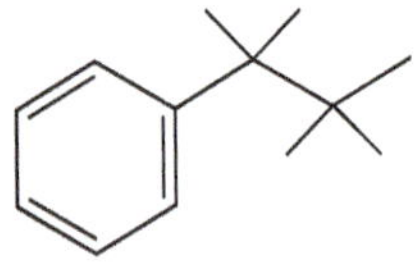

65. **a.** HNO_3 / H_2SO_4 → NO_2 ; H_2, Pd/C → NH_2 ; CH_3I excess, K_2CO_3 (see page 405 of the text) → $^+N(CH_3)_3$ I^-

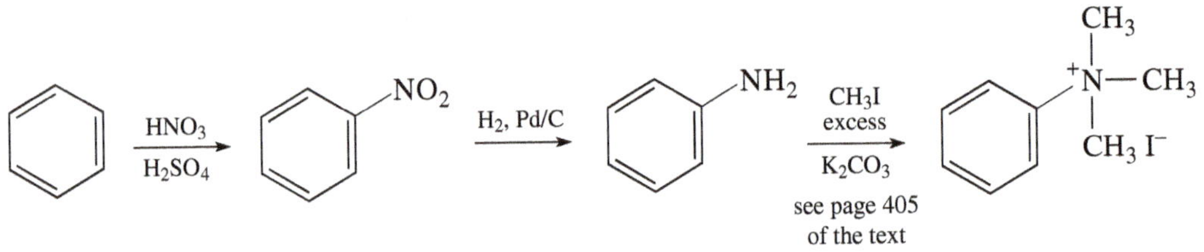

b. CH_3Cl / $AlCl_3$ → CH_3 ; HNO_3 / H_2SO_4 → CH_3, NO_2 ; H_2, Pd/C → CH_3, NH_2 ; $NaNO_2$, HCl, 0 °C → CH_3, $N^+{\equiv}N$ Cl^- ; Cu_2O, $Cu(NO_3)_2$, H_2O **or** H_3O^+, Δ → CH_3, OH ; HNO_3 / H_2SO_4 → O_2N, CH_3, OH

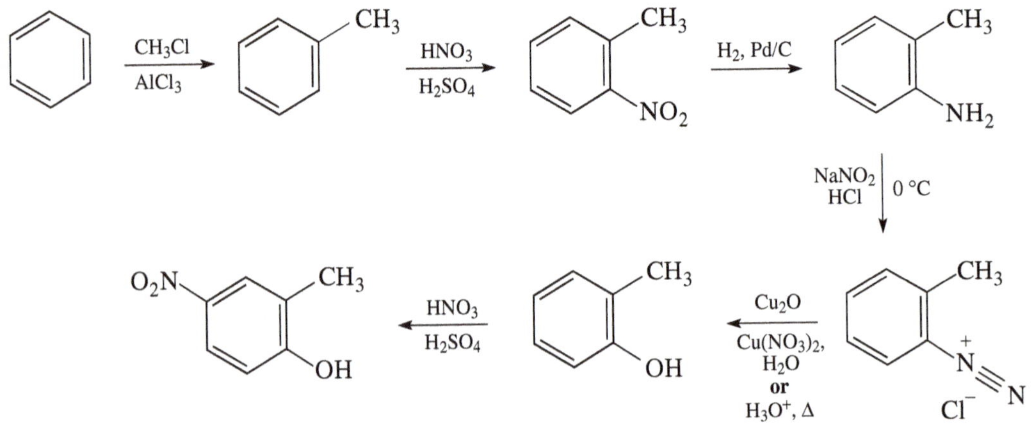

c. Benzene $\xrightarrow[AlCl_3]{C_6H_5CH_2Cl}$ $C_6H_5CH_2C_6H_5$ $\xrightarrow[FeCl_3]{Cl_2}$ $p\text{-}ClC_6H_4CH_2C_6H_5$

d. Benzene $\xrightarrow[AlCl_3]{CH_3Cl}$ $C_6H_5CH_3$ $\xrightarrow[h\nu]{Br_2}$ $C_6H_5CH_2Br$ $\xrightarrow{CH_3O^-}$ $C_6H_5CH_2OCH_3$

e. Benzene $\xrightarrow[FeBr_3]{Br_2}$ C_6H_5Br $\xrightarrow[H_2SO_4]{HNO_3}$ $p\text{-}O_2NC_6H_4Br$ $\xrightarrow[\Delta]{^-NH_2}$ $p\text{-}O_2NC_6H_4NH_2$

f. Benzene $\xrightarrow[H_2SO_4]{HNO_3}$ $C_6H_5NO_2$ $\xrightarrow[FeBr_3]{Br_2}$ $m\text{-}BrC_6H_4NO_2$ $\xrightarrow{H_2, Pd/C}$ $m\text{-}BrC_6H_4NH_2$ $\xrightarrow[0\ °C]{NaNO_2, HCl}$ $m\text{-}BrC_6H_4\overset{+}{N}{\equiv}N\ Cl^-$ $\xrightarrow{KI}$ $m\text{-}BrC_6H_4I$

g. Benzene $\xrightarrow[FeBr_3]{Br_2}$ C_6H_5Br $\xrightarrow[FeBr_3]{Br_2}$ $p\text{-}BrC_6H_4Br$ $\xrightarrow[Et_2O]{Mg}$ $p\text{-}BrMgC_6H_4MgBr$ $\xrightarrow{D_2O}$ $p\text{-}DC_6H_4D$

h. Benzene $\xrightarrow[FeBr_3]{Br_2}$ C_6H_5Br $\xrightarrow[H_2SO_4]{HNO_3}$ $p\text{-}O_2NC_6H_4Br$ $\xrightarrow[2.\ HO^-]{1.\ CH_3NH_2,\ \Delta}$ $p\text{-}O_2NC_6H_4NHCH_3$

i. Benzene $\xrightarrow[H_2SO_4]{HNO_3}$ $C_6H_5NO_2$ $\xrightarrow[FeBr_3]{Br_2}$ $m\text{-}BrC_6H_4NO_2$

66. **a.** The halogens withdraw electrons inductively and donate electrons by resonance. Because they all deactivate the benzene ring toward electrophilic aromatic substitution, we know that their electron-withdrawing effect is stronger than their electron-donating effect. Therefore, an *ortho*-halo-substituted benzoic acid is a stronger acid than benzoic acid.

b. Because fluorine is the weakest deactivator of the halogens (Table 18.1), we know that overall it donates electrons better than the other halogens. Therefore, *ortho*-fluorobenzoic acid is the weakest of the *ortho*-halo-substituted benzoic acids.

c. The smaller the halogen, the more electronegative it is and, therefore, the better it is at withdrawing electrons inductively. The smaller the halogen, the better it is at donating electrons by resonance because a 2*p* orbital of carbon overlaps a 2*p* orbital of a halogen better than a 3*p* orbital of a halogen and overlaps a 3*p* orbital better than a 4*p* orbital. Therefore, Br does not withdraw electrons as well as Cl, and Br does not donate electrons as well as Cl, so their pK_a values are similar.

67. **a.** The weaker the base attached to the acyl group, the stronger its electron-withdrawing ability; therefore, the easier it is to form the tetrahedral intermediate. (*para*-Chlorophenol is a stronger acid than phenol, so the conjugate base of *para*-chlorophenol is a weaker base than the conjugate base of phenol, etc.)

O=C(CH3)O–C6H4–Cl > O=C(CH3)O–C6H5 > O=C(CH3)O–C6H4–CH3 > O=C(CH3)O–C6H11

D A C B

b. The tetrahedral intermediate collapses by eliminating the OR group. The lower the basicity of the OR group, the easier it is to eliminate.

O=C(CH3)O–C6H4–Cl > O=C(CH3)O–C6H5 > O=C(CH3)O–C6H4–CH3 > O=C(CH3)O–C6H11

Thus, the rate of both formation of the tetrahedral intermediate and collapse of the tetrahedral intermediate is decreased by increasing the basicity of the OR group.

68. **a.**

benzene + *para*-xylene $\xrightarrow{^{+}NO_2}$ nitrobenzene (minor product) + 1,4-dimethyl-2-nitrobenzene (major product)

CH3, CH3, NO2, NO2, CH3, CH3

b. The major product is 1,4-dimethyl-2-nitrobenzene.
para-Xylene is more reactive than benzene toward electrophilic aromatic substitution. The methyl groups activate the ortho positions, so all four positions on the ring are activated. Attack on any one of the four leads to the same compound.

69. *meta*-Xylene reacts more rapidly. In *meta*-xylene, both methyl groups activate the same position, whereas in *para*-xylene, each methyl group activates a different position. Therefore, *meta*-xylene is more highly activated.

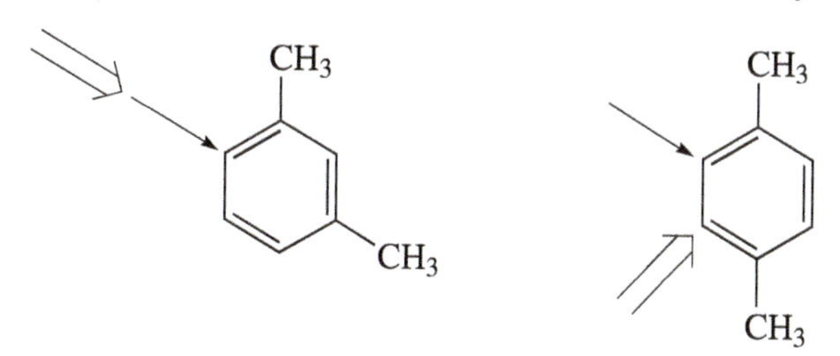

70. The reactions in part **a** and part **b** are intramolecular Friedel–Crafts acylation reactions.

a. Drawing the mechanism can help you determine the product of the reaction.

O Cl: $AlCl_3$ O Cl—$AlCl_3$ O $AlCl_4^-$ + HB$^+$ H O :B O

b. The mechanism in part **a** can help you determine the product.

O

c. —CH_2—

71. The ^{1}H NMR spectrum is the spectrum of butylbenzene.
The benzene ring protons at ~7.2 ppm that integrate to 5 protons indicate a monosubstitutied benzene ring.
The two triplets and two multiplets indicate a straight-chain butyl group.

—$CH_2CH_2CH_2CH_3$

Therefore, the acyl chloride has a straight-chain propyl group and a carbonyl group that will be reduced to a methylene group.

$CH_3CH_2CH_2C(=O)Cl$

72. **a.** 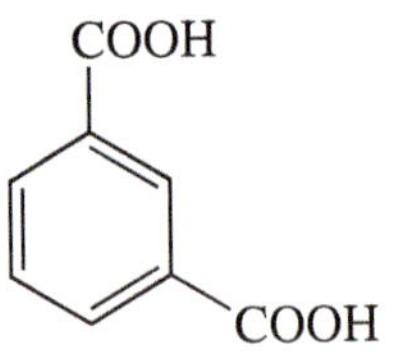 (benzene with COOH and COOH, meta) **b.** 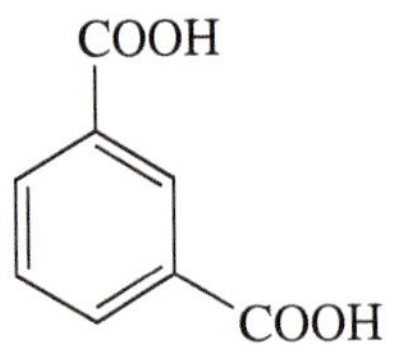 (benzene with COOH and COOH, meta) **c.** 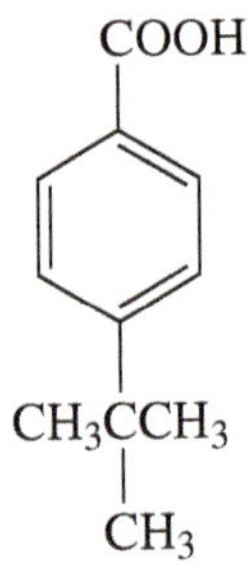 (benzene with COOH and $C(CH_3)_3$, para)

COOH COOH

COOH COOH

COOH

$CH_3\overset{|}{C}CH_3$

CH_3

73. Recall that the highest frequency signal is given by the most deshielded proton, that is, the one with the least electron density (the signal farthest to the left on the spectrum).

a. The oxygen withdraws electrons inductively from the carbon bonded to the hydrogens being compared, so the oxygen-containing compound has the signal at the highest frequency.

$$CH_3O-CH_2\underline{CH_3}$$

b. The oxygen donates electrons by resonance, so the oxygen-containing compound has the signal at the lowest frequency.

$$CH_3\ddot{O}-CH=CH_2 \longleftrightarrow CH_3\overset{+}{O}=CH-\overset{-}{\ddot{C}}H_2$$

Therefore, the non-oxygen-containing compound has the signal at the highest frequency.

$$CH_3CH=\underline{CH_2}$$

74. generation of the electrophile

$$+ \; H-F \longrightarrow \; + \; F^-$$

reaction of benzene with the electrophile

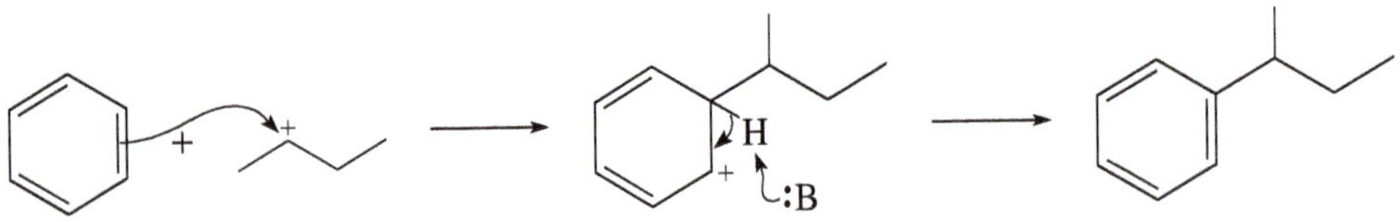

H

:B

75.

a. 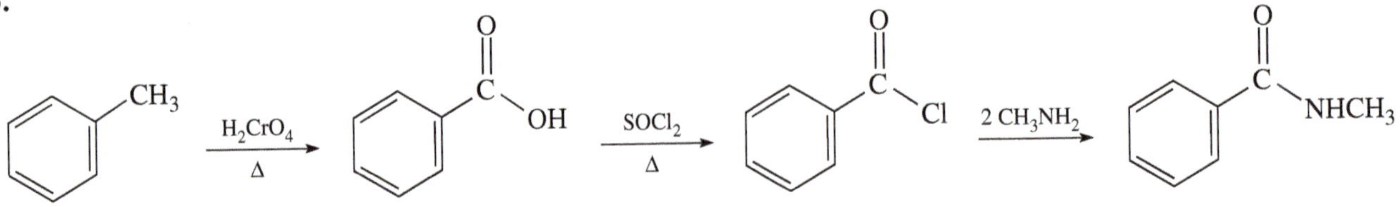 CH_3 $\xrightarrow[\Delta]{H_2CrO_4}$ C(=O)OH $\xrightarrow[\Delta]{SOCl_2}$ C(=O)Cl $\xrightarrow{2\ CH_3NH_2}$ C(=O)NHCH$_3$

b. 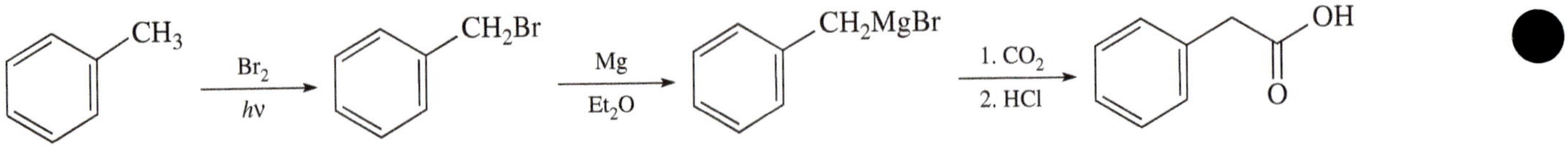 CH_3 $\xrightarrow[h\nu]{Br_2}$ CH_2Br $\xrightarrow[Et_2O]{Mg}$ CH_2MgBr $\xrightarrow[2.\ HCl]{1.\ CO_2}$ OH, O

76. The greater the electron-withdrawing ability of the para substituent, the greater the K_{eq} for hydrate formation. (See Section 16.9 in the text.)

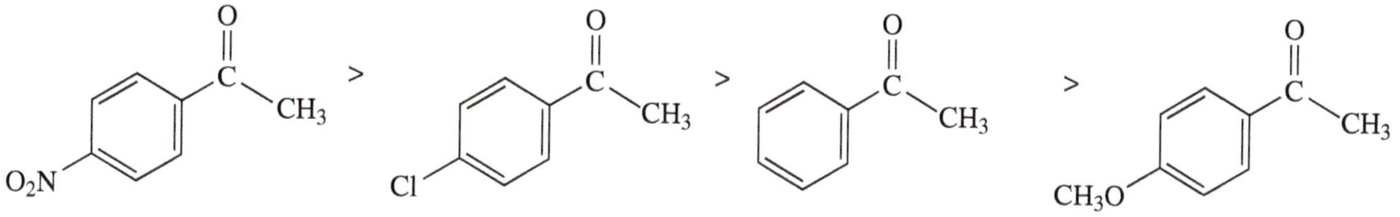

p-nitroacetophenone (O_2N, $C(=O)CH_3$) > p-chloroacetophenone (Cl, $C(=O)CH_3$) > acetophenone ($C(=O)CH_3$) > p-methoxyacetophenone (CH_3O, $C(=O)CH_3$)

77. The original compound was 1,3-dibromobenzene. 1,4-Dibromobenzene would form only one product, and 1,2-dibromobenzene would form only two.

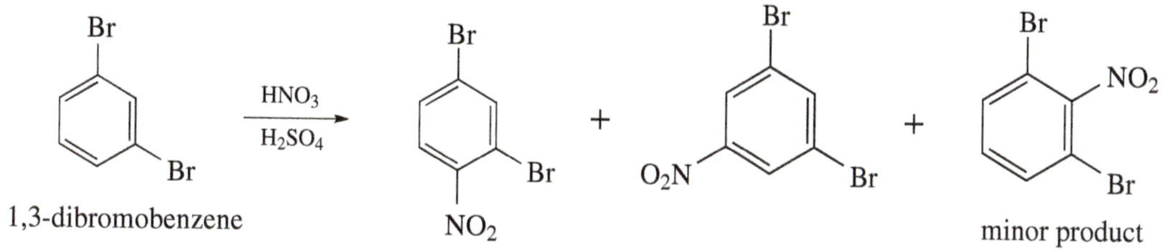

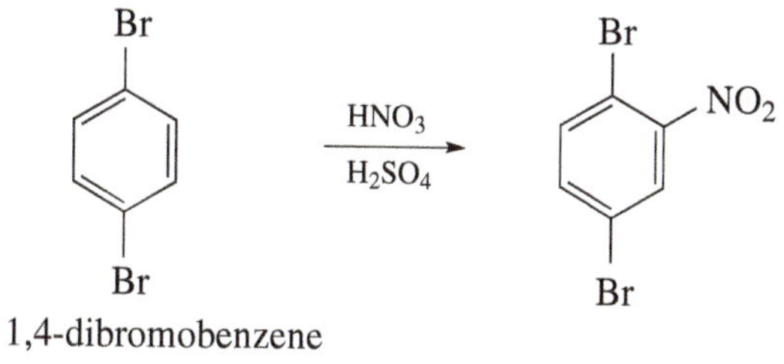

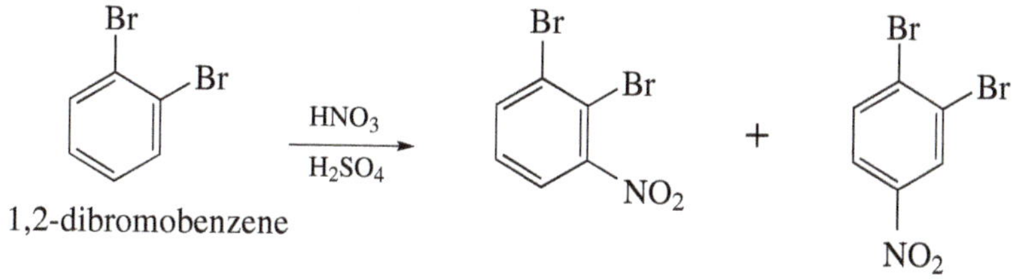

1,3-dibromobenzene $\xrightarrow[H_2SO_4]{HNO_3}$ (Br, Br, NO_2) + (O_2N, Br, Br) + (Br, NO_2, Br) minor product

1,4-dibromobenzene $\xrightarrow[H_2SO_4]{HNO_3}$ (Br, NO_2, Br)

1,2-dibromobenzene $\xrightarrow[H_2SO_4]{HNO_3}$ (Br, Br, NO_2) + (Br, Br, NO_2)

78. **a.** acetophenone (O) → ethylbenzene

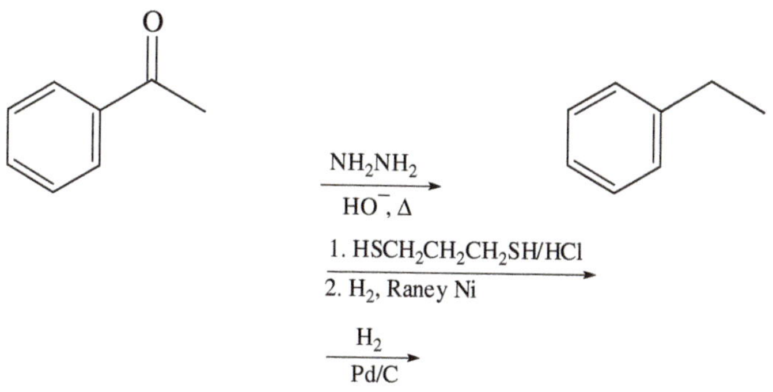

$\xrightarrow[HO^-, \Delta]{NH_2NH_2}$

$\xrightarrow[2.\ H_2,\ Raney\ Ni]{1.\ HSCH_2CH_2CH_2SH/HCl}$

$\xrightarrow[Pd/C]{H_2}$

b.

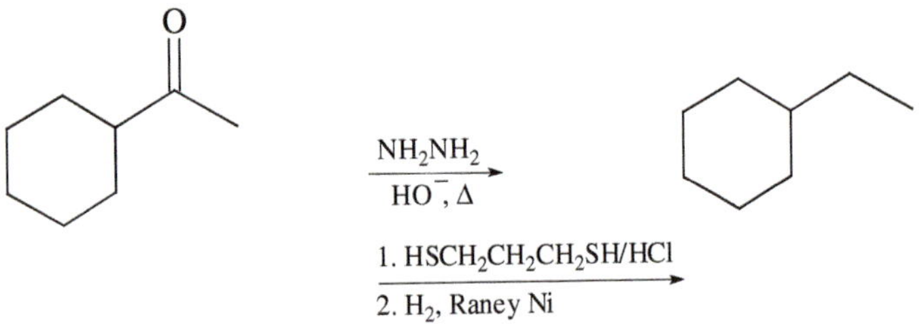

NH_2NH_2 / HO^-, Δ

1. $HSCH_2CH_2CH_2SH/HCl$
2. H_2, Raney Ni

The reason there are three ways to carry out the reaction in part **a** but only two ways to carry it out in part **b** is because only a carbonyl group adjacent to a benzene ring can be reduced to a methylene group by H_2, Pd/C.

79. **a.** HCl adds to the alkene, forming a secondary carbocation that undergoes a 1,2-hydride shift to form a tertiary carbocation. The tertiary carbocation is an electrophile that can add either to the double bond in a second molecule of the reactant (in an intermolecular reaction) or to the benzene ring in the same molecule of the reactant (in an intramolecular reaction).

The intramolecular reaction is favored because it forms a stable five-membered ring. (See Section 9.16 in the text.) After the electrophile adds to the benzene ring, a base (B:) in the reaction mixture removes a proton and the aromaticity of the benzene ring is restored.

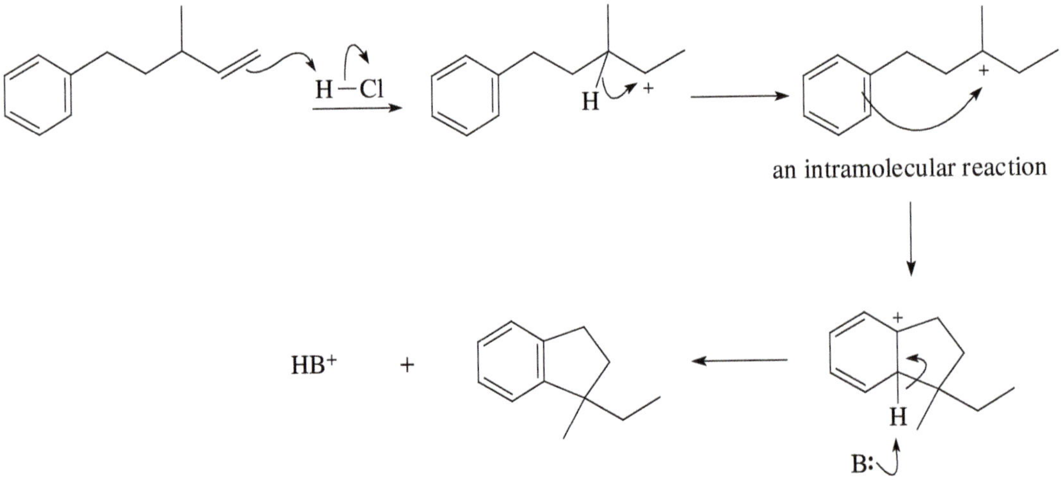

H—Cl

H +

+

an intramolecular reaction

HB^+ +

+

H

B:

b. As in part **a**, an electrophile is formed that can react in either an intermolecular reaction or an intramolecular reaction. Seeing that the product of the reaction has two benzene rings and that there are twice as many carbons in the product as in the reactant indicates that two reactant molecules react in an intermolecular reaction. In this case, the intermolecular reaction is favored, because the intramolecular reaction would lead to a highly strained three-membered ring. The electrophile that is formed in the intermolecular reaction can add to the benzene ring in an intramolecular reaction to form a stable five-membered ring.

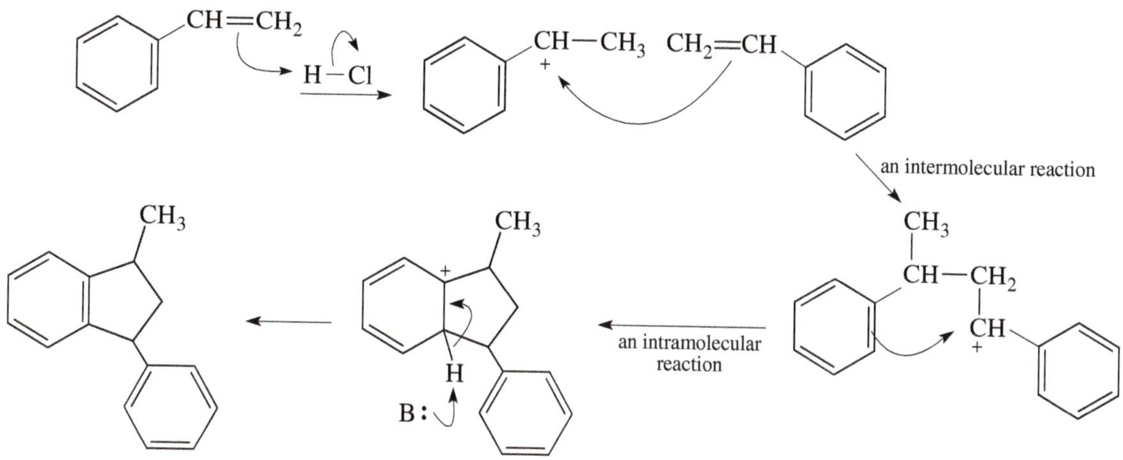

$CH{=}CH_2$ + $H{-}Cl$ → $\overset{+}{C}H{-}CH_3$ + $CH_2{=}CH$

an intermolecular reaction

CH_3, $CH{-}CH_2$, $\overset{+}{C}H$

an intramolecular reaction

CH_3, H, B:

CH_3

80. **a.** Br + $B(OR)_2$ $\xrightarrow[HO^-]{PdL_2}$

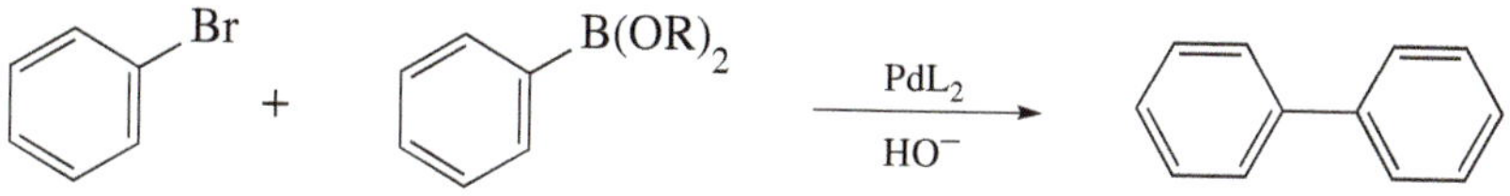

b. + Cl (acyl chloride) $\xrightarrow[2.\ H_2O]{1.\ AlCl_3}$ O, C

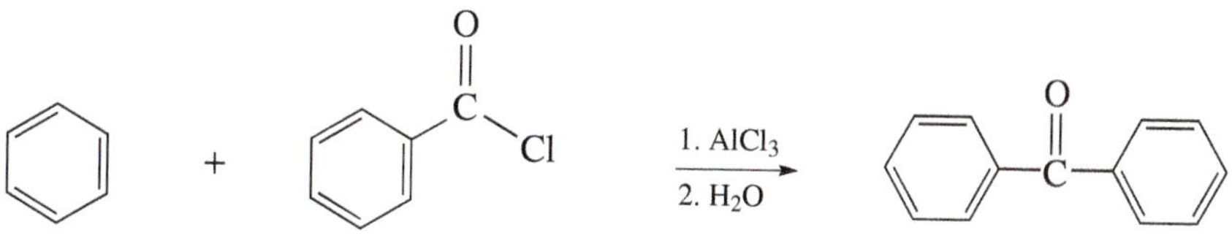

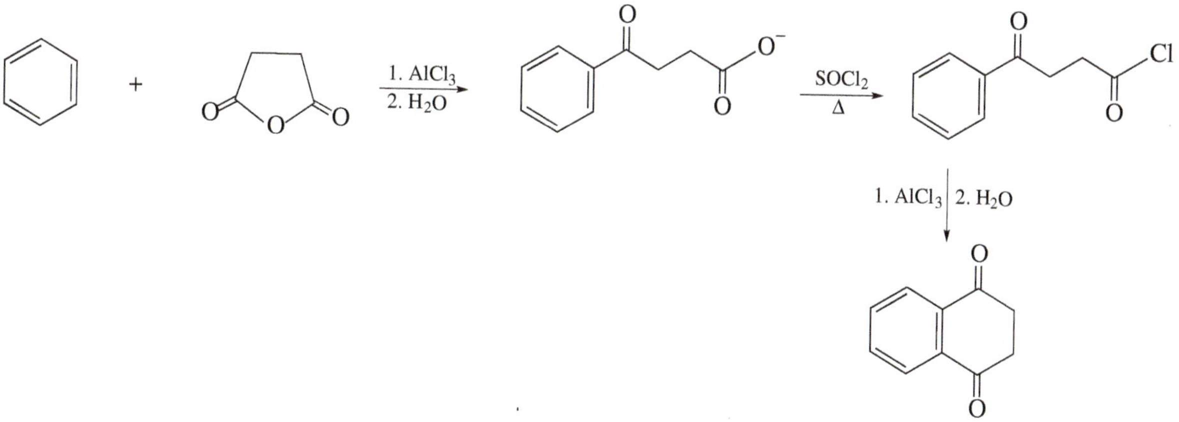

c. + O, O, O $\xrightarrow[2.\ H_2O]{1.\ AlCl_3}$ O, O^-, O $\xrightarrow[\Delta]{SOCl_2}$ O, Cl, O

1. $AlCl_3$ 2. H_2O

O, O

d.

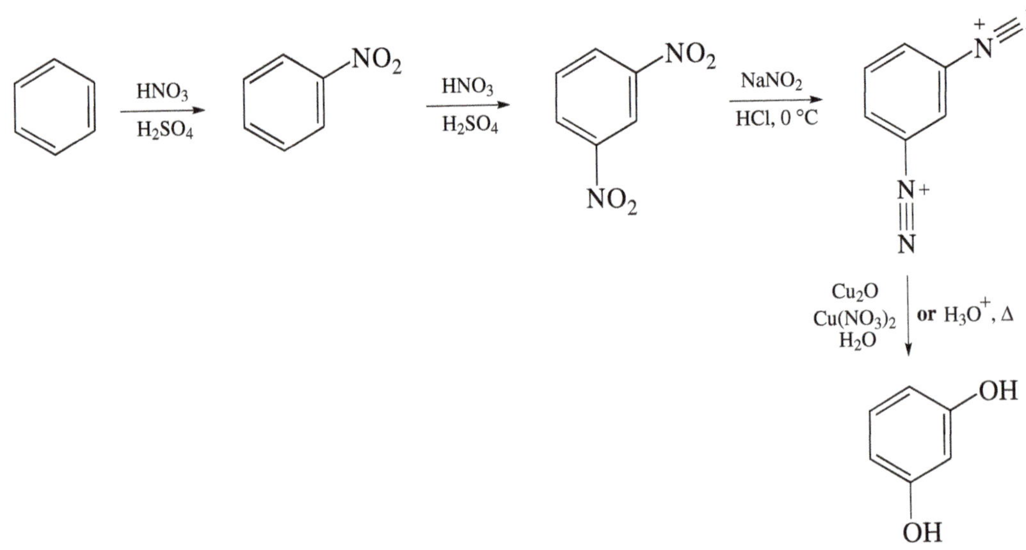

$\xrightarrow[H_2SO_4]{HNO_3}$ NO_2 $\xrightarrow[H_2SO_4]{HNO_3}$ NO_2, NO_2 $\xrightarrow[HCl,\ 0\ °C]{NaNO_2}$ $\overset{+}{N}\equiv N$, $N^+\equiv N$

$\downarrow$ Cu_2O, $Cu(NO_3)_2$, H_2O **or** H_3O^+, Δ

OH, OH

81.

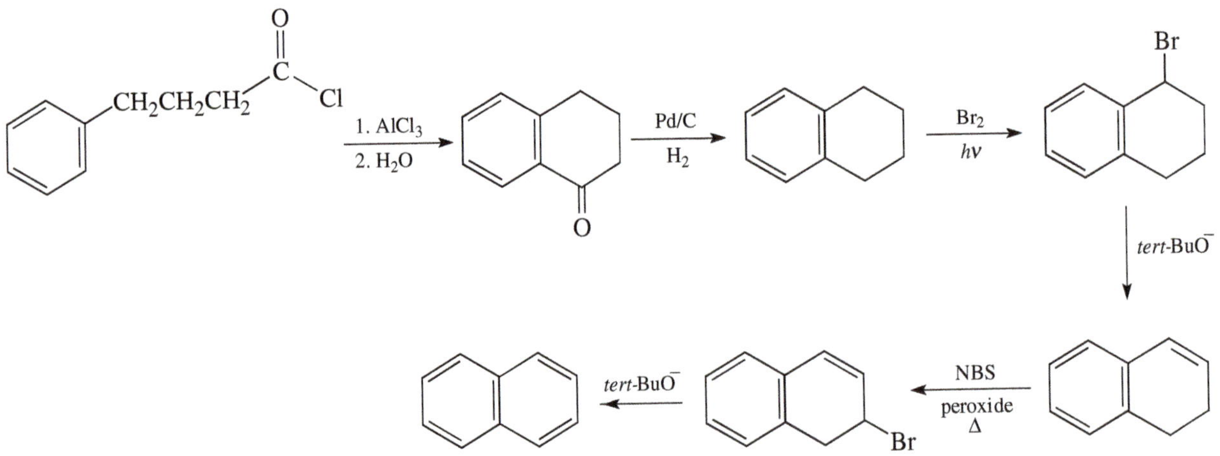

$CH_2CH_2CH_2C(=O)Cl$ $\xrightarrow[2.\ H_2O]{1.\ AlCl_3}$ O $\xrightarrow[H_2]{Pd/C}$ $\xrightarrow[h\nu]{Br_2}$ Br

$\downarrow$ *tert*-BuO$^-$

$\xleftarrow{tert\text{-}BuO^-}$ Br $\xleftarrow[\text{peroxide},\ \Delta]{NBS}$

82. The carbocation formed by putting an electrophile at the ortho or para positions can be stabilized by resonance electron donation from the phenyl substituent.

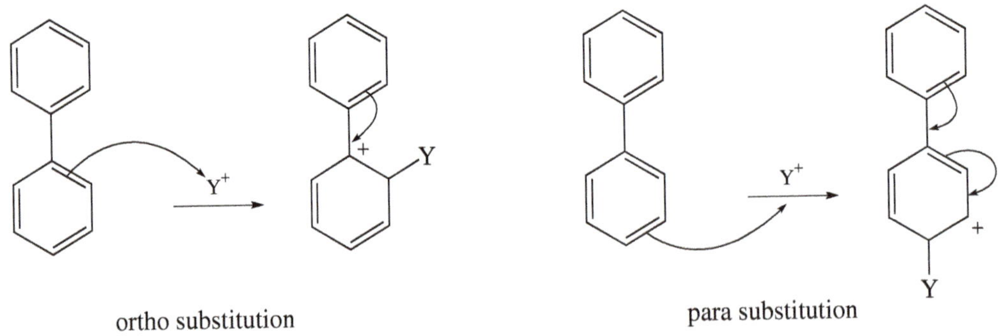

$\xrightarrow{Y^+}$ +, Y

ortho substitution

$\xrightarrow{Y^+}$ +, Y

para substitution

The carbocation formed by putting an electrophile at the meta position cannot be stabilized by resonance electron donation from the phenyl substituent.

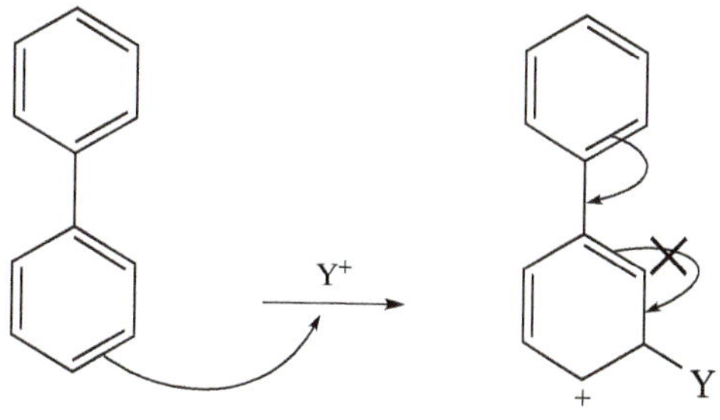

Y^+ Y +

meta substitution

83. **1.** $\xrightarrow[FeBr_3]{Br_2}$ Br $\xrightarrow[H_2SO_4]{HNO_3}$ O_2N Br $\xrightarrow[\Delta]{CH_3O^-}$ O_2N OCH_3 $\xrightarrow{H_2, Pd/C}$ H_2N OCH_3

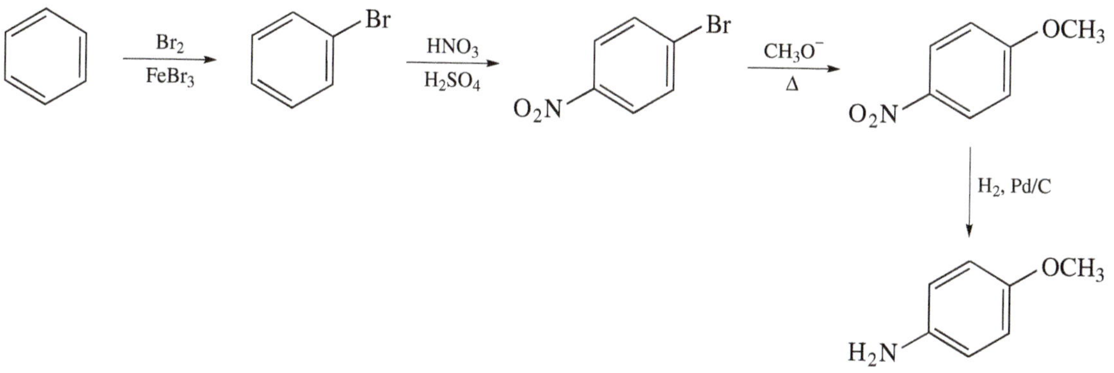

2. $\xrightarrow[H_2SO_4]{HNO_3}$ NO_2 $\xrightarrow{H_2, Pd/C}$ NH_2 $\xrightarrow[NaNO_2, 0\ °C]{HCl}$ $\overset{+}{N}{\equiv}N$ Cl^- $\xrightarrow{CH_3OH \mid HCl, \Delta}$ OCH_3 $\xrightarrow[H_2SO_4]{HNO_3}$ O_2N OCH_3 $\xrightarrow{H_2, Pd/C}$ H_2N OCH_3

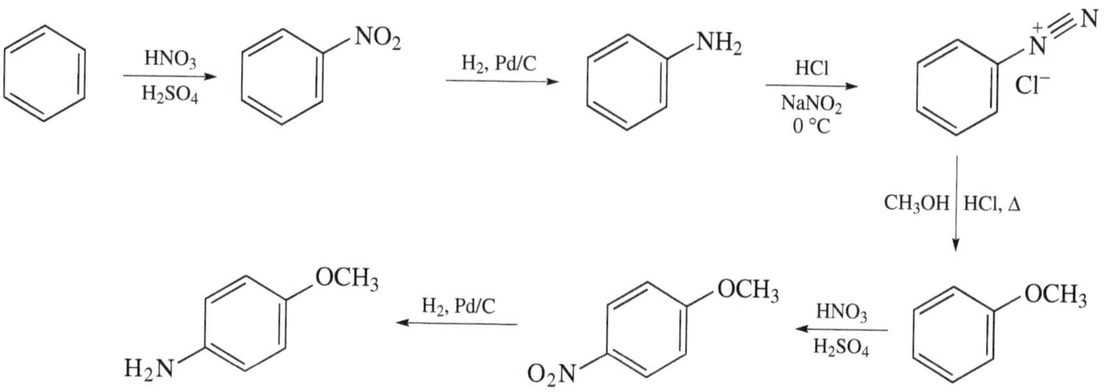

84. Synthesizing benzaldehyde from benzene would be easy if formyl chloride could be used. However, this compound is unstable and must be generated in situ via a Gatterman–Koch reaction (see page 884 in the text), or it can be synthesized via a Friedel–Crafts alkylation reaction. Conversion of the methyl ketone to a carboxylic acid is called a haloform reaction. (See Chapter 17, Problem 58.)

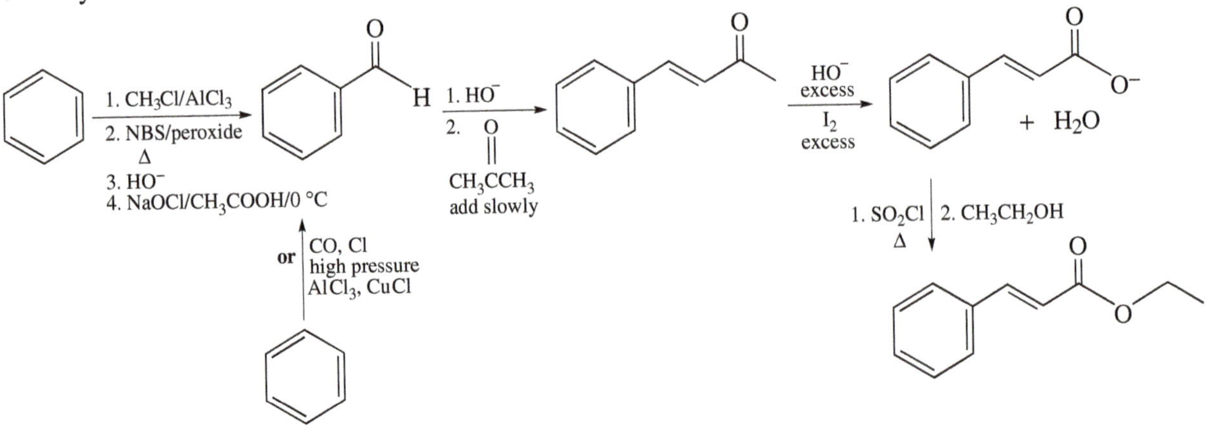

1. $CH_3Cl/AlCl_3$
2. NBS/peroxide, Δ
3. HO^-
4. $NaOCl/CH_3COOH/0\ °C$

or: CO, Cl; high pressure; $AlCl_3$, CuCl

1. HO^-
2. CH_3CCH_3 (O), add slowly

HO^- excess, I_2 excess

$+ H_2O$

1. SO_2Cl, Δ 2. CH_3CH_2OH

85.

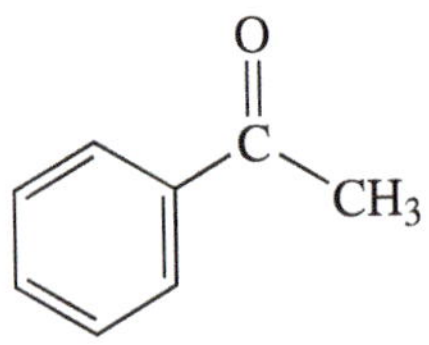

A: $C_6H_5C(=O)CH_3$

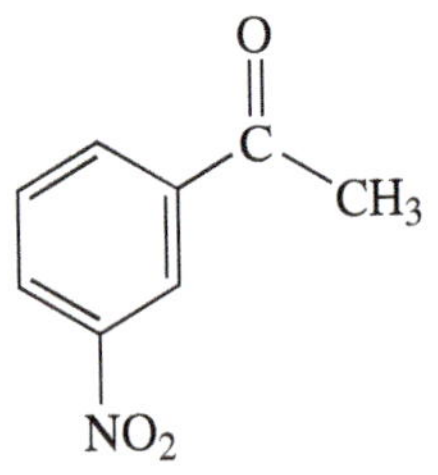

B: $C_6H_4(NO_2)C(=O)CH_3$

C: $C_6H_5CH_2CH_3$

D: C_6H_5COOH

E: $C_6H_5CHBrCH_3$

F: $C_6H_5CH(OCH_3)CH_3$ + $C_6H_5CH{=}CH_2$

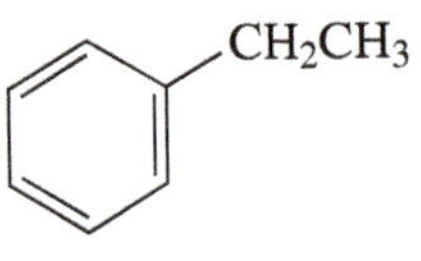

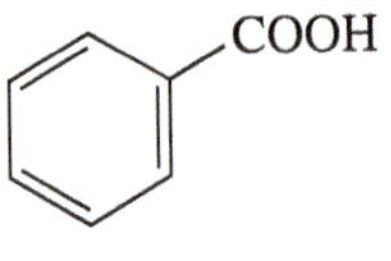

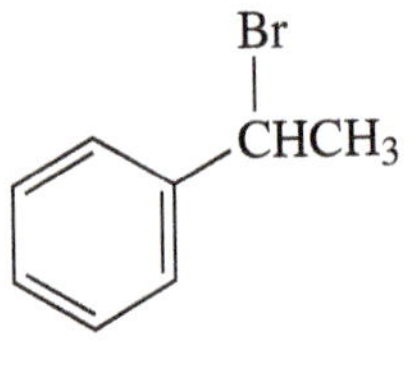

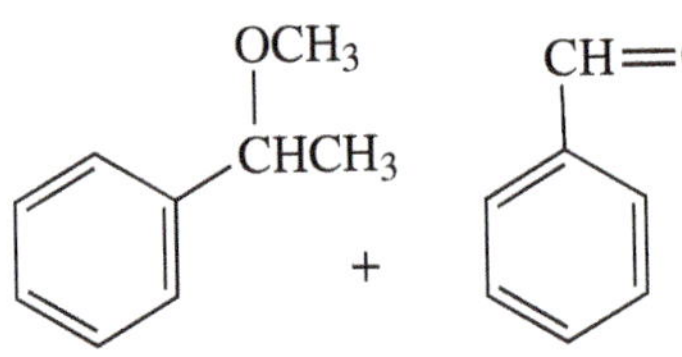

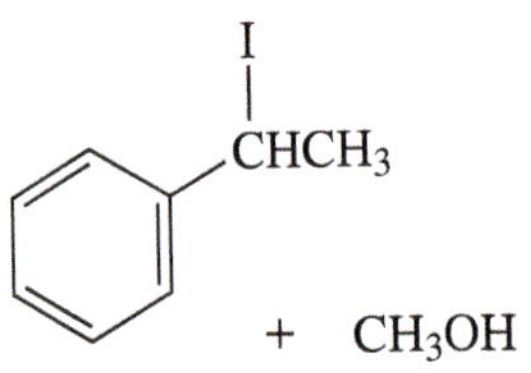

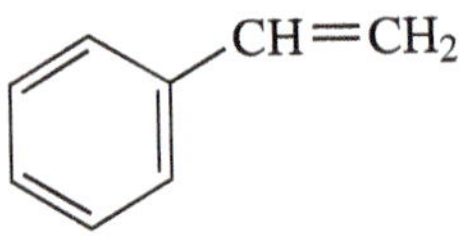

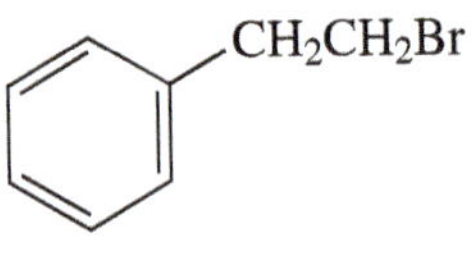

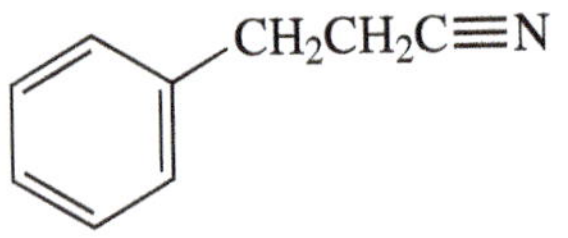

G: $C_6H_5CHICH_3$ + CH_3OH

H: $C_6H_5CH{=}CH_2$

I: $C_6H_5CH_2CH_2Br$

J: $C_6H_5CH_2CH_2C{\equiv}N$

86. **a.** The hydroxy-substituted carbocation intermediate is more stable because the positive charge can be stabilized by resonance electron donation from the OH group.

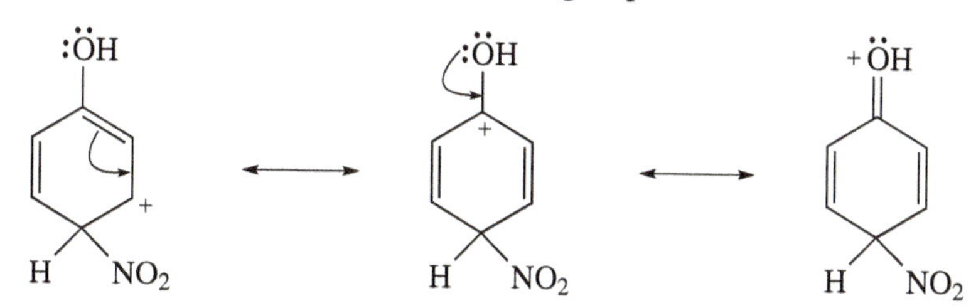

b. The carbanion with the negative charge meta to the nitro group is more stable because a negative charge in this position can be delocalized onto the nitro group but a negative charge ortho to the nitro group position cannot.

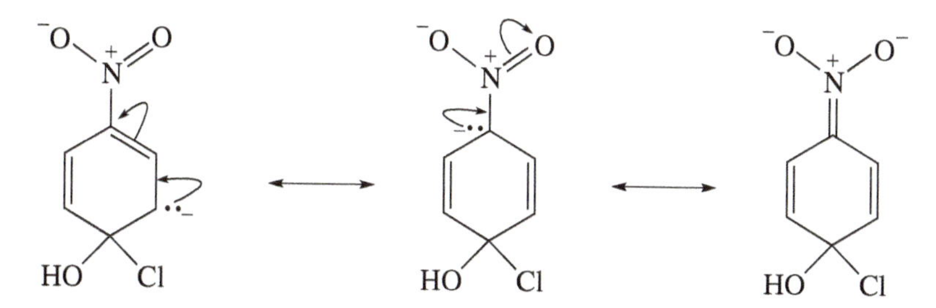

87. **a.** The chloro substituent primarily withdraws electrons inductively. (It only minimally donates electrons by resonance.) The closer it is to the COOH group, the more it withdraws electrons from the OH bond and the stronger the acid. Therefore, the ortho isomer is the strongest acid and the para isomer is the weakest acid.

b. The nitro substituent withdraws electrons inductively. It also withdraws electrons by resonance, if it is ortho or para to the COOH group. Therefore, the ortho and para isomers are the strongest acids, and the ortho isomer is a stronger acid than the para isomer because of greater inductive electron withdrawal from the closer position.

c. The amino substituent primarily donates electrons by resonance, but it can donate electrons by resonance to the COOH group only if it is ortho or para to it. Therefore, the meta isomer is the strongest acid. Seeing that the ortho isomer is the weakest acid tells us that resonance electron donation to the COOH group is more efficient from the ortho position.

88. **a.**

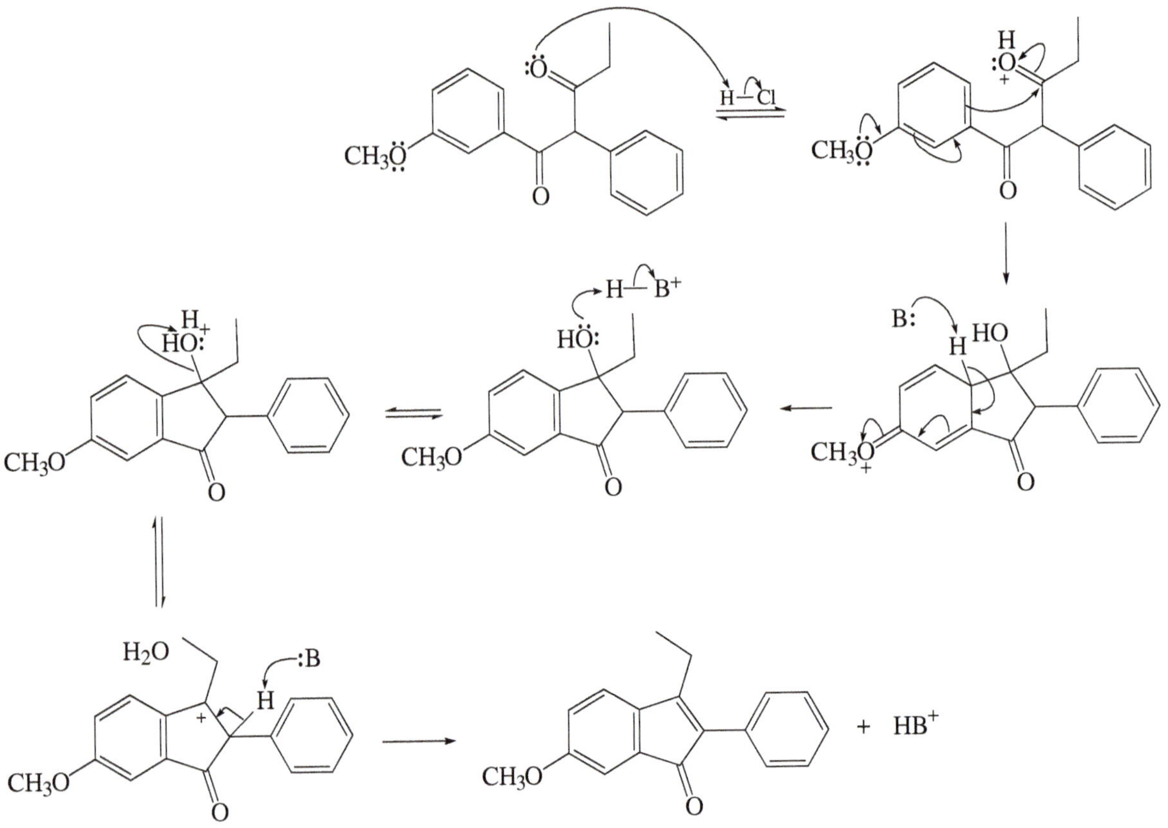

b.

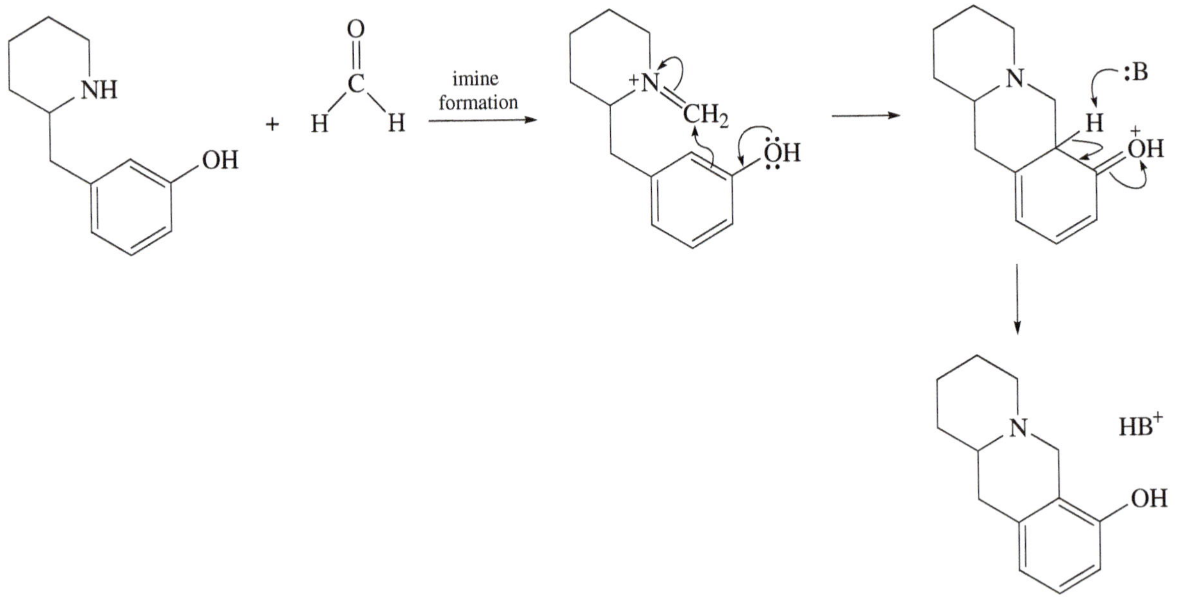

89. A fluoro substituent is more electronegative than a chloro substituent. Therefore, nucleophilic attack on the carbon bearing the fluoro substituent is easier than nucleophilic attack on the carbon bearing the chloro substituent. In addition, the smaller fluoro substituent provides less steric hindrance to attack by the nucleophile.

A fluoro substituent is a stronger base than a chloro substituent, so elimination of the halogen in the second step of the reaction is harder for a fluoro-substituted benzene than for a chloro-substituted benzene.

The fact that the fluoro-substituted compound is more reactive tells you that attack of the nucleophile on the aromatic ring is the rate-determining step of the reaction.

90. The signal at 7.1–7.3 ppm that integrates to 5 protons indicates a monosubstituted benzene ring. The doublet at 1.3 ppm that integrates to 6 protons and the septet at 2.9 ppm that integrates to 1 proton indicate an isopropyl group. Therefore, compound **A** is isopropylbenzene.

91. **a.** benzene $\xrightarrow[AlCl_3]{CH_3Cl}$ toluene (CH_3) $\xrightarrow[FeBr_3]{Br_2}$ p-bromotoluene (CH_3, Br) $\xrightarrow[PdL_2,\ HO^-]{CH_2{=}C(CH_3)B(OR)_2}$ p-methyl-substituted benzene with $CH_3C{=}CH_2$

or

benzene $\xrightarrow[AlCl_3]{CH_3CHClCH_3}$ CH_3CHCH_3-substituted benzene $\xrightarrow[h\nu]{Br_2}$ CH_3CBrCH_3-substituted benzene $\xrightarrow{tert\text{-}BuO^-}$ $CH_3C{=}CH_2$-substituted benzene $\xrightarrow[AlCl_3]{CH_3Cl}$ p-CH_3, $CH_3C{=}CH_2$-substituted benzene

b. benzene $\xrightarrow[FeBr_3]{Br_2}$ bromobenzene $\xrightarrow[H_2SO_4]{HNO_3}$ p-bromonitrobenzene (Br, NO_2) $\xrightarrow[FeBr_3]{Br_2}$ 1,2-dibromo-4-nitrobenzene (Br, Br, NO_2) $\xrightarrow[\Delta]{CH_3O^-}$ OCH_3, Br, NO_2-substituted benzene

c. benzene $\xrightarrow[2.\ H_2O]{1.\ (CH_3)_2CHCCl\ (C{=}O),\ AlCl_3}$ $C_6H_5C(=O)CH(CH_3)_2$ $\xrightarrow{H_2,\ Pd/C\ \text{or}\ H_2NNH_2,\ HO^-,\ \Delta}$ $C_6H_5CH_2CH(CH_3)_2$ $\xrightarrow{1.\ CH_3CCl\ (C{=}O),\ AlCl_3\quad 2.\ H_2O}$ p-$CH_3C(=O)$–C_6H_4–$CH_2CH(CH_3)_2$ $\xrightarrow{H_2SO_4}$ $CH_3C(=O)$, SO_3H, $CH_2CH(CH_3)_2$-substituted benzene

d. benzene $\xrightarrow[\text{2. } H_2O]{\text{1. } CH_3\overset{O}{\overset{\|}{C}}Cl,\ AlCl_3}$ $CH_3C(=O)$–benzene $\xrightarrow[FeBr_3]{Br_2}$ $CH_3C(=O)$–benzene–Br

$\xrightarrow{H_2,\ Pd/C}$ CH_3CH_2–benzene–Br $\xrightarrow[AlCl_3]{(CH_3)_3CCl}$ CH_3CH_2–benzene(–$C(CH_3)_3$)–Br

92. **a.** The first three compounds will not show a carbonyl stretch at 1700 cm^{-1}, and the bottom four will show this absorption band. The first three can be distinguished from one another by the presence or absence of the indicated absorption bands.

CH_2OH (cyclohexane ring)	CH_2OH (benzene ring)	CH_2OCH_3 (benzene ring)
band at 3300 cm^{-1} no band at 1600 cm^{-1}	band at 3300 cm^{-1} band at 1600 cm^{-1}	no band at 3300 cm^{-1} band at 1600 cm^{-1}

The last four compounds all have an absorption band at 1700 cm^{-1}.
They can be distinguished by the presence or absence of the indicated absorption bands.

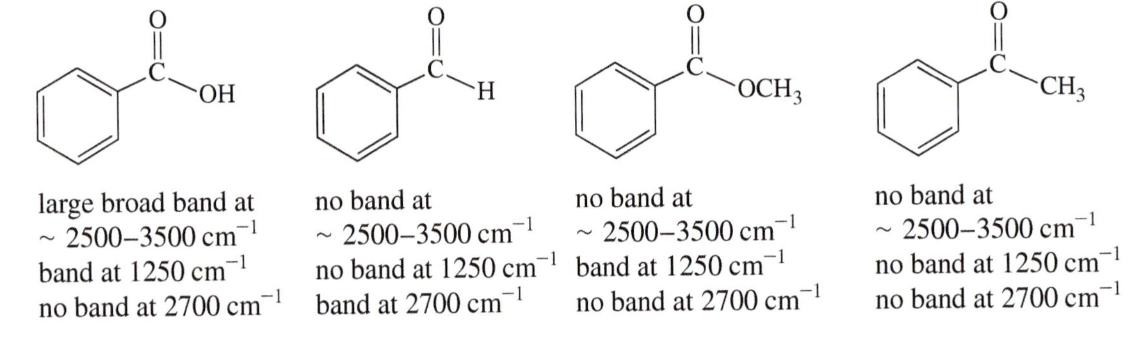

b. This is the only compound without the characteristic benzene ring hydrogens at ~ 7–8 ppm.

CH_2OH (cyclohexane ring)

Only two compounds will have two signals other than the signals for the benzene ring hydrogens. They can be distinguished by integration (2:3 versus 2:1) or by the two sharp singlets for the ether versus the somewhat broader singlet for the hydrogen bonded to oxygen.

CH_2OCH_3 (benzene ring) CH_2OH (benzene ring)

Each of the following four compounds has only one signal (a singlet) in addition to the benzene ring hydrogens. The four can be distinguished by the position of the singlet.

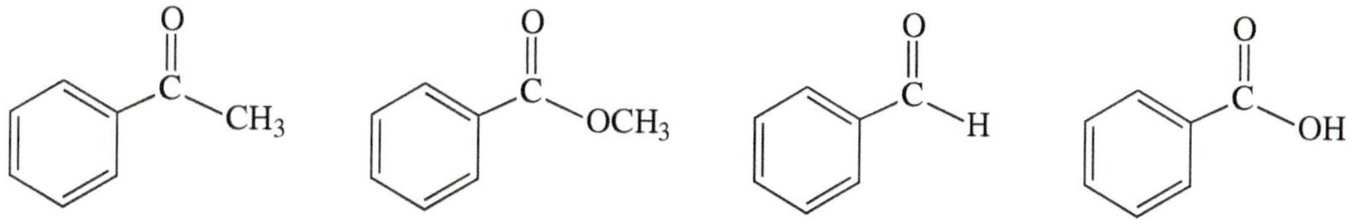

~2 ppm ~4 ppm ~9–10 ppm ~10–12 ppm

93.

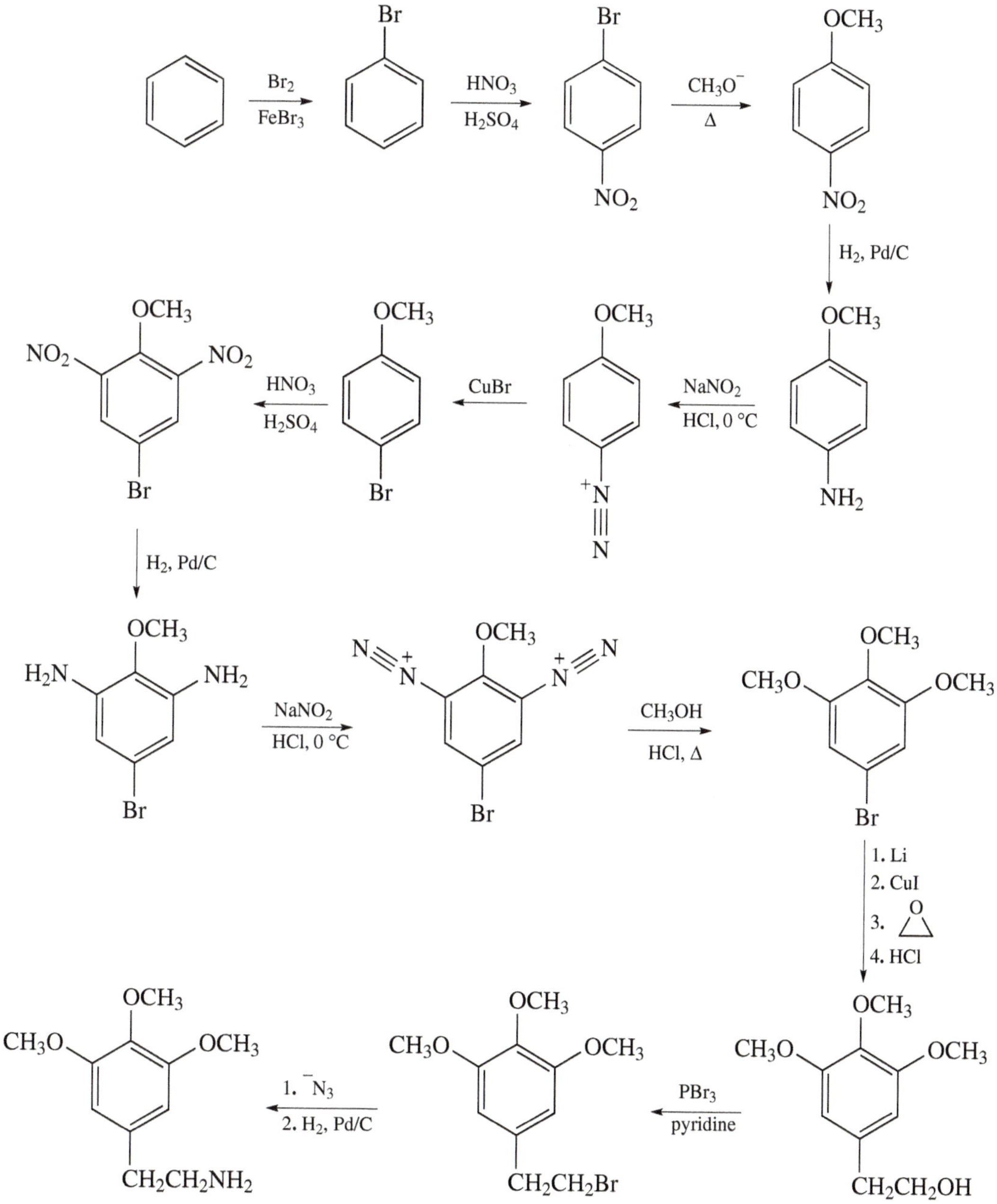

For the last reaction, see Problem 17 on page 405.

94. The configuration of the asymmetric center in the reactant will be retained only if the asymmetric center undergoes two successive S_N2 reactions. In the first step of the reaction, the NH_2 group is converted to a good leaving group. The first S_N2 reaction involves the carboxylate ion that is closest to the carbon attached to the leaving group. The product of the reaction is a strained three-membered ring that reacts with the other carboxylate ion in a second S_N2 reaction, forming a more stable five-membered ring.

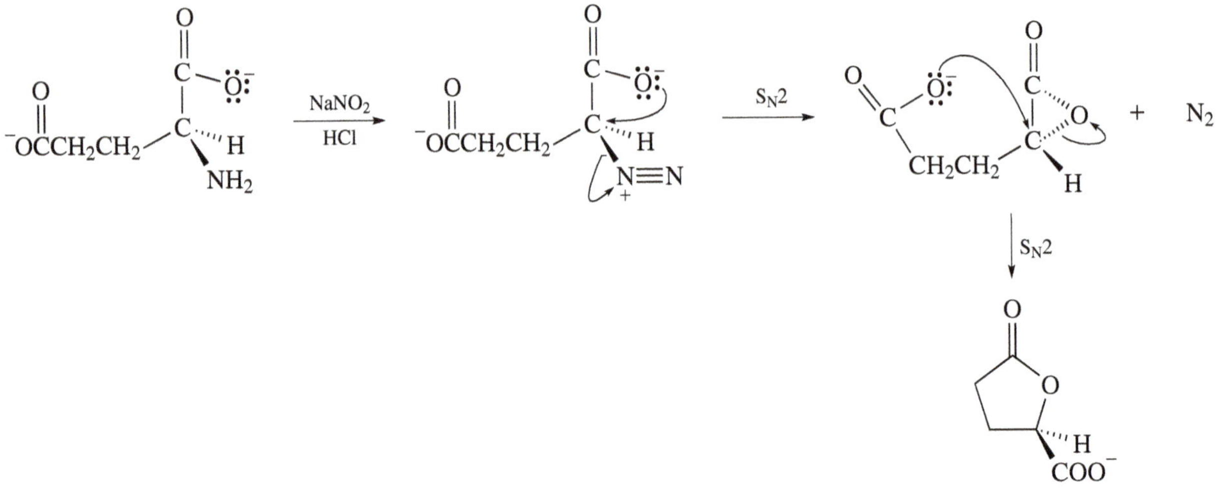

NaNO$_2$ / HCl; S_N2; + N_2; S_N2

95. a. + HB^+

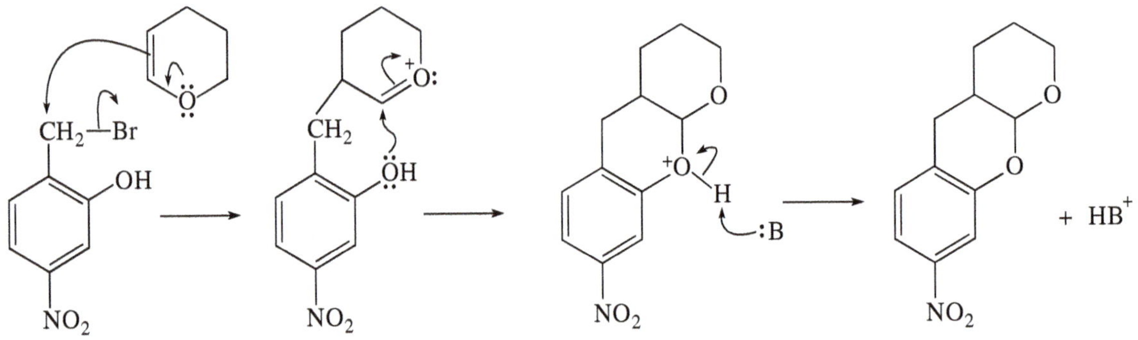

b.

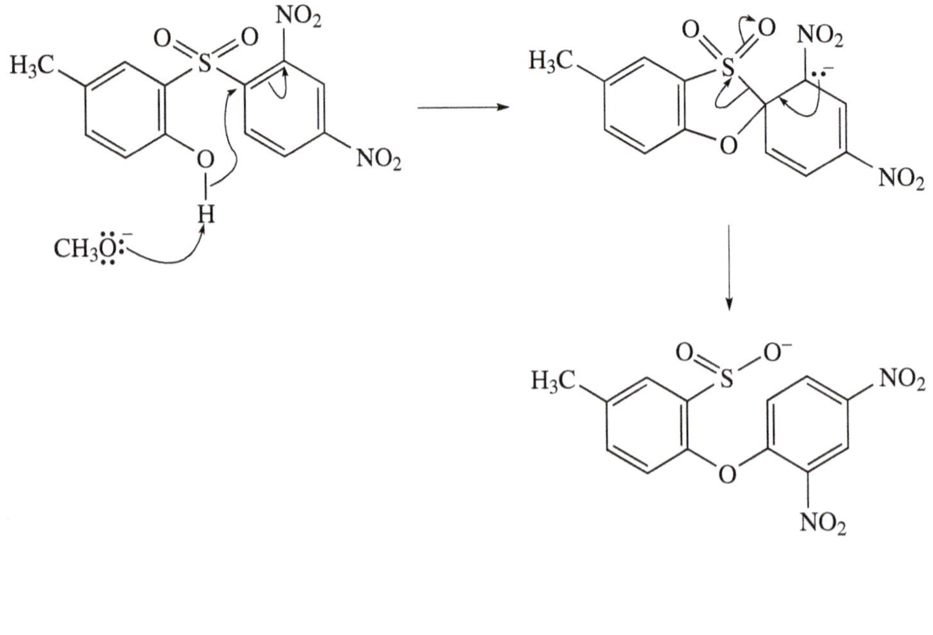

96.

Benzene $\xrightarrow[FeBr_3]{Br_2}$ bromobenzene $\xrightarrow[2.\ CuI]{1.\ Li}$ $(C_6H_5)_2CuLi$ $\xrightarrow[2.\ HCl]{1.\ \text{ethylene oxide}}$ $C_6H_5CH_2CH_2OH$ $\xrightarrow{PBr_3/\text{pyridine}}$ $C_6H_5CH_2CH_2Br$ $\xrightarrow{Mg,\ Et_2O}$ $C_6H_5CH_2CH_2MgBr$

CH_3CHO $\xrightarrow[2.\ HCl]{1.\ CH_3CH_2MgBr}$ $CH_3CH(OH)CH_2CH_3$ $\xrightarrow[CH_3COOH,\ 0\ °C]{NaOCl}$ $CH_3COCH_2CH_3$

$C_6H_5CH_2CH_2MgBr$ + $CH_3COCH_2CH_3$ $\longrightarrow$ $C_6H_5CH_2CH_2C(O^-)(CH_3)CH_2CH_3$ $\xrightarrow{HCl}$ $C_6H_5CH_2CH_2C(OH)(CH_3)CH_2CH_3$

97. The unsplit signal at ~7.1 ppm suggests that all the benzene ring hydrogens in the product are chemically equivalent. The triplet at 1.2 ppm and quartet at 2.6 ppm indicate an ethyl group.

$CH_3CH_2-C_6H_5 \xrightarrow[AlCl_3]{CH_3CH_2Cl} CH_3CH_2-C_6H_4-CH_2CH_3$ (para)

98. **a.** The alkyl diazonium ion is very unstable. Loss of N_2 and a 1,2-hydride shift forms a *tert*-butyl carbocation, which can undergo either substitution or elimination.

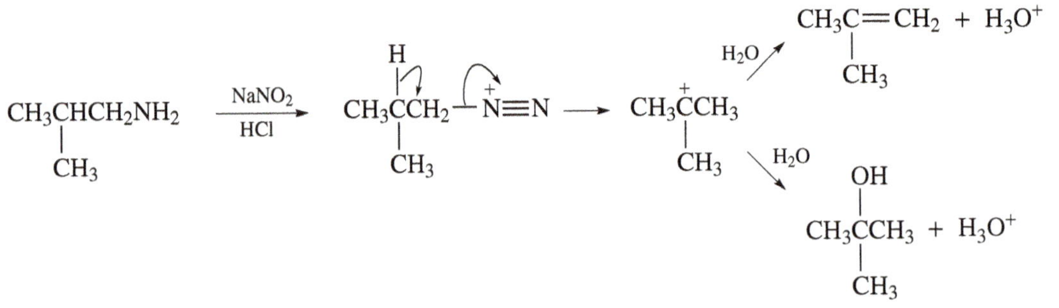

$$CH_3CH(CH_3)CH_2NH_2 \xrightarrow[HCl]{NaNO_2} CH_3CH(CH_3)CH_2-\overset{+}{N}\equiv N \longrightarrow CH_3\overset{+}{C}(CH_3)CH_3$$

$$CH_3\overset{+}{C}(CH_3)CH_3 \xrightarrow{H_2O} CH_3C(CH_3)=CH_2 + H_3O^+$$

$$CH_3\overset{+}{C}(CH_3)CH_3 \xrightarrow{H_2O} CH_3C(OH)(CH_3)CH_3 + H_3O^+$$

b. The cation formed from the diazonium ion will undergo a pinacol-like rearrangement. (See Chapter 10, Problem 92.)

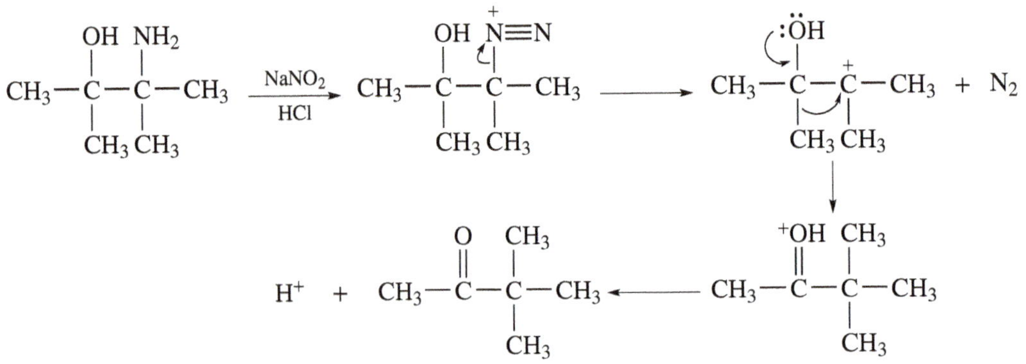

$$CH_3-C(OH)(CH_3)-C(NH_2)(CH_3)-CH_3 \xrightarrow[HCl]{NaNO_2} CH_3-C(OH)(CH_3)-C(\overset{+}{N}\equiv N)(CH_3)-CH_3 \longrightarrow CH_3-C(\ddot{O}H)(CH_3)-\overset{+}{C}(CH_3)-CH_3 + N_2$$

$$\longrightarrow CH_3-C(={}^{+}OH)-C(CH_3)_2-CH_3 \longrightarrow H^+ + CH_3-C(=O)-C(CH_3)_2-CH_3$$

99. A chloro group is a better leaving group than the ammonium group, so the product is formed without hydroxide ion catalysis.

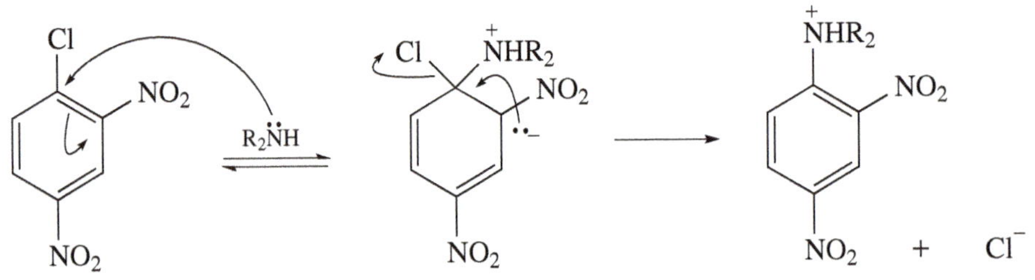

$$\text{1-chloro-2,4-dinitrobenzene} \underset{}{\overset{R_2\ddot{N}H}{\rightleftharpoons}} \text{adduct (Cl, }\overset{+}{N}HR_2\text{; } NO_2\text{, } NO_2\text{)} \longrightarrow \text{2,4-dinitro-}\overset{+}{N}HR_2\text{-benzene} + Cl^-$$

A methoxy group is a poorer leaving group than the ammonium group, so the ammonium group is eliminated, reforming starting materials. If hydroxide ion is added to the solution, hydroxide ion converts the ammonium group into an amino group. Because the amino group is a poorer leaving group than the methoxy group, the methoxy group will be eliminated.

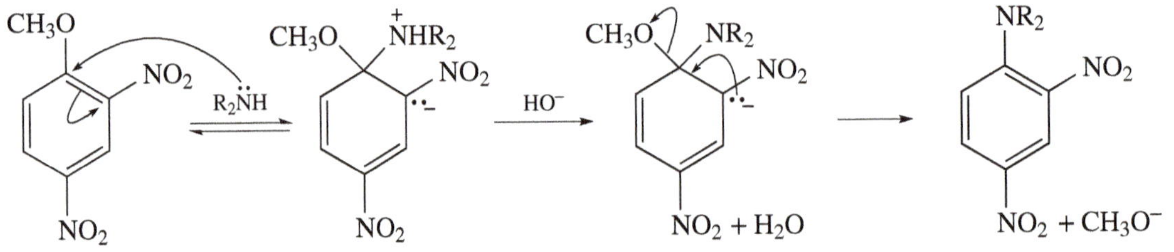

$$\text{1-methoxy-2,4-dinitrobenzene} \overset{R_2\ddot{N}H}{\rightleftharpoons} \text{adduct (}CH_3O\text{, }\overset{+}{N}HR_2\text{; } NO_2\text{, } NO_2\text{)} \xrightarrow{HO^-} \text{adduct (}CH_3O\text{, }NR_2\text{; } NO_2\text{, } NO_2\text{)} + H_2O \longrightarrow \text{2,4-dinitro-}NR_2\text{-benzene} + CH_3O^-$$

100.

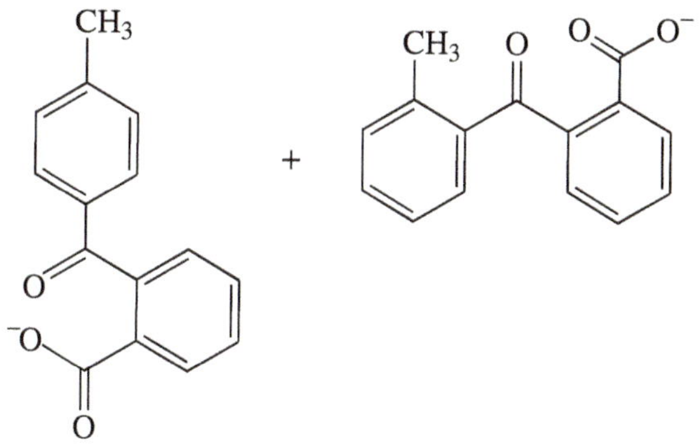

101. **a.**

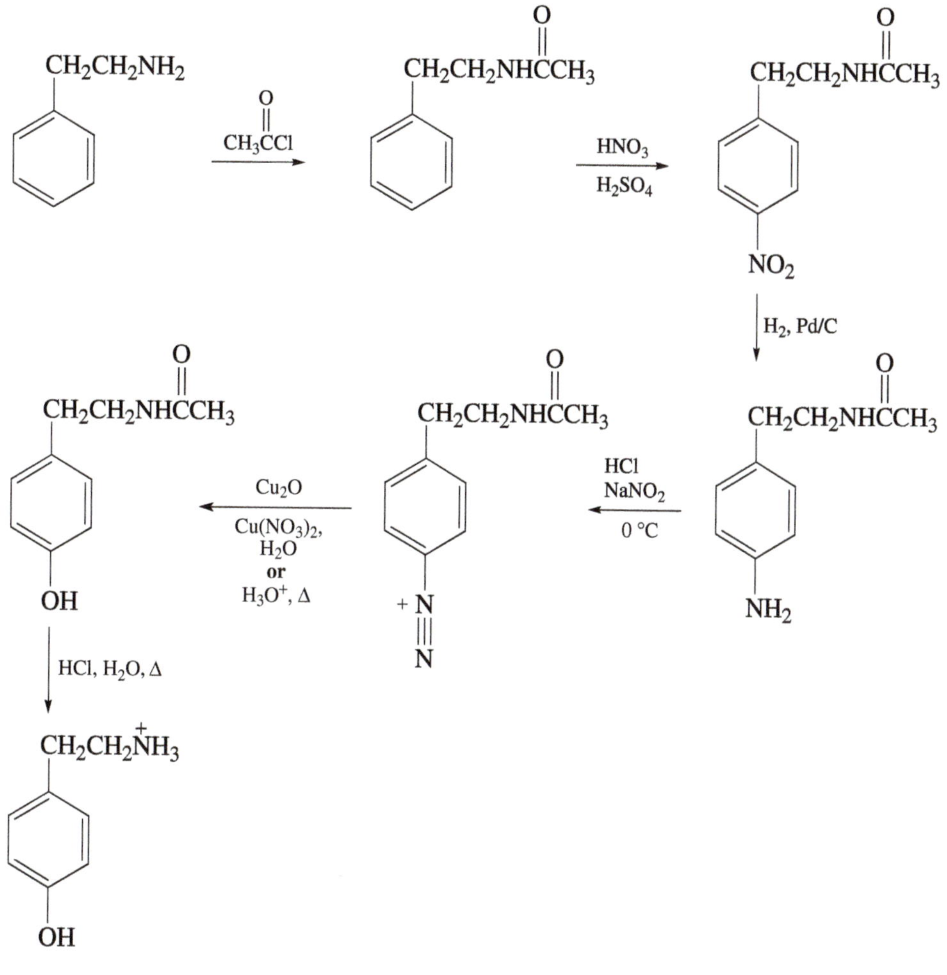

b. The NH_2 group must be protected before the compound reacts with nitric acid.

c. $C_6H_5CH_2CH_2Cl$ $\xrightarrow[\text{2. } H_2, \text{ Pd/C}]{\text{1. } ^-N_3}$ $C_6H_5CH_2CH_2NH_2$

$C_6H_5CH_2CH_2Cl$ $\xrightarrow[\text{2. HCl, } H_2O, \Delta \text{; 3. } HO^-]{\text{1. phthalimide anion}}$ $C_6H_5CH_2CH_2NH_2$

d. $C_6H_5CH_2Cl$ $\xrightarrow{^-C{\equiv}N}$ $C_6H_5CH_2C{\equiv}N$ $\xrightarrow[\text{Raney Ni}]{H_2}$ $C_6H_5CH_2CH_2NH_2$

e. C_6H_5CHO $\xrightarrow[\text{2. } H_3O^+]{\text{1. } NaBH_4}$ $C_6H_5CH_2OH$ $\xrightarrow[\text{pyridine}]{SOCl_2}$ $C_6H_5CH_2Cl$ $\xrightarrow{^-C{\equiv}N}$ $C_6H_5CH_2C{\equiv}N$ $\xrightarrow[\text{Raney Ni}]{H_2}$ $C_6H_5CH_2CH_2NH_2$

102. a. benzene $\xrightarrow[AlCl_3]{CH_3Cl}$ toluene (CH_3) $\xrightarrow[H_2SO_4]{HNO_3}$ o-nitrotoluene (CH_3, NO_2) $\xrightarrow[\Delta]{H_2CrO_4}$ (COOH, NO_2) $\xrightarrow[\text{Pd/C}]{H_2}$ (COOH, NH_2) $\xrightarrow[0\ °C]{NaNO_2,\ HCl}$ (COOH, $\overset{+}{N}{\equiv}N$ Cl^-) $\xrightarrow[\text{or } H_3O^+, \Delta]{Cu_2O,\ Cu(NO_3)_2,\ H_2O}$ (COOH, OH) $\xrightarrow{CH_3COCCH_3\ (\text{O O})}$ (COOH, $OC(=O)CH_3$)

b.

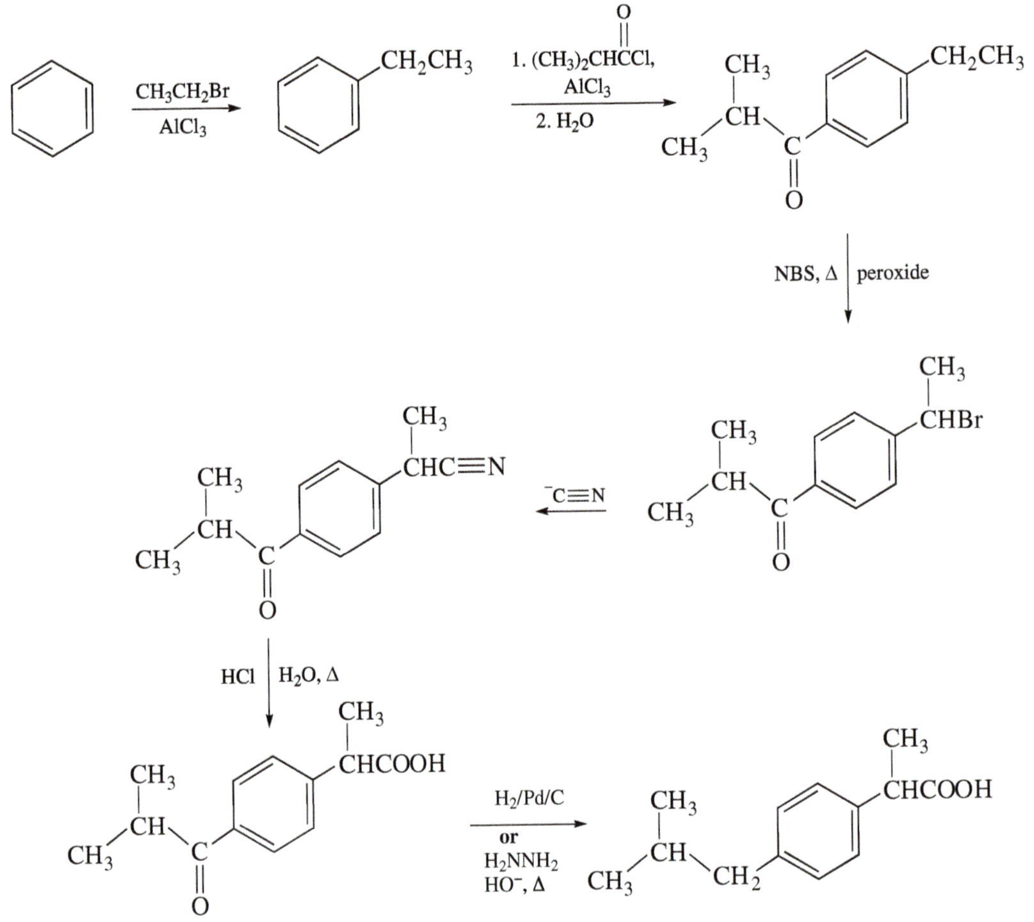

benzene $\xrightarrow[AlCl_3]{CH_3CH_2Br}$ ethylbenzene (CH_2CH_3) $\xrightarrow[\text{2. } H_2O]{\text{1. } (CH_3)_2CHCCl\ (C{=}O),\ AlCl_3}$ $(CH_3)_2CH{-}C({=}O){-}C_6H_4{-}CH_2CH_3$

$\xrightarrow{NBS,\ \Delta,\ \text{peroxide}}$ $(CH_3)_2CH{-}C({=}O){-}C_6H_4{-}CHBr(CH_3)$

$\xrightarrow{^{-}C{\equiv}N}$ $(CH_3)_2CH{-}C({=}O){-}C_6H_4{-}CH(CH_3)C{\equiv}N$

$\xrightarrow{HCl,\ H_2O,\ \Delta}$ $(CH_3)_2CH{-}C({=}O){-}C_6H_4{-}CH(CH_3)COOH$

$\xrightarrow{H_2/Pd/C\ \textbf{or}\ H_2NNH_2,\ HO^-,\ \Delta}$ $(CH_3)_2CH{-}CH_2{-}C_6H_4{-}CH(CH_3)COOH$

c.

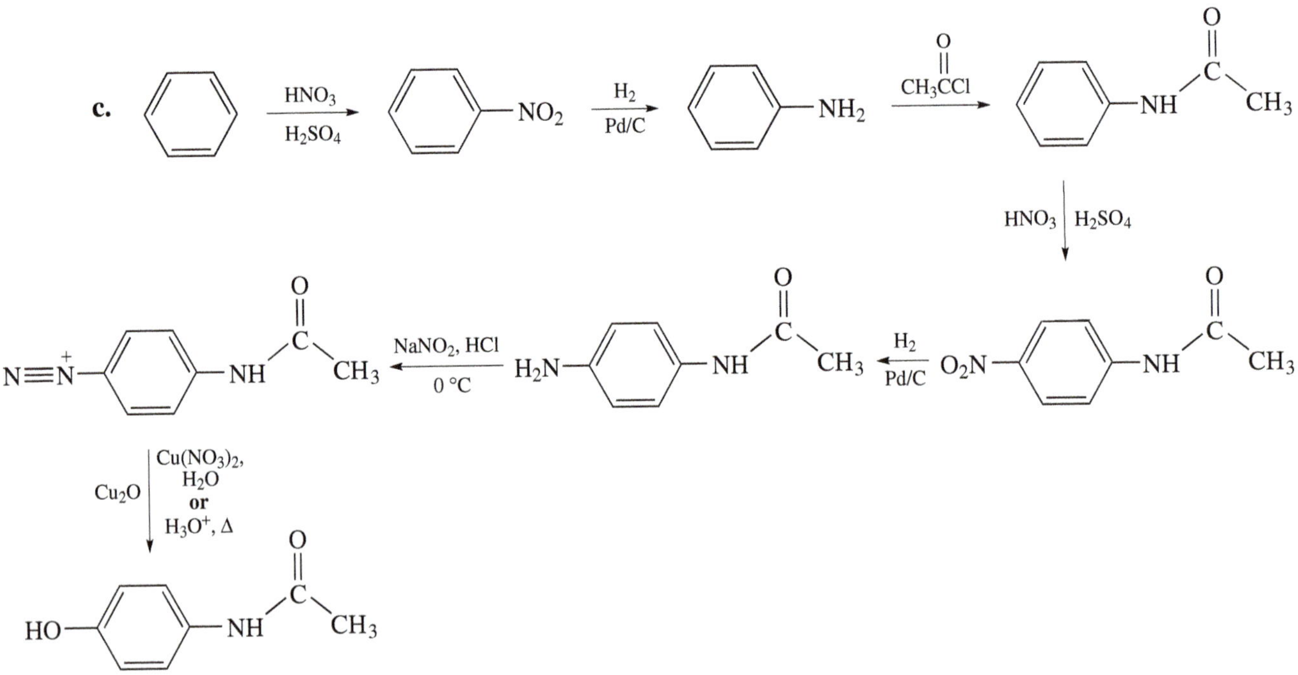

benzene $\xrightarrow[H_2SO_4]{HNO_3}$ $C_6H_5{-}NO_2$ $\xrightarrow[Pd/C]{H_2}$ $C_6H_5{-}NH_2$ $\xrightarrow{CH_3CCl\ (C{=}O)}$ $C_6H_5{-}NH{-}C({=}O){-}CH_3$

$\xrightarrow{HNO_3,\ H_2SO_4}$ $O_2N{-}C_6H_4{-}NH{-}C({=}O){-}CH_3$

$\xrightarrow[Pd/C]{H_2}$ $H_2N{-}C_6H_4{-}NH{-}C({=}O){-}CH_3$

$\xrightarrow[0\ ^\circ C]{NaNO_2,\ HCl}$ $N{\equiv}\overset{+}{N}{-}C_6H_4{-}NH{-}C({=}O){-}CH_3$

$\xrightarrow{Cu_2O,\ Cu(NO_3)_2,\ H_2O\ \textbf{or}\ H_3O^+,\ \Delta}$ $HO{-}C_6H_4{-}NH{-}C({=}O){-}CH_3$

103. **a.**

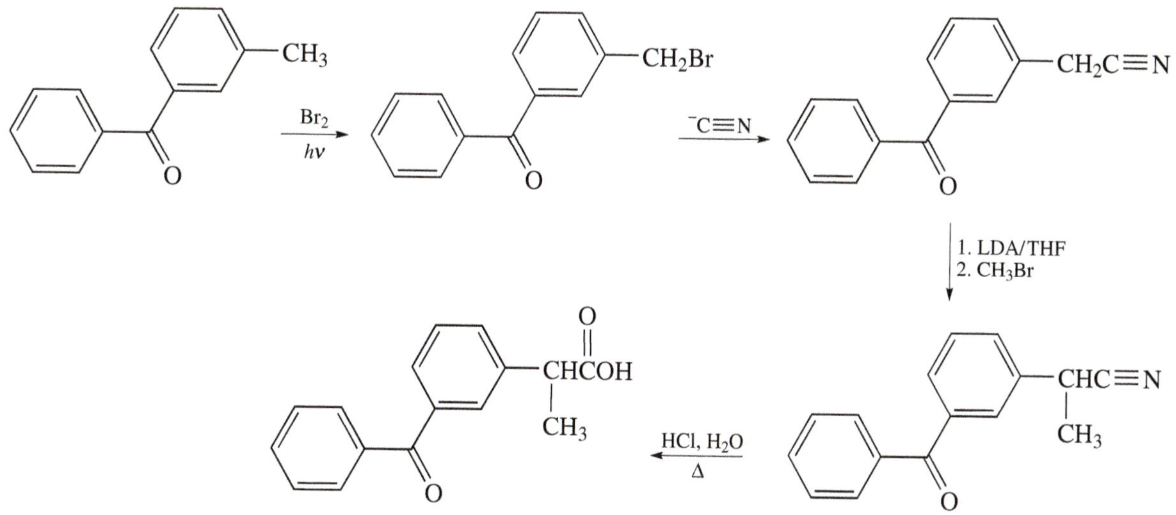

Br_2, $h\nu$; $^-C\equiv N$; 1. LDA/THF 2. CH_3Br; HCl, H_2O, Δ

CH_3 → CH_2Br → $CH_2C\equiv N$ → $CHC\equiv N$ (CH_3) → $CHCOH$ (=O, CH_3)

b. The substituent in ibuprofen is placed on the ring by a Friedel–Crafts acylation reaction. A Friedel–Crafts reaction cannot be done in the synthesis of ketoprofen, because the benzene rings are deactivated and deactivated rings cannot undergo Friedel–Crafts reactions.

104.

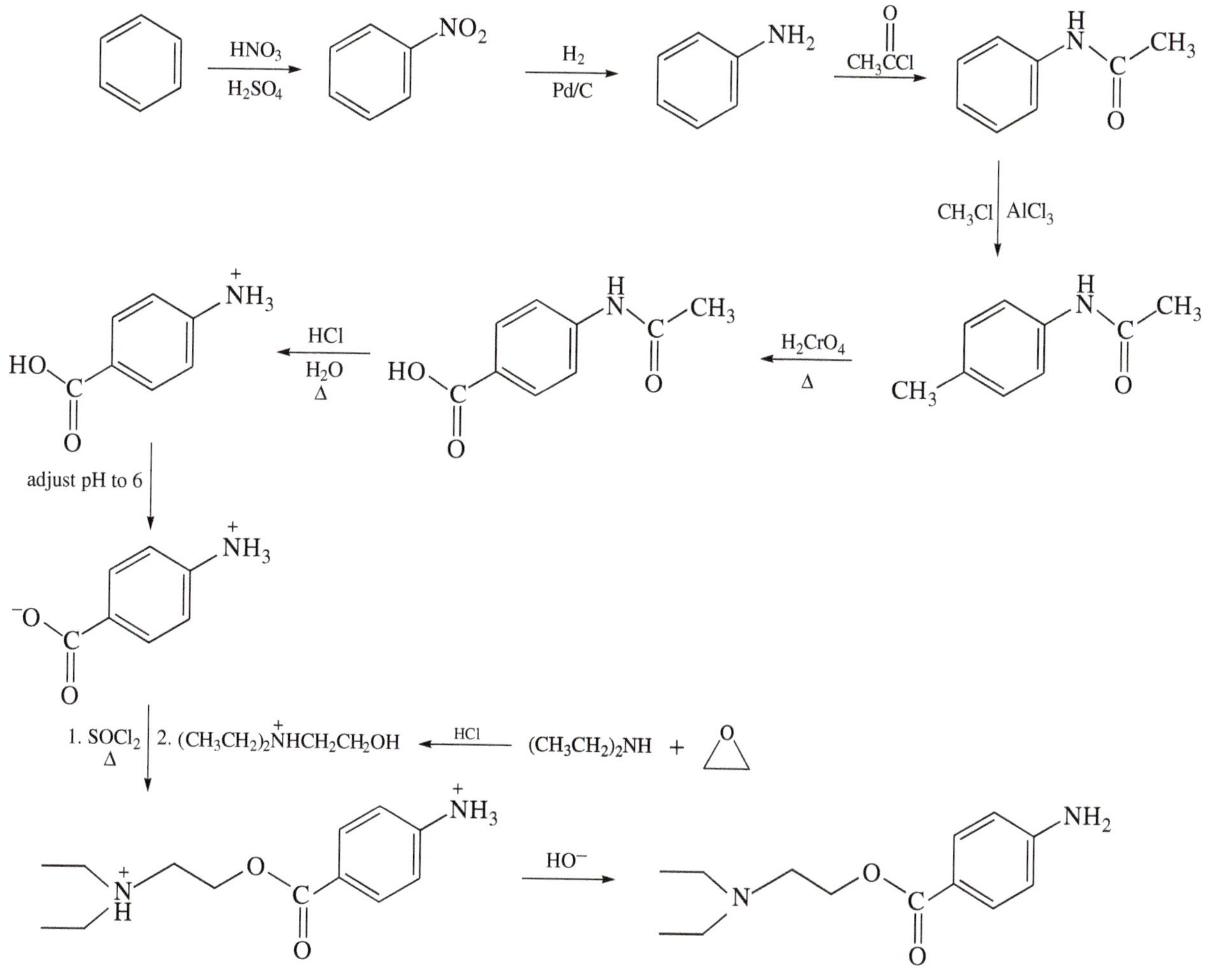

HNO_3/H_2SO_4 → NO_2; H_2, Pd/C → NH_2; CH_3CCl (=O) → $NHC(=O)CH_3$; CH_3Cl, $AlCl_3$ → CH_3, $NHC(=O)CH_3$; H_2CrO_4, Δ → HOC(=O), $NHC(=O)CH_3$; HCl, H_2O, Δ → HOC(=O), $\overset{+}{N}H_3$

adjust pH to 6 → $^-OC(=O)$, $\overset{+}{N}H_3$

1. $SOCl_2$, Δ 2. $(CH_3CH_2)_2\overset{+}{N}HCH_2CH_2OH$ ← HCl ← $(CH_3CH_2)_2NH$ + ethylene oxide

→ $\overset{+}{N}H$ ester, $\overset{+}{N}H_3$; HO^- → N ester, NH_2

105. Notice that the Friedel–Crafts reaction can be done on a ring with a sulfonic acid group because its deactivation is offset by the activating amide group. The sulfonic acid group is put on the ring to block the para position and thereby force the methyl groups to go to the ortho positions. Notice that the Friedel–Crafts alkylation reaction can be done because the deactivating effect of the meta director is offset by the activating effect of the amide. Also notice that because some (or all) of the amide may be hydrolyzed when the sulfonic acid group is removed, the acyl chloride is added to reform the amide.

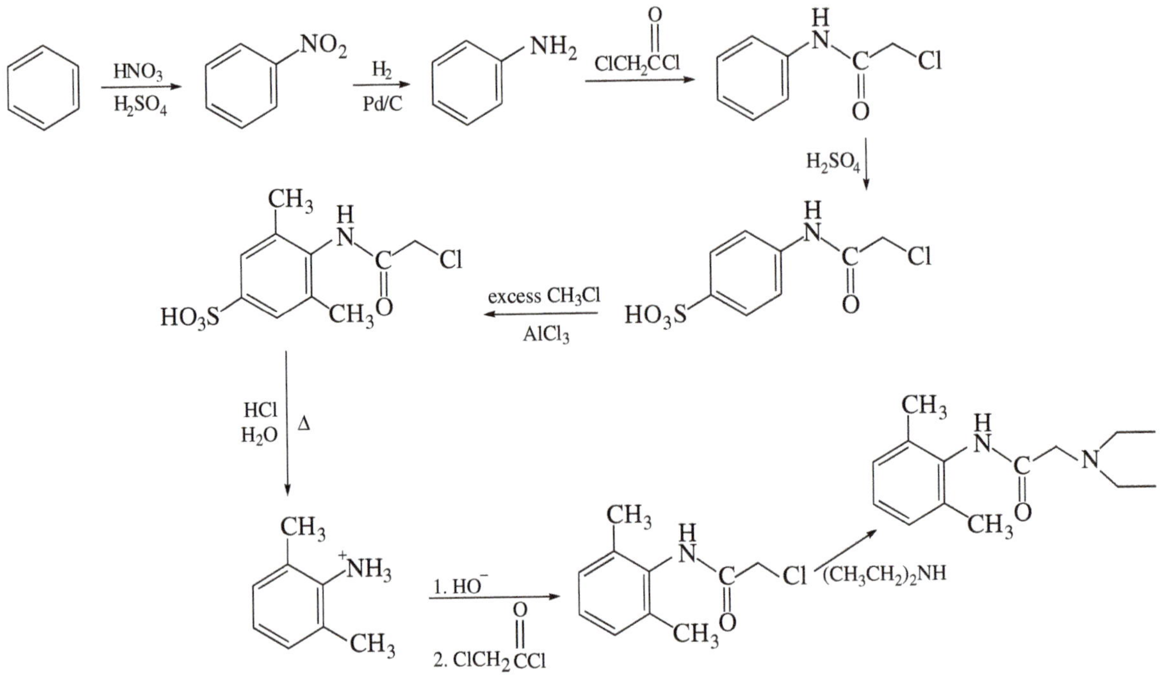

106. Notice that thionyl chloride can replace the OH group of a carboxylic acid and the OH group of a sulfonic acid with a Cl.

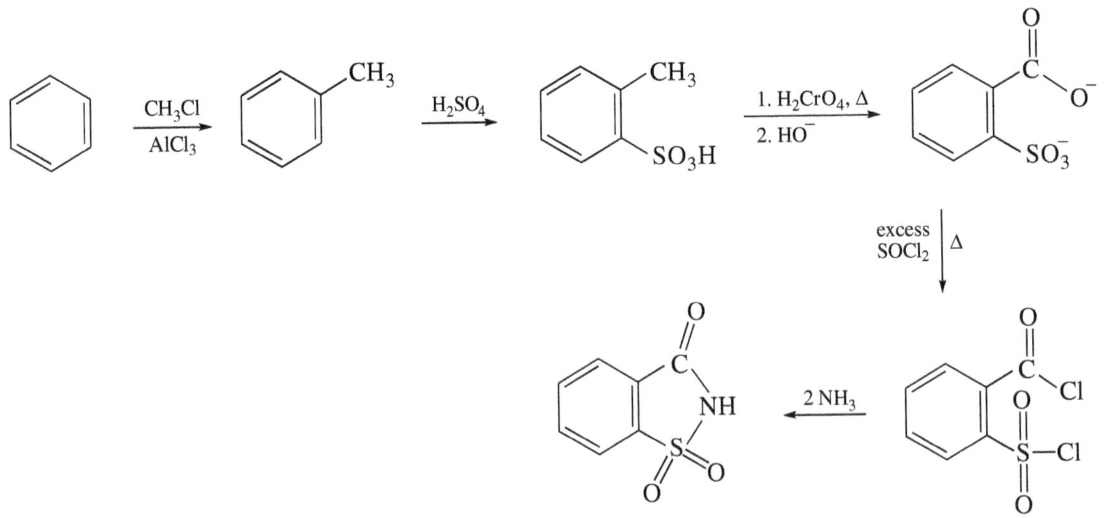

Chapter 18 Practice Test

1. Name the following:

a. 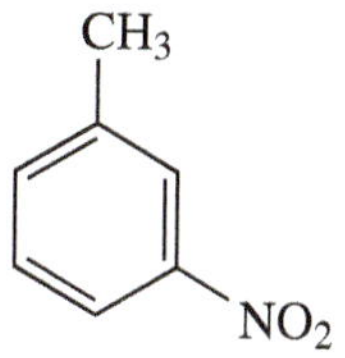 CH_3, NO_2

b. 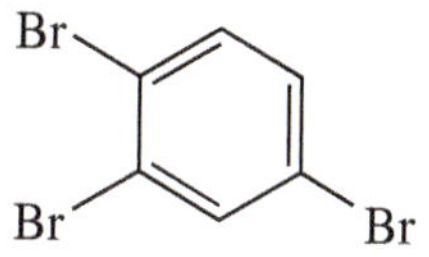 Br, Br, Br

c. 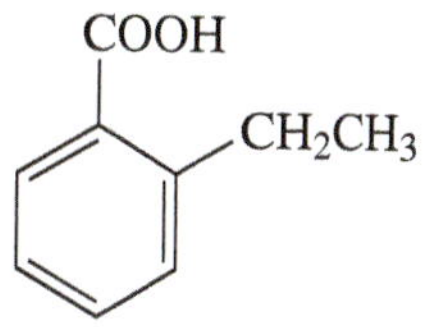 COOH, CH_2CH_3

d. 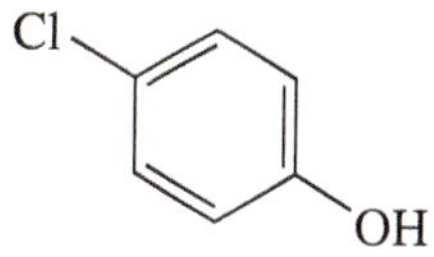 Cl, OH

2. Rank the following compounds from most reactive to least reactive toward reaction with $Br_2/FeBr_3$:

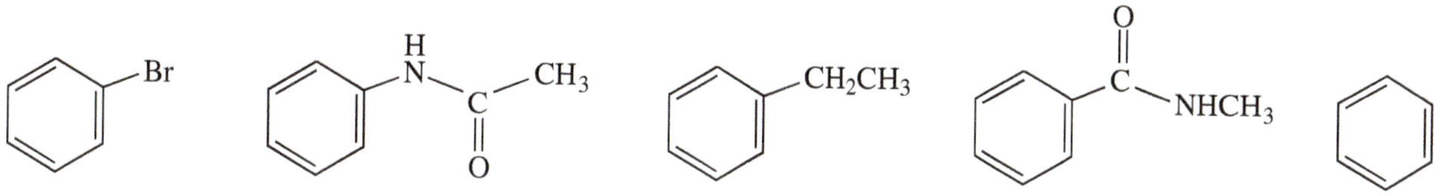

Br; $\mathrm{NH{-}C(=O){-}CH_3}$; CH_2CH_3; $\mathrm{C(=O){-}NHCH_3}$; benzene

3. For each of the following pairs, circle the stronger acid:

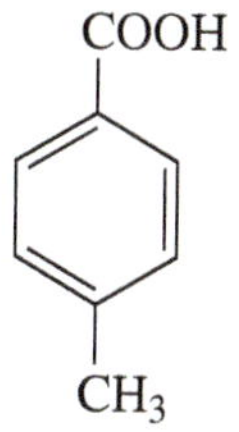

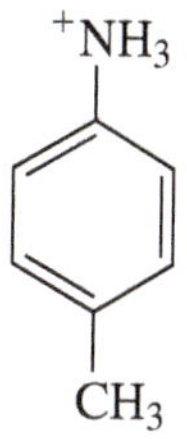

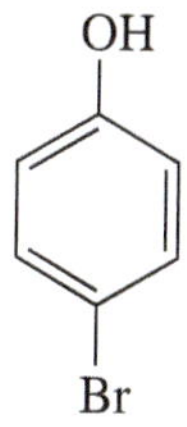

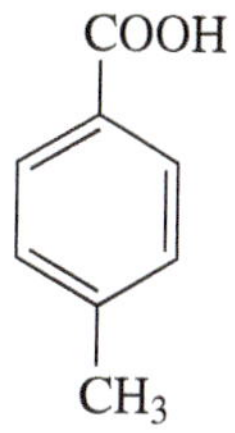

a. 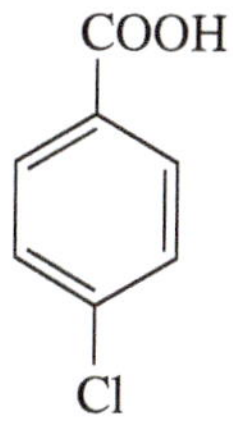 COOH / Cl or COOH / CH_3

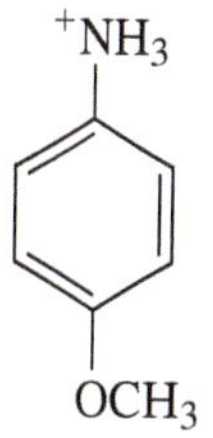

b. 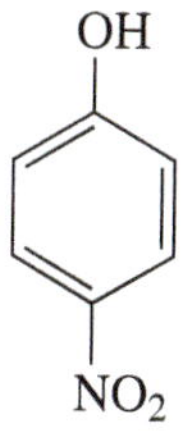 OH / NO_2 or OH / Br

c. $^{+}NH_3$ / CH_3 or $^{+}NH_3$ / OCH_3

d. COOH / CH_3 or COOH

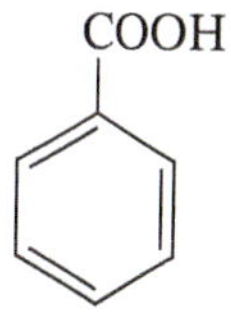

4. a. Which is more reactive in a nucleophilic substitution reaction, *para*-bromonitrobenzene or *para*-bromoethylbenzene?

b. Which is more reactive in an electrophilic substitution reaction, *para*-bromonitrobenzene or *para*-bromoethylbenzene?

5. What acid anhydride would you use in the synthesis of propylbenzene?

6. Draw the mechanism for the formation of the nitronium ion from nitric acid and sulfuric acid.

7. Draw the major product(s) of the following reactions:

a. 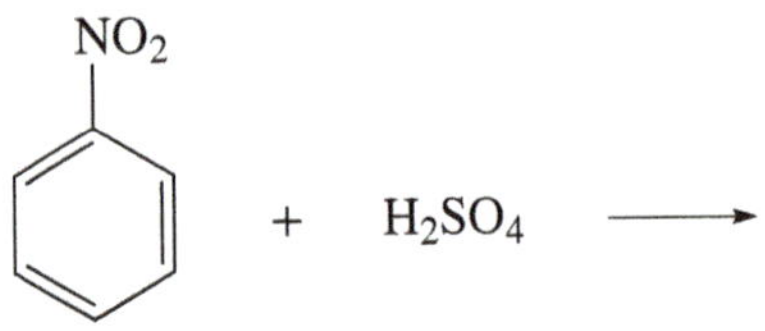 (nitrobenzene, NO_2) + H_2SO_4 ⟶

b. 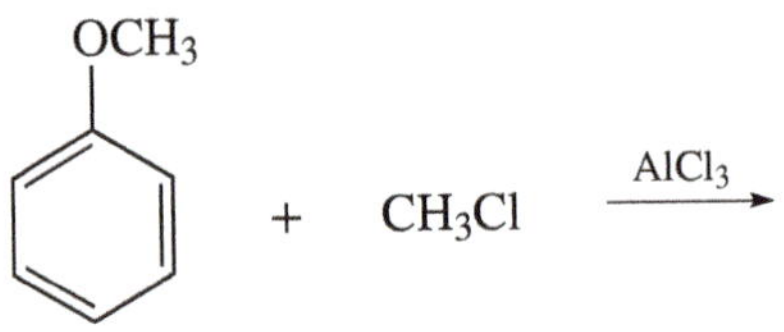 (anisole, OCH_3) + CH_3Cl $\xrightarrow{AlCl_3}$

c. 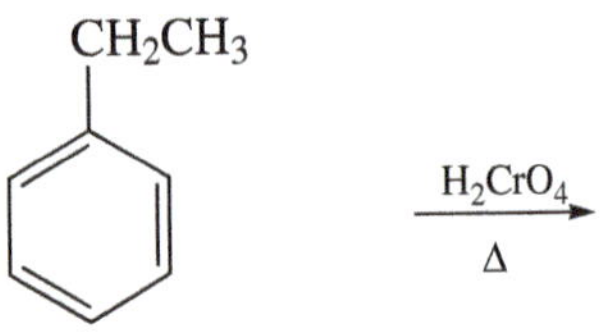 (ethylbenzene, CH_2CH_3) $\xrightarrow[\Delta]{H_2CrO_4}$

d. 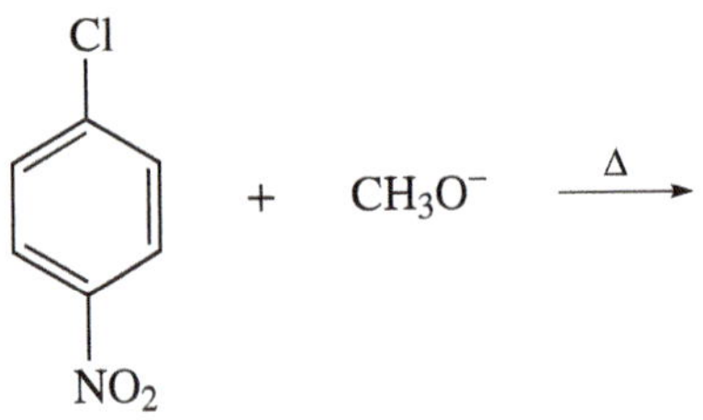 (1-chloro-4-nitrobenzene, Cl and NO_2) + CH_3O^- $\xrightarrow{\Delta}$

e. 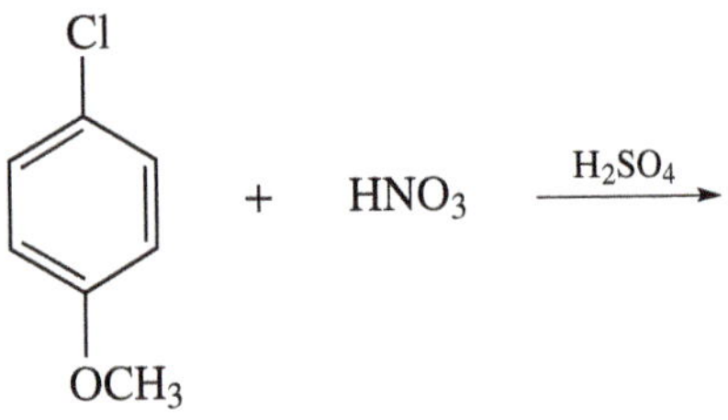 (1-chloro-4-methoxybenzene, Cl and OCH_3) + HNO_3 $\xrightarrow{H_2SO_4}$

f. 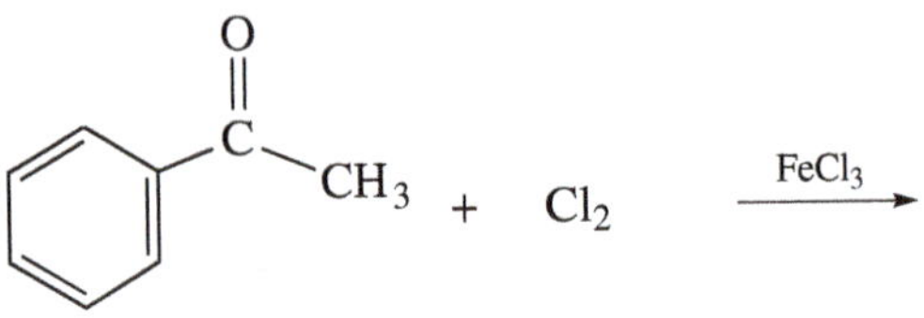 (acetophenone, $C_6H_5C(=O)CH_3$) + Cl_2 $\xrightarrow{FeCl_3}$

8. Indicate whether each of the following statements is true or false:

a. Benzoic acid is more reactive than benzene toward electrophilic aromatic substitution. T F

b. *para*-Chlorobenzoic acid is more acidic than *para*-methoxybenzoic acid. T F

c. A CH=CH_2 group is a meta director. T F

d. *para*-Nitroaniline is more basic than *para*-chloroaniline. T F

9. Draw the resonance contributors for the carbocation intermediate that is formed when benzene reacts with an electrophile (Y^+).

CHAPTER 19
More About Amines • Reactions of Heterocyclic Compounds

Important Terms

furan	a five-membered ring aromatic compound with an oxygen ring atom.
heteroatom	an atom other than a carbon or a hydrogen.
heterocyclic compound (heterocycle)	a cyclic compound in which one or more of the ring atoms are heteroatoms.
imidazole	a five-membered ring aromatic compound with two nitrogen ring atoms.
ligation	the sharing of nonbonded electrons with a metal.
porphyrin ring system	a compound that consists of four pyrrole rings joined by one-carbon bridges.
purine	a pyrimidine ring fused to an imidazole ring.
pyrimidine	a benzene ring with nitrogens at the 1- and 3-positions.
pyrrole	a five-membered ring aromatic compound with a nitrogen ring atom.
thiophene	a five-membered ring aromatic compound with a sulfur ring atom.

Solutions to Problems

1.

a. 2,2-dimethylazacyclopropane or 2,2-dimethylaziridine

b. 4-ethylazacyclohexane 4-ethylpiperidine

c. 2-methylthiacyclopropane or 2-methylthiirane

d. 3-methylazacyclobutane or 3-methylazetidine

e. 2,3-dimethyloxacyclopentane or 2,3-dimethyltetrahydrofuran

f. 2-ethyloxacyclobutane or 2-ethyloxetane

2. Solved in the text.

3. The oxygen in morpholine withdraws electrons inductively, which make protonated morpholine the stronger acid. Recall that inductive electron withdrawal increases acidity (Section 2.7).

$\mathrm{p}K_\mathrm{a} = 9.28$ ⇌ morpholine + H^+

$\mathrm{p}K_\mathrm{a} = 11.12$ ⇌ piperidine + H^+

4. **a.**

b. The conjugate acid of 3-quinuclidinone has a lower $\mathrm{p}K_\mathrm{a}$ than the conjugate acid of morpholine because the sp^2 oxygen of 3-quinuclidinone is more electronegative than the sp^3 oxygen of morpholine (Section 2.6) and the oxygen is closer to the nitrogen. So we know that its $\mathrm{p}K_\mathrm{a}$ is less than 9.

conjugate acid of 3-quinuclidinone $\mathrm{p}K_\mathrm{a} = 7.46$

conjugate acid of morpholine $\mathrm{p}K_\mathrm{a} = 9.28$

c. The conjugate acid of 3-chloroquinuclidine has a lower $\mathrm{p}K_\mathrm{a}$ than the conjugate acid of 3-bromoquinuclidine because chlorine is more electronegative than bromine, so it is better at withdrawing electrons inductively.

conjugate acid of 3-chloroquinuclidine

conjugate acid of 3-bromoquinuclidine

5. a. 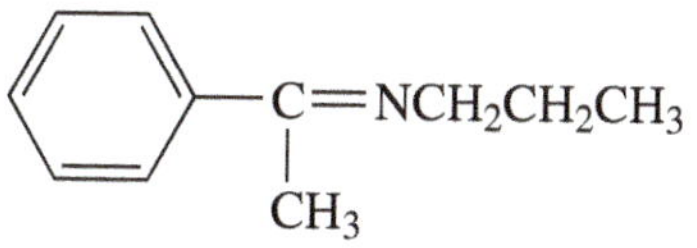 C=$NCH_2CH_2CH_3$ 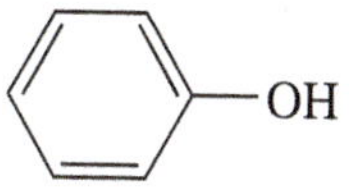 | CH_3 c. OH

b. 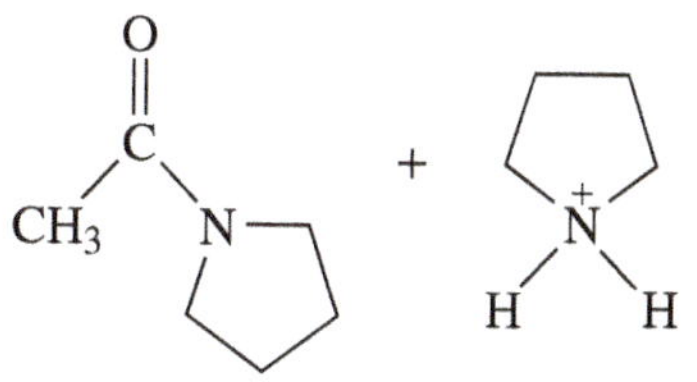 O ‖ C CH_3 N + $\overset{+}{N}$ H H d. 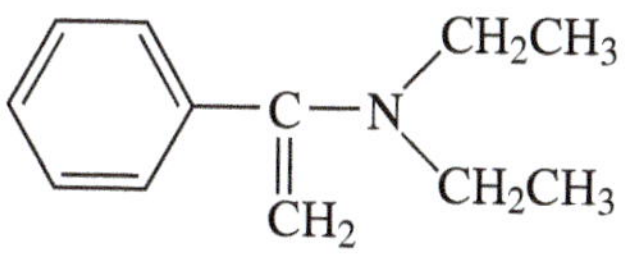 C—N CH_2CH_3 CH_2CH_3 ‖ CH_2

6. 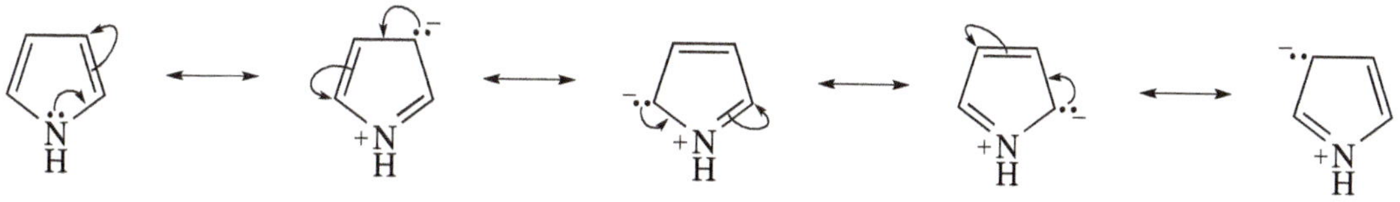 $\ddot{N}$ H ⟷ +N H ⟷ +N H ⟷ +N H ⟷ +N H

7. Pyrrole and its conjugate base are both aromatic. Cyclopentadiene does not become aromatic until it loses a proton. It is the drive for the nonaromatic compound to become a stable aromatic compound that causes cyclopentadiene to be a stronger acid than pyrrole.

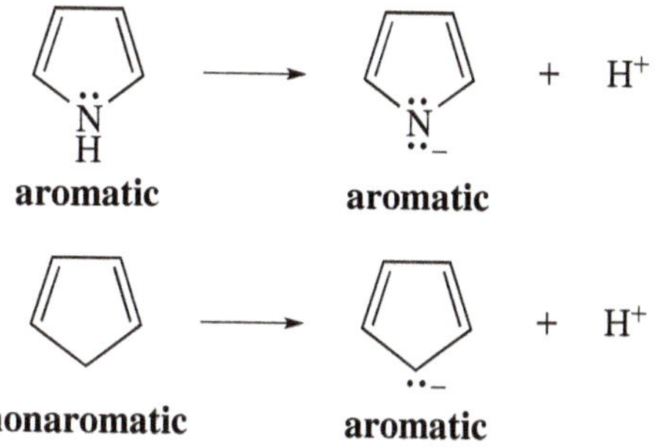

$\ddot{N}$ H → $\ddot{N}^-$ + H^+

aromatic **aromatic**

→ + H^+

nonaromatic **aromatic**

8. 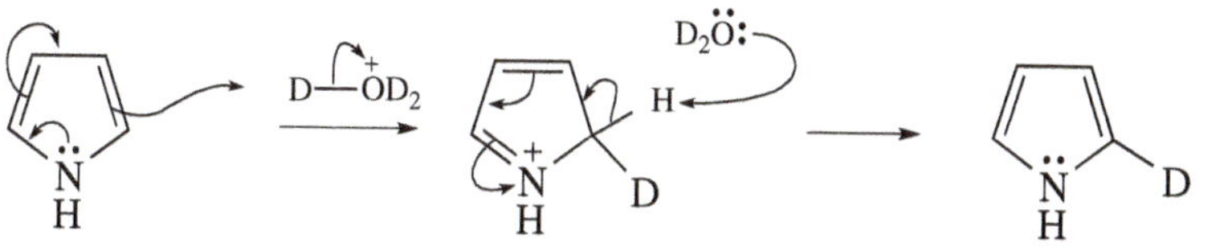 $\ddot{N}$ H → D—$\overset{+}{O}D_2$ → +N H D $D_2\ddot{O}$: H → $\ddot{N}$ H D

2-deuteriopyrrole

9. Solved in the text.

10. Pyridine will act as an amine with the alkyl bromide, forming a quaternary ammonium salt.

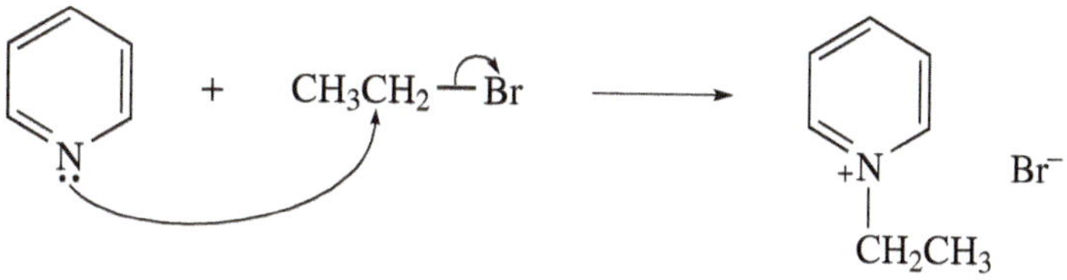

$\ddot{N}$ + CH_3CH_2—Br → +N | CH_2CH_3 Br^-

11. The first reaction is a two-step nucleophilic aromatic substitution reaction (S_NAr).

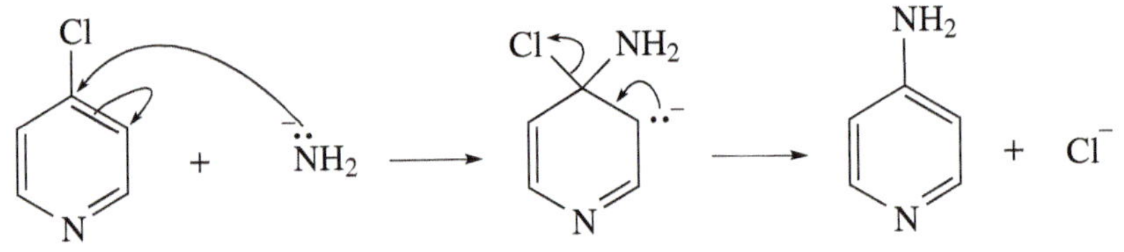

Cl NH$_2$ NH$_2$ Cl NH$_2$ N N N + Cl$^-$

The second reaction is a one-step nucleophilic substitution reaction (S_N2).

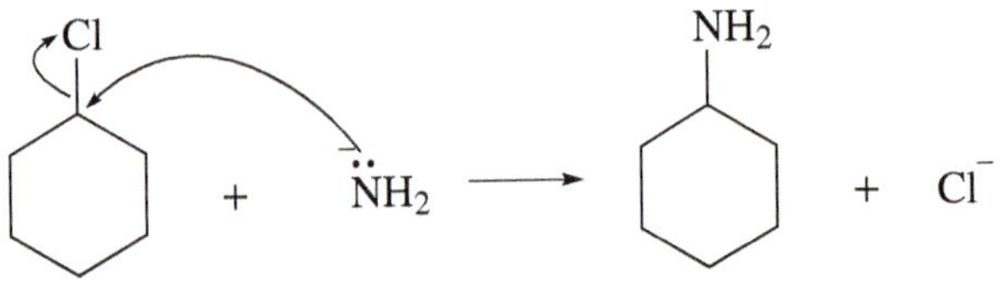

Cl NH$_2$ NH$_2$ + Cl$^-$

12. **a.** N + HO$^-$ → N H OH ⇌ (HO$^-$ / H$_2$O) N H O$^-$ + H—OH

The hydride ion is a better leaving group than O^{2-}.

HO$^-$ + N H O ← (H$_2$O) N O ⟷ N O$^-$ + H$_2$ + HO$^-$

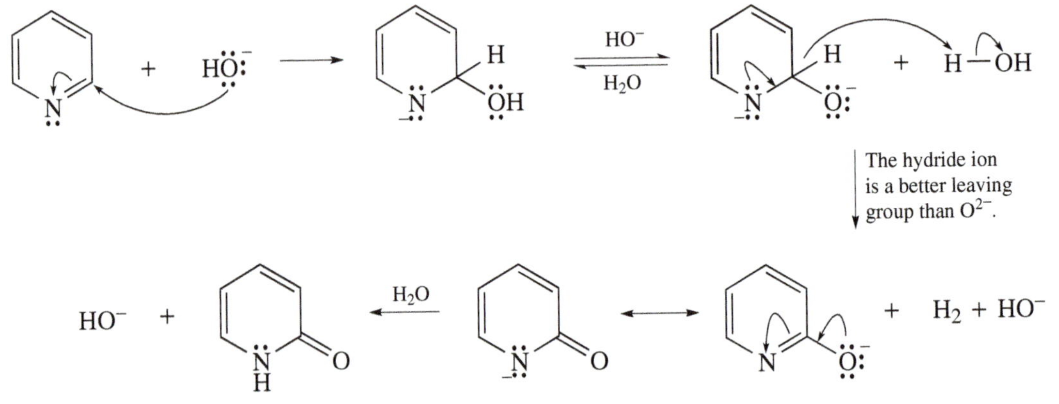

2-pyridone

b. 4-Pyridone is also formed because nucleophilic addition of hydroxide ion can take place at the 4-position as well as at the 2-position. It proceeds by the same mechanism as the one shown in part **a** for the formation of 2-pyridone.

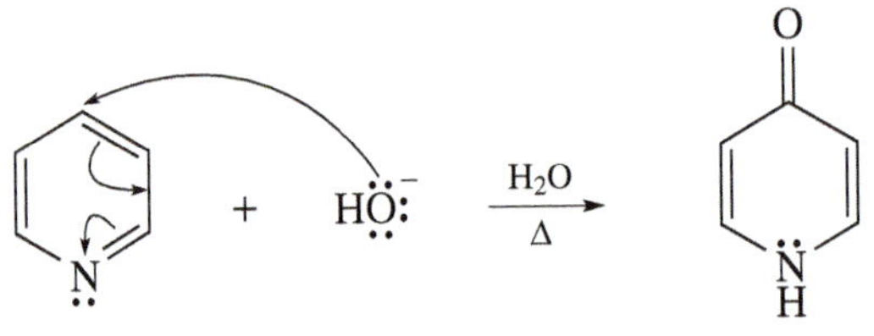

N + HO$^-$ → (H$_2$O, Δ) O N H

4-pyridone

13. It is easiest to remove a proton from the methyl group of the *N*-alkylated pyridine because the electrons left behind when the proton is removed can be delocalized onto the positively charged nitrogen. (A positively charged nitrogen more readily accepts the delocalized electrons than does a neutral nitrogen.)

It is easier to remove a proton from 4-methylpyridine than from 3-methylpyridine because in the former, the electrons left behind when the proton is removed can be delocalized onto the electronegative nitrogen. In contrast, in 3-methylpyridine the electrons can be delocalized onto only carbons.

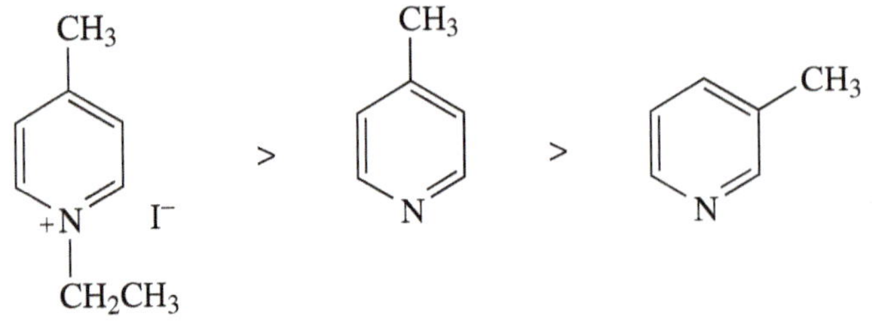

CH$_3$ +N I$^-$ CH$_2$CH$_3$ > CH$_3$ N > CH$_3$ N

14. There are three possible sites for electrophilic substitution: C-2, C-4, and C-5. To determine the major product, compare the relative stabilities of the carbocations formed in the first step of the reaction.

Substitution at C-2

:N NCH$_3$ $\xrightarrow{Br-Br-FeBr_3}$:N NCH$_3$ (Br) ⟷ :N NCH$_3$ (Br) ⟷ :N NCH$_3$ (Br)

Substitution at C-2 forms an intermediate with three resonance contributors; all the atoms in one contributor have complete octets and a positive charge on N; one contributor has a carbon with an incomplete octet, and one has a nitrogen with an incomplete octet.

Substitution at C-4

:N NCH$_3$ $\xrightarrow{Br-Br-FeBr_3}$:N NCH$_3$ (Br) ⟷ :N NCH$_3$ (Br)

Substitution at C-4 forms an intermediate with two resonance contributors; all the atoms in one contributor have complete octets and a positive charge on N; one contributor has a carbon with an incomplete octet.

Substitution at C-5

:N NCH$_3$ $\xrightarrow{Br-Br-FeBr_3}$:N NCH$_3$ (Br) ⟷ :N NCH$_3$ (Br) ⟷ :N NCH$_3$ (Br)

Substitution at C-5 forms an intermediate with three resonance contributors; all the atoms in one contributor have complete octets and a positive charge on N; two contributors have a carbon with an incomplete octet.

Substitution at C-4 forms the least stable intermediate because it has only two of the three resonance contributors that the others have. Substitution at C-5 forms the most stable intermediate because a positively charged carbon with an incomplete octet is more stable than a positively charged nitrogen with an incomplete octet.

Therefore, the major product of the reaction is 5-bromo-*N*-methylimidazole.

Br

N NCH$_3$

5-bromo-*N*-methylimidazole

15. Pyrrole and imidazole are both more reactive than benzene because each reacts with an electrophile to form a carbocation intermediate that is stabilized by resonance electron donation into the ring by a nitrogen atom.

Pyrrole is more reactive than imidazole because the second nitrogen atom of imidazole cannot donate electrons into the ring by resonance but can withdraw electrons from the ring inductively.

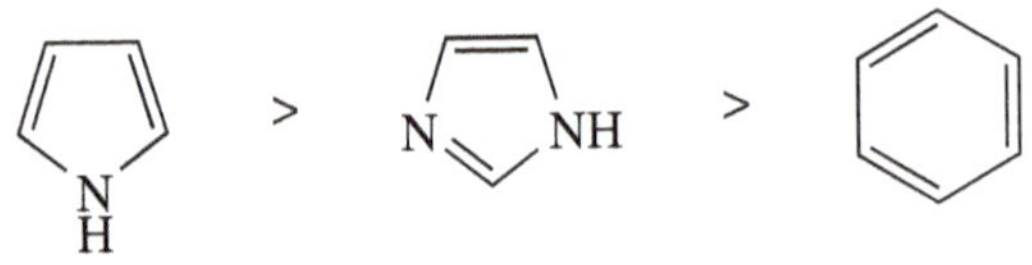

pyrrole > imidazole > benzene

16. Imidazole forms intermolecular hydrogen bonds, whereas *N*-methylimidazole cannot form hydrogen bonds because it does not have a hydrogen bonded to a nitrogen.

Because the hydrogen bonds have to be broken in order for the compound to boil, imidazole has a higher boiling point.

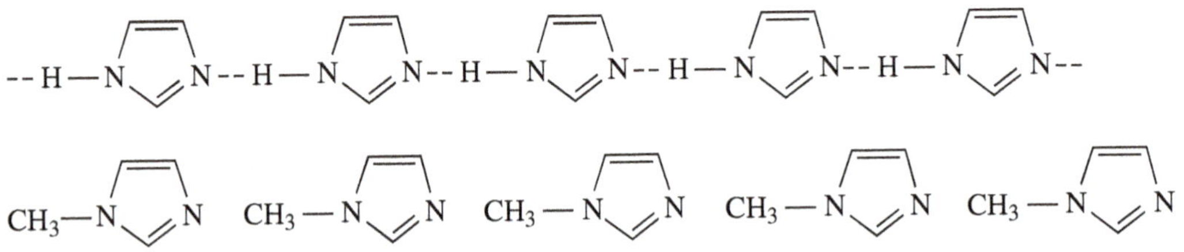

17. The second nitrogen in imidazole, onto which the electrons left behind when the proton is removed can be delocalized, causes imidazole to be a stronger acid than pyrrole because nitrogen is more electronegative than carbon.

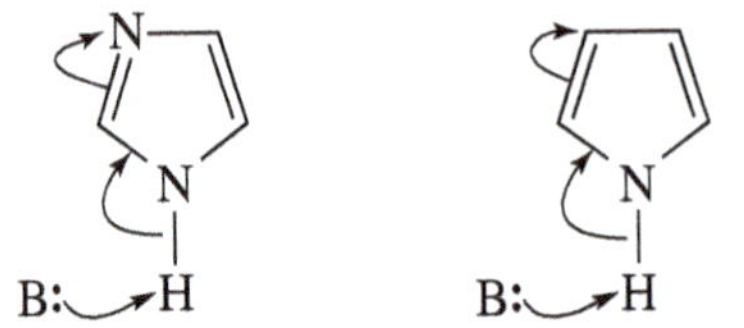

imidazole $pK_a = 14.4$

pyrrole $pK_a = 17$

18. For the derivation of the equation used in this problem, see Special Topic I in this *Study Guide and Solutions Manual.*

$$\text{fraction of imidazole in the acidic form} = \frac{[H^+]}{K_a + [H^+]}$$

$$pH = 7.4;\ [H^+] = 4.0 \times 10^{-8}$$

$$pK_a = 6.8;\ K_a = 1.58 \times 10^{-7} \text{ (see Table 20.1)}$$

$$\text{fraction of imidazole in the acidic form} = \frac{4.0 \times 10^{-8}}{1.58 \times 10^{-7} + 4.0 \times 10^{-8}}$$

$$= \frac{4.0 \times 10^{-8}}{1.98 \times 10^{-7}}$$

$$= 0.20$$

$$\text{percent of imidazole in the acidic form} = 20\%$$

19.

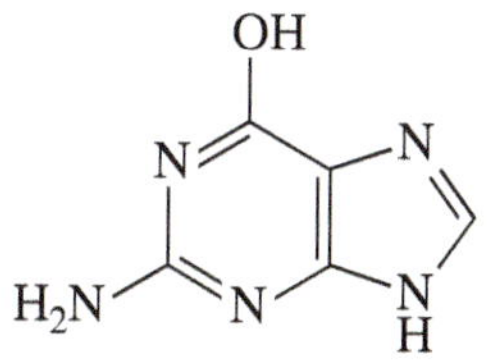

enol form of guanine

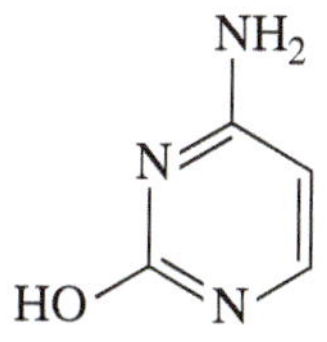

enol form of cytosine

20. The second nitrogen in protonated pyrimidine (that withdraws electrons inductively) causes it to be a stronger acid than protonated pyridine.

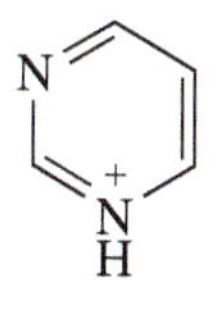

protonated pyrimidine
$pK_a = 1.0$

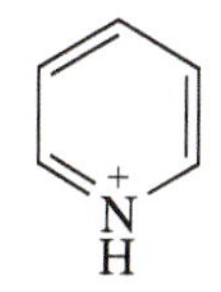

protonated pyridine
$pK_a = 5.2$

21.
a. 2-methylazacyclobutane or 2-methylazetidine
b. 2,3-dimethylazacyclohexane or 2,3-dimethylpiperidine
c. 3-chloropyrrole
d. 2-ethyl-5-methylazacyclohexane or 2-ethyl-5-methylpiperidine

22.

a.

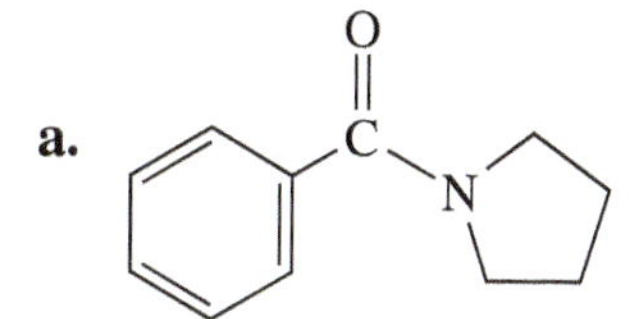

b.

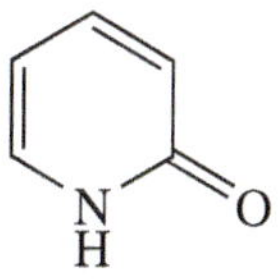

c. $CH_3CH_2CH_2CH_2CH_2NH_2$

d.

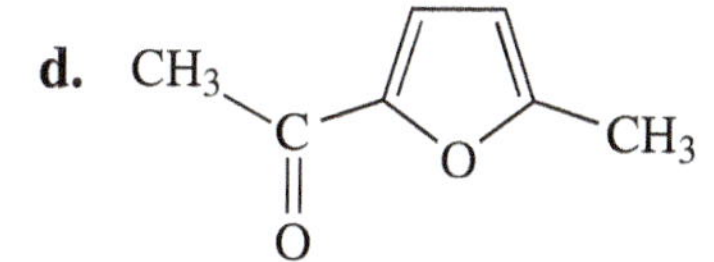

e.

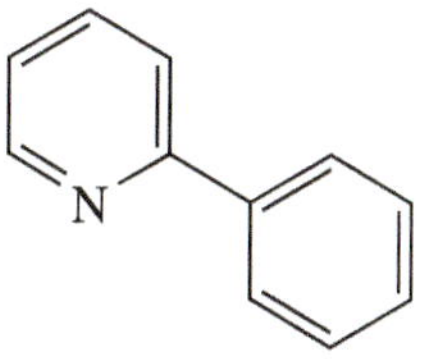

f. $\overset{+}{N}H_2CH_2CH_2CH_3$ Br^-

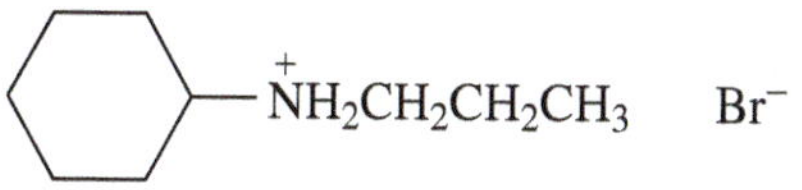

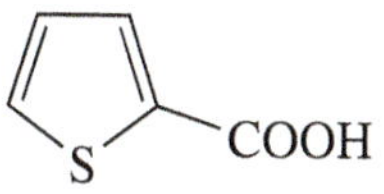

g.

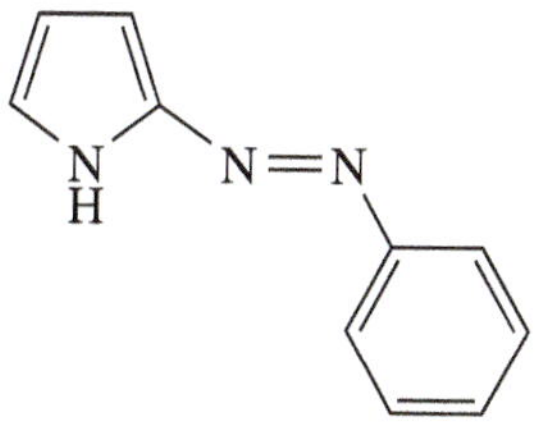

h. $CH_2CH_2CH_2CH_3$

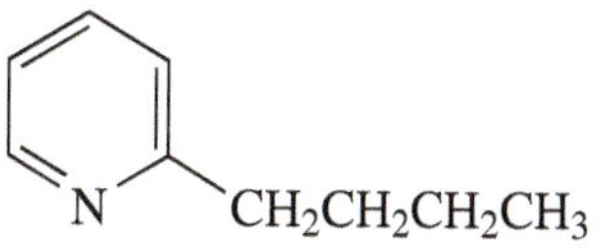

i. COOH

23. A > B > C > D > E > F > G

A B C D E F G

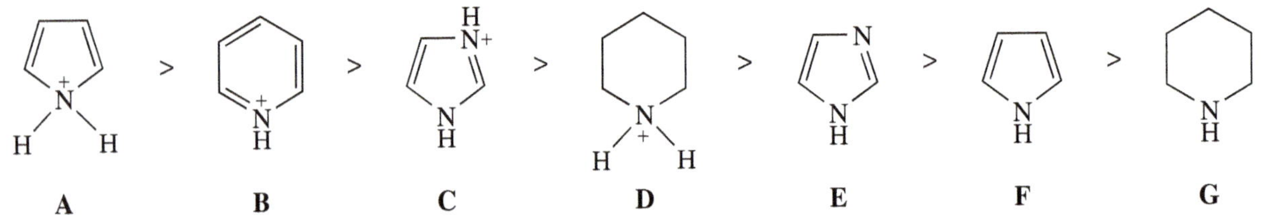

A is the most acidic, because it becomes aromatic when it loses a proton.

B, **C**, and **D** are the next most acidic, because in all three, the proton is bonded to a positively charged nitrogen.

B and **C** are more acidic than **D**, because in **B** and **C**, the proton to be lost is bonded to an sp^2 nitrogen, which is more electronegative than the sp^3 nitrogen in **D**.

B is more acidic than **C**, because the uncharged nitrogen in **C** can donate electrons by resonance to the positively charged nitrogen, which stabilizes the protonated form since the positive charge is shared by two nitrogens.

Neutral compounds **E**, **F**, and **G** are the least acidic. **E** and **F** are more acidic than **G**, because **E** and **F** lose a proton from an sp^2 nitrogen, whereas a proton is lost from a less electronegative sp^3 nitrogen in **G**.

E is more acidic than **F**, because the negative charge on **E**'s conjugate base can be delocalized onto the second nitrogen.

24. The compound on the right (the compound shown below) is easier to decarboxylate, because the electrons left behind when CO_2 is removed can be delocalized onto nitrogen, forming a stable neutral species as a result of proton donation to nitrogen. The electrons left behind when the other compound loses CO_2 cannot be delocalized.

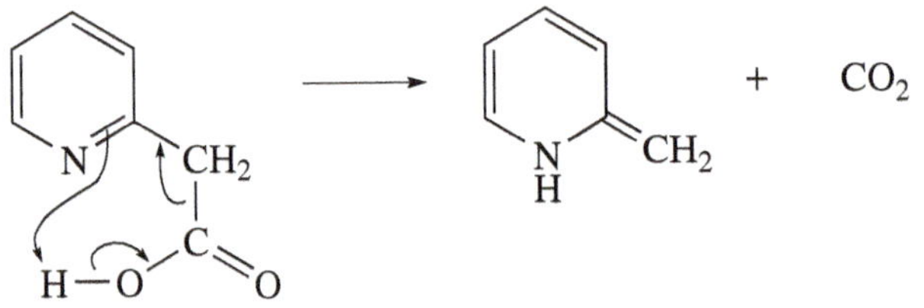

$\longrightarrow$ + CO_2

25. The *N*-, *O*-, and *S*-substituted benzenes have the same relative reactivity toward electrophilic aromatic substitution as the *N*-, *O*-, and *S*-containing five-membered heterocyclic rings and for the same reason.

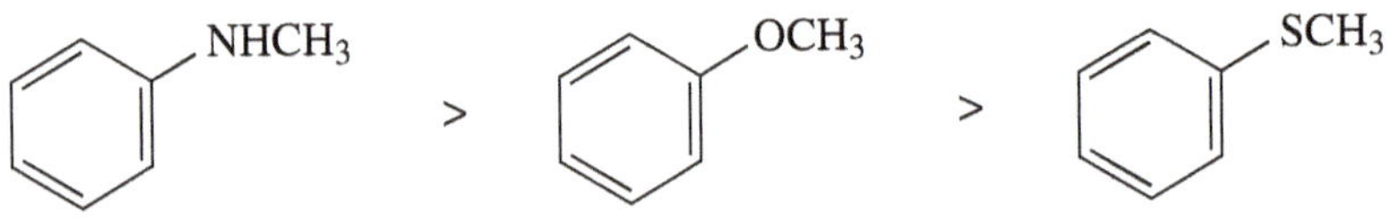

$NHCH_3$ > OCH_3 > SCH_3

The *N*-substituted benzene is more reactive than the *O*-substituted benzene, because nitrogen is more effective than oxygen at donating electrons into the benzene ring since it is less electronegative than oxygen. (Recall that electrophilic substitution is aided by electron donation to the ring because it stabilizes the carbocation formed in the rate-limiting step.)

The *S*-substituted benzene is the least reactive because the lone-pair electrons of sulfur are in a 3*p* orbital, whereas the lone-pair electrons of nitrogen and oxygen are in a 2*p* orbital. Electron delocalization by overlap of the 3*p* orbital of sulfur and the 2*p* orbital of carbon is less effective than electron delocalization by overlap of the 2*p* orbital of nitrogen and the 2*p* orbital of carbon.

26. By comparing the carbocation intermediates formed in the rate-determining step when the amino-substituted compound undergoes electrophilic substitution at C-3 and C-4, you can see that the carbocation formed in the case of substitution at C-3 is more stable; one of its resonance contributors is relatively stable because it is the only one that does not have an atom with an incomplete octet. As a result, the amino-substituted compound undergoes electrophilic substitution predominantly at C-3. Therefore, the keto-substituted compound is the one that undergoes electrophilic substitution predominantly at C-4.

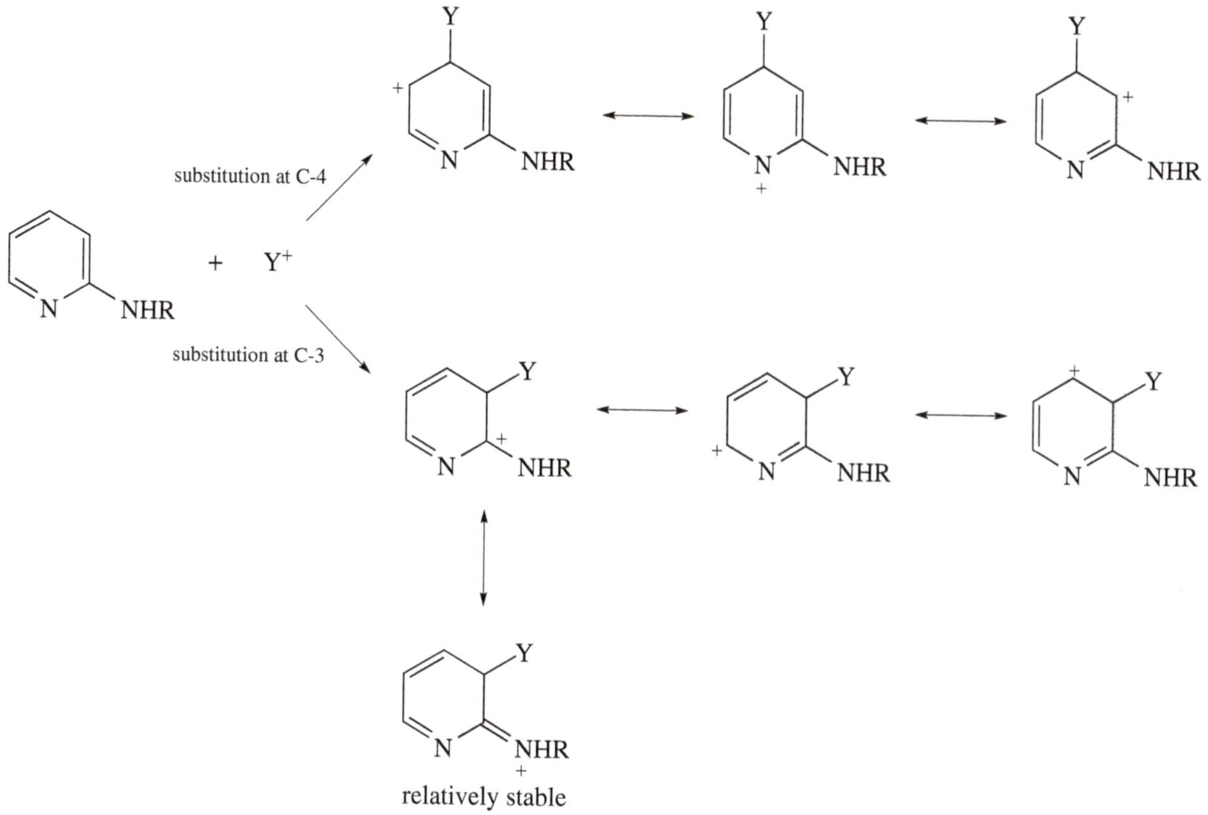

substitution at C-4

+ Y^+

substitution at C-3

relatively stable

27. **a.** The Lewis acid, $AlCl_3$, complexes with nitrogen, causing the aziridine ring to open when it is attacked by the nucleophilic benzene ring. The ring will open in the direction that puts the partial positive charge on the more substituted carbon because it is more stable than the transition state that would have the positive charge on the less substituted carbon.

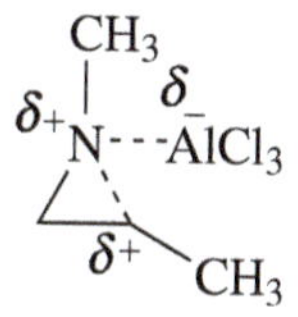

Therefore, the major and minor products are those shown.

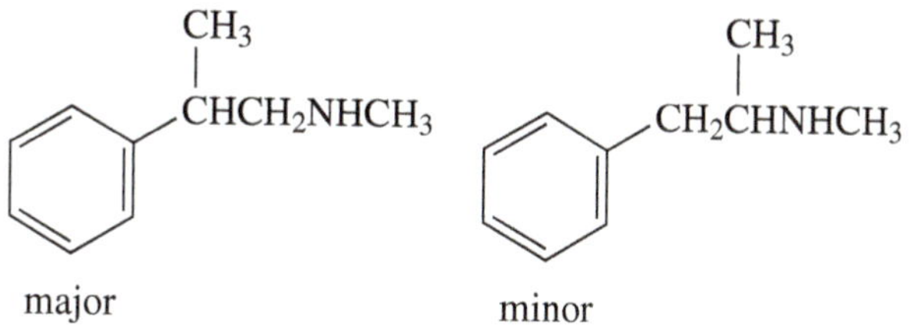

major minor

b. Yes, epoxides can undergo similar reactions.

28. The electrophile adds preferentially to the 2-position.

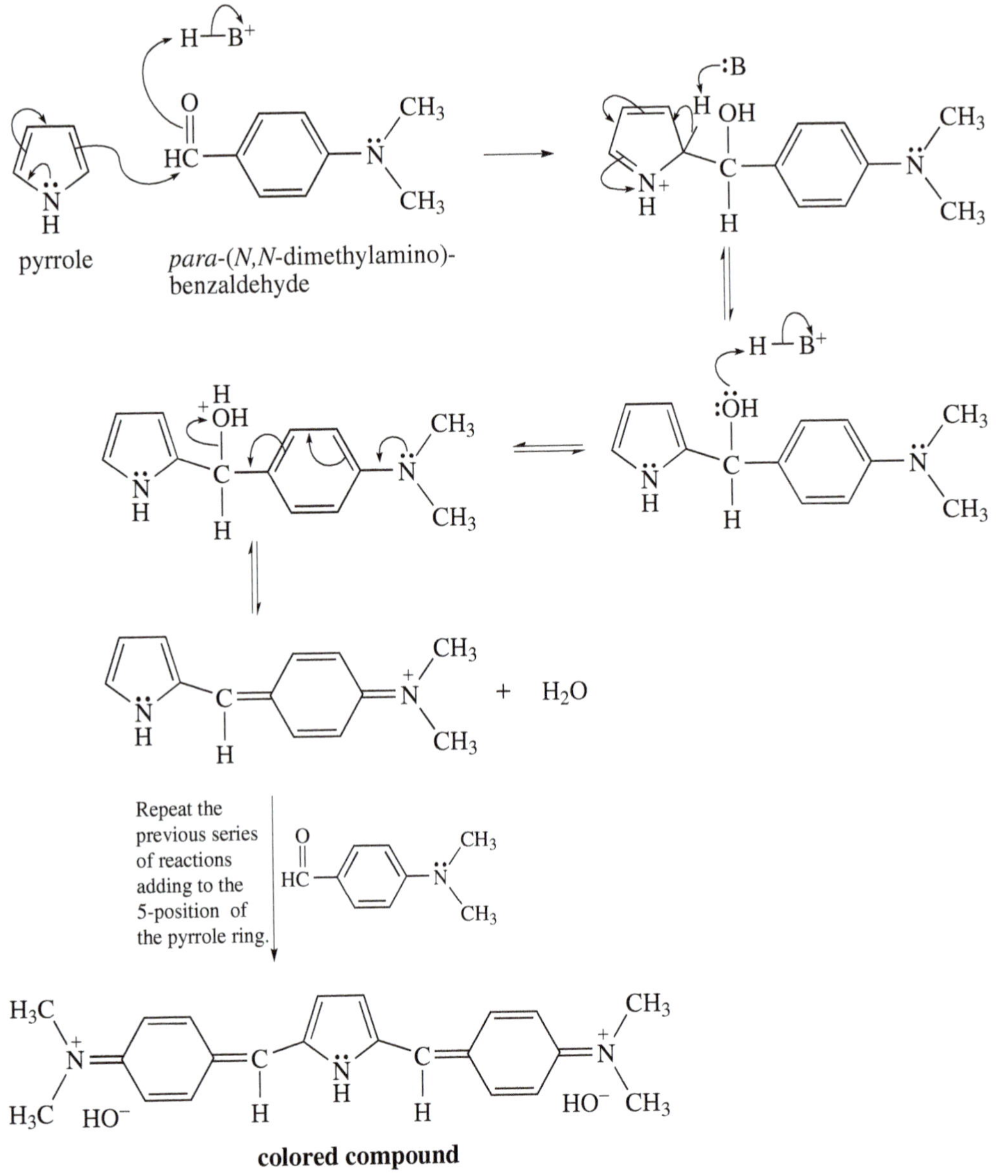

pyrrole

para-(*N*,*N*-dimethylamino)-benzaldehyde

Repeat the previous series of reactions adding to the 5-position of the pyrrole ring.

$+ H_2O$

colored compound

29. Oxygen is the negative end of the dipole in both compounds. Tetrahydrofuran has the greater dipole moment because the effect of the electron-withdrawing oxygen in furan is decreased by the ability of oxygen to donate electrons by resonance into the ring.

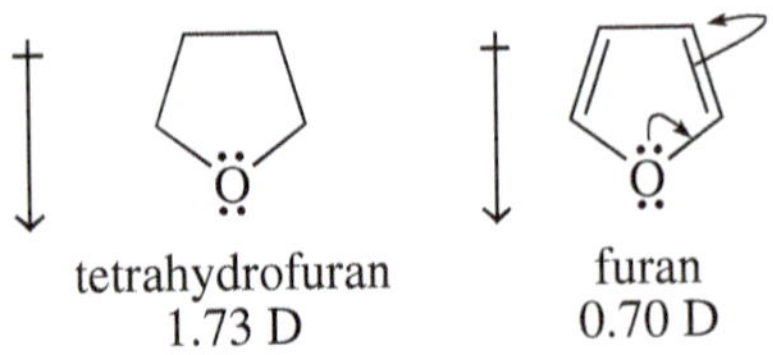

tetrahydrofuran
1.73 D

furan
0.70 D

30. **a.** 4-chloro-3-isopropylpyridine
b. 2-isopropyl-3-methylazacyclopropane
c. 4-ethyl-2,2,3-trimethyloxacyclobutane

31. The products of the reaction show that the hydrogen is removed from the β-carbon that is bonded to the most hydrogens. Therefore, we can conclude that the transition state has a carbanion-like one. (See page 417 in the text.)

32. **a.** $CH_3N(CH_3)OH$ + $CH_2{=}CHCH_3$

b. 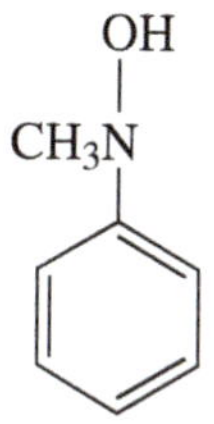 $CH_3N(OH)C_6H_5$ + $CH_2{=}CHCH_3$

c. $CH_2{=}CH_2$ + $N(CH_3)(OH)CH_2CH(CH_3)CH_3$

d. $CH_2{=}CHCH_2CH_2CH_2CH_2N(OH)CH_3$

The carbon of the methyl group is the β-carbon that is bonded to the most hydrogens.

33. **a.** The increase in the electron density of the ring as a result of resonance electron donation by oxygen causes pyridine-*N*-oxide to be more reactive toward electrophilic aromatic substitution than pyridine because the extra electron density stabilizes the carbocation intermediate.

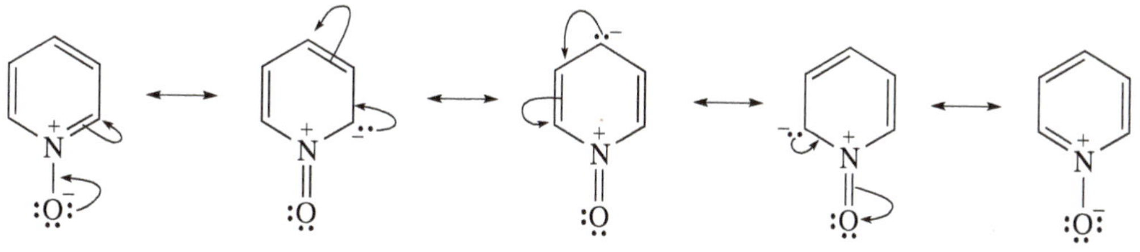

b. In Section 18.13, we saw that substituents that are able to donate electrons by resonance into the ring are ortho/para directors. Therefore, pyridine-*N*-oxide will undergo electrophilic aromatic substitution at the 2- and 4-positions. Because the 2-positions are somewhat sterically hindered, pyridine-*N*-oxide undergoes electrophilic substitution primarily at the 4-position.

34. Pyrrolidine is a saturated nonaromatic compound, whereas pyrrole and pyridine are unsaturated aromatic compounds. The C-2 hydrogens of pyrrolidine are at δ 2.82 ppm, about where one would expect the signal for hydrogens bonded to an sp^3 carbon adjacent to an electron-withdrawing amino group.

The C-2 hydrogens of pyrrole and pyridine are expected to be at a higher frequency because of diamagnetic anisotropy (Section 14.8). Because the nitrogen of pyrrole donates electrons into the ring and the nitrogen of pyridine withdraws electrons from the ring, the C-2 hydrogens of pyrrole are in an environment with a greater electron density, so they should show a signal at a lower frequency relative to the C-2 hydrogens of pyridine. Thus, the C-2 hydrogens of pyrrole are at δ 6.42 ppm, and the C-2 hydrogens of pyridine are at δ 8.50 ppm.

35. A UV spectrum results from the π electron system. The lone-pair electrons on the nitrogen in aniline are delocalized into the benzene ring and, therefore, are part of the π system. Protonation of aniline removes two electrons from the π system. This has a significant effect on its UV spectrum.

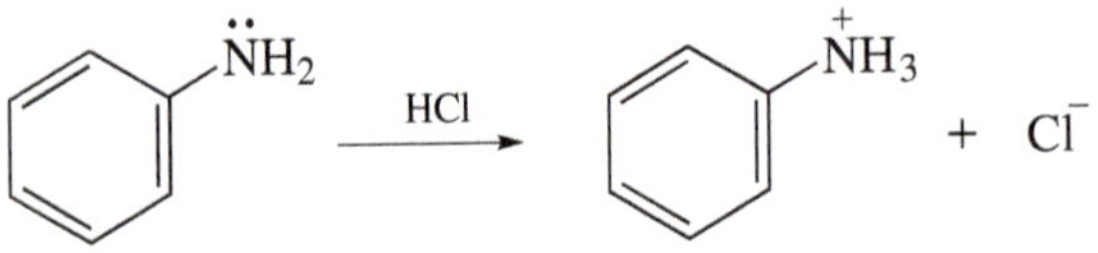

$$C_6H_5\ddot{N}H_2 \xrightarrow{HCl} C_6H_5\overset{+}{N}H_3 + Cl^-$$

The lone-pair electrons on the nitrogen atom in pyridine are sp^2 electrons and thus are not part of the π system. Protonation of pyridine, therefore, does not remove any electrons from the π system and has only a minor effect on the UV spectrum.

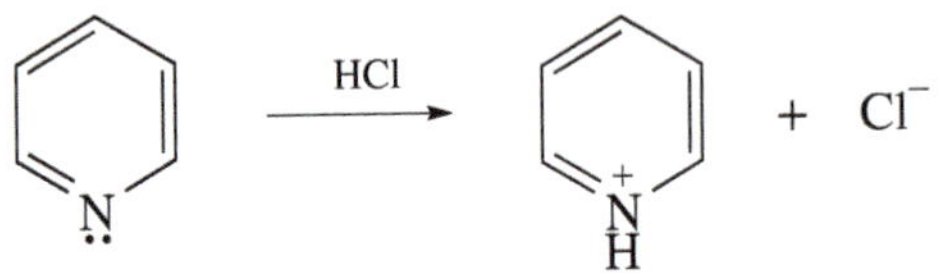

$$C_5H_5\ddot{N} \xrightarrow{HCl} C_5H_5\overset{+}{N}H + Cl^-$$

36. When ammonia loses a proton, the electrons left behind remain on nitrogen.

$$\ddot{N}H_3 + :B \longrightarrow {}^{-}\!:\!\ddot{N}H_2 + HB^+$$

When pyrrole loses a proton, the electrons left behind can be delocalized onto the four ring carbons. Electron delocalization stabilizes the anion and makes it easier to form. Recall that stabilizing the base increases the acidity of its conjugate acid.

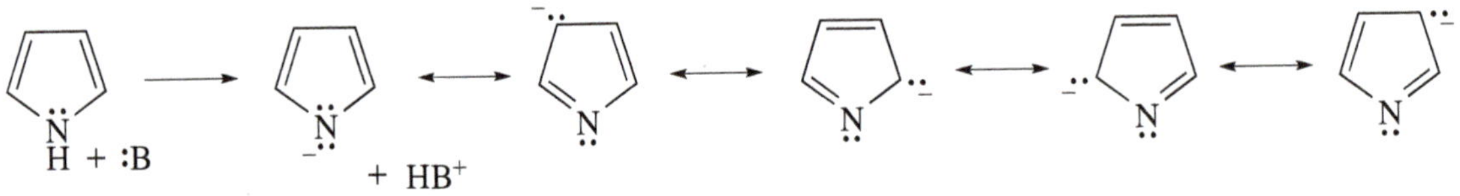

Pyrrole (N–H) + :B ⟶ pyrrolide anion + HB^+ ⟷ ⟷ ⟷ ⟷ (resonance contributors with the negative charge on N and on each of the four ring carbons)

37. The electrophile adds preferentially to the 2-position.

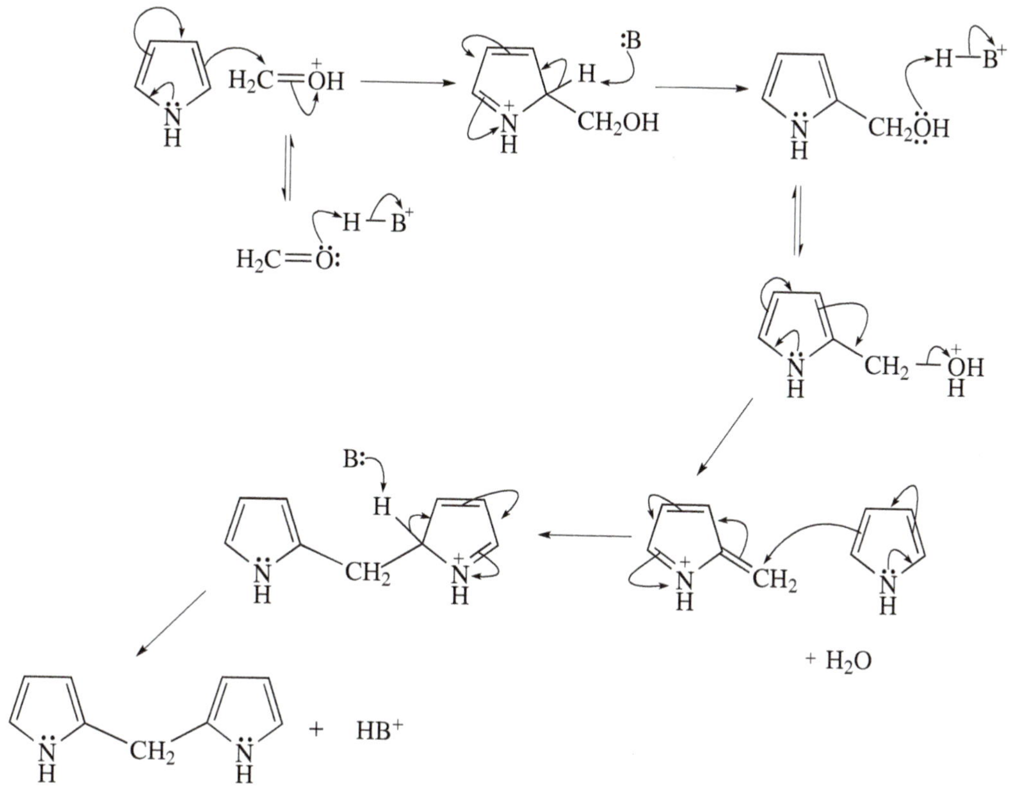

$H_2C{=}\overset{+}{O}H$ $H_2C{=}\ddot{O}$: H—B^+ :B CH_2OH $CH_2\ddot{O}H$ CH_2—$\overset{+}{O}H$ H B: CH_2 + H_2O + HB^+

38. Before we can answer the questions, we must figure out the mechanism of the reaction. Once the mechanism is known, it will be relatively easy to determine how a change in a reactant will affect the product. The mechanism is shown below.

Propenal, an α,β-unsaturated aldehyde, undergoes a conjugate addition reaction with aniline. This is followed by an intramolecular electrophilic aromatic substitution reaction. Dehydration of the alcohol results in 1,2-dihydroquinoline, which is oxidized to quinoline by nitrobenzene.

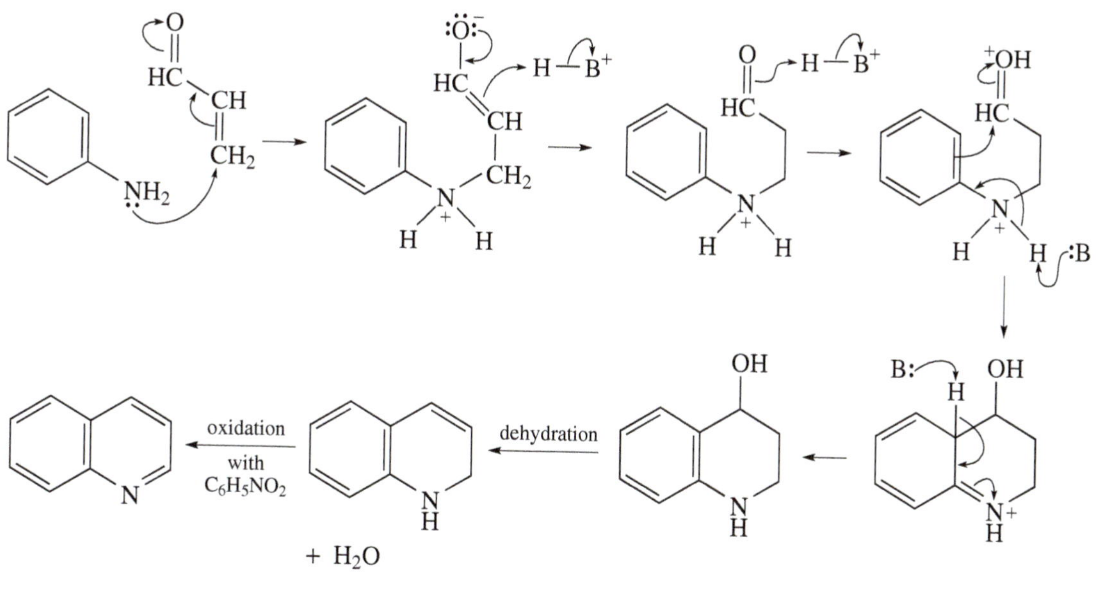

HC CH CH_2 NH_2 H—B^+ :B OH oxidation with $C_6H_5NO_2$ dehydration + H_2O

a. 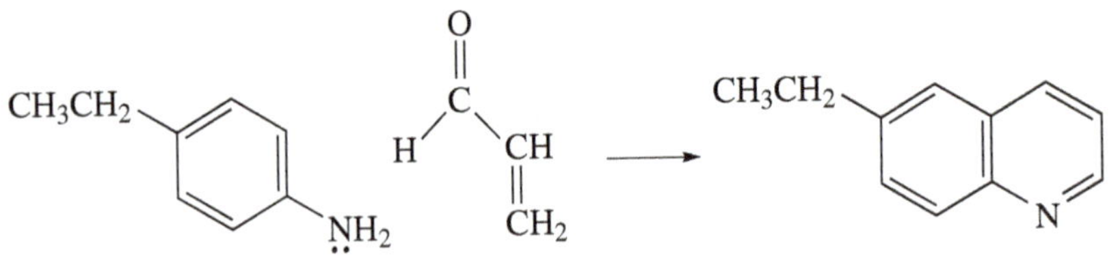 CH_3CH_2 $\ddot{N}H_2$ O C H CH CH CH_2 ⟶ CH_3CH_2 N

b. $\ddot{N}H_2$ O C CH_3 CH CH CH_2CH_3 ⟶ CH_3 N CH_2CH_3

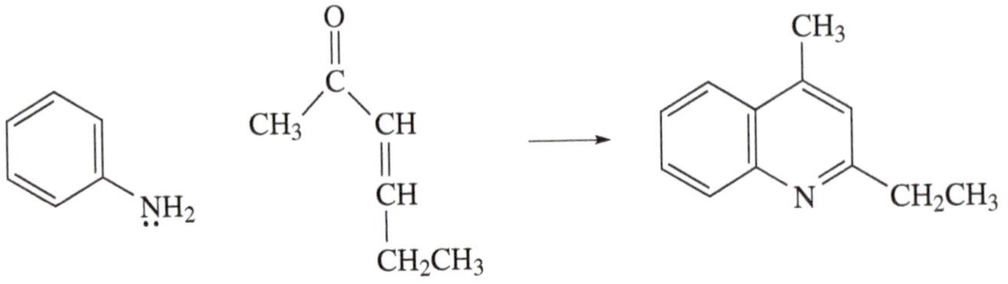

c.

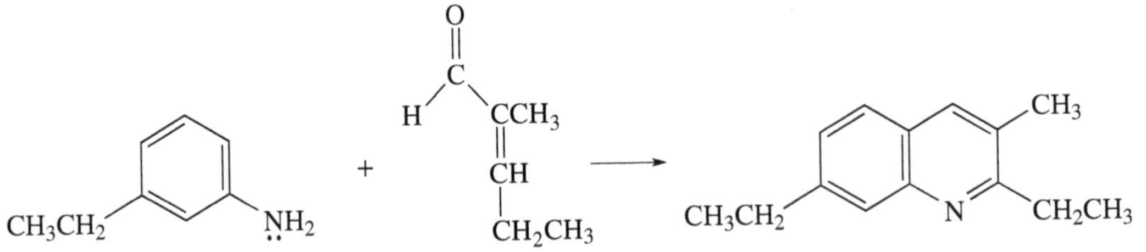

CH_3CH_2 $\ddot{N}H_2$ + O C H CCH_3 CH CH_2CH_3 ⟶ CH_3 CH_3CH_2 N CH_2CH_3

meta-ethylaniline

2-methyl-2-pentenal

2,7-diethyl-3-methylquinoline

39. **a.** The electrophile adds to the 3-position.

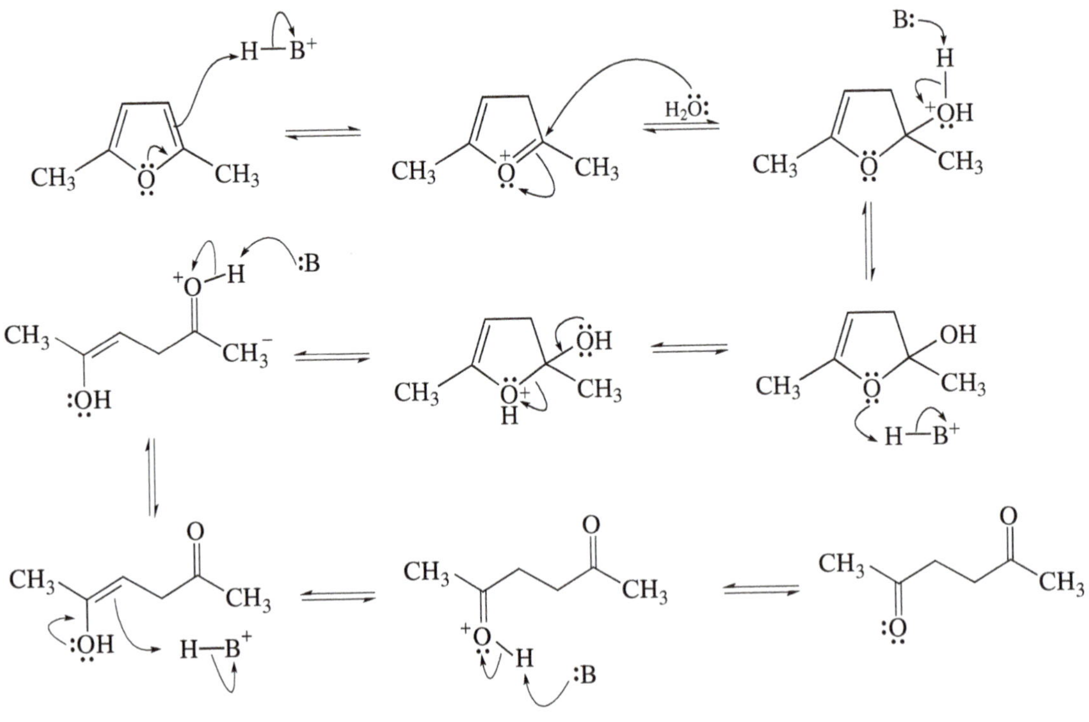

H—B^+

CH_3 $\ddot{O}$ CH_3 ⇌ CH_3 $\overset{+}{O}$ CH_3 ⇌ $H_2\ddot{O}$: ⇌ B: H $^+\ddot{O}H$ CH_3 $\ddot{O}$ CH_3

^+O—H :B

CH_3 :$\ddot{O}H$ CH_3^- ⇌ CH_3 $\ddot{O}^+$ H $\ddot{O}H$ CH_3 ⇌ CH_3 $\ddot{O}$ OH CH_3 H—B^+

O

CH_3 :$\ddot{O}H$ CH_3 H—B^+ ⇌ CH_3 $^+\ddot{O}$ H :B O CH_3 ⇌ CH_3 :$\ddot{O}$ O CH_3

b. The electrophile adds preferentially to the 2-position.

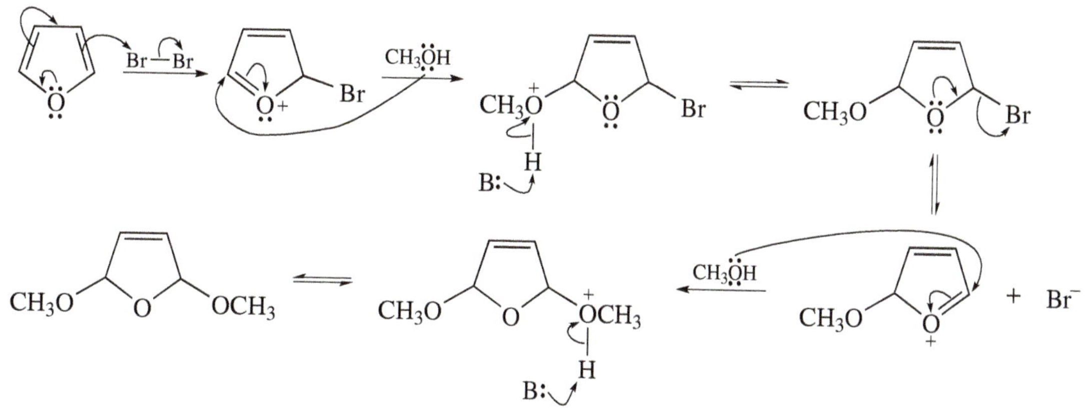

Br—Br → O+ Br $CH_3\ddot{O}H$ → $CH_3\overset{+}{O}$ O Br ⇌ CH_3O O Br
H B:

CH_3O O OCH_3 ⇌ CH_3O O $\overset{+}{O}CH_3$ ← $CH_3\ddot{O}H$ CH_3O $\overset{+}{O}$ + Br^-
H B:

40. **a.** O C CH_3 O → (HNO_3) O C CH_3 O_2N O

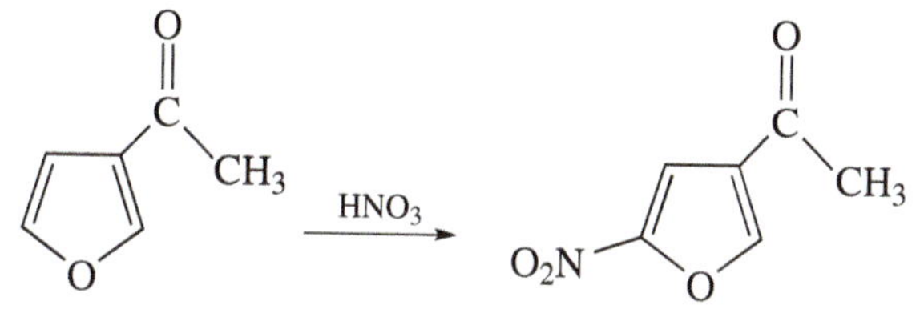

You can see why the nitro substituent goes to this position by examining the relative stabilities of the possible carbocation intermediates.

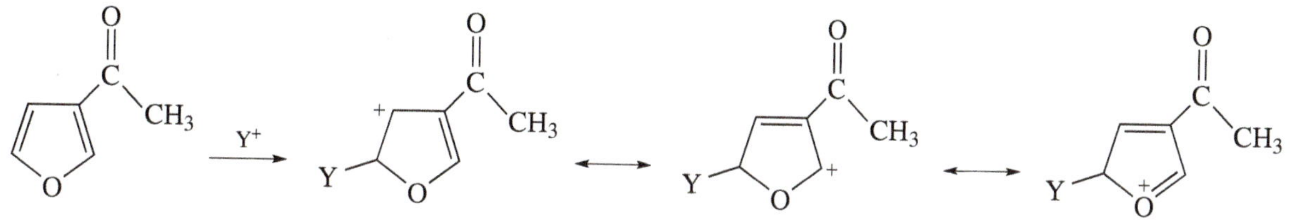

O C CH_3 O → (Y^+) O C CH_3 Y O + ⟷ O C CH_3 Y O + ⟷ O C CH_3 Y $\overset{+}{O}$

The carbocation has three resonance contributors.

O C CH_3 O → (Y^+) Y O C CH_3 + O ⟷ Y O C CH_3 $\overset{+}{O}$

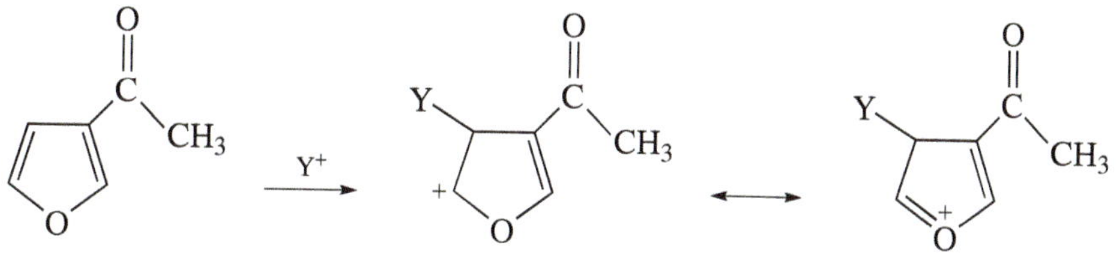

The carbocation has two resonance contributors.

O C CH_3 O → (Y^+) O C + CH_3 O Y ⟷ O C CH_3 + O Y ⟷ O C CH_3 $\overset{+}{O}$ Y

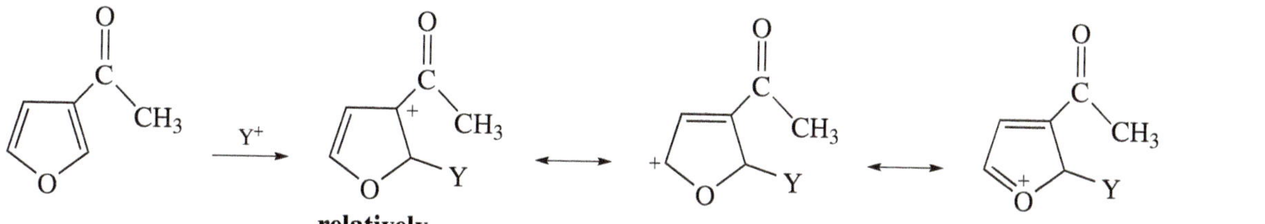

relatively unstable

The carbocation has three resonance contributors, but the first one is relatively unstable because the positive charge is on the carbon attached to the electron-withdrawing substituent.

b. 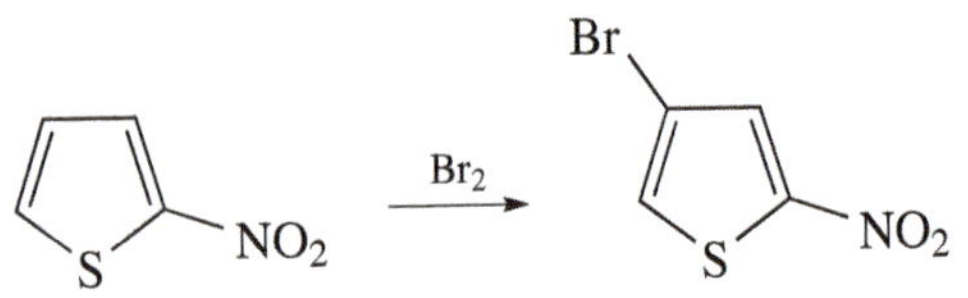 $\xrightarrow{Br_2}$

NO_2 S Br S NO_2

You can understand why the bromo substituent goes to this position by examining the relative stabilities of the possible carbocation intermediates.

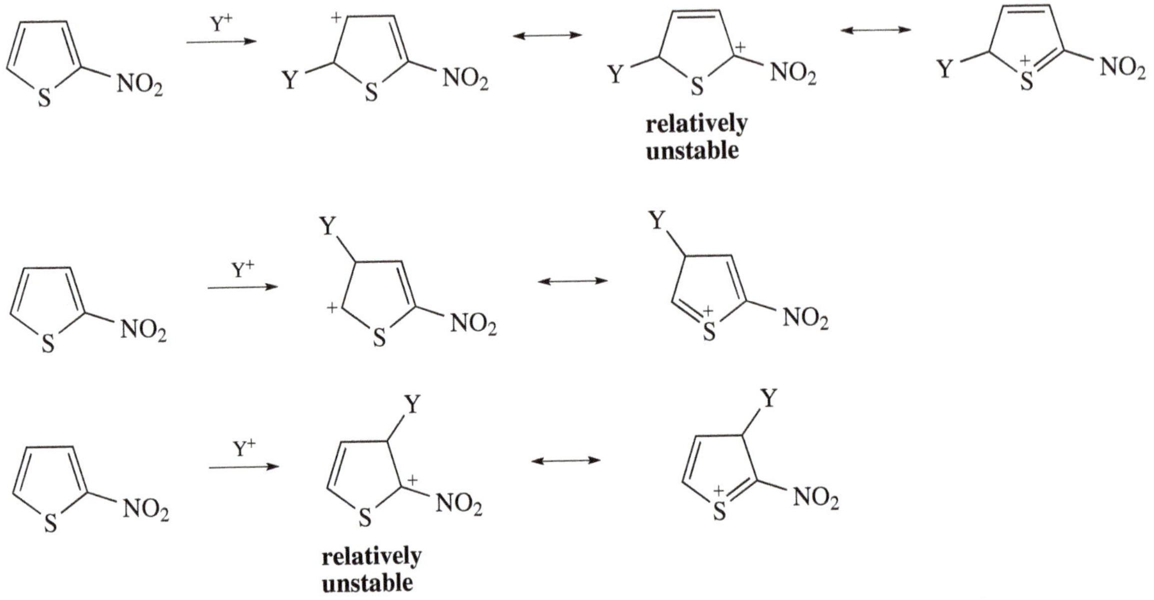

Y^+ NO_2 Y NO_2 Y NO_2 Y NO_2

relatively unstable

Y^+ Y NO_2 Y NO_2

Y^+ Y NO_2 Y NO_2

relatively unstable

c. 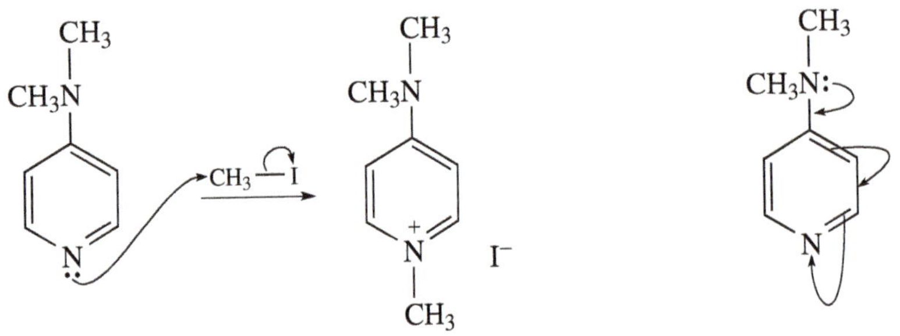 CH_3 CH_3N CH_3—I CH_3 CH_3N CH_3 I^- CH_3 CH_3N

The other nitrogen is not alkylated, because its lone pair is delocalized into the pyridine ring. Therefore, it is not available to react with the alkyl halide.

d. 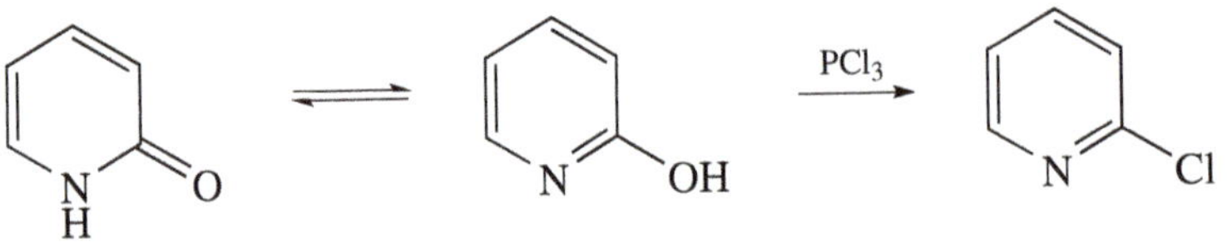 N H O ⇌ N OH $\xrightarrow{PCl_3}$ N Cl

PCl_3 substitutes a Cl for an OH, as it does in alcohols and carboxylic acids.

e. In the first step, hydroxide ion removes the most acidic hydrogen from the compound. A hydrogen bonded to the C-4 methyl group is the most acidic hydrogen because the electrons left behind when the proton is removed can be delocalized onto the positively charged nitrogen of the pyridine ring.

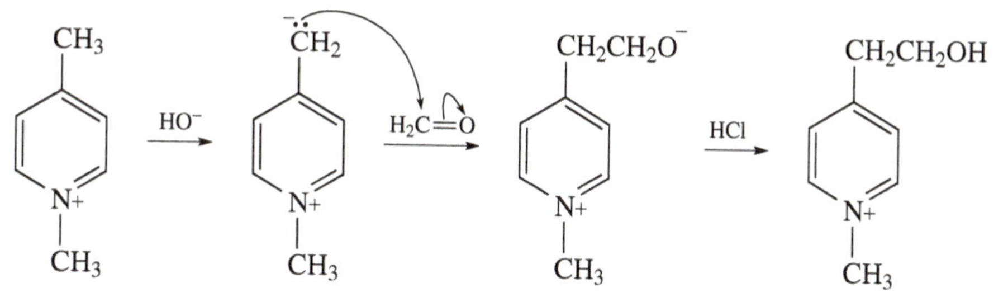

CH_3 / CH_3 (N+) $\xrightarrow{HO^-}$ $\ddot{C}H_2^-$ / CH_3 (N+) $\xrightarrow{H_2C=O}$ $CH_2CH_2O^-$ / CH_3 (N+) $\xrightarrow{HCl}$ CH_2CH_2OH / CH_3 (N+)

f. pyrrole (N–H) + CH_3CH_2–MgBr ⟶ pyrrolide anion ($\ddot{N}^-$) + CH_3CH_3 + Mg^{2+} + Br^-

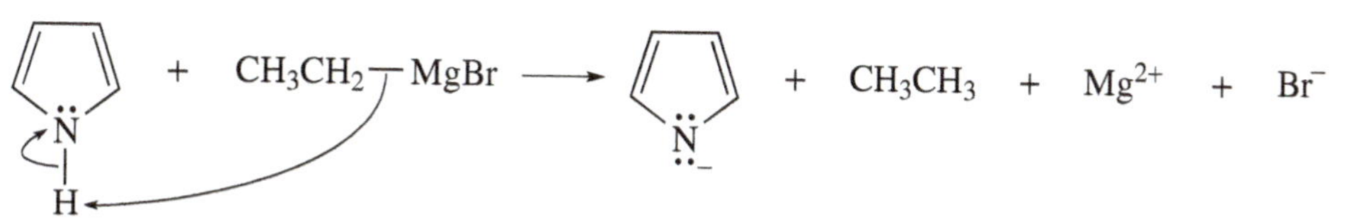

41.

$C_6H_5NHNH_2$ + $O{=}C(CH_3){-}C_6H_5$ ⇌ $C_6H_5NHN{=}C(CH_3){-}C_6H_5$

⇌ $C_6H_5\ddot{N}H{-}NH{-}C({=}CH_2){-}C_6H_5$ ⟶ (H, $CH_2{-}C(C_6H_5){=}NH$, =NH; :B, H–B$^+$) ⟶ ($CH_2{-}C(C_6H_5){=}\ddot{N}H$, NH_2; H–B$^+$)

⇌ ($CH_2{-}C(C_6H_5){=}\overset{+}{N}H_2$, $\ddot{N}H_2$) ⟶ (C_6H_5, NH_2, $\overset{+}{N}H$–H, :B) ⇌ (C_6H_5, $\ddot{N}H_2$, $\ddot{N}H$; H–B$^+$)

⇌ (C_6H_5, $\overset{+}{N}H_3$, $\ddot{N}H$) ⇌ (B: H, C_6H_5, $\overset{+}{N}H$) ⇌ (C_6H_5, NH)

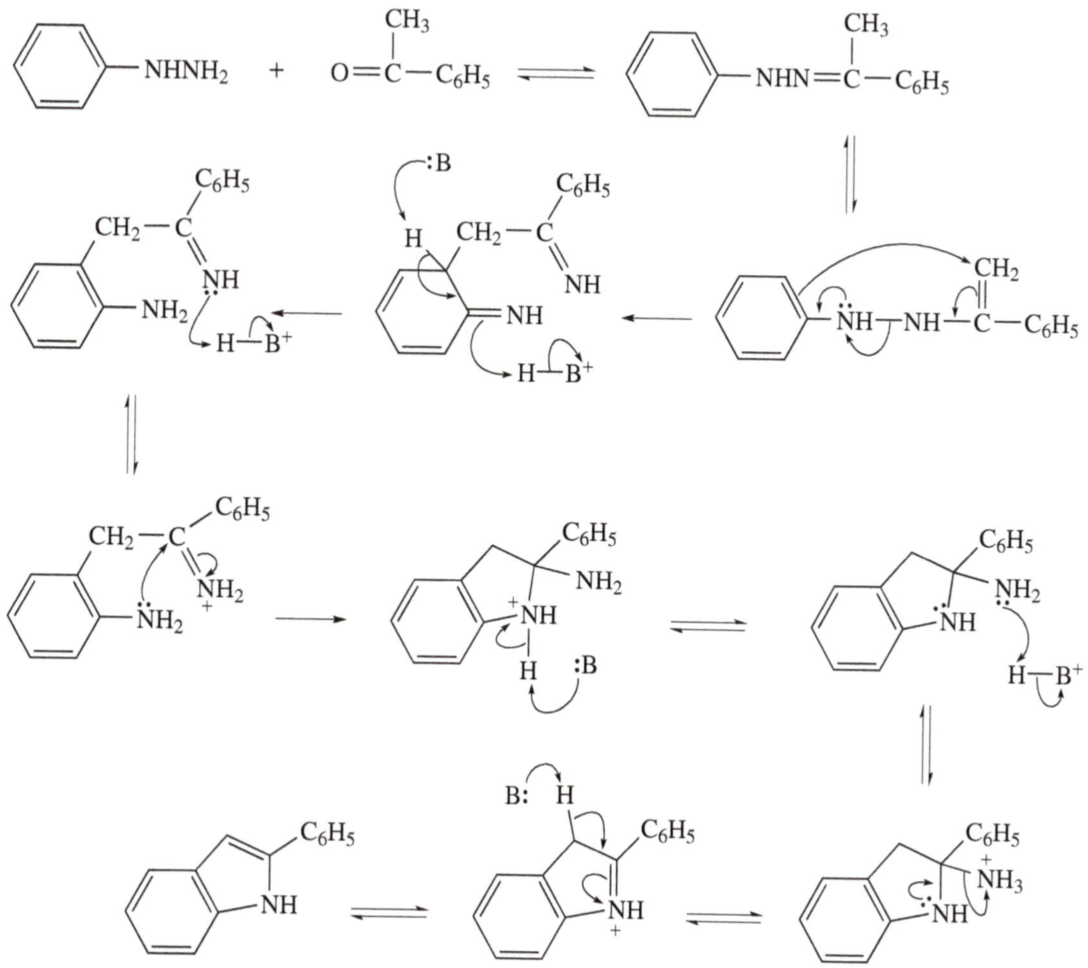

42. **a.** C_6H_5–NHNH$_2$ + $CH_3\overset{O}{\overset{\|}{C}}CH_2CH_3$

b. C_6H_5–NHNH$_2$ + $CH_3CH_2CH_2\overset{O}{\overset{\|}{C}}H$

c. C_6H_5–NHNH$_2$ + cyclohexanone (=O)

43. HC=Ö: ; BF_3 ; $HC=\overset{+}{O}\overset{-}{B}F_3$; N ; H ; B: ; H ; $\overset{-}{O}BF_3$; N^+ ; H ; CH ; C_6H_5 ; $H-B^+$; :$\overset{-}{O}BF_3$; N ; H ; CH ; C_6H_5 ; H ; $\overset{+}{O}\overset{-}{B}F_3$; N ; H ; CH ; C_6H_5 ; N^+ ; H ; CH ; C_6H_5 ; N ; H ; N ; H ; CH ; H ; N^+ ; H ; C_6H_5 ; :B ; N ; H ; CH ; N ; H ; C_6H_5

Repeat two more times.

N ; H ; CH ; C_6H_5 ; N ; H ; CH ; C_6H_5 ; N ; H ; CH ; C_6H_5 ; N ; H

1. Repeat with benzaldehyde.
2. Instead of using another molecule of pyrrole, do an intramolecular reaction with the pyrrole at the end of the chain.

tetraphenylporphyrin

44. **a.** thiophene (S) $\xrightarrow[SnCl_4]{CH_3CH_2Cl}$ 2-ethylthiophene (S) $\xrightarrow{HNO_3,\ H_2SO_4}$ (S) NO_2

b. pyridine (N) $\xrightarrow[H_2SO_4,\ \Delta]{HNO_3}$ (N) NO_2 $\xrightarrow[Pd/C]{H_2}$ (N) NH_2 $\xrightarrow[Fe_2Br_3]{Br_2}$ Br (N) NH_2

Chapter 19 Practice Test

1. Give two names for the following compounds:

a. H_3C … N H … CH_3 b. N … CH_3 c. CH_2CH_3 … O d. Br … N H

2. Draw the product of the following reactions:

a. $CH_3C(=O)Cl$ + 2 $CH_3CH_2NH_2$ ⟶

b. cyclohexanone + piperidine $\xrightarrow{\text{trace acid}}$

c. pyridine + Br_2 $\xrightarrow[300\ °C]{FeBr_3}$

d. pyridine + CH_3I ⟶

e. furan + Cl_2 ⟶

3. Which is a stronger acid?

a. piperidinium (N⁺, H, H) or morpholinium (O, N⁺, H, H)

b. 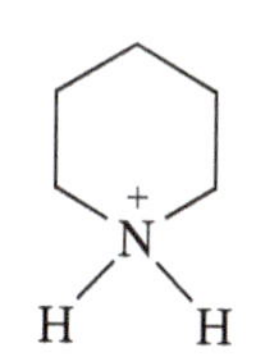 or 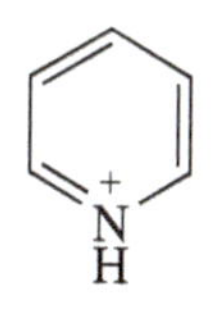

c. 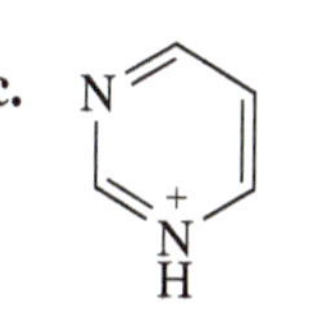 or 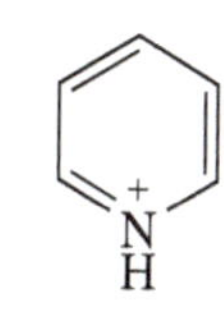

d.  or 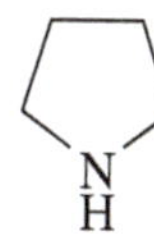

4. Indicate whether each of the following statements is true or false:

a. 4-Chloropyridine is more reactive toward nucleophilic aromatic substitution than is 3-chloropyrrole. T F

b. Pyrrole is more reactive toward electrophilic aromatic substitution than is furan. T F

c. Pyrrole is more reactive toward electrophilic aromatic substitution than is benzene T F

d. Pyridine is more reactive toward electrophilic aromatic substitution than is benzene. T F

5. Show how the following compound can be prepared from the given starting material:

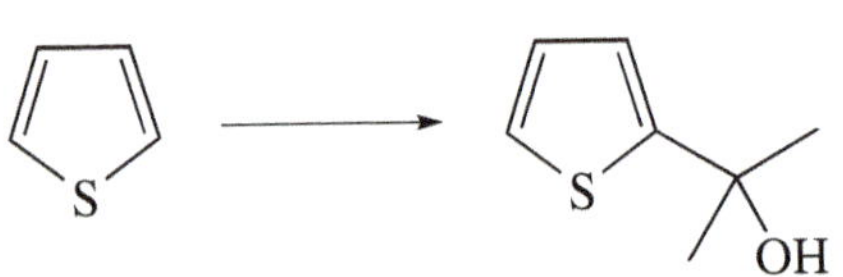

CHAPTER 20
The Organic Chemistry of Carbohydrates

Important Terms

aldaric acid	the compound that results when the aldehyde and primary alcohol groups of an aldose are oxidized to carboxylic acids.
alditol	the compound that results when the carbonyl group of an aldose is reduced to an alcohol.
aldonic acid	the compound that results when the aldehyde group of an aldose is oxidized to a carboxylic acid.
aldose	a polyhydroxy aldehyde.
amino sugar	a sugar in which one of the OH groups is replaced by an NH_2 group.
anomeric carbon	the carbon in a cyclic sugar that is the carbonyl carbon in the straight-chain form.
anomeric effect	preference for the axial position by substituents bonded to the anomeric carbon.
anomers	two cyclic sugars that differ in configuration only at the anomeric carbon (the carbon that is the carbonyl carbon in the straight-chain form).
bioorganic compound	an organic compound found in a living system.
carbohydrate	a sugar, a saccharide. Naturally occurring carbohydrates have the D-configuration.
complex carbohydrate	a carbohydrate that contains two or more sugar molecules linked together.
deoxy sugar	a sugar in which one of the OH groups has been replaced by a hydrogen.
disaccharide	a compound containing two sugar molecules linked together.
enediol rearrangement	a base-catalyzed reaction that interconverts monosaccharides.
epimerization	changing the configuration of a carbon by removing a proton and then reprotonating it.
epimers	monosaccharides that differ in configuration at only one carbon.
furanose	a five-membered ring sugar.
furanoside	a five-membered ring glycoside.
glycoprotein	a protein that is covalently bonded to an oligosaccharide.
glycoside	the acetal of a sugar.

N-glycoside — a glycoside with a nitrogen instead of an oxygen at the glycosidic linkage.

glycosidic bond — the bond between the anomeric carbon of a sugar and an OR or NHR group.

α-1,4′-glycosidic linkage — a glycosidic linkage between the C-1 of one sugar and the C-4 of a second sugar with the oxygen atom at C-1 in the axial position.

α-1,6′-glycosidic linkage — a glycosidic linkage between the C-1 of one sugar and the C-6 of a second sugar with the oxygen atom at C-1 in the axial position.

β-1,4′-glycosidic linkage — a glycosidic linkage between the C-1 of one sugar and the C-4 of a second sugar with the oxygen atom at C-1 in the equatorial position.

Haworth projection — a way to show the structure of a sugar in which the five- and six-membered rings are represented as being flat.

heptose — a monosaccharide with seven carbons.

hexose — a monosaccharide with six carbons.

ketose — a polyhydroxy ketone.

Kiliani–Fischer synthesis — a method used to increase the number of carbons in an aldose by one, resulting in the formation of a pair of C-2 epimers.

molecular recognition — the recognition of one molecule by another as a result of specific interactions.

monosaccharide — a single sugar molecule.

mutarotation — a slow change in optical rotation to an equilibrium value.

nonreducing sugar — a sugar that cannot be oxidized by reagents such as Ag^+ or Br_2. Nonreducing sugars are not in equilibrium with the open-chain aldose or ketose.

oligosaccharide — 3 to 10 sugar molecules linked by glycosidic bonds.

oxocarbenium ion — an ion in which the positive charge is shared by a carbon and an oxygen.

pentose — a monosaccharide with five carbons.

polysaccharide — a compound containing 10 or more sugar molecules linked together.

pyranose — a six-membered ring sugar.

pyranoside — a six-membered ring glycoside.

reducing sugar a sugar that can be oxidized by reagents such as Ag^+ or Br_2. Reducing sugars are in equilibrium with the open-chain aldose or ketose forms.

simple carbohydrate a single sugar molecule.

tetrose a monosaccharide with four carbons.

triose a monosaccharide with three carbons.

Wohl degradation a method used to shorten an aldose by one carbon.

Solutions to Problems

1. D-Ribose is an aldopentose.
D-Sedoheptulose is a ketoheptose.
D-Mannose is an aldohexose.

2. Notice that an L-sugar is the mirror image of a D-sugar.

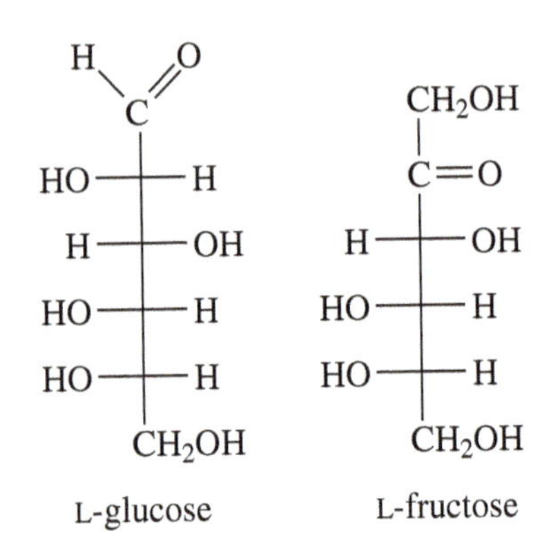

3. First determine whether the structure represents *R*-glyceraldehyde or *S*-glyceraldehyde. Then, because *R*-glyceraldehyde = D-glyceraldehyde and *S*-glyceraldehyde = L-glyceraldehyde, you can answer the question.

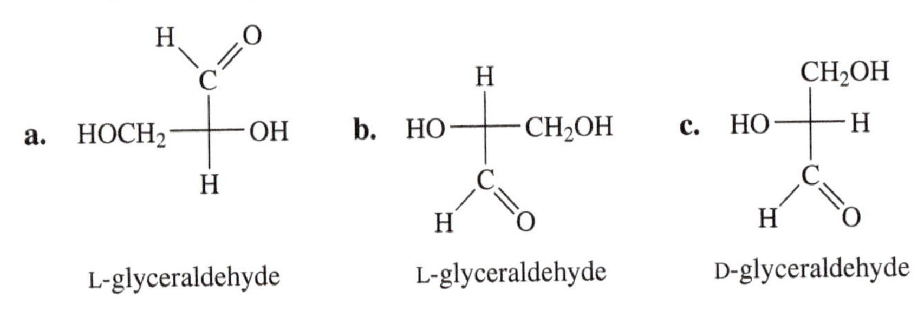

4. **a.** enantiomers because they are mirror images
b. diastereomers because the configuration of one asymmetric center is the same in both and the configuration of the other asymmetric center is the opposite in both

5. **a.** D-ribose **b.** L-talose **c.** L-allose **d.** L-ribose

6. **a.** D-glucose = (2*R*,3*S*,4*R*,5*R*)-2,3,4,5,6-pentahydroxyhexanal

b. Because D-mannose is the C-2 epimer of D-glucose, the systematic name of D-mannose can be obtained just by changing the configuration of the C-2 carbon in the systematic name of D-glucose.

D-mannose = (2*S*,3*S*,4*R*,5*R*)-2,3,4,5,6-pentahydroxyhexanal

c. D-Galactose is the C-4 epimer of D-glucose. Therefore, each of its carbon atoms, except C-4, has the same configuration as it has in D-glucose.

D-galactose = (2*R*,3*S*,4*S*,5*R*)-2,3,4,5,6-pentahydroxyhexanal

d. L-Glucose is the mirror image of D-glucose, so each carbon in L-glucose has the opposite configuration to that in D-glucose.

L-glucose = (2*S*,3*R*,4*S*,5*S*)-2,3,4,5,6-pentahydroxyhexanal

7. D-psicose

8.

a. A ketoheptose has four asymmetric centers ($2^4 = 16$ stereoisomers).

b. An aldoheptose has five asymmetric centers ($2^5 = 32$ stereoisomers).

c. A ketotriose does not have an asymmetric center; therefore, it has no stereoisomers.

9.

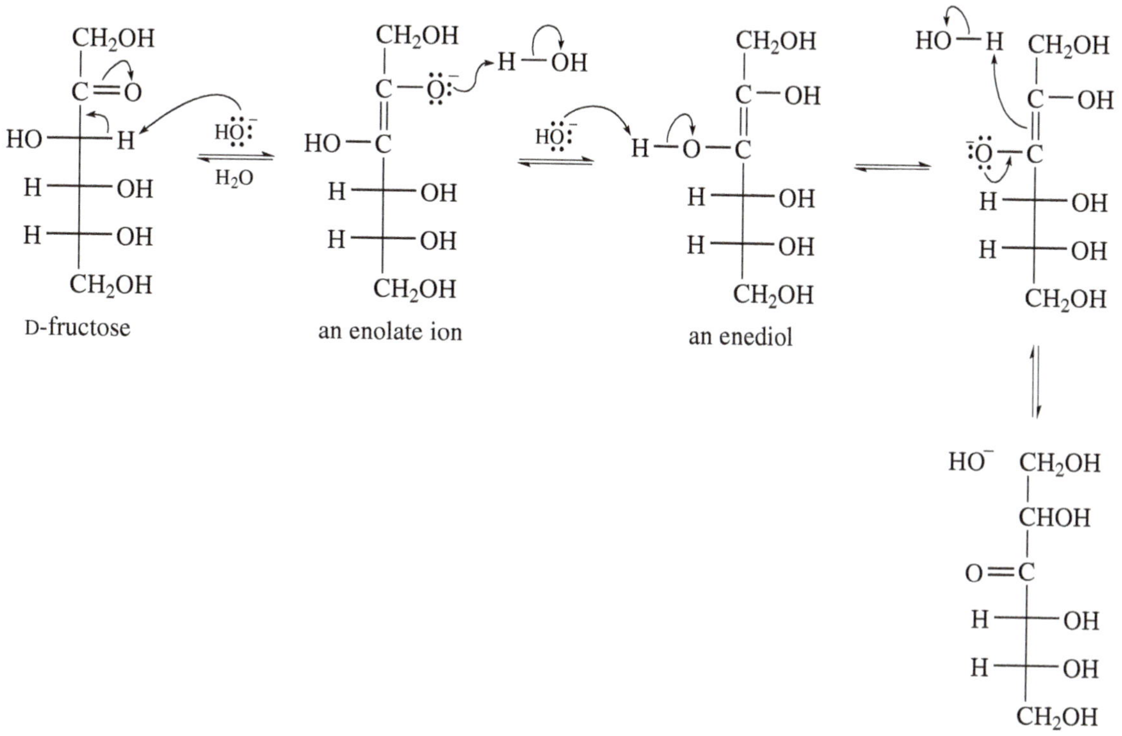

D-fructose

an enolate ion

an enediol

10. Removal of an α-hydrogen creates an enol that can enolize back to the ketone (using the OH at C-2) or can enolize to an aldehyde (using the OH at C-1). The aldehyde has a new asymmetric center (indicated by an *); one of the epimers is D-glucose, and the other is D-mannose.

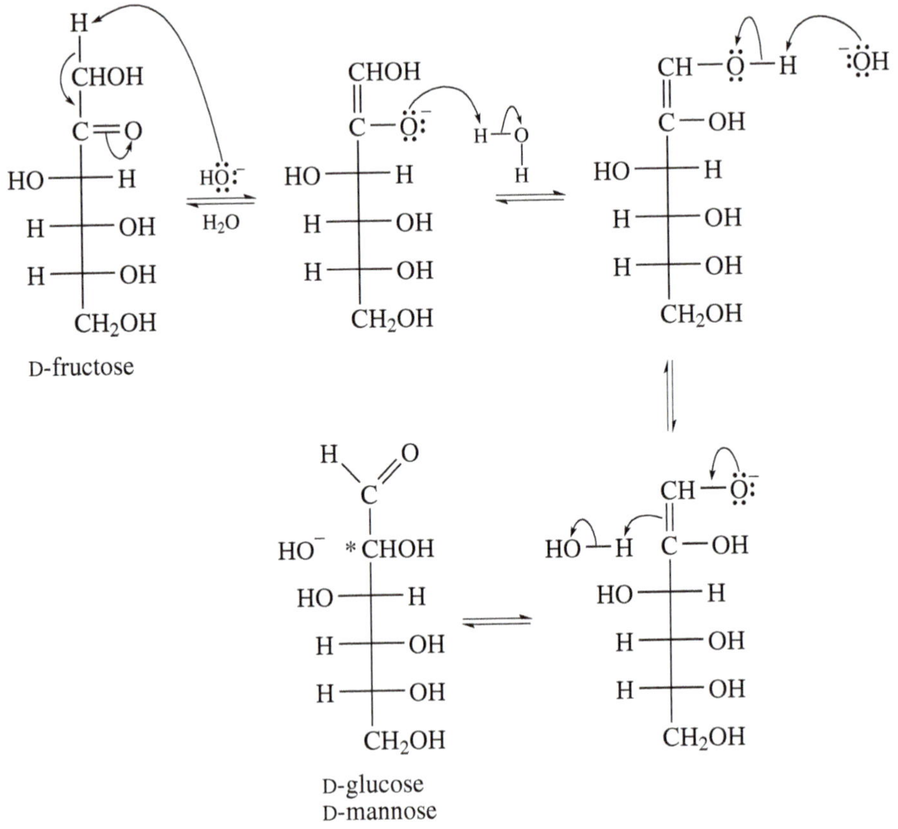

D-fructose

D-glucose
D-mannose

11. D-talose and D-galactose in addition to D-tagatose and D-sorbose.

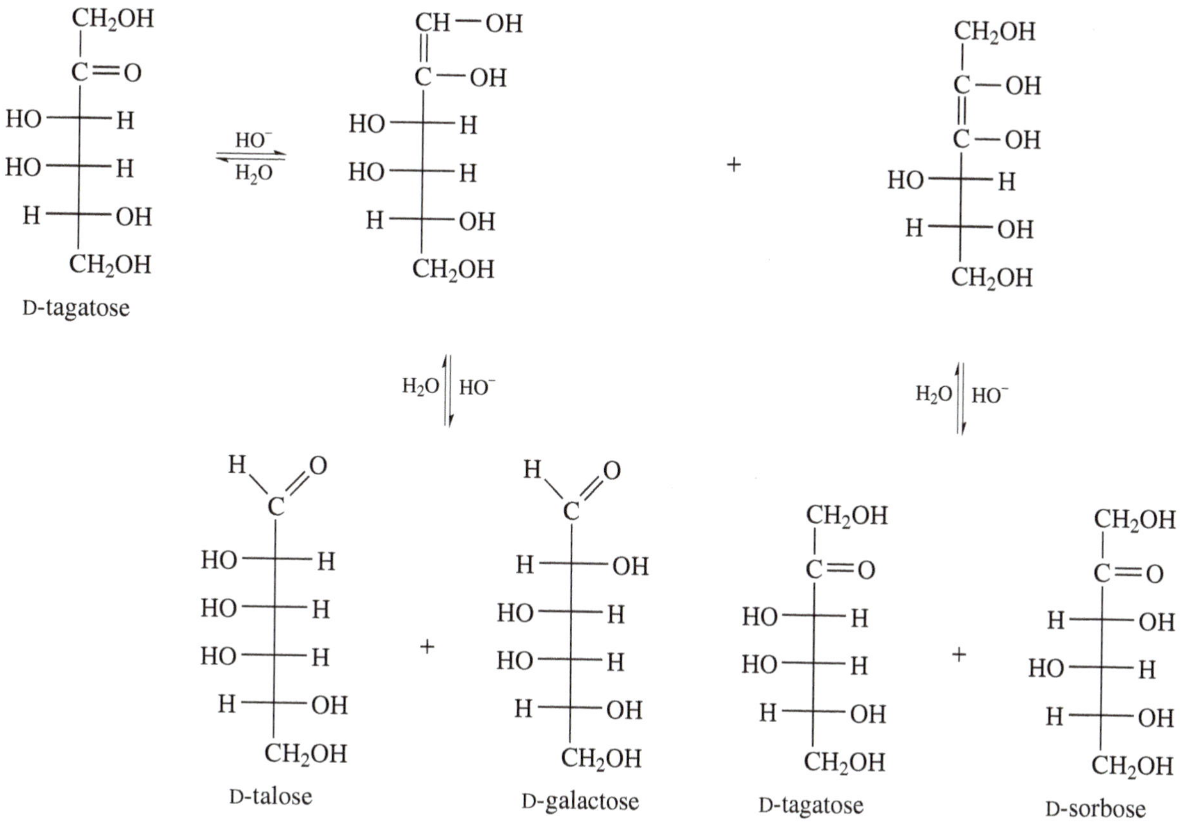

CH₂OH / C=O / HO—H / HO—H / H—OH / CH₂OH
D-tagatose

$\underset{H_2O}{\overset{HO^-}{\rightleftharpoons}}$

CH—OH / C—OH / HO—H / HO—H / H—OH / CH₂OH + CH₂OH / C—OH / C—OH / HO—H / H—OH / CH₂OH

H₂O ⇅ HO⁻

D-talose + D-galactose   D-tagatose + D-sorbose

12. **a.** When D-idose is reduced, D-iditol is formed.

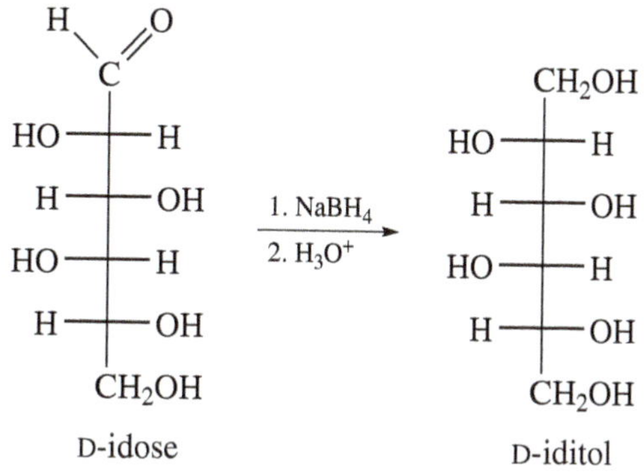

D-idose $\xrightarrow[2.\ H_3O^+]{1.\ NaBH_4}$ D-iditol

b. When D-sorbose is reduced, C-2 becomes an asymmetric center, so both D-iditol and the C-2 epimer of D-iditol (D-gulitol) are formed.

D-sorbose $\xrightarrow[\text{2. } H_3O^+]{\text{1. NaBH}_4}$ D-iditol + D-gulitol

13. **a.** **1.** D-Altrose is reduced to the same alditol that D-talose is reduced to. The easiest way to answer this question is to draw D-talose and its alditol. Then draw the monosaccharide with the same configuration at C-2, C-3, C-4, and C-5 as D-talose, reversing the functional groups at C-1 and C-6. (Put the primary alcohol group at the top and the aldehyde group at the bottom.) When this structure is rotated 180° in the plane of the paper, the monosaccharide can be identified.

D-talose $\xrightarrow[\text{2. } H_3O^+]{\text{1. NaBH}_4}$ the alditol of D-talose $\xleftarrow[\text{2. } H_3O^+]{\text{1. NaBH}_4}$ ⟷ (rotate 180°) D-altrose

2. L-Gulose is reduced to the same alditol that D-glucose is reduced to.

D-glucose $\xrightarrow[\text{2. } H_3O^+]{\text{1. NaBH}_4}$ the alditol of D-glucose $\xleftarrow[\text{2. } H_3O^+]{\text{1. NaBH}_4}$ ⟷ (rotate 180°) L-gulose

3. L-Galactose is reduced to the same alditol that D-galactose is reduced to.

D-galactose $\xrightarrow[2.\ H_3O^+]{1.\ NaBH_4}$ the alditol of D-galactose $\xleftarrow[2.\ H_3O^+]{1.\ NaBH_4}$ (rotate 180°) L-galactose

b. 1. The ketohexose (D-tagatose) with the same configuration at C-3, C-4, and C-5 as D-talose forms the same alditol (D-talitol) that D-talose forms. The other alditol (D-galactitol) that is formed has the opposite configuration at C-2.

D-tagatose $\xrightarrow[2.\ H_3O^+]{1.\ NaBH_4}$ D-talitol + D-galactitol

2. Similarly, the ketohexose (D-psicose) with the same configuration at C-3, C-4, and C-5 as D-allose forms the same alditol that D-allose forms. The other alditol (D-altritol) that is formed has the opposite configuration at C-2.

D-psicose $\xrightarrow[2.\ H_3O^+]{1.\ NaBH_4}$ D-allitol + D-altritol

14. **a.** L-gulose (Note the similarity of this problem to Problem 13. a. 2.)

D-glucose D-glucaric acid L-gulose

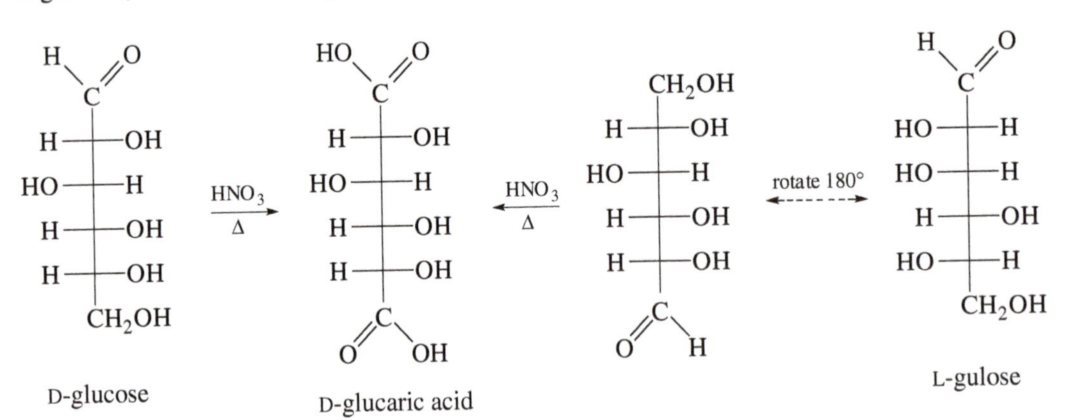

b. L-Gularic acid because it is also the oxidation product of L-gulose.

c. D-allose and L-allose, D-altrose and D-talose, L-altrose and L-talose, D-galactose and L-galactose

15. The monosaccharides are the aldoses with one additional carbon that are C-2 epimers and whose other asymmetric centers have the same configuration as the given aldose.

a. D-gulose and D-idose **b.** L-xylose and L-lyxose

16. The monosaccharides are the aldoses whose C-1 carbon has been removed, whose C-2 carbon has been converted to an aldehyde, and whose other asymmetric centers have the same configuration as the given aldose.

a. D-allose and D-altrose **b.** D-glucose and D-mannose **c.** L-allose and L-altrose

17. Solved in the text.

18.

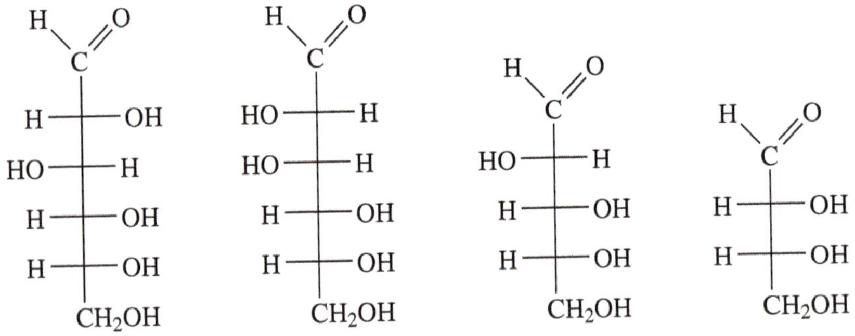

D-glucose **A** D-mannose **B** D-arabinose **C** D-erythrose **D**

19. The hemiacetals in **a** and **b** have one asymmetric center; therefore, each has two stereoisomers. The hemi-acetals in **c** and **b** have two asymmetric centers; therefore, each has four stereoisomers.

a. Solved in the text.

b. H_3C O OH H_3C O OH O H_3C OH O OH H_3C

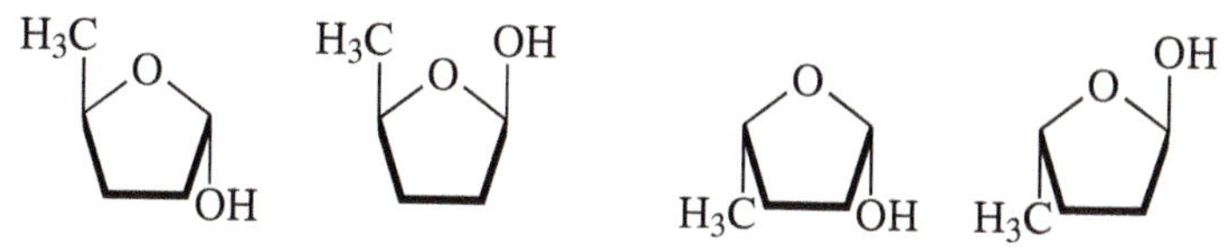

c. O OH H O H OH

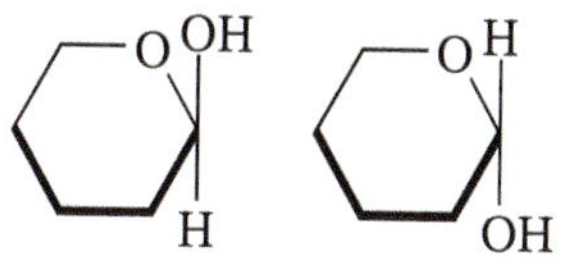

d. $CH_3CH_2CH_2$ O OH $CH_3CH_2CH_2$ OH O O $CH_3CH_2CH_2$ OH OH O $CH_3CH_2CH_2$

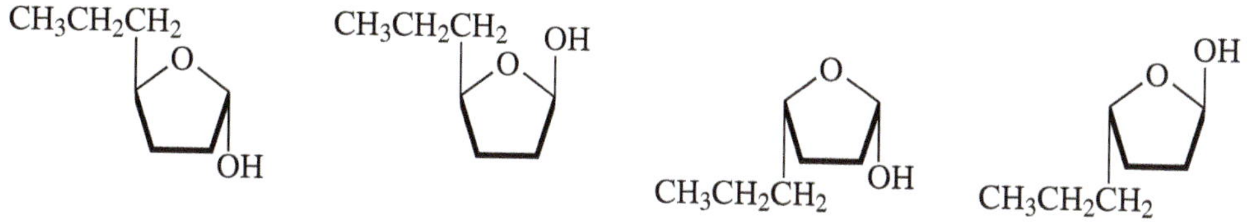

20. **a.** CH_2OH HO O OH OH OH

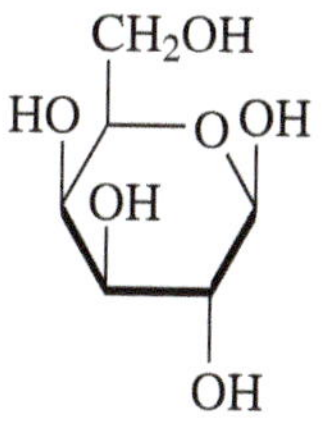

b. O CH_2OH OH HO HO OH

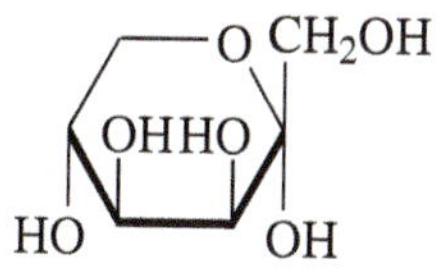

c. CH_2OH O OH HO OH OH CH_2OH O HO HO OH OH

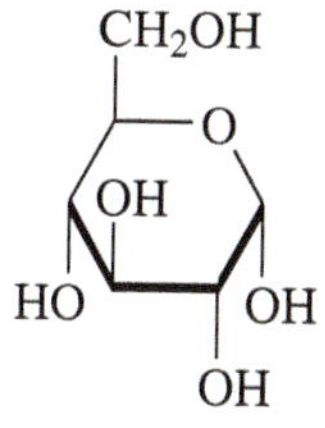

α-D-glucose

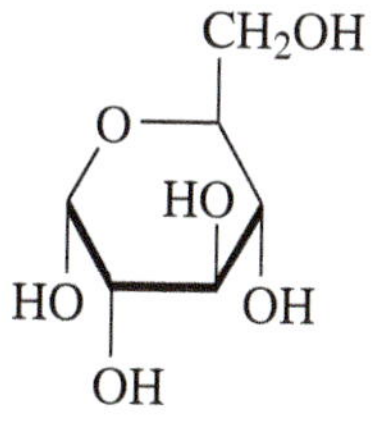

α-L-glucose
the mirror image of α-D-glucose

21. CH_2OH HO H O OH OH OH

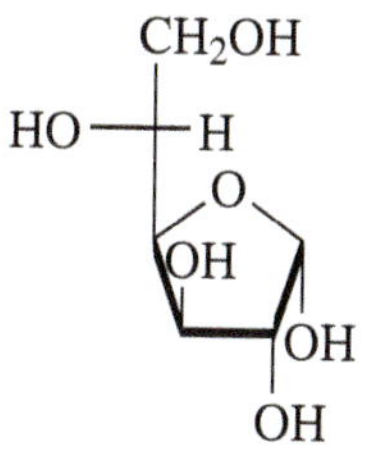

22.

D-erythrose α-D-erythrofuranose β-D-erythrofuranose

23. First recall that in the chair conformation, an α-anomer has the anomeric OH group in the axial position and a β-anomer has the anomeric OH group in the equatorial position. Then recall that glucose has all of its OH groups in equatorial positions. Now this question can be answered easily.

All the OH groups in β-D-glucose are in equatorial positions. Because β-D-mannose is a C-2 epimer of β-D-glucose, the C-2 OH group of β-D-mannose is the only one in the axial position.

24. **a.** D-Idose differs in configuration from D-glucose at C-2, C-3, and C-4. Therefore, the OH groups at C-2, C-3, and C-4 in β-D-idose are in axial positions.

b. D-Allose is a C-3 epimer of D-glucose. Therefore, the OH group at C-3 is in the axial position and because it is the α-anomer, the OH group at C-1 (the anomeric carbon) is also in the axial position.

25.

$\xrightarrow[\text{HCl}]{CH_3CH_2OH}$ +

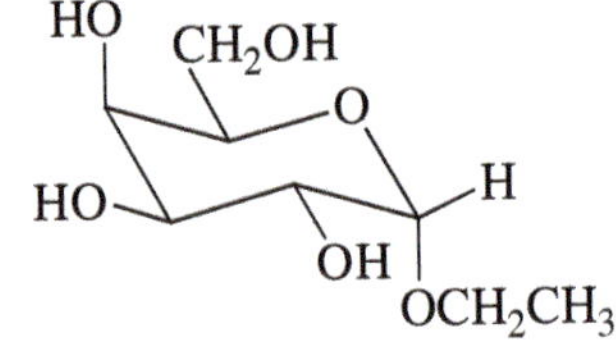

β-D-galactose ethyl β-D-galactoside ethyl α-D-galactoside

26. If more than a trace amount of acid is used, the amine that acts as a nucleophile when it forms the *N*-glycoside becomes protonated, and a protonated amine is not a nucleophile.

27. Solved in the text.

28. **a.** α-D-talose (or α-D-talopyranose) (reducing; it is a hemiacetal)

b. methyl α-D-galactoside (or methyl α-D-galactopyranoside) (nonreducing; it is an acetal)

c. ethyl β-D-psicoside (or ethyl β-D-psicofuranoside) (nonreducing; it is an acetal)

29.

$$\begin{aligned} \text{specific rotation of glucose} + \text{specific rotation of fructose} &= -22.0 \\ +52.7 + \text{specific rotation of fructose} &= -22.0 \\ \text{specific rotation of fructose} &= -22.0 + (-52.7) \\ \text{specific rotation of fructose} &= -74.7 \end{aligned}$$

30. **a.** Both sugars in the disaccharide are glucose.

b. It is a 1,6′-glycosidic linkage.

31. **a.** Amylose has α-1,4′-glycosidic linkages, whereas cellulose has β-1,4′-glycosidic linkages.

b. Amylose has α-1,4′-glycosidic linkages, whereas amylopectin has both α-1,4′-glycosidic linkages and α-1,6′-glycosidic linkages.

c. Glycogen and amylopectin have the same kind of linkages, but glycogen has a higher frequency of α-1,6′-glycosidic linkages.

d. Cellulose has a hydroxy group at C-2, whereas chitin has an *N*-acetyl amino group at that position.

32. A proton is more easily lost from the C-3 OH group because the electrons that are left behind when the proton is removed are delocalized onto an oxygen. When a proton is removed from the C-2 OH group, the electrons that are left behind are delocalized onto a carbon. Because oxygen is more electronegative than carbon, a negatively charged oxygen is more stable than a negatively charged carbon. Recall that the more stable the base, the stronger its conjugate acid.

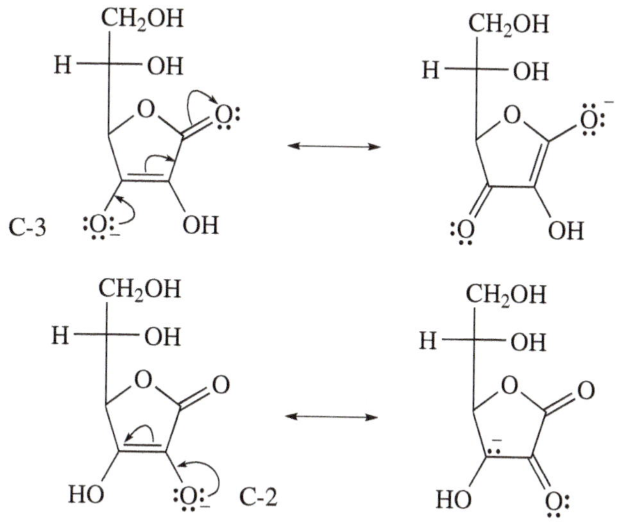

33. **a.** People with type O blood can receive blood only from other people with type O blood because types A, B, and AB blood have sugar components that type O blood does not have.

b. People with type AB blood can give blood only to other people with type AB blood. They cannot give blood to people with A, B, or O blood because type AB blood has sugar components that types A, B, and O blood do not have.

34.

a.
COOH
H—OH
HO—H
HO—H
H—OH
COOH

b.
COO^-
H—OH
HO—H
HO—H
H—OH
CH_2OH

c.
CH_2OH
H—OH
HO—H
HO—H
H—OH
CH_2OH

d. CH_3O, CH_2OCH_3, O, CH_3O, CH_3O, OCH_3

e.

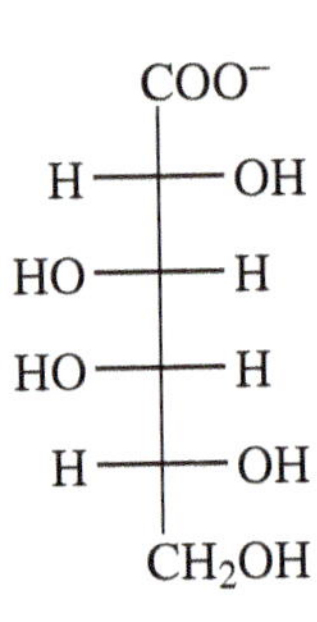

f.

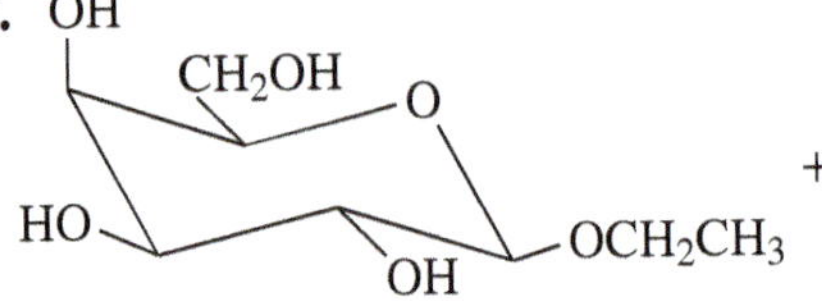

+

OH, CH_2OH, O, HO, OH, OCH_2CH_3

g.
HC=O
HO—H
HO—H
H—OH
CH_2OH

35. C-2 epimer = D-mannose
C-3 epimer = D-allose
C-4 epimer = D-galactose
C-5 epimer = L-idose

36. **a.** D-lyxose **b.** D-talose **c.** D-psicose

37.

D-threose (HC=O; HO—H; H—OH; CH_2OH) $\xrightarrow{\text{Kiliani–Fischer synthesis}}$ sugar A, D-xylose (HC=O; H—OH; HO—H; H—OH; CH_2OH) + sugar B, D-lyxose (HC=O; HO—H; HO—H; H—OH; CH_2OH) $\xrightarrow{HNO_3}$ (COOH; H—OH; HO—H; H—OH; COOH) optically inactive + (COOH; HO—H; HO—H; H—OH; COOH) optically active

38. **a.** D-ribose and L-ribose, D-arabinose and L-arabinose, D-xylose and L-xylose, D-lyxose and L-lyxose

b. D-ribose and D-arabinose, D-xylose and D-lyxose, L-ribose and L-arabinose, L-xylose and L-lyxose

c. D-arabinose, L-arabinose, D-lyxose, and L-lyxose

39. **a.** 2*R*,3*S*,4*R*,5*R* **b.** 2*R*,3*S*,4*S*,5*R* **c.** 2*R*,3*R*,4*R* **d.** 2*R*,3*S*,4*R* **e.** 3*R*,4*S*,5*R*

40.

methyl α-D-ribofuranose

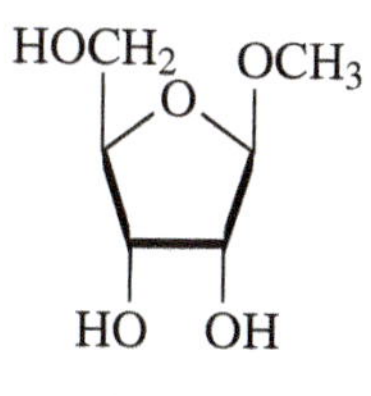

methyl β-D-ribofuranose

methyl α-D-ribopyranose

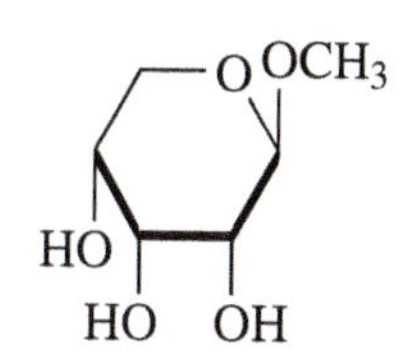

methyl β-D-ribopyranose

41. **a.** methyl β-D-sorboside **b.** ethyl β-D-guloside **c.** methyl α-D-idoside

42. A monosaccharide with a molecular weight of 150 must have five carbons (5 Cs = 60, 5 Os = 80, and 10 Hs = 10 for a total of 150). All aldopentoses are optically active. Therefore, the compound must be a ketopentose. The following is the only ketopentose that is not optically active.

CH_2OH
H—C—OH
C=O
H—C—OH
CH_2OH

43.

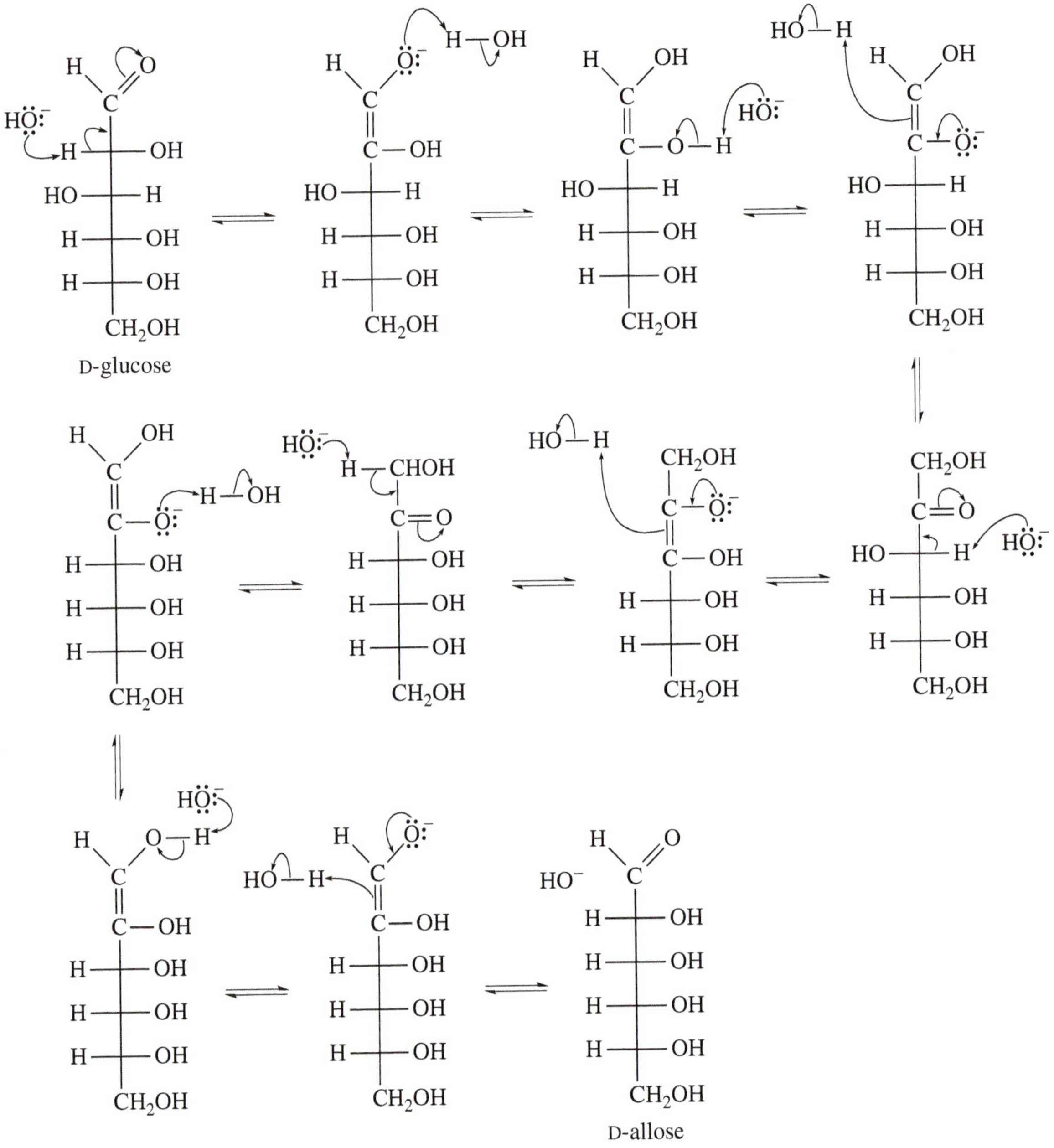

D-glucose

D-allose

44.

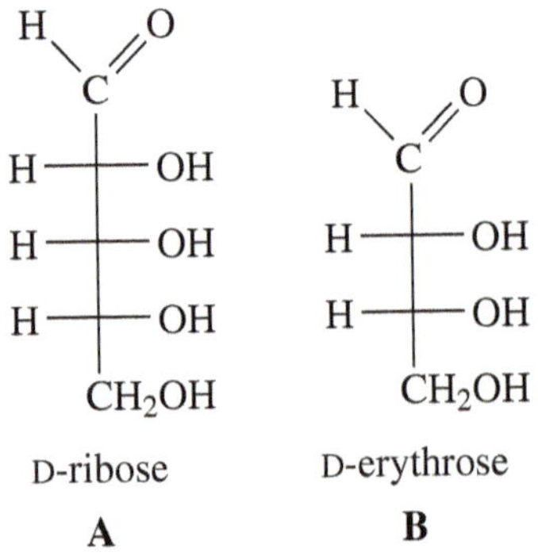

D-ribose

A

D-erythrose

B

45.

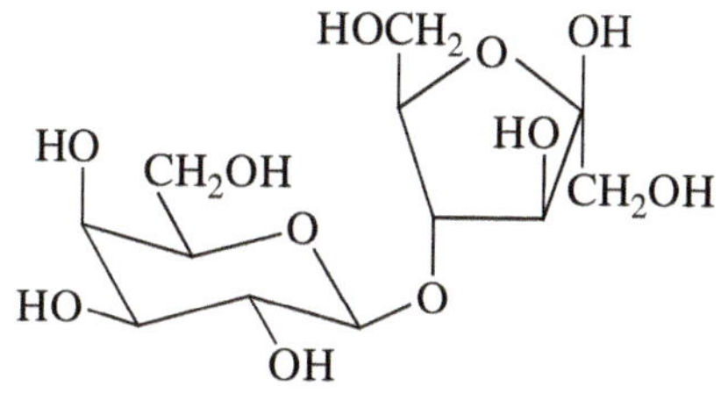

46. The hexose is D-altrose.

1. Knowing that (+)-glyceraldehyde is D-glyceraldehyde gives the configuration at C-5.
2. Knowing that the second Wohl degradation gives D-erythrose gives the configuration at C-4.
3. Knowing that the first Wohl degradation followed by oxidation gives an optically inactive aldaric acid gives the configuration at C-3.
4. Knowing that the original hexose forms an optically active aldaric acid gives the configuration at C-2.

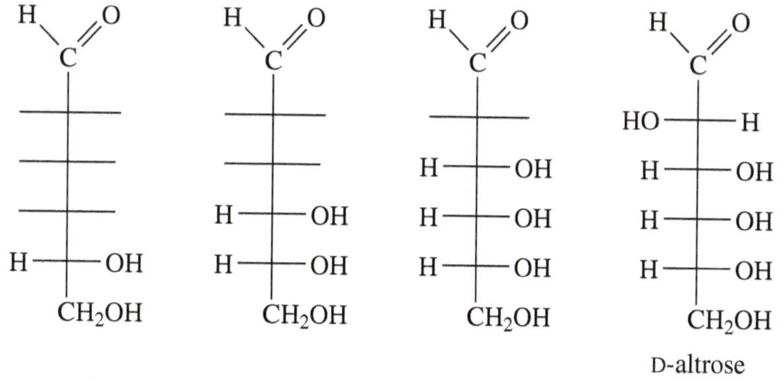

D-altrose

47.

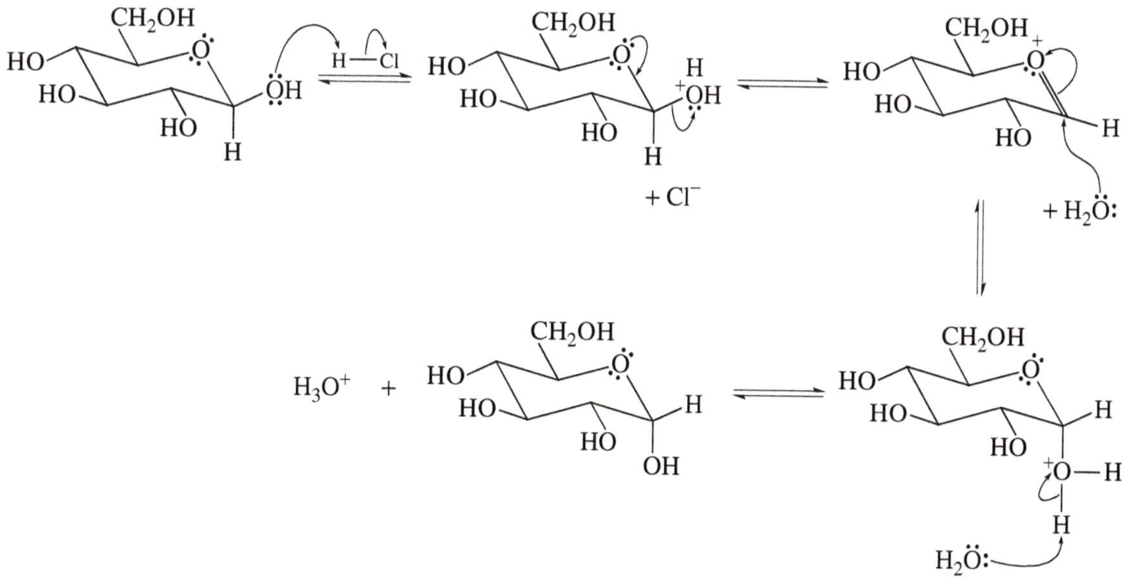

48. There are two possible structures for an optically active five-carbon alditol formed from a Wohl degradation of a D-hexose. Notice that the three asymmetric centers in the structure on the left have the same configuration as the equivalent asymmetric centers in D-glucose. Therefore, the OH groups in the structure on the left are all in equatorial positions.

	CH_2OH				CH_2OH	
HO	—	H		HO	—	H
H	—	OH		*HO	—	H
H	—	OH		H	—	OH
	CH_2OH				CH_2OH	

All OH groups are in equatorial positions. Only the starred OH group is in an axial position.

The unknown aldohexose has only one OH group in an axial position. Therefore, the aldohexose that gives the alditol on the left needs to have the OH on C-2 pointing to the left (so it will be axial). The other aldohexose has an axial hydrogen, so it needs to have the OH on C-2 pointing to the right (so it will not be axial). The two possible aldoses are D-mannose and D-galactose:

	HC=O				HC=O	
HO	—	H		H	—	OH
HO	—	H		HO	—	H
H	—	OH		*HO	—	H
H	—	OH		H	—	OH
	CH_2OH				CH_2OH	
	D-mannose				D-galactose	

The two possibilities can be distinguished by reducing the two aldoses to alditols and then measuring their optical rotation in a polarimeter. The alditol of D-mannose will be optically active and the alditol of D-galactose will be optically inactive.

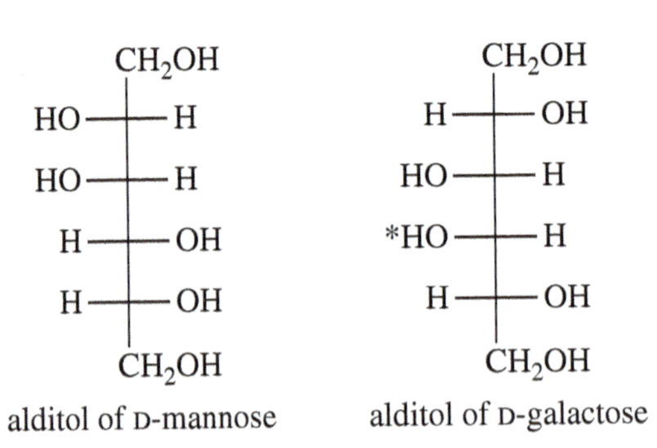

alditol of D-mannose alditol of D-galactose

49. The hydrogen that is bonded to the anomeric carbon is the hydrogen that has its signal at the highest frequency, because it is the only hydrogen that is bonded to a carbon that is bonded to two oxygens. So the two anomeric hydrogens (one from the α-anomer and one from the β-anomer) are responsible for the two high-frequency doublets.

50.

the β-D-glucuronide

the α-D-glucuronide

51.

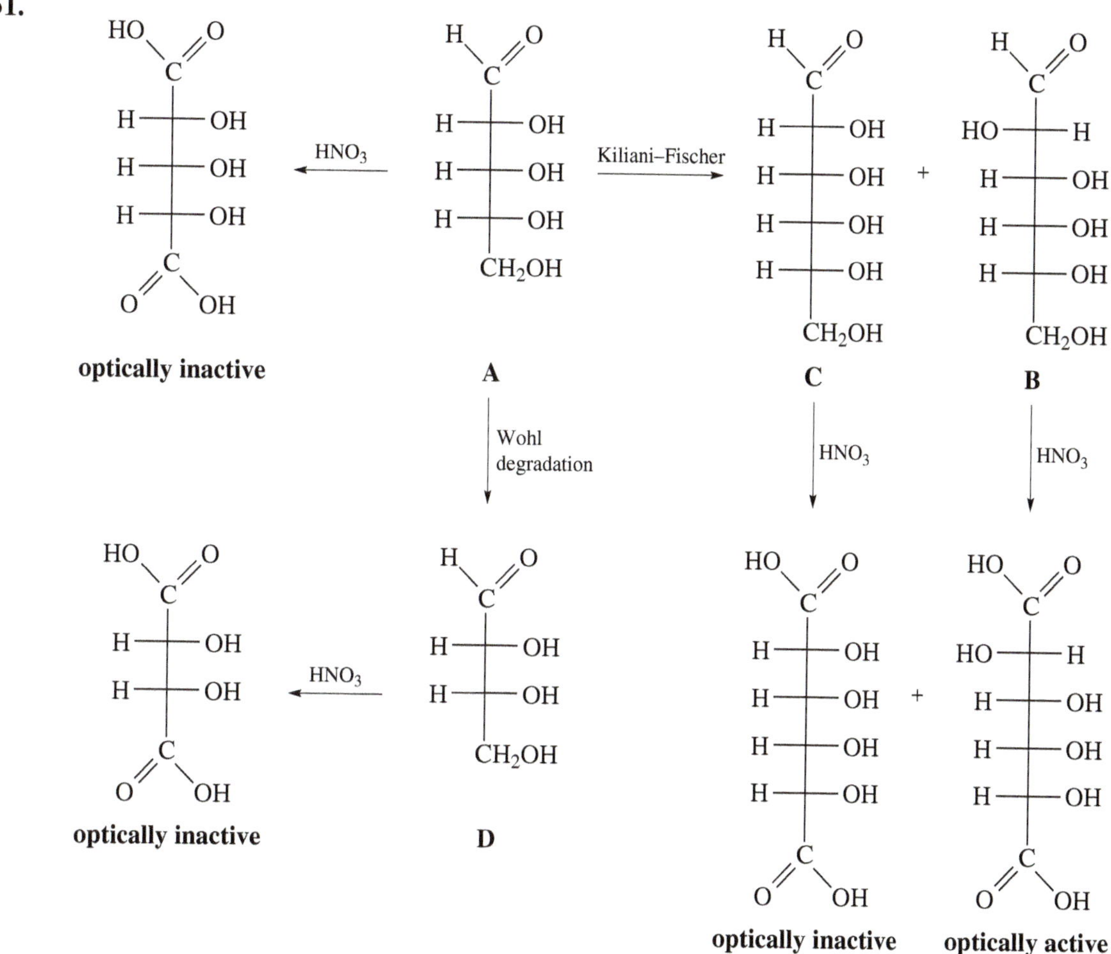

52.

1. As Fischer did, we can narrow our search to eight aldohexoses because there are eight pairs of enantiomers. First, we need to find an aldopentose that forms (+)-galactose as a product of a Kiliani–Fischer synthesis. That sugar is the one known as (−)-lyxose. The Kiliani–Fischer synthesis on (−)-lyxose yields two sugars with melting points that show them to be the sugars known as (+)-galactose and (+)-talose. Now we know that (+)-galactose and (+)-talose are C-2 epimers. The eight aldohexoses are sugars 1 and 2, 3 and 4, 5 and 6, or 7 and 8. (See page 954 of the text.)

2. When (+)-galactose and (+)-talose react with HNO_3, (+)-galactose forms an optically inactive aldaric acid and (+)-talose forms an optically active aldaric acid. Therefore, (+)-galactose and (+)-talose are sugars 1 and 2 or 7 and 8. Because (+)-galactose is the one that forms the optically inactive aldaric acid, it is either sugar 1 or 7.

3. To determine the structure of (+)-galactose, we can go back to (−)-lyxose, the sugar that forms sugars 7 and 8 by a Kiliani–Fischer synthesis, and oxidize it with HNO_3. Finding that the aldaric acid is optically active allows us to conclude that (+)-galactose is sugar 7, because the aldopentose that leads to sugars 1 and 2 would give an optically inactive aldaric acid.

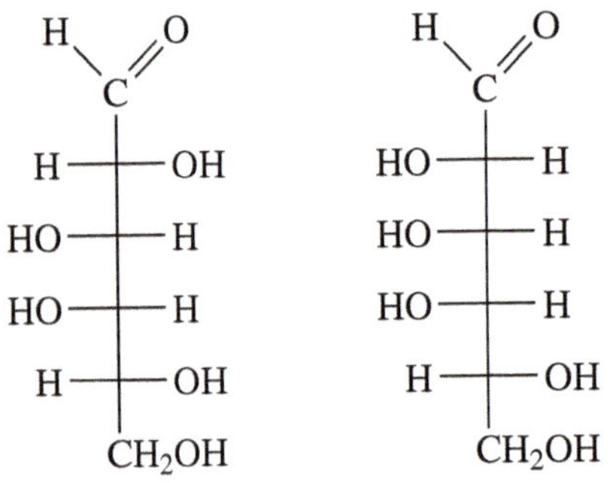

7
D-galactose

8
D-talose

53. **D-Arabinose**. The only D-aldopentoses that are oxidized to optically active aldaric acids are D-arabinose and D-lyxose. A Wohl degradation of D-arabinose forms D-erythrose, whereas a Wohl degradation of D-lyxose forms D-threose. Because D-erythrose forms an optically inactive aldaric acid but D-threose does not, we can conclude that the D-aldopentose is D-arabinose.

54.

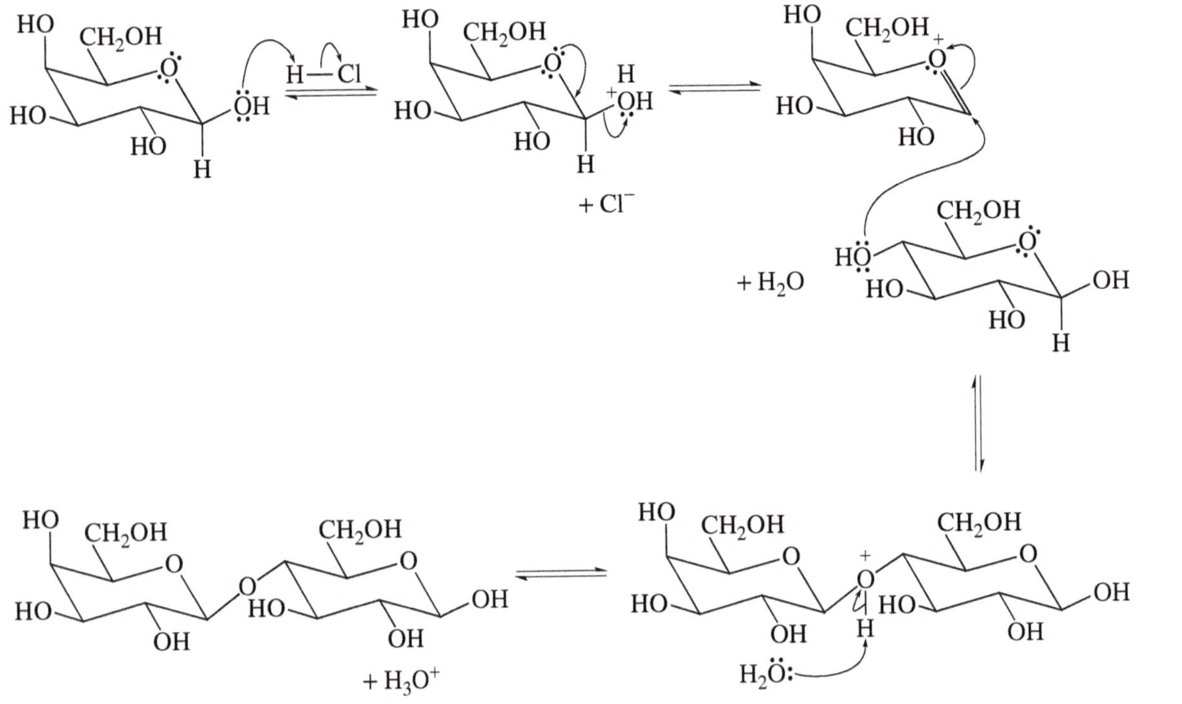

$+ Cl^-$

$+ H_2O$

$+ H_3O^+$

lactose

55.

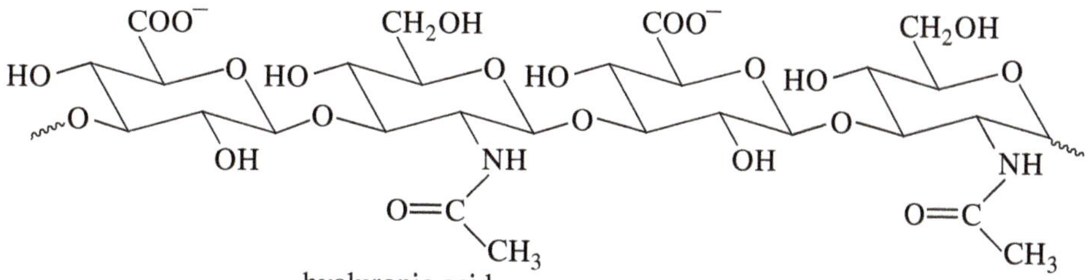

COO⁻ CH₂OH COO⁻ CH₂OH
HO O HO O HO O HO O
O O O O
OH NH OH NH
O=C O=C
CH₃ CH₃

hyaluronic acid

56. She can take a sample of one of the sugars and oxidize it with nitric acid to an aldaric acid or reduce it with sodium borohydride to an alditol. If the product is optically active, the sugar was D-lyxose; if the product is not optically active, the sugar was D-xylose.

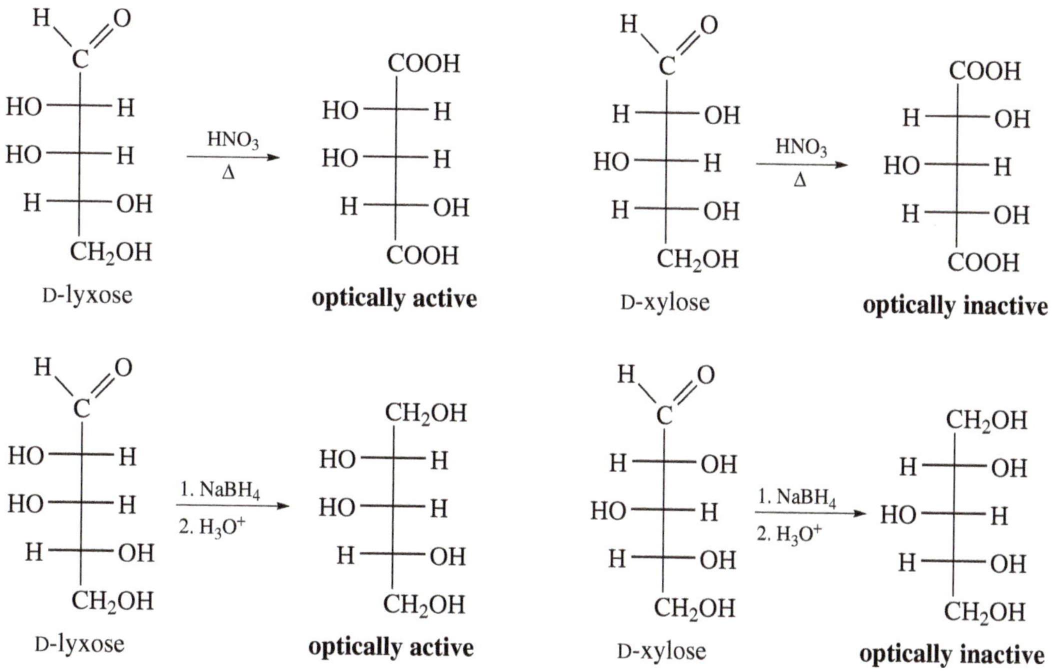

D-lyxose $\xrightarrow[\Delta]{HNO_3}$ **optically active**

D-xylose $\xrightarrow[\Delta]{HNO_3}$ **optically inactive**

D-lyxose $\xrightarrow[2.\ H_3O^+]{1.\ NaBH_4}$ **optically active**

D-xylose $\xrightarrow[2.\ H_3O^+]{1.\ NaBH_4}$ **optically inactive**

57.

H—Cl H₂Ö: NH₂ O—H :B H—B⁺ NH₂ OH NH₃⁺ :ÖH H :B

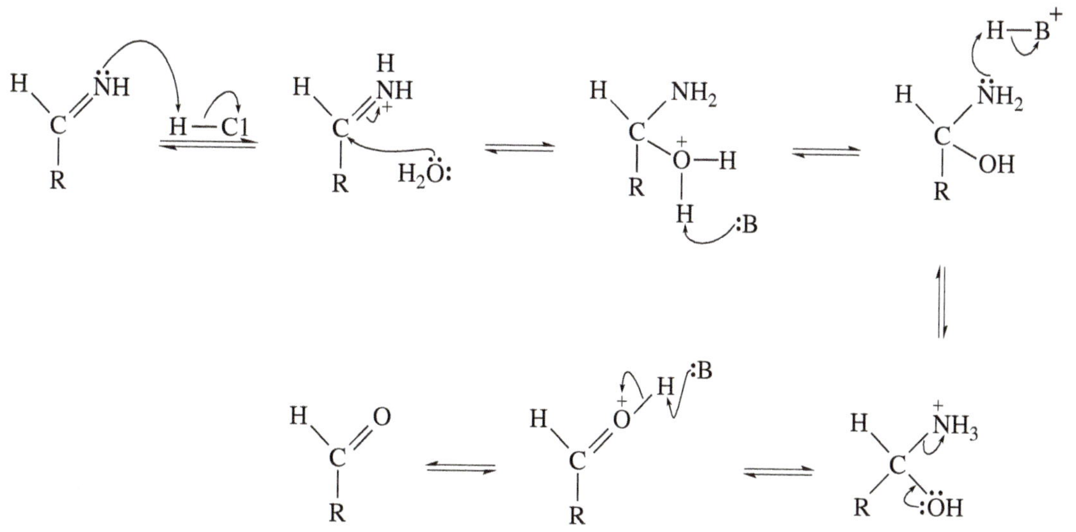

58.

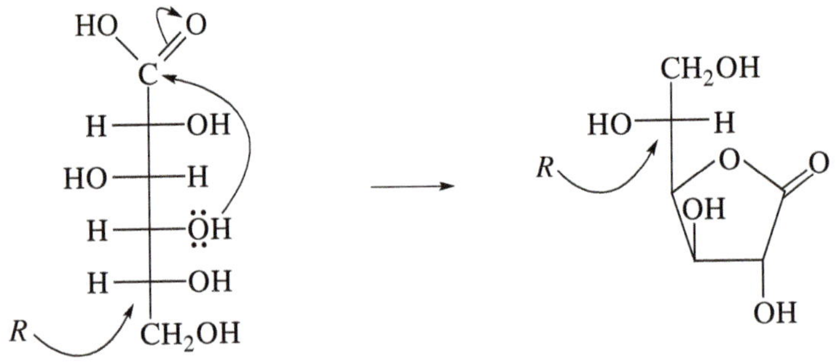

HO, O, C, H—OH, HO—H, H—ÖH, H—OH, R, CH_2OH → CH_2OH, HO—H, R, O, O, OH, OH

aldonic acid of
D-glucose

59.

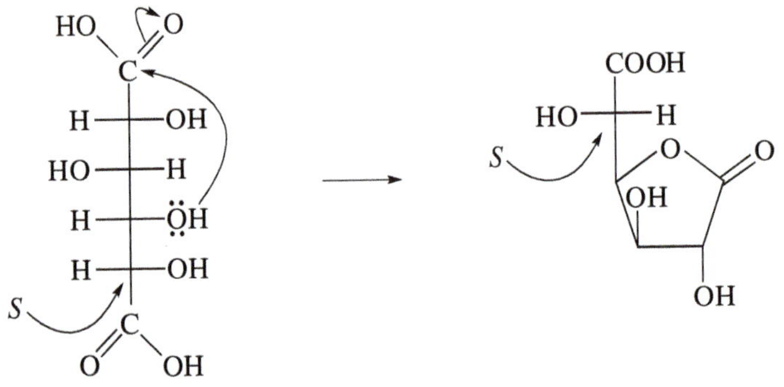

HO, O, C, H—OH, HO—H, H—ÖH, H—OH, S, C, O, OH → COOH, HO—H, S, O, O, OH, OH

aldaric acid of
D-glucose

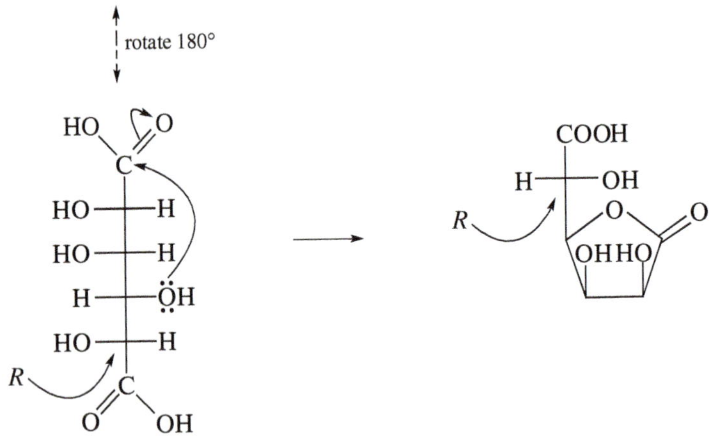

rotate 180°

HO, O, C, HO—H, HO—H, H—ÖH, HO—H, R, C, O, OH → COOH, H—OH, R, O, O, OH HO

60.

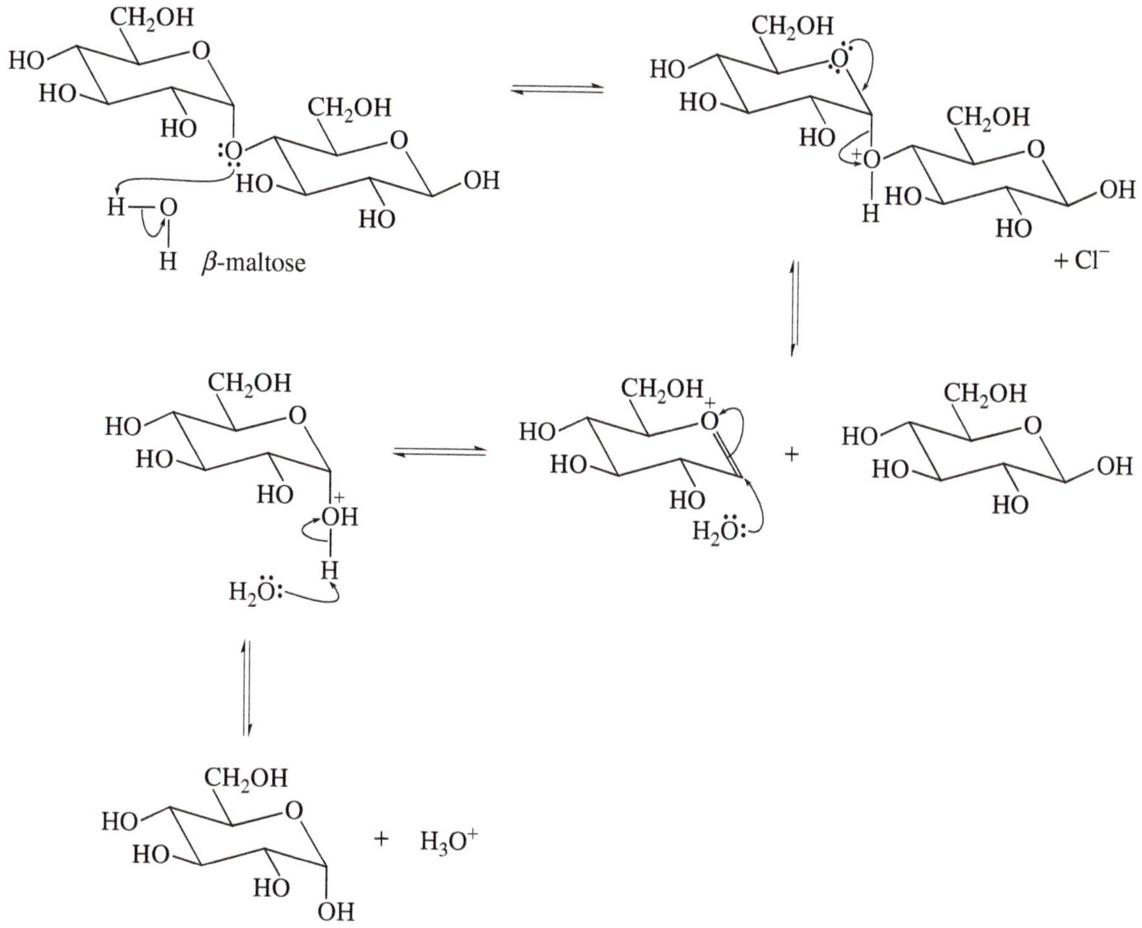

The β-anomer will also be formed when H_2O adds to the top of the plane of the oxocarbenium ion.

61. **10** aldaric acids

Each of the following pairs forms the same aldaric acid:

D-allose and L-allose	L-altrose and L-talose
D-galactose and L-galactose	D-gulose and L-glucose
D-altrose and D-talose	L-gulose and D-glucose

Therefore, 12 aldohexoses form 6 aldaric acids. The other 4 aldohexoses each form a distinctive aldaric acid, and $6 + 4 = 10$.

62. We know the hexose is a ketohexose because it does not react with Br_2; we know it is a 2-ketohexose because if it were a 3-ketohexose, it would not be able to form a hemiacetal (because the hemiacetal would have an unstable four-membered ring), and, therefore, would not undergo mutarotation.

Knowing that the hexose is oxidized by Tollens reagent to the aldonic acids D-talonic acid and D-galactonic acid tells us that the aldonic acids and the ketohexose have the same configuration at C-3, C-4, and C-5. Therefore, the hexose is D-tagatose.

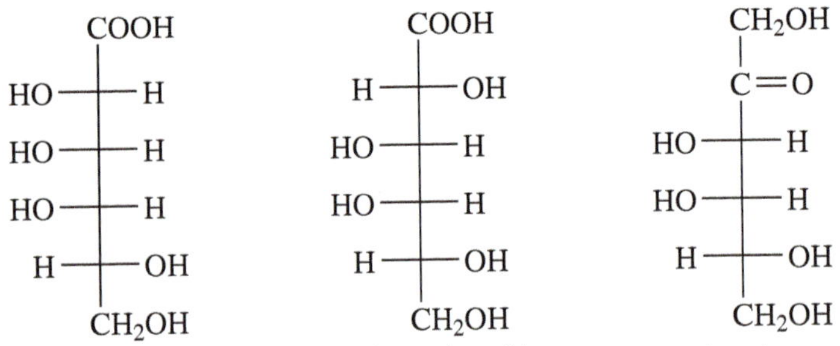

COOH
HO—H
HO—H
HO—H
H—OH
CH_2OH
D-talonic acid

COOH
H—OH
HO—H
HO—H
H—OH
CH_2OH
D-galactonic acid

CH_2OH
C=O
HO—H
HO—H
H—OH
CH_2OH
D-tagatose

63.

H DÖ:⁻
CHOH
C=O
HO—H
H—OH
H—OH
CH_2OH

⇌

D—OD
CHOH
C—Ö:⁻
HO—H
H—OH
H—OH
CH_2OH

⇌

H DÖ:⁻
CDOH
C=O
HO—H
H—OH
H—OH
CH_2OH

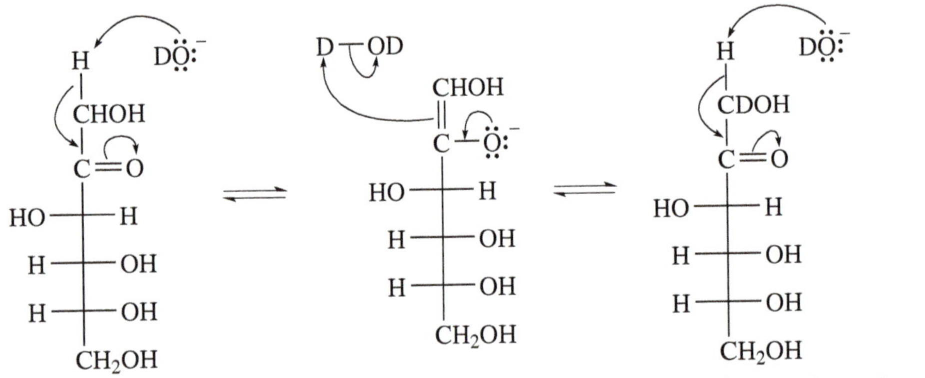

D-fructose with one deuterium

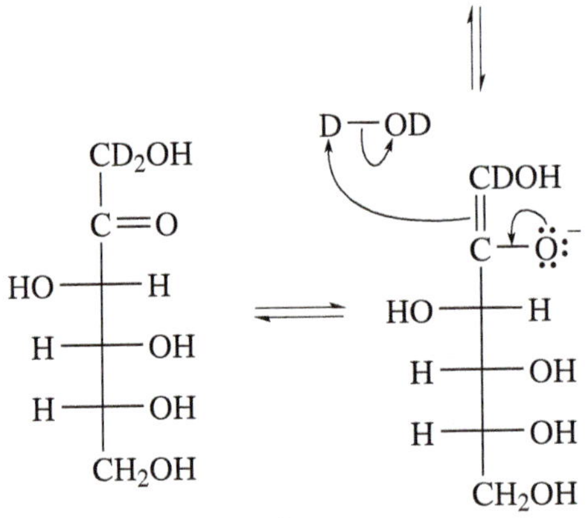

⇅

D—OD
CDOH
C—Ö:⁻
HO—H
H—OH
H—OH
CH_2OH

⇌

CD_2OH
C=O
HO—H
H—OH
H—OH
CH_2OH

D-fructose with two deuteriums

64.

a. HO CH_2OH HO O HO OH

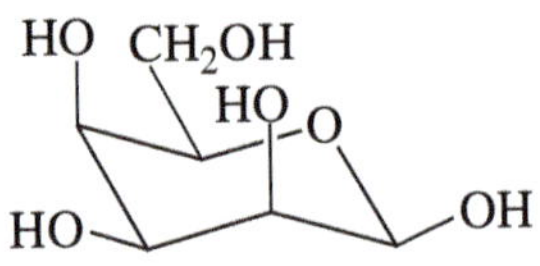

b. HO CH_2OH HO O HO OH

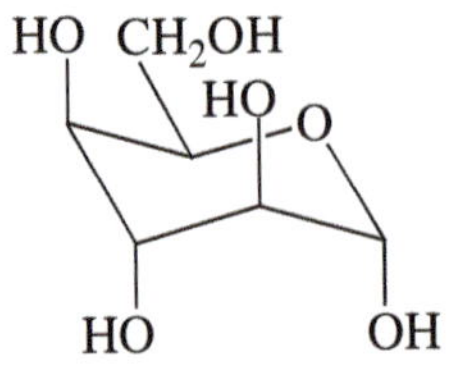

c. OH O HO HO CH_2OH OH

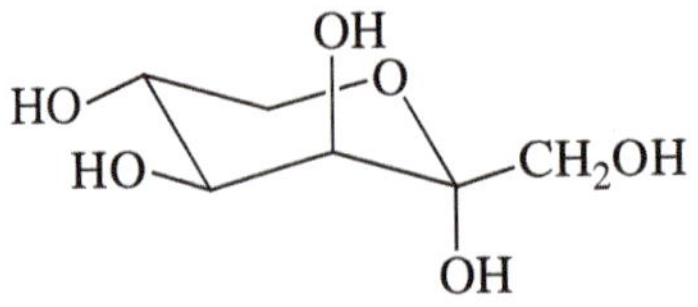

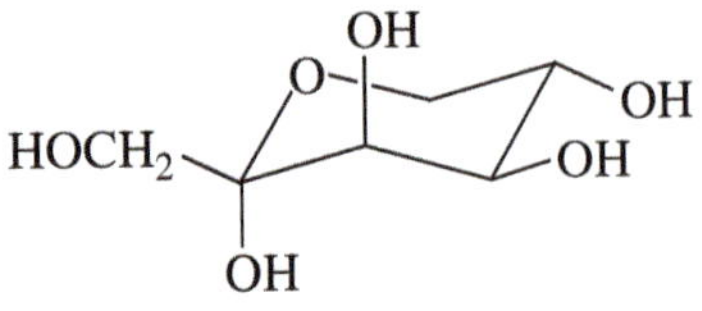

d. $HOCH_2$ O OH CH_2OH OH OH

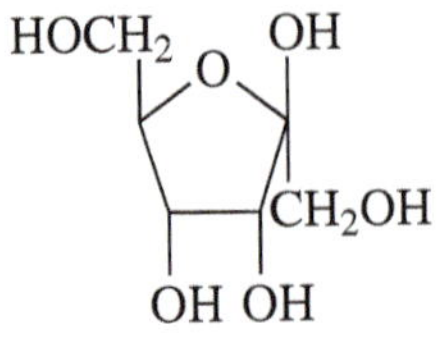

e. $HOCH_2$ OH O OH HO OH

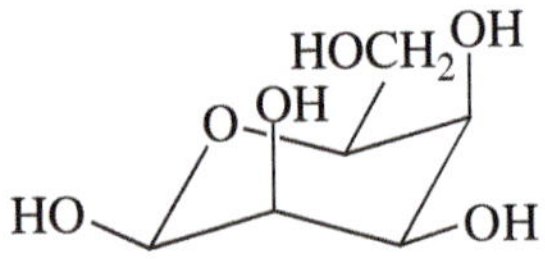

f. OH O OH $HOCH_2$ OH OH

65. Let A = the fraction of D-glucose in the α-form and B = the fraction of D-glucose in the β-form.

$$A + B = 1$$
$$B = 1 - A$$

specific rotation of A = 112.2
specific rotation of B = 18.7
specific rotation of the equilibrium mixture = 52.7
specific rotation of the mixture = specific rotation of A × fraction of D-glucose in the α-form + specific rotation of B × fraction of D-glucose in the β-form

$$52.7 = 112.2\,A + (1 - A)\,18.7$$
$$52.7 = 112.2\,A + 18.7 - 18.7\,A$$
$$34.0 = 93.5\,A$$
$$A = 0.36$$
$$B = 0.64$$

This calculation shows that 36% is in the α-form and 64% is in the β-form. This agrees with the values given in Section 20.10.

66. Let A = the fraction of D-galactose in the α-form and B = the fraction of D-galactose in the β-form.

$$A + B = 1$$
$$B = 1 - A$$

specific rotation of A = 150.7
specific rotation of B = 52.8
specific rotation of the equilibrium mixture = 80.2
specific rotation of the mixture = specific rotation of A × fraction of galactose in the α-form + specific rotation of B × fraction of galactose in the β-form

$$80.2 = 150.7\,A + 52.8\,(1 - A)$$
$$80.2 = 150.7\,A + 52.8 - 52.8\,A$$
$$27.4 = 97.9\,A$$
$$A = .280$$

Therefore, 28% is α-D-galactose and 72% is β-D-galactose.

67.

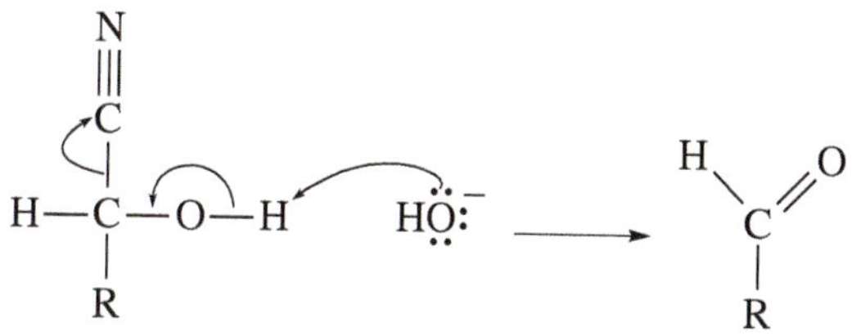

N≡C–C(H)(R)–O–H + HO^- ⟶ H–C(=O)–R

68. Silver oxide increases the leaving tendency of the iodide ion from methyl iodide, thereby allowing the nucleophilic substitution reaction to take place with the poorly nucleophilic alcohol groups. Because mannose is missing the methyl substituent on the oxygen at C-6, the disaccharide must be formed using the C-6 OH group of mannose and the anomeric carbon of galactose.

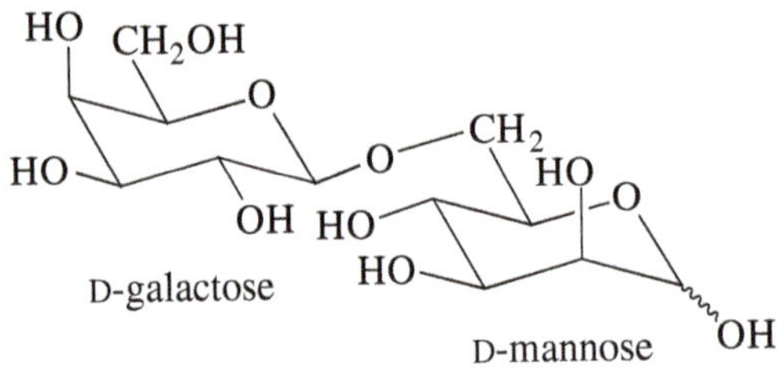

D-galactose D-mannose

69. D-Altrose most likely exists as a furanose because:
(1) the furanose is particularly stable since all the large substituents are trans to each other, and
(2) the pyranose has two of its OH groups in the unstable axial position.

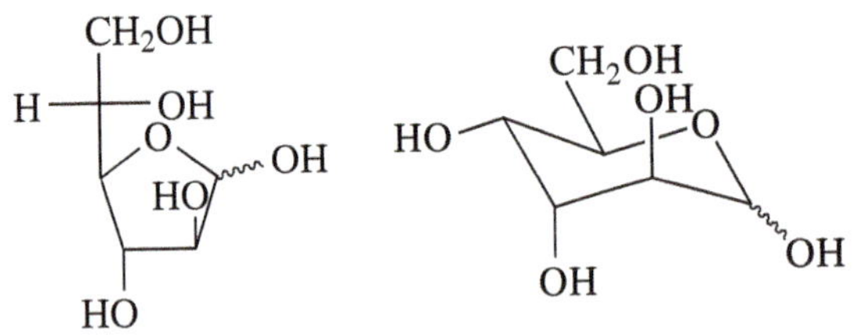

D-altrofuranose D-altropyranose

70. From its molecular formula and the fact that only glucose is formed when trehalose is hydrolyzed, we know that it is a disaccharide. Trehalose can be a nonreducing sugar only if the anomeric carbon of one glucose is connected to the anomeric carbon of the other glucose. Because it can be hydrolyzed by maltase, its glycosidic linkage must have the same geometry as an α-1,4′-glycosidic linkage. Therefore, the OH group attached to the anomeric carbon of the glucose on the left must be axial, and the OH group attached to the anomeric carbon of the glucose on the right must be equatorial. (Notice that the glucose on the right is drawn upside down and is reversed horizontally.)

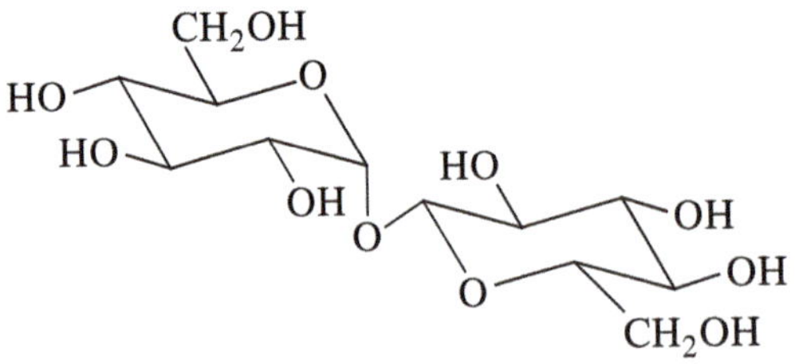

71.

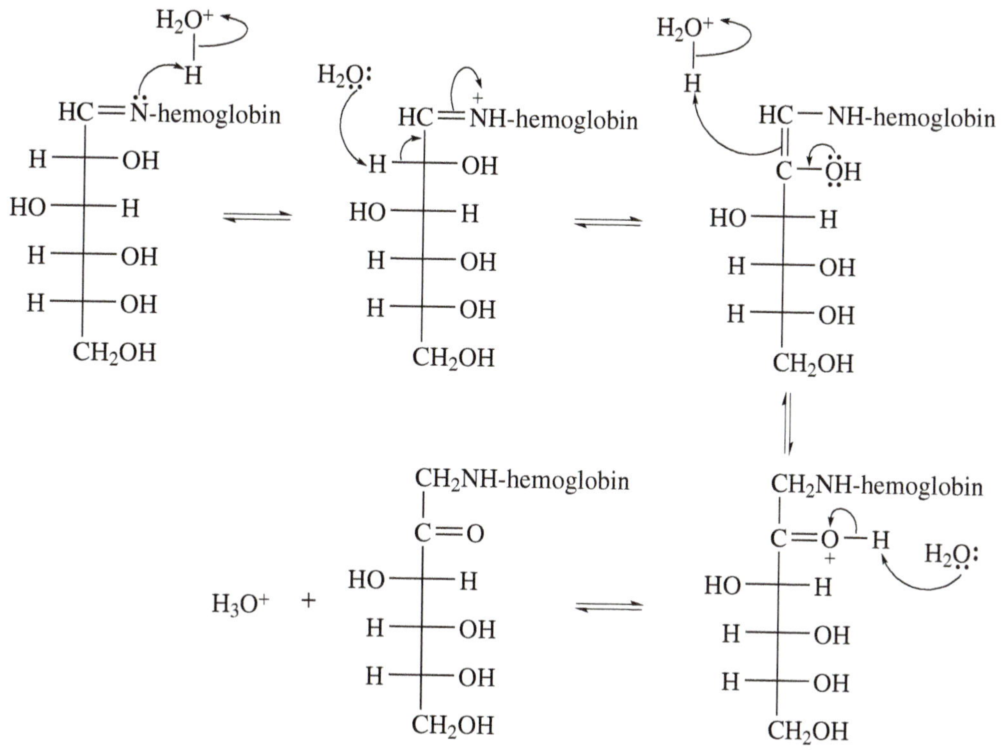

HC=N-hemoglobin, H—OH, HO—H, H—OH, H—OH, CH_2OH (protonated by H_3O^+) ⇌ HC=$\overset{+}{N}$H-hemoglobin, H—OH, HO—H, H—OH, H—OH, CH_2OH (H removed by H_2O:) ⇌ HC—NH-hemoglobin, C—OH, HO—H, H—OH, H—OH, CH_2OH (protonated by H_3O^+) ⇌ CH_2NH-hemoglobin, C=$\overset{+}{O}$—H, HO—H, H—OH, H—OH, CH_2OH (H removed by H_2O:) ⇌ H_3O^+ + CH_2NH-hemoglobin, C=O, HO—H, H—OH, H—OH, CH_2OH

72. Because all the glucose units have six-membered rings, the 5-position is never methylated.
2,3,4,6-tetra-*O*-methyl-D-glucose has only its 1-position in an acetal linkage.
2,4,6-tri-*O*-methyl-D-glucose has its 1- and 3-positions in an acetal linkage.
2,3,4-tri-*O*-methyl-D-glucose has its 1- and 6-positions in an acetal linkage.
2,4-di-*O*-methyl-D-glucose has its 1-, 3-, and 6-positions in an acetal linkage.

1,6′-linkage **3,6′-linkage** **1,3′-linkage** **1,6′-linkage**

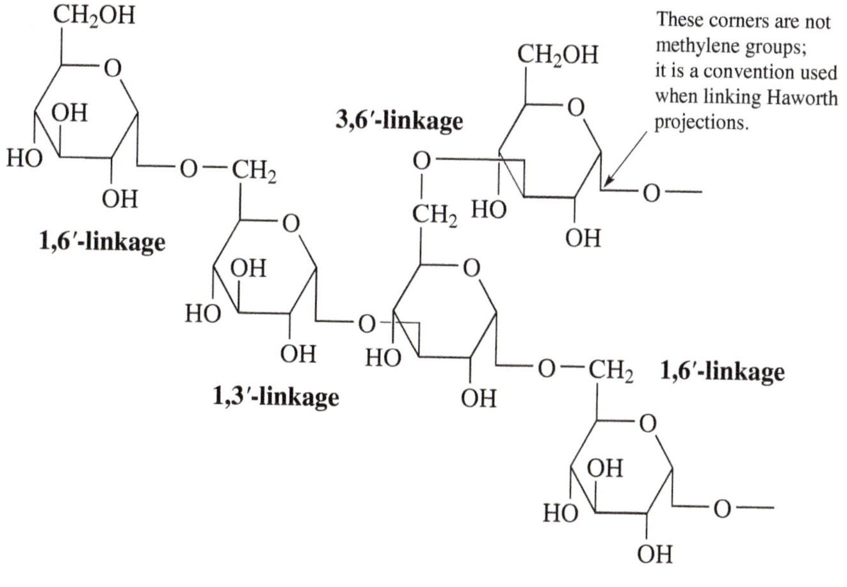

These corners are not methylene groups; it is a convention used when linking Haworth projections.

73. In the case of D-idose, the chair conformer with both the OH substituent at C-1 and the CH_2OH substituent in axial positions (which is necessary for the formation of the anhydro form) has the OH substituents at C-2, C-3, and C-4 in equatorial positions. Thus, this is a relatively stable conformer because three of the five large substituents are in equatorial positions, where there is more room for a substituent.

In the case of D-glucose, the chair conformer with both the OH substituent at C-1 and the CH_2OH substituent in axial positions has the OH substituents at C-2, C-3, and C-4 in axial positions. This is a relatively unstable conformer because all the large substituents are in axial positions and will have unfavorable 1,3-diaxial interactions.

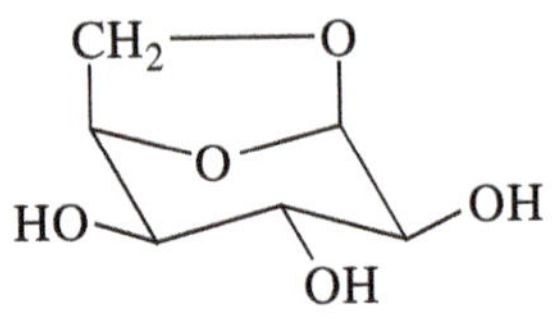

anhydro form of D-idose

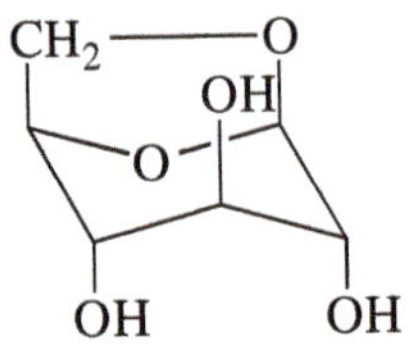

anhydro form of D-glucose

Therefore, a large percentage of D-idose but only a small percentage of D-glucose exists in the anhydro form at 100 °C.

Chapter 20 Practice Test

1. Draw the product(s) of the following reactions:

a.

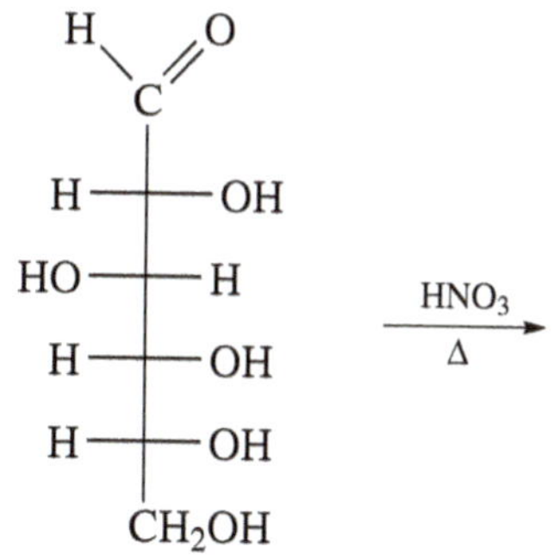

H–C(=O)
H—OH
HO—H
H—OH
H—OH
CH_2OH

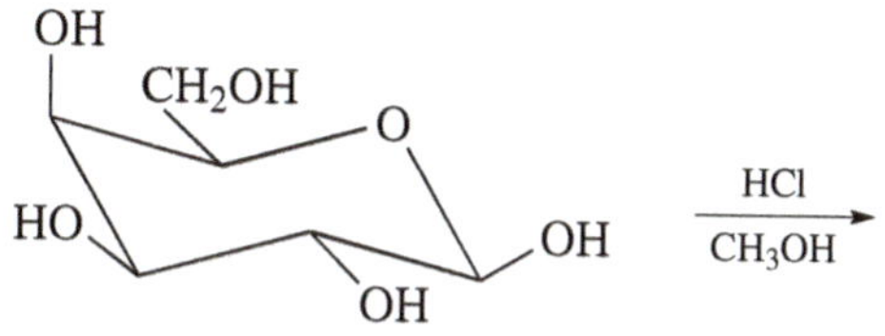

$\xrightarrow[\Delta]{HNO_3}$

b.

OH, CH_2OH, O, HO, OH, OH

$\xrightarrow[CH_3OH]{HCl}$

c.

H–C(=O)
HO—H
HO—H
H—OH
CH_2OH

1. $HC{\equiv}N$
2. $H_2/Pd/BaSO_4$
3. HCl, H_2O

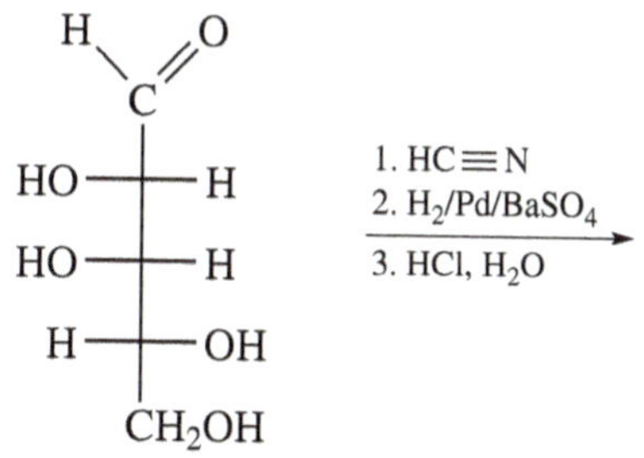

d.

H–C(=O)
H—OH
H—OH
H—OH
CH_2OH

\+ Br_2 $\xrightarrow{H_2O}$

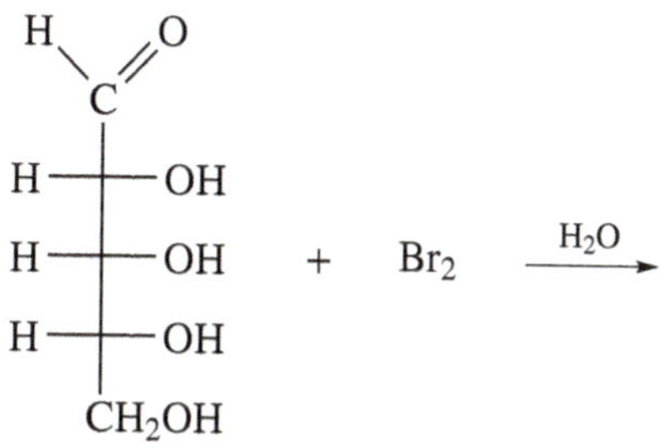

2. Indicate whether each of the following statements is true or false:

a. Glycogen contains α-1,4′- and β-1,6′-glycosidic linkages. T F

b. D-Mannose is a C-1 epimer of D-glucose. T F

c. D-Glucose and L-glucose are anomers. T F

d. D-Erythrose and D-threose are diastereomers. T F

d. Wohl degradations of D-glucose and D-gulose form the same aldopentose. T F

3. Which of the following sugars form an optically active aldaric acid?

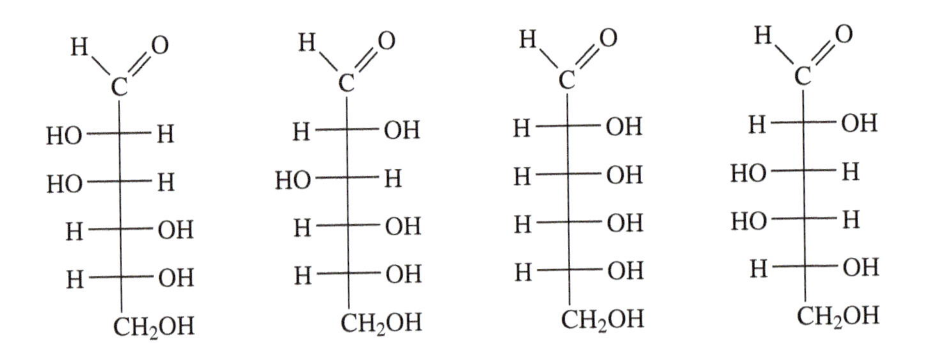

4. When crystals of D-fructose are dissolved in a basic aqueous solution, two aldohexoses are obtained. Identify the aldohexoses.

5. Draw three tetroses that form the same enol.

6. D-Talose and ________ are reduced to the same alditol.

7. What is the main structural difference between amylose and cellulose?

8. What aldohexoses are formed from a Kiliani–Fischer synthesis starting with D-xylose?

9. What aldohexose is the C-3 epimer of D-glucose?

10. Draw the most stable chair conformer of β-D-allose, a C-3 epimer of β-D-glucose.

Chapter 21
Amino Acids, Peptides, and Proteins

Important Terms

amino acid — an α-amino carboxylic acid. Naturally occurring amino acids have the L-configuration.

D-amino acid — the configuration of an amino acid drawn in a Fischer projection with the carboxyl group on top, the hydrogen on the left, and the amino group on the right.

L-amino acid — the configuration of an amino acid drawn in a Fischer projection with the carboxyl group on top, the hydrogen on the right, and the amino group on the left.

amino acid analyzer — an instrument that automates the ion-exchange separation of amino acids.

anion-exchange resin — a resin that binds anions.

antiparallel β-pleated sheet — a type of secondary structure in which the adjacent hydrogen-bonded polypeptide chains in a β-pleated sheet run in opposite directions.

automated solid-phase peptide synthesis — an automated technique that synthesizes a polypeptide (in the C-terminal to N-terminal direction) while its C-terminal amino acid is attached to a solid support.

cation-exchange resin — a resin that binds cations.

coil conformation (loop conformation) — the part of a protein that is highly ordered but not in an α-helix or β-pleated sheet.

C-terminal amino acid — the terminal amino acid of a peptide (or protein); it has a free carboxyl group.

denaturation — the destruction of the highly organized secondary and tertiary structure of a protein.

dipeptide — two amino acids linked by an amide bond.

disulfide — a compound with an S—S bond.

disulfide bridge — a disulfide (S—S) bond formed by two cysteine residues in a polypeptide or protein.

Edman's reagent — phenyl isothiocyanate; the reagent used to determine the N-terminal amino acid of a polypeptide.

electrophoresis — a technique that separates amino acids on the basis of their pI values.

endopeptidase — an enzyme that hydrolyzes a peptide bond that is not at the end of a polypeptide chain.

essential amino acid — an amino acid that humans must obtain from their diet because they cannot synthesize it at all or cannot synthesize it in adequate amounts.

exopeptidase an enzyme that hydrolyzes a peptide bond at the end of a polypeptide chain.

fibrous protein a water-insoluble protein; its polypeptide chains are arranged in bundles.

globular protein a water-soluble protein; it tends to have a roughly spherical shape.

α-helix the backbone of a polypeptide coiled in a right-handed spiral with hydrogen bonding occurring within the helix.

hydrophobic interactions interactions between nonpolar groups. These interactions increase stability by decreasing the amount of structured water, thereby increasing entropy.

interchain disulfide bridge a disulfide bridge between two cysteine residues in different polypeptide chains.

intrachain disulfide bridge a disulfide bridge between two cysteine residues in the same polypeptide chain.

ion-exchange chromatography a technique that uses a column packed with an insoluble resin to separate compounds on the basis of their charge and polarity.

isoelectric point (pI) the pH at which there is no net charge on an amino acid.

kinetic resolution separating enantiomers based on the difference in their rate of reaction with an enzyme.

loop conformation (coil conformation) see coil conformation.

N-terminal amino acid the terminal amino acid of a polypeptide (or protein); it has a free amino group.

oligopeptide 3 to 10 amino acids linked by amide bonds.

paper chromatography a technique that separates amino acids based on polarity.

parallel β-pleated sheet a type of secondary structure in which the adjacent hydrogen-bonded polypeptide chains in a β-pleated sheet run in the same direction.

partial hydrolysis a technique that hydrolyzes only some of the peptide bonds in a polypeptide.

peptidase an enzyme that catalyzes the hydrolysis of a peptide bond.

peptide a polymer of amino acids linked by amide bonds.

peptide bond the amide bond that links the amino acids in a polypeptide or protein.

β-pleated sheet a type of secondary structure in which the backbone of a polypeptide extends in a zigzag structure with hydrogen bonding between neighboring chains.

polypeptide many amino acids linked by amide bonds.

primary structure the sequence of amino acids and the location of the disulfide bridges in a polypeptide or protein.

protein a naturally occurring polymer of 40 to 4000 amino acids linked by amide bonds.

quaternary structure a description of the way in which the individual polypeptide chains in a protein with more than one chain are arranged with respect to one other.

secondary structure a description of the conformation of the backbone of a polypeptide or protein.

side chain the substituent attached to the α-carbon of an amino acid.

structural protein a protein that gives strength to a biological structure.

subunit a polypeptide chain of protein that has more than one polypeptide chain.

tertiary structure a description of the three-dimensional arrangement of all the atoms in a protein.

thin-layer chromatography a technique that separates compounds on the basis of their polarity.

tripeptide three amino acids linked by amide bonds.

zwitterion a compound with a negative charge and a positive charge on nonadjacent atoms.

Solutions to Problems

1. **a.** When the imidazole ring is protonated, the double-bonded nitrogen is the one that accepts the proton. (Its lone pair is in an sp^2 orbital.)

$$\text{:N}\diagdown\text{:NH} \xrightarrow{\text{HCl}} \text{HN}^+\text{:NH} + \text{Cl}^-$$

The lone-pair electrons on the single-bonded nitrogen (π electrons) are delocalized and, therefore, are not available to be protonated.

b. The lone-pair electrons on the double-bonded nitrogen are protonated because the lone-pair electrons on the other nitrogens are delocalized and, therefore, cannot be protonated. In addition, protonation of the double-bonded nitrogen leads to a highly resonance-stabilized conjugate acid.

2. **a.** D-alanine is (*R*)-alanine **b.** D-asparate is (*R*)-aspartate

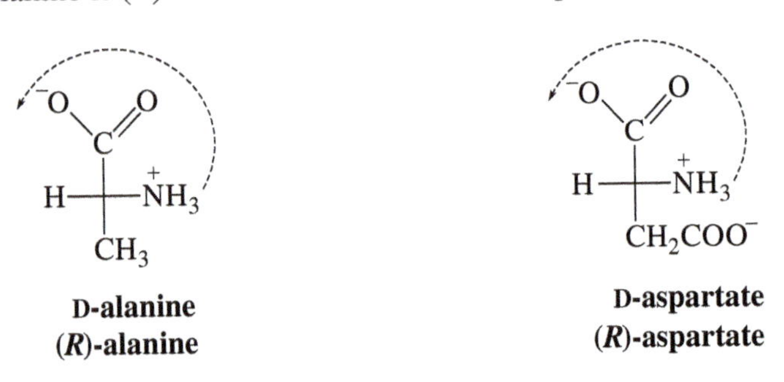

c. The α-carbons of all the D-amino acids except cysteine have the *R*-configuration. Similarly, the α-carbons of all the L-amino acids except cysteine have the *S*-configuration.

In all the amino acids except cysteine, the amino group has the highest priority and the carboxyl group has the second-highest priority. In cysteine, the thiomethyl group has a higher priority than the carboxylate group because sulfur has a greater atomic number than oxygen, causing the counterclockwise arrow to be a clockwise arrow.

D-cysteine
(*S*)-cysteine

3. Solved in the text.

4. Isoleucine is the only other amino acid that has more than one asymmetric center. Like threonine, it has two asymmetric centers.

$$CH_3CH_2\overset{*}{C}H(CH_3)-\overset{*}{C}H(^{+}NH_3)-COO^{-}$$

isoleucine

5. Alanine exists predominately as a zwitterion in an aqueous solution with pH > 2.34 and pH < 9.69.

6. The electron-withdrawing $^{+}NH_3$ substituent on the α-carbon increases the acidity of the carboxyl group.

7. **a.** Solved in the text.

b. (imidazol-4-yl)$CH_2CH(^{+}NH_3)COO^{-}$

c. $H_2NC(=O)CH_2CH_2CH(^{+}NH_3)COO^{-}$

d. $H_3\overset{+}{N}CH_2CH_2CH_2CH_2CH(^{+}NH_3)COO^{-}$

e. $H_2NC(=^{+}NH_2)NHCH_2CH_2CH_2CH(^{+}NH_3)COO^{-}$

f. $HO-C_6H_4-CH_2CH(^{+}NH_3)COO^{-}$

8. **a.** $HOC(=O)CH_2CH_2CH(^{+}NH_3)C(=O)OH$

b. $HOC(=O)CH_2CH_2CH(^{+}NH_3)COO^{-}$

c. $^{-}OC(=O)CH_2CH_2CH(^{+}NH_3)COO^{-}$

d. $^{-}OC(=O)CH_2CH_2CH(NH_2)COO^{-}$

9. **a.** The carboxyl group of the aspartic acid side chain is a stronger acid than the carboxyl group of the glutamic acid side chain. The carboxyl group of the aspartic acid side chain is closer to the electron-withdrawing protonated amino group.

b. The protonated lysine side chain is a stronger acid than the protonated arginine side chain. The protonated arginine side chain has less of a tendency to lose a proton because its positive charge is delocalized over three nitrogens.

10. **a.** asparagine $pI = \frac{2.02 + 8.84}{2} = \frac{10.86}{2} = 5.43$

b. arginine $pI = \frac{9.04 + 12.48}{2} = \frac{21.52}{2} = 10.76$

c. serine $pI = \frac{2.21 + 9.15}{2} = \frac{11.36}{2} = 5.68$

d. aspartate $pI = \frac{2.09 + 3.86}{2} = \frac{5.95}{2} = 2.98$

11. **a.** aspartate (pI = 2.98)

b. arginine (pI = 10.76)

c. Aspartate because its pI is lower than that of glutamate. Therefore, pH 6.20 is farther away from the pH at which it has no net charge. The farther an amino acid with no net charge has to "move" to get to the given pH value, the more charged it will become.

d. Methionine because at pH = 6.20, methionine is farther away from the pH at which it has no net charge (pI of methionine = 5.75, pI of glycine = 5.97).

12. Tyrosine and cysteine each have two groups that are neutral in their acidic forms and negatively charged in their basic forms. Unlike other amino acids that have similarly ionizing groups, the pK_a value of one of the two similarly ionizing groups in tyrosine and in cysteine is close to the pK_a value of the group that ionizes differently. Therefore, the group that ionizes differently cannot be ignored in calculating the pI.

13. **a.** $< 25\%$ **b.** $> 75\%$

14. For the amino acid to have no net charge, the two amino groups must have a +1 charge between them in order to cancel out the −1 charge of the carboxylate group. Because they are positively charged in their acidic forms and neutral in their basic forms, the sum of their charges will be +1 at the midpoint of their pK_a values.

15. $CH_3CH(CH_3)CHO$ (O=C(H)–CH(CH₃)CH₃)

The R group of the aldehyde is the same as the R group of the amino acid.

16. Leucine and isoleucine both have C_4H_9 side chains and, therefore, have the same polarity. Consequently, the spots for both amino acids appear at the same place on the chromatographic plate. Therefore, the chromatographic plate has one less spot than the number of amino acids.

17. Because the amino acid analyzer contains a cation-exchange resin (it binds cations), the less positively charged the amino acid is, the less tightly it is bound to the column. Using buffer solutions of increasingly higher pH to elute the column causes the amino acids bound to the column to become increasingly deprotonated (less positively charged) so they can be released from the column.

18. Cation-exchange chromatography releases amino acids in order of their pI values. The amino acid with the lowest pI is released first because at a given pH, it is the amino acid with the highest concentration of negative charge, and negatively charged molecules are not bound by the negatively charged resin. The relatively nonpolar resin releases polar amino acids before nonpolar amino acids.

a. Asp (pI = 2.98) is more negative at pH = 4 than is Ser (pI = 5.68).

b. Ser is more polar than Ala.

c. Val is more polar than Leu.

d. Tyr is more polar than Phe.

19. A column containing an anion-exchange resin releases amino acids in reverse order of their pI values (the opposite of what happens in a cation-exchange resin). The amino acid with the highest pI is released first because, at a given pH, it will be the amino acid with the highest concentration of positive charge.

His > Val > Ser > Asp

20. The first equivalent of ammonia will react with the acidic proton of the carboxylic acid to form an ammonium ion, which is not nucleophilic and, therefore, cannot substitute for Br. Thus, a second equivalent of ammonia is needed for the desired nucleophilic substitution reaction.

21. **a.** The following reactions show that pyruvic acid forms alanine, oxaloacetic acid forms aspartate, and α-ketoglutarate forms glutamate.

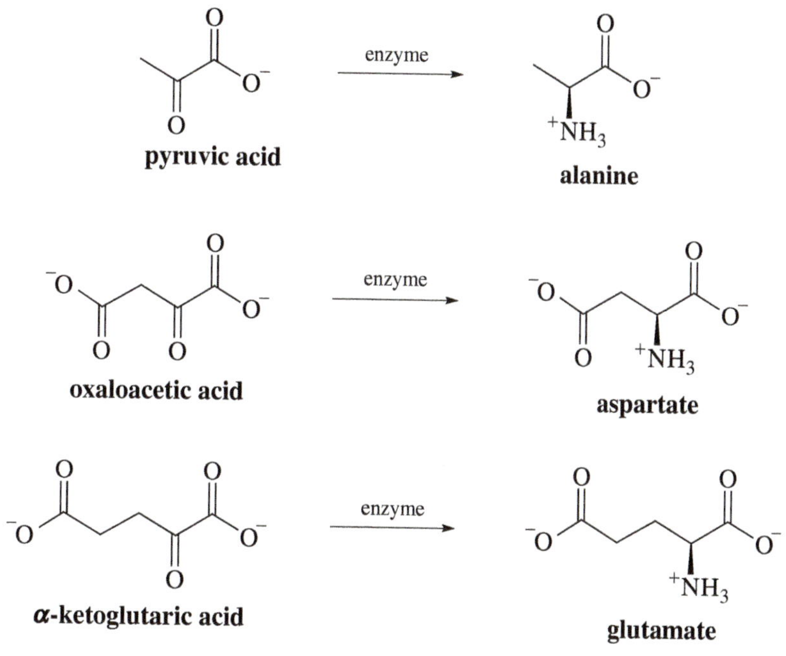

O, O, O^-, enzyme, O, O^-, $^+NH_3$

pyruvic acid → **alanine**

^-O, O, O^-, O, O, enzyme, ^-O, O, O^-, O, $^+NH_3$

oxaloacetic acid → **aspartate**

O, O, ^-O, O^-, O, enzyme, O, O, ^-O, O^-, $^+NH_3$

α-ketoglutaric acid → **glutamate**

If reductive amination (page 764) is carried out in the cell, only the L-isomer of each amino acid will be formed.

b. Recall that imine formation is best carried at a pH about 1.5 units lower than the pK_a of the protonated ammonium ion (Section 16.8), that is, at about pH = 8. Therefore, the carboxyl groups will be in their basic forms. If reductive amination is carried out in the laboratory, both the D- and L-isomer (a racemic mixture) of each amino acid will be formed.

pyruvic acid → (1. NH_3, trace acid; 2. H_2, Pd/C) → alanine + alanine

oxaloacetic acid → (1. NH_3, trace acid; 2. H_2, Pd/C) → aspartate + aspartate

α-ketoglutaric acid → (1. NH_3, trace acid; 2. H_2, Pd/C) → glutamate + glutamate

22. Notice that the R group attached to the Br is the same as the R group attached to the α-carbon of the amino acid.

R—Br corresponds to R—CH($^+NH_3$)COO$^-$

a. leucine **b.** methionine

23. As in Problem 22, the R group attached to the Br is the same as the R group attached to the α-carbon of the desired amino acid.

a. 4-bromo-1-butanamine **b.** benzyl bromide

24. Notice that the R group attached to the carbonyl group of the aldehyde is the same as the R group attached to the α-carbon of the amino acid.

R—CH(=O) corresponds to R—CH($^+NH_3$)COO$^-$

a. alanine **b.** isoleucine **c.** leucine

25. Convert the amino acids into esters using $SOCl_2$ followed by ethanol. Then treat the esters with pig liver esterase. Because the enzyme hydrolyzes only esters of L-amino acids, the products will be the L-amino acid, ethanol, and the ester of the D-amino acid. These compounds can be readily separated. After they are separated, the D-amino acid can be obtained by acid-catalyzed hydrolysis of the ester of the D-amino acid. This separation technique is called a kinetic resolution because the enantiomers are separated (resolved) as a result of reacting at different rates in the enzyme-catalyzed reaction.

26.

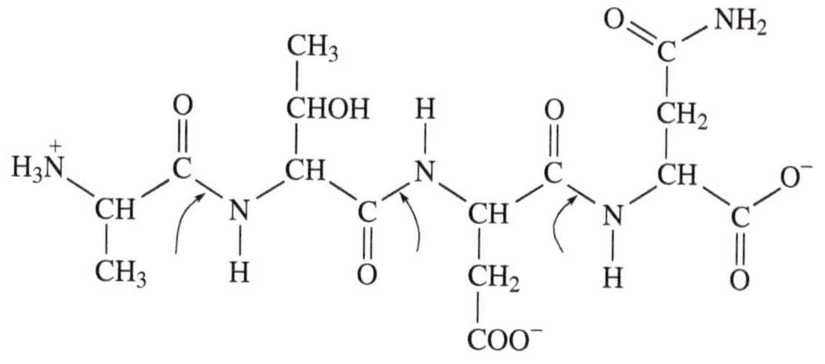

The peptide bonds are indicated by arrows.

27.

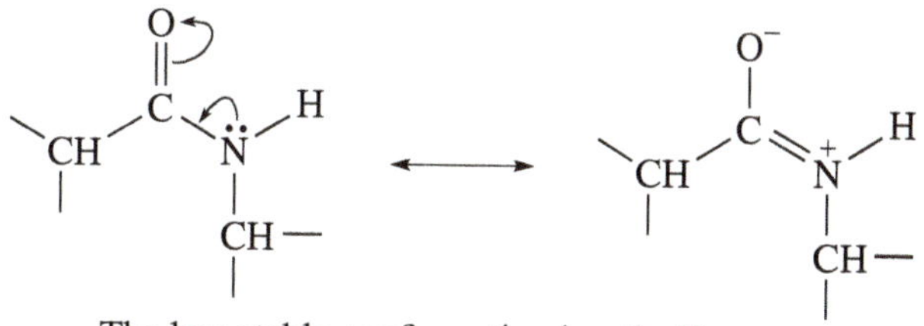

The less stable configuration has the R groups on the same side of the double bond.

28. The bonds on either side of the α-carbon can freely rotate. In other words, the bond between the α-carbon and the carbonyl carbon and the bond between the α-carbon and the nitrogen (the bonds indicated by arrows) can freely rotate. The bond between the C and N (the peptide bond) cannot rotate because it has partial double-bond character.

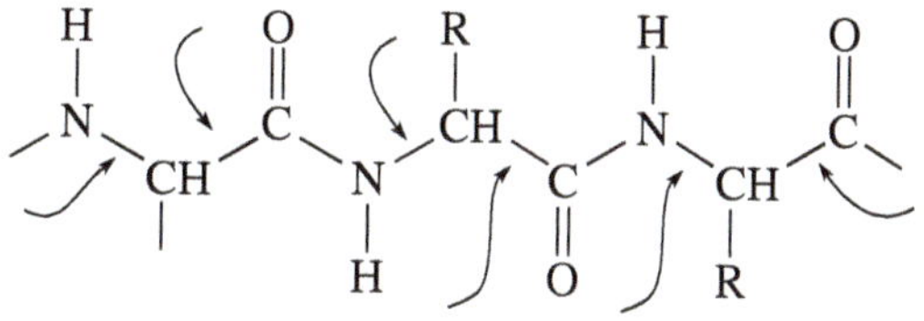

29. **a.**

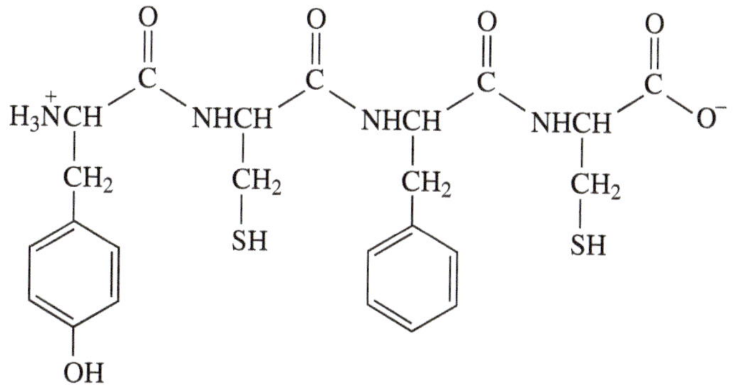

b.

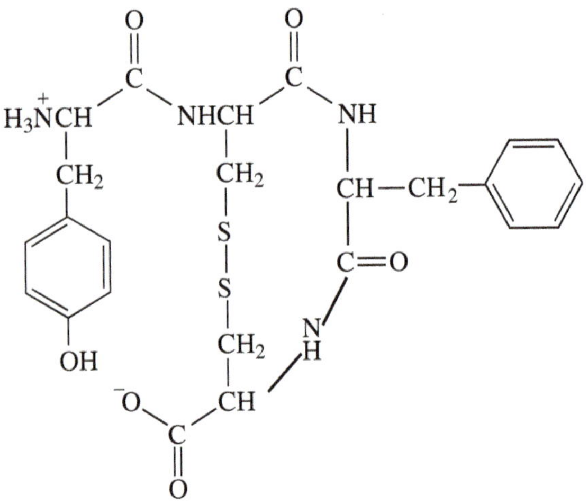

30.

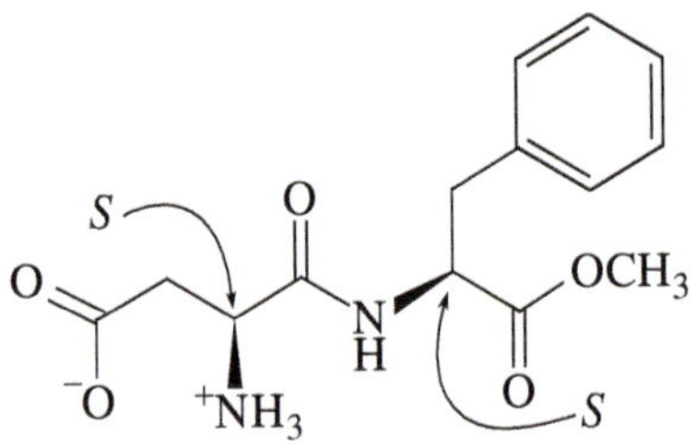

31. **a.** glutamate, cysteine, and glycine

b. In forming the amide bond between glutamate and cysteine, the amino group of cysteine reacts with the γ-carboxyl group of glutamate rather than with its α-carboxyl group.

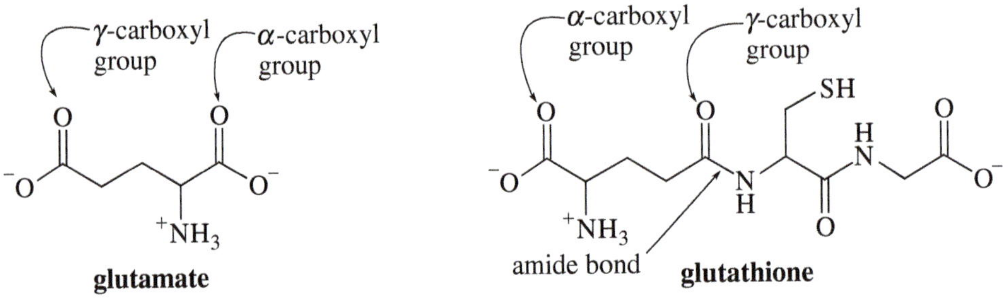

32. **Leu-Val** and **Val-Val** will be formed because the amino group of leucine is protected, so leucine cannot react with a carboxyl group (that is, leucine cannot be the C-terminal amino acid). The amino group of valine can react equally easily with the carboxyl group of leucine and the carboxyl group of valine.

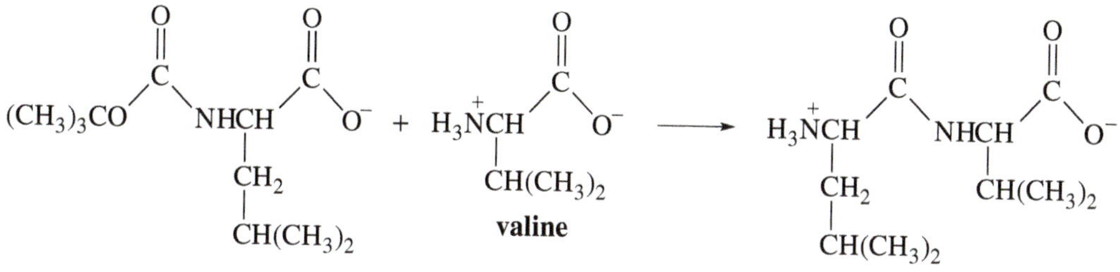

$(CH_3)_3CO$ ... $NHCH$ ... O^- + $H_3\overset{+}{N}CH$ ... O^- ⟶ $H_3\overset{+}{N}CH$ ... $NHCH$ ... O^-

CH_2 | $CH(CH_3)_2$ | CH_2 | $CH(CH_3)_2$

$CH(CH_3)_2$ | valine | $CH(CH_3)_2$

N-protected leucine **valine** **Leu-Val**

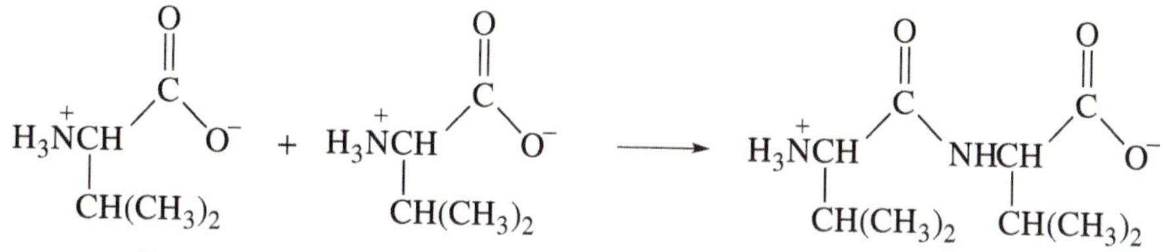

$H_3\overset{+}{N}CH$ ... O^- + $H_3\overset{+}{N}CH$ ... O^- ⟶ $H_3\overset{+}{N}CH$ ... $NHCH$ ... O^-

$CH(CH_3)_2$ | $CH(CH_3)_2$ | $CH(CH_3)_2$ $CH(CH_3)_2$

valine **valine** **Val-Val**

33. If valine's carboxyl group is activated with thionyl chloride, the OH group of serine, as well as the NH_2 group of serine, would react readily with the very reactive acyl chloride, forming both an ester (with its OH group) and an amide (with its NH_2 group).

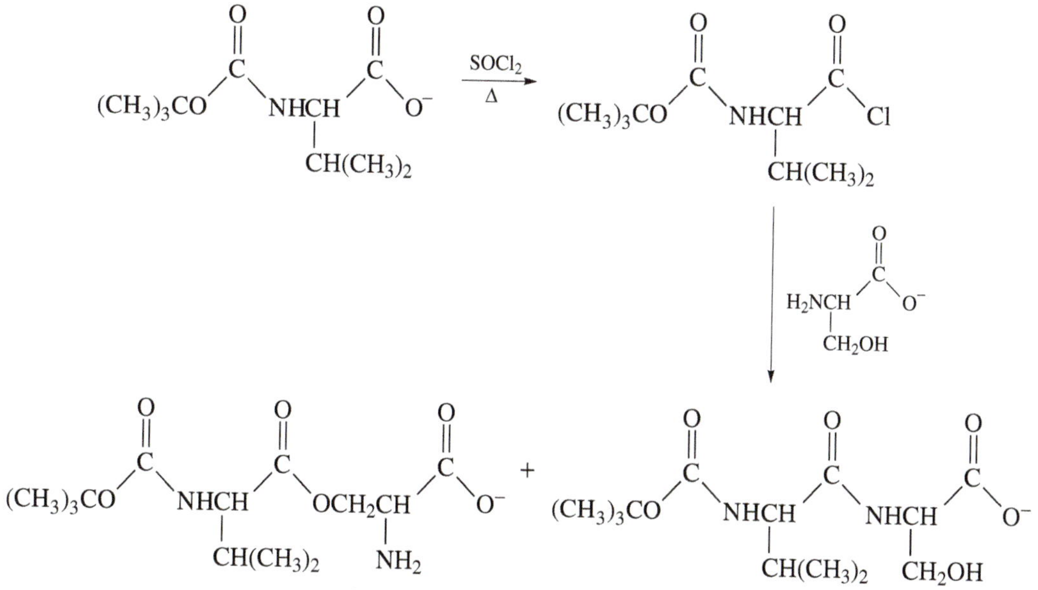

$(CH_3)_3CO$ ... $NHCH$ ... O^- $\xrightarrow[\Delta]{SOCl_2}$ $(CH_3)_3CO$ ... $NHCH$ ... Cl

$CH(CH_3)_2$ | $CH(CH_3)_2$

H_2NCH ... O^- (CH_2OH)

$(CH_3)_3CO$ ... $NHCH$ ... OCH_2CH ... O^- + $(CH_3)_3CO$ ... $NHCH$ ... $NHCH$ ... O^-

$CH(CH_3)_2$ NH_2 | $CH(CH_3)_2$ CH_2OH

ester **amide**

If valine's carboxyl group is activated with DCCD, an imidate will be formed. Because an imidate is less reactive than an acyl chloride, the imidate will react with the more reactive NH_2 group in preference to the less reactive OH group.

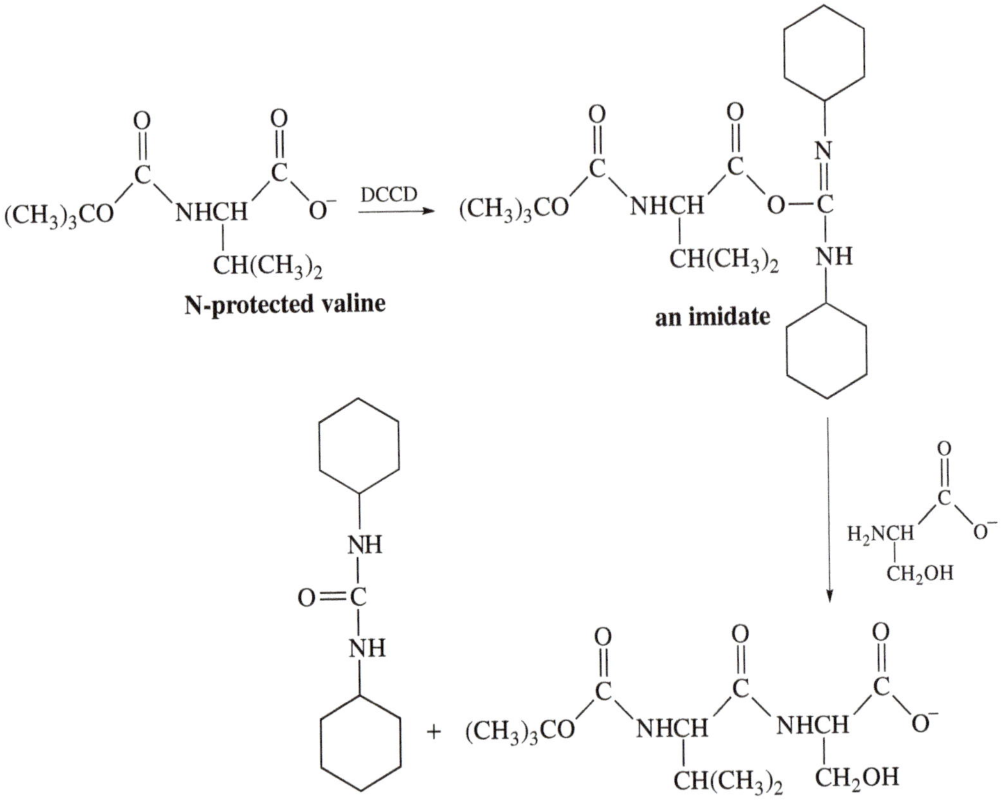

$(CH_3)_3COC(=O)NHCH(CH(CH_3)_2)CO_2^-$ $\xrightarrow{\text{DCCD}}$ $(CH_3)_3COC(=O)NHCH(CH(CH_3)_2)C(=O)O{-}C(=N{-}C_6H_{11})NH{-}C_6H_{11}$

N-protected valine

an imidate

$\xrightarrow{H_2NCH(CH_2OH)CO_2^-}$

$C_6H_{11}NH{-}C(=O){-}NHC_6H_{11}$ + $(CH_3)_3COC(=O)NHCH(CH(CH_3)_2)C(=O)NHCH(CH_2OH)CO_2^-$

34. $(CH_3)_3COC(=O)OC(=O)OC(CH_3)_3$ + $H_2NCH(CH_2CH(CH_3)_2)CO_2^-$ $\longrightarrow$ $(CH_3)_3COC(=O)NHCH(CH_2CH(CH_3)_2)CO_2^-$

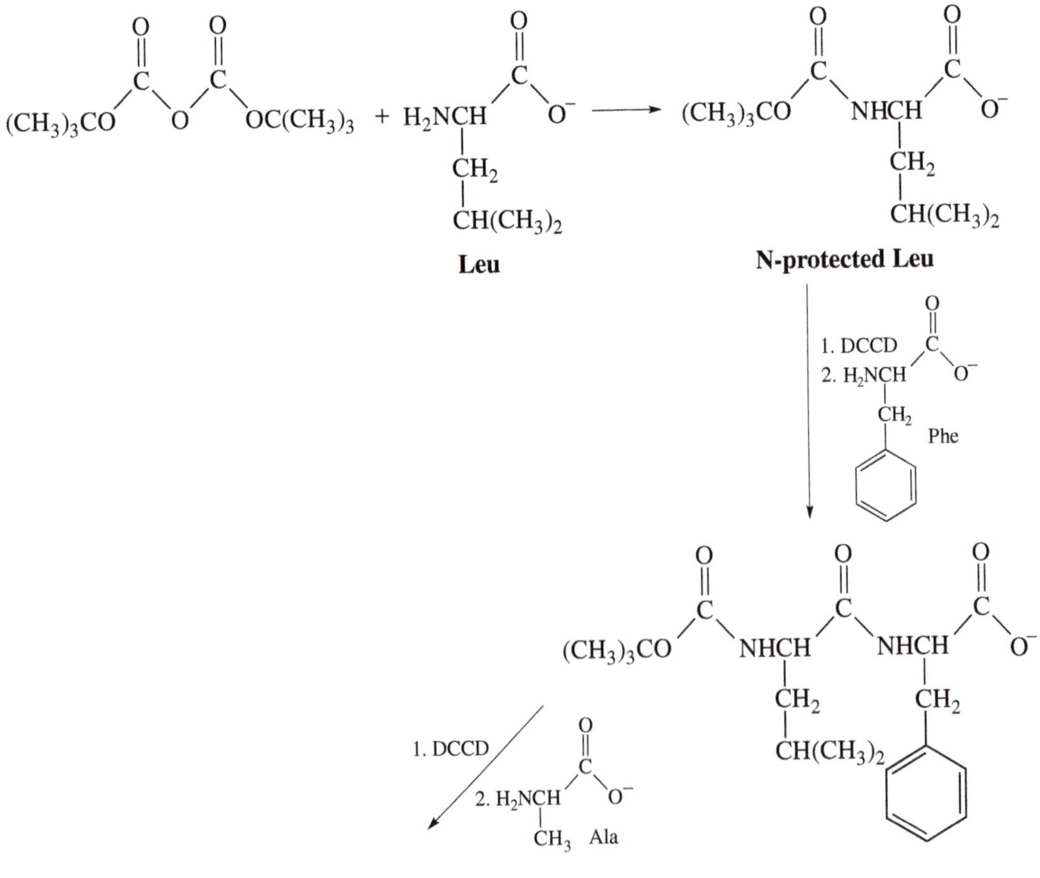

Leu

N-protected Leu

1. DCCD
2. $H_2NCH(CH_2C_6H_5)CO_2^-$ Phe

$(CH_3)_3COC(=O)NHCH(CH_2CH(CH_3)_2)C(=O)NHCH(CH_2C_6H_5)CO_2^-$

1. DCCD
2. $H_2NCH(CH_3)CO_2^-$ Ala

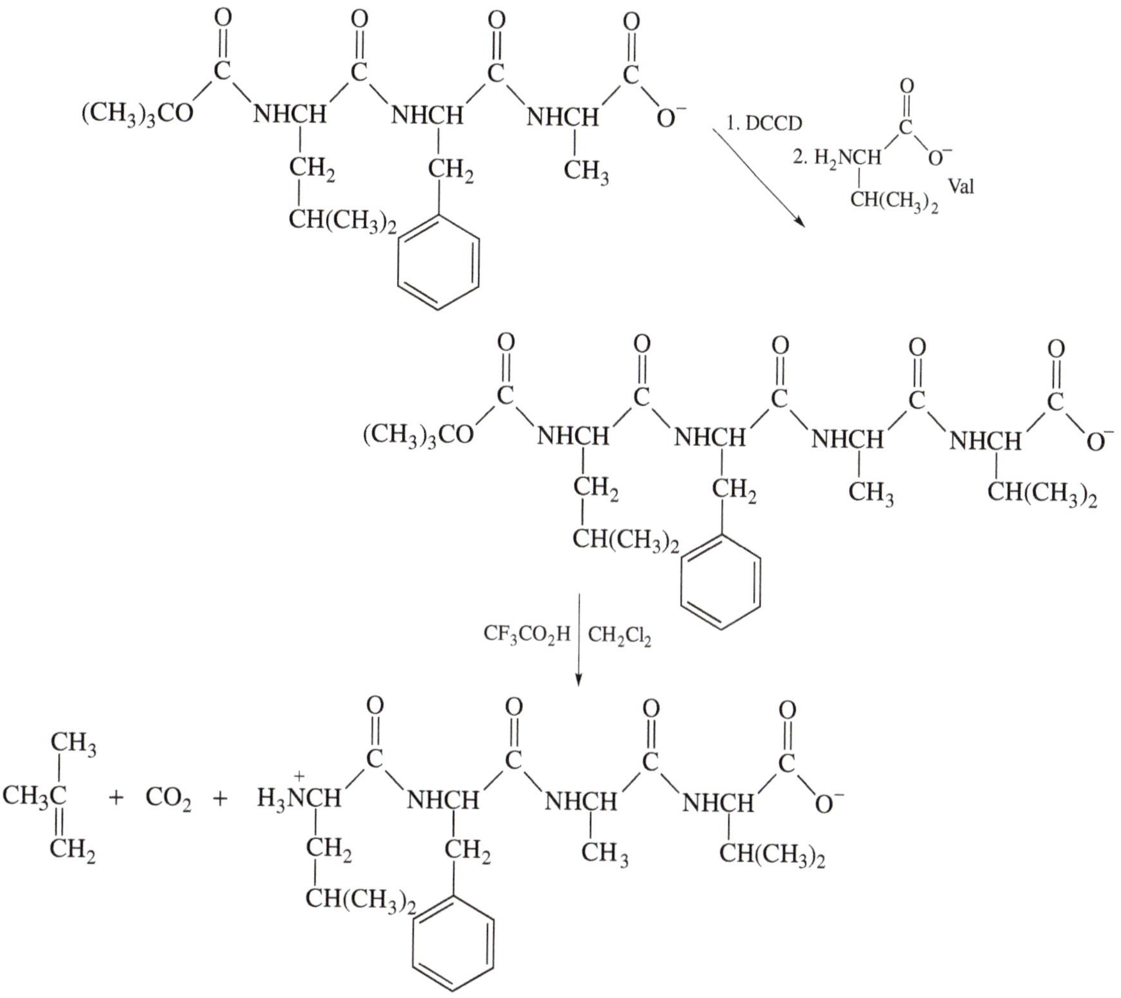

1. DCCD
2. $H_2NCH(CH(CH_3)_2)CO_2^-$ Val

CF_3CO_2H | CH_2Cl_2

$(CH_3)_2C{=}CH_2$ + CO_2 + $H_3\overset{+}{N}$-Leu-Phe-Ala-Val-O^-

Leu **Phe** **Ala** **Val**

35. Notice that the number of steps is one less than the number of amino acids in the peptide.

a. 5.8%

2	**3**	**4**	**5**	**6**	**7**	**8**	**9**
70%	49%	34%	24%	17%	12%	8.2%	5.8%

b. 4.4%

2	**3**	**4**	**5**	**6**	**7**	**8**	**9**	**10**	**11**	**12**	**13**	**14**	**15**
80%	64%	51%	41%	33%	26%	21%	17%	13%	11%	8.6%	6.9%	5.5%	4.4%

36.

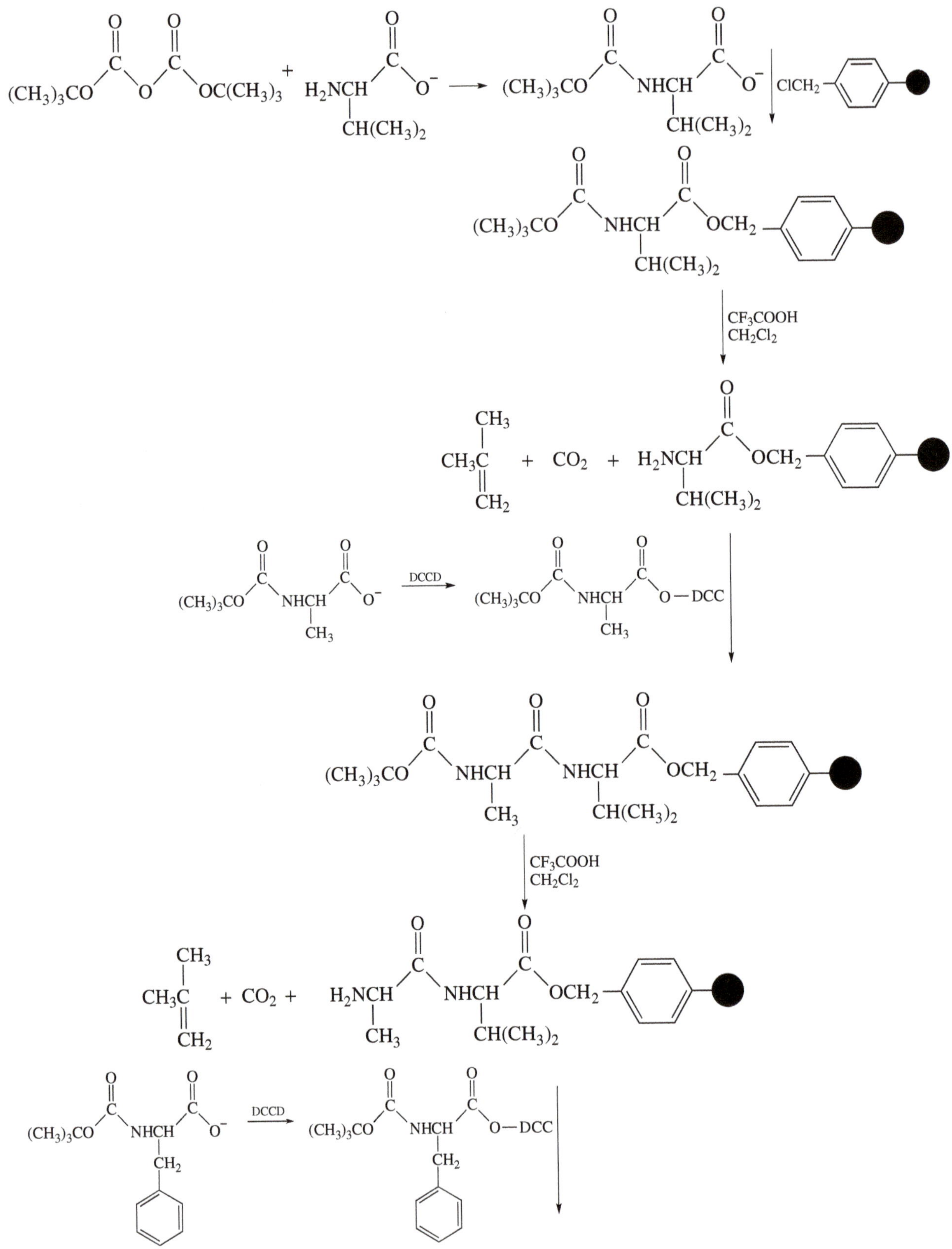

$(CH_3)_3COC(=O)OC(=O)OC(CH_3)_3$ + $H_2NCH(CH(CH_3)_2)CO^-$ (C=O) → $(CH_3)_3COC(=O)NHCH(CH(CH_3)_2)C(=O)O^-$

$ClCH_2$–(benzene ring)–● ↓

$(CH_3)_3COC(=O)NHCH(CH(CH_3)_2)C(=O)OCH_2$–(benzene ring)–●

↓ CF_3COOH, CH_2Cl_2

$CH_3C(CH_3)=CH_2$ + CO_2 + $H_2NCH(CH(CH_3)_2)C(=O)OCH_2$–(benzene ring)–●

$(CH_3)_3COC(=O)NHCH(CH_3)C(=O)O^-$ $\xrightarrow{DCCD}$ $(CH_3)_3COC(=O)NHCH(CH_3)C(=O)O{-}DCC$ ↓

$(CH_3)_3COC(=O)NHCH(CH_3)C(=O)NHCH(CH(CH_3)_2)C(=O)OCH_2$–(benzene ring)–●

↓ CF_3COOH, CH_2Cl_2

$CH_3C(CH_3)=CH_2$ + CO_2 + $H_2NCH(CH_3)C(=O)NHCH(CH(CH_3)_2)C(=O)OCH_2$–(benzene ring)–●

$(CH_3)_3COC(=O)NHCH(CH_2C_6H_5)C(=O)O^-$ $\xrightarrow{DCCD}$ $(CH_3)_3COC(=O)NHCH(CH_2C_6H_5)C(=O)O{-}DCC$ ↓

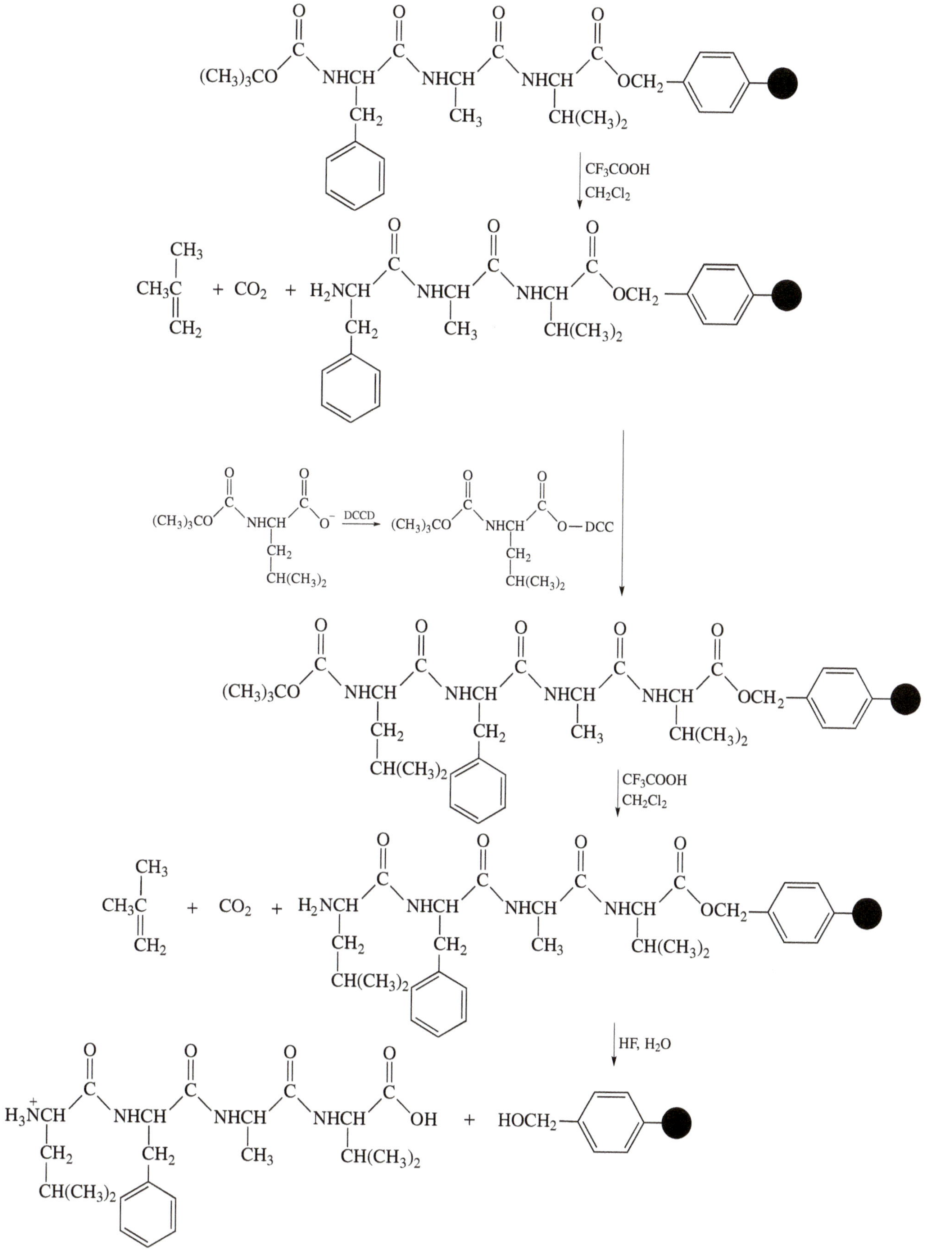

$(CH_3)_3CO-C(=O)-NHCH(CH_2C_6H_5)-C(=O)-NHCH(CH_3)-C(=O)-NHCH(CH(CH_3)_2)-C(=O)-OCH_2-C_6H_4-$●

$\downarrow$ CF_3COOH, CH_2Cl_2

$CH_3C(CH_3)=CH_2$ + CO_2 + $H_2NCH(CH_2C_6H_5)-C(=O)-NHCH(CH_3)-C(=O)-NHCH(CH(CH_3)_2)-C(=O)-OCH_2-C_6H_4-$●

$(CH_3)_3CO-C(=O)-NHCH(CH_2CH(CH_3)_2)-C(=O)-O^-$ $\xrightarrow{DCCD}$ $(CH_3)_3CO-C(=O)-NHCH(CH_2CH(CH_3)_2)-C(=O)-O-DCC$

$\downarrow$

$(CH_3)_3CO-C(=O)-NHCH(CH_2CH(CH_3)_2)-C(=O)-NHCH(CH_2C_6H_5)-C(=O)-NHCH(CH_3)-C(=O)-NHCH(CH(CH_3)_2)-C(=O)-OCH_2-C_6H_4-$●

$\downarrow$ CF_3COOH, CH_2Cl_2

$CH_3C(CH_3)=CH_2$ + CO_2 + $H_2NCH(CH_2CH(CH_3)_2)-C(=O)-NHCH(CH_2C_6H_5)-C(=O)-NHCH(CH_3)-C(=O)-NHCH(CH(CH_3)_2)-C(=O)-OCH_2-C_6H_4-$●

$\downarrow$ HF, H_2O

$H_3\overset{+}{N}CH(CH_2CH(CH_3)_2)-C(=O)-NHCH(CH_2C_6H_5)-C(=O)-NHCH(CH_3)-C(=O)-NHCH(CH(CH_3)_2)-C(=O)-OH$ + $HOCH_2-C_6H_4-$●

37. It is an S_N2 reaction followed by dissociation of a proton.

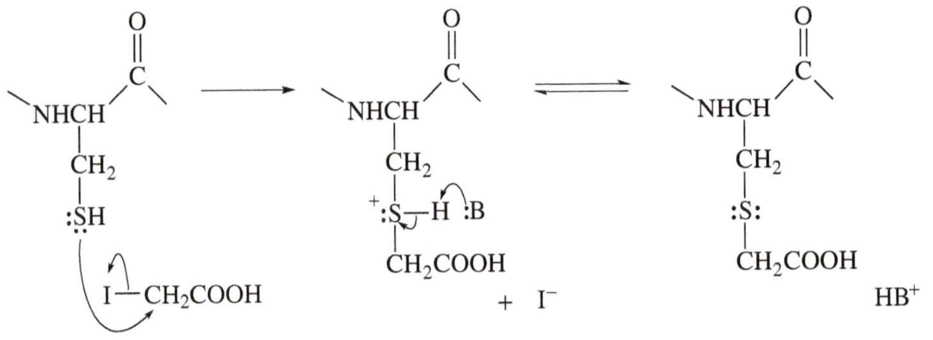

38. Because insulin has two peptide chains, treatment with Edman's reagent would release two PTH-amino acids in approximately equal amounts.

39. Knowing that the N-terminal amino acid is Gly, look for a peptide fragment that contains Gly.

"Fragment 6" tells you that the second amino acid is Arg.

"Fragment 5" tells you that the next two are Ala-Trp or Trp-Ala.

"Fragment 4" tells you that Glu is next to Ala, so the third and fourth amino acids must be Trp-Ala and the fifth is Glu.

"Fragment 7" tells you that the sixth amino acid is Leu.

"Fragment 8" tells you that the next two are Met-Pro or Pro-Met.

"Fragment 3" tells you that Pro is next to Val, so the seventh and eighth amino acids must be Met-Pro and the ninth is Val.

"Fragment 2" tells you that the last amino acid is Asp.

Gly-Arg-Trp-Ala-Glu-Leu-Met-Pro-Val-Asp

40. Cysteine can react with cyanogen bromide, but the sulfur would not be positively charged, so it would be a poor leaving group. In addition, the lactone would not be formed because it would have a strained four-membered ring. Without lactone formation, the imine would not be formed, so cleavage cannot occur.

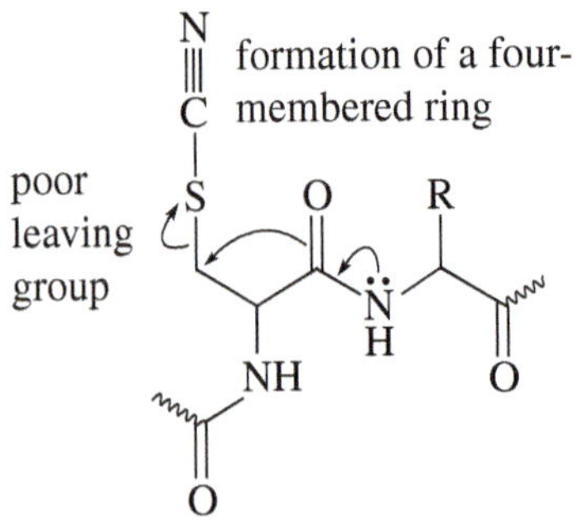

formation of a four-membered ring

poor leaving group

41. **a.** His-Lys Leu-Val-Glu-Pro-Arg Ala-Gly-Ala

b. Leu-Gly-Ser-Met-Phe-Pro-Tyr Gly-Val

42. Solved in the text.

43. The data from treatment with Edman's reagent and carboxypeptidase A identify the first and last amino acids.

Leu ___ ___ ___ ___ ___ ___ Ser

The data from cleavage with cyanogen bromide identify the position of Met and identify the other amino acids in the pentapeptide and tripeptide but not their order.

cleavage with cyanogen bromide

Leu (Arg, Lys, Tyr) ___ ___ ___ Met | (Arg, Phe) ___ ___ Ser

The data from treatment with trypsin put the remaining amino acids in the correct positions.

Leu Tyr Lys | Arg | Met Phe Arg | Ser

44. **a.** Trypsin cleaves at Arg and Lys. There are two possible primary structures:

Val-Gly-Asp-Lys-Leu-Glu-Pro-Ala-Arg-Ala-Leu-Gly-Asp

or

Leu-Glu-Pro-Ala-Arg-Val-Gly-Asp-Lys-Ala-Leu-Gly-Asp

The two possible primary structures can be distinguished by Edman's reagent. Edman's reagent will release Val in one case and Leu in the other.

b. Trypsin cleaves at Arg and Lys. There are two possible primary structures:

Ala-Glu-Pro-Arg-Ala-Met-Gly-Lys-Val-Leu-Gly-Glu

or

Ala-Met-Gly-Lys-Ala-Glu-Pro-Arg-Val-Leu-Gly-Glu

The two possible primary structures can be distinguished by treatment with cyanogen bromide. Cyanogen bromide will cleave one of the possible polypeptides into two hexamers and the other into a dimer and a decamer.

45. **a.** 74 amino acids/3.6 amino acids per turn of the helix = 20.6 turns of the helix
20.6×5.4 Å = **110 Å** in an α-helix (5.4 Å is the repeat distance of the α-helix.)

b. 74 amino acids $\times$ 3.5 Å = **260 Å** in fully extended polypeptide chain

46. It would fold so that its nonpolar residues are on the outside of the protein in contact with the nonpolar membrane and its polar residues are on the inside of the protein.

47. A protein folds to maximize the number of polar groups on the surface of the protein and the number of nonpolar groups on the inside of the protein.

a. A cigar-shaped protein has the greatest surface area to volume ratio, so it has the highest percentage of polar amino acids.

b. A subunit of a hexamer has the smallest percentage of polar amino acids, because part of the surface of the subunit can be on the inside of the hexamer and, therefore, have nonpolar amino acids on its surface.

48. Lemon juice contains citric acid. Some of the side chains of the enzyme will become protonated in an acidic solution. This will change the charge of the group (for example, a negatively charged aspartate, when protonated, becomes neutral; a neutral lysine, when protonated, becomes positively charged). Because the shape of an enzyme is determined by the interaction of the side chains, changing the charges of the side chains causes the enzyme to undergo a conformational change that leads to denaturation. When the enzyme is denatured, it loses its ability to catalyze the reaction that causes apples to turn brown.

49. **a.** pH = 9.69 **b.** pH = (its pI) **c.** pH = 2.34

50. **a.** Val-Arg-Gly-Met-Arg-Ala Ser

b. Ser-Phe-Lys-Met Pro-Ser-Ala-Asp

c. Arg Ser-Pro-Lys Lys Ser-Glu-Gly

51. **a.** The pK_a of 6.0 indicates that the amino acid is histidine.

b. Because it requires two equivalents of hydroxide ion to get to a neutral pH, the amino acid has two acidic groups. The pK_a of 9.8 indicates that the amino acid is aspartic acid.

c. The pK_a values indicate that the amino acid is most likely alanine. However, it could be isoleucine.

52. As an amino acid moves from a solution with a pH equal to its pI to a more basic solution, the amino acid becomes more and more negatively charged. Because asparagine has a lower pI than leucine, in a solution of pH = 7.3, asparagine has moved farther from its pI than has leucine. Asparagine, therefore, will have a higher percentage of negative charge at pH = 7.3.

53.

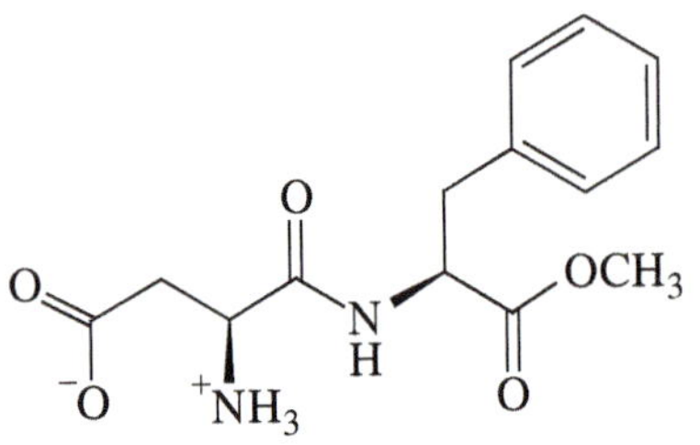

O, O, $^-$O, $^+NH_3$, N H, OCH_3, O

54. **a.** $HO-C(=O)-CH_2CH(^+NH_3)-C(=O)-OH$

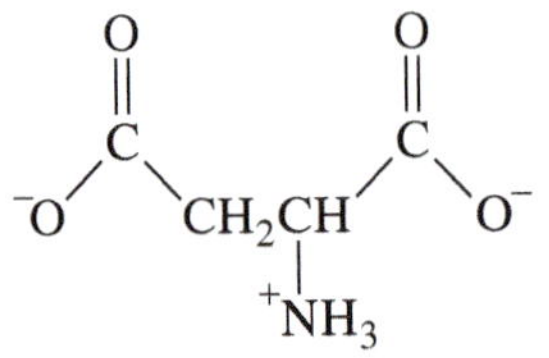

b. $HO-C(=O)-CH_2CH(^+NH_3)-C(=O)-O^-$

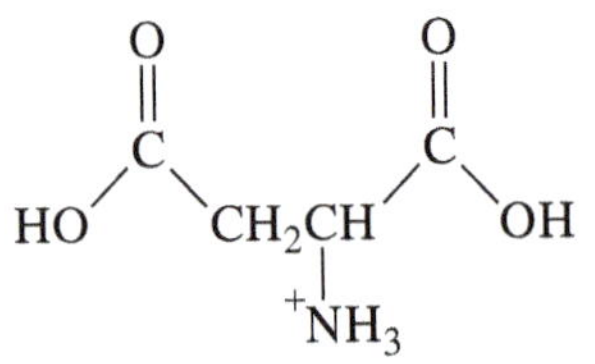

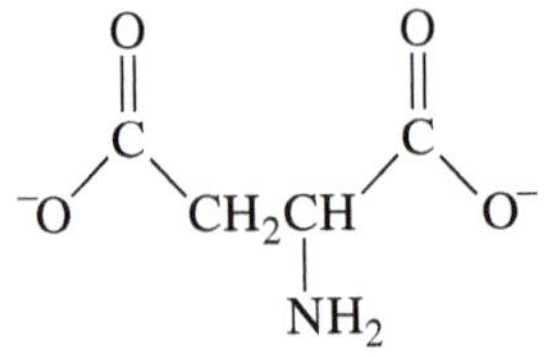

c. $^-O-C(=O)-CH_2CH(^+NH_3)-C(=O)-O^-$

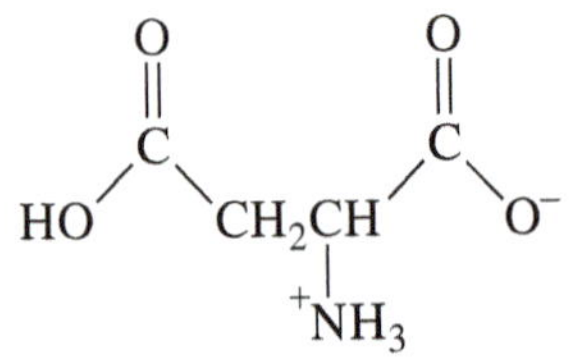

d. $^-O-C(=O)-CH_2CH(NH_2)-C(=O)-O^-$

55. $C_6H_5CH_2C(=O)C(=O)OH \xrightarrow[\text{2. } H_2,\ Pd/C]{\text{1. excess } NH_3,\ \text{trace acid}} C_6H_5CH_2CH(\overset{+}{N}H_3)C(=O)O^-$

56. The student is correct. At the pI, the total of the positive charges on the tripeptide's amino groups must be one to balance the one negative charge of the carboxylate group. When the pH of the solution is equal to the pK_a of a lysine residue, the three lysine groups each have one-half of a positive charge for a total of one and one-half positive charges. Therefore, the solution must be more basic than this to have just one positive charge.

57. **a.** $HO\text{—}C_6H_4\text{—}CH_2C(=O)H$ **b.** $CH_3CH(CH_3)CH_2C(=O)H$ **c.** $H_2N\text{—}C(=\overset{+}{N}H_2)\text{—}NHCH_2CH_2CH_2C(=O)H$

58. In a solution of pH = 5, **His** will have an overall positive charge and **Glu** will have an overall negative charge. **His**, therefore, will migrate to the cathode, and **Glu** will migrate to the anode.

Ser is more polar than **Thr** (both have OH groups, but **Thr** has an additional carbon).

Thr is more polar than **Met** (**Met** has $CH_3SCH_2CH_2$ instead of $CH_3CH(OH)$). **Met** is more polar than **Leu** (**Leu** has four carbons instead of the sulfur and three carbons that **Met** has).

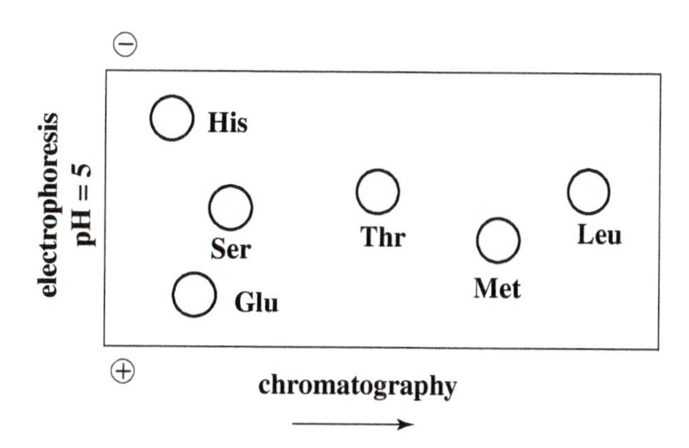

59. We know that "Fragment 3" (Leu-Pro-Phe) is at the C-terminal end of the polypeptide. Now the only question we need to answer is which of the two fragments obtained by cleavage with cyanogen bromide (which begin with Gly and end with Met) is nearest the C-terminal end of the polypeptide. "Fragment 4", obtained from trypsin cleavage, gives the answer. The Met nearest the C-terminal end must be preceded by Arg or Lys. Therefore, the polypeptide has the following sequence:

Gly-Leu-Tyr-Phe-Lys-Ser-Met-Gly-Leu-Tyr-Lys-Val-Ile-Arg-Met-Leu-Pro-Phe

60. An amino acid is insoluble in diethyl ether (a relatively nonpolar solvent) because an amino acid exists as a highly polar zwitterion at neutral pH. Carboxylic acids and amines are less polar because they either are neutral or have a single charge depending on the extent of dissociation in diethyl ether.

61. You would (correctly) expect serine and cysteine to have lower pK_a values than alanine, because a hydroxymethyl and a thiomethyl group are more electron withdrawing than a methyl group. Because oxygen is more electronegative than sulfur, you would expect serine to have a lower pK_a than cysteine. The fact that cysteine has a lower pK_a than serine can be explained by stabilization of serine's carboxyl proton by hydrogen bonding to the β-OH group of serine, which causes it to have less of a tendency to be removed by a base.

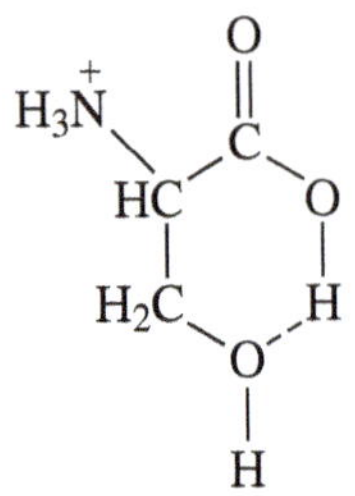

$H_3\overset{+}{N}$–HC(–C(=O)–O–H)–H_2C–O–H

62. Each compound has two groups that can act as a buffer, one amino group and one carboxyl group. Therefore, the compound in higher concentration (0.2 M glycine) is a more effective buffer.

63. Groups that are not fully charged at the given pH are shown in the form that predominates at that pH. For example, tyrosine has a pK_a = 9.11, so at pH = 7, it is shown in its acidic form (with its proton).

a. $H_3\overset{+}{N}CH(C=O)NHCH(C=O)NHCH(C=O)NHCH(C=O)NHCH(C=O)NHCH(C=O)OH$

Side chains: $(CH_2)_4{}^+NH_3$; CH_2OH; CH_2COOH; CH_2SH; CH_2–imidazolium (NH, $^+$N–H); CH_2–C$_6$H$_4$–OH

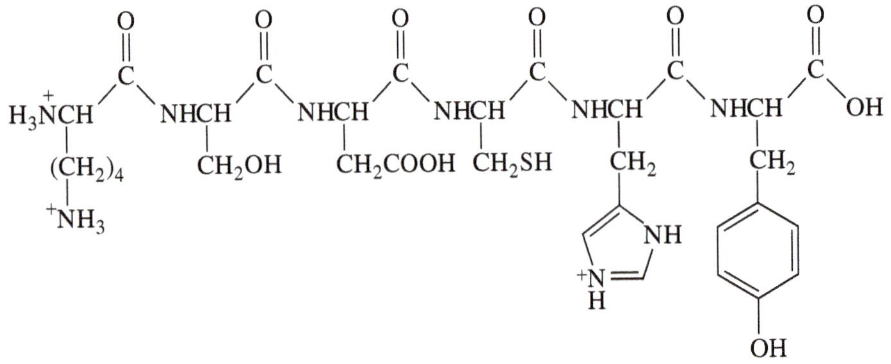

b. $H_3\overset{+}{N}CH(C=O)NHCH(C=O)NHCH(C=O)NHCH(C=O)NHCH(C=O)NHCH(C=O)O^-$

Side chains: $(CH_2)_4{}^+NH_3$; CH_2OH; CH_2COO^-; CH_2SH; CH_2–imidazolium (NH, $^+$N–H); CH_2–C$_6$H$_4$–OH

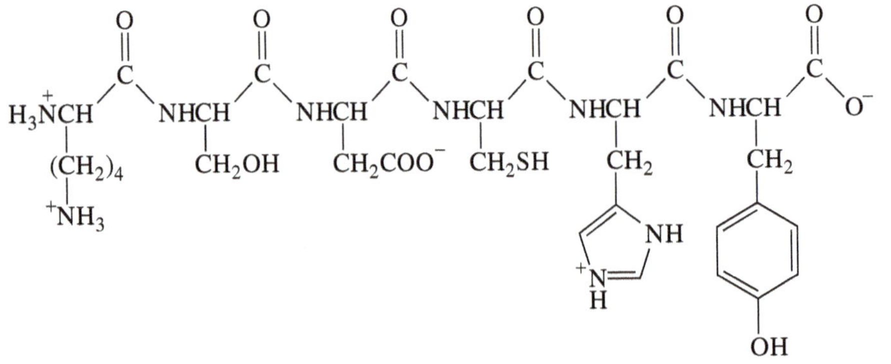

c. $H_2NCH(C{=}O)$–$NHCH(C{=}O)$–$NHCH(C{=}O)$–$NHCH(C{=}O)$–$NHCH(C{=}O)$–$NHCH(C{=}O)O^-$; side chains: $(CH_2)_4NH_3^+$, CH_2OH, CH_2COO^-, CH_2SH, CH_2-imidazole (NH, N=), CH_2-phenyl-OH

d. $H_2NCH(C{=}O)$–$NHCH(C{=}O)$–$NHCH(C{=}O)$–$NHCH(C{=}O)$–$NHCH(C{=}O)$–$NHCH(C{=}O)O^-$; side chains: $(CH_2)_4NH_2$, CH_2OH, CH_2COO^-, CH_2S^-, CH_2-imidazole (NH, N=), CH_2-phenyl-O^-

64. —NHCH(C=O)–$(CH_2)_4$–NH_2 (lysine) + maleic anhydride ⟶ —NHCH(C=O)–$(CH_2)_4$–NH–(tetrahedral intermediate with O^-) ⟶ —NHCH(C=O)–$(CH_2)_4$–NH–C(=O)–CH=CH–COO^-

lysine

maleic anhydride

65. a. When the polypeptide is treated with maleic anhydride, lysine reacts with maleic anhydride (see Problem 64), but the amino group of arginine is not sufficiently nucleophilic to react with maleic anhydride. Therefore, trypsin will cleave only at arginine residues because the enzyme no longer recognizes lysine residues.

b. Four fragments will be obtained from the polypetide. Remember that trypsin will not cleave the Arg-Pro bond.

c. The N-terminal end of each fragment will be positively charged because of the $^+NH_3$ group.
The C-terminal end will be negatively charged because of the COO^- group.
Arginine residues will be positively charged.
Aspartate and glutamate residues will be negatively charged.
Lysine residues will be negatively charged because they are attached to the maleic acid group.

The least negatively charged fragment will be eluted first, and the most negatively charged fragment will be eluted last:

A > D > C > B

A Gly-Ala-Asp-Ala-Leu-Pro-Gly-Ile-Leu-Val-Arg overall charge = 0

\+ − +−

B Asp-Val-Gly-Lys-Val-Glu-Val-Phe-Glu-Ala-Gly-Arg overall charge = −3

+− − − − +−

C Ala-Glu-Phe-Lys-Glu-Pro-Arg overall charge = −2

\+ − − − + −

D Leu-Val-Met-Lys-Val-Glu-Gly-Arg-Pro-Val-Ala-Ala-Gly-Leu-Trp overall charge = −1

\+ − − + −

66. First, mark off where the chains would be cleaved by chymotrypsin (C-side of Phe, Trp, Tyr).

Val-Met-Tyr | Ala-Cys-Ser-Phe | Ala-Glu-Ser

Ser-Cys-Phe | Lys-Cys-Trp | Lys-Tyr | Cys-Phe | Arg-Cys-Ser

Then from the fragments given, you can determine where the disulfide bridges are in the original intact peptide. For example, "Fragment 2" has two Phe, two Cys, and one Ser. Therefore, the first and fourth fragments of the second row must be connected by a disulfide bond. Fragment "5" provides the evidence for the disulfide bond between the two chains.

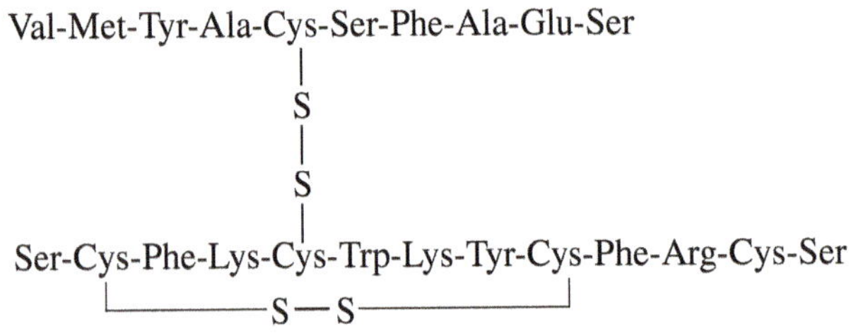

```
Val-Met-Tyr-Ala-Cys-Ser-Phe-Ala-Glu-Ser
                |
                S
                |
                S
                |
Ser-Cys-Phe-Lys-Cys-Trp-Lys-Tyr-Cys-Phe-Arg-Cys-Ser
    |_______________S—S_____________________|
```

67. The methyl ester of phenylalanine rather than phenylalanine itself should be added in the second peptide bond-forming step because if esterification of phenylalanine is done after amide bond formation, both the carboxyl group of phenylalanine and the γ-carboxyl group of aspartate can be esterified. Both the carboxyl group of phenylalanine and the γ-carboxyl group of aspartate can be activated by DCCD. Therefore, some product will be obtained in which the amide bond is formed with the γ-carboxyl group of aspartate rather than with the α-carboxyl group.

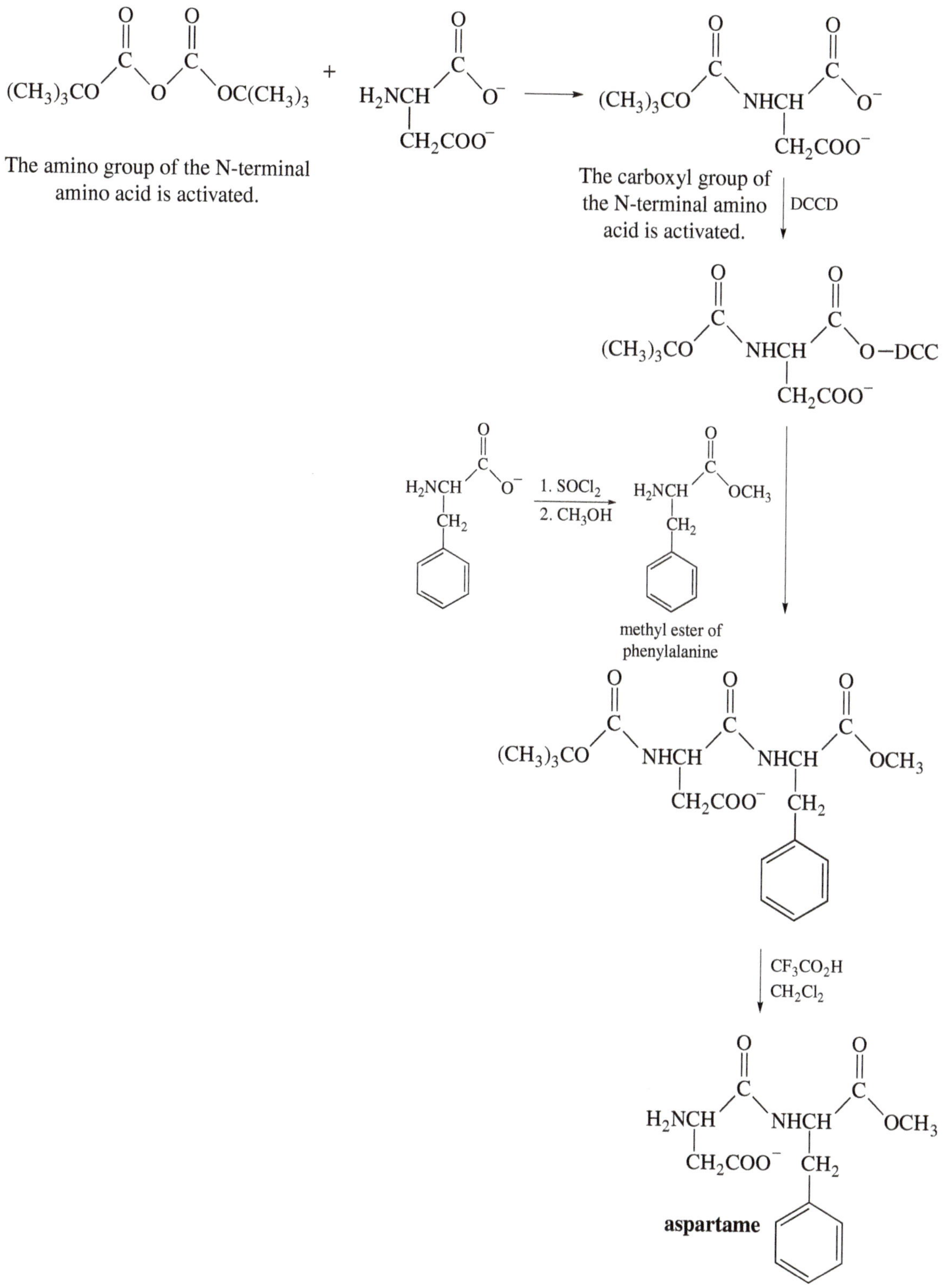

$(CH_3)_3CO-C(=O)-O-C(=O)-OC(CH_3)_3$ + $H_2NCH(CH_2COO^-)-C(=O)-O^-$ ⟶ $(CH_3)_3CO-C(=O)-NHCH(CH_2COO^-)-C(=O)-O^-$

The amino group of the N-terminal amino acid is activated.

The carboxyl group of the N-terminal amino acid is activated.

↓ DCCD

$(CH_3)_3CO-C(=O)-NHCH(CH_2COO^-)-C(=O)-O{-}DCC$

$H_2NCH(CH_2C_6H_5)-C(=O)-O^-$ —1. $SOCl_2$, 2. CH_3OH⟶ $H_2NCH(CH_2C_6H_5)-C(=O)-OCH_3$

methyl ester of phenylalanine

$(CH_3)_3CO-C(=O)-NHCH(CH_2COO^-)-C(=O)-NHCH(CH_2C_6H_5)-C(=O)-OCH_3$

↓ CF_3CO_2H, CH_2Cl_2

$H_2NCH(CH_2COO^-)-C(=O)-NHCH(CH_2C_6H_5)-C(=O)-OCH_3$

aspartame

68. **a.** $RCHO \xrightarrow[\text{trace acid}]{NH_3} RCH{=}NH_2^+$ (intermediate I) $\xrightarrow{^-C{\equiv}N} RCH(NH_2)C{\equiv}N$ (intermediate II) $\xrightarrow[H_2O,\ \Delta]{HCl} RCH(^+NH_3)COOH$

b. $CH_3CH(CH_3)CH_2CHO$ (3-methylbutanal) $\longrightarrow CH_3CH(CH_3)CH_2CH(^+NH_3)COOH$ (leucine)

c. $CH_3CH_2CH(CH_3)CHO$ (2-methylbutanal) $\longrightarrow CH_3CH_2CH(CH_3){-}CH(^+NH_3)COO^-$ (isoleucine)

69. Ser-Glu-Leu-Trp-Lys-Ser-Val-Glu-His-Gly-Ala-Met

From the experiment with carboxypeptidase A, we know that the C-terminal amino acid is Met.

"Fragment 12" tells us the amino acid adjacent to Met is Ala.

"Fragment 5" tells us the next amino acid is Gly.

"Fragment 2" tells us the next amino acid is His.

"Fragment 7" tells us the next amino acid is Glu.

"Fragment 10" tells us the next amino acid is Val.

"Fragment 3" tells us the next amino acid is Ser.

"Fragment 9" tells us the next amino acid is Lys.

"Fragment 1" tells us the next amino acid is Trp.

"Fragment 8" tells us the next amino acid is Leu.

"Fragment 11" tells us the next amino acid is Glu.

"Fragment 4" tells us the next (first) amino acid is Ser.

70. **a.** x^y = the number of amino acids

y = the number of spots in the peptide for each amino acid

$20^8 = 25{,}600{,}000{,}000$

b. $20^{100} = 1.3 \times 10^{130}$

71. The pK_a of the carboxylic acid group of the amino acid is lower than the pK_a of the carboxylic acid group of the dipeptide because the positively charged ammonium group of the amino acid is more strongly electron withdrawing than the amide group of the peptide. This causes the amino acid to be a stronger acid and, therefore, have a lower pK_a.

$$H_3\overset{+}{N}CH_2C(=O)OH \qquad H_3\overset{+}{N}CH_2C(=O)NHCH_2C(=O)OH$$

(OH of the amino acid: lower pK_a)

The pK_a of the ammonium group of the dipeptide is lower than the pK_a of the ammonium group of the amino acid because the amide group of the dipeptide is more strongly electron withdrawing than the carboxylate group of the amino acid.

$$H_3\overset{+}{N}CH_2C(=O)O^- \qquad H_3\overset{+}{N}CH_2C(=O)NHCH_2C(=O)O^-$$

($H_3\overset{+}{N}$ of the dipeptide: lower pK_a)

72. Finding that there is one less spot than the number of amino acids tells you that the spots for two of the amino acids superimpose. Because leucine and isoleucine have identical polarities, they are good candidates for being the amino acids that migrate to the same location.

Val (pI = 5.97), Trp (pI = 5.89), and Met (pI = 5.75) can be ordered based on their pI values, because the one with the greatest pI will be the one with the greatest amount of positive charge at pH = 5. (See Problem 52.)

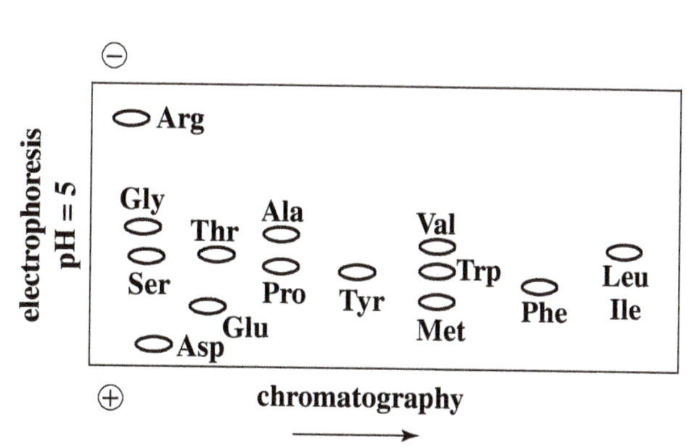

73.

$$(CH_3)_3CO{-}C({=}O){-}O{-}C({=}O){-}OC(CH_3)_3 + H_2\ddot{N}CH(R){-}C({=}O){-}O^- \longrightarrow (CH_3)_3CO{-}C(\ddot{O}:^-)(H{-}\overset{+}{N}H{-}CH(R)COO^-){-}O{-}C({=}O){-}OC(CH_3)_3 \quad B:$$

$$\downarrow$$

$$(CH_3)_3CO{-}C({=}O){-}NHCH(R){-}C({=}O){-}O^- + (CH_3)_3CO{-}C({=}O){-}O^- \longleftarrow (CH_3)_3CO{-}C(\ddot{O}:^-)(NH{-}CHCOO^-(R)){-}O{-}C({=}O){-}OC(CH_3)_3 \quad HB^+$$

74. Oxidation of dithiothreitol is an intramolecular reaction, so it occurs with a larger rate constant than the oxidation of 2-mercaptoethanol, which is an intermolecular reaction. The reverse reduction reaction should occur with about the same rate constant in both cases.

Increasing the rate of the oxidation reaction while keeping the rate of the reduction reaction constant is responsible for the greater equilibrium constant because $K_{eq} = k_1/k_{-1}$, where k_1 is the rate constant for oxidation and k_{-1} is the rate constant for reduction.

$$\text{dithiothreitol (HO, HO, SH, SH)} \underset{k_{-1}}{\overset{k_1}{\rightleftharpoons}} \text{cyclic disulfide (HO, HO, S–S)}$$

$$2\ HOCH_2CH_2SH \underset{k_{-1}}{\overset{k_1}{\rightleftharpoons}} HOCH_2CH_2S{-}SCH_2CH_2OH$$

75. a.

$$(CH_3)_2CHCH_2{-}C({=}O){-}OH \xrightarrow[2.\ H_2O]{1.\ Br_2,\ PBr_3} (CH_3)_2CHCH(Br){-}C({=}O){-}OH \xrightarrow[\text{excess}]{NH_3} (CH_3)_2CHCH(\overset{+}{N}H_3){-}C({=}O){-}O^- + \overset{+}{N}H_4$$

valine

b.

$$(CH_3)_2CH{-}CH{=}O \xrightarrow[\text{trace acid}]{NH_3} (CH_3)_2CH{-}CH{=}NH \xrightarrow[HCl]{^-C{\equiv}N} (CH_3)_2CHCH(\overset{+}{N}H_3){-}C{\equiv}N \xrightarrow[\Delta]{HCl,\ H_2O} (CH_3)_2CHCH(\overset{+}{N}H_3){-}C({=}O){-}OH$$

an aldehyde — an imine — valine

c.
$$(CH_3)_2CHC(=O)COOH \xrightarrow[\text{trace acid}]{NH_3\ \text{excess}} (CH_3)_2CHC(=NH)COO^- \xrightarrow[Pd/C]{H_2} (CH_3)_2CHCH(NH_2)COO^-$$

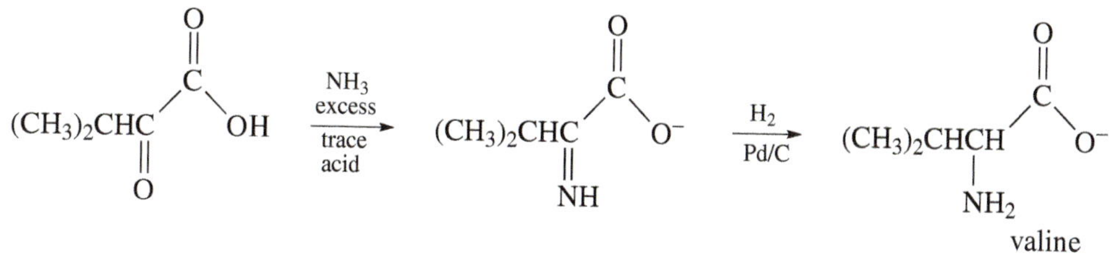

valine

d.
$$C_2H_5OC(=O)-CH(Br)-COC_2H_5(=O) + \text{potassium phthalimide } (N^-K^+) \longrightarrow \text{phthalimido}-N-CH(COOC_2H_5)_2$$

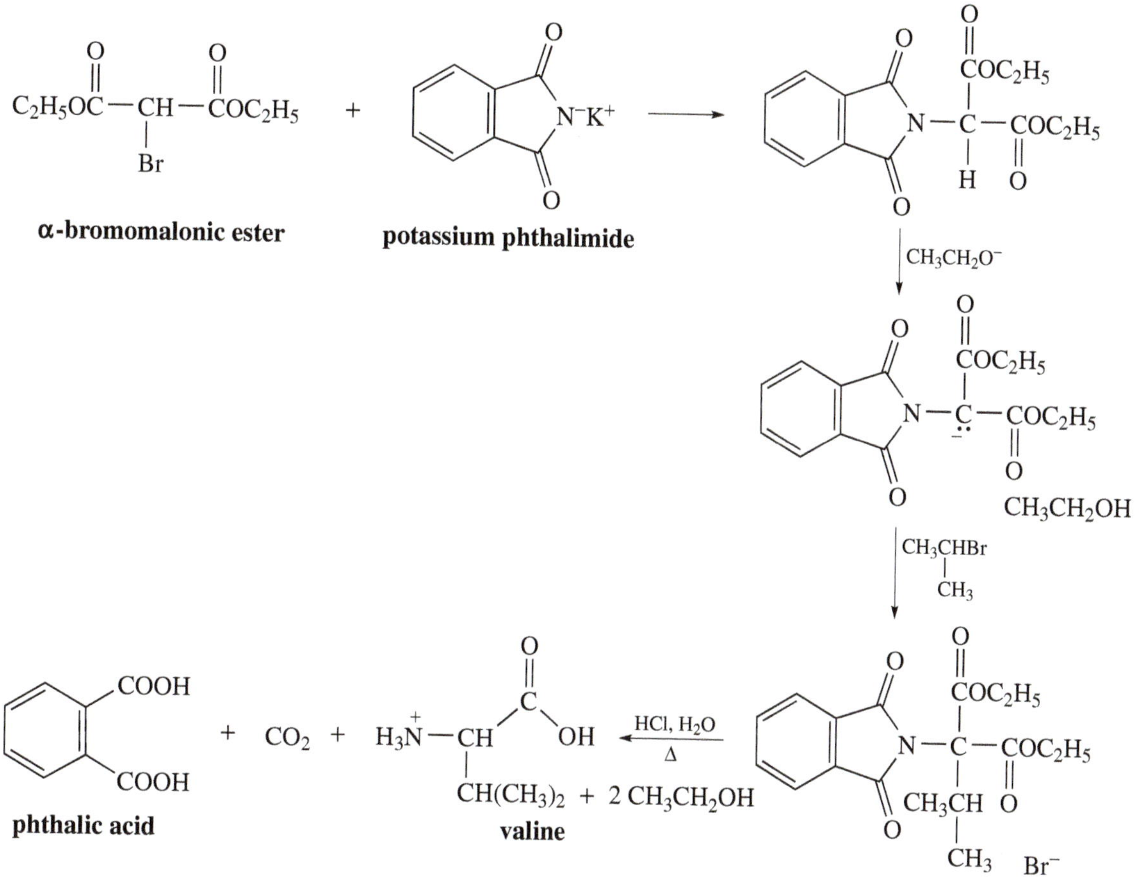

α-bromomalonic ester **potassium phthalimide**

$$\xrightarrow{CH_3CH_2O^-} \text{phthalimido}-N-\ddot{C}^-(COOC_2H_5)_2 + CH_3CH_2OH$$

$$\xrightarrow{CH_3CHBr(CH_3)} \text{phthalimido}-N-C(CH(CH_3)_2)(COOC_2H_5)_2 + Br^-$$

$$\xrightarrow[\Delta]{HCl,\ H_2O} \text{phthalic acid } (C_6H_4(COOH)_2) + CO_2 + H_3\overset{+}{N}-CH(CH(CH_3)_2)-COOH + 2\ CH_3CH_2OH$$

phthalic acid **valine**

e.
$$C_2H_5OC(=O)-CH(NHCOCH_3)-COC_2H_5(=O) \xrightarrow{CH_3CH_2O^-} C_2H_5OC(=O)-\ddot{C}^-(NHCOCH_3)-COC_2H_5(=O) + CH_3CH_2OH$$

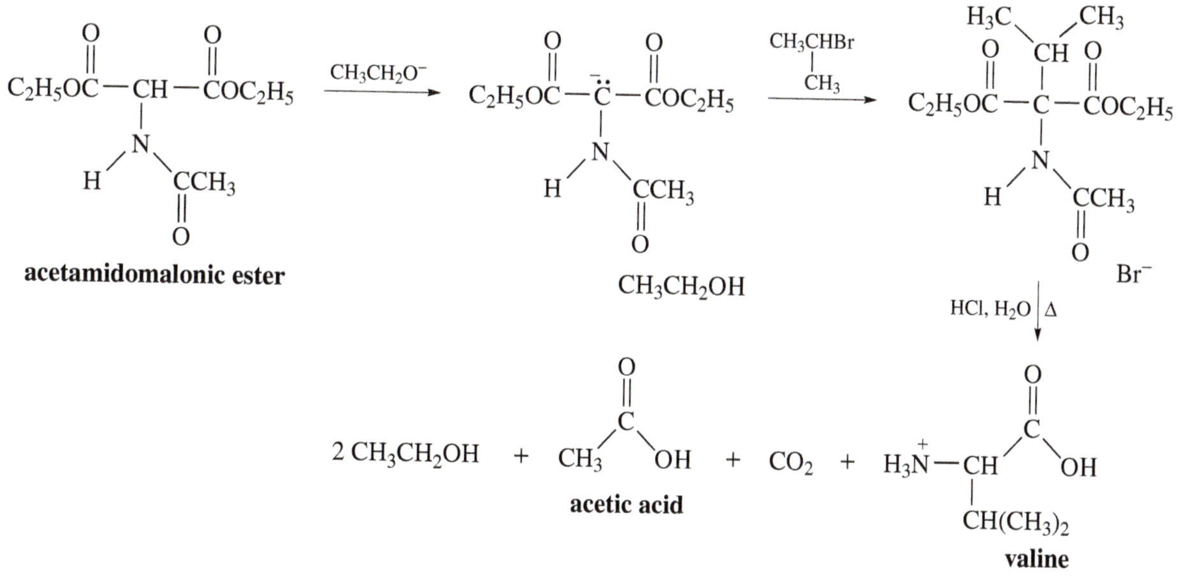

acetamidomalonic ester

$$\xrightarrow{CH_3CHBr(CH_3)} C_2H_5OC(=O)-C(CH(CH_3)_2)(NHCOCH_3)-COC_2H_5(=O) + Br^-$$

$$\xrightarrow[\Delta]{HCl,\ H_2O} 2\ CH_3CH_2OH + CH_3COOH + CO_2 + H_3\overset{+}{N}-CH(CH(CH_3)_2)-COOH$$

acetic acid **valine**

76. **a.** **1.** Tyr-Gly-Gly-Phe-Met-Thr-Ser-Gly-Lys
Ser-Gln-Thr-Pro-Leu-Val-Thr-Leu-Phe-Lys-
Asn-Ala-Ile-Ile-Lys, Asn-Ala-Tyr-Lys, Lys, and Gly-Glu

2. Tyr-Gly-Gly-Phe-Met
Thr-Ser-Gly-Lys-Ser-Gln-Thr-Pro-Leu-Val-Thr-Leu-Phe-Lys-Asn-Ala-Ile-Ile-Lys-Asn-Ala-
Tyr-Lys-Lys-Gly-Glu

3. Tyr
Gly-Gly-Phe
Met-Thr-Ser-Glu-Lys-Ser-Gln-Thr-Pro-Leu-Val-Thr-Leu-Phe
Lys-Asn-Ala-Ile-Ile-Lys-Asn-Ala-Tyr
Lys-Lys-Gly-Glu

b. N-terminal end: Tyr-Gly-Gly-Phe-Met
C-terminal end: Tyr-Lys-Lys-Gly-Glu or Tyr-Lys-Lys-Glu-Gly

77. Because the native enzyme has four disulfide bridges, we know that the denatured enzyme has eight cysteine residues. The first cysteine has a one in seven chance of forming a disulfide bridge with the correct cysteine. The first cysteine of the next pair has a one in five chance, and the first cysteine of the third pair has a one in three chance.

$$\frac{1}{7} \times \frac{1}{5} \times \frac{1}{3} = 0.0095$$

If disulfide bridge formation were entirely random, the recovered enzyme should have 0.95% of its original activity. The fact that the enzyme the chemist recovered had 80% of its original activity supports his hypothesis that disulfide bridges form after the minimum energy conformation of the protein has been achieved. In other words, disulfide bridge formation is not random, but is determined by the tertiary structure of the protein.

78. The acid protonates the lone pair that are sp^2 electrons. Formation of the tetrahedral intermediate (the two arrows that represent electron flow to the right) and collapse of the tetrahedral intermediate (the three arrows that represent electron flow to the left) are shown together in the second step.

a thiazoline

a PTH-amino acid

79. The spot marked with an **X** is the peptide that is different in the normal and mutant polypeptide. The spot is closer to the cathode and farther to the right, indicating that the substituted amino acid in the mutant has a greater pI and is less polar.

The fingerprints are those of hemoglobin (normal) and sickle-cell hemoglobin (mutant). In sickle-cell hemoglobin, a glutamate in the normal polypeptide has been substituted with a valine. This agrees with our observation that the substituted amino acid is less negative and more nonpolar. (See the discussion of sickle-cell anemia on page 1171 of the text.)

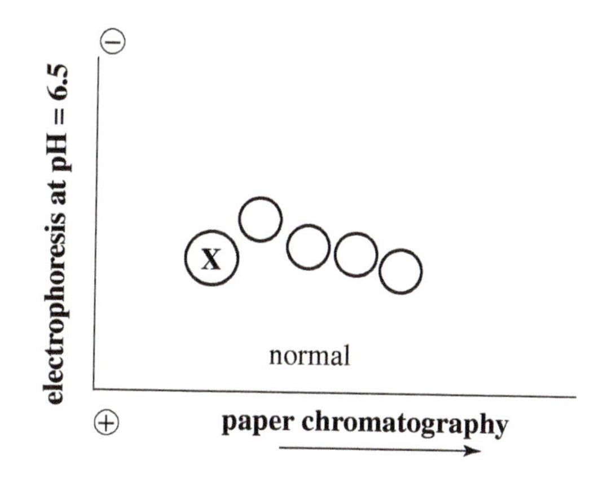

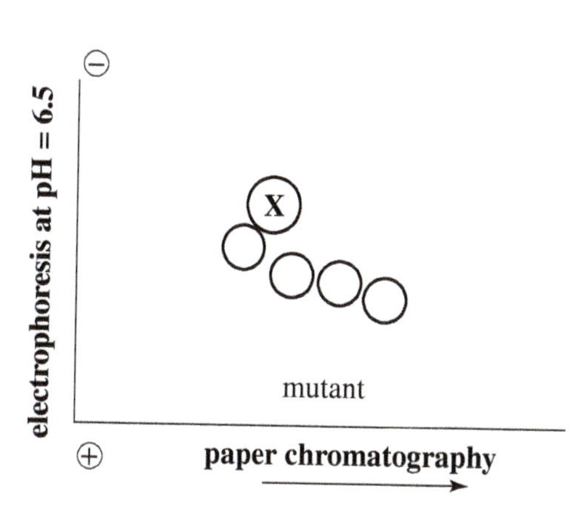

80. **a.** Acid-catalyzed hydrolysis indicates the peptide contains 12 amino acids.

__ __ __ __ __ __ __ __ __ __ __ __

b. Treatment with Edman's reagent indicates that Val is the N-terminal amino acid.

Val __ __ __ __ __ __ __ __ __ __ __ __

c. Treatment with carboxypeptidase A indicates that Ala is the C-terminal amino acid.

Val __ __ __ __ __ __ __ __ __ __ __ __ Ala

d. Treatment with cyanogen bromide indicates that Met is the fifth amino acid with Arg, Gly, Ser in an unknown order in positions 2, 3, and 4.

Val __ __ __ Met __ __ __ __ __ __ Ala
Arg, Gly, Ser

e. Treatment with trypsin indicates that Arg is the third amino acid, Ser is second, Gly is fourth, Tyr is sixth, and Lys is seventh. Because Lys is in the terminal fragment, cleavage did not occur at Lys, so Pro must be at lysine's cleavage site, but we don't know whether Lys-Pro comes before of after Phe and Ser.

Val Ser Arg Gly Met Tyr Lys __ __ __ __ Ala
Lys-Pro, Phe, Ser

f. Treatment with chymotrypsin indicates that Phe is the tenth amino acid and Ser is the eleventh.

Val Ser Arg Gly Met Tyr Lys Lys Pro Phe Ser Ala

Chapter 21 Practice Test

1. Draw the structure of the following amino acids at pH = 7:

   a. glutamic acid **b.** lysine **c.** isoleucine **d.** arginine **e.** asparagine

2. Draw the form of histidine that predominates at:

   a. pH = 1 **b.** pH = 4 **c.** pH = 8 **d.** pH = 11

3. Answer the following:

   a. Alanine has a pI = 6.02, and serine has a pI = 5.68. Which will have the highest concentration of positive charge at pH = 5.50?

   b. Which amino acid is the only one that does not have an asymmetric center?

   c. Which are the two most nonpolar amino acids?

   d. Which amino acid has the lowest pI?

4. Why does the carboxyl group of alanine have a lower pK_a than the carboxyl group of propanoic acid?

   $CH_3CH(^{+}NH_3)C(=O)OH$ $pK_a = 2.2$ alanine

   $CH_3CH_2C(=O)OH$ $pK_a = 4.7$ propanoic acid

5. Indicate whether each of the following statements is true or false:

   a. A cigar-shaped protein has a greater percentage of polar residues than a spherical protein. T F

   b. Naturally occurring amino acids have the L-configuration. T F

   c. There is free rotation about a peptide bond. T F

6. What compound is obtained from mild oxidation of cysteine?

7. Define each of the following:

   a. the primary structure of a protein

   b. the tertiary structure of a protein

   c. the quaternary structure of a protein

8. Identify the spots.

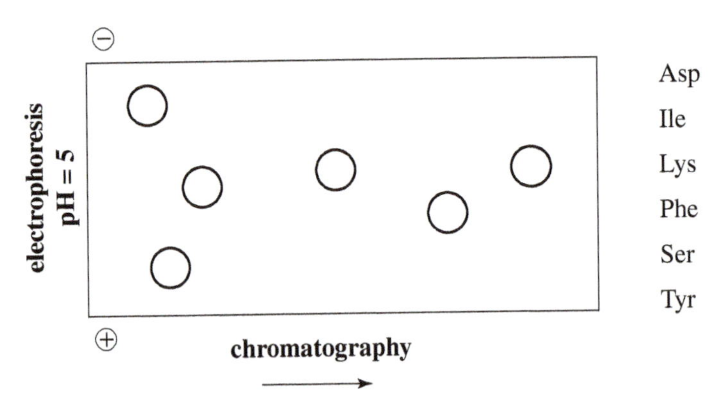

9. Calculate the pI of the following amino acids:

a. phenylalanine (pK_as = 2.16, 9.18) **b.** arginine (pK_as = 2.17, 9.04, 12.48)

10. From the following information, determine the primary sequence of the decapeptide:

a. Acid hydrolysis gives Ala, 2 Arg, Gly, His, Ile, Lys, Met, Phe, Ser.

b. Reaction with Edman's reagent liberates Ala.

c. Reaction with carboxypeptidase A liberates Ile.

d. Reaction with cyanogen bromide (cleaves on the C-side of Met)

1. Gly, 2 Arg, Ala, Met, Ser

2. Lys, Phe, Ile, His

e. Reaction with trypsin (cleaves on the C-side of Arg and Lys)

1. Arg, Gly

2. Ile

3. Phe, Lys, Met, His

4. Arg, Ser, Ala

f. Reaction with thermolysin (cleaves on the N-side of Leu, Ile, Phe, Trp, Tyr)

1. Lys, Phe

2. 2 Arg, Ser, His, Gly, Ala, Met

3. Ile

11. Describe how acetaldehyde can be converted to alanine.

12. Draw the mechanism for the conversion of a thiol to a disulfide in a basic solution of Br_2.

CHAPTER 22
Catalysis in Organic Reactions and in Enzymatic Reactions

Important Terms

acid catalyst	a catalyst that increases the rate of a reaction by donating a proton.
active site	a pocket or cleft in an enzyme where the substrate is bound.
acyl-enzyme intermediate	an amino acid residue of an enzyme that has been acylated while catalyzing a reaction.
amino acid side chain	the substituent on the α-carbon of an amino acid.
base catalyst	a catalyst that increases the rate of a reaction by removing a proton.
catalyst	a substance that increases the rate of a reaction without itself being changed or consumed in the overall reaction.
covalent catalysis (nucleophilic catalysis)	catalysis that occurs as a result of a nucleophile forming a covalent bond with one of the reactants.
effective molarity	the concentration of the reagent that would be required in an intermolecular reaction for it to have the same rate as an intramolecular reaction.
electrophilic catalyst	an electrophile that facilitates a reaction.
electrostatic catalysis	the stabilization of a charge by an opposite charge.
enzyme	a protein that is a catalyst.
general-acid catalysis	catalysis in which a proton is transferred to the reactant during the slow step of the reaction.
general-base catalysis	catalysis in which a proton is removed from the reactant during the slow step of the reaction.
induced fit model	a model that describes the specificity of an enzyme for its substrate: the shape of the active site does not become completely complementary to the shape of the substrate until after the enzyme has bound the substrate.
intramolecular catalysis	catalysis in which the catalyst that facilitates the reaction is part of the molecule undergoing reaction.
lock-and-key model	a model that describes the specificity of an enzyme for its substrate: the substrate fits the enzyme like a key fits into a lock.
metal-ion catalysis	catalysis in which the species that facilitates the reaction is a metal ion.

molecular recognition the recognition of one molecule by another as a result of specific interactions (for example, the specificity of an enzyme for its substrate).

nucleophilic catalysis (covalent catalysis) catalysis that occurs as a result of a nucleophile forming a covalent bond with one of the reactants.

nucleophilic catalyst a catalyst that increases the rate of a reaction by acting as a nucleophile.

pH-activity profile or pH-rate profile a plot of the activity of an enzyme as a function of the pH of the reaction mixture.

relative rate the relative rate is obtained by dividing the actual rate constant by the rate constant of the slowest reaction in the group being compared.

site-specific mutagenesis a technique that substitutes one amino acid of a protein for another.

specific-acid catalysis catalysis in which the proton is fully transferred to the reactant before the slow step of the reaction.

specific-base catalysis catalysis in which the proton is completely removed from the reactant before the slow step of the reaction.

substrate the reactant of an enzyme-catalyzed reaction.

Solutions to Problems

1. A catalyst increases the rate of a reaction by decreasing the difference in energy between the reactant and the transition state of the rate-limiting step. That is, it decreases the height of the energy hill of the rate-limiting step.

The following parameters would be different for a reaction carried out in the presence of a catalyst: $\Delta H^{\ddagger}$, E_a, $\Delta S^{\ddagger}$, $\Delta G^{\ddagger}$, k_{rate}. These are the parameters that reflect the difference in energy between the reactant and the transition state.

The other factors do not change because they reflect the difference in energy between the reactant and product, which is not affected by catalysis.

2. Notice that **(1)** and **(2)** have only the first part of the mechanism for the acid-catalyzed hydrolysis of an ester (because the final product of the reaction is a tetrahedral compound), **(3)** has only the second part (because the initial reactant is a tetrahedral compound), and **(4)** has both the first and second parts.

Mechanism for acid-catalyzed ester hydrolysis: first part

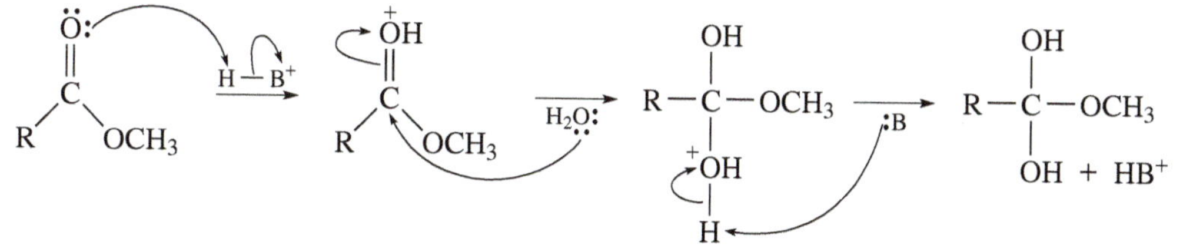

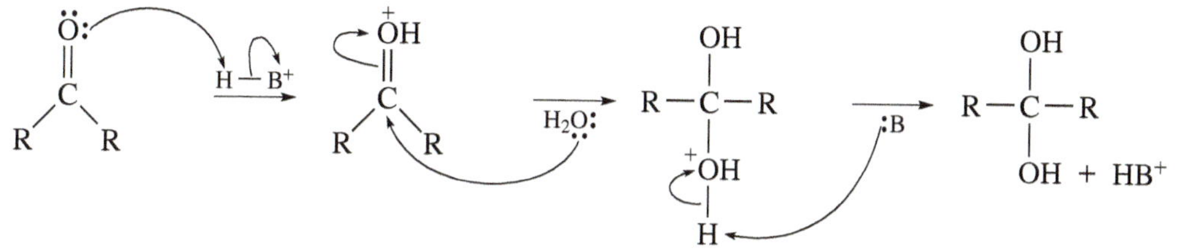

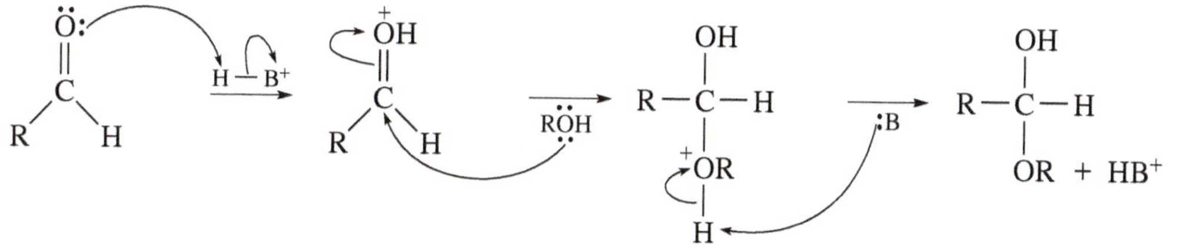

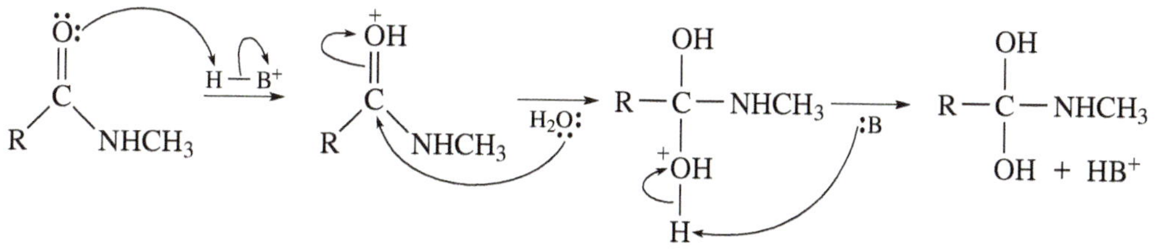

a. **Similarities:** the first step is protonation of the carbonyl compound, the second step is addition of a nucleophile to the protonated carbonyl compound, and the third step is loss of a proton.

b. **Differences:** the carbonyl compound that is used as the starting material; the nucleophile used in **(2)** is an alcohol rather than water.

Mechanism for acid-catalyzed ester hydrolysis: second part

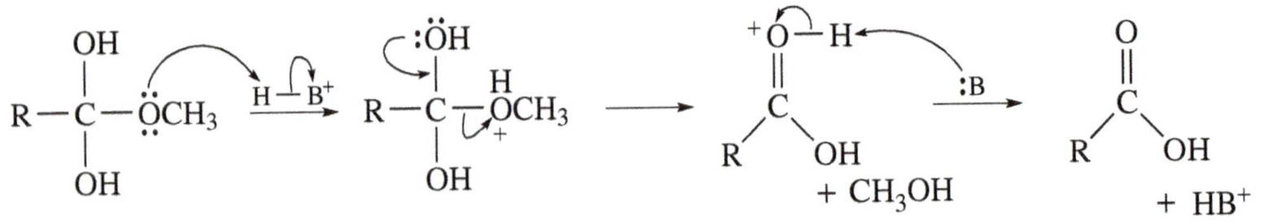

(3)

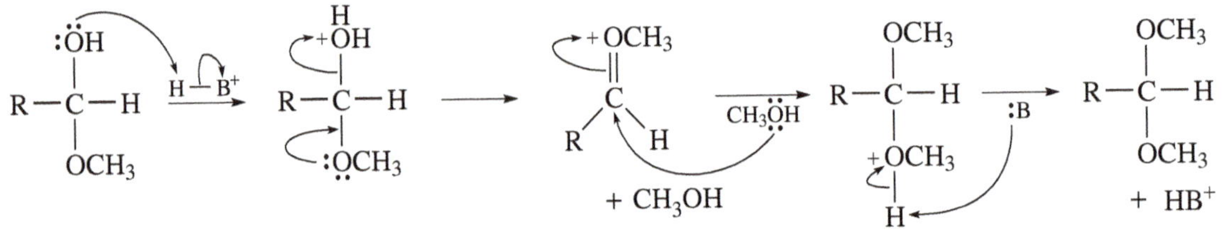

(4)

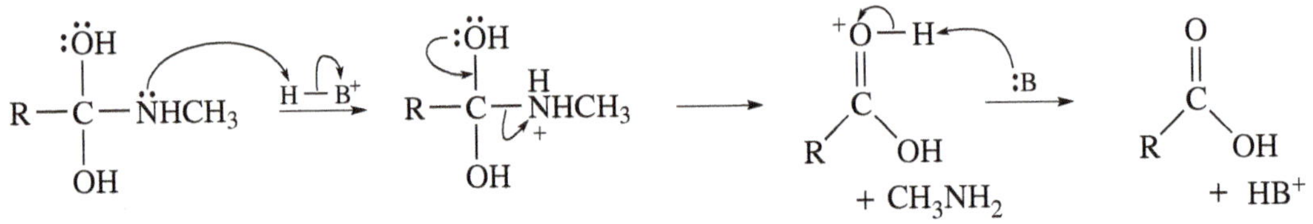

a. **Similarities:** the first step is protonation of the tetrahedral intermediate, and the second step is elimination of a group from the tetrahedral intermediate. In two of the three reactions, the third step is loss of a proton.

b. **Differences:** the group that is protonated—O in the first two and N in the third—and the group that is eliminated from the tetrahedral intermediate. The third step in acetal formation is not loss of a proton because the intermediate does not have a proton to lose. Instead, the third step is addition of a nucleophile, and the fourth step is loss of a proton.

3. Both slow steps are specific-acid catalyzed; the compound is protonated before the slow step in each case.

4. **a.** In specific-acid catalysis, the proton is added to the reactant before the O—C bond forms.

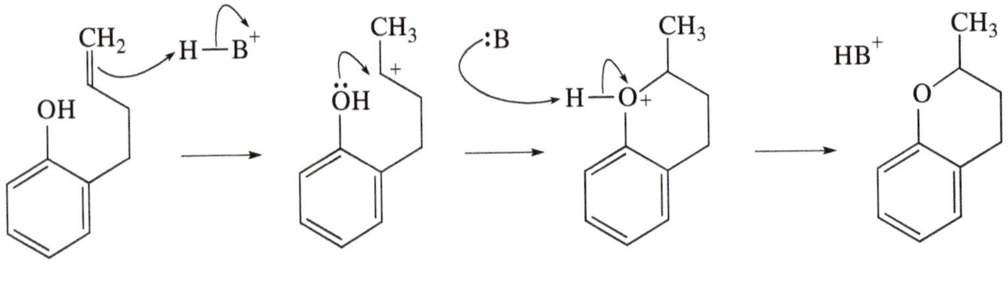

b. In general-acid catalysis, the proton is added to the reactant and the O—C bond forms in the same step.

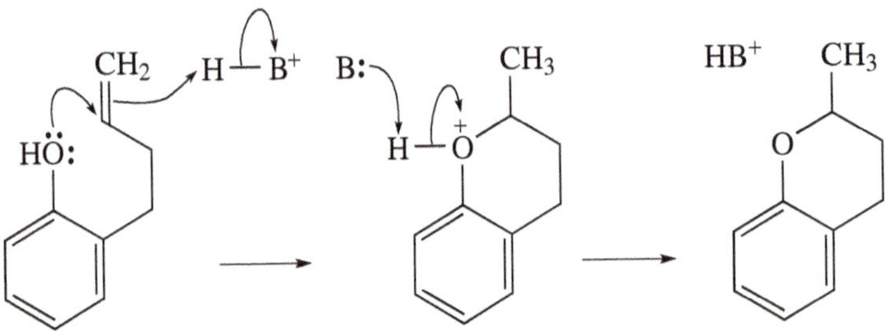

5. Solved in the text.

6. **a.** In specific-base catalysis, the proton is removed from the reactant before the O—C bond forms.

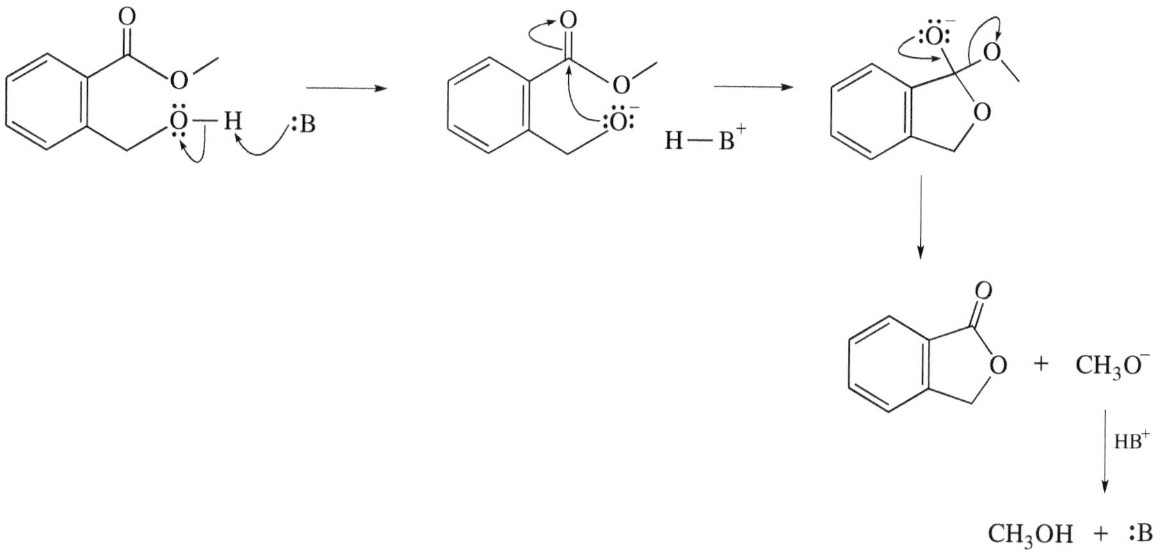

$CH_3O^- \xrightarrow{HB^+} CH_3OH + :B$

b. In general-base catalysis, the proton is removed from the reactant and the O—C bond forms in the same step.

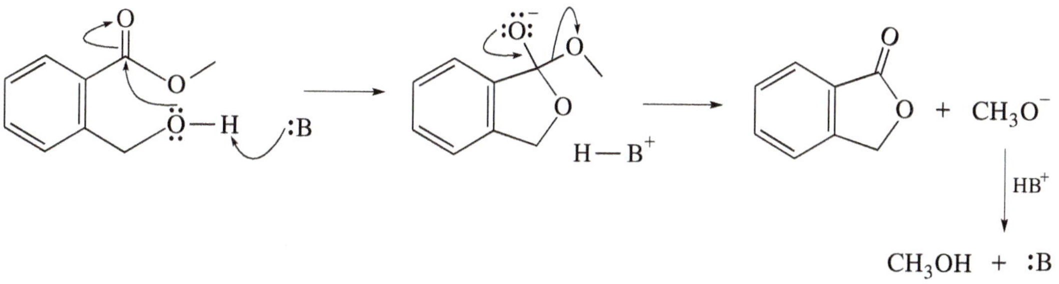

$CH_3O^- \xrightarrow{HB^+} CH_3OH + :B$

7. The metal ion catalyzes the decarboxylation reaction by complexing with the negatively charged oxygen of the carboxylate group and carbonyl oxygen of the β-keto group, thereby making it easier for the carbonyl oxygen to accept the electrons that are left behind when CO_2 is eliminated.

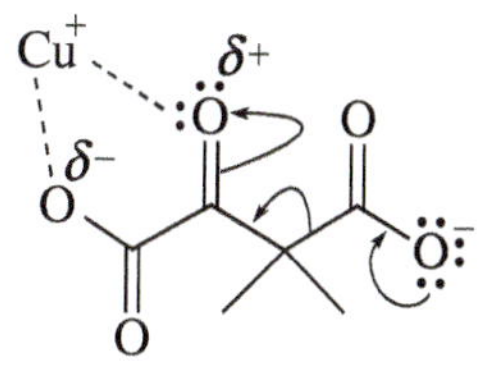

Because acetoacetate and the monoethyl ester of dimethyloxaloacetate do not have a negatively charged oxygen on one carbon and a carbonyl group on an adjacent carbon with which to form a complex, a metal ion does not catalyze decarboxylation of these compounds.

8. Co^{2+} can catalyze the reaction in three different ways. It can complex with the reactant, increasing the susceptibility of the carbonyl group to nucleophilic addition. It can also complex with water, increasing the tendency of water to lose a proton, resulting in a better nucleophile for hydrolysis. And it can complex with the leaving group, decreasing its basicity and thereby making it a better leaving group.

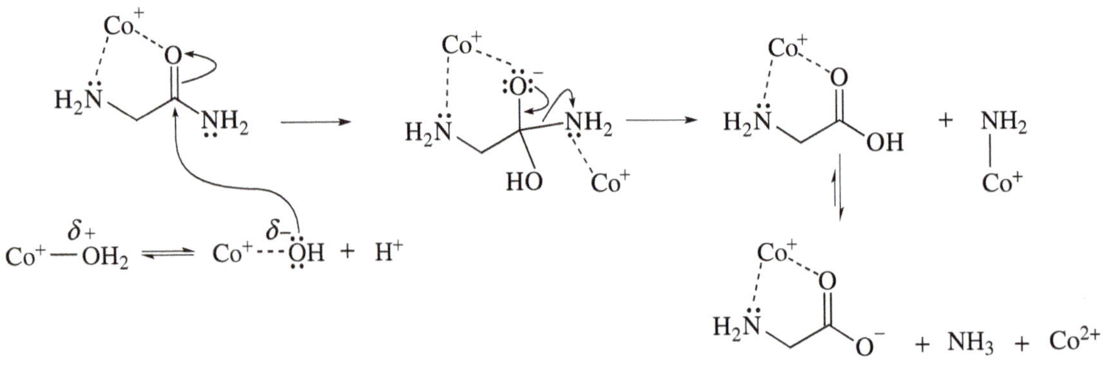

9. Because the reacting groups in the trans isomer are pointed in opposite directions, they cannot react in an intramolecular reaction. Because they can react only via an intermolecular pathway, they will have approximately the same rate of reaction as they would have if the reacting groups were in separate molecules. Consequently, the relative rate would be expected to be close to one.

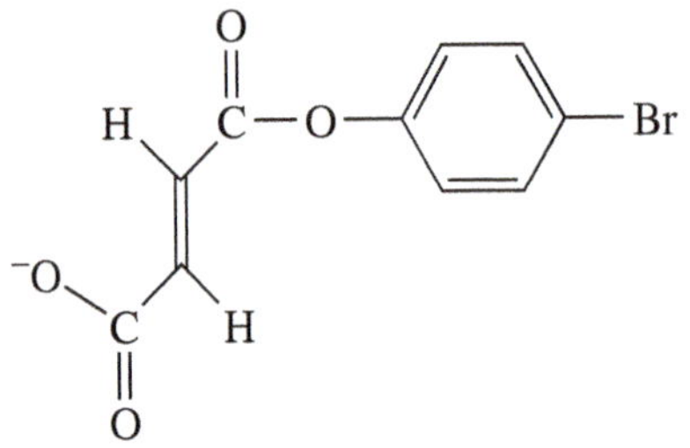

10. The nucleophile can attack the back side of either of the two ring carbons to which the sulfur is bonded in the intermediate, thereby forming two trans products. There are two nucleophiles (water and ethanol), so a total of four products will be formed.

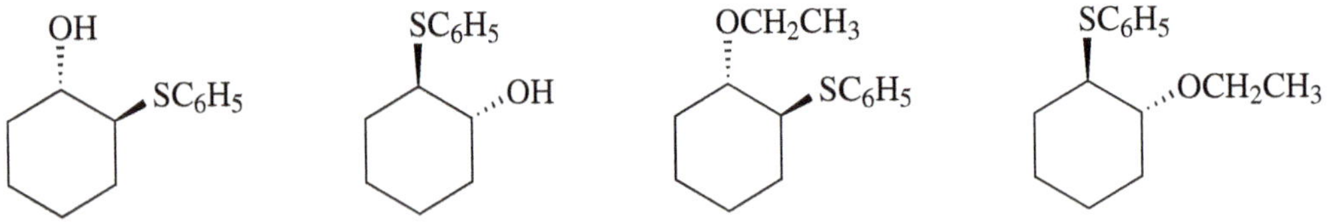

11. Solved in the text.

12. The tetrahedral intermediate has two leaving groups, a carboxylate ion and a phenolate ion. When there are no nitro substituents on the benzene ring, the carboxylate ion is a weaker base (a better leaving group) than the phenolate ion, so the tetrahedral intermediate reforms the ester. Thus, the reaction proceeds through general-base catalyzed hydrolysis of the ester.

In contrast, when there are nitro groups on the benzene ring, the 2,4-dinitrophenolate ion is a weaker base (a better leaving group) than the carboxylate ion, so the tetrahedral intermediate forms the anhydride. Thus, the reaction proceeds through hydrolysis of the anhydride formed by nucleophilic catalysis.

13. If the *ortho*-carboxyl substituent acts as an intramolecular base catalyst, ^{18}O will be incorporated into acetic acid and not into salicylic acid.

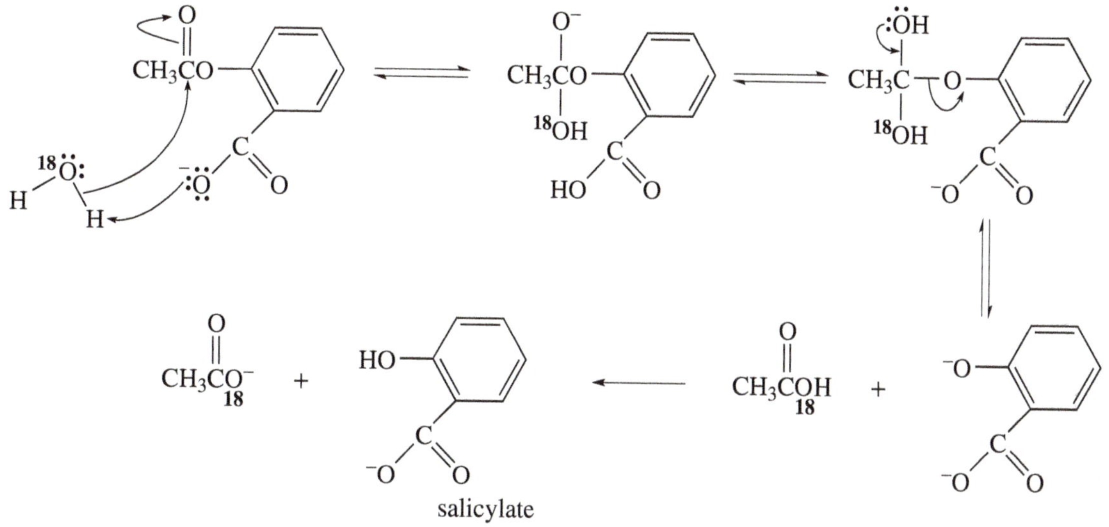

O
CH3CO—
^{18}Ö:
H H
:Ö: C O
⇌
O⁻
CH3CO—
^{18}OH
C
HO O
⇌
:ÖH
CH3C—O—
^{18}OH
C
⁻O O
⇌
O
CH3CO⁻
18
+
HO—
C
⁻O O
salicylate
⟵
O
CH3COH
18
+
⁻O—
C
⁻O O

If the *ortho*-carboxyl substituent acts as an intramolecular nucleophilic catalyst, ^{18}O would be incorporated into both salicylic acid (if water adds to the carbonyl group attached to the benzene ring) and acetic acid (if water adds to the carbonyl group attached to the methyl group).

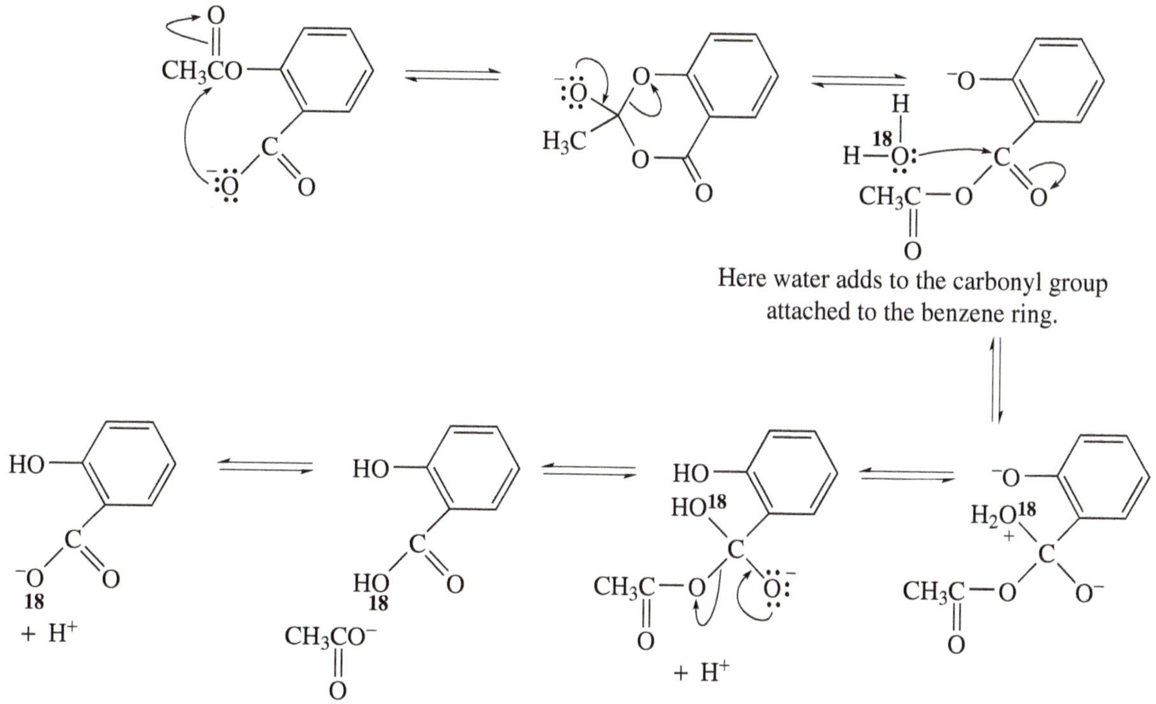

O
CH3CO—
C
:Ö: O
⇌
:Ö:
O—
H3C
O—
O
⇌
⁻O—
H
18
H—Ö:
C
CH3C—O O
O
Here water adds to the carbonyl group attached to the benzene ring.
⇌
HO—
C
⁻O O
18
+ H⁺
⇌
HO—
C
HO O
18
CH3CO⁻
O
⇌
HO—
HO18
C
CH3C—O :Ö:⁻
O
+ H⁺
⇌
⁻O—
H2O18
+
C
CH3C—O O⁻
O

Not all of the salicylic acid would contain ^{18}O, because the anhydride intermediate has two different carbonyl groups. If water adds to the benzyl carbonyl group (above), salicylic acid will contain ^{18}O. If, however, water adds to the acetyl carbonyl group (below), acetic acid will contain ^{18}O.

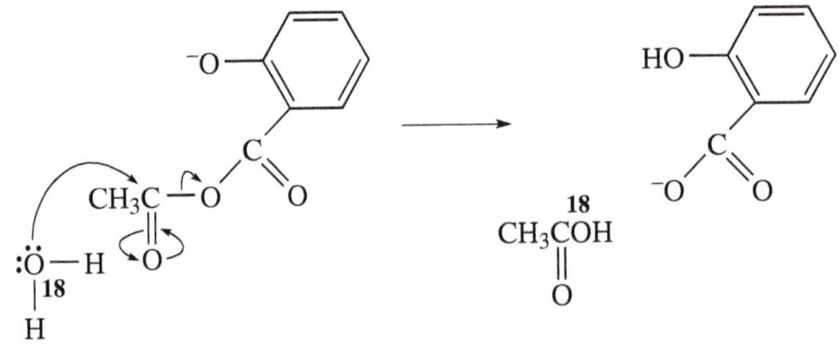

Here water adds to the carbonyl group attached to the methyl group.

14. Solved in the text.

15. **2**, **3**, and **4** are bases, so they can help remove a proton.

16. Ser-Ala-Leu is more readily cleaved by carboxypeptidase A, because the nonpolar isobutyl substituent of phenylalanine is more attracted to the hydrophobic pocket of the enzyme than is the negatively charged substituent of aspartate.

17. Glu 270 adds to the carbonyl group of the ester, forming a tetrahedral intermediate. Collapse of the tetrahedral intermediate is most likely catalyzed by an acidic group of the enzyme, which increases the leaving ability of the RO group. (Perhaps the HO substituent of tyrosine is close enough in the esterase to be the catalyst.) The group that donates the proton can then act as a base catalyst to remove a proton from water as it hydrolyzes the anhydride.

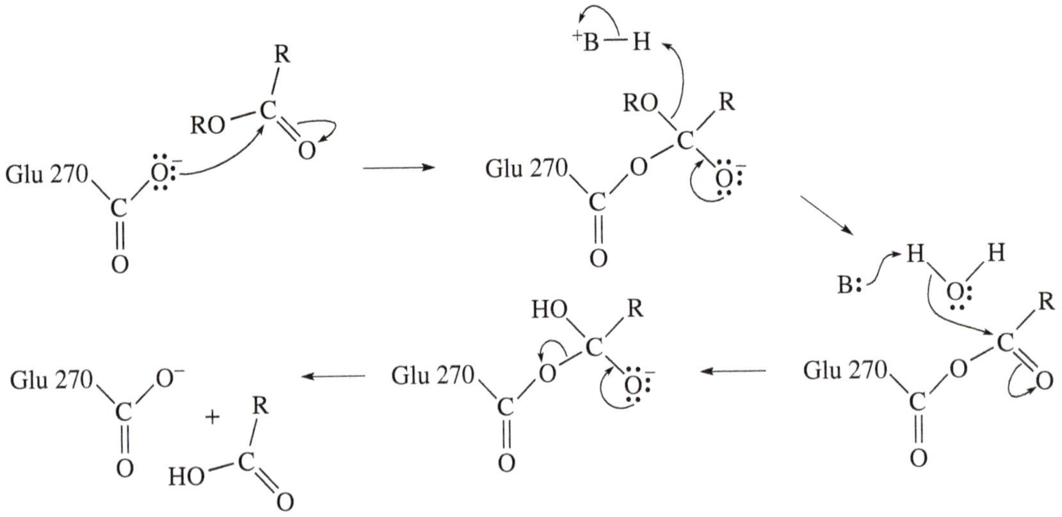

18. Because arginine extends farther into the binding pocket, it must be the one that forms direct hydrogen bonds. Lysine, which is shorter, needs the mediation of a water molecule in order to engage in hydrogen bond formation with aspartate.

arginine | lysine

end of arginine | end of lysine

19. The side chains of D-Arg and D-Lys are not positioned to bind correctly at the active site. They would, however, be able to bind at a mirror image of the active site.

20. NAM will contain ^{18}O because it is the ring that undergoes nucleophilic attack by $H_2^{18}O$.

NAM | NAG

21. **a.** Because the catalytic group is an acid catalyst, it will be active in its acidic form and inactive in its basic form. The pH at the midpoint of the curve corresponds to the pK_a of the catalytic group because pH = pK_a when $[HA] = [A^-]$. (See page 70 of the text.)

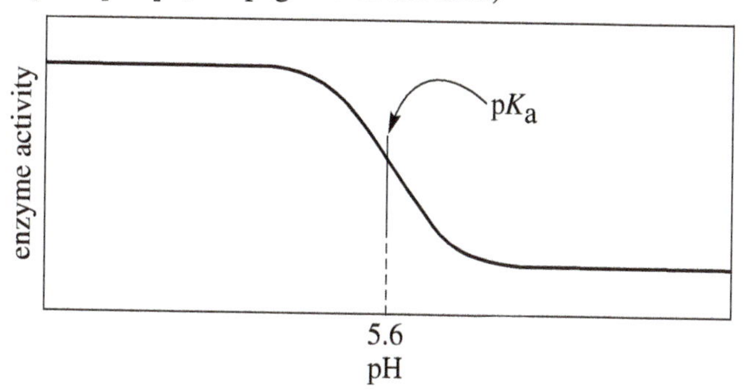

b. Because the catalytic group is a base catalyst, it will be inactive in its acidic form and active in its basic form.

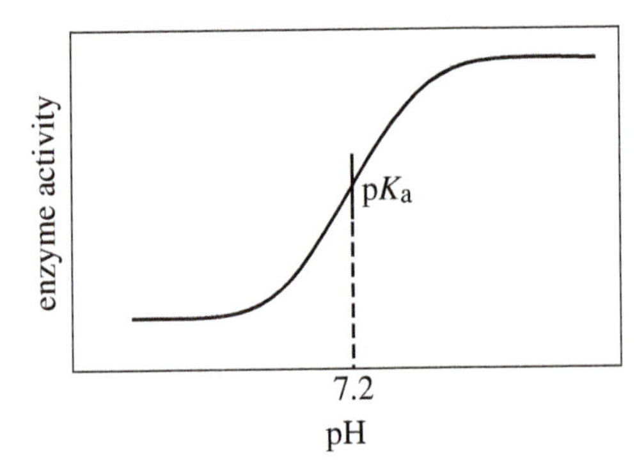

22. The ascending leg of the pH-rate profile is due to a group that is a general-base catalyst because the rate is at a maximum when the group is in its basic form. From the description of the mechanism, we know that amino acid is histidine.

The descending leg of the pH-rate profile is due to a group that is a general-acid catalyst because the rate is at a maximum when the group is in its acidic form. From the description of the mechanism, we know that amino acid is lysine.

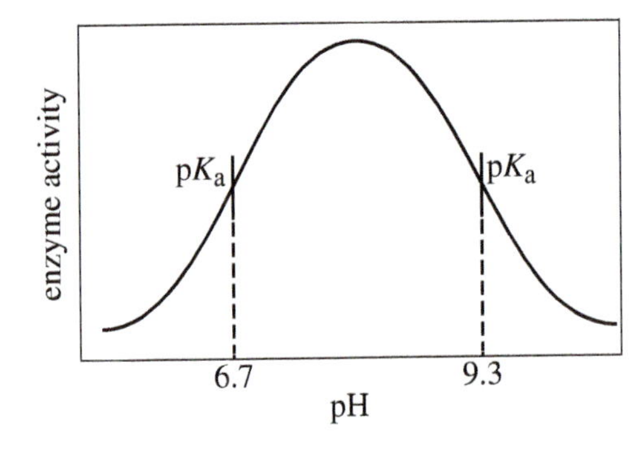

23. In the absence of an enzyme, D-fructose is in equilibrium with D-glucose and D-mannose as a result of an enediol rearrangement (Section 20.5). Both C-2 epimers are formed, because a new asymmetric center is formed at C-2 and it can have either the *R* or the *S* configuration. Enzyme-catalyzed reactions are typically highly stereoselective—the enzyme catalyzes the formation of a single stereoisomer. Therefore, D-fructose is in equilibrium with only D-glucose in the presence of the enzyme that catalyzes the enediol rearrangement.

24.

$CH_2OPO_3^{2-}$, C=O, HO—C—H, H—C—O—H :ÖH⁻, H—C—OH, $CH_2OPO_3^{2-}$ ⟶ $CH_2OPO_3^{2-}$, C—Ö:⁻, HO—C—H, H—O—H + H, C=O, H—C—OH, $CH_2OPO_3^{2-}$ ⟶ $CH_2OPO_3^{2-}$, C=O, CH_2OH + HO^-

25. The positively charged nitrogen of the protonated imine can accept the electrons that are left behind when the C3—C4 bond breaks.

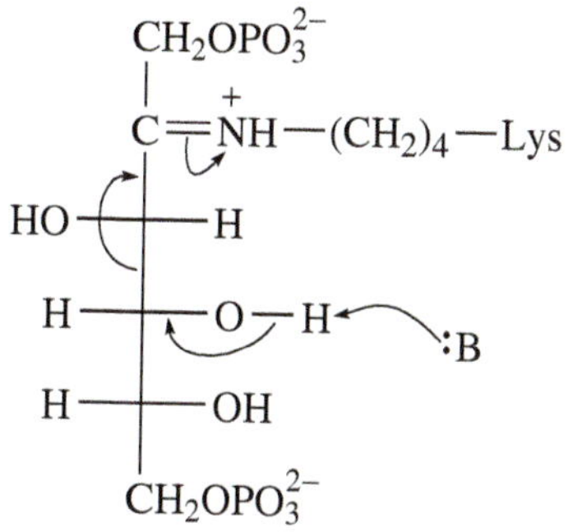

$CH_2OPO_3^{2-}$ / $C{=}\overset{+}{N}H{-}(CH_2)_4{-}Lys$ / HO—H / H—O—H ←:B / H—OH / $CH_2OPO_3^{2-}$

In the absence of the protonated imine, the electrons would be delocalized onto a neutral oxygen. (See Problem 24.) The neutral oxygen is not as electron withdrawing as the positively charged nitrogen of the protonated imine. In other words, imine formation makes it easier to break the C3—C4 bond.

26. Because **2** is a primary amine, it can form an imine. Notice that **1** cannot form an imine because the lone pair on the NH_2 group is delocalized onto the oxygen, so this NH_2 group is not a nucleophile. The N in **3** is not nucleophilic because its lone pair is delocalized (the lone pair is part of indole's π cloud).

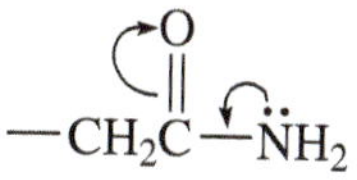

$-CH_2C(=O)-\ddot{N}H_2$

27. In order to break the C3—C4 bond, the carbonyl group has to be at the 2-position as it is in fructose so that it can accept the electrons that result from C3—C4 bond cleavage; the carbonyl group at the 1-position in glucose cannot accept these electrons. Therefore, glucose must isomerize to a ketose before the reaction can occur.

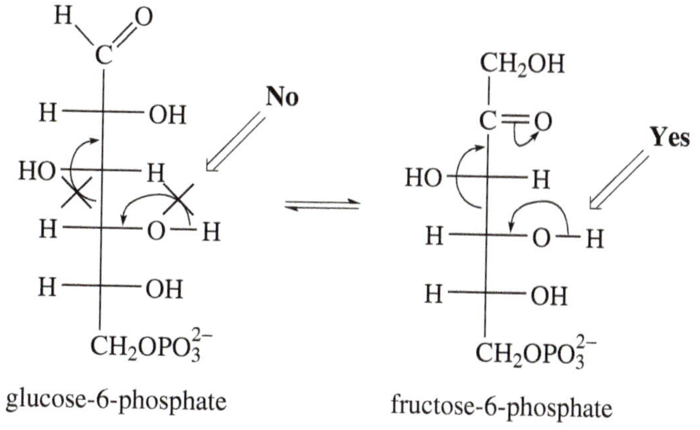

glucose-6-phosphate (No) ⇌ fructose-6-phosphate (Yes)

28. Cysteine residues react with iodoacetic acid because a thiol is a good nucleophile and iodine is a good leaving group. If a cysteine residue is at the active site of an enzyme, adding a substituent to the sulfur in this way could interfere with the enzyme's being able to bind the substrate or it could interfere with positioning the tyrosine residue that catalyzes the reaction. Adding a substituent to cysteine might also cause a conformational change in the enzyme that could destroy its activity.

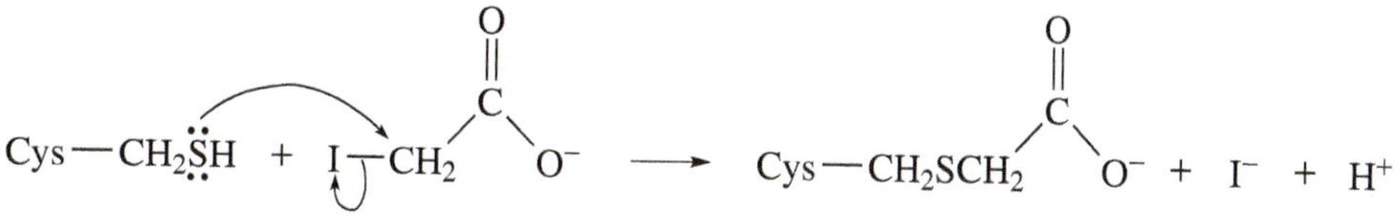

$$Cys{-}CH_2\ddot{S}H + I{-}CH_2C(=O)O^- \longrightarrow Cys{-}CH_2SCH_2C(=O)O^- + I^- + H^+$$

29. The following compound eliminates HBr more rapidly because the negatively charged oxygen is in position to act as an intramolecular general-base catalyst.

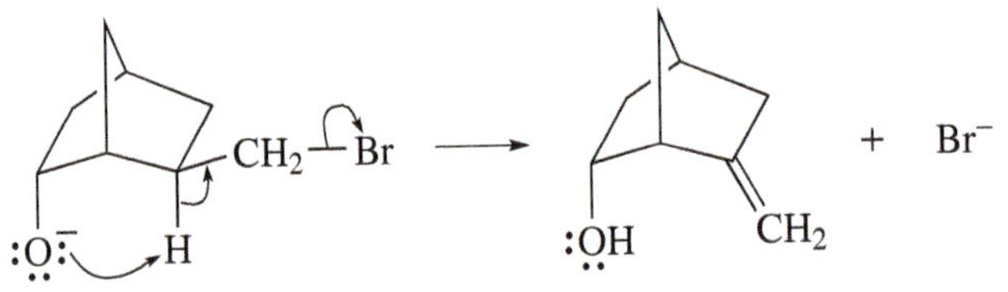

30. The following compound will form an anhydride more rapidly because it forms a five-membered-ring anhydride, which is less strained than the seven-membered-ring anhydride formed by the other compound. The greater stability of the five-membered-ring product causes the transition state leading to its formation to be more stable than the transition state leading to formation of the seven-membered-ring product.

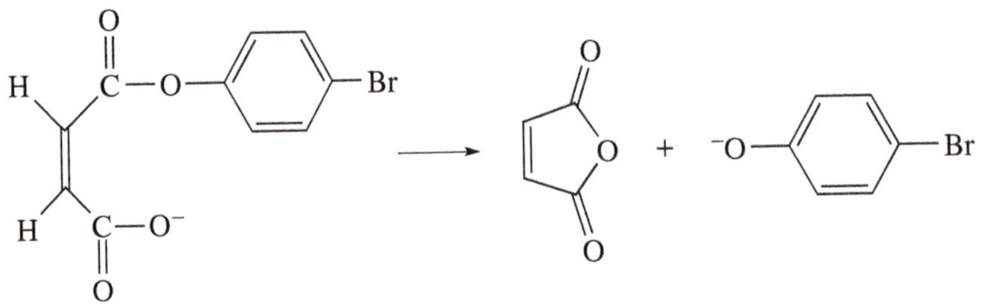

31.

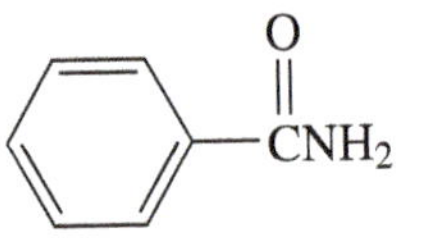

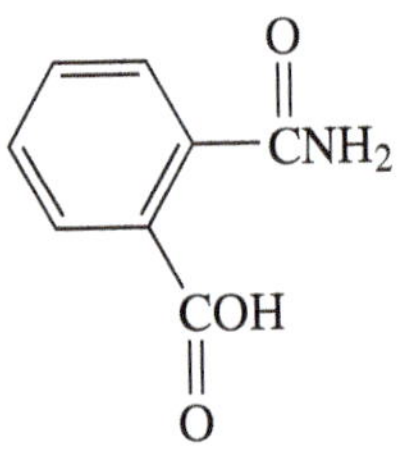

ortho-carboxybenzamide

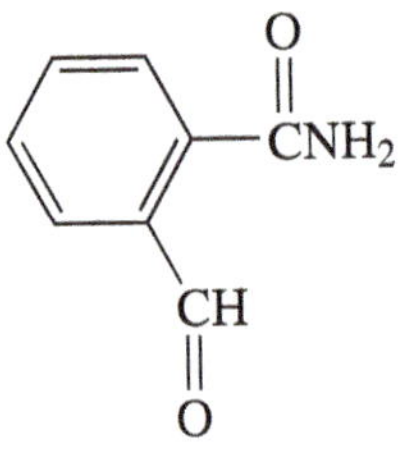

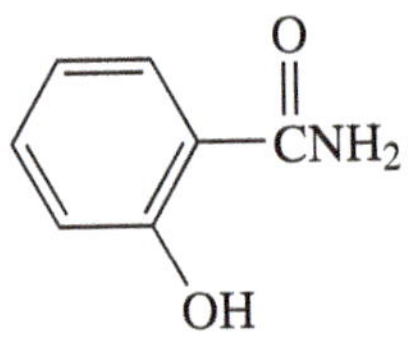

ortho-hydroxybenzamide

In order to hydrolyze an amide, the NH_2 group in the tetrahedral intermediate has to leave in preference to the less basic OH group. This can happen if the NH_2 group is protonated because $^+NH_3$ is a weaker base and, therefore, easier to eliminate than OH. Of the four compounds, two (*ortho*-carboxybenzamide and *ortho*-hydroxybenzamide) have substituents that can protonate the NH_2 by acting as acid catalysts in a hydrolysis reaction carried out at pH = 3.5.

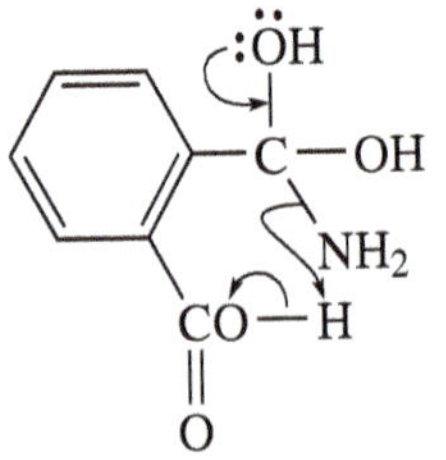

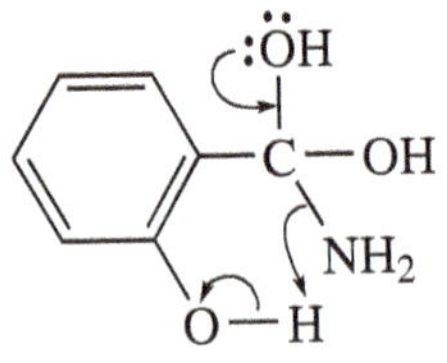

Because the carboxy group withdraws electrons from the ring by resonance and the OH group donates electrons to the ring by resonance, formation of the tetrahedral intermediate will be faster for the carboxy-substituted compound.

The carboxy group is a stronger acid than the OH group (and with a $pK_a = 4.2$, 83% will be in the acidic form at pH = 3.5). Therefore, a larger fraction of the tetrahedral intermediate will be protonated than in the case of the hydroxy-substituted compound, so collapse of the tetrahedral intermediate will be faster. Therefore, the *ortho*-carboxybenzamide has the faster rate of hydrolysis.

32. **a.**  $CH_3CH_2\ddot{S}CH_2CH_2-Cl$ intramolecular nucleophilic catalysis

b. 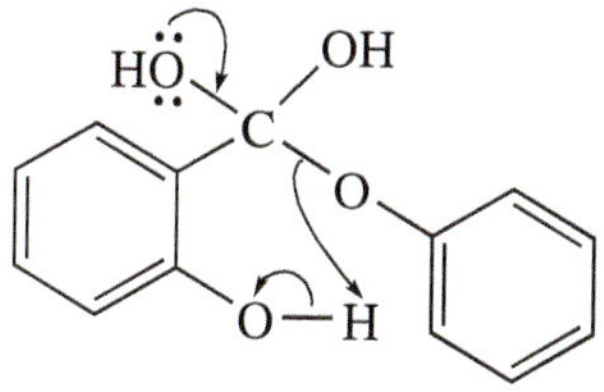 intramolecular general-acid catalysis
The OH substituent is protonating the leaving group as it departs, causing it to be a weaker base and, therefore, a better leaving group.

33. If the *ortho*-carboxylate ion is acting as a general-base catalyst, the kinetic isotope effect will be greater than 1.0 (see Problem 91 in Chapter 9), because an O—H (or O—D) bond is broken in the slow step of the reaction and an O—D bond is stronger than an O—H bond and, therefore, is harder to break.

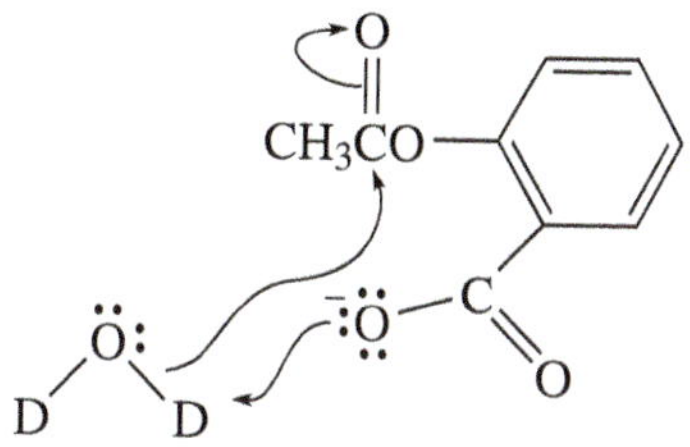

If the *ortho*-carboxylate ion is acting as a nucleophilic catalyst, the kinetic isotope effect will be about 1.0, because an O—H (or O—D) bond is not broken in the slow step of the reaction.

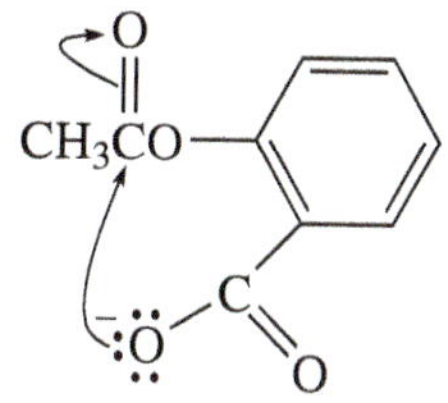

34.

$$\frac{\text{rate constant for the catalyzed reaction}}{\text{rate constant for the uncatalyzed reaction}} \quad \frac{1.5 \times 10^6\ M^{-1}\,s^{-1}}{0.6\ M^{-1}\,s^{-1}} = 2.5 \times 10^6$$

35. Co^{2+} can catalyze the hydrolysis reaction by complexing with three nitrogen atoms in the substrate as well as with water. Complexation increases the acidity of water, thereby providing a better nucleophile (metal-bound hydroxide) for the hydrolysis reaction. Complexation with three nitrogens locks the nucleophile into the correct position for addition to the carbonyl carbon.

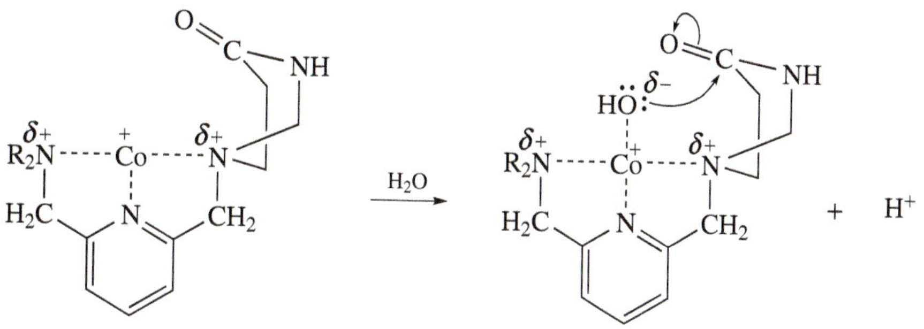

$\xrightarrow{H_2O}$ + H^+

R_2N H_2N N O O^- H_2C N CH_2 ⟵ HO $\ddot{O}$: H +N R_2N---- Co^+--- N H_2C N CH_2

36. Class I aldolases use a protonated imine as a site to which electrons can be delocalized (see Problem 25). Class II aldolases use a metal ion complexed with the carbonyl group as a site to which electrons can be delocalized.

$CH_2OPO_3^{2-}$ C=O--Zn^{2+} ($\delta+$) HO—H H—O—H :Ö—(benzene ring)—CH_2— H—OH $CH_2OPO_3^{2-}$ ⟶ $CH_2OPO_3^{2-}$ C—Ö:—Zn^+ ($\delta-$) HO—C—H H C=O H—O—(benzene ring)—CH_2— H—OH $CH_2OPO_3^{2-}$

37. Knowing that the reaction is much slower if the N is replaced by a CH tells us that the N must be acting as a catalyst. We see that the N is in position to be an intramolecular nucleophilic catalyst.

CH_3—N: Cl Cl ⟶ CH_3—$\overset{+}{N}$ Cl ⟶ (HÖ:$^-$) CH_3—N: OH Cl ⟶ CH_3—$\overset{+}{N}$ OH ⟶ (HÖ:$^-$) CH_3—N: OH OH

38. In the absence of morpholine, the first step of the reaction is addition of water to the **ester** because addition of water to the aldehyde would form a hydrate that would be in equilibrium with the aldehyde.

Two mechanisms can account for the observed catalysis by morpholine. In the presence of morpholine, the first step of the reaction (in both mechanisms) is addition of morpholine to the **aldehyde**. The reaction of morpholine with the aldehyde is a faster reaction than the reaction of water with the ester because morpholine is a better nucleophile than water and an aldehyde is more susceptible to nucleophilic addition than an ester.

In one mechanism, the negatively charged aldehyde oxygen (which is a much better nucleophile than water) is the nucleophile that adds to the ester. The tetrahedral intermediate collapses to form a lactone. Iminium ion formation followed by imine hydrolysis forms the final product.

H
C=O
OCH3
C
O
HN O
N O
O
:O:
OCH3
:N O
O
O
+ CH3OH
O
C
H
O−
C
O
HN O
H2O
imine
hydrolysis
H
C=N+ O
O−
C
O

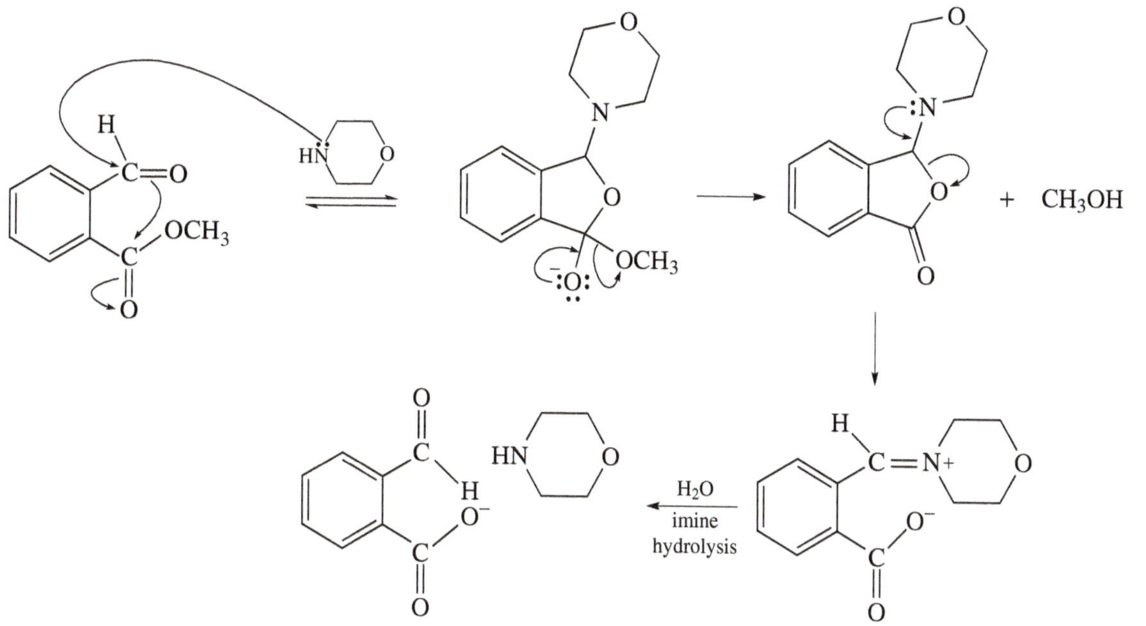

In the other mechanism, the reaction of morpholine with the aldehyde forms an iminium ion. The positively charged iminium ion makes it easier for water to add to the ester by stabilizing the negative charge that develops on the oxygen. Collapse of the tetrahedral intermediate followed by imine hydrolysis forms the final product.

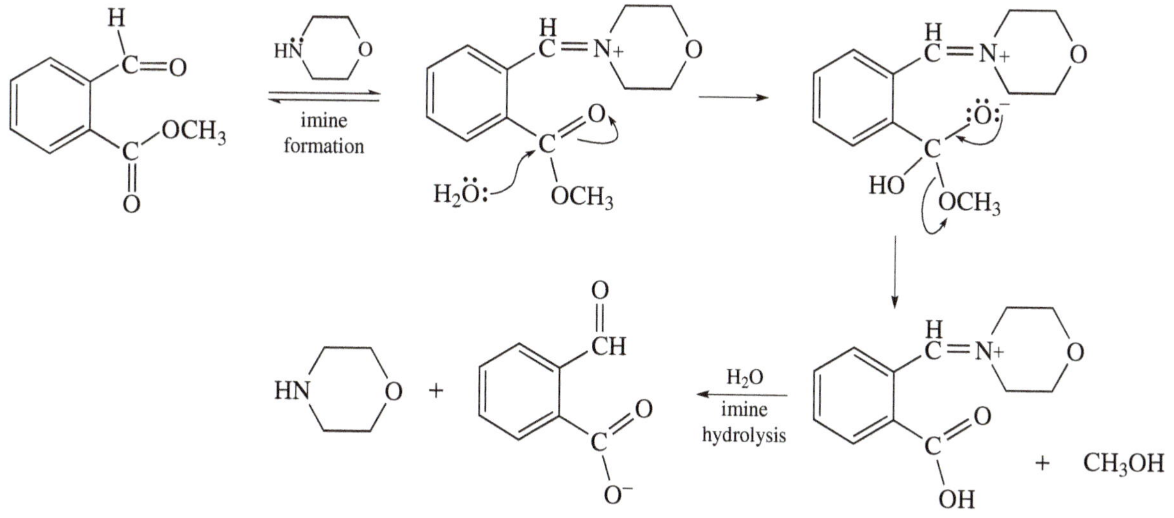

H
C=O
OCH3
C
O
HN O
imine
formation
H
C=N+ O
C=O
H2Ö:
OCH3
H
C=N+ O
:Ö:−
C
HO
OCH3
O
CH
HN O +
C=O
O−
H2O
imine
hydrolysis
H
C=N+ O
C=O
OH
+ CH3OH

39.

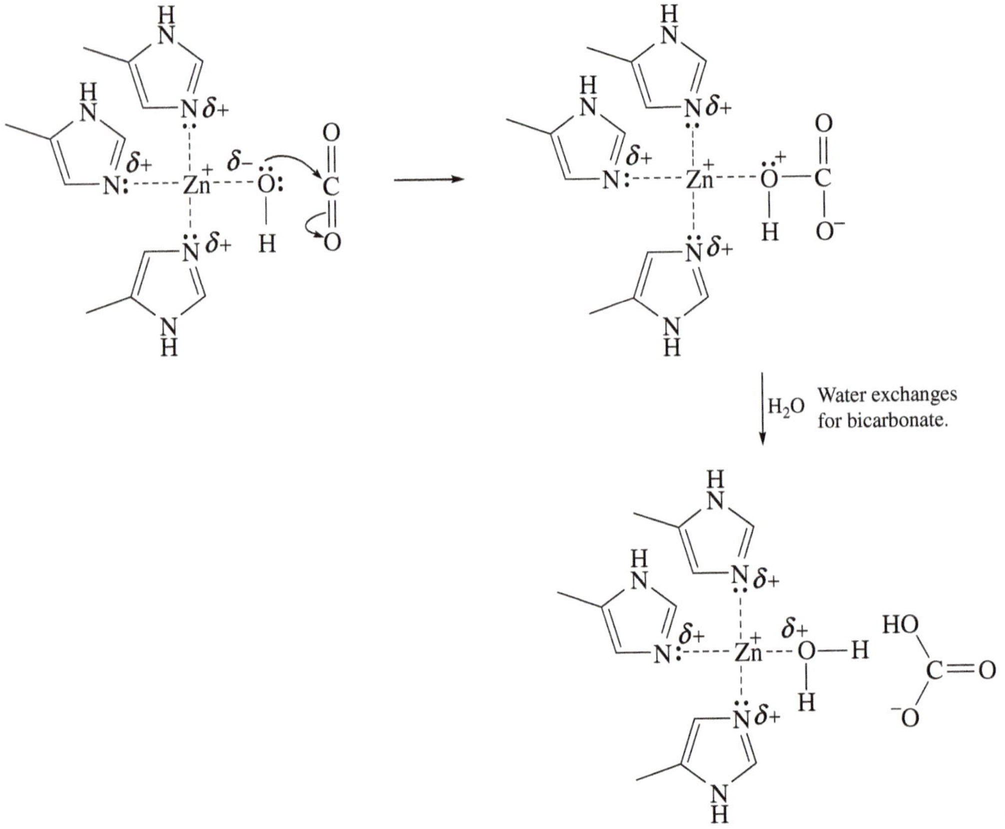

H₂O
Water exchanges for bicarbonate.

40. At pH = 12, the nucleophile is hydroxide ion. Addition of hydroxide ion to the carbonyl group is faster in **A** because the negative charge on the oxygen that is created in the tetrahedral intermediate is stabilized by the positively charged nitrogen.

A faster **B**

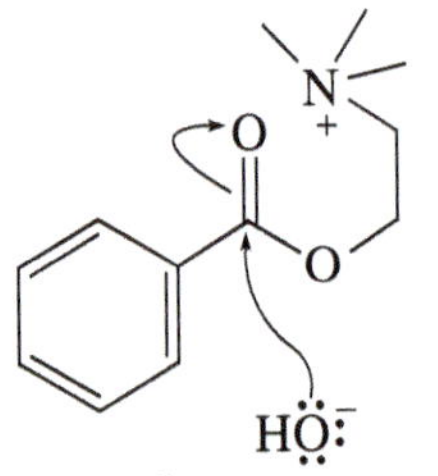

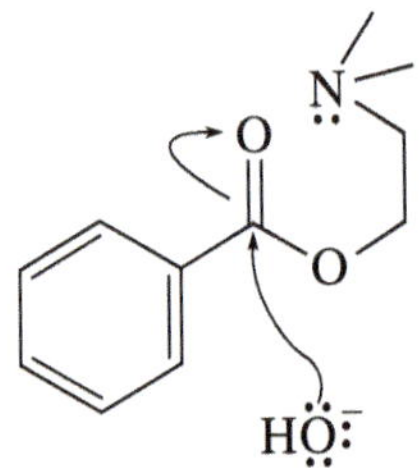

At pH = 8, the nucleophile is water because the concentration of hydroxide ion is only 1×10^{-6} M. Addition of water to the carbonyl group is faster in **B** because the amino group can act as a general-base catalyst to make water a better nucleophile.

A **B** 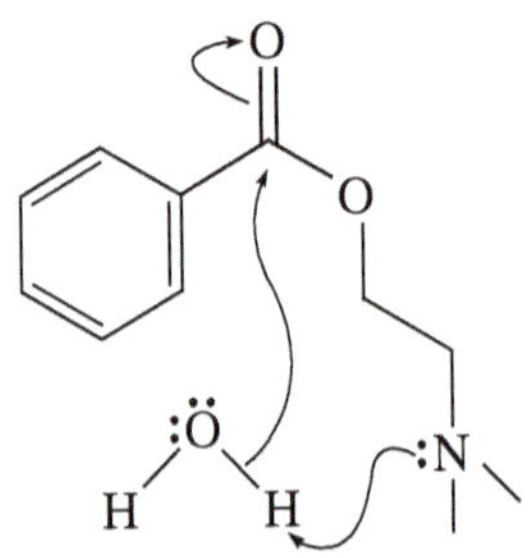 faster

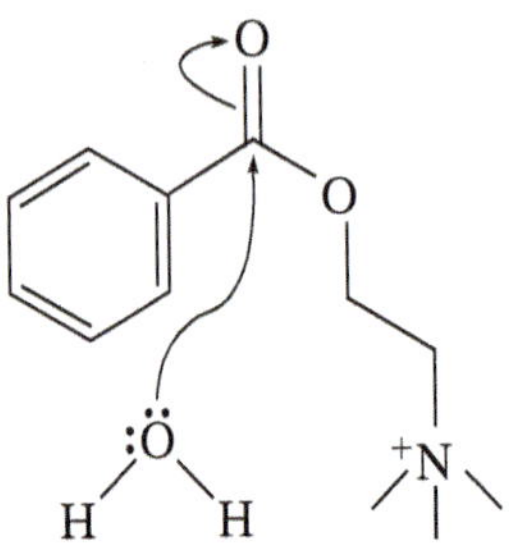

41. **a.** The cis reactants each undergo an S_N2 reaction. Because the acetate displaces the tosylate group by back-side attack, each cis reactant forms a trans product.

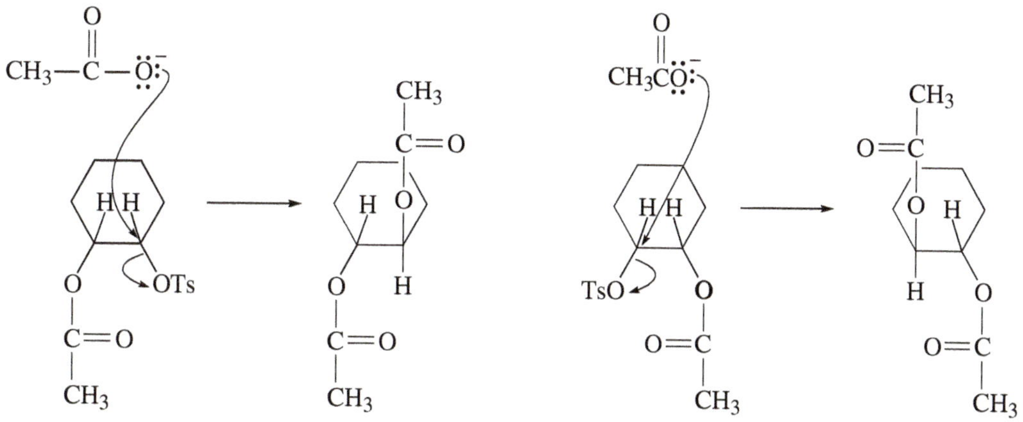

b. The acetate group in a trans reactant is positioned to be able to displace the tosylate leaving group by an intramolecular S_N2 reaction. Acetate ion then attacks in a second S_N2 reaction from the back side of the group it displaces, so trans products are formed. Because both trans reactants form the same intermediate, they both form the same products. Because the acetate ion can attack either of the carbons in the intermediate equally easily, a racemic mixture will be formed.

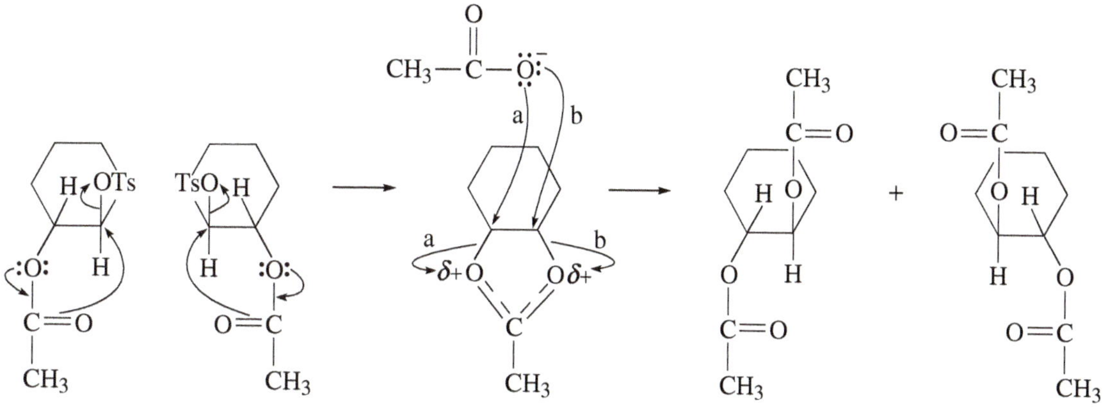

c. The trans reactants are more reactive because the tosylate leaving group is displaced in an intramolecular reaction, and the resulting positively charged cis intermediate is considerably more reactive than the neutral cis reactants. Thus, a trans isomer undergoes two successive S_N2 reactions, each of which is faster than the one step S_N2 reaction that a cis isomer undergoes.

42. Reduction of the imine linkage with sodium borohydride causes fructose to become permanently attached to the enzyme because the hydrolyzable imine bond has been lost. Acid-catalyzed hydrolysis removes the two phosphate groups and hydrolyzes the two peptide bonds, so the radioactive fragment that is isolated after hydrolysis is the lysine residue (covalently attached to fructose) that originally formed the imine.

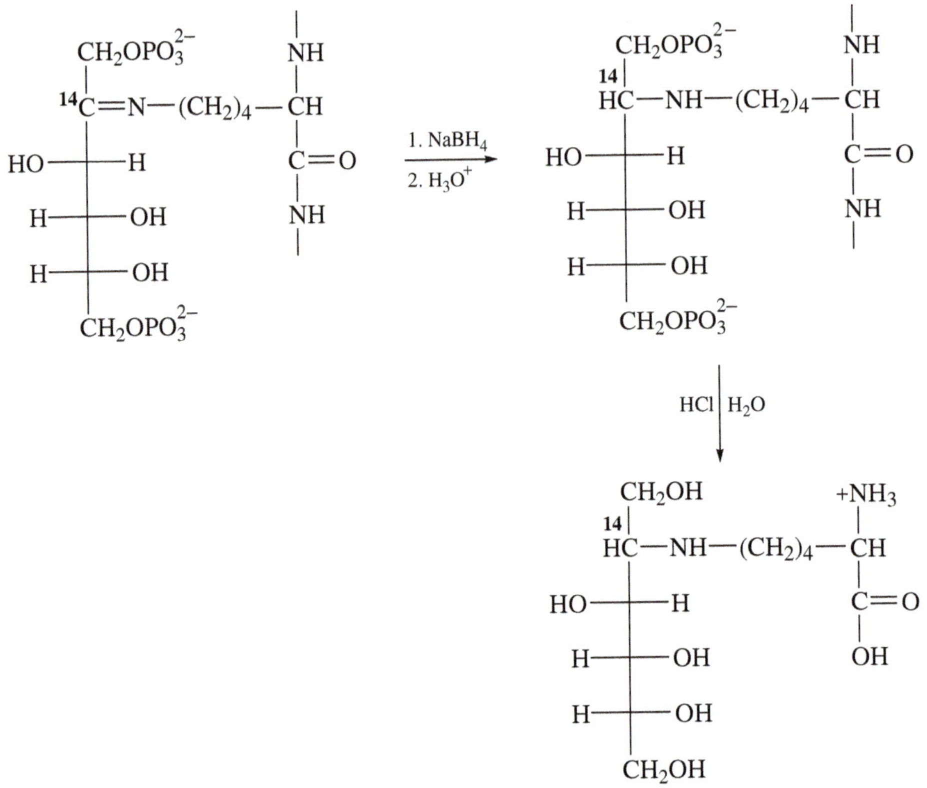

43. **a.** 3-Amino-2-oxindole catalyzes the decarboxylation of an α-keto acid by first forming an imine, which increases the acidity of the hydrogen on the α-carbon. Now the compound can be decarboxylated because the electrons left behind are delocalized onto an oxygen.

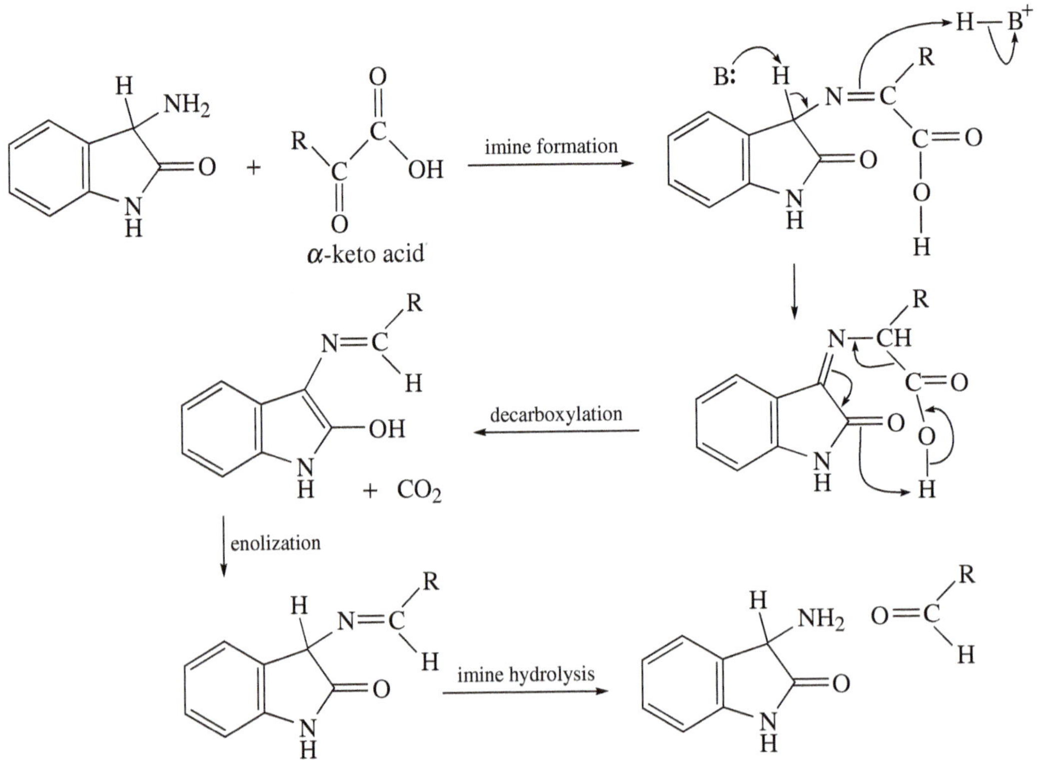

b. 3-Aminoindole would not be as effective a catalyst, because the electrons left behind when CO_2 is eliminated cannot be delocalized onto an electronegative atom.

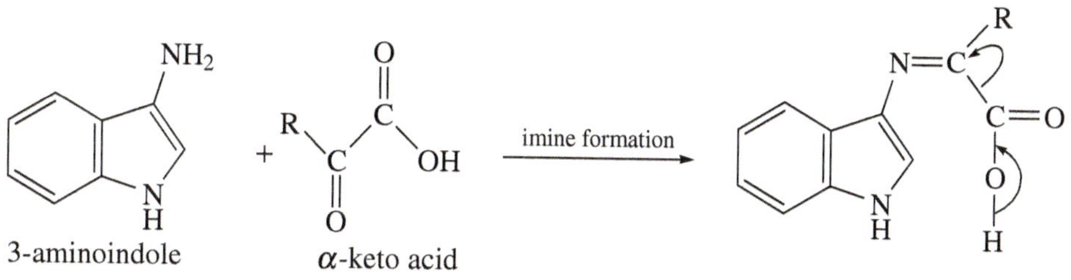

3-aminoindole + α-keto acid → (imine formation)

44. **a.** Intramolecular nucleophilic attack on an alkyl halide occurs more rapidly than intermolecular attack on an alkyl halide, because the reacting groups are tethered in the former, which makes it more likely that they will find each other to react. The intramolecular reaction is followed by another relatively rapid S_N2 reaction, because the strain in the three-membered ring causes it to break easily.

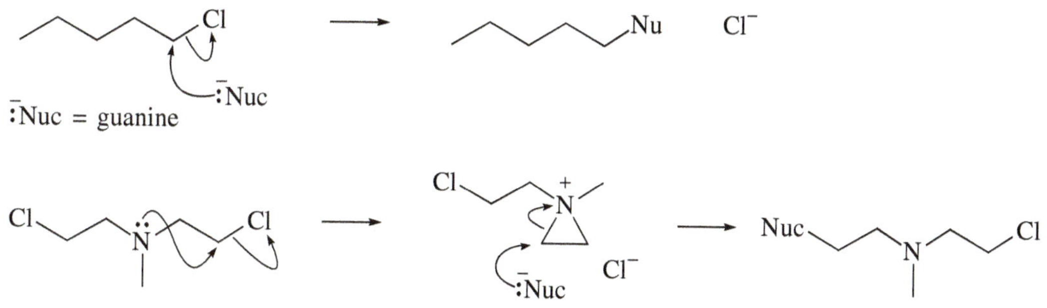

:Nuc⁻ = guanine

b.

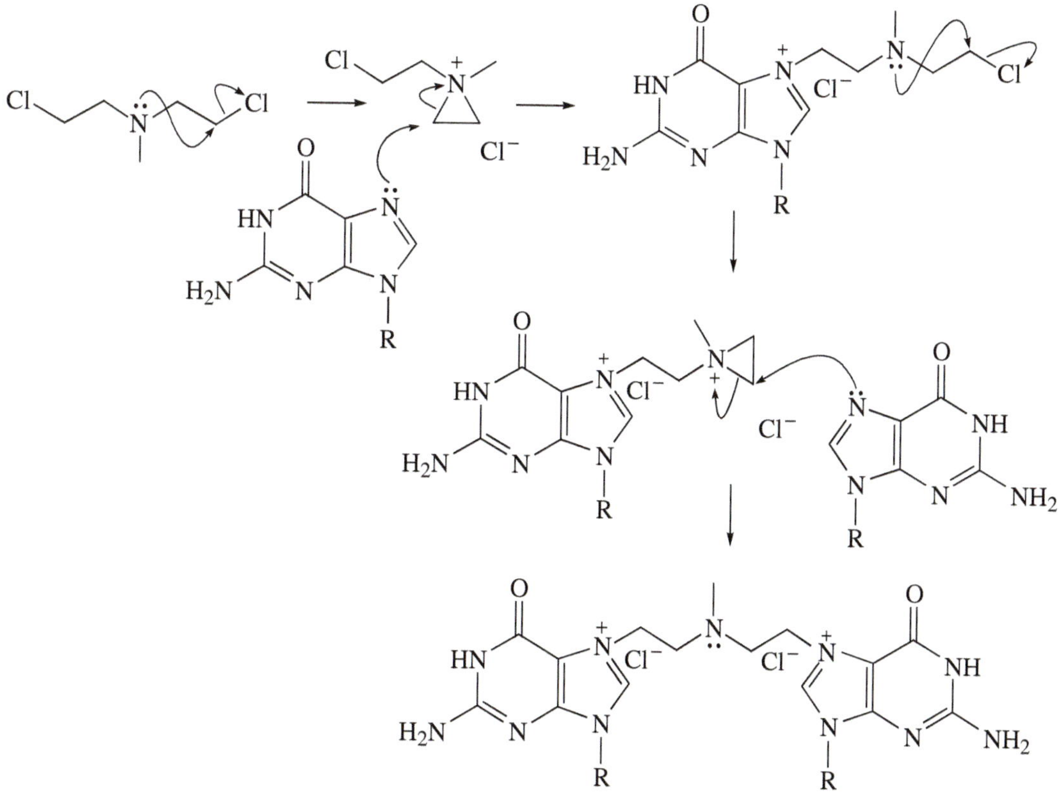

45.

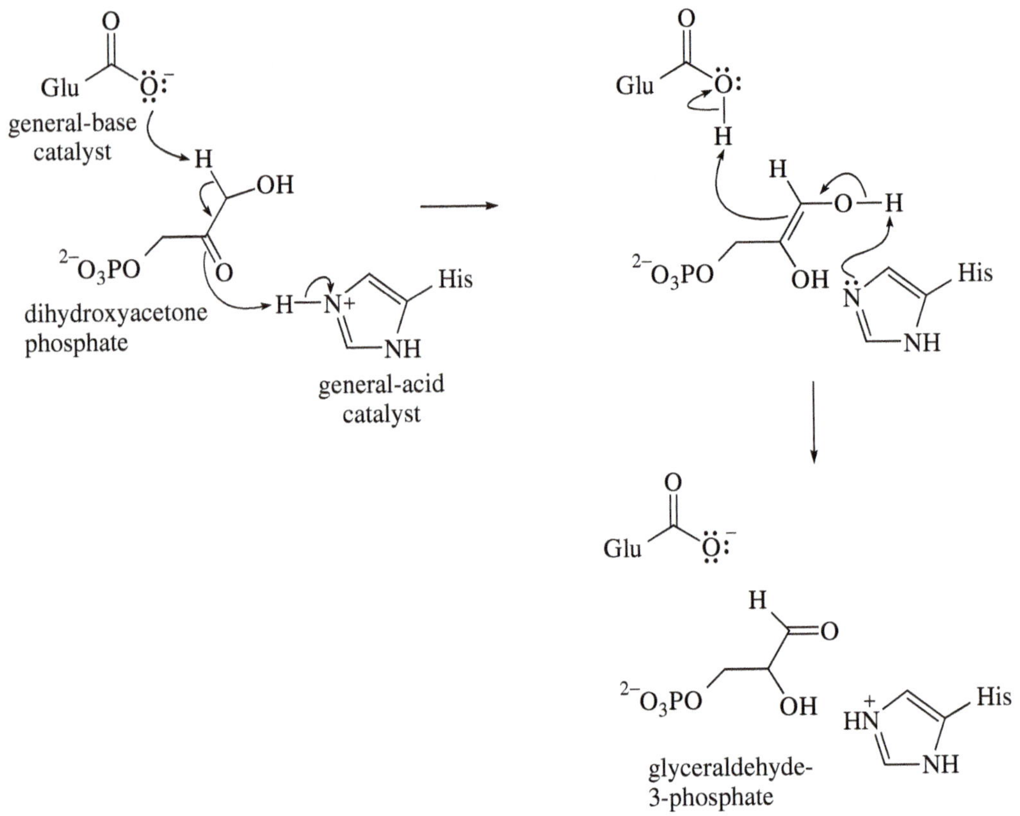
Glu
general-base catalyst
H
OH
$^{2-}O_3PO$
O
dihydroxyacetone phosphate
H—N+
His
NH
general-acid catalyst
Glu
H
H
O—H
$^{2-}O_3PO$
OH
N
His
NH
Glu
H
O
$^{2-}O_3PO$
OH
HN
His
NH
glyceraldehyde-3-phosphate

Chapter 22 Practice Test

1. Indicate whether each of the following statements is true or false:

   a. A catalyst increases the equilibrium constant of a reaction. T F

   b. An acid catalyst donates a proton to the substrate, and a base catalyst removes a proton from the substrate. T F

   c. The reactant of an enzyme-catalyzed reaction is called a substrate. T F

   d. Complexing with a metal ion increases the pK_a of water. T F

2. In lysozyme, glutamate 35 is a catalyst that is active in its acid form. Explain its catalytic function.

3. a. Draw a pH-rate profile for an enzyme that has one catalytic group at the active site with a $pK_a = 5.7$ that functions as a general-base catalyst.

   b. Draw a pH-rate profile for an enzyme that has one catalytic group at the active site with a $pK_a = 5.7$ that functions as a general-acid catalyst.

4. Show the curved mechanism arrows for the first step in the mechanism for chymotrypsin.

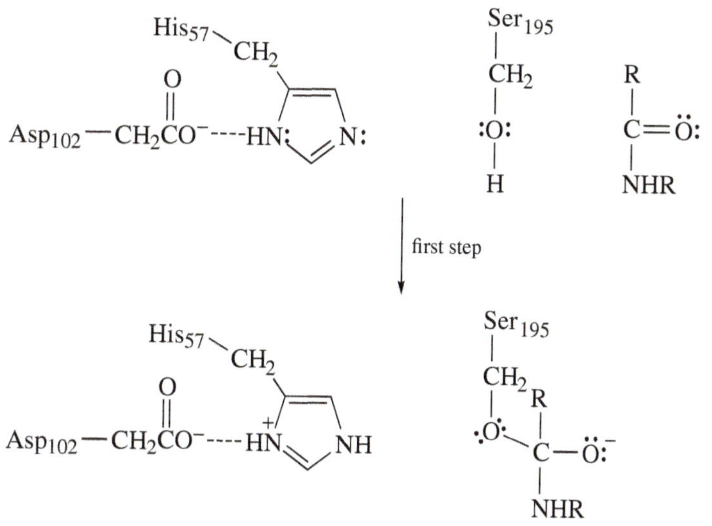

His57—CH2
Asp102—CH2CO⁻----HN: N:
Ser195—CH2—:O:—H
R—C=Ö:—NHR

first step

His57—CH2
Asp102—CH2CO⁻----HN+ NH
Ser195—CH2—:O:—C(R)(NHR)—Ö:⁻

   a. What kind of catalyst is histidine in this step?

   b. What kind of catalyst is serine in this step?

   c. How does aspartate catalyze the reaction?

CHAPTER 23
The Organic Chemistry of the Coenzymes, Compounds Derived from Vitamins

Important Terms

biotin — the coenzyme required by enzymes that catalyze carboxylation of a carbon adjacent to an ester or a keto group.

coenzyme — a cofactor that is an organic molecule.

coenzyme A — a thiol used by biological organisms to form thioesters.

coenzyme B_{12} — the coenzyme required by enzymes that catalyze certain rearrangement reactions.

cofactor — an organic molecule or a metal ion that an enzyme needs to catalyze a reaction.

competitive inhibitor — a compound that inhibits an enzyme by competing with the substrate for binding at the active site.

dehydrogenase — an enzyme that carries out an oxidation reaction by removing hydrogen from the substrate.

electron sink — a site to which electrons can be delocalized.

flavin adenine dinucleotide (FAD) — a coenzyme required in certain oxidation reactions. It is reduced to $FADH_2$, which is a coenzyme required in certain reduction reactions.

heterocyclic compound — a cyclic compound in which one or more of the ring atoms is an atom other than carbon.

lipoate — a coenzyme required in certain oxidation reactions.

mechanism-based inhibitor (suicide inhibitor) — an inhibitor that inactivates an enzyme by undergoing part of the normal catalytic mechanism.

molecular recognition — the recognition of one molecule by another as a result of specific interactions.

nicotinamide adenine dinucleotide (NAD^+) — a coenzyme required in certain oxidation reactions. It is reduced to NADH, which is a coenzyme required in certain reduction reactions.

nicotinamide adenine dinucleotide phosphate ($NADP^+$) — a coenzyme that is reduced to NADPH, which is a coenzyme required in certain reduction reactions.

nucleotide — a heterocycle attached in the β-position to the anomeric carbon of a phosphorylated ribose or 2-deoxyribose.

pyridoxal phosphate the coenzyme required by enzymes that catalyze certain transformations of amino acids.

suicide inhibitor (mechanism-based inhibitor) a compound that inactivates an enzyme by undergoing part of its normal catalytic mechanism.

tetrahydrofolate (THF) the coenzyme required by enzymes that catalyze a reaction that donates a group containing a single carbon to its substrate.

thiamine pyrophosphate (TPP) the coenzyme required by enzymes that catalyze a reaction that transfers an acyl group from one species to another.

transamination a reaction in which an amino group is transferred from one compound to another.

transimination the reaction of a primary amine with an imine to form a new imine and a primary amine derived from the original imine.

vitamin a substance needed in small amounts for normal body function that the body cannot synthesize or cannot synthesize in adequate amounts.

vitamin KH_2 the coenzyme required by the enzyme that catalyzes the carboxylation of glutamate side chains.

Solutions to Problems

1. The metal ion (Zn^{2+}) makes the carbonyl carbon more susceptible to nucleophilic addition, increases the nucleophilicity of water by making it more like a hydroxide ion, and stabilizes the negative charge on the transition state.

2. The alcohol is oxidized to a ketone.

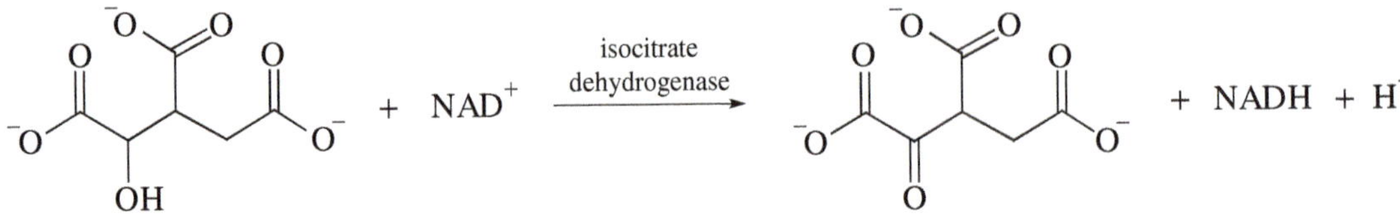

isocitrate + NAD^+ —(isocitrate dehydrogenase)→ α-ketoglutarate precursor (ketone) + NADH + H^+

3. The ketone is reduced to an alcohol.

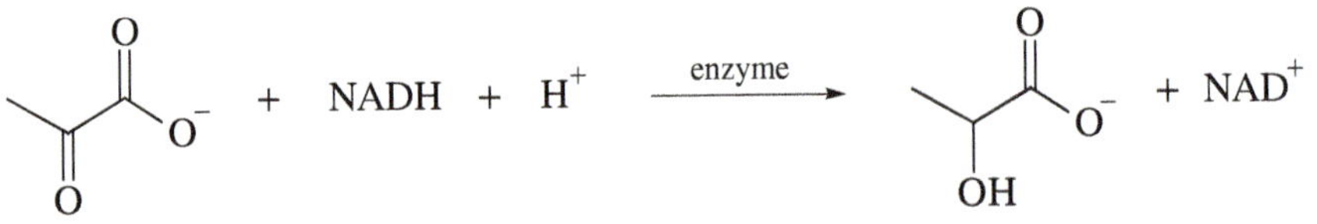

pyruvate + NADH + H^+ —(enzyme)→ lactate + NAD^+

4. **a.** FAD has seven conjugated double bonds (indicated by *).

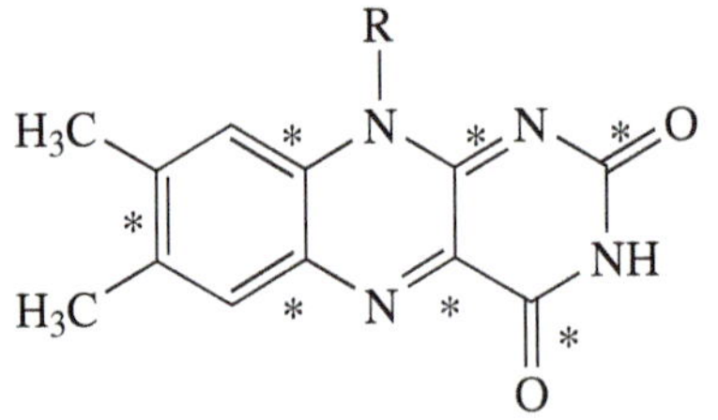

b. $FADH_2$ has three conjugated double bonds (on the left of the molecule). It also has three conjugated double bonds (on the right of the molecule) when the amides are written with a double bond between nitrogen and carbon.

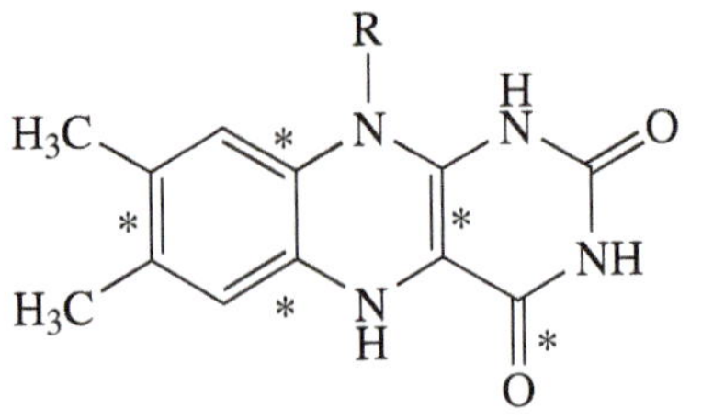

5. The nitrogen that is the stronger base is the one most apt to be protonated. N-1 is a weaker base than N-5 because the π electrons that belong to N-1 are delocalized onto an oxygen when a nucleophile adds to C-10a. The π electrons that belong to N-5 end up on nitrogen when a nucleophile adds to C-4a. Nucleophilic addition, therefore, occurs at the position (4a) that results in the stronger base (N-5) being protonated.

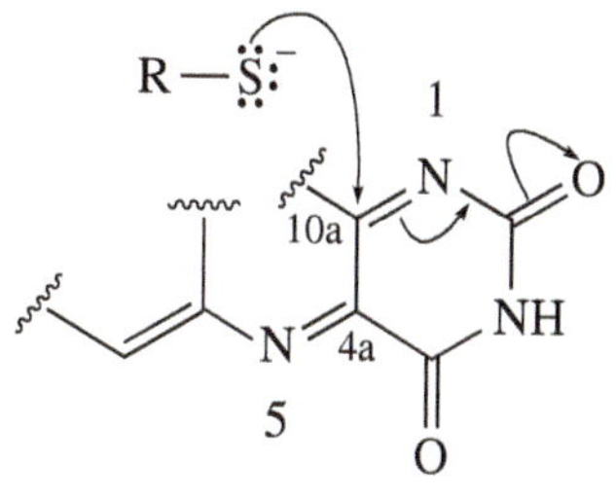

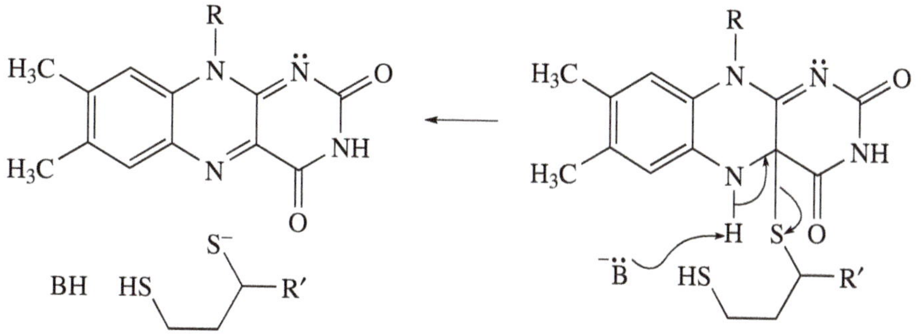

6. Solved in the text.

7. When a proton is removed from the methyl group at C-8, the electrons that are left behind can be delocalized onto the oxygen at the 2-position or onto the oxygen at the 4-position.

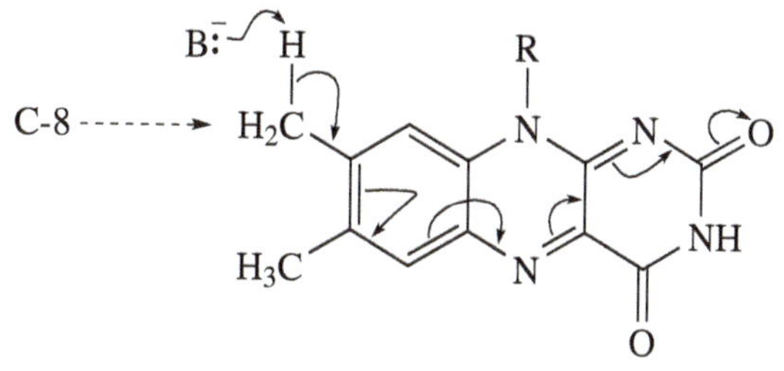

When a proton is removed from the methyl group at C-7, the electrons that are left behind can be delocalized only onto carbons that, being less electronegative than oxygen, are less able to accommodate the electrons.

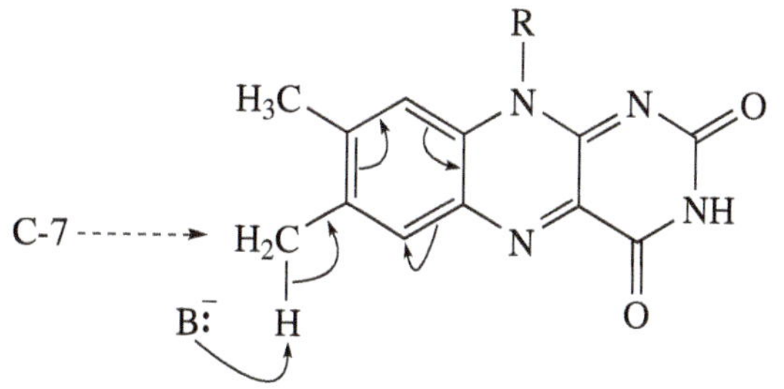

8.

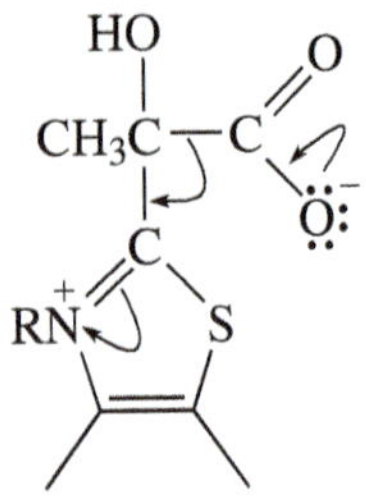

pyruvate-TPP intermediate

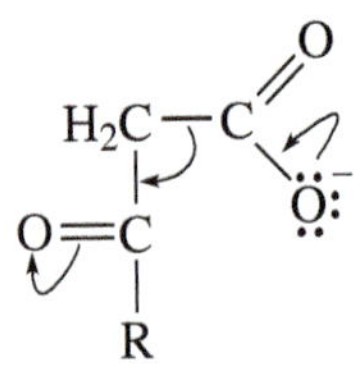

a β-keto acid

9. Notice that the only difference in the mechanisms for pyruvate decarboxylase and acetolactate synthase is the species to which the two-carbon fragment is transferred—a proton in the case of pyruvate carboxylase and pyruvate in the case of acetolactate synthase.

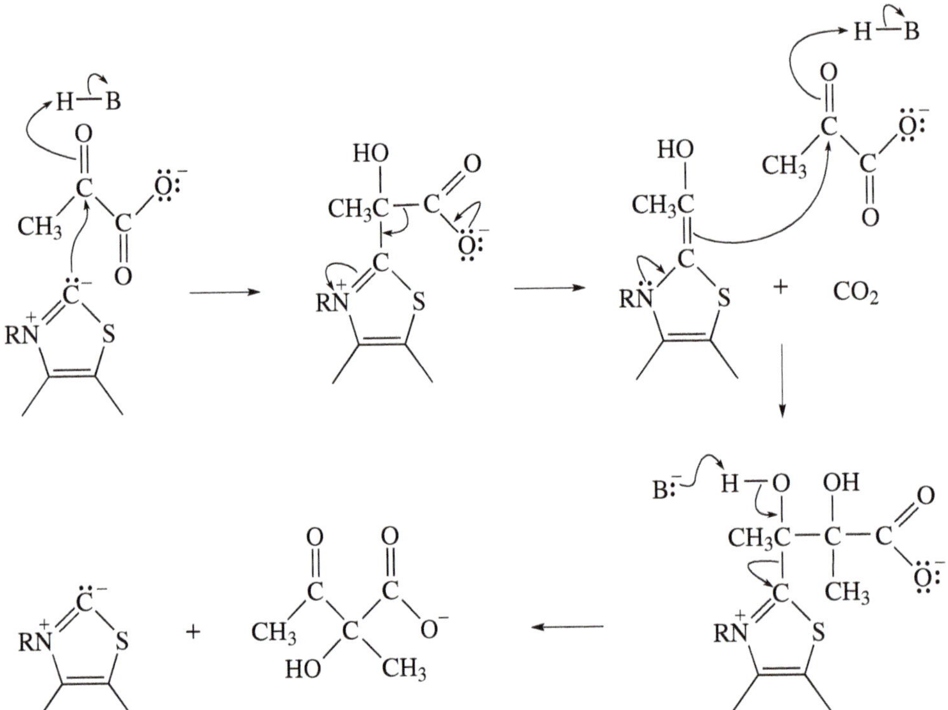

acetolactate

10. Notice that the only difference in this reaction and the one in Problem 9 is that the species to which the two-carbon fragment is transferred has an ethyl group in place of the methyl group.

H—B O C CH$_3$ C O Ö:$^-$ C:$^-$ RN$^+$ S → HO CH_3C—C O Ö:$^-$ C RN$^+$ S → HO CH_3CH_2 C O C O Ö:$^-$ H—B CH_3C C RN S + CO_2

↓

B:$^-$ H—O OH CH_3C—C—C O Ö:$^-$ C CH_2CH_3 RN$^+$ S → RN$^+$ C:$^-$ S + O O C C CH_3 C O$^-$ HO CH_2CH_3

α-aceto-α-hydroxybutyrate

11. Solved in the text.

12. **a.** CH_3 C O **b.** CH_3 C O **c.** $HOCH_2$ C O

13. The compound on the right is more easily decarboxylated because the electrons left behind when CO_2 is eliminated are delocalized onto the positively charged nitrogen of the pyridine ring. The electrons left behind if the other compound is decarboxylated cannot be delocalized.

CH_2CH_2 C O Ö:$^-$ N$^+$ H CH_2 C O Ö:$^-$ N$^+$ H

14. The α-keto group that accepts the amino group from pyridoxamine is converted into an amino group.

a.

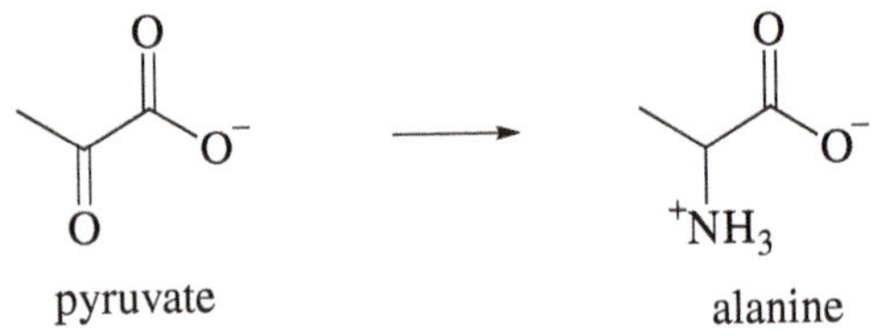

pyruvate → alanine

b.

oxaloacetate → aspartate

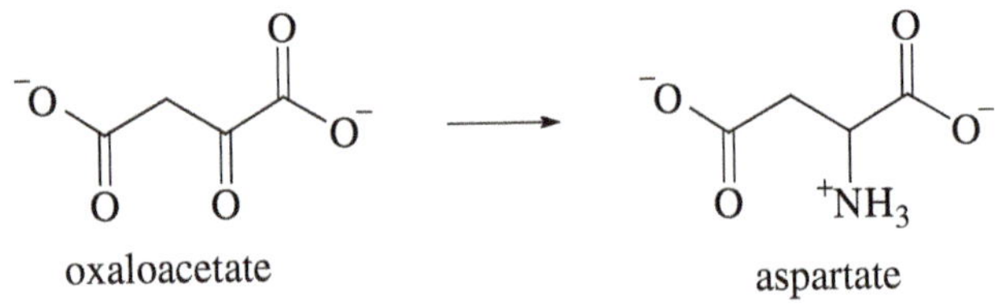

15. The first step (after transimination) in all amino acid transformations catalyzed by PLP is removal of a substituent from the α-carbon of the amino acid. The electrons left behind when the substituent is removed are delocalized onto the positively charged nitrogen of the pyridine ring. If the ring nitrogen were not protonated, it would be less attractive to the electrons. In other words, it would be a less effective electron sink.

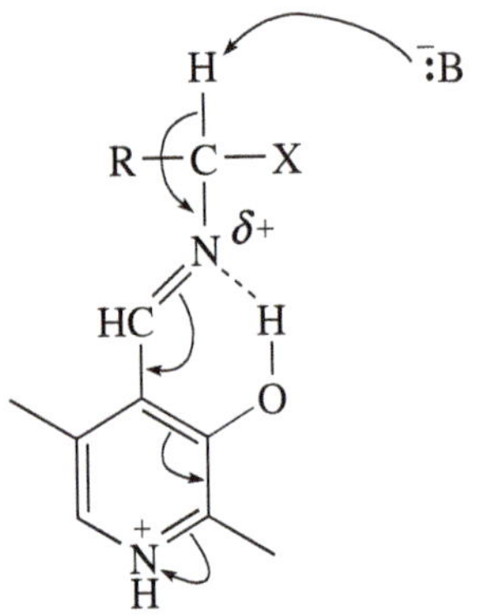

16. The hydrogen of the OH substituent forms a hydrogen bond with the nitrogen of the imine linkage. (See Problem 15.) This puts a partial positive charge on the nitrogen, which makes it easier for the amino acid to add to the imine carbon in the transimination reaction that attaches the amino acid to the coenzyme. It also makes it easier to remove a substituent from the α-carbon of the amino acid. If the OH substituent is replaced by an OCH_3 substituent, a proton is no longer available to form the hydrogen bond.

17. For clarity, the substituents are not put on the three bonds attached to the pyridinium ring in Problems 17 and 18.

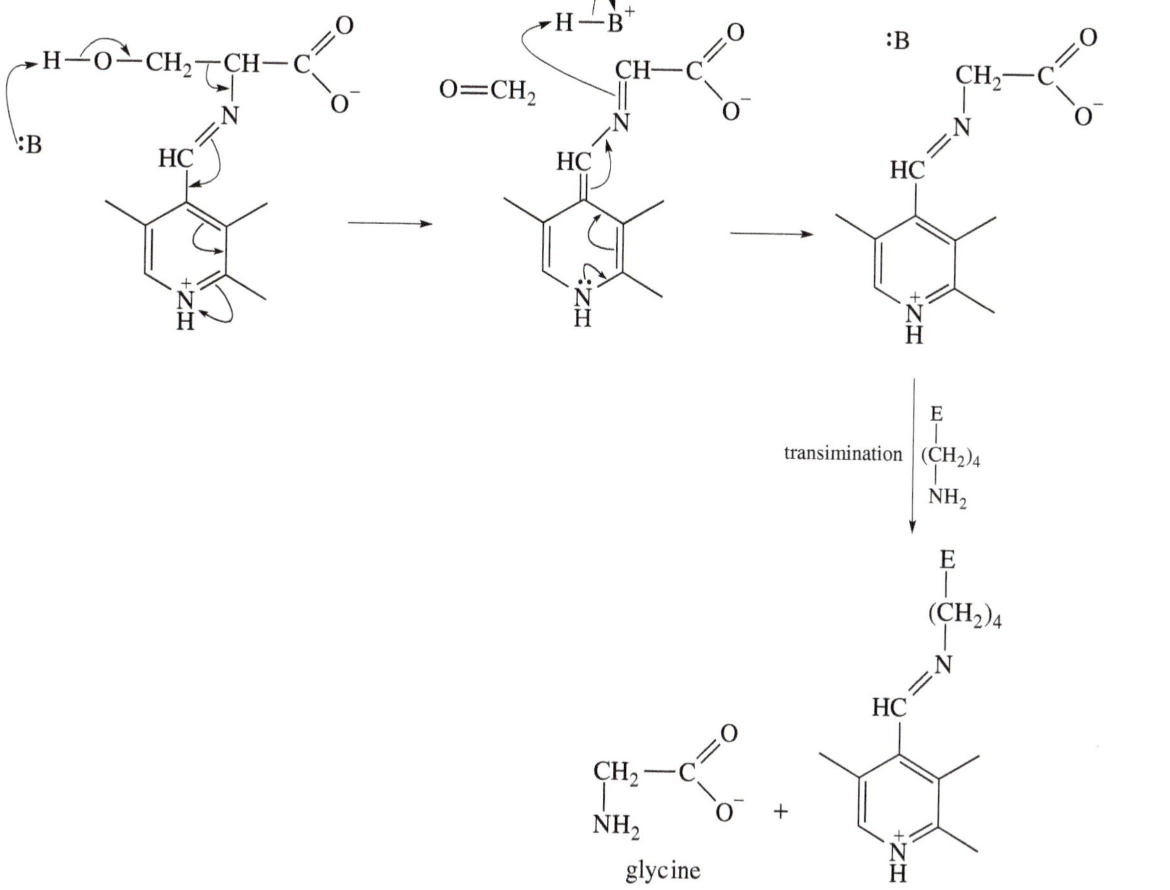

H—O—CH$_2$—CH—C(=O)O$^-$:B HC N

O=CH$_2$ H—B$^+$ CH—C(=O)O$^-$ N HC

:B CH$_2$—C(=O)O$^-$ N HC

transimination E—(CH$_2$)$_4$—NH$_2$

E (CH$_2$)$_4$ N HC

CH$_2$—C(=O)O$^-$ NH$_2$ glycine +

18. In the first step, a proton is removed from the α-carbon; in the next step, the leaving group is eliminated from the β-carbon. Transimination produces an enamine. The enamine tautomerizes to an imine that is hydrolyzed to the final product.

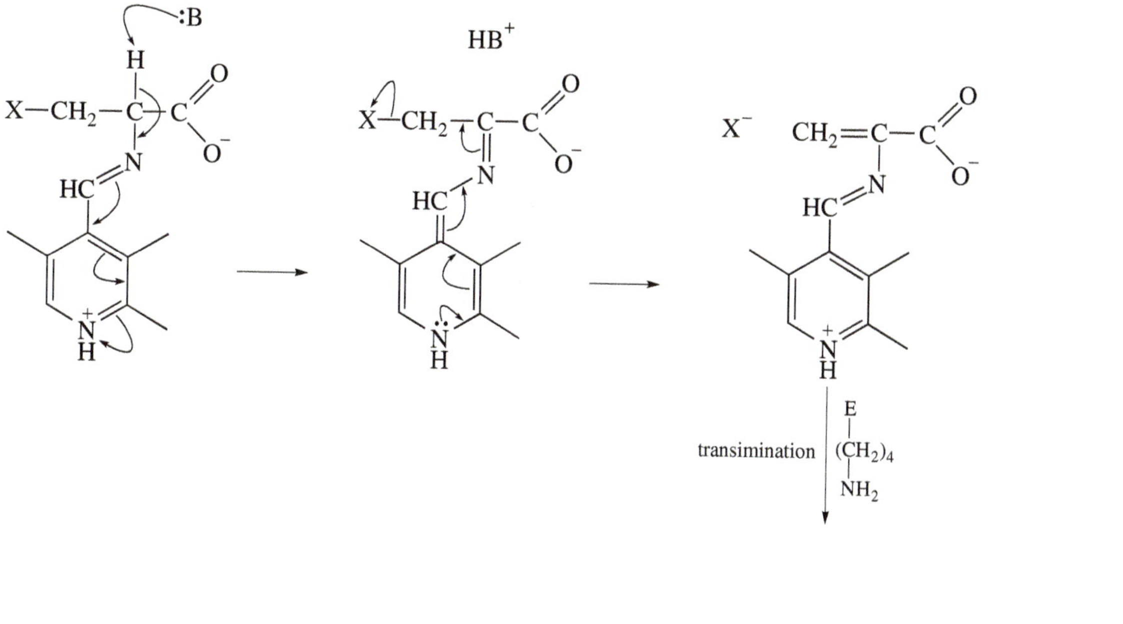

:B H X—CH$_2$—C—C(=O)O$^-$ N HC

HB$^+$ X—CH$_2$—C—C(=O)O$^-$ N HC

X$^-$ CH$_2$=C—C(=O)O$^-$ N HC

transimination E—(CH$_2$)$_4$—NH$_2$

$CH_3C(=NH)CO_2^-$ $\xrightleftharpoons{\text{tautomerization}}$ $CH_2{=}C(NH_2){-}CO_2^-$ + pyridinium imine (HC=N–$(CH_2)_4$–E)

an enamine

$CH_3C(=NH)CO_2^-$ $\xrightarrow[\text{H}_2\text{O}]{\text{HCl}}$ (imine hydrolysis) $CH_3C(=O)CO_2^-$ + $\overset{+}{N}H_4$

19. The mechanism is the same as that shown in the text for dioldehydrase. The tetrahedral intermediate that is formed as a result of the coenzyme B_{12}–catalyzed isomerization is unstable and loses ammonia to give acetaldehyde, the final product of the reaction.

AdCH2–Co(III) + HO–CH2–CH(H)–NH2 → AdĊH2 Co(II) + substrate → AdCH3 Co(II) + HO–CH2–ĊH–NH2 → AdCH2–H + ·CH2–CH(OH)–NH2 (Co(II)) → AdĊH2 Co(II) + H–CH2–CH(OH)–:NH2 → AdCH2–Co(III)

H–CH2–CH(OH)–:NH2 + H–B+ ⇌ H–CH2–CH(OH)–NH3+ + :B ⇌ H–CH2–CH=O+–H + :NH3 ⇌ CH_3CHO + $^+NH_4$

20.

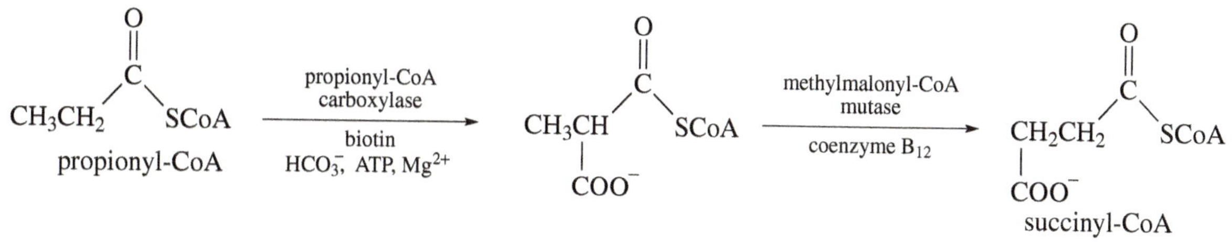

$$CH_3CH_2C(=O)SCoA \xrightarrow[\text{biotin, } HCO_3^-, \text{ ATP, } Mg^{2+}]{\text{propionyl-CoA carboxylase}} CH_3CH(COO^-)C(=O)SCoA \xrightarrow[\text{coenzyme } B_{12}]{\text{methylmalonyl-CoA mutase}} {}^-OOCCH_2CH_2C(=O)SCoA$$

propionyl-CoA → succinyl-CoA

21. They differ only in the circled part of the molecule.

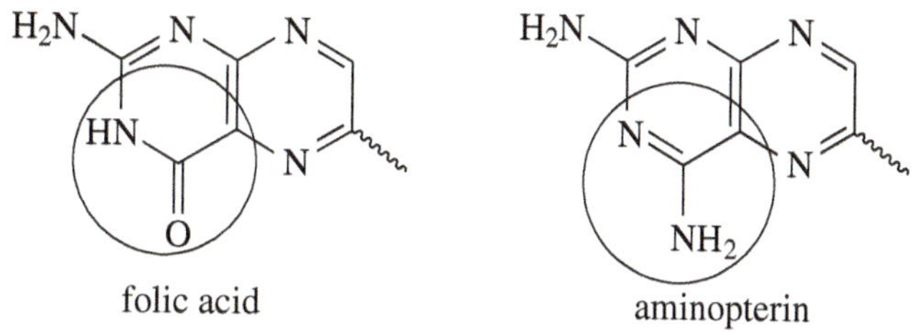

folic acid (circled: HN, C=O) aminopterin (circled: N, NH_2)

22. The methyl group in thymidine comes from the methylene group of N^5,N^{10}-methylene-THF, followed by the addition of a hydride ion from the coenzyme. The methylene group of N^5,N^{10}-methylene-THF comes from the CH_2OH group of serine by means of a PLP-catalyzed C_α—C_β cleavage. (See page 1082 in the text.)

23. Two thiol groups are oxidized in each of the two reactions that overall reduces vitamin K epoxide to vitamin KH_2. Because dihydrolipoate has two thiol groups, each thiol oxidation involves an intramolecular reaction. When thiols such as ethanethiol or propanethiol are used, each thiol oxidation involves an intermolecular reaction. These thiols react more slowly than dihydrolipoate, because the two thiol groups are not in the same molecule. Therefore, they have to find each other to react.

24.
a. thiamine pyrophosphate
b. FAD oxidizes dihydrolipoate back to lipoate.
c. NAD^+ oxidizes $FADH_2$ back to FAD.
d.

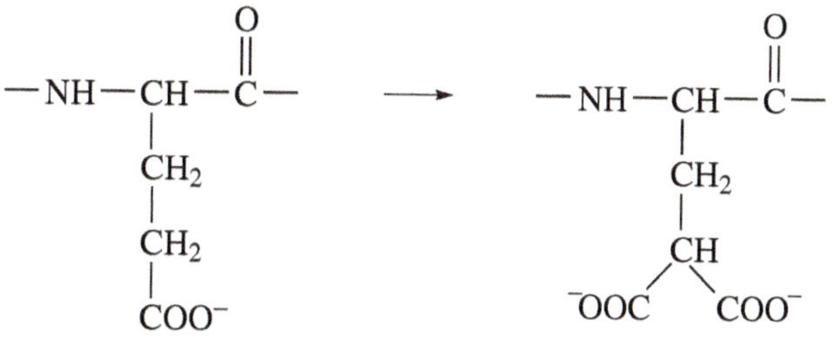

$$—NH—CH(CH_2CH_2COO^-)—C(=O)— \longrightarrow —NH—CH(CH_2CH(COO^-)_2)—C(=O)—$$

e. thiamine pyrophosphate and pyridoxal phosphate
f. Thiamine pyrophosphate is used for the decarboxylation of α-keto acids. Pyridoxal phosphate is used for the decarboxylation of amino acids.
g. biotin and vitamin KH_2
h. Biotin carboxylates a carbon adjacent to a carbonyl group (that is, an α-carbon). Vitamin KH_2 carboxylates the γ-carbon of a glutamate.

25.
a. thiamine pyrophosphate, carboxybiotin, pyridoxal phosphate
b. NAD^+, $NADP^+$, FAD
c. vitamin KH_2
d. *N*-substituted tetrahydrofolate

26. **a.** acetyl-CoA carboxylase; biotin
b. dihydrolipoyl dehydrogenase; FAD
c. methylmalonyl-CoA mutase; coenzyme B_{12}
d. aspartate transaminase; pyridoxal phosphate
e. propionyl-CoA carboxylase; biotin

27. **a.** thiamine pyrophosphate, lipoate, coenzyme A, FAD, NAD^+, indicated by numbers 1–5 in the mechanism below
b. The mechanism is exactly the same as that for the pyruvate dehydrogenase system.

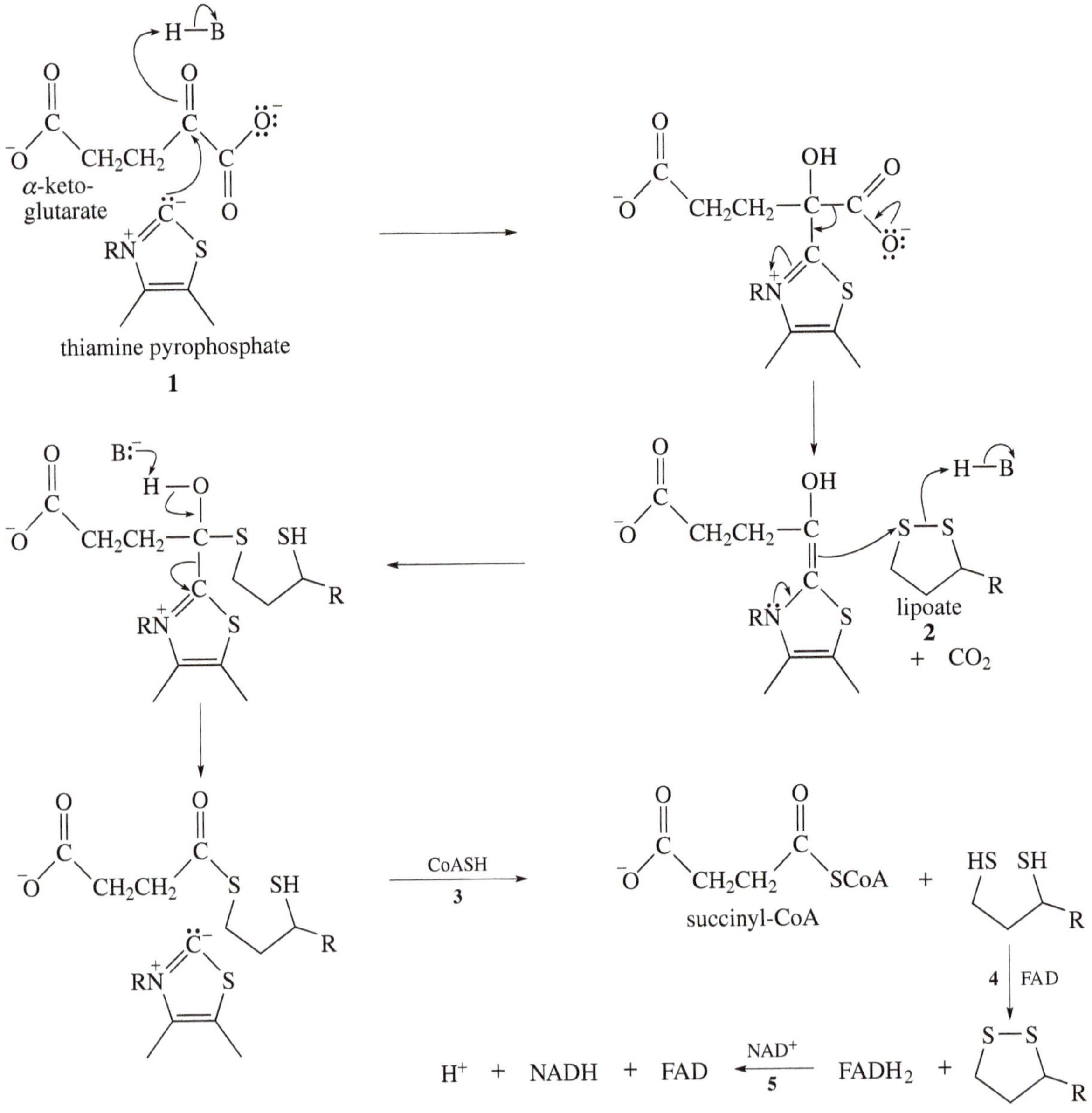

α-keto-glutarate

thiamine pyrophosphate
1

lipoate
2
+ CO_2

CoASH
3

succinyl-CoA

4 FAD

H^+ + NADH + FAD $\xleftarrow[5]{NAD^+}$ $FADH_2$ +

28. The product of transamination of an amino acid has a carbonyl group in place of the amino group.

a. $CH_3CH(CH_3)COCOO^-$ derived from Val $\quad$ $CH_3CH(CH_3)CH_2COCOO^-$ derived from Leu $\quad$ $CH_3CH_2CH(CH_3)COCOO^-$ derived from Ile

b. $CH_3CH(CH_3)COSCoA$ $\quad$ $CH_3CH(CH_3)CH_2COSCoA$ $\quad$ $CH_3CH_2CH(CH_3)COSCoA$

c. The reaction catalyzed by α-keto acid dehydrogenase is identical to the reaction catalyzed by the pyruvate dehydrogenase complex (Section 23.3). Therefore, they both require the same coenzymes: thiamine pyrophosphate, lipoate, coenzyme A, FAD, NAD^+.

d. The disease can be treated by a diet low in branched-chain amino acids.

29. $CH_3C(=O)-$ $\quad$ $HOCH_2C(=O)-$ $\quad$ $^-OOCCH_2CH_2C(=O)-$

$CH_3CH(CH_3)C(=O)-$ $\quad$ $CH_3CH(CH_3)CH_2C(=O)-$ $\quad$ $CH_3CH_2CH(CH_3)C(=O)-$

30. $AdCH_2-Co(III)$ + $CH_3-CH(COO^-)-C(=O)SCoA$ $\longrightarrow$ $AdCH_2\cdot$ Co(II) + $CH_3-CH(COO^-)-C(=O)SCoA$ $\longrightarrow$ $AdCH_3$ Co(II) + $\cdot CH_2-CH(COO^-)-C(=O)SCoA$

$\downarrow$

$\cdot CH_2-C\cdot(COO^-)H$ ... $AdCH_2-H$ Co(II) + $CH_2-\dot{C}(COO^-)-H$ with $C=O$, SCoA $\longrightarrow$ $CH_2(C(=O)SCoA)-CH_2COO^-$ + $AdCH_2\cdot$ Co(II) $\longrightarrow$ $AdCH_2-Co(III)$

31. The products of the reaction are shown below.

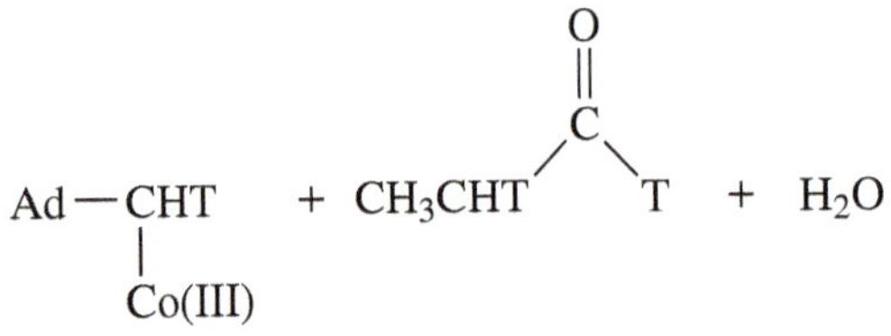

$$\underset{\displaystyle Co(III)}{Ad-CHT} + CH_3CHT-\overset{\displaystyle O}{\overset{\|}{C}}-T + H_2O$$

The following mechanism explains the products:

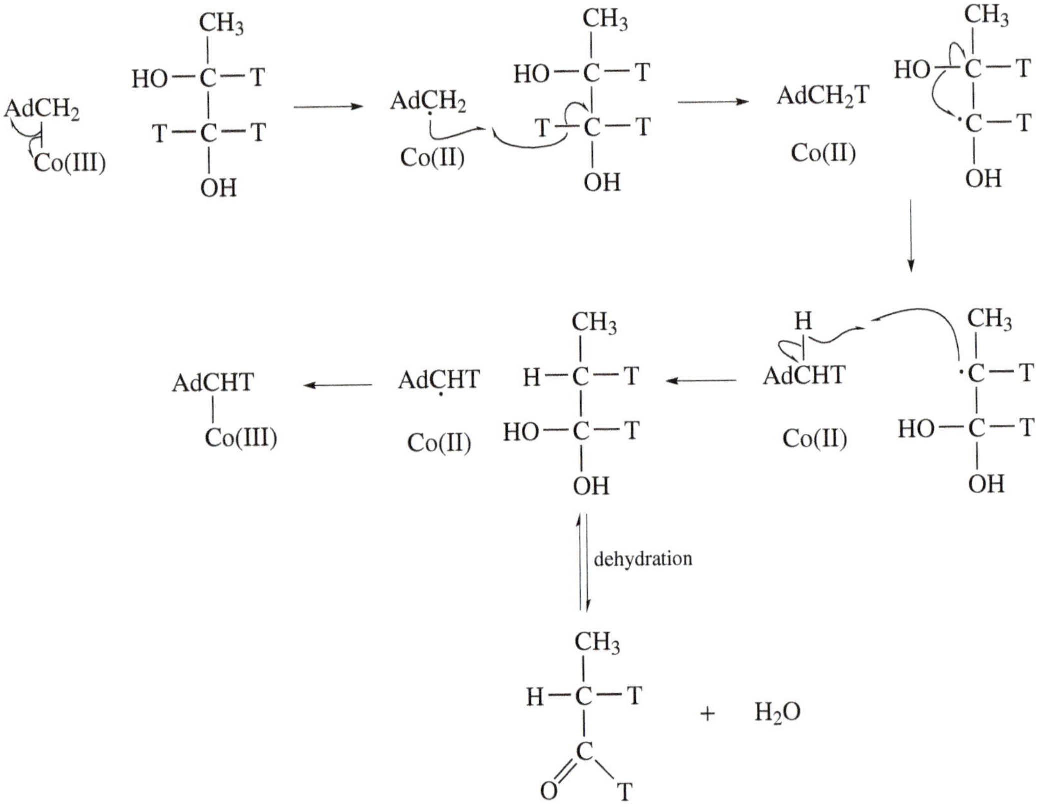

$$AdCH_2\text{–}Co(III) + HO-C(CH_3)(T)-C(T)(T)-OH \longrightarrow Ad\dot{C}H_2 \;\; Co(II) + HO-C(CH_3)(T)-C(T)(T)-OH \longrightarrow AdCH_2T \;\; Co(II) + HO-C(CH_3)(T)-\dot{C}(T)-OH$$

$$\downarrow$$

$$\underset{\displaystyle Co(III)}{AdCHT} \longleftarrow Ad\dot{C}HT \;\; Co(II) + H-C(CH_3)(T)-C(T)(OH)-OH \longleftarrow \overset{\displaystyle H}{AdCHT} \;\; Co(II) + \dot{C}(CH_3)(T)-C(T)(OH)-OH$$

$$\updownarrow \text{dehydration}$$

$$H-C(CH_3)(T)-C(=O)-T + H_2O$$

If there is only a limited amount of coenzyme, the second H of the coenzyme can be replaced by T in a subsequent round of catalysis, which means that the final product will be:

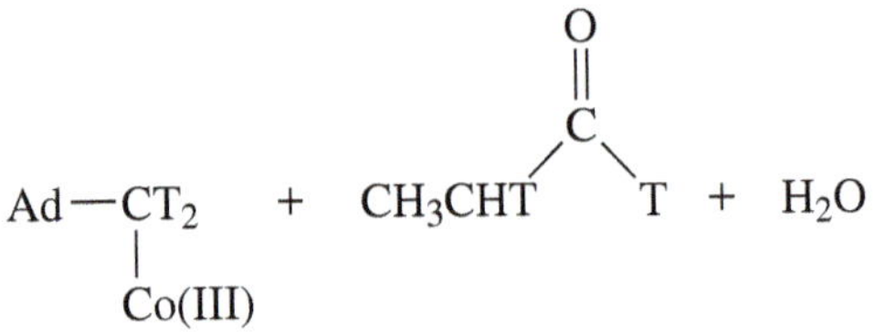

$$\underset{\displaystyle Co(III)}{Ad-CT_2} + CH_3CHT-\overset{\displaystyle O}{\overset{\|}{C}}-T + H_2O$$

32.

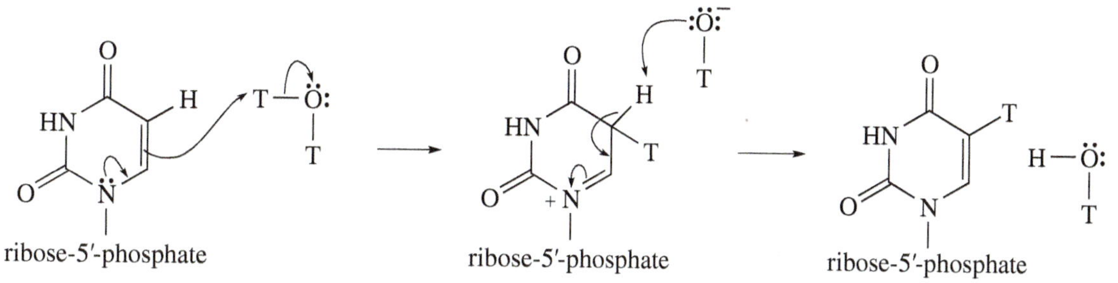

33. For clarity, the substituents are not put on the three bonds attached to the pyridinium ring in Problems 33–36. Each reaction in Problems 33–35 starts after the amino acid has undergone transimination with the lysine-bound coenzyme.

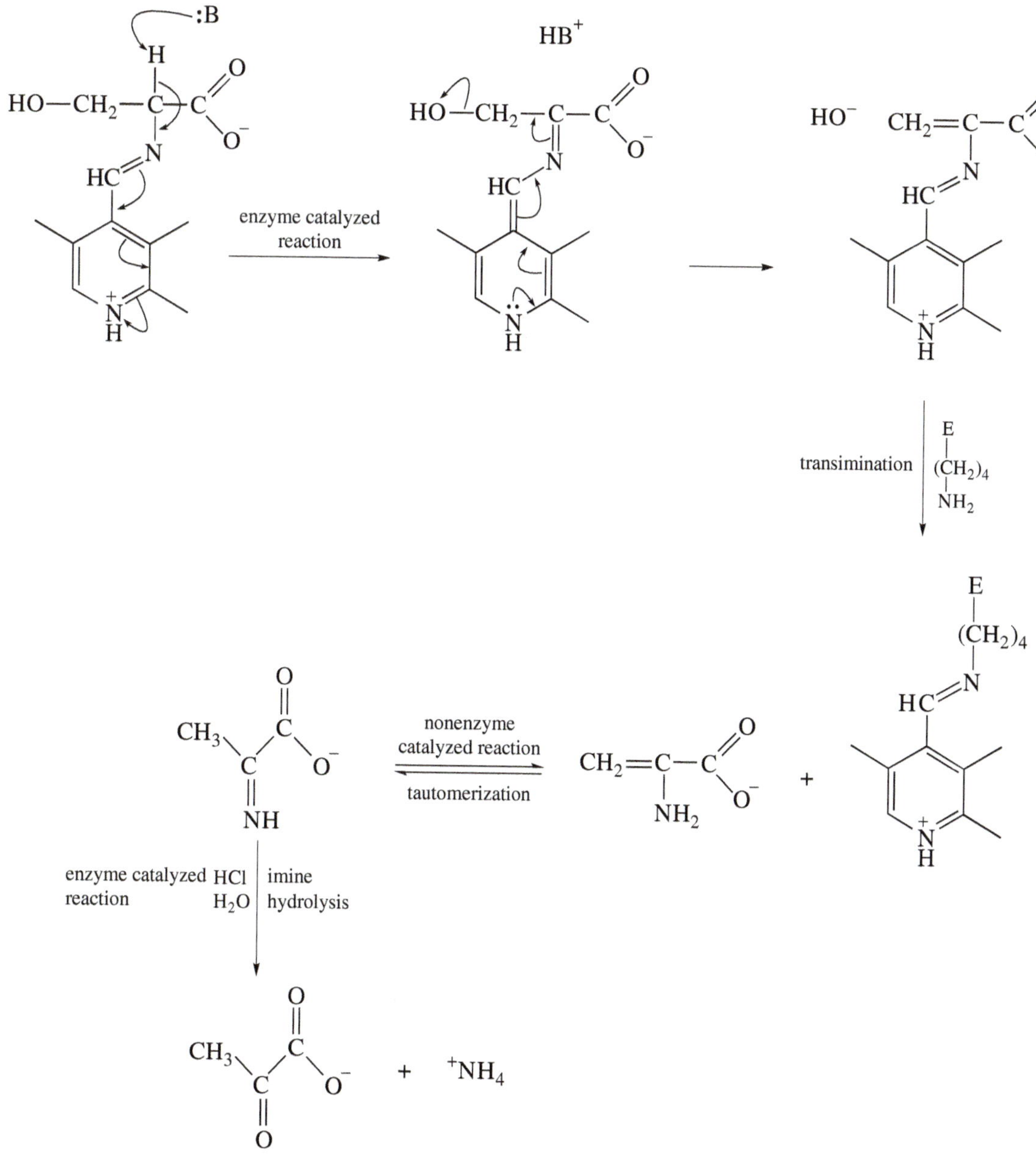

34.

:B

$X{-}CH_2{-}CH(N{=}CH{-}\text{pyridinium})){-}CO_2^-$

$X{-}CH_2{-}C({=}N{-}CH{=}\text{pyridine}){-}CO_2^-$

X^- $CH_2{=}C(N{=}CH{-}\text{pyridinium}){-}CO_2^-$ H—B$^+$

:Y$^-$

$Y{-}CH_2{-}CH(N{=}CH{-}\text{pyridinium}){-}CO_2^-$

transimination E—$(CH_2)_4$—NH_2

E—$(CH_2)_4$—N=CH—pyridinium + $Y{-}CH_2{-}CH(NH_2){-}CO_2^-$

35.

$$X{-}CH_2{-}CH_2{-}\underset{\underset{N=CH\text{-PLP}^{+}}{|}}{\overset{H}{\overset{|}{C}}}{-}COO^- \xrightarrow{:B} X{-}CH_2{-}CH(H){-}\underset{\underset{N{-}HC=\text{PLP}}{\|}}{C}{-}COO^- + HB^+ \xrightarrow{B:} X^- \; CH_2{=}CH{-}\underset{\underset{N}{\|}}{C}{-}COO^- \;\; H{-}B^+$$

$$\downarrow$$

$$:B \quad CH_3CH{=}\underset{\underset{N=HC\text{-PLP}^{+}}{|}}{C}{-}COO^-$$

$$\underset{\text{transimination}}{\overset{E{-}(CH_2)_4{-}NH_2}{\rightleftharpoons}} \quad CH_3CH{=}\underset{\underset{NH_2}{|}}{C}{-}COO^- \; + \; E{-}(CH_2)_4{-}N{=}HC\text{-PLP}^{+}$$

$$\Big\updownarrow \text{ tautomerization}$$

$$CH_3CH_2{-}\underset{\underset{NH}{\|}}{C}{-}COO^- \xrightarrow[\text{imine hydrolysis}]{HCl,\ H_2O} CH_3CH_2{-}\underset{\underset{O}{\|}}{C}{-}COO^- \; + \; {}^{+}NH_4$$

36.

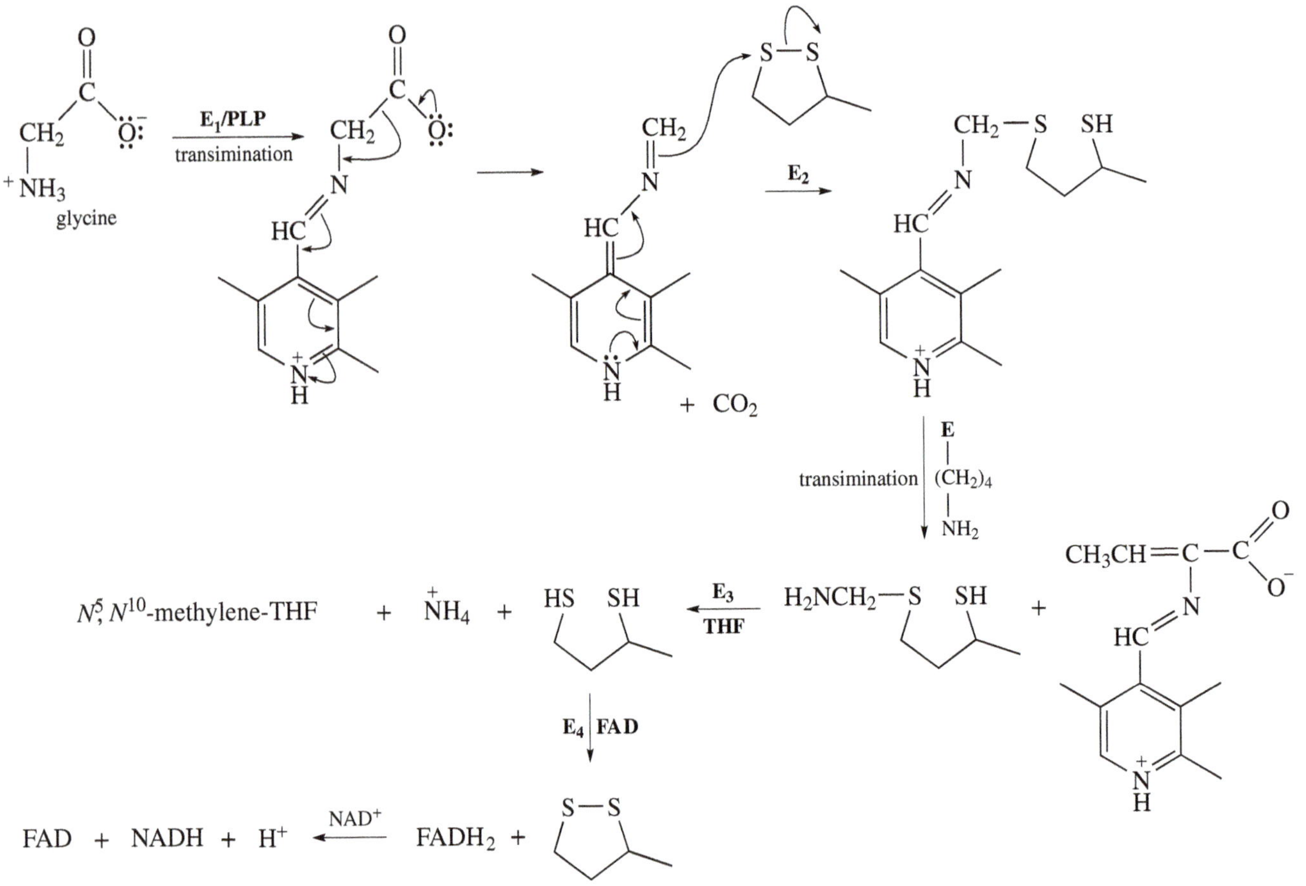

glycine

E_1/PLP transimination

E_2

+ CO_2

transimination E–$(CH_2)_4$–NH_2

N^5,N^{10}-methylene-THF + $\overset{+}{N}H_4$ + [dihydrolipoate] $\xleftarrow[\text{THF}]{E_3}$ H_2NCH_2–S SH + $CH_3CH{=}C{-}CO_2^-$ (PLP imine)

E_4 FAD

FAD + NADH + H^+ $\xleftarrow{NAD^+}$ $FADH_2$ + [lipoate]

37. Nonenzyme-bound FAD is a stronger oxidizing agent than NAD^+, so NAD^+ cannot oxidize $FADH_2$ to FAD. When the enzyme binds FAD, it makes it a weaker oxidizing agent. Therefore, when bound to the enzyme, FAD is a weaker oxidizing agent than NAD^+, so NAD^+ can oxidize enzyme-bound $FADH_2$ to enzyme-bound FAD.

38.

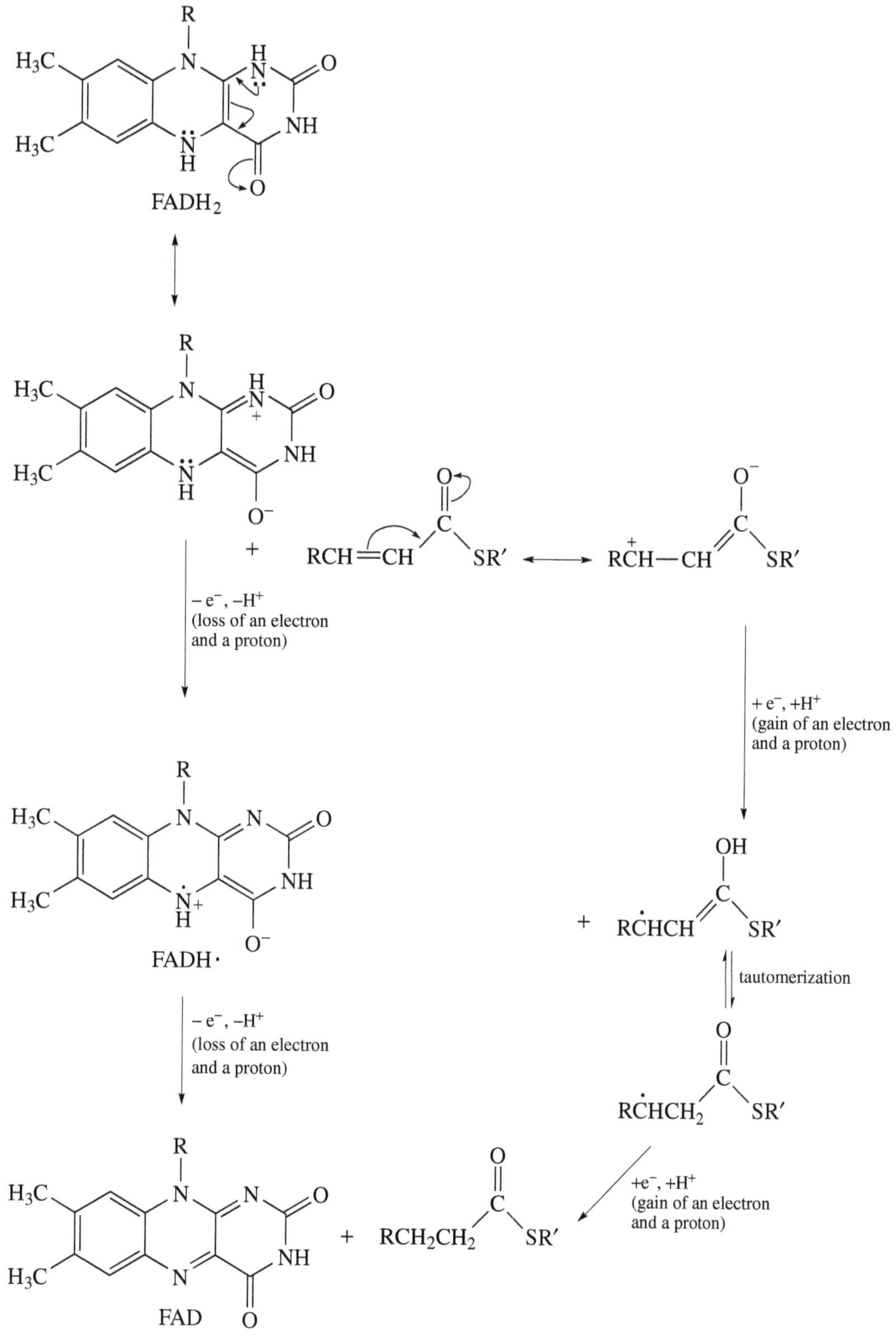

R; H_3C; H_3C; N; $\ddot{N}$H; NH; O; O; $\ddot{N}$; H

$FADH_2$

↕

R; H_3C; H_3C; N; H; $\overset{+}{N}$; O; NH; $\ddot{N}$; H; O^-

$+ \; RCH{=}CH\text{—}C({=}O)\text{—}SR' \longleftrightarrow \overset{+}{R}CH\text{—}CH{=}C(O^-)\text{—}SR'$

$-e^-, -H^+$
(loss of an electron
and a proton)

$+e^-, +H^+$
(gain of an electron
and a proton)

R; H_3C; H_3C; N; N; O; NH; $\dot{N}^+$; H; O^-

FADH·

$+ \; R\dot{C}HCH{=}C(OH)\text{—}SR'$

tautomerization

$R\dot{C}HCH_2\text{—}C({=}O)\text{—}SR'$

$-e^-, -H^+$
(loss of an electron
and a proton)

$+e^-, +H^+$
(gain of an electron
and a proton)

R; H_3C; H_3C; N; N; O; NH; N; O

FAD

$+ \; RCH_2CH_2\text{—}C({=}O)\text{—}SR'$

Chapter 23 Practice Test

1. What two coenzymes put carboxyl groups on their substrates?

2. What are the three one-carbon groups that tetrahydrofolate coenzymes put on their substrates?

3. Show the mechanism for NADPH reducing its substrate.

4. Draw the structure of the compound obtained when the following amino acid undergoes transamination:

$$CH_3CH_2CH(CH_3)CH(^+NH_3)C(=O)O^-$$

5. What is the first step in the reaction of the substrate with coenzyme B_{12} in an enzyme-catalyzed reaction that requires coenzyme B_{12}?

6. What coenzyme is required for each of the following enzyme-catalyzed reactions?

$$CH_3CH_2C(=O)SCoA \xrightarrow{\text{enzyme}} CH_3CH(COO^-)C(=O)SCoA \xrightarrow{\text{enzyme}} CH_2(COO^-)CH_2C(=O)SCoA$$

7. Show the enzyme-catalyzed reaction that requires vitamin KH_2 as a coenzyme.

8. Draw the product of the enzyme-catalyzed reaction that requires biotin and whose substrate is acetyl-CoA.

9. **a.** Other than the substrate, enzyme, and coenzyme, what three additional reagents are needed by a reaction that requires biotin as a coenzyme?

b. What is the function of each of these reagents?

10. What is the function of FAD in the pyruvate dehydrogenase system?

11. Draw the structures of the two products obtained from the following transamination reaction:

$$HO-C_6H_4-CH_2CH(^+NH_3)C(=O)O^- + {}^-OC(=O)CH_2CH_2C(=O)C(=O)O^- \xrightarrow{\text{transamination}}$$

12. Why is NADPH needed in order to convert uridines to thymidines?

13. Indicate whether each of the following statements is true or false:

a. Vitamin B_1 is the only water-insoluble vitamin that has a coenzyme function. T F

b. $FADH_2$ is a reducing agent. T F

c. Thiamine pyrophospate is vitamin B_6. T F

d. Cofactors that are organic molecules are called coenzymes. T F

e. Vitamin K is a water-soluble vitamin. T F

f. Lipoic acid is covalently bound to its enzyme by an amide linkage. T F

CHAPTER 24
The Organic Chemistry of the Metabolic Pathways

Important Terms

acyl adenylate $R{-}C(=O){-}O{-}P(=O)(O^-){-}O{-}$adenosine

acyl phosphate $R{-}C(=O){-}O{-}P(=O)(O^-){-}O^-$

acyl pyrophosphate $R{-}C(=O){-}O{-}P(=O)(O^-){-}O{-}P(=O)(O^-){-}O^-$

allosteric activator/inhibitor a compound that activates/inhibits an enzyme by binding to a site on the enzyme other than the active site.

anabolism the reactions living organisms carry out that result in the synthesis of complex biomolecules from simple precursor molecules.

catabolism the reactions living organisms carry out to provide energy and simple precursor molecules for synthesis.

citric acid cycle a series of reactions that convert the acetyl group of acetyl-CoA into two molecules of CO_2 and a molecule of CoASH.

fatty acid a long, straight-chain carboxylic acid.

feedback inhibitor a compound that inhibits a step at the beginning of the pathway for its biosynthesis.

gluconeogenesis the synthesis of glucose from pyruvate.

glycolysis the series of reactions that converts glucose to two molecules of pyruvate.

high-energy bond a bond that releases a great deal of energy when it is broken.

metabolism reactions living organisms carry out to obtain the energy they need and to synthesize the compounds they require.

β-oxidation a repeating series of four reactions that convert a fatty acyl-CoA molecule into molecules of acetyl-CoA.

oxidative phosphorylation — the fourth stage of catabolism in which NADH and $FADH_2$ are oxidized back to NAD^+ and FAD: for each NADH that is oxidized, 2.5 ATPs are formed; for each $FADH_2$ that is oxidized, 1.5 ATPs are formed.

phosphoanhydride bond — the bond that holds two phosphoric acid molecules together.

phosphoryl transfer reaction — the transfer of a phosphate group from one compound to another.

regulatory enzyme — an enzyme that catalyzes an irreversible reaction near the beginning of a pathway, thereby allowing independent control over degradation and synthesis.

Solutions to Problems

1. Solved in the text.

2. **a.** The pK_a values of the three OH groups of ADP are 0.9, 2.8, and 6.8. At pH 7.4, two of the groups will be in their basic forms, giving ADP two negative charges.

We can determine the fraction of the group with a pK_a of 6.8 that will be in its basic form at pH 7.4 using the method shown in Problem 1.

$$\text{fraction of group present in basic form} = \frac{K_a}{K_a + [H^+]} = \frac{1.6 \times 10^{-7}}{1.6 \times 10^{-7} + 4.0 \times 10^{-8}}$$

$$= \frac{1.6 \times 10^{-7}}{1.6 \times 10^{-7} + 0.4 \times 10^{-7}} = 0.8$$

$$\text{total negative charge on ADP} = 2.0 + 0.8 = 2.8$$

b. The pK_a values of the alkyl phosphate are 1.9 and 6.7. At pH 7.4, the OH group with a pK_a of 1.9 will account for one negative charge. We need to calculate the fraction of the group with a pK_a value of 6.7 that will be negatively charged at pH 7.4.

$$\text{fraction of group present in basic form} = \frac{K_a}{K_a + [H^+]} = \frac{2.0 \times 10^{-7}}{2.0 \times 10^{-7} + 0.4 \times 10^{-7}} = 0.8$$

$$\text{total negative charge on the alkyl phosphate} = 1.0 + 0.8 = 1.8$$

3.

CH_2OH–CHOH–CH_2O–H (:B) + ATP (^-O–P(=O)(O^-)–O–P(=O)(O^-)–O–P(=O)(O^-)–O–Ad) ⟶ CH_2OH–CHOH–$CH_2OPO_3^{2-}$ + ADP + H^+

glycerol **glycerol-3-phosphate** **ADP**

4.

1 OH, C, H, $HOCH_2$ 3, $CH_2OPO_3^{2-}$ 2 or HO–C–H with CH_2OH (top) and $CH_2OPO_3^{2-}$ (bottom)

(*R*)-glycerol-3-phosphate

5. The resonance contributor on the right shows that the β-carbon of the α,β-unsaturated carbonyl compound has a partial positive charge. The nucleophilic OH group, therefore, is attracted to the β-carbon.

RCH=CH–C(=O)–SCoA ⟷ $\overset{+}{\text{RCH}}$(β)–CH(α)=C(O^-)–SCoA

6. Because palmitic acid has 16 carbons and the acyl group of acetyl-CoA has 2 carbons, 8 molecules of acetyl-CoA are formed from 1 molecule of palmitic acid.

7. seven: one mole of NADH is obtained from each round of β-oxidation.

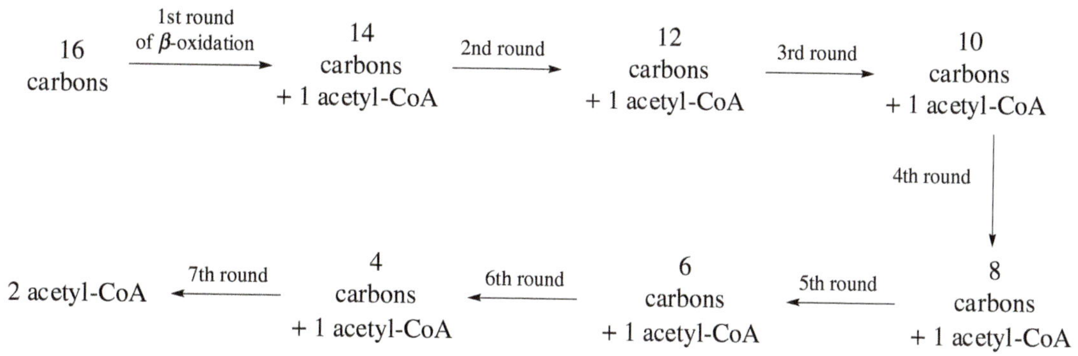

16 carbons → (1st round of β-oxidation) 14 carbons + 1 acetyl-CoA → (2nd round) 12 carbons + 1 acetyl-CoA → (3rd round) 10 carbons + 1 acetyl-CoA → (4th round) 8 carbons + 1 acetyl-CoA → (5th round) 6 carbons + 1 acetyl-CoA → (6th round) 4 carbons + 1 acetyl-CoA → (7th round) 2 acetyl-CoA

8. The OH group attacks the γ-phosphorus of ATP.

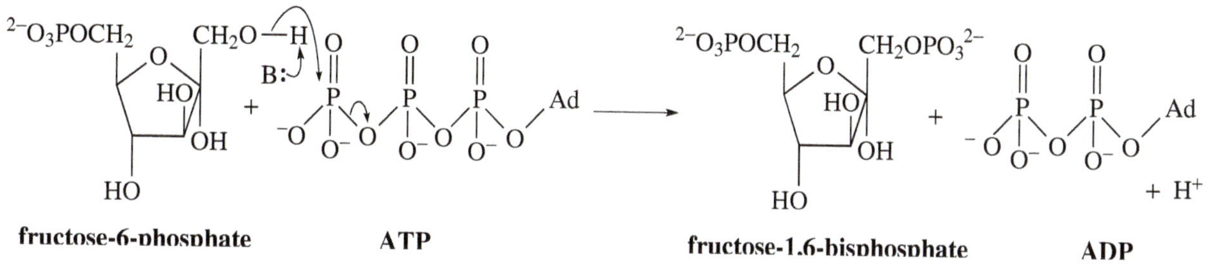

fructose-6-phosphate + ATP → fructose-1,6-bisphosphate + ADP + H^+

9. The reaction that follows the oxidation of glyceraldehyde-3-phosphate to 1,3-bisphosphoglycerate (the conversion of 1,3-bisphosphoglycerate to 3-phosphoglycerate) is highly exergonic. Therefore, as 1,3-bisphosphoglycerate is converted to 3-phosphoglycerate, glyceraldehyde-3-phosphate will be converted to 1,3-bisphosphoglycerate to replenish it.

glyceraldehyde-3-phosphate ⇌ 1,3-bisphosphoglycerate ⇌ 3-phosphoglycerate

10. two; each molecule of D-glucose is converted to two molecules of glyceraldehyde-3-phosphate, and each molecule of glyceraldehyde-3-phosphate requires one molecule of NAD^+ for it to be converted to one molecule of pyruvate.

11. acetaldehyde reductase

12. a ketone

13. thiamine pyrophosphate

14.

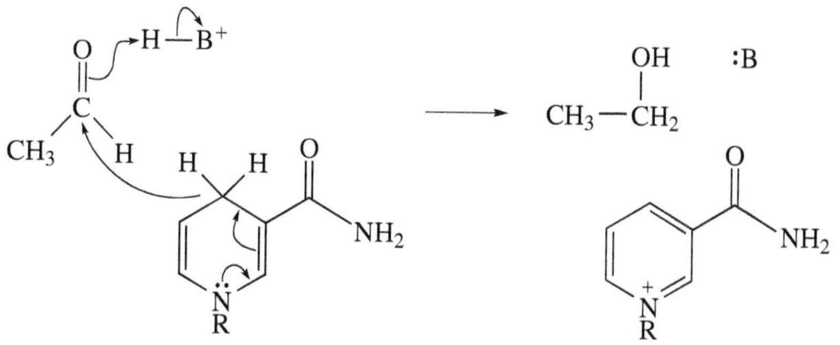

$CH_3CH=O$ + NADH (H–B$^+$) → CH_3-CH_2OH + NAD^+ + :B

15. 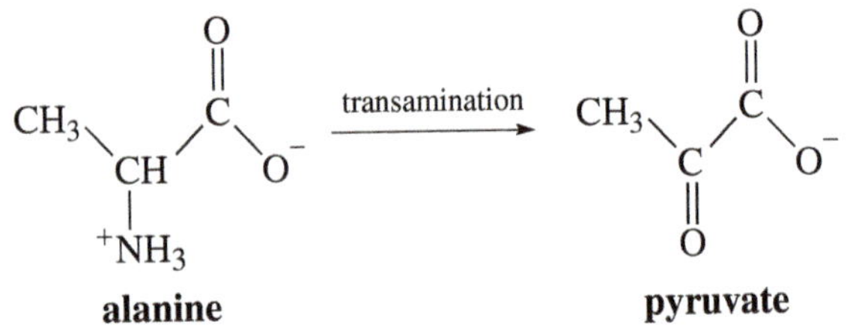 $CH_3CH(^{+}NH_3)COO^-$ —transamination→ CH_3COCOO^-

alanine **pyruvate**

16. Protonated histidine (pK_a = 6.0) is not strong enough an acid to fully protonate the OH group to make it a good leaving group (H_2O) that would be able to leave in the first step of the elimination reaction, which is required for an E1 reaction. Therefore, it is an E2 reaction with protonated histidine acting as a general-acid catalyst to protonate the OH group as it departs.

17. a secondary alcohol

18. citrate and isocitrate (and the alkene intermediate generated during the conversion of citrate to isocitrate)

19. 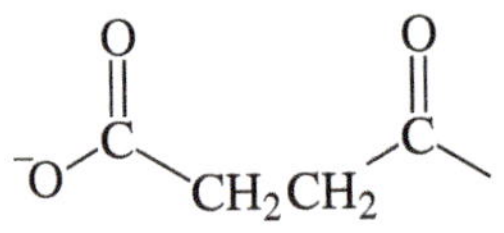 $^-OOCCH_2CH_2COCH_3$

20.

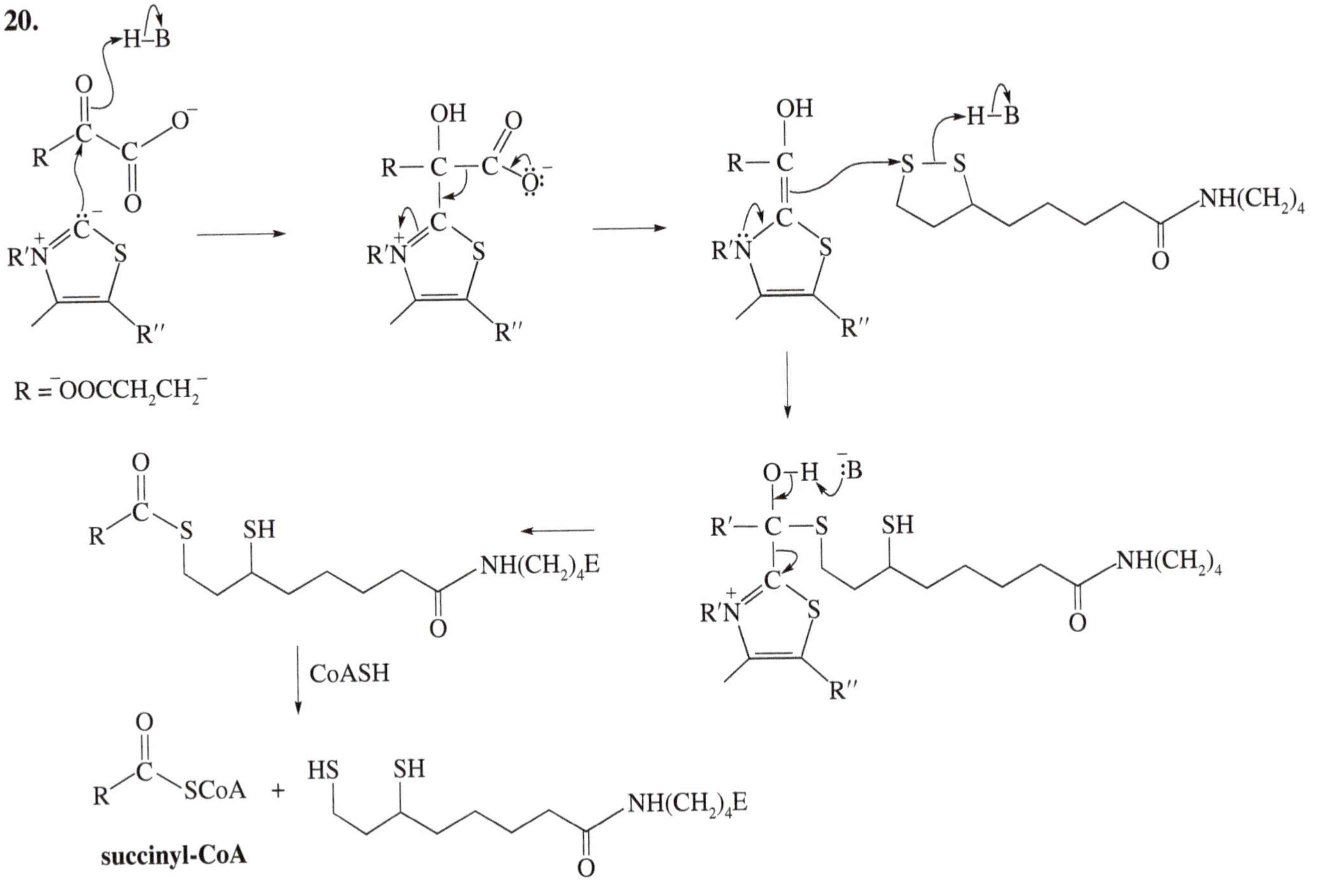

H–B

R, O, O^-, C, C, O

R'N, C, S, R''

OH, R–C–C, O, :Ö:⁻

R'N, C, S, R''

OH, R–C, H–B, S–S, $NH(CH_2)_4$, O

R'N, C, S, R''

$R = {}^-OOCCH_2CH_2$

O–H :B⁻

R'–C–S, SH, $NH(CH_2)_4$, O

R'N, C, S, R''

R, C, O, S, SH, $NH(CH_2)_4E$, O

CoASH

R, C, O, SCoA + HS, SH, $NH(CH_2)_4E$, O

succinyl-CoA

21. **a.** Succinyl-CoA synthetase: this enzyme catalyzes a reaction of succinyl-CoA, but it is named for the reverse reaction, which is the synthesis of succinyl-CoA.

b. aldolase: it is a retroaldol reaction, but it is named for an aldol addition; phosphoglycerate kinase (a kinase puts a phosphoryl group on the substrate); pyruvate kinase

22. **a.** The conversion of one molecule of glycerol to dihydroxyacetone phosphate consumes one molecule of ATP. The conversion of dihydroxyacetone phosphate to pyruvate produces two molecules of ATP. Therefore, one molecule of ATP is obtained from the conversion of one molecule of glycerol to pyruvate.

b. One NADH is formed from the conversion of one molecule of glycerol to dihydroxyacetone phosphate, and one NADH is formed from the conversion of dihydroxyacetone phosphate to pyruvate. Each NADH forms 2.5 ATP in the fourth stage of catabolism. So when the fourth stage of catabolism is included, six molecules ($2.5 + 2.5 + 1 = 6$) of ATP are obtained from the conversion of one molecule of glycerol to pyruvate.

23. **a.** glycerol kinase **b.** phosphatidic acid phosphatase

24. The first step is an S_N2 reaction; the second step is a nucleophilic acyl substitution reaction.

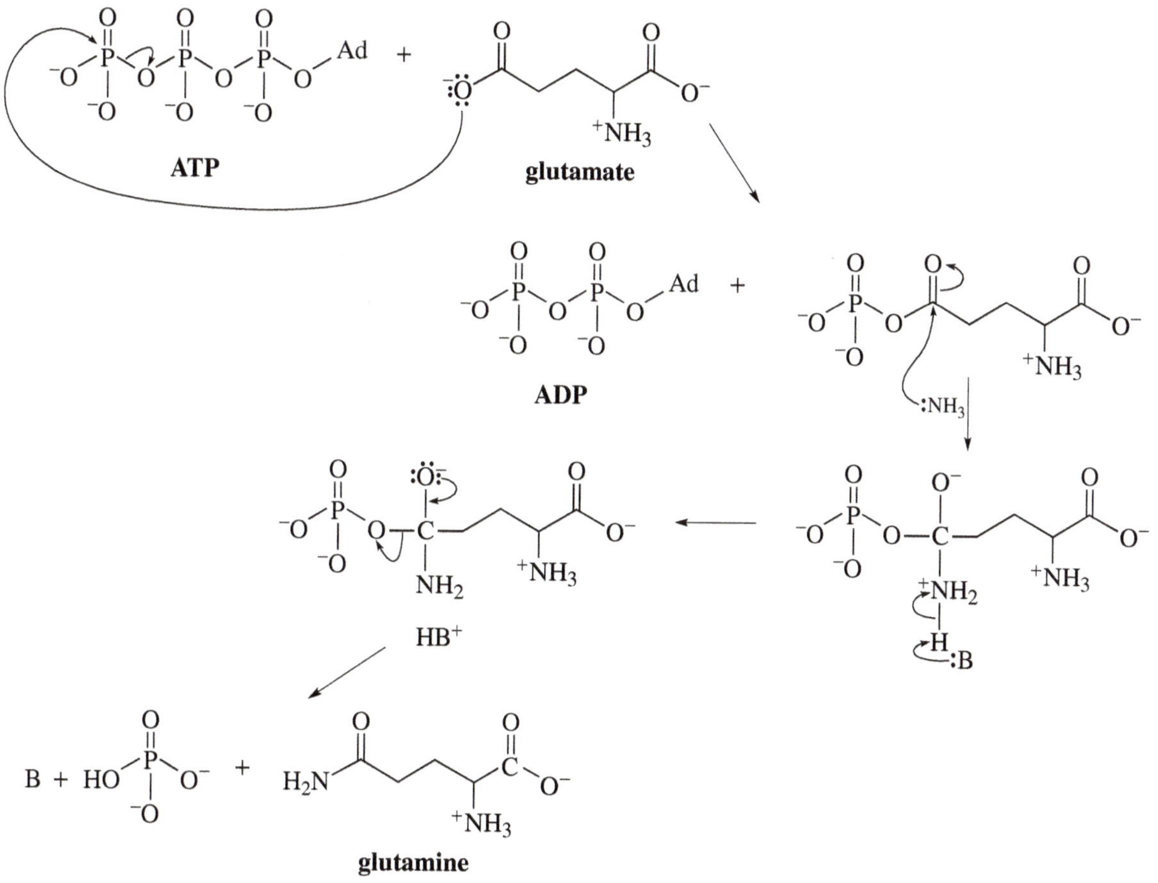

25. **a.** catabolic **b.** catabolic

26. The OH group attacks the g-phosphorus of ATP.

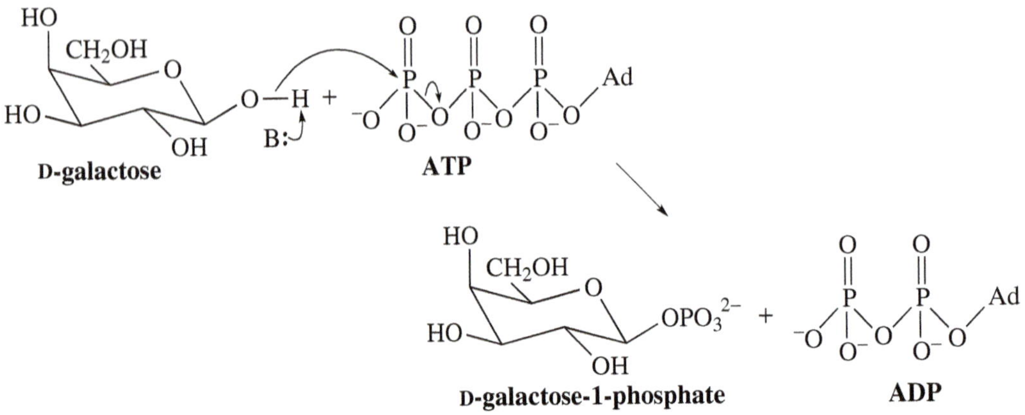

HO, CH_2OH, O, HO, OH, O—H + B:, O, O, O, P, P, P, Ad, ^-O, O^-, O, O^-, O, O^-, O

D-galactose **ATP**

HO, CH_2OH, O, HO, OH, OPO_3^{2-} + O, O, P, P, Ad, ^-O, O^-, O, O^-, O

D-galactose-1-phosphate **ADP**

27. The hydrogen on the α-carbon.

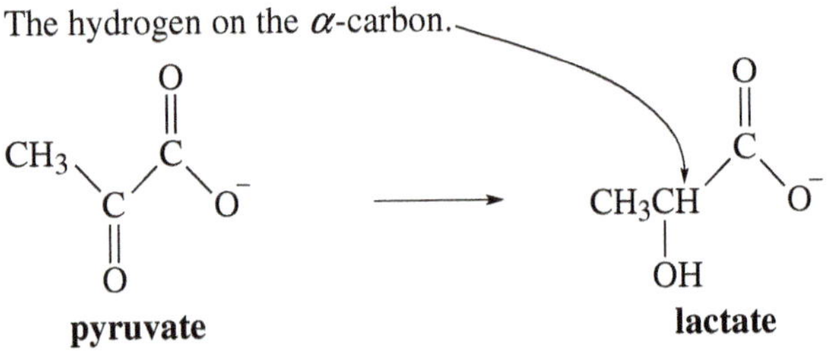

CH_3, C, C, O, O^-, O → CH_3CH, C, O, O^-, OH

pyruvate **lactate**

28.

$^+NH_3$, ^-OOC, S, CH_3 + O, O, O, P, P, P, ^-O, O^-, O, O^-, O, O^-, O, NH_2, N, N, N, N, O, OH OH

↓

O, O, O, P, P, P, ^-O, O^-, O, O^-, O, O^-, O^- + $^+NH_3$, ^-OOC, ^+S, CH_3, NH_2, N, N, N, N, O, OH OH

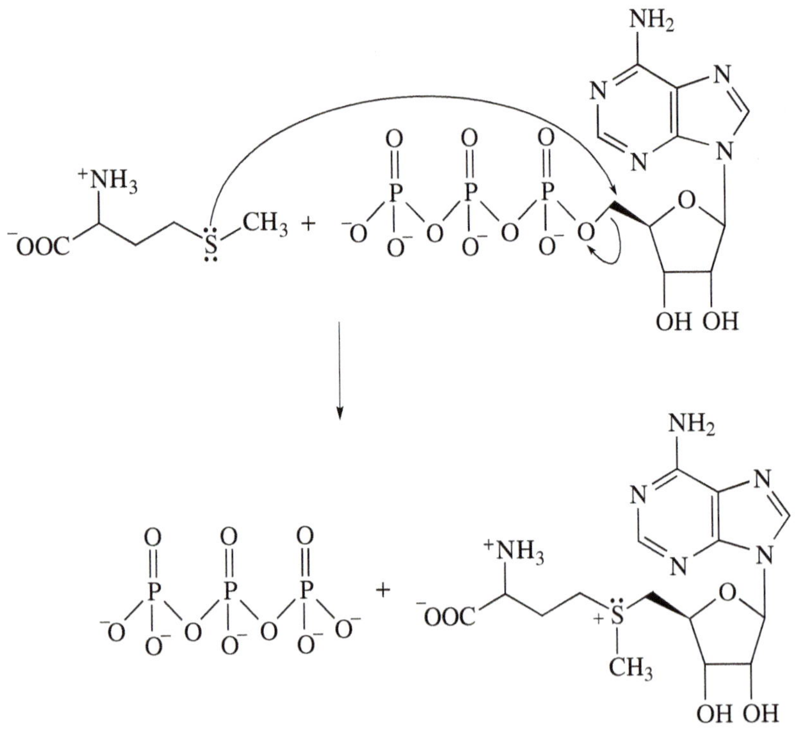

29. **a.** reactions 1 and 3 **b.** reactions 2, 5, and 8 **c.** reaction 6 **d.** reaction 9

30. the conversion of citrate to isocitrate
the conversion of fumarate to (*S*)-malate

31. **a.** the conversion of pyruvate to acetyl-CoA
b. It is catalyzed by the pyruvate dehydrogenase complex, a group of 3 enzymes and 5 coenzymes (pages 1076–1077 in the text).

32.

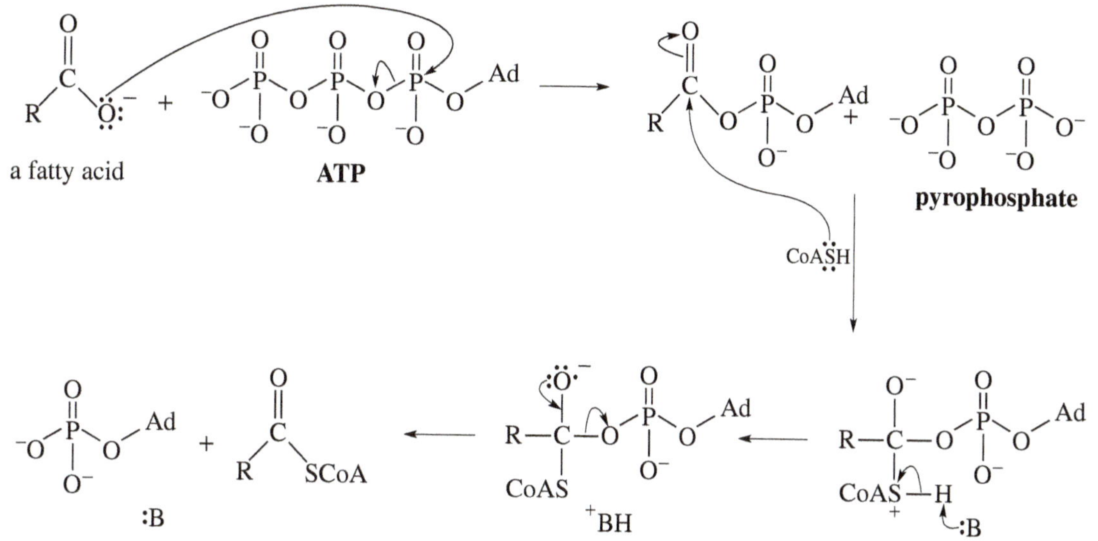

a fatty acid + **ATP** → acyl adenylate + **pyrophosphate**

CoASH

:B + acyl-CoA ← tetrahedral intermediate ($^+$BH) ← tetrahedral intermediate (:B)

33.

α-ketoglutarate —reduction→

34. The label will be on the phosphate group that is attached to the enzyme (phosphoglycerate mutase) that catalyzes the isomerization of 3-phosphoglycerate to 2-phosphoglycerate.

35. If you examine the mechanism for the isomerism of glucose-6-phosphate to fructose-6-phosphate on page 1056 of the text, you can see that C-1 in D-glucose is also C-1 in D-fructose.

Now if you examine the mechanism for the aldolase-catalyzed cleavage of fructose-1,6-bisphosphate to form dihydroxyacetone phosphate glyceraldehyde-3-phosphate on page 1058 of the text, you can see which carbons in D-glucose correspond to the carbons in dihydroxyacetone phosphate and D-glyceraldehyde-3-phosphate.

Then if you look at the bottom of page 1109, you will see how the carbons in dihydroxyacetone phosphate correspond to the carbons in D-glyceraldehyde-3-phosphate.

D-glucose-6-phosphate → **D-fructose-1,6-bisphosphate** → **dihydroxyacetone phosphate** → **D-glyceraldehyde-3-phosphate**

D-glyceraldehyde-3-phosphate → **D-glyceraldehyde-3-phosphate** → **pyruvate**

3 and 4	COO^-
2 and 5	$C=O$
1 and 6	CH_3

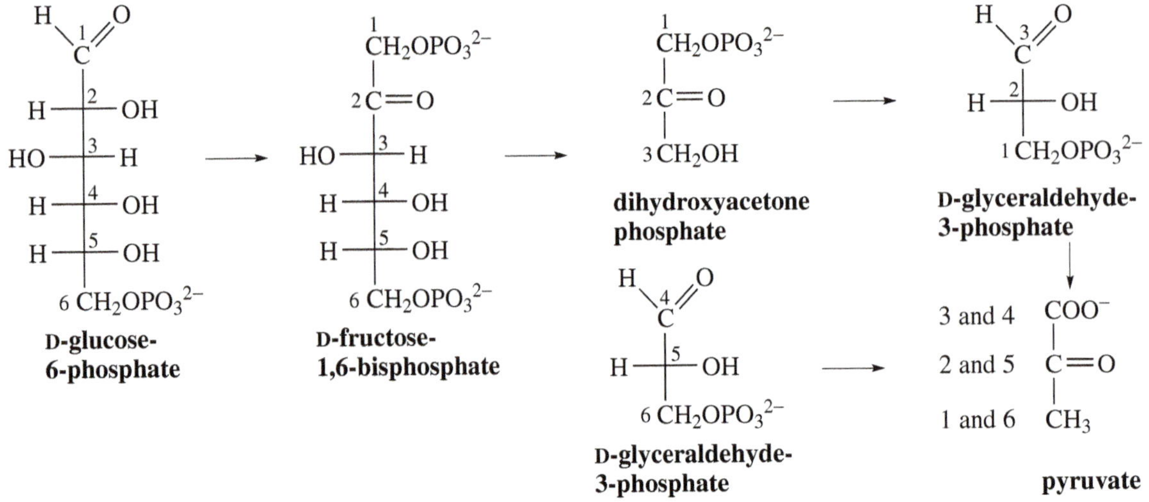

Finally, we see how the carbons in D-glyceraldehyde-3-phosphate correspond to the carbons in pyruvate. Therefore, both C-3 and C-4 of glucose become a carboxyl group in pyruvate.

36. Pyruvate loses its carboxyl group when it is converted to ethanol. Because the carboxyl group is C-3 or C-4 of glucose, half of the ethanol molecules contain C-1 and C-2 of glucose and the other half contain C-5 and C-6 of glucose.

37. At the beginning of a fast, blood glucose levels would be normal.
After a 24-hour fast, blood glucose levels would be very low because both dietary glucose and stored glucose (glycogen) have been depleted and glucose cannot be synthesized because of the deficiency of fructose-1,6-bisphosphatase.

38. The conversion of pyruvate to lactate is a reversible reaction. Lactate can be converted back to pyruvate by oxidation.

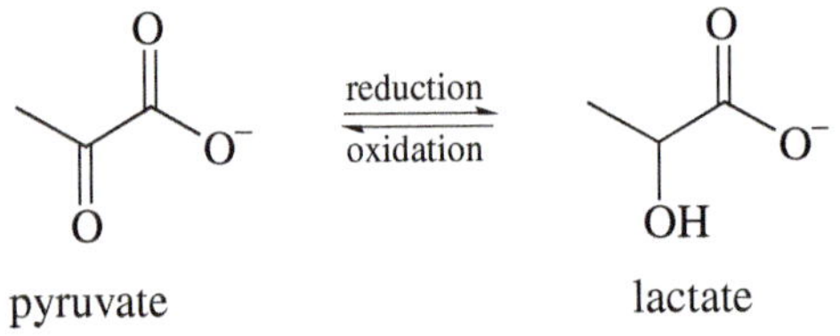

pyruvate ⇌ lactate (reduction / oxidation)

The conversion of pyruvate to acetaldehyde is not a reversible reaction because it is a decarboxylation. The CO_2 cannot be put back onto acetaldehyde.

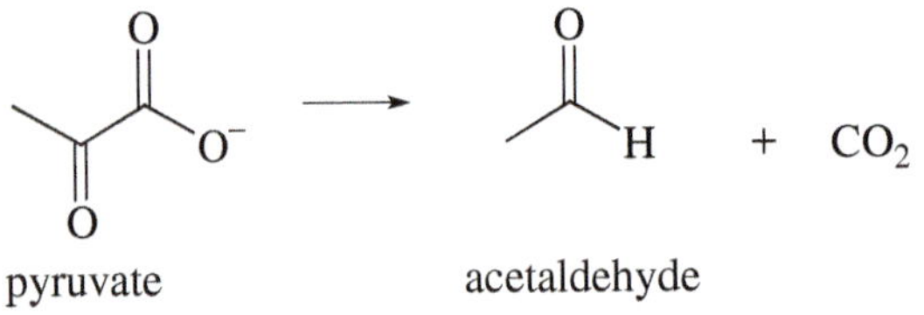

pyruvate → acetaldehyde + CO_2

39. The β-oxidation of a molecule of a 16-carbon fatty acyl-CoA forms 8 molecules of acetyl-CoA.

40. Each molecule of acetyl-CoA forms 2 molecules of CO_2. Therefore, the 8 molecules of acetyl-CoA obtained from a molecule of a 16-carbon fatty acyl-CoA will form 16 molecules of CO_2.

41. No ATP is formed from β-oxidation.

42. Each molecule of acetyl-CoA that is cleaved from the 16-carbon fatty acyl-CoA forms 1 molecule of $FADH_2$ and 1 molecule of NADH. Because a 16-carbon fatty acyl-CoA undergoes 7 cleavages, 7 molecules of $FADH_2$ and 7 molecules of NADH are formed from the 16-carbon fatty acyl-CoA.

43. Because each NADH forms 2.5 molecules of ATP and each $FADH_2$ forms 1.5 molecules of ATP in oxidative phosphorylation, the 7 molecules of NADH form 17.5 molecules of ATP and the 7 molecules of $FADH_2$ form 10.5 molecules of ATP. Therefore, 28 molecules of ATP are formed.

44. We have seen that each molecule of acetyl-CoA that enters the citric acid cycle forms 10 molecules of ATP (Section 24.10). A molecule of a 16-carbon fatty acid will form 8 molecules of acetyl-CoA. These will form 80 molecules of ATP. When these are added to the number of ATP molecules formed from the NADH and $FADH_2$ generated in β-oxidation (80 + 28), we see that 108 molecules of ATP are formed from complete metabolism of a 16-carbon saturated fatty acyl-CoA.

45. Each molecule of glucose, while being converted to 2 molecules of pyruvate, forms 2 molecules of ATP and 2 molecules of NADH.

The 2 molecules of pyruvate form 2 molecules of NADH while being converted to 2 molecules of acetyl-CoA.

Each molecule of acetyl-CoA that enters the citric acid cycle forms 3 molecules of NADH, 1 molecule of $FADH_2$ and 1 molecule of ATP. Therefore, the 2 molecules of acetyl-CoA obtained from glucose form 6 molecules of NADH, 2 molecules of $FADH_2$, and 2 molecules of ATP.

Therefore, each molecule of glucose forms 4 molecules of ATP, 10 molecules of NADH (2 + 2 + 6), and 2 molecules of $FADH_2$.

Because each NADH forms 2.5 molecules of ATP and each $FADH_2$ forms 1.5 molecules of ATP, 1 molecule of glucose forms 4 + (10 × 2.5) + (2 × 1.5) molecules of ATP. That is, each molecule of glucose forms 32 molecules of ATP.

46. Pyruvate can be converted to alanine (transamination), oxaloacetate (carboxylation), lactate (reduction), and acetyl-CoA (by the pyruvate dehydrogenase complex).

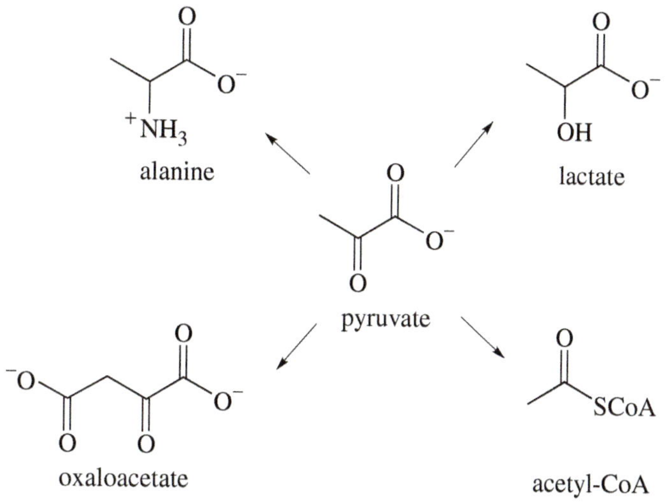

alanine
lactate
pyruvate
oxaloacetate
acetyl-CoA

47. The conversion of propionyl-CoA to methylmalonyl-CoA requires biotin (vitamin H).
The conversion of methylmalonyl-CoA to succinyl-CoA requires coenzyme B_{12} (vitamin B_{12}).

48. In Problem 35, we saw how the carbons in D-glyceraldehyde-3-phosphate correspond to the carbons in pyruvate.

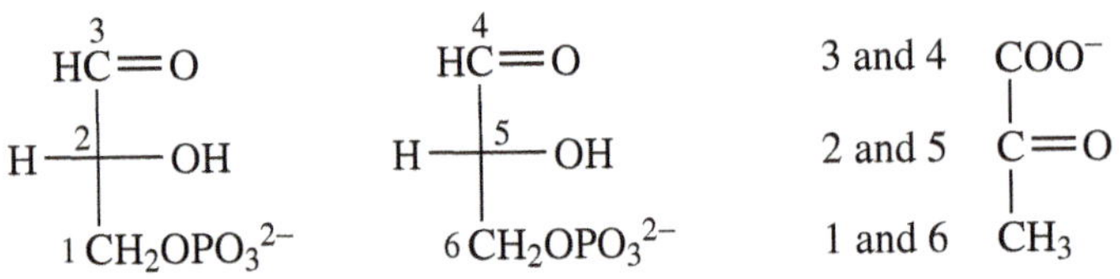

D-glyceraldehyde-3-phosphate	D-glyceraldehyde-3-phosphate	carbons	pyruvate
3 HC=O	4 HC=O	3 and 4	COO^-
H–2–OH	H–5–OH	2 and 5	C=O
1 $CH_2OPO_3^{2-}$	6 $CH_2OPO_3^{2-}$	1 and 6	CH_3

D-glyceraldehyde-3-phosphate **pyruvate**

Now we can answer the questions. The label in pyruvate is indicated by *.

a. COO^-–C=O–*CH_3

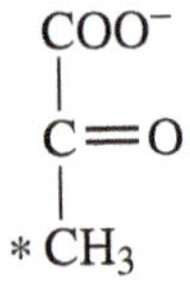

b. COO^-–*C=O–CH_3

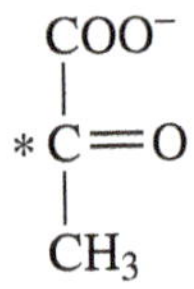

c. *COO^-–C=O–CH_3

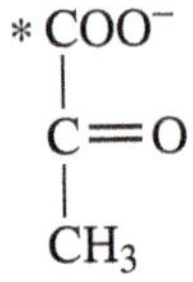

d. *COO^-–C=O–CH_3

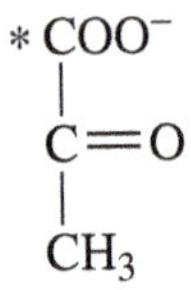

e. COO^-–*C=O–CH_3

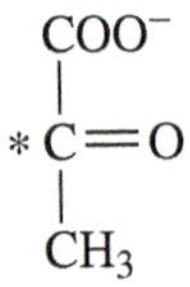

f. COO^-–C=O–*CH_3

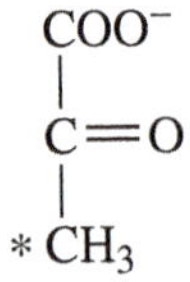

49.

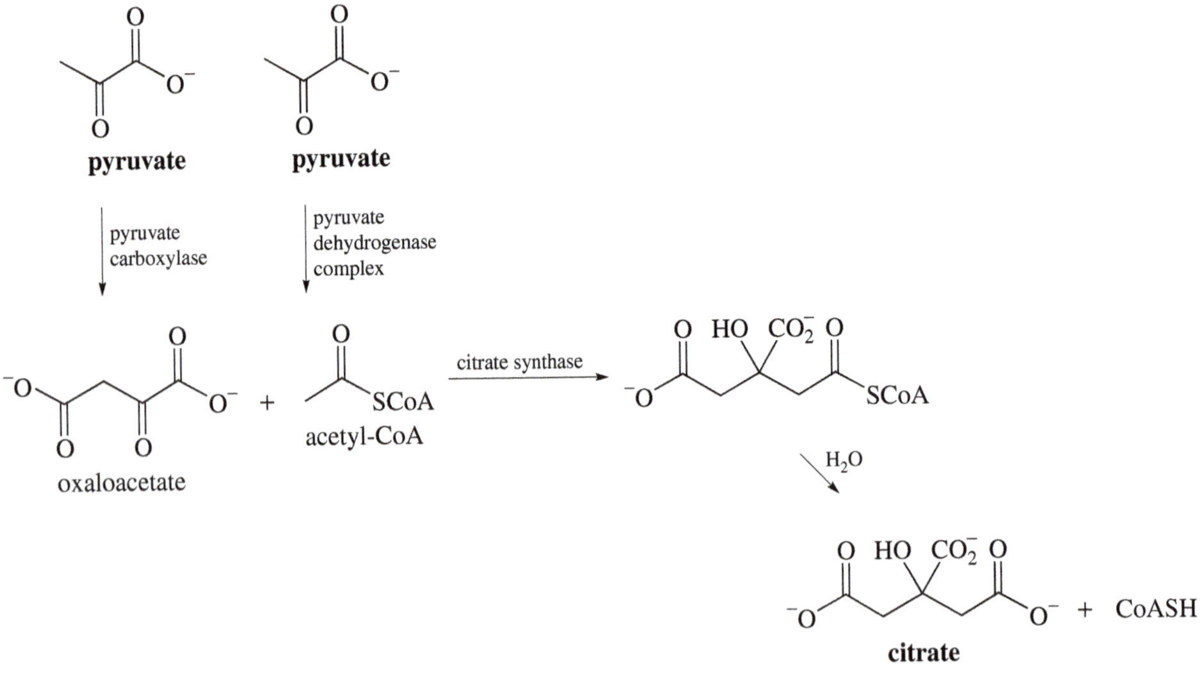

pyruvate → (pyruvate carboxylase) → oxaloacetate

pyruvate → (pyruvate dehydrogenase complex) → acetyl-CoA

oxaloacetate + acetyl-CoA (SCoA) → (citrate synthase) → citryl-CoA (O HO CO_2^- O, SCoA)

→ (H_2O) → citrate (O HO CO_2^- O) + CoASH

50. A Claisen condensation between two molecules of acetyl-CoA forms acetoacetyl-CoA that, when hydrolyzed, forms acetoacetate.

Claisen condensation

acetyl-CoA

hydrolysis

$CoASH$ + **acetoacetate** ← **acetoacetyl-CoA** + $CoASH$

Acetoacetate can undergo decarboxylation to form acetone, or it can be reduced to 3-hydroxybutyrate.

acetoacetate → (decarboxylation, H^+) → CO_2 + enol → (enolization) → **acetone**

acetoacetate + NAD—H → (reduction) → NAD^+ → (H^+) → **3-hydroxybutyrate**

51. The mechanism for the conversion of fructose-1,6-bisphosphate to glyceraldehyde-3-phosphate and dihydroxyacetone phosphate is shown on page 1058 of the text.

The mechanism for the conversion of dihydroxyacetone phosphate to glyceraldehyde-3-phosphate is shown on page 1109 of the text.

From these mechanisms, you can see that the label (*) was at C-1 in glyceraldehyde-3-phosphate.

aldolase, see page 1058 of the text

triosephosphate isomerase, see page 1109 of the text

52. **a.** UDP-galactose and UDP-glucose are C-4 epimers. NAD⁺ oxidizes the C-4 OH group of UDP-galactose to a ketone. When NADH reduces the ketone back to an OH, it attacks the sp^2 carbon from above the plane, forming the C-4 epimer of the starting material.

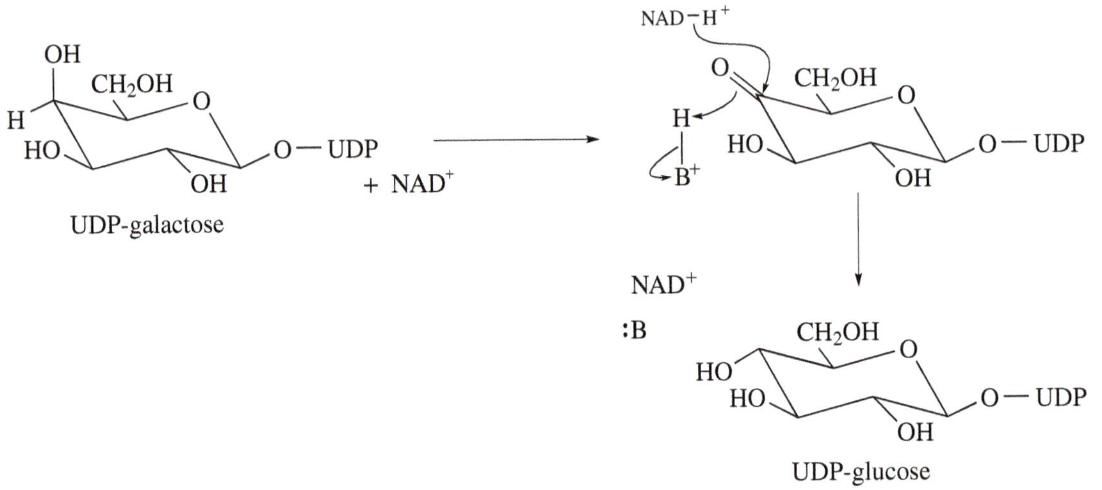

UDP-galactose + NAD⁺ → (NAD–H⁺, :B⁺) → NAD⁺, :B + UDP-glucose

UDP-galactose

UDP-glucose

b. The enzyme is called an epimerase because it converts a compound into an epimer (in this case, a C-4 epimer).

53. Because the compound that would react in the second step with the activated carboxylic acid group is excluded from the incubation mixture, the reaction between the carboxylate ion and ATP will come to equilibrium.

If radioactively labeled pyrophosphate is put into the incubation mixture, ATP will become radioactive if the mechanism involves attack on the α-phosphorus because pyrophosphate is a reactant in the reverse reaction that forms ATP.

ATP will not become radioactive if the mechanism involves attack on the β-phosphorus because pyrophosphate is not a reactant in the reverse reaction that forms ATP. (In other words, because pyrophosphate is not a product of the reaction, it cannot become incorporated into ATP in the reverse reaction.)

attack on the α-phosphorus

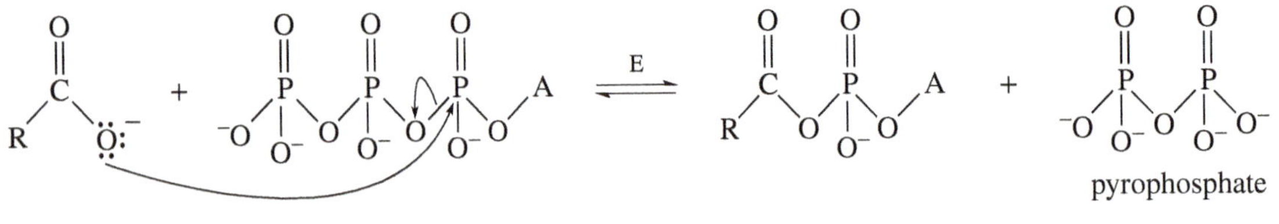

pyrophosphate

attack on the β-phosphorus

AMP

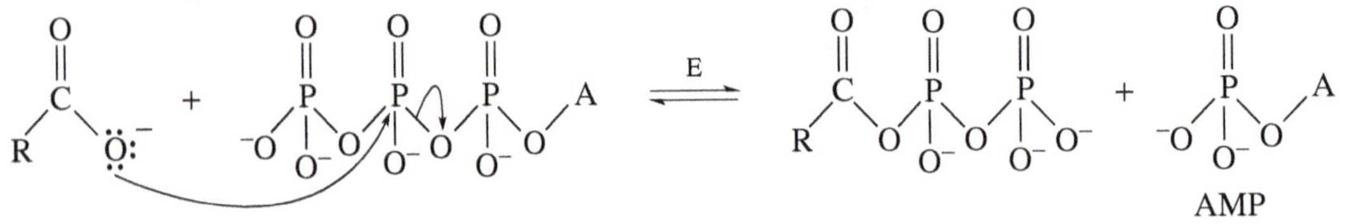

54. If radioactive AMP is added to the reaction mixture, the results will be the opposite. If the mechanism involves attack on the α-phosphorus, ATP will not become radioactive because AMP is not a reactant in the reverse reaction that forms ATP. If the mechanism involves attack on the β-phosphorus, ATP will become radioactive because AMP is a reactant in the reverse reaction that forms ATP.

Chapter 24 Practice Test

1. Draw a structure for the following:

a. the intermediate formed when a nucleophile (RO^-) attacks the γ-phosphorus of ATP

b. an acyl pyrophosphate

c. pyrophosphate

2. Fill in the six blanks in the following scheme:

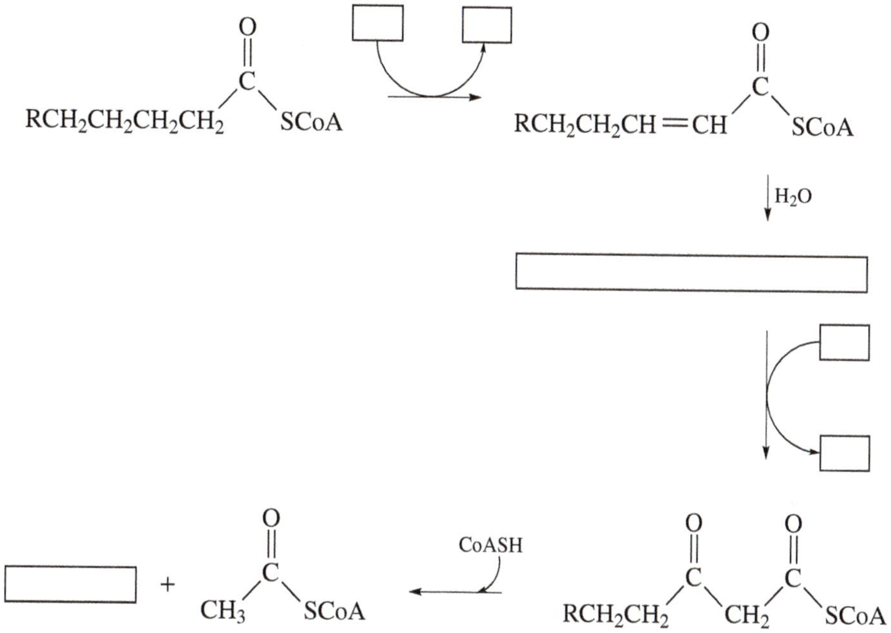

$RCH_2CH_2CH_2CH_2C(=O)SCoA$ → (☐ → ☐) → $RCH_2CH_2CH{=}CHC(=O)SCoA$

↓ H_2O

☐

↓ (☐ → ☐)

$RCH_2CH_2C(=O)CH_2C(=O)SCoA$ → (CoASH) → ☐ + $CH_3C(=O)SCoA$

3. Which of the following are not citric acid cycle intermediates: fumarate, acetate, citrate?

4. Which provide energy to the cell: anabolic reactions or catabolic reactions?

5. What compounds are formed when proteins undergo the first stage of catabolism?

6. What compound is formed when fatty acids undergo the second stage of catabolism?

7. Indicate whether each of the following statements is true or false.

a. Each molecule of $FADH_2$ forms 2.5 molecules of ATP in the fourth stage of catabolism. T F

b. $FADH_2$ is oxidized to FAD. T F

c. NAD^+ is oxidized to NADH. T F

d. Acetyl-CoA is a citric acid cycle intermediate. T F

Chapter 25
The Organic Chemistry of Lipids

Important Terms

angular methyl group	a methyl substituent at the 10- or 13-position of a steroid ring system.
carotenoid	compounds responsible for the red and orange colors of fruits, vegetables, and fall leaves.
cholesterol	a steroid that is the precursor of all other steroids.
detergent	a sodium or potassium salt of a benzenesulfonic acid.
dimethylallyl pyrophosphate	a precursor needed for the biosynthesis of terpenes, which is biosynthesized from isopentenyl pyrophosphate.
diterpene	a terpene that contains 20 carbons.
fat	a triester of glycerol that exists as a solid at room temperature.
fatty acid	a long straight-chain carboxylic acid.
hormone	a chemical messenger that stimulates or inhibits some process in target tissues.
isopentenyl pyrophosphate	the starting material for the biosynthesis of terpenes.
lipid	a water-insoluble compound found in a living system.
lipid bilayer	two layers of phosphoacylglycerols arranged so that their polar heads are on the outside and their nonpolar fatty acid chains are on the inside.
membrane	the material that surrounds the cell to isolate its contents.
micelle	a spherical aggregation of molecules, each with a long hydrophobic tail and a polar head, arranged so that the polar head points to the outside of the sphere.
mixed triacylglycerol	a triacylglycerol in which the fatty acid components are different.
monoterpene	a terpene that contains 10 carbons.
oil	a triester of glycerol that exists as a liquid at room temperature.
phosphatidic acid	a phosphoglyceride with one OH group of glycerol esterified with phosphoric acid.
phosphoglyceride (phosphoacylglycerol)	formed when two OH groups of glycerol form esters with fatty acids and the terminal OH group is part of a phosphodiester.
phospholipid	a lipid that contains a phosphate group.
polyunsaturated fatty acid	a fatty acid with more than one double bond.

prostaglandin a carboxylic acid, derived from arachidonic acid, that is responsible for a variety of physiological functions.

protein prenylation the process of attaching farnesyl and geranylgeranyl groups to proteins to allow them to become anchored to membranes.

saponification hydrolysis of a fat under basic conditions.

sesquiterpene a terpene that contains 15 carbons.

simple triacylglycerol a triacylglycerol in which the three fatty acid components are the same.

soap a sodium or potassium salt of a fatty acid.

sphingolipid a lipid that contains sphingosine.

squalene a triterpene that is a precursor of steroid molecules.

steroid a class of compounds that contains a steroid ring system.

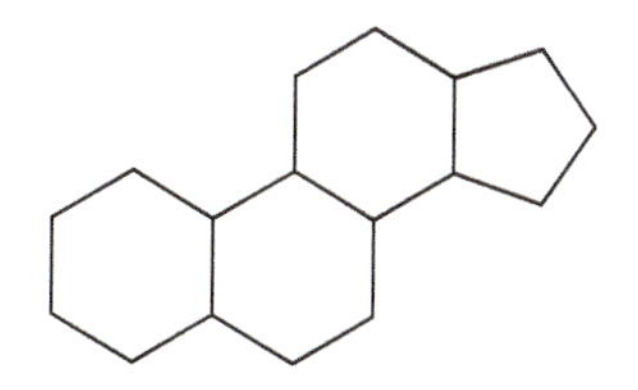

steroid ring system

α-substituent a substituent on the opposite side of a steroid ring system from the angular methyl groups (the bottom face as typically drawn).

β-substituent a substituent on the same side of a steroid ring system as the angular methyl groups (the top face as typically drawn).

terpene a lipid isolated from a plant that contains carbon atoms in multiples of five.

terpenoid a terpene that contains oxygen.

tetraterpene a terpene that contains 40 carbons.

trans fused two rings fused together such that if one ring is considered to be two substituents of the other ring, the substituents would be opposite sides of the first ring.

triacylglycerol the compound formed when the three OH groups of glycerol are esterified with fatty acids.

triterpene a terpene that contains 30 carbons.

wax an ester formed from a long straight-chain carboxylic acid and a long straight-chain alcohol.

Solutions to Problems

1. **a.** Stearic acid has the higher melting point, because it has two more methylene groups (giving it a greater surface area and, therefore, stronger London dispersion forces) than palmitic acid.

b. Stearic acid has the higher melting point because stearic acid is saturated, so its molecules can pack closer together than can the molecules of oleic acid that has a double bond.

c. Oleic acid has the higher melting point because it has two more methylene groups than palmitoleic acid (see part **a**).

d. Oleic acid has the higher melting point, because it has one cis double bond, whereas linoleic acid has two cis double bonds. The greater the number of double bonds, the harder for the molecules to pack closely together.

2. Glyceryl tripalmitate has a higher melting point, because the carboxylic acid components are saturated and can, therefore, pack more closely together than the cis-unsaturated carboxylic acid components of glyceryl tripalmitoleate.

3. To be optically inactive, the fat must have a plane of symmetry. In other words, the fatty acid residues at C-1 and C-3 must be identical. Therefore, stearic acid must form the esters at C-1 and C-3.

$$
\begin{array}{l}
CH_2-O-\overset{\overset{\displaystyle O}{\|}}{C}-(CH_2)_{16}CH_3 \\
| \\
CH-O-\overset{\overset{\displaystyle O}{\|}}{C}-(CH_2)_{10}CH_3 \\
| \\
CH_2-O-\overset{\overset{\displaystyle O}{\|}}{C}-(CH_2)_{16}CH_3
\end{array}
$$

4. To be optically active, the fat must not have a plane of symmetry. Therefore, the two stearic acid residues must be attached to adjacent alcohol groups (either C-1 and C-2 or C-2 and C-3).

$$
\begin{array}{l}
CH_2-O-\overset{\overset{\displaystyle O}{\|}}{C}-(CH_2)_{16}CH_3 \\
| \\
CH-O-\overset{\overset{\displaystyle O}{\|}}{C}-(CH_2)_{16}CH_3 \\
| \\
CH_2-O-\overset{\overset{\displaystyle O}{\|}}{C}-(CH_2)_{10}CH_3
\end{array}
$$

5. Solved in the text.

6. The identities of R^1 and R^2 have no effect on the configuration of the asymmetric center at C-2 because:
the group with priority #1 is always the carboxyl group on C-2;
the group with priority #2 is always the group that contains the phosphorus;
the group with priority #3 is always the group that contains the other carboxyl group;
the group with priority #4 is always the H on C-2.

7. Because the interior of a membrane is nonpolar and the surface of a membrane is polar, integral proteins have a higher percentage of nonpolar amino acids.

8. The bacteria could synthesize phosphoacylglycerols with more saturated fatty acids. These triacylglycerols would pack more tightly in the lipid bilayer and, therefore, would have higher melting points and be less fluid.

9. Membranes must be kept in a semifluid state to allow transport across them. Cells closer to the hoof of an animal are going to be in a colder average environment than cells closer to the body. Therefore, the cells closer to the hoof have a higher degree of unsaturation to give them a lower melting point so that the membranes will not solidify at the colder temperature.

10.

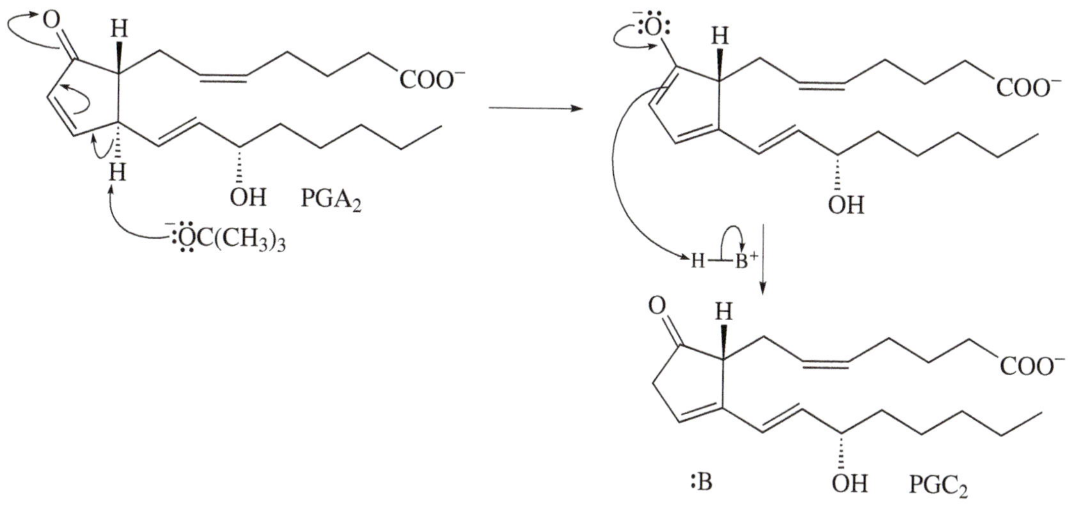

O H COO⁻ OH PGA_2 $^{-}$:ÖC$(CH_3)_3$

:Ö: H COO⁻ OH H—B⁺

O H COO⁻ :B OH PGC_2

11. Solved in the text.

12.

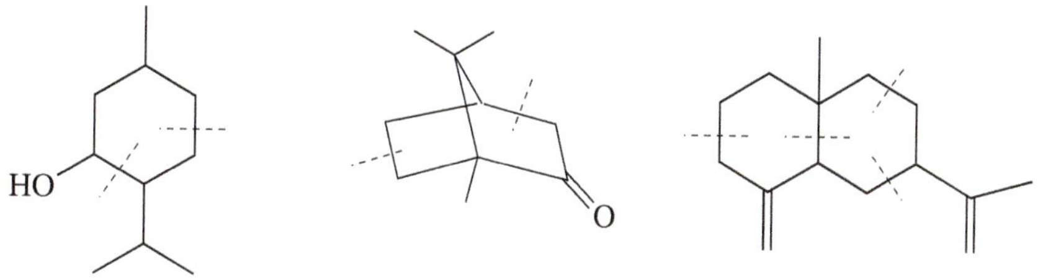

HO

menthol **camphor** **β-selinene**

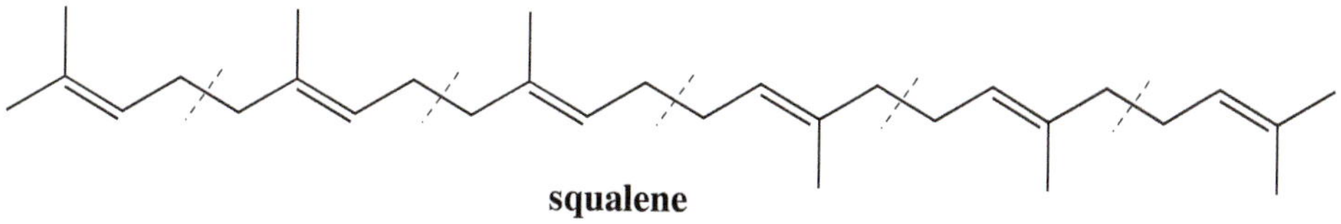

squalene

13. The fact that the tail-to-tail linkage occurs in the exact center of the molecule suggests that the two halves are synthesized (in a head-to-tail fashion) and then joined in a tail-to-tail linkage.

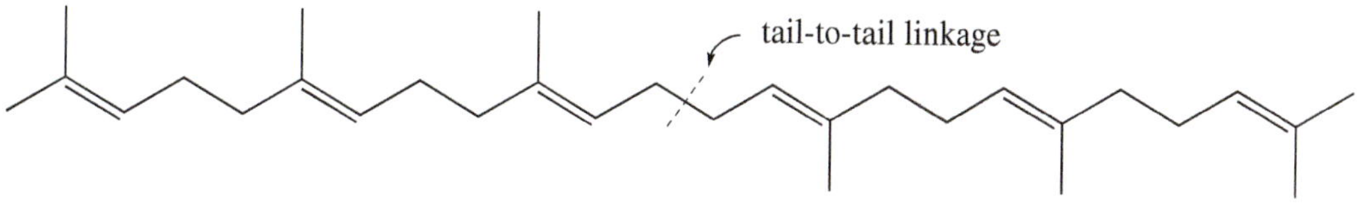

tail-to-tail linkage

14. Squalene, lycopene, and β-carotene are all synthesized in the same way. In each case, two halves are synthesized (in a head-to-tail fashion) and then joined in a tail-to-tail linkage.

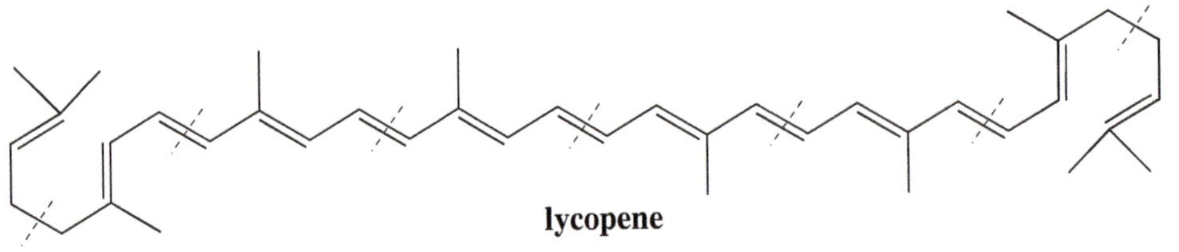

lycopene

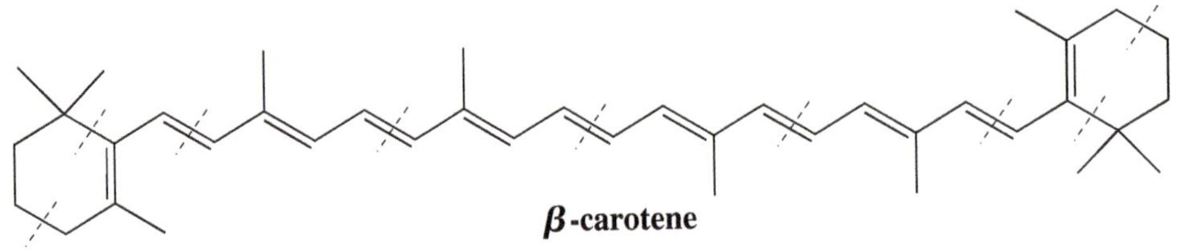

$\boldsymbol{\beta}$**-carotene**

15.

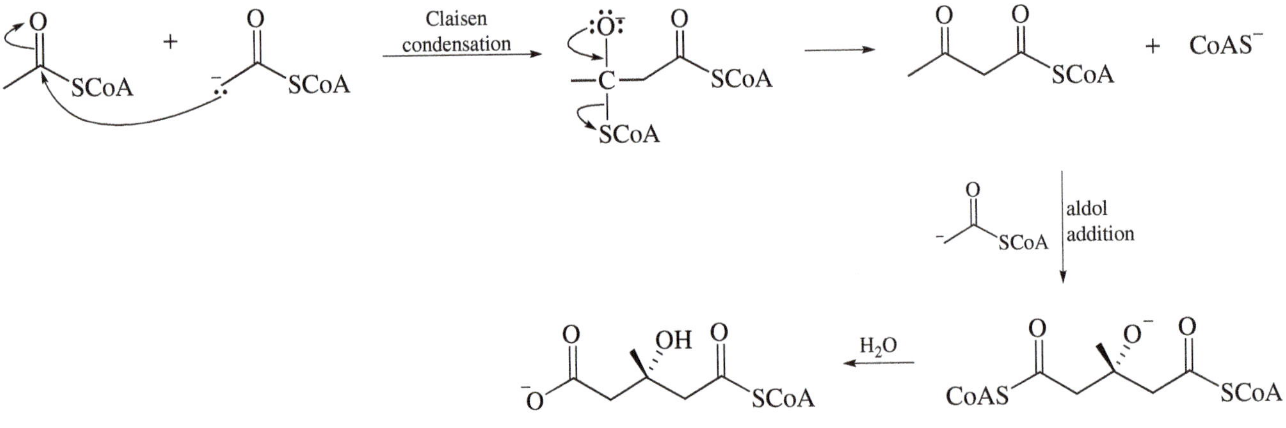

Claisen condensation

SCoA + CoAS$^-$

aldol addition

H_2O

hydroxymethylglutaryl-CoA

16. One equivalent of NADPH is required to reduce the thioester to an aldehyde. The second equivalent of NADPH is needed to reduce the aldehyde to an alcohol.

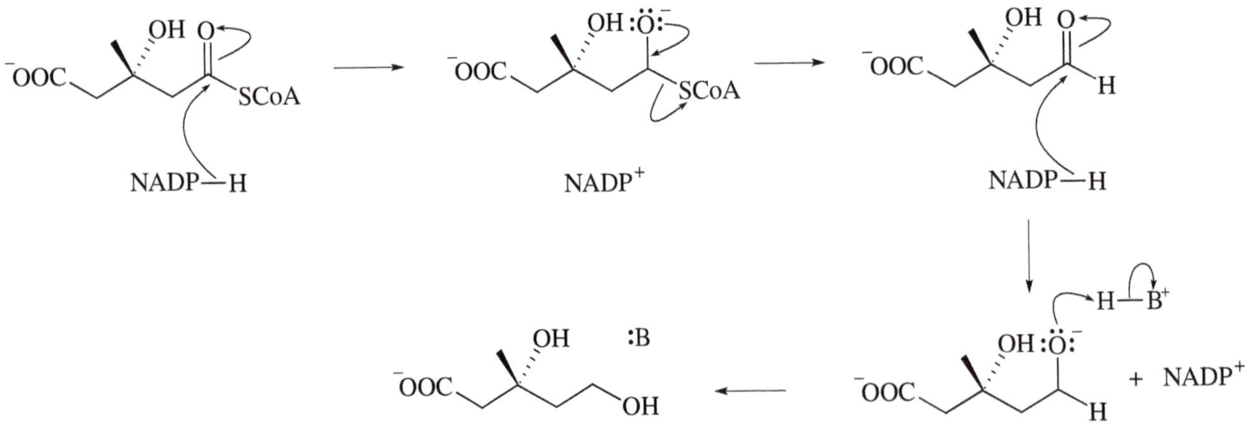

NADP—H

NADP$^+$

NADP—H

+ NADP$^+$

17.

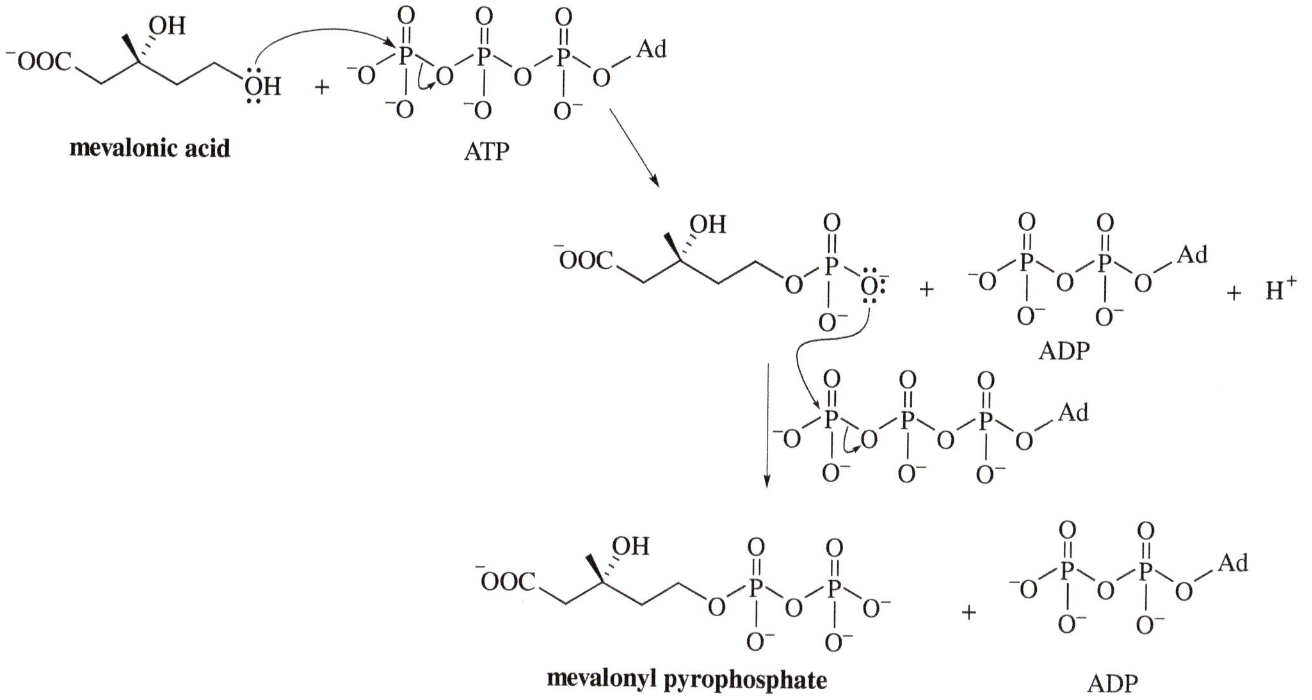
OH
ÖH
mevalonic acid
ATP
Ad
ADP
H+
mevalonyl pyrophosphate
ADP

18. Solved in the text.

19.

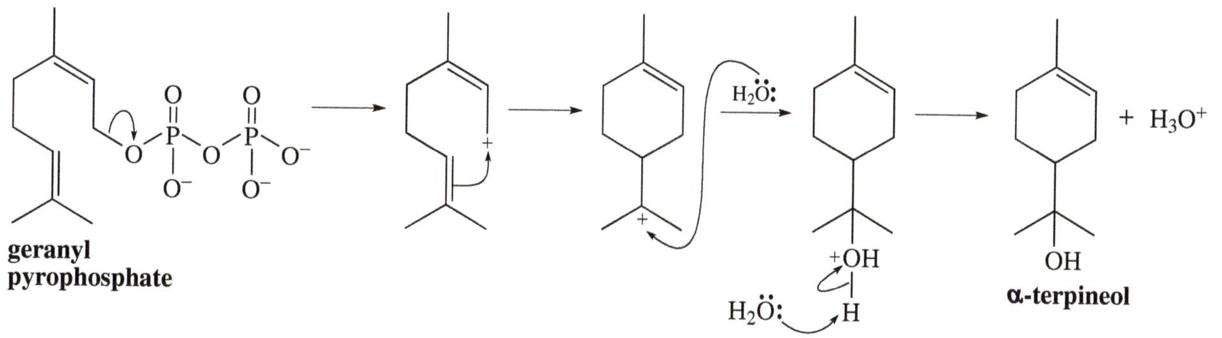
geranyl pyrophosphate
H2Ö:
H3O+
+OH
H
OH
α-terpineol

20.

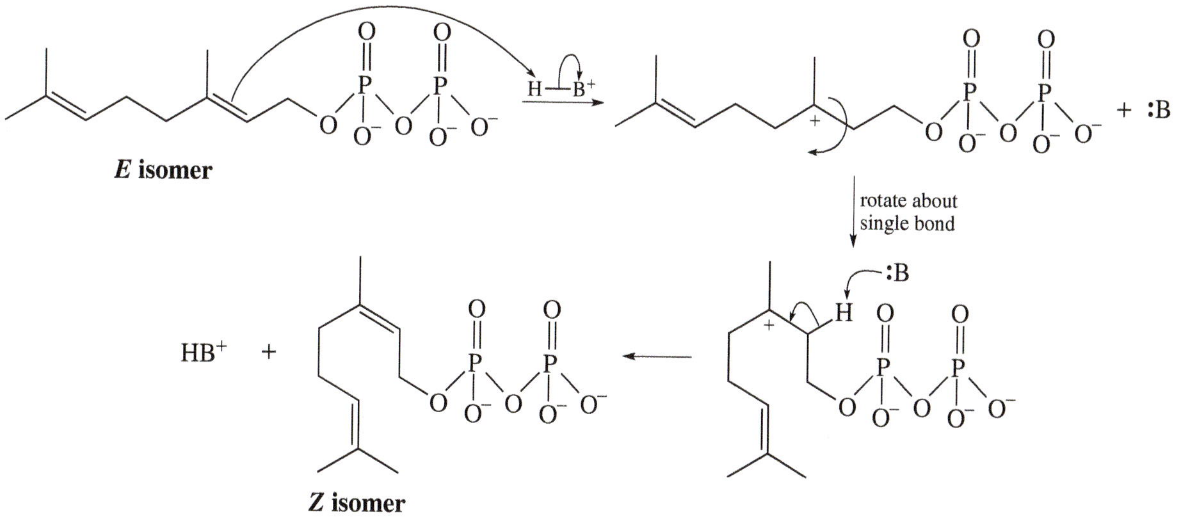
E isomer
H—B+
:B
rotate about single bond
HB+
Z isomer

21. It tells you that the reaction is an S_N1 reaction because the fluoro-substituted carbocation is less stable than the non-fluoro-substituted carbocation (due to the strongly electron-withdrawing fluoro substituent), so it would form more slowly.

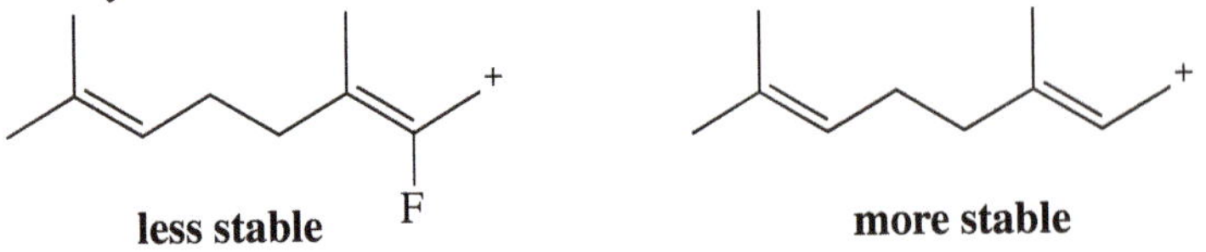

F

less stable **more stable**

If the reaction had been an S_N2 reaction, the fluoro-substituted compound would have reacted more rapidly than the non-fluoro-substituted carbocation because the electron-withdrawing fluoro substituent would make the compound more susceptible to nucleophilic attack.

22.

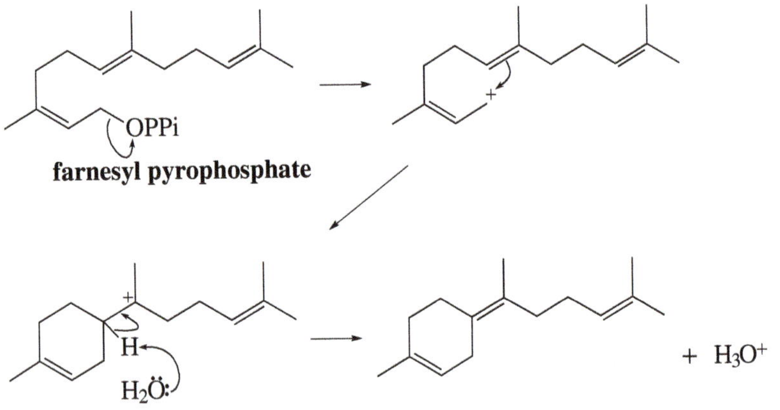

OPPi

farnesyl pyrophosphate

H

$H_2\ddot{O}$:

+ H_3O^+

23. Solved in the text.

24. Because acetyl-CoA is converted into malonyl-CoA (see Section 25.8), mevalonyl pyrophosphate will contain three labeled carbons, which means that the α-terpineol (juniper oil) will contain six labeled carbons.

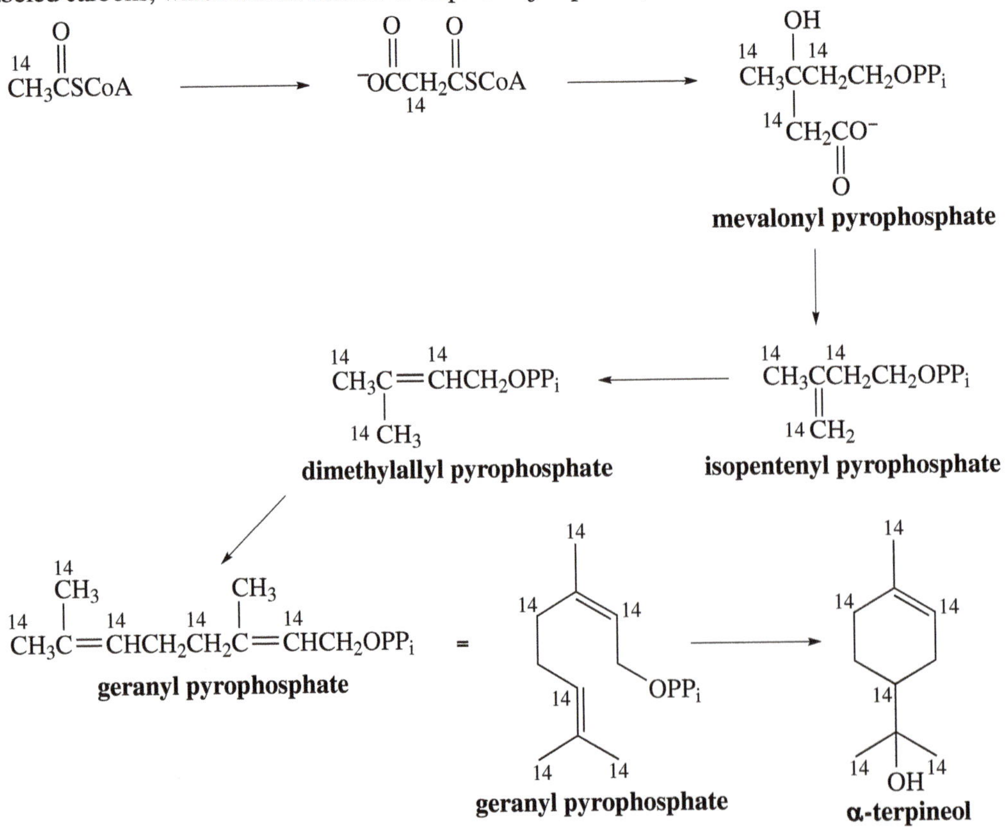

$^{14}CH_3C(=O)SCoA \longrightarrow {}^{-}OC(=O)\,^{14}CH_2C(=O)SCoA \longrightarrow$ $^{14}CH_3\,^{14}C(OH)(CH_2CH_2OPP_i)\,^{14}CH_2CO^-$ (C=O)

mevalonyl pyrophosphate

$^{14}CH_3\,^{14}C(=^{14}CH_2)CH_2CH_2OPP_i$

isopentenyl pyrophosphate

$^{14}CH_3\,^{14}C(^{14}CH_3)=CHCH_2OPP_i$

dimethylallyl pyrophosphate

$^{14}CH_3\,^{14}C(^{14}CH_3)=^{14}CHCH_2CH_2\,^{14}C(CH_3)=^{14}CHCH_2OPP_i$

geranyl pyrophosphate

=

14 14 14 14 OPP$_i$ 14 14

geranyl pyrophosphate

14 14 14 14 14 OH 14

α-terpineol

25. There are two 1,2-hydride shifts and two 1,2-methyl shifts. The last step is elimination of a proton.

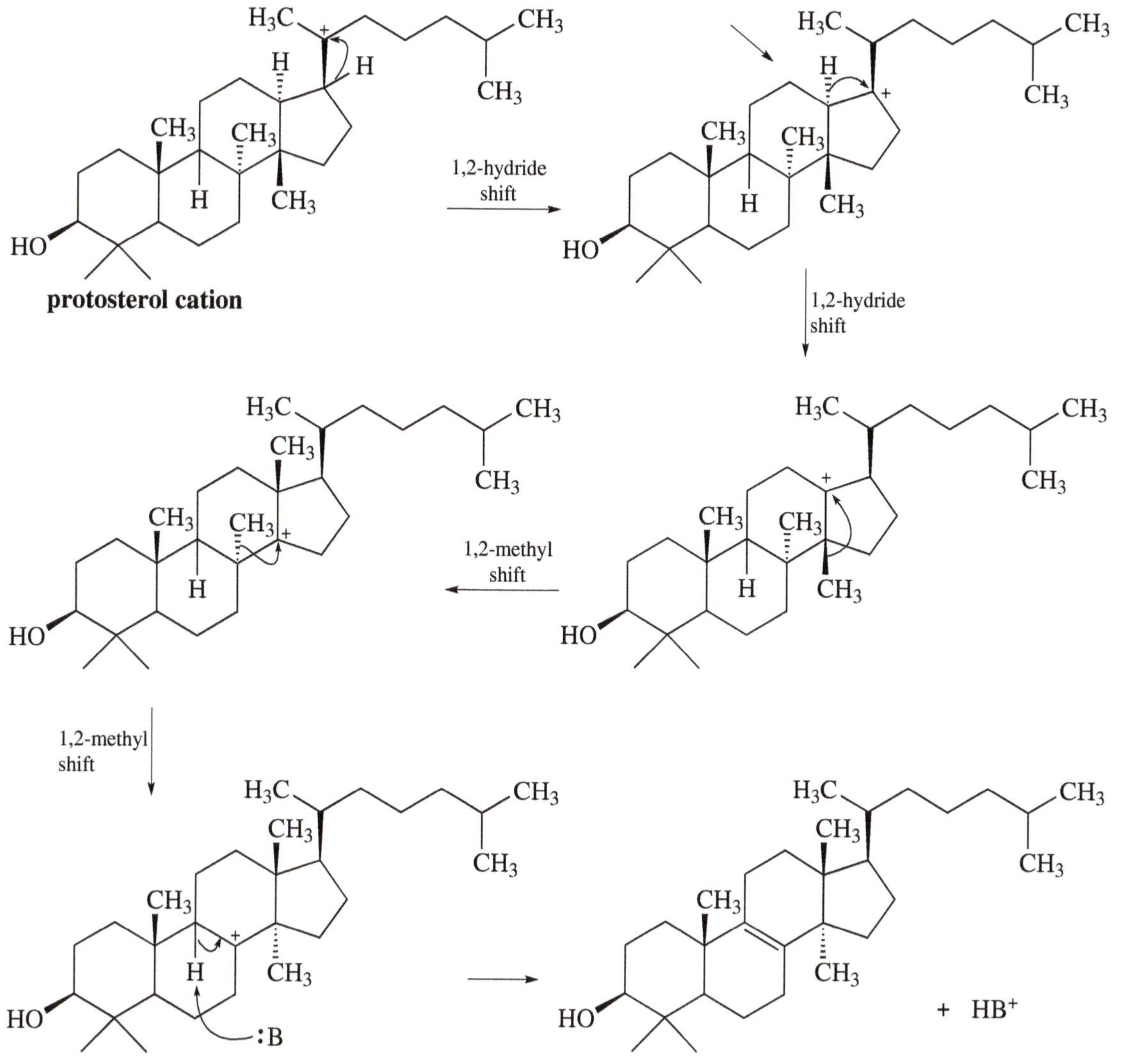

protosterol cation

1,2-hydride shift

1,2-hydride shift

1,2-methyl shift

1,2-methyl shift

+ HB^+

26. A ***β*-hydrogen** at C-5 means that the A and B rings are **cis** fused; an ***α*-hydrogen** at C-5 means that they are **trans** fused.

27. Because the OH substituent is on the **same side** of the steroid ring system as the angular methyl groups, it is a ***β*-substituent**.

28.

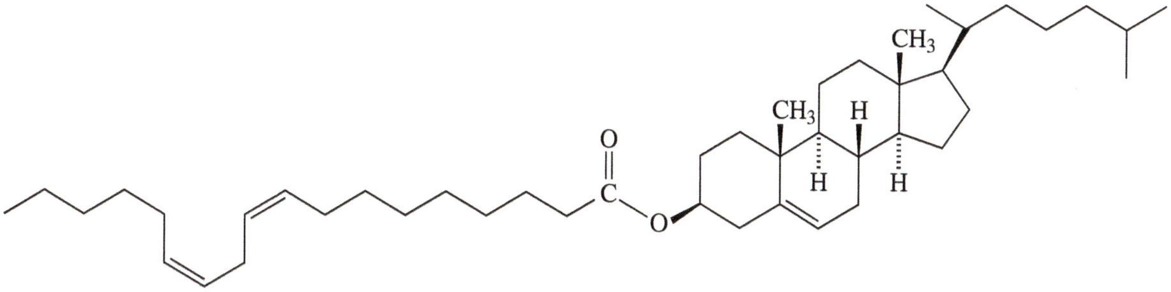

29. The hemiacetal is formed by reaction of the primary alcohol with the aldehyde.

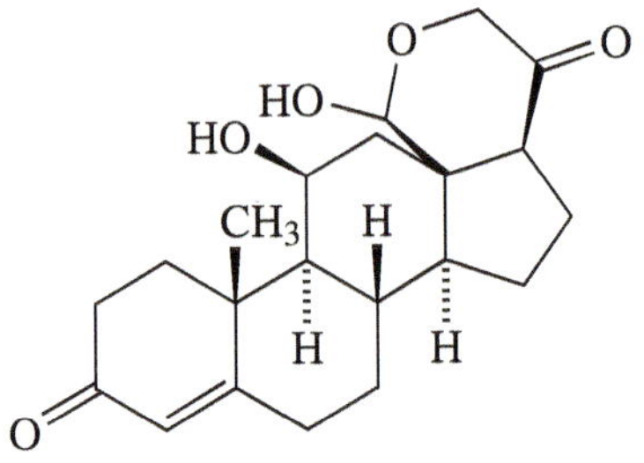

30. **a.** Chenodeoxycholic acid is missing the OH group on the C ring (at the top of the molecule) that cholic acid has.

b. Notice that because the methyl and hydrogen at the juncture of the A and B rings are on the same side, we know that the A and B rings are cis fused. The three OH groups are all in axial positions.

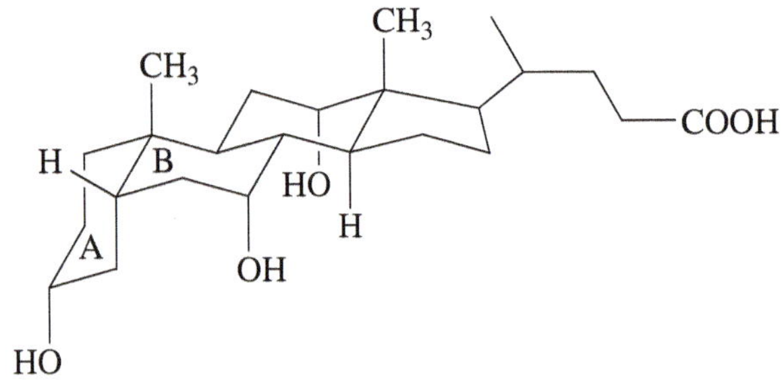

31.

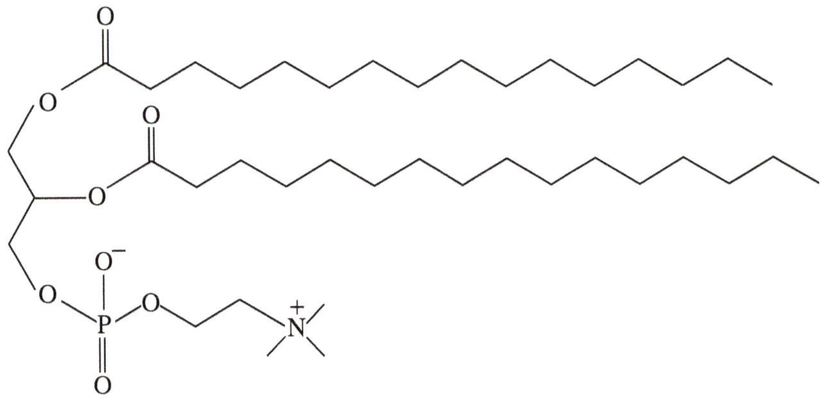

32. If stearic acid were at C-1 and C-3, the fat would not be optically active.
Therefore, stearic acid must be attached to adjacent OH groups (C-1 and C-2 or C-2 and C-3).

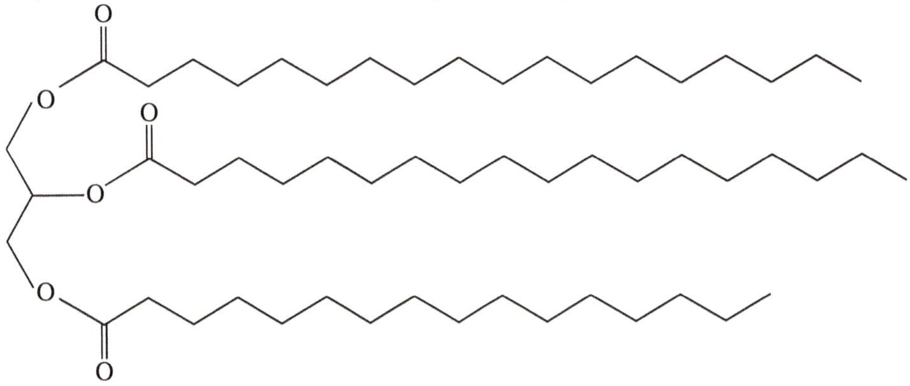

33. All triacylglycerols do not have the same number of asymmetric centers. If the carboxylic acid components at C-1 and C-3 of glycerol are not identical, the triacylglycerol has one asymmetric center (C-2). If the carboxylic acid components at C-1 and C-3 of glycerol are identical, the triacylglycerol has no asymmetric centers.

34. **a.** There are three triacylglycerols in which one of the fatty acid components is lauric acid and two are myristic acid. Myristic acid can be at C-1 and C-3 of glycerol, in which case the triacylglycerol does not have any asymmetric centers. If myristic acid is at C-1 and C-2 of glycerol, C-2 is an asymmetric center, and consequently, the compound can have a pair of enantiomers.

b. There are six triacylglycerols in which one of the fatty acid components is lauric acid, one is myristic acid, and one is palmitic acid. The three possible arrangements are shown below (with the fatty acid components abbreviated as L, M, and P). Because each has an asymmetric center, each can exist as a pair of enantiomers for a total of six triacylglycerols.

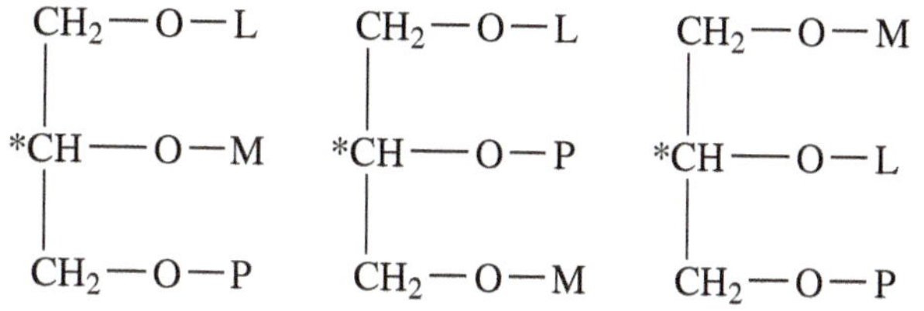

$$\begin{array}{lll} CH_2{-}O{-}L & CH_2{-}O{-}L & CH_2{-}O{-}M \\ {}^{*}CH{-}O{-}M & {}^{*}CH{-}O{-}P & {}^{*}CH{-}O{-}L \\ CH_2{-}O{-}P & CH_2{-}O{-}M & CH_2{-}O{-}P \end{array}$$

35.

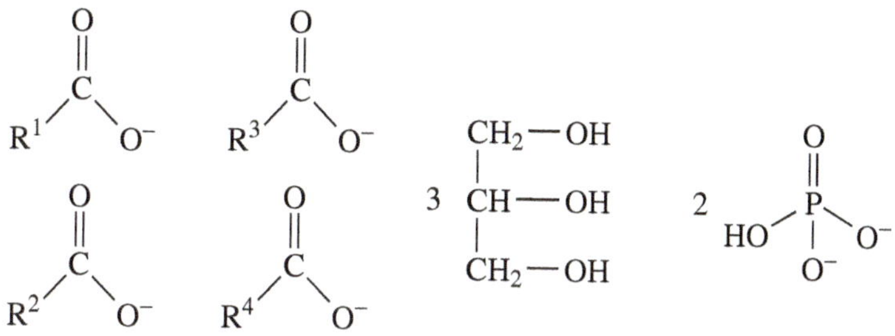

$$R^1{-}C(=O){-}O^- \quad R^3{-}C(=O){-}O^- \quad R^2{-}C(=O){-}O^- \quad R^4{-}C(=O){-}O^-$$

$$3\ \begin{array}{l} CH_2{-}OH \\ CH{-}OH \\ CH_2{-}OH \end{array} \qquad 2\ HO{-}P(=O)(O^-){-}O^-$$

36.

$$\begin{array}{l} CH_2{-}O{-}C(=O){-}CH_3 \\ CH{-}O{-}C(=O){-}CH_3 \\ CH_2{-}O{-}C(=O){-}CH_3 \end{array}$$

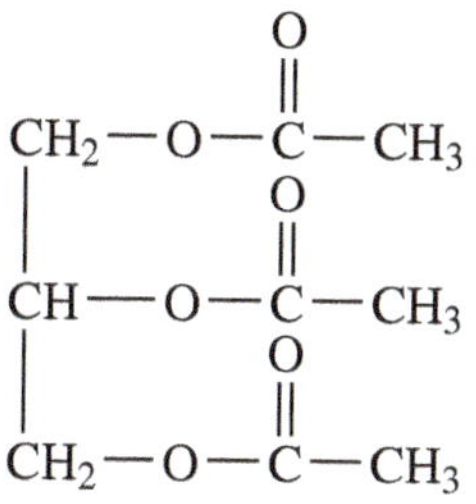

The structure to the left has a molecular formula = $C_9H_{14}O_6$ and a molecular weight = 218.

Subtracting 218 from the total molecular weight gives the molecular weight contribution from the methylene (CH_2) groups in the triacylglycerol.

$$722 - 218 = 504$$

Dividing 504 by the molecular weight contribution from a methylene group (14) gives the number of methylene groups.

$$\frac{504}{14} = 36$$

Because there are 36 methylene groups, each fatty acid in the triacylglycerol has 12 methylene groups.

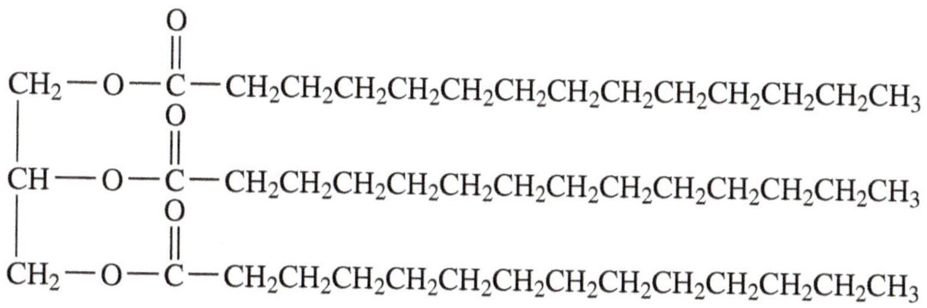

$$\begin{array}{l} CH_2{-}O{-}C(=O){-}CH_2CH_2CH_2CH_2CH_2CH_2CH_2CH_2CH_2CH_2CH_2CH_2CH_3 \\ CH{-}O{-}C(=O){-}CH_2CH_2CH_2CH_2CH_2CH_2CH_2CH_2CH_2CH_2CH_2CH_2CH_3 \\ CH_2{-}O{-}C(=O){-}CH_2CH_2CH_2CH_2CH_2CH_2CH_2CH_2CH_2CH_2CH_2CH_2CH_3 \end{array}$$

nutmeg

37.

a. **b.** 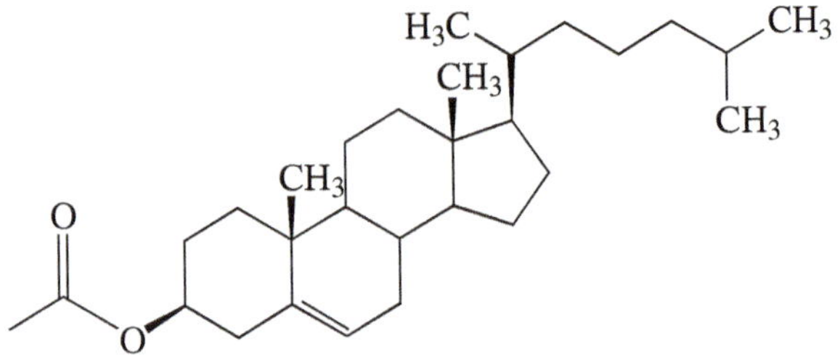 **c.** **d.** **e.**

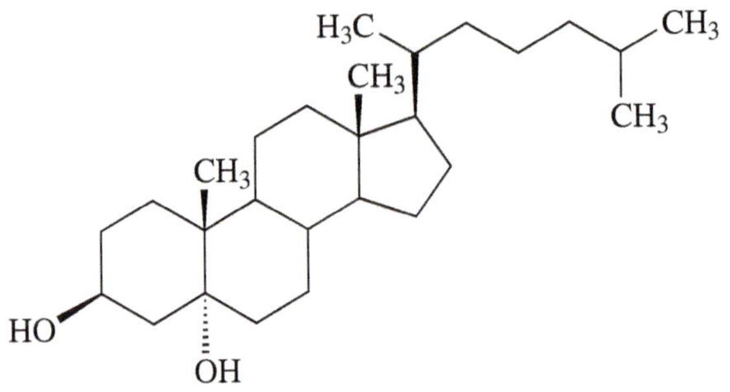

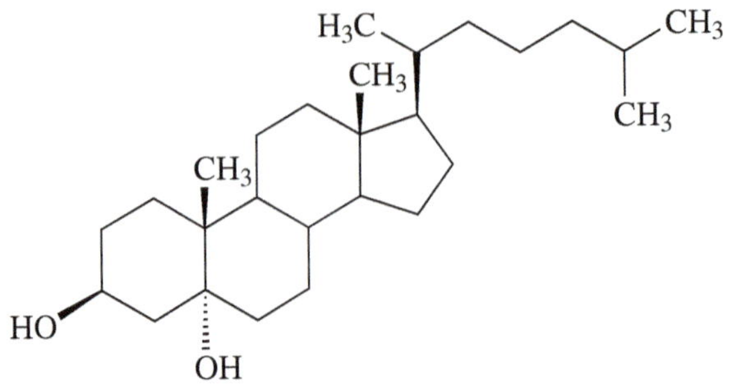

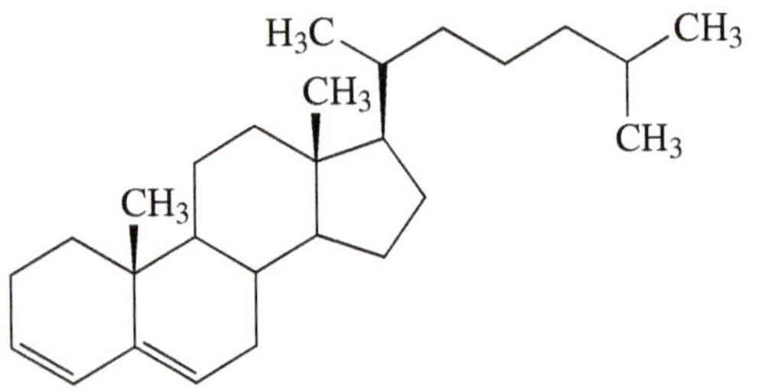

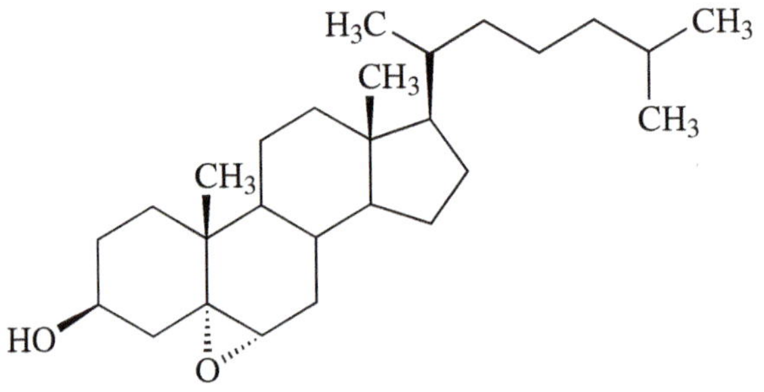

38. **a.** Starting with mevalonyl pyrophosphate, you can trace the location of the label in the compounds that lead to the formation of geranyl pyrophosphate. Geranyl pyrophosphate is converted to citronellal.

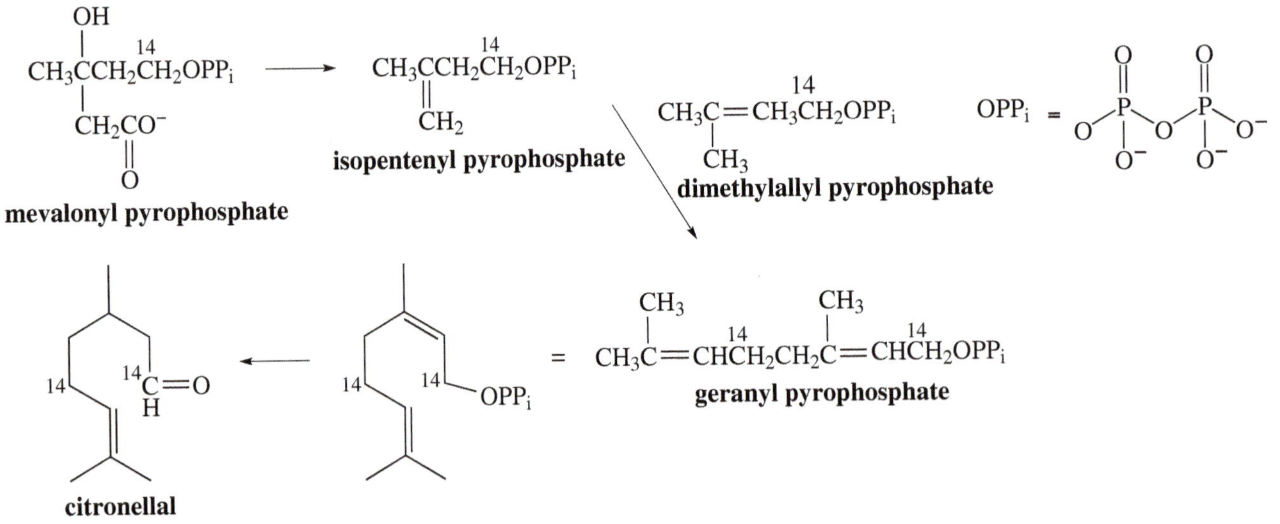

mevalonyl pyrophosphate → **isopentenyl pyrophosphate** → **dimethylallyl pyrophosphate** → **geranyl pyrophosphate** → **citronellal**

b. The label is lost from sample B when mevalonic pyrophosphate loses CO_2 to form isopentenyl pyrophosphate, so none of the carbons will be labeled in citronellal.

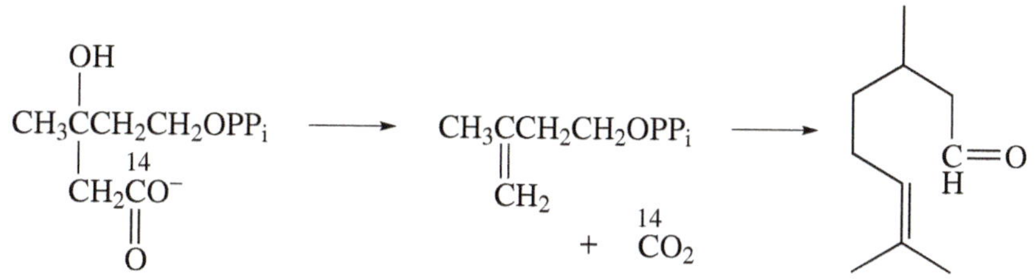

$CH_3C(OH)(CH_2\overset{14}{C}O_2^-)CH_2CH_2OPP_i \longrightarrow CH_3C(=CH_2)CH_2CH_2OPP_i + {}^{14}CO_2 \longrightarrow$ citronellal

c. Because the methyl groups are equivalent in the carbocation that is formed as an intermediate when isopentenyl pyrophosphate is converted to dimethylallyl pyrophosphate, either of the methyl groups can be labeled in dimethylallyl pyrophosphate. This means that either of the two methyl groups can be labeled in geranyl pyrophosphate and in citronellal.

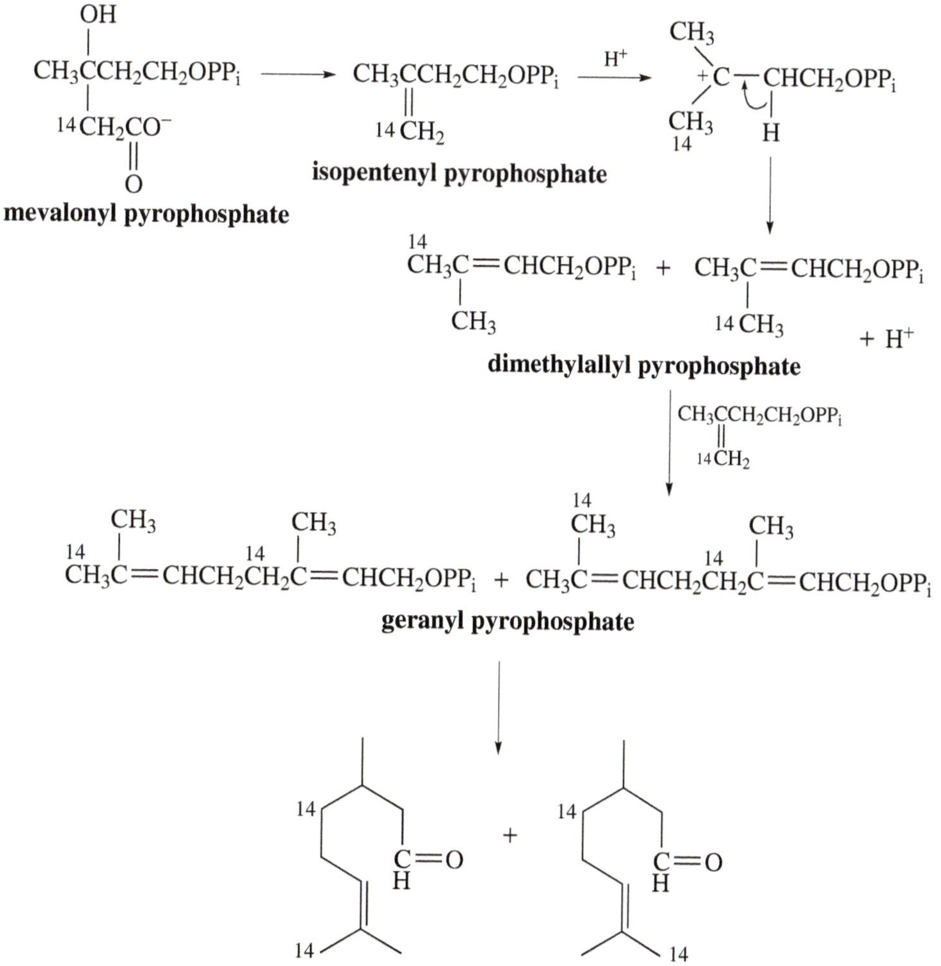

$CH_3C(OH)(^{14}CH_2CO_2^-)CH_2CH_2OPP_i$ **mevalonyl pyrophosphate** $\longrightarrow CH_3C(=^{14}CH_2)CH_2CH_2OPP_i$ **isopentenyl pyrophosphate** $\xrightarrow{H^+} (CH_3)(^{14}CH_3)C^+{-}CH(H)CH_2OPP_i$

$\longrightarrow CH_3\overset{14}{C}=CHCH_2OPP_i$ (with CH_3) $+ CH_3C=CHCH_2OPP_i$ (with $^{14}CH_3$) $+ H^+$

dimethylallyl pyrophosphate

$\xrightarrow{CH_3C(=^{14}CH_2)CH_2CH_2OPP_i}$

$^{14}CH_3C(CH_3)=CHCH_2CH_2\overset{14}{C}(CH_3)=CHCH_2OPP_i + CH_3C(^{14}CH_3)=CHCH_2CH_2C(CH_3)=CHCH_2OPP_i$

geranyl pyrophosphate

$\longrightarrow$ citronellal (14C-labeled) + citronellal (14C-labeled)

citronellal

39. + $CH_2=O$

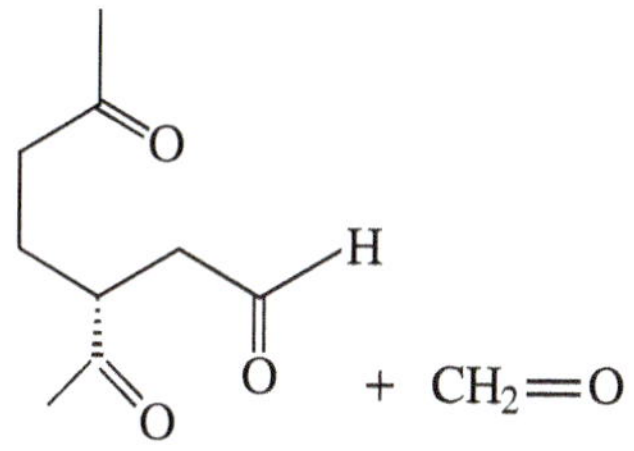

40.

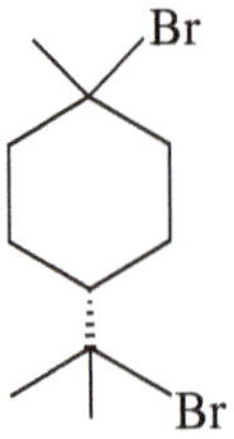

Br

Br

This compound is achiral.

41. **a.**

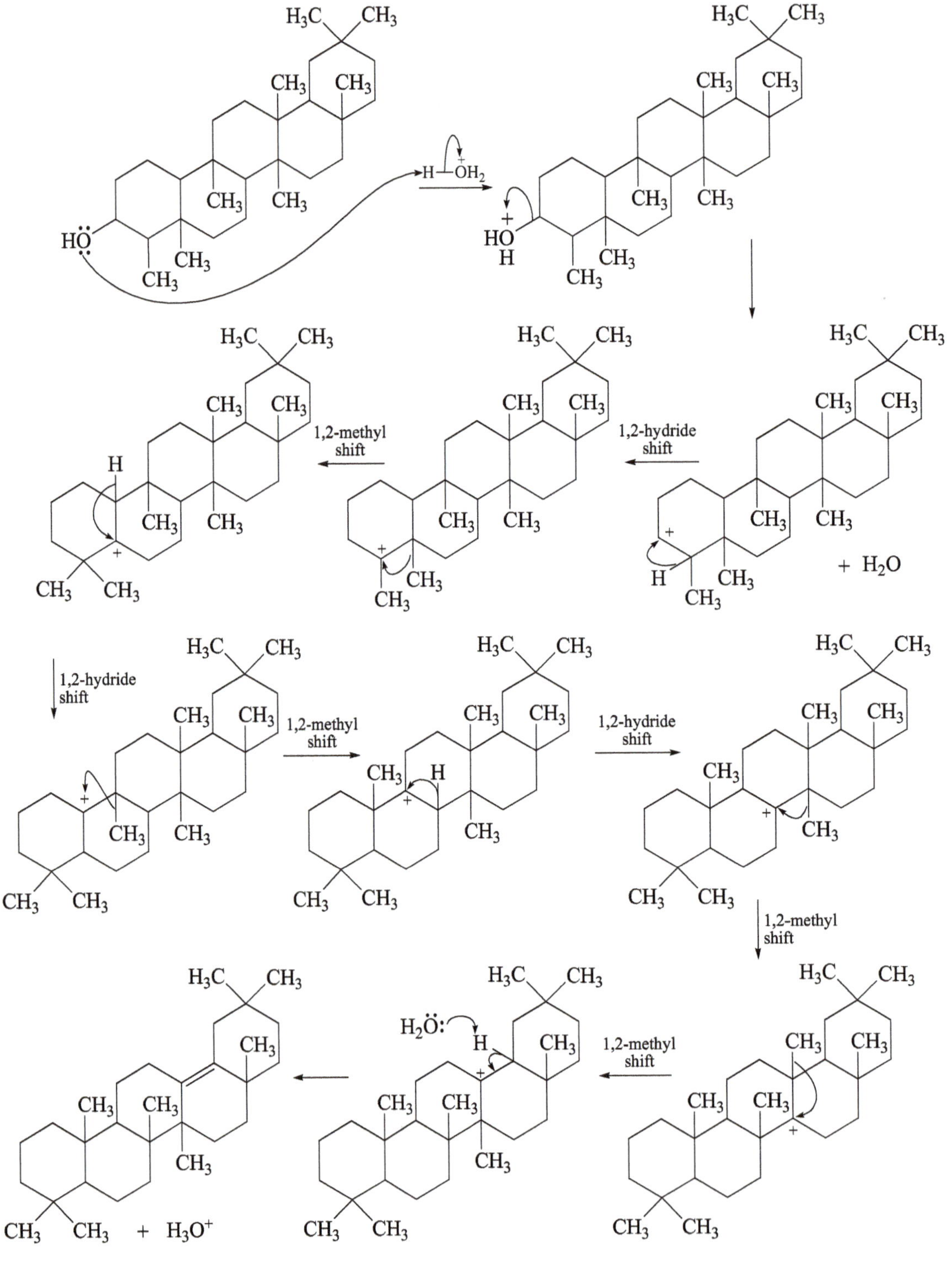

H_3C CH_3 CH_3 CH_3 CH_3 CH_3 HÖ CH_3 CH_3

H—$\overset{+}{O}H_2$

H_3C CH_3 CH_3 CH_3 CH_3 CH_3 $\overset{+}{HO}$ H CH_3 CH_3

1,2-methyl shift

1,2-hydride shift

$+ H_2O$

1,2-hydride shift

1,2-methyl shift

1,2-hydride shift

1,2-methyl shift

1,2-methyl shift

$H_2\ddot{O}$:

$+ H_3O^+$

b. It has 30 carbons, so it is a triterpene.

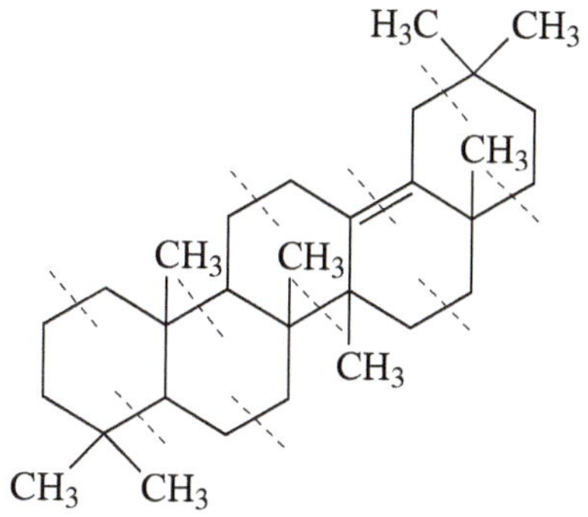

H_3C CH_3 CH_3 CH_3 CH_3 CH_3 CH_3 CH_3

42. First, draw the starting material (geranyl pyrophosphate) so that its structure is as similar as possible to that of the target molecule.

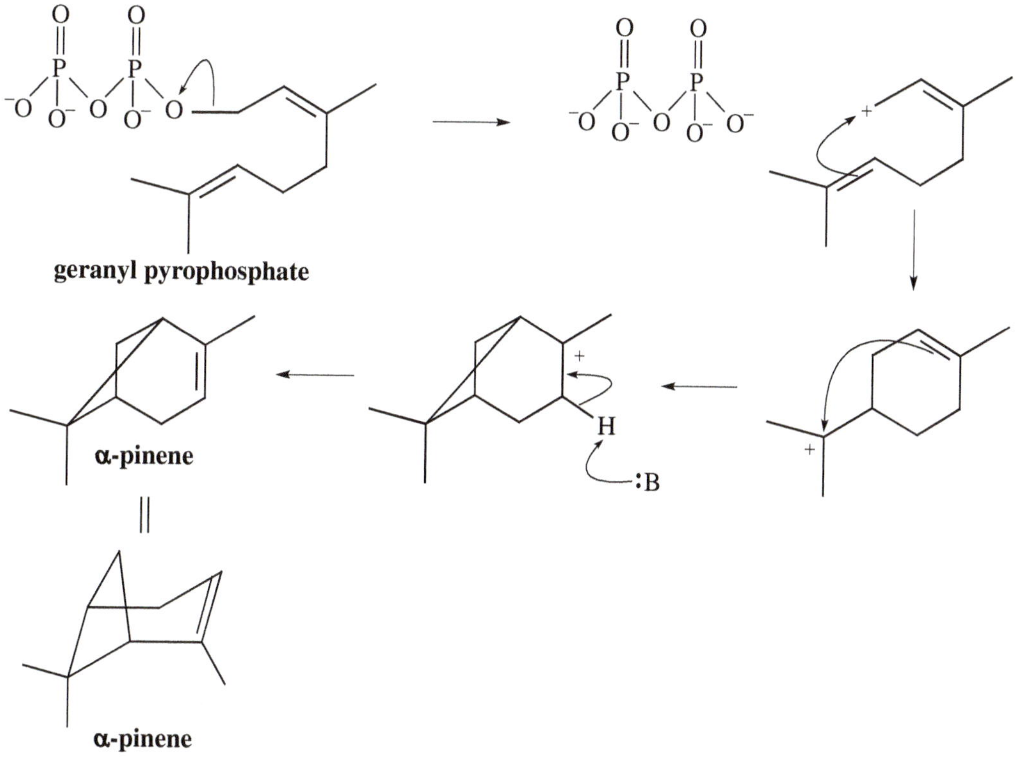

geranyl pyrophosphate

:B

α-pinene

α-pinene

43.

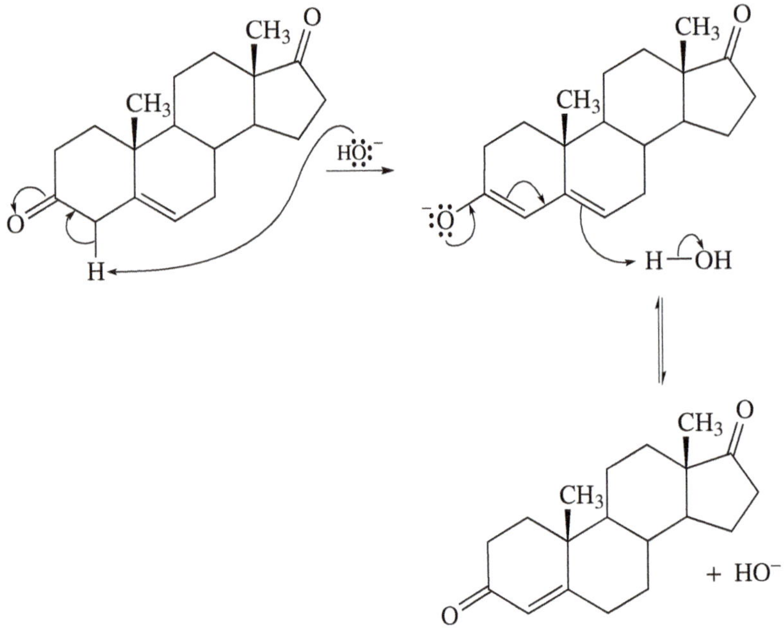

44. The OH groups will react only if they are in equatorial positions, because introduction of bulky axial substituents would decrease the stability of the molecule.

In the case of 5α-cholestane-3β,7β-diol, the two OH groups are on the same side of the ring system as the angular methyl group, which means that they are in equatorial positions. Both OH groups react with ethyl chloroformate.

In the case of 5α-cholestane-3β,7α-diol, only one of the OH groups is on the same side of the ring system as the angular methyl group. The other is on the opposite side of the ring, which means that it is in an axial position. Only the OH group that is in the equatorial position reacts with ethyl chloroformate.

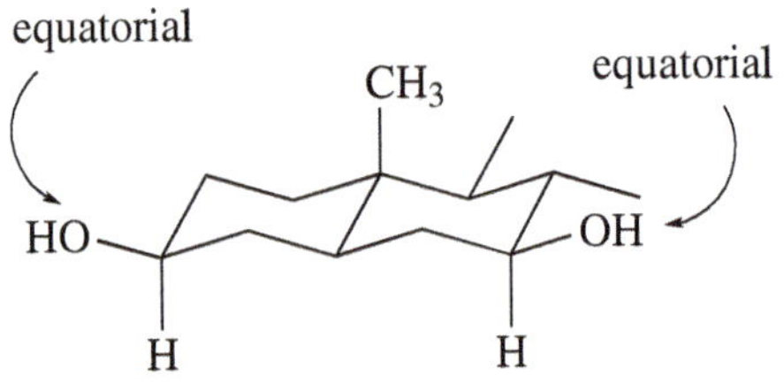

5α-cholestane-3β,7β-diol

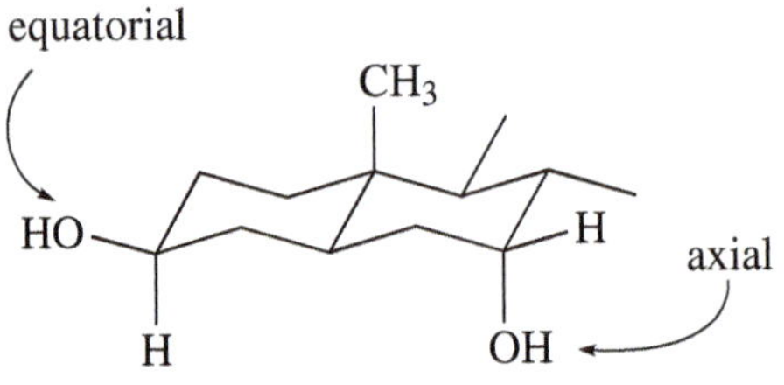

5α-cholestane-3β,7α-diol

45. The mechanism for the conversion of isopentenyl pyrophosphate to geranyl pyrophospate to farnesyl pyrophosphate is shown in the text on pages 1143 and 1145. Now we need to show how farnesyl pyrophosphate is converted to eudesmol.

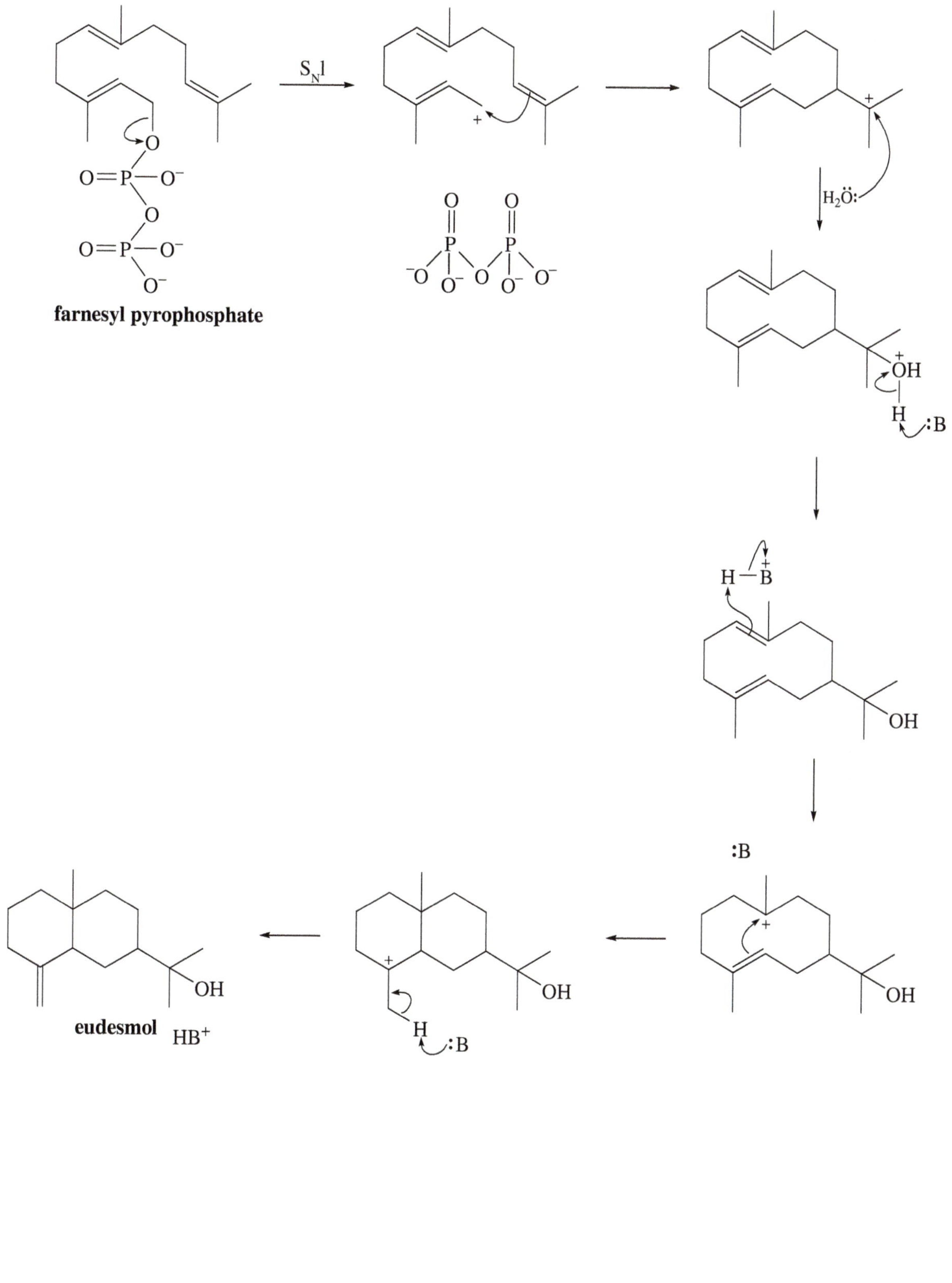

S_N1

farnesyl pyrophosphate

$H_2\ddot{O}$:

eudesmol HB^+

46. + $H_2C{=}O$ (H–C(=O)–H)

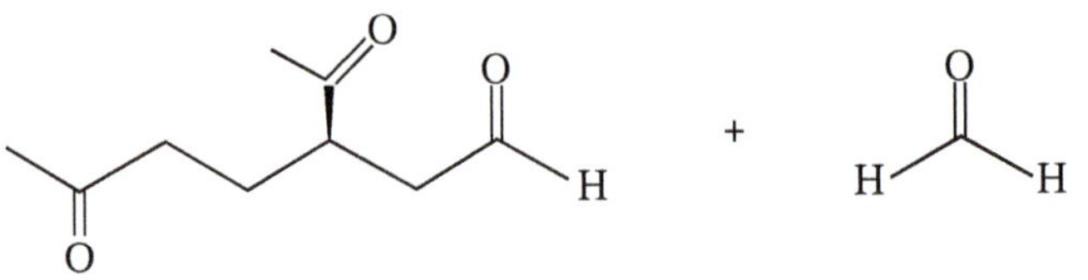

47. There are 14 possible structures for compound **A** and 4 possible structures for compound **B**.

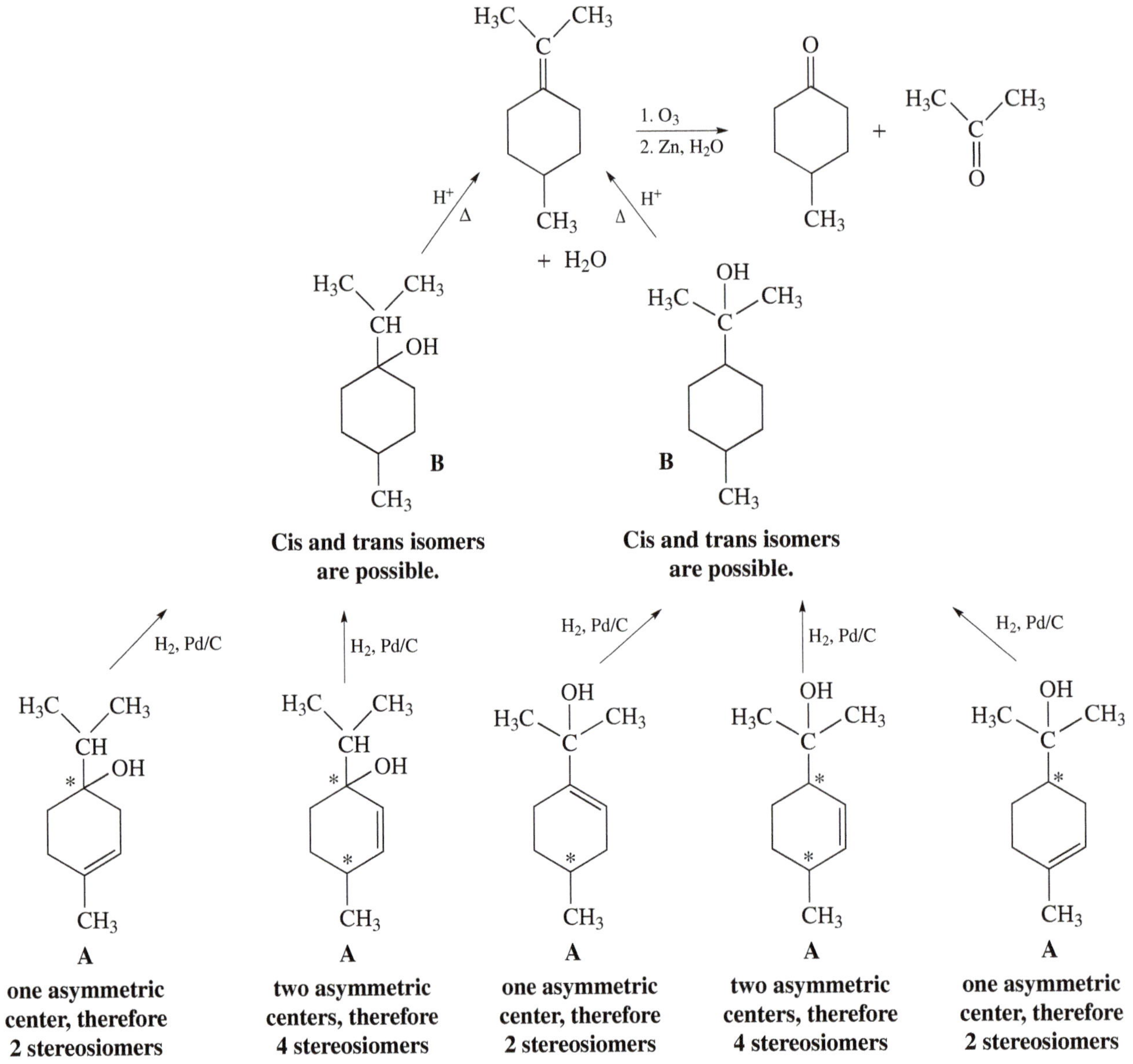

1. O_3
2. Zn, H_2O

H^+ Δ

CH_3 + H_2O

H_3C CH_3 CH OH

CH_3

B

OH H_3C C CH_3

CH_3

B

Cis and trans isomers are possible.

Cis and trans isomers are possible.

H_2, Pd/C

A

one asymmetric center, therefore 2 stereosiomers

A

two asymmetric centers, therefore 4 stereosiomers

A

one asymmetric center, therefore 2 stereosiomers

A

two asymmetric centers, therefore 4 stereosiomers

A

one asymmetric center, therefore 2 stereosiomers

48.

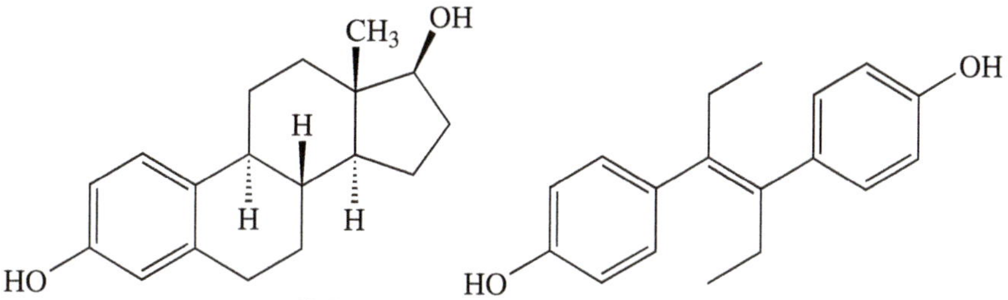

estradiol

DES

49. A sesquiterpene is synthesized from farnesyl pyrophosphate. The carbons that would be labeled in farnesyl pyrophosphate, if it is synthesized from acetate with a ^{14}C-labeled carbonyl carbon, are indicated by an asterisk. (See Problem 23.)

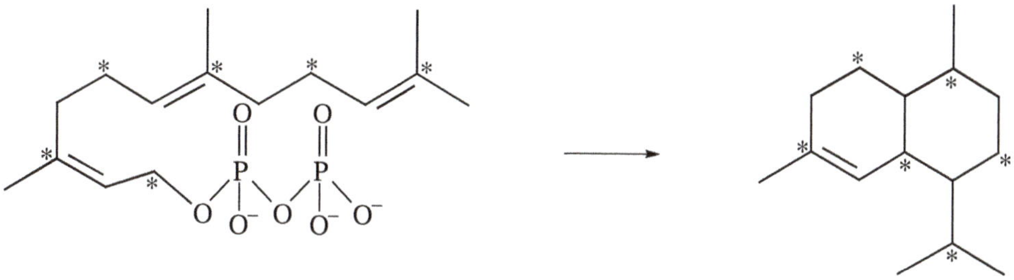

farnesyl pyrophosphate

Chapter 25 Practice Test

1. Mark off the isoprene units in squalene.

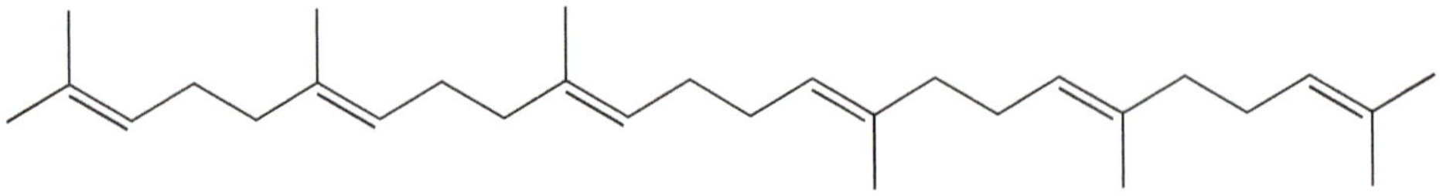

squalene

2. Draw the mechanism for the following reaction:

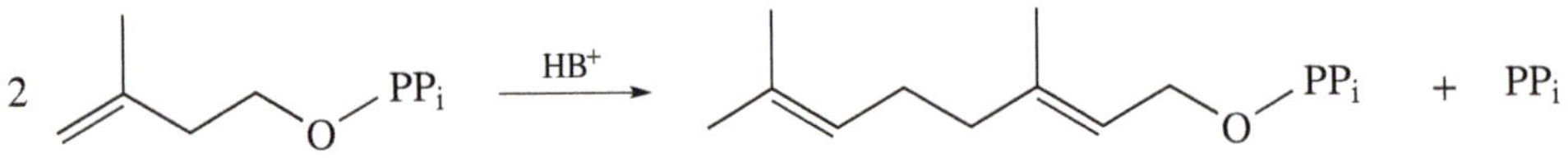

2 (O–PP$_i$) $\xrightarrow{HB^+}$ (O–PP$_i$) + PP$_i$

(PP$_i$ = pyrophosphate)

3. Draw structures of the following:

a. a phospholipid that contains ethanolamine

b. a wax

4. Indicate whether each of the following statements is true or false:

a. Cholesterol is the precursor of all other steroids. T F

b. The double bonds in unsaturated fats are conjugated. T F

c. Fats have a higher percentage of saturated fatty acids than do oils. T F

d. A saturated fatty acid has a lower melting point than an unsaturated fatty acid with the same number of carbons. T F

e. Lipids are insoluble in water. T F

f. A diterpene contains 20 carbons. T F

g. Vitamin A is not a coenzyme. T F

5. How many carbon atoms are in a triterpene?

6. What size is the ring in a prostaglandin?

7. What is the starting material for the synthesis of prostaglandins?

CHAPTER 26
The Chemistry of the Nucleic Acids

Important Terms

anticodon the three bases at the bottom of the middle loop in tRNA.

base a nitrogen-containing heterocyclic compound (a purine or a pyrimidine) found in DNA and RNA.

codon a sequence of three bases in mRNA that specifies the amino acid to be incorporated into a protein.

deamination a hydrolysis reaction that results in the removal of ammonia.

deoxyribonucleic acid (DNA) a polymer of deoxyribonucleotides containing all the genetic information of an organism.

deoxyribonucleotide a nucleotide where the sugar component is D-2-deoxyribose.

dinucleotide two nucleotides linked by a phosphodiester bond.

double helix the term used to describe the secondary structure of DNA.

exon a stretch of bases in DNA that are a portion of a gene.

gene a segment of DNA that encodes a protein.

gene therapy a technique that inserts a synthetic gene into the DNA of an organism that is defective in that gene.

genetic code the amino acid specified by each three-base sequence of mRNA.

genetic engineering recombinant DNA technology where DNA segments are inserted into DNA in a host cell and allowed to replicate.

human genome the total DNA of a human cell.

major groove the wider and deeper of the two alternating grooves in DNA.

minor groove the narrower and more shallow of the two alternating grooves in DNA.

nucleic acid a chain of five-membered ring sugars linked by phosphodiester groups with each sugar bearing a heterocyclic amine at the anomeric carbon in the β-position. The two kinds of nucleic acids are DNA and RNA.

nucleoside a heterocyclic base (purine or pyrimidine) bonded to the anomeric carbon of a sugar (D-ribose or D-2-deoxyribose) in the β-position.

nucleotide a nucleoside with one of its OH groups bonded to a phosphate group via a phosphoester linkage.

oligonucleotide	3 to 10 nucleotides linked by phosphodiester groups.
phosphodiester	a species in which two of the OH groups of phosphoric acid have been converted to OR groups.
polynucleotide	many nucleotides linked by phosphodiester groups.
primary structure	the sequence of bases in a nucleic acid.
pyrosequencing	an automated technique used to sequence DNA; it detects the identity of the base that adds to the DNA primer.
recombinant DNA	DNA that has been incorporated into a host cell.
replication	the synthesis of identical copies of DNA.
replication fork	the position on DNA where replication begins.
restriction endonuclease	an enzyme that cleaves DNA at a specific base sequence.
restriction fragment	a fragment that is formed when DNA is cleaved by a restriction endonuclease.
ribonucleic acid (RNA)	a polymer of ribonucleotides.
ribonucleotide	a nucleotide where the sugar component is D-ribose.
ribosomal RNA (rRNA)	the structural component of ribosomes, the particles on which protein synthesis takes place.
RNA splicing	the step in RNA processing that cuts out nonsense bases and splices informational pieces together.
semiconservative replication	the mode of replication that results in a daughter molecule of DNA having one of the original DNA strands in addition to a newly synthesized strand.
sense strand	the strand in DNA that is not read during transcription; it has the same sequence of bases as the synthesized mRNA strand, except that the mRNA has Us in place of the Ts in DNA.
stacking interactions	weak attractive forces between the mutually induced dipoles of adjacent pairs of bases in DNA.
stop codon	a codon that signals "stop protein synthesis here."
template strand	the strand in DNA that is read during transcription.
transcription	the synthesis of mRNA from a DNA blueprint.
transfer RNA (tRNA)	a single-stranded RNA molecule that carries an amino acid to be incorporated into a protein.
translation	the synthesis of a protein from an mRNA blueprint.

Solutions to Problems

1. The most basic atom in the ring is protonated. In the case of a purine, this is the nitrogen at the 7-position. In the next step, the bond between the heterocyclic base and the sugar breaks, with the anomeric carbocation being stabilized by the ring oxygen's nonbonding electrons.

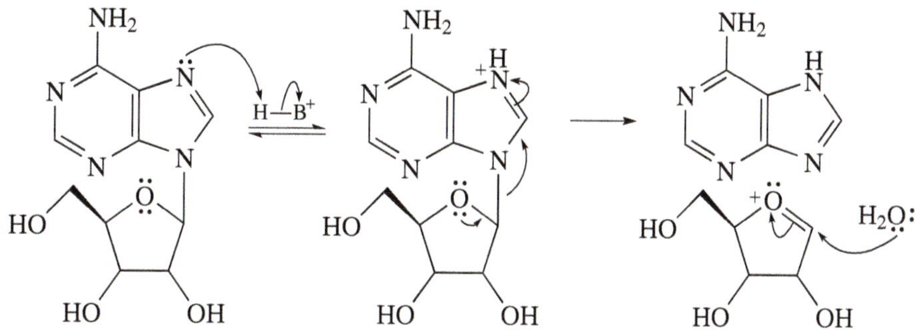

NH2 N N N N HO O HO OH H—B+ NH2 N +N H N N HO O HO OH → NH2 N H N N N HO +O HO OH H2O:

The mechanism is exactly the same for pyrimidines, except that the initial protonation takes place at the 3-position.

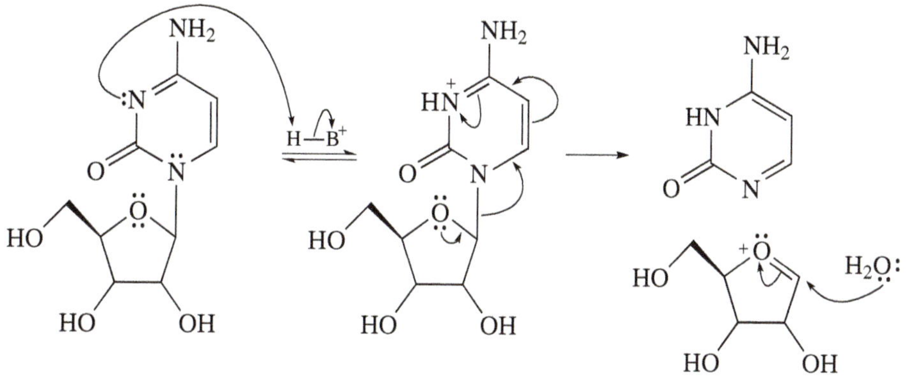

NH2 :N O N HO O HO OH H—B+ NH2 +HN O N HO O HO OH → NH2 HN O N HO +O HO OH H2O:

2. a.

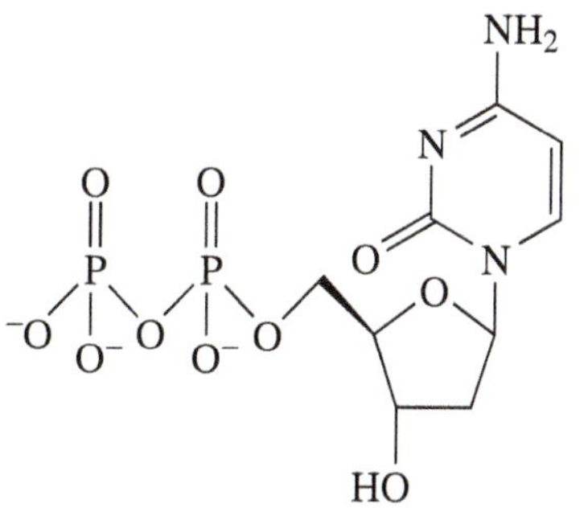

NH2 N O N O P O- O- O O P O- O O HO

dCDP

b.

O CH3 HN O N O P O- O- O O P O- O O P O- O O HO

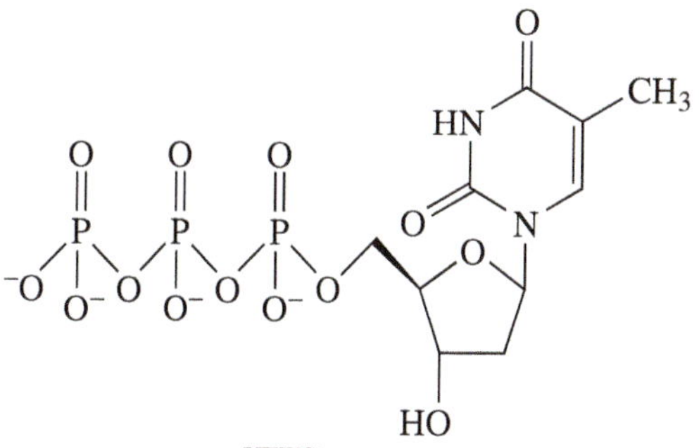

dTTP

c.

dUMP

e.

guanosine 5′-triphosphate
GTP

d.

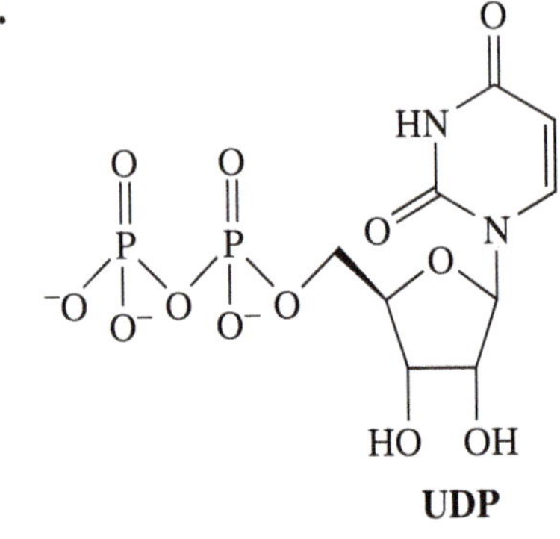

UDP

f.

adenosine 3′-monophosphate
AMP

3. A hydrogen bond acceptor is indicated by **A** (it is an atom with a lone pair); a hydrogen bond donor is indicated by **D** (it is an atom with a hydrogen attached to a nitrogen).
The **A** and **D** designations show that the maximum number of hydrogen bonds that can form are two between thymine and adenine and three between cytosine and guanine. Notice that uracil and thymine have the same **A** and **D** designations.

uracil **thymine** **adenine**

cytosine **guanine**

4. If the bases existed in the enol form, no hydrogen bonds could form between the bases unless one of the bases were shifted vertically.

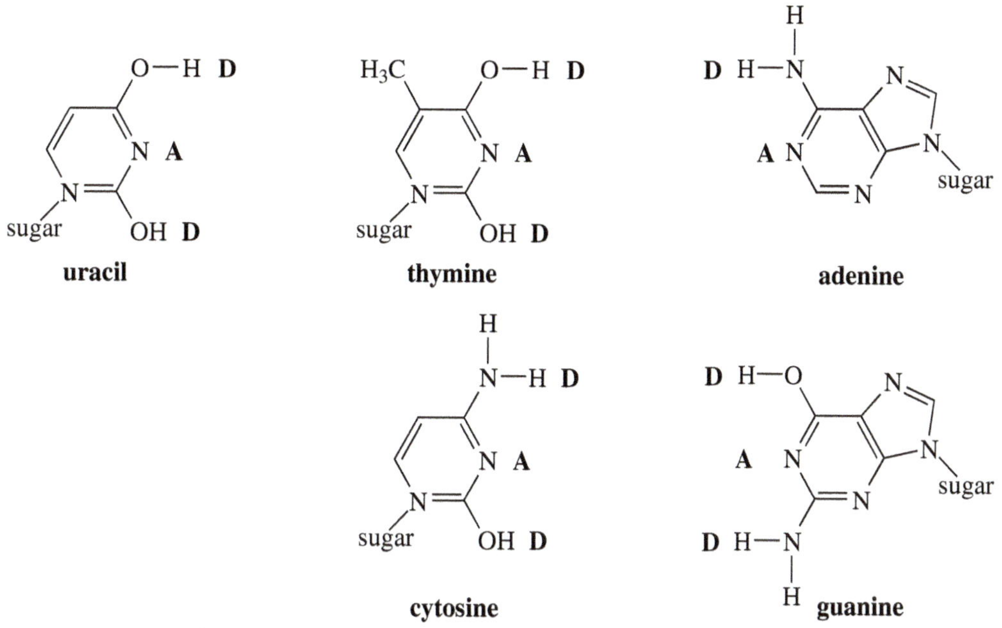

O—H **D**
N **A**
N
sugar
OH **D**
uracil

H_3C
O—H **D**
N **A**
N
sugar
OH **D**
thymine

H
D H—N
N
A N
N
N
sugar
adenine

H
N—H **D**
N **A**
N
sugar
OH **D**
cytosine

D H—O
N
A N
N
N
sugar
D H—N
H
guanine

5. **a.** 3′—C—C—T—G—T—T—A—G—A—C—G—5′
b. guanine

6. Notice that when a nucleophile attacks the phosphorus of a diester, the π bond breaks. However, when a nucleophile attacks the phosphorus of an anhydride, a σ bond breaks instead of the π bond.

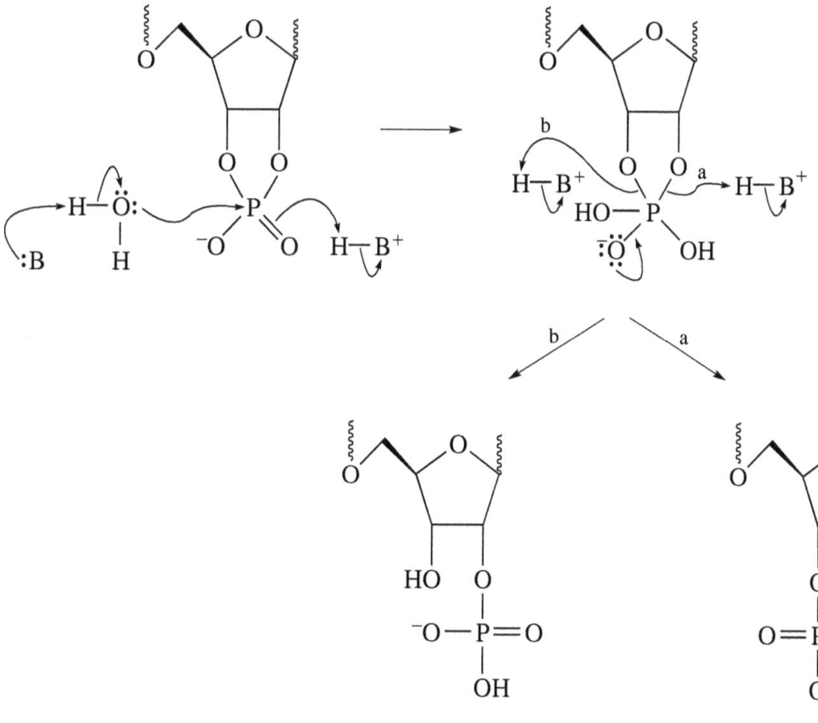

:B H—Ö: H O O P ⁻O O H—B⁺ → b H—B⁺ O O a H—B⁺ HO—P :Ö: OH

b a

HO O ⁻O—P=O OH

O OH O=P—O⁻ OH

7.

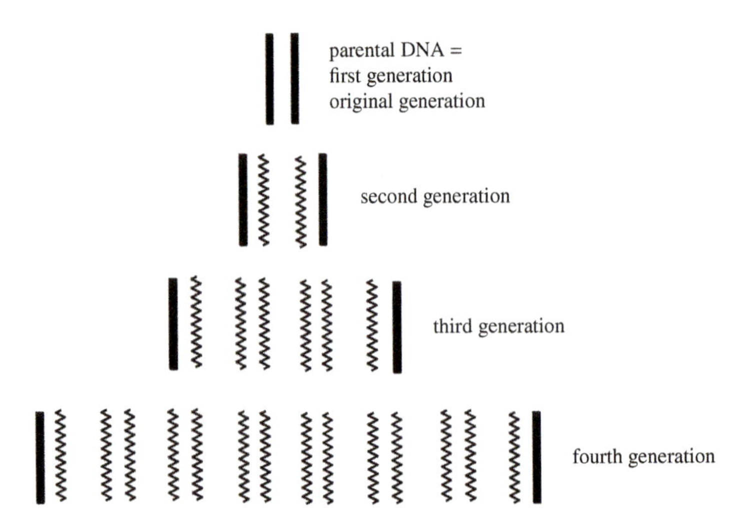

8. Thymine and uracil differ only in that thymine has a methyl substituent that uracil does not have (thymine is 5-methyluracil). Because thymine and uracil both have the same groups in the same positions that can participate in hydrogen bonding, they both call for incorporation of the same purine. Because thymine and uracil form one hydrogen bond with guanine and two with adenine, they both incorporate adenine to maximize hydrogen bonding.

9. Because methionine is known to be the first base incorporated into the heptapeptide, the mRNA sequence is read beginning at AUG, since that is the only codon that codes for methionine.

Met-Asp-Pro-Val-Ile-Lys-His

10. Met-Asp-Pro-Leu-Leu-Asn

11. It does not cause protein synthesis to stop because the sequence UAA does not occur within a triplet. The reading frame causes the triplets to be AU**U** and **AA**A. In other words, the U is at the end of a triplet and the next triplet starts with AA.

12. A change in the third base of a codon is least likely to cause a mutation because the third base is variable for many amino acids. For example, CUU, CUC, CUA, and CUG all code for leucine.

13. The sequence of bases in the template strand of DNA specifies the sequence of bases in mRNA, so the bases in the template strand and the bases in mRNA are complementary. Therefore, the sequence of bases in the sense strand of DNA is identical to the sequence of bases in mRNA, except that wherever there is a U in mRNA, there is a T in the sense strand of DNA.

5′—G—C—A—T—G—G—A—C—C—C—C—G—T—T—A—T—T—A—A—A—C—A—C—3′

14.

	Met	Asp	Pro	Val	Ile	Lys	His
<u>codons</u>	AUG	GAU	CCU	GUU	AUU	AAA	CAU
		GAC	CCC	GUC	AUC	AAG	CAC
			CCA	GUA	AUA		
			CCG	GUG			

anticodons							
	CAU	AUC	AGG	AAC	AAU	UUU	AUG
		GUC	GGG	GAC	GAU	CUU	GUG
			UGG	UAC	UAU		
			CGG	CAC			

Notice that the anticodons are written in the $5' \rightarrow 3'$ direction. For example, the anticodon of AUG is written as CAU (not UAC).

codon 5′ A U G 3′

anticodon 3′ U A C 5′

15.

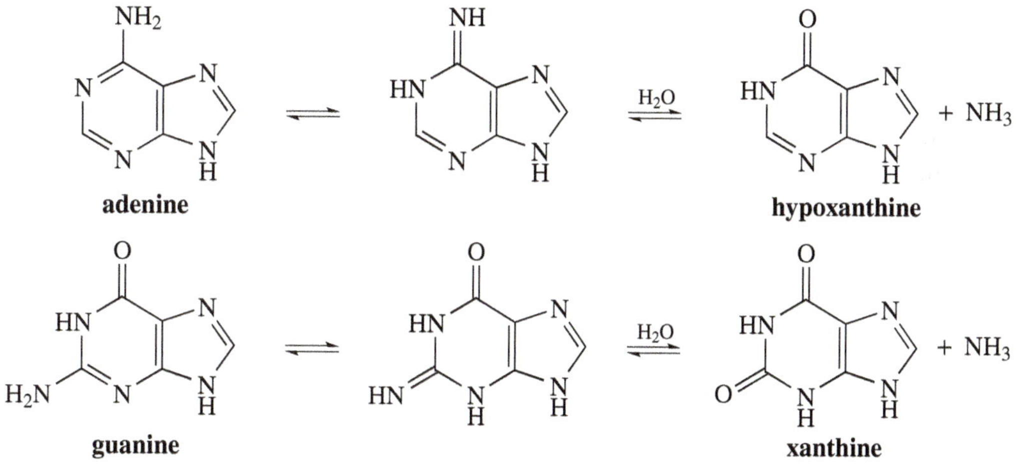

adenine $\rightleftharpoons$ $\overset{H_2O}{\rightleftharpoons}$ hypoxanthine + NH_3

guanine $\rightleftharpoons$ $\overset{H_2O}{\rightleftharpoons}$ xanthine + NH_3

16. Deamination involves hydrolyzing an imine to form a carbonyl group and ammonia.

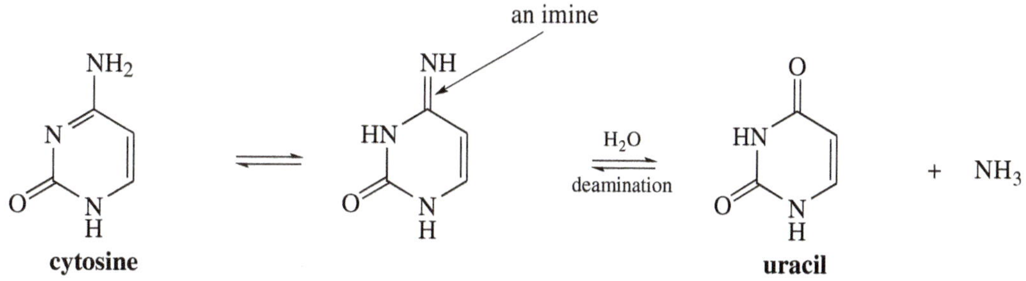

an imine

cytosine $\rightleftharpoons$ $\overset{H_2O}{\underset{\text{deamination}}{\rightleftharpoons}}$ uracil + NH_3

Thymine does not have an amino substituent on the ring, which means that it cannot form an imine and, therefore, cannot be deaminated.

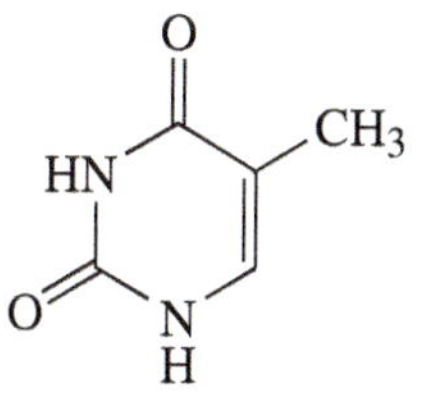

thymine

17. **B** is the only sequence that has a chance of being recognized by a restriction endonuclease because it is the only one that has the same sequence of bases in the 5′ → 3′ direction that the complementary strand has in the 5′ → 3′ direction.

$$^{5'}A—C—G—C—G—T^{3'}$$
$$^{3'}T—G—C—G—C—A^{5'}$$

18. Lys-Val-Gly-Tyr-Pro-Gly-Met-Val-Val

19. 5′—GAC—CAC—CAT—TCC—GGG—GTA—GCC—AAC—TTT—3′

20. 5′—AAA—GTT—GGC—TAC—CCC—GGA—ATG—GTG—GTC—3′

21. A segment of DNA with 18 base pairs has 36 bases. If there are 7 cytosines, then there are 7 guanines. This accounts for 14 bases. Therefore, 22 (36 − 14 = 22) are adenines and thymines.

a. 11 thymines **b.** 7 guanines

22.
a. guanosine 3′-monophosphate
b. cytidine 5′-diphosphate
c. 2′-deoxyadenosine 5′-monophosphate
d. thymidine

23. The third base in each codon has some variability.

mRNA 5′-GG(UCA or G)UC(UCA or G)CG(UCA or G)GU(UCA or G)CA(U or C)GA(A or G)-3′
or AG(U or C) AG(A or G)

DNA <u>template</u> 3′-CC(AGT or C)AG(AGT or C)GC(AGT or C)CA(AGT or C)GT(A or G)CT(T or C)-5′
or TC(A or G) TC(T or C)

<u>sense</u> 5′-GG(TCA or G)TC(TCA or G)CG(TCA or G)GT(TCA or G)CA(T or C)GA(A or G)-3′
or AG(T or C) AG(A or G)

Notice that Ser and Arg are two of three amino acids that can be specified by six different codons.

24.

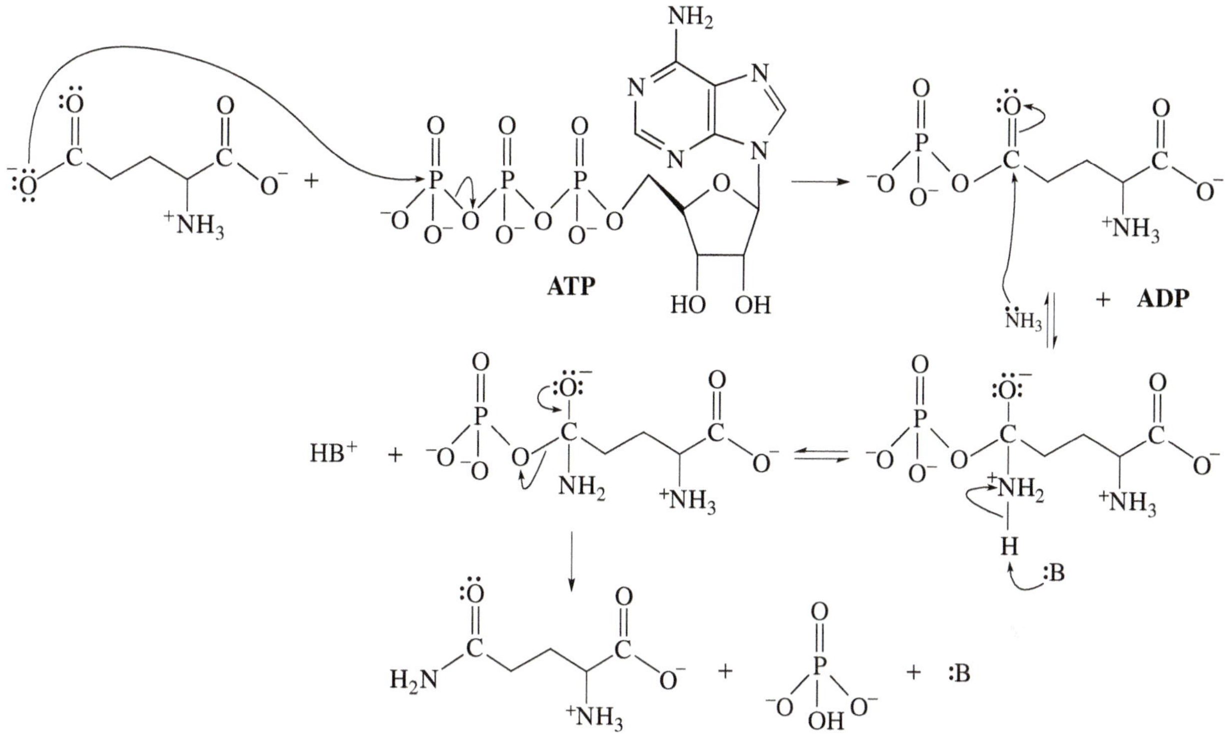

ATP

+ ADP

HB$^+$ + :B

25. **a.** Ile **b.** Asp **c.** Val **d.** Val

26. The ribosome, the particle on which protein synthesis occurs, has a binding site for the growing peptide chain and a binding site for the next amino acid to be incorporated into the chain.

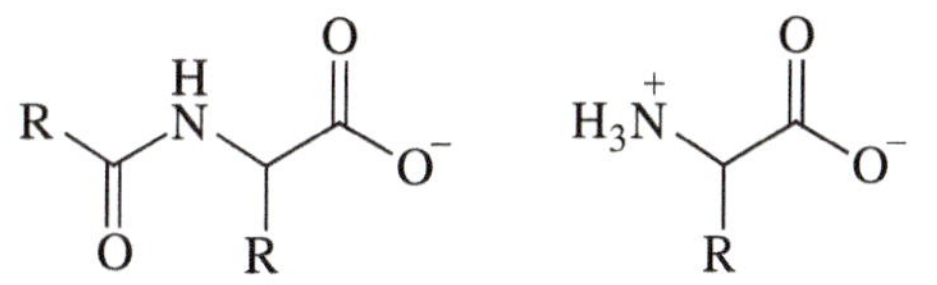

peptide binding site **amino acid binding site**

In protein synthesis, all peptide bonds are formed by the reaction of an amino acid with a peptide, except the first peptide bond, which must be formed by the reaction of two amino acids. Therefore, for the synthesis of the first peptide bond, the first (N-terminal) amino acid must have a peptide bond that will fit into the peptide binding site. The formyl group of *N*-formylmethionine provides the peptide group that will be recognized by the peptide binding site for formation of the first peptide bond.

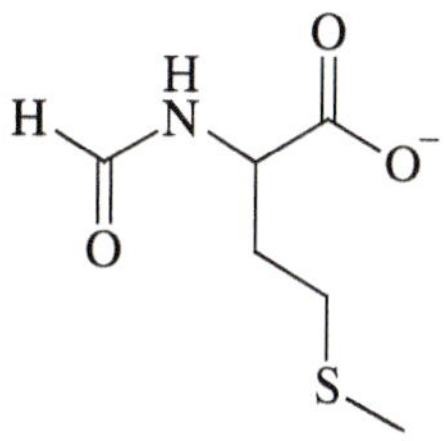

***N*-formylmethionine**

27.

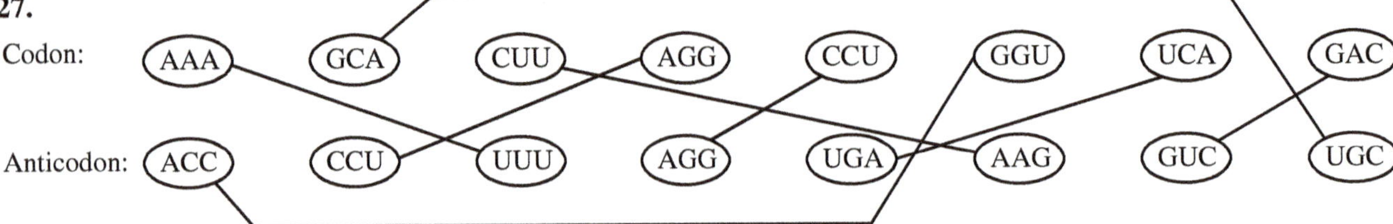

28. **a.**

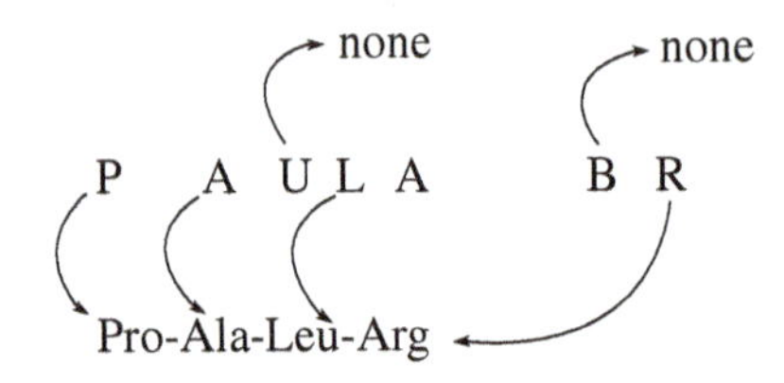

b. mRNA CC(UCA or G)GC(UCA or G)CU(UCA or G)CG(UCA or G)

or UU(A or G) AG(A or G)

c. DNA (sense strand) CC(TCA or G)GC(TCA or G)CT(TCA or G)CG(TCA or G)

or TT(A or G) AG(A or G)

Notice that because mRNA is complementary to the template strand of DNA, which is complementary to the sense strand, mRNA and the sense strand of DNA have the same sequence of bases (except DNA has a T where RNA has a U). Also notice that Leu and Arg are each specified by six codons.

29. **a.** CC and GG **c.** CA and TG

CA and TG are formed in equal amounts because A pairs with T and C pairs with G.

5' C A 3'
3' G T 5'

(Remember that the dinucleotides are written in the $5' \longrightarrow 3'$ direction. For example, part **f** is not a correct answer because it has A pairing with A and T pairing with T.)

30. AZT is incorporated into DNA when the 3′-OH group of the last nucleotide incorporated into the growing chain of DNA attacks the α-phosphorus of AZT-triphosphate instead of the α-phosphorus of a normal nucleotide. When AZT is thus incorporated into DNA, DNA synthesis stops because AZT does not have a 3′-OH group that can react with another nucleotide.

31.

UUU = phenylalanine
GGG = glycine
UUG = leucine
UGU = cysteine
GUU = valine
GGU = glycine
GUG = valine
UGG = tryptophan

32. The number of different codons using four nucleotides:

for a two-letter code: $(4)^2 = 16$
for a three-letter code: $(4)^3 = 64$
for a four-letter code: $(4)^4 = 256$

Because 20 amino acids must be specified, a two-letter code does not provide enough codons.
A three-letter code provides enough codons for all the amino acids and also provides the necessary stop codons.
A four-letter code provides many more codons than are needed.

33. In the first step of the reaction, the imidazole ring of one histidine acts as a general-base catalyst, removing a proton from the 2′-OH group to make it a better nucleophile. In the second step, the imidazole ring of the other histidine acts as a general-acid catalyst, protonating the leaving group to make it a weaker base and, therefore, a better leaving group. In the third and fourth steps, the roles of the two imidazole rings are reversed.

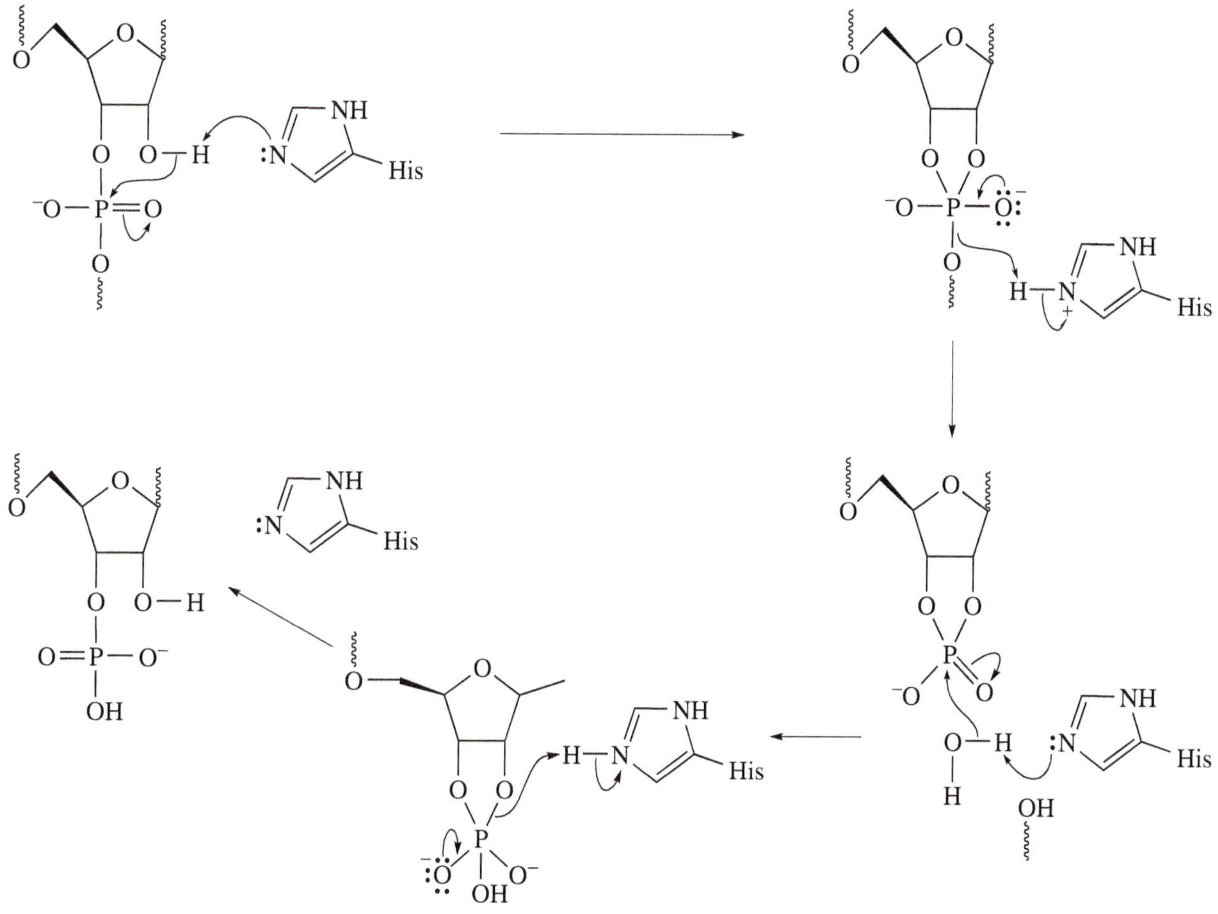

34. The normal and mutant peptides have the following base sequence in their mRNA:

normal: CA(AG) UA(UC) GG(UCAG) AC(UCAG) CG(UCAG) UA(UC) GU(UCAG)
mutant: CA(AG) UC(UCAG) GA(AG) CC(UCGA) GG(UCGA) AC(UCAG)

a. The middle nucleotide (A) in the second triplet was deleted. This means that an A was deleted in the sense strand of DNA or a T was deleted in the template strand of DNA.

b. The mRNA for the mutant peptide has an unused 3′-terminal two-letter code, U(UCAG). The last amino acid in the octapeptide of the normal fragment is leucine, so its last triplet is UU(AG) or CU(UCAG).

This means that the triplet for the last amino acid in the mutant is U(UCAG)(UC) and that the last amino acid in the mutant is one of the following: Phe, Ser, Tyr, or Cys.

35. If deamination does not occur, the mRNA sequence will be:

AUG–UCG–CUA–AUC, which will code for the following tetrapeptide:
Met-Ser-Leu-Ile

Deamination of a cytosine results in a uracil.
If the cytosines are deaminated, the mRNA sequence will be:

AUG–UUG–UUA–AUU, which will code for the following tetrapeptide:
Met-Leu-Leu-Ile

The only cytosine that would change the amino acid that is incorporated into the peptide is the first one. Therefore, this is the cytosine that could cause the most damage to an organism if it were deaminated.

36. 5-Bromouracil is incorporated into DNA in place of thymine because of their similar size. Thymine exists primarily in the keto form and pairs with adenine via two hydrogen bonds. 5-Bromouracil exists primarily in the enol form. The enol cannot form any hydrogen bonds with adenine, but it can form two hydrogen bonds with guanine. Therefore, 5-bromouracil pairs with guanine. Because 5-bromouracil causes guanine to be incorporated instead of adenine into newly synthesized DNA strands, it causes mutations.

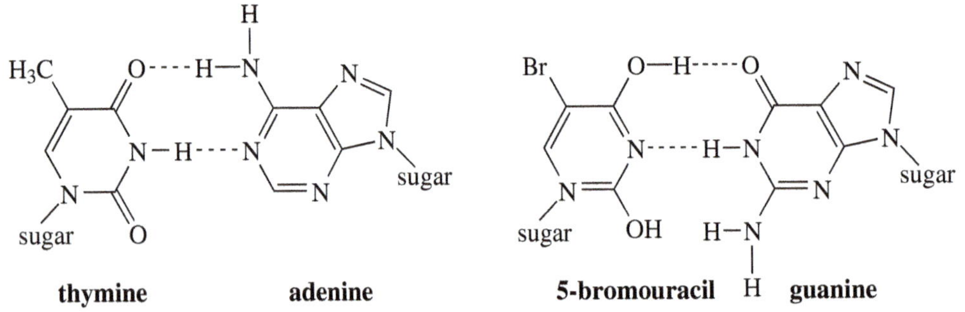

thymine **adenine** **5-bromouracil** **guanine**

37. In an acidic environment, nitrite ion is protonated to nitrous acid. We have seen that the nitrosonium ion is formed from nitrous acid (Section 18.20).

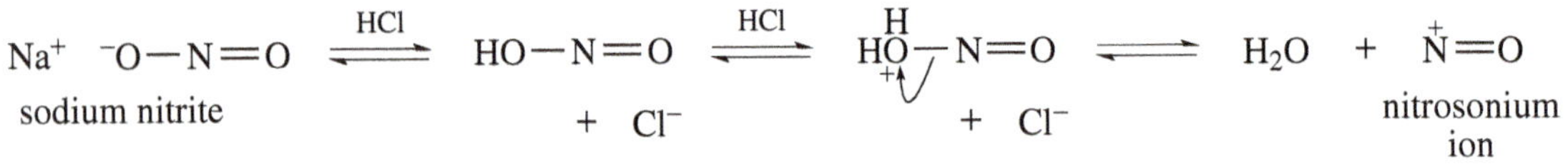

$$Na^+ \ {}^-O{-}N{=}O \underset{}{\overset{HCl}{\rightleftharpoons}} HO{-}N{=}O + Cl^- \overset{HCl}{\rightleftharpoons} H\overset{+}{O}H{-}N{=}O + Cl^- \rightleftharpoons H_2O + \overset{+}{N}{=}O$$

sodium nitrite; nitrosonium ion

The nitrosonium ion reacts with a primary amino group to form a diazonium ion, which can be displaced by water (Section 18.18).

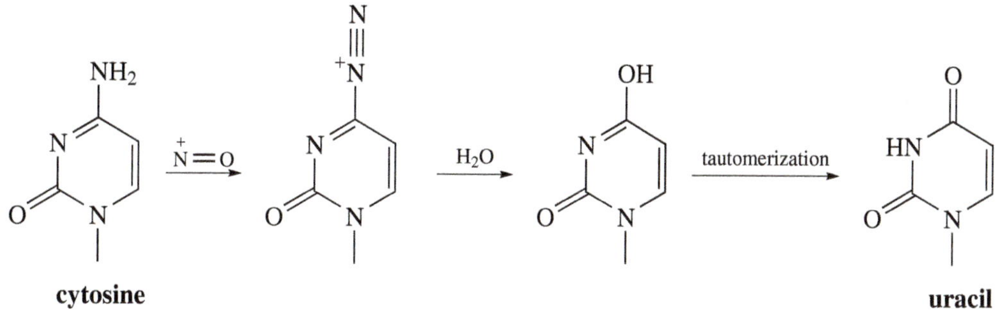

cytosine

uracil

38. It requires energy to break the hydrogen bonds that hold the two chains together, so an enormous amount of energy would be required to unravel the chain completely.
As the new nucleotides that are incorporated into the growing chain form hydrogen bonds with the parent chain, energy is released, and this energy can be used to unwind the next part of the double helix.

39. Ca^{2+} decreases the pK_a of water, forming calcium-bound hydroxide ion, which is a better nucleophile than water. The positively charged nitrogen (of the guanidinium group) of arginine stabilizes the negatively charged oxygen formed when the P=O π bond of the diester breaks and therefore makes it easier to form. Glutamic acid protonates the OR oxygen, thereby making it a weaker base and, therefore, a better leaving group when the P=O π bond reforms.

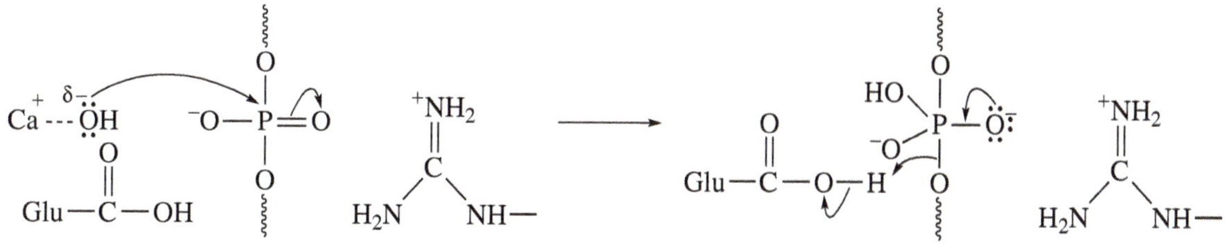

Chapter 26 Practice Test

1. Is the following compound dTMP, UMP, dUMP, or dUTP?

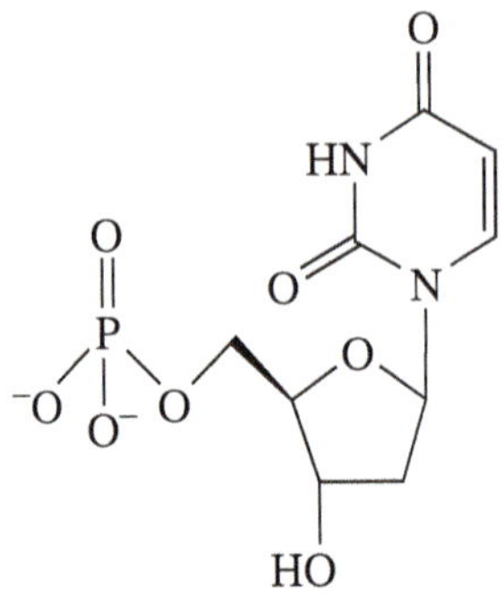

2. If one of the strands of DNA has the following sequence of bases running in the 5′ → 3′ direction, what is the sequence of bases in the complementary strand?

5′—A—C—T—T—G—C—A—T—3′

3. What base is closest to the 5′-end in the complementary strand?

4. Indicate whether each of the following statements is true or false:

a.	Guanine and cytosine are purines.	T	F
b.	The 3′-OH group allows RNA to be easily cleaved.	T	F
c.	The number of As in DNA is equal to the number of Ts.	T	F
d.	rRNA carries the amino acid that will be incorporated into a protein.	T	F
e.	The template strand of DNA is the one transcribed to form RNA.	T	F
f.	The 5′-end of DNA has a free OH group.	T	F
g.	The synthesis of proteins from an RNA blueprint is called transcription.	T	F
h.	A nucleotide consists of a base and a sugar.	T	F
i.	RNA contains Ts, and DNA contains Us.	T	F

5. Which of the following base sequences would most likely be recognized by a restriction endonuclease?

1. ACGCGT **3.** ACGGCA **5.** ACATCGT

2. ACGGGT **4.** ACACGT **6.** CCAACC

6. What would be the sequence of bases in the mRNA obtained from the following segment of DNA?

sense strand
5′—G–C–A–T–G–G–A–C–C–C–C–G–T—3′
3′—C–G–T–A–C–C–T–G–G–G–G–C–A—5′
template strand

7. Which of the following pairs of dinucleotides occur in equal amounts in DNA?

(Remember, nucleotides are always written in the $5' \rightarrow 3'$ direction.)

CA and GT CG and AT
CG and GG CA and TG

CHAPTER 27
Synthetic Polymers

Important Terms

addition polymer (chain-growth polymer) — a polymer that is made by adding monomers to the growing end of a chain.

alpha olefin — a monosubstituted ethylene.

alternating copolymer — a copolymer in which two monomers alternate.

anionic polymerization — a chain-growth polymerization where the initiator is a nucleophile and the propagation site is an anion.

aramide — an aromatic polyamide.

atactic polymer — a polymer in which the substituents are randomly oriented on the extended carbon chain.

biodegradable polymer — a polymer that can be degraded by microorganisms.

biopolymer — a polymer that is synthesized in nature.

block copolymer — a copolymer in which there are blocks of each kind of monomer within the polymer chain.

cationic polymerization — a chain-growth polymerization where the initiator is an electrophile and the propagation site is a cation.

chain-growth polymer (addition polymer) — a polymer that is made by adding monomers to the growing end of a chain.

chain transfer — a reaction in which a growing polymer chain reacts with a molecule XY in a manner that allows X· to terminate the chain, leaving behind Y· to initiate a new chain.

condensation polymer (step-growth polymer) — a polymer that is made by combining two molecules while removing a small molecule (usually water or an alcohol).

conducting polymer — a polymer that can conduct electricity down its backbone.

copolymer — a polymer formed using two or more different monomers.

cross-linking — connecting polymer chains with covalent bonds.

crystallites — regions of a polymer in which the chains are highly ordered.

elastomer — a polymer that can stretch and then revert back to its original shape.

epoxy resin	a resin that is formed by mixing a low-molecular-weight prepolymer with a compound that forms a cross-linked polymer.
graft copolymer	a copolymer that contains branches of a polymer of one monomer grafted onto the backbone of a polymer made from another monomer.
head-to-tail addition	the head of one molecule is added to the tail of another molecule.
homopolymer	a polymer that contains only one kind of monomer.
isotactic polymer	a polymer in which all the substituents are on the same side of the extended carbon chain.
living polymer	a nonterminated chain-growth polymer that remains active. Therefore, the polymerization reaction can continue upon addition of more monomer.
materials science	the science of creating new materials that have practical applications.
monomer	a repeating unit in a polymer.
oriented polymer	a polymer obtained by stretching polymer chains and putting them back together in a parallel fashion.
plasticizer	an organic molecule that dissolves in a polymer and allows the polymer chains to slide by each other.
polyamide	a polymer in which monomers are connected by amide groups.
polycarbonate	a polymer in which the monomers are connected by carbonate groups.
polyester	a polymer in which monomers are connected by ester groups.
polymer	a large molecule made by linking monomers together.
polymer chemistry	the field of chemistry that deals with synthetic polymers; part of the larger discipline known as materials science.
polymerization	the process of linking up monomers to form a polymer.
polyurethane	a polymer in which monomers are connected by urethane groups.
propagating site	the reactive end of a chain-growth polymer.
radical polymerization	a chain-growth polymerization where the initiator is a radical and the propagation site is a radical.
random copolymer	a copolymer with a random distribution of monomers.
ring-opening polymerization	a chain-growth polymerization that involves opening the ring of the monomer.

step-growth polymer (condensation polymer) a polymer that is made by combining two molecules while removing a small molecule (usually water or an alcohol).

syndiotactic polymer a polymer in which the substituents regularly alternate on both sides of the fully extended carbon chain.

synthetic polymer a polymer that is not synthesized in nature.

thermoplastic polymer a polymer that has both ordered crystalline regions and amorphous non-crystalline regions.

thermosetting polymers cross-linked polymers that, after they are hardened, cannot be remelted by heating.

urethane (carbamate) a compound with a carbonyl group that is both an amide and an ester.

$$RO-C(=O)-NHR$$

vinyl polymer a polymer in which the monomer is ethylene or a substituted ethylene.

vulcanization increasing the flexibility of rubber by heating it with sulfur.

Ziegler–Natta catalyst an aluminum-titanium initiator that controls the stereochemistry of a polymer.

Solutions to Problems

1. **a.** $CH_2{=}CHCl$ **b.** 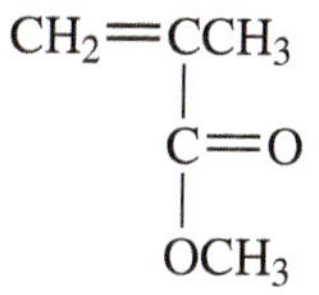 $CH_2{=}C(CH_3)-C(=O)OCH_3$ **c.** $CF_2{=}CF_2$

2. Poly(vinyl chloride) would be more apt to contain head-to-head linkages, because a chloro substituent is less able (compared with a phenyl substituent) to stabilize the growing end of the polymer chain by electron delocalization.

3. 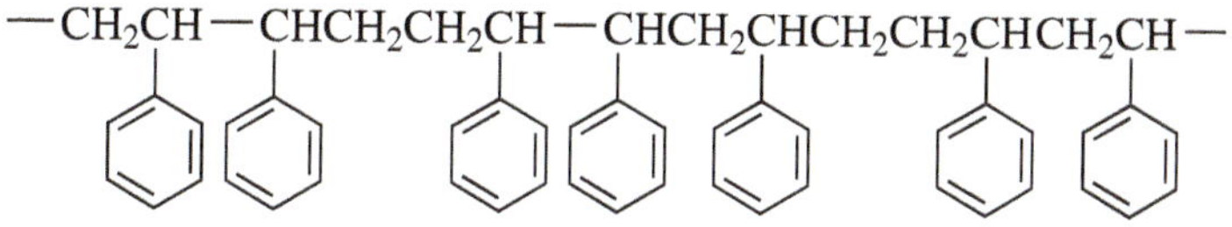 $-CH_2CH-CHCH_2CH_2CH-CHCH_2CHCH_2CH_2CHCH_2CH-$ (each CH bears a phenyl group)

4.

$$HO-OH \longrightarrow 2\ HO\cdot$$

$$HO\cdot + CH_2{=}CHCl \longrightarrow HO-CH_2\underset{Cl}{CH}\cdot$$

$$HO-CH_2\underset{Cl}{CH}\cdot + CH_2{=}CHCl \longrightarrow HO-CH_2\underset{Cl}{CH}CH_2\underset{Cl}{CH}\cdot$$

$$HO-CH_2\underset{Cl}{CH}CH_2\underset{Cl}{CH}\cdot + CH_2{=}CHCl \longrightarrow HO-CH_2\underset{Cl}{CH}CH_2\underset{Cl}{CH}CH_2\underset{Cl}{CH}\cdot$$

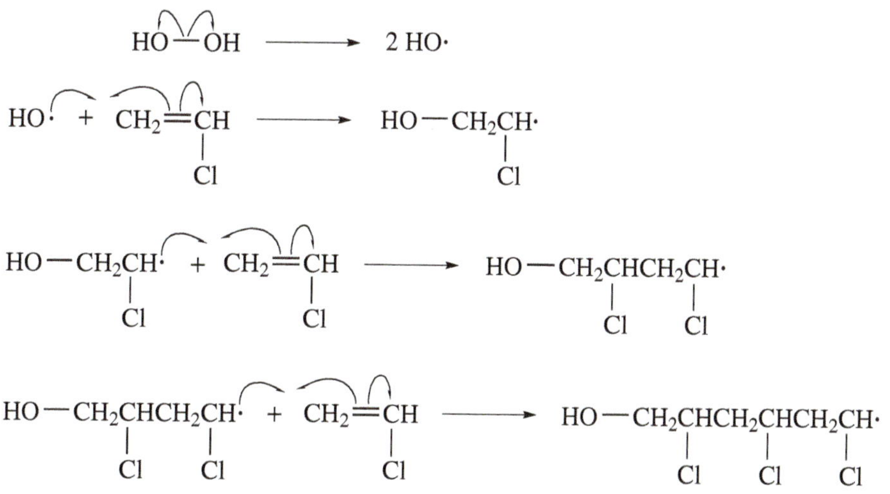

5. Beach balls are made from more highly branched polyethylene because branching increases the flexibility of the polymer.

6.

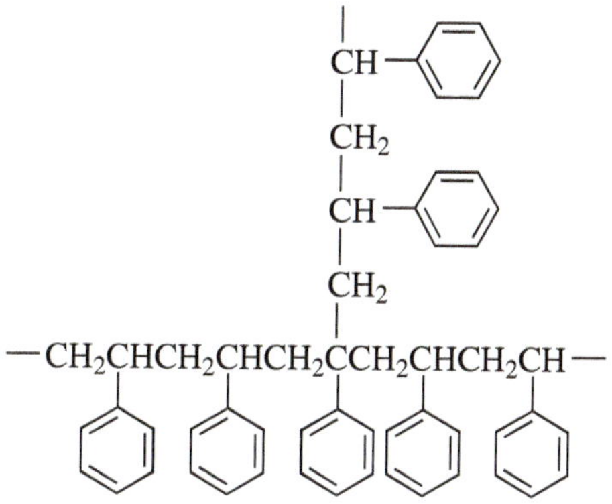

$-CH_2CHCH_2CHCH_2CCH_2CHCH_2CH-$ (the five CH/C carbons each bear a phenyl group; the quaternary C also bears a branch $-CH_2CH(C_6H_5)CH_2CH(C_6H_5)-$)

7. Decreasing ability to undergo cationic polymerization is in the same order as decreasing stability of the carbocation intermediate. (Electron donation increases the stability of the carbocation.)

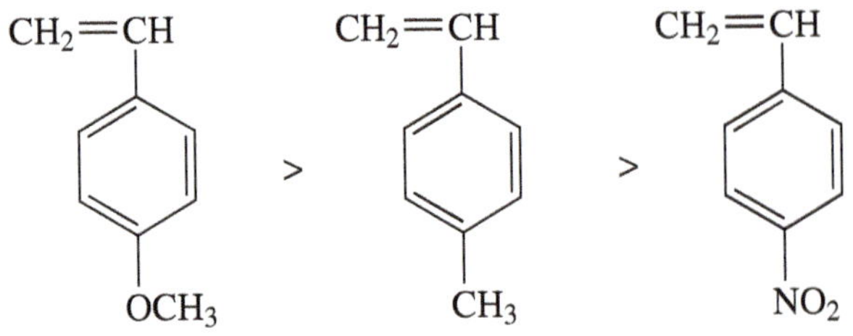

$CH_2{=}CH$–C_6H_4–OCH_3 > $CH_2{=}CH$–C_6H_4–CH_3 > $CH_2{=}CH$–C_6H_4–NO_2

donates electrons by resonance (OCH₃) ... **withdraws electrons by resonance** (NO₂)

$CH_2{=}CH{-}OCH_3 > CH_2{=}CHCH_3 > CH_2{=}CH{-}\overset{O}{\overset{\|}{C}}CH_3$

donates electrons by resonance ... **withdraws electrons by resonance**

$CH_2{=}CCH_3$–C_6H_5 > $CH_2{=}CH$–C_6H_5

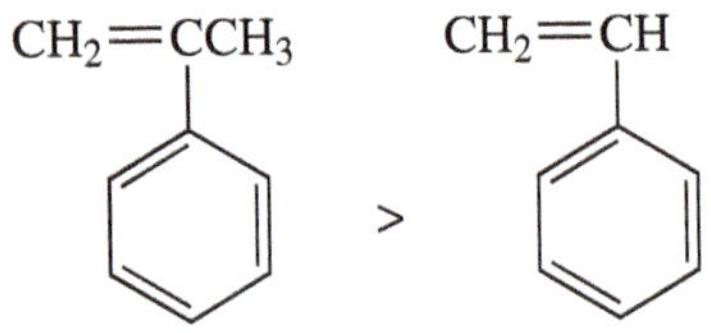

A tertiary benzylic cation is more stable than a secondary benzylic cation.

8. Decreasing ability to undergo anionic polymerization is in the same order as decreasing stability of the carbanion intermediate. (Electron withdrawal increases the stability of the carbanion.)

a. $CH_2{=}CH$–C_6H_4–NO_2 > $CH_2{=}CH$–C_6H_4–CH_3 > $CH_2{=}CH$–C_6H_4–OCH_3

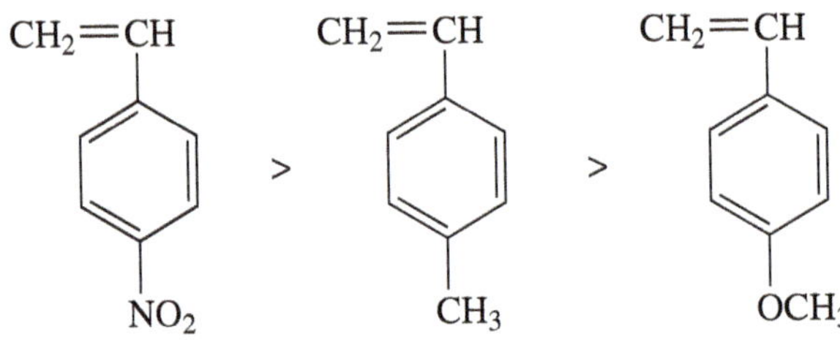

withdraws electrons by resonance (NO₂) ... **donates electrons by resonance** (OCH₃)

b. $CH_2{=}CHC{\equiv}N > CH_2{=}CHCl > CH_2{=}CHCH_3$

withdraws electrons by resonance ($CH_2{=}CHC{\equiv}N$); **withdraws electrons inductively** ($CH_2{=}CHCl$)

9. Methyl methacrylate does not undergo cationic polymerization because the carbocation propagating site would be very unstable, since the ester group is strongly electron withdrawing.

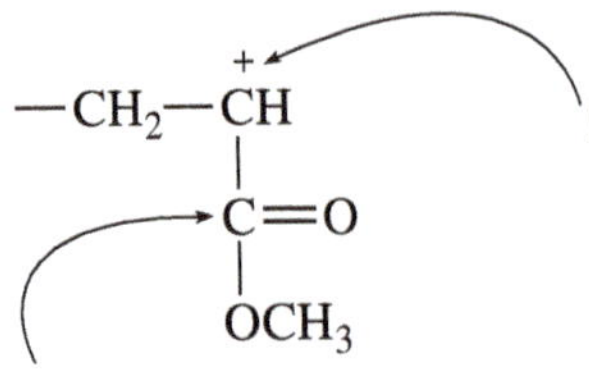

$$-CH_2-\overset{+}{C}H-C(=O)OCH_3$$

propagating site ($\overset{+}{C}H$)

a strongly electron-withdrawing group ($C{=}O$ / OCH_3)

10. **a.** $RO^- + (CH_3)_2C(O)CH_2 \longrightarrow RO{-}CH_2C(CH_3)_2O^-$

$RO{-}CH_2C(CH_3)_2O^- + (CH_3)_2C(O)CH_2 \longrightarrow RO{-}CH_2C(CH_3)_2OCH_2C(CH_3)_2O^-$

$RO{-}CH_2C(CH_3)_2OCH_2C(CH_3)_2O^- + (CH_3)_2C(O)CH_2 \longrightarrow RO{-}CH_2C(CH_3)_2OCH_2C(CH_3)_2OCH_2C(CH_3)_2O^-$

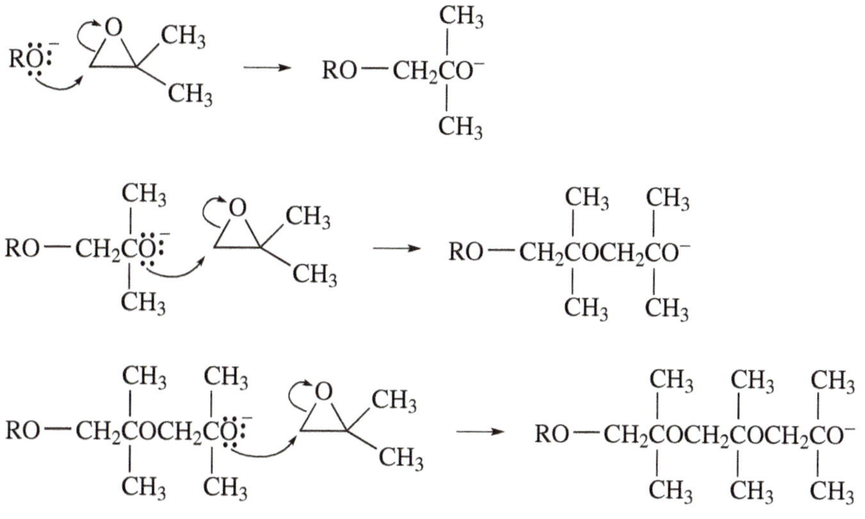

b. epoxide + $H{-}B^+ \longrightarrow$ protonated epoxide ($H\overset{+}{O}$:)

$H\overset{+}{O}$-epoxide + epoxide $\longrightarrow HOCH_2C(CH_3)_2{-}\overset{+}{O}$-epoxide

$HOCH_2C(CH_3)_2{-}\overset{+}{O}$-epoxide + epoxide $\longrightarrow HOCH_2C(CH_3)_2OCH_2C(CH_3)_2{-}\overset{+}{O}$-epoxide

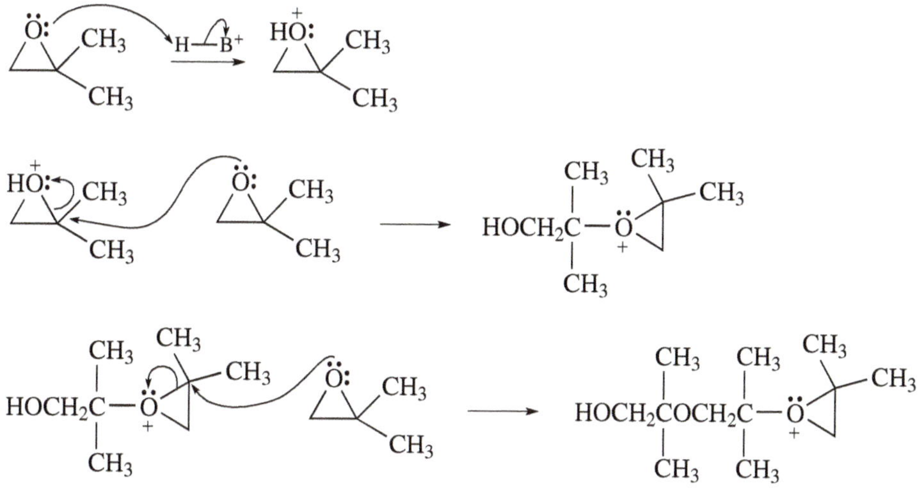

11. In anionic polymerization, nucleophilic attack occurs at the less substituted carbon because it is less sterically hindered; in cationic polymerization, nucleophilic attack occurs at the more substituted carbon because the ring opens to give the more stable partial carbocation (Section 10.7).

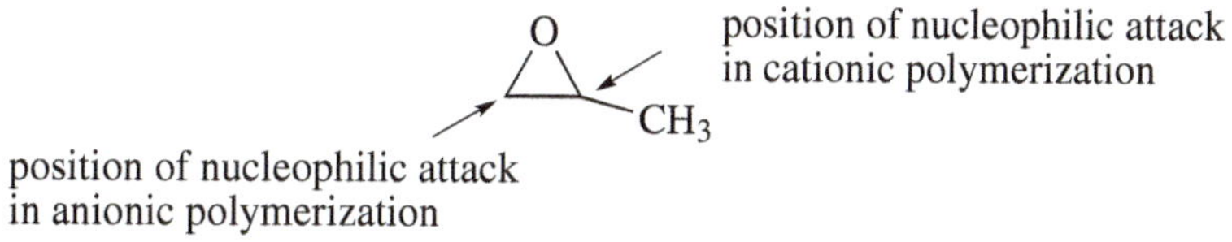

position of nucleophilic attack in cationic polymerization

position of nucleophilic attack in anionic polymerization

12. **a.** $CH_2{=}C(CH_3)CH_3 + BF_3 + H_2O$

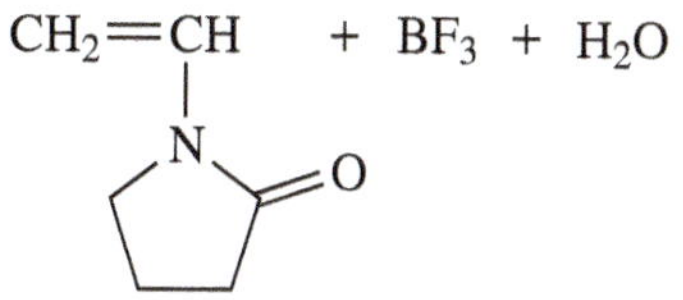

c. ethylene oxide $+ CH_3O^-$

b. $CH_2{=}CH{-}N$(2-pyrrolidinone) $+ BF_3 + H_2O$

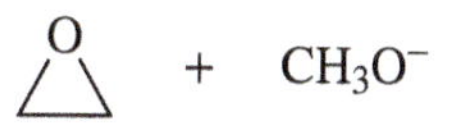

d. $CH_2{=}CH{-}COCH_3$ (with C=O) $+ BuLi$

13. 3,3-Dimethyloxacyclobutane undergoes cationic polymerization by the following mechanism:

H_3C, H_3C, $\overset{+}{O}H$ + H_3C, H_3C, $\ddot{O}$: $\longrightarrow$

$$H_3C,\ H_3C\ \text{(ring)}\ \overset{+}{O}-CH_2\underset{CH_3}{\overset{CH_3}{C}}CH_2OH \quad + \quad H_3C,\ H_3C\ \text{(ring)}\ \ddot{O}:$$

$$\longrightarrow H_3C,\ H_3C\ \text{(ring)}\ \overset{+}{O}-CH_2\underset{CH_3}{\overset{CH_3}{C}}CH_2OCH_2\underset{CH_3}{\overset{CH_3}{C}}CH_2OH \quad + \quad H_3C,\ H_3C\ \text{(ring)}\ \ddot{O}:$$

$$\longrightarrow H_3C,\ H_3C\ \text{(ring)}\ \overset{+}{O}-CH_2\underset{CH_3}{\overset{CH_3}{C}}CH_2OCH_2\underset{CH_3}{\overset{CH_3}{C}}CH_2OCH_2\underset{CH_3}{\overset{CH_3}{C}}CH_2OH$$

14.

15. **a.** $CH_2{=}CH{-}CH{=}CH_2 \quad CH_2{=}CH{-}CH{=}CH_2 \quad CH_2{=}CH{-}CH{=}CH_2$

$$\downarrow$$

$$-CH_2-CH{=}CH-CH_2-CH_2-CH{=}CH-CH_2-CH_2-CH{=}CH-CH_2-$$

b. $CH_2{=}\underset{CH=CH_2}{CH} \quad CH_2{=}\underset{CH=CH_2}{CH} \quad CH_2{=}\underset{CH=CH_2}{CH}$

$$\downarrow$$

$$-CH_2-\underset{CH=CH_2}{CH}-CH_2-\underset{CH=CH_2}{CH}-CH_2-\underset{CH=CH_2}{CH}-$$

16. **a.** $-NHCH_2CH_2CH_2\overset{O}{\overset{\|}{C}}NHCH_2CH_2CH_2\overset{O}{\overset{\|}{C}}-$

b. $-NH(CH_2)_4NH\overset{O}{\overset{\|}{C}}CH_2CH_2\overset{O}{\overset{\|}{C}}NH(CH_2)_4NH\overset{O}{\overset{\|}{C}}CH_2CH_2\overset{O}{\overset{\|}{C}}-$

17. $-\overset{O}{\overset{\|}{C}}(CH_2)_4\overset{O}{\overset{\|}{C}}\left[NH(CH_2)_6NH-\overset{O}{\overset{\|}{C}}(CH_2)_4\overset{O}{\overset{\|}{C}}\right]_n NH(CH_2)_6NH-$

$$\downarrow H_2SO_4$$

$$n\ HO\overset{O}{\overset{\|}{C}}(CH_2)_4\overset{O}{\overset{\|}{C}}OH \quad + \quad n\ H_3\overset{+}{N}(CH_2)_6\overset{+}{N}H_3$$

18. They hydrolyze to form salts of dicarboxylic acids and diols.

$-\overset{O}{\overset{\|}{C}}-C_6H_4-\overset{O}{\overset{\|}{C}}\left[OCH_2-C_6H_{10}-CH_2O-\overset{O}{\overset{\|}{C}}-C_6H_4-\overset{O}{\overset{\|}{C}}\right]_n OCH_2-C_6H_{10}-CH_2O-$

Kodel

$$\downarrow NaOH$$

$$n\ Na^+\ {}^-O\overset{O}{\overset{\|}{C}}-C_6H_4-\overset{O}{\overset{\|}{C}}O^-\ Na^+ \quad + \quad n\ HOCH_2-C_6H_{10}-CH_2OH$$

$-\overset{O}{\overset{\|}{C}}-C_6H_4-\overset{O}{\overset{\|}{C}}\left[OCH_2CH_2O-\overset{O}{\overset{\|}{C}}-C_6H_4-\overset{O}{\overset{\|}{C}}\right]_n OCH_2CH_2O-$

Dacron

$$\downarrow NaOH$$

$$n\ Na^+\ {}^-O\overset{O}{\overset{\|}{C}}-C_6H_4-\overset{O}{\overset{\|}{C}}O^-\ Na^+ \quad + \quad n\ HOCH_2CH_2OH$$

19. a.

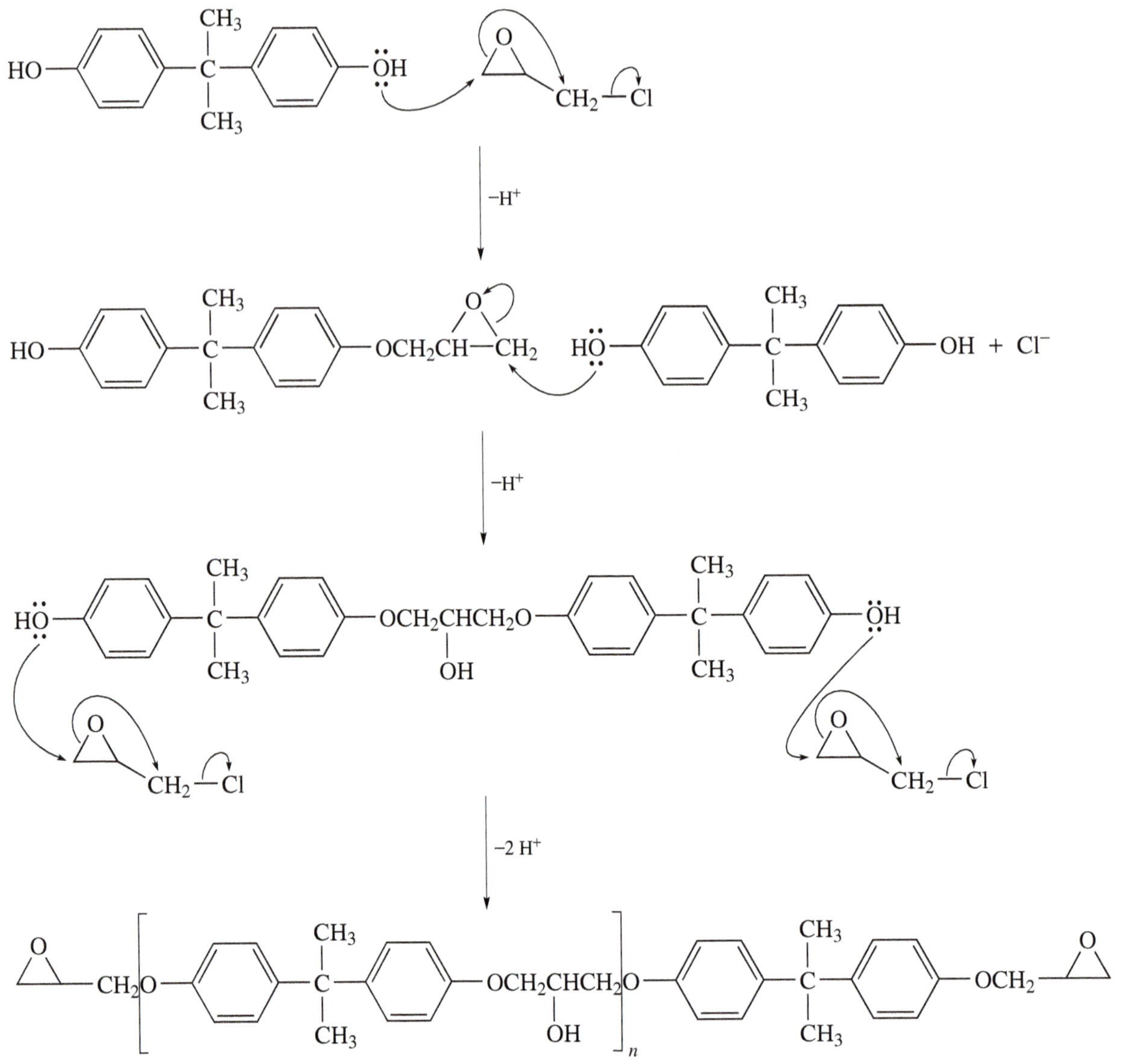

HO—C$_6$H$_4$—C(CH$_3$)$_2$—C$_6$H$_4$—ÖH + (epoxide)—CH$_2$—Cl

$-H^+$

HO—C$_6$H$_4$—C(CH$_3$)$_2$—C$_6$H$_4$—OCH$_2$CH—CH$_2$ (epoxide O) + HÖ—C$_6$H$_4$—C(CH$_3$)$_2$—C$_6$H$_4$—OH + Cl$^-$

$-H^+$

HÖ—C$_6$H$_4$—C(CH$_3$)$_2$—C$_6$H$_4$—OCH$_2$CHCH$_2$O—C$_6$H$_4$—C(CH$_3$)$_2$—C$_6$H$_4$—ÖH (CH carries OH) + 2 (epoxide)—CH$_2$—Cl

$-2\ H^+$

(epoxide)—CH$_2$[O—C$_6$H$_4$—C(CH$_3$)$_2$—C$_6$H$_4$—OCH$_2$CHCH$_2$]$_n$O—C$_6$H$_4$—C(CH$_3$)$_2$—C$_6$H$_4$—OCH$_2$—(epoxide) (CH in brackets carries OH)

b.

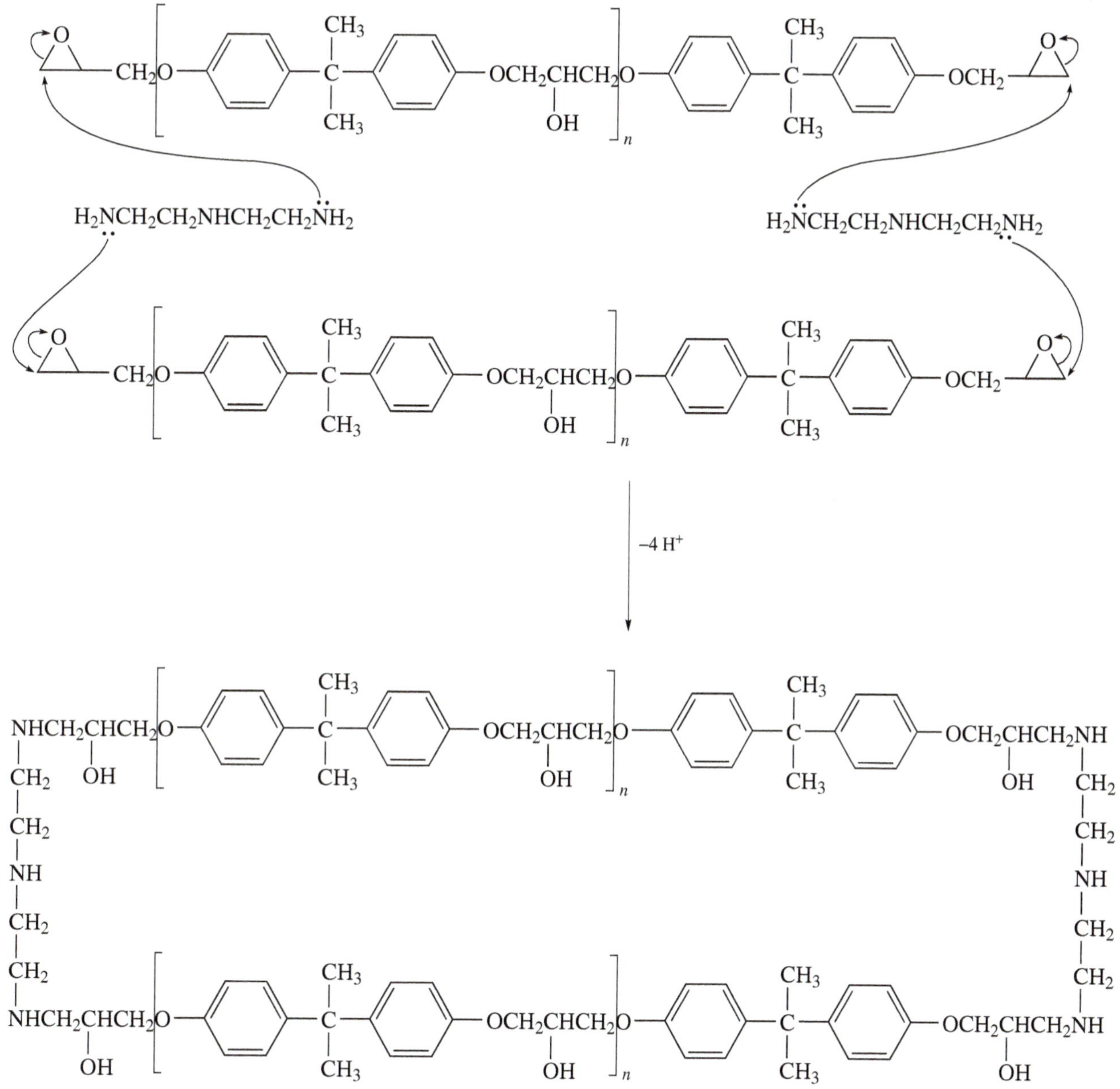

$H_2NCH_2CH_2NHCH_2CH_2\ddot{N}H_2$

$H_2\ddot{N}CH_2CH_2NHCH_2CH_2NH_2$

$-4\ H^+$

20. The polymer is stiffer because the OH on the middle carbon of glycerol cross-links the polymer chains.

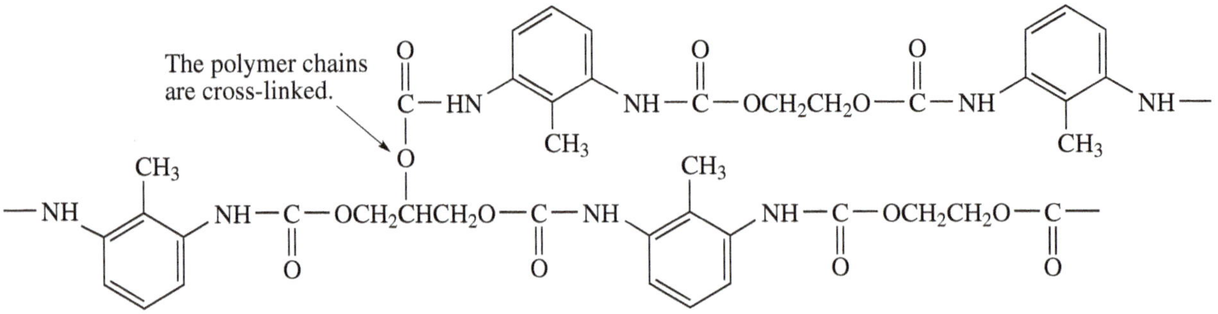

The polymer chains are cross-linked.

$$-\text{NH}-\text{C}_6\text{H}_3(\text{CH}_3)-\text{NH}-\overset{\text{O}}{\overset{\|}{\text{C}}}-\text{HN}-\text{C}_6\text{H}_3(\text{CH}_3)-\text{NH}-\overset{\text{O}}{\overset{\|}{\text{C}}}-\text{OCH}_2\text{CH}_2\text{O}-\overset{\text{O}}{\overset{\|}{\text{C}}}-\text{NH}-\text{C}_6\text{H}_3(\text{CH}_3)-\text{NH}-$$

$$-\text{NH}-\text{C}_6\text{H}_3(\text{CH}_3)-\text{NH}-\underset{\text{O}}{\underset{\|}{\text{C}}}-\text{OCH}_2\text{CHCH}_2\text{O}-\underset{\text{O}}{\underset{\|}{\text{C}}}-\text{NH}-\text{C}_6\text{H}_3(\text{CH}_3)-\text{NH}-\underset{\text{O}}{\underset{\|}{\text{C}}}-\text{OCH}_2\text{CH}_2\text{O}-\underset{\text{O}}{\underset{\|}{\text{C}}}-$$

21. Formation of a protonated imine between formaldehyde and one amino group, followed by reaction with a second amino group, accounts for formation of the linkage that holds the monomers together.

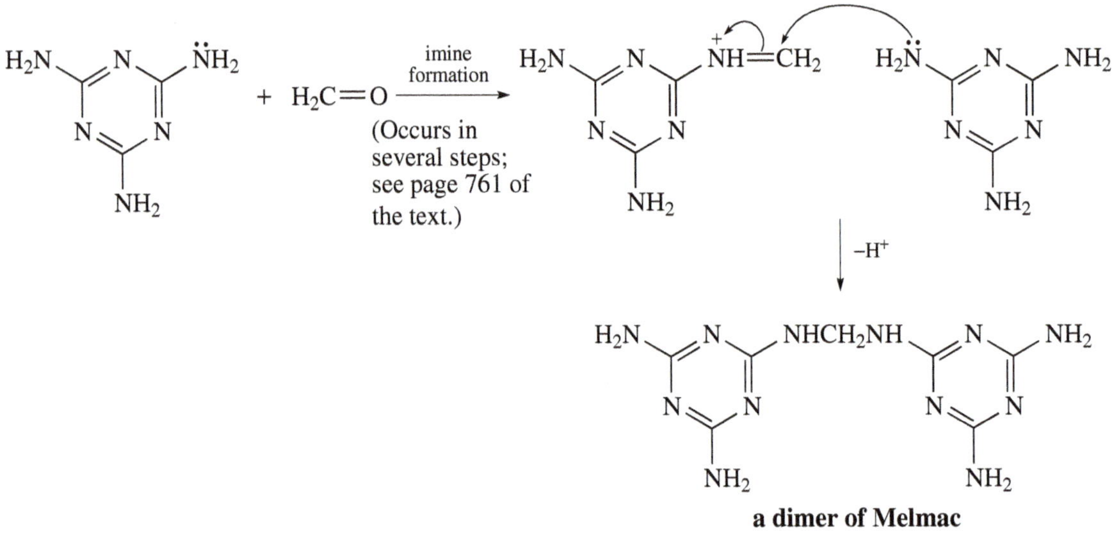

melamine (H_2N, NH_2, NH_2 on the triazine ring) + $H_2C{=}O$ $\xrightarrow{\text{imine formation}}$ $\overset{+}{\text{NH}}{=}\text{CH}_2$ (on the triazine ring, with two NH_2 groups) + $H_2\ddot{N}$ (on a second melamine)

(Occurs in several steps; see page 761 of the text.)

$\downarrow -H^+$

H_2N, NH_2 (triazine)–$NHCH_2NH$–(triazine) NH_2, NH_2

a dimer of Melmac

22.

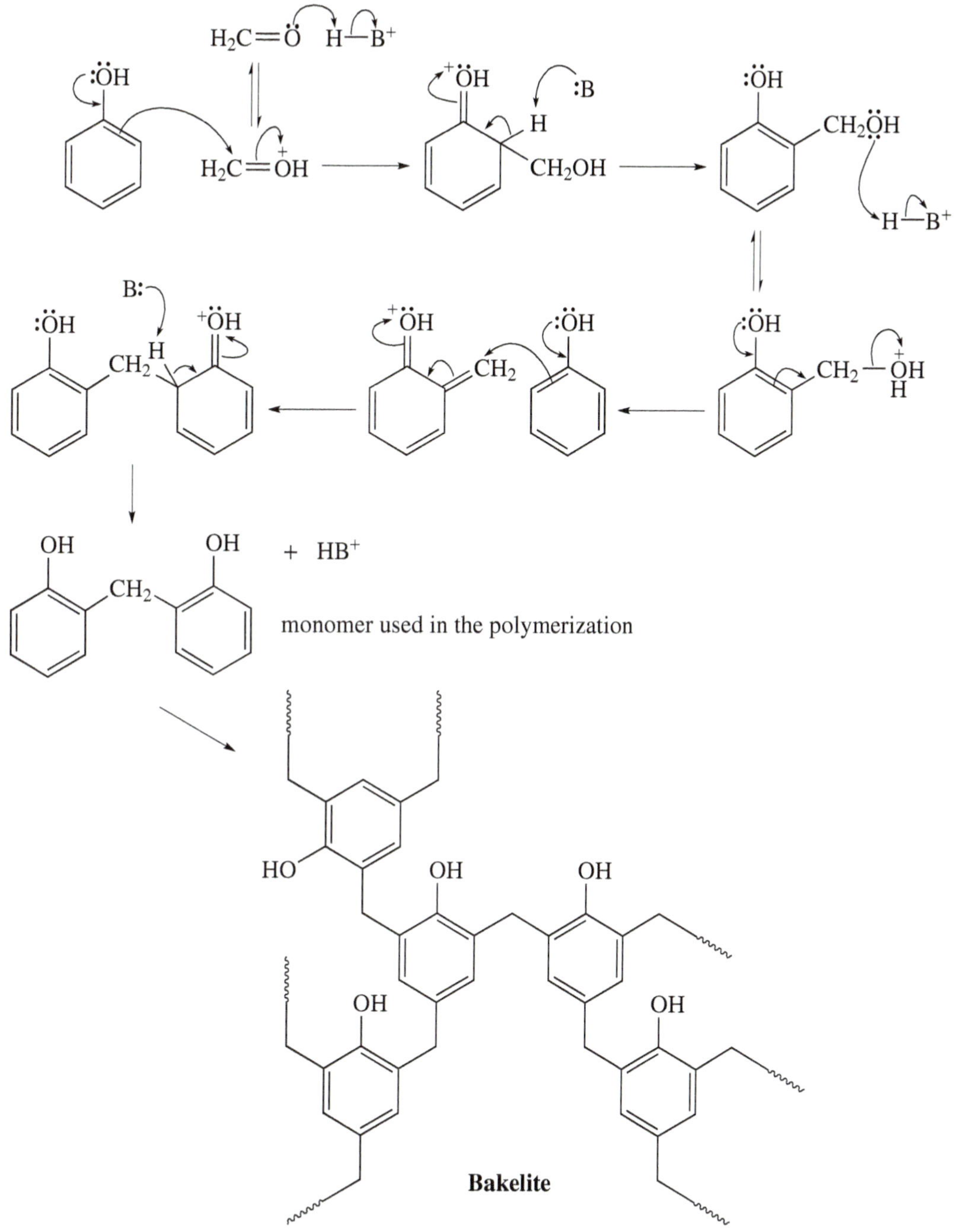
monomer used in the polymerization
Bakelite

23. **a.**

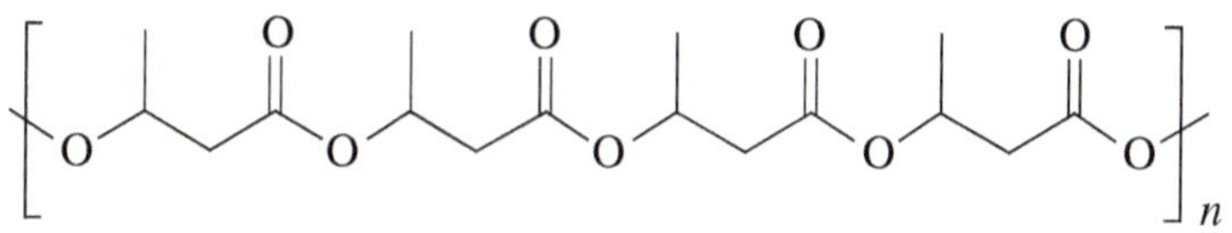

b.

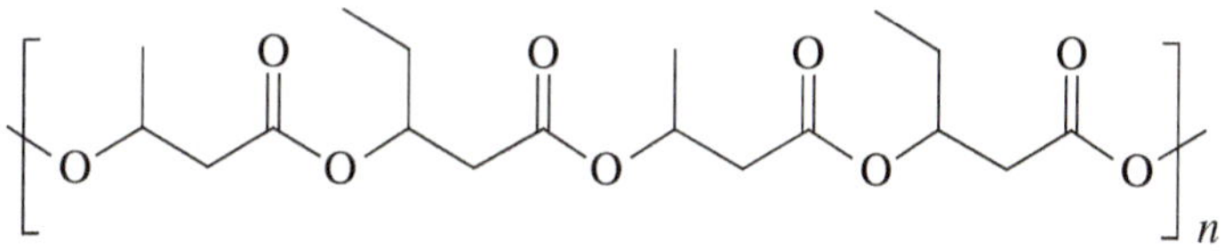

24. **a.** 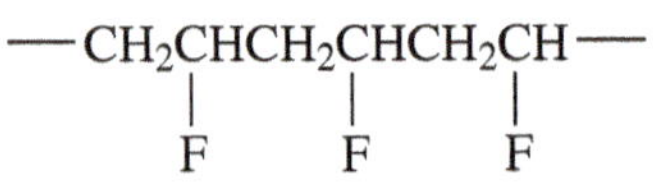 $—CH_2\underset{F}{\underset{|}{C}}HCH_2\underset{F}{\underset{|}{C}}HCH_2\underset{F}{\underset{|}{C}}H—$ chain-growth polymer

b. 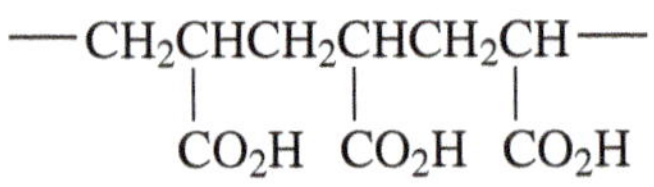 $—CH_2\underset{CO_2H}{\underset{|}{C}}HCH_2\underset{CO_2H}{\underset{|}{C}}HCH_2\underset{CO_2H}{\underset{|}{C}}H—$ chain-growth polymer

c. $—O(CH_2)_5\overset{O}{\overset{\|}{C}}O(CH_2)_5\overset{O}{\overset{\|}{C}}O(CH_2)_5\overset{O}{\overset{\|}{C}}—$ 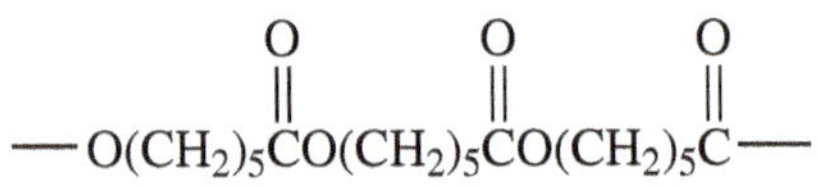 step-growth polymer

d. 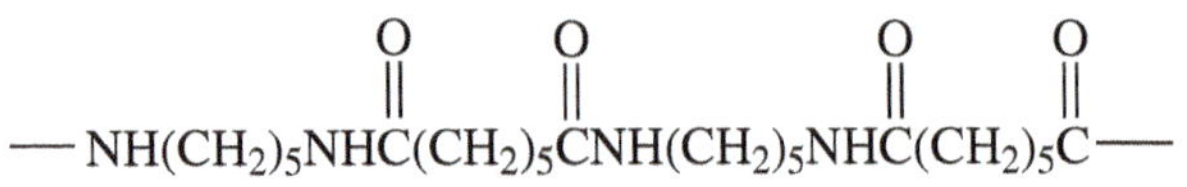 $—NH(CH_2)_5NH\overset{O}{\overset{\|}{C}}(CH_2)_5\overset{O}{\overset{\|}{C}}NH(CH_2)_5NH\overset{O}{\overset{\|}{C}}(CH_2)_5\overset{O}{\overset{\|}{C}}—$ step-growth polymer

e. 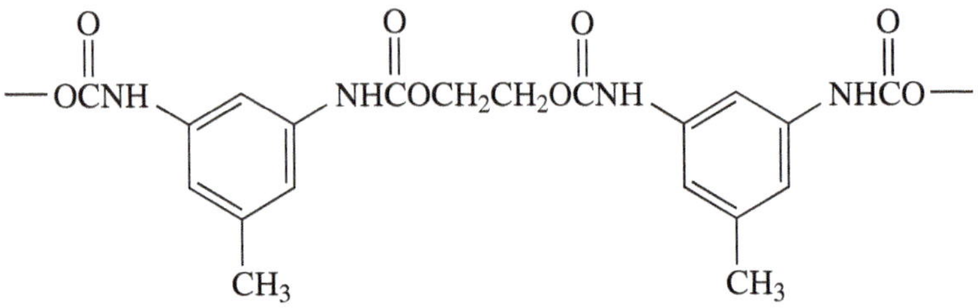 $—O\overset{O}{\overset{\|}{C}}NH$–(benzene ring, CH_3)–$NH\overset{O}{\overset{\|}{C}}OCH_2CH_2O\overset{O}{\overset{\|}{C}}NH$–(benzene ring, CH_3)–$NH\overset{O}{\overset{\|}{C}}O—$ step-growth polymer

25. **a.** 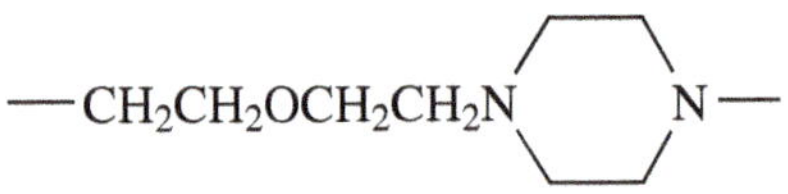 $—CH_2CH_2OCH_2CH_2N$ (ring) $N—$

b. 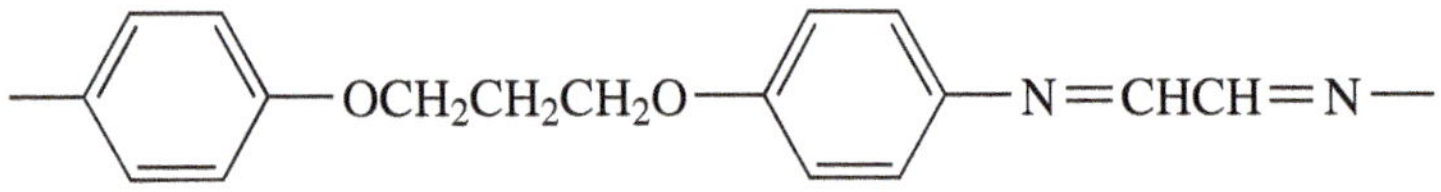 —(benzene ring)—$OCH_2CH_2CH_2O$—(benzene ring)—$N{=}CHCH{=}N—$

c. 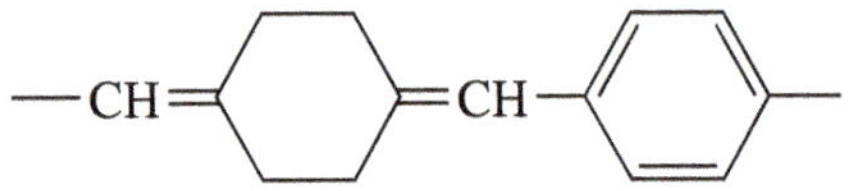 $—CH{=}$(ring)${=}CH$—(benzene ring)—

26. 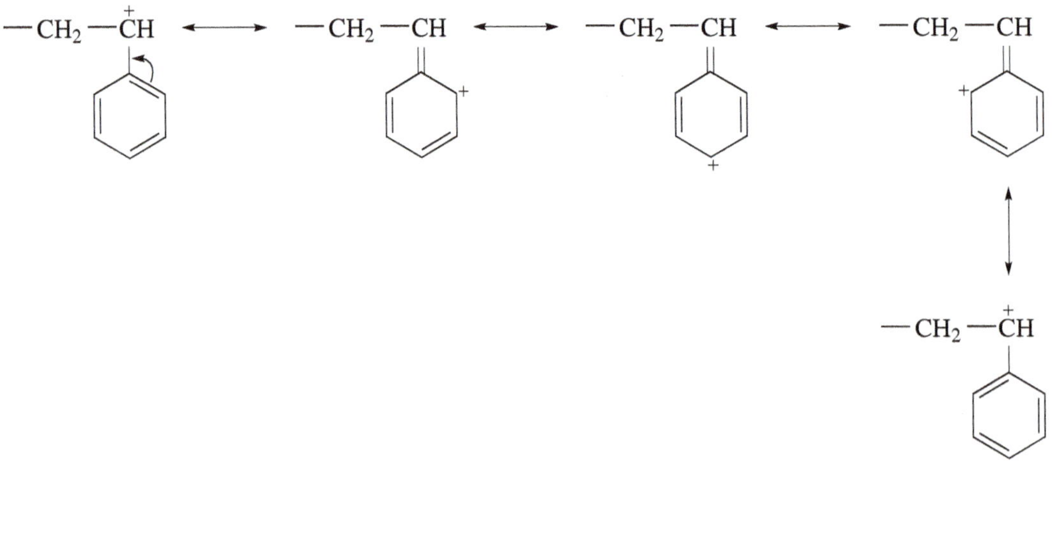 $—CH_2—\overset{+}{C}H$ ⟷ $—CH_2—CH$ ⟷ $—CH_2—CH$ ⟷ $—CH_2—CH$ ⟷ $—CH_2—\overset{+}{C}H$

27. **a.** $CH_2{=}CHCH_2CH_3$

b. (2-methyloxirane: epoxide ring with CH_3 substituent)

c. $CH_2{=}CH$–(4-pyridyl)

d. $ClSO_2$–C_6H_4–SO_2Cl + $H_2N(CH_2)_6NH_2$

e. $CH_2{=}C(CH_3)CH{=}CH_2$

f. $HO(CH_2)_5COH$ (with C=O)

g. $CH_2{=}CCH_3$ with phenyl on the C

h. $HOC(=O)$–C_6H_4–$C(=O)OH$ + $HOCH_2CH_2OH$

a, **b**, **c**, **e**, and **g** are chain-growth polymers.

d, **f**, and **h** are step-growth polymers.

28. Whether a polymer is isotactic, syndiotactic, or atactic depends on whether the substituents are all on one side of the carbon chain, alternate on both sides of the chain, or are random with respect to the chain. Because a polymer of isobutylene has two identical substituents on each carbon in the chain, different configurations are not possible.

$$-CH_2C(CH_3)_2CH_2C(CH_3)_2CH_2C(CH_3)_2CH_2C(CH_3)_2-$$

polymer of isobutylene

29. **a.** $CH_3OCH_2CH(CH_3)OCH_2CH(CH_3)OCH_2CH(CH_3)OCH_2CH(CH_3)O-$

b. $-CH_2C(CH_3)(CHCH_3C_6H_5)-CH_2C(CH_3)(CHCH_3C_6H_5)-CH_2C(CH_3)(CHCH_3C_6H_5)-CH_2C(CH_3)(CHCH_3C_6H_5)-$

c. $-CH_2CH(COCH_3)-CH_2CH(COCH_3)-CH_2CH(COCH_3)-CH_2CH(COCH_3)-$ (each $COCH_3$ with C=O)

d. $-CH_2CH(OCH_3)-CH_2CH(OCH_3)-CH_2CH(OCH_3)-CH_2CH(OCH_3)-$

30. **a.** Because it is a polyamide (and not a polyester), it is a nylon.

b. 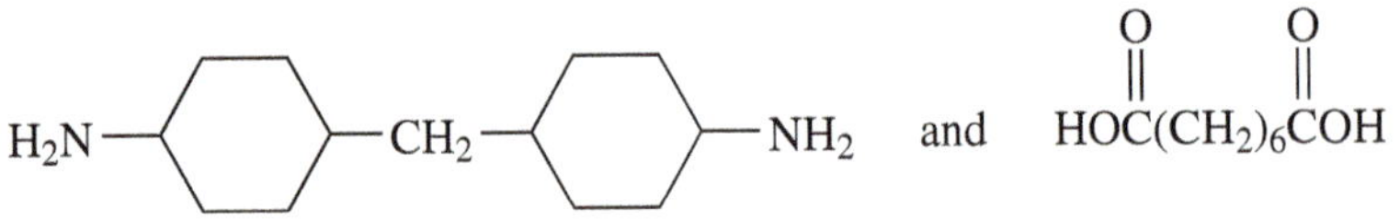 H_2N–(cyclohexyl)–CH_2–(cyclohexyl)–NH_2 and $HOC(=O)(CH_2)_6C(=O)OH$

31. Each ester group in the compound reacts with HCl/CH_3OH in the same way.

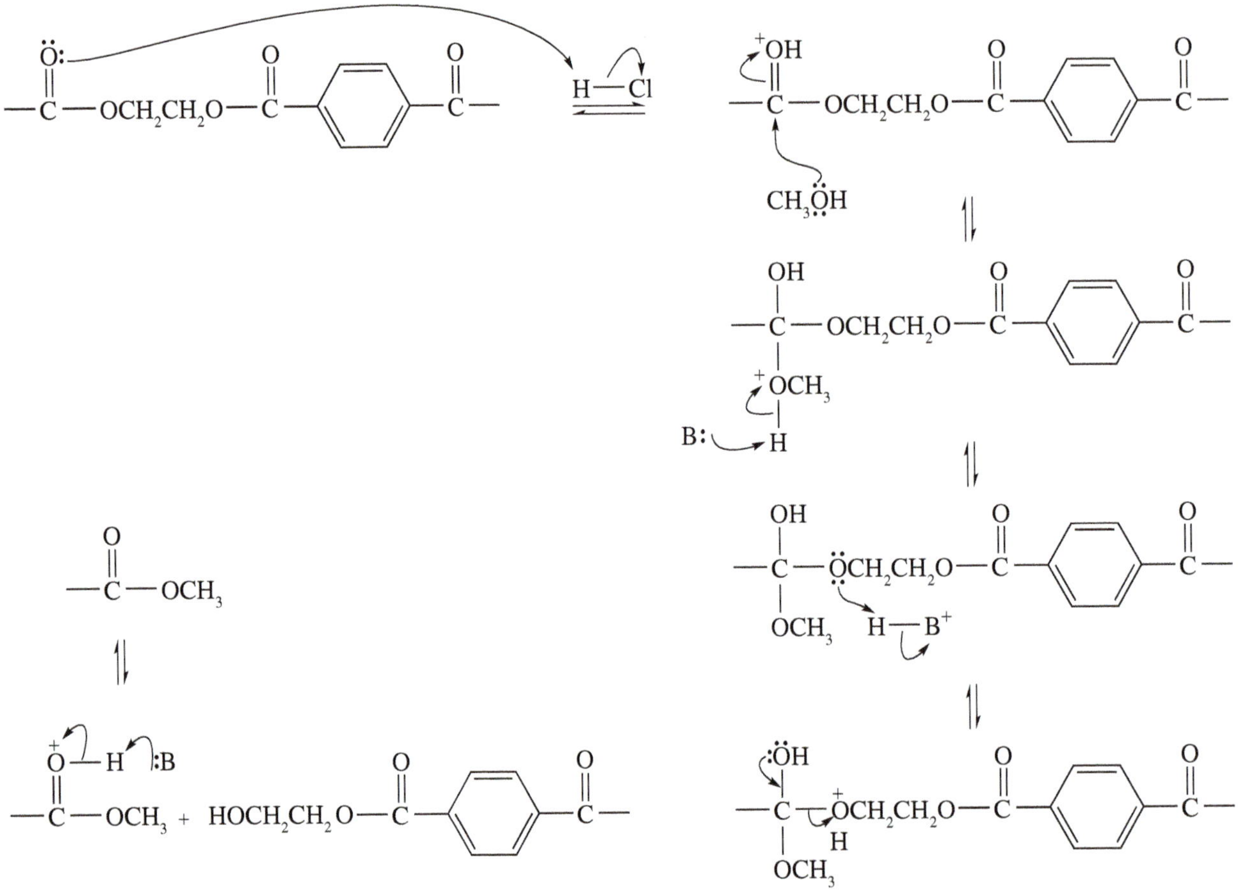

$$-C(=\ddot{O})-OCH_2CH_2O-C(=O)-C_6H_4-C(=O)- \;\; \overset{H-Cl}{\rightleftharpoons} \;\; -C(=\overset{+}{O}H)-OCH_2CH_2O-C(=O)-C_6H_4-C(=O)-$$

$CH_3\ddot{O}H$

$$\rightleftharpoons \;\; -C(OH)(\overset{+}{O}CH_3H)-OCH_2CH_2O-C(=O)-C_6H_4-C(=O)- \quad B:$$

$$\rightleftharpoons \;\; -C(OH)(OCH_3)-\ddot{O}CH_2CH_2O-C(=O)-C_6H_4-C(=O)- \quad H-B^+$$

$$\rightleftharpoons \;\; -C(:\ddot{O}H)(OCH_3)-\overset{+}{O}(H)CH_2CH_2O-C(=O)-C_6H_4-C(=O)-$$

$$\rightleftharpoons \;\; -C(=\overset{+}{O}-H)-OCH_3 \; :B \;\; + \;\; HOCH_2CH_2O-C(=O)-C_6H_4-C(=O)-$$

$$\rightleftharpoons \;\; -C(=O)-OCH_3$$

32. A copolymer is composed of more than one kind of monomer. Because the initially formed carbocation can rearrange, two different monomers (the unrearranged carbocation and the rearranged carbocation) are involved in formation of the polymer.

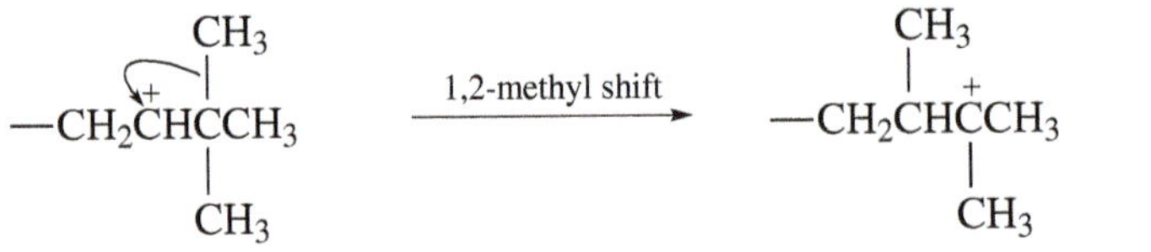

$$-CH_2\overset{+}{C}HC(CH_3)_2CH_3 \xrightarrow{\text{1,2-methyl shift}} -CH_2CH(CH_3)\overset{+}{C}(CH_3)CH_3$$

unrearranged monomer **rearranged monomer**

33. The polymer in the flask that contained a high-molecular-weight polymer and little material of intermediate molecular weight was formed by a chain-growth mechanism, whereas the polymer in the flask that contained mainly material of intermediate molecular weight was formed by a step-growth mechanism.

In a chain-growth mechanism, monomers are added to the growing end of a chain. This means that at any one time, there will be polymeric chains and monomers.

Step-growth polymerization is not a chain reaction; any two monomers can react. Therefore, high-molecular-weight material will not be formed until the end of the reaction when pieces of intermediate molecular weight combine.

34. **a.** Vinyl alcohol is unstable; it tautomerizes to acetaldehyde.

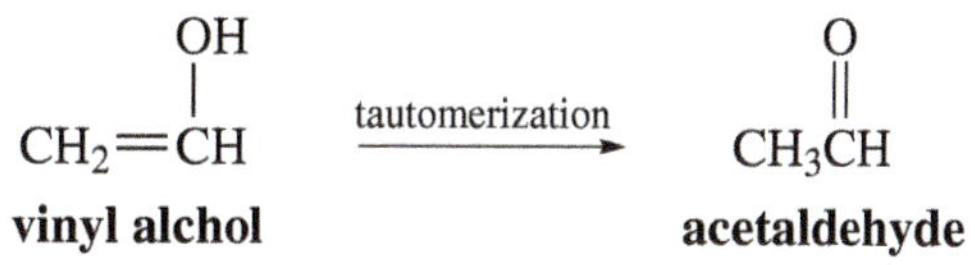

$$CH_2{=}CH(OH) \xrightarrow{\text{tautomerization}} CH_3CH({=}O)$$

vinyl alchol → **acetaldehyde**

b. It is not a true polyester. It has ester groups as substituents **on** the backbone of the chain, but it does not have ester groups **within** the backbone of the polymer chain. A true polyester has ester groups within the backbone of the polymer chain.

35. Each of the following five carbocations can add the growing end of the polymer chain.

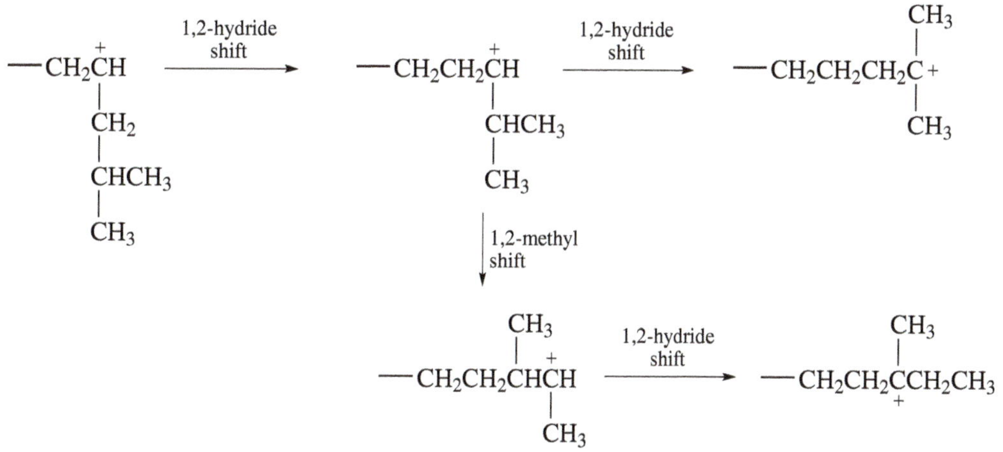

$$—CH_2\overset{+}{C}H(CH_2CH(CH_3)CH_3) \xrightarrow[\text{shift}]{\text{1,2-hydride}} —CH_2CH_2\overset{+}{C}H(CH(CH_3)CH_3) \xrightarrow[\text{shift}]{\text{1,2-hydride}} —CH_2CH_2CH_2\overset{+}{C}(CH_3)_2$$

$$—CH_2CH_2\overset{+}{C}H(CH(CH_3)CH_3) \xrightarrow[\text{shift}]{\text{1,2-methyl}} —CH_2CH_2CH(CH_3)\overset{+}{C}HCH_3 \xrightarrow[\text{shift}]{\text{1,2-hydride}} —CH_2CH_2\overset{+}{C}(CH_3)CH_2CH_3$$

36. Bu—CH_2CH(—$CH_2C_6H_5$)—N(CH_3)—CH_2CH(—$CH_2C_6H_5$)—N(CH_3)—CH_2CH(—$CH_2C_6H_5$)—N(CH_3)—

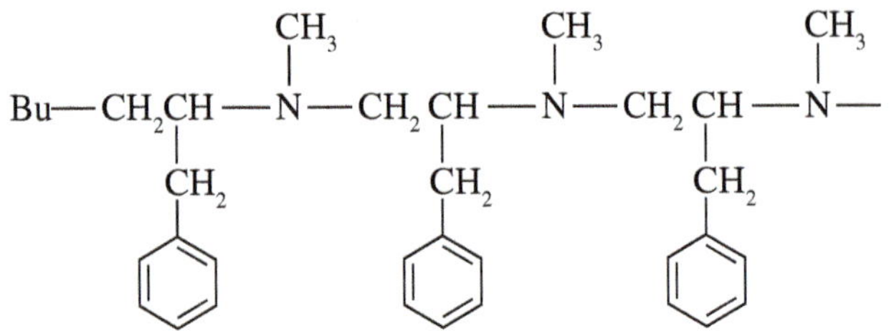

37. Because 1,4-divinylbenzene has substituents on both ends of the benzene ring that can engage in polymerization, the polymer chains can become cross-linked, which increases the rigidity of the polymer.

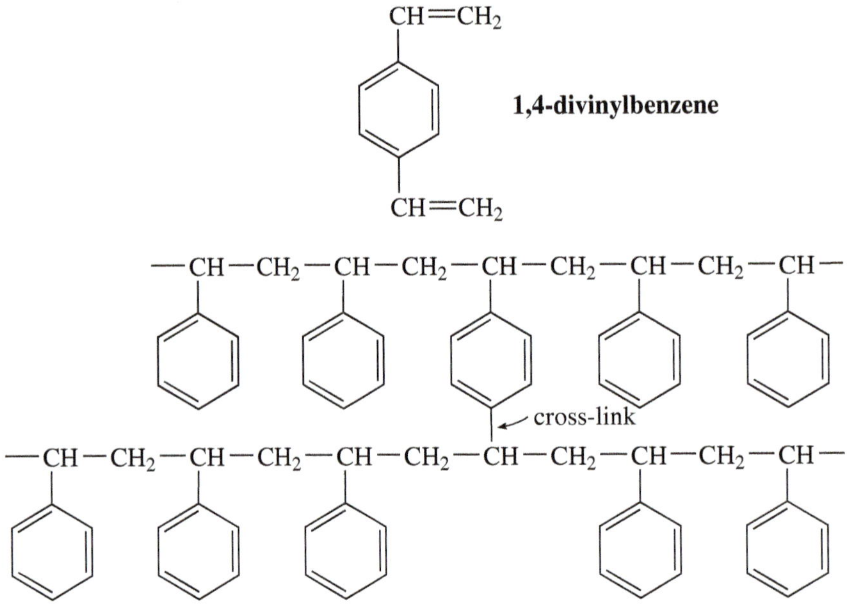

$CH{=}CH_2$ (top) and $CH{=}CH_2$ (bottom) on a benzene ring: **1,4-divinylbenzene**

—CH—CH_2—CH—CH_2—CH—CH_2—CH—CH_2—CH—

cross-link

—CH—CH_2—CH—CH_2—CH—CH_2—CH—CH_2—CH—CH_2—CH—

38. The OH group on the middle carbon of glycerol allows for cross-linking during polymerization. Glyptal gets its strength from this cross-linking.

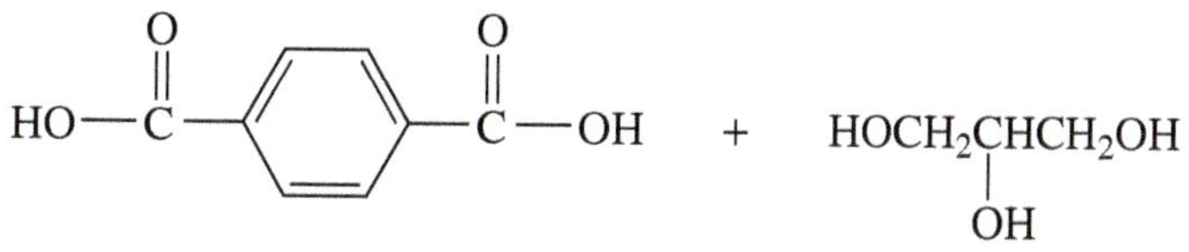

$$HO-\overset{O}{\overset{\|}{C}}-C_6H_4-\overset{O}{\overset{\|}{C}}-OH \quad + \quad HOCH_2\underset{OH}{\underset{|}{C}}HCH_2OH$$

terephthalic acid **glycerol**

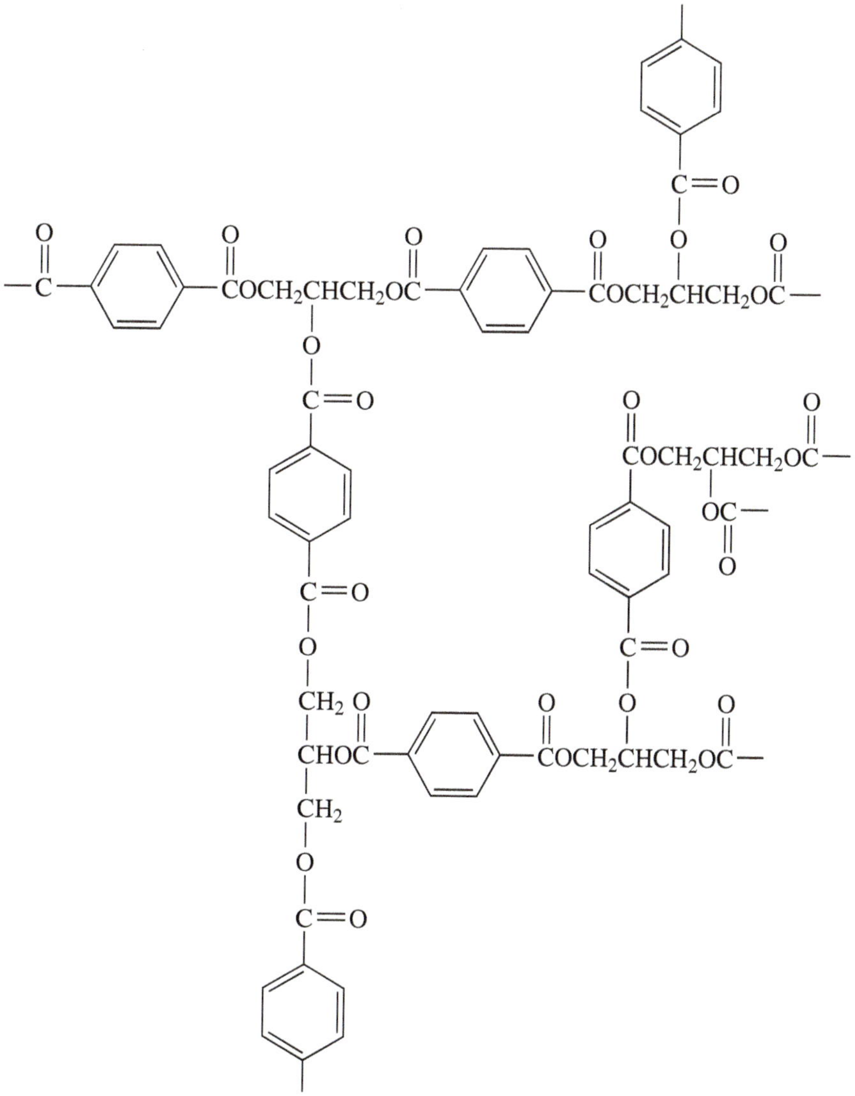

39. The five-membered ring and the four-membered ring react to form the species that polymerizes to form the alternating copolymer.

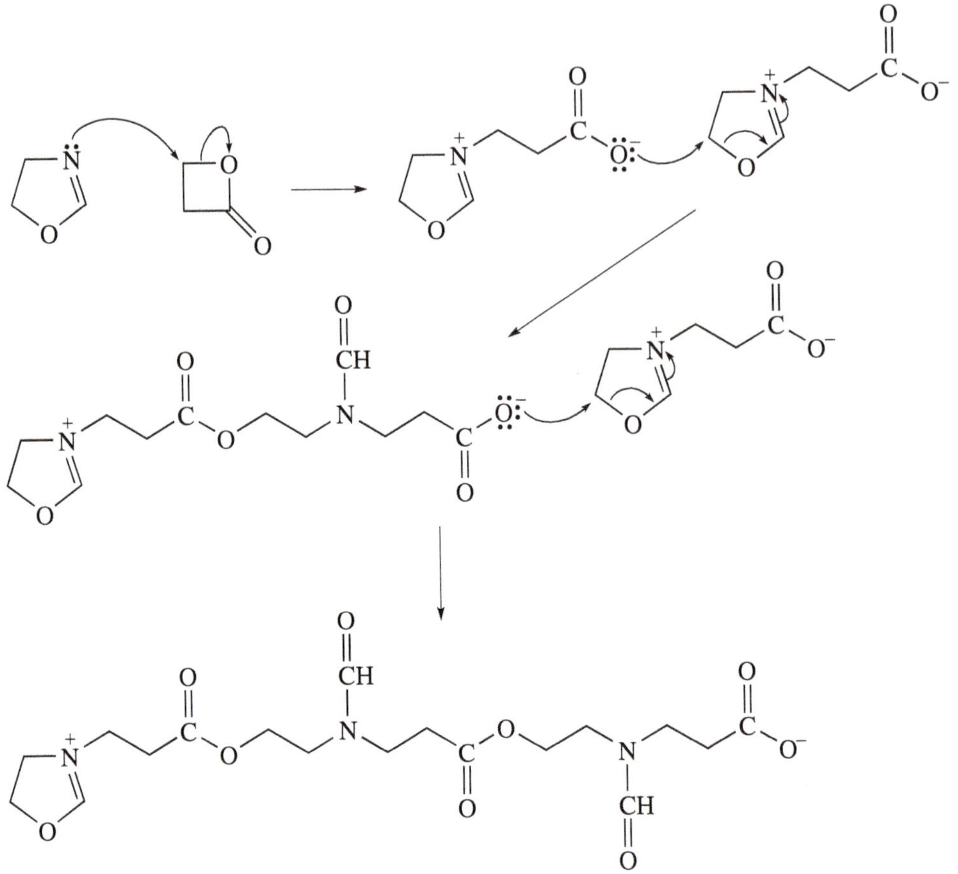

40. Both compounds can form esters via intramolecular or intermolecular reactions. The product of the intramolecular reaction is a lactone; the intermolecular reaction leads to a polymer.

5-Hydroxypentanoic acid reacts intramolecularly to form a six-membered-ring lactone, whereas 6-hydroxyhexanoic acid reacts intramolecularly to form a seven-membered-ring lactone.

5-hydroxypentanoic acid

6-hydroxyhexanoic acid

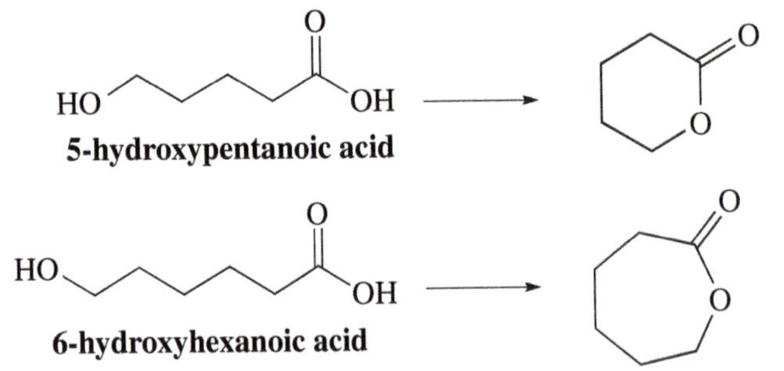

The compound that forms the most polymer is the one that forms the least lactone because the two reactions compete with each other.

The six-membered-ring lactone is more stable and, therefore, has a more stable transition state for its formation, compared to a seven-membered-ring lactone. Because it is easier for 5-hydroxypentanoic acid to form the six-membered-ring lactone than for 6-hydroxyhexanoic acid to form the seven-membered-ring lactone, **6-hydroxyhexanoic acid** forms more polymer.

41. Rubber contains cis double bonds. Ozone (O_3), which is present in the air, oxidizes double bonds to carbonyl groups, which destroys the polymer chain. Polyethylene does not contain double bonds, so it is not air-oxidized.

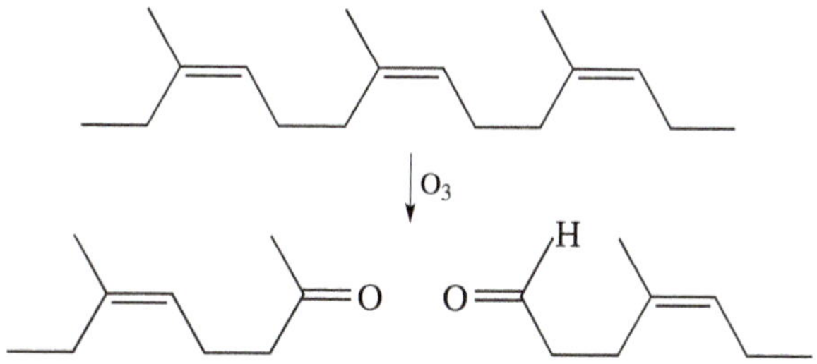

O_3

H

O O

42. $CH_2{=}CH{-}CH{=}O$ $-CH_2{-}CH{=}CH{-}O-$

acrolein **repeating unit**

$CH_2{=}CH(CH{=}O)$ $-CH_2{-}CH(CH{=}O)-$

acrolein **repeating unit**

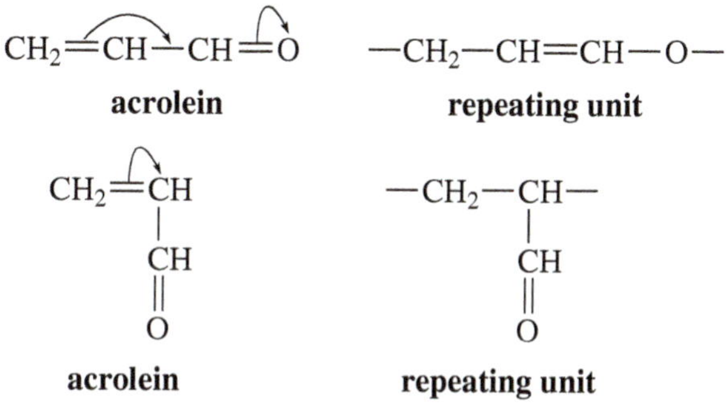

43. The plasticizer that keeps vinyl soft and pliable can vaporize over time, causing the polymer to become brittle. For this reason, high-boiling materials are preferred over low-boiling materials as plasticizers.

44. **a.** Because the negative charge on the propagation site can be delocalized onto the carbonyl oxygen, the polymer is best prepared by anionic polymerization.

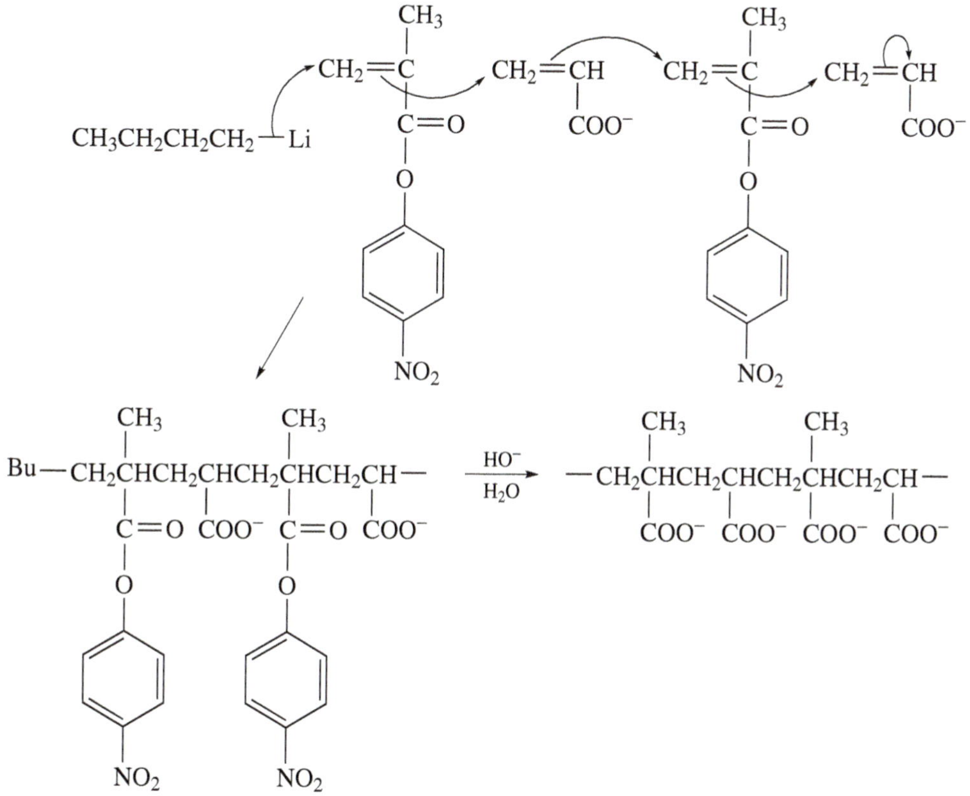

$CH_3CH_2CH_2CH_2{-}Li$ $CH_2{=}C(CH_3)(C{=}O{-}O{-}C_6H_4{-}NO_2)$ $CH_2{=}CH(COO^-)$ $CH_2{=}C(CH_3)(C{=}O{-}O{-}C_6H_4{-}NO_2)$ $CH_2{=}CH(COO^-)$

$Bu{-}CH_2\underset{C=O\,(O{-}C_6H_4{-}NO_2)}{\overset{CH_3}{C}}HCH_2\underset{COO^-}{CH}CH_2\underset{C=O\,(O{-}C_6H_4{-}NO_2)}{\overset{CH_3}{C}}HCH_2\underset{COO^-}{CH}- \xrightarrow[H_2O]{HO^-} -CH_2\underset{COO^-}{\overset{CH_3}{C}}HCH_2\underset{COO^-}{CH}CH_2\underset{COO^-}{\overset{CH_3}{C}}HCH_2\underset{COO^-}{CH}-$

NO2 NO2

b. The carboxyl substituent is in position to remove a proton from water, making water a better nucleophile.

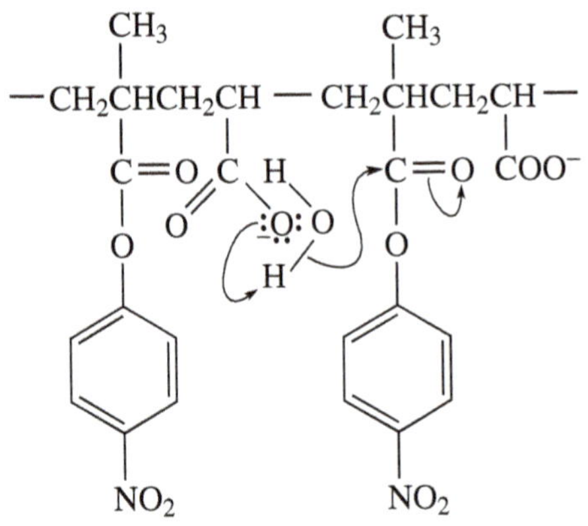

45. Hydrolysis converts the ester substituents into alcohol substituents, which can react with ethylene oxide to graft a polymer of ethylene oxide onto the backbone of the alternating copolymer of styrene and vinyl acetate.

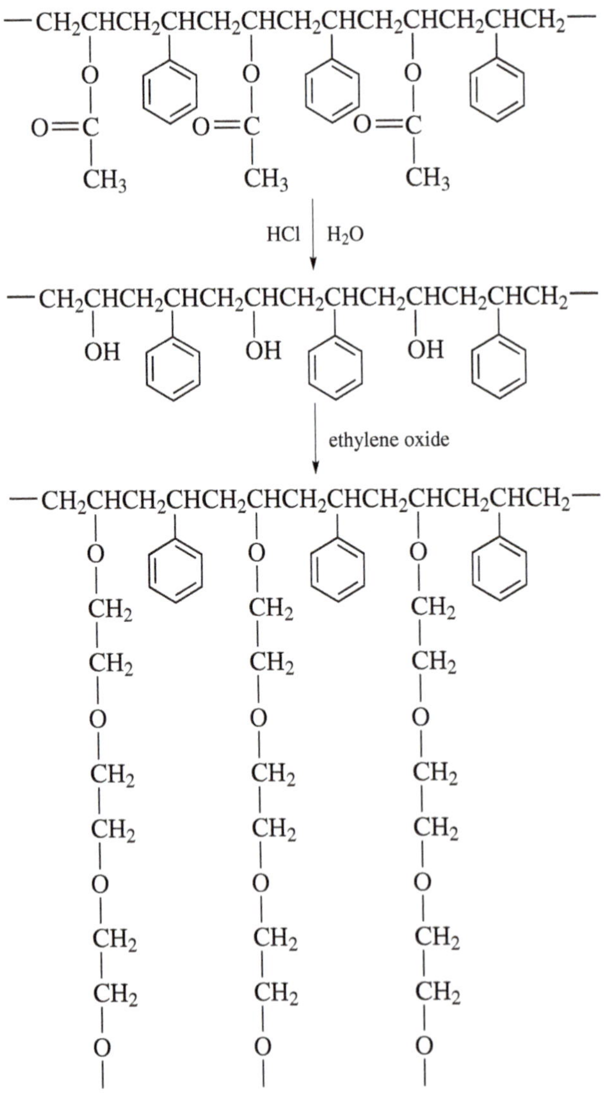

46. The desired polymer is an alternating copolymer of ethylene and 1,2-dibromoethylene.

$CH_2{=}CH_2$ $\quad$ $\underset{Br}{CH}{=}\underset{Br}{CH}$

47. **a.** $H\overset{+}{O}{=}C(H)(H) \quad :\ddot{O}{=}C(H)(H) \longrightarrow HO{-}CH_2{-}\overset{+}{O}{=}C(H)(H) \quad :\ddot{O}{=}C(H)(H)$

$\downarrow$

$HO{-}CH_2{-}O{-}CH_2{-}\overset{+}{O}{=}C(H)(H) \quad :\ddot{O}{=}C(H)(H)$

$\downarrow$

$HO{-}CH_2{-}O{-}CH_2{-}O{-}CH_2{-}\overset{+}{O}{=}C(H)(H)$

$\downarrow$

$HO{-}CH_2{-}O{-}CH_2{-}O{-}CH_2{-}O{-}CH_2{-}O{-}CH_2{-}$

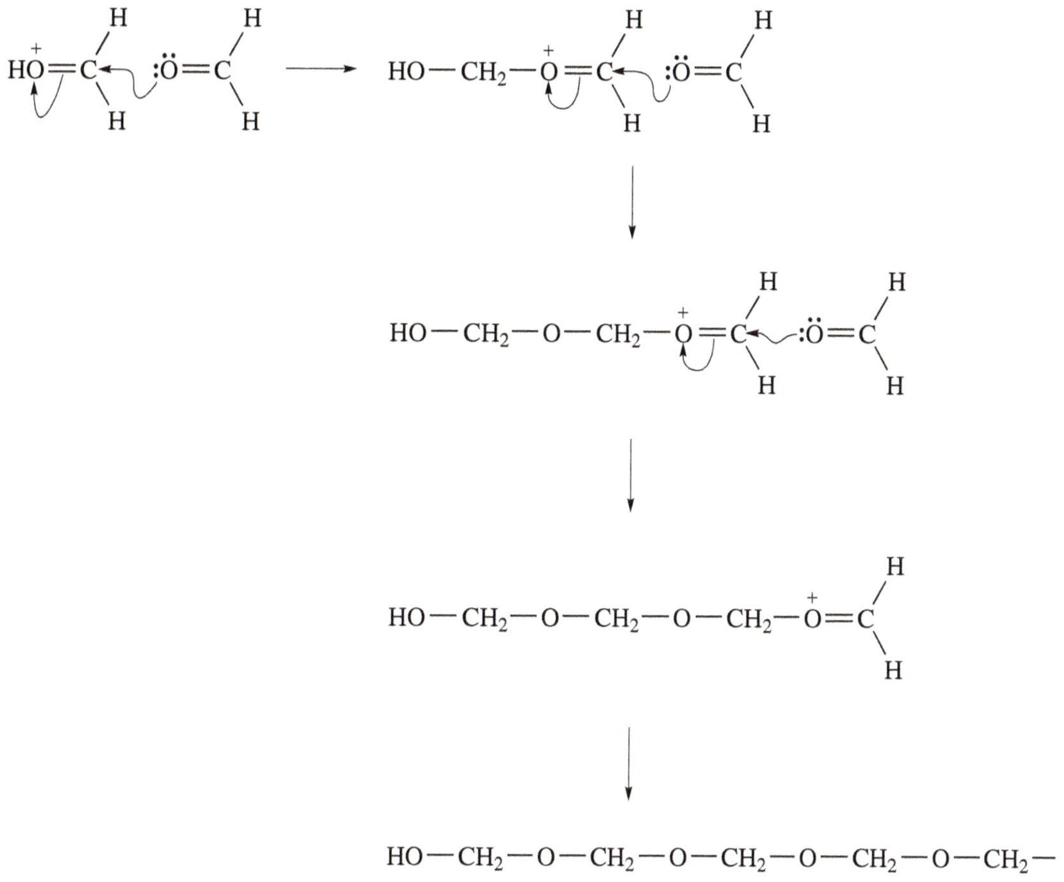

Delrin

b. Delrin is a chain-growth polymer because it is made by adding monomers to the end of a growing chain.

Chapter 27 Practice Test

1. Draw a short segment of the polymer obtained from each of the following monomers. In each case, tell whether it is a chain-growth polymer or a step-growth polymer.

a. $CH_2{=}CHCOOH$

b. 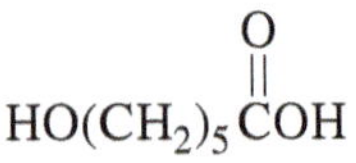 $HO(CH_2)_5\overset{\overset{\large O}{\|}}{C}OH$

c. $CH_2{=}CCl_2$

d. 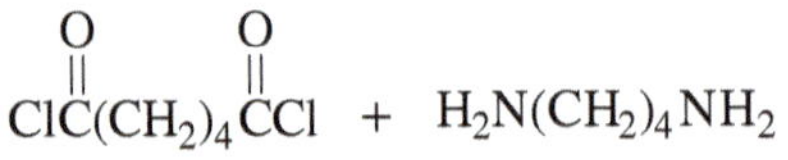 $Cl\overset{\overset{\large O}{\|}}{C}(CH_2)_4\overset{\overset{\large O}{\|}}{C}Cl + H_2N(CH_2)_4NH_2$

2. Draw the structure of the monomer or monomers used to synthesize the following polymers. In each case, tell whether the polymer is a chain-growth polymer or a step-growth polymer.

a. 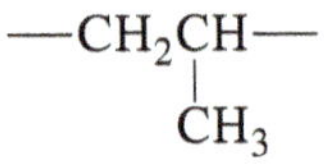 $-CH_2\underset{\underset{\large CH_3}{|}}{C}H-$

b. $-CH_2\overset{\overset{\large CH_3}{|}}{C}{=}CHCH_2-$

c. $-CH_2\underset{\underset{\large CH_3}{|}}{C}HO-$

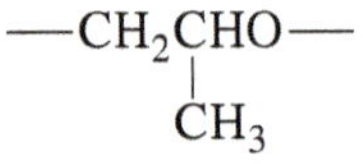

d. 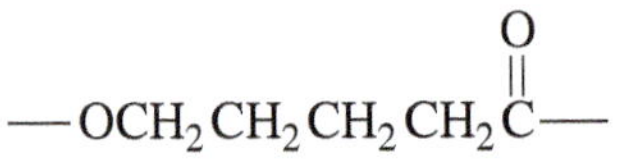 $-OCH_2CH_2CH_2CH_2\overset{\overset{\large O}{\|}}{C}-$

3. Draw short segments of the polymers obtained from the following compounds under the given reaction conditions.

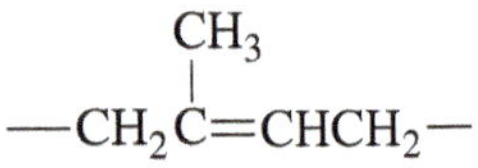

a. 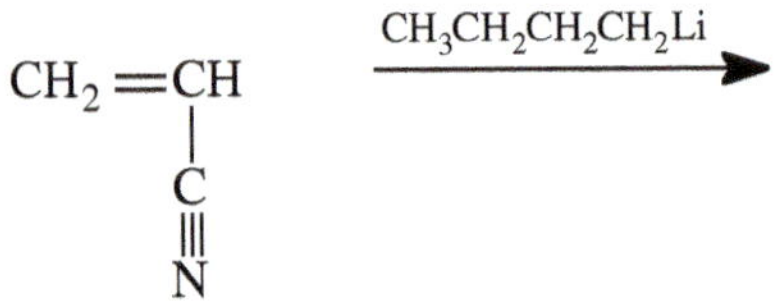 $CH_2{=}\underset{\underset{\large C\equiv N}{|}}{C}H \xrightarrow{CH_3CH_2CH_2CH_2Li}$

b. $CH_2{=}\underset{\underset{\large \underset{\large CH_3}{|}}{C=O}}{|}}{C}H \xrightarrow{BF_3,\ H_2O}$

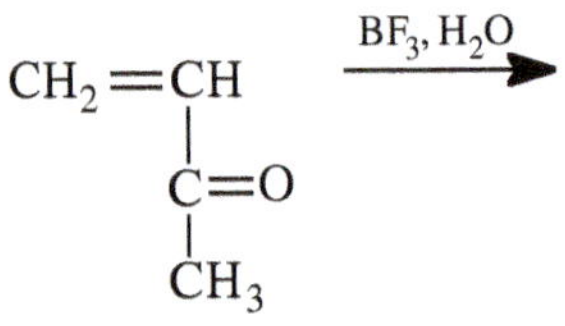

c. $H_2C\overset{O}{—}CHCH_2CH_3 \xrightarrow{CH_3O^-}$ (epoxide: three-membered ring of H_2C, O, and $CHCH_2CH_3$)

4. Explain why $CH_2{=}CCl_2$ does not form an isotactic, syndiotactic, or atactic polymer.

CHAPTER 28
Pericyclic Reactions

Important Terms

antarafacial bond formation formation and/or cleavage of two σ bonds that occur on opposite sides of the π system of the reactant.

antarafacial rearrangement a rearrangement where the migrating group moves to the opposite face of the π system.

antibonding π molecular orbital a molecular orbital that results when two parallel atomic orbitals with opposite phases interact. Electrons in an antibonding orbital decrease bond strength.

antisymmetric molecular orbital a molecular orbital in which the left half is not a mirror image of the right half but would be if one-half of the MO were turned upside down.

bonding π molecular orbital a molecular orbital that results when two parallel atomic orbitals with the same phase interact. Electrons in a bonding orbital increase bond strength.

Claisen rearrangement a [3,3] sigmatropic rearrangement of an allyl vinyl ether.

conrotatory ring closure a ring closure that achieves head-to-head overlap of *p* orbitals by rotating the orbitals in the same direction.

conservation of orbital symmetry theory a theory that explains the relationship between the structure and stereochemistry of the reactant, the conditions under which a pericyclic reaction takes place, and the stereochemistry of the product.

Cope rearrangement a [3,3] sigmatropic rearrangement of a 1,5-diene.

cycloaddition reaction a reaction in which two π-bond-containing molecules react to form a cyclic compound.

disrotatory ring closure a ring closure that achieves head-to-head overlap of *p* orbitals by rotating the orbitals in opposite directions.

electrocyclic reaction an intramolecular reaction in which a new σ bond is formed between the ends of a conjugated system.

excited state a description of which orbitals the electrons of an atom or a molecule occupy when an electron in the ground state has been moved to a higher-energy orbital.

frontier orbital analysis an analysis that determines the outcome of a pericyclic reaction using frontier orbitals.

frontier orbitals the HOMO and the LUMO of the two reacting species in a pericyclic reaction.

frontier orbital theory a theory that, like the conservation of orbital symmetry, explains the relationship between reactant, product, and reaction conditions in a pericyclic reaction.

ground state a description of which orbitals the electrons of an atom or a molecule occupy when they are all in their lowest-energy orbitals.

highest occupied molecular orbital (HOMO) the molecular orbital of highest energy that contains an electron.

linear combination of atomic orbitals (LCAO) the combination of atomic orbitals to produce a molecular orbital.

lowest unoccupied molecular orbital (LUMO) the molecular orbital of lowest energy that does not contain an electron.

molecular orbital (MO) theory a theory that describes a model in which the electrons occupy orbitals as they do in atoms but with the orbitals extending over the entire molecule.

pericyclic reaction a concerted reaction that occurs as a result of a cyclic reorganization of electrons.

photochemical reaction a reaction that takes place when a reactant absorbs light.

polar reaction the reaction between a nucleophile and an electrophile.

radical reaction a reaction in which a new bond is formed using one electron from one reactant and one electron from another reactant.

selection rules the rules that determine the outcome of a pericyclic reaction.

sigmatropic rearrangement a reaction in which a σ bond is broken in the reactant, a new σ bond is formed in the product, and the π bonds rearrange.

suprafacial bond formation formation and/or cleavage of two σ bonds that occur on the same side of the π system of the reactant.

suprafacial rearrangement a rearrangement where the migrating group remains on the same face of the π system.

symmetric molecular orbital a molecular orbital in which the left half is a mirror image of the right half.

symmetry-allowed pathway a pathway that leads to overlap of in-phase orbitals.

symmetry-forbidden pathway a pathway that leads to overlap of out-of-phase orbitals.

thermal reaction a reaction that takes place without the reactant having to absorb light.

Woodward–Hoffmann rules a series of selection rules for pericyclic reactions.

Solutions to Problems

1. **a.** electrocyclic reaction
   b. sigmatropic rearrangement
   c. cycloaddition reaction
   d. cycloaddition reaction

2. **a.** bonding orbitals = ψ_1, ψ_2, ψ_3; antibonding orbitals = ψ_4, ψ_5, ψ_6
   b. ground-state HOMO = ψ_3; ground-state LUMO = ψ_4
   c. excited-state HOMO = ψ_4; excited-state LUMO = ψ_5
   d. symmetric orbitals = ψ_1, ψ_3, ψ_5; antisymmetric orbitals = ψ_2, ψ_4, ψ_6
   e. The HOMO and LUMO have opposite symmetries.

3. **a.** eight molecular orbitals **b.** ψ_4 **c.** seven nodes (There is also a node that passes through the nuclei—that is, through the centers of the *p* orbitals.)

4. **a.** 1,3-Pentadiene has two conjugated π bonds, so it has the same molecular orbital description as 1,3-butadiene, a compound that also has two conjugated π bonds. (See Figure 28.2 on page 1216 of the text.)
   b. The π bonds in 1,4-pentadiene are isolated, so its molecular orbital description is the same as ethene, a compound with an isolated π bond. (See Figure 28.1 on page 1216 of the text.)
   c. 1,3,5-Heptatriene has three conjugated π bonds, so it has the same molecular orbital description as 1,3,5-hexatriene, a compound that also has three conjugated π bonds. (See Figure 28.3 on page 1217 of the text.)
   d. 1,3,5,8-Nonatetraene has three conjugated π bonds and an isolated π bond. The three conjugated π bonds are described by Figure 28.3 and the isolated π bond by Figure 28.1.

5. **a.** 2, 4, or 6 conjugated double bonds 3, 5, or 7 conjugated double bonds

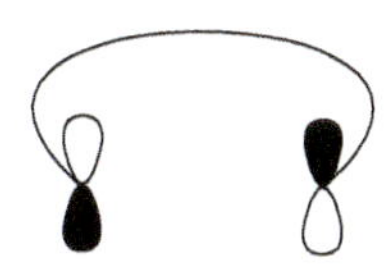

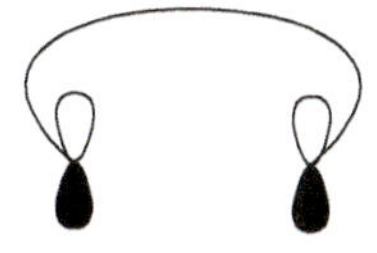

   b. Under thermal conditions, electrocyclic ring closure involves the ground-state HOMO of the polyene (a compound with several double bonds). If the polyene has an even number of double bonds, conrotatory ring closure will result in in-phase overlap of the terminal *p* orbitals in the HOMO. If the polyene has an odd number of double bonds, disrotatory ring closure will result in in-phase overlap of the terminal *p* orbitals in the HOMO.

   Under photochemical conditions, electrocyclic ring closure involves the excited-state HOMO of the polyene, which has the opposite symmetry of the ground-state HOMO and, therefore, requires the opposite mode of ring closure to that required under thermal conditions.

6. **a.** (2*E*,4*Z*,6*Z*,8*E*)-2,4,6,8-Decatetraene has an even number of conjugated π bonds (4). Therefore, under thermal conditions, ring closure will be conrotatory.
   b. The substituents point in opposite directions, and conrotatory ring closure of such substituents will cause them to be trans in the ring-closed product.

c. Under photochemical conditions, a compound with four conjugated π bonds undergoes disrotatory ring closure.

d. Because ring closure is disrotatory and the substituents point in opposite directions, the product will have the cis configuration.

7. a. correct b. correct c. correct

8. 1. a. conrotatory b. trans 2. a. disrotatory b. cis

9. Solved in the text.

10. The reaction of maleic anhydride with 1,3-butadiene involves three π bonds in the reacting system. Such a reaction under thermal conditions involves suprafacial ring closure.

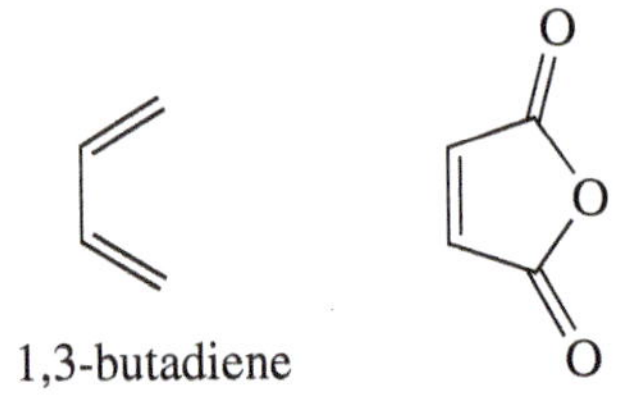

1,3-butadiene

The reaction of maleic anhydride with ethylene involves two π bonds in the reacting system, and such a reaction under thermal conditions involves antarafacial ring closure, which cannot occur in the formation of a four-membered ring.

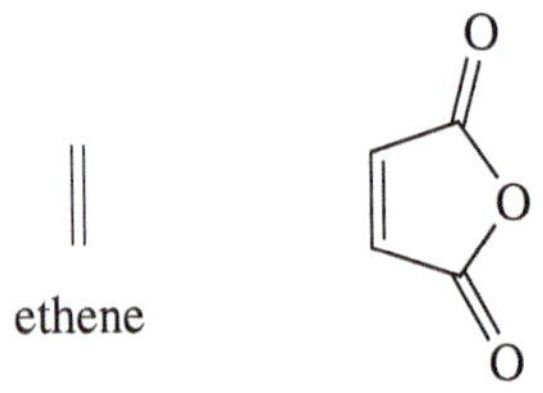

ethene

11. Yes. Because we are dealing with formation of a small ring, ring closure must be suprafacial. Under photochemical conditions, suprafacial ring closure requires an even number of π bonds in the reacting system. Therefore, a concerted reaction will occur but it will use only one of the π bonds of 1,3-butadiene.

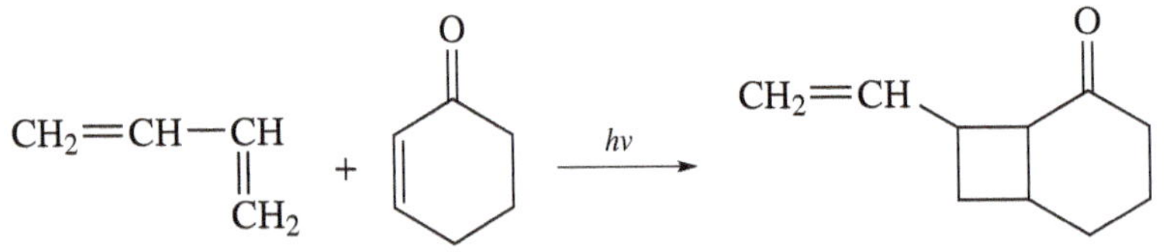

$CH_2{=}CH{-}CH{=}CH_2$ + cyclohexenone $\xrightarrow{h\nu}$ $CH_2{=}CH$–(bicyclic cyclobutane-fused cyclohexanone)

12. a. 1. [1,7] sigmatropic rearrangement
2. [1,5] sigmatropic rearrangement
3. [5,5] sigmatropic rearrangement
4. [3,3] sigmatropic rearrangement

b. 1.

2.

3.

4.

13. **a.** and **b.**

14. If a nondeuterated reactant had been used, the product would be identical to the reactant. Therefore, the rearrangement would not have been detectable.

15. A suprafacial rearrangement can take place under photochemical conditions if there are an even number of electron pairs in the reacting system. Therefore, a 1,3-hydrogen shift occurs involving four electrons.

a [1,3] sigmatropic migration of deuterium

A suprafacial rearrangement can take place under thermal conditions if an odd number of electron pairs are in the reacting system. Therefore, a 1,5-hydrogen shift occurs involving six electrons.

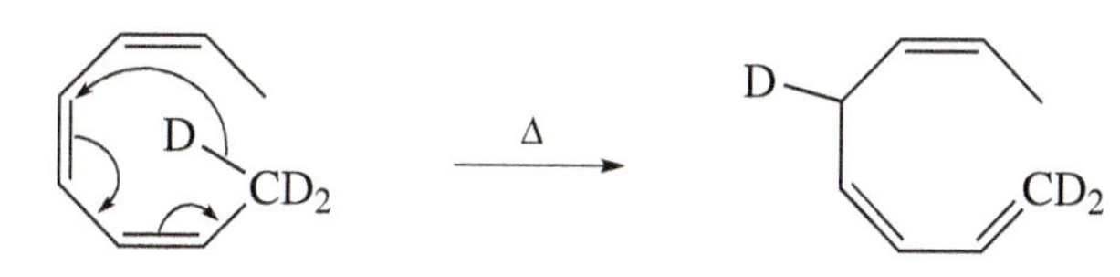

a [1,5] sigmatropic migration of deuterium

16. Solved in the text.

17. [1,3] Sigmatropic migrations of hydrogen cannot occur under thermal conditions because the four-membered transition state does not allow the required antarafacial rearrangement.

[1,3] Sigmatropic migrations of carbon can occur under thermal conditions because carbon can achieve the required antarafacial rearrangement by using both lobes of its *p* orbital when it migrates.

18. **a.** Because 1,3-migration of carbon requires carbon to migrate using both lobes of its *p* orbital (it involves an even number of pairs of electrons, so it takes place by an antarafacial pathway), migration occurs with inversion of configuration.

b. Because 1,5-migration of carbon requires carbon to migrate using only one lobe of its *p* orbital (it involves an odd number of pairs of electrons, so it takes place by a suprafacial pathway), migration occurs with retention of configuration.

19. Because the [1,7] sigmatropic rearrangement takes place under thermal conditions and involves an even number (4) of pairs of electrons, migration of hydrogen involves antarafacial rearrangement. Because the cyclic transition state involves eight ring atoms, antarafacial rearrangement is possible.

20. Because the reactant (provitamin D_3) has an odd number (3) of conjugated π bonds and reacts under photochemical conditions, ring closure is conrotatory. The methyl and hydrogen substituents point in opposite directions in provitamin D_3. Conrotatory ring closure causes substituents that point in opposite directions in the reactant to be trans in the product.

21. Chorismate mutase catalyzes a [3,3] sigmatropic Claisen rearrangement.

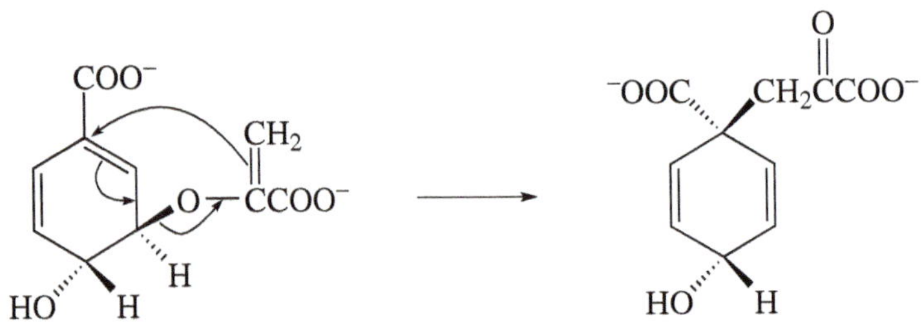

22. Were you able to convince yourself that TE-AC is valid?

23. **a.**

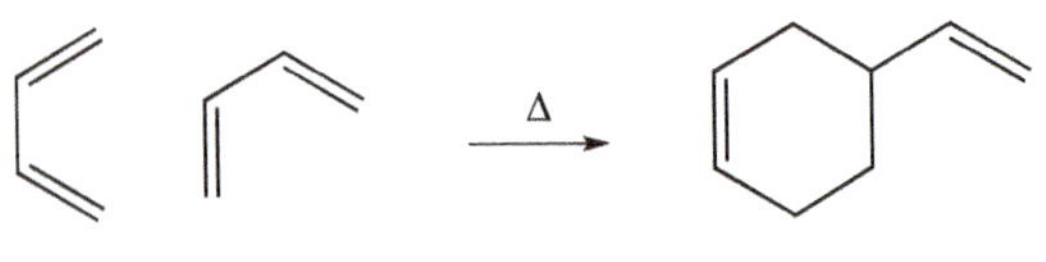

b.

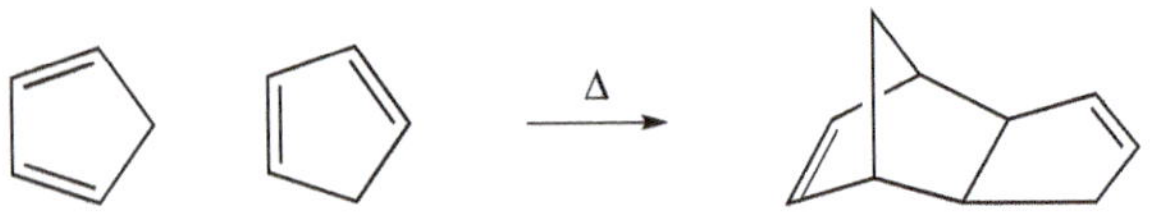

c.

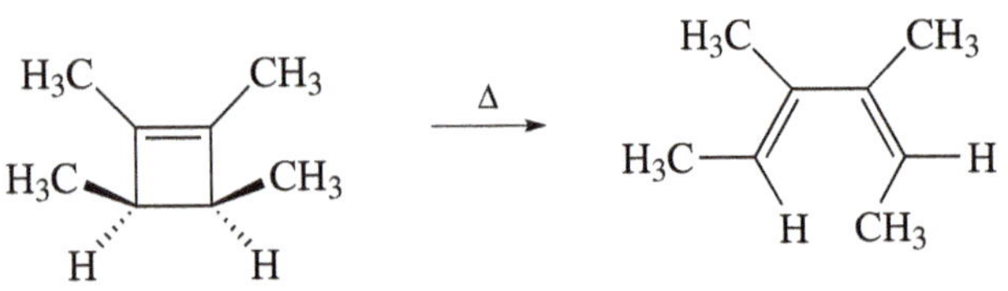

d. 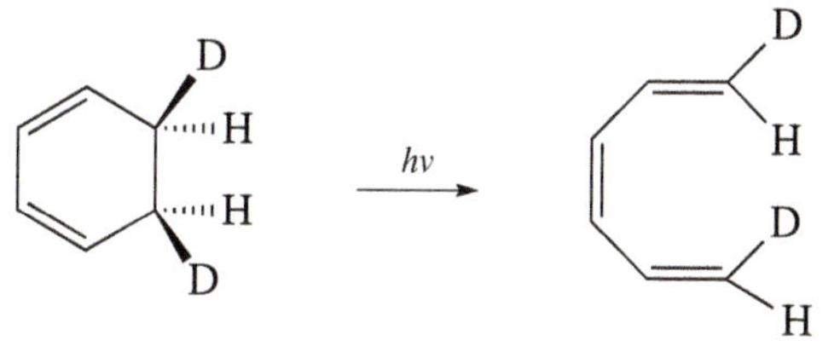 D, H, H, D — $h\nu$ → D, H, D, H

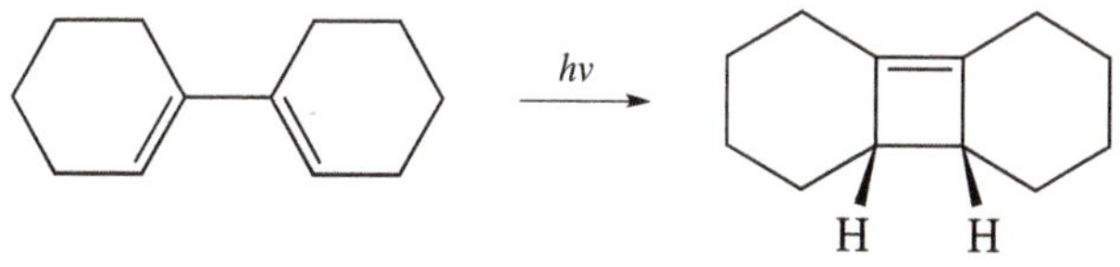

e. 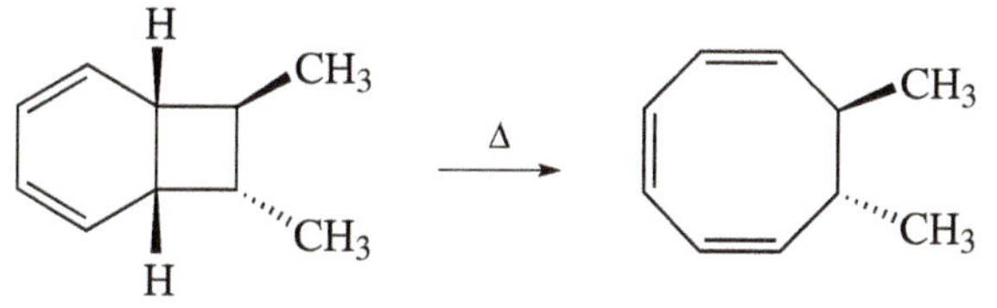 H, CH_3, CH_3, H — Δ → CH_3, CH_3

f. 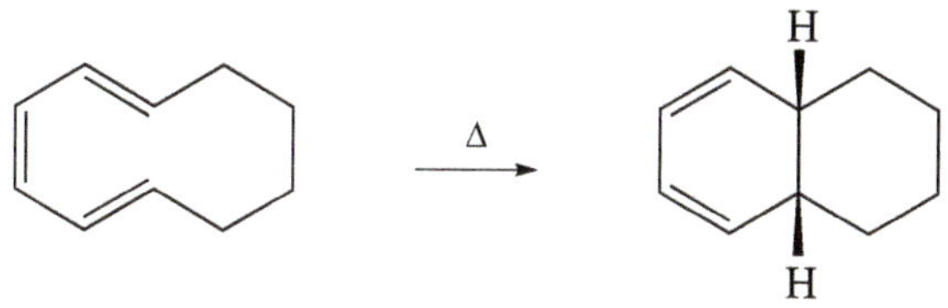 Δ → H, H

g. $h\nu$ → H H

h. H_3C — Δ → H_3C

24. Because the compound has an odd number of π bonds, it undergoes disrotatory ring closure under thermal conditions and conrotatory ring closure under photochemical conditions.

In the compounds in which the two methyl substituents point in opposite directions, the substituents will be cis in the ring-closed product when ring closure is disrotatory and trans in the ring-closed product when ring closure is conrotatory.

In the compounds in which the two methyl substituents point in the same direction, the substituents will be trans in the ring-closed product when ring closure is disrotatory and cis in the ring-closed product when ring closure is conrotatory.

a. 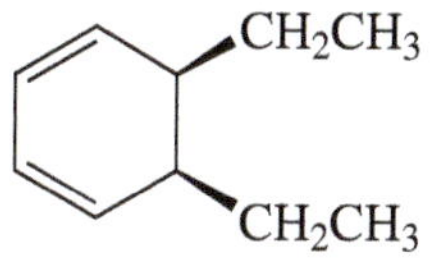 CH_2CH_3, CH_2CH_3

b. 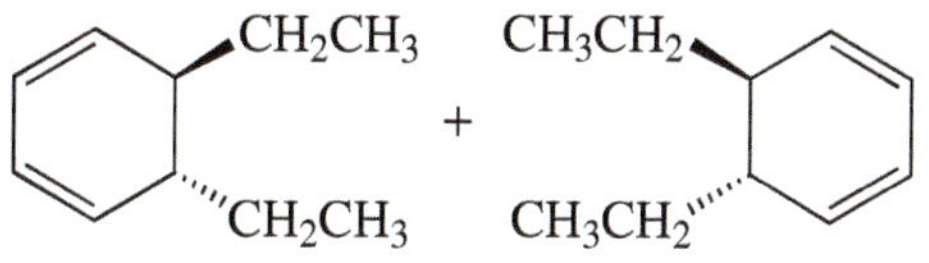 CH_2CH_3, CH_2CH_3 + CH_3CH_2, CH_3CH_2

c. 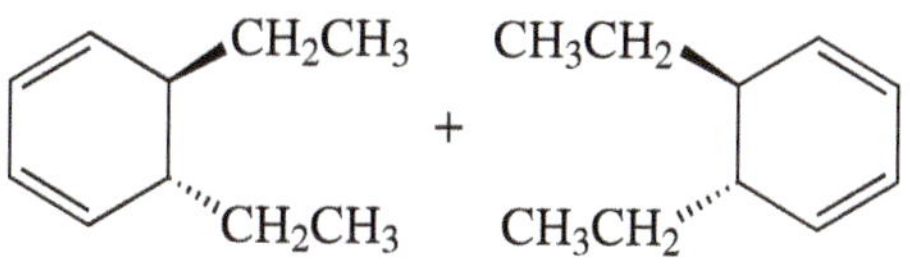 CH_2CH_3, CH_2CH_3 + CH_3CH_2, CH_3CH_2

d. 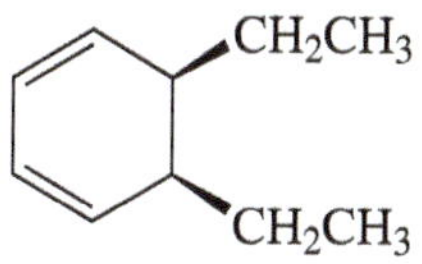 CH_2CH_3, CH_2CH_3

25. The hydrogens that end up at the ring juncture in the first reaction point in opposite directions in the reactant. Because ring closure is disrotatory (odd number of π bonds, thermal conditions), the hydrogens in the ring-closed product are cis. (See Table 28.2 on page 1224 of the text.)

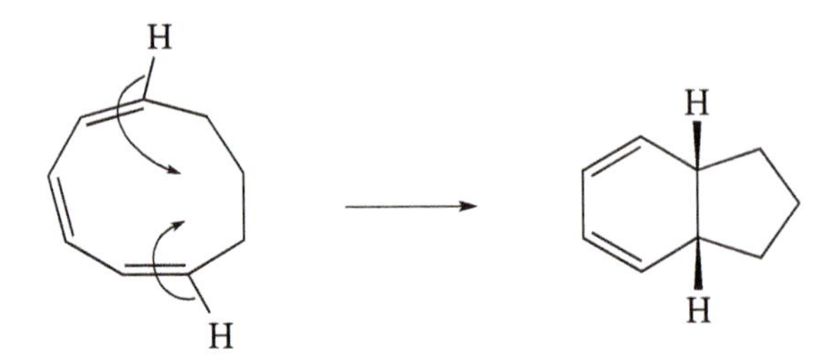

In the second example, the hydrogens that end up at the ring juncture point in the same direction in the reactant. Ring closure is still disrotatory, so the hydrogens in the ring-closed product are trans.

H H H H

26. CH_2 $\|$ CH_2 $\xrightarrow{\Delta}$ $\xrightarrow[Pd/C]{H_2}$

27. The first step is similar to a Cope rearrangement; the second step is tautomerization of the enol.

OH OH O

28. **1.** Because the compound has an even number of π bonds, it undergoes conrotatory ring closure under thermal conditions and disrotatory ring closure under photochemical conditions. Because the two methyl substituents point in opposite directions, they will be trans in the ring-closed product when ring closure is conrotatory and cis in the ring-closed product when ring closure is disrotatory.

a. CH_3 CH_3 $\xrightarrow{\Delta}$ CH_3 CH_3 + H_3C H_3C

b. CH_3 CH_3 $\xrightarrow{h\nu}$ CH_3 CH_3

2. Because the compound has an even number of π bonds, it will undergo conrotatory ring closure under thermal conditions and disrotatory ring closure under photochemical conditions. Because the two methyl substituents point in the same direction, they will be cis in the ring-closed product when ring closure is conrotatory and trans in the ring-closed product when ring closure is disrotatory.

a.

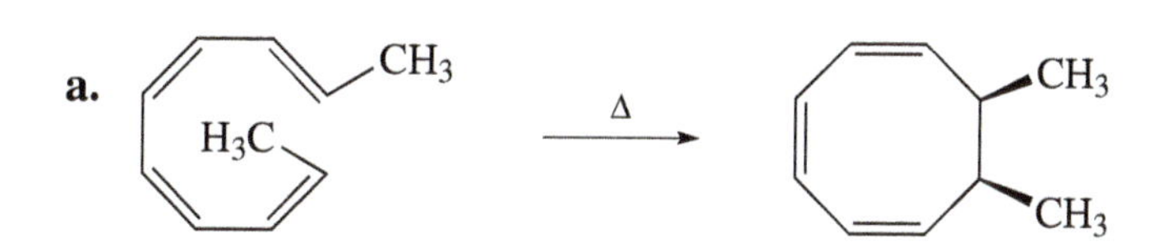

b. $\xrightarrow{h\nu}$

29. a. $\xrightarrow{h\nu}$

b. $\xrightarrow{\Delta}$

c. $\xrightarrow{\Delta}$

tautomerization

d. $\xrightarrow{\Delta}$

e. $\xrightarrow{\Delta}$ tautomerization

30. **a.** This is a [1,5] sigmatropic rearrangement, so it can take place by a concerted mechanism under thermal conditions.

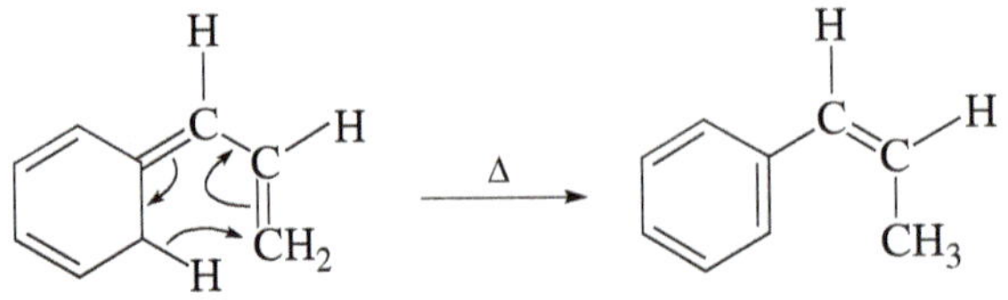

b. This is a [1,3] sigmatropic rearrangement, so it can take place by a concerted mechanism under photochemical conditions.

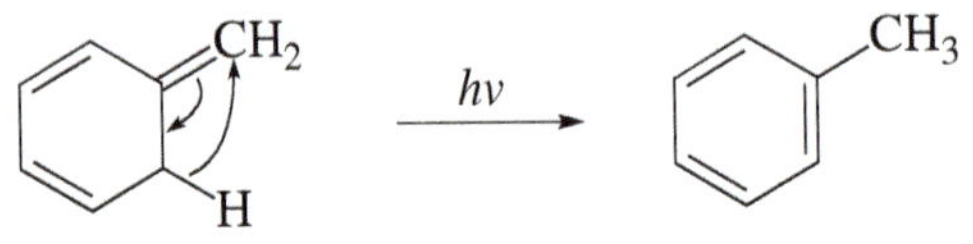

31. At first glance, it is surprising that the isomerization of Dewar benzene (a highly-strained and unstable molecule) to benzene (a stable aromatic compound) is so slow. However, the isomerization requires conrotatory ring opening, which is symmetry forbidden under thermal conditions. The reaction, therefore, cannot take place by a concerted pathway and must take place by a much slower stepwise process.

32. **B** is the product. Because the reaction is a [1,3] sigmatropic rearrangement, antarafacial ring closure is required. Carbon, therefore, must migrate using both lobes of its *p* orbital. This means that the configuration of the migrating carbon will undergo inversion. The configuration of the migrating carbon has been inverted in **B** (the H's are cis to each other but were trans to each other in the reactant) and retained in **A**.

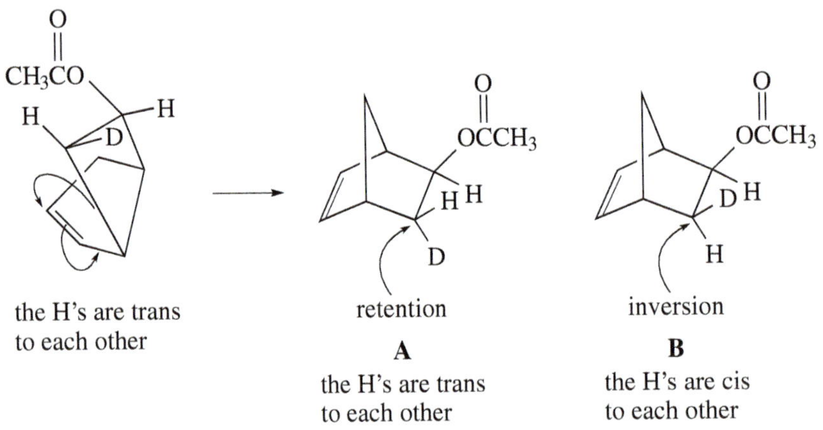

the H's are trans to each other

retention

A

the H's are trans to each other

inversion

B

the H's are cis to each other

33. Hydrogen cannot undergo a [1,3] sigmatropic rearrangement, because it cannot migrate by an antarafacial pathway that is required for a sigmatropic rearrangement involving an even number of pairs of electrons under thermal conditions. Carbon can undergo a [1,3] sigmatropic rearrangement because it can migrate by a suprafacial pathway if it uses both lobes of its *p* orbital. Therefore, the first compound can undergo only a 1,3-methyl group migration.

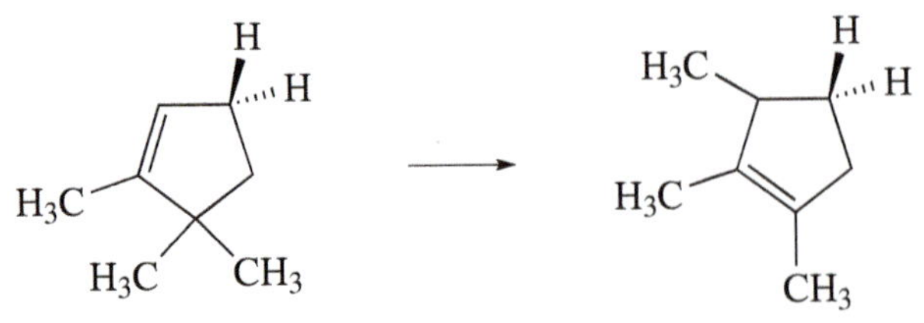

The second compound can undergo the 1,3-methyl group migration that the first compound undergoes, and the *sec*-butyl group can also undergo a [1,3] sigmatropic rearrangement. The migrating *sec*-butyl group will have its configuration inverted due to the antarafacial nature of the rearrangement.

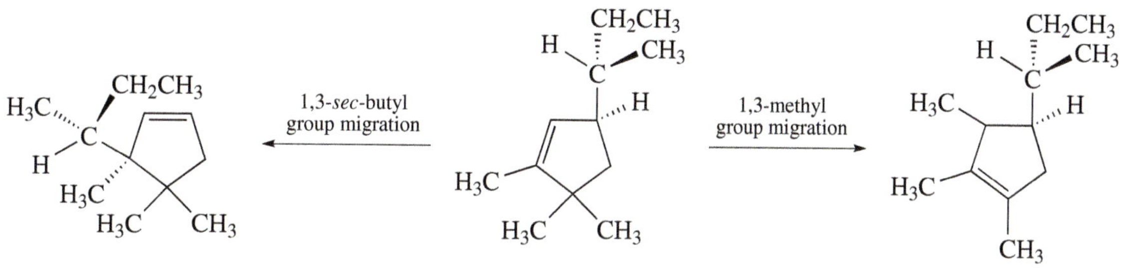

1,3-*sec*-butyl group migration

1,3-methyl group migration

34. An infrared absorption band is indicative of a carbonyl group. A [3,3] sigmatropic rearrangement of the reactant leads to a compound with two enolic groups. Tautomerization of the enols results in keto groups. The keto carbonyl groups give the absorbance at 1715 cm^{-1}.

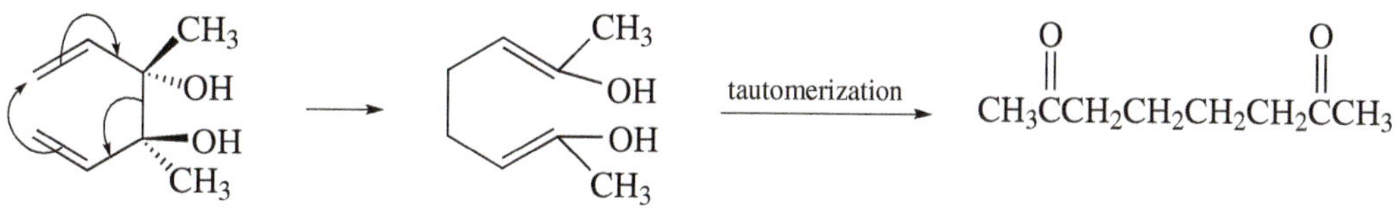

tautomerization

$CH_3\overset{O}{\overset{\|}{C}}CH_2CH_2CH_2CH_2\overset{O}{\overset{\|}{C}}CH_3$

35. The reaction is a [1,7] sigmatropic rearrangement. Because the reaction involves four pairs of electrons, antarafacial rearrangement occurs. Therefore, when H migrates, because it is above the plane of the reactant molecule, it ends up below the plane of the product molecule. When D migrates, because it is below the plane of the reactant molecule, it ends up above the plane of the product molecule.

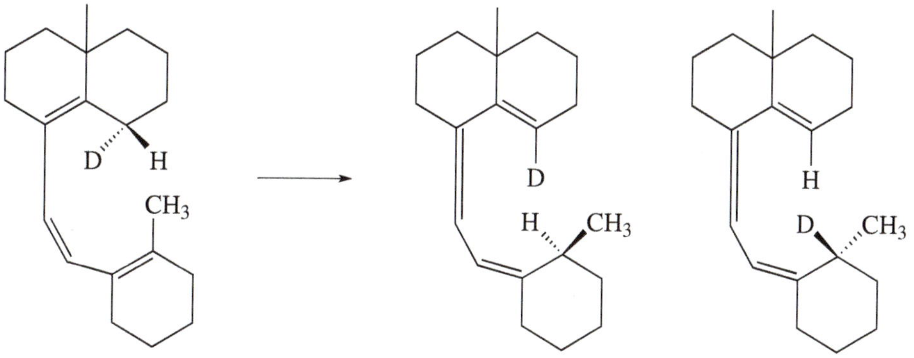

36.

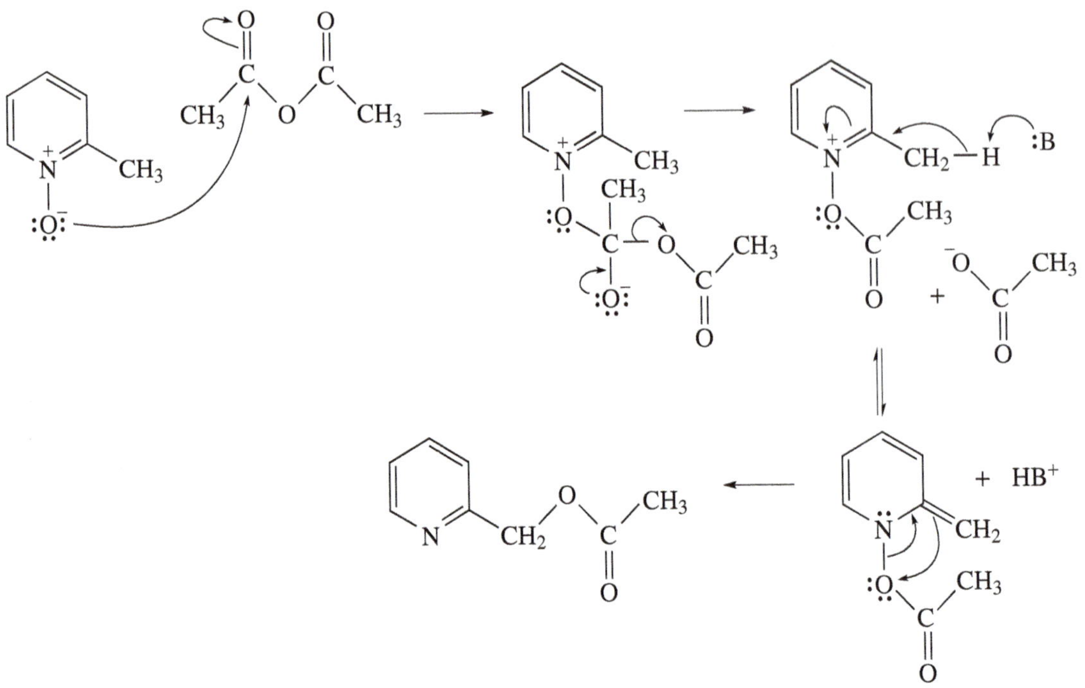

37.

a. [3,3] → tautomerization →

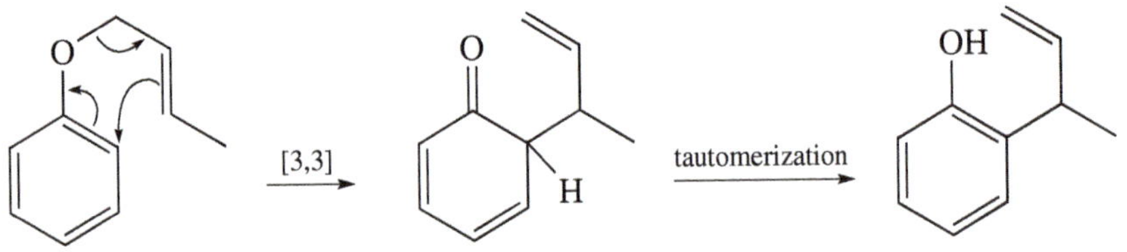

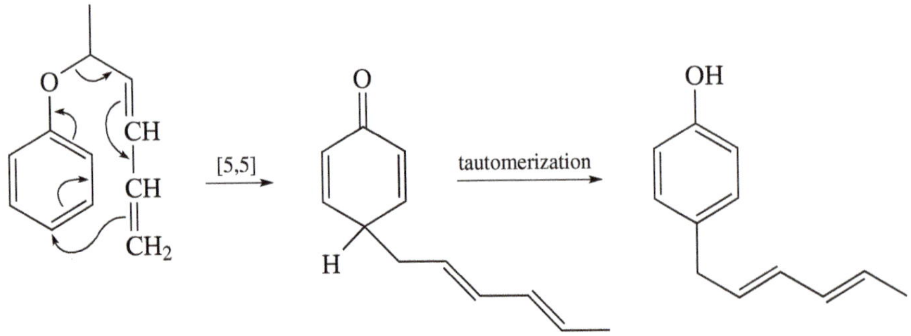

b. [3,3] → tautomerization →

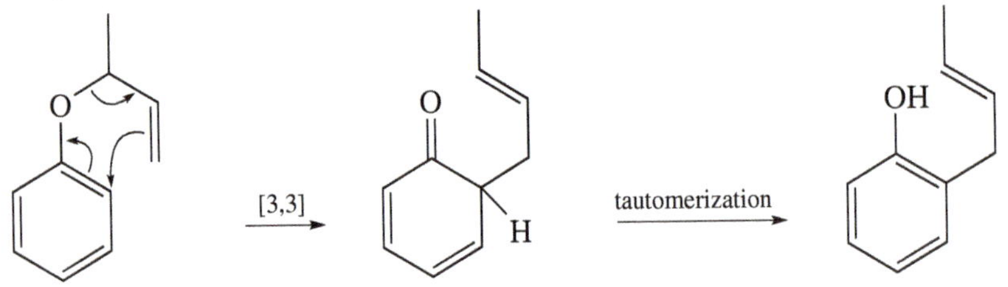

c. [5,5] → tautomerization →

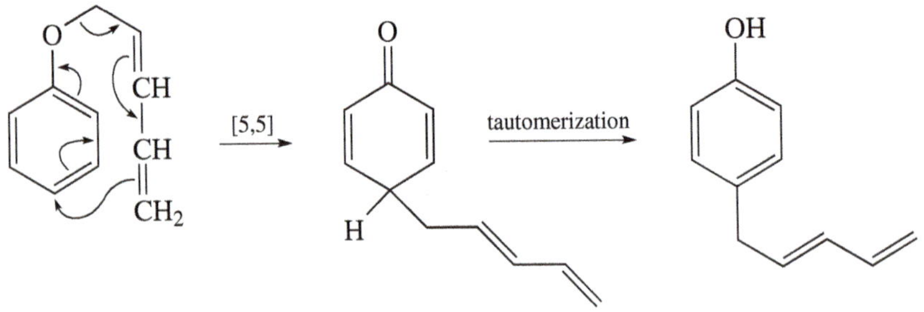

d. [5,5] → tautomerization →

38. **a.**

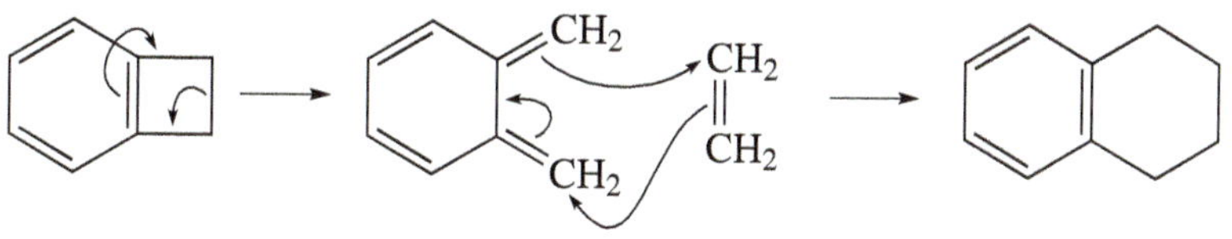

CH_2 CH_2 CH_2 CH_2

b.

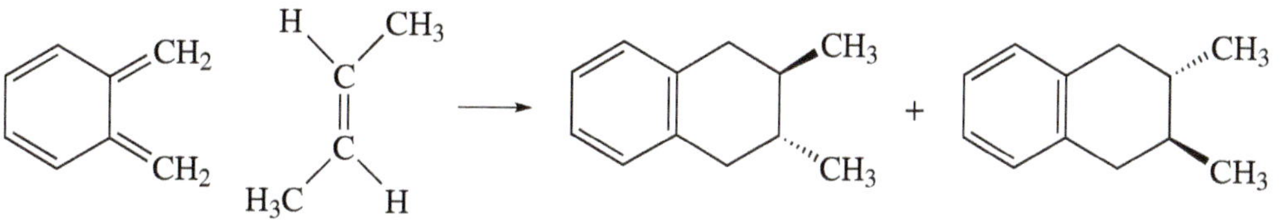

CH_2 CH_2 H CH_3 C C H_3C H CH_3 CH_3 + CH_3 CH_3

39. Disrotatory ring closure of (2*E*,4*Z*,6*Z*)-octatriene leads to the trans isomer, which can exist as a pair of enantiomers. One enantiomer is formed if the "top lobes" of the *p* orbitals rotate toward each other, and the other enantiomer is formed if the "bottom lobes" of the *p* orbitals rotate toward each other.

CH_3 H CH_3 H Δ CH_3 CH_3 + CH_3 CH_3

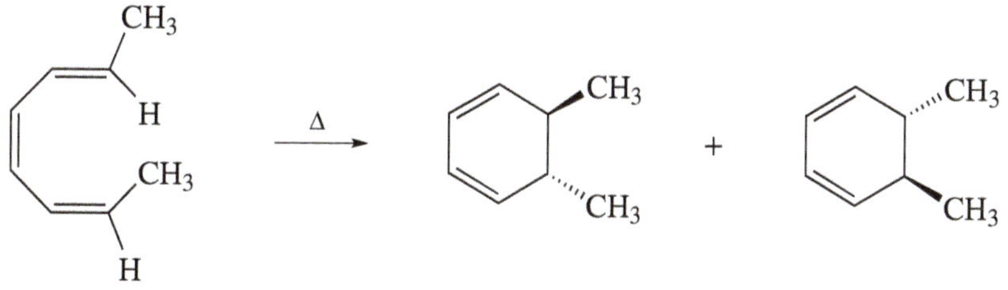

(2*E*, 4*Z*, 6*Z*)-octatriene

In contrast, disrotatory ring closure of (2*E*,4*Z*,6*E*)-octatriene leads to the cis isomer, which is a meso compound and, consequently, does not have a nonsuperimposable mirror image. Therefore, the same compound is formed from the "top lobes" of the *p* orbitals rotating toward each other and from the "bottom lobes" of the *p* orbitals rotating toward each other.

CH_3 H H CH_3 Δ CH_3 CH_3

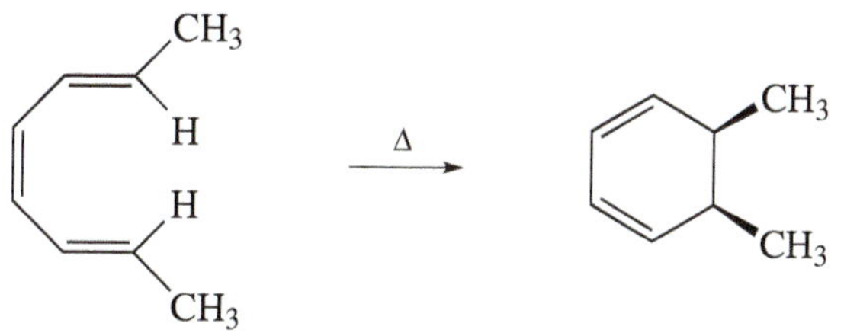

(2*E*, 4*Z*, 6*E*)-octatriene

40. Under thermal conditions, a compound with two π bonds undergoes conrotatory ring closure. Conrotatory ring closure that results in a ring-closed compound with the substituents cis to each other requires that the substituents point in the same direction in the reactant. Therefore, the product with the methyl substituents pointing in the same direction is obtained in 99% yield.

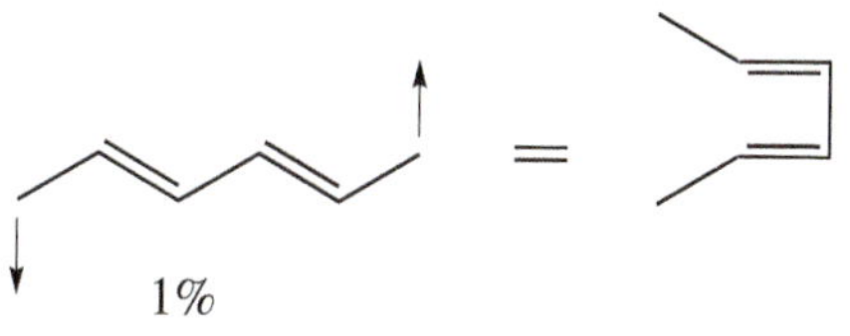

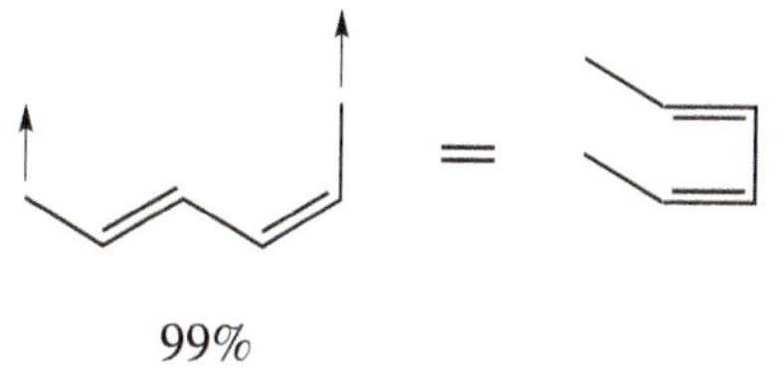

1% 99%

Methyl groups point in opposite directions. **Methyl groups point in the same directions.**

41. Because the reactant has two π bonds, electrocyclic ring closure is conrotatory under thermal conditions. Two different compounds, **X** and **Y**, can be formed because conrotatory ring closure can occur in either a clockwise or counterclockwise direction.

both clockwise → **X**

both counterclockwise → **Y**

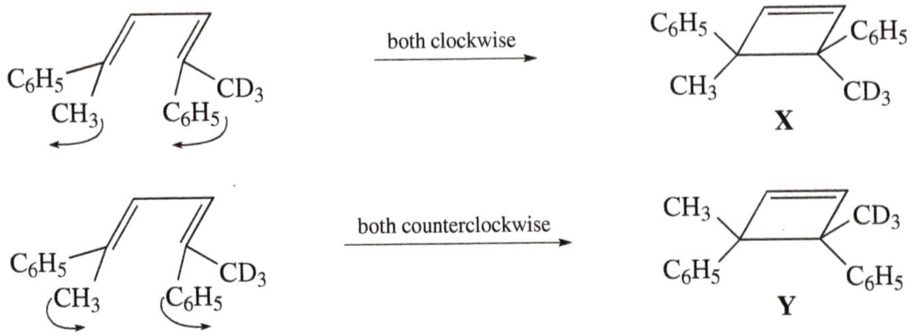

Each of the compounds (**X** and **Y**) can undergo a conrotatory ring-opening reaction in either a clockwise or counterclockwise direction to form either **A** or **B**. **A** and **B** are the only isomers that can be formed; formation of **C** and **D** requires disrotatory ring closure.

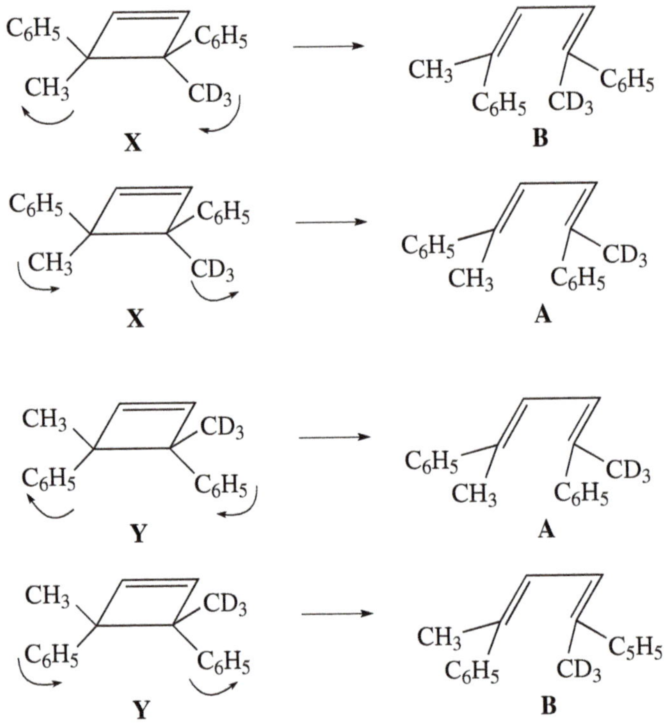

X → **B**

X → **A**

Y → **A**

Y → **B**

42.

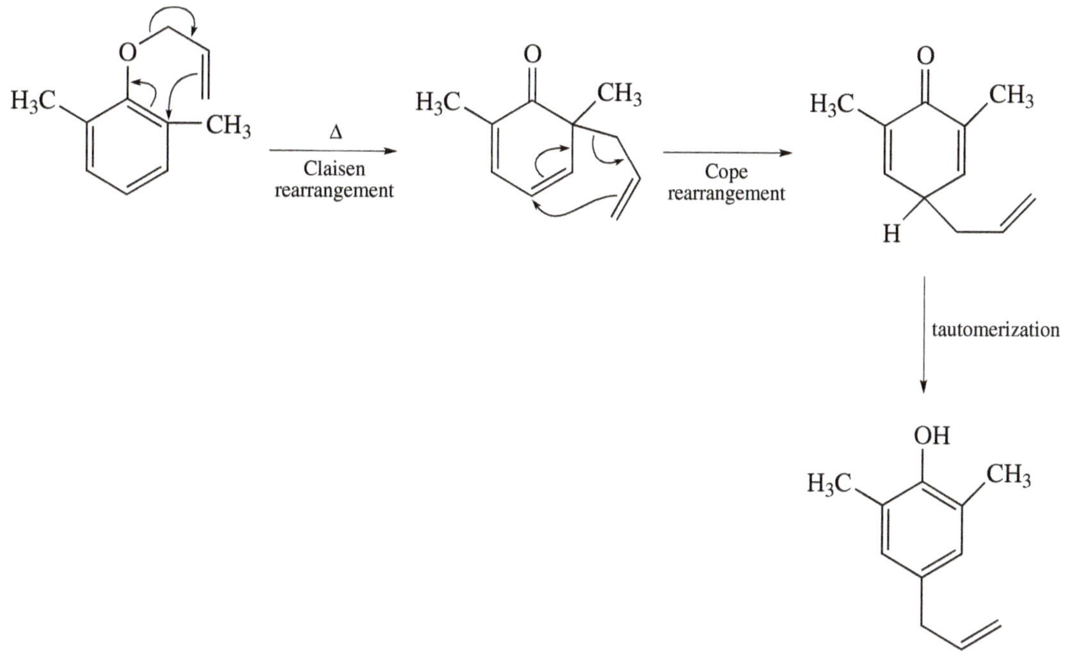
H3C
O
CH3
Δ
Claisen
rearrangement
Cope
rearrangement
H
tautomerization
OH

43. Because the compounds that undergo ring closure to give **A** and **B** have two π bonds, ring opening of **A** and **B** under thermal conditions is conrotatory. Because the hydrogens in **A** and **B** are cis, they must point in the same direction in the ring-opened product. To have the two hydrogens pointing in the same direction, one of the double bonds in the ring-opened compound must be cis and the other must be trans.

An eight-membered ring is too small to accommodate conjugated double bonds with one cis and the other trans, so **A** will not be able to undergo a ring-opening reaction under thermal conditions. A 10-membered ring can accommodate a trans double bond, so **B** is able to undergo a ring-opening reaction under thermal conditions.

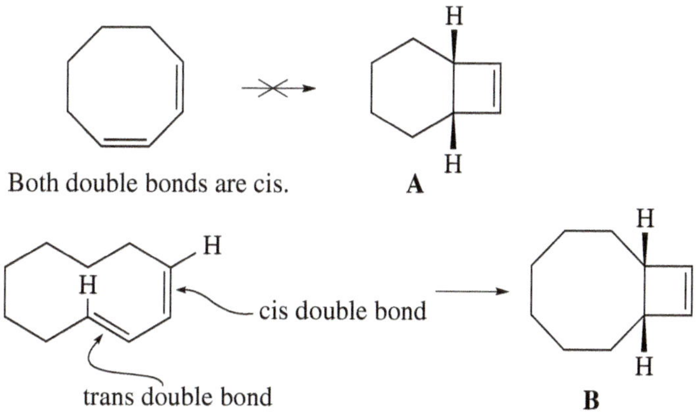
H
H
Both double bonds are cis.
A
H
H
cis double bond
trans double bond
H
H
B

44. The compound undergoes a 1,5-hydrogen shift of D or a 1,5-hydrogen shift of H. In each case, an unstable nonaromatic intermediate is formed that undergoes a subsequent 1,5-hydrogen shift to form an aromatic product.

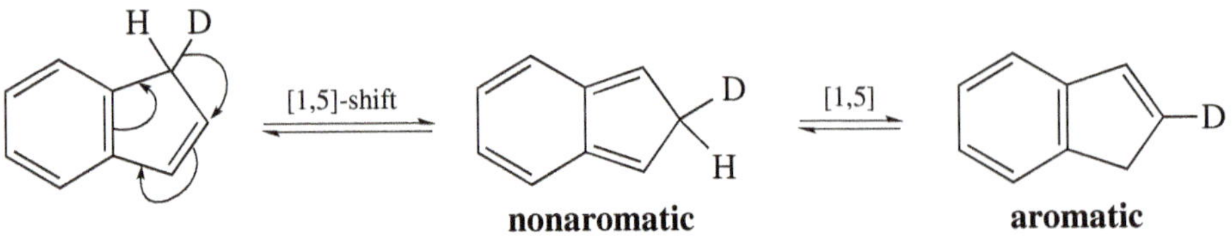

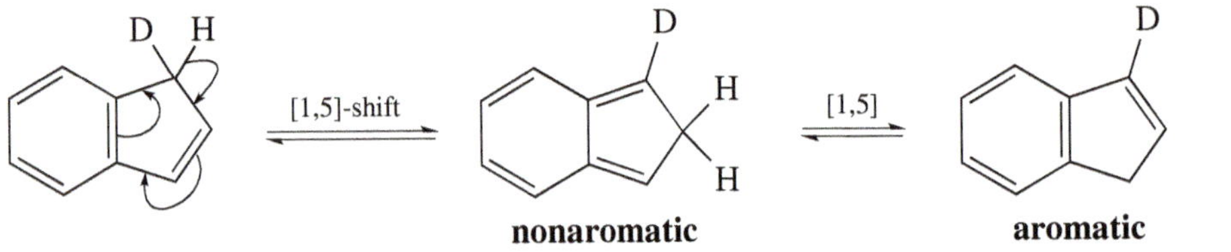

H D

[1,5]-shift

D

H

[1,5]

D

nonaromatic

aromatic

D H

[1,5]-shift

D

H

H

[1,5]

D

nonaromatic

aromatic

45. Because the ring-opened compound formed in the first step has three conjugated π bonds involved in an electrocyclic reaction, conrotatory ring opening of the reactant will occur under photochemical conditions, and the trans hydrogens in the reactant require that the hydrogens point in the opposite direction in the ring-opened compound. Thermal electrocyclic ring closure of a three π bond system is disrotatory in step two, and disrotatory ring closure of a compound with hydrogens that point in opposite directions will cause those hydrogens to be cis in the ring-closed product.

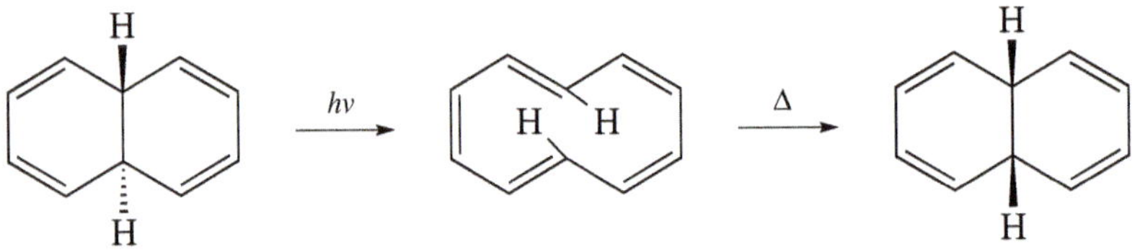

H

H

$h\nu$

H H

Δ

H

H

46. A Diels–Alder reaction is followed by a reverse Diels–Alder reaction that eliminates CO_2. Loss of a stable gas molecule (CO_2) and formation of a stable aromatic product provide the driving force for the second step.

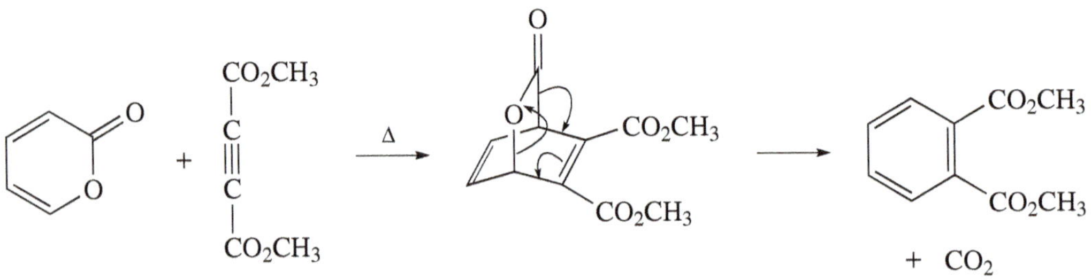

O

O

+

CO_2CH_3

C

C

CO_2CH_3

Δ

O

O

CO_2CH_3

CO_2CH_3

CO_2CH_3

CO_2CH_3

+ CO_2

47. H ÖH H—Cl ⇌ H OH (H, +) Cl⁻ → H + + H_2O

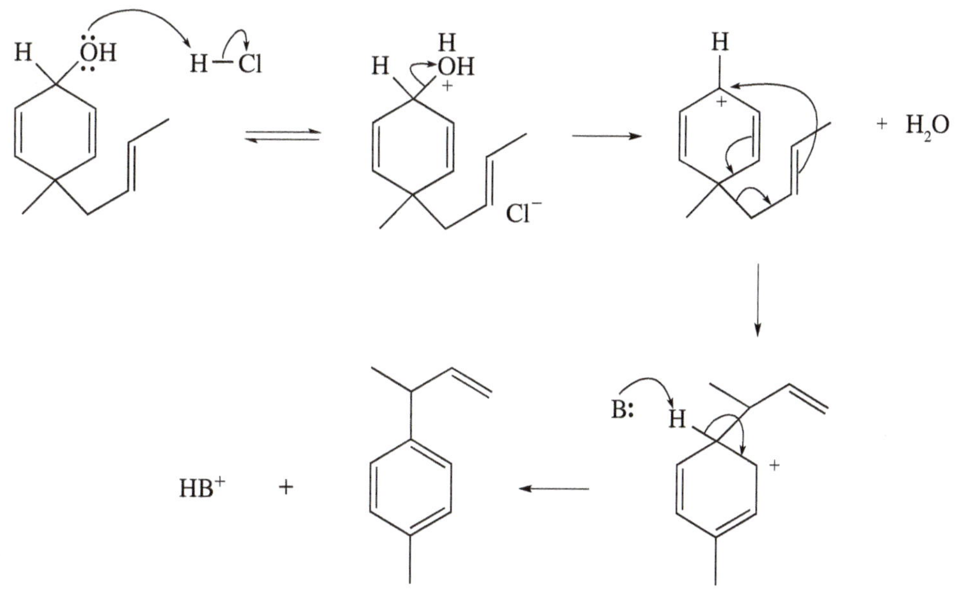

↓

B: H +

HB⁺ + ←

Chapter 28 Practice Test

1. How many molecule orbitals does 1,3,5-hexatriene produce?

2. Which of the molecular orbitals of 1,3,5-hexatriene are symmetric?

3. Which of the molecular orbitals of 1,3,5-hexatriene are antisymmetric?

4. Which are bonding molecular orbitals?

5. A compound with three double bonds undergoes __________ ring closure under thermal conditions.

6. A compound with two double bonds undergoes __________ ring closure under photochemical conditions.

7. A compound with three pairs of electrons in the reacting system undergoes __________ rearrangement under thermal conditions.

8. A compound with two pairs of electrons in the reacting system undergoes __________ rearrangement under photochemical conditions.

9. [1,5] Sigmatropic hydrogen migration involves _______ pairs of electrons.

10. A [2+2] cycloaddition reaction occurs only under __________ conditions.

Answers to Chapter 1 Practice Test

1. **a.** a carbon–fluorine bond **b.** one **c.** a carbon–hydrogen bond in ethane **d.** the bond angle in ammonia

2. $^{+}CH_3$ sp^2 $^{-}CH_3$ sp^3 $^{\cdot}CH_3$ sp^2

3. 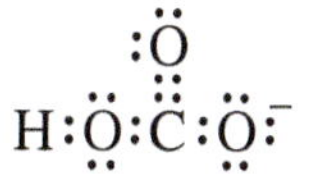 $H{:}\ddot{O}{:}C{:}\ddot{O}{:}^{-}$ (with $:\ddot{O}$ double-bonded above the C)

4. CH_2Cl_2 (circled: CH_3CH_3) CH_3Cl $H_2C{=}O$ (circled: CCl_4)

5. $^{+}NH_4$

6. **a.** $^{+}CH_3$ **b.** $H{:}^{-}$ **c.** $:\ddot{Br}\cdot$ **d.** CH_3CH_3

7. $CH_3CH_2CH_2CH{=}CH_2$ **or** $CH_3CH_2CH{=}CHCH_3$ **or** $CH_3CHCH{=}CH_2$ (with CH_3 on the second carbon)

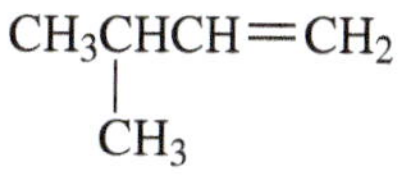

8. 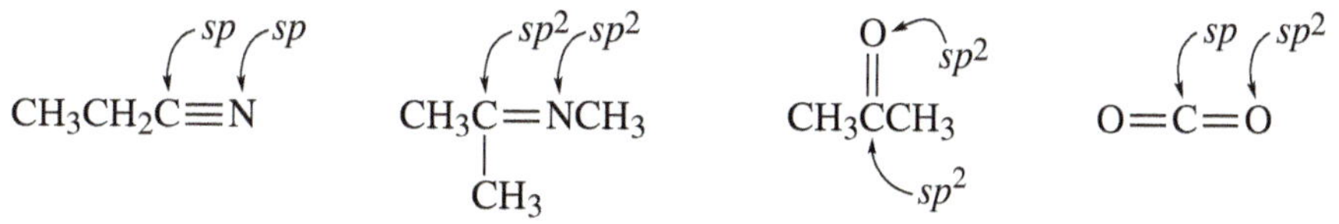 $CH_3CH_2C{\equiv}N$ (C: *sp*, N: *sp*) $CH_3C{=}NCH_3$ with CH_3 on C (C: *sp2*, N: *sp2*) CH_3CCH_3 with O double-bonded to C (O: *sp2*, C: *sp2*) $O{=}C{=}O$ (O: *sp*... C: *sp*, O: *sp2*)

9. **a.** $1s^2\,2s^2\,2p_x 2p_y$ **b.** $1s^2\,2s\,2p_x\,2p_y\,2p_z$ **c.** $1s^2\,2sp^3\,2sp^3\,2sp^3\,2sp^3$

10. **a.** 109.5° **b.** 180° **c.** 120° **d.** 104.5°

11. 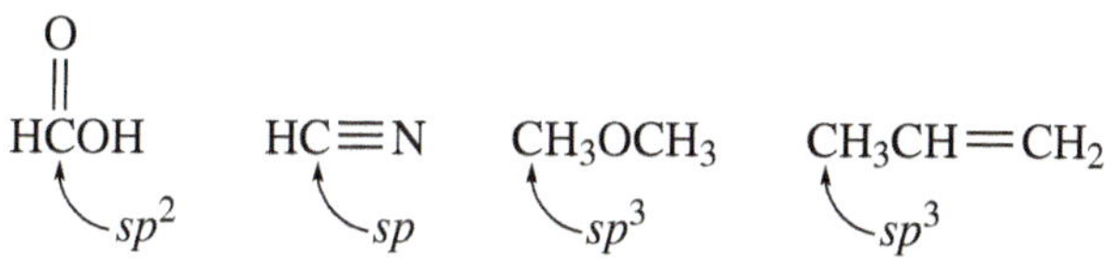 $HCOH$ with O double-bonded to C (sp^2) $HC{\equiv}N$ (*sp*) CH_3OCH_3 (sp^3) $CH_3CH{=}CH_2$ (sp^3)

12.
a. A pi bond is stronger than a sigma bond. F
b. A triple bond is shorter than a double bond. T
c. The oxygen–hydrogen bonds in water are formed by the overlap of an sp^2 orbital of oxygen with an *s* orbital of hydrogen. F
d. A double bond is stronger than a single bond. T
e. A tetrahedral carbon has bond angles of 107.5°. F

Answers to Chapter 2 Practice Test

1. **a.** 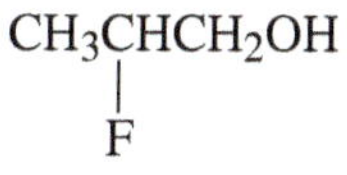 CH_3CHCH_2OH with F on the second carbon **b.** HI **c.** 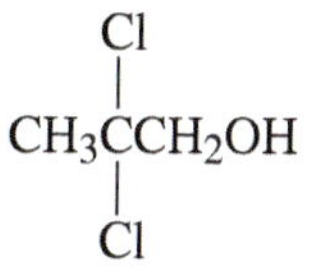 CH_3CCH_2OH with two Cl on the second carbon **d.** NH_3

2. **a.** $CH_3CH_2NH_2$ **b.** F^{-}

3. 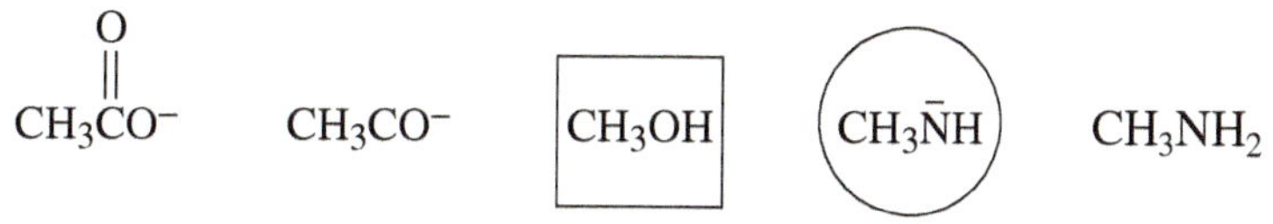 CH_3CO^{-} with O double-bonded to C CH_3CO^{-} (boxed: CH_3OH) (circled: $CH_3\bar{N}H$) CH_3NH_2

4. CH_3COO^- CH_3CH_2OH CH_3OH $CH_3CH_2\overset{+}{N}H_3$

5. **a.** $CH_3NH_2 + H_2O \rightleftharpoons CH_3\overset{+}{N}H_3 + HO^-$ **b.** reactants

6. **a.** $^-NH_2$ **b.** $^+NH_4$

7. 6.2

8. **a.** $CH_3OH + {}^+NH_4 \rightleftharpoons CH_3\overset{+}{O}H_2 + NH_3$ **b.** reactants

9. between 8 and 9

10. CH_3CH_2OH **2** $CH_3CH_2NH_2$ **3** CH_3CH_2SH **1** $CH_3CH_2CH_3$ **4**

11. formic acid/sodium formate

12.

a.	HO^- is a stronger base than $^-NH_2$.	F
b.	A Lewis acid is a compound that accepts a share in a pair of electrons.	T
c.	CH_3CH_3 is more acidic than $H_2C{=}CH_2$.	F
d.	The weaker the acid, the more stable the conjugate base.	F
e.	The larger the pK_a, the weaker the acid.	T
f.	The weaker the base, the more stable it is.	T

Answers to Chapter 3 Practice Test

1. **a.** 3-methyloctane **b.** 2-methyl-1-heptanol **c.** 3-octanol

2. **a.** (Newman projection: front CH_2CH_3, H, H; back H, H, CH_2CH_3) **b.** (Newman projection: CH_3CH_2 CH_2CH_3; H, H; H, H) **c.** (Newman projection: CH_2CH_3, CH_3CH_2, H; H, H, H)

3. **a.** *sec*-butyl chloride, 2-chlorobutane **c.** cyclopentyl bromide, bromocyclopentane

b. isohexyl alcohol, 4-methyl-1-pentanol

4. **a.** $CH_3CH_2CH_2CH_2CH_2Br$ **1** $CH_3CH_2CH_2Br$ **3** $CH_3CH_2CH_2CH_2Br$ **2**

b. $CH_3CH_2CH_2CH_2CH_3$ **3** $CH_3CH_2CH_2CH_2OH$ **1** $CH_3CH_2CH_2CH_2Cl$ **2**

c. $CH_3C(CH_3)_2{-}C(CH_3)_2CH_3$ **3** $CH_3CH_2CH_2CH_2CH_2CH_2CH_2CH_3$ **1** $CH_3CH(CH_3)CH_2CH_2CH_2CH_2CH_3$ **2**

5. **a.** 6-methyl-3-heptanol
b. 3-ethoxyheptane
c. 1-bromo-3-methylcyclopentane
d. 1,4-dichloro-5-methylheptane

6.

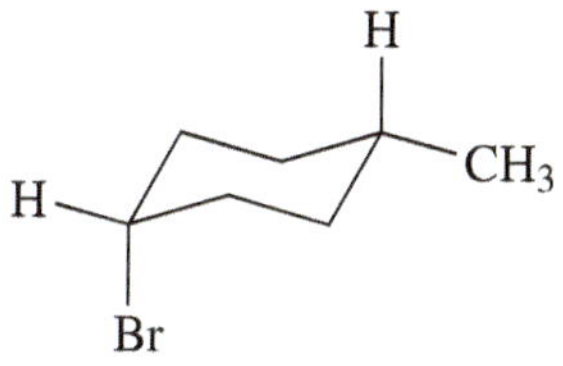

H

H— —CH_3

Br

7. *cis*-1-isopropyl-3-methylcyclohexane

8. **a.** butyl alcohol
b. 1-butanol
c. hexane
d. pentylamine
e. ethyl alcohol

9. **a.** isopentyl bromide
1-bromo-3-methylbutane
b. isopentyl alcohol
3-methyl-1-butanol
c. isopentylamine
3-methyl-1-butanamine

10. **a.**

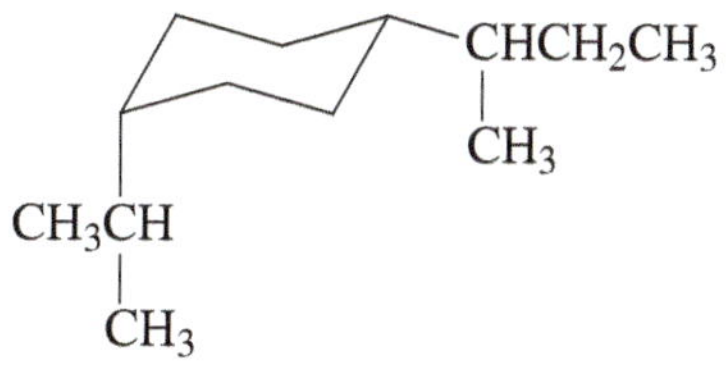

$CHCH_2CH_3$
|
CH_3

CH_3CH
|
CH_3

b.

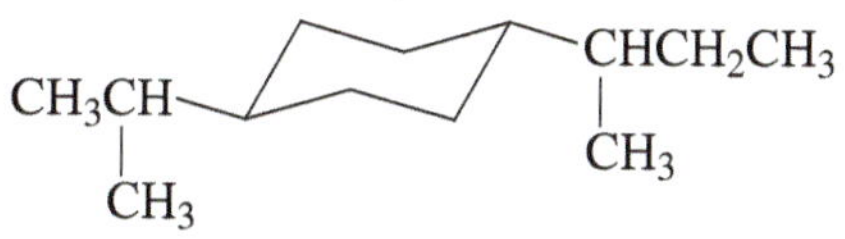

CH_3CH— —$CHCH_2CH_3$
| |
CH_3 CH_3

c.

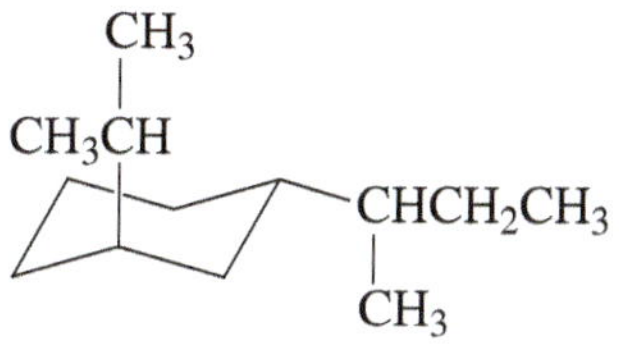

CH_3
|
CH_3CH

$CHCH_2CH_3$
|
CH_3

11. **a.** 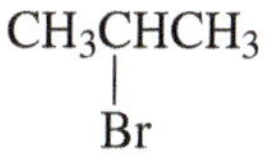 CH_3CHCH_3
|
Br

b. $CH_3CH_2NHCH_3$

c. 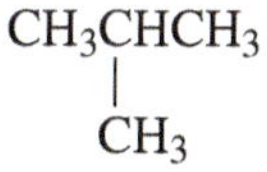 CH_3CHCH_3 (with CH_3 on the central carbon) or CH_3CCH_3 (with CH_3 above and below the central carbon) or CH_4 or $CH_3C—CCH_3$ (with CH_3 above and below each central carbon)

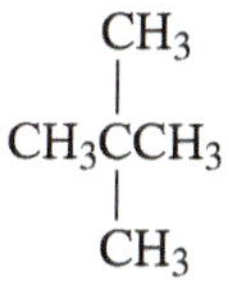

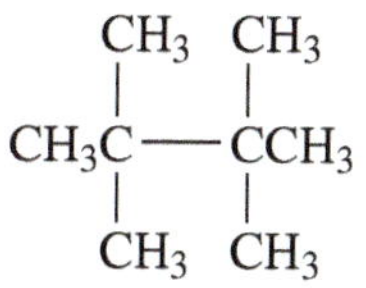

d. 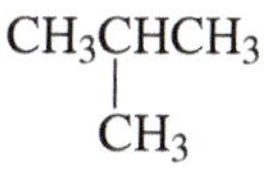 CH_3CHCH_3
|
CH_3

e. $CH_3CH_2CH_2OH$ CH_3CHOH (with CH_3 below) $CH_3CH_2OCH_3$

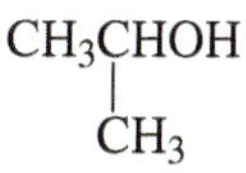

12. **a.** 2,5-dimethylheptane
b. 7-bromo-2-heptanol
c. 2-chloro-4-heptanol
d. 4-bromo-2-chloro-1-methylcyclohexane
e. 1-butoxy-2-methylpentane
f. 3-methyl-*N*-propyl-1-pentanamine

Answers to Chapter 4 Practice Test

1. **a.** a pair of enantiomers **b.** a pair of enantiomers

2.

$-\overset{O}{\overset{\Vert}{C}}CH_3$	$-C{=}CH_2$	$-Cl$	$-C{\equiv}N$
2	**4**	**1**	**3**

3. $\dfrac{-4.8}{0.80 \times 2} = -3.0$

4.

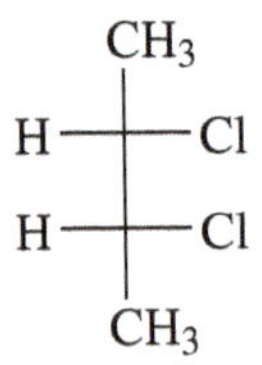

CH_3
H — — Cl
H — — Cl
CH_3

5. $CH_3CH_2CH_2CH_2Cl$ $CH_3CH_2\underset{Cl}{CH}CH_3$ $CH_3\underset{CH_3}{CH}CH_2Cl$ $CH_3\overset{CH_3}{\underset{Cl}{C}}CH_3$

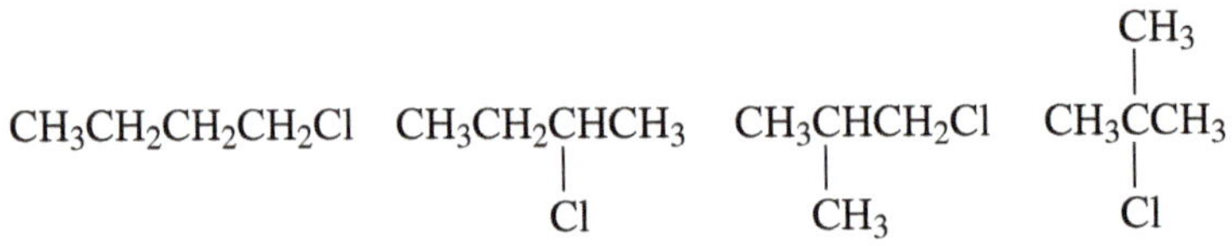

6. **a.** *Z* **b.** *Z*

7. **a.** HO OH HO OH HO OH

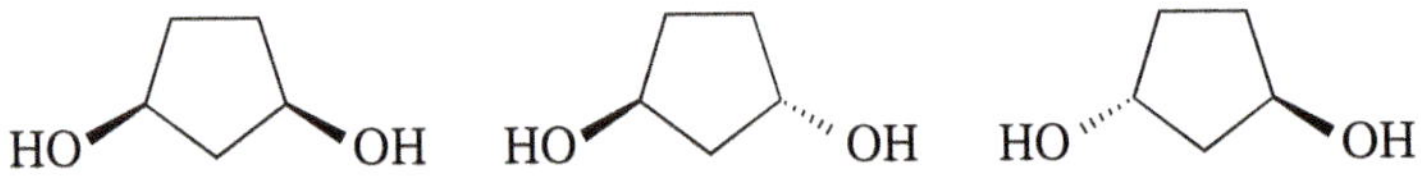

b.

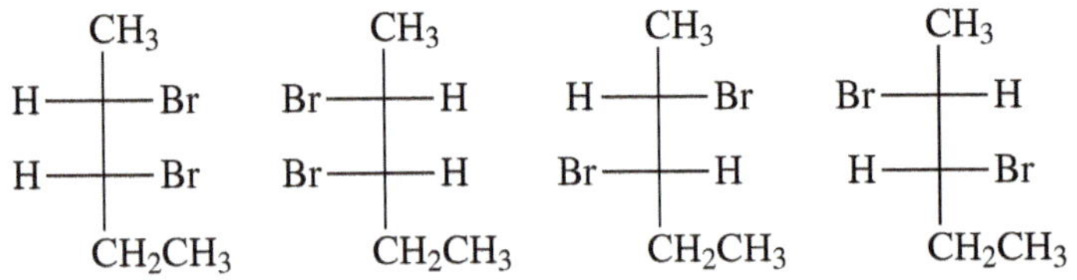

CH_3 / H — Br / H — Br / CH_2CH_3 CH_3 / Br — H / Br — H / CH_2CH_3 CH_3 / H — Br / Br — H / CH_2CH_3 CH_3 / Br — H / H — Br / CH_2CH_3

or

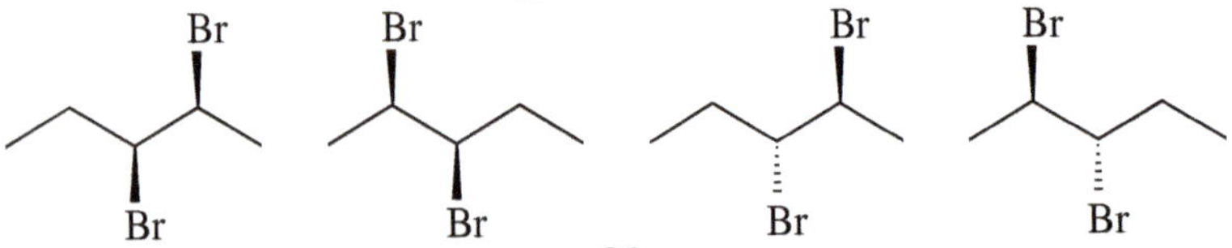

Br Br Br Br
Br Br Br Br

or

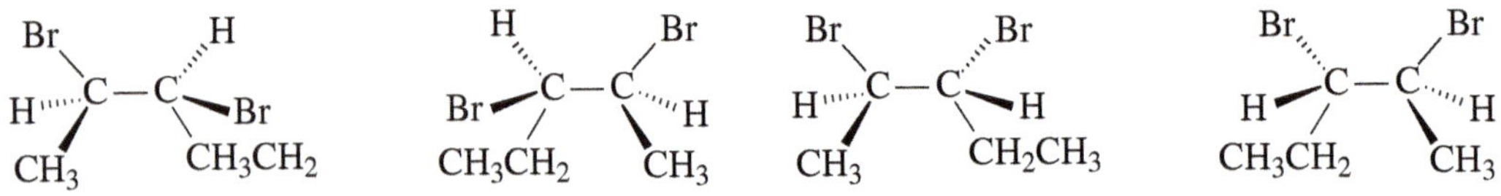

Br H H C—C Br CH_3 CH_3CH_2 H Br Br C—C H CH_3CH_2 CH_3 Br Br H C—C H CH_3 CH_2CH_3 Br Br H C—C H CH_3CH_2 CH_3

c. no stereoisomers

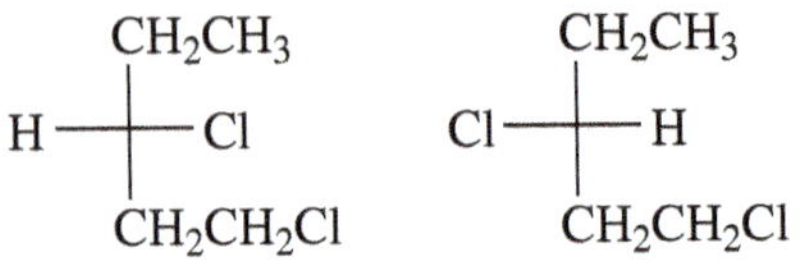

d. CH_2CH_3 / H — Cl / CH_2CH_2Cl CH_2CH_3 / Cl — H / CH_2CH_2Cl **or** Cl Cl Cl Cl

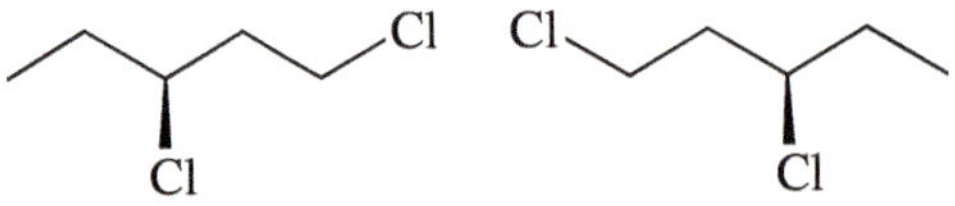

or CH_2CH_3 C H $ClCH_2CH_2$ Cl CH_2CH_3 H C CH_2CH_2Cl Cl

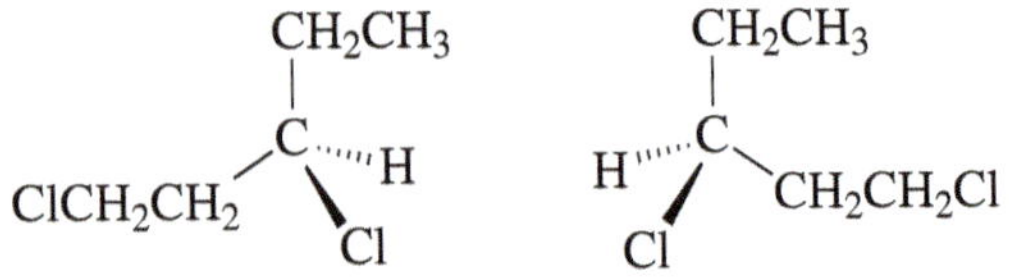

e.

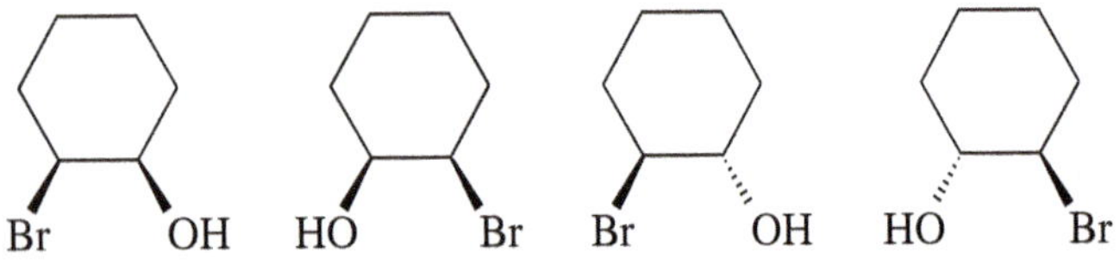

Br OH HO Br Br OH HO Br

f. 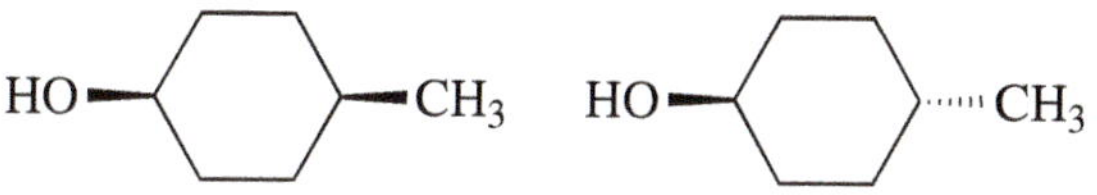 HO–(cyclohexane)–CH_3 HO–(cyclohexane)····CH_3

8. 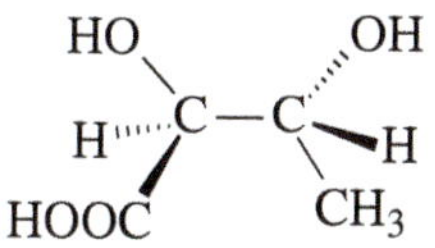 HO, OH; H····C—C····H; HOOC, CH_3

9. (−)-2-Methylbutanoic acid has the *S* configuration.

10. −16

11. a. 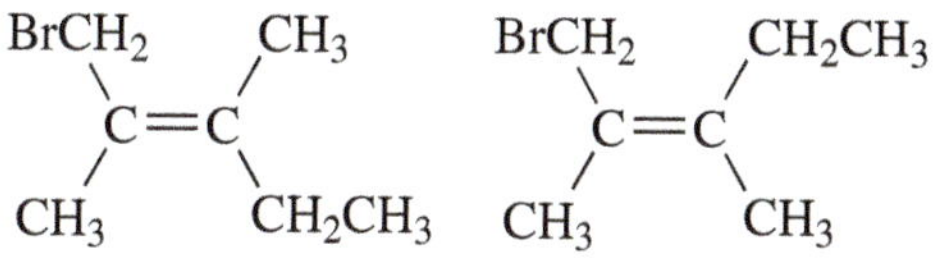 $BrCH_2(CH_3)C{=}C(CH_3)(CH_2CH_3)$ ***E*** $BrCH_2(CH_3)C{=}C(CH_2CH_3)(CH_3)$ ***Z***

b. 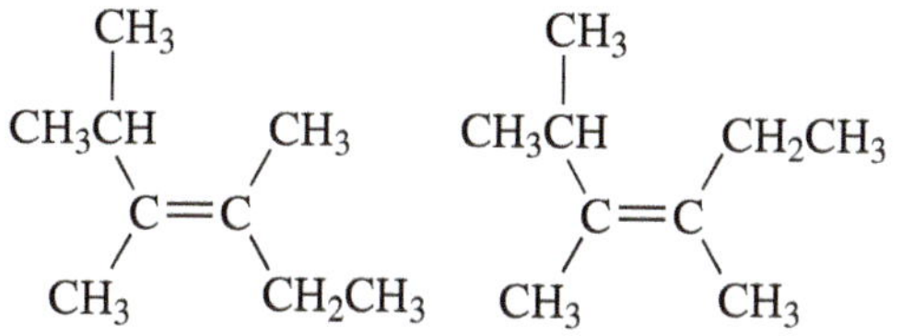 $(CH_3)_2CH(CH_3)C{=}C(CH_3)(CH_2CH_3)$ ***E*** $(CH_3)_2CH(CH_3)C{=}C(CH_2CH_3)(CH_3)$ ***Z***

12. 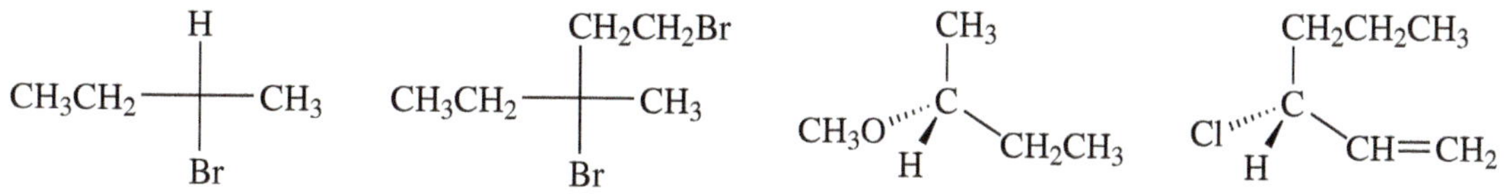 CH_3CH_2—C(H)(Br)—CH_3 CH_3CH_2—C(CH_2CH_2Br)(Br)—CH_3 CH_3O····C(CH_3)(H)CH_2CH_3 Cl····C($CH_2CH_2CH_3$)(H)$CH{=}CH_2$

13. a. identical

b. 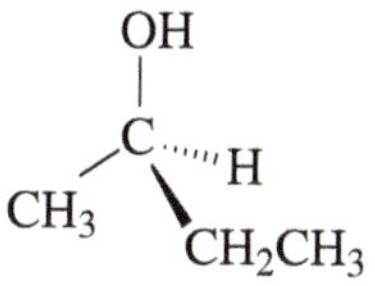 OH, C····H, CH_3, CH_2CH_3

c. CH_3; H—C—Cl; H—C—OH; CH_2CH_3

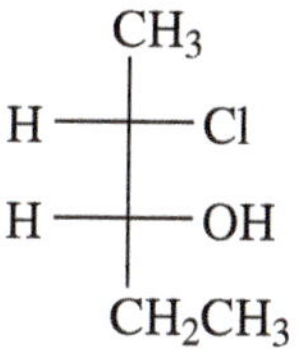

d. 1. 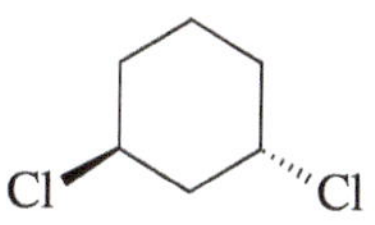 Cl–(cyclohexane)····Cl or Cl····(cyclohexane)–Cl

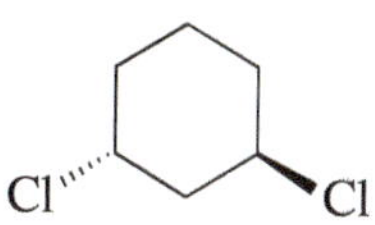

2. 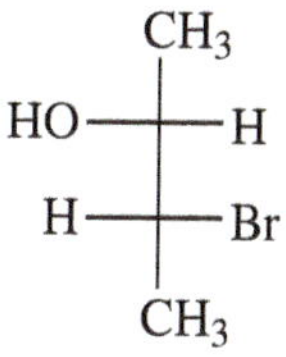 CH_3; HO—C—H; H—C—Br; CH_3 or CH_3; H—C—OH; Br—C—H; CH_3

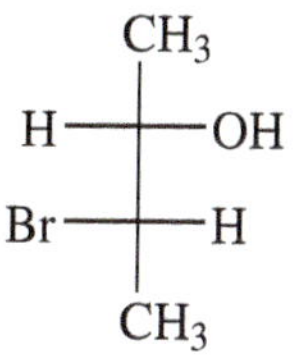

14. a. Diastereomers have the same melting points. F
b. 3-Chloro-2,3-dimethylpentane has two asymmetric centers. F
c. Meso compounds do not rotate the plane of polarization of plane-polarized light. T
d. 2,3-Dichloropentane has a stereoisomer that is a meso compound. F
e. All compounds with the *R* configuration are dextrorotatory. F
f. A compound with three asymmetric centers can have a maximum of nine stereoisomers. F

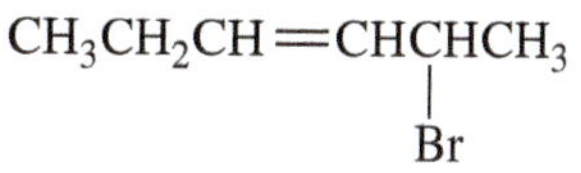

15. 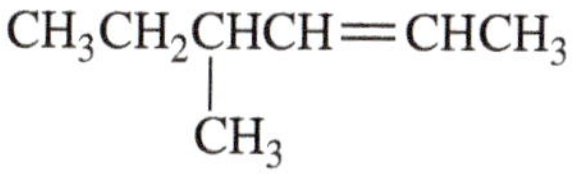 $CH_3CH_2CH(CH_3)CH{=}CHCH_3$ and $CH_3CH_2CH{=}CHCH(Br)CH_3$

Answers to Chapter 5 Practice Test

1. **a.** 4-methyl-1-hexene **c.** 7-methyl-3-nonene

b. 4-bromocyclopentene **d.** 4-chloro-3-methylcyclohexene

2. **a.** CH_3 (cyclohexene ring with CH_3 on the double-bond carbon) **b.** H_3C, H / C=C / H, CH_3 (H₃C and H on one carbon; H and CH₃ on the other, trans)

3. **a.** 2-pentene **c.** 3-methyl-2-pentene

b. 3-methyl-1-hexene **d.** 1-methylcyclohexene

4.

a. Increasing the energy of activation increases the rate of the reaction. F
b. Decreasing the entropy of the products compared to the entropy of the reactants makes the equilibrium constant more favorable. F
c. An exergonic reaction is one with a $-\Delta G°$. T
d. An alkene is an electrophile. F
e. The higher the energy of activation, the more slowly the reaction takes place. T
f. Another name for *trans*-2-butene is (*Z*)-2-butene. F
g. A reaction with a negative $\Delta G°$ has an equilibrium constant greater than one. T
h. Increasing the free energy of the reactants increases the rate of the reaction. T
i. Increasing the free energy of the products increases the rate of the reaction. F
j. The magnitude of a rate constant is not dependent on the concentration of the reactants. T
k. 2,3-Dimethyl-2-pentene is more stable than 3,4-dimethyl-2-pentene. T

5. B

6. **a.** $CH_2{=}CHCH_2OH$ **c.** CH_3CH_2, $CH_2CH_2CH_3$ / C=C / H, H (cis)

b. (cyclohexene ring with CH_3) **d.** $CH_2{=}CHBr$

7. 5 **8.** $CH_3CH{=}CH_2 + H{-}\ddot{\underset{..}{Cl}}{:} \rightleftharpoons CH_3\overset{+}{C}H{-}CH_3 + {:}\ddot{\underset{..}{Cl}}{:}^- \longrightarrow CH_3CH(\ddot{\underset{..}{Cl}}{:}){-}CH_3$

9. a negative $\Delta H°$, a positive $\Delta S°$, a high temperature

10. **a.** 4 kcal/mol **b.** 35 °C **c.** one reactant forms two products

11.

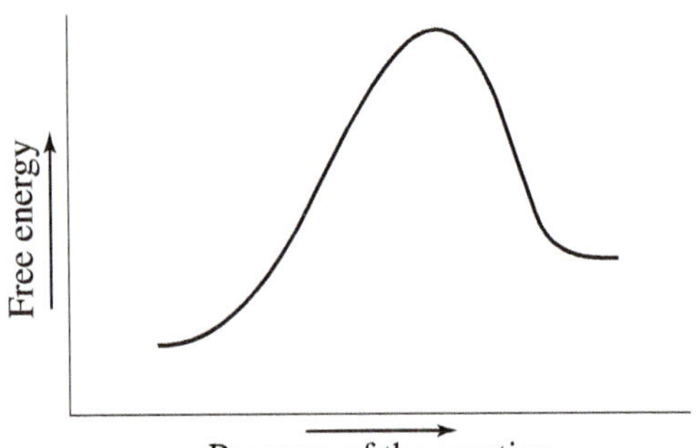

Answers to Chapter 6 Practice Test

1. **a.** 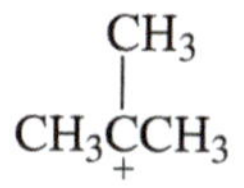 $CH_3\overset{CH_3}{\underset{+}{|}}CCH_3$ (carbocation: $(CH_3)_3C^+$) **b.** $CH_3CH_2\overset{+}{C}HCH_3$ **c.** $CH_3CH_2\overset{+}{C}H_2$

2. $CH_3CH_2CH_2CH{=}CH_2$

3. **a.** 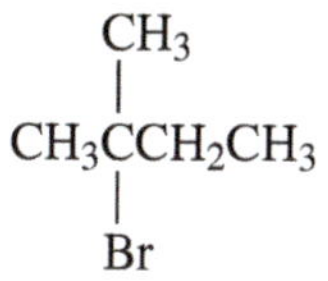 $CH_3C(CH_3)(Br)CH_2CH_3$

d. 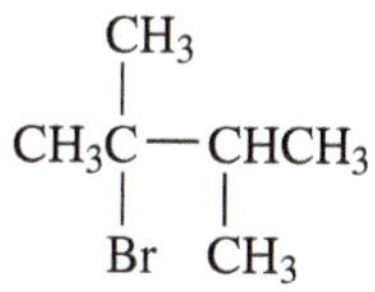 $CH_3C(CH_3)(Br){-}CH(CH_3)CH_3$

b. 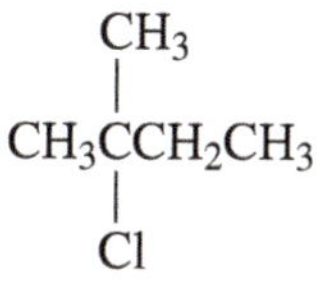 $CH_3C(CH_3)(Cl)CH_2CH_3$

e. 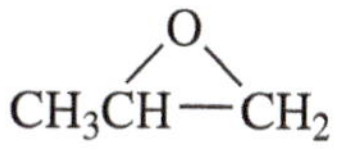 $CH_3CH{-}CH_2$ (epoxide, with O bridging the two carbons)

c. 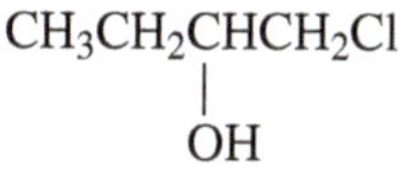 $CH_3CH_2CH(OH)CH_2Cl$

f. 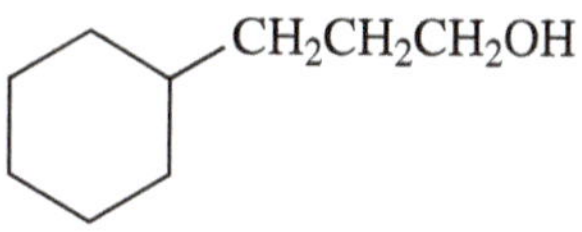 cyclohexane–$CH_2CH_2CH_2OH$

4. **a.** 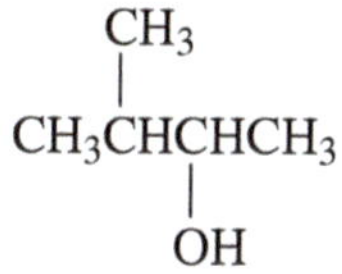 $CH_3CH(CH_3)CH(OH)CH_3$ **b.** 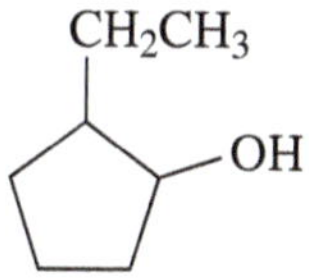 cyclopentane ring with CH_2CH_3 and adjacent OH

5. **a.** 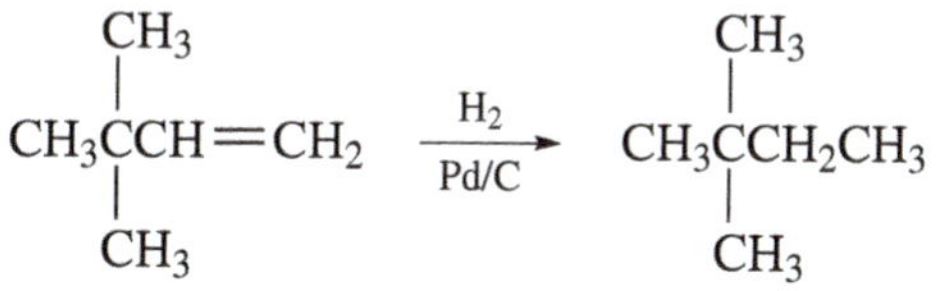 $CH_3C(CH_3)_2CH{=}CH_2 \xrightarrow[Pd/C]{H_2} CH_3C(CH_3)_2CH_2CH_3$

c. 1-methylcyclohexene (ring with CH_3) $\xrightarrow{HBr}$ cyclohexane with CH_3 and Br on the same carbon

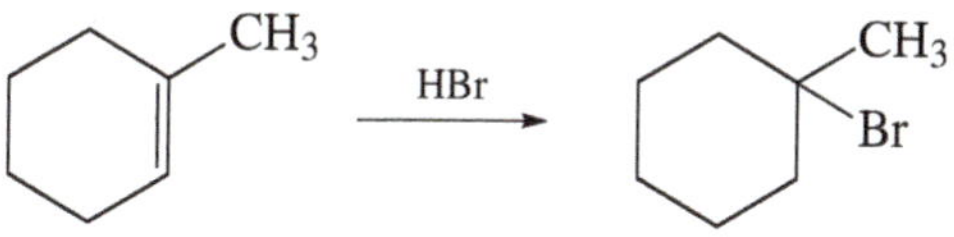

b. 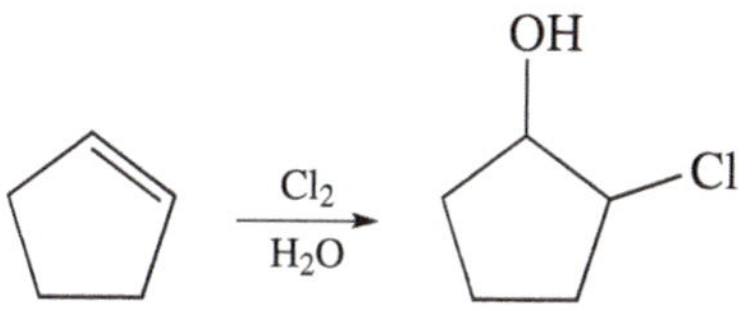 cyclopentene $\xrightarrow[H_2O]{Cl_2}$ cyclopentane with OH and adjacent Cl

d. 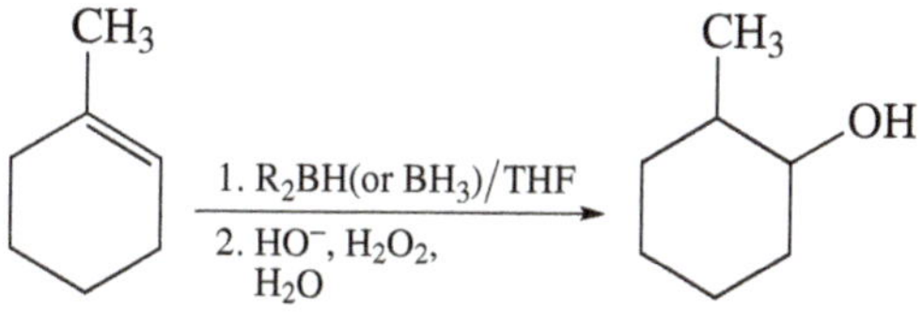 1-methylcyclohexene (ring with CH_3) $\xrightarrow[2.\ HO^-,\ H_2O_2,\ H_2O]{1.\ R_2BH(\text{or } BH_3)/THF}$ cyclohexane with CH_3 and adjacent OH

6. 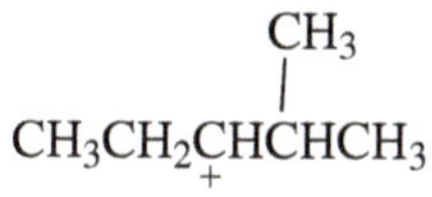 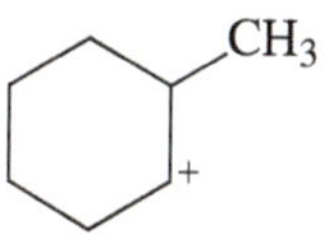 $CH_3CH_2\overset{+}{C}(CH_3)CHCH_3$ (as printed: $CH_3CH_2\underset{+}{C}HCHCH_3$ with CH_3 on the charged carbon) cyclohexyl cation with CH_3 (+ on ring carbon adjacent to CH_3)

7.
a. The addition of Br_2 to 1-butene to form 1,2-dibromobutane is a concerted reaction. F
b. The reaction of 1-butene with HCl forms 1-chlorobutane as the major product. F
c. The reaction of HBr with 3-methylcyclohexene is more highly regioselective than is the reaction of HBr with 1-methylcyclohexene. F
d. The reaction of an alkene with a carboxylic acid forms an epoxide. F
e. A catalyst increases the equilibrium constant of a reaction. F
f. The addition of HBr to 3-methyl-2-pentene is a stereospecific reaction. F
g. The addition of HBr to 3-methyl-2-pentene is a stereoselective reaction. F
h. The addition of HBr to 3-methyl-2-pentene is a regioselective reaction. T

8.
a. 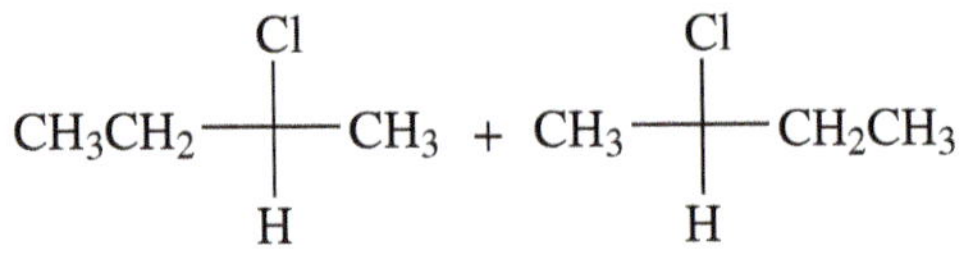 $CH_3CH_2C(Cl)(H)CH_3$ + $CH_3C(Cl)(H)CH_2CH_3$

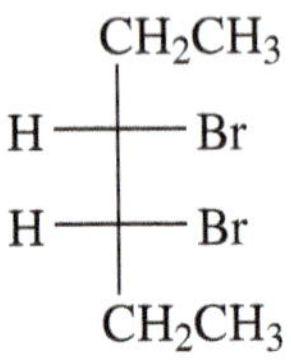

b. 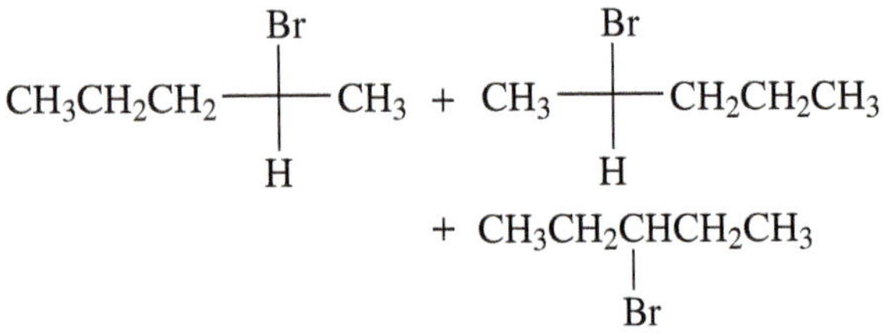 $CH_3CH_2CH_2C(Br)(H)CH_3$ + $CH_3C(Br)(H)CH_2CH_2CH_3$ + $CH_3CH_2CHCH_2CH_3$ (Br on C-3)

c. Fischer projection: top CH_2CH_3; H—C—Br; H—C—Br; bottom CH_2CH_3

d. 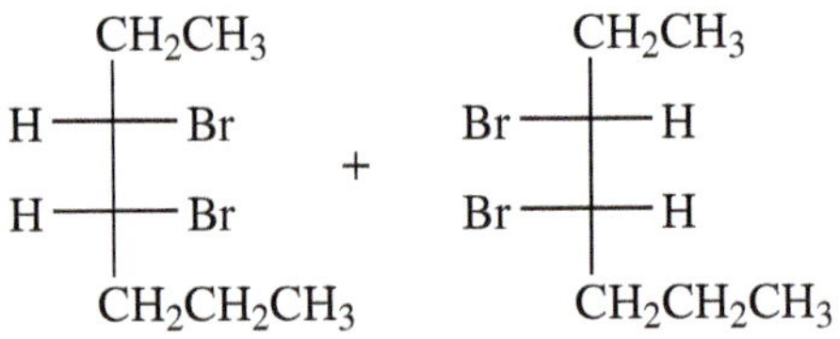 Fischer projections: top CH_2CH_3; H—C—Br; H—C—Br; bottom $CH_2CH_2CH_3$ + top CH_2CH_3; Br—C—H; Br—C—H; bottom $CH_2CH_2CH_3$

9.
a. 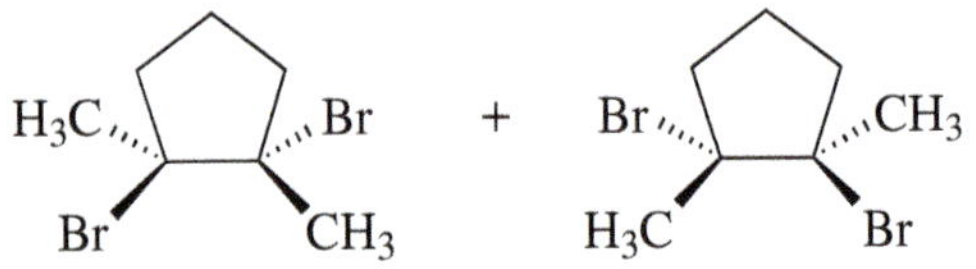 H₃C, Br, Br, CH₃ substituted cyclopentane + Br, CH₃, H₃C, Br substituted cyclopentane

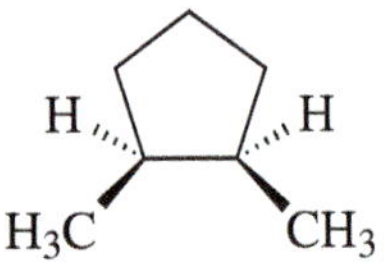

b. 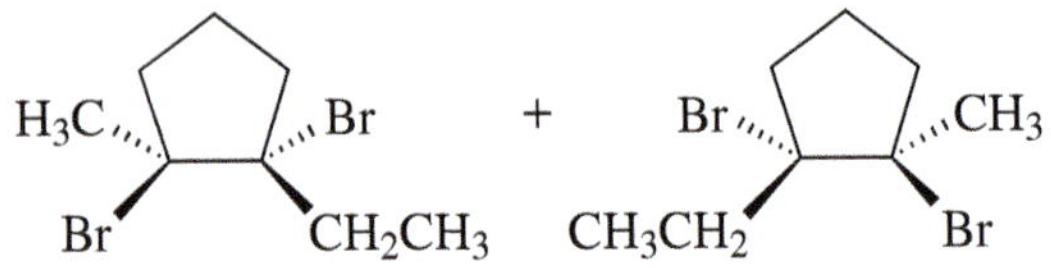 H₃C, Br, Br, CH_2CH_3 substituted cyclopentane + Br, CH₃, CH_3CH_2, Br substituted cyclopentane

c. H, H, H_3C, CH_3 substituted cyclopentane

d. 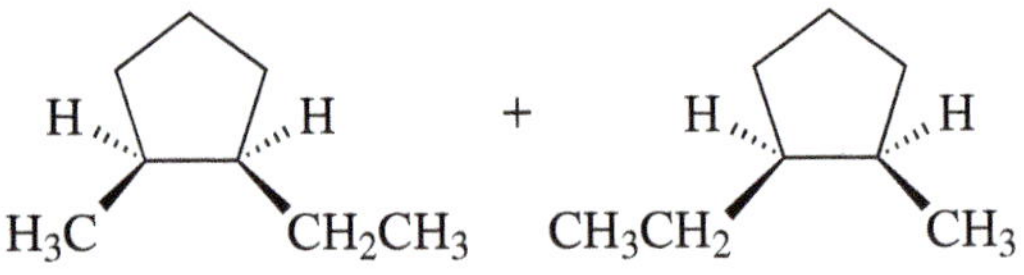 H, H, H_3C, CH_2CH_3 substituted cyclopentane + H, H, CH_3CH_2, CH_3 substituted cyclopentane

Answers to Chapter 7 Practice Test

1.
a. **1.** R_2BH/THF **2.** HO^-, H_2O_2, H_2O
b. H_2/Lindlar catalyst

2.
a. 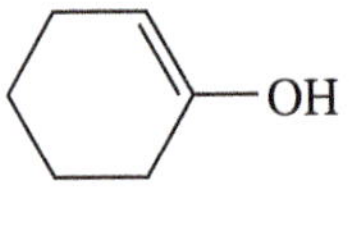 1-cyclohexenol (cyclohexene with OH on the double-bond carbon)
b. 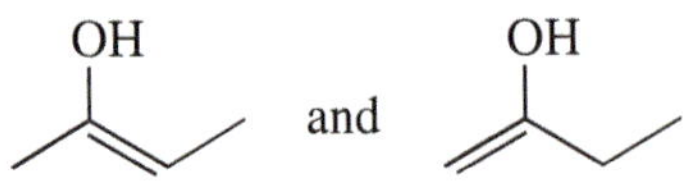 OH enol structures, and OH enol structures

3.
a. $CH_3CH_2CHC{\equiv}CCH_2CHCH_3$ (with CH_3 on C-3 and C-7)
b. 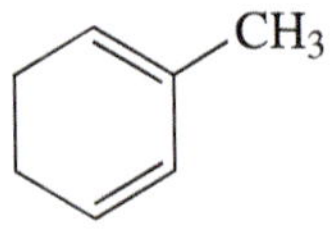 1-methyl-1,4-cyclohexadiene (CH_3)

4.
a. A terminal alkyne is more stable than an internal alkyne. F
b. Propyne is more reactive than propene toward reaction with HBr. F
c. 1-Butyne is more acidic than 1-butene. T

d. An sp^2 carbon is more electronegative than an sp^3 carbon. T
e. The reactions of internal alkynes are more regioselective than the reactions of terminal alkynes. F
f. Alkenes are more reactive than alkynes. T

5. **a.** 1-bromo-5-methyl-3-hexyne **b.** 5-methyl-3-hexyn-1-ol

6. **a.** $CH_3CH_2CH_2C{\equiv}CH$ **b.** $CH_3CH_2C{\equiv}CCH_2CH_3$

7.

NH_3	$CH_3{\equiv}CH$	CH_3CH_3	H_2O	$CH_3CH{=}CH_2$
3	**2**	**5**	**1**	**4**

8. $CH_3CH_2CH_2C(=O)CH_2CH_3$ (3-methylcyclohexanone: O= ring with CH_3)

9. **a.** $CH_3CH_2C{\equiv}CH \xrightarrow{^-NH_2} CH_3CH_2C{\equiv}C^- \xrightarrow{CH_3CH_2Br} CH_3CH_2C{\equiv}CCH_2CH_3 \xrightarrow{H_2/Pt} CH_3CH_2CH_2CH_2CH_2CH_3$

b. $CH_3CH_2C{\equiv}CH \xrightarrow{1.\ ^-NH_2\ \ 2.\ CH_3CH_2Br} CH_3CH_2C{\equiv}CCH_2CH_3 \xrightarrow[\text{or Na, } NH_3(\text{liq}),\ -78\ ^\circ C]{H_2/\text{Lindlar catalyst}} CH_3CH_2CH{=}CHCH_2CH_3 \xrightarrow{HBr} CH_3CH_2CH_2CH(Br)CH_2CH_3$

c. $CH_3CH_2C{\equiv}CH \xrightarrow{^-NH_2} CH_3CH_2C{\equiv}C^- \xrightarrow{CH_3CH_2Br} CH_3CH_2C{\equiv}CCH_2CH_3 \xrightarrow{H_2O,\ H_2SO_4} CH_3CH_2C(=O)CH_2CH_2CH_3$

Answers to Chapter 8 Practice Test

1. **a.** $C_6H_5\overset{+}{C}H_2$ (benzyl cation)

b. $CH_3\bar{C}HC{\equiv}CH$

c. $CH_3\bar{C}HC(=O)CH_3$

d. $CH_2{=}CH\overset{+}{C}H_2$

e. cyclohexenyl–$\overset{+}{C}HCH_3$

2. **a.** $CH_3CH{=}CH{-}\ddot{O}CH_3 \longleftrightarrow CH_3\bar{C}H{-}CH{=}\overset{+}{O}CH_3$

b. $CH_3CH{=}CH{-}CH{=}CH{-}\overset{+}{C}H_2 \longleftrightarrow CH_3CH{=}CH{-}\overset{+}{C}H{-}CH{=}CH_2$

$\updownarrow$

$CH_3\overset{+}{C}H{-}CH{=}CH{-}CH{=}CH_2$

c. $^{-}CH_2{-}CH{=}CH{-}\overset{O}{\overset{\|}{C}}H \longleftrightarrow CH_2{=}CH{-}\overset{-}{C}H{-}\overset{O}{\overset{\|}{C}}H \longleftrightarrow CH_2{=}CH{-}CH{=}\overset{O^-}{\overset{|}{C}}H$

3. $CH_3\overset{CH_3}{\overset{|}{\underset{+}{C}}}CH_2CH{=}CH_2$ $CH_2{=}CHCH_2CH{=}CH_2$ $CH_3CH_2NHCH_2CH{=}CHCH_3$

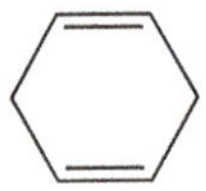

4. a. 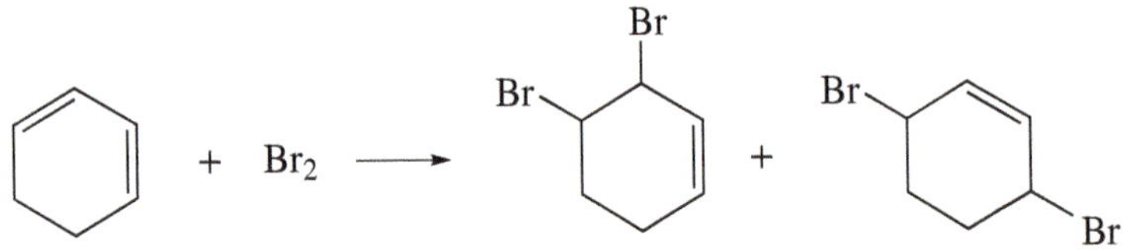 + Br_2 ⟶ Br Br + Br Br

b. 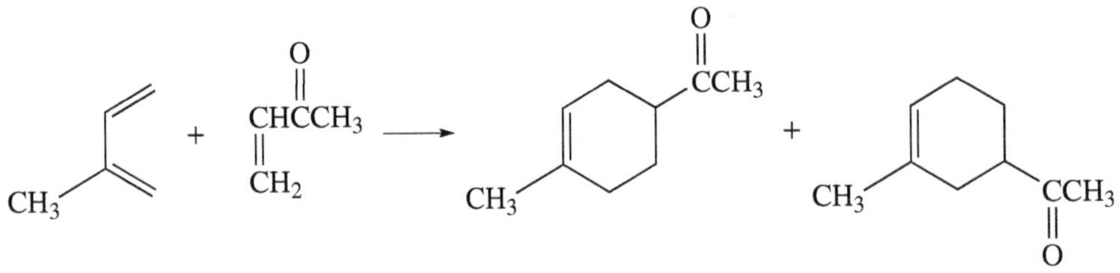 CH_3 + $\overset{O}{\overset{\|}{CHCCH_3}}$ $\|$ CH_2 ⟶ CH_3 $\overset{O}{\overset{\|}{C}}CH_3$ + CH_3 $\underset{O}{\underset{\|}{C}}CH_3$

5. $CH_3\overset{O}{\overset{\|}{C}}OH$ and $CH_3\overset{O^-}{\overset{|}{C}}{=}\overset{+}{O}H$

6. 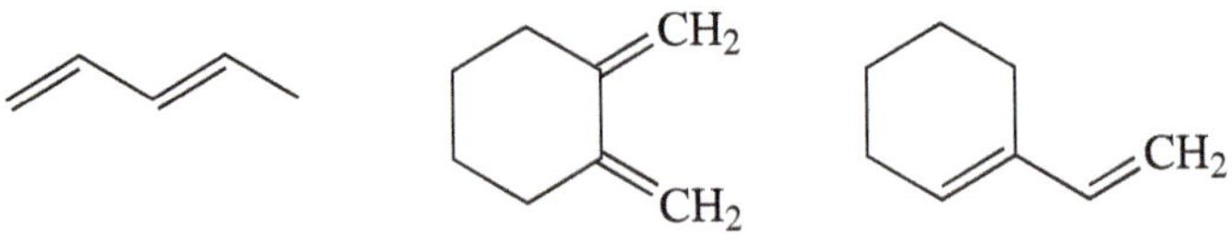 CH_2 CH_2 CH_2

7. a. 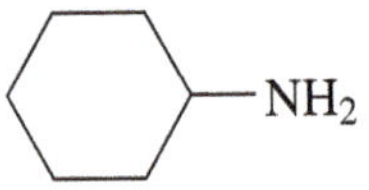 NH_2 **b.** 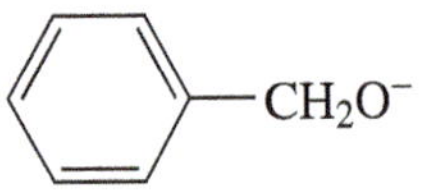 CH_2O^-

8. a. 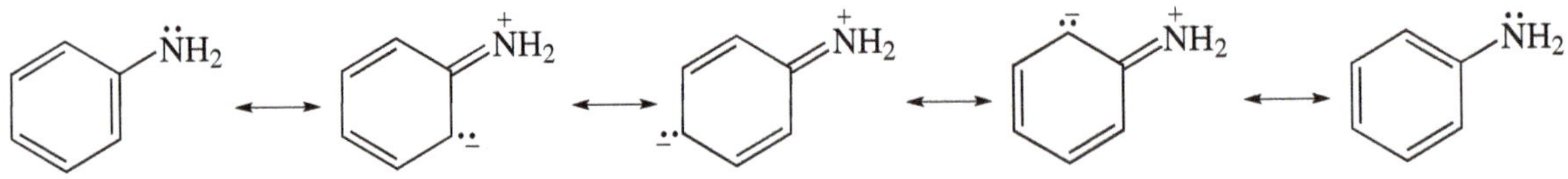 $\ddot{N}H_2$ ⟷ $\overset{+}{N}H_2$ ⟷ $\overset{+}{N}H_2$ ⟷ $\overset{+}{N}H_2$ ⟷ $\ddot{N}H_2$

b. 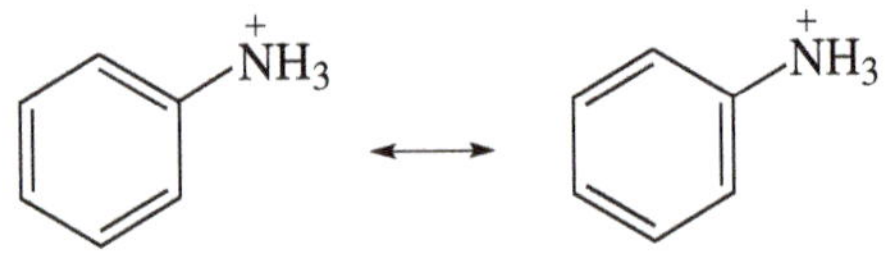 $\overset{+}{N}H_3$ ⟷ $\overset{+}{N}H_3$

c. 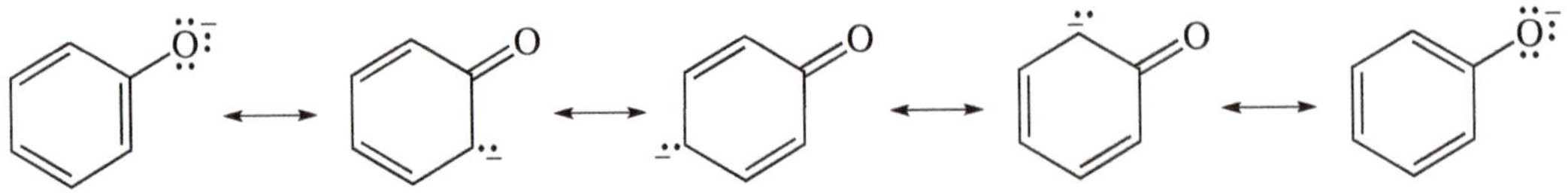 $\ddot{O}{:}^-$ ⟷ O ⟷ O ⟷ O ⟷ $\ddot{O}{:}^-$

9. a. 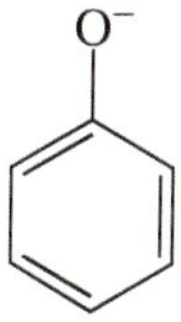 (phenoxide ion: O^- on benzene ring) b. 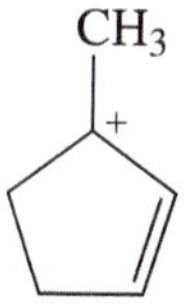 (cyclopentene ring with CH_3 and positive charge)

10. a. A compound with four conjugated double bonds has four molecular orbitals. F

b. ψ_1 and ψ_2 are symmetric molecular orbitals. F

c. If ψ_3 is the HOMO in the ground state, ψ_4 will be the HOMO in the excited state. T

d. If ψ_3 is the LUMO, ψ_4 will be the HOMO. F

e. If the ground-state HOMO is symmetric, the ground-state LUMO will be antisymmetric. T

f. A single bond formed by an *sp*^{2}—*sp*^{2} overlap is longer than a single bond formed by an *sp*^{2}—*sp*^{3} overlap. F

g. The thermodynamically controlled product is the major product obtained when the reaction is carried out under mild conditions. F

h. 1,3-Hexadiene is more stable than 1,4-hexadiene. T

11. 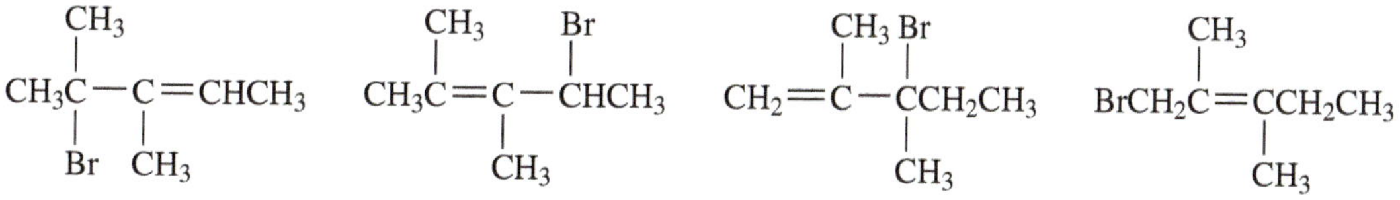 $CH_3C(CH_3)(Br)-C(CH_3)=CHCH_3$ $CH_3C(CH_3)=C(CH_3)-CH(Br)CH_3$ $CH_2=C(CH_3)-C(Br)(CH_3)CH_2CH_3$ $BrCH_2C(CH_3)=C(CH_3)CH_2CH_3$

12. 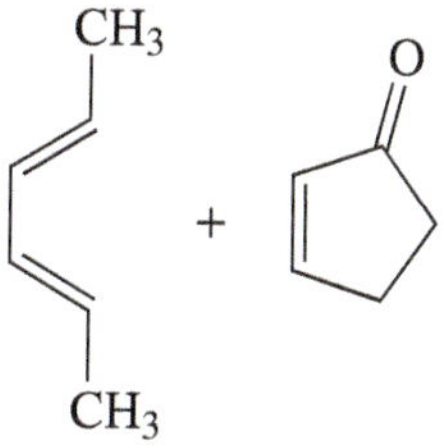 CH_3–diene–CH_3 + cyclopentenone

13. 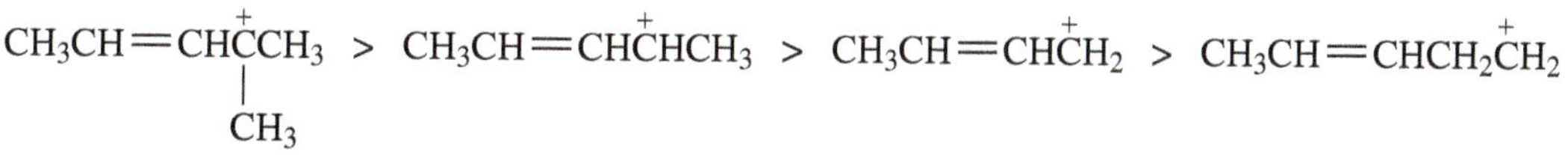 $CH_3CH=CH\overset{+}{C}(CH_3)CH_3$ > $CH_3CH=CH\overset{+}{C}HCH_3$ > $CH_3CH=CH\overset{+}{C}H_2$ > $CH_3CH=CHCH_2\overset{+}{C}H_2$

14. a. 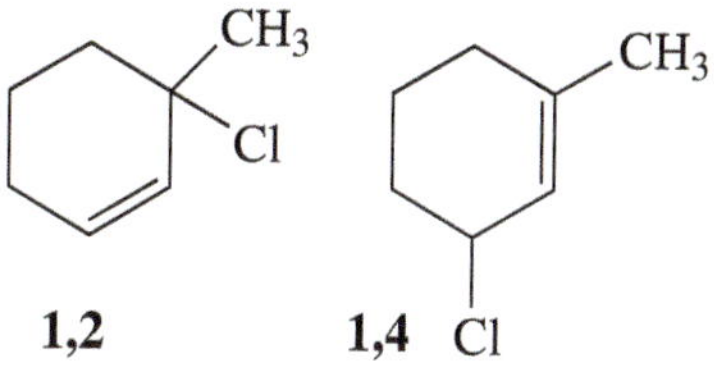 CH_3, Cl (1,2) CH_3, Cl (1,4) b. CH_3, Cl

1,2 1,4

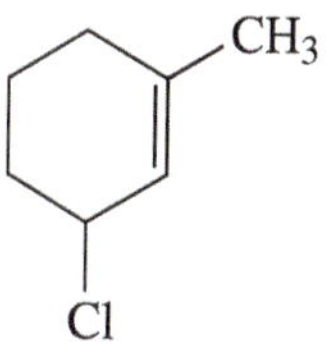

15. 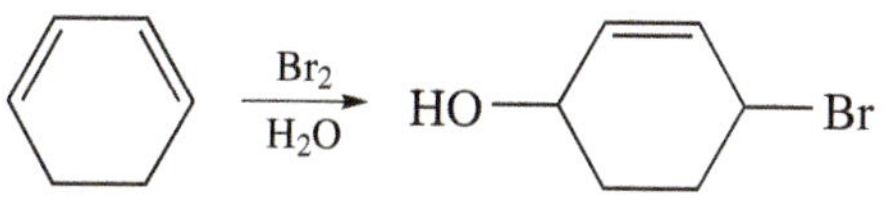 cyclohexadiene $\xrightarrow[H_2O]{Br_2}$ HO–cyclohexene–Br

16. 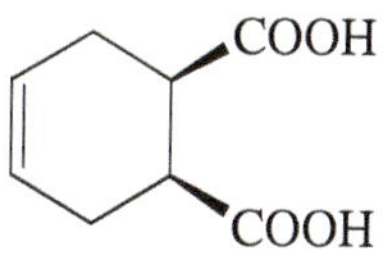 cyclohexene with two COOH

17. **a.** $CH_2{=}CH{-}C(CH_3)(Cl){-}CH_3$ — product of kinetic control

$ClCH_2{-}CH_2{=}C(CH_3){-}CH_3$ — product of thermodynamic control

b. 3-bromo-3-methylcyclohexene (ring carbon bearing CH_3 and Br) — product of kinetic control

1-methyl-3-bromocyclohexene (CH_3 on the double-bond carbon, Br on ring) — product of thermodynamic control

18. cyclopropenyl cation (+) ; cyclopentadienyl anion (−)

19. cycloheptatrienyl cation (+)

20. **a.** $C_6H_5{-}S(=O)_2{-}OH$ **c.** $C_6H_5{-}COOH$ **e.** $C_6H_5{-}\overset{+}{N}H_3$

b. cyclopentadiene (CH_2 with H, H) **d.** cyclohexene with OH on the double-bond carbon **f.** $O_2N{-}C_6H_4{-}OH$ (para)

Answers to Chapter 9 Practice Test

1. **a.** $C_6H_5{-}CH(Br)CH_2CH_3$ **b.** $CH_3CH{=}CHCH(Br)CH_3$

2. **a.** $CH_3CH_2CH(CH_3)Br$ **b.** $C_6H_5{-}CH_2Br$

3. **a.** 1-methoxy-2-methylcyclohexane (OCH_3 and CH_3 as drawn) + 1-methylcyclohexene

b. $C_6H_5{-}CH(OCH_3)CH_2CH_3$ (R and S) + $C_6H_5{-}CH{=}CHCH_3$ (cis and trans)

c. HO—C(CH₃)(H)(CH₂CH₃) + $CH_3CH{=}CHCH_3$ (cis and trans)

d. $H{-}C(CH_3)(CH_2CH_3){-}NHCH_3$

mainly a substitution product because the amine is a relatively weak base

e. $CH_3CH{=}CHCH_2OCH_3$ + $CH_3CH(OCH_3)CH{=}CH_2$ + $CH_2{=}CHCH{=}CH_2$

because you do not whether the reactant is cis or trans, you can tell whether the product is cis or trans

R and S

4.

a. Increasing the concentration of the nucleophile favors an S_N1 reaction over an S_N2 reaction. F

b. Ethyl iodide is more reactive than ethyl chloride in an S_N2 reaction. T

c. In an S_N1 reaction, the product with the retained configuration is obtained in greater yield. F

d. The rate of a substitution reaction in which none of the reactants is charged will increase if the polarity of the solvent is increased. T

e. An S_N2 reaction is a two-step reaction. F

f. The pK_a of a carboxylic acid is greater in water than it is in a less polar solvent. F

g. 4-Bromo-1-butanol forms a cyclic ether faster than does 3-bromo-1-propanol. T

5. **a.** CH_3O^- **b.** CH_3S^-

6.

a. $CH_3CH_2CH_2Cl + HO^-$

b. $CH_3CH_2CH_2I + HO^-$

c. $CH_3CH_2CH_2Br + HO^-$

d. 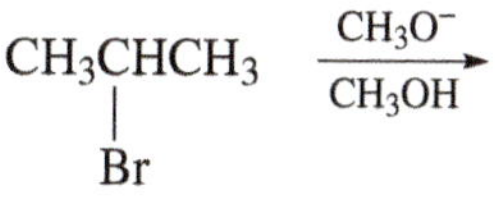 $CH_3CH(Br)CH_3 \xrightarrow[CH_3OH]{CH_3O^-}$

e. $BrCH_2CH_2CH_2CH_2NHCH_3$

7. All are aprotic solvents except ethanol.

8.

a. The rate of the reaction would increase.

b. The rate of the reaction would increase.

c. The rate of the reaction would decrease.

d. The pK_a would decrease.

e. The pK_a would decrease.

9. **a.** 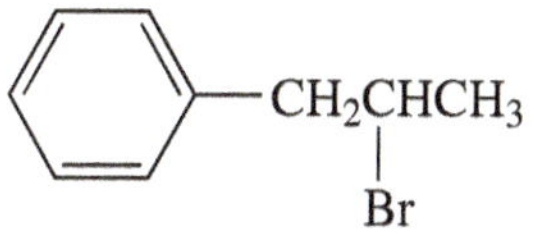 $C_6H_5CH_2CH(Br)CH_3$ **b.** 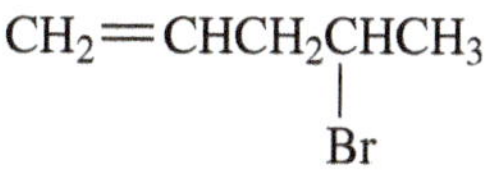 $CH_2{=}CHCH_2CH(Br)CH_3$

10. 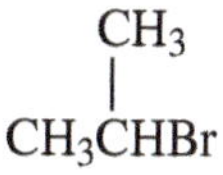 $CH_3C(CH_3)HBr$

11. 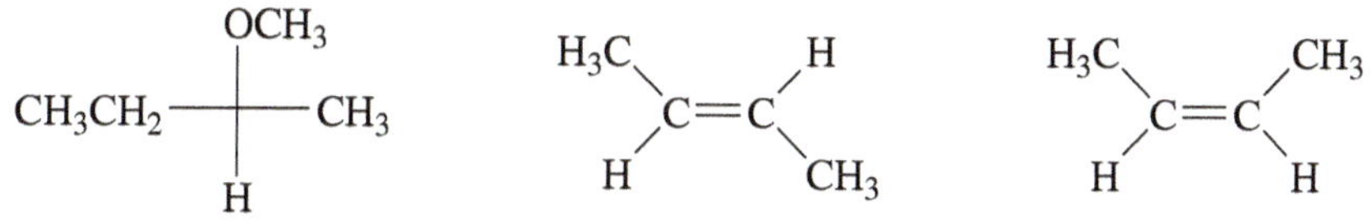 CH_3CH_2—C(OCH$_3$)(H)—CH_3 (Fischer projection: OCH_3 top, H bottom, CH_3CH_2 left, CH_3 right)

$H_3C(H)C{=}C(H)CH_3$ (trans) — major

$H_3C(H)C{=}C(H)CH_3$ (cis) — minor

12. **a.** 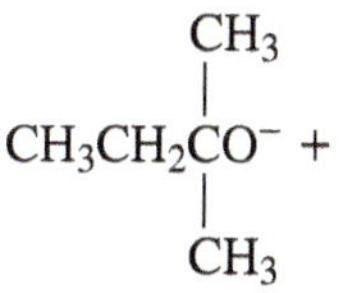 $CH_3CH_2C(CH_3)_2O^- + CH_3CH_2CH_2Br$ **b.** 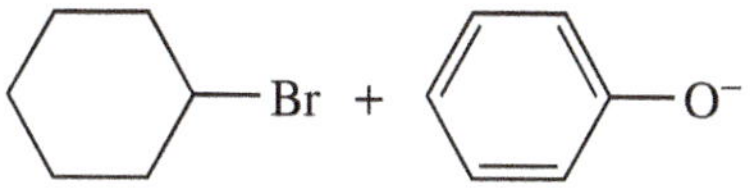 cyclohexyl—Br + phenyl—O^- **c.** $CH_3CH_2CH_2O^- + CH_3Br$

13.

a. 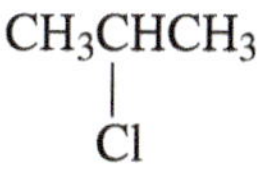 $CH_3CH(Cl)CH_3 + HO^-$

b. $CH_3CH_2CH_2I + HO^-$

c. $CH_3CH_2CH_2Br + HO^-$

d. 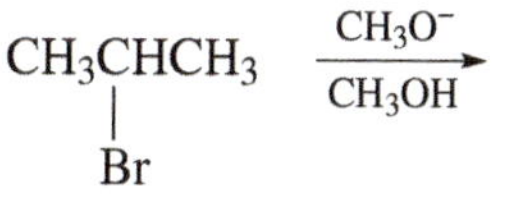 $CH_3CH(Br)CH_3 \xrightarrow[CH_3OH]{CH_3O^-}$

e. 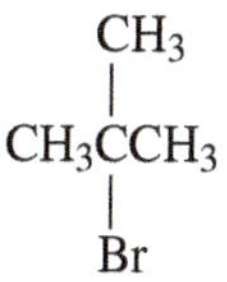 $CH_3C(CH_3)(Br)CH_3 + HO^-$

14. **a.** 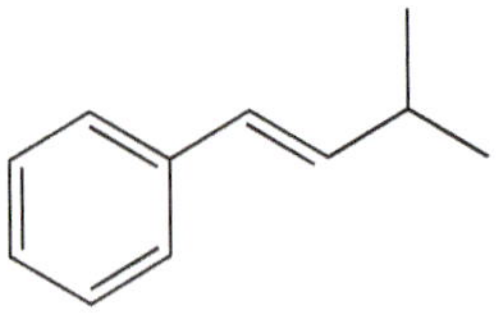 **b.**  **c.** 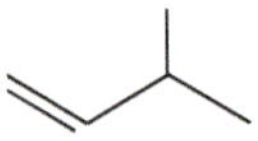 **d.**

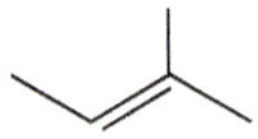

15. *cis*-1-bromo-2-methylcyclohexane

Answers to Chapter 10 Practice Test

1. HBr **2.** $SOCl_2$

3.

a. $HOCH_2C(CH_2CH_3)(OCH_3)CH_2CH_3$

$$\begin{array}{c} CH_2CH_3 \\ | \\ HOCH_2CCH_2CH_3 \\ | \\ OCH_3 \end{array}$$

b. $CH_3OCH_2C(CH_2CH_3)(OH)CH_2CH_3$

$$\begin{array}{c} CH_2CH_3 \\ | \\ CH_3OCH_2CCH_2CH_3 \\ | \\ OH \end{array}$$

4.

a. $CH_3CH(CH_3)CH_2N(CH_3)_2$ + $CH_2{=}CHCH_2CH_3$

b. 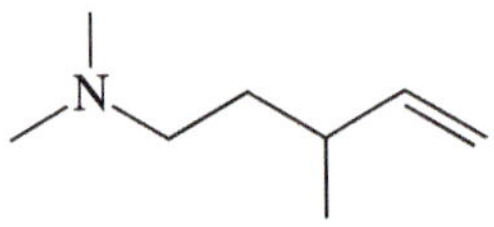 **c.**

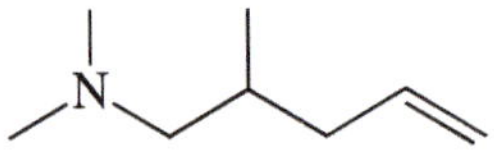

5.

a. $CH_3CH_2C(CH_3){=}C(CH_3)CH_3$ (printed as $CH_3CH_2C{=}CCH_3$ with CH_3 above and below)

b. $CH_3CH_2CH_2C(CH_3){=}C(CH_3)CH_3$ (printed as $CH_3CH_2CH_2C{=}CCH_3$ with CH_3 above and below)

c. $CH_3CH_2CH{=}CHCH_3$

d.

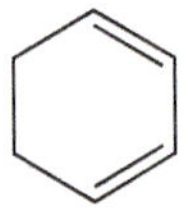

c. will form both trans and cis but more trans.

6.

a.	Tertiary alcohols are easier to dehydrate than secondary alcohols.	T
b.	Alcohols are more acidic than thiols.	F
c.	Alcohols have higher boiling points than thiols.	T
d.	The acid-catalyzed dehydration of a primary alcohol is an S_N1 reaction.	F
e.	The Hofmann elimination reaction is an E2 reaction.	T

7.

a. $CH_3CH_2C(CH_3)_2{-}I$ + CH_3OH (printed as $CH_3CH_2C{-}I$ with CH_3 above and below)

b. 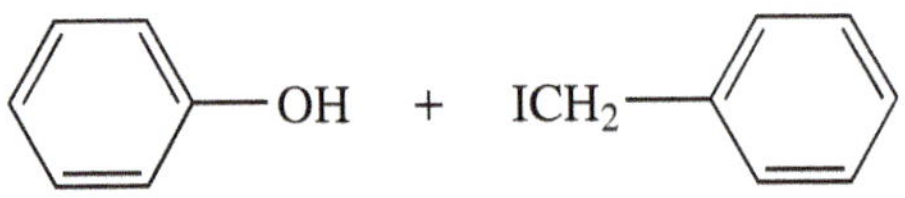 $C_6H_5{-}OH$ + $ICH_2{-}C_6H_5$

8.

a. $CH_3CH_2CH_2NHCH_2CH_3$

b. $CH_3CH_2CH_2Cl$

c. $CH_3CH_2CH_2CH{=}CHCH_3$ + $CH_3CH_2CH{=}CHCH_2CH_3$

d. 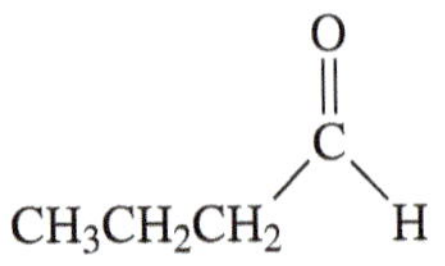 $CH_3CH_2CH_2C(=O)H$

e. 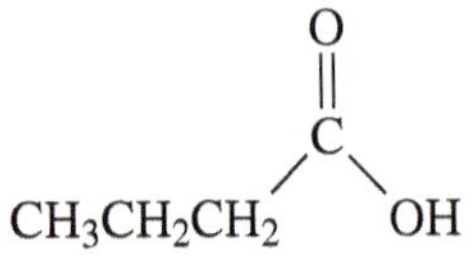 $CH_3CH_2CH_2C(=O)OH$

Answers to Chapter 11 Practice Test

1. an organocadmium compound

2. 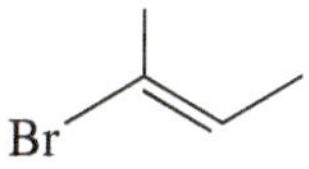 Br and Br

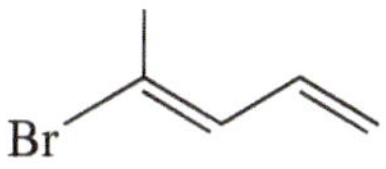

3. **a.** 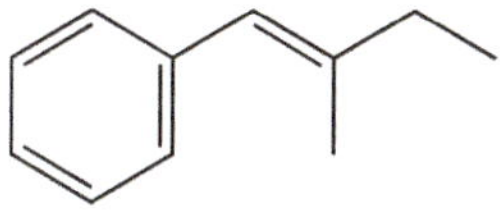 **b.** 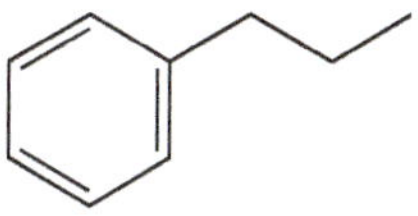 **c.**

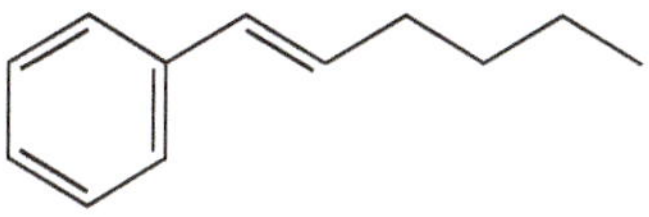

4. **a.** 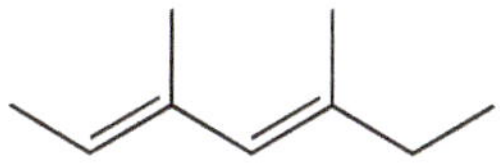 **b.** 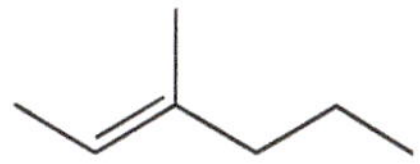 **c.**

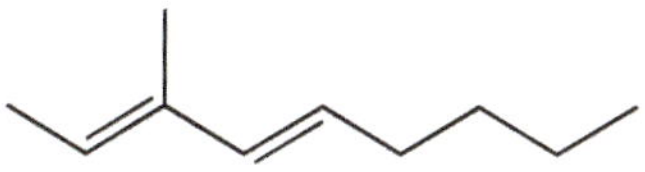

5. **a.** $CH_2{=}CH_2$ +

b. $CH_2{=}CH_2$ +

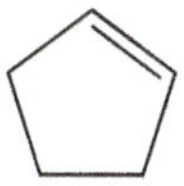

c. $CH_2{=}CH_2$ +

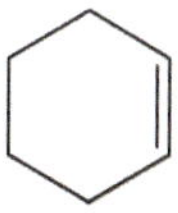

6. **a.** 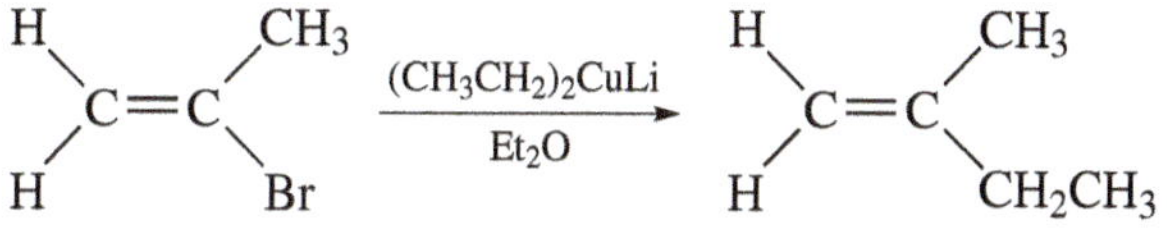 $(H)_2C{=}C(CH_3)(Br)$ $\xrightarrow[Et_2O]{(CH_3CH_2)_2CuLi}$ $(H)_2C{=}C(CH_3)(CH_2CH_3)$

b. 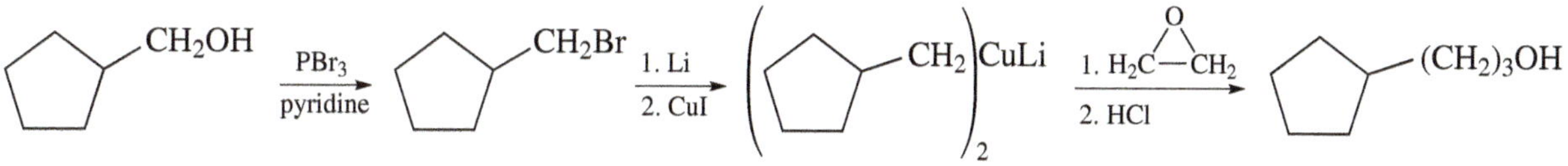 CH_2OH $\xrightarrow[\text{pyridine}]{PBr_3}$ CH_2Br $\xrightarrow[\text{2. CuI}]{\text{1. Li}}$ $(\ \ CH_2)_2CuLi$ $\xrightarrow[\text{2. HCl}]{\text{1. } H_2C{-}CH_2 \text{ (O)}}$ $(CH_2)_3OH$

7. **a.** $CH_3CH_2CH_2CH_2CH_2OH$ **b.** 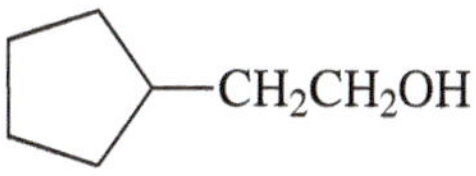 $-CH_2CH_2OH$

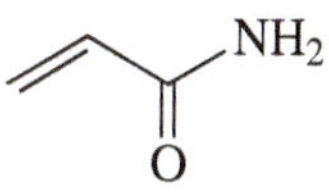

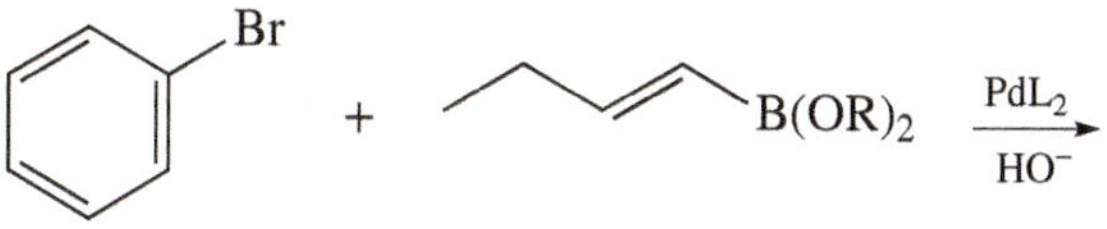

8. 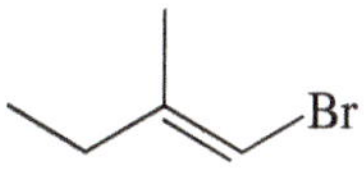 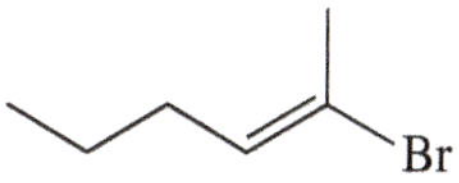 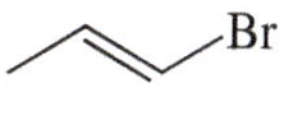 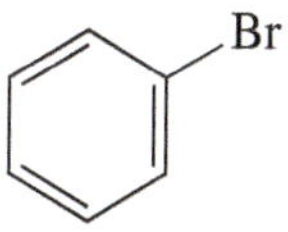 Br Br Br Br

9. **a.** **1.** 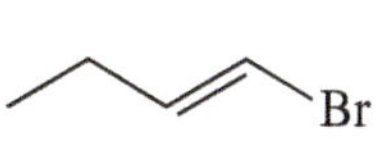 Br + NH_2 O **2.** Br +

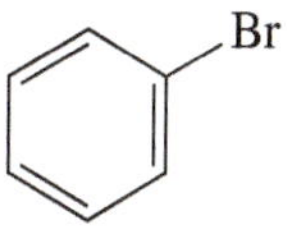

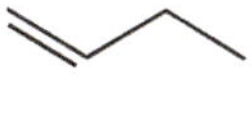

b. **1.**  Br NH_2 O + $B(OR)_2$ $\xrightarrow[HO^-]{PdL_2}$

2. Br + $B(OR)_2$ $\xrightarrow[HO^-]{PdL_2}$

10. **a.**  **b.**

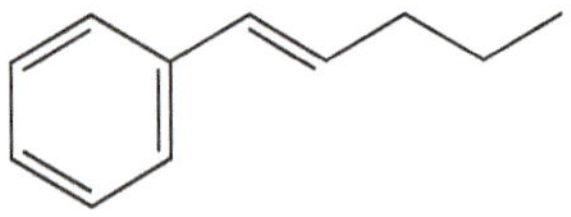

11. 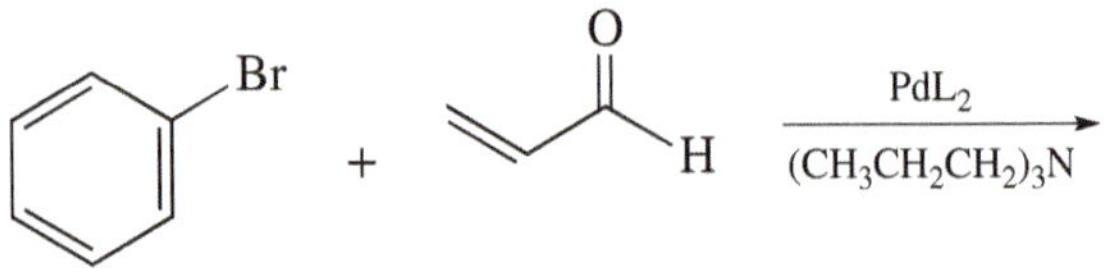 Br + O H $\xrightarrow[(CH_3CH_2CH_2)_3N]{PdL_2}$

Answers to Chapter 12 Practice Test

1. **a.** 4 **b.** 4 **c.** 5 **d.** 4 **e.** 3

2. $CH_3CH_3 + \cdot Cl \longrightarrow CH_3\dot{C}H_2 + HCl$

3.

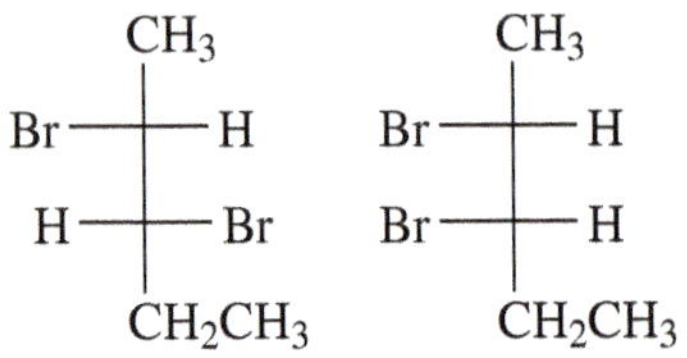

CH$_3$ / Br—H / H—Br / CH$_2$CH$_3$

CH$_3$ / Br—H / Br—H / CH$_2$CH$_3$

4. $CH_3CH_3 + Cl\cdot \longrightarrow CH_3\dot{C}H_2 + HCl \qquad \Delta H^\circ = 101 - 103 = -2\text{ kcal/mol}$

$CH_3\dot{C}H_2 + Cl_2 \longrightarrow CH_3CH_2Cl + Cl\cdot \qquad \Delta H^\circ = 58 - 85 = -27\text{ kcal/mol}$

5.

$\dot{C}H_2CH_2CH_2CH{=}CH$ **3**

$CH_3CH_2\dot{C}HCH{=}CH_2$ **1**

$CH_2CH_2CH_2CH{=}\dot{C}H$ **4**

$CH_3\dot{C}HCH_2CH{=}CH_2$ **2**

6. **a.** 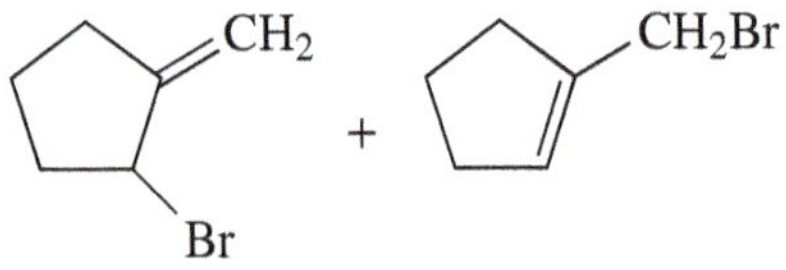 CH$_2$ / Br + CH$_2$Br

b. 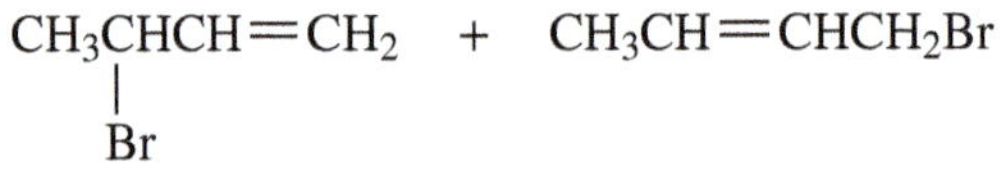 $CH_3CHCH{=}CH_2$ (Br on C-2) + $CH_3CH{=}CHCH_2Br$

7. **a.**

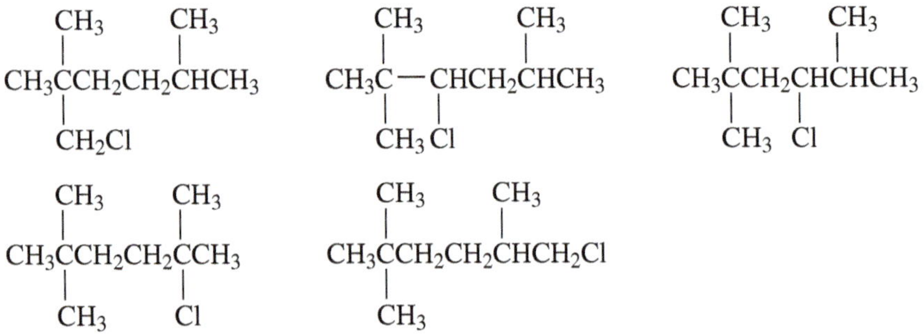

$CH_3CCH_2CH_2CHCH_3$ (CH$_3$ and CH$_2$Cl on C-2; CH$_3$ on C-5)

$CH_3C{-}CHCH_2CHCH_3$ (CH$_3$, CH$_3$ on C-2; Cl on C-3; CH$_3$ on C-5)

$CH_3CCH_2CHCHCH_3$ (CH$_3$, CH$_3$ on C-2; CH$_3$ on C-4; Cl on C-5)

$CH_3CCH_2CH_2CCH_3$ (CH$_3$, CH$_3$ on C-2; CH$_3$, Cl on C-5)

$CH_3CCH_2CH_2CHCH_2Cl$ (CH$_3$, CH$_3$ on C-2; CH$_3$ on C-5)

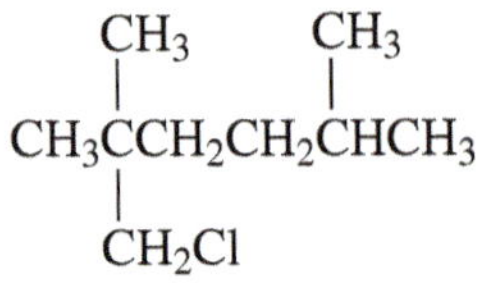

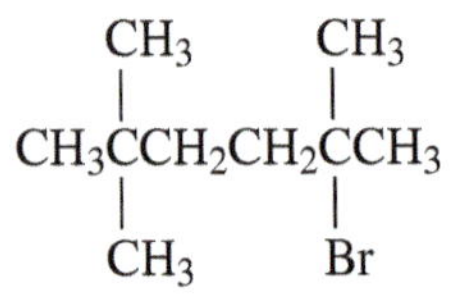

b. $CH_3CCH_2CH_2CHCH_3$ (CH$_3$ and CH$_2$Cl on C-2; CH$_3$ on C-5)

c. $CH_3CCH_2CH_2CCH_3$ (CH$_3$, CH$_3$ on C-2; CH$_3$, Br on C-5)

8. 23%

9.

a. 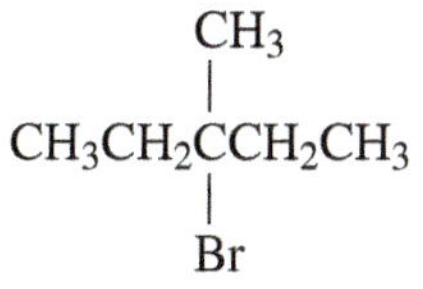 $CH_3CH_2C(CH_3)(Br)CH_2CH_3$

$$\begin{array}{c} CH_3 \\ | \\ CH_3CH_2CCH_2CH_3 \\ | \\ Br \end{array}$$

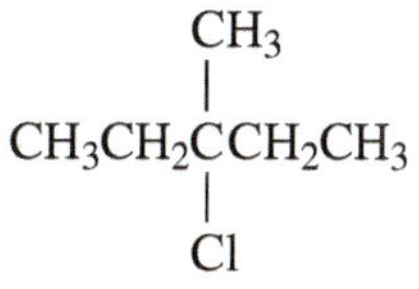

b. 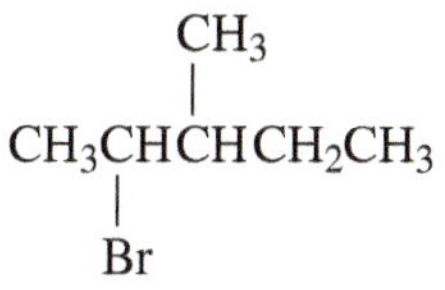 $CH_3CH(Br)CH(CH_3)CH_2CH_3$

$$\begin{array}{c} \quad CH_3 \\ \quad | \\ CH_3CHCHCH_2CH_3 \\ | \quad \\ Br \quad \end{array}$$

c. $CH_3CH_2C(CH_3)(Cl)CH_2CH_3$

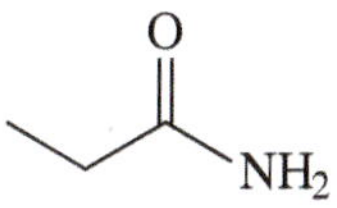

$$\begin{array}{c} CH_3 \\ | \\ CH_3CH_2CCH_2CH_3 \\ | \\ Cl \end{array}$$

d. 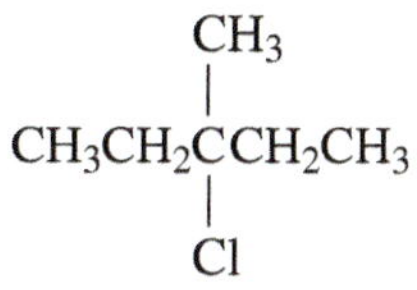 $CH_3CH_2C(CH_3)(Cl)CH_2CH_3$

$$\begin{array}{c} CH_3 \\ | \\ CH_3CH_2CCH_2CH_3 \\ | \\ Cl \end{array}$$

10. b

Answers to Chapter 13 Practice Test

1.

a. 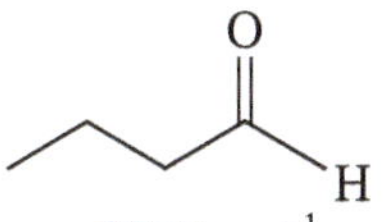 $CH_3CH_2CH_2CHO$ (skeletal: butanal, C=O with H) ~2770 cm^{-1}

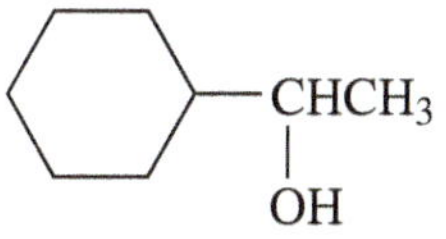

b. $CH_3CH_2CONH_2$ (skeletal: propanamide, C=O with NH_2) ~3300 cm^{-1}

c. $CH_3CH_2CH_2CH_2OH$ ~3600−3200 cm^{-1}

d. $C_6H_{11}CH(OH)CH_3$ (cyclohexyl–CHCH₃ with OH) ~1380 cm^{-1}

e. 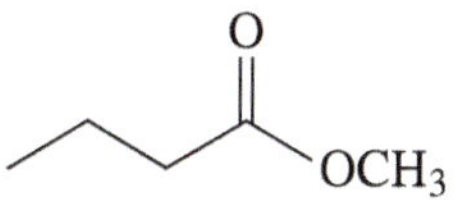 $CH_3CH_2CH_2COOCH_3$ (skeletal: C=O with OCH_3) ~1050 or ~1250 cm^{-1}

f. $CH_3CH_2CH{=}CHCH_3$ ~1600 cm^{-1}, ~3100 cm^{-1} $\qquad$ $CH_3CH_2C{\equiv}CCH_3$ ~2100 cm^{-1}

g. $CH_3CH_2C{\equiv}CH$ ~3300 cm^{-1}

2.

a.	The O—H stretch of a concentrated solution of an alcohol occurs at a higher frequency than the O—H stretch of a dilute solution.	F
b.	Light of 2 nm is of higher energy than light of 3 nm.	T
c.	It takes more energy for a bending vibration than for a stretching vibration.	F
d.	Propyne will not have an absorption band at 3100 cm^{-1} because there is no change in the dipole moment.	F
e.	The M + 2 peak of an alkyl chloride is half the height of the M peak.	F

3.

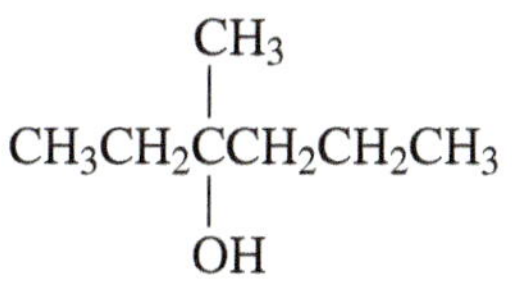

$$\begin{array}{c} CH_3 \qquad\qquad \\ | \qquad\qquad \\ CH_3CH_2CCH_2CH_2CH_3 \\ | \qquad\qquad \\ OH \qquad\qquad \end{array}$$

4.

$$\text{Absorbance} = \text{molar absorptivity} \times \text{concentration} \times \text{length of light path (in cm)}$$
$$0.75 = \text{molar absorptivity} \times 3.8 \times 10^{-4} \times 1$$
$$\text{molar absorptivity} = 2000\ M^{-1}\ cm^{-1}$$

5. **a.** OH OH

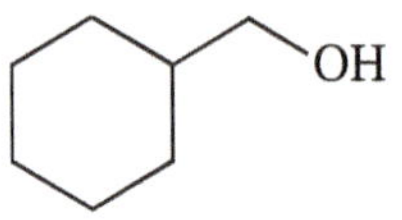

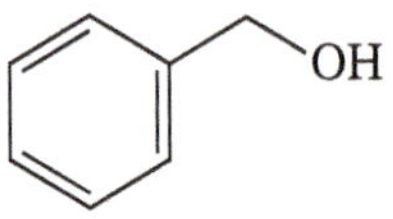

~3100 cm^{-1}
~1500 cm^{-1}

d. O OCH$_3$

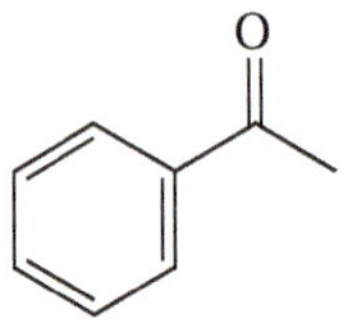

~1700 cm^{-1} 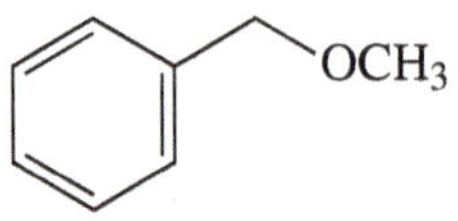 ~1050 cm^{-1}

b. O H O

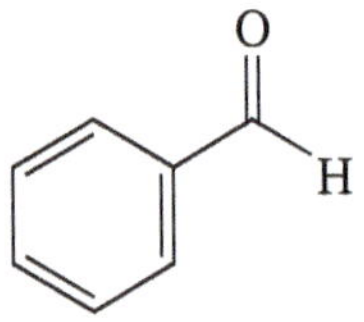

~2700 cm^{-1} 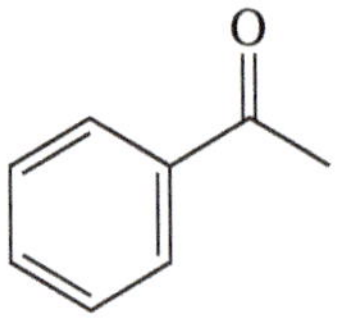 ~2900 cm^{-1}
~1380 cm^{-1}

e.

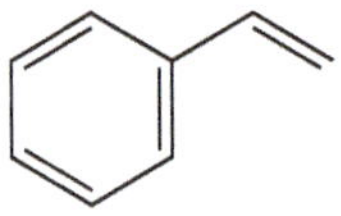

~3100 cm^{-1} 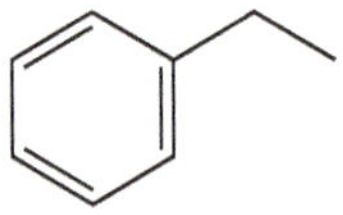 ~2900 cm^{-1}
~1380 cm^{-1}

c. O OH O OCH$_3$

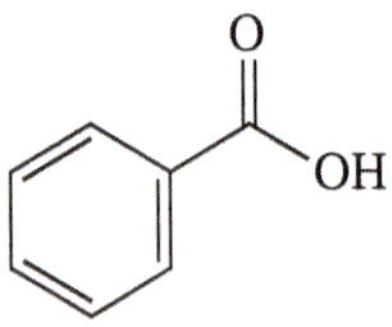

~3300–2500 cm^{-1} 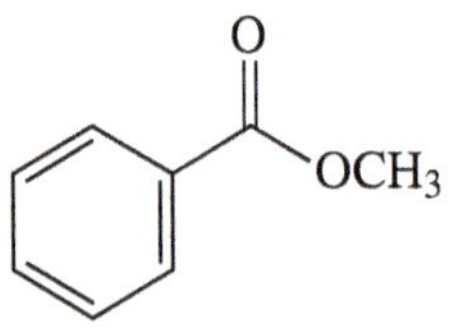 ~2900 cm^{-1}
~1380 cm^{-1}
~1050 cm^{-1}

6. **a.** **b.** H N **c.** NH$_2$

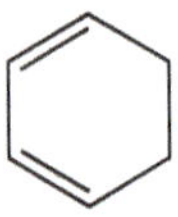

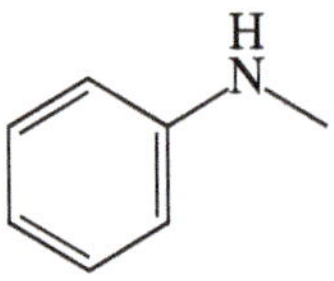

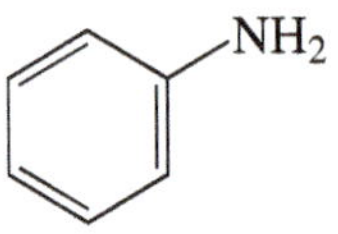

7.

$$\text{absorbance} = \text{molar absorbtivity} \times \text{concentration} \times \text{length of light path (in cm)}$$
$$0.76 = 1200 \times \text{concentration}$$
$$\text{concentration} = 6.3 \times 10^{-4}\ M$$

8. $\frac{60}{13} = 4$ with 8 left over $C_4H_{12} = C_3H_8O$ $CH_3CH_2CH_2OH$ and CH_3CHCH_3 (with OH on the central carbon)

9. A bond between a carbon and an atom of similar electronegativity breaks homolytically, whereas a bond between a carbon and a more electronegative atom breaks heterolytically.

Answers to Chapter 14 Practice Test

1. $CH_3CH_2CH_2C(=O)CH_3$ **4** $CH_3CH_2CHClCH_2CH_3$ **3** $CH_2{=}CHC(=O)H$ **4**

nitrobenzene ($C_6H_5NO_2$) **3** 1,3-dichlorobenzene **3** $CH_3CH(CH_3)CH_2CH(CH_3)CH_3$ **3**

2.

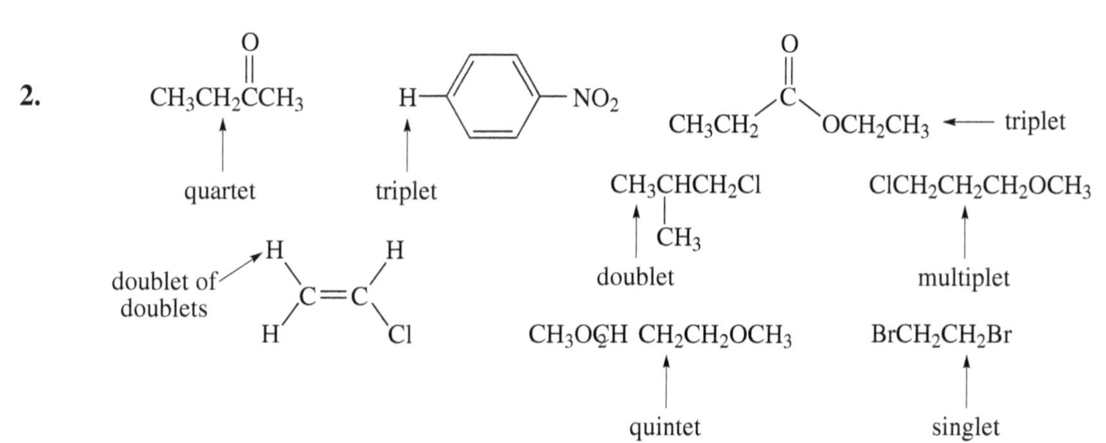

3. $CH_3C(=O)OCH_2CH_3$ **3** signals — The signal at the highest frequency (farthest downfield) is a quartet.

$CH_3CH_2C(=O)OCH_3$ **3** signals — The signal at the highest frequency (farthest downfield) is a singlet.

$HC(=O)OCH_2CH_2CH_3$ **4** signals

4. **a.** The signals on the right of an NMR spectrum are deshielded compared to the signals on the left. F

b. Dimethyl ketone has the same number of signals in its ^1H NMR spectrum as in its ^{13}C NMR spectrum. F

c. In the ^1H NMR spectrum of the compound shown below, the lowest-frequency signal (the one farthest upfield) is a singlet and the highest-frequency signal (the one farthest downfield) is a doublet. T

$O_2N-C_6H_4-CH_3$

d. The greater the frequency of the signal, the greater its chemical shift in ppm. T

5. **a.** $CH_3CH_2CH_2Cl$ (triplet) **3 signals**

b. $CH_3CH_2C(=O)OCH_3$ (singlet) **3 signals**

c. $CH_3CH(Br)CH_3$ (septet) **2 signals**

6. **a.** $CH_3CH_2CH_2Cl$ **3 signals** (triplet)

b. $CH_3CH_2C(=O)OCH_3$ **4 signals** (singlet)

c. $CH_3CH(Br)CH_3$ **2 signals** (doublet)

Answers to Chapter 15 Practice Test

1. **a.** $CH_3C(=O)OCH_3$ **b.** $CH_3C(=O)O-C_6H_5$ **c.** $CH_3C(=O)O-C_6H_4-NO_2$ (para) **d.** $CH_3C(=O)Cl$

2. **a.** *N*-ethylpentanamide **c.** methyl 4-phenylbutanoate

b. 3-methylpentanoic acid **d.** ethanoic propanoic anhydride

3. **a.** $CH_3CH_2C(=O)OC(=O)CH_2CH_3$

b. $CH_3C(=O)Cl + H_2O \longrightarrow CH_3C(=O)OH + HCl$

any reaction in which a reactant is cleaved as a result of reaction with water

c. $CH_3C(=O)OCH_3 + CH_3CH_2OH \overset{HCl}{\rightleftharpoons} CH_3C(=O)OCH_2CH_3 + CH_3OH$

d. $CH_3C(=O)OCH_3 + CH_3NH_2 \xrightarrow{\Delta} CH_3C(=O)NHCH_3 + CH_3OH$

4. **a.** $CH_3C(=O)OH$ **b.** $CH_3C(=O)OCH_3$ **c.** $CH_3C(=O)NH_2$ **d.** $CH_3C(=O)OH$

5. **a.** $CH_3CH_2C(=O)OH$ + $\overset{+}{N}H_4$

b. $CH_3CH_2C(=O)O^-$ + CH_3CH_2OH

c. $CH_3CH_2CH_2C(=O)OH$ + $CH_3CH_2C(=O)OH$

d. $CH_3CH_2C(=O)NHCH_2CH_3$ + Cl^- + $CH_3CH_2\overset{+}{N}H_3$

e. $C_6H_{11}C(=O)OCH_2CH_3$ (cyclohexyl) + CH_3OH

6. **a.** $CH_3CH_2CH_2Br \xrightarrow{^-C\equiv N} CH_3CH_2CH_2C\equiv N \xrightarrow[\text{Raney Ni}]{H_2} CH_3CH_2CH_2CH_2NH_2$

b. $CH_3CH_2CH_2Br \xrightarrow{^-C\equiv N} CH_3CH_2CH_2C\equiv N \xrightarrow[\Delta]{HCl,\ H_2O} CH_3CH_2CH_2C(=O)OH$

$\xrightarrow{SOCl_2} CH_3CH_2CH_2C(=O)Cl \xrightarrow{CH_3C(=O)O^-} CH_3CH_2CH_2C(=O)OC(=O)CH_3$

7. **a.** $CH_3CH_2CH_2C(=O)OCH_2CH_2CH_3$

b. $CH_3CH_2C(=O)OH$ + $CH_3\overset{+}{N}H_3$

c. $CH_3CH_2C(=O)OH$ + HO—C_6H_5 (phenol)

d. $CH_3CH_2CH_2\overset{+}{N}H_3$

e. $CH_3C(=O)OCH_2CH_3$ + $CH_3C(=O)OH$

Answers to Chapter 16 Practice Test

1. **a.** 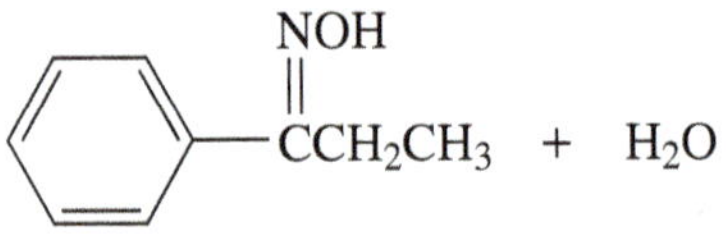
NOH
CCH₂CH₃ + H₂O

d. 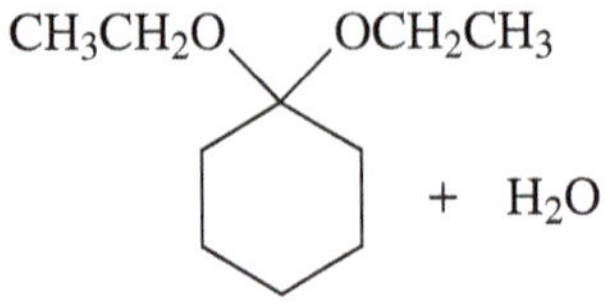
CH₃CH₂O
OCH₂CH₃
+ H₂O

g. 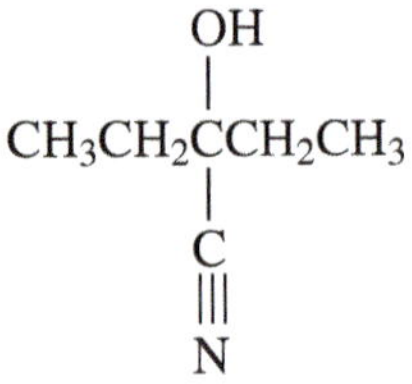
OH
CH₃CH₂CCH₂CH₃
C
N

b. 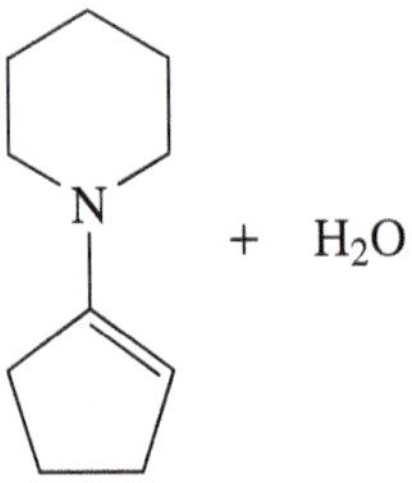
N
+ H₂O

e. 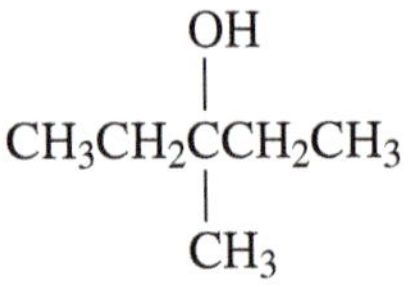
OH
CH₃CH₂CCH₂CH₃
CH₃

h. 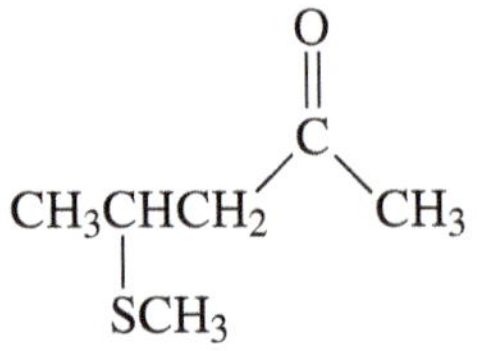
O
C
CH₃CHCH₂
CH₃
SCH₃

c. 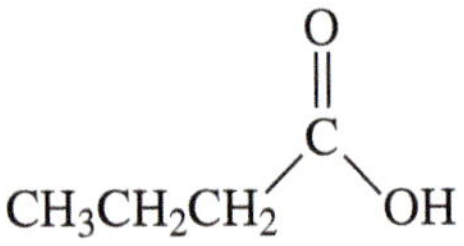
O
C
CH₃CH₂CH₂
OH

f. 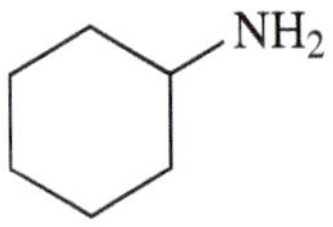
NH₂

i. 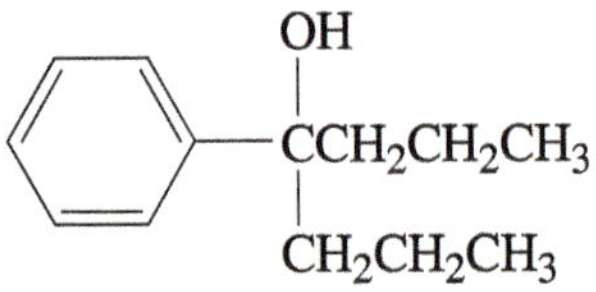
OH
CCH₂CH₂CH₃
CH₂CH₂CH₃

2. 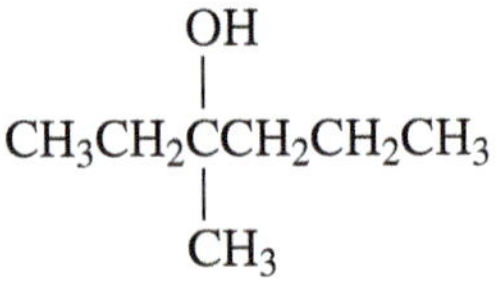
OH
CH₃CH₂CCH₂CH₂CH₃
CH₃

3. **a.** ethyl 4-hydroxyhexanoate **b.** 4-oxoheptanal **c.** 4-formylhexanamide

4. **a.** 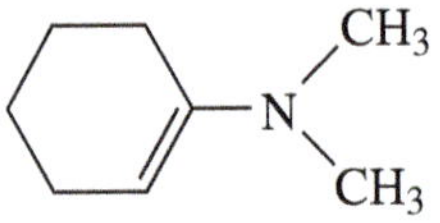
CH₃
N
CH₃

c. 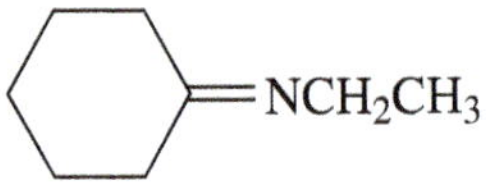
NCH₂CH₃

e. 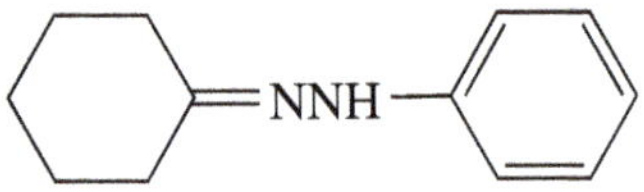
NNH

b. 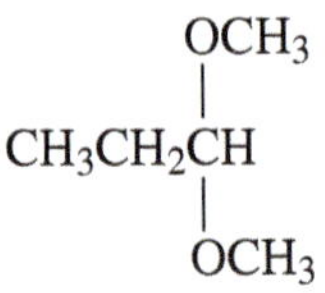
OCH₃
CH₃CH₂CH
OCH₃

d. 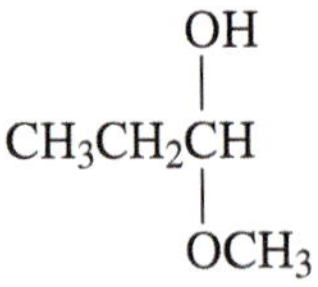
OH
CH₃CH₂CH
OCH₃

5. **a.** butanal **b.** 2-pentanone

6. **a.** 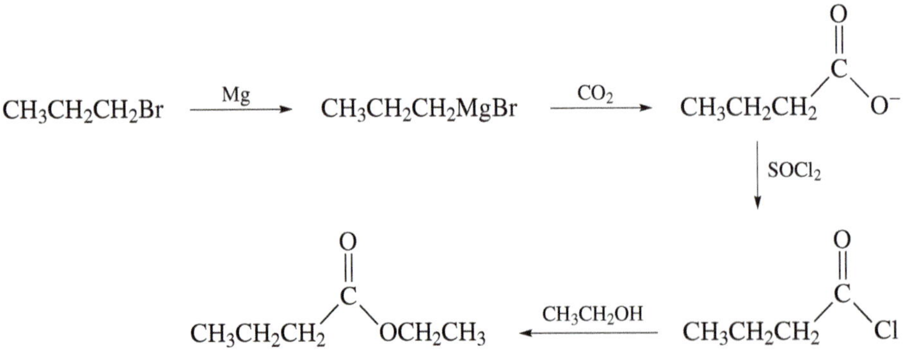
CH₃CH₂CH₂Br
Mg
CH₃CH₂CH₂MgBr
CO₂
O
C
CH₃CH₂CH₂
O⁻
SOCl₂
O
C
CH₃CH₂CH₂
OCH₂CH₃
CH₃CH₂OH
O
C
CH₃CH₂CH₂
Cl

b. $CH_3C(=O)OCH_3$ (O, C, CH_3, OCH_3) $\xrightarrow[CH_3MgBr]{\text{excess}}$ $CH_3\overset{O^-}{\underset{CH_3}{C}}CH_3$ $\xrightarrow{H_3O^+}$ $CH_3\overset{OH}{\underset{CH_3}{C}}CH_3$

c. CO_2 $\xrightarrow{CH_3MgBr}$ $CH_3\overset{O}{\overset{\|}{C}}O^-$ $\xrightarrow[H_3O^+]{LiAlH_4}$ CH_3CH_2OH

d. cyclohexanone (=O) $\xrightarrow[2.\ H_3O^+]{1.\ LiAlH_4}$ cyclohexanol (OH) $\xrightarrow[\text{pyridine}]{SOCl_2}$ chlorocyclohexane (Cl) $\xrightarrow{^-C{\equiv}N}$ cyclohexane–$C{\equiv}N$

Answers to Chapter 17 Practice Test

1. $CH_3\overset{O}{\overset{\|}{C}}CH_2\overset{O}{\overset{\|}{C}}OCH_3$ — **2** $CH_3\overset{O}{\overset{\|}{C}}CH_2\overset{O}{\overset{\|}{C}}CH_3$ — **1** $CH_3\overset{O}{\overset{\|}{C}}CH_3$ — **3**

2. **a.** $CH_3\overset{OH}{C}{=}CH\overset{O}{\overset{\|}{C}}CH_3$ **b.** $CH_3\overset{O}{\overset{\|}{C}}CH_2\overset{O}{\overset{\|}{C}}OCH_3$

3. **a.** $CH_3CH_2\overset{O}{\overset{\|}{C}}CH_3$ + CO_2

b. $CH_3CH_2\overset{Br}{\underset{Br}{C}}\overset{O}{\overset{\|}{C}}H$ + 2 Br^-

c. $CH_3CH_2CH_2\overset{O}{\overset{\|}{C}}\underset{CH_2CH_3}{CH}\overset{O}{\overset{\|}{C}}OCH_3$

d. $CH_3\underset{O}{\underset{\|}{C}}\overset{CH_2CH_2\overset{O}{\overset{\|}{C}}CH_3}{CH}\underset{O}{\underset{\|}{C}}CH_3$

e. cyclopentanone with O and Br (2-bromocyclopentanone)

f. cyclohexanone with O; Cl, Cl; Cl, Cl (2,2,6,6-tetrachlorocyclohexanone)

g. $CH_3CH_2CH_2CH_2\overset{O}{\overset{\|}{C}}OH$

4. **a.** $2\ CH_3CH_2C(=O)H \xrightleftharpoons{HO^-} CH_3CH_2CH(OH)CH(CH_3)C(=O)H$

b. $2\ CH_3CH_2C(=O)H \xrightleftharpoons{HO^-} CH_3CH_2CH(OH)CH(CH_3)C(=O)H \xrightarrow[\Delta]{H_2SO_4} CH_3CH_2CH{=}C(CH_3)C(=O)H$

c. $2\ CH_3CH_2C(=O)OCH_3 \xrightarrow[2.\ H_3O^+]{1.\ CH_3O^-} CH_3CH_2C(=O)CH(CH_3)C(=O)OCH_3 + CH_3OH$

d. $CH_3OC(=O)CH_2CH_2CH_2CH_2CH_2C(=O)OCH_3 \xrightarrow[2.\ H_3O^+]{1.\ CH_3O^-}$ (cyclohexanone ring bearing $C(=O)OCH_3$)

e. $CH_3CH_2OC(=O)CH_2C(=O)OCH_2CH_3 \xrightarrow[2.\ CH_3CH_2Br\quad 3.\ H^+,\ H_2O,\ \Delta]{1.\ CH_3CH_2O^-} CH_3CH_2CH_2C(=O)OH$

f. $CH_3C(=O)CH_2C(=O)OCH_2CH_3 \xrightarrow[2.\ CH_3CH_2Br\quad 3.\ H^+,\ H_2O,\ \Delta]{1.\ CH_3CH_2O^-} CH_3CH_2CH_2C(=O)CH_3$

5.

$CH_3CH(CH_3)CH_2CH_2CH(OH)CH(CH_2CH_3)C(=O)H$

$CH_3CH(CH_3)CH_2CH_2CH(OH)CH(CH_2CH(CH_3)CH_3)C(=O)H$

$CH_3CH_2CH_2CH(OH)CH(CH_2CH_3)C(=O)H$

$CH_3CH_2CH_2CH(OH)CH(CH_2CH(CH_3)CH_3)C(=O)H$

6. a. $CH_3CH_2CH_2-\overset{O}{\overset{\|}{C}}-OCH_3$

b. $CH_3O-\overset{O}{\overset{\|}{C}}-CH_2CH_2CH_2CH_2CH_2-\overset{O}{\overset{\|}{C}}-OCH_3$

Answers to Chapter 18 Practice Test

1. a. *meta*-methylnitrobenzene
1-methyl-3-nitrobenzene

b. 1,2,4-tribromobenzene

c. *ortho*-ethylbenzoic acid
2-ethylbenzoic acid

d. *para*-chlorophenol
4-chlorophenol

2. (benzene ring)–Br: 4

(benzene ring)–N(H)–C(=O)–CH$_3$: 1

(benzene ring)–CH$_2$CH$_3$: 2

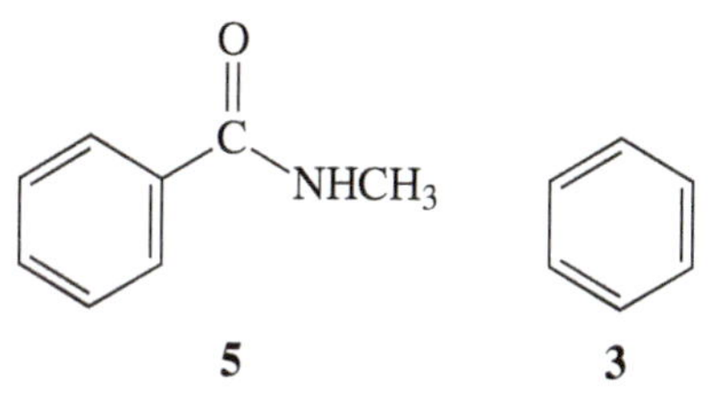

3. a. COOH (top), Cl (bottom)

b. OH (top), NO$_2$ (bottom)

c. $^+NH_3$ (top), CH$_3$ (bottom)

d. COOH

4. a. *para*-bromonitrobenzene

b. *para*-bromoethylbenzene

5. $CH_3CH_2-\overset{O}{\overset{\|}{C}}-O-\overset{O}{\overset{\|}{C}}-CH_2CH_3$

6. $H\ddot{O}-NO_2 + H-OSO_3H \rightleftharpoons H\overset{H}{\overset{|}{\overset{+}{O}}}-NO_2 \rightleftharpoons H_2O + {}^+NO_2$

7. a. NO$_2$ (position 1), SO$_3$H (position 3)

b. OCH$_3$, CH$_3$ (ortho) + OCH$_3$, CH$_3$ (para)

c. COOH

d. OCH$_3$ (top), NO$_2$ (bottom)

e. Cl, NO$_2$, OCH$_3$

f. $\overset{O}{\overset{\|}{C}}CH_3$, Cl

8. **a.** Benzoic acid is more reactive than benzene toward electrophilic aromatic substitution. F
b. *para*-Chlorobenzoic acid is more acidic than *para*-methoxybenzoic acid. T
c. A $CH{=}CH_2$ group is a meta director. F
d. *para*-Nitroaniline is more basic than *para*-chloroaniline. F

9.

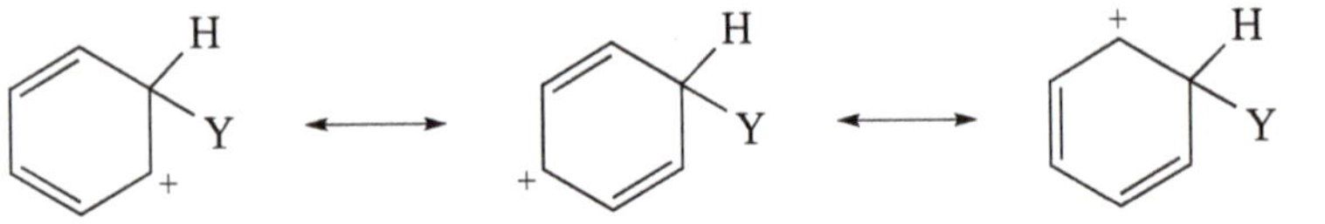

Answers to Chapter 19 Practice Test

1. **a.** 2,4-dimethyl pyrrolidine
2,4-dimethylazacyclopentane

b. *N*-methylpiperidine
N-methylazacyclohexane

c. 3-ethyltetrahydrofuran
3-ethyloxacyclopentane

d. 3-bromopiperidine
3-bromoazacyclohexane

2. **a.** 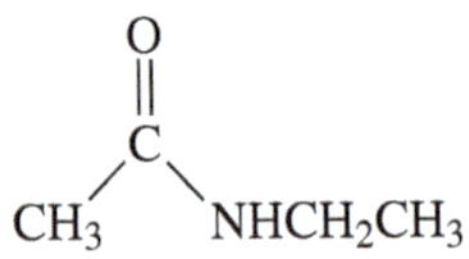 $CH_3C(=O)NHCH_2CH_3$ + $CH_3CH_2\overset{+}{N}H_3$ Cl^-

b. 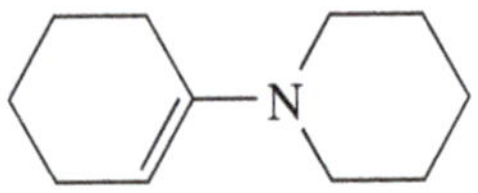 + H_2O

c.

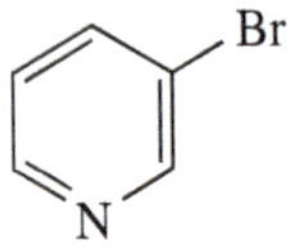

d.

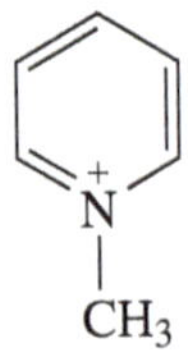

e.

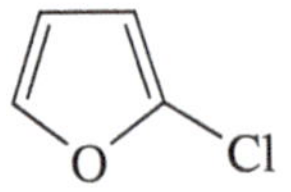

3. **a.** 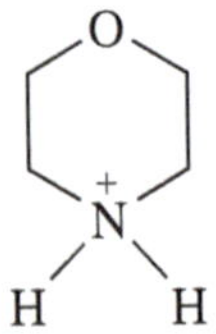 **b.** 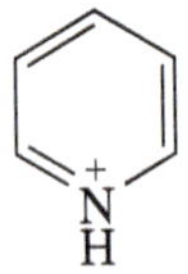 **c.** 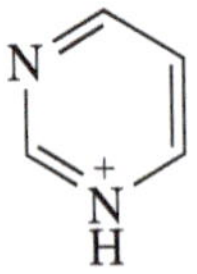 **d.**

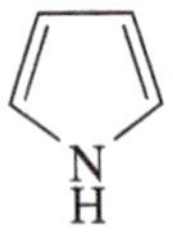

4. **a.** 4-Chloropyridine is more reactive toward nucleophilic aromatic substitution than is 3-chloropyrrole. T
b. Pyrrole is more reactive toward electrophilic aromatic substitution than is furan. T
c. Pyrrole is more reactive toward electrophilic aromatic substitution than is benzene. T
d. Pyridine is more reactive toward electrophilic aromatic substitution than is benzene. F

5. 1. $SnCl_4$ 2. H_2O → 1. CH_3MgBr 2. H_2O →

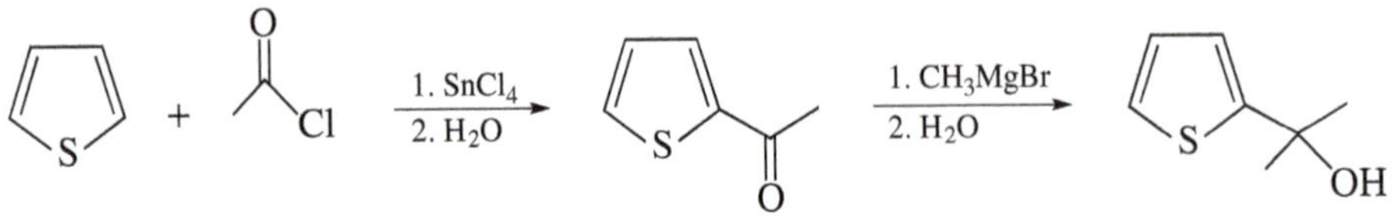

Answers to Chapter 20 Practice Test

1.

a.

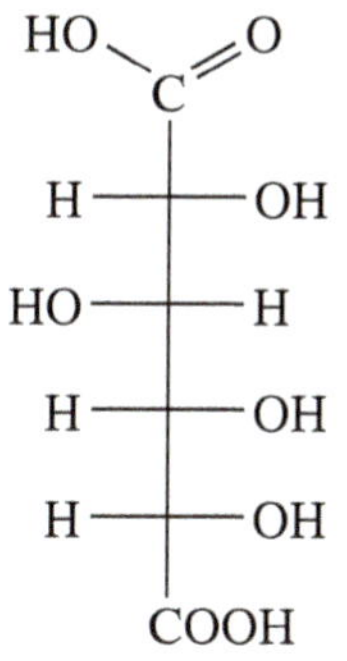

HO–C=O

H–OH

HO–H

H–OH

H–OH

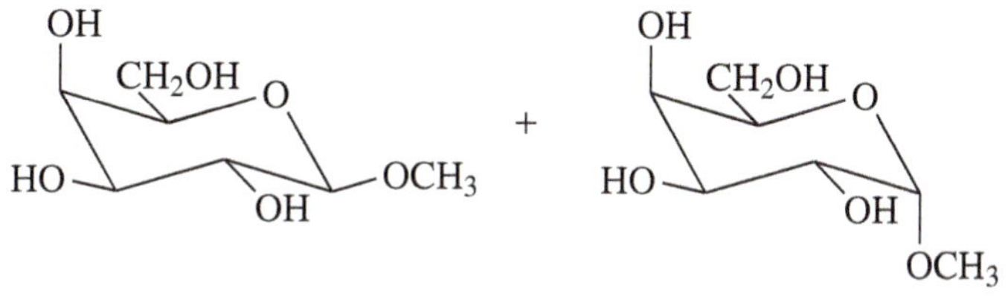

COOH

c.

H–C=O

H–OH

HO–H

HO–H

H–OH

CH_2OH

\+

H–C=O

HO–H

HO–H

HO–H

H–OH

CH_2OH

b.

OH, CH_2OH, O, HO, OH, OCH_3

\+

OH, CH_2OH, O, HO, OH, OCH_3

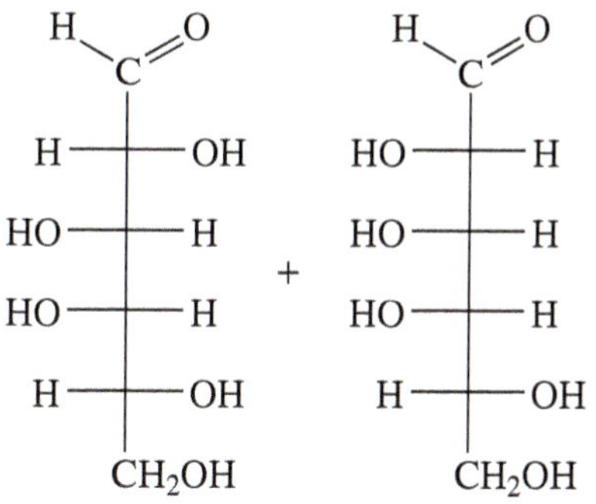

d.

HO–C=O

H–OH

H–OH

H–OH

CH_2OH

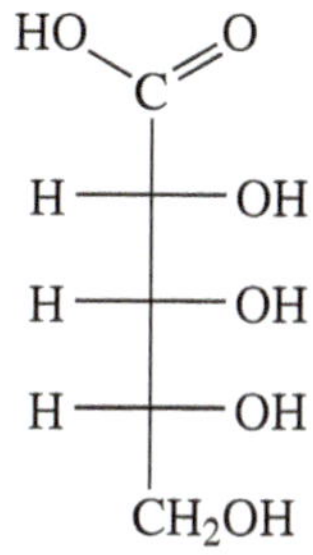

2.

a. Glycogen contains α-1,4′ and β-1,6′-glycosidic linkages. F
b. D-Mannose is a C-1 epimer of D-glucose. F
c. D-Glucose and L-glucose are anomers. F
d. D-Erythrose and D-threose are diastereomers. T
e. Wohl degradations of D-glucose and D-gulose form the same aldopentose. F

3.

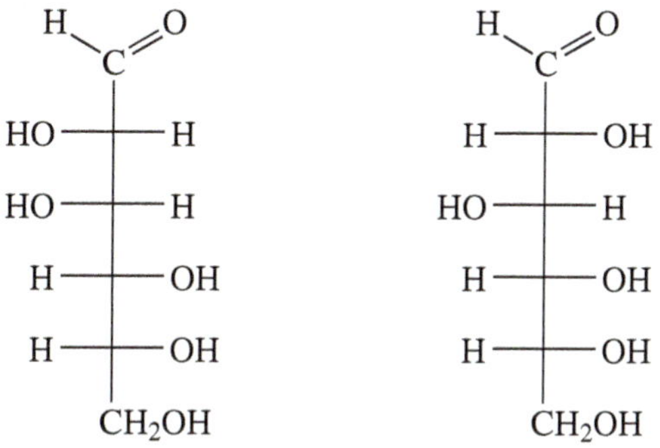

H–C=O

HO–H

HO–H

H–OH

H–OH

CH_2OH

H–C=O

H–OH

HO–H

H–OH

H–OH

CH_2OH

4. D-mannose and D-glucose

5.

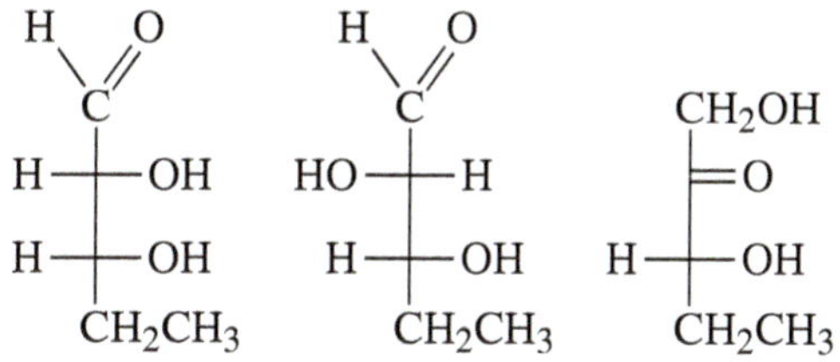

H–C=O

H–OH

H–OH

CH_2CH_3

H–C=O

HO–H

H–OH

CH_2CH_3

CH_2OH

=O

H–OH

CH_2CH_3

6. D-altrose

7. Amylose has α-1,4′-glycosidic linkages, whereas cellulose has β-1,4′-glycosidic linkages.

8. D-gulose and D-idose

9. D-allose

10. (chair structure) CH_2OH, O, HO, OH, OH, HO

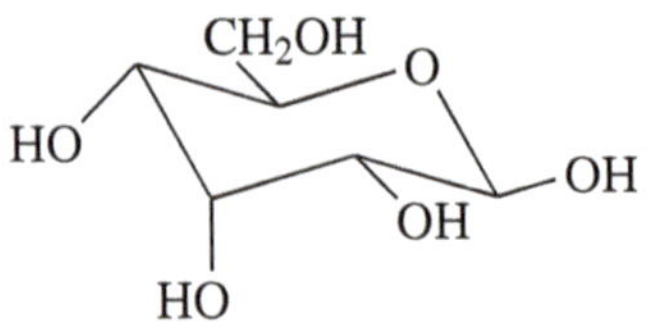

Answers to Chapter 21 Practice Test

1.

a. $^{-}O-\overset{O}{\overset{\|}{C}}-CH_2CH_2\underset{^{+}NH_3}{\underset{|}{C}H}-\overset{O}{\overset{\|}{C}}-O^{-}$

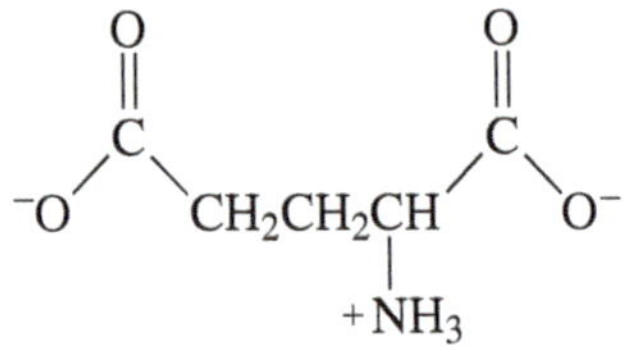

b. $H_3\overset{+}{N}CH_2CH_2CH_2CH_2\underset{^{+}NH_3}{\underset{|}{C}H}-\overset{O}{\overset{\|}{C}}-O^{-}$

c. $CH_3CH_2\overset{CH_3}{\overset{|}{C}}HCH(^{+}NH_3)-\overset{O}{\overset{\|}{C}}-O^{-}$

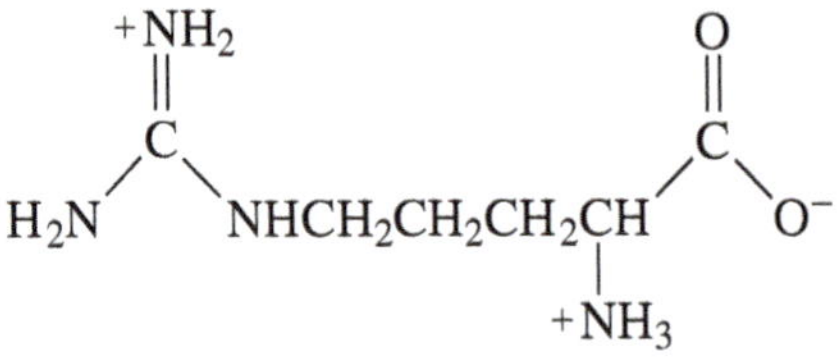

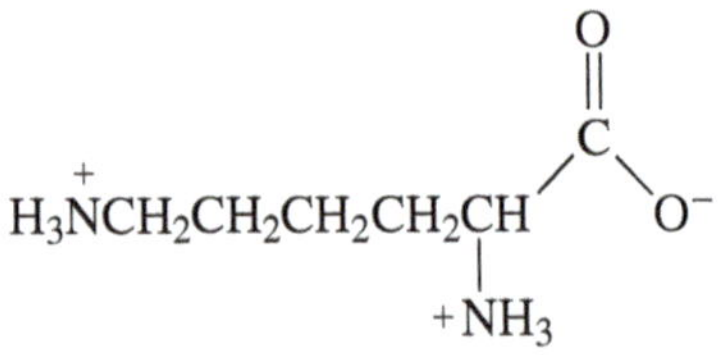

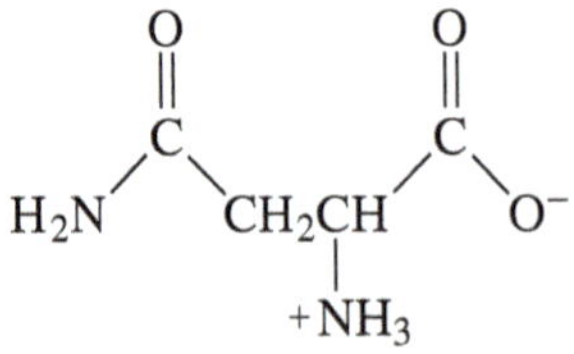

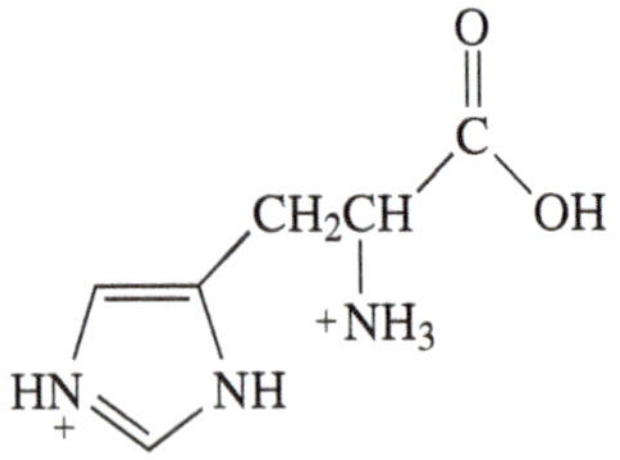

d. $H_2N-\overset{^{+}NH_2}{\overset{\|}{C}}-NHCH_2CH_2CH_2\underset{^{+}NH_3}{\underset{|}{C}H}-\overset{O}{\overset{\|}{C}}-O^{-}$

e. $H_2N-\overset{O}{\overset{\|}{C}}-CH_2\underset{^{+}NH_3}{\underset{|}{C}H}-\overset{O}{\overset{\|}{C}}-O^{-}$

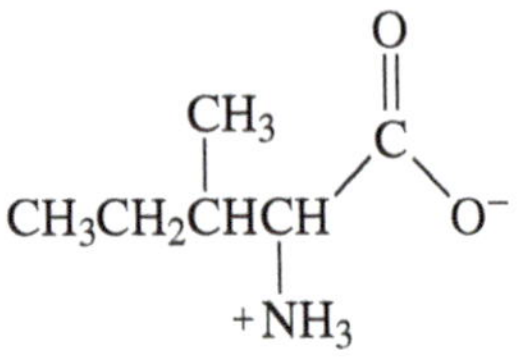

2.

a. (imidazole ring, $H\overset{+}{N}$ and NH)$-CH_2\underset{^{+}NH_3}{\underset{|}{C}H}-\overset{O}{\overset{\|}{C}}-OH$

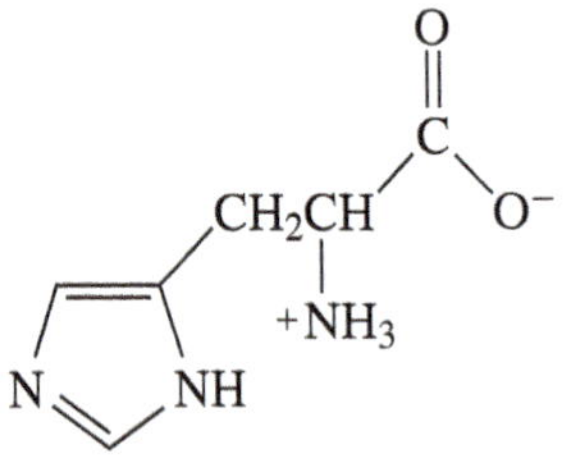

b. (imidazole ring, $H\overset{+}{N}$ and NH)$-CH_2\underset{^{+}NH_3}{\underset{|}{C}H}-\overset{O}{\overset{\|}{C}}-O^{-}$

c. (imidazole ring, N and NH)$-CH_2\underset{^{+}NH_3}{\underset{|}{C}H}-\overset{O}{\overset{\|}{C}}-O^{-}$

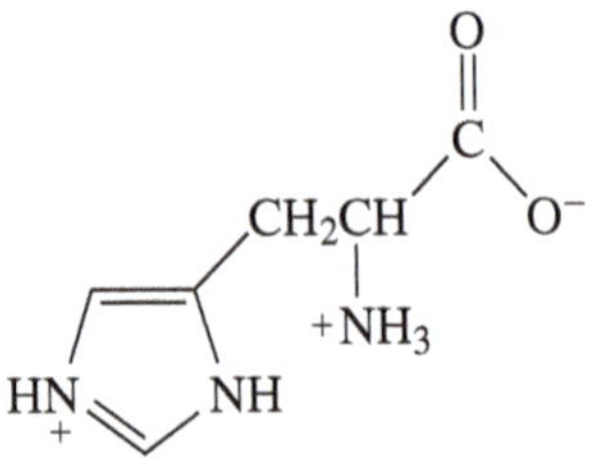

d. (imidazole ring, N and NH)$-CH_2\underset{NH_2}{\underset{|}{C}H}-\overset{O}{\overset{\|}{C}}-O^{-}$

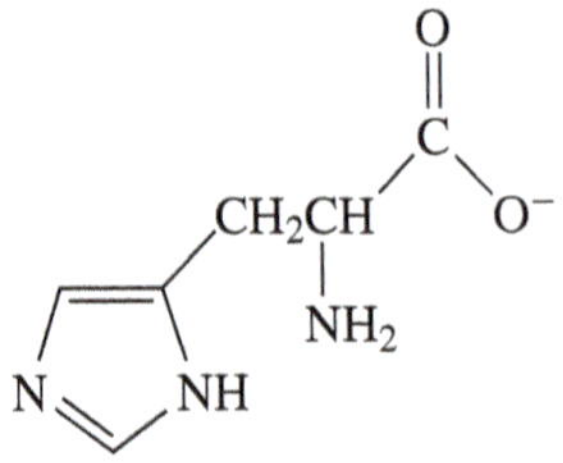

3.

a. alanine, because it is farther away from its pI

b. glycine

c. leucine and isoleucine

d. aspartic acid

4. The electron-withdrawing protonated amino group causes the carboxyl group of alanine to have a lower pK_a.

5. **a.** A cigar-shaped protein has a greater percentage of polar residues than a spherical protein. T

b. Naturally occurring amino acids have the L-configuration. T

c. There is free rotation about a peptide bond. F

6. 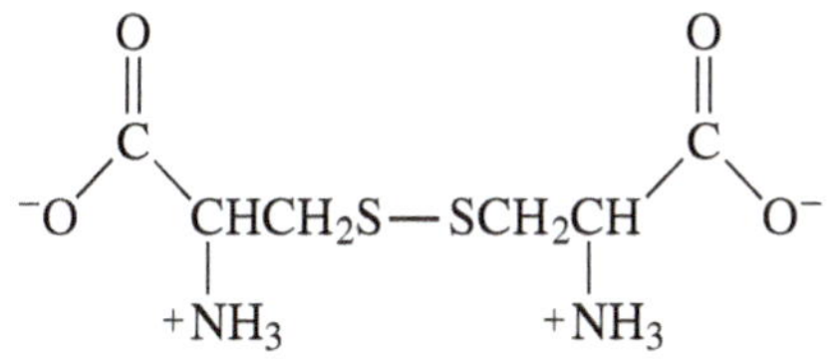 $^{-}O(O{=})C{-}CH(^{+}NH_3)CH_2S{-}SCH_2CH(^{+}NH_3){-}C({=}O)O^{-}$

7. **a.** the sequence of the amino acids and the location of the disulfide bonds in the protein

b. the three-dimensional arrangement of all the atoms in the protein

c. a description of the way the subunits of an oligomer are arranged in space

8.

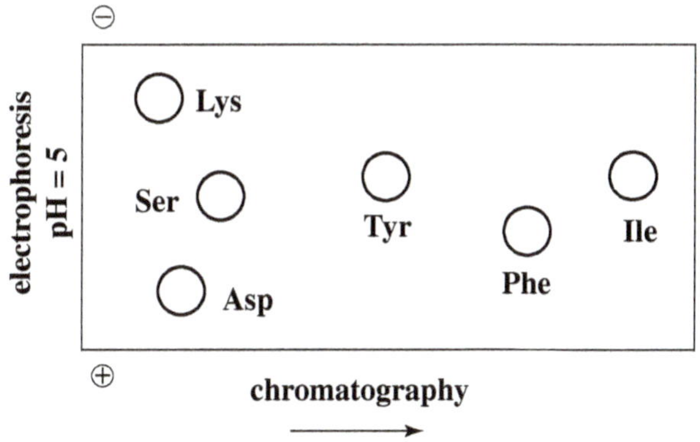

⊖

electrophoresis pH = 5

Lys

Ser

Tyr

Ile

Phe

Asp

⊕

chromatography →

9. **a.** $\frac{2.16 + 9.18}{2} = \frac{11.34}{2} = 5.67$ **b.** $\frac{9.04 + 12.48}{2} = \frac{21.52}{2} = 10.76$

10. Ala Ser Arg Gly Arg Met His Phe Lys Ile

11. 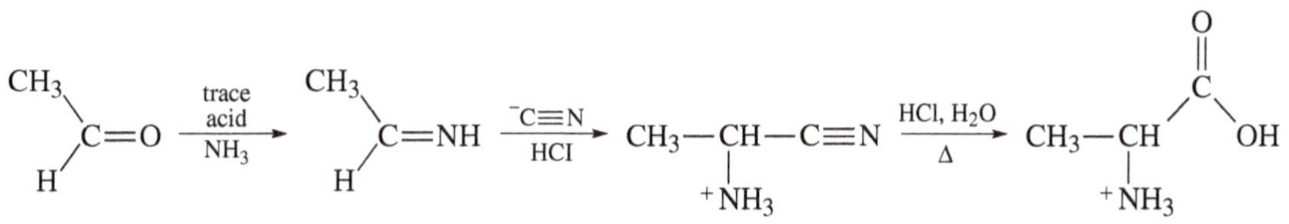 $CH_3(H)C{=}O \xrightarrow[NH_3]{\text{trace acid}} CH_3(H)C{=}NH \xrightarrow[HCl]{^{-}C{\equiv}N} CH_3{-}CH(^{+}NH_3){-}C{\equiv}N \xrightarrow[\Delta]{HCl,\ H_2O} CH_3{-}CH(^{+}NH_3){-}C({=}O)OH$

12. 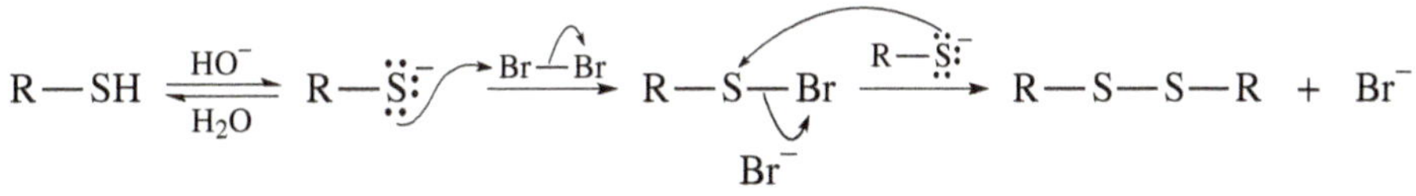 $R{-}SH \underset{H_2O}{\overset{HO^{-}}{\rightleftharpoons}} R{-}\ddot{S}{:}^{-} \xrightarrow{Br{-}Br} R{-}S{-}Br + Br^{-} \xrightarrow{R{-}\ddot{S}{:}^{-}} R{-}S{-}S{-}R + Br^{-}$

Answers to Chapter 22 Practice Test

1.
a. A catalyst increases the equilibrium constant of a reaction. F
b. An acid catalyst donates a proton to the substrate, and a base catalyst removes a proton from the substrate. T
c. The reactant of an enzyme-catalyzed reaction is called a substrate. T
d. Complexing with a metal ion increases the pK_a of water. F

2. It protonates the leaving group to make it a better leaving group.

3. **a.**

rate
pH 5.7

b.

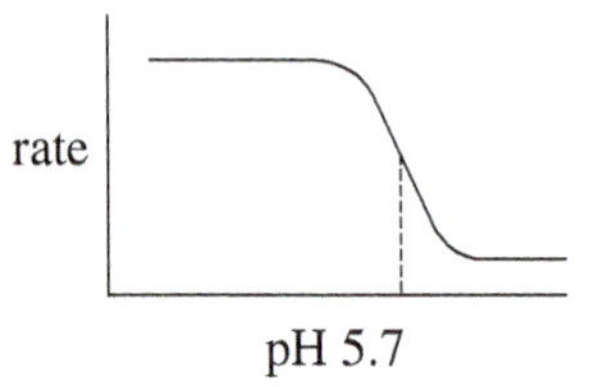

4.

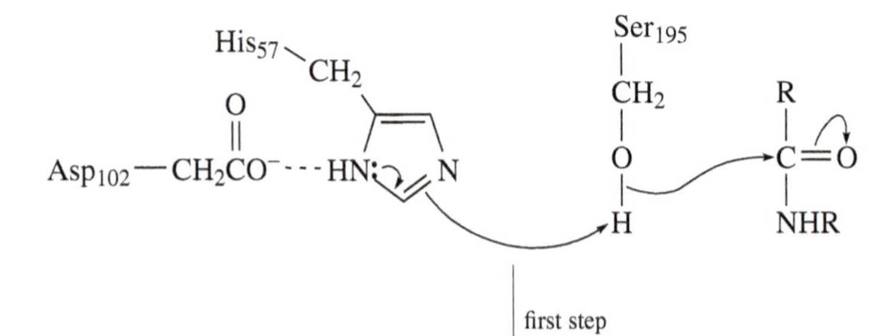

first step

His57
CH2
O
Asp102—CH2CO−---HN+
NH

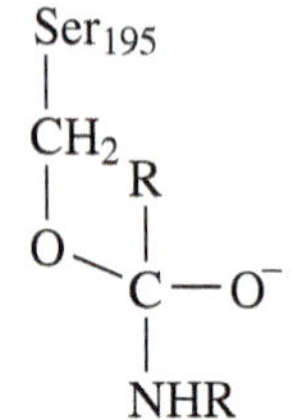

a. base catalyst

b. nucleophilic catalyst

c. It stabilizes the positive charge on histidine.

Answers to Chapter 23 Practice Test

1. biotin and vitamin KH_2

2. methyl (CH_3), methylene (CH_2), formyl $(HC{=}O)$

3.

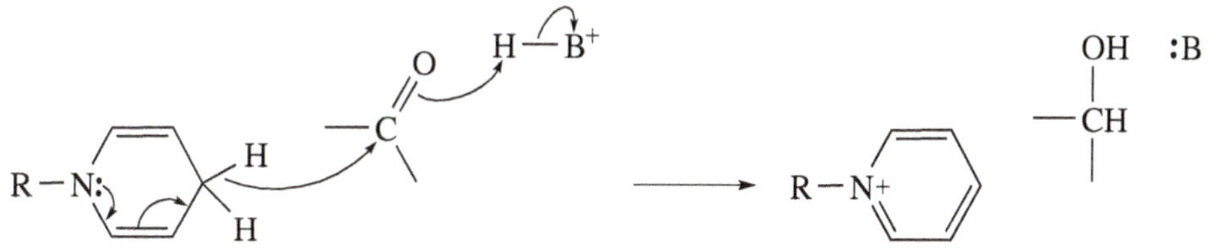

R—N: H H O C H—B⁺ → R—N⁺ OH :B —CH

4.

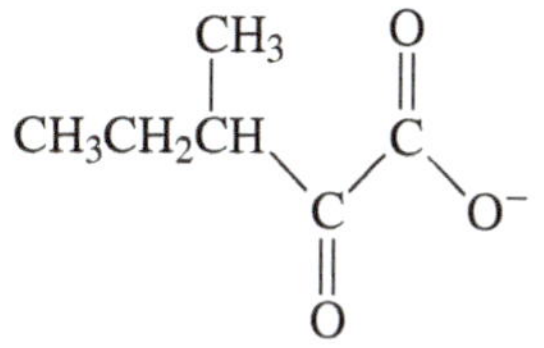

$CH_3CH_2CH(CH_3)C(=O)C(=O)O^-$

5. The first step is removing the hydrogen atom that is going to change places with a group on an adjacent carbon.

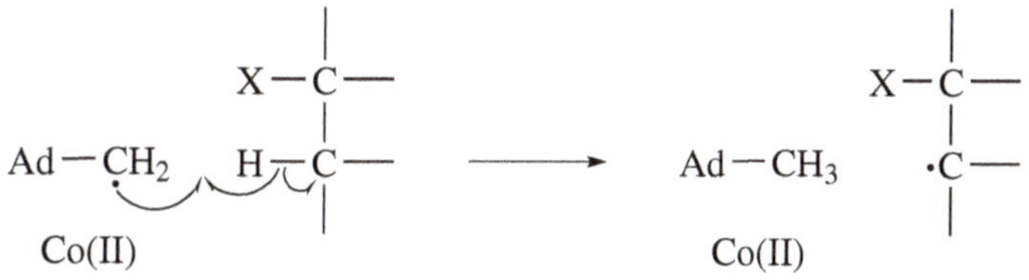

Ad—CH₂ H—C— X—C— Co(II) → Ad—CH₃ •C— X—C— Co(II)

6. first coenzyme = biotin
second coenzyme = coenzyme B_{12}

7.

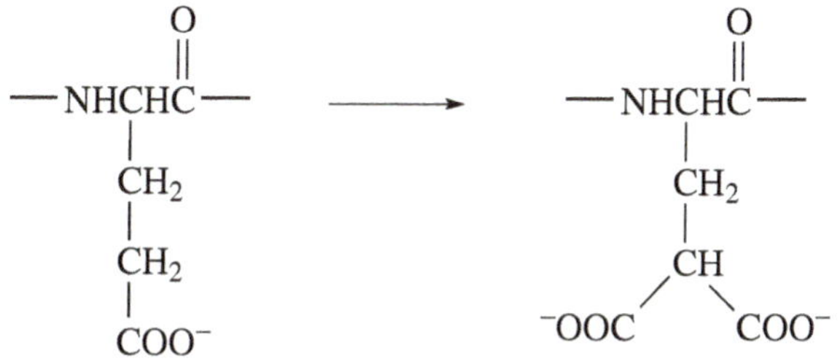

—NHCHC(=O)— with $CH_2CH_2COO^-$ side chain → —NHCHC(=O)— with $CH_2CH(COO^-)_2$ side chain

8.

$^-OOC{-}CH_2{-}C(=O)SCoA$

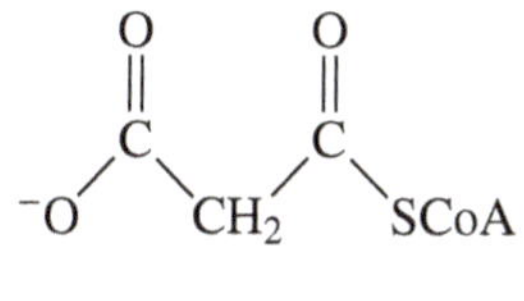

9. **a.** ATP, Mg^{2+}, HCO_3^-

b. ATP activates bicarbonate (HCO_3^-) by putting a good leaving group (phosphate) on it. Mg^{2+} complexes with ATP in order to reduce the negative charge on ATP so that it can react with a nucleophile.

HCO_3^- is the source of the COO^- group that is put on the reactant.

10. FAD oxidizes dihydrolipoate to lipoate.

11. $HO-C_6H_4-CH_2-C(=O)-COO^-$ + $^-OOC-CH_2CH_2CH(^+NH_3)-COO^-$

12. In the process of converting uridines to thymidines, tetrahydrofolate is oxidized to dihydrofolate. NADPH is used to reduce dihydrofolate back to tetrahydrofolate.

13.

a. Vitamin B_1 is the only water-insoluble vitamin that has a coenzyme function. F
b. $FADH_2$ is a reducing agent. T
c. Thiamine pyrophosphate is vitamin B_6. F
d. Cofactors that are organic molecules are called coenzymes. T
e. Vitamin K is a water-soluble vitamin. F
f. Lipoic acid is covalently bound to its enzyme by an amide linkage. T

Answers to Chapter 24 Practice Test

1. **a.** $RO-P(=O)(O^-)-O^-$ **b.** $R-C(=O)-O-P(=O)(O^-)-O-P(=O)(O^-)-O^-$ **c.** $^-O-P(=O)(O^-)-O-P(=O)(O^-)-O^-$

2. $RCH_2CH_2CH_2CH_2-C(=O)-SCoA$ $\xrightarrow{FAD \rightarrow FADH_2}$ $RCH_2CH_2CH{=}CH-C(=O)-SCoA$

$\xrightarrow{H_2O}$ $RCH_2CH_2CH(OH)-CH_2-C(=O)-SCoA$

$\xrightarrow{NAD^+ \rightarrow NADH, H^+}$ $RCH_2CH_2-C(=O)-CH_2-C(=O)-SCoA$

$\xrightarrow{CoASH}$ $RCH_2CH_2-C(=O)-SCoA$ + $CH_3-C(=O)-SCoA$

3. acetate

4. catabolic reactions

5. amino acids

6. acetyl-CoA

7.
a. Each molecule of $FADH_2$ forms 2.5 molecules of ATP in the fourth stage of catabolism. F
b. $FADH_2$ is oxidized to FAD. T
c. NAD^+ is oxidized to NADH. F
d. Acetyl-CoA is a citric acid cycle intermediate. F

Answers to Chapter 25 Practice Test

1.

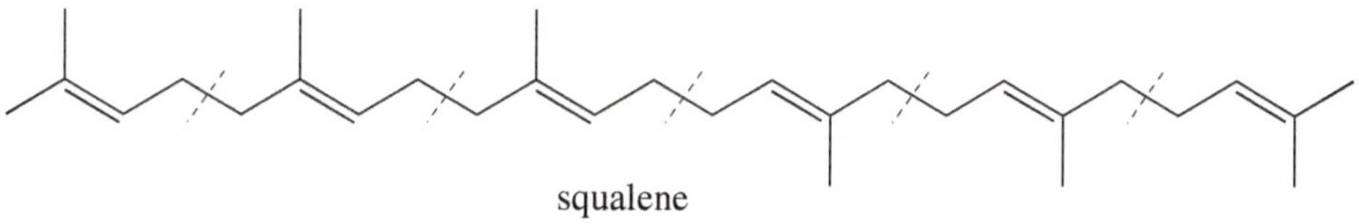

squalene

2.

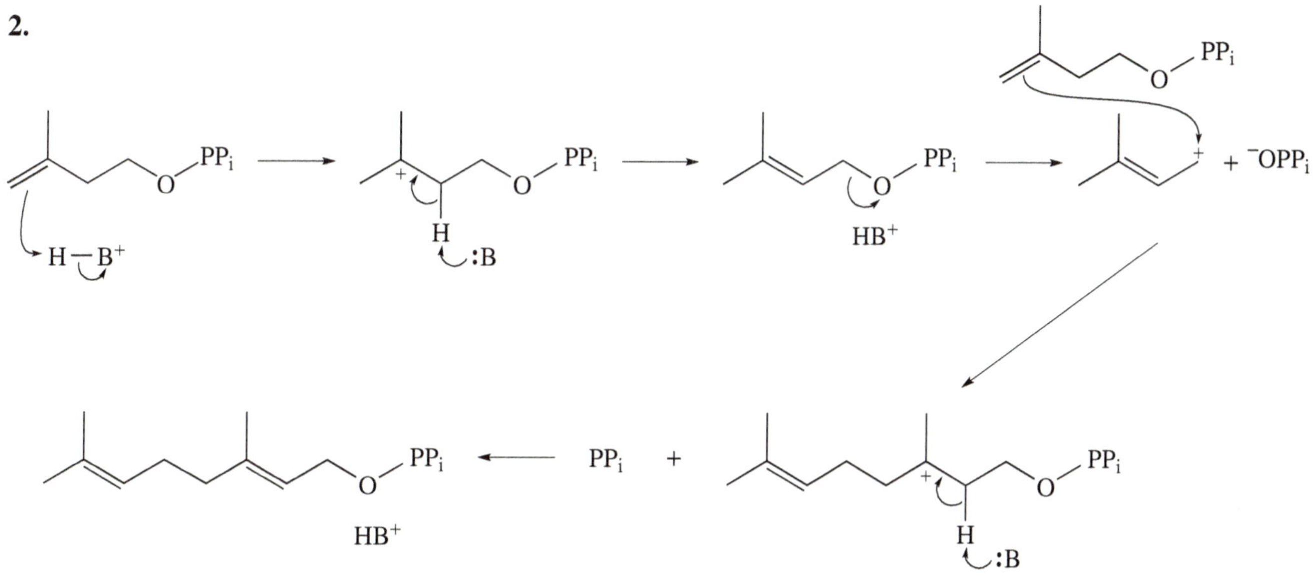

PP_i PP_i PP_i PP_i + $^-OPP_i$

H—B^+ H :B HB^+

PP_i PP_i + PP_i

HB^+ H :B

3. **a.**

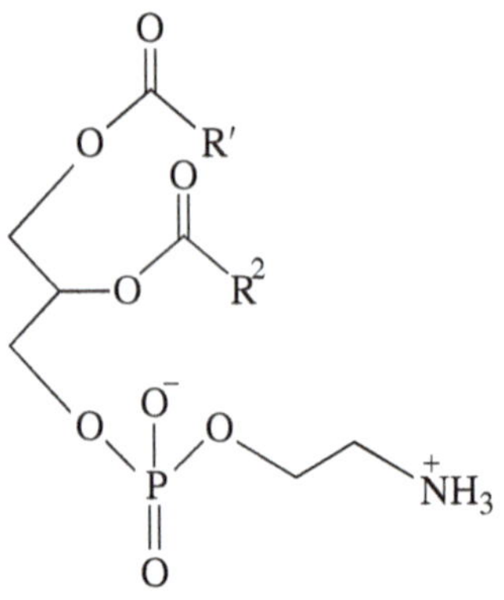

b. 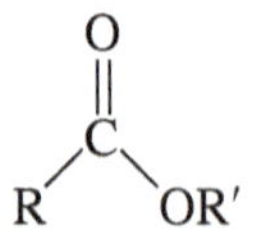 R and R′ are long, straight chains.

4.

a. Cholesterol is the precursor of all other steroids. T
b. The double bonds in unsaturated fats are conjugated. F
c. Fats have a higher percentage of saturated fatty acids than do oils. T
d. A saturated fatty acid has a lower melting point than an unsaturated fatty acid with the same number of carbons. F
e. Lipids are insoluble in water. T
f. A diterpene contains 20 carbons. T
g. Vitamin A is not a coenzyme. T

5. 30

6. 5-membered ring

7. arachidonic acid

Answers to Chapter 26 Practice Test

1. dUMP

2. 5′—A—T—G—C—A—A—G—T—3′

3. A

4.

a. Guanine and cytosine are purines. F
b. The 3′-OH group allows RNA to be easily cleaved. F
c. The number of As in DNA is equal to the number of Ts. T
d. rRNA carries the amino acid that will be incorporated into a protein. F
e. The template strand of DNA is the one transcribed to form RNA. T
f. The 5′-end of DNA has a free OH group. F
g. The synthesis of proteins from an RNA blueprint is called transcription. F
h. A nucleotide consists of a base and a sugar. F
i. RNA contains Ts, and DNA contains Us. F

5. only #1

6. 5′—G—C—A—U—G—G—A—C—C—C—C—G—U—3′

7. CA and TG

Answers to Chapter 27 Practice Test

1. **a.** $-CH_2CHCH_2CHCH_2CH-$ (each CH bearing a COOH substituent)

COOH COOH COOH

b. $-O(CH_2)_5\overset{O}{\overset{\|}{C}}-O(CH_2)_5\overset{O}{\overset{\|}{C}}-O(CH_2)_5\overset{O}{\overset{\|}{C}}-$

c. $-CH_2CCl_2CH_2CCl_2CH_2CCl_2-$

(Cl above and Cl below each C)

d. $-\overset{O}{\overset{\|}{C}}(CH_2)_4\overset{O}{\overset{\|}{C}}-NH(CH_2)_4NH-\overset{O}{\overset{\|}{C}}(CH_2)_4\overset{O}{\overset{\|}{C}}-NH(CH_2)_4NH-$

2. **a.** $CH_2{=}CHCH_3$

b. $CH_2{=}\underset{}{\overset{CH_3}{\overset{|}{C}}}-CH{=}CH_2$

c. 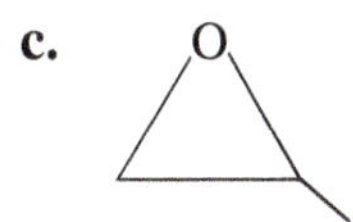

d. $HO(CH_2)_4\overset{O}{\overset{\|}{C}}OH$

3. **a.** $CH_3CH_2CH_2CH_2CH_2CHCH_2CHCH_2CH-$ (each CH bearing a $C{\equiv}N$ substituent)

C≡N C≡N C≡N

b. $CH_3CHCH_2CHCH_2CH-$ (each CH bearing a $C{=}O$ group with CH_3)

C=O C=O C=O

CH_3 CH_3 CH_3

c. $CH_3O-CH_2CHOCH_2CHOCH_2CHO-$ (each CH bearing a CH_2CH_3 substituent)

CH_2CH_3 CH_2CH_3 CH_2CH_3

4. because the two substituents attached to the carbon are identical

Answers to Chapter 28 Practice Test

1. 6

2. ψ_1, ψ_3, and ψ_5

3. ψ_2, ψ_4, and ψ_6

4. ψ_1, ψ_2, and ψ_3

5. A compound with three double bonds undergoes disrotatory ring closure under thermal conditions.

6. A compound with two double bonds undergoes disrotatory ring closure under photochemical conditions.

7. A compound with three pairs of electrons in the reacting system undergoes suprafacial rearrangement under thermal conditions.

8. A compound with two pairs of electrons in the reacting system undergoes suprafacial rearrangement under photochemical conditions.

9. [1,5] Sigmatropic hydrogen migration involves three pairs of electrons.

10. A [2+2] cycloaddition reaction occurs only under photochemical conditions.